Aligned on Class 11 & 12 Syllabus

Mathematics

1978-2014

Contains Topic-wise:

IIT-JEE (1978-2012)
AIEEE (2002-12)
JEE Main & Advanced 2013

Years 37 JEE ADVANCED +13 Years JEE MAIN

Topic Wise Solved Papers

22 Topics
9 Types
3100+ Qns.

Must for JEE Main, Advanced, BITSAT & Other State Exams

Mamta Batra

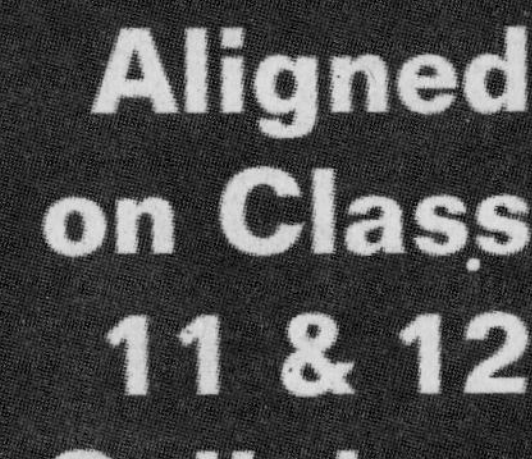

• **Head Office :** B-32, Shivalik Main Road, Malviya Nagar, New Delhi-110017
• **Sales Office :** B-48, Shivalik Main Road, Malviya Nagar, New Delhi-110017
Tel. : 011-26691021 / 26691713

Price : ₹ 420

Typeset by Disha DTP Team

For further information about the books from DISHA,
Log on to **www.dishapublication.com** or email to **info@dishapublication.com**

Index

CHAPTERS BASED ON CLASS 11 SYLLABUS*

CHAPTERS BASED ON CLASS 12 SYLLABUS*

* *The chapters have been divided as per the Class 11 & 12 syllabus followed by the NCERT books. Some of the chapters which are split in the class 11 & 12 syllabus in NCERT have been combined here. There might be certain topics/ chapters which are not covered in NCERT but are a part of JEE Advanced/IIT-JEE syllabus.*

JEE ADVANCED 2014 MATHEMATICS

PAPER - 1

SECTION - I

This section contains **10 multiple choice questions.** Each question has 4 choices (a), (b), (c) and (d) out of which **ONLY ONE or MORE THAN ONE are correct.**

1. Let $f : [a, b] \to [1, \infty)$ be a continuous function and let $g : R \to R$ be defined as

$$g(x) = \begin{cases} 0, & \text{if } x < a, \\ \int_a^x f(t)\,dt, & \text{if } a \le x \le b; \text{ then} \\ \int_a^b f(t)\,dt, & \text{if } x > b. \end{cases}$$

(a) $g(x)$ is continuous but not differentiable at a
(b) $g(x)$ is differentiable on R
(c) $g(x)$ is continuous but not differentiable at b
(d) $g(x)$ is continuous and differentiable at either (a) or (b) but not both

2. For every pair of continuous functions $f, g : [0, 1] \to R$ such that $\max\{f(x) : x \in [0,1]\} = \max\{g(x) : x \in [0,1]\}$, the correct statement(s) is (are):

(a) $(f(c))^2 + 3f(c) = (g(c))^2 + 3g(c)$ for some $c \in [0, 1]$
(b) $(f(c))^2 + f(c) = (g(c))^2 + 3g(c)$ for some $c \in [0, 1]$
(c) $(f(c))^2 + 3f(c) = (g(c))^2 + g(c)$ for some $c \in [0, 1]$
(d) $(f(c))^2 = (g(c))^2$ for some $c \in [0, 1]$

3. Let M be a 2×2 symmetric matrix with integer entries. Then M is invertible if

(a) The first column of M is the transpose of the second row of M
(b) The second row of M is the transpose of the first column of M
(c) M is a diagonal matrix with non-zero entries in the main diagonal
(d) The product of entries in the main diagonal of M is not the square of an integer

4. Let $\vec{x}$, $\vec{y}$ and $\vec{z}$ be three vectors each of magnitude $\sqrt{2}$ and the angle between each pair of them is $\frac{\pi}{3}$. If $\vec{a}$ is a non-zero vector perpendicular to $\vec{x}$ and $\vec{y} \times \vec{z}$ and $\vec{b}$ is a non-zero vector perpendicular to $\vec{y}$ and $\vec{z} \times \vec{x}$, then

(a) $\vec{b} = \left(\vec{b} \cdot \vec{z}\right)\left(\vec{z} - \vec{x}\right)$

(b) $\vec{a} = \left(\vec{a} \cdot \vec{y}\right)\left(\vec{y} - \vec{z}\right)$

(c) $\vec{a} \cdot \vec{b} = -\left(\vec{a} \cdot \vec{y}\right)\left(\vec{b} \cdot \vec{z}\right)$

(d) $\vec{a} = -\left(\vec{a} \cdot \vec{y}\right)\left(\vec{z} - \vec{y}\right)$

5. From a point $P(\lambda, \lambda, \lambda)$, perpendicular PQ and PR are drawn respectively on the lines $y = x, z = 1$ and $y = -x, z = -1$. If P is such that $\angle QPR$ is a right angle, then the possible value(s) of λ is/(are)

(a) $\sqrt{2}$ (b) 1
(c) -1 (d) $-\sqrt{2}$

6. Let M and N be two 3×3 matrices such that $MN = NM$. Further, if $M \neq N^2$ and $M^2 = N^4$, then

(a) determinant of $(M^2 + MN^2)$ is 0
(b) there is 3×3 non-zero matrix U such that $(M^2 + MN^2)U$ is the zero matrix
(c) determinant of $(M^2 + MN^2) \ge 1$
(d) for a 3×3 matrix U, if $(M^2 + MN^2)U$ equals the zero matrix then U is the zero matrix

7. Let $f : (0, \infty) \to R$ be given by $f(x) \int_{\frac{1}{x}}^{x} e^{-\left(t+\frac{1}{t}\right)} \frac{dt}{t}$. Then

(a) $f(x)$ is monotonically increasing on $[1, \infty)$
(b) $f(x)$ is monotonically decreasing on $(0, 1)$
(c) $f(x) + f\left(\frac{1}{x}\right) = 0$, for all $x \in (0, \infty)$
(d) $f(2^x)$ is an odd function of x on R

8. Let $f: \left(-\frac{\pi}{2}, \frac{\pi}{2}\right) \to R$ be given by $f(x) = (\log(\sec x + \tan x))^3$.

Then

(a) $f(x)$ is an odd function

(b) $f(x)$ is non-one function

(c) $f(x)$ is an onto function

(d) $f(x)$ is an even function

9. A circle S passes through the point (0, 1) and is orthogonal to the circles $(x-1)^2 + y^2 = 16$ and $x^2 + y^2 = 1$. Then

(a) radius of S is 8

(b) radius of S is 7

(c) centre of S is $(-7, 1)$

(d) centre of S is $(-8, 1)$

10. Let $a \in R$ and let $f: R \to R$ be given by $f(x) = x^5 - 5x + a$. Then

(a) $f(x)$ has three real roots if $a > 4$

(b) $f(x)$ has only real root if $a > 4$

(c) $f(x)$ has three real roots if $a < -4$

(d) $f(x)$ has three real roots if $-4 < a < 4$

SECTION - II

This section contains **10 questions.** Each question, when worked out will result in one integer from 0 to 9 *(both inclusive)*.

11. The slope of the tangent to the curve $(y - x^5)^2 = x(1 + x^2)^2$ at the point (1, 3) is

12. Let $f: [0, 4\pi] \to [0, \pi]$ be defined by $f(x) = \cos^{-1}(\cos x)$. The number of points $x \in [0, 4\pi]$ satisfying the equation

$$f(x) = \frac{10 - x}{10} \text{ is}$$

13. The largest value of non-negative integer a for which

$$\lim_{x \to 1} \left\{ \frac{-ax + \sin(x-1) + a}{x + \sin(x-1) - 1} \right\}^{\frac{1-x}{1-\sqrt{x}}} = \frac{1}{4} \text{ is}$$

14. Let $f: R \to R$ and $g: R \to R$ be respectively given by $f(x) = |x| + 1$ and $g(x) = x^2 + 1$. Define $h: R \to R$ by

$$h(x) = \begin{cases} \max \{f(x), g(x)\} & \text{if } x \le 0, \\ \min \{f(x), g(x)\} & \text{if } x > 0. \end{cases}$$

The number of points at which $h(x)$ is not differentiable is

15. For a point P in the plane, let $d_1(P)$ and $d_2(P)$ be the distance of the point P form the lines $x - y = 0$ and $x + y = 0$ respectively. The area of the region R consisting of all points P lying in the first quadrant of the plane and satisfying $2 \le d_1(P) + d_2(P) \le 4$, is

16. Let $n_1 < n_2 < n_3 < n_4 < n_5$ be positive integers such that $n_1 + n_2 + n_3 + n_4 + n_5 = 20$. Then the number of such distinct arrangements $(n_1, n_2, n_3, n_4, n_5)$ is

17. The value of $\int_0^1 4x^3 \left\{ \frac{d^2}{dx^2}\left(1 - x^2\right)^5 \right\} dx$ is

18. Let $\vec{a}$, $\vec{b}$ and $\vec{c}$ be three non-coplanar unit vectors such that the angle between every pair of them is $\frac{\pi}{3}$. If $\vec{a} \times \vec{b} + \vec{b} \times \vec{c} = p\vec{a} + q\vec{b} + r\vec{c}$, where p, q and r are scalars, then the value of $\frac{p^2 + 2q^2 + r^2}{q^2}$ is

19. Let a, b, c be positive integers such that $\frac{b}{a}$ is an integer. If a, b, c are in geometric progression and the arithmetic mean of a, b, c is $b + 2$, then the value of $\frac{a^2 + a - 14}{a + 1}$ is

20. Let $n \ge 2$ be an integer. Take n distinct points on a circle and join each pair of points by a line segment. Colour the line segment joining every pair of adjacent points by blue and the rest by red. If the number of red and blue line segments are equal, then the value of n is

PAPER - 2

SECTION - I

This section contains **10 multiple choice questions**. Each question has four choices (a), (b), (c) and (d) out of which **ONLY ONE** option is correct.

1. Six cards and six envelopes are numbered 1, 2, 3, 4, 5, 6 and cards are to be placed in envelopes so that each envelope contains exactly one card and no card is placed in the envelope bearing the same number and moreover the card numbered 1 is always placed in envelope numbered 2. Then the number of ways it can be done is
(a) 264 (b) 265
(c) 53 (d) 67

2. In a triangle the sum of two sides is x and the product of the same sides is y. If $x^2 - c^2 = y$, where c is the third side of the triangle, then the ratio of the in radius to the circum-radius of the triangle is
(a) $\frac{3y}{2x(x+c)}$ (b) $\frac{3y}{2c(x+c)}$
(c) $\frac{3y}{4x(x+c)}$ (d) $\frac{3y}{4c(x+c)}$

3. The common tangents to the circle $x^2 + y^2 = 2$ and the parabola $y^2 = 8x$ touch the circle at the points P, Q and the parabola at the points R, S. Then the area of the quadrilateral $PQRS$ is
(a) 3 (b) 6
(c) 9 (d) 15

4. Three boys and two girls stand in a queue. The probability, that the number of boys ahead of every girl is at least one more than the number of girls ahead of her, is
(a) $\frac{1}{2}$ (b) $\frac{1}{3}$
(c) $\frac{2}{3}$ (d) $\frac{3}{4}$

5. The quadratic equation $p(x) = 0$ with real coefficients has purely imaginary roots. Then the equation $p(p(x)) = 0$ has
(a) one purely imaginary roots
(b) all real roots
(c) two real and two purely imaginary roots
(d) neither real nor purely imaginary roots

6. For $x \in (0, \pi)$, the equation $\sin x + 2\sin 2x - \sin 3x = 3$ has
(a) infinitely many solutions
(b) three solutions
(c) one solution
(d) no solution

7. The following integral $\int_{\frac{\pi}{4}}^{\frac{\pi}{2}} (2\operatorname{cosec} x)^{17}\, dx$ is equal to
(a) $\int_0^{\log(1+\sqrt{2})} 2\left(e^u + e^{-u}\right)^{16} du$
(b) $\int_0^{\log(1+\sqrt{2})} \left(e^u + e^{-u}\right)^{17} du$
(c) $\int_0^{\log(1+\sqrt{2})} \left(e^u - e^{-u}\right)^{17} du$
(d) $\int_0^{\log(1+\sqrt{2})} 2\left(e^u - e^{-u}\right)^{16} du$

8. Coefficient of x^{11} in the expansion of $(1 + x^2)^4(1 + x^3)^7(1 + x^4)^{12}$ is
(a) 1051 (b) 1106
(c) 1113 (d) 1120

9. Let $f : [0, 2] \to R$ be a function which is continuous on $[0, 2]$ and is differentiable on $(0, 2)$ with $f(0) = 1$. Let
$F(x) = \int_0^{x^2} f\left(\sqrt{t}\right) dt$ for $x \in [0,2]$. If $F'(x) = f'(x)$ for all $x \in (0,2)$, then $F(2)$ equals
(a) $e^2 - 1$ (b) $e^4 - 1$
(c) $e - 1$ (d) e^4

10. The function $y = f(x)$ is the solution of the differential equation
$\frac{dy}{dx} + \frac{xy}{x^2 - 1} = \frac{x^4 + 2x}{\sqrt{1 - x^2}}$ in $(-1, 1)$ satisfying $f(0) = 0$. Then
$\int_{-\frac{\sqrt{3}}{2}}^{\frac{\sqrt{3}}{2}} f(x)\, d(x)$ is
(a) $\frac{\pi}{3} - \frac{\sqrt{3}}{2}$ (b) $\frac{\pi}{3} - \frac{\sqrt{3}}{4}$
(c) $\frac{\pi}{6} - \frac{\sqrt{3}}{4}$ (d) $\frac{\pi}{6} - \frac{\sqrt{3}}{2}$

SECTION - II

This section contains 3 paragraphs, each describing theory, experiments, data etc. 6 questions related to the three paragraphs with two questions on each paragraph. Each question has only one correct answer among the four given options (a), (b), (c) and (d).

Paragraph for Questions 11 and 12

Box 1 contains three cards bearing numbers 1, 2, 3; box 2 contains five cards bearing numbers 1, 2, 3, 4, 5; and box 3 contains seven cards bearing numbers 1, 2, 3, 4, 5, 6, 7. A card is drawn from each of the boxes. Let x_i be number on the card drawn from the ith box, $i = 1, 2, 3$.

11. The probability that $x_1 + x_2 + x_3$ is odd, is

(a) $\frac{29}{105}$ (b) $\frac{53}{105}$

(c) $\frac{57}{105}$ (d) $\frac{1}{2}$

12. The probability that x_1, x_2, x_3 are in an arithmetic progression, is

(a) $\frac{9}{105}$ (b) $\frac{10}{105}$

(c) $\frac{11}{105}$ (d) $\frac{7}{105}$

Paragraph For Questions 13 and 14

Let a, r, s, t be nonzero real numbers. Let P $(at^2, 2at)$, Q, R $(ar^2, 2ar)$ and S $(as^2, 2as)$ be distinct points on the parabola $y^2 = 4ax$. Suppose that PQ is the focal chord and lines QR and PK are parallel, where K is the point $(2a, 0)$

13. The value of r is

(a) $-\frac{1}{t}$ (b) $\frac{t^2+1}{t}$

(c) $\frac{1}{t}$ (d) $\frac{t^2-1}{t}$

14. If $st = 1$, then the tangent at P and the normal at S to the parabola meet at a point whose ordinate is

(a) $\frac{\left(t^2+1\right)^2}{2t^3}$ (b) $\frac{a\left(t^2+1\right)^2}{2t^3}$

(c) $\frac{a\left(t^2+1\right)^2}{t^3}$ (d) $\frac{a\left(t^2+2\right)^2}{t^3}$

Paragraph For Questions 15 and 16

Given that for each $a \in (0, 1)$, $\lim\limits_{h \to 0^+} \int_h^{1-h} t^{-a}(1-t)^{a-1}\,dt$ exists. Let this limit be $g(a)$. In addition, it is given that the function $g(a)$ is differentiable on $(0, 1)$.

15. The value of $g\left(\frac{1}{2}\right)$ is

(a) π (b) 2π

(c) $\frac{\pi}{2}$ (d) $\frac{\pi}{4}$

16. The value of $g'\left(\frac{1}{2}\right)$ is

(a) $\frac{\pi}{2}$ (b) π

(c) $-\frac{\pi}{2}$ (d) 0

SECTION - III

This section contains 4 questions, each having two matching lists. Choice for the correct combination of elements from List-I and List-II are given as options (a), (b), (c) and (d), out of which one is correct.

17.

	List - I		List - II
P.	The number of polynomials $f(x)$ with non-negative integer coefficients of degree ≤ 2, satisfying $f(0) = 0$ and $\int_0^1 f(x)\,dx = 1$, is	**1.**	8
Q.	The number of points in the interval $\left[-\sqrt{13}, \sqrt{13}\right]$ at which $f(x) = \sin(x^2) + \cos(x^2)$ attains its maximum value, is	**2.**	2
R.	$\int_{-2}^{2} \frac{3x^2}{\left(1+e^x\right)}\,dx$ equals	**3.**	4
S.	$\frac{\left(\int_{-\frac{1}{2}}^{\frac{1}{2}} \cos 2x \log\left(\frac{1+x}{1-x}\right) dx\right)}{\left(\int_{0}^{\frac{1}{2}} \cos 2x \log\left(\frac{1+x}{1-x}\right) dx\right)}$	**4.**	0

	P	Q	R	S
(a)	3	2	4	1
(b)	2	3	4	1
(c)	3	2	1	4
(d)	2	3	1	4

18. **List - I** **List - II**

P. Let $y(x)=\cos\left(3\cos^{-1}x\right)$, $x\in[-1,1]$, $x\neq\pm\frac{\sqrt{3}}{2}$. Then $\frac{1}{y(x)}\left\{\left(x^2-1\right)\frac{d^2y(x)}{dx^2}+x\frac{dy(x)}{dx}\right\}$ equals — **1.** 1

Q. Let $A_1, A_2, ..., A_n\ (n>2)$ be the vertices of a regular polygon of n sides with its centre at the origin. Let $\vec{a_k}$ be the position vector of the point A_k, $k = 1, 2, ..., n$. If $\left|\sum_{k=1}^{n-1}\left(\vec{a_k}\times\vec{a_{k+1}}\right)\right| = \left|\sum_{k=1}^{n-1}\left(\vec{a_k}.\vec{a_{k+1}}\right)\right|$, then the minimum value of n is — **2.** 2

R. If the normal from the point $P(h, 1)$ on the ellipse $\frac{x^2}{6}+\frac{y^2}{3}=1$ is perpendicular to the line $x+y=8$, then the value of h is — **3.** 8

S. Number of positive solutions satisfying the equation $\tan^{-1}\left(\frac{1}{2x+1}\right)+\tan^{-1}\left(\frac{1}{4x+1}\right)=\tan^{-1}\left(\frac{2}{x^2}\right)$ is — **4.** 9

	P	Q	R	S
(a)	4	3	2	1
(b)	2	4	3	1
(c)	4	3	1	2
(d)	2	4	1	3

19. Let $f_1: R\to R$, $f_2:[0,\infty)\to R$, $f_3: R\to R$ and $f_4: R\to[0,\infty)$ be defined by $f_1(x)=\begin{cases}|x| & \text{if } x<0,\\ e^x & \text{if } x\geq 0;\end{cases}$

$f_2(x)=x^2$; $f_3(x)=\begin{cases}\sin x & \text{if } x<0,\\ x & \text{if } x\geq 0;\end{cases}$

and $f_4(x)=\begin{cases}f_2(f_1(x)) & \text{if } x<0,\\ f_2(f_1(x))-1 & \text{if } x\geq 0.\end{cases}$

	List-I		List-II
P.	f_4 is	**1.**	Onto but not one-one
Q.	f_3 is	**2.**	Neither continuous nor one-one
R.	$f_2 o f_1$ is	**3.**	Differentiable but not one-one
S.	f_2 is	**4.**	Continuous and one-one

	P	Q	R	S
(a)	3	1	4	2
(b)	1	3	4	2
(c)	3	1	2	4
(d)	1	3	2	4

20. Let $z_k=\cos\left(\frac{2k\pi}{10}\right)+i\sin\left(\frac{2k\pi}{10}\right)$; $k = 1, 2, ..., 9$..

	List-I		List-II
P.	For each z_k there exists as z_j such that $z_k.\ z_j=1$	**1.**	True
Q.	There exists a $k\in\{1,2,...,9\}$ such that $z_1.z=z_k$ has no solution z in the set of complex numbers	**2.**	False
R.	$\frac{\lvert 1-z_1\rvert\lvert 1-z_2\rvert....\lvert 1-z_9\rvert}{10}$ equals	**3.**	1
S.	$1-\sum_{k=1}^{9}\cos\left(\frac{2k\pi}{10}\right)$ equals	**4.**	2

	P	Q	R	S
(a)	1	2	4	3
(b)	2	1	3	4
(c)	1	2	3	4
(d)	2	1	4	3

SOLUTIONS

Paper - 1

1. (a,c) $g(x)$ may be discontinuous at $x = a$ or $x = b$.
Let us check the continuity of $g(x)$ at $x = a$ and $x = b$

$$\lim_{x\to a^-} g(x) = 0$$

$$\lim_{x\to a^+} g(x) = \lim_{x\to a^+} \int_a^x f(t)\,dt = \int_a^a f(t)\,dt = 0$$

$$g(a) = \int_a^a f(t)\,dt = 0$$

$\therefore$ $g(x)$ is continuous at $x = a$

Also $\lim_{x\to b^-} g(x) = \lim_{x\to b^-} \int_a^x f(t)\,dt = \int_a^b f(t)\,dt$

$$\lim_{x\to b^+} g(x) = \int_a^b f(t)\,dt$$

$$g(b) = \int_a^b f(t)\,dt$$

$\therefore$ $g(x)$ is continuous at $x = b$
Hence $g(x)$ is continuous $\forall x \in R$

Now $g'(x) = \begin{cases} 0, & x < a \\ f(x), & a \le x \le b \\ 0, & x > b \end{cases}$

$g'(a^-) = 0$ and $g'(a^+) = f(a)$

$g'(b^-) = f(b)$ and $g'(b^+) = 0$

As $f(a), f(b) \in [1, \infty)$ $\therefore f(a), f(b) \ne 0$

Hence $g'(a^-) \ne g'(a^+)$ and $g'(b^-) \ne g'(b^+)$

$\therefore$ g is not differentiable at a and b.

2. (a, d) Let f and g be maximum at c_1 and c_2 respectively,
$c_1, c_2 \in (0,1)$
Then, $f(c_1) = g(c_2)$
Let $h(x) = f(x) - g(x)$
Then, $h(c_1) = f(c_1) - g(c_1) > 0$
and $h(c_2) = f(c_2) - g(c_2) < 0$

$\therefore$ $h(x) = 0$ has atleast one root in (c_1, c_2)
$C \in (c_1, c_2)$
i.e. for
$f(c) = g(c)$
which gives A and D are correct options.

3. (c,d) Let $M = \begin{bmatrix} a & b \\ b & c \end{bmatrix}$ where a, b, c are integers.

M is invertible if $\begin{vmatrix} a & b \\ b & c \end{vmatrix} \ne 0 \Rightarrow ac \ne b^2$

Then $\begin{bmatrix} a \\ b \end{bmatrix} = \begin{bmatrix} b \\ c \end{bmatrix} \Rightarrow a = b = c \Rightarrow ac = b^2$.

$\therefore$ (a) is not correct.

If $\begin{bmatrix} b & c \end{bmatrix} = \begin{bmatrix} a & b \end{bmatrix} \Rightarrow b = a = c$
$\Rightarrow ac = b^2$
$\therefore$ (b) is not correct.

If $M = \begin{bmatrix} a & 0 \\ 0 & c \end{bmatrix}$, then $|M| = ac \ne 0$
$\therefore$ M is invertible.
(c) is correct

As $ac \ne (\text{integer})^2 \Rightarrow ac \ne b^2$
$\therefore$ (d) is correct.

4. (a, b, c) $|\vec{x}| = |\vec{y}| = |\vec{z}| = \sqrt{2}$

Angle between each pair is $\frac{\pi}{3}$

$$\vec{a} = \lambda\left[\vec{x} \times \left(\vec{y} \times \vec{z}\right)\right]$$

$$= \lambda\left[\left(\vec{x}\cdot\vec{z}\right)\vec{y} - \left(\vec{x}\cdot\vec{y}\right)\vec{z}\right]$$

$$= \lambda\left[\left(\sqrt{2}.\sqrt{2}\cos\frac{\pi}{3}\right)\vec{y} - \left(\sqrt{2}.\sqrt{2}\cos\frac{\pi}{3}\right)\vec{z}\right]$$

$$= \lambda\left(\vec{y} - \vec{z}\right)$$

$$\vec{b} = \mu\left[\vec{y} \times \left(\vec{z} \times \vec{x}\right)\right]$$

$$= \mu\left[\left(\vec{y}.\vec{x}\right)\vec{z} - \left(\vec{y}.\vec{z}\right)\vec{x}\right]$$

$$= \mu\left[\left(\sqrt{2}.\sqrt{2}.\cos\frac{\pi}{3}\right)\vec{z} - \left(\sqrt{2}.\sqrt{2}.\cos\frac{\pi}{3}\right)\vec{x}\right]$$

$$= \mu\left(\vec{z} - \vec{x}\right)$$

Now $\vec{b}.\vec{z} = \mu\left[\vec{z}.\vec{z} - \vec{x}.\vec{z}\right]$

$= \mu(2-1) = \mu$

$\therefore \vec{b} = \left(\vec{b}.\vec{z}\right)\left(\vec{z} - \vec{x}\right)$ is correct

Also $\vec{a}.\vec{y} = \lambda\left(\vec{y}.\vec{y} - \vec{z}.\vec{y}\right) = \lambda(2-1) = \lambda$

$\therefore \vec{a} = \left(\vec{a}.\vec{y}\right)\left(\vec{y} - \vec{z}\right)$ is also correct

$$\vec{a}.\vec{b} = \lambda\mu\left(\vec{y}.\vec{z} - \vec{y}.\vec{x} - \vec{z}.\vec{z} + \vec{z}.\vec{x}\right)$$

$$= \lambda\mu(1-1-2+1) = -\lambda\mu = -\left(\vec{a}.\vec{y}\right)\left(\vec{b}.\vec{z}\right)$$

$\therefore$ (c) is correct.

$$-\left(\vec{a}.\vec{y}\right)\left(\vec{z} - \vec{y}\right) = \lambda\left(\vec{z} - \vec{y}\right) = -\vec{a}$$

(d) is not correct.

5. (c) Lines are $x = y, z = 1$

or $\dfrac{x-0}{1} = \dfrac{y-0}{1} = \dfrac{z-1}{0} = \alpha$ ...(1)

and $y = -x, z = -1$

or $\dfrac{x-0}{-1} = \dfrac{y-0}{1} = \dfrac{z+1}{0} = \beta$ ...(2)

Let $Q(\alpha, \alpha, 1)$ and $R(-\beta, \beta, -1)$

Direction ratios of PQ are $\lambda - \alpha, \lambda - \alpha, \lambda - 1$

and direction ratios of PR are $\lambda + \beta, \lambda - \beta, \lambda + 1$

$\because PQ$ is perpendicular to line (1)

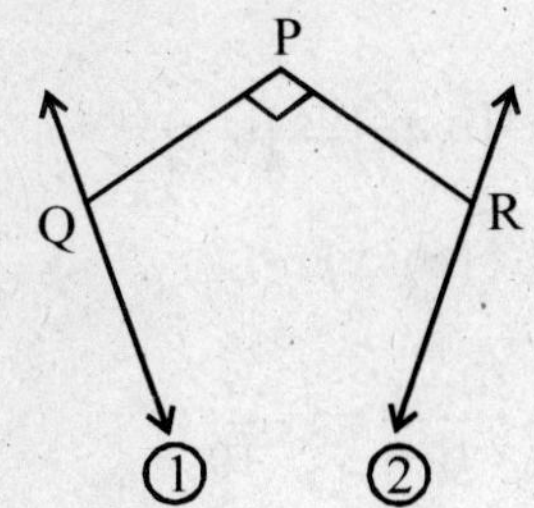

$\therefore -(\lambda + \beta) + \lambda - \beta = 0$

$\Rightarrow \beta = 0$

$\therefore$ dr's of PQ are $0, 0, \lambda - 1$

and dr's of PR are $\lambda, \lambda, \lambda + 1$

$\because \angle QPR = 90° \Rightarrow (\lambda - 1)(\lambda + 1) = 0 \Rightarrow \lambda = 1$ or -1

But for $\lambda = 1$, we get point Q itself

$\therefore$ we take $\lambda = -1$

6. (a,b) Given $MN = NM$, $M \neq N^2$ and $M^2 = N^4$.

Then $M^2 = N^4 \Rightarrow (M + N^2)(M - N^2) = 0$

$\Rightarrow$ (i) $M + N^2 = 0$ and $M - N^2 \neq 0$

(ii) $|M + N^2| = 0$ and $|M - N^2| = 0$

In each case $|M + N^2| = 0$

$\therefore |M^2 + MN^2| = |M|\,|M + N^2| = 0$

$\therefore$ (a) is correct and (c) is not correct.

Also we know if $|A| = 0$, then there can be many matrices U, such that $AU = 0$

$\therefore (M^2 + MN^2)U = 0$ will be true for many values of U.

Hence (b) is correct.

Again if $AX = 0$ and $|A| = 0$, then X can be non-zero.

$\therefore$ (d) is not correct.

7. (a, c, d) $f(x) = \displaystyle\int_{1/x}^{x} e^{-\left(t + \frac{1}{t}\right)} \frac{dt}{t}$

$$\therefore f'(x) = \frac{e^{-\left(x + \frac{1}{x}\right)}}{x} + \frac{x}{x^2} e^{-\left(\frac{1}{x} + x\right)} = \frac{2}{x} e^{-\left(x + \frac{1}{x}\right)}$$

For $x \in [1, \infty)$, $f'(x) > 0$

$\therefore f$ is monotonically increasing on $[1, \infty)$

(a) is correct.

For $x \in (0, 1)$, $f'(x) > 0$

$\therefore$ (b) is not correct

$$f(x) + f\left(\frac{1}{x}\right) = \int_{1/x}^{x} e^{-\left(t + \frac{1}{t}\right)} \frac{dt}{t} + \int_{x}^{1/x} e^{-\left(t + \frac{1}{t}\right)} \frac{dt}{t} = 0$$

$\therefore$ (c) is not correct.

Replacing x by 2^x in $f(x) + f\left(\dfrac{1}{x}\right) = 0$

We get $f(2^x) + f\left(2^{-x}\right) = 0$

or $f(2^x) = -f(2^{-x})$

$\therefore$ $f(2^x)$ is an odd function.

$\therefore$ (d) is correct.

8. (a, b, c) $f:\left[-\frac{\pi}{2},\frac{\pi}{2}\right]\longrightarrow R$

$f(x)=[\log(\sec x+\tan x)]^3$
$f(-x)=[\log(\sec x-\tan x)]^3$

$$=\left[\log\left(\frac{(\sec x-\tan x)(\sec x+\tan x)}{\sec x+\tan x}\right)\right]^3$$

$$=\left[\log\left(\frac{1}{\sec x+\tan x}\right)\right]^3=\left[-\log(\sec x+\tan x)\right]^3$$

$$=-\left[\log(\sec x+\tan x)\right]^3=-f(x)$$

$\therefore f$ is an odd function.
(a) is correct and (d) is not correct.
Also

$$f'(x)=3\left[\log(\sec x+\tan x)\right]^2.\frac{\sec x\tan x+\sec^2 x}{\sec x+\tan x}$$

$$=3\sec x\left[\log(\sec x+\tan x)\right]^2>0\ \forall x\in\left(-\frac{\pi}{2},\frac{\pi}{2}\right)$$

$\therefore f$ is increasing on $\left(-\frac{\pi}{2},\frac{\pi}{2}\right)$

We know that strictly increasing function is one one.
$\therefore f$ is one one
$\therefore$ (b) is correct.

Also $\lim\limits_{x\to\frac{\pi}{2}^-}\left[\log(\sec x+\tan x)\right]^3\to\infty$

and $\lim\limits_{x\to\frac{\pi}{2}^+}\left[\log(\sec x+\tan x)\right]^3\to-\infty$

$\therefore$ Range of $f=(-\infty,\infty)=R$
$\therefore f$ is an onto function.
$\therefore$ (c) is correct.

9. (b, c) Let the equation of circle be
$x^2+y^2+2gx+2fy+c=0$
It passes through (0, 1)
$\therefore 1+2f+c=0$ ...(i)
This circle is orthogonal to $(x-1)^2+y^2=16$
i.e. $x^2+y^2-2x-15=0$
and $x^2+y^2-1=0$
$\therefore$ We should have
$2g(-1)+2f(0)=c-15$
or $2g+c-15=0$ ...(ii)
and $2g(0)+2f(0)=c-1$
or $c=1$ ...(iii)
Solving (i), (ii) and (iii), we get
$c=1, g=7, f=-1$

$\therefore$ Required circle is
$x^2+y^2+14x-2y+1=0$
With centre $(-7, 1)$ and radius = 7
$\therefore$ (b) and (c) are correct options.

10. (b,d) $f(x)=x^5-5x+a$

$f(x)=0\Rightarrow x^5-5x+a=0$
$\Rightarrow a=5x-x^5=g(x)$
$\Rightarrow g(x)=0$ when $x=0,\ 5^{1/4}\ -5^{1/4}$

and $g'(x)=0\Rightarrow x=1,-1$
Also $g(-1)=-4$ and $g(1)=4$
$\therefore$ graph of $g(x)$ will be as shown below.
From graph

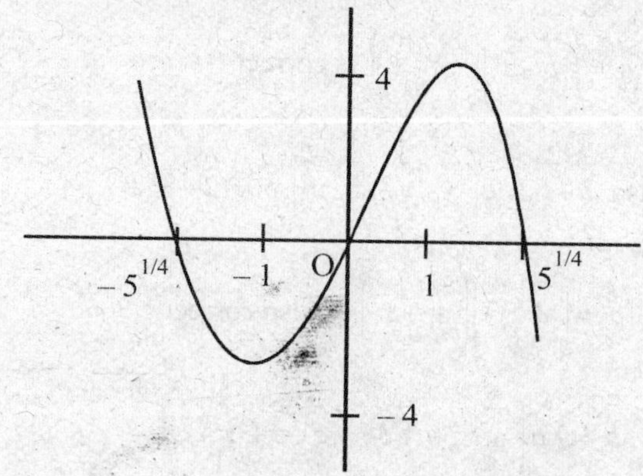

if $a\in(-4,-4)$
then $g(x)=a$ or
$f(x)=0$ has 3 real roots
If $a>4$ or $a<-4$
then $f(x)=0$ has only one real root.
$\therefore$ (b) and (d) are the correct options.

11. (8) $\left(y-x^5\right)^2=x\left(1+x^2\right)^2$

$$2\left(y-x^5\right)\left(\frac{dy}{dx}-5x^4\right)=\left(1+x^2\right)^2+2x\left(1+x^2\right).2x$$

At point (1, 3)

$$2(3-1)\left(\frac{dy}{dx}-5\right)=(1+1)^2+2(1+1).2$$

$$\Rightarrow\frac{dy}{dx}=8$$

12. (3) We have $f:[0,4\pi]\to[0,\pi]$

$f(x)=\cos^{-1}(\cos x)$

and $g(x)=\frac{10-x}{10}=1-\frac{x}{10}$

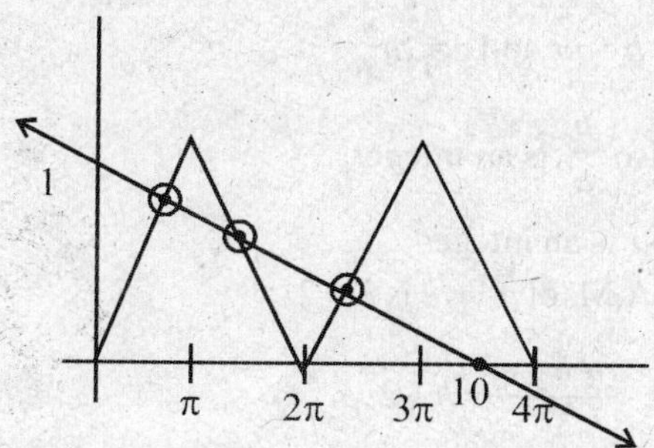

The graph of $y = f(x)$ and $y = g(x)$ are as follows.
Clearly $f(x) = g(x)$ has 3 solutions.

13. (2) $\lim_{x\to 1}\left\{\frac{-ax+\sin(x-1)+a}{x+\sin(x-1)-1}\right\}^{\frac{1-x}{1-\sqrt{x}}} = \frac{1}{4}$

$$\Rightarrow \lim_{x\to 1}\left\{\frac{a(1-x)+\sin(x-1)}{(x-1)+\sin(x-1)}\right\}^{1+\sqrt{x}}$$

$$\Rightarrow \lim_{x\to 1}\left\{\frac{-a+\frac{\sin(x-1)}{x-1}}{1+\frac{\sin(x-1)}{x-1}}\right\}^{1+\sqrt{x}}$$

$$\Rightarrow \left(\frac{-a+1}{2}\right)^2 = \frac{1}{4}$$

$\Rightarrow a = 0$ or 2

∴ Largest value of a is 2.

14. (3) $f(x) = |x|+1 = \begin{cases} x+1, & x \ge 0 \\ -x+1, & x < 0 \end{cases}$

$g(x) = x^2 + 1$

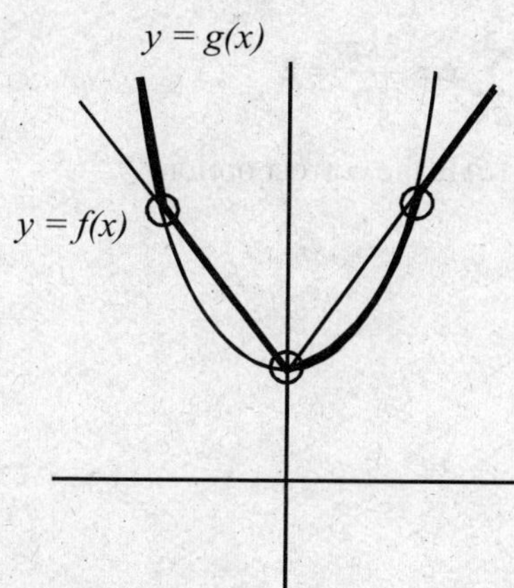

From graph there are 3 points at which $h(x)$ is not differentiable.

15. (6) Let the point P be (x, y)

Then $d_1(P) = \left|\frac{x-y}{\sqrt{2}}\right|$ and $d_2(P) = \left|\frac{x+y}{\sqrt{2}}\right|$

For P lying in first quadrant $x > 0, y > 0$.
Also $2 \le d_1(P) + d_2(P) \le 4$

$$\Rightarrow 2 \le \left|\frac{x-y}{\sqrt{2}}\right| + \left|\frac{x+y}{\sqrt{2}}\right| \le 4$$

If $x > y$, then $2 \le \frac{x-y+x+y}{\sqrt{2}} \le 4$

or $\sqrt{2} \le x \le 2\sqrt{2}$

If $x < y$, then

$$2 \le \frac{y-x+x+y}{\sqrt{2}} \le 4$$

or $\sqrt{2} \le y \le 2\sqrt{2}$

The required region is the shaded region in the figure given below.

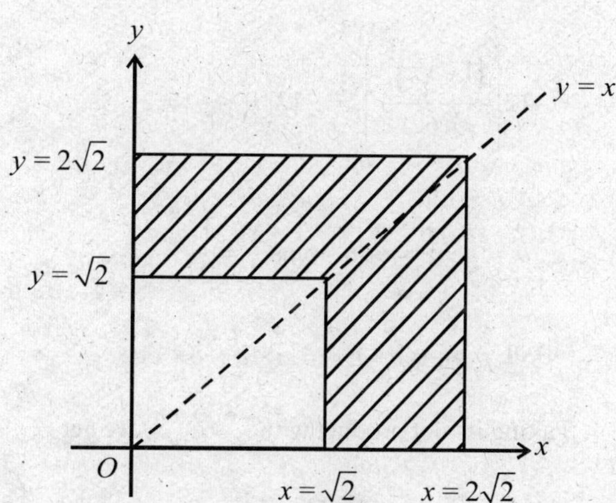

∴ Required area $= (2\sqrt{2})^2 = (\sqrt{2})^2 = 8 - 2 = 6$ sq units.

16. (7) ∵ n_1, n_2, n_3, n_4 and n_5 are positive integers such that $n_1 < n_2 < n_3 < n_4 < n_5$

Then for $n_1 + n_2 + n_3 + n_4 + n_5 = 20$

If n_1, n_2, n_3, n_4 take minimum values 1, 2, 3, 4 respectively then n_5 will be maximum 10.

∴ Corresponding to $n_5 = 10$, there is only one solution $n_1 = 1, n_2 = 2, n_3 = 3, n_4 = 4$.

Corresponding to $n_5 = 9$, we can have, only solution $n_1 = 1, n_2 = 2, n_3 = 3, n_4 = 5$ i.e., one solution

Corresponding to $n_5 = 8$, we can have, only solution
$n_1 = 1, n_2 = 2, n_3 = 3, n_4 = 6$
or $n_1 = 1, n_2 = 2, n_3 = 4, n_4 = 4$
i.e., 2 solution

For $n_5 = 7$, we can have
$n_1 = 1, n_2 = 1, n_3 = 4, n_4 = 6$
or $n_1 = 1, n_2 = 3, n_3 = 4, n_4 = 5$ *i.e.* 2 solutions
For $n_5 = 6$, we can have
$n_1 = 2, n_2 = 3, n_3 = 4, n_4 = 5$
i.e., one solution
Thus there can be 7 solutions.

17. (2) $\int_0^1 4x^3 \left[\frac{d^2}{dx^2}(1-x^2)^5\right] dx$

$$= 4x^3\left[\frac{d}{dx}(1-x^2)^5\right]\Big|_0^1 - \int_0^1\left[\frac{d}{dx}(1-x^2)^5\right].12x^2dx$$

$$= -12x^2(1-x^2)^5\Big|_0^1 + \int_0^1 (1-x^2)^5 .24xdx$$

$$= -12\int_0^1 (1-x^2)^5 .(-2x)dx$$

$$= -12\left(\frac{(1-x^2)^6}{6}\right)_0^1 = -12\left(0-\frac{1}{6}\right) = 2$$

18. (4) $\vec{a}.\vec{b} = \vec{b}.\vec{c} = \vec{c}.\vec{a} = \cos\frac{\pi}{3} = \frac{1}{2}$

Given $p\vec{a} + q\vec{b} + r\vec{c} = \vec{a}\times\vec{b} + \vec{b}\times\vec{c}$

Taking its dot product with $\vec{a}, \vec{b}, \vec{c}$, we get

$$p + \frac{1}{2}q + \frac{1}{2}r = \left[\vec{a}\ \ \vec{b}\ \ \vec{c}\right] \quad ...(1)$$

$$\frac{1}{2}p + q + \frac{1}{2}r = 0 \quad ...(2)$$

$$\frac{1}{2}p + \frac{1}{2}q + r = \left[\vec{a}\ \ \vec{b}\ \ \vec{c}\right] \quad ...(3)$$

From (1) and (3), $p = r$ Using (2) $q = -p$

$$\therefore \frac{p^2+2q^2+r^2}{q^2} = \frac{p^2+2p^2+p^2}{p^2} = 4$$

19. (4) $\because a, b, c$ are in G.P
$\therefore b = ar$ and $c = ar^2$

Also $\frac{b}{a}$ is an integer

$\Rightarrow r$ is an integer
$\because$ A.M. of a, b, c is $b + 2$

$$\Rightarrow \frac{a+b+c}{3} = b+2$$

$$\Rightarrow \quad a + ar + ar^2 = 3ar + 6$$

$$\Rightarrow \quad a(r^2 - 2r + 1) = 6$$

$$\Rightarrow \quad a(r-1)^2 = 6$$

$\because a$ and r are integers
$\therefore$ The only possible values of a and r can be 6 and 2 respectively.

Then $\frac{a^2+a-14}{a+1} = \frac{36+6-14}{6+1} = \frac{28}{7} = 4$

20. (5) Number of adjacent lines $= n$
Number of non adjacent lines $= {}^nC_2 - n$

$$\therefore {}^nC_2 - n = n \Rightarrow \frac{n(n-1)}{2} - 2n = 0$$

$$\Rightarrow n^2 - 5n = 0 \Rightarrow n = 0 \text{ or } 5$$

But $n \geq 2 \Rightarrow n = 5$
S(4) : $1, Z_1, Z_2, ... Z_9$ are 10th roots of unity.
$\therefore Z^{10} - 1 = 0$
From equation $1 + Z_1 + Z_2 + ... + Z_9 = 0$
$\text{Re}(1) + \text{Re}(Z_1) + \text{Re}(Z_2) + ... + \text{Re}(Z_9) = 0$
$\Rightarrow \text{Re}(Z_1) + \text{Re}(Z_2) + ... \text{Re}(Z_9) = -1$

$$\Rightarrow \sum_{K=1}^{9}\cos\frac{2k\pi}{10} = -1$$

$$\Rightarrow 1 - \sum_{K=1}^{9}\cos\frac{2k\pi}{10} = 2$$

Hence (c) is the correct option.

Paper - 2

1. (c) $\because$ Card numbered 1 is always placed in envelope numbered 2, we can consider two cases.

Case I: Card numbered 2 is placed in envelope numbered 1.

Then it is dearrangement of 4 objects, which can be done in

$$4!\left(1-\frac{1}{1!}+\frac{1}{2!}-\frac{1}{3!}+\frac{1}{4!}\right) = 9 \text{ ways}$$

Case II: Card numbered 2 is not placed in envelope numbered 1.

Then it is dearrangement of 5 objects, which can be done in

$$5!\left(1-\frac{1}{1!}+\frac{1}{2!}-\frac{1}{3!}+\frac{1}{4!}-\frac{1}{5!}\right) = 44 \text{ ways}$$

$\therefore$ Total ways $= 44 + 9 = 53$

2. (b) Let two sides of Δ be a and b.

Then $a + b = x$ and $ab = y$

Also given $x^2 - c^2 = y$, where c is the third side of Δ.

$$\Rightarrow (a+b)^2 - c^2 = ab \Rightarrow a^2+b^2-c^2 = -ab$$

$$\Rightarrow \frac{a^2+b^2-c^2}{2ab} = -\frac{1}{2} \Rightarrow \cos c = -\frac{1}{2} \Rightarrow c = 120°$$

$$\therefore \frac{r}{R} = \frac{\Delta}{s}\times\frac{4\Delta}{abc} \text{ where } \Delta = \text{area of triangle}$$

$$\Rightarrow \frac{r}{R} = \frac{4\Delta^2}{\frac{(a+b+c)}{2}abc} = \frac{8\times\left(\frac{1}{2}ab\sin c\right)^2}{(a+b+c)abc}$$

$$\Rightarrow \frac{2a^2b^2\sin^2 120°}{(a+b+c)abc} = \frac{2ab\times\frac{3}{4}}{(x+c)c} = \frac{3y}{2c(x+c)}$$

3. (d)

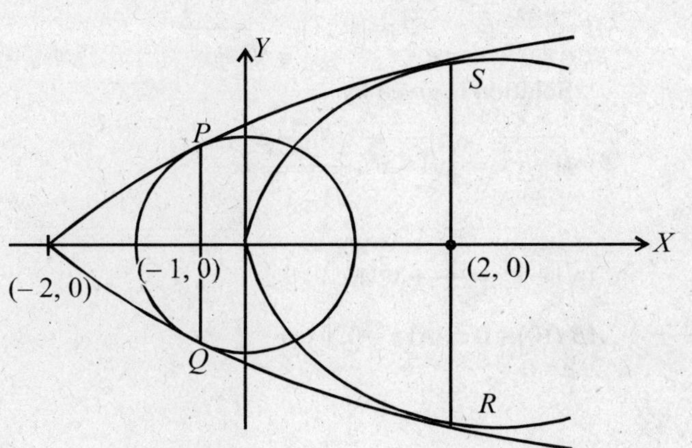

Let the tangent to

$y^2 = 8x$ be

$$y = mx + \frac{2}{m}$$

If it is common tangent to parabola and circle, then

$y = mx + \frac{2}{m}$ is a tangent to $x^2 + y^2 = 2$

$$\therefore \left|\frac{\frac{2}{m}}{\sqrt{m^2+1}}\right| = \sqrt{2} \Rightarrow \frac{4}{m^2(1+m^2)} = 2$$

$\Rightarrow m^4 + m^2 - 2 = 0$

$\Rightarrow (m^2 + 2)(m^2 - 1) = 0$

$\Rightarrow m = 1$ or -1

$\therefore$ Required tangents are $y = x + 2$ and $y = -x - 2$

Their common point is $(-2, 0)$

$\therefore$ Tangents are drawn from $(-2, 0)$

$\therefore$ Chord of contact PQ to circle is

$x.(-2) + y.0 = 2$ or $x = -1$

and Chord of contact RS to parabola is

$y.0 = 4(x - 2)$ or $x = 2$

Hence coordinates of P and Q are

$(-1, 1)$ and $(-1, -1)$

Also coordinates of R and S are

$(2, -4)$ and $(2, 4)$

$\therefore$ Area of trapezium $PQRS$ is $\frac{1}{2}(2+8)\times 3 = 15$

4. (a) According to given condition, we can have the following cases

(I) GGBBB

(II) BGGBB

(III) GBGBB

(IV) BGBGB

(V) GBBGB

i.e., the two girls can occupy two of the first three places (case I, II, III) or second and fourth (case IV) or first and fourth (case V) places.

Thus favourable cases are

$= 3 \times 2! \times 3! + 2 \times 2! \times 3! = 60$

Total ways in which 5 persons can be seated $= 5! = 120$

$$\therefore \text{Required probability} = \frac{60}{120} = \frac{1}{2}$$

5. (d) Quadratic equation with real coefficients and purely imaginary roots can be considered as

$p(x) = x^2 + a = 0$ where $a > 0$ and $a \in R$

The $p[p(x)] = 0$

$\Rightarrow (x^2 + a)^2 + a = 0$

$\Rightarrow x^4 + 2ax^2 + (a^2 + a) = 0$

$\Rightarrow x^2 = \dfrac{-2a \pm \sqrt{4a^2 - 4a^2 - 4a}}{2}$

$\Rightarrow x^2 = -a \pm \sqrt{a}\, i$

$\Rightarrow x = \sqrt{-a \pm \sqrt{a}\, i} = \alpha \pm i\beta$ where $\alpha, \beta \neq 0$

$\therefore p[p(x)] = 0$ has complex roots which are neither purely real nor purely imaginary.

6. (d) $\sin x + 2\sin 2x - \sin 3x = 3$

$\Rightarrow \sin x + 4\sin x \cos x - 3\sin x + 4\sin^3 x = 3$

$\Rightarrow \sin x\,(-2 + 2\cos x + 4\sin^2 x) = 3$

$\Rightarrow \sin x\,(-2 + 2\cos x + 4 - 4\cos^2 x) = 3$

$\Rightarrow 2 + 2\cos x - 4\cos^2 x = \dfrac{3}{\sin x}$

$\Rightarrow 2 - \left(4\cos^2 x - 2.2\cos x.\dfrac{1}{2} + \dfrac{1}{4}\right) + \dfrac{1}{4} = \dfrac{3}{\sin x}$

$\Rightarrow \dfrac{9}{4} - \left(2\cos x - \dfrac{1}{2}\right)^2 = \dfrac{3}{\sin x}$

LHS $\leq \dfrac{9}{4}$ and RHS ≥ 3

$\therefore$ Equation has no solution.

7. (a) Let $I = \displaystyle\int_{\pi/4}^{\pi/2} (2\operatorname{cosec} x)^{17}\, dx$

$= \displaystyle\int_{\pi/4}^{\pi/2} (\operatorname{cosec} x + \cot x + \operatorname{cosec} x - \cot x)^{16}\, 2\operatorname{cosec} x\, dx$

$I = 2\displaystyle\int_{\pi/4}^{\pi/2} \left(\operatorname{cosec} x + \cot x + \frac{1}{\operatorname{cosec} x + \cot x}\right)^{16} . \operatorname{cosec} x\, dx$

Let $\operatorname{cosec} x + \cot x = e^u$

$\Rightarrow \left(-\operatorname{cosec} x \cot x - \operatorname{cosec}^2 x\right) dx = e^u du$

$\Rightarrow -\operatorname{cosec} x\, dx = du$

Also at $x = \dfrac{\pi}{4},\ u = ln\left(\sqrt{2} + 1\right)$

at $x = \dfrac{\pi}{2}, u = ln\, 1 = 0$

$\therefore I = -2 \displaystyle\int_{ln(\sqrt{2}+1)}^{0} \left(e^u + e^{-u}\right)^{16} du$

$= 2\displaystyle\int_{0}^{ln(\sqrt{2}+1)} \left(e^u + e^{-u}\right)^{16} du$

8. (c) Coeff. of x^{11} in exp. of $\left(1+x^2\right)^4 \left(1+x^3\right)^7 \left(1+x^4\right)^{12}$

= (Coeff. of x^a) × (Coeff. of x^b) × (Coeff. of x^c)

Such that $a + b + c = 11$

Here $a = 2m,\ b = 3n,\ c = 4p$

$\therefore 2m + 3n + 4p = 11$

Case I : $m = 0, n = 1, p = 2$

Case II : $m = 1, n = 3, p = 0$

Case III : $m = 2, n = 1, p = 1$

Case IV : $m = 4, n = 1, p = 0$

$\therefore$ Required coeff.

$= {}^4C_0 \times {}^7C_1 \times {}^{12}C_2 + {}^4C_1 \times {}^7C_3 \times {}^{12}C_0$

$+ {}^4C_2 \times {}^7C_1 \times {}^{12}C_1 + {}^4C_4 \times {}^7C_1 \times {}^{12}C_0$

$= 462 + 140 + 504 + 7 = 1113$

9. (b) $F(x) = \displaystyle\int_0^{x^2} f\left(\sqrt{t}\right) dt$ for $x \in [0, 2]$

$\Rightarrow F'(x) = f(x).2x$

Now $F'(x) = f'(x)\ \forall x \in (0, 2)$

$\Rightarrow f(x).2x = f'(x)$

$\Rightarrow \dfrac{f'(x)}{f(x)} = 2x$

$\Rightarrow ln f(x) = x^2 + c$

$\Rightarrow f(x) = e^{x^2 + c} = e^{x^2}.\, e^c$

As $f(0) = 1 \Rightarrow 1 = e^c$

$\therefore f(x) = e^{x^2}$

So $F(x) = \displaystyle\int_0^{x^2} e^x dx = e^{x^2} - 1$

$\therefore F(2) = e^4 - 1$

10. (b) Given D.E. can be written as

$\dfrac{dy}{dx} - \dfrac{x}{1-x^2} y = \dfrac{x^4 + 2x}{\sqrt{1-x^2}}$

If $e^{\int \frac{-x}{1-x^2} dx} = e^{\frac{+1}{2}\log\left(1-x^2\right)} = \sqrt{1-x^2}$

$\therefore$ Solution is given by

$y\sqrt{1-x^2} = \displaystyle\int \sqrt{1-x^2}.\frac{x^4 + 2x}{\sqrt{1-x^2}} dx$

$y\sqrt{1-x^2} = \dfrac{x^5}{5} + x^2 + c$

$AB\, f(0) = 0 \Rightarrow$ At $x = 0, y = 0$

$\therefore c = 0$

$$\therefore \quad y\sqrt{1-x^2} = \frac{x^5}{5}+x^2$$

or $\quad y = f(x) = \dfrac{\frac{x^5}{5}+x^2}{\sqrt{1-x^2}}$

$$\therefore \quad I = \int_{-\frac{\sqrt{3}}{2}}^{\frac{\sqrt{3}}{2}} \frac{\frac{x^5}{5}+x^2}{\sqrt{1-x^2}}dx$$

$$= 2\int_0^{\frac{\sqrt{3}}{2}} \frac{x^2}{\sqrt{1-x^2}}dx \quad \left(\because \frac{x^5}{\sqrt{1-x^2}} \text{ is odd}\right)$$

put $x = \sin\theta \Rightarrow dx = \cos\theta\, d\theta$

$$\therefore I = 2\int_0^{\frac{\pi}{3}} \sin^2\theta d\theta = \int_0^{\frac{\pi}{3}} (1-\cos 2\theta)\, d\theta$$

$$= \left(\theta - \frac{\sin 2\theta}{2}\right)_0^{\frac{\pi}{3}} = \frac{\pi}{3} - \frac{\sqrt{3}}{4}$$

11. (b) $x_1 + x_2 + x_3$ will be odd
If two are even and one is odd
or all three are odd.
$\therefore$ Required probability
$= P(EEO) + P(EOE) + P(OEE) + P(OQO)$

$$= \frac{1}{3}\times\frac{2}{5}\times\frac{4}{7}+\frac{1}{3}\times\frac{3}{5}\times\frac{3}{7}+\frac{2}{3}\times\frac{2}{5}\times\frac{3}{7}+\frac{2}{3}\times\frac{3}{5}\times\frac{4}{7}$$

$$= \frac{8+9+12+24}{105} = \frac{53}{105}$$

12. (c) If x_1, x_2, x_3 are in AP then $2x_2 = x_1 + x_3$
$\because$ LHS is even, x_1 & x_3 can be both even or both odd.
x_1 and x_3 both can be even in $1 \times 3 = 3$ ways
x_1 and x_3 both can be odd in $2 \times 4 = 8$ ways
$\therefore$ Total favourable ways $= 3 + 8 = 11$
Also one number from each box can be drawn in $3 \times 5 \times 7$ ways
$\therefore$ Total ways $= 105$

Hence required probability $= \dfrac{11}{105}$

13. (d) $\because$ PQ is a focal chord, $Q\left(\dfrac{a}{t^2}, \dfrac{-2a}{t}\right)$
Also $QR \parallel PK \Rightarrow m_{QR} = m_{PK}$

$$\Rightarrow \frac{-\frac{2a}{t}-2ar}{\frac{a}{t^2}-ar^2} = \frac{0-2at}{2a-at^2}$$

$$\Rightarrow \frac{-2a\left(\frac{1}{t}+r\right)}{a\left(\frac{1}{t}+r\right)\left(\frac{1}{t}-r\right)} = \frac{-2at}{a\left(2-t^2\right)}$$

$$\Rightarrow 2-t^2 = t\left(\frac{1}{t}-r\right)$$

$$\left[\because r \neq -\frac{1}{t} \text{ otherwise } Q \text{ will coincide with } R\right]$$

$$\Rightarrow 2-t^2 = 1-tr \Rightarrow r = \frac{t^2-1}{t}$$

14. (b) Tangent at P is

$$ty = x + at^2 \qquad ....(i)$$

Normal at S

$$sx + y = 2as + as^3 \qquad ....(ii)$$

But given $st = 1 \Rightarrow s = \dfrac{1}{t}$

$$\therefore \quad \frac{x}{t} + y = \frac{2a}{t} + \frac{a}{t^3}$$

$$\Rightarrow xt^2 = yt^3 = 2at^2 + a$$

Putting value of x from equation (i) in above equation we get

$$\Rightarrow t^2\left(ty - at^2\right) + yt^3 = 2at^2 + a$$

$$\Rightarrow \left(t^3+t^3\right)y - at^4 = 2at^2 + a$$

$$\Rightarrow 2t^3 y = a\left(t^4 + 2t^2 + 1\right)$$

$$y = \frac{a\left(t^4+2t^2+1\right)}{2t^3} = \frac{a\left(t^2+1\right)^2}{2t^3}$$

15. (a)

$$g(a) = \lim_{h\to 0^+} \int_h^{1-h} t^{-a}(1-t)^{a-1}\, dt$$

$$\therefore \; g\left(\frac{1}{2}\right) = \lim_{h\to 0^+} \int_h^{1-h} t^{-1/2}(1-t)^{-1/2}\, dt$$

$$= \lim_{h\to 0^+} \int_h^{1-h} \frac{1}{\sqrt{t(1-t)}}dt$$

$$= \lim_{h\to 0^+} \int_h^{1-h} \frac{1}{\sqrt{\left(\frac{1}{2}\right)^2 - \left(t-\frac{1}{2}\right)^2}}dt$$

$$= \lim_{h\to 0^+} \left[\sin^{-1}\left(\frac{t-\frac{1}{2}}{\frac{1}{2}}\right)\right]_h^{1-h} = \lim_{h\to 0^+} \sin^{-1}(2t-1)\Big]_h^{1-h}$$

$= \lim_{h\to 0^+}\left[\sin^{-1}(1-2h)-\sin^{-1}(2h-1)\right]$

$= \frac{\pi}{2}-\left(-\frac{\pi}{2}\right)=\frac{\pi}{2}$

16. (d) $g(a)=\lim_{h\to 0^+}\int_h^{1-h} t^{-a}(1-t)^{a-1}\,dt$

$g(a)=\lim_{h\to 0^+}\int_h^{1-h}(1-t)^{-a}t^{a-1}\,dt$

$\left(\text{using}\int_a^b f(x)\,dx=\int_a^b f(a+b-x)\,dx\right)$

Also $g(1-a)=\lim_{h\to 0^+}\int_h^{1-h} t^{a-1}(1-t)^{-a}\,dt$

Thus $g(a)=g(1-a)$

$\Rightarrow g'(a)=-g'(1-a)\Rightarrow g'(a)+g'(1-a)=0$

Putting $a=\frac{1}{2}$ we get $g'\left(\frac{1}{2}\right)+g'\left(\frac{1}{2}\right)=0$

or $g'\left(\frac{1}{2}\right)=0$

17. (d) $P(2)$ Let $f(x)=ax^2+bx+c$
where $a, b, c\ge 0$ and a, b, c are integers.
$\because f(0)=0\Rightarrow c=0$
$\therefore f(x)=ax^2+bx$

Also $\int_0^1 f(x)\,dx=1$

$\Rightarrow\left[\frac{ax^3}{3}+\frac{bx^2}{2}\right]_0^1=1$

$\Rightarrow\frac{a}{3}+\frac{b}{2}=1$

$\Rightarrow 2a+3b=6$
$\because$ a and b are integers
$a=0$ and $b=2$
or $a=3$ and $b=0$
$\therefore$ There are only 2 solutions.
$Q(3)$ $f(x)=\sin x^2+\cos x^2$

$f(x)$ is max. $\sqrt{2}$ at $x^2=\frac{\pi}{4}$ or $\frac{9\pi}{4}$

$\Rightarrow x=\pm\frac{\sqrt{\pi}}{2}$ or $\pm\frac{3\sqrt{\pi}}{2}\in\left[-\sqrt{13},\sqrt{13}\right]$

$\therefore$ There are four points.

$R(1)$ $I=\int_{-2}^{2}\frac{3x^2}{1+e^x}dx=\int_{-2}^{2}\frac{3x^2}{1+e^{-x}}dx$

$\left[\text{Using}\int_a^b f(x)\,dx=\int_a^b f(a+b-x)\,dx\right]$

$=\int_{-2}^{2}\frac{3x^2e^x}{1+e^x}dx$

$2I=\int_{-2}^{2}\frac{3x^2\left(1+e^x\right)}{1+e^x}dx=\int_{-2}^{2}3x^2\,dx$

$2I=\left(x^3\right)_{-2}^{2}=8-(-8)=16$

$\Rightarrow I=8$

$S(4)$ $\dfrac{\int_{-1/2}^{1/2}\cos 2x\log\left(\frac{1+x}{1-x}\right)dx}{\int_0^{1/2}\cos 2x\log\left(\frac{1+x}{1-x}\right)dx}=0$

$\because$ Numerator = 0, function being odd.
Hence option (d) is correct sequence.

18. (a) $P(4)$ $y=\cos\left(3\cos^{-1}x\right)$

$y=\cos\left[\cos^{-1}\left(4x^3-3x\right)\right]$
$y=4x^3-3x$

$\Rightarrow\frac{dy}{dx}=12x^2-3$ and $\frac{d^2y}{dx^2}=24x$

$\therefore\ \frac{1}{y}\left\{\left(x^2-1\right)\frac{d^2y}{dx}+x\frac{dy}{dx}\right\}$

$=\frac{1}{4x^3-3x}\left\{\left(x^2-1\right)24x+x\left(12x^2-3\right)\right\}$

$=\frac{3x}{4x^3-3x}\left\{8x^2-8+4x^2-1\right\}$

$=\frac{3x\left\{12x^2-9\right\}}{4x^3-3x}=\frac{9\left\{4x^3-3x\right\}}{4x^3-3x}=9$

$Q(3)$ $\because A_1, A_2, A_3, \ldots A_n$ are the vertices of a regular polygon of n sides with its centre at origin and $\vec{a}_1,\vec{a}_2$, ... $\vec{a}_n$ are their position vectors.

$\therefore\quad\left|\vec{a}_1\right|=\left|\vec{a}_2\right|=\ldots=\left|\vec{a}_n\right|=\lambda$

Then $\vec{a_k} \times \vec{a_{k+1}} = \lambda^2 \sin\frac{2\pi}{n}$

and $\vec{a_k} \times \vec{a_{k+1}} = \lambda^2 \cos\frac{2\pi}{n}$

Hence given equation reduces to

$$(n-1)\lambda^2 \sin\frac{2\pi}{n} = (n-1)\lambda^2 \cos\frac{2\pi}{n}$$

$$\Rightarrow \tan\frac{2\pi}{n} = 1 \Rightarrow \frac{2\pi}{n} = \frac{\pi}{4} \Rightarrow n = 8$$

R.(2) Normal from *P*(*h*, 1) on $\frac{x^2}{6} + \frac{y^2}{3} = 1$ is

$$\frac{x-h}{h/6} = \frac{y-1}{1/3}$$

$\Rightarrow 2(x-h) = h(y-1)$

$\Rightarrow 2x - hy - h = 0$

It is perpendicular to $x + y = 8$

$\therefore \frac{2}{h} \times -1 = -1 \Rightarrow h = 2$

S.(1) $\tan^{-1}\left(\frac{1}{2x+1}\right) + \tan^{-1}\left(\frac{1}{4x+1}\right) = \tan^{-1}\left(\frac{2}{x^2}\right)$

$$\Rightarrow \tan^{-1}\left(\frac{\frac{1}{2x+1} + \frac{1}{4x+1}}{1 - \frac{1}{2x+1}\cdot\frac{1}{4x+1}}\right) = \tan^{-1}\left(\frac{2}{x^2}\right)$$

$$\Rightarrow \tan^{-1}\left(\frac{4x+1+2x+1}{8x^2+6x+1-1}\right) = \tan^{-1}\left(\frac{2}{x^2}\right)$$

$$\Rightarrow \tan^{-1}\left(\frac{6x+2}{8x^2+6x}\right) = \tan^{-1}\left(\frac{2}{x^2}\right)$$

$$\Rightarrow \tan^{-1}\left(\frac{3x+1}{4x^2+3x}\right) = \tan^{-1}\left(\frac{2}{x^2}\right)$$

$$\Rightarrow \frac{3x+1}{4x^2+3x} = \frac{2}{x^2}$$

$\Rightarrow 3x^2 - 7x - 6 = 0$ (for $x > 0$)

$\Rightarrow x = 3$ or $-\frac{2}{3}$ (rejected as $x > 0$)

$\therefore$ Only one +ve solution is there

Hence (a) is the correct option.

19. (d) $P(1)$: $f_4(x) = \begin{cases} x^2, & x < 0 \\ e^{2x} - 1, & x \geq 0 \end{cases}$

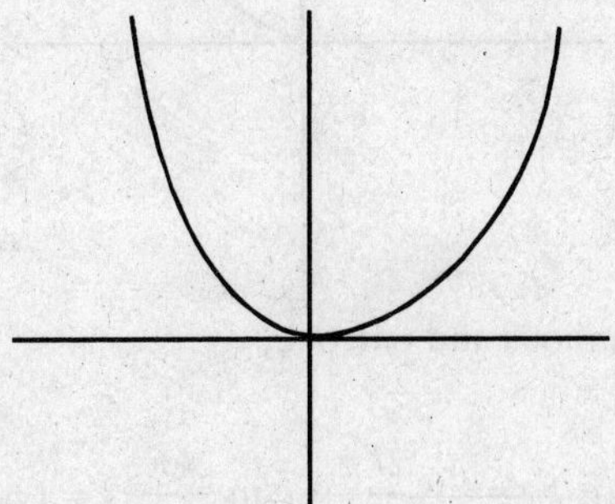

Range of $f_4 = [0, \infty)$

$\therefore f_4$ is onto

From graph f_4 is not one one.

$Q(3)$: $f_3(x) = \begin{cases} \sin x, & x < 0 \\ x, & x \geq 0 \end{cases}$

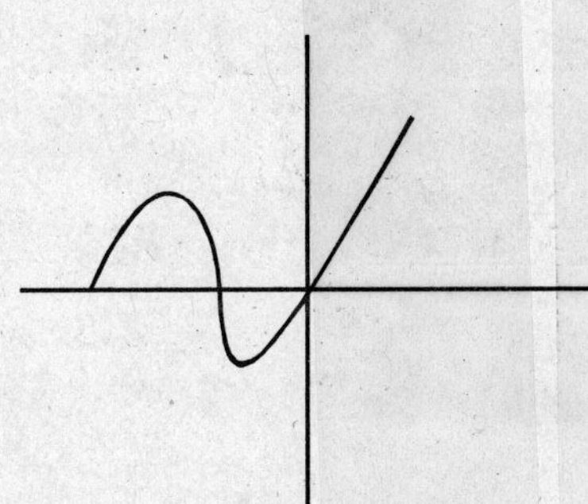

From graph *f* is differentiable but not one one.

$R(2)$: $f_2 o f_1(x) = \begin{cases} x^2, & x < 0 \\ e^{2x}, & x \geq 0 \end{cases}$

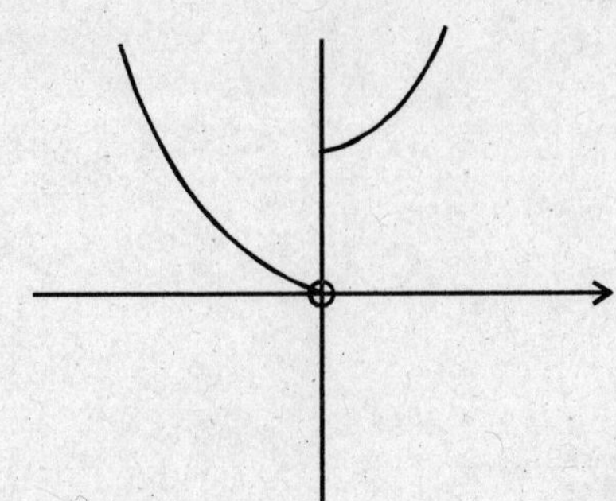

From graph $f_2 o f_1$ is neither continuous nor one one.

$S(4)$: $f_2(x) = x^2, x \in [0, \infty)$

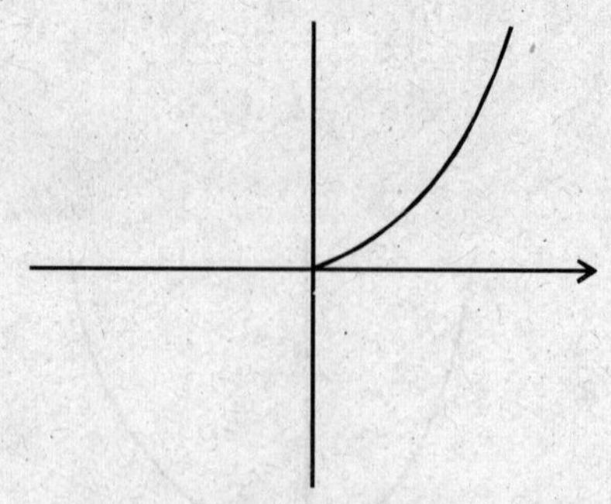

It is continuous and one one.

20. (c) $P(1)$: $z_k = \cos\frac{2k\pi}{10} + i\sin\frac{2k\pi}{10}$, $k = 1$ to 9

$\therefore\ z_k = e^{i\frac{2k\pi}{10}}$

Now $z_k.z_j = 1 \Rightarrow z_j = \frac{1}{z_k} = e^{-i\frac{2k\pi}{10}} = \overline{z_k}$

We know if z_k is 10th root of unity so will be $\overline{z}_k$.

$\therefore$ For every z_k, there exist $z_i = \overline{z}_k$

Such that $z_k.z_j = z_k.\overline{z}_k = 1$

Hence the statement is true.

$Q(2)$: $z_1 = z_k \Rightarrow z = \frac{z_k}{z_1}$ for $z_1 \neq 0$

$\therefore$ We can always find a solution to $z_1.z = z_k$

Hence the statement is false.

$R(3)$: We know $z^{10} - 1 = (z-1)(z-z_1)....(z-z_9)$

$\Rightarrow (z-z_1)(z-z_2).....(z-z_9) = \frac{z^{10}-1}{z-1}$

$= 1 + z + z^2 + ... z^9$

For z = 1 we get

$(1-z_1)(1-z_2).....(1-z_9) = 10$

$\therefore\ \frac{|1-z_1||1-z_2|.....|1-z_9|}{10} = 1$

JEE MAIN 2014 MATHEMATICS

(Held on 6th April-2014)

1. If $X=\{4^n-3n-1 : n\in N\}$ and $Y=\{9(n-1): n\in N\}$, where N is the set of natural numbers, then $X\cup Y$ is equal to:

(a) X (b) Y

(c) N (d) $Y-X$

2. If z is a complex number such that $|z|\ge 2$, then the minimum value of $\left|z+\frac{1}{2}\right|$:

(a) is strictly greater than $\frac{5}{2}$

(b) is strictly greater than $\frac{3}{2}$ but less than $\frac{5}{2}$

(c) is equal to $\frac{5}{2}$

(d) lie in the interval (1, 2)

3. If $a\in R$ and the equation

$-3(x-[x])^2+2(x-[x])+a^2=0$

(where $[x]$ denotes the greatest integer $\le x$) has no integral solution, then all possible values of a lie in the interval:

(a) $(-2,-1)$ (b) $(-\infty,-2)\cup(2,\infty)$

(c) $(-1,0)\cup(0,1)$ (d) $(1,2)$

4. Let α and β be the roots of equation $px^2+qx+r=0$, $p\ne 0$. If p, q, r are in A.P. and $\frac{1}{\alpha}+\frac{1}{\beta}=4$, then the value of $|\alpha-\beta|$ is:

(a) $\frac{\sqrt{34}}{9}$ (b) $\frac{2\sqrt{13}}{9}$

(c) $\frac{\sqrt{61}}{9}$ (d) $\frac{2\sqrt{17}}{9}$

5. If $\alpha,\beta\ne 0$, and $f(n)=\alpha^n+\beta^n$ and

$$\begin{vmatrix} 3 & 1+f(1) & 1+f(2) \\ 1+f(1) & 1+f(2) & 1+f(3) \\ 1+f(2) & 1+f(3) & 1+f(4) \end{vmatrix}$$

$=K(1-\alpha)^2(1-\beta)^2(\alpha-\beta)^2$, then K is equal to:

(a) 1 (b) –1

(c) $\alpha\beta$ (d) $\frac{1}{\alpha\beta}$

6. If A is an 3×3 non-singular matrix such that $AA'=A'A$ and $B=A^{-1}A'$, then BB' equals:

(a) B^{-1} (b) $(B^{-1})'$

(c) $I+B$ (d) I

7. If the coefficents of x^3 and x^4 in the expansion of $(1+ax+bx^2)(1-2x)^{18}$ in powers of x are both zero, then (a, b) is equal to:

(a) $\left(14,\frac{272}{3}\right)$ (b) $\left(16,\frac{272}{3}\right)$

(c) $\left(16,\frac{251}{3}\right)$ (d) $\left(14,\frac{251}{3}\right)$

8. If $(10)^9+2(11)^1(10^8)+3(11)^2(10)^7+\ldots\ldots$ $+10(11)^9=k(10)^9$, then k is equal to:

(a) 100 (b) 110

(c) $\frac{121}{10}$ (d) $\frac{441}{100}$

9. Three positive numbers form an increasing G. P. If the middle term in this G.P. is doubled, the new numbers are in A.P. then the common ratio of the G.P. is:

(a) $2-\sqrt{3}$ (b) $2+\sqrt{3}$

(c) $\sqrt{2}+\sqrt{3}$ (d) $3+\sqrt{2}$

10. $\lim\limits_{x\to 0}\frac{\sin\left(\pi\cos^2 x\right)}{x^2}$ is equal to:

(a) $-\pi$ (b) π

(c) $\frac{\pi}{2}$ (d) 1

11. If g is the inverse of a function f and $f'(x)=\frac{1}{1+x^5}$, then $g'(x)$ is equal to:

(a) $\frac{1}{1+\{g(x)\}^5}$ (b) $1+\{g(x)\}^5$

(c) $1+x^5$ (d) $5x^4$

12. If f and g are differentiable functions in [0, 1] satisfying $f(0)=2=g(1), g(0)=0$ and $f(1)=6$, then for some $c \in]0,1[$

(a) $f'(c)=g'(c)$ (b) $f'(c)=2g'(c)$
(c) $2f'(c)=g'(c)$ (d) $2f'(c)=3g'(c)$

13. If $x=-1$ and $x=2$ are extreme points of $f(x)=\alpha\log|x|+\beta x^2+x$ then

(a) $\alpha=2, \beta=-\frac{1}{2}$ (b) $\alpha=2, \beta=\frac{1}{2}$
(c) $\alpha=-6, \beta=\frac{1}{2}$ (d) $\alpha=-6, \beta=-\frac{1}{2}$

14. The integral $\int\left(1+x-\frac{1}{x}\right)e^{x+\frac{1}{x}}dx$ is equal to

(a) $(x+1)e^{x+\frac{1}{x}}+c$ (b) $-xe^{x+\frac{1}{x}}+c$
(c) $(x-1)e^{x+\frac{1}{x}}+c$ (d) $xe^{x+\frac{1}{x}}+c$

15. The integral $\int_0^{\pi}\sqrt{1+4\sin^2\frac{x}{2}-4\sin\frac{x}{2}}\,dx$ equals:

(a) $4\sqrt{3}-4$ (b) $4\sqrt{3}-4-\frac{\pi}{3}$
(c) $\pi-4$ (d) $\frac{2\pi}{3}-4-4\sqrt{3}$

16. The area of the region described by $A=\{(x,y): x^2+y^2\le 1 \text{ and } y^2\le 1-x\}$ is:

(a) $\frac{\pi}{2}-\frac{2}{3}$ (b) $\frac{\pi}{2}+\frac{2}{3}$
(c) $\frac{\pi}{2}+\frac{4}{3}$ (d) $\frac{\pi}{2}-\frac{4}{3}$

17. Let the population of rabbits surviving at time t be governed by the differential equation $\frac{dp(t)}{dt}=\frac{1}{2}p(t)-200$. If $p(0)=100$, then $p(t)$ equals:

(a) $600-500\,e^{t/2}$ (b) $400-300\,e^{-t/2}$
(c) $400-300\,e^{t/2}$ (d) $300-200\,e^{-t/2}$

18. Let PS be the median of the triangle vertices $P(2, 2)$, $Q(6,-1)$ and $R(7, 3)$. The equation of the line passing through $(1,-1)$ and parallel to PS is:

(a) $4x+7y+3=0$ (b) $2x-9y-11=0$
(c) $4x-7y-11=0$ (d) $2x+9y+7=0$

19. Let a, b, c and d be non-zero numbers. If the point of intersection of the lines $4ax+2ay+c=0$ and $5bx+2by+d=0$ lies in the fourth quadrant and is equidistant from the two axes then

(a) $3bc-2ad=0$ (b) $3bc+2ad=0$
(c) $2bc-3ad=0$ (d) $2bc+3ad=0$

20. The locus of the foot of perpendicular drawn from the centre of the ellipse $x^2+3y^2=6$ on any tangent to it is

(a) $(x^2+y^2)^2=6x^2+2y^2$
(b) $(x^2+y^2)^2=6x^2-2y^2$
(c) $(x^2-y^2)^2=6x^2+2y^2$
(d) $(x^2-y^2)^2=6x^2-2y^2$

21. Let C be the circle with centre at (1, 1) and radius = 1. If T is the circle centred at $(0, y)$, passing through origin and touching the circle C externally, then the radius of T is equal to

(a) $\frac{1}{2}$ (b) $\frac{1}{4}$
(c) $\frac{\sqrt{3}}{\sqrt{2}}$ (d) $\frac{\sqrt{3}}{2}$

22. The slope of the line touching both the parabolas $y^2=4x$ and $x^2=-32y$ is

(a) $\frac{1}{8}$ (b) $\frac{2}{3}$
(c) $\frac{1}{2}$ (d) $\frac{3}{2}$

23. The image of the line $\frac{x-1}{3}=\frac{y-3}{1}=\frac{z-4}{-5}$ in the plane $2x-y+z+3=0$ is the line:

(a) $\frac{x-3}{3}=\frac{y+5}{1}=\frac{z-2}{-5}$
(b) $\frac{x-3}{-3}=\frac{y+5}{-1}=\frac{z-2}{5}$
(c) $\frac{x+3}{3}=\frac{y-5}{1}=\frac{z-2}{-5}$
(d) $\frac{x+3}{-3}=\frac{y-5}{-1}=\frac{z+2}{5}$

24. The angle between the lines whose direction cosines satisfy the equations $l+m+n=0$ and $l^2+m^2+n^2$ is

(a) $\frac{\pi}{6}$ (b) $\frac{\pi}{2}$
(c) $\frac{\pi}{3}$ (d) $\frac{\pi}{4}$

25. If $\left[\vec{a}\times\vec{b}\;\; \vec{b}\times\vec{c}\;\; \vec{c}\times\vec{a}\right]=\lambda\left[\vec{a}\,\vec{b}\,\vec{c}\right]^2$ then λ is equal to

(a) 0 (b) 1
(c) 2 (d) 3

26. Let A and B be two events such that $P\left(\overline{A \cup B}\right) = \frac{1}{6}$, $P\left(\overline{A \cap B}\right) = \frac{1}{4}$ and $P\left(\overline{A}\right) = \frac{1}{4}$, where $\overline{A}$ stands for the complement of the event A. Then the events A and B are

(a) independent but not equally likely.
(b) independent and equally likely.
(c) mutually exclusive and independent.
(d) equally likely but not independent.

27. The variance of first 50 even natural numbers is

(a) 437 (b) $\frac{437}{4}$
(c) $\frac{833}{4}$ (d) 833

28. Let $f_k(x) = \frac{1}{k}\left(\sin^k x + \cos^k x\right)$ where $x \in R$ and $k \geq 1$. Then $f_4(x) - f_6(x)$ equals

(a) $\frac{1}{4}$ (b) $\frac{1}{12}$
(c) $\frac{1}{6}$ (d) $\frac{1}{3}$

29. A bird is sitting on the top of a vertical pole 20 m high and its elevation from a point O on the ground is $45°$. It flies off horizontally straight away from the point O. After one second, the elevation of the bird from O is reduced to $30°$. Then the speed (in m/s) of the bird is

(a) $20\sqrt{2}$ (b) $20\left(\sqrt{3}-1\right)$
(c) $40\left(\sqrt{2}-1\right)$ (d) $40\left(\sqrt{3}-\sqrt{2}\right)$

30. The statement $\sim\left(p \leftrightarrow \sim q\right)$ is:

(a) a tautology (b) a fallacy
(c) eqivalent to $p \leftrightarrow q$ (d) equivalent to $\sim p \leftrightarrow q$

SOLUTIONS

1. (b) $4^n - 3n - 1 = (1+3)^n - 3n - 1$

$= [{}^nC_0 + {}^nC_1.3 + {}^nC_2.3^2 + ... + {}^nC_n 3^n] - 3n - 1$

$= 9\,[{}^nC_2 + {}^nC_3.3 + ... + {}^nC_n.3^{n-2}]$

$\therefore 4^n - 3n - 1$ is a multiple of 9 for all n.

$\therefore X = \{x : x \text{ is a multiple of } 9\}$

Also, $Y = \{9(n-1) : n \in \mathbf{N}\} = \{\text{All multiples of } 9\}$

Clearly $X \subset Y$. $\therefore X \cup Y = Y$

2. (d) We know minimum value of $|Z_1 + Z_2|$ is $||Z_1| - |Z_2||$

Thus minimum value of $\left|Z + \frac{1}{2}\right|$ is $\left||Z| - \frac{1}{2}\right|$

$\leq \left|Z + \frac{1}{2}\right| \leq |Z| + \frac{1}{2}$

Since, $|Z| \geq 2$ therefore $2 - \frac{1}{2} < \left|Z + \frac{1}{2}\right| < 2 + \frac{1}{2}$

$\Rightarrow \frac{3}{2} < \left|Z + \frac{1}{2}\right| < \frac{5}{2}$

3. (c) Consider $-3(x - [x])^2 + 2[x - [x]) + a^2 = 0$

$\Rightarrow 3\{x\}^2 - 2\{x\} - a^2 = 0 \quad (\because x - [x] = \{x\})$

$\Rightarrow 3\left(\{x\}^2 - \frac{2}{3}\{x\}\right) = a^2, a \neq 0$

$\Rightarrow a^2 = 3\{x\}\left(\{x\} - \frac{2}{3}\right)$

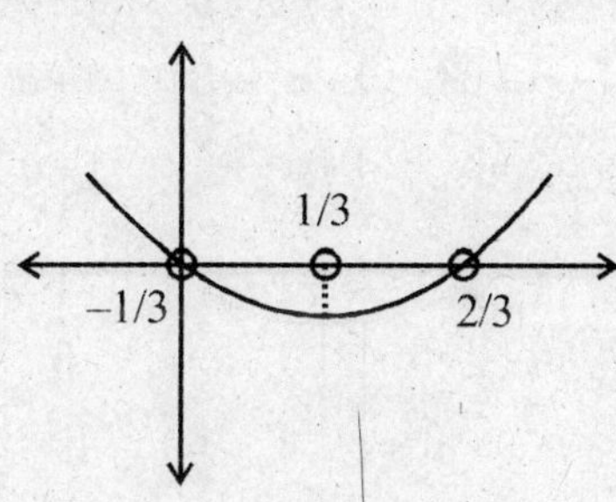

Now, $\{x\} \in (0,1)$ and $\frac{-2}{3} \leq a^2 < 1$ (by graph)

Since, x is not an integer

$\therefore \quad a \in (-1, 1) - \{0\}$

$\Rightarrow \quad a \in (-1, 0) \cup (0, 1)$

4. (b) Let p, q, r are in AP

$\Rightarrow 2q = p + r$ ...(i)

Given $\frac{1}{\alpha} + \frac{1}{\beta} = 4$

$\Rightarrow \quad \frac{\alpha + \beta}{\alpha\beta} = 4$

We have $\alpha + \beta = -q/p$ and $\alpha\beta = \frac{r}{p}$

$\Rightarrow \quad \frac{-\frac{q}{p}}{\frac{r}{p}} = 4 \Rightarrow q = -4r$ ....(ii)

From (i), we have

$2(-4r) = p + r$

$p = -9r$

$q = -4r$

$r = r$

Now $|\alpha - \beta| = \sqrt{(\alpha+\beta)^2 - 4\alpha\beta}$

$= \sqrt{\left(\frac{-q}{p}\right)^2 - \frac{4r}{p}}$

$= \frac{\sqrt{q^2 - 4pr}}{|p|}$

$= \frac{\sqrt{16r^2 + 36r^2}}{|-9r|}$

$= \frac{2\sqrt{13}}{9}$

5. (a) Consider

$$\begin{vmatrix} 3 & 1+f(1) & 1+f(2) \\ 1+f(1) & 1+f(2) & 1+f(3) \\ 1+f(2) & 1+f(3) & 1+f(4) \end{vmatrix}$$

$$= \begin{vmatrix} 1+1+1 & 1+\alpha+\beta & 1+\alpha^2+\beta^2 \\ 1+\alpha+\beta & 1+\alpha^2+\beta^2 & 1+\alpha^3+\beta^3 \\ 1+\alpha^2+\beta^2 & 1+\alpha^3+\beta^3 & 1+\alpha^4+\beta^4 \end{vmatrix}$$

$$= \begin{vmatrix} 1 & 1 & 1 \\ 1 & \alpha & \beta \\ 1 & \alpha^2 & \beta^2 \end{vmatrix} \times \begin{vmatrix} 1 & 1 & 1 \\ 1 & \alpha & \beta \\ 1 & \alpha^2 & \beta^2 \end{vmatrix}$$

$$= \begin{vmatrix} 1 & 1 & 1 \\ 1 & \alpha & \beta \\ 1 & \alpha^2 & \beta^2 \end{vmatrix}^2$$

$= [(1-\alpha)(1-\beta)(\alpha-\beta)]^2$

So, $\boxed{k=1}$

6. (d) $BB' = B(A^{-1}A')' = B(A')'(A^{-1})' = BA(A^{-1})'$

$= (A^{-1}A')(A(A^{-1})')$

$= A^{-1}A.A'.(A^{-1})'$ $\quad\{\text{as } AA' = A'A\}$

$= I(A^{-1}A)'$

$= I.I = I^2 = I$

7. (b) Consider $(1 + ax + bx^2)(1 - 2x)^{18}$

$= (1 + ax + bx^2)\,[{}^{18}C_0 - {}^{18}C_1(2x) + {}^{18}C_2(2x)^2 - {}^{18}C_3(2x)^3 + {}^{18}C_4(2x)^4 - ....]$

Coeff of $x^3 = {}^{18}C_3(-2)^3 + a.(-2)^2.\,{}^{18}C_2 + b(-2).\,{}^{18}C_1 = 0$

Coeff. of $x^3 = -{}^{18}C_3.8 + a \times 4.\,{}^{18}C_2 - 2b \times 18 = 0$

$= -\frac{18\times17\times16}{6}.8 + \frac{4a + 18\times17}{2} - 36b = 0$

$= -51 \times 16 \times 8 + a \times 36 \times 17 - 36b = 0$

$= -34 \times 16 + 51a - 3b = 0$

$= 51a - 3b = 34 \times 16 = 544$

$= 51a - 3b = 544$ ....(i)

Only option number (b) satisfies the equation number (i)

8. (a) Let $10^9 + 2.(11)(10)^8 + 3(11)^2(10)^7 + ... + 10(11)^9 = k(10)^9$

Let $x = 10^9 + 2.(11)(10)^8 + 3(11)^2(10)^7 + ... + 10(11)^9$

Multiplied by $\frac{11}{10}$ on both the sides

$\frac{11}{10}x = 11.10^8 + 2.(11)^2.(10)^7 + ... + 9(11)^9 + 11^{10}$

$x\left(1 - \frac{11}{10}\right) = 10^9 + 11(10)^8 + 11^2 \times (10)^7 + ... + 11^9 - 11^{10}$

$\Rightarrow -\frac{x}{10} = 10^9\left[\frac{\left(\frac{11}{10}\right)^{10} - 1}{\frac{11}{10} - 1}\right] - 11^{10}$

$\Rightarrow -\frac{x}{10} = (11^{10} - 10^{10}) - 11^{10} = -10^{10}$

$\Rightarrow x = 10^{11} = k.10^9$ Given

$\Rightarrow k = 100$

9. (b) Let a, ar, ar^2 are in G.P.

According to the question

$a, 2ar, ar^2$ are in A.P.

$\Rightarrow 2 \times 2ar = a + ar^2$

$\Rightarrow 4r = 1 + r^2$

$\Rightarrow r^2 - 4r + 1 = 0$

$r = \frac{4 \pm \sqrt{16-4}}{2} = 2 \pm \sqrt{3}$

Since $r > 1$

$\therefore\ r = 2 - \sqrt{3}$ is rejected

Hence, $r = 2 + \sqrt{3}$

10. (b) Consider $\lim_{x\to0} \frac{\sin(\pi\cos^2 x)}{x^2}$

$= \lim_{x\to0} \frac{\sin\left[\pi(1-\sin^2 x)\right]}{x^2}$

$= \lim_{x\to0} \sin\frac{(\pi - \pi\sin^2 x)}{x^2}$ $\quad[\because \sin(\pi-\theta) = \sin\theta]$

$= \lim_{x\to0} \sin\frac{(\pi\sin^2 x)}{\pi\sin^2 x} \times \frac{\pi\sin^2 x}{x^2}$

$= \lim_{x\to0} 1 \times \pi\left(\frac{\sin x}{x}\right)^2 = \pi$

11. (b) Since $f(x)$ and $g(x)$ are inverse of each other

$$\therefore \quad g'(f(x)) = \frac{1}{f'(x)}$$

$$\Rightarrow \quad g'(f(x)) = 1 + x^5 \quad \left(\because \; f'(x) = \frac{1}{1+x^5}\right)$$

Here $x = g(y)$

$$\therefore \quad g'(y) = 1 + [g(y)]^5$$

$$\Rightarrow \quad g'(x) = 1 + (g(x))^5$$

12. (b) Since, f and g both are continuous function on [0, 1] and differentiable on (0, 1) then $\exists c \in (0,1)$ such that

$$f'(c) = \frac{f(1) - f(0)}{1} = \frac{6-2}{1} = 4$$

and $g'(c) = \dfrac{g(1) - g(0)}{1} = \dfrac{2-0}{1} = 2$

Thus, we get

$$f'(c) = 2g'(c)$$

13. (a) Let $f(x) = \alpha \log|x| + \beta x^2 + x$

Differentiate both side,

$$f'(x) = \frac{\alpha}{x} + 2\beta x + 1$$

Since $x = -1$ and $x = 2$ are extreme points therefore $f'(x) = 0$ at these points.

Put $x = -1$ and $x = 2$ in $f'(x)$, we get

$-\alpha - 2\beta + 1 = 0 \Rightarrow \alpha + 2\beta = 1$ ...(i)

$\dfrac{\alpha}{2} + 4\beta + 1 = 0 \Rightarrow \alpha + 8\beta = -2$ ...(ii)

On solving (i) and (ii), we get

$$6\beta = -3 \Rightarrow \beta = -\frac{1}{2} \quad \therefore \quad \alpha = 2$$

14. (d) Let $I = \int\left(1 + x - \dfrac{1}{x}\right)e^{x+1/x}dx$

$$= \int e^{x+1/x}dx + \int\left(x - \frac{1}{x}\right)e^{x+1/x}dx$$

$$= x.e^{x+1/x} - \int x\left(1 - \frac{1}{x^2}\right)e^{x+1/x}dx$$

$$+ \int\left(x - \frac{1}{x}\right)e^{x+1/x}dx$$

$$= x.e^{x+\frac{1}{x}} - \int\left(x - \frac{1}{x}\right)e^{x+1/x}dx$$

$$+ \int\left(x - \frac{1}{x}\right)e^{x+1/x}dx$$

$$= xe^{x+\frac{1}{x}} + C$$

15. (b) Let $I = \displaystyle\int_0^{\pi}\sqrt{1 + 4\sin^2\frac{x}{2} - 4\sin\frac{x}{2}}\,dx$

$$= \int_0^{\pi}\left|2\sin\frac{x}{2} - 1\right|dx$$

$$= \int_0^{\pi/3}\left(1 - 2\sin\frac{x}{2}\right)dx + \int_{\pi/3}^{\pi}\left(2\sin\frac{x}{2} - 1\right)dx$$

$$\left[\because \; \sin\frac{x}{2} = \frac{1}{2} \Rightarrow \frac{x}{2} = \frac{\pi}{6} \Rightarrow x = \frac{\pi}{3}, \frac{x}{2} = \frac{5\pi}{6} \Rightarrow x = \frac{5\pi}{3}\right]$$

$$= \left[x + 4\cos\frac{x}{2}\right]_0^{\pi/3} + \left[-4\cos\frac{x}{2} - x\right]_{\pi/3}^{\pi}$$

$$= \frac{\pi}{3} + 4\frac{\sqrt{3}}{2} - 4 + \left(0 - \pi + 4\frac{\sqrt{3}}{2} + \frac{\pi}{3}\right)$$

$$= 4\sqrt{3} - 4 - \frac{\pi}{3}$$

16. (c) Given curves are $x^2 + y^2 = 1$ and $y^2 = 1 - x$. Intersecting points are $x = 0, 1$

Area of shaded portion is the required area.

So, Required Area = Area of semi-circle + Area bounded by parabola

$$= \frac{\pi r^2}{2} + 2\int_0^1\sqrt{1-x}\,dx$$

$$= \frac{\pi}{2} + 2\int_0^1\sqrt{1-x}\,dx \qquad (\because \text{ radius of circle} = 1)$$

$$= \frac{\pi}{2} + 2\left[\frac{(1-x)^{3/2}}{-3/2}\right]_0^1 = \frac{\pi}{2} - \frac{4}{3}(-1) = \frac{\pi}{2} + \frac{4}{3} \text{ Sq. unit}$$

17. (c) Given differential equation is

$$\frac{dp(t)}{dt} = \frac{1}{2}p(t) - 200$$

By separating the variable, we get

$$dp(t) = \left[\frac{1}{2}p(t) - 200\right]dt$$

$$\Rightarrow \quad \frac{dp(t)}{\frac{1}{2}p(t) - 200} = dt$$

Integrate on both the sides,

$$\int\frac{d(p(t))}{\frac{1}{2}p(t) - 200} = \int dt$$

Let $\frac{1}{2}p(t)-200=s$

$\Rightarrow \quad \frac{dp(t)}{2}=ds$

So, $\int\frac{d\,p(t)}{\left(\frac{1}{2}p(t)-200\right)}=\int dt$

$\Rightarrow \quad \int\frac{2ds}{s}=\int dt$

$\Rightarrow \quad 2\log s=t+c$

$\Rightarrow \quad 2\log\left(\frac{p(t)}{2}-200\right)=t+c$

$\Rightarrow \quad \frac{p(t)}{2}-200=e^{\frac{1}{2}}k$

Using given condition $p(t)=400-300\,e^{t/2}$

18. (d) Let P, Q, R, be the vertices of ΔPQR

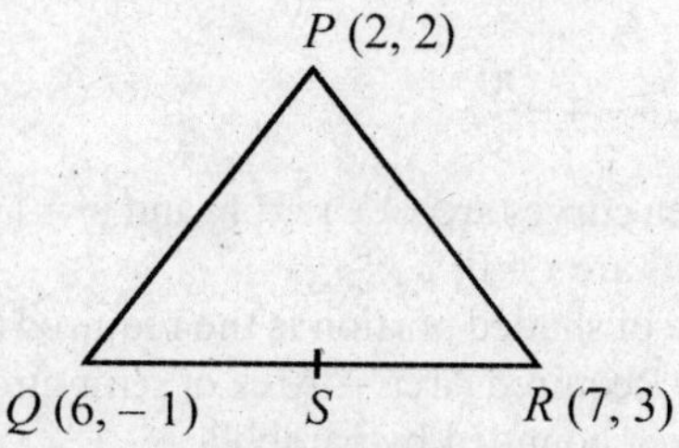

Since PS is the median

S is mid-point of QR

So, $S=\left(\frac{7+6}{2},\frac{3-1}{2}\right)=\left(\frac{13}{2},1\right)$

Now, slope of $PS=\frac{2-1}{2-\frac{13}{2}}=-\frac{2}{9}$

Since, required line is parallel to PS therefore slope of required line = slope of PS Now, eqn of line passing through $(1,-1)$ and having slope $-\frac{2}{9}$ is

$y-(-1)=-\frac{2}{9}(x-1)$

$9y+9=-2x+2 \Rightarrow 2x+9y+7=0$

19. (a) Given lines are

$4ax+2ay+c=0$

$5bx+2by+d=0$

The point of intersection will be

$\frac{x}{2ad-2bc}=\frac{-y}{4ad-5bc}=\frac{1}{8ab-10ab}$

$\Rightarrow \quad x=\frac{2(ad-bc)}{-2ab}=\frac{bc-ad}{ab}$

$\Rightarrow \quad y=\frac{5bc-4ad}{-2ab}=\frac{4ad-5bc}{2ab}$

$\because$ Point of intersection is in fourth quadrant so x is positive and y is negative.

Also distance from axes is same

So $x=-y$ ($\because$ distance from x-axis is $-y$ as y is negative)

$\frac{bc-ad}{ab}=\frac{5bc-4ad}{2ab}$

$\Rightarrow \quad 3bc-2ad=0$

20. (a) Given eq^n of ellipse can be written as

$\frac{x^2}{6}+\frac{y^2}{2}=1$

$\Rightarrow \quad a^2=6, b^2=2$

Now, equation of any variable tangent is

$y=mx\pm\sqrt{a^2m^2+b^2}$ ...(i)

where m is slope of the tangent

So, equation of perpendicular line drawn from centre to tangent is

$y=\frac{-x}{m}$ ...(ii)

Eliminating m, we get

$(x^4+y^4+2x^2y^2)=a^2x^2+b^2y^2$

$\Rightarrow \quad (x^2+y^2)^2=a^2x^2+b^2y^2$

$\Rightarrow \quad \boxed{(x^2+y^2)^2=6x^2+2y^2}$

21. (b)

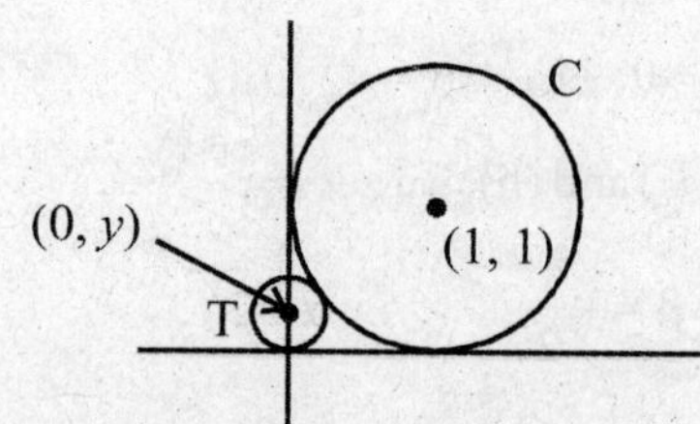

Equation of circle $C\equiv(x-1)^2+(y-1)^2=1$

Radius of T = $|y|$

T touches C externally

therefore,

Distance between the centres = sum of their radii

$\Rightarrow \quad \sqrt{(0-1)^2+(y-1)^2}=1+|y|$

$\Rightarrow \quad (0-1)^2+(y-1)^2=(1+|y|)^2$

$\Rightarrow \quad 1+y^2+1-2y=1+y^2+2|y|$

$2|y|=1-2y$

If $y>0$ then $2y=1-2y \Rightarrow y=\frac{1}{4}$

If $y<0$ then $-2y=1-2y \Rightarrow 0=1$ (not possible)

$\therefore \quad y=\frac{1}{4}$

22. (c) Given parabolas are

$y^2 = 4x$ ...(i)

$x^2 = -32y$ ...(ii)

Let m be slope of common tangent

Equation of tangent of parabola (1)

$$y = mx + \frac{1}{m} \quad \text{(i)}$$

Equation of tangent of parabola (2)

$y = mx + 8m^2$ ...(ii)

(i) and (ii) are identical

$$\Rightarrow \frac{1}{m} = 8m^2$$

$$\Rightarrow m^3 = \frac{1}{8}$$

$$\Rightarrow \boxed{m = \frac{1}{2}}$$

Alternative method:

Let tangent to $y^2 = 4x$ be $y = mx + \frac{1}{m}$

Since this is also tangent to $x^2 = -32y$

$$\therefore \quad x^2 = -32\left(mx + \frac{1}{m}\right)$$

$$\Rightarrow x^2 + 32mx + \frac{32}{m} = 0$$

Now, D = 0

$$(32)^2 - 4\left(\frac{32}{m}\right) = 0$$

$$\Rightarrow m^3 = \frac{4}{32} \Rightarrow m = \frac{1}{2}$$

23. (c)

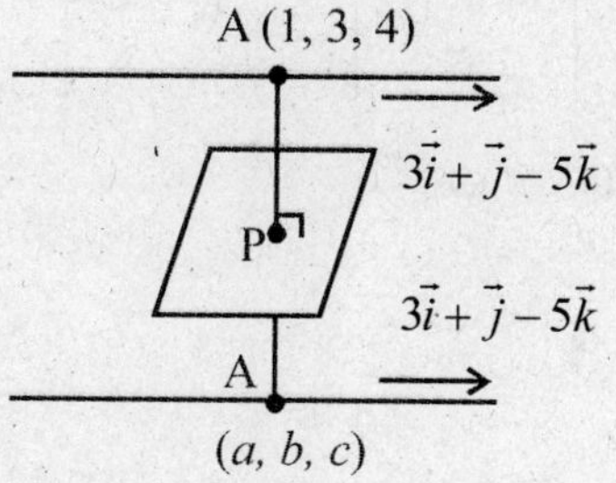

A (1, 3, 4)

$3\vec{i} + \vec{j} - 5\vec{k}$

P

$3\vec{i} + \vec{j} - 5\vec{k}$

A

(a, b, c)

$$\frac{a-1}{2} = \frac{b-3}{-1} = \frac{c-4}{1} = \lambda \text{(let)}$$

$$\Rightarrow a = 2\lambda + 1$$

$$b = 3 - \lambda$$

$$c = 4 + \lambda$$

$$P = \left(\frac{a+1}{2}, \frac{b+3}{2}, \frac{c+4}{2}\right)$$

$$= \left(\lambda + 1, \frac{6-\lambda}{2}, \frac{\lambda+8}{2}\right)$$

$$\therefore \quad 2(\lambda+1) - \frac{6-\lambda}{2} + \frac{\lambda+8}{2} + 3 = 0$$

$$3\lambda + 6 = 0 \Rightarrow \lambda = -2$$

$$a = -3, b = 5, c = 2$$

Required line is $\frac{x+3}{3} = \frac{y-5}{1} = \frac{z-2}{-5}$

24. (c) Given

$l + m + n = 0$ and $l^2 = m^2 + n^2$

Now, $(-m-n)^2 = m^2 + n^2$

$$\Rightarrow mn = 0$$

$$\Rightarrow m = 0 \text{ or } n = 0$$

If $m = 0$ then $l = -n$

We know $l^2 + m^2 + n^2 = 1$

$$\Rightarrow n = \pm\frac{1}{\sqrt{2}}$$

i.e. $(l_1, m_1, n_1) = \left(-\frac{1}{\sqrt{2}}, 0, \frac{1}{\sqrt{2}}\right)$

If $n = 0$ then $l = -m$

$l^2 + m^2 + n^2 = 1 \Rightarrow 2m^2 = 1$

$$\Rightarrow m = \pm\frac{1}{\sqrt{2}}$$

Let $m = \frac{1}{\sqrt{2}}$

$$\Rightarrow l = -\frac{1}{\sqrt{2}} \text{ and } n = 0$$

$$(l_2, m_2, n_2) = \left(-\frac{1}{\sqrt{2}}, \frac{1}{\sqrt{2}}, 0\right)$$

$$\therefore \quad \cos\theta = \frac{1}{2}$$

$$\Rightarrow \theta = \frac{\pi}{3}$$

25. (b) L.H.S

$= (\vec{a}\times\vec{b}).[(\vec{b}\times\vec{c})\times(\vec{c}\times\vec{a})]$

$= (\vec{a}\times\vec{b}).[(\vec{b}\times\vec{c}.\vec{a})\vec{c} - (\vec{b}\times\vec{c}.\vec{c})\vec{a}]$

$= (\vec{a}\times\vec{b}).[[\vec{b}\,\vec{c}\,\vec{a}]\vec{c}] \qquad [\because \vec{b}\times\vec{c}.\vec{c} = 0]$

$= [\vec{a}\;\vec{b}\,\vec{c}].(\vec{a}\times\vec{b}.\vec{c}) = [\vec{a}\,\vec{b}\,\vec{c}]^2$

$[\vec{a}\times\vec{b} \quad \vec{b}\times\vec{c} \quad \vec{c}\times\vec{a}] = [\vec{a}\,\vec{b}\,\vec{c}]^2$

So $\lambda = 1$

26. (a) Given

$$P(\overline{A \cup B}) = \frac{1}{6} \Rightarrow P(A \cup B) = 1 - \frac{1}{6} = \frac{5}{6}$$

$$P(\bar{A}) = \frac{1}{4} \Rightarrow P(A) = 1 - \frac{1}{4} = \frac{3}{4}$$

We know

$$P(A \cup B) = P(A) + P(B) - P(A \cap B)$$

$$\Rightarrow \quad \frac{5}{6} = \frac{3}{4} + P(B) - \frac{1}{4} \quad \left(\because P(A \cap B) = \frac{1}{4}\right)$$

$$\Rightarrow \quad P(B) = \frac{1}{3}$$

$\because$ $P(A) \neq P(B)$ so they are not equally likely.

Also $P(A) \times P(B) = \frac{3}{4} \times \frac{1}{3} = \frac{1}{4} = P(A \cap B)$

So A & B are independent.

27. (d) First 50 even natural numbers are 2, 4, 6 ....., 100

$$\text{Variance} = \frac{\sum x_i^2}{N} - (\bar{x})^2$$

$$\Rightarrow \quad \sigma^2 = \frac{2^2 + 4^2 + ... + 100^2}{50} - \left(\frac{2+4+...+100}{50}\right)^2$$

$$= \frac{4(1^2 + 2^2 + 3^2 + ... + 50^2)}{50} - (51)^2$$

$$= 4\left(\frac{50 \times 51 \times 101}{50 \times 6}\right) - (51)^2 = 3434 - 2601$$

$$\Rightarrow \quad \sigma^2 = 833$$

28. (b) Let $f_k(x) = \frac{1}{k}(\sin^k x + \cos^k x)$

Consider

$$f_4(x) - f_6(x) = \frac{1}{4}(\sin^4 x + \cos^4 x) - \frac{1}{6}(\sin^6 x + \cos^6 x)$$

$$= \frac{1}{4}[1 - 2\sin^2 x \cos^2 x] - \frac{1}{6}[1 - 3\sin^2 x \cos^2 x]$$

$$= \frac{1}{4} - \frac{1}{6} = \frac{1}{12}$$

29. (b) Let the speed be y m/sec.

Let AC be the vertical pole of height 20 m.

Let O be the point on the ground such that $\angle AOC = 45°$

Let $OC = x$

Time $t = 1$ s

From ΔAOC, $\tan 45° = \frac{20}{x}$ ...(i)

and from ΔBOD, $\tan 30° = \frac{20}{x+y}$ ...(ii)

From (i) and (ii), we have $x = 20$ and $\frac{1}{\sqrt{3}} = \frac{20}{x+y}$

$$\Rightarrow \quad \frac{1}{\sqrt{3}} = \frac{20}{20+y} \Rightarrow 20 + y = 20\sqrt{3}$$

So, $y = 20(\sqrt{3} - 1)$

i.e., speed $= 20(\sqrt{3} - 1)$ m/s

30. (c)

p	q	$\sim q$	$p \leftrightarrow \sim q$	$\sim(p \leftrightarrow \sim q)$
F	F	T	F	T
F	T	F	T	F
T	F	T	T	F
T	T	F	F	T

Clearly equivalent to $p \leftrightarrow q$

Chapter 1 Trigonometric Functions & Equations

Section-A JEE Advanced/ IIT-JEE

(A) FILL IN THE BLANKS :

1. Suppose $\sin^3 x \sin 3x = \sum_{m=0}^{n} C_m \cos mx$ is an identity in x, where $C_0, C_1, ...C_n$ are constants, and $C_n \neq 0$. then the value of n is ________ **(1981 - 2 Marks)**

2. The solution set of the system of equations $x + y = \frac{2\pi}{3}$, $\cos x + \cos y = \frac{3}{2}$, where x and y are real, is ________. **(1987 - 2 Marks)**

3. The set of all x in the interval $[0, \pi]$ for which $2\sin^2 x - 3 \sin x + 1 \geq 0$, is ________. **(1987 - 2 Marks)**

4. The sides of a triangle inscribed in a given circle subtend angles α, β and γ at the centre. The minimum value of the arithmetic mean of $\cos\left(\alpha + \frac{\pi}{2}\right)$, $\cos\left(\beta + \frac{\pi}{2}\right)$ and $\cos\left(\gamma + \frac{\pi}{2}\right)$ is equal to ________ **(1987 - 2 Marks)**

5. The value of
$\sin\frac{\pi}{14}\sin\frac{3\pi}{14}\sin\frac{5\pi}{14}\sin\frac{7\pi}{14}\sin\frac{9\pi}{14}\sin\frac{11\pi}{14}\sin\frac{13\pi}{14}$ is equal to ________ **(1991 - 2 Marks)**

6. If $K = \sin(\pi/18)\sin(5\pi/18)\sin(7\pi/18)$, then the numerical value of K is ________. **(1993 - 2 Marks)**

7. If $A > 0, B > 0$ and $A + B = \pi/3$, then the maximum value of $\tan A \tan B$ is________ . **(1993 - 2 Marks)**

8. General value of θ satisfying the equation $\tan^2\theta + \sec 2\theta = 1$ is________. **(1996 - 1 Mark)**

9. If $\cos(x - y), \cos x$ and $\cos(x + y)$ are in H.P., then $\cos x \sec\left(\frac{y}{2}\right) =$ ________ **(1997C - 2 Marks)**

10. The real roots of the equation $\cos^7 x + \sin^4 x = 1$ in the interval $(-\pi, \pi)$ are ..., ..., and ________. **(1997 - 2 Marks)**

(B) TRUE / FALSE :

1. If $\tan A = (1 - \cos B)/\sin B$, then $\tan 2A = \tan B$. **(1983 - 1 Mark)**

2. There exists a value of θ between 0 and 2π that satisfies the equation $\sin^4\theta - 2\sin^2\theta - 1 = 0$. **(1984 - 1 Mark)**

(C) MCQ's WITH ONE CORRECT ANSWER :

1. If $\tan\theta = -\frac{4}{3}$, then $\sin\theta$ is **(1979)**

   (a) $-\frac{4}{5}$ but not $\frac{4}{5}$ (b) $-\frac{4}{5}$ or $\frac{4}{5}$
   (c) $\frac{4}{5}$ but not $-\frac{4}{5}$ (d) None of these.

2. If $\alpha + \beta + \gamma = 2\pi$, then **(1979)**

   (a) $\tan\frac{\alpha}{2} + \tan\frac{\beta}{2} + \tan\frac{\gamma}{2} = \tan\frac{\alpha}{2}\tan\frac{\beta}{2}\tan\frac{\gamma}{2}$
   (b) $\tan\frac{\alpha}{2}\tan\frac{\beta}{2} + \tan\frac{\beta}{2}\tan\frac{\gamma}{2} + \tan\frac{\gamma}{2}\tan\frac{\alpha}{2} = 1$
   (c) $\tan\frac{\alpha}{2} + \tan\frac{\beta}{2} + \tan\frac{\gamma}{2} = -\tan\frac{\alpha}{2}\tan\frac{\beta}{2}\tan\frac{\gamma}{2}$
   (d) None of these.

3. Given $A = \sin^2\theta + \cos^4\theta$ then for all real values of θ

   (a) $1 \leq A \leq 2$ (b) $\frac{3}{4} \leq A \leq 1$ **(1980)**
   (c) $\frac{13}{16} \leq A \leq 1$ (d) $\frac{3}{4} \leq A \leq \frac{13}{16}$

4. The equation $2\cos^2\frac{x}{2}\sin^2 x = x^2 + x^{-2}$; $0 < x \leq \frac{\pi}{2}$ has

   (a) no real solution (b) one real solution
   (c) more than one solution (d) none of these **(1980)**

5. The general solution of the trigonometric equation $\sin x + \cos x = 1$ is given by : **(1981 - 2 Marks)**

   (a) $x = 2n\pi$; $n = 0, \pm1, \pm2 ...$
   (b) $x = 2n\pi + \pi/2$; $n = 0, \pm1, \pm2...$
   (c) $x = n\pi + (-1)^n \frac{\pi}{4} - \frac{\pi}{4}$ $n = 0, \pm1, \pm2...$
   (d) none of these

6. The value of the expression $\sqrt{3}\operatorname{cosec} 20° - \sec 20°$ is equal to *(1988 - 2 Marks)*
(a) 2 (b) $2 \sin 20°/\sin 40°$
(c) 4 (d) $4 \sin 20°/\sin 40°$

7. The general solution of $\sin x - 3 \sin 2x + \sin 3x = \cos x - 3 \cos 2x + \cos 3x$ is *(1989 - 2 Marks)*
(a) $n\pi + \frac{\pi}{8}$ (b) $\frac{n\pi}{2} + \frac{\pi}{8}$
(c) $(-1)^n \frac{n\pi}{2} + \frac{\pi}{8}$ (d) $2n\pi + \cos^{-1}\frac{3}{2}$

8. The equation $(\cos p - 1)x^2 + (\cos p)x + \sin p = 0$ In the variable x, has real roots. Then p can take any value in the interval *(1990 - 2 Marks)*
(a) $(0, 2\pi)$ (b) $(-\pi, 0)$
(c) $\left(-\frac{\pi}{2}, \frac{\pi}{2}\right)$ (d) $(0, \pi)$

9. Number of solutions of the equation $\tan x + \sec x = 2\cos x$ lying in the interval $[0, 2\pi]$ is : *(1993 - 1 Mark)*
(a) 0 (b) 1
(c) 2 (d) 3

10. Let $0 < x < \frac{\pi}{4}$ then $(\sec 2x - \tan 2x)$ equals *(1994)*
(a) $\tan\left(x - \frac{\pi}{4}\right)$ (b) $\tan\left(\frac{\pi}{4} - x\right)$
(c) $\tan\left(x + \frac{\pi}{4}\right)$ (d) $\tan^2\left(x + \frac{\pi}{4}\right)$

11. Let n be a positive integer such that $\sin\frac{\pi}{2n} + \cos\frac{\pi}{2n} = \frac{\sqrt{n}}{2}$. Then *(1994)*
(a) $6 \le n \le 8$ (b) $4 < n \le 8$
(c) $4 \le n \le 8$ (d) $4 < n < 8$

12. If ω is an imaginary cube root of unity then the value of $\sin\left\{(\omega^{10} + \omega^{23})\pi - \frac{\pi}{4}\right\}$ is *(1994)*
(a) $-\frac{\sqrt{3}}{2}$ (b) $-\frac{1}{\sqrt{2}}$
(c) $\frac{1}{\sqrt{2}}$ (d) $\frac{\sqrt{3}}{2}$

13. $3(\sin x - \cos x)^4 + 6(\sin x + \cos x)^2 + 4(\sin^6 x + \cos^6 x) =$ *(1995S)*
(a) 11 (b) 12
(c) 13 (d) 14

14. The general values of θ satisfying the equation $2\sin^2\theta - 3\sin\theta - 2 = 0$ is *(1995S)*
(a) $n\pi + (-1)^n \pi/6$ (b) $n\pi + (-1)^n \pi/2$
(c) $n\pi + (-1)^n 5\pi/6$ (d) $n\pi + (-1)^n 7\pi/6$

15. $\sec^2\theta = \frac{4xy}{(x+y)^2}$ is true if and only if *(1996 - 1 Mark)*
(a) $x + y \ne 0$ (b) $x = y, x \ne 0$
(c) $x = y$ (d) $x \ne 0, y \ne 0$

16. In a triangle PQR, $\angle R = \pi/2$. If $\tan(P/2)$ and $\tan(Q/2)$ are the roots of the equation $ax^2 + bx + c = 0$ $(a \ne 0)$ then. *(1999 - 2 Marks)*
(a) $a + b = c$ (b) $b + c = a$
(c) $a + c = b$ (d) $b = c$

17. Let $f(\theta) = \sin\theta(\sin\theta + \sin 3\theta)$. Then $f(\theta)$ is *(2000S)*
(a) ≥ 0 only when $\theta \ge 0$ (b) ≤ 0 for all real θ
(c) ≥ 0 for all real θ (d) ≤ 0 only when $\theta \le 0$

18. The number of distinct real roots of $\begin{vmatrix} \sin x & \cos x & \cos x \\ \cos x & \sin x & \cos x \\ \cos x & \cos x & \sin x \end{vmatrix} = 0$ in the interval $-\frac{\pi}{4} \le x \le \frac{\pi}{4}$ is *(2001S)*
(a) 0 (b) 2
(c) 1 (d) 3

19. The maximum value of $(\cos\alpha_1).(\cos\alpha_2)\ldots(\cos\alpha_n)$, under the restrictions $0 \le \alpha_1, \alpha_2, \ldots, \alpha_n \le \frac{\pi}{2}$ and $(\cot\alpha_1).(\cot\alpha_2)\ldots(\cot\alpha_n) = 1$ is *(2001S)*
(a) $1/2^{n/2}$ (b) $1/2^n$
(c) $1/2n$ (d) 1

20. If $\alpha + \beta = \pi/2$ and $\beta + \gamma = \alpha$, then $\tan\alpha$ equals *(2001S)*
(a) $2(\tan\beta + \tan\gamma)$ (b) $\tan\beta + \tan\gamma$
(c) $\tan\beta + 2\tan\gamma$ (d) $2\tan\beta + \tan\gamma$

21. The number of integral values of k for which the equation $7\cos x + 5\sin x = 2k + 1$ has a solution is *(2002S)*
(a) 4 (b) 8
(c) 10 (d) 12

22. Given both θ and ϕ are acute angles and $\sin\theta = \frac{1}{2}$, $\cos\phi = \frac{1}{3}$, then the value of $\theta + \phi$ belongs to *(2004S)*
(a) $\left(\frac{\pi}{3}, \frac{\pi}{2}\right]$ (b) $\left(\frac{\pi}{2}, \frac{2\pi}{3}\right)$
(c) $\left(\frac{2\pi}{3}, \frac{5\pi}{6}\right]$ (d) $\left(\frac{5\pi}{6}, \pi\right]$

23. $\cos(\alpha-\beta)=1$ and $\cos(\alpha+\beta)=1/e$ where $\alpha, \beta \in [-\pi, \pi]$. Pairs of α, β which satisfy both the equations is/are *(2005S)*

(a) 0 (b) 1
(c) 2 (d) 4

24. The values of $\theta \in (0, 2\pi)$ for which $2\sin^2\theta - 5\sin\theta + 2 > 0$, are *(2006 - 3M, –1)*

(a) $\left(0, \frac{\pi}{6}\right) \cup \left(\frac{5\pi}{6}, 2\pi\right)$ (b) $\left(\frac{\pi}{8}, \frac{5\pi}{6}\right)$

(c) $\left(0, \frac{\pi}{8}\right) \cup \left(\frac{\pi}{6}, \frac{5\pi}{6}\right)$ (d) $\left(\frac{41\pi}{48}, \pi\right)$

25. Let $\theta \in \left(0, \frac{\pi}{4}\right)$ and $t_1 = (\tan\theta)^{\tan\theta}$, $t_2 = (\tan\theta)^{\cot\theta}$, $t_3 = (\cot\theta)^{\tan\theta}$ and $t_4 = (\cot\theta)^{\cot\theta}$, then *(2006 - 3M, –1)*

(a) $t_1 > t_2 > t_3 > t_4$ (b) $t_4 > t_3 > t_1 > t_2$
(c) $t_3 > t_1 > t_2 > t_4$ (d) $t_2 > t_3 > t_1 > t_4$

26. The number of solutions of the pair of equations

$2\sin^2\theta - \cos 2\theta = 0$

$2\cos^2\theta - 3\sin\theta = 0$

in the interval $[0, 2\pi]$ is *(2007 - 3 Marks)*

(a) zero (b) one (c) two (d) four

D MCQ's WITH ONE OR MORE THAN ONE CORRECT

1. $\left(1+\cos\frac{\pi}{8}\right)\left(1+\cos\frac{3\pi}{8}\right)\left(1+\cos\frac{5\pi}{8}\right)\left(1+\cos\frac{7\pi}{8}\right)$ is equal to *(1984 - 3 Marks)*

(a) $\frac{1}{2}$ (b) $\cos\frac{\pi}{8}$

(c) $\frac{1}{8}$ (d) $\frac{1+\sqrt{2}}{2\sqrt{2}}$

2. The expression $3\left[\sin^4\left(\frac{3\pi}{2}-\alpha\right)+\sin^4(3\pi+\alpha)\right] - 2\left[\sin^6\left(\frac{\pi}{2}+\alpha\right)+\sin^6(5\pi-\alpha)\right]$ is equal to

(a) 0 (b) 1 *(1986 - 2 Marks)*
(c) 3 (d) $\sin 4\alpha + \cos 6\alpha$
(e) none of these

3. The number of all possible triplets (a_1, a_2, a_3) such that $a_1 + a_2\cos(2x) + a_3\sin^2(x) = 0$ for all x is *(1987 - 2 Marks)*

(a) zero (b) one
(c) three (d) infinite
(e) none

4. The values of θ lying between $\theta = 0$ and $\theta = \pi/2$ and satisfying the equation *(1988 - 2 Marks)*

$$\begin{vmatrix} 1+\sin^2\theta & \cos^2\theta & 4\sin 4\theta \\ \sin^2\theta & 1+\cos^2\theta & 4\sin 4\theta \\ \sin^2\theta & \cos^2\theta & 1+4\sin 4\theta \end{vmatrix} = 0 \text{ are}$$

(a) $7\pi/24$ (b) $5\pi/24$ (c) $11\pi/24$ (d) $\pi/24$.

5. Let $2\sin^2 x + 3\sin x - 2 > 0$ and $x^2 - x - 2 < 0$ (x is measured in radians). Then x lies in the interval *(1994)*

(a) $\left(\frac{\pi}{6}, \frac{5\pi}{6}\right)$ (b) $\left(-1, \frac{5\pi}{6}\right)$

(c) $(-1, 2)$ (d) $\left(\frac{\pi}{6}, 2\right)$

6. The minimum value of the expression $\sin\alpha + \sin\beta + \sin\gamma$, where α, β, γ are real numbers satisfying $\alpha + \beta + \gamma = \pi$ is

(a) positive (b) zero *(1995)*
(c) negative (d) –3

7. The number of values of x in the interval $[0, 5\pi]$ satisfying the equation $3\sin^2 x - 7\sin x + 2 = 0$ is *(1998 - 2 Marks)*

(a) 0 (b) 5
(c) 6 (d) 10

8. Which of the following number(s) is/are rational? *(1998 - 2 Marks)*

(a) $\sin 15°$ (b) $\cos 15°$
(c) $\sin 15° \cos 15°$ (d) $\sin 15° \cos 75°$

9. For a positive integer n, let *(1999 - 3 Marks)*

$$f_n(\theta) = \left(\tan\frac{\theta}{2}\right)(1+\sec\theta)(1+\sec 2\theta)(1+\sec 4\theta)....(1+\sec 2^n\theta).$$

Then

(a) $f_2\left(\frac{\pi}{16}\right) = 1$ (b) $f_3\left(\frac{\pi}{32}\right) = 1$

(c) $f_4\left(\frac{\pi}{64}\right) = 1$ (d) $f_5\left(\frac{\pi}{128}\right) = 1$

10. If $\frac{\sin^4 x}{2} + \frac{\cos^4 x}{3} = \frac{1}{5}$, then *(2009)*

(a) $\tan^2 x = \frac{2}{3}$ (b) $\frac{\sin^8 x}{8} + \frac{\cos^8 x}{27} = \frac{1}{125}$

(c) $\tan^2 x = \frac{1}{3}$ (d) $\frac{\sin^8 x}{8} + \frac{\cos^8 x}{27} = \frac{2}{125}$

11. For $0 < \theta < \frac{\pi}{2}$, the solution (s) of

$$\sum_{m=1}^{6} \operatorname{cosec}\left(\theta + \frac{(m-1)\pi}{4}\right)\operatorname{cosec}\left(\theta + \frac{m\pi}{4}\right) = 4\sqrt{2}$$

is (are) *(2009)*

(a) $\frac{\pi}{4}$ (b) $\frac{\pi}{6}$

(c) $\frac{\pi}{12}$ (d) $\frac{5\pi}{12}$

12. Let $\theta, \varphi \in [0, 2\pi]$ be such that $2\cos\theta(1-\sin\varphi) = \sin^2\theta$ $\left(\tan\frac{\theta}{2}+\cot\frac{\theta}{2}\right)\cos\varphi - 1, \tan(2\pi-\theta) > 0$ and $-1 < \sin\theta < -\frac{\sqrt{3}}{2}$, then φ cannot satisfy *(2012)*

(a) $0 < \varphi < \frac{\pi}{2}$ (b) $\frac{\pi}{2} < \varphi < \frac{4\pi}{3}$

(c) $\frac{4\pi}{3} < \varphi < \frac{3\pi}{2}$ (d) $\frac{3\pi}{2} < \varphi < 2\pi$

13. The number of points in $(-\infty\ \infty)$, for which $x^2 - x\sin x - \cos x = 0$, is *(JEE Adv. 2013)*

(a) 6 (b) 4

(c) 2 (d) 0

14. Let $f(x) = x\sin\pi x, x > 0$. Then for all natural numbers n, $f'(x)$ vanishes at *(JEE Adv. 2013)*

(a) A unique point in the interval $\left(n, n+\frac{1}{2}\right)$

(b) A unique point in the interval $\left(n+\frac{1}{2}, n+1\right)$

(c) A unique point in the interval $(n, n+1)$

(d) Two points in the interval $(n, n+1)$

E SUBJECTIVE PROBLEMS :

1. If $\tan\alpha = \frac{m}{m+1}$ and $\tan\beta = \frac{1}{2m+1}$, find the possible values of $(\alpha+\beta)$. *(1978)*

2. (a) Draw the graph of $y = \frac{1}{\sqrt{2}}(\sin x + \cos x)$ from $x = -\frac{\pi}{2}$ to $x = \frac{\pi}{2}$.

(b) If $\cos(\alpha+\beta) = \frac{4}{5}$, $\sin(\alpha-\beta) = \frac{5}{13}$, and α, β lies between 0 and $\frac{\pi}{4}$, find $\tan 2\alpha$. *(1979)*

3. (a) If two sides of a triangle and the included angle are given by $a = (1+\sqrt{3})$ cm, $b = 2$ cm, $C = 60°$, find the other two angles and the third side.

(b) Prove that $5\cos\theta + 3\cos\left(\theta+\frac{\pi}{3}\right) + 3$ lies between -4 and 10. *(1979)*

4. Given $\alpha+\beta-\gamma = \pi$, prove that $\sin^2\alpha + \sin^2\beta - \sin^2\gamma = 2\sin\alpha\sin\beta\cos\gamma$ *(1980)*

5. Given $A = \left\{x : \frac{\pi}{6} \le x \le \frac{\pi}{3}\right\}$ and $f(x) = \cos x - x(1+x)$; find $f(A)$. *(1980)*

6. For all θ in $[0, \pi/2]$ show that, $\cos(\sin\theta) \ge \sin(\cos\theta)$. *(1981 - 4 Marks)*

7. Without using tables, prove that $(\sin 12°)(\sin 48°)(\sin 54°) = \frac{1}{8}$. *(1982 - 2 Marks)*

8. Find the coordinates of the points of intersection of the curves $y = \cos x$, $y = \sin 3x$ if $-\frac{\pi}{2} \le x \le \frac{\pi}{2}$. *(1982 - 3 Marks)*

9. Show that $16\cos\left(\frac{2\pi}{15}\right)\cos\left(\frac{4\pi}{15}\right)\cos\left(\frac{8\pi}{15}\right)\cos\left(\frac{16\pi}{15}\right) = 1$ *(1983 - 2 Marks)*

10. Find all the solution of $4\cos^2 x\sin x - 2\sin^2 x = 3\sin x$ *(1983 - 2 Marks)*

11. Find the values of $x \in (-\pi, +\pi)$ which satisfy the equation $8^{(1+|\cos x|+|\cos^2 x|+|\cos^3 x|+......)} = 4^3$ *(1984 - 2 Marks)*

12. Prove that $\tan\alpha + 2\tan 2\alpha + 4\tan 4\alpha + 8\cot 8\alpha = \cot\alpha$ *(1988 - 2 Marks)*

13. ABC is a triangle such that $\sin(2A+B) = \sin(C-A) = -\sin(B+2C) = \frac{1}{2}$.

If A, B and C are in arithmetic progression, determine the values of A, B and C. *(1990 - 5 Marks)*

14. If $\exp\{(\sin^2 x + \sin^4 x + \sin^6 x + \ldots\ldots\ldots\ldots \infty)\ln 2\}$ satisfies the equation $x^2 - 9x + 8 = 0$, find the value of $\frac{\cos x}{\cos x + \sin x}$, $0 < x < \frac{\pi}{2}$. *(1991 - 4 Marks)*

15. Show that the value of $\frac{\tan x}{\tan 3x}$, wherever defined never lies between $\frac{1}{3}$ and 3. *(1992 - 4 Marks)*

16. Determine the smallest positive value of x (in degrees) for which $\tan(x+100°) = \tan(x+50°)\tan(x)\tan(x-50°)$. *(1993 - 5 Marks)*

17. Find the smallest positive number p for which the equation $\cos(p\sin x) = \sin(p\cos x)$ has a solution $x \in [0, 2\pi]$. *(1995 - 5 Marks)*

18. Find all values of θ in the interval $\left(-\frac{\pi}{2}, \frac{\pi}{2}\right)$ satisfying the equation $(1-\tan\theta)(1+\tan\theta)\sec^2\theta + 2^{\tan^2\theta} = 0$. *(1996 - 2 Marks)*

19. Prove that the values of the function $\frac{\sin x\cos 3x}{\sin 3x\cos x}$ do not lie between $\frac{1}{3}$ and 3 for any real x. *(1997 - 5 Marks)*

20. Prove that $\sum_{k=1}^{n-1}(n-k)\cos\frac{2k\pi}{n} = -\frac{n}{2}$, where $n \ge 3$ is an integer. *(1997 - 5 Marks)*

21. In any triangle ABC, prove that *(2000 - 3 Marks)*

$\cot\frac{A}{2} + \cot\frac{B}{2} + \cot\frac{C}{2} = \cot\frac{A}{2}\cot\frac{B}{2}\cot\frac{C}{2}$.

22. Find the range of values of t for which $2\sin t = \frac{1-2x+5x^2}{3x^2-2x-1}$, $t \in \left[-\frac{\pi}{2}, \frac{\pi}{2}\right]$. *(2005 - 2 Marks)*

F MATCH THE FOLLOWING :

Each question contains statements given in two columns, which have to be matched. The statements in Column-I are labelled A, B, C and D, while the statements in Column-II are labelled p, q, r, s and t. Any given statement in Column-I can have correct matching with ONE OR MORE statement(s) in Column-II. The appropriate bubbles corresponding to the answers to these questions have to be darkened as illustrated in the following example :

If the correct matches are A-p, s and t; B-q and r; C-p and q; and D-s then the correct darkening of bubbles will look like the given.

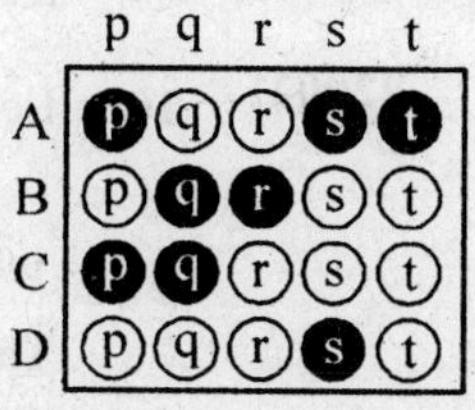

1. In this questions there are entries in columns 1 and 2. Each entry in column 1 is related to exactly one entry in column 2. Write the correct letter from column 2 against the entry number in column 1 in your answer book.

$\dfrac{\sin 3\alpha}{\cos 2\alpha}$ is **(1992 - 2 Marks)**

Column I	Column II
(A) positive	(p) $\left(\dfrac{13\pi}{48}, \dfrac{14\pi}{48}\right)$
(B) negative	(q) $\left(\dfrac{14\pi}{48}, \dfrac{18\pi}{48}\right)$
	(r) $\left(\dfrac{18\pi}{48}, \dfrac{23\pi}{48}\right)$
	(s) $\left(0, \dfrac{\pi}{2}\right)$

I INTEGER VALUE CORRECT TYPE :

1. The number of all possible values of θ where $0 < \theta < \pi$, for which the system of equations

$$(y+z)\cos 3\theta = (xyz)\sin 3\theta$$

$$x\sin 3\theta = \frac{2\cos 3\theta}{y} + \frac{2\sin 3\theta}{z}$$

$$(xyz)\sin 3\theta = (y+2z)\cos 3\theta + y\sin 3\theta$$

have a solution (x_0, y_0, z_0) with $y_0 z_0 \neq 0$, is **(2010)**

2. The number of values of θ in the interval, $\left(-\dfrac{\pi}{2}, \dfrac{\pi}{2}\right)$ such that $\theta \neq \dfrac{n\pi}{5}$ for n = 0, ±1, ±2 and $\tan\theta = \cot 5\theta$ as well as $\sin 2\theta = \cos 4\theta$ is **(2010)**

3. The maximum value of the expression

$$\frac{1}{\sin^2\theta + 3\sin\theta\cos\theta + 5\cos^2\theta}$$ is **(2010)**

4. Two parallel chords of a circle of radius 2 are at a distance $\sqrt{3}+1$ apart. If the chords subtend at the center, angles of $\dfrac{\pi}{k}$ and $\dfrac{2\pi}{k}$, where $k > 0$, then the value of $[k]$ is **(2010)**

[Note : $[k]$ denotes the largest integer less than or equal to k]

5. The positive integer value of $n > 3$ satisfying the equation

$$\frac{1}{\sin\left(\frac{\pi}{n}\right)} = \frac{1}{\sin\left(\frac{2\pi}{n}\right)} + \frac{1}{\sin\left(\frac{3\pi}{n}\right)}$$ is **(2011)**

Section-B JEE Main/ AIEEE

1. The period of $\sin^2 \theta$ is **[2002]**
(a) π^2 (b) π
(c) 2π (d) $\pi/2$

2. The number of solution of $\tan x + \sec x = 2\cos x$ in $[0, 2\pi)$ is **[2002]**
(a) 2 (b) 3
(c) 0 (d) 1

3. Which one is not periodic **[2002]**
(a) $|\sin 3x| + \sin^2 x$ (b) $\cos\sqrt{x} + \cos^2 x$
(c) $\cos 4x + \tan^2 x$ (d) $\cos 2x + \sin x$

4. Let α, β be such that $\pi < \alpha - \beta < 3\pi$.
If $\sin\alpha + \sin\beta = -\dfrac{21}{65}$ and $\cos\alpha + \cos\beta = -\dfrac{27}{65}$, then the value of $\cos\dfrac{\alpha-\beta}{2}$ **[2004]**
(a) $\dfrac{-6}{65}$ (b) $\dfrac{3}{\sqrt{130}}$
(c) $\dfrac{6}{65}$ (d) $-\dfrac{3}{\sqrt{130}}$

5. If $u = \sqrt{a^2\cos^2\theta + b^2\sin^2\theta} + \sqrt{a^2\sin^2\theta + b^2\cos^2\theta}$ then the difference between the maximum and minimum values of u^2 is given by **[2004]**
(a) $(a-b)^2$ (b) $2\sqrt{a^2+b^2}$
(c) $(a+b)^2$ (d) $2(a^2+b^2)$

6. A line makes the same angle θ, with each of the x and z axis. If the angle β, which it makes with y-axis, is such that $\sin^2\beta = 3\sin^2\theta$, then $\cos^2\theta$ equals **[2004]**
(a) $\dfrac{2}{5}$ (b) $\dfrac{1}{5}$
(c) $\dfrac{3}{5}$ (d) $\dfrac{2}{3}$

7. The number of values of x in the interval $[0, 3\pi]$ satisfying the equation $2\sin^2 x + 5\sin x - 3 = 0$ is **[2006]**
(a) 4 (b) 6
(c) 1 (d) 2

8. If $0 < x < \pi$ and $\cos x + \sin x = \dfrac{1}{2}$, then $\tan x$ is **[2006]**
(a) $\dfrac{(1-\sqrt{7})}{4}$ (b) $\dfrac{(4-\sqrt{7})}{3}$
(c) $-\dfrac{(4+\sqrt{7})}{3}$ (d) $\dfrac{(1+\sqrt{7})}{4}$

9. Let **A** and **B** denote the statements
A : $\cos\alpha + \cos\beta + \cos\gamma = 0$
B : $\sin\alpha + \sin\beta + \sin\gamma = 0$
If $\cos(\beta-\gamma) + \cos(\gamma-\alpha) + \cos(\alpha-\beta) = -\dfrac{3}{2}$, then : **[2009]**
(a) **A** is false and **B** is true (b) both **A** and **B** are true
(c) both **A** and **B** are false (d) **A** is true and **B** is false

10. Let $\cos(\alpha+\beta) = \dfrac{4}{5}$ and $\sin(\alpha-\beta) = \dfrac{5}{13}$, where $0 \le \alpha, \beta \le \dfrac{\pi}{4}$.
Then $\tan 2\alpha$ = **[2010]**
(a) $\dfrac{56}{33}$ (b) $\dfrac{19}{12}$
(c) $\dfrac{20}{7}$ (d) $\dfrac{25}{16}$

11. If $A = \sin^2 x + \cos^4 x$, then for all real x : **[2011]**
(a) $\dfrac{13}{16} \le A \le 1$ (b) $1 \le A \le 2$
(c) $\dfrac{3}{4} \le A \le \dfrac{13}{16}$ (d) $\dfrac{3}{4} \le A \le 1$

12. The possible values of $\theta \in (0, \pi)$ such that $\sin(\theta) + \sin(4\theta) + \sin(7\theta) = 0$ are : **[2011 RS]**
(a) $\dfrac{\pi}{4}, \dfrac{5\pi}{12}, \dfrac{\pi}{2}, \dfrac{2\pi}{3}, \dfrac{3\pi}{4}, \dfrac{8\pi}{9}$
(b) $\dfrac{2\pi}{9}, \dfrac{\pi}{4}, \dfrac{\pi}{2}, \dfrac{2\pi}{3}, \dfrac{3\pi}{4}, \dfrac{35\pi}{36}$
(c) $\dfrac{2\pi}{9}, \dfrac{\pi}{4}, \dfrac{\pi}{2}, \dfrac{2\pi}{3}, \dfrac{3\pi}{4}, \dfrac{8\pi}{9}$
(d) $\dfrac{2\pi}{9}, \dfrac{\pi}{4}, \dfrac{4\pi}{9}, \dfrac{\pi}{2}, \dfrac{3\pi}{4}, \dfrac{8\pi}{9}$

13. In a ΔPQR, If $3\sin P + 4\cos Q = 6$ and $4\sin Q + 3\cos P = 1$, then the angle R is equal to : **[2012]**
(a) $\dfrac{5\pi}{6}$ (b) $\dfrac{\pi}{6}$ (c) $\dfrac{\pi}{4}$ (d) $\dfrac{3\pi}{4}$

14. ABCD is a trapezium such that AB and CD are parallel and BC $\perp$ CD. If $\angle ADB = \theta$, BC $= p$ and CD $= q$, then AB is equal to : **[JEE M 2013]**
(a) $\dfrac{(p^2+q^2)\sin\theta}{p\cos\theta + q\sin\theta}$ (b) $\dfrac{p^2+q^2\cos\theta}{p\cos\theta + q\sin\theta}$
(c) $\dfrac{p^2+q^2}{p^2\cos\theta + q^2\sin\theta}$ (d) $\dfrac{(p^2+q^2)\sin\theta}{(p\cos\theta + q\sin\theta)^2}$

15. The expression $\dfrac{\tan A}{1-\cot A} + \dfrac{\cot A}{1-\tan A}$ can be written as : **[JEE M 2013]**
(a) sinA cosA + 1 (b) secA cosecA + 1
(c) tanA + cotA (d) secA + cosecA

Solutions & Explanations

Section-A : JEE Advanced/ IIT-JEE

A 1. 6 2. ϕ 3. $\left[0, \frac{\pi}{6}\right] \cup \left\{\frac{\pi}{2}\right\} \cup \left[\frac{5\pi}{6}, \pi\right]$ 4. $\frac{-\sqrt{3}}{2}$

5. $\frac{1}{64}$ 6. $\frac{1}{8}$ 7. $\frac{1}{3}$ 8. $n\pi, n\pi \pm \frac{\pi}{3}$ 9. $\pm\sqrt{2}$ 10. $-\frac{\pi}{2}, \frac{\pi}{2}, 0$

B 1. T 2. F

C 1. (b) 2. (a) 3. (b) 4. (a) 5. (c) 6. (c) 7. (b) 8. (d)
9. (c) 10. (b) 11. (d) 12. (c) 13. (c) 14. (d) 15. (b) 16. (a)
17. (c) 18. (c) 19. (a) 20. (c) 21. (b) 22. (b) 23. (d) 24. (a)
25. (b) 26. (c)

D 1. (c) 2. (b) 3. (d) 4. (a, c) 5. (d) 6. (c) 7. (c) 8. (c)
9. (a,b,c,d) 10. (a, b) 11. (c, d) 12. (a, c, d) 13. (c) 14. (b, c)

E 1. $n\pi + \frac{\pi}{4}$ 2. (b) $\frac{56}{33}$ 3. (a) $45°, 75°, \sqrt{6}$ cm 5. $\left[\frac{1}{2} - \frac{\pi}{3}\left(1 + \frac{\pi}{3}\right), \frac{\sqrt{3}}{2} - \frac{\pi}{6}\left(1 + \frac{\pi}{6}\right)\right]$

8. $\left(\frac{\pi}{4}, \frac{1}{\sqrt{2}}\right), \left(\frac{\pi}{8}, \cos\frac{\pi}{8}\right), \left(-\frac{3\pi}{8}, \cos\frac{3\pi}{8}\right)$ 10. $n\pi, n\pi + (-1)^n \frac{\pi}{10}, n\pi + (-1)^n \left(\frac{-3\pi}{10}\right)$

11. $\pm\frac{\pi}{3}, \pm\frac{2\pi}{3}$ 13. $45°, 60°, 75°$ 14. $\frac{\sqrt{3}-1}{2}$ 16. $30°$ 17. $\frac{\pi\sqrt{2}}{4}$ 18. $\pm\frac{\pi}{3}$

22. $\left[-\frac{\pi}{2}, \frac{-\pi}{10}\right] \cup \left[\frac{3\pi}{10}, \frac{\pi}{2}\right]$ 23. 3 24. 3 25. 2 26. 3

F 1. (A) - r, (B) - p

I 1. 3 2. 3 3. 2 4. 3 5. 7

Section-B : JEE Main/ AIEEE

1. (b) 2. (b) 3. (b) 4. (d) 5. (a) 6. (c) 7. (a) 8. (c)
9. (b) 10. (a) 11. (d) 12. (d) 13. (b) 14. (a) 15. (b)

Section-A JEE Advanced/ IIT-JEE

A FILL IN THE BLANKS :

1. Given $\sin^3 x . \sin 3x \sum_{m=0}^{n} C_m \cos mx$

$$\sin^3 x \sin 3x = \frac{1}{4}[3\sin x - \sin 3x]\sin 3x$$

$$= \frac{1}{4}\left[\frac{3}{2}.2\sin x.\sin 3x - \sin^2 3x\right]$$

$$= \frac{1}{4}\left[\frac{3}{2}(\cos 2x - \cos x) - \frac{1}{2}(1 - \cos x)\right]$$

$$= \frac{1}{8}\left[\cos 6x + 3\cos 2x - 3\cos x - 1\right]$$

We observe that on LHS 6 is the max value of m.

$\therefore \quad n = 6$

2. The equations are $x + y = 2\pi/3$ ...(i)

$\cos x + \cos y = 3/2$ ...(ii)

From eq. (ii) $2\cos\frac{x+y}{2}\cos\frac{x-y}{2}=\frac{3}{2}$

$\Rightarrow\ 2\cos\frac{\pi}{3}\cos\frac{x-y}{2}=\frac{3}{2}$ [Using eq. (i)]

$\Rightarrow\ 2.\frac{1}{2}\cos\frac{x-y}{2}=\frac{3}{2}\ \Rightarrow\ \cos\frac{x-y}{2}=\frac{3}{2}>1$

Which has no solution.

$\therefore$ The solution of given equations is imaginary.

3. We have $2\sin^2 x-3\sin x+1\geq 0$

$\Rightarrow\ (2\sin x-1)(\sin x-1)\geq 0$

$\Rightarrow\ \left(\sin x-\frac{1}{2}\right)(\sin x-1)\geq 0\ \Rightarrow\ \sin x\leq\frac{1}{2}$ or $\sin x\geq 1$

But we know that $\sin x\leq 1$ and $\sin x\geq 0$ for $x\in[0,\pi]$

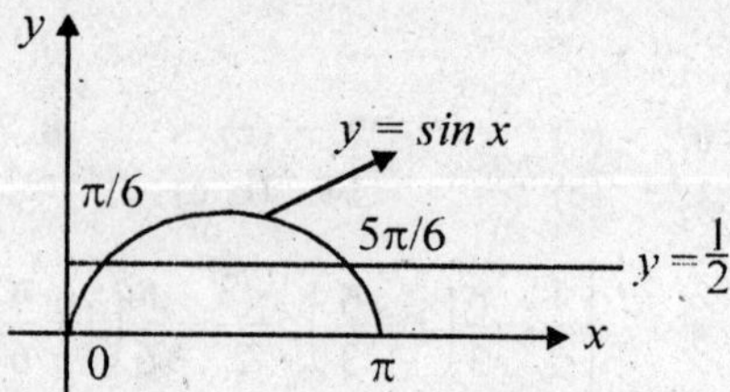

$\Rightarrow$ either $\sin x=1$ or $0\leq\sin x\leq\frac{1}{2}$

$\Rightarrow$ either $x=\pi/2$ or $x\in[0,\pi/6]\cup[5\pi/6,\pi]$

Combining, we get $x\in[0,\pi/6]\cup\{\pi/2\}\cup[5\pi/6,\pi]$

4. Let ΔABC be inscribed in circle with centre at O such that $\angle AOB=\alpha, \angle BOC=\beta$ and $\angle COA=\gamma$.

Also $\alpha+\beta+\gamma=2\pi$

$\Rightarrow\ \frac{\alpha}{2}+\frac{\beta}{2}+\frac{\gamma}{2}=\pi$ ....(1)

Now A.M. of $\cos(\alpha+\pi/2)$, $\cos(\beta+\pi/2)$ and $\cos(\gamma+\pi/2)$]

$=\frac{1}{3}[\cos(\alpha+\pi/2)+\cos(\beta+\pi/2)+\cos(\gamma+\pi/2)]$

$=\frac{1}{3}[-\sin\alpha-\sin\beta-\sin\gamma]$ [Using $\cos(\pi/2+\theta)=-\sin\theta$]

$=-\frac{1}{3}[\sin\alpha+\sin\beta+\sin\gamma]$

$=-\frac{1}{3}\left[2\sin\frac{\alpha+\beta}{2}\cos\frac{\alpha-\beta}{2}+2\sin\frac{\gamma}{2}\cos\frac{\gamma}{2}\right]$

$=-\frac{1}{3}\left[2\sin\left(\pi-\frac{\gamma}{2}\right)\cos\frac{\alpha-\beta}{2}+2\sin\frac{\gamma}{2}\cos\frac{\gamma}{2}\right]$

[Using eq.(1)]

$=-\frac{1}{3}\left[2\sin\frac{\gamma}{2}\cos\frac{\alpha-\beta}{2}+2\sin\frac{\gamma}{2}\cos\frac{\gamma}{2}\right]$

$=-\frac{2}{3}\sin\frac{\gamma}{2}\left[\cos\frac{\alpha-\beta}{2}+\cos\left(\pi-\frac{\alpha+\beta}{2}\right)\right]$

[Using eq.(1)]

$=-\frac{2}{3}\sin\frac{\gamma}{2}\left[\cos\frac{\alpha-\beta}{2}-\cos\frac{\alpha+\beta}{2}\right]$

$=-\frac{2}{3}\sin\frac{\gamma}{2}\left[2\sin\frac{\alpha}{2}\sin\frac{\beta}{2}\right]=-\frac{4}{3}\sin\frac{\alpha}{2}\sin\frac{\beta}{2}\sin\frac{\gamma}{2}$

Now the arithmatic will be least when $\sin\alpha/2\sin\beta/2\sin\gamma/2$ is max. and it is max when $\sin\alpha/2=\sin\beta/2=\sin\gamma/2$

$\Rightarrow\ \alpha=\beta=\gamma$ ....(2)

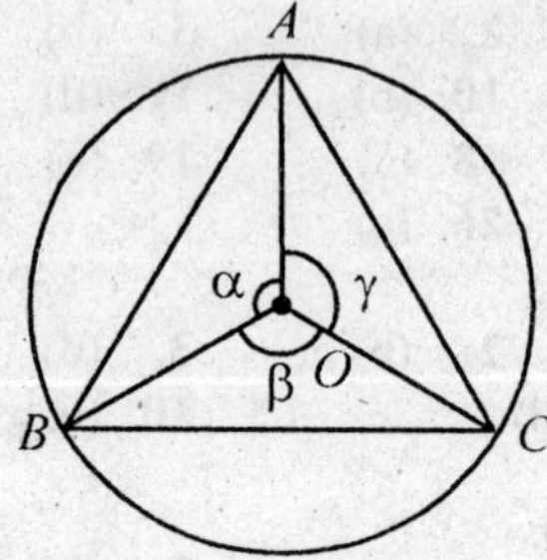

Using (1) and (2), we get $\alpha=\beta=\gamma=120°$

Thus min value of A.M. $=-\frac{4}{3}\sin^3 120°$

$=-\frac{4}{3}\times\left(\frac{\sqrt{3}}{2}\right)^3=-\frac{\sqrt{3}}{2}$

ALTERNATE SOLUTION

We know that A.M. $\geq$ G.M.

$\Rightarrow$ Min value of AM. is obtained when AM = GM

$\Rightarrow$ The quantities whose AM is being taken are equal.

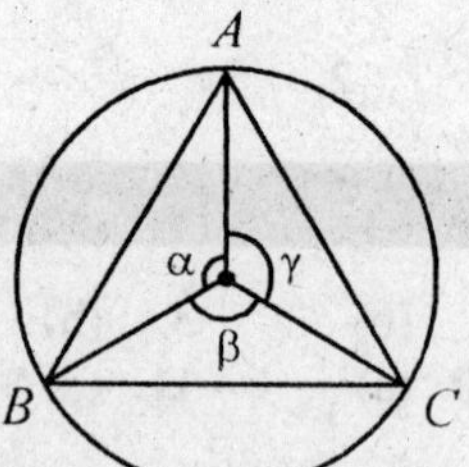

i.e., $\cos\left(\alpha+\frac{\pi}{2}\right)=\cos\left(\beta+\frac{\pi}{2}\right)=\cos\left(\gamma+\frac{\pi}{2}\right)$

$\Rightarrow\ \sin\alpha=\sin\beta=\sin\gamma$

Also $\alpha+\beta+\gamma=360°\Rightarrow\alpha=\beta=\gamma=120°=2\pi/3$

$\therefore$ Min value of A.M.

$=\dfrac{\cos\left(\frac{2\pi}{3}+\frac{\pi}{2}\right)+\cos\left(\frac{2\pi}{3}+\frac{\pi}{2}\right)+\cos\left(\cos\frac{2\pi}{3}+\frac{\pi}{2}\right)}{3}$

$=\dfrac{-\sin\frac{2\pi}{3}}{3}=-\frac{\sqrt{3}}{2}$

5. $\sin\frac{\pi}{14}\sin\frac{3\pi}{14}\sin\frac{5\pi}{14}\sin\frac{7\pi}{14}\sin\frac{9\pi}{14}\sin\frac{11\pi}{14}\sin\frac{13\pi}{14}$

$$=\left(\sin\frac{\pi}{14}\sin\frac{3\pi}{14}\sin\frac{5\pi}{14}\right)^2$$

$$=\left[\cos\left(\frac{\pi}{2}-\frac{\pi}{14}\right)\cos\left(\frac{\pi}{2}-\frac{3\pi}{14}\right)\cos\left(\frac{\pi}{2}-\frac{5\pi}{14}\right)\right]^2$$

$$=\left[\cos\frac{3\pi}{7}\cos\frac{2\pi}{7}\cos\frac{\pi}{7}\right]^2$$

$$=\left(\frac{1}{8\sin\pi/7}\sin\frac{8\pi}{7}\right)^2=\left(\frac{\sin(\pi+\pi/7)}{8\sin\pi/7}\right)^2$$

$$=\left(\frac{-\sin\pi/7}{8\sin\pi/7}\right)^2=\left(\frac{1}{8}\right)^2=\frac{1}{64}$$

6. $K=\sin\frac{\pi}{18}\sin\frac{5\pi}{18}\sin\frac{7\pi}{18}$

$$=\cos\left(\frac{\pi}{2}-\frac{\pi}{18}\right)\cos\left(\frac{\pi}{2}-\frac{5\pi}{18}\right)\cos\left(\frac{\pi}{2}-\frac{7\pi}{18}\right)$$

$$=\cos\frac{\pi}{9}\cos\frac{2\pi}{9}\cos\frac{4\pi}{9}$$

$$=\frac{1}{2^3\sin\frac{\pi}{9}}.\sin\frac{8\pi}{9}$$ **NOTE THIS STEP**

{Using $\cos\alpha\cos2\alpha\cos2^2\alpha\,.........\cos2^{n-1}\alpha$

$$=\frac{1}{2^n\sin\alpha}.\sin(2^n\alpha)]$$

$$=\frac{1}{8\sin\pi/9}.\sin\pi/9=\frac{1}{8}$$

7. We have $A, B>0$ and $A+B=\frac{\pi}{3}$

To find max value of $\tan A\tan B$.

Let $z=\tan A\tan B=\tan A\tan(\pi/3-A)$

$$=\tan A\left[\frac{\sqrt{3}-\tan A}{1+\sqrt{3}\tan A}\right]$$

Let $\tan A=x$ then $z=\frac{\sqrt{3}x-x^2}{1+\sqrt{3}x}$

$$\Rightarrow\frac{dz}{dx}=\frac{(\sqrt{3}-2x)(1+\sqrt{3}x)-\sqrt{3}(\sqrt{3}x-x^2)}{(1+\sqrt{3}x)^2}$$

For max. value of z, put $\frac{dz}{dx}=0$

$$\Rightarrow\frac{-\sqrt{3}x^2-2x+\sqrt{3}}{(1+\sqrt{3}x)^2}=\Rightarrow\sqrt{3}x^2+2x-\sqrt{3}=0$$

$$\Rightarrow x=1/\sqrt{3},-\sqrt{3}$$

As A and $B>0$ and $A+B=\frac{\pi}{3}$

$\Rightarrow$ A and B both are +ve acute angles

$\therefore\ x=\frac{1}{\sqrt{3}}(x\neq -ve)$

$$\frac{d^2z}{dx^2}=\frac{(-2\sqrt{3}x-2)(1+\sqrt{3}x)^2-2\sqrt{3}(1+\sqrt{3}x)(-\sqrt{3}x^2-2x+\sqrt{3})}{(1+\sqrt{3}x)^4}$$

$$\therefore\ \left.\frac{d^2z}{dx^2}\right|_{x=\frac{1}{\sqrt{3}}}=\frac{(-4)(2)^2-0}{(2)^4}=-ve$$

$\Rightarrow$ z is max at $x=\frac{1}{\sqrt{3}}$

i.e., at $\tan A=\frac{1}{\sqrt{3}}\Rightarrow A=30°$

$\Rightarrow B=30°$

$\therefore$ Max value of $z=\tan30°\tan30°$

$$=\frac{1}{\sqrt{3}}.\frac{1}{\sqrt{3}}=\frac{1}{3}$$

ALTERNATE SOLUTION

$A+B=\pi/3\ \Rightarrow\tan(A+B)=\sqrt{3}$

$$\Rightarrow\frac{\tan A+\tan B}{1-\tan A\tan B}=\sqrt{3}\Rightarrow\frac{\tan A+\frac{y}{\tan A}}{1-y}=\sqrt{3}$$

[where $y=\tan A\tan B$]

$\Rightarrow\tan^2A+\sqrt{3}\,(y-1)\tan A+y=0$

For real value of $\tan A$, $3(y-1)^2-4y\geq0$

$\Rightarrow 3y^2-10y+3\geq0\Rightarrow(y-3)\,(y-\frac{1}{3})\geq0$

$\Rightarrow\ y\leq\frac{1}{3}$ or ≥3

But $A, B>0$ and $A+B=\pi/3\Rightarrow A, B<\pi/3$

$\Rightarrow\ \tan A\tan B<3$

$\therefore\ y\leq\frac{1}{3},$ i.e., max. value of y is 1/3.

8. $\tan^2\theta+\sec2\theta=1$

$t^2+\frac{1+t^2}{1-t^2}=1$ where $t=\tan\theta$

$\therefore\ t^2(t^2-3)=0\ \therefore\ \tan\theta=0,\pm\sqrt{3}$ etc.

which means $\theta=n\pi$ and $\theta=n\pi\pm\pi/3$

9. As given that $\cos(x-y)$, $\cos x$ and $\cos(x+y)$ are in H.P.

$$\therefore\ \frac{2}{\cos x}=\frac{1}{\cos(x-y)}+\frac{1}{\cos(x+y)}$$

$\Rightarrow \frac{2}{\cos x}=\frac{\cos(x+y)+\cos(x-y)}{\cos(x+y)\cos(x-y)}$

$\Rightarrow \frac{2}{\cos x}=\frac{2\cos(x)\cos y}{\cos^2 x-\sin^2 y}$

$\Rightarrow \cos^2 x-\sin^2 y=\cos^2 x\cos y$

$\Rightarrow \cos^2 x(1-\cos y)=\sin^2 y$

$\Rightarrow \cos^2 x=\frac{(1-\cos^2 y)}{1-\cos y}=1+\cos y$

$\Rightarrow \cos^2 x=2\cos^2(y/2)$

$\Rightarrow \cos^2 x\sec^2 y/2=2$

$\Rightarrow \cos x\sec(y/2)=\pm\sqrt{2}$

10. $\cos^7 x=1-\sin^4 x=(1-\sin^2 x)(1+\sin^2 x)$

$=\cos^2 x(1+\sin^2 x)$

$\therefore \cos x=0$ or $x=\pi/2,-\pi/2$

or $\cos^5 x=1+\sin^2 x$ or $\cos^5 x-\sin^2 x=1$

Now maximum value of each $\cos x$ or $\sin x$ is 1.

Hence the above equation will hold when

$\cos x=1$ and $\sin x=0$. Both these imply $x=0$

Hence $x=-\frac{\pi}{2},\frac{\pi}{2},0$

B TRUE / FALSE :

1. $\tan A=\frac{1-\cos B}{\sin B}=\frac{2\sin^2 B/2}{2\sin B/2\cos/2}=\tan B/2$

Hence $\tan 2A=\frac{2\tan A}{1-\tan^2 A}=\frac{2\tan B/2}{1-\tan^2 B/2}=\tan B$

$\therefore$ Statement is true.

2. Given equation is $\sin^4\theta-2\sin^2\theta-1=0$

Here, $D=4+4=8$

$\therefore \sin^2\theta=\frac{2\pm\sqrt{8}}{2}=1\pm\sqrt{2}.$

But $\sin^2\theta$ can not be –ve

$\therefore \sin^2\theta=\sqrt{2}+1$

But as $-1\le\sin\theta\le 1$ $\therefore \sin^2\theta\ne\sqrt{2}+1$

Thus there is no value of θ which satisfy the given equation.

$\therefore$ Statement is false.

C MCQ's WITH ONE CORRECT ANSWER :

1. **(b)** $\tan\theta=\frac{-4}{3}\Rightarrow\theta\in$ II quad or IV quad.

$\therefore 0<\sin\theta<1$ or $-1<\sin\theta<0$

$\therefore \sin\theta$ may be $\frac{4}{5}$ or $-\frac{4}{5}$

2. **(a)** $\alpha+\beta+\gamma=2\pi\Rightarrow\frac{\alpha}{2}+\frac{\beta}{2}+\frac{\gamma}{2}=\pi$

$\therefore \tan\left(\frac{\alpha}{2}+\frac{\beta}{2}\right)=\tan\left(\pi-\frac{\gamma}{2}\right)=-\tan\frac{\gamma}{2}$

$\Rightarrow \frac{\tan\alpha/2+\tan\beta/2}{1-\tan\alpha/2\tan\beta/2}=-\tan\gamma/2$

$\Rightarrow \tan\alpha/2+\tan\beta/2+\tan\gamma/2$

$=\tan\alpha/2\tan\beta/2\tan\gamma/2$

3. **(b)** Given that $A=\sin^2\theta+\cos^4\theta$

$\Rightarrow A=\left(\frac{1-\cos 2\theta}{2}\right)+\left(\frac{1+\cos 2\theta}{2}\right)^2$

$=\frac{1}{4}[2-2\cos 2\theta+1+2\cos 2\theta+\cos^2 2\theta]$

$=\frac{1}{4}\left(3+\frac{1+\cos 4\theta}{2}\right)=\frac{1}{4}\left(\frac{7+\cos 4\theta}{2}\right)$

$=\frac{7+\cos 4\theta}{8}$

Now $-1\le\cos 4\theta\le 1$, real values of θ

$\Rightarrow 6\le 7+\cos 4\theta\le 8$

$\Rightarrow \frac{6}{8}\le\frac{7+\cos 4\theta}{8}\le\frac{8}{8}\Rightarrow\frac{3}{4}\le A\le 1$

ALTERNATE SOLUTION

$A=\sin^2\theta+\cos^4\theta=\sin^2\theta+(1-\sin^2\theta)^2$

$=\sin^4\theta-\sin^2\theta+1$

$\Rightarrow A=\left(\sin^2\theta-\frac{1}{2}\right)^2+\frac{3}{4}$

But $0\le\left(\sin^2\theta-\frac{1}{2}\right)^2\le\frac{1}{4}$

$\therefore \frac{3}{4}\le\left(\sin^2\theta-\frac{1}{2}\right)^2+\frac{3}{4}\le 1$ or $\frac{3}{4}\le A\le 1$

4. **(a)** The given equation is

$2\cos^2\left(\frac{x}{2}\right)\sin^2 x=x^2+\frac{1}{x^2}$ where $0<x\le\frac{\pi}{2}$

LHS $=2\cos^2\frac{x}{2}\sin^2 x=(1+\cos x)\sin^2 x$

$\because 1+\cos x<2$ and $\sin^2 x\le 1$ for $0<x\le\frac{\pi}{2}$

$\therefore (1+\cos x)\sin^2 x<2$

And R.H.S. $=x^2+\frac{1}{x^2}\ge 2$ $\therefore$ For $0<x\le\frac{\pi}{2}$,

given equation is not possible for any real value of x.

5. **(c)** $\sin x + \cos x = 1$

$\Rightarrow \frac{1}{\sqrt{2}}\sin x + \frac{1}{\sqrt{2}}\cos = \frac{1}{\sqrt{2}}$

$\Rightarrow \sin x \cos\frac{\pi}{4} + \cos x \sin\frac{\pi}{4} = \sin\frac{\pi}{4}$

$\Rightarrow \sin(x + \pi/4) = \sin \pi/4$

$\Rightarrow x + \pi/4 = n\pi + (-1)^n \pi/4, n \in Z$ (the set of integers)

$\Rightarrow x = n\pi + (-1)^n \pi/4 - \pi/4$

where n = 0, ± 1, ± 2, ...

6. **(c)** The given expression is $\sqrt{3}\,\text{cosec}\,20° - \sec 20°$

$= \frac{\sqrt{3}}{\sin 20°} - \frac{1}{\cos 20°} = \frac{\sqrt{3}\cos 20° - \sin 20°}{\sin 20° \cos 20°}$

$= 4\left[\frac{\frac{\sqrt{3}}{2}\cos 20° - \frac{1}{2}\sin 20°}{2\sin 20° \cos 20°}\right]$

$= 4\left[\frac{\sin 60° \cos 20° - \cos 60° \sin 20°}{\sin(2 \times 20°)}\right]$

$= \frac{4\sin(60° - 20°)}{\sin 40°} = \frac{4\sin 40°}{\sin 40°} = 4$

7. **(b)** The given equation is

$\sin x - 3\sin 2x + \sin 3x = \cos x - 3\cos 2x + \cos 3x$

$\Rightarrow 2\sin 2x\cos x - 3\sin 2x = 2\cos 2x\cos x - 3\cos 2x$

$\Rightarrow \sin 2x(2\cos x - 3) = \cos 2x(2\cos x - 3)$

$\Rightarrow \sin 2x = \cos 2x$ (as $\cos x \neq 3/2$)

$\Rightarrow \tan 2x = 1$

$\Rightarrow 2x = n\pi + \pi/4$

$\Rightarrow x = \frac{n\pi}{2} + \frac{\pi}{8}$

8. **(d)** The given equation is

$(\cos p - 1)x^2 + (\cos p)x + \sin p = 0$

For this equation to have real roots $D \geq 0$

$\Rightarrow \cos^2 p - 4\sin p(\cos p - 1) \geq 0$

$\Rightarrow \cos^2 p - 4\sin p\cos p + 4\sin^2 p + 4\sin p - 4\sin^2 p \geq 0$

$\Rightarrow (\cos p - 2\sin p)^2 + 4\sin p(1 - \sin p) \geq 0$

For every real value of p $(\cos p - 2\sin p)^2 \geq 0$ and $1 - \sin p \geq 0$

$\therefore D \geq 0, \forall p \in (0, \pi)$

9. **(c)** The given eq is, $\tan x + \sec x = 2\cos x$

$\Rightarrow \frac{\sin x}{\cos x} + \frac{1}{\cos x} = 2\cos x$

$\Rightarrow \sin x + 1 = 2\cos^2 x$

$\Rightarrow 2\sin^2 x + \sin x - 1 = 0$

$\Rightarrow (2\sin x - 1)(\sin x + 1) = 0 \Rightarrow \sin x = \frac{1}{2}, -1$

$\Rightarrow x = \frac{\pi}{6}, \frac{5\pi}{6}, \frac{3\pi}{2} \in [0, 2\pi]$

But for $x = \frac{3\pi}{2}$, given eq. is not defined,

$\therefore$ only 2 solutions.

10. **(b)** $\sec 2x - \tan 2x = \frac{1 - \sin 2x}{\cos 2x} = \frac{1 - \cos 2\left(\frac{\pi}{4} - x\right)}{\sin 2\left(\frac{\pi}{4} - x\right)}$

$= \frac{2\sin^2\left(\frac{\pi}{4} - x\right)}{2\sin\left(\frac{\pi}{4} - x\right)\cos\left(\frac{\pi}{4} - x\right)} = \tan\left(\frac{\pi}{4} - x\right)$

11. **(d)** $\sin\frac{\pi}{2n} + \cos\frac{\pi}{2n} = \frac{\sqrt{n}}{2}$

$\Rightarrow \sin^2\frac{\pi}{2n} + \cos^2\frac{\pi}{2n} + 2\sin\frac{\pi}{2n}\cos\frac{\pi}{2n} = \frac{n}{4}$

$\Rightarrow 1 + \sin\frac{\pi}{n} = \frac{n}{4} \Rightarrow \sin\frac{\pi}{n} = \frac{n-4}{4}$

For $n = 2$ the given equation is not satisfied. Considering

$n > 1$ and $n \neq 2, 0 < \sin\frac{\pi}{n} < 1 \Rightarrow 0 < \frac{n-4}{4} < 1$

$\Rightarrow 4 < n < 8.$

12. **(c)** $\sin\left\{(\omega^{10} + \omega^{23})\pi - \frac{\pi}{4}\right\} = \sin\left\{(\omega + \omega^2)\pi - \frac{\pi}{4}\right\}$

$= \sin\left(-\pi - \frac{\pi}{4}\right)$

$= -\sin\left(\pi + \frac{\pi}{4}\right) = \sin \pi/4 = 1/\sqrt{2}$

13. **(c)** $3(\sin x - \cos x)^4 + 6(\sin x + \cos x)^2 + 4(\sin^6 x + \cos^6 x)$

$= 3(1 - \sin 2x)^2 + 6(1 + \sin 2x)$

$+ 4[(\sin^2 x + \cos^2 x)^3 - 3\sin^2 x\cos^2 x(\sin^2 x + \cos^2 x)]$

$= 3 - 6\sin 2x + 3\sin^2 2x + 6 + 6\sin 2x + 4\left[1 - \frac{3}{4}\sin^2 2x\right]$

$= 13 + 3\sin^2 2x - 3\sin^2 2x = 13$

14. **(d)** The given equation is

$2\sin^2\theta - 3\sin\theta - 2 = 0$

$\Rightarrow (2\sin\theta + 1)(\sin\theta - 2) = 0$

$\Rightarrow \sin\theta = -\frac{1}{2}$ [$\because \sin\theta - 2 = 0$ is not possible]

$\Rightarrow \sin\theta = \sin(-\pi/6) = \sin(7\pi/6)$

$\Rightarrow \theta = n\pi + (-1)^n(-\pi/6) = n\pi + (-1)^n 7\pi/6$

$\Rightarrow$ Thus, $\theta = n\pi + (-1)^n 7\pi/6$

15. **(b)** We have $\sec^2\theta = \dfrac{4xy}{(x+y)^2}$

But $\sec^2\theta \geq 1 \Rightarrow \dfrac{4xy}{(x+y)^2} \geq 1$

$\Rightarrow 4xy \geq x^2 + y^2 + 2xy$

$\Rightarrow x^2 + y^2 - 2xy \leq 0 \Rightarrow (x-y)^2 \leq 0$

$\Rightarrow x - y = 0$ [as perfect square of real number can never be negative]

Also then $x \neq 0$ as then $\sec^2\theta$ will become indeterminate.

16. **(a)** Given that in ΔPQR, $\angle R = \pi/2$

$\Rightarrow \angle P + \angle Q = \pi/2 \Rightarrow \dfrac{\angle P}{2} + \dfrac{\angle Q}{2} = \dfrac{\pi}{4}$

Also tan $P/2$ and tan $Q/2$ are roots of the equation $ax^2 + bx + c = 0\ (a \neq 0)$

$\therefore \tan P/2 + \tan Q/2 = -\dfrac{b}{a}; \tan P/2 \tan Q/2 = c/a$

Now consider,

$$\tan\left(\frac{P+Q}{2}\right) = \frac{\tan P/2 + \tan Q/2}{1 - \tan P/2 \tan Q/2}$$

$\Rightarrow \tan\dfrac{\pi}{4} = \dfrac{-b/a}{1 - c/a} \Rightarrow 1 - \dfrac{c}{a} = -\dfrac{b}{a}$

$\Rightarrow a - c = -b \Rightarrow a + b = c$

17. **(c)** $f(\theta) = \sin\theta(\sin\theta + \sin 3\theta)$

$= (\sin\theta + 3\sin\theta - 4\sin^3\theta).\sin\theta$

$= (4\sin\theta - 4\sin^3\theta)\sin\theta = \sin^2\theta\,(4 - 4\sin^2\theta)$

$= 4\sin^2\theta\,(1 - \sin^2\theta)$

$= 4\sin^2\theta\cos^2\theta = (2\sin\theta\cos\theta)^2 = (\sin 2\theta)^2 \geq 0$

which is true for all θ.

18. **(c)** To simplify the det. Let $\sin x = a$; $\cos x = b$ the equation becomes

$\begin{vmatrix} a & b & b \\ b & a & b \\ b & b & a \end{vmatrix} = 0$ Operating $C_2 - C_1$; $C_3 - C_2$ we get

$$\begin{vmatrix} a & b-a & 0 \\ b & a-b & b-a \\ b & 0 & a-b \end{vmatrix} = 0$$

$\Rightarrow a(a-b)^2 - (b-a)[b(a-b) - b(b-a)] = 0$

$\Rightarrow a(a-b)^2 - 2b(b-a)(a-b) = 0$

$\Rightarrow (a-b)^2(a-2b) = 0 \Rightarrow (a = b)$ or $a = 2b$

$\Rightarrow \dfrac{a}{b} = 1$ or $\dfrac{a}{b} = 2$

$\Rightarrow \tan x = 1$ or $\tan x = 2$. But we have $-\pi/4 \leq x \leq \pi/4$

$\Rightarrow \tan(-\pi/4) \leq \tan x \leq \tan(\pi/4) \Rightarrow -1 \leq \tan x \leq 1$

$\therefore \tan x = 1 \Rightarrow x = \pi/4$ $\therefore$ Only one real root is there.

19. **(a)** We are given that

$(\cot\alpha_1).(\cot\alpha_2)\ldots.(\cot\alpha_n) = 1$

$\Rightarrow (\cos\alpha_1)(\cos\alpha_2)\ldots.(\cos\alpha_n)$

$= (\sin\alpha_1)(\sin\alpha_2)\ldots.(\sin\alpha_n)$ ....(1)

Let $y = (\cos\alpha_1)(\cos\alpha_2)\ldots,(\cos\alpha_n)$ (to be max.)

Squaring both sides, we get

$y^2 = (\cos^2\alpha_1)(\cos^2\alpha_2)\ldots.(\cos^2\alpha_n)$

$= \cos\alpha_1\sin\alpha_1\cos\alpha_2\sin\alpha_2\ldots.\cos\alpha_n\sin\alpha_n$ (Using (1))

$= \dfrac{1}{2^n}[\sin 2\alpha_1\sin 2\alpha_2\ldots.\sin 2\alpha_n]$

As $0 \leq \alpha_1, \alpha_2, \ldots.\alpha_n \leq \pi/2$

$\therefore 0 \leq 2\alpha_1, 2\alpha_2, \ldots..2\alpha_n \leq \pi$

$\Rightarrow 0 \leq \sin 2\alpha_1, \sin 2\alpha_2, \ldots..\sin 2\alpha_n \leq 1$

$\therefore y^2 \leq \dfrac{1}{2^n}.1 \Rightarrow y \leq \dfrac{1}{2^{n/2}}$

$\therefore$ Max. value of y is $1/2^{n/2}$.

20. **(c)** Given that $\alpha + \beta = \pi/2$.

$\Rightarrow \alpha = \pi/2 - \beta$

$\Rightarrow \tan\alpha = \tan(\pi/2 - \beta) = \cot\beta = \dfrac{1}{\tan\beta}$

$\Rightarrow \tan\alpha\tan\beta = 1 \Rightarrow 1 + \tan\alpha\tan\beta = 2.$

$\therefore \tan(\alpha - \beta) = \dfrac{\tan\alpha - \tan\beta}{1 + \tan\alpha\tan\beta}$

$\Rightarrow \tan\gamma = \dfrac{\tan\alpha - \tan\beta}{2}$

$\Rightarrow 2\tan\gamma = \tan\alpha - \tan\beta$

$\Rightarrow \tan\alpha = 2\tan\gamma + \tan\beta$

21. **(b)** We know that

$\Rightarrow -\sqrt{a^2+b^2} \leq a\cos\theta + b\sin\theta \leq \sqrt{a^2+b^2}$

NOTE THIS STEP

$\Rightarrow -\sqrt{74} \leq 7\cos x + 5\sin x \leq \sqrt{74}$

$\Rightarrow -\sqrt{74} \leq 2k+1 \leq \sqrt{74} \Rightarrow -8.6 \leq 2k+1 \leq 8.6$

$\Rightarrow -4.8 \leq k \leq 3.8$

(considering only integral values)

$\Rightarrow k$ can take 8 integral values.

22. **(b)** Given that $\sin\theta = 1/2$ and $\cos\phi = 1/3$ and θ and ϕ both are acute angles

$\therefore \theta = \pi/6$ and $0 < \dfrac{1}{3} < \dfrac{1}{2}$

or $\cos\pi/2 < \cos\phi < \cos\pi/3$ or $\pi/3 < \phi < \pi/2$

$\therefore \dfrac{\pi}{3} + \dfrac{\pi}{6} < \theta + \phi < \dfrac{\pi}{2} + \dfrac{\pi}{6}$ or $\dfrac{\pi}{2} < \theta + \phi < \dfrac{2\pi}{3}$

$\Rightarrow \theta + \phi \in \left(\dfrac{\pi}{2}, \dfrac{2\pi}{3}\right)$

23. (d) Given that $\cos(\alpha-\beta)=1$ and $\cos(\alpha+\beta)=1/e$ where $\alpha,\beta\in[-\pi,\pi]$

Now, $\cos(\alpha-\beta)=1 \Rightarrow \alpha-\beta=0 \Rightarrow \alpha=\beta$

Now, $\cos(\alpha+\beta)=1/e \Rightarrow \cos 2\alpha=1/e$

$\because 0<1/e<1$ and $2\alpha\in[-2\pi,2\pi]$

$\Rightarrow$ There will be two values of 2α satisfying $\cos 2\alpha=1/e$ in $[0,2\pi]$ and two in $[-2\pi,0]$.

$\Rightarrow$ There will be four values of α in $[-\pi,\pi]$ and correspondingly four values of β. Hence there are four sets of (α,β).

ALTERNATE SOLUTION

Given $\cos(\alpha-\beta)=1$, $\cos(\alpha+\beta)=1/e$ and $\alpha,\beta\in[-\pi,\pi]$

$\cos(\alpha-\beta)=1 \Rightarrow \alpha-\beta=0, 2\pi, -2\pi$

$\Rightarrow \alpha-\beta=0 \quad [\because \alpha,\beta\in[-\pi,\pi]]$

$\Rightarrow \alpha=\beta$

$\therefore \cos 2\alpha=1/e$

No. of solutions of above will be no.of pts of intersection of the curves $y=\cos 2\alpha$ and $y=1/e$ $\alpha,\beta\in[-\pi,\pi]$

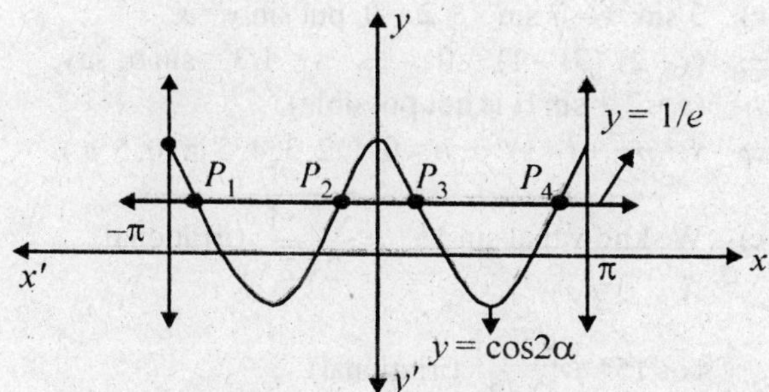

From graph it is clear that there are 4 solutions, corresponding to four points of intersection.

24. (a) $2\sin^2\theta-5\sin\theta+2>0$

$\Rightarrow (\sin\theta-2)(2\sin\theta-1)>0$

$\Rightarrow \sin\theta<\frac{1}{2} \quad [\because -1\le\sin\theta\le1]$

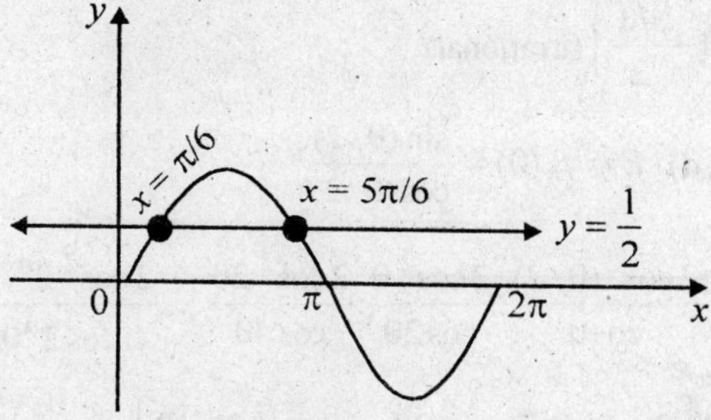

From graph, we get

$$x\in\left(0,\frac{\pi}{6}\right)\cup\left(\frac{5\pi}{6},2\pi\right)$$

25. (b) $\because \theta\in\left(0,\frac{\pi}{4}\right)\Rightarrow \tan\theta<1$ and $\cot\theta>1$

Let $\tan\theta=1-x$ and $\cot\theta=1+y$

Where $x, y>0$ and are very small, then

$\therefore t_1=(1-x)^{1-x}, t_2=(1-x)^{1+y}, t_3=(1+y)^{1-x}, t_4=(1+y)^{1+y}$

Clearly, $t_4>t_3$ and $t_1>t_2$ also, $t_3>t_1$ **NOTE THIS STEP**

Thus $t_4>t_3>t_1>t_2$.

26. (c) $2\sin^2\theta-\cos 2\theta=0 \Rightarrow 1-2\cos 2\theta=0$

$\Rightarrow \cos 2\theta=\frac{1}{2} \Rightarrow 2\theta=\frac{\pi}{3},\frac{5\pi}{3},\frac{7\pi}{3},\frac{11\pi}{3}$

$\Rightarrow \theta=\frac{\pi}{6},\frac{5\pi}{6},\frac{7\pi}{6},\frac{11\pi}{6}$ ....(1)

where $\theta\in[0,2\pi]$

Also $2\cos^2\theta-3\sin\theta=0$

$\Rightarrow 2\sin^2\theta+3\sin\theta-2=0$

$\Rightarrow (2\sin\theta-1)(\sin\theta+2)=0 \Rightarrow \sin\theta=\frac{1}{2}$

$[\because \sin\theta\ne-2]$

$\Rightarrow \theta=\frac{\pi}{6},\frac{5\pi}{6}$ ....(2)

where $\theta\in[0,2\pi]$

Combining (1) and (2), we get $\theta=\frac{\pi}{6},\frac{5\pi}{6}$

$\therefore$ Two solutions are there.

(D) MCQ's WITH ONE OR MORE THAN ONE CORRECT :

1. (c) We have,

$(1+\cos\pi/8)(1+\cos3\pi/8)(1+\cos5\pi/8)(1+\cos7\pi/8)$

$=(1+\cos\pi/8)(1+\cos3\pi/8)(1+\cos(\pi-3\pi/8))(1+\cos(\pi-\pi/8))$

$=(1+\cos\pi/8)(1+\cos3\pi/8)(1-\cos3\pi/8)(1-\cos\pi/8)$

$=(1-\cos^2\pi/8)(1-\cos^2 3\pi/8)$

$=\sin^2\pi/8\,\sin^2 3\pi/8$

$=\frac{1}{4}.[2\sin\pi/8\,\sin(\pi/2-\pi/8)]^2$

$=\frac{1}{4}.[2\sin\pi/8\,\cos\pi/8)]^2$

$=\frac{1}{4}.\sin^2\pi/4=\frac{1}{4}\times\frac{1}{2}=\frac{1}{8}$

$\therefore$ (c) is the correct answer.

2. (b) The given expression is

$$= 3\left[\sin^4\left(\frac{3\pi}{2}-\alpha\right)+\sin^4(3\pi+\alpha)\right]$$
$$-2\,[\sin^6(\pi/2+\alpha)+\sin^6(5\pi-\alpha)]$$

$$= 3\left[\cos^4\alpha+\sin^4\alpha\right]-2\left[\cos^6\alpha+\sin^6\alpha\right]$$

$$= 3\left[\left(\cos^2\alpha+\sin^2\alpha\right)^2-2\sin^2\alpha\cos^2\alpha\right]$$
$$-2\left[(\cos^2\alpha+\sin^2\alpha)^3-3\cos^2\alpha\sin^2\alpha\,(\cos^2\alpha+\sin^2\alpha)\right]$$

$$= 3\left[1-2\sin^2\alpha\cos^2\alpha\right]-2\left[1-3\cos^2\alpha\sin^2\alpha\right]$$

$$= 3-6\sin^2\alpha\cos^2\alpha-2+6\sin^2\alpha\cos^2\alpha=1$$

3. (d) Since $a_1 + a_2 \cos 2x + a_3 \sin^2 x = 0$ for all x

Putting $x = 0$ and $x = \pi/2$, we get $a_1 + a_2 = 0$ ....(1)
and $a_1 - a_2 + a_3 = 0$ ....(2)

$\Rightarrow a_2 = -a_1$ and $a_3 = -2a_1$

$\therefore$ The given equation becomes

$a_1 - a_1 \cos 2x - 2a_1 \sin^2 x = 0, \ \forall x$

$\Rightarrow a_1 (1 - \cos 2x - 2\sin^2 x) = 0, \forall x$

$\Rightarrow a_1 (2\sin^2 x - 2\sin^2 x) = 0, \forall x$

The above is satisfied for all values of a_1.
Hence infinite number of triplets $(a_1, -a_1, -2a_1)$ are possible.

4. (a, c) We have

$$\begin{vmatrix} 1+\sin^2\theta & \cos^2\theta & 4\sin 4\theta \\ \sin^2\theta & 1+\cos^2\theta & 4\sin 4\theta \\ \sin^2\theta & \cos^2\theta & 1+4\sin 4\theta \end{vmatrix} = 0$$

Operating $C_1 \to C_1 + C_2$

$$\begin{vmatrix} 2 & \cos^2\theta & 4\sin 4\theta \\ 2 & 1+\cos^2\theta & 4\sin 4\theta \\ 1 & \cos^2\theta & 1+4\sin 4\theta \end{vmatrix} = 0$$

Operating $R_1 \to R_1 - R_2$; $R_2 \to R_2 - R_3$

$$\begin{vmatrix} 0 & -1 & 0 \\ 1 & 1 & -1 \\ 1 & \cos^2\theta & 1+4\sin 4\theta \end{vmatrix} = 0$$

Expanding along R_1 we get

$$[1 + 4\sin 4\theta + 1] = 0$$

$$\Rightarrow 2(1+2\sin 4\theta+1)=0 \Rightarrow \sin 4\theta = -\frac{1}{2}$$

$$\Rightarrow 4\theta = \pi+\pi/6 \quad \text{or} \quad 2\pi-\pi/6$$

$$\Rightarrow 4\theta = 7\pi/6 \quad \text{or} \quad 11\pi/6$$

$$\Rightarrow \theta = 7\pi/24 \quad \text{or} \quad 11\pi/24$$

5. (d) $2\sin^2 x + 3\sin x - 2 > 0$

$(2\sin x - 1)(\sin x + 2) > 0$

$\Rightarrow 2\sin x - 1 > 0 \quad (\because -1 \le \sin x \le 1)$

$\Rightarrow \sin x > 1/2 \Rightarrow x \in (\pi/6, 5\pi/6)$ ....(1)

Also $x^2 - x - 2 < 0$

$\Rightarrow (x-2)(x+1) < 0 \Rightarrow -1 < x < 2$ ....(2)

Combining (1) and (2) $x \in (\pi/6, 2)$.

6. (c)

$$\sin\alpha+\sin\beta+\sin\gamma = 2\sin\frac{\alpha+\beta}{2}\cos\frac{\alpha-\beta}{2}+2\sin\frac{\gamma}{2}\cos\frac{\gamma}{2}$$

$$= 2\sin\left(\frac{\pi}{2}-\frac{\gamma}{2}\right)\cos\frac{\alpha-\beta}{2}+2\sin\left(\frac{\pi}{2}-\frac{\alpha+\beta}{2}\right)\cos\frac{\gamma}{2}$$

$$= 2\cos\frac{\gamma}{2}\left[\cos\frac{\alpha-\beta}{2}+\cos\frac{\alpha+\beta}{2}\right]$$

$= 2\cos\alpha/2\ \cos\beta/2\ \cos\gamma/2$

$\therefore$ Each $\cos\alpha/2$, $\cos\beta/2$, $\cos\gamma/2$ lies between -1 and 1,

therefore $-1 \le \cos\alpha/2, \cos\beta/2, \cos\gamma/2 \le 1$

$\Rightarrow -2 \le 2\cos\alpha/2, \cos\beta/2, \cos\gamma/2 \le 2$

$\Rightarrow -2 \le \cos\alpha+\cos\beta+\cos\gamma \le 2$

$\therefore$ min value $= -2$.

7. (c) $3\sin^2 x - 7\sin x + 2 = 0$, put $\sin x = s$

$\Rightarrow (s-2)(3s-1) = 0 \quad \Rightarrow \quad s = 1/3 = \sin\alpha$, say,

$(s = 2 = \sin\alpha$ is not possible)

$\Rightarrow x = n\pi + (-1)^n\alpha, n = 0, 1, 2, 3, 4, 5$ in $(0, 5\pi)$.

8. (c) We know that $\sin 15° = \dfrac{\sqrt{3}-1}{2\sqrt{2}}$ (irrational)

$\cos 15° = \dfrac{\sqrt{3}+1}{2\sqrt{2}}$ (irrational)

$\sin 15° . \cos 15° = \dfrac{1}{2}(2\sin 15° \cos 15°)$

$= \dfrac{1}{2}\sin 30° = \dfrac{1}{4}$ (rational)

$\sin 15° \cos 75° = \sin 15° \cos(90 - 15°)$

$= \sin 15° \sin 15° = \sin^2 15° = \dfrac{1}{2}(1+\cos 30°)$

$= \dfrac{1}{2}\left(1-\dfrac{\sqrt{3}}{2}\right)$ (irrational)

9. (a, b, c, d) $E = f_n(\theta) = \dfrac{\sin(\theta/2)}{\cos(\theta/2)}$

$$\left[\frac{2\cos^2(\theta/2)}{\cos\theta}.\frac{2\cos^2\theta}{\cos 2\theta}.\frac{2\cos^2 2\theta}{\cos 4\theta}.....\frac{2\cos^2 2^{n-1}\theta}{\cos 2^n\theta}\right]$$

$$= \frac{\sin\theta}{\cos\theta}\left[\frac{2\cos^2\theta}{\cos 2\theta}.\frac{2\cos^2 2\theta}{\cos 4\theta}....\frac{2\cos^2 2^{n-1}\theta}{\cos 2^n\theta}\right] = \tan 2^n\theta$$

$$n = 2, \theta = \frac{\pi}{16}$$

$f_2\left(\frac{\pi}{16}\right) = \tan 4.\frac{\pi}{16} = \tan\frac{\pi}{4} = 1.$

Similarly, $f_3\left(\frac{\pi}{32}\right), f_4\left(\frac{\pi}{64}\right)$ *and* $f_5\left(\frac{\pi}{128}\right)$ *is* $\tan\frac{\pi}{4} = 1.$

10. **(a, b)** Given that

$$\frac{\sin^4 x}{2} + \frac{\cos^4 x}{3} = \frac{1}{5} \Rightarrow 3\sin^4 x + 2\cos^4 x = \frac{6}{5}$$

$$\Rightarrow \sin^4 x + 2[\sin^4 x + \cos^4 x] = \frac{6}{5}$$

$$\Rightarrow \sin^4 x + 2[1 - 2\sin^2 x \cos^2 x] = \frac{6}{5}$$

$$\Rightarrow \sin^4 x + 2 - 4\sin^2 x(1 - \sin^2 x) = \frac{6}{5}$$

$$\Rightarrow 5\sin^4 x - 4\sin^2 x + 2 - \frac{6}{5} = 0$$

$$\Rightarrow 25\sin^4 x - 20\sin^2 x + 4 = 0$$

$$\Rightarrow (5\sin^2 x - 2)^2 = 0 \Rightarrow \sin^2 x = \frac{2}{5}$$

$$\Rightarrow \cos^2 x = \frac{3}{5} \text{ and } \tan^2 x = \frac{2}{3}$$

Also $\frac{\sin^8 x}{8} + \frac{\cos^8 x}{27} = \frac{2}{625} + \frac{3}{625} = \frac{5}{625} = \frac{1}{125}$

11. **(c, d)** We have

$$\sum_{m=1}^{6} \operatorname{cosec}\left[\theta + \frac{(m-1)\pi}{4}\right]\operatorname{cosec}\left[\theta + \frac{m\pi}{4}\right] = 4\sqrt{2}$$

$$\Rightarrow \sum_{m=1}^{6} \frac{\sin\frac{\pi}{4}}{\sin\left[\theta + \frac{(m-1)\pi}{4}\right]\sin\left[\theta + \frac{m\pi}{4}\right]} = 4$$

$$\Rightarrow \sum_{m=1}^{6} \frac{\sin\left[\left(\theta + \frac{m\pi}{4}\right) - \left(\theta + \frac{(m-1)\pi}{4}\right)\right]}{\sin\left(\theta + \frac{(m-1)\pi}{4}\right)\sin\left(\theta + \frac{m\pi}{4}\right)} = 4$$

$$\Rightarrow \sum_{m=1}^{6} \frac{\left[\begin{array}{l}\sin\left(\theta + \frac{m\pi}{4}\right)\cos\left(\theta + \frac{(m-1)\pi}{4}\right) \\ -\cos\left(\theta + \frac{m\pi}{4}\right)\sin\left(\theta + \frac{(m-1)\pi}{4}\right)\end{array}\right]}{\sin\left(\theta + \frac{(m-1)\pi}{4}\right)\sin\left(\theta + \frac{m\pi}{4}\right)} = 4$$

$$\Rightarrow \sum_{m=1}^{6}\left[\cot\left(\theta + \frac{(m-1)\pi}{4}\right) - \cot\left(\theta + \frac{m\pi}{4}\right)\right] = 4$$

$$\Rightarrow \left[\cot\theta - \cot\left(\theta + \frac{\pi}{4}\right)\right] + \left[\cot\left(\theta + \frac{\pi}{4}\right) - \cot\left(\theta + \frac{2\pi}{4}\right)\right]$$

$$+ \ldots + \left[\cot\left(\theta + \frac{5\pi}{4}\right) - \cot\left(\theta + \frac{6\pi}{4}\right)\right] = 4$$

$$\Rightarrow \cot\theta - \cot\left(\theta + \frac{3\pi}{2}\right) = 4 \Rightarrow \cot\theta + \tan\theta = 4$$

$$\Rightarrow \cos^2\theta + \sin^2\theta = 4\sin\theta\cos\theta$$

$$\Rightarrow \sin 2\theta = \frac{1}{2} \Rightarrow 2\theta = \frac{\pi}{6} \text{ or } \frac{5\pi}{6}$$

$$\Rightarrow \theta = \frac{\pi}{12} \text{ or } \frac{5\pi}{12}$$

12. **(a, c, d)**
As $\tan(2\pi - \theta) > 0 \Rightarrow -\tan\theta > 0 \Rightarrow \tan\theta < 0$
$\therefore$ $\theta \in$ II or IV quadrant ...(1)

And $-1 < \sin\theta < \frac{-\sqrt{3}}{2}$

$\Rightarrow \theta \in$ III or IV quadrant ...(2)
Also $\theta \in [0, 2\pi]$ ...(3)
Combining above three equations (1), (2) and (3) ;
$\theta \in$ IV quadrant and more precisely $\frac{3\pi}{2} < \theta < \frac{5\pi}{3}$

$\left(\because -1 < \sin\theta < -\frac{\sqrt{3}}{2}\right)$

Now,
$2\cos\theta\,(1 - \sin\phi) = \sin^2\theta\,(\tan\theta/2 + \cot\theta/2)\cos\phi - 1$

$\Rightarrow 2\cos\theta(1 - \sin\phi) = \sin^2\theta \times \frac{1}{\sin\theta/2\ \cos\theta/2}.\cos\phi - 1$

$\Rightarrow 2\cos\theta(1 - \sin\phi) = 2\sin\theta\cos\phi - 1$

$\Rightarrow 2\cos\theta + 1 = 2\sin(\theta + \phi)$

$\because$ $\theta \in \left(\frac{3\pi}{2}, \frac{5\pi}{3}\right)$ $\therefore 2\cos\theta + 1 \in (1, 2)$

$\Rightarrow$ $1 < 2\sin(\theta + \phi) < 2$

$\Rightarrow \frac{1}{2} < \sin(\theta + \phi) < 1 \Rightarrow \theta + \phi \in \left(\frac{\pi}{6}, \frac{5\pi}{6}\right)$

or $\theta + \phi \in \left(2\pi + \frac{\pi}{6}, 2\pi + \frac{5\pi}{6}\right)$ or $\left(\frac{13\pi}{6}, \frac{17\pi}{6}\right)$

But $\theta \in \left(\frac{3\pi}{2}, \frac{5\pi}{3}\right)$ (as $\theta + \phi \in [0, 4\pi]$)

$$\Rightarrow \frac{13\pi}{6} < \theta + \phi < \frac{17\pi}{6} \Rightarrow \frac{13\pi}{6} - \theta < \phi < \frac{17\pi}{6} - \theta$$

$$\Rightarrow \frac{13\pi}{6} - \theta_{max} < \phi < \frac{17\pi}{6} - \theta_{min}$$

$$\Rightarrow \frac{13\pi}{6} - \frac{5\pi}{3} < \phi < \frac{17\pi}{6} - \frac{3\pi}{2} \Rightarrow \frac{\pi}{2} < \phi < \frac{4\pi}{3}$$

13. **(c)** Let $f(x) = x^2 - x\sin x - \cos x$
$\therefore$ $f'(x) = 2x - x\cos x = x(2 - \cos x)$
$\therefore$ f is increasing on $(0, \infty)$ and decreasing on $(-\infty, 0)$

Also $\lim_{x\to\infty} f(x)=\infty$, $\lim_{x\to-\infty} f(x)=\infty$ and $f(o)=-1$

$\therefore$ $y=f(x)$ meets x-axis twice.

i.e., $f(x)=0$ has two points in $(-\infty,\infty)$.

14. **(b, c)** We have $f(x)=x\sin\pi x, x>0$

$\Rightarrow f'(x)=\sin\pi x+x\pi\cos\pi x$

$f'(x)=0 \Rightarrow \tan\pi x=-\pi x$

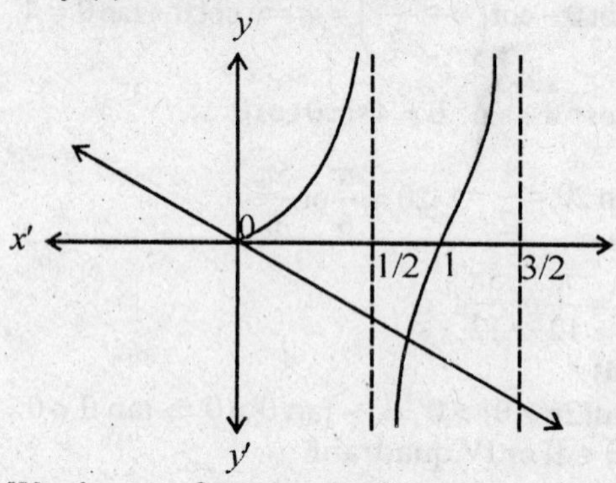

We observe, from graph of $y=\tan\pi x$ and $y=-\pi x$ that they intersect at unique point in the intervals

$(n, n+1)$ and $\left(n+\frac{1}{2}, n+1\right)$

E SUBJECTIVE PROBLEMS :

1. We know $\tan(\alpha+\beta)=\dfrac{\tan\alpha+\tan\beta}{1-\tan\alpha\tan\beta}$

$$\Rightarrow \tan(\alpha+\beta)=\frac{\frac{m}{m+1}+\frac{1}{2m+1}}{1-\frac{m}{m+1}\cdot\frac{1}{2m+1}}=\frac{2m^2+2m+1}{2m^2+2m+1}=1$$

$\Rightarrow \alpha+\beta=n\pi+\pi/4$ where $n\in Z$.

2. **(a)** Given: $y=\frac{1}{\sqrt{2}}(\sin x+\cos x)=\sin\left(x+\frac{\pi}{4}\right)$ ...(1)

Now, to draw the graph of $y=\sin\left(x+\frac{\pi}{4}\right)$, we first draw the graph of $y=\sin x$ and then on shifting it by $-\frac{\pi}{4}$, we will obtain the required graph as shown in figure given below.

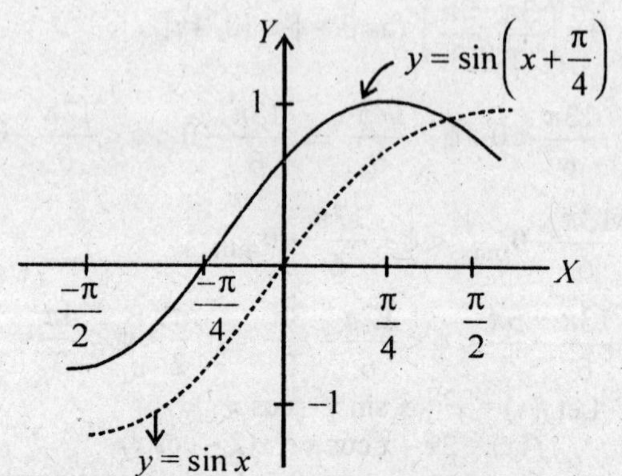

(b) $\cos(\alpha+\beta)=\frac{4}{5}$

$\Rightarrow \tan(\alpha+\beta)=\frac{3}{4}, 0<\alpha,\beta<\frac{\pi}{4}$

$\sin(\alpha-\beta)=\frac{5}{13}\Rightarrow \tan(\alpha-\beta)=\frac{5}{12}$

$\therefore \tan 2\alpha=\tan[(\alpha+\beta)+(\alpha-\beta)]$

$$=\frac{\tan(\alpha+\beta)+\tan(\alpha-\beta)}{1-\tan(\alpha+\beta)\tan(\alpha-\beta)}$$

$$=\frac{\frac{3}{4}+\frac{5}{12}}{1-\frac{3}{4}\times\frac{5}{12}}=\frac{9+5}{12}\times\frac{16}{11}=\frac{56}{33}$$

3. **(a)** We have $a=(\sqrt{3}+1)$ cm

$b=2$ cm

$\angle C=60°$

By Cosine Law

$$\cos C=\frac{a^2+b^2-c^2}{2ab}\Rightarrow \cos 60°=\frac{(\sqrt{3}+1)^2+4-c^2}{2(\sqrt{3}+1)\times 2}$$

$$\Rightarrow \frac{1}{2}=\frac{4+2\sqrt{3}+4-c^2}{4(\sqrt{3}+1)}$$

$\Rightarrow c^2=6\Rightarrow c=\sqrt{6}$ cm.

Now, $\cos B=\dfrac{c^2+a^2-b^2}{2ca}=\dfrac{6+4+2\sqrt{3}-4}{2\sqrt{6}(\sqrt{3}+1)}$

$$=\frac{\sqrt{3}(\sqrt{3}+1)}{\sqrt{6}(\sqrt{3}+1)}=\frac{1}{\sqrt{2}}$$

$\Rightarrow \angle B=45°$. Further $\angle A=\pi-(45°+60°)=75°$

ALTERNATE SOLUTION

Given that in ΔABC $a=(\sqrt{3}+1)$ cm.

$b=2$ cm, $\angle C=60°$

$\therefore A+B=120°$ ....(1)

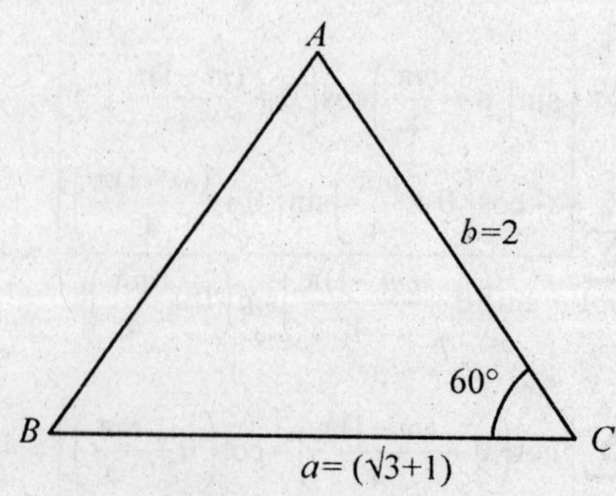

Also using Napier's analogy

$$\tan\frac{A-B}{2}=\frac{a-b}{a+b}\cot C/2$$

$$=\left(\frac{\sqrt{3}+1-2}{\sqrt{3}+1+2}\right)\cot 30^\circ=\frac{\sqrt{3}-1}{\sqrt{3}\ (\sqrt{3}+1)}\times\sqrt{3}$$

$$=\frac{\sqrt{3}-1}{\sqrt{3}+1}=\tan 15^\circ \Rightarrow A-B=30^\circ \quad ....(2)$$

Solving (1) and (2), $A=75^\circ$ and $B=45^\circ$

Then by sine law, $\frac{a}{\sin A}=\frac{b}{\sin B}=\frac{c}{\sin C}$

$$\Rightarrow c=\frac{b\sin C}{\sin B}=\frac{2\sin 60^\circ}{\sin 45^\circ}=2\times\frac{\sqrt{3}}{2}\times\frac{\sqrt{2}}{1}=\sqrt{6}\text{ cm}$$

(b) Let $y=5\cos\theta+3\cos\left(\theta+\frac{\pi}{3}\right)+3$

$$=5\cos\theta+3\cos\theta\cos\frac{\pi}{3}-3\sin\theta\sin\frac{\pi}{3}+3$$

$$=5\cos\theta+\frac{3}{2}\cos\theta-\frac{3\sqrt{3}}{2}\sin\theta+3$$

$$=\frac{13}{2}\cos\theta-\frac{3\sqrt{3}}{2}\sin\theta+3$$

Now, $-\sqrt{\left(\frac{13}{2}\right)^2+\left(\frac{3\sqrt{3}}{2}\right)^2}\le\frac{13}{2}\cos\theta-\frac{3\sqrt{3}}{2}\sin\theta$

$$\le\sqrt{\left(\frac{13}{2}\right)^2+\left(\frac{3\sqrt{3}}{2}\right)^2}$$

$$\Rightarrow -7\le\frac{13}{2}\cos\theta-\frac{3\sqrt{3}}{2}\sin\theta\le 7$$

$$\Rightarrow -4\le\frac{13}{2}\cos\theta-\frac{3\sqrt{3}}{2}\sin\theta+3\le 10$$

$\therefore$ y lies between -4 and 10.

4. Given $\alpha+\beta-\gamma=\pi$ and to prove that

$\sin^2\alpha+\sin^2\beta-\sin^2\gamma=2\sin\alpha\sin\beta\cos\gamma$

L.H.S. $=\sin^2\alpha+\sin^2\beta-\sin^2\gamma$

[Using $\sin^2 A-\sin^2 B=\sin(A+B)\sin(A-B)$]

$=\sin^2\alpha+\sin(\beta+\gamma)\sin(\beta-\gamma)$

$=\sin^2\alpha+\sin(\beta+\gamma)\sin(\pi-\alpha)$ $\quad(\because \alpha+\beta-\gamma=\pi)$

$=\sin^2\alpha+\sin(\beta+\gamma)\sin\alpha$

$=\sin\alpha(\sin\alpha+\sin(\beta+\gamma))$

$=\sin\alpha[\sin[\pi-(\beta-\gamma)]+\sin(\beta+\gamma)]$

$=\sin\alpha[\sin(\beta-\gamma)+\sin(\beta+\gamma)]$

$=\sin\alpha[2\sin\beta\cos\gamma]$

$=2\sin\alpha\sin\beta\cos\gamma=$ R.H.S.

5. $A=\left\{x:\frac{\pi}{6}\le x\le\frac{\pi}{3}\right\}$

$f(x)=\cos x-x(1+x)$

$f'(x)=-\sin x-1-2x<0,\ \forall x\in A$

$\therefore f$ is a decreasing function.

$\therefore$ as $\frac{\pi}{6}\le x\le\frac{\pi}{3}\Rightarrow f\left(\frac{\pi}{3}\right)\le f(x)\le\left(\frac{\pi}{6}\right)$

$$\Rightarrow \cos\frac{\pi}{3}-\frac{\pi}{3}\left(1+\frac{\pi}{3}\right)\le f(x)\le\cos\frac{\pi}{6}-\frac{\pi}{6}\left(1+\frac{\pi}{6}\right)$$

$$\therefore f(A)=\left[\frac{1}{2}-\frac{\pi}{3}\left(1+\frac{\pi}{3}\right),\frac{\sqrt{3}}{2}-\frac{\pi}{6}\left(1+\frac{\pi}{6}\right)\right]$$

6. We have

$$\cos\theta+\sin\theta=\sqrt{2}\left[\frac{1}{\sqrt{2}}\cos\theta+\frac{1}{\sqrt{2}}\sin\theta\right]$$

$$=\sqrt{2}\sin(\pi/4+\theta)$$

$\therefore \cos\theta+\sin\theta\le\sqrt{2}<\pi/2$ $\left(\because \begin{matrix}\sqrt{2}=1.414\\ \pi/2=1.57\end{matrix}\right)$

$\therefore \cos\theta+\sin\theta<\pi/2$

$\Rightarrow \cos\theta<\pi/2-\sin\theta$ $\quad$.....(1)

As $\theta\in[0,\pi/2]$ in which $\sin\theta$ increases.

$\therefore$ Taking sin on both sides of eq. (1), we get

$\sin(\cos\theta)<\sin(\pi/2-\sin\theta)$

$\sin(\cos\theta)<\cos(\sin\theta)$

$\Rightarrow \cos(\sin\theta)>\sin(\cos\theta)$ $\quad$....(1)

Hence the result.

ALTERNATE SOLUTION

We know that for $x\in[0,\pi/2]$

$\sin x\le x$ ....(1)

For $x=\cos\theta$, above can be written as

$\sin(\cos\theta)\le\cos\theta$ $\quad$....(2)

Also in I quad $\cos\theta$ is a decreasing function

$\therefore$ Replacing x by θ in (1) and then taking cosine on both sides, we get

$\cos(\sin\theta)\ge\cos\theta$ $\quad$....(3)

From (2) and (3) we get

$\cos(\sin\theta)\ge\cos\theta\ge\sin(\cos\theta)$

$\Rightarrow \cos(\sin\theta)>\sin(\cos\theta)$

Hence the result.

7. L.H.S.

$$=\sin 12^\circ\sin 48^\circ\sin 54^\circ=\frac{1}{2}[2\sin 12^\circ\cos 42^\circ]\sin 54^\circ$$

$$=\frac{1}{2}\sin^2 54^\circ-\frac{1}{2}\sin 54^\circ=\frac{1}{4}[2\sin^2 54^\circ-\sin 54^\circ]$$

Now we know that $\sin 54^\circ=\frac{1+\sqrt{5}}{4}$

$$\therefore \text{We get, }=\frac{1}{4}\left[2\left(\frac{1+\sqrt{5}}{4}\right)^2-\left(\frac{1+\sqrt{5}}{4}\right)\right]$$

$$=\frac{1}{4}\left[2\left(\frac{1+5+2\sqrt{5}}{16}\right)-\left(\frac{1+\sqrt{5}}{4}\right)\right]$$

$$=\frac{1}{4}\times\frac{1}{8}\left[6+2\sqrt{5}-2-2\sqrt{5}\right]$$

$$=\frac{1}{32}\times 4=\frac{1}{8}=\text{R.H.S.}$$

8. To find the pts of intersection of the curves

$y = \cos x$; $y = \sin 3x$

Consider $\sin 3x = \cos x$

$\Rightarrow \cos x = \cos(\pi/2 - 3x)$

$x = 2n\pi \pm (\pi/2 - 3x)$

Taking '+' sign, $x = 2n\pi + \pi/2 - 3x$

$\Rightarrow 4x = 2n\pi + \pi/2 \Rightarrow x = n\pi/2 + \pi/8$

For $n = 0, x = \pi/8$

$n = -1, x = -\pi/2 + \pi/8 = -3\pi/8$

For other value of n x is rejected as $\frac{-\pi}{2} \le x \le \frac{\pi}{2}$

Taking '–' sign, $x = 2n\pi - \pi/2 + 3x$

$\Rightarrow -2x = 2n\pi - \pi/2 \Rightarrow x = -n\pi + \pi/4$

For $n = 0$; $x = \pi/4$

For other values of n, x is rejected as $-\frac{\pi}{2} \le x \le \frac{\pi}{2}$

For three pts of intersection are there

$x = \pi/4,\ \pi/8,\ -3\pi/8$

$\Rightarrow y = \cos\pi/4, \cos\pi/8, \cos(-3\pi/8)$

$\therefore$ Req. pts are $(\pi/4, 1/\sqrt{2}), (\pi/8, \cos\pi/8)$, $(-3\pi/8, \cos 3\pi/8)$.

9. We know that,

$\cos A \cos 2A \cos 4A \ldots \cos 2^n A$

$$= \frac{1}{2^{n+1}\sin A}\sin.(2^{n+1}A)$$

$$\therefore 16\cos\frac{2\pi}{15}\cos 2\left(\frac{2\pi}{15}\right)\cos 2^2\left(\frac{2\pi}{15}\right)\cos 2^3\left(\frac{2\pi}{15}\right)$$

$$= 16\,\frac{\sin(2^4 A)}{2^4 \sin A} \quad \text{(where } A = 2\pi/15)$$

$$= 16\,\frac{\sin(32\pi/15)}{16\sin 2\pi/15} = \frac{\sin(32\pi/15)}{\sin(2\pi + 2\pi/15)}$$

$$= \frac{\sin(32\pi/15)}{\sin(32\pi/15)} = 1$$

10. Given eq. is,

$4\cos^2 x \sin x - 2\sin^2 x = 3\sin x$

$\Rightarrow 4\cos^2 x \sin x - 2\sin^2 x - 3\sin x = 0$

$\Rightarrow 4(1 - \sin^2 x)\sin x - 2\sin^2 x - 3\sin x = 0$

$\Rightarrow \sin x[4\sin^2 x + 2\sin x - 1] = 0$

$\Rightarrow$ either $\sin x = 0$ or $4\sin^2 x + 2\sin x - 1 = 0$

If $\sin x = 0 \Rightarrow x = n\pi$

$\Rightarrow$ If $4\sin^2 x + 2\sin x - 1 = 0$

$$\sin x = \frac{-1 \pm \sqrt{5}}{4}$$

If $\sin x = \frac{-1 \pm \sqrt{5}}{4} = \sin 18° = \sin\frac{\pi}{10}$

then $x = nx + (-1)^4 \frac{\pi}{10}$

If $\sin x = -\left(\frac{\sqrt{5}+1}{4}\right) = \sin(-54°) = \sin\left(\frac{-3\pi}{10}\right)$

then $x = n\pi + (-1)^n\left(\frac{-3\pi}{10}\right)$

Hence, $x = n\pi,\ n\pi + (-1)^n\frac{\pi}{10}$ or $n\pi + (-1)^n\left(\frac{-3\pi}{10}\right)$

where n is some integer

11. The given equation is

$$8^{(1+|\cos x|+|\cos^2 x|+|\cos^3 x|+....)} = 4^3$$

$$\Rightarrow 2^{3(1+|\cos x|+|\cos^2 x|+|\cos^3 x|+....)} = 2^6$$

$\Rightarrow 3(1 + |\cos x| + |\cos^2 x| + |\cos^3 x| + ...) = 6$

$\Rightarrow 1 + |\cos x| + |\cos^2 x| + |\cos^3 x| + ... = 2$

$\Rightarrow \frac{1}{1 - |\cos x|} = 2$ **NOTE THIS STEP**

$\Rightarrow 1 - \cos x = 1/2 \Rightarrow |\cos x| = \frac{1}{2}$

$\Rightarrow x = \pi/3, -\pi/3, 2\pi/3, -2\pi/3, \ldots$

The values of $x \in (-\pi, \pi)$ are $\pm\pi/3, \pm 2\pi/3$.

12. We know that $\tan 2\alpha = \frac{2\tan\alpha}{1 - \tan^2\alpha}$

$\Rightarrow \frac{1 - \tan^2\alpha}{\tan\alpha} = 2\cot 2\alpha$

$\Rightarrow \cot\alpha - \tan\alpha = 2\cot 2\alpha$

Now we have to prove

$\tan\alpha + 2\tan 2\alpha + 4\tan 4\alpha + 8\cot 8\alpha = \cot\alpha$

LHS

$\tan\alpha + 2\tan 2\alpha + 4\tan 4\alpha + 4(2\cot 2.4\alpha)$

$= \tan\alpha + 2\tan 2\alpha + 4\tan 4\alpha + 4(\cot 4\alpha - \tan 4\alpha)$ [Using (1)]

$= \tan\alpha + 2\tan 2\alpha + 4\tan 4\alpha + 4\cot 4\alpha - 4\tan 4\alpha$

$= \tan\alpha + 2\tan 2\alpha + 2(2\cot 2.2\alpha)$

$= \tan\alpha + 2\tan 2\alpha + 2(\cot\alpha - \tan 2\alpha)$

$= \tan\alpha + 2\tan 2\alpha + 2(2\cot 2\alpha - \tan 2\alpha)$ [Using (1)]

$= \tan\alpha + 2\cot 2\alpha$

$= \tan\alpha + (\cot\alpha - \tan\alpha)$ [Using (1)]

$= \cot\alpha = \text{RHS}$.

13. Given that in ΔABC, A, B and C are in A.P.

$\therefore A + C = 2B$

also $A + B + C = 180°$

$\Rightarrow B + 2B = 180° \Rightarrow B = 60°$

$\Rightarrow$ Also given that,

$\sin(2A + B) = \sin(C - A) = -\sin(B + 2C) = \frac{1}{2}$

$\Rightarrow \sin(2A + 60°) = \sin(C - A) = -\sin(60 + 2C) = \frac{1}{2}$ ...(1)

From eq. (1), we have

$\sin(2A+60°)=\frac{1}{2} \Rightarrow 2A+60°=30°,\ 150°$

but A can not be –ve

$\therefore\ 2A+60°=150°$

$\Rightarrow 2A=90° \Rightarrow \quad A=45°$

Again from (1) $\sin(60°+2C)=-\frac{1}{2}$

$\Rightarrow 60°+2C=210°$ or $330°$

$\Rightarrow C=75°$ or $135°$

Also from (1) $\sin(C-A)=\frac{1}{2}$

$\Rightarrow C-A=30°, 150°$

For $A=45°, C=75°$ or $195°$ (not possible)

$\therefore\ C=75°$

Hence we have $A=45°, B=60°, C=75°$.

14. Let $y=\exp[\sin^2 x+\sin^4 x+\sin^6 x+....\infty]\ln 2$

$=e^{\ln \cdot 2^{\sin^2 x+\sin^4 x+\sin^6 x+....\infty}}$

$=2^{\sin^2 x+\sin^4 x+\sin^6 x+....\infty}$

$=2^{\frac{\sin^2 x}{1-\sin^2 x}}=2^{\tan^2 x}$

As y satisfies the eq.

$x^2-9x+8=0 \quad \therefore y^2-9y+8=0$

$\Rightarrow (y-1)(y-8)=0 \Rightarrow y=1, 8$

$\Rightarrow 2^{\tan^2 x}=1$ or $2^{\tan^2 x}=8$

$\Rightarrow \tan^2 x=0$ or $\tan^2 x=3$

$\Rightarrow \tan x=0$ or $\tan x=\sqrt{3}, -\sqrt{3}$

$\Rightarrow x=0$ or $x=\pi/3,\ 2\pi/3$

But given that $0<x<\pi/2$

$\Rightarrow x=\pi/3$

Hence $\frac{\cos x}{\cos x+\sin x}=\frac{1}{1+\tan x}=\frac{1}{1+\sqrt{3}}=\frac{\sqrt{3}-1}{2}$

15. Let $y=\frac{\tan x}{\tan 3x} \Rightarrow y=\frac{\tan x(1-3\tan^2 x)}{3\tan x-\tan^3 x}$

$\Rightarrow 3y-3\tan^2 x=1-3\tan^2 x$

$\Rightarrow (y-3)\tan^2 x=3y-1$

$\Rightarrow \tan^2 x=\frac{3y-1}{y-3}$

$\Rightarrow \frac{3y-1}{y-3}>0$ (L.H.S. being a prefect square)

$\Rightarrow \frac{(3y-1)(y-3)}{(y-3)^2}>0 \Rightarrow (3y-1)(y-3)>0$

$\Rightarrow$ [number line: $-\infty$ … +ve … $1/3$ … –ve … 3 … +ve] $\Rightarrow y<\frac{1}{3}$ or $y>3$

Thus y never lies between $\frac{1}{3}$ and 3.

16. Given that,

$\tan(x+100°)=\tan(x+50°)\tan x\tan(x-50°)$

$\Rightarrow \frac{\tan(x+100°)}{\tan x}=\tan(x+50°)\tan(x-50°)$

$\Rightarrow \frac{\sin(x+100°)\cos x}{\cos(x+100°)\sin x}=\frac{\sin(x+50°)\sin(x-50°)}{\cos(x+50°)\cos(x-50°)}$

$\Rightarrow \frac{\sin(2x+100°)+\sin 100°}{\sin(2x+100°)-\sin 100°}=\frac{\cos 100°-\cos 2x}{\cos 100°+\cos 2x}$

Applying componendo and dividendo, we get

$\Rightarrow \frac{2\sin(2x+100°)}{2\sin 100°}=\frac{2\cos 100°}{-2\cos 2x}$

$\Rightarrow 2\sin(2x+100°)\cos 2x=-2\sin 100°\cos 100°$

$\Rightarrow \sin(4x+100°)+\sin 100°=-\sin 200°$

$\Rightarrow \sin(4x+10°+90°)+\sin(90°+10°)=-\sin(180+20°)$

$\Rightarrow \cos(4x+10°)+\cos 10°=\sin 20°$

$\Rightarrow \cos(4x+10°)=\sin 20°-\cos 10°$

$\Rightarrow \cos(4x+10°)=\sin 20°-\sin 80°$

$=-2\cos 50°\sin 30°=-2\cos 50°.\frac{1}{2}$

$=-\cos 50°=\cos 130°$

$\Rightarrow 4x+10°=130°$

$\Rightarrow x=30°$

17. Given that $\cos\theta=\sin\phi$

where $\theta=p\sin x,\ \phi=p\cos x$

Above is possible when both $\theta=\phi=\frac{\pi}{4}$

or $\theta=\phi=\frac{5\pi}{4}$

$\therefore\ p\sin x=\frac{\pi}{4}$ or $p\sin x=\frac{5\pi}{4}$

and $p\cos x=\frac{\pi}{4}$ or $p\cos x=\frac{5\pi}{4}$

Squaring and adding, $p^2=\frac{\pi^2}{16}.2$ or $\frac{25\pi^2}{16}.2$

$\therefore\ p=\frac{\pi}{4}\sqrt{2}$ only for least positive value

or $p=\frac{\pi}{4}\sqrt{2}$

ALTERNATE SOLUTION

$\cos(p\sin x)=\sin(p\cos x)$

$\Rightarrow \cos(p\sin x)=\cos(\pi/2-p\cos x)$

$\Rightarrow p\sin x=2n\pi\pm(\pi/2-p\cos x)$

$\Rightarrow p\sin x\pm p\cos x=2n\pi\pm\pi/2$

$\Rightarrow \sqrt{2}p(\sin x\cos\pi/4\pm\cos x\sin\pi/4)=2n\pi\pm\pi/2$

$\Rightarrow \sqrt{2}p\sin(x\pm\pi/4)=2n\pi+\pi/2$ ....(1)

Differentiating (1) with respect to x, we get

$$\sqrt{2}\frac{dp}{dx}\sin\ (x\pm\pi/4)+\sqrt{2}\,p\cos\ (x\ \pm\pi/4)=0$$

$$\Rightarrow\quad \frac{dp}{dx}=-p\cot\ (x\pm\pi/4)$$

For smallest value of p, $\frac{dp}{dx}=0$

$\Rightarrow$ $\cot\ (x\ \pm\pi/4)=0$

$\Rightarrow$ $\cot\ (x+\pi/4)=0$ or $\cot\ (x-\pi/4)=0$

$\Rightarrow$ $x=\pi/4$ or $5\pi/4$ or $x=3\pi/4$ or $7\pi/4$

Now, $p=\frac{1}{\sqrt{2}\sin\ (x\pm\pi/4)}(2n\pi\pm\pi/2)$ [from (1)]

For smallest +ve value of p, $n=0$

$$\therefore\quad p=\frac{1}{\sqrt{2}\sin\ (x\pm\pi/4)}(\pm\pi/2)$$

Now for $x=\pi/4 \Rightarrow p=\frac{\pi}{2\sqrt{2}}$; $x=5\pi/4$

$$\Rightarrow\quad p=\frac{-\pi}{2\sqrt{2}}$$

$x=3\pi/4 \Rightarrow p=\frac{-\pi}{2\sqrt{2}}$; $x=7\pi/4 \Rightarrow p=\frac{\pi}{2\sqrt{2}}$

Thus the smallest positive value of $p=\frac{\pi}{2\sqrt{2}}=\frac{\pi\sqrt{2}}{4}$

18. Given : $(1-\tan\theta)\ (1+\tan\theta)\sec^2\theta+2^{\tan^2\theta}=0$

or $(1-\tan^2\theta)\ (1+\tan^2\theta)+2^{\tan^2\theta}=0$

Let us put $\tan^2\theta=t$

$\therefore$ $(1-t)(1+t)+2^t=0$

or $1-t^2+2^t=0$

It is clearly satisfied by $t=3$.

as $-8+8=0$ $\therefore$ $\tan^2\theta=3$

$\therefore$ $p=\pm\pi/3$ in the given interval.

9. Let $y=\frac{\sin x\cos 3x}{\sin 3x\cos x}=\frac{\tan x}{\tan 3x}$

We have $y=\frac{\tan x}{\tan 3x}=\frac{\tan x(1-3\tan^2 x)}{3\tan x-\tan^3 x}$

$$=\frac{1-3\tan^2 x}{3-\tan^2 x}$$

(the expression is not defined if $\tan x=0$)

$\Rightarrow$ $3y-(\tan^2 x)\,y=1-3\tan^2 x$

$\Rightarrow$ $3y-1=(y-3)\tan^2 x$

$$\Rightarrow\quad \tan^2 x=\frac{3y-1}{y-3}=\frac{(3y-1)\,(y-3)}{(y-3)^2}$$

Since $\tan^2 x>0$, we get

$(3y-1)\ (y-3)>0$

$$\Rightarrow\quad \left(y-\frac{1}{3}\right)(y-3)>0$$

$$\Rightarrow\quad y<\frac{1}{3}\quad\text{or}\quad y>3$$

This shows that y cannot lie between $\frac{1}{3}$ and 3.

20. Expanding the sigma on putting $k=1, 2, 3, ..., n$

$$S=(n-1)\cos\frac{2\pi}{n}+(n-2)\cos 2.\frac{2\pi}{n}+......$$

$$+1.\cos\ (n-1)\cos\frac{2\pi}{n}\quad ...(1)$$

We know that $\cos\theta=\cos\ (2\pi-\theta)$

Replacing each angle θ by $2\pi-\theta$ in (1), we get

$$S=(n-1)\cos\ (n-1)\frac{2\pi}{n}+(n-2)\cos\ (n-2)\ \frac{2\pi}{n}+......$$

$$+1.\cos\frac{2\pi}{n}\ \text{by (1)}\quad .....(2)$$

Add terms in (1) and (2) having the same angle and take n common

$$\therefore 2S=n\left[\cos\frac{2\pi}{n}+\cos\frac{4\pi}{n}+\cos\frac{6\pi}{n}+....+\cos\ (n-1)\frac{2\pi}{n}\right]$$

Angles are in A.P. of $d=\frac{2\pi}{n}$

$$2S=n\left[\frac{\sin(n-1)\frac{\pi}{n}}{\sin\frac{\pi}{n}}\cos\frac{\frac{2\pi}{n}+(n-1)\frac{2\pi}{n}}{2}\right]$$

NOTE THIS STEP

$=n.1\cos\pi=-n$

$\because \sin\ (\pi-\theta)=\sin\theta$

$\therefore$ $S=-n/2$

21. We have, $A+B+C=\pi$

$$\Rightarrow\quad \frac{A}{2}+\frac{B}{2}+\frac{C}{2}=\frac{\pi}{2}\Rightarrow\frac{A}{2}+\frac{B}{2}=\frac{\pi}{2}-\frac{C}{2}$$

$$\text{or}\quad \cot\left(\frac{A}{2}+\frac{B}{2}\right)=\cot\left(\frac{\pi}{2}-\frac{C}{2}\right)$$

$$\Rightarrow\quad \frac{\cot\frac{A}{2}.\cot\frac{B}{2}-1}{\cot\frac{A}{2}+\cot\frac{B}{2}}=\tan\frac{C}{2}$$

$$\Rightarrow\quad \cot\frac{A}{2}+\cot\frac{B}{2}+\cot\frac{C}{2}=\cot\frac{A}{2}.\cot\frac{B}{2}.\cot\frac{C}{2}$$

22. Given that,

$$2\sin t=\frac{1-2x+5x^2}{3x^2-2x-1},\quad t\in[-\pi/2,\pi/2]$$

This can be written as

$(6 \sin t - 5) x^2 + 2(1 - 2 \sin t) x - (1 + 2 \sin t) = 0$

For given equation to hold, x should be some real number, therefore above equation should have real roots i.e., $D \geq 0$

$\Rightarrow 4(1 - 2 \sin t)^2 + 4(6 \sin t - 5)(1 + 2 \sin t) \geq 0$

$\Rightarrow 16 \sin^2 t - 8 \sin t - 4 \geq 0$

$\Rightarrow (4 \sin^2 t - 2 \sin t - 1) \geq 0$

$\Rightarrow 4\left(\sin t - \frac{\sqrt{5}+1}{4}\right)\left(\sin t + \frac{\sqrt{5}-1}{4}\right) \geq 0$

$\Rightarrow \sin t \leq -\left(\frac{\sqrt{5}-1}{4}\right)$ or $\sin t \geq \frac{\sqrt{5}+1}{4}$

$\Rightarrow \sin t \leq \sin(-\pi/10)$ or $\sin t \geq \sin(3\pi/10)$

$\Rightarrow t \leq -\pi/10$ or $t \geq 3\pi/10$

(Note that $\sin x$ is an increasing function from $-\pi/2$ to $\pi/2$)

$\therefore$ range of t is $[-\pi/2, -\pi/10] \cup [3\pi/10, \pi/2]$.

(F) MATCH THE FOLLOWING :

1. (i) If $\frac{13\pi}{48} < \alpha < \frac{14\pi}{48} \Rightarrow \frac{13\pi}{16} < 3\alpha < \frac{14\pi}{16}$

and $\frac{13\pi}{24} < 2\alpha < \frac{14\pi}{24}$

$\Rightarrow 3\alpha \in II$ quad and $2\alpha \in II$ quad

$\Rightarrow \sin 3\alpha = +$ ve

$\cos 2\alpha = -$ ve $\quad \therefore \frac{\sin 3\alpha}{\cos 2\alpha} = -ve$

$\therefore$ (B) corresponds to (p).

If $\alpha \in \left(\frac{14\pi}{48}, \frac{18\pi}{48}\right) \Rightarrow \frac{14\pi}{16} < 3\alpha < \frac{18\pi}{16}$

and $\frac{14\pi}{24} < 2\alpha < \frac{18\pi}{24}$

$\Rightarrow 3\alpha \in II$ or III quad and $2\alpha \in II$ quad

$\Rightarrow$ Nothing can be said about the sign of $\frac{\sin 3\alpha}{\cos 2\alpha}$ over this interval.

If $\alpha \in \left(\frac{18\pi}{48}, \frac{23\pi}{48}\right)$ then $\frac{18\pi}{16} < 3\alpha < \frac{23\pi}{16}$

and $\frac{18\pi}{24} < 2\alpha < \frac{23\pi}{24}$

$\Rightarrow 3\alpha \in III$ quad and $2\alpha \in II$ quad

$\Rightarrow \sin 3\alpha = -$ ve, $\cos 2a = -$ ve

$\frac{\sin 3\alpha}{\cos 2\alpha} = +ve$

$\therefore$ (A) corresponds to (r)

If $\alpha \in (0, \pi/2)$

$\Rightarrow 0 < 3\alpha < 3\pi/2$ and $2 < 2\alpha < \pi$

$\Rightarrow$ Nothing can be said about the sign of $\frac{\sin 3\alpha}{\cos 2\alpha}$ over the given interval.

(I) INTEGER VALUE CORRECT TYPE :

1. **3** The given equations are

$xyz \sin 3\theta = (y + z) \cos 3\theta$ —(1)

$xyz \sin 3\theta = 2z \cos 3\theta + 2y \sin 3\theta$ —(2)

$xyz \sin 3\theta = (y + 2z) \cos 3\theta + y \sin 3\theta$ —(3)

Operating (1) – (2) and (3) – (1), we get

$(\cos 3\theta - 2\sin\theta) y - (\cos 3\theta) z = 0$

and $\sin 3\theta\, y + (\cos 3\theta) z = 0$

which is homogeneous system of linear equation. But $y \neq 0, z \neq 0$

$\therefore \frac{\cos 3\theta - 2\sin 3\theta}{\sin 3\theta} = -\frac{\cos 3\theta}{\cos 3\theta} \Rightarrow \cos 3\theta = \sin 3\theta$

$\Rightarrow \tan 3\theta = 1 \Rightarrow 3\theta = n\pi + \frac{\pi}{4}$

$\Rightarrow \theta = (4n+1)\frac{\pi}{12}, n \in Z$

For $\theta \in (0, \pi)$

$\theta = \frac{\pi}{12}, \frac{5\pi}{12}, \frac{3\pi}{4}$

$\therefore$ Three such solutions are possible.

2. $\tan\theta = \cot 5\theta, \theta \neq \frac{n\pi}{5}$

$\Rightarrow \cos\theta \cos 5\theta - \sin 5\theta \sin\theta = 0$

$\Rightarrow \cos 6\theta = 0$

$\Rightarrow 6\theta = \frac{-5\pi}{2}, \frac{-3\pi}{2}, \frac{-\pi}{2}, \frac{\pi}{2}, \frac{3\pi}{2}, \frac{5\pi}{2}$

$\Rightarrow \theta = \frac{-5\pi}{12}, \frac{-\pi}{4}, \frac{-\pi}{12}, \frac{\pi}{12}, \frac{\pi}{4}, \frac{5\pi}{12}$

Again $\sin 2\theta = \cos 4\theta = 1 - 2\sin^2 2\theta$

$\Rightarrow 2\sin^2 2\theta + \sin 2\theta - 1 = 0$

$\Rightarrow \sin 2\theta = -1, \frac{1}{2} \Rightarrow 2\theta = \frac{-\pi}{2}, \frac{\pi}{6}, \frac{5\pi}{6}$

$\Rightarrow \theta = \frac{-\pi}{4}, \frac{\pi}{12}, \frac{5\pi}{12}$

So common solutions are $\theta = \frac{-\pi}{4}, \frac{\pi}{12}$ and $\frac{5\pi}{12}$

$\therefore$ Number of solutions = 3.

3. Let $f(\theta) = \frac{1}{g(\theta)}$

Where $g(\theta) = \sin^2\theta + 3\sin\theta\cos\theta + 5\cos^2\theta$

Clearly f is maximum when g is minimum

Now $g(\theta)=\frac{1-\cos 2\theta}{2}+\frac{3}{2}\sin 2\theta+\frac{5}{2}(1+\cos 2\theta)$

$=3+2\cos 2\theta+\frac{3}{2}\sin 2\theta \geq 3+\left(-\sqrt{4+\frac{9}{4}}\right)$

$\therefore g_{\min}=3-\frac{5}{2}=\frac{1}{2} \Rightarrow f_{\max}=2.$

4. 3

From figure, we get

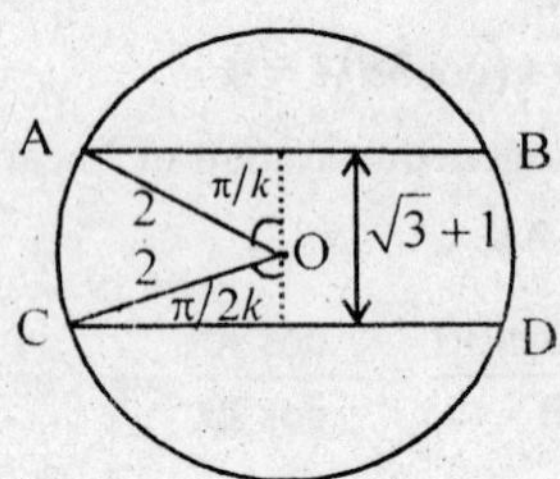

$2\cos\frac{\pi}{k}+2\cos\frac{\pi}{2k}=\sqrt{3}+1$

$\Rightarrow 2\times 2\cos^2\frac{\pi}{2k}+2\cos\frac{\pi}{2k}-2=\sqrt{3}+1$

$\Rightarrow 4\cos^2\frac{\pi}{2k}+2\cos\frac{\pi}{2k}-(3+\sqrt{3})=0$

$\Rightarrow \cos\frac{\pi}{2k}=\frac{-2\pm\sqrt{4+16(3+\sqrt{3})}}{8}=\frac{-1\pm\sqrt{13+4\sqrt{3}}}{4}$

$=\frac{-1\pm(2\sqrt{3}+1)}{4}=\frac{\sqrt{3}}{2}$ or $-\left(\frac{\sqrt{3}+1}{2}\right)$

As $\frac{\pi}{2k}$ is an acute angle, $\cos\frac{\pi}{2k}=\frac{\sqrt{3}}{2}=\cos\frac{\pi}{6}$

$\Rightarrow k=3$

5. 7

We have

$\frac{1}{\sin\frac{\pi}{n}}-\frac{1}{\sin\frac{3\pi}{n}}=\frac{1}{\sin\frac{2\pi}{n}}$

$\Rightarrow \frac{\sin\frac{3\pi}{n}-\sin\frac{\pi}{n}}{\sin\frac{\pi}{n}\sin\frac{3\pi}{n}}=\frac{1}{\sin\frac{2\pi}{n}}$

$\Rightarrow \frac{2\cos\frac{2\pi}{n}\sin\frac{\pi}{n}}{\sin\frac{\pi}{n}\sin\frac{3\pi}{n}}=\frac{1}{\sin\frac{2\pi}{n}}$

$\Rightarrow 2\sin\frac{2\pi}{n}\cos\frac{2\pi}{n}=\sin\frac{3\pi}{n}$

$\Rightarrow \sin\frac{4\pi}{n}-\sin\frac{3\pi}{n}=0$

$\Rightarrow 2\cos\frac{7\pi}{2n}\sin\frac{\pi}{2n}=0 \Rightarrow \cos\frac{7\pi}{2n}=0$ or $\sin\frac{\pi}{2n}=0$

$\Rightarrow \frac{7\pi}{2n}=(2k+1)\frac{\pi}{2}$ or $\frac{\pi}{2n}=2k\pi$ where $k\in Z$

$\Rightarrow n=\frac{7}{2k+1}$ or $n=\frac{1}{4k}$

($n=\frac{1}{4k}$ not possible for any integral value of k)

As $n>3$, for $k=0$, we get $n=7$.

Section-B JEE Main/ AIEEE

1. (b) $\sin^2\theta=\frac{1-\cos 2\theta}{2}$; Period $=\frac{2\pi}{2}=\pi$

2. (b) The given equation is tanx + secx = 2 cos x;

$\Rightarrow \sin x+1=2\cos^2 x \Rightarrow \sin x+1=2(1-\sin^2 x);$

$\Rightarrow 2\sin^2 x+\sin x-1=0;$

$\Rightarrow (2\sin x-1)(\sin x+1)=0 \Rightarrow \sin x=\frac{1}{2},-1.;$

$\Rightarrow x=30°, 150°, 270°.$

3. (b) $\because \cos\sqrt{x}$ is non periodic

$\therefore \cos\sqrt{x}+\cos^2 x$ can not be periodic.

4. (d) $\pi<\alpha-\beta<3\pi$

$\Rightarrow \frac{\pi}{2}<\frac{\alpha-\beta}{2}<\frac{3\pi}{2} \Rightarrow \cos\frac{\alpha-\beta}{2}<0$

$\sin\alpha+\sin\beta=-\frac{21}{65}$

$\Rightarrow 2\sin\frac{\alpha+\beta}{2}\cos\frac{\alpha-\beta}{2}=-\frac{21}{65}$ .....(1)

$\cos\alpha+\cos\beta=-\frac{27}{65}$

$\Rightarrow 2\cos\frac{\alpha+\beta}{2}\cos\frac{\alpha-\beta}{2}=-\frac{27}{65}$ .....(2)

Square and add (1) and (2)

$$4\cos^2\frac{\alpha-\beta}{2}=\frac{(21)^2+(27)^2}{(65)^2}=\frac{1170}{65\times 65}$$

$$\therefore \cos^2\frac{\alpha-\beta}{2}=\frac{9}{130}\Rightarrow \cos\frac{\alpha-\beta}{2}=-\frac{3}{\sqrt{130}}$$

ALTERNATE SOLUTION

Given that $\sin\alpha+\sin\beta=\frac{21}{65}$ ....(1)

$\cos\alpha+\cos\beta=\frac{-27}{65}$ ....(2)

Squaring and adding equation (1) and (2) we get

$$\sin^2\alpha+\sin^2\beta+2\sin\alpha\sin\beta+\cos^2\alpha$$

$$+\cos^2\beta+2\cos\alpha\cos\beta=\left(\frac{-21}{65}\right)^2+\left(\frac{-27}{65}\right)^2$$

$$\Rightarrow 2+2(\cos\alpha\cos\beta+\sin\alpha\sin\beta)=\frac{1170}{4225}$$

$$\Rightarrow 2[1+\cos(\alpha-\beta)]=\frac{1170}{4425}$$

$$\Rightarrow 4\cos^2\left(\frac{\alpha-\beta}{2}\right)=\frac{1170}{4425}\Rightarrow \cos^2\frac{\alpha-\beta}{2}=\frac{9}{130}$$

$$\Rightarrow \cos\frac{\alpha-\beta}{2}=\frac{-3}{\sqrt{130}}$$

Negative sign is taken because $\frac{\pi}{2}<\frac{\alpha-\beta}{2}<\frac{3\pi}{2}$

5. (a) $u^2=a^2+b^2+2\sqrt{\begin{array}{l}(a^4+b^4)\cos^2\theta\sin^2\theta\\ +a^2b^2(\cos^4\theta+\sin^4\theta)\end{array}}$ ... (1)

Now $(a^4+b^4)\cos^2\theta\sin^2\theta+a^2b^2(\cos^4\theta+\sin^4\theta)$

$=(a^4+b^4)\cos^2\theta\sin^2\theta+a^2b^2(1-2\cos^2\theta\sin^2\theta)$

$=(a^4+b^4-2a^2b^2)\cos^2\theta\sin^2\theta+a^2b^2$

$=(a^2-b^2)^2.\frac{\sin^2 2\theta}{4}+a^2b^2$ ...(2)

$\because 0\le\sin^2 2\theta\le 1$

$\Rightarrow 0\le(a^2-b^2)^2\frac{\sin^2 2\theta}{4}\le\frac{(a^2-b^2)^2}{4}$

$\Rightarrow a^2b^2\le(a^2-b^2)^2\frac{\sin^2 2\theta}{4}+a^2b^2$

$\le(a^2-b^2)^2.\frac{1}{4}+a^2b^2$ ....(3)

$\therefore$ from (1), (2) and (3)

Minimum value of $u^2=a^2+b^2+2\sqrt{a^2b^2}=(a+b)^2$

Maximum value of u^2

$=a^2+b^2+2\sqrt{(a^2-b^2)^2.\frac{1}{4}+a^2b^2}$

$=a^2+b^2+\frac{2}{2}\sqrt{(a^2+b^2)^2}=2(a^2+b^2)$

$\therefore$ Max value - Min value

$=2(a^2+b^2)-(a+b^2)=(a-b)^2$

6. (c) The direction cosines of the line are $\cos\theta$, $\cos\beta$, $\cos\theta$

$\therefore \cos^2\theta+\cos^2\beta+\cos^2\theta=1$

$\Rightarrow 2\cos^2\theta=\sin^2\beta=3\sin^2\theta$ (given)

$\Rightarrow 2\cos^2\theta=3-3\cos^2\theta$

$\therefore \cos^2\theta=\frac{3}{5}$

7. (a)

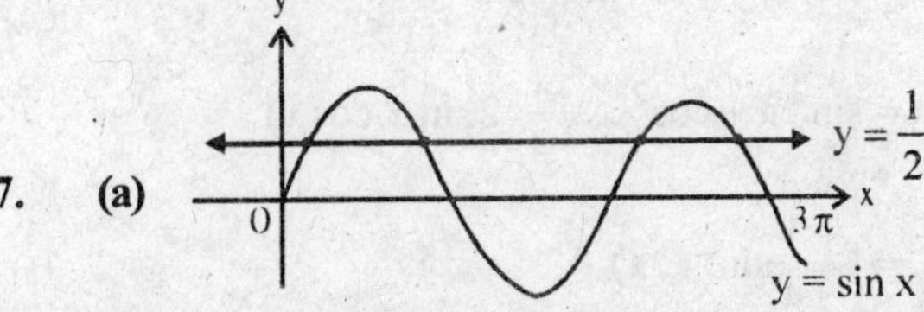

$2\sin^2 x+5\sin x-3=0$

$\Rightarrow (\sin x+3)(2\sin x-1)=0$

$\Rightarrow \sin x=\frac{1}{2}$ and $\sin x\ne -3$

$\therefore$ In $[0,3\pi]$, x has 4 values.

8. (c) $\cos x+\sin x=\frac{1}{2}\Rightarrow 1+\sin 2x=\frac{1}{4}$

$\Rightarrow \sin 2x=-\frac{3}{4}$, so x is obtuse and

$\frac{2\tan x}{1+\tan^2 x}=-\frac{3}{4}\Rightarrow 3\tan^2 x+8\tan x+3=0$

$\therefore \tan x=\frac{-8\pm\sqrt{64--36}}{6}=-\frac{-4\pm\sqrt{7}}{3}$

as $\tan x<0$ $\therefore \tan x=\frac{-4-\sqrt{7}}{3}$

9. (b) We have

$\cos(\beta-\gamma)+\cos(\gamma-\alpha)+\cos(\alpha-\beta)=-\frac{3}{2}$

$\Rightarrow 2[\cos(\beta-\gamma)+\cos(\gamma-\alpha)+\cos(\alpha-\beta)]+3=0$

$\Rightarrow 2[\cos(\beta-\gamma)+\cos(\gamma-\alpha)+\cos(\alpha-\beta)]$

$+\sin^2\alpha+\cos^2\alpha+\sin^2\beta+\cos^2\beta+\sin^2\gamma+\cos^2\alpha=0$

$\Rightarrow$ $[\sin^2\alpha + \sin^2\beta + \sin^2\gamma + 2\sin\alpha\sin\beta + 2\sin\beta\sin\gamma$
$+ 2\sin\gamma\sin\alpha] + [\cos^2\alpha + \cos^2\beta + \cos^2\gamma + 2\cos\alpha\cos\beta$
$+ 2\cos\beta\cos\gamma + 2\cos\gamma\cos\alpha] = 0$

$\Rightarrow$ $[\sin\alpha + \sin\beta + \sin\gamma]^2 + (\cos\alpha + \cos\beta + \cos\gamma)^2 = 0$

$\Rightarrow$ $\sin\alpha + \sin\beta + \sin\gamma = 0$ and $\cos\alpha + \cos\beta + \cos\gamma = 0$

$\therefore$ A and B both are true.

10. **(a)** $\cos(\alpha+\beta) = \frac{4}{5} \Rightarrow \tan(\alpha+\beta) = \frac{3}{4}$

$\sin(\alpha-\beta) = \frac{5}{13} \Rightarrow \tan(\alpha-\beta) = \frac{5}{12}$

$$\tan 2\alpha = \tan[(\alpha+\beta)+(\alpha-\beta)] = \frac{\frac{3}{4}+\frac{5}{12}}{1-\frac{3}{4}.\frac{5}{12}} = \frac{56}{33}$$

11. **(d)** $A = \sin^2 x + \cos^4 x$

$= \sin^2 x + \cos^2 x(1-\sin^2 x)$

$= \sin^2 x + \cos^2 x - \frac{1}{4}(2\sin x.\cos x)^2$

$= 1 - \frac{1}{4}\sin^2(2x)$

Now $0 \le \sin^2(2x) \le 1$

$\Rightarrow 0 \ge -\frac{1}{4}\sin^2(2x) \ge -\frac{1}{4}$

$\Rightarrow 1 \ge 1 - \frac{1}{4}\sin^2(2x) \ge 1 - \frac{1}{4} \Rightarrow 1 \ge A \ge \frac{3}{4}$

12. **(d)** $\sin 4\theta + 2\sin 4\theta\cos 3\theta = 0$

$\sin 4\theta(1+2\cos 3\theta) = 0$

$\sin 4\theta = 0$ or $\cos 3\theta = -\frac{1}{2}$

$4\theta = n\pi; n \in I$ or $3\theta = 2n\pi \pm \frac{2\pi}{3}, n \in I$

$\theta = \frac{\pi}{4}, \frac{\pi}{2}, \frac{3\pi}{4}$ or $\theta = \frac{2\pi}{9}, \frac{8\pi}{9}, \frac{4\pi}{9}$ $[\because \theta, \in (0,\pi)]$

13. **(b)** Given $3\sin P + 4\cos Q = 6$ ...(i)
$4\sin Q + 3\cos P = 1$ ...(ii)
Squaring and adding (i) & (ii) we get
$9\sin^2 P + 16\cos^2 Q + 24\sin P\cos Q$
$+ 16\sin^2 Q + 9\cos^2 P + 24\sin Q\cos P = 36 + 1 = 37$
$\Rightarrow$ $9(\sin^2 P + \cos^2 P) + 16(\sin^2 Q + \cos^2 Q)$
$+ 24(\sin P\cos Q + \cos P\sin Q) = 37$
$\Rightarrow$ $9 + 16 + 24\sin(P+Q) = 37$
$[\because \sin^2\theta + \cos^2\theta = 1$ and $\sin A\cos B + \cos A\sin B = \sin(A+B)]$

$\Rightarrow$ $\sin(P+Q) = \frac{1}{2} \Rightarrow P+Q = \frac{\pi}{6}$ or $\frac{5\pi}{6}$

$\Rightarrow$ $R = \frac{5\pi}{6}$ or $\frac{\pi}{6}$ $(\because P+Q+R=\pi)$

If $R = \frac{5\pi}{6}$ then $0 < P, Q < \frac{\pi}{6}$

$\Rightarrow$ $\cos Q < 1$ and $\sin P < \frac{1}{2}$

$\Rightarrow$ $3\sin P + 4\cos Q < \frac{11}{2}$ which is not true.

So $R = \frac{\pi}{6}$

14. **(a)** From Sine Rule

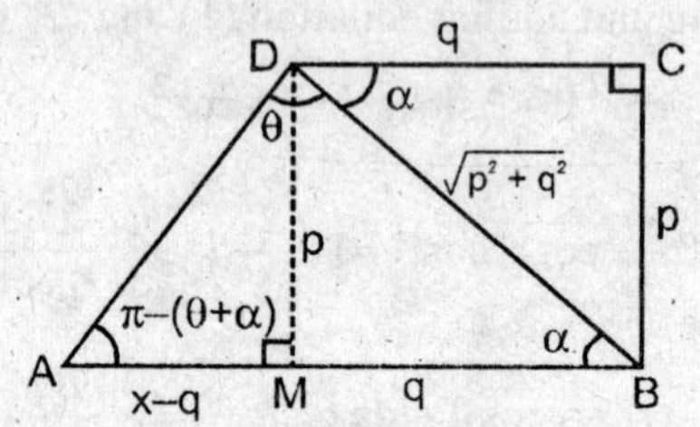

$$\frac{AB}{\sin\theta} = \frac{\sqrt{p^2+q^2}}{\sin(\pi-(\theta+\alpha))}$$

$$AB = \frac{\sqrt{p^2+q^2}\sin\theta}{\sin\theta\cos\alpha + \cos\theta\sin\alpha}$$

$$= \frac{(p^2+q^2)\sin\theta}{q\sin\theta + p\cos\theta}$$

$$\left(\because \cos\alpha = \frac{q}{\sqrt{p^2+q^2}} \text{ and } \sin\alpha = \frac{p}{\sqrt{p^2+q^2}}\right)$$

15. **(b)** Given expression can be written as

$$\frac{\sin A}{\cos A} \times \frac{\sin A}{\sin A - \cos A} + \frac{\cos A}{\sin A} \times \frac{\cos A}{\cos A - \sin A}$$

$$\left(\because \tan A = \frac{\sin A}{\cos A} \text{ and } \cot A = \frac{\cos A}{\sin A}\right)$$

$$= \frac{1}{\sin A - \cos A}\left\{\frac{\sin^3 A - \cos^3 A}{\cos A\sin A}\right\}$$

$$= \frac{\sin^2 A + \sin A\cos A + \cos^2 A}{\sin A\cos A} = 1 + \sec A \operatorname{cosec} A$$

Chapter 2 Complex Numbers

Section-A JEE Advanced/ IIT-JEE

A FILL IN THE BLANKS :

1. If the expression *(1987 - 2 Marks)*

$$\frac{\left[\sin\left(\frac{x}{2}\right)+\cos\left(\frac{x}{2}\right)+i\tan(x)\right]}{\left[1+2\,i\sin\left(\frac{x}{2}\right)\right]}$$

is real, then the set of all possible values of x is ....

2. For any two complex numbers z_1, z_2 and any real number a and b. *(1988 - 2 Marks)*
$|az_1 - bz_2|^2 + |bz_1 + az_2|^2 =$ ....

3. If a, b, c, are the numbers between 0 and 1 such that the points $z_1 = a + i, z_2 = 1 + bi$ and $z_3 = 0$ form an equilateral triangle, then $a =$ ....and $b =$ .... *(1989 - 2 Marks)*

4. $ABCD$ is a rhombus. Its diagonals AC and BD intersect at the point M and satisfy $BD = 2AC$. If the points D and M represent the complex numbers $1 + i$ and $2 - i$ respectively, then A represents the complex number ....or.......... *(1993 - 2 Marks)*

5. Suppose Z_1, Z_2, Z_3 are the vertices of an equilateral triangle inscribed in the circle $|Z| = 2$. If $Z_1 = 1 + i\sqrt{3}$ then $Z_2 =$ ...., $Z_3 =$ .... *(1994 - 2 Marks)*

6. The value of the expression
$1\cdot(2-\omega)(2-\omega^2)+2\cdot(3-\omega)(3-\omega^2)+....+(n-1).(n-\omega)(n-\omega^2)$, where ω is an imaginary cube root of unity, is..... *(1996 - 2 Marks)*

B TRUE / FALSE :

1. For complex number $z_1 = x_1 + iy_1$ and $z_2 = x_2 + iy_2$, we write $z_1 \cap z_2$, if $x_1 \le x_2$ and $y_1 \le y_2$. Then for all complex numbers z with $1 \cap z$, we have $\frac{1-z}{1+z} \cap 0$. *(1981 - 2 Marks)*

2. If the complex numbers, Z_1, Z_2 and Z_3 represent the vertices of an equilateral triangle such that
$|Z_1| = |Z_2| = |Z_3|$ then $Z_1 + Z_2 + Z_3 = 0$. *(1984 - 1 Mark)*

3. If three complex numbers are in A.P. then they lie on a circle in the complex plane. *(1985 - 1 Mark)*

4. The cube roots of unity when represented on Argand diagram form the vertices of an equilateral triangle. *(1988 - 1 Mark)*

C MCQ's WITH ONE CORRECT ANSWER :

1. If the cube roots of unity are $1, \omega, \omega^2$, then the roots of the equation $(x-1)^3 + 8 = 0$ are *(1979)*
(a) $-1, 1+2\omega, 1+2\omega^2$ (b) $-1, 1-2\omega, 1-2\omega^2$
(c) $-1, -1, -1$ (d) None of these

2. The smallest positive integer n for which *(1980)*
$\left(\frac{1+i}{1-i}\right)^n = 1$ is
(a) $n = 8$ (b) $n = 16$
(c) $n = 12$ (d) none of these

3. The complex numbers $z = x + iy$ which satisfy the equation $\left|\frac{z-5i}{z+5i}\right| = 1$ lie on *(1981 - 2 Marks)*
(a) the x-axis
(b) the straight line $y = 5$
(c) a circle passing through the origin
(d) none of these

4. If $z = \left(\frac{\sqrt{3}}{2}+\frac{i}{2}\right)^5 + \left(\frac{\sqrt{3}}{2}-\frac{i}{2}\right)^5$, then *(1982 - 2 Marks)*
(a) $\text{Re}(z) = 0$ (b) $\text{Im}(z) = 0$
(c) $\text{Re}(z) > 0, \text{Im}(z) > 0$ (d) $\text{Re}(z) > 0, \text{Im}(z) < 0$

5. The inequality $|z-4| < |z-2|$ represents the region given by *(1982 - 2 Marks)*
(a) $\text{Re}(z) \ge 0$ (b) $\text{Re}(z) < 0$
(c) $\text{Re}(z) > 0$ (d) none of these

6. If $z = x + iy$ and $\omega = (1 - iz)/(z - i)$, then $|\omega| = 1$ implies that, in the complex plane, *(1983 - 1 Mark)*
(a) z lies on the imaginary axis
(b) z lies on the real axis
(c) z lies on the unit circle
(d) None of these

7. The points $z_1, z_2, z_3\ z_4$ in the complex plane are the vertices of a parallelogram taken in order if and only if *(1983 - 1 Mark)*

(a) $z_1+z_4=z_2+z_3$ (b) $z_1+z_3=z_2+z_4$
(c) $z_1+z_2=z_3+z_4$ (d) None of these

8. If a, b, c and u, v, w are complex numbers representing the vertices of two triangles such that $c=(1-r)a+rb$ and $w=(1-r)u+rv$, where r is a complex number, then the two triangles *(1985 - 2 Marks)*

(a) have the same area (b) are similar
(c) are congruent (d) none of these

9. If $\omega\ (\neq 1)$ is a cube root of unity and $(1+\omega)^7=A+B\omega$ then A and B are respectively *(1995S)*

(a) 0, 1 (b) 1, 1
(c) 1, 0 (d) −1, 1

11. Let z and ω be two non zero complex numbers such that $|z|=|\omega|$ and $\text{Arg}\, z+\text{Arg}\,\omega=\pi$, then z equals *(1995S)*

(a) ω (b) $-\omega$
(c) $\overline{\omega}$ (d) $-\overline{\omega}$

12. Let z and ω be two complex numbers such that $|z|\le 1$, $|\omega|\le 1$ and $|z+i\omega|=|z-i\overline{\omega}|=2$ then z equals *(1995S)*

(a) 1 or i (b) i or $-i$
(c) 1 or −1 (d) i or −1

13. For positive integers n_1, n_2 the value of the expression $(1+i)^{n_1}+(1+i^3)^{n_1}+(1+i^5)^{n_2}+(1+i^7)^{n_2}$, where $i=\sqrt{-1}$ is a real number if and only if *(1996 - 1 Marks)*

(a) $n_1=n_2+1$ (b) $n_1=n_2-1$
(c) $n_1=n_2$ (d) $n_1>0, n_2>0$

14. If $i=\sqrt{-1}$, then $4+5\left(-\frac{1}{2}+\frac{i\sqrt{3}}{2}\right)^{334}+3\left(-\frac{1}{2}+\frac{i\sqrt{3}}{2}\right)^{365}$ is equal to *(1999 - 2 Marks)*

(a) $1-i\sqrt{3}$ (b) $-1+i\sqrt{3}$
(c) $i\sqrt{3}$ (d) $-i\sqrt{3}$

15. If $\arg(z)<0$, then $\arg(-z)-\arg(z)=$ *(2000S)*

(a) π (b) $-\pi$
(c) $-\frac{\pi}{2}$ (d) $\frac{\pi}{2}$

16. If z_1, z_2 and z_3 are complex numbers such that *(2000S)*

$$|z_1|=|z_2|=|z_3|=\left|\frac{1}{z_1}+\frac{1}{z_2}+\frac{1}{z_3}\right|=1,\ \text{then}\ |z_1+z_2+z_3|\ \text{is}$$

(a) equal to 1 (b) less than 1
(c) greater than 3 (d) equal to 3

17. Let z_1 and z_2 be n^{th} roots of unity which subtend a right angle at the origin. Then n must be of the form *(2001S)*

(a) $4k+1$ (b) $4k+2$
(c) $4k+3$ (d) $4k$

18. The complex numbers z_1, z_2 and z_3 satisfying $\frac{z_1-z_3}{z_2-z_3}=\frac{1-i\sqrt{3}}{2}$ are the vertices of a triangle which is *(2001S)*

(a) of area zero
(b) right-angled isosceles
(c) equilateral
(d) obtuse-angled isosceles

19. For all complex numbers z_1, z_2 satisfying $|z_1|=12$ and $|z_2-3-4i|=5$, the minimum value of $|z_1-z_2|$ is *(2002S)*

(a) 0 (b) 2
(c) 7 (d) 17

20. If $|z|=1$ and $\omega=\frac{z-1}{z+1}$ (where $z\neq -1$), then $\text{Re}(\omega)$ is *(2003S)*

(a) 0 (b) $-\frac{1}{|z+1|^2}$
(c) $\left|\frac{z}{z+1}\right|.\frac{1}{|z+1|^2}$ (d) $\frac{\sqrt{2}}{|z+1|^2}$

21. If $\omega\ (\neq 1)$ be a cube root of unity and $(1+\omega^2)^n=(1+\omega^4)^n$, then the least positive value of n is *(2004S)*

(a) 2 (b) 3
(c) 5 (d) 6

22. The locus of z which lies in shaded region (excluding the boundaries) is best represented by *(2005S)*

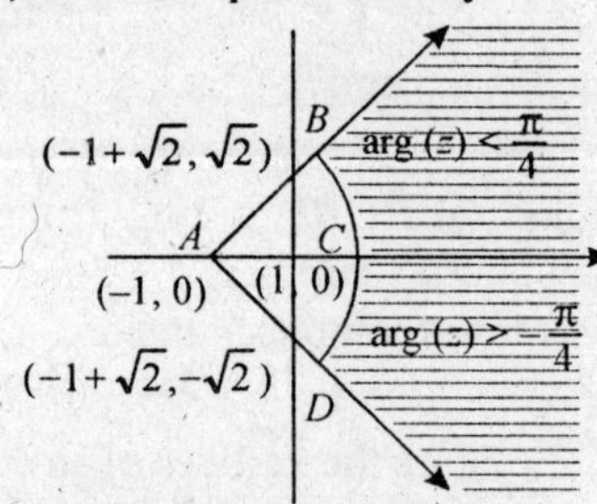

(a) $z:|z+1|>2$ and $|\arg(z+1)|<\pi/4$
(b) $z:|z-1|>2$ and $|\arg(z-1)|<\pi/4$
(c) $z:|z+1|<2$ and $|\arg(z+1)|<\pi/2$
(d) $z:|z-1|<2$ and $|\arg(z+1)|<\pi/2$

23. a, b, c are integers, not all simultaneously equal and ω is cube root of unity $(\omega\neq 1)$, then minimum value of $|a+b\omega+c\omega^2|$ is *(2005S)*

(a) 0 (b) 1
(c) $\frac{\sqrt{3}}{2}$ (d) $\frac{1}{2}$

24. Let $\omega=-\frac{1}{2}+i\frac{\sqrt{3}}{2}$, then the value of the det.

$$\begin{vmatrix}1 & 1 & 1\\ 1 & -1-\omega^2 & \omega^2\\ 1 & \omega^2 & \omega^4\end{vmatrix}\ \text{is}$$ *(2002 - 2 Marks)*

(a) 3ω (b) $3\omega(\omega-1)$
(c) $3\omega^2$ (d) $3\omega(1-\omega)$

25. If $\frac{w-\bar{w}z}{1-z}$ is purely real where $w=\alpha+i\beta$, $\beta\neq 0$ and $z\neq 1$, then the set of the values of z is *(2006 - 3M, –1)*

(a) $\{z:|z|=1\}$ (b) $\{z:z=\bar{z}\}$
(c) $\{z:z\neq 1\}$ (d) $\{z:|z|=1, z\neq 1\}$

26. A man walks a distance of 3 units from the origin towards the north-east (N 45° E) direction. From there, he walks a distance of 4 units towards the north-west (N 45° W) direction to reach a point P. Then the position of P in the Argand plane is *(2007 -3 marks)*

(a) $3e^{i\pi/4}+4i$ (b) $(3-4i)e^{i\pi/4}$
(c) $(4+3i)e^{i\pi/4}$ (d) $(3+4i)e^{i\pi/4}$

27. If $|z|=1$ and $z\neq\pm 1$, then all the values of $\frac{z}{1-z^2}$ lie on

(a) a line not passing through the origin *(2007 -3 marks)*
(b) $|z|=\sqrt{2}$
(c) the x-axis
(d) the y-axis

28. A particle P starts from the point $z_0=1+2i$, where $i=\sqrt{-1}$. It moves horizontally away from origin by 5 units and then vertically away from origin by 3 units to reach a point z_1. From z_1 the particle moves $\sqrt{2}$ units in the direction of the vector $\hat{i}+\hat{j}$ and then it moves through an angle $\frac{\pi}{2}$ in anticlockwise direction on a circle with centre at origin, to reach a point z_2. The point z_2 is given by *(2008)*

(a) $6+7i$ (b) $-7+6i$
(c) $7+6i$ (d) $-6+7i$

29. Let $z=\cos\theta+i\sin\theta$. Then the value of $\sum_{m=1}^{15}\text{Im}(z^{2m-1})$ at $\theta=2°$ is *(2009)*

(a) $\frac{1}{\sin 2°}$ (b) $\frac{1}{3\sin 2°}$
(c) $\frac{1}{2\sin 2°}$ (d) $\frac{1}{4\sin 2°}$

30. Let $z=x+iy$ be a complex number where x and y are integers. Then the area of the rectangle whose vertices are the roots of the equation

$$z\bar{z}^3+\bar{z}z^3=350$$

is *(2009)*

(a) 48 (b) 32
(c) 40 (d) 80

31. Let z be a complex number such that the imaginary part of z is non-zero and $a=z^2+z+1$ is real. Then a cannot take the value *(2012)*

(a) -1 (b) $\frac{1}{3}$ (c) $\frac{1}{2}$ (d) $\frac{3}{4}$

32. Let complex numbers α and $\frac{1}{\bar{\alpha}}$ lie on circles $(x-x_0)^2+(y-y_0)^2=r^2$ and $(x-x_0)^2+(y-y_0)^2=4r^2$, respectively. If $z_0=x_0+iy_0$ satisfies the equation $2|z_0|^2=r^2+2$, then $|\alpha|=$ *(JEE Adv. 2013)*

(a) $\frac{1}{\sqrt{2}}$ (b) $\frac{1}{2}$
(c) $\frac{1}{\sqrt{7}}$ (d) $\frac{1}{3}$

D MCQ's WITH ONE OR MORE THAN ONE CORRECT :

1. If $z_1=a+ib$ and $z_2=c+id$ are complex numbers such that $|z_1|=|z_2|=1$ and $\text{Re}(z_1\bar{z}_2)=0$, then the pair of complex numbers $w_1=a+ic$ and $w_2=b+id$ satisfies – *(1985 - 2 Marks)*

(a) $|w_1|=1$ (b) $|w_2|=1$
(c) $\text{Re}(w_1\bar{w}_2)=0$ (d) none of these

2. Let z_1 and z_2 be complex numbers such that $z_1\neq z_2$ and $|z_1|=|z_2|$. If z_1 has positive real part and z_2 has negative imaginary part, then $\frac{z_1+z_2}{z_1-z_2}$ may be *(1986 - 2 Marks)*

(a) zero (b) real and positive
(c) real and negative (d) purely imaginary
(e) none of these.

3. If z_1 and z_2 are two nonzero complex numbers such that $|z_1+z_2|=|z_1|+|z_2|$, then Arg z_1 – Arg z_2 is equal to *(1987 - 2 Marks)*

(a) $-\pi$ (b) $-\frac{\pi}{2}$
(c) 0 (d) $\frac{\pi}{2}$
(e) π

4. The value of $\sum_{k=1}^{6}\left(\sin\frac{2\pi k}{7}-i\cos\frac{2\pi k}{7}\right)$ is *(1987 - 2 Marks)*

(a) -1 (b) 0
(c) $-i$ (d) i
(e) None

5. If ω is an imaginary cube root of unity, then $(1+\omega-\omega^2)^7$ equals *(1998 - 2 Marks)*

(a) 128ω (b) -128ω
(c) $128\omega^2$ (d) $-128\omega^2$

6. The value of the sum $\sum_{n=1}^{13}(i^n+i^{n+1})$, where $i=\sqrt{-1}$, equals

(a) i (b) $i-1$ *(1998 - 2 Marks)*
(c) $-i$ (d) 0

7. If $\begin{vmatrix} 6i & -3i & 1 \\ 4 & 3i & -1 \\ 20 & 3 & i \end{vmatrix} = x+iy$, then *(1998 - 2 Marks)*

(a) $x=3, y=2$ (b) $x=1, y=3$
(c) $x=0, y=3$ (d) $x=0, y=0$

8. Let z_1 and z_2 be two distinct complex numbers and let $z=(1-t)z_1+tz_2$ for some real number t with $0<t<1$. If Arg (w) denotes the principal argument of a non-zero complex number w, then *(2010)*

(a) $|z-z_1|+|z-z_2|=|z_1-z_2|$
(b) $\text{Arg}(z-z_1)=\text{Arg}(z-z_2)$
(c) $\begin{vmatrix} z-z_1 & \bar{z}-\bar{z}_1 \\ z_2-z_1 & \bar{z}_2-\bar{z}_1 \end{vmatrix}$
(d) $\text{Arg}(z-z_1)=\text{Arg}(z_2-z_1)$

9. Let $w=\dfrac{\sqrt{3}+i}{2}$ and $P=\{w^n : n=1,2,3,...\}$. Further $H_1 = \left\{z\in\mathbb{C} : \text{Re}\, z>\frac{1}{2}\right\}$ and $H_2=\left\{z\in\mathbb{C} : \text{Re}\, z<\frac{-1}{2}\right\}$, where c is the set of all complex numbers. If $z_1 \in P\cap H_1, z_2\in P\cap H_2$ and O represents the origin, then $\angle z_1Oz_2=$ *(JEE Adv. 2013)*

(a) $\frac{\pi}{2}$ (b) $\frac{\pi}{6}$
(c) $\frac{2\pi}{3}$ (d) $\frac{5\pi}{6}$

E SUBJECTIVE PROBLEMS :

1. Express $\dfrac{1}{1-\cos\theta+2i\sin\theta}$ in the form $x+iy$. *(1978)*
2. If $x=a+b, y=a\gamma+b\beta$ and $z=a\beta+b\gamma$ where γ and β are the complex cube roots of unity, show that $xyz=a^3+b^3$. *(1978)*
3. If $x+iy=\sqrt{\dfrac{a+ib}{c+id}}$, prove that $(x^2+y^2)^2=\dfrac{a^2+b^2}{c^2+d^2}$. *(1979)*
4. It is given that n is an odd integer greater than 3, but n is not a multiple of 3. Prove that x^3+x^2+x is a factor of $(x+1)^n-x^n-1$. *(1980)*
5. Find the real values of x and y for which the following equation is satisfied $\dfrac{(1+i)x-2i}{3+i}+\dfrac{(2-3i)y+i}{3-i}=i$ *(1980)*
6. Let the complex number z_1, z_2 and z_3 be the vertices of an equilateral triangle. Let z_0 be the circumcentre of the triangle. Then prove that $z_1^2+z_2^2+z_3^2=3z_0^2$. *(1981 - 4 Marks)*
7. Prove that the complex numbers z_1, z_2 and the origin form an equilateral triangle only if $z_1^2+z_2^2-z_1z_2=0$. *(1983 - 3 Marks)*
8. If $1, a_1, a_2 ......, a_{n-1}$ are the n roots of unity, then show that $(1-a_1)(1-a_2)(1-a_3)....(1-a_{n-1})=n$ *(1984 - 2 Marks)*
9. Show that the area of the triangle on the Argand diagram formed by the complex numbers z, iz and $z+iz$ is $\frac{1}{2}|z|^2$. *(1986 - 2½ Marks)*
10. Complex numbers z_1, z_2, z_3 are the vertices A, B, C respectively of an isosceles right angled triangle with right angle at C. Show that $(z_1-z_2)^2=2(z_1-z_3)(z_3-z_2)$. *(1986 - 2½ Marks)*
11. Let $Z_1=10+6i$ and $Z_2=4+6i$. If Z is any complex number such that the argument of $\dfrac{(Z-Z_1)}{(Z-Z_2)}$ is $\dfrac{\pi}{4}$, then prove that $|Z-7-9i|=3\sqrt{2}$. *(1990 - 4 Marks)*
12. If $iz^3+z^2-z+i=0$, then show that $|z|=1$. *(1995 - 5 Marks)*
13. If $|Z|\le 1, |W|\le 1$, show that $|Z-W|^2\le(|Z|-|W|)^2+(Arg\,Z-Arg\,W)^2$ *(1995 - 5 Marks)*
14. Find all non-zero complex numbers Z satisfying $\bar{Z}=iZ^2$. *(1996 - 2 Marks)*
15. Let $\bar{b}z+b\bar{z}=c$, $b\ne 0$, be a line in the complex plane, where $\bar{b}$ is the complex conjugate of b. If a point z_1 is the reflection of a point z_2 through the line, then show that $c=\bar{z}_1b+z_2\bar{b}$. *(1997C - 5 Marks)*
16. Let z_1 and z_2 be roots of the equation $z^2+pz+q=0$, where the coefficients p and q may be complex numbers. Let A and B represent z_1 and z_2 in the complex plane. If $\angle AOB=\alpha\ne 0$ and $OA=OB$, where O is the origin, prove that $p^2=4q\cos^2\left(\dfrac{\alpha}{2}\right)$. *(1997 - 5 Marks)*
17. For complex numbers z and w, prove that $|z|^2w-|w|^2z=z-w$ if and only if $z=w$ or $z\bar{w}=1$. *(1999 - 10 Marks)*
18. Let a complex number α, $\alpha\ne 1$, be a root of the equation $z^{p+q}-z^p-z^q+1=0$, where p, q are distinct primes. Show that either $1+\alpha+\alpha^2+....+\alpha^{p-1}=0$ or $1+\alpha+\alpha^2+...+\alpha^{q-1}=0$, but not both together. *(2002 - 5 Marks)*
19. If z_1 and z_2 are two complex numbers such taht $|z_1|<1<|z_2|$ then prove that $\left|\dfrac{1-z_1\bar{z}_2}{z_1-z_2}\right|<1$. *(2003 - 2 Marks)*
20. Prove that there exists no complex number z such that $|z|<\dfrac{1}{3}$ and $\sum\limits_{r=1}^{n}a_rz^r=1$ where $|a_r|<2$. *(2003 - 2 Marks)*
21. Find the centre and radius of circle given by $\left|\dfrac{z-\alpha}{z-\beta}\right|=k, k\ne 1$ where, $z=x+iy$, $\alpha=\alpha_1+i\alpha_2$, $\beta=\beta_1+i\beta_2$ *(2004 - 2 Marks)*
22. If one the vertices of the square circumscribing the circle $|z-1|=\sqrt{2}$ is $2+\sqrt{3}\,i$. Find the other vertices of the square. *(2005 - 4 Marks)*

F MATCH THE FOLLOWING :

Each question contains statements given in two columns, which have to be matched. The statements in Column-I are labelled A, B, C and D, while the statements in Column-II are labelled p, q, r, s and t. Any given statement in Column-I can have correct matching with ONE OR MORE statement(s) in Column-II. The appropriate bubbles corresponding to the answers to these questions have to be darkened as illustrated in the following example :
If the correct matches are A-p, s and t; B-q and r; C-p and q; and D-s then the correct darkening of bubbles will look like the given.

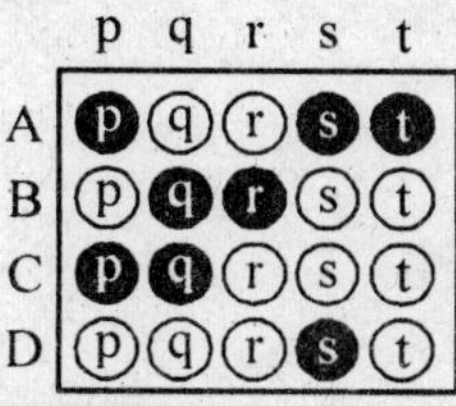

1. $z \neq 0$ is a complex number **(1992 - 2 Marks)**

Column I	Column II
(A) $\text{Re}\, z = 0$	(p) $\text{Re}\, z^2 = 0$
(B) $\text{Arg}\, z = \frac{\pi}{4}$	(q) $\text{Im}\, z^2 = 0$
	(r) $\text{Re}\, z^2 = \text{Im}\, z^2$

2. Match the statements in **Column-1** with those in **Column -II.** **(2010)**
[**Note :** Here z takes values in the complex plane and Im z and Re z denote , respectively, the imaginary part and the real part of z.]

Column-I	Column-II
(A) The set of points z satisfying $\|z - i\|z\|\| = \|z + i\|z\|\|$ is contained in or equal to	(p) an ellipse with eccentricity $\frac{4}{5}$
	(q) the set of points z satisfying $\text{Im}\, z = 0$
(B) The set of points z satisfying $\|z+4\|+\|z-4\| = 10$ is contained in or equal to	(r) the set of points z satisfying $\|\text{Im}\, z\| \leq 1$
(C) If $\|w\| = 2$, then the set of points $z = w - \frac{1}{w}$ is contained in or equal to	(s) the set of points z satisfying $\|\text{Re}\, z\| \leq 2$
(D) If $\|w\| = 1$, then the set of points $z = w + \frac{1}{w}$ is contained in or equal to.	(t) the set of points z satisfying $\|z\| \leq 3$

G COMPREHENSION BASED QUESTIONS :

PASSAGE-I

Let A, B, C be three sets of complex numbers as defined below

$$A = \{z : \text{Im}\, z \geq 1\}$$

$$B = \{z : |z - 2 - i| = 3\}$$

$$C = \{z : \text{Re}((1-i)z) = \sqrt{2}\}$$

1. The number of elements in the set $A \cap B \cap C$ is **(2008)**
(a) 0 (b) 1
(c) 2 (d) ∞

2. Let z be any point in $A \cap B \cap C$.
Then, $|z + 1 - i|^2 + |z - 5 - i|^2$ lies between **(2008)**
(a) 25 and 29 (b) 30 and 34
(c) 35 and 39 (d) 40 and 44

3. Let z be any point $A \cap B \cap C$ and let w be any point satisfying $|w - 2 - i| < 3$. Then, $|z| - |w| + 3$ lies between
(a) –6 and 3 (b) –3 and 6 **(2008)**
(c) –6 and 6 (d) –3 and 9

PASSAGE-II

Let $S = S_1 \cap S_2 \cap S_3$, where

$$S_1 = \{z \in \mathbb{C} : |z| < 4\},\ S_2 = \left\{z \in \mathbb{C} : \text{Im}\left[\frac{z - 1 + \sqrt{3}i}{1 - \sqrt{3}i}\right] > 0\right\}$$

and $S_3 = \{z \in \mathbb{C} : \text{Re}\, z > 0\}$.

4. Area of $S =$ **(JEE Adv. 2013)**
(a) $\frac{10\pi}{3}$ (b) $\frac{20\pi}{3}$
(c) $\frac{16\pi}{3}$ (d) $\frac{32\pi}{3}$

5. $\min_{z \in S} |1-3i-z| =$ *(JEE Adv. 2013)*

(a) $\frac{2-\sqrt{3}}{2}$

(b) $\frac{2+\sqrt{3}}{2}$

(c) $\frac{3-\sqrt{3}}{2}$

(d) $\frac{3+\sqrt{3}}{2}$

I INTEGER VALUE CORRECT TYPE :

1. If z is any complex number satisfying $|z-3-2i| \le 2$, then the minimum value of $|2z-6+5i|$ is *(2011)*

2. Let $\omega = e^{\frac{i\pi}{3}}$, and a, b, c, x, y, z be non-zero complex numbers such that *(2011)*

$$a+b+c=x$$
$$a+b\omega+c\omega^2=y$$
$$a+b\omega^2+c\omega=z$$

Then the value of $\frac{|x|^2+|y|^2+|z|^2}{|a|^2+|b|^2+|c|^2}$ is

Section-B JEE Main/ AIEEE

1. z and w are two nonzero complex numbers such that $|z|=|w|$ and Arg z + Arg w = π then z equals **[2002]**
(a) $\overline{\omega}$ (b) $-\overline{\omega}$ (c) ω (d) $-\omega$

2. If $|z-4|<|z-2|$, its solution is given by **[2002]**
(a) $Re(z)>0$ (b) $Re(z)<0$
(c) $Re(z)>3$ (d) $Re(z)>2$

3. The locus of the centre of a circle which touches the circle $|z-z_1|=a$ and $|z-z_2|=b$ externally (z, z_1 & z_2 are complex numbers) will be **[2002]**
(a) an ellipse (b) a hyperbola
(c) a circle (d) none of these

4. If z and ω are two non-zero complex numbers such that $|z\omega|=1$ and $Arg(z)-Arg(\omega)=\frac{\pi}{2}$, then $\bar{z}\omega$ is equal to **[2003]**
(a) $-i$ (b) 1 (c) -1 (d) i.

5. Let Z_1 and Z_2 be two roots of the equation $Z^2+aZ+b=0$, Z being complex. Further, assume that the origin, Z_1 and Z_2 form an equilateral triangle. Then **[2003]**
(a) $a^2=4b$ (b) $a^2=b$
(c) $a^2=2b$ (d) $a^2=3b$

6. If $\left(\frac{1+i}{1-i}\right)^x=1$ then **[2003]**
(a) $x=2n+1$, where n is any positive integer
(b) $x=4n$, where n is any positive integer
(c) $x=2n$, where n is any positive integer
(d) $x=4n+1$, where n is any positive integer.

7. Let z and w be complex numbers such that $\bar{z}+i\bar{w}=0$ and arg $zw=\pi$. Then arg z equals **[2004]**
(a) $\frac{5\pi}{4}$ (b) $\frac{\pi}{2}$ (c) $\frac{3\pi}{4}$ (d) $\frac{\pi}{4}$

8. If $z=x-iy$ and $z^{\frac{1}{3}}=p+iq$, then $\left(\frac{x}{p}+\frac{y}{q}\right)\Big/(p^2+q^2)$ is equal to **[2004]**
(a) -2 (b) -1 (c) 2 (d) 1

9. If $|z^2-1|=|z|^2+1$, then z lies on **[2004]**
(a) an ellipse (b) the imaginary axis
(c) a circle (d) the real axis

10. If the cube roots of unity are **1**, ω, ω^2 then the roots of the equation $(x-1)^3+8=0$, are **[2005]**
(a) $-1, -1+2\omega, -1-2\omega^2$
(b) $-1, -1, -1$
(c) $-1, 1-2\omega, 1-2\omega^2$
(d) $-1, 1+2\omega, 1+2\omega^2$

11. If z_1 and z_2 are two non- zero complex numbers such that $|z_1+z_2|=|z_1|+|z_2|$, then arg z_1 − arg z_2 is equal to **[2005]**
(a) $\frac{\pi}{2}$ (b) $-\pi$
(c) 0 (d) $\frac{-\pi}{2}$

12. If $\omega=\frac{z}{z-\frac{1}{3}i}$ and $|\omega|=1$, then z lies on **[2005]**
(a) an ellipse (b) a circle
(c) a straight line (d) a parabola

13. The value of $\sum_{k=1}^{10}\left(\sin\frac{2k\pi}{11}+i\cos\frac{2k\pi}{11}\right)$ is **[2006]**
(a) i (b) 1
(c) -1 (d) $-i$

14. If $z^2 + z + 1 = 0$, where z is complex number, then the value of

$\left(z+\frac{1}{z}\right)^2+\left(z^2+\frac{1}{z^2}\right)^2+\left(z^3+\frac{1}{z^3}\right)^2+.........+\left(z^6+\frac{1}{z^6}\right)^2$ is **[2006]**

(a) 18 (b) 54
(c) 6 (d) 12

15. If $|z+4| \le 3$, then the maximum value of $|z+1|$ is **[2007]**

(a) 6 (b) 0
(c) 4 (d) 10

16. The conjugate of a complex number is $\frac{1}{i-1}$ then that complex number is **[2008]**

(a) $\frac{-1}{i-1}$ (b) $\frac{1}{i+1}$ (c) $\frac{-1}{i+1}$ (d) $\frac{1}{i-1}$

17. Let R be the real line. Consider the following subsets of the plane $R \times R$:

$S=\{(x, y): y=x+1 \text{ and } 0<x<2\}$

$T=\{(x, y): x-y \text{ is an integer}\}$,

Which one of the following is true? **[2008]**

(a) Neither S nor T is an equivalence relation on R
(b) Both S and T are equivalence relation on R
(c) S is an equivalence relation on R but T is not
(d) T is an equivalence relation on R but S is not

18. The number of complex numbers z such that $|z-1|=|z+1|=|z-i|$ equals **[2010]**

(a) 1 (b) 2
(c) ∞ (d) 0

19. Let α, β be real and z be a complex number. If $z^2+\alpha z+\beta=0$ has two distinct roots on the line $\text{Re } z=1$, then it is necessary that : **[2011]**

(a) $\beta \in (-1, 0)$ (b) $|\beta| = 1$
(c) $\beta \in (1, \infty)$ (d) $\beta \in (0, 1)$

20. If $\omega(\neq 1)$ is a cube root of unity, and $(1+\omega)^7 = A + B\omega$. Then (A, B) equals **[2011]**

(a) (1, 1) (b) (1, 0)
(c) (–1, 1) (d) (0, 1)

21. If $\omega \neq 1$ is the complex cube root of unity and matrix $H=\begin{bmatrix}\omega & 0\\ 0 & \omega\end{bmatrix}$, then H^{70} is equal to – **[2011 RS]**

(a) 0 (b) –H
(c) H^2 (d) H

22. If $z \neq 1$ and $\frac{z^2}{z-1}$ is real, then the point represented by the complex number z lies : **[2012]**

(a) either on the real axis or on a circle passing through the origin.
(b) on a circle with centre at the origin
(c) either on the real axis or on a circle not passing through the origin.
(d) on the imaginary axis.

23. If z is a complex number of unit modulus and argument θ, then $\arg\left(\frac{1+z}{1+\bar{z}}\right)$ equals: **[JEE M 2013]**

(a) $-\theta$ (b) $\frac{\pi}{2}-\theta$
(c) θ (d) $\pi-\theta$

Solutions & Explanations

Section-A : JEE Advanced/ IIT-JEE

A 1. $2n\pi,\ n\pi+\frac{\pi}{4}$ 2. $(a^2+b^2)(|z_1|^2+|z_2|^2)$ 3. $2-\sqrt{3},\ 2-\sqrt{3}$

4. $3-\frac{i}{2}$ or $1-\frac{3}{2}i$ 5. $-2,\ 1-i\sqrt{3}$ 6. $\frac{1}{4}n(n-1)(n^2+3n+4)$

B 1. T 2. T 3. F 4. T

C

1. (b)	2. (d)	3. (a)	4. (b)	5. (d)	6. (b)
7. (b)	8. (b)	9. (d)	10. (b)	11. (d)	12. (c)
13. (d)	14. (c)	15. (a)	16. (a)	17. (d)	18. (c)
19. (b)	20. (a)	21. (b)	22. (a)	23. (b)	24. (b)
25. (d)	26. (d)	27. (d)	28. (d)	29. (d)	30. (a)
31. (d)	32. (c)				

D

1. (a, b, c)	2. (a, d)	3. (c)	4. (d)	5. (d)	6. (b)
7. (d)	8. (a, c, d)	9. (c, d)			

E 1. $\left(\frac{1}{5+3\cos\theta}\right)+\left(\frac{-2\cot\theta/2}{5+3\cos\theta}\right)i$ 5. $x=3, y=-1$ 14. $i,\ \frac{\pm\sqrt{3}}{2}-\frac{i}{2}$ 21. centre $=\frac{\alpha-k^2\beta}{1-k^2}$,

radius $=\frac{k}{|1-k^2|}|\alpha-\beta|$

22. $(1-\sqrt{3})+i,\ -i\sqrt{3},\ (\sqrt{3}+1)-i$

F 1. (A) - q ; (B) - p 2. (A) - q, r ; (B) - p; (C) - p, s, t; (D) - q, r, s, t

G 1. (b) 2. (c) 3. (d) 4. (b) 5. (c)

I 1. 5 2. 3

Section-B : JEE Main/ AIEEE

1. (b)	2. (c)	3. (b)	4. (a)	5. (d)	6. (b)
7. (c)	8. (a)	9. (b)	10. (c)	11. (c)	12. (c)
13. (d)	14. (d)	15. (a)	16. (c)	17. (d)	18. (a)
19. (c)	20. (a)	21. (d)	22. (a)	23. (c)	

Section-B JEE Main/ AIEEE

A. FILL IN THE BLANKS :

1. Let $z=\frac{\sin x/2+\cos x/2+i\tan x}{1+2i\sin x/2}$

$$=\frac{(\sin x/2+\cos x/2+i\tan x)(1-2i\sin x/2)}{(1+2i\sin x/2)(1-2i\sin x/2)}$$

$$=\frac{[\sin x/2+\cos x/2+2\sin x/2\tan x)+i(\tan x-2\sin^2 x/2-2\sin x/2\cos x/2)]}{(1+4\sin^2 x/2)}$$

But ATQ, $I_m(z)=0$ (as z is real)

$$\Rightarrow \tan x-2\sin\frac{x}{2}\left(\sin\frac{x}{2}+\cos\frac{x}{2}\right)=0$$

$$\Rightarrow \frac{\sin x}{\cos x}-2\sin^2 x/2-2\sin x/2\cos x/2=0$$

$$\Rightarrow \frac{\sin x}{\cos x}-(1-\cos x)-\sin x=0$$

$$\Rightarrow \sin x\left[\frac{1}{\cos x}-1\right]-[1-\cos x]=0$$

$$\Rightarrow \left(\frac{1-\cos x}{\cos x}\right)\sin x-[1-\cos x]=0$$

$\Rightarrow \quad (1-\cos x)\left(\frac{\sin x}{\cos x}-1\right)=0$

$\Rightarrow \quad \cos x = 1 \quad \Rightarrow \quad x = 2n\pi$ and

$\tan x = 1 \quad \Rightarrow \quad x = n\pi + \pi/4$

$\therefore \quad x = 2n\pi,\ n\pi + \pi/4$ **Ans.**

2. **KEY CONCEPT :** $|z|^2 = z\bar{z}$

$|az_1 - bz_2|^2 + |bz_1 + az_2|^2$

$= (az_1 - bz_2)\overline{(az_1 - bz_2)} + (bz_1 + az_2)\overline{(bz_1 + az_2)}$

$= (az_1 - bz_2)(a\bar{z}_1 - b\bar{z}_2) + (bz_1 + az_2)(b\bar{z}_1 + a\bar{z}_2)$

$= a^2|z_1|^2 + b^2|z_2|^2 - abz_1\bar{z}_2 - ab\bar{z}_1 z_2$

$+ b^2|z_1|^2 + a^2|z_2|^2 + ab\bar{z}_1 z_2 + abz_1\bar{z}_2$

$= \quad (a^2+b^2)(|z_1|^2 + |z_2|^2)$

ALTERNATE SOLUTION :

$|az_1 - bz_2|^2 + |bz_1 + az_2|^2$

$= \quad a^2|z_1|^2 + b^2|z_2|^2 - 2ab\,\mathrm{Re}(z_1\bar{z}_2) + b^2|z_1|^2$

$+ a^2|z_2|^2 + 2ab\,\mathrm{Re}(z_1\bar{z}_2)$

$= (a^2+b^2)(|z_1|^2 + |z_2|^2)$

3. **KEY CONCEPT :** $|z_1 - z_2|$ = distance between two points represented by z_1 and z_2.

As $z_1 = a + i$, $z_2 = 1 + bi$ and $z_3 = 0$ form an equilateral triangle, therefore

$$|z_1 - z_3| = |z_2 - z_3| = |z_1 - z_2|$$

$\Rightarrow \quad |a+i| = |1+bi| = |(a-1) + i(1-b)|$

$\Rightarrow \quad a^2 + 1 = 1 + b^2 = (a-1)^2 + (1-b)^2$

$\Rightarrow \quad a^2 = b^2 = a^2 + b^2 - 2a - 2b + 1$

$\Rightarrow \quad a = b$ ....(1)

$(\because\ a, b > 0 \quad \therefore \quad a \neq -b)$ and

$b^2 - 2a - 2b + 1 = 0$

or $\quad a^2 - 2a - 2b + 1 = 0$ ....(2)

$\Rightarrow \quad a^2 - 2a - 2a + 1 = 0 \qquad [\because a = b]$

$\Rightarrow \quad a^2 - 4a + 1 = 0$

$\Rightarrow \quad a = \frac{4 \pm 2\sqrt{3}}{2} = 2 \pm \sqrt{3}$ But $0 < a, b < 1$

$\therefore \quad a = 2 - \sqrt{3}$ also $b = 2 - \sqrt{3}$

ALTERNATE SOLUTION :

Since z_1, z_2 and z_3 form an equilateral Δ

$\Rightarrow z_1^2 + z_2^2 + z_3^2 = z_1 z_2 + z_2 z_3 + z_3 z_1$

or $(a+i)^2 + (1+ib)^2 + (0)^2 = (a+i)(1+ib) + 0 + 0$

$\Rightarrow a^2 - 1 + 2ai + 1 - b^2 + 2ib = a + i(ab+1) - b$

$\Rightarrow (a^2 - b^2) + 2i(a+b) = (a-b) + i(ab+1)$.

on comparing $a^2 - b^2 = a - b$ and $2(a+b) = ab + 1$

$\Rightarrow (a-b)(a+b-1) = 0$ and $2(a+b) = ab + 1$

$\Rightarrow (a = b$ or $a + b = 1)$ and $2(a+b) = ab + 1$

If $a = b \Rightarrow 2(2a) = a^2 + 1$

or $a^2 - 4a + 1 = 0$

$\Rightarrow \quad a = \frac{4 \pm \sqrt{16-4}}{2} = 2 \pm \sqrt{3}$ If $a + b = 1$,

$\Rightarrow 2 = a(1-a) + 1 \Rightarrow a^2 - a + 1 = 0$

$\Rightarrow a = \frac{1 \pm \sqrt{1-4}}{2}$, but a and $b \in R$

$\therefore$ only solution when $a = b$, $0 < a; b < 1$

$\Rightarrow a = b = 2 - \sqrt{3}$

4. **KEY CONCEPT :**

If PB is obtained on rotating PA about P, through an angle θ in anticlockwise direction then

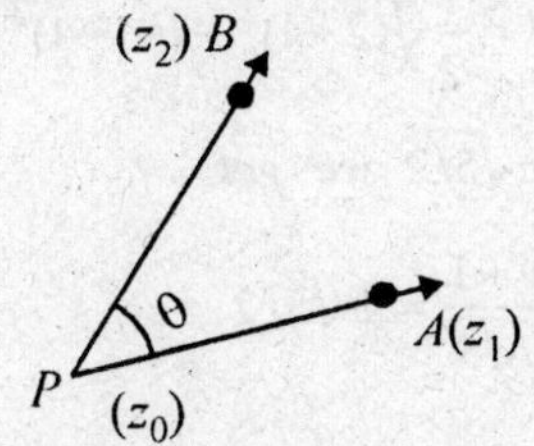

$$\frac{z_2 - z_0}{|z_2 - z_0|} = \frac{z_1 - z_0}{|z_1 - z_0|} e^{i\theta}$$

If rotation is about origin and magnitude of PA and PB are equal then $z_2 = z_1 e^{i\theta}$.

If angle of rotaion is $\pi/2$ then $z_2 = z_1 i$

If angle of rotation is $2\pi/3$ then $z_2 = z_1 \omega$ where ω is complex cube root of unity.

Let $z_1 = 1 + i$, $z_2 = 2 - i$ and z_3 represent point A.

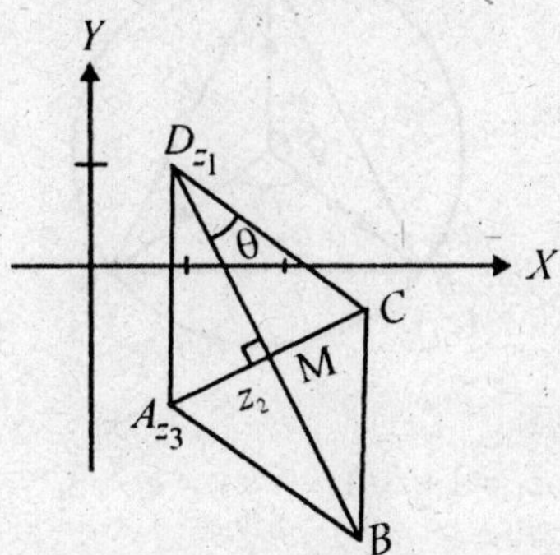

Possible position of A can be obtained by rotating D about M, through an angle of $\pi/2$ either anticlockwise or clockwise .

$$\frac{z_3 - z_2}{|z_3 - z_2|} = \frac{z_1 - z_2}{|z_1 - z_2|} e^{\pm i\frac{\pi}{2}} \Rightarrow \frac{z_3 - z_2}{z_1 - z_2} = \frac{|z_3 - z_2|}{|z_1 - z_2|} e^{\pm i\frac{\pi}{2}}$$

$\Rightarrow \quad \frac{z_3 - z_2}{z_1 - z_2} = \frac{1}{2} i$ or $\frac{-1}{2} i$

$\Rightarrow \quad \frac{z_3 - 2 + i}{2i - 1} = \frac{i}{2}$ or $-\frac{i}{2}$

$\Rightarrow \quad z_3 = 1 - \frac{3}{2}i \quad or \quad 3 - \frac{i}{2}$

So, A represents the complex numbers $3 - \frac{i}{2}$ or $1 - \frac{3}{2}i$

ALTERNATE SOLUTION :

If we see the problem as in co-ordinate geometry we have $D \equiv (1,1)$ and $M \equiv (2, -1)$

We know that diagonals of rhombus bisect each other at 90°

$\therefore$ AC passes through M and is $\perp$ to BD

$\therefore$ Eq. of AC in symmetric form can be written as

$$\frac{x-2}{2/\sqrt{5}} = \frac{y+1}{1/\sqrt{5}} = r$$

Now for pt. A, as

$$AM = \frac{1}{2}DM = \frac{1}{2}\sqrt{(2-1)^2 + (-1-1)^2} = \sqrt{5}/2$$

Putting $r = \pm\sqrt{5}/2$ we get,

$$\frac{x-2}{2/\sqrt{5}} = \frac{y+1}{1/\sqrt{5}} = \pm\sqrt{5}/2$$

$\Rightarrow \quad x = \pm 1 + 2, y = \pm\frac{1}{2} - 1$

$\Rightarrow \quad x = 3$ or $1, y = \frac{-1}{2}$ or $\frac{-3}{2}$

$\therefore$ Pt. A is $3 - i/2$ or $1 - (3/2)i$

5. Let z_1, z_2, z_3 be the vertices A, B and C respectively of equilateral ΔABC, inscribed in a circle $|z| = 2$, centre (0, 0) rasius = 2

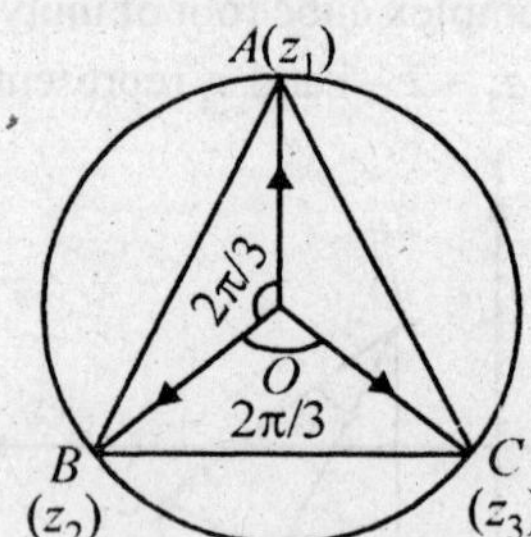

Given $\quad z_1 = 1 + i\sqrt{3} \qquad z_2 = e^{\frac{2\pi i}{3}} z_1$

$$= \left(\cos\frac{2\pi}{3} + i\sin\frac{2\pi}{3}\right)(1 + i\sqrt{3}) = \frac{-1-3}{2} = -2$$

and $z_3 = e^{4(\pi/3)i} z_1$

$$= \left(\cos\frac{4\pi}{3} + i\sin\frac{4\pi}{3}\right)(1 + i\sqrt{3})$$

$$= \left(\frac{-1-i\sqrt{3}}{2}\right)(1 + i\sqrt{3}) = \frac{-1-2i\sqrt{3}+3}{2} = 1 - i\sqrt{3}$$

6. rth term of the given series,

$$= r[(r+1) - \omega](r+1) - \omega^2]$$

$$= r[(r+1)^2 - (\omega + \omega^2)(r+1) + \omega^3]$$

$$= r[(r+1)^2 - (-1)(r+1) + 1]$$

$$= r[(r^2 + 3r + 3] = r^3 + 3r^2 + 3r$$

Thus, sum of the given series,

$$= \sum_{r=1}^{(n-1)} (r^3 + 3r^2 + 3r)$$

$$= \frac{1}{4}(n-1)^2 n^2 + 3.\frac{1}{6}(n-1)\ (n)\ (2n-1)$$

$$+ 3.\frac{1}{2}(n-1)n$$

$$= (n-1)\ (n)\left[\frac{1}{4}(n-1)n + \frac{1}{2}(2n-1) + \frac{3}{2}\right]$$

$$= \frac{1}{4}(n-1)\ n[n^2 - n + 4n - 2 + 6]$$

$$= \frac{1}{4}(n-1)n[n^2 + 3n + 4]$$

ALTERNATE SOLUTION :

Given expression $= \sum_{n=2}^{n}(n^3 - 1) = \sum_{n=1}^{n}(n^3 - 1)$

$$= \left[\left(\frac{n(n+1)}{2}\right)^2 - n\right] \quad \textbf{NOTE THIS STEP}$$

$$= \frac{n^2(n^2 + 2n + 1) - 4n}{4} = \frac{1}{4}n(n^3 + 2n^2 + n - 4)$$

$$= \frac{1}{4}n(n-1)[n^2 + 3n + 4]$$

(B) TRUE / FALSE :

1. Let $z = x + iy$

then $1 \cap z \Rightarrow 1 \le x \,\&\, 0 \le y$ (by def.)

Consider

$$\frac{1-z}{1+z} = \frac{1-(x+iy)}{1+(x+iy)} = \frac{(1-x)-iy}{(1+x)+iy} \times \frac{(1+x)-iy}{(1+x)-iy}$$

$$= \frac{1-x^2-y^2}{(1+x)^2+y^2} - \frac{iy(1-x+1+x)}{(1+x)^2+y^2}$$

$$= \frac{1-x^2-y^2}{(1+x)^2+y^2} - \frac{2iy}{(1+x)^2+y^2}$$

$$\frac{1-z}{1+z} \cap 0 \quad \Rightarrow \quad \frac{1-x^2-y^2}{(1+x)^2+y^2} \le 0$$

and $\frac{-2y}{(1+x)^2+y^2} \le 0$

$\Rightarrow$ $1-x^2-y^2 \le 0$ and $-2y \le 0$

$\Rightarrow$ $x^2+y^2 \ge 1$ and $y \ge 0$ which is true as

$x \ge 1$ & $y \ge 0$

$\therefore$ The given statement is true $\forall\ z \in C$.

2. As $|z_1| = |z_2| = |z_3|$

$\therefore$ z_1, z_2, z_3 are equidistant from origin. Hence O is the circumcentre of ΔABC.

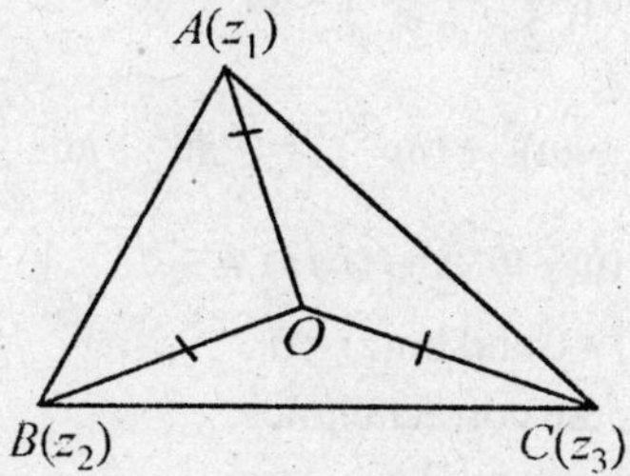

But according to question ΔABC is equilateral and we know that in an equilateral Δ circumcentre and centriod coincide.

$\therefore$ Centriod of $\Delta ABC = 0$

$\Rightarrow$ $\frac{z_1+z_2+z_3}{3} = 0 \Rightarrow z_1+z_2+z_3 = 0$

$\therefore$ Statement is true.

3. If z_1, z_2, z_3 are in A.P. then, $\frac{z_1+z_3}{2} = z_2$

$\Rightarrow$ z_2 is mid pt. of line joining z_1 and z_3.

$\Rightarrow$ z_1, z_2, z_3 lie on a st. line

$\therefore$ Given statement is false

4. Cube roots of unity are $1, \frac{-1+i\sqrt{3}}{2}, \frac{-1-i\sqrt{3}}{2}$

Let these be represented by the points A, B and C respectively in Argand plane. Then

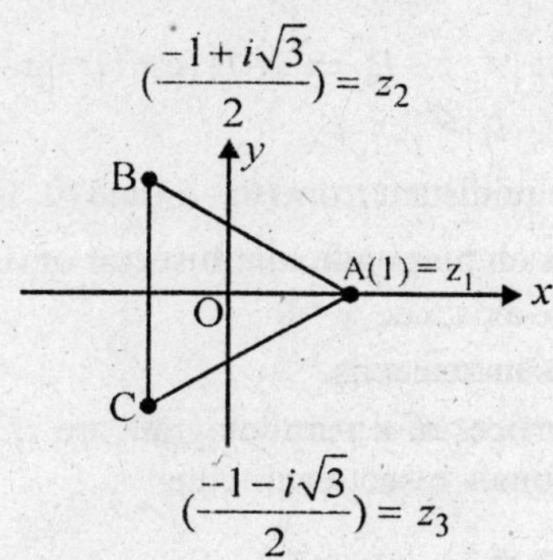

$$AB = \left|\frac{-1+i\sqrt{3}}{2} - 1\right| = \left|\frac{-3+i\sqrt{3}}{2}\right|$$

$$= \sqrt{\left(\frac{-3}{2}\right)^2 + \left(\frac{\sqrt{3}}{2}\right)^2} = \sqrt{\frac{9}{4}+\frac{3}{4}} = \sqrt{\frac{12}{4}} = \sqrt{3}$$

Similarly, $BC = CA = \sqrt{3}$

$\therefore$ ΔABC is an equilateral triangle

$\therefore$ Statement is True

ALTERNATE SOLUTION-1 :

Let point A, B, C represent z_1, z_2, z_3 respectively

Then $\angle CBA = \arg\left(\frac{z_1-z_2}{z_3-z_2}\right) = \arg\left(\frac{\frac{3-\sqrt{3}i}{2}}{-\sqrt{3}i}\right)$

$$= \arg\left(\frac{\sqrt{3}-i}{-2i}\right) = \arg\left(\frac{1}{2}+\frac{\sqrt{3}}{2}i\right) = \pi/3$$

Also $\angle CAB = \arg\left(\frac{z_3-z_1}{z_2-z_1}\right) = \arg\left(\frac{\frac{-3-\sqrt{3}i}{2}}{\frac{-3+i\sqrt{3}}{2}}\right)$

$$= \arg\left(\frac{\sqrt{3}+i}{\sqrt{3}-i}\right) = \arg\left(\frac{1+i\sqrt{3}}{2}\right) = \pi/3$$

$\therefore$ ΔABC is an equilateral Δ

$\therefore$ Statement is true.

ALTERNATE SOLUTION-2 :

The cube roots of unity are,

$z_1 = 1 = \cos 0 + i \sin 0$

$$z_2 = \frac{-1+i\sqrt{3}}{2} = \cos\frac{2\pi}{3} + i\sin\frac{2\pi}{3}$$

$$z_3 = \frac{-1-i\sqrt{3}}{2} = \cos\frac{4\pi}{3} + i\sin\frac{4\pi}{3}$$

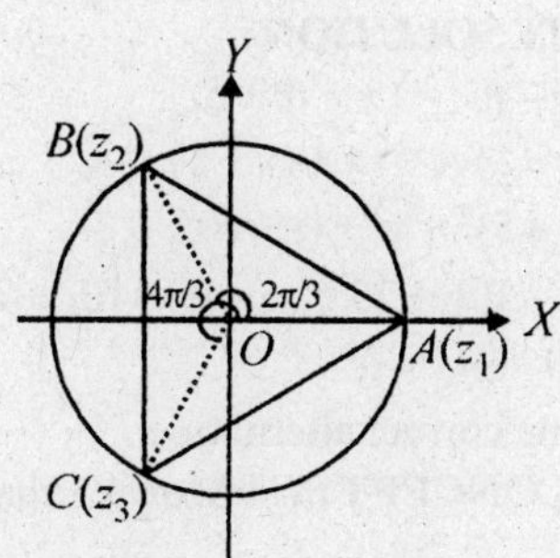

As $|z_1| = |z_2| = |z_3| = 1$.

$\therefore$ All z_1, z_2, z_3 lie on a circle of unit radius.

Clearly $\angle AOB = 2\pi/3 \Rightarrow \angle ACB = \pi/3$

$\angle BOC = 4\pi/3 - 2\pi/3 = 2\pi/3 \Rightarrow \angle BAC = \pi/3$

$\therefore$ ΔABC is an equilateral Δ. Hence the given statement is true.

ALTERNATE SOLUTION-3 :

$\because$ Cube roots of unity are $1, \frac{-1+i\sqrt{3}}{2}, \frac{-1-\sqrt{3}}{2}$

$\therefore$ Vertices of triangle are

$$A(1,0), B\left(\frac{-1}{2}, \frac{\sqrt{3}}{2}\right), C\left(\frac{-1}{2}, \frac{-\sqrt{3}}{2}\right)$$

$\Rightarrow AB = BC = CA \quad \therefore \Delta$ is equilateral.

C MCQ's WITH ONE CORRECT ANSWER :

1. (b) $(x-1)^3 + 8 = 0$

$\Rightarrow (x-1)^3 = -8 = (-2)^3$

$\Rightarrow x - 1 = -2$ or -2ω or $-2\omega^2$

$\Rightarrow x = -1, 1-2\omega, 1-2\omega^2$

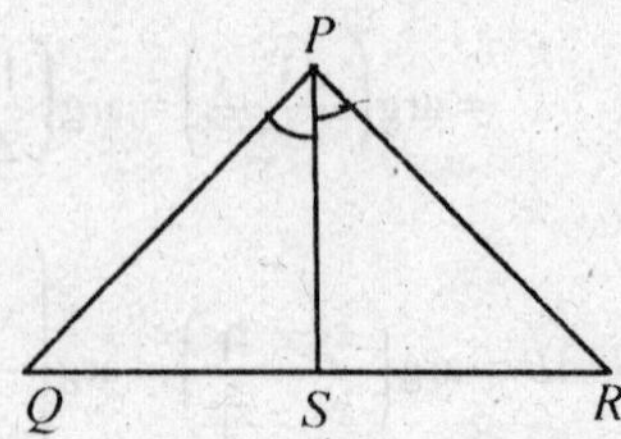

2. (d) $\frac{1+i}{1-i} = \frac{(1+i)^2}{(1-i)(1+i)} = \frac{1-1+2i}{2} = i$

Now $i^n = 1 \Rightarrow$ the smallest positive integral value of n should be 4.

3. (a) **KEY CONCEPT :** If $|z-z_1| = |z-z_2|$ then locus of z is the line, perpendicular bisector of line-segment joining z_1 and z_2.

$z = x + iy$

$\Rightarrow |z-5i| = |z+5i|$

$\Rightarrow z$ remains equidistant from $z_1 = 5i$ and $z_2 = -5i$.

$\therefore$ z lies on perpendicular bisector of line segment joining z_1 and z_2

which is clearly the axis of x i.e. $y = 0$

$\therefore$ 'a' is the correct alternative.

ALTERNATE SOLUTION :

ATQ $|x+iy-5i| = |x+iy+5i|$

$\Rightarrow |x+(y-5)i| = |x+(y+5)i|$

$\Rightarrow x^2 + (y-5)^2 = x^2 + (y+5)^2$

$\Rightarrow x^2 + y^2 - 10y + 25 = x^2 + y^2 + 10y + 25$

$\Rightarrow 20y = 0 \Rightarrow y = 0$

$\therefore$ 'a' is the correct alternative.

4. (b) **KEY CONCEPT :** (D' Moivre's thm)

$(\cos\theta + i\sin\theta)^n = \cos n\theta + i\sin n\theta$, n is any rational no.

$$z = \left(\frac{\sqrt{3}}{2} + \frac{i}{2}\right)^5 + \left(\frac{\sqrt{3}}{2} - \frac{i}{2}\right)^5$$

we have $\frac{\sqrt{3}}{2} + \frac{i}{2} = \cos\pi/6 + i\sin\pi/6$ and

$\frac{\sqrt{3}}{2} - \frac{i}{2} = \cos\pi/6 - i\sin\pi/6$

$\therefore \; z = (\cos\pi/6 + i\sin\pi/6)^5 + (\cos\pi/6 - i\sin\pi/6)^5$

$= \cos 5\pi/6 + i\sin 5\pi/6 + \cos 5\pi/6 - i\sin 5\pi/6^6$

$= 2\cos 5\pi/6 = -\sqrt{3}$

$\therefore$ Re $(z) < 0$ and Im $(z) = 0$

$\therefore$ (b) is the correct choice.

ALTERNATE SOLUTION :

$$\left(\frac{\sqrt{3}}{2} + \frac{i}{2}\right) = -i\left(\frac{-1}{2} + \frac{i\sqrt{3}}{2}\right) = i\omega$$

$$\frac{\sqrt{3}}{2} - \frac{i}{2} = i\left(\frac{-1}{2} - \frac{i\sqrt{3}}{2}\right) = i\omega^2$$

$\therefore \quad z = (-i\omega)^5 + (i\omega^2)^5 = -i\omega^2 + i\omega$

$= i(\omega - \omega^2) = i(i\sqrt{3}) = -\sqrt{3}$

$\Rightarrow$ Re$(z) < 0$ and Im$(z) = 0$

$\therefore$ (b) is the correct choice.

5. (d) $|z-4| < |z-2|$

$\Rightarrow |(x-4) + iy| < |(x-2) + iy|$

$\Rightarrow (x-4)^2 + y^2 < (x-2)^2 + y^2$

$\Rightarrow -8x + 16 < -4x + 4 \quad \Rightarrow 4x - 12 > 0$

$\Rightarrow x > 3 \Rightarrow$ Re$(z) > 3$

6. (b) $|\omega| = 1 \Rightarrow \left|\frac{1-iz}{z-i}\right| = 1$

$\Rightarrow |1-iz| = |z-i|$

$\Rightarrow |1-i(x+iy)| = |x+iy-i|$

$\Rightarrow |(y+1) - ix| = |x + i(y-1)|$

$\Rightarrow x^2 + (y+1)^2 = x^2 + (y-1)^2$

$\Rightarrow 4y = 0 \quad \Rightarrow \quad y = 0$

$\Rightarrow$ z lies on real axis

ALTERNATE SOLUTION :

$|\omega| = 1 \Rightarrow \left|\frac{1-iz}{z-i}\right| = 1$

$\Rightarrow |1-iz| = |z-i| \Rightarrow |-i||z+i| = |z-i|$

$\Rightarrow |z-(-i)| = |z-i|$

$\Rightarrow$ z is equidistant from $(0,-1)$ and $(0, 1)$

$\Rightarrow$ z lies on perpendicular bisector of $(0,-1)$ and $(0, 1)$, i.e., x-axis, i.e., $y = 0$

$\therefore$ z lies on real axis.

7. (b) If vertices of a parallelogram are z_1, z_2, z_3, z_4 then as diagonals bisect each other

$\therefore \quad \frac{z_1+z_3}{2} = \frac{z_2+z_4}{2} \quad \Rightarrow z_1 + z_3 = z_2 + z_4$

8. (b) Let ABC be the Δ with vertices a, b, c and PQR be the Δ with vertices u, v, w.

Then $c = (1-r)\,a + rb$

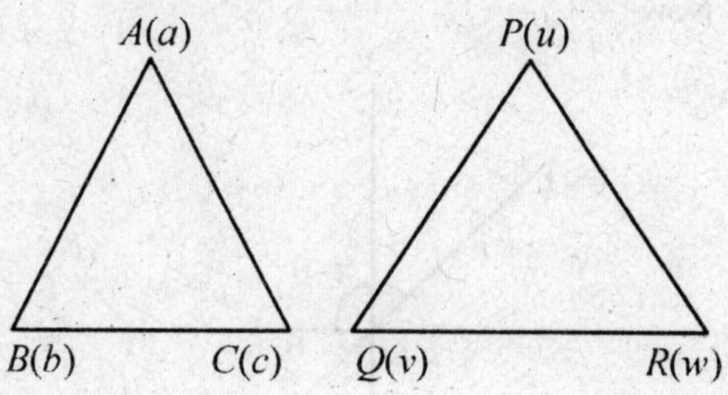

$\Rightarrow \quad c-a=r(b-a) \quad \Rightarrow \dfrac{c-a}{b-a}=r \quad$ ....(1)

$\Rightarrow \quad w=(1-r)\,u+rv \quad \Rightarrow \dfrac{w-u}{v-u}=r \quad$ ....(2)

From (1) and (2) $\left|\dfrac{c-a}{b-a}\right|=\left|\dfrac{w-u}{v-u}\right|$ and

$\arg\left(\dfrac{c-a}{b-a}\right)=\arg\left(\dfrac{w-u}{v-u}\right)$

$\Rightarrow \quad \dfrac{AC}{AB}=\dfrac{PR}{PQ}$ and $\angle CAB=\angle RPQ$

$\Rightarrow \quad \Delta ABC \sim \Delta PQR$

ALTERNATE SOLUTION :

KEY CONCEPT : Two Δ's with vertices a, b, c and u, v, w are similar if

$$\begin{vmatrix} a & u & 1 \\ b & v & 1 \\ c & w & 1 \end{vmatrix}=0$$

Here we have

$$\begin{vmatrix} a & u & 1 \\ b & v & 1 \\ c & w & 1 \end{vmatrix}$$

$$=\begin{vmatrix} a & u & 1 \\ b & v & 1 \\ c-(1-r)a-rb & w-(1-r)u-rw & 1-(1-r)-r \end{vmatrix}$$

$[R_3 \to R_3-(1-r)R_1-rR_2]$

But given that

$c=(1-r)\,a+rb$ and $w=(1-r)\,u+rw$

$\therefore$ Determinant $=0$

$\Rightarrow$ The two Δ's are similar.

9. **(d)** Let $z_1=\sin x+i\cos 2x$; $z_2=\cos x-i\sin 2x$

Then ATQ $\overline{z_1}=z_2$

$\Rightarrow \sin x-i\cos 2x=\cos x-i\sin 2x$

$\Rightarrow \sin x=\cos x$ and $\cos 2x=\sin 2x$

[Equating real and imaginary parts.]

$\Rightarrow \tan x=1$ and $\tan 2x=1$

$\Rightarrow x=\dfrac{\pi}{4}$ and $x=\dfrac{\pi}{8}$

which is not possible at the same time.

$\therefore$ There is no value of x for which above both are satisfied simultaneously.

$\therefore$ (d) is correct answer

10. **(b)** $(1+\omega)^7=A+B\omega$

$\Rightarrow (-\omega^2)^7=A+B\omega \qquad (\because 1+\omega+\omega^2=0)$

$\Rightarrow -\omega^{14}=A+B\omega$

$\Rightarrow -\omega^2=A+B\omega \qquad (\because \omega^3=1)$

$\Rightarrow 1+\omega=A+B\omega \Rightarrow A=1, B=1$

11. **(d)** $\because |z|=|\omega|$ and $\arg z=\pi-\arg\omega$

Let $\omega=re^{i\theta}$ then $z=re^{i(\pi-\theta)}$

$\Rightarrow z=re^{i\pi}\cdot e^{-i\theta}$

$=(re^{-i\theta})(\cos\pi+i\sin\pi)=\overline{\omega}(-1)=-\overline{\omega}$

12. **(c)** We have $2=|z+i\omega|\le|z|+|\omega|$

$[\because |z_1+z_2|\le|z_1|+|z_2|]$

$\therefore \quad |z|+|\omega|\ge 2$ ....(1)

But given that $|z|\le 1$ and $|\omega|\le 1$

$\Rightarrow |z|+|\omega|\le 2$ ....(2)

From (1) and (2), $|z|=|\omega|=1$

Also $|z+i\omega|=|z-i\overline{\omega}|$

$\Rightarrow |z+i\omega|^2=|z-i\overline{\omega}|^2$

$\Rightarrow (z+i\omega)\overline{(z+i\omega)}=(z-i\overline{\omega})\overline{(z-i\overline{\omega})}$

NOTE THIS STEP

$\Rightarrow (z+i\omega)(\overline{z}-i\overline{\omega})=(\overline{z}+i\omega)(z-i\overline{\omega})$

$\Rightarrow z\overline{z}+i\omega\overline{z}-iz\overline{\omega}+\omega\overline{\omega}=\overline{z}z-i\overline{z}\,\overline{\omega}+i\omega z+\omega\overline{\omega}$

$\Rightarrow \omega\overline{z}-z\overline{\omega}+\overline{z}\,\overline{\omega}-\omega z=0$

$\Rightarrow (\overline{z}-z)(\omega+\overline{\omega})=0 \Rightarrow z=\overline{z}$ or $\omega=-\overline{\omega}$

$\Rightarrow \text{Im}(z)=0$ or $\text{Re}(\omega)=0$

Also $|z|=1, |\omega|=1$

$\Rightarrow z=1$ or -1 and $\omega=i$ or $-i$.

ALTERNATE SOLUTION :

Given that $|z+i\omega|=|z-i\overline{\omega}|$

$\Rightarrow |z-(-i\omega)|=|z-\overline{(-i\omega)}|$

$\Rightarrow$ z lies on perpendicular bisector of the line segment joining $(-i\omega)$ and $\overline{(-i\omega)}$, which is real axis, $(-i\omega)$ and $\overline{(-i\omega)}$ being mirror images of each other.

$\therefore \text{Im}(z)=0.$

If $z=x$ then $|z|\le 1 \Rightarrow x^2\le 1 \Rightarrow -1\le x\le 1$

$\therefore$ (c) is the correct option.

13. (d) $(1+i)^{n_1}+(1-i)^{n_1}+(1+i)^{n_2}+(1-i)^{n_2}$

$$=\left[{}^{n_1}C_0-{}^{n_1}C_1 i+{}^{n_1}C_2 i^2+{}^{n_1}C_3 i^3+..............{}^{n_1}C_{n_1}(i)^{n_1}\right]$$

$$+\left[{}^{n_1}C_0-{}^{n_1}C_1 i+{}^{n_1}C_2 i^2-{}^{n_1}C_3 i^3+...+(-1)^{n_1}\,{}^{n_1}C_{n_1}(i)^{n_1}\right]$$

$$+\left[{}^{n_2}C_0-{}^{n_2}C_1 i+{}^{n_2}C_2 i^2+{}^{n_2}C_3 i^3+...+{}^{n_2}C_{n_2}(i)^{n_2}\right]$$

$$+\left[{}^{n_2}C_0-{}^{n_2}C_2 i+{}^{n_2}C_2 i^2-{}^{n_2}C_3 i^3+......+(-1)^{n_2}\,{}^{n_2}C_2(i)^{n_2}\right]$$

$$=2\left[\begin{array}{l}{}^{n_1}C_0+{}^{n_1}C_2 i^2+{}^{n_1}C_4 i^4+.......+\left(\frac{n_1+1}{2}\right)^{th} or \left(\frac{n_1}{2}\right)^{th}\\ \text{term as } n_1 \text{ is odd or even respectively.}\end{array}\right]$$

$$+2\left[\begin{array}{l}{}^{n_2}C_0+{}^{n_2}C_2 i^2+{}^{n_2}C_4 i^4+.......+\left(\frac{n_2+1}{2}\right)^{th} or \left(\frac{n_2}{2}\right)^{th}\\ \text{term as } n_2 \text{ is odd or even respectively.}\end{array}\right]$$

$$=2\left[\begin{array}{l}{}^{n_1}C_0-{}^{n_1}C_2+{}^{n_1}C_4+...........+\left(\frac{n_1+1}{2}\right)^{th} or \left(\frac{n_1}{2}\right)^{th}\\ \text{term as } n_1 \text{ is odd or even respectively.}\end{array}\right]$$

$$+2\left[\begin{array}{l}{}^{n_2}C_0-{}^{n_2}C_2+{}^{n_2}C_4-...........+\left(\frac{n_2+1}{2}\right)^{th} or \left(\frac{n_2}{2}\right)^{th}\\ \text{term as } n_2 \text{ is odd or even respectively.}\end{array}\right]$$

This is a real number irrespective of the values of n_1 and n_2.

ALTERNATE SOLUTION :

$(1+i)^{n_1}+(1+i^3)^{n_1}+(1+i^5)^{n_2}+(1+i^7)^{n_2}$

$$=(1+i)^{n_1}+(1-i)^{n_1}+(1+i)^{n_2}+(1-i)^{n_2}$$

Using $1+i=\sqrt{2}\,(\cos\pi/4+i\sin\pi/4)$

and $1-i=\sqrt{2}\,(\cos\pi/4-i\sin\pi/4)$

We get the given expression as

$$=(\sqrt{2})^{n_1}\left[\cos\frac{n_1\pi}{4}+i\sin\frac{n_1\pi}{4}\right]$$

$$+(\sqrt{2})^{n_2}\left[\cos\frac{n_2\pi}{4}+i\sin\frac{n_2\pi}{4}\right]$$

$$+(\sqrt{2})^{n_2}\left[\cos\frac{n_2\pi}{4}-i\sin\frac{n_2\pi}{4}\right]$$

$$=(\sqrt{2})^{n_1}\left[2\cos\frac{n_1\pi}{4}\right]+(\sqrt{2})^{n_2}\left[2\cos\frac{n_2\pi}{4}\right]$$

= real number irrespective the values of n_1 and n_2

$\therefore$ (*d*) is the most appropriate answer.

14. (c) $E=4+5(\omega)^{334}+3(\omega)^{365}=4+5\omega+3\omega^2$

$$=1+2\omega+3(1+\omega+\omega^2)=1+(-1+i\sqrt{3})=i\sqrt{3}$$

15. (a) $\arg(z)<0$ (given) $\Rightarrow \arg(z)=-\theta$

Now

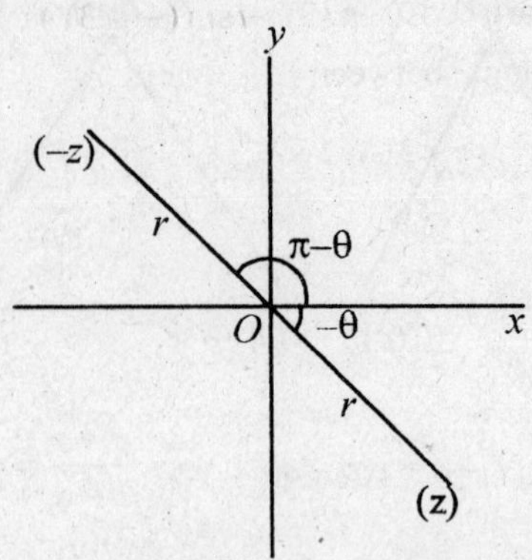

$z=r\cos(-\theta)+i\sin(-\theta)=r[\cos(\theta)-i\sin(\theta)]$

Again $-z=-r\,[\cos(\theta)-i\sin(\theta)]$

$$=r\,[\cos(\pi-\theta)+i\sin(\pi-\theta)]$$

$\therefore\quad \arg(-z)=\pi-\theta;$

Thus $\arg(-z)-\arg(z)=\pi-\theta-(-\theta)=\pi-\theta+\theta=\pi$

ALTERNATE SOLUTION :

$\arg(-z)-\arg(z)=\arg\left(\frac{-z}{z}\right)=\arg(-1)=\pi$

16. (a) $|z_1|=|z_2|=|z_3|=1$ (given)

Now, $|z_1|=1 \Rightarrow |z_1|^2=1 \Rightarrow z_1\bar{z}_1=1$

Similarly $z_2\bar{z}_2=1,\ z_3\bar{z}_3=1$

Now, $\left|\frac{1}{z_1}+\frac{1}{z_2}+\frac{1}{z_3}\right|=1 \Rightarrow |\bar{z}_1+\bar{z}_2+\bar{z}_3|=1$

$\Rightarrow \left|\overline{z_1+z_2+z_3}\right|=1$ **NOTE THIS STEP**

$\Rightarrow |z_1+z_2+z_3|=1$

17. (d) Let $z=(1)^{1/n}=(\cos 2k\pi+i\sin 2k\pi)^{1/n}$

$$z=\cos\frac{2k\pi}{n}+i\sin\frac{2k\pi}{n}, k=0,\ 1,2,.....,n-1.$$

Let $z_1=\cos\left(\frac{2k_1\pi}{n}\right)+i\sin\left(\frac{2k_1\pi}{n}\right)$

and $z_2=\cos\left(\frac{2k_2\pi}{n}\right)+i\sin\frac{2k_2\pi}{n}$

be the two values of z. s.t. they subtend $\angle$ of 90° at origin.

$\therefore\quad \frac{2k_1\pi}{n}-\frac{2k_2\pi}{n}=\pm\frac{\pi}{2}\Rightarrow 4(k_1-k_2)=\pm n$

As k_1 and k_2 are integers and $k_1\neq k_2$.

$\therefore\quad n=4k,\ k\in I$

18. (c) $\frac{z_1-z_3}{z_2-z_3}=\frac{1-i\sqrt{3}}{2}$

$\Rightarrow \arg\left(\frac{z_1-z_3}{z_2-z_3}\right)=\arg\left(\frac{1-i\sqrt{3}}{2}\right)$

$\Rightarrow \arg(\cos(-\pi/3) + i\sin(-\pi/3))$

$\Rightarrow$ angle between $z_1 - z_3$ and $z_2 - z_3$ is 60°.

and $\left|\frac{z_1 - z_3}{z_2 - z_3}\right| = \left|\frac{1 - i\sqrt{3}}{2}\right|$

$\Rightarrow \left|\frac{z_1 - z_3}{z_2 - z_3}\right| = 1 \Rightarrow |z_1 - z_3| = |z_2 - z_3|$

NOTE THIS STEP

$\Rightarrow$ The Δ with vertices z_1, z_2 and z_3 is isosceles with vertical $\angle 60°$. Hence rest of the two angles should also be 60° each.

$\Rightarrow$ Req. Δ is an equilateral Δ.

19. (b) We know that,

$|z_1 - z_2| \geq ||z_1| - |z_2||$ ....(1)

Here $|z_1| = 12$

and $|z_2 - 3 - 4i| = 5$

but $|z_2 - (3+4i)| \geq ||z_2| - |3+4i||$

$\Rightarrow 5 \geq |z_2| - 5 \Rightarrow |z_2| \leq 10$

Also from (i) $|z_1 - z_2|$ will have least value when $|z_2|$ has greatest value i.e. 10

$\therefore \quad |z_1 - z_2| \geq 12 - 10 = 2$

Thus min. value of $|z_1 - z_2|$ is 2.

ALTERNATE SOLUTION:

$|z_1| = 12 \Rightarrow z_1$ lies on a circle with centre (0, 0) and radius 12 units, and $|z_2 - 3 - 4i| = 5 \Rightarrow z_2$ lies on a circle with centre (3, 4) and radius 5 units.

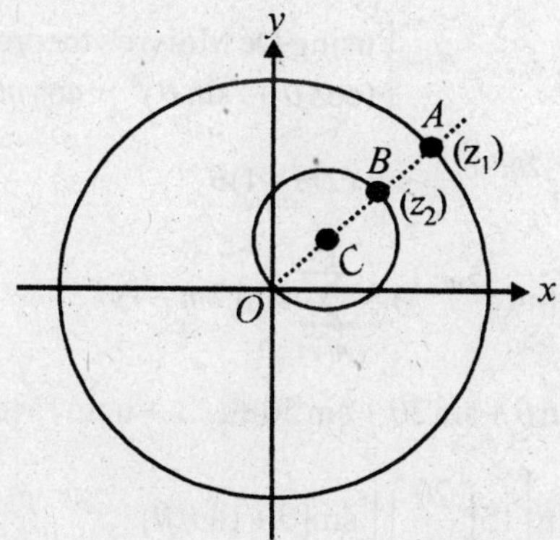

From fig. it is clear that $|z_1 - z_2|$ i.e., distance between z_1 and z_2 will be min when they lie at A and B resp. i.e., O, C, B, A are collinear as shown. Then $z_1 - z_2 = AB = OA - OB = 12 - 2(5) = 2$. As above is the min, value, we must have $|z_1 - z_2| \geq 2$.

20. (a) Given that $|z| = 1$ and $\omega = \frac{z-1}{z+1} (z \neq -1)$

Now we know that $z\bar{z} = |z|^2$

$\Rightarrow \quad z\bar{z} = 1 \qquad$ (for $|z| = 1$)

$\therefore \quad \omega = \left(\frac{z-1}{z+1}\right) \times \frac{(\bar{z}+1)}{(\bar{z}+1)} = \frac{z\bar{z} + z - \bar{z} - 1}{z\bar{z} + z + \bar{z} + 1} = \frac{2iy}{2+2x}$

[$\because z\bar{z} = 1$ and taking $z = x + iy$ so that $z + \bar{z} = 2x$ *and* $z - \bar{z} = 2iy$]

$\Rightarrow \quad \text{Re}(\omega) = 0$

ALTERNATE SOLUTION:

$\omega = \frac{z-1}{z+1} \Rightarrow z = \frac{1+\omega}{1-\omega}; |z| = 1 \Rightarrow \left|\frac{1+\omega}{1-\omega}\right| = 1$

$\Rightarrow |\omega + 1| = |\omega - 1| \Rightarrow \omega$ is equidistant from (1, 0) and (– 1, 0) and hence must lie on perpendicular bisector of line segment joining (1, 0) and (– 1, 0), i.e., imaginary axis $\Rightarrow \omega$ is purely imaginary $\Rightarrow \text{Re}(\omega) = 0$.

21. (b) $(1 + \omega^2)^n = (1 + \omega^4)^n$

$\Rightarrow (-\omega)^n = (1 + \omega)^n = (-\omega^2)^n \Rightarrow \omega^n = 1 \Rightarrow n = 3$

22. (a) Here we observe that.

$AB = AC = AD = 2$

$\therefore$ BCD is an arc of a circle with centre at A and radius 2. Shaded region is outer (exterior) part of this sector $ABCDA$.

$\therefore$ For any pt. z on arc BCD we should have

$|z - (-1)| = 2$

and for shaded region, $|z + 1| > 2$ ....(i)

For shaded region we also have

$-\pi/4 < \arg(z+1) < \pi/4$

or $\quad |\arg(z+1)| < \pi/4$ ...(ii)

Combining (i) and (ii), (a) is the correct option.

23. (b) Given that a, b, c are integers not all equal, ω is cube root of unity $\neq 1$, then

$|a + b\omega + c\omega^2|$

$= \left|a + b\left(\frac{-1+i\sqrt{3}}{2}\right) + c\left(\frac{-1-i\sqrt{3}}{2}\right)\right|$

$= \left|\left(\frac{2a-b-c}{2}\right) + i\left(\frac{b\sqrt{3} - c\sqrt{3}}{2}\right)\right|$

$= \frac{1}{2}\sqrt{(2a-b-c)^2 + 3(b-c)^2}$

$= \frac{1}{2}\sqrt{4a^2 + b^2 + c^2 - 4ab + 2bc - 4ac + 3b^2 + 3c^2 - 6bc}$

$= \sqrt{a^2 + b^2 + c^2 - ab - bc - ca}$

$= \sqrt{\frac{1}{2}[(a-b)^2 + (b-c)^2 + (c-a)^2]}$

R.H.S. will be min. when $a = b = c$, but we cannot take $a = b = c$ as per question.

$\therefore$ The min value is obtained when any two are zero and third is a minimum magnitude integer i.e. 1.

Thus $b = c = 0$, $a = 1$ gives us the minimum value 1.

ALTERNATE SOLUTION:

We observe that, $\overline{(a + b\omega + c\omega^2)} = (a + \overline{b\omega} + \overline{c\omega^2})$

$= (a + b\omega^2 + c\omega)$

$\Rightarrow \ |a+b\omega+c\omega^2|^2 = |(a+b\omega+c\omega^2)|\ |(a+b\omega^2+c\omega)|$

$(\because |z|^2 = z.\bar{z})$

$= \ |(a+b\omega+c\omega^2)(a+b\omega^2+c\omega)|$

$= \ |a^2+b^2+c^2-ab-bc-ca|$

$= \ \frac{1}{2}[(a-b)^2+(b-c)^2+c-a)^2]$

$\Rightarrow \ |a+b\omega+c\omega^2| = \sqrt{\frac{1}{2}[(a-b)^2+(b-c)^2+(c-a)^2]}$

Now same reason as given in previous method.

24. **(b)** Operating $R_1+R_2+R_3$, we get

$$\begin{vmatrix} 3 & 0 & 0 \\ 1 & -1-\omega^2 & \omega^2 \\ 1 & \omega^2 & \omega 4 \end{vmatrix} = 3[-\omega-1-\omega] = 3(\omega^2-\omega)$$

25. **(d)** $\because \frac{w - wz}{l - z}$ is purely real

$\therefore \overline{\left(\frac{w-\bar{w}z}{1-z}\right)} = \left(\frac{w-\bar{w}z}{1-z}\right) \Rightarrow \frac{\bar{w}-w\bar{z}}{1-\bar{z}} = \frac{w-\bar{w}z}{1-z}$

$\Rightarrow \ \bar{w}-\bar{w}z-w\bar{z}+wz\bar{z} = w-w\bar{z}-\bar{w}z+\bar{w}z\bar{z}$

$\Rightarrow \ w-\bar{w} = (w-\bar{w})|z|^2$

$\Rightarrow \ |z|^2 = 1 \quad (\because w = \alpha+i\beta \text{ and } \beta \neq 0)$

$\Rightarrow \ |z| = 1$ also given $z \neq 1$

$\therefore$ The required set is $\{z : |z|=1, z \neq 1\}$

$= 3\omega\ (\omega-1)$

26. **(d)** $\overrightarrow{OP} = \overrightarrow{OA} + \overrightarrow{AP}$

$\Rightarrow \ \overrightarrow{OP} = \overrightarrow{OA} + \overrightarrow{OB}$

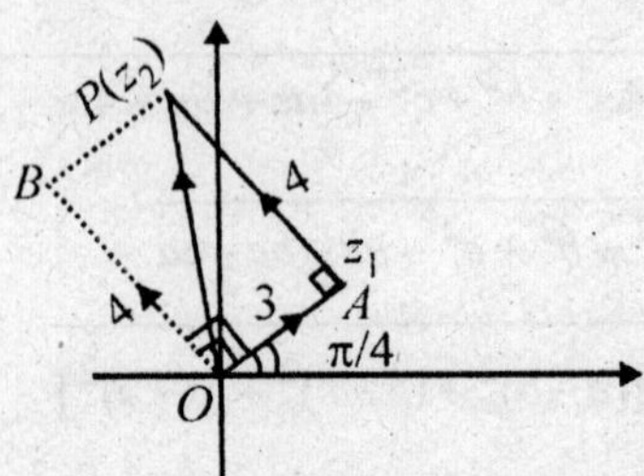

$\Rightarrow \ \overrightarrow{OP} = 3e^{i\pi/4} + 4e^{i(\pi/2+\pi/4)}$ **NOTE THIS STEP**

$= 3e^{i\pi/4} + 4e^{i\pi/2}.e^{i\pi/4}$

$= 3e^{i\pi/4} + 4ie^{i\pi/4} = e^{i\pi/4}(3+4i).$

27. **(d)** Given $|z| = 1$ and $z \neq \pm 1$

To find locus of $\omega = \frac{z}{1-z^2}$

We have $\omega = \frac{z}{1-z^2} = \frac{z}{z\bar{z}-z^2}$

$[\because |z| = 1 \Rightarrow |z|^2 = z\bar{z} = 1]$

$= \frac{1}{\bar{z}-z}$ = purely imaginary number

$\therefore \ \omega$ must lie on y – axis.

ALTERNATE SOLUTION :

$\because |z| = 1$ let $z = e^{i\theta}$

then $\omega = \frac{e^{i\theta}}{1-e^{2i\theta}} = \frac{e^{i\theta}\ .\ e^{-i\theta}}{e^{-i\theta}-e^{2i\theta}.\ e^{-i\theta}} = \frac{i}{2\sin\theta}$

$\Rightarrow$ Re $(\omega) = 0 \Rightarrow x = 0 \ \therefore$ locus of ω is y - axis.

28. **(d)** The initial position of point is $Z_0 = 1+2i$

$\therefore \ Z_1 = (1+5)+(2+3)i = 6+5i$

Now Z_1 is moved through a distance of $\sqrt{2}$ units in the direction $\hat{i}+\hat{j}$. (i.e. by $1+i$)

$\therefore$ It becomes $Z_1' = Z_1 + (1+i) = 7+6i$

Now OZ_1' is rotated through an angle $\frac{\pi}{2}$ in anticlockwise direction, therefore $Z_2 = iZ_1' = -6+7i$

29. **(d)** $z = \cos\theta + i\sin\theta$

$\Rightarrow \ z^{2m-1} = (\cos\theta + i\sin\theta)^{2m-1}$

$= \cos(2m-1)\theta + i\sin(2m-1)\theta$

$\left[\begin{array}{l}\text{using De Moivre's theorem} \\ (\cos\theta + i\sin\theta)^n = \cos n\theta + i\sin n\theta\end{array}\right]$

$\therefore \ \text{Im}(z^{2m-1}) = \sin(2m-1)\theta$

$\therefore \ \sum_{m=1}^{15} \text{Im}(z^{2m-1}) = \sum_{m=1}^{15} \sin(2m-1)\theta$

$= \sin\theta + \sin 3\theta + \sin 5\theta + \ldots\ldots + \text{upto } 15 \text{ terms}$

$= \frac{\sin\left[15\left(\frac{2\theta}{2}\right)\right].\sin[\theta + 14\times\theta]}{\sin\theta}$

$\left[\begin{array}{l}\text{Using } \sin\alpha + \sin(\alpha+\beta) + \sin(\alpha+2\beta) + \ldots n \text{ terms} \\ = \frac{\sin(n\beta/2).\sin[\alpha+(n-1)\beta/2]}{\sin(\beta/2)}\end{array}\right]$

$= \frac{\sin 15\theta.\sin 15\theta}{\sin\theta} = \frac{\sin 30°.\sin 30°}{\sin 2°}$

$= \frac{1}{4\sin 2°}$

30. **(a)** Given $z = x+iy$ where x and y are integer

Also $z\bar{z}^3 + \bar{z}z^3 = 350$

$\Rightarrow \ |z|^2(\bar{z}^2 + z^2) = 350$

$\Rightarrow (x^2+y^2)(x^2-y^2)=175$

$\Rightarrow (x^2+y^2)(x^2-y^2)=25\times 7$ ...(i)

or $(x^2+y^2)(x^2-y^2)=35\times 5$ ...(ii)

$\because$ x and y are integers,

$\therefore x^2+y^2=25$ and $x^2-y^2=7$ [From eq (i)]

$\Rightarrow x^2=16$ and $y^2=9$

$\Rightarrow x=\pm 4$ and $y=\pm 3$

$\therefore$ Vertices of rectangle are $(4,3),(4,-3),(-4,-3),(-4,3)$.

So, area of rectangle $=8\times 6=48$ sq. units

Now from eq. (ii)

or $x^2+y^2=35$ and $x^2-y^2=5$

$\Rightarrow x^2=20$, which is not possible for any integral value of x

31. **(d)** $\because$ $Im(z)\neq 0 \Rightarrow z$ is non real

and equation $z^2+z+(1-a)=0$ will have non real roots, if $D<0$

$\Rightarrow 1-4(1-a)<0 \Rightarrow 4a<3 \Rightarrow a<\frac{3}{4}$

$\therefore$ a can not take the value $\frac{3}{4}$

32. **(c)** As α lies on the circle $(x-x_0)^2+(y-y_0)^2=r^2$

$\therefore |\alpha-z_0|^2=r^2$

$\Rightarrow (\alpha-z_0)(\bar{\alpha}-\bar{z}_0)=r^2$

$\Rightarrow \alpha\bar{\alpha}-\alpha\bar{z}_0-\bar{\alpha}z_0+z_0\bar{z}_0=r^2$

$\Rightarrow |\alpha|^2+|z_0|^2-\alpha\bar{z}_0-\bar{\alpha}z_0=r^2$ (i)

Also $\frac{1}{\bar{\alpha}}$ lies on the circle $(x-x_0)^2+(y-y_0)^2=4r^2$

$\therefore \left|\frac{1}{\bar{\alpha}}-z_0\right|^2=4r^2$

$\Rightarrow \left(\frac{1}{\bar{\alpha}}-z_0\right)\left(\frac{1}{\alpha}-\bar{z}_0\right)=4r^2$

$\Rightarrow \frac{1}{\alpha\bar{\alpha}}-\frac{z_0}{\alpha}-\frac{\bar{z}_0}{\bar{\alpha}}+z_0\bar{z}_0=4r^2$

$\Rightarrow \frac{1}{|\alpha|^2}-\frac{z_0\bar{\alpha}}{|\alpha|^2}-\frac{\bar{z}_0\alpha}{|\alpha|^2}+|z_0|^2=4r^2$

$\Rightarrow 1+|\alpha|^2|z_0|^2-z_0\bar{\alpha}-\bar{z}_0\alpha=4r^2|\alpha|^2$ (ii)

Subtracting eq^n (i) from (ii) we get

$1-|\alpha|^2+|z_0|^2(|\alpha|^2-1)=r^2(4|\alpha|^2-1)$

or $(|\alpha|^2-1)(|z_0|^2-1)=r^2(4|\alpha|^2-1)$

Using $|z_0|^2=\frac{r^2+2}{2}$ we get

$(|\alpha|^2-1)\frac{r^2}{2}=r^2(4|\alpha|^2-1)$

$\Rightarrow |\alpha|^2-1=8|\alpha|^2-2 \Rightarrow |\alpha|=\frac{1}{\sqrt{7}}$

(D) MCQ's WITH ONE OR MORE THAN ONE CORRECT :

1. **(a, b, c)** Given that $z_1=a+ib$ and $z_2=c+id$ and $|z_1|=|z_2|=1$

$\therefore$ Let $z_1=\cos\theta+i\sin\theta=a+ib$

and $z_2=\cos\phi+i\sin\phi=c+id$

$\Rightarrow a=\cos\theta;\ b=\sin\theta;$

$c=\cos\phi;\ d=\sin\phi$

Also $w_1=a+ic=\cos\theta+i\cos\phi$

$w_2=b+id=\sin\theta+i\sin\phi$

ATQ Re $(z_1\bar{z}_2)=0$

$\Rightarrow$ Re$[(\cos\theta+i\sin\theta)\overline{(\cos\phi+i\sin\phi)}]=0$

$\Rightarrow$ Re$[(\cos\theta+i\sin\theta)(\cos\phi-i\sin\phi)]=0$

$\Rightarrow$ Re$[\cos(\theta-\phi)+i\sin(\theta-\phi)]=0$

$\Rightarrow \cos(\theta-\phi)=0 \Rightarrow \theta-\phi=\pi/2$

$\Rightarrow \theta=\pi/2+\phi$

$\therefore w_1=\cos(\pi/2+\phi)+i\cos\phi=-\sin\phi+i\cos\phi;$

$w_2=\sin(\pi/2+\phi)+i\sin\phi=\cos\phi+i\sin\phi$

$\therefore |w_1|=1;\ |w_2|=1$

Also

Re $(w_1\overline{w_2})=$ Re $[(-\sin\phi+i\cos\phi)(\cos\phi-i\sin\phi)]$

$\Rightarrow$ Re $[(-\sin\phi\cos\phi+\sin\phi\cos\phi)+i(\cos^2\phi+\sin^2\phi)]$

$\Rightarrow$ Re $(w_1\overline{w}_2)=0$

ALTERNATE SOLUTION:

$z_1=a+ib$ and $z_2=c+id$.

ATQ $|zi|^2=|z_2|^2=1$

$\Rightarrow a^2+b^2=1$ and $c^2+d^2=1$. ....(1)

Also Re $(z_1\ \bar{z}_2)=0 \Rightarrow ac+bd=0$

$\Rightarrow \frac{a}{b}=\frac{-d}{c}=\alpha\ (say)$ ....(2)

From (1) and (2), we get

$b^2\alpha^2+b^2=c^2\alpha^2+c^2 \Rightarrow b^2=c^2;$

Similarly $a^2=d^2$

$\therefore |\omega_1|=\sqrt{a^2+c^2}=\sqrt{c^2+d^2}=1$

and $|\omega_2|=\sqrt{b^2+d^2}=\sqrt{c^2+d^2}=1$

Also Re $(\omega_1\bar{\omega}_2)=ab+cd=(b\alpha)b+c(-c\alpha)$

$=\alpha(b^2-c^2)=0$

2. **(a,d)** Let $z_1=a+ib,\ a>0$ and $b\in R;\ z_2=c+id,\ d<0, c\in R$

then $|z_1|=|z_2| \Rightarrow a_2+b_2=c_2+d_2$

$$\Rightarrow a^2-c^2=d^2-b^2 \quad ....(1)$$

Now, $\frac{z_1+z_2}{z_1-z_2}=\frac{(a+c)+i(b+d)}{(a-c)+i(b-d)}$

$$= \frac{[(a^2-c^2)+(b^2-d^2)]+i[(a-c)(b+d)-(a+c)(b-d)]}{(a-c)^2+(b-d)^2}$$

$$= \frac{i[(a-c)(b+d)-(a+c)(b-d)]}{(a-c)^2+(b-d)^2} \quad \text{[Using (1)]}$$

= purely imaginary number or zero in case $a+c=b+d=0$.

3. **(c)** Let $z_1=r_1(\cos\theta_1+i\sin\theta_1)$

and $z_2=r_2(\cos\theta_2+i\sin\theta_2)$

where $r_1=|z_1|, r_2=|z_2|, \theta_1=\arg(z_1), \theta_2=\arg(z_2)$

$\therefore \quad z_1+z_2=r_1(\cos\theta_1+i\sin\theta_1)+r_2(\cos\theta_2+i\sin\theta_2)$

$= (r_1\cos\theta_1+r_2\cos\theta_2)+i(r_1\sin\theta_1+r_2\sin\theta_2)$

$= r_1^2\cos^2\theta_1+r_2^2\cos^2\theta_2+2r_1r_2\cos\theta_1\cos\theta_2$

$\quad +r_1^2\sin^2\theta_1+r_2^2\sin^2\theta_2+2r_1r_2\sin\theta_1\sin\theta_2$

$=r_1^2+r_2^2+2r_1r_2\cos(\theta_1-\theta_2)$

and $|z_1|+|z_2|=r_1+r_2$

Since $|z_1+z_2|=|z_1|+|z_2|$ (given)

$\Rightarrow |z_1+z_2|^2=|z_1|^2+|z_2|^2+2|z_1||z_2|$

$\Rightarrow r_1^2+r_2^2+2r_1r_2\cos(\theta_1-\theta_2)=r_1^2+r_2^2+2r_1r_2$

$\Rightarrow \cos(\theta_1-\theta_2)=1 \Rightarrow \theta_1-\theta_2=0$

$\Rightarrow \text{Arg}(z_1)=\arg(z_2)$

4. **(d)** Let $z=\cos\frac{2\pi}{7}+i\sin\frac{2\pi}{7}$

Then by DeMoivre's theorem, we have

$$z^k=\cos\frac{2\pi k}{7}+i\sin\frac{2\pi k}{7}$$

Now, $\sum_{k=1}^{6}\left(\sin\frac{2\pi k}{7}-i\cos\frac{2\pi k}{7}\right)$

$$=\sum_{k=1}^{6}(-i)\left(\cos\frac{2\pi k}{7}+i\sin\frac{2\pi k}{7}\right)$$

$$=(-i)\sum_{k=1}^{6}z^k=-i\,z\frac{(1-z^6)}{1-z}=-i\left(\frac{z-z^7}{1-z}\right)$$

$$=(-i)\left(\frac{z-1}{1-z}\right)=[\text{Using } z^7=\cos 2\pi+i\sin 2\pi=1]$$

$$=i\left(\frac{1-z}{1-z}\right)=i$$

5. **(d)** We have $(1+\omega+\omega^2)^7=(-\omega^2-\omega^2)^7$

$(-2)^7(\omega^2)^7=-128\omega^{14}=-128\omega^2$

6. **(b)** $\sum_{i=1}^{13}(i^n+i^{n+1})=\sum_{i=1}^{13}i^n(1+i)=(1+i)\sum_{i=1}^{13}i^n$

This forms a G.P.

Sum of G.P. $=i(1+i)\frac{(1-i^{13})}{1-i}=i-1$ as $i^{13}=i$

7. **(d)** Taking $-3i$ common from C_2, we get

$$-3i\begin{vmatrix}6i & 1 & 1\\ 4 & -1 & -1\\ 20 & i & i\end{vmatrix}=0 \quad (\because C_2\equiv C_3)$$

$\Rightarrow x=0,\ y=0$

8. **(a,c,d)** Given that $z=(1-t)z_1+t\,z_2$ where $0<t<1$

$\Rightarrow z=\frac{(1-t)z_1+tz_2}{(1-t)+t}$

$\Rightarrow$ z divides the join of z_1 and z_2 internally in the ratio $t:(1-t)$.

$\therefore z_1, z$ and z_2 are collinear

$t \quad 1-t$

$z_1 \quad z \quad z_2$

$\Rightarrow |z-z_1|+|z-z_2|=|z_1-z_2|$

Also $z=(1-t)z_1+t\,z_2$

$\Rightarrow \frac{z-z_1}{z_2-z_1}=t=$ purely real number

$\therefore \arg\left(\frac{z-z_1}{z_2-z_1}\right)=0 \Rightarrow \arg(z-z_1)=\arg(z_2-z_1)$

Also $\frac{z-z_1}{z_2-z_1}=t\Rightarrow\frac{\bar{z}-\bar{z}_1}{\bar{z}_2-\bar{z}_1}=t$

$\Rightarrow \frac{z-z_1}{z_2-z_1}=\frac{\bar{z}-\bar{z}_1}{\bar{z}_2-\bar{z}_1}$

$\Rightarrow (z-z_1)(\bar{z}_2-\bar{z}_1)=(\bar{z}-\bar{z}_1)(z_2-z_1)$

$\Rightarrow \begin{vmatrix}z-z_1 & \bar{z}-\bar{z}_1\\ z_2-z_1 & \bar{z}_2-\bar{z}_1\end{vmatrix}=0$

9. **(c, d)** $\text{w}=\frac{\sqrt{3}+\text{i}}{2}=\cos\frac{\pi}{6}+\text{i}\sin\frac{\pi}{6}$

and $\text{w}^\text{n}=\cos\frac{\text{n}\pi}{6}+\text{i}\sin\frac{\text{n}\pi}{6}$

$\therefore$ P contains all those points which lie on unit circle and have arguments $\frac{\pi}{6},\frac{2\pi}{6},\frac{3\pi}{6}$ and so on.

As $z_1\in P\cap H_1$ and $z_2\in P\cap H_2$, therefore z_1 and z_2 can have possible positions as shown in the figure.

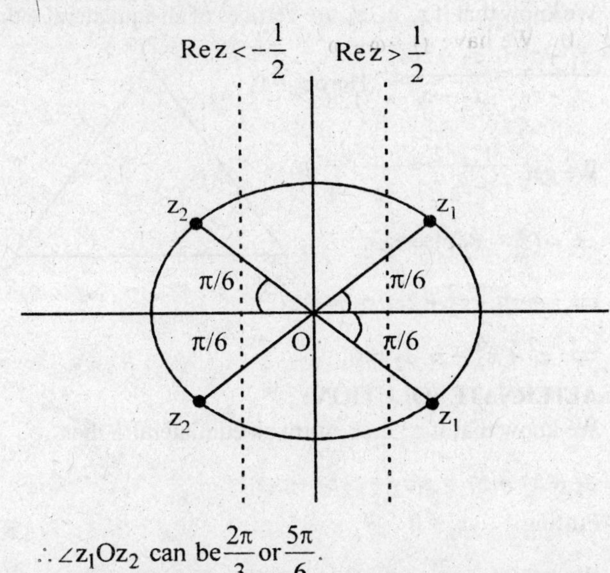

$\therefore \angle z_1Oz_2$ can be $\frac{2\pi}{3}$ or $\frac{5\pi}{6}$.

(E) SUBJECTIVE PROBLEMS :

1. $\frac{1}{1-\cos\theta+2i\sin\theta}$

$$= \frac{1}{2\sin^2\theta/2+4i\sin\theta/2\cos\theta/2} = \frac{1}{2\sin\theta/2}$$

$$\left[\frac{\sin\theta/2-2i\cos\theta/2}{(\sin\theta/2+2i\cos\theta/2)(\sin\theta/2-2i\cos\theta/2)}\right]$$

$$= \frac{1}{2\sin\theta/2}\left[\frac{\sin\theta/2-2i\cos\theta/2}{(\sin^2\theta/2+4\cos^2\theta/2)}\right]$$

$$= \frac{1}{2\sin\theta/2}\left[\frac{2\sin\theta/2-4i\cos\theta/2}{1-\cos\theta+4+4\cos\theta}\right]$$

$$= \frac{2}{2\sin\theta/2}\left[\frac{2\sin\theta/2-2i\cos\theta/2}{5+3\cos\theta}\right]$$

$$= \left(\frac{1}{5+3\cos\theta}\right)+\left(\frac{-2\cot\theta/2}{5+3\cos\theta}\right)i$$

which is of the form $X+iY$.

2. As β and γ are the complex cube roots of unity therefore, let $\beta=\omega$ and $\gamma=\omega^2$ so that $\omega+\omega^2+1=0$ and $\omega^3=1$.
Then $xyz=(a+b)(a\omega^2+b\omega)(a\omega+b\omega^2)$
$=(a+b)(a^2\omega^3+ab\omega^4+ab\omega^2+b^2\omega^3)$
$=(a+b)(a^2+ab\omega+ab\omega^2+b^2)$ (using $\omega^3=1$)
$=(a+b)(a^2+ab(\omega+\omega^2)+b^2)$
$=(a+b)(a^2-ab+b^2)$ (using $\omega+\omega^2=-1$)
$=a^3+b^3$ Hence proved.

3. Given $x+iy=\sqrt{\frac{c+ib}{c+id}}$

$$\Rightarrow (x+iy)^2=\frac{a+ib}{c+id} \quad ....(1)$$

Taking conjugate on both sides, we get

$$(x-iy)^2=\frac{a-ib}{c-id} \quad ....(2)$$

Multiply (1) and (2), we get

$$(x^2+y^2)^2=\frac{a^2+b^2}{c^2+d^2}$$

4. Given that n is an odd integer >3 and n is not a multiple of 3.
Let $p(x)=(x+1)^n-x^n-1$ and $q(x)=x^3+x^2+x$
$=x(x^2+x+1)=x(x-\omega)(x-\omega^2)$
where ω and ω^2 are cube roots of unity.
Clearly $0,\omega,\omega^2$ are zeros of the polynomial $q(x)$.
Now, $p(0)=1^n-0-1=0 \Rightarrow 0$ is a zero of $p(x)$.
$p(\omega)=(\omega+1)^n-\omega^n-1 \Rightarrow (-\omega^2)^n-\omega^n-1$
$= -\omega^{2n}-\omega^n-1$ [$\because$ n is odd]
$= -(\omega^n+\omega^{2n})-1$
$= -(-1)-1$ [$\because \omega^n+\omega^{2n}+1=0$ if $n\neq 3m$] $=0$
$\Rightarrow$ ω is a zero of $p(x)$.
Also, $p(\omega^2)=(\omega^2+1)^n-(\omega^2)^n-1 \Rightarrow (-\omega)^n-\omega^{2n}-1$
$=-\omega^n-\omega^{2n}-1=-(1+\omega^n+\omega^{2n})=0$ [for $n\neq 3m$, $m\in N$]
$\Rightarrow$ ω^2 is a zero of $p(x)$.
$\because$ $0,\omega,\omega^2$ are zeroes of $p(x)$
$\therefore$ $x, x-\omega, x-\omega^2$ are factors of $p(x)$.
$\Rightarrow$ $x(x-\omega)(x-\omega^2)$ is a factor of $p(x)$
$\Rightarrow$ x^3+x^2+x is a factor of $p(x)$.

5. $\frac{(1+i)x-2i}{3+i}+\frac{(2-3i)y+i}{3-i}=i$
$\Rightarrow$ $(4+2i)x-6i-2+(9-7i)y+3i-1=10i$
$\Rightarrow$ $(4x+9y-3)+(2x-7y-3)i=10i$
$\Rightarrow$ $4x+9y-3=0$ and $2x-7y-3=10$
On solving these two, we get $x=3$, $y=-1$

6. Let the vertices of ΔABC be represented by z_1, z_2 and z_3, and its circumcentre by z_0.

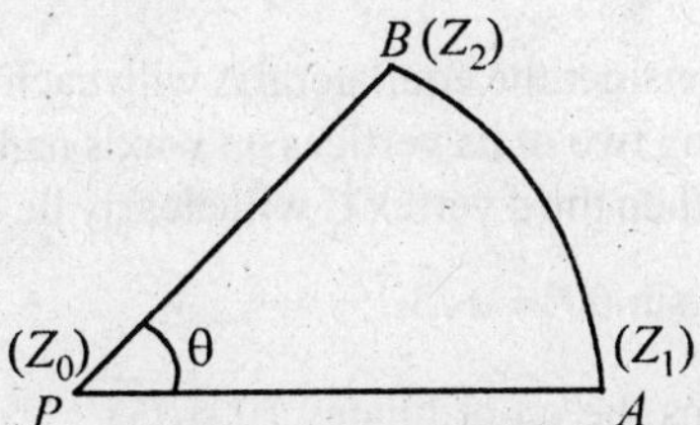

Now we know that if PA is rotated, about P, in anticlockwise sense, by an angle θ then we write

$z_2-z_0=(z_1-z_0)e^{i\theta}$ **NOTE THIS STEP**

Similarly in ΔABC, by rotation in anticlockwise sense about A and B we get

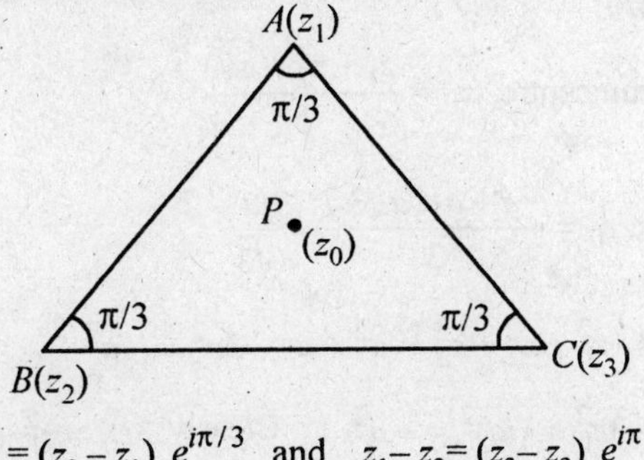

$z_3-z_1=(z_2-z_1)e^{i\pi/3}$ and $z_1-z_2=(z_3-z_2)e^{i\pi/3}$

Dividing, we get $\dfrac{z_3 - z_1}{z_1 - z_2} = \dfrac{z_2 - z_1}{z_3 - z_2}$

$\Rightarrow \ (z_3 - z_1)(z_3 - z_2) = -(z_2 - z_1)^2$

$\Rightarrow \ z_3^2 - z_2 z_3 - z_1 z_3 + z_1 z_2 + z_2^2 + z_1^2 - 2z_1 z_2 = 0$

$\Rightarrow \ z_1^2 + z_2^2 + z_3^2 = z_1 z_2 + z_2 z_3 + z_3 z_1$ ....(1)

Now in an equilateral Δ circumcentre coincides with centriod. Therefore

$$z_0 = \frac{z_1 + z_2 + z_3}{3}$$

Squaring, $9z_0^2 = z_1^2 + z_2^2 + z_3^2 + 2z_1 z_2 + 2z_2 z_3 + 2z_3 z_1$

$\Rightarrow \ \dfrac{9z_0^2 - (z_1^2 + z_2^2 + z_3^2)}{2} = z_1 z_2 + z_2 z_3 + z_3 z_1$ ....(2)

From (1) and (2) we get,

$$z_1^2 + z_2^2 + z_3^2 = \frac{9z_0^2 - (z_1^2 + z_2^2 + z_3^2)}{2}$$

$\Rightarrow \ 2(z_1^2 + z_2^2 + z_3^2) + (z_1^2 + z_2^2 + z_3^2) = 9z_0^2$

$\Rightarrow \ 3(z_1^2 + z_2^2 + z_3^2) = 9z_0^2 \Rightarrow z_1^2 + z_2^2 + z_3^2 = 3z_0^2$

ALTERNATE SOLUTION :

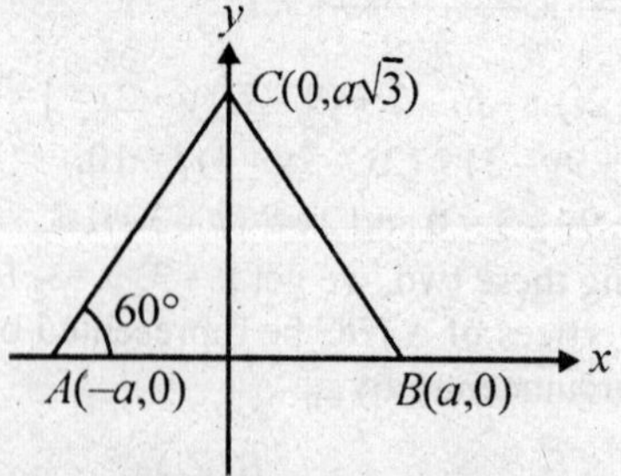

Let us consider the equilateral Δ with each side of length $2a$ and having two of its vertices on x-axis namely $A(-a,0)$ and $B(a, 0)$, then third vertex C will clearly lie on y-axis s.t.

$OC = 2a \sin 60° = a\sqrt{3}$

$\therefore$ C has the co-ordinates $(0, a\sqrt{3})$.

Now in the form of complex numbers if A, B and C are represented by z_1, z_2, z_3 then

$z_1 = -a \ ; \ z_2 = a \ ; \ z_3 = a\sqrt{3}\, i$

As in an equilateral Δ, centriod and circumcentre coincide, we get

Circumcentre, $z_0 = \dfrac{z_1 + z_2 + z_3}{3}$

$\Rightarrow \ z_0 = \dfrac{-a + a + a\sqrt{3}\, i}{3} = \dfrac{ia}{\sqrt{3}}$

Now, $z_1^2 + z_2^2 + z_3^2 = a^2 + a^2 - 3a^2 = -a^2$

and $3z_0^2 = (ia)^2 = -a^2 \ \therefore$ Clearly $3z_0^2 = z_1^2 + z_2^2 + z_3^2$

7. We know that if z_1, z_2, z_3 are vertices of an equilateral Δ then

$\dfrac{z_1 - z_2}{z_3 - z_2} = \dfrac{z_3 - z_1}{z_2 - z_1}$ Here $z_3 = 0$,

We get $\dfrac{z_1 - z_2}{-z_2} = \dfrac{-z_1}{z_2 - z_1}$

$\Rightarrow -(z_1 - z_2)^2 = z_1 z_2$

$\Rightarrow \ -z_1^2 - z_2^2 + 2z_1 z_2 = z_1 z_2$

$\Rightarrow \ z_1^2 + z_2^2 - z_1 z_2 = 0.$

ALTERNATE SOLUTION :

We know that if z_1, z_2, z_3 form an equilateral Δ then

$$z_1^2 + z_2^2 + z_3^2 = z_1 z_2 + z_2 z_3 + z_3 z_1$$

Putting $z_3 = 0$

We get $z_1^2 + z_2^2 = z_1 z_2$ or $z_1^2 + z_2^2 - z_1 z_2 = 0$

8. $1, a_1, a_2, \ldots\ldots a_{n-1}$ are the n roots of unity. Clearly above n values are roots of eq. $x^n - 1 = 0$

Therefore we must have (by factor theorem)

$x^n - 1 = (x - 1)(x - a_1)(x - a_2) \ldots(x - a_{n-1})$ ....(1)

$\Rightarrow \ \dfrac{x^n - 1}{x - 1} = (x - a_1)(x - a_2) \ldots(x - a_{n-1})$ ....(2)

ALTERNATE SOLUTION-1 :

LHS of eq. (2) is the sum

$1 + x + x^2 + \ldots + x^{n-1}$

$\therefore$ We get

$1 + x + x^2 + \ldots\ldots + x^{n-1} = (x - a_1)(x - a_2) \ldots (x - a_{n-1})$

Putting $x = 1$, we get

$1 + 1 + 1 + \ldots\ldots n$ times $= (1 - a_1)(1 - a_2) \ldots(1 - a_{n-1})$

$\Rightarrow \ (1 - a_1)(1 - a_2) \ldots(1 - a_{n-1}) = n$ Proved.

ALTERNATE SOLUTION-2 :

Taking limit as $x \to 1$ on both sides of eq. (2) we get

$$\lim_{x \to 1} \left(\frac{x^n - 1}{x - 1} \right) = \lim_{x \to 1} (x - a_1)(x - a_2) \ldots (x - a_{n-1})$$

$\Rightarrow \ n.1^{n-1} = (1 - a_1)(1 - a_2) \ldots (1 - a_{n-1})$

$\Rightarrow \ (1 - a_1)(1 - a_2) \ldots (1 - a_{n-1}) = n$

ALTERNATE SOLUTION-3 :

Differentiating both sides of eq. (1), we get

$nx^{n-1} = (x - a_1)(x - a_2) \ldots (x - a_{n-1}) + (x - 1)(x - a_2) \ldots(x - a_{n-1}) + \ldots + (x - 1)(x - a_1) \ldots (x - a_{n-2})$

For $x = 1$, we get

$n = (1 - a_1)(1 - a_2) \ldots (1 - a_{n-1})$

[All the terms except first contain $(x - 1)$ and hence become zero for $x = 1$] Proved.

9. Let $A = z = x + iy$, $B = iz = -y + ix$,

$C = z + iz = (x - y) + i(x + y)$

Now, area of

$$\Delta ABC = \frac{1}{2} \begin{vmatrix} x & y & 1 \\ -y & x & 1 \\ x - y & x + y & 1 \end{vmatrix}$$

Operating $R_2 - R_1$, $R_3 - R_1$, we get

$$\Delta = \frac{1}{2}\begin{vmatrix} x & y & 1 \\ -y-x & x-y & 0 \\ -y & x & 0 \end{vmatrix}$$

$$= \frac{1}{2}|x(-y-x)+y(x-y)|$$

$$= \frac{1}{2}|-xy-x^2+xy-y^2| = \frac{1}{2}|-x^2-y^2|$$

$$= \frac{1}{2}|x^2+y^2| = \frac{1}{2}|z|^2 \text{ Hence Proved.}$$

ALTERNATE SOLUTION :

Let $A(z)$, $B(iz)$, $C(z+iz)$

We know that iz is obtained by rotating OA through an angle 90°, also pt $z+iz$ can be obtained by completing the parallelogram two of whose adjacent sides are OA and OB.

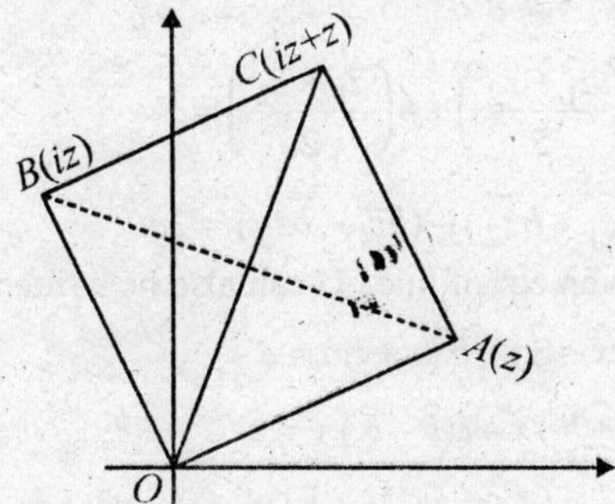

From Argand diagram, it is clear that

Ar (ΔABC) = ar (ΔOAB)

$$= \frac{1}{2} \times OA \times OB \qquad [\because rt\ \angle ed\ at\ pt\ 0]$$

$$= \frac{1}{2}|z| \times |z| = \frac{1}{2}|z|^2$$

[**NOTE :** $OA = |z|, OB = |z|$]

Hence Proved.

10. $\because$ $BC = AC$ and $\angle C = \pi/2$, we have by rotation about C in anti clockwise sense

$$CB = CA\ e^{i\pi/2}$$

$$\Rightarrow z_2 - z_3 = (z_1 - z_3)^{i\pi/2} \Rightarrow z_2 - z_3 = i(z_1 - z_3)$$

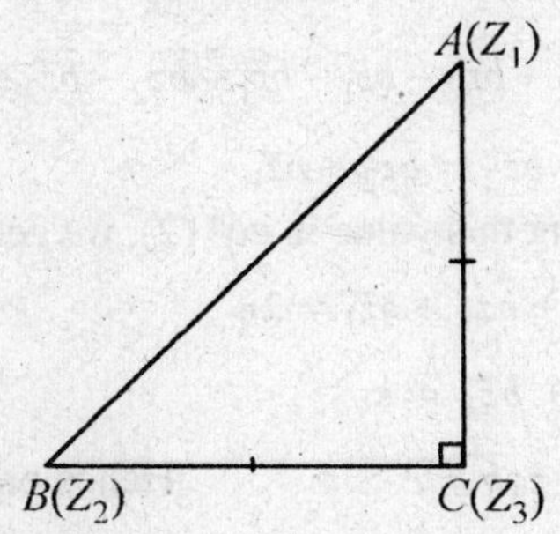

Squaring both sides, we get

$$(z_2 - z_3)^2 = -(z_1 - z_3)^2$$

$$\Rightarrow z_2^2 + z_3^2 - 2z_2 z_3 = -z_1^2 - z_3^2 + 2z_1 z_3$$

$$\Rightarrow z_1^2 + z_2^2 - 2z_1 z_2 = 2z_1 z_3 + 2z_2 z_3 - 2z_3^2 - 2z_1 z_2$$

$$\Rightarrow (z_1 - z_2)^2 = 2[(z_1 z_3 - z_3^2) - (z_1 z_2 - z_2 z_3)]$$

$$\Rightarrow (z_1 - z_2)^2 = 2(z_1 - z_3)(z_3 - z_2)$$

11. We are given that $z_1 = 10 + 6i$ and $z_2 = 4 + 6i$

Also $\arg\left(\frac{z - z_1}{z - z_2}\right) = \frac{\pi}{4}$

$$\Rightarrow \arg(z - z_1) - \arg(z - z_2) = \frac{\pi}{4} \quad \textbf{NOTE THIS STEP}$$

$$\Rightarrow \arg((x+iy) - (10+6i)) - \arg((x+iy) - (4+6i)) = \frac{\pi}{4}$$

$$\Rightarrow \arg[(x-10) + i(y-6)] - \arg[(x-4) + i(y-6)] = \frac{\pi}{4}$$

$$\Rightarrow \tan^{-1}\left(\frac{y-6}{x-10}\right) - \tan^{-1}\left(\frac{y-6}{x-4}\right) = \frac{\pi}{4}$$

$$\Rightarrow \tan^{-1}\left(\frac{\frac{y-6}{x-10} - \frac{y-6}{x-4}}{1 + \frac{(y-6)^2}{(x-4)(x-10)}}\right) = \frac{\pi}{4}$$

$$\Rightarrow \frac{(x-4)(y-6) - (x-10)(y-6)}{(x-4)(x-10) + (y-6)^2} = \tan\frac{\pi}{4}$$

$$\Rightarrow (x - 4 - x + 10)(y-6) = (x-4)(x-10) + (y-6)^2$$

$$\Rightarrow 6y - 36 = x^2 + y^2 - 14x - 12y + 40 + 36$$

$$\Rightarrow x^2 + y^2 - 14x - 18y + 112 = 0$$

$$\Rightarrow (x^2 - 14x + 49) + (y^2 - 18y + 81) = 18$$

$$\Rightarrow (x-7)^2 + (y-9)^2 = (3\sqrt{2})^2$$

$$\Rightarrow |(x+iy) - (7+9i)| = 3\sqrt{2}$$

$$\Rightarrow |z - (7+9i)| = 3\sqrt{2}. \qquad \text{Hence Proved.}$$

12. Dividing through out by i and knowing that $\frac{1}{i} = -i$, we get

$z^3 - iz^2 + iz + 1 = 0$

or $z^2(z - i) + i(z - i) = 0$ as $1 = -i^2$

or $(z - i)(z^2 + i) = 0$ $\therefore z = i$ or $z^2 = -i$

$\therefore$ $|z| = |i| = 1$ or $|z^2| = |z|^2 = |-i| = 1 \Rightarrow |z| = 1$

Hence in either case $|z| = 1$

13. Let $Z = r_1(\cos\theta_1 + i\sin\theta_1)$

and $W = r_2(\cos\theta_2 + i\sin\theta_2)$

We have $|Z| = r_1$, $|W| = r_2$, Arg $Z = \theta_1$ and Arg $W = \theta_2$

Since $|Z| \le 1, |W| \le 1$, it follows that $r_1 \le$ *and* $r_2 \le 1$

We have $Z - W = (r_1 \cos\theta_1 - r_2 \cos\theta_2)$

$$+i(r_1 \sin\theta_1 - r_2 \sin\theta_2)$$

$$|Z-W|^2 = (r_1 \cos\theta_1 - r_2\cos\theta_2)^2 + (r_1\sin\theta_1 - r_2\sin\theta_2)^2$$

$=$

$$r_1^2\cos^2\theta_1 + r_2^2\cos^2\theta_2 - 2r_1r_2\cos\theta_1\cos\theta_2 + r_1^2\sin^2\theta_1$$

$$+r_2^2\sin^2\theta_2 - 2r_1r_2\sin\theta_1\sin\theta_2$$

$$= r_1^2(\cos^2\theta_1 + \sin^2\theta_1) + r_2^2(\cos^2\theta_2 + \sin^2\theta_2)$$

$$-2r_1r_2(\cos\theta_1\cos\theta_2 + \sin\theta_1\sin\theta_2)$$

$$= r_1^2 + r_2^2 - 2r_1r_2\cos(\theta_1 - \theta_2)$$

$$= (r_1 - r_2)^2 + 2r_1r_2[1 - \cos(\theta_1 - \theta_2)]$$

$$= (r_1 - r_2)^2 + 4r_1r_2\sin^2\left(\frac{\theta_1 - \theta_2}{2}\right)$$

$$= |r_1 - r_2|^2 + 4r_1r_2\left|\sin\left(\frac{\theta_1-\theta_2}{2}\right)\right|^2$$

$$\le |r_1 - r_2|^2 + 4\left|\sin\left(\frac{\theta_1-\theta_2}{2}\right)\right| \quad [\because r_1, r_2 \le 1]$$

But $|\sin\theta| \le |\theta| \ \forall\ \theta \in R$ **NOTE THIS STEP**

Therefore,

$$|Z-W|^2 \le |r_1 - r_2|^2 + 4\left|\frac{\theta_1-\theta_2}{2}\right|^2$$

$$\le |r_1 - r_2|^2 + |\theta_1 - \theta_2|^2$$

Thus $|Z-W|^2 \le (|Z| - |W|)^2 + (\text{Arg } Z - \text{Arg } W)^2$

14. Let $z = x + iy$ then $\bar{z} = iz^2$

$\Rightarrow x - iy = i(x^2 - y^2 + 2ixy)$

$\Rightarrow x - iy = i(x^2 - y^2) - 2xy$

$\Rightarrow x(1 + 2y) = 0;\ x^2 - y^2 + y = 0$

$\Rightarrow x = 0$ or $y = -\dfrac{1}{2}$

$\Rightarrow x = 0,\ y = 0,\ 1$

or $\quad y = -\dfrac{1}{2},\ x = \pm\dfrac{\sqrt{3}}{2}$

For non zero complex number z

$$x = 0, y = 1;\ x = \frac{\sqrt{3}}{2},\ y = -\frac{1}{2};\ x = \frac{-\sqrt{3}}{2},\ y = -\frac{1}{2}$$

$$\therefore \quad z = i, \frac{\sqrt{3}}{2} - \frac{i}{2}, -\frac{\sqrt{3}}{2} - \frac{i}{2}$$

15.

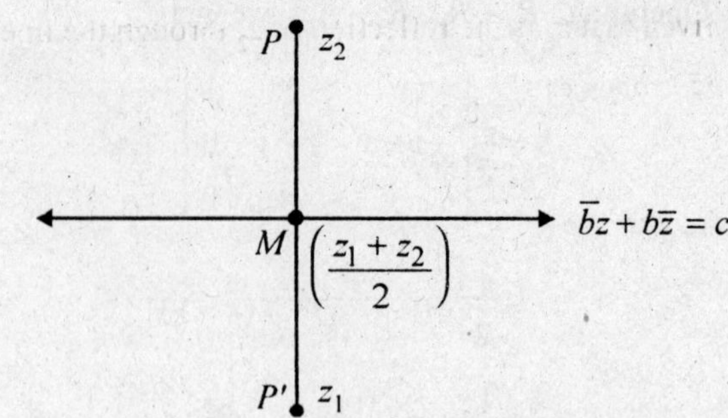

Since z_1 is reflection of point z_2, through the line

$\bar{b}z + b\bar{z} = c$ ....(1)

Therefore the mid point of z_1 and z_2 i.e. $\dfrac{z_1+z_2}{2}$ must lie on line (1), thus

$$\bar{b}\left(\frac{z_1+z_2}{2}\right) + b\left(\frac{\overline{z_1+z_2}}{2}\right) = c$$

$$\Rightarrow \bar{b}\left(\frac{z_1+z_2}{2}\right) + b\left(\frac{\bar{z}_1+\bar{z}_2}{2}\right) = c$$

$\Rightarrow (\bar{b}z_1 + b\bar{z}_2) + (b\bar{z}_1 + \bar{b}z_2) = 2c$ ....(2)

Now given eq^n of line (1) can also be written as

$$b(x - iy) + \bar{b}(x + iy) = c$$

or $\quad (b + \bar{b})x - i(b - \bar{b})y = c$

$\Rightarrow$ slope of line $= \dfrac{b+\bar{b}}{i(b-\bar{b})} = m_1$ (say)

Also slope of line joining $z_1\ (x_1 + iy_1)$ and $z_2\ (x_2 + iy_2)$ is

$$m_2 = \frac{y_2 - y_1}{x_2 - x_1} = \frac{\frac{z_2 - \bar{z}_2}{2i} - \frac{z_1 - \bar{z}_1}{2i}}{\frac{z_2 + \bar{z}_2}{2} - \frac{z_1 + \bar{z}_1}{2}} = \frac{z_2 - \bar{z}_2 - z_1 + \bar{z}_1}{i(z_2 + \bar{z}_2 - z_1 - \bar{z}_1)}$$

As the above two lines are perpendicular to each other

$\therefore \quad m_1 \times m_2 = -1$

$$\Rightarrow \frac{b+\bar{b}}{i(b-b)} \times \frac{z_2 - \bar{z}_2 - z_1 + \bar{z}_1}{i(z_2 + \bar{z}_2 - z_1 - \bar{z}_1)} = -1$$

$$\Rightarrow \frac{b+\bar{b}}{(b-\bar{b})} \times \frac{z_2 - \bar{z}_2 - z_1 + \bar{z}_1}{z_2 + \bar{z}_2 - z_1 - \bar{z}_1} = 1$$

$$\Rightarrow bz_2 - b\bar{z}_2 - bz_1 + b\bar{z}_1 + \bar{b}z_2 - \bar{b}\bar{z}_2 - \bar{b}z_1 + \bar{b}\bar{z}_1$$

$$= bz_2 + b\bar{z}_2 - bz_1 - b\bar{z}_1 - \bar{b}z_2 - \bar{b}\bar{z}_2 + \bar{b}z_1 + \bar{b}\bar{z}_1$$

$$\Rightarrow \bar{b}z_1 + b\bar{z}_2 = \bar{b}z_2 + b\bar{z}_1$$

Substituting this value in eq^n (2), we get

$$\bar{b}z_2 + b\bar{z}_1 + \bar{b}z_2 + b\bar{z}_1 = 2c$$

$$\Rightarrow \bar{b}z_2 + b\bar{z}_1 = c$$

$\Rightarrow \bar{z}_1 b + z_2\bar{b} = c$ Hence proved.

ALTERNATE SOLUTION :

Given that z_1 is the reflection of z_2 through the line

$b\bar{z} + \bar{b}z = c$ ....(1)

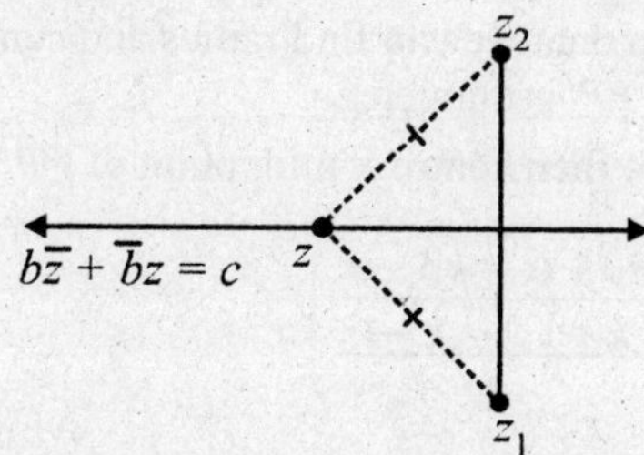

therefore for any arbitrary pt. z on the line, we must have

$|z - z_1| = |z - z_2| \Rightarrow |z - z_1|^2 = |z - z_2|^2$

$\Rightarrow (z - z_1)\overline{(z - z_1)} = (z - z_2)\overline{(z - z_2)}$

$\Rightarrow (z - z_1)(\bar{z} - \bar{z}_1) = (z - z_2)(\bar{z} - \bar{z}_2)$

$\Rightarrow (\overline{z_2} - \overline{z_1})z + (z_2 - z_1)\bar{z} + z_1\overline{z_1} - z_2.\overline{z_2} = 0$ ....(2)

But the above eq^n represents the given line. Thus eq^{n}s (1) and (2) must be identical.

$\therefore$ Comparing eq^{n}s (1) and (2) we get

$$\frac{\bar{b}}{\bar{z}_2 - \bar{z}_1} = \frac{b}{z_2 - z_1} = \frac{c}{z_2\bar{z}_2 - z_1\bar{z}_1} = \lambda \text{ (say)}$$

$\Rightarrow \bar{b} = \lambda(\overline{z_2} - \overline{z_1}),\ b = \lambda(z_2 - z_1)$

$c = \lambda(z_2\bar{z}_2 - z_1\bar{z}_1)$

Now, $\bar{z}_1 b + z_2\bar{b} = \bar{z}_1(\lambda(z_2 - z_1)) + z_2\lambda(\overline{z_2} - \overline{z_1})$

$= \lambda(\overline{z_1}z_2 - z_1\overline{z_1} + z_2\overline{z_2} - z_2\overline{z_1})$

$= \lambda(z_2\overline{z_2} - z_1\overline{z_1}) = c$ Hence proved.

16. $z^2 + pz + q = 0$

$z_1 + z_2 = -p,\ z_1z_2 = q$

By rotation through α in anticlockwise direction

$z_2 = z_1 e^{i\alpha}$ ....(1)

$$\frac{z_2}{z_1} = \frac{e^{i\alpha}}{1} = \frac{\cos\alpha + i\sin\alpha}{1}$$

Add 1 in both sides to get $z_1 + z_2 = -p$

$$\therefore \frac{z_1 + z_2}{z_1} = \frac{1 + \cos\alpha + i\sin\alpha}{1} = 2\cos\frac{\alpha}{2}\left[\cos\frac{\alpha}{2} + i\sin\frac{\alpha}{2}\right]$$

or $\dfrac{(z_2 + z_1)}{z_1} = 2\cos\dfrac{\alpha}{2}e^{i\alpha/2}$

On squaring $(z_2 + z_1)^2 = 4\cos^2(\alpha/2)z_1^2.e^{i\alpha}$

$= 4\cos^2\dfrac{\alpha}{2}z_1^2.\dfrac{z_2}{z_1} = 4\cos^2\dfrac{\alpha}{2}z_1z_2$

or $p^2 = 4q\cos^2\dfrac{\alpha}{2}$

17. Given that z and w are two complex numbers.

To prove $|z|^2 w - |w|^2 z = z - w \Leftrightarrow z = w$ or $z\overline{w} = 1$

First let us consider

$|z|^2 w - |w|^2 z = z - w$ ....(1)

$\Rightarrow z(1 + |w|^2) = w(1 + |z|^2)$

$\Rightarrow \dfrac{z}{w} = \dfrac{1 + |z|^2}{1 + |w|^2}$ = a real number

$\Rightarrow \overline{\left(\dfrac{z}{w}\right)} = \dfrac{z}{w} \Rightarrow \dfrac{\bar{z}}{\bar{w}} = \dfrac{z}{w}$

$\Rightarrow \bar{z}w = z\bar{w}$ ....(2)

Again from equation (1),

$z\bar{z}w - w\bar{w}z = z - w$

$z(\bar{z}w - 1) - w(\bar{w}z - 1) = 0$

$z(z\bar{w} - 1) - w(z\bar{w} - 1) = 0$ (Using equation (2))

$\Rightarrow (z\bar{w} - 1)(z - w) = 0 \Rightarrow z\bar{w} = 1$ *or* $z = w$

Conversely if $z = w$ then

L.H.S. of $(1) = |w|^2 w - |w|^2 w = 0$

R.H.S. of $(1) = w - w = 0$

$\therefore$ (1) holds

Also if $z\bar{w} = 1$ then

L.H.S. of $(1) = z\bar{z}w - w\bar{w}z$

$= zz\bar{w} - w\bar{w}z = z - w =$ R.H.S. Hence proved.

18. The given equation can be written as

$(z^p - 1)(z^q - 1) = 0$

$\therefore\ z = (1)^{1/p}$ or $(1)^{1/q}$ ....(1)

where p and q are distinct prime numbers.

Hence both the equations will have distinct roots and as $z \neq 1$, both will not be simultaneously zero for any value of z given by equations in (1)

NOTE THIS STEP

Also $1 + \alpha + \alpha^2 + ... + \alpha^{p-1} = \dfrac{1 - \alpha^p}{1 - \alpha} = 0\ (\alpha \neq 1)$

or $1 + \alpha + \alpha^2 + ... + \alpha^p = \dfrac{1 - \alpha^q}{1 - \alpha} = 0\ (\alpha \neq 1)$

Because of (1) either $\alpha^p = 1$ and if $\alpha^q = 1$ but not both simultaneously as p and q are distinct primes.

19. Given that $|z_1| < 1 < |z_2|$

Then $\left|\dfrac{1 - z_1\bar{z}_2}{z_1 - z_2}\right| < 1$ is true

if $|1 - z_1\bar{z}_2| < |z_1 - z_2|$ is true

if $|1 - z_1\bar{z}_2|^2 < |z_1 - z_2|^2$ is true

if $(1 - z_1\bar{z}_2)\overline{(1 - z_1\bar{z}_2)} < (z_1 - z_2)\overline{(z_1 - z_2)}$ is true

if $(1 - z_1\bar{z}_2)(1 - \bar{z}_1z_2) < (z_1 - z_2)(\bar{z}_1 - \bar{z}_2)$

if $1 - z_1\bar{z}_2 - \bar{z}_1z_2 + z_1\bar{z}_1 z_2\bar{z}_2 < z_1\bar{z}_1 - z_1\bar{z}_2$ $-\bar{z}_1z_2 + z_2\bar{z}_2$ is true

if $1 + |z_1|^2|z_2|^2 < |z_1|^2 + |z_2|^2$ is true

if $(1 - |z_1|^2)(1 - |z_2|^2) < 0$ is true.

which is obviously true

as $|z_1|<1<|z_2|$

$\Rightarrow |z_1|^2<1<|z_2|^2$

$\Rightarrow |1-|z_1|^2>0$ and $(1-|z_2|^2)<0$ Hence proved.

20. Let us consider, $\sum_{r=1}^{n} a_r z^r = 1$ where $|a_r|<2$

$\Rightarrow a_1 z + a_2 z^2 + a_3 z^3 + ... + a_n z^n = 1$

$\Rightarrow |a_1 z + a_2 z^2 + a_3 z^3 + ... + a_n z^n| = 1$ ....(1)

But we know that $|z_1+z_2| \le |z_1| + |z_2|$

$\therefore$ Using its generalised form, we get

$|a_1 z + a_2 z^2 + a_3 z^3 + ... + a_n z^n|$

$\le |a_1 z| + |a_2 z^2| + ... + |a_n z^n|$

$\Rightarrow 1 \le |a_1||z| + |a_2||z^2| + |a_3||z^3| + ... + |a_n||z^n|$ (Using eq^n (1))

But given that $|a_r|<2 \forall r = 1(1)^n$

$\therefore 1 < 2[|z| + |z|^2 + |z|^3 + ... + |z|^n]$ [Using $|z^n| = |z|^n$]

$\Rightarrow 1 < 2\left[\frac{|z|(1-|z|^n)}{1-|z|}\right] \Rightarrow 2\left[\frac{|z|-|z|^{n+1}}{1-|z|}\right] > 1$

$\Rightarrow 2[|z|-|z|^{n+1}] > 1-|z|$ ($\because 1-|z|>0$ as $|z|<1/3$)

$\Rightarrow [|z|-|z|^{n+1}] > \frac{1}{2} - \frac{1}{2}|z| \Rightarrow \frac{3}{2}|z| > \frac{1}{2} + |z|^{n+1}$

$\Rightarrow |z| > \frac{1}{3} + \frac{2}{3}|z|^{n+1} \Rightarrow |z| > \frac{1}{3}$

which is a contradiction as given that $|z|<\frac{1}{3}$

$\therefore$ There exist no such complex number.

21. We are given that

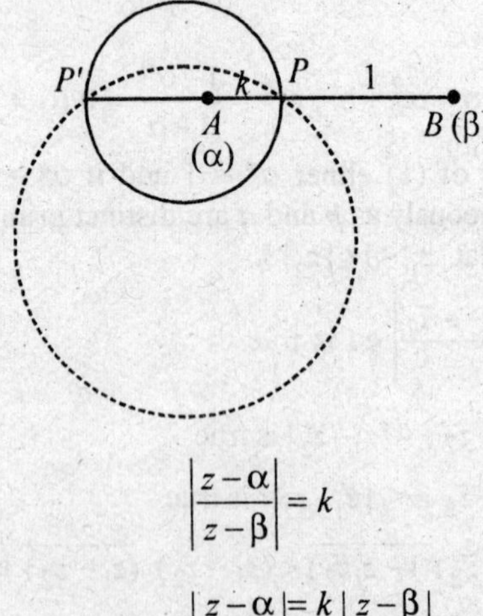

$$\left|\frac{z-\alpha}{z-\beta}\right| = k$$

$\Rightarrow |z-\alpha| = k|z-\beta|$

Let pt. A represents complex number α and B that of β, and P represents z. then $|z-\alpha| = k|z-\beta|$

$\Rightarrow$ z is the complex number whose distance from A is k times its distance from B.

i.e. $PA = k\ PB$

$\Rightarrow$ P divides AB in the ratio $k:1$ internally or externally (at P').

Then $P\left(\frac{k\beta+\alpha}{k+1}\right)$ and $P'\left(\frac{k\beta-\alpha}{k-1}\right)$

Now through PP' there can pass a number of circles, but with given data we can find radius and centre of that circle for which PP' is diameter.

And hence then centre = mid. point of PP'

$$= \left(\frac{\frac{k\beta+\alpha}{k+1} + \frac{k\beta-\alpha}{k-1}}{2}\right)$$

$$= \frac{k^2\beta + k\alpha - k\beta - \alpha + k^2\beta - k\alpha + k\beta - \alpha}{2(k^2-1)}$$

$$= \frac{k^2\beta - \alpha}{k^2-1} = \frac{\alpha - k^2\beta}{1-k^2}$$

Also radius

$$= \frac{1}{2}|PP'| = \frac{1}{2}\left|\frac{k\beta+\alpha}{k+1} - \frac{k\beta-\alpha}{k-1}\right|$$

$$= \frac{1}{2}\left|\frac{k^2\beta + k\alpha - k\beta - \alpha - k^2\beta + k\alpha - k\beta + \alpha}{k^2-1}\right|$$

$$= \frac{k|\alpha-\beta|}{|1-k^2|}$$

ALTERNATE SOLUTION:

The given circle is $\left|\frac{z-\alpha}{z-\beta}\right| = k,\ k \ne 1$ where

$z = x+iy,\ \alpha = \alpha_1 + i\alpha_2,\ \beta = \beta_1 + i\beta_2$

We know that complex equation of a circle in its standard form is

$z\bar{z} - a\bar{z} - \bar{a}z + a\bar{a} - r^2 = 0$ ....(1)

where a is the centre and r is the radius.

Let us reduce given equation of circle into its standard form

$$\left|\frac{z-\alpha}{z-\beta}\right| = k$$

$\Rightarrow |z-\alpha| = k|z-\beta|$

Squaring on both sides, we get

$|z-\alpha|^2 = k^2|z-\beta|^2$

$\Rightarrow (z-\alpha)\overline{(z-\alpha)} = k^2(z-\beta)\overline{(z-\beta)}$

$\Rightarrow (z-\alpha)(\bar{z}-\bar{\alpha}) = k^2(z-\beta)(\bar{z}-\bar{\beta})$

$\Rightarrow z\bar{z} - \alpha\bar{z} - \bar{\alpha}z + \alpha\bar{\alpha} = k^2(z\bar{z} - \beta\bar{z} - \bar{\beta}z + \beta\bar{\beta})$

$\Rightarrow (1-k^2)z\bar{z} - (\alpha - k^2\beta)\bar{z}$

$- (\bar{\alpha} - k^2\bar{\beta})z + \alpha\bar{\alpha} - k^2\beta\bar{\beta} = 0$

$$\Rightarrow z\bar{z}-\left(\frac{\alpha-k^2\beta}{1-k^2}\right)\bar{z}-\left(\frac{\bar{\alpha}-k^2\bar{\beta}}{1-k^2}\right)z+\left(\frac{\alpha\bar{\alpha}-k^2\beta\bar{\beta}}{1-k^2}\right)=0$$

$$\Rightarrow z\bar{z}-\left(\frac{\alpha-k^2\beta}{1-k^2}\right)\bar{z}-\overline{\left(\frac{\alpha-k^2\beta}{1-k^2}\right)}z+\left(\frac{\alpha\bar{\alpha}-k^2\beta\bar{\beta}}{1-k^2}\right)=0 \quad ....(2)$$

On comparing (1) and (2), we get

$$a=\frac{\alpha-k^2\beta}{1-k^2} \text{ and } a\bar{a}-r^2=\frac{\alpha\bar{\alpha}-k^2\beta\bar{\beta}}{1-k^2}$$

$\therefore$ Centre is $a=\dfrac{\alpha-k^2\beta}{1-k^2}$ and r is given by

$$\left(\frac{\alpha-k^2\beta}{1-k^2}\right)\overline{\left(\frac{\alpha-k^2\beta}{1-k^2}\right)}-r^2=\frac{\alpha\bar{\alpha}-k^2\beta\bar{\beta}}{1-k^2}$$

$$\Rightarrow r^2=\left(\frac{\alpha-k^2\beta}{1-k^2}\right)\left(\frac{\bar{\alpha}-k^2\bar{\beta}}{1-k^2}\right)-\frac{\alpha\bar{\alpha}-k^2\beta\bar{\beta}}{1-k^2}$$

$$=\frac{\alpha\bar{\alpha}-k^2\alpha\bar{\beta}-k^2\bar{\alpha}\beta+k^4\beta\bar{\beta}-\alpha\bar{\alpha}+k^2\beta\bar{\beta}+k^2\alpha\bar{\alpha}-k^4\beta\bar{\beta}}{(1-k^2)^2}$$

$$=\frac{k^2\alpha\bar{\alpha}+k^2\beta\bar{\beta}-k^2\alpha\bar{\beta}+k^2\bar{\alpha}\beta}{(1-k^2)^2}$$

$$=\frac{k^2[\alpha\bar{\alpha}+\beta\bar{\beta}-\alpha\bar{\beta}-\bar{\alpha}\beta]}{(1-k^2)^2}=\frac{k^2(\alpha-\beta)(\bar{\alpha}-\bar{\beta})}{(1-k^2)^2}$$

$$=\frac{k^2(\alpha-\beta)\overline{(\alpha-\beta)}}{(1-k^2)^2}=\frac{k^2|\alpha-\beta|^2}{(1-k^2)^2}$$

$\therefore$ Radius $r=\dfrac{k}{|1-k^2|}|\alpha-\beta|$

22. The given circle is $|z-1|=\sqrt{2}$ where z_0=1 is the centre and $\sqrt{2}$ is radius of circle. z_1 is one of the vertex of square inscribed in the given circle.

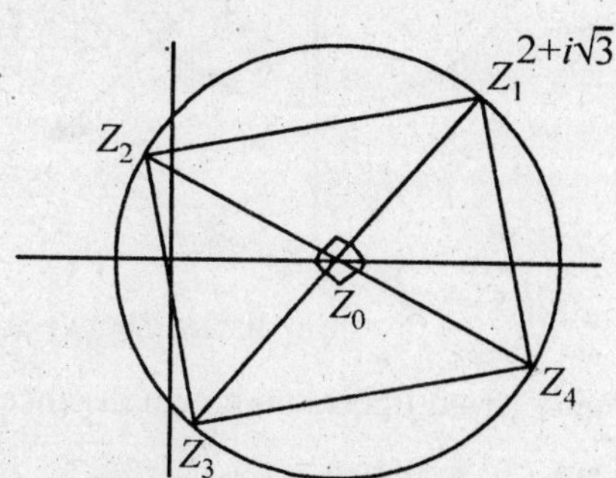

Clearly z_2 can be obtained by rotating z_1 by an $\angle 90°$ in anticlockwise sense, about centre z_0

Thus, $z_2-z_0=(z_1-z_0)e^{i\pi/2}$

or $\quad z_2-1=(2+i\sqrt{3}-1)i \Rightarrow z_2=i-\sqrt{3}+1$

$z_2=(1-\sqrt{3})+i$

Again rotating z_2 by 90° about z_0 we get

$z_3-z_0=(z_2-z_0)\,i$

$\Rightarrow z_3-1=[(1-\sqrt{3})+i-1]\,i=-\sqrt{3}i-1 \Rightarrow z_3=-i\sqrt{3}$

and similarly $1=(-i\sqrt{3}-1)\,i=\sqrt{3}-i$

$\Rightarrow z_4=(\sqrt{3}+1)-i$

Thus the remaining vertices are

$(1-\sqrt{3})+i, -i\sqrt{3}, (\sqrt{3}+1)-i$

ALTERNATE SOLUTION :

Obtaining $z_2=(1-\sqrt{3})+i$ as in previous method, we observe that z_0 is mid point of z_1 and z_3; and z_2 and z_4.

$\therefore \dfrac{z_1+z_3}{2}=z_0 \Rightarrow \dfrac{2+i\sqrt{3}+z_3}{2}=1$

$\Rightarrow z_3=-i\sqrt{3}$

and $\dfrac{z_2+z_4}{2}=z_0 \Rightarrow \dfrac{(1-\sqrt{3})i+z_4}{2}=1$

$\Rightarrow z_4=(\sqrt{3}+1)-i.$

F MATCH THE FOLLOWING :

1. $z\neq 0$ Let $z=a+ib$
Re $(z)=0 \Rightarrow z=ib \Rightarrow z^2=-b^2$
$\therefore Im(z)^2=0$
$\therefore$ (A) corresponds to (q)

Arg $z=\dfrac{\pi}{4} \Rightarrow a=b \Rightarrow z=a+ia$
$z^2=a^2-a^2+2ia^2$
$z^2=2ia^2 \Rightarrow$ Re $(z)^2=0$
$\therefore$ (B) corresponds to (p).

2. (A) $\rightarrow$ (q, r) $\big|z-i|z|\big|=\big|z+i|z|\big|$

$\Rightarrow z$ is equidistant from two points $(0,|z|)$ and $(0,-|z|)$ which lie on imaginary axis.

$\therefore z$ must lie on real axis $\Rightarrow$ Im$(z)=0$ also $|I_m(z)|\leq 1$

(B) $\rightarrow$ p

Sum of distances of z from two fined points (–4, 0) and (4, 0) is 10 which is greater than 8.

$\therefore z$ traces an ellipse with $2a=10$ and $2ae=8$

$\Rightarrow e=4/5$

(C) $\rightarrow$ (p, s, t)

Let $\omega=2(\cos\theta+i\sin\theta)$

then $z=\omega-\dfrac{1}{\omega}=2(\cos\theta+i\sin\theta)-\dfrac{1}{2}(\cos\theta-i\sin\theta)$

$\Rightarrow x+iy=\dfrac{3}{2}\cos\theta+i\dfrac{5}{2}\sin\theta$

Here $|z| = \sqrt{\frac{9+25}{4}} = \sqrt{\frac{34}{4}} \le 3$ and $|R_e(z)| \le 2$

Also $x = \frac{3}{2}\cos\theta, y = \frac{5}{2}\sin\theta \Rightarrow \frac{4x^2}{9} + \frac{4y^2}{25} = 1$

Which is an ellipse with $e = \sqrt{1 - \frac{9}{25}} = \frac{4}{5}$

(D) → (q, r, s, t)

Let $\omega = \cos\theta + i\sin\theta$ then $z = 2\cos\theta \Rightarrow \text{Im } z = 0$

Also $|z| \le 3$ and $|\text{Im}(z)| \le 1, |R_e(z)| \le 2$

G COMPREHENSION BASED QUESTIONS :

1-3. We have $A = \{z : \text{Im}(z) \ge 1\} = \{(x, y) : y \ge 1\}$

Clearly A is the set of all points lying on or above the line $y = 1$ in cartesian plane.

$B = \{z : |z - 2 - i| = 3\} = \{(x, y) : (x-2)^2 + (y-1)^2 = 9\}$

$\Rightarrow B$ is the set of all points lying on the boundary of the circle with centre (2, 1) and radius 3.

$C = \{z : \text{Re}[(1-i)z] = \sqrt{2}\} = \{(x, y) : x + y = \sqrt{2}\}$

$\Rightarrow$ C is the set of all points lying on the straight line represented by $x + y = \sqrt{2}$.

Graphically, the three sets are represented as shown below :

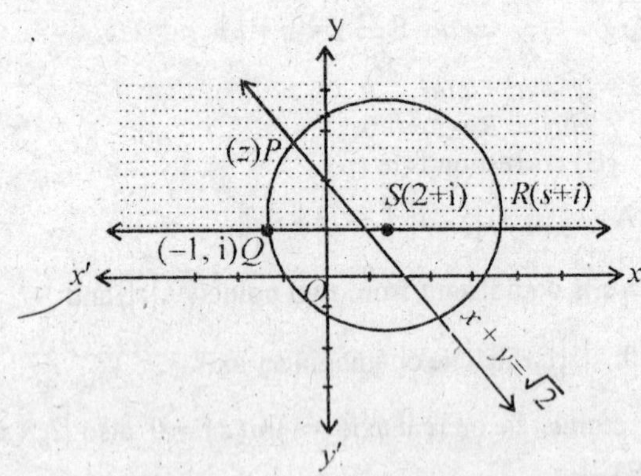

1. (b) From graph $A \cap B \cap C$ consists of only one point P [the common point of the region $y \ge 1, (x-2)^2 + (y-1)^2 = 9$ and $x + y = \sqrt{2}$]

$\therefore n(A \cap B \cap C) = 1$

2. (c) As z is a point of $A \cap B \cap C$

$\Rightarrow z$ represents the point P

$\therefore |z + 1 - i|^2 + |z - 5 - i|^2$

$\Rightarrow |z - (-1 + i)|^2 + |z - (5 + i)|^2$

$\Rightarrow PQ^2 + PR^2 = QR^2 = 6^2 = 36$

which lies between 35 and 39

$\therefore$ (c) is correct option.

3. (d) Given that $|w - 2 - i| < 3$

$\Rightarrow$ Distance between w and $2 + i$ i.e. S is smaller than 3.

$\Rightarrow w$ is a point lying inside the circle with centre S and radius 3.

$\Rightarrow$ Distance between z (i.e. the point P) and w should be smaller than 6 (the diameter of the circle)

i.e. $|z - w| < 6$

But we know that $||z| - |w|| < |z - w|$

$\Rightarrow ||z| - |w|| < 6 \Rightarrow -6 < |z| - |w| < 6$

$-3 < |z| - |w| + 3 < 9$

4. (b) Area of shaded region

$= \frac{\pi}{4} \times 4^2 + \frac{\pi \times 4^2 \times 60°}{360°}$

$= 4\pi + \frac{8\pi}{3} = \frac{20\pi}{3}$

5. (c) $S_1 : x^2 + y^2 < 16$

$S_2 : \text{Im}\left[\frac{(x-1) + i(y+\sqrt{3})}{1 - i\sqrt{3}}\right] > 0$

$\Rightarrow \sqrt{3}(x-1) + (y + \sqrt{3}) > 0 \Rightarrow y + \sqrt{3}x > 0$

$S_3 : x > 0$

Then $S : S_1 \cap S_2 \cap S_3$ is as shown in the figure given below.

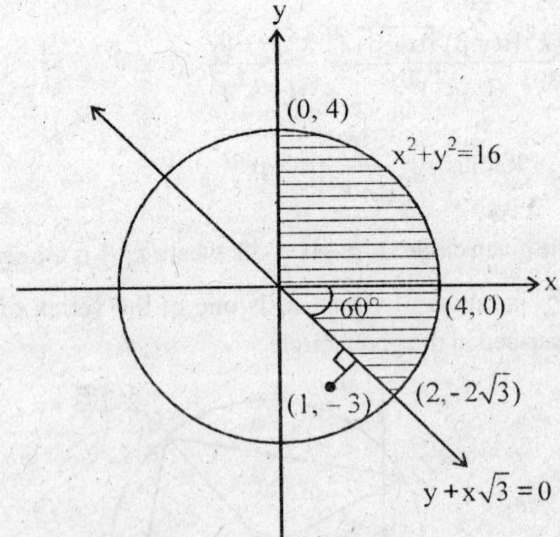

$\therefore \min_{z \in S} |1 - 3i - z| =$ min distance between z and (1, –3)

Clearly (from figure) minimum distance between $z \in S$ and (1, –3) from line $y + x\sqrt{3} = 0$

i.e. $\left|\frac{\sqrt{3} - 3}{\sqrt{3} + 1}\right| = \frac{3 - \sqrt{3}}{2}$

I INTEGER VALUE CORRECT TYPE :

1. **5**

Given $|z-3-2i| \le 2$

which represents a circular region with centre (3, 2) and radius 2.

Now $|2z-6+5i| = 2\left|z-\left(3-\frac{5}{2}i\right)\right|$

$= 2 \times$ distance of z from P (where Z lies in or on the circle)

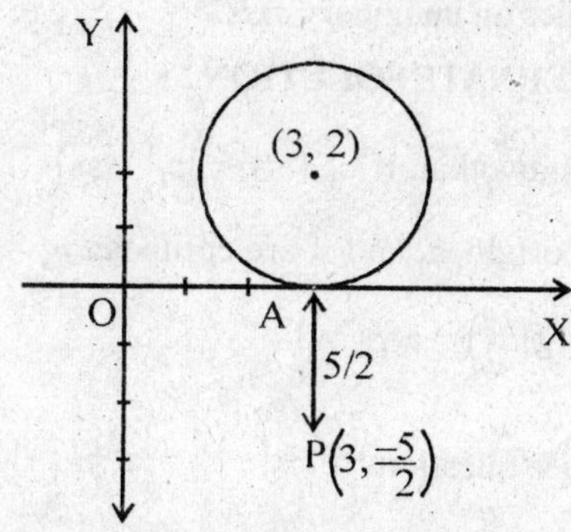

Also min distance of z from $P = \frac{5}{2}$

$\therefore$ Minimum value of $|2z-6+5i| = 5$

Alternative

$\left|z-3+\frac{5}{2}i\right| = \left|z-3-2i+2i+\frac{5}{2}i\right|$

$= \left|(z-3-2i)+\frac{9}{2}i\right|$

$\ge \left||z-3-2i|-\frac{9}{2}\right|$ (using $|z_1+z_2| \ge ||z_1|-|z_2||$)

$\ge \left|2-\frac{9}{2}\right| = \frac{5}{2}$ $\therefore \left|z-3+\frac{5}{2}i\right| \ge \frac{5}{2}$

or $|2z-6+5i| \ge 5$

2. **3**

The expression may not attain integral value for all a, b, c. If we consider a = b = c then

$x = 3a, y = a(1+\omega+\omega^2) = a(1+i\sqrt{3})$

$Z = a(1+\omega^2+\omega) = a(1+i\sqrt{3})$

$\Rightarrow |x|^2+|y|^2+|z|^2 = 9|a|^2+4|a|^2+4|a|^2 = 17|a|^2$

$\Rightarrow \frac{|x|^2+|y|^2+|z|^2}{|a|^2+|b|^2+|c|^2} = \frac{17}{3}$ (which is not an integer)

Note : However if $\omega = e^{i\left(2\frac{\pi}{3}\right)}$, then the value of expression can be evaluated as follows

$\frac{|x|^2+|y|^2+|z|^2}{|a|^2+|b|^2+|c|^2} = \frac{x\bar{x}+y\bar{y}+z\bar{z}}{|a|^2+|b|^2+|c|^2}$

$$= \frac{(a+b+c)(\bar{a}+\bar{b}+\bar{c})+(a+b\omega+c\omega^2)(\bar{a}+\bar{b}\omega^2+\bar{c}\omega)+(a+b\omega^2+c\omega)(\bar{a}+\bar{b}\omega+\bar{c}\omega^2)}{|a|^2+|b|^2+|c|^2}$$

$$= \frac{3|a|^2+3|b|^2+3|c|^2+(a\bar{b}+\bar{a}b+b\bar{c}+\bar{b}c+a\bar{c}+\bar{a}c)(1+\omega+\omega^2)}{|a|^2+|b|^2+|c|^2}$$

$= 3$ $(\because 1+\omega+\omega^2 = 0)$

Section-B JEE Main/ AIEEE

1. **(b)** Let $|z| = |\omega| = r$

$\therefore z = re^{i\theta}$, $\omega = re^{i\phi}$ where $\theta + \phi = \pi$.

$\therefore z = re^{i(\pi-\phi)} = re^{i\pi}.e^{-i\phi} = -re^{-i\phi} = -\bar{\omega}.[\because \bar{\omega} = re^{-i\phi}]$

2. **(c)** Given $|z-4| < |z-2|$ Let $z = x+iy$

$\Rightarrow |(x-4)+iy)| < |(x-2)+iy|$

$\Rightarrow (x-4)^2+y^2 < (x-2)^2+y^2$

$\Rightarrow x^2-8x+16 < x^2-4x+4 \Rightarrow 12 < 4x$

$\Rightarrow x > 3 \Rightarrow \text{Re}(z) > 3$

3. **(b)** Let the circle be $|z-z_0| = r$. Then according to given conditions $|z_0-z_1| = r+a$ and $|z_0-z_2| = r+b$. Eliminating r, we get $|z_0-z_1|-|z_0-z_2| = a-b$.

$\therefore$ Locus of centre z_0 is $|z-z_1|-|z-z_2| = a-b$, which represents a hyperbola.

4. **(a)** $|\bar{z}\omega| = |\bar{z}||\omega| = |z||\omega| = |z\omega| = 1$

$\text{Arg}(\bar{z}\omega) = \arg(\bar{z})+\arg(\omega) = -\arg(z)+\arg\omega$

$= -\frac{\pi}{2}$ $\therefore \bar{z}\omega = -1$

ALTERNATE SOLUTION :

Let $z = r_1e^{i\theta}$ and $w = r_2e^{i\phi}$ $\therefore \bar{z} = r_1e^{-i\theta}$

Now $|z\omega| = 1 \Rightarrow \left|r_1r_2e^{i(\theta+\phi)}\right| = 1 \Rightarrow r_1r_2 = 1$

Also arg (z) – arg $(\omega) = \frac{\pi}{2} \Rightarrow \theta-\phi = \frac{\pi}{2}$

Now $\bar{z}\omega = r_1e^{-i\theta}.r_2e^{i\phi} = r_1r_2e^{-i(\theta-\phi)} = e^{-\frac{i\pi}{2}} = -1$

5. (d) $z^2 + az + b = 0$; $z_1 + z_2 = -a \,\&\, z_1 z_2 = b$

$0, z_1, z_2$ form an equilateral Δ

$\therefore 0^2 + z_1^2 + z_2^2 = 0.z_1 + z_1.z_2 + z_2.0$

(for an equilateral triangle,

$z_1^2 + z_2^2 + z_3^2 = z_1z_2 + z_2z_3 + z_3z_1$)

$\Rightarrow z_1^2 + z_2^2 = z_1z_2 \Rightarrow (z_1+z_2)^2 = 3z_1z_2$

$\therefore a^2 = 3b$

6. (b) $\left(\frac{1+i}{1-i}\right)^x = 1 \Rightarrow \left[\frac{(1+i)^2}{1-i^2}\right]^x = 1$

$\left(\frac{1+i^2+2i}{1+1}\right)^x = 1 \Rightarrow (i)^x = 1;\ \therefore x = 4n;\ \ n \in I^+$

7. (c) $\arg zw = \pi \Rightarrow \arg z + \arg w = \pi$...(1)

$\bar{z} + i\bar{w} = 0 \Rightarrow \bar{z} = -i\bar{w}$

$\therefore z = iw \Rightarrow \arg z = \frac{\pi}{2} + \arg w$

$\Rightarrow \arg z = \frac{\pi}{2} + \pi - \arg z$ (from(1)) $\therefore \arg z = \frac{3\pi}{4}$

8. (a) $z^{\frac{1}{3}} = p + iq \Rightarrow z = p^3 + (iq)^3 + 3p(iq)(p+iq)$

$\Rightarrow x - iy = p^3 - 3pq^2 + i(3p^2q - q^3)$

$\therefore x = p^3 - 3pq^2 \Rightarrow \frac{x}{p} = p^2 - 3q^2$

$y = q^3 - 3p^2q \Rightarrow \frac{y}{q} = q^2 - 3p^2$

$\therefore \frac{x}{p} + \frac{y}{q} = -2p^2 - 2q^2 \ \therefore \left(\frac{x}{p}+\frac{y}{q}\right)\Big/(p^2+q^2) = -2$

9. (b) $|z^2 - 1| = |z|^2 + 1 \Rightarrow |z^2-1|^2 = (z\bar{z}+1)^2$

$\Rightarrow (z^2-1)(\bar{z}^2-1) = (z\bar{z}+1)^2$

$\Rightarrow z^2\bar{z}^2 - z^2 - \bar{z}^2 + 1 = z^2\bar{z}^2 + 2z\bar{z} + 1$

$\Rightarrow z^2 + 2z\bar{z} + \bar{z}^2 = 0 \Rightarrow (z+\bar{z})^2 = 0 \Rightarrow z = -\bar{z}$

$\Rightarrow z$ is purely imaginary

ALTERNATE SOLUTION 1 :

Let $z = r(\cos\theta + i\sin\theta)$

Then $|z^2 - 1| = |r^2(\cos 2\theta + i\sin 2\theta) - 1|$

$= \sqrt{r^4 - 2r^2\cos 2\theta + 1}$ and $|z^2-1|^2 = (|z|^2+1)^2$

$\Rightarrow r^4 - 2r^2\cos 2\theta + 1 = r^4 + 2r^2 + 1$

$\Rightarrow 2\cos^2\theta = 0 \Rightarrow \cos\theta = \pm\frac{\pi}{2}$

$\therefore z$ lies on imaginary axis.

ALTERNATE SOLUTION 2 :

We know that, if $|z_1 + z_2| = |z_1| + |z_2|$

then origin, z_1 and z_2 are collinear

$\Rightarrow \arg(z_1) = \arg(z_2)$

As per question $\left|z^2 + (-1)\right| = \left|z^2\right| + |-1|$

$\Rightarrow \arg\left(z^2\right) = \arg(-1) \Rightarrow 2\arg(z) = \pi \Rightarrow \arg(z) = \frac{\pi}{2}$

$\Rightarrow z$ lies on imaginary axis.

10. (c) $(x-1)^3 + 8 = 0 \Rightarrow (x-1) = (-2)\,(1)^{1/3}$

$\Rightarrow x - 1 = -2$ or -2ω or $-2\omega^2$

or $x = -1$ or $1 - 2\omega$ or $1 - 2\omega^2$.

11. (c) $|z_1 + z_2| = |z_1| + |z_2| \Rightarrow z_1$ and z_2 are collinear and are to the same side of origin; hence $\arg z_1 - \arg z_2 = 0$.

12. (c) As given $w = \frac{z}{z - \frac{1}{3}i} \Rightarrow |w| = \frac{|z|}{|z - \frac{1}{3}i|} = 1$

$\Rightarrow |z| = \left|z - \frac{1}{3}i\right|$

$\Rightarrow$ distance of z from origin and point $\left(0, \frac{1}{3}\right)$ is same hence z lies on bisector of the line joining points (0, 0) and (0, 1/3).

Hence z lies on a straight line.

13. (d) $\sum_{k=1}^{10}\left(\sin\frac{2k\pi}{11}+i\cos\frac{2k\pi}{11}\right)$

$$=i\sum_{k=1}^{10}\left(\cos\frac{2k\pi}{11}-i\sin\frac{2k\pi}{11}\right)$$

$$=i\sum_{k=1}^{10}e^{-\frac{2k\pi}{11}i}=i\left\{\sum_{k=0}^{10}e^{-\frac{2k\pi}{11}i}-1\right\}$$

$$=i\left[1+e^{-\frac{2\pi}{11}i}+e^{-\frac{4\pi}{11}i}+\text{....11 terms}\right]-i$$

$$=i\left[\frac{1-\left(e^{-\frac{2\pi}{11}i}\right)^{11}}{1-e^{-\frac{2\pi}{11}i}}\right]-i=i\left[\frac{1-e^{-2\pi i}}{1-e^{-\frac{2\pi}{11}i}}\right]-i$$

$$=i\times 0-i \qquad [\because e^{-2\pi i}=1]$$

$$=-i$$

14. (d) $z^2+z+1=0 \Rightarrow z=\omega$ or ω^2

So, $z+\frac{1}{z}=\omega+\omega^2=-1$

$z^2+\frac{1}{z^2}=\omega^2+\omega=-1,\ z^3+\frac{1}{z^3}=\omega^3+\omega^3=2$

$z^4+\frac{1}{z^4}=-1,\ z^5+\frac{1}{z^5}=-1$ and $z^6+\frac{1}{z^6}=2$

$\therefore$ The given sum $=1+1+4+1+1+4$

$=12$

15. (a) z lies on or inside the circle with centre (–4, 0) and radius 3 units.

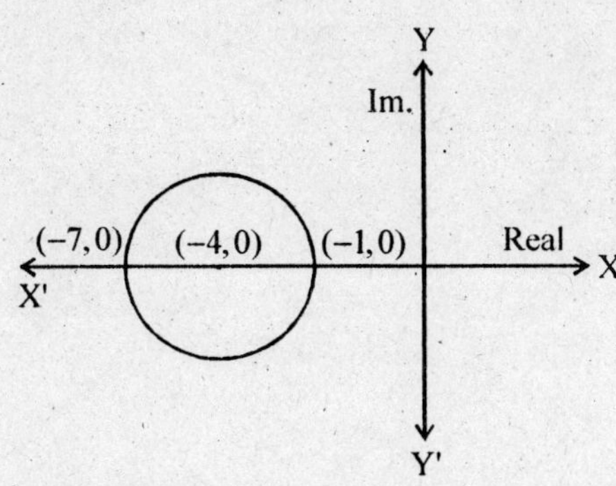

From the Argand diagram maximum value of $|z+1|$ is 6

ALTERNATE SOLUTION :

$|z+1|=|z+4-3|\le|z+4|+|-3|\le|3|+|-3|$

$\Rightarrow |z+1|\le 6 \Rightarrow |z+1|_{max}=6$

16. (c) $\left(\frac{1}{i-1}\right)=\frac{1}{-i-1}=\frac{-1}{i+1}$

17. (d) Given $S=\{(x,y): y=x+1 \text{ and } 0<x<2\}$

$\because x\ne x+1$ for any $x\in(0,2)\Rightarrow(x,x)\notin S$

$\therefore$ S is not reflexive.

Hence S in not an equivalence relation.

Also $T=\{x,y): x-y$ is an integer$\}$

$\because x-x=0$ is an integer $\forall x\in R$

$\therefore$ T is reflexive.

If $x-y$ is an integer then $y-x$ is also an integer

$\therefore$ T is symmetric

If $x-y$ is an integer and $y-z$ is an integer then $(x-y)+(y-z)=x-z$ is also an integer.

$\therefore$ T is transitive

Hence T is an equivalence relation.

18. (a) Let $z=x+iy$

$|z-1|=|z+1|(x-1)^2+y^2=(x+1)^2+y^2$

$\Rightarrow \text{Re } z=0 \quad\Rightarrow x=0$

$|z-1|=|z-i|(x-1)^2+y^2=x^2+(y-1)^2$

$\Rightarrow x=y$

$|z+1|=|z-i|(x+1)^2+y^2=x^2+(y-1)^2$

Only (0, 0) will satisfy all conditions.

$\Rightarrow$ Number of complex number $z=1$

19. (c) $\because$ Real part of roots is 1

Let roots are $1+pi,\ 1+q$

$\therefore$ sum of roots.

$=1+pi+1+qi=-\alpha$ which is real

$\Rightarrow q=-p$ or root are

$1+pi$ and $1-pi$ product of roots

$=1+p^2=\beta\in(1,\infty)$

$p\ne 0$ as roots are distinct.

20. (a) $(1+\omega)^7=A+B\omega$

$(-\omega^2)^7=A+B\omega$

$-\omega^2=A+B\omega$

$1+\omega=A+B\omega$

$\Rightarrow A=1, B=1.$

21. (d) $H^2 = \begin{bmatrix} \omega & 0 \\ 0 & \omega \end{bmatrix}\begin{bmatrix} \omega & 0 \\ 0 & \omega \end{bmatrix} = \begin{bmatrix} \omega^2 & 0 \\ 0 & \omega^2 \end{bmatrix}$

If $H^k = \begin{bmatrix} \omega^k & 0 \\ 0 & \omega \end{bmatrix}$ then $H^{k+1} = \begin{bmatrix} \omega^{k+1} & 0 \\ 0 & \omega^{k+1} \end{bmatrix}$

So by principal of mathematical induction,

$$H^{70} = \begin{bmatrix} \omega^{70} & 0 \\ 0 & \omega^{70} \end{bmatrix} = \begin{bmatrix} \omega^{69}\omega & 0 \\ 0 & \omega^{69}\omega \end{bmatrix} = \begin{bmatrix} \omega & 0 \\ 0 & \omega \end{bmatrix} = H$$

22. (a) Let $z = x + iy$

$\therefore\ z^2 = x^2 - y^2 + 2ixy$

Now $\frac{z^2}{z-1}$ is real $\Rightarrow \text{Im}\left(\frac{z^2}{z-1}\right) = 0$

$\Rightarrow \text{Im}\left(\frac{x^2 - y^2 + 2ixy}{(x-1)+iy}\right) = 0$

$\Rightarrow \text{Im}\ [(x^2 - y^2 + 2ixy)\ (x-1) - iy)] = 0$

$\Rightarrow 2xy\ (x-1) - y\ (x^2 - y^2) = 0$

$\Rightarrow y(x^2 + y^2 - 2x) = 0$

$\Rightarrow y = 0;\ x^2 + y^2 - 2x = 0$

$\therefore$ z lies either on real axis or on a circle through origin.

Alternative

Since we know $z = \bar{z}$ iff z is real.

Therefore, $\frac{z^2}{z-1} = \frac{\bar{z}^2}{\bar{z}-1}$

$\Rightarrow\ z\bar{z}z - z^2 = z.\bar{z}.\bar{z} - \bar{z}^2 \Rightarrow |z|^2.z - z^2 = |z|^2.\bar{z} - \bar{z}^2$

$\Rightarrow\ |z|^2(z-\bar{z}) - (z-\bar{z})(z+\bar{z}) = 0$

$\Rightarrow\ (z-\bar{z})\left(|z|^2 - (z+\bar{z})\right) = 0$

Either $z - \bar{z} = 0$ or $|z|^2 - (z+\bar{z}) = 0$

Either $z = \bar{z} \Rightarrow$ real axis

or $|z|^2 = z + \bar{z} \Rightarrow z\bar{z} - z - \bar{z} = 0$

represents a circle passing through origin.

23. (c) Given $|z| = 1$, $\arg z = \theta$

As we know, $\bar{z} = \frac{1}{z}$

$\therefore\ \arg\left(\frac{1+z}{1+\bar{z}}\right) = \arg\left(\frac{1+z}{1+\frac{1}{z}}\right) = \arg(z) = \theta.$

Chapter 3 Quadratic Equation and Inequations (Inequalities)

Section-A JEE Advanced/ IIT-JEE

A FILL IN THE BLANKS :

1. The coefficient of x^{99} in the polynomial $(x-1)(x-2)....(x-100)$ is .... *(1982 - 2 Marks)*

2. If $2+i\sqrt{3}$ is a root of the equation $x^2+px+q=0$, where p and q are real, then $(p, q) = ($.................. , ...$)$. *(1982 - 2 Marks)*

3. If the product of the roots of the equation $x^2 - 3kx + 2e^{2\ln k} - 1 = 0$ is 7, then the roots are real for $k =$ .... *(1984 - 2 Marks)*

4. If the quadratic equations $x^2+ax+b=0$ and $x^2+bx+a=0$ $(a \neq b)$ have a common root, then the numerical value of $a+b$ is .... *(1986 - 2 Marks)*

5. The solution of equation $\log_7 \log_5(\sqrt{x+5}+\sqrt{x})=0$ is .... *(1986 - 2 Marks)*

6. If $x<0, y<0, x+y+\dfrac{x}{y}=\dfrac{1}{2}$ and $(x+y)\dfrac{x}{y}=-\dfrac{1}{2}$, then $x=$ .... and $y=$ ...... *(1990 - 2 Marks)*

7. Let n and k be positive such that $n \geq \dfrac{k(k+1)}{2}$. The number of solutions $(x_1, x_2, ..., x_k)$, $x_1 \geq 1$, $x_2 \geq 2$, ..., $x_k \geq k$, all integers, satisfying $x_1+x_2+....+x_k=n$, is .... *(1996 - 2 Marks)*

8. The sum of all the real roots of the equation $|x-2|^2+|x-2|-2=0$ is .... *(1997 - 2 Marks)*

B TRUE / FALSE :

1. For every integer $n>1$, the inequality $(n!)^{1/n} < \dfrac{n+1}{2}$ holds. *(1981 - 2 Marks)*

2. The equation $2x^2+3x+1=0$ has an irrational root. *(1983 - 1 Mark)*

3. If $a<b<c<d$, then the roots of the equation $(x-a)(x-c)+2(x-b)(x-d)=0$ are real and distinct. *(1984 - 1 Mark)*

4. If $n_1, n_2, ...n_p$ are p positive integers, whose sum is an even number, then the number of odd integers among them is odd. *(1985 - 1 Mark)*

5. If $P(x)=ax^2+bx+c$ and $Q(x)=-ax^2+dx+c$, where $ac \neq 0$, then $P(x)Q(x)=0$ has at least two real roots. *(1985 - 1 Mark)*

6. If x and y are positive real numbers and m, n are any positive integers, then $\dfrac{x^n y^m}{(1+x^{2n})(1+y^{2m})} > \dfrac{1}{4}$ *(1989 - 1 Mark)*

C MCQ's WITH ONE CORRECT ANSWER :

1. If ℓ, m, n are real, $\ell \neq m$, then the roots by the equation: $(\ell-m)x^2-5(\ell+m)x-2(\ell-m)=0$ are *(1979)*
   (a) Real and equal (b) Complex
   (c) Real and unequal (d) None of these.

2. The equation $x+2y+2z=1$ and $2x+4y+4z=9$ have *(1979)*
   (a) Only one solution
   (b) Only two solutions
   (c) Infinite number of solutions
   (d) None of these.

3. If x, y and z are real and different and $u=x^2+4y^2+9z^2-6yz-3zx-2xy$, then u is always. *(1979)*
   (a) non negative (b) zero
   (c) non positive (d) none of these

4. Let $a>0$, $b>0$ and $c>0$. Then the roots of the equation $ax^2+bx+c=0$ *(1979)*
   (a) are real and negative (b) have negative real parts
   (c) both (a) and (b) (d) none of these

5. Both the roots of the equation $(x-b)(x-c)+(x-a)(x-c)+(x-a)(x-b)=0$ are always *(1980)*
   (a) positive (b) real
   (c) negative (d) none of these.

6. The least value of the expression $2\log_{10}x - \log_x(0.01)$, for $x>1$, is *(1980)*
   (a) 10 (b) 2
   (c) -0.01 (d) none of these.

7. If (x^2+px+1) is a factor of (ax^3+bx+c), then *(1980)*
   (a) $a^2+c^2=-ab$ (b) $a^2-c^2=-ab$
   (c) $a^2-c^2=ab$ (d) none of these

8. The number of real solutions of the equation $|x|^2-3|x|+2=0$ is *(1982 - 2 Marks)*
   (a) 4 (b) 1
   (c) 3 (d) 2

9. Two towns A and B are 60 km apart. A school is to be built to serve 150 students in town A and 50 students in town B. If the total distance to be travelled by all 200 students is to be as small as possible, then the school should be built at *(1982 - 2 Marks)*
(a) town B (b) 45 km from town A
(c) town A (d) 45 km from town B

10. If p, q, r are any real numbers, then *(1982 - 2 Marks)*
(a) $\max(p, q) < \max(p, q, r)$
(b) $\min(p, q) = \frac{1}{2}(p+q-|p-q|)$
(c) $\max(p, q) < \min(p, q, r)$
(d) none of these

11. The largest interval for which $x^{12} - x^9 + x^4 - x + 1 > 0$ is *(1982 - 2 Marks)*
(a) $-4 < x \le 0$ (b) $0 < x < 1$
(c) $-100 < x < 100$ (d) $-\infty < x < \infty$

12. The equation $x - \frac{2}{x-1} = 1 - \frac{2}{x-1}$ has *(1984 - 2 Marks)*
(a) no root (b) one root
(c) two equal roots (d) infinitely many roots

13. If $a^2 + b^2 + c^2 = 1$, then $ab + bc + ca$ lies in the interval *(1984 - 2 Marks)*
(a) $[\frac{1}{2}, 2]$ (b) $[-1, 2]$
(c) $[-\frac{1}{2}, 1]$ (d) $[-1, \frac{1}{2}]$

14. If $\log_{0.3}(x-1) < \log_{0.09}(x-1)$, then x lies in the interval – *(1985 - 2 Marks)*
(a) $(2, \infty)$ (b) $(1, 2)$
(c) $(-2, -1)$ (d) none of these

15*. If α and β are the roots of $x^2 + px + q = 0$ and α^4, β^4 are the roots of $x^2 - rx + s = 0$, then the equation $x^2 - 4qx + 2q^2 - r = 0$ has always *(1989 - 2 Marks)*
(a) two real roots
(b) two positive roots
(c) two negative roots
(d) one positive and one negative root

* Question has more than one correct option.

16. Let a, b, c be real numbers, $a \neq 0$. If α is a root of $a^2x^2 + bx + c = 0$. β is the root of $a^2x^2 - bx - c = 0$ and $0 < \alpha < \beta$, then the equation $a^2x^2 + 2bx + 2c = 0$ has a root γ that always satisfies *(1989 - 2 Marks)*
(a) $\gamma = \frac{\alpha+\beta}{2}$ (b) $\gamma = \alpha + \frac{\beta}{2}$
(c) $\gamma = \alpha$ (d) $\alpha < \gamma < \beta$.

17. The number of solutions of the equation $\sin(e)^x = 5^x + 5^{-x}$ is *(1990 - 2 Marks)*
(a) 0 (b) 1
(c) 2 (d) Infinitely many

18. Let α, β be the roots of the equation $(x-a)(x-b) = c, c \neq 0$. Then the roots of the equation $(x-\alpha)(x-\beta) + c = 0$ are *(1992 - 2 Marks)*
(a) a, c (b) b, c
(c) a, b (d) $a + c, b + c$

19. The number of points of intersection of two curves $y = 2\sin x$ and $y = 5x^2 + 2x + 3$ is *(1994)*
(a) 0 (b) 1
(c) 2 (d) ∞

20. If p, q, r are +ve and are in $A.P.$, the roots of quadratic equation $px^2 + qx + r = 0$ are all real for *(1994)*
(a) $\left|\frac{r}{p} - 7\right| \ge 4\sqrt{3}$ (b) $\left|\frac{p}{r} - 7\right| \ge 4\sqrt{3}$
(c) all p and r (d) no p and r

21. Let $p, q \in \{1, 2, 3, 4\}$. The number of equations of the form $px^2 + qx + 1 = 0$ having real roots is *(1994)*
(a) 15 (b) 9
(c) 7 (d) 8

22. The equation $\sqrt{x+1} - \sqrt{x-1} = \sqrt{4x-1}$ has *(1997C - 2 Marks)*
(a) no solution (b) one solution
(c) two solutions (d) more than two solutions

23. If the roots of the equation $x^2 - 2ax + a^2 + a - 3 = 0$ are real and less than 3, then *(1999 - 2 Marks)*
(a) $a < 2$ (b) $2 \le a \le 3$
(c) $3 < a \le 4$ (d) $a > 4$

24. If α and β $(\alpha < \beta)$ are the roots of the equation $x^2 + bx + c = 0$, where $c < 0 < b$, then *(2000S)*
(a) $0 < \alpha < \beta$ (b) $\alpha < 0 < \beta < |\alpha|$
(c) $\alpha < \beta < 0$ (d) $\alpha < 0 < |\alpha| < \beta$

25. If a, b, c, d are positive real numbers such that $a + b + c + d = 2$, then $M = (a+b)(c+d)$ satisfies the relation *(2000S)*
(a) $0 \le M \le 1$ (b) $1 \le M \le 2$
(c) $2 \le M \le 3$ (d) $3 \le M \le 4$

26. If $b > a$, then the equation $(x-a)(x-b) - 1 = 0$ has *(2000S)*
(a) both roots in (a, b)
(b) both roots in $(-\infty, a)$
(c) both roots in $(b, +\infty)$
(d) one root in $(-\infty, a)$ and the other in $(b, +\infty)$

27. For the equation $3x^2 + px + 3 = 0, p > 0$, if one of the root is square of the other, then p is equal to *(2000S)*
(a) 1/3 (b) 1
(c) 3 (d) 2/3

28. If $a_1, a_2, \ldots, a_n$ are positive real numbers whose product is a fixed number c, then the minimum value of $a_1 + a_2 + \ldots + a_{n-1} + 2a_n$ is *(2002S)*
(a) $n(2c)^{1/n}$ (b) $(n+1)c^{1/n}$
(c) $2nc^{1/n}$ (d) $(n+1)(2c)^{1/n}$

29. The set of all real numbers x for which $x^2 - |x+2| + x > 0$, is **(2002S)**

(a) $(-\infty, -2) \cup (2, \infty)$ (b) $(-\infty, -\sqrt{2}) \cup (\sqrt{2}, \infty)$

(c) $(-\infty, -1) \cup (1, \infty)$ (d) $(\sqrt{2}, \infty)$

30. If $\alpha \in \left(0, \frac{\pi}{2}\right)$ then $\sqrt{x^2+x} + \frac{\tan^2 \alpha}{\sqrt{x^2+x}}$ is always greater than or equal to **(2003S)**

(a) $2\tan\alpha$ (b) 1

(c) 2 (d) $\sec^2\alpha$

31. For all 'x', $x^2 + 2ax + 10 - 3a > 0$, then the interval in which 'a' lies is **(2004S)**

(a) $a < -5$ (b) $-5 < a < 2$

(c) $a > 5$ (d) $2 < a < 5$

32. If one root is square of the other root of the equation $x^2 + px + q = 0$, then the relation between p and q is **(2004S)**

(a) $p^3 - q(3p-1) + q^2 = 0$ (b) $p^3 - q(3p+1) + q^2 = 0$

(c) $p^3 + q(3p-1) + q^2 = 0$ (d) $p^3 + q(3p+1) + q^2 = 0$

33. Let a, b, c be the sides of a triangle where $a \neq b \neq c$ and $\lambda \in R$. If the roots of the equation $x^2 + 2(a+b+c)x + 3\lambda(ab+bc+ca) = 0$ are real, then **(2006 - 3M, –1)**

(a) $\lambda < \frac{4}{3}$ (b) $\lambda > \frac{5}{3}$

(c) $\lambda \in \left(\frac{1}{3}, \frac{5}{3}\right)$ (d) $\lambda \in \left(\frac{4}{3}, \frac{5}{3}\right)$

34. Let α, β be the roots of the equation $x^2 - px + r = 0$ and $\frac{\alpha}{2}$, 2β be the roots of the equation $x^2 - qx + r = 0$. Then the value of r is **(2007 -3 marks)**

(a) $\frac{2}{9}(p-q)(2q-p)$ (b) $\frac{2}{9}(q-p)(2p-q)$

(c) $\frac{2}{9}(q-2p)(2q-p)$ (d) $\frac{2}{9}(2p-q)(2q-p)$

35. Let p and q be real numbers such that $p \neq 0$, $p^3 \neq q$ and $p^3 \neq -q$. If α and β are nonzero complex numbers satisfying $\alpha + \beta = -p$ and $\alpha^3 + \beta^3 = q$, then a quadratic equation having $\frac{\alpha}{\beta}$ and $\frac{\beta}{\alpha}$ as its roots is **(2010)**

(a) $(p^3+q)x^2 - (p^3+2q)x + (p^3+q) = 0$

(b) $(p^3+q)x^2 - (p^3-2q)x + (p^3+q) = 0$

(c) $(p^3-q)x^2 - (5p^3-2q)x + (p^3-q) = 0$

(d) $(p^3-q)x^2 - (5p^3+2q)x + (p^3-q) = 0$

36. Let (x_0, y_0) be the solution of the following equations

$$(2x)^{\ell n 2} = (3y)^{\ell n 3}$$

$$3^{\ell n x} = 2^{\ell n y}$$

Then x_0 is **(2011)**

(a) $\frac{1}{6}$ (b) $\frac{1}{3}$ (c) $\frac{1}{2}$ (d) 6

37. Let α and β be the roots of $x^2 - 6x - 2 = 0$, with $\alpha > \beta$. If $a_n = \alpha^n - \beta^n$ for $n \geq 1$, then the value of $\frac{a_{10} - 2a_8}{2a_9}$ is **(2011)**

(a) 1 (b) 2 (c) 3 (d) 4

38. A value of b for which the equations

$$x^2 + bx - 1 = 0$$

$$x^2 + x + b = 0$$

have one root in common is **(2011)**

(a) $-\sqrt{2}$ (b) $-i\sqrt{3}$ (c) $i\sqrt{5}$ (d) $\sqrt{2}$

(D) MCQ's WITH ONE OR MORE THAN ONE CORRECT :

1. For real x, the function $\frac{(x-a)(x-b)}{x-c}$ will assume all real values provided **(1984 - 3 Marks)**

(a) $a > b > c$ (b) $a < b < c$

(c) $a > c > b$ (d) $a < c < b$

2. If S is the set of all real x such that $\frac{2x-1}{2x^3+3x^2+x}$ is positive, then S contains **(1986 - 2 Marks)**

(a) $\left(-\infty, -\frac{3}{2}\right)$ (b) $\left(-\frac{3}{2}, -\frac{1}{4}\right)$

(c) $\left(-\frac{1}{4}, \frac{1}{2}\right)$ (d) $\left(\frac{1}{2}, 3\right)$

(e) none of these

3. If a, b and c are distinct positive numbers, then the expression $(b+c-a)(c+a-b)(a+b-c) - abc$ is **(1986 - 2 Marks)**

(a) positive (b) negative

(c) non-positive (d) non-negative

(e) none of these

4. If a, b, c, d and p are distinct real numbers such that $(a^2+b^2+c^2)p^2 - 2(ab+bc+cd)p + (b^2+c^2+d^2) \leq 0$ then a, b, c, d **(1987 - 2 Marks)**

(a) are in A. P. (b) are in G.P.

(c) are in H. P. (d) satisfy $ab = cd$

(e) satisfy none of these

5. The equation $x^{3/4(\log_2 x)^2+\log_2 x-5/4}=\sqrt{2}$ has (1989 - 2 Marks)
(a) at least one real solution
(b) exactly three solutions
(c) exactly one irrational solution
(d) complex roots.

6. The product of n positive numbers is unity Then their sum is (1991 - 2 Marks)
(a) a positive integer (b) divisible by n
(c) equal to $n+\frac{1}{n}$ (d) never less than n

7. Number of divisor of the form $4n+2\ (n\geq 0)$ of the integer 240 is (1998 - 2 Marks)
(a) 4 (b) 8
(c) 10 (d) 3

8. If $3^x=4^{x-1}$, then $x=$ (JEE Adv. 2013)
(a) $\frac{2\log_3 2}{2\log_3 2-1}$ (b) $\frac{2}{2-\log_2 3}$
(c) $\frac{1}{1-\log_4 3}$ (d) $\frac{2\log_2 3}{2\log_2 3-1}$

E SUBJECTIVE PROBLEMS :

1. Solve for x: $4^x-3^{x-\frac{1}{2}}=3^{x+\frac{1}{2}}-2^{2x-1}$ (1978)

2. If $(m,n)=\dfrac{(1-x^m)(1-x^{m-1})..........(1-x^{m-n+1})}{(1-x)(1-x^2)............(1-x^n)}$ (1978)
where m and n are positive integers ($n\leq m$), show that $(m,n+1)=(m-1,n+1)+x^{m-n-1}(m-1,n)$.

3. Solve for x: $\sqrt{x+1}-\sqrt{x-1}=1$. (1978)

4. Solve the following equation for x: (1978)
$2\log_x a+\log_{ax}a+3\log_{a^2x}a=0,\ a>0$

5. Show that the square of $\dfrac{\sqrt{26-15\sqrt{3}}}{5\sqrt{2}-\sqrt{38+5\sqrt{3}}}$ is a rational number. (1978)

6. Sketch the solution set of the following system of inequalities:
$x^2+y^2-2x\geq 0;\ 3x-y-12\leq 0;\ y-x\leq 0;\ y\geq 0.$ (1978)

7. Find all integers x for which (1978)
$(5x-1)<(x+1)^2<(7x-3)$.

8. If α, β are the roots of $x^2+px+q=0$ and γ, δ are the roots of $x^2+rx+s=0$, evaluate $(\alpha-\gamma)(\alpha-\delta)(\beta-\gamma)(\beta-\delta)$ in terms of p, q, r and s.
Deduce the condition that the equations have a common root. (1979)

9. Show that for any triangle with sides a, b and c
$3(ab+bc+ca)<(a+b+c)^2<4(bc+ca+ab)$
when are the first two expressions equal? (1979)

10. Given $n^4<10^n$ for a fixed positive integer $n\geq 2$, prove that $(n+1)^4<10^{n+1}$. (1980)

11. Let $y=\sqrt{\dfrac{(x+1)(x-3)}{(x-2)}}$ (1980)
Find all the real values of x for which y takes real values.

12. For what values of m, does the system of equations
$3x+my=m$
$2x-5y=20$
has solution satisfying the conditions $x>0,\ y>0$. (1980)

13. Find the solution set of the system (1980)
$x+2y+z=1;$
$2x-3y-w=2;$
$x\geq 0;\ y\geq 0;\ z\geq 0;\ w\geq 0.$

14. Show that the equation $e^{\sin x}-e^{-\sin x}-4=0$ has no real solution. (1982 - 2 Marks)

15. mn squares of euqal size are arranged to from a rectangle of dimension m by n, where m and n are natural numbers. Two squares will be called 'neighbours' if they have exactly one common side. A natural number is written in each square such that the number written in any square is the arithmetic mean of the numbers written in its neighbouring squares. Show that this is possible only if all the numbers used are equal. (1982 - 5 Marks)

16. If one root of the quadratic equation $ax^2+bx+c=0$ is equal to the n-th power of the other, then show that
$(ac^n)^{\frac{1}{n+1}}+(a^nc)^{\frac{1}{n+1}}+b=0$ (1983 - 2 Marks)

17. Find all real values of x which satisfy $x^2-3x+2>0$ and $x^2-2x-4\leq 0$ (1983 - 2 Marks)

18. Solve for x; $(5+2\sqrt{6})^{x^2-3}+(5-2\sqrt{6})^{x^2-3}=10$ (1985 - 5 Marks)

19. For $a\leq 0$, determine all real roots of the equation $x^2-2a|x-a|-3a^2=0$ (1986 - 5 Marks)

20. Find the set of all x for which $\dfrac{2x}{(2x^2+5x+2)}>\dfrac{1}{(x+1)}$ (1987 - 3 Marks)

21. Let α_1,α_2 and β_1,β_2 be the roots of $ax^2+bx+c=0$ and $px^2+qx+r=0$ respectively. If the system of equations $\alpha_1y+\alpha_2z=0$ and $\beta_1y+\beta_2z=0$ has a nontrivial solution, then prove that $\dfrac{b^2}{q^2}=\dfrac{ac}{pr}$. (1987 - 3 Marks)

22. Solve $|x^2+4x+3|+2x+5=0$ (1988 - 5 Marks)

23. Let a, b, c be real. If $ax^2+bx+c=0$ has two real roots α and β, where $\alpha<-1$ and $\beta>1$, then show that $1+\dfrac{c}{a}+\left|\dfrac{b}{a}\right|<0$. (1995 - 5 Marks)

24. Find the set of all solutions of the equation
$2^{|y|}-|2^{y-1}-1|=2^{y-1}+1$ (1997C - 3 Marks)

25. Let S be a square of unit area. Consider any quadrilateral which has one vertex on each side of S. If a, b, c, and d denote the lengths of the sides of the quadrilateral, prove that $2 \le a^2+b^2+c^2+d^2 \le 4$. *(1997 - 5 Marks)*

26. If α, β are the roots of $ax^2 + bx + c = 0$, $(a \ne 0)$ and $\alpha+\delta, \beta+\delta$ are the roots of $Ax^2 + Bx + C = 0$, $(A \ne 0)$ for some constant δ, then prove that $\frac{b^2-4ac}{a^2} = \frac{B^2-4AC}{A^2}$. *(2000 - 4 Marks)*

27. Let a, b, c be real numbers with $a \ne 0$ and let α, β be the roots of the equation $ax^2 + bx + c = 0$. Express the roots of $a^3x^2 + abcx + c^3 = 0$ in terms of α, β. *(2001 - 4 Marks)*

28. If $x^2 + (a-b)x + (1-a-b) = 0$ where $a, b \in R$ then find the values of a for which equation has unequal real roots for all values of b. *(2003 - 4 Marks)*

29. If a, b, c are positive real numbers. Then prove that $(a+1)^7(b+1)^7(c+1)^7 > 7^7 a^4b^4c^4$ *(2004 - 4 Marks)*

30. Let a and b be the roots of the equation $x^2 - 10cx - 11d = 0$ and those of $x^2 - 10ax - 11b = 0$ are c, d then the value of $a+b+c+d$, when $a \ne b \ne c \ne d$, is. *(2006 - 6M)*

(H) ASSERTION & REASON TYPE QUESTIONS :

1. Let a, b, c, p, q be real numbers. Suppose α, β are the roots of the equation $x^2 + 2px + q = 0$ and $\alpha, \frac{1}{\beta}$ are the roots of the equation $ax^2 + 2bx + c = 0$, where $\beta^2 \notin \{-1, 0, 1\}$

STATEMENT - 1 : $(p^2-q)(b^2-ac) \ge 0$

and

STATEMENT - 2 : $b \ne pa$ or $c \ne qa$ *(2008)*

(a) STATEMENT - 1 is True, STATEMENT - 2 is True; STATEMENT - 2 is a correct explanation for STATEMENT - 1

(b) STATEMENT - 1 is True, STATEMENT - 2 is True; STATEMENT - 2 is **NOT** a correct explanation for STATEMENT - 1

(c) STATEMENT - 1 is True, STATEMENT - 2 is False

(d) STATEMENT - 1 is False, STATEMENT - 2 is True

(I) INTEGER VALUE CORRECT TYPE :

1. Let (x, y, z) be points with integer coordinates satisfying the system of homogeneous equations :

$$3x - y - z = 0$$
$$-3x + z = 0$$
$$-3x + 2y + z = 0$$

Then the number of such points for which $x^2 + y^2 + z^2 \le 100$ is *(2009)*

2. The smallest value of k, for which both the roots of the equation

$$x^2 - 8kx + 16(k^2 - k + 1) = 0$$

are real, distinct and have values at least 4, is *(2009)*

3. The minimum value of the sum of real numbers a^{-5}, a^{-4}, $3a^{-3}$, 1, a^8 and a^{10} where $a > 0$ is *(2011)*

4. The number of distinct real roots of

$$x^4 - 4x^3 + 12x^2 + x - 1 = 0 \text{ is}$$ *(2011)*

Section-B JEE Main/ AIEEE

1. If $\alpha \ne \beta$ but $\alpha^2 = 5\alpha - 3$ and $\beta^2 = 5\beta - 3$ then the equation having α/β and β/α as its roots is **[2002]**

(a) $3x^2 - 19x + 3 = 0$ (b) $3x^2 + 19x - 3 = 0$
(c) $3x^2 - 19x - 3 = 0$ (d) $x^2 - 5x + 3 = 0$.

2. Difference between the corresponding roots of $x^2+ax+b=0$ and $x^2+bx+a=0$ is same and $a \ne b$, then **[2002]**

(a) $a+b+4=0$ (b) $a+b-4=0$
(c) $a-b-4=0$ (d) $a-b+4=0$

3. Product of real roots of the equation $t^2x^2+|x|+9=0$ **[2002]**

(a) is always positive
(b) is always negative
(c) does not exist
(d) none of these

4. If p and q are the roots of the equation $x^2+px+q=0$, then **[2002]**

(a) $p=1, q=-2$ (b) $p=0, q=1$
(c) $p=-2, q=0$ (d) $p=-2, q=1$

5. If a, b, c are distinct +ve real numbers and $a^2+b^2+c^2=1$ then $ab+bc+ca$ is **[2002]**

(a) less than 1 (b) equal to 1
(c) greater than 1 (d) any real no.

6. If the sum of the roots of the quadratic equation $ax^2 + bx + c = 0$ is equal to the sum of the squares of their reciprocals, then $\frac{a}{c}, \frac{b}{a}$ and $\frac{c}{b}$ are in **[2003]**

(a) Arithmetic - Geometric Progression
(b) Arithmetic Progression
(c) Geometric Progression
(d) Harmonic Progression.

7. The value of 'a' for which one root of the quadratic equation $(a^2 - 5a + 3)x^2 + (3a-1)x + 2 = 0$ is twice as large as the other is **[2003]**

(a) $-\frac{1}{3}$ (b) $\frac{2}{3}$
(c) $-\frac{2}{3}$ (d) $\frac{1}{3}$

8. The number of real solutions of the equation $x^2 - 3|x| + 2 = 0$ is

(a) 3 (b) 2
(c) 4 (d) 1

9. The real number x when added to its inverse gives the minimum value of the sum at x equal to **[2003]**

(a) –2 (b) 2
(c) 1 (d) –1

10. Let two numbers have arithmetic mean 9 and geometric mean 4. Then these numbers are the roots of the quadratic equation **[2004]**

(a) $x^2 - 18x - 16 = 0$ (b) $x^2 - 18x + 16 = 0$
(c) $x^2 + 18x - 16 = 0$ (d) $x^2 + 18x + 16 = 0$

11. If $(1-p)$ is a root of quadratic equation $x^2+px+(1-p)=0$ then its root are **[2004]**
(a) $-1, 2$ (b) $-1, 1$
(c) $0, -1$ (d) $0, 1$

12. If one root of the equation $x^2+px+12=0$ is 4, while the equation $x^2+px+q=0$ has equal roots, then the value of 'q' is **[2004]**
(a) 4 (b) 12
(c) 3 (d) $\frac{49}{4}$

13. In a triangle PQR, $\angle R=\frac{\pi}{2}$. If $\tan\left(\frac{P}{2}\right)$ and $-\tan\left(\frac{Q}{2}\right)$ are the roots of $ax^2+bx+c=0$, $a\neq 0$ then **[2005]**
(a) $a=b+c$ (b) $c=a+b$
(c) $b=c$ (d) $b=a+c$

14. If both the roots of the quadratic equation $x^2-2kx+k^2+k-5=0$ are less than 5, then k lies in the interval **[2005]**
(a) $(5,6]$ (b) $(6,\infty)$
(c) $(-\infty,4)$ (d) $[4,5]$

15. If the roots of the quadratic equation $x^2+px+q=0$ are $\tan 30°$ and $\tan 15°$, respectively, then the value of $2+q-p$ is **[2006]**
(a) 2 (b) 3
(c) 0 (d) 1

16. All the values of m for which both roots of the equation $x^2-2mx+m^2-1=0$ are greater than -2 but less then 4, lie in the interval **[2006]**
(a) $-2<m<0$ (b) $m>3$
(c) $-1<m<3$ (d) $1<m<4$

17. If x is real, the maximum value of $\frac{3x^2+9x+17}{3x^2+9x+7}$ is **[2006]**
(a) $\frac{1}{4}$ (b) 41
(c) 1 (d) $\frac{17}{7}$

18. If the difference between the roots of the equation $x^2+ax+1=0$ is less than $\sqrt{5}$, then the set of possible values of a is **[2007]**
(a) $(3,\infty)$ (b) $(-\infty,-3)$
(c) $(-3,3)$ (d) $(-3,\infty)$.

DIRECTIONS : *Q. No. 19 is Assertion-Reason type questions. Each of these questions contains two statements: Statement-1(Assertion) and Statement-2(Reason). Each of these questions also has four alternative choices, only one of which is the correct answer. You have to select the correct choice.*

19. Statement-1 : For every natural number $n\geq 2$,
$$\frac{1}{\sqrt{1}}+\frac{1}{\sqrt{2}}+.........+\frac{1}{\sqrt{n}}>\sqrt{n}.$$
Statement-2 : For every natural number $n\geq 2$,
$$\sqrt{n(n+1)}<n+1.$$
(a) Statement -1 is false, Statement-2 is true **[2008]**
(b) Statement -1 is true, Statement-2 is true; Statement -2 is a correct explanation for Statement-1
(c) Statement -1 is true, Statement-2 is true; Statement -2 is not a correct explanation for Statement-1
(d) Statement -1 is true, Statement-2 is false

20. The quadritic equations $x^2-6x+a=0$ and $x^2-cx+6=0$ have one root in common. The other roots of the first and second equations are integers in the ratio 4 : 3. Then the common root is **[2009]**
(a) 1 (b) 4
(c) 3 (d) 2

21. If the roots of the equation $bx^2+cx+a=0$ be imaginary, then for all real values of x, the expression $3b^2x^2+6bcx+2c^2$ is : **[2009]**
(a) less than $4ab$ (b) greater than $-4ab$
(c) less than $-4ab$ (d) greater than $4ab$

22. If $\left|z-\frac{4}{z}\right|=2$, then the maximum value of $|Z|$ is equal to :
(a) $\sqrt{5}+1$ (b) 2 **[2009]**
(c) $2+\sqrt{2}$ (d) $\sqrt{3}+1$

23. If α and β are the roots of the equation $x^2-x+1=0$, then $\alpha^{2009}+\beta^{2009}=$ **[2010]**
(a) -1 (b) 1
(c) 2 (d) -2

24. Sachin and Rahul attempted to solve a quadratic equation. Sachin made a mistake in writing down the constant term and ended up in roots (4,3). Rahul made a mistake in writing down coefficient of x to get roots (3,2). The correct roots of equation are : **[2011 RS]**
(a) 6, 1 (b) 4, 3
(c) $-6, -1$ (d) $-4, -3$

25. Let for $a\neq a_1\neq 0$,
$f(x)=ax^2+bx+c$ $g(x)=a_1x^2+b_1x+c_1$
and $p(x)=f(x)-g(x)$.
If $p(x)=0$ only for $x=-1$ and $p(-2)=2$, then the value of $p(2)$ is : **[2011 RS]**
(a) 3 (b) 9
(c) 6 (d) 18

26. The equation $e^{\sin x}-e^{-\sin x}-4=0$ has : ***[2012]***
(a) infinite number of real roots
(b) no real roots
(c) exactly one real root (d) exactly four real roots

27. The real number k for which the equation, $2x^3+3x+k=0$ has two distinct real roots in [0, 1]
(a) lies between 1 and 2 **[JEE M 2013]**
(b) lies between 2 and 3
(c) lies between -1 and 0
(d) does not exist.

28. The number of values of k, for which the system of equations : **[JEE M 2013]**
$(k+1)x+8y=4k$
$kx+(k+3)y=3k-1$
has no solution, is
(a) infinite (b) 1
(c) 2 (d) 3

29. If the equations $x^2+2x+3=0$ and $ax^2+bx+c=0$, $a,b,c\in$ R, have a common root, then $a:b:c$ is **[JEE M 2013]**
(a) $1:2:3$ (b) $3:2:1$
(c) $1:3:2$ (d) $3:1:2$

Solutions & Explanations

Section-A : JEE Advanced/ IIT-JEE

A **1.** –5050 **2.** –4, 7 **3.** 2 **4.** 1 **5.** 4

6. –1/4, –1/4 **7.** $\dfrac{\left[k+\left(n-\dfrac{k(k+1)}{2}\right)-1\right]!}{\left[n-\dfrac{k(k+1)}{2}\right]!\,(k-1)!}$ **8.** 4

B **1.** T **2.** F **3.** T **4.** F **5.** T **6.** F

C

1. (c)	**2.** (d)	**3.** (a)	**4.** (c)	**5.** (b)	**6.** (b)	**7.** (c)	**8.** (a)
9. (c)	**10.** (b)	**11.** (d)	**12.** (a)	**13.** (c)	**14.** (a)	**15.** (a) (d)	**16.** (d)
17. (a)	**18.** (c)	**19.** (a)	**20.** (b)	**21.** (c)	**22.** (a)	**23.** (a)	**24.** (b)
25. (a)	**26.** (d)	**27.** (c)	**28.** (a)	**29.** (b)	**30.** (a)	**31.** (b)	**32.** (a)
33. (a)	**34.** (d)	**35.** (b)	**36.** (c)	**37.** (c)	**38.** (c)		

D **1.** (c, d) **2.** (a, d) **3.** (b) **4.** (b) **5.** (a,b,c) **6.** (d) **7.** (a) **8.** (a, b, c)

E **1.** $\frac{3}{2}$ **3.** $\frac{5}{4}$ **4.** $a^{-1/2}, a^{-4/3}$ **7.** 3

8. $q(r-p)^2 - p(r-p)(s-q)+(s-q)^2$; $(q-s)^2 = (r-p)(ps-qr)$ **11.** $[-1,2)\cup[3,\infty)$

12. $m\in\left(-\infty, \dfrac{-15}{2}\right)\cup(30,\infty)$ **13.** $x=1,\ y=0,\ z=0,\ w=0$ **17.** $[-1,1)\cup(2,4]$

18. $\pm 2, \pm\sqrt{2}$ **19.** $\{a\pm a\sqrt{2}, -a\pm a\sqrt{6}\}$ **20.** $(-2,-1)\cup\left(-\dfrac{2}{3}, -\dfrac{1}{2}\right)$

22. $-4, -1-\sqrt{3}$ **24.** $\{-1\}\cup[1,\infty)$ **27.** $\alpha^2\beta,\ \alpha\beta^2$ **28.** $a>1$ **30.** 1210

H **1.** (b)

I **1.** 7 **2.** 2 **3.** 8 **4.** 2

Section-B : JEE Main/ AIEEE

1. (a)	**2.** (a)	**3.** (a)	**4.** (a)	**5.** (a)	**6.** (d)	**7.** (b)	**8.** (c)
9. (c)	**10.** (b)	**11.** (c)	**12.** (d)	**13.** (b)	**14.** (c)	**15.** (b)	**16.** (c)
17. (b)	**18.** (c)	**19.** (b)	**20.** (d)	**21.** (b)	**22.** (a)	**23.** (b)	**24.** (a)
25. (d)	**26.** (b)	**27.** (d)	**28.** (b)	**29.** (a)			

Section-A JEE Advanced/ IIT-JEE

A FILL IN THE BLANKS :

1. Given polynomial :

$(x-1)(x-2)(x-3)\ldots(x-100)$

$= x^{100} - (1+2+3+\ldots+100)x^{99} + (\ldots)x^{98} + \ldots$

Here coeff. of $x^{99} = -(1+2+3+\ldots+100)$

$$= \frac{-100\times 101}{2} = -5050.$$

2. As p and q are real;and one root is $2+i\sqrt{3}$, other should be $2-i\sqrt{3}$

Then $p = -$ (sum of roots) $= -4$,

$q =$ product of roots $= 4+3 = 7$.

3. The given equation is $x^2 - 3kx + 2e^{2\ln k} - 1 = 0$

Or $x^2 - 3kx + (2k^2 - 1) = 0$

Here product of roots $= 2k^2 - 1$

$\therefore\ 2k^2 - 1 = 7 \Rightarrow\ k^2 = 4\ \Rightarrow k = 2, -2$

Now for real roots we must have $D \geq 0$

$\Rightarrow 9k^2 - 4(2k^2 - 1) \geq 0 \Rightarrow k^2 + 4 \geq 0$

Which is true for all k. Thus $k = 2, -2$

But for $k = -2$, $\ln k$ is not defined

$\therefore$ Rejecting $k = -2$, we get $k = 2$.

4. According to the question, $x^2 + ax + b = 0$ and $x^2 + bx + a = 0$ $(a \neq b)$ have a common root, let it be α then we have

$$\alpha^2 + a\alpha + b = 0 \quad \text{.... (i)}$$

$$\alpha^2 + b\alpha + a = 0 \quad \text{.... (ii)}$$

$$\Rightarrow \frac{\alpha^2}{a^2 - b^2} = \frac{\alpha}{b - a} = \frac{1}{b - a} \Rightarrow \frac{\alpha^2}{a + b} = \frac{\alpha}{-1} = \frac{1}{-1}$$

$\Rightarrow \alpha^2 = -(a+b); \alpha = 1 \Rightarrow a + b = -1$

$\therefore$ Numerical value of $a + b = -1$.

5. $\log_7 \log_5 \left(\sqrt{x+5} + \sqrt{x}\right) = 0$

$\Rightarrow \log_5 \left(\sqrt{x+5} + \sqrt{x}\right) = 1$ **NOTE THIS STEP**

$\Rightarrow \sqrt{x+5} + \sqrt{x} = 5 \Rightarrow x + 5 = 25 + x - 10\sqrt{x}$

$\Rightarrow 2 = \sqrt{x} \Rightarrow x = 4$ which satisfies the given equation.

6. Given $x < 0$, $y < 0$

$$x + y + \frac{x}{y} = \frac{1}{2} \text{ and } (x+y).\frac{x}{y} = -\frac{1}{2}$$

Let $x + y = a$ and $\frac{x}{y} = b$ .... (1)

$\therefore$ We get $a + b = \frac{1}{2}$ and $ab = -\frac{1}{2}$

Solving these two, we get $a + \left(-\frac{1}{2a}\right) = \frac{1}{2}$

$\Rightarrow 2a^2 - a - 1 = 0 \Rightarrow a = 1, -1/2$

$\Rightarrow b = -1/2, 1$

$\therefore$ (1) $\Rightarrow x + y = 1$ and $\frac{x}{y} = -\frac{1}{2}$

or $x + y = \frac{-1}{2}$ and $\frac{x}{y} = 1$ But $x, y < 0$

$\therefore x + y < 0 \Rightarrow x + y = \frac{-1}{2}$ and $\frac{x}{y} = 1$

On solving, we get $x = -1/4$ and $y = -1/4$.

7. We have

$x_1 + x_2 + \ldots\ldots\ldots\ldots + x_k = n$ .... (1)

where $x_1 \geq 1, x_2 \geq 2, x_3 \geq 3, \ldots\ldots\ldots, x_k \geq k$; all integers

Let $y_1 = x_1 - 1, y_2 = x_2 - 2, \ldots\ldots\ldots, y_k = x_k - k$

so that $y_1, y_2, \ldots\ldots\ldots, y_k \geq 0$

Substituting the values of $x_1, x_2, \ldots\ldots\ldots x_k$ in equation .. (1)

We get $y_1 + y_2 + \ldots y_k = n - (1 + 2 + 3 \ldots + k)$

$$= n - \frac{k(k+1)}{2} \quad \text{.... (2)}$$

Now keeping in mind that number of solutions of the equation

$\alpha + 2\beta + 3\gamma + \ldots\ldots q\theta = n$

for $\alpha, \beta, \gamma, \ldots\ldots\ldots \theta \in \mathrm{I}$ and each is ≥ 0, is given by coeff of x^n in

$(1 + x + x^2 + \ldots\ldots.)(1 + x^2 + x^4 + \ldots\ldots..)$

$(1 + x^3 + x^6 + \ldots\ldots.)\ldots\ldots.(1 + x^q + x^{2q} + \ldots\ldots\ldots..)$

We find that no. of solutions of equation (2)

$= \text{coeff of } x^{n - \frac{k(k+1)}{2}}$ in $(1 + x + x^2 + \ldots\ldots\ldots)^k$

NOTE THIS STEP

$= \text{coeff of } x^{n - \frac{k(k+1)}{2}}$ in $(1 - x)^{-k}$

$= \text{coeff of } x^{n - \frac{k(k+1)}{2}}$ in $(1 + {}^kC_1 x + {}^{k+1}C_2 x^2$

$$+ {}^{k+2}C_3 x^3 + \ldots\ldots\ldots\ldots) = {}^{k + \left(n - \frac{k(k+1)}{2}\right) - 1}C_{n - \frac{k(k+1)}{2}}$$

$$= \frac{\left[k + \left(n - \frac{k(k+1)}{2}\right) - 1\right]!}{\left[n - \frac{k(k+1)}{2}\right]!(k-1)!}.$$

8. $|x - 2|^2 + |x - 2| - 2 = 0$

Case 1. $x \geq 2$

$\Rightarrow (x-2)^2 + (x-2) - 2 = 0$

$\Rightarrow x^2 - 3x = 0 \Rightarrow x(x-3) = 0$

$\Rightarrow x = 0, 3$ (0 is rejected as $x \geq 2$)

$\Rightarrow x = 3$ ....(1)

Case 2. $x < 2$

$\{-(x-2)\}^2 - (x-2) - 2 = 0$

$\Rightarrow x^2 + 4 - 4x - x = 0 \Rightarrow (x-1)(x-4) = 0$

$\Rightarrow x = 1, 4$ (4 is rejected as $x < 2$)

$\Rightarrow x = 1$ ....(2)

Therefore, the sum of the roots is $3 + 1 = 4$.

B TRUE / FALSE :

1. Consider n numbers, namely 1, 2, 3, 4, ...n.
KEY CONCEPT : Now using A.M. > G.M. for distinct numbers, we get

$$\frac{1+2+3+4....+n}{n} > (1.2.3.4..n)^{1/n}$$

$$\Rightarrow \frac{n(n+1)}{2n} > (n!)^{1/n} \Rightarrow (n!)^{1/n} < \frac{n+1}{2} \therefore True.$$

ALTERNATE SOLUTION :
The given inequality can be written as

$n! < \left(\frac{n+1}{2}\right)^n$ for $n > 1$.

Let us use mathematical induction to check the validity of given inequality.

For $n = 2$, we have $2! < \left(\frac{3}{2}\right)^2 = \frac{9}{4}$ which is true

$\therefore$ Inequality is valid for $n = 2$.

Let it be valid for $n = k$ then $k! < \left(\frac{k+1}{2}\right)^k$ ....(1)

Consider :

$(k+1)! = (k+1)\,k! < (k+1)\left(\frac{k+1}{2}\right)^k$ (Using (1))

Now we will try to check

$(k+1)\left(\frac{k+1}{2}\right)^k < \left(\frac{k+2}{2}\right)^{k+1}$ ....(2)

Which is equivalent to write $2 < \left(\frac{k+2}{k+1}\right)^{k+1}$ ....(3)

Now, $\left(\frac{k+2}{k+1}\right)^{k+1} = \left(1+\frac{1}{k+1}\right)^{k+1}$

$$= 1+(k+1)\frac{1}{k+1}+\frac{(k+1)k}{2!}\left(\frac{1}{k+1}\right)^2+....$$

(Using Binomial expansion)

$\therefore \left(\frac{k+2}{k+1}\right)^{k+1} > 2 \Rightarrow$ (3) holds and hence (2) holds

$\Rightarrow (k+1)! < \left(\frac{k+2}{2}\right)^{k+1}$. Thus statement is true for $n = k+1$.

Hence, by Principle of Mathematical Induction given statement is true, $\forall n > 1$.

2. $2x^2+3x+1=0 \Rightarrow x=-1,-1/2$ both are rational
$\therefore$ Statement is FALSE.

ALTERNATE SOLUTION :
If coefficients of a quadratic equation are rational and D is a perfect square, roots are rational.
As coefficients are rational and $D = 9 - 8 = 1$ is perfect square
$\therefore$ roots must be rational.
$\therefore$ Statement is FALSE.

3. The given equation is

$$(x-a)(x-c)+2(x-b)(x-d)=0$$

$$\Rightarrow 3x^2-(a+c+2b+2d)x+(ac+2bd)=0$$

For nature of roots, we find discriminant

$$D=(a+c+2b+2d)^2-12(ac+2bd)$$

$$=[(a+2d)-(c+2b)]^2+4(a+2d)(c+2b)-12(ac+2bd)$$

$$=[(a+2d)-(c+2b)]^2+8ab+8cd-8ac-8bd$$

$=[(a+2d)-(c+2b)]^2+8(c-b)(d-a)>0$. as given that $a<b<c<d \Rightarrow$ Statement is TRUE.

ALTERNATE SOLUTION :
$f(x)=(x-a)(x-c)+2(x-b)(x-d)$.
$f(a)=+ve$
$f(b)=-ve$
$f(c)=-ve$
$f(d)=+ve$
$\therefore$ There exists two real and distinct roots one in the interval (a, b) and other in (c, d). Hence, (True).

4. Consider $N = n_1+n_2+n_3+....+n_p$, where N is an even number. Let k numbers among these p numbers be odd, then $p-k$ are even numbers. Now sum of $(p-k)$ even numbers is even and for N to be an even number, sum of k odd numbers must be even which is possible only when k is even.
$\therefore$ The given statement is false.

5. We have $P(x) = ax^2+bx+c$
for which $D_1 = b^2-4ac$ ....(1)
and $Q(x) = -ax^2+dx+c$
for which $D_2 = d^2+4ac$ .... (2)
Given that $ac \neq 0$ $\therefore$ Following two cases are possible.
If $ac > 0$ then from eq. (2), D_2 is +ve $\Rightarrow$ Q(x) has real roots.
If $ac < 0$ then from eq. (1), D_1 is +ve $\Rightarrow$ P(x) has real roots.
Thus $P(x)\,Q(x) = 0$ has at least two real roots.
$\therefore$ Given statement is true.

ALTERNATE SOLUTION :
$P(x).Q(x) = (ax^2+bx+c)(-ax^2+bx+c)$
$\Rightarrow D_1 = b^2-4ac$ and $D_2 = b^2+4ac$
clearly, $D_1+D_2 = 2b^2 \geq 0$
$\therefore$ atleast one of D_1 and D_2 is (+ ve). Hence, atleast real roots.
Thus, (True)

$$\Rightarrow n\left(\sum_{i=1}^{n} x_i^2\right) \geq \left(\sum_{i=1}^{n} x_i\right)^2$$

6. As x and y are positive real numbers and m and n are positive integers

$\therefore \quad \frac{1+x^{2n}}{2} \geq (1 \times x^{2n})^{1/2}$ and $\frac{1+y^{2m}}{2} \geq (1 \times y^{2m})^{1/2}$

{For two +ve numbers A.M. $\geq$ G.M.}

$$\Rightarrow \left(\frac{1+x^{2n}}{2}\right) \geq x^n \quad ....(1)$$

and $\left(\frac{1+y^{2m}}{2}\right) \geq y^m \quad ....(2)$

Multiplying (1) and (2), we get

$$\frac{(1+x^{2n})(1+y^{2m})}{4} \geq x^n y^m \Rightarrow \frac{1}{4} \geq \frac{x^n y^m}{(1+x^{2n})(1+y^{2m})}$$

Hence the statement is false.

ALTERNATE SOLUTION :

$$\frac{x^n y^m}{(1+x^{2n})(1+y^{2m})} = \frac{1}{(x^n+x^{-n})(y^m+y^{-m})}$$

But $\quad x^n + \frac{1}{x^n} \geq 2$ and $y^m + \frac{1}{y^m} \geq 2$

$\therefore \quad (x^n+x^{-n})(y^m+y^{-m}) \geq 4$

$$\Rightarrow \frac{1}{(x^n+x^{-n})(y^m+y^{-m})} \leq \frac{1}{4}.$$

(C) MCQ's WITH ONE CORRECT ANSWER :

1. (c) ℓ, m, n are real, $\ell \neq m$
Given equation is

$$(\ell - m)x^2 - 5(\ell + m)x - 2(\ell - m) = 0$$

$$D = 25(\ell+m)^2 + 8(\ell-m)^2 > 0, \ell, m \in R$$

$\therefore$ Roots are real and unequal.

2. (d) The given equations are
$x + 2y + 2z = 1 \quad ....(1)$
and $2x + 4y + 4z = 9 \quad ....(2)$
Subtracting (1) × (2) from (2), we get 0 =7 (not possible)
$\therefore$ No solution.

3. (a) $u = x^2 + 4y^2 + 9z^2 - 6yx - 3zx - 2xy$

$$= \frac{1}{2}[2x^2 + 8y^2 + 18z^2 - 12yz - 6zx - 4xy]$$

$$= \frac{1}{2}[(x^2 - 4xy + 4y^2) + (4y^2 + 9z^2 - 12yz) + (x^2 + 9z^2 - 6zx)]$$

$$= \frac{1}{2}[(x-2y)^2 + (2y-3z)^2 + (3z-x)^2] \geq 0$$

$\therefore$ u is always non-negative.

4. (c) As $a, b, c > 0$, a, b, c should be real (note that order relation is not defined in the set of complex numbers)
$\therefore$ Roots of equation are either real or complex conjugate.
Let α, β be the roots of $ax^2 + bx + c = 0$, then

$$\alpha + \beta = -\frac{b}{a} = -ve, \quad \alpha\beta = \frac{c}{a} = +ve$$

$\Rightarrow$ Either both α, β are – ve (if roots are real) or both α, β have – ve real parts (if roots are complex conjugate)

5. (b) The given equation is

$$(x-b)(x-c) + (x-a)(x-c) + (x-a)(x-b) = 0$$

$$\Rightarrow 3x^2 - 2(a+b+c)x + (ab+bc+ca) = 0$$

Discriminant $= 4(a+b+c)^2 - 12(ab+bc+ca)$

$$= 4[a^2 + b^2 + c^2 - ab - bc - ca]$$

$$= 2[(a-b)^2 + (b-c)^2 + (c-a)^2] \geq 0 \quad \forall\, a, b, c$$

$\therefore$ Roots of given equation are always real.

6. (b) Let $y = 2\log_{10} x - \log_x 0.01$

$$= 2\log_{10} x - \frac{\log_{10} 0.01}{\log_{10} x} = 2\log_{10} x + \frac{2}{\log_{10} x}$$

$$= 2\left[\log_{10} x + \frac{2}{\log_{10} x}\right]$$

[Here $x > 1 \Rightarrow \log_{10} x > 0$]
Now since sum of a real + ve number and its reciprocal is always greater than or equal to 2.
$\therefore \quad y \geq 2 \times 2 \Rightarrow y \geq 4$
$\therefore$ Least value of y is 4.

7. (c) As $(x^2 + px + 1)$ is a factor of $ax^3 + bx + c$, we can assume that zeros of $x^2 + px + 1$ are α, β and that of $ax^3 + bx + c$ be α, β, γ so that
$\alpha + \beta = -p \quad .... (i)$
$\alpha\beta = 1 \quad .... (ii)$
and $\alpha + \beta + \gamma = 0 \quad .... (iii)$

$$\alpha\beta + \beta\gamma + \gamma\alpha = \frac{b}{a} \quad .... (iv)$$

$$\alpha\beta\gamma = \frac{-c}{a} \quad .... (v)$$

Solving (ii) and (v) we get $\gamma = -c/a$.
Also from (i) and (iii) we get $\gamma = p$
$\therefore \quad p = \gamma = -c/a$
Using equations (i), (ii) and (iv) we get

$$1 + \gamma(-p) = \frac{b}{a}$$

$$\Rightarrow 1 + \left(-\frac{c}{a}\right)\left(\frac{c}{a}\right) = \frac{b}{a} \quad \text{(using } \gamma = p = -c/a\text{)}$$

$$\Rightarrow 1 - \frac{c^2}{a^2} = \frac{b}{a} \Rightarrow a^2 - c^2 = ab$$

$\therefore$ (c) is the correct answer.

8. **(a)** $|x|^2 - 3|x| + 2 = 0$

Case I : $x < 0$ then $|x| = -x$

$\Rightarrow$ $x^2 + 3x + 2 = 0 \Rightarrow (x+1)(x+2) = 0$

$x = -1, -2$ (both acceptable as < 0)

Case II: $x > 0$ then $|x| = x$

$\Rightarrow$ $x^2 - 3x + 2 = 0$

$\Rightarrow$ $(x-1)(x-2) = 0$

$x = 1, 2$ (both acceptable as > 0)

$\therefore$ There are 4 real solutions.

9. **(c)** Let the distance of school from $A = x$

$\therefore$ The distance of the school form $B = 60 - x$

Total distance covered by 200 students

$= 2[150x + 50(60 - x)] = 2[100x + 3000]$

This is min., when $x = 0$

$\therefore$ school should be built at town A.

10. **(b)** if $p = 5, q = 3, r = 2$

$\max(p, q) = 5$; $\max(p, q, r) = 5$

$\Rightarrow$ $\max(p, q) = \max(p, q, r)$

$\therefore$ (a) is not true. Similarly we can show that (c) is not true.

Also $\min(p,q) = \frac{1}{2}(p + q - |p - q|)$

Let $p < q$ then LHS $= p$

and R.H.S. $= \frac{1}{2}(p + q - q + p) = p$

Similarly, we can prove that (*b*) is true for $q < p$ too.

11. **(d)** Given expression $x^{12} - x^9 + x^4 - x + 1 = f(x)$ (say)

For $x < 0$ put $x = -y$ where $y > 0$

then we get $f(x) = y^{12} + y^9 + y^4 + y + 1 > 0$ for $y > 0$

For $0 < x < 1$, $x^9 < x^4 \Rightarrow -x^9 + x^4 > 0$

Also $1 - x > 0$ and $x^{12} > 0$

$\Rightarrow$ $x^{12} - x^9 + x^4 + 1 - x > 0 \Rightarrow f(x) > 0$

For $x > 1$

$f(x) = x(x^3 - 1)(x^8 + 1) + 1 > 0$

So $f(x) > 0$ for $-\infty < x < \infty$.

12. **(a)** Given equation is $x - \frac{2}{x-1} = 1 - \frac{2}{x-1}$

Clearly $x \neq 1$ for the given eq. to be defined. If $x - 1 \neq 0$, we can cancel the common term $\frac{-2}{x-1}$ on both sides to get $x = 1$, but it is not possible. So given eq. has no roots.

$\therefore$ (*a*) is the correct answer.

13. **(c)** Given that $a^2 + b^2 + c^2 = 1$ ....(1)

We know $(a + b + c)^2 \geq 0$

$\Rightarrow$ $a^2 + b^2 + c^2 + 2ab + 2bc + 2ca \geq 0$

$\Rightarrow$ $2(ab + bc + ca) \geq -1$ [Using (1)]

$\Rightarrow$ $ab + bc + ca \geq -1/2$ ....(2)

Also we know that

$\frac{1}{2}\left[(a-b)^2 + (b-c)^2 + (c-a)^2\right] \geq 0$

$\Rightarrow$ $a^2 + b^2 + c^2 - ab - bc - ca \geq 0$

$\Rightarrow$ $ab + bc + ca \leq 1$ [Using (1)]

$\Rightarrow$ $ab + bc + ca \leq 1$ ....(3)

Combining (2) and (3), we get

$-1/2 \leq ab + bc + ca \leq 1$

$\therefore$ $ab + bc + ca \in [-1/2, 1]$

$\therefore$ (c) is the correct answer.

14. **(a)** First of all for $\log(x-1)$ to be defined, $x - 1 > 0$

$\Rightarrow$ $x > 0$ ....(1)

Now, $\log_{0.3}(x-1) < \log_{0.09}(x-1)$

$\Rightarrow$ $\log_{0.3}(x-1) < \log_{(0.3)^2}(x-1)$

$\Rightarrow$ $\log_{0.3}(x-1) < \frac{1}{2}\log_{0.3}(x-1)$

$\Rightarrow$ $2\log_{0.3}(x-1) < \log_{0.3}(x-1)$

$\Rightarrow$ $\log_{0.3}(x-1)^2 < \log_{0.3}(x-1)$

$\Rightarrow$ $(x-1)^2 > (x-1)$ **NOTE THIS STEP**

[The inequality is reversed since base lies between 0 and 1]

$\Rightarrow$ $(x-1)^2 - (x-1) > 0 \Rightarrow (x-1)(x-2) > 0$ ....(2)

Combining (1) and (2) we get $x > 2$

$\therefore$ $x \in (2, \infty)$

15. **(d)** Given α, β are the roots of

$x^2 + px + q = 0 \Rightarrow \alpha + \beta = -p$

and $\alpha\beta = q$ ...(1)

α^4, β^4 are the roots of $x^2 - rx + s = 0$

$\Rightarrow \alpha^4 + \beta^4 = r \;\&\; \alpha^4 . \beta^4 = s$ ...(2)

Now, $x^2 - 4qx + 2q^2 - r = 0$

$\Rightarrow (x - 2q)^2 - 4q^2 + 2q^2 - r = 0$

$\Rightarrow (x - 2q)^2 = 2q^2 + r$

$(x - 2\alpha\beta)^2 = 2\alpha^2\beta^2 + \alpha^4 + \beta^4$ [From (1) and (2)]

$\Rightarrow$ $(x - 2\alpha\beta)^2 = (\alpha^2 + \beta^2)^2 \Rightarrow x = 2\alpha\beta \pm (\alpha^2 + \beta^2)$

$\Rightarrow$ $x = \alpha^2 + \beta^2 + 2\alpha\beta, -\alpha^2 - \beta^2 + 2\alpha\beta$

$\Rightarrow$ $x = (\alpha + \beta)^2, -(\alpha - \beta)^2$

Hence, the equation has one positive and one negative root.

ALTERNATE SOLUTION : From the first method,

$q = \alpha\beta, r = \alpha^4 + \beta^4$ ...(3)

Product of the roots of the equation

$x^2 - 4qx + (2q^2 - r) = 0$

$= 2q^2 - r = 2\alpha^2\beta^2 - \alpha^4 - \beta^4 = -(\alpha^2 - \beta^2)^2$ [From (3)]

$= -$ (positive quantity) $= -$ ve quantity

$\Rightarrow$ one root is positive and other is negative.

16. **(d)** **KEY CONCEPT :** If $f(\alpha)$ and $f(\beta)$ are of opposite signs then there must lie a value γ between α and β such that $f(\gamma)=0$.

a, b, c are real numbers and $a \neq 0$.
As α is a root of $a^2x^2+bx+c=0$
$\therefore\ a^2\alpha^2+b\alpha+c=0$ ....(1)
Also β is a root of $a^2x^2-bx-c=0\ \therefore$
$a^2\beta^2-b\beta-c=0$ .... (2)
Now, let $f(x)=a^2x^2+2bx+2c$
Then $f(\alpha)=a^2\alpha^2+2b\alpha+2c$
$=a^2\alpha^2+2(b\alpha+c)$
$=a^2\alpha^2+2(-a^2\alpha^2)$ [Using eq. (1)]
$=-a^2\alpha^2$.
and $f(\beta)=a^2\beta^2+2b\beta+2c$
$=a^2\beta^2+2(b\beta+c)$
$=a^2\beta^2+2(a^2\beta^2)$ [Using eq. (2)]
$=3a^2\beta^2>0$.
Since $f(\alpha)$ and $f(\beta)$ are of opposite signs and γ is a root of equation $f(x)=0$

NOTE THIS STEP

$\therefore$ γ must lie between α and β
Thus $\alpha<\gamma<\beta$. $\therefore$ (d) is the correct option.

17. **(a)** The given eq. is $\sin(e^x)=5^x+5^{-x}$
We know 5^x and 5^{-x} both are +ve real numbers using AM $\geq$ GM

$\therefore\ 5^x+\frac{1}{5^x}\geq 2 \Rightarrow 5^x+5^{-x}\geq 2$

$\therefore$ R.H.S. of given eq. ≥ 2

While $\sin e^x \in[-1,1]$ i.e. LHS $\in[-1,1]$

$\therefore$ The equation is not possible for any real value of x.
Hence (a) is the correct answer.

18. **(c)** α, β are roots of the equation $(x-a)(x-b)=c, c\neq 0$
$\therefore$ $(x-a)(x-b)-c=(x-\alpha)(x-\beta)$
$\Rightarrow$ $(x-\alpha)(x-\beta)+c=(x-a)(x-b)$
$\Rightarrow$ roots of $(x-\alpha)(x-\beta)+c=0$ are a and b.
$\therefore$ (c) is the correct option.

19. **(a)** We have

$$y=5x^2+2x+3=5\left(x+\frac{1}{5}\right)^2+\frac{14}{5}>2,\ \forall x\in R$$

while $y=2\sin x\leq 2, \forall x\in R$

$\Rightarrow$ The two curves do not meet at all.

20. **(b)** For real roots $q^2-4pr\geq 0$

$\Rightarrow \left(\frac{p+r}{2}\right)^2-4pr\geq 0$ ($\because$ p, q, r are in A.P.)

$\Rightarrow p^2+r^2-14pr\geq 0 \Rightarrow \frac{p^2}{r^2}-14\frac{p}{r}+1\geq 0$

$\Rightarrow \left(\frac{p}{r}-7\right)^2-48\geq 0 \Rightarrow \left|\frac{p}{r}-7\right|\geq 4\sqrt{3}$

21. **(c)** For the equation $px^2+qx+1=0$ to have real roots

$D\geq 0 \qquad \Rightarrow q^2\geq 4p$

If $p=1$ then $q^2\geq 4 \Rightarrow q=2,3,4$

If $p=2$ then $q^2\geq 8 \Rightarrow q=3,4$

If $p=3$ then $q^2\geq 12 \Rightarrow q=4$

If $p=4$ then $q^2\geq 16 \Rightarrow q=4$

$\therefore$ No. of req. equations = 7.

22. **(a)** The given eq^n is $\sqrt{x+1}-\sqrt{x-1}=\sqrt{4x-1}$
Squaring both sides, we get

$x+1+x-1-2\sqrt{x^2-1}=4x-1$

$\Rightarrow -2\sqrt{x^2-1}=2x-1$

Again squaring both sides, we get
$\Rightarrow 4(x^2-1)=4x^2-4x+1\Rightarrow -4x=-5$
$\Rightarrow x=5/4$
Substituting this value of x in given eq^n, we get

$$\sqrt{\frac{5}{4}+1}-\sqrt{\frac{5}{4}-1}=\sqrt{4\times\frac{5}{4}-1}$$

$\Rightarrow \frac{3}{2}-\frac{1}{2}=2$(not satisfied)

$\therefore$ $x=5/4$ is not a sol^n of given eq^n. $\Rightarrow$ the given eq^n has no solution.
$\therefore$ (a) is the correct option.

23. **(a)** **KEY CONCEPT :** If both roots of a quadratic equation $ax^2+bx+c=0$ are less than k then $af(k)>0, D\geq 0, \alpha+\beta<2k$.

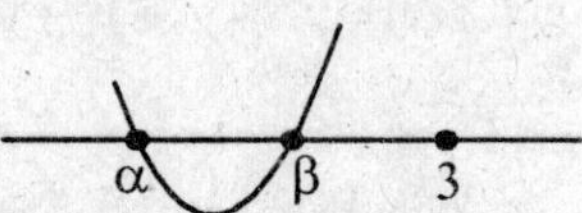

$f(x)=x^2-2ax+a^2+a-3=0$,
$f(3)>0, \alpha+\beta<6, D\geq 0$.
$\Rightarrow a^2-5a+6>0, a<3, -4a+12\geq 0$
$\Rightarrow a<2$ or $a>3, a<3, a<3 \Rightarrow a<2$.

24. **(b)** Given $c<0<b$ and $\alpha+\beta=-b$ ....(1)
$\alpha\beta=c$ ....(2)
From (2), $c<0\Rightarrow\alpha\beta<0\Rightarrow$ either α is -ve or β is -ve and second ;quantity is positive.
from (1), $b>0\Rightarrow -b<0\Rightarrow\alpha+\beta<0\Rightarrow$ the sum is negative
$\Rightarrow$ modules of nengative quantity is > modulus of positive quantity but $\alpha<\beta$ is given. Therefore, it is clear that α is negative and β is positive and modulus of α is greater than modulus of $\beta \Rightarrow \alpha<0<\beta<|\alpha|$

25. **(a)** As A.M. $\geq$ G.M. for positive real numbers, we get

$$\frac{(a+b)+(c+d)}{2}\geq\sqrt{(a+b)(c+d)}\Rightarrow M\leq 1$$

(Putting values)
Also $(a+b)(c+d)>0$ [$\therefore$ a,b,c,d > 0]
$\therefore$ $0\leq M\leq 1$

26. (d) KEY CONCEPT : $f(x) = ax^2 + bx + c$, represents an upward parabola if $a > 0$ and downward parabola if $a < 0$.

The given equation is $(x - a)(x - b) - 1 = 0, b > a$.

or $x^2 - (a + b)x + ab - 1 = 0$

Let $f(x) = x^2 - (a + b)x + ab - 1$

Since coeff. of x^2 i.e. $1 > 0$, $\therefore$ it represents upward parabola, intersecting x - axis at two points. (corresponding to two real roots, D being +ve). Also $f(a) = f(b) = -1 \Rightarrow$ curve is below x-axis at a and b

$\Rightarrow$ a and b both lie between the roots.

Thus the graph of given eq^n is as shown.

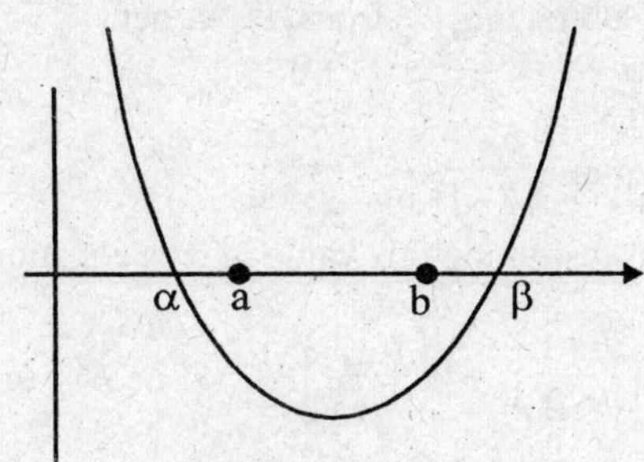

from graph it is clear that one root of the equation lies in $(-\infty, a)$ and other in (b, ∞).

27. (c) Let α, α^2 be the roots of $3x^2 + px + 3$.

$\therefore$ $\alpha + \alpha^2 = -p/3$ and $\alpha^3 = 1$

$\Rightarrow (\alpha - 1)(\alpha^2 + \alpha + 1) = 0 \Rightarrow \alpha = 1$ or $\alpha^2 + \alpha = -1$

If $\alpha = 1, p = -6$ which is not possible as $p > 0$

If $\alpha^2 + \alpha = -1 \Rightarrow -p/3 = -1 \Rightarrow p = 3$.

28. (a) We have

$$\frac{(a_1 + a_2 + ... + a_{n-1} + 2a_n)}{n} \geq (a_1 a_2 ... a_{n-1} 2a_n)^{1/n}$$

[Using A.M. ≥ G.M.]

$\Rightarrow a_1 + a_2 + a_3 + ... + a_{n-1} + 2a_n \geq n(2c)^{1/n}$

29. (b) For $x < -2$, $|x + 2| = -(x + 2)$ and the inequality becomes

$x^2 + x + 2 + x > 0 \quad \Rightarrow (x + 1)^2 + 1 > 0$

which is valid $\forall \; x \in R$ but $x < -2$

$\therefore \quad x \in (-\infty, -2)$ ....(1)

For $x \geq 2, |x + 2| = x + 2$ and the inequality becomes

$x^2 - x - 2 + x > 0 \Rightarrow x^2 > 2 \Rightarrow x > \sqrt{2}$ or $x < -\sqrt{2}$

i.e., $x \in (-\infty, -\sqrt{2}) \cup (\sqrt{2}, \infty)$

but $x \geq -2 \Rightarrow x \in [-2, -\sqrt{2}) \cup (\sqrt{2}, \infty)$ ....(2)

From (1) and (2)

$x \in (-\infty, -2) \cup [-2, -\sqrt{2}) \cup (\sqrt{2}, \infty)$

$\Rightarrow \quad x \in (-\infty, -\sqrt{2}) \cup (\sqrt{2}, \infty)$

30. (a) Let $a = \sqrt{x^2 + x}$ and $b = \dfrac{\tan^2 \alpha}{\sqrt{x^2 + x}}$

then using AM ≥ GM, we get $\dfrac{a + b}{2} \geq \sqrt{ab}$

$\Rightarrow \quad a + b \geq 2\sqrt{ab}$

$\Rightarrow \quad \sqrt{x^2 + x} + \dfrac{\tan^2 \alpha}{\sqrt{x^2 + x}} \geq 2\sqrt{\tan^2 \alpha} = 2\tan\alpha$

$[\because \alpha \in (0, \pi/2)]$

31. (b) KEY CONCEPT : $f(x) = ax^2 + bx + c$ has same sign as that of a if $D < 0$.

$x^2 + 2ax + 10 - 3a > 0 \, \forall x$

$\Rightarrow \quad D < 0 \Rightarrow 4a^2 - 4(10 - 3a) < 0$

$\Rightarrow \quad a^2 + 3a - 10 < 0$

$\Rightarrow \quad (a + 5)(a - 2) < 0 \Rightarrow a \in (-5, 2)$

32. (a) $x^2 + px + q = 0$

Let roots be α and α^2

$\Rightarrow \quad \alpha + \alpha^2 = -p, \alpha\alpha^2 = q \Rightarrow \alpha = q^{1/3}$

$\therefore \quad (q)^{1/3} + (q^{1/3})^2 = -p$

Taking cube of both sides, we get

$q + q^2 + 3q(q^{1/3} + q^{2/3}) = -p^3$

$\Rightarrow \quad q + q^2 - 3pq = -p^3$

$\Rightarrow \quad p^3 + q^2 - q(3p - 1) = 0$

33. (a) $\because$ a, b, c are sides of a triangle and $a \neq b \neq c$

$\therefore \quad |a - b| < |c| \Rightarrow a^2 + b^2 - 2ab < c^2$

Similarly, we have

$b^2 + c^2 - 2bc < a^2$; $c^2 + a^2 - 2ca < b^2$

On adding, we get

$a^2 + b^2 + c^2 < 2(ab + bc + ca)$

$\Rightarrow \quad \dfrac{a^2 + b^2 + c^2}{ab + bc + ca} < 2$ ....(1)

$\because$ Roots of the given equation are real

$\therefore \quad (a + b + c)^2 - 3\lambda(ab + bc + ca) \geq 0$

$\Rightarrow \quad \dfrac{a^2 + b^2 + c^2}{ab + bc + ca} \geq 3\lambda - 2$ ....(2)

From (1) and (2), we get $3\lambda - 2 < 2 \Rightarrow \lambda < \dfrac{4}{3}$.

34. (d) As α, β are the roots of $x^2 - px + r = 0$

$\therefore \quad \alpha + \beta = p$ ....(1)

and $\alpha\beta = r$ ....(2)

Also $\dfrac{\alpha}{2}, 2\beta$ are the roots of $x^2 - qx + r = 0$

$\therefore \quad \dfrac{\alpha}{2} + 2\beta = q$ or $\alpha + 4\beta = 2q$ ....(3)

Solving (1) and (3) for α and β ,we get

$\beta = \frac{1}{3}(2q-p)$ and $\alpha = \frac{2}{3}(2q-q)$

Substituting values of α and β, in equation (2),

we get $\frac{2}{9}(2p-q)(2q-p)=r.$

35. **(b)** Given that $\alpha+\beta=-p$ and $\alpha^3+\beta^3=q$

$\Rightarrow (\alpha+\beta)^3-3\alpha\beta(\alpha+\beta)=q$

$\Rightarrow -p^3-3\alpha\beta(-p)=q \Rightarrow \alpha\beta=\frac{p^3+q}{3p}$

Now for required quadratic equation,

sum of roots $=\frac{\alpha}{\beta}+\frac{\beta}{\alpha}=\frac{\alpha^2+\beta^2}{\alpha\beta}$

$$=\frac{(\alpha+\beta)^2-2\alpha\beta}{\alpha\beta}=\frac{p^2-2\left(\frac{p^3+q}{3p}\right)}{\frac{p^3+q}{3p}}$$

$$=\frac{3p^3-2p^3-2q}{p^3+q}=\frac{p^3-2q}{p^3+q}$$

and product of roots $=\frac{\alpha}{\beta}.\frac{\beta}{\alpha}=1$

$\therefore$ Required equation is

$$x^2-\left(\frac{p^3-2q}{p^3+q}\right)x+1=0$$

or $(p^3+q)x^2-(p^3-2q)x+(p^3+q)=0$

36. **(c)** We have $(2x)^{\ell n2}=(3y)^{\ell n3}$

$\Rightarrow \ell n2.\ \ell n2x=\ell n3.\ \ell n3y$

$\Rightarrow \ell n2.\ \ell n2x=\ell n3.\ (\ell n3+\ell ny)$ ...(1)

Also given $3^{\ell nx}=2^{\ell ny}$

$\Rightarrow \ell nx.\ \ell n3=\ell ny.\ \ell n2$

$\Rightarrow \ell ny=\frac{\ell nx.\ell n3}{\ell n2}$

Substituting this value of ℓny in equation (1), we get

$\ell n2.\ \ell n2x=\ell n3\left[\ell n3+\frac{\ell nx.\ell n3}{\ell n2}\right]$

$\Rightarrow (\ell n2)^2\ \ell n2x=(\ell n3)^2\ \ell n2+(\ell n3)^2\ \ell nx$

$\Rightarrow (\ell n2)^2\ \ell n2x=(\ell n3)^2(\ell n2+\ell nx)$

$\Rightarrow (\ell n2)^2\ \ell n2x-(\ell n3)^2\ \ell n2x=0$

$\Rightarrow [(\ell n2)^2-(\ell n3)^2]\ \ell n2x=0$

$\Rightarrow \ell n2x=0$

$\Rightarrow 2x=1$ or $x=\frac{1}{2}$

37. **(c)** $\because \alpha, \beta$ are the roots of $x^2-6x-2=0$

$\therefore \alpha^2-6\alpha-2=0$

$\Rightarrow \alpha^{10}-6\alpha^9-2\alpha^8=0$

$\Rightarrow \alpha^{10}-2\alpha^8=6\alpha^9$ ...(1)

Similarly $\beta^{10}-2\beta^8=6\beta^9$ ...(2)

From equation (1) and (2)

$\alpha^{10}-\beta^{10}-2(\alpha^8-\beta^8)=6(\alpha^9-\beta^9)$

$\Rightarrow a_{10}-2a_8=6a_9$

$\Rightarrow \frac{a_{10}-2a_8}{2a_9}=3$

38. **(b)** Let α be the common root of given equations, then

$\alpha^2+b\alpha-1=0$ ...(1)

and $\alpha^2+\alpha+b=0$ ...(2)

Subtracting (2) from (1), we get

$(b-1)\alpha-(b+1)=0$

or $\alpha=\frac{b+1}{b-1}$

Substituting this value of α in equation (1), we get

$\left(\frac{b+1}{b-1}\right)^2+b\left(\frac{b+1}{b-1}\right)-1=0$ or $b^3+3b=0$

$\Rightarrow b=0,\ i\sqrt{3},-i\sqrt{3}$

(D) MCQ's WITH ONE OR MORE THAN ONE CORRECT :

1. **(c,d)** Let $y=\frac{(x-a)(x-b)}{(x-c)}$

$\Rightarrow (x-c)y=x^2-(a+b)x+ab$

$\Rightarrow x^2-(a+b+y)x+ab+cy=0$

Here, $\Delta=(a+b+y)^2-4(ab+cy)$

$=y^2+2y(a+b-2c)+(a-b)^2.$

Since x is real and y assumes all real values.

$\therefore$ $\Delta\ge 0$ for all real values of y

$\Rightarrow$ $y^2+2y(a+b-2c)+(a-b)^2\ge 0$

Now we know that the sign of a quad is same as of coeff of y^2 provided its descriminant

$B^2-4AC<0$

This will be so if,

$4(a+b-2c)^2-4(a-b)^2<0$

or $4(a+b-2c+a-b)(a+b-2c-a+b)<0$

$\Rightarrow$ $16(a-c)(b-c)<0$

$\Rightarrow$ $16(c-a)(c-b)<0$ ....(1)

Now,

If $a<b$ then from inequation (1), we get $c\in(a,b)$

$\Rightarrow$ $a<c<b$

or If $a>b$ then from inequation (1) we get, $c\in(a,b)$

$\Rightarrow b<c<a$ or $a>c>b$

Thus, we observe that both (c) and (d) are the correct answer.

2. (a, d) **KEY CONCEPT :** Wavy curve method :

Let $f(x)=(x-\alpha_1)(x-\alpha_2)...(x-\alpha_n)$

To find sign of $f(x)$, plot α_1, α_2, ... α_n on number line in ascending order of magnitude. Starting from right extreme put + ve, –ve signs alternately. $f(x)$ is positive in the intervals having + ve sign and negative in the intervals having –ve sign.

We have,

$$f(x)=\frac{2x-1}{2x^3+3x^2+x}=\frac{2x-1}{x(2x+1)(x+1)}$$

NOTE THIS STEP : Critical points are $x = 1/2, 0, -1/2, -1$

On number line by wavy method, we have

$$\overset{+}{-\infty}\quad -1\quad \overset{-}{}\quad -1/2\quad \overset{+}{}\quad 0\quad \overset{-}{}\quad 1/2\quad \overset{+}{}\quad \infty$$

For $f(x)>0$, when

$x\in(-\infty,-1)\cup(-1/2,0)\cup(1/2,\infty)$

Clearly S contains $(-\infty,-3/2)$ and $(1/2, 3)$

3. (b) Given that a, b, c are distinct +ve numbers. The expression whose sign is to be checked is $(b+c-a)(c+a-b)(a+b-c)-abc$.

As this expression is symmetric in a, b, c, without loss of generality, we can assume that $a<b<c$.

Then $c-a=+$ ve and $c-b=+$ ve

$\therefore\quad b+c-a=+$ ve and $c+a-b=+$ ve

But $a+b-c$ may be + ve or – ve.

Case I : If $a+b-c=+$ ve then we can say that a, b, c, are such that sum of any two is greater than the 3rd. Consider

$x=a+b-c,\quad y=b+c-a,\quad z=c+a-b$

then x, y, z all are + ve.

and then $a=\dfrac{x+z}{2}, b=\dfrac{y+x}{2}, c=\dfrac{z+y}{2}$

Now we know that A.M. > G.M. for distinct real numbers

$$\therefore\quad \frac{x+y}{2}>\sqrt{xy}, \frac{y+z}{2}>\sqrt{yz}, \frac{z+x}{2}>\sqrt{zx}$$

$$\Rightarrow\left(\frac{x+y}{2}\right)\left(\frac{y+z}{2}\right)\left(\frac{z+x}{2}\right)>xyz$$

$$\Rightarrow abc>(a+b-c)(b+c-a)(c+a-b)$$

$$\Rightarrow (b+c-a)(c+a-b)(a+b-c)-abc<0$$

Case II : If $a+b-c=$ -ve then

$(b+c-a)(c+a-b)(a+b-c)-abc$

$=(+\text{ve})(+\text{ve})(-\text{ve})-(+\text{ve})$

$=(-\text{ve})-(+\text{ve})=(-\text{ve})$

$\Rightarrow (b+c-a)(c+a-b)(a+b-c)<abc$

Hence in either case given expression is –ve.

4. (b) Given that a, b, c, d, p are real and distinct numbers such that

$$(a^2+b^2+c^2)p^2-2(ab+bc+cd)p+(b^2+c^2+d^2)\le 0$$

$$\Rightarrow (a^2p^2+b^2p^2+c^2p^2)-(2abp+2bcp+2cdp)+(b^2+c^2+d^2)\le 0$$

$$\Rightarrow (a^2p^2-2abp+b^2)+(b^2p^2-2bcp+c^2)+(c^2p^2-2cdp+d^2)\le 0$$

$$\Rightarrow (ap-b)^2+(bp-c)^2+(cp-d)^2\le 0$$

Being sum of perfect squares, LHS can never be –ve, therefore the only possibility is

$(ap-b)^2+(bp-c)^2+(cp-d)^2=0$

Which is possible only when each term is zero individually i.e.

$ap-b=0;\ bp-c=0;\ cp-d=0$

$$\Rightarrow \frac{b}{a}=p;\frac{c}{b}=p;\frac{d}{c}=p\Rightarrow \frac{b}{a}=\frac{c}{b}=\frac{d}{c}=p$$

$\Rightarrow\ a, b, c, d$ are in G.P.

5. (a, b, c) The given equation is, $x^{\frac{3}{4}(\log_2 x)^2+\log_2 x-\frac{5}{4}}=\sqrt{2}$

For $x>0$, taking log on both sides to the base x, we get

$$\frac{3}{4}(\log_2 x)^2+(\log_2 x)-\frac{5}{4}=\log_x\sqrt{2}=\frac{1}{2}\log_x 2$$

NOTE THIS STEP

Let $\log_2 x=y$, then we get, $\dfrac{3}{4}y^2+y-\dfrac{5}{4}=\dfrac{1}{2y}$

$$\Rightarrow 3y^3+4y^2-5y-2=0$$

$$\Rightarrow (y-1)(y+2)(3y+1)=0\Rightarrow y=1,-2,-1/3$$

$$\Rightarrow \log_2 x=1,-2,-1/3\Rightarrow x=2,2^{-2},2^{-1/3}$$

$\Rightarrow x=2,\dfrac{1}{4},\dfrac{1}{2^{1/3}}$ (All accepted as > 0)

$\therefore$ There are three real solution in which one is irrational.

6. (d) Let $x_1, x_2, ..., x_n$ be the n +ve numbers

According to the question,

$x_1x_2x_3....x_n=1$ ....(1)

We know for +ve no.'s A.M. ≥ G.M.

$$\Rightarrow \frac{x_1+x_2+...+x_n}{n}\ge\sqrt[n]{x_1x_2...x_n}$$

$\Rightarrow \dfrac{x_1+x_2+...+x_n}{n}\ge 1$ [Using eq. (1)]

$\Rightarrow x_1+x_2+...+x_n\ge n$.

7. (a) We have $240=2^4.3.5$.

Divisors of 240 are

1, 2, 4, 8, 16
3, 6, 12, 24, 48
5, 10, 20, 40, 80,
15, 30, 60, 120, 240

Out of these divisors just 4 divisors viz., 2, 6, 10, 30 are of the form 4n +2.

8. **(a, b, c)**

$3^x = 4^{x-1} \Rightarrow x\log 3 = 2(x-1)\log 2$

$\Rightarrow x = \dfrac{2\log 2}{2\log 2 - \log 3}$

$\Rightarrow x = \dfrac{2\log_3 2}{2\log_3 2 - 1} = \dfrac{2}{2-\log_2 3}$

Also $x = \dfrac{1}{1-\frac{1}{2}\log_2 3} = \dfrac{1}{1-\log_4 3}$

(E) SUBJECTIVE PROBLEMS :

1. $4^x - 3^{x-1/2} = 3^{x+1/2} - \dfrac{(2^2)^x}{2}$

$\Rightarrow 4^x - \dfrac{3^x}{\sqrt{3}} = 3^x\sqrt{3} - \dfrac{4^x}{2}$

$\Rightarrow \dfrac{3}{2}\cdot 4^x = 3^x\left(\sqrt{3} + \dfrac{1}{\sqrt{3}}\right) \Rightarrow \dfrac{3}{2}\cdot 4^x = 3^x \dfrac{4}{\sqrt{3}}$

$\Rightarrow \dfrac{4^{x-1}}{4^{1/2}} = \dfrac{3^{x-1}}{\sqrt{3}} \Rightarrow 4^{x-3/2} = 3^{x-3/2}$

$\Rightarrow \left(\dfrac{4}{3}\right)^{x-3/2} = 1 \Rightarrow x - \dfrac{3}{2} = 0$

$\Rightarrow x = 3/2$

2. RHS $= (m-1, n+1) + x^{m-n-1}(m-1, n)$

$= \dfrac{(1-x^{m-1})(1-x^{m-2})\dots(1-x^{m-n-1})}{(1-x)(1-x^2)\dots(1-x^{n+1})}$

$+x^{m-n-1}\left[\dfrac{(1-x^{m-1})(1-x^{m-2})\dots(1-x^{m-n})}{(1-x)(1-x^2)\dots(1-x^n)}\right]$

$= \dfrac{(1-x^{m-1})(1-x^{m-2})\dots(1-x^{m-n})}{(1-x)(1-x^2)\dots(1-x^n)}$

$\left[\dfrac{1-x^{m-n-1}}{1-x^{n+1}} + x^{m-n-1}\right]$

$\left[\dfrac{1-x^{m-n-1}+x^{m-n-1}-x^m}{1-x^{n+1}}\right]$

$= \dfrac{(1-x^m)(1-x^{m-1})\dots(1-x^{m-n})}{(1-x)(1-x^2)\dots(1-x^n)(1-x^{n+1})}$

$= (m, n+1) =$ L.H.S. Hence Proved

3. $\sqrt{x+1} = 1 + \sqrt{x-1}$

Squaring both sides, we get

$x+1 = 1 + x - 1 + 2\sqrt{x-1} \Rightarrow 1 = 2\sqrt{x-1}$.

$\Rightarrow 1 = 4(x-1)$

$\Rightarrow x = 5/4$

4. Given $a > 0$, so we have to consider two cases : $a \neq 1$ and $a = 1$. Also it is clear that $x > 0$ and $x \neq 1$, $ax \neq 1$, $a^2x \neq 1$.

Case I : If $a > 0, \neq 1$

then given equation can be simplified as

$\dfrac{2}{\log_a x} + \dfrac{1}{1+\log_a x} + \dfrac{3}{2+\log_a x} = 0$

Putting $\log_a x = y$, we get

$2(1+y)(2+y) + y(2+y) + 3y(1+y) = 0$

$\Rightarrow 6y^2 + 11y + 4 = 0 \Rightarrow y = -4/3$ and $-1/2$

$\Rightarrow \log_a x = -4/3$ and $\log_a x = -1/2$

$\Rightarrow x = a^{-4/3}$ and $x = a^{-1/2}$

Case II : If $a = 1$ then equation becomes

$2\log_x 1 + \log_x 1 + 3\log_x 1 = 6\log_x 1 = 0$

which is true $\forall\ x > 0, \neq 1$

Hence solution is if $a = 1, x > 0, \neq 1$

if $a > 0, \neq 1$; $x = a^{-1/2}, a^{-4/3}$

5. Let $x = \dfrac{\sqrt{26-15\sqrt{3}}}{5\sqrt{2} - \sqrt{38+5\sqrt{3}}}$

$\Rightarrow x^2 = \dfrac{26-15\sqrt{3}}{50+38+5\sqrt{3}-10\sqrt{76+10\sqrt{3}}}$

$\Rightarrow x^2 = \dfrac{26-15\sqrt{3}}{88+5\sqrt{3}-10\sqrt{75+1+10\sqrt{3}}}$

$\Rightarrow x^2 = \dfrac{26-15\sqrt{3}}{88+5\sqrt{3}-10\sqrt{(5\sqrt{3})^2+(1)^2+2\times 5\sqrt{3}\times 1}}$

$\Rightarrow x^2 = \dfrac{26-15\sqrt{3}}{88+5\sqrt{3}-10\sqrt{(5\sqrt{3}+1)^2}}$

$\Rightarrow x^2 = \dfrac{26-15\sqrt{3}}{88+5\sqrt{3}-10(5\sqrt{3}+1)}$

$\Rightarrow x^2 = \dfrac{26-15\sqrt{3}}{88+5\sqrt{3}-50\sqrt{3}-10} = \dfrac{26-15\sqrt{3}}{78-45\sqrt{3}}$

$= \dfrac{26-15\sqrt{3}}{3(26-15\sqrt{3})} = \dfrac{1}{3}$, which is a rational number.

6. $x^2 + y^2 - 2x \geq 0 \Rightarrow x^2 - 2x + 1 + y^2 \geq 1$

$\Rightarrow (x-1)^2 + y^2 \geq 1$ which represents the boundary and exterior region of the circle with centre at (1,0) and radius as 1.

For $3x - y \leq 12$, the corresponding equation is $3x - y = 12$; any two points on it can be taken as (4, 0), (2, – 6). Also putting (0, 0) in given inequation, we get $0 \leq 12$ which is true.

$\therefore$ given inequation represents that half plane region of line $3x - y = 12$ which contains origin.

For $y \le x$, the corresponding equation $y = x$ has any two points on it as (0, 0) and (1, 1). Also putting (2, 1) in the given inequation, we get $1 \le 2$ which is true, so $y \le x$ represents that half plane which contains the points (2, 1). $y \ge 0$ represents upper half cartesian plane.
Combining all we find the solution set as the shaded region in the graph.

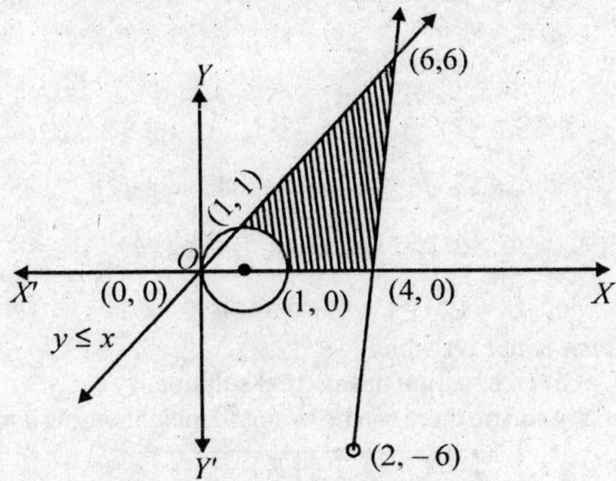

7. There are two parts of this question
$(5x-1) < (x+1)^2$ and $(x+1)^2 < (7x-3)$
Taking first part

$(5x-1) < (x+1)^2 \Rightarrow 5x-1 < x^2+2x+1$

$\Rightarrow x^2-3x+2 > 0 \Rightarrow (x-1)(x-2) > 0$

(using wavy method) [+ on $(-\infty, 1)$, – on $(1, 2)$, + on $(2, +\infty)$]

$\Rightarrow x < 1$ or $x > 2$ ....(1)

Taking second part

$(x+1)^2 < (7x-3) \Rightarrow x^2-5x+4 < 0$

$\Rightarrow (x-1)(x-4) < 0$

(using wavy method) [+ on $(-\infty, 1)$, – on $(1, 4)$, + on $(4, +\infty)$]

$\Rightarrow 1 < x < 4$ ....(2)

Combining (1) and (2) [taking common solution], we get $2 < x < 4$ but x is an integer therefore $x = 3$.

8. $\because$ α, β are the roots of $x^2+px+q=0$

$\therefore$ $\alpha+\beta=-p, \quad \alpha\beta=q$

$\because$ γ, δ are the roots of $x^2+rx+s=0$

$\therefore$ $\gamma+\delta=-r, \gamma\delta=s$

Now, $(\alpha-\gamma)(\alpha-\delta)(\beta-\gamma)(\beta-\delta)$

$= [\alpha^2-(\gamma+\delta)\alpha+\gamma\delta][\beta^2-(\gamma+\delta)\beta+\gamma\delta]$

$= [\alpha^2+r\alpha+s][\beta^2+r\beta+s]$

[$\because \alpha, \beta$ are roots of $x^2+px+q=0$

$\therefore \alpha^2+p\alpha+q=0$ and $\beta^2+p\beta+q=0$]

$= [(r-p)\alpha+(s-q)][(r-p)\beta+(s-q)]$

$= (r-p)^2\alpha\beta+(r-p)(s-q)(\alpha+\beta)+(s-q)^2$

$= q(r-p)^2-p(r-p)(s-q)+(s-q)^2$

Now if the equations $x^2+px+q=0$ and $x^2+rx+s=0$ have a common root say α, then $\alpha^2+p\alpha+q=0$ and $\alpha^2+r\alpha+s=0$

$$\Rightarrow \frac{\alpha^2}{ps-qr}=\frac{\alpha}{q-s}=\frac{1}{r-p}$$

$$\Rightarrow \alpha^2=\frac{ps-qr}{r-p} \text{ and } \alpha=\frac{q-s}{r-p}$$

$\Rightarrow (q-s)^2=(r-p)(ps-qr)$ which is the required condition.

9. We know that for sides a, b, c of a Δ

$(a-b)^2 \ge 0$

$\Rightarrow$ $a^2+b^2 \ge 2ab$ ....(1)

Similarly $b^2+c^2 \ge 2bc$ ....(2)

$c^2+a^2 \ge 2ca$ ....(3)

Adding the three inequations, we get

$2(a^2+b^2+c^2) \ge 2(ab+bc+ca)$

$\Rightarrow a^2+b^2+c^2 \ge ab+bc+ca$

Adding 2 $(ab+bc+ca)$ to both sides, we get

$(a+b+c)^2 \ge 3(ab+bc+ca)$

or $3(ab+bc+ca) \le (a+b+c)^2$ ....(A)

Also $c < a+b$ (triangle inequality)

$\Rightarrow$ $c^2 < ac+bc$ ....(4)

Similarly $b^2 < ab+bc$ ....(5)

$a^2 < ab+ca$ ....(6)

Adding (4), (5) and (6), we get

$a^2+b^2+c^2 < 2(ab+bc+ca)$

Adding 2 $(ab+bc+ca)$ to both sides, we get

$\Rightarrow$ $(a+b+c)^2 < 4(ab+bc+ca)$ ....(B)

Combining (A) and (B), we get

$3(ab+bc+ca) \le (a+b+c)^2 < 4(ab+bc+ca)$

First two expressions will be equal for $a = b = c$.

10. Given that $n^4 < 10^n$ for a fixed +ve integer $n \ge 2$.
To prove that $(n+1)^4 < 10^{n+1}$
Proof: Since $n^4 < 10^n \Rightarrow 10n^4 < 10^{n+1}$ ....(1)
So it is sufficient to prove that $(n+1)^4 < 10n^4$

$$\text{Now } \left(\frac{n+1}{n}\right)^4 = \left(1+\frac{1}{n}\right)^4 \le \left(1+\frac{1}{2}\right)^4 \quad [\because n \ge 2]$$

$$= \frac{81}{16} < 10$$

$\Rightarrow (n+1)^4 < 10n^4$ .... (2)

From (1) and (2), $(n+1)^4 < 10^{n+1}$

11. $y=\sqrt{\dfrac{(x+1)(x-3)}{(x-2)}}$

y will take all real values if $\dfrac{(x+1)(x-3)}{(x-2)}\geq 0$

By wavy method

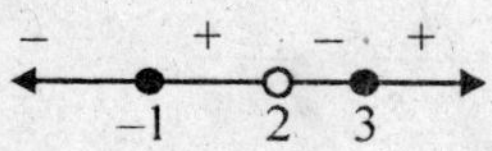

$-\quad +\quad -\quad +$

$-1\quad 2\quad 3$

$x\in[-1,2)\cup[3,\infty)$

[2 is not included as it makes denominator zero, and hence y an undefined number.]

12. The given equations are $3x+my-m=0$ and $2x-5y-20=0$

Solving these equations by cross product method, we get

$$\frac{x}{-20m-5m}=\frac{y}{-2m+60}=\frac{1}{-15-2m}$$ **NOTE THIS STEP**

$$\Rightarrow\quad x=\frac{25m}{2m+15},\ y=\frac{2m-60}{2m+15}$$

For $x>0\Rightarrow\dfrac{25m}{2m+15}>0$

$$\Rightarrow m<-\frac{15}{2}\text{ or } m>0 \qquad ....(1)$$

[using wavy method]

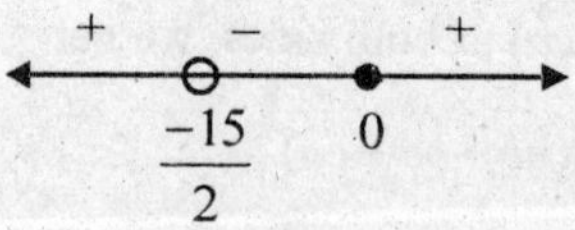

$+\quad -\quad +$

$\frac{-15}{2}\quad 0$

For $y>0\Rightarrow\dfrac{2(m-30)}{2m+15}>0$

$$\Rightarrow m<-\frac{15}{2}\text{ or } m>30 \qquad ....(2)$$

[using wavy method]

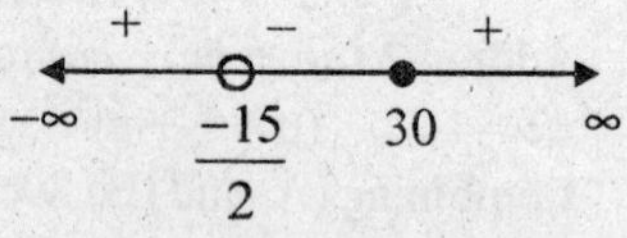

$+\quad -\quad +$

$-\infty\quad \frac{-15}{2}\quad 30\quad \infty$

Combining (1) and (2), we get the common values of m as follows :

$$m<-\frac{15}{2}\text{ or } m>30 \quad\therefore\ m\in\left(-\infty,\frac{-15}{2}\right)\cup(30,\infty)$$

13. The given system is

$x+2y+z=1$ ....(1)

$2x-3y-\omega=2$ ....(2)

where $x, y, z,\ \omega\geq 0$

Multiplying eqn. (1) by 2 and subtracting from (2), we get

$$7y+2z+\omega=0\Rightarrow\ \omega=-(7y+2z)$$

Now if $y, z>0$, $\omega<0$ (not possible)

If $y=0, z=0$ then $x=1$ and $\omega=0$.

$\therefore$ The only solution is $x=1, y=0, z=0, \omega=0$.

14. $e^{\sin x}-e^{-\sin x}-4=0$

Let $e^{\sin x}=y$ then $e^{-\sin x}=1/y$

$\therefore$ Equation becomes, $y-\dfrac{1}{y}-4=0$

$\Rightarrow\ y^2-4y-1=0$

$\Rightarrow\ y=2+\sqrt{5}, 2-\sqrt{5}$

But y is real +ve number,

$\therefore\ y\neq 2-\sqrt{5}\Rightarrow y=2+\sqrt{5}$

$\Rightarrow\ e^{\sin x}=2+\sqrt{5}\Rightarrow\ \sin x=\log_e(2+\sqrt{5})$

But $2+\sqrt{5}>e\Rightarrow\ \log_e(2+\sqrt{5})>\log_e e$

$\Rightarrow\ \log_e(2+\sqrt{5})>1$ Hence, $\sin x>1$

Which is not possible.

$\therefore$ Given equation has no real solution.

15. For any square there can be at most 4, neighbouring squares.

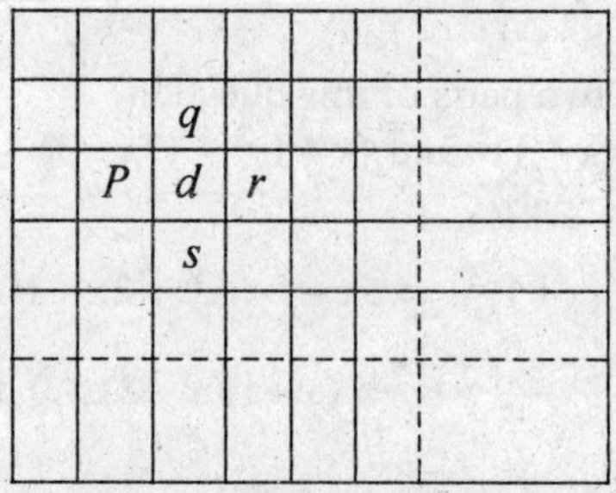

	q		
P	d	r	
	s		

Let for a square having largest number d, p, q, r, s be written then

According to the question,

$$p+q+r+s=4d$$

$$\Rightarrow\quad (d-p)+(d-q)+(d-r)+(d-s)=0$$

Sum of four +ve numbers can be zero only if these are zero individually

$\therefore\ d-p=0=d-q=d-r=d-s$

$\Rightarrow\quad p=q=r=s=d$

$\Rightarrow$ all the numbers written are same.

Hence Proved.

16. Let α, β be the roots of eq. $ax^2+bx+c=0$

According to the question,

$\beta=\alpha^n$

Also $\alpha+\beta=-b/a$; $\alpha\beta=c/a$

$$\alpha\beta=\frac{c}{a}\Rightarrow\alpha.\alpha^n=\frac{c}{a}\Rightarrow\alpha=\left(\frac{c}{a}\right)^{\frac{1}{n+1}}$$

then $\alpha+\beta=-b/a\Rightarrow\alpha+\alpha^n=\dfrac{-b}{a}$

$$\text{or}\quad\left(\frac{c}{a}\right)^{\frac{1}{n+1}}+\left(\frac{c}{a}\right)^{\frac{n}{n+1}}=\frac{-b}{a}$$

$$\Rightarrow\quad a.\left(\frac{c}{a}\right)^{\frac{1}{n+1}}+a.\left(\frac{c}{a}\right)^{\frac{n}{n+1}}+b=0$$

$\Rightarrow\ a^{\frac{n}{n+1}}c^{\frac{1}{n+1}} + a^{\frac{1}{n+1}}c^{\frac{n}{n+1}} + b = 0$

$\Rightarrow\ (a^n c)^{\frac{1}{n+1}} + (ac^n)^{\frac{1}{n+1}} + b = 0$

Hence Proved.

17. $x^2 - 3x + 2 > 0,\quad x^2 - 3x - 4 \le 0$

$\Rightarrow (x-1)(x-2) > 0$ and $(x-4)(x+1) < 0$

-∞ –1 1 2 4 ∞

$\Rightarrow\ x \in (-\infty, 1) \cup (2, \infty)$ and $x \in [-1, 4]$

$\therefore$ Common solution is $[-1, 1) \cup (2, 4]$

18. The given equation is

$(5+2\sqrt{6})^{x^2-3} + (5-2\sqrt{6})^{x^2-3} = 10$ ....(1)

Let $(5+2\sqrt{6})^{x^2-3} = y$ ....(2)

then $(5-2\sqrt{6})^{x^2-3} = \left(\frac{(5-2\sqrt{6})(5+2\sqrt{6})}{5+2\sqrt{6}}\right)^{x^2-3}$

$= \left(\frac{25-24}{5+2\sqrt{6}}\right)^{x^2-3} = \left(\frac{1}{5+2\sqrt{6}}\right)^{x^2-3} = \frac{1}{y}$ (Using (2))

$\therefore$ The given equation (1) becomes $y + \frac{1}{y} = 10$

$\Rightarrow\ y^2 - 10y + 1 = 0$

$\Rightarrow\ y = \frac{10 \pm \sqrt{100-4}}{2} = \frac{10 \pm 4\sqrt{6}}{2}$

$\Rightarrow\ y = 5 + 2\sqrt{6}$ or $5 - 2\sqrt{6}$

Consider, $y = 5 + 2\sqrt{6}$

$\Rightarrow\ (5+2\sqrt{6})^{x^2-3} = (5+2\sqrt{6})$

$\Rightarrow\ x^2 - 3 = 1 \Rightarrow x^2 = 4 \Rightarrow x = \pm 2$

Again consider

$y = 5 - 2\sqrt{6} = \frac{1}{5+2\sqrt{6}} = (5+2\sqrt{6})^{-1}$

$\Rightarrow\ (5+2\sqrt{6})^{x^2-3} = (5+2\sqrt{6})^{-1} \Rightarrow x^2 - 3 = -1$

$\Rightarrow\ x^2 = 2$

$\Rightarrow\ x = \pm\sqrt{2}$

Hence the solutions are $2, -2, \sqrt{2}, -\sqrt{2}$.

19. The given equation is,

$x^2 - 2a|x-a| - 3a^2 = 0$

Here two cases are possible.

Case I : $x - a > 0$ then $|x-a| = x - a$

$\therefore$ Eq. becomes

$x^2 - 2a(x-a) - 3a^2 = 0$

or $x^2 - 2ax - a^2 = 0 \Rightarrow x = \frac{2a \pm \sqrt{4a^2 + 4a^2}}{2}$

$\Rightarrow x = a \pm a\sqrt{2}$

Case II : $x - a < 0$ then $|x-a| = -(x-a)$

$\therefore$ Eq. becomes

$x^2 + 2a(x-a) - 3a^2 = 0$

or $x^2 + 2ax - 5a^2 = 0$

$\Rightarrow\ x = \frac{-2a \pm \sqrt{4a^2 + 20a^2}}{2}$

$\Rightarrow\ x = \frac{-2a \pm 2a\sqrt{6}}{2}$

$x = -a \pm a\sqrt{6}$

Thus the solution set is $\{a \pm a\sqrt{2}, -a \pm a\sqrt{6}\}$

20. We are given $\frac{2x}{2x^2+5x+2} > \frac{1}{x+1}$

$\Rightarrow\ \frac{2x}{2x^2+5x+2} - \frac{1}{x+1} > 0$

$\Rightarrow\ \frac{2x^2+2x-2x^2-5x-2}{(2x^2+5x+2)(x+1)} > 0$

$\Rightarrow\ \frac{-3x-2}{(2x+1)(x+1)(x+2)} > 0$

$\Rightarrow\ \frac{(3x+2)}{(x+1)(x+2)(2x+1)} < 0$

$\Rightarrow\ \frac{(3x+2)(x+1)(x+2)(2x+1)}{(x+1)^2(x+2)^2(2x+1)^2} < 0$

$\Rightarrow\ (3x+2)(x+1)(x+2)(2x+1) < 0$ ....(1)

NOTE THIS STEP : Critical pts are $x = -2/3, -1, -2, -1/2$

On number line

+ – + – +

–∞ –2 –1 –2/3 –1/2 ∞

Clearly Inequality (1) holds for,

$x \in (-2, -1) \cup (-2/3, -1/2)$

$[as\ x \neq -2, -1, -2/3, -1/2]$

21. We are given that α_1, α_2 are the roots of

$ax^2 + bx + c = 0$

$\therefore \quad \alpha_1+\alpha_2=-\frac{b}{a};\alpha_1\alpha_2=\frac{c}{a}$ ....(1)

and β_1, β_2 are the roots of $px^2+qx+r=0$

$\therefore \quad \beta_1+\beta_2=-\frac{q}{p};\beta_1\beta_2=\frac{r}{p}$ ....(2)

The system of equations, $\alpha_1 y+\alpha_2 z=0$
and $\beta_1 y+\beta_2 z=0$ has a non trivial solution.

$\therefore$ We must have $\begin{vmatrix}\alpha_1 & \beta_1\\ \alpha_2 & \beta_2\end{vmatrix}=0$

NOTE THIS STEP :

$\Rightarrow \alpha_1\beta_2-\alpha_2\beta_1=0 \Rightarrow \frac{\alpha_1}{\alpha_2}=\frac{\beta_1}{\beta_2}$

By componendo and dividendo, we get

$\frac{\alpha_1+\alpha_2}{\alpha_1-\alpha_2}=\frac{\beta_1+\beta_2}{\beta_1-\beta_2}$

$\Rightarrow (\alpha_1+\alpha_2)(\beta_1-\beta_2)=(\alpha_1-\alpha_2)(\beta_1+\beta_2)$

$\Rightarrow (\alpha_1+\alpha_2)^2[(\beta_1+\beta_2)^2-4\beta_1\beta_2]$

$=[(\alpha_1+\alpha_2)^2-4\alpha_1\alpha_2](\beta_1+\beta_2)^2$

Using equations (1) and (2) we get

$$\frac{b^2}{a^2}\left[\frac{q2}{p^2}-\frac{4r}{p}\right]=\frac{q^2}{p^2}\left[\frac{b^2}{a^2}-\frac{4c}{a}\right]$$

$$\Rightarrow \frac{b^2q2}{a^2p^2}-\frac{4b^2r}{a^2p}=\frac{q^2b^2}{p^2a^2}-\frac{4cq^2}{ap^2} \Rightarrow \frac{-4b^2r}{a^2p}=\frac{-4cq^2}{ap^2}$$

$$\Rightarrow \frac{b^2r}{a}=\frac{sq^2}{p} \Rightarrow \frac{b^2}{q^2}=\frac{ac}{pr} \quad \text{Hence Proved}$$

22. The Given equation is,
$|x^2+4x+3|+2x+5=0$
Now there can be two cases.

Case I : $x^2+4x+3\ge 0 \Rightarrow (x+1)(x+3)\ge 0$

$\Rightarrow x\in(-\infty,-3]\cup[-1,\infty)$ ....(i)

Then given equation becomes,

$\Rightarrow x^2+6x+8=0$

$\Rightarrow (x+4)(x+2)=0 \Rightarrow x=-4,-2$

But $x=-2$ does not satisfy (i), hence rejected

$\therefore \quad x=-4$ is the sol.

Case II : $x^2+4x+3<0$

$\Rightarrow (x+1)(x+3)<0$

$\Rightarrow x\in(-3,-1)$ ....(ii)

Then given equation becomes,

$-(x^2+4x+3)+2x+5=0$

$\Rightarrow -x^2-2x+2=0 \Rightarrow x^2+2x-2=0$

$\Rightarrow x=\frac{-2\pm\sqrt{4+8}}{2}$

$\Rightarrow x=-1+\sqrt{3},-1-\sqrt{3}$

Out of which $x=-1-\sqrt{3}$ is sol.

Combining the two cases we get the solutions of given equation as $x=-4,-1-\sqrt{3}$.

23. Given that for $a, b, c\in R$, $ax^2+bx+c=0$ has two real roots α and β, where $\alpha<-1$ and $\beta>1$. There may be two cases depending upon value of a, as shown below.
In each of cases (i) and (ii) $af(-1)<0$ and $af(1)<0$

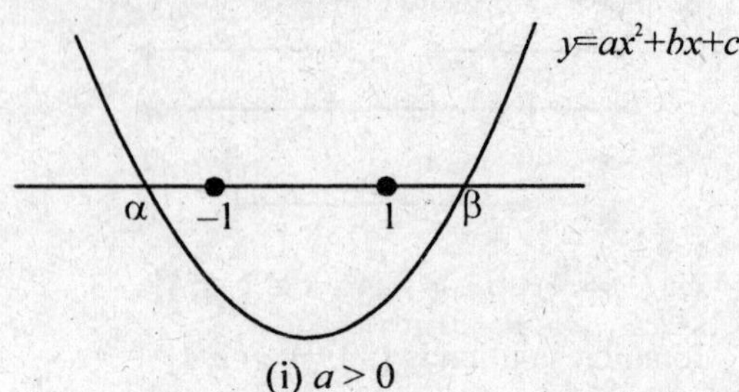

(i) $a>0$

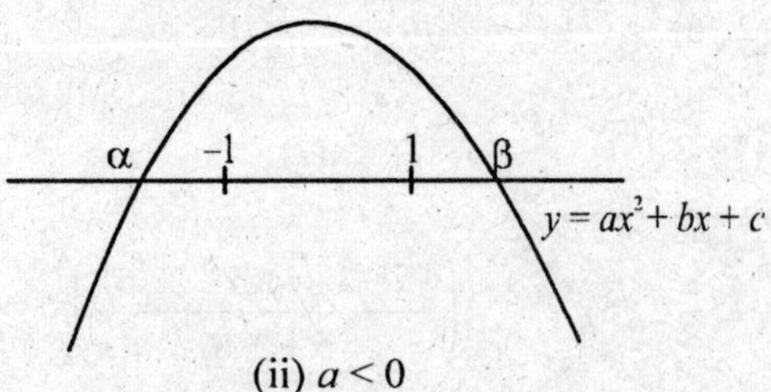

(ii) $a<0$

$\Rightarrow a(a-b+c)<0$ and $a(a+b+c)<0$

Dividing by a^2 (> 0), we get

$$1-\frac{b}{a}+\frac{c}{a}<0 \quad \text{....(1)}$$

and $$1+\frac{b}{a}+\frac{c}{a}<0 \quad \text{....(2)}$$

Combining (1) and (2) we get

$1+\left|\frac{b}{a}\right|+\frac{c}{a}<0$ **or** $1+\frac{c}{a}+\left|\frac{b}{a}\right|<0$ **Hence Proved.**

24. The given equation is,

$2^{|y|}-|2^{y-1}-1|=2^{y-1}+1$

On the basis of absolute values involved here ($|y|$ and $|2^{y-1}-1|$), there are two critical pts 0 and 1. So we shall consider three cases, when y lies in three different intervals namely $(-\infty,0)$, $[0,1]$, $(1,\infty)$

Case I : $y\in(-\infty,0)$ then

$|y|=-y$ and $|2^{y-1}-1|=1-2^{y-1}$

$\therefore$ The given equation becomes
$2^{-y}-1+2^{y-1}=2^{y-1}+1$

$\Rightarrow 2^{-y}=2 \Rightarrow y=-1\in(-\infty,0)$

Case II : $y \in [0,1]$

If $y = 0$ we get $1-\left|\frac{1}{2}-1\right| = \frac{1}{2}+1$

$$1-\frac{1}{2}=\frac{1}{2}+1 \qquad \text{(not satisfied)}$$

$\therefore$ $y = 0$ is not a sol^n

If $y = 1$ we get $2-|2^0-1| = 2^0+1$

$\Rightarrow$ $2 = 2$ (satisfied)

$\therefore$ $y = 1$ is a sol^n.

If $y \in (0, 1)$ then $|y| = y$ and $|2^{y-1}-1| = 1-2^{y-1}$

$\therefore$ The eq^n becomes

$2^y - 1 + 2^{y-1} = 2^{y-1} + 1 \Rightarrow 2^y = 2$

$\Rightarrow$ $y = 1 \notin (0,1)$

$\therefore$ $y = 1$ is the only sol^n in this case.

Case III : $y \in (1, \infty)$

then $|y| = 1, |2^{y-1}-1| = 2^{y-1}-1$

the given eq^n becomes, $2^y - 2^{y-1} + 1 = 2^{y-1} + 1$

$\Rightarrow$ $2^y - 2^y = 0$

which is satisfied for all real values of y but $y \in (1, \infty)$

$\therefore$ $(1, \infty)$ is the sol^n in this case.

Combining all the cases, we get the sol^n as $y \in \{-1\} \cup [1, \infty)$

25. $a^2 = p^2 + s^2, b^2 = (1-p)^2 + q^2$

$c^2 = (1-q)^2 + (1-r)^2$, $d^2 = r^2 + (1-s)^2$

$\therefore$ $a^2 + b^2 + c^2 + d^2 = \{p^2 + (1-p)^2\} + \{q^2 - (1-q)^2\}$
$+ \{r^2\} + (1-r)^2\} + \{s^2 + (1-s)^2\}$

where p, q, r, s all vary in the interval [0, 1].

Now consider the function

$$y^2 = x^2 + (1-x)^2, \ 0 \le x \le 1,$$

$$2y\frac{dy}{dx} = 2x - 2(1-x) = 0$$

$\Rightarrow$ $x = \frac{1}{2}$ which $\frac{d^2y}{dx^2} = 4$ i.e.+ive

Hence y is minimum at $x = \frac{1}{2}$ and its minimum

value is $\frac{1}{4}+\frac{1}{4}=\frac{1}{2}$.

Clearly value is maximum at the end pts which is 1.

$\therefore$ Minimum value of $a^2 + b^2 + c^2 + d^2 = \frac{1}{2}+\frac{1}{2}+\frac{1}{2}+\frac{1}{2} = 2$

and maximum value is $1 + 1 + 1 + 1 = 4$. Hence proved.

ALTERNATE SOLUTION :

$$x^2 + y^2 \le (x+y)^2 \le 1 \quad if \ x + y = 1$$

Here $x = p, y = 1-p$ $\therefore$ $x + y = 1$

$\therefore$ $a^2 + b^2 + c^2 + d^2 \le 1 + 1 + 1 + 1 = 4$

Again $x^2 + y^2 = (x+y)^2 - 2xy = 1 - 2xy$

NOTE THIS STEP : $\therefore$ Minimum of $(x^2+y^2) = 1-2$ (maximum of xy).

KEY CONCEPT : Now we know that product of two quantities xy is maximum when the quantities are equal provided their sum is constant.

Here $x + y = p + 1 - p = 1$ = constant.

$\therefore$ xy is maximum when $\frac{x}{1}=\frac{y}{1}=\frac{x+y}{2}=\frac{1}{2}$

$\therefore$ $x = 1/2, y = 1/2$

Minimum of $x^2 + y^2 = 1 - 2.\frac{1}{2}.\frac{1}{2} = 1 - \frac{1}{2} = \frac{1}{2}$

$\therefore$ Minimum value of

$$a^2 + b^2 + c^2 + d^2 = \frac{1}{2}+\frac{1}{2}+\frac{1}{2}+\frac{1}{2} = 2$$

$\therefore$ $2 \le a^2 + b^2 + c^2 + d^2 \le 4$.

26. We know that,

$$(\alpha-\beta)^2 = [(\alpha+\delta)-(\beta+\delta)]^2$$

$\Rightarrow$ $(\alpha+\beta)^2 - 4\alpha\beta = (\alpha+\delta+\beta+\delta)^2 - 4(\alpha+\delta)(\beta+\delta)$

$\Rightarrow$ $\frac{b^2}{a^2}-\frac{4c}{a}=\frac{B^2}{A^2}-\frac{4C}{A} \Rightarrow \frac{4ac-b^2}{a^2}=\frac{4AC-B^2}{A^2}$

[Here $\alpha+\beta = -\frac{b}{a}$, $\alpha\beta = \frac{c}{a}$,

$(\alpha+\delta)(\beta+\delta)$

$= -\frac{B}{A}$ and $(\alpha+\delta)(\beta+\delta) = \frac{C}{A}$]

Hence proved.

27. $\alpha+\beta = \frac{-b}{a}$, $\alpha\beta = \frac{c}{a}$

Roots of the equation $a^3x^2 + abcx + c^3 = 0$ are

$$x = \frac{-abc \pm \sqrt{(abc)^2 - 4a^3c^3}}{2a^3}$$

$$= \frac{\left(-\frac{b}{a}\right)\left(\frac{c}{a}\right) \pm \sqrt{\left(\frac{b}{a}\right)^2\left(\frac{c}{a}\right)^2 - 4\left(\frac{c}{a}\right)^3}}{2}$$

$$= \frac{(\alpha+\beta)(\alpha\beta) \pm \sqrt{(\alpha+\beta)^2(\alpha\beta)^2 - 4(\alpha\beta)^3}}{2}$$

$$= (\alpha\beta)\frac{((\alpha+\beta) \pm \sqrt{(\alpha-\beta)^2}}{2}$$

$$= (\alpha\beta)\frac{((\alpha+\beta) \pm (\alpha-\beta))}{2} = \alpha^2\beta, \alpha\beta^2$$

Let γ and δ be the required roots. Then

$\gamma = \alpha^2\beta$ and $\delta = \alpha\beta^2$.

ALTERNATE SOLUTION :

$ax^2 + bx + c = 0$ has roots α and β. (given)

$\Rightarrow \alpha + \beta = -b/a$ and $\alpha\beta = c/a$

Now, $a^3x^2 + abcx + c^3 = 0$

Divides the equation by c^2, we get

$$\frac{a^3}{c^2}x^2 + \frac{abcx}{c^2} + \frac{c^3}{c^2} = 0,\quad a\left(\frac{ax}{c}\right)^2 + b\left(\frac{ax}{c}\right) + c = 0$$

$\Rightarrow \frac{ax}{c} = \alpha, \beta$ are the roots

$\Rightarrow x = \frac{c}{a}\alpha, \frac{c}{a}\beta$ are the roots

$\Rightarrow x = \alpha\,\beta\,\alpha,\ \alpha\,\beta\,\beta$ are the roots

$\Rightarrow x = \alpha^2\beta,\ \alpha\beta^2$ are the roots

ALTERNATE SOLUTION :

Divide the equation by a^3, we get

$$x^2 + \frac{b}{a}.\frac{c}{a}.x + \left(\frac{c}{a}\right)^3 = 0$$

$\Rightarrow x^2 - (\alpha + \beta).(\alpha\beta)x + (\alpha\beta)^3 = 0$

$\Rightarrow x^2 - \alpha^2\beta x - \alpha\beta^2 x + (\alpha\beta)^3 = 0$

$\Rightarrow x(x - \alpha^2\beta) - \alpha\beta^2(x - \alpha^2\beta) = 0$

$\Rightarrow (x - \alpha^2\beta)(x - \alpha\beta^2) = 0$

$\Rightarrow x = \alpha^2\beta, \alpha\beta^2$ which is the required answer.

28. The given equation is,

$x^2 + (a - b)x + (1 - a - b) = 0$

$a, b \in R$

For this eq^n to have unequal real roots $\forall\, b$

$D > 0$

$\Rightarrow (a - b)^2 - 4(1 - a - b) > 0$

$\Rightarrow a^2 + b^2 - 2ab - 4 + 4a + 4b > 0$

$\Rightarrow b^2 + b(4 - 2a) + a^2 + 4a - 4 > 0$

Which is a quadratic expression in b, and it will be true $\forall\, b \in R$ if discriminant of above eq^n less than zero.

i.e., $(4 - 2a)^2 - 4(a^2 + 4a - 4) < 0$

$\Rightarrow (2 - a)^2 - (a^2 + 4a - 4) < 0$

$\Rightarrow 4 - 4a + a^2 - a^2 - 4a + a < 0$

$\Rightarrow -8a + 8 < 0$

$\Rightarrow a > 1$

29. Given that a, b, c are positive real numbers. To prove that $(a+1)^7(b+1)^7(c+1)^7 > 7^7 a^4b^4c^4$

Consider L.H.S. $= (1+a)^7.(1+b)^7.(1+c)^7$

$= [(1+a)(1+b)(1+c)]^7$

$[1 + a + b + c + ab + bc + ca + abc]^7$

$> [a + b + c + ab + bc + ca + abc]^7$ ....(1)

Now we know that AM ≥ GM using it for +ve no's a, b, c, ab, bc, ca and abc, we get

$$\frac{a+b+c+ab+bc+ca+abc}{7} \geq (a^4b^4c^4)^{1/7}$$

$\Rightarrow (a+b+c+ab+bc+ca+abc)^7 \geq 7^7(a^4b^4c^4)a$

From (1) and (2), we get

$[(1+a)(1+b)(1+c)]^7 > 7^7a^4b^4c^4$

Hence Proved.

30. Roots of $x^2 - 10cx - 11d = 0$ are a and b

$\Rightarrow a + b = 10c$ and $ab = -11d$

Similarly c and d are the roots of $x^2 - 10ax - 11b = 0$

$\Rightarrow c + d = 10a$ and $cd = -11b$

$\Rightarrow a + b + c + d = 10(a + c)$ and $abcd = 121\,bd$

$\Rightarrow b + d = 9(a + c)$ and $ac = 121$

Also we have $a^2 - 10\,ac - 11d = 0$ and $c^2 - 10ac - 11b = 0$

$\Rightarrow a^2 + c^2 - 20ac - 11(b + d) = 0$

$\Rightarrow (a + c)^2 - 22 \times 121 - 99(a + c) = 0$

$\Rightarrow a + c = 121$ or -22

For $a + c = -22$, we get $a = c$

∴ rejecting this value we have $a + c = 121$

∴ $a + b + c + d = 10(a + c) = 1210$

(H) ASSERTION & REASON TYPE QUESTIONS :

1. **(b)** As $a, b, c, p, q \in R$ and the two given equations have exactly one common root

⇒ Either both equations have real roots or both eqations have imaginary roots

⇒ Either $\Delta_1 \geq 0$ and $\Delta_2 \geq 0$ or $\Delta_1 \leq 0$ and $\Delta_2 \leq 0$

$\Rightarrow p^2 - q \geq 0$ and $b^2 - ac \geq 0$

or $p^2 - q \leq 0$ and $b^2 - ac \leq 0$

$\Rightarrow (p^2 - q)(b^2 - ac) \geq 0$

∴ Statement 1 is true.

Also we have $\alpha\beta = q$ and $\frac{\alpha}{\beta} = \frac{c}{a}$

$\therefore \frac{\alpha\beta}{\alpha/\beta} = \frac{q}{c} \times a \Rightarrow \beta^2 = \frac{qa}{c}$

As $\beta \neq 1$ or $-1 \Rightarrow \beta^2 \neq 1 \Rightarrow \frac{qa}{c} \neq 1$ or $c \neq qa$

Again, as exactly one root α is common, and $\beta \neq 1$

$\therefore \alpha + \beta \neq \alpha + \frac{1}{\beta} \Rightarrow \frac{-2b}{a} \neq -2p$

$\Rightarrow b \neq ap$

∴ Statement 2 is correct.

But Statement 2 is not a correct explanation of Statement 1.

(I) INTEGER VALUE CORRECT TYPE :

1. The given system of equations is

$3x - y - z = 0$

$-3x + z = 0$

$-3x + 2y + z = 0$

Let $x = p$ where p is an integer then $y = 0$ and $z = 3p$

But $x^2 + y^2 + z^2 \leq 100 \Rightarrow p^2 + 9p^2 \leq 100$

$\Rightarrow p^2 \leq 10 \Rightarrow p = 0, \pm 1, \pm 2 \pm 3$

i.e. p can take 7 different values.

∴ Number of points (x, y, z) are 7.

2. The given equation is

$$x^2 - 8kx + 16(k^2 - k + 1) = 0$$

$\because$ Both the roots are real and distinct

$\therefore$ $D > 0 \Rightarrow (8k)^2 - 4 \times 16(k^2 - k + 1) > 0$

$\Rightarrow$ k > 1 ...(i)

$\because$ Both the roots are greater than or equal to 4

$\therefore$ $\alpha + \beta > 8$ and $f(4) \ge 0$

$\Rightarrow$ $k > 1$ ...(ii)

and $16 - 32k + 16(k^2 - k + 1) \ge 0$

$\Rightarrow k^2 - 3k + 2 \ge 0 \Rightarrow (k-1)(k-2) \ge 0$

$\Rightarrow k \in (-\infty, 1] \cup [2, \infty)$ ...(iii)

Combining (i), (ii) and (iii), we get $k \ge 2$ or the smallest value of $k = 2$.

3. **8**

$\because a > 0, \therefore a^{-5}, a^{-4}, 3a^{-3}, 1, a^8, a^{10} > 0$

Using AM $\ge$ GM for positive real numbers we get

$$\frac{\frac{1}{a^5} + \frac{1}{a^4} + \frac{1}{a^3} + \frac{1}{a^3} + \frac{1}{a^3} + 1 + a^8 + a^{10}}{8} \ge$$

$$\left(\frac{1}{a^5} \cdot \frac{1}{a^4} \cdot \frac{1}{a^3} \cdot \frac{1}{a^3} \cdot \frac{1}{a^3} \cdot 1 \cdot a^8 \cdot a^{10}\right)^{\frac{1}{8}}$$

$$\Rightarrow \frac{1}{a^5} + \frac{1}{a^4} + \frac{3}{a^3} + 1 + a^8 + a^{10} \ge 8(1)^{\frac{1}{8}}$$

4. **2**

We have $x^4 - 4x^3 + 12x^2 + x - 1 = 0$

$\Rightarrow x^4 - 4x^3 + 6x^2 - 4x + 1 + 6x^2 + 5x - 2 = 0$

$\Rightarrow (x-1)^4 + 6x^2 + 5x - 2 = 0$

$\Rightarrow (x-1)^4 = -6x^2 - 5x + 2$

To solve the above polynomial, it is equivalent to find the intersection points of the curves $y = (x-1)^4$ and

$y = -6x^2 - 5x + 2$ or $y = (x-1)^4$ and $\left(x + \frac{5}{12}\right)^2 = -\frac{1}{6}\left(y - \frac{73}{24}\right)$

The graph of above two curves as follows.
Clearly they have two points of intersection.
Hence the given polynomial has two real roots.

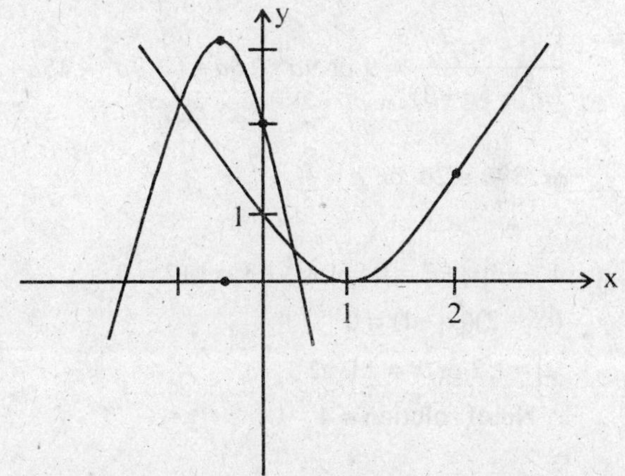

Section-B JEE Main/ AIEEE

1. (a) We have $\alpha^2 = 5\alpha - 3$ and $\beta^2 = 5\beta - 3$;
$\Rightarrow \alpha$ & β are roots of equation, $x^2 = 5x - 3$
or $x^2 - 5x + 3 = 0$ $\therefore \alpha + \beta = 5$ and $\alpha\beta = 3$

Thus, the equation having $\frac{\alpha}{\beta}$ & $\frac{\beta}{\alpha}$ as its roots is

$$x^2 - x\left(\frac{\alpha}{\beta} + \frac{\beta}{\alpha}\right) + \frac{\alpha\beta}{\alpha\beta} = 0$$

$$\Rightarrow x^2 - x\left(\frac{\alpha^2 + \beta^2}{\alpha\beta}\right) + 1 = 0 \text{ or } 3x^2 - 19x + 3 = 0$$

2. (a) Let α, β and γ, δ be the roots of the equations $x^2 + ax + b = 0$ and $x^2 + bx + a = 0$ respectively.
$\therefore \alpha + \beta = -a, \alpha\beta = b$ and $\gamma + \delta = -b, \gamma\delta = a.$
Given $|\alpha - \beta| = |\gamma - \delta| \Rightarrow (\alpha - \beta)^2 = (\gamma - \delta)^2$
$\Rightarrow (\alpha + \beta)^2 - 4\alpha\beta = (\gamma + \delta)^2 - 4\gamma\delta$
$\Rightarrow a^2 - 4b = b^2 - 4a \Rightarrow (a^2 - b^2) + 4(a - b) = 0$
$\Rightarrow a + b + 4 = 0$ $(\because a \neq b)$

3. (a) Product of real roots $= \frac{9}{t^2} > 0, \forall t \in R$

$\therefore$ Product of real roots is always positive.

4. (a) $p + q = -p$ and $pq = q \Rightarrow q(p-1) = 0$
$\Rightarrow q = 0$ or $p = 1.$
If $q = 0$, then $p = 0$. i.e. $p = q$
$\therefore p = 1$ and $q = -2.$

5. (a) $\because (a-b)^2 + (b-c)^2 + (c-a)^2 > 0$
$\Rightarrow 2(a^2 + b^2 + c^2 - ab - bc - ca) > 0$
$\Rightarrow 2 > 2(ab + bc + ca) \Rightarrow ab + bc + ca < 1$

6. (d) $ax^2 + bx + c = 0, \ \alpha + \beta = \frac{-b}{a}, \alpha\beta = \frac{c}{a}$

As for given conditon, $\alpha + \beta = \frac{1}{\alpha^2} + \frac{1}{\beta^2}$

$$\alpha + \beta = \frac{\alpha^2 + \beta^2}{\alpha^2\beta^2} \Rightarrow -\frac{b}{a} = \frac{\frac{b^2}{a^2} - \frac{2c}{a}}{\frac{c^2}{a^2}}$$

On simplification $2a^2c = ab^2 + bc^2$

$\Rightarrow \frac{2a}{b} = \frac{c}{a} + \frac{b}{c} \quad \Rightarrow \frac{c}{a}, \frac{a}{b}, \frac{b}{c}$ are in A.P.

$\therefore \frac{a}{c}, \frac{b}{a}, \& \frac{c}{b}$ are in H.P.

7. (b) Let the roots of given equation be α and 2α then

$$\alpha + 2\alpha = 3\alpha = \frac{1-3a}{a^2-5a+3}$$

$$\& \ \alpha.2\alpha = 2\alpha^2 = \frac{2}{a^2-5a+3} \Rightarrow \alpha = \frac{1-3a}{3(a^2-5a+3)}$$

$$\therefore 2\left[\frac{1}{9}\frac{(1-3a)^2}{(a^2-5a+3)^2}\right] = \frac{2}{a^2-5a+3}$$

$$\frac{(1-3a)^2}{(a^2-5a+3)} = 9 \text{ or } 9a^2 - 6a + 1 = 9a^2 - 45a + 27$$

or $39a = 26$ or $a = \frac{2}{3}$

8. (c) $x^2 - 3|x| + 2 = 0 \Rightarrow |x|^2 - 3|x| + 2 = 0$

$(|x|-2)(|x|-1) = 0$

$|x| = 1, 2$ or $x = \pm 1, \pm 2$

$\therefore$ No. of solution = 4

9. (c) $y = x + \frac{1}{x}$ or $\frac{dy}{dx} = 1 - \frac{1}{x^2}$

For max. or min., $1 - \frac{1}{x^2} = 0 \Rightarrow x = \pm 1$

$$\frac{d^2y}{dx^2} = \frac{2}{x^3} \Rightarrow \left(\frac{d^2y}{dx^2}\right)_{x=2} = 2 \text{ (+ve minima)}$$

$\therefore x = 1$

10. (b) Let two numbers be a and b then $\frac{a+b}{2} = 9$ and

$\sqrt{ab} = 4$

$\therefore$ Equation with roots a and b is

$x^2 - (a+b)x + ab = 0 \Rightarrow x^2 - 18x + 16 = 0$

11. (c) Let the second root be α.

Then $\alpha + (1-p) = -p \Rightarrow \alpha = -1$

Also $\alpha.(1-p) = 1-p$

$\Rightarrow (\alpha - 1)(1-p) = 0 \Rightarrow p = 1 [\because \alpha = -1]$

$\therefore$ Roots are $\alpha = -1$ and $p - 1 = 0$

12. (d) 4 is a root of $x^2 + px + 12 = 0$

$\Rightarrow 16 + 4p + 12 = 0 \Rightarrow p = -7$

Now, the equation $x^2 + px + q = 0$ has equal roots.

$\therefore p^2 - 4q = 0 \Rightarrow q = \frac{p^2}{4} = \frac{49}{4}$

13. (b) $\tan\left(\frac{P}{2}\right), \tan\left(\frac{Q}{2}\right)$ are the roots of $ax^2 + bx + c = 0$

$$\tan\left(\frac{P}{2}\right) + \tan\left(\frac{Q}{2}\right) = -\frac{b}{a}, \quad \tan\left(\frac{P}{2}\right) \cdot \tan\left(\frac{Q}{2}\right) = \frac{c}{a}$$

$$\frac{\tan\left(\frac{P}{2}\right) + \tan\left(\frac{Q}{2}\right)}{1 - \tan\left(\frac{P}{2}\right)\tan\left(\frac{Q}{2}\right)} = \tan\left(\frac{P}{2} + \frac{Q}{2}\right) = 1$$

$$\Rightarrow \frac{-\frac{b}{a}}{1 - \frac{c}{a}} = 1 \Rightarrow -\frac{b}{a} = \frac{a}{a} - \frac{c}{a}$$

$\Rightarrow -b = a - c$ or $c = a + b$.

14. (c) both roots are less than 5

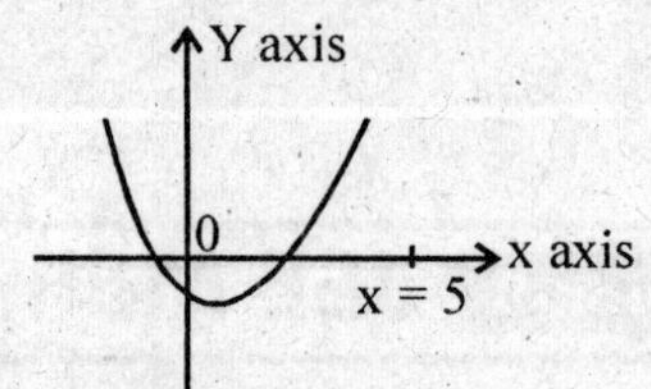

then (i) Discriminant ≥ 0

(ii) $p(5) > 0$

(iii) $\frac{\text{Sum of roots}}{2} < 5$

Hence (i) $4k^2 - 4(k^2 + k - 5) \geq 0$

$4k^2 - 4k^2 - 4k + 20 \geq 0$

$4k \leq 20 \Rightarrow k \leq 5$

(ii) $f(5) > 0; 25 - 10k + k^2 + k - 5 > 0$

or $k^2 - 9k + 20 > 0$

or $k(k-4) - 5(k-4) > 0$

or $(k-5)(k-4) > 0$

$\Rightarrow k \in (-\infty, 4) \cup (-\infty, 5)$

(ii) $\frac{\text{Sum of roots}}{2} = -\frac{b}{2a} = \frac{2k}{2} < 5$

The interection of (i), (ii) & (iii) gives $k \in (-\infty, 4)$.

15. (b) $x^2 + px + q = 0$

Sum of roots $= \tan 30° + \tan 15° = -p$

Product of roots = $\tan 30^\circ . \tan 15^\circ = q$

$$\tan 45^\circ = \frac{\tan 30^\circ + \tan 15^\circ}{1 - \tan 30^\circ . \tan 15^\circ} = \frac{-p}{1-q} = 1$$

$\Rightarrow -p = 1 - q \Rightarrow q - p = 1 \quad \therefore 2 + q - p = 3$

16. (c) Equation $x^2 - 2mx + m^2 - 1 = 0$

$(x-m)^2 - 1 = 0$ or $(x-m+1)(x-m-1) = 0$

$x = m-1, m+1$

$m - 1 > -2$ and $m + 1 < 4$

$\Rightarrow m > -1$ and $m < 3$ or, $-1 < m < 3$

17. (b) $y = \dfrac{3x^2 + 9x + 17}{3x^2 + 9x + 7}$

$3x^2(y-1) + 9x(y-1) + 7y - 17 = 0$

$D \geq 0 \because$ x is real

$81(y-1)^2 - 4 \times 3(y-1)(7y-17) \geq 0$

$\Rightarrow (y-1)(y-41) \leq 0 \Rightarrow 1 \leq y \leq 41$

$\therefore$ Max value of y is 41

ALTERNATE SOLUTION :

Given $f(x) = \dfrac{3x^2 + 9x + 17}{3x^2 + 9x + 7}$

$\Rightarrow f(x) = 1 + \dfrac{10}{3x^2 + 9x + 7}$

Clearly $f(x)$ is maximum when $g(x) = 3x^2 + 9x + 7$ is min.

Here $g(x) = 3\left(x^2 + 3x + \dfrac{9}{4}\right) + 7 - \dfrac{27}{4}$

$= 3\left(x + \dfrac{3}{2}\right)^2 + \dfrac{1}{4}$

which is minimum when $x = \dfrac{-3}{2}$

$$\therefore f_{max} = 1 + \frac{10}{3 \times \frac{9}{4} - 9 \times \frac{3}{2} + 7} = 1 + \frac{10 \times 4}{27 - 54 + 28} = 41$$

18. (c) Let α and β are roots of the equation $x^2 + ax + 1 = 0$

$\alpha + \beta = -a$ and $\alpha\beta = 1$

given $|\alpha - \beta| < \sqrt{5} \Rightarrow \sqrt{(\alpha+\beta)^2 - 4\alpha\beta} < \sqrt{5}$

$\left(\because (\alpha-\beta)^2 = (\alpha+\beta)^2 - 4\alpha\beta\right)$

$\Rightarrow \sqrt{a^2 - 4} < \sqrt{5} \Rightarrow a^2 - 4 < 5$

$\Rightarrow a^2 - 9 < 0 \Rightarrow a^2 < 9 \Rightarrow -3 < a < 3 \Rightarrow a \in (-3, 3)$

19. (b) Statement 2 is $\sqrt{n(n+1)} < n+1, n \geq 2$

$\Rightarrow \sqrt{n} < \sqrt{n+1}, n \geq 2$ which is true

$\Rightarrow \sqrt{2} < \sqrt{3} < \sqrt{4} < \sqrt{5} < \text{-------} \sqrt{n}$

Now $\sqrt{2} < \sqrt{n} \Rightarrow \dfrac{1}{\sqrt{2}} > \dfrac{1}{\sqrt{n}}$

$\sqrt{3} < \sqrt{n} \Rightarrow \dfrac{1}{\sqrt{3}} > \dfrac{1}{\sqrt{n}}; \quad \sqrt{n} \leq \sqrt{n} \Rightarrow \dfrac{1}{\sqrt{n}} \geq \dfrac{1}{\sqrt{n}}$

Also $\dfrac{1}{\sqrt{1}} > \dfrac{1}{\sqrt{n}}$ $\therefore$ Adding all, we get

$$\frac{1}{\sqrt{1}} + \frac{1}{\sqrt{2}} + \frac{1}{\sqrt{3}} + \text{..........} + \frac{1}{n} > \frac{n}{\sqrt{n}} = \sqrt{n}$$

Hence both the statements are correct and statement 2 is a correct explanation of statement-1.

20. (d) Let the roots of equation $x^2 - 6x + a = 0$ be α and 4β and that of the equation $x^2 - cx + 6 = 0$ be α and 3β. Then

$\alpha + 4\beta = 6; \quad 4\alpha\beta = a$

and $\alpha + 3\beta = c; \quad 3\alpha\beta = 6$

$\Rightarrow a = 8$

$\therefore$ The equation becomes $x^2 - 6x + 8 = 0$

$\Rightarrow (x-2)(x-4) = 0 \Rightarrow$ roots are 2 and 4

$\Rightarrow \alpha = 2, \beta = 1 \quad \therefore$ Common root is 2.

21. (b) Given that roots of the equation $bx^2 + cx + a = 0$ are imaginary

$\therefore c^2 - 4ab < 0$ ....(i)

Let $y = 3b^2x^2 + 6\,bc\,x + 2c^2$

$\Rightarrow 3b^2x^2 + 6\,bc\,x + 2c^2 - y = 0$

As x is real, $D \geq 0$

$\Rightarrow 36\,b^2c^2 - 12\,b^2\,(2c^2 - y) \geq 0$

$\Rightarrow 12\,b^2\,(3\,c^2 - 2\,c^2 + y) \geq 0$

$\Rightarrow c^2 + y \geq 0 \quad \Rightarrow y \geq -c^2$

But from eqn. (i), $c^2 < 4ab$ or $-c^2 > -4ab$

$\therefore$ we get $y \geq -c^2 > -4ab$

$\Rightarrow y > -4\,ab$

22. (a) Given that $\left|z - \dfrac{4}{z}\right| = 2$

Now $|z| = \left|z - \dfrac{4}{z} + \dfrac{4}{-z}\right| \leq \left|z - \dfrac{4}{z}\right| + \dfrac{4}{|z|}$

$\Rightarrow |z| \leq 2 + \dfrac{4}{|z|} \quad \Rightarrow |z|^2 - 2|z| - 4 \leq 0$

$\Rightarrow \left(|z| - \dfrac{2+\sqrt{20}}{2}\right)\left(|z| - \dfrac{2-\sqrt{20}}{2}\right) \leq 0$

$\Rightarrow \left(|z| - (1+\sqrt{5})\right)\left(|z| - (1-\sqrt{5})\right) \leq 0$

$\Rightarrow \left(-\sqrt{5}+1\right) \leq |z| \leq \left(\sqrt{5}+1\right)$

$\Rightarrow |z|_{max} = \sqrt{5} + 1$

23. (b) $x^2 - x + 1 = 0 \Rightarrow x = \frac{1 \pm \sqrt{1-4}}{2}$

$x = \frac{1 \pm \sqrt{3}i}{2}$

$\alpha = \frac{1}{2} + i\frac{\sqrt{3}}{2} = -\omega^2 \qquad \beta = \frac{1}{2} - \frac{i\sqrt{3}}{2} = -\omega$

$\alpha^{2009} + \beta^{2009} = (-\omega^2)^{2009} + (-\omega)^{2009} = -\omega^2 - \omega = 1$

24. (a) Let the correct equation be $ax^2 + bx + c = 0$

Now Sachin's equation $\Rightarrow ax^2 + bx + c' = 0$
Roots found by Sachin's are 4 and 3

Rahul's equation, $ax^2 + b'x + c = 0$
Roots found by Rahul's are 3 and 2

$-\frac{b}{a} = 7$ ....(i)

$\frac{c}{a} = 6$ ...(ii)

From (i) and (ii), roots of the correct equation

$x^2 - 7x + 6 = 0$ are 6 and 1.

25. (d) $P(x) = 0$

$\Rightarrow f(x) = g(x)$

$\Rightarrow ax^2 + bx + c = a_1x^2 + b_1x + c_1$

$\Rightarrow (a - a_1)x^2 + (b - b_1)x + (c - c_1) = 0.$

It has only one solution, $x = -1$

$\Rightarrow b - b_1 = a - a_1 + c - c_1$ ...(1)

vertex $= (-1, 0)$

$\Rightarrow \frac{b - b_1}{2(a - a_1)} = -1$

$\Rightarrow b - b_1 = 2(a - a_1)$ ...(2)

Now $P(-2) = 2$
$\Rightarrow f(-2) - g(-2) = 2$
$\Rightarrow 4a - 2b + c - 4a_1 + 2b_1 - c_1 = 2$
$\Rightarrow 4(a - a_1) - 2(b - b_1) + (c - c_1) = 2$ ...(3)
From equations, (1), (2) and (3)

$a - a_1 = c - c_1 = \frac{1}{2}(b - b_1) = 2$

Now, $P(2) = f(2) - g(2)$

$= 4(a - a_1) + 2(b - b_1) + (c - c_1)$

$= 8 + 8 + 2 = 18$

26. (b) Given equation is $e^{\sin x} - e^{-\sin x} - 4 = 0$
Put $e^{\sin x} = t$ in the given equation, we get
$t^2 - 4t - 1 = 0$

$\Rightarrow t = \frac{4 \pm \sqrt{16+4}}{2} = \frac{4 \pm \sqrt{20}}{2} = \frac{4 \pm 2\sqrt{5}}{2} = 2 \pm \sqrt{5}$

$\Rightarrow e^{\sin x} = 2 \pm \sqrt{5} \; (\because t = e^{\sin x})$

$\Rightarrow e^{\sin x} = 2 - \sqrt{5}$ and $e^{\sin x} = 2 + \sqrt{5}$

$\Rightarrow e^{\sin x} = 2 - \sqrt{5} < 0$ and $\sin x = \ln(2 + \sqrt{5}) > 1$

So rejected So, rejected
Hence given equation has no solution.
$\therefore$ The equation has no real roots.

27. (d) $f(x) = 2x^3 + 3x + k$
$f'(x) = 6x^2 + 3 > 0 \quad \forall x \in R \quad (\because x^2 > 0)$
$\Rightarrow f(x)$ is strictly increasing function
$\Rightarrow f(x) = 0$ has only one real root, so two roots are not possible.

28. (b) From the given system, we have

$\frac{k+1}{k} = \frac{8}{k+3} \neq \frac{4k}{3k-1}$ ($\because$ System has no solution)

$\Rightarrow k^2 + 4k + 3 = 8k$
$\Rightarrow k = 1, 3$
If $k = 1$

then $\frac{8}{1+3} \neq \frac{4.1}{2}$ which is false

And if $k = 3$

then $\frac{8}{6} \neq \frac{4.3}{9-1}$ which is true, therefore $k = 3$

Hence for only one value of k. System has no solution.

29. (a) Given equations are
$x^2 + 2x + 3 = 0$ ...(i)
$ax^2 + bx + c = 0$ ...(ii)
Roots of equation (i) are imaginary roots.
According to the question (ii) will also have both roots same as (i). Thus

$\frac{a}{1} = \frac{b}{2} = \frac{c}{3} = \lambda$ (say) $\Rightarrow a = \lambda, b = 2\lambda, c = 3\lambda$

Hence, required ratio is 1 : 2 : 3

Chapter 4

Permutations and Combinations

Section-A JEE Advanced/ IIT-JEE

(A) FILL IN THE BLANKS :

1. In a certain test, a_i students gave wrong answers to atleast i questions, where $i = 1, 2, \ldots, k$. No student gave more than k wrong answers. The total number of wrong answers given is .... ***(1982 - 2 Marks)***

2. The side AB, BC and CA of a triangle ABC have 3, 4 and 5 interior points respectively on them. The number of triangles that can be constructed using these interior points as vertices is .... ***(1984 - 2 Marks)***

3. Total number of ways in which six '+' and four '–' signs can be arranged in a line such that no two '–' signs occur together is .... ***(1988 - 2 Marks)***

4. There are four balls of different colours and four boxes of colours, same as those of the balls. The number of ways in which the balls, one each in a box, could be placed such that

(B) TRUE / FALSE :

1. The product of any r consecutive natural numbers is always divisible by $r!$. ***(1985 - 1 Mark)***

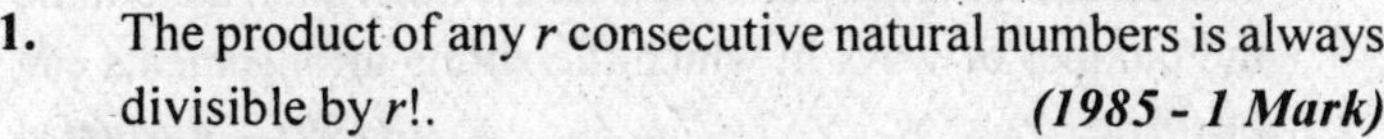

(C) MCQ's WITH ONE CORRECT ANSWER :

1. ${}^nC_{r-1} = 36, {}^nC_r = 84$ and ${}^nC_{r+1} = 126$, then r is : ***(1979)***

(a) 1 (b) 2
(c) 3 (d) None of these.

2. Ten different letters of an alphabet are given words with five letters are formed from three given letters. Then the number of words which have at least one letter repeated are ***(1982 - 2 Marks)***

(a) 69760 (b) 30240
(c) 99748 (d) none of these

3. The value of the expression ${}^{47}C_4 + \sum_{j=1}^{5} {}^{52-j}C_3$ is equal to ***(1982 - 2 Marks)***

(a) ${}^{47}C_5$ (b) ${}^{52}C_5$
(c) ${}^{52}C_4$ (d) none of these

4. Eight chairs are numbered 1 to 8. Two women and three men wish to occupy one chair each. First the women choose the chairs from amongst the chairs marked 1 to 4 ; and then the men select the chairs from amongst the remaining. The number of possible arrangements is ***(1982 - 2 Marks)***

(a) ${}^6C_3 \times {}^4C_2$ (b) ${}^4P_2 \times {}^4P_3$
(c) ${}^4C_2 + {}^4P_3$ (d) none of these

5. A five-digit numbers divisible by 3 is to be formed using the numerals 0, 1, 2, 3, 4 and 5, without repetition. The total number of ways this can be done is ***(1989 - 2 Marks)***

(a) 216 (b) 240
(c) 600 (d) 3125

6. How many different nine digit numbers can be formed from the number 223355888 by rearranging its digits so that the odd digits occupy even positions ? ***(2000S)***

(a) 16 (b) 36
(c) 60 (d) 180

7. Let T_n denote the number of triangles which can be formed using the vertices of a regular polygon of n sides. If $T_{n+1} - T_n = 21$, then n equals ***(2001S)***

(a) 5 (b) 7
(c) 6 (d) 4

8. The number of arrangements of the letters of the word BANANA in which the two N's do not appear adjacently is ***(2002S)***

(a) 40 (b) 60
(c) 80 (d) 100

9. A rectangle with sides of length $(2m - 1)$ and $(2n - 1)$ units is divided into squares of unit length by drawing parallel lines as shown in the diagram, then the number of rectangles possible with odd side lengths is ***(2005S)***

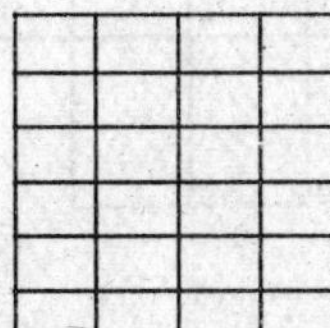

(a) $(m + n - 1)^2$ (b) 4^{m+n-1}
(c) m^2n^2 (d) $m(m + 1)n(n + 1)$

10. If the LCM of p, q is $r^2t^4s^2$, where r, s, t are prime numbers and p, q are the positive integers then the number of ordered pair (p, q) is *(2006 - 3M, –1)*

(a) 252 (b) 254
(c) 225 (d) 224

11. The letters of the word COCHIN are permuted and all the permutations are arranged in an alphabetical order as in an English dictionary. The number of words that appear before the word COCHIN is *(2007 -3 marks)*

(a) 360 (b) 192
(c) 96 (d) 48

12. The number of seven digit integers, with sum of the digits equal to 10 and formed by using the digits 1, 2 and 3 only, is *(2009)*

(a) 55 (b) 66
(c) 77 (d) 88

13. The total number of ways in which 5 balls of different colours can be distributed among 3 persons so that each person gets at least one ball is *(2012)*

(a) 75 (b) 150
(c) 210 (d) 243

D MCQ's WITH ONE OR MORE THAN ONE CORRECT :

1. An n-digit number is a positive number with exactly n digits. Nine hundred distinct n- digit numbers are to be formed using only the three digits 2, 5 and 7. The smallest value of n for which this is possible is *(1998 - 2 Marks)*

(a) 6 (b) 7
(c) 8 (d) 9

E SUBJECTIVE PROBLEMS :

1. Six X's have to be placed in the squares of figure below in such a way that each row contains at least one X. In how many different ways can this be done. *(1978)*

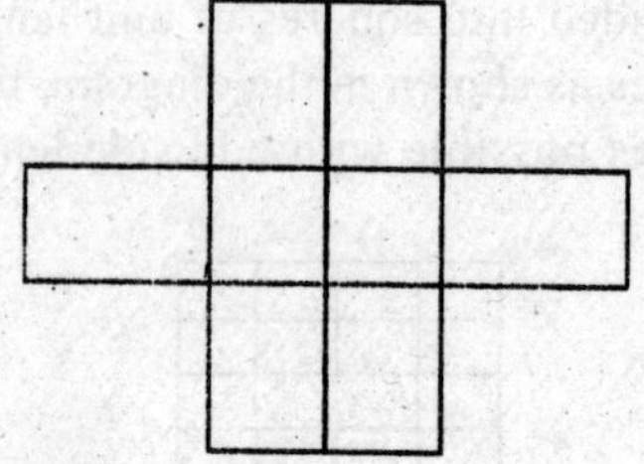

2. In how many ways can a pack of 52 cards be

(i) divided equally among four players in order

(ii) formed into 4 groups of 13 cards each.

(iii) In 4 sets, three of them having 17 cards each and the fourth just 1 card? *(1979)*

3. Five balls of different colours are to be placed in there boxes of different size. Each box can hold all five. In how many different ways can we place the balls so that no box remains empty ? *(1981 - 4 Marks)*

4. m men and n women are to be seated in a row so that no two women sit together. If $m > n$, then show that the number of ways in which they can be seated is $\dfrac{m!(m+1)!}{(m-n+1)!}$ *(1983 - 2 Marks)*

5. 7 relatives of a man comprises 4 ladies and 3 gentlemen; his wife has also 7 relatives; 3 of them are ladies and 4 gentlemen. In how many ways can they invite a dinner party of 3 ladies and 3 gentlemen so that there are 3 of man's relatives and 3 of the wife's relatives? *(1985 - 5 Marks)*

6. A box contains two white balls, three black balls and four red balls. In how many ways can three balls be drawn from the box if at least one black ball is to be included in the draw? *(1986 - 2½ Marks)*

7. A student is allowed to select atmost n books from a collection of $(2n + 1)$ books. If the total number of ways in which he can select at least one book is 63, find the value of n. *(1987 - 3 Marks)*

8. Eighteen guests have to be seated, half on each side of a long table. Four particular guests desire to sit on one particular side and three others on the other side. Determine the number of ways in which the sitting arrangements can be made. *(1991 - 4 Marks)*

9. A committee of 12 is to be formed from 9 women and 8 men. In how many ways this can be done if at least five women have to be included in a committee? In how many of these committees *(1994 - 4 Marks)*

(a) The women are in majority?
(b) The men are in majority?

10. Prove by permutation or otherwise $\dfrac{(n^2)!}{(n!)^n}$ is an integer $(n \in I^+)$. *(2004 - 2 Marks)*

11. If total number of runs scored in n matches is $\left(\dfrac{n+1}{4}\right)(2^{n+1} - n - 2)$ where $n > 1$, and the runs scored in the kth match are given by k. 2^{n+1-k}, where $1 \le k \le n$. Find n. *(2005 - 2 Marks)*

(F) MATCH THE FOLLOWING :

Each question contains statements given in two columns, which have to be matched. The statements in Column-I are labelled A, B, C and D, while the statements in Column-II are labelled p, q, r, s and t. Any given statement in Column-I can have correct matching with ONE OR MORE statement(s) in Column-II. The appropriate bubbles corresponding to the answers to these questions have to be darkened as illustrated in the following example :

If the correct matches are A-p, s and t; B-q and r; C-p and q; and D-s then the correct darkening of bubbles will look like the given.

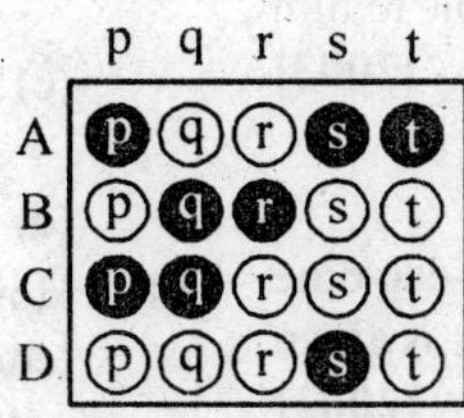

1. Consider all possible permutations of the letters of the word ENDEANOEL. Match the Statements / Expressions in **Column I** with the Statements / Expressions in **Column II** and indicate your answer by darkening the appropriate bubbles in the 4 × 4 matrix given in the ORS. ***(2008)***

Column I	Column II
(A) The number of permutations containing the word ENDEA is	(p) 5!
(B) The number of permutations in which the letter E occurs in the first and the last positions is	(q) 2 × 5!
(C) The number of permutations in which none of the letters D, L, N occurs in the last five positions is	(r) 7 × 5!
(D) The number of permutations in which the letters A, E, O occur only in odd positions is	(s) 21 × 5!

(G) COMPREHENSION BASED QUESTIONS :

PASSAGE - 1

Let a_n denote the number of all *n*-digit positive integers formed by the digits 0, 1 or both such that no consecutive digits in them are 0. Let b_n = the number of such *n*-digit integers ending with digit 1 and c_n = the number of such *n*–digit integers ending with digit 0. ***(2012)***

1. The value of b_6 is
(a) 7 (b) 8
(c) 9 (d) 11

2. Which of the following is correct?
(a) $a_{17} = a_{16} + a_{15}$ (b) $c_{17} \neq c_{16} + c_{15}$
(c) $b_{17} \neq b_{16} + c_{16}$ (d) $a_{17} = c_{17} + b_{16}$

(I) INTEGER VALUE CORRECT TYPE :

1. Consider the set of eight vectors $V = \{a\hat{i} + b\hat{j} + c\hat{k} : a, b, c \in \{-1, 1\}\}$. Three non-coplanar vectors can be chosen from V in 2^p ways. Then p is ***(JEE Adv. 2013)***

Section-B JEE Main/ AIEEE

1. Total number of four digit odd numbers that can be formed using 0, 1, 2, 3, 5, 7 (using repetition allowed) are **[2002]**
(a) 216 (b) 375
(c) 400 (d) 720

2. Number greater than 1000 but less than 4000 is formed using the digits 0, 1, 2, 3, 4 (repetition allowed). Their number is **[2002]**
(a) 125 (b) 105
(c) 375 (d) 625

3. Five digit number divisible by 3 is formed using 0, 1, 2, 3, 4, 6 and 7 without repetition. Total number of such numbers are **[2002]**
(a) 312 (b) 3125
(c) 120 (d) 216

4. The sum of integers from 1 to 100 that are divisible by 2 or 5 is **[2002]**
(a) 3000 (b) 3050
(c) 3600 (d) 3250

5. If ${}^nC_r$ denotes the number of combination of n things taken r at a time, then the expression ${}^nC_{r+1}+{}^nC_{r-1}+2\times{}^nC_r$ equals [2003]

(a) ${}^{n+1}C_{r+1}$ (b) ${}^{n+2}C_r$

(c) ${}^{n+2}C_{r+1}$ (d) ${}^{n+1}C_r$.

6. A student is to answer 10 out of 13 questions in an examination such that he must choose at least 4 from the first five questions. The number of choices available to him is [2003]

(a) 346 (b) 140

(c) 196 (d) 280

7. The number of ways in which 6 men and 5 women can dine at a round table if no two women are to sit together is given by [2003]

(a) 7×5 (b) 6×5

(c) 30 (d) 5×4

8. How many ways are there to arrange the letters in the word GARDEN with vowels in alphabetical order

(a) 480 (b) 240 [2004]

(c) 360 (d) 120

9. The number of ways of distributing 8 identical balls in 3 distinct boxes so that none of the boxes is empty is

(a) ${}^8C_3$ (b) 21 [2004]

(c) 3^8 (d) 5

10. If the letters of the word SACHIN are arranged in all possible ways and these words are written out as in dictionary, then the word SACHIN appears at serial number [2005]

(a) 601 (b) 600

(c) 603 (d) 602

11. At an election, a voter may vote for any number of candidates, not greater than the number to be elected. There are 10 candidates and 4 are of be selected, if a voter votes for at least one candidate, then the number of ways in which he can vote is [2006]

(a) 5040 (b) 6210

(c) 385 (d) 1110

12. The set $S = \{1, 2, 3, ..., 12\}$ is to be partitioned into three sets A, B, C of equal size. Thus $A \cup B \cup C = S$, $A\cap B = B\cap C = A\cap C = \phi$. The number of ways to partition S is [2007]

(a) $\frac{12!}{(4!)^3}$ (b) $\frac{12!}{(4!)^4}$

(c) $\frac{12!}{3!(4!)^3}$ (d) $\frac{12!}{3!(4!)^4}$

DIRECTIONS : *Q. No. 13 is Assertion-Reason type questions. Each of these questions contains two statements: Statement-1(Assertion) and Statement-2(Reason). Each of these questions also has four alternative choices, only one of which is the correct answer. You have to select the correct choice.* [2008]

13. In a shop there are five types of ice-creams available. A child buys six ice-creams.

Statement-1 : The number of different ways the child can buy the six ice-creams is ${}^{10}C_5$.

Statement -2 : The number of different ways the child can buy the six ice-creams is equal to the number of different ways of arranging 6 A's and 4 B's in a row.

(a) Statement -1 is false, Statement-2 is true

(b) Statement -1 is true, Statement-2 is true; Statement -2 is a correct explanation for Statement-1

(c) Statement -1 is true, Statement-2 is true; Statement -2 is not a correct explanation for Statement-1

(d) Statement -1 is true, Statement-2 is false

14. How many different words can be formed by jumbling the letters in the word MISSISSIPPI in which no two S are adjacent?

(a) $8.{}^6C_4.{}^7C_4$ (b) $6.7.{}^8C_4$

(c) $6.8.{}^7C_4$ (d) $7.{}^6C_4.{}^8C_4$

15. From 6 different novels and 3 different dictionaries, 4 novels and 1 dictionary are to be selected and arranged in a row on a shelf so that the dictionary is always in the middle. Then the number of such arrangement is: [2009]

(a) at least 500 but less than 750

(b) at least 750 but less than 1000

(c) at least 1000

(d) less than 500

16. There are two urns. Urn A has 3 distinct red balls and urn B has 9 distinct blue balls. From each urn two balls are taken out at random and then transferred to the other. The number of ways in which this can be done is [2010]

(a) 36 (b) 66

(c) 108 (d) 3

17. **Statement-1:** The number of ways of distributing 10 identical balls in 4 distinct boxes such that no box is empty is ${}^9C_3$.

Statement-2: The number of ways of choosing any 3 places from 9 different places is ${}^9C_3$. [2011]

(a) Statement-1 is true, Statement-2 is true; Statement-2 is not a correct explanation for Statement-1.

(b) Statement-1 is true, Statement-2 is false.

(c) Statement-1 is false, Statement-2 is true.

(d) Statement-1 is true, Statement-2 is true; Statement-2 is a correct explanation for Statement-1.

18. Assuming the balls to be identical except for difference in colours, the number of ways in which one or more balls can be selected from 10 white, 9 green and 7 black balls is : [2012]

(a) 880 (b) 629 (c) 630 (d) 879

19. Let T_n be the number of all possible triangles formed by joining vertices of an n-sided regular polygon. If $T_{n+1} - T_n = 10$, then the value of n is :

(a) 7 (b) 5 [JEE M 2013]

(c) 10 (d) 8

Solutions & Explanations

Section-A : JEE Advanced/ IIT-JEE

A 1. $a_1 + a_2 + ... + a_k$ 2. 205 3. 35 4. 9

B 1. T

C 1. (c) 2. (a) 3. (c) 4. (d) 5. (a) 6. (c) 7. (b) 8. (a) 9. (c) 10. (c) 11. (c) 12. (c) 13. (b)

D 1. (b)

E 1. 26 2. (i) $\frac{52!}{(13!)^4}$ (ii) $\frac{52!}{(13!)^4 \times 4!}$ (iii) $\frac{52!}{(17!)^3 \times 3!}$ 3. 300 5. 485 6. 64 7. 3 8. ${}^{11}C_5 \times 9! \times 9!$ 9. 6062, (a) 2702, (b) 1008 11. 7

F 1. (A) - p; (B) - s; (C) - q; (D) - q

G 1. (b) 2. (a)

I 1. 5

Section-B : JEE Main/ AIEEE

1. (d) 2. (c) 3. (d) 4. (b) 5. (c) 6. (c) 7. (a) 8. (c) 9. (b) 10. (a) 11. (c) 12. (a) 13. (a) 14. (d) 15. (c) 16. (c) 17. (d) 18. (d) 19. (b)

Section-A JEE Advanced/ IIT-JEE

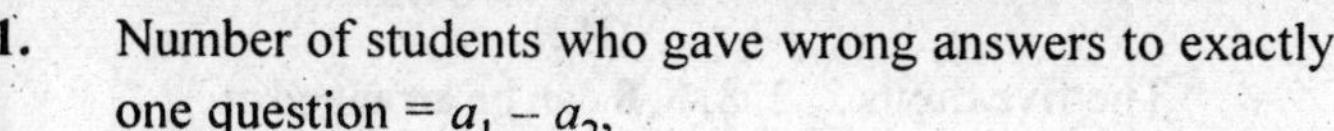

A FILL IN THE BLANKS :

1. Number of students who gave wrong answers to exactly one question $= a_1 - a_2$,
Two questins $= a_2 - a_3$
Three questions $= a_3 - a_4$, $k-1$ question $= a_{k-1} - a_k$, $k_{\text{question}} = a_k$
$\therefore$ Total number of wrong answers
$= 1\,(a_1 - a_2) + 2\,(a_2 - a_3) + 3\,(a_3 - a_4) + ... (k-1)\,(a_{k-1} - a_k) + k\,a_k$
$= a_1 + a_2 + a_3 + ...a_k$

2. We have total $3 + 4 + 5 = 12$ points out of which 3 fall on one line, 4 on other line and 5 on still other line. So number of Δ's that can be formed using 12 such points are
$= {}^{12}C_3 - {}^3C_3 - {}^4C_3 - {}^5C_3$
$$= \frac{12 \times 11 \times 10}{6} - 1 - 4 - \frac{5 \times 4}{2 \times 1} = 220 - 15 = 205$$

3. '+' signs can be put in a row in 1 way, creating 7 ticked places to keep '–' sign so that no two '–' signs occur together
√ + √ + √ + √ + √ + √ + √
Out of these 7 places 4 can be chosen in ${}^7C_4$ ways.
$\therefore$ Required no. of arrangements are
$$= {}^7C_4 = {}^7C_3 = \frac{7.\,6.\,5}{3.2.1} = 35$$

4. **KEY CONCEPT :** We know that number of dearrangements of n objects
$$= n!\left[1 - \frac{1}{1!} + \frac{1}{2!} - \frac{1}{3!} + \frac{1}{4!} - ..... \frac{1}{n!}\right]$$
$\therefore$ No. of ways of putting all the 4 balls into boxes of different colour
$$= 4!\left[1 - \frac{1}{1!} + \frac{1}{2!} - \frac{1}{3!} + \frac{1}{4!}\right] = 4!\left(\frac{1}{2} - \frac{1}{6} + \frac{1}{24}\right)$$
$$= 24\left(\frac{12 - 4 + 1}{24}\right) = 9$$

B TRUE / FALSE :

1. Consider $\frac{(n+1)(n+2)...(n+r)}{r!}$

$= \frac{1.2.3...(n-1)n\,(n+1)(n+2)...(n+r)}{1.2.3...n.r!}$

$= \frac{(n+r)!}{n!r!} = {}^{n+r}C_r$

= some integral value

$\Rightarrow$ (n + 1) (n + 2) ...(n + r) is divisible by r!

Thus given statement is true.

C MCQ's WITH ONE CORRECT ANSWER :

1. (c) ${}^nC_{r-1} = 36, {}^nC_r = 84, {}^nC_{r+1} = 126$

KEY CONCEPT : We know that

$$\frac{{}^nC_{r-1}}{{}^nC_r} = \frac{r}{n-r+1}$$

$\Rightarrow \frac{36}{84} = \frac{r}{n-r+1}$

$\Rightarrow \frac{r}{n-r+1} = \frac{3}{7} \Rightarrow 3n - 10r + 3 = 0$ ....(1)

Also $\frac{{}^nC_r}{{}^nC_{r+1}} = \frac{r+1}{n-r} = \frac{84}{126} = \frac{2}{3}$

$\Rightarrow 2n - 5r - 3 = 0$ ....(2)

Solving (1) and (2), we get $n = 9$ and $r = 3$.

2. (a) Total number of words that can be formed using 5 letters out of 10 given different letters
= 10 × 10 × 10 × 10 × 10 (as letters can repeat)
= 1,00,000
Number of words that can be formed using 5 different letters out of 10 different letters
= ${}^{10}P_5$ (none can repeat)

$= \frac{10!}{5!} = 30,240$

$\therefore$ Number of words in which at least one letter is repeated
= total words–words with none of the letters repeated
= 1,00,000 – 30,240
= 69760

3. (c) ${}^{47}C_4 + \sum_{j=1}^{5} {}^{52-j}C_3 = {}^{47}C_4 + {}^{51}C_3 + {}^{50}C_3$

$+ {}^{49}C_3 + {}^{48}C_3 + {}^{47}C_3$

$= {}^{51}C_3 + {}^{50}C_3 + {}^{49}C_3 + {}^{48}C_3 + ({}^{47}C_3 + {}^{47}C_4)$

[Using ${}^nC_r + {}^nC_{r+1} = {}^{n+1}C_{r+1}$]

$= {}^{51}C_3 + {}^{50}C_3 + {}^{49}C_3 + ({}^{48}C_3 + {}^{48}C_4)$

$= {}^{51}C_3 + {}^{50}C_3 + ({}^{49}C_3 + {}^{49}C_4)$

$= {}^{51}C_3 + ({}^{50}C_3 + {}^{50}C_4)$

$= {}^{51}C_3 + {}^{51}C_4 = {}^{52}C_4$

4. (d) 1 2 3 4 5 6 7 8

Two women can choose two chairs out of 1, 2, 3, 4, in ${}^4C_2$ ways and can arrange themselves in 2! ways. Three men can choose 3 chairs out of 6 remaining chairs in ${}^6C_3$ ways and can arrange themselves in 3! ways

$\therefore$ Total number of possible arrangements are

${}^4C_2 \times 2! \times {}^6C_3 \times 3! = {}^4P_2 \times {}^6P_3$

5. (a) **KEY CONCEPT :** We know that a number is divisible by 3 if the sum of its digits is divisibly by 3.
Now out of of 0, 1, 2, 3, 4, 5 if we take 1, 2, 3, 4, 5 or 0, 1, 2, 4, 5 then the 5 digit numbers will be divisible by 3.
Case I : Number of 5 digit numbers formed using the digits 1, 2, 3, 4, 5 = 5! = 120
Case II : Taking 0, 1, 2, 4, 5 if we make 5 digit number then
I place can be filled in = 4 ways (0 can not come at I place)
II place can be filled in = 4 ways
III place can be filled in = 3 ways
IV place can be filled in = 2 ways
V place can be filled in = 1 ways
$\therefore$ Total numbers are = 4 × 4! = 96
Thus total numbers divisible by 3 are = 120 + 96 = 216

6. (c) $X - X - X - X - X$. The four digits 3, 3, 5, 5 can be arranged at (–) places in $\frac{4!}{2!2!} = 6$ ways.
The five digits 2, 2, 8, 8, 8 can be arranged at

(X) places in $\frac{5!}{2!3!} = 10$ ways.

Total no. of arrangements = 6 × 10 = 60 ways.

7. (b) **KEY CONCEPT :** A regular polygon of n sides has n vertices, no more than two of which are collinear. Out of these n points ${}^nC_3$ Δ's can be formed.

$\therefore$ $T_n = {}^nC_3$; $T_{n+1} = {}^{n+1}C_3$

As per question,

$T_{n+1} - T_n = 21 \Rightarrow {}^{n+1}C_3 - {}^nC_3 = 21$

$\frac{(n+1)n(n-1)}{3.2.1} - \frac{n(n-1)(n-2)}{3.2.1} = 21$

$\Rightarrow \quad n(n-1)(n+1-n+2) = 126$

$\Rightarrow \quad n(n-1) = 42$

$\Rightarrow \quad n(n-1) = 7 \times 6 \Rightarrow n = 7.$

8. **(a)** Total number of ways of arranging the letters of the word BANANA is $\frac{6!}{2!3!} = 60$. Number of words in which 2 N's come together is $\frac{5!}{3!} = 20$.

Hence the required number $= 60 - 20 = 40$

9. **(c)** Along horizontal side one unit can be taken in $= (2m-1)$ ways

3 units can be taken in $= (2m-3)$ ways

5 units can be taken in $= (2m-5)$ ways

and so on

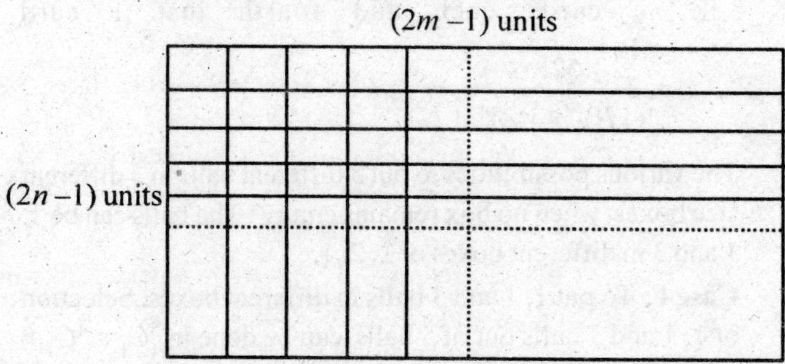

$\therefore$ The number of ways of selecting a side horizontally is

$(2m-1)+(2m-3)+(2m-5)+ \ldots +3+1 = m^2$

Similarly the number of ways of selecting a side vertically is

$(2n-1)+(2n-3)+(2n-5)+ \ldots +3+1 = n^2$

$\therefore$ total number of rectangles $= m^2 \times n^2 = m^2n^2$.

ALTERNATE SOLUTION :

1 2 3 4 2m –1 2m

1 2 3 4 2n –1 2n

If we see the blocks in terms of lines then there are $2m$ vertical lines and $2n$ horizontal lines. To form the required rectangle we must select two horizontal lines, one even numbered (out of 2, 4, ...$2n$) and one odd numbered (out of 1, 3....$2n$–1) and similarly two vertical lines. The number of rectangles is

${}^mC_1 \cdot {}^mC_1 \cdot {}^nC_1 \cdot {}^nC_1 = m^2n^2$

10. **(c)** Let the two sides be A and B. Assume that four particular guests wish to sit on side A. Four guests who wish to sit on side A can be accommodated on nine chairs in ${}^9P_4$ ways and three guests who wish to sit on side B can be accommodated in ${}^9P_3$ ways.

Now, the remaining guests are left who can sit on 11 chairs on both the sides of the table in (11!) ways. Hence, the total number of ways in which 18 persons can be seated $= {}^9P_4 \times {}^9P_3 \times (11)!$.

11. **(c)** The letter of word *COCHIN* in alphabetic order are C, C, H, I, N, O.

Fixing first letter C and keeping C at second place, rest 4 can be arranged in 4! ways.

Similarly the words starting with CH, CI, CN are 4! in each case.

Then fixing first two letters as *CO* next four places when filled in alphabetic order give the word *COCHIN*.

$\therefore$ Numbers of words coming before *COCHIN* are $4 \times 4! = 4 \times 24 = 96$

12. **(c)** We have to form 7 digit numbers, using the digits 1, 2 and 3 only, such that the sum of the digits in a number = 10.

This can be done by taking 2, 2, 2, 1, 1, 1, 1, or by taking 2, 3, 1, 1, 1, 1, 1.

$\therefore$ Number of ways $= \frac{7!}{3!4!} + \frac{7!}{5!} = 77.$

13. **(b)** $\because$ Each person gets at least one ball.

$\therefore$ 3 Persons can have 5 balls in the following systems

Person	I	II	III
No. of balls	1	1	3

or

Person	I	II	III
No. of balls	1	2	2

The number of ways to distribute the balls in first system

$= {}^5C_1 \times {}^4C_1 \times {}^3C_3$

Also 3, persons having 1, 1 and 3 balls can be arranged in $\frac{3!}{2!}$ ways.

$\therefore$ No. of ways to distribute 1, 1, 3 balls to the three persons

$= {}^5C_1 \times {}^4C_1 \times {}^3C_3 \times \frac{3!}{2!} = 60$

Similarly the total no. of ways to distribute 1, 2, 2 balls to the three persons

$$= {}^5C_1 \times {}^4C_2 \times {}^2C_2 \times \frac{3!}{2!} = 90$$

$\therefore$ The required number of ways = 60 + 90 = 150

Alternative method

Key Concept : The number of ways in which n different things can be distributed into r different groups (when blank groups are not allowed) is

$$r^n - {}^rC_1\,(r-1)^n + {}^rC_2\,(r-2)^n - \ldots\ldots\ldots + (-1)^{r-1}\,{}^rC_{r-1}$$

Required number of ways

$$= 3^5 - {}^3C_1\,(3-1)^5 + {}^3C_2\,(3-2)^5 = 243 - 96 + 3 = 150$$

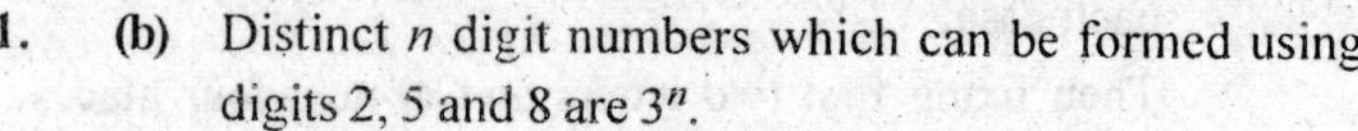

D MCQ's WITH ONE OR MORE THAN ONE CORRECT :

1. **(b)** Distinct n digit numbers which can be formed using digits 2, 5 and 8 are 3^n.

We have to find n so that

$3^n \geq 900 \Rightarrow 3^{n-2} \geq 100$

$\Rightarrow n-2 \geq 5 \Rightarrow n \geq 7$. So the least value of n is 7.

E SUBJECTIVE PROBLEMS :

1. As all the X's are identical, the question is of selection of 6 squares from 8 squares, so that no row remains empty. Here R_1 has 2 squares, R_2 has 4 squares and R_3 has 2 squares. The selection scheme is as follows :

	R_1	R_2	R_3
	1	4	1
or	1	3	2
or	2	3	1
or	2	2	2

$\therefore$ Number of selections are

$${}^2C_1 \times {}^4C_4 \times {}^2C_1 + {}^2C_1 \times {}^4C_3 \times {}^2C_2 + {}^2C_2 \times {}^4C_3 \times {}^2C_1 + {}^2C_2 \times {}^4C_2 \times {}^2C_2$$

$$= 4 + 8 + 8 + 6 = 26$$

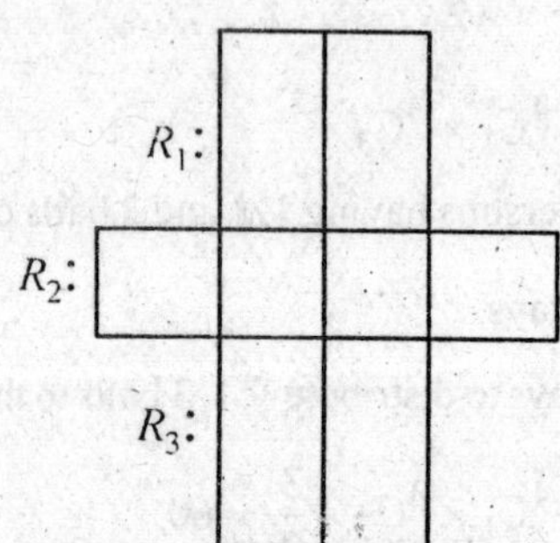

2. **KEY CONCEPT :** If $4n$ different things are to be equally distributed amongst 4 persons the number of ways are $\frac{4n!}{(n!)^4}$ but if $4n$ things are to form 4 equal groups then the number of ways are $\frac{(4n)!}{(n!)^4 \times 4!}$

(i) Number of ways of dividing 52 cards equally among 4 players $= \frac{52!}{(13!)^4}$

(ii) Number of ways to form 4 groups of 13 cards each

$$= \frac{52!}{(13!)^4 \times 4!}$$

(iii) Number of ways to form 4 sets, three of them having 17 cards each and fourth just 1 card

$$= \frac{52!}{(17!)^3 \times 3! \times 1!}$$

3. The various possibilities to put 5 different balls in 3 different size boxes, when no box remains empty : The balls can be 1, 1 and 3 in different boxes or 2, 2, 1.

Case I : To put 1, 1 and 3 balls in different boxes. Selection of 1, 1 and 3 balls out of 5 balls can be done in ${}^5C_1 \times {}^4C_1 \times {}^3C_3$ ways and then 1, 1, 3 can permute (as defferent size boxes) in 3 ! ways.

$\therefore$ No. of ways

$= {}^5C_1 \times {}^4C_1 \times {}^3C_3 \times 3! = 5 \times 4 \times 1 \times 6 = 120$

Case II : To put 2, 2 and 1 ball in different boxes. Selection of 2, 2 and 1 balls out of 5 balls can be done in ${}^5C_2 \times {}^3C_2 \times {}^1C_1$ ways

and then 2, 2, 1 can permute (different boxes) in 3 ! ways

$\therefore$ No. of ways

$= {}^5C_1 \times {}^3C_2 \times {}^1C_1 \times 3! = 10 \times 3 \times 1 \times 6 = 180$

Combining case I and II, total number of required ways are = 120 + 180 = 300.

4. m men can be seated in m! ways creating $(m+1)$ places for ladies to sit.

n ladies out of $(m+1)$ places (as $n < m$) can be seated in ${}^{m+1}P_n$ ways

$\therefore$ Total ways $= m! \times {}^{m+1}P_n$

$$= m! \times \frac{(m+1)!}{(m+1-n)!} = \frac{(m+1)!\,m!}{(m-n+1)!}$$

5. There are four possibilities:

(i) 3 ladies from husband's side and 3 gentlemen from wife's side.

No. of ways in this case

$= {}^4C_3 \times {}^4C_3 = 4 \times 4 = 16$

(ii) 3 gentlemen from husband's side and 3 ladies from wife's side.

No. of ways in this case $= {}^3C_3 \times {}^3C_3 = 1 \times 1 = 1$

(iii) 2 ladies and one gentlemen from husband's side and one lady and 2 gentlemen from wife's side.

No. of ways in this case

$= ({}^4C_2 \times {}^3C_1) \times ({}^3C_1 \times {}^4C_2) = 6 \times 3 \times 3 \times 6 = 324$

(iv) One lady and 2 gentlemen from husband's side and 2 ladies and one gentlemen from wife's side.

No. of ways in this case

$= ({}^4C_1 \times {}^3C_2) \times ({}^3C_2 \times {}^4C_1) = 4 \times 3 \times 3 \times 4 = 144$

Hence the total no. of ways are

$= 16 + 1 + 324 + 144 = 485$

6. Number of ways of drawing at least one black ball

= 1 black and 2 other or 2 black and 1 other or 3 black

$= {}^3C_1 \times {}^6C_2 + {}^3C_2 \times {}^6C_1 + {}^3C_3 = 3 \times 15 + 3 \times 6 + 1$

$= 45 + 18 + 1 = 64$

ALTERNATE SOLUTION :

Number of ways of drawing at least one black ball

= Total ways– No. of ways of drawing no black ball

$= {}^9C_3 - {}^6C_3 = 84 - 20 = 64$

7. Number of ways in which a student can select at least one and atmost n books out of $(2n+1)$ books is equal to

$= {}^{2n+1}C_1 + {}^{2n+1}C_2 + {}^{2n+1}C_3 + ... + {}^{2n+1}C_n$

$$= \frac{1}{2}[2.{}^{2n+1}C_1 + 2.{}^{2n+1}C_2 + 2.{}^{2n+1}C_3 + ... + 2.{}^{2n+1}C_n]$$

$$= \frac{1}{2}[({}^{2n+1}C_1 + {}^{2n+1}C_{2n}) + ({}^{2n+1}C_2 + {}^{2n+1}C_{2n-1})$$

$$+ ({}^{2n+1}C_3 + {}^{2n+1}C_{2n-2}) + ... + ({}^{2n+1}C_n + {}^{2n+1}C_{n+1})]$$

[Using ${}^nC_r = {}^nC_{n-r}$]

$$= \frac{1}{2}[{}^{2n+1}C_1 + {}^{2n+1}C_2 + {}^{2n+1}C_3 + ... + {}^{2n+1}C_n$$

$$+ {}^{2n+1}C_{n+1} + {}^{2n+1}C_{n+2} + ... + {}^{2n+1}C_{2n}]$$

$$= \frac{1}{2}[{}^{2n+1}C_0 + {}^{2n+1}C_1 + {}^{2n+1}C_2 + ... + {}^{2n+1}C_{2n+1} - 1 - 1]$$

$$= \frac{1}{2}[2^{2n+1} - 2] = 2^{2n} - 1$$

ATQ, $2^{2n} - 1 = 63 \Rightarrow 2^{2n} = 64 = 2^6$

$\Rightarrow 2n = 6 \Rightarrow n = 3$

8. Out of 18 guests half i.e. 9 to be seated on side A and rest 9 on side B. Now out of 18 guests, 4 particular guests desire to sit on one particular side say side A and other 3 on other side B. Out of rest $18 - 4 - 3 = 11$ guests we can select 5 more for side A and rest 6 can be seated on side B. Selection of 5 out of 11 can be done in ${}^{11}C_5$ ways and 9 guests on each sides of table can be seated in $9! \times 9!$ ways. Thus there are total ${}^{11}C_5 \times 9! \times 9!$ arrangements.

9. Given that there are 9 women and 8 men. A committee of 12 is to be formed including at least 5 women.

This can be doen in the following ways.

= $5W$ and $7M$ or $6W$ and $6M$

or $7W$ and $5M$

or $8W$ and $4M$

or $9W$ and $3M$

No. of ways of forming committee is

$= {}^9C_5 \times {}^8C_7 + {}^9C_6 \times {}^8C_6 + {}^9C_7 \times {}^8C_5 + {}^9C_8 \times {}^8C_4 + {}^9C_9 \times {}^8C_3$

$$= \frac{9.8.7.6}{4.3.2.1} \times 8 + \frac{9.8.7}{3.2.1} \times \frac{8.7}{2.1} + \frac{9.8}{2.1} \times \frac{8.7.6}{3.2.1}$$

$$+ 9 \times \frac{8.7.6.5}{4.3.2.1} + 1 \times \frac{8.7.6}{3.2.1}$$

$= 126 \times 8 + 84 \times 28 + 36 \times 56 + 9 \times 70 + 56 = 6062$ ways.

(a) The women are in majority in

$2016 + 630 + 56 = 2702$ ways.

(b) The men are in majority in 1008 ways.

10. Let there be n sets of different objects each set containing n identical objects [eg (1, 1, 1 ... 1 (n times)), (2, 2, 2 ..., 2 (n times) ... ($n, n, n ... n$ (n times))]

Then the no. of ways in which these $n \times n = n^2$ objects can be arranged in a row

$$= \frac{(n^2)!}{n!n!...n!} = \frac{(n^2)!}{(n!)^n}$$

But these number of ways should be a natural number.

Hence $\frac{(n^2)!}{(n!)^n}$ is an integer. $(n \in I^+)$

11. Given that

runs scored in kth match $= k.2^{n+1-k}$, $1 \le k \le n$

and runs scored in n matches

$$= \frac{n+1}{4}(2^{n+1} - n - 2)$$

$$\therefore \sum_{k=1}^{n} k.2^{n+1-k} = \frac{n+1}{4}(2^{n+1} - n - 2)$$

$$\Rightarrow 2^{n+1}\left[\sum_{k=1}^{n} \frac{k}{2^k}\right] = \frac{n+1}{4}(2^{n+1} - n - 2)$$

$$\Rightarrow 2^{n+1}\left[\frac{1}{2} + \frac{2}{2^2} + \frac{3}{2^3} + ... + \frac{n}{2^n}\right]$$

$$= \frac{n+1}{4}(2^{n+1} - n - 2) \quad ....(i)$$

Let $S = \frac{1}{2} + \frac{2}{2^2} + \frac{3}{2^3} + ... + \frac{n}{2^n}$

$\frac{1}{2}S = \frac{1}{2^2} + \frac{2}{2^3} + ... + \frac{n-1}{2^n} + \frac{n}{2^{n+1}}$

Subtracting the above two, we get

$\frac{1}{2}S = \frac{1}{2} + \frac{1}{2^2} + \frac{1}{2^3} + ... + \frac{1}{2^n} - \frac{n}{2^{n+1}}$.

$\Rightarrow \frac{1}{2}S = \frac{\frac{1}{2}\left(1 - \frac{1}{2^n}\right)}{1 - \frac{1}{2}} - \frac{n}{2^{n+1}}$

$\Rightarrow S = 2\left[1 - \frac{1}{2^n} - \frac{n}{2^{n+1}}\right]$

$\therefore$ Equation (i) becomes

$2.2^{n+1}\left[1 - \frac{1}{2^n} - \frac{n}{2^{n+1}}\right] = \frac{n+1}{4}\left[2^{n+1} - n - 2\right]$

$\Rightarrow 2.\left[2^{n+1} - 2 - n\right] = \frac{n+1}{4}\left[2^{n+1} - 2 - n\right]$

$\Rightarrow \frac{n+1}{4} = 2 \Rightarrow n = 7$

F MATCH THE FOLLOWING :

1. (A)-p; (B)-s; (C)-q; (D)-q

(A) For the permutations containing the word ENDEA we consider 'ENDEA' as single letter. Then we have total ENDEA, N, O, E, L i.e. 5 letters which can be arranged in 5! ways.

$\therefore$ (A) $\to$ (p)

(B) If E occupies the first and last position, the middle 7 positions can be filled by N, D, E, A, N, O, L.

These can be arranged in $\frac{7!}{2!} = 21 \times 5!$ ways.

$\therefore$ (B) $\to$ (s)

(C) If none of the letters D, L, N occur in the last five positions then we should arrange D, D, L, N at first four positions and rest five i.e. E, E, E, A,O at last five positions. This can be done in

$\frac{4!}{2!} \times \frac{5!}{3!}$ ways .

(C) $\to$ (q)

(D) As per question A, E, E, E, O can be arranged at 1st, 3rd, 5th, 7th and 9th positions and rest N, D, N, L at rest 4 positions. This can be done in

$\frac{5!}{3!} \times \frac{4!}{2!}$ ways $= 2 \times 5!$ ways

(D) $\to$ (q)

G COMPREHENSION BASED QUESTIONS :

1. (b) $\because$ a_n = number of all n digit +ve integers formed by the digits 0, 1 or both such that no consecutive digits in them are 0.

and b_n = number of such n digit integers ending with 1

c_n = number of such n digit integers ending with 0

Clearly, $a_n = b_n + c_n$ ($\because$ a_n can end with 0 or 1)

Also $b_n = a_{n-1}$

and $c_n = a_{n-2}$ [$\because$ if last digit is 0, second last has to be 1]

$\therefore$ We get $a_n = a_{n-1} + a_{n-2}$, $n \geq 3$

Also $a_1 = 1$, $a_2 = 2$,

By this recurring formula

$a_3 = a_2 + a_1 = 3$

$a_4 = a_3 + a_2 = 3 + 2 = 5$

$a_5 = a_4 + a_3 = 5 + 3 = 8$

Also $b_6 = a_5 = 8$

2. (a) By recurring formula, $a_{17} = a_{16} + a_{15}$ is correct

Also $c_{17} \neq c_{16} + c_{15}$

$\Rightarrow a_{15} \neq a_{14} + a_{13}$ ($\because$ $c_n = a_{n-2}$)

$\therefore$ Incorrect

Similarly, other parts are also incorrect.

I INTEGER VALUE CORRECT TYPE :

1. 5 Given 8 vectors are

(1, 1, 1), (–1, –1, –1); (–1, 1, 1), (1, –1, –1); (1, –1, 1), (–1, 1, –1); (1, 1, –1), (–1, –1, 1)

These are 4 diagonals of a cube and their opposites.

For 3 non coplanar vectors first we select 3 groups of diagonals and its opposite in ${}^4C_3$ ways. Then one vector from each group can be selected in 2 × 2 × 2 ways.

$\therefore$ Total ways = ${}^4C_3 \times 2 \times 2 \times 2 = 32 = 2^5$

$\therefore$ $p = 5$

Section-B JEE Main/ AIEEE

1. (d) Required number of numbers
$= 5 \times 6 \times 6 \times 4 = 36 \times 20 = 720.$

2. (c) Required number of numbers $= 3 \times 5 \times 5 \times 5 = 375$

3. (d) We know that a number is divisible by 3 only when the sum of the digits is divisible by 3. The given digits are 0, 1, 2, 3, 4, 5.
Here the possible number of combinations of 5 digits out of 6 are ${}^5C_4 = 5$, which are as follows–
$1+2+3+4+5 = 15 = 3 \times 5$
$0+2+3+4+5 = 14$ (not divisible by 3)
$0+1+3+4+5 = 13$ (not divisible by 3)
$0+1+2+4+5 = 12 = 3 \times 4$
$0+1+2+3+5 = 11$ (not divisible by 3)
$0+1+2+3+4 = 10$ (not divisible by 3)
Thus the number should contain the digits 1, 2, 3, 4, 5 or the digits 0, 1, 2, 4, 5.
Taking 1, 2, 3, 4, 5, the 5 digit numbers are = 5! = 120
Taking 0, 1, 2, 4, 5, the 5 digit numbers are = 5! – 4! = 96
$\therefore$ Total number of numbers = 120 + 96 = 216

4. (b) Required sum
$= (2+4+6+ ... +100) + (5+10+15+ ... +100)$
$-(10+20+ ... +100)$
$= 2550 + 1050 - 530 = 3050.$

5. (c) ${}^nC_{r+1} + {}^nC_{r-1} + 2\,{}^nC_r = {}^nC_{r-1} + {}^nC_r + {}^nC_r + {}^nC_{r+1}$
$= {}^{n+1}C_r + {}^{n+1}C_{r+1} = {}^{n+2}C_{r+1}$

6. (c) As for given question two cases are possible.
i) Selecting 4 out of first five question and 6 out of remaining 8 question $= {}^5C_4 \times {}^8C_6 = 140$ choices.
ii) Selecting 5 out of first five question and 5 out of remaining 8 questions $= {}^5C_5 \times {}^8C_5 = 56$ choices.
Therefore, total number of choices
$= 140 + 56 = 196.$

7. (a) No. of ways in which 6 men can be arranged at a round table = (6 - 1)! = 5!
Now women can be arranged in ${}^6P_5 = 6!$ Ways.
Total Number of ways = 6! × 5!

8. (c) Total number of arrangements of letters in the word GARDEN = 6 ! = 720 there are two vowels A and E, in half of the arrangements A preceeds E and other half A follows E.
So, vowels in alphabetical order in $\frac{1}{2} \times 720 = 360$

9. (b) We know that the number of ways of distributing n identical items among r persons, when each one of them receives at least one item is ${}^{n-1}C_{r-1}$
$\therefore$ The required number of ways
$= {}^{8-1}C_{3-1} = {}^7C_2 = \frac{7!}{2!5!} = \frac{7 \times 6}{2 \times 1} = 21$

10. (a) Alphabetical order is
A, C, H, I, N, S
No. of words starting with A – 5!
No. of words starting with C – 5!
No. of words starting with H – 5!
No. of words starting with I – 5!
No. of words starting with N – 5!
SACHIN – 1
$\therefore$ sachin appears at serial no 601

11. (c) ${}^{10}C_1 + {}^{10}C_2 + {}^{10}C_3 + {}^{10}C_4$
$= 10 + 45 + 120 + 210 = 385$

12. (a) Set S = {1, 2, 3, ... 12}
$A \cup B \cup C = S,\ A \cap B = B \cap C = A \cap C = \phi$
$\therefore$ The number of ways to partition
$= {}^{12}C_4 \times {}^8C_4 \times {}^4C_4 = \frac{12!}{4!8!} \times \frac{8!}{4!4!} \times \frac{4!}{4!0!} = \frac{12!}{(4!)^3}$

13. (a) The given situation in statement 1 is equivalent to find the non negative integral solutions of the equation
$x_1 + x_2 + x_3 + x_4 + x_5 = 6$
which is coeff. of x^6 in the expansion of
$(1 + x + x^2 + x^3 + ...\infty)^5$ = coeff. of x^6 in $(1-x)^{-5}$
= coeff. of x^6 in $1 + 5x + \frac{5.6}{2!}x^2$.........
$= \frac{5 \cdot 6 \cdot 7 \cdot 8 \cdot 9 \cdot 10}{6!} = \frac{10!}{6!4!} = {}^{10}C_6$
$\therefore$ Statement 1 is wrong.
Number of ways of arranging 6A's and 4B's in a row
$= \frac{10!}{6!4!} = {}^{10}C_6$ which is same as the number of ways the child can buy six icecreams.
$\therefore$ Statement 2 is true.

14. (d) First let us arrange M, I, I, I, I, P, P
Which can be done in $\frac{7!}{4!2!}$ ways
√M√I√I√I√I√P√P√

Now 4 S can be kept at any of the ticked places in $^8C_4$ ways so that no two S are adjacent.
Total required ways

$$=\frac{7!}{4!2!}\,{}^8C_4=\frac{7!}{4!2!}\,{}^8C_4=7\times{}^6C_4\times{}^8C_4$$

15. **(c)** 4 novels, out of 6 novels and 1 dictionary out of 3 can be selected in $^6C_4\times{}^3C_1$ ways
Then 4 novels with one dictionary in the middle can be arranged in 4! ways.

$\therefore$ Total ways of arrangement $={}^6C_4\times{}^3C_1\times 4!=1080$

16. **(c)** Total number of ways $={}^3C_2\times{}^9C_2$

$$=3\times\frac{9\times 8}{2}=3\times 36=108$$

17. **(d)** The number of ways of distributing 10 identical balls in 4 distinct boxes such that no box empty is same as the number of ways of selecting $(r-1)$ places out of $(n-1)$ different places, that is $^{n-1}C_{r-1}$.

Hence required number of ways $={}^{10-1}C_{4-1}={}^9C_3$

$\therefore$ Both statements are correct and second statement is a correct explanation of statement -1.

18. **(d)** Number of white balls = 10
Number of green balls = 9
and Number of black balls = 7
$\therefore$ Required probability $=(10+1)(9+1)(7+1)$
$-1=11.10.8-1=879$

[$\because$ The total number of ways of selecting one or more items from p identical items of one kind, q identical items of second kind; r identical items of third kind is $(p+1)(q+1)(r+1)-1$]

19. **(b)** We know,
$T_n={}^nC_3,\ T_{n+1}={}^{n+1}C_3$
ATQ, $T_{n+1}-T_n={}^{n+1}C_3-{}^nC_3=10$
$\Rightarrow {}^nC_2=10$
$\Rightarrow n=5.$

Chapter 5 Mathematical Induction and Binomial Theorem

Section-A JEE Advanced/ IIT-JEE

A FILL IN THE BLANKS :

1. The larger of $99^{50} + 100^{50}$ and 101^{50} is .... **(1982 - 2 Marks)**
2. The sum of the coefficients of the plynomial $(1+x-3x^2)^{2163}$ is .... **(1982 - 2 Marks)**
3. If $(1+ax)^n = 1+8x+24x^2+....$ then $a =$ ...... and $n =$ .... **(1983 - 2 Marks)**
4. Let n be positive integer. If the coefficients of 2nd, 3rd, and 4th terms in the expansion of $(1+x)^n$ are in A.P., then the value of n is .... **(1994 - 2 Marks)**
5. The sum of the rational terms in the expansion of $(\sqrt{2}+3^{1/5})^{10}$ is .... **(1997 - 2 Marks)**

C MCQ's WITH ONE CORRECT ANSWER :

1. Given positive integers $r > 1$, $n > 2$ and that the coefficient of $(3r)$th and $(r+2)$th terms in the binomial expansion of $(1+x)^{2n}$ are equal . Then **(1983 - 1 Mark)**
   (a) $n = 2r$ (c) $n = 2r+1$
   (c) $n = 3r$ (d) none of these
2. The coefficient of x^4 in $\left(\frac{x}{2}-\frac{3}{x^2}\right)^{10}$ is **(1983 - 1 Mark)**
   (a) $\frac{405}{256}$ (b) $\frac{504}{259}$
   (c) $\frac{450}{263}$ (d) none of these
3. The expression $\left(x+(x^3-1)^{\frac{1}{2}}\right)^5 + \left(x-(x^3-1)^{\frac{1}{2}}\right)^5$ is a polynimial of degree **(1992 - 2 Marks)**
   (a) 5 (b) 6
   (c) 7 (d) 8
4. If in the expansion of $(1+x)^m(1-x)^n$, the coefficients of x and x^2 are 3 and -6 respectively, then m is **(1999 - 2 Marks)**
   (a) 6 (b) 9
   (c) 12 (d) 24
5. For $2 \le r \le n$, $\binom{n}{r}+2\binom{n}{r-1}+\binom{n}{r-2}=$ **(2000S)**
   (a) $\binom{n+1}{r-1}$ (b) $2\binom{n+1}{r+1}$
   (c) $2\binom{n+2}{r}$ (d) $\binom{n+2}{r}$
6. In the binomial expansion of $(a-b)^n$, $n \ge 5$, the sum of the 5th and 6th terms is zero. Then a/b equals **(2001S)**
   (a) $(n-5)/6$ (b) $(n-4)/5$
   (c) $5/(n-4)$ (d) $6/(n-5)$
7. The sum $\sum_{i=0}^{m}\binom{10}{i}\binom{20}{m-i}$, (where $\binom{p}{q}=0$ if $p < q$) is maximum when m is **(2002S)**
   (a) 5 (b) 10
   (c) 15 (d) 20
8. Coefficient of t^{24} in $(1+t^2)^{12}(1+t^{12})(1+t^{24})$ is **(2003S)**
   (a) ${}^{12}C_6+3$ (b) ${}^{12}C_6+1$
   (c) ${}^{12}C_6$ (d) ${}^{12}C_6+2$
9. If ${}^{n-1}C_r = (k^2-3)\,{}^nC_{r+1}$, then $k \in$ **(2004S)**
   (a) $(-\infty,-2]$ (b) $[2,\infty)$
   (c) $[-\sqrt{3},\sqrt{3}]$ (d) $(\sqrt{3},2]$
10. The value of
    $\binom{30}{0}\binom{30}{10}-\binom{30}{1}\binom{30}{11}+\binom{30}{2}\binom{30}{12}.....+\binom{30}{20}\binom{30}{30}$ is where $\binom{n}{r}={}^nC_r$ **(2005S)**
    (a) $\binom{30}{10}$ (b) $\binom{30}{15}$
    (c) $\binom{60}{30}$ (d) $\binom{31}{10}$
11. For r = 0, 1, …, 10, let A_r, B_r and C_r denote, respectively, the coefficient of x^r in the expansions of $(1+x)^{10}$, $(1+x)^{20}$ and $(1+x)^{30}$. Then $\sum_{r=1}^{10} A_r(B_{10}B_r - C_{10}A_r)$ is equal to **(2010)**
    (a) $B_{10}-C_{10}$ (b) $A_{10}(B^2{}_{10}C_{10}A_{10})$
    (c) 0 (d) $C_{10}-B_{10}$

D MCQ's WITH ONE OR MORE THAN ONE CORRECT :

1. If C_r stands for ^nC_r, then the sum of the series

$$\frac{2\left(\frac{n}{2}\right)!\left(\frac{n}{2}\right)!}{n!}[C_0^2 - 2C_1^2 + 3C_2^2 - \text{.........} + (-1)^n (n+1)C_n^2],$$

where n is an even positive integer, is equal to **(1986 - 2 Marks)**

(a) 0 (b) $(-1)^{n/2}(n+1)$

(c) $(-1)^{n/2}(n+2)$ (d) $(-1)^n n$

(e) none of these.

2. If $a_n = \sum_{r=0}^{n} \frac{1}{^nC_r}$, then $\sum_{r=0}^{n} \frac{r}{^nC_r}$ equals **(1998 - 2 Marks)**

(a) $(n-1)a_n$ (b) na_n

(c) $\frac{1}{2}na_n$ (d) None of the above

E SUBJECTIVE PROBLEMS :

1. Given that **(1979)**

$C_1 + 2C_2x + 3C_3x^2 + \text{............} + 2n\,C_{2n}x^{2n-1} = 2n(1+x)^{2n-1}$

where $C_r = \frac{(2n)!}{r!(2n-r)!}$ $\quad r = 0, 1, 2, \text{..................}, 2n$

Prove that

$C_1^2 - 2C_2^2 + 3C_3^2 - \text{......................} - 2nC_{2n}^2 = (-1)^n n\,C_n.$

2. Let $u_1 = 1, u_2 = 1$ and $u_{n+2} = u_{n+1} + u_n,\ \ n \geq 1.$

Use mathematical induction to show that

$$u_n = \frac{1}{\sqrt{5}}\left[\left(\frac{1+\sqrt{5}}{2}\right)^n - \left(\frac{1-\sqrt{5}}{2}\right)^n\right] \text{ for all } n > 1.$$

(1981 - 4 Marks)

3. Prove that $7^{2n} + (2^{3n-3})(3^{n-1})$ is divisible by 25 for any natural number n. **(1982 - 5 Marks)**

4. If $(1+x)^n = C_0 + C_1x + C_2x^2 + \text{......} + C_nx^n$ then show that the sum of the products of the C_i's taken two at a time, represented by $\sum\limits_{0 \le i < j \le n}\sum C_iC_j$ is equal to $2^{2n-1} - \frac{(2n)!}{2(n!)^2}$

(1983 - 3 Marks)

5. Use mathematical Induction to prove : If n is any odd positive integer, then $n(n^2-1)$ is divisible by 24. **(1983 - 2 Marks)**

6. If p be a natural number then prove that $p^{n+1} + (p+1)^{2n-1}$ is divisible by $p^2 + p + 1$ for every positive integer n. **(1984 - 4 Marks)**

7. Given $s_n = 1 + q + q^2 + \text{......} + q^n$;

$$S_n = 1 + \frac{q+1}{2} + \left(\frac{q+1}{2}\right)^2 + \text{.....} + \left(\frac{q+1}{2}\right)^n, q \neq 1 \text{ Prove that}$$

$$^{n+1}C_1 + {}^{n+1}C_2s_1 + {}^{n+1}C_3s_2 + \text{.....} + {}^{n+1}C_ns_n = 2^n S_n$$

(1984 - 4 Marks)

8. Use method of mathematical induction $2.7^n + 3.5^n - 5$ is divisible by 24 for all $n > 0$ **(1985 - 5 Marks)**

9. Using mathematical induction prove that

$$\sum_{k=0}^{n} k^2\ {}^nC_k = n(n+1)\ 2^{n-2} \text{ for } n > 1 \quad \textbf{(1986 - 5 Marks)}$$

10. Prove by mathematical induction that – **(1987 - 3 Marks)**

$$\frac{(2n)!}{2^{2n}(n!)^2} \leq \frac{1}{(3n+1)^{1/2}} \text{ for all positive Integers n.}$$

11. Let $R = (5\sqrt{5}+11)^{2n+1}$ and $f = R - [R]$, where [ ] denotes the greatest integer function. Prove that $Rf = 4^{2n+4}$. **(1988 - 5 Marks)**

12. Using mathematical induction, prove that **(1989 - 3 Marks)**

$$^mC_0\,{}^nC_k + {}^mC_1\,{}^nC_{k-1} + \text{.............}\ {}^mC_k\,{}^nC_0 = {}^{(m+n)}C_k,$$

where m, n, k are positive integers, and $^pC_q = 0$ for $p < q$.

13. Prove that **(1989 - 5 Marks)**

$$C_0 - 2^2C_1 + 3^2C_2 - \text{...........} + (-1)^n(n+1)^2C_n = 0,$$

$n > 2$, where $C_r = {}^nC_r$.

14. Prove that $\frac{n^7}{7} + \frac{n^5}{5} + \frac{2n^3}{3} - \frac{n}{105}$ is an integer for every positive integer n. **(1990 - 2 Marks)**

15. Using induction or otherwise, prove that for any non-negative integers m, n, r and k, **(1991 - 4 Marks)**

$$\sum_{m=0}^{k}(n-m)\frac{(r+m)!}{m!} = \frac{(r+k+1)!}{k!}\left[\frac{n}{r+1} - \frac{k}{r+2}\right]$$

16. If $\sum_{r=0}^{2n} a_r(x-2)^r = \sum_{r=0}^{2n} b_r(x-3)^r$ and $a_k = 1$ for all $k \geq n$, then show that $b_n = {}^{2n+1}C_{n+1}$ **(1992 - 6 Marks)**

17. Let $p \geq 3$ be an integer and α, β be the roots of $x^2 - (p+1)x + 1 = 0$ using mathematical induction show that $\alpha^n + \beta^n$.

(i) is an integer and (ii) is not divisible by p

(1992 - 6 Marks)

18. Using mathematical induction, prove that

$\tan^{-1}(1/3)+\tan^{-1}(1/7)+.....\tan^{-1}\{1/(n^2+n+1)\}$

$=\tan^{-1}\{n/(n+2)\}$ *(1993 - 5 Marks)*

19. Prove that $\sum_{r=1}^{k}(-3)^{r-1}\ {}^{3n}C_{2r-1}=0$, where $k=(3n)/2$ and n is an even positive integer. *(1993 - 5 Marks)*

20. If x is not an integral multiple of 2π use mathematical induction to prove that : *(1994 - 4 Marks)*

$$\cos x+\cos 2x+........+\cos nx=\cos\frac{n+1}{2}x\sin\frac{nx}{2}\operatorname{cosec}\frac{x}{2}$$

21. Let n be a positive integer and *(1994 - 5 Marks)*
$(1+x+x^2)^n=a_0+a_1x+..........+a_{2n}x^{2n}$
Show that $a_0^2-a_1^2+a_2^2\ ....+a_{2n}^2=a_n$

22. Using mathematical induction prove that for every integer $n\geq 1$, $(3^{2n}-1)$ is divisible by 2^{n+2} but not by 2^{n+3}. *(1996 - 3 Marks)*

23. Prove that $\frac{3!}{2(n+3)}=\sum_{r=0}^{n}(-1)^r\left(\frac{{}^nC_r}{{}^{r+3}C_r}\right)$ *(1997C - 5 Marks)*

24. Let $0<A_i<\pi$ for $i=1,2....,n$. Use mathematical induction to prove that

$$\sin A_1+\sin A_2...+\sin A_n\leq n\sin\left(\frac{A_1+A_2+.....+A_n}{n}\right)$$

where ≥ 1 is a natural number.
{You may use the fact that
$p\sin x+(1-p)\sin y\leq\sin[px+(1-p)y]$,
where $0\leq p\leq 1$ and $0\leq x,y\leq\pi$.} *(1997 - 5 Marks)*

25. Let p be a prime and m a positive integer. By mathematical induction on m, or otherwise, prove that whenever r is an integer such that p does not divide r, p divides ${}^{mp}C_r$, *(1998 - 8 Marks)*
[Hint: You may use the fact that $(1+x)^{(m+1)p}=(1+x)^p(1+x)^{mp}$]

26. Let n be any positive integer. Prove that *(1999 - 10 Marks)*

$$\sum_{k=0}^{m}\frac{\binom{2n-k}{k}}{\binom{2n-k}{n}}\cdot\frac{(2n-4k+1)}{(2n-2k+1)}2^{n-2k}=\frac{\binom{n}{m}}{\binom{2n-2m}{n-m}}2^{n-2m}$$

for each non-be gatuve integer $m\leq n$. $\left(\text{Here}\binom{p}{q}={}^pC_q\right)$.

27. For any positive integer m,n (with $n\geq m$), let $\binom{n}{m}={}^nC_m$.

Prove that

$\binom{n}{m}+\binom{n-1}{m}+\binom{n-2}{m}+.....+\binom{m}{m}=\binom{n+1}{m+2}$ Hence or otherwise, prove that

$$\binom{n}{m}+2\binom{n-1}{m}+3\binom{n-2}{m}+.....+(n-m+1)\binom{m}{m}=\binom{n+2}{m+2}.$$

(2000 - 6 Marks)

28. For every positive integer n, prove that $\sqrt{(4n+1)}<\sqrt{n}+\sqrt{n+1}<\sqrt{4n+2}$. Hence or otherwise, prove that $[\sqrt{n}+\sqrt{(n+1)}]=[\sqrt{4n+1}]$, where $[x]$ denotes the greatest integer not exceeding x. *(2000 - 6 Marks)*

29. Let a, b, c be positive real numbers such that $b^2-4ac>0$ and let $\alpha_1=c$. Prove by induction that

$\alpha_{n+1}=\frac{a\alpha_n^2}{(b^2-2a(\alpha_1+\alpha_2+...+\alpha_n)}$ is well – defined and

$\alpha_{n+1}<\frac{\alpha_n}{2}$ for all $n=1,2,...$ (Here, 'well – defined' means that the denominator in the expression for α_{n+1} is not zero.) *(2001 - 5 Marks)*

30. Use mathematical induction to show that $(25)^{n+1}-24n+5735$ is divisible by $(24)^2$ for all $n=1,2,........$ *(2002 - 5 Marks)*

31. Prove that

$$2^k\binom{n}{0}\binom{n}{k}-2^{k-1}\binom{n}{1}\binom{n-1}{k-1}+2^{k-2}\binom{n}{2}\binom{n-2}{k-2}-.....(-1)^k\binom{n}{k}\binom{n-k}{0}=\binom{n}{k}.$$

(2003 - 2 Marks)

32. A coin has probability p of showing head when tossed. It is tossed n times. Let p_n denote the probability that no two (or more) consecutive heads occur. Prove that $p_1=1$, $p_2=1-p^2$ and $p_n=(1-p).\ p_{n-1}+p(1-p)p_{n-2}$ for all $n\geq 3$.
Prove by induction on n, that $p_n=A\alpha^n+B\beta^n$ for all $n\geq 1$, where α and β are the roots of quadratic equation

$x^2-(1-p)x-p(1-p)=0$ and $A=\frac{p^2+\beta-1}{\alpha\beta-\alpha^2}$, $B=\frac{p^2+\alpha-1}{\alpha\beta-\beta^2}$.

(2000 - 5 Marks)

I INTEGER VALUE CORRECT TYPE :

1. The coefficients of three consecutive terms of $(1+x)^{n+5}$ are in the ratio 5 : 10 : 14. Then n = *(JEE Adv. 2013)*

Section-B JEE Main/ AIEEE

1. The coefficients of x^p and x^q in the expansion of $(1+x)^{p+q}$ are
(a) equal [2002]
(b) equal with opposite signs
(c) reciprocals of each other
(d) none of these

2. If the sum of the coefficients in the expansion of $(a+b)^n$ is 4096, then the greatest coeficient in the expansion is
(a) 1594 (b) 792 [2002]
(c) 924 (d) 2924

3. The positive integer just greater than $(1+0.0001)^{10000}$ is
(a) 4 (b) 5 [2002]
(c) 2 (d) 3

4. r and n are positive integers $r > 1$, $n > 2$ and coefficient of $(r+2)^{\text{th}}$ term and $3r^{\text{th}}$ term in the expansion of $(1+x)^{2n}$ are equal, then n equals [2002]
(a) $3r$ (b) $3r+1$
(c) $2r$ (d) $2r+1$

5. If $a_n = \sqrt{7+\sqrt{7+\sqrt{7+\ldots\ldots}}}$ haing n radical signs then by methods of mathematical induction which is true
(a) $a_n > 7\ \forall\ n \geq 1$ (b) $a_n < 7\ \forall\ n \geq 1$ [2002]
(c) $a_n < 4\ \forall\ n \geq 1$ (d) $a_n < 3\ \forall\ n \geq 1$

6. If x is positive, the first negative term in the expansion of $(1+x)^{27/5}$ is [2003]
(a) 6th term (b) 7th term
(c) 5th term (d) 8th term.

7. The number of integral terms in the expansion of $(\sqrt{3}+\sqrt[8]{5})^{256}$ is [2003]
(a) 35 (b) 32
(c) 33 (d) 34

8. Let $S(K) = 1+3+5\ldots+(2K-1) = 3+K^2$. Then which of the following is true [2004]
(a) Principle of mathematical induction can be used to prove the formula
(b) $S(K) \Rightarrow S(K+1)$
(c) $S(K) \not\Rightarrow S(K+1)$
(d) $S(1)$ is correct

9. The coefficient of the middle term in the binomial expansion in powers of x of $(1+\alpha x)^4$ and of $(1-\alpha x)^6$ is the same if α equals [2004]
(a) $\frac{3}{5}$ (b) $\frac{10}{3}$
(c) $\frac{-3}{10}$ (d) $\frac{-5}{3}$

10. The coefficient of x^n in expansion of $(1+x)(1-x)^n$ is
(a) $(-1)^{n-1}n$ (b) $(-1)^n(1-n)$ [2004]
(c) $(-1)^{n-1}(n-1)^2$ (d) $(n-1)$

11. The value of ${}^{50}C_4 + \sum\limits_{r=1}^{6} {}^{56-r}C_3$ is [2005]
(a) ${}^{55}C_4$ (b) ${}^{55}C_3$
(c) ${}^{56}C_3$ (d) ${}^{56}C_4$

12. If $A = \begin{bmatrix} 1 & 0 \\ 1 & 1 \end{bmatrix}$ and $I = \begin{bmatrix} 1 & 0 \\ 0 & 1 \end{bmatrix}$, then which one of the following holds for all $n \geq 1$, by the principle of mathematical induction [2005]
(a) $A^n = nA-(n-1)\text{I}$
(b) $A^n = 2^{n-1}A-(n-1)\text{I}$
(c) $A^n = nA+(n-1)\text{I}$
(d) $A^n = 2^{n-1}A+(n-1)\text{I}$

13. If the coefficient of x^7 in $\left[ax^2+\left(\frac{1}{bx}\right)\right]^{11}$ equals the coefficient of x^{-7} in $\left[ax-\left(\frac{1}{bx^2}\right)\right]^{11}$, then a and b satisfy the relation [2005]
(a) $a-b=1$ (b) $a+b=1$
(c) $\frac{a}{b}=1$ (d) $ab=1$

14. If x is so small that x^3 and higher powers of x may be neglected, then $\dfrac{(1+x)^{\frac{3}{2}}-\left(1+\frac{1}{2}x\right)^3}{(1-x)^{\frac{1}{2}}}$ may be approximated as [2005]
(a) $1-\frac{3}{8}x^2$ (b) $3x+\frac{3}{8}x^2$
(c) $-\frac{3}{8}x^2$ (d) $\frac{x}{2}-\frac{3}{8}x^2$

15. If the expansion in powers of x of the function $\frac{1}{(1-ax)(1-bx)}$ is $a_0 + a_1x + a_2x^2 + a_3x^3$...... then a_n is **[2006]**

(a) $\frac{b^n - a^n}{b-a}$ (b) $\frac{a^n - b^n}{b-a}$

(c) $\frac{a^{n+1} - b^{n+1}}{b-a}$ (d) $\frac{b^{n+1} - a^{n+1}}{b-a}$

16. For natural numbers m, n if $(1-y)^m(1+y)^n = 1 + a_1y + a_2y^2 + ...$ and $a_1 = a_2 = 10$, then (m,n) is **[2006]**

(a) $(20,45)$ (b) $(35,20)$

(c) $(45,35)$ (d) $(35,45)$

17. In the binomial expansion of $(a-b)^n$, $n \geq 5$, the sum of 5th and 6th terms is zero, then a/b equals **[2007]**

(a) $\frac{n-5}{6}$ (b) $\frac{n-4}{5}$

(c) $\frac{5}{n-4}$ (d) $\frac{6}{n-5}$.

18. The sum of the series **[2007]**

$^{20}C_0 - {}^{20}C_1 + {}^{20}C_2 - {}^{20}C_3 + ... -..... + {}^{20}C_{10}$ is

(a) 0 (b) $^{20}C_{10}$

(c) $-^{20}C_{10}$ (d) $\frac{1}{2}\,{}^{20}C_{10}$

DIRECTIONS : *Q. No. 19 is Assertion-Reason type questions. Each of these questions contains two statements: Statement-1(Assertion) and Statement-2(Reason). Each of these questions also has four alternative choices, only one of which is the correct answer. You have to select the correct choice.*

19. Statement-1 : $\sum_{r=0}^{n}(r+1)\,{}^nC_r = (n+2)2^{n-1}$.

Statement-2 : $\sum_{r=0}^{n}(r+1)\,{}^nC_r x^r = (1+x)^n + nx(1+x)^{n-1}$.

[2008]

(a) Statement -1 is false, Statement-2 is true

(b) Statement -1 is true, Statement-2 is true; Statement -2 is a correct explanation for Statement-1

(c) Statement -1 is true, Statement-2 is true; Statement -2 is not a correct explanation for Statement-1

(d) Statement -1 is true, Statement-2 is false

20. The remainder left out when $8^{2n} - (62)^{2n+1}$ is divided by 9 is : **[2009]**

(a) 2 (b) 7

(c) 8 (d) 0

21. Let $S_1 = \sum_{j=1}^{10} j(j-1)^{10}C_j$, $S_2 = \sum_{j=1}^{10} j^{10}C_j$ and $S_3 = \sum_{j=1}^{10} j^2\,{}^{10}C_j$.

Statement -1 : $S_3 = 55 \times 2^9$.

Statement - 2: $S_1 = 90 \times 2^8$ and $S_2 = 10 \times 2^8$. **[2010]**

(a) Statement -1 is true, Statement -2 is true ; Statement -2 is **not** a correct explanation for Statement -1.

(b) Statement -1 is true, Statement -2 is false.

(c) Statement -1 is false, Statement -2 is true .

(d) Statement - 1 is true, Statement 2 is true ; Statement -2 is a correct explanation for Statement -1.

22. The coefficient of x^7 in the expansion of $(1-x-x^2+x^3)^6$ is **[2011]**

(a) –132 (b) –144

(c) 132 (d) 144

23. Statement - 1 : For each natural number n, $(n+1)^7 - 1$ is divisible by 7.

Statement - 2 : For each natural number $n, n^7 - n$ is divisible by 7. **[2011 RS]**

(a) Statement-1 is true, Statement-2 is true; Statement-2 is a correct explanation for Statement-1.

(b) Statement-1 is true, Statement-2 is true; Statement-2 is NOT a correct explanation for Statement-1

(c) Statement-1 is true, Statement-2 is false

(d) Statement-1 is false, Statement-2 is true

24. If n is a positive integer , then $(\sqrt{3}+1)^{2n} - (\sqrt{3}-1)^{2n}$ is : ***[2012]***

(a) an irrational number

(b) an odd positive integer

(c) an even positive integer

(d) a rational number other than positive integers

25. The term independent of x in expansion of

$\left(\frac{x+1}{x^{2/3} - x^{1/3} + 1} - \frac{x-1}{x - x^{1/2}}\right)^{10}$ is **[JEE M 2013]**

(a) 4 (b) 120

(c) 210 (d) 310

Solutions & Explanations

Section-A : IIT-JEE

A	1. $(101)^{50}$	2. -1	3. $a=2, n=4$	4. 7	5. 41	
C	1. (a)	2. (a)	3. (c)	4. (c)	5. (d)	6. (b)
	7. (c)	8. (d)	9. (d)	10. (a)	11. (d)	
D	1. (c)	2. (c)				
I	1. 6					

Section-B : AIEEE

1. (a)	2. (c)	3. (d)	4. (c)	5. (b)	6. (d)
7. (c)	8. (b)	9. (c)	10. (b)	11. (d)	12. (a)
13. (d)	14. (c)	15. (d)	16. (d)	17. (b)	18. (d)
19. (b)	20. (a)	21. (b)	22. (b)	23. (a)	24. (a)
25. (c)					

Section-A JEE Advanced/ IIT-JEE

A FILL IN THE BLANKS :

1. Consider $(101)^{50}-\{(99)^{50}+(100)^{50}\}$
$=(100+1)^{50}-(100-1)^{50}-(100)^{50}$
$=(100)^{50}[(1+0.01)^{50}-(1-0.01)^{50}-1]$
$=(100)^{50}[2({}^{50}C_1(0.01)+{}^{50}C_3(0.01)^3+....)-1]$
$=(100)^{50}[2({}^{50}C_3(0.01)^3+....)]$
>0
$\therefore\ (101)^{50}>(99)^{50}+(100)^{50}\quad \therefore (101)^{50}$ is greater.

2. If we put $x=1$ in the expansion of $(1+x-3x^2)^{2163}=A_0+A_1x+A_2x^2+...$ we will get the sum of coefficients of given polynomial, which clearly comes to be -1.

3. $(1+ax)^n=1+8x+24x^2+...$

$\Rightarrow (1+ax)^n=1+nxa+\frac{n(n-1)}{2!}.a^2x^2+...$

$=1+8x+24x^2+...$

Comparing like powers of x we get

$nax=8x \Rightarrow na=8$ ....(1)

$\frac{n(n-1)a^2}{2}=24 \Rightarrow n(n-1)a^2=48$ ....(2)

Solving (1) and (2), $n=4,\ a=2$

4. We know that for a +ve integer n
$(1+x)^n={}^nC_0+{}^nC_1x+{}^nC_2x^2+......+{}^nC_nx^n$
ATQ coefficients of 2nd, 3nd, and 4th terms are in A.P.
i.e. ${}^nC_1, {}^nC_2, {}^nC_3$ are in A.P.
$\Rightarrow 2.{}^nC_2={}^nC_1+{}^nC_3$

$\Rightarrow 2\times\frac{n(n-1)}{2}=n+\frac{n(n-1)(n-2)}{3!}$

$\Rightarrow n-1=1+\frac{n^2-3n+2}{6} \Rightarrow n^2-9n+14=0$

$\Rightarrow (n-7)(n-2)=0 \Rightarrow n=7$ or 2

But for the existance of 4th term, n = 7.

5. Let T_{r+1} be the general term in the expansion of $(\sqrt{2}+3^{1/5})^{10}$

$\therefore\ T_{r+1}={}^{10}C_r(\sqrt{2})^{10-r}.(3^{1/5})^r.(0\le r\le 10)$

$=\frac{10!}{r!(10-r)!}.2^{5-r/2}.3^{r/5}$

Let T_{r+1} will be rational if $2^{5-r/2}$ and $3^{r/5}$ are rational numbers.

$\Rightarrow 5-\frac{r}{2}$ and $\frac{r}{5}$ are integers.

$\Rightarrow$ r = 0 and r = 10 $\Rightarrow T_1$ and T_{11} are rational terms.

$\Rightarrow$ Sum of T_1 and $T_{11}={}^{10}C_0 2^{5-0}.3^0+{}^{10}C_{10}2^{5-5}.3^2$
$=1.32.1+1.1.9=32+9=41$

C MCQ's WITH ONE CORRECT ANSWER :

1. (a) Given that r and n are +ve integers such that r > 1, n > 2
Also in the expansion of $(1+x)^{2n}$
coeff. of $(3r)^{th}$ term = coeff. of $(r+2)^{th}$ term
$\Rightarrow {}^{2n}C_{3r-1}={}^{2n}C_{r+1}$
[using ${}^nC_x={}^nC_y \Rightarrow x=y$ or $x+y=n$]
$\Rightarrow 3r-1=r+1$ or $3r-1+r+1=2n$
$\Rightarrow r=1$ or $2r=n$
But r > 1 $\therefore$ n = 2r

2. (a) General term in the expansion $\left(\frac{x}{2}-\frac{3}{x^2}\right)^{10}$ is

$$T_{r+1} = {}^{10}C_r\left(\frac{x}{2}\right)^{10-r}\left(\frac{-3}{x^2}\right)^r = {}^{10}C_r x^{10-3r}\frac{(-1)^r 3^r}{2^{10-r}}$$

For coeff of x^4, we should have

$10-3r=4 \Rightarrow r=2$

$\therefore$ Coeff of $x^4 = {}^{10}C_2 \frac{(-1)^2 3^2}{2^8} = \frac{405}{256}$

3. (c) The given expression is

$(x+\sqrt{x^3-1})^5+(x-\sqrt{x^3-1})^5$

We know by binomial theorem, that

$(x + a)^n + (x - a)^n = 2\ [{}^nC_0x^n + {}^nC_2x^{n-2}a^2 + {}^nC_4x^{n-4}a^4 + ...]$

$\therefore$ The given expression is equal to

$2\ [{}^5C_0x^5 + {}^5C_2x^3(x^3 - 1) + {}^5C_4x\ (x^3 - 1)^2]$

Max. power of x involved here is 7, also only +ve integral powers of x are involved, therefore given expression is a polynomial of degree 7.

4. (c) We have $(1 + x)^m\ (1-x)^n$

$$\left[1+mx+\frac{m(m-1)}{2!}x^2+....\right]\left[1-nx+\frac{n(n-1)}{2!}x^2-..\right]$$

$$=1+(m-n)x+\left[\frac{m(m-1)}{2}+\frac{n(n-1)}{2}-mn\right]x^2+....$$

ATQ, $m-n=3$ ....(1)

and $\frac{1}{2}m(m-1)+\frac{1}{2}n(n-1)-mn=-6$

$\Rightarrow m^2+n^2-2mn-(m+n)=-12$

$\Rightarrow (m-n)^2-(m+n)=-12$

$\Rightarrow m+n=9+12=21$ ....(2)

From (1) and (2), we get

$m=12$

5. (d) $\binom{n}{r}+2\binom{n}{r-1}+\binom{n}{r-2}$

$$=\left[\binom{n}{r}+\binom{n}{r-1}\right]+\left[\binom{n}{r-1}+\binom{n}{r-2}\right]$$

NOTE THIS STEP : $\binom{n+1}{r}+\binom{n+1}{r-1}=\binom{n+2}{r}$

$[\because\ {}^nC_r+{}^nC_{r-1}={}^{n+1}C_r]$

6. (b) $(a-b)^n,\ \geq 5$

In binomial expansion of above $T_5+T_6=0$

$\Rightarrow\ {}^nC_4\,a^{n-4}\,b^{4n}\,C_5\,a^{n-5}\,b^5=0$

$$\Rightarrow \frac{{}^nC_4}{{}^nC_5}.\frac{a}{b}=1 \Rightarrow \frac{4+1}{n-4}.\frac{a}{b}=1$$

$$\left[\text{Using } \frac{{}^nC_r}{{}^nC_{r+1}}=\frac{r+1}{n-r}\right]$$

$$\Rightarrow \frac{a}{b}=\frac{n-4}{5}$$

7. (c) $\sum_{i=0}^{m}{}^{10}C_i{}^{20}C_{m-i}={}^{10}C_0{}^{20}C_m+{}^{10}C_1{}^{20}C_{m-1}$

$+{}^{10}C_2{}^{20}C_{m-2}+.....+{}^{10}C_m{}^{20}C_0$

$=$ Coeff of x^m in the expansion of product $(1+x)^{10}(1+x)^{20}$

$=$ Coeff of x^m in the expansion of product $(1+x)^{30}$

$={}^{30}C_m$

To get max. value of given sum, ${}^{30}C_m$ should be max. which is so when $m = 30/2 = 15$.

$$\left[\text{Using the fact that max }({}^nC_r)=\begin{cases}{}^nC_{n/2}\text{ if } n \text{ is even}\\ {}^nC_{\frac{n+1}{2}}\text{ if } n \text{ is odd}\end{cases}\right]$$

8. (d) $(1+t^2)^{12}\,(1+t^{12})\,(1+t^{24})$

$=(1+t^{12}+t^{24}+t^{36})\,(1+t^2)^{12}$

$\therefore$ Coeff. of t^{24}

$= 1\times$ Coeff. of t^{24} in $(1+t^2)^{12}+1\times$ Coeff. of t^{12} in $(1+t^2)^{12}+1\times$ constant term in $(1+t^2)^{12}$

$={}^{12}C_{12}+{}^{12}C_6+{}^{12}C_0=1+{}^{12}C_6+1={}^{12}C_6+2$

9. (d) ${}^{n-1}C_r={}^nC_{r+1}(k^2-3)$

$$\Rightarrow k^2-3=\frac{{}^{n-1}C_r}{{}^nC_{r+1}}=\frac{r+1}{n}$$

Since $0\leq r\leq n-1$

$$\Rightarrow 1\leq r+1\leq n \Rightarrow \frac{1}{n}\leq\frac{r+1}{n}\leq 1 \Rightarrow \frac{1}{n}\leq k^2-3\leq 1$$

$$\Rightarrow 3+\frac{1}{n}\leq k^2\leq 4 \Rightarrow \sqrt{3+\frac{1}{n}}\leq k\leq 2$$

as $n\to\infty$

$\Rightarrow \sqrt{3}<k\leq 2 \Rightarrow k\in(\sqrt{3},2]$

10. (a) To find

${}^{30}C_0{}^{30}C_{10}-{}^{30}C_1{}^{30}C_{11}+{}^{30}C_2{}^{30}C_{12}-....+{}^{30}C_{20}{}^{30}C_{30}$

We know that

$(1+x)^{30}={}^{30}C_0+{}^{30}C_1x+{}^{30}C_2x^2+....+{}^{30}C_{20}x^{20}+....{}^{30}C_{30}x^{30}$ ....(1)

$(x-1)^{30}={}^{30}C_0x^{30}-{}^{30}C_1x^{29}+....+{}^{30}C_{10}x^{20}-{}^{30}C_{11}x^{19}+{}^{30}C_{12}x^{18}+....{}^{30}C_{30}x^0$ ....(2)

Multiplying eq^n (1) and (2), we get

$(x^2-1)^{30}=(\ )\times(\ )$

Equating the coefficients of x^{20} on both sides, we get

$${}^{30}C_{10} = {}^{30}C_0\,{}^{30}C_{10} - {}^{30}C_1\,{}^{30}C_{11} + {}^{30}C_2\,{}^{30}C_{12} - .... + {}^{30}C_{20}\,{}^{30}C_{30}$$

$\therefore$ Req. value is ${}^{30}C_{10}$

11. **(d)** Clearly $A_r = {}^{10}C_r,\ B_r = {}^{20}C_r,\ C_r = {}^{30}C_r$

$$\text{Now}\sum_{r=1}^{10} {}^{10}C_r\left({}^{20}C_{10}\,{}^{20}C_r - {}^{30}C_{10}\,{}^{10}C_r\right)$$

$$= {}^{20}C_{10}\sum_{r=1}^{10}{}^{10}C_r\,{}^{20}C_r - {}^{30}C_{10}\sum_{r=1}^{10}{}^{10}C_r \times {}^{10}C_r$$

$$= {}^{20}C_{10}\left({}^{10}C_1\,{}^{20}C_1 + {}^{10}C_2\,{}^{20}C_2 + ... + {}^{10}C_{10}\,{}^{20}C_{10}\right)$$

$$- {}^{30}C_{10}\left({}^{10}C_1 \times {}^{10}C_1 + {}^{10}C_2 \times {}^{10}C_2 + ... + {}^{10}C_{10}\,{}^{10}C_{10}\right) ....(1)$$

Now expanding $(1+x)^{10}$ and $(1+x)^{20}$ by binomial theorem and comparing the coefficients of x^{20} in their product, on both sides, we get

$${}^{10}C_0\,{}^{20}C_0 + {}^{10}C_1\,{}^{20}C_1 + {}^{10}C_2\,{}^{20}C_2 + ... + {}^{10}C_{10}\,{}^{20}C_{10}$$

$= \text{coeff of } x^{20} \text{ in } (1+x)^{30}$

$= {}^{30}C_{20} = {}^{30}C_{10}$

$$\therefore\ {}^{10}C_1\,{}^{20}C_1 + {}^{10}C_2\,{}^{20}C_2 + ... + {}^{10}C_{10}\,{}^{20}C_{10} = {}^{30}C_{10} - 1$$

Again expending $(1+x)^{10}$ and $(x+1)^{10}$ by binomial theorem and comparing the coefficients of x^{10} in their product on both sides, we get

$$\therefore\ \left({}^{10}C_0\right)^2\left({}^{10}C_1\right)^2 + \left({}^{10}C_2\right)^2 + ... + \left({}^{10}C_{10}\right)^2 =$$

coeff of x^{10} in $(1+x)^{20} = {}^{20}C_{10}$

$$\therefore\ \left({}^{10}C_1\right)^2 + \left({}^{10}C_2\right)^2 + ... + \left({}^{10}C_{10}\right)^2 = {}^{20}C_{10} - 1$$

Substituting these values in equation (1), we get

$$= {}^{20}C_{10}\left({}^{30}C_{10} - 1\right) - {}^{30}C_{10}\left({}^{20}C_{10} - 1\right)$$

$$= {}^{30}C_{10} - {}^{20}C_{10} = C_{10} - B_{10}$$

(D) MCQ's WITH ONE OR MORE THAN ONE CORRECT :

1. **(c)** $\because$ n is even, let $n = 2m$ then

$$\text{LHL} = S = \frac{2.m!m!}{(2m)!}[C_0^2 - 2C_1^2 + 3C_2^2.....$$

$$+(-1)^{2m}(2m+1)C_{2m}^2....] \quad ....(1)$$

$$= \frac{2.m!.m!}{(2m)!}C_{2m}^2 - 2C_{2m-1}^2 + 3C_{2m-2}^2 - .....$$

$$+(-1)^{2m}(2m+1)C_0^2 \qquad [\text{Using } C_r = C_{n-r}]$$

$$\Rightarrow\ S = \frac{2.m!m!}{(2m)!}[(2m+1)C_0^2 - 2mC_1^2$$

$$+(2m-1)C_2^2...... - 2C_{2m-1}^2 + C_{2m}^2] \quad ....(2)$$

Adding (1) and (2):

$$2S = 2\frac{m!m!}{(2m)!}[2m+2][C_0^2 - C_1^2 + C_2^2 + ... + C_{2m}^2]$$

Now keeping in mind that

$$C_0^2 - C_1^2 + C_2^2 - ...... + C_n^2 = (-1)^{n/2}\,{}^nC_{n/2}$$

If n is even we get

$$S = \frac{m!m!}{(2m)!}(2m+2)\,[(-1)^m\,{}^{2m}C_m] = \left(2\frac{n}{2}+2\right)(-1)^{n/2}$$

$$= (-1)^{n/2}(n+2)$$

ALTERNATE SOLUTION

We have

$$C_0^2 - 2C_1^2 + 3C_2^2 - .... + (-1)^n\,(n+1)C_n^2$$

$$= [C_0^2 - C_1^2 + C_2^2 - .... + (-1)^n C_n^2]$$

$$-[C_1^2 - 2C_2^2 + 3C_3^2 - ....(-1)^n\,nC_n^2] \quad ....(1)$$

$= X - Y$

where $\quad X = C_0^2 - C_1^2 + C_2^2 - .... + (-1)^n C_n^2$

and $\quad Y = C_1^2 - 2C_2^2 + 3C_3^2 - .... + (-1)^n\,nC_n^2$

Evaluation of X :

$$(1+x)^n = C_0 + C_1x + C_2x^2 + ... + C_nx^n$$

$$(x-1)^n = C_0x^n - C_1x^{n-1} + C_2x^{n-2} - .... + (-1)^n C_n$$

Multiplying both sides, we get

$$(1+x^n)(x-1)^n = (C_0 + C_1x + C_2x^2 + ... + C_nx^n)$$

$$(C_0x^n - C_1x^{n-1} + C_2x^{n-2} .... + (-1)^n C_n)$$

$$\Rightarrow\ (-1)^n(1-x^2)^n = (\ )(\ )$$

Now $C_0^2 - C_1^2 + C_2^2 - ....$ is the coefficient of x^n in the above product in RHS.

$\therefore$ Comparing the co-efficients of x^n on both sides, we get

NOTE THIS STEP :

$$X = C_0^2 - C_1^2 + C_2^2 - .... = (-1)^{n/2}\frac{n!}{\frac{n}{2}!\frac{n}{2}!}$$

Evaluation of Y : Consider

$$(1+x)^{2n} = C_0 + C_1x + C_2x^2 + ... + C_{2n}x^{2n}$$

Differentiating the above with respect to x we get

$$2n(1+x)^{2n-1} = C_1 + 2C_2x + 3C_3x^2 + ... + 2n\,C_{2n}x^{2n-1} \quad ....(a)$$

Again we have,

$$\left(1-\frac{1}{x}\right)^{2n} = C_0 - C_1.\frac{1}{x} + C_2.\frac{1}{x^2} - .... + C_{2n}.\frac{1}{x^{2n}} \quad ....(b)$$

Multiplying (a) and (b), we get

$$2n(1+x)^{2n-1}\left(1-\frac{1}{x}\right)^{2n}$$

$$= (C_1+2C_2x+3C_3x^2+....+2nC_{2n}x^{2n-1})$$

$$\times\left(C_0-C_1.\frac{1}{x}+C_2.\frac{1}{x^2}....+C_{2n}.\frac{l}{x^{2n}}\right)$$

The coefficient of $1/x$ in RHS is,

$$-(C_1^2-2C_2^2+3C_3^2-....-2nC_{2n}^2)$$

Also coefficient of $1/x$ in LHS is,

$$= \text{coeff of } \frac{1}{x} in \frac{2n(1+x)^{2n-1}(x-1)^{2n}}{x^{2n}}$$

$$= \text{coeff of } \frac{1}{x} \text{ in } \frac{2n}{x^{2n}}(1-x^2)^{2n-1}(x-1)$$

$$= \text{coeff of } x^{2n-1} \text{in } 2n(1-x^2)^{2n-1}(1-x)$$

$$= \text{coeff of } (x^2)^{n-1} \text{ in } 2n(1-x^2)^{2n-1} \times \text{ coeff of } x\, n(1-x)$$

$$= 2n(-1)^{n-1}.{}^{2n-1}C_{n-1}(-1)$$

$$= (-1)^n.2n.\frac{(2n-1)!}{(n-1)!n!} = (-1)^{n-1}\frac{2n!}{n!\,n!}.n$$

$\therefore$ Comparing the coefficients of $\frac{1}{x}$ on both sides, we get

$$C_1^2-2C_2^2+3C_3^2-....-2nC_{2n}^2 = (-1)^{n-1}\frac{2n!}{n!n!}.n$$

$$\Rightarrow Y = (-1)^{n/2-1}\frac{n!}{\frac{n}{2}!\frac{n}{2}!}.\frac{n}{2} \text{ (Replacing } n \text{ by } n/2)$$

Substituting the values of X and Y in eq. (i), we get

$$C_0^2-2C_1^2+3C_2^2-....+(-1)^n(n+1)C_n^2$$

$$= (-1)^{n/2}\frac{n!}{\frac{n}{2}!\frac{n}{2}!}\left(1+\frac{n}{2}\right) = (-1)^{n/2}\frac{n+2}{2}\frac{n!}{\frac{n}{2}!\frac{n}{2}!}$$

$\therefore$ Sum of the series,

$$= \frac{2\left(\frac{n}{2}\right)!\left(\frac{n}{2}\right)!}{n!}[C_0^2-2C_1^2+3C_2^2-....+(-1)^n(n+1)C_n^2]$$

NOTE THIS STEP :

$$= \frac{2.\left(\frac{n}{2}\right)!\left(\frac{n}{2}\right)!}{n!}\left[(-1)^{n/2}\frac{n+2}{2}.\frac{n!}{\frac{n}{2}!\frac{n}{2}!}\right] = (-1)^{n/2}(n+2)$$

2. (c) Let $b = \sum_{r=0}^{n}\frac{r}{^nC_r} = \sum_{r=0}^{n}\frac{n-(n-r)}{^nC_r}$

$$= na_n - \sum_{r=0}^{n}\frac{n-r}{^nC_{n-r}} \qquad [\because {}^nC_r = {}^nC_{n-r}]$$

$$= na_n - b$$

$$\Rightarrow \quad 2b = na_n \Rightarrow b = \frac{n}{2}a_n$$

(E) SUBJECTIVE PROBLEMS :

1. Given that

$$C_1+2C_2x+3C_3x^2+....+2nC_{2n}x^{2n-1} = 2n(1+x)^{2n-1} \quad ....(1)$$

where $\quad C_r = \frac{2n!}{r!(2n-r)!}$

Integrating both sides with respect to x, under the limits 0 to x, we get

$$[C_1x+C_2x^2+C_3x^3+....+C_{2n}x^{2n}]_0^x = [(1+x)^{2n}]_0^x$$

$$\Rightarrow C_1x+C_2x^2+C_3x^3+....+C_{2n}x^{2n} = (1+x)^{2n}-1$$

$$\Rightarrow C_0+C_1x+C_2x^2+C_3x^3+....+C_{2n}x^{2n} = (1+x)^{2n} \quad ....(2)$$

Changing x *by* $-\frac{1}{x}$, we get

$$\Rightarrow C_0-\frac{C_1}{x}+\frac{C_2}{x^2}-\frac{C_3}{x^3}+....+(-1)^{2n}\frac{C_{2n}}{x^{2n}} = \left(1-\frac{1}{x}\right)^{2n}$$

$$\Rightarrow C_0x^{2n}-C_1x^{2n-1}+C_2x^{2n-2}-C_3x^{2n-3}$$
$$+....+C_{2n} = (x-1)^{2n} \quad ....(3)$$

Multiplying eqn. (1) and (3) and equating the coefficients of x^{2n-1} on both sides, we get

$$-C_1^2+2C_2^2-3C_3^2+....+2nC_{2n}^2$$

$$= \text{coeff.of } x^{2n-1} \; in\, 2n(x-1)(x^2-1)^{2n-1}$$

$$= 2n\,[\text{coeff. of } x^{2n-2} \text{ in } (x^2-1)^{2n-1}$$
$$- \text{coeff. of } x^{2n-1} \text{ in } (x^2-1)^{2n-1}]$$

$$= 2n\,[{}^{2n-1}C_{n-1}(-1)^{n-1}-0]$$

$$= (-1)^{n-1}.2n\,{}^{2n-1}C_{n-1}$$

$$\Rightarrow C_1^2-2C_2^2+3C_3^2+....+2nC_{2n}^2$$

$$= (-1)^n.2n\,{}^{2n-1}C_{n-1} = (-1)^n n.\left(\frac{2n}{n}.{}^{2n-1}C_{n-1}\right)$$

$$= (-1)^n n.{}^{2n}C_n = (-1)^n n.C_n. \qquad (\because {}^{2n}C_n = C_n)$$

Hence Proved.

2. Given $u_1 = 1$, $u_2 = 1$, $u_{n+2} = u_{n+1} + u_n$, $n \geq 1$

$$u_n = \frac{1}{\sqrt{5}}\left[\left(\frac{1+\sqrt{5}}{2}\right)^n - \left(\frac{1-\sqrt{5}}{2}\right)^n\right]$$

$$\therefore \quad u_2 = \frac{1}{\sqrt{5}}\left[\left(\frac{1+\sqrt{5}}{2}\right)^2 - \left(\frac{1-\sqrt{5}}{2}\right)^2\right]$$

$$= \frac{1}{\sqrt{5}}\left(\frac{1+\sqrt{5}}{2}+\frac{1-\sqrt{5}}{2}\right)\left(\frac{1+\sqrt{5}}{2}-\frac{1-\sqrt{5}}{2}\right)$$

$$= \frac{1}{\sqrt{5}}[1.\sqrt{5}] = 1$$

$\therefore$ u_n is true for $n = 2$ (as given $u_2 = 1$).

Let it be true for $n \geq k > 2$ then

$$u_k = \frac{1}{\sqrt{5}}\left[\left(\frac{1+\sqrt{5}}{2}\right)^k - \left(\frac{1-\sqrt{5}}{2}\right)^k\right]$$

.... (Induction Hypothesis)

Consider $U_{k+1} = u_k + u_{k-1}$ [Using $u_{n+2} = u_{n+1} + u_n$]

$$\therefore \quad u_{k+1} = \frac{1}{\sqrt{5}}\left[\left(\frac{1+\sqrt{5}}{2}\right)^k - \left(\frac{1-\sqrt{5}}{2}\right)^k\right]$$

$$+\frac{1}{\sqrt{5}}\left[\left(\frac{1+\sqrt{5}}{2}\right)^{k-1} - \left(\frac{1-\sqrt{5}}{2}\right)^{k-1}\right]$$

$$= \frac{1}{\sqrt{5}}\left[\left(\frac{1+\sqrt{5}}{2}\right)^k + \left(\frac{1+\sqrt{5}}{2}\right)^{k-1}\right]$$

$$-\frac{1}{\sqrt{5}}\left[\left(\frac{1-\sqrt{5}}{2}\right)^k + \left(\frac{1-\sqrt{5}}{2}\right)^{k-1}\right]$$

$$= \frac{1}{\sqrt{5}}\left[\left(\frac{1+\sqrt{5}}{2}\right)^{k-1}\left(\frac{1+\sqrt{5}}{2}+1\right)\right.$$

$$\left.-\left(\frac{1-\sqrt{5}}{2}\right)^{k-1}\left\{\frac{1-\sqrt{5}}{2}+1\right\}\right]$$

$$= \frac{1}{\sqrt{5}}\left[\left(\frac{1+\sqrt{5}}{2}\right)^{k-1}\left\{\frac{6+2\sqrt{5}}{4}\right\}\right.$$

$$\left.-\left(\frac{1-\sqrt{5}}{2}\right)^{k-1}\left\{\frac{6-2\sqrt{5}}{4}\right\}\right]$$

$$= \frac{1}{\sqrt{5}}\left[\left(\frac{1+\sqrt{5}}{2}\right)^{k-1}\left\{\frac{(\sqrt{5})^2+1^2+2\sqrt{5}}{4}\right\}\right.$$

$$\left.-\left(\frac{1-\sqrt{5}}{2}\right)^{k-1}\left(\frac{1^2+(\sqrt{5})^2-2\sqrt{5}}{4}\right)\right]$$

$$= \frac{1}{\sqrt{5}}\left[\left(\frac{1+\sqrt{5}}{2}\right)^{k+1} - \left(\frac{1-\sqrt{5}}{2}\right)^{k+1}\right]$$

$\therefore$ u_{k+1} is also true.

Hence by principle of mathematical induction u_n is true $\forall\ n > 1$

3. $P(n): 7^{2n} + 2^{3n-3}, 3^{n-1}$ is divisible by 25 $\forall n \in N$.

Let us prove it by Mathematical Induction :

$P(1): 7^2 + 2^0.3^0 = 49 + 1 = 50$ which is divisible by 25.

$\therefore$ $P(1)$ is true.

Let $P(k)$ be true that is $7^{2k} + 2^{3k-3}, 3^{k-1}$ is divisible by 25.

$\Rightarrow$ $7^{2k} + 2^{3k-3}.3^{k-1} = 25m$ where $m \in Z$.

$\Rightarrow$ $2^{3k-3}.\ 3^{k-1} = 25m - 7^{2k}$ ....(1)

Consider $P(k+1)$:

$7^{2(k+1)} + 2^{3(k+1)-3}.3^{k+1-1} = 7^{2k}.7^2 + 2^{3k}.3^k$

$= 49.7^{2k} + 2^3.3.2^{3k-3}.3^{k-1} = 49.7^{2k} + 24(25m - 7^{2k})$

(Using IH eq. (1))

$= 49.7^{2k} + 24 \times 25m - 24 \times 7^{2k}$

$= 25.7^{2k} + 24 \times 25m = 25(7^{2k} + 24m)$

$= 25 \times$ some integral value which is divisible by 25.

$\therefore$ $P(k+1)$ is also true.

Hence by the principle of mathematical induction

$P(n)$ is true $\forall\ n \in Z$.

4. $S = \sum\sum C_i C_j$

$0 \leq i < j \leq n$

NOTE THIS STEP

$\Rightarrow$ $S = C_0(C_1 + C_2 + C_3 + ... + C_n) + C_1(C_2 + C_3 + ... + C_n) + C_2(C_3 + C_4 + C_5 + ... C_n) + ... C_{n-1}(C_n)$

$\Rightarrow$ $S = C_0(2^n - C_0) + C_1(2^n - C_0 - C_1) + C_2(2^n - C_0 - C_1 - C_2) + ... + C_{n-1}(2^n - C_0 - C_1 .... C_{n-1}) + C_n(2^n - C_0 - C_1 .... C_n)$

$\Rightarrow$ $S = 2^n(C_0 + C_1 + C_2 + ... + C_{n-1} + C_n) - (C_0^2 + C_1^2 + C_2^2 + ... + C_n^2) - S$

$$\Rightarrow \quad 2S = 2^n.2^n - \frac{2n!}{(n!)^2} = 2^{2n} - \frac{2n!}{(n!)^2}$$

$$\Rightarrow \quad S = 2^{2n-1} - \frac{2n!}{2(n!)^2}$$

5. $P(n): n(n^2-1)$ is divisible by 24 for n odd +ve integer.

For $n = 2m-1$, it can be restated as

$P(m): (2m-1)(4m^2 - 4m) = 4m(m-1)(2m-1)$

is divisible by 24 $\forall\ m \in N$

$\Rightarrow$ $P(m): m(m-1)(2m-1)$ is divisible by 6 $\forall\ m \in N$.

Here $P(1) = 0$, divisible by 6.

$\therefore$ $P(1)$ is true.

Let it be true for $m = k$, i.e.,

$k(k-1)(2k-1) = 6p$

$\Rightarrow$ $2k^3 - 3k^2 + k = 6p$ ....(1)

Consider

$P(k+1): k(k+1)(2k+1) = 2k^3 + 3k^2 + k$

$= 6p + 3k^2 + 3k^2$ (Using (1)

$= 6(p + k^2)$

$\Rightarrow$ divisible by 6

$\therefore$ $P(k+1)$ is also true.

Hence $P(m)$ is true $\forall\ m \in N$.

6. $P(n) : P^{n+1} + (p+1)^{2n-1}$ is divisible by $p^2 + p + 1$
For $n = 1$, $P(1) : p^2 + p + 1$ which is divisible by $p^2 + p + 1$.
$\therefore$ $P(1)$ is true.
Let $P(k)$ be true, i.e.,
$p^{k+1} + (p+1)^{2k-1}$ is divisible by $p^2 + p + 1$
$\Rightarrow p^{k+1} + (p+1)^{2k-1} = (p^2 + p + 1)\, m$ ....(1)
Consider $P(k+1) : p^{k+2} + (p+1)^{2k+1}$
$= p \,.\, p^{k+1} + (p+1)^{2k-1} .\, (p+1)^2$
$= p\,[m\,(p^2 + p + 1) - (p+1)^{2k-1}] + (p+1)^{2k-1}(p+1)^2$
$= p\,(p^2 + p + 1)m - p\,(p+1)^{2k-1} + (p+1)^{2k-1}\,(p^2 + 2p + 1)$
$= p\,(p^2 + p + 1)m + (p+1)^{2k-1}(p^2 + p + 1)$
$= (p^2 + p + 1)\,[mp + (p+1)^{2k-1}]$
$= (p^2 + p + 1)$ some integral value
$\therefore$ divisible by $p^2 + p + 1$ $\therefore P(k+1)$ is also true.
Hence by principle of mathematical induction $P(n)$ is true $\forall\, n \in N$.

7. We have $s_n = \dfrac{1 - q^{n+1}}{1 - q}$ ....(1)

and $S_n = \dfrac{1 - \left(\dfrac{q+1}{2}\right)^{n+1}}{1 - \left(\dfrac{q+1}{2}\right)} = \dfrac{2^{n+1} - (q+1)^{n+1}}{2^n (1-q)}$ ....(2)

Now, ${}^{n+1}C_1 + {}^{n+1}C_2 s_1 + {}^{n+1}C_3 s_2 + ... + {}^{n+1}C_{n+1} s_n$

$= \dfrac{1}{1-q}[{}^{n+1}C_1(1-q) + {}^{n+1}C_2(1-q^2) + {}^{n+1}C_3(1-q^3) + ... +$
$+ .... + {}^{n+1}C_n(1 - q^{n+1})]$ Using (1)

$= \dfrac{1}{1-q}\Big[\left({}^{n+1}C_1 + {}^{n+1}C_2 + ... + {}^{n+1}C_{n+1}\right)$
$- \left({}^{n+1}C_1 q + {}^{n+1}C_2 q^2 + ... + {}^{n+1}C_{n+1} q^{n+1}\right)$

$= \dfrac{1}{1-q}\left[2^{n+1} - 1) - \{(1+q)^{n+1} - 1\}\right]$

$= \dfrac{2^{n+1} - (1+q)^{n+1}}{(1-q)} = 2^n S_n$ [Using eq. (2)]

8. Let $A_n = 2.7^n + 3.5^n - 5$
Then $A_1 = 2.7 + 3.5 - 5 = 14 + 15 - 5 = 24$.
Hence A_1 is divisible by 24.
Now assume that A_m is divisible by 24 so that we may write
$A_m = 2.7^m + 3.5^m - 5 = 24k,\ k \in N$ ....(1)
Then $A_{m+1} - A_m = 2\,(7^{m+1} - 7^m) + 3\,(5^{m+1} - 5^m) - 5 + 5$
$= 2.7^m\,(7-1) + 3.5^m\,(5-1) = 12.\,(7^m + 5^m)$
Since 7^m and 5^m are odd integers $\forall\ m \in N$, their sum must be an even integer, say
$7^m + 5^m = 2p, p \in N$.
Hence $A_{m+1} - A_m = 12.2\,p = 24\,p$
or $A_{m+1} = A_m + 24p = 24k + 24p$ [by (1)]
Hence A_{m+1} is divisible by 24.
It follows by mathematical induction that A_n is divisible by 24 for all $n \in N$.

9. Let $P(n) : \sum_{k=0}^{n} k^2\, {}^nC_k = n(n+1)2^{n-2}$ for $n > 1$

$P(2) : \text{LHS} = \sum_{k=0}^{2} k^2\, {}^2C_k = 0 + {}^2C_1 = 6$

$\text{RHS} = 2(2+1)\,.\,2^{2-2} = 6$
$\therefore$ LHS = RHS
$\therefore$ $P(2)$ is true.
Let $P(m)$ be true i.e.

$$\sum_{k=0}^{m} k^2\, {}^mC_k = m(m+1).2^{m-2} \quad ....(1)$$

Consider $P(m+1)$:

$$\text{LHS} = \sum_{k=0}^{m+1} k^2\, {}^{m+1}C_k = \sum_{k=0}^{m+1} k^2\left({}^mC_k + {}^mC_{k-1}\right)$$

$$= \sum_{k=0}^{m} k^2\, {}^mC_k + \sum_{k=1}^{m+1} k^{2\,m}C_{k-1}$$

[As $I\Sigma$ is meaning less for $k = m + I$ and II for $k = 0$]

$$= m\,(m+1)\,2^{\,m-2} + \sum_{k=1}^{m+1} k^2\, {}^mC_{k-1} \quad \text{[Using (1)]}$$

$$= m\,(m+1)\,2^{m-2} + \sum_{p=0}^{m} (p+1)^2\, {}^mC_p$$

[Substituting $k - 1 = $ p]

$$= \text{m}\,(\text{m}+1)\,2^{\text{m}-2} + \sum_{p=0}^{m} (p)^2\, {}^mC_p + 2\sum_{p=0}^{m} p\, {}^mC_p + \sum_{p=0}^{m} {}^mC_p$$

$= m\,(m+1)2^{m-2} + m(m+1)2^{m-2} + 2X + 2^m$ ....(2)
[Using eq. (1) for I sigma, assuming X for II sigma and III sigma is sum of binomial coefficients in the expansion $(1+x)^m$]
Now we have
$X = {}^mC_1 + 2\,.\,{}^mC_2 + 3\,.\,{}^mC_3 + ... + m\,.\,{}^mC_m$
Consider
$(1+x)^n = {}^nC_0 + {}^nC_1\, x + {}^nC_2 x^2 + ... + {}^nC_n x^n$
Differentiating both sides with respect to x, we get
$n\,(1+x)^{n-1} = {}^nC_1 + 2x.\, {}^nC_2 + 3\, x^2\, {}^nC_3 + ... + n\, x^{n-1}\, {}^nC_n$
For $x = 1$, we get
$n\,.2^{n-1} = {}^nC_1 + 2.\, {}^nC_2 + 3.\, {}^nC_3 + ... + n\,.{}^nC_n$

Using this relationship, we get

$X = m.2^{m-1}$

$\therefore$ Eq. (2) becomes, LHS

$= m(m+1)2^{m-2} + m(m+1)2^{m-2} + 2m.2^{m-1} + 2^m.$

$= 2m(m+1)2^{m-2} + 2m.2^{m-1} + 2^m$

$= m(m+1)2^{m-1} + 2m.2^{m-1} + 2.2^{m-1}$

$= (m^2+3m+2).2^{m-1} = (m+1)(m+2)\cdot 2^{m-1}$

$\Rightarrow$ $P(m+1)$ is also true.

Hence by Principle of Mathematical Induction we conclude that $P(n)$ is true $\forall\, n \in N$.

10. Let $P(n): \dfrac{(2n)!}{2^{2n}(n!)^2} \le \dfrac{1}{(3n+1)^{1/2}}$

For $n=1$, $P(1): \dfrac{2!}{2^2(1!)^2} \le \dfrac{1}{(3+1)^{1/2}} \Rightarrow \dfrac{1}{4} \le \dfrac{1}{2}$

$\Rightarrow \dfrac{1}{2} \le \dfrac{1}{2}$ which is true for $n=1$

Assume that P(k) is true, then

$$P(k): \frac{(2k)!}{2^{2k}(k!)^2} \le \frac{1}{(3k+1)^{1/2}} \quad ....(1)$$

For $n = k+1$,

$$\frac{[2(k+1)]!}{2^{2(k+1)}[(k+1)!]^2} = \frac{(2k+2)!}{2^{2k+2}[(k+1)!]^2}$$

$$= \frac{(2k+2)(2k+1)(2k)!}{4.2^{2k}(k+1)^2(k!)^2}$$

$$\le \frac{(2k+2)(2k+1)}{4(k+1)^2}\cdot\frac{1}{(3k+1)^{1/2}}$$

[Using Induction hypothesis (1)]

$$= \frac{(2k+1)}{2.(k+1)(3k+1)^{1/2}}$$

Thus, $\dfrac{[2(k+1)]!}{2^{2(k+1)}[(k+1)!]^2} \le \dfrac{(2k+1)}{2(k+1)(3k+1)^{1/2}}$ ....(2)

In order to prove $P(k+1)$, it is sufficient to prove that

$$\frac{(2k+1)}{2(k+1)(3k+1)^{1/2}} \le \frac{1}{(3k+4)^{1/2}} \quad ....(3)$$

Squaring eq. (3), we get

$$\frac{(2k+1)^2}{4(k+1)^2(3k+1)} \le \frac{1}{3k+4}$$

$\Rightarrow (2k+1)^2(3k+4) - 4(k+1)^2(3k+1) \le 0$

$\Rightarrow (4k^2+4k+1)(3k+4) - 4(k^2+2k+1)(3k+1) \le 0$

$\Rightarrow (12k^3+28k^2+19k+4) - (12k^3+28k^2+20k+4) \le 0$

$\Rightarrow -k \le 0$

which is true.

Hence from (2) and (3), we get

$$\frac{(2k+2)!}{2^{2k+2}[(k+1)!]^2} \le \frac{1}{(3k+4)1/2}$$

Hence the above inequation is true for $n = k+1$ and by the principle of induction it is true for all $n \in N$.

11. We have $5\sqrt{5} - 11 = \dfrac{4}{5\sqrt{5}+11} < 1$

Therefore $0 < 5\sqrt{5} - 11 < 1$

This gives us $0 < (5\sqrt{5}-11)^{2n+1} < 1$ for every positive integer n.

Also $(5\sqrt{5}+11)^{2n+1} - (5\sqrt{5}-11)^{2n+1}$

$= 2[{}^{2n+1}C_1(5\sqrt{5})^{2n}.11 + {}^{2n+1}C_3(5\sqrt{5})^{2n-2}.11^3 + \; ....+ {}^{2n+1}C_{2n+1}11^{2n+1}]$

$= 2[{}^{2n+1}C_1(125)^n.11 + {}^{2n+1}C_3(125)^{n-1}.11^3 + \; ....+ {}^{2n+1}C_{2n+1}11^{2n+1}]$

$= 2k$ ....(1)

where k is some positive integer.

Let $F = (5\sqrt{5}-11)^{2n+1}$

Then equation (1) becomes

$R - F = 2k$

$\Rightarrow [R] + R - [R] - F = 2k \Rightarrow [R] + f - F = 2k$

$\Rightarrow f - F = 2k - [R] \Rightarrow f - F$ is an integer.

But $0 \le f < 1$ and $0 < F < 1$

Therefore $-1 < f - F < 1$

Since $f - F$ is an integer, we must have $f - F = 0$

$\Rightarrow f = F.$

Now, $Rf = RF = (5\sqrt{5}+11)^{2n+1}(5\sqrt{5}-11)^{2n+1}$

$= [(5\sqrt{5})^2 - 12]^{2n+1} = 4^{2n+1}$

12. Let the given statement be

$P(m,n): {}^mC_0{}^nC_k + {}^mC_1{}^nC_{k-1} + ... + {}^mC_k{}^nC_0 = {}^{m+n}C_k$

where $m, n, k \in N$ and ${}^pC_q = 0$ for $p < q$.

As k is a positive integer and ${}^pC_q = 0$ for $p < q$.

$\therefore$ k must be a positive integer less than or equal to the smaller of m and n,

We have $k = 1$, when $m = n = 1$

$\therefore$ $P(1,1)$ is ${}^1C_0{}^1C_1 + {}^1C_1{}^1C_0 = {}^2C_1 \Rightarrow 1+1 = 2.$

Thus $P(1,1)$ is true.

Now let us assume that $P(m, n)$ holds good for any fixed value of m and n i.e.

$${}^mC_0{}^nC_k + {}^mC_1{}^nC_{k-1} + ... + {}^mC_k{}^nC_0 = {}^{m+n}C_k \quad ....(1)$$

Then $P(m+1, n+1)$ will be

$${}^{m+1}C_0\,{}^{n+1}C_k + {}^{m+1}C_1\,{}^{n+1}C_{k-1} + ... + {}^{m+1}C_k\,{}^{n+1}C_0$$

$$= {}^{m+n+2}C_k \quad ....(2)$$

Consider LHS

$$= {}^{m+1}C_0\,{}^{n+1}C_k + {}^{m+1}C_1\,{}^{n+1}C_{k-1} + ... + {}^{m+1}C_k\,{}^{n+1}C_0$$

$$= 1.({}^nC_{k-1} + {}^nC_k) + ({}^mC_0 + {}^mC_1)({}^nC_{k-2} + {}^nC_{k-1})$$

$$+ ({}^mC_1 + {}^mC_2)({}^nC_{k-3} + {}^nC_{k-2}) + ... + ({}^mC_{k-1} + {}^mC_k).1$$

$$= ({}^nC_{k-1} + {}^mC_1{}^nC_{k-2} + {}^mC_2\,{}^nC_{k-3} + ... + {}^mC_{k-1}\,{}^nC_0)$$

$$+ ({}^nC_k + {}^mC_1{}^nC_{k-1} + {}^mC_2\,{}^nC_{k-2} + ... + {}^mC_{k-1}\,{}^nC_1 + {}^mC_k)$$

$$+ ({}^mC_0\,{}^nC_{k-2} + {}^mC_1\,{}^nC_{k-3} + ... + {}^mC_{k-2}{}^nC_0)$$

$$+ ({}^mC_0{}^nC_{k-1} + {}^mC_1\,{}^nC_{k-2} + {}^mC_2{}^nC_{k-3} + ... + {}^mC_{k-2}{}^nC_1 + {}^mC_{k-1})$$

$$= {}^{m+n}C_{k-1} + {}^{m+n}C_k + {}^{m+n}C_{k-2} + {}^{m+n}C_{k-1} \quad \text{[Using (1)]}$$

$$= {}^{m+n+1}C_k + {}^{m+n+1}C_{k-1} = {}^{m+n+2}C_k$$

Hence the theorem holds for the next integers $m+1$ and $n+1$. Then by mathematical induction the statement $P(m, n)$ holds for all positive integral values of m and n.

13. We know that

$(1-x)^n = C_0 - C_1x + C_2x^2 - C_3x^3 + ... + (-1)^n C_nx^n$

Multiplying both sides by x, we get

$x(1-x)^n = C_0x - C_1x^2 + C_2x^3 - C_3x^4 + ... + (-1)^n C_nx^{n+1}$

Differentiating both sides w.r. to x, we get

$(1-x)^n - nx(1-x)^{n-1}$

$= C_0 - 2C_1x + 3C_2x^2 - 4C_3x^3 + ... + (-1)^n (n+1) C_nx^n$

Again multiplying both sides by x, we get

$x(1-x)^n - nx^2(1-x)^{n-1}$

$= C_0x - 2C_1x^2 + 3C_2x^3 - 4C_3x^4 + ... + (-1)^n (n+1) C_nx^{n+1}$

Differentiating above with respect to x, we get

$(1-x)^n - nx(1-x)^{n-1} - 2nx(1-x)^{n-1} + nx^2(n-1)(1-x)^{n-2}$

$= C_0 - 2^2C_1x + 3^2C_2x^2 - 4^2C_3x^3 + ... + (-1)^n (n+1)^2 C_nx^n$

Substituting $x = 1$, in above, we get

$0 = C_0 - 2^2C_1 + 3^2C_2 - 4^2C_3 + ... + (-1)^n (n+1)^2 C_n$

Hence Proved.

14. We have

$$P(n): \frac{n^7}{7} + \frac{n^5}{5} + \frac{2n^3}{3} - \frac{n}{105} \text{ is an integer, } \forall n \in N$$

$$P(1): \frac{1}{7} + \frac{1}{5} + \frac{2}{2} - \frac{1}{105}$$

$$= \frac{15+21+70-1}{105} = \frac{105}{105} = 1 \text{ an integer}$$

$\therefore$ $P(1)$ is true

Let $P(k)$ be true i.e.

$$\frac{k^7}{7} + \frac{k^5}{5} + \frac{2k^3}{3} - \frac{k}{105} \text{ is an integer}$$

$$\Rightarrow \frac{k^7}{7} + \frac{k^5}{5} + \frac{2k^3}{3} - \frac{k}{105} = m, \text{(say)}$$

$$m \in N \quad ....(1)$$

Consider $P(k+1)$:

$$= \frac{(k+1)^7}{7} + \frac{(k+1)^5}{5} + \frac{2(k+1)^3}{3} - \frac{(k+1)}{105}$$

$$= \left(\frac{k^7 + 7k^6 + 21k^5 + 35k^4 + 35k^3 + 21k^2 + 7k + 1}{7}\right)$$

$$+ \left(\frac{k^5 + 5k^4 + 10k^3 + 10k^2 + 5k + 1}{5}\right)$$

$$+ 2\left(\frac{k^3 + 3k^2 + 3k + 1}{3}\right) - \left(\frac{k+1}{105}\right)$$

$$= \left(\frac{k^7}{7} + \frac{k^5}{5} + \frac{2k^3}{3} - \frac{k}{105}\right)$$

$$+ [k^6 + 3k^5 + 5k^4 + 5k^3 + 3k^2 + k + k^4$$

$$+ 2k^3 + 2k^2 + k + 2k^2 + 2k] + \left(\frac{1}{7} + \frac{1}{5} + \frac{2}{3} - \frac{1}{105}\right)$$

$= m +$ some integral value $+ 1$

$=$ some integral value

$\therefore$ $P(k+1)$ is also true.

Hence $P(n)$ is true $\forall\, n \in N$, (by the Principle of Mathematical Induction.)

15. Let $P(k) = \displaystyle\sum_{m=0}^{k} \frac{(n-m)(r+m)!}{m!} = \frac{(r+k+1)!}{k!}\left[\frac{n}{r+1} - \frac{k}{r+2}\right]$

For $k = 1$, we will have two terms, on LHS, in sigma for $m = 0$ and $m = 1$, so that

$$LHS = (n-0)\frac{r!}{0!} + (n-1)\frac{(r+1)!}{1!}$$

and $$RHS = \frac{(r+2)!}{1!}\left[\frac{n}{r+1} - \frac{1}{r+2}\right]$$

Hence LHS = RHS for $k = 1$.

Now let the formula holds for $k = s$, that is let

$$\sum_{m=0}^{s} \frac{(n-m)\,(r+m)!}{m!} = \frac{(r+s+1)!}{s!}\left(\frac{n}{r+1} - \frac{s}{r+2}\right) \quad ...(1)$$

Let us add next term corresponding to $m = s+1$ i.e.

adding $\dfrac{(n-s-1)(r+s+1)!}{(s+1)!}$ to both sides, we get

$$\sum_{m=0}^{s+1} \frac{(n-m)\,(r+m)!}{m!} = \frac{(r+s+1)!}{s!}\left[\frac{n}{r+1} - \frac{s}{r+2}\right]$$

$$+ \frac{(n-s-1)(r+s+1)!}{(s+1)!}$$

$$= \frac{(r+s+1)!}{(s+1)!}\left[\frac{(s+1)n}{r+1} - \frac{s(s+1)}{r+2} + n - s - 1\right]$$

$$= \frac{(r+s+1)!}{(s+1)!}\left[n\left\{\frac{s+1}{r+1}+1\right\} - (s+1)\left\{\frac{s}{r+2}+1\right\}\right]$$

$$= \frac{(r+s+2)\,(r+s+1)!}{(s+1)!}\left[\frac{n}{r+1} - \frac{s+1}{r+2}\right]$$

Hence the formula holds for $k = s+1$ and so by the induction principle, the formula holds for all natural numbers k.

16. Given that

$$\sum_{r=0}^{2n} a_r (x-2)^r = \sum_{r=0}^{2n} b_r (x-3)^r \quad ....(1)$$

and $a_k = 1, \forall\, k \geq n$

To prove $b_n = {}^{2n+1}C_{n+1}$

In the given equation (1) let us put $x - 3 = y$ so that $x - 2 = y + 1$ and we get

$$\sum_{r=0}^{2n} a_r (1+y)^r = \sum_{r=0}^{2n} b_r (y)^r$$

$\Rightarrow a_0 + a_1(1+y) + ...a_{n-1} + (1+y)^{n-1}(1+y)^n + (1+y)^{n+1} + ...(1+y)^{2n}$

$$= \sum_{r=0}^{2n} b_r y^r \quad \text{[Using } a_k = 1, \forall\, k \geq n]$$

Equating the coefficients of y^n on both sides we get

NOTE THIS STEP:

$\Rightarrow {}^nC_n + {}^{n+1}C_n + {}^{n+2}C_n + ... + {}^{2n}C_n = b_n$

$\Rightarrow ({}^{n+1}C_{n+1} + {}^{n+1}C_n) + {}^{n+2}C_n + ... + {}^{2n}C_n = b_n$

[Using ${}^nC_n = {}^{n+1}C_{n+1} = 1$]

$\Rightarrow b_n = {}^{n+2}C_{n+1} + {}^{n+2}C_n + ... + {}^{2n}C_n$

[Using ${}^mC_r + {}^mC_{r-1} = {}^{m+1}C_r$]

Combining the terms in similar way, we get

$\Rightarrow b_n = {}^{2n}C_{n+1} + {}^{2n}C_n \Rightarrow b_n = {}^{2n+1}C_{n+1}$

Hence Proved

17. Since α, β are the roots of $x^2 - (p+1)x + 1 = 0$

$\therefore \quad \alpha + \beta = p + 1;\ \alpha\beta = 1$

Here $p \geq 3$ and $p \in Z$

(i) To prove that $\alpha^n + \beta^n$ is an integer.

Let us consider the statement, "$\alpha^n + \beta^n$ is an integer."

Then for $n = 1$, $\alpha + \beta = p + 1$ which is an integer, p being an integer.

$\therefore$ Statement is true for $n = 1$

Let the statement be true for $n \leq k$, i.e., $\alpha^k + \beta^k$ is an integer

Then,

$$\alpha^{k+1} + \beta^{k+1} = \alpha^k.\alpha + \beta^k.\beta$$

$$= \alpha(\alpha^k + \beta^k) + \beta(\alpha^k + \beta^k) - \alpha\beta^k - \alpha^k\beta$$

$$= (\alpha+\beta)(\alpha^k + \beta^k) - \alpha\beta(\alpha^{k-1} + \beta^{k-1}$$

$$= (\alpha+\beta)(\alpha^k + \beta^k) - (\alpha^{k-1} + \beta^{k-1}) \quad .....(1)$$

[as $\alpha\beta = 1$]

= difference of two integers = some integral value

$\Rightarrow$ Statement is true for $n = k + 1$.

$\therefore$ By the principle of mathematical induction the given statement is true for $\forall\, n \in N$.

(ii) Let R_n be the remainder of $\alpha^n + \beta^n$ when divided by p where $0 \leq R_n \leq p - 1$

Since $\alpha + \beta = p + 1 \quad \therefore \quad R_1 = 1$

Also $\alpha^2 + \beta^2 = (\alpha+\beta)^2 - 2\alpha\beta = (p+1)^2 - 2$

$= p^2 + 2p - 1 = p(p+1) + p - 1$

$\therefore \quad R_2 = p - 1$

Also from equation (1) of previous part (i), we have

$$\alpha^{n+1} + \beta^{n+1} = (p+1)(\alpha^n + \beta^n) - (\alpha^{n-1} + \beta^{n-1})$$

$$= p(\alpha^n + \beta^n) + (\alpha^n + \beta^n) - (\alpha^{n-1} + \beta^{n-1})$$

$\Rightarrow R_{n+1}$ is the remainder of $R_n - R_{n-1}$ when divided by p

$\therefore$ We observe that $R_2 - R_1 = p - 1 - 1$

$\therefore \quad R_3 = p - 2$

Similarly, R_4 is the remainder when $R_3 - R_2$ is divided by p where

$R_3 - R_2 = p - 2 - p + 1 = -1 = -p + (p-1) \quad \therefore \quad R_4 = p - 1$

$R_4 - R_3 = p - 1 - p + 1 = 1 \quad \therefore \quad R_5 = 1$

$R_5 - R_4 = 1 - p + 1 = -p + 2 \quad \therefore \quad R_6 = p - 2$

It is evident for above that the remainder is either 1 or $p - 1$ or $p - 2$.

Since $p \geq 3$, so none is divisible by p.

ALTERNATE SOLUTION

For part (ii)

Let $P(n)$: "$\alpha^n + \beta^n$ is not divisible by p"

$\therefore$ Then for $n = 1$

$\alpha + \beta = p + 1$, which is not divisible by p

$\therefore \quad P(1)$ is true .... (1)

For $n = 2$,

$\alpha^2 + \beta^2 = p^2 + 2p - 1 = p(p+1) + (p-1)$

which is not divisible by p.

$\therefore \quad P(2)$ is true .....(2)

For $n = 3$

$\alpha^3 + \beta^3 = p^3 + 3p^2 - 2 = p(p^2 + 3p - 1) + p - 2$

which is not divisible by p (as $p \geq 3$) .....(3)

Let $f(k) = \alpha^k + \beta^k$

$f(k+1) = \alpha^{k+1} + \beta^{k+1}$

$= (\alpha+\beta)(\alpha^k + \beta^k) - \alpha\beta(\alpha^{k-1} + \beta^{k-1})$

$= (p+1)f(k) - f(k-1)$

$f(k+2) = (p+1)f(k+1) - f(k)$

$= (p+1)\,[(p+1)f(k) - f(k-1)] - f(k)$

$= p\,[(p+2)f(k) - f(k-1)] - f(k-1)$

$\Rightarrow$ If $f(k-1)$ is not divisible by p, so is $f(k+2)$ .... (4)

Using (1), (2), (3) and (4) we see that $f(4)$, $f(5)$ .... are not divisible by p.

$\therefore \quad P(n)$ is true $\forall\, n \in N$.

18. To prove

$$P(n):\ \tan^{-1}\left(\frac{1}{3}\right) + \tan^{-1}\left(\frac{1}{7}\right) + ...\tan^{-1}\left(\frac{1}{n^2+n+1}\right)$$

$$= \tan^{-1}\left(\frac{n}{n+2}\right)$$

For $n = 1$, LHS $= \tan^{-1}\frac{1}{3}$;

RHS $= \tan^{-1}\frac{1}{3} \Rightarrow$ LHS = RHS.

$\therefore$ $P(1)$ is true.

Let $P(k)$ be true, i.e.

$$\tan^{-1}\left(\frac{1}{3}\right) + \tan^{-1}\left(\frac{1}{7}\right) + \ldots \tan^{-1}\left(\frac{1}{k^2+k+1}\right)$$

$$= \tan^{-1}\left(\frac{k}{k+2}\right)$$

Consider $P(k+1)$

$$\tan^{-1}\frac{1}{3} + \tan^{-1}\frac{1}{7} + \ldots \tan^{-1}\left(\frac{1}{k^2+k+1}\right)$$

$$+\tan^{-1}\left(\frac{1}{(k+1)^2+(k+1)+1}\right)$$

$$= \tan^{-1}\left[\frac{k+1}{(k+1)+2}\right]$$

$$\text{LHS} = \tan^{-1}\left[\frac{k}{(k+2}\right] + \tan^{-1}\left(\frac{1}{(k^2+3k+3)}\right)$$

[Using equation (1)]

$$= \tan^{-1}\left[\frac{\frac{k}{(k+2} + \frac{1}{k^2+3k+3}}{1-\left(\frac{k}{k+2}\right)\left(\frac{1}{k^2+3k+3}\right)}\right]$$

$$= \tan^{-1}\left[\frac{k^3+3k^2+3k+k+2}{k^3+3k^2+3k+2k^2+6k+6-k}\right]$$

$$= \tan^{-1}\left[\frac{k^3+3k^2+4k+2}{k^3+5k^2+8k+6}\right]$$

$$= \tan^{-1}\left[\frac{(k+1)(k^2+2k+2)}{(k+3)(k^2+2k+2)}\right] = \tan^{-1}\left(\frac{k+1}{k+3}\right) = RHS$$

$\therefore$ $P(k+1)$ is also true.

Hence by the principle of mathematical induction $P(n)$ is true for every natural number.

19. To evaluate $\sum_{r=1}^{k}(-3)^{r-1}\,{}^{3n}C_{2r-1}$ where $k = \frac{3n}{2}$ and n is +ve even interger.

Let $n = 2m$, where $m \in z^+$ $\therefore$ $k = \frac{3(2m)}{2} = 3m$

$$\therefore \sum_{r=1}^{k}(-3)^{r-1}\,{}^{3n}C_{2r-1} = \sum_{r=1}^{3m}(-3)^{r-1}\,{}^{6m}C_{2r-1}$$

$$= {}^{6m}C_1 - 3.{}^{6m}C_3 + 3^2\,{}^{6m}C_5 - \ldots\ldots \quad \text{....(1)}$$

Now we know that

$$(1+a)^{6m} - (1-a)^{6m}$$
$$= 2[{}^{6m}C_1 a + {}^{6m}C_3 a^3 + {}^{6m}C_5 a^5 + \ldots..] \quad \text{....(2)}$$

Keeping in mind the form of RHS in equation (1) and in equation (2)

We put $a = i\sqrt{3}$ in equation (2) to get

$$(1+i\sqrt{3})^{6m} - (1+i\sqrt{3})^{6m}$$

$$= 2[{}^{6m}C_1 i\sqrt{3} - {}^{6m}C_3 i3\sqrt{3} + {}^{6m}C_5 i3^2\sqrt{3}\ldots.]$$

$$\Rightarrow (1+i\sqrt{3})^{6m} - (1-i\sqrt{3})^{6m}$$

$$= 2\sqrt{3}\,i[{}^{6m}C_1 - 3.{}^{6m}C_3 + 3^2\,{}^{6m}C_5\ldots.] \quad \text{....(3)}$$

But $1 + i\sqrt{3} = 2(\cos\pi/3 + i\sin\pi/3)$

$$\therefore (1+i\sqrt{3})^{6m} = 2^{6m}(\cos\pi/3 + i\sin\pi/3)^{6m}$$

NOTE THIS STEP

$$= 2^{6m}\left(\cos\frac{6m\pi}{3} + i\sin\frac{6m\pi}{3}\right)$$

[Using D' Moivre's thm.]

Similarly,

$$(1-i\sqrt{3})^{6m} = 2^{6m}\left(\cos\frac{6m\pi}{3} - i\sin\frac{6m\pi}{3}\right)$$

$$\therefore (1+i\sqrt{3})^{6m} - (1-i\sqrt{3})^{6m} = 2^{6m}.2\sin 2m\pi = 0$$

Substituting the above in equation (3) we get

$${}^{6m}C_1 - 3.{}^{6m}C_3 + 3^2\,{}^{6m}C_5 - \ldots.. = 0$$

$$\Rightarrow \sum_{r=1}^{k}(-3)^{r-1}\,{}^{3n}C_{2r-1} = 0.$$

Hence Proved

20. Let $P(n): \cos x + \cos 2x + \ldots. + \cos nx$

$$= \cos\frac{n+1}{2}x\sin\frac{nx}{2}\ \mathrm{cosec}\ \frac{x}{2} \quad \text{....(1)}$$

where x is not an integral multiple of 2π.

For $n = 1$ $P(1)$: L.H.S. $= \cos x$

$$R.H.S. = \cos\frac{1+1}{2}x\sin\frac{x}{2}\ \mathrm{cosec}\ \frac{x}{2} = \cos x$$

L.H.S. = R.H.S.

$\Rightarrow$ $P(1)$ is true.

Let $P(k)$ be true i.e.

$\cos x + \cos 2x + \ldots. + \cos kx$

$$= \cos\frac{k+1}{2}x\sin\frac{kx}{2}\ \text{cosec}\ \frac{x}{2} \quad \text{....(2)}$$

Consider $P(k+1)$:

$\cos x + \cos 2x + \ldots. + \cos kx + \cos(k+1)x$

$$= \cos\left(\frac{k+2}{2}\right)x\sin\frac{(k+1)x}{2}\ \text{cosec}\ \frac{x}{2}$$

L.H.S. $[\cos x + \cos 2x + ... + \cos kx + \cos (k+1)x$

$= \cos\left(\frac{k+1}{2}\right)x \sin \text{cosec}\frac{kx}{2}\frac{x}{2} + \cos(k+1)x$

[Using (2)]

$= \left[\cos\left(\frac{k+1}{2}\right)x \sin\frac{kx}{2} + \cos (k+1)x \sin\frac{x}{2}\right] \text{cosec}\frac{x}{2}$

$= \frac{1}{2}\left[2\cos\frac{(k+1)x}{2}\sin\frac{kx}{2} + 2\cos (k+1)x \sin\frac{x}{2}\right] \text{cosec}\frac{x}{2}$

$= \frac{1}{2}[\sin\left(\frac{2k+1}{2}\right)x - \sin\frac{x}{2}$

$+ \sin\left(xk + \frac{3x}{2}\right) - \sin\left(xk + \frac{x}{2}\right)]\text{cosec}\frac{x}{2}$

$= \frac{1}{2}\left[\sin\left(xk + \frac{3x}{2}\right) - \sin\frac{x}{2}\right]\text{cosec}\frac{x}{2}$

$= \frac{1}{2}\left[2\cos\frac{(k+2)x}{2}\sin\frac{(k+1)}{2}\right]\text{cosec}\frac{x}{2}$

$= \cos\frac{(k+2)x}{2}\sin\frac{(k+1)x}{2}\text{cosec}\frac{x}{2} = R.H.S.$

$\therefore$ $P(k+1)$ is also true.

Hence by the principle of mathematical induction

$P(n)$ is true $\forall n \in N$.

21. Given that,

$(1+x+x^2)^n = a_0 + a_1x + ... + a_{2n}x^{2n}$ ....(1)

where n is a +ve integer.

Replacing x by $-\frac{1}{x}$ in eq^n (1), we get

$\left(1 - \frac{1}{x} + \frac{1}{x^2}\right)^n = a_0 - \frac{a_1}{x} + \frac{a_2}{x^2} - \frac{a_3}{x^3} + ... + \frac{a_{2n}}{x^{2n}}$ ....(2)

Multiplying eq.'s (1) and (2) :

$\frac{(1+x+x^2)^n (x^2 - x + 1)^n}{x^{2n}}$

$= (a_0 + a_1x + ... + a_{2n}x^{2n})\left(a_0 - \frac{a_1}{x} + \frac{a_2}{x^2} + ... + \frac{a_n}{x^{2n}}\right)$

Equating the constant terms on both sides we get

$a_0^2 - a_1^2 + a_2^2 - a_3^2 + ... + a_{2n}^2 =$ constant term in the expansion

of $\frac{[(1+x+x^2)(1-x+x^2)]^n}{x^{2n}}$

= Coeff. of x^{2n} in the expansion of $(1+x^2+x^4)^n$

But replacing x by x^2 in eq's (1), we have

$(1+x^2+x^4)^n = a_0 + a_1x^2 + ... + a_{2n}(x^2)^{2n}$

$\therefore$ Coeff of $x^{2n} = a_n$

Hence we obtain, $a_0^2 - a_1^2 + a_2^2 - a_3^2 + ... + a_{2n}^2 = a_n$

22. For $n = 1$, $3^{2^n} - 1 = 3^{2^1} - 1 = 9 - 1 = 8$ which is divisible by $2^{n+2} = 2^3 = 8$ but is not divisible by $2^{n+3} = 2^4 = 16$

Therefore, the result is true for $n = 1$.

Assume that the result is true for $n = k$. That is, assume that $3^{2^k} - 1$ is divisible by 2^{k+2} but is not divisible by 2^{k+3},

Since $3^{2^k} - 1$ is divisble by 2^{k+2} but not by 2^{k+3}, we can write $3^{2^k} - 1 = (m)\,2^{k+2}$

where m must be an odd positive integer, for otherwise $3^{2k} - 1$ will become divisible by 2^{k+3}.

For $n = k+1$, we have

$3^{2^{k+1}} - 1 = 3^{2^k.2} - 1 = \left(3^{2^k}\right)^2 - 1$

$= (m.2^{k+2} + 1)^2 - 1$ [Using (1)]

$= m^2.(2^{k+2})^2 + 2m.2^{k+2} + 1 - 1$

$= m^2.2^{2k+4} + m.2^{k+3} = 2^{k+3}(m^2.2^{k+1} + m.)$

$\Rightarrow$ $3^{2^{k+1}} - 1$ is divisible by 2^{k+3}.

But $3^{2^{k+1}} - 1$ is not divisible by 2^{k+4} for otherwise we must have 2 divides $m^2.\,2^{k+1} + m$. But this is not possible as m is odd.

Thus, the result is true for $n = k+1$.

23. To prove that,

$$\frac{3!}{2(n+3)} = \sum_{r=0}^{n}(-1)^r\left[\frac{^nC_r}{^{r+3}C_r}\right]$$

Let us consider

$$\frac{^nC_r}{^{r+3}C_r} = \frac{n!}{r!(n-r)!}\frac{r!3!}{r!(r+3)!} = \frac{n!}{(n-r)!}\frac{3!}{(r+3)!}$$

$$= \frac{(n+3)!3!}{(n+3)(n+2)(n+1)(n-r)!(r+3)!}$$

$$= \frac{3!}{(n+3)(n+2)(n+1)}\,{}^{n+3}C_{r+3}$$

Now, R.H.S.

$$\sum_{r=0}^{n}(-1)^r\left[\frac{^nC_r}{^{r+3}C_r}\right]$$

$$= \sum_{r=0}^{n}(-1)^r\frac{3!}{(n+3)(n+2)(n+1)}\,{}^{n+3}C_{r+3}$$

$$= \frac{3!}{(n+3)(n+2)(n+1)}$$

$$\left[{}^{n+3}C_0 - {}^{n+3}C_1 + {}^{n+3}C_2 - \sum_{r=-3}^{n}(-1)^{r+3}\,{}^{n+3}C_{r+3}\right]$$

$$= \frac{3!}{(n+3)(n+2)(n+1)}\left[1-n-3+\frac{1}{2}(n+3)(n+2)-0\right]$$

$$\left(\because \sum_{r=0}^{n}(-1)^r\ {}^nC_r = 0\right)$$

$$= \frac{3!(n+2)}{(n+3)(n+2)(n+1)}\left[\frac{(n+3)}{2}-1\right]$$

$$= \frac{3!(n+2)(n+1)}{2(n+3)(n+2)(n+1)}\frac{3!}{2(n+2)} = \text{L.H.S.}$$ **Hence Proved**

ALTERNATE SOLUTION

We know that, $(1+x)^n = \sum_{r=0}^{n} {}^nC_r\, x^r$

Integrating both sides we get

$$\frac{(1+x)^{n+1}}{n+1} = \sum_{r=0}^{n} {}^nC_r\, \frac{x^{r+1}}{r+1} + k_1$$

Put $x = 0$ we get $k_1 = \frac{1}{n+1}$

$\therefore$ We get,

$$\frac{(1+x)^{n+1}}{n+1} - \frac{1}{n+1} = \sum_{r=0}^{n} {}^nC_r\, \frac{x^{r+1}}{r+1}$$

Integrating the above again w.r. to x

$$\frac{(1+x)^{n+2}}{(n+1)(n+2)} - \frac{x}{n+1} = \sum_{r=0}^{n} {}^nC_r\, \frac{x^{r+2}}{(r+1)(r+2)} + k_2$$

Put $x = 0$, we get, $k_2 = \frac{1}{(n+1)(n+2)}$

$\therefore$ We get,

$$\frac{(1+x)^{n+2}}{(n+1)(n+2)} - \frac{x}{n+1} - \frac{1}{(r+1)(n+2)}$$

$$= \sum_{r=0}^{n} {}^nC_r\, \frac{x^{r+2}}{(r+1)(r+2)}$$

Integrating again w.r. to x, we get

$$\frac{(1+x)^{n+3}}{(n+1)(n+2)(n+3)} - \frac{x^2}{2(n+1)} - \frac{x}{(r+1)(n+2)}$$

$$= \sum_{r=0}^{n} {}^nC_r\, \frac{x^{r+3}}{(r+1)(r+2)(r+3)} + k_3$$

Put $x = 0$ to get,

$$k_3 = \frac{1}{(n+1)(n+2)(n+3)}$$

Thus we get

$$\frac{(1+x)^{n+3}}{(n+1)(n+2)(n+3)} - \frac{x^2}{2(n+1)} - \frac{x}{(n+1)(n+2)} - \frac{1}{(n+1)(n+2)(n+3)}$$

$$\sum_{r=0}^{n} {}^nC_r\, \frac{x^{r+3}}{(r+1)(r+2)(r+3)}$$

Substituting $x = -1$ in the above, we get

$$-\frac{1}{2(n+1)} - \frac{1}{(n+1)(n+2)} - \frac{1}{(n+1)(n+2)(n+3)}$$

$$= \sum_{r=0}^{n} \frac{{}^nC_r(-1)^{r+3}}{(r+1)(r+2)(r+3)}$$

$$\Rightarrow \frac{-(n+2)(n+3)+2(n+3)-2}{2(n+3)(n+2)(n+1)}$$

$$\sum_{r=0}^{n}(-1)^3(-1)^r\, \frac{{}^nC_r\, r!}{(r+1)(r+2)(r+3)r!}$$

$$\Rightarrow \frac{-n^2-5n-6-2n+6-2}{2(n+3)(n+2)(n+1)} = -\frac{1}{3!}$$

$$\sum_{r=0}^{n}(-1)^r\, \frac{{}^nC_r\, r!3!}{(r+3)!}$$

$$\Rightarrow \frac{-(n^2+3n+2)}{2(n+3)(n^2+3n+2)} = -\frac{1}{3!}\sum_{r=0}^{n}(-1)^r\, \frac{{}^nC_r}{{}^{r+3}C_r}$$

$$\Rightarrow \frac{3!}{2(n+3)} = \sum_{r=0}^{n}(-1)^r\left[\frac{{}^nC_r}{{}^{r+3}C_r}\right]$$

Hence Proved.

24. For $n = 1$, the inequalitity becomes

$\sin A_1 \le \sin A_1$,

which is clearly true.

Assume that the inequality holds for $n = k$ where k is some positive integer. That is, assume that

$$\sin A_1 + \sin A_2 + ... + \sin A_k \le k\sin\left(\frac{A_1 + A_2 + ... + A_k}{k}\right) \quad ....(1)$$

for same positive integer k.

We shall now show that the result holds for $n = k + 1$ that is, we show that

$$\sin A_1 + \sin A_2 + ... + \sin A_k + \sin A_{k+1}$$

$$\le (k+1)\sin\left(\frac{A_1 + A_2 + ... + A_{k+1}}{k+1}\right) \quad ....(2)$$

L.H.S. of (2)

$= \sin A_1 + \sin A_2 + ... + \sin A_k + \sin A_{k+1}$

$\le k \sin\left(\frac{A_1 + A_2 + ... + A_k}{k}\right) + \sin A_{k+1}$

[Induction assumption]

$= (k+1)\left[\frac{k}{k+1}\sin\alpha + \frac{1}{k+1}\sin A_{k+1}\right];$

where $\alpha + \frac{A_1 + A_2 + ...A_k}{k}$

$\therefore$ L.H.S. of

(2) $\le (k+1)\left[\left(1-\frac{k}{k+1}\right)\sin\alpha + \frac{1}{k+1}\sin A_{k+1}\right]$

$\le (k+1)\sin\left\{\left(1-\frac{k}{k+1}\right)\alpha + \frac{1}{k+1}A_{k+1}\right\}$

[Using the fact $p\sin x + (1-p)\sin y \le \sin[px + (1-p)y]$ for $0 \le p \le 1, 0 \le x, y \le \pi$]

$= (k+1)\sin\left\{\frac{k}{k+1}\left(\frac{A_1 + A_2 + ...A_k}{k}\right) + \frac{1}{k+1}A_{k+1}\right\}$

$= (k+1)\sin\left(\frac{A_1 + A_2 + ...A_{k+1}}{k+1}\right\}$

Thus, the inequality holds for $n = k + 1$. Hence, by the principle of mathematical induction the inequality holds for all $n \in N$.

25. We know that ${}^nC_r = \frac{n}{r}\,{}^{n-1}C_{r-1}$

$\therefore \quad {}^{mp}C_r = \frac{mp}{r}\,{}^{mp-1}C_{r-1} = \left[\frac{m.{}^{mp-1}C_{r-1}}{r}\right]p$

Now, L.H.S is an integer

$\Rightarrow$ RHS must be an integer

But p and z are coprime (given)

$\therefore$ r must divide $m.\,{}^{mp-1}C_{r-1}$

or $\frac{m.\,{}^{mp-1}C_{r-1}}{r}$ is an integer.

$\Rightarrow \frac{{}^{mp}C_r}{p}$ is an integer or ${}^{mp}C_r$ is divisible by p.

ALTERNATE SOLUTION

(Proof by Induction)

As r does not divide p and

For $m = 1$, ${}^{mp}C_r = {}^pC_r = \frac{p}{r}\,{}^{p-1}C_{r-1}$

As ${}^pC_r$ is an integer, $\frac{p}{r}\,{}^{p-1}C_{r-1}$ must be an integer, but p and r are coprime.

$\therefore$ ${}^{p-1}C_{r-1}$ must be divisible by r

or $\frac{{}^{p-1}C_{r-1}}{r}$ should be an integer.

$\Rightarrow \frac{{}^pC_r}{p}$ should be an integer.

$\Rightarrow$ p divides ${}^pC_r$ ....(1)

$\therefore$ $P(1)$ is true

For $m = k$,

Let $P(k)$ be true, i.e.,

p divides ${}^{kp}C_r$ (Induction Hypothesis) ....(2)

Consider

$P(k+1)$: ${}^{(k+1)p}C_r$ is divisible by p.

To prove it let us consider

$(1+x)^{(k+1)p} = (1+x)^{kp}(1+x)^p$

$\Rightarrow {}^{(k+1)p}C_0 + {}^{(k+1)p}C_1x + {}^{(k+1)p}C_2x + ... {}^{(k+1)p}C_{(k+1)p}x^{(k+1)p}$

$= [{}^{kp}C_0 + {}^{kp}C_1x + {}^{kp}C_2x^2 + ... + {}^{kp}C_{kp}x^{kp}]$

$\times\ {}^pC_0 + {}^pC_1x + {}^pC_2x^2 + ... + {}^pC_px^p]$

NOTE THIS STEP

Equating the coeff's of x^r on both sides we get

${}^{(k+1)p}C_r = {}^{kp}C_0\,{}^pC_r + {}^{kp}C_1\,{}^pC_{r-1} + {}^{kp}C_2\,{}^pC_{r-2} + ... + {}^{kp}C_r\,{}^pC_0$

$= {}^pC_r + {}^{kp}C_1\,{}^pC_{r-1} + {}^{kp}C_2\,{}^pC_{r-2} + ... + {}^{kp}C_r$

Form (1) and (2) RHS is divisible by p and so is LHS.

$\therefore$ $P(k+1)$ is also true.

Hence by the principle of Mathematical Induction given statement is true $\forall$ m.

26. Let $P(m) = \sum_{k=0}^{m} \frac{\binom{2n-k}{k}^{(2n-4k+1)}}{\binom{2n-k}{n}^{(2n-2k+1)}} 2^{n-2k}$

$= \frac{\binom{n}{m}}{\binom{2n-2m}{n-m}}.2^{n-2m}$ ....(1)

For $m = 0$,

LHS $= \frac{\binom{2n}{0}}{\binom{2n}{n}}.\frac{2n+1}{2n+1}.2^n = \frac{1}{\binom{2n}{n}}2^n,$

R.H.S. $= \frac{\binom{n}{0}}{\binom{2n}{n}}.2^n = \frac{1}{\binom{2n}{n}}2^n = L.H.S$

[$\because m = 0 \Rightarrow k = 0$]

$\therefore$ $P(0)$ holds true. Now assuming $P(m)$

L.H.S. of $P(m+1)$ = L.H.S. of

$$P(m)+\frac{\binom{2n-m-1}{m+1}}{\binom{2n-m-1}{n}}\cdot\frac{(2n-4m-3)}{(2n-2m-1)}.2^{n-2m-2}$$

$$=\frac{n!(n-m)!}{m!(2n-2m)!}.2^{n-2m}$$

$$+\frac{n!(n-m-1)!(2n-4m-3)}{(m+1)!(2n-2m-2)!(2n-2m-1)}.2^{n-2m-2}$$

$$=\frac{n!(n-m-1)!2^{n-2m-2}}{(m+1)!(2n-2m-1)!}$$

$$\times\left\{\frac{(n-m).4\,(m+1)}{(2n-2m)}+(2n-4m-3)\right\}$$

$$=\frac{n!(n-m-1)!2^{n-2m-2}\,(2n-2m-1)}{(m+1)!(2n-2m-1)!}$$

$$=\frac{n!(n-m-1)!2^{n-2m-2}}{(m+1)!(2n-2m-2)!}=\frac{\binom{n}{m+1}}{\binom{2n-2m-2}{n-m-1}}.2^{n-2m-2}$$

= R.H.S. of $P(m+1)$.

Hence by mathematical induction, result follows for all $0 \le m \le n$.

27. Given that for positive integers m and n such that $n \ge m$, then to prove that

$${}^nC_m + {}^{n-1}C_m + {}^{n-2}C_m + ... + {}^mC_m = {}^{n+1}C_{m+1}$$

L.H.S. ${}^mC_m + {}^{m+1}C_m + {}^{m+2}C_m + ... + {}^{n-1}C_m + {}^nC_m$

[writing L.H.S. in reverse order]

$= ({}^{m+1}C_{m+1} + {}^{m+1}C_m) + {}^{m+2}C_m + ... + {}^{n-1}C_m + {}^nC_m$

$[\because {}^mC_m = {}^{m+1}C_{m+1}]$

$= ({}^{m+2}C_{m+1} + {}^{m+2}C_m) + {}^{m+3}C_m + ... + {}^nC_m$

$[\because {}^nC_{r+1} + {}^nC_r = {}^{n+1}C_{r+1}]$

$= {}^{m+3}C_{m+1} + {}^{m+3}C_m + ... + {}^nC_m$

Combining in the same way we get

$= {}^nC_{m+1} + {}^nC_m = {}^{n+1}C_{m+1}$ = R.H.S.

Again we have to prove

${}^nC_m + 2\,{}^{n-1}C_m + 3\,{}^{n-2}C_m + ... + (n-m+1)\,{}^mC_m$

$= {}^{n+2}C_{m+2}$

$= [{}^nC_m + {}^{n-1}C_m + {}^{n-2}C_m + ... + {}^mC_m] + [{}^{n-1}C_m + {}^{n-2}C_m + ... + {}^mC_m] + [{}^{n-2}C_m + ... + {}^mC_m] + ... + [{}^mC_m]$

[$n - m + 1$ bracketed terms]

$= {}^{n+1}C_{m+1} + {}^nC_{m+1}\ {}^{n-1}C_{m+1} .... + {}^{m+1}C_{m+1}$

[using previous result.]

$= {}^{n+2}C_{m+2}$

[Replacing n by $n+1$ and m by $m+1$ in the previous result.]

= R.H.S.

ALTERNATE SOLUTION : We know that coeff. of x^r in the binomial expansion of $(1+x)^n$ is ${}^nC_r$

$\therefore\ {}^nC_m + {}^{n-1}C_m + {}^{n-2}C_m + ... + {}^mC_m$

= Coeff. of x^m in the expansion of s

$[(1+x)^n + (1+x)^{n-1} + (1+x)^{n-2} + ... + (1+x)^m]$

= Coeff. of x^m in

$[(1+x)^m + (1+x)^{m-1} + (1+x)^{m-2} + ... + (1+x)^n]$

(writing in reserve order)

= Coeff. of x^m in

$$\left[(1+x)^m\frac{\{(1+x)^{n-m+1}-1\}}{1+x-1}\right][\textit{Sum of G.P.}]$$

$$= \text{Coeff. of } x^m \text{ in } \frac{[(1+x)^m\{(1+x)^{n-m+1}-1\}]}{x}$$

= Coeff. of x^{m+1} in $(1+x)^m\,[(1+x)^{n-m+1}-1]$

= Coeff. of x^{m+1} in $[(1+x)^{n+1}-(1+x)^m]$

$= {}^{n+1}C_{m+1} - 0$

[As there is no term containing x^{m+1} in the expansion of $(1+x)^m$]

$= {}^{n+1}C_{m+1}$

Now to prove

${}^nC_m + 2\,{}^{n-1}C_m + 3.\,{}^{n-2}C_m + ... + (n-m+1)\,{}^mC_m$

$= {}^{n+2}C_{m+2}$

Let us consider

$S = (1+x)^n + 2(1+x)^{n-1} + 3(1+x)^{n-2} + ... + (n-m+1)(1+x)^m$ ....(1)

$(1+x)S = (1+x)^{n+1} + 2(1+x)^n + 3(1+x)^{n-1} + ... + (n-m+1)(1+x)^{m+1}$ ....(2)

Substracting (i) from (ii) we get

$xS = (1+x)^{n+1} + (1+x)^n + (1+x)^{n-1} + (1+x)^{n-2} + ... + (1+x)^m + (n-m+1)(1+x)^m$

$$=\frac{(1+x)^{n+1}\left[1-\left(\frac{1}{1+x}\right)^{n+1-m}\right]}{1-\frac{1}{1+x}}-(n-m+1)\,(1+x)^m$$

$$=\frac{(1+x)^{n+1}[(1+x)^{n+1-m}-1]}{(1+x)^{n-m}.\,x}-(n-m+1)\,(1+x)^m$$

$$=\frac{(1+x)^{m+1}[(1+x)^{n+1-m}-1]}{x}-(n-m+1)\,(1+x)^m$$

$$\Rightarrow\ S=\frac{(1+x)^{n+2}-(1+x)^{m+1}}{x^2}-\frac{(n-m+1)\,(1+x)^m}{x}$$

Now,

${}^nC_m + 2.\,{}^{n-1}C_m + 3.\,{}^{n-2}C_m + ... + (n-m+1)\,{}^mC_m$

= Coeff. of x^m in S

= Coeff of x^m in

$$\left[\frac{(1+x)^{n+2}-(1+x)^{m+1}}{x^2}-\frac{(n-m+1)\,(1+x)^m}{x}\right]$$

= Coeff. of

x^{m+2} in $[(1+x)^{n+2}-(1+x)^{m+1}]-(n-m+1)$

Coeff. of x^{m+1} in $(1+x)^m = {}^{n+2}C_{m+2} - 0 - 0\ (n-m+1)$
$= {}^{n+2}C_{m+2}$

Hence proved

28. For $n > 0, \sqrt{4n+1} > 0, \sqrt{n}+\sqrt{n+1} > 0$ *and* $\sqrt{4n+2} > 0$

Now, $\sqrt{4n+1} < \sqrt{n}+\sqrt{n+1} < \sqrt{4n+2}$ to be proved.

I. To prove $\sqrt{4n+1} < \sqrt{n}+\sqrt{n+1}$

Squaring both sides in $\sqrt{4n+1} < \sqrt{n}+\sqrt{n+1}$

$\Rightarrow \quad 4n+1 < n+n+1+2\sqrt{n(n+1)}$

$\Rightarrow \quad 2n < 2\sqrt{n(n+1)} \Rightarrow n < \sqrt{n(n+1)}$ which is true.

II. To prove $\sqrt{n}+\sqrt{n+1} < \sqrt{4n+2}$

Squaring both sides,

$n+n+1+2\sqrt{n(n+1)} < 4n+2$

$\Rightarrow \quad 2\sqrt{n(n+1)} < 2n+1$ Squaring again

$4[n(n+1)] < 4n^2+1+4n$ or $0 < 1$ which is true

Hence $\sqrt{4n+1} < \sqrt{n}+\sqrt{n+1} < \sqrt{4n+2}$

Further to prove $[\sqrt{n}+\sqrt{n+1}] = [\sqrt{4n+1}]$, we have to prove that there is no positive integer which lies between $\sqrt{4n+1}$ and $\sqrt{4n+2}$ or $[\sqrt{4n+1}] = [\sqrt{4n+2}]$. Using Mathematical induction.

We have to check $[\sqrt{4n+1}] = [\sqrt{4n+2}]$ for $n = 1$

$[\sqrt{5}] = [\sqrt{6}] \Rightarrow 2 = 2$, which is true

Assume for $n = k$ (arbitrary)

i.e., $[\sqrt{4k+1}] = [\sqrt{4k+2}]$ To prove for $n = k+1$

To check $[\sqrt{4k+5}] = [\sqrt{4k+6}]$ since $k \ge 0$

Here $4k+5$ is an odd number and $4k+6$ is even number. Their greatest integer will be different iff $4k+6$ is a perfect square that is $4k+6 = r^2$

$\Rightarrow \quad k = \frac{r^2}{4} - \frac{6}{4}, \frac{6}{4}$ is not integer. But k has to be integer.

So $4k+6$ cannot be perfect square.

$\Rightarrow \quad [\sqrt{4k+5}] = [\sqrt{4k+6}]$

By Sandwich theorem

$\Rightarrow \quad [\sqrt{n}+\sqrt{n+1}] = [\sqrt{4n+1}]$

29. We have a, b, c the +ve real number s.t. $b^2 - 4ac > 0$; $\alpha_1 = c$.

$$P(n): \alpha_{n+1} = \frac{a\alpha_n^2}{b^2 - 2a(\alpha_1+\alpha_2+....\alpha_n)}$$

is well defined and $\alpha_{n+1} < \frac{\alpha_n}{2}, \forall n = 1, 2, ...$

For $n = 1$, $\alpha_2 = \frac{a\alpha_1^2}{b^2 - 2a\alpha_1} = \frac{ac^2}{b^2 - 2ac}$

Now, $b^2 - 4ac > 0 \Rightarrow b^2 - 2ac > 2ac > 0$

$\therefore \quad \alpha_2$ is well defined (as denomination is not zero)

Also $\left[\begin{array}{l} \because \ b^2 - 2ac > 2ac \\ \Rightarrow \ \frac{1}{b^2-2ac} < \frac{1}{2ac} \end{array}\right] \Rightarrow \frac{\alpha_2}{c} < \frac{1}{2} \Rightarrow \frac{\alpha_2}{\alpha_1} < \frac{1}{2}$

$\therefore \quad P(n)$ is true for $n = 1$.

Let the statement be true for $1 \le n \le k$ *i.e.,*

$\alpha_{k+1} = \frac{a\alpha_k^2}{b^2 - 2a(\alpha_1+\alpha_2+....\alpha_k)}$ is well defined

and $\alpha_{k+1} < \frac{\alpha_k}{2}$

Now, we will prove that $P(k+1)$ is also true

i.e., $\alpha_{k+2} = \frac{a\alpha_{k+1}^2}{b^2 - 2a(\alpha_1+\alpha_2+....+\alpha_k+\alpha_{k+1})}$ is well defined and $\alpha_{k+2} < \frac{\alpha_{k+1}}{2}$.

We have

$\alpha_1 = c, \alpha_2 < \frac{c}{2}, \alpha_3 < \frac{\alpha_2}{2} < \frac{c}{2^2}, \alpha_4 < \frac{\alpha_3}{2} < \frac{c}{2^3}, ...$ (by IH)

Now, $(\alpha_1 + \alpha_2 + ...\alpha_k + \alpha_{k+1} < c + \frac{c}{2} + \frac{c}{2^k} + ... + \frac{c}{2^k}$

$= \frac{c\left(1 - \frac{1}{2^{k+1}}\right)}{1 - 1/2} = 2c\left(1 - \frac{1}{2^{k+1}}\right) < 2c$

$\therefore \quad \alpha_1 + \alpha_2 + ... + \alpha_{k+1} < 2c$

$\Rightarrow \quad -2a(\alpha_1 + \alpha_2 + ... + \alpha_{k+1}) > -4ac$

$\Rightarrow \quad b^2 - 2a(\alpha_1 + \alpha_2 + ... + \alpha_{k+1}) > b^2 - 4ac > 0$

$\therefore \quad \alpha_{k+2}$ is well defined. Again by IH we have

$\alpha_{k+1} < \frac{\alpha_k}{2} \Rightarrow 2\alpha_{k+1} < \alpha_k$

$\Rightarrow \quad 4\alpha_{k+1}^2 < \alpha_k^2$ [*As by def.* α_{k+1}, α_k *are* +ve]

$\Rightarrow \quad 4\alpha_{k+1} < \frac{\alpha_k^2}{\alpha_{k+1}}$

$\Rightarrow \quad 4\alpha_{k+1} < \frac{b^2 - 2a(\alpha_1+\alpha_2+....+\alpha_k)}{a}$

$\Rightarrow \quad 4a\alpha_{k+1} < b^2 - 2a(\alpha_1+\alpha_2+....+\alpha_k)$

$\Rightarrow \quad 2a\alpha_{k+1} < b^2 - 2a(\alpha_1+\alpha_2+....+\alpha_k+\alpha_{k+1})$

$\Rightarrow \quad \dfrac{a\alpha_{k+1}^2}{b^2 - 2a(\alpha_1 + \alpha_2 + ... + \alpha_{k+1})} < \dfrac{1}{2}$

$\Rightarrow \quad \dfrac{a\alpha_{k+1}}{b^2 - 2a(\alpha_1 + \alpha_2 + ... + \alpha_{k+1})} < \dfrac{\alpha_{k+1}}{2}$

$\Rightarrow \quad \alpha_{k+2} < \dfrac{\alpha_{k+1}}{2}$

$\therefore$ $P(k+1)$ is also true.

Thus by the Principle of Mathematical Induction the Statement $P(n)$ is true $\forall n \in N$.

30. Let $P(n) : (25)^{n+1} - 24n + 5735$

For n = 1.

P(1) : $625 - 24 + 5735 = 6336 = (24)^2 \times (11)$, which is divisible by 24^2. Hence P(1) is true

Let $P(k)$ be true, where $k \geq 1$

$\Rightarrow \ (25)^{k+1} - 24k + 5735 = (24)^2 \lambda$ where $\lambda \in N$

For $n = k+1$, $P(k+1) : (25)^{k+2} - 24(k+1) + 5735$

$= 25[(25)^{k+1} - 24k + 5735]$

$+ 25.24.k - (25)(5735) + 5735 - 24(k+1)$

$= 25(24)^2 \lambda + (24)^2 k - 5735 \times 24 - 24$

$= 25(24)^2 \lambda + (24)^2 k - (24)(5736)$

$= 25(24)^2 \lambda + (24)^2 k - (24)^2 (239)$,

$= (24)^2 [25\lambda + k - 239]$

which is divisible by $(24)^2$.

Hence, by the method of mathematical induction result is true $\forall n \in N$.

31. To prove that

$2^k\, {}^nC_0\, {}^nC_k - 2^{k-1}\, {}^nC_1\, {}^{n-1}C_{k-1} + 2^{k-2}\, {}^nC_2\, {}^{n-2}C_{k-2}$

$- ..... + (-1)^k\, {}^nC_k\, {}^{n-k}C_0 = {}^nC_k$

LHS of above equation can be written as

$$\sum_{r=0}^{k} (-1)^r 2^{k-r}\, {}^nC_r\, {}^{n-r}C_{k-r}$$

$$= \sum_{r=0}^{k} (-1)^r 2^{k-r} \frac{n!}{r!(n-r)!} \frac{(n-r)!}{(k-r)!(n-k)!}$$

$$= \sum_{r=0}^{k} (-1)^r 2^{k-r} \frac{n!k!}{r!k!(n-k)!(k-r)!}$$

$$= \sum_{r=0}^{k} (-1)^r \frac{2^k}{2^r} . \frac{n!}{k!(n-k)!} \frac{k!}{r!(k-r)!}$$

$$= 2^k\, {}^nC_k \sum_{r=0}^{k} (-1/2)^r \frac{k!}{r!(k-r)!}$$

$$= 2^k\, {}^nC_k \sum_{r=0}^{k} {}^kC_r (-1/2)^r = 2^k\, {}^nC_k (1 - 1/2)^k$$

$= 2^k\, {}^nC_k \dfrac{1}{2^k} = {}^nC_k$ = R.H.S. **Hence Proved**

32. We have $\alpha + \beta = 1 - p$ and $\alpha\beta = -p(1-p)$

For $n = 1$, $p_n = p_1 = 1$

Also, $A\alpha^n + B\beta^n = A\alpha + B\beta = \dfrac{(p^2 + \beta - 1)\alpha}{\alpha\beta - \alpha^2}$

$+ \dfrac{(p^2 + \alpha - 1)\beta}{\alpha\beta - \beta^2} = \dfrac{p^2 + \beta - 1}{\beta - \alpha} + \dfrac{p^2 + \alpha - 1}{\alpha - \beta}$

$= \dfrac{p^2 + \beta - 1 - p^2 - \alpha + 1}{\beta - \alpha} = \dfrac{\beta - \alpha}{\beta - \alpha} = 1$

For n = 2, $p_2 = 1 - p^2$

Also, $A\alpha^n + B\beta^n = A\alpha^2 + B\beta^2$

$= \dfrac{(p^2 + \beta - 1)\alpha^2}{\alpha\beta - \alpha^2} + \dfrac{(p^2 + \alpha - 1)\beta^2}{\alpha\beta - \alpha^2}$

which is true for $n = 2$

Now let result is true for $k < n$ where $n \geq 3$.

$P_n = (1-p)P_{n-1} + p(1-p)P_{n-2}$

$= (1-p)(A\alpha^{n-1} + B\beta^{n-1}) + p(1-p)(A\alpha^{n-2} + B\beta^{n-2})$

$= A\alpha^{n-2}\{(1-p)\alpha + p(1-p)\} + B\beta^{n-2}\{(1-p)\beta - p(1-p)$

$= A\alpha^{n-2}\{(\alpha+\beta)\alpha - \alpha\beta\}$

$+ B\beta^{n-2}\{(\alpha+\beta)\beta - \alpha\beta\}$ by (1)

$= A\alpha^{n-2}\{\alpha^2 + \beta\alpha - \alpha\beta\} + B\beta^{n-2}\{\alpha\beta + \beta^2 - \alpha\beta\}$

$= A\alpha^{n-2}(\alpha^2) + B\beta^{n-2}(\beta^2) = A\alpha^n + B\beta^n$

This is true for n. Hence by principle of mathematical induction, the result holds good for all $n \in N$.

I INTEGER VALUE CORRECT TYPE :

1. **6** Let the coefficients of three consecutive terms of $(1+x)^{n+5}$ be

${}^{n+5}C_{r-1}$, ${}^{n+5}C_r$, ${}^{n+5}C_{r+1}$, then we have

${}^{n+5}C_{r-1} : {}^{n+5}C_r : {}^{n+5}C_{r+1} = 5 : 10 : 14$

$\dfrac{{}^{n+5}C_{r-1}}{{}^{n+5}C_r} = \dfrac{5}{10} \Rightarrow \dfrac{r}{n+6-r} = \dfrac{1}{2}$

or $n - 3r + 6 = 0$ ...(1)

Also $\dfrac{{}^{n+1}C_r}{{}^{n+5}C_{r+1}} = \dfrac{10}{14} \Rightarrow \dfrac{r+1}{n-r+5} = \dfrac{5}{7}$

or $5n - 12r + 18 = 0$ (2)

Solving (1) and (2) we get $n = 6$.

Section-B JEE Main/ AIEEE

1. (a) We have $t_{p+1} = {}^{p+q}C_p x^p$ and $t_{q+1} = {}^{p+q}C_q x^q$
${}^{p+q}C_p = {}^{p+q}C_q$.[Remember ${}^nC_r = {}^nC_{n-r}$]

2. (c) We have $2^n = 4096 = 2^{12} \Rightarrow n = 12$; the greatest coeff = coeff of middle term. So middle term

$= t_7$; $t_7 = t_{6+1} \Rightarrow$ coeff of $t_7 = {}^{12}C_6 = \frac{12!}{6!6!} = 924$.

3. (d) $(1+0.0001)^{10000} = \left(1+\frac{1}{n}\right)^n, n = 10000$

$= 1 + n.\frac{1}{n} + \frac{n(n-1)}{2!}\frac{1}{n^2} + \frac{n(n-1)(n-2)}{3!}\frac{1}{n^3} + ...$

$= 1 + 1 + \frac{1}{2!}\left(1-\frac{1}{n}\right) + \frac{1}{3!}\left(1-\frac{1}{n}\right) + \left(1-\frac{2}{n}\right) + ...$

$< 1 + \frac{1}{1!} + \frac{1}{2!} + \frac{1}{3!} + ... + \frac{1}{(9999)!}$

$= 1 + \frac{1}{1!} + \frac{1}{2!} + ...\infty = e < 3$

4. (c) $t_{r+2} = {}^{2n}C_{r+1}\, x^{r+1}; t_{3r} = {}^{2n}C_{3r-1}\, x^{3r-1}$
Given ${}^{2n}C_{r+1} = {}^{2n}C_{3r-1}$;
$\Rightarrow {}^{2n}C_{2n-(r+1)} = {}^{2n}C_{3r-1}$
$\Rightarrow 2n - r - 1 = 3r - 1 \Rightarrow 2n = 4r \Rightarrow n = 2r$

5. (b) $a_1 = \sqrt{7} < 7$. Let $a_m < 7$

Then $a_{m+1} = \sqrt{7 + a_m} \Rightarrow a^2_{m+1} = 7 + a_m < 7 + 7 < 14$.

$\Rightarrow a_{m+1} < \sqrt{14} < 7$; So by the principle of mathematical induction $a_n < 7 \;\forall\, n$.

6. (d) $T_{r+1} = \frac{n(n-1)(n-2).........(n-r+1)}{r!}(x)^r$

For first negative term, $n - r + 1 < 0 \Rightarrow r > n + 1$

$\Rightarrow r > \frac{32}{5} \therefore r = 7. \left(\because n = \frac{27}{5}\right)$

Therefore, first negative term is T_8.

7. (c) $T_{r+1} = {}^{256}C_r(\sqrt{3})^{256-r}(\sqrt[8]{5})^r = {}^{256}C_r(3)^{\frac{256-r}{2}}(5)^{r/8}$

Terms will be integral if $\frac{256-r}{2}$ & $\frac{r}{8}$ both are +ve integer, which is so if r is an integral multiple of 8. As $0 \le r \le 256$

$\therefore$ r = 0,8,16,24,........256 , total 33 values.

8. (b) $S(k) = 1+3+5+...+(2k-1) = 3 + k^2$

$S(1): 1 = 3 + 1$, which is not true

$\because S(1)$ is not true.

$\therefore$ P.M.I cannot be applied

Let $S(k)$ is true, i.e.

$1 + 3 + 5.... + (2k-1) = 3 + k^2$

$\Rightarrow 1 + 3 + 5.... + (2k-1) + 2k + 1$

$= 3 + k^2 + 2k + 1 = 3 + (k+1)^2$

$\therefore S(k) \Rightarrow S(k+1)$

9. (c) The middle term in the expansion of
$(1+\alpha x)^4 = T_3 = {}^4C_2(\alpha x)^2 = 6\alpha^2x^2$
The middle term in the expansion of
$(1-\alpha x)^6 = T_4 = {}^6C_3(-\alpha x)^3 = -20\alpha^3x^3$
According to the question

$6\alpha^2 = -20\alpha^3 \Rightarrow \alpha = -\frac{3}{10}$

10. (b) Coeff. of x^n in $(1+x)(1-x)^n$
= coeff of x^n in

$(1+x)(1 - {}^nC_1x + {}^nC_2x^2 - .... + (-1)^{nn}C_nx^n)$

$= (-1)^{n\,n}C_n + (-1)^{n-1\,n}C_{n-1} = (-1)^n + (-1)^{n-1}.n$

$= (-1)^n(1-n)$

ALTERNATE SOLUTION

Coeff of x^n in $(1+x)(1-x)^n$

= Coeff of x^n in $(1-x)^n$ + Coeff of x^{n-1} in $(1-x)^n$

$= (-1)^n\,{}^nC_n + (-1)^{n-1}\,{}^nC_{n-1} = (-1)^n 1 + (-1)^{n-1}n$

$= (-1)^n[1-n]$

11. (d) ${}^{50}C_4 + \sum_{r=1}^{6} {}^{56-r}C_3$

$\Rightarrow {}^{50}C_4 + \left[\begin{matrix}{}^{55}C_3 + {}^{54}C_3 + {}^{53}C_3 + {}^{52}C_3 \\ + {}^{51}C_3 + {}^{50}C_3\end{matrix}\right]$

We know $\left[{}^nC_r + {}^nC_{r-1} = {}^{n+1}C_r\right]$

$\Rightarrow ({}^{50}C_4 + {}^{50}C_3)$

$+ {}^{51}C_3 + {}^{52}C_3 + {}^{53}C_3 + {}^{54}C_3 + {}^{55}C_3$

$\Rightarrow ({}^{51}C_4 + {}^{51}C_3) + {}^{52}C_3 + {}^{53}C_3 + {}^{54}C_3 + {}^{55}C_3$

Proceeding in the same way, we get

$\Rightarrow {}^{55}C_4 + {}^{55}C_3 = {}^{56}C_4$.

12. (a) We observe that

$A^2=\begin{bmatrix}1&0\\2&1\end{bmatrix}, A^3=\begin{bmatrix}1&0\\3&1\end{bmatrix}$ and we can prove by induction that $A^n=\begin{bmatrix}1&0\\n&1\end{bmatrix}$

Now $nA-(n-1)I=\begin{bmatrix}n&0\\n&n\end{bmatrix}-\begin{bmatrix}n-1&0\\0&n-1\end{bmatrix}$

$=\begin{bmatrix}1&0\\n&1\end{bmatrix}=A^n$

$\therefore nA-(n-1)I=A^n$

13. (d) T_{r+1} in the expansion

$$\left[ax^2+\frac{1}{bx}\right]^{11}={}^{11}C_r(ax^2)^{11-r}\left(\frac{1}{bx}\right)^r$$

$={}^{11}C_r(a)^{11-r}(b)^{-r}(x)^{22-2r-r}$

For the Coefficient of x^7, we have

$\Rightarrow 22-3r=7 \Rightarrow r=5$

$\therefore$ Coefficient of $x^7={}^{11}C_5(a)^6(b)^{-5}$ ...(1)

Again T_{r+1} in the expansion

$$\left[ax-\frac{1}{bx^2}\right]^{11}={}^{11}C_r(ax^2)^{11-r}\left(-\frac{1}{bx^2}\right)^r$$

$={}^{11}C_r(a)^{11-r}(-1)^r\times(b)^{-r}(x)^{-2r}(x)^{11-r}$

For the Coefficient of x^{-7}, we have

Now $11-3r=-7 \Rightarrow 3r=18 \Rightarrow r=6$

$\therefore$ Coefficient of $x^{-7}={}^{11}C_6a^5\times1\times(b)^{-6}$

$\therefore$ Coefficient of x^7 = Coefficient of x^{-7}

$\Rightarrow {}^{11}C_5(a)^6(b)^{-5}={}^{11}C_6a^5\times(b)^{-6} \Rightarrow ab=1.$

14. (c) $\because$ x^3 and higher powers of x may be neglected

$$\therefore \frac{(1+x)^{\frac{3}{2}}-\left(1+\frac{x}{2}\right)^3}{\left(1-x^{\frac{1}{2}}\right)}$$

$$=(1-x)^{\frac{-1}{2}}\left[\left(1+\frac{3}{2}x+\frac{\frac{3}{2}\cdot\frac{1}{2}}{2!}x^2\right)-\left(1+\frac{3x}{2}+\frac{3.2}{2!}\frac{x^2}{4}\right)\right]$$

$$=\left[1+\frac{x}{2}+\frac{\frac{1}{2}\cdot\frac{3}{2}}{2!}x^2\right]\left[\frac{-3}{8}x^2\right]=\frac{-3}{8}x^2$$

(as x^3 and higher powers of x can be neglected)

15. (d) $(1-ax)^{-1}(1-bx)^{-1}$

$=(1+ax+a^2x^2+...)(1+bx+b^2x^2+...)$

$\therefore$ Coefficient of x^n

$x^n=b^n+ab^{n-1}+a^2b^{n-2}+.......+a^{n-1}b+a^n$

{which is a G.P. with $r=\frac{a}{b}$

$$\therefore \text{ Its sum is }=\frac{b^n\left[1-\left(\frac{a}{b}\right)^{n+1}\right]}{1-\frac{a}{b}}\}=\frac{b^{n+1}-a^{n+1}}{b-a}$$

$$\therefore a_n=\frac{b^{n+1}-a^{n+1}}{b-a}$$

16. (d) $(1-y)^m(1+y)^n$

$=[1-{}^mC_1y+{}^mC_2y^2-......]$

$[1+{}^nC_1y+{}^nC_2y^2+.....]$

$$=1+(n-m)+\left\{\frac{m(m-1)}{2}+\frac{n(n-1)}{2}-mn\right\}y^2+.....$$

$\therefore a_1=n-m=10$

and $a_2=\frac{m^2+n^2-m-n-2mn}{2}=10$

So, $n-m=10$ and $(m-n)^2-(m+n)=20$

$\Rightarrow m+n=80$

$\therefore m=35, n=45$

17. (b) $T_{r+1}=(-1)^r.{}^nC_r(a)^{n-r}.(b)^r$ is an expansion of $(a-b)^n$

$\therefore$ 5th term $=t_5=t_{4+1}$

$=(-1)^4.{}^nC_4(a)^{n-4}.(b)^4={}^nC_4.a^{n-4}.b^4$

6th term $=t_6=t_{5+1}=(-1)^5\,{}^nC_5(a)^{n-5}(b)^5$

Given $t_5+t_6=0$

$\therefore$ ${}^nC_4.a^{n-4}.b^4+(-{}^nC_5.a^{n-5}.b^5)=0$

$$\Rightarrow \frac{n!}{4!(n-4)!}\cdot\frac{a^n}{a^4}.b^4-\frac{n!}{5!(n-5)!}\cdot\frac{a^nb^5}{a^5}=0$$

$$\Rightarrow \frac{n!.a^nb^4}{4!(n-5)!.a^4}\left[\frac{1}{(n-4)}-\frac{b}{5.a}\right]=0$$

or, $\frac{1}{n-4}-\frac{b}{5a}=0 \Rightarrow \frac{a}{b}=\frac{n-4}{5}$

18. (d) We know that, $(1+x)^{20}={}^{20}C_0+{}^{20}C_1x+{}^{20}C_2x^2+......$ ${}^{20}C_{10}x^{10}+.....{}^{20}C_{20}x^{20}$

Put $x=-1$, $(0)={}^{20}C_0-{}^{20}C_1+{}^{20}C_2-{}^{20}C_3+......+{}^{20}C_{10}$ $-{}^{20}C_{11}....+{}^{20}C_{20}$

$\Rightarrow 0=2[{}^{20}C_0-{}^{20}C_1+{}^{20}C_2-{}^{20}C_3+.....-{}^{20}C_9]+{}^{20}C_{10}$

$\Rightarrow {}^{20}C_{10}=2[{}^{20}C_0-{}^{20}C_1+{}^{20}C_2-{}^{20}C_3+......-{}^{20}C_9+{}^{20}C_{10}]$

$\Rightarrow {}^{20}C_0-{}^{20}C_1+{}^{20}C_2-{}^{20}C_3+....+{}^{20}C_{10}=\frac{1}{2}\,{}^{20}C_{10}$

19. (b) We have

$$\sum_{r=0}^{n}(r+1)^nC_rx^r = \sum_{r=0}^{n} r.^nC_rx^r + \sum_{r=0}^{n} {}^nC_rx^r$$

$$= \sum_{r=1}^{n} r\cdot\frac{n}{r}\,{}^{n-1}C_{r-1}x^r + (1+x)^n$$

$$= nx\sum_{r=1}^{n} {}^{n-1}C_{r-1}x^{r-1} + (1+x)^n$$

$= nx\,(1+x)^{n-1} + (1+x)^n =$ RHS

∴ Statement 2 is correct.

Putting $x = 1$, we get

$$\sum_{r=0}^{n}(r+1)^nC_r = n\cdot 2^{n-1} + 2^n = (n+2)\cdot 2^{n-1}.$$

∴ Statement 1 is also true and statement 2 is a correct explanation for statement 1.

20. (a) $(8)^{2n} - (62)^{2n+1}$

$= (64)^n - (62)^{2n+1} = (63+1)^n - (63-1)^{2n+1}$

$$= \left[{}^nC_0(63)^n + {}^nC_1(63)^{n-1} + {}^nC_2(63)^{n-2} + \ldots\ldots + {}^nC_{n-1}(63) + {}^nC_n\right]$$

$$= \left[{}^{2n+1}C_0(63)^{2n+1} - {}^{2n+1}C_1(63)^{2n} + {}^{2n+1}C_2(63)^{2n-1} - \ldots\ldots + (-1)^{2n+1}\,{}^{2n+1}C_{2+1}\right]$$

$$= 63\times\left[{}^nC_0(63)^{n-1} + {}^nC_1(63)^{n-2} + {}^nC_2(63)^{n-3} + \ldots\ldots\right] + 1$$

$$- 63\times\left[{}^{2n+1}C_0(63)^{2n} - {}^{2n+1}C_1(63)^{2n-1} + \ldots\ldots\right] + 1$$

⇒ 63 × some integral value + 2

⇒ $8^{2n} - (62)^{2n+1}$ when divided by 9 leaves 2 as the remainder.

21. (b) $S_2 = \sum_{j=1}^{10} j\ {}^{10}C_j = \sum_{j=1}^{10} 10\ {}^9C_{j-1}$

$$= 10\left[{}^9C_0 + {}^9C_1 + {}^9C_2 + \ldots + {}^9C_9\right] = 10.2^9$$

22. (b) $(1-x-x^2+x^3)^6 = [(1-x) - x^2(1-x)]^6$

$= (1-x)^6(1-x^2)^6$

$= (1-6x+15x^2-20x^3+15x^4-6x^5+x^6)$

$\times (1-6x^2+15x^4-20x^6+15x^8-6x^{10}+x^{12})$

Coefficient of $x^7 = (-6)(-20) + (-20)(15) + (-6)(-6)$

$= -144$

23. (a) **Statement 2 :** $P(n) : n^7 - n$ is divisible by 7

Put $n = 1$, $1 - 1 = 0$ is divisible by 7, which is true

Let $n = k$, $P(k) : k^7 - k$ is divisible by 7, true

Put $n = k + 1$

∴ $P(k+1) : (k+1)^7 - (k+1)$ is div. by 7

$P(k+1) : k^7 + {}^7C_1k^6 + {}^7C_2k^2 + \ldots\ldots + {}^7C_6k + 1 - k - 1$, is div. by 7.

$P(k+1) : (k^7 - k) + ({}^7C_1k^6 + {}^7C_2k^5 + \ldots\ldots + {}^7C_6k)$ is div. by 7.

Since 7 is coprime with 1, 2, 3, 4, 5, 6.

So ${}^7C_1, {}^7C_2, \ldots\ldots {}^7C_6$ are all divisible by 7

∴ $P(k+1)$ is divisible by 7

Hence $P(n) : n^7 - n$ is divisible by 7

Statement 1 : $n^7 - n$ is divisible by 7

⇒ $(n+1)^7 - (n+1)$ is divisible by 7

⇒ $(n+1)^7 - n^7 - 1 + (n^7 - n)$ is divisible by 7

⇒ $(n+1)^7 - n^7 - 1$ is divisible by 7

Hence both Statements 1 and 2 are correct and Statement 2 is the correct explanation of Statement-1.

24. (a) $(\sqrt{3}+1)^{2n} - (\sqrt{3}-1)^{2n}$

$$= \left[(\sqrt{3}+1)^2\right]^n - \left[(\sqrt{3}-1)^2\right]^n$$

$$= (4+2\sqrt{3})^n - (4-2\sqrt{3})^n$$

$$= 2^n\left[(2+\sqrt{3})^n - (2-\sqrt{3})^n\right]$$

$$= 2^n\times 2\left[{}^nC_1\,2^{n-1}\sqrt{3} + {}^nC_3\cdot 2^{n-3}3\sqrt{3} + \ldots\right]$$

$$= 2^{n+1}\sqrt{3}\left[{}^nC_1\cdot 2^{n-1} + {}^nC_3 2^{n-3}\cdot 3 + \ldots\right]$$

$= \sqrt{3}\times$ Some integer

∴ irrational number

25. (c) Given expression can be written as

$$\left((x^{1/3}+1) - \left(\frac{\sqrt{x}+1}{\sqrt{x}}\right)\right)^{10} = \left(x^{1/3} + 1 - 1 - \frac{1}{\sqrt{x}}\right)^{10}$$

$= (x^{1/3} - x^{-1/2})^{10}$

General term $= T_{r+1} = {}^{10}C_r(x^{1/3})^{10-r}(-x^{-1/2})^r$

$$= {}^{10}C_r\, x^{\frac{10-r}{3}}\cdot(-1)^r\cdot x^{-\frac{r}{2}} = {}^{10}C_r(-1)^r\cdot x^{\frac{10-r}{3}-\frac{r}{2}}$$

Term will be independent of x when $\frac{10-r}{3} - \frac{r}{2} = 0$

⇒ $r = 4$

So, required term $= T_5 = {}^{10}C_4 = 210$

Chapter 6

Sequences and Series

Section-A JEE Advanced/ IIT-JEE

A FILL IN THE BLANKS :

1. The sum of integers from 1 to 100 that are divisible by 2 or 5 is .... *(1984 - 2 Marks)*

2. The solution of the equation $\log_7 \log_5(\sqrt{x+5}+\sqrt{x}) = 0$ is .... *(1986 - 2 Marks)*

3. The sum of the first n terms of the series $1^2+2.2^2+3^2+2.4^2+5^2+2.6^2+$......... is $n\ (n+1)^2/2$, when n is even. When n is odd, the sum is .... *(1988 - 2 Marks)*

4. Let the harmonic mean and geometric mean of two positive numbers be the ratio 4 : 5. Then the two number are in the ratio .... *(1992 - 2 Marks)*

5. For any odd integer $n \geq 1$, $n^3-(n-1)^3+...+(-1)^{n-1}\ 1^3 =$ .... *(1996 - 1 Mark)*

6. Let x be the arithmetic mean and y, z be the two geometric means between any two positive numbers. Then $\frac{y^3+z^3}{xyz} =$ .... *(1997C - 2 Marks)*

7. Let p and q be roots of the equation $x^2-2x+A=0$ and let r and s be the roots of the equation $x^2 - 18x + B = 0$. If $p<q<r<s$ are in arithmetic progression, then $A =$ .... and $B =$............. *(1997 - 2 Marks)*

C MCQ's WITH ONE CORRECT ANSWER :

1. If x, y and z are pth, qth and rth terms respectively of an A.P. and also of a G.P., then $x^{y-z}y^{z-x}z^{x-y}$ is equal to :
   (a) xyz (b) 0
   (c) 1 (d) None of these

2. The third term of a geometric progression is 4. The product of the first five terms is *(1982 - 2 Marks)*
   (a) 4^3 (b) 4^5
   (c) 4^4 (d) none of these

3. The rational number, which equals the number $2.\overline{357}$ with recurring decimal is *(1983 - 1 Mark)*
   (a) $\frac{2355}{1001}$ (b) $\frac{2379}{997}$
   (c) $\frac{2355}{999}$ (d) none of these

4. If a, b, c are in G.P., then the equations $ax^2+2bx+c=0$ and $dx^2+2ex+f=0$ have a common root if $\frac{d}{a}, \frac{e}{b}, \frac{f}{c}$ are in — *(1985 - 2 Marks)*
   (a) A.P. (b) G.P.
   (c) H.P. (d) none of these

5. Sum of the first n terms of the series $\frac{1}{2}+\frac{3}{4}+\frac{7}{8}+\frac{15}{16}+$............. is equal to *(1988 - 2 Marks)*
   (a) 2^n-n-1 (b) $1-2^{-n}$
   (c) $n+2^{-n}-1$ (d) 2^n+1.

6. The number $\log_2 7$ is *(1990 - 2 Marks)*
   (a) an integer (b) a rational number
   (c) an irrational number (d) a prime number

7. If $ln(a+c)$, $ln\ (a-c)$, $ln\ (a-2b+c)$ are in A.P., then *(1994)*
   (a) c, b, c are in A.P. (b) a^2, b^2, c^2 are in A.P.
   (c) a, b, c are in G.P. (d) a, b, c are in H.P.

8. Let $a_1, a_2, ... a_{10}$ be in A, P, and $h_1, h_2,....h_{10}$ be in H.P. If $a_1 = h_1 = 2$ and $a_{10} = h_{10} = 3$, then a_4h_7 is *(1999 - 2 Marks)*
   (a) 2 (b) 3
   (c) 5 (d) 6

9. The harmonic mean of the roots of the equation $\left(5+\sqrt{2}\right)x^2-\left(4+\sqrt{5}\right)x+8+2\sqrt{5}=0$ is *(1999 - 2 Marks)*
   (a) 2 (b) 4
   (c) 6 (d) 8

10. Consider an infinite geometric series with first term a and common ratio r. If its sum is 4 and the second term is 3/4, then *(2000S)*
    (a) $a=\frac{4}{7}, r=\frac{3}{7}$ (b) $a=2, r=\frac{3}{8}$
    (c) $a=\frac{3}{2}, r=\frac{1}{2}$ (d) $a=3, r=\frac{1}{4}$

11. Let α, β be the roots of $x^2-x+p=0$ and γ, δ be the roots of $x^2-4x+q=0$. If α, β, γ, δ are in G.P., then the integral values of p and q respectively, are *(2001S)*
    (a) $-2, -32$ (b) $-2, 3$
    (c) $-6, 3$ (d) $-6, -32$

12. Let the positive numbers a, b, c, d be in A.P. Then abc, abd, acd, bcd are *(2001S)*

(a) NOT in A.P./G.P./H.P. (b) in A.P.

(c) in G.P. (d) in H.P.

13. If the sum of the first $2n$ terms of the A.P. 2, 5, 8, ..., is equal to the sum of the first n terms of the A.P. 57, 59, 61, ..., then n equals *(2001S)*

(a) 10 (b) 12

(c) 11 (d) 13

14. Suppose a, b, c are in A.P. and a^2, b^2, c^2 are in G.P. if $a<b<c$ and $a+b+c=\frac{3}{2}$, then the value of a is *(2002S)*

(a) $\frac{1}{2\sqrt{2}}$ (b) $\frac{1}{2\sqrt{3}}$

(c) $\frac{1}{2}-\frac{1}{\sqrt{3}}$ (d) $\frac{1}{2}-\frac{1}{\sqrt{2}}$

15. An infinite G.P. has first term 'x' and sum '5', then x belongs to *(2004S)*

(a) $x<-10$ (b) $-10<x<0$

(c) $0<x<10$ (d) $x>10$

16. In the quadratic equation $ax^2+bx+c=0, \Delta=b^2-4ac$ and $\alpha+\beta, \alpha^2+\beta^2, \alpha^3+\beta^3$, are in G.P. where α, β are the root of $ax^2+bx+c=0$, then *(2005S)*

(a) $\Delta \neq 0$ (b) $b\Delta=0$

(c) $c\Delta=0$ (d) $\Delta=0$

17. In the sum of first n terms of an A.P. is cn^2, then the sum of squares of these n terms is *(2009)*

(a) $\frac{n(4n^2-1)c^2}{6}$ (b) $\frac{n(4n^2+1)c^2}{3}$

(c) $\frac{n(4n^2-1)c^2}{3}$ (d) $\frac{n(4n^2+1)c^2}{6}$

18. Let $a_1, a_2, a_3, ...$ be in harmonic progression with $a_1=5$ and $a_{20}=25$. The least positive integer n for which $a_n<0$ is *(2012)*

(a) 22 (b) 23 (c) 24 (d) 25

D MCQ's WITH ONE OR MORE THAN ONE CORRECT :

1. If the first and the $(2n-1)$st terms of an A.P., a G.P. and an H.P. are equal and their n-th terms are a, b and c respectively, then *(1988 - 2 Marks)*

(a) $a=b=c$ (b) $a \geq b \geq c$

(c) $a+c=b$ (d) $ac-b^2=0$.

2. For $0<\phi<\pi/2$, if

$$x=\sum_{n=0}^{\infty}\cos^{2n}\phi,\ y=\sum_{n=0}^{\infty}\sin^{2n}\phi,\ z=\sum_{n=0}^{\infty}\cos^{2n}\phi\sin^{2n}\phi$$

then: *(1993 - 2 Marks)*

(a) $xyz=xz+y$ (b) $xyz=xy+z$

(c) $xyz=x+y+z$ (d) $xyz=yz+x$

3. Let n be an odd integer. If $\sin n\theta=\sum_{r=0}^{n} b_r \sin^r\theta$, for every value of θ, then *(1998 - 2 Marks)*

(a) $b_0=1, b_1=3$ (b) $b_0=0, b_1=n$

(c) $b_0=-1, b_1=n$ (d) $b_0=0, b_1=n^2-3n+3$

4. Let T_r be the rth term of an A.P., for $r=1, 2, 3, ...$ If for some positive integers m, n we have

$T_m=\frac{1}{n}$ and $T_n=\frac{1}{m}$, then T_{mn} equals *(1998 - 2 Marks)*

(a) $\frac{1}{mn}$ (b) $\frac{1}{m}+\frac{1}{n}$

(c) 1 (d) 0

5. If $x>1, y>1, z>1$ are in G.P., then $\frac{1}{1+\ln x}, \frac{1}{1+\ln y}, \frac{1}{1+\ln z}$ are in *(1998 - 2 Marks)*

(a) A.P. (b) H.P.

(c) GP (d) None of these

6. For a positive integer n, let

$a(n)=1+\frac{1}{2}+\frac{1}{3}+\frac{1}{4}+....\frac{1}{(2^n)-1}$. Then *(1999 - 3 Marks)*

(a) $a(100) \leq 100$ (b) $a(100)>100$

(c) $a(200) \leq 100$ (d) $a(200)>100$

7. A straight line through the vertex P of a triangle PQR intersects the side QR at the point S and the circumcircle of the triangle PQR at the point T. If S is not the centre of the circumcircle, then *(2008)*

(a) $\frac{1}{PS}+\frac{1}{ST}<\frac{2}{\sqrt{QS \times SR}}$ (b) $\frac{1}{PS}+\frac{1}{ST}>\frac{2}{\sqrt{QS \times SR}}$

(c) $\frac{1}{PS}+\frac{1}{ST}<\frac{4}{QR}$ (d) $\frac{1}{PS}+\frac{1}{ST}>\frac{4}{QR}$

8. Let $S_n=\sum_{k=1}^{n}\frac{n}{n^2+kn+k^2}$ and $T_n=\sum_{k=0}^{n-1}\frac{n}{n^2+kn+k^2}$ for $n=1, 2, 3, ...$ Then, *(2008)*

(a) $S_n<\frac{\pi}{3\sqrt{3}}$ (b) $S_n>\frac{\pi}{3\sqrt{3}}$

(c) $T_n<\frac{\pi}{3\sqrt{3}}$ (d) $T_n>\frac{\pi}{3\sqrt{3}}$

9. Let $S_n=\sum_{k=1}^{4n}(-1)^{\frac{k(k+1)}{2}}k^2$. Then S_n can take value(s) *(JEE Adv. 2013)*

(a) 1056 (b) 1088

(c) 1120 (d) 1332

(E) SUBJECTIVE PROBLEMS :

1. The harmonic mean of two numbers is 4. Their arithmetic mean A and the geometric mean G satisfy the relation.
$2A+G^2=27$
Find the two numbers. **(1979)**

2. The interior angles of a polygon are in arithmetic progression. The smallest angle is 120°, and the common difference is 5°, Find the number of sides of the polygon. **(1980)**

3. If a_1, a_2............, a_n are in arithmetic progression, where $a_i > 0$ for all i, Show that

$$\frac{1}{\sqrt{a_1}+\sqrt{a_2}}+\frac{1}{\sqrt{a_2}+\sqrt{a_3}}+........+\frac{1}{\sqrt{a_{n-1}}+\sqrt{a_n}}=\frac{n-1}{\sqrt{a_1}+\sqrt{a_n}}$$

(1982 - 2 Marks)

4. Does there exist a geometric progression containing 27, 8 and 12 as three of its terms ? If it exits, how many such progressions are possible ? **(1982 - 3 Marks)**

5. Find three numbers a, b, c, between 2 and 18 such that
(i) their sum is 25
(ii) the numbers 2, a, b sare consecutive terms of an A.P. and
(iii) the numbers b, c, 18 are consecutive terms of a G.P. **(1983 - 2 Marks)**

6. If $a>0$, $b>0$ and $c>0$, prove that

$$(a+b+c)\left(\frac{1}{a}+\frac{1}{b}+\frac{1}{c}\right)\geq 9$$ **(1984 - 2 Marks)**

7. If n is a natural number such that
$n=p_1^{\alpha_1}\cdot p_2^{\alpha_2}\cdot p_3^{\alpha_3}........p_k^{\alpha_k}$ and $p_1, p_2, ..., p_k$ are distinct primes, then show that $\ln n \geq k \ln 2$ **(1984 - 2 Marks)**

8. Find the sum of the series :

$$\sum_{r=0}^{n}(-1)^r\,{}^nC_r\left[\frac{1}{2^r}+\frac{3^r}{2^{2r}}+\frac{7^r}{2^{3r}}+\frac{15^r}{2^{4r}}......\text{ up to } m \text{ terms}\right]$$

(1985 - 5 Marks)

9. The sum of the squares of three distinct real numbers, which are in G. P., is S^2. If their sum is aS, show that $a^2 \in \left(\frac{1}{3},1\right)\cup(1,3)$. **(1986 - 5 Marks)**

10. Solve for x the following equation : **(1987 - 3 Marks)**

$$\log_{(2x+3)}(6x^2+23x+21)=4-\log_{(3x+7)}(4x^2+12x+9)$$

11. If $\log_3 2$, $\log_3(2^x-5)$, and $\log_3\left(2^x-\frac{7}{2}\right)$ are in arithmetic progression, determine the value of x. **(1990 - 4 Marks)**

12. Let p be the first of the n arithmetic means between two numbers and q the first of n harmonic means between the same numbers. Show that q does not lie between p and $\left(\frac{n+1}{n-1}\right)^2 p$. **(1991 - 4 Marks)**

13. If S_1, S_2, S_3, ..., S_n are the sums of infinite geometric series whose first terms are 1, 2, 3, ..., n and whose common ratios are $\frac{1}{2},\frac{1}{3},\frac{1}{4}$, ... $\frac{1}{n+1}$ respectively, then find the values of $S_1^2+S_2^2+S_3^2+..............+S_{2n-1}^2$ **(1991 - 4 Marks)**

14. The real numbers x_1, x_2, x_3 satisfying the equation $x^3-x^2+\beta x+\gamma=0$ are in AP. Find the intervals in which β and γ lie. **(1996 - 3 Marks)**

15. Let a, b, c, d be real numbers in G.P. If u, v, w, satisfy the system of equations **(1999 - 10 Marks)**

$$u+2v+3w=6$$
$$4u+5v+6w=12$$
$$6u+9v=4$$

then show that the roots of the equation

$$\left(\frac{1}{u}+\frac{1}{v}+\frac{1}{w}\right)x^2+[(b-c)^2+(c-a)^2+(d-b)^2]\,x+u+v+w=0$$

and $20x^2+10(a-d)^2x-9=0$ are reciprocals of each other.

16. The fourth power of the common difference of an arithmatic progression with integer entries is added to the product of any four consecutive terms of it. Prove that the resulting sum is the square of an integer. **(2000 - 4 Marks)**

17. Let $a_1, a_2, ..., a_n$ be positive real numbers in geometric progression. For each n, let A_n, G_n, H_n be respectively, the arithmetic mean, geometric mean, and harmonic mean of $a_1, a_2, ..., a_n$. Find an expression for the geometric mean of $G_1, G_2, ..., G_n$ in terms of $A_1, A_2, ..., A_n, H_1, H_2, ..., H_n$. **(2001 - 5 Marks)**

18. Let a, b be positive real numbers. If a, A_1, A_2, b are in arithmetic progression, a, G_1, G_2, b are in geometric progression and a, H_1, H_2, b are in harmonic progression, show that $\frac{G_1G_2}{H_1H_2}=\frac{A_1+A_2}{H_1+H_2}=\frac{(2a+b)(a+2b)}{9ab}$. **(2002 - 5 Marks)**

19. If a, b, c are in A.P., a^2, b^2, c^2 are in H.P., then prove that either $a=b=c$ or a, b, $-\frac{c}{2}$ form a G.P. **(2003 - 4 Marks)**

20. If $a_n=\frac{3}{4}-\left(\frac{3}{4}\right)^2+\left(\frac{3}{4}\right)^3+.....(-1)^{n-1}\left(\frac{3}{4}\right)^n$ and $b_n=1-a_n$, then find the least natural number n_0 such that $b_n > a_n \forall n \geq n_0$. **(2006 - 6M)**

(G) COMPREHENSION BASED QUESTIONS :

PASSAGE - 1

Let V_r denote the sum of first r terms of an arithmetic progression (A.P.) whose first term is r and the common difference is $(2r - 1)$. Let $T_r = V_{r+1} - V_r - 2$ and $Q_r = T_{r+1} - T_r$ for $r = 1, 2, ...$

1. The sum $V_1 + V_2 + ... + V_n$ is *(2007 - 4 marks)*
   (a) $\frac{1}{12}n(n+1)(3n^2 - n + 1)$
   (b) $\frac{1}{12}n(n+1)(3n^2 + n + 2)$
   (c) $\frac{1}{2}n(2n^2 - n + 1)$
   (d) $\frac{1}{3}(2n^3 - 2n + 3)$
2. T_r is always *(2007 - 4 marks)*
   (a) an odd number (b) an even number
   (c) a prime number (d) a composite number
3. Which one of the following is a correct statement ? *(2007 - 4 marks)*
   (a) $Q_1, Q_2, Q_3, ...$ are in A.P. with common difference 5
   (b) $Q_1, Q_2, Q_3, ...$ are in A.P. with common difference 6
   (c) $Q_1, Q_2, Q_3, ...$ are in A.P. with common difference 11
   (d) $Q_1 = Q_2 = Q_3 = ...$

PASSAGE - 2

Let A_1, G_1, H_1 denote the arithmetic, geometric and harmonic means, respectively, of two distinct positive numbers. For $n \geq 2$, Let A_{n-1} and H_{n-1} have arithmetic, geometric and harmonic means as A_n, G_n, H_n respectively.

4. Which one of the following statements is correct ? *(2007 - 4 marks)*
   (a) $G_1 > G_2 > G_3 > ...$
   (b) $G_1 < G_2 < G_3 < ...$
   (c) $G_1 = G_2 = G_3 = ...$
   (d) $G_1 < G_3 < G_5 < ...$ and $G_2 > G_4 > G_6 > ....$
5. Which one of the following statements is correct ? *(2007 - 4 marks)*
   (a) $A_1 > A_2 > A_3 > ...$
   (b) $A_1 < A_2 < A_3 < ...$
   (c) $A_1 > A_3 > A_5 > ...$ and $A_2 < A_4 < A_6 < ...$
   (d) $A_1 < A_3 < A_5 < ...$ and $A_2 > A_4 > A_6 > ...$
6. Which one of the following statements is correct ? *(2007 - 4 marks)*
   (a) $H_1 > H_2 > H_3 > ...$
   (b) $H_1 < H_2 < H_3 < ...$
   (c) $H_1 > H_3 > H_5 > ...$ and $H_2 < H_4 < H_6 < ...$
   (d) $H_1 < H_3 < H_6 < ...$ and $H_2 > H_4 > H_6 > ...$

(H) ASSERTION & REASON TYPE QUESTIONS :

1. Suppose four distinct positive numbers a_1, a_2, a_3, a_4 are in G.P. Let $b_1 = a_1$, $b_2 = b_1 + a_2$, $b_3 = b_2 + a_3$ and $b_4 = b_3 + a_4$.
   STATEMENT - 1 : The numbers b_1, b_2, b_3, b_4 are neither in A.P. nor in G.P.
   and
   STATEMENT - 2 : The numbers b_1, b_2, b_3, b_4 are in H.P. *(2008)*
   (a) STATEMENT - 1 is True, STATEMENT - 2 is True; STATEMENT - 2 is a correct explanation for STATEMENT - 1
   (b) STATEMENT - 1 is True, STATEMENT - 2 is True; STATEMENT - 2 is **NOT** a correct explanation for STATEMENT - 1
   (c) STATEMENT - 1 is True, STATEMENT - 2 is False
   (d) STATEMENT - 1 is False, STATEMENT - 2 is True

(I) INTEGER VALUE CORRECT TYPE :

1. Let S_k, $k = 1, 2, ... , 100$, denote the sum of the infinite geometric series whose first term is $\frac{k-1}{k!}$ and the common ratio is $\frac{1}{k}$. Then the value of $\frac{100^2}{100!} + \sum_{k=1}^{100} |(k^2 - 3k + 1)S_k|$ is *(2010)*
2. Let $a_1, a_2, a_3, ..., a_{11}$ be real numbers satisfying $a_1 = 15$, $27 - 2a_2 > 0$ and $a_k = 2a_{k-1} - a_{k-2}$ for $k = 3, 4, ...11$. if $\frac{a_1^2 + a_2^2 + ... + a_{11}^2}{11} = 90$, then the value of $\frac{a_1 + a_2 + ... + a_{11}}{11}$ is equal to *(2010)*
3. Let $a_1, a_2, a_3 .....a_{100}$ be an arithmetic progression with $a_1 = 3$ and $S_p = \sum_{i=1}^{p} a_i, 1 \leq p \leq 100$. For any integer n with $1 \leq n \leq 20$, let $m = 5n$. If $\frac{S_m}{S_n}$ does not depend on n, then a_2 is *(2011)*
4. A pack contains n cards numbered from 1 to n. Two consecutive numbered cards are removed from the pack and the sum of the numbers on the remaining cards is 1224. If the smaller of the numbers on the removed cards is k, then k – 20 = *(JEE Adv. 2013)*

Section-B JEE Main/ AIEEE

1. If 1, $\log_9(3^{1-x}+2)$, $\log_3(4.3^x-1)$ are in A.P. then x equals **[2002]**
(a) $\log_3 4$ (b) $1-\log_3 4$
(c) $1-\log_4 3$ (d) $\log_4 3$

2. l, m, n are the pth, q^{th} and r^{th} term of a G. P. all positive, then $\begin{vmatrix} \log l & p & 1 \\ \log m & q & 1 \\ \log n & r & 1 \end{vmatrix}$ equals
(a) -1 (b) 2 **[2002]**
(c) 1 (d) 0

3. The value of $2^{1/4}.\,4^{1/8}.\,8^{1/16}$ ... ∞ is **[2002]**
(a) 1 (b) 2
(c) 3/2 (d) 4

4. Fifth term of a GP is 2, then the product of its 9 terms is **[2002]**
(a) 256 (b) 512
(c) 1024 (d) none of these

5. Sum of infinite number of terms of GP is 20 and sum of their square is 100. The common ratio of GP is **[2002]**
(a) 5 (b) 3/5
(c) 8/5 (d) 1/5

6. $1^3-2^3+3^3-4^3+...+9^3=$ **[2002]**
(a) 425 (b) -425
(c) 475 (d) -475

7. The sum of the series **[2003]**

$\frac{1}{1.2}-\frac{1}{2.3}+\frac{1}{3.4}$ .... up to ∞ is equal to

(a) $\log_e\left(\frac{4}{e}\right)$ (b) $2\log_e 2$
(c) $\log_e 2-1$ (d) $\log_e 2$

8. If $S_n=\sum_{r=0}^{n}\frac{1}{^nC_r}$ and $t_n=\sum_{r=0}^{n}\frac{r}{^nC_r}$, then $\frac{t_n}{S_n}$ is equal to

(a) $\frac{2n-1}{2}$ (b) $\frac{1}{2}n-1$ **[2004]**
(c) $n-1$ (d) $\frac{1}{2}n$

9. Let T_r be the rth term of an A.P. whose first term is a and common difference is d. If for some positive integers $m, n, m \neq n$, $T_m=\frac{1}{n}$ and $T_n=\frac{1}{m}$, then $a-d$ equals

(a) $\frac{1}{m}+\frac{1}{n}$ (b) 1 **[2004]**
(c) $\frac{1}{mn}$ (d) 0

10. The sum of the first n terms of the series

$1^2+2.2^2+3^2+2.4^2+5^2+2.6^2+...$

is $\frac{n(n+1)^2}{2}$ when n is even. When n is odd the sum is

(a) $\left[\frac{n(n+1)}{2}\right]^2$ (b) $\frac{n^2(n+1)}{2}$ **[2004]**
(c) $\frac{n(n+1)^2}{4}$ (d) $\frac{3n(n+1)}{2}$

11. The sum of series $\frac{1}{2!}+\frac{1}{4!}+\frac{1}{6!}+$..... is **[2004]**

(a) $\frac{(e^2-2)}{e}$ (b) $\frac{(e-1)^2}{2e}$
(c) $\frac{(e^2-1)}{2e}$ (d) $\frac{(e^2-1)}{2}$

12. If the coefficients of rth, $(r+1)$th, and $(r+2)$th terms in the the binomial expansion of $(1+y)^m$ are in A.P., then m and r satisfy the equation **[2005]**

(a) $m^2-m(4r-1)+4r^2-2=0$
(b) $m^2-m(4r+1)+4r^2+2=0$
(c) $m^2-m(4r+1)+4r^2-2=0$
(d) $m^2-m(4r-1)+4r^2+2=0$

13. If $x=\sum_{n=0}^{\infty}a^n$, $y=\sum_{n=0}^{\infty}b^n$, $z=\sum_{n=0}^{\infty}c^n$ where a, b, c are in A.P and $|a|<1, |b|<1, |c|<1$ then x, y, z are in **[2005]**
(a) G.P.
(b) A.P.
(c) Arithmetic - Geometric Progression
(d) H.P.

14. The sum of the series **[2005]**

$1+\frac{1}{4.2!}+\frac{1}{16.4!}+\frac{1}{64.6!}+$ ....ad inf. is

(a) $\frac{e-1}{\sqrt{e}}$ (b) $\frac{e+1}{\sqrt{e}}$
(c) $\frac{e-1}{2\sqrt{e}}$ (d) $\frac{e+1}{2\sqrt{e}}$

15. Let a_1, a_2, a_3........... be terms on A.P. If

$$\frac{a_1+a_2+..........a_p}{a_1+a_2+..........+a_q}=\frac{p^2}{q^2},\ p\neq q\text{, then }\frac{a_6}{a_{21}}\text{ equals}$$

(a) $\frac{41}{11}$ (b) $\frac{7}{2}$ **[2006]**

(c) $\frac{2}{7}$ (d) $\frac{11}{41}$

16. If $a_1, a_2, ..., a_n$ are in H.P., then the expression $a_1a_2+a_2a_3+.........+a_{n-1}a_n$ is equal to **[2006]**

(a) $n(a_1-a_n)$ (b) $(n-1)(a_1-a_n)$

(c) na_1a_n (d) $(n-1)a_1a_n$

17. The sum of series $\frac{1}{2!}-\frac{1}{3!}+\frac{1}{4!}-$....... upto infinity is

(a) $e^{-\frac{1}{2}}$ (b) $e^{+\frac{1}{2}}$ **[2007]**

(c) e^{-2} (d) e^{-1}

18. In a geometric progression consisting of positive terms, each term equals the sum of the next two terms. Then the common ratio of its progression is equals **[2007]**

(a) $\sqrt{5}$ (b) $\frac{1}{2}(\sqrt{5}-1)$

(c) $\frac{1}{2}(1-\sqrt{5})$ (d) $\frac{1}{2}\sqrt{5}$.

19. The first two terms of a geometric progression add up to 12. the sum of the third and the fourth terms is 48. If the terms of the geometric progression are alternately positive and negative, then the first term is **[2008]**

(a) –4 (b) –12

(c) 12 (d) 4

20. The sum to infinite term of the series

$1+\frac{2}{3}+\frac{6}{3^2}+\frac{10}{3^3}+\frac{14}{3^4}+...$ is **[2009]**

(a) 3 (b) 4

(c) 6 (d) 2

21. A person is to count 4500 currency notes. Let a_n denote the number of notes he counts in the nth minute. If $a_1 = a_2 = ... = a_{10} = 150$ and $a_{10}, a_{11}, ...$ are in an AP with common difference –2, then the time taken by him to count all notes is **[2010]**

(a) 34 minutes (b) 125 minutes

(c) 135 minutes (d) 24 minutes

22. A man saves ₹ 200 in each of the first three months of his service. In each of the subsequent months his saving increases by ₹ 40 more than the saving of immediately previous month. His total saving from the start of service will be ₹ 11040 after **[2011]**

(a) 19 months (b) 20 months

(c) 21 months (d) 18 months

23. Let a_n be the nth term of an A.P. If $\sum_{r=1}^{100} a_{2r}=\alpha$ and $\sum_{r=1}^{100} a_{2r-1}=\beta$, then the common difference of the A.P. is **[2011 RS]**

(a) $\alpha-\beta$ (b) $\frac{\alpha-\beta}{100}$

(c) $\beta-\alpha$ (d) $\frac{\alpha-\beta}{200}$

24. **Statement-1:** The sum of the series 1 + (1 + 2 + 4) + (4 + 6 + 9) + (9 + 12 + 16) + ... + (361 + 380 + 400) is 8000.

Statement-2: $\sum_{k=1}^{n}\left(k^3-(k-1)^3\right)=n^3$, for any natural number *n*. ***[2012]***

(a) Statement-1 is false, Statement-2 is true.

(b) Statement-1 is true, statement-2 is true; statement-2 is a correct explanation for Statement-1.

(c) Statement-1 is true, statement-2 is true; statement-2 is **not** a correct explanation for Statement-1.

(d) Statement-1 is true, statement-2 is false.

25. The sum of first 20 terms of the sequence 0.7, 0.77, 0.777,....., is **[JEE M 2013]**

(a) $\frac{7}{81}(179-10^{-20})$ (b) $\frac{7}{9}(99-10^{-20})$

(c) $\frac{7}{81}(179+10^{-20})$ (d) $\frac{7}{9}(99+10^{-20})$

Solutions & Explanations

Section-A : JEE Advanced/ IIT-JEE

A 1. 3050 2. 4 3. $\frac{n^2(n+1)}{2}$ 4. 4 : 1 or 1 : 4 5. $\frac{1}{4}(n+1)^2(2n-1)$

6. 2 7. –3, 77

C 1. (c) 2. (b) 3. (c) 4. (a) 5. (c) 6. (c) 7. (d)
8. (d) 9. (b) 10. (d) 11. (a) 12. (d) 13. (c) 14. (d)
15. (c) 16. (c) 17. (c) 18. (d)

D 1. (a, b, d) 2. (b, c) 3. (b) 4. (c) 5. (b) 6. (a, d) 7. (b, d)
8. (a, d) 9. (a, d)

E 1. 3 and 6 or 6 and 3 2. 9 4. yes, infinite 5. 5, 8, 12 8. $\frac{2^{mn}-1}{2^{mn}(2^n-1)}$ 10. $-\frac{1}{4}$

11. 3 13. $\frac{n(2n+1)(4n+1)-3}{3}$ 14. $\beta \in \left(-\infty, \frac{1}{3}\right], \gamma \in \left[-\frac{1}{27}, \infty\right)$

17. $G=(A_1 A_2 .....A_n H_1 H_2 ....H_n)^{1/2n}$. 20. 6

G 1. (b) 2. (d) 3. (b) 4. (c) 5. (a) 6. (b)

H 1. (c)

I 1. (3) 2. (0) 3. (9) 4. (5)

Section-B : JEE Main/ AIEEE

1. (b) 2. (d) 3. (b) 4. (b) 5. (b) 6. (a) 7. (a) 8. (d)
9. (d) 10. (b) 11. (b) 12. (c) 13. (d) 14. (d) 15. (d) 16. (d)
17. (d) 18. (b) 19. (b) 20. (a) 21. (a) 22. (c) 23. (b) 24. (b)
25. (c)

Section-A JEE Advanced/ IIT-JEE

(A) FILL IN THE BLANKS :

1. The sum of integers from 1 to 100 that are divisible by 2 or = sum of integers from 1 to 100 divisible by 2 + sum of integers from 1 to 100 divisible by 5 – sum of integers from 1 to 100 divisible by 10

$$=(2+4+6+...+100)+(5+10+15+...+100)-(10+20+...+100)$$

$$=\frac{50}{2}[2\times 2+49\times 2]+\frac{20}{2}[2\times 5+19\times 5]$$

$$-\frac{10}{2}[2\times 10+9\times 10]$$

$$=2550+1050-550=3050$$

2. The given equation is

$$\log_7 \log_5 (\sqrt{x+5}+\sqrt{x})=0$$

$$\Rightarrow \log_5 (\sqrt{x+5}+\sqrt{x})=1$$

$$\Rightarrow \sqrt{x+5}+\sqrt{x}=5 \Rightarrow \sqrt{x+5}=5-\sqrt{x}$$

Squaring both sides

$$\Rightarrow x+5=25-10\sqrt{x}+x \Rightarrow 10\sqrt{x}=20$$

$$\Rightarrow \sqrt{x}=2 \Rightarrow x=4$$

3. When n is odd, let $n=2m+1$

$\therefore$ The req. sum

$$=1^2+2.2^2+3^2+2.4^2+...+2(2m)^2+(2m+1)^2$$

$$=\Sigma(2m+1)^2+4[1^2+2^2+3^2+...+m^2]$$

$$=\frac{(2m+1)(2m+2)(4m+2+1)}{6}+\frac{4m(m+1)(2m+1)}{6}$$

$$=\frac{(2m+1)(m+1)}{6}[2(4m+3)+4m]$$

$$=\frac{(2m+1)(2m+2)(6m+3)}{6}=\frac{(2m+1)^2(2m+2)}{2}$$

$$=\frac{n^2(n+1)}{2} \quad [\because 2m+1=n)$$

ALTERNATE SOLUTION :

$\because$ n is odd, last term = n^2

$\therefore$ req. sum

$= [1^2 + 2.2^2 + 3^2 + 2.4^2 + ... + 2(n-1)^2] + n^2$

$= \frac{(n-1)n^2}{2} + n^2$ [Using sum for $(n-1)$ to be even]

$= \frac{n^2(n+1)}{2}$

4. Let a and b be two positive numbers.

Then, $H.M. = \frac{2ab}{a+b}$ and $G.M. = \sqrt{ab}$

ATQ $HM : GM = 4 : 5$

$\therefore \quad \frac{2ab}{(a+b)\sqrt{ab}} = \frac{4}{5}$

$\Rightarrow \frac{2\sqrt{ab}}{a+b} = \frac{4}{5} \Rightarrow \frac{a+b+2\sqrt{ab}}{a+b-2\sqrt{ab}} = \frac{5+4}{5-4}$

$\Rightarrow \frac{(\sqrt{a}+\sqrt{b})^2}{(\sqrt{a}-\sqrt{b})^2} = \frac{9}{1} \Rightarrow \frac{\sqrt{a}+\sqrt{b}}{\sqrt{a}-\sqrt{b}} = 3, -3$

$\Rightarrow \frac{2\sqrt{a}}{2\sqrt{b}} = \frac{3+1}{3-1}, \frac{-3+1}{-3-1}$

$\Rightarrow \frac{\sqrt{a}}{\sqrt{b}} = 2, \frac{1}{2} \Rightarrow \frac{a}{b} = 4, \frac{1}{4}$

$a : b = 4 : 1$ or $1 : 4$

ALTERNATE SOLUTION:

Let for two + ve no.'s a and b, $\frac{a}{b} = m$

then $G = \sqrt{ab} = b\sqrt{m}$ and $H = \frac{2ab}{a+b} = \frac{2b\sqrt{m}}{b+bm}$

$\therefore \quad \frac{H}{G} = \frac{4}{5} \Rightarrow \frac{2\sqrt{m}}{m+1} = \frac{4}{5} \Rightarrow 5\sqrt{m} = 2m+2$

$\Rightarrow 2m - 5\sqrt{m} + 2 = 0 \Rightarrow (\sqrt{m}-2)(\sqrt{m}-\frac{1}{2}) = 0$

$\Rightarrow m = 4$ or $1/4$ $\therefore$ req. ratio = 4 : 1 or 1 : 4.

5. Since n is an odd integer, $(-1)^{n-1} = 1$ and $n-1, n-3, n-5$, ... are even integers.

We have

$n^3 - (n-1)^3 + (n-2)^3 - (n-3)^3 + ... + (-1)^{n-1}1^3$

$= n^3 + (n-1)^3 + (n-2)^3 + ... + 1^3 - 2[(n-1)^3 + (n-3)^3 + ... + 2^3]$

$= [n^3 + (n-1)^3 + (n-2)^3 + ... + 1^3]$

$-2 \times 2^3 \left[\left(\frac{n-1}{2}\right)^3 + \left(\frac{n-3}{2}\right)^3 + ... + 1^3\right]$

[$\because$ n – 1, n – 3.........are even integers]

Here the first square bracket contain the sum of cubes of 1st n natural numbers. Whereas the second square bracket contains the sum of the cubes of natural numbers from 1 to $\left(\frac{n-1}{2}\right)$, where $n-1, n-3$, ... are even integers. Using the formula for sum of cubes of 1st n natural numbers we get the summation

$= \left[\frac{n(n+1)}{2}\right]^2 - 16\left[\left\{\frac{1}{2}\left(\frac{n-1}{2}\right)\left(\frac{n-1}{2}+1\right)\right\}\right]^2$

$= \frac{1}{4}n^2(n+1)^2 - 16\frac{(n-1)^2(n+1)^2}{16 \times 4}$

$= \frac{1}{4}(n+1)^2[n^2 - (n-1)^2] = \frac{1}{4}(n+1)^2(2n-1)$

6. Let x be the A.M. between a and b and y, z be the G.M.'s between a and b where a and b are +ve.

Then a, x, b are in A.P. $\Rightarrow x = \frac{a+b}{2}$ and a, y, z, b are in G.P.

$y = ar$, $z = ar^2$ where $r = \sqrt[3]{\frac{b}{a}}$ also, $yz = ab$

Now, $\quad \frac{y^3+z^3}{xyz} = \frac{a^3r^3 + a^3r^6}{\left(\frac{a+b}{2}\right)ab} = \frac{2a^2r^3(1+r^3)}{b(a+b)}$

$= \frac{a^3.\frac{b}{a} + a^3\frac{b^2}{a^2}}{ab\left(\frac{a+b}{2}\right)} = \frac{2(a^2b+ab^2)}{(a^2b+ab^2)} = 2$

7. It is given

$p + q = 2, \quad pq = A$

and $\quad r + s = 18, \quad rs = B$

and it is given that p, q, r, are in A.P.

Therefore, let $p = a - 3s$, $q = a - d$, $r = a + d$ and $s = a + 3d$.

As $\quad p < q < r < s$,

we have $\quad d > 0$

Now, $\quad 2 = p + q = a - 3d + a - d = 2a - 4d$

$\Rightarrow a - 2d = 1$ ....(1)

Again $18 = r + s = a + d + a + 3d$

$\Rightarrow 18 = 2a + 4d \Rightarrow 9 = a + 2d$. ....(2)

Subtracting (1) from (2)

$\Rightarrow 8 = 4d$

$\Rightarrow 2 = d$ Putting in (2) we obtain a = 5

$\therefore \quad p = a - 3d = 5 - 6 = -1, \quad q = a - d = 5 - 2 = 3$

$r = a + d = 5 + 2 = 7, \quad s = a + 3d = 5 + 6 = 11$

Therefore, $A = pq = -3$ and $B = rs = 77$.

(C) MCQ's WITH ONE CORRECT ANSWER :

1. (c) $\because$ x, y, z are the pth, q^{th} and r^{th} terms of an AP.

$\therefore x = A + (p-1)D; y = A + (q-1)D;$

$z = A + (r-1)D$

$\Rightarrow x - y = (p-q)D; \; y - z = (q-r)D$

$z - x = (r-p)D$ ....(1)

where A is the first term and D is the common difference.

Also x, y, z are the pth, q^{th}, and r^{th} terms of a GP.

$\therefore x = AR^{p-1}, y = AR^{q-1}, z = AR^{r-1}$

$\therefore x^{y-z} y^{z-x} z^{x-y} = (AR^{p-1})^{y-z} (AR^{q-1})^{z-x} (AR^{r-1})^{x-y}$

$= A^{y-z+z-x+x-y} R^{(p-1)(y-z)+(q-1)(z-x)+(r-1)(x-y)}$

$= A^0 r^{(p-1)(q-r)D+(q-1)(r-p)D+(r-1)(p-q)D}$

$= A^0 R^0 = 1$

2. **(b)** $ar^2 = 4$ ....(1)

$a.\ ar.\ ar^2.\ ar^3.\ ar^4 = a^5 r^{10} = (ar^2)^5 = 4^5.$

3. **(c)** $2.\overline{357} = 2 + .357 + 0.000357 + ...\infty$

$$= 2 + \frac{357}{10^3} + \frac{357}{10^6} + ...\infty$$

$$= 2 + \frac{\frac{357}{10^3}}{1 - \frac{1}{10^3}} = 2 + \frac{357}{999} = \frac{2355}{999}$$

ALTERNATE SOLUTION :

Let $x = 2.\overline{357} \Rightarrow 1000x = 2357.\overline{357}$

On subtracting, we get

$$999x = 2355 \quad \Rightarrow \quad x = \frac{2355}{999}$$

4. **(a)** a, b, c are in G.P.

$b^2 = ac$ ....(1)

$ax^2 + 2bx + c = 0$

and $dx^2 + 2ex + f = 0$ have a common root

Let it be α, then $a\alpha^2 + 2b\alpha + c = 0$

$d\alpha^2 + 2e\alpha + f = 0$

$$\Rightarrow \frac{\alpha^2}{2(bf - ec)} = \frac{\alpha}{cd - af} = \frac{1}{2(ae - bd)}$$

$$\Rightarrow \alpha^2 = \frac{bf - ce}{ae - bd}; \ \alpha = \frac{cd - af}{2\,(ae - bd)}$$

Substituting the value of α, we get

$$\frac{(cd - af)^2}{4\,(ae - bd)^2} = \frac{bf - ce}{ae - bd}$$

$\Rightarrow (cd - af)^2 = 4\,(ae - bd)\,(bf - ce)$

Dividing both sides by a^2c^2 we get

$$\left(\frac{d}{a} - \frac{f}{c}\right)^2 = 4\left(\frac{e}{c} - \frac{bd}{ac}\right)\left(\frac{bf}{ac} - \frac{e}{a}\right)$$

$$\left(\frac{d}{a} - \frac{f}{c}\right)^2 = 4\left(\frac{e}{c} - \frac{d}{b}\right)\left(\frac{f}{b} - \frac{e}{a}\right) \quad \text{[Using eq. (1)]}$$

$$\Rightarrow \frac{d^2}{a^2} + \frac{f^2}{c^2} - \frac{2df}{ac} = \frac{4ef}{cb} - \frac{4e^2}{ac} - \frac{4df}{b^2} + \frac{4de}{ab}$$

$$\Rightarrow \frac{d^2}{a^2} + \frac{f^2}{c^2} + \frac{4e^2}{b^2} + 2\frac{d}{a}.\frac{f}{c} - 4\frac{e}{b}.\frac{f}{c} - 4.\frac{d}{a}.\frac{e}{b} = 0$$

[Using eq.(1)]

$$\Rightarrow \left(\frac{d}{a} + \frac{f}{c} - 2\frac{e}{b}\right)^2 = 0 \Rightarrow \frac{d}{a} + \frac{f}{c} = \frac{2e}{b}$$

$\Rightarrow \frac{d}{a}, \frac{e}{b}, \frac{f}{c}$ are in A.P.

5. **(c)** Let $S = \frac{1}{2} + \frac{3}{4} + \frac{7}{8} + \frac{15}{16} + ...n$ terms

$$= \left(1 - \frac{1}{2}\right) + \left(1 - \frac{1}{4}\right) + \left(1 - \frac{1}{8}\right) + \left(1 - \frac{1}{16}\right)....\text{n terms}$$

$$= (1 + 1 + 1 + ...n \text{ terms}) - \left(\frac{1}{2} + \frac{1}{2^2} + \frac{1}{2^3} + \frac{1}{2^4} + ... + \frac{1}{2^n}\right)$$

$$= n - \left[\frac{\frac{1}{2}\left(1 - \frac{1}{2^n}\right)}{1 - \frac{1}{2}}\right] = n - 1 + 2^{-n}$$

6. **(c)** We know that $\log_2 4 = 2$ and $\log_2 8 = 3$

$\therefore \log_2 7$ lies between 2 and 3

$\therefore \log_2 7$ is either rational or irrational but not integer or prime number.

If possible let $\log_2 7 = \frac{p}{q}$ (a rational number)

$\Rightarrow 2^{p/q} = 7 \Rightarrow 2^p = 7^q$

$\Rightarrow$ even number = odd number

$\therefore$ We get a contradiction, so assumption is wrong. Hence $\log_2 7$ must be an irrational number.

7. **(d)** In $(a + c)$, In $(a - c)$, In $(a - 2b + c)$ are in A.P.

$\Rightarrow a + c, a - c, a - 2b + c$ are in G.P.

$\Rightarrow (c - a)^2 = (a + c)\,(a - 2b + c)$

$\Rightarrow (c - a)^2 = (a + c)^2 - 2b\,(a + c)$

$\Rightarrow 2b\,(a + c) = (a + c)^2 - (c - a)^2$

$$\Rightarrow 2b\,(a + c) = 4ac \Rightarrow b = \frac{2ac}{a + c}$$

$\Rightarrow a, b, c$ are in H.P.

8. **(d)** $a_1 = h_1 = 2, a_{10} = h_{10} = 3$

$3 = a_{10} = 2 + 9d \Rightarrow d = 1/9$

$\therefore a_4 = 2 + 3d = 7/3$

$$3 = h_{10} \Rightarrow \frac{1}{3} = \frac{1}{h_{10}} = \frac{1}{2} + 9D \quad \therefore D = -\frac{1}{54}$$

$$\frac{1}{h_7} = \frac{1}{2} + 6D = \frac{1}{2} - \frac{1}{9} = \frac{7}{18} \qquad \therefore a_4 h_7 = \frac{7}{3} \times \frac{18}{7} = 6.$$

9. **(b)** $$\frac{1}{H} = \frac{1}{2}\left(\frac{1}{\alpha} + \frac{1}{\beta}\right) = \frac{\alpha + \beta}{2\alpha\beta} = \frac{(4 + \sqrt{5})}{(5 + \sqrt{2})}.\frac{(5 + \sqrt{2})}{(8 + 2\sqrt{5})}.\frac{1}{2} = \frac{1}{4}$$

$\therefore H = 4.$

10. **(d)** Sum = 4 and second term = 3/4, it is given that first term is a and common ratio r

$$\Rightarrow \frac{a}{1 - r} = 4 \text{ and } ar = 3/4 \Rightarrow r = \frac{3}{4a}$$

$$\text{Therefore, } \frac{a}{1 - \frac{3}{4a}} = 4 \Rightarrow \frac{4a^2}{4a - 3} = 4$$

or $a^2 - 4a + 3 = 0 \Rightarrow (a - 1)\,(a - 3) = 0$

$\Rightarrow a = 1$ or 3

When $a = 1, r = 3/4$ and when $a = 3, r = 1/4$

11. **(a)** α, β are the roots of $x^2 - x + p = 0$

$\therefore \quad \alpha + \beta = 1 \quad$ ....(1)

$\alpha\beta = p \quad$ ...(2)

γ, δ are the roots of $x^2 - 4x + q = 0$

$\therefore \quad \gamma + \delta = 4 \quad$ ....(3)

$\gamma\delta = q \quad$ ....(4)

$\alpha, \beta, \gamma, \delta$ are in G.P.

$\therefore$ Let $\alpha = a;\ \beta = ar,\ \gamma = ar^2,\ \delta = ar^3$.

Substituting these values in equations (1), (2), (3) and (4), we get

$a + ar = 1 \quad$ ....(5)

$a^2r = p \quad$ ....(6)

$ar^2 + ar^3 = 4 \quad$ ....(7)

$a^2r^5 = q \quad$ ....(8)

Dividing (7) by (5) we get

$$\frac{ar^2(1+r)}{a(1+r)} = \frac{4}{1} \Rightarrow r^2 = 4 \Rightarrow r = 2, -2$$

$$(5) \quad \Rightarrow a = \frac{1}{1+r} = \frac{1}{1+2} \text{ or } \frac{1}{1-2} = \frac{1}{3} \text{ or} -1$$

As p is an integer (given), r is also an integer (2 or – 2).

$\therefore (6) \Rightarrow a \neq \frac{1}{3}$. Hence $a = -1$ and $r = -2$.

$\therefore \quad p = (-1)^2 \times (-2) = -2$

$q = (-1)^2 \times (-2)^5 = -32$

12. **(d)** a, b, c, d are in A.P.

$\therefore \quad d, c, b, a$ are also in A.P.

$\Rightarrow \quad \frac{d}{abcd}, \frac{c}{abcd}, \frac{b}{abcd}, \frac{a}{abcd}$ are also in A.P.

$\Rightarrow \quad \frac{1}{abc}, \frac{1}{abd}, \frac{1}{acd}, \frac{1}{bcd}$ are in A.P.

$\Rightarrow \quad abc, abd, acd, bcd$ are in H.P.

13. **(c)** ATQ $2 + 5 + 8 + ... 2n$ terms $= 57 + 59 + 61 + ...\ n$ terms

$$\Rightarrow \quad \frac{2n}{2}[4 + (2n-1)\ 3] = \frac{n}{2}[114 + (n-1)\ 2]$$

$\Rightarrow \quad 6n + 1 = n + 56 \Rightarrow 5n = 55 \Rightarrow n = 11$

14. **(d)** Given that a, b, c are in A.P.

$\Rightarrow \quad 2b = a + c$

$\Rightarrow \quad$ but given $a + b + c = 3/2$

$\Rightarrow \quad 3b = 3/2$

$\Rightarrow \quad b = 1/2$ and then $a + c = 1$

Again a^2, b^2, c^2, are in G.P. $\Rightarrow b^4 = a^2c^2$

$$\Rightarrow \quad b^2 = \pm\ ac \Rightarrow ac = \frac{1}{4} \text{ or } -\frac{1}{4}$$

and $\quad a + c = 1 \quad$ ....(1)

Considering $a + c = 1$ and $ac = 1/4$

$\Rightarrow \quad (a-c)^2 = 1 - 1 = 0 \ \Rightarrow a = c$ but $a \neq c$ as given that $a < b < c$

$\therefore \quad$ We consider $a + c = 1$ and $ac = -1/4$

$\Rightarrow \quad (a-c)^2 = 1 + 1 = 2 \Rightarrow a - c = \pm\sqrt{2}$

but $a < c \Rightarrow a - c = -\sqrt{2} \quad$ ....(2)

Solving (1) and (2) we get $a = \frac{1}{2} - \frac{1}{\sqrt{2}}$

15. **(c)** $\frac{x}{1-r} = 5 \Rightarrow r = 1 - \frac{x}{5}$

Since G.P. contains infinite terms

$\therefore \quad -1 < r < 1$

$$\Rightarrow \quad -1 < 1 - \frac{x}{5} < 1 \Rightarrow -2 < -\frac{x}{5} < 0$$

$$\Rightarrow \quad -10 < x < 0. \Rightarrow 0 < \frac{x}{5} < 2$$

$\Rightarrow \quad 0 < x < 10.$

16. **(c)** In the quadratic equation $ax^2 + bx + c = 0$

$$\Delta = b^2 - 4ac \text{ and } \alpha + \beta = -\frac{b}{a}, \alpha\beta = \frac{c}{a}$$

$$\alpha^2 + \beta^2 = (\alpha + \beta)^2 - 2\alpha\beta$$

$$= \frac{b^2}{a^2} - \frac{2c}{a} = \frac{b^2 - 2ac}{a^2}$$

$$\text{and} \quad \alpha^3 + \beta^3 = -\frac{b^3}{a^3} - \frac{3c}{a}\left(-\frac{b}{a}\right) = -\left(\frac{b^3 - 3abc}{a^3}\right)$$

ATQ $\quad \alpha + \beta, \alpha^2 + \beta^2, \alpha^3 + \beta^3$ are in G.P.

$$\Rightarrow \quad -\frac{b}{a}, \frac{b^2 - 2ac}{a^2}, -\frac{(b^3 - 3\ abc)}{a^3} \text{ are in G.P.}$$

$$\Rightarrow \quad \left(\frac{b^2 - 2\ ac}{a^2}\right)^2 = \frac{b}{a}\left(\frac{b^3 - 3\ abc}{a^3}\right)$$

$\Rightarrow \quad b^4 + 4a^2c^2 - 4ab^2c = b^4 - 3ab^2c$

$\Rightarrow \quad 4a^2c^2 - ab^2c = 0 \Rightarrow ac\,\Delta = 0$

$\Rightarrow \quad c\Delta = 0 \quad (\because$ In quadratic $a \neq 0)$

ALTERNATE SOLUTION :

ATQ $\quad (\alpha^2 + \beta^2)^2 = (\alpha + \beta)\ (\alpha^3 + \beta^3)$

$\Rightarrow \quad \alpha^4 + \beta^4 + 2\alpha^2\beta^2 = \alpha^4 + \beta^4 + \alpha\beta^3 + \beta\alpha^3$

$\Rightarrow \quad \alpha\beta\ (\alpha^2 + \beta^2 - 2\alpha\beta) = 0$

$\Rightarrow \quad \alpha\beta\ (\alpha - \beta)^2 = 0$

$\Rightarrow \quad \alpha\beta = 0$ or $(\alpha - \beta) = 0$

$\Rightarrow \quad \frac{c}{a} = 0$ or $\alpha = \beta$

$\Rightarrow \quad c = 0$ or $\Delta = 0$ (equal roots)

$\therefore \quad c\Delta = 0$ is the correct option.

ALTERNATE SOLUTION :

ATQ $\ \alpha + \beta, \alpha^2 + \beta^2$ and $\alpha^3 + \beta^3$ are in G.P.

We observe that the above form in G.P. if $\beta = 0$ i.e. one root is zero which happens when $c = 0$. Again $2\alpha, 2\alpha^2, 2\alpha^3$ are also in G.P. (for $\alpha = \beta$)

$\therefore \quad$ G.P. is formed when roots are equal i.e. $\Delta = 0$

Therefore we get either $c = 0$ or $\Delta = 0$

$\Rightarrow \quad c\Delta = 0.$

17. (c) Given that for an A.P, $S_n = cn^2$

Then $T_n = S_n - S_{n-1} = cn^2 - c(n-1)^2$

$= (2n-1)c$

$\therefore$ Sum of squares of n terms of this A.P

$= \sum T_n^2 = \sum (2n-1)^2 . c^2$

$= c^2\left[4\sum n^2 - 4\sum n + n\right]$

$= c^2\left[\frac{4n(n+1)(2n+1)}{6} - \frac{4n(n+1)}{2} + n\right]$

$= c^2 n\left[\frac{2(2n^2+3n+1)-6(n+1)+3}{3}\right]$

$= c^2 n\left[\frac{4n^2-1}{3}\right] = \frac{n(4n^2-1)c^2}{3}$

18. (d) $\because a_1, a_2, a_3, ...$ are in H.P.

$\therefore \frac{1}{a_1}, \frac{1}{a_2}, \frac{1}{a_3}$ .... are in A.P.

$\therefore \frac{1}{a_1} = \frac{1}{5}$ and $\frac{1}{a_{20}} = \frac{1}{25}$

$\frac{1}{a_1} + 19d = \frac{1}{a_{20}} \Rightarrow \frac{1}{5} + 19d = \frac{1}{25} \Rightarrow d = \frac{-4}{475}$

Now $\frac{1}{a_n} = \frac{1}{5} + (n-1)\left(\frac{-4}{475}\right)$

Clearly $a_n < 0$ if $\frac{1}{a_n} < 0 \Rightarrow \frac{1}{5} - \frac{4n}{475} + \frac{4}{475} < 0$

$\Rightarrow -4n < -99$ or $n > \frac{99}{4} = 24\frac{3}{4}$ $\therefore n \geq 25$

Hence least value of n is 25.

D MCQ's WITH ONE OR MORE THAN ONE CORRECT :

1. **(a,b,d)** Let x be the first term and y the $(2n-1)$th terms of AP, GP and HP whose nth terms are a, b, c respectively.
For AP, $y = x + (2n-2)d$

$\Rightarrow d = \frac{y-x}{2(n-1)}$

$\therefore a = x + (n-1)d = x + \frac{1}{2}(y-x) = \frac{1}{2}(x+y)$ ....(1)

For G.P. $y = xr^{2n-2} \Rightarrow r = \left(\frac{y}{x}\right)^{\frac{1}{2n-2}}$

$\therefore b = xr^{n-1} = x.\left(\frac{y}{x}\right)^{1/2} = \sqrt{xy}$ ....(2)

For H.P. $\frac{1}{y} = \frac{1}{x} + (2n-2)d_1$

$\Rightarrow d_1 = \frac{x-y}{2xy(n-1)}$

$\therefore \frac{1}{c} = \frac{1}{x} + (n-1)d_1 = \frac{1}{x} + \frac{x-y}{2xy}$

$\frac{1}{c} = \frac{x+y}{2xy} \Rightarrow c = \frac{2xy}{x+y}$ ....(3)

Thus from (1), (2) and (3), a, b, c are A.M., G.M. and H.M. respectively of x and y.

NOTE THIS STEP

We know that $AM \geq GM \geq HM$

$\therefore a \geq b \geq c$

Also$(G.M.)^2 = (A.M.)\ (H.M.)$

$\Rightarrow b^2 = ac$

2. **(b, c)** We have for $0 < \phi < \frac{\pi}{2}$

$x = \sum_{n=0}^{\infty} \cos^{2n}\phi = 1 + \cos^2\phi + \cos^4\phi + ...\infty$

$\frac{1}{1-\cos^2\phi} = \frac{1}{\sin^2\phi}$ ....(1)

[Using sum of infinite G.P. $\cos^2\alpha$ being < 1]

$y = \sum_{n=0}^{\infty} \sin^{2n}\phi = 1 + \sin^2\phi + \sin^4\phi + ...\infty$

$= \frac{1}{1-\sin^2\phi} = \frac{1}{\cos^2\phi}$ ....(2)

$z = \sum_{n=0}^{\infty} \cos^{2n}\phi \sin^{2n}\phi$

$= 1 + \cos^2\phi\sin^2\phi + \cos^4\phi\sin^4\phi + ...\infty$

$= \frac{1}{1-\cos^2\phi\sin^2\phi}$ ....(3)

Substituting the values of $\cos^2\phi$ and $\sin^2\phi$ in (3), from (1) and (2), we get

$z = \frac{1}{1-\frac{1}{x}.\frac{1}{y}} \Rightarrow z = \frac{xy}{xy-1} \Rightarrow xyz - z = xy \Rightarrow xyz = xy + z.$

Also, $x + y + z = \frac{1}{\cos^2\phi} + \frac{1}{\sin^2\phi} + \frac{1}{1-\cos^2\phi\sin^2\phi}$

$$\frac{[\sin^2\phi(1-\cos^2\phi\sin^2\phi) + \cos^2\phi(1-\cos^2\phi\sin^2\phi) + \cos^2\phi\sin^2\phi]}{\cos^2\phi\sin^2\phi(1-\cos^2\phi\sin^2\phi)}$$

$$= \frac{(\sin^2\phi+\cos^2\phi)(1-\cos^2\phi\sin^2\phi) + \cos^2\phi\sin^2\phi}{\cos^2\phi\sin^2\phi(1-\cos^2\phi\sin^2\phi)}$$

$$= \frac{1}{\cos^2\phi\sin^2\phi(1-\cos^2\phi\sin^2\phi)} = xyz$$

Thus (b) and (c) both are correct.

3. **(b)** Putting $\theta = 0$, we get $b_0 = 0$

$\therefore \quad \sin n\theta = \sum_{r=1}^{n} b_r \sin^r \theta \Rightarrow \frac{\sin n\theta}{\sin\theta} = \sum_{r=1}^{n} b_r (\sin\theta)^{r-1}$

$= b_1 + b_2 \sin\theta + b_3 \sin^2\theta + ... + b_n \sin^{n-1}\theta$

Taking limit as $\theta \to 0$, we obtain

$\lim_{\theta\to 0} \frac{\sin n\theta}{\sin\theta} = b_1 \Rightarrow b_1 = n.$

4. **(c)** $T_m = a + (m-1)\,d = 1/n$ and $t_n = a + (n-1)d = 1/m$

$\Rightarrow (m-n)\,d = 1/n - 1/m = (m-n)/mn \Rightarrow d = 1/mn$

$\Rightarrow a = \frac{1}{mn}$

$\therefore \quad t_{mn} = a + (mn-1)\,d = \frac{1}{mn} + (mn-1)\frac{1}{mn}$

$= \frac{1}{mn} + 1 - \frac{1}{mn} = 1$

5. **(b)** Let the common ratio of the G.P. be r.
Then $y = xr$ and $z = xr^2$
$\ln y = \ln x + \ln r$ and $\ln z = \ln x + 2\ln r$
Putting $A = 1 + \ln x$, $D = \ln r$ in given numbers
Then

$\frac{1}{1+\ln x} = \frac{1}{A},$

$\frac{1}{1+\ln y} = \frac{1}{1+\ln xr} = \frac{1}{1+\ln x + \ln r},$

$= \frac{1}{A+D}$ and $\frac{1}{1+\ln z} = \frac{1}{1+\ln x + 2\ln r} = \frac{1}{A+2D}$

Therefore, $\frac{1}{1+\ln x}, \frac{1}{1+\ln y}, \frac{1}{1+\ln z}$ are in H.P.

ALTERNATE SOLUTION :
NOTE THIS STEP
If x, y, z are in G.P. $(x, y, z > 1)$; $\log x, \log y, \log z$ will be in A.P.

$\Rightarrow$ $1 + \log x, 1 + \log y, 1 + \log z$ will also be in A.P.

$\Rightarrow \frac{1}{1+\log x}, \frac{1}{1+\log y}, \frac{1}{1+\log z}$ will be in H.P.

6. **(a, d)** We have

$a(n) = 1 + \frac{1}{2} + \frac{1}{3} + \frac{1}{4} + \frac{1}{5} + ... + \frac{1}{2^n - 1}$

$= 1 + \left(\frac{1}{2} + \frac{1}{3}\right) + \left(\frac{1}{4} + \frac{1}{5} + \frac{1}{6} + \frac{1}{7}\right) + \left(\frac{1}{8} + ... + \frac{1}{15}\right) + ... + \left(\frac{1}{2^{n-1}} + ... + \frac{1}{2^n - 1}\right)$

$< 1 + \left(\frac{1}{2} + \frac{1}{2}\right) + \left(\frac{1}{4} + \frac{1}{4} + ... + \frac{1}{4}\right) + \left(\frac{1}{8} + \frac{1}{8} + ... + \frac{1}{8}\right) + .. + \left(\frac{1}{2^{n-1}} + .. + \frac{1}{2^{n-1}}\right)$

$< 1 + \frac{2}{2} + \frac{4}{4} + \frac{8}{8} + ... + \frac{2^{n-1}}{2^{n-1}} = 1 + 1 + ... + 1 = n$

Thus, $a(100) < 100$
Also

$a(n) = 1 + \frac{1}{2} + \left(\frac{1}{3} + \frac{1}{4}\right) + \left(\frac{1}{5} + \frac{1}{6} + \frac{1}{7} + \frac{1}{8}\right) + ... + \left(\frac{1}{2^{n-1}+1} + .. + \frac{1}{2^n}\right) - \frac{1}{2^n}$

$> 1 + \frac{1}{2} + \left(\frac{1}{4} + \frac{1}{4}\right) + \left(\frac{1}{8} + \frac{1}{8} + ... + \frac{1}{8}\right) + ... + \left(\frac{1}{2^n} + .. + \frac{1}{2^n}\right) - \frac{1}{2^n}$

$= 1 + \frac{1}{2} + \frac{2}{4} + \frac{4}{8} + ... + \frac{2^{n-1}}{2^n} - \frac{1}{2^n}$

$= 1 + \underbrace{\left(\frac{1}{2} + \frac{1}{2} + \frac{1}{2} + ... + \frac{1}{2}\right)}_{n \text{ times}} - \frac{1}{2^n}$

$= 1 + \frac{n}{2} - \frac{1}{2^n} = \left(1 - \frac{1}{2^n}\right) + \frac{n}{2}$

Thus, $a(200) > \left(1 - \frac{1}{2^{200}}\right) + \frac{200}{2} > 100,$

i.e. $a(200) > 100$.

7. **(b, d)**

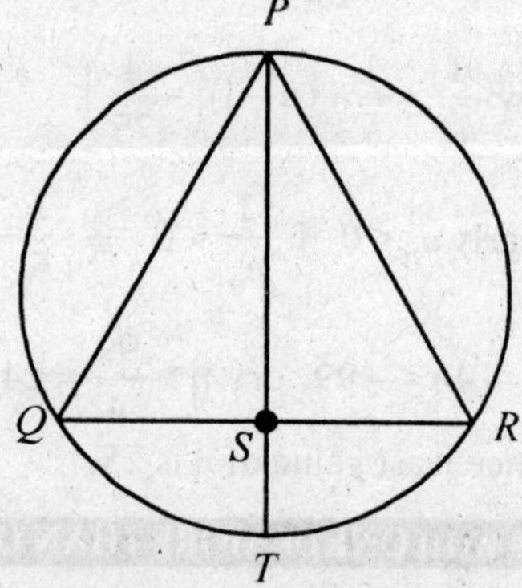

We know by geometry $PS \times ST = QS \times SR$ ...(1)

$\because$ S is not the centre of circulm circle,

$PS \neq ST$

And we know that for two unequal real numbers. H.M. < G.M.

$\Rightarrow \frac{2}{\frac{1}{PS} + \frac{1}{ST}} < \sqrt{PS \times ST} \Rightarrow \frac{1}{PS} + \frac{1}{ST} > \frac{2}{\sqrt{PS \times ST}}$

$\Rightarrow \frac{1}{PS} + \frac{1}{ST} > \frac{2}{\sqrt{QS \times SR}}$ [using eqn (1)] ...(2)

$\therefore$ (b) is the correct option.

Also $\sqrt{QS \times SR} < \frac{QS + SR}{2}$ $(\because GM < AM)$

$\Rightarrow \frac{1}{\sqrt{QS \times SR}} > \frac{2}{QR} \Rightarrow \frac{2}{\sqrt{QS \times SR}} > \frac{4}{QR}$ ...(3)

From equations (2) and (3) we get $\frac{1}{PS}+\frac{1}{ST}>\frac{4}{QR}$

$\therefore$ (d) is also the correct option.

8. **(a,d)** We have $S_n=\sum_{k=1}^{n}\frac{n}{n^2+kn+k^2}$

and $T_n=\sum_{k=0}^{n-1}\frac{n}{n^2+kn+k^2}$ $n=1,2,3........$

For $n=1$ we get

$S_1=\frac{1}{1+1+1}=\frac{1}{3}=0.3$ and $T_1=\frac{1}{1+0}=1$

Also $\frac{\pi}{3\sqrt{3}}=\frac{\pi\sqrt{3}}{9}=\frac{3.14\times1.73}{9}=0.34\times1.73=0.58$

$\therefore S_1<\frac{\pi}{3\sqrt{3}}<T_1$, $\therefore S_n<\frac{\pi}{3\sqrt{3}}$ and $T_n>\frac{\pi}{3\sqrt{3}}$

9. **(a,d)** $S_n=-1^2-2^2+3^2+4^2-5^2-6^2+....$

$=(3^2+7^2+11^2+...)+(4^2+8^2+12^2+...)$

$-(1^2+5^2+9^2+...)-(2^2+6^2+10^2+...)$

$$=\sum_{r=1}^{n}(4r-1)^2+\sum_{r=1}^{n}(4r)^2-\sum_{r=1}^{n}(4r-3)^2-\sum_{r=1}^{n}(4r-2)^2$$

$$=\left[\sum_{r=1}^{n}(4r-1)^2-(4r-3)^2\right]+4\left[\sum_{r=1}^{n}(2r)^2-(2r-1)^2\right]$$

$$=8\sum_{r=1}^{n}(2r-1)+4\sum_{r=1}^{n}(4r-1)$$

$$=8\left[2\frac{n(n+1)}{2}-n\right]+4\left[4\frac{n(n+1)}{2}-n\right]$$

$=8n^2+8n^2+4n=16n^2+4n$

For $n=8$, $16n^2+4n=1056$

and for $n=9$, $16n^2+4n=1332$

(E) SUBJECTIVE PROBLEMS :

1. Let the two numbers be a and b, then

$\frac{2ab}{a+b}=4...(1);\frac{a+b}{2}=A;\sqrt{ab}=G$

Also $2A+G^2=27\Rightarrow a+b+ab=27$ ....(2)

Putting $ab=27-(a+b)$ in eqn. (1), we get

$\frac{54-2(a+b)}{a+b}=4\Rightarrow a+b=9$ then $ab=27-9=18$

Solving the two we get $a=6, b=3$ or $a=3, b=6$, which are the required numbers.

2. Let there be n sides in the polygon.

Then by geometry, sum of all n interior angles of polygon $=(n-2)\times180°$

Also the angles are in A.P. with the smallest angle $=120°$, common difference $=5°$

$\therefore$ Sum of all interior angles of polygon

$=\frac{n}{2}[2\times120+(n-1)\times5]$

Thus we should have

$\frac{n}{2}[2\times120+(n-1)\times5]=(n-2)\times180$

$\Rightarrow \frac{n}{2}[5n+235]=(n-2)\times180$

$\Rightarrow 5n^2+235n=360n-720$

$\Rightarrow 5n^2-125n+720=0\Rightarrow n^2-25n+144=0$

$\Rightarrow (n-16)(n-9)=0\Rightarrow n=16, 9$

Also if $n=16$ then 16th angle $=120+15\times5=195°>180°$

$\therefore$ not possible.

Hence $n=9$.

3. $a_1, a_2.... a_n$ are in A.P. $\forall\ a_i>0$

$\therefore$ $a_1-a_2=a_2-a_3=...=a_{n-1}-a_n=d$ (a constant) ....(1)

Now we have to prove

$$\frac{1}{\sqrt{a_1}+\sqrt{a_2}}+\frac{1}{\sqrt{a_2}+\sqrt{a_3}}+.....+\frac{1}{\sqrt{a_{n-1}}+\sqrt{a_n}}=\frac{n-1}{\sqrt{a_1}+\sqrt{a_n}}$$

$$\text{L.H.S.}=\frac{1}{\sqrt{a_1}+\sqrt{a_2}}+\frac{1}{\sqrt{a_2}+\sqrt{a_3}}+.....+\frac{1}{\sqrt{a_{n-1}}+\sqrt{a_n}}$$

$$=\frac{\sqrt{a_1}-\sqrt{a_2}}{a_1-a_2}+\frac{\sqrt{a_2}-\sqrt{a_3}}{a_2-a_3}+.....+\frac{\sqrt{a_{n-1}}-\sqrt{a_n}}{a_{n-1}-a_n}$$

$$=\frac{1}{d}[\sqrt{a_1}-\sqrt{a_2}+\sqrt{a_2}-\sqrt{a_3}+.....+\sqrt{a_{n-1}}-\sqrt{a_n}]$$

(Using equation (1))

$$=\frac{1}{d}[\sqrt{a_1}-\sqrt{a_n}]$$

$$=\frac{a_1-a_n}{d(\sqrt{a_1}+\sqrt{a_n})}=\frac{(n-1)d}{d\sqrt{a_1}+\sqrt{a_n}}=\frac{n-1}{\sqrt{a_1}+\sqrt{a_n}}$$

= RHS. Hence Proved.

4. If possible let for a G.P.

$T_p=27=AR^{p-1}$ ....(1)

$T_q=8=AR^{q-1}$ ....(2)

$T_r=12=AR^{r-1}$ ....(3)

From (1) and (2)

$R^{p-q}=\frac{27}{8}\Rightarrow R^{p-q}=(3/2)^3$ ....(4)

From (2) and (3):

$R^{q-r}=\frac{8}{12}\Rightarrow R^{q-r}=(3/2)^{-1}$ ....(5)

From (4) and (5):

$R=3/2; p-q=3;\ q-r=-1$

$p-2q+r=4;\quad p,q,r\in N$ ....(6)

As there can be infinite natural numbers for p, q and r to satisfy equation (6)

$\therefore$ There can be infinite G.P's.

5. $2<a, b, c<18\qquad a+b+c=25$ ....(1)

2, a, b are in AP $\Rightarrow 2a=b+2$

$\Rightarrow 2a-b=2$ ....(2)

b, c, 18 are in GP $\Rightarrow c^2=18b$ ....(3)

From (2) $\Rightarrow a=\frac{b+2}{2}$

(1) $\Rightarrow \frac{b+2}{2}+b+c=25 \Rightarrow 3b=48-2c$

(3) $\Rightarrow c^2=6\ (48-2c) \Rightarrow c^2+12c-288=0$

$\Rightarrow\ c=12, -24$ (rejected)

$\Rightarrow\ a=5,\ b=8,\ c=12$

6. Given that $a, b, c>0$

We know for +ve numbers A.M. ≥ G.M.

∴ For +ve numbers a, b, c we get

$$\frac{a+b+c}{3} \geq \sqrt[3]{abc} \quad ....(1)$$

Also for +ve numbers $\frac{1}{a}, \frac{1}{b}, \frac{1}{c}$, we get

$$\frac{\frac{1}{a}+\frac{1}{b}+\frac{1}{c}}{3} \geq \sqrt[3]{\frac{1}{abc}} \quad ....(2)$$

Multiplying in eqs (1) and (2) we get

$$\left(\frac{a+b+c}{3}\right)\left(\frac{\frac{1}{a}+\frac{1}{b}+\frac{1}{c}}{3}\right) \geq \sqrt[3]{abc} \times \frac{1}{\sqrt[3]{abc}}$$

$$\Rightarrow \quad (a+b+c)\left(\frac{1}{a}+\frac{1}{b}+\frac{1}{c}\right) \geq 9 \qquad \textbf{Proved.}$$

7. Given that $n=p_1^{\alpha_1}.\ p_2^{\alpha_2}.\ p_3^{\alpha_3}....\ p_k^{\alpha_k}$ ....(1)

Where $n \in N$ and $p_1, p_2, p_3, ...p_k$ are distinct prime numbers.

Taking log on both sides of eq. (1), we get

$\log n=\alpha_1 \log p_1+\alpha_2 \log p_2+....+\alpha_k \log p_k$ ....(2)

Since every prime number is such that

$p_i \geq 2$

∴ $\log_e p_i \geq \log_e 2$ ....(3)

$\forall\ i=1\,(1)\,k$

Using (2) and (3) we get

$\log n \geq \alpha_1 \log 2+\alpha_2 \log 2+\alpha_3 \log 2+....+\alpha_k \log 2$

$\Rightarrow\ \log n \geq (\alpha_1+\alpha_2+....+\alpha_k) \log 2$

$\Rightarrow\ \log n \geq k \log 2$ **Proved.**

8. The given series is

$$\sum_{r=0}^{n}(-1)^r\,{}^nC_r\left[\frac{1}{2^r}+\frac{3^r}{2^{2r}}+\frac{7^r}{2^{3r}}+\frac{15^r}{2^{4r}}+....\text{ up to } m \text{ terms}\right]$$

$$\sum_{r=0}^{n}(-1)^r\,{}^nC_r\left[\left(\frac{1}{2}\right)^r+\left(\frac{3}{4}\right)^r+\left(\frac{7}{8}\right)^r+\left(\frac{15}{16}\right)^r+....\text{to } m \text{ terms}\right]$$

Now, $\sum_{r=0}^{n}(-1)^r\,{}^nC_r\left(\frac{1}{2}\right)^r=1-{}^nC_1.\frac{1}{2}+{}^nC_2.\frac{1}{2^2}-{}^nC_3.\frac{1}{2^3}+...$

$$=\left(1-\frac{1}{2}\right)^n=\frac{1}{2^n}$$

Similarly, $\sum_{r=0}^{n}(-1)^r\,{}^nC_r\left(\frac{3}{4}\right)^r=\left(1-\frac{3}{4}\right)^n=\frac{1}{4^n}$ etc.

Hence the given series is,

$$=\frac{1}{2^n}+\frac{1}{4^n}+\frac{1}{8^n}+\frac{1}{16^n}+....\text{to } m \text{ terms}$$

$$=\frac{\frac{1}{2^n}\left(1-\left(\frac{1}{2^n}\right)^m\right)}{1-\frac{1}{2^n}} \qquad \text{[Summing the G.P.]}$$

$$=\frac{2^{mn}-1}{2^{mn}(2^n-1)}$$

9. Let the three distinct real numbers be $\frac{\alpha}{r}, \alpha, \alpha r$.

Since sum of squares of three numbers be S^2.

$$\therefore \quad \frac{\alpha^2}{r^2}+\alpha^2+a^2r^2=S^2$$

$$\text{or} \quad \frac{\alpha^2(1+r^2+r^4)}{r^2}=S^2 \quad ....(1)$$

Sum of numbers is aS

$$\therefore \quad \frac{\alpha(1+r+r^2)}{r}=aS \quad ....(2)$$

Dividing eq. (1) by the square of (2), we get

$$\frac{\alpha^2(1+r^2+r^4)}{r^2}\times\frac{r^2}{\alpha^2(1+r+r^2)^2}=\frac{S^2}{a^2S^2}$$

$$\frac{(1+2r^2+r^4)-r^2}{(1+r+r^2)^2}=\frac{1}{a^2},\ \frac{(1+r+r^2)(1-r+r^2)}{(1+r+r^2)^2}=\frac{1}{a^2}$$

$$\frac{1-r+r^2}{1+r+r^2}=\frac{1}{a^2} \Rightarrow a^2r^2-a^2r+a^2=1+r+r^2$$

$$\Rightarrow\ (a^2-1)\,r^2-(a^2+1)\,r+(a^2-1)=0$$

$$\Rightarrow\ r^2+\left(\frac{1+a^2}{1-a^2}\right)r+1=0 \quad ....(3)$$

For real values of r, $D \geq 0$

$$\Rightarrow\ \left(\frac{1+a^2}{1-a^2}\right)^2-4 \geq 0$$

$$\Rightarrow\ 1+2a^2+a^2-4+8a^2-4a^4 \geq 0$$

$$\Rightarrow\ 3a^4-10a^2+3 \leq 0 \Rightarrow (3a^2-1)(a^2-3) \leq 0$$

$$\Rightarrow\ \left(a^2-\frac{1}{3}\right)(a^2-3) \leq 0$$

Clearly the above inequality holds for

$$\frac{1}{3} \leq a^2 \leq 3$$

But from eq. (3), $a \neq 1$

$\therefore \quad a^2 \in \left(\frac{1}{3}, 1\right) \cup (1, 3)$

10. The given equation is

$\log_{(2x+3)}(6x^2+23+21)$

$= 4 - \log_{3x+7}(4x^2+12x+9)$

$\Rightarrow \log_{(2x+3)}(6x^2+23x+21)$
$+ \log_{(3x+7)}(4x^2+12x+9) = 4$

$\Rightarrow \log_{(2x+3)}(2x+3)(3x+7) + \log_{(3x+7)}(2x+3)^2\, x = 4$

$\Rightarrow 1 + \log_{(2x+3)}(3x+7) + 2\log_{(3x+7)}(2x+3) = 4$

[Using $\log ab = \log a + \log b$ and $\log a^n = n\log a$]

NOTE THIS STEP

$$\Rightarrow \log_{(2x+3)}(3x+7) + \frac{2}{\log_{(2x+3)}(3x+7)} = 3$$

[Using $\log_a b = \frac{1}{\log_b a}$]

Let $\log_{(2x+3)}(3x+7) = y$ ....(1)

$\Rightarrow y + \frac{2}{y} = 3 \Rightarrow y^2 - 3y + 2 = 0$

$\Rightarrow (y-1)(y-2) = 0 \Rightarrow y = 1, 2$

Substituting the values of y in (1), we get

$\Rightarrow \log_{(2x+3)}(3x+7) = 1$ and $\log_{(2x+3)}(3x+7) = 2$

$\Rightarrow 3x+7 = 2x+3$ and $3x+7 = (2x+3)^2$

$\Rightarrow x = -4$ and $4x^2+9x+2 = 0$

$\Rightarrow x = -4$ and $(x+2)(4x+1) = 0$

$\Rightarrow x = -4$ and $x = -2, x = -\frac{1}{4}$

As $\log_a x$ is defined for $x > 0$ and $a > 0$ ($a \neq 1$), the possible value of x should satisfy all of the following inequalities :

$\Rightarrow 2x+3 > 0$ and $3x+7 > 0$

Also $2x+3 \neq 1$ and $3x+7 \neq 1$

Out of $x = -4, x = -2$ and $x = -\frac{1}{4}$ only $x = -\frac{1}{4}$ satisfies the above inequalities.

So only solution is $x = -\frac{1}{4}$.

11. Given that $\log_3 2, \log_3(2^x - 5), \log_3(2^x - 7/2)$ are in A.P.

$\Rightarrow 2\log_3(2^x-5) = \log_3 2 + \log_3(2^x - 7/2)$

$\Rightarrow (2^x-5)^2 = 2\left(2^x - \frac{7}{2}\right)$

$\Rightarrow (2^x)^2 - 10.2^x + 25 - 2.2^x + 7 = 0$

$\Rightarrow (2^x)^2 - 12.2^x + 32 = 0$

Let $2^x = y$, then we get,

$y^2 - 12y + 32 = 0 \Rightarrow (y-4)(y-8) = 0$

$\Rightarrow y = 4$ or $8 \Rightarrow 2^x = 2^2$ or $2^3 \Rightarrow x = 2$ or 3

But for $\log_3(2^x - 5)$ and $\log_3(2^x - 7/2)$ to be defined

$2^x - 5 > 0$ and $2^x - 7/2 > 0$

$\Rightarrow 2^x > 5$ and $2^x > 7/2$

$\Rightarrow 2^x > 5$

$\Rightarrow x \neq 2$ and therefore $x = 3$.

12. Let a and b be two numbers and $A_1, A_2, A_3, \ldots A_n$ be n A.M's between a and b.

Then a, $A_1, A_2, \ldots A_n$, b are in A.P.

There are $(n+2)$ terms in the series, so that

$a + (n+1)d = b \Rightarrow d = \frac{b-a}{n+1}$

$\therefore \quad A_1 = a + \frac{b-a}{n+1} = \frac{an+b}{n+1}$

$\therefore$ ATQ $p = \frac{an+b}{n+1}$ ....(1)

The first H.M. between a and b, when nHM's are inserted between a and b can be obtained by replacing a by $\frac{1}{a}$ and b by $\frac{1}{b}$ in eq. (1) and then taking its reciptocal.

Therefore, $q = \dfrac{1}{\dfrac{\left(\frac{1}{a}\right)n + \frac{1}{b}}{n+1}} = \frac{(n+1)ab}{bn+a}$

$\therefore \quad q = \frac{(n+1)ab}{a+bn}$ ....(2)

We have to prove that q cannot lie between p *and* $\frac{(n+1)^2}{(n-1)^2}p$.

Now, $n+1 > n-1 \Rightarrow \frac{n+1}{n-1} > 1$

$\Rightarrow \left(\frac{n+1}{n-1}\right)^2 > 1$ or $p\left(\frac{n+1}{n-1}\right)^2 > p$

$\Rightarrow p < p\left(\frac{n+1}{n-1}\right)^2$ ....(3)

Now to prove the given, we have to show that q is less than p.

For this, let, $\frac{p}{q} = \frac{(na+b)(nb+a)}{(n+1)^2 ab}$

$\Rightarrow \frac{p}{q} - 1 = \frac{n(a^2+b^2) + ab(n^2+1) - (n+1)^2 ab}{(n+1)^2 ab}$

$\Rightarrow \frac{p}{q} - 1 = \frac{n(a^2+b^2-2ab)}{(n+1)^2 ab}$

$\Rightarrow \frac{p}{q} - 1 = \frac{n}{(n+1)^2}\left(\frac{a-b}{\sqrt{ab}}\right)^2$

$= \frac{n}{(n+1)^2}\left(\sqrt{\frac{a}{b}} - \sqrt{\frac{b}{a}}\right)^2 \Rightarrow \frac{p}{q} - 1 > 0$

$\Rightarrow$ (provided a and b and hence p and q are +ve)

$p > q$ ....(4)

From 3 and (4), we get,

$$q < p < \left(\frac{n+1}{n+1}\right)^2 p$$

$\therefore$ q can not lie between p and $\left(\frac{n+1}{n+1}\right)^2 p$, if a and b are +ve numbers.

ALTERNATE SOLUTION :

After getting equations (1) and (2) as in the previous method, substitute $b = p(n+1) - an$ [from (1)] in equation (2) to get

$aq + nq\,[p\,(n+1) - an] = (n+1)\,a[p(n+1) - an]$

$\Rightarrow\; a^2 n\,(n+1) + a\,[q\,(1-n^2) - p(n+1)^2] + npq(n+1) = 0$

$\Rightarrow\; na^2 - [(n+1)p + (n-1)q]\;a + npq = 0$

$\Rightarrow\; D \geq 0 \qquad (\because a \text{ is real})$

$\Rightarrow\; [n+1)\,p + (n-1)q]^2 - 4n^2pq \geq 0$

$\Rightarrow\; (n-1)^2q^2 + \{2(n^2-1) - 4n^2\}\,pq + (n+1)^2p^2 \geq 0$

$$\Rightarrow\; q^2 - 2\frac{n^2+1}{(n-1)^2}pq + \left(\frac{n+1}{n-1}\right)^2 p^2 \geq 0$$

$$\Rightarrow\; \left[q - p\left(\frac{n+1}{n-1}\right)^2\right][q-p] \geq 0$$

[On factorizing by discriminent method] $\Rightarrow$ q can not lie between p and $p\left(\frac{n+1}{n-1}\right)^2$.

13. ATQ we have,

$$S_1 = 1 + \frac{1}{2} + \left(\frac{1}{2}\right)^2 + \left(\frac{1}{2}\right)^3 + \ldots.\infty$$

$$S_2 = 2 + 2.\frac{1}{3} + 2\left(\frac{1}{3}\right)^2 + \ldots.\infty$$

....

....

$$S_3 = 3 + 3.\frac{1}{4} + 3\left(\frac{1}{4}\right)^2 + \ldots.\infty$$

$$S_n = n + n.\frac{1}{n+1} + n\left(\frac{1}{n+1}\right)^2 + \ldots.\infty$$

$$\Rightarrow\; S_1 = \frac{1}{1-\frac{1}{2}} = 2 \qquad \left[\text{Using } S_\infty = \frac{a}{1-r}\right]$$

$$S_2 = \frac{2}{1-\frac{1}{3}} = 3, \qquad S_3 = \frac{3}{1-\frac{1}{4}} = 4,$$

$$S_n = \frac{n}{1-\frac{1}{n+1}} = (n+1)$$

$\therefore\; S_1^2 + S_2^2 + S_3^2 + \ldots. + S_{2n-1}^2$

$= 2^2 + 3^2 + 4^2 + \ldots.. + (n+1)^2 + \ldots.. + (2n)^2$

NOTE THIS STEP :

$$\sum_{r=1}^{2n} r^2 - 1 = \frac{2n(2n+1)(4n+1)}{6} - 1^2$$

$$= \frac{n(2n+1)(4n+1) - 3}{3}$$

14. Since x_1, x_2, x_3 are in A.P. Therefore, let $x_1 = a - d, x_2 = a$ and $x_3 = a + d$ and x_1, x_2, x_3 are the roots of $x^3 - x^2 + \beta x + \gamma = 0$

We have $\Sigma\alpha = a - d + a + a + d = 1$ ....(1)

$\Sigma\alpha\beta = (a-d)\,a + a\,(a+d) + (a-d)\,(a+d) = \beta$ ....(2)

$\alpha\,\beta\,\gamma = (a-d)\,a\,(a+d) = -\gamma$ ....(3)

From (1), we get, $3a = 1 \Rightarrow a = 1/3$

From (2), we get, $3a^2 - d^2 = \beta$

$\Rightarrow 3\,(1/3)^2 - d^2 = \beta \;\Rightarrow\; 1/3 - \beta = d^2$

(**NOTE :** In this equation we have two variables β and d but we have only one equation. So at first sight it looks that this equation cannot be solved but we know that $d^2 \geq 0\;\forall\; d \in R$ then ;β can be solved).

$$\Rightarrow\; \frac{1}{3} - \beta \geq 0 \qquad \because d^2 \geq 0$$

$$\Rightarrow\; \beta \leq \frac{1}{3} \;\Rightarrow\; \beta \in (-\infty, 1/3]$$

From (3), $a\,(a^2 - d^2) = -\gamma$

$$\Rightarrow\; \frac{1}{3}\left(\frac{1}{9} - d^2\right) = -\gamma \;\Rightarrow\; \frac{1}{27} - \frac{1}{3}d^2 = -\gamma$$

$$\Rightarrow\; \gamma + \frac{1}{27} = \frac{1}{3}d^2 \;\Rightarrow\; \gamma + \frac{1}{27} \geq 0$$

$$\Rightarrow\; \gamma \geq -\frac{1}{27} \;\Rightarrow\; \gamma \in \left[-\frac{1}{27}, \infty\right)$$

Hence $\beta \in (-\infty, 1/3)$ and $\gamma \in [-1/27, \infty]$

15. Solving the system of equations, $u + 2v + 3w = 6$, $4u + 5v + 6w = 12$ and $6u + 9v = 4$

we get $u = -1/3,\; v = 2/3,\; w = 5/3$

$$\therefore\; u + v + w = 2,\; \frac{1}{u} + \frac{1}{v} + \frac{1}{w} = -\frac{9}{10}$$

Let r be the common ratio of the G.P., a, b, c, d. Then $b = ar$, $c = ar^2$, $d = ar^3$.

Then the first equation

$$\left(\frac{1}{u} + \frac{1}{v} + \frac{1}{w}\right)x^2 + [(b-c)^2 + (c-a)^2 + (d-b)^2]x + (u+v+w) = 0$$

becomes

$$-\frac{9}{10}x^2 + [(ar - ar^2)^2 + (ar^2 - a)^2 + (ar^3 - ar)^2]\,x + 2 = 0$$

i.e., $9x^2 - 10a^2\,(1-r)^2\,[r^2 + (r+1)^2 + r^2\,(r+1)^2]x - 20 = 0$

i.e., $9x^2 - 10a^2\,(1-r)^2\,(r^4 + 2r^3 + 3r^2 + 2r + 1)\,x - 20 = 0$

i.e., $9x^2 - 10a^2\,(1-r)^2\,(1 + r + r^2)^2\;x - 20 = 0$,

i.e., $9x^2 - 10a^2\,(1-r^3)^2\;x - 20 = 0$ ....(1)

The second equation is,

$20\,x^2 + 10\,(a - ar^3)^2\,x - 9 = 0$

i.e., $20\,x^2 + 10\,a^2\,(1-r^3)^2 x - 9 = 0$ ....(2)

Since (2) can be obtained by the substitution $x \to 1/x$, equations (1) and (2) have reciprocal roots.

16. Let $a-3d$, $a-d$, $a+d$ and $a+3d$ be any four consecutive terms of an A.P. with common difference 2d. $\because$ Terms of A.P. are integers, $2d$ is also an integer.

Hence $P=(2d)^4+(a-3d)(a-d)(a+d)(a+3d)$

$=16d^4+(a^2-9d^2)(a^2-d^2)=(a^2-5d^2)^2$

Now, $a^2-5d^2=a^2-9d^2+4d^2$

$=(a-3d)(a+3d)+(2d)^2=$ some integer

Thus, $P=$ square of an integer.

17. Given that $a_1, a_2, ...a_n$ are +ve real no's in G.P.

$$\left.\begin{aligned} a_1 &= a \\ a_2 &= ar \\ \therefore \quad a_3 &= ar^2 \\ &\vdots \\ a_n &= ar^{n-1} \end{aligned}\right\} \begin{aligned} &\text{As } a_1, a_2, ...a_n \text{ are} +ve \\ &\therefore \quad r>0 \end{aligned}$$

A_n is A.M. of $a_1, a_2, ..., a_n$

$$\therefore \quad A_n=\frac{a_1+a_2+....+a_n}{n}=\frac{a+ar+....+ar^{n-1}}{n}$$

$$A_n=\frac{a(1-r^n)}{n(1-r)} \quad .... (1) \text{ (For } r \neq 1)$$

G_n is G.M. of $a_1, a_2, ..., a_n$

$$\therefore \quad G_n=\sqrt[n]{a_1, a_2...a_n}=\sqrt[n]{a.ar.ar^2....ar^{n-1}}$$

$$=\sqrt[n]{a^n, r^{\frac{n(n-1)}{2}}}=ar^{\frac{(n-1)}{2}}$$

$$G_n=ar^{\frac{(n-1)}{2}} \quad .... (2) \ (r \neq 1)$$

H_n is H.M. of $a_1, a_2, ..., a_n$

$$\therefore \quad H_n=\frac{n}{\frac{1}{a_1}+\frac{1}{a_2}+....+\frac{1}{a_n}}=\frac{n}{\frac{1}{a}+\frac{1}{ar}+....+\frac{1}{ar^{n-1}}}$$

$$=\frac{n}{\frac{1}{a}\frac{\left(\frac{1}{r^n}-1\right)}{\left(\frac{1}{r}-1\right)}}=\frac{n}{\frac{1}{a}\left(\frac{1-r^n}{r^n}\right).\frac{r}{1-r}}$$

$$H_n=\frac{anr^{n-1}(1-r)}{(1-r^n)} \quad (r \neq 1) \quad ...(3)$$

We also observe that

$$A_n H_n=\frac{a(1-r^n)}{(n(1-r))}\times\frac{anr^{n-1}(1-r)}{(1-r^n)}=a^n r^{n-1}=G_n^2$$

$$\therefore \quad A_n H_n=G_n^2 \quad ....(4)$$

$\therefore$ Now, G.M. of $G_1, G_2, ... G_n$ is

$$G=\sqrt[n]{G_1\ G_2\ ....G_n}$$

$$G=\sqrt[n]{\sqrt{A_1 H_1}\ \sqrt{A_2 H_2}...\sqrt{A_n H_n}}$$

[Using equation (4)

$$G=(A_1A_2....A_nH_1H_2....H_n)^{1/2n} \quad ...(5)$$

If $r=1$ then

$$A_n=G_n=H_n=a$$

Also $A_n H_n=G_n^2$

$\therefore$ For $r=1$ also, equation (5) holds.

Hence we get,

$$G=(A_1A_2....A_nH_1H_2....H_n)^{1/2n}$$

18. Clearly $A_1+A_2=a+b$

$$\frac{1}{H_1}+\frac{1}{H_2}=\frac{1}{a}+\frac{1}{b}$$

$$\Rightarrow \quad \frac{H_1+H_2}{H_1H_2}=\frac{a+b}{ab}=\frac{A_1+A_2}{G_1G_2}$$

$$\Rightarrow \quad \frac{G_1G_2}{H_1H_2}=\frac{A_1+A_2}{H_1+H_2}$$

$$\text{Also } \frac{1}{H_1}=\frac{1}{a}+\frac{1}{3}\left(\frac{1}{b}-\frac{1}{a}\right) \Rightarrow H_1=\frac{3ab}{2b+a}$$

$$\frac{1}{H_2}=\frac{1}{a}+\frac{2}{3}\left(\frac{1}{b}-\frac{1}{a}\right) \Rightarrow H_2=\frac{3ab}{2a+b}$$

$$\Rightarrow \quad \frac{A_1+A_2}{H_1+H_2}=\frac{a+b}{3ab\left(\frac{1}{2b+a}+\frac{1}{2a+b}\right)}$$

$$=\frac{(2b+a)(2a+b)}{9ab}$$

19. Given that a, b, c are in A.P.

$\Rightarrow \ 2b=a+c$ ....(1)

and a^2, b^2, c^2 are in H.P.

$$\Rightarrow \quad \frac{1}{b^2}-\frac{1}{a^2}=\frac{1}{c^2}-\frac{1}{b^2}$$

$$\Rightarrow \quad \frac{(a-b)(a+b)}{b^2a^2}=\frac{(b-c)(b+c)}{b^2c^2}$$

$\Rightarrow \ ac^2+bc^2=a^2b+a^2c \quad [\because a-b=b-c]$

$\Rightarrow \ ac(c-a)+b(c-a)(c+a)=0$

$\Rightarrow \ (c-a)(ab+bc+ca)=0$

$\Rightarrow$ either $c-a=0$ or $ab+bc+ca=0$

$\Rightarrow$ either $c=a$ or $(a+c)b+ca=0$

and then from (i) $2b^2+ca=0$

Either $a=b=c$ or $b^2=a\left(\frac{-c}{2}\right)$

i.e. a, b, – c/2 are in G.P. **Hence Proved.**

20. $$a_n=\frac{3}{4}-\left(\frac{3}{4}\right)^2+\left(\frac{3}{4}\right)^3+....+(-1)^{n-1}\left(\frac{3}{4}\right)^n$$

$$=\frac{\frac{3}{4}\left(1-\left(-\frac{3}{4}\right)^n\right)}{1+\frac{3}{4}}=\frac{3}{7}\left(1-\left(-\frac{3}{4}\right)^n\right)$$

$b_n=1-a_n$ and $b_n>a_n \ \forall \ n \geq n_0$

$\therefore \quad 1-a_n>a_n \Rightarrow 2a_n<1$

$\Rightarrow \frac{6}{7}\left[1-\left(-\frac{3}{4}\right)^n\right]<1 \Rightarrow -\left(-\frac{3}{4}\right)^n<\frac{1}{6}$

$\Rightarrow (-3)^{n+1}<2^{2n-1}$

For n to be even, inequality always holds. For n to be odd, it holds for $n \geq 7$.

$\therefore$ The least natural no., for which it holds is 6 ($\because$ it holds for every even natural no.)

G COMPREHENSION BASED QUESTIONS :

1. (b) $V_1+V_2+....+V_n=\sum_{r=1}^{n} V_r=\sum_{r=1}^{n}\left(r^3-\frac{r^2}{2}+\frac{r}{2}\right)$

$= \Sigma n^3-\frac{\Sigma n^2}{2}+\frac{\Sigma n}{2}$

$= \frac{n^2(n+1)^2}{4}-\frac{n(n+1)(2n+1)}{12}+\frac{n(n+1)}{4}$

$= \frac{n(n+1)}{4}\left[n(n+1)-\frac{2n+1}{3}+1\right]$

$= \frac{n(n+1)(3n^2+n+2)}{12}$

2. (d) $T_r=V_{r+1}-V_r-2$

$= \left[(r+1)^3-\frac{(r+1)^2}{2}+\frac{r+1}{2}\right]-\left[r^3-\frac{r^2}{2}+\frac{r}{2}\right]-2$

$= 3r^2+2r+1$

$T_r=(r+1)(3r-1)$

For each r, T_r has two different factors other than 1 and itself.

$\therefore$ T_r is always a composite number.

3. (b) $\therefore Q_{r+1}-Q_r=T_{r+2}-T_{r+1}-(T_{r+1}-T_r)$

$= T_{r+2}-2T_{r+1}+T_r$

$= (r+3)(3r+5)-2(r+2)(3r+2)+(r+1)(3r-1)$

$\because Q_{r+1}-Q_r=6(r+1)+5-6r-5=6$ (constant)

$\therefore Q_1, Q_2, Q_3, ...$ are in AP with common difference 6.

4. (c) Given $A_1=\frac{a+b}{2}, G_1=\sqrt{ab}, \ H_1=\frac{2ab}{a+b}$

also $A_n=\frac{A_{n-1}+H_{n-1}}{2}, G_n=\sqrt{A_{n-1}H_{n-1}}$

$H_n=\frac{2A_{n-1}H_{n-1}}{A_{n-1}+H_{n-1}}$

$\Rightarrow G_n^2=A_nH_n \Rightarrow A_nH_n=A_{n-1}H_{n-1}$

Similarly we can prove

$A_nH_n=A_{n-1}H_{n-1}=A_{n-2}H_{n-2}=....=A_1H_1$

$\Rightarrow A_nH_n=ab \ \therefore$

$\therefore G_1^2=G_2^2=G_3^2 ....=ab$

$\Rightarrow G_1=G_2=G_3 ....=\sqrt{ab}$

5. (a) We have $A_n=\frac{A_{n-1}+H_{n-1}}{2}$

$\therefore A_n-A_{n-1}=\frac{A_{n-1}+H_{n-1}}{2}-A_{n-1}$

$= \frac{H_{n-1}-A_{n-1}}{2}<0 \ (\because A_{n-1}>H_{n-1})$

$\Rightarrow A_n<A_{n-1}$ or $A_{n-1}>A_n$

$\therefore$ We can conclude that $A_1>A_2>A_3>....$

6. (b) We have $A_nH_n=ab \Rightarrow H_n=\frac{ab}{A_n}$

$\because \frac{1}{A_{n-1}}<\frac{1}{A_n} \Rightarrow H_{n-1}<H_n \quad \therefore H_1<H_2<H_3<.......$

H ASSERTION & REASON TYPE QUESTIONS :

1. (c) Given a_1, a_2, a_3, a_4 are in GP.

Then b_1, b_2, b_3, b_4 are the numbers

$a_1, a_1+a_2, a_1+a_2+a_3, \ a_1+a_2+a_3+a_4$

or $a, a+ar, \ a+ar+ar^2, \ a+ar+ar^2+ar^3$

Clearly above numbers are neither in AP nor in G.P. and hence statement 1 is true.

Also $\frac{1}{a}, \frac{1}{a+ar}, \frac{1}{a+ar+ar^2}, \frac{1}{a+ar+ar^2+ar^3}$ are not in A.P.

$\therefore b_1, b_2, b_3, b_4$ are not in H.P.

$\therefore$ Statement 2 is false.

I INTEGER VALUE CORRECT TYPE :

1. 3 Using $S_\infty=\frac{a}{1-r}$, we get

$$S_k=\begin{cases}\dfrac{\frac{k-1}{k!}}{1-\frac{1}{k!}}, & k\neq 1\\ 0, & k=1\\ \dfrac{1}{(k-1)!}, & k\geq 2\end{cases}$$

Now $\sum_{k=1}^{100}|(k^2-3k+1)S_k|=\sum_{k=2}^{100}|(k^2-3k+1)|\frac{1}{(k-1)!}$

$=|-1|+\sum_{k=3}^{100}\frac{(k^2-1)+1-3(k-1)-2}{(k-1)!}$ as $k^2-3k+1>0 \forall k\geq 3$

$= 1+\sum_{k=3}^{100}\left(\frac{1}{(k-3)!}-\frac{1}{(k-1)!}\right)$

$$= 1+\left(1-\frac{1}{2!}\right)+\left(\frac{1}{1!}-\frac{1}{3!}\right)+\left(\frac{1}{2!}-\frac{1}{4!}\right)+....$$

$$+\left(\frac{1}{96!}-\frac{1}{98!}\right)+\left(\frac{1}{97!}-\frac{1}{99!}\right)$$

$$= 3-\frac{1}{98!}-\frac{1}{99!}=3-\frac{9900}{100!}-\frac{100}{100!}$$

$$= 3-\frac{10000}{100!}=3-\frac{(100)^2}{100!}$$

$$\therefore \frac{100^2}{100!}+3\sum_{k=1}^{100}\left|\left(k^2-3k+1\right)S_k\right|=3.$$

2. (0) Given that $a_k = 2a_{k-1}-a_{k-2}$

$\Rightarrow \frac{a_{k-2}+a_k}{2}=a_{k-1}, 3 \le k \le 11$

$\Rightarrow a_1, a_2, a_3, ..., a_{11}$ are in AP.

If a is the first term and D the commen difference then

$a_1^2+a_2^2+...+a_{11}^2=990$

$\Rightarrow 11a^2+d^2\left(1^2+2^2+...+10^2\right)+2ad(1+2+...+10)$

$=990$

$\Rightarrow 11a^2+\frac{10\times11\times21}{6}d^2+2ad\times\frac{10\times11}{2}=990$

$\Rightarrow a^2+35d^2+150d=90$

Using $a = 15$, we get

$35d^2+150d+135=0$ or $7d^2+30d+27=0$

$\Rightarrow (d+3)(7d+9)=0 \Rightarrow d=-3$ or $-9/7$

then $a_2=15-3=12$ or $15-\frac{9}{7}=\frac{96}{7}>\frac{27}{2}$

$\therefore d \ne -9/7$

Hence $\frac{a_1+a_2+...+a_{11}}{11}=\frac{\frac{11}{2}[2\times15+10(-3)]}{11}=0$

3. (9)

We have $\frac{S_m}{S_n}=\frac{S_{5n}}{S_n}=\frac{\frac{5n}{2}[2\times3+(5n-1)d]}{\frac{n}{2}[6+(n-1)d]}$

$=\frac{5[(6-d)+5nd]}{(6-d)+nd]}$

which will be independent of n if $d = 6$ or $d = 0$

For a proper A.P. we take $d = 6$

then $a_2 = 3 + 6 = 9$

4. (5) Let $k, k + 1$ be removed from pack.

$\therefore (1+2+3+...+n)-(k+k+1)=1224$

$\frac{n(n+1)}{2}-2k=1225$

$k=\frac{n(n+1)-2450}{4}$

for $n=50$, $k=25$

$\therefore\ k-20=5$

Section-B JEE Main/ AIEEE

1. (b) $1, \log_9(3^{1-x}+2), \log_3(4.3^x-1)$ are in A.P.

$\Rightarrow 2\log_9(3^{1-x}+2)=1+\log_3(4.3^x-1)$

$\Rightarrow \log_3(3^{1-x}+2)=\log_3 3+\log_3(4.3^x-1)$

$\Rightarrow \log_3(3^{1-x}+2)=\log_3[3(4.3^x-1)]$

$\Rightarrow 3^{1-x}+2=3(4.3^x-1)$

$\Rightarrow 3.3^{-x}+2=12.3^x-3.$

Put $3^x=t$

$\Rightarrow \frac{3}{t}+2=12t-3$ or $12t^2-5t-3=0;$

Hence $t=-\frac{1}{3},\frac{3}{4} \Rightarrow 3^x=\frac{3}{4}$ (as $3^x \ne -ve$)

$\Rightarrow x=\log_3\left(\frac{3}{4}\right)$ or $x=\log_3 3-\log_3 4$

$\Rightarrow x=1-\log_3 4$

2. (d) $l=AR^{p-1} \Rightarrow \log l=\log A+(p-1)\log R$

$m=AR^{q-1} \Rightarrow \log m=\log A+(q-1)\log R$

$n=AR^{r-1} \Rightarrow \log n=\log A+(r-1)\log R$

Now,

$$\begin{vmatrix}\log l & p & 1\\ \log m & q & 1\\ \log n & r & 1\end{vmatrix}=\begin{vmatrix}\log A+(p-1)\log R & p & 1\\ \log A+(q-1)\log R & q & 1\\ \log A+(r-1)\log R & r & 1\end{vmatrix}$$

Operating $C_1-(\log R)C_2+(\log R-\log A)C_3$

$$=\begin{vmatrix}0 & p & 1\\ 0 & q & 1\\ 0 & r & 1\end{vmatrix}=0$$

3. (b) The product is $P=2^{1/4}.2^{2/8}.2^{3/16}$ ....

$=2^{1/4+2/8+3/16+.........\infty}$

Now let $S=\frac{1}{4}+\frac{2}{8}+\frac{3}{16}+........\infty$ ......(1)

$\frac{1}{2}S=\frac{1}{8}+\frac{2}{16}+........\infty$ ......(2)

Subtracting (2) from (1)

$\Rightarrow \frac{1}{2}S=\frac{1}{4}+\frac{1}{8}+\frac{1}{16}+........\infty$

or $\frac{1}{2}S=\frac{1/4}{1-1/2}=\frac{1}{2}\Rightarrow S=1$

$\therefore P=2^S=2$

4. **(b)** $ar^4=2$

$a\times ar\times ar^2\times ar^3\times ar^4\times ar^5\times ar^6\times ar^7\times ar^8$

$=a^9r^{36}=(ar^4)^9=2^9=512$

5. **(b)** Let a = first term of G.P. and r = common ratio of G.P.; Then G.P. is a, ar, ar^2

Given $S_\infty=20\Rightarrow \frac{a}{1-r}=20\Rightarrow a=20(1-r)$... (i)

Also $a^2+a^2r^2+a^2r^4+...$ to ∞ =100

$\Rightarrow \frac{a^2}{1-r^2}=100\Rightarrow a^2=100(1-r)(1+r)$... (ii)

From (i), $a^2=400(1-r)^2$;

From (ii), we get $100(1-r)(1+r)=400(1-r)^2$

$\Rightarrow 1+r=4-4r\Rightarrow 5r=3\Rightarrow r=3/5$.

6. **(a)** $1^3-2^3+3^3-4^3+........+9^3$

$=1^3+2^3+3^3+.......+9^3-2(2^3+4^3+6^3+8^3)$

$=\left[\frac{9\times10}{2}\right]^2-2.2^3\left[1^3+2^3+3^3+4^3\right]$

$=(45)^2-16.\left[\frac{4\times5}{2}\right]^2=2025-1600=425$

7. **(a)** $\frac{1}{1.2}-\frac{1}{2.3}+\frac{1}{3.4}..............\infty$

$|T_n|=\frac{1}{n(n+1)}=\left(\frac{1}{n}-\frac{1}{n+1}\right)$

$S=T_1-T_2+T_3-T_4+T_5...........\infty$

$=\left(\frac{1}{1}-\frac{1}{2}\right)-\left(\frac{1}{2}-\frac{1}{3}\right)+\left(\frac{1}{3}-\frac{1}{4}\right)-\left(\frac{1}{4}-\frac{1}{5}\right)....$

$=1-2\left[\frac{1}{2}-\frac{1}{3}+\frac{1}{4}-\frac{1}{5}..............\infty\right]$

$=1-2[-\log(1+1)+1]=2\log2-1=\log\left(\frac{4}{e}\right)$.

8. **(d)** $S_n=\frac{1}{^nC_0}+\frac{1}{^nC_1}+\frac{1}{^nC_2}+....+\frac{1}{^nC_n}$

$t_n=\frac{0}{^nC_0}+\frac{1}{^nC_1}+\frac{2}{^nC_2}+....+\frac{n}{^nC_n}$

$t_n=\frac{n}{^nC_n}+\frac{n-1}{^nC_{n-1}}+\frac{n-2}{^nC_{n-2}}+....+\frac{0}{^nC_0}$

Add, $2t_n=(n)\left[\frac{1}{^nC_0}+\frac{1}{^nC_1}+....\frac{1}{^nC_n}\right]=nS_n$

$\therefore \frac{t_n}{S_n}=\frac{n}{2}$

9. **(d)** $T_m=a+(m-1)d=\frac{1}{n}$............(1)

$T_n=a+(n-1)d=\frac{1}{m}$..............(2)

(1) - (2) $\Rightarrow (m-n)d=\frac{1}{n}-\frac{1}{m}\Rightarrow d=\frac{1}{mn}$

From (1) $a=\frac{1}{mn}\Rightarrow a-d=0$

10. **(b)** If n is odd, the required sum is

$1^2+2.2^2+3^2+2.4^2+......+2.(n-1)^2+n^2$

$=\frac{(n-1)(n-1+1)^2}{2}+n^2$

[$\because$ $(n-1)$ is even

$\therefore$ using given formula for the sum of $(n-1)$ terms.]

$=\left(\frac{n-1}{2}+1\right)n^2=\frac{n^2(n+1)}{2}$

11. **(b)** We know that $e=1+\frac{1}{1!}+\frac{1}{2!}+\frac{1}{3!}+.......$

and $e^{-1}=1-\frac{1}{1!}+\frac{1}{2!}-\frac{1}{3!}+.......$

$\therefore e+e^{-1}=2\left[1+\frac{1}{2!}+\frac{1}{4!}+....\right]$

$\therefore \frac{1}{2!}+\frac{1}{4!}+\frac{1}{6!}+......=\frac{e+e^{-1}}{2}-1$

$=\frac{e^2+1-2e}{2e}=\frac{(e-1)^2}{2e}$

12. (c) Given ${}^mC_{r-1},\ {}^mC_r,\ {}^mC_{r+1}$ are in A.P.

$$2\,{}^mC_r = {}^mC_{r-1} + {}^mC_{r+1}$$

$$\Rightarrow\ 2 = \frac{{}^mC_{r-1}}{{}^mC_r} + \frac{{}^mC_{r+1}}{{}^mC_r} = \frac{r}{m-r+1} + \frac{m-r}{r+1}$$

$$\Rightarrow\ m^2 - m(4r+1) + 4r^2 - 2 = 0.$$

13. (d) $x = \sum_{n=0}^{\infty} a^n = \frac{1}{1-a}$ $\quad a = 1 - \frac{1}{x}$

$y = \sum_{n=0}^{\infty} b^n = \frac{1}{1-b}$ $\quad b = 1 - \frac{1}{y}$

$z = \sum_{n=0}^{\infty} c^n = \frac{1}{1-c}$ $\quad c = 1 - \frac{1}{z}$

a, b, c are in A.P. OR $2b = a + c$

$$2\left(1 - \frac{1}{y}\right) = 1 - \frac{1}{x} + 1 - \frac{1}{y}$$

$\frac{2}{y} = \frac{1}{x} + \frac{1}{z}$ $\Rightarrow$ x, y, z are in H.P.

14. (d) $\frac{e^x + e^{-x}}{2} = 1 + \frac{x^2}{2!} + \frac{x^4}{4!} + \frac{x^6}{6!}$............

Putting $x = \frac{1}{2}$ we get

$$1 + \frac{1}{4.2!} + \frac{1}{16.4!} + \frac{1}{64.6!} + ...$$

$$\infty = \frac{e^{\frac{1}{2}} + e^{\frac{-1}{2}}}{2} = \frac{\sqrt{e} + \frac{1}{\sqrt{e}}}{2} = \frac{e+1}{2\sqrt{e}}$$

15. (d) $\frac{\frac{p}{2}[2a_1 + (p-1)d]}{\frac{q}{2}[2a_1 + (q-1)d]} = \frac{p^2}{q^2} \Rightarrow \frac{2a_1 + (p-1)d}{2a_1 + (q-1)d} = \frac{p}{q}$

$$\frac{a_1 + \left(\frac{p-1}{2}\right)d}{a_1 + \left(\frac{q-1}{2}\right)d} = \frac{p}{q}$$

For $\frac{a_6}{a_{21}}$, $p = 11, q = 41 \Rightarrow \frac{a_6}{a_{21}} = \frac{11}{41}$

16. (d) $\frac{1}{a_2} - \frac{1}{a_1} = \frac{1}{a_3} - \frac{1}{a_2} =$ .... $= \frac{1}{a_n} - \frac{1}{a_{n-1}} = d$ (say)

Then $a_1a_2 = \frac{a_1 - a_2}{d}$, $a_2a_3 = \frac{a_2 - a_3}{d}$,

..........., $a_{n-1}a_n = \frac{a_{n-1} - a_n}{d}$

$\therefore\ a_1a_2 + a_2a_3 + ... + a_{n-1}a_n$

$$= \frac{a_1 - a_2}{d} + \frac{a_2 - a_3}{d} + ... + \frac{a_{n-1} - a_n}{d}$$

$$= \frac{1}{d}[a_1 - a_2 + a_2 - a_3 + ... + a_{n-1} - a_n] = \frac{a_1 - a_n}{d}$$

Also, $\frac{1}{a_n} = \frac{1}{a_1} + (n-1)d$

$$\Rightarrow \frac{a_1 - a_n}{a_1a_n} = (n-1)d \Rightarrow \frac{a_1 - a_n}{d} = (n-1)a_1a_n$$

Which is the required result.

17. (d) We know that $e^x = 1 + x + \frac{x^2}{2!} + \frac{x^3}{3!} + ...\infty$

Put $x = -1$

$\therefore\ e^{-1} = 1 - 1 + \frac{1}{2!} - \frac{1}{3!} + \frac{1}{4!}$........$\infty$

$\therefore\ e^{-1} = \frac{1}{2!} - \frac{1}{3!} + \frac{1}{4!} - \frac{1}{5!}$........$\infty$

18. (b) Let the series a, ar, ar^2, ... are in geometric progression. given, $a = ar + ar^2$

$\Rightarrow\ 1 = r + r^2 \quad \Rightarrow r^2 + r - 1 = 0$

$\Rightarrow\ r = \frac{-1 \pm \sqrt{1 - 4 \times -1}}{2} \Rightarrow \quad r = \frac{-1 \pm \sqrt{5}}{2}$

$\Rightarrow\ r = \frac{\sqrt{5} - 1}{2}$ [$\because$ terms of G.P. are positive

$\therefore$ r should be positive]

19. (b) As per question,

$a + ar = 12$ ...(1)

$ar^2 + ar^3 = 48$ ...(2)

$\Rightarrow\ \frac{ar^2(1+r)}{a(1+r)} = \frac{48}{12} \Rightarrow r^2 = 4, \Rightarrow r = -2$

($\because$ terms are = + ve and –ve alternately)

$\Rightarrow\ a = -12$

20. (a) We have

$S = 1 + \frac{2}{3} + \frac{6}{3^2} + \frac{10}{3^3} + \frac{14}{3^4} + ...\infty$ ....(1)

Multiplying both sides by $\frac{1}{3}$ we get

$\frac{1}{3}S = \frac{1}{3} + \frac{2}{3^2} + \frac{6}{3^3} + \frac{10}{3^4} + ...\infty$ ....(2)

Subtracting eqn. (2) from eqn. (1) we get

$$\frac{2}{3}S = 1 + \frac{1}{3} + \frac{4}{3^2} + \frac{4}{3^3} + \frac{4}{3^4} + ...\infty$$

$\Rightarrow \dfrac{2}{3}S=\dfrac{4}{3}+\dfrac{4}{3^2}+\dfrac{4}{3^3}+\dfrac{4}{3^4}+........\infty$

$\Rightarrow \dfrac{2}{3}S=\dfrac{\frac{4}{3}}{1-\frac{1}{3}}=\dfrac{4}{3}\times\dfrac{3}{2} \Rightarrow S=3$

21. **(a)** Till 10th minute number of counted notes = 1500

$3000=\dfrac{n}{2}[2\times148+(n-1)(-2)]=n[148-n+1]$

$n^2-149n+3000=0$

$\Rightarrow n=125,24$

But $n=125$ is not possible

$\therefore$ total time = 24 + 10 = 34 minutes.

22. **(c)** Let required number of months = n

$\therefore 200\times3+(240+280+320+...+(n-3)^{th}\text{ term})=11040$

$\Rightarrow \dfrac{n-3}{2}[2\times240+(n-4)\times40]=11040-600$

$\Rightarrow (n-3)[240+20n-80]=10440$

$\Rightarrow (n-3)(20n+160)=10440$

$\Rightarrow (n-3)(n+8)=522$

$\Rightarrow n^2+5n-546=0$

$\Rightarrow (n+26)(n-21)=0$

$\therefore n=21$

23. **(b)** Let A.P. be $a, a+d, a+2d,.........$

$a_2+a_4+...........+a_{200}=\alpha$

$\Rightarrow \dfrac{100}{2}[2(a+d)+(100-1)d]=\alpha$ ....(i)

and $a_1+a_3+a_5+.........+a_{199}=\beta$

$\Rightarrow \dfrac{100}{2}[2a+(100-1)d]=\beta$ ....(ii)

On solving (i) and (ii), we get

$d=\dfrac{\alpha-\beta}{100}$

24. **(b)** nth term of the given series

$=T_n=(n-1)^2+(n-1)n+n^2$

$=\dfrac{((n-1)^3-n^3)}{(n-1)-n}=n^3-(n-1)^3$

$\Rightarrow S_n=\sum_{k=1}^{n}\left[k^3-(k-1)^3\right]\Rightarrow 8000=n^3$

$\Rightarrow$ $n=20$ which is a natural number.

Now, put $n=1,2,3,....20$

$T_1=1^3-0^3$

$T_2=2^3-1^3$

$\vdots$

$T_{20}=20^3-19^3$

Now, $T_1+T_2+---+T_{20}=S_{20}$

$\Rightarrow S_{20}=20^3-0^3=8000$

Hence, both the given statements are true and statement 2 supports statement 1.

25. **(c)** Given sequence can be written as

$\dfrac{7}{10}+\dfrac{77}{100}+\dfrac{777}{10^3}+......+$ up to 20 terms

$=7\left[\dfrac{1}{10}+\dfrac{11}{100}+\dfrac{111}{10^3}+......+\text{up to 20 terms}\right]$

Multiply and divide by 9

$=\dfrac{7}{9}\left[\dfrac{9}{10}+\dfrac{99}{100}+\dfrac{999}{1000}+......\text{up to 20 terms}\right]$

$=\dfrac{7}{9}\left[\left(1-\dfrac{1}{10}\right)+\left(1-\dfrac{1}{10^2}\right)+\left(1-\dfrac{1}{10^3}\right)+......\text{up to 20 terms}\right]$

$=\dfrac{7}{9}\left[20-\dfrac{\frac{1}{10}\left(1-\left(\frac{1}{10}\right)^{20}\right)}{1-\frac{1}{10}}\right]$

$=\dfrac{7}{9}\left[\dfrac{179}{9}+\dfrac{1}{9}\left(\dfrac{1}{10}\right)^{20}\right]=\dfrac{7}{81}[179+(10)^{-20}]$

Chapter 7

Straight Lines and Pair of Straight Lines

Section-A JEE Advanced/ IIT-JEE

A FILL IN THE BLANKS :

1. The area enclosed within the curve $|x|+|y|=1$ is .... *(1981 - 2 Marks)*
2. $y = 10^x$ is the reflection of $y = \log_{10} x$ in the line whose equation is .... *(1982 - 2 Marks)*
3. The set of lines $ax+by+c = 0$, where $3a + 2b + 4c = 0$ is concurrent at the point .... *(1982 - 2 Marks)*
4. Given the points A (0, 4) and B (0, –4), the equation of the locus of the point $P(x, y)$ such that $|AP-BP|=6$ is .... *(1983 - 1 Mark)*
5. If a, b and c are in A.P., then the straight line $ax+by+c=0$ will always pass through a fixed point whose coordinates are .... *(1984 - 2 Marks)*
6. The orthocentre of the triangle formed by the lines $x + y = 1$, $2x + 3y = 6$ and $4x - y + 4 = 0$ lies in quadrant number .... *(1985 - 2 Marks)*
7. Let the algebraic sum of the perpendicular distances from the points (2, 0), (0, 2) and (1, 1) to a variable straight line be zero; then the line passes through a fixed point whose cordinates are .... *(1991 - 2 Marks)*
8. The vertices of a triangle are $A(-1,-7)$, $B(5, 1)$ and $C(1, 4)$. The equation of the bisector of the angle $\angle ABC$ is .... *(1993 - 2 Marks)*

B TRUE / FALSE :

1. The straight line $5x + 4y = 0$ passes through the point of intersection of the straight lines $x + 2y - 10 = 0$ and $2x+y+5=0$. *(1983 - 1 Mark)*
2. The lines $2x + 3y + 19 = 0$ and $9x + 6y - 17 = 0$ cut the coordinate axes in concyclic points. *(1988 - 1 Mark)*

C MCQ's WITH ONE CORRECT ANSWER :

1. The points $(-a, -b)$, (0, 0), (a, b) and (a^2, ab) are : *(1979)*
   (a) Collinear
   (b) Vertices of a parallelogram
   (c) Vertices of a rectangle
   (d) None of these
2. The point (4, 1) undergoes the following three transformations successively. *(1980)*
   (i) Reflection about the line $y=x$.
   (ii) Translation through a distance 2 units along the positive direction of x-axis.
   (iii) Rotation through an angle p/4 about the origin in the counter clockwise direction.
   Then the final position of the point is given by the coordinates.
   (a) $\left(\frac{1}{\sqrt{2}}, \frac{7}{\sqrt{2}}\right)$ (b) $(-\sqrt{2}, 7\sqrt{2})$
   (c) $\left(-\frac{1}{\sqrt{2}}, \frac{7}{\sqrt{2}}\right)$ (d) $(\sqrt{2}, 7\sqrt{2})$
3. The straight lines $x+y=0$, $3x+y-4=0$, $x+3y-4=0$ form a triangle which is *(1983 - 1 Mark)*
   (a) isosceles (b) equilateral
   (c) right angled (d) none of these
4. If $P=(1, 0)$, $Q=(-1, 0)$ and $R=(2, 0)$ are three given points, then locus of the point S satisfying the relation $SQ^2+SR^2=2SP^2$, is *(1988 - 2 Marks)*
   (a) a straight line parallel to x-axis
   (b) a circle passing through the origin
   (c) a circle with the centre at the origin
   (d) a straigth line parallel to y-axis.
5. Line L has intercepts a and b on the coordinate axes. When the axes are rotated through a given angle, keeping the origin fixed, the same line L has intercepts p and q, then *(1990 - 2 Marks)*
   (a) $a^2+b^2 = p^2+q^2$ (b) $\frac{1}{a^2}+\frac{1}{b^2}=\frac{1}{p^2}+\frac{1}{q^2}$
   (c) $a^2+p^2 = b^2+q^2$ (d) $\frac{1}{a^2}+\frac{1}{p^2}=\frac{1}{b^2}+\frac{1}{q^2}$
6. If the sum of the distances of a point from two perpendicular lines in a plane is 1, then its locus is *(1992 - 2 Marks)*
   (a) square (b) circle
   (c) straight line (d) two intersecting lines
7. The locus of a variable point whose distance from (–2, 0) is 2/3 times its distance from the line $x = -\frac{9}{2}$ is *(1994)*
   (a) ellipse (b) parabola
   (c) hyperbola (d) none of these
8. The equations to a pair of opposite sides of parallelogram are $x^2 - 5x + 6 = 0$ and $y^2 - 6y + 5 = 0$, the equations to its diagonals are *(1994)*
   (a) $x+4y=13, y=4x-7$ (b) $4x+y=13, 4y=x-7$
   (c) $4x+y=13, y=4x-7$ (d) $y-4x=13, y+4x=7$
9. The orthocentre of the triangle formed by the lines $xy = 0$ and $x+y=1$ is *(1995S)*
   (a) $\left(\frac{1}{2}, \frac{1}{2}\right)$ (b) $\left(\frac{1}{3}, \frac{1}{3}\right)$
   (c) (0,0) (d) $\left(\frac{1}{4}, \frac{1}{4}\right)$

10. Let PQR be a right angled isosceles triangle, right angled at $P(2, 1)$. If the equation of the line QR is $2x+y=3$, then the equation representing the pair of lines PQ and PR is *(1999 - 2 Marks)*

(a) $3x^2-3y^2+8xy+20x+10y+25=0$
(b) $3x^2-3y^2+8xy-20x-10y+25=0$
(c) $3x^2-3y^2+8xy+10x+15y+20=0$
(d) $3x^2-3y^2-8xy-10x-15y-20=0$

11. If x_1, x_2, x_3 as well as y_1, y_2, y_3, are in G.P. with the same common ratio, then the points (x_1, y_1), (x_2, y_2) and (x_3, y_3). *(1999 - 2 Marks)*

(a) lie on a straight line (b) lie on an ellipse
(c) lie on a circle (d) are vertices of a triangle

12. Let PS be the median of the triangle with vertices $P(2, 2)$, $Q(6, -1)$ and $R(7, 3)$. The equation of the line passing through $(1, -1)$ and parallel to PS is *(2000S)*

(a) $2x-9y-7=0$ (b) $2x-9y-11=0$
(c) $2x+9y-11=0$ (d) $2x+9y+7=0$

13. The incentre of the triangle with vertices $(1, \sqrt{3})$, $(0, 0)$ and $(2, 0)$ is *(2000S)*

(a) $\left(1, \frac{\sqrt{3}}{2}\right)$ (b) $\left(\frac{2}{3}, \frac{1}{\sqrt{3}}\right)$
(c) $\left(\frac{2}{3}, \frac{\sqrt{3}}{2}\right)$ (d) $\left(1, \frac{1}{\sqrt{3}}\right)$

14. The number of integer values of m, for which the x-coordinate of the point of intersection of the lines $3x+4y=9$ and $y=mx+1$ is also an integer, is *(2001S)*

(a) 2 (b) 0
(c) 4 (d) 1

15. Area of the parallelogram formed by the lines $y=mx$, $y=mx+1$, $y=nx$ and $y=nx+1$ equals *(2001S)*

(a) $|m+n|/(m-n)^2$ (b) $2/|m+n|$
(c) $1/(|m+n|)$ (d) $1/(|m-n|)$

16. Let $0<\alpha<\frac{\pi}{2}$ be fixed angle. If

$P=(\cos\theta, \sin\theta)$ and $Q=(\cos(\alpha-\theta), \sin(\alpha-\theta))$, then Q is obtained from P by *(2002S)*

(a) clockwise rotation around origin through an angle α
(b) anticlockwise rotation around origin through an angle α
(c) reflection in the line through origin with slope $\tan\alpha$
(d) reflection in the line through origin with slope $\tan(\alpha/2)$

17. Let $P=(-1, 0)$, $Q=(0, 0)$ and $R=(3, 3\sqrt{3})$ be three points. Then the equation of the bisector of the angle PQR is

(a) $\frac{\sqrt{3}}{2}x+y=0$ (b) $x+\sqrt{3}y=0$ *(2002S)*
(c) $\sqrt{3}x+y=0$ (d) $x+\frac{\sqrt{3}}{2}y=0$

18. A straight line through the origin O meets the parallel lines $4x+2y=9$ and $2x+y+6=0$ at points P and Q respectively. Then the point O divides the segemnt PQ in the ratio

(a) $1:2$ (b) $3:4$ *(2002S)*
(c) $2:1$ (d) $4:3$

19. The number of intergral points (integral point means both the coordinates should be integer) exactly in the interior of the triangle with vertices (0,0), (0,21) and (21,0), is *(2003S)*

(a) 133 (b) 190
(c) 233 (d) 105

20. Orthocentre of triangle with vertices (0, 0), (3, 4) and (4, 0) is

(a) $\left(3, \frac{5}{4}\right)$ (b) $(3, 12)$ *(2003S)*
(c) $\left(3, \frac{3}{4}\right)$ (d) $(3, 9)$

21. Area of the triangle formed by the line $x+y=3$ and angle bisectors of the pair of straight lines $x^2-y^2+2y=1$ is *(2004S)*

(a) 2 sq. units (b) 4 sq. units
(c) 6 sq. units (d) 8 sq. units

22. Let $O(0, 0)$, $P(3, 4)$, $Q(6, 0)$ be the vertices of the triangles OPQ. The point R inside the triangle OPQ is such that the triangles OPR, PQR, OQR are of equal area. The coordinates of R are *(2007 -3 marks)*

(a) $\left(\frac{4}{3}, 3\right)$ (b) $\left(3, \frac{2}{3}\right)$
(c) $\left(3, \frac{4}{3}\right)$ (d) $\left(\frac{4}{3}, \frac{2}{3}\right)$

23. A straight line L through the point (3, –2) is inclined at an angle 60° to the line $\sqrt{3}x+y=1$. If L also intersects the x-axis, then the equation of L is *(2011)*

(a) $y+\sqrt{3}x+2-3\sqrt{3}=0$
(b) $y-\sqrt{3}x+2+3\sqrt{3}=0$
(c) $\sqrt{3}y-x+3+2\sqrt{3}=0$
(d) $\sqrt{3}y+x-3+2\sqrt{3}=0$

(D) MCQ's WITH ONE OR MORE THAN ONE CORRECT :

1. Three lines $px+qy+r=0$, $qx+ry+p=0$ and $rx+py+q=0$ are concurrent if *(1985 - 2 Marks)*

(a) $p+q+r=0$
(b) $p^2+q^2+r^2=qr+rp+pq$
(c) $p^3+q^3+r^3=3pqr$
(d) none of these.

2. The points $\left(0, \frac{8}{3}\right)$, $(1, 3)$ and $(82, 30)$ are vertices of

(a) an obtuse angled triangle *(1986 - 2 Marks)*
(b) an acute angled triangle
(c) a right angled triangle
(d) an isosceles triangle
(e) none of these.

3. All points lying inside the triangle formed by the points (1, 3), (5, 0) and (–1, 2) satisfy ***(1986 - 2 Marks)***
(a) $3x+2y \geq 0$ (b) $2x+y-13 \geq 0$
(c) $2x-3y-12 \leq 0$ (d) $-2x+y \geq 0$
(e) none of these.

4. A vector $\vec{a}$ has components $2p$ and 1 with respect to a rectangular cartesian system. This system is rotated through a certain angle about the origin in the counter clockwise sense. If, with respect to the new system, $\vec{a}$ has components $p+1$ and 1, then ***(1986 - 2 Marks)***
(a) $p=0$ (b) $p=1$ or $p=-\frac{1}{3}$
(c) $p=-1$ or $p=\frac{1}{3}$ (d) $p=1$ or $p=-1$
(e) none of these .

5. If (P(1, 2), Q(4, 6), R(5, 7) and S(a, b) are the vertices of a parallelogram PQRS, then ***(1998 - 2 Marks)***
(a) $a=2, b=4$ (b) $a=3, b=4$
(c) $a=2, b=3$ (d) $a=3, b=5$

6. The diagonals of a parallelogram PQRS are along the lines $x+3y=4$ and $6x-2y=7$. Then PQRS must be a. ***(1998 - 2 Marks)***
(a) rectangle (b) square
(c) cyclic quadrilateral (d) rhombus.

7. If the vertices P, Q, R of a triangle PQR are rational points, which of the following points of the triangle PQR is (are) always rational point(s)? ***(1998 - 2 Marks)***
(a) centroid (b) incentre
(c) circumcentre (d) orthocentre
(A rational point is a point both of whose co-ordinates are rational numbers.)

8. Let L_1 be a strainght line passing through the origin and L_2 be the straight line $x+y=1$. If the intercepts made by the circle $x^2+y^2-x+3y=0$ on L_1 and L_2 are equal, then which of the following equations can represent L_1? ***(1999 - 3 Marks)***
(a) $x+y=0$ (b) $x-y=0$
(c) $x+7y=0$ (d) $x-7y=0$

9. For a > b > c > 0, the distance between (1, 1) and the point of intersection of the lines ax + by + c = 0 and bx + ay + c = 0 is less than $2\sqrt{2}$. Then ***(JEE Adv. 2013)***
(a) a + b – c > 0 (b) a – b + c < 0
(c) a – b + c > 0 (d) a + b – c < 0

E SUBJECTIVE PROBLEMS :

1. A straight line segment of length ℓ moves with its ends on two mutually perpendicular lines. Find the locus of the point which divides the line segment in the ratio 1 : 2. ***(1978)***

2. The area of a triangle is 5. Two of its vertices are A (2, 1) and B (3, –2). The third vertex C lies on $y=x+3$. Find C. ***(1978)***

3. One side of a rectangle lies along the line $4x+7y+5=0$. Two of its vertices are (–3, 1) and (1, 1). Find the equations of the other three sides. ***(1978)***

4. (a) Two vertices of a triangle are (5, –1) and (–2, 3). If the orthocentre of the triangle is the origin, find the coordinates of the third point.
(b) Find the equation of the line which bisects the obtuse angle between the lines $x-2y+4=0$ and $4x-3y+2=0$. ***(1979)***

5. A straight line L is perpendicular to the line $5x-y=1$. The area of the triangle formed by the line L and the coordinate axes is 5. Find the equation of the line L. ***(1980)***

6. The points (1, 3) and (5, 1) are two opposite vertices of a rectangle. The other two vertices lie on the line $y=2x+c$. Find c and the remaining vertices. ***(1981 - 4 Marks)***

7. The end A, B of a straight line segment of constant length c slide upon the fixed rectangular axes OX, OY respectively. If the rectangle OAPB be completed, then show that the locus of the foot of the perpendicular drawn from P to AB is
$x^{\frac{2}{3}}+y^{\frac{2}{3}}=c^{\frac{2}{3}}$ ***(1983 - 2 Marks)***

8. The vertices of a triangle are $[at_1t_2, a(t_1+t_2)]$, $[at_2t_3, a(t_2+t_3)]$, $[at_3t_1, a(t_3+t_1)]$. Find the orthocentre of the triangle. ***(1983 - 3 Marks)***

9. The coordinates of A, B, C are (6, 3), (–3, 5), (4, – 2) respectively, and P is any point (x, y). Show that the ratio of the area of the triangles Δ PBC and ΔABC is $\left|\frac{x+y-2}{7}\right|$ ***(1983 - 2 Marks)***

10. Two equal sides of an isosceles triangle are given by the equations $7x-y+3=0$ and $x+y-3=0$ and its third side passes through the point (1, –10). Determine the equation of the third side. ***(1984 - 4 Marks)***

11. One of the diameters of the circle circumscribing the rectangle ABCD is $4y=x+7$. If A and B are the points (–3, 4) and (5, 4) respectively, then find the area of rectangle. ***(1985 - 3 Marks)***

12. Two sides of a rhombus ABCD are parallel to the lines $y=x+2$ and $y=7x+3$. If the diagonals of the rhombus intersect at the point (1, 2) and the vertex A is on the y-axis, find possible co-ordinates of A. ***(1985 - 5 Marks)***

13. The equations of the perpendicular bisectors of the sides AB and AC of a triangle ABC are $x-y+5=0$ and $x+2y=0$, respectively. If the point A is (1, –2), find the equation of the line BC. ***(1986 - 5 Marks)***

14. Lines $L_1 \equiv ax+by+c=0$ and $L_2 \equiv lx+my+n=0$ intersect at the point P and make an angle θ with each other. Find the equation of a line L different from L_2 which passes through P and makes the same angle θ with L_1. ***(1988 - 5 Marks)***

15. Let ABC be a triangle with AB = AC. If D is the midpoint of BC, E is the foot of the perpendicular drawn from D to AC and F the mid-point of DE, prove that AF is perpendicular to BE. ***(1989 - 5 Marks)***

16. Straight lines $3x + 4y = 5$ and $4x - 3y = 15$ intersect at the point A. Points B and C are chosen on these two lines such that $AB = AC$. Determine the possible equations of the line BC passing through the point (1, 2). *(1990 - 4 Marks)*

17. A line cuts the x-axis at $A(7, 0)$ and the y-axis at $B(0, -5)$. A variable line PQ is drawn perpendicular to AB cutting the x-axis in P and the y-axis in Q. If AQ and BP intersect at R, find the locus of R. *(1990 - 4 Marks)*

18. Find the equation of the line passing through the point (2, 3) and making intercept of length 2 units between the lines $y + 2x = 3$ and $y + 2x = 5$. *(1991 - 4 Marks)*

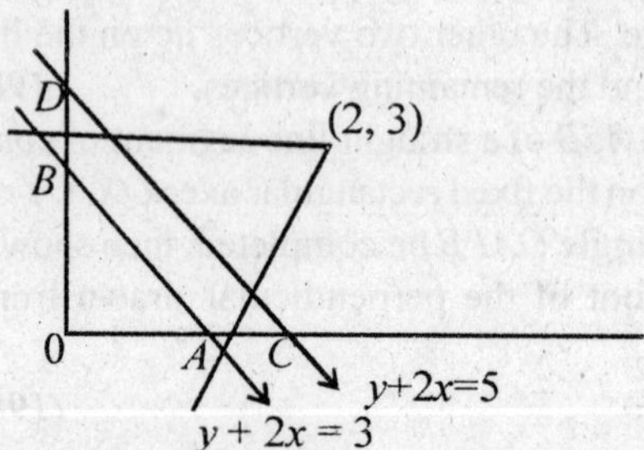

19. Show that all chords of the curve $3x^2 - y^2 - 2x + 4y = 0$, which subtend a right angle at the origin, pass through a fixed point. Find the coordinates of the point. *(1991 - 4 Marks)*

20. Determine all values of α for which the point (α, α^2) lies inside the triangle formed by the lines *(1992 - 6 Marks)*

$$2x + 3y - 1 = 0$$
$$x + 2y - 3 = 0$$
$$5x - 6y - 1 = 0$$

21. Tagent at a point P_1 {other than (0, 0)} on the curve $y = x^3$ meets the curve again at P_2. The tangent at P_2 meets the curve at P_3, and so on. Show that the abscissae of $P_1, P_2, P_3 .... P_n$, form a G.P. Also find the ratio.

[area $(\Delta P_1, P_2, P_3)$]/[area$(P_2 P_3, P_4)$] *(1993 - 5 Marks)*

22. A line through $A(-5, -4)$ meets the line $x + 3y + 2 = 0$, $2x + y + 4 = 0$ and $x - y - 5 = 0$ at the points B, C and D respectively. If $(15/AB)^2 + (10/AC)^2 = (6/AD)^2$, find the equation of the line. *(1993 - 5 Marks)*

23. A rectangle $PQRS$ has its side PQ parallel to the line $y = mx$ and vertices P, Q and S on the lines $y = a$, $x = b$ and $x = -b$, respectively. Find the locus of the vertex R. *(1996 - 2 Marks)*

24. A variable straight line of slope 4 intersects the hyperbola $xy = 1$ at two points. Find the locus of the point which divides the line segment between these two points in the ratio 1 : 2. *(1997C - 5 Marks)*

25. Using co-ordinate geometry, prove that the three altitudes of any triangle are concurrent. *(1998 - 8 Marks)*

26. For points $P = (x_1, y_1)$ and $Q = (x_2, y_2)$ of the co-ordinate plane, a new distance $d(P, Q)$ is defined by $d(P, Q) = |x_1 - x_2| + |y_1 - y_2|$. Let $O = (0, 0)$ and $A = (3, 2)$. Prove that the set of points in the first quadrant which are equidistant (with respect to the new distance) from O and A consists of the union of a line segment of finite length and an infinite ray. Sketch this set in a labelled diagram. *(2000 - 10 Marks)*

27. Let ABC and PQR be any two triangles in the same plane. Assume that the prependiculars from the points A, B, C to the sides QR, RP, PQ respectively are concurrent. Using vector methods or otherwise, prove that the prependuculars from P, Q, R to BC, CA, AB respectively are also concurrent. *(2000 - 10 Marks)*

28. Let a, b, c be real numbers with $a^2 + b^2 + c^2 = 1$. Show that the equation

$$\begin{vmatrix} ax - by - c & bx + ay & cx + a \\ bx + ay & -ax + by - c & cy + b \\ cx + a & cy + b & -ax - by + c \end{vmatrix} = 0 \text{ represents a}$$

straight line. *(2001 - 6 Marks)*

29. A straight line L through the origin meets the lines $x + y = 1$ and $x + y = 3$ at P and Q respectively. Through P and Q two straight lines L_1 and L_2 are drawn, parallel to $2x - y = 5$ and $3x + y = 5$ respectively. Lines L_1 and L_2 intersect at R. Show that the locus of R, as L varies, is a straight line. *(2002 - 5 Marks)*

30. A straight line L with negative slope passes through the point (8, 2) and cuts the positive coordinate axes at points P and Q. Find the absolute minimum value of $OP + OQ$, as L varies, where O is the origin. *(2002 - 5 Marks)*

31. The area of the triangle formed by the intersection of a line parallel to x-axis and passing through $P(h, k)$ with the lines $y = x$ and $x + y = 2$ is $4h^2$. Find the locus of the point P. *(2005 - 2 Marks)*

(H) ASSERTION & REASON TYPE QUESTIONS :

1. Lines $L_1 : y - x = 0$ and $L_2 : 2x + y = 0$ intersect the line $L_3 : y + 2 = 0$ at P and Q, respectively. The bisector of the acute angle between L_1 and L_2 intersects L_3 at R.

STATEMENT-1 : The ratio $PR : RQ$ equals $2\sqrt{2} : \sqrt{5}$.

because

STATEMENT-2 : In any triangle, bisector of an angle divides the triangle into two similar triangles. *(2007 - 3 marks)*

(a) Statement-1 is True, Statement-2 is True; Statement-2 is not a correct explanation for Statement-1

(b) Statement-1 is True, Statement-2 is True; Statement-2 is NOT a correct explanation for Statement-1

(c) Statement-1 is True, Statement-2 is False

(d) Statement-1 is False, Statement-2 is True.

Section-B JEE Main/ AIEEE

1. A triangle with vertices (4, 0), (–1, –1), (3, 5) is [2002]
   (a) isosceles and right angled
   (b) isosceles but not right angled
   (c) right angled but not isosceles
   (d) neither right angled nor isoceles

2. Locus of mid point of the portion between the axes of $x \cos\alpha + y \sin\alpha = p$ whre p is constant is [2002]
   (a) $x^2+y^2=\frac{4}{p^2}$ (b) $x^2+y^2=4p^2$
   (c) $\frac{1}{x^2}+\frac{1}{y^2}=\frac{2}{p^2}$ (d) $\frac{1}{x^2}+\frac{1}{y^2}=\frac{4}{p^2}$

3. If the pair of lines $ax^2+2hxy+by^2+2gx+2fy+c=0$ intersect on the y-axis then [2002]
   (a) $2fgh=bg^2+ch^2$ (b) $bg^2 \neq ch^2$
   (c) $abc=2fgh$ (d) none of these

4. The pair of lines represented by $3ax^2+5xy+(a^2-2)y^2=0$ are perpendicular to each other for [2002]
   (a) two values of a
   (b) $\forall a$
   (c) for one value of a
   (d) for no values of a

5. A square of side a lies above the x-axis and has one vertex at the origin. The side passing through the origin makes an angle $\alpha\left(0<\alpha<\frac{\pi}{4}\right)$ with the positive direction of x-axis. The equation of its diagonal not passing through the origin is [2003]
   (a) $y(\cos\alpha+\sin\alpha)+x(\cos\alpha-\sin\alpha)=a$
   (b) $y(\cos\alpha-\sin\alpha)-x(\sin\alpha-\cos\alpha)=a$
   (c) $y(\cos\alpha+\sin\alpha)+x(\sin\alpha-\cos\alpha)=a$
   (d) $y(\cos\alpha+\sin\alpha)+x(\sin\alpha+\cos\alpha)=a$.

6. If the pair of straight lines $x^2-2pxy-y^2=0$ and $x^2-2qxy-y^2=0$ be such that each pair bisects the angle between the other pair, then [2003]
   (a) $pq=-1$ (b) $p=q$
   (c) $p=-q$ (d) $pq=1$.

7. Locus of centroid of the triangle whose vertices are $(a\cos t, a\sin t), (b\sin t, -b\cos t)$ and (1, 0), where t is a parameter, is [2003]
   (a) $(3x+1)^2+(3y)^2=a^2-b^2$
   (b) $(3x-1)^2+(3y)^2=a^2-b^2$
   (c) $(3x-1)^2+(3y)^2=a^2+b^2$
   (d) $(3x+1)^2+(3y)^2=a^2+b^2$.

8. If x_1, x_2, x_3 and y_1, y_2, y_3 are both in G.P. with the same common ratio, then the points $(x_1, y_1), (x_2, y_2)$ and (x_3, y_3) [2003]
   (a) are vertices of a triangle
   (b) lie on a straight line
   (c) lie on an ellipse
   (d) lie on a circle.

9. If the equation of the locus of a point equidistant from the point (a_1, b_1) and (a_2, b_2) is $(a_1-b_2)x+(a_1-b_2)y+c=0$, then the value of \`c\` is [2003]
   (a) $\sqrt{a_1^2+b_1^2-a_2^2-b_2^2}$
   (b) $\frac{1}{2}(a_2^2+b_2^2-a_1^2-b_1^2)$
   (c) $a_1^2-a_2^2+b_1^2-b_2^2$
   (d) $\frac{1}{2}(a_1^2+a_2^2+b_1^2+b_2^2)$.

10. Let $A(2, -3)$ and $B(-2, 3)$ be vertices of a triangle ABC. If the centroid of this triangle moves on the line $2x+3y=1$, then the locus of the vertex C is the line [2004]
    (a) $3x-2y=3$ (b) $2x-3y=7$
    (c) $3x+2y=5$ (d) $2x+3y=9$

11. The equation of the straight line passing through the point (4, 3) and making intercepts on the co-ordinate axes whose sum is –1 is [2004]
    (a) $\frac{x}{2}-\frac{y}{3}=1$ and $\frac{x}{-2}+\frac{y}{1}=1$
    (b) $\frac{x}{2}-\frac{y}{3}=-1$ and $\frac{x}{-2}+\frac{y}{1}=-1$
    (c) $\frac{x}{2}+\frac{y}{3}=1$ and $\frac{x}{2}+\frac{y}{1}=1$
    (d) $\frac{x}{2}+\frac{y}{3}=-1$ and $\frac{x}{-2}+\frac{y}{1}=-1$

12. If the sum of the slopes of the lines given by $x^2-2cxy-7y^2=0$ is four times their product c has the value [2004]
    (a) –2 (b) –1
    (c) 2 (d) 1

13. If one of the lines given by $6x^2-xy+4cy^2=0$ is $3x+4y=0$, then c equals [2004]
    (a) –3 (b) –1
    (c) 3 (d) 1

14. The line parallel to the x- axis and passing through the intersection of the lines $ax + 2by + 3b = 0$ and $bx - 2ay - 3a = 0$, where $(a, b) \neq (0, 0)$ is **[2005]**

(a) below the x - axis at a distance of $\frac{3}{2}$ from it

(b) below the x - axis at a distance of $\frac{2}{3}$ from it

(c) above the x - axis at a distance of $\frac{3}{2}$ from it

(d) above the x - axis at a distance of $\frac{2}{3}$ from it

15. If a vertex of a triangle is (1, 1) and the mid points of two sides through this vertex are (–1, 2) and (3, 2) then the centroid of the triangle is **[2005]**

(a) $\left(-1, \frac{7}{3}\right)$ (b) $\left(\frac{-1}{3}, \frac{7}{3}\right)$

(c) $\left(1, \frac{7}{3}\right)$ (d) $\left(\frac{1}{3}, \frac{7}{3}\right)$

16. A straight line through the point A (3, 4) is such that its intercept between the axes is bisected at A. Its equation is

(a) $x + y = 7$ (b) $3x - 4y + 7 = 0$ **[2006]**

(c) $4x + 3y = 24$ (d) $3x + 4y = 25$

17. If (a, a^2) falls inside the angle made by the lines $y = \frac{x}{2}$, $x > 0$ and $y = 3x$, $x > 0$, then a belong to **[2006]**

(a) $\left(0, \frac{1}{2}\right)$ (b) $(3, \infty)$

(c) $\left(\frac{1}{2}, 3\right)$ (d) $\left(-3, -\frac{1}{2}\right)$

18. Let A (h, k), B(1, 1) and C (2, 1) be the vertices of a right angled triangle with AC as its hypotenuse. If the area of the triangle is 1square unit, then the set of values which 'k' can take is given by **[2007]**

(a) $\{-1, 3\}$ (b) $\{-3, -2\}$

(c) $\{1, 3\}$ (d) $\{0, 2\}$

19. Let $P = (-1, 0)$, $Q = (0, 0)$ and $R = (3, 3\sqrt{3})$ be three point. The equation of the bisector of the angle PQR is **[2007]**

(a) $\frac{\sqrt{3}}{2}x + y = 0$ (b) $x + \sqrt{3}y = 0$

(c) $\sqrt{3}x + y = 0$ (d) $x + \frac{\sqrt{3}}{2}y = 0$.

20. If one of the lines of $my^2 + (1 - m^2)xy - mx^2 = 0$ is a bisector of the angle between the lines $xy = 0$, then m is

(a) 1 (b) 2 **[2007]**

(c) –1/2 (d) –2.

21. The perpendicular bisector of the line segment joining P (1, 4) and Q(k, 3) has y-intercept –4. Then a possible value of k is **[2008]**

(a) 1 (b) 2

(c) –2 (d) –4

22. The shortest distance between the line $y - x = 1$ and the curve $x = y^2$ is : **[2009]**

(a) $\frac{2\sqrt{3}}{8}$ (b) $\frac{3\sqrt{2}}{5}$

(c) $\frac{\sqrt{3}}{4}$ (d) $\frac{3\sqrt{2}}{8}$

23. The lines $p(p^2 + 1)x - y + q = 0$ and $(p^2 + 1)^2x + (p^2 + 1)y + 2q = 0$ are perpendicular to a common line for : **[2009]**

(a) exactly one values of p

(b) exactly two values of p

(c) more than two values of p

(d) no value of p

24. Three distinct points A, B and C are given in the 2-dimensional coordinates plane such that the ratio of the distance of any one of them from the point (1, 0) to the distance from the point (–1, 0) is equal to $\frac{1}{3}$. Then the circumcentre of the triangle ABC is at the point: **[2009]**

(a) $\left(\frac{5}{4}, 0\right)$ (b) $\left(\frac{5}{2}, 0\right)$

(c) $\left(\frac{5}{3}, 0\right)$ (d) (0, 0)

25. The line L given by $\frac{x}{5} + \frac{y}{b} = 1$ passes through the point (13, 32). The line K is parallel to L and has the equation $\frac{x}{c} + \frac{y}{3} = 1$. Then the distance between L and K is **[2010]**

(a) $\sqrt{17}$ (b) $\frac{17}{\sqrt{15}}$

(c) $\frac{23}{\sqrt{17}}$ (d) $\frac{23}{\sqrt{15}}$

26. The lines $L_1 : y - x = 0$ and $L_2 : 2x + y = 0$ intersect the line $L_3 : y + 2 = 0$ at P and Q respectively. The bisector of the acute angle between L_1 and L_2 intersects L_3 at R.

Statement-1: The ratio $PR : RQ$ equals $2\sqrt{2} : \sqrt{5}$

Statement-2: In any triangle, bisector of an angle divides the triangle into two similar triangles. **[2011]**

(a) Statement-1 is true, Statement-2 is true; Statement-2 is **not** a correct explanation for Statement-1.

(b) Statement-1 is true, Statement-2 is false.

(c) Statement-1 is false, Statement-2 is true.

(d) Statement-1 is true, Statement-2 is true; Statement-2 is a correct explanation for Statement-1.

27. The lines $x+y=|a|$ and $ax-y=1$ interesect each other in the first quadrant. Then the set of all possible values of a is the interval : **[2011 RS]**

(a) $(0,\infty)$ (b) $[1,\infty)$

(c) $(-1,\infty)$ (d) $(-1,1)$

28. If A $(2,-3)$ and B $(-2,1)$ are two vertices of a triangle and third vertex moves on the line $2x+3y=9$, then the locus of the centroid of the triangle is : **[2011 RS]**

(a) $x-y=1$ (b) $2x+3y=1$

(c) $2x+3y=3$ (d) $2x-3y=1$

29. There are 10 points in a plane, out of these 6 are collinear. If N is the number of triangles formed by joining these points. Then : **[2011 RS]**

(a) $N \le 100$ (b) $100 < N \le 140$

(c) $140 < N \le 190$ (d) $N > 190$

30. If the line $2x+y=k$ passes through the point which divides the line segment joining the points (1,1) and (2,4) in the ratio 3 :2, then k equals : **[2012]**

(a) $\frac{29}{5}$ (b) 5 (c) 6 (d) $\frac{11}{5}$

31. A ray of light along $x+\sqrt{3}y=\sqrt{3}$ gets reflected upon reaching x-axis, the equation of the reflected ray is **[JEE M 2013]**

(a) $y=x+\sqrt{3}$ (b) $\sqrt{3}y=x-\sqrt{3}$

(c) $y=\sqrt{3}x-\sqrt{3}$ (d) $\sqrt{3}y=x-1$

32. The x-coordinate of the incentre of the triangle that has the coordinates of mid points of its sides as (0, 1) (1. 1) and (1, 0) is : **[JEE M 2013]**

(a) $2+\sqrt{2}$ (b) $2-\sqrt{2}$

(c) $1+\sqrt{2}$ (d) $1-\sqrt{2}$

Solutions & Explanations

Section-A : JEE Advanced/ IIT-JEE

A 1. 2 sq. units 2. $y = x$ 3. $\left(\frac{3}{4}, \frac{1}{2}\right)$ 4. $\frac{y^2}{9} - \frac{x^2}{7} = 1$ 5. $(1, -2)$ 6. first quadrant

7. $(1, 1)$ 8. $x - 7y + 2 = 0$

B 1. T 2. T

C 1. (a) 2. (c) 3. (a) 4. (d) 5. (b) 6. (a)
7. (a) 8. (c) 9. (c) 10. (b) 11. (a) 12. (d)
13. (d) 14. (a) 15. (d) 16. (d) 17. (c) 18. (b)
19. (b) 20. (c) 21. (a) 22. (c) 23. (b)

D 1. (a, b, c) 2. (e) 3. (a, c) 4. (b) 5. (c) 6. (d)
7. (a,c, d) 8. (b, c) 9. (a)

E 1. $9x^2 + 36y^2 = 4\ell^2$ 2. $\left(\frac{-3}{2}, \frac{3}{2}\right)$ or $\left(\frac{7}{2}, \frac{13}{2}\right)$

3. $4x + 7y - 11 = 0$, $7x - 4y - 3 = 0$; $7x - 4y + 25 = 0$ 4. (a) $(-4, -7)$ (b) $(4 - \sqrt{5})x + (2\sqrt{5} - 3)y - (4\sqrt{5} - 2) = 0$

5. $x + 5y - 5\sqrt{2} = 0$ or $x + 5y + 5\sqrt{2} = 0$ 6. $c = -4, (4, 4), (2, 0)$

8. $(-a,\ a(t_1 + t_2 + t_3) + at_1t_2t_3)$ 10. $x - 3y - 31 = 0$ or $3x + y + 7 = 0$

11. 32 sq. units 12. $(0, 0)$ or $(0, 5/2)$

13. $14x + 23y - 40 = 0$ 14. $(a^2 + b^2)(lx + my + n) - 2(al + bm)(ax + by + c) = 0$

16. $x - 7y + 13 = 0$ or $7x + y - 9 = 0$ 17. $x^2 + y^2 - 7x + 5y = 0$

18. $3x + 4y - 18 = 0$ or $x - 2 = 0$ 19. $(1, -2)$

20. $\alpha \in \left(-\frac{3}{2}, -1\right) \cup \left(\frac{1}{2}, 1\right)$ 21. $\frac{1}{64}$ sq units

22. $2x + 3y + 22 = 0$ 23. $x(m^2 - 1) - ym + (m^2 + 1)b + am = 0$

24. $16x^2 + 10xy + y^2 = 2$

30. 18 31. $y = 2x + 1$ or $y = -2x + 1$

H 1. (c)

Section-B : JEE Main/ AIEEE

1. (a) 2. (d) 3. (a) 4. (a) 5. (a) 6. (a) 7. (c) 8. (b) 9. (b) 10. (d) 11. (a) 12. (c)
13. (a) 14. (a) 15. (c) 16. (c) 17. (c) 18. (a) 19. (c) 20. (a) 21. (d) 22. (d) 23. (a) 24. (a)
25. (c) 26. (b) 27. (b) 28. (b) 29. (a) 30. (c) 31. (b) 32. (b)

Section-A JEE Advanced/ IIT-JEE

(A) FILL IN THE BLANKS :

1. $|x|+|y|=1$

The curve represents four lines

$x+y=1, x-y=1, -x+y=1, -x-y=1$

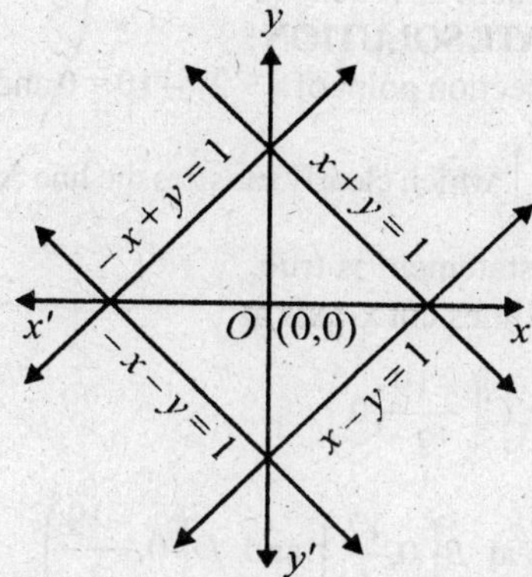

which enclose a square of side = distance between opp. sides

$x+y=1$ and $x+y=-1$

$$=\frac{1+1}{\sqrt{1+1}}=\sqrt{2}$$

KEY CONCEPT:

Perpendicular distance between two parallel lines :

$ax+by+c_1=0$

$ax+by+c_2=0$

is given by $\left|\frac{c_1-c_2}{\sqrt{a^2+b^2}}\right|=p$

$\therefore$ Req. area = $(\text{side})^2$ = 2 sq. units.

2. As $y=\log_{10}x$ can be obtained by replacing x by y and y by x in $y=10^x$

$\therefore$ The line of reflection is $y=x$.

3. Given that $3a+2b+4c=0$

$\Rightarrow \frac{3}{4}a+\frac{1}{2}b+c=0$

$\Rightarrow$ The set of lines $ax+by+c=0$ passes through the point (3/4, 1/2).

4. $|AP-BP|=6$

KEY CONCEPT:

We know that locus of a point, difference of whose distances from two fixed points is constant, is hyperbola with the fixed points as focii and the difference of distances as length of transverse axis.

Thus, $ae=4$ and $2a=6$

$\Rightarrow a=3, e=4/3$

$\Rightarrow b^2=9\left(\frac{16}{9}-1\right)=7 \quad \therefore$ Equation is $\frac{y^2}{9}-\frac{x^2}{7}=1$

(foci being on y-axis, it is vertical hyperbola)

ALTERNATE SOLUTION:

$|AP-BP|=6 \Rightarrow AP-BP=\pm 6$

$\Rightarrow AP=BP\pm 6$

$\Rightarrow \sqrt{x^2+(y-4)^2}=\sqrt{x^2+(y+4)^2}\pm 6$

$\Rightarrow x^2+y^2-8y+16=x^2+y^2+8y+16+36 \pm 12\sqrt{x^2+(y+4)^2}$

$\Rightarrow -16y-36=\pm 12\sqrt{x^2+(y+4)^2}$

$\Rightarrow 4y+9=\mp 3\sqrt{x^2+(y+4)^2}$

$\Rightarrow 16y^2+72y+81=9(x^2+y^2+8y+16)$

$\Rightarrow 7y^2-9x^2=63 \Rightarrow \frac{y^2}{9}-\frac{x^2}{7}=1$

5. If a, b, c are in A.P. then

$a+c=2b \Rightarrow a-2b+c=0$

$\Rightarrow ax+by+c=0$ passes through (1, –2).

6. **First quadrant.**

The equations of sides of triangle ABC are

$AB : x+y=1$

$BC : 2x+3y=6$

$CA : 4x-y=-4$

Solving these pairwise we get the vertices of Δ as follows $A(-3/5, 8/5)\ B(-3, 4)\ C(-3/7, 16/7)$

Now AD is line $\perp^{\text{lar}}$ to BC and passes through A. Any line perpendicular to BC is

$3x-2y+\lambda=0$

As it passes through $A(-3/5, 8/5)$

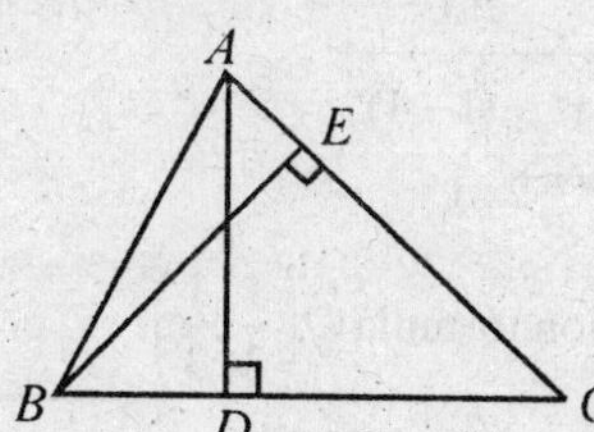

$\therefore \frac{-9}{5}-\frac{16}{5}+\lambda=0$

$\Rightarrow \lambda=5$

$\therefore$ Equation of altitude AD is

$3x-2y+5=0$ ....(1)

Any line perpendicular to side AC is

$x+4y+\mu=0$

As it passes through point $B(-3, 4)$

$\therefore -3+16+\mu=0 \Rightarrow \mu=-13$

$\therefore$ Equation of altitude BE is $x+4y-13=0$ ....(2)

Now orthocentre is the point of intersection of equations (1) and (2) (AD and BE)

Solving (1) and (2), we get $x=3/7, y=22/7$

As both the co-ordinates are positive, orthocentre lies in first quadrant.

7. Let the variable line be $ax + by + c = 0$ ....(1)

Then $\perp^{lar}$ distance of line from (0, 2) $= \frac{2a+c}{\sqrt{a^2+b^2}} = p_1$

$\perp^{lar}$ distance of line from (0, 2) $= \frac{2b+c}{\sqrt{a^2+b^2}} = p_2$

$\perp^{lar}$ distance of line from (1, 1) $= \frac{a+b+c}{\sqrt{a^2+b^2}} = p_3$

ATQ $p_1 + p_2 + p_3 = 0$

$\Rightarrow \frac{2a+c+2b+c+a+b+c}{\sqrt{a^2+b^2}} = 0$

$\Rightarrow 3a+3b+3c=0$

$\Rightarrow a+b+c=0$ ....(2)

From (1) and (2), we can say variable line (1) passes through the fixed point (1, 1).

8. Let BD be the bisector of $\angle ABC$.

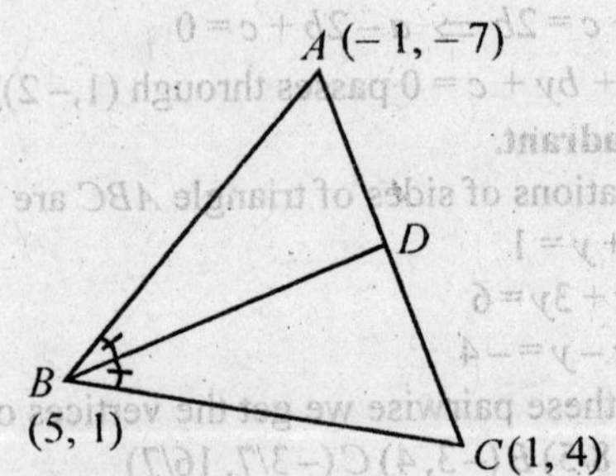

NOTE THIS STEP:

Then $AD : DC = AB : BC$

And $AB = \sqrt{(5+1)^2 + (1+7)^2} = 10$

$BC = \sqrt{(5-1)^2 + (1-4)^2} = 5$

$\therefore \quad AD : DC = 2 : 1$

$\therefore$ By section formula $D\left(\frac{1}{3}, \frac{1}{3}\right)$

Therefore equation of BD is

$y - 1 = \frac{1/3 - 1}{1/3 - 5}(x-5) \Rightarrow y - 1 = \frac{-2/3}{-14/3}(x-5)$

$\Rightarrow 7y - 7 = x - 5 \Rightarrow x - 7y + 2 = 0$

B TRUE / FALSE :

1. **KEY CONCEPT:**

Three lines $a_1x + b_1y + c_1 = 0$, $a_2x + b_2y + c_2 = 0$ and $a_3x + b_3y + c_3 = 0$ are concurrent if

$$\begin{vmatrix} a_1 & b_1 & c_1 \\ a_2 & b_2 & c_2 \\ a_3 & b_2 & c_2 \end{vmatrix} = 0$$

Statement can be restated as– 'three lines are concurrent' for which

$$\begin{vmatrix} 5 & 4 & 0 \\ 1 & 2 & -10 \\ 2 & 1 & 5 \end{vmatrix} = 0$$

LHS on expanding gives $5(10+10) - 4(5+20)$
$= 100 - 100 = 0 = \text{RHS}$

$\therefore$ Statement is True.

ALTERNATE SOLUTION :

Intersection point of $x + 2y - 10 = 0$ and $2x + y + 5 = 0$ is $\left(\frac{-20}{3}, \frac{25}{3}\right)$ which clearly satisfies the line $5x + 4y = 0$. Hence the given statement is true.

2. The given lines cut x-axis at

$A\left(\frac{17}{9}, 0\right), C\left(\frac{-19}{2}, 0\right)$

and y-axis at $B\left(0, \frac{17}{6}\right)$ and $D\left(0, \frac{-19}{3}\right)$.

Now A, B, C, D are concyclic if for AC and BD intersecting at O we have

$AO \times OC = BO \times OD$

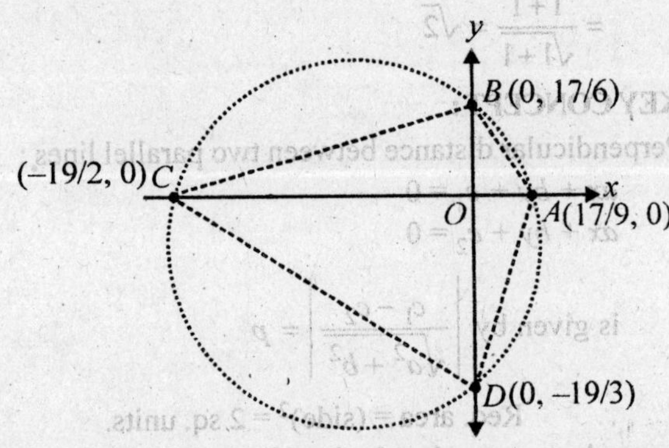

or, $\frac{AO}{BO} = \frac{OD}{OC}$ if $\frac{17/9}{17/6} = \frac{-19/3}{-19/2}$ i.e. $\frac{2}{3} = \frac{2}{3}$ which is true.

$\therefore$ The given statement is true.

C MCQ's WITH ONE CORRECT ANSWER :

1. (a) The given points are $A(-a, -b)$, $B(0, 0)$, $C(a, b)$ and $D(a^2, ab)$.

Slope of $AB = \frac{b}{a}$ = slope of BC = slope of BD

$\therefore$ A, B, C, D are collinear.

ALTERNATE SOLUTION:

Let $A = (-a, -b)$, $B = (0, 0)$, $C = (a, b)$ and $D = (a^2, ab)$

Area of triangle formed by the points A, B, C is given by

$$\Delta_1 = \frac{1}{2}\begin{vmatrix} -a & -b & 1 \\ 0 & 0 & 1 \\ a & b & 1 \end{vmatrix} = \frac{1}{2}[(-ab+ab)] = 0$$

[Expanding along R_2]

Clearly, points A, B, C are collinear. .....(1)

Similarly, area of triangle formed by the points B, C, D is given by

$$\Delta_2 = \frac{1}{2}\begin{vmatrix} 0 & 0 & 1 \\ a & b & 1 \\ a^2 & ab & 1 \end{vmatrix} = \frac{1}{2}\left[1(a^2b - a^2b)\right] = 0$$

[Expanding along R_1]

Clearly, points B, C, D are collinear ......(2)

From results (1) and (2) it is clear that the points A, B, C, D are collinear.

2. **(c)** Reflection about the line $y = x$, changes the point (4, 1) to (1, 4).

On translation of (1, 4) through a distance of 2 units along +ve direction of x-axis the point becomes (1 + 2, 4), i.e., (3, 4).

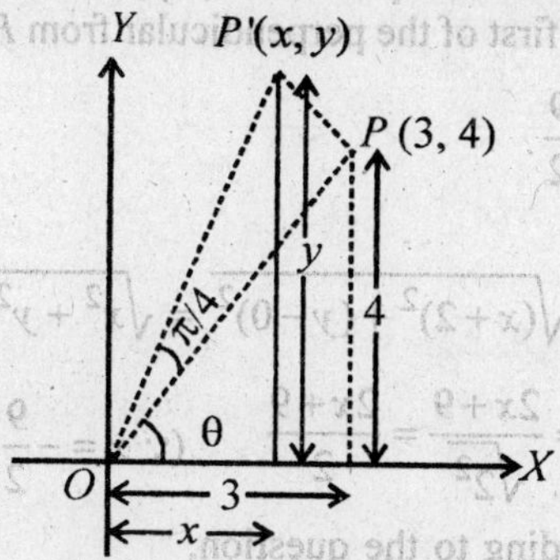

On rotation about origin through an angle $\pi/4$ the point P takes the position P' such that $OP = OP'$

Also $OP = 5 = OP'$ and $\cos\theta = \frac{3}{5}$, $\sin\theta = \frac{4}{5}$

Now, $x = OP'\cos\left(\frac{\pi}{4} + \theta\right)$

$$= 5\left(\cos\frac{\pi}{4}\cos\theta - \sin\frac{\pi}{4}\sin\theta\right)$$

$$= 5\left(\frac{3}{5\sqrt{2}} - \frac{4}{5\sqrt{2}}\right) = -\frac{1}{\sqrt{2}}$$

$$y = OP'\sin\left(\frac{\pi}{4} + \theta\right)$$

$$= 5\left(\sin\frac{\pi}{4}\cos\theta + \cos\frac{\pi}{4}\sin\theta\right)$$

$$= 5\left(\frac{3}{5\sqrt{2}} + \frac{4}{5\sqrt{2}}\right) = \frac{7}{\sqrt{2}} \quad \therefore \quad P' = \left(-\frac{1}{\sqrt{2}}, \frac{7}{\sqrt{2}}\right)$$

3. **(a)** Solving the given equations of lines pairwise, we get the vertices of Δ as

$A(-2, 2)\ B(2, -2), C(1, 1)$

Then $AB = \sqrt{16 + 16} = 4\sqrt{2}$

$BC = \sqrt{1 + 9} = \sqrt{10}$

$CA = \sqrt{9 + 1} = \sqrt{10}$ $\quad \therefore \Delta$ is isosceles.

ALTERNATE SOLUTION :

Angle between AB and AC is

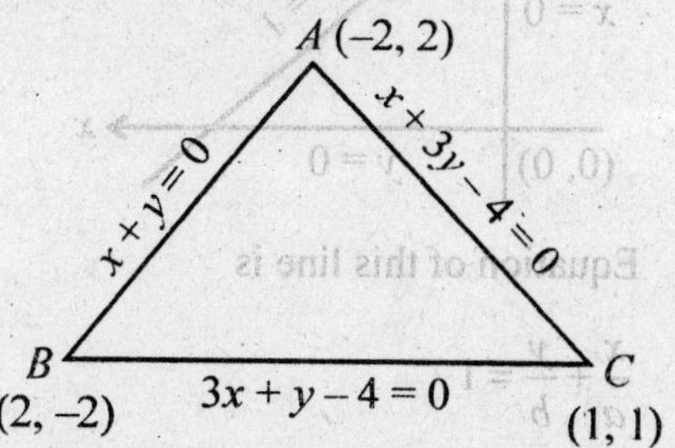

$$\angle A = \tan^{-1}\left|\frac{-1 + 1/3}{1 + 1/3}\right| = \tan^{-1}\left(\frac{1}{2}\right)$$

Angle between AB and BC is

$$\angle B = \tan^{-1}\left|\left(\frac{-1 + 3}{1 + 3}\right)\right| = \tan^{-1}\left(\frac{1}{2}\right)$$

Two $\angle's$ are equal (and not 60°)

$\therefore$ Δ is isosceles.

Hence (a) is the correct alternative.

4. **(a)** We have

$P = (1, 0), Q = (-1, 0), R = (2, 0)$

Let $S = (x, y)$

ATQ $SQ^2 + SR^2 = 2SP^2$

$\Rightarrow (x + 1)^2 + y^2 + (x - 2)^2 + y^2 = 2[(x - 1)^2 + y^2]$

$\Rightarrow 2x^2 + 2y^2 - 2x + 5 = 2x^2 + 2y^2 - 4x + 2$

$\Rightarrow 2x + 3 = 0 \Rightarrow x = -3/2$

Which is a straight line parallel to y-axis.

5. **(b)** As L has intercepts a and b on axes, equation of L is

$$\frac{x}{a} + \frac{y}{b} = 1 \qquad \text{....... (1)}$$

Let x and y axes be rotated through an angle θ in anticlockwise direction.

In new system intercepts are p and q, therefore equation of L becomes

$$\frac{x}{p} + \frac{y}{q} = 1 \qquad \text{....... (2)}$$

KEY CONCEPT :

As the origin is fixed in rotation, the distance of line from origin in both the cases should be same.

$$\therefore \quad \text{We get } d = \left|\frac{1}{\sqrt{\frac{1}{a^2} + \frac{1}{b^2}}}\right| = \left|\frac{1}{\sqrt{\frac{1}{p^2} + \frac{1}{q^2}}}\right|$$

$$\Rightarrow \frac{1}{a^2} + \frac{1}{b^2} = \frac{1}{p^2} + \frac{1}{q^2}$$

$\therefore$ (b) is the correct answer.

ALTERNATE SOLUTION : The given line has intercepts a and b on the co-ordinate axes.

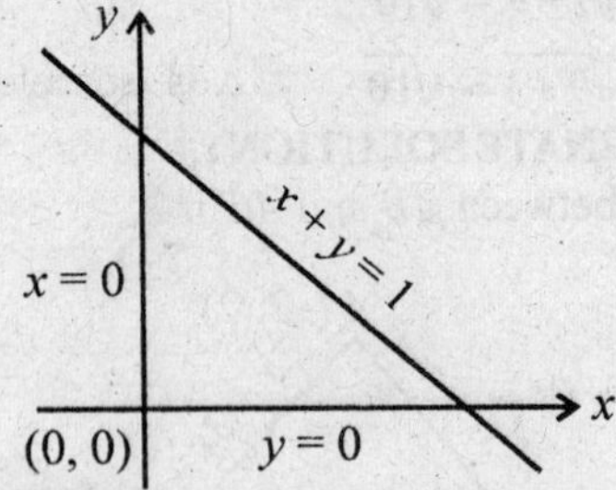

$\therefore$ Equation of this line is

$$\frac{x}{a}+\frac{y}{b}=1 \qquad \text{........(1)}$$

KEY CONCEPT : Now, we know that if co-ordinate axes are rotated by an angle θ, then transformed equation can be obtained by replacing x by $x\cos\theta - y\sin\theta$ and y by $x\sin\theta + y\cos\theta$.

Therefore transformed form of equation (1) is

$$\frac{x\cos\theta-y\sin\theta}{a}+\frac{x\sin\theta+y\cos\theta}{b}=1$$

$$\Rightarrow x\left(\frac{\cos\theta}{a}+\frac{\sin\theta}{b}\right)+y\left(\frac{\cos\theta}{b}-\frac{\sin\theta}{a}\right)=1$$

$$\Rightarrow \frac{x}{\left(\frac{ab}{b\cos\theta+a\sin\theta}\right)}+\frac{y}{\left(\frac{ab}{a\cos\theta-b\sin\theta}\right)}=1$$

$\therefore$ Intercepts of line on rotated (new) axes are

$$\frac{ab}{b\cos\theta+a\sin\theta},\ \frac{ab}{a\cos\theta-b\sin\theta}$$

But ATQ these are given to be p and q.

$\therefore$ We must have

$$p=\frac{ab}{b\cos\theta+a\sin\theta},\ q=\frac{ab}{a\cos\theta-b\sin\theta}$$

$$\Rightarrow \frac{1}{a}\cos\theta+\frac{1}{b}\sin\theta=\frac{1}{p} \qquad \text{........(i)}$$

$$\Rightarrow \frac{1}{b}\cos\theta-\frac{1}{a}\sin\theta=\frac{1}{q} \qquad \text{........(ii)}$$

Squaring and adding (i) and (ii), we get

$$\left(\frac{1}{a}\cos\theta+\frac{1}{b}\sin\theta\right)^2+\left(\frac{1}{b}\cos\theta-\frac{1}{a}\sin\theta\right)^2=\frac{1}{p^2}+\frac{1}{q^2}$$

$$\left(\frac{1}{a^2}+\frac{1}{b^2}\right)\cos^2\theta+\left(\frac{1}{a^2}+\frac{1}{b^2}\right)\sin^2\theta$$

$$=\frac{1}{p^2}+\frac{1}{q^2} \Rightarrow \frac{1}{a^2}+\frac{1}{b^2}=\frac{1}{p^2}+\frac{1}{q^2}$$

$\therefore$ (b) is the correct answer.

6. (a) Let the two perpendicular lines be the co-ordinate axes. Let (x, y) be the point sum of whose distances from two axes is 1 then we must have

$|x|+|y|=1$

or, $\pm x \pm y = 1$

These are the four lines

$x+y=1, x-y=1, \ -x+y=1, -x-y=1$

Any two adjacent sides are perpendicular to each other. Also each line is equidistant from origin. Therefore figure formed is a square.

7. (a) If variable point is P and $S(-2, 0)$ then $PS=\frac{2}{3}PM$

where PM is the perpendicular distance of point P from given line $x=-9/2$

$\therefore$ By definition P describes an ellipse. $\left(e=\frac{2}{3}<1\right)$

ALTERNATE SOLUTION :

Let the variable point be $P(x, y)$ and let $S(-2, 0)$. Let M be the first of the perpendicular from P to the line $x=-\frac{9}{2}$.

Now,

$$SP=\sqrt{(x+2)^2+(y-0)^2}=\sqrt{x^2+y^2+4x+4}$$

$$PM=\frac{2x+9}{\sqrt{2^2}}=\frac{2x+9}{2} \qquad \left(\because x=-\frac{9}{2}\Rightarrow 2x+9=0\right)$$

According to the question,

$$SP=\frac{2}{3}PM \Rightarrow 9SP^2=4PM^2$$

$$\Rightarrow 9x^2+9y^2+36x+36=4x^2+36x+81$$

$$\Rightarrow 5x^2+9y^2=45 \Rightarrow \frac{x^2}{9}+\frac{y^2}{5}=1$$

which is the equation of an ellipse.

8. (c) The sides of parallelogram are $x=2, x=3, y=1, y=5$.

D (2, 5) C (3, 5)

A (2, 1) B (3, 1)

$\therefore$ Diagonal AC is $\frac{y-1}{5-1}=\frac{x-2}{3-2}$ or $y=4x-7$

Equation diagonal BD is $\frac{x-2}{3-2}=\frac{y-5}{1-5}$ or $4x+y=13$

9. (c) The lines by which Δ is formed are $x=0$, $y=0$ and $x+y=1$.

KEY CONCEPT : Clearly, it is right Δ and we know that in a rightΔ orthocentre coincides with the vertex at which right ∠ is formed.

∴ Orthocentre is (0, 0).

10. (b) Let m be the slope of PQ then

$$\tan 45° = \left|\frac{m-(-2)}{1+m(-2)}\right|$$

$$\Rightarrow 1 = \left|\frac{m+2}{1-2m}\right| \Rightarrow \pm 1 = \frac{m+2}{1-2m}$$

$\Rightarrow m+2 = 1-2m$ or $-1+2m = m+2$

$\Rightarrow m = -1/3$ or $m = 3$

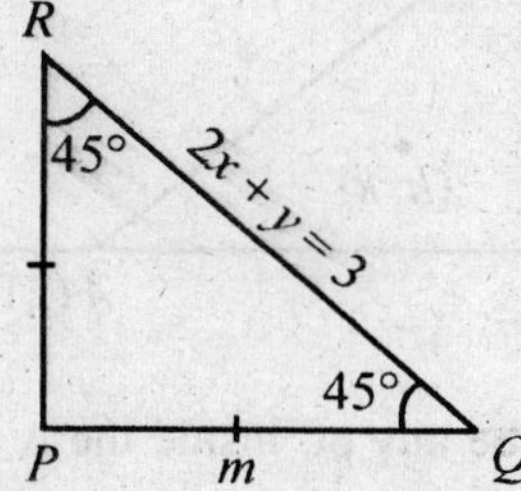

As PR also makes ∠ 45° with RQ.

∴ The above two values of m are for PQ and PR.

∴ Equation of PQ, $y-1 = -\frac{1}{3}(x-2)$

$\Rightarrow 3y-3 = -x+2 \quad \Rightarrow x+3y-5=0$

and equation of PR is $\Rightarrow 3x-y-5=0$

∴ Combined equation of PQ and PR is

$(x-3y-5)(3x-y-5)=0$

$\Rightarrow 3x^2-3y^2+8xy-20x-10y+25=0$

11. (a) $x_2 = x_1 r, x_3 = x_1 r^2$ and so is $y_2 = y_1 r, y_3 = y_1 r^2$

$$\Delta = \begin{vmatrix} x_1 & y_1 & 1 \\ x_2 & y_2 & 1 \\ x_3 & y_3 & 1 \end{vmatrix} = r.r^2 \begin{vmatrix} x_1 & y_1 & 1 \\ x_1 & y_1 & 1 \\ x_1 & y_1 & 1 \end{vmatrix} = 0$$

Hence the points lie on a line, i.e., they are collinear.

12. (d) S is the midpoint of Q and R

$$\text{Therefore, } S \equiv \left(\frac{7+6}{2}, \frac{3-1}{2}\right) = \left(\frac{13}{2}, 1\right)$$

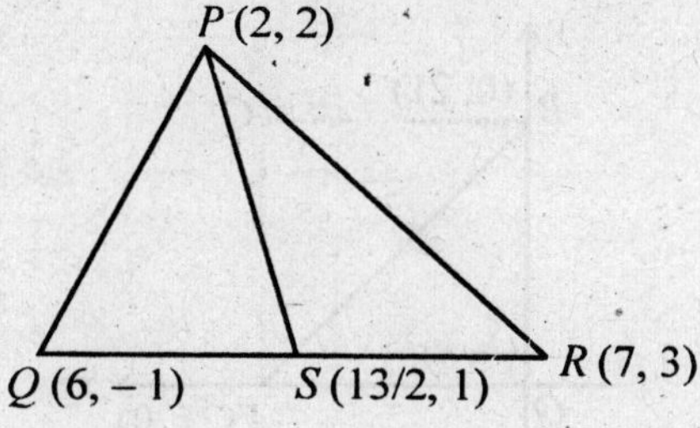

Now slope of $PS = m = \frac{2-1}{2-13/2} = -\frac{2}{9}$

Now equation of the line passing through (1, – 1) and parallel to PS is

$$y+1 = -\frac{2}{9}(x-1) \text{ or } 2x+9y+7=0$$

13. (d) Here AB = BC = CA = 2. So, it is an equilaterial triangle and the incentre coincides with centroid. Therefore,

$$\text{Incentre} = \left(\frac{0+1+2}{3}, \frac{0+0+\sqrt{3}}{3}\right) = \left(1, \frac{1}{\sqrt{3}}\right)$$

14. (a) Intersection of $3x+4y=9$ and $y=mx+1$.

For x co-ordinate

$3x+4(mx+1)=9 \Rightarrow (3+4m)x=5$

$$x = \frac{5}{3+4m}$$

For x to be an integer $3+4m$ should be a divisor of 5 i.e., 1, – 1, 5 or – 5.

$3+4m=1 \Rightarrow m=-1/2$ (not integer)

$3+4m=-1 \Rightarrow m=-1$ (integer)

$3+4m=5 \Rightarrow m=1/2$ (not an integer)

$3+4m=-5 \Rightarrow m=-2$ (integer)

∴ There are 2 integral values of m.

∴ (a) is the correct alternative.

15. (d)

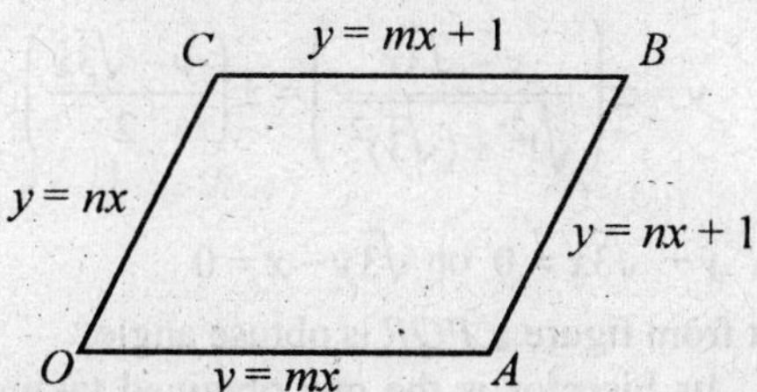

The vertices, $O(0,0), A\left(\frac{1}{m-n}, \frac{m}{m-n}\right), B(0,1)$

$Ar(\|^{gm} OABC) = 2\,Ar(\Delta OAB)$

$$= 2\frac{1}{2}\left|\left[0\left(\frac{m}{m-n}-1\right)+\frac{1}{m-n}(1-0)+0\left(0-\frac{m}{m-n}\right)\right]\right|$$

$$= \frac{1}{|m-n|}$$

16. (d) Clearly $OP = OQ = 1$ and $\angle QOP = \alpha - \theta - \theta = \alpha - 2\theta$.

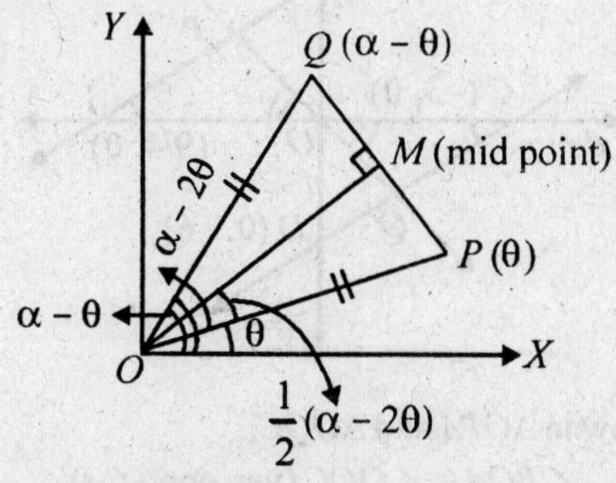

The bisector of $\angle QOP$ will be a perpendicular to PQ and also bisect it. Hence Q is reflection of P in the line OM which makes an angle $\angle MOP + \angle POX$ with x-axis, i.e.,

$$\frac{1}{2}(\alpha - 2\theta) + \theta = \alpha/2.$$

So that slope of OM is $\tan \alpha/2$.

17. (c) $\tan\theta = \sqrt{3} \Rightarrow \theta = 60° \Rightarrow \angle PQR = 120°$

$\Rightarrow$ bisector will have slope tan 120°

$\Rightarrow$ equation of bisector is $\sqrt{3}x + y = 0$

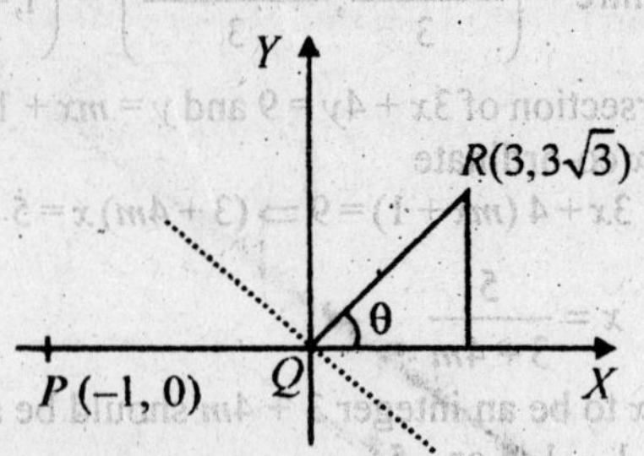

ALTERNATE SOLUTION:

Equation of PQ is $y = 0$

Equation of QR is $y = \frac{3\sqrt{3}x}{3}$

$\therefore$ Angle bisector of $\angle PQR$ is given by

$$y = \pm\left(\frac{y - \sqrt{3}x}{\sqrt{1^2 + (\sqrt{3})^2}}\right) = \pm\left(\frac{y - \sqrt{3}x}{2}\right)$$

$\Rightarrow y + \sqrt{3}x = 0$ or $\sqrt{3}y - x = 0$

But from figure $\angle PQR$ is obtuse angle.

$\therefore$ Its bisector is the one obtained taking +ve sign

i.e. $\sqrt{3}x + y = 0$

18. (b) The given lines are

$2x + y = 9/2$ ....(1)

and $2x + y = -6$ ....(2)

Signs of constants on R.H.S. show that two lines lie on opp. sides of origin. Let any line through origin meets these lines in P and Q respectively then req. ratio is $OP : OQ$

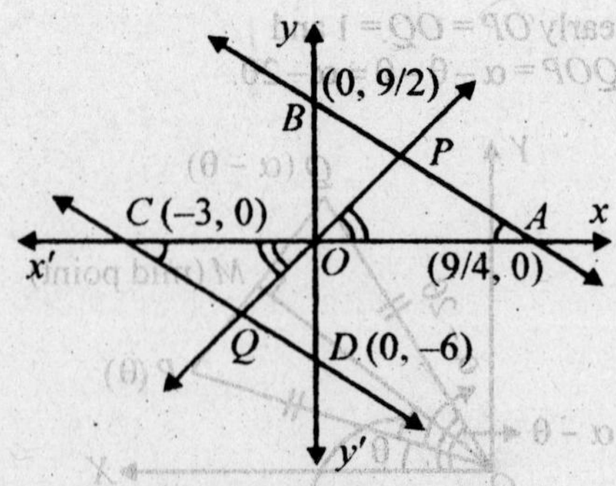

Now in ΔOPA and ΔOQC,

$\angle POA = \angle QOC$ (ver. opp. $\angle$'s)

$\angle PAO = \angle OCQ$ (alt. int. $\angle$'s)

$\therefore$ $\Delta OPA \sim \Delta OQC$ (by AA similarly)

$\therefore$ $\frac{OP}{OQ} = \frac{OA}{OC} = \frac{9/4}{3} = \frac{3}{4}$

$\therefore$ Req. ratio is 3 : 4.

ALTERNATE SOLUTION : In particular, PQ can be taken perpendicular to the two parallel lines then required ratio = ratio of perpendicular distances of lines from origin

$$= OP : OQ = \left|\frac{9}{\sqrt{20}}\right| : \left|\frac{-6}{\sqrt{5}}\right| = 3 : 4$$

19. (b) Equation of AB is $\frac{x}{21} + \frac{y}{21} = 1$ or $x + y = 21$

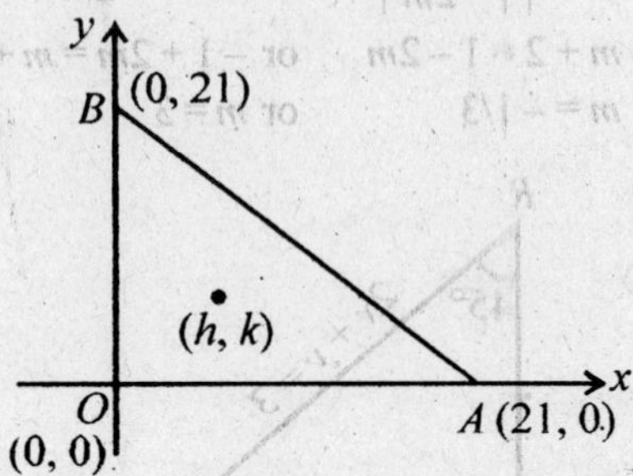

Let (h, k) be any pt. inside the Δ OAB then we must have

$h < 21, h > 0$ ....(i)

$k < 21, k > 0$ ....(ii)

and $h + k < 21$ ....(iii)

For integral values of (h, k) satisfying (i), (ii) and (iii) simultaneously, let

	Total no. of pts
$h = 1$ then $k = 1, 2, 3, \ldots 19$	19
$h = 2$ then $k = 1, 2, 3, \ldots 18$	18
$h = 3$ then $k = 1, 2, 3, \ldots 17$	17
: :	:
: :	1
$h = 19$ then $k = 1$	

$\therefore$ Total no. of integral points

$$= 19 + 18 + 17 + \ldots + 1 = \frac{19 \times 20}{2} = 190$$

ALTERNATE SOLUTION:

Total no. of points within the square $OABC$

$= 20 \times 20 = 400$

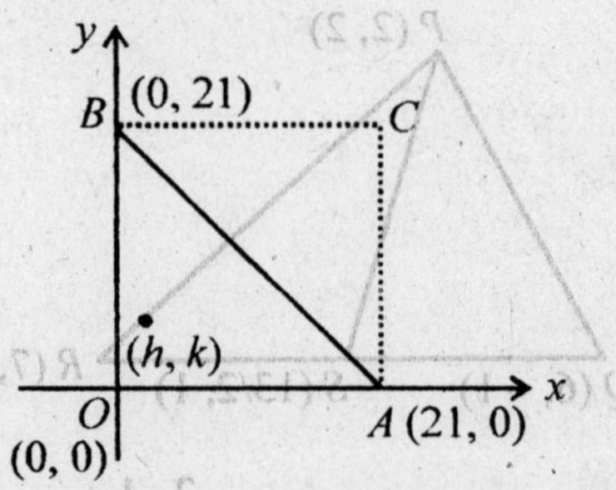

Points on line $AB = 20$ ((1, 1), (2, 2), ... (20, 20))

$\therefore$ Points within ΔOBC and ΔABC = 400 – 20 = 380

By symmetry points within ΔOAB $= \frac{380}{2} = 190$

20. (c) KEY CONCEPT : We know that orthocentre is the meeting point of altitudes of a Δ.

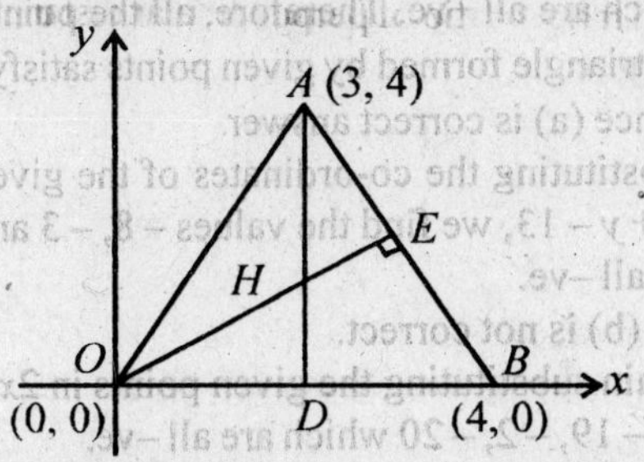

Equation of alt. AD

⇒ line parallel to y-axis through (3, 4)

⇒ $x = 3$ .... (1)

Similarly eq^n of $OE \perp AB$ is

$$y = -\frac{3-4}{4-0}x$$

⇒ $y = x/4$ .... (2)

Solving (1) and (2), we get orthocentre as (3, 3/4).

21. (a) $x^2 - y^2 + 2y = 1$

⇒ $x = \pm(y-1)$

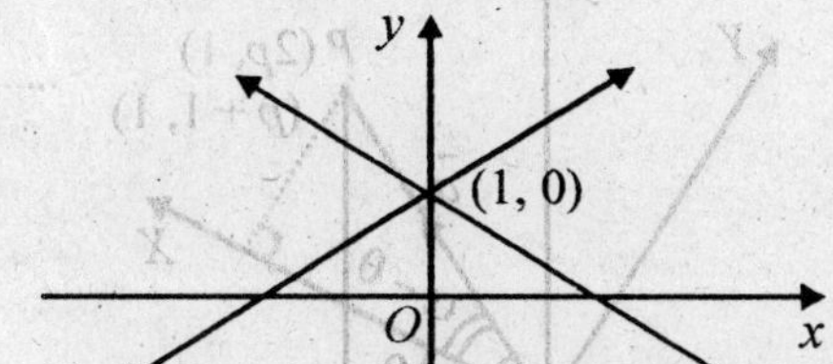

Bisectors of above lines are x = 0 and y = 1.

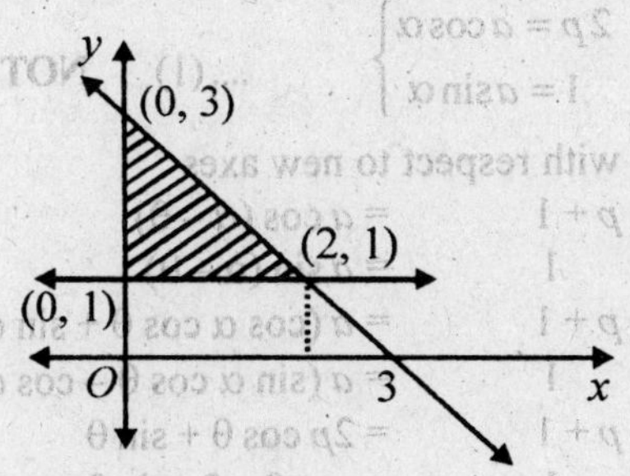

So area between x = 0, y = 1 and x + y = 3 is shaded region shown in figure.

Area $= \frac{1}{2} \times 2 \times 2 = 2$ sq. units.

22. (c) ∵ $Ar(\Delta OPR) = Ar(\Delta PQR) = Ar(\Delta OQR)$

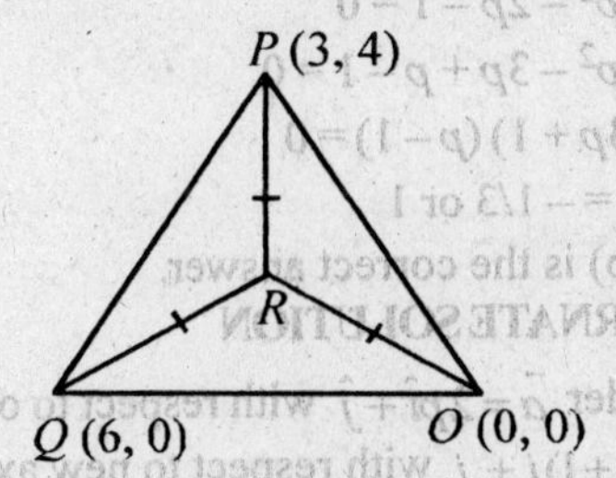

∴ By simply geometry

R should be the centroid of ΔPQO

$$\Rightarrow R\left(\frac{3+6+0}{3}, \frac{4+0+0}{3}\right) = \left(3, \frac{4}{3}\right)$$

ALTERNATE SOLUTION :

Here $OP = PQ = 5$

⇒ ΔOPQ is an isosceles Δ.

∴ R must lie on angle bisector of $\angle P$.

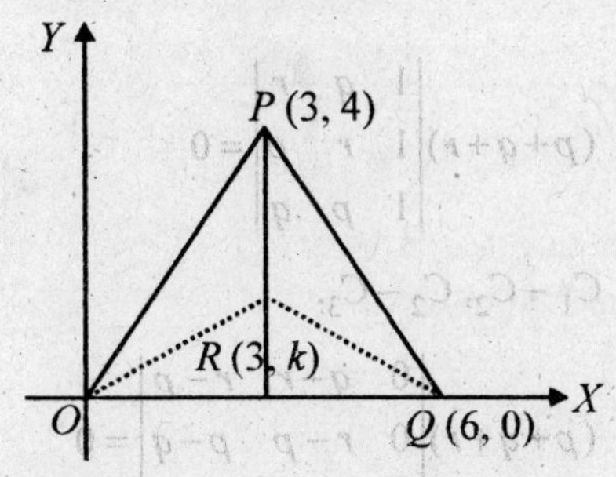

Let $R(3, k)$

Now $Ar(\Delta ORQ) = \frac{1}{3} Ar(\Delta OPQ)$

$$\Rightarrow \frac{1}{2} \times 6 \times k = \frac{1}{3} \times \frac{1}{2} \times 6 \times 4 \Rightarrow k = \frac{4}{3} \quad \therefore R\left(3, \frac{4}{3}\right)$$

23. (b) Let the slope of line L be m.

Then $\left|\frac{m+\sqrt{3}}{1-\sqrt{3}m}\right| = \sqrt{3}$

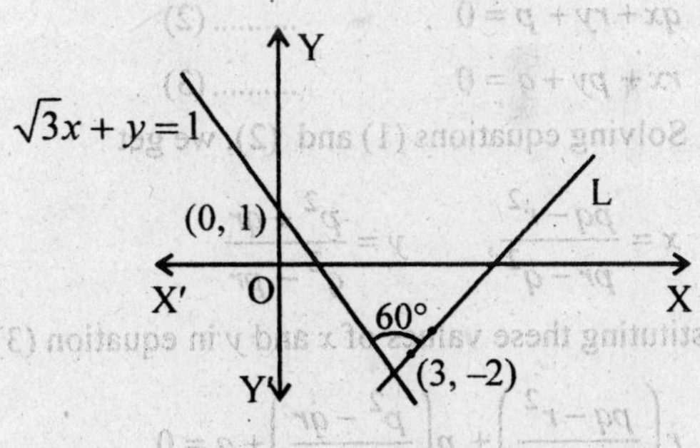

$\Rightarrow m + \sqrt{3} = \pm(\sqrt{3} - 3m)$

$\Rightarrow 4m = 0$ or $2m = 2\sqrt{3} \Rightarrow m = 0$ or $m = \sqrt{3}$

∵ L intersects x-axis, ∴ $m = \sqrt{3}$

∴ Equation of L is $y + 2 = \sqrt{3}\,(x - 3)$

or $\sqrt{3}\,x - y - (2 + 3\sqrt{3}) = 0$

D MCQ's WITH ONE OR MORE THAN ONE CORRECT :

1. (a, b, c)

For concurrency of three lines

$px + qy + r = 0$; $qx + ry + p = 0$;

$rx + py + q = 0$

We must have,

$$\begin{vmatrix} p & q & r \\ q & r & p \\ r & p & q \end{vmatrix} = 0 \quad \textbf{NOTE THIS STEP}$$

$$\Rightarrow C_1+C_2+C_3, \begin{vmatrix} p+q+r & q & r \\ p+q+r & r & p \\ p+q+r & p & q \end{vmatrix} = 0$$

$$\Rightarrow (p+q+r)\begin{vmatrix} 1 & q & r \\ 1 & r & p \\ 1 & p & q \end{vmatrix} = 0$$

$\Rightarrow C_1 - C_2, C_2 - C_3,$

$$\Rightarrow (p+q+r)\begin{vmatrix} 0 & q-r & r-p \\ 0 & r-p & p-q \\ 1 & p & q \end{vmatrix} = 0$$

$\Rightarrow (p+q+r)(pq-q^2-rp+rq-r^2+pr+pr-p^2)=0$

$\Rightarrow (p+q+r)(p^2+q^2+r^2-pq-pr-rq)=0$

$\Rightarrow p^3+q^3+r^3-3pqr=0$

It is clear that a, b, c are correct options.

ALTERNATE SOLUTION :

KEY CONCEPT : If the point of intersection of any two lines of the system of three lines satisfies the third equation, then the three lines are concurrent.

Given lines are:

$px+qy+r=0$ .... (1)

$qx+ry+p=0$ .... (2)

$rx+py+q=0$ .... (3)

Solving equations (1) and (2), we get

$$x = \frac{pq-r^2}{pr-q^2}, \quad y = \frac{p^2-qr}{q^2-pr}$$

Substituting these values of x and y in equation (3), we get

$$r\left(\frac{pq-r^2}{pr-q^2}\right)+p\left(\frac{p^2-qr}{q^2-pr}\right)+q=0$$

$\Rightarrow p^3+q^3+r^3-3pqr=0$ .... (4)

$\Rightarrow (p+q+r)(p^2+q^2+r^2-pq-qr-rp)=0$ .... (5)

From equations (4) and (5), we obtain

$p^3+q^3+r^3=3pqr,\ p+q+r=0,$

$p^2+q^2+r^2=pq+qr+rp$

2. **(e)** Let $A(0, 8/3)$, $B(1, 3)$ and $C(82, 30)$.

Now, slope of line $AB = \frac{3-8/3}{1-0} = \frac{1}{3}$

Slope of line $BC = \frac{30-3}{82-1} = \frac{27}{81} = \frac{1}{3}$

$\Rightarrow AB \parallel BC$ and B is common point.

$\Rightarrow A, B, C$ are collinear. **NOTE THIS STEP**

3. **(a, c)** Substituting the co-ordinates of the points (1, 3), (5, 0) and (– 1, 2) in $3x + 2y$, we obtain the value 8, 15 and 1 which are all +ve. Therefore, all the points lying inside the triangle formed by given points satisfy $3x + 2y \geq 0$. Hence (a) is correct answer.

Substituting the co-ordinates of the given points in $2x + y - 13$, we find the values – 8, – 3 and – 13 which are all –ve.

So, (b) is not correct.

Again substituting the given points in $2x - 3y - 12$ we get – 19, – 2, – 20 which are all –ve.

It follows that all points lying inside the triangle formed by given points satisfy $2x - 3y - 12 \leq 0$.

So, (c) is the correct answer.

Finally substituting the co-ordinates of the given points in $-2x + y$, we get 1, – 10 and 4 which are not all +ve.

So, (d) is not correct.

Hence, (a) and (c) are the correct answers.

4. **(b)** Let $\vec{a}$ makes an angle α with positive x-axis.

Let the original co-ordinate system be rotated through angle θ and $|\vec{a}| = a$.

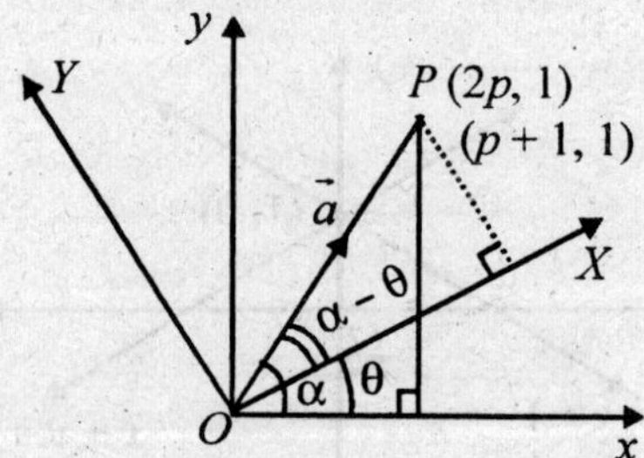

Then w.r. to original axis.

$$\left.\begin{aligned} 2p &= a\cos\alpha \\ 1 &= a\sin\alpha \end{aligned}\right\} \quad .... (1) \quad \textbf{NOTE THIS STEP}$$

And with respect to new axes

$p+1 = a\cos(\alpha-\theta)$

$1 = a\sin(\alpha-\theta)$

$\Rightarrow p+1 = a(\cos\alpha\cos\theta+\sin\alpha\sin\theta)$

$1 = a(\sin\alpha\cos\theta-\cos\alpha\sin\theta)$

$\Rightarrow p+1 = 2p\cos\theta+\sin\theta$ .... (i)

$1 = \cos\theta-2p\sin\theta$ .... (ii)

Using equation (1)

Squaring and adding (i) and (ii), we get

$(p+1)^2+1^2 = 4p^2\cos^2\theta+\sin^2\theta+4p\sin\theta\cos\theta+\cos^2\theta+4p^2\sin^2\theta-4p\sin\theta\cos\theta$

$\Rightarrow p^2+2p+2=4p^2+1$

$\Rightarrow 3p^2-2p-1=0$

$\Rightarrow 3p^2-3p+p-1=0$

$\Rightarrow (3p+1)(p-1)=0$

$\Rightarrow p=-1/3$ or 1

$\therefore$ (b) is the correct answer.

ALTERNATE SOLUTION

Consider $\vec{a}=2p\hat{i}+\hat{j}$ with respect to original axes and $a=(p+1)i+j$ with respect to new axes.

Now, as length of vector will remain the same

$\therefore \quad |\vec{a}| = \sqrt{(2p)^2 + 1} = \sqrt{(p+1)^2 + 1^2}$

$\Rightarrow p^2 + 2p + 2 = 4p^2 + 1$

$\Rightarrow 3p^2 - 2p - 1 = 0$

$\Rightarrow p = 1$ or $- 1/3$

$\therefore$ (b) is the correct answer.

5. **(c)** *PQRS* will represent a parallelogram if and only if the mid-point of *PR* is same as that of the mid-point of *QS*. That is, if and only if

$$\frac{1+5}{2} = \frac{4+a}{2} \text{ and } \frac{2+7}{2} = \frac{6+b}{2}$$

$\Rightarrow a = 2$ and $b = 3$.

6. **(d)** Slope of $x + 3y = 4$ is $- 1/3$ and slope of $6x - 2y = 7$ is 3. Therefore, these two lines are perpendicular which shows that both diagonals are perpendicular. Hence *PQRS* must be a rhombus.

7. **(a, c, d)**
Since the co-ordinates of in the centre depend on lengths of side of Δ. $\therefore$ it can have irrational coordinates

8. **(b, c)** We know that length of intercept made by a circle on a line is given by $= 2\sqrt{r^2 - p^2}$

where $p = \perp$ distance of line from the centre of the circle.

Here circle is $x^2 + y^2 - x + 3y = 0$ with centre $\left(\frac{1}{2}, \frac{-3}{2}\right)$

and radius $= \frac{\sqrt{10}}{2}$

$L_1 :\ y = mx$ (any line through origin)
$L_2 :\ x + y - 1 = 0$ (given line)
ATQ circle makes equal intercepts on L_1 and L_2

$$\Rightarrow 2\sqrt{\frac{10}{4} - \frac{\left(\frac{m}{2} + \frac{3}{2}\right)^2}{m^2 + 1}} = 2\sqrt{\frac{10}{4} - \frac{\left(\frac{1}{2} - \frac{3}{2} - 1\right)^2}{2}}$$

$$\Rightarrow \frac{\left(\frac{m+3}{2}\right)^2}{m^2 + 1} = 2$$

$\Rightarrow m^2 + 6m + 9 = 8m^2 + 8$

$\Rightarrow 7m^2 - 6m - 1 = 0$

$\Rightarrow 7m^2 - 7m + m - 1 = 0$

$\Rightarrow (7m + 1)(m - 1) = 0$

$\Rightarrow m = 1, -1/7$

$\therefore$ The req. line L_1 is $y = x$ or $y = -\frac{x}{7}$,

i.e., $x - y = 0$ or $x + 7y = 0$.

9. **(a)** The intersection point of two lines is $\left(\frac{-c}{a+b}, \frac{-c}{a+b}\right)$

Distance between (1, 1) and $\left(\frac{-c}{a+b}, \frac{-c}{a+b}\right) < 2\sqrt{2}$

$$\Rightarrow 2\left(1 + \frac{c}{a+b}\right)^2 < 8 \Rightarrow 1 + \frac{c}{a+b} < 2$$

$\Rightarrow a + b - c > 0$

(E) SUBJECTIVE PROBLEMS :

1. Let P (x, y) divides line segment AB in the ratio 1 : 2, so that AP = $\ell/3$ and BP = $2\ell/3$ where AB = ℓ.
Then PN = x and PM = y
Let $\angle$ PAM = θ = $\angle$ BPN

In ΔPMA, $\sin\theta = \frac{y}{\ell/3} = \frac{3y}{\ell}$

In ΔPNB, $\cos\theta = \frac{x}{2\ell/3} = \frac{3x}{2\ell}$

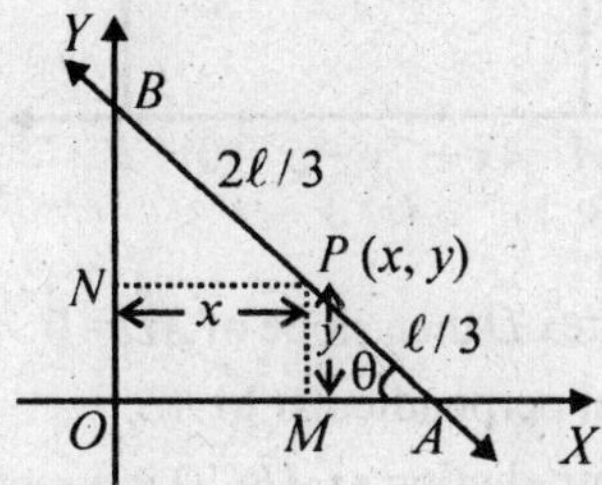

Now, $\sin^2\theta + \cos^2\theta = 1$

$\Rightarrow \frac{9y^2}{\ell^2} + \frac{9x^2}{4\ell^2} = 1 \Rightarrow 9x^2 + 36y^2 = 4\ell^2$

ALTERNATE SOLUTION

Let P (x, y) divides the line segment AB in the ratio 1 : 2. Let A (a, 0) and B (0, b)
Then by section formula

$x = \frac{a \times 2 + 1 \times 0}{2+1}$

and $y = \frac{0 \times 2 + b \times 1}{2+1}$

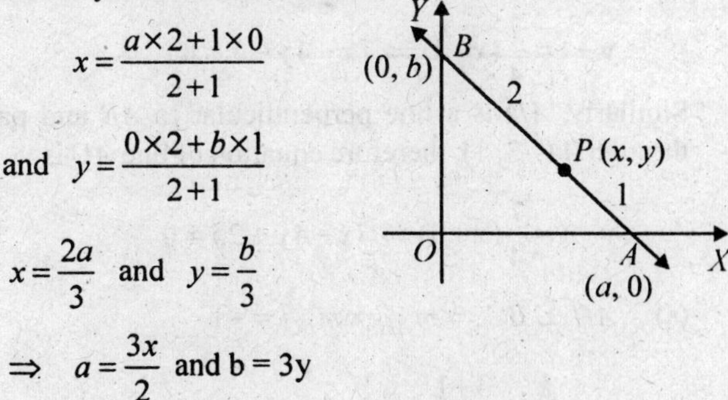

$x = \frac{2a}{3}$ and $y = \frac{b}{3}$

$\Rightarrow a = \frac{3x}{2}$ and b = 3y

Also $OA^2 + OB^2 = AB^2 \Rightarrow a^2 + b^2 = \ell^2$

$\Rightarrow \left(\frac{3x}{2}\right)^2 + (3y)^2 + \ell^2$

$\Rightarrow 9x^2 + 36y^2 = 4\ell^2$ which the required locus.

2. As *C* lies on the line $y = x + 3$, let the co-ordinates of *C* be $(\lambda, \lambda + 3)$. Also *A* (2, 1), *B* (3, –2).
Then area of ΔABC is given by

$$\frac{1}{2}\begin{vmatrix} 2 & 1 & 1 \\ 3 & -2 & 1 \\ \lambda & \lambda+3 & 1 \end{vmatrix} = \pm 5$$

$\Rightarrow |2(-2-\lambda-3)-1(3-\lambda)(3\lambda+9+2\lambda)|=10$

$\Rightarrow |-2\lambda-10-3+\lambda+5\lambda+9|=10 \Rightarrow |4\lambda-4|=10$

$\Rightarrow 4\lambda-4=10$ or $4\lambda-4=-10$

$\Rightarrow \lambda=7/2$ or $\lambda=-3/2$

$\therefore$ Coordinates of C are $\left(\frac{7}{2},\frac{13}{2}\right)$ or $\left(\frac{-3}{2},\frac{3}{2}\right)$

3. Let side AB of rectangle $ABCD$ lies along $4x+7y+5=0$.

As $(-3, 1)$ lies on the line, let it be vertex A.

Now $(1, 1)$ is either vertex C or D.

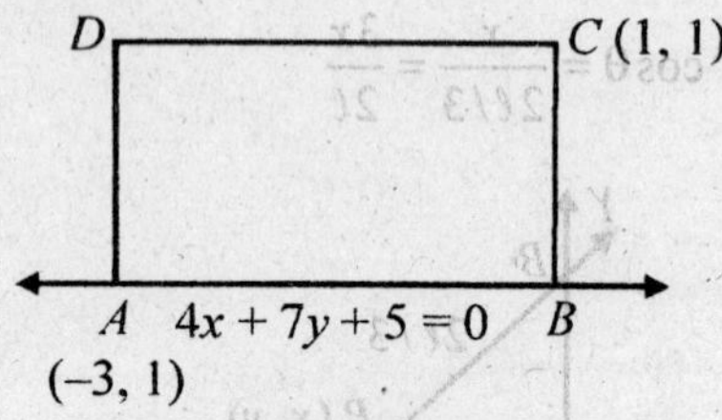

If $(1, 1)$ is vertex D then slope of $AD = 0$

$\Rightarrow$ AD is not perpendicular to AB.

But it is a contradiction as $ABCD$ is a rectangle.

$\therefore$ (1, 1) are the co-ordinates of vertex C.

CD is a line parallel to AB and passing through C, therefore equation of CD is

$$y-1=-\frac{4}{7}(x-1) \Rightarrow 4x+7y-11=0$$

Also BC is a line perpendicular to AB and passing through C, therefore equation of BC is

$$y-1=\frac{7}{4}(x-1) \Rightarrow 7x-4y-3=0$$

Similarly, AD is a line perpendicular to AB and passing through $A(-3, 1)$, therefore equation of line AD is

$$y-1=\frac{7}{4}(x+3) \Rightarrow 7x-4y+25=0$$

4. (a) $AH \perp BC \Rightarrow m_{AH} \times m_{BC} = -1$

$$\Rightarrow \frac{k}{h}\times\frac{3+1}{-2-5}=-1$$

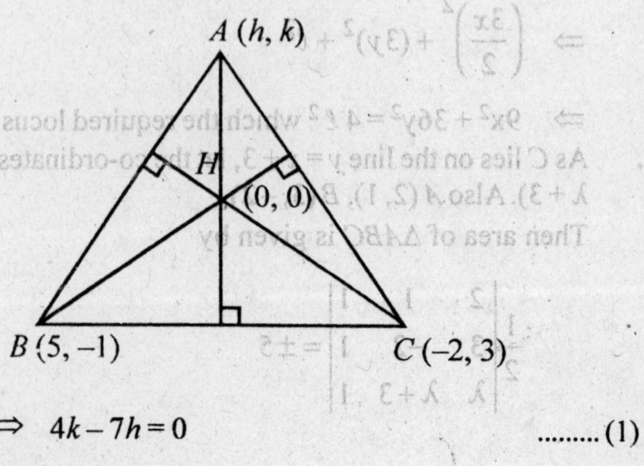

$\Rightarrow 4k-7h=0$ .... (1)

Also, $BH \perp AC$

$$\Rightarrow \frac{-1}{5}\times\frac{3-k}{-2-h}=-1 \Rightarrow 3-k=-10-5h$$

$\Rightarrow 5h-k+13=0$ .... (2)

Solving (1) and (2), we get $h=-4, k=-7$

$\therefore$ Third vertex is $(-4, -7)$.

(b) The given lines are $x-2y+4=0$ .... (1)

and $4x-3y+2=0$ ......(2)

Both the lines have constant terms of same sign.

$\therefore$ The equation of bisectors of the angles between the given lines are

$$\frac{x-2y+4}{\sqrt{1+4}}=\pm\frac{4x-3y+2}{\sqrt{16+9}}$$

Here $a_1a_2+b_1b_2>0$ therefore, taking +ve sign on RHS, we get obtuse angle bisector as

$$(4-\sqrt{5})x+(2\sqrt{5}-3)y-(4\sqrt{5}-2)=0 \quad ....(3)$$

5. The given line is $5x-y=1$

$\therefore$ The equation of line L which is perpendicular to the given line is $x+5y=\lambda$. This line meets co-ordinate axes at $A(\lambda, 0)$ and $B(0, \lambda/5)$.

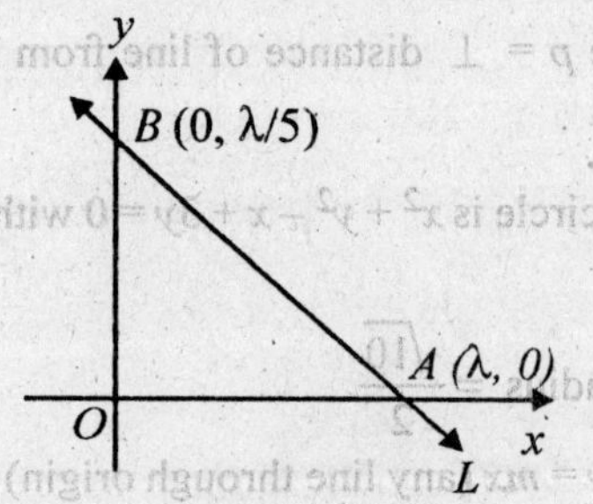

$\therefore$ Area of $\Delta OAB = \frac{1}{2}\times OA\times OB$

$$\Rightarrow 5=\frac{1}{2}\times\lambda\times\frac{\lambda}{5} \Rightarrow \lambda^2=5^2\times2 \Rightarrow \lambda=\pm5\sqrt{2}$$

$\therefore$ The equation of line L is $x+5y-5\sqrt{2}=0$

or $x+5y+5\sqrt{2}=0$.

6. Let $ABCD$ be a rectangle and co-ordinates of its opp. vertices A and C are $(1, 3)$ and $(5, 1)$ resp.

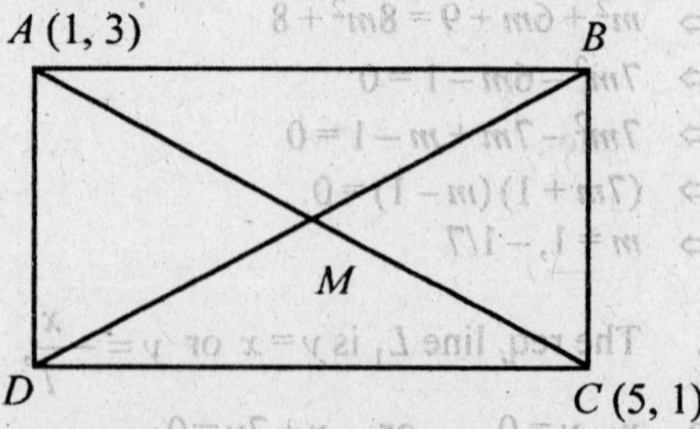

Given that other two vertices B and D lie on $y=2x+c$

$\therefore$ Eq. of diagonal BD is $2x-y+c=0$

Now we know that diagonals of a rectangle bisect each other.

$\therefore$ Mid point of AC will lie on BD

$$M\left(\frac{1+5}{2},\frac{3+1}{2}\right)=(3,2)$$

As (3, 2) lies on $2x - y + c = 0$

$\therefore \quad 2 \times 3 - 2 + c = 0 \Rightarrow c = -4$

$\therefore$ Eq. of BD is $2x - y - 4 = 0$

For which $m = 2 \Rightarrow \tan\theta = 2 \Rightarrow \cos\theta = 1/\sqrt{5}$;

$\sin\theta = 2/\sqrt{5}$.

Now we know that if a line passes though (x_1, y_1) has inclination θ then its symmetric eq. is

$$\frac{x - x_1}{\cos\theta} = \frac{y - y_1}{\sin\theta} = r$$

where r is the distance between (x, y) and (x_1, y_1)

$\therefore$ Eq. of BD in symm. form can be written as

$$\frac{x-3}{1/\sqrt{5}} = \frac{y-2}{2/\sqrt{5}} = r \quad \text{.......(1)}$$

Now length of diagonal

$$AC = \sqrt{4^2 + 2^2} = \sqrt{20} = 2\sqrt{5}$$

As diagonals of a rectangle are equal in length

$\therefore \quad BD = 2\sqrt{5}$

Also M is the mind point of BD

$\therefore \quad BM = MD = \sqrt{5}$.

In order to find B, we use symm. form of line BD with $r = \sqrt{5}$

$$\therefore \frac{x-3}{1/\sqrt{5}} = \frac{y-2}{2/\sqrt{5}} = \sqrt{5}$$

$\Rightarrow x = 1 + 3, y = 2 + 2 \Rightarrow B(4, 4)$

Similarly for point D (being in opp. direction of B) we use $r = -\sqrt{5}$, then we get

$$\frac{x-3}{1/\sqrt{5}} = \frac{y-2}{2/\sqrt{5}} = -\sqrt{5} \Rightarrow x = -1 + 3, y = -2 + 2$$

$\Rightarrow x = 2, y = 0 \qquad \therefore D(2, 0)$

Hence $c = -4$ and remaining vertices are (4, 4) and (2, 0).

7. Here eq. of AB is $\dfrac{x}{c\cos\alpha} + \dfrac{y}{c\sin\alpha} = 1$

$\Rightarrow x\sin\alpha + y\cos\alpha = c\sin\alpha\cos\alpha \quad \text{........(1)}$

Eq. of PN (perpendicular to AB and through P)

$y - c\sin\alpha = \cot\alpha\,(x - c\cos\alpha)$

$\Rightarrow x\cos\alpha - y\sin\alpha = c(\cos^2\alpha - \sin^2\alpha) \quad \text{.........(2)}$

N is intersection point of (1) and (2),

Multiplying (1) by sin α and (2) by cos α and subtracting.

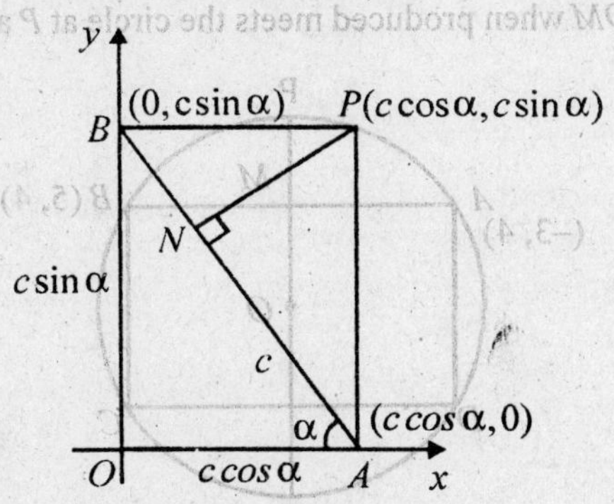

We get, $x = c\cos^3\alpha$, $y = c\sin^3\alpha$

$$\therefore \quad \cos\alpha = \left(\frac{x}{c}\right)^{\frac{1}{3}}, \ \sin\alpha = \left(\frac{y}{c}\right)^{\frac{1}{3}}$$

$$\therefore \quad \text{Locus of N (x, y) is, } \left(\frac{x}{c}\right)^{\frac{2}{3}} + \left(\frac{y}{c}\right)^{\frac{2}{3}} = 1$$

$$\Rightarrow \quad x^{\frac{2}{3}} + y^{\frac{2}{3}} = c^{\frac{2}{3}}$$

ALTERNATE SOLUTION

From figure,

x = OA – AL

= c cos α – AN cos α

= c cos α – (AP sin α.) cos α

= c cos α – c sin α . sin α cos α

= c cos α (1 – sin² α)

= c cos³ α

y = OB – MB

= c sin α – BN sin α

= c sin α – BP cos α sin α

= c sin α – c cos α . cos α sin α

= c sin α (1 – cos² α) = c sin³ α

$$\therefore \text{ Locus of (x, y) is } \left(\frac{x}{c}\right)^{\frac{2}{3}} + \left(\frac{y}{c}\right)^{\frac{2}{3}} = 1 \text{ or } x^{\frac{2}{3}} + y^{\frac{2}{3}} = c^{\frac{2}{3}}$$

8. **KEY CONCEPT :** We know orthocentre of Δ is meeting point of altitudes.

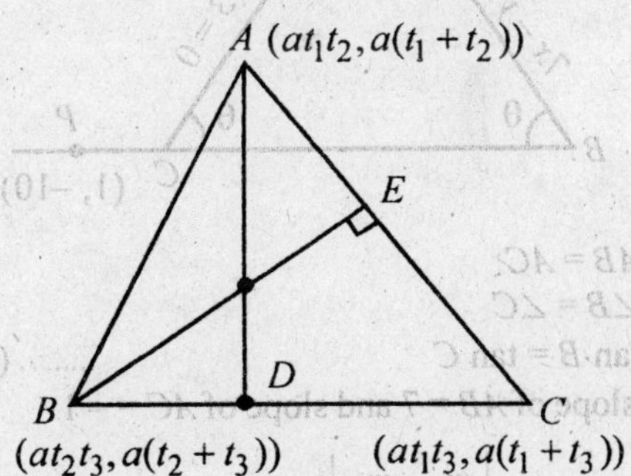

$$\text{Slope of } BC = \frac{a(t_1+t_3) - a(t_2+t_3)}{at_1t_3 - at_2t_3}$$

$$= \frac{a(t_1 + t_3 - t_2 - t_3)}{at_3(t_1 - t_2)} = \frac{1}{t_3}$$

$\therefore$ Slope of $AD = -t_3$

$\therefore$ Eq. of AD,

$y - a(t_1 + t_2) = -t_3(x - at_1t_2)$

or $\quad xt_3 + y = at_1t_2t_3 + a(t_1 + t_2) \quad \text{......(1)}$

Similarly, by symm. equation of BE is

$\Rightarrow xt_1 + y = at_1t_2t_3 + a(t_2 + t_3) \quad \text{......(2)}$

Solving (1) and (2), we get

$x = -a$

$y = a(t_1 + t_2 + t_3) + at_1t_2t_3$

$\therefore$ Orthocentre $H(-a, a(t_1 + t_2 + t_3) + at_1t_2t_3)$

9. Area of $\Delta ABC = \frac{1}{2}\begin{vmatrix} 6 & 3 & 1 \\ -3 & 5 & 1 \\ 4 & -2 & 1 \end{vmatrix}$

$= \frac{1}{2}[6(7)+3(5)+4(-2)] = \frac{49}{2}$

Area of $\Delta PBC = \frac{1}{2}\begin{vmatrix} x & y & 1 \\ -3 & 5 & 1 \\ 4 & -2 & 1 \end{vmatrix}$

$= \frac{1}{2}(7x+7y-14) - \frac{7}{2}|x+y-2|$

$ATQ, \frac{Ar(\Delta PBC)}{Ar(\Delta ABC)} = \frac{\frac{7}{2}|x+y-2|}{\frac{49}{2}} = \left|\frac{x+y-2}{7}\right|$

10. Let equations of equal sides AB and AC of isosceles ΔABC are

$7x-y+3=0$ ....(1)

and $x+y-3=0$ ....(2)

The third side BC of Δ passes through the point $(1,-10)$. Let its slope be m.

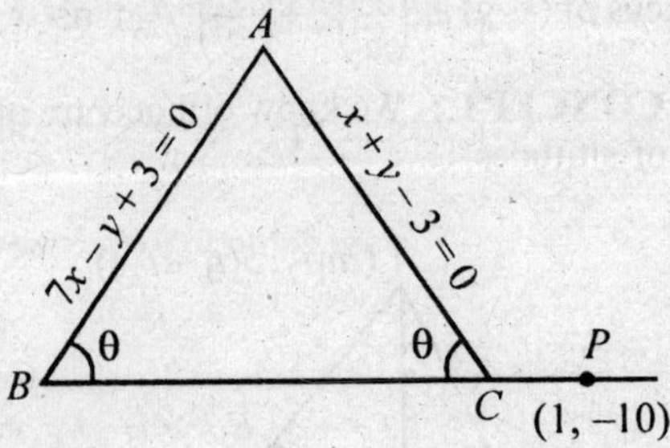

As $AB = AC$

$\therefore \angle B = \angle C$

$\Rightarrow \tan B = \tan C$ ....(3)

Now slope of $AB = 7$ and slope of $AC = -1$

Using $\tan\theta = \left|\frac{m_1 - m_2}{1+m_1 m_2}\right|$, we get

$\tan B = \left|\frac{7-m}{1+7m}\right|$ and $\tan C = \left|\frac{-1-m}{1-m}\right|$

From eq. (3), we get

$\left|\frac{7-m}{1+7m}\right| = \left|\frac{-1-m}{1-m}\right|$

$\Rightarrow \frac{7-m}{1+7m} = \pm\left(\frac{-1-m}{1-m}\right)$

Taking '+' sign, we get

$(7-m)(1-m) = -(1+m)(1+7m)$

$\Rightarrow 7-8m+m^2+7m^2+8m+1=0$

$\Rightarrow 8m^2+8=0 \Rightarrow m^2+1=0$

It has not real solution.

Taking '–' sign, we get

$(7-m)(1-m) = (1+m)(1+7m)$

$\Rightarrow 7-8m+m^2-7m^2-8m-1=0$

$\Rightarrow -6m^2-16m+6=0 \Rightarrow 3m^2+8m-3=0$

$\Rightarrow (3m-1)(m+3)=0 \Rightarrow m = 1/3, -3$

$\therefore$ The required line is

$y+10 = \frac{1}{3}(x-1)$ or $y+10 = -3(x-1)$

i.e. $x-3y-31=0$ or $3x+y+7=0$.

ALTERNATE SOLUTION

The given equations of equal sides are

$7x-y+3=0$ ....(1)

$x+y-3=0$ ....(2)

Let m be the slope of line which passes through the point $(1,-10)$.

Thus equation of line will be :

$y+10 = m(x-1)$ ....(3)

This line with slope m will be parallel to the line whose equations are the angle bisector of lines given by (1) and (2).

Therefore equations of two bisectors of the angle formed by (1) and (2) are given by

$$\frac{7x-y+3}{\sqrt{7^2+(-1)^2}} = \pm\frac{x+y-3}{\sqrt{1^2+1^2}}$$

or $\frac{7x-y+3}{\sqrt{50}} = \pm\frac{x+y-3}{\sqrt{2}}$ or $7x-y+3 = \pm 5(x+y-3)$

Taking +ve sign,

$7x-y+3 = 5x+5y-15$ or $2x-6y-18=0$

$\Rightarrow x-3y-9=0$ ...(4)

Taking – ve sign,

$7x-y+3 = -5x-5y+15$

$\Rightarrow 12x+4y-12=0$

$\Rightarrow 3x+y-4=0$ ...(5)

Slopes of (4) and (5) are given by -3, $1/3$

Thus the required equations of the third side are given by

$y+10 = \frac{1}{3}(x-1)$

or $3y+30 = x-1$ or $x+3y+31=0$

and $y+10 = -3(x-1)$

or $y+10 = -3x+3$ or $3x+y+7=0$

11. Let O be the centre of the circle. M is the mid point of AB. Then

$OM \perp AB$

Let OM when produced meets the circle at P and Q.

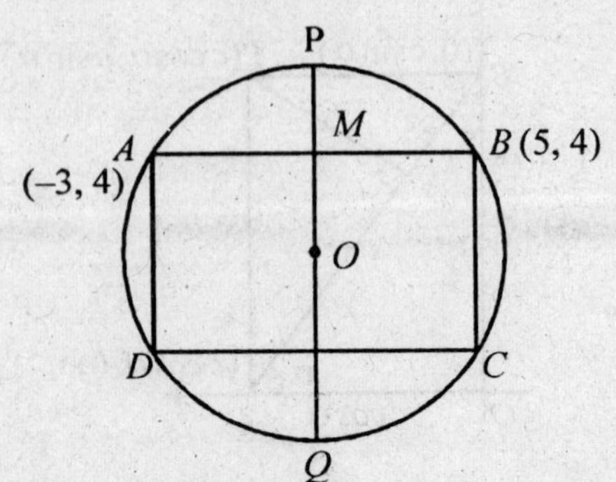

$\therefore$ PQ is a diameter perpendicular to AB and passing through M.

$$M=\left(\frac{-3+5}{2},\frac{4+4}{2}\right)=(1,4)$$

Slope of $AB=\frac{4-4}{5+3}=0$

$\therefore$ PQ, being perpendicular to AB, is a line parallel to y-axis passing through (1, 4).

$\therefore$ Its equation is

$x=1$ .... (1)

Also eq. of one of the diameter given is

$4y=x+7$ .... (2)

Solving (1) and (2), we get co-ordinates of centre O

$O(1,2)$

Also let co-ordinates of D be (α, β)

Then O is mid point of BD, therefore

$$\left(\frac{\alpha+5}{2},\frac{\beta+4}{2}\right)=(1,2) \Rightarrow \alpha=-3, \beta=0$$

$\therefore$ $D(-3,0)$

Using the distance formula we get

$$AD=\sqrt{(-3+3)^2+(4-0)^2}=4$$

$$AB=\sqrt{(5+3)^2+(4-4)^2}=8$$

$\therefore$ Area of rectangle $ABCD=AB\times AD=8\times 4$
$= 32$ square units.

12. A being on y-axis, may be chosen as (0, a).
The diagonals intersect at P (1, 2).

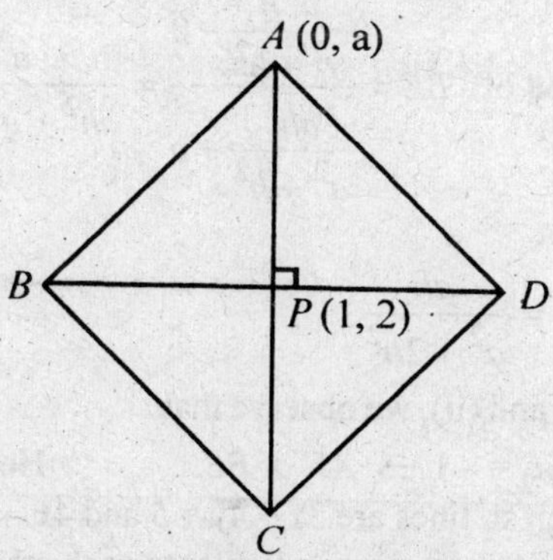

Again we know that diagonals will be parallel to the angle bisectors of the two sides

$y=x+2$ and $y=7x+3$

i.e., $\frac{x-y+2}{\sqrt{2}}=\pm\frac{7x-y+3}{5\sqrt{2}}$

$\Rightarrow$ $5x-5y+10=\pm(7x-y+3)$

$\Rightarrow$ $2x+4y-7=0$ and $12x-6y+13=0$

$m_1=-1/2$ $\qquad m_2=2$

Let diagonal d_1 be parallel to $2x+4y-7=0$ and diagonal d_2 be parallel to $12x-6y+13=0$. The vertex A could be on any of the two diagonals. Hence slope of AP is either $-1/2$ or 2.

$\Rightarrow$ $\frac{2-a}{1-0}=2$ or $\frac{-1}{2}$

$\Rightarrow$ $a=0$ or $\frac{5}{2}$

$\therefore$ A is (0, 0) or (0, 5/2)

13. Eq. of $\perp^{lar}$ bisector of side AB is

$x-y+5=0$ .... (1)

$m=1$

$\therefore$ Slope of $AB=-1$

$\therefore$ Eq. of side AB is,

$y+2=-1(x-1)$

$\Rightarrow$ $x+y+1=0$ .... (2)

Eq. of perpendicular bisector of side AC is

$x+2y=0$

$m=-1/2$

Slope of $AC=2$

$\therefore$ Eq. of side AC is,

$y+2=2(x-1) \Rightarrow 2x-y-4=0$

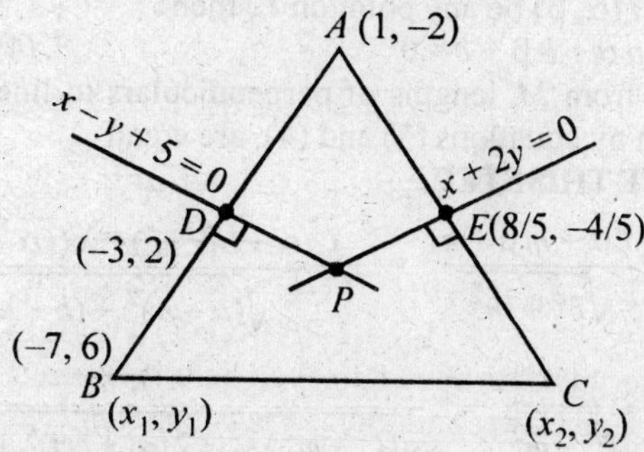

Now D is point of intersection of (1) and (2)

$\therefore$ Solving (1) and (2), we get

$D(-3,2)$

Similarly E is point of intersection of (3) and (4).

$\therefore$ Solving (3) and (4), we get

$$E\left(\frac{8}{5},\frac{4}{5}\right)$$

Now let $B(x_1, y_1)$ and $C(x_2, y_2)$

As D is mid point of AB, therefore

$$\left(\frac{x_1+1}{2},\frac{y_1-2}{2}\right)=(-3,2)$$

$\Rightarrow$ $x_1=-7, y_1=6$ $\qquad \therefore B(-7,6)$

As E is mid point of AC, therefore

$$\frac{x_2+1}{2}=\frac{8}{5} \text{ and } \frac{y_2-2}{2}=\frac{-4}{5}$$

$\Rightarrow$ $x_2=\frac{11}{5}$ and $y_2=\frac{2}{5}$

$\therefore$ $C(11/5, 2/5)$

Therefore equation of side BC is

$$y-6=\frac{2/5-6}{11/5+7}(x+7)$$

$$y-6=\frac{-28}{46}(x+7)$$

or $23y-138=-14x-98$ or $14x+23y-40=0$

14. Let the equation of other line L, which passes through the point of intersection P of lines

$L_1 \equiv ax + by + c = 0$ .... (1)

and $L_2 \equiv \ell x + my + n = 0$ .... (2)

be $L_1 + \lambda L_2 = 0$ **NOTE THIS STEP**

i.e. $(ax + by + c) + \lambda(\ell x + my + n) = 0$ .... (3)

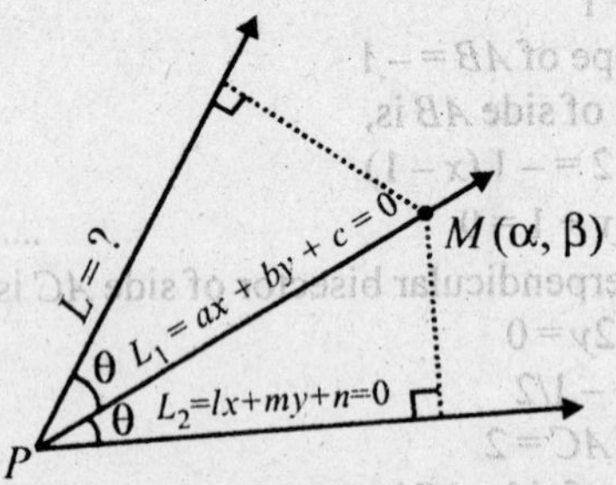

From figure it is clear that L_1 is the bisector of the angle between the lines given by (2) and (3) [i.e. L_2 and L]

Let $M(\alpha, \beta)$ be any point on L_1 then

$a\alpha + b\beta + c = 0$ ...... (4)

Also from M, lengths of perpendiculars to lines L and L_2 given by equations (3) and (4), are equal

NOTE THIS STEP:

$$\frac{\ell\alpha + m\beta + n}{\sqrt{\ell^2 + m^2}} = \pm \frac{(a\alpha + b\beta + c) + \lambda(l\alpha + m\beta + n)}{\sqrt{(a+\lambda)^2 + (b + \lambda m)^2}}$$

$$\Rightarrow \frac{1}{\sqrt{\ell^2 + m^2}} = \pm \frac{\lambda}{\sqrt{(\ell^2 + m^2)\lambda^2 + 2(a\ell + bm)\lambda + (a^2 + b^2)}}$$ [Using 4]

$$\Rightarrow (\ell^2 + m^2)\lambda^2 + 2(a\ell + bm)\lambda + (a^2 + b^2) = \lambda^2(\ell^2 + m^2)$$

$$\Rightarrow \lambda = \frac{a^2 + b^2}{2(a\ell + bm)}$$

Substituting this value of λ in eq. (3), we get L as

$$(ax + by + c) - \frac{(a^2 + b^2)}{2(a\ell + bm)}(\ell x + my + n) = 0$$

$$\Rightarrow (a^2 + b^2)(\ell x + my + n) - 2(a\ell + bm)(ax + by + c) = 0$$

15. Let BC be taken as x-axis with origin at D, the mid-point of BC, and DA will be y-axis.

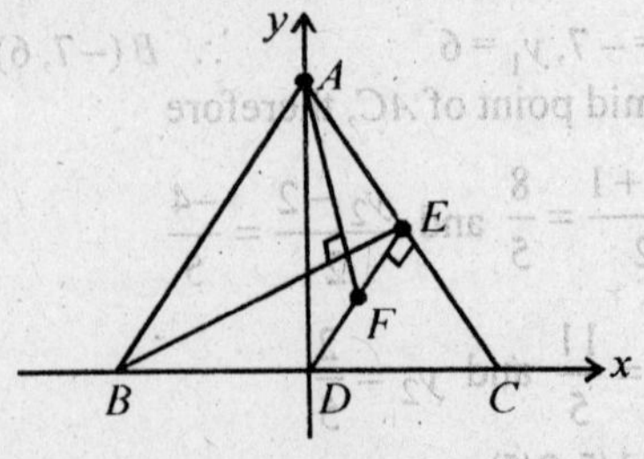

$AB = AC.$

Let $BC = 2a$, then the co-ordinates of B and C are $(-a, 0)$ and $(a, 0)$.

Let $DA = h$, so that co-ordinates of A are $(0, h)$.

NOTE THIS STEP:

Then equation of AC is $\frac{x}{a} + \frac{y}{h} = 1$ .... (1)

And equation of $DE \perp$ to AC and passing through origin is

$$\frac{x}{h} - \frac{y}{a} = 0 \Rightarrow x = \frac{hy}{a}$$ .... (2)

Solving (1) and (2) we get the co-ordinates of pt E as follows

$$\frac{hy}{a^2} + \frac{y}{h} = 1 \Rightarrow h^2 y + a^2 y = a^2 h$$

$$\Rightarrow y = \frac{a^2 h}{a^2 + h^2} \Rightarrow x = \frac{ah^2}{a^2 h^2}$$

$$\therefore E\left(\frac{ah^2}{a^2 + h^2}, \frac{a^2 h}{a^2 + h^2}\right)$$

Since F is mid pt. of DE, therefore, its co-ordinates are

$$F\left(\frac{ah^2}{2(a^2 + h^2)}, \frac{a^2 h}{2(a^2 + h^2)}\right)$$

$$\therefore \text{Slope of } AF = \frac{h - \frac{a^2 h}{2(a^2 + h^2)}}{0 - \frac{ah^2}{2(a^2 + h^2)}} = \frac{2h(a^2 + h^2) - a^2 h}{-ah^2}$$

$$\Rightarrow m_1 = -\frac{a^2 + 2h^2}{ah}$$ .... (i)

$$\text{And slope of } BE = \frac{\frac{a^2 h}{a^2 + h^2} - 0}{\frac{ah^2}{a^2 + h^2} + a} = \frac{a^2 h}{ah^2 + a^3 + ah^2}$$

$$\Rightarrow m_2 = \frac{ah}{a^2 + 2h^2}$$ .... (ii)

From (i) and (ii), we observe that

$m_1 m_2 = -1 \Rightarrow AF \perp BE.$ **Hence Proved.**

16. The given st. lines are $3x + 4y = 5$ and $4x - 3y = 15$. Clearly these st. lines are perpendicular to each other $(m_1 m_2 = -1)$, and intersect at A. Now B and C are pts on these lines such that $AB = AC$ and BC passes through $(1, 2)$

From fig. it is clear that

$\angle B = \angle C = 45°$

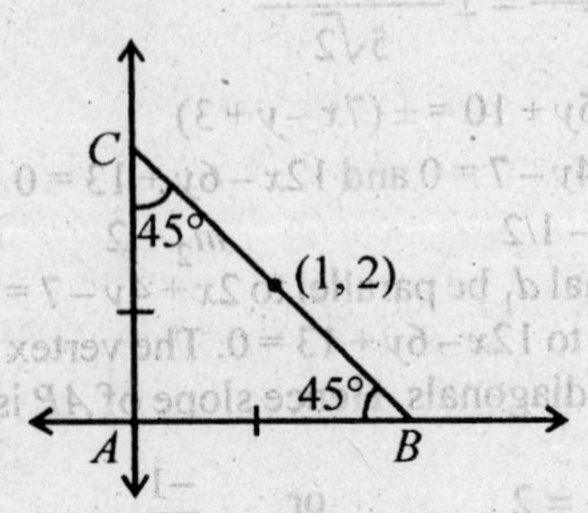

Let slope of BC be m. Then using

$$\tan B = \left|\frac{m_1 - m_2}{1 + m_1 m_2}\right|, \text{ we get}$$

$$\tan 45° = \left|\frac{m + 3/4}{1 - \frac{3}{4}m}\right|$$

$\Rightarrow \quad 4m + 3 = \pm(4 - 3m)$

$\Rightarrow \quad 4m + 3 = 4 - 3m$ or $4m + 3 = -4 + 3m$

$\Rightarrow \quad m = 1/7$ or $m = -7$

$\therefore \quad$ Eq. of BC is, $y - 2 = \frac{1}{7}(x - 1)$

or $\quad y - 2 = -7(x - 1)$

$\Rightarrow \quad 7y - 14 = x - 1$ or $y - 2 = -7x + 7$

$\Rightarrow \quad x - 7y + 13 = 0$ or $7x + y - 9 = 0$

ALTERNATE SOLUTION

As BC makes equal $\angle$'s with AB and AC,

$\therefore$ BC must be parallel to one of the $\angle$ bisectors of AB and AC.

$\angle$ Bisectors of $4x - 3y - 15 = 0$ and $3x + 4y - 5 = 0$ are

$$\frac{4x - 3y - 15}{5} = \pm\frac{3x + 4y - 5}{5}$$

$\Rightarrow \quad 4x - 3y - 15 = 3x + 4y - 5$

and $4x - 3y - 15 = -3x - 4y + 5$

$\Rightarrow \quad x - 7y - 10 = 0 \quad$ ....(1)

and $7x + y - 20 = 0 \quad$ ....(2)

Now the line parallel to (1) and passing through (1, 2) is

$$y - 2 = \frac{1}{7}(x - 1) \Rightarrow x - 7y + 13 = 0$$

And the line parallel to (2) and passing through (1, 2) is

$y - 2 = -7(x - 1)$

$\Rightarrow \quad 7x + y - 9 = 0$

$\therefore \quad$ The required lines are

$x - 7y + 13 = 0$ and $7x + y - 9 = 0$.

17. Eq. of the line AB is

$$\frac{x}{7} - \frac{y}{5} = 1 \qquad [A(7, 0), B(0, -5)]$$

$\Rightarrow \quad 5x - 7y - 35 = 0$

NOTE THIS STEP:

Eq. of line $PQ \perp AB$ is $7x + 5y + \lambda = 0$ which meets axes of x and y at pts $P(-\lambda/7, 0)$ and $Q(0, -\lambda/5)$ resp.

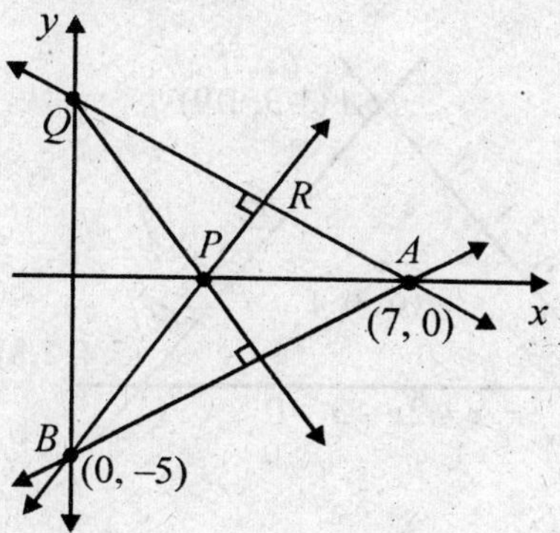

Eq. of AQ is,

$$\frac{x}{y} + \frac{y}{-\lambda/5} = 1 \Rightarrow \lambda x - 35y - 7\lambda = 0 \qquad ....(2)$$

Eq. of BP is,

$$\frac{-7x}{\lambda} - \frac{y}{5} = 1 \Rightarrow 35x + \lambda y + 5\lambda = 0 \qquad ....(3)$$

Locus of R the pt. of intersection of (2) and (3) can be obtained by eliminating λ from these eq.'s, as follows

$$35x + (5 + y)\left(\frac{35y}{x - 7}\right) = 0$$

$\Rightarrow \quad 35x(x - 7) + 35y(5 + y) = 0 \Rightarrow x^2 + y^2 - 7x + 5y = 0$

18. Let the equation of line through A which makes an intercept of 2 units between.

$2x + y = 3 \quad$ ....(1)

and $2x + y = 5 \quad$ ....(2)

be $\dfrac{x - 2}{\cos\theta} = \dfrac{y - 3}{\sin\theta}$ **NOTE THIS STEP**

Let $AP = r$ then $AQ = r + 2$

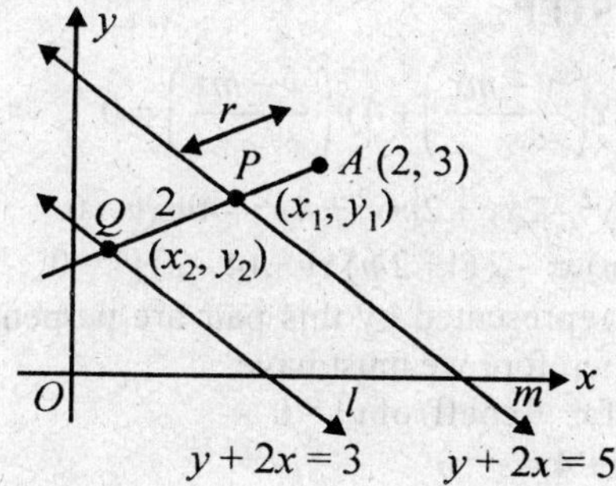

Then for pt $P(x_1, y_1)$,

$$\frac{x_1 - 2}{\cos\theta} = \frac{y_1 - 3}{\sin\theta} = r \Rightarrow \frac{2(x_1 - 2) + (y_1 - 3)}{2\cos\theta + \sin\theta} = r$$

$$\left(\text{Using } \frac{a_1}{a_2} = \frac{b_1}{b_2} = \frac{\lambda a_1 + \mu b_1}{\lambda a_2 + \mu b_2}\right)$$

$$\Rightarrow \quad \frac{(2x_1 + y_1) - 7}{2\cos\theta + \sin\theta} = r \Rightarrow \frac{5 - 7}{2\cos\theta + \sin\theta} = r$$

[Using $2x_1 + y_1 = 5$ as $P(x_1, y_1)$ lies on $2x + y = 5$]

$$\frac{-2}{2\cos\theta + \sin\theta} = r \qquad \text{...(i)}$$

For pt $Q(x_2, y_2)$,

$$\frac{x_2 - 2}{\cos\theta} = \frac{y_2 - 3}{\sin\theta} = r + 2$$

$$\Rightarrow \quad \frac{2(x_2 - 2) + (y_2 - 3)}{2\cos\theta + \sin\theta} = r + 2$$

$$\Rightarrow \quad \frac{-4}{2\cos\theta + \sin\theta} = r + 2 \qquad \text{...(ii)}$$

(ii) – (i)

$$\Rightarrow \quad \frac{-2}{2\cos\theta + \sin\theta} = 2$$

$\Rightarrow$ $2\cos\theta + \sin\theta = -1$ ... (3)

$\Rightarrow$ $2\cos\theta = -(1 + \sin\theta)$

Squaring on both sides, we get

$\Rightarrow$ $4\cos^2\theta = 1 + 2\sin\theta + \sin^2\theta$

$\Rightarrow$ $5\sin^2\theta + 2\sin\theta - 3 = 0$

$\Rightarrow$ $5\sin^2\theta + 5\sin\theta - 3\sin\theta - 3 = 0$

$\Rightarrow$ $(5\sin\theta - 3)(\sin\theta + 1) = 0 \Rightarrow \sin\theta = 3/5, -1$

$\Rightarrow$ $\cos\theta = -4/5, 0$ [Using eq. (3)]

$\therefore$ The required equation is either

$$\frac{x-2}{-4/5} = \frac{y-3}{3/5} \text{ or } \frac{x-2}{0} = \frac{y-3}{-1}$$

$\Rightarrow$ either $3x - 6 = -4y + 12$ or $x - 2 = 0$

$\Rightarrow$ either $3x + 4y - 18 = 0$ or $x - 2 = 0$

19. The given curve is

$3x^2 - y^2 - 2x + 4y = 0$ ... (1)

Let $y = mx + c$ be the chord of curve (1) which subtends an $\angle$ of 90° at origin.

Then the combined eq. of lines joining points of intersection of curve (1) and chord $y = mx + c$ to the origin, can be obtained by making the eq. of curve homogeneous with the help of eq. of chord, as follows.

NOTE THIS STEP :

$$3x^2 - y^2 - 2x\left(\frac{y-mx}{c}\right) + 4y\left(\frac{y-mx}{c}\right) = 0$$

$\Rightarrow$ $3cx^2 - cy^2 - 2xy + 2mx^2 + 4y^2 - 4mxy = 0$

$\Rightarrow$ $(3c + 2m)x^2 - 2(1 + 2m)xy + (4 - c)y^2 = 0$

As the lines represented by this pair are perpendicular to each other, therefore we must have

coeff. of x^2 + coeff. of $y^2 = 0$

$\Rightarrow$ $3c + 2m + 4 - c = 0$

$\Rightarrow$ $-2 = m.1 + c$

Which on comparision with eq. of chord, implies that $y = mx + c$ passes though $(1, -2)$.

Hence the family of chords must pass through $(1, -2)$.

20. **KEY CONCEPT :** Points (x_1, y_1) and (x_2, y_2) lie on the same or on opposite side of line $ax + by + c = 0$ according as $ax_1 + by_1 + c$ and $ax_2 + y_2 + c$ are of same or of opposite sign, that is as

$$\frac{ax_1 + by_1 + c}{ax_2 + by_2 + c} > 0 \text{ or } < 0.$$

The rough graph of the sides of Δ are shown as in figure

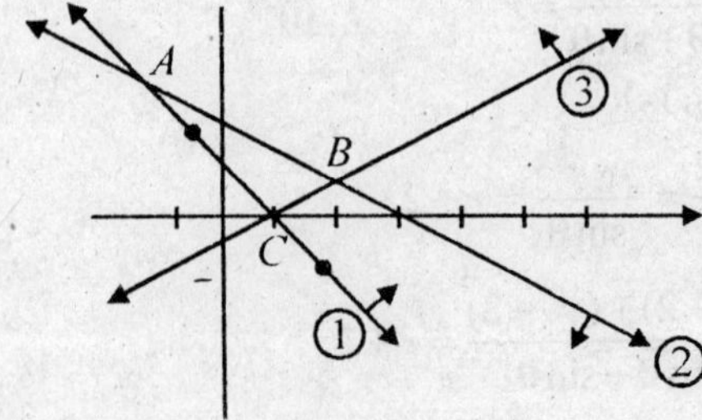

Where

$2x + 3y - 1 = 0$ ... (1)

$x + 2y - 3 = 0$ ... (2)

$5x - 6y - 1 = 0$ ... (3)

For the pt. (α, α^2) to lie inside the Δ we must have the following inequations satisfied by pt (α, α^2).

NOTE THIS STEP

$2x + 3y - 1 > 0$ (pt and (0, 0) are on opp. sides)

$x + 2y - 3 < 0$ (pt and (0, 0) are on same sides)

NOTE THIS STEP

$5x - 6y - 1 < 0$ (pt and (0, 0) are on opp. sides)

Substituting $x = \alpha$ and $y = \alpha^2$, we get

$2\alpha - 3\alpha^2 - 1 > 0,\quad \alpha + 2\alpha^2 - 3 < 0$

$5\alpha - 6\alpha^2 - 1 < 0$

or $3\alpha^2 + 2\alpha - 1 > 0$ or $(3\alpha - 1)(\alpha + 1) > 0$

$\Rightarrow$ $\alpha \in (-\infty, -1) \cup \left(\frac{1}{3}, \infty\right)$ and $2\alpha^2 + \alpha - 3 < 0$

$\Rightarrow$ $(2\alpha + 3)(\alpha - 1) < 0 \Rightarrow \alpha \in \left(-\frac{3}{2}, 1\right)$

and $6\alpha^2 - 5\alpha + 1 < 0$

$\Rightarrow$ $(2\alpha - 1)(3\alpha - 1) > 0 \Rightarrow \alpha \in \left(-\infty, \frac{1}{3}\right) \cup \left(\frac{1}{2}, \infty\right)$

Thus the required value of α is the common solution of the above inequations which can be determined by the following number line

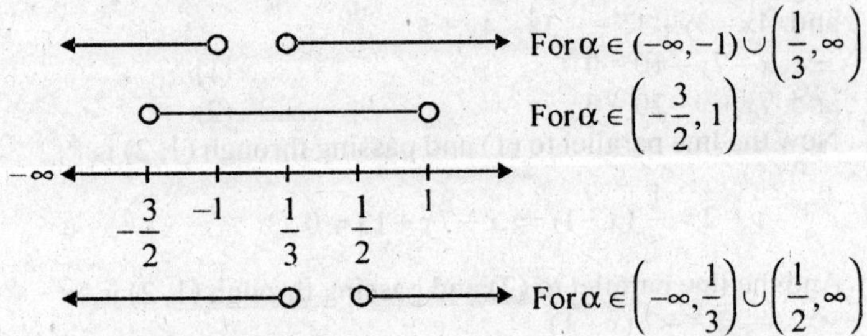

The required common solution clearly is

$$\left(-\frac{3}{2}, -1\right) \cup \left(\frac{1}{2}, 1\right)$$

ALTERNATE SOLUTION :

The points of intersection of given lines are

$$A\left(\frac{1}{3}, \frac{1}{9}\right), B(-7, 5), C\left(\frac{5}{4}, \frac{7}{8}\right)$$

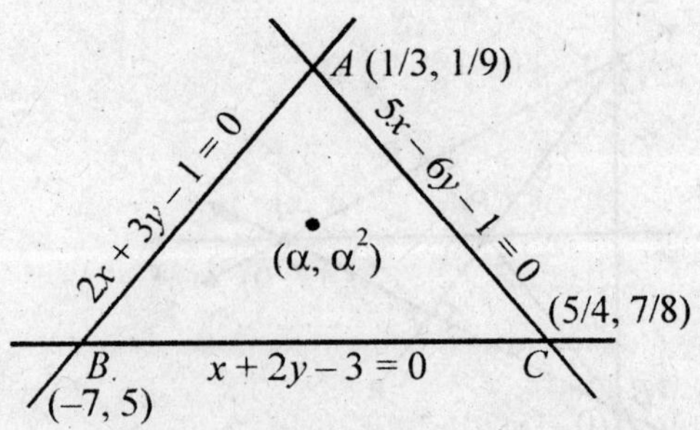

If (α, α^2) lies inside the Δ formed by the given lines, then $\left(\frac{1}{3},\frac{1}{9}\right)$ and (α, α^2) lie on the same side of the line $x+2y-3=0$

$$\Rightarrow \frac{\alpha+2\alpha^2-3}{\frac{1}{3}+\frac{2}{9}-3}>0 \Rightarrow 2\alpha^2+\alpha-3<0 \ldots (1)$$

Similarly $\left(\frac{5}{4},\frac{7}{8}\right)$ and (α, α^2) lie on the same side of the line $2x+3y-1=0$.

$$\Rightarrow \frac{2\alpha+3\alpha^2-1}{\frac{10}{4}+\frac{21}{8}-1}>0 \Rightarrow 3\alpha^2+2\alpha-1>0 \ldots (2)$$

$(-7, 5)$ and (α, α^2) lie on the same side of the line $5x-6y-1=0$.

$$\Rightarrow \frac{5\alpha+6\alpha^2-1}{-35-30-1}>0 \Rightarrow 6\alpha^2-5\alpha+1>0 \ldots (3)$$

Now common solution of (1), (2) and (3) can be obtained as in the previous method,

$$\therefore \quad \alpha \in \left(-\frac{3}{2},-1\right) \cup \left(\frac{1}{2},1\right)$$

21. The given curve is

$y = x^3 \qquad \ldots (1)$

Let the pt, P_1 be (t, t^3), $t \neq 0$

Then slope of tangent at P_1 is

$$= \frac{dy}{dx} = (3x^2)_{x=t} = 3t^2$$

$\therefore$ Equation of tangent at P_1 is

$y-t^3 = 3t^2(x-t) \Rightarrow y = 3t^2x-2t^3$

$\Rightarrow 3t^2x - y - 2t^3 = 0 \qquad \ldots (2)$

Now this tangent meets the curve again at P_2 which can be obtained by solving (1) and (2)

i.e., $3t^2x - x^2 - 2t^3 = 0$

or $x^3 - 3t^2x + 2t^3 = 0$

$(x-t)^2(x+2t) = 0 \Rightarrow x = -2t$ as $x = t$ is for P_1

$\therefore \quad y = -8t^3$

Hence pt P_2 is $(-2t, -8t^3) = (t_1, t_1^3)$ say.

Similarly, we can find that tangent at P_2 which meets the curve again at $P_3(2t_1, -8t_1^3)$ i.e., $(4t, 64t^3)$.

Similarly, $P_4 \equiv (-8t, -512t^3)$ and so on.

We observe that abscissae of pts. $P_1, P_2, P_3 \ldots$ are $t, -2t, 4t, \ldots$ which form a GP with common ratio -2. Also ordinates of these pts. $t^3, -8t^3, 64t^3, \ldots$ also form a GP with common ratio -8.

$$\text{Now, } \frac{Ar(\Delta P_1P_2P_3)}{Ar(\Delta P_2P_3P_4)} = \frac{\begin{vmatrix} 1 & t & t^3 \\ 1 & -2t & -8t^3 \\ 1 & 4t & 64t^3 \end{vmatrix}}{\begin{vmatrix} 1 & -2t & -8t^3 \\ 1 & 4t & -64t^3 \\ 1 & -8t & -512t^3 \end{vmatrix}}$$

$$= \frac{t^4\begin{vmatrix} 1 & 1 & 1 \\ 1 & -2 & -8 \\ 1 & -4 & 64 \end{vmatrix}}{(-2)(-8)t^4\begin{vmatrix} 1 & 1 & 1 \\ 1 & -2 & -8 \\ 1 & 4 & -64 \end{vmatrix}}$$

$= \frac{1}{16}$ sq. units.

22. Let θ be the inclination of line through $A(-5,-4)$. Therefore equation of this line is

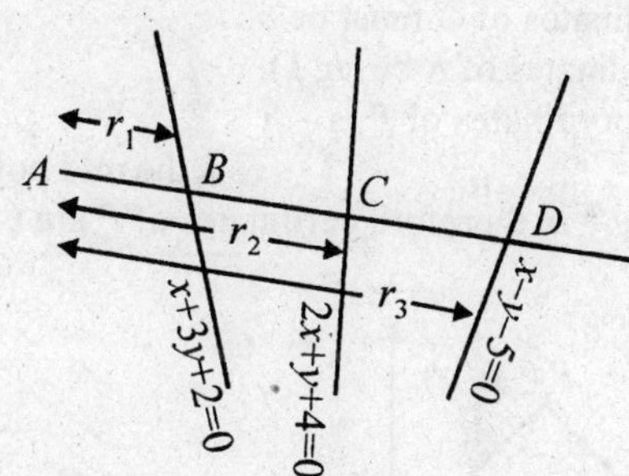

$$\frac{x+5}{\cos\theta} = \frac{y+4}{\sin\theta} = r_1, r_2, r_3$$

$\Rightarrow B(r_1\cos\theta-5, r_1\sin\theta-4)$

$C(r_2\cos\theta-5, r_2\sin\theta-4)$

$D(r_3\cos\theta-5, r_3\sin\theta-4)$

But B lies on $x+3y+2=0$. therefore

$r_1\cos\theta - 5 + 3r_1\sin\theta - 12 + 2 = 0$

$$\Rightarrow r_1 = \frac{15}{\cos\theta+3\sin\theta} = AB$$

$$\Rightarrow \frac{15}{AB} = \cos\theta + 3\sin\theta \qquad \ldots (1)$$

As C lies on $2x+y+4=0$, therefore

$2(r_2\cos\theta-5)+(r_2\sin\theta-4)+4=0$

$$\Rightarrow r_2 = \frac{10}{2\cos\theta+\sin\theta} = AC$$

$$\Rightarrow \frac{10}{AC} = 2\cos\theta+\sin\theta \qquad \ldots (2)$$

Similarly D lines on $x-y-5=0$, therefore

$r_3\cos\theta - 5 - r_3\sin\theta + 4 - 5 = 0$

$\Rightarrow \quad r_3 = \dfrac{6}{\cos\theta - \sin\theta} = AD$

$\Rightarrow \quad \dfrac{6}{AD} = \cos\theta - \sin\theta \qquad \ldots(3)$

Now, ATQ, $\left(\dfrac{15}{AB}\right)^2 + \left(\dfrac{10}{AC}\right)^2 = \left(\dfrac{6}{AD}\right)^2$

$\Rightarrow \quad (\cos\theta + 3\sin\theta)^2 + (2\cos\theta + \sin\theta)^2$

$= (\cos\theta - \sin\theta)^2$ [Using (1), (2) and (3)]

$\Rightarrow \quad 4\cos^2\theta + 9\sin^2\theta + 12\sin\theta\cos\theta = 0$

$\Rightarrow \quad 2\cos\theta + 3\sin\theta = 0$

$\Rightarrow \quad \tan\theta = -\dfrac{2}{3}$

$\therefore$ Equation of req. line is

$y + 4 = -\dfrac{2}{3}(x+5)$

$3y + 12 = -2x - 10$

$\Rightarrow \quad 2x + 3y + 22 = 0$

23. Let the co-ordinates of Q be (b, α) and that of S be $(-b, \beta)$. Let PR and SQ intersect each other at G.

$\because$ G is the mid pt of SQ.

($\because$ diagonals of a rectangle bisect each other)

$\therefore$ x co-ordinates of G must be a.

Let the co-ordinates of R be (h, k).

$\therefore$ The x-coordinates of P is $-h$

($\because$ G is the mid point of PR)

As P lies on $y = a$, therefore cordinates of P are $(-h, a)$.

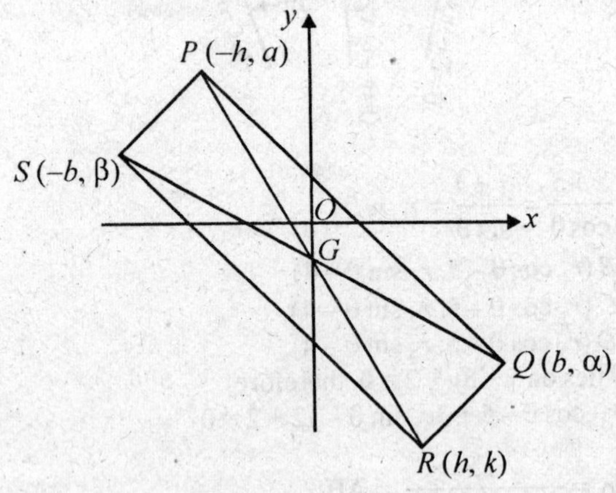

$\because$ PQ is parallel to $y = mx$,

Slope of $PQ = m$

$\therefore \quad \dfrac{\alpha - a}{b + h} = m \Rightarrow \alpha = a + m(b+h) \qquad \ldots(1)$

Also $RQ \perp PQ \Rightarrow$

Slope of $RQ = \dfrac{-1}{m}$

$\therefore \quad \dfrac{k - \alpha}{h - b} = \dfrac{-1}{m} \Rightarrow \alpha = k + \dfrac{1}{m}(h-b) \qquad \ldots(2)$

From (1) and (2) we get

$a + m(b+h) = k + \dfrac{1}{m}(h-b)$

$\Rightarrow \quad am + m^2(b+h) = km + (h-b)$

$\Rightarrow \quad (m^2-1)h - mk + b(m^2+1) + am = 0$

$\therefore$ Locus of vertex $R(h, k)$ is

$(m^2-1)x - my + b(m^2+1) + am = 0.$

24. Equation of line with slope 4 is

$y = 4x + C \qquad \ldots(1)$

Let it meets the given curve $xy = 1$

at $\quad A\left(a, \dfrac{1}{a}\right)$ and $B\left(b, \dfrac{1}{b}\right)$

Then A and B must satisfy eq^n (1)

$\therefore$ We get, $\dfrac{1}{a} = 4a + C$ and $\dfrac{1}{b} = 4b + C$

On subtracting, we get

$\dfrac{1}{a} - \dfrac{1}{b} = 4(a-b)$

$\Rightarrow \quad \dfrac{b-a}{ab} = 4(a-b)$

For two different pts $a \neq b$

$\therefore$ we get $\qquad 4ab + 1 = 0 \qquad \ldots(2)$

Now let $P(x_1, y_1)$ be the pt. on AB which divides AB in the ratio 1 : 2.

$\therefore \quad x_1 = \dfrac{2a+b}{3}$ and $y_1 = \dfrac{\frac{2}{a} + \frac{1}{b}}{3}$

$\Rightarrow \quad 2a + b = 3x_1$ and $3y_1 ab = a + 2b$

$\Rightarrow \quad 2a + b = 3x_1 \qquad \ldots(4)$

and $a + 2b = -\dfrac{3}{4}y_1 \qquad \ldots(5)$

(using eq^n (2))

or $\quad 4a + 8b = -3y_1$

Solving eq^{n}s (4) and (5) we get

$a = \dfrac{8x_1 + y_1}{4}$ and $b = \dfrac{-(2x_1 + y_1)}{2}$

Substituting these values in eq^n (2), we get

$-(8x_1 + y_1)(2x_1 + y_1) + 2 = 0 \Rightarrow 16x_1^2 + 10x_1y_1 + y_1^2 = 2$

$\therefore$ locus of (x_1, y_1) is,

$16x^2 + 10xy + y^2 = 2$

25. Choose one side $BC = 2a$ along x-axis with its mid-point O as origin so that the points B and C are $(-a, 0)$ and $(a, 0)$ respectively. Let the third vertex be the point (h, k), AD, BE and CF are three altitudes whose equations are

$AD : x = h$ or $x + y - h = 0 \qquad \ldots(1)$

$BE : y - 0 = -\dfrac{(h-a)}{k}(x+a)$

or $\quad x(h-a) + ky + a(h-a) = 0 \qquad \ldots(2)$

CF : Replace a by $-a$ in (2)

or $\quad x(h+a) + ky - a(h+a) = 0 \qquad \ldots(3)$

The three lines will be concurrent if $\Delta = 0$.

$$\text{or} \quad \begin{vmatrix} 1 & 0 & -h \\ h-a & k & a(h-a) \\ h+a & k & -a(h+a) \end{vmatrix}$$

Apply $R_3 - R_2$ and two rows become identical

$\therefore \quad \Delta = 0$

$\therefore$ concurrent.

ALTERNATE SOLUTION:

Let $A(x_1, y_1)$, $B(x_2, y_2)$, $C(x_3, y_3)$ be the vertices of ΔABC

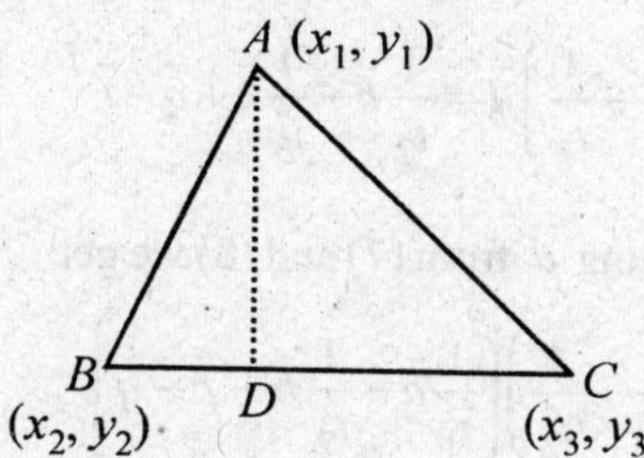

Then equation of alt. AD is

$$y - y_1 = -\left[\frac{x_2 - x_3}{y_2 - y_3}\right](x - x_1)$$

or $(x - x_1)(x_2 - x_3) + (y - y_1)(y_2 - y_3) = 0$ ... (1)

Similarly equations of other two attitudes are

$(x - x_2)(x_3 - x_1) + (y - y_2)(y_3 - y_1) = 0$ ...(2)

and $(x - x_3)(x_1 - x_2) + (y - y_3)(y_1 - y_2) = 0$ ...(3)

Now, above three lines will be concurrent if

$$\begin{vmatrix} x_2 - x_3 & y_2 - y_3 & -x_1(x_2 - x_3) - y_1(y_2 - y_3) \\ x_3 - x_1 & y_3 - y_1 & -x_2(x_3 - x_1) - y_2(y_3 - y_1) \\ x_1 - x_2 & y_1 - y_2 & -x_3(x_1 - x_2) - y_3(y_1 - y_2) \end{vmatrix} = 0$$

On L.H.S.

Operating $R_1 + R_2 + R_3$, R_1 becomes row of zeros.

$\therefore$ Value of determinant $= 0 =$ R.H.S.

Hence the altitudes are concurrent.

26. Let $P = (h, k)$ be a general point in the first quadrant such that

$$d(P, A) = d(P, O)$$

$\Rightarrow |h - 3| + |k - 2| = |h| + |k| = h + k$ ... (1)

[h and k are + ve, pt (h, k) being in I quadrant.]

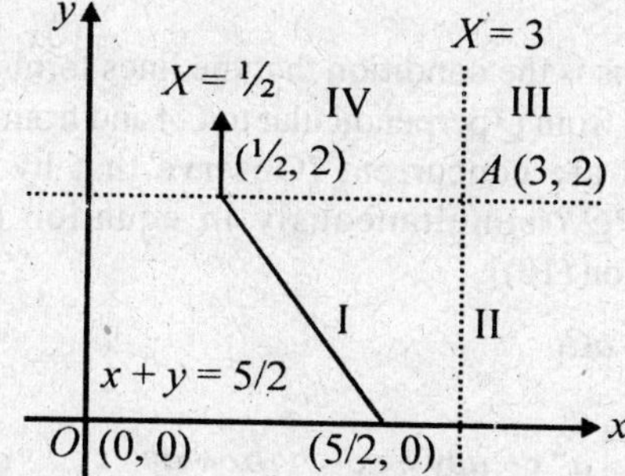

If $h < 3, k < 2$ then (h, k) lies in region I.

It $h > 3, k < 2$, (h, k) lies in region II.

If $h > 3, k > 2$ (h, k) lies in region III.

If $h < 3, k > 2$ (h, k) lies in region IV.

In region I, eq. (1)

$\Rightarrow 3 - h + 2 - k = h + k$

$\Rightarrow h + k = \frac{5}{2}$

In region II, eq. (1) becomes

$\Rightarrow h - 3 + 2 - k = h + k$

$\Rightarrow k = -\frac{1}{2}$ not possible.

In region III, eq. (1) becomes

$\Rightarrow h - 3 + k - 2 = h + k$

$\Rightarrow -5 = 0$ not possible.

In region IV, eq. (1) becomes

$\Rightarrow 3 - h + k - 2 = h + k$

$\Rightarrow h = 1/2$

$\Rightarrow$ Hence required set consists of line segment $x + y = 5/2$ of finite length as shown in the first region and the ray $x = 1/2$ in the fourth region.

27. Let the co-ordinates of the vertices of the ΔABC be $A(a_1, b_1)$, $B(a_2, b_2)$ and $C(a_3, b_3)$ and co-ordinates of the vertices of the ΔPQR be

$P(x_1, y_1)$, $B(x_2, y_2)$ and $R(x_3, y_3)$

$$\text{Slope of } QR = \frac{y_2 - y_3}{x_2 - x_3}$$

$\Rightarrow$ Slope of straight line perpendicular to

$$QR = -\frac{x_2 - x_3}{y_2 - y_3}$$

Equation of straight line passing through $A(a_1, b_1)$ and perpendicular to QR is

$$y - b_1 = -\frac{x_2 - x_3}{y_2 - y_3}(x - a_1)$$

$\Rightarrow (x_2 - x_3)x + (y_2 - y_3)y - a_1(x_2 - x_3) - b_1(y_2 - y_3) = 0$ ... (1)

Similarly equation of straight line from B and perpendicular to RP is

$(x_3 - x_1)x + (y_3 - y_1)y - a_2(x_3 - x_1) - b_2(y_3 - y_1) = 0$ ... (2)

and eq^n of straight line from C and perpendicular to PQ is

$(x_1 - x_2)x + (y_1 - y_2)y - a_3(x_1 - x_2) - b_3(y_1 - y_2) = 0$ ... (3)

As straight lines (1), (2) and (3) are given to be concurrent, we should have

$$\begin{vmatrix} x_2 - x_3 & y_2 - y_3 & a_1(x_2 - x_3) + b_1(y_2 - y_3) \\ x_3 - x_1 & y_3 - y_1 & a_2(x_3 - x_1) + b_2(y_3 - y_1) \\ x_1 - x_2 & y_1 - y_2 & a_1(x_1 - x_2) + b_3(y_1 - y_2) \end{vmatrix} = 0 \quad \text{... (4)}$$

Operating $R_1 \to R_1 + R_2 + R_3$, we get

$$\begin{vmatrix} 0 & 0 & S \\ x_3 - x_1 & y_3 - y_1 & a_2(x_3 - x_1) + b_2(y_3 - y_1) \\ x_1 - x_2 & y_1 - y_2 & a_3(x_1 - x_2) + b_3(y_1 - y_2) \end{vmatrix} = 0$$

where

$[S=a_1(x_2-x_3)+b_1(y_2-y_3)+a_2(x_3-x_1)+b_2(y_3-y_1)+a_3(x_1-x_2)+b_3(y_1-y_2)]$

Expanding along R_1

$\Rightarrow\ [(x_3-x_1)(y_1-y_2)-(x_1-x_2)(y_3-y_1)]\,S=0$

$$\Rightarrow\ \left[\frac{y_1-y_2}{x_1-x_2}-\frac{y_3-y_1}{x_3-x_1}\right]S=0$$

$\Rightarrow\ [m_{PQ}-m_{PR}]\,S=0$

$\Rightarrow\ S=0$

$[m_{PQ}=m_{PR}\Rightarrow PQ\parallel PR$ which is not possible in $\Delta PQR]$

$\Rightarrow\ a_1(x_2-x_3)+b_1(y_2-y_3)+a_2(x_3-x_1)+b_2(y_3-y_1)+a_3(x_1-x_2)+b_3(y_1-y_2)=0$ ... (5)

$\Rightarrow\ x_1(a_3-a_2)+y_1(b_3-b_2)+x_2(a_1-a_3)+y_2(b_1-b_3)+x_3(a_2-a_1)+y_3(b_2-b_1)=0$

(Rearranging the equation (5))

But above condition shows

NOTE THIS STEP

$$\begin{vmatrix} a_3-a_2 & b_3-b_2 & x_1(a_3-a_2)+y_1(b_3-b_2)\\ a_1-a_3 & b_1-b_3 & x_2(a_1-a_3)+y_2(b_1-b_3)\\ a_2-a_1 & b_2-b_1 & x_3(a_2-a_1)+y_3(b_2-b_1)\end{vmatrix}=0$$

[Using the fact that as (4) $\Leftrightarrow$ (5) in the same way (6) $\Leftrightarrow$ (7)]

Clearly equation (7) shows that lines through P and perpendicular to BC, through Q and perpendicular to AB are concurrent. **Hence Proved.**

ALTERNATE SOLUTION

Let the position vectors of A, B, C be $\vec{a},\vec{b}$ and $\vec{c}$ respectively and that of P, Q, R be $\vec{p},\vec{q}$ and $\vec{r}$ respectively.

Let $\vec{h}$ be the position vector of the orthocentre H of ΔPQR.

We have $HP\perp QR$.

Equation of straight line passing through A and perpendicular of QR i.e., parallel to $\vec{HP}=\vec{p}-\vec{h}$ is

$\vec{r}=\vec{a}+t_1(\vec{p}-\vec{h})$ ... (1)

where t_1 is a parameter.

Similarly, equation of straight line through B and perpendicular to RP is

$\vec{r}=\vec{b}+t_2(\vec{q}-\vec{h})$ ... (2)

Again equation of straight line through C and perpendicular to PQ is

$\vec{r}=\vec{c}+t_3(\vec{r}-\vec{h})$ ... (3)

If the lines (1), (2) and (3) are concurrent, then there exist a point D with position vector $\vec{d}$ which lies on all of them, that is for some values of t_1, t_2 and t_3 which implies that

$\frac{1}{t_1}\vec{d}=\frac{1}{t_1}\vec{a}+\vec{p}-\vec{h}$ ...(4)

$\frac{1}{t_2}\vec{d}=\frac{1}{t_2}\vec{b}+\vec{q}-\vec{h}$ ...(5)

$\frac{1}{t_3}\vec{d}=\frac{1}{t_2}\vec{c}+\vec{r}-\vec{h}$ ...(6)

From (4) and (5) we get

$$\left(\frac{1}{t_1}-\frac{1}{t_2}\right)\vec{d}=\frac{1}{t_1}\vec{a}-\frac{1}{t_2}\vec{b}+\vec{p}-\vec{q}\quad ...(7)$$

and form (5) and (6) we get

$$\left(\frac{1}{t_1}-\frac{1}{t_3}\right)\vec{d}=\frac{1}{t_2}\vec{b}-\frac{1}{t_3}\vec{c}+\vec{q}-\vec{r}\quad ...(8)$$

Eliminating $\vec{d}$ from (7) and (8) we get

$$\left(\frac{1}{t_2}-\frac{1}{t_3}\right)\left[\frac{1}{t_1}\vec{a}-\frac{1}{t_2}\vec{b}+\vec{p}-\vec{q}\right]=\left(\frac{1}{t_1}-\frac{1}{t_2}\right)\left[\frac{1}{t_2}\vec{b}-\frac{1}{t_3}\vec{c}+\vec{q}-\vec{r}\right]$$

$\Leftrightarrow\ (t_3-t_2)[t_2\vec{a}-t_1\vec{b}+t_1t_2(\vec{p}\vec{q})]=(t_3-t_1)[t_3\vec{b}+t_2\vec{c}+t_2t_3(\vec{q}-\vec{r})]$

$\Leftrightarrow\ t_2(t_3-t_2)\vec{a}+t_2(t_1-t_3)\vec{b}+t_2(t_2-t_1)\vec{c}+t_1t_2(t_3-t_2)\vec{p}+t_2^2(t_1-t_3)\vec{q}+t_2t_3(t_2-t_1)\vec{r}=\vec{0}$

Thus (1), (2) and (3) are concurrent is equivalent to say that there exist scalars t_1, t_2, t_3 such that

$(t_2-t_3)\vec{a}+(t_3-t_1)\vec{b}+(t_1-t_2)\vec{c}+t(t_2-t_3)\vec{p}+t_2(t_3-t_1)\vec{q}+t_3(t_1-t_2)\vec{r}=\vec{0}$ ...(9)

Dividing by $t_1t_2t_3$ we get

$$\left(\frac{1}{t_2}-\frac{1}{t_3}\right)\vec{p}+\left(\frac{1}{t_3}-\frac{1}{t_1}\right)\vec{q}+\left(\frac{1}{t_1}-\frac{1}{t_2}\right)\vec{r}+\frac{1}{t_1}\left(\frac{1}{t_2}-\frac{1}{t_3}\right)\vec{a}+\frac{1}{t_2}\left(\frac{1}{t_3}-\frac{1}{t_1}\right)\vec{b}+\frac{1}{t_3}\left(\frac{1}{t_1}-\frac{1}{t_2}\right)\vec{c}=0$$

.... (10)

But this is the condition that the lines form P perpendicular to BC, from Q perpendicular to CA and from R perpendicular to AB are concurrent (Observe that by changing ABC and PQR simultaneously in equation (9), we obtain equation (10)). Hence proved.

28. $C_1\to aC_1$

$$\Delta=\frac{1}{a}\begin{vmatrix} a^2x-aby-ac & bx+ay & cx+a\\ abx+a^2y & -ax+by-c & cy+b\\ acx+a^2 & cy+b & -ax-by+c\end{vmatrix}$$

Applying $C_1 \to C_1 + bC_2 + cC_3$

$$\Delta = \frac{1}{a}\begin{vmatrix} (a^2+b^2+c^2)x & ay+bx & cx+a \\ (a^2+b^2+c^2)y & by-c-ax & cy+b \\ (a^2+b^2+c^2) & b+cy & -ax-by+c \end{vmatrix}$$

$$= \frac{1}{a}\begin{vmatrix} x & ay+bx & cx+a \\ y & by-c-ax & b+cy \\ 1 & b+cy & c-ax-by \end{vmatrix},$$

as $a^2+b^2+c^2=1$

$C_2 \to C_2 - bC_1$ and $C_3 \to C_3 - cC_1$

then $\Delta = \frac{1}{a}\begin{vmatrix} x & ay & a \\ y & -c-ax & b \\ 1 & cy & -ax-by \end{vmatrix}$

$$= \frac{1}{ax}\begin{vmatrix} x^2 & axy & ax \\ y & -c-ax & b \\ 1 & cy & -ax-by \end{vmatrix}$$

$R_1 \to R_1 + yR_2 + R_3$

$$\Delta = \frac{1}{ax}\begin{vmatrix} x^2+y^2+1 & 0 & 0 \\ y & -c-ax & b \\ 1 & cy & -ax-by \end{vmatrix}$$

On expanding along R_1

$$\Delta = \frac{(x^2+y^2+1)}{ax} ax(ax+by+c)$$

$$= (x^2+y^2+1)(ax+by+c)$$

Given $\Delta = 0$

$\Rightarrow ax + by + c = 0$, which represents a straight line.

$[\because x^2+y^2+1 \neq 0$, being + ve].

29. The line $y = mx$ meets the given lines in

$P\left(\frac{1}{m+1}, \frac{m}{m+1}\right)$ and $Q\left(\frac{3}{m+1}, \frac{3m}{m+1}\right)$

Hence equation of L_1 is

$$y - \frac{m}{m+1} = 2\left(x - \frac{1}{m+1}\right)$$

$\Rightarrow y - 2x - 1 = -\frac{3}{m+1}$ ...(1)

and that of L_2 is

$$y - \frac{3m}{m+1} = -3\left(x - \frac{3}{m+1}\right)$$

$\Rightarrow y + 3x - 3 = \frac{6}{m+1}$ ...(2)

From (1) and (2)

$$\frac{y-2x-1}{y+3x-3} = -\frac{1}{2}$$

$\Rightarrow x - 3y + 5 = 0$ which is a straight line.

30. Let the equation of the line be

$(y-2) = m(x-8)$ where $m < 0$

$\Rightarrow P \equiv \left(8 - \frac{2}{m}, 0\right)$ and $Q \equiv (0, 2-8m)$

Now, $OP + OQ = \left|8 - \frac{2}{m}\right| + |2 - 8m|$

$$= 10 + \frac{2}{-m} + 8(-m) \geq 10 + 2\sqrt{\frac{2}{-m} \times 8(-m)} \geq 18$$

31. A line passing through $P(h, k)$ and parallel to x-axis is

$y = k$. ...(1)

The other two lines given are

$y = x$ ...(2)

and $x + y = 2$ ...(3).

Let ABC be the Δ formed by the points of intersection of the lines (1), (2) and (3), as shown in the figure.

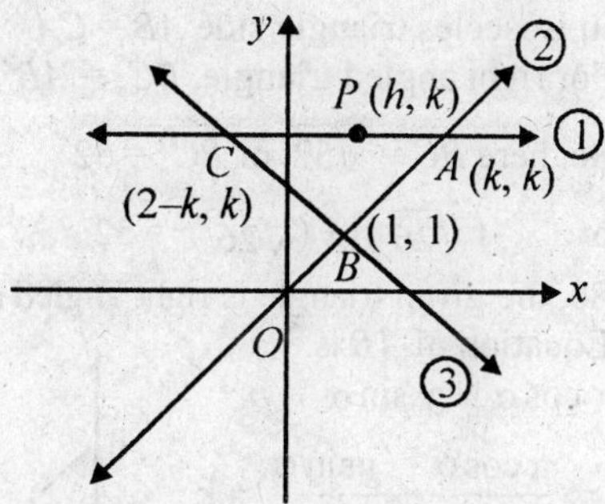

Then $A(k, k)$, $B(1, 1)$, $C(2-k, k)$

$\therefore$ Area of $\Delta ABC = \frac{1}{2}\begin{vmatrix} k & k & 1 \\ 1 & 1 & 1 \\ 2-k & k & 1 \end{vmatrix} = 4h^2$

Operating $C_1 - C_2$ we get

$$\frac{1}{2}\begin{vmatrix} 0 & k & 1 \\ 0 & 1 & 1 \\ 2-2k & k & 1 \end{vmatrix} = 4h^2$$

$\Rightarrow \frac{1}{2}|(2-2k)(k-1)| = 4h^2$

$\Rightarrow (k-1)^2 = 4h^2$

$\Rightarrow k - 1 = 2h$ or $k - 1 = -2h$

$\Rightarrow k = 2h + 1$ or $k = -2h + 1$

$\therefore$ Locus of (h, k) is, $y = 2x + 1$ or $y = -2x + 1$.

(H) ASSERTION & REASON TYPE QUESTIONS :

1. **(c)** Point of intersection of L_1 and L_2 is $A(0, 0)$.
Also $P(-2, -2)$, $Q(1, -2)$

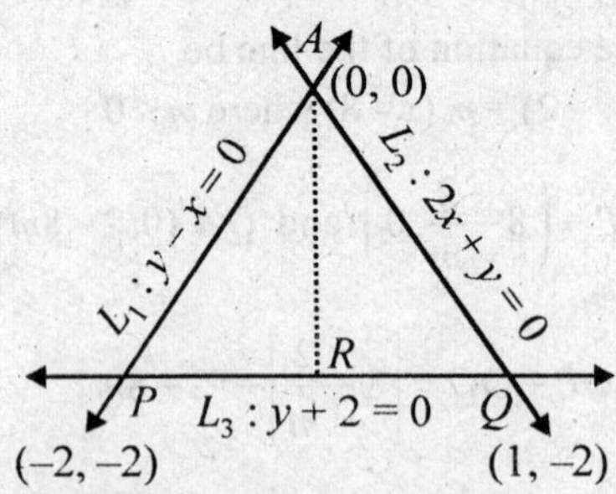

$\because$ AR is the bisector of $\angle PAQ$, therefore R divides PQ in the same ratio as $AP : AQ$.

Thus $PR : RQ = AP : AQ = 2\sqrt{2} : \sqrt{5}$

$\therefore$ Statement-1 is true.

Statement-2 is clearly false.

Section-B JEE Main/ AIEEE

1. **(a)** $AB = \sqrt{(4+1)^2 + (0+1)^2} = \sqrt{26}$;

$BC = \sqrt{(3+1)^2 + (5+1)^2} = \sqrt{52}$

$CA = \sqrt{(4-3)^2 + (0-5)^2} = \sqrt{26}$;

In isosceles triangle side $AB = CA$
For right angled triangle, $BC^2 = AB^2 + AC^2$

So, here $BC = \sqrt{52}$ or $BC^2 = 52$

or $(\sqrt{26})^2 + (\sqrt{26})^2 = 52$

So, the given triangle is right angled and also isosceles

2. **(d)** Equation of AB is

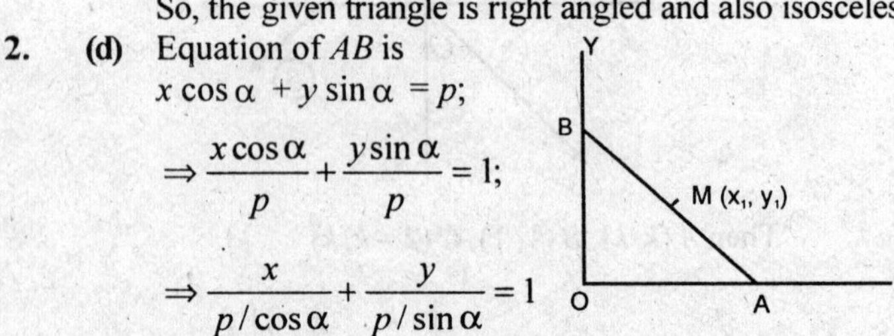

$x\cos\alpha + y\sin\alpha = p$;

$\Rightarrow \dfrac{x\cos\alpha}{p} + \dfrac{y\sin\alpha}{p} = 1$;

$\Rightarrow \dfrac{x}{p/\cos\alpha} + \dfrac{y}{p/\sin\alpha} = 1$

So co-ordinates of A and B are

$\left(\dfrac{p}{\cos\alpha}, 0\right)$ and $\left(0, \dfrac{p}{\sin\alpha}\right)$;

So coordinates of midpoint of AB are

$\left(\dfrac{p}{2\cos\alpha}, \dfrac{p}{2\sin\alpha}\right) = (x_1, y_1)(let)$;

$x_1 = \dfrac{p}{2\cos\alpha}$ & $y_1 = \dfrac{p}{2\sin\alpha}$;

$\Rightarrow \cos\alpha = p/2x_1$ and $\sin\alpha = p/2y_1$;

$\cos^2\alpha + \sin^2\alpha = 1 \Rightarrow \dfrac{p^2}{4}\left(\dfrac{1}{x_1^2} + \dfrac{1}{y_1^2}\right) = 1$

Locus of (x_1, y_1) is $\dfrac{1}{x^2} + \dfrac{1}{y^2} = \dfrac{4}{p^2}$.

3. **(a)** Put $x = 0$ in the given equation
$\Rightarrow by^2 + 2fy + c = 0$.
For unique point of intersection $f^2 - bc = 0$
$\Rightarrow af^2 - abc = 0$.
Since $abc + 2fgh - af^2 - bg^2 - ch^2 = 0$
$\Rightarrow 2fgh - bg^2 - ch^2 = 0$

4. **(a)** $3a + a^2 - 2 = 0 \Rightarrow a^2 + 3a - 2 = 0$.;

$\Rightarrow a = \dfrac{-3 \pm \sqrt{9+8}}{2} = \dfrac{-3 \pm \sqrt{17}}{2}$

5. **(a)** Co-ordinates of $A = (a\cos\alpha, a\sin\alpha)$

Equation of OB,

$y = \tan\left(\dfrac{\pi}{4} + \alpha\right)x$

$CA \perp^r$ to OB

$\therefore$ slope of $CA = -\cot\left(\dfrac{\pi}{4} + \alpha\right)$

Equation of CA

$y - a\sin\alpha = -\cot\left(\dfrac{\pi}{4} + \alpha\right)(x - a\cos\alpha)$

$\Rightarrow (y - a\sin\alpha)\left(\tan\left(\dfrac{\pi}{4} + \alpha\right)\right) = (a\cos\alpha - x)$

$\Rightarrow (y - a\sin\alpha)\left(\dfrac{\tan\dfrac{\pi}{4} + \tan\alpha}{1 - \tan\dfrac{\pi}{4}\tan\alpha}\right)(a\cos\alpha - x)$

$\Rightarrow (y - a\sin\alpha)(1 + \tan\alpha) = (a\cos\alpha - x)(1 - \tan\alpha)$

$\Rightarrow (y - a\sin\alpha)(\cos\alpha + \sin\alpha) = (a\cos\alpha - x)(\cos\alpha - \sin\alpha)$

$\Rightarrow y(\cos + \sin\alpha) - a\sin\alpha\cos\alpha - a\sin^2\alpha$

$= a\cos^2\alpha - a\cos\alpha\sin\alpha - x(\cos\alpha - \sin\alpha$

$\Rightarrow y(\cos\alpha+\sin\alpha)+x(\cos\alpha-\sin\alpha)=a$

$y(\sin\alpha+\cos\alpha)+x(\cos\alpha-\sin\alpha)=a.$

6. **(a)** Equation of bisectors of second pair of straight lines is, $qx^2+2xy-qy^2=0$ ....(1)

It must be identical to the first pair

$x^2-2pxy-y^2=0$ ....(2)

from (1) and (2) $\frac{q}{1}=\frac{2}{-2p}=\frac{-q}{-1}\Rightarrow pq=-1$.

7. **(c)** $x=\frac{a\cos t+b\sin t+1}{3}\Rightarrow a\cos t+b\sin t=3x-1$

$y=\frac{a\sin t-b\cos t}{3}\Rightarrow a\sin t-b\cos t=3y$

Squaring & adding, $(3x-1)^2+(3y)^2=a^2+b^2$

8. **(b)** Taking co-ordinates as

$\left(\frac{x}{r},\frac{y}{r}\right);(x,y)\,\&\,(xr,yr)$.

Then slope of line joining

$\left(\frac{x}{r},\frac{y}{r}\right),(x,y)=\frac{y\left(1-\frac{1}{r}\right)}{x\left(1-\frac{1}{r}\right)}=\frac{y}{x}$.

and slope of line joining (x, y) and (xr, yr)

$=\frac{y(r-1)}{x(r-1)}=\frac{y}{x}$

$\therefore m_1=m_2$

$\Rightarrow$ Points lie on the straight line.

9. **(b)** $(x-a_1)^2+(y-b_1)^2=(x-a_2)^2+(y-b_2)^2$

$(a_1-a_2)x+(b_1-b_2)y$

$+\frac{1}{2}(a_2^2+b_2^2-a_1^2-b_1^2)=0$

$c=\frac{1}{2}(a_2^2+b_2^2-a_1^2-b_1^2)$

10. **(d)** Let the vertex C be (h, k), then the centroid of ΔABC is $\left(\frac{2-2+h}{3},\frac{-3+1+k}{3}\right)$

or $\left(\frac{h}{3},\frac{-2+k}{3}\right)$. It lies on $2x+3y=1$

$\Rightarrow\frac{2h}{3}-2+k=1\Rightarrow 2h+3k=9$

= Locus of C is $2x+3y=9$

11. **(a)** Let the required line be $\frac{x}{a}+\frac{y}{b}=1$ ....(1)

then $a+b=-1$ ....(2)

(1) passes through (4, 3), $\Rightarrow\frac{4}{a}+\frac{3}{b}=1$

$\Rightarrow 4b+3a=ab$ ....(3)

Eliminating b from (2) and (3), we get

$a^2-4=0\Rightarrow a=\pm2\Rightarrow b=-3$ or 1

$\therefore$ Equations of straight lines are

$\frac{x}{2}+\frac{y}{-3}=1$ or $\frac{x}{-2}+\frac{y}{1}=1$

12. **(c)** Let the lines be $y=m_1x$ and $y=m_2x$ then

$m_1+m_2=-\frac{2c}{7}$ and $m_1m_2=-\frac{1}{7}$

Given $m_1+m_2=4\,m_1m_2$

$\Rightarrow\frac{2c}{7}=-\frac{4}{7}\Rightarrow c=2$

13. **(a)** $3x+4y=0$ is one of the lines of the pair

$6x^2-xy+4cy^2=0$, Put $y=-\frac{3}{4}x$,

we get $6x^2+\frac{3}{4}x^2+4c\left(-\frac{3}{4}x\right)^2=0$

$\Rightarrow 6+\frac{3}{4}+\frac{9c}{4}=0\Rightarrow c=-3$

14. **(a)** The line passing through the intersection of line $ax+2by=3b=0$ and $bx-2ay-3a=0$ is

$ax+2by+3b+\lambda(bx-2ay-3a)=0$

$\Rightarrow(a+b\lambda)x+(2b-2a\lambda)y+3b-3\lambda a=0$

As this line is parallel to x-axis.

$\therefore a+b\lambda=0\Rightarrow\lambda=-a/b$

$\Rightarrow\quad ax+2by+3b-\frac{a}{b}(bx-2ay-3a)=0$

$\Rightarrow\quad ax+2by+3b-ax+\frac{2a^2}{b}y+\frac{3a^2}{b}=0$

$y\left(2b+\frac{2a^2}{b}\right)+3b+\frac{3a^2}{b}=0$

$y\left(\frac{2b^2+2a^2}{b}\right)=-\left(\frac{3b^2+3a^2}{b}\right)$

$y=\frac{-3(a^2+b^2)}{2(b^2+a^2)}=\frac{-3}{2}$

So it is 3/2 units below x-axis.

15. (c) Vertex of tringle is (1, 1) and midpoint of sides through this vertex is (– 1, 2) and (3, 2)

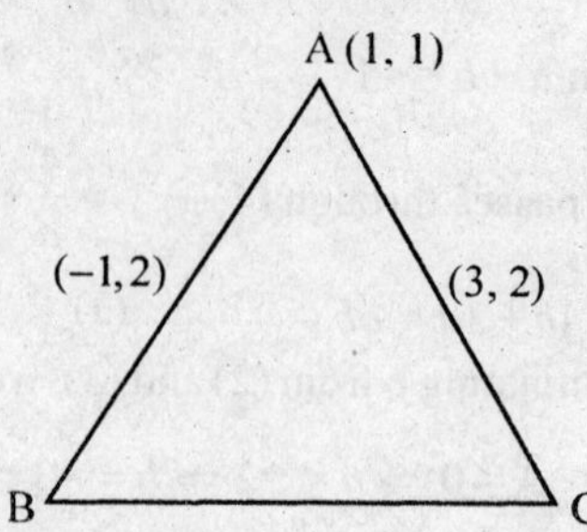

⇒ vertex B and C come out to be (– 3, 3) and (5, 3)

$\therefore$ Centroid is $\frac{1-3+5}{3}, \frac{1+3+5}{3} \Rightarrow \left(1, \frac{7}{3}\right)$

16. (c)

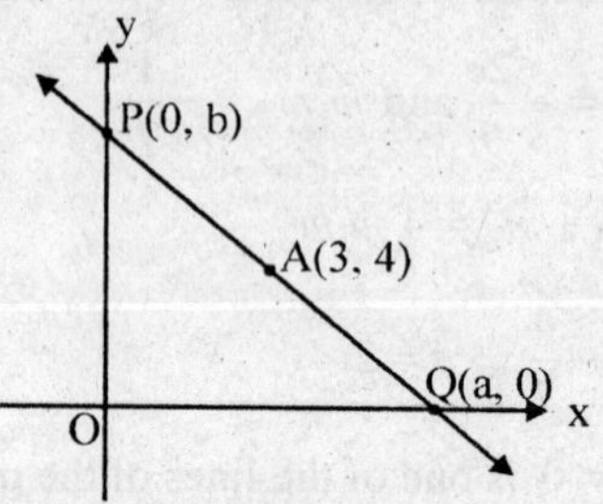

$\because A$ is the mid point of PQ, therefore

$\frac{a+0}{2} = 3, \quad \frac{0+b}{2} = 4 \Rightarrow a = 6, b = 8$

$\therefore$ Equation of line is $\frac{x}{6} + \frac{y}{8} = 1$ or $4x + 3y = 24$

17. (c) Clearly for point P,

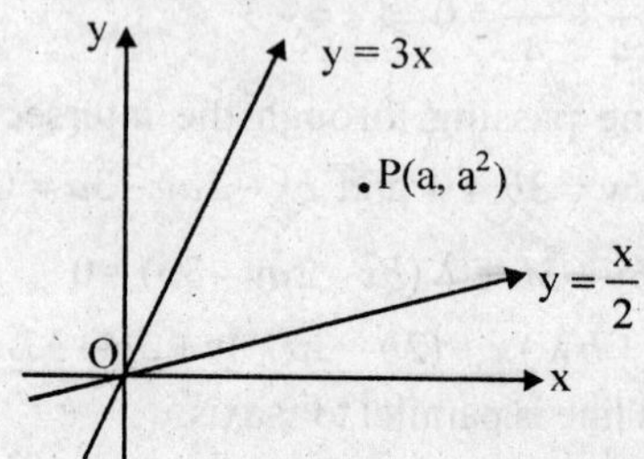

$a^2 - 3a < 0$ **and** $a^2 - \frac{a}{2} > 0 \Rightarrow \frac{1}{2} < a < 3$

18. (a) Given : The vertices of a right angled triangle A(l, k), B(1, 1) and C(2, 1) and Area of ΔABC = 1 square unit

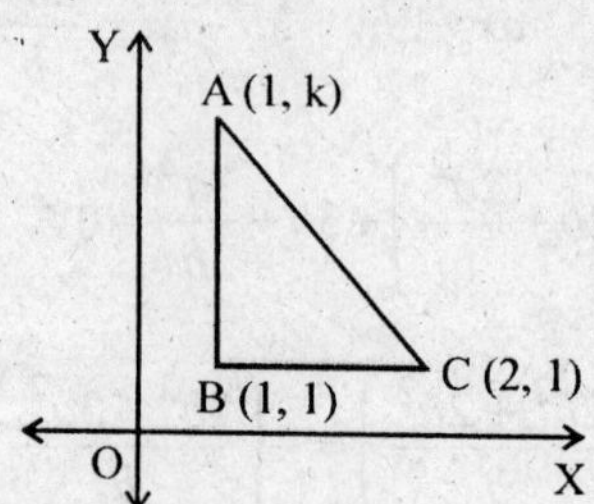

We know that, area of right angled triangle

$= \frac{1}{2} \times BC \times AB = 1 = \frac{1}{2}(1)|(k-1)|$

$\Rightarrow \pm(k-1) = 2 \Rightarrow k = -1, 3$

19. (c) Given : The coordinates of points P, Q, R are (–1, 0), (0, 0), $(3, 3\sqrt{3})$ respectively.

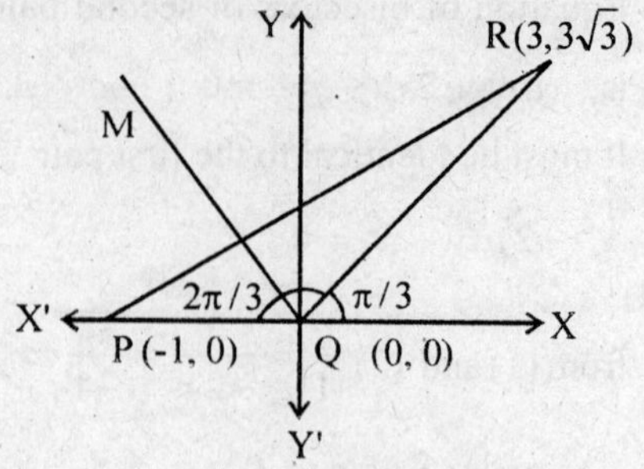

Slope of QR $= \frac{y_2 - y_1}{x_2 - x_1} = \frac{3\sqrt{3}}{3}$

$\Rightarrow \quad \tan\theta = \sqrt{3} \Rightarrow \theta = \frac{\pi}{3} \Rightarrow \angle RQX = \frac{\pi}{3}$

$\therefore \quad \angle RQP = \pi - \frac{\pi}{3} = \frac{2\pi}{3}$

Let QM bisects the $\angle PQR$,

$\therefore$ Slope of the line QM $= \tan\frac{2\pi}{3} = -\sqrt{3}$

$\therefore$ Equation of line QM is $(y-0) = -\sqrt{3}(x-0)$

$\Rightarrow \quad y = -\sqrt{3}x \Rightarrow \sqrt{3}x + y = 0$

20. (a) Equation of bisectors of lines, xy = 0 are $y = \pm x$

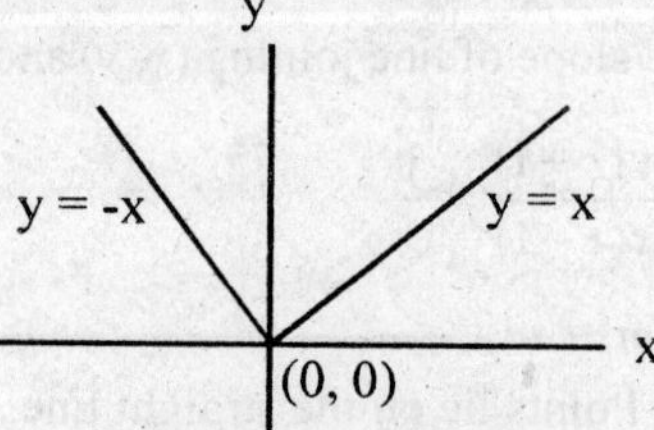

$\therefore$ Put $y = \pm x$ in the given equation

$my^2 + (1-m^2)xy - mx^2 = 0$

$\therefore \quad mx^2 + (1-m^2)x^2 - mx^2 = 0$

$\Rightarrow \quad 1 - m^2 = 0 \Rightarrow m = \pm 1$

21. (d) Slope of $PQ = \frac{3-4}{k-1} = \frac{-1}{k-1}$

$\therefore$ Slope of perpendicular bisector of PQ = (k – 1)

Also mid point of PQ $\left(\frac{k+1}{2}, \frac{7}{2}\right)$.

$\therefore$ Equation of perpendicular bisector is

$$y - \frac{7}{2} = (k-1)\left(x - \frac{k+1}{2}\right)$$

$\Rightarrow \quad 2y - 7 = 2(k-1)x - (k^2 - 1)$

$\Rightarrow \quad 2(k-1)x - 2y + (8 - k^2) = 0$

$\therefore \quad$ y-intercept $= -\frac{8-k^2}{-2} = -4$

$\Rightarrow \quad 8 - k^2 = -8$ or $k^2 = 16 \Rightarrow k = \pm 4$

22. **(d)** Let (a^2, a) be the point of shortest distance on $x = y^2$
Then distance between (a^2, a) and line $x - y + 1 = 0$ is given by

$$D = \frac{a^2 - a + 1}{\sqrt{2}} = \frac{1}{\sqrt{2}}\left[\left(a - \frac{1}{2}\right)^2 + \frac{3}{4}\right]$$

It is min when $a = \frac{1}{2}$ and $D_{min} = \frac{3}{4\sqrt{2}} = \frac{3\sqrt{2}}{8}$

23. **(a)** If the lines $p(p^2+1)x - y + q = 0$
and $(p^2+1)^2 x + (p^2+1)y + 2q = 0$
are perpendicular to a common line then these lines must be parallel to each other,

$\therefore \quad m_1 = m_2 \quad \Rightarrow -\frac{p(p^2+1)}{-1} = -\frac{(p^2+1)^2}{p^2+1}$

$\Rightarrow (p^2+1)(p+1) = 0$

$\Rightarrow p = -1$

$\therefore$ p can have exactly one value.

24. **(a)** Given that

$P(1, 0), Q(-1, 0)$ and $\frac{AP}{AQ} = \frac{BP}{BQ} = \frac{CP}{CQ} = \frac{1}{3}$

$\Rightarrow 3AP = AQ$

Let $A = (x, y)$ then

$3AP = AQ \Rightarrow 9\,AP^2 = AQ^2$

$\Rightarrow 9(x-1)^2 + 9y^2 = (x+1)^2 + y^2$

$\Rightarrow 9x^2 - 18x + 9 + 9y^2 = x^2 + 2x + 1 + y^2$

$\Rightarrow 8x^2 - 20x + 8y^2 + 8 = 0$

$\Rightarrow x^2 + y^2 - \frac{5}{3}x + 1 = 0 \qquad$ ....(1)

$\therefore$ A lies on the circle given by eq (1). As B and C also follow the same condition, they must lie on the same circle.

$\therefore$ Centre of circumcircle of ΔABC

= Centre of circle given by (1) = $\left(\frac{5}{4}, 0\right)$

25. **(c)** Slope of line L = $-\frac{b}{5}$

Slope of line K = $-\frac{3}{c}$

Line L is parallel to line k.

$\Rightarrow \frac{b}{5} = \frac{3}{c} \Rightarrow bc = 15$

(13, 32) is a point on L.

$\therefore \frac{13}{5} + \frac{32}{b} = 1 \Rightarrow \frac{32}{b} = -\frac{8}{5}$

$\Rightarrow b = -20 \Rightarrow c = -\frac{3}{4}$

Equation of K: $y - 4x = 3 \Rightarrow 4x - y + 3 = 0$

Distance between L and K = $\frac{|52 - 32 + 3|}{\sqrt{17}} = \frac{23}{\sqrt{17}}$

26. **(b)**

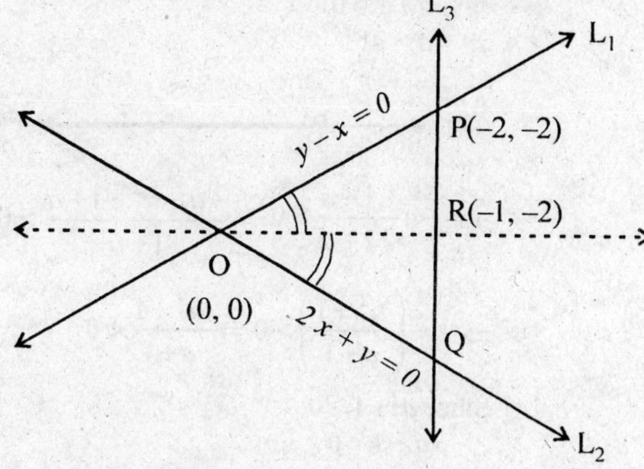

$L_1 : y - x = 0$

$L_2 : 2x + y = 0$

$L_3 : y + 2 = 0$

On solving the equation of line L_1 and L_2 we get their point of intersection (0, 0) i.e., origin O.

On solving the equation of line L_1 and L_3, we get $P = (-2, -2)$.

Similarly, we get $Q = (-1, -2)$

We know that bisector of an angle of a triangle, divide the opposite side the triangle in the ratio of the sides including the angle [Angle Bisector Theorem of a Triangle]

$$\therefore \frac{PR}{RQ} = \frac{OP}{OQ} = \frac{\sqrt{(-2)^2 + (-2)^2}}{\sqrt{(-1)^2 + (-2)^2}} = \frac{2\sqrt{2}}{\sqrt{5}}$$

27. **(b)** $x + y = |a|$

and $ax - y = 1$

Case I : If $a > 0$

$x + y = a$ ...... (1)

$ax - y = 1$ ...... (2)

On adding equation (1) and (2), we get

$x(1+a) = 1 + a \Rightarrow x = 1$

$y = a - 1$

It is in first quadrant

so = $a - 1 \geq 0$

$\Rightarrow a \geq 1$

$\Rightarrow a \in [1, \infty)$

Case II : If $a < 0$

$x + y = -a$ ....(3)

$ax - y = 1$ ....(4)

On adding equation (3) and (4), we get

$x(1+a) = 1 - a$

$x = \frac{1-a}{1+a} > 0 \Rightarrow \frac{a-1}{a+1} < 0$

Since $a-1<0$

$\therefore \quad a+1>0$

$\Rightarrow \quad a>-1$

..............(5)

$$y=-a-\frac{1-a}{1+a}>0 = \frac{-a-a^2-1+a}{1+a}>0$$

$$\Rightarrow \quad -\left(\frac{a^2+1}{a+1}\right)>0 \Rightarrow \frac{a^2+1}{a+1}<0$$

Since $a^2+1>0$

$\therefore \quad a+1<0$

$\Rightarrow \quad a<-1$

..............(6)

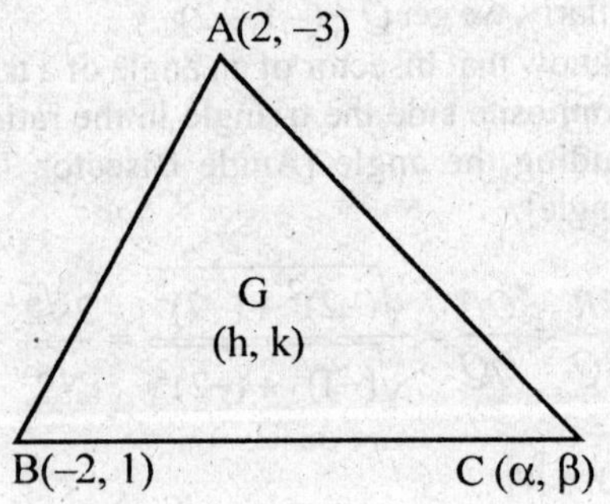

From (5) and (6), $a \in \phi$

Hence Case-II is not possible.

So, correct answer is $a \in [1, \infty)$

28. (b)

$$\alpha = 3h$$

$$\beta - 2 = 3k$$

$$\beta = 3k+2$$

Third vertex (α, β) lies on the line

$$2x+3y=9$$

$$2\alpha+3\beta=9$$

$$2(3h)+3(3k+2)=9$$

$$2h+3k=1$$

$$2x+3y=1$$

29. (a) Number of required triangles $= {}^{10}C_3 - {}^{6}C_3$

$$=\frac{10\times9\times8}{6}-\frac{6\times5\times4}{6}=120-20=100$$

30. (c) Let the joining points be A (1,1) and B (2,4). Let point C divides line AB in the ratio 3 : 2. So, by section formula we have

$$C=\left(\frac{3\times2+2\times1}{3+2},\frac{3\times4+2\times1}{3+2}\right)=\left(\frac{8}{5},\frac{14}{5}\right)$$

Since Line $2x+y=k$ passes through $C\left(\frac{8}{5},\frac{14}{5}\right)$

$\therefore \quad C$ satisfies the equation $2x+y=k$.

$$\Rightarrow \quad \frac{2+8}{5}+\frac{14}{5}=k \Rightarrow k=6$$

31. (b) Suppose B(0, 1) be any point on given line and co-ordinate of A is $(\sqrt{3}, 0)$. So, equation of

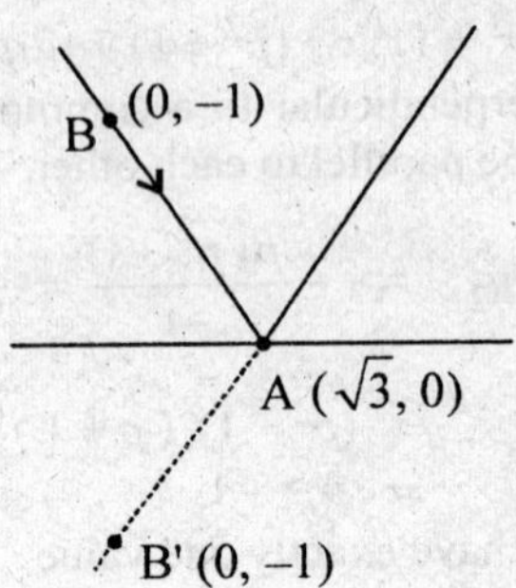

Reflected Ray is $\dfrac{-1-0}{0-\sqrt{3}}=\dfrac{y-0}{x-\sqrt{3}}$

$$\Rightarrow \quad \sqrt{3}y = x-\sqrt{3}$$

32. (b) From the figure, we have

$a=2, b=2\sqrt{2}, c=2$

$x_1=0, x_2=0, x_3=2$

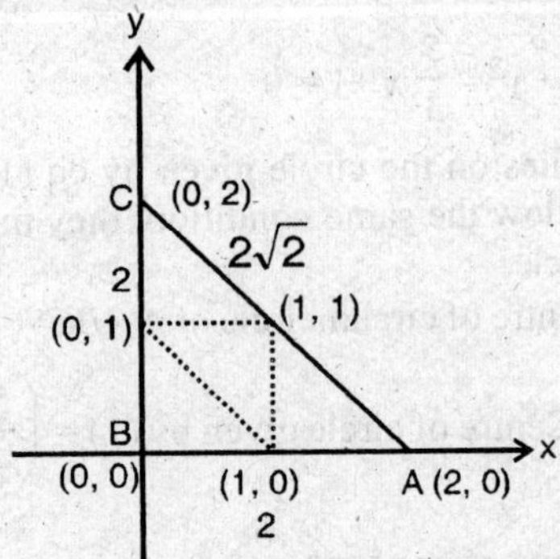

Now, x-co-ordinate of incentre is given as

$$\frac{ax_1+bx_2+cx_3}{a+b+c}$$

$$\Rightarrow \ x\text{-coordinate of incentre} = \frac{2\times0+2\sqrt{2}.0+2.2}{2+2+2\sqrt{2}}$$

$$=\frac{2}{2+\sqrt{2}}=2-\sqrt{2}$$

Chapter 8

Circle

Section-A JEE Advanced/ IIT-JEE

A FILL IN THE BLANKS :

1. If A and B are points in the plane such that $PA/PB = k$ (constant) for all P on a given circle, then the value of k cannot be equal to .... *(1982 - 2 Marks)*
2. The points of intersection of the line $4x - 3y - 10 = 0$ and the circle $x^2 + y^2 - 2x + 4y - 20 = 0$ are .... and .... *(1983 - 2 Marks)*
3. The lines $3x - 4y + 4 = 0$ and $6x - 8y - 7 = 0$ are tangents to the same circle. The radius of this circle is .... *(1984 - 2 Marks)*
4. Let $x^2 + y^2 - 4x - 2y - 11 = 0$ be a circle. A pair of tangents from the point (4, 5) with a pair of radii form a quadrilateral of area .... *(1985 - 2 Marks)*
5. From the origin chords are drawn to the circle $(x - 1)^2 + y^2 = 1$. The equation of the locus of the mid-points of these chords is .... *(1985 - 2 Marks)*
6. The equation of the line passing through the points of intersection of the circles $3x^2 + 3y^2 - 2x + 12y - 9 = 0$ and $x^2 + y^2 + 6x + 2y - 15 = 0$ is .... *(1986 - 2 Marks)*
7. From the point $A(0, 3)$ on the circle $x^2 + 4x + (y - 3)^2 = 0$, a chord AB is drawn and extended to a point M such that $AM = 2AB$. The equation of the locus of M is .... *(1986 - 2 Marks)*
8. The area of the triangle formed by the tangents from the point (4, 3) to the the circle $x^2 + y^2 = 9$ and the line joining their points of contact is .... *(1987 - 2 Marks)*
9. If the circle $C_1 : x^2 + y^2 = 16$ intersects another circle C_2 of radius 5 in such a manner that common chord is of maximum length and has a slope equal to 3/4, then the coordinates of the centre of C_2 are .... *(1988 - 2 Marks)*
10. The area of the triangle formed by the positive x-axis and the normal and the tangent to the circle $x^2 + y^2 = 4$ at $(1, \sqrt{3})$ is, ... *(1989 - 2 Marks)*
11. If a circle passes through the points of intersection of the coordinate axes with the lines $\lambda x - y + 1 = 0$ and $x - 2y + 3 = 0$, then the value of λ = ... *(1991 - 2 Marks)*
12. The equation of the locus of the mid-points of the circle $4x^2 + 4y^2 - 12x + 4y + 1 = 0$ that subtend an angle of $2\pi/3$ at its centre is .... *(1993 - 2 Marks)*
13. The intercept on the line y = x by the circle $x^2 + y^2 - 2x = 0$ is AB. Equation of the circle with AB as a diameter is .... *(1996 - 1 Mark)*
14. Two vertices of an equilateral triangle are (–1, 0) and (1, 0), and its third vertex lies above the x-axis. The equation of its circumcircle is.................. *(1997C - 2 Marks)*
15. For each natural number k, let C_k denote the circle with radius k centimetres and centre at the origin. On the circle C_k, α-particle moves k centimetres in the counter-clockwise direction. After completing its motion on C_k, the particle moves to C_{k+1} in the radial direction. The motion of the particle continues in this manner. The particle starts at (1, 0). If the particle crosses the positive direction of the x-axis for the first time on the circle C_n then n = ... *(1997 - 2 Marks)*
16. The chords of contact of the pair of tangents drawn from each point on the line $2x + y = 4$ to circle $x^2 + y^2 = 1$ pass through the point .... *(1997 - 2 Marks)*

B TRUE / FALSE :

1. No tangent can be drawn from the point (5/2, 1) to the circumcircle of the triangle with vertices $(1, \sqrt{3})$ $(1, -\sqrt{3})$, $(3, -\sqrt{3})$. *(1985 - 1 Mark)*
2. The line $x + 3y = 0$ is a diameter of the circle $x^2 + y^2 - 6x + 2y = 0$. *(1989 - 1 Mark)*

C MCQ's WITH ONE CORRECT ANSWER :

1. A square is inscribed in the circle $x^2 + y^2 - 2x + 4y + 3 = 0$. Its sides are parallel to the coordinate axes. The one vertex of the square is *(1980)*

   (a) $(1 + \sqrt{2}, -2)$ (b) $(1 - \sqrt{2}, -2)$
   (c) $(1, -2 + \sqrt{2})$ (d) none of these

2. Two circles $x^2+y^2=6$ and $x^2+y^2-6x+8=0$ are given. Then the equation of the circle through their points of intersection and the point (1, 1) is *(1980)*
(a) $x^2+y^2-6x+4=0$ (b) $x^2+y^2-3x+1=0$
(c) $x^2+y^2-4y+2=0$ (d) none of these

3. The centre of the circle passing through the point (0, 1) and touching the curve $y=x^2$ at (2, 4) is *(1983 - 1 Mark)*
(a) $\left(\frac{-16}{5},\frac{27}{10}\right)$ (b) $\left(\frac{-16}{7},\frac{53}{10}\right)$
(c) $\left(\frac{-16}{5},\frac{53}{10}\right)$ (d) none of these

4. The equation of the circle passing through (1, 1) and the points of intersection of $x^2+y^2+13x-3y=0$ and $2x^2+2y^2+4x-7y-25=0$ is *(1983 - 1 Mark)*
(a) $4x^2+4y^2-30x-10y-25=0$
(b) $4x^2+4y^2+30x-13y-25=0$
(c) $4x^2+4y^2-17x-10y+25=0$
(d) none of these

5. The locus of the mid-point of a chord of the circle $x^2+y^2=4$ which subtends a right angle at the origin is *(1984 - 2 Marks)*
(a) $x+y=2$ (b) $x^2+y^2=1$
(c) $x^2+y^2=2$ (d) $x+y=1$

6. If a circle passes through the point (a, b) and cuts the circle $x^2+y^2=k^2$ orthogonally, then the equation of the locus of its centre is *(1988 - 2 Marks)*
(a) $2ax+2by-(a^2+b^2+k^2)=0$
(b) $2ax+2by-(a^2-b^2+k^2)=0$
(c) $x^2+y^2-3ax-4by+(a^2+b^2-k^2)=0$
(d) $x^2+y^2-2ax-3by+(a^2-b^2-k^2)=0$.

7. If the two circles $(x-1)^2+(y-3)^2=r^2$ and $x^2+y^2-8x+2y+8=0$ intersect in two distinct points, then *(1989 - 2 Marks)*
(a) $2<r<8$ (b) $r<2$
(c) $r=2$ (d) $r>2$

8. The lines $2x-3y=5$ and $3x-4y=7$ are diameters of a circle of area 154 sq. units. Then the equation of this circle is *(1989 - 2 Marks)*
(a) $x^2+y^2+2x-2y=62$
(b) $x^2+y^2+2x-2y=47$
(c) $x^2+y^2-2x+2y=47$
(d) $x^2+y^2-2x+2y=62$

9. The centre of a circle passing through the points (0, 0), (1, 0) and touching the circle $x^2+y^2=9$ is *(1992 - 2 Marks)*
(a) $\left(\frac{3}{2},\frac{1}{2}\right)$ (b) $\left(\frac{1}{2},\frac{3}{2}\right)$
(c) $\left(\frac{1}{2},\frac{1}{2}\right)$ (d) $\left(\frac{1}{2},-2^{\frac{1}{2}}\right)$

10. The locus of the centre of a circle, which touches externally the circle $x^2+y^2-6x-6y+14=0$ and also touches the y-axis, is given by the equation: *(1993 - 1 Marks)*
(a) $x^2-6x-10y+14=0$ (b) $x^2-10x-6y+14=0$
(c) $y^2-6x-10y+14=0$ (d) $y^2-10x-6y+14=0$

11. The circles $x^2+y^2-10x+16=0$ and $x^2+y^2=r^2$ intersect each other in two distinct points if *(1994)*
(a) $r<2$ (b) $r>8$
(c) $2<r<8$ (d) $2\le r\le 8$

12. The angle between a pair of tangents drawn from a point P to the circle $x^2+y^2+4x-6y+9\sin^2\alpha+13\cos^2\alpha=0$ is 2α. The equation of the locus of the point P is *(1996 - 1 Mark)*
(a) $x^2+y^2+4x-6y+4=0$ (b) $x^2+y^2+4x-6y-9=0$
(c) $x^2+y^2+4x-6y-4=0$ (d) $x^2+y^2+4x-6y+9=0$

13. If two distinct chords, drawn from the point (p, q) on the circle $x^2+y^2=px+qy$ (where $pq\neq 0$) are bisected by the x-axis, then *(1999 - 2 Marks)*
(a) $p^2=q^2$ (b) $p^2=8q^2$
(c) $P^2<8q^2$ (d) $p^2>8q^2$.

14. The triangle PQR is inscribed in the circle $x^2+y^2=25$. If Q and R have co-ordinates (3,4) and (–4, 3) respectively, then $\angle QPR$ is equal to *(2000S)*
(a) $\frac{\pi}{2}$ (b) $\frac{\pi}{3}$
(c) $\frac{\pi}{4}$ (d) $\frac{\pi}{6}$

15. If the circles $x^2+y^2+2x+2ky+6=0$, $x^2+y^2+2ky+k=0$ intersect orthogonally, then k is *(2000S)*
(a) 2 or $-\frac{3}{2}$ (b) -2 or $-\frac{3}{2}$
(c) 2 or $\frac{3}{2}$ (d) -2 or $\frac{3}{2}$

16. Let AB be a chord of the circle $x^2+y^2=r^2$ subtending a right angle at the centre. Then the locus of the centroid of the triangle PAB as P moves on the circle is *(2001S)*
(a) a parabola (b) a circle
(c) an ellipse (d) a pair of straight lines

17. Let PQ and RS be tangents at the extremities of the diameter PR of a circle of radius r. If PS and RQ intersect at a point X on the circumference of the circle, then $2r$ equals *(2001S)*
(a) $\sqrt{PQ.RS}$ (b) $(PQ+RS)/2$
(c) $2PQ.RS/(PQ+RS)$ (d) $\sqrt{(PQ^2+RS^2)}/2$

18. If the tangent at the point P on the circle $x^2+y^2+6x+6y=2$ meets a straight line $5x-2y+6=0$ at a point Q on the y-axis, then the length of PQ is *(2002S)*
(a) 4 (b) $2\sqrt{5}$
(c) 5 (d) $3\sqrt{5}$.

19. The centre of circle inscribed in square formed by the lines $x^2 - 8x + 12 = 0$ and $y^2 - 14y + 45 = 0$, is *(2003S)*

(a) $(4, 7)$ (b) $(7, 4)$
(c) $(9, 4)$ (d) $(4, 9)$

20. If one of the diameters of the circle $x^2 + y^2 - 2x - 6y + 6 = 0$ is a chord to the circle with centre (2, 1), then the radius of the circle is *(2004S)*

(a) $\sqrt{3}$ (b) $\sqrt{2}$
(c) 3 (d) 2

21. A circle is given by $x^2 + (y-1)^2 = 1$, another circle C touches it externally and also the x-axis, then the locus of its centre is *(2005S)*

(a) $\{(x, y) : x^2 = 4y\} \cup \{(x, y) : y \le 0\}$
(b) $\{(x, y) : x^2 + (y-1)^2 = 4\} \cup \{(x, y) : y \le 0\}$
(c) $\{(x, y) : x^2 = y\} \cup \{(0, y) : y \le 0\}$
(d) $\{(x, y) : x^2 = 4y\} \cup \{(0, y) : y \le 0\}$

22. Tangents drawn from the point $P(1, 8)$ to the circle $x^2 + y^2 - 6x - 4y - 11 = 0$ touch the circle at the points A and B. The equation of the circumcircle of the triangle PAB is *(2009)*

(a) $x^2 + y^2 + 4x - 6y + 19 = 0$
(b) $x^2 + y^2 - 4x - 10y + 19 = 0$
(c) $x^2 + y^2 - 2x + 6y - 29 = 0$
(d) $x^2 + y^2 - 6x - 4y + 19 = 0$

23. The circle passing through the point $(-1, 0)$ and touching the y-axis at $(0, 2)$ also passes through the point. *(2011)*

(a) $\left(-\frac{3}{2}, 0\right)$ (b) $\left(-\frac{5}{2}, 2\right)$
(c) $\left(-\frac{3}{2}, \frac{5}{2}\right)$ (d) $(-4, 0)$

24. The locus of the mid-point of the chord of contact of tangents drawn from points lying on the straight line $4x - 5y = 20$ to the circle $x^2 + y^2 = 9$ is *(2012)*

(A) $20(x^2 + y^2) - 36x + 45y = 0$
(B) $20(x^2 + y^2) + 36x - 45y = 0$
(C) $36(x^2 + y^2) - 20x + 45y = 0$
(D) $36(x^2 + y^2) + 20x - 45y = 0$

(D) MCQ's WITH ONE OR MORE THAN ONE CORRECT :

1. The equations of the tangents drawn from the origin to the circle $x^2 + y^2 - 2rx - 2hy + h^2 = 0$, are *(1988 - 2 Marks)*

(a) $x = 0$ (b) $y = 0$
(c) $(h^2 - r^2)x - 2rhy = 0$ (d) $(h^2 - r^2)x + 2rhy = 0$

2. The number of common tangents to the circles $x^2 + y^2 = 4$ and $x^2 + y^2 - 6x - 8y = 24$ is *(1998 - 2 Marks)*

(a) 0 (b) 1
(c) 3 (d) 4

3. If the circle $x^2 + y^2 = a^2$ intersects the hyperbola $xy = c^2$ in four points $P(x_1, y_1)$, $Q(x_2, y_2)$, $R(x_3, y_3)$, $S(x_4, y_4)$, then *(1998 - 2 Marks)*

(a) $x_1 + x_2 + x_3 + x_4 = 0$ (b) $y_1 + y_2 + y_3 + y_4 = 0$
(c) $x_1x_2x_3x_4 = c^4$ (d) $y_1y_2y_3y_4 = c^4$

4. Circle(s) touching x-axis at a distance 3 from the origin and having an intercept of length $2\sqrt{7}$ on y-axis is (are) *(JEE Adv. 2013)*

(a) $x^2 + y^2 - 6x + 8y + 9 = 0$
(b) $x^2 + y^2 - 6x + 7y + 9 = 0$
(c) $x^2 + y^2 - 6x - 8y + 9 = 0$
(d) $x^2 + y^2 - 6x - 7y + 9 = 0$

(E) SUBJECTIVE PROBLEMS :

1. Find the equation of the circle whose radius is 5 and which touches the circle $x^2 + y^2 - 2x - 4y - 20 = 0$ at the point (5, 5). *(1978)*

2. (a) Find the equation of a circle which passes through the point (2, 0) and whose centre is the limit of the point of intersection of the lines

$3x + 5y = 1$
$(2 + c)x + 5c^2y = 1$ as c tends to 1.

(b) A tent is made in the form of a frustum A of a right circular cone surmounted by another right circular cone B. The diameter of the ends of the frustum A are 8 m and 4m, its height is 3m and the height of the cone B is 2m. Find the area of the canvas required. *(1979)*

3. Let A be the centre of the circle $x^2 + y^2 - 2x - 4y - 20 = 0$. Suppose that the tangents at the points $B(1, 7)$ and $D(4, -2)$ on the circle meet at the point C. Find the area of the quadrilateral $ABCD$. *(1981 - 4 Marks)*

4. Find the equations of the circle passing through (–4, 3) and touching the lines $x + y = 2$ and $x - y = 2$. *(1982 - 3 Marks)*

5. Through a fixed point (h, k) secants are drawn to the circle $x^2 + y^2 = r^2$. Show that the locus of the mid-points of the secants intercepted by the circle is $x^2 + y^2 = hx + ky$. *(1983 - 5 Marks)*

6. The abscissa of the two points A and B are the roots of the equation $x^2 + 2ax - b^2 = 0$ and their ordinates are the roots of the equation $x^2 + 2px - q^2 = 0$. Find the equation and the radius of the circle with AB as diameter. *(1984 - 4 Marks)*

7. Lines $5x + 12y - 10 = 0$ and $5x - 12y - 40 = 0$ touch a circle C_1 of diameter 6. If the centre of C_1 lies in the first quadrant, find the equation of the circle C_2 which is concentric with C_1 and cuts intercepts of length 8 on these lines. *(1986 - 5 Marks)*

8. Let a given line L_1 intersects the x and y axes at P and Q, respectively. Let another line L_2, perpendicular to L_1, cut the x and y axes at R and S, respectively. Show that the locus of the point of intersection of the lines PS and QR is a circle passing through the origin. *(1987 - 3 Marks)*

9. The circle $x^2 + y^2 - 4x - 4y + 4 = 0$ is inscribed in a triangle which has two of its sides along the co-ordinate axes. The locus of the circumcentre of the triangle is $x + y - xy + k(x^2 + y^2)^{1/2} = 0$. Find k. *(1987 - 4 Marks)*

10. Let $S \equiv x^2 + y^2 + 2gx + 2fy + c = 0$ be a given circle. Find the locus of the foot of the perpendicular drawn from the origin upon any chord of S which subtends a right angle at the origin. *(1988 - 5 Marks)*

11. If $\left(m_i, \frac{1}{m_i}\right)$, $m_i > 0, i = 1, 2, 3, 4$ are four distinct points on a circle, then show that $m_1 m_2 m_3 m_4 = 1$ *(1989 - 2 Marks)*

12. A circle touches the line $y = x$ at a point P such that $OP = 4\sqrt{2}$, where O is the origin. The circle contains the point $(-10, 2)$ in its interior and the length of its chord on the line $x + y = 0$ is $6\sqrt{2}$. Determine the equation of the circle. *(1990 - 5 Marks)*

13. Two circles, each of radius 5 units, touch each other at $(1, 2)$. If the equation of their common tangent is $4x + 3y = 10$, find the equation of the circles. *(1991 - 4 Marks)*

14. Let a circle be given by $2x(x-a) + y(2y-b) = 0, (a \neq 0, b \neq 0)$. Find the condition on a and b if two chords, each bisected by the x- axis, can be drawn to the circle from $\left(a, \frac{b}{2}\right)$. *(1992 - 6 Marks)*

15. Consider a family of circles passing through two fixed points $A(3,7)$ and $B(6, 5)$. Show that the chords in which the circle $x^2 + y^2 - 4x - 6y - 3 = 0$ cuts the members of the family are concurrent at a point. Find the coordinate of this point. *(1993 - 5 Marks)*

16. Find the coordinates of the point at which the circles $x^2 + y^2 - 4x - 2y = -4$ and $x^2 + y^2 - 12x - 8y = -36$ touch each other. Also find equations common tangents touching the circles in the distinct points. *(1993 - 5 Marks)*

17. Find the intervals of values of a for which the line $y + x = 0$ bisects two chords drawn from a point $\left(\frac{1+\sqrt{2}a}{2}, \frac{1-\sqrt{2}a}{2}\right)$ to the circle $2x^2 + 2y^2 - (1+\sqrt{2}a)x - (1-\sqrt{2}a)y = 0$. *(1996 - 5 Marks)*

18. A circle passes through three points A, B and C with the line segment AC as its diameter. A line passing through A intersects the chord BC at a point D inside the circle. If angles DAB and CAB are α and β respectively and the distance between the point A and the mid point of the line segment DC is d, prove that the area of the circle is

$$\frac{\pi d^2 \cos^2 \alpha}{\cos^2 \alpha + \cos^2 \beta + 2 \cos \alpha \cos \beta \cos(\beta - \alpha)}$$

(1996 - 5 Marks)

19. Consider a curve $ax^2 + 2hxy + by^2 = 1$ and a point P not on the curve. A line drawn from the point P intersects the curve at points Q and R. If the product $PQ.\ PR$ is independent of the slope of the line, then show that the curve is a circle. *(1997C - 5 Marks)*

20. Let C be any circle with centre $(0, \sqrt{2})$. Prove that at the most two rational points can be there on C. (A rational point is a point both of whose coordinates are rational numbers.) *(1997 - 5 Marks)*

21. C_1 and C_2 are two concentric circles, the radius of C_2 being twice that of C_1. From a point P on C_2, tangents PA and PB are drawn to C_1. Prove that the centroid of the triangle PAB lies on C_1. *(1998 - 8 Marks)*

22. Let T_1, T_2 be two tangents drawn from $(-2, 0)$ onto the circle $C: x^2 + y^2 = 1$. Determine the circles touching C and having T_1, T_2 as their pair of tangents. Further, find the equations of all possible common tangents to these circles, when taken two at a time. *(1999 - 10 Marks)*

23. Let $2x^2 + y^2 - 3xy = 0$ be the equation of a pair of tangents drawn from the origin O to a circle of radius 3 with centre in the first quadrant. If A is one of the points of contact, find the length of OA. *(2001 - 5 Marks)*

24. Let C_1 and C_2 be two circles with C_2 lying inside C_1. A circle C lying inside C_1 touches C_1 internally and C_2 externally. Identify the locus of the centre of C. *(2001 - 5 Marks)*

25. For the circle $x^2 + y^2 = r^2$, find the value of r for which the area enclosed by the tangents drawn from the point $P(6, 8)$ to the circle and the chord of contact is maximum. *(2003 - 2 Marks)*

26. Find the equation of circle touching the line $2x + 3y + 1 = 0$ at $(1, -1)$ and cutting orthogonally the circle having line segment joining $(0, 3)$ and $(-2, -1)$ as diameter. *(2004 - 4 Marks)*

27. Circles with radii 3, 4 and 5 touch each other externally. If P is the point of intersection of tangents to these circles at their points of contact, find the distance of P from the points of contact. *(2005 - 2 Marks)*

G COMPREHENSION BASED QUESTIONS :

PASSAGE-1

$ABCD$ is a square of side length 2 units. C_1 is the circle touching all the sides of the square $ABCD$ and C_2 is the circumcircle of square $ABCD$. L is a fixed line in the same plane and R is a fixed point.

1. If P is any point of C_1 and Q is another point on C_2, then $\frac{PA^2 + PB^2 + PC^2 + PD^2}{QA^2 + QB^2 + QC^2 + QD^2}$ is equal to *(2006 - 5M, –2)*

   (a) 0.75 (b) 1.25
   (c) 1 (d) 0.5

2. If a circle is such that it touches the line L and the circle C_1 externally, such that both the circles are on the same side of the line, then the locus of centre of the circle is *(2006 - 5M, –2)*

   (a) ellipse (b) hyperbola
   (c) parabola (d) pair of straight line

3. A line L' through A is drawn parallel to BD. Point S moves such that its distances from the line BD and the vertex A are equal. If locus of S cuts L' at T_2 and T_3 and AC at T_1, then area of $\Delta T_1 T_2 T_3$ is *(2006 - 5M, –2)*

   (a) $\frac{1}{2}$ sq. units (b) $\frac{2}{3}$ sq. units
   (c) 1 sq. units (d) 2 sq. units

PASSAGE-2

A circle C of radius 1 is inscribed in an equilateral triangle PQR. The points of contact of C with the sides PQ, QR, RP are D, E, F, respectively. The line PQ is given by the equation $\sqrt{3}x+y-6=0$ and the point D is $\left(\frac{3\sqrt{3}}{2}, \frac{3}{2}\right)$. Further, it is given that the origin and the centre of C are on the same side of the line PQ.

4. The equation of circle C is *(2008)*

(a) $(x-2\sqrt{3})^2+(y-1)^2=1$

(b) $(x-2\sqrt{3})^2+\left(y+\frac{1}{2}\right)^2=1$

(c) $(x-\sqrt{3})^2+(y+1)^2=1$

(d) $(x-\sqrt{3})^2+(y-1)^2=1$

5. Points E and F are given by *(2008)*

(a) $\left(\frac{\sqrt{3}}{2}, \frac{3}{2}\right), (\sqrt{3}, 0)$ (b) $\left(\frac{\sqrt{3}}{2}, \frac{1}{2}\right), (\sqrt{3}, 0)$

(c) $\left(\frac{\sqrt{3}}{2}, \frac{3}{2}\right), \left(\frac{\sqrt{3}}{2}, \frac{1}{2}\right)$ (d) $\left(\frac{3}{2}, \frac{\sqrt{3}}{2}\right), \left(\frac{\sqrt{3}}{2}, \frac{1}{2}\right)$

6. Equations of the sides QR, RP are *(2008)*

(a) $y=\frac{2}{\sqrt{3}}x+1,\ y=-\frac{2}{\sqrt{3}}x-1$

(b) $y=\frac{1}{\sqrt{3}}x,\ y=0$

(c) $y=\frac{\sqrt{3}}{2}x+1,\ y=-\frac{\sqrt{3}}{2}x-1$

(d) $y=\sqrt{3}x,\ y=0$

PASSAGE-3

A tangent PT is drawn to the circle $x^2+y^2=4$ at the point $P(\sqrt{3},1)$. A straight line L, perpendicular to PT is a tangent to the circle $(x-3)^2+y^2=1$. *(2012)*

7. A possible equation of L is

(a) $x-\sqrt{3}y=1$ (b) $x+\sqrt{3}y=1$

(c) $x-\sqrt{3}y=-1$ (d) $x+\sqrt{3}y=5$

8. A common tangent of the two circles is

(a) $x=4$ (b) $y=2$

(c) $x+\sqrt{3}y=4$ (d) $x+2\sqrt{2}y=6$

(H) ASSERTION & REASON TYPE QUESTIONS :

1. Tangents are drawn from the point (17, 7) to the circle $x^2+y^2=169$.

STATEMENT-1 : The tangents are mutually perpendicular.

because

STATEMENT-2 : The locus of the points from which mutually perpendicular tangents can be drawn to the given circle is $x^2+y^2=338$. ***(2007 -3 marks)***

(a) Statement-1 is True, statement-2 is True; Statement-2 is a correct explanation for Statement-1.

(b) Statement-1 is True, Statement-2 is True; Statement-2 is NOT a correct explanation for Statement-1

(c) Statement-1 is True, Statement-2 is False

(d) Statement-1 is False, Statement-2 is True.

2. Consider $L_1: 2x+3y+p-3=0$

$L_2: 2x+3y+p+3=0$

where p is a real number, and $C: x^2+y^2+6x-10y+30=0$

STATEMENT - 1 : If line L_1 is a chord of circle C, then line L_2 is not always a diameter of circle C

and

STATEMENT - 2 : If line L_1 is a diameter of circle C, then line L_2 is not a chord of circle C. ***(2008)***

(a) Statement - 1 is True, Statement - 2 is True; Statement - 2 **is** a correct explanation for Statement - 1

(b) Statement - 1 is True, Statement - 2 is True; Statement - 2 **is NOT** a correct explanation for Statement - 1

(c) Statement - 1 is True, Statement - 2 is False

(d) Statement - 1 is False, Statement - 2 is True

(I) INTEGER VALUE CORRECT TYPE :

1. The centres of two circles C_1 and C_2 each of unit radius are at a distance of 6 units from each other. Let P be the mid point of the line segement joining the centres of C_1 and C_2 and C be a circle touching circles C_1 and C_2 externally. If a common tangent to C_1 and C passing through P is also a common tangent to C_2 and C, then the radius of the circle C is ***(2009)***

2. The straight line $2x-3y=1$ divides the circular region $x^2+y^2 \le 6$ into two parts.

If $S=\left\{\left(2,\frac{3}{4}\right),\left(\frac{5}{2},\frac{3}{4}\right),\left(\frac{1}{4},-\frac{1}{4}\right),\left(\frac{1}{8},\frac{1}{4}\right)\right\}$ then the number of points (s) in S lying inside the smaller part is ***(2011)***

Section-B JEE Main/ AIEEE

1. If the chord $y = mx + 1$ of the circle $x^2+y^2=1$ subtends an angle of measure 45° at the major segment of the circle then value of m is [2002]

(a) $2 \pm \sqrt{2}$ (b) $-2 \pm \sqrt{2}$

(c) $-1 \pm \sqrt{2}$ (d) none of these

2. The centres of a set of circles, each of radius 3, lie on the circle $x^2+y^2=25$. The locus of any point in the set is [2002]

(a) $4 \le x^2+y^2 \le 64$ (b) $x^2+y^2 \le 25$

(c) $x^2+y^2 \ge 25$ (d) $3 \le x^2+y^2 \le 9$

3. The centre of the circle passing through (0, 0) and (1, 0) and touching the circle $x^2+y^2=9$ is [2002]

(a) $\left(\frac{1}{2}, \frac{1}{2}\right)$ (b) $\left(\frac{1}{2}, -\sqrt{2}\right)$

(c) $\left(\frac{3}{2}, \frac{1}{2}\right)$ (d) $\left(\frac{1}{2}, \frac{3}{2}\right)$

4. The equation of a circle with origin as a centre and passing through equilateral triangle whose median is of length $3a$ is [2002]

(a) $x^2+y^2=9a^2$ (b) $x^2+y^2=16a^2$

(c) $x^2+y^2=4a^2$ (d) $x^2+y^2=a^2$

5. If the two circles $(x-1)^2 + (y-3)^2 = r^2$ and $x^2 + y^2 - 8x + 2y + 8 = 0$ intersect in two distinct point, then [2003]

(a) $r > 2$ (b) $2 < r < 8$

(c) $r < 2$ (d) $r = 2$.

6. The lines $2x - 3y = 5$ and $3x - 4y = 7$ are diameters of a circle having area as 154 sq.units.Then the equation of the circle is [2003]

(a) $x^2 + y^2 - 2x + 2y = 62$

(b) $x^2 + y^2 + 2x - 2y = 62$

(c) $x^2 + y^2 + 2x - 2y = 47$

(d) $x^2 + y^2 - 2x + 2y = 47$.

7. If a circle passes through the point (a, b) and cuts the circle $x^2 + y^2 = 4$ orthogonally, then the locus of its centre is [2004]

(a) $2ax - 2by - (a^2 + b^2 + 4) = 0$

(b) $2ax + 2by - (a^2 + b^2 + 4) = 0$

(c) $2ax - 2by + (a^2 + b^2 + 4) = 0$

(d) $2ax + 2by + (a^2 + b^2 + 4) = 0$

8. A variable circle passes through the fixed point $A(p, q)$ and touches x-axis . The locus of the other end of the diameter through A is [2004]

(a) $(y-q)^2 = 4px$ (b) $(x-q)^2 = 4py$

(c) $(y-p)^2 = 4qx$ (d) $(x-p)^2 = 4qy$

9. If the lines $2x + 3y + 1 = 0$ and $3x - y - 4 = 0$ lie along diameter of a circle of circumference 10π, then the equation of the circle is [2004]

(a) $x^2 + y^2 + 2x - 2y - 23 = 0$

(b) $x^2 + y^2 - 2x - 2y - 23 = 0$

(c) $x^2 + y^2 + 2x + 2y - 23 = 0$

(d) $x^2 + y^2 - 2x + 2y - 23 = 0$

10. Intercept on the line $y = x$ by the circle $x^2 + y^2 - 2x = 0$ is AB. Equation of the circle on AB as a diameter is [2004]

(a) $x^2 + y^2 + x - y = 0$ (b) $x^2 + y^2 - x + y = 0$

(c) $x^2 + y^2 + x + y = 0$ (d) $x^2 + y^2 - x - y = 0$

11. If the circles $x^2 + y^2 + 2ax + cy + a = 0$ and $x^2 + y^2 - 3ax + dy - 1 = 0$ intersect in two distinct points P and Q then the line $5x + by - a = 0$ passes through P and Q for [2005]

(a) exactly one value of a

(b) no value of a

(c) infinitely many values of a

(d) exactly two values of a

12. A circle touches the x- axis and also touches the circle with centre at (0,3) and radius 2. The locus of the centre of the circle is [2005]

(a) an ellipse (b) a circle

(c) a hyperbola (d) a parabola

13. If a circle passes through the point (a, b) and cuts the circle $x^2 + y^2 = p^2$ orthogonally, then the equation of the locus of its centre is [2005]

(a) $x^2 + y^2 - 3ax - 4by + (a^2 + b^2 - p^2) = 0$

(b) $2ax + 2by - (a^2 - b^2 + p^2) = 0$

(c) $x^2 + y^2 - 2ax - 3by + (a^2 - b^2 - p^2) = 0$

(d) $2ax + 2by - (a^2 + b^2 + p^2) = 0$

14. If the pair of lines $ax^2 + 2(a+b)xy + by^2 = 0$ lie along diameters of a circle and divide the circle into four sectors such that the area of one of the sectors is thrice the area of another sector then **[2005]**

(a) $3a^2 - 10ab + 3b^2 = 0$

(b) $3a^2 - 2ab + 3b^2 = 0$

(c) $3a^2 + 10ab + 3b^2 = 0$

(d) $3a^2 + 2ab + 3b^2 = 0$

15. If the lines $3x - 4y - 7 = 0$ and $2x - 3y - 5 = 0$ are two diameters of a circle of area 49π square units, the equation of the circle is **[2006]**

(a) $x^2 + y^2 + 2x - 2y - 47 = 0$

(b) $x^2 + y^2 + 2x - 2y - 62 = 0$

(c) $x^2 + y^2 - 2x + 2y - 62 = 0$

(d) $x^2 + y^2 - 2x + 2y - 47 = 0$

16. Let C be the circle with centre (0, 0) and radius 3 units. The equation of the locus of the mid points of the chords of the circle C that subtend an angle of $\frac{2\pi}{3}$ at its center is

(a) $x^2 + y^2 = \frac{3}{2}$ (b) $x^2 + y^2 = 1$ **[2006]**

(c) $x^2 + y^2 = \frac{27}{4}$ (d) $x^2 - y^2 = \frac{9}{4}$

17. Consider a family of circles which are passing through the point (– 1, 1) and are tangent to x-axis. If (h, k) are the coordinate of the centre of the circles, then the set of values of k is given by the interval **[2007]**

(a) $-\frac{1}{2} \le k \le \frac{1}{2}$ (b) $k \le \frac{1}{2}$

(c) $0 \le k \le \frac{1}{2}$ (d) $k \ge \frac{1}{2}$

18. The point diametrically opposite to the point $P(1, 0)$ on the circle $x^2 + y^2 + 2x + 4y - 3 = 0$ is **[2008]**

(a) $(3, -4)$ (b) $(-3, 4)$

(c) $(-3, -4)$ (d) $(3, 4)$

19. The differential equation of the family of circles with fixed radius 5 units and centre on the line $y = 2$ is

(a) $(x-2)y'^2 = 25 - (y-2)^2$

(b) $(y-2)y'^2 = 25 - (y-2)^2$

(c) $(y-2)^2 y'^2 = 25 - (y-2)^2$

(d) $(x-2)^2 y'^2 = 25 - (y-2)^2$

20. If P and Q are the points of intersection of the circles $x^2 + y^2 + 3x + 7y + 2p - 5 = 0$ and $x^2 + y^2 + 2x + 2y - p^2 = 0$ then there is a circle passing through P, Q and (1, 1) for: **[2009]**

(a) all except one value of p

(b) all except two values of p

(c) exactly one value of p

(d) all values of p

21. The circle $x^2 + y^2 = 4x + 8y + 5$ intersects the line $3x - 4y = m$ at two distinct points if **[2010]**

(a) $-35 < m < 15$ (b) $15 < m < 65$

(c) $35 < m < 85$ (d) $-85 < m < -35$

22. The two circles $x^2 + y^2 = ax$ and $x^2 + y^2 = c^2$ $(c > 0)$ touch each other if **[2011]**

(a) $|a| = c$ (b) $a = 2c$

(c) $|a| = 2c$ (d) $2|a| = c$

23. The equation of the circle passing through the point (1, 0) and (0, 1) and having the smallest radius is - **[2011 RS]**

(a) $x^2 + y^2 - 2x - 2y + 1 = 0$

(b) $x^2 + y^2 - x - y = 0$

(c) $x^2 + y^2 + 2x + 2y - 7 = 0$

(d) $x^2 + y^2 + x + y - 2 = 0$

24. The length of the diameter of the circle which touches the x-axis at the point (1,0) and passes through the point (2,3) is: **[2012]**

(a) $\frac{10}{3}$ (b) $\frac{3}{5}$ (c) $\frac{6}{5}$ (d) $\frac{5}{3}$

25. The circle passing through (1, –2) and touching the axis of x at (3, 0) also passes through the point **[JEE M 2013]**

(a) $(-5, 2)$ (b) $(2, -5)$

(c) $(5, -2)$ (d) $(-2, 5)$

Solutions & Explanations

Section-A : JEE Advanced/ IIT-JEE

A 1. 1 2. (4, 2), (–2, –6) 3. $\frac{3}{4}$ 4. 8 sq. units 5. $x^2+y^2-x=0$ 6. $10x-3y-18=0$

7. $x^2+y^2+8x-6y+9=0$ 8. $\frac{192}{25}$ 9. $\left(-\frac{9}{5},\frac{12}{5}\right)$ or $\left(\frac{9}{5},-\frac{12}{5}\right)$ 10. $2\sqrt{3}$ sq. units

11. 2 12. $16x^2+16y^2-48x+16y+31=0$ 13. $x^2+y^2-x-y=0$ 14. $x^2+y^2-\frac{2y}{\sqrt{3}}-1=0$

15. 7 16. $\left(\frac{1}{2},\frac{1}{4}\right)$

B 1. *T* 2. *T*

C 1. (d) 2. (b) 3. (c) 4. (b) 5. (c) 6. (a) 7. (a) 8. (c) 9. (d)
10. (d) 11. (c) 12. (d) 13. (d) 14. (c) 15. (a) 16. (b) 17. (a) 18. (c)
19. (a) 20. (c) 21. (d) 22. (b) 23. (d) 24. (a)

D 1. (a, c) 2. (b) 3. (a,b,c,d) 4. (a, c)

E 1. $x^2+y^2-18x-16y+120=0$ 2. (a) $25x^2+25y^2-20x+2y-60=0$, (b) $2(3\sqrt{13}+2\sqrt{2})\pi\ m^2$

3. 75 sq. units 4. $x^2+y^2+2(10\pm\sqrt{54})x+55\pm\sqrt{54}=0$

6. $x^2+y^2+2ax+2py-b^2-q^2=0, \sqrt{a^2+p^2+b^2+q^2}$ 7. $x^2+y^2-10x-4y+4=0$

9. $k=1$ 10. $x^2+y^2+gx+fy+\frac{c}{2}=0$ 12. $x^2+y^2+18x-2y+32=0$

13. $x^2+y^2+6x+2y-15=0$ and $x^2+y^2-10x-10y+25=0$ 14. $a^2>2b^2$ 15. $\left(2,\frac{23}{3}\right)$

16. $\left(\frac{14}{5},\frac{8}{5}\right)$, $y=0$ and $7y-24x+16=0$ 17. $a\in\,]-\infty,-2[\,\cup\,]2,\infty[$

22. $(x-4)^2+y^2=3^2$ and $\left(x+\frac{4}{3}\right)^2+y^2=\left(\frac{1}{3}\right)^2$; $y=\pm\frac{5}{\sqrt{39}}\left(x+\frac{4}{5}\right)$ 23. $3(3+\sqrt{10})$

24. ellipse 25. 5 16. $2x^2+2y^2-10x-5y+1=0$ 27. $\sqrt{5}$

G 1. (a) 2. (b) 3. (c) 4. (d) 5. (a) 6. (d) 7. (a) 8. (d)

H 1. (a) 2. (c)

I 1. 8 2. 2

Section-B : JEE Main/ AIEEE

1. (c) 2. (a) 3. (b) 4. (c) 5. (b) 6. (d) 7. (b) 8. (d) 9. (d)
10. (d) 11. (b) 12. (d) 13. (d) 14. (d) 15. (d) 16. (d) 17. (d) 18. (c)
19. (c) 20. (a) 21. (a) 22. (a) 23. (b) 24. (a) 25. (c)

Section-A JEE Advanced/ IIT-JEE

(A) FILL IN THE BLANKS :

1. As P lies on a circle and A and B two points in the plane such that $\frac{PA}{PB} = k$

Then k can be any real number except 1 as otherwise P will lie on perpendicular bisector of AB which is a line.

2. For point of intersection of line

$4x - 3y - 10 = 0$ ...(1)

and circle $x^2 + y^2 - 2x + 4y - 20 = 0$ ...(2)

Solving (1) and (2), we get

$$\left(\frac{3y+10}{4}\right)^2 + y^2 - 2\left(\frac{3y+10}{4}\right) + 4y - 20 = 0$$

$\Rightarrow 9y^2 + 60y + 100 + 16y^2 - 24y - 80 + 64y - 320 = 0$

$\Rightarrow 25y^2 + 100y - 300 = 0$

$\Rightarrow y^2 + 4y - 12 = 0$

$\Rightarrow y = 2, -6$

$\Rightarrow x = 4, -2$

$\therefore$ Points are (4, 2) and (–2, –6)

3. Let $3x - 4y + 4 = 0$ be the tangent at point A and $6x - 8y - 7 = 0$ be the tangent of point B of the circle.

As the two tangents parallel to each other

$\therefore$ AB should be the diameter of circle.

$\therefore$ AB = distance between parallel lines

$3x - 4y + 4 = 0$ and $6x - 8y - 7 = 0$

= distance between $6x - 8y + 8 = 0$ and $6x - 8y - 7 = 0$

$$= \left|\frac{8+7}{\sqrt{36+64}}\right| = \frac{15}{10} = \frac{3}{2}$$

$\therefore$ radius of circle $= \frac{1}{2}(AB) = \frac{3}{4}$

4. 8 *sq.* units

KEY CONCEPT:

Length of tangent from a point (x_1, y_1) to a circle $x^2 + y^2 + 2gx + 2fy + c = 0$ is given by

$\sqrt{x_1^2 + y_1^2 + 2gx_1 + 2fy_1 + c}$

The equation of circle is,

$x^2 + y^2 - 4x - 2y - 11 = 0$

It's centre is (2, 1), radius

$= \sqrt{4+1+11} = 4 = BC$

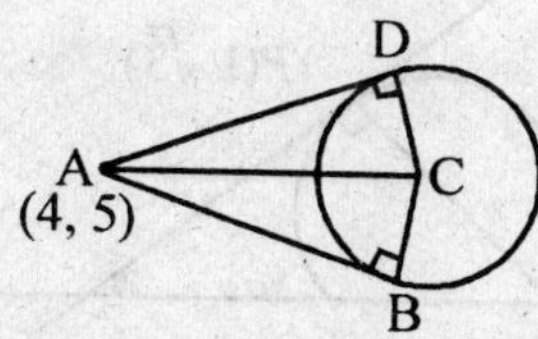

length of tangent from the *pt.* (4, 5) is

$= \sqrt{16+25-16-10-11} = \sqrt{4} = 2 = AB$

$\therefore$ Area of quad. $ABCD$

$= 2$ (Area of ΔABC) $= 2 \times \frac{1}{2} \times AB \times BC$

$= 2 \times \frac{1}{2} \times 2 \times 4 = 8$ sq. units.

5. The equation of given circle is

$(x-1)^2 + y^2 = 1$

or $x^2 + y^2 - 2x = 0$ ...(1)

KEY CONCEPT : We know that equation of chord of curve $S = 0$, whose mid point is (x_1, y_1) is given by $T = S_1$ where T is tangent to curve $S = 0$ at (x_1, y_1).

$\therefore$ If (x_1, y_1) is the mid point of chord of given circle (1), then equation of chord is

$xx_1 + yy_1 - (x + x_1) = x_1^2 + y_1^2 - 2x_1$

$\Rightarrow (x_1 - 1)x + y_1 y + x_1 - x_1^2 - y_1^2 = 0$

At it passes through origin, we get

$x_1 - x_1^2 - y_1^2 = 0$

or $x_1^2 + y_1^2 - x_1 = 0$

$\therefore$ locus of (x_1, y_1) is $x^2 + y^2 - x = 0$

ALTERNATE SOLUTION

The given circle is $(x-1)^2 + y^2 = 1$

Centre (1, 0) radius = 1

Consider chord OA of this circle with $M(x_1, y_1)$ as mid pt of OA.

Let co-ordinates of point A, on circle, be

$(1 + \cos\theta, \sin\theta)$

[Using parametric equation of circle as $x = h + r\cos\theta$, $y = k + r\sin\theta$ with centre (h, k) and radius r.]

As $M(x_1, y_1)$ is mid pt of OA, using mid pt formula we get

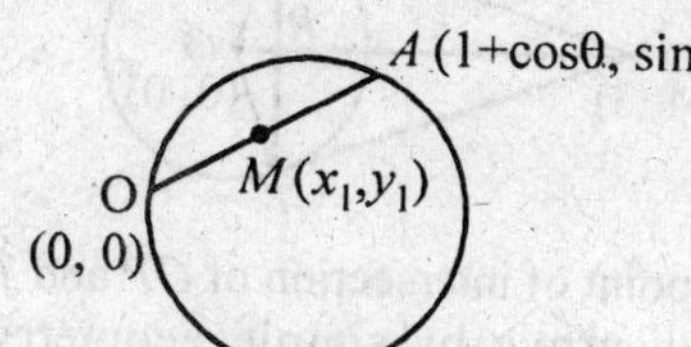

$$x_1 = \frac{1+\cos\theta}{2}, y_1 = \frac{\sin\theta}{2}$$

$\Rightarrow \cos\theta = 2x_1 - 1, \sin\theta = 2y_1$

Squaring and adding, we get

$(2x_1 - 1)^2 + (2y_1)^2 = 1$

$\Rightarrow 4x_1^2 - 4x_1 + 1 + 4y_1^2 = 1$

$\Rightarrow x_1^2 + y_1^2 - x_1 = 0$

$\therefore$ locus of (x_1, y_1) is $x^2 + y^2 - x = 0$

6. The equation of two circles are

$$x^2 + y^2 - \frac{2}{3}x + 4y - 3 = 0 \quad ...(1)$$

and $x^2 + y^2 + 6x + 2y - 15 = 0$ ...(2)

Now we know eq. of common chord of two circles $S_1 = 0$ and $S_2 = 0$ is $S_1 - S_2 = 0$

$\Rightarrow\ 6x + \frac{2}{3}x + 2y - 4y - 15 + 3 = 0$

$\Rightarrow\ \frac{20x}{3} - 2y - 12 = 0$

$\Rightarrow\ 10x - 3y - 18 = 0$

7. The equation of circle is,
$x^2 + y^2 + 4x - 6y + 9 = 0$ ...(1)

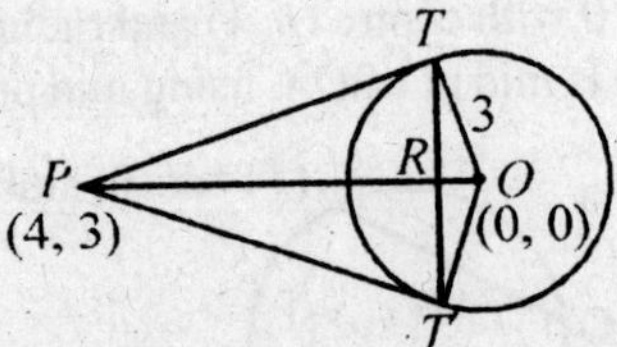

$AM = 2AB$

$\Rightarrow\ AB = BM$

Let the co-ordinates of M be (h, k)
Then B is mid pt of AM

$\therefore\ B\left(\frac{0+h}{2}, \frac{3+k}{2}\right) = \left(\frac{h}{2}, \frac{k+3}{2}\right)$

As B lies on circle (1),

$\therefore\ \left(\frac{h}{2}\right)^2 + \left(\frac{k+3}{2}\right)^2 + 4\times\frac{h}{2} - 6\left(\frac{k+3}{2}\right) + 9 = 0$

$\Rightarrow\ h^2 + k^2 + 6k + 9 + 8h - 12k - 36 + 36 = 0$

$\Rightarrow\ h^2 + k^2 + 8h - 6k + 9 = 0$

$\therefore$ locus of (h, k) is ,
$x^2 + y^2 + 8x - 6y + 9 = 0$

8. From $P(4, 3)$ two tangents PT and PT' are drawn to the circle $x^2 + y^2 = 9$ with O (0, 0) as centre and $r = 3$.
To find the area of $\Delta PTT'$.

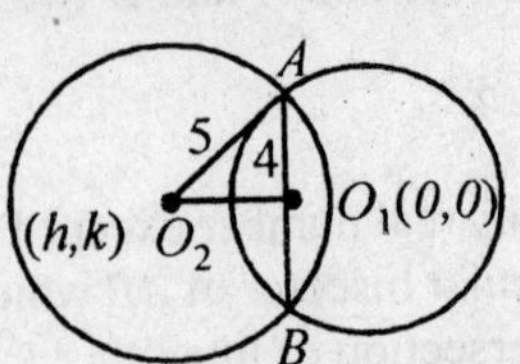

Let R be the point of intersection of OP and TT'.
Then we can prove by simple geometry that OP is perpendicular bisector of TT'.
Equation of chord of contact TT' is
$4x + 3y = 9$
Now, OR = length of the perpendicular from O to TT' is

$= \left|\frac{4\times 0 + 3\times 0 - 9}{\sqrt{4^2 + +3^2}}\right| = \frac{9}{5}.$

OT = radius of circle = 3

$\therefore\ TR = \sqrt{OT^2 - OR^2} = \sqrt{9 - \frac{81}{25}} = \frac{12}{5}$

Again $OP = \sqrt{(4-0)^2 + (3-0)^2} = 5$

$\therefore\ PR = OP - OR = 5 - \frac{9}{5} = \frac{16}{5}$

Area of the triangle

$PTT' = PR \times TR = \frac{16}{5} \times \frac{12}{5} = \frac{192}{25}$

9. We have $C_1 : x^2 + y^2 = 16$, Centre O_1 (0, 0) radius $= 4$. C_2 is another circle with radius 5, let its centre O_2 be (h, k).

Now the common chord of circles C_1 and C_2 is of maximum length when chord is diameter of smaller circle C_1, and then it passes through centre O_1 of circle C_1. Given that slope of this chord is 3/4.

$\therefore$ Equation of AB is,

$y = \frac{3}{4}x \Rightarrow 3x - 4y = 0$ ...(1)

In right ΔAO_1O_2,

$O_1O_2 = \sqrt{5^2 - 4^2} = 3$

Also $O_1O_2 = \perp^{lar}$ distance from (h, k) to (1)

$\Rightarrow\ 3 = \left|\frac{3h - 4k}{\sqrt{3^2 + 4^2}}\right|$ **NOTE THIS STEP**

$\Rightarrow\ \pm 3 = \frac{3h - 4k}{5}$

$\Rightarrow\ 3h - 4k \pm 15 = 0$ ...(2)

Again $AB \perp O_1O_2$

$\Rightarrow\ m_{AB} \times m_{O_1O_2} = -1$

$\Rightarrow\ \frac{3}{4}\times\frac{k}{h} = -1 \Rightarrow 4h + 3k = 0$ ...(3)

Solving, $3h - 4k + 15 = 0$ and $4h + 3k = 0$
We get $h = -9/5$, $k = 12/5$
Again solving $3h - 4k - 15 = 0$ and $4h + 3k = 0$
We get $h = 9/5$, $k = -12/5$

Thus the required centre is $\left(\frac{-9}{5}, \frac{12}{5}\right)$ or $\left(\frac{9}{5}, \frac{-12}{5}\right)$.

10. Tangent at $P(1, \sqrt{3})$ to the circle $x^2 + y^2 = 4$ is

$x \cdot 1 + y \cdot \sqrt{3} = 4$

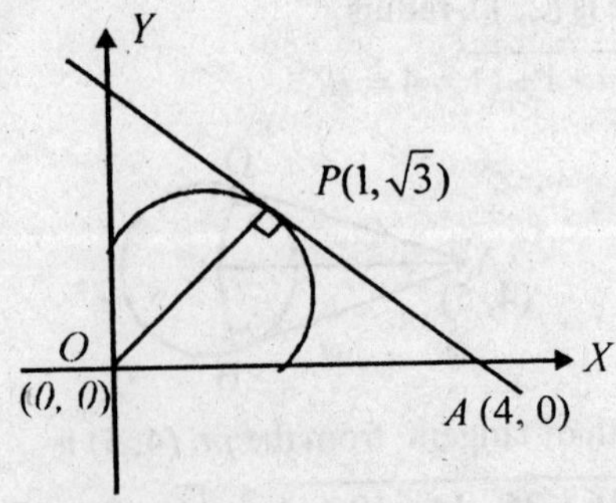

It meets x-axis at A (4, 0),

$\therefore \quad OA = 4$

Also OP = radius of circle = 2

$\therefore \quad PA = \sqrt{4^2 - 2^2} = \sqrt{12}$

$\therefore \quad$ Area of $\Delta OPA = \frac{1}{2} \times OP \times PA = \frac{1}{2} \times 2 \times \sqrt{12}$

$= 2\sqrt{3}$ sq. units

11. The given lines are $\lambda x - y + 1 = 0$ and $x - 2y + 3 = 0$ which meet x-axis at $A\left(-\frac{1}{\lambda}, 0\right)$ and $B(-3, 0)$ and y-axis at $C(0, 1)$ and $D\left(0, \frac{3}{2}\right)$ respectively.

Then we must have,

$OA \times OB = OC \times OD$ **NOTE THIS STEP**

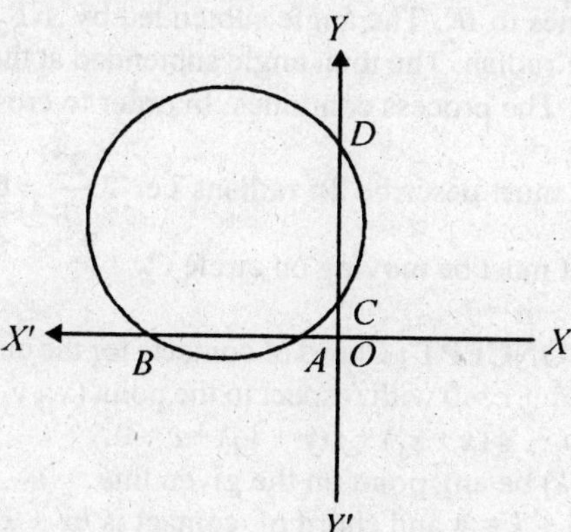

$\Rightarrow \left(-\frac{1}{\lambda}\right)(-3) = 1 \times \frac{3}{2}$

$\Rightarrow \lambda = 2$

ALTERNATE SOLUTION

The given lines $\lambda x - y + 1 = 0$ and $x - 2y + 3 = 0$ meet x-axis at A$\left(\frac{-1}{\lambda}, 0\right)$ and $B(-3, 0)$ and y-axis at $C(0, 1)$ and $D\left(0, \frac{3}{2}\right)$ respectively.

Let the equation of circle through $ABCD$ be

$x^2 + y^2 + 2gx + 2fy + c = 0$

As it passes through (–3, 0), therefore

$9 - 6g + c = 0 \quad \ldots(1)$

As it passes through (0, 1), therefore

$1 + 2f + c = 0 \quad \ldots(2)$

As it passes through (0, 3/2), therefore

$\frac{9}{4} + 3f + c = 0 \quad \ldots(3)$

Solving the above three equations, we get

$f = -5/4, c = 3/2, g = 7/4$

$\therefore$ The circle becomes

$x^2 + y^2 + \frac{7}{2}x - \frac{5}{2}y + \frac{3}{2} = 0$

or $2x^2 + 2y^2 + 7x - 5y + 3 = 0$

It passes through $(-1/\lambda, 0)$ also, therefore

$\frac{2}{\lambda^2} - \frac{7}{\lambda} + 3 = 0, \quad \lambda \neq 0$

$\Rightarrow 3\lambda^2 - 7\lambda + 2 = 0 \Rightarrow (3\lambda - 1)(\lambda - 2) = 0$

$\Rightarrow \lambda = \frac{1}{3}$ or $\lambda = 2$

But for $\lambda = \frac{1}{3}$ two lines coincide.

$\therefore$ We must have $\lambda = 2$.

12. The given circle is,

$4x^2 + 4y^2 - 12x + 4y + 1 = 0$

or $x^2 + y^2 - 3x + y + \frac{1}{4} = 0$ with centre $\left(\frac{3}{2}, -\frac{1}{2}\right)$

and $r = \sqrt{\frac{9}{4} + \frac{1}{4} - \frac{1}{4}} = \frac{3}{2}$

Let $M(h, k)$ be the mid pt. of the chord AB of the given circle, then $CM \perp AB$. $\angle ACB = 120°$.

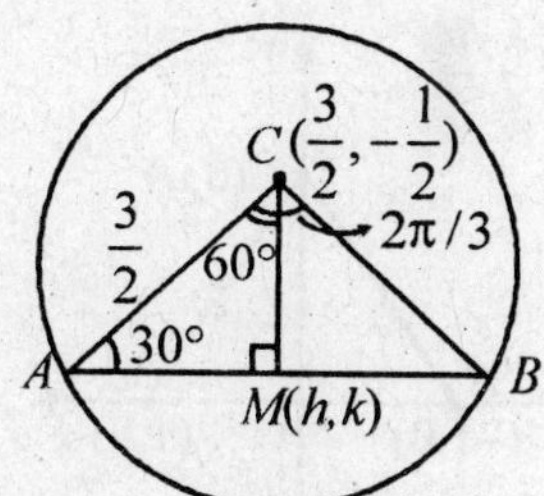

In ΔACM,

$\angle ACM = \frac{1}{2}\angle ACB = 60°$ and $\angle A = 30°$

$\therefore \quad \sin A = \frac{CM}{AC}$

$\sin 30° = \frac{\sqrt{(h - 3/2)^2 + (k + 1/2)^2}}{3/2}$

$\Rightarrow \left(\frac{3}{4}\right)^2 = \left(h - \frac{3}{2}\right)^2 + \left(k + \frac{1}{2}\right)^2$

$\Rightarrow h^2 + k^2 - 3h + k + \frac{9}{4} + \frac{1}{4} - \frac{9}{16} = 0$

$\Rightarrow 16h^2 + 16k^2 - 48h + 16k + 31 = 0$

$\therefore$ locus of (h, k) is

$16x^2 + 16y^2 - 48x + 16y + 31 = 0$.

13. Equation of any circle passing through the point of intersection of $x^2 + y^2 - 2x = 0$ and $y = x$ is

$x^2 + y^2 - 2x + \lambda(y - x) = 0$ **NOTE THIS STEP**

or $x^2 + y^2 - (2 + \lambda)x + \lambda y = 0$

Its centre is $\left(\frac{2+\lambda}{2}, \frac{-\lambda}{2}\right)$

For AB to be the diameter of the required circle, the centre must lie on AB. That is,

$$\frac{2+\lambda}{2} = -\frac{\lambda}{2} \Rightarrow \lambda = -1$$

Thus, equation of required circle is
$x^2+y^2-2x-y+x=0$
or $x^2+y^2-x-y=0$

ALTERNATE SOLUTION

KEY CONCEPT : If (x_1, y_1) and (x_2, y_2) are the end points of diameter of a circle then equation of circle is
$(x-x_1)(x-x_2)+[(y-y_1)(y-y_2)]=0.$

Points of intersection of line $y = x$ and circle $x^2+y^2-2x=0$ can be obtained by solving these two equations i.e.
$x^2+y^2-2x=0$
$x=0, 1$
$\Rightarrow y=0, 1$
$\therefore A(0, 0), B(1, 1)$

Now equation of circle with AB as diameter is
$x(x-1)+y(y-1)=0$
$\Rightarrow x^2+y^2-x-y=0$

14. Let ABC be the given equilateral Δ then C must lie on y-axis. Let $(0, a)$. Also $AC = AB$

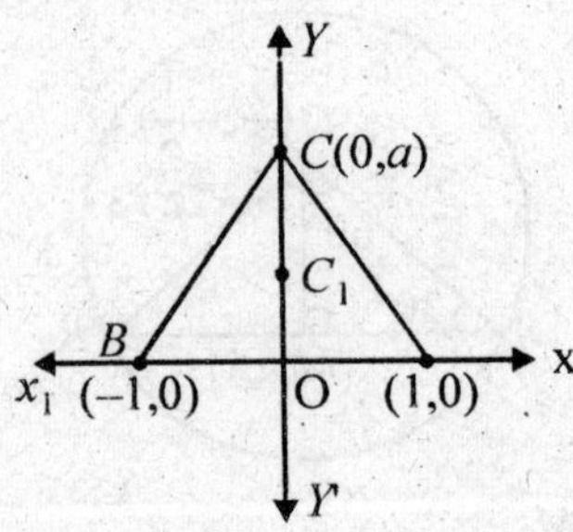

$\Rightarrow \sqrt{1+a^2} = \sqrt{2^2+0^2}$
$\Rightarrow 1+a^2=4$
$\Rightarrow a=\sqrt{3}$
$\therefore C(0, \sqrt{3})$

Then centroid of ΔABC is $\left(0, \frac{1}{\sqrt{3}}\right)$

But in an equilateral Δ circumcenter coincides with centroid

$\therefore$ Circumcenter is $\left(0, \frac{1}{\sqrt{3}}\right)$

Also radius of circumcircle $= C_1 B$

$$= \sqrt{(1-0)^2 + \left(0-\frac{1}{\sqrt{3}}\right)^2} = \sqrt{1+\frac{1}{3}} = \frac{2}{\sqrt{3}}$$

$\therefore$ Equation of circumcircle is

$$(x-0)^2 + \left(y-\frac{1}{\sqrt{3}}\right)^2 = \left(\frac{2}{\sqrt{3}}\right)^2$$

$$\Rightarrow x^2+y^2-\frac{2y}{\sqrt{3}}+\frac{1}{3}=\frac{4}{3}$$

$$\Rightarrow x^2+y^2-\frac{2y}{\sqrt{3}}-1=0$$

15.

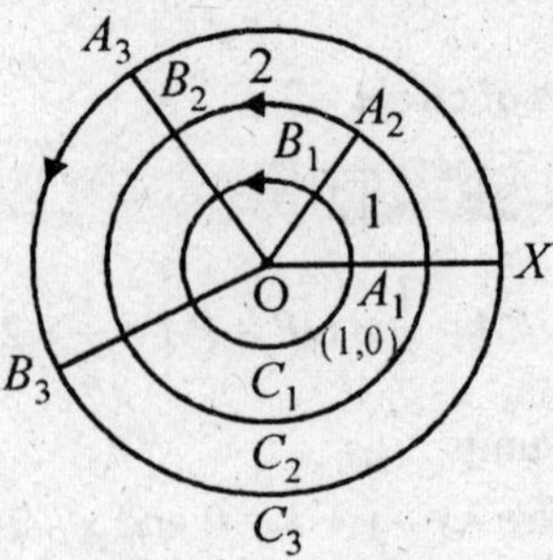

The radius of circle C_1 is 1 cm, C_2 is 2 cm and soon. It starts from A_1 (1, 0) on C_1, moves a distance of 1 cm on C_1 to come to B_1. The angle subtended by A_1B_1 at the centre will be $\frac{1}{r} = \theta$ radians, i.e. 1 radian.

From B_1 it moves along radius, OB_1 and comes to A_2 on circle C_2 of radius 2. From A_2 it moves on C_2 a distance 2 cm and comes to B_2. The angle subtended by A_2B_2 is again as before 1 radian. The total angle subtended at the centre is 2 radians. The process continues. In order to cross the x-axis again it must describe 2π radians i.e. $2.\frac{22}{7} = 6.7$ radians.

Hence it must be moving on circle C_7
$\therefore n=7$

16. **KEY CONCEPT :** Chord of contacts for the circle $x^2+y^2+2gx+2fy+c=0$ with respect to the point (x_1, y_1) is given by $xx_1+yy_1+g(x+x_1)+f(y+y_1)+c=0.$

Let (h, k) be any point on the given line
$\therefore$ $2h+k=4$ and chord of contact is $hx+ky=1$
or $hx+(4-2h)y=1$
or $(4y-1)+h(x-2y)=0$
$P+\lambda Q=0$. It passes through the intersection of $P=0$ and $Q=0$ i.e. $\left(\frac{1}{2}, \frac{1}{4}\right)$.

(B) TRUE / FALSE :

1. The circle passes through the points $A(1, \sqrt{3}), B(1, -\sqrt{3})$ and $C(3, -\sqrt{3})$.

Here line AB is parallel to y-axis and BC is parallel to x-axis, there $\angle ABC = 90°$
$\therefore$ AC is a diameter of circle.
$\therefore$ Eq. of circle is
$(x-1)(x-3)+(y-\sqrt{3})(y+\sqrt{3})=0$
$\Rightarrow x^2+y^2-4x=0$ ...(1)

Let us check the position of pt (5/2, 1) with. respect to the circle (1), we get

$$S_1 = \frac{25}{4}+1-10<0$$

$\therefore$ Point lies inside the circle.
$\therefore$ No tangent can be drawn to the given circle from point (5/2, 1).
$\therefore$ Given statement is true.

2. The centre of the circle $x^2+y^2-6x+2y=0$ is $(3,-1)$ Which lies on the line $x+3y=0$

$\therefore$ The statement is true.

C MCQ's WITH ONE CORRECT ANSWER :

1. **(d)** The given circle is $x^2+y^2-2x+4y+3=0$. Centre $(1,-2)$. Lines through centre $(1,-2)$ and parallel to axes are $x=1$ and $y=-2$.

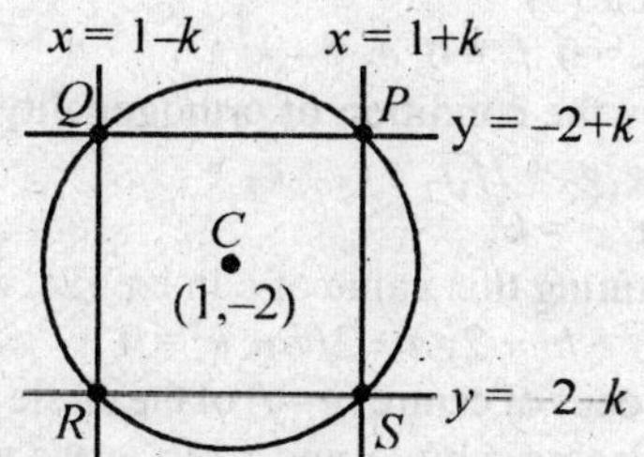

Let the side of square be $2k$.

Then sides of square are $x=1-k$ and $x=1+k$ and $y=-2-k$ and $y=-2+k$

$\therefore$ Co-ordinates of P, Q, R, S are $(1+k,-2+k)$, $(1-k,-2+k)$, $(1-k,-2-k)$, $(1+k,-2-k)$ respectively.

Also $P(1+k,-2+k)$ lies on circle

$\therefore (1+k)^2+(-2+k)^2-2(1+k)+4(-2+k)+3=0$

$\Rightarrow 2k^2=2$

$\Rightarrow k=1$ or -1

If $k=1$

$P(2,-1), Q(0,-1), R(0,-3), S(2,-3)$

If $k=-1$

$P(0,-3), Q(2,-3), R(2,-1), S(0,-1)$

2. **(b)** The circle through points of intersection of the two circles $x^2+y^2-6=0$ and $x^2+y^2-6x+8=0$ is

$(x^2+y^2-6)+\lambda(x^2+y^2-6x+8)=0$

As it passes through $(1, 1)$

$(1+1-6)+\lambda(1+1-6+8)=0$

$\Rightarrow \lambda=\frac{4}{4}=1$

$\therefore$ The required circle is

$2x^2+2y^2-6x+2=0$

or, $x^2+y^2-3x+1=0$

3. **(c)** **KEY CONCEPT** : Equation of tangent at a point (x_1, y_1) of a curve can be obtained by replacing x^2 by xx_1, y^2 by yy_1, x by

$\frac{1}{2}(x+x_1)$ and y by $\frac{1}{2}(y+y_1)$.

If the circle touches $y=x^2$ at $(2, 4)$ then the tangent at $(2, 4)$ to $y=x^2$ will be same as tangent to circle at $[(2, 4)]$ which is

$\frac{1}{2}(y+4)=2x.$

or $4x-y=4$ ...(1)

Let eq. of circle be

$x^2+y^2+2gx+2fx+c=0$

with centre $(-g,-f)$

As it passes through $(0, 1)$

$\Rightarrow 1+2f+c=0$

$\therefore c=-1-2f$ ...(2)

Also the eq. of tangent of circle at $(2, 4)$ is

$x.2+y.4+g(x+2)+f(y+4)+c=0$

Using eq. (2) it becomes

$(2+g)x+(4+f)y+2g+2f-1=0$ ...(3)

But (1) and (3) represent the same line

$\therefore \frac{2+g}{4}=\frac{4+f}{-1}=\frac{2g+2f-1}{-4}$

$\Rightarrow -2-g=16+4f$ and $2+g=-2g-2f+1$

$\Rightarrow g+4f+18=0$ ...(1)

$3g+2f+1=0$

Solving we get $g=16/5, f=-53/10$

$\therefore$ Centre $(-16/5, 53/10)$.

ALTERNATE SOLUTION

Let $C(h, k)$ be the centre of circle touching $x^2=y$ at B $(2, 4)$. Then equation of common tangent at B is

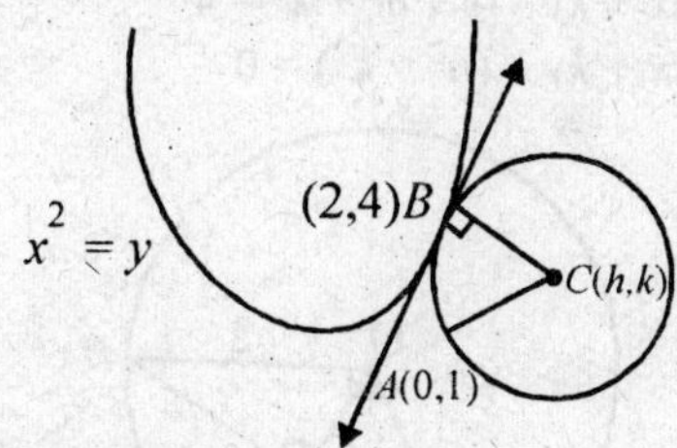

NOTE THIS STEP

$2.x=\frac{1}{2}(y+4)$ i.e., $4x-y=4$

Radius is perpendicular to this tangent

$\therefore 4\left(\frac{k-4}{h-2}\right)=-1\Rightarrow 4k=18$ ...(1)

Also $AC=BC$

$\Rightarrow h^2+(k-1)^2=(h-2)^2+(k-4)^2$

$\Rightarrow 4h+6k=19$ ...(2)

Solving (1) and (2)

we get the centre as $\left(-\frac{16}{5},\frac{53}{10}\right)$.

4. **(b)** **KEY CONCEPT**

Circle through pts. of intersection of two circles $S_1=0$ and $S_2=0$ is $S_1+\lambda S_2=0$

$\therefore$ Req. circle is,

$(x^2+y^2+13x-3y)+\lambda(x^2+y^2+2x-\frac{7}{2}y-\frac{25}{2})=0$

$\Rightarrow (1+\lambda)x^2+(1+\lambda)y^2+(13+2\lambda)$

$x+\left(-3-\frac{7}{2}\lambda\right)y-\frac{25\lambda}{2}=0$

As it passes through $(1, 1)$

$\therefore 1+\lambda+1+\lambda+13+2\lambda-3-\frac{7\lambda}{2}-\frac{25\lambda}{2}=0$

$\Rightarrow -12\lambda + 12 = 0 \Rightarrow \lambda = 1$

$\therefore$ Req. circle is,

$$2x^2 + 2y^2 + 15x - \frac{13y}{2} - \frac{25}{2} = 0$$

or $4x^2 + 4y^2 + 30x - 13y - 25 = 0$

5. (c) Let AB be the chord with its mid pt $M(h, k)$.

As $\angle AOB = 90°$

$\therefore \quad AB = \sqrt{2^2 + 2^2} = 2\sqrt{2}.$

$\therefore \quad AM = \sqrt{2}$

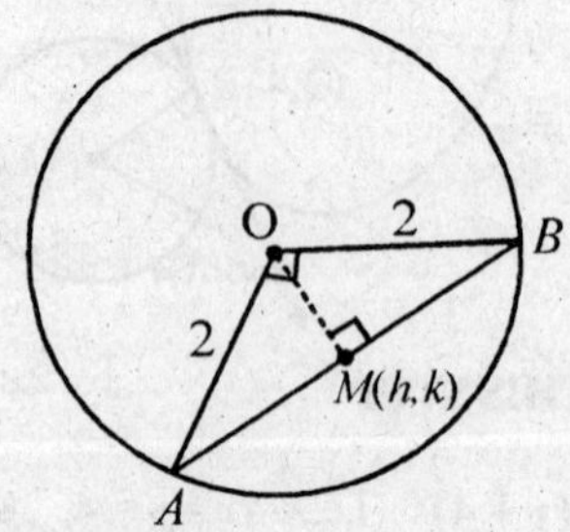

NOTE THIS STEP

By prop. of rt. Δ

$AM = MB = OM$

$\therefore \quad OM = \sqrt{2} \Rightarrow h^2 + k^2 = 2$

$\therefore$ locus of (h, k) is $x^2 + y^2 = 2$

ALTERNATE SOLUTION

Let (h, k) be the mid pt of chord of circle $x^2 + y^2 = 4$ then eq. of chord AB is given by $T = S_1$

i.e., $hx + ky - 4 = h^2 + k^2 - 4$

$\Rightarrow hx + ky - (h^2 + k^2) = 0 \quad \ldots(1)$

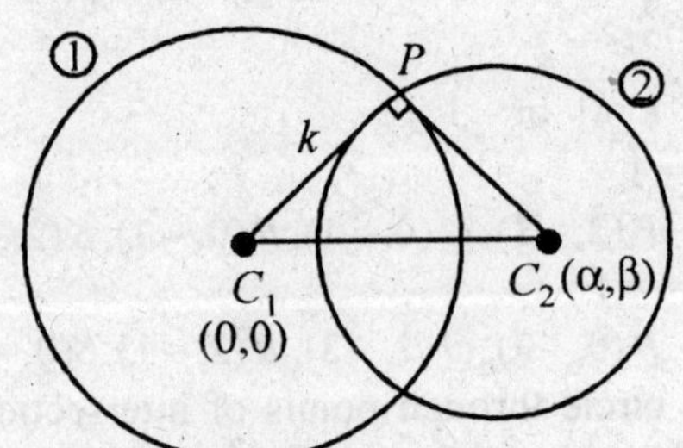

Now combined equation of lines joining pts of intersection of chord AB and circle

$$x^2 + y^2 = 4 \quad \ldots(2)$$

to origin can be obtained by making the equation of circle homogeneous with the help of eq. of chord. Thus making (2) homogeneous with the help of (1) we get

$$x^2 + y^2 - 4\left(\frac{hx + ky}{h^2 + k^2}\right)^2 = 0 \quad \ldots(3)$$

Above represents joint equations of OA and OB. But $\angle$ between OA and $OB = 90°$.

NOTE THIS STEP

$\therefore$ Coeff. of x^2 + Coeff. of $y^2 = 0$

$$\Rightarrow 1 - \frac{4h^2}{(h^2 + k^2)^2} + 1 - \frac{4k^2}{(h^2 + k^2)^2} = 0$$

$$\Rightarrow \frac{4(h^2 + k^2)}{(h^2 + k^2)^2} = 2 \Rightarrow h^2 + k^2 = 2$$

$\therefore$ locus of (h, k) is $x^2 + y^2 = 2$

6. (a) KEY CONCEPT

Two circles $x^2 + y^2 + 2g_1x + 2f_1y + c_1 = 0$ and $x^2 + y^2 + 2g_2x + 2f_2y + c_2 = 0$ are orthogonal iff $2g_1g_2 + 2f_1f_2 = c_1 + c_2$

(a) Let the required circle be,

$$x^2 + y^2 + 2gx + 2fy + c = 0 \quad \ldots(1)$$

As it passes through (a, b), we get,

$$a^2 + b^2 + 2ag + 2bf + c = 0 \quad \ldots(2)$$

Also (1) is orthogonal with the circle,

$$x^2 + y^2 = k^2 \quad \ldots(3)$$

For circle (1)

$g_1 = g, f_1 = f, c_1 = c$

For circle (3)

$g_2 = 0, f_2 = 0, c_2 = -k^2$

$\therefore$ By the condition of orthogonality,

$2g_1g_2 + 2f_1f_2 = c_1 + c_2$

We get, $c = k^2$

Substituting this value of c in eq. (2), we get

$a^2 + b^2 + 2ga + 2fb + k^2 = 0$

$\therefore$ Locus of centre $(g - f)$ of the circle can be obtained by replacing g by $-x$ and f by $-y$ we get

$a^2 + b^2 - 2ax - 2by + k^2 = 0$

i.e. $2ax + 2by - (a^2 + b^2 + k^2) = 0$

ALTERNATE SOLUTION

$$x^2 + y^2 = k^2 \quad \ldots(1)$$

is the given circle with centre C_1 (0, 0) and radius $C_1P = k$.

Let the centre of circle (2) which cuts (1) orthogonally be $C_2(\alpha, \beta)$.

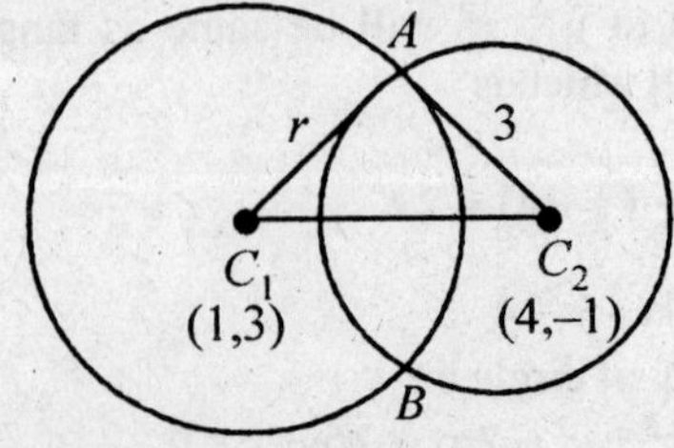

As it passes through (a, b), radius

$$C_2P = \sqrt{(\alpha - a)^2 + (\beta - b)^2}$$

Now as (1) and (2) are orthogonal, therefore

$$C_1P^2 + C_2P^2 = C_1C_2^2$$

$\Rightarrow k^2 + (\alpha - a)^2 + (\beta - b)^2 = \alpha^2 + \beta^2$

$\Rightarrow k^2 + a^2 + b^2 - 2a\alpha - 2b\beta = 0$

$\therefore$ locus of (α, β) is

$k^2 + a^2 + b^2 - 2ax - 2by = 0$

or $2ax + 2by - (a^2 + b^2 + k^2) = 0$

7. (a) We have two circles $(x - 1)^2 + (y - 3)^2 = r^2$

Centre (1, 3), radius $= r$

and $x^2 + y^2 - 8x + 2y + 8 = 0$

Centre (4, – 1), radius $= \sqrt{16 + 1 - 8} = 3$

As the two circles intersect each other in two distinct points we should have

$C_1 C_2 < r_1 + r_2$ and $C_1 C_2 > |r_1 - r_2|$

$\Rightarrow C_1 C_2 < r + 3$ and $C_1 C_2 < |r_1 - r_2|$

$\Rightarrow \sqrt{9+16} < r+3$ and $5 > |r-3|$

$\Rightarrow 5 < r+3 \quad \Rightarrow \quad |r-3| < 5$

$\Rightarrow r > 2 \ldots$ (i) $\quad \Rightarrow \quad -5 < r-3 < 5$

$\Rightarrow -2 < r < 8 \ldots$ (ii)

Combining (i) and (ii), we get

$2 < r < 8$

8. (c) As $2x - 3y - 5 = 0$ and $3x - 4y - 7 = 0$ are diameters of circles.

$\therefore$ Centre of circle is solution of these two eq. 's, i.e.

$$\frac{x}{21-20} = \frac{y}{-15+14} = \frac{1}{-8+9}$$

$\Rightarrow x = 1, y = -1$

$\therefore C(1, -1)$

Also area of circle, $\pi r^2 = 154$

$\Rightarrow r^2 = \frac{154}{22} \times 7 = 49 \Rightarrow r = 7$

$\therefore$ Equation of required circle is

$(x-1)^2 + (y+1)^2 = 7^2$

$\Rightarrow x^2 + y^2 - 2x + 2y = 47$

9. (d) Let P be the pt. of contact of two circles and $P(3\cos\theta, 3\sin\theta)$ parametric pt. with respect to circle $x^2 + y^2 = 9$.

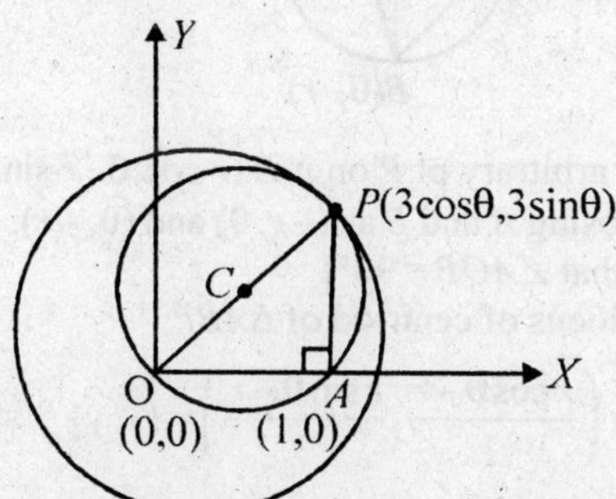

We know that pt. of contact of two circles lies on the line joining their centres therefore, centre of smaller circle must lie on OP, or OP is diameter of smaller circle.

Therefore equation of circle is

$x(x - 3\cos\theta) + y(y - 3\sin\theta) = 0$

or $x^2 + y^2 - 3x\cos\theta - 3y\sin\theta = 0$

As the circle passes through (1, 0), we get

$1 - 3\cos\theta = 0$

$\Rightarrow \cos\theta = \frac{1}{3}$

$\Rightarrow \sin\theta = \pm\sqrt{1 - \frac{1}{9}} = \frac{\pm 2\sqrt{2}}{3}$

$\therefore P(1, \pm 2\sqrt{2})$

Hence centre of circle is $\left(\frac{1}{2}, \pm\sqrt{2}\right)$

ALTERNATE SOLUTION 1 :

With respect to the fig. given in previous method, the centre of the circle passing through the points (0, 0) and (1, 0) has coordiantes $\left(\frac{1}{2}, a\right)$ for some real value of a as centre lies on perpendicular bisector of chord joining (0, 0) and (1, 0). Also any circle touching $x^2 + y^2 = 9$, must have its centre on a line passing through the origin. Let $P(x, y)$ be the pt of contact of two circles. Then OP is diameter of the smaller circle and hence mid pt. of OP = centre of smaller circle

$\Rightarrow \frac{0+x}{2} = \frac{1}{2}, \frac{0+y}{2} = a$

$\Rightarrow x = 1, \quad y = 2a$

But $(1, 2a)$ must lie on the circle

$x^2 + y^2 = 9$

$\therefore 1 + 4a^2 = 9$

$\Rightarrow 4a^2 = 8 \quad \Rightarrow \quad a = \pm\sqrt{2}$

$\therefore$ The required centres are $\left(\frac{1}{2}, \pm\sqrt{2}\right)$

ALTERNATE SOLUTION 2 :

Let the equation of the circle be

$x^2 + y^2 + 2gx + 2fy + c = 0.$

As this circle passes through (0, 0) and (1, 0) we get $c = 0, 1 + 2g = 0$

$\Rightarrow g = -\frac{1}{2}$

According to the question this circle touches the given circle $x^2 + y^2 = 9$

$\therefore$ 2 × radius of required circle = radius of given circle

$\Rightarrow 2\sqrt{g^2 + f^2} = 3$

$\Rightarrow g^2 + f^2 = \frac{9}{4}$

$\Rightarrow \frac{1}{4} + f^2 = \frac{9}{4}$

$\Rightarrow f^2 = 2$

$\Rightarrow f = \pm\sqrt{2}$

$\therefore$ The centre is $\left(\frac{1}{2}, \sqrt{2}\right), \left(\frac{1}{2}, -\sqrt{2}\right)$.

10. (d) The given circle is $x^2 + y^2 - 6x + 14 = 0$, centre (3, 3), radius = 2

Let (h, k) be the centre of touching circle. Then radius of touching circle = h [as it touches y-axis also]

$\therefore$ Distance between centres of two circles = sum of the radii of two circles

$\Rightarrow \sqrt{(h-3)^2 + (k-3)^2} = 2 + h$

$\Rightarrow (h-3)^2 + (k-3)^2 = (2+h)^2$

$\Rightarrow h^2 - 6h + 9 + k^2 - 6k + 9 = 4 + 4h + h^2$

$\Rightarrow k^2 - 10h - 6k + 14 = 0$

$\therefore$ locus of (h, k) is $y^2 - 10x - 6y + 14 = 0$

11. (c) Centres and radii of two circles are C_1 (5, 0); $3 = r_1$ and C_2 (0, 0); $r = r_2$

As circles intersect each other in two distinct points,

$$|r_1 - r_2| < C_1C_2 < r_1 + r_2$$

$$\Rightarrow |r-3| < 5 < r+3$$

$$\Rightarrow 2 < r < 8$$

12. (d) Centre of the circle

$x^2 + y^2 + 4x - 6y + 9\sin^2\alpha + 13\cos^2\alpha = 0$

is $C(-2, 3)$ and its radius is

$$\sqrt{2^2 + (-3)^2 - 9\sin^2\alpha - 13\cos^2\alpha}$$

$$= \sqrt{4 + 9 - 9\sin^2\alpha - 13\cos^2\alpha} = 2\sin\alpha$$

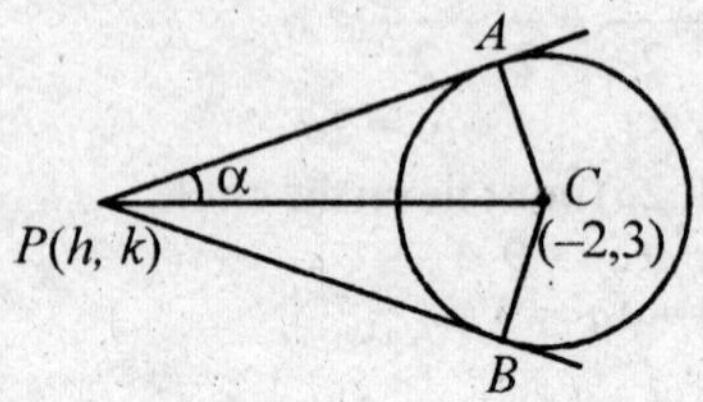

Let $P(h, k)$ be any point on the locus. The $\angle APC = \alpha$

Also $\angle PAC = \dfrac{\pi}{2}$

That is, triangle APC is a right triangle.

Thus, $\sin\alpha = \dfrac{AC}{PC} = \dfrac{2\sin\alpha}{\sqrt{(h+2)^2 + (k-3)^2}}$

$$\Rightarrow \sqrt{(h+2)^2 + (k-3)^2} = 2$$

$$\Rightarrow (h+2)^2 + (k-3)^2 = 4$$

or $h^2 + k^2 + 4h - 6k + 9 = 0$

Thus required equation of the locus is

$$x^2 + y^2 + 4x - 6y + 9 = 0$$

13. (d) The given equation of the circle is

$$x^2 + y^2 - px - qy = 0, pq \neq 0$$

Let the chord drawn from (p, q) is bisected by x-axis at point $(x_1, 0)$.

Then equation of chord is

NOTE THIS STEP

$$xx_1 - \frac{p}{2}(x + x_1) - \frac{q}{2}(y + 0) = x_1^2 - px_1 \quad (\text{using } T = S_1)$$

As it passes through (p, q), therefore,

$$px_1 - \frac{p}{2}(p + x_1) - \frac{q^2}{2} = x_1^2 - px_1$$

$$\Rightarrow x_1^2 - \frac{3}{2}px_1 + \frac{p^2}{2} + \frac{q^2}{2} = 0$$

$$\Rightarrow 2x_1^2 - 3px_1 + p^2 + q^2 = 0$$

As through (p,q) two distinct chords can be drawn.

$\therefore$ Roots of above equation be real and distinct, i.e., $D > 0$.

$$\Rightarrow 9p^2 - 4 \times 2(p^2 + q^2) > 0$$

$$\Rightarrow p^2 > 8q^2$$

14. (c) O is the point at centre and P is the point at circumference. Therefore, angle QOR is double the angle QPR.

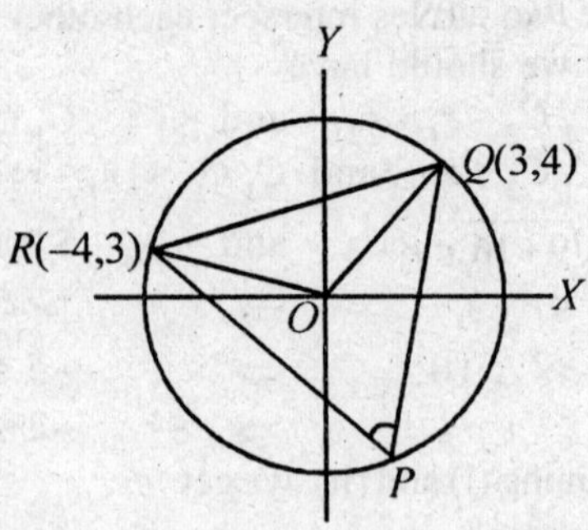

So, it sufficient to find the angle QOR. Now slope of $OQ = 4/3$

Slope of $OR = -3/4$

Again $m_1 m_2 = -1$

Therefore, $\angle QOR = 90°$ which implies that $\angle QPR = 45°$.

15. (a) $2g_1g_2 + 2f_1f_2 = c_1 + c_2$

(formula for orthogonal intersection of two cricles)

$$\Rightarrow 2(1)(0) + 2(k)(k) = 6 + k$$

$$\Rightarrow 2k^2 - k - 6 = 0$$

$$\Rightarrow k = -3/2, 2$$

16. (b) $x^2 + y^2 = r^2$ is a circle with centre at (0, 0) and radius r units.

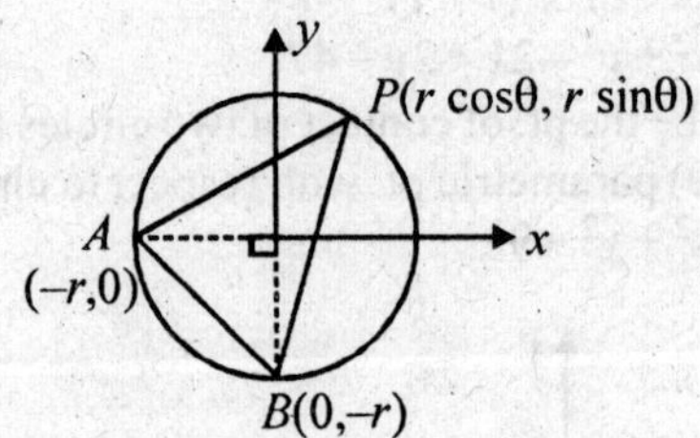

Any arbitrary pt P on it is $(r\cos\theta, r\sin\theta)$

Choosing A and B as $(-r, 0)$ and $(0, -r)$.

[So that $\angle AOB = 90°$]

For locus of centriod of ΔABP

$$\left(\frac{r\cos\theta - r}{3}, \frac{r\sin\theta - r}{3}\right) = (x, y)$$

$$\Rightarrow r\cos\theta - r = 3x$$

$$r\sin\theta - r = 3y$$

$$\Rightarrow r\cos\theta = 3x + r$$

$$r\sin\theta = 3y + r$$

Squaring and adding $(3x + r)^2 + (3y + r)^2 = r^2$ which is a circle.

17. (a) Let $\angle RPS = \theta$

$\angle XPQ = 90° - \theta$

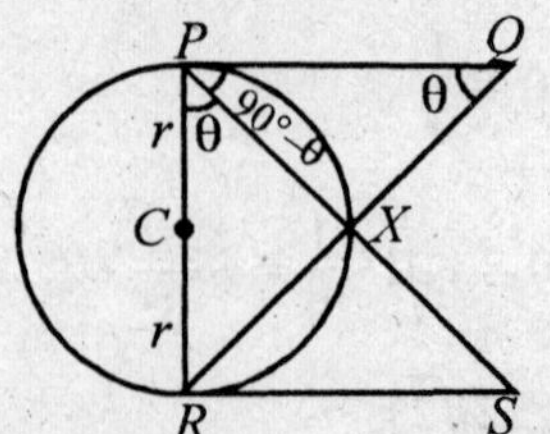

$\therefore$ $\angle PQX = \theta$ $\quad (\because \angle PXQ = 90°)$

$\therefore$ $\Delta PRS \sim \Delta QPR$ $\quad$ (AA similarity)

$\therefore \quad \frac{PR}{QP} = \frac{RS}{PR}$

$\Rightarrow \quad PR^2 = PQ \cdot RS$

$\Rightarrow \quad PR = \sqrt{PQ.RS} \Rightarrow 2r = \sqrt{PQ.RS}$

18. **(c)** Line $5x - 2y + 6 = 0$ is intersected by tangent at P to circle $x^2 + y^2 + 6x + 6y - 2 = 0$ on y-axis at $Q\,(0, 3)$.
In other words tangent passes through (0, 3)

$\therefore \quad PQ$ = length of tangent to circle from (0, 3)

$= \sqrt{0+9+0+18-2}$

$= \sqrt{25} = 5$

19. **(a)** $x^2 - 8x + 12 = 0 \quad \Rightarrow (x-6)(x-2) = 0$

$y^2 - 14y + 45 = 0 \quad \Rightarrow (y-5)(y-9) = 0$

Thus sides of square are
$x = 2, x = 6, y = 5, y = 9$
Then centre of circle inscribed in square will be

$\left(\frac{2+6}{2}, \frac{5+9}{2}\right) = (4, 7).$

20. **(c)** The given circle is $x^2 + y^2 - 2x - 6y + 6 = 0$ with centre $C(1, 3)$ and radius $= \sqrt{1+9-6} = 2$. Let AB be one of its diameter which is the chord of other circle with centre at $C_1\,(2, 1)$.

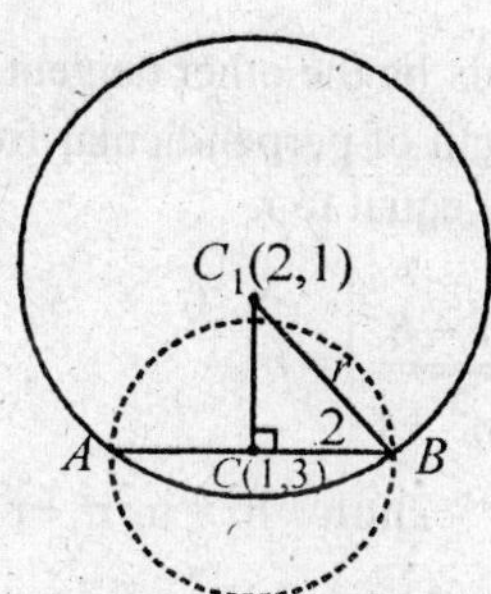

Then in ΔC_1CB,

$C_1B^2 = CC_1^2 + CB^2$

$r^2 = [(2-1)^2 + (1-3)^2] + (2)^2$

$\Rightarrow \quad r^2 = 1 + 4 + 4$

$\Rightarrow \quad r^2 = 9$

$\Rightarrow \quad r = 3.$

21. **(d)** Let the centre of circle C be (h, k). Then as this circle touches axis of x, its radius $= |k|$

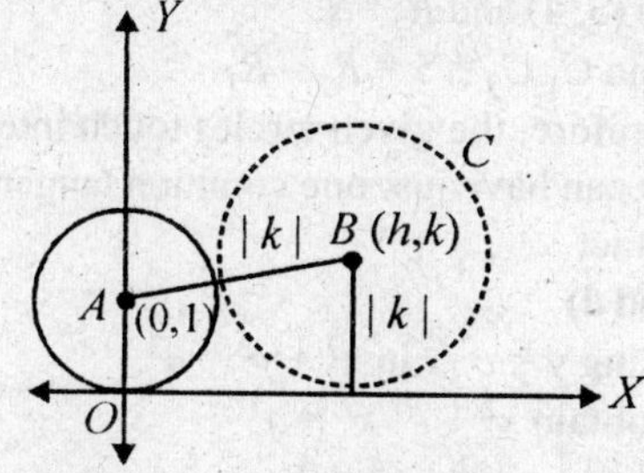

Also it touches the given circle $x^2 + (y-1)^2 = 1$, centre (0, 1) radius 1, externally
Therefore, the distance between centres = sum of radii

$\Rightarrow \quad \sqrt{(h-0)^2 + (k-1)^2} = 1 + |k|$

$\Rightarrow \quad h^2 + k^2 - 2k + 1 = (1 + |k|)^2$

$\Rightarrow \quad h^2 + k^2 - 2k + 1 = 1 + 2|k| + k^2$

$\Rightarrow \quad h^2 = 2k + 2|k|$

$\therefore$ Locus of (h, k) is, $x^2 = 2y + 2|y|$
Now if $y > 0$, it becomes $x^2 = 4y$
and if $y \le 0$, it becomes $x = 0$

$\therefore$ Combining the two, the required locus is
$\{(x, y) : x^2 = 4y\} \cup \{(0, y) : y \le 0\}$

22. **(b)** Tangents PA and PB are drawn from the point P (1, 3) to circle $x^2 + y^2 - 6x - 4y - 11 = 0$ with centre C (3, 2)

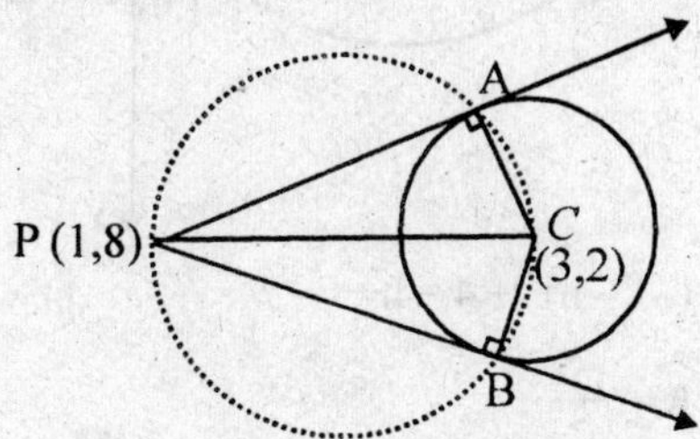

Clearly the circumcircle of ΔPAB will pass through C and as $\angle A = 90°$, PC must be a diameter of the circle.

$\therefore$ Equation of required circle is

$(x-1)(x-3) + (y-8)(y-2) = 0$

$\Rightarrow \quad x^2 + y^2 - 4x - 10y + 19 = 0$

ALTERNATIVE SOLUTION :

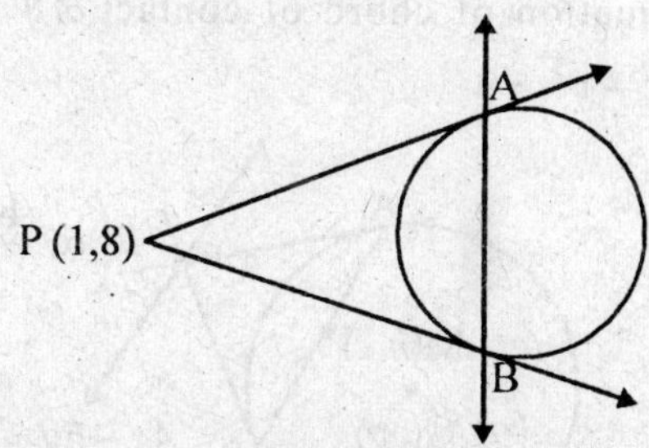

Equation of chord of contact i.e. AB is

$x.1 + y.8 - 3(x+1) - 2(y+8) - 11 = 0$

or $\quad x - 3y + 15 = 0$

Now equation of circle passing through intersection point of circle $x^2 + y^2 - 6x - 4y - 11 = 0$ and of line $x - 3y + 15 = 0$ is given by $S + \lambda L = 0$

i.e. $(x^2 + y^2 - 6x - 4y - 11) + \lambda(x - 3y + 15) = 0$

As this circle passes through P (1, 8) also, we get

$(1 + 64 - 6 - 32 - 11) + \lambda\,(1 - 24 + 15) = 0$

$\Rightarrow \quad \lambda = 2$

$\therefore$ The required circle is

$(x^2 + y^2 - 6x - 4y - 11) + 2(x - 3y + 15) = 0$

or $x^2 + y^2 - 4x - 10y + 19 = 0$

23. **(d)** Let centre of the circle be (h, 2) then radius = |h|

$\therefore$ Equation of circle becomes $(x-h)^2+(y-2)^2=h^2$

As it passes through (–1, 0)

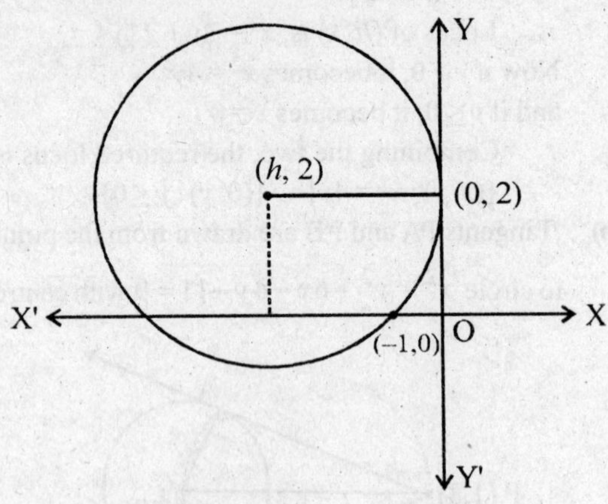

$\Rightarrow (-1-h)^2+4=h^2$

$\Rightarrow h=\frac{-5}{2}$

$\therefore$ Centre $\left(\frac{-5}{2},2\right)$ and $r=\frac{5}{2}$

Distance of centre from (–4, 0) is $\frac{5}{2}$

$\therefore$ It lies on the circle.

3. **(a)** Any point P on line $4x-5y=20$ is $\left(\alpha,\frac{4\alpha-20}{5}\right)$.

Equation of chord of contact AB to the circle $x^2+y^2=9$

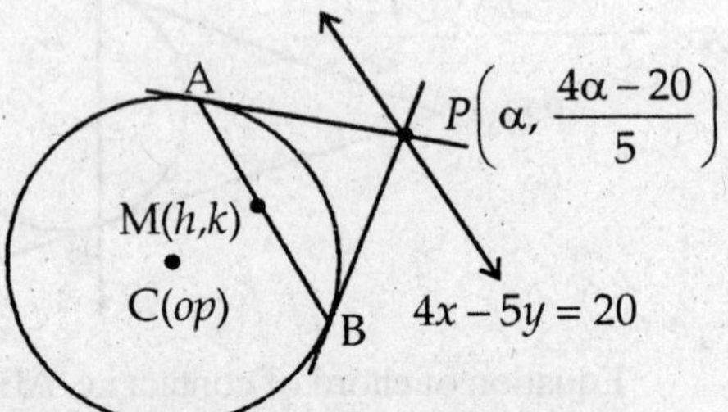

drawn from point $P\left(\alpha,\frac{4\alpha-20}{5}\right)$ is

$$x.\alpha+y.\left(\frac{4\alpha-20}{5}\right)=9 \quad ....(1)$$

Also the equation of chord AB whose mid point is (h, k) is

$$hx+ky=h^2+k^2 \quad ....(2)$$

$\because$ Equations (1) and (2) represent the same line, therefore

$$\frac{h}{\alpha}=\frac{k}{\frac{4\alpha-20}{5}}=\frac{h^2+k^2}{9}$$

$\Rightarrow \quad 5k\alpha=4h\alpha-20h$ and $9h=\alpha\left(h^2+k^2\right)$

$\Rightarrow \quad \alpha=\frac{20h}{4h-5k}$ and $\alpha=\frac{9h}{h^2+k^2}$

$\Rightarrow \quad \frac{20h}{4h-5k}=\frac{9h}{h^2+k^2} \Rightarrow 20(h^2+k^2)=9(4h-5k)$

$\therefore$ Locus of (h, k) is

$$20\left(x^2+y^2\right)-36x+45y=0$$

D MCQ's WITH ONE OR MORE THAN ONE CORRECT

1. **(a, c)** The given circle is $x^2+y^2-2rx-2hy+h^2=0$ with centre (r, h) and radius = r.

Clearly circle touches y-axis so one of its tangent is $x=0$.

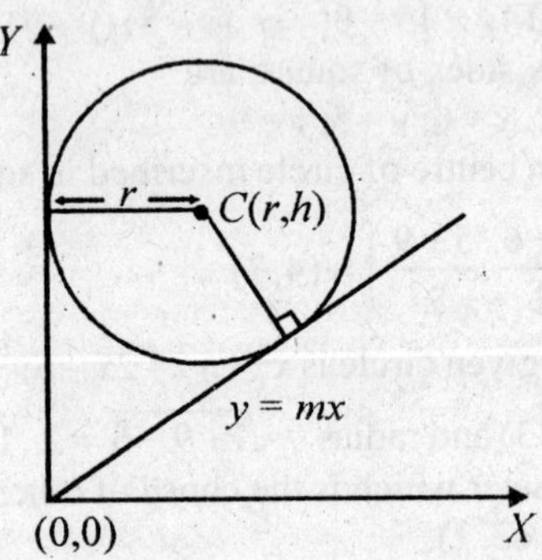

Let y = mx be the other tangent through origin. Then length of perpendicular from C (r, h) to y = mx should be equal to r.

$\therefore \quad \left|\frac{mr-h}{\sqrt{m^2+1}}\right|=r$

$\Rightarrow \quad m^2r^2-2mrh+h^2=m^2r^2+r^2$

$\Rightarrow \quad m=\frac{h^2-r^2}{2rh}$

$\therefore \quad$ Other tangent is $y=\frac{h^2-r^2}{2rh}x$

or $\quad (h^2-r^2)x-2rhy=0$

2. **(b)** $x^2+y^2=4$ (given)

Centre $C_1\equiv(0,0)$ and $R_1=2$.

Also for circle $x^2+y^2-6x-8y-24=0$

$C_2\equiv(3,4)$ and $R_2=7$.

Again $C_1C_2=5=R_2-R_1$

Therefore, the given circles touch internally such that they can have just one common tangent at the point of contact.

3. **(a, b, c and d)**

Putting $y=c^2/x$ in $x^2+y^2=a^2$,

we obtain $x^2+c^4/x^2=a^2$

$\Rightarrow \quad x^4-a^2x^2+c^4=0 \quad ...(1)$

As x_1, x_2, x_3 and x_4 are roots of (1),

$\Rightarrow \quad x_1+x_2+x_3+x_4=0$ and $x_1x_2x_3x_4=c^4$

Similarly, forming equation in y, we get

$y_1+y_2+y_3+y_4=0$ and $y_1y_2y_3y_4=c^4$.

4. **(a, c)** There can be two possibilites for the given circle as shown in the figure

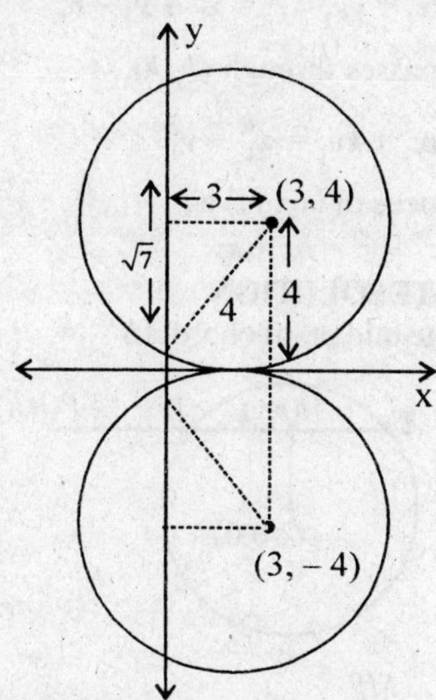

$\therefore$ The equations of circles can be

$(x-3)^2+(y-4)^2=4^2$

or $(x-3)^2+(y+4)^2=4^2$

i.e. $x^2+y^2-6x-8y+9=0$

or $x^2+y^2-6x+8y+9=0$

E) SUBJECTIVE PROBLEMS :

1. The given circle is

$x^2+y^2-2x-4y-20=0$

whose centre is (1, 2) and radius = 5

Radius of required circle is also 5.

Let its centre be $C_2(\alpha,\beta)$. Both the circles touch each other at $P(5,5)$.

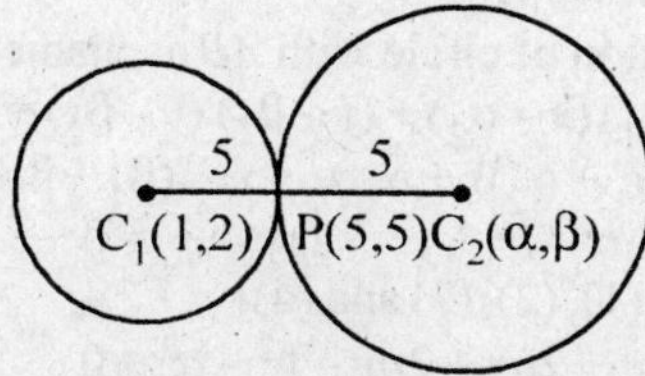

It is clear from figure that $P(5,5)$ is the mid-point of C_1C_2. Therefore, we should have

$$\frac{1+\alpha}{2}=5 \text{ and } \frac{2+\beta}{2}=5$$

$\Rightarrow$ $\alpha=9$ and $\beta=8$

$\therefore$ Centre of required circle is (9, 8) and equation of required circle is

$(x-9)^2+(y-8)^2=5^2$

$\Rightarrow$ $x^2+y^2-18x-16y+120=0$

2. (a) The point of intersection of the lines $3x+5y=1$ and $(2+c)x+5c^2y=1$ is

$$\left(\frac{1-c^2}{2+c-3c^2},\frac{c-1}{5(2+c-3c^2)}\right)$$

If (h, k) is the centre of the required circle then

$$h=\lim_{c\to 1}\frac{1-c^2}{2+c-3c^2}=\lim_{c\to 1}\frac{(1-c)(1+c)}{(1-c)(2+3c)}$$

$$=\lim_{c\to 1}\frac{1+c}{2+3c}=\frac{2}{5}$$

and $k=\lim_{c\to 1}\frac{c-1}{5(2+c-3c^2)}=\lim_{c\to 1}\frac{-1}{5(2+3c)}=\frac{-1}{25}$

$\therefore$ Centre is $\left(\frac{2}{5},\frac{-1}{25}\right)$

Also circle passes through (2, 0)

$\therefore$ radius $=\sqrt{\left(2-\frac{2}{5}\right)^2+\left(\frac{1}{25}\right)^2}=\frac{\sqrt{1601}}{25}$

$\therefore$ Equation of required circle is

$$\left(x-\frac{2}{5}\right)^2+\left(y+\frac{1}{25}\right)^2=\frac{1601}{625}$$

$\Rightarrow$ $25x^2+25y^2-20x+2y-60=0$

(b) Slant height of frustum A

$=\sqrt{(4-2)^2+3^2}=\sqrt{13}$ m

$\therefore$ Curved surface area of frustum

$=\pi(4+2)\sqrt{13}=6\sqrt{13}\pi\,m^2$

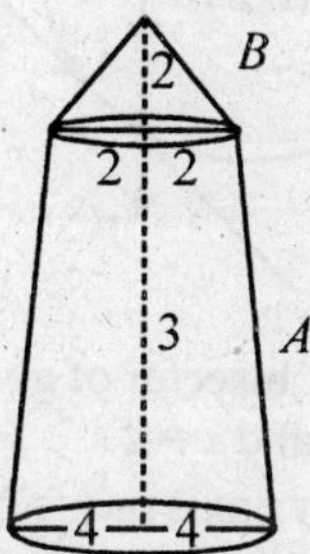

Also slant height of cone $B=\sqrt{2^2+2^2}=2\sqrt{2}$

$\therefore$ Curved surface area of cone

$=\pi\times 2\times 2\sqrt{2}=4\sqrt{2}\pi\,m^2$

$\therefore$ Area of canvas required $=6\sqrt{13}\pi+4\sqrt{2}\pi$

$=2(3\sqrt{13}+2\sqrt{2})\pi\,m^2$

3. The eq. of circle is

$x^2+y^2-2x-4y-20=0$

Centre (1, 2), radius $=\sqrt{1+4+20}=5$

Using eq. of tangent at (x_1, y_1) of

$x^2+y^2+2gx_1+2fy_1+c=0$ is

$xx_1+yy_1+g(x+x_1)f(y+y_1)+c=0$

Eq. of tangent at (1, 7) is

$x.1+y.7-(x+1)-2(y+7)-20=0$

$\Rightarrow$ $y-7=0$ ...(1)

Similarly eq. of tangent at (4, – 2) is

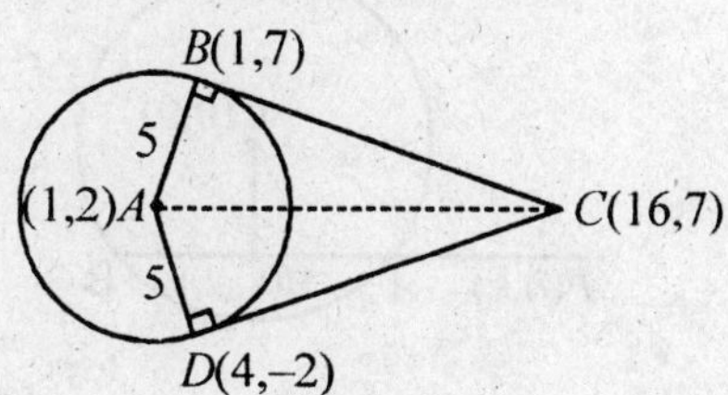

$4x-2y-(x+4)-2(y-2)-20=0$

$\Rightarrow\ 3x-4y-20=0 \qquad \ldots(2)$

For pt C, solving (1) and (2), we get

$x=16, y=7$

$\therefore\ C(16,7).$

Now, clearly ar (quad $ABCD$) = 2 Ar (rt ΔABC)

$= 2\times\frac{1}{2}\times AB\times BC = AB\times BC$

where AB = radius of circle = 5

and BC = length of tangent from C to circle

$= \sqrt{16^2+7^2-32-28-20}$

$= \sqrt{225}$

$= 15$

$\therefore\ ar$ (quad $ABCD$) = 5 × 15 = 75 sq. units.

4. Given st. lines are $x+y=2$

$x-y=2$

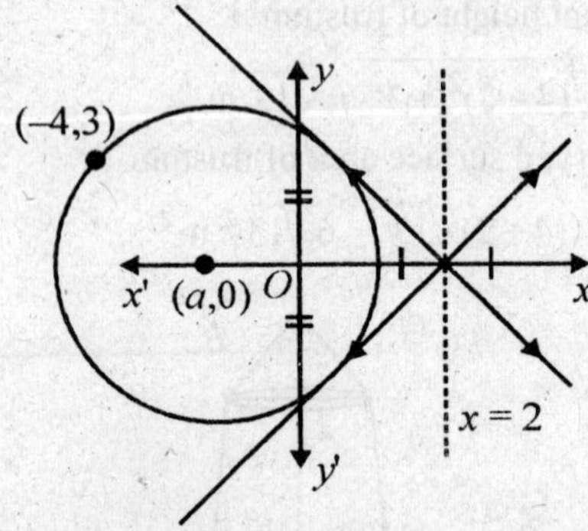

As centre lies on ∠ bisector of given equations (lines) which are the lines $y=0$ and $x=2$.

$\therefore$ Centre lies on x axis or $x=2$.

But as it passes through (– 4, 3), i.e., II quadrant.

$\therefore$ Centre must lie on x-axis

Let it be $(a, 0)$ then distance between $(a, 0)$ and $(-4, 3)$ is = length of ⊥ lar distance from $(a, 0)$ to $x+y-2=0$

$\Rightarrow\ (a+4)^2+(0-3)^2=\left(\frac{a-2}{\sqrt{2}}\right)^2$

$\Rightarrow\ 2(a^2+8a+16+9)=a^2-4a+4$

$\Rightarrow\ a^2+20a+46=0 \Rightarrow\ a=-10\pm\sqrt{54}$

$\therefore$ Equation of circle is

$\Rightarrow\ [x+(10\pm\sqrt{54}\,)]^2+y^2=[-(10\pm\sqrt{54}\,)+4]^2+3^2$

$\Rightarrow\ x^2+y^2+2(10\pm\sqrt{54}\,)x+8(10\pm\sqrt{54}\,)-25=0$

$\Rightarrow\ x^2+y^2+2(10\pm\sqrt{54}\,)x+55\pm\sqrt{54}=0.$

5. Equation of chord whose mid point is given is

$T=S_1$

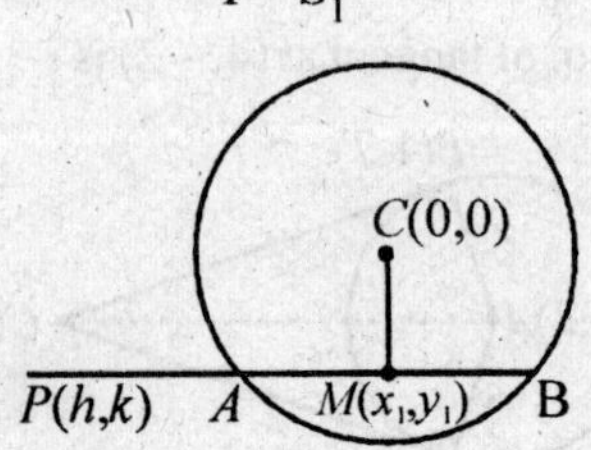

[Consider (x_1, y_1) be mid pt. of AB]

$\Rightarrow\ xx_1+yy_1-r^2=x_1^2+y_1^2-r^2$

As it passes through (h, k),

$\therefore\ hx_1+ky_1=x_1^2+y_1^2$

$\therefore$ locus of (x_1, y_1) is,

$x^2+y^2=hx+ky$

ALTERNATE SOLUTION

Let M be the mid pt. of chord AB

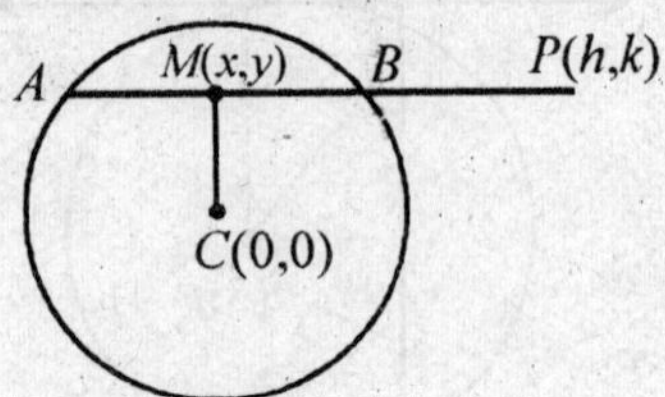

Then $CM \perp MP$

$\Rightarrow\ m_{CM}\times m_{MP}=-1$

$\Rightarrow\ \frac{y}{x}\times\frac{k-y}{h-x}=-1$

$\Rightarrow\ ky-y^2=-hx+x^2$

$\Rightarrow\ x^2+y^2=hx+ky$ which is req. locus.

6. Let the two points be

$A=(\alpha_1, \beta_1)$ and $B=(\alpha_2, \beta_2)$

Thus α_1, α_2 are roots of

$x^2+2ax-b^2=0$

$\therefore\ \alpha_1+\alpha_2=-2a \qquad \ldots(1)$

$\alpha_1\alpha_2=-b^2 \qquad \ldots(2)$

β_1, β_2 are roots of $x^2+2px-q^2=0$

$\therefore\ \beta_1+\beta_2=-2p \qquad \ldots(3)$

$\beta_1\beta_2=-q^2 \qquad \ldots(4)$

Now equation of circle with AB as diameter is

$(x-\alpha_1)(x-\alpha_2)+(y-\beta_1)(y-\beta_2)=0$

$\Rightarrow\ x^2-(\alpha_1+\alpha_2)x+\alpha_1\alpha_2+y^2-(\beta_1+\beta_2)y+\beta_1\beta_2=0$

$\Rightarrow\ x^2+2ax-b^2+y^2+2py-q^2=0$

[Using eq. (1), (2), (3) and (4)]

$\Rightarrow\ x^2+y^2+2ax+2py-b^2-q^2=0$

Which is the equation of required circle, with its centre $(-a, -p)$ and radius

$=\sqrt{a^2+p^2+b^2+q^2}$

7. Let equation of tangent PAB be $5x+12y-10=0$ and that of PXY be

$5x-12y-40=0$

Now let centre of circles C_1 and C_2 be $C(h, k)$.

Let $CM \perp PAB$ then CM = radius of $C_1 = 3$

Also C_2 makes an intercept of length 8 units on $PAB \Rightarrow AM=4$

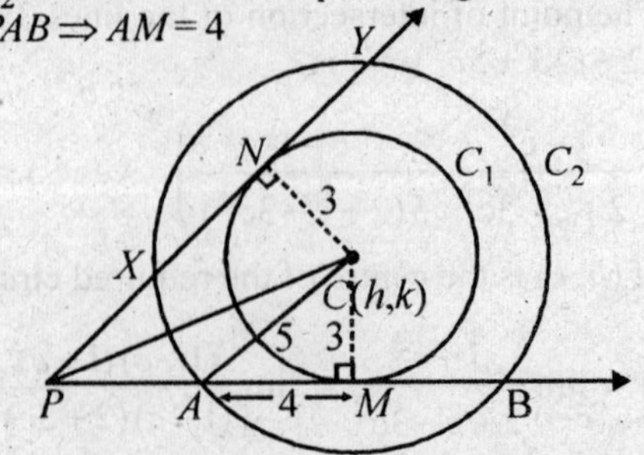

Then in ΔAMC, we get

$AC = \sqrt{4^2 + 3^2} = 5$

$\therefore$ Radius of C_2 is $= 5$ units

Also, as $5x + 12y - 10 = 0$ ... (1)

and $5x - 12y - 40 = 0$ ... (2)

are tangents to C_1, length of perpendicular from C to $AB = 3$ units

$\therefore$ We get $\frac{5h + 12k - 10}{13} = \pm 3$

$\Rightarrow$ $5h + 12k - 49 = 0$ ... (i)

or $5h + 12k + 29 = 0$ ... (ii)

Similarly, $\frac{5h - 12k - 40}{13} = \pm 3$

$\Rightarrow$ $5h - 12k - 79 = 0$ ... (iii)

or $5h - 12k - 1 = 0$ ... (iv)

As C lies in first quadrant

$\therefore$ h, k are $+ve$

$\therefore$ Eq. (ii) is not possible.

Solving (i) and (iii), we get

$h = 64/5, k = -5/4$

This is also not possible.

Now solving (i) and (iv), we get $h = 5, k = 2$.

Thus centre for C_2 is (5, 2) and radius 5.

Hence, equation of C_2 is

$(x - 5)^2 + (y - 2)^2 = 5^2$

$\Rightarrow$ $x^2 + y^2 - 10x - 4y + 4 = 0$

ALTERNATE SOLUTION

If C be the centre of the circle C_1 then it lies on the bisectors of the given lines which are

$$\frac{5x + 12y - 10}{13} = \pm \frac{5x - 12y - 40}{13}$$

$\Rightarrow$ $x = 5$ ...(i)

or $y = -5/4$ ...(ii)

Since, centre lies in 1st quadrent therefore, (ii) is not possible.

$\therefore$ Let the centre be $(5, k)$, then its perpendicular distance from each of lines will be 3

i.e., $\frac{25 + 12k - 10}{13} = \pm 3$ and $\frac{25 - 12k - 40}{13} = \pm 3$

$\Rightarrow$ $k = 2$ or $k = -ve$ and $k = -ve$ in both cases

$\therefore$ The only possibility for k is the value 2.

$\therefore$ Centre is (5, 2)

Now the circle C_2 cuts off intercepts 8 from these lines and if its radius be r, then

$r^2 = 3^2 + 4^2 = 5^2 \Rightarrow r = 5$

$\therefore$ Circle C_2 is,

$(x - 5)^2 + (y - 2)^2 = 5^2$

i.e. $x^2 + y^2 - 10x - 4y + 4 = 0$

8. Let the equation of L_1 be

$x \cos \alpha + y \sin \alpha = p_1$

Then any line perpendicular to L_1 is

$x \sin \alpha - y \cos \alpha = p_2$

where p_2 is a variable.

Then L_1 meets x-axis at P ($p_1 \sec \alpha$, 0) and y-axis at Q (0, $p_1 \operatorname{cosec} \alpha$).

Similarly L_2 meets x-axis at R ($p_2 \operatorname{cosec} \alpha$, 0) and y-axis at S (0, $- p_2 \sec \alpha$).

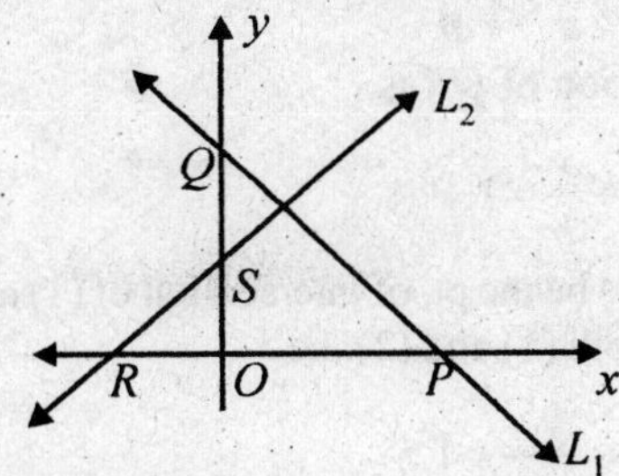

Now equation of PS is,

$$\frac{x}{p_1 \sec \alpha} + \frac{y}{-p_2 \sec \alpha} = 1$$

$\Rightarrow$ $\frac{x}{p_1} - \frac{y}{p_2} = \sec \alpha$ ...(1)

Similarly, equation of QR is,

$\Rightarrow$ $\frac{x}{p_2 \operatorname{cosec} \alpha} + \frac{y}{p_1 \operatorname{cosec} \alpha} = 1$

$\Rightarrow$ $\frac{x}{p_2} + \frac{y}{p_1} = \operatorname{cosec} \alpha$ ...(2)

Locus of point of intersection of PS and QR can be obtained by eliminating the variable p_2 from (1) and (2)

i.e. $\left(\frac{x}{p_1} - \sec \alpha\right)\frac{x}{y} + \frac{y}{p_1} = \operatorname{cosec} \alpha$

[Substituting the value of $\frac{1}{p_2}$ from (1) in (2)]

$\Rightarrow$ $(x - p_1 \sec \alpha) x + y^2 = p_1 y \operatorname{cosec} \alpha$

$\Rightarrow$ $x^2 + y^2 - p_1 x \sec \alpha - p_1 y \operatorname{cosec} \alpha = 0$

which is a circle through origin.

ALTERNATE SOLUTION

Let L_1 be the line $\frac{x}{a} + \frac{y}{b} = 1$ which meets x-axis at $P(a, 0)$ and y-axis at Q (0, b).

As L_1 is fixed, a and b are fixed.

Now any variable line L_2, perpendicular to L_1 is given by

$\frac{x}{b} - \frac{y}{a} = k$

where k is variable.

This line L_2, meets x-axis at R (bk, 0) and y-axis at S(0, $- ak$)

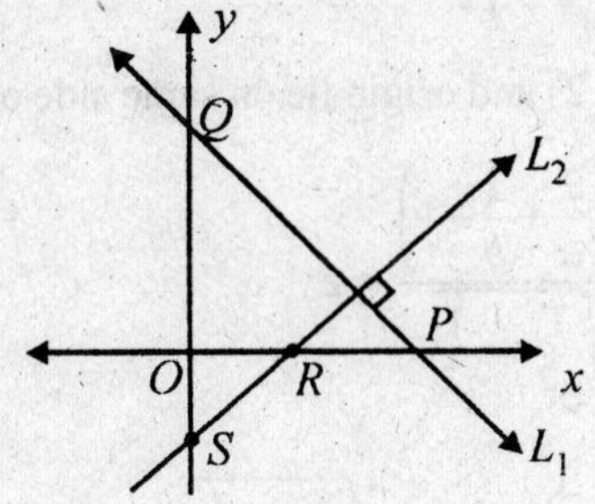

$\therefore$ Equation of PS is

$$\frac{x}{a}+\frac{y}{-ak}=1$$

and equation of QR is

$$\frac{x}{bk}+\frac{y}{b}=1$$

Let (x_1, y_1) be the pt. of intersection of (1) and (2) then (x_1, y_1) lies on both (1) and (2), i.e.

$$\frac{x_1}{a}+\frac{y_1}{-ak}=1 \quad \ldots(1)$$

$$\frac{x_1}{bk}+\frac{y_1}{b}=1 \quad \ldots(2)$$

To find the locus of (x_1, y_1), we eliminate k from (1) and (2)

Operating (1) $\times \frac{x_1}{b}+(2)\times\frac{y_1}{b}$, we get

$$\frac{x_1^2}{ab}+\frac{y_1^2}{ab}=\frac{x_1}{b}+\frac{y_1}{a}$$

$$\Rightarrow x_1^2+y_1^2-(ax_1+by_1)=0$$

$\therefore$ locus of (x_1, y_1) is $x^2+y^2-ax-by=0$

Which is a circle through origin.

9. The given circle is $x^2+y^2-4x-4y+4=0$.

This can be re-written as

$$(x-2)^2+(y-2)^2=4$$

Which has centre C (2, 2) and radius 2.

Let the eq. of third side AB of ΔOAB is $\frac{x}{a}+\frac{y}{b}=1$ such that $A(a, 0)$ and $B(0, b)$

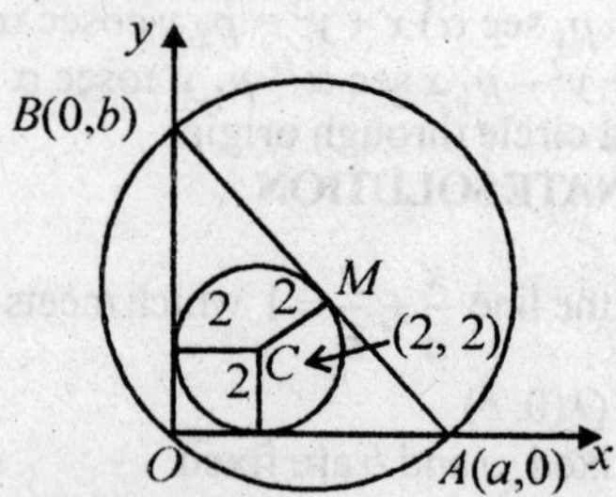

Length of perpendicular form (2, 2) on AB = radius = CM = 2

$$\therefore \frac{\left|\frac{2}{a}+\frac{2}{b}-1\right|}{\sqrt{\frac{1}{a^2}+\frac{1}{b^2}}}=2$$

Since (2, 2) and origin lie on same side of AB

$$\therefore \frac{-\left(\frac{2}{a}+\frac{2}{b}-1\right)}{\sqrt{\frac{1}{a^2}+\frac{1}{b^2}}}=2$$

$$\Rightarrow \frac{2}{a}+\frac{2}{b}-1=-2\sqrt{\frac{1}{a^2}+\frac{1}{b^2}} \quad \ldots(1)$$

Since $\angle AOB=\pi/2$.

Hence, AB is the diameter of the circle passing through ΔOAB, mid point of AB is the centre of the circle i.e. $(a/2, b/2)$

Let centre be $(h, k)\equiv\left(\frac{a}{2},\frac{b}{2}\right)$

then $a=2h$, $b=2k$.

Substituting the values of a and b in (1), we get

$$\frac{2}{2h}+\frac{2}{2k}-1=-2\sqrt{\frac{1}{4h^2}+\frac{1}{4k^2}}$$

$$\Rightarrow \frac{1}{h}+\frac{1}{k}-1=-\sqrt{\frac{1}{h^2}+\frac{1}{k^2}}$$

$$\Rightarrow h+k-hk+\sqrt{h^2+k^2}=0$$

$\therefore$ Locus of $M(h, k)$ is,

$$x+y-xy+\sqrt{x^2+y^2}=0 \quad \ldots(2)$$

Comparing it with given equation of locus of circumcentre of Δ i.e.

$$x+y-xy+k\sqrt{x^2+y^2}=0 \quad \ldots(3)$$

We get, $k=1$

10. The equation of given circle is

$$x^2+y^2+2gx+2fy+c=0 \quad \ldots(1)$$

Let $P(h, k)$ be the foot of perpendicular drawn from origin to the chord LM of circle S.

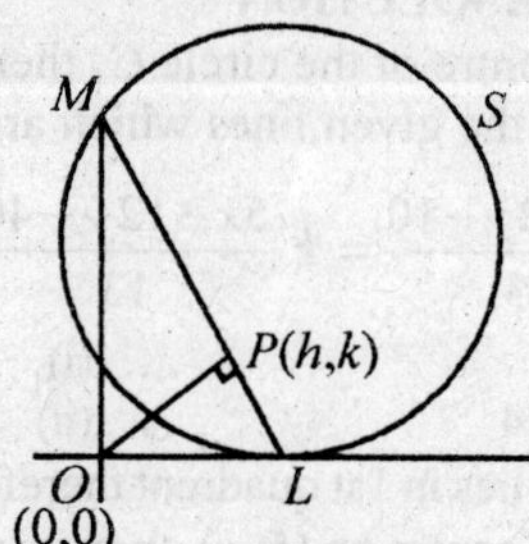

Slope of $OP=\frac{k}{h}$

$\therefore$ Slope of $LM=-\frac{h}{k}$

$\therefore$ Eq. of LM is :

$$y-k=-\frac{h}{k}(x-h)$$

$$\Rightarrow ky-k^2=-hx+h^2$$

$$\Rightarrow hx+ky=h^2+k^2 \quad \ldots(2)$$

Now the combined equations of lines joining the points of intersection of (1) and (2) to origin can be obtained by making eq. (1) homogeneous with the help of (2) as follows :

$$x^2+y^2+(2gx+2fy)\left(\frac{hx+ky}{h^2+k^2}\right)+c\left(\frac{hx+ky}{h^2+k^2}\right)^2=0$$

$$\Rightarrow (h^2+k^2)^2(x^2+y^2)+(h^2+k^2)(2gx+2fy)(hx+ky)+c(hx+ky)^2=0$$

$\Rightarrow$ $(h^2+k^2)^2(x^2+y^2)+(h^2+k^2)[\{2ghx^2+2fky^2+(2gk+2fh)xy\}]+c(h^2x^2+k^2y^2+2hkxy)=0$

$\Rightarrow$ $[(h^2+k^2)^2+2gh(h^2+k^2)+ch^2]x^2+[(2gk+2fh)(h^2+k^2)+2hkc]xy+[(h^2+k^2)^2+2fk(h^2+k^2)+ck^2]y^2=0$ ...(3)

Above represents the combined eq. of OL and OM.

But $\angle LOM=90°$

$\therefore$ From eq. (3), we must have coeff. of x^2 + coeff. of $y^2=0$

$\Rightarrow$ $(h^2+k^2)^2+2gh(h^2+k^2)+ch^2+(h^2+k^2)^2+2fk(h^2+k^2)+ck^2=0$

$\Rightarrow$ $h^2+k^2+gh+fk+\dfrac{c}{2}=0$

$\therefore$ Locus of (h, k) is $x^2+y^2+gx+fy+\dfrac{c}{2}=0$

11. Given that $\left(m_i,\dfrac{1}{m_i}\right)$, $m_i>0$, i = 1, 2, 3, 4 are four distinct points on a circle.

Let the equation of circle be

$x^2+y^2+2gx+2fy+c=0$

As the point $\left(m,\dfrac{1}{m}\right)$ lies on it, therefore, we have

$$m^2+\frac{1}{m^2}+2gm+\frac{2f}{m}+c=0$$

$\Rightarrow$ $m^4+2gm^3+cm^2+2fm+1=0$

Since m_1, m_2, m_3, m_4 are roots of this equation, therefore product of roots = 1

$\Rightarrow$ $m_1m_2m_3m_4=1$

12. Let AB be the length of chord intercepted by circle on $y+x=0$

Let CM be perpendicular to AB from centre $C(h, k)$.

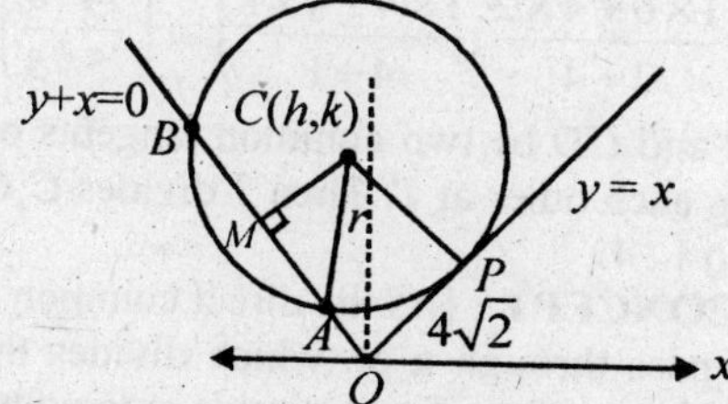

Also $y-x=0$ and $y+x=0$ are perpendicular to each other.

$\therefore$ $OPCM$ is rectangle.

$\therefore$ $CM=OP=4\sqrt{2}$.

Let r be the radius of cirlce.

Also $AM=\dfrac{1}{2}AB=\dfrac{1}{2}\times 6\sqrt{2}=3\sqrt{2}$

$\therefore$ In ΔCAM, $AC^2=AM^2+MC^2$

$\Rightarrow$ $r^2=(3\sqrt{2})^2+(4\sqrt{2})^2$

$\Rightarrow$ $r^2=(5\sqrt{2})^2$

$\Rightarrow$ $r=5\sqrt{2}$.

Again $y=x$ is tangent to the circle at P

$\therefore$ $CP=r$

$\Rightarrow$ $\left|\dfrac{h-k}{\sqrt{2}}\right|=5\sqrt{2}$

$\Rightarrow$ $h-k=\pm 10$ ...(1)

Also $CM=4\sqrt{2}$

$\Rightarrow$ $\left|\dfrac{h+k}{\sqrt{2}}\right|=4\sqrt{2}$

$\Rightarrow$ $h+k=\pm 8$ ...(2)

Solving four sets of eq's given by (1) and (2), we get the possible centres as

$(9,-1), (1,-9), (-1, 9), (-9, 1)$

$\therefore$ Possible circles are

$(x-9)^2+(y+1)^2-50=0$

$(x-1)^2+(y+9)^2-50=0$

$(x+1)^2+(y-9)^2-50=0$

$(x+9)^2+(y-1)^2-50=0$

But the pt $(-10, 2)$ lies inside the circle.

$\therefore$ $S_1<0$ which is satisfied only for $(x+9)^2+(y-1)^2-50=0$

$\therefore$ The required eq. of circle is $x^2+y^2+18x-2y+32=0$.

13. Let t be the common tangent given by

$4x+3y=10$ ...(1)

Common pt of contact being $P(1, 2)$

Let A and B be the centres of the circles, required. Clearly, AB is the line perpendicular to t and passing through $P(1, 2)$.

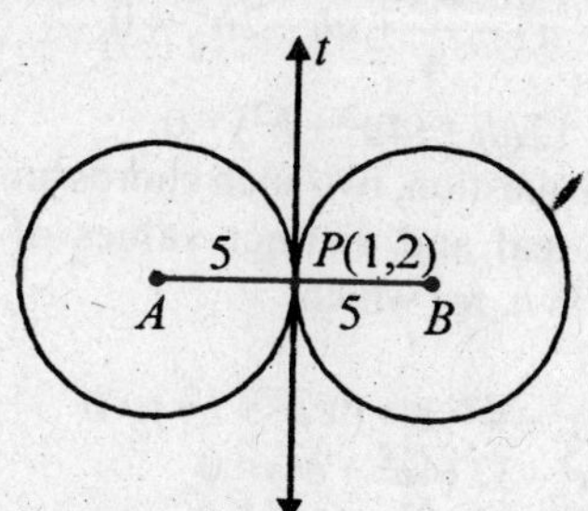

Therefore eq. of AB is

$$\frac{x-1}{4/5}=\frac{x-2}{3/5}=r\quad\left[\begin{array}{l}\text{As slope of } t \text{ is} = -4/3\\ \therefore \text{ slope of } AB \text{ is} = 3/4 = \tan\theta\\ \therefore \cos\theta = 4/5; \sin\theta = 3/5\end{array}\right.$$

For pt A, $r=-5$ and for pt B, $r=5$, we get

$$\frac{x-1}{4/5}=\frac{y-2}{3/5}=-5,5\left(\begin{array}{l}\text{radius of each circle}\\ \text{being } 5,\ AP=PB=5\end{array}\right)$$

$\Rightarrow$ For pt A $x=-4+1, y=-3+2$

and For pt B $x=4+1, y=3+2$

$\therefore$ $A(-3,-1)$ $B(5, 5)$.

$\therefore$ Eq.'s of required circles are

$(x+3)^2+(y+1)^2=5^2$

and $(x-5)^2+(y-5)^2=5^2$

$$\Rightarrow\quad \left.\begin{array}{r}x^2+y^2+6x+2y-15=0\\ \text{and } x^2+y^2-10x-10y+25=0\end{array}\right\}$$

14. The given circle is

$2x(x-a)+y(2y-b)=0$ $(a, b\neq 0)$

$\Rightarrow$ $2x^2+2y^2-2ax-by=0$ ....(1)

Let us consider the chord of this circle which passes through the pt $\left(a, \frac{b}{2}\right)$ and whose mid pt. lies on x-axis.

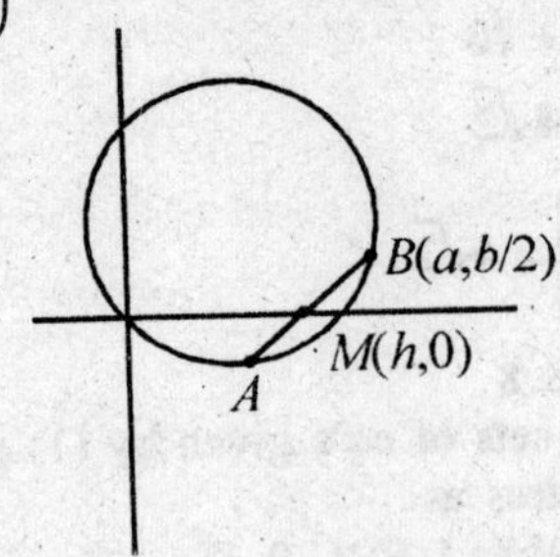

Let (h, 0) be the mid point of the chord, then eq. of chord can be obtained by

$T = S_1$

i.e., $2xh + 2y.0 - a(x+h) - \frac{b}{2}(y+0) = 2h^2 - 2ah$

$\Rightarrow \quad (2h-a)x - \frac{b}{2}y + ah - 2h^2 = 0$

This chord passes through $\left(a, \frac{b}{2}\right)$, therefore

$(2h-a)a - \frac{b}{2}.\frac{b}{2} + ah - 2h^2 = 0$

$\Rightarrow \quad 2ah - a^2 - \frac{b^2}{4} + ah - 2h^2 = 0$

$\Rightarrow \quad 8h^2 - 12ah + (4a^2 + b^2) = 0$

As given in question, two such chords are there, so we should have two real and distinct values of h from the above quadratic in h, for which

$D > 0$

$\Rightarrow \quad (12a)^2 - 4 \times 8 \times (4a^2 + b^2) > 0$

$\Rightarrow \quad 144a^2 - 32(4a^2 + b^2) > 0$

$\Rightarrow \quad 144a^2 - 128a^2 - 32b^2 > 0$

$\Rightarrow \quad 16a^2 > 32b^2$

$\Rightarrow \quad a^2 > 2b^2$

15. Let the family of circles, passing through A (3, 7) and B (6, 5), be

$x^2 + y^2 + 2gx + 2fy + c = 0$

As it passes through (3, 7)

$\therefore \quad 9 + 49 + 6g + 14f + c = 0$

or, $\quad 6g + 14f + c + 58 = 0 \quad \ldots(1)$

As it passes through (6, 5)

$\therefore \quad 36 + 25 + 12g + 10f + c = 0$

$12g + 10f + c + 61 = 0 \quad \ldots(2)$

(2) – (1) gives,

$6g - 4f + 3 = 0 \Rightarrow g = \frac{4f-3}{6}$

Substituting the value of g in equation (1), we get

$4f - 3 + 14f + c + 58 = 0$

$\Rightarrow \quad 18f + 55 + c = 0 \Rightarrow c = -18f - 55$

Thus the family is

$x^2 + y^2 + \left(\frac{4f-3}{3}\right)x + 2fy - (18f + 55) = 0$

Members of this family are cut by the circle

$x^2 + y^2 - 4x - 6y - 3 = 0$

$\therefore$ Equation of family of chords of intersection of above two circles is

$S_1 - S_2 = 0$

$\Rightarrow \quad \left(\frac{4f-3}{3} + 4\right)x + (2f+6)y - 18f + 52) = 0$

which can be written as

$(3x + 6y - 52) + f\left(\frac{4}{3}x + 2y - 18\right) = 0$

which represents the family of lines passing through the pt. of intersection of the lines

$3x + 6y - 52 = 0$ and $4x + 6y - 54 = 0$

Solving which we get $x = 2$ and $y = 23/3$.

Thus the required pt. of intersection is $\left(2, \frac{23}{3}\right)$

16. The given circles are

$x^2 + y^2 - 4x - 2y = -4$

and $x^2 + y^2 - 12x - 8y = -36$

i.e., $x^2 + y^2 - 4x - 2y + 4 = 0 \quad \ldots(1)$

$x^2 + y^2 - 12x - 8y + 36 = 0 \quad \ldots(2)$

with centres $C_1(2, 1)$ and $C_2(6, 4)$ and radii 1 and 4 respectively.

Also $C_1C_2 = 5$

As $\quad r_1 + r_2 = C_1C_2$

$\Rightarrow$ Two circles touch each other externally, at P.

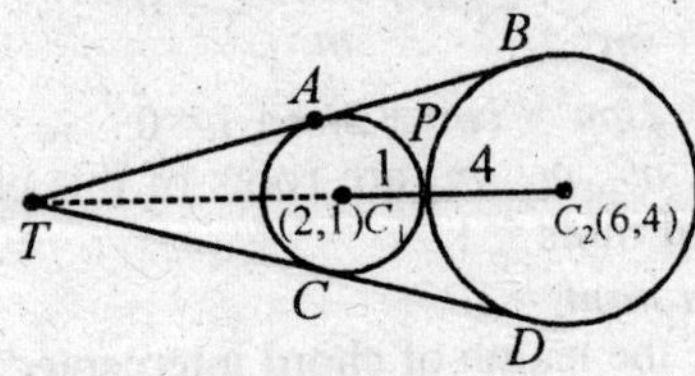

Clearly, P divides C_1C_2 in the ratio 1 : 4

$\therefore$ Co-ordinates of P are

$$\left(\frac{1\times6+4\times2}{1+4}, \frac{1\times4+4\times1}{4+1}\right) = \left(\frac{14}{5}, \frac{8}{5}\right)$$

Let AB and CD be two common tangents of given circles, meeting each other at T. Then T divides C_1C_2 externally in the ratio 1 : 4.

KEY CONCEPT : [As the direct common tangents of two circles pass through a pt. which divides the line segment joining the centres of two circles externally in the ratio of their radii.]

Hence, $T \equiv \left(\frac{1\times6-4\times2}{1-4}, \frac{1\times4-4\times1}{1-4}\right) = \left(\frac{2}{3}, 0\right)$

Let m be the slope of the tangent, then equation of tangent through (2/3, 0) is

$y - 0 = m\left(x - \frac{2}{3}\right) \Rightarrow y - mx + \frac{2}{3}m = 0$

Now, length of perpendicular from (2, 1), to the above tangent is radius of the circle

$$\therefore \quad \left|\frac{1 - 2m + \frac{2}{3}m}{\sqrt{m^2+1}}\right| = 1$$

$\Rightarrow \quad (3-4m)^2 = 9(m^2+1) \Rightarrow \quad 9-24m+16m^2 = 9m^2+9$

$\Rightarrow \quad 7m^2 - 24m = 0 \Rightarrow \quad m = 0, \dfrac{24}{7}$

Thus the equations of the tangents are $y = 0$ and $7y - 24x + 16 = 0$.

17. Let the given point be

$(p, \bar{p}) = \left(\dfrac{1+\sqrt{2}\,a}{2}, \dfrac{1-\sqrt{2}\,a}{2}\right)$ and the equation of the circle becomes

$x^2 + y^2 - px - \bar{p}y = 0$

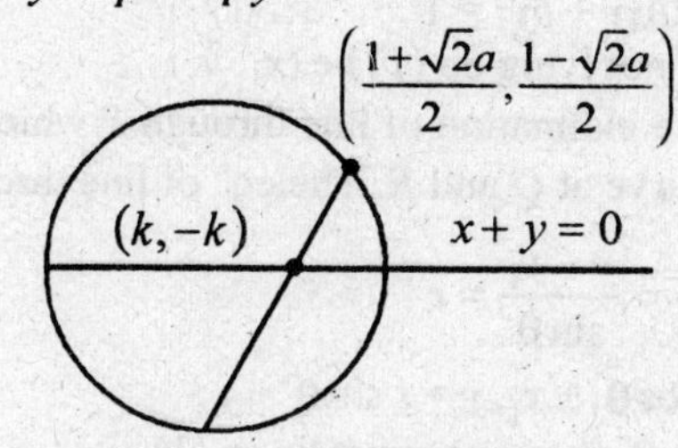

Since the chord is bisected by the line $x + y = 0$, its mid-point can be chosen as $(k, -k)$. Hence the equation of the chord by $T = S_1$ is

$$kx - ky - \frac{p}{2}(x+k) - \frac{\bar{p}}{2}(y-k) = k^2 + k^2 - pk + \bar{p}k$$

It passes through $A\ (p, \bar{p})$

$$\therefore \quad kp - k\bar{p} - \frac{p}{2}(p+k) - \frac{\bar{p}}{2}(\bar{p}-k) = 2k^2 - pk + \bar{p}k$$

or $\quad 3k(p - \bar{p}) = 4k^2 + (p^2 + \bar{p}^2) \qquad \ldots(1)$

Put $p - \bar{p} = a\sqrt{2}, p^2 - \bar{p}^2 = 2.\dfrac{(1+2a^2)}{4} = \dfrac{1+2a^2}{2} \qquad \ldots(2)$

Hence, from (1) by the help of (2), we get

$$4k^2 - 3\sqrt{2}\,ak + \frac{1}{2}(1+2a^2) = 0 \qquad \ldots(3)$$

Since, there are two chords which are bisected by $x + y = 0$, we must have two real values of k from (3)

$\therefore \quad \Delta > 0$

or $\quad 18a^2 - 8(1+2a^2) > 0$

or, $\quad a^2 - 4 > 0$

or, $\quad (a+2)(a-2) > 0$

$\therefore \quad a < -2$ or > 2

$\therefore \quad a \in (-\infty, -2) \cup (2, \infty)$

or $\quad a \in]-\infty, -2[\cup]2, \infty[$

18. Let r be the radius of circle, then $AC = 2r$

Since, AC is the diameter

$\therefore \quad \angle ABC = 90°$

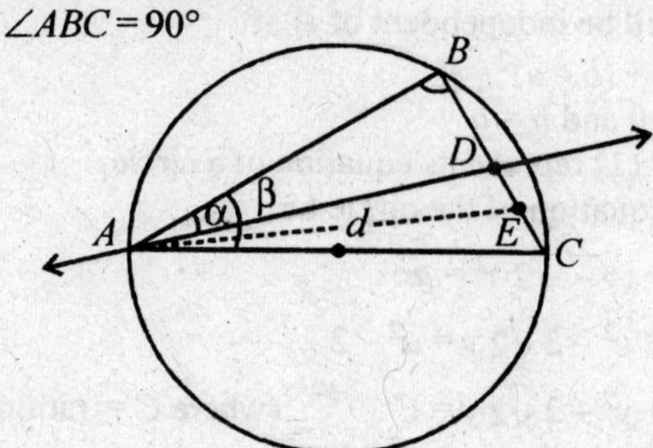

$\therefore$ In ΔABC

$BC = 2r \sin\beta$

$AB = 2r\cos\beta$

In rt $\angle ed'\ \Delta ABC$

$BD = AB\tan\alpha = 2r\cos\beta\tan\alpha$

$AD = AB\sec\alpha = 2r\cos\beta\sec\alpha$

$\therefore \quad DC = BC - BD = 2r\sin\beta - 2r\cos\beta\tan\alpha$

Now since E is the mid point of DC

$$\therefore \quad DE = \frac{DC}{2} = \frac{2r\sin\beta - 2r\cos\beta\tan\alpha}{2}$$

$\Rightarrow \quad DE = r\sin\beta - r\cos\beta\tan\alpha$

Now in ΔADC, AE is the median

NOTE THIS STEP

$\therefore \quad 2(AE^2 + DE^2) = AD^2 + AC^2$

$\Rightarrow \quad 2[d^2 + r^2(\sin\beta - \cos\beta\tan\alpha)^2] = 4r^2\cos^2\beta\sec^2\alpha + 4r^2$

$$\Rightarrow \quad d^2 + \frac{r^2(\sin\beta\cos\alpha - \cos\beta\sin\alpha)^2]}{\cos^2\alpha} = 2r^2\cos^2\beta\sec^2\alpha + 2r^2$$

$$\Rightarrow \quad d^2 + \frac{r^2\sin^2(\beta-\alpha)}{\cos^2\alpha} = 2r^2\cos^2\beta\sec^2\alpha + 2r^2$$

$$\Rightarrow \quad d^2 = r^2\left[2\cos^2\beta\sec^2\alpha + 2 - \frac{\sin^2(\beta-\alpha)}{\cos^2\alpha}\right]$$

$\Rightarrow \quad d^2\cos^2\alpha = r^2(2\cos^2\beta + 2\cos^2\alpha - \sin^2(\beta-\alpha))$

$= r^2[\cos^2\alpha + \cos^2\beta + \cos^2\alpha + \cos^2\beta - 1 + \cos^2(\beta-\alpha)]$

$$= r^2\left[\cos^2\alpha + \cos^2\beta + \frac{1+\cos 2\alpha}{2} + \frac{1+\cos 2\beta}{2} - 1 + \cos^2(\beta-\alpha)\right]$$

$$= r^2\left[\cos^2\alpha + \cos^2\beta + \frac{1}{2}(\cos 2\alpha + \cos 2\beta) + \cos^2(\beta-\alpha)\right]$$

$= r^2[\cos^2\alpha + \cos^2\beta + \cos(\alpha+\beta)\cos(\beta-\alpha) + \cos^2(\beta-\alpha)$

$= r^2[\cos^2\alpha + \cos^2\beta + \cos(\beta-\alpha)(\cos(\alpha+\beta) + \cos(\beta-\alpha))]$

$= r^2[\cos^2\alpha + \cos^2\beta + 2\cos\alpha\cos\beta\cos(\beta-\alpha)$

$$\Rightarrow \quad r^2 = \frac{d^2\cos^2\alpha}{\cos^2\alpha + \cos^2\beta + 2\cos\alpha\cos\beta\cos(\beta-\alpha)}$$

$\Rightarrow$ Area of circle,

$$\pi r^2 = \frac{\pi d^2\cos^2\alpha}{\cos^2\alpha + \cos^2\beta + 2\cos\alpha\cos\beta\cos(\beta-\alpha)}$$

ALTERNATE SOLUTION

Let the radius of circle be r then $OA = OC = r$ and since AC is diameter,

$\therefore \quad AC = 2r$

Considering AC along x-axis and with centre O as orign we get co-ordinates of A and C as $(-r, 0)$ and $(r, 0)$ respectively.

$\angle ABC = 90°$

Also $\angle BAC = \beta \Rightarrow \angle BOC = 2\beta$

then co-ordinates of B are $(r\cos 2\beta, r\sin 2\beta)$

Slope of $AD = \tan(\beta - \alpha)$

Let co-ordinates of pt. D be (x, y), then equation of AD is

$y - 0 = \tan(\beta - \alpha)(x + r)$

or $y = \tan(\beta - \alpha)(x + r)$ ...(1)

Equation of BC is

$$y = \frac{r\sin 2\beta}{r\cos 2\beta - r}(x - r)$$

or $$y = \frac{2\sin\beta\cos\beta}{-2\sin^2\beta}(x - r)$$

$y = -\cot\beta\,(x - r)$ ...(2)

Now D is pt of intersection of (1) and (2). Substituting value of y from (2) in (1) we get

$\tan(\beta - \alpha)(x + r) = -\cot\beta\,(x - r)$

$\Rightarrow x[\tan(\beta - \alpha) + \cot\beta] = r[\cot\beta - \tan(\beta - \alpha)]$

$$\Rightarrow x\left[\frac{\sin(\beta-\alpha)}{\cos(\beta-\alpha)} + \frac{\cos\beta}{\sin\beta}\right] = r\left[\frac{\cos\beta}{\sin\beta} - \frac{\sin(\beta-\alpha)}{\cos(\beta-\alpha)}\right]$$

$\Rightarrow x[\cos(\beta - \alpha)\cos\beta + \sin(\beta - \alpha)\sin\beta]$
$= r[\cos(\beta - \alpha)\cos\beta - \sin\beta\sin\beta - \alpha)]$

$\Rightarrow x\cos(\beta - \alpha - \beta) = r\cos(\beta - \alpha + \beta)$

$$\Rightarrow x = \frac{r\cos(2\beta - \alpha)}{\cos\alpha}$$

$$\Rightarrow y = -\frac{\cos\beta}{\sin\beta}\left[\frac{r\cos(2\beta-\alpha)}{\cos\alpha} - r\right]$$

$$= -\frac{r\cos\beta}{\sin\beta}\left[\frac{\cos(2\beta-\alpha) - \cos\alpha}{\cos\alpha}\right]$$

$$= -\frac{r\cos\beta}{\sin\beta}\frac{2\sin\beta\sin(\alpha-\beta)}{\cos\alpha}$$

$$= \frac{-2r\cos\beta\sin(\alpha-\beta)}{\cos\alpha}$$

$\therefore$ Co-ordinates of D are

$$\left(\frac{r\cos(2\beta-\alpha)}{\cos\alpha}, \frac{-2r\cos\beta\sin(\alpha-\beta)}{\cos\alpha}\right)$$

Thus co-ordinates of E are

$$\left(\frac{r\cos(2\beta-\alpha) + r\cos\alpha}{2\cos\alpha}, \frac{-r\cos\beta\sin(\alpha-\beta)}{\cos\alpha}\right)$$

or, $$\left(\frac{r\cos\beta\cos(\beta-\alpha)}{\cos\alpha}, \frac{r\cos\beta\sin(\alpha-\beta)}{\cos\alpha}\right)$$

Since, $AD = d$, we get

$$d^2 = r^2\left[\frac{\cos\beta\cos(\beta-\alpha)}{\cos\alpha} + 1\right]^2 + \left[\frac{\cos\beta\sin(\beta-\alpha)}{\cos\alpha}\right]^2$$

$$= \frac{r^2}{\cos^2\alpha}[\cos^2\beta\cos^2(\beta-\alpha) + \cos^2\alpha$$
$$+ 2\cos\beta\cos(\beta-\alpha)\cos\alpha + \cos^2\beta\sin^2(\beta-\alpha)]$$

$$= \frac{r^2}{\cos^2\alpha}[\cos^2\beta + \cos^2\alpha + 2\cos\alpha\cos\beta\cos(\beta-\alpha)]$$

$$\Rightarrow r^2 = \frac{d^2\cos^2\alpha}{\cos^2\beta + \cos^2\alpha + 2\cos\alpha\cos\beta\cos(\beta-\alpha)}$$

$\therefore$ Area of circle $= \pi r^2$

$$= \frac{\pi d^2\cos^2\alpha}{\cos^2\beta + \cos^2\alpha + 2\cos\alpha\cos\beta\cos(\beta-\alpha)}$$

19. The given curve is

$ax^2 + 2hxy + by^2 = 1$ ...(1)

Let the pt P not lying on (1) be (x_1, y_1)

Let θ be the inclination of line through P which intersects the given curve at Q and R. The eq^n of line through P is

$$\frac{x - x_1}{\cos\theta} = \frac{y - y_1}{\sin\theta} = r$$

$\Rightarrow x = r\cos\theta + x_1, y = r\sin\theta + y_1$

For pt. Q or R, above pt. must lie on (1)

$\Rightarrow a[x_1 + r\cos\theta]^2 + 2h[x_1 + r\cos\theta]$
$[y_2 + r\sin\theta] + b[y_1 + r\sin\theta]^2 = 1$

$\Rightarrow (a\cos^2\theta + 2h\sin\theta\cos\theta + b\sin^2\theta)r^2 +$
$2(ax_1\cos\theta\, hx_1\sin\theta + hy_1\cos\theta + by_1\sin\theta)r$
$+ (ax_1^2 + 2hx_1y_1 + by_1^2 - 1) = 0$

It is quadratic in r giving two values of r as PQ and PR.

$$\text{Therefore product of roots} = \frac{\text{constant term}}{\text{coeff. of } r^2}$$

$$\Rightarrow PQ.PR = \frac{ax_1^2 + 2hx_1y_1 + by_1^2 - 1}{a\cos^2\theta + 2h\sin\theta\cos\theta + b\sin^2\theta}$$

Here $ax_1^2 + 2hx_1y_1 + by_1^2 - 1 \neq 0$ as (x_1, y_1) does not lie on (1).

Here denominator

$= a\cos^2\theta + 2h\sin\theta\cos\theta + b^2\sin^2\theta$

$= a + 2h\sin\theta\cos\theta + (b - a)\sin^2\theta$

$= a + \sin\theta[2h\cos\theta + (b - a)\sin\theta]$

$= a + \sin\theta . \sqrt{4h^2 + (b-a)^2}$

$[\cos\theta\sin\phi + \sin\theta\cos\phi]$

where, $\tan\phi = \dfrac{b-a}{2h}$

$= a + \sqrt{4h^2 + (b-a)^2}\sin\theta\sin(\theta + \phi)$

which will be independent of θ if

$4h^2 + (b - a)^2 = 0$

$\Rightarrow h = 0$ and $b = a$

$\Rightarrow$ Eq^n (1) represents equation of a circle.

20. Let the equation of the circle be

$x^2 + (y - \sqrt{2})^2 = a^2$

$\Rightarrow x^2 + y^2 - 2\sqrt{2}y = a^2 - 2$

$\Rightarrow x^2 + y^2 - 2\sqrt{2}y = C$ (where C = rational no.)

Let $(x_1, y_1), (x_2, y_2), (x_3, y_3)$ be 3 rational pts on the circle i.e.

$$x_1^2 + y_1^2 - 2\sqrt{2}y_1 = C \quad \ldots(1)$$

$$x_2^2 + y_2^2 - 2\sqrt{2}y_2 = C \quad \ldots(2)$$

$$x_3^2 + y_3^2 - 2\sqrt{2}y_3 = C \quad \ldots(3)$$

As R.H.S. is same in all the three equations we get

$$x_1^2 + y_1^2 - 2\sqrt{2}y_1 = x_2^2 + y_2^2 - 2\sqrt{2}y_2$$

$$= x_3^2 + y_3^2 - 2\sqrt{2}y_3$$

Comparing irrational parts, we get

$$y_1 = y_2 = y_3 \quad \ldots(4)$$

Comparing rational parts, we get

$$x_1^2 + y_1^2 = x_2^2 + y_2^2 = x_3^2 + y_3^2$$

Using (4) we get, $x_1^2 = x_2^2 = x_3^2$

Thus there are only two possible values of abscissa $\pm x_1$ and only one value of ordinate y_1.

$\Rightarrow$ There could be only 2 rational points (x_1, y_1) and $(-x_1, y_1)$

ALTERNATE SOLUTION

Given C is the circle with centre at $(0, \sqrt{2})$ and radius r (say)

then $C \equiv x^2 + (y - \sqrt{2})^2 = r^2$

$\Rightarrow (y - \sqrt{2})^2 = (r^2 - x^2)$

$\Rightarrow y - \sqrt{2} = \pm\sqrt{r^2 - x^2}$

$\Rightarrow y = \sqrt{2} \pm \sqrt{r^2 - x^2} \quad \ldots(1)$

The only rational value which y can have is 0. Suppose the possible value of x for which y is 0 is x_1. Certainly $-x_1$ will also give the value of y as 0 (from (1)). Thus, at the most, there are two rational pts which satisfy the eq^n of C.

21. Let $P(h, k)$ be on C_2

$\therefore \quad h^2 + k^2 = 4r^2$

Chord of contact of P w.r.t. C_1 is

$hx + ky = r^2$

It intersects C_1,

$x^2 + y^2 = a^2$ in A and B.

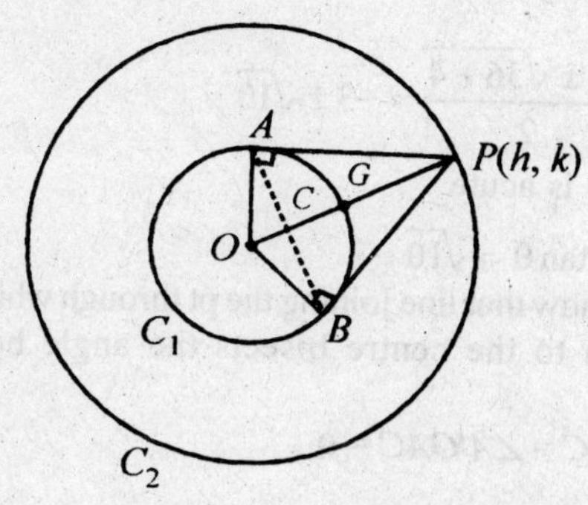

Eliminating y, we get,

$$x^2 + \left(\frac{r^2 - hx}{k}\right)^2 = r^2$$

or, $x^2(h^2 + k^2) - 2r^2hx + r^4 - r^2k^2 = 0$

or, $x^2.4r^2 - 2r^2hx + r^2(r^2 - k^2) = 0$

$$\therefore \quad x_1 + x_2 = \frac{2r^2h}{4r^2} = \frac{h}{2}, y_1 + y_2 = \frac{k}{2}$$

If (x, y) be the centroid of ΔPAB, then

$$3x = x_1 + x_2 + h = \frac{h}{2} + h = \frac{3h}{2}$$

$\therefore \quad x = \frac{h}{2}$ or $h = 2x$ and

Similarly $k = 2y$

Putting in (1) we get

$4x^2 + 4y^2 = 4r^2$

$\therefore$ Locus is

$x^2 + y^2 = r^2$ i.e., C_1

ALTERNATE SOLUTION

Let r be the radius of C_1 and $2r$ that of C_2, then

$OA = OB = r$

and $OP = 2r$.

Since PA and PB are tangents to C_1, therefore

$\angle OAP = \angle OBP = 90°$

Let OP meet C_1 at G.

Let $\angle OPA = \theta$ then

$$\sin\theta = \frac{OA}{OP} = \frac{r}{2r} = \frac{1}{2}$$

$\Rightarrow \theta = 30°$

$\Rightarrow \angle AOP = 60°$

In ΔOAC

$$\cos 60° = \frac{OC}{OA}$$

$$\Rightarrow \frac{1}{2} = \frac{OC}{r}$$

$\Rightarrow OC = r/2$

$$\therefore \quad PC = 2r - \frac{r}{2} = \frac{3r}{2}$$

Also $CG = r - OC = r - \frac{r}{2} = \frac{r}{2}$

Clearly $CG = \frac{1}{3}PC$

$\Rightarrow$ G is centriod of ΔABP.

22. The given circle is $x^2 + y^2 = 1 \quad \ldots(1)$

Centre $O(0, 0)$ radius $= 1$

Let T_1 and T_2 be the tangents drawn from $(-2, 0)$ to the circle (1).

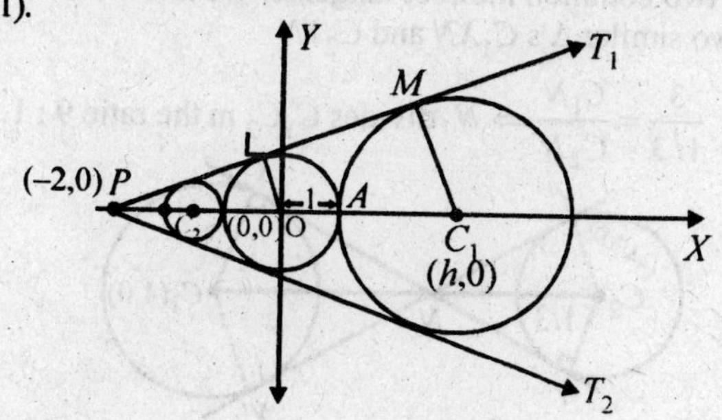

Let m be the slope of tangent then equations of tangents are

$$y-0=m(x+2)$$

or, $mx-y+2m=0$ ...(2)

As it is tangent to circle (1) length of $\perp$ lar from (0, 0) to (2) = radius of (1)

$$\Rightarrow \left|\frac{2m}{\sqrt{m^2+1}}\right|=1$$

$$\Rightarrow 4m^2=m^2+1 \Rightarrow m=\pm 1/\sqrt{3}$$

$\therefore$ The two tangents are $x+\sqrt{3}y+2=0 (T_1)$ and $x-\sqrt{3}y+2=0(T_2)$

Now any other circle touching (1) and T_1, T_2 is such that its centre lies on x-axis.

Let $(h, 0)$ be the centre of such circle, then from fig.

$OC_1=OA+AC_1 \Rightarrow |h|=1+|AC_1|$

But AC_1 = $\perp$ lar distance of $(h, 0)$ to tangent

$$\Rightarrow |h|=1+\left|\frac{h+2}{2}\right| \Rightarrow |h|-1=\left|\frac{h+2}{2}\right|$$

Squaring,

$$h^2-2|h|+1=\frac{h^2+4h+4}{4}$$

$\Rightarrow 4h^2\pm 8h+4=h^2+4h+4$

'+' $\Rightarrow 3h^2=-4h \quad \Rightarrow h=-4/3$

'–' $\Rightarrow 3h^2=12h \quad \Rightarrow h=4$

Thus centres of circles are (4, 0), $\left(-\frac{4}{3},0\right)$.

$\therefore$ Radius of circle with centre (4, 0) is = 4 – 1 = 3 and radius of circle with centre $\left(\frac{-4}{3},0\right)$ is $=\frac{4}{3}-1=\frac{1}{3}$

$\therefore$ The two possible circles are

$(x-4)^2+y^2=3^2$ ...(3)

And $\left(x+\frac{4}{3}\right)+y^2=\left(\frac{1}{3}\right)^2$ ...(4)

Now, common tangents of (1) and (3). Since (1) and (3) are two touching circles they have three common tangents T_1, T_2 and $x = 1$ (clear from fig.)

Similarly common tangents of (1) and (4) are T_1, T_2 and $x=-1$.

For the circles (3) and (4) there will be four common tangents of which two are direct common tangents.

XY and $x'y'$ and two are indirect common tangents. Let us find two common indirect tangents. We know that

In two similar Δ's C_1XN and C_2YN

$$\frac{3}{1/3}=\frac{C_1N}{C_2N} \Rightarrow N \text{ divides } C_1C_2 \text{ in the ratio } 9:1.$$

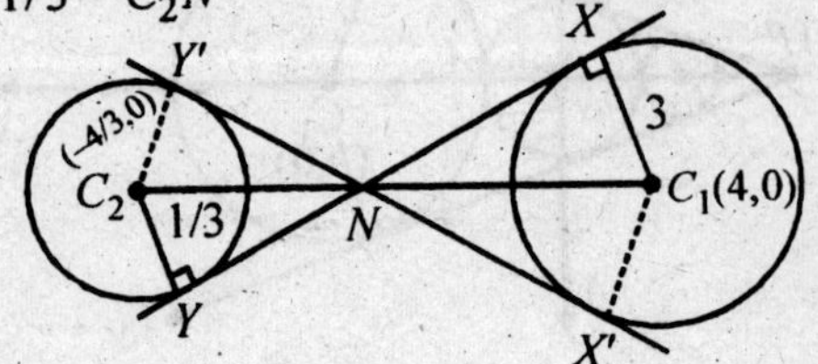

Clearly N lies on x-axis.

$$\therefore \quad N=\left(\frac{9\times(-4/3)+1\times 4}{10},0\right)=\left(\frac{-4}{5},0\right)$$

Any line through N is

$$y=m\left(x+\frac{4}{5}\right)$$

or $5mx-5y+4m=0$

If it is tangent to (3) then

$$\left|\frac{20m+4m}{\sqrt{25m^2+25}}\right|=3$$

$$\Rightarrow 24m=15\sqrt{m^2+1} \Rightarrow 64m^2=25m^2+25$$

$$\Rightarrow 39m^2=25 \Rightarrow m=\pm 5/\sqrt{39}$$

$\therefore$ Required tangents are

$$y=\pm\frac{5}{\sqrt{39}}\left(x+\frac{4}{5}\right)$$

23. The equation $2x^2-3xy+y^2=0$ represents pair of tangents OA and OA'.

Let angle between these to tangents be 2θ.

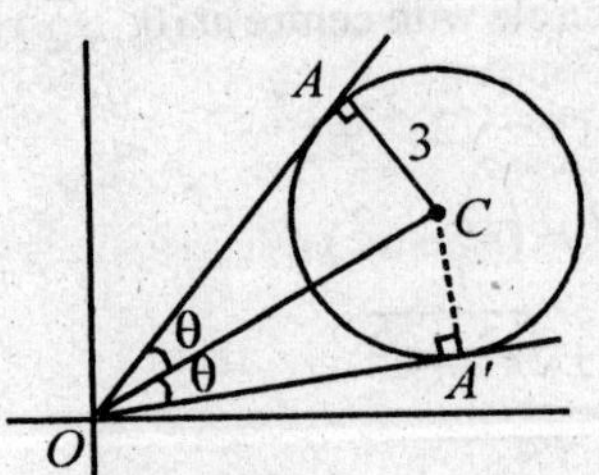

$$\text{Then, } \tan 2\theta=\frac{2\sqrt{\left(\frac{-3}{2}\right)^2-2\times 1}}{2+1}$$

$$\left[\text{Using } \tan\theta=\frac{2\sqrt{h^2-ab}}{a+b}\right]$$

$$\frac{2\tan\theta}{1-\tan^2\theta}=\frac{1}{3} \Rightarrow \tan^2\theta+6\tan\theta-1=0$$

$$\tan\theta=\frac{-6\pm\sqrt{36+4}}{2}=-3\pm\sqrt{10}$$

As θ is acute

$$\therefore \quad \tan\theta=\sqrt{10}-3$$

Now we know that line joining the pt through which tangents are drawn to the centre bisects the angle between the tangents,

$\therefore \quad \angle AOC=\angle A'OAC=\theta$

In ΔAOC,

$$\tan\theta=\frac{3}{OA} \quad \Rightarrow OA=\frac{3}{\sqrt{10}-3}\times\frac{\sqrt{10}+3}{\sqrt{10}+3}$$

$$\therefore \quad OA=3(3+\sqrt{10}).$$

24. Let equation of C_1 be $x^2 + y^2 = r_1^2$ and of C_2 be $(x-a)^2 + (y-b)^2 = r_2^2$

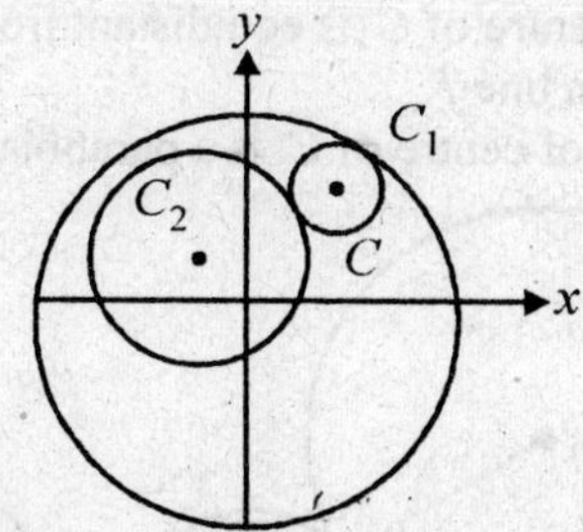

Let centre of C be (h, k) and radius be r, then by the given conditions.

$\sqrt{(h-a)^2+(k-b)^2} = r + r_2$ and $\sqrt{h^2+k^2} = r_1 - r$

$\Rightarrow \sqrt{(h-a)^2+(k-b)^2} + \sqrt{h^2+k^2} = r_1 + r_2$

Required locus is

$$\sqrt{(x-a)^2+(y-b)^2} + \sqrt{x^2+y^2} = r_1 + r_2,$$

which represents an ellipse whose foci are at (a, b) and $(0, 0)$.

[$\because PS + PS' =$ constant $\Rightarrow$ locus of P is an ellipse with foci at S and S']

25. The given circle is $x^2 + y^2 = r^2$

From pt. (6, 8) tangents are drawn to this circle.

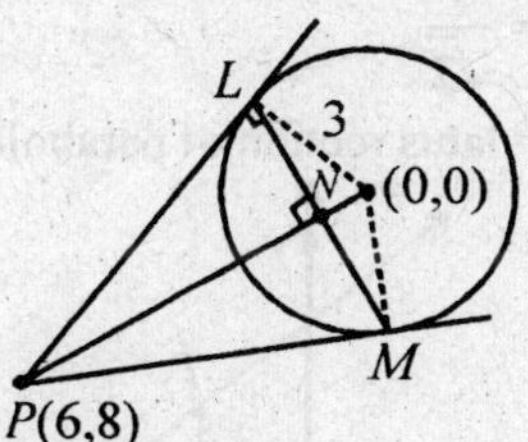

Then length of tangent

$$PL = \sqrt{6^2+8^2-r^2} = \sqrt{100-r^2}$$

Also equation of chord of contact LM is $6x + 8y - r^2 = 0$

PN = length of $\perp^{lar}$ from P to LM

$$= \frac{36+64-r^2}{\sqrt{36+64}} = \frac{100-r^2}{10}$$

Now in rt. ΔPLN, $LN^2 = PL^2 - PN^2$

$$= (100-r^2) - \frac{(100-r^2)^2}{100} = \frac{(100-r^2)r^2}{100}$$

$$\Rightarrow \quad LN = \frac{r\sqrt{100-r^2}}{10}$$

$$\therefore \quad LM = \frac{r\sqrt{100-r^2}}{5} \qquad (\because LM = 2\,LN)$$

$$\therefore \quad \text{Area of } \Delta PLM = \frac{1}{2} \times LM \times PN$$

$$= \frac{1}{2} \times \frac{r\sqrt{100-r^2}}{5} \times \frac{100-r^2}{10}$$

$$= \frac{1}{100}\left[r(100-r^2)^{\frac{3}{2}}\right]$$

For max value of area, we should have

$$\frac{dA}{dr} = 0$$

$$\Rightarrow \frac{1}{100}\left[(100-r^2)^{\frac{3}{2}} + r.\frac{3}{2}(100-r^2)^{\frac{1}{2}}(-2r)\right] = 0$$

$$\Rightarrow \quad (100-r^2)^{\frac{1}{2}}[100-r^2-3r^2] = 0$$

$\Rightarrow \quad r = 10$ or $r = 5$

But $r = 10$ gives length of tangent $PL = 0$

$\therefore \quad r \neq 10$

Hence, $r = 5$

26. We are given that line $2x + 3y + 1 = 0$ touches a circle $S = 0$ at $(1, -1)$.

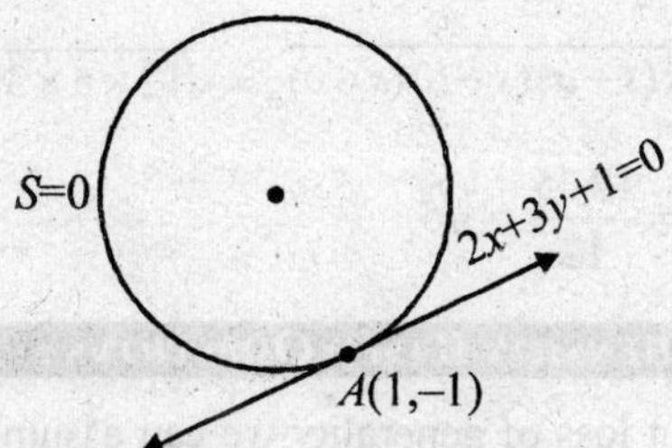

So, eq^n of this circle can be given by

$$(x-1)^2 + (y+1)^2 + \lambda(2x+3y+1) = 0.$$

[**Note :** $(x-1)^2 + (y+1)^2 = 0$ represents a pt. circle with centre at $(1, -1)$].

or $x^2 + y^2 + 2x(\lambda - 1) + y(3\lambda + 2) + (\lambda + 2) = 0$ ...(1)

But given that this circle is orthogonal to the circle, the extremities of whose diameter are (0, 3) and (– 2, – 1) i.e.

$$x(x+2) + (y-3)(y+1) = 0$$

$$x^2 + y^2 + 2x - 2y - 3 = 0 \qquad \text{.......... (2)}$$

Applying the condition of orthogonality for (1) and (2), we get

$$2(\lambda-1).1 + 2\left(\frac{3\lambda+2}{2}\right).(-1) = \lambda + 2 + (-3)$$

$$[2g_1g_2 + 2f_1f_2 = c_1 + c_2]$$

$$\Rightarrow \quad 2\lambda - 2 - 3\lambda - 2 = \lambda - 1$$

$$\Rightarrow \quad 2\lambda = -3 \Rightarrow \lambda = \frac{-3}{2}$$

Substituting this value of λ in eq^n (1) we get the required circle as

$$x^2 + y^2 - 5x - \frac{5}{2}y + \frac{1}{2} = 0$$

or, $\quad 2x^2 + 2y^2 - 10x - 5y + 1 = 0$

27. Given these circles with centres at C_1, C_2 and C_3 and with radii 3, 4 and 5 respectively, The three circles touch each other externally as shown in the figure.

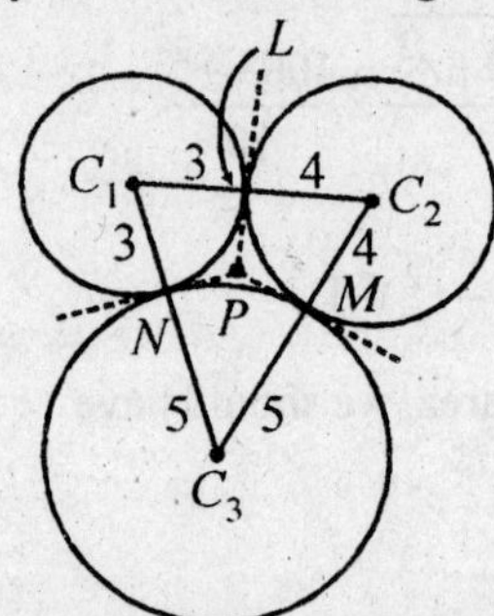

P is the point of intersection of the three tangents drawn at the pts of contacts, L, M and N. Since lengths of tangents to a circle from a point are equal, we get

$PL = PM = PN$

Also $PL \perp C_1C_2, PM \perp C_2C_3, PN \perp C_1C_3$

($\because$ tangent is perpendicular to the radius at pt. of contact)

Clearly P is the incentre of $\Delta C_1C_2C_3$ and its distance from pt.of contact i.e., PL is the radius of incircle of $\Delta C_1C_2C_3$.

In $\Delta C_1C_2C_3$ sides are

$a = 3+4=7, b=4+5=9, c=5+3=8$

$\therefore \quad s = \frac{a+b+c}{2} = 12$

$\therefore \quad \Delta = \sqrt{s(s-a)(s-b)(s-c)} = \sqrt{12\times5\times3\times4} = 12\sqrt{5}$

$\therefore \quad r = \frac{\Delta}{s} = \frac{12\sqrt{5}}{12} = \sqrt{5}$

G COMPREHENSION BASED QUESTIONS :

1. (a) Without loss of generality we can assume the square $ABCD$ with its vertices $A(1, 1)$, $B(-1, 1)$, $C(-1,-1)$, $D(1,-1)$

P to be the point $(0, 1)$ and Q as $(\sqrt{2}, 0)$.

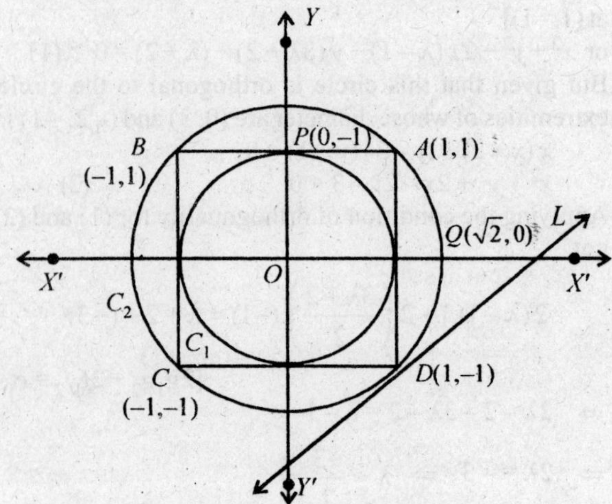

Then, $\frac{PA^2+PB^2+PC^2+PD^2}{QA^2+QB^2+QC^2+QD^2}$

$= \frac{1+1+5+5}{2[(\sqrt{2}-1)^2+1]+2((\sqrt{2}+1)^2+1]} = \frac{12}{16} = 0.75$

2. (b) Let C' be the said circle touching C_1 and L, so that C_1 and C' are on the same side of L. Let us draw a line T parallel to L at a distance equal to the radius of circle C_1, on opposite side of L.

Then the centre of C' is equidistant from the centre of C_1 and from line T.

$\Rightarrow$ locus of centre of C' is a parabola.

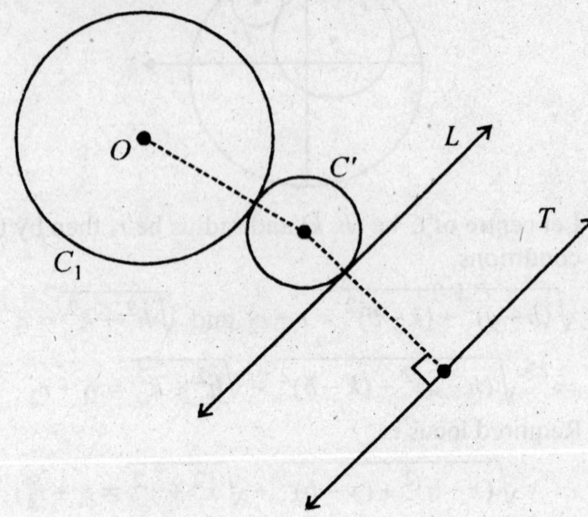

3. (c) Since S is equidistant form A and line BD, it traces a parabola. Clearly, AC is the axis, $A(1, 1)$ is the focus and $T_1\left(\frac{1}{2},\frac{1}{2}\right)$ is the vertex of parabola.

$AT_1 = \frac{1}{\sqrt{2}}$.

T_2T_3 = latus rectum of parabola

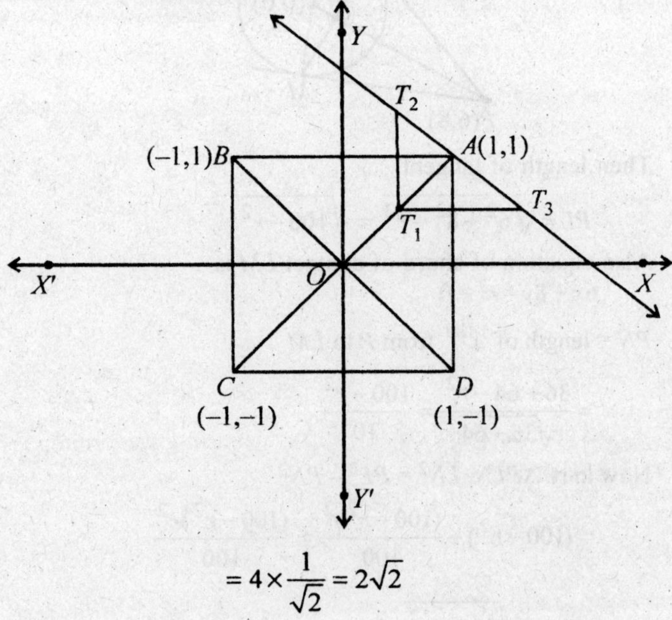

$= 4\times\frac{1}{\sqrt{2}} = 2\sqrt{2}$

$\therefore \quad$ Area $(\Delta T_1T_2T_3) = \frac{1}{2}\times\frac{1}{\sqrt{2}}\times2\sqrt{2} = \frac{1}{2} = 1$ sq. units

4. (d) Slope of $CD = \frac{1}{\sqrt{3}}$

$\therefore$ Parametric equation of CD is

$$\frac{x-\frac{3\sqrt{3}}{2}}{\frac{\sqrt{3}}{2}}=\frac{y-\frac{3}{2}}{\frac{1}{2}}=\pm 1$$

$\therefore$ Two possible coordinates of C are

$$\left(\frac{\sqrt{3}}{2}+\frac{3\sqrt{3}}{2},\frac{1}{2}+\frac{3}{2}\right) \text{ or } \left(\frac{-\sqrt{3}}{2}+\frac{3\sqrt{3}}{2},-\frac{1}{2}+\frac{3}{2}\right)$$

i.e. $(2\sqrt{3},2)$ or $(\sqrt{3},1)$

As (0, 0) and C lie on the same side of PQ

$\therefore$ $(\sqrt{3},1)$ should be the coordinates of C.

NOTE THIS STEP : Remember (x_1,y_1) and (x_2,y_2) lie on the same or opposite side of a line ax + by + c = 0

according as $\frac{ax_1+by_1+c}{ax_2+by_2+c}>0$ or <0.

$\therefore$ Equation of the circle is

$(x-\sqrt{3})^2+(y-1)^2=1$

5. (a) ΔPQR is an equilateral triangle, the incentre C must coincide with centriod of ΔPQR and D, E, F must concide with the mid points of sides PQ, QR and RP respectively.

Also $\angle CPD = 30° \Rightarrow PD = \sqrt{3}$

Writing the equation of side PQ in symmetric form we get,

$$\frac{x-\frac{3\sqrt{3}}{2}}{-\frac{1}{2}}=\frac{y-\frac{3}{2}}{\frac{\sqrt{3}}{2}}=\mp\sqrt{3}$$

$$\therefore \text{ Coordinates of P} = \left(\frac{\sqrt{3}}{2}+\frac{3\sqrt{3}}{2},\frac{-3}{2}+\frac{3}{2}\right)$$

$$=(2\sqrt{3},\ 0) \text{ and}$$

$$\text{coordinates of Q} = \left(\frac{-\sqrt{3}}{2}+\frac{3\sqrt{3}}{2},\frac{3}{2}+\frac{3}{2}\right)=(\sqrt{3},3)$$

Let coordinates of R be (α,β), then using the formula for centriod of Δ we get

$\frac{\sqrt{3}+2\sqrt{3}+\alpha}{3}=\sqrt{3}$ and $\frac{3+0+\beta}{3}=1$

$\Rightarrow \alpha=0$ and $\beta=0$

$\therefore$ Coordinates of R = (0, 0)

Now coordinates of E = mid point of QR = $\left(\frac{\sqrt{3}}{2},\frac{3}{2}\right)$

and coordinates of F = mid point of PR = $(\sqrt{3},0)$

6. (d) Equation of side QR is $y=\sqrt{3}x$ and equation of side RP is $y=0$

Paragraph 3

Given the implicit function $y^3-3y+x=0$

For $x\in(-\infty,-2)\cup(2,\infty)$ it is y = f (x) real valued differentiable function.

and for $x\in(-2,2)$ it is y = g(x) real valued differentiable function.

7. (a) Equation of tangent PT to the circle $x^2+y^2=4$ at the point $P(\sqrt{3},1)$ is $x\sqrt{3}+y=4$.

Let the line L, perpendicular to tangent PT be

$x-y\sqrt{3}+\lambda=0$

As it is tangent to the circle $(x-3)^2+y^2=1$

$\therefore$ length of perpendicular from centre of circle to the tangent = radius of circle.

$\Rightarrow \left|\frac{3+\lambda}{2}\right|=1 \Rightarrow \lambda=-1$ or -5

$\therefore$ Equation of L can be $x-\sqrt{3}y=1$ or $x-\sqrt{3}y=5$

8. (d)

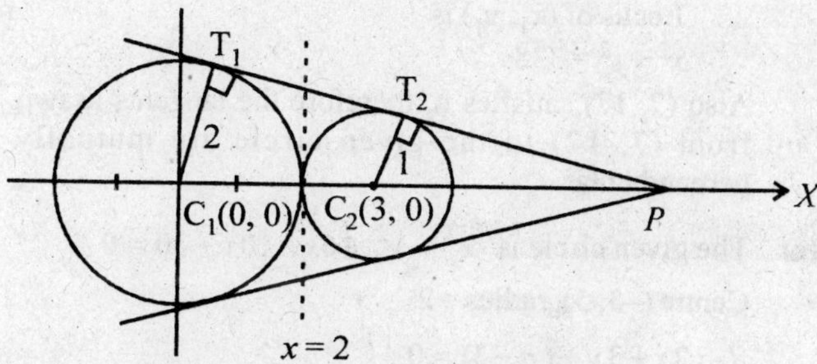

From the figure it is clear that the intersection point of two direct common tangents lies on x-axis.

Also $\Delta PT_1C_1 \sim \Delta PT_2C_2$

$\Rightarrow PC_1 : PC_2 = 2:1$

or P divides C_1C_2 in the ratio 2 : 1 externally

$\therefore$ Coordinates of P are (6, 0)

Let the equation of tangent through P be

$y=m(x-6)$

As it touches $x^2+y^2=4$

$$\therefore \left|\frac{6m}{\sqrt{m^2+1}}\right|=2 \Rightarrow 36m^2=4(m^2+1)$$

$$\Rightarrow m=\pm\frac{1}{2\sqrt{2}}$$

$\therefore$ Equations of common tangents are

$$y=\pm\frac{1}{2\sqrt{2}}(x-6)$$

Also $x=2$ is the common tangent to the two circles.

(H) ASSERTION & REASON TYPE QUESTIONS :

1. (a) Equation of director circle of the given circle $x^2+y^2=169$ is $x^2+y^2=2\times 169=338$.

We know from every point on director circle, the tangents drawn to given circle are perpendicular to each other.

Here (17, 7) lies on director circle.

$\therefore$ The tangent from (17, 7) to given circle are mutually perpendicular.

ALTERNATE SOLUTION

Let (x_1, y_1) be the point from which the tangents drawn to the circle $x^2+y^2=169$ are mutually perpendicular.

Then combined equation of tangents is given by

$$SS_1=T^2$$

$$\Rightarrow (x^2+y^2+169)(x_1^2+y_1^2-169)=(xx_1+yy_1-169)^2$$

As lines represented by this equation are mutually perpendicular.

$\therefore$ Coeff. of x^2 + Coeff. of $y^2=0$

$$x_1^2+y_1^2-169-x_1^2+x_1^2+y_1^2-169-y_1^2=0$$

$$\Rightarrow x_1^2+y_1^2=338$$

$\therefore$ Locus of (x_1, y_1) is

$$x^2+y^2=338$$

Also (7, 17) satisfies it, therefore the tangents drawn from (7, 17) to the given circle are mutually perpendicular.

2. (c) The given circle is $x^2+y^2+6x-10y+30=0$

Centre (–3, 5), radius = 2

$L_1: 2x+3y+(p-3)=0$;

$L_2: 2x+3y+p+3=0$

Clearly $L_1 \parallel L_2$

Distance between L_1 and L_2

$$=\left|\frac{p+3-p+3}{\sqrt{2^2+3^2}}\right|=\frac{6}{\sqrt{13}}<2$$

$\Rightarrow$ If one line is a chord of the given circle, other line may or may not the diameter of the circle.

$\therefore$ Statement 1 is true and statement 2 is false.

(I) INTEGER VALUE CORRECT TYPE :

1. (8) Let r be the radius of required circle.

Clearly, in ΔC_1CC_2, $C_1C=C_2C=r+1$

and P is mid point of C_1C_2

$\therefore$ $CP \perp C_1C_2$

Also $PM \perp CC_1$

Now $\Delta PMC_1 \sim \Delta CPC_1$ (by AA similarity)

$$\therefore \frac{MC_1}{PC_1}=\frac{PC_1}{CC_1}$$

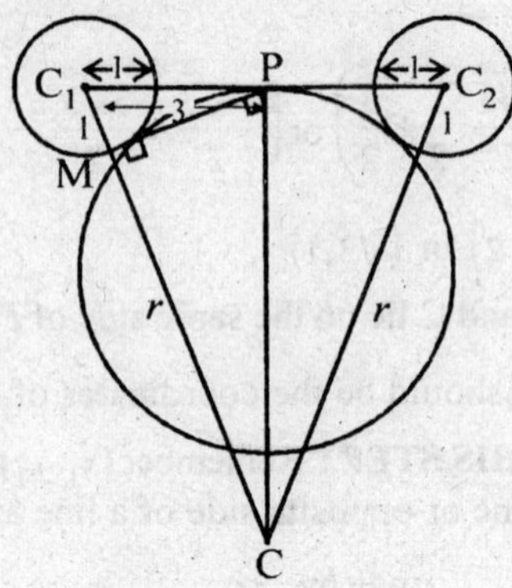

$$\Rightarrow \frac{1}{3}=\frac{3}{r+1} \Rightarrow r+1=9 \Rightarrow r=8.$$

2. (2)

The smaller region of circle is the region given by

$x^2+y^2 \le 6$ ...(1)

and $2x-3y \ge 1$ ...(2)

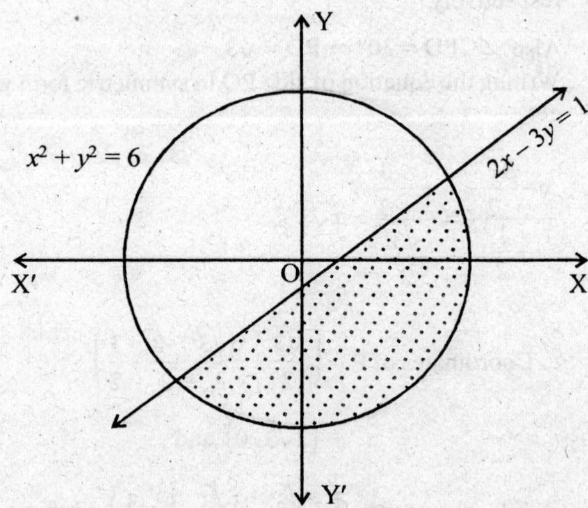

We observe that only two points $\left(2, \frac{3}{4}\right)$ and $\left(\frac{1}{4}, -\frac{1}{4}\right)$ satisfy both the inequations (1) and (2)

$\therefore$ 2 points in S lie inside the smaller part.

Section-B JEE Main/ AIEEE

1. (c) Equation of circle $x^2+y^2=1=(1)^2$
$\Rightarrow x^2+y^2=(y-mx)^2 \Rightarrow x^2=m^2x^2-2\,mxy;$
$\Rightarrow x^2(1-m^2)+2mxy=0$. Which represents the pair of lines between which the angle is 45°.

$$\tan 45=\pm\frac{2\sqrt{m^2-0}}{1-m^2}=\frac{\pm 2m}{1-m^2};$$

$\Rightarrow 1-m^2=\pm 2m \Rightarrow m^2\pm 2m-1=0$

$$\Rightarrow m=\frac{-2\pm\sqrt{4+4}}{2}=\frac{-2\pm 2\sqrt{2}}{2}=-1\pm\sqrt{2}.$$

2. (a) For any point P (x, y) in the given circle,

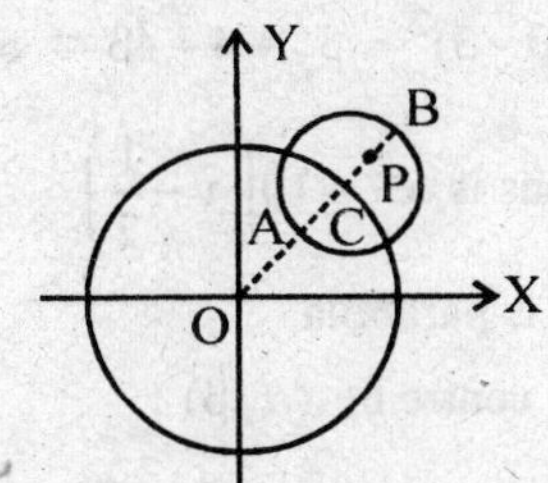

we should have

$OA\le OP\le OB \Rightarrow (5-3)\le\sqrt{x^2+y^2}\le 5+3$

$\Rightarrow 4\le x^2+y^2\le 64$

3. (b) Let the required circle be
$x^2+y^2+2gx+2fy+c=0$
Since it passes through (0, 0) and (1, 0)

$\Rightarrow c=0$ and $g=-\frac{1}{2}$

Points (0, 0) and (1, 0) lie inside the circle $x^2+y^2=9$, so two circles touch internally
$\Rightarrow c_1c_2=r_1-r_2$

$\therefore \sqrt{g^2+f^2}=3-\sqrt{g^2+f^2}\Rightarrow\sqrt{g^2+f^2}=\frac{3}{2}$

$\Rightarrow f^2=\frac{9}{4}-\frac{1}{4}=2 \qquad \therefore f=\pm\sqrt{2}.$

Hence, the centres of required circle are

$\left(\frac{1}{2},\sqrt{2}\right)$ or $\left(\frac{1}{2},-\sqrt{2}\right)$

4. (c) Let ABC be an equilateral triangle, whose median is AD.

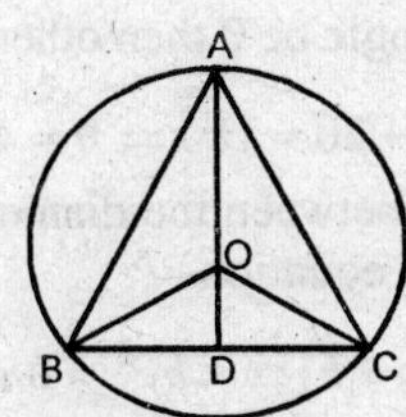

Given $AD=3a$.
In ΔABD, $AB^2=AD^2+BD^2$;
$\Rightarrow x^2=9a^2+(x^2/4)$ where $AB=BC=AC=x$.

$\frac{3}{4}x^2=9a^2\Rightarrow x^2=12a^2.$

In ΔOBD, $OB^2=OD^2+BD^2$

$\Rightarrow r^2=(3a-r)^2+\frac{x^2}{4}$

$\Rightarrow r^2=9a^2-6ar+r^2+3a^2;\Rightarrow 6ar=12a^2$

$\Rightarrow r=2a$

So equation of circle is $x^2+y^2=4a^2$

5. (b) $|r_1-r_2|<C_1C_2$ for intersection

$\Rightarrow r-3<5\Rightarrow r<8$ ...(1)

and $r_1+r_2>C_1C_2$, $r+3>5\Rightarrow r>2$ ...(2)

From (1) and (2), $2<r<8$.

6. (d) $\pi r^2=154\Rightarrow r=7$
For centre on solving equation
$2x-3y=5$ & $3x-4y=7$ we get $x=1, y=-1$
$\therefore$ centre $=(1,-1)$
Equation of circle, $(x-1)^2+(y+1)^2=7^2$
$x^2+y^2-2x+2y=47$

7. (b) Let the variable circle is
$x^2+y^2+2gx+2fy+c=0$ ......(1)
It passes through (a, b)
$\therefore a^2+b^2+2ga+2fb+c=0$ ......(2)

(1) cuts $x^2+y^2=4$ orthogonally
$\therefore 2(g\times 0+f\times 0)=c-4\Rightarrow c=4$

$\therefore$ from (2) $a^2+b^2+2ga+2fb+4=0$
$\therefore$ Locus of centre $(-g,-f)$ is
$a^2+b^2-2ax-2by+4=0$
or $2ax+2by=a^2+b^2+4$

8. (d) Let the variable circle be
$x^2+y^2+2gx+2fy+c=0$ ....(1)
$\therefore p^2+q^2+2gp+2fq+c=0$ ....(2)
Circle (1) touches x-axis,
$\therefore g^2-c=0\Rightarrow c=g^2$. From (2)
$p^2+q^2+2gp+2fq+g^2=0$ ....(3)

Let the other end of diameter through (p, q) be (h, k), then

$$\frac{h+p}{2}=-g \text{ and } \frac{k+q}{2}=-f$$

Put in (3)

$$p^2+q^2+2p\left(-\frac{h+p}{2}\right)+2q\left(-\frac{k+q}{2}\right)+\left(\frac{h+p}{2}\right)^2=0$$

$$\Rightarrow h^2+p^2-2hp-4kq=0$$

$\therefore$ locus of (h, k) is $x^2+p^2-2xp-4yq=0$

$$\Rightarrow (x-p)^2=4qy$$

9. (d) Two diameters are along

$2x+3y+1=0$ and $3x-y-4=0$

solving we get centre $(1,-1)$

circumference $=2\pi r=10\pi$

$\therefore r=5.$

Required circle is, $(x-1)^2+(y+1)^2=5^2$

$$\Rightarrow x^2+y^2-2x+2y-23=0$$

10. (d) Solving $y=x$ and the circle

$x^2+y^2-2x=0$, we get

$x=0, y=0$ and $x=1, y=1$

$\therefore$ Extremities of diameter of the required circle are $(0, 0)$ and $(1, 1)$. Hence, the equation of circle is

$(x-0)(x-1)+(y-0)(y-1)=0$

$$\Rightarrow x^2+y^2-x-y=0$$

11. (b) $s_1=x^2+y^2+2ax+cy+a=0$

$s_2=x^2+y^2-3ax+dy-1=0$

Equation of common chord of circles s_1 and s_2 is given by $s_1-s_2=0$

$$\Rightarrow 5ax+(c-d)y+a+1=0$$

Given that $5x+by-a=0$ passes through P and Q

$\therefore$ The two equations should represent the same line

$$\Rightarrow \frac{a}{1}=\frac{c-d}{b}=\frac{a+1}{-a} \Rightarrow a+1=-a^2$$

$a^2+a+1=0$

No real value of a.

12. (d) Equation of circle with centre $(0, 3)$ and radius 2 is

$x^2+(y-3)^2=4$

Let locus of the variable circle is (α, β)

$\because$ It touches x - axis.

$\therefore$ It's equation is $(x-\alpha)^2+(y+\beta)^2=\beta^2$

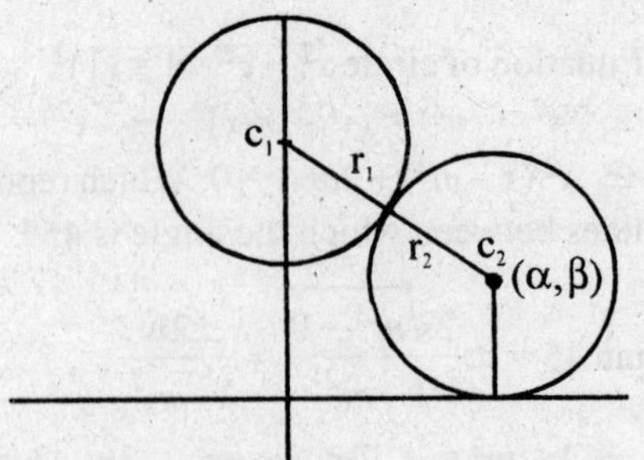

Circle touch externally $\Rightarrow c_1c_2=r_1+r_2$

$\therefore \sqrt{\alpha^2+(\beta-3)^2}=2+\beta$

$\alpha^2+(\beta-3)^2=\beta^2+4+4\beta \Rightarrow \alpha^2=10(\beta-1/2)$

$\therefore$ Locus is $x^2=10\left(y-\frac{1}{2}\right)$

Which is parabola.

13. (d) Let the centre be (α, β)

$\because$ It cuts the circle $x^2+y^2=p^2$ orthogonally

$\therefore$ Using $2g_1g_2+2f_1f_2=c_1+c_2$, we get

$2(-\alpha)\times 0+2(-\beta)\times 0=c_1-p^2$

$c_1=p^2$

Let equation of circle is

$x^2+y^2-2\alpha x-2\beta y+p^2=0$

It passes through

$(a,b) \Rightarrow a^2+b^2-2\alpha a-2\beta b+p^2=0$

$\therefore$ Locus of (α, β) is

$\therefore 2ax+2by-(a^2+b^2+p^2)=0.$

14. (d)

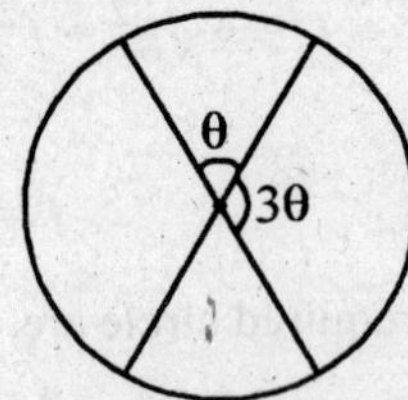

As per question area of one sector = 3 area of another sector

$\Rightarrow$ angle at centre by one sector $=3\times$ angle at centre by another sector

Let one angle be θ then other $=3\theta$

Clearly $\theta+3\theta=180 \Rightarrow \theta=45°$

$\therefore$ Angle between the diameters represented by combined equation

$ax^2+2(a+b)xy+by^2=0$ is $45°$

$\therefore$ Using $\tan\theta = \dfrac{2\sqrt{h^2 - ab}}{a+b}$

we get $\tan 45^\circ = \dfrac{2\sqrt{(a+b)^2 - ab}}{a+b}$

$\Rightarrow 1 = \dfrac{2\sqrt{a^2 + b^2 + ab}}{a+b} \Rightarrow (a+b)^2 = 4\left(a^2 + b^2 + ab\right)$

$\Rightarrow a^2 + b^2 + 2ab = 4a^2 + 4b^2 + 4ab$

$\Rightarrow 3a^2 + 3b^2 + 2ab = 0$

15. (d) Point of intersection of $3x - 4y - 7 = 0$ and $2x - 3y - 5 = 0$ is $(1, -1)$ which is the centre of the circle and radius = 7

$\therefore$ Equation is $(x-1)^2 + (y+1)^2 = 49$

$\Rightarrow x^2 + y^2 - 2x + 2y - 47 = 0$

16. (d) Let $M(h, k)$ be the mid point of chord AB where

$\angle AOB = \dfrac{2\pi}{3}$

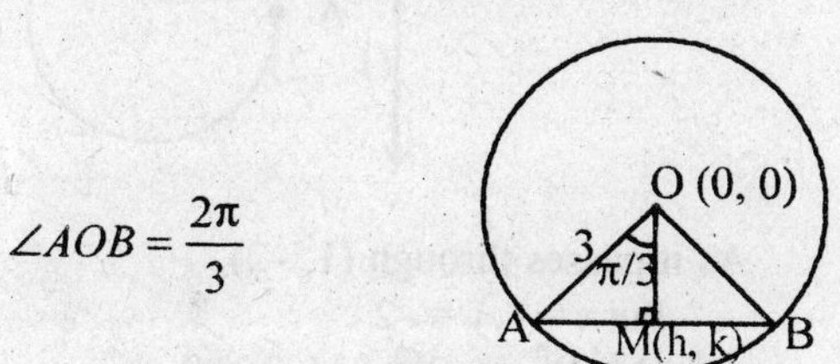

$\therefore \angle AOM = \dfrac{\pi}{3}$. Also $OM = 3\cos\dfrac{\pi}{3} = \dfrac{3}{2}$

$\Rightarrow \sqrt{h^2 + k^2} = \dfrac{3}{2} \Rightarrow h^2 + k^2 = \dfrac{9}{4}$

$\therefore$ Locus of (h, k) is $x^2 + y^2 = \dfrac{9}{4}$

17. (d) Equation of circle whose centre is (h, k) i.e $(x - h)^2 + (y - k)^2 = k^2$

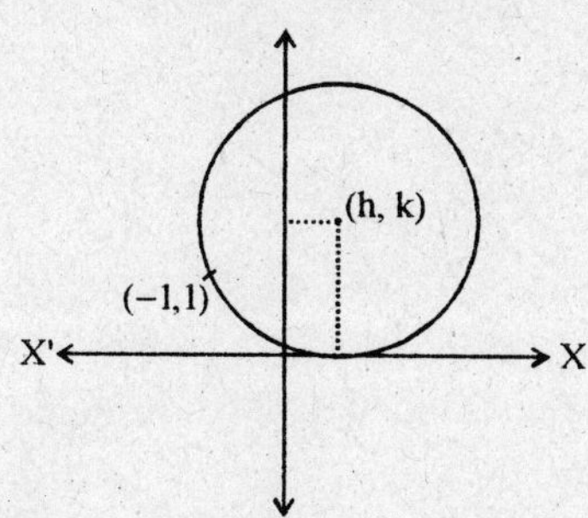

(radius of circle = k because circle is tangent to x-axis)
Equation of circle passing through $(-1, +1)$

$\therefore (-1 - h)^2 + (1 - k)^2 = k^2$

$\Rightarrow 1 + h^2 + 2h + 1 + k^2 - 2k = k^2$

$\Rightarrow h^2 + 2h - 2k + 2 = 0$

$D \geq 0$

$\therefore (2)^2 - 4 \times 1.(-2k + 2) \geq 0$

$\Rightarrow 4 - 4(-2k + 2) \geq 0 \Rightarrow 1 + 2k - 2 \geq 0 \Rightarrow k \geq \dfrac{1}{2}$

18. (c) The given circle is $x^2 + y^2 + 2x + 4y - 3 = 0$

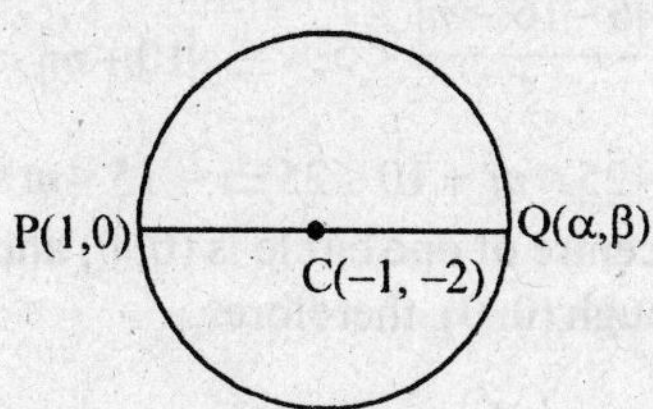

Centre $(-1, -2)$
Let $Q(\alpha, \beta)$ be the point diametrically opposite to the point $P(1, 0)$,

then $\dfrac{1+\alpha}{2} = -1$ and $\dfrac{0+\beta}{2} = -2$

$\Rightarrow \alpha = -3, \beta = -4$

So, Q is $(-3, -4)$

19. (c) Let the centre of the circle be $(h, 2)$

$\therefore$ Equation of circle is

$(x - h)^2 + (y - 2)^2 = 25$ ...(1)

Differentiating with respect to x, we get

$2(x - h) + 2(y - 2)\dfrac{dy}{dx} = 0$

$\Rightarrow x - h = -(y - 2)\dfrac{dy}{dx}$

Substituting in equation (1) we get

$(y-2)^2\left(\dfrac{dy}{dx}\right)^2 + (y-2)^2 = 25$

$\Rightarrow (y-2)^2 (y')^2 = 25 - (y-2)^2$

20. (a) The given circles are

$S_1 \equiv x^2 + y^2 + 3x + 7y + 2p - 5 = 0$ ....(1)

$S_2 \equiv x^2 + y^2 + 2x + 2y - p^2 = 0$ ....(2)

$\therefore$ Equation of common chord PQ is $S_1 - S_2 = 0$

$\Rightarrow L \equiv x + 5y + p^2 + 2p - 5 = 0$

$\Rightarrow$ Equation of circle passing through P and Q is $S_1 + \lambda L = 0$

$\Rightarrow (x^2 + y^2 + 3x + 7y + 2p - 5) + \lambda (x + 5y + p^2 + 2p - 5) = 0$

As it passes through (1, 1), therefore

$\Rightarrow (7 + 2p) + \lambda (2p + p^2 + 1) = 0$

$\Rightarrow \lambda = -\dfrac{2p+7}{(p+1)^2}$

which does not exist for $p = -1$

21. (a) Circle $x^2 + y^2 - 4x - 8y - 5 = 0$

Centre = (2, 4), Radius = $\sqrt{4 + 16 + 5} = 5$

If circle is intersecting line $3x - 4y = m$, at two distinct points.

$\Rightarrow$ length of perpendicular from centre to the line < radius

$$\Rightarrow \frac{|6-16-m|}{5} < 5 \Rightarrow |10+m| < 25$$

$$\Rightarrow -25 < m + 10 < 25 \Rightarrow -35 < m < 15$$

22. (a) As centre of one circle is (0, 0) and other circle passes through (0, 0), therefore

Also $C_1\left(\frac{a}{2}, 0\right)$ $C_2(0, 0)$

$r_1 = \frac{a}{2}$ $r_2 = C$

$C_1C_2 = r_1 - r_2 = \frac{a}{2}$

$\Rightarrow C - \frac{a}{2} = \frac{a}{2}$

$\Rightarrow C = a$

If the two circles touch each other, then they must touch each other internally.

23. (b) Circle whose diametric end points are (1,0) and (0,1) will be of smallest radius. Equation of this smallest circle is $(x-1)(x-0)+(y-0)(y-1)=0$

$\Rightarrow x^2 + y^2 - x - y = 0$

24. (a) Let centre of the circle be (1,h)

[$\because$ circle touches x-axis at (1,0)]

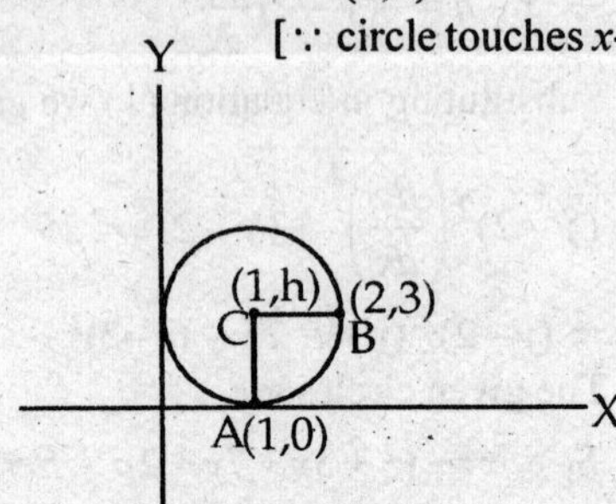

Let the circle passes through the point B (2,3)

$\therefore$ $CA = CB$ (radius)

$\Rightarrow$ $CA^2 = CB^2$

$\Rightarrow$ $(1-1)^2 + (h-0)^2 = (1-2)^2 + (h-3)^2$

$\Rightarrow$ $h^2 = 1 + h^2 + 9 - 6h$

$\Rightarrow$ $h = \frac{10}{6} = \frac{5}{3}$

Thus, diameter is $2h = \frac{10}{3}$.

25. (c) Since circle touches x-axis at (3, 0)

$\therefore$ The equation of circle be

$(x-3)^2 + (y-0)^2 + \lambda y = 0$

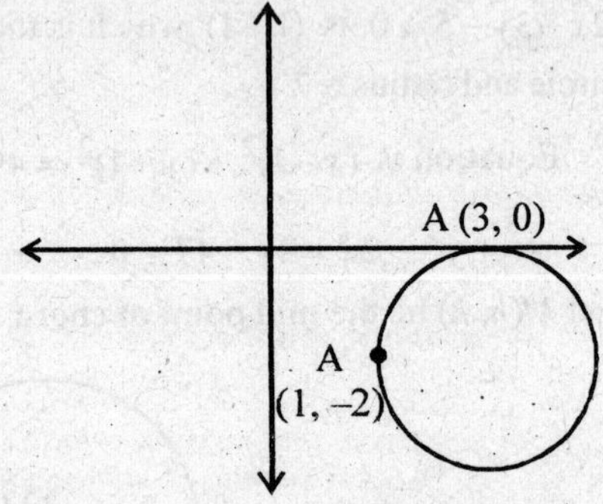

As it passes through (1, –2)

$\therefore$ Put $x = 1, y = -2$

$\Rightarrow$ $(1-3)^2 + (-2)^2 + \lambda(-2) = 0$

$\Rightarrow$ $\lambda = 4$

$\therefore$ equation of circle is $(x-3)^2 + y^2 - 8 = 0$

Now, from the options (5, –2) satisfies equation of circle.

❐ ❐

Chapter 9 Conic Sections

Section-A JEE Advanced/ IIT-JEE

A FILL IN THE BLANKS :

1. The point of intersection of the tangents at the ends of the latus rectum of the parabola $y^2 = 4x$ is....... *(1994 - 2 Marks)*

2. An ellipse has eccentricity $\frac{1}{2}$ and one focus at the point $P\left(\frac{1}{2}, 1\right)$. Its one directrix is the common tangent, nearer to the point P, to the circle $x^2 + y^2 = 1$ and the hyperbola $x^2 - y^2 = 1$. The equation of the ellipse, in the standard form, is........... *(1996 - 2 Marks)*

3. An ellipse has OB as a semi minor axis. F, F' are its foci, and the angle FBF' is a right angle. Then the eccentricity of the ellipse is .... *(1997C - 2 Marks)*

C MCQ's WITH ONE CORRECT ANSWER :

1. The equation $\frac{x^2}{1-r} - \frac{y^2}{1+r} = 1$, $r > 1$ represents *(1981 - 2 Marks)*
   (a) an ellipse (b) a hyperbola
   (c) a circle (d) none of these

2. Each of the four inequalties given below defines a region in the xy plane. One of these four regions does not have the following property. For any two points (x_1, y_1) and (x_2, y_2) in the region, the point $\left(\frac{x_1 + x_2}{2}, \frac{y_1 + y_2}{2}\right)$ is also in the region. The inequality defining this region is *(1981 - 2 Marks)*
   (a) $x^2 + 2y^2 \leq 1$ (b) Max $\{|x|, |y|\} \leq 1$
   (c) $x^2 - y^2 \leq 1$ (d) $y^2 - x \leq 0$

3. The equation $2x^2 + 3y^2 - 8x - 18y + 35 = k$ represents *(1994)*
   (a) no locus if $k > 0$ (b) an ellipse if $k < 0$
   (c) a point if $k = 0$ (d) a hyperbola if $k > 0$

4. Let E be the ellipse $\frac{x^2}{9} + \frac{y^2}{4} = 1$ and C be the circle $x^2 + y^2 = 9$. Let P and Q be the points (1, 2) and (2, 1) respectively. Then *(1994)*
   (a) Q lies inside C but outside E
   (b) Q lies outside both C and E
   (c) P lies inside both C and E
   (d) P lies inside C but outside E

5. Consider a circle with its centre lying on the focus of the parabola $y^2 = 2px$ such that it touches the directrix of the parabola. Then a point of intersection of the circle and parabola is *(1995S)*
   (a) $\left(\frac{p}{2}, p\right)$ or $\left(\frac{p}{2}, -p\right)$ (b) $\left(\frac{p}{2}, -\frac{p}{2}\right)$
   (c) $\left(-\frac{p}{2}, p\right)$ (d) $\left(-\frac{p}{2}, -\frac{p}{2}\right)$

6. The radius of the circle passing through the foci of the ellipse $\frac{x^2}{16} + \frac{y^2}{9} = 1$, and having its centre at (0, 3) is *(1995S)*
   (a) 4 (b) 3
   (c) $\sqrt{\frac{1}{2}}$ (d) $\frac{7}{2}$

7. Let P ($a \sec\theta$, $b \tan\theta$) and Q ($a \sec\phi$, $b \tan\phi$), where $\theta + \phi = \pi/2$, be two points on the hyperbola $\frac{x^2}{a^2} - \frac{y^2}{b^2} = 1$. If (h, k) is the point of intersection of the normals at P and Q, then k is equal to *(1999 - 2 Marks)*
   (a) $\frac{a^2 + b^2}{a}$ (b) $-\left(\frac{a^2 + b^2}{a}\right)$
   (c) $\frac{a^2 + b^2}{b}$ (d) $-\left(\frac{a^2 + b^2}{b}\right)$

8. If $x = 9$ is the chord of contact of the hyperbola $x^2 - y^2 = 9$, then the equation of the corresponding pair of tangents is *(1999 - 2 Marks)*
   (a) $9x^2 - 8y^2 + 18x - 9 = 0$ (b) $9x^2 - 8y^2 - 18x + 9 = 0$
   (c) $9x^2 - 8y^2 - 18x - 9 = 0$ (d) $9x^2 - 8y^2 + 18x + 9 = 0$

9. The curve described parametrically by $x = t^2 + t + 1$, $y = t^2 - t + 1$ represents *(1999 - 2 Marks)*
(a) a pair of straight lines (b) an ellipse
(c) a parabola (d) a hyperbola

10. If $x + y = k$ is normal to $y^2 = 12x$, then k is *(2000S)*
(a) 3 (b) 9
(c) −9 (d) −3

11. If the line $x - 1 = 0$ is the directrix of the parabola $y^2 - kx + 8 = 0$, then one of the values of k is *(2000S)*
(a) 1/8 (b) 8
(c) 4 (d) 1/4

12. The equation of the common tangent touching the circle $(x-3)^2 + y^2 = 9$ and the parabola $y^2 = 4x$ above the x-axis is *(2001S)*
(a) $\sqrt{3}y = 3x + 1$ (b) $\sqrt{3}y = -(x+3)$
(c) $\sqrt{3}y = x + 3$ (d) $\sqrt{3}y = -(3x+1)$

13. The equation of the directrix of the parabola $y^2 + 4y + 4x + 2 = 0$ is *(2001S)*
(a) $x = -1$ (b) $x = 1$
(c) $x = -3/2$ (d) $x = 3/2$

14. If $a > 2b > 0$ then the positive value of m for which $y = mx - b\sqrt{1+m^2}$ is a common tangent to $x^2 + y^2 = b^2$ and $(x - a)^2 + y^2 = b^2$ is *(2002S)*
(a) $\dfrac{2b}{\sqrt{a^2 - 4b^2}}$ (b) $\dfrac{\sqrt{a^2 - 4b^2}}{2b}$
(c) $\dfrac{2b}{a - 2b}$ (d) $\dfrac{b}{a - 2b}$

15. The locus of the mid-point of the line segment joining the focus to a moving point on the parabola $y^2 = 4ax$ is another parabola with directrix *(2002S)*
(a) $x = -a$ (b) $x = -a/2$
(c) $x = 0$ (d) $x = a/2$

16. The equation of the common tangent to the curves $y^2 = 8x$ and $xy = -1$ is *(2002S)*
(a) $3y = 9x + 2$ (b) $y = 2x + 1$
(c) $2y = x + 8$ (d) $y = x + 2$

17. The area of the quadrilateral formed by the tangents at the end points of latus rectum to the ellipse $\dfrac{x^2}{9} + \dfrac{y^2}{5} = 1$, is *(2003S)*
(a) 27/4 sq. units (b) 9 sq. units
(c) 27/2 sq. units (d) 27 sq. units

18. The focal chord to $y^2 = 16x$ is tangent to $(x - 6)^2 + y^2 = 2$, then the possible values of the slope of this chord, are *(2003S)*
(a) {−1, 1} (b) {−2, 2}
(c) {−2, −1/2} (d) {2, −1/2}

19. For hyperbola $\dfrac{x^2}{\cos^2\alpha} - \dfrac{y^2}{\sin^2\alpha} = 1$ which of the following remains constant with change in 'α' *(2003S)*
(a) abscissae of vertices (b) abscissae of foci
(c) eccentricity (d) directrix

20. If tangents are drawn to the ellipse $x^2 + 2y^2 = 2$, then the locus of the mid-point of the intercept made by the tangents between the coordinate axes is *(2004S)*
(a) $\dfrac{1}{2x^2} + \dfrac{1}{4y^2} = 1$ (b) $\dfrac{1}{4x^2} + \dfrac{1}{2y^2} = 1$
(c) $\dfrac{x^2}{2} + \dfrac{y^2}{4} = 1$ (d) $\dfrac{x^2}{4} + \dfrac{y^2}{2} = 1$

21. The angle between the tangents drawn from the point (1, 4) to the parabola $y^2 = 4x$ is *(2004S)*
(a) π/6 (b) π/4
(c) π/3 (d) π/2

22. If the line $2x + \sqrt{6}\,y = 2$ touches the hyperbola $x^2 - 2y^2 = 4$, then the point of contact is *(2004S)*
(a) $(-2, \sqrt{6})$ (b) $(-5, 2\sqrt{6})$
(c) $\left(\dfrac{1}{2}, \dfrac{1}{\sqrt{6}}\right)$ (d) $\left(4, -\sqrt{6}\right)$

23. The minimum area of triangle formed by the tangent to the $\dfrac{x^2}{a^2} + \dfrac{y^2}{b^2} = 1$ & coordinate axes is *(2005S)*
(a) ab sq. units (b) $\dfrac{a^2 + b^2}{2}$ sq. units
(c) $\dfrac{(a+b)^2}{2}$ sq. units (d) $\dfrac{a^2 + ab + b^2}{3}$ sq. units

24. Tangent to the curve $y = x^2 + 6$ at a point (1, 7) touches the circle $x^2 + y^2 + 16x + 12y + c = 0$ at a point Q. Then the coordinates of Q are *(2005S)*
(a) (−6, −11) (b) (−9, −13)
(c) (−10, −15) (d) (−6, −7)

25. The axis of a parabola is along the line $y = x$ and the distances of its vertex and focus from origin are $\sqrt{2}$ and $2\sqrt{2}$ respectively. If vertex and focus both lie in the first quadrant, then the equation of the parabola is *(2006 - 3M, –1)*
(a) $(x + y)^2 = (x - y - 2)$ (b) $(x - y)^2 = (x + y - 2)$
(c) $(x - y)^2 = 4(x + y - 2)$ (d) $(x - y)^2 = 8(x + y - 2)$

26. A hyperbola, having the transverse axis of length 2 sin θ, is confocal with the ellipse $3x^2 + 4y^2 = 12$. Then its equation is *(2007 - 3 marks)*
(a) $x^2\operatorname{cosec}^2\theta - y^2\sec^2\theta = 1$ (b) $x^2\sec^2\theta - y^2\operatorname{cosec}^2\theta = 1$
(c) $x^2\sin^2\theta - y^2\cos^2\theta = 1$ (d) $x^2\cos^2\theta - y^2\sin^2\theta = 1$

27. Let a and b be non-zero real numbers. Then, the equation $(ax^2 + by^2 + c)(x^2 - 5xy + 6y^2) = 0$ represents *(2008)*
(a) four straight lines, when $c = 0$ and a, b are of the same sign.
(b) two straight lines and a circle, when $a = b$, and c is of sign opposite to that of a
(c) two straight lines and a hyperbola, when a and b are of the same sign and c is of sign opposite to that of a
(d) a circle and an ellipse, when a and b are of the same sign and c is of sign opposite to that of a

28. Consider a branch of the hyperbola

$$x^2 - 2y^2 - 2\sqrt{2}x - 4\sqrt{2}y - 6 = 0$$

with vertex at the point A. Let B be one of the end points of its latus rectum. If C is the focus of the hyperbola nearest to the point A, then the area of the triangle ABC is ***(2008)***

(a) $1-\sqrt{\dfrac{2}{3}}$ (b) $\sqrt{\dfrac{3}{2}}-1$

(c) $1+\sqrt{\dfrac{2}{3}}$ (d) $\sqrt{\dfrac{3}{2}}+1$

29. The line passing through the extremity A of the major axis and extremity B of the minor axis of the ellipse

$$x^2 + 9y^2 = 9$$

meets its auxiliary circle at the point M. Then the area of the triangle with vertices at A, M and the origin O is ***(2009)***

(a) $\dfrac{31}{10}$ (b) $\dfrac{29}{10}$

(c) $\dfrac{21}{10}$ (d) $\dfrac{27}{10}$

30. The normal at a point P on the ellipse $x^2 + 4y^2 = 16$ meets the x - axis at Q. If M is the mid point of the line segment PQ, then the locus of M intersects the latus rectums of the given ellipse at the points ***(2009)***

(a) $\left(\pm\dfrac{3\sqrt{5}}{2}, \pm\dfrac{2}{7}\right)$ (b) $\left(\pm\dfrac{3\sqrt{5}}{2}, \pm\sqrt{\dfrac{19}{4}}\right)$

(c) $\left(\pm 2\sqrt{3}, \pm\dfrac{1}{7}\right)$ (d) $\left(\pm 2\sqrt{3}, \pm\dfrac{4\sqrt{3}}{7}\right)$

31. The locus of the orthocentre of the triangle formed by the lines

$$(1+p)x - py + p(1+p) = 0,$$
$$(1+q)x - qy + q(1+q) = 0,$$

and $y = 0$, where $p \neq q$, is ***(2009)***

(a) a hyperbola (b) a parabola

(c) an ellipse (d) a straight line

32. Let $P(6, 3)$ be a point on the hyperbola $\dfrac{x^2}{a^2} - \dfrac{y^2}{b^2} = 1$. If the normal at the point P intersects the x-axis at $(9, 0)$, then the eccentricity of the hyperbola is ***(2011)***

(a) $\sqrt{\dfrac{5}{2}}$ (b) $\sqrt{\dfrac{3}{2}}$ (c) $\sqrt{2}$ (d) $\sqrt{3}$

33. Let (x, y) be any point on the parabola $y^2 = 4x$. Let P be the point that divides the line segment from $(0, 0)$ to (x, y) in the ratio 1 : 3. Then the locus of P is ***(2011)***

(a) $x^2 = y$ (b) $y^2 = 2x$

(c) $y^2 = x$ (d) $x^2 = 2y$

34. The ellipse $E_1 : \dfrac{x^2}{9} + \dfrac{y^2}{4} = 1$ is inscribed in a rectangle R whose sides are parallel to the coordinate axes. Another ellipse E_2 passing through the point $(0, 4)$ circumscribes the rectangle R. The eccentricity of the ellipse E_2 is ***(2012)***

(a) $\dfrac{\sqrt{2}}{2}$ (b) $\dfrac{\sqrt{3}}{2}$ (c) $\dfrac{1}{2}$ (d) $\dfrac{3}{4}$

D MCQ's WITH ONE OR MORE THAN ONE CORRECT :

1. The number of values of c such that the straight line $y = 4x + c$ touches the curve $(x^2/4) + y^2 = 1$ is ***(1998 - 2 Marks)***

(a) 0 (b) 1

(c) 2 (d) infinite.

2. If $P = (x, y)$, $F_1 = (3, 0)$, $F_2 = (-3, 0)$ and $16x^2 + 25y^2 = 400$, then $PF_1 + PF_2$ equals ***(1998 - 2 Marks)***

(a) 8 (b) 6

(c) 10 (d) 12

3. On the ellipse $4x^2 + 9y^2 = 1$, the points at which the tangents are parallel to the line $8x = 9y$ are ***(1999 - 3 Marks)***

(a) $\left(\dfrac{2}{5}, \dfrac{1}{5}\right)$ (b) $\left(-\dfrac{2}{5}, \dfrac{1}{5}\right)$

(c) $\left(-\dfrac{2}{5}, -\dfrac{1}{5}\right)$ (d) $\left(\dfrac{2}{5}, -\dfrac{1}{5}\right)$

4. The equations of the common tangents to the parabola $y = x^2$ and $y = -(x-2)^2$ is/are ***(2006 - 5M, –1)***

(a) $y = 4(x-1)$ (b) $y = 0$

(c) $y = -4(x-1)$ (d) $y = -30x - 50$

5. Let a hyperbola passes through the focus of the ellipse $\dfrac{x^2}{25} + \dfrac{y^2}{16} = 1$. The transverse and conjugate axes of this hyperbola coincide with the major and minor axes of the given ellipse, also the product of eccentricities of given ellipse and hyperbola is 1, then ***(2006 - 5M, –1)***

(a) the equation of hyperbola is $\dfrac{x^2}{9} - \dfrac{y^2}{16} = 1$

(b) the equation of hyperbola is $\dfrac{x^2}{9} - \dfrac{y^2}{25} = 1$

(c) focus of hyperbola is $(5, 0)$

(d) vertex of hyperbola is $(5\sqrt{3}, 0)$

6. Let $P(x_1, y_1)$ and $Q(x_2, y_2)$, $y_1 < 0$, $y_2 < 0$, be the end points of the latus rectum of the ellipse $x^2 + 4y^2 = 4$. The equations of parabolas with latus rectum PQ are ***(2008)***

(a) $x^2 + 2\sqrt{3}y = 3 + \sqrt{3}$ (b) $x^2 - 2\sqrt{3}y = 3 + \sqrt{3}$

(c) $x^2 + 2\sqrt{3}y = 3 - \sqrt{3}$ (d) $x^2 - 2\sqrt{3}y = 3 - \sqrt{3}$

7. In a triangle ABC with fixed base BC, the vertex A moves such that

$$\cos B+\cos C=4\sin^2\frac{A}{2}.$$

If a, b and c denote the lengths of the sides of the triangle opposite to the angles A, B and C, respectively, then *(2009)*
(a) $b+c=4a$
(b) $b+c=2a$
(c) locus of point A is an ellipse
(d) locus of point A is a pair of straight lines

8. The tangent PT and the normal PN to the parabola $y^2=4ax$ at a point P on it meet its axis at points T and N, respectively. The locus of the centroid of the triangle PTN is a parabola whose *(2009)*
(a) vertex is $\left(\frac{2a}{3},0\right)$ (b) directrix is $x=0$
(c) latus rectum is $\frac{2a}{3}$ (d) focus is $(a, 0)$

9. An ellipse intersects the hyperbola $2x^2-2y^2=1$ orthogonally. The eccentricity of the ellipse is reciprocal of that of the hyperbola. If the axes of the ellipse are along the coordinate axes, then *(2009)*
(a) equation of ellipse is $x^2+2y^2=2$
(b) the foci of ellipse are $(\pm1,0)$
(c) equation of ellipse is $x^2+2y^2=4$
(d) the foci of ellipse are $(\pm\sqrt{2},0)$

10. Let A and B be two distinct points on the parabola $y^2=4x$. If the axis of the parabola touches a circle of radius r having AB as its diameter, then the slope of the line joining A and B can be *(2010)*
(a) $-\frac{1}{r}$ (b) $\frac{1}{r}$
(c) $\frac{2}{r}$ (d) $-\frac{2}{r}$

11. Let the eccentricity of the hyperbola $\frac{x^2}{a^2}-\frac{y^2}{b^2}=1$ be reciprocal to that of the ellipse $x^2+4y^2=4$. If the hyperbola passes through a focus of the ellipse, then *(2011)*
(a) the equation of the hyperbola is $\frac{x^2}{3}-\frac{y^2}{2}=1$
(b) a focus of the hyperbola is $(2, 0)$
(c) the eccentricity of the hyperbola is $\sqrt{\frac{5}{3}}$
(d) the equation of the hyperbola is $x^2-3y^2=3$

12. Let L be a normal to the parabola $y^2=4x$. If L passes through the point $(9, 6)$, then L is given by *(2011)*
(a) $y-x+3=0$ (b) $y+3x-33=0$
(c) $y+x-15=0$ (d) $y-2x+12=0$

13. Tangents are drawn to the hyperbola $\frac{x^2}{9}-\frac{y^2}{4}=1$, parallel to the straight line $2x-y=1$. The points of contact of the tangents on the hyperbola are *(2012)*
(a) $\left(\frac{9}{2\sqrt{2}},\frac{1}{\sqrt{2}}\right)$ (b) $\left(-\frac{9}{2\sqrt{2}},-\frac{1}{\sqrt{2}}\right)$
(c) $\left(3\sqrt{3},-2\sqrt{2}\right)$ (d) $\left(-3\sqrt{3},2\sqrt{2}\right)$

E SUBJECTIVE PROBLEMS :

1. Suppose that the normals drawn at three different points on the parabola $y^2=4x$ pass through the point (h, k). Show that $h>2$. *(1981 - 4 Marks)*

2. A is a point on the parabola $y^2=4ax$. The normal at A cuts the parabola again at point B. If AB subtends a right angle at the vertex of the parabola. find the slope of AB. *(1982 - 5 Marks)*

3. Find the equation of the normal to the curve $x^2=4y$ which passes through the point $(1, 2)$. *(1984 - 4 Marks)*

4. Three normals are drawn from the point (c, 0) to the curve $y^2=x$. Show that c must be greater than 1/2. One normal is always the x-axis. Find c for which the other two normals are perpendicular to each other. *(1991 - 4 Marks)*

5. Through the vertex O of parabola $y^2=4x$, chords OP and OQ are drawn at right angles to one another. Show that for all positions of P, PQ cuts the axis of the parabola at a fixed point. Also find the locus of the middle point of PQ. *(1994 - 4 Marks)*

6. Show that the locus of a point that divides a chord of slope 2 of the parabola $y^2=4x$ internally in the ratio 1: 2 is a parabola. Find the vertex of this parabola. *(1995 - 5 Marks)*

7. Let 'd' be the perpendicular distance from the centre of the ellipse $\frac{x^2}{a^2}+\frac{y^2}{b^2}=1$ to the tangent drawn at a point P on the ellipse. If F_1 and F_2 are the two foci of the ellipse, then show that $(PF_1-PF_2)^2=4a^2\left(1-\frac{b^2}{d^2}\right)$. *(1995 - 5 Marks)*

8. Points A, B and C lie on the parabola $y^2=4ax$. The tangents to the parabola at A, B and C, taken in pairs, intersect at points P, Q and R. Determine the ratio of the areas of the triangles ABC and PQR. *(1996 - 3 Marks)*

9. From a point A common tangents are drawn to the circle $x^2+y^2=a^2/2$ and parabola $y^2=4ax$. Find the area of the quadrilateral formed by the common tangents, the chord of contact of the circle and the chord of contact of the parabola. *(1996 - 2 Marks)*

10. A tangent to the ellipse $x^2+4y^2=4$ meets the ellipse $x^2+2y^2=6$ at P and Q. Prove that the tangents at P and Q of the ellipse $x^2+2y^2=6$ are at right angles. *(1997 - 5 Marks)*

11. The angle between a pair of tangents drawn from a point P to the parabola $y^2=4ax$ is 45°. Show that the locus of the point P is a hyperbola. *(1998 - 8 Marks)*

12. Consider the family of circles $x^2 + y^2 = r^2, 2 < r < 5$. If in the first quadrant, the common taingent to a circle of this family and the ellipse $4x^2 + 25y^2 = 100$ meets the co-ordinate axes at A and B, then find the equation of the locus of the mid-point of AB. ***(1999 - 10 Marks)***

13. Find the co-ordinates of all the points P on the ellipse $\frac{x^2}{a^2}+\frac{y^2}{b^2}=1$, for which the area of the triangle PON is maximum, where O denotes the origin and N, the foot of the perpendicular from O to the tangent at P. ***(1999 - 10 Marks)***

14. Let ABC be an equilateral triangle inscribed in the circle $x^2 + y^2 = a^2$. Suppose perpendiculars from A, B, C to the major axis of the ellipse $\frac{x^2}{a^2}+\frac{y^2}{b^2}=1, (a > b)$ meets the ellipse respectively, at P, Q, R. so that P, Q, R lie on the same side of the major axis as A, B, C respectively. Prove that the normals to the ellipse drawn at the points P, Q and R are concurrent. ***(2000 - 7 Marks)***

15. Let C_1 and C_2 be respectively, the parabolas $x^2 = y - 1$ and $y^2 = x - 1$. Let P be any point on C_1 and Q be any point on C_2. Let P_1 and Q_1 be the reflections of P and Q, respectively, with respect to the line $y = x$. Prove that P_1 lies on C_2, Q_1 lies on C_1 and $PQ \geq \min\{PP_1, QQ_1\}$. Hence or otherwise determine points P_0 and Q_0 on the parabolas C_1 and C_2 respectively such that $P_0Q_0 \leq PQ$ for all pairs of points (P,Q) with P on C_1 and Q on C_2. ***(2000 - 10 Marks)***

16. Let P be a point on the ellipse $\frac{x^2}{a^2}+\frac{y^2}{b^2}=1, 0<b<a$. Let the line parallel to y–axis passing through P meet the circle $x^2 + y^2 = a^2$ at the point Q such that P and Q are on the same side of x–axis. For two positive real numbers r and s, find the locus of the point R on PQ such that $PR : RQ = r : s$ as P varies over the ellipse. ***(2001 - 4 Marks)***

17. Prove that, in an ellipse, the perpendicular from a focus upon any tangent and the line joining the centre of the ellipse to the point of contact meet on the corresponding directrix. ***(2002 - 5 Marks)***

18. Normals are drawn from the point P with slopes m_1, m_2, m_3 to the parabola $y^2 = 4x$. If locus of P with $m_1 m_2 = \alpha$ is a part of the parabola itself then find α. ***(2003 - 4 Marks)***

19. Tangent is drawn to parabola $y^2 - 2y - 4x + 5 = 0$ at a point P which cuts the directrix at the point Q. A point R is such that it divides QP externally in the ratio $1/2 : 1$. Find the locus of point R. ***(2004 - 4 Marks)***

20. Tangents are drawn from any point on the hyperbola $\frac{x^2}{9}-\frac{y^2}{4}=1$ to the circle $x^2 + y^2 = 9$. Find the locus of mid-point of the chord of contact. ***(2005 - 4 Marks)***

21. Find the equation of the common tangent in 1st quadrant to the circle $x^2 + y^2 = 16$ and the ellipse $\frac{x^2}{25}+\frac{y^2}{4}=1$. Also find the length of the intercept of the tangent between the coordinate axes. ***(2005 - 4 Marks)***

(F) MATCH THE FOLLOWING :

DIRECTION (for Qs. 1-3) : *Each question contains statements given in two columns, which have to be matched. The statements in Column-I are labelled A, B, C and D, while the statements in Column-II are labelled p, q, r, s and t. Any given statement in Column-I can have correct matching with ONE OR MORE statement(s) in Column-II. The appropriate bubbles corresponding to the answers to these questions have to be darkened as illustrated in the following example :*
If the correct matches are A-p, s and t; B-q and r; C-p and q; and D-s then the correct darkening of bubbles will look like the given.

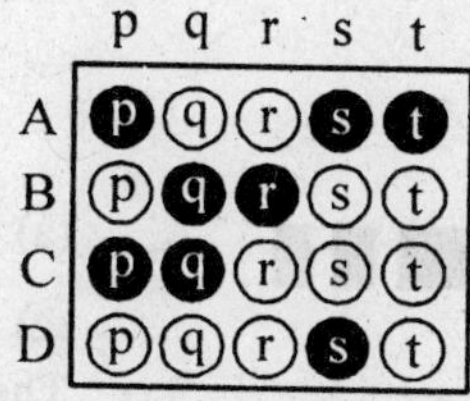

1. Match the following : (3, 0) is the pt. from which three normals are drawn to the parabola $y^2 = 4x$ which meet the parabola in the points P, Q and R. Then ***(2006 - 6M)***

Column I	***Column II***
(A) Area of ΔPQR	(p) 2
(B) Radius of circumcircle of ΔPQR	(q) 5/2
(C) Centroid of ΔPQR	(r) (5/2, 0)
(D) Circumcentre of ΔPQR	(s) (2/3, 0)

2. Match the statements in Column I with the properties in Column II and indicate your answer by darkening the appropriate bubbles in the 4 × 4 matrix given in the ORS. **(2007 -6 marks)**

Column I	*Column II*
(A) Two intersecting circles	(p) have a common tangent
(B) Two mutually external circles	(q) have a common normal
(C) Two circles, one strictly inside the other	(r) do not have a common tangent
(D) Two branches of a hyperbola	(s) do not have a common normal

(2009)

3. Match the conics in **Column I** with the statements/expressions in **Column II.**

Column I	*Column II*
(A) Circle	(p) The locus of the point (h,k) for which the line $hx + ky = 1$ touches the circle $x^2 + y^2 = 4$
(B) Parabola	(q) Points z in the complex plane satisfying $\|z+2\|-\|z-2\|=\pm 3$
(C) Ellipse	(r) Points of the conic have parametric representation $x=\sqrt{3}\left(\frac{1-t^2}{1+t^2}\right),\ y=\frac{2t}{1+t^2}$
(D) Hyperbola	(s) The eccentricity of the conic lies in the interval $1 \le x < \infty$
	(t) Points z in the complex plane satisfying $\text{Re }(z+1)^2 = \|z\|^2 + 1$

DIRECTION (for Qs. 4):

4. A line $L : y = mx + 3$ meets y-axis at $E(0, 3)$ and the arc of the parabola $y^2 = 16x$, $0 \le y \le 6$ at the point $F(x_0, y_0)$. The tangent to the parabola at $F(x_0, y_0)$ intersects the y-axis at $G(0, y_1)$. The slope m of the line L is chosen such that the area of the triangle EFG has a local maximum. **(JEE Adv. 2013)**

Match List I with List II and select the correct answer using the code given below the lists :

List I	**List II**
P. $m=$	1. $\frac{1}{2}$
Q. Maximum area of ΔEFG is	2. 4
R. $y_0 =$	3. 2
S. $y_1 =$	4. 1

Codes:

	P	Q	R	S
(a)	4	1	2	3
(b)	3	4	1	2
(c)	1	3	2	4
(d)	1	3	4	2

G COMPREHENSION BASED QUESTIONS :

Passage 1

Consider the circle $x^2 + y^2 = 9$ and the parabola $y^2 = 8x$. They intersect at P and Q in the first and the fourth quadrants, respectively. Tangents to the curcle at P and Q intersect the x-axis at R and tangents to the parabola at P and Q intersect the x-axis at S. **(2007 -4 marks)**

1. The ratio of the areas of the triangles PQS and PQR is

(a) $1:\sqrt{2}$ (b) $1:2$

(c) $1:4$ (d) $1:8$

2. The radius of the circumcircle of the triangle PRS is **(2007 -4 marks)**

(a) 5 (b) $3\sqrt{3}$

(c) $3\sqrt{2}$ (d) $2\sqrt{3}$

3. The radius of the incircle of the triangle PQR is **(2007 -4 marks)**

(a) 4 (b) 3

(c) $\frac{8}{3}$ (d) 2

Passage 2

The circle $x^2 + y^2 - 8x = 0$ and hyperbola $\frac{x^2}{9} - \frac{y^2}{4} = 1$ intersect at the points A and B. **(2010)**

4. Equation of a common tangent with positive slope to the circle as well as to the hyperbola is

(a) $2x-\sqrt{5}y-20=0$ (b) $2x-\sqrt{5}y+4=0$

(c) $3x-4y+8=0$ (d) $4x-3y+4=0$

5. Equation of the circle with AB as its diameter is

(a) $x^2+y^2-12x+24=0$ (b) $x^2+y^2+12x+24=0$

(c) $x^2+y^2+24x-12=0$ (d) $x^2+y^2-24x-12=0$

Passage 3

Tangents are drawn from the point P(3, 4) to the ellipse $\frac{x^2}{9}+\frac{y^2}{4}=1$ touching the ellipse at points A and B. ***(2010)***

6. The coordinates of A and B are

(a) (3,0) and (0,2)

(b) $\left(-\frac{8}{5},\frac{2\sqrt{161}}{15}\right)$ and $\left(-\frac{9}{5},\frac{8}{5}\right)$

(c) $\left(-\frac{8}{5},\frac{2\sqrt{161}}{15}\right)$ and (0, 2)

(d) (3,0) and $\left(-\frac{9}{5},\frac{8}{5}\right)$

7. The orthocenter of the triangle PAB is

(a) $\left(5,\frac{8}{7}\right)$ (b) $\left(\frac{7}{5},\frac{25}{8}\right)$

(c) $\left(\frac{11}{5},\frac{8}{5}\right)$ (d) $\left(\frac{8}{25},\frac{7}{5}\right)$

8. The equation of the locus of the point whose distances from the point P and the line AB are equal, is

(a) $9x^2+y^2-6xy-54x-62y+241=0$

(b) $x^2+9y^2+6xy-54x+62y-241=0$

(c) $9x^2+9y^2-6xy-54x-62y-241=0$

(d) $x^2+y^2-2xy+27x+31y-120=0$

Let PQ be a focal chord of the parabola $y^2=4ax$. The tangents to the parabola at P and Q meet at a point lying on the line $y=2x+a$, $a>0$.

9. Length of chord PQ is ***(JEE Adv. 2013)***

(a) $7a$ (b) $5a$

(c) $2a$ (d) $3a$

10. If chord PQ subtends an angle θ at the vertex of $y^2=4ax$, then $\tan\theta =$ ***(JEE Adv. 2013)***

(a) $\frac{2}{3}\sqrt{7}$ (b) $\frac{-2}{3}\sqrt{7}$

(c) $\frac{2}{3}\sqrt{5}$ (d) $\frac{-2}{3}\sqrt{5}$

(H) ASSERTION & REASON TYPE QUESTIONS :

1. **STATEMENT-1 :** The curve $y=\frac{-x^2}{2}+x+1$ is symmetric with respect to the line $x=1$. because

STATEMENT-2 : A parabola is symmetric about its axis. ***(2007 -3 marks)***

(a) Statement-1 is True, Statement-2 is True; Statement-2 is a correct explanation for Statement-1

(b) Statement-1 is True, Statement-2 is True; Statement-2 is NOT a correct explanation for Statement-1

(c) Statement-1 is True, Statement-2 is False

(d) Statement-1 is False, Statement-2 is True.

(I) INTEGER VALUE CORRECT TYPE :

1. The line $2x+y=1$ is tangent to the hyperbola $\frac{x^2}{a^2}-\frac{y^2}{b^2}=1$. If this line passes through the point of intersection of the nearest directrix and the x-axis, then the eccentricity of the hyperbola is ***(2010)***

2. Consider the parabola $y^2=8x$. Let Δ_1 be the area of the triangle formed by the end points of its latus rectum and the point $P\left(\frac{1}{2},2\right)$ on the parabola and Δ_2 be the area of the triangle formed by drawing tangents at P and at the end points of the latus rectum. Then $\frac{\Delta_1}{\Delta_2}$ is ***(2011)***

3. Let S be the focus of the parabola $y^2=8x$ and let PQ be the common chord of the circle $x^2+y^2-2x-4y=0$ and the given parabola. The area of the triangle PQS is ***(2012)***

4. A vertical line passing through the point (h, 0) intersects the ellipse $\frac{x^2}{4}+\frac{y^2}{3}=1$ at the points P and Q. Let the tangents to the ellipse at P and Q meet at the point R. If $\Delta(h)$ = area of the triangle PQR, $\Delta_1=\max\limits_{1/2\le h\le 1}\Delta(h)$ and $\Delta_2=\min\limits_{1/2\le h\le 1}\Delta(h)$, then $\frac{8}{\sqrt{5}}\Delta_1-8\Delta_2=$ ***(JEE Adv. 2013)***

Section-B JEE Main/ AIEEE

1. Two common tangents to the circle $x^2+y^2=2a^2$ and parabola $y^2=8ax$ are [2002]

(a) $x=\pm(y+2a)$ (b) $y=\pm(x+2a)$

(c) $x=\pm(y+a)$ (d) $y=\pm(x+a)$

2. The normal at the point $(bt_1^2, 2bt_1)$ on a parabola meets the parabola again in the point $(bt_2^2, 2bt_2)$, then

(a) $t_2=t_1+\frac{2}{t_1}$ (b) $t_2=-t_1-\frac{2}{t_1}$ [2003]

(c) $t_2=-t_1+\frac{2}{t_1}$ (d) $t_2=t_1-\frac{2}{t_1}$

3. The foci of the ellipse $\frac{x^2}{16}+\frac{y^2}{b^2}=1$ and the hyperbola $\frac{x^2}{144}-\frac{y^2}{81}=\frac{1}{25}$ coincide. Then the value of b^2 is

(a) 9 (b) 1 [2003]

(c) 5 (d) 7

4. If $a \neq 0$ and the line $2bx+3cy+4d=0$ passes through the points of intersection of the parabolas $y^2=4ax$ and $x^2=4ay$, then [2004]

(a) $d^2+(3b-2c)^2=0$ (b) $d^2+(3b+2c)^2=0$

(c) $d^2+(2b-3c)^2=0$

(d) $d^2+(2b+3c)^2=0$

5. The eccentricity of an ellipse, with its centre at the origin, is $\frac{1}{2}$. If one of the directrices is $x=4$, then the equation of the ellipse is: [2004]

(a) $4x^2+3y^2=1$ (b) $3x^2+4y^2=12$

(c) $4x^2+3y^2=12$ (d) $3x^2+4y^2=1$

6. Let P be the point $(1, 0)$ and Q a point on the locus $y^2=8x$. The locus of mid point of PQ is [2005]

(a) $y^2-4x+2=0$ (b) $y^2+4x+2=0$

(c) $x^2+4y+2=0$ (d) $x^2-4y+2=0$

7. The locus of a point $P(\alpha, \beta)$ moving under the condition that the line $y=\alpha x+\beta$ is a tangent to the hyperbola $\frac{x^2}{a^2}-\frac{y^2}{b^2}=1$ is [2005]

(a) an ellipse (b) a circle

(c) a parabola (d) a hyperbola

8. An ellipse has OB as semi minor axis, F and F' its focii and the angle FBF' is a right angle. Then the eccentricity of the ellipse is [2005]

(a) $\frac{1}{\sqrt{2}}$ (b) $\frac{1}{2}$

(c) $\frac{1}{4}$ (d) $\frac{1}{\sqrt{3}}$

9. The locus of the vertices of the family of parabolas $y=\frac{a^3x^2}{3}+\frac{a^2x}{2}-2a$ is [2006]

(a) $xy=\frac{105}{64}$ (b) $xy=\frac{3}{4}$

(c) $xy=\frac{35}{16}$ (d) $xy=\frac{64}{105}$

10. In an ellipse, the distance between its foci is 6 and minor axis is 8. Then its eccentricity is [2006]

(a) $\frac{3}{5}$ (b) $\frac{1}{2}$

(c) $\frac{4}{5}$ (d) $\frac{1}{\sqrt{5}}$

11. Angle between the tangents to the curve $y=x^2-5x+6$ at the points (2, 0) and (3, 0) is [2006]

(a) π (b) $\frac{\pi}{2}$

(c) $\frac{\pi}{6}$ (d) $\frac{\pi}{4}$

12. For the Hyperbola $\frac{x^2}{\cos^2\alpha}-\frac{y^2}{\sin^2\alpha}=1$, which of the following remains constant when α varies = ? [2007]

(a) abscissae of vertices (b) abscissae of foci

(c) eccentricity (d) directrix.

13. The equation of a tangent to the parabola $y^2=8x$ is $y=x+2$. The point on this line from which the other tangent to the parabola is perpendicular to the given tangent is [2007]

(a) (2,4) (b) (–2,0)

(c) (–1,1) (d) (0,2)

14. The normal to a curve at $P(x, y)$ meets the x-axis at G. If the distance of G from the origin is twice the abscissa of P, then the curve is a [2007]

(a) circle (b) hyperbola

(c) ellipse (d) parabola.

15. A focus of an ellipse is at the origin. The directrix is the line $x = 4$ and the eccentricity is $\frac{1}{2}$. Then the length of the semi-major axis is **[2008]**

(a) $\frac{8}{3}$ (b) $\frac{2}{3}$ (c) $\frac{4}{3}$ (d) $\frac{5}{3}$

16. A parabola has the origin as its focus and the line $x = 2$ as the directrix. Then the vertex of the parabola is at **[2008]**

(a) $(0,2)$ (b) $(1,0)$
(c) $(0,1)$ (d) $(2,0)$

17. The ellipse $x^2+4y^2=4$ is inscribed in a rectangle aligned with the coordinate axes, which in turn is inscribed in another ellipse that passes through the point (4, 0). Then the equation of the ellipse is : **[2009]**

(a) $x^2+12y^2=16$ (b) $4x^2+48y^2=48$
(c) $4x^2+64y^2=48$ (d) $x^2+16y^2=16$

18. If two tangents drawn from a point P to the parabola $y^2 = 4x$ are at right angles, then the locus of P is **[2010]**

(a) $2x+1=0$ (b) $x=-1$
(c) $2x-1=0$ (d) $x=1$

19. Equation of the ellipse whose axes are the axes of coordinates and which passes through the point (–3, 1) and has eccentricity $\sqrt{\frac{2}{5}}$ is **[2011]**

(a) $5x^2+3y^2-48=0$ (b) $3x^2+5y^2-15=0$
(c) $5x^2+3y^2-32=0$ (d) $3x^2+5y^2-32=0$

20. The equation of the hyperbola whose foci are (– 2, 0) and (2, 0) and eccentricity is 2 is given by : **[2011 RS]**

(a) $x^2-3y^2=3$ (b) $3x^2-y^2=3$
(c) $-x^2+3y^2=3$ (d) $-3x^2+y^2=3$

21. **Statement-1 :** An equation of a common tangent to the parabola $y^2=16\sqrt{3}x$ and the ellipse $2x^2+y^2=4$ is $y=2x+2\sqrt{3}$

Statement-2 : If the line $y=mx+\frac{4\sqrt{3}}{m}, (m\neq 0)$ is a common tangent to the parabola $y^2=16\sqrt{3}x$ and the ellipse $2x^2 + y^2 = 4$, then m satisfies $m^4+2m^2=24$ **[2012]**

(a) Statement-1 is false, Statement-2 is true.
(b) Statement-1 is true, statement-2 is true; statement-2 is a correct explanation for Statement-1.
(c) Statement-1 is true, statement-2 is true; statement-2 is **not** a correct explanation for Statement-1.
(d) Statement-1 is true, statement-2 is false.

22. An ellipse is drawn by taking a diameter of the circle $(x-1)^2 + y^2 = 1$ as its semi-minor axis and a diameter of the circle $x^2+(y-2)^2=4$ is semi-major axis. If the centre of the ellipse is at the origin and its axes are the coordinate axes, then the equation of the ellipse is : **[2012]**

(a) $4x^2+y^2=4$ (b) $x^2+4y^2=8$
(c) $4x^2+y^2=8$ (d) $x^2+4y^2=16$

23. The equation of the circle passing through the foci of the ellipse $\frac{x^2}{16}+\frac{y^2}{9}=1$, and having centre at (0, 3) is **[JEE M 2013]**

(a) $x^2+y^2-6y-7=0$
(b) $x^2+y^2-6y+7=0$
(c) $x^2+y^2-6y-5=0$
(d) $x^2+y^2-6y+5=0$

24. **Given :** A circle, $2x^2+2y^2=5$ and a parabola, $y^2=4\sqrt{5}x$.

Statement-1 : An equation of a common tangent to these curves is $y=x+\sqrt{5}$.

Statement-2 : If the line, $y=mx+\frac{\sqrt{5}}{m}$ $(m\neq 0)$ is their common tangent, then m satisfies $m^4-3m^2+2=0$. **[JEE M 2013]**

(a) Statement-1 is true; Statement-2 is true; Statement-2 is a correct explanation for Statement-1.
(b) Statement-1 is true; Statement-2 is true; Statement-2 is not a correct explanation for Statement-1.
(c) Statement-1 is true; Statement-2 is false.
(d) Statement-1 is false; Statement-2 is true.

Solutions & Explanations

Section-A : JEE Advanced/ IIT-JEE

A **1.** $(-1, 0)$ **2.** $\dfrac{\left(x-\frac{1}{3}\right)^2}{\left(\frac{1}{3}\right)^2}+\dfrac{(y-1)^2}{\left(\frac{1}{2\sqrt{3}}\right)^2}=1$ **3.** $\dfrac{1}{\sqrt{2}}$

C **1.** (d) **2.** (c) **3.** (c) **4.** (d) **5.** (a) **6.** (a) **7.** (d)
8. (b) **9.** (c) **10.** (b) **11.** (c) **12.** (c) **13.** (d) **14.** (a)
15. (c) **16.** (d) **17.** (d) **18.** (a) **19.** (b) **20.** (a) **21.** (c)
22. (d) **23.** (a) **24.** (d) **25.** (d) **26.** (a) **27.** (b) **28.** (b)
29. (d) **30.** (c) **31.** (d) **32.** (b) **33.** (c) **34.** (c)

D **1.** (c) **2.** (c) **3.** (b, d) **4.** (a, b) **5.** (a, c) **6.** (b, c) **7.** (b, c)
8. (a, d) **9.** (a, b) **10.** (c, d) **11.** (b, d) **12.** (a, b, d) **13.** (a, b)

E **2.** $m=\pm\sqrt{2}$. **3.** $x+y-3=0$ **4.** $c=\dfrac{3}{4}$ **5.** $y^2=2(x-4)$

6. $\left(\dfrac{2}{9},\dfrac{8}{9}\right)$ **8.** $2:1$ **9.** $\dfrac{15a^2}{4}$ **12.** $\dfrac{25}{x^2}+\dfrac{4}{y^2}=4$

13. $\left(\dfrac{a^2}{\sqrt{a^2+b^2}},\dfrac{b^2}{\sqrt{a^2+b^2}}\right)$ **16.** $\dfrac{x^2}{a^2}+\dfrac{y^2(r+s)^2}{(bs+ar)^2}=1$ **18.** $\alpha=2$

19. $(x-1)(y-1)^2+4=0$ **20.** $\dfrac{x^2}{9}-\dfrac{y^2}{4}=\left(\dfrac{x^2+y^2}{9}\right)^2$ **21.** $y=-\dfrac{2}{\sqrt{3}}x+4\sqrt{\dfrac{7}{3}}$; $\dfrac{14}{\sqrt{3}}$

F **1.** (A)-(p), (B)-(q), (C)-(s), (D)-(r) **2.** (A)-p, q; (B)-p, q ; (C)-q, r ; (D)-q, r
3. (A)-p; (B)-s, t ; (C)-r ; (D)-q, s **4.** (a)

G **1.** (c) **2.** (b) **3.** (d) **4.** (b) **5.** (a) **6.** (d) **7.** (c) **8.** (a) **9.** (b) **10.** (d)

H **1.** (a)

I **1.** (2) **2.** (2) **3.** (4) **4.** (9)

Section-B : JEE Main/ AIEEE

1. (b) **2.** (b) **3.** (d) **4.** (d) **5.** (b) **6.** (a) **7.** (d)
8. (a) **9.** (a) **10.** (a) **11.** (b) **12.** (b) **13.** (b) **14.** (b)
15. (a) **16.** (b) **17.** (a) **18.** (b) **19.** (d) **20.** (b) **21.** (b)
22. (d) **23.** (a) **24.** (b)

Section-A JEE Advanced/ IIT-JEE

(A) FILL IN THE BLANKS :

1. Given parabola is $y^2 = 4x$; $a = 1$

Extremities of latus rectum are $(1, 2)$ and $(1, -2)$ tangent to $y^2 = 4x$ at $(1, 2)$ is $y.2 = 2(x+1)$

i.e. $y = x + 1$ ...(1)

Similarly tangent at $(1, -2)$ is, $y = -x - 1$ ...(2)

Intersection pt. of these tangents can be obtained by solving (1) and (2), which is $(-1, 0)$.

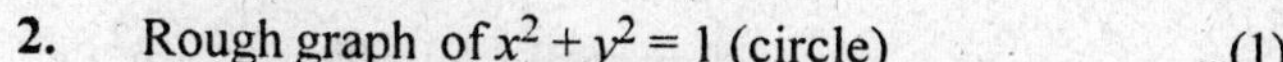

2. Rough graph of $x^2 + y^2 = 1$ (circle) ...(1)

and $x^2 - y^2 = 1$ (hyperbola) ...(2)

is as shown below.

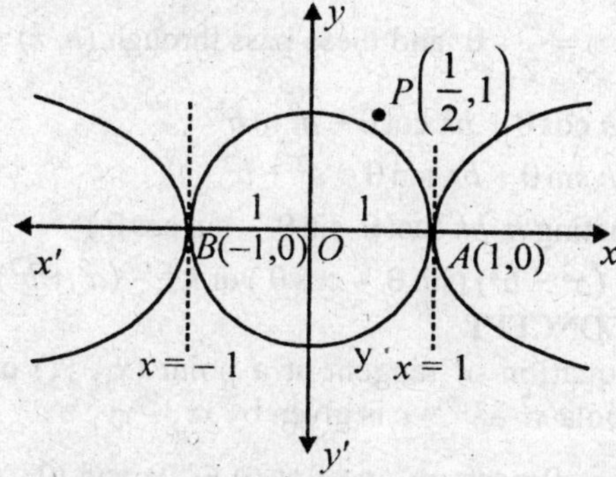

It is clear from graph that there are two common tangents to the curves (1) and (2) namely $x = 1$ and $x = -1$ out of which $x = 1$ is nearer to pt. P.

Hence directrix of required ellipse is $x - 1 = 0$

Also $e = 1/2$, focus $(1/2, 1)$ then equation of ellipse is given by

$$(x - 1/2)^2 + (y-1)^2 = \frac{1}{4}(x-1)^2$$

$$\text{or } 4\left(x^2 - x + \frac{1}{4} + y^2 - 2y + 1\right) = x^2 - 2x + 1$$

$$\Rightarrow 4x^2 - 4x + 1 + 4y^2 - 8y + 4 - x^2 + 2x - 1 = 0$$

$$\Rightarrow 3x^2 - 2x + 4y^2 - 8y + 4 = 0$$

$$\Rightarrow 3\left(x^2 - \frac{2}{3}x + \frac{1}{9}\right) + 4(y^2 - 2y + 1) = \frac{3}{9}$$

$$\Rightarrow 3\left(x - \frac{1}{3}\right)^2 + 4(y-1)^2 = \frac{1}{3}$$

$$\Rightarrow 9(x - 1/3)^2 + 12(y-1)^2 = 1$$

$$\Rightarrow \frac{(x-1/3)^2}{(1/3)^2} + \frac{(y-1)^2}{(1/2\sqrt{3})^2} = 1$$

which is the standard equation of the ellipse.

3.

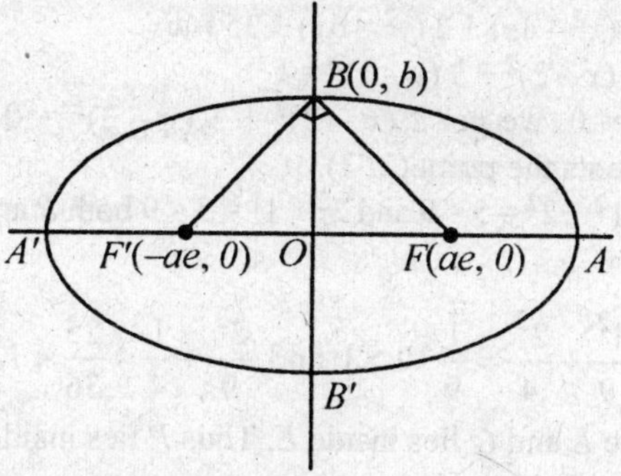

$$\Rightarrow m_{BF} \times m_{BF'} = -1$$

$$\Rightarrow \frac{0-b}{ae-0} \times \frac{0-b}{-ae-0} = -1$$

$$\Rightarrow \frac{b^2}{-a^2e^2} = -1$$

$$\Rightarrow b^2 = a^2e^2 \Rightarrow a^2(1-e^2) = a^2e^2$$

$$\Rightarrow 2e^2 = 1 \Rightarrow e = 1/\sqrt{2}$$

[Note: $e \neq -1/\sqrt{2}$ as the eccentricity of ellipse is $0 < e < 1$**]**

(C) MCQ's WITH ONE CORRECT ANSWER :

1. **(d)** Given that $\frac{x^2}{1-r} - \frac{y^2}{1+r} = 1$, $r > 1$

As $r > 1$

$\therefore$ $1 - r < 0$ and $1 + r > 0$

$\therefore$ Let $1 - r = -a^2$, $1 + r = b^2$

then we get

$$\frac{x^2}{-a^2} - \frac{y^2}{b^2} = 1 \quad \Rightarrow \frac{x^2}{a^2} + \frac{y^2}{b^2} = -1$$

which is not possible for any real values of x and y.

2. **(c)** (a) $x^2 + 2y^2 \leq 1$ represents interior region of an ellipse where on taking any two pts the mid pt of that segment will also lie inside that ellipse

(b) Max $\{|x|, |y|\} \leq 1$

$\Rightarrow |x| \leq 1, |y| \leq 1 \Rightarrow -1 \leq x \leq 1$ and $-1 \leq y \leq 1$ which represents the interior region of a square with its sides $x = \pm 1$ and $y = \pm 1$ in which for any two pts, their mid pt also lies inside the region.

(c) $x^2 - y^2 \geq 1$ represents the exterior region of hyperbola in which if we take two points $(2, 0)$ and $(-2, 0)$ then their mid pt $(0, 0)$ does not lie in the same region (as shown in the figure.)

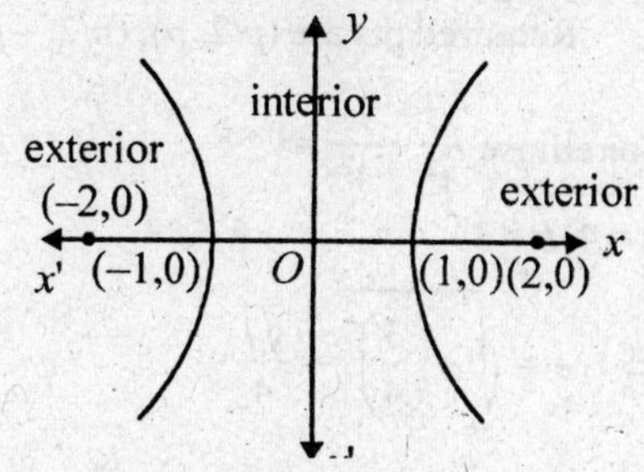

(d) $y^2 \le x$ represents interior region of parabola in which for any two pts, their mid point also lie inside the region.

3. **(c)** We have $2x^2 + 3y^2 - 8x - 18y + 35 = k$

$\Rightarrow\ 2(x^2 - 4x) + 3(y^2 - 6y) + 35 = k$

$\Rightarrow\ 2(x-2)^2 + 3(y-3)^2 = k$

For $k = 0$, we get $2(x-2)^2 + 3(y-3)^2 = 0$ which represents the point (2, 3).

4. **(d)** Since $1^2 + 2^2 = 5 < 9$ and $2^2 + 1^1 = 5 < 9$ both P and Q lie inside C.

Also $\frac{1^2}{9} + \frac{2^2}{4} = \frac{1}{9} + 1 > 1$ and $\frac{2^2}{9} + \frac{1}{4} = \frac{25}{36} < 1$, P lies outside E and Q lies inside E. Thus P lies inside C but outside E.

5. **(a)** The focus of parabola $y^2 = 2px$ is $\left(\frac{p}{2}, 0\right)$ and directrix $x = -p/2$

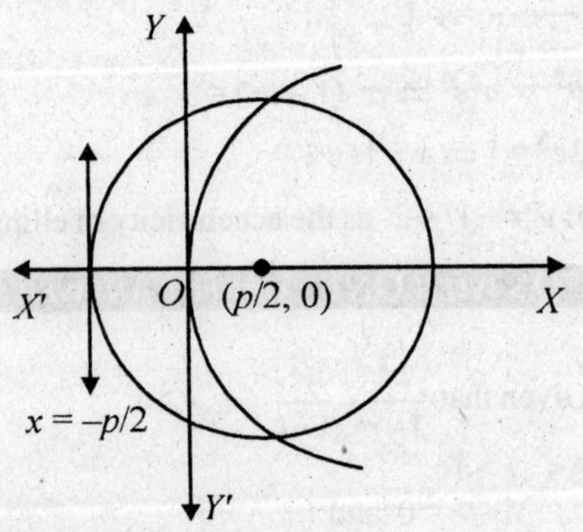

NOTE THIS STEP

In the figure, we have supposed that $p > 0$]

$\therefore$ Centre of circle is $\left(\frac{p}{2}, 0\right)$ and radius $= \frac{p}{2} + \frac{p}{2} = p$

$\therefore$ Equation of circle is $\left(x - \frac{p}{2}\right)^2 + y^2 = p^2$

For pts of intersection of $y^2 = 2px$ ...(i)

and $4x^2 + 4y^2 - 4px - 3p^2 = 0$ ...(ii)

can be obtained by solving (i) and (ii) as follows

$4x^2 + 8px - 4px - 3p^2 = 0$

$4x^2 + 4px - 3p^2 = 0$

$(2x + 3p)(2x - p) = 0$

$\Rightarrow\ x = \frac{-3p}{2}, \frac{p}{2}$

$\Rightarrow\ y^2 = -3p^2$ (not possible), p^2

$\Rightarrow\ y = \pm p$

$\therefore$ Required pts are $(p/2, p), (p/2, -p)$

6. **(a)** For ellipse $\frac{x^2}{4^2} + \frac{y^2}{3^2} = 1$

$a = 4, b = 3$

$\Rightarrow\ e = \sqrt{1 - \left(\frac{3}{4}\right)^2} = \frac{\sqrt{7}}{4}$

$\therefore$ Foci are $(\sqrt{7}, 0)$ and $(-\sqrt{7}, 0)$

Centre of circle is at (0, 3) and it passes through $(\pm\sqrt{7}, 0)$, therefore radius of circle

$= \sqrt{(\sqrt{7})^2 + (3)^2} = 4$

7. **(d)** **KEY CONCEPT:**

Equation of the normal to the hyperbola $\frac{x^2}{a^2} - \frac{y^2}{b^2} = 1$ at the point $(a \sec\alpha, b \tan\alpha)$ is given by $ax \cos\alpha + by \cot\alpha = a^2 + b^2$

Normals at θ, ϕ are $\begin{cases} ax\cos\theta + by\cot\theta = a^2 + b^2 \\ ax\cos\theta + by\cot\phi = a^2 + b^2 \end{cases}$

where $\phi = \frac{\pi}{2} - \theta$ and these pass through (h, k)

$\therefore\ ah\cos\theta + bk\cot\theta = a^2 + b^2$

$ah\sin\theta + bk\tan\theta = a^2 + b^2$

Eliminating h, $bk(\cot\theta\ \sin\theta - \tan\cos\theta)$

$= (a^2 + b^2)(\sin\theta - \cos\theta)$ or $k = -(a^2 + b^2)/b$

8. **(b)** **KEY CONCEPT**

The equation of tangent at a point (x_1, y_1) on the hyperbola $x^2 - y^2 = c$ is given by $xx_1 - yy_1 = c$

Chord $x = 9$ meets $x^2 - y^2 = 9$ at $(9, 6\sqrt{2})$ and $(9, -6\sqrt{2})$ at which tangents are

$9x - 6\sqrt{2}y = 9$ and $9x + 6\sqrt{2}y = 9$

or $3x - 2\sqrt{2}y - 3 = 0$ and $3x + 2\sqrt{2}y - 3 = 0$

$\therefore$ Combined equation of tangents is

$(3x - 2\sqrt{2}y - 3)(3x + 2\sqrt{2}y - 3) = 0$

or $9x^2 - 8y^2 - 18x + 9 = 0$

ALTERNATE SOLUTION

Let a pair of tangents be drawn from pt (x_1, y_1) to hyperbola $x^2 - y^2 = 9$

Then chord of contact will be

$xx_1 - yy_1 = 9$ ...(1) (Using T = 0)

But given chord of contact is

$x = 9$ ...(2)

As equation (1) and (2) represent same line, these equations should be identical and hence

$\frac{x_1}{1} = \frac{-y_1}{0} = \frac{9}{9}$

$\Rightarrow\ x_1 = 1, y_1 = 0$

$\therefore$ Equation of pair of tangents drawn from (1, 0) to $x^2 - y^2 = 9$ is

$(x^2 - y^2 - 9)(1^2 - 0^2 - 9) = (x.1 - y.0 - 9)^2$

(Using $SS_1 = T^2$)

$\Rightarrow\ (x^2 - y^2 - 9)(-8) = (x - 9)^2$

$\Rightarrow\ -8x^2 + 8y^2 + 72 = x^2 - 18x + 81$

$\Rightarrow\ 9x^2 - 8y^2 - 18x + 9 = 0$

9. (c) KEY CONCEPT

The equation $ax^2 + 2hxy + by^2 + 2gx + 2fy + c = 0$ represents a parabola if $\Delta \neq 0$ and $h^2 = ab$ where $\Delta = abc + 2fgh - af^2 - bg^2 - ch^2$

Now we have $x = t^2 + t + 1$ and $y = t^2 - t + 1$

$\frac{x+y}{2} = t^2 + 1, \frac{x-y}{2} = t$ (Adding and subtracting values of x and y)

Eliminating t,

$2(x+y) = (x-y)^2 + 4$ .... (1)

$\Rightarrow x^2 - 2xy + y^2 - 2x - 2y + 4 = 0$ .... (2)

Here, $a = 1, h = -1, b = 1, g = -1, f = -1, c = 4$

$\therefore \Delta \neq 0$. and $h^2 = ab$

Hence the given curve represents a parabola.

10. (b) $y = mx + c$ is normal to the parabola $y^2 = 4ax$ if $c = -2am - am^3$

Here $m = -1$, $c = k$ and $a = 3$

$\therefore \quad c = k = -2(3)(-1) - 3(-1)^3 = 9$

11. (c) KEY CONCEPT : The directrix of the parabola $y^2 = 4a(x - x_1)$ is given by $x = x_1 - a$.

$y^2 = kx - 8 \Rightarrow y^2 = k\left(x - \frac{8}{k}\right)$

Directrix of parabola is $x = \frac{8}{k} - \frac{k}{4}$;

Now, $x = 1$ also coincides with $x = \frac{8}{k} - \frac{k}{4}$

On comparision, $\frac{8}{k} - \frac{k}{4} = 1$, or $k^2 - 4k - 32 = 0$

On solving we get $k = 4$

12. (c) KEY CONCEPT

The line $y = mx + \frac{a}{m}$ is the tangent to the parabola $y^2 = 4ax$ at the point $\left(\frac{a}{m^2}, \frac{2a}{m}\right)$, m being slope of the tangent.

Let the equation of tangent to $y^2 = 4x$ be $y = mx + \frac{1}{m}$ where m is the slope of the tangent.

If it is tangent to the circle $(x-3)^2 + y^2 = 9$ then length of perpendicular to tangent from centre (3, 0) should be equal to the radius 3.

$\therefore \frac{3m + \frac{1}{m}}{\sqrt{m^2+1}} = 3$

$\Rightarrow 9m^2 + \frac{1}{m^2} + 6 = 9m^2 + 9 \Rightarrow m = \pm\frac{1}{\sqrt{3}}$

$\therefore$ Tangents are $x - y\sqrt{3} + 3 = 0$ and $x + y\sqrt{3} + 3 = 0$ out of which $x - y\sqrt{3} + 3 = 0$ meets the parabola at $(3, 2\sqrt{3})$ i.e., above x-axis.

ALTERNATE SOLUTION

Let at pt (x_1, y_1) of parabola $y^2 = 4x$, equation of tangent is $yy_1 = 2(x + x_1)$

i.e., $2x - yy_1 + 2x_1 = 0$ ....(1)

As it is tangent to the circle $(x-3)^2 + y^2 = 9$

$\therefore$ Length of $\perp$ from (3, 0) to (1) is 3

$\therefore \left|\frac{6 + 2x_1}{\sqrt{4 + y_1^2}}\right| = 3$

$\Rightarrow 36 + 24x_1 + 4x_1^2 = 9(4 + y_1^2)$

Also $y_1^2 = 4x_1 \quad \therefore$ We get

$4x_1^2 + 24x_1 + 36 = 36 + 36x_1$

$x_1^2 + 6x_1 - 9x_1 = 0$

$x_1 = 0, 3 \Rightarrow y_1 = 0, \pm 2\sqrt{3}$

As tangent is above x-axis, $y_1 \neq 0, y_1 \neq -2\sqrt{3}$

$\therefore \quad x_1 = 3$ and $y_1 = 2\sqrt{3}$

Equation is $2x - 2\sqrt{3}\,y + 6 = 0$

substituting values in (1)

$\Rightarrow \sqrt{3}\,y = x + 3$

13. (d) $y^2 + 4y + 4x + 2 = 0$

$y^2 + 4y + 4 = -4x + 2$

$(y+2)^2 = -4(x - 1/2)$

It is of the form $Y^2 = -4AX$

whose directrix is given by $X = A$

$\therefore$ Req. equation is $x - 1/2 = 1 \Rightarrow x = 3/2$.

14. (a) Given that $a > 2b > 0$ and $m > 0$

Also $y = mx - b\sqrt{1+m^2}$ ...(1)

is tangent to $x^2 + y^2 = b^2$ ...(2)

as well as to $(x-a)^2 + y^2 = b^2$ ...(3)

$\because$ (1) is tangent to (3)

$\therefore \left|\frac{am - b\sqrt{1+m^2}}{\sqrt{m^2+1}}\right| = b$

[length of perpendicular from $(a, 0)$ to (1) = radius b]

$\Rightarrow am - b\sqrt{1+m^2} = \pm b\sqrt{1+m^2}$

$\Rightarrow am - 2b\sqrt{1+m^2} = 0$

or $am = 0$ (not possible as $a, m > 0$)

$\Rightarrow a^2m^2 = 4b^2(1+m^2)$

$\Rightarrow m^2 = \frac{4b^2}{a^2 - 4b^2}$

$\Rightarrow m = \frac{2b}{\sqrt{a^2 - 4b^2}}$ $(\because m > 0)$

ALTERNATE SOLUTION

$\because \quad a > 2b > 0$

Therefore, the two given circles do not intersect each other.

$\therefore$ Common tangents can be direct or indirect. As $y = \pm b$ are direct common tangets

$\therefore$ Given line should be indirect common tanget.

Clearly line $y = mx - b\sqrt{1+m^2}$ will pass from point $(a/2, 0)$ (mid-point of the centres of the circles)

$\Rightarrow \quad m = \dfrac{2b}{\sqrt{a^2 - 4b^2}}$

15. **(c)** If (h, k) is the mid point of line joining focus $(a, 0)$ and $Q\,(at^2, 2at)$ on parabola then

$h = \dfrac{a+at^2}{2}, k = at$

Eliminating t, we get $2h = a + a\left(\dfrac{k^2}{a^2}\right)$

$\Rightarrow \quad k^2 = a\,(2h - a)$

$\Rightarrow \quad k^2 = 2a\,(h - a/2)$

$\therefore \quad$ Locus of (h, k) is $y^2 = 2a\,(x - a/2)$

whose directrix is $(x - a/2) = -\dfrac{a}{2}$

$\Rightarrow \quad x = 0$

ALTERNATE SOLUTION

Let the moving point be $P(\alpha, \beta)$ and the mid-point of the focus $(a, 0)$ and P be $M\,(h, k)$.

So $h = \dfrac{\alpha + a}{2}, \; k = \dfrac{\beta}{2}$

$\Rightarrow \quad \alpha = 2h - a, \beta = 2k$

As $P(\alpha, \beta)$ lies on $y^2 = 4ax$

$\Rightarrow \quad \beta^2 = 4a\alpha$

$\Rightarrow \quad 4k^2 = 4a\,(2h - a) \Rightarrow k^2 = a\,(2h - a)$

$\Rightarrow \quad$ Locus of M is $y^2 = 2a\left(x - \dfrac{a}{2}\right)$

which is a parabolas having directrix

$x - \dfrac{a}{2} = -\dfrac{a}{2} \Rightarrow x = 0$

16. **(d)** The given curves are

$y^2 = 8x \quad$ ...(1)

and $\quad xy = -1 \quad$ ...(2)

If m is the slope of tangent to (1), then eq^n of tangent is

$y = mx + 2/m$

If this tangent is also a tangent to (2), then putting value of y in curve (2)

$x\left(mx + \dfrac{2}{m}\right) = -1$

$\Rightarrow \quad mx^2 + \dfrac{2}{m}x + 1 = 0$

$\Rightarrow \quad m^2x^2 + 2x + m = 0$

We should get repeated roots for the eq^n (condition of tangency)

$\Rightarrow \quad D = 0$

$\therefore \quad (2)^2 - 4m^2.m = 0$

$\Rightarrow \quad m^3 = 1 \qquad \Rightarrow m = 1$

Hence required tangent is $y = x + 2$

17. **(d)** **KEY CONCEPT**

The end point of latus rectum of ellipse $\dfrac{x^2}{a^2} + \dfrac{y^2}{b^2} = 1$ in first quadrant is $\left(ae, \dfrac{b^2}{a}\right)$ and the tangent at this point intersects x-axis at $\left(\dfrac{a}{e}, 0\right)$ and y-axis at (0, a).

The given ellipse is $\dfrac{x^2}{9} + \dfrac{y^2}{5} = 1$

Then $a^2 = 9, b^2 = 5$

$\Rightarrow \quad e = \sqrt{1 - \dfrac{5}{9}} = \dfrac{2}{3}$

$\therefore \quad$ end point of latus rectum in first quadrent is $L\,(2, 5/3)$

Equation of tangent at L is $\dfrac{2x}{9} + \dfrac{y}{3} = 1$

It meets x-axis at $A\,(9/2, 0)$ and y-axis at $B\,(0, 3)$

$\therefore \quad$ Area of ΔOAB $= \dfrac{1}{2} \times \dfrac{9}{2} \times 3 = \dfrac{27}{4}$

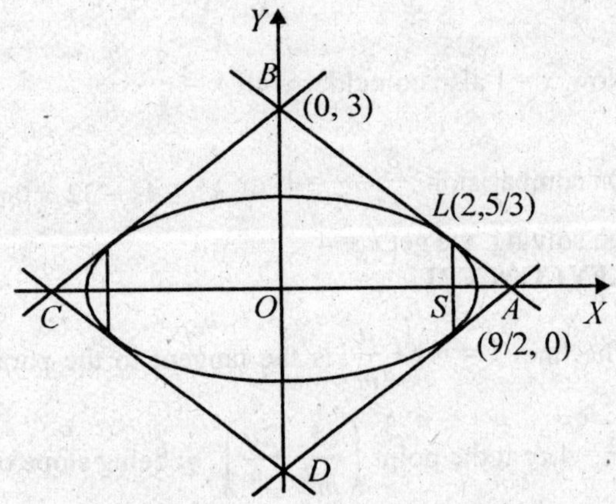

By symmetry area of quadrilateral

$= 4 \times (\text{Area } \Delta OAB) = 4 \times \dfrac{27}{4} = 27$ sq. units.

18. **(a)** For parabola $y^2 = 16x$, focus $\equiv (4, 0)$. Let m be the slope of focal chord then eq^n is

$y = m\,(x - 4) \qquad$ ...(1)

But given that above is a tangent to the circle

$(x - 6)^2 + y^2 = 2$

With Centre, $C\,(6, 0)$, $r = \sqrt{2}$

$\therefore \quad$ Length of $\perp^{lar}$ from (6, 0) to (1) = r

$\Rightarrow \quad \dfrac{6m - 4m}{\sqrt{m^2 + 1}} = \sqrt{2}$

$\Rightarrow \quad 2m = \sqrt{2(m^2 + 1)}$

$\Rightarrow \quad 2m^2 = m^2 + 1$

$\Rightarrow \quad m^2 = 1 \Rightarrow m = \pm 1$

19. **(b)** The given eq^n of hyperbola is

$$\frac{x^2}{\cos^2\alpha}-\frac{y^2}{\sin^2\alpha}=1$$

$\Rightarrow$ $a=\cos\alpha, b=\sin\alpha$

$\Rightarrow$ $e=\sqrt{1+\frac{b^2}{a^2}}=\sqrt{1+\tan^2\alpha}=\sec\alpha$

$\Rightarrow$ $ae=1$

$\therefore$ foci $(\pm 1, 0)$

$\therefore$ foci remain constant with respect to α.

20. **(a)** Any tangent to ellipse $\frac{x^2}{2}+\frac{y^2}{1}=1$ is

$$\frac{x\cos\theta}{\sqrt{2}}+y\sin\theta=1$$

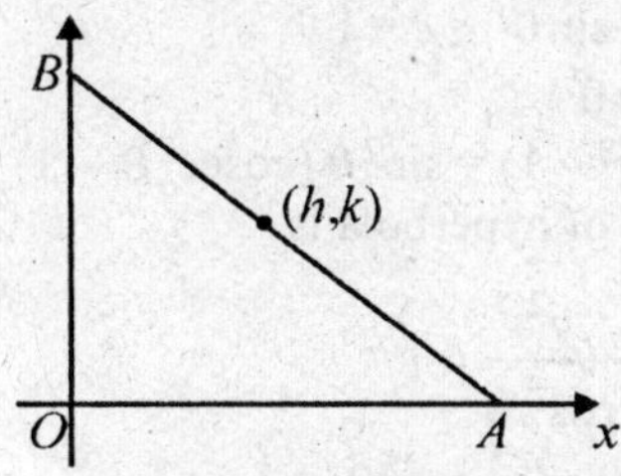

$\therefore$ $A(\sqrt{2}\sec\theta, 0)$; $B(0, \operatorname{cosec}\theta)$

$\Rightarrow$ $2h=\sqrt{2}\sec\theta$ and $2k=\operatorname{cosec}\theta$

(Using mid pt. formula)

$\Rightarrow$ $\cos\theta=\frac{1}{\sqrt{2}h}$ and $\sin\theta=\frac{1}{2k}$

$\Rightarrow$ $\left(\frac{1}{\sqrt{2}h}\right)^2+\left(\frac{1}{2k}\right)^2=1$

$\Rightarrow$ $\frac{1}{2h^2}+\frac{1}{4k^2}=1$

Required locus,

$$\frac{1}{2x^2}+\frac{1}{4y^2}=1$$

21. **(c)** $y=mx+1/m$

Above tangent passes through (1, 4)

$\Rightarrow$ $4=m+1/m \Rightarrow m^2-4m+1=0$

Now angle between the lines is given by

$$\tan\theta=\left|\frac{m_1-m_2}{1+m_1m_2}\right|=\frac{\sqrt{(m_1+m_2)^2-4m_1m_2}}{1+m_1m_2}$$

$$=\frac{\sqrt{16-4}}{1+1}=\sqrt{3}\Rightarrow\theta=\frac{\pi}{3}$$

ALTERNATE SOLUTION

The combined equation of tangents drawn from (1, 4) to the parabola $y^2=4x$ is

$(y^2-4x)(4^2-4\times 1)=[y\times 4-2(x+1)]^2$

[Using $SS_1=T^2$]

$\Rightarrow$ $12(y^2-4x)=4[2y-x-1]^2$

$\Rightarrow$ $3(y^2-4x)=4y^2+x^2+1-4xy+2x-4y$

$\Rightarrow$ $x^2+y^2-4xy+14x-4y+1=0$

Now, we know angle between two lines, given by $ax^2+2hxy+by^2+2gx+2fy+c=0$ is

$$\theta=\tan^{-1}\left(\frac{2\sqrt{h^2-ab}}{a+b}\right)$$

$\therefore$ The req. angle is $\theta=\tan^{-1}\left(\frac{2\sqrt{2^2-1\times 1}}{1+1}\right)$

$$=\tan^{-1}(\sqrt{3})=\pi/3$$

22. **(d)** Equation of tangent to hyperbola $x^2-2y^2=4$ at any point (x_1, y_1) is

$xx_1-2yy_1=4$

Comparing with $2x+\sqrt{6}\,y=2$ or $4x+2\sqrt{6}\,y=4$

$\Rightarrow x_1=4$ and $-2y_1=2\sqrt{6}\Rightarrow(4,-\sqrt{6})$ is the required point.

ALTERNATE SOLUTION

KEY CONCEPT

We know that $y=mx+c$ is tangent to hyperbola $\frac{x^2}{a^2}-\frac{y^2}{b^2}=1$ if $c^2=a^2m^2-b^2$ and point of contact is $\left(\frac{-a^2m}{c},\frac{-b^2}{c}\right)$

We are given that $y=-\frac{2}{\sqrt{6}}x+\frac{2}{\sqrt{6}}$ touches

$$\frac{x^2}{4}-\frac{y^2}{2}=1$$

where $m=-2/\sqrt{6}$, $c=2/\sqrt{6}$, $a^2=4$, $b^2=2$

$\therefore$ Point of contact is

$$\left(\frac{-4(-2/\sqrt{6})}{2/\sqrt{6}},\frac{-2}{2/\sqrt{6}}\right)=(4,-\sqrt{6})$$

23. **(a)** Any tangent to the ellipse $\frac{x^2}{a^2}+\frac{y^2}{b^2}=1$ at $P(a\cos\theta, b\sin\theta)$ is $\frac{x\cos\theta}{a}+\frac{y\sin\theta}{b}=1$

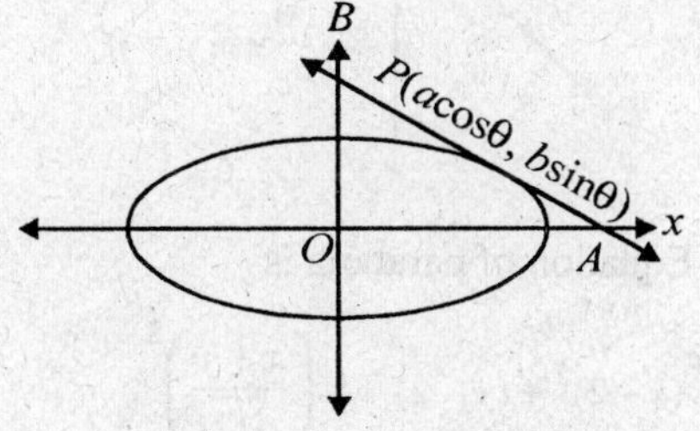

It meets co-ordinate axes at $A\,(a\sec\theta\,,\,0)$ and $B\,(0,\,b\,\text{cosec}\,\theta\,)$

$\therefore$ Area of $\Delta OAB = \frac{1}{2}\times\ a\sec\theta\times b\,\text{cosec}\,\theta$

$\Rightarrow\quad \Delta = \frac{ab}{\sin 2\theta}$

For Δ to be min, $\sin 2\theta$ should be max. and we know max value of $\sin 2\theta = 1$

$\therefore\quad \Delta_{max} = ab$ sq. units.

24. (d) The given curve is $y = x^2 + 6$

Equation of tangent at (1, 7) is

$\frac{1}{2}(y+7) = x.1 + 6$

$\Rightarrow\quad 2x - y + 5 = 0$ ...(1)

As given this tangent (1) touches the circle

$x^2 + y^2 + 16x + 12y + c = 0$ at Q

Centre of circle = (– 8, – 6).

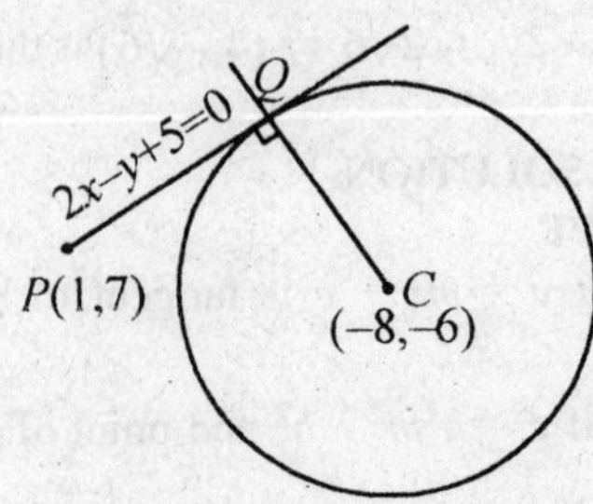

Then equation of CQ which is perpendicular to (1) and passes through (– 8, – 6) is

$y + 6 = -\frac{1}{2}(x+8)$

$\Rightarrow\quad x + 2y + 20 = 0$ ...(2)

Now Q is pt. of intersection of (1) and (2)

$\therefore$ Solving eq^n (1) & (2) we get

$x = -6,\ y = -7$

$\therefore$ Req. pt. is (– 6, – 7).

25. (d) Since, distance of vertex from origin is $\sqrt{2}$ and focus is $2\sqrt{2}$

$\therefore$ Vertex is (1, 1) and focus is (2, 2), directrix $x + y = 0$

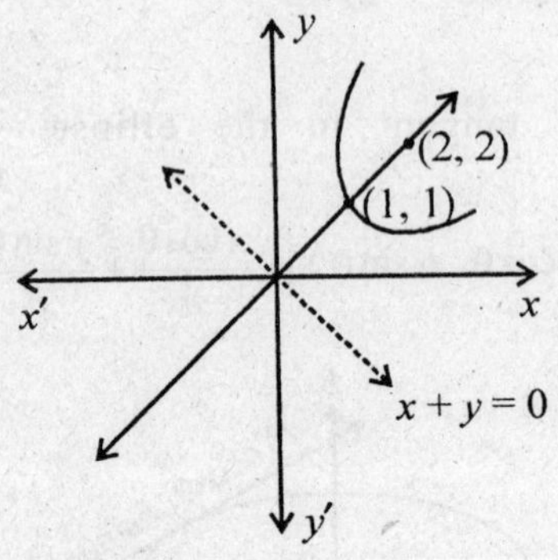

$\therefore$ Equation of parabola is

$(x-2)^2 + (y-2)^2 = \left(\frac{x+y}{\sqrt{2}}\right)^2$

$\Rightarrow 2(x^2 - 4x + 4) + 2(y^2 - 4y + 4) = x^2 + y^2 + 2xy$

$\Rightarrow\quad x^2 + y^2 - 2xy = 8\,(x + y - 2)$

$\Rightarrow\quad (x - y)^2 = 8\,(x + y - 2)$

26. (a) The length of transverse axis = $2\sin\theta = 2a$

$\Rightarrow\quad a = \sin\theta$

Also for ellipse $3x^2 + 4y^2 = 12$

or $\quad \frac{x^2}{4} + \frac{y^2}{3} = 1,\ a^2 = 4,\ b^2 = 3$

$e = \sqrt{1-\frac{b^2}{a^2}} = \sqrt{1-\frac{3}{4}} = \frac{1}{2}$

$\therefore$ Focus of ellipse $= \left(2\times\frac{1}{2}, 0\right) \Rightarrow (1, 0)$

As hyperbola is confocal with ellipse, focus of hyperbola = (1, 0)

$\Rightarrow\quad ae = 1 \Rightarrow \sin\theta\ \times e = 1$

$\Rightarrow\quad e = \text{cosec}\,\theta$

$\therefore\quad b^2 = a^2(e^2 - 1) = \sin^2\theta\,(\text{cosec}^2\theta - 1) = \cos^2\theta$

$\therefore$ Equation of hyperbola is

$\frac{x^2}{\sin^2\theta} - \frac{y^2}{\cos^2\theta} = 1$

or, $\quad x^2\text{cosec}^2\theta - y^2\sec^2\theta = 1$

27. (b) $x^2 - 5xy + 6y^2 = 0$ represents a pair of straight lines given by $x - 3y = 0$ and $x - 2y = 0$.

Also $ax^2 + by^2 + c = 0$ will represent a circle if $a = b$ and c is of sign opposite to that of a.

28. (b) The given hyperbola is

$x^2 - 2y^2 - 2\sqrt{2}x - 4\sqrt{2}y - 6 = 0$

$\Rightarrow (x^2 - 2\sqrt{2}x + 2) - 2(y^2 + 2\sqrt{2}y + 2) = 6 + 2 - 4$

$\Rightarrow (x - \sqrt{2})^2 - 2(y + \sqrt{2})^2 = 4$

$\Rightarrow \frac{(x-\sqrt{2})^2}{2^2} - \frac{(y+\sqrt{2})^2}{(\sqrt{2})^2} = 1$

$\therefore\ a = 2,\ b = \sqrt{2}\ \Rightarrow\ e = \sqrt{1+\frac{2}{4}} = \sqrt{\frac{3}{2}}$

Clearly ΔABC is a right triangle.

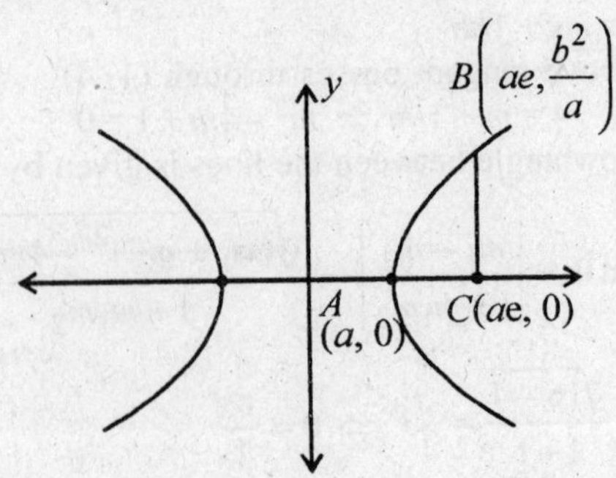

$\therefore$ Ar (ΔABC) $= \frac{1}{2}\times \text{AC}\times\text{BC} = \frac{1}{2}(ae - a)\times\frac{b^2}{a}$

$= \frac{1}{2}(e-1)\times b^2 = \frac{1}{2}\left(\sqrt{\frac{3}{2}} - 1\right)\times 2 = \sqrt{\frac{3}{2}} - 1$

29. (d) The given ellipse is $x^2+9y^2=9$ or $\frac{x^2}{3^2}+\frac{y^2}{1^2}=1$

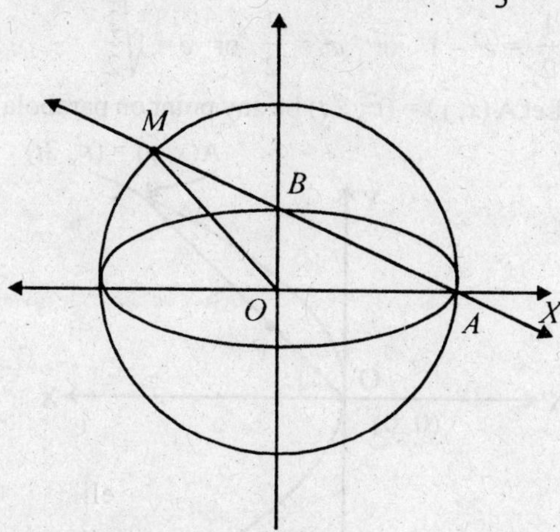

So, that $A(3,0)$ and $B(0,1)$

$\therefore$ Equation of AB is $\frac{x}{3}+\frac{y}{1}=1$

or $x+3y-3=0$ (1)

Also auxillary circle of given ellipse is

$x^2+y^2=9$ (2)

Solving equation (1) and (2), we get the point M where line AB meets the auxillary circle.

Putting $x=3-3y$ from eq^n(1) in eq^n(2)

we get $(3-3y)^2+y^2=9$

$\Rightarrow 9-18y+9y^2+y^2=9$

$\Rightarrow 10y^2-18y=0$

$\Rightarrow y=0, \frac{9}{5}$

$\Rightarrow x=3, \frac{-12}{5}$

Clearly M $\left(\frac{-12}{5},\frac{9}{5}\right)$

$\therefore$ Area of $\Delta OAM=\frac{1}{2}\begin{vmatrix}0 & 0 & 1\\ 3 & 0 & 1\\ \frac{-12}{5} & \frac{9}{5} & 1\end{vmatrix}=\frac{27}{10}$

ALTERNATE SOLUTION

The given ellipse is $\frac{x^2}{3^2}+\frac{y^2}{1^2}=1$, so that

$OA=3, OB=1$ (refer to fig given in privious solution)

Let $\angle OAM=\theta=\angle OMA$ $(\because OA=OM)$

In ΔOAB, $\tan\theta=\frac{1}{3}$

Also in ΔOAM, $\angle AOM=180-2\theta$

So that $\sin\angle AOM=\sin(180-2\theta)=\sin 2\theta$

$$=\frac{2\tan\theta}{1+\tan^2\theta}=\frac{2\times\frac{1}{3}}{1+\frac{1}{9}}=\frac{3}{5}$$

$\therefore$ $\text{Ar}(\Delta OAM)=\frac{1}{2}\times OA^2\times\sin\angle AOM$

$\left[\Delta=\frac{1}{2}r^2\sin\theta \text{ for isosceles triangle}\right]$

$$=\frac{1}{2}\times 3^2\times\frac{3}{5}=\frac{27}{10}$$

30. (c) The given ellipse is $\frac{x^2}{4^2}+\frac{y^2}{2^2}=1$

such that $a^2=16$ and $b^2=4$

$\therefore$ $e^2=1-\frac{4}{16}=\frac{3}{4} \Rightarrow e=\frac{\sqrt{3}}{2}$

Let $P(4\cos\theta, 2\sin\theta)$ be any point on the ellipse, then equation of normal at P is

$4x\sin\theta-2y\cos\theta=12\sin\theta\cos\theta$

$\Rightarrow \frac{x}{3\cos\theta}-\frac{y}{6\sin\theta}=1$

$\therefore$ Q, the point where normal at P meets x-axis, has coordinates $(3\cos\theta, 0)$

$\therefore$ Mid point of PQ is $M\left(\frac{7\cos\theta}{2},\sin\theta\right)$

For locus of point M we consider

$x=\frac{7\cos\theta}{2}$ and $y=\sin\theta$

$\Rightarrow \cos\theta=\frac{2x}{7}$ and $\sin\theta=y$

$\Rightarrow \frac{4x^2}{49}+y^2=1$ ...(1)

Also the latus rectum of given ellipse is

$x=\pm ae=\pm 4\times\frac{\sqrt{3}}{2}=\pm 2\sqrt{3}$ or $x=\pm 2\sqrt{3}$ ...(2)

Solving equations (1) and (2), we get

$\frac{4\times 12}{49}+y^2=1 \Rightarrow y^2=\frac{1}{49}$ or $y=\pm\frac{1}{7}$

$\therefore$ The required points are $\left(\pm 2\sqrt{3},\pm\frac{1}{7}\right)$.

31. (d) The triangle is formed by the lines

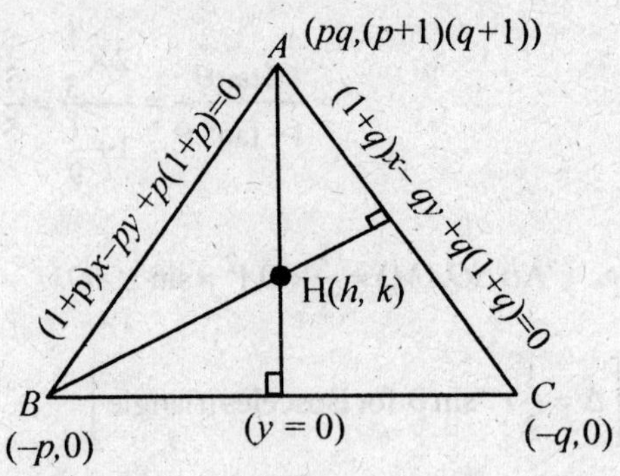

$AB: (1+p)x - py + p(1+p) = 0$

$AC: (1+q)x - qy + q(1+q) = 0$

$BC: y = 0$

So that the vertices are

$A(pq, (p+1)(q+1)),\ B(-p, 0), C(-q, 0)$

Let $H(h, k)$ be the orthocentre of ΔABC. Then as $AH \perp BC$ and passes through $A(pq, (p+1)(q+1))$

The eq^n of AH is $x = pq$

$\therefore \quad h = pq \quad ...(1)$

Also BH is perpendicular to AC

$\therefore \quad m_1 m_2 = -1 \Rightarrow \frac{k-0}{h+p} \times \frac{1+q}{q} = -1$

$\Rightarrow \quad \frac{k}{pq+p} \times \frac{1+q}{q} = -1 \quad$ (using eq^n (1))

$\Rightarrow \quad k = -pq \quad ...(2)$

From (1) and (2) we observe $h + k = 0$

$\therefore$ Locus of (h, k) is $x + y = 0$ which is a straight line.

32. (b) For hyperbola $\frac{x^2}{a^2} - \frac{y^2}{b^2} = 1$, we have

$\frac{2x}{a^2} - \frac{2y}{b^2}\frac{dy}{dx} = 0 \Rightarrow \frac{dy}{dx} = \frac{b^2 x}{a^2 y}$

$\therefore$ Slope of normal at P (6, 3)

$= -\frac{1}{\left(\frac{dy}{dx}\right)_{(6,3)}} = -\frac{3a^2}{6b^2}$

$\therefore$ Equation of normal is

$\frac{y-3}{x-6} = -\frac{3a^2}{6b^2}$

As it intersects x-axis at (9, 0)

$\therefore \quad \frac{0-3}{9-6} = \frac{-3a^2}{6b^2} \Rightarrow a^2 = 2b^2 \quad ...(1)$

Also for hyperbola, $b^2 = a^2(e^2 - 1)$

Using $a^2 = 2b^2$; we get

$b^2 = 2b^2(e^2 - 1)$

$\frac{1}{2} = e^2 - 1$ or $e^2 = \frac{3}{2}$ or $e = \sqrt{\frac{3}{2}}$

33. (c) Let A $(x, y) = (t^2, 2t)$ be any point on parabola $y^2 = 4x$.

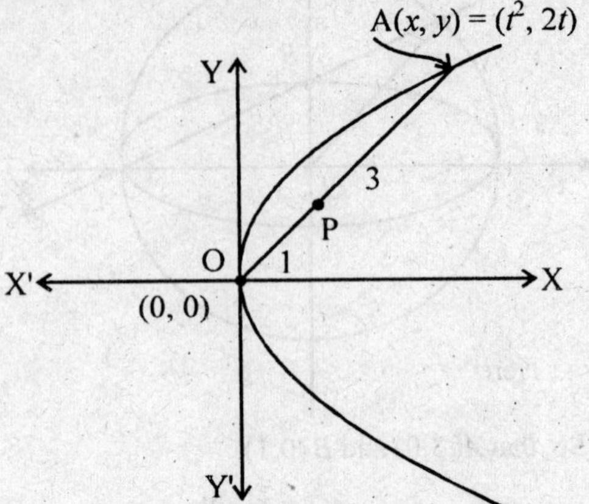

Let P (h, k) divides OA in the ratio 1 : 3

Then $(h, k) = \left(\frac{t^2}{4}, \frac{2t}{4}\right)$

$\Rightarrow h = \frac{t^2}{4}$ and $k = \frac{t}{2}$

$\Rightarrow h = k^2$

$\therefore$ locus of P (h, k) is $x = y^2$.

34. (c) As rectangle $ABCD$ circumscribed the ellipse

$\frac{x^2}{9} + \frac{y^2}{4} = 1$

$\therefore \quad A = (3, 2)$

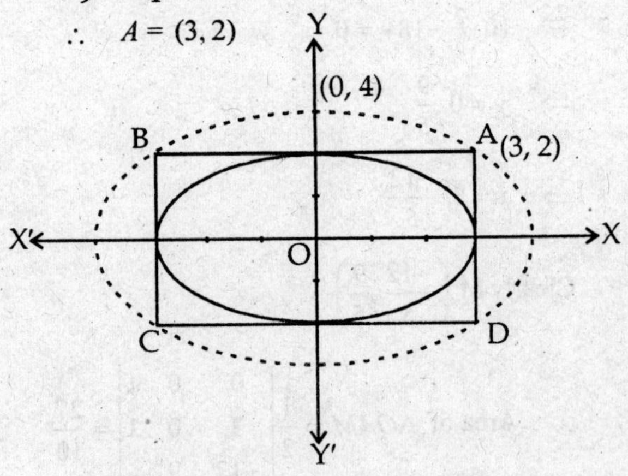

Let the ellipse circumscribing the rectangle $ABCD$ is

$\frac{x^2}{a^2} + \frac{y^2}{b^2} = 1$

Given that it passes through $(a, 4)$

$\therefore \ b^2 = 16$

Also it passes through A (3, 2)

$\therefore \quad \frac{9}{a^2} + \frac{4}{16} = 1 \Rightarrow a^2 = 12$

$\therefore \quad e = \sqrt{1 - \frac{12}{16}} = \sqrt{\frac{1}{4}} = \frac{1}{2}$

(D) MCQ's WITH ONE OR MORE THAN ONE CORRECT :

1. (c) The given curve is $\frac{x^2}{4}+\frac{y^2}{1}=1$ (an ellipse) and given line is $y=4x+c$.
We know that $y = mx + c$ touches the ellipse
$$\frac{x^2}{a^2}+\frac{y^2}{b^2}=1 \text{ if } c=\pm\sqrt{a^2m^2+b^2}$$
Here $c=\pm\sqrt{4\times16+1}=\pm\sqrt{65}$
$\therefore$ two values of c exist

2. (c) The ellipse can be written as, $\frac{x^2}{25}+\frac{y^2}{16}=1$
Here $a^2 = 25$, $b^2 = 16$, but $b^2 = a^2(1-e^2)$
$\Rightarrow$ $16/25=1-e^2$
$\Rightarrow$ $e^2=1-16/25=9/25 \Rightarrow e=3/5$
Foci of the ellipse are $(+ae, 0) = (\pm 3, 0)$, i.e., F_1 and F_2
$\therefore$ We have $PF_1 + PF_2 = 2a = 10$ for every point P on the ellipse.

3. (b,d) Let (x_1, y_1) be the point at which tangents to ellipse $4x^2 + 9y^2 = 1$ are parallel to $8x = 9y$, then

slope of tangent at $(x_1, y_1)=\frac{8}{9}$
$$\Rightarrow \left(\frac{dy}{dx}\right)_{(x_1,y_1)}=\frac{8}{9} \quad ...(1)$$
Differentiating equation of ellipse w.r.t. to x we get
$$8x+18y\frac{dy}{dx}=0$$
$$\Rightarrow \left(\frac{dy}{dx}\right)_{(x_1,y_1)}=\frac{-8x_1}{18y_1}=\frac{-4x_1}{9y_1}$$
Substituting in equation (1) we get
$$\frac{-4x_1}{9y_1}=\frac{8}{9}\Rightarrow -x_1=2y_1 \quad ...(2)$$
Also (x_1, y_1) being point of contact must lie on curve
$$\therefore \quad 4x_1^2+9y_1^2=1$$
$$\Rightarrow \quad 4.4y_1^2+9y_1^2=1 \qquad \text{(Using (2))}$$
$$\Rightarrow \quad y_1^2=\frac{1}{25} \qquad \Rightarrow y_1=\pm\frac{1}{5}$$
$$\Rightarrow \quad x_1=\mp\frac{2}{5}$$
Thus the required points are $\left(\frac{-2}{5},\frac{1}{5}\right)$ and $\left(\frac{2}{5},\frac{-1}{5}\right)$

ALTERNATE SOLUTION

Let $y=\frac{8}{9}x+C$ be the tangent to $\frac{x^2}{1/4}+\frac{y^2}{1/9}=1$

where $C=\pm\sqrt{a^2m^2+b^2}=\pm\sqrt{\frac{1}{4}\times\frac{64}{81}+\frac{1}{9}}=\pm\frac{5}{9}$

and pts of contact are $\left(\frac{-a^2m}{c},\frac{b^2}{c}\right)=\left(\frac{2}{5},\frac{-1}{5}\right)$

or $\left(\frac{-2}{5},\frac{1}{5}\right)$

4. (a,b) if $y = mx + c$ is tangent to $y = x^2$ then $x^2 - mx - c = 0$ has equal roots
$$\Rightarrow \quad m^2+4c=0\Rightarrow c=-\frac{m^2}{4}$$
$$\therefore \quad y=mx-\frac{m^2}{4} \text{ is tangent to } y=x^2$$
$\therefore$ This is also tangent to $y=-(x-2)^2$
$$\Rightarrow \quad mx-\frac{m^2}{4}=-x^2+4x-4$$
$$\Rightarrow \quad x^2+(m-4)x+\left(4-\frac{m^2}{4}\right)=0 \text{ has equal roots}$$
$$\Rightarrow \quad m^2-8m+16=-m^2+16\Rightarrow m=0,4$$
$\Rightarrow$ $y = 0$ or $y = 4x - 4$ are the tangents.

5. (a,c) For the given ellipse $\frac{x^2}{25}+\frac{y^2}{16}=1$
$$e=\sqrt{1-\frac{16}{25}}=\frac{3}{5}$$
$$\Rightarrow \quad \text{Eccentricity of hyperbola}=\frac{5}{3}$$
Let the hyperbola be $\frac{x^2}{A^2}-\frac{y^2}{B^2}=1$ then
$$B^2=A^2\left(\frac{25}{9}-1\right)=\frac{16}{9}A^2$$
$\therefore$ $\frac{x^2}{A^2}-\frac{9y^2}{16A^2}=1$ As it passes through focus of ellipse i.e. $(3, 0)$
$\therefore$ we get
$A^2 = 9 \Rightarrow B^2 = 16$

$\therefore$ Equation of hyperbola is $\frac{x^2}{9}-\frac{y^2}{16}=1$, focus of hyperbola is $(5, 0)$, vertex of hyperbola is $(3, 0)$.

6. (b,c)

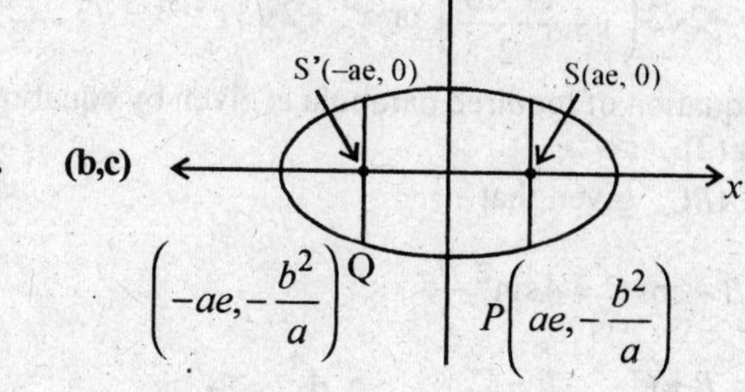

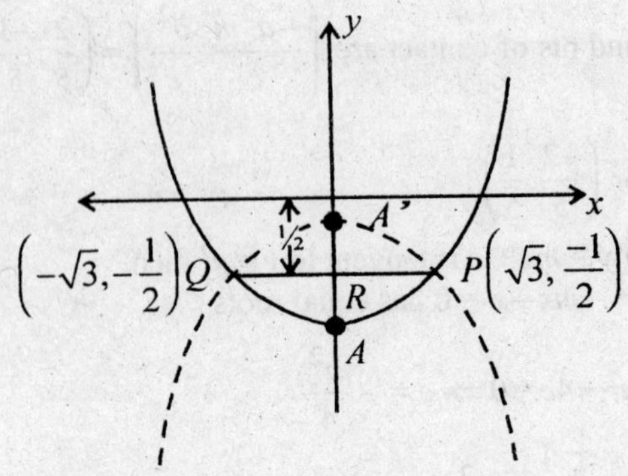

Given ellipse is $x^2+4y^2=4$

or $\frac{x^2}{2^2}+\frac{y^2}{1}=1 \Rightarrow a=2, b=1$

$\therefore \quad e=\sqrt{1-\frac{1}{4}}=\frac{\sqrt{3}}{2} \quad \therefore \quad ae=\sqrt{3}$

As per question $P\equiv(ae,-b^2/a)=\left(\sqrt{3},-\frac{1}{2}\right)$

$Q\equiv(-ae,-b^2/a)=\left(-\sqrt{3},-\frac{1}{2}\right)$

$\therefore \quad PQ=2\sqrt{3}$

Now if PQ is the length of latus rectum to be found,

then $PQ=4a=2\sqrt{3} \Rightarrow a=\frac{\sqrt{3}}{2}$

Also as PQ is horizontal, parabola with PQ as latus rectum can be upward parabola (with vertex at A) or down ward parabola (with vertex at A').

For upward parabola,

$AR=a=\frac{\sqrt{3}}{2}$ $\therefore$ Coordinates of $A=\left(0,-\left(\frac{\sqrt{3}+1}{2}\right)\right)$

So equation of upward parabola is given by

$$x^2=2\sqrt{3}\left(y+\frac{\sqrt{3}+1}{2}\right) \text{ or } x^2-2\sqrt{3}y=3+\sqrt{3} \quad ...(1)$$

For downward parabola $A'R=a=\frac{\sqrt{3}}{2}$

$\therefore$ Coordinates of $A'=\left(0,-\left(\frac{1-\sqrt{3}}{2}\right)\right)$

So equation of downward parabola is given by

$$x^2=-2\sqrt{3}\left(y+\frac{1-\sqrt{3}}{2}\right) \text{ or } x^2+2\sqrt{3}y=3-\sqrt{3} \quad ...(2)$$

$\therefore$ the equation of required parabola is given by equation (1) or (2).

7. **(b,c)** In ΔABC, given that

$$\cos B+\cos C=4\sin^2\frac{A}{2}$$

$$\Rightarrow 2\cos\frac{B+C}{2}\cos\frac{B-C}{2}-4\sin^2\frac{A}{2}=0$$

$$\Rightarrow 2\sin\frac{A}{2}\left[\cos\frac{B-C}{2}-2\sin\frac{A}{2}\right]=0$$

$$\Rightarrow \sin\frac{A}{2}=0 \text{ or } \cos\frac{B-C}{2}-2\cos\frac{B+C}{2}=0$$

But in a triangle $\sin\frac{A}{2}\neq 0$

$$\therefore \quad \cos\frac{B-C}{2}-2\cos\frac{B+C}{2}=0$$

$$\Rightarrow \frac{\cos\left(\frac{B+C}{2}\right)}{\cos\left(\frac{B-C}{2}\right)}=\frac{1}{2}$$

Applying componendo and dividendo, we get

$$\frac{\cos\left(\frac{B+C}{2}\right)+\cos\left(\frac{B-C}{2}\right)}{\cos\left(\frac{B+C}{2}\right)-\cos\left(\frac{B-C}{2}\right)}=\frac{1+2}{1-2}=-3$$

$$\Rightarrow \frac{2\cos\frac{B}{2}\cos\frac{C}{2}}{-2\sin\frac{B}{2}\sin\frac{C}{2}}=-3$$

$$\Rightarrow \tan\frac{B}{2}\tan\frac{C}{2}=\frac{1}{3}$$

$$\Rightarrow \sqrt{\frac{(s-a)(s-c)}{s(s-b)}}\sqrt{\frac{(s-a)(s-b)}{s(s-c)}}=\frac{1}{3}$$

$$\Rightarrow \frac{s-a}{s}=\frac{1}{3} \text{ or } 2s=3a$$

$$\Rightarrow a+b+c=3a \quad \text{or} \quad b+c=2a$$

i.e. $AC+AB=$ constant

($\because$ Base $BC=a$ is given to be constant)

$\Rightarrow$ A moves on an ellipse.

8. **(a,d)** Let $P(at^2, 2at)$ be any point on the parabola

$$y^2=4ax.$$

Then tangent to parabola at P is $y=\frac{x}{t}+at$

which meets the axis of parabola *i.e* x-axis at $T(-at^2, 0)$.

Also normal to parabola at P is $tx+y=2at+at^3$

which meets the axis of parabola at $N(2a+at^2, 0)$

Let $G(x, y)$ be the centriod of ΔPTN, then

$$x=\frac{at^2-at^2+2a+at^2}{3} \text{ and } y=\frac{2at}{3}$$

$$\Rightarrow x=\frac{2a+at^2}{3} \text{ and } y=\frac{2at}{3}$$

Eliminating t from above, we get the locus of centriod G as

$$3x = 2a + a\left(\frac{3y}{2a}\right)^2 \Rightarrow y^2 = \frac{4a}{3}\left(x - \frac{2}{3}a\right)$$

which is a parabola with vertex $\left(\frac{2a}{3}, 0\right)$, directrix as

$x - \frac{2a}{3} = -\frac{a}{3}$ or $x = \frac{a}{3}$, latus rectum as $\frac{4a}{3}$ and focus as $(a, 0)$.

9. **(a,b)** The given hyperbola is

$$x^2 - y^2 = \frac{1}{2} \quad ...(1)$$

which is a rectangular hyperbola (i.e. $a = b$)

$\therefore \quad e = \sqrt{2}.$

Let the ellipse be $\frac{x^2}{a^2} + \frac{y^2}{b^2} = 1$

Its eccentricity $= \frac{1}{\sqrt{2}}$

$\therefore \quad b^2 = a^2\left(1 - \frac{1}{2}\right) \Rightarrow b^2 = \frac{a^2}{2}$

So, the equation of ellipse becomes

$$x^2 + 2y^2 = a^2 \quad ...(2)$$

Let the hyperbola (1) and ellipse (2) intersect each other at $P(x_1, y_1)$.

Then slope of hyperbola (1) at P is given by

$$m_1 = \left(\frac{dy}{dx}\right)_{(x_1, y_1)} = \frac{x_1}{y_1}$$

and that of ellipse (2) at P is

$$m_2 = \left(\frac{dy}{dx}\right)_{(x_1, y_1)} = -\frac{x_1}{2y_1}$$

As the two curves intersect orthogonally,

$\therefore \quad m_1 m_2 = -1$

$\Rightarrow \quad \frac{x_1}{y_1} \cdot \left(-\frac{x_1}{2y_1}\right) = -1$

$\Rightarrow \quad x_1^2 = 2y_1^2 \quad ...(i)$

Also $P(x_1, y_1)$ lies on $x^2 - y^2 = \frac{1}{2}$

$\therefore \quad x_1^2 - y_1^2 = \frac{1}{2} \quad ...(ii)$

Solving (i) and (ii), we get $y_1^2 = \frac{1}{2}$ and $x_1^2 = 1$

Also $P(x_1, y_1)$ lies on ellipse $x^2 + 2y^2 = a^2$

$\therefore \quad x_1^2 + 2y_1^2 = a^2 \Rightarrow 1 + 1 = a^2$ or $a^2 = 2$

$\therefore$ The required ellipse is $x^2 + 2y^2 = 2$ whose foc are $(\pm ae, 0) = \left(\pm\sqrt{2} \times \frac{1}{\sqrt{2}}, 0\right) = (\pm 1, 0)$

10. **(c,d)** Given parabola $y^2 = 4x$

Let $A(t_1^2, 2t_1)$ and $B(t_2^2, 2t_2)$

Then centre of circle drawn with AB as diameter is

$$\left(\frac{t_1^2 + t_2^2}{2}, t_1 + t_2\right)$$

As circle touches x-axis

$\therefore r = |t_1 + t_2| \Rightarrow t_1 + t_2 = \pm r$

Also slope of AB $= \frac{2(t_2 - t_1)}{t_2^2 - t_1^2} = \frac{2}{t_2 + t_1} = \pm\frac{2}{r}$

11. **(b, d)** For $x^2 + 4y^2 = 4$ or $\frac{x^2}{4} + \frac{y^2}{1} = 1$

$$e = \sqrt{1 - \frac{1}{4}} = \frac{\sqrt{3}}{2}$$

As per question, $\sqrt{1 + \frac{b^2}{a^2}} = \frac{2}{\sqrt{3}} \Rightarrow \frac{b}{a} = \frac{1}{\sqrt{3}}$

focus of ellipse is $(\pm\sqrt{3}, 0)$

As hyperbola passes through $(\pm\sqrt{3}, 0)$

$\therefore \frac{3}{a^2} = 1$ or $a = \sqrt{3}$

$\therefore b = 1$ and focus of hyperbola $(\pm 2, 0)$

$\therefore$ Equation of hyperbola is $\frac{x^2}{3} - \frac{y^2}{1} = 1$

or $x^2 - 3y^2 = 3$

12. **(a, b, d)**

The equation of normal to $y^2 = 4x$ is $y = mx - 2m - m^3$

As it passes through (9, 6)

$\therefore 6 = 9m - 2m - m^3$

$\Rightarrow m^3 - 7m + 6 = 0 \Rightarrow (m - 1)(m^2 + m - 6) = 0$

$\Rightarrow (m - 1)(m + 3)(m - 2) = 0 \Rightarrow m = 1, 2, -3$

$\therefore$ Normal is $y = x - 3$ or $y = 2x - 12$ or $y = -3x + 33$

$\therefore$ a, b, d are the correct option.

13. **(a, b)**

KEY CONCEPT : If slope of tangent is m, then equations of tangents to hyperbola $\frac{x^2}{a^2} - \frac{y^2}{b^2} = 1$ are

$y = mx \pm \sqrt{a^2m^2 - b^2}$ with the points of contact

$$\left(\frac{\pm a^2 m}{\sqrt{a^2m^2 - b^2}}, \frac{\pm b^2}{\sqrt{a^2m^2 - b^2}}\right)$$

$\because$ Tangent to hyperbola $\frac{x^2}{9}-\frac{y^2}{4}=1$ is parallel to $2x-y=1$, therefore slope of tangent = 2

$\therefore$ Points of contact are $\left(\frac{\pm 9\times 2}{\sqrt{9\times 4-4}}, \frac{\pm 4}{\sqrt{9\times 4-4}}\right)$

i.e. $\left(\frac{9}{2\sqrt{2}}, \frac{1}{\sqrt{2}}\right)$ and $\left(\frac{-9}{2\sqrt{2}}, \frac{-1}{\sqrt{2}}\right)$

Alternative : Let P(3 sec θ, 2 tan θ) be any point on hyperbola $\frac{x^2}{9}-\frac{y^2}{4}=1$ at which equation of tangent is $\frac{x\sec\theta}{3}-\frac{y\tan\theta}{2}=1$

As tangent is parallel to $2x-y=1$

$\therefore$ slope of tangent = 2

$\Rightarrow \frac{2\sec\theta}{3\tan\theta}=2 \Rightarrow \sin\theta=\frac{1}{3}$

If $\theta\in$ I quad, $\sec\theta=\frac{3}{2\sqrt{2}}$ and $\tan\theta=\frac{1}{2\sqrt{2}}$

$\therefore$ point of contact is $\left(\frac{9}{2\sqrt{2}}, \frac{1}{\sqrt{2}}\right)$

If $\theta\in$ II quad, $\sec\theta=\frac{-3}{2\sqrt{2}}$ and $\tan\theta=\frac{-1}{2\sqrt{2}}$

$\therefore$ point of contact is $\left(\frac{-9}{2\sqrt{2}}, \frac{-1}{\sqrt{2}}\right)$

(E) SUBJECTIVE PROBLEMS :

1. The equation of a normal to the parabola $y^2 = 4ax$ in its slope form is given by

$y=mx-2am-am^3$

$\therefore$ Eq. of normal to $y^2=4x$, is

$y=mx-2m-m^3$ ...(1)

Since the normal drawn at three different points on the parabola pass through (h, k), it must satisfy the equation (1)

$\therefore\ k=mh-2m-m^3$

$\Rightarrow\ m^3-(h-2)m+k=0$

This cubic eq. in m has three different roots say m_1, m_2, m_3

$\therefore\ m_1+m_2+m_3=0$ ...(2)

$m_1m_2+m_2m_3+m_3m_1=-(h-2)$ ...(3)

Now, $(m_1+m_2+m_3)^2=0$ [Squaring (2)]

$\Rightarrow\ m_1^2+m_2^2+m_3^2=-2(m_1m_2+m_2m_3+m_3m_1)$

$\Rightarrow\ m_1^2+m_2^2+m_3^2=2(h-2)$ [Using (3)]

Since LHS of this equation is he sum of perfect squares, therefore it is + ve

$\therefore\ h-2>0\Rightarrow h>2$ Proved

2. Parabola $y^2=4ax$.

Let at any pt A equation of normal is

$y=mx-2am-am^3$. ...(1)

Combined equation of OA and OB can be obtained by making equation of parabola homogeneous with the help of normal.

$\therefore$ Combined eq. of OA and OB is

$y^2=4ax\left(\frac{mx-y}{2am+am^3}\right)$

[From eqn. (1) using $\frac{mx-y}{2am+am^3}=1$]

$y^2=\frac{4x(mx-y)}{2m+m^3}$

$\Rightarrow\ 4mx^2-4xy-(2m+m^3)y^2=0$

But angle between the lines represented by this pair is 90°.

$\Rightarrow$ coeff. of x^2 + coeff of $y^2=0$

$\Rightarrow\ 4m-2m-m^3=0$

$\Rightarrow\ m^3-2m=0$

$\Rightarrow\ m=0, \sqrt{2}, -\sqrt{2}$

But for m = 0 eq. of normal becomes y = 0 which does not intersect the parabola at any other point.

$\therefore\ m=\pm\sqrt{2}$

ALTERNATE SOLUTION

Let the co-ordinates of pts A and B be

$A(at_1^2, 2at_1)$, $B(at_2^2, 2at_2)$

The equation of the normal at A is $y-2at_1=-t_1(x-at_1^2)$

As it passes through B,

$2at_2-2at_1=-t_1(at_2^2-at_1^2)$

$\Rightarrow\ 2a(t_2-t_1)=-at_1(t_2+t_1)(t_2-t_1)$

$\Rightarrow\ 2=-t_1(t_2+t_1)$ ...(1)

Since OA and OB are perpendicular to each other

$\therefore\ -t_1t_2=4$ or $t_2=-4/t_1$

Substituting this value of t_2 in eq. (1) we get

$2=-t_1\left(\frac{-4}{t_1}+t_1\right)=4-t_1^2$

or, $t_1^2=2\Rightarrow t_2=-4/(\pm\sqrt{2})=\mp 2\sqrt{2}$

$\therefore$ Slope of $AB=\frac{2}{t_1+t_2}=\frac{-2t_1}{2}$ [Using (1)]

$=\pm\sqrt{2}$

3. The equation of curve is $y=x^2/4$

then $\frac{dy}{dx}=\frac{x}{2}$

Let the pt. of contact be (h, k) then it lies on the curve $y=x^2/4$

$\therefore\ k=h^2/4$ ..(1)

Slope of normal $=\frac{-1}{\left(\frac{dy}{dx}\right)_{h,k}}=\frac{-1}{h/2}=\frac{-2}{h}$

$\therefore$ Equation of normal is $y-k=\frac{-2}{h}(x-h)$

$\Rightarrow\ hy-hk=-2k+2h$

$\Rightarrow\ 2x+hy-2h+hk$ ....(2)

As normal passes through (1, 2)

$\therefore \quad 2+2h=2h+hk$

$\Rightarrow \quad hk=2 \quad$ ...(3)

Solving equations (1) and (3), we get

$$h.\frac{h^2}{4}=2 \Rightarrow h^3=8$$

$\Rightarrow \quad h=2$

$\Rightarrow \quad k=1 \quad$ [from (3)]

$\therefore$ Equation of normal is.

$2x+2y=4+2$ or $x+y-3=0$

4. Given parabola is $y^2=x$.

KEY CONCEPT

We know that from any pt normal to $y^2=4ax$ is given by

$y=mx-2am-am^3$

Here, $a=\frac{1}{4}$,

$\therefore$ Normal is $y=mx-\frac{m}{2}-\frac{m^3}{4}$

As per question this normal passes through $(c, 0)$ therefore, we get

$$mc-\frac{m}{2}-\frac{m^3}{4}=0 \quad ...(1)$$

$$\Rightarrow m\left[c-\frac{1}{2}-\frac{m^2}{4}\right]=0$$

$$\Rightarrow m=0 \text{ or } m^2=4\left(c-\frac{1}{2}\right)$$

$m=0$ shows normal is $y=0$ i.e. x-axis is always a normal.

Also $m^2 \geq 0 \Rightarrow 4\left(c-\frac{1}{2}\right) \geq 0$

$\Rightarrow \quad c \geq 1/2$

At $c=\frac{1}{2}$, from (1) $m=0$

$\therefore$ for other real values of m, $c>1/2$

Now for other two normals to be perpendicular to each other, we must have $m_1.m_2=-1$

Or in other words, if m_1, m_2 are roots of $\frac{m^2}{4}+\frac{1}{2}-c=0$ then product of roots $=-1$

$$\Rightarrow \frac{\left(\frac{1}{2}-c\right)}{1/4}=-1 \Rightarrow \frac{1}{2}-c=-\frac{1}{4}$$

$\Rightarrow \quad c=3/4$

5. Let the equation of chord OP be $y=mx$.

Then eq^n of chord OQ will be $y=-\frac{1}{m}x \quad [\because OQ \perp OP]$

P is pt. of intersection of $y=mx$ and $y^2=4x$.

Solving the two we get $P\left(\frac{4}{m^2}, \frac{4}{m}\right)$

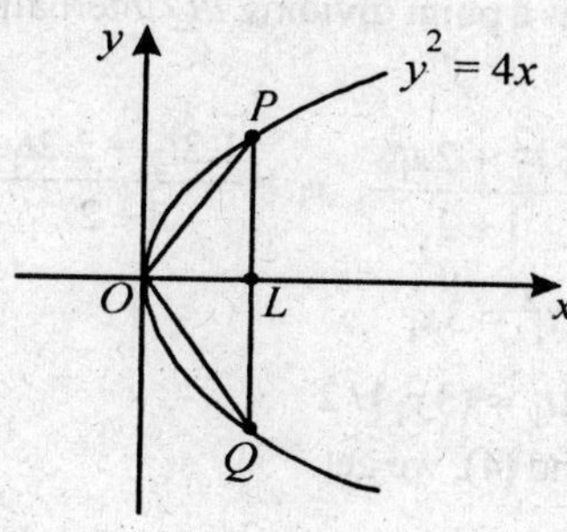

Q is pt. of intersection of $y=-\frac{1}{m}x$ and $y^2=4x$.

Solving the two we get $Q(4m^2, -4m)$

Now eq. of PQ is,

$$y+4m=\frac{\frac{4}{m}+4m}{\frac{4}{m^2}-4m^2}(x-4m^2)$$

$$\Rightarrow y+4m=\frac{m}{1-m^2}(x-4m^2)$$

$\Rightarrow \quad (1-m^2)y+4m-4m^3=mx-4m^3$

$\Rightarrow \quad mx-(1-m^2)y-4m=0$

This line meets x-axis where $y=0$ i.e. $x=4$

$\Rightarrow \quad OL=4$, which is constant as independent of m. Again let (h, k) be the mid pt. of PQ, then

$$h=\frac{4m^2+\frac{4}{m^2}}{2} \text{ and } k=\frac{\frac{4}{m}-4m}{2}$$

$$\Rightarrow h=2\left(m^2+\frac{1}{m^2}\right) \text{ and } k=2\left(\frac{1}{m}-m\right)$$

$$\Rightarrow h=2\left[\left(\frac{1}{m}-m\right)+2\right] \Rightarrow h=2\left[\frac{k^2}{4}+2\right]$$

$\Rightarrow \quad 2h=k^2+8 \Rightarrow k^2=2(h-4)$

$\therefore$ Locus of (h, k) is $y^2=2(x-4)$

6. Let $P(t_1^2, 2t_1)$ and $Q(t_2^2, 2t_2)$ be the ends of the chord PQ of the parabola

$y^2=4x \quad$ ...(1)

$$\therefore \text{ Slope of chord } PQ=\frac{2t_2-2t_1}{t_2^2-t_1^2}=2$$

$\Rightarrow \quad t_2+t_1=1 \quad$ ...(2)

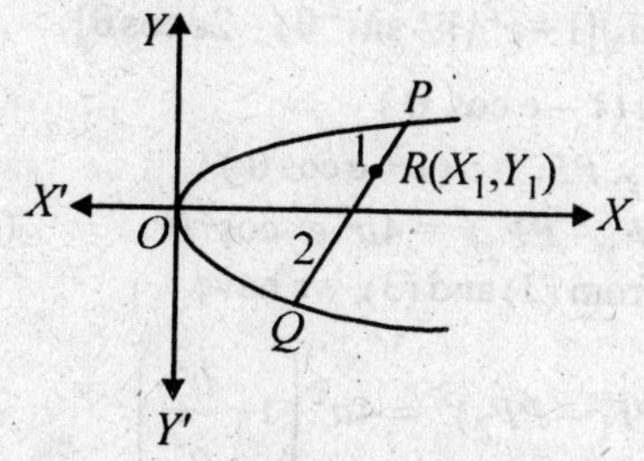

If $R(x_1, y_1)$ is a point dividing PQ internally in the ratio 1 : 2, then

$$x_1 = \frac{1.t_2^2 + 2.t_1^2}{1+2},\ y_1 = \frac{1.2t_2 + 2.2t_1}{1+2}$$

$\Rightarrow \quad t_2^2 + 2t_1^2 = 3x_1$ ...(3)

and $t_2 + 2t_1 = (3y_1)/2$ ...(4)

From (2) and (4), we get

$$t_1 = \frac{3}{2}y_1 - 1, t_2 = 2 - \frac{3}{2}y_1$$

Substituting in (3), we get

$$\left(2 - \frac{3}{2}y_1\right)^2 + 2\left(\frac{3}{2}y_1 - 1\right)^2 = 3x_1$$

$$\Rightarrow \quad (9/4)y_1^2 - 4y_1 = x_1 - 2$$

$$\left(y_1 - \frac{8}{9}\right)^2 = \left(\frac{4}{9}\right)\left(x_1 - \frac{2}{9}\right)$$

$\therefore$ Locus of the point $R(x_1, y_1)$ is

$(y - 8/9)^2 = (4/9)(x - 2/9)$

which is a parabola having vertex at the point (2/9, 8/9).

7. Equation to the tangent at the point $P(a\cos\theta, b\sin\theta)$ on $x^2/a^2 + y^2/b^2 = 1$ is

$$\frac{x}{a}\cos\theta + \frac{y}{b}\sin\theta = 1 \quad ...(1)$$

$\therefore$ d = perpendicular distance of (1) from the centre (0, 0) of the ellipse

$$= \frac{1}{\sqrt{\frac{1}{a^2}\cos^2\theta + \frac{1}{b^2}\sin^2\theta}} = \frac{(ab)}{\sqrt{b^2\cos^2\theta + a^2\sin^2\theta}}$$

$$\therefore \quad 4a^2\left(1 - \frac{b^2}{d^2}\right) = 4a^2\left\{1 - \frac{b^2\cos^2\theta + a^2\sin^2\theta}{a^2}\right\}$$

$= 4(a^2 - b^2)\cos^2\theta = 4a^2e^2\cos^2\theta$ ....(2)

The coordinates of focii F_1 and F_2 are

$F_1 = (ae, 0)$ and $F_2 = (-ae, 0)$

$$\therefore \quad PF_1 = \sqrt{[(a\cos\theta - ae)^2 + (b\sin\theta)^2]}$$

$$= \sqrt{[(a^2(\cos\theta - e)^2 + (b\sin\theta)^2]}$$

$$= \sqrt{[(a^2(\cos\theta - e)^2 + a^2(1-e^2)\sin^2\theta)]}$$

[Using $b^2 = a^2(1 - e^2)$]

$$= a\sqrt{[1 + e^2(1 - \sin^2\theta) - 2e\cos\theta]}$$

$= a(1 - e\cos\theta)$

Similarly, $PF_2 = a(1 + e\cos\theta)$

$\therefore \quad (PF_1 - PF_2)^2 = 4a^2e^2\cos^2\theta$ ...(3)

Hence from (2) and (3), we have

$$(PF_1 - PF_2)^2 = 4a^2\left(1 - \frac{b^2}{a^2}\right)$$

8. Let the three points on the parabola $y^2 = 4ax$ be $A(at_1^2, 2at_1)$, $B(at_2^2, 2at_2)$ and $C(at_3^2, 2at_3)$.

Then using the fact that equation of tangent to $y^2 = 4ax$ at $(at^2, 2at)$ is $y = \frac{x}{t} + at$, we get equations of tangents at A, B and C as folllows

$$y = \frac{x}{t_1} + at_1 \quad ...(1)$$

$$y = \frac{x}{t_2} + at_2 \quad ...(2)$$

$$y = \frac{x}{t_3} + at_3 \quad ...(3)$$

Solving the above equations pair wise we get the pts.

$P(at_1t_2, a(t_1 + t_2))$

$Q(at_2t_3, a(t_2 + t_3))$

$R(at_3t_1, a(t_3 + t_1))$

$$\text{Now, area of } \Delta ABC = \frac{1}{2}\begin{vmatrix} 1 & at_1^2 & 2at_1 \\ 1 & at_2^2 & 2at_2 \\ 1 & at_3^2 & 2at_3 \end{vmatrix}$$

$$= a^2\begin{vmatrix} 1 & t_1 & t_1^2 \\ 1 & t_2 & t_2^2 \\ 1 & t_3 & t_3^2 \end{vmatrix}$$

$$= \left|a^2(t_1 - t_2)(t_2 - t_3)(t_3 - t_1)\right| \quad ...(4)$$

$$\text{Also area of } \Delta PQR = \frac{1}{2}\begin{vmatrix} 1 & at_1t_2 & a(t_1 + t_2) \\ 1 & at_2t_3 & a(t_2 + t_3) \\ 1 & at_3t_1 & a(t_3 + t_1) \end{vmatrix}$$

$$= \frac{a^2}{2}\begin{vmatrix} 1 & t_1t_2 & t_1 + t_2 \\ 1 & t_2t_3 & t_2 + t_3 \\ 1 & t_3t_1 & t_3 + t_1 \end{vmatrix}$$

$$= \frac{a^2}{2}\begin{vmatrix} 0 & (t_1 - t_3)t_2 & t_1 - t_3 \\ 0 & (t_2 - t_1)t_3 & t_2 - t_1 \\ 1 & t_3t_1 & t_3 + t_1 \end{vmatrix}$$

$[R_1 \to R_1 - R_3$ and $R_2 \to R_2 - R_3]$

Expanding along C_1,

$$= \left|\frac{a^2}{2}(t_1 - t_3)(t_2 - t_1)(t_2 - t_3)\right|$$

$$= \left|\frac{a^2}{2}(t_1 - t_2)(t_2 - t_3)(t_3 - t_1)\right| \quad ...(5)$$

From equation (4) and (5), we get

$$\frac{Ar(\Delta ABC)}{Ar(\Delta PQR)} = \frac{a^2 |(t_1-t_2)(t_2-t_3)(t_3-t_1)|}{\frac{a^2}{2}|(t_1-t_2)(t_2-t_3)(t_3-t_1)|} = \frac{2}{1}$$

$\therefore$ The required ratio is 2 : 1

9. **KEY CONCEPT :** Equation of any tangent to the parabola $y^2 = 4ax$ is $y = mx + a/m$

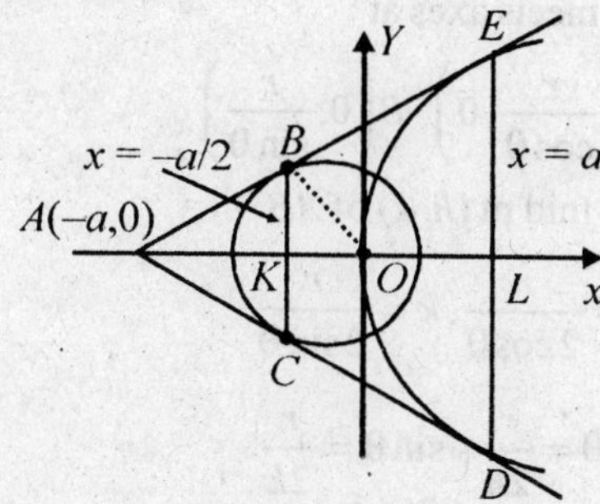

This line will touch the circle $x^2 + y^2 = a^2/2$

if $\frac{a}{m} = \pm\frac{a}{\sqrt{2}}\sqrt{m^2+1}$ $[c = \pm r\sqrt{1+m^2}]$

$\Rightarrow \frac{a^2}{m^2} = \frac{a^2}{2}(m^2+1)$

$\Rightarrow 2 = m^4 + m^2 \Rightarrow m^4 + m^2 - 2 = 0$

$\Rightarrow (m^2+2)(m^2-1) = 0 \Rightarrow m = 1, -1$

Thus the two tangents (common one) are

$y = x + a$ and $y = -x - a$

These two intersect each other at $(-a, 0)$

The chord of contact at $A(-a, 0)$ for the circle $x^2 + y^2 = a^2/2$ is $(-a.x) + 0.y = a^2/2$ i.e., $x = -a/2$

and the chord of contact at $A(-a,0)$ for the parabola $y^2 = 4ax$ is

$0.y = 2a(x-a)$ i.e., $x = a$

Note that DE is latus recturn of parabola $y^2 = 4ax$, therefore its lengths is $4a$.

Chords of contact are clearly parallel to each other, so req. quadrilateral is a trapezium.

$$Ar\text{ (trap } BCDE) = \frac{1}{2}(BC + DE) \times KL$$

$$= \frac{1}{2}(a+4a)\left(\frac{3a}{2}\right) = \frac{15a^2}{4}$$

10. The given ellipses are

$$\frac{x^2}{4} + \frac{y^2}{1} = 1 \quad ...(1)$$

and $\frac{x^2}{6} + \frac{y^2}{3} = 1$ ...(2)

Then the equation of tangent to (1) at any point $T(2\cos\theta, \sin\theta)$ is given by

$$\frac{x.2\cos\theta}{4} + \frac{y.\sin\theta}{1} = 1$$

or $\frac{x\cos\theta}{2} + y\sin\theta = 1$ ...(3)

Let this tangent meet the ellipse (2) at P and Q.

Let the tangents drawn to ellipse (2) at P and Q meet each other at $R(x_1, y_1)$

Then PQ is chord of contact of ellipse (2) with respect to the pt $R(x_1, y_1)$ and is given by

$$\frac{xx_1}{6} + \frac{yy_1}{3} = 1 \quad ...(4)$$

Clearly equations (3) and (4) represent the same lines and hence should be identical. Therefore comparing the cofficients, we get

$$\frac{\frac{\cos\theta}{2}}{\frac{x_1}{6}} = \frac{\sin\theta}{\frac{y_1}{3}} = \frac{1}{1}$$

$\Rightarrow x_1 = 3\cos\theta, y_1 = 3\sin\theta \Rightarrow x_1^2 + y_1^2 = 9$

$\Rightarrow$ Locus of (x_1, y_1) is $x^2 + y^2 = 9$

which is the director circle of the ellipse $\frac{x^2}{6} + \frac{y^2}{3} = 1$ and

KEY CONCEPT : We know that the director circle is the locus of intersection point of the tangents which are at right $\angle$.

Thus tangents at P and Q are at right $\angle$'s.

11. Let $P(e, f)$ be any point on the locus. Equation of pair of tangents from $P(e, f)$ to the parabola $y^2 = 4ax$ is

$[fy - 2a(x+e)]^2 = (f^2 - 4ae)(y^2 - 4ax)$ $[T^2 = SS_1]$

Here, a = coefficient of $x^2 = 4a^2$ ...(1)

$2h$ = coefficient of $xy = -4af$ ...(2)

and b = coefficient of $y^2 = f^2 - (f^2 - 4ae) = 4ae$ ...(3)

If they include an angle 45°, then

$$1 = \tan 45° = \frac{2\sqrt{h^2 - ab}}{a+b}$$

or, $(a+b)^2 = 4(h^2 - ab)$

or, $(4a^2 + 4ae)^2 = 4[4a^2f^2 - (4a^2)(4ae)]$

or, $(a+e)^2 = f^2 - 4ae$ or $e^2 + 6ae + a^2 - f^2 = 0$

or $(e+3a)^2 - f^2 = 8a^2$

Hence the required locus is $(x+3a)^2 - y^2 = 8a^2$, which is a hyperbola.

ALTERNATE SOLUTION

Let pt P be (x_1, y_1) from which a pair of tangents is drawn to parabola $y^2 = 4ax$, where the angle between two tangent is 45°.

Let any one of these tangents be $y = mx + a/m$ where m is the slope of tangent. As it passes through (x_1, y_1) we get

$y_1 = mx_1 + a/m \Rightarrow m^2x_1 - my_1 + a = 0$

which is quadratic in m.

If its roots are m_1 and m_2 then these must be the slopes of two tangents drawn from P to parabola.

Also $m_1 + m_2 = \frac{y_1}{x_1}, m_1 m_2 = \frac{a}{x_1}$

$$m_1 - m_2 = \sqrt{(m_1+m_2)^2 - 4m_1m_2} = \sqrt{\left(\frac{y_1}{x_1}\right)^2 - \frac{4a}{x_1}}$$

$$\Rightarrow \quad (m_1 - m_2)^2 = \frac{y_1^2 - 4ax_1}{x_1^2}$$

Now, angle between tangents, as given, is 45°

$$\therefore \quad \tan 45° = \left|\frac{m_1 - m_2}{1 + m_1 m_2}\right| \Rightarrow 1 = \frac{(m_1 - m_2)^2}{(1+m_1m_2)^2}$$

$$\Rightarrow \quad \left(1 + \frac{a}{x_1}\right)^2 = \frac{y_1^2 - 4ax_1}{x_1^2} \Rightarrow (x_1 + a)^2 = y_1^2 - 4ax_1$$

$$\Rightarrow \quad x_1^2 + 6ax_1 + a^2 - y_1^2 = 0 \Rightarrow (x_1 + 3a)^2 - y_1^2 = 8a^2$$

$$\Rightarrow \quad \frac{(x_1+3a)^2}{8a^2} - \frac{y_1^2}{8a^2} = 1$$

$\therefore$ Locus of (x_1, y_1) is $\frac{(x+3a)^2}{8a^2} - \frac{y^2}{8a^2} = 1$

Which is the equation of hyperbola.

12. Let any point P on ellipse $4x^2 + 25y^2 = 100$ be $(5 \cos \theta, 2 \sin \theta)$. So equation of tangent to the ellipse at P will be

$$\frac{x\cos\theta}{5} + \frac{y \sin\theta}{2} = 1$$

Tangent (1) also touches the circle $x^2 + y^2 = r^2$, so its distance from origin must be r.

Tangent (2) intersects the coordinate axes at $A\left(\frac{5}{\cos\theta}, 0\right)$ and $B\left(0, \frac{2}{\sin\theta}\right)$ respectively. Let $M(h, k)$ be the midpoint of line segment AB. Then by mid point formula

$$h = \frac{5}{2\cos\theta}, k = \frac{1}{\sin\theta}$$

$$\Rightarrow \quad \cos\theta = \frac{5}{2h}, \quad \sin\theta = \frac{1}{k}$$

$$\Rightarrow \quad \cos^2\theta + \sin^2\theta = \frac{25}{4h^2} + \frac{1}{k^2}$$

Hence locus of $M(h, k)$ is $\frac{25}{x^2} + \frac{4}{y^2} = 4$

Locus is independent of r.

ALTERNATE SOLUTION

Equation of any tangent to circle $x^2 + y^2 = r^2$ is

$x \cos\theta + y \sin\theta = r$ ...(1)

If it is tangent to ellipse $4x^2 + 25y^2 = 100$ at (x_1, y_1). Then its equation should be

$4xx_1 + 25yy_1 = 100$ ...(2)

As (1) and (2) represent the same line, these should be identical.

$$\therefore \quad \frac{4x_1}{\cos\theta} = \frac{25y_1}{\sin\theta} = \frac{100}{r}$$

$$\Rightarrow \quad x_1 = \frac{25\cos\theta}{r}, y_1 = \frac{4\sin\theta}{r}$$

$$\Rightarrow \quad \cos\theta = \frac{rx_1}{25}, \sin\theta = \frac{ry_1}{4} \quad ...(3)$$

Now (1) meets axes at

$$A\left(\frac{r}{\cos\theta}, 0\right) \; B\left(0, \frac{r}{\sin\theta}\right)$$

$\therefore$ For mid pt (h, k) of AB

$$h = \frac{r}{2\cos\theta}, k = \frac{r}{2\sin\theta}$$

But $\cos\theta = \frac{r}{2h}, \sin\theta = \frac{r}{2k}$ ...(4)

From (3) and (4),

$$x_1 = \frac{25}{2h}, \quad y_1 = \frac{4}{2k}$$

But (x_1, y_1) lies on ellipse (being pt of contact)

$$\therefore \quad 4x_1^2 + 25y_1^2 = 100$$

$$\Rightarrow \quad 4 \times \frac{625}{4h^2} + 25 \times \frac{16}{4k^2} = 100 \Rightarrow \frac{25}{h^2} + \frac{4}{k^2} = 4$$

$\therefore$ Locus of (h, k) is $\frac{25}{x^2} + \frac{4}{y^2} = 4$

13. The ellipse is $\frac{x^2}{a^2} + \frac{y^2}{b^2} = 1$ ...(1)

Since this ellipse is symmetrical in all four quadrants, either there exists no such P or four points, one in each qudrant. Without loss of generality we can assume that $a > b$ and P lies in first quadrant.

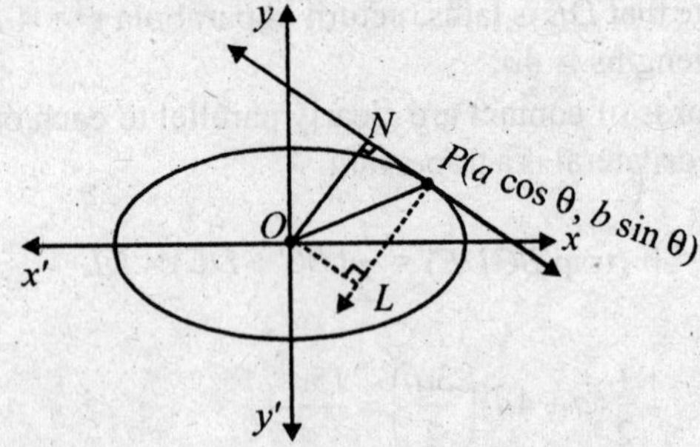

Let $P(a\cos\theta, b\sin\theta)$ then equation of tangent is

$$\frac{x}{a}\cos\theta + \frac{y}{b}\sin\theta = 1$$

$$\therefore \quad ON = \frac{ab}{\sqrt{b^2\cos^2\theta + a^2\sin^2\theta}}$$

Equation of ON is, $\frac{x}{b}\sin\theta - \frac{y}{a}\cos\theta = 0$

Equation of normal at P is
$ax \sec\theta - by \operatorname{cosec}\theta = a^2 - b^2$

$$\therefore OL = \frac{a^2 - b^2}{\sqrt{a^2 \sec^2\theta + b^2 \operatorname{cosec}^2\theta}} = \frac{(a^2 - b^2)\sin\theta\cos\theta}{\sqrt{a^2 \sin^2 + b^2 \cos^2\theta}}$$

and $NP = OL$

$$\therefore \quad NP = \frac{(a^2 - b^2)\sin\theta\cos\theta}{\sqrt{a^2 \sin^2\theta + b^2 \cos^2\theta}}$$

$$\therefore \quad Z = \text{Area of } OPN = \frac{1}{2} \times ON \times NP$$

$$= \frac{1}{2} ab(a^2 - b^2) \frac{\sin\theta\cos\theta}{a^2 \sin^2\theta + b^2 \cos^2\theta}$$

Let $u = \dfrac{a^2 \sin^2\theta + b^2 \cos^2\theta}{\sin\theta\cos\theta} = a^2 \tan\theta + b^2 \cot\theta$

$\dfrac{du}{d\theta} = a^2 \sec^2\theta - b^2 \operatorname{cosec}^2\theta = 0 \Rightarrow \tan\theta = b/a$

$\left(\dfrac{d^2u}{d\theta^2}\right)_{\tan^{-1} b/a} > 0$, u is minimum at $\theta = \tan^{-1} b/a$

So Z is maximum at $\theta = \tan^{-1} b/a$

$$\therefore \quad P \text{ is } \left(\frac{a^2}{\sqrt{a^2 + b^2}}, \frac{b^2}{\sqrt{a^2 + b^2}}\right)$$

By symmetry, we have four such points

$$\left(\pm\frac{a^2}{\sqrt{a^2 + b^2}}, \pm\frac{b^2}{\sqrt{a^2 + b^2}}\right)$$

14. Let A, B, C be the point on circle whose coordinates are $A = [a\cos\theta, a\sin\theta]$

$$B = \left[a\cos\left(\theta + \frac{2\pi}{3}\right), a\sin\left(\theta + \frac{2\pi}{3}\right)\right]$$

and $C = \left[a\cos\left(\theta + \dfrac{4\pi}{3}\right), a\sin\left(\theta + \dfrac{4\pi}{3}\right)\right]$

Further, $P[a\cos\theta, b\sin\theta]$ (Given)

$$Q = \left[a\cos\left(\theta + \frac{2\pi}{3}\right), b\sin\left(\theta + \frac{2\pi}{3}\right)\right]$$

and $R = \left[a\cos\left(\theta + \dfrac{4\pi}{3}\right), b\sin\left(\theta + \dfrac{4\pi}{3}\right)\right]$

It is given that P, Q, R are on the same side of x-axis as A, B, C.

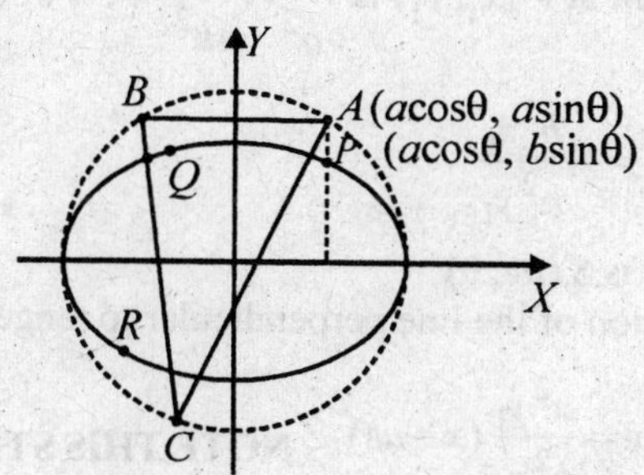

So required normals to the ellipse are

$ax \sec\theta - by \operatorname{cosec}\theta = a^2 - b^2$ ...(1)

$ax\sec\left(\theta + \dfrac{2\pi}{3}\right) - by\operatorname{cosec}\left(\theta + \dfrac{2\pi}{3}\right) = a^2 - b^2$ ...(2)

$ax\sec\left(\theta + \dfrac{4\pi}{3}\right) - by\operatorname{cosec}\left(\theta + \dfrac{4\pi}{3}\right) = a^2 - b^2$ ...(3)

Now, above three normals are concurrent

$\Rightarrow \quad \Delta = 0$

$$\text{where } \Delta = \begin{vmatrix} \sec\theta & \operatorname{cosec}\theta & 1 \\ \sec\left(\theta + \frac{2\pi}{3}\right) & \operatorname{cosec}\left(\theta + \frac{2\pi}{3}\right) & 1 \\ \sec\left(\theta + \frac{4\pi}{3}\right) & \operatorname{cosec}\left(\theta + \frac{4\pi}{3}\right) & 1 \end{vmatrix}$$

Multiplying and dividing the different rows R_1, R_2 and R_3 by $\sin\theta\cos\theta$,

$$\sin\left(\theta + \frac{2\pi}{3}\right)\cos\left(\theta + \frac{2\pi}{3}\right)$$

and $\sin\left(\theta + \dfrac{4\pi}{3}\right)\cos\left(\theta + \dfrac{4\pi}{3}\right)$ respectively, we get

$$\Delta = \frac{1}{\sin\theta\cos\theta\sin\left(\theta + \frac{2\pi}{3}\right)\cos\left(\theta + \frac{2\pi}{3}\right)\sin\left(\theta + \frac{4\pi}{3}\right)\cos\left(\theta + \frac{4\pi}{3}\right)} \times$$

$$\begin{vmatrix} \sin\theta & \cos\theta & \sin 2\theta \\ \sin\left(\theta + \frac{2\pi}{3}\right) & \cos\left(\theta + \frac{2\pi}{3}\right) & \sin\left(2\theta + \frac{4\pi}{3}\right) \\ \sin\left(\theta - \frac{2\pi}{3}\right) & \cos\left(\theta - \frac{2\pi}{3}\right) & \sin\left(2\theta - \frac{4\pi}{3}\right) \end{vmatrix} = 0$$

[Operating $R_2 \to R_2 + R_3$ and simplyfing R_2 we get $R_2 \equiv R_1$]

15. Given that

$C_1 : x^2 = y - 1$; $\quad C_2 : y^2 = x - 1$

Let $P\ (x_1, x_1^2 + 1)$ on C_1 and $Q\ (y_2^2 + 1, y_2)$ on C_2.

Now the reflection of pt P in the line $y = x$ can be obtained by interchanging the values of abscissa and ordiante.

Thus reflection of pt. $P\ (x_1, x_1^2 + 1)$ is $P_1(x_1^2 + 1, x_1)$

and reflection of pt. $Q\ (y_2^2 + 1, y_2)$ is $Q_1(y_2, y_2^2 + 1)$

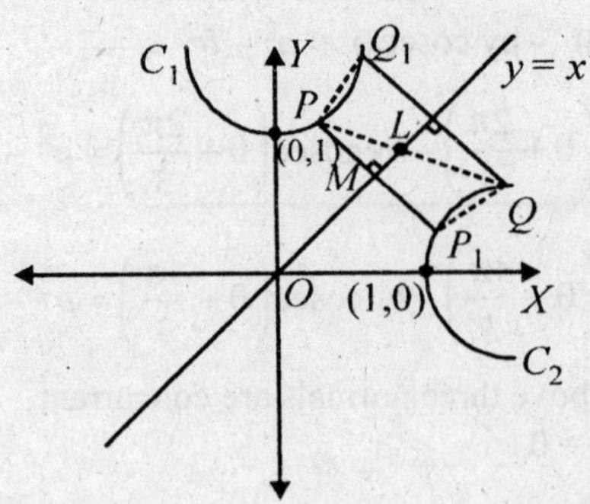

It can be seen clearly that P_1 lies on C_2 and Q_1 on C_1. Now PP_1 and QQ_1 both are perpendicular to mirror line $y = x$. Also M is mid pt. of PP_1 ($\because$ P_1 is morror image of P in $y = x$)

$\therefore \quad PM = \frac{1}{2}PP_1$

In rt ΔPML,

$PL > PM$

$\Rightarrow \quad PL > \frac{1}{2}PP_1$ ...(i)

Similarly,

$LQ > \frac{1}{2}QQ_1$ ...(ii)

Adding (i) and (ii) we get

$$PL + LQ > \frac{1}{2}(PP_1 + QQ_1)$$

$$\Rightarrow \quad PQ > \frac{1}{2}(PP_1 + QQ_1)$$

$\Rightarrow$ PQ is more than the mean of PP_1 and QQ_1

$\Rightarrow$ $PQ \geq \min(PP_1, QQ_1)$ **NOTE THIS STEP**

Let $\min(PP_1, QQ_1) = PP_1$

then $PQ^2 \geq PP_1^2 = (x_1^2 + 1 - x_1)^2 + (x_1^2 + 1 - x_1)^2$

$= 2(x_1^2 + 1 - x_1)^2 = f(x_1)$

$\Rightarrow \quad f'(x_1) = 4(x_1^2 + 1 - x_1)(2x_1 - 1)$

$$= 4\left(\left(x_1 - \frac{1}{2}\right)^2 + \frac{3}{4}\right)(2x_1 - 1)$$

$\therefore \quad f'(x_1) = 0$ when $x_1 = \frac{1}{2}$

Also $f'(x_1) < 0$ if $x_1 < \frac{1}{2}$

and $f'(x_1) > 0$ if $x_1 > \frac{1}{2}$

$\Rightarrow \quad f(x_1)$ is min when $x_1 = \frac{1}{2}$

Thus if at $x_1 = \frac{1}{2}$ pt P is P_0 on C_1

$$P_0\left(\frac{1}{2}, \left(\frac{1}{2}\right)^2 + 1\right) = \left(\frac{1}{2}, \frac{5}{4}\right)$$

Similarly Q_0 on C_2 will be image of P_0 with respect to $y = x$

$$\therefore \quad Q_0\left(\frac{5}{4}, \frac{1}{2}\right)$$

16. Let the co-ordinates of P be $(a\cos\theta, b\sin\theta)$ then co-ordinates of Q are $(a\cos\theta, a\sin\theta)$

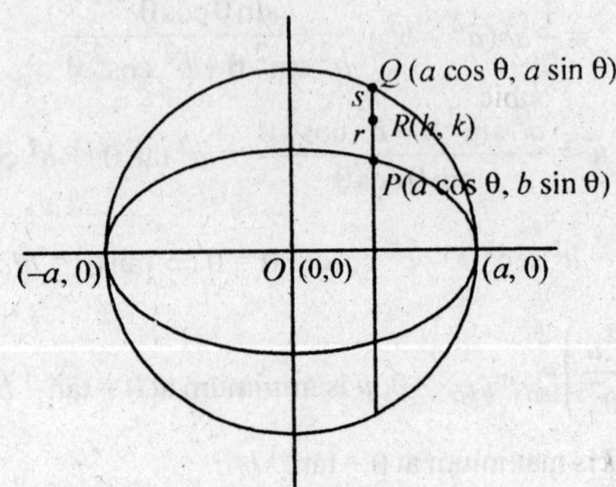

As $R(h, k)$ divides PQ in the ratio $r : s$, then

$$h = \frac{s.(a\cos\theta) + r(a\cos\theta)}{(r+s)} = a\cos\theta$$

$$\Rightarrow \quad \cos\theta = \frac{h}{a}$$

$$k = \frac{s(b\sin\theta) + r(a\sin\theta)}{(r+s)} = \frac{\sin\theta(bs + ar)}{(r+s)}$$

$$\Rightarrow \quad \sin\theta = \frac{k(r+s)}{(bs+ar)} \quad \because \cos^2\theta + \sin^2\theta = 1$$

$$\therefore \quad \frac{h^2}{a^2} + \frac{k^2(r+s)^2}{(bs+ar)^2} = 1.$$

Hence locus of R is $\frac{x^2}{a^2} + \frac{y^2(r+s)^2}{(bs+ar)^2} = 1$ which is equation of an ellipse.

17. Let the ellipse be $\frac{x^2}{a^2} + \frac{y^2}{b^2} = 1$ and O be the centre.

Tangent at $P(x_1, y_1)$ is $\frac{xx_1}{a^2} + \frac{yy_1}{b^2} - 1 = 0$ whose

slope $= -\frac{b^2 x_1}{a^2 y_1}$

Focus is $S(ae, 0)$.

Equation of the line perpendicular to tangent at P is

$$y = \frac{a^2 y_1}{b^2 x_1}(x - ae) \quad \textbf{NOTE THIS STEP} \quad ....(1)$$

Equation of OP is $y = \frac{y_1}{x_1}x$ ...(2)

(1) and (2) intersect $\Rightarrow \frac{y_1}{x_1}x = \frac{a^2 y_1}{b^2 x_1}(x - ae)$

$\Rightarrow\ x(a^2 - b^2) = a^3 e \Rightarrow x.\, a^2e^2 = a^3 e$

$\Rightarrow\ x = a/e$

Which is the corresponding directrix.

18. Let P be the pt. (h, k). Then eqn of normal to parabola $y^2 = 4x$ from point (h, k), if m is the slope of normal, is

$y = mx - 2m - m^3$

As it passes through (h, k), therefore

$mh - k - 2m - m^3 = 0$

or, $m^3 + (2 - h) + k = 0$ ...(1)

Whcih is cubic in m, giving three values of m say m_1, m_2 and m_3. Then $m_1 m_2 m_3 = -k$ (from eq^n) but given that $m_1 m_2 = \alpha$

$\therefore$ We get $m_3 = -\frac{k}{\alpha}$

But m_3 must satisfy eq^n (1)

$\therefore\ \frac{-k^3}{\alpha^3} + (2 - h)\left(\frac{-k}{\alpha}\right) + k = 0$

$\Rightarrow\ k^2 - 2\alpha^2 - h\alpha^2 - \alpha^3 = 0$

$\therefore$ Locus of $P(h, k)$ is $y^2 = \alpha^2 x + (\alpha^3 - 2\alpha^2)$

But ATQ, locus of P is a part of parabola $y^2 = 4x$, therefore comparing the two, we get $\alpha^2 = 4$ and $\alpha^3 - 2\alpha^2 = 0$

$\Rightarrow \alpha = 2$

ALTERNATE SOLUTION

Let $P(t_1^2, 2t_1)$ and $Q(t_2^2, 2t_2)$ be two pts on parabola $y^2 = 4x$. Then equations of normal at P and Q are respectively.

$xt_1 + y = 2t_1 + t_1^3$ ...(1)

and $xt_2 + y = 2t_2 + t_2^3$ ...(2)

Pt. of intersection of normals at P and Q can be obtained by solving (1) and (2) as $(2 + t_1^2 + t_2^2 + t_1 t_2, -t_1 t_2(t_1 + t_2))$

ATQ, this pt. of intersection lies on parabola $y^2 = 4x$, therefore we get

$t_1^2 t_2^2 [t_1 + t_2]^2 = 8 + 4[(t_1 + t_2)^2 - t_1 t_2]$

$\Rightarrow\ (t_1 + t_2)^2[t_1^2 t_2^2 - 4] + 4(t_1 t_2 - 2) = 0$

$\Rightarrow\ (t_1 t_2 - 2)[(t_1 + t_2)^2 (t_1 t_2 + 2) + 4] = 0$

$\Rightarrow\ t_1 t_2 - 2 = 0$

$\Rightarrow\ t_1 t_2 = 2$

$\Rightarrow\ (-t_1)(-t_2) = 2$

$\Rightarrow$ (slope of normal at P) (slope of normal at Q) = 2

$\Rightarrow\ m_1 m_2 = 2$

$\Rightarrow\ \alpha = 2$ $(\because$ ATQ $m_1 m_2 = \alpha)$

19. The given eq^n of parabola is

$y^2 - 2y - 4x + 5 = 0$ ...(1)

$\Rightarrow\ (y - 1)^2 = 4(x - 1)$

Any parametric point on this parabola is

$P(t^2 + 1, 2t + 1)$ **NOTE THIS STEP**

Differentiating (1) w.r. to x, we get

$2y\frac{dy}{dx} - 2\frac{dy}{dx} - 4 = 0 \Rightarrow \frac{dy}{dx} = \frac{2}{y - 1}$

$\therefore$ Slope of tangent to (1) at point P $(t^2 + 1, 2t + 1)$ is

$m = \frac{2}{2t} = \frac{1}{t}$

$\therefore$ Eq^n of tangent at P $(t^2 + 1, 2t + 1)$ is

$y - (2t + 1) = \frac{1}{t}(x - t^2 - 1)$

$\Rightarrow\ yt - 2t^2 - t = x - t^2 = 1$

$\Rightarrow\ x - yt + (t^2 + t - 1) = 0$ ...(2)

Now directrix of given parabola is

$(x - 1) = -1 \Rightarrow x = 0$

Tangent (2) meets directix at $Q\left(0, \frac{t^2 + t - 1}{t}\right)$

Let pt. R be (h, k)

ATQ, R divides the line joing QP in the ratio $\frac{1}{2} : 1$ i.e., $1 : 2$ externally.

$\therefore\ (h, k) = \left(\frac{1(1 + t^2) - 0}{-1}, \frac{t + 2t^2 - 2t^2 - 2t + 2}{-t}\right)$

$\Rightarrow\ h = -(1 + t^2)$ and $k = \frac{t - 2}{t}$

$\Rightarrow\ t^2 = -1 - h$ and $t = \frac{2}{1 - k}$

Eliminating t, we get $\left(\frac{2}{1 - k}\right)^2 = -1 - h$

$\Rightarrow\ 4 = -(1 - k)^2(1 - h)$

$\Rightarrow\ (h - 1)(k - 1)^2 + 4 = 0$

$\therefore$ Locus of R (h, k) is $(x - 1)(y - 1)^2 + 4 = 0$

20. Any pt on the hyperbola $\frac{x^2}{9} - \frac{y^2}{4} = 1$ is $(3\,\text{sce}\,\theta, 2\tan\theta)$

Then, equation of chord of contact to the circle $x^2 + y^2 = 9$, with respect to the pt. $(3\sec\theta, 2\tan\theta)$ is

$(3\sec\theta)x + (2\tan\theta)y = 9$ ...(i)

If (h, k) be the mid point of chord of contact then equation of chord of contact will be

$hx + ky - 9 = h^2 + k^2 - 9$ $(T = S_1)$

or, $hx + ky = h^2 + k^2$ ...(ii)

But equations (i) and (ii) represent the same st. line and hence should be identical, therefore, we get

$\frac{3\sec\theta}{h} = \frac{2\tan\theta}{k} = \frac{9}{h^2 + k^2}$

$\Rightarrow\ \sec\theta = \frac{3h}{h^2 + k^2},\ \tan\theta = \frac{9k}{2(h^2 + k^2)}$

$\sec^2\theta - \tan^2\theta = 1$

$\Rightarrow \quad \frac{9h^2}{(h^2+k^2)^2} - \frac{81k^2}{4(h^2+k^2)^2} = 1$

$\Rightarrow \quad 4h^2 - 9k^2 = \frac{4}{9}(h^2+k^2)^2$

or, $\frac{h^2}{9} - \frac{k^2}{4} = \left(\frac{h^2+k^2}{9}\right)^2$

$\therefore$ Locus of (h, k) is $\frac{x^2}{9} - \frac{y^2}{4} = \left(\frac{x^2+y^2}{9}\right)^2$

21. Let the common tangent to circle $x^2 + y^2 = 16$ and ellipse $x^2/25 + y^2/4 = 1$ be

$y = mx + \sqrt{25m^2+4}$ **NOTE THIS STEP** ..(i)

As it is tangent to circle $x^2 + y^2 = 16$, we should have

$\frac{\sqrt{25m^2+4}}{\sqrt{m^2+1}} = 4$

[Using : length of perpendicular from (0,0) to (1) = 4]

$\Rightarrow \quad 25m^2 + 4 = 16m^2 + 16 \Rightarrow 9m^2 = 12$

$\Rightarrow \quad m = \frac{-2}{\sqrt{3}}$

[Leaving + ve sign to consider tangent in I quadrant]

$\therefore$ Equation of common tangent is

$y = -\frac{2}{\sqrt{3}}x + \sqrt{25.\frac{4}{3}+4}$

$\Rightarrow \quad y = -\frac{2}{\sqrt{3}}x + 4\sqrt{\frac{7}{3}}$

This tangent meets the axes at $A(2\sqrt{7}, 0)$ and $B\left(0, 4\sqrt{\frac{7}{3}}\right)$

$\therefore$ Length of intercepted portion of tangent between axes

$= AB = \sqrt{(2\sqrt{7}) + \left(4\sqrt{\frac{7}{3}}\right)^2} = 14/\sqrt{3}$

(F) MATCH THE FOLLOWING :

1. Let $y = mx - 2m - m^3$ be the equation of normal to $y^2 = 4x$. As it passes through (3, 0), we get $m = 0, 1, -1$

Then three points on parabola are given by $(m^2, -2m)$ for $m = 0, 1, -1$

$\therefore$ $P(0, 0), Q(1, 2), R(1, -2)$

$\therefore$ Area of $\Delta PQR = \frac{1}{2} \times \begin{vmatrix} 0 & 0 & 1 \\ 1 & 2 & 1 \\ 1 & -2 & 1 \end{vmatrix} = 2$ sq. units

Radius of circum-circle,

$R = \frac{abc}{4\Delta} = \frac{\sqrt{5}\times\sqrt{5}\times 4}{4\times 2} = \frac{5}{2}$ **NOTE THIS STEP**

(where, a, b, c are the sides of ΔPQR)

Centroid of $\Delta PQR = \left(\frac{2}{3}, 0\right)$

Circumcentre $= \left(\frac{5}{2}, 0\right)$

Thus, **(A) - (p); (B) - (q); (C) – (s); (D) – (r)**

2. (A) - p, q

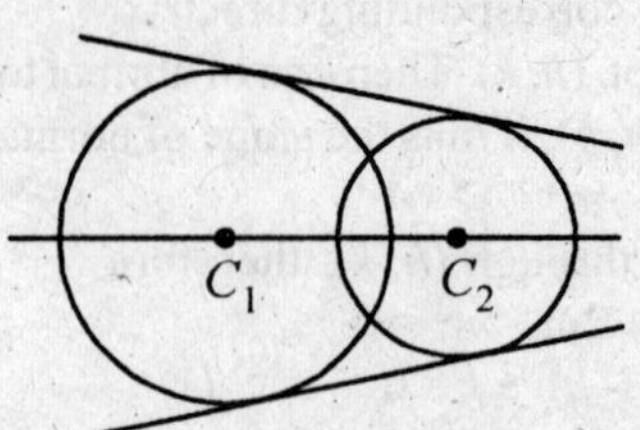

It is clear from the figure that two intersecting circles have a common tangent and a common normal joining the centres

(B) - p, q

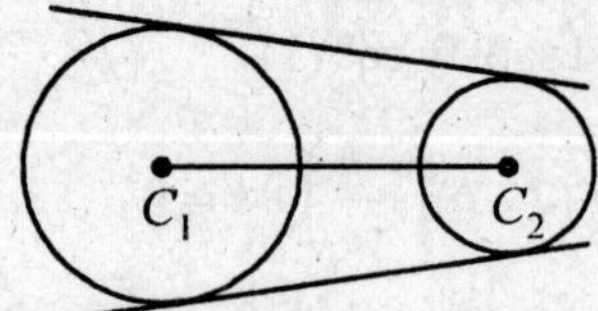

(C) – q, r

Two circle when one is completely inside the other have a common normal C_1C_2 but no common tangent.

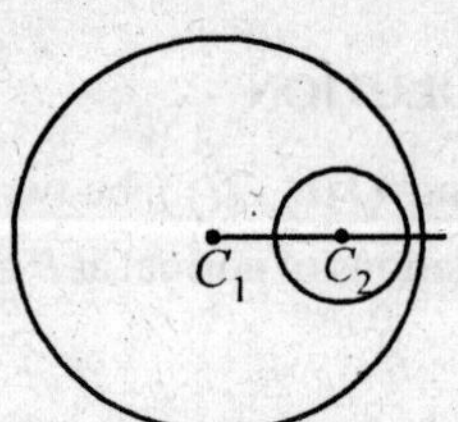

(D) – q, r

Two branches of hyperbola have no common tangent but have a common normal joining S_1S_2.

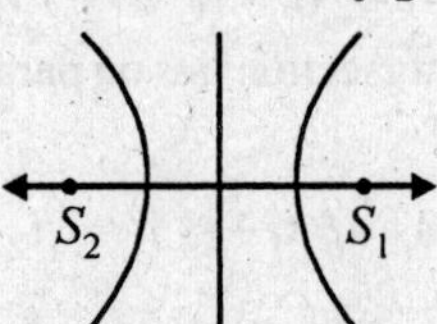

Matrix Match

(A) – p, q; (B) – p, q; (C) – q, r; (D) – q, r

3. $A \rightarrow p;\ B \rightarrow s,t;\ C \rightarrow r;\ D \rightarrow q,s$

(p) As the line $hx + ky = 1$, touches the circle $x^2 + y^2 = 4$

$\therefore$ Length of perpendicular from centre (0 , 0) of circle to line = radius of the circle

$\Rightarrow \quad \frac{1}{\sqrt{h^2+k^2}} = 2 \Rightarrow h^2 + k^2 = \frac{1}{4}$

Locus of (h, k) is $x^2 + y^2 = \frac{1}{4}$, which is a circle.

(q) We know that if $|z-z_1|-|z-z_2|=k$ where $|k|<|z_1-z_2|$ then z traces a hyperbola.
Here $|z+2|-|z-2|=\pm 3$
$\therefore$ Locus of z is a hyperbola.

(r) We have $x=\sqrt{3}\left(\frac{1-t^2}{1+t^2}\right), y=\frac{2t}{1+t^2}$

$\Rightarrow \frac{x}{\sqrt{3}}=\frac{1-t^2}{1+t^2}$ and $y=\frac{2t}{1+t^2}$

On squaring and adding, we get

$$\frac{x^2}{3}+y^2=\frac{(1-t^2)^2+4t^2}{(1+t^2)^2}=1 \text{ or } \frac{x^2}{3}+\frac{y^2}{1}=1$$

which is the equation of an ellipse.

(s) We know eccentricity for a parabola = 1
for an ellipse < 1
for a hyperbola > 1

$\therefore$ The conics whose eccentricity lies in $1 \le x < \infty$ are parabola and hyperbola.

(t) Let $z=x+iy$ then

Re $[(x+1)+iy]^2 = x^2+y^2+1$

$\Rightarrow (x+1)^2-y^2=x^2+y^2+1$

$\Rightarrow y^2=x$, which is a parabola.

4. **(a)** Equation of tangent to $y^2=16x$ at F (x_0, y_0)
$yy_0=8(x+x_0)$

$\Rightarrow G\left(0, \frac{8x_0}{y_0}\right)$

Area of $\Delta EFG=\frac{1}{2}\times(3-y_1)\times x_0$

$$A=\frac{1}{2}x_0\left(3-\frac{8x_0}{y_0}\right)$$

$$A=\frac{1}{2}\times\frac{y_0^2}{16}\left(3-\frac{y_0}{2}\right)=\frac{1}{32}\left(3y_0^2-\frac{y_0^3}{2}\right)$$

$$\frac{dA}{dy_0}=\frac{1}{32}\left(6y_0-\frac{3y_0^2}{2}\right)$$

$$\frac{dA}{dy_0}=0 \Rightarrow y_0=4 \Rightarrow x_0=1$$

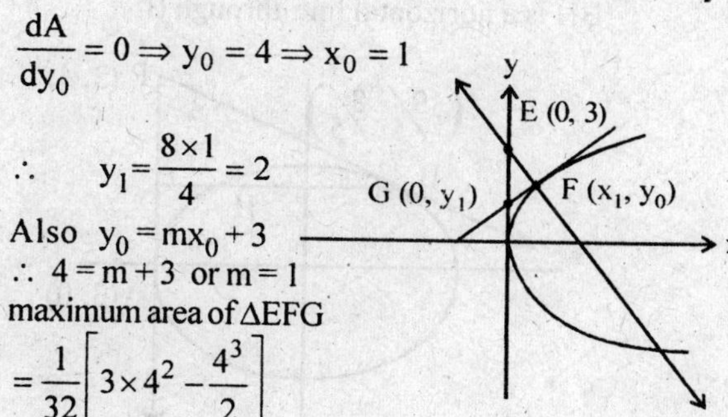

$\therefore y_1=\frac{8\times 1}{4}=2$

Also $y_0=mx_0+3$
$\therefore 4=m+3$ or $m=1$
maximum area of ΔEFG

$$=\frac{1}{32}\left[3\times 4^2-\frac{4^3}{2}\right]$$

$$=\frac{1}{32}[48-32]=\frac{1}{2}$$

$\therefore$ (P)→(4), (Q)→(1), (R)→(2), (S)→(3)

G COMPREHENSION BASED QUESTIONS :

For Qs. 1-3 :

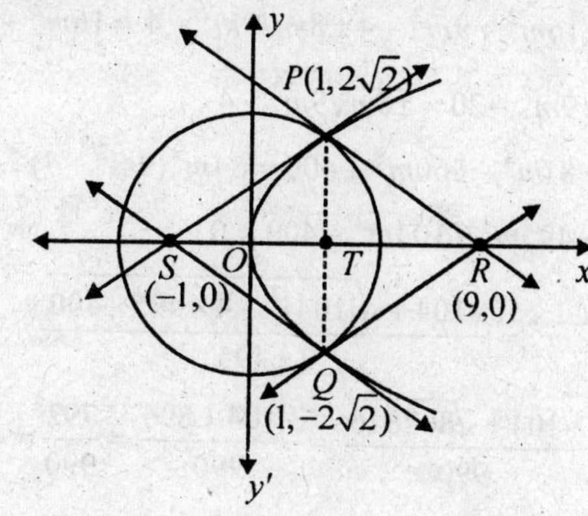

1. **(c)** $\frac{Ar\,\Delta PQS}{Ar\,\Delta PQR}=\frac{\frac{1}{2}PQ\times ST}{\frac{1}{2}PQ\times TR}=\frac{ST}{TR}=\frac{2}{8}=\frac{1}{4}a$

2. **(b)** For ΔPRS,

$$Ar(\Delta PRS)=\Delta=\frac{1}{2}\times SR\times PT=\frac{1}{2}\times 10\times 2\sqrt{2}$$

$\therefore \Delta=10\sqrt{2}, a=PS=2\sqrt{3}$

$b=PR=6\sqrt{2}, c=SR=10$

$\therefore$ Radius of circumference

$$=R=\frac{abc}{4\Delta}=\frac{2\sqrt{3}\times 6\sqrt{2}\times 10}{4\times 10\sqrt{2}}=3\sqrt{3}$$

3. **(d)** Radius of incircle

$$=\frac{\text{area of }\Delta PQR}{\text{semi perimeter of }\Delta PQR}=\frac{\Delta}{s}$$

We have $a=PR=6\sqrt{2}$, $b=QP=PR=6\sqrt{2}$
$c=PQ=4\sqrt{2}$

and $\Delta=\frac{1}{2}\times PQ\times TR=16\sqrt{2}$

$$\therefore s=\frac{6\sqrt{2}+6\sqrt{2}+4\sqrt{2}}{2}=8\sqrt{2}$$

$$\therefore r=\frac{16\sqrt{2}}{8\sqrt{2}}=2$$

4. **(b)** We known that tangent to hyperbola $\frac{x^2}{a^2}-\frac{y^2}{b^2}=1$ is $y=mx\pm\sqrt{a^2m^2-b^2}$

$\therefore$ The tangent to hyperbola $\frac{x^2}{9}-\frac{y^2}{4}=1$ is

$y=mx\pm\sqrt{9m^2-4}$ where $m>0$

As it is also the tangent to the circle

$x^2 + y^2 - 8x = 0$

With centre (4 ,0) and radius 4

$\therefore \left|\frac{4m \pm \sqrt{9m^2 - 4}}{\sqrt{m^2+1}}\right| = 4$

$\Rightarrow 16m^2 + 9m^2 - 4 \pm 8m\sqrt{9m^2-4} = 16m^2 + 16$

$\Rightarrow 9m^2 - 20 = \pm 8m\sqrt{9m^2-4}$

$\Rightarrow 81m^4 - 360m^2 + 400 = 64m^2(9m^2-4)$

$\Rightarrow 495m^4 + 104m^2 - 400 = 0$

$\Rightarrow m^2 = \frac{-104 + \sqrt{(104)^2 + 4 \times 495 \times 400}}{2 \times 495}$

$= \frac{-104 + \sqrt{802816}}{990} = \frac{-104 + 896}{990} = \frac{792}{990} = \frac{4}{5}$

$\Rightarrow m = \frac{2}{\sqrt{5}}$ as $m > 0$

$\therefore$ *the tangent is* $y = \frac{2}{\sqrt{5}}m + \frac{4}{\sqrt{5}}$

or $2x - \sqrt{5}y + 4 = 0$

ALTERNATIVE :

Any tangent to $\frac{x^2}{9} - \frac{y^2}{4} = 1$ is $\frac{x\sec\alpha}{3} - \frac{y\tan\alpha}{2} = 1$

It touches circle with center (4,0) and radius = 4

$\therefore \frac{\frac{4\sec\alpha - 3}{3}}{\sqrt{\frac{\sec^2\alpha}{9} + \frac{\tan^2\alpha}{4}}} = 4$

$\Rightarrow 16\sec^2\alpha - 24\sec\alpha + 9 = 144\left(\frac{\sec^2\alpha}{9} + \frac{\tan^2\alpha}{4}\right)$

$\Rightarrow 12\sec^2\alpha + 8\sec - 15 = 0 \Rightarrow \sec\alpha = \frac{5}{6}$ or $\frac{-3}{2}$

but $\sec\alpha = \frac{5}{6} < 1$ is not possible

$\therefore \sec\alpha = -3/2 \Rightarrow \tan\alpha = \pm\frac{\sqrt{5}}{2}$

$\therefore$ slope of tangent $= \frac{2\sec\alpha}{3\tan\alpha} = \frac{2(-3/2)}{3(-\sqrt{5}/2)}$

(for +ve value of m)

$= \frac{2}{\sqrt{5}}$

$\therefore$ Equation of tangent is $\frac{-x}{2} + \frac{y\sqrt{5}}{4} = 1$

or $2x - \sqrt{5}y + 4 = 0$

5. (a) The intersection points of given circle

$x^2 + y^2 - 8x = 0$ ...(1)

and hyperbola $4x^2 - 9y^2 - 36 = 0$ ...(2)

can be obtained by solving these equations Substituting value of y^2 from eqn (1) in eqn (2), we get

$4x^2 - 9(8x - x^2) = 36 \Rightarrow 13x^2 - 72x - 36 = 0$

$\Rightarrow x = 6, \frac{-6}{13} \Rightarrow y^2 = 12, \frac{-48}{13}\frac{-36}{169}$ *(not possible)*

$\therefore (6, 2\sqrt{3})$ and $(6, -2\sqrt{3})$ are points of intersection.

So eqn of required circle is

$(x-6)(x-6) + (y - 2\sqrt{3})(y + 2\sqrt{3}) = 0$

$\Rightarrow x^2 + 36 - 12x + y^2 - 12 = 0$

$\Rightarrow x^2 + y^2 - 12x + 24 = 0$

6. (d) Tangent to $\frac{x^2}{3^2} + \frac{y^2}{2^2} = 1$ at the point $(3\cos\theta, 2\sin\theta)$ is

$\frac{x\cos\theta}{3} + \frac{y\sin\theta}{2} = 1$

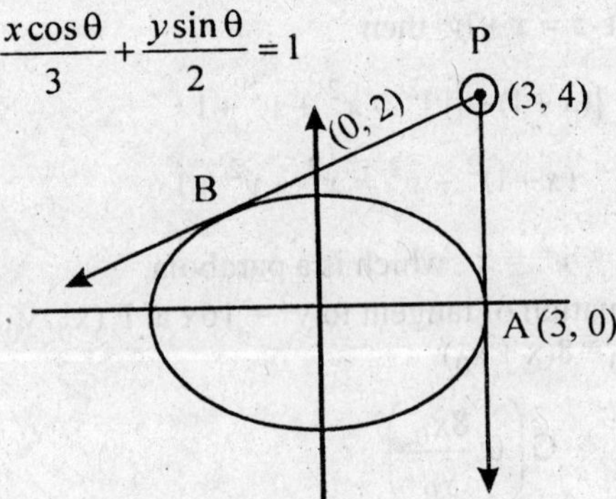

As it passes throught (3,4), we get

$\cos\theta + 2\sin\theta = 1$

$\Rightarrow 4\sin^2\theta = 1 + \cos^2\theta - 2\cos\theta$

$\Rightarrow 5\cos^2\theta - 2\cos\theta - 3 = 0$

$\Rightarrow \cos\theta = 1, -\frac{3}{5} \Rightarrow \sin\theta = 0, \frac{4}{5}$

$\therefore$ Required points are A (3,0) and B$\left(-\frac{9}{5}, \frac{8}{5}\right)$

7. (c) Let H be the orthocentre of ΔPAB, then as $BH \perp AP$, BH is a horizontal line through B.

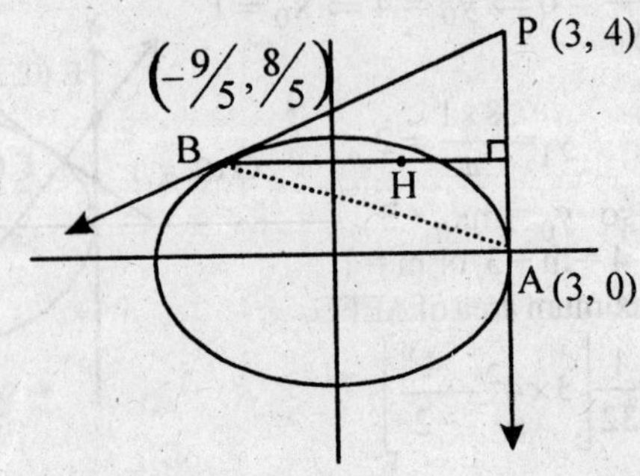

$\therefore$ y- cordinate of B = 8/5

Let H has coordinater $(\alpha, 8/5)$

Then slope of PH $= \dfrac{\frac{8}{5}-4}{\alpha-3} = \dfrac{-12}{5(\alpha-3)}$

and slope of AB $= \dfrac{\frac{8}{5}-0}{\frac{-9}{5}-3} = \dfrac{8}{-24} = \dfrac{-1}{3}$

But PH $\perp$ AB $\Rightarrow \dfrac{-12}{5(\alpha-3)} \times \left(\dfrac{-1}{3}\right) = -1$

$\Rightarrow 4 = -5\alpha + 15$ or $\alpha = 11/5$

Hence $H\left(11/5, 8/5\right)$.

8. **(a)** Clearly the moving point traces a parabola with focus at $P(3,4)$ and directrix as

AB: $\dfrac{y-0}{x-3} = \dfrac{-1}{3}$ or $x+3y-3=0$

$\therefore$ Equation of parabola is

$$(x-3)^2+(y-4)^2 = \frac{(x+3y-3)^2}{10}$$

or $9x^2+y^2-6xy-54x-62y+241=0$

9. **(b)** $PQ = \sqrt{\left(at^2-\dfrac{a}{t^2}\right)^2+\left(2at+\dfrac{2a}{t}\right)^2}$

$= a\sqrt{\left(t+\dfrac{1}{t}\right)^2\left(t-\dfrac{1}{t}\right)^2+4\left(t+\dfrac{1}{t}\right)^2}$

$= a\left(t+\dfrac{1}{t}\right)\sqrt{\left(t-\dfrac{1}{t}\right)^2+4}$

$= a\left(t+\dfrac{1}{t}\right)^2 = 5a$

10. **(d)** As PQ is the focal chord of $y^2 = 4ax$

$\therefore$ Coordinates of P and Q can be taken as

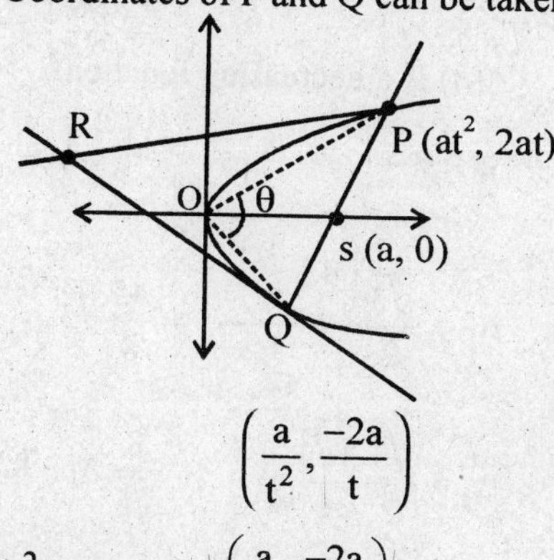

$\left(\dfrac{a}{t^2}, \dfrac{-2a}{t}\right)$

$P(at^2, 2at)$ and $Q\left(\dfrac{a}{t^2}, \dfrac{-2a}{t}\right)$

Tangents at P and Q are

$y = \dfrac{x}{t} + at$ and $y = -xt - \dfrac{a}{t}$

which intersect each other at $R\left(-a, a\left(t-\dfrac{1}{t}\right)\right)$

As R lies on the $y = 2x + a,\ a > 0$

$\therefore a\left(t-\dfrac{1}{t}\right) = -2a + a \Rightarrow t - \dfrac{1}{t} = -1$

$\Rightarrow t + \dfrac{1}{t} = \sqrt{5}$

Now, $m_{OP} = \dfrac{2}{t}$ and $m_{OQ} = -2t$

$\therefore \tan\theta = \dfrac{\frac{2}{t}+2t}{1-4} = \dfrac{2\left(t+\frac{1}{t}\right)}{-3} = \dfrac{2\sqrt{5}}{-3}$

(H) ASSERTION & REASON TYPE QUESTIONS :

1. **(a)** The given curve is $y = -\dfrac{x^2}{2} + x + 1$

or $(x-1)^2 = -2(y-3/2)$

which is a parabola, so should be symmetric with respect to its axis $x - 1 = 0$

$\therefore$ Both the statements are true and statement 2 is a correct explanation for statement 1.

(I) INTEGER VALUE CORRECT TYPE :

1. **(2)** Intersection point of nearest directrix $x = \dfrac{a}{e}$ and x-axis is $\left(\dfrac{a}{e}, 0\right)$

As $2x + y = 1$ passes through $\left(\dfrac{a}{e}, 0\right)$

$\therefore \dfrac{2a}{e} = 1 \Rightarrow a = \dfrac{e}{2}$

Also $y = -2x + 1$ is a tangent to $\dfrac{x^2}{a^2} - \dfrac{y^2}{b^2} = 1$

$\therefore 1 = a^2(-2)^2 - b^2 \Rightarrow 4a^2 - b^2 = 1$

$\Rightarrow 4a^2 - a^2(e^2-1) = 1$

$\Rightarrow 4\times\dfrac{e^2}{4} - \dfrac{e^2}{4}(e^2-1) = 1$

$\Rightarrow 4e^2 - e^4 + e^2 = 4 \Rightarrow e^4 - 5e^4 + 4 = 0$

$\Rightarrow e^2 = 4$ *as* $e > 1$ for hyperbola.

$\Rightarrow e = 2$

2. (2)

$\Delta_1 = \text{Area of } \Delta PLL' = \frac{1}{2}\times 8\times\frac{3}{2} = 6$

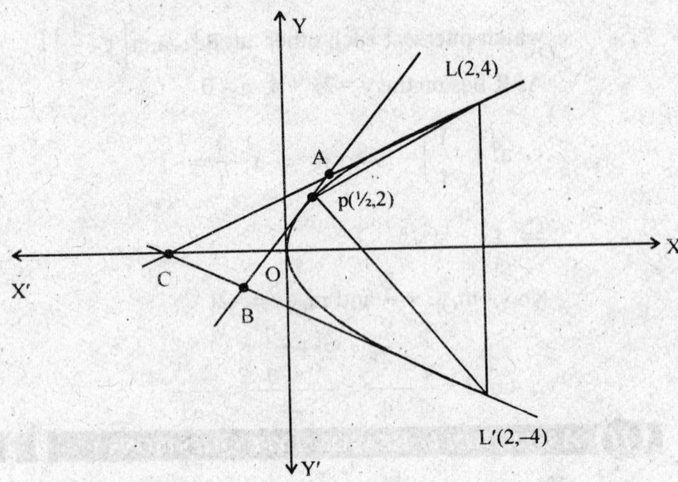

Equation of AB, $y = 2x+1$ Equation of AC, $y = x+2$
Equation of BC, $-y = x+2$ Solving above equations we get
A (1, 3), B (–1, –1), C (–2, 0)

$\therefore \Delta_2 = \frac{1}{2}\begin{vmatrix} 1 & 3 & 1 \\ -1 & -1 & 1 \\ -2 & 0 & 1 \end{vmatrix} = 3 \therefore \frac{\Delta_1}{\Delta_2} = 2$

3. (4) We observe both parabola $y^2 = 8x$ and circle $x^2 + y^2 - 2x - 4y = 0$ pass through origin

$\therefore$ One end of common chord PQ is origin. Say P(0, 0)

Let Q be the point $(2t^2, 4t)$, then it will satisfy the equation of circle.

$\therefore \quad 4t^4 + 16t^2 - 4t^2 - 16t = 0$

$\Rightarrow \quad t^4 + 3t^2 - 4t = 0 \Rightarrow t(t^3 + 3t - 4) = 0$

$\Rightarrow \quad t(t-1)(t^2 + t - 4) = 0 \Rightarrow t = 0 \text{ or } 1$

For $t = 0$, we get point P, therefore $t = 1$ gives point Q as (2, 4).

We also observe here that P(0, 0) and Q(2, 4) are end points of diameter of the given circle and focus of the parabola is the point S(2, 0).

$\therefore$ Area of $\Delta PQS = \frac{1}{2}\times PS\times QS = \frac{1}{2}\times 2\times 4 = 4$ sq. units

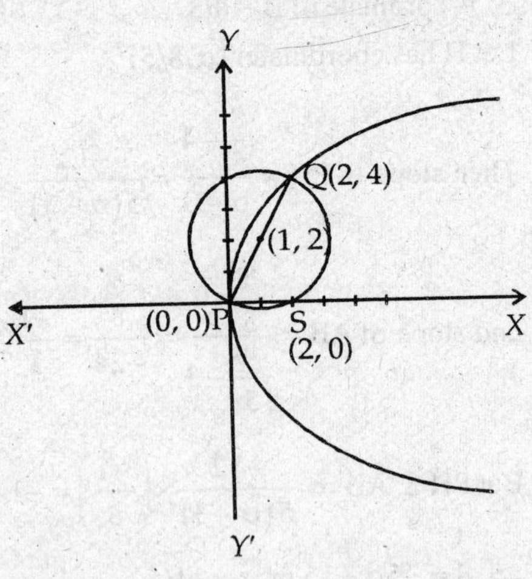

4. (9) Vertical line $x = h$, meets the ellipse $\frac{x^2}{4} + \frac{y^2}{3} = 1$ at

$P\left(h, \frac{\sqrt{3}}{2}\sqrt{4-h^2}\right)$ and $Q\left(h, \frac{-\sqrt{3}}{2}\sqrt{4-h^2}\right)$

By symmetry, tangents at P and Q will meet each other at x-axis.

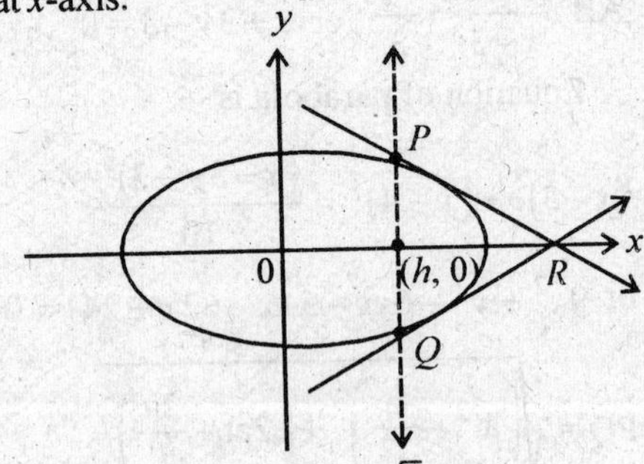

Tangent at P is $\frac{xh}{4} + \frac{y\sqrt{3}}{6}\sqrt{4-h^2} = 1$

which meets x-axis at $R\left(\frac{4}{h}, 0\right)$

Area of $\Delta PQR = \frac{1}{2}\times\sqrt{3}\sqrt{4-h^2}\times\left(\frac{4}{h} - h\right)$

i.e., $\Delta(h) = \frac{\sqrt{3}}{2}\frac{(4-h^2)^{3/2}}{h}$

$\frac{d\Delta}{dh} = -\sqrt{3}\left[\frac{\sqrt{4-h^2}\,(h^2+2)}{h^2}\right] < 0$

$\therefore$ $\Delta(h)$ is a decreasing function.

$\therefore \quad \frac{1}{2} \le h \le 1 \Rightarrow \Delta_{max} = \Delta\left(\frac{1}{2}\right)$ and $\Delta_{min} = \Delta(1)$

$\therefore \quad \Delta_1 = \frac{\sqrt{3}}{2}\frac{\left(4 - \frac{1}{4}\right)^{3/2}}{\frac{1}{2}} = \frac{45}{8}\sqrt{5}$

$\Delta_2 = \frac{\sqrt{3}}{2}\frac{3\sqrt{3}}{1} = \frac{9}{2} \quad \therefore \frac{8}{\sqrt{5}}\Delta_1 - 8\Delta_2 = 45 - 36 =$

Section-B JEE Main/ AIEEE

1. (b) Any tangent to the parabola $y^2 = 8ax$ is

$$y = mx + \frac{2a}{m} \quad ...(i)$$

If (i) is a tangent to the circle, $x^2 + y^2 = 2a^2$ then,

$$\sqrt{2}a = \pm\frac{2a}{m\sqrt{m^2+1}}$$

$\Rightarrow m^2(1+m^2) = 2 \Rightarrow (m^2+2)(m^2-1) = 0 \Rightarrow m = \pm 1.$

So from (i), $y = \pm(x+2a)$.

2. (b) Equation of the normal to a parabola $y^2 = 4bx$ at point $\left(bt_1^2, 2bt_1\right)$ is

$$y = -t_1 x + 2bt_1 + bt_1^3$$

As given, it also passes through $\left(bt_2^2, 2bt_2\right)$ then

$$2bt_2 = -t_1 bt_2^2 + 2bt_1 + bt_1^3$$

$$2t_2 - 2t_1 = -t_1\left(t_2^2 - t_1^2\right) = -t_1(t_2+t_1)(t_2-t_1)$$

$$\Rightarrow 2 = -t_1(t_2+t_1) \Rightarrow t_2 + t_1 = -\frac{2}{t_1}$$

$$\Rightarrow t_2 = -t_1 - \frac{2}{t_1}$$

3. (d) $\frac{x^2}{144} - \frac{y^2}{81} = \frac{1}{25}$

$$a = \sqrt{\frac{144}{25}}, b = \sqrt{\frac{81}{25}}, \quad e = \sqrt{1+\frac{81}{144}} = \frac{15}{12} = \frac{5}{4}$$

$\therefore$ Foci $= (\pm 3, 0)$

$\therefore$ foci of ellipse = foci of hyperbola

$\therefore$ for ellipse $ae = 3$ but $a = 4$,

$\therefore$ $e = \frac{3}{4}$

Then $b^2 = a^2(1-e^2) \Rightarrow b^2 = 16\left(1-\frac{9}{16}\right) = 7$

4. (d) Solving equations of parabolas

$y^2 = 4ax$ and $x^2 = 4ay$

we get (0, 0) and (4a, 4a)

Substituting in the given equation of line

$2bx + 3cy + 4d = 0$,

we get $d = 0$ and $2b + 3c = 0 \Rightarrow d^2 + (2b+3c)^2 = 0$

5. (b) $e = \frac{1}{2}$. Directrix, $x = \frac{a}{e} = 4$

$$\therefore a = 4 \times \frac{1}{2} = 2 \qquad \therefore b = 2\sqrt{1-\frac{1}{4}} = \sqrt{3}$$

Equation of elhipe is $\frac{x^2}{4} + \frac{y^2}{3} = 1 \Rightarrow 3x^2 + 4y^2 = 12$

6. (a) $P = (1, 0)$ $Q = (h, k)$ Such that $K^2 = 8h$

Let (α, β) be the midpoint of PQ

$$\alpha = \frac{h+1}{2}, \qquad \beta = \frac{k+0}{2}$$

$$2\alpha - 1 = h \qquad 2\beta = k.$$

$$(2\beta)^2 = 8(2\alpha - 1) \Rightarrow \beta^2 = 4\alpha - 2$$

$$\Rightarrow y^2 - 4x + 2 = 0.$$

7. (d) Tangent to the hyperbola $\frac{x^2}{a^2} - \frac{y^2}{b^2} = 1$ is

$$y = mx \pm \sqrt{a^2m^2 - b^2}$$

Given that $y = \alpha x + \beta$ is the tangent of hyperbola

$\Rightarrow \quad m = \alpha$ and $a^2m^2 - b^2 = \beta^2$

$\therefore \quad a^2\alpha^2 - b^2 = \beta^2$

Locus is $a^2x^2 - y^2 = b^2$ which is hyperbola.

8. (a) $\because \angle FBF' = 90° \Rightarrow FB^2 + F'B^2 = FF'^2$

$$\therefore \left(\sqrt{a^2e^2+b^2}\right)^2 + \left(\sqrt{a^2e^2+b^2}\right)^2 = (2ae)^2$$

$$\Rightarrow 2(a^2e^2 + b^2) = 4a^2e^2 \Rightarrow e^2 = \frac{b^2}{a^2}$$

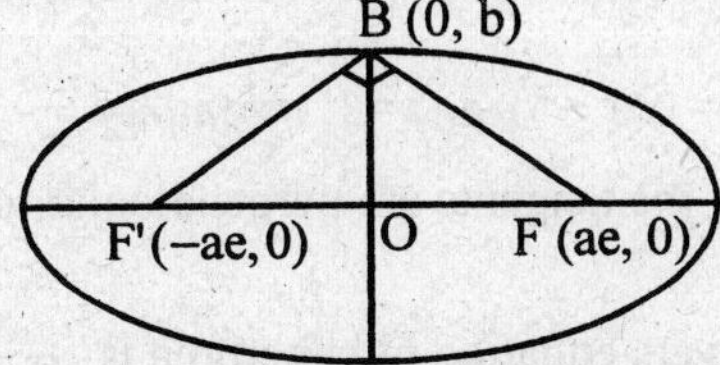

Also $e^2 = 1 - b^2/a^2 = 1 - e^2$

$$\Rightarrow 2e^2 = 1, \; e = \frac{1}{\sqrt{2}}.$$

ALTERNATE SOLUTION

$m_{BF} = \dfrac{b}{-ae}$

$m_{BF'} = \dfrac{b}{ae}$

$m_{BF} \times m_{BF'} = -1$

$\Rightarrow b^2 = a^2e^2 \Rightarrow a^2(1-e^2) = a^2e^2$

$\Rightarrow 2e^2 = 1$ or $e = \dfrac{1}{\sqrt{2}}$

9. (a) Given parabola is $y = \dfrac{a^3x^2}{3} + \dfrac{a^2x}{2} - 2a$

$\Rightarrow y = \dfrac{a^3}{3}\left(x^3 + \dfrac{3}{2a}x + \dfrac{9}{16a^2}\right) - \dfrac{3a}{16} - 2a$

$\Rightarrow y + \dfrac{35a}{16} = \dfrac{a^3}{3}\left(x + \dfrac{3}{4a}\right)^2$

$\therefore$ Vertex of parabola is $\left(\dfrac{-3}{4a}, \dfrac{-35a}{16}\right)$

To find locus of this vertex,

$x = \dfrac{-3}{4a}$ and $y = \dfrac{-35a}{16}$

$\Rightarrow a = \dfrac{-3}{4x}$ and $a = -\dfrac{16y}{35}$

$\Rightarrow \dfrac{-3}{4x} = \dfrac{-16y}{35} \Rightarrow 64xy = 105$

$\Rightarrow xy = \dfrac{105}{64}$ which is the required locus.

10. (a) $2ae = 6 \Rightarrow ae = 3$; $2b = 8 \Rightarrow b = 4$

$b^2 = a^2(1-e^2)$; $16 = a^2 - a^2e^2 \Rightarrow a^2 = 16 + 9 = 25$

$\Rightarrow a = 5 \qquad \therefore e = \dfrac{3}{a} = \dfrac{3}{5}$

11. (b) $\dfrac{dy}{dx} = 2x - 5 \therefore m_1 = (2x-5)_{(2,0)} = -1$,

$m_2 = (2x-5)_{(3,0)} = 1 \Rightarrow m_1m_2 = -1$

i.e. the tangents are perpendicular to each other.

12. (b) Given, equation of hyperbola is $\dfrac{x^2}{\cos^2\alpha} - \dfrac{y^2}{\sin^2\alpha} = 1$

We know that the equation of hyperbola is

$\dfrac{x^2}{a^2} - \dfrac{y^2}{b^2} = 1$ Here, $a^2 = \cos^2\alpha$ and $b^2 = \sin^2\alpha$

We know that, $b^2 = a^2(e^2 - 1)$

$\Rightarrow \sin^2\alpha = \cos^2\alpha(e^2 - 1)$

$\Rightarrow \sin^2\alpha + \cos^2\alpha = \cos^2\alpha . e^2$

$\Rightarrow e^2 = 1 + \tan^2\alpha = \sec^2\alpha \Rightarrow e = \sec\alpha$

$\therefore ae = \cos\alpha . \dfrac{1}{\cos\alpha} = 1$

Co-ordinates of foci are $(\pm ae, 0)$ i.e. $(\pm 1, 0)$

Hence, abscissae of foci remain constant when α varies.

13. (b) Parabola $y^2 = 8x$

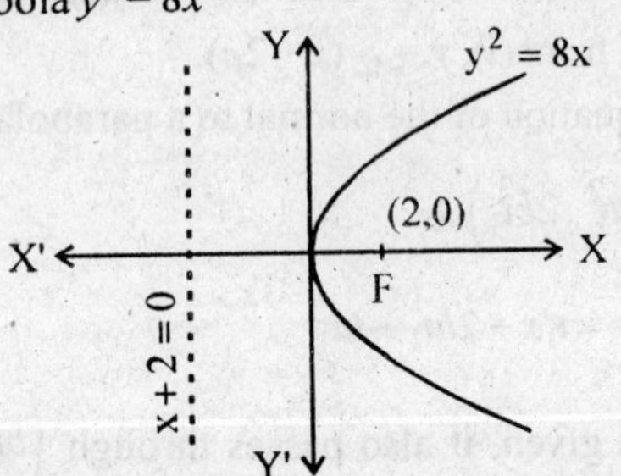

We know that the locus of point of intersection of two perpendicular tangents to a parabola is its directrix. Point must be on the directrix of parabola

$\because$ equation of directrix $x + 2 = 0 \Rightarrow x = -2$

Hence the point is (–2, 0)

14. (b) Equation of normal at $P(x, y)$ is $Y - y = -\dfrac{dx}{dy}(X - x)$

Coordinate of G at X axis is $(X, 0)$ (let)

$\therefore 0 - y = -\dfrac{dx}{dy}(X - x) \Rightarrow y\dfrac{dy}{dx} = X - x$

$\Rightarrow X = x + y\dfrac{dy}{dx} \therefore$ Co-ordinate of $G\left(x + y\dfrac{dy}{dx}, 0\right)$

Given distance of G from origin = twice of the abscissa of P.

$\because$ distance cannot be –ve, therefore abscissa x should be +ve

$\therefore x + y\dfrac{dy}{dx} = 2x \Rightarrow y\dfrac{dy}{dx} = x \Rightarrow ydy = xdx$

On Integrating $\Rightarrow \dfrac{y^2}{2} = \dfrac{x^2}{2} + c_1 \Rightarrow x^2 - y^2 = -2c_1$

$\therefore$ the curve is a hyperbola

15. (a) Perpendicular distance of directrix from focus

$= \dfrac{a}{e} - ae = 4$

$\Rightarrow a\left(2 - \dfrac{1}{2}\right) = 4$

$\Rightarrow a = \dfrac{8}{3}$

$\therefore$ Semi major axis = 8/3

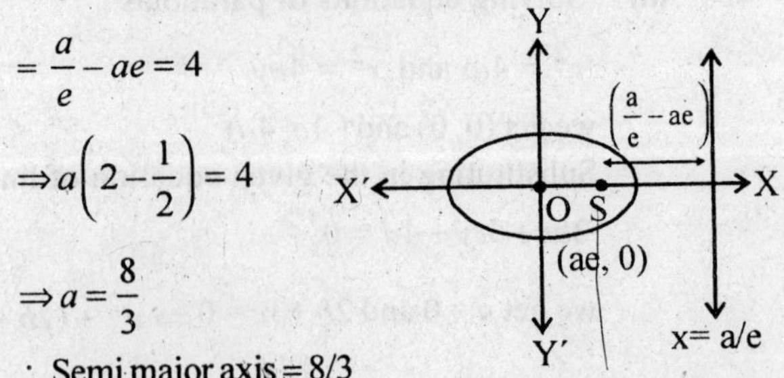

16. (b) Vertex of a parabola is the mid point of focus and the point where directrix meets the axis of the parabola.
Here focus is $O(0, 0)$ and directxix meets the axis at $B(2, 0)$
$\therefore$ Vertex of the parabola is $(1, 0)$

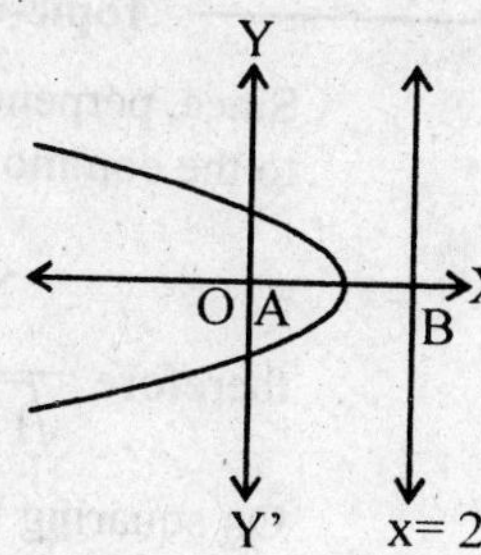

17. (a) The given ellipse is $\frac{x^2}{4}+\frac{y^2}{1}=1$

So $A = (2, 0)$ and $B= (0, 1)$
If $PQRS$ is the rectangle in which it is inscribed, then $P = (2, 1)$.

Let $\frac{x^2}{a^2}+\frac{y^2}{b^2}=1$

be the ellipse circumscribing the rectangle $PQRS$. Then it passes through $P\ (2,1)$

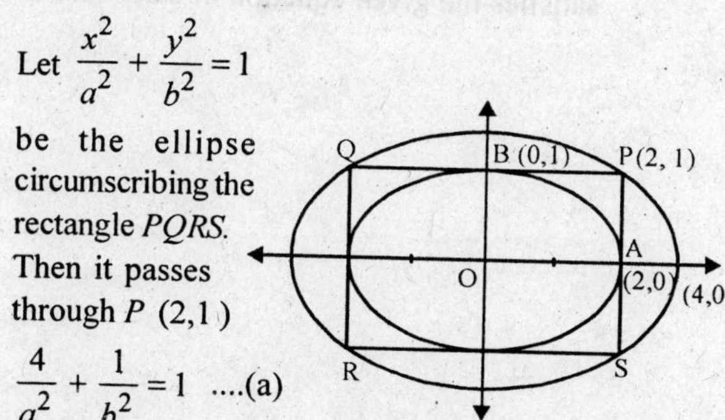

$\therefore\ \frac{4}{a^2}+\frac{1}{b^2}=1$ ....(a)

Also, given that, it passes through (4, 0)

$\therefore\ \frac{16}{a^2}+0=1 \Rightarrow a^2=16$

$\Rightarrow\ b^2 = 4/3$ [substituting $a^2 = 16$ in eq^n (a)]

$\therefore$ The required ellipse is $\frac{x^2}{16}+\frac{y^2}{4/3}=1$ or $x^2+12y^2=16$

18. (b) The locus of perpendicular tangents is directrix i.e., $x = -1$

19. (d) Let the ellipse be $\frac{x^2}{a^2}+\frac{y^2}{b^2}=1$

It passes through $(-3, 1)$ so $\frac{9}{a^2}+\frac{1}{b^2}=1$ ...(i)

Also, $b^2 = a^2(1-2/5)$

$\Rightarrow 5b^2 = 3a^2$ ...(ii)

Solving (i) and (ii) we get $a^2 = \frac{32}{3}, b^2 = \frac{32}{5}$

So, the equation of the ellipse is $3x^2+5y^2=32$

20. (b) $ae = 2$
$e = 2$
$\therefore\ a = 1$

$b^2 = a^2\left(e^2-1\right)$

$b^2 = 1(4-1)$

$b^2 = 3$

Equation of hyperbola, $\frac{x^2}{a^2}-\frac{y^2}{b^2}=1$

$\Rightarrow \frac{x^2}{1}-\frac{y^2}{3}=1$

$3x^2-y^2=3$

21. (b) Given equation of ellipse is $2x^2+y^2=4$

$\Rightarrow \frac{2x^2}{4}+\frac{y^2}{4}=1 \Rightarrow \frac{x^2}{2}+\frac{y^2}{4}=1$

Equation of tangent to the ellipse $\frac{x^2}{2}+\frac{y^2}{4}=1$ is

$y = mx \pm\sqrt{2m^2+4}$ ...(1)

($\because$ equation of tangent to the ellipse $\frac{x^2}{a^2}+\frac{y^2}{b^2}=1$ is $y = mx + c$ where $c=\pm\sqrt{a^2m^2+b^2}$)

Now, Equation of tangent to the parabola $y^2=16\sqrt{3}x$ is $y = mx+\frac{4\sqrt{3}}{m}$ ...(2)

($\because$ equation of tangent to the parabola $y^2 = 4ax$ is $y = mx+\frac{a}{m}$)

On comparing (1) and (2), we get

$\frac{4\sqrt{3}}{m}=\pm\sqrt{2m^2+4}$

Squaring on both the sides, we get
$16(3) = (2m^2+4)m^2$
$\Rightarrow 48 = m^2(2m^2+4) \Rightarrow\ 2m^4+4m^2-48=0$
$\Rightarrow m^4+2m^2-24=0 \Rightarrow\ (m^2+6)(m^2-4)=0$
$\Rightarrow m^2 = 4\ (\because m^2 \neq -6) \Rightarrow\ m = \pm 2$

$\Rightarrow$ Equation of common tangents are $y=\pm 2x\pm 2\sqrt{3}$
Thus, statement-1 is true.
Statement-2 is obviously true.

22. (d) Equation of circle is $(x-1)^2+y^2=1$
$\Rightarrow$ radius = 1 and diameter = 2
$\therefore$ Length of semi-minor axis is 2.
Equation of circle is $x^2+(y-2)^2=4=(2)^2$
$\Rightarrow$ radius = 2 and diameter = 4
$\therefore$ Length of semi major axis is 4
We know, equation of ellipse is given by

$$\frac{x^2}{(\text{Major axis})^2}+\frac{y^2}{(\text{Minor axis})^2}=1$$

$\Rightarrow \frac{x^2}{(4)^2}+\frac{y^2}{(2)^2}=1 \Rightarrow \frac{x^2}{16}+\frac{y^2}{4}=1$

$\Rightarrow x^2+4y^2=16$

23. (a) From the given equation of ellipse, we have

$a=4, b=3, e=\sqrt{1-\frac{9}{16}}$

$\Rightarrow e=\frac{\sqrt{7}}{4}$

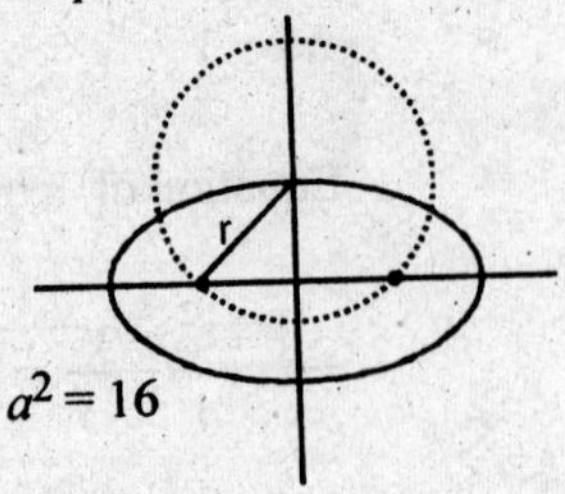

Now, radius of this circle $= a^2 = 16$

$\Rightarrow$ Focii $= (\pm\sqrt{7}, 0)$

Now equation of circle is $(x-0)^2+(y-3)^2=16$

$x^2+y^2-6y-7=0$

24. (b) Let common tangent be

$y = mx + \frac{\sqrt{5}}{m}$

Since, perpendicular distance from centre of the circle to the common tangent is equal to radius of the circle,

therefore $\dfrac{\frac{\sqrt{5}}{m}}{\sqrt{1+m^2}} = \sqrt{\frac{5}{2}}$

On squaring both the side, we get

$m^2(1+m^2)=2$

$\Rightarrow m^4+m^2-2=0$

$\Rightarrow (m^2+2)(m^2-1)=0$

$\Rightarrow m=\pm 1 \qquad (\because m \neq \pm\sqrt{2})$

$y=\pm(x+\sqrt{5})$, both statements are correct as $m=\pm 1$ satisfies the given equation of statement-2.

Chapter 10

Functions

Section-A JEE Advanced/ IIT-JEE

(A) FILL IN THE BLANKS :

1. The values of $f(x) = 3\sin\left(\sqrt{\frac{\pi^2}{16} - x^2}\right)$ lie in the interval .... *(1983 - 1 Mark)*

2. For the function $f(x) = \frac{x}{1+e^{1/x}}, \quad x \neq 0$
$= 0 \quad , \quad x = 0$
the derivative from the right, $f'(0+)$ = ..., and the derivative from the left, $f'(0-)$ = ... *(1983 - 2 Marks)*

3. The domain of the function $f(x) = \sin^{-1}(\log_2 \frac{x^2}{2})$ is given by .... *(1984 - 2 Marks)*

4. Let A be a set of n distinct elements. Then the total number of distinct functions from A to A is .... and out of these .... are onto functions. *(1985 - 2 Marks)*

5. If $f(x) = \sin \ln\left(\frac{\sqrt{4-x^2}}{1-x}\right)$, then domain of $f(x)$ is .... and its range is .... *(1985 - 2 Marks)*

6. There are exactly two distinct linear functions, ..., and .... which map [– 1, 1] onto [0, 2]. *(1989 - 2 Marks)*

7. If f is an even function defined on the interval (-5, 5), then four real values of x satisfying the equation $f(x) = f\left(\frac{x+1}{x+2}\right)$ are ...., ..., ..., and .... *(1996 - 1 Mark)*

8. If $f(x) = \sin^2 x + \sin^2\left(x+\frac{\pi}{3}\right) + \cos x \cos\left(x+\frac{\pi}{3}\right)$ and $g\left(\frac{5}{4}\right) = 1$, then (gof) (x) = ... *(1996 - 2 Marks)*

(B) TRUE / FALSE :

1. If $f(x) = (a - x^n)^{1/n}$ where $a > 0$ and n is a positive integer, then $f[f(x)] = x$. *(1983 - 1 Mark)*

2. The function $f(x) = \frac{x^2+4x+30}{x^2-8x+18}$ is not one -to -one. *(1983 - 1 Mark)*

3. If $f_1(x)$ and $f_2(x)$ are defined on domains D_1 and D_2 respectively, then $f_1(x) + f_2(x)$ is defined on $D_1 \cup D_2$. *(1988 - 1 Mark)*

(C) MCQ's WITH ONE CORRECT ANSWER :

1. Let R be the set of real numbers. If $f: R \to R$ is a function defined by $f(x) = x^2$, then f is : *(1979)*
   (a) Injective but not surjective
   (b) Surjective but not injective
   (c) Bijective
   (d) None of these.

2. The entire graphs of the equation $y = x^2 + kx - x + 9$ is strictly above the x-axis if and only if *(1979)*
   (a) $k < 7$ (b) $-5 < k < 7$
   (c) $k > -5$ (d) None of these.

3. Let $f(x) = |x - 1|$. Then *(1983 - 1 Mark)*
   (a) $f(x^2) = (f(x))^2$ (b) $f(x+y) = f(x) + f(y)$
   (c) $f(|x|) = |f(x)|$ (d) None of these

4. If x satisfies $|x-1| + |x-2| + |x-3| \geq 6$, then *(1983 - 1 Mark)*
   (a) $0 \leq x \leq 4$ (b) $x \leq -2$ or $x \geq 4$
   (c) $x \leq 0$ or $x \geq 4$ (d) None of these

5. If $f(x) = \cos(\ln x)$, then $f(x)f(y) - \frac{1}{2}\left[f\left(\frac{x}{y}\right) + f(xy)\right]$ has the value *(1983 - 1 Mark)*
   (a) –1 (b) 1/2
   (c) –2 (d) none of these

6. The domain of definition of the function $y = \frac{1}{\log_{10}(1-x)} + \sqrt{x+2}$ is *(1983 - 1 Mark)*
   (a) (–3, –2) excluding –2.5 (b) [0, 1] excluding 0.5
   (c) [–2, 1) excluding 0 (d) none of these

7. Which of the following functions is periodic? *(1983 - 1 Mark)*

(a) $f(x) = x - [x]$ where [x] denotes the largest integer less than or equal to the real number x

(b) $f(x) = \sin\frac{1}{x}$ for $x \neq 0$, $f(0) = 0$

(c) $f(x) = x \cos x$

(d) none of these

8. Let $f(x)$ = sin$x$ and g(x) = ln | x |. If the ranges of the composition functions fog and gof are R_1 and R_2 respectively, then *(1994 - 2 Marks)*

(a) $R_1 = \{u : -1 \le u < 1\}$, $R_2 = \{v : -\infty < v < 0\}$

(b) $R_1 = \{u : -\infty < u < 0\}$, $R_2 = \{v : -1 \le v \le 0\}$

(c) $R_1 = \{u : -1 < u < 1\}$, $R_2 = \{v : -\infty < v < 0\}$

(d) $R_1 = \{u : -1 \le u \le 1\}$, $R_2 = \{v : -\infty < v \le 0\}$

9. Let $f(x) = (x+1)^2 - 1,\ x \ge -1$. Then the set $\{x : f(x) = f^{-1}(x)\}$ is *(1995)*

(a) $\left\{0, -1, \frac{-3+i\sqrt{3}}{2}, \frac{-3-i\sqrt{3}}{2}\right\}$

(b) $\{0, 1, -1\}$

(c) $\{0, -1\}$

(d) empty

10. The function $f(x) = |px - q| + r\,|\,x\,|,\ x \in (-\infty, \infty)$ where $p > 0, q > 0, r > 0$ assumes its minimum value only on one point if *(1995)*

(a) $p \neq q$ (b) $r \neq q$

(c) $r \neq p$ (d) $p = q = r$

11. Let $f(x)$ be defined for all $x > 0$ and be continuous. Let $f(x)$ satisfy $f\left(\frac{x}{y}\right) = f(x) - f(y)$ for all x, y and $f(e) = 1$. Then *(1995S)*

(a) $f(x)$ is bounded (b) $f\left(\frac{1}{x}\right) \to 0$ as $x \to 0$

(c) $x f(x) \to 1$ as $x \to 0$ (d) $f(x) = \ln x$

12. If the function f: $[1, \infty) \to [1, \infty)$ is defined by $f(x) = 2^{x(x-1)}$, then $f^{-1}(x)$ is *(1999 - 2 Marks)*

(a) $\left(\frac{1}{2}\right)^{x(x-1)}$ (b) $\frac{1}{2}(1 + \sqrt{1 + 4\log_2 x})$

(c) $\frac{1}{2}(1 - \sqrt{1 + 4\log_2 x})$ (d) not defined

13. Let $f : R \to R$ be any function. Define $g : R \to R$ by $g(x) = |f(x)|$ for all x. Then g is *(2000S)*

(a) onto if f is onto

(b) one-one if f is one-one

(c) continuous if f is continuous

(d) differentiable if f is differentiable.

14. The domain of definition of the function $f(x)$ given by the equation $2^x + 2^y = 2$ is *(2000S)*

(a) $0 < x \le 1$ (b) $0 \le x \le 1$

(c) $-\infty < x \le 0$ (d) $-\infty < x < 1$

15. Let $g(x) = 1 + x - [x]$ and $f(x) = \begin{cases} -1, & x < 0 \\ 0, & x = 0 \\ 1, & x > 0 \end{cases}$. Then for all x, $f(g(x))$ is equal to *(2001S)*

(a) x (b) 1

(c) $f(x)$ (d) $g(x)$

16. If f:$[1, \infty) \to [2, \infty)$ is given by $f(x) = x + \frac{1}{x}$ then $f^{-1}(x)$ equals *(2001S)*

(a) $(x + \sqrt{x^2 - 4})/2$ (b) $x/(1 + x^2)$

(c) $\left(x - \sqrt{x^2 - 4}\right)/2$ (d) $1 + \sqrt{x^2 - 4}$

17. The domain of definition of $f(x) = \frac{\log_2(x+3)}{x^2 + 3x + 2}$ is *(2001S)*

(a) $R \setminus \{-1, -2\}$ (b) $(-2, \infty)$

(c) $R \setminus \{-1, -2, -3\}$ (d) $(-3, \infty) \setminus \{-1, -2\}$

18. Let $E = \{1, 2, 3, 4\}$ and $F = \{1, 2\}$. Then the number of onto functions from E to F is *(2001S)*

(a) 14 (b) 16

(c) 12 (d) 8

19. Let $f(x) = \frac{\alpha x}{x+1}, x \neq -1$. Then, for what value of α is $f(f(x)) = x$? *(2001S)*

(a) $\sqrt{2}$ (b) $-\sqrt{2}$

(c) 1 (d) -1

20. Suppose $f(x) = (x + 1)^2$ for $x \ge -1$. If $g(x)$ is the function whose graph is the reflection of the graph of $f(x)$ with respect to the line $y = x$, then $g(x)$ equals *(2002S)*

(a) $-\sqrt{x} - 1, x \ge 0$ (b) $\frac{1}{(x+1)^2}, x > -1$

(c) $\sqrt{x+1}, x \ge -1$ (d) $\sqrt{x} - 1, x \ge 0$

21. Let function $f : R \to R$ be defined by $f(x) = 2x + \sin x$ for $x \in R$, then f is *(2002S)*

(a) one-to-one and onto

(b) one-to-one but NOT onto

(c) onto but NOT one-to-one

(d) neither one-to-one nor onto

22. If $f : [0, \infty) \longrightarrow [0, \infty)$, and $f(x) = \frac{x}{1+x}$ then f is *(2003S)*

(a) one-one and onto

(b) one-one but not onto

(c) onto but not one-one

(d) neither one-one nor onto

23. Domain of definition of the function $f(x) = \sqrt{\sin^{-1}(2x) + \frac{\pi}{6}}$ for real valued x, is *(2003S)*

(a) $\left[-\frac{1}{4}, \frac{1}{2}\right]$ (b) $\left[-\frac{1}{2}, \frac{1}{2}\right]$

(c) $\left(-\frac{1}{2}, \frac{1}{9}\right)$ (d) $\left[-\frac{1}{4}, \frac{1}{4}\right]$

24. Range of the function $f(x)=\dfrac{x^2+x+2}{x^2+x+1}; x \in R$ is **(2003S)**

(a) $(1, \infty)$ (b) $(1,11/7]$
(c) $(1, 7/3]$ (d) $(1, 7/5]$

25. If $f(x)=x^2+2bx+2c^2$ and $g(x)=-x^2-2cx+b^2$ such that min $f(x) >$ max $g(x)$, then the relation between b and c, is **(2003S)**

(a) no real value of b & c (b) $0 < c < b\sqrt{2}$
(c) $|c| < |b|\sqrt{2}$ (d) $|c| > |b|\sqrt{2}$

26. If $f(x) = \sin x + \cos x$, $g(x) = x^2 - 1$, then $g(f(x))$ is invertible in the domain **(2004S)**

(a) $\left[0, \dfrac{\pi}{2}\right]$ (b) $\left[-\dfrac{\pi}{4}, \dfrac{\pi}{4}\right]$
(c) $\left[-\dfrac{\pi}{2}, \dfrac{\pi}{2}\right]$ (d) $[0, \pi]$

27. If the functions f(x) and g(x) are defined on $R \to R$ such that

$$f(x)=\begin{cases}0, & x \in \text{rational}\\ x, & x \in \text{irrational}\end{cases}; \quad g(x)=\begin{cases}0, & x \in \text{irrational}\\ x, & x \in \text{rational}\end{cases} \text{ then}$$

$(f-g)(x)$ is **(2005S)**
(a) one-one & onto
(b) neither one-one nor onto
(c) one-one but not onto
(d) onto but not one-one

28. X and Y are two sets and $f: X \to Y$. If $\{f(c) = y; c \subset X, y \subset Y\}$ and $\{f^{-1}(d) = x; d \subset Y, x \subset X\}$, then the true statement is **(2005S)**

(a) $f(f^{-1}(b)) = b$ (b) $f^{-1}(f(a)) = a$
(c) $f(f^{-1}(b)) = b, b \subset y$ (d) $f^{-1}(f(a)) = a, a \subset x$

29. If $F(x)=\left(f\left(\dfrac{x}{2}\right)\right)^2+\left(g\left(\dfrac{x}{2}\right)\right)^2$ where $f''(x) = -f(x)$ and $g(x) = f'(x)$ and given that $F(5) = 5$, then $F(10)$ is equal to

(a) 5 (b) 10 **(2006 - 3M, –1)**
(c) 0 (d) 15

30. Let $f(x)=\dfrac{x}{(1+x^n)^{1/n}}$ for $n \geq 2$ and

$g(x) = \underbrace{(fofo...of)}_{f \text{ occurs } n \text{ times}}(x)$. Then $\int x^{n-2} g(x)dx$ equals.

(2007 -3 marks)

(a) $\dfrac{1}{n(n-1)}(1+nx^n)^{1-\frac{1}{n}}+K$ (b) $\dfrac{1}{n-1}(1+nx^n)^{1-\frac{1}{n}}+K$

(c) $\dfrac{1}{n(n+1)}(1+nx^n)^{1+\frac{1}{n}}+K$ (d) $\dfrac{1}{n+1}(1+nx^n)^{1+\frac{1}{n}}+K$

31. Let f, g and h be real-valued functions defined on the interval [0, 1] by $f(x)=e^{x^2}+e^{-x^2}$, $g(x)=xe^{x^2}+e^{-x^2}$ and $h(x)=x^2e^{x^2}+e^{-x^2}$. If a, b and c denote, respectively, the absolute maximum of f, g and h on [0, 1], then **(2010)**

(a) $a = b$ and $c \neq b$ (b) $a = c$ and $a \neq b$
(c) $a \neq b$ and $c \neq b$ (d) $a = b = c$

32. Let $f(x) = x^2$ and $g(x) = \sin x$ for all $x \in R$. Then the set of all x satisfying (f o g o g o f) (x) = (g o g o f) (x), where (f o g) (x) $= f(g(x))$, is

(a) $\pm\sqrt{n\pi}, n \in \{0,1,2,....\}$

(b) $\pm\sqrt{n\pi}, n \in \{1,2,....\}$

(c) $\dfrac{\pi}{2}+2n\pi, n \in \{...-2,-1,0,1,2....\}$

(d) $2n\pi, n \in \{...-2,-1,0,1,2,....\}$

33. The function f : [0, 3] $\to$ [1, 29], defined by $f(x) = 2x^3 - 15x^2 + 36x + 1$, is

(a) one–one and onto (b) onto but not one–one
(c) one–one but not onto
(d) neither one–one nor onto

D MCQ's WITH ONE OR MORE THAN ONE CORRECT :

1. If $y=f(x)=\dfrac{x+2}{x-1}$ then **(1984 - 3 Marks)**

(a) $x = f(y)$
(b) $f(1) = 3$
(c) y increases with x for $x < 1$
(d) f is a rational function of x

2. Let $g(x)$ be a function defined on [– 1, 1]. If the area of the equilateral triangle with two of its vertices at (0,0) and $[x, g(x)]$ is $\dfrac{\sqrt{3}}{4}$, then the function $g(x)$ is **(1989 - 2 Marks)**

(a) $g(x) = \pm\sqrt{1-x^2}$ (b) $g(x) = \sqrt{1-x^2}$
(c) $g(x) = -\sqrt{1-x^2}$ (d) $g(x) = \sqrt{1+x^2}$

3. If $f(x) = \cos[\pi^2]x + \cos[-\pi^2]x$, where $[x]$ stands for the greatest integer function, then **(1991 - 2 Marks)**

(a) $f\left(\dfrac{\pi}{2}\right)=-1$ (b) $f(\pi) = 1$

(c) $f(-\pi) = 0$ (d) $f\left(\dfrac{\pi}{4}\right)=1$

4. If $f(x) = 3x - 5$, then $f^{-1}(x)$ **(1998 - 2 Marks)**

(a) is given by $\dfrac{1}{3x-5}$

(b) is given by $\dfrac{x+5}{3}$

(c) does not exist because f is not one-one
(d) does not exist because f is not onto.

5. If $g(f(x)) = |\sin x|$ and $f(g(x)) = (\sin\sqrt{x})^2$, then

(a) $f(x) = \sin^2 x$, $g(x) = \sqrt{x}$ **(1998 - 2 Marks)**
(b) $f(x) = \sin x$, $g(x) = |x|$
(c) $f(x) = x^2$, $g(x) = \sin\sqrt{x}$
(d) f and g cannot be determined.

6. Let $f: (0, 1) \to R$ be defined by $f(x) = \frac{b-x}{1-bx}$, where b is a constant such that $0 < b < 1$. Then
(a) f is not invertible on (0, 1)
(b) $f \neq f^{-1}$ on (0, 1) and $f'(b) = \frac{1}{f'(0)}$
(c) $f = f^{-1}$ on (0, 1) and $f'(b) = \frac{1}{f'(0)}$
(d) f^{-1} is differentiable (0, 1)

7. Let $f: (-1, 1) \to IR$ be such that $f(\cos 4\theta) = \frac{2}{2-\sec^2\theta}$ for $\theta \in \left(0, \frac{\pi}{4}\right) \cup \left(\frac{\pi}{4}, \frac{\pi}{2}\right)$. Then the value (s) of $f\left(\frac{1}{3}\right)$ is (are)
(a) $1-\sqrt{\frac{3}{2}}$ (b) $1+\sqrt{\frac{3}{2}}$
(c) $1-\sqrt{\frac{2}{3}}$ (d) $1+\sqrt{\frac{2}{3}}$

8. The function $f(x) = 2|x| + |x+2| - ||x+2| - 2|x||$ has a local minimum or a local maximum at $x =$ *(JEE Adv. 2013)*
(a) -2 (b) $\frac{-2}{3}$
(c) 2 (d) $\frac{2}{3}$

E SUBJECTIVE PROBLEMS :

1. Find the domain and range of the function $f(x) = \frac{x^2}{1+x^2}$. Is the function one-to-one? *(1978)*
2. Draw the graph of $y = |x|^{1/2}$ for $-1 \le x \le 1$. *(1978)*
3. If $f(x) = x^9 - 6x^8 - 2x^7 + 12x^6 + x^4 - 7x^3 + 6x^2 + x - 3$, find $f(6)$. *(1979)*
4. Consider the following relations in the set of real numbers R.
$R = \{(x, y); x \in R, y \in R, x^2 + y^2 \le 25\}$
$R' = \left\{(x, y) : x \in R, y \in R, y \ge \frac{4}{9}x^2\right\}$
Find the domain and range of $R \cap R'$. Is the relation $R \cap R'$ a function? *(1979)*
5. Let A and B be two sets each with a finite number of elements. Assume that there is an injective mapping from A to B and that there is an injective mapping from B to A. Prove that there is a bijective mapping from A to B. *(1981 - 2 Marks)*
6. Let $f(x+y) = f(x) f(y)$ for all x and y. Suppose $f(5) = 2$ and $f'(0) = 3$. Find $f'(5)$. *(1981 - 2 Marks)*
7. Let f be a one-one function with domain $\{x, y, z\}$ and range $\{1, 2, 3\}$. It is given that exactly one of the following statements is true and the remaining two are false $f(x) = 1, f(y) \neq 1, f(z) \neq 2$ determine $f^{-1}(1)$. *(1982 - 3 Marks)*
8. Let R be the set of real numbers and $f: R \longrightarrow R$ be such that for all x and y in R $|f(x) - f(y)| \le |x-y|^3$. Prove that $f(x)$ is a constant. *(1988 - 2 Marks)*
9. Find the natural number 'a' for which $\sum_{k=1}^{n} f(a+k) = 16(2^n - 1)$, where the function '$f$' satisfies the relation $f(x+y) = f(x) f(y)$ for all natural numbers x, y and further $f(1) = 2$. *(1992 - 6 Marks)*
10. Let $\{x\}$ and $[x]$ denotes the fractional and integral part of a real number x respectively. Solve $4\{x\} = x + [x]$. *(1994 - 4 Marks)*
11. A function $f: IR \to IR$, where IR is the set of real numbers, is defined by $f(x) = \frac{\alpha x^2 + 6x - 8}{\alpha + 6x - 8x^2}$. Find the interval of values of α for which f is onto. Is the function one-to-one for $\alpha = 3$? Justify your answer. *(1996 - 5 Marks)*
12. Let $f(x) = Ax^2 + Bx + C$ where A, B, C are real numbers. Prove that if $f(x)$ is an integer whenever x is an integer, then the numbers $2A$, $A + B$ and C are all integers. Conversely, prove that if the numbers $2A$, $A+B$ and C are all integers then $f(x)$ is an integer whenever x is an integer. *(1998 - 8 Marks)*

F MATCH THE FOLLOWING :

Each question contains statements given in two columns, which have to be matched. The statements in Column-I are labelled A, B, C and D, while the statements in Column-II are labelled p, q, r, s and t. Any given statement in Column-I can have correct matching with ONE OR MORE statement(s) in Column-II. The appropriate bubbles corresponding to the answers to these questions have to be darkened as illustrated in the following example :
If the correct matches are A-p, s and t; B-q and r; C-p and q; and D-s then the correct darkening of bubbles will look like the given.

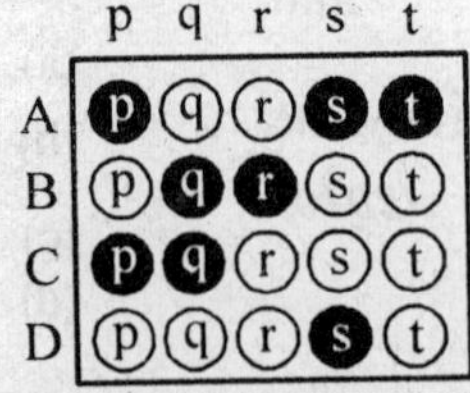

1. Let the function defined in column 1 have domain $\left(-\frac{\pi}{2}, \frac{\pi}{2}\right)$ and range $(-\infty, \infty)$ *(1992 - 2 Marks)*

Column I	Column II
(A) $1+2x$	(p) onto but not one-one
(B) $\tan x$	(q) one- one but not onto
	(r) one- one and onto
	(s) neither one-one nor onto

2. Let $f(x)=\dfrac{x^2-6x+5}{x^2-5x+6}$ *(2007 -6 marks)*

Match of expressions/statements in Column I with expressions/statements in Column II and indicate your answer by darkening the appropriate bubbles in the 4 × 4 matrix given in the ORS.

Column I	*Column II*
(A) If $-1<x<1$, then $f(x)$ satisfies	(p) $0<f(x)<1$
(B) If $1<x<2$, then $f(x)$ satisfies	(q) $f(x)<0$
(C) If $3<x<5$, then $f(x)$ satisfies	(r) $f(x)>0$
(D) If $x>5$, then $f(x)$ satisfies	(s) $f(x)<1$

Section-B JEE Main/ AIEEE

1. The domain of $\sin^{-1}[\log_3(x/3)]$ is **[2002]**
(a) $[1,9]$ (b) $[-1,9]$
(c) $[-9,1]$ (d) $[-9,-1]$

2. The function $f(x)=\log\left(x+\sqrt{x^2+1}\right)$, is **[2003]**
(a) neither an even nor an odd function
(b) an even function
(c) an odd function
(d) a periodic function.

3. Domain of definition of the function $f(x)=\dfrac{3}{4-x^2}+\log_{10}(x^3-x)$, is **[2003]**
(a) $(-1,0)\cup(1,2)\cup(2,\infty)$ (b) $(a,2)$
(c) $(-1,0)\cup(a,2)$ (d) $(1,2)\cup(2,\infty)$.

4. If $f: R\to R$ satisfies $f(x+y)=f(x)+f(y)$, for all $x, y\in R$ and $f(1)=7$, then $\sum_{r=1}^{n} f(r)$ is **[2003]**
(a) $\dfrac{7n(n+1)}{2}$ (b) $\dfrac{7n}{2}$
(c) $\dfrac{7(n+1)}{2}$ (d) $7n+(n+1)$.

5. A function f from the set of natural numbers to integers defined by **[2003]**

$$f(n)=\begin{cases}\dfrac{n-1}{2}, \text{when n is odd}\\ -\dfrac{n}{2}, \text{when n is even}\end{cases}$$ is

(a) neither one -one nor onto
(b) one-one but not onto
(c) onto but not one-one
(d) one-one and onto both.

6. The range of the function $f(x)={}^{7-x}P_{x-3}$ is **[2004]**
(a) $\{1,2,3,4,5\}$ (b) $\{1,2,3,4,5,6\}$
(c) $\{1,2,3,4,\}$ (d) $\{1,2,3,\}$

7. If $f: R\to S$, defined by
$f(x)=\sin x-\sqrt{3}\cos x+1$, is onto, then the interval of S is **[2004]**
(a) $[-1,3]$ (b) $[-1,1]$
(c) $[0,1]$ (d) $[0,3]$

8. The graph of the function $y=f(x)$ is symmetrical about the line $x=2$, then **[2004]**
(a) $f(x)=-f(-x)$ (b) $f(2+x)=f(2-x)$
(c) $f(x)=f(-x)$ (d) $f(x+2)=f(x-2)$

9. The domain of the function $f(x)=\dfrac{\sin^{-1}(x-3)}{\sqrt{9-x^2}}$ is **[2004]**
(a) $[1,2]$ (b) $[2,3)$
(c) $[1,2]$ (d) $[2,3]$

10. Let $f:(-1,1)\to B$, be a function defined by $f(x)=\tan^{-1}\dfrac{2x}{1-x^2}$, then f is both one - one and onto when B is the interval **[2005]**
(a) $\left(0,\frac{\pi}{2}\right)$ (b) $\left[0,\frac{\pi}{2}\right)$
(c) $\left[-\frac{\pi}{2},\frac{\pi}{2}\right]$ (d) $\left(-\frac{\pi}{2},\frac{\pi}{2}\right)$

11. A function is matched below against an interval where it is supposed to be increasing. Which of the following pairs is incorrectly matched? **[2005]**

	Interval	*Function*
(a)	$(-\infty, \infty)$	$x^3 - 3x^2 + 3x + 3$
(b)	$[2, \infty)$	$2x^3 - 3x^2 - 12x + 6$
(c)	$\left(-\infty, \frac{1}{3}\right]$	$3x^2 - 2x + 1$
(d)	$(-\infty, -4)$	$x^3 + 6x^2 + 6$

12. A real valued function $f(x)$ satisfies the functional equation

$$f(x-y) = f(x)f(y) - f(a-x)f(a+y)$$

where a is a given constant and $f(0) = 1, f(2a - x)$ is equal to **[2005]**

(a) $-f(x)$ (b) $f(x)$

(c) $f(a) + f(a-x)$ (d) $f(-x)$

13. The largest interval lying in $\left(\frac{-\pi}{2}, \frac{\pi}{2}\right)$ for which the function,

$f(x) = 4^{-x^2} + \cos^{-1}\left(\frac{x}{2} - 1\right) + \log(\cos x)$, is defined, is **[2007]**

(a) $\left[-\frac{\pi}{4}, \frac{\pi}{2}\right)$ (b) $\left[0, \frac{\pi}{2}\right)$

(c) $[0, \pi]$ (d) $\left(-\frac{\pi}{2}, \frac{\pi}{2}\right)$

14. Let $f: N \rightarrow Y$ be a function defined as $f(x) = 4x + 3$ where $Y = \{y \in N : y = 4x + 3 \text{ for some } x \in N\}$.
Show that f is invertible and its inverse is **[2008]**

(a) $g(y) = \frac{3y+4}{3}$ (b) $g(y) = 4 + \frac{y+3}{4}$

(c) $g(y) = \frac{y+3}{4}$ (d) $g(y) = \frac{y-3}{4}$

15. Let $f(x) = (x+1)^2 - 1, x \geq -1$

Statement -1 : The set $\{x : f(x) = f^{-1}(x) = \{0, -1\}$

Statement-2 : f is a bijection. **[2009]**

(a) Statement-1 is true, Statement-2 is true. Statement-2 is not a correct explanation for Statement-1.

(b) Statement-1 is true, Statement-2 is false.

(c) Statement-1 is false, Statement-2 is true.

(d) Statement-1 is true, Statement-2 is true. Statement-2 is not a correct explanation for Statement-1.

16. For real x, let $f(x) = x^3 + 5x + 1$, then **[2009]**

(a) f is onto R but not one-one

(b) f is one-one and onto R

(c) f is neither one-one nor onto R

(d) f is one-one but not onto R

17. The domain of the function $f(x) = \frac{1}{\sqrt{|x| - x}}$ is **[2011]**

(a) $(0, \infty)$ (b) $(-\infty, 0)$

(c) $(-\infty, \infty) - \{0\}$ (d) $(-\infty, \infty)$

18. Let f be a function defined by $f(x) = (x-1)^2 + 1, (x \geq 1)$.

Statement - 1 :

The set $\{x : f(x) = f^{-1}(x)\} = \{1, 2\}$.

Statement - 2:

f is a bijection and $f^{-1}(x) = 1 + \sqrt{x-1}, x \geq 1$. **[2011 RS]**

(a) Statement-1 is true, Statement-2 is true; Statement-2 is a correct explanation for Statement-1.

(b) Statement-1 is true, Statement-2 is true; Statement-2 is NOT a correct explanation for Statement-1.

(c) Statement-1 is true, Statement-2 is false.

(d) Statement-1 is false, Statement-2 is true.

Solutions & Explanations

Section-A : JEE Advanced/ IIT-JEE

A 1. $\left[0, \frac{3}{\sqrt{2}}\right]$ 2. 0, 1 3. $[-2, -1]\cup[1, 2]$ 4. $n^n, n!$

5. $(-2, 1), [-1, 1]$ 6. $x+1$ and $-x+1$ 7. $\frac{-3\pm\sqrt{5}}{2}, \frac{3\pm\sqrt{5}}{2}$ 8. 1

B 1. T 2. T 3. F

C 1. (d) 2. (b) 3. (d) 4. (c) 5. (d) 6. (c) 7. (a)
8. (d) 9. (b) 10. (c) 11. (d) 12. (b) 13. (a) 14. (d)
15. (a) 16. (d) 17. (d) 18. (a) 19. (b) 20. (a) 21. (c)
22. (d) 23. (b) 24. (a) 25. (d) 26. (d) 27. (c) 28. (c)
29. (a) 30. (a) 31. (d) 32. (a) 33. (b)

D 1. (a, d) 2. (b, c) 3. (a, c) 4. (b) 5. (a) 6. (a, b) 7. (a, b)
8. (a, b)

E 1. R, [0, 1) ; f is not one to one 3. 3

5. domain = $\{x : x \in R, 16x^4 + 81x^2 - 2025 \le 0\}$; range = $\{y : y \in R, y \ge \frac{4x^2}{9}\}$; $R \cap R'$ is not a function.

6. 6 7. y 9. $a=3$ 10. $-\frac{5}{3}, 0, \frac{5}{3}$ 11. $2 \le \alpha \le 14$, No

F 1. (A) - q; (B) - r 2. (A) -p, r, s ; (B) - q, s ; (C) - q, s ; (D) -p, r, s

Section-B : JEE Main/ AIEEE

1. (a) 2. (c) 3. (a) 4. (a) 5. (d) 6. (d) 7. (a)
8. (b) 9. (b) 10. (d) 11. (c) 12. (a) 13. (b) 14. (d)
15. (b) 16. (b) 17. (b) 18. (a)

Section-A JEE Advanced/ IIT-JEE

(A) FILL IN THE BLANKS :

1. For the given function to be defined

$$\frac{\pi^2}{16} - x^2 \ge 0 \Rightarrow -\pi/4 \le x \le \pi/4$$

$\therefore D_f = [-\pi/4, \pi/4]$

Now, for $x \in [-\pi/4, \pi/4]$, $\sqrt{\pi^2/16 - x^2} \in [0, \pi/4]$

and sine function increases on $[0, \pi/4]$

$$\therefore 0 = \sin 0 \le \sin\sqrt{\frac{\pi^2}{16} - x^2} \le \sin \pi/4 = 1/\sqrt{2}$$

$$\Rightarrow 0 \le 3\sin\sqrt{\frac{\pi^2}{16} - x^2} \le 3/\sqrt{2}$$

$\therefore f(x) = [0, 3/\sqrt{2}]$

2. $f(x) = \begin{cases} \dfrac{x}{1+e^{1/x}}, & x \ne 0 \\ 0, & x = 0 \end{cases}$

$$f'(0^+) = \lim_{h\to0} \frac{f(0+h) - f(0)}{h}$$

$$= \lim_{h\to0} \frac{\frac{h}{1+e^{1/h}}}{h} = \lim_{h\to0} \frac{1}{1+e^{1/h}}$$

$$= \lim_{h\to0} \frac{e^{-1/h}}{e^{-1/h}+1} = \frac{0}{1} = 0$$

$$f'(0^-) = \lim_{h\to0} \frac{f(0) - f(0-h)}{h}$$

$$= \lim_{h\to0} \frac{0 - \frac{-h}{1+e^{-1/h}}}{h} = \lim_{h\to0} \frac{1}{1+e^{-1/h}} = 1$$

Thus $f'(0^+) = 0$ and $f'(0^-) = 1$

3. To find domain of function $f(x)=\sin^{-1}\left(\log_2 \frac{x^2}{2}\right)$

For $f(x)$ to be defined we should have

$$-1\le \log_2\left(\frac{x^2}{2}\right)\le 1$$

NOTE THIS STEP :

$\Rightarrow \quad 2^{-1}\le \frac{x^2}{2}\le 2^1 \;\Rightarrow\; 1\le x^2\le 4$

$\Rightarrow -2\le x\le -1$ or $1\le x\le 2$

$\Rightarrow x\in[-2,-1]\cup[1,2]$

4. Set A has n distinct elements.
Then to define a function from A to A, we need to associate each element of set A to any one the n elements of set A. So total number of functions from set A to set A is equal to the number of ways of doing n jobs where each job can be done in n ways. The total number such ways is $n\times n\times n\times \ldots \times n$ (n - times).
Hence the total number of functions from A to A is n^n.
Now for an onto function from A to A, we need to associate each element of A to one and only one element of A. So total number of onto functions from set A to A is equal to number of ways of arranging n elements in range (set A) keeping n elements fixed in domain (set A). n elements can be arranged in $n!$ ways.
Hence, the total number of functions from A to A is $n!$.

5. The given function is,

$$f(x)=\sin\left[\ell n\left(\frac{\sqrt{4-x^2}}{1-x}\right)\right]$$

For ℓn to be defined $\frac{\sqrt{4-x^2}}{1-x}>0$

$\Rightarrow 1-x>0$

Also $4-x^2>0$

$\Rightarrow x<1$ and $-2<x<2$

Combining these two inequalities, we get $x\in(-2,1)$

$\therefore$ Domain of f is $(-2, 1)$

Also $\sin\theta$ always lies in $[-1, 1]$.

$\therefore$ Range of f is $[-1, 1]$

6. **KEY CONCEPT :** Every linear function is either strictly increasing or strictly decreasing. If $f(x)=ax+b$, $D_f=[p,q]$, $R_f=[m,n]$
Then $f(p)=m$ and $f(q)=n$, if $f(x)$ is strictly increasing and $f(p)=n, f(q)=m$, If $f(x)$ is strictly decreasing function.
Let $f(x)=ax+b$ be the linear function which maps $[-1, 1]$ onto $[0, 2]$. then $f(-1)=0$ and $f(1)=2$
or $f(-1)=2$ and $f(1)=0$
Depending upon $f(x)$ is increasing or decreasing respectively.

$\Rightarrow \quad -a+b=0$ and $a+b=2$ ....(1)

or $\quad -a+b=2$ and $a+b=0$ ....(2)

Solving (1), we get $a=1, b=1$.

Solving (2), we get $a=-1, b=1$

Thus there are only two functions i.e., $x+1$ and $-x+1$.

7. Given that $f(x)=f\left(\frac{x+1}{x+2}\right)$ and f is an even function

$\therefore f(x)=f(-x)=f\left(\frac{-x+1}{-x+2}\right)$

$\Rightarrow x=\frac{-x+1}{-x+2} \qquad \Rightarrow x^2-3x+1=0$

$\Rightarrow x=\frac{3\pm\sqrt5}{2}$

Also $f(x)=f\left(\frac{x+1}{x+2}\right)=f(-x)$

$\Rightarrow \frac{x+1}{x+2}=-x \Rightarrow x^2+3x+1=0$

$\Rightarrow x=\frac{-3\pm\sqrt5}{2}$

$\therefore$ Four values of x are

$\frac{3+\sqrt5}{2}, \frac{3-\sqrt5}{2}, \frac{-3+\sqrt5}{2}, \frac{-3-\sqrt5}{2}$

8. $f(x)=\sin^2 x+\sin^2\left(x+\frac{\pi}{3}\right)+\cos x\cos\left(x+\frac{\pi}{3}\right)$

$\Rightarrow f(x)=\sin^2 x+\left[\sin\left(x+\frac{\pi}{3}\right)\right]^2+\cos x\cos\left(x+\frac{\pi}{3}\right)$

$\Rightarrow f(x)=\sin^2 x+\frac14(\sin x+\sqrt3\cos x)^2$

$+\frac12\cos x(\cos x-\sqrt3\sin x)$

$=\frac54(\sin^2 x+\cos^2 x)=\frac54$

$\therefore \quad (gof)\,x=g\,[f(x)]=g\,(5/4)=1$

ALTERNATE SOLUTION :

$f'(x)=2\sin x\cos x+2\sin\left(x+\frac{\pi}{3}\right)\cos\left(x+\frac{\pi}{3}\right)$

$-\sin x\cos\left(x+\frac{\pi}{3}\right)-\cos x\sin\left(x+\frac{\pi}{3}\right)$

$=\sin 2x+\sin(2\pi+2\pi/3)-\sin(2x+\pi/3)$

$=2\,[\sin(2x+\pi/3)].\ \frac12-\sin\left(2x+\frac{\pi}{3}\right)=0$

$\Rightarrow \quad f(x)=\text{constant}=k,$

Also $f(0)=5/4$

$\therefore \ f(x)=5/4,\ \forall x\in R$

$\therefore \ (gof)(x) \ =g\,[f(x)]=g\,(5/4)=1$

B TRUE / FALSE :

1. $f(x)=(a-x^n)^{1/n}$, $a>0$, n is + ve integer

$f(f(x))=f[(a-x^n)^{1/n}]=[a-\{(a-x^n)^{1/n}\}^n]^{1/n}$

$=(a-a+x^n)^{1/n}=x$

2. **KEY CONCEPT :** A function is one-one if it is strictly increasing or strictly decreasing, other wise it is many one.

$$f(x) = \frac{x^2 + 4x + 30}{x^2 - 8x + 18}$$

$$\Rightarrow \; f'(x) = \frac{-12[x^2 + 2x - 26]}{(x^2 - 8x + 18)^2}$$

$$\Rightarrow \; f'(x) = \frac{-12(x - 3\sqrt{3} + 1)(x + 3\sqrt{3} + 1)}{(x^2 - 8x + 18)^2}$$

$\Rightarrow$ $f(x)$ increases on $\left(-3\sqrt{3} - 1, 3\sqrt{3} - 1\right)$ and decreases otherwise.

$\Rightarrow$ $f(x)$ is many one.

3. We know that sum of any two functions is defined only on the points where both f_1 as well as f_2 are defined that is $f_1 + f_2$ is defined on $D_1 \cap D_2$.

$\therefore$ The given statement is false.

C MCQ's WITH ONE CORRECT ANSWER :

1. (d) $f(x) = x^2$ is many one as $f(1) = f(-1) = 1$

Also f is into as – ve real number have no pre-image.

$\therefore$ F is neither injective nor surjective.

2. (b) $y = x^2 + (k-1)x + 9 = \left(x + \frac{k+1}{2}\right)^2 + 9 - \left(\frac{k-1}{2}\right)^2$

For entire graph to be above x-axis, we should have

$$9 - \left(\frac{k-1}{2}\right)^2 > 0$$

+ – + on number line with points –5 and 7

$\Rightarrow$ $k^2 - 2k - 35 < 0$

$\Rightarrow$ $(k-7)(k+5) < 0$

i.e, $-5 < k < 7$

3. (d) $f(x) = |x - 1| = \begin{cases} -x + 1, & x < 1 \\ x - 1, & x \geq 1 \end{cases}$

Consider $f(x^2) = (f(x))^2$

If it is true it should be $\forall\ x$

$\therefore$ Put $x = 2$

LHS $= f(2^2) = |4 - 1| = 3$

RHS $= (f(2))^2 = 1$

$\therefore$ (a) is not correct

Consider $f(x + y) = f(x) + f(y)$

Put $x = 2, y = 5$ we get

$f(7) = 6; f(2) + f(5) = 1 + 4 = 5$

$\therefore$ (b) is not correct

Consider $f(|x|) = |f(x)|$

Put $x = -5$ then $f(|-5|) = f(5) = 4$

$|f(-5)| = |-5 - 1| = 6$

$\therefore$ (c) is not correct.

Hence (d) is the correct alternative.

4. (c) $|x - 1| + |x - 2| + |x - 3|) \geq 6$

Consider $f(x) = |x - 1| + |x - 2| + |x - 3|$

$$f(x) = \begin{cases} 6 - 3x, & x < 1 \\ 4 - x, & 1 \leq x < 2 \\ x, & 2 \leq x < 3 \\ 3x - 6, & x \geq 3 \end{cases}$$

NOTE THIS STEP :

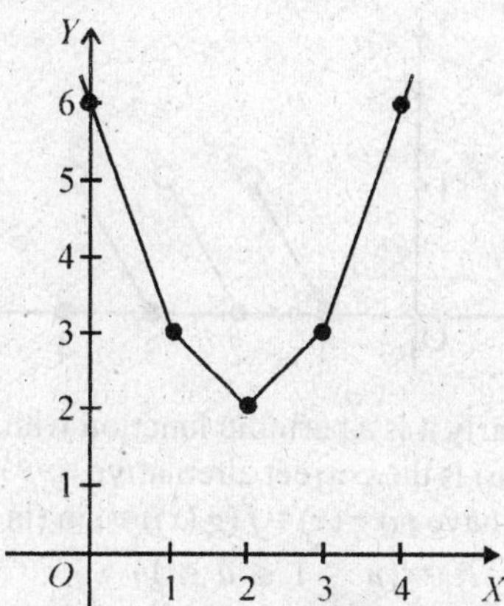

Graph of $f(x)$ shows $f(x) \geq 6$ for $x \leq 0$ or $x \geq 4$

5. (d) $f(x) = \cos(\log x)$

$$\therefore \quad f(x)f(y) - \frac{1}{2}\left[f\left(\frac{x}{y}\right) + f(xy)\right]$$

$$= \cos(\log x)\cos(\log y) - \frac{1}{2}[\cos(\log x - \log y) + \cos(\log x + \log y)]$$

$$= \cos(\log x)\cos(\log y) - \frac{1}{2}[2\cos(\log x)\cos(\log y)]$$

$= 0$

6. (c) $y = \frac{1}{\log_{10}(1 - x)} + \sqrt{x + 2}$

$y = f(x) + g(x)$

NOTE THIS STEP : Then domain of given function is $D_f \cap D_g$

Now, for domain of $f(x) = \frac{1}{\log_{10}(1 - x)}$

We know it is defined only when $1 - x > 0$ and $1 - x \neq 1$

$\Rightarrow x < 1$ and $x \neq 0$

$\therefore$ $D_f = (-\infty, 1) - \{0\}$

For domain of $g(x) = \sqrt{x + 2}$

$x + 2 \geq 0$

$\Rightarrow$ $x \geq -2$

$\therefore$ $D_g = [-2, \infty)$

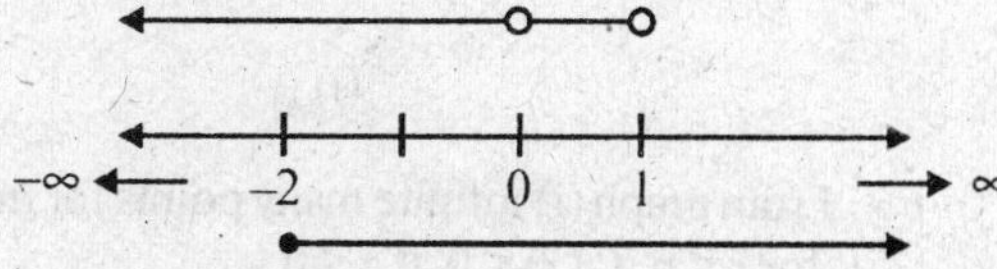

$\therefore$ Common domain is $[-2, 1) - \{0\}$

7. (a) $f(x)=x-[x]=\begin{cases} .... \\ x-1, & 1\le x<2 \\ x-2, & 2\le x<3 \\ x-3, & 3\le x<4 \\ .... \end{cases}$

graph of function is

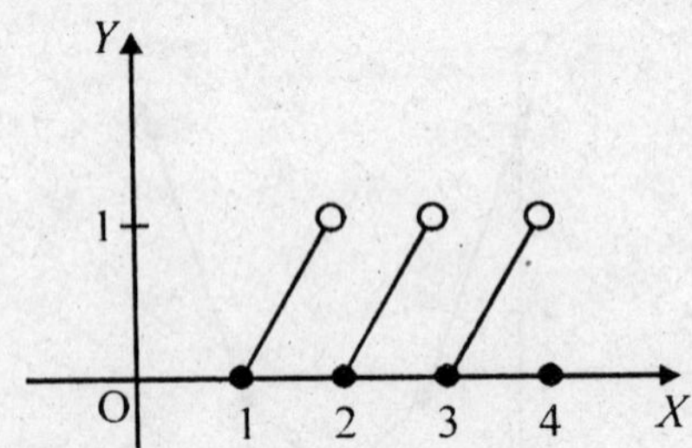

Clearly it is a periodic function with period 1.
$\therefore$ (a) is the correct alternative.

8. (d) We have $fog(x)=f(g(x))=\sin(\ln|x|)$

$\therefore \quad R_1=\{u:-1\le u\le 1\}$

$(\because -1\le \sin\theta\le 1, \forall\theta)$

Also $gof(x)=g(f(x))=\ln|\sin x|$

$\because \quad 0\le|\sin x|\le 1$

$\therefore \quad -\infty<\ln|\sin x|\le 0$

$\therefore \quad R_2=\{v:-\infty<v\le 0\}$

9. (c) $f(x)=f^{-1}(x)\Rightarrow fof(x)=x$

$[(x+1)^2-1+1]^2-1=x$

$\Rightarrow \quad (x+1)^4=x+1$

$\therefore \quad x=0$ or -1

$\therefore$ Req. set is $\{0,-1\}$

10. (c) $f(x)=|px-q|+r|x|$

$=\begin{cases}-px+q-rx, & x\le 0\\ -px+q+rx, & 0<x\le q/p\\ px-q+rx, & q/p<x\end{cases}$

For $r=p, f'(x)<0$ if $x<0$
$=0$ if $0<x<q/p$
>0 if $x>q/p$

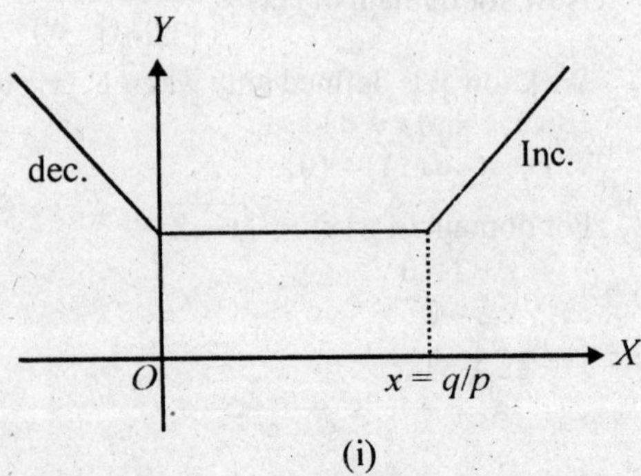

(i)

From graph (i) infinite many points for min value of $f(x)$
for $r<p, f'(x)<0$ if $x\le 0$
<0 if $0<x\le q/p$
>0 if $x<q/p$

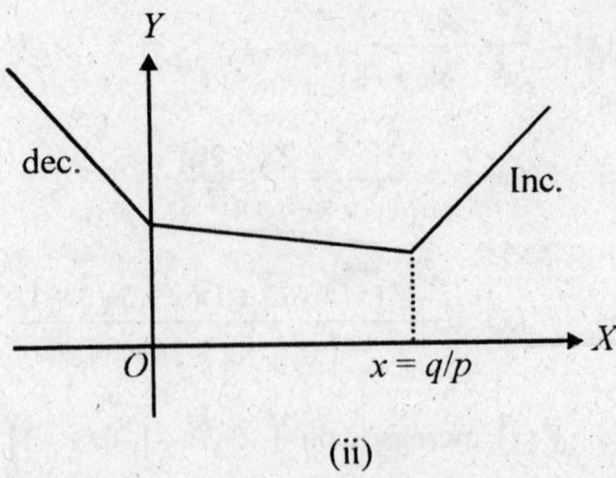

(ii)

From graph (ii) only pt. of min of $f(x)$ at $x=q/p$
For $r>p, f'(x)<0$ if $x\le 0$
>0 if $0<x$

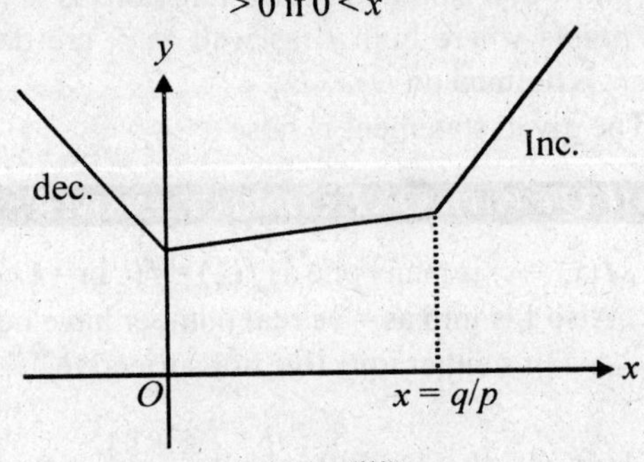

(iii)

From graph (iii) only one pt. of min of $f(x)$ at $x=0$

11. (d) $f(x)$ is continuous and defined for all $x>0$ and

$f\left(\frac{x}{y}\right)=f(x)-f(y)$

Also $f(e)=1$
$\Rightarrow$ Clearly $f(x)=\ell n\,x$ which satisfies all these properties
$\therefore f(x)=\ell n\,x$

12. (b) Let $y=2^{x(x-1)}$
$\Rightarrow x^2-x-\log_2 y=0;$

$x=\frac{1}{2}\left(1\pm\sqrt{1+4\log_2 y}\right)$

Since x is + ive, we choose only + out of $\pm$
(for $y\ge 1, \log_2 y\ge 0$)

$\therefore \; x=\frac{1}{2}\left(1+\sqrt{1+4\log_2 y}\right)$

or $\quad f^{-1}(x)=\frac{1}{2}\left(1+\sqrt{1+4\log_2 x}\right)$

13. (c) Let $h(x)=|x|$ then
$g(x)=|f(x)|=h(f(x))$
Since composition of two continuous functions is continuous, therefore g is continuous if f is continuous.

14. (d) It is given that

$2^x+2^y=2 \; \forall \; x,y\in R$

but $2^x, 2^y>0 \; \forall \; x,y\in R$
Therefore, $2^x=2-2^y<2$
$\Rightarrow 0<2^x<2\Rightarrow x<1$
Hence domain $=(-\infty, 1)$

15. (b) $g(x) = 1 + x - [x]$;

$$f(x) = \begin{cases} -1, & x < 0 \\ 0, & x = 0 \\ 1, & x > 0 \end{cases}$$

For integral values of x; $g(x) = 1$
For $x < 0$; (but not integral value) $x - [x] > 0 \Rightarrow g(x) > 1$
For $x > 0$; (but not integral value) $x - [x] > 0 \Rightarrow g(x) > 1$
$\therefore g(x) \geq 1, \forall x$ $\qquad \therefore f(g(x)) = 1, \forall x$

16. (a) $f(x) = x + \dfrac{1}{x} = y \Rightarrow x^2 - yx + 1 = 0$

$$\Rightarrow x = \frac{y \pm \sqrt{y^2 - 4}}{2}$$

$$\therefore x = \frac{y + \sqrt{y^2 - 4}}{2} \quad (\because x \geq 1 \text{ and } y \geq 2)$$

$$\therefore f^{-1}(x) = \frac{x + \sqrt{x^2 - 4}}{2}$$

NOTE THIS STEP :

17. (d) For domain of $f(x) = \dfrac{\log_2(x+3)}{x^2 + 3x + 2}$
$x^2 + 3x + 2 \neq 0$ and $x + 3 > 0$
$\Rightarrow x \neq -1, -2$ and $x > -3$
$\therefore D_f = (-3, \infty) - \{-1, -2\}$

18. (a) From E to F we can define, in all, $2 \times 2 \times 2 \times 2 = 16$ functions (2 options for each element of E) out of which 2 are into, when all the elements of E either map to 1 or to 2.
$\therefore$ No. of onto functions $= 16 - 2 = 14$

19. (d) $f(x) = \dfrac{\alpha x}{x+1}, x \neq -1$

$$f(f(x)) = x \Rightarrow \frac{\alpha\left(\dfrac{\alpha x}{x+1}\right)}{\dfrac{\alpha x}{x+1} + 1} = x$$

$$\Rightarrow \frac{\alpha^2 x}{(\alpha+1)x + 1} = x$$

$\Rightarrow (\alpha + 1)x^2 + (1 - \alpha^2)x = 0$ ...(1)
$\Rightarrow \alpha + 1 = 0$ and $1 - \alpha^2 = 0$
$\Rightarrow \alpha = -1$

20. (d) Given that $f(x) = (x+1)^2, x \geq -1$
Now if $g(x)$ is the reflection of $f(x)$ in the line $y = x$ then it can be obtained by interchanging x and y in $f(x)$
i.e., $y = (x+1)^2$ changes to $x = (y+1)^2$
$\Rightarrow y + 1 = \sqrt{x}$

$\left[y + 1 \neq -\sqrt{x}, \text{since } y \geq -1 \right.$ as in figure.

$\Rightarrow y = \sqrt{x} - 1$ defined $\forall x \geq 0$

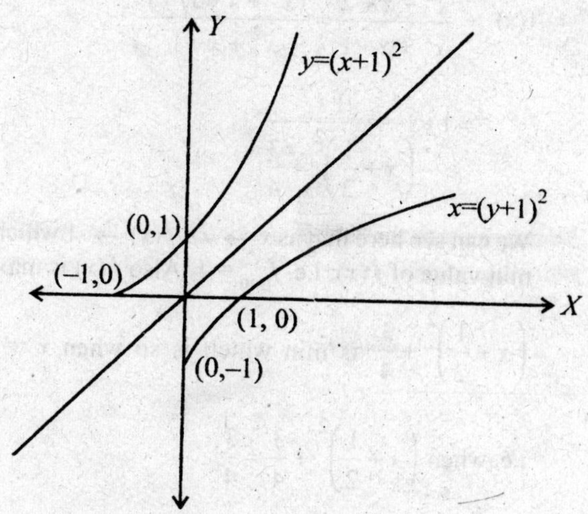

$\therefore g(x) = \sqrt{x} - 1 \quad \forall x \geq 0$

21. (a) Given that
$f(x) = 2x + \sin x, \ x \in R$
$\Rightarrow f'(x) = 2 + \cos x$
But $-1 \leq \cos x \leq 1$
$\Rightarrow 1 \leq 2 + \cos x \leq 3$
$\Rightarrow 1 \leq 2 + \cos x \leq 3$
$\therefore f'(x) > 0, \forall x \in R$
$\Rightarrow f(x)$ is strictly increasing and hence one-one
Also as $x \to \infty$, $f(x) \to \infty$ and $x \to -\infty$, $f(x) \to -\infty$
$\therefore$ Range of $f(x) = R =$ domain of $f(x) \Rightarrow f(x)$ is onto.
Thus, $f(x)$ is one-one and onto.

22. (b) Given that $f: [0, \infty) \to [0, \infty)$

Such that $f(x) = \dfrac{x}{x+1}$

Then $f'(x) = \dfrac{1 + x - x}{(1+x)^2} = \dfrac{1}{(1+x)^2} > 0 \, \forall x$

$\therefore f$ is an increasing function $\Rightarrow f$ is one-one.
Also, $D_f = [0, \infty)$

And for range let $\dfrac{x}{1+x} = y \Rightarrow x = \dfrac{y}{1-y}$
$x \geq 0 \Rightarrow 0 \leq y < 1$
$\therefore R_f = [0, 1) \neq$ Co-domain
$\therefore f$ is not onto.

23. (a) For $f(x) = \sqrt{\sin^{-1}(2x) + \dfrac{\pi}{6}}$ to be defined and real
if $\sin^{-1} 2x + \pi/6 \geq 0$

$\Rightarrow \sin^{-1} 2x \geq -\dfrac{\pi}{6}$ ...(1)

But we know that
$-\pi/2 \leq \sin^{-1} 2x \leq \pi/2$ ...(2)
Combining (1) and (2), we get

$$-\frac{\pi}{6} \leq \sin^{-1} 2x \leq \frac{\pi}{2}$$

$\Rightarrow \sin(-\pi/6) \leq 2x \leq \sin(\pi/2) \Rightarrow -1/2 \leq 2x \leq 1$

$\Rightarrow -1/4 \leq x \leq 1/2 \ \therefore D_f = \left[-\dfrac{1}{4}, \dfrac{1}{2}\right]$

24. (c) We have

$$f(x) = \frac{x^2+x+2}{x^2+x+1} = \frac{(x^2+x+1)+1}{x^2+x+1}$$

$$= 1+\frac{1}{\left(x+\frac{1}{2}\right)^2+\frac{3}{4}}$$

We can see here that as $x \to \infty$, $f(x) \to 1$ which is the min value of $f(x)$. i.e. $f_{min} = 1$. Also $f(x)$ is max when $\left(x+\frac{1}{2}\right)^2+\frac{3}{4}$ is min which is so when $x = -1/2$

i.e. when $\left(x+\frac{1}{2}\right)^2+\frac{3}{4}=\frac{3}{4}$.

$\therefore f_{max} = 1+\frac{1}{3/4} = 7/3$

$\therefore R_f = (1, 7/3]$

25. (d) We have

$f(x) = x^2+2bx+2c^2$; $g(x) = -x^2-2cx+b^2$

$\Rightarrow f(x) = (x+b)^2+2c^2-b^2$

and $g(x) = -(x+c)^2+b^2+c^2$

$\Rightarrow f_{min} = 2c^2-b^2$ and $g_{max} = b^2+c^2$

For $f_{min} > g_{max} \Rightarrow 2c^2-b^2 > b^2+c^2$

$\Rightarrow c^2 > 2b^2 \Rightarrow |c| > |b|\sqrt{2}$

26. (b) $f(x) = \sin x + \cos x$, $g(x) = x^2-1$

$\Rightarrow g(f(x)) = (\sin x+\cos x)^2 - 1 = \sin 2x$

Clearly $g(f(x))$ is invertible in $-\frac{\pi}{2} \le 2x \le \frac{\pi}{2}$

$[\because \sin\theta$ is invertible when $-\pi/2 \le \theta \le \pi/2]$

$\Rightarrow -\frac{\pi}{4} \le x \le \frac{\pi}{4}$

27. (a) We are given that

$f: R \to R$ such that $f(x) = \begin{cases} 0, & x \in \text{rational} \\ x, & x \in \text{irrational} \end{cases}$

$g: R \to R$ such that $g(x) = \begin{cases} 0, & x \in \text{irrational} \\ x, & x \in \text{rational} \end{cases}$

$\therefore (f-g): R \to R$ such that

$(f-g)(x) = \begin{cases} -x, \text{ if } & x \in \text{rational} \\ x, \text{ if } & x \in \text{irrational} \end{cases}$

Since $f-g: R \to R$ for any x there is only one value of $(f(x)-g(x))$ whether x is rational or irrational. Moreover as $x \in R$, $f(x)-g(x)$ also belongs to R. Therefore, $(f-g)$ is one-one onto.

28. (d) Given that X and Y are two sets and $f: X \to Y$.

$\{f(c) = y; c \subset X, y \subset Y\}$ and

$\{f^{-1}(d) = x : d \subset Y, x \subset X\}$

The pictorial representation of given information is as shown:

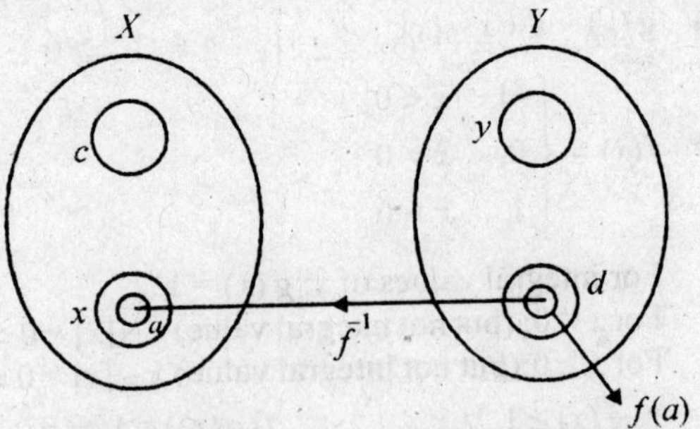

Since $f^{-1}(d) = x \Rightarrow f(x) = d$ Now if $a \subset x$

$\Rightarrow f(a) \subset f(x) = d \Rightarrow f^{-1}[f(a)] = a$

$\therefore f^{-1}(f(a)) = a$, $a \subset x$ is the correct option.

29. (a) $F(x) = \left(f\left(\frac{x}{2}\right)\right)^2 + \left(g\left(\frac{x}{2}\right)\right)^2$

$\Rightarrow F'(x) = 2f\left(\frac{x}{2}\right).f'\left(\frac{x}{2}\right).\frac{1}{2}+2g\left(\frac{x}{2}\right).g'\left(\frac{x}{2}\right).\frac{1}{2}$

$= f\left(\frac{x}{2}\right).f'\left(\frac{x}{2}\right)+f'\left(\frac{x}{2}\right).f''\left(\frac{x}{2}\right)$

$[\because g(x) = f'(x) \Rightarrow g'(x) = f''(x)]$

$= f\left(\frac{x}{2}\right).f'\left(\frac{x}{2}\right)-f'\left(\frac{x}{2}\right)f\left(\frac{x}{2}\right)$

$= 0 \qquad [\because f''(x) = -f(x)]$

$\Rightarrow F(x)$ is a constant function.

$\therefore F(x) = F(5) = 5 \ \forall x \in R \Rightarrow F(10) = 5$

30. (a) Given $f(x) = \frac{x}{(1+x^n)^{1/n}}$ for $n \ge 2$

$\therefore fof(x) = f[f(x)] = f\left[\frac{x}{(1+x^n)^{1/n}}\right]$

$$= \frac{\frac{x}{(1+x^n)^{1/n}}}{\left[1+\frac{x^n}{1+x^n}\right]^{1/n}} = \frac{x}{(1+2x^n)^{1/n}}$$

Further, $fofof(x) = \frac{x}{(1+3x^n)^{1/n}}$

Proceeding in the similar manner, we get

$g(x) = fofof.....of(x) = \frac{x}{(1+nx^n)^{1/n}}$

(f occurs n times)

Now, $\int x^{n-2} g(x)dx = \int \frac{x^{n-1}}{(1+nx^n)^{1/n}} dx$

Let $1+nx^n = t \Rightarrow n^2x^{n-1}\,dx = dt$

$\therefore$ Integral becomes $= \frac{1}{n^2}\int t^{-1/n} dt = \frac{1}{n^2}.\frac{t^{-\frac{1}{n}+1}}{-\frac{1}{n}+1}+K$

$$= \frac{1}{n}.\frac{t^{1-1/n}}{n-1}+K = \frac{(1+nx^n)^{1-1/n}}{n(n-1)}+K$$

31. **(d)** $f(x)=e^{x^2}+e^{-x^2} \Rightarrow f'(x)=2x\left(e^{x^2}-e^{-x^2}\right)\geq 0,$

$\forall x \in [0,1]$

$\therefore$ f (x) is an increasing function on $[0,1]$

Hence $f_{max}=f(1)=e+\frac{1}{e}=a$

$g(x)=xe^{x^2}+e^{-x^2}$

$\Rightarrow g'(x)=(2x^2+1)e^{x^2}-2xe^{-x^2}\geq 0, \forall x\in[0,1]$

$\therefore$ g (x) is an increasing function on $[0,1]$

$\therefore$ $g_{\max}=g(1)=e+\frac{1}{e}=b$

$h(x)=x^2e^{x^2}+e^{-x^2}$

$h'(x)=2x\left[e^{x^2}(1+x^2)-e^{-x^2}\right]\geq 0, \forall x\in[0,1]$

$\therefore$ h (x) is an increasing function on $[0,1]$

$\therefore$ $h_{\max}=h(1)=e+\frac{1}{e}=c$

Hence $a=b=c$.

32. **(a)** Given that $f(x)=x^2$ and $g(x)=\sin x, \forall x\in R$

Then (gof) (x) = $\sin x^2$

$\Rightarrow$ (gogof) (x) = $\sin(\sin x^2)$

$\Rightarrow$ (fogogof) (x) = $\sin^2(\sin x^2)$

As given that (fogogof) (x) = (gogof) (x)

$\Rightarrow \sin^2(\sin x^2)=\sin(\sin x^2)$

$\Rightarrow \sin(\sin x^2)=0, 1$

$\Rightarrow \sin x^2=n\pi$ or $\left((4n+1)\frac{\pi}{2}\right.$ where $n\in Z$

$\Rightarrow \sin x^2=0 \quad \because \sin x^2\in[-1,1]$

$\Rightarrow x^2=n\pi$

$\Rightarrow x=\pm\sqrt{n\pi}$ where $n\in W$

33. **(b)** We have $f(x)=2x^3-15x^2+36x+1$

$\Rightarrow \quad f'(x)=6x^2-30x+36$

$= 6(x^2-5x+6)$

$= 6(x-2)(x-3)$

$\because$ $f'(x)>0 \ \forall x\in[0,2]$ and $f'(x)<0 \ \forall x\in[2,3]$

$\therefore$ $f(x)$ is increasing on [0, 2] and decreasing on [2, 3]

$\therefore$ $f(x)$ is many one on [0, 3]

Also $f(0)=1, f(2)=29, f(3)=28$

$\therefore$ Global min = 1 and Global max = 29

i.e., Range of $f=[1,29]$ = codomain

$\therefore f$ is onto.

D MCQ's WITH ONE OR MORE THAN ONE CORRECT :

1. **(a, d)**

Given that $f(x)=y=\frac{x+2}{x-1}$

Let us check each option one by one.

(a) $f(x)=\frac{x+2}{x-1}=y\Rightarrow x=f(y)$

$\therefore$ (a) is correct

(b) $f(1)\neq 3$ as function is not defined for $x=1$

$\therefore$ (b) is not correct.

(c) $f'(x)=\frac{x-1-x-2}{(x-1)^2}=\frac{-3}{(x-1)^2}$

$\therefore$ $f'(x)<0$, if $x\neq 1$

$\Rightarrow$ $f(x)$ is decreasing if $x\neq 1$

$\therefore$ (c) is not correct.

(d) $f(x)=\frac{x+2}{x-1}$, which is a rational function of x.

2. **(b, c)**

As (0, 0) and (x, g (x)) are two vertices of an equilateral triangle; therefore, length of the side of Δ is

$=\sqrt{(x-0)^2+(g(x)-0)^2}=\sqrt{x^2+(g(x))^2}$

$\therefore$ The area of equilateral $\Delta=\frac{\sqrt{3}}{4}(x^2+(g(x))^2)$

ATQ, this area $=\frac{\sqrt{3}}{4}$

$\therefore$ We get $\frac{\sqrt{3}}{4}(x^2+(g(x))^2)=\frac{\sqrt{3}}{4}$

$\Rightarrow (g(x))^2=1-x^2$

$\Rightarrow g(x)=\pm\sqrt{1-x^2}$

$\therefore$ (b), (c) are the correct answers as (a) is not a function ($\because$ image of x is not unique)

3. **(a, c)**

$f(x)=\cos[\pi^2]x+\cos[-\pi^2]x$

We know $9<\pi^2<10$ and $-10<-\pi^2<-9$

$\Rightarrow [\pi^2]=9$ and $[-\pi^2]=-10$

$\Rightarrow \therefore f(x)=\cos 9x+\cos(-10x)$

$f(x)=\cos 9x+\cos 10x$

Let us check each option one by one.

(a) $f\left(\frac{\pi}{2}\right)=\cos\frac{9\pi}{2}+\cos 5\pi=-1$ **(true)**

(b) $f(\pi)=\cos 9\pi+\cos 10\pi=-1+1=0$ **(false)**

(c) $f(-\pi)=\cos(-9\pi)+\cos(-10\pi)$
$=\cos 9\pi+\cos 10\pi=-1+1=0$ **(true)**

(d) $f\left(\frac{\pi}{4}\right)=\cos\frac{9\pi}{4}+\cos\frac{5\pi}{2}$

$=\cos\left(2\pi+\frac{\pi}{4}\right)+0=\frac{1}{\sqrt{2}}$ **(false)**

Thus (a) and (c) are the correct options.

4. **(b)** $f(x) = 3x - 5$ (given), which is strictly increasing on R, so $f^{-1}(x)$ exists.

Let $y = f(x) = 3x - 5$

$$\Rightarrow y + 5 = 3x \Rightarrow x = \frac{y+5}{3} \quad ...(1)$$

and $y = f(x) \Rightarrow x = f^{-1}(y)$ ...(2)

From (1) and (2):

$$f^{-1}(y) = \frac{y+5}{3} = f^{-1}(x) = \frac{x+5}{3}$$

5. **(a)** Let us check each option one by one.

(a) $f(x) = \sin^2 x$ and $g(x) = \sqrt{x}$

Now, $fog = f(g(x)) = f(\sqrt{x}) = \sin^2 \sqrt{x} = (\sin\sqrt{x})^2$

and $gof(x) = g(f(x)) = g(\sin^2 x) = \sqrt{\sin^2 x} = |\sin x|$

(a) is true.

(b) $f(x) = \sin x$, $g(x) = |x|$

$fog(x) = f(g(x)) = f(|x|) = \sin|x| \neq (\sin\sqrt{x})^2$

$\therefore$ (b) is not true

(c) $f(x) = x^2$, $g(x) = \sin\sqrt{x}$

$fog(x) = f[g(x)] = f(\sin\sqrt{x}) = (\sin\sqrt{x})^2$

and $(gof)(x) = g[f(x)] = g(x^2) = \sin\sqrt{x^2}$
$= \sin|x| \neq |\sin x|$

$\therefore$ (c) is not true.

6. **(a, b)** We have $f(x) = \dfrac{b-x}{1-bx}$, $0 < b < 1$

Let $f(x_1) = f(x_2)$

$$\Rightarrow \frac{b-x_1}{1-bx_1} = \frac{b-x_2}{1-bx_2}$$

$\Rightarrow b - b^2x_2 - x_1 + bx_1x_2 = b - x_2 - b^2x_1 + bx_1x_2$

$\Rightarrow x_2(1-b^2) = x_1(1-b^2)$

$\Rightarrow x_1 = x_2$ as $1 - b^2 \neq 0$

$\therefore f$ is one one.

Also $\dfrac{b-x}{1-bx} = y$

$\Rightarrow b - x = y - bxy$

$\Rightarrow (by - 1)x = y - b$

$$\Rightarrow x = \frac{y-b}{by-1}$$

For $y = \dfrac{1}{b}$, x is not defined

$\therefore f$ is not onto.

$\therefore f$ is not invertible

$$\text{Also } f'(x) = \frac{-1(1-bx)-(-b)(b-x)}{(1-bx)^2} = \frac{b^2-1}{(1-bx)^2}$$

$\therefore f'(b) = \dfrac{1}{b^2-1}$ and $f'(0) = b^2 - 1 \Rightarrow f'(b) = \dfrac{1}{f'(0)}$

Hence a and b are the correct options.

7. **(a, b)**

$$\text{Given } f(\cos 4\theta) = \frac{2}{2-\sec^2\theta} = \frac{2\cos^2\theta}{2\cos^2\theta - 1}$$

$$= \frac{1+\cos 2\theta}{\cos 2\theta} = 1 + \frac{1}{\cos 2\theta}$$

$$\text{Let } \cos 4\theta = \frac{1}{3} \Rightarrow 2\cos^2 2\theta - 1 = \frac{1}{3}$$

$$\Rightarrow \cos 2\theta = \pm\sqrt{\frac{2}{3}}$$

$$\therefore f(\cos 4\theta) = 1 \pm \sqrt{\frac{3}{2}} \text{ or } f\left(\frac{1}{3}\right) = 1 \pm \sqrt{\frac{3}{2}}$$

8. **(a, b)** For $f(x) = 2|x| + |x+2| - ||x+2| - 2|x||$

the critical points can be obtained by solving $|x| = 0$, $|x+2| = 0$ and $||x+2| - 2|x|| = 0$

we get $x = 0, -2, 2, -\dfrac{2}{3}$

Then we can write

$$f(x)\begin{cases} -2x-4, & x \le -2 \\ 2x+4, & -2 < x \le -\frac{2}{3} \\ -4x, & -\frac{2}{3} < x \le 0 \\ 4x, & 0 < x \le 2 \\ 2x+4, & x > 2 \end{cases}$$

The graph of $y = f(x)$ is as follows

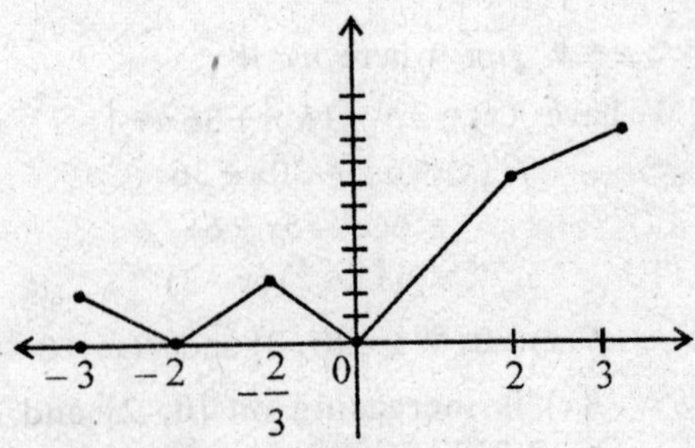

From graph $f(x)$ has local minimum at –2 and 0 and local maximum at $-\dfrac{2}{3}$

E) SUBJECTIVE PROBLEMS :

Since $f(x)$ is defined and real for all real values of x, therefore domain of f is R.

Also $\frac{x^2}{1+x^2} \geq 0$, for all $x \in R$

and $\frac{x^2}{1+x^2} < 1$ ($\because x^2 < 1+x^2$) for all $x \in R$

$\therefore \quad 0 \leq \frac{x^2}{1+x^2} < 1 \Rightarrow 0 \leq f(x) < 1$

$\Rightarrow$ Range of $f = [0, 1)$

Also since $f(1) = f(-1) = 1/2$

$\therefore \quad f$ is not one-to-one.

$y = |x|^{1/2}, -1 \leq x \leq 1$

$\Rightarrow \quad y = \sqrt{-x}$ if $-1 \leq x \leq 0 = \sqrt{x}$ if $0 \leq x \leq 1$

$\Rightarrow \quad y^2 = -x$ if $-1 \leq x \leq 0$ and $y^2 = x$ if $0 \leq x \leq 1$

[Here y should be taken always + ve, as by definition y is a + ve square root].

Clearly $y^2 = -x$ represents upper half of left handed parabola (upper half as y is + ve)

and $y^2 = x$ represents upper half of right handed parabola.

Therefore the resulting graph is as shown below :

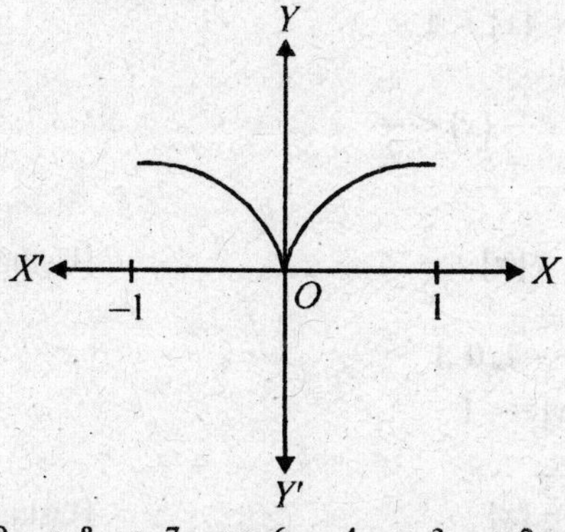

$f(x) = x^9 - 6x^8 - 2x^7 + 12x^6 + x^4 - 7x^3 + 6x^2 + x - 3$

Then $f(6) = 6^9 - 6 \times 6^8 - 2 \times 6^7 + 12 \times 6^6 + 6^4 - 7 \times 6^3 + 6 \times 6^2 + 6 - 3$

$= 6^9 - 6^9 - 2 \times 6^7 + 2 \times 6^7 + 6^4 - 7 \times 6^3 + 6^3 + 6 - 3 = 3$

$R = [(x, y); x \in R, y \in R, x^2 + y^2 \leq 25]$ which represents all the points inside and on the circle $x^2 + y^2 = 5^2$, with centre (0, 0) and radius = 5

$$R' = \left\{(x, y) : x \in R, y \in R, y \geq \frac{4}{9}x^2\right\}$$

which represents all the points inside and on the upward parabola $x^2 \leq \frac{9}{4}y$.

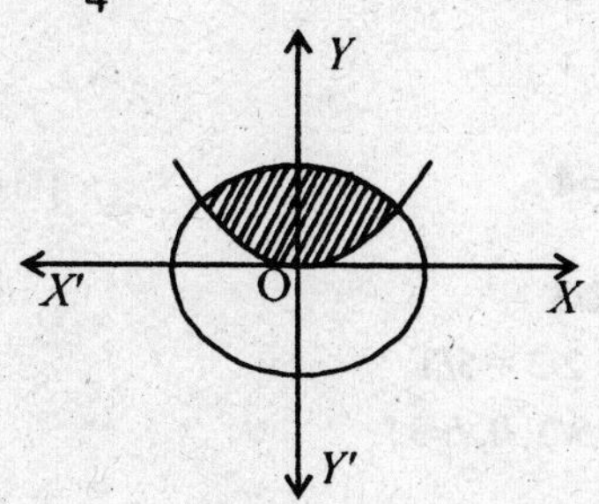

Thus $R \cap R'$ = The set of all points in shaded region.

For domain of $R \cap R'$.

$x^2 + y^2 \leq 25$

$\Rightarrow \quad x^2 \leq 25 - y^2$ ....(1)

and $y \geq \frac{4}{9}x^2 \Rightarrow \frac{16x^4}{81} \leq y^2$

$\Rightarrow \quad -\frac{16x^4}{81} \geq -y^2$

$\Rightarrow \quad 25 - \frac{16x^4}{81} \geq 25 - y^2$ ....(2)

$\therefore$ Combining (1) and (2) $x^2 \leq 25 - \frac{16}{81}x^4$

$\Rightarrow \quad 16x^4 + 81x^2 - 2025 \leq 0$

$\therefore$ Domain of $R \cap R'$ =

$\{x : x \in R, 16x^4 + 81x^2 - 2025 \leq 0\}$ and range of $R \cap R'$

$= \{y : y \in R, y \geq \frac{4x^2}{9}, 16x^4 + 81x^2 - 2025 \leq 0\}$

$R \cap R'$ is not a function because image of an element is not unique, e.g., (0, 1), (0, 2), (0,3) ..... $\in R \cap R'$.

5. As there is an injective maping from A to B, each element of A has unique image in B. Similarly as there is an injective mapping from B to A, each element of B has unique image in A. So we can conclude that each element of A has unique image in B and each element of B has unique image in A or in other words there is one to one mapping from A to B. Thus there is bijective mapping from A to B.

6. Given that $\forall x, y; f(x+y) = f(x)f(y)$

$\therefore \quad f(0+0) = f(0).f(0) = [f(0)]^2$

$\Rightarrow \quad f(0) = [f(0)]^2$

$\Rightarrow \quad f(0) = 1$ Now, given that

$f'(0) = 3$

$\Rightarrow \quad \lim_{h\to 0} \frac{f(0+h) - f(0)}{h} = 3$

$\Rightarrow \quad \lim_{h\to 0} \frac{f(0)f(h) - f(0)}{h} = 3$

[Using $f(x+y) = f(x).f(y)$]

$\Rightarrow \quad \lim_{h\to 0} \frac{f(h) - 1}{h} = 3$ ...(1) [Using $f(0) = 1$]

Now, $f'(5) = \lim_{h\to 0} \frac{f(5+h) - f(5)}{h}$

$= \lim_{h\to 0} \frac{f(5)f(h) - f(5)}{h}$

{Using prop. $f(x+y) = f(x)f(y)$}

$= \lim_{h\to 0} \frac{2f(h) - 2}{h}$ [Using $f(5) = 2$]

$= 2\lim_{h\to 0} \frac{f(h) - 1}{h} = 2 \times 3 = 6$ [Using eq. (1)]

7. f is one one function,

$D_f = \{x, y, z\}; R_f = \{1, 2, 3\}$

Eaxctly one of the following is true :

$f(x) = 1, f(y) \neq 1, f(z) \neq 2$

To determine $f^{-1}(1)$:

Case I: $f(x) = 1$ is true.

$\Rightarrow f(y) \neq 1, f(z) \neq 2$ are false.

$\Rightarrow f(y) = 1, f(z) = 2$ are true.

But $f(x) = 1, f(y) = 1$ are true, is not possible as f is one to one.

$\therefore$ This case is not possible.

Case II: $f(y) \neq 1$ is true.

$\Rightarrow f(x) = 1$ and $f(z) \neq 2$ are false

$\Rightarrow f(x) \neq 1$ and $f(z) = 2$ are true

Thus, $f(x) \neq 1, f(y) \neq 1, f(z) = 2$

$\Rightarrow$ Either $f(x)$ or $f(y) = 2$. So, f is not one to one

$\therefore$ This case is also not possible.

$\therefore f(z) \neq 2$ is true

$\therefore f(x) = 1$ and $f(y) \neq 1$ are false.

$\Rightarrow f(x) \neq 1$ and $f(y) = 1$ are true.

$\Rightarrow f^{-1}(1) = y$

8. Since $|f(x) - f(y)| \leq |x - y|^3$ is true $\forall\ x, y \in R$

We have for $x \neq y$, $\dfrac{|f(x) - f(y)|}{|x - y|} \leq |x - y|^2$

$$\Rightarrow \lim_{y \to x}\left|\frac{f(x) - f(y)}{x - y}\right| \leq \lim_{y \to x}|x - y|^2$$

$$\Rightarrow \left|\lim_{y \to x}\frac{f(x) - f(y)}{x - y}\right| \leq 0$$

$\Rightarrow |f'(x)| \leq 0$

$\Rightarrow f'(x) = 0$

$\Rightarrow f(x)$ is a constant function. Hence Proved.

9. Given that $f(x + y) = f(x) f(y)\ \forall\ x,\ y \in N$ and $f(1) = 2$

To find 'a' such that,

$$\sum_{k=1}^{n} f(a + k) = 16(2^n - 1) \quad \text{...(1)}$$

For this we start with

$f(1) = 2$ ...(2)

Then $f(2) = f(1 + 1) = f(1) f(1)$

$\Rightarrow f(2) = 2^2$ Using (2)

Similarly we get, $f(3) = 2^3$,

$f(4) = 2^4, \ldots, f(n) = 2^n$

Now eq. (1) can be written as

$f(a + 1) + f(a + 2) + f(a + 3) + \ldots + f(a + n) = 16\,(2^n - 1)$

$\Rightarrow f(a)f(1) + f(a)f(2) + f(a)f(3) + \ldots + f(a)f(n) = 16\,(2^n - 1)$

$\Rightarrow f(a)f(1) + f(a)f(2) + f(a)f(3) + \ldots\ldots + f(a)f(n) = 16\,(2^n - 1)$

$\Rightarrow f(a)\,[f(1) + f(2) + f(3) + \ldots + f(n)] = 16\,[2^n - 1]$

$\Rightarrow f(a)\,[2 + 2^2 + 2^3 + \ldots + 2^n] = 16\,[2^n - 1]$

$$\Rightarrow f(a)\left[\frac{2(2^n - 1)}{2 - 1}\right] = 16[2^n - 1]$$

$\Rightarrow f(a) = 8 = 2^3 = f(3) \Rightarrow a = 3$

10. Given that $4\{x\} = x + [x]$

Where $[x]$ = greatest integer $\leq x$

$\{x\}$ = fractional part of x

$\therefore x = [x] + \{x\}$ for any $x \in R$

$\therefore$ Given eq^n becomes

$4\{x\} = [x] + \{x\} + [x]$

$\Rightarrow 3\{x\} = 2[x]$

$\Rightarrow [x] = \dfrac{3}{2}\{x\}$ ....(1)

Now $-1 < \{x\} < 1$

$\Rightarrow -\dfrac{3}{2} < \dfrac{3}{2}\{x\} < \dfrac{3}{2}$

$\Rightarrow -\dfrac{3}{2} < [x] < \dfrac{3}{2}$ [Using eqn (1)]

$\Rightarrow [x] = -1, 0, 1$

If $[x] = -1$

$\Rightarrow -1 = \dfrac{3}{2}\{x\}$ [Using eqn (1)]

$\Rightarrow \{x\} = -\dfrac{2}{3}$

$\therefore x = [x] + \{x\}$

$\Rightarrow x = -1 + (-2/3) = -5/3$

If $[x] = 0$

$\Rightarrow \dfrac{3}{2}\{x\} = 0$ [Using eq^n (1)]

$\Rightarrow \{x\} = 0$

$\therefore x = 0 + 0 = 0$

If $[x] = 1$

$\Rightarrow \dfrac{3}{2}\{x\} = 1$ [Using eq^n (1)]

$\Rightarrow \{x\} = 2/3$

$\Rightarrow x = 1 + 2/3 = 5/3$

Thus, $x = -5/3, 0, 5/3$

11. Let us put $y = \frac{\alpha x^2 + 6x - 8}{\alpha + 6x - 8x^2}$

$\Rightarrow (\alpha + 6x - 8x^2)y = \alpha x^2 + 6x - 8$

$\Rightarrow (\alpha + 8y)x^2 + 6(1-y)x - (8 + \alpha y) = 0$

Since x is real, $D \geq 0$

$\Rightarrow 36(1-y)^2 + 4(\alpha + 8y)(8 + \alpha y) \geq 0$

$\Rightarrow 9(1 - 2y + y^2) + [8\alpha + (64 + \alpha^2)y + 8\alpha y^2] \geq 0$

$\Rightarrow y^2(9 + 8\alpha) + y(46 + \alpha^2) + (9 + 8\alpha) \geq 0$ ...(1)

For (1) to hold for each $y \in R$, $9 + 8\alpha > 0$

and $(46 + \alpha^2)^2 - 4(9 + 8\alpha)^2 \leq 0$

$\Rightarrow \alpha > -9/8$

and $[46 + \alpha^2 - 2(9 + 8\alpha)][46 + \alpha^2 + 2(9 + 8\alpha)] \leq 0$

$\Rightarrow \alpha > -9/8$

and $(\alpha^2 - 16\alpha + 28)(\alpha^2 + 16\alpha + 64) \leq 0$

$\Rightarrow \alpha > -9/8$

and $(\alpha - 2)(\alpha - 14)(\alpha + 8)^2 \leq 0$

$\Rightarrow \alpha > -8/9$

and $(\alpha - 2)(\alpha - 14) \leq 0$ $[\because (\alpha + 8)^2 \geq 0]$

$\Rightarrow \alpha > -8/9$ and $2 \leq \alpha \leq 14$

$\Rightarrow 2 \leq \alpha \leq 14$

Thus, $f(x) = \frac{\alpha x^2 + 6x - 8}{\alpha + 6x - 8x^2}$ will be onto if $2 \leq \alpha \leq 14$.

When $\alpha = 3$

$$f(x) = \frac{3x^2 + 6x - 8}{3 + 6x - 8x^2}$$

In this case $f(x) = 0$ implies, $3x^2 + 6x - 8 = 0$

$$\Rightarrow x = \frac{-6 \pm \sqrt{36 + 96}}{6} = \frac{-6 \pm \sqrt{132}}{6} = \frac{-6 \pm 2\sqrt{33}}{6}$$

$$= \frac{1}{3}(-3 \pm \sqrt{33})$$

This shows that

$$f\left[\frac{1}{3}(-3 + \sqrt{33})\right] = f\left[\frac{1}{3}(-3 - \sqrt{33})\right] = 0$$

Therefore, f is not one-to-one at $\alpha = 3$.

12. Suppose $f(x) = Ax^2 + Bx + C$ is an integer whenever x is an integer.

$\therefore$ $f(0), f(1), f(-1)$ are integers

$\Rightarrow$ $C, A + B + C, A - B + C$ are integers.

$\Rightarrow$ $C, A + B, A - B$ are integers

$\Rightarrow$ $C, A + B, (A + B) + (A - B) = 2A$ are integers.

Conversely suppose $2A$, $A + B$ and C are integers.

Let x be any integer.

We have

$$f(x) = Ax^2 + Bx + C$$

$$= 2A\left[\frac{x(x-1)}{2}\right] + (A + B)\,x + C$$

Since x is an integer x, $x(x - 1)/2$ is an integer.

Also $2A$, $A + B$ and C are integers.

We get $f(x)$ is an integer for all integer x.

F MATCH THE FOLLOWING :

1. (A) $f(x) = 1 + 2x$,

$D_f = (-\pi/2, \pi/2)$

The given function represents a straight line so it is one one.

But $f_{min} = 1 - \pi = f\left(-\frac{\pi}{2}\right)$

$f_{max} = 1 + \pi = f\left(\frac{\pi}{2}\right)$

$\therefore$ Range $f = (1 - \pi, 1 + \pi) \in (-\infty, \infty)$

$\therefore$ f is not onto. Hence (A) $\to$ (q).

(B) $f(x) = \tan x$

It is an increasing function on $(-\pi/2, \pi/2)$ and its range $= (-\infty, \infty) =$ co-domain of f.

$\therefore$ f is one one onto.

$\therefore$ (B) $\to$ r

2. We have $f(x) = \frac{x^2 - 6x + 5}{x^2 - 5x + 6} = \frac{(x-5)(x-1)}{(x-2)(x-3)}$

(A) If $-1 < x < 1$ then $f(x) = \frac{(-ve)(-ve)}{(-ve)(-ve)} = +ve$

$\therefore$ $f(x) > 0$ (r)

Also $f(x) - 1 = \frac{-x-1}{x^2 - 5x + 6} = -\frac{(x+1)}{(x-2)(x-3)}$

For $-1 < x < 1$, $f(x) - 1 = \frac{-(+ve)}{(-ve)(-ve)} = -ve$

$\Rightarrow f(x) - 1 < 0 \Rightarrow f(x) < 1$ (s)

$\therefore$ $0 < f(x) < 1$ (p)

(B) If $1 < x < 2$ then $f(x) = \frac{(-ve)(+ve)}{(-ve)(-ve)} = -ve$

$\therefore$ $f(x) < 0$ (q) and so $f(x) < 1$ (s)

(C) If $3 < x < 5$ then $f(x) = \frac{(-ve)(+ve)}{(+ve)(+ve)} = -ve$

$\therefore$ $f(x) < 0$ (q) and so $f(x) < 1$ (s)

(D) For $x > 5$, $f(x) > 0$ (r)

Also $f(x) - 1 = \frac{-(x+1)}{(x-2)(x-3)} < 0$

For $x > 5$, $f(x) < 1$ (s)

$\therefore$ $0 < f(x) < 1$ (p)

Section-B JEE Main/ AIEEE

(a) $f(x)=\sin^{-1}\left(\log_3\left(\frac{x}{3}\right)\right)$ exists

if $-1 \le \log_3\left(\frac{x}{3}\right) \le 1 \Leftrightarrow 3^{-1} \le \frac{x}{3} \le 3^1$

$\Leftrightarrow 1 \le x \le 9$ or $x \in [1, 9]$

(c) $f(x) = \log(x+\sqrt{x^2+1})$

$$f(-x) = \log\left\{-x+\sqrt{x^2+1}\right\} = \log\left\{\frac{-x^2+x^2+1}{x+\sqrt{x^2+1}}\right\}$$

$= -\log(x+\sqrt{x^2+1}) = -f(x)$

$\Rightarrow f(x)$ is an odd function.

(a) $f(x) = \dfrac{3}{4-x^2} + \log_{10}(x^3-x)$

$4-x^2 \ne 0;\ x^3-x>0;$

$x \ne \pm\sqrt{4}$ and $-1<x<0$ or $1<x<\infty$

$\therefore D = (-1, 0)\cup(1,\infty)-\left\{\sqrt{4}\right\}$

$D = (-1, 0)\cup(1, 2)\cup(2,\infty)$.

(a) $f(x+y) = f(x)+f(y)$.

Function should be $f(x) = mx$

$f(1) = 7;\ \therefore m = 7,\ f(x) = 7x$

$$\sum_{r=1}^{n} f(r) = 7\sum_{1}^{n} r = \frac{7n(n+1)}{2}$$

(d) We have $f : N \to I$

If x and y are two even natural numbers,

then $f(x) = f(y) \Rightarrow \frac{-x}{2} = \frac{-y}{2} \Rightarrow x = y$

Again if x and y are two odd natural numbers then

$f(x) = f(y) \Rightarrow \frac{x-1}{2} = \frac{y-1}{2} \Rightarrow x = y$

$\therefore f$ is onto.

Also each negative integer is an image of even natural number and each positive integer is an image of odd natural number.

$\therefore f$ is onto.

Hence f is one one and onto both.

6. (d) ${}^{7-x}P_{x-3}$ is defined if

$7-x \ge 0, x-3 \ge 0$ and $7-x \ge x-3$

$\Rightarrow 3 \le x \le 5$ and $x \in I$

$\therefore x = 3, 4, 5$

$\therefore f(3) = {}^{7-3}P_{3-3} = {}^4P_0 = 1$

$\therefore f(4) = {}^{7-4}P_{4-3} = {}^3P_1 = 3$

$\therefore f(5) = {}^{7-5}P_{5-3} = {}^2P_2 = 2$

Hence range = {1, 2, 3}

7. (a) $f(x)$ is onto $\therefore S =$ range of $f(x)$

Now $f(x) = \sin x - \sqrt{3}\cos x + 1 = 2\sin\left(x-\frac{\pi}{3}\right)+1$

$\because -1 \le \sin\left(x-\frac{\pi}{3}\right) \le 1$

$-1 \le 2\sin\left(x-\frac{\pi}{3}\right)+1 \le 3$

$\therefore f(x) \in [-1, 3] = S$

ALTERNATE SOLUTION :

We know that

$-\sqrt{a^2+b^2} \le a\sin\theta + b\cos\theta \le \sqrt{a^2+b^2}$

$\therefore -2 \le \sin x - \sqrt{3}\cos x \le 2$

$\Rightarrow -1 \le \sin x - \sqrt{3}\cos x + 1 \le 3$

$\therefore f(x) \in [-1, 3]$

(b) Let us consider a graph symm. with respect to line $x = 2$ as shown in the figure.

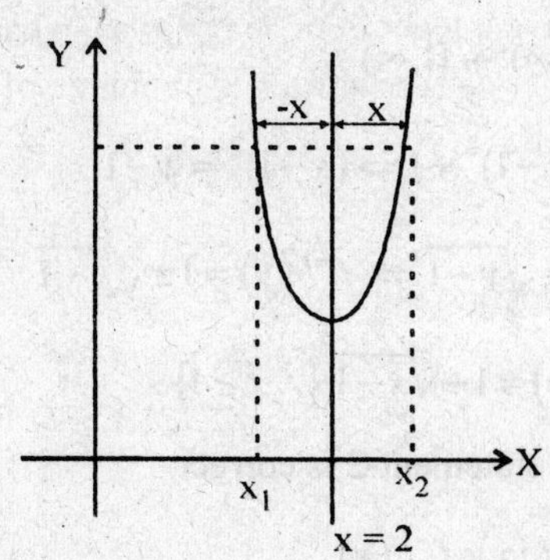

From the figure

$f(x_1) = f(x_2)$, where $x_1 = 2 - x$ and $x_2 = 2 + x$

$\therefore f(2-x) = f(2+x)$

(b) $f(x) = \dfrac{\sin^{-1}(x-3)}{\sqrt{9-x^2}}$ is defined

if (i) $-1 \le x-3 \le 1 \Rightarrow 2 \le x \le 4$

and (ii) $9 - x^2 > 0 \Rightarrow -3 < x < 3$

Taking common solution of (i) and (ii),

we get $2 \le x < 3$ $\therefore$ Domain = [2, 3)

(d) Given $f(x) = \tan^{-1}\left(\dfrac{2x}{1-x^2}\right) = 2\tan^{-1}x$ for $x \in (-1, 1)$

If $x \in (-1, 1) \Rightarrow \tan^{-1} x \in \left(\dfrac{-\pi}{4}, \dfrac{\pi}{4}\right)$

$\Rightarrow 2\tan^{-1} x \in \left(\dfrac{-\pi}{2}, \dfrac{\pi}{2}\right)$

Clearly, range of $f(x) = \left(-\dfrac{\pi}{2}, \dfrac{\pi}{2}\right)$

For f to be onto, codomain = range

$\therefore$ Co-domain of function $= B = \left(-\dfrac{\pi}{2}, \dfrac{\pi}{2}\right)$.

(c) Clearly function $f(x) = 3x^2 - 2x + 1$ is increasing when

$f'(x) = 6x - 2 \ge 0 \Rightarrow \quad x \in [1/3, \infty)$

$\therefore f(x)$ is incorrectly matched with $\left(-\infty, \dfrac{1}{3}\right)$

12. (a) $f(2a-x) = f(a-(x-a))$

$= f(a)f(x-a) - f(0)f(x) = f(a)f(x-a) - f(x)$

$= -f(x)$

$[\because x = 0, y = 0,\ f(0) = f^2(0) - f^2(a)$

$\Rightarrow f^2(a) = 0 \Rightarrow f(a) = 0]$

$\Rightarrow f(2a-x) = -f(x)$

13. (b) $f(x) = 4^{-x^2} + \cos^{-1}\left(\dfrac{x}{2} - 1\right) + \log(\cos x)$

$f(x)$ is defined if $-1 \le \left(\dfrac{x}{2} - 1\right) \le 1$ and $\cos x > 0$

or $\quad 0 \le \dfrac{x}{2} \le 2$ and $-\dfrac{\pi}{2} < x < \dfrac{\pi}{2}$

or $\quad 0 \le x \le 4$ and $-\dfrac{\pi}{2} < x < \dfrac{\pi}{2}$

$\therefore \quad x \in \left[0, \dfrac{\pi}{2}\right)$

14. (d) Clearly f is one one and onto, so invertible

Also $f(x) = 4x + 3 = y$

$\Rightarrow x = \dfrac{y-3}{4}$

$\therefore\ g(y) = \dfrac{y-3}{4}$

15. (b) Given that $f(x) = (x+1)^2 - 1,\ x \ge -1$

Clearly $D_f = [-, \infty)$ but co-demain is not given. Therefore $f(x)$ need not be necessarily onto.

But if $f(x)$ is onto then as $f(x)$ is one one also, $(x + 1)$ being something +ve, $f^{-1}(x)$ will exist where

$(x+1)^2 - 1 = y$

$\Rightarrow x + 1 = \sqrt{y+1}$ (+ve square root as $x + 1 \ge 0$)

$\Rightarrow x = -1 + \sqrt{y+1}$

$\Rightarrow f^{-1}(x) = \sqrt{x+1} - 1$

Then $\quad f(x) = f^{-1}(x)$

$\Rightarrow (x+1)^2 - 1 = \sqrt{x+1} - 1$

$\Rightarrow (x+1)^2 = \sqrt{x+1} \Rightarrow (x+1)^4 = (x+1)$

$\Rightarrow (x+1)[(x+1)^3 - 1] = 0 \Rightarrow x = -1, 0$

$\therefore$ The statement-1 is correct but statement-2 is false.

16. **(b)** Given that $f(x) = x^3 + 5x + 1$

$\therefore$ $f'(x) = 3x^2 + 5 > 0, \ \forall x \in R$

$\Rightarrow$ $f(x)$ is strictly increasing on R

$\Rightarrow$ $f(x)$ is one one

$\therefore$ Being a polynomial $f(x)$ is cont. and inc.

on R with $\lim\limits_{x \to \infty} f(x) = -\infty$,

and $\lim\limits_{x \to \infty} f(x) = \infty$

$\therefore$ Range of $f = (-\infty, \infty) = R$

Hence f is onto also. So, f is one one and onto R.

17. **(b)** $f(x) = \dfrac{1}{\sqrt{|x| - x}}$, define if $|x| - x > 0$

$\Rightarrow |x| > x, \Rightarrow x < 0$

Hence domain of $f(x)$ is $(-\infty, 0)$

18. **(a)** $f(x) = (x-1)^2 + 1, x \geq 1$

Since f is a bijective function

$\therefore$ $f : [1, \infty) \to [1, \infty)$

$\Rightarrow$ $y = (x-1)^2 + 1 \Rightarrow (x-1)^2 = y - 1$

$\Rightarrow$ $x = 1 \pm \sqrt{y-1} \Rightarrow f^{-1}(y) = 1 \pm \sqrt{y-1}$

$\Rightarrow$ $f^{-1}(x) = 1 + \sqrt{x-1} \ \{\therefore x \geq 1\}$

Hence statement-2 is correct

Now $f(x) \ f^{-1}(x)$

$\Rightarrow f(x) = x \Rightarrow (x-1)^2 + 1 = x$

$\Rightarrow$ $x^2 - 3x + 2 = 0 \Rightarrow x = 1, 2$

Hence statement-1 is correct

Chapter 11 Limits, Continuity and Differentiability

Section-A JEE Advanced/ IIT-JEE

A FILL IN THE BLANKS :

1. Let $f(x)=\begin{cases}(x-1)^2\sin\dfrac{1}{(x-1)}-|x| & \text{if } x\neq 1\\ -1, & \text{if } x=1\end{cases}$

be a real-valued function. Then the set of points where $f(x)$ is not differentiable is .... *(1981 - 2 Marks)*

2. Let $f(x)=\begin{cases}\dfrac{(x^3+x^2-16x+20)}{(x-2)^2}, & \text{if } x\neq 2\\ k, & \text{if } x=2\end{cases}$

If $f(x)$ is continuous for all x, then $k=$ .... *(1981 - 2 Marks)*

3. A discontinuous function $y=f(x)$ satisfying $x^2+y^2=4$ is given by $f(x)=$ .... *(1982 - 2 Marks)*

4. $\lim\limits_{x\to 1}(1-x)\tan\dfrac{\pi x}{2}=$ .... *(1984 - 2 Marks)*

5. If $f(x)=\sin x,\ x\neq n\pi,\ n=0,\pm 1,\pm 2,\pm 3,$............
$= 2$, otherwise

and $g(x)=x^2+1,\ x\neq 0, 2$
$=4, x=0$
$=5, x=2,$

then $\lim\limits_{x\to 0} g[f(x)]$ is .... *(1986 - 2 Marks)*

6. $\lim\limits_{x\to -\infty}\left[\dfrac{x^4\sin\left(\frac{1}{x}\right)+x^2}{(1+|x|^3)}\right]=$ .... *(1987 - 2 Marks)*

7. If $f(9)=9,\ f'(9)=4$, then $\mathop{Lim}\limits_{x\to 9}\dfrac{\sqrt{f(x)}-3}{\sqrt{x}-3}$ equals........... *(1988 - 2 Marks)*

8. ABC is an isosceles triangle inscribed in a circle of radius r. If $AB = AC$ and h is the altitude from A to BC then the triangle ABC has perimeter $P=2(\sqrt{2hr-h^2})+\sqrt{2hr})$ and area $A=$ .... also $\lim\limits_{h\to 0}\dfrac{A}{P^3}=$ .... *(1989 - 2 Marks)*

9. $\mathop{Lt}\limits_{x\to\infty}\left(\dfrac{x+6}{x+1}\right)^{x+4}=$ .... *(1990 - 2 Marks)*

10. Let $f(x)=x\,|x|$. The set of points where $f(x)$ is twice differentiable is .... *(1992 - 2 Marks)*

11. Let $f(x)=[x]\sin\left(\dfrac{\pi}{[x+1]}\right)$, where [•] denotes the greatest integer function. The domain of f is... and the points of discontinuity of f in the domain are..... *(1996 - 2 Marks)*

12. $\lim\limits_{x\to 0}\left(\dfrac{1+5x^2}{1+3x^2}\right)^{1/x^2}=$ ... *(1996 - 1 Mark)*

13. $\lim\limits_{x\to 0}\dfrac{\int_0^{x^2}\cos t^2\,dt}{x\sin x}=$ .... *(1997C - 2 Marks)*

14. Let $F(x)=f(x)\,g(x)\,h(x)$ for all real x, where $f(x)$, $g(x)$ and $h(x)$ are differentiable functions. At some point x_0, $F'(x_0)=21F(x_0)$, $f'(x_0)=4f(x_0)$, $g'(x_0)=-7g(x_0)$ and $h'(x_0)=kh(x_0)$. Then $k=$...... *(1997C - 2 Marks)*

15. $\lim\limits_{h\to 0}\dfrac{\ln(1+2h)-2\ln(1+h)}{h^2}=$ ...... *(1997C - 2 Marks)*

16. Let $f(x)$ be a continuous function defined for $1\le x\le 3$. If $f(x)$ takes rational values for all x and $f(2)=10$, then $f(1.5)=$............ *(1997 - 2 Marks)*

B TRUE / FALSE :

1. If $\mathop{Lt}\limits_{x\to a}[f(x)g(x)]$ exists then both $\mathop{Lt}\limits_{x\to a}f(x)$ and $\mathop{Lt}\limits_{x\to a}g(x)$ exist. *(1981 - 2 Marks)*

C MCQ's WITH ONE CORRECT ANSWER :

1. If $f(x)=\sqrt{\dfrac{x-\sin x}{x+\cos^2 x}}$, then $\lim\limits_{x\to\infty}f(x)$ is *(1979)*

(a) 0 (b) ∞

(c) 1 (d) None of these

1. For a real number y, let $[y]$ denotes the greatest integer less than or equal to y: Then the function $f(x)=\dfrac{\tan(\pi[x-\pi])}{1+[x]^2}$ is *(1981 - 2 Marks)*
   (a) discontinuous at some x
   (b) continuous at all x, but the derivative $f'(x)$ does not exist for some x
   (c) $f'(x)$ exists for all x, but the second derivative $f'(x)$ does not exist for some x
   (d) $f'(x)$ exists for all x

2. There exist a function $f(x)$, satisfying $f(0)=1, f'(0)=-1$, $f(x)>0$ for all x, and *(1982 - 2 Marks)*
   (a) $f''(x)>0$ for all x (b) $-1<f''(x)<0$ for all x
   (c) $-2\le f''(x)\le -1$ for all (d) $f''(x)<-2$ for all x

3. If $G(x)=-\sqrt{25-x^2}$ then $\lim\limits_{x\to 1}\dfrac{G(x)-G(1)}{x-1}$ has the value *(1983 - 1 Mark)*
   (a) $\dfrac{1}{24}$ (b) $\dfrac{1}{5}$
   (c) $-\sqrt{24}$ (d) none of these

4. If $f(a)=2,\ f'(a)=1,\ g(a)=-1,\ g'(a)=2$, then the value of $\lim\limits_{x\to a}\dfrac{g(x)f(a)-g(a)f(x)}{x-a}$ is *(1983 - 1 Mark)*
   (a) -5 (b) $\dfrac{1}{5}$
   (c) 5 (d) none of these

5. The function $f(x)=\dfrac{\ln(1+ax)-\ln(1-bx)}{x}$ is not defined at $x=0$. The value which should be assigned to f at $x=0$ so that it is continuous at $x=0$, is *(1983 - 1 Mark)*
   (a) $a-b$ (b) $a+b$
   (c) $\ln a-\ln b$ (d) none of these

6. $\lim\limits_{n\to\infty}\left\{\dfrac{1}{1-n^2}+\dfrac{2}{1-n^2}+....+\dfrac{n}{1-n^2}\right\}$ is equal to *(1984 - 2 Marks)*
   (a) 0 (b) $-\dfrac{1}{2}$
   (c) $\dfrac{1}{2}$ (d) none of these

7. If $f(x)=\dfrac{\sin[x]}{[x]},\ [x]\neq 0$ *(1985 - 2 Marks)*
   $=0,\quad [x]=0$
   Where $[x]$ denotes the greatest integer less than or equal to x. then $\lim\limits_{x\to 0}f(x)$ equals –
   (a) 1 (b) 0
   (c) -1 (d) none of these

8. Let $f: R\to R$ be a differentiable function and $f(1)=4$. Then the value of $\lim\limits_{x\to 1}\displaystyle\int_4^{f(x)}\frac{2t}{x-1}dt$ is *(1990 - 2 Marks)*
   (a) $8f'(1)$ (b) $4f'(1)$
   (c) $2f'(1)$ (d) $f'(1)$

10. Let [.] denote the greatest integer function and $f(x)=[\tan^2 x]$, then: *(1993 - 1 Mark)*
   (a) $\lim\limits_{x\to 0}f(x)$ does not exist
   (b) $f(x)$ is continuous at $x=0$
   (c) $f(x)$ is not differentiable at $x=0$
   (d) $f'(0)=1$

11. The function $f(x)=[x]\cos\left(\dfrac{2x-1}{2}\right)\pi$, [.] denotes the greatest integer function, is discontinuous at *(1995S)*
   (a) All x (b) All integer points
   (c) No x (d) x which is not an integer

12. $\lim\limits_{n\to\infty}\dfrac{1}{n}\sum\limits_{r=1}^{2n}\dfrac{r}{\sqrt{n^2+r^2}}$ equals *(1997 - 2 Marks)*
   (a) $1+\sqrt{5}$ (b) $-1+\sqrt{5}$
   (c) $-1+\sqrt{2}$ (d) $1+\sqrt{2}$.

13. The function $f(x)=[x]^2-[x^2]$ (where $[y]$ is the greatest integer less than or equal to y), is discontinuous at *(1999 - 2 Marks)*
   (a) all integers
   (b) all integers except 0 and 1
   (c) all integers except 0
   (d) all integers except 1

14. The function $f(x) = (x^2-1)\,|x^2-3x+2|+\cos(|x|)$ is NOT differentiable at *(1999 - 2 Marks)*
   (a) -1 (b) 0
   (c) 1 (d) 2

15. $\lim\limits_{x\to 0}\dfrac{x\tan 2x-2x\tan x}{(1-\cos 2x)^2}$ is *(1999 - 2 Marks)*
   (a) 2 (b) -2
   (c) 1/2 (d) $-1/2$

16. For $x\in R$, $\lim\limits_{x\to\infty}\left(\dfrac{x-3}{x+2}\right)^x=$ *(2000S)*
   (a) e (b) e^{-1}
   (c) e^{-5} (d) e^5

17. $\lim\limits_{x\to 0}\dfrac{\sin(\pi\cos^2 x)}{x^2}$ equals *(2001S)*
   (a) $-\pi$ (b) π
   (c) $\pi/2$ (d) 1

18. The left-hand derivative of $f(x) = [x] \sin(\pi x)$ at $x = k$, k an integer, is (2001S)
(a) $(-1)^k(k-1)\pi$ (b) $(-1)^{k-1}(k-1)\pi$
(c) $(-1)^k k\pi$ (d) $(-1)^{k-1}k\pi$

19. Let $f: R \to R$ be a function defined by $f(x) = \max\{x, x^3\}$. The set of all points where $f(x)$ is NOT differentiable is (2001S)
(a) $\{-1, 1\}$ (b) $\{-1, 0\}$
(c) $\{0, 1\}$ (d) $\{-1, 0, 1\}$

20. Which of the following functions is differentiable at $x = 0$? (2001S)
(a) $\cos(|x|) + |x|$ (b) $\cos(|x|) - |x|$
(c) $\sin(|x|) + |x|$ (d) $\sin(|x|) - |x|$

21. The domain of the derivative of the function

$$f(x) = \begin{cases} \tan^{-1} x & \text{if } |x| \le 1 \\ \frac{1}{2}(|x|-1) & \text{if } |x| > 1 \end{cases}$$ is (2002S)

(a) $R - \{0\}$ (b) $R - \{1\}$
(c) $R - \{-1\}$ (d) $R - \{-1, 1\}$

22. The integer n for which $\lim_{x \to 0} \frac{(\cos x - 1)(\cos x - e^x)}{x^n}$ is a finite non-zero number is (2002S)
(a) 1 (b) 2
(c) 3 (d) 4

23. Let $f: R \to R$ be such that $f(1) = 3$ and $f'(1) = 6$. Then $\lim_{x \to 0} \left(\frac{f(1+x)}{f(1)}\right)^{1/x}$ equals (2002S)
(a) 1 (b) $e^{1/2}$
(c) e^2 (d) e^3

24. If $\lim_{x \to 0} \frac{((a-n)nx - \tan x)\sin nx}{x^2} = 0$, where n is nonzero real number, then a is equal to (2003S)
(a) 0 (b) $\frac{n+1}{n}$
(c) n (d) $n + \frac{1}{n}$

25. $\lim_{h \to 0} \frac{f(2h+2+h^2) - f(2)}{f(h-h^2+1) - f(1)}$, given that $f'(2) = 6$ and $f'(1) = 4$
(a) does not exist (b) is equal to $-3/2$
(c) is equal to $3/2$ (d) is equal to 3 (2003S)

26. If (x) is differentiable and strictly increasing function, then the value of $\lim_{x \to 0} \frac{f(x^2) - f(x)}{f(x) - f(0)}$ is (2004S)
(a) 1 (b) 0
(c) -1 (d) 2

27. The function given by $y = ||x| - 1|$ is differentiable for all real numbers except the points (2005S)
(a) $\{0, 1, -1\}$ (b) ± 1
(c) 1 (d) -1

28. If $f(x)$ is continuous and differentiable function an $f(1/n) = 0 \ \forall \ n \ge 1$ and $n \in I$, then (2005S
(a) $f(x) = 0, x \in (0, 1]$
(b) $f(0) = 0, f'(0) = 0$
(c) $f(0) = 0 = f'(0), \ x \in (0, 1]$
(d) $f(0) = 0$ and $f'(0)$ need not to be zero

29. The value of $\lim_{x \to 0} \left((\sin x)^{1/x} + (1+x)^{\sin x}\right)$, where $x > 0$ i
(a) 0 (b) -1 (2006 - 3M, –
(c) 1 (d) 2

30. Let $f(x)$ be differentiable on the interval $(0, \infty)$ such th $f(1) = 1$, and $\lim_{t \to x} \frac{t^2 f(x) - x^2 f(t)}{t - x} = 1$ for each $x > 0$. The $f(x)$ is (2007 - 3 mark
(a) $\frac{1}{3x} + \frac{2x^2}{3}$ (b) $\frac{-1}{3x} + \frac{4x^2}{3}$
(c) $\frac{-1}{x} + \frac{2}{x^2}$ (d) $\frac{1}{x}$

31. $\lim_{x \to \frac{\pi}{4}} \frac{\int_2^{\sec^2 x} f(t)\,dt}{x^2 - \frac{\pi^2}{16}}$ equals (2007 - 3 mark
(a) $\frac{8}{\pi} f(2)$ (b) $\frac{2}{\pi} f(2)$
(c) $\frac{2}{\pi} f\left(\frac{1}{2}\right)$ (d) $4f(2)$

32. Let $g(x) = \frac{(x-1)^n}{\log \cos^m (x-1)}$; $0 < x < 2$, m and n are integer $m \ne 0$, $n > 0$, and let p be the left hand derivative of $|x -$ at $x = 1$. If $\lim_{x \to 1^+} g(x) = p$, then (200
(a) $n = 1, m = 1$ (b) $n = 1, m = -1$
(c) $n = 2, m = 2$ (d) $n > 2, m = n$

33. If $\lim_{x \to 0} \left[1 + x \ln(1+b^2)\right]^{1/x} = 2b \sin^2 \theta, b > 0$ and $\theta \in (-\pi, \pi$ then the value of θ is (201
(a) $\pm\frac{\pi}{4}$ (b) $\pm\frac{\pi}{3}$
(c) $\pm\frac{\pi}{6}$ (d) $\pm\frac{\pi}{2}$

34. If $\lim_{x\to\infty}\left(\frac{x^2+x+1}{x+1}-ax-b\right)=4$, then *(2012)*

(a) $a=1, b=4$ (b) $a=1, b=-4$
(c) $a=2, b=-3$ (d) $a=2, b=3$

35. Let $f(x)=\begin{cases} x^2\left|\cos\frac{\pi}{x}\right|, & x\neq 0 \\ 0, & x=0 \end{cases}$, $x\in R$ then f is *(2012)*

(a) differentiable both at $x=0$ and at $x=2$
(b) differentiable at $x=0$ but not differentiable at $x=2$
(c) not differentiable at $x=0$ but differentiable at $x=2$
(d) differentiable neither at $x=0$ nor at $x=2$

36. Let $\alpha(a)$ and $\beta(a)$ be the roots of the equation $\left(\sqrt[3]{1+a}-1\right)x^2+\left(\sqrt{1+a}-1\right)x+\left(\sqrt[6]{1+a}-1\right)=0$ where $a>-1$. Then $\lim_{a\to 0^+}\alpha(a)$ and $\lim_{a\to 0^+}\beta(a)$ are *(2012)*

(a) $-\frac{5}{2}$ and 1 (b) $-\frac{1}{2}$ and -1
(c) $-\frac{7}{2}$ and 2 (d) $-\frac{9}{2}$ and 3

(D) MCQ's WITH ONE OR MORE THAN ONE CORRECT :

1. If $x+|y|=2y$, then y as a function of x is *(1984 - 3 Marks)*
(a) defined for all real x
(b) continuous at $x=0$
(c) differentiable for all x
(d) such that $\frac{dy}{dx}=\frac{1}{3}$ for $x<0$

2. If $f(x)=x(\sqrt{x}-\sqrt{x+1})$, then– *(1985 - 2 Marks)*
(a) $f(x)$ is continuous but not differentiable at $x=0$
(b) $f(x)$ is differentiable at $x=0$
(c) $f(x)$ is not differentiable at $x=0$
(d) none of these

3. The function $f(x)=1+|\sin x|$ is *(1986 - 2 Marks)*
(a) continuous nowhere
(b) continuous everywhere
(c) differentiable nowhere
(d) not differentiable at $x=0$
(e) not differentiable at infinite number of points.

4. Let $[x]$ denote the greatest integer less than or equal to x. If $f(x)=[x\sin\pi x]$, then $f(x)$ is *(1986 - 2 Marks)*
(a) continuous at $x=0$ (b) continuous in $(-1, 0)$
(c) differentiable at $x=1$ (d) differentiable in $(-1, 1)$
(e) none of these

5. The set of all points where the function $f(x)=\frac{x}{(1+|x|)}$ is differentiable, is *(1987 - 2 Marks)*
(a) $(-\infty,\infty)$ (b) $[0,\infty)$
(c) $(-\infty,0)\cup(0,\infty)$ (d) $(0,\infty)$
(e) None

6. The function $f(x)=\begin{cases} |x-3|, & x\geq 1 \\ \frac{x^2}{4}-\frac{3x}{2}+\frac{13}{4}, & x<1 \end{cases}$ is *(1988 - 2 Marks)*
(a) continous at $x=1$ (b) differentiable at $x=1$
(c) continous at $x=3$ (d) differentiable at $x=3$.

7. If $f(x)=\frac{1}{2}x-1$, then on the interval $[0,\pi]$ *(1989 - 2 Marks)*
(a) $\tan[f(x)]$ and $1/f(x)$ are both continuous
(b) $\tan[f(x)]$ and $1/f(x)$ are both discontinuous
(c) $\tan[f(x)]$ and $f^{-1}(x)$ are both continuous
(d) $\tan[f(x)]$ is continuous but $1/f(x)$ is not.

8. The value of $\lim_{x\to 0}\frac{\sqrt{\frac{1}{2}(1-\cos 2x)}}{x}$ *(1991 - 2 Marks)*
(a) 1 (b) -1
(c) 0 (d) none of these

9. The following functions are continuous on $(0,\pi)$. *(1991 - 2 Marks)*
(a) $\tan x$
(b) $\int_0^x t\sin\frac{1}{t}\,dt$
(c) $\begin{cases} 1, & 0<x\leq\frac{3\pi}{4} \\ 2\sin\frac{2}{9}x, & \frac{3\pi}{4}<x<\pi \end{cases}$
(d) $\begin{cases} x\sin x, & 0<x\leq\frac{\pi}{2} \\ \frac{\pi}{2}\sin(\pi+x), & \frac{\pi}{2}<x<\pi \end{cases}$

10. Let $f(x)=\begin{cases} 0, & x<0 \\ x^2, & x\geq 0 \end{cases}$ then for all x *(1994)*
(a) f' is differentiable (b) f is differentiable
(c) f' is continuous (d) f is continuous

11. Let $g(x)=xf(x)$, where $f(x)=\begin{cases} x\sin\frac{1}{x}, & x\neq 0 \\ 0, & x=0 \end{cases}$. At $x=0$ *(1994)*
(a) g is differentiable but g' is not continuous
(b) g is differentiable while f is not
(c) both f and g are differentiable
(d) g is differentiable and g' is continuous

12. The function $f(x)=\max\{(1-x),(1+x),2\}$, $x\in(-\infty,\infty)$ is *(1995)*
(a) continuous at all points
(b) differentiable at all points
(c) differentiable at all points except at $x=1$ and $x=-1$
(d) continuous at all points except at $x=1$ and $x=-1$, where it is discontinuous

13. Let $h(x) = \min\{x, x^2\}$, for every real number of x, Then *(1998 - 2 Marks)*
(a) h is continuous for all x
(b) h is differentiable for all x
(c) $h'(x) = 1$, for all $x > 1$
(d) h is not differentiable at two values of x.

14. $\lim_{x\to 1} \frac{\sqrt{1-\cos 2(x-1)}}{x-1}$ *(1998 - 2 Marks)*
(a) exists and it equals $\sqrt{2}$
(b) exists and it equals $-\sqrt{2}$
(c) does not exist because $x - 1 \to 0$
(d) does not exist because the left hand limit is not equal to the right hand limit.

15. If $f(x) = \min\{1, x^2, x^3\}$, then *(2006 - 5M, –1)*
(a) $f(x)$ is continuous $\forall x \in R$
(b) $f(x)$ is continuous and differentiable everywhere.
(c) $f(x)$ is not differentiable at two points
(d) $f(x)$ is not differentiable at one point

16. Let $L = \lim_{x\to 0} \frac{a - \sqrt{a^2 - x^2} - \frac{x^2}{4}}{x^4}, a > 0$.
If L is finite, then *(2009)*
(a) $a = 2$ (b) $a = 1$
(c) $L = \frac{1}{64}$ (d) $L = \frac{1}{32}$

17. Let $f: \mathbf{R} \to \mathbf{R}$ be a function such that $f(x+y) = f(x) + f(y)$, $\forall x, y \in \mathbf{R}$. If $f(x)$ is differentiable at $x = 0$, then *(2011)*
(a) $f(x)$ is differentiable only in a finite interval containing zero
(b) $f(x)$ is continuous $\forall x \in R$
(c) $f'(x)$ is constant $\forall x \in R$
(d) $f(x)$ is differentiable except at finitely many points.

18. If $f(x) = \begin{cases} -x - \frac{\pi}{2}, & x \le -\frac{\pi}{2} \\ -\cos x, & -\frac{\pi}{2} < x \le 0 \\ x - 1, & 0 < x \le 1 \\ \ell n x, & x > 1 \end{cases}$, then *(2011)*
(a) $f(x)$ is continuous at $x = -\frac{\pi}{2}$
(b) $f(x)$ is not differentiable at $x = 0$
(c) $f(x)$ is differentiable at $x = 1$
(d) $f(x)$ is differentiable at $x = -\frac{3}{2}$

19. For every integer n, let a_n and b_n be real numbers. Let function $f: IR \to IR$ be given by *(2012)*
$$f(x) = \begin{cases} a_n + \sin \pi x, & \text{for } x \in [2n, 2n+1] \\ b_n + \cos \pi x, & \text{for } x \in (2n-1, 2n) \end{cases}$$
for all integers n. If f is continuous, then which of the following hold(s) for all n?
(a) $a_{n-1} - b_{n-1} = 0$ (b) $a_n - b_n = 1$
(c) $a_n - b_{n+1} = 1$ (d) $a_{n-1} - b_n = -1$

20. For $a \in \mathbb{R}$ (the set of all real numbers), $a \neq -1$, *(JEE Adv. 2013)*
$$\lim_{n\to\infty} \frac{(1^a + 2^a + ... + n^a)}{(n+1)^{a-1}[(na+1)+(na+2)+...+(na+n)]} = \frac{1}{60}.$$
Then $a =$
(a) 5 (b) 7
(c) $\frac{-15}{2}$ (d) $\frac{-17}{2}$

E SUBJECTIVE PROBLEMS :

1. Evaluate $\lim_{x\to a} \frac{\sqrt{a+2x} - \sqrt{3x}}{\sqrt{3a+x} - 2\sqrt{x}}, (a \neq 0)$ *(1978)*

2. $f(x)$ is the integral of $\frac{2\sin x - \sin 2x}{x^3}, x \neq 0$, find $\lim_{x\to 0} f'(x)$ *(1979)*

3. Evaluate : $\lim_{h\to 0} \frac{(a+h)^2 \sin(a+h) - a^2 \sin a}{h}$ *(1980)*

4. Let $f(x+y) = f(x) + f(y)$ for all x and y. If the function $f(x)$ is continuous at $x = 0$, then show that $f(x)$ is continuous at all x. *(1981 - 2 Marks)*

5. Use the formula $\lim_{x\to 0} \frac{a^x - 1}{x} = \ln a$ to find
$\lim_{x\to 0} \frac{2^x - 1}{(1+x)^{1/2} - 1}$ *(1982 - 2 Marks)*

6. Determine the values a, b, c for which the function
$$f(x) = \begin{cases} \frac{\sin(a+1)x + \sin x}{x}, & x < 0 \\ c, & x = 0 \\ \frac{(x+bx^2)^{1/2} - x^{1/2}}{bx^{3/2}}, & x > 0 \end{cases}$$
is continuous at $x = 0$ *(1982 - 3 Marks)*

7. Let $f(x) = \begin{cases} 1 + x, & 0 \le x \le 2 \\ 3 - x, & 2 \le x \le 3 \end{cases}$ *(1983 - 2 Marks)*
Determine the form of $g(x) = f[f(x)]$ and hence find the points of discontinuity of g, if any

8. Let $f(x) = \begin{cases} \frac{x^2}{2}, & 0 \le x < 1 \\ 2x^2 - 3x + \frac{3}{2}, & 1 \le x \le 2 \end{cases}$ *(1983 - 2 Marks)*
Discuss the continuity of f, f' and f'' on $[0, 2]$.

9. Let $f(x) = x^3 - x^2 + x + 1$ and

$$g(x) = \max\{f(t); 0 \le t \le x\}, \ 0 \le x \le 1 \qquad \textbf{(1985 - 5 Marks)}$$
$$= 3 - x \qquad 0 \le x \le 2$$

Discuss the continuity and differentiability of the function g(x) in the interval (0, 2).

10. Let $f(x)$ be defined in the interval [–2, 2] such that

$$f(x) = \begin{cases} -1, -2 \le x \le 0 \\ x - 1, 0 < x \le 2 \end{cases}$$

and $g(x) = f(|x|) + |f(x)|$

Test the differentiability of $g(x)$ in (–2, 2). ***(1986 - 5 Marks)***

11. Let $f(x)$ be a continuous and $g(x)$ be a discontinuous function. prove that $f(x) + g(x)$ is a discontinuous function. ***(1987 - 2 Marks)***

12. Let $f(x)$ be a function satisfying the condition $f(-x) = f(x)$ for all real x. If $f'(0)$ exists, find its value. ***(1987 - 2 Marks)***

13. Let g(x) be a polynomial of degree one and $f(x)$ be defined by ***(1987 - 6 Marks)***

$$f(x) = \begin{cases} g(x) & x \le 0 \\ \left[\dfrac{(1+x)}{(2+x)}\right]^{1/x} & x > 0 \end{cases}$$

Find the continuous fonction $f(x)$ satsfying. $f'(1) = f(-1)$.

14. Find the values of a and b so that the function

$$f(x) = \begin{cases} x + a\sqrt{2}\sin x, & 0 \le x < \pi/4 \\ 2x\cot x + b & \pi/4 \le x \le \pi/2 \\ a\cos 2x - b\sin x & \pi/2 < x \le \pi \end{cases}$$

is continuous for $0 \le x \le \pi$. ***(1989 - 2 Marks)***

15. Draw a graph of the function $y = [x] + |1 - x|, -1 \le x \le 3$. Determine the points, if any, where this function is not differentiable. ***(1989 - 4 Marks)***

16. Let $f(x) = \begin{cases} \dfrac{1 - \cos 4x}{x^2}, & x < 0 \\ a, & x = 0 \\ \dfrac{\sqrt{x}}{\sqrt{16 + \sqrt{x}} - 4}, & x > 0 \end{cases}$ ***(1990 - 4 Marks)***

Determine the value of a, if possible, so that the function is continuous at $x = 0$

17. A function $f: R \to R$ satisfies the equation $f(x + y) = f(x) f(y)$ for all x, y in R and $f(x) \ne 0$ for any x in R. Let the function be differentiable at $x = 0$ and $f'(0) = 2$. Show that $f'(x) = 2f(x)$ for all x in R. Hence, determine $f(x)$. ***(1990 - 4 Marks)***

18. Find $\lim_{x \to 0} \{\tan(\pi/4 + x)\}^{1/x}$ ***(1993 - 2 Marks)***

19. Let $f(x) = \begin{cases} \{1 + |\sin x|\}^{a/|\sin x|} & ; \ -\dfrac{\pi}{6} < x < 0 \\ b & ; \ x = 0 \\ e^{\tan 2x / \tan 3x} & ; \ 0 < x < \dfrac{\pi}{6} \end{cases}$

(1994 - 4 Marks)

Determine a and b such that $f(x)$ is continuous at $x = 0$

20. Let $f\left(\dfrac{x+y}{2}\right) = \dfrac{f(x) + f(y)}{2}$ for all real x and y. If $f'(0)$ exists and equals – 1 and $f(0) = 1$, find $f(2)$. ***(1995 - 5 Marks)***

21. Let $f(x) = \begin{cases} xe^{-\left(\frac{1}{|x|} + \frac{1}{x}\right)}, & x \ne 0 \\ 0, & x = 0 \end{cases}$ ***(1997C - 5 Marks)***

Test whether

(a) $f(x)$ is continuous at $x = 0$

(b) $f(x)$ is differentiable at $x = 0$

22. Determine the values of x for which the following functio fails to be continuous or differentiable: ***(1997 - 5 Marks)***

$$f(x) = \begin{cases} 1 - x, & x < 1 \\ (1 - x)(2 - x), & 1 \le x \le 2 \\ 3 - x, & x > 2 \end{cases}$$ Justify your answer.

23. Let $f(x), x \ge 0$, be a non–negative continuous function, an let $F(x) = \int_0^x f(t)\,dt, x \ge 0$. If for some $c > 0, f(x) \le cF(x)$ for a $x \ge 0$, then show that $f(x) = 0$ for all $x \ge 0$. ***(2001 - 5 Marks)***

24. Let $\alpha \in R$. Prove that a function $f: R \to R$ is differentiable a α if and only if there is a function $g: R \to R$ which i continuous at α and satisfies $f(x) - f(\alpha) = g(x)(x - \alpha)$ for a $x \in R$. ***(2001 - 5 Marks)***

25. Let $f(x) = \begin{cases} x + a & \text{if } x < 0 \\ |x - 1| & \text{if } x \ge 0, \end{cases}$ and ***(2002 - 5 Marks)***

$$g(x) = \begin{cases} x + 1 & \text{if } x < 0 \\ (x - 1)^2 + b & \text{if } x \ge 0, \end{cases}$$ where a and b ar non-negative real numbers. Determine the composite functio $g \, o \, f$. If $(g \, o \, f)(x)$ is continuous for all real x, determine th values of a and b. Further, for these values of a and b, is g f differentiable at $x = 0$? Justify your answer.

26. If a function $f: [-2a, 2a] \to R$ is an odd function such tha $f(x) = f(2a - x)$ for $x \in [a, 2a]$ and the left hand derivative a $x = a$ is 0 then find the left hand derivative at $x = -a$.

(2003 - 2 Marks)

27. $f'(0) = \lim_{n \to \infty} nf\left(\dfrac{1}{n}\right)$ and $f(0) = 0$. Using this fin

$$\lim_{n \to \infty}\left((n+1)\frac{2}{\pi}\cos^{-1}\left(\frac{1}{n}\right) - n\right), \left|\cos^{-1}\frac{1}{n}\right| < \frac{\pi}{2}$$

(2004 - 2 Marks)

28. If $|c| \le \frac{1}{2}$ and $f(x)$ is a differentiable function at $x = 0$ given

by $f(x) = \begin{cases} b\sin^{-1}\left(\frac{c+x}{2}\right), & -\frac{1}{2} < x < 0 \\ \frac{1}{2}, & x = 0 \\ \frac{e^{ax/2}-1}{x}, & 0 < x < \frac{1}{2} \end{cases}$.

Find the value of 'a' and prove that $64\ b^2 = 4 - c^2$ *(2004 - 4 Marks)*

29. If $f(x-y) = f(x)\cdot g(y) - f(y)\cdot g(x)$ and $g(x-y) = g(x)\cdot g(y) - f(x)\cdot f(y)$ for all $x, y \in R$. If right hand derivative at $x = 0$ exists for $f(x)$. Find derivative of $g(x)$ at $x = 0$ *(2005 - 4 Marks)*

F MATCH THE FOLLOWING :

Each question contains statements given in two columns, which have to be matched. The statements in Column-I are labelled A, B, C and D, while the statements in Column-II are labelled p, q, r, s and t. Any given statement in Column-I can have correct matching with ONE OR MORE statement(s) in Column-II. The appropriate bubbles corresponding to the answers to these questions have to be darkened as illustrated in the following example :

If the correct matches are A-p, s and t; B-q and r; C-p and q; and D-s then the correct darkening of bubbles will look like the given.

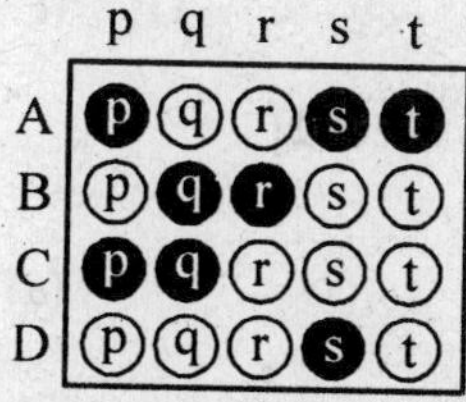

1. In this questions there are entries in columns I and II. Each entry in column I is related to exactly one entry in column II. Write the correct letter from column II against the entry number in column I in your answer book. *(1992 - 2 Marks)*

Column I	*Column II*
(A) $\sin(\pi[x])$	(p) differentiable everywhere
(B) $\sin(\pi(x-[x]))$	(q) nowhere differentiable
	(r) not differentiable at 1 and – 1

2. In the following $[x]$ denotes the greatest integer less than or equal to x.

Match the functions in **Column I** with the properties in **Column II** and indicate your answer by darkening the appropriate bubbles in the 4 × 4 matrix given in the ORS. *(2007 - 6 marks)*

Column I	*Column II*
(A) $x\,|x|$	(p) continuous in (–1, 1)
(B) $\sqrt{|x|}$	(q) differentiable in (–1, 1)
(C) $x + [x]$	(r) strictly increasing in (–1, 1)
(D) $|x-1| + |x+1|$	(s) not differentiable at least at one point in (–1, 1)

I INTEGER VALUE CORRECT TYPE :

1. Let $f : [1, \infty) \to [2, \infty)$ be a differentiable function such that $f(1) = 2$. If $6\int_1^x f(t)dt = 3xf(x) - x^3$ for all $x \ge 1$, then the value of $f(2)$ is *(2011)*

Section-B JEE Main/ AIEEE

1. $\lim_{x\to 0}\frac{\sqrt{1-\cos 2x}}{\sqrt{2}x}$ is [2002]

(a) 1 (b) –1
(c) zero (d) does not exist

2. $\underset{x\to\infty}{\text{Lim}}\left(\frac{x^2+5x+3}{x^2+x+3}\right)^x$ [2002]

(a) e^4 (b) e^2
(c) e^3 (d) 1

3. Let $f(x)=4$ and $f'(x)=4$. Then $\lim_{x\to 2}\frac{xf(2)-2f(x)}{x-2}$ is given by [2002]

(a) 2 (b) –2
(c) –4 (d) 3

4. $\lim_{n\to\infty}\frac{1^p+2^p+3^p+.....+n^p}{n^{p+1}}$ is [2002]

(a) $\frac{1}{p+1}$ (b) $\frac{1}{1-p}$
(c) $\frac{1}{p}-\frac{1}{p-1}$ (d) $\frac{1}{p+2}$

5. $\lim_{x\to 0}\frac{\log x^n-[x]}{[x]}, n\in N$, ([x] denotes greatest integer less than or equal to x) [2002]

(a) has value -1 (b) has value 0
(c) has vlaue 1 (d) does not exist

6. If $f(1)=1, f^1(1)=2$, then $\lim_{x\to 1}\frac{\sqrt{f(x)}-1}{\sqrt{x}-1}$ is [2002]

(a) 2 (b) 4
(c) 1 (d) 1/2

7. f is defined in [-5, 5] as [2002]

$f(x)=x$ if x is rational
$=-x$ if x is irrational. Then

(a) $f(x)$ is continuous at every x, except $x=0$
(b) $f(x)$ is discontinuous at every x, except $x=0$
(c) $f(x)$ is continuous everywhere
(d) $f(x)$ is discontinuous everywhere

8. $f(x)$ and $g(x)$ are two differentiable functions on [0, 2] such that $f''(x)-g''(x)=0$, $f'(1)=2g'(1)=4f(2)=3g(2)=9$ then $f(x)-g(x)$ at $x=3/2$ is [2002]

(a) 0 (b) 2
(c) 10 (d) 5

9. If $f(x+y)=f(x).f(y)\forall x.y$ and $f(5)=2, f'(0)=3$, then $f'(5)$ is [2002]

(a) 0 (b) 1
(c) 6 (d) 2

10. $\lim_{n\to\infty}\frac{1+2^4+3^4+...n^4}{n^5}-\lim_{n\to\infty}\frac{1+2^3+3^3+...n^3}{n^5}$ [2003]

(a) $\frac{1}{5}$ (b) $\frac{1}{30}$
(c) Zero (d) $\frac{1}{4}$

11. If $\lim_{x\to 0}\frac{\log(3+x)-\log(3-x)}{x}=k$, the value of k is [2003]

(a) $-\frac{2}{3}$ (b) 0 (c) $-\frac{1}{3}$ (d) $\frac{2}{3}$

12. The value of $\lim_{x\to 0}\frac{\int_0^{x^2}\sec^2 t\,dt}{x\sin x}$ is [2003]

(a) 0 (b) 3
(c) 2 (d) 1

13. Let $f(a)=g(a)=k$ and their nth derivatives $f^n(a), g^n(a)$ exist and are not equal for some n. Further if

$$\lim_{x\to a}\frac{f(a)g(x)-f(a)-g(a)f(x)+f(a)}{g(x)-f(x)}=4$$

then the value of k is [2003]

(a) 0 (b) 4
(c) 2 (d) 1

14. $\lim_{x\to\frac{\pi}{2}}\frac{\left[1-\tan\left(\frac{x}{2}\right)\right][1-\sin x]}{\left[1+\tan\left(\frac{x}{2}\right)\right][\pi-2x]^3}$ is [2003]

(a) ∞ (b) $\frac{1}{8}$
(c) 0 (d) $\frac{1}{32}$

15. If $f(x)=\begin{cases} xe^{-\left(\frac{1}{|x|}+\frac{1}{x}\right)} & , x\neq 0 \\ 0 & , x=0 \end{cases}$ then f(x) is

(a) discontinuous every where **[2003]**
(b) continuous as well as differentialble for all x
(c) continuous for all x but not differentiable at $x=0$
(d) neither differentiable nor continuous at $x=0$

16. If $\lim\limits_{x\to\infty}\left(1+\frac{a}{x}+\frac{b}{x^2}\right)^{2x}=e^2$, then the values of a and b, are

(a) $a=1$ and $b=2$ (b) $a=1, b\in \boldsymbol{R}$ **[2004]**
(c) $a\in \boldsymbol{R}, b=2$ (d) $a\in \boldsymbol{R}, b\in \boldsymbol{R}$

17. Let $f(x)=\dfrac{1-\tan x}{4x-\pi}, x\neq\dfrac{\pi}{4}, x\in\left[0,\dfrac{\pi}{2}\right]$. If $f(x)$ is continuous in $\left[0,\dfrac{\pi}{2}\right]$, then $f\left(\dfrac{\pi}{4}\right)$ is **[2004]**

(a) -1 (b) $\dfrac{1}{2}$
(c) $-\dfrac{1}{2}$ (d) 1

18. $\lim\limits_{n\to\infty}\left[\frac{1}{n^2}\sec^2\frac{1}{n^2}+\frac{2}{n^2}\sec^2\frac{4}{n^2}..........+\frac{1}{n}\sec^2 1\right]$ equals **[2005]**

(a) $\dfrac{1}{2}\sec 1$ (b) $\dfrac{1}{2}\operatorname{cosec} 1$
(c) $\tan 1$ (d) $\dfrac{1}{2}\tan 1$

19. Let α and β be the distinct roots of $ax^2+bx+c=0$, then $\lim\limits_{x\to\alpha}\dfrac{1-\cos(ax^2+bx+c)}{(x-\alpha)^2}$ is equal to **[2005]**

(a) $\dfrac{a^2}{2}(\alpha-\beta)^2$ (b) 0
(c) $\dfrac{-a^2}{2}(\alpha-\beta)^2$ (d) $\dfrac{1}{2}(\alpha-\beta)^2$

20. Suppose $f(x)$ is differentiable at $x = 1$ and $\lim\limits_{h\to 0}\dfrac{1}{h}f(1+h)=5$, then $f'(1)$ equals **[2005]**

(a) 3 (b) 4
(c) 5 (d) 6

21. Let f be differentiable for all x. If $f(1)=-2$ and $f'(x)\geq 2$ for $x\in[1,6]$, then **[2005]**

(a) $f(6)\geq 8$ (b) $f(6)<8$
(c) $f(6)<5$ (d) $f(6)=5$

22. If f is a real valued differentiable function satisfying $|f(x)-f(y)|\leq(x-y)^2, x,y\in R$ and $f(0)=0$, then $f(1)$ equals **[2005]**

(a) -1 (b) 0
(c) 2 (d) 1

23. Let $f:R\to R$ be a function defined by $f(x)=\min\{x+1,|x|+1\}$, Then which of the following is true?

(a) $f(x)$ is differentiable everywhere **[2007]**
(b) $f(x)$ is not differentiable at $x=0$
(c) $f(x)\geq 1$ for all $x\in R$
(d) $f(x)$ is not differentiable at $x=1$

24. The function $f:R/\{0\}\to R$ given by **[2007]**

$$f(x)=\frac{1}{x}-\frac{2}{e^{2x}-1}$$

can be made continuous at $x=0$ by defining $f(0)$ as

(a) 0 (b) 1
(c) 2 (d) -1

25. Let $f(x)=\begin{cases}(x-1)\sin\dfrac{1}{x-1} & \text{if } x\neq 1\\ 0 & \text{if } x=1\end{cases}$ **[2008]**

Then which one of the following is true?

(a) f is neither differentiable at $x=0$ nor at $x=1$
(b) f is differentiable at $x=0$ and at $x=1$
(c) f is differentiable at $x=0$ but not at $x=1$
(d) f is differentiable at $x=1$ but not at $x=0$

26. Let f : **R** → **R** be a positive increasing function with $\lim\limits_{x\to\infty}\dfrac{f(3x)}{f(x)}=1$

Then $\lim\limits_{x\to\infty}\dfrac{f(2x)}{f(x)}=$ **[2010]**

(a) $\dfrac{2}{3}$ (b) $\dfrac{3}{2}$
(c) 3 (d) 1

27. $\lim\limits_{x\to 2}\left(\dfrac{\sqrt{1-\cos\{2(x-2)\}}}{x-2}\right)$ **[2011]**

(a) equals $\sqrt{2}$ (b) equals $-\sqrt{2}$
(c) equals $\dfrac{1}{\sqrt{2}}$ (d) does not exist

28. The values of p and q for which the function **[2011]**

$$f(x)=\begin{cases}\dfrac{\sin(p+1)x+\sin x}{x}, & x<0\\ q, & x=0\\ \dfrac{\sqrt{x+x^2}-\sqrt{x}}{x^{3/2}}, & x>0\end{cases}$$ is continuous for all x in R, are

(a) $p=\frac{5}{2}, q=\frac{1}{2}$ (b) $p=-\frac{3}{2}, q=\frac{1}{2}$

(c) $p=\frac{1}{2}, q=\frac{3}{2}$ (d) $p=\frac{1}{2}, q=-\frac{3}{2}$

29. Define $f(x)$ as the product of two real functions

$f_1(x)=x, x\in R$, and $f_2(x)=\begin{cases}\sin\frac{1}{x}, & \text{if } x\neq 0\\ 0, & \text{if } x=0\end{cases}$

as follows : **[2011 RS]**

$$f(x)=\begin{cases}f_1(x).f_2(x), & \text{if } x=0\\ 0 & \text{if } x=0\end{cases}$$

Statement - 1 : $f(x)$ is continuous on R.

Statement - 2 : $f_1(x)$ and $f_2(x)$ are continuous on R.

(a) Statement -1 is true, Statement-2 is true; Statement-2 is a correct explanation for Statement-1.
(b) Statement-1 is true, Statement-2 is true; Statement-2 is NOT a correct explanation for Statement-1
(c) Statement-1 is true, Statement-2 is false
(d) Statement-1 is false, Statement-2 is true

30. Let $f: R \to [0,\infty)$ be such that $\lim_{x\to 5} f(x)$ exists and

$$\lim_{x\to 5}\frac{(f(x))^2-9}{\sqrt{|x-5|}}=0$$

Then $\lim_{x\to 5} f(x)$ equals :

(a) 0 (b) 1
(c) 2 (d) 3

31. If function $f(x)$ is differentiable at $x = a$,

then $\lim_{x\to a}\dfrac{x^2 f(a)-a^2 f(x)}{x-a}$ is : **[2011 RS]**

(a) $-a^2 f'(a)$ (b) $a f(a)-a^2 f'(a)$

(c) $2af(a)-a^2 f'(a)$ (d) $2af(a)+a^2 f'(a)$

32. If $f: R \to R$ is a function defined by $f(x) = [x]\cos\left(\frac{2x-1}{2}\right)\pi$, where $[x]$ denotes the greatest integer function, then f is . **[2012]**

(a) continuous for every real x.
(b) discontinuous only at $x = 0$
(c) discontinuous only at non-zero integral values of x.
(d) continuous only at $x = 0$.

33. Consider the function, $f(x)=|x-2|+|x-5|, x\in R$.

Statement-1 : $f'(4)=0$

Statement-2 : f is continuous in [2,5], differentiable in (2,5) and $f(2)=f(5)$. **[2012]**

(a) Statement-1 is false, Statement-2 is true.
(b) Statement-1 is true, statement-2 is true; statement-2 is a correct explanation for Statement-1.
(c) Statement-1 is true, statement-2 is true; statement-2 is **not** a correct explanation for Statement-1.
(d) Statement-1 is true, statement-2 is false.

34. $\lim_{x\to 0}\dfrac{(1-\cos 2x)(3+\cos x)}{x\tan 4x}$ is equal to **[JEE M 2013]**

(a) $-\frac{1}{4}$ (b) $\frac{1}{2}$
(c) 1 (d) 2

Solutions & Explanations

Section-A : JEE Advanced/ IIT-JEE

A

1. {0} 2. $k = 7$ 3. $f(x) = \sqrt{4-x^2}, -2 \le x \le 0 = -\sqrt{4-x^2}, 0 \le x \le 2$

4. $\frac{2}{\pi}$ 5. 1 6. –1 7. 4 8. $\sqrt{2rh-h^2}, \frac{1}{128r}$ 9. e^5

10. $R-\{0\}$ 11. $(-\infty, -1) \cup [0, \infty)$, $I-\{0\}$ where I is the set of ingeger except $n = -1$

12. e^2 13. 1 14. 24 15. –1 16. 10

B

1. F

C

1. (c) 2. (d) 3. (a) 4. (d) 5. (c) 6. (b) 7. (b)
8. (d) 9. (a) 10. (b) 11. (c) 12. (b) 13. (d) 14. (d)
15. (c) 16. (c) 17. (b) 18. (a) 19. (d) 20. (d) 21. (d)
22. (c) 23. (c) 24. (d) 25. (d) 26. (c) 27. (a) 28. (b)
29. (c) 30. (a) 31. (a) 32. (c) 33. (d) 34. (b) 35. (b)
36. (b)

D

1. (a, b, d) 2. (b) 3. (b, d, e) 4. (a, b, d) 5. (a) 6. (a, b, c)
7. (b) 8. (d) 9. (b, c) 10. (b, c, d) 11. (a, b) 12. (a, c)
13. (a, c, d) 14. (d) 15. (a, d) 16. (a, c) 17. (b, c) 18. (a, b, c, d)
19. (b, d) 20. (b, d)

E

1. $\frac{2}{3\sqrt{3}}$ 2. 1 3. $a^2 \cos a + 2a \sin a$ 5. $2 \ln 2$

6. $a = -\frac{3}{2}, b \in R-\{0\}, c = \frac{1}{2}$ 7. $g(x) = \begin{cases} 2+x, & 0 \le x \le 1 \\ 2-x, & 1 < x \le 2 \\ 4-x, & 2 < x \le 3 \end{cases}$ discontinuity at $x = 1, 2$

8. f and f' are continuous and f'' is discontinuous on [0, 2]

9. cont. on (0, 2) and differentiable on $(0, 2) - \{1\}$ 10. not differentiable at $x = 1$ 12. $f'(0) = 0$

13. $f(x) = \begin{cases} \left(-\frac{2}{3}\log\frac{3}{2}+\frac{1}{9}\right)x, & x \le 0 \\ \left(\frac{1+x}{2+x}\right)^{1/x}, & x > 0 \end{cases}$ 14. $a = \frac{\pi}{6}, b = -\frac{\pi}{12}$ 15. $x = 0, 1, 2, 3$

16. $a = 8$ 17. $f(x) = e^{2x}$ 18. e^2 19. $a = \frac{2}{3}, b = e^{2/3}$ 20. $f(2) = -1$

21. continuous at $x = 0$, not differentiable at $x = 0$ 22. f is continuous and differentiable at all points except at $x = 2$.

25. $g(f(x)) = \begin{cases} x+a+1. & \text{if } x < -a \\ (x+a-1)^2+b & \text{if } a \le x < 0 \\ x^2+b & \text{if } 0 \le x \le 1 \\ (x-2)^2+b & \text{if } x > 1 \end{cases}$, $a = 1, b = 0$, gof is differentiable at $x = 0$ 26. 0

27. $\frac{\pi-2}{\pi}$ 28. 1 29. 0

F

1. (A) - p, (B) - r 2. (A) - p, q, r ; (B) - p, s ; (C) - r, s ; (D) - p, q

I

1. 6

Section-B : JEE Main/ AIEEE

1. (d)	2. (a)	3. (c)	4. (a)	5. (d)	6. (a)	7. (b)
8. (d)	9. (c)	10. (a)	11. (d)	12. (d)	13. (b)	14. (d)
15. (c)	16. (b)	17. (c)	18. (d)	19. (a)	20. (c)	21. (a)
22. (b)	23. (a)	24. (b)	25. (c)	26. (d)	27. (d)	28. (b)
29. (c)	30. (d)	31. (c)	32. (a)	33. (c)	34. (d)	

Section-A JEE Advanced/ IIT-JEE

A FILL IN THE BLANKS :

1. Given $f(x)=\begin{cases}(x-1)^2\sin\frac{1}{x-1}-|x|, x\neq 1\\ -1, \quad x=1\end{cases}$

We know that $|x|$ is not differentiable at $x=0$

$\therefore\ (x-1)^2\sin\frac{1}{x-1}-|x|$ is not differentiable at $x=0$.

At all other values of x, $f(x)$ is differentiable.

$\therefore$ The req. set of points is $\{0\}$.

2. It will be continuous at $x=2$ if

$$\lim_{x\to 2} f(x)=f(2)$$

$$\Rightarrow \lim_{x\to 2}\frac{x^3+x^2-16x+20}{(x-2)^2}=k$$

$$\Rightarrow k=\lim_{x\to 2}\frac{(x-2)^2(x+5)}{(x-2)^2}$$

$$=\lim_{x\to 2}(x+5)=7$$

$\therefore\ k=7$

3. $f(x)=\sqrt{4-x^2}, -2\le x\le 0$

$=-\sqrt{4-x^2}, 0\le x\le 2$

By choosing any arcs of circle $x^2-y^2=4$, we can define a discontinuous function, one of which is

$$f(x)=\begin{cases}\sqrt{4-x^2}, -2\le x\le 0\\ -\sqrt{4-x^2}, 0\le x\le 2\end{cases}$$

4. **KEY CONCEPT**

(L' Hospital rule)

$$\lim_{x\to a}\frac{f(x)}{g(x)}=\lim_{x\to a}\frac{f'(x)}{g'(x)}$$

if $\frac{f(a)}{g(a)}$ is of the form $\frac{0}{0}$ or $\frac{\infty}{\infty}$ or $0\times\infty$.

$\lim_{x\to 1}(1-x)\tan\frac{\pi x}{2}$ [form $0\times\infty$]

$=\lim_{x\to 1}\frac{1-x}{\cot(\pi x/2)}$ $\left[\text{Form}\frac{0}{0}\right]$

$=\lim_{x\to 1}\frac{-1}{\frac{-\pi}{2}\operatorname{cosec}^2\left(\frac{\pi x}{2}\right)}$ [Applying *L'* Hospital's rule]

$=\frac{2}{\pi}$.

ALTERNATE SOLUTION

$$\lim_{x\to 1}(1-x)\tan\frac{\pi x}{2}$$

Let us put $x=1+y$ so that when $x\to 1$; $y\to 0$

$\therefore$ We get $\lim_{y\to 0}[1-(1+y)]\tan\left[\frac{\pi}{2}(1+y)\right]$

$$=\lim_{y\to 0} y\cot\left(\frac{\pi y}{2}\right)$$

$$=\lim_{y\to 0}\frac{\left(\frac{\pi y}{2}\right)}{\tan\left(\frac{\pi y}{2}\right)}\times\frac{2}{\pi}=1\times\frac{2}{\pi}=\frac{2}{\pi}$$

5. Given that,

$f(x)=\sin x, x\neq n\pi, n=0,\pm1,\pm2,...$

$=2$, otherwise

And $g(x)=x^2+1, x\neq 0,2$

$=4, x=0$

$=5, x=2$

Then $\lim_{x\to 0} g[f(x)]=\lim_{x\to 0} g(\sin x)$

$$\lim_{x\to 0}(\sin^2 x+1)=1$$

6. $\lim_{x\to-\infty}\left[\frac{x^4\sin\left(\frac{1}{x}\right)+x^2}{(1+|x|^3)}\right]$

Let $L = \lim_{x\to-\infty} \frac{x^3}{1+|x|^3}\left[x\sin\left(\frac{1}{x}\right)+\frac{1}{x}\right]$

$$= \lim_{x\to-\infty} \frac{x^3}{|x|^3}\left[\frac{1}{1+\frac{1}{|x|x^2}}\right]\left[x\sin\left(\frac{1}{x}\right)+\frac{1}{x}\right] \quad ...(1)$$

$$= \lim_{x\to-\infty} \frac{x^3}{|x|^3}.1 = \lim_{x\to-\infty} \frac{x^3}{-x^3} = -1$$

7. Given that $f(9)=9, f'(9)=4$

Then,

$$\lim_{x\to 9} \frac{\sqrt{f(x)}-3}{\sqrt{x}-3} = \lim_{x\to 9} \frac{(\sqrt{f(x)}-3)(\sqrt{f(x)}+3)}{(\sqrt{x}-3)(\sqrt{x}+3)}$$

$$\lim_{x\to 9} \frac{\sqrt{x}+3}{\sqrt{f(x)}+3}$$

$$= \lim_{x\to 9} \frac{f(x)-9}{x-9}.\left[\frac{3+3}{3+3}\right]$$

$$= \lim_{x\to 9} \frac{f(x)-f(9)}{x-9}.1 = f'(9) = 4$$

8. In ΔABC, $AB = AC$

$AD \perp BC$ (D is mid pt of BC)

Let r = radius of circumcircle

$\therefore$ $OA = OB = OC = r$

Now $BD = \sqrt{BO^2 - OD^2} = \sqrt{r^2-(h-r)^2}$

$= \sqrt{2rh-h^2}$

$\therefore$ $BC = 2\sqrt{2rh-h^2}$

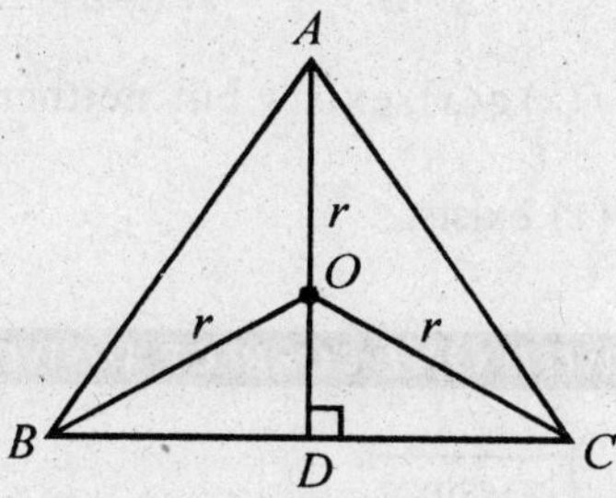

$\therefore$ Area of $\Delta ABC = \frac{1}{2}\times BC \times AD = h\sqrt{2rh-h^2}$

Also $\lim_{h\to 0} \frac{A}{P^3} = \lim_{h\to 0} \frac{h\sqrt{2rh-h^2}}{8(\sqrt{2rh-h^2}+\sqrt{2rh})^3}$

$$= \lim_{h\to 0} \frac{h^{3/2}\sqrt{2r-h}}{8h^{3/2}(\sqrt{2r-h})+\sqrt{2r})^3}$$

$$= \lim_{h\to 0} \frac{\sqrt{2r-h}}{8(2r-h+\sqrt{2r})^3}$$

$$= \frac{\sqrt{2r}}{8(\sqrt{2r}+\sqrt{2r})^3} = \frac{\sqrt{2r}}{8.8.2r.\sqrt{2r}} = \frac{1}{128r}$$

9. $\lim_{x\to\infty}\left(\frac{x+6}{x+1}\right)^{x+4} = \lim_{x\to\infty}\left\{\left[1+\frac{5}{x+1}\right]^{\frac{x+1}{5}}\right\}^{5\left(\frac{x+4}{x+1}\right)}$

[Using $\lim_{x\to\infty}\left(1+\frac{1}{x}\right)^x = e$]

$$e^{\lim_{x\to\infty} 5\left(\frac{x+4}{x+1}\right)} = e^{5\lim_{x\to\infty}\left(\frac{1+4/x}{1+1/x}\right)} = e^5$$

ALTERNATE SOLUTION

$$\lim_{x\to\infty}\left(\frac{x+6}{x+1}\right)^{x+4} = \lim_{x\to\infty}\left(\frac{1+\frac{6}{x}}{1+\frac{1}{x}}\right)^{x+4}$$

$$= \lim_{x\to\infty}\frac{\left(1+\frac{6}{x}\right)^x}{\left(1+\frac{1}{x}\right)^x}.\left(\frac{1+\frac{6}{x}}{1+\frac{1}{x}}\right)^4$$

[Using $\lim_{x\to\infty}\left(1+\frac{\lambda}{x}\right)^x = e^{\lambda}$]

$$= \frac{e^6}{e}.\left(\frac{1}{1}\right)^4 = e^5$$

10. We have,

$$f(x) = x|x| = \begin{cases} -x^2, & x<0 \\ x^2, & x\ge 0 \end{cases}$$

$$f'(x) = \begin{cases} -2x, & x<0 \\ 2x, & x\ge 0 \end{cases}$$

$$f''(x) = \begin{cases} -2, & x<0 \\ 2, & x\ge 0 \end{cases}$$

Clearly $f''(x)$ exists at every pt. except at $x = 0$

Thus $f(x)$ is twice differentiable on $R - \{0\}$.

11. Thus function is not defined for those values of x for which $[x + 1] = 0$. In other words it means that

$0 \le x+1 < 1$ or $-1 \le x < 0$ ....(1)

Hence the function is defined outside the region given by (1).

Required domain is $]-\infty, -1[\cup[0,\infty]$

Now, consider integral values of x say $x = n$

$$\text{R.H.L.} = \lim_{h\to 0}[n+h]\sin\frac{\pi}{[n+1+h]} = n\sin\frac{\pi}{(n+1)}$$

$$\text{L.H.L.} = \lim_{h\to 0}[n-h]\sin\frac{\pi}{[n+1-h]} = (n-1)\frac{\pi}{n}$$

Clearly RHL $\neq$ LHL. Hence the given function is not continuous for integral values of n $(n \neq 0, -1)$.

At $x = 0, f(0) = 0,$

$$\lim f(0+h) = \lim_{h\to 0}[h]\sin\frac{\pi}{[h+1]} = 0$$

The function is not defined for $x < 0$. Hence we cannot find $\lim f(0-h)$. Thus $f(x)$ is continuous at $x = 0$. Hence the points of discontinuity are given by $I - \{0\}$ where I is set of integers n except $n = -1$

12. KEY CONCEPT

$$\lim_{x\to 0}[f(x)]^{g(x)} = e^{\lim_{x\to 0} g(x)\log f(x)}$$

$$\lim_{x\to 0}\left(\frac{1+5x^2}{1+3x^2}\right)^{1/x^2} = e^{\lim_{x\to 0}\frac{1}{x^2}\log\left[\frac{1+5x^2}{1+3x^2}\right]}$$

$$= e^{\lim_{x\to 0}\left[5.\frac{\log(1+5x^2)}{5x^2} - 3.\frac{\log(1+3x^2)}{3x^2}\right]}$$

$$= e^{5-3} = e^2$$

13. KEY CONCEPT

$$\left[\frac{d}{dx}\int_{f_1(x)}^{f_2(x)} f(t)dt = f[f_2(x)]f_2'(x) - f[f_1(x)]f_1'(x)\right]$$

$$\lim_{x\to 0}\frac{d\left[\int_0^{x^2}\cos t^2 dt\right]}{\frac{d}{dx}(x\sin x)} \qquad \left[\frac{0}{0}\text{ form}\right]$$

$$= \lim_{x\to 0}\frac{\cos(x^2)^2 . \frac{d}{dx}(x^2)}{x\cos x + \sin x} \text{ [Applying } L \text{ Hospital's rule]}$$

$$= \lim_{x\to 0}\frac{2x\cos x^4}{x\cos x + \sin x} = \lim_{x\to 0}\frac{2\cos x^4}{\cos x + \frac{\sin x}{x}}$$

$$= \frac{2\cos 0}{\cos 0 + 1} = \frac{2}{2} = 1$$

14. $F(x) = f(x)\, g(x)\, h(x),\ \forall\, x \in R$

$f(x), g(x), h(x)$ are differentiable functions, therefore

$F'(x) = f'(x)\, g(x)\, h(x) + f(x)\, g'(x)\, h(x) + f(x)\, g(x)\, h'(x)$

At $x = x_0$

$$F(x_0) = f'(x_0)\, g(x_0)\, h(x_0) + f(x_0)\, g'(x_0)\, h(x_0) + f(x_0)\, g(x_0)\, h'(x_0)$$

Using the given values of $F'(x_0), f'(x_0), g'(x_0)$ and $h'(x_0)$; we get

$$21F(x_0) = 4f(x_0)\, g(x_0)\, h(x_0) - 7f(x_0)\, g(x_0)\, h(x_0) + kh(x_0)\, f(x_0)\, g(x_0)$$

$\Rightarrow\ 21 = 4 - 7 + k \qquad (\because F(x_0) = f(x_0)\, g(x_0)\, h(x_0))$

$\Rightarrow\ k = 24$

15. $$\lim_{h\to 0}\frac{\ln(1+2h) - 2\ln(1+h)}{h^2} = \lim_{h\to 0}\frac{\ln\left[\frac{(1+2h)}{1+2h+h^2}\right]}{h^2}$$

$$= \lim_{h\to 0}\frac{\ln\left[1+\frac{-h^2}{1+2h+h^2}\right]}{\frac{-h^2}{1+2h+h^2}} \times \frac{\frac{-h^2}{1+2h+h^2}}{h^2}$$

$$= 1.\lim_{h\to 0}\frac{-1}{1+2h+h^2} \quad \left[\text{Using } \lim_{x\to 0}\frac{\log(1+x)}{x} = 1\right]$$

$= -1$

16. Since $f(x)$ is given continuous on the closed bounded interval $[1, 3]$, $f(x)$ is bounded and assumes all the values lying in the interval $[m, M]$ where

$$m = \min f(x) \text{ and } M = \max f(x)$$

$$1 \le x \le 3 \Rightarrow f(1) \le f(x) \le (3)$$

If $m \gtrless M$, then $f(x)$ must assume all the irrational values lying in the $[m, M]$. But since $f(x)$ takes only rational values, we must have $m = M$ i.e., $f(x)$ must be a constant function. As $f(2) = 10$, we get

$$f(x) = 10\ \forall\ x \in [1, 3]$$

$$f(1.5) = 10$$

B TRUE / FALSE :

1. Consider $f(x) = \frac{|x-a|}{x-a}, g(x) = \frac{x-a}{|x-a|}$

then $\lim_{x\to a}(f(x)g(x))$ exists but neither $\lim_{x\to a} f(x)$ nor $\lim_{x\to a} g(x)$ exists.

C MCQ's WITH ONE CORRECT ANSWER :

1. **(c)** $f(x) = \sqrt{\frac{x - \sin x}{x + \cos^2 x}}$;

$$\Rightarrow \lim_{x\to\infty} f(x) = \lim_{x\to\infty}\sqrt{\frac{1-\frac{\sin x}{x}}{1+\frac{\cos^2 x}{x}}} = \sqrt{\frac{1-0}{1+0}} = 1$$

2. **(d)** $f(x) = \frac{\tan(\pi[x-\pi])}{1+[x]^2}$

By def. $[x-\pi]$ is an integer whatever be the value of x. And so $\pi[x-\pi]$ is an integral multiple of π. Consequently

$\tan(\pi[x-\pi]) = 0, \forall x.$

And since $1+[x]^2 \neq 0$ for any x, we conclude that $f(x)=0$.

Thus $f(x)$ is constant function and so, it is continuous and differentiable any no. of times, that is $f'(x), f''(x), f'''(x),\ldots$ all exist for every x, their value being 0 at every pt. x. Hence, out of all the alternatives only (d) is correct.

3. (a) $f(x) = e^{-x}$ is one such function.

Here $f(0)=1, f'(0)=-1, f(x)>0, \forall x.$

$\therefore\ f''(x)>0\ \forall x$

4. (d) $\lim\limits_{x\to 1} \dfrac{-\sqrt{25-x^2}-(-\sqrt{24})}{x-1}$

$$= \lim_{x\to 1}\frac{\sqrt{24}-\sqrt{25-x^2}}{x-1}\times\frac{\sqrt{24}+\sqrt{25-x^2}}{\sqrt{24}+\sqrt{25-x^2}}$$

$$= \lim_{x\to 1}\frac{x^2-1}{(x-1)[\sqrt{24}+\sqrt{25-x^2}]}$$

$$= \lim_{x\to 1}\frac{x+1}{[\sqrt{24}+\sqrt{25-x^2}]}$$

$$= \frac{2}{2\sqrt{24}} = \frac{1}{2\sqrt{6}}$$

5. (c) $\lim\limits_{x\to a}\dfrac{g(x)f(a)-g(a)f(x)}{x-a}$

$$= \lim_{h\to 0}\frac{g(a+h)f(a)-g(a)f(a+h)}{h} \quad [\text{For } x=a+h]$$

$$= \lim_{h\to 0}\frac{g(a+h)f(a)-g(a)f(a)+g(a)f(a)-g(a)f(a+h)}{h}$$

$$= \lim_{h\to 0} f(a)\left[\frac{g(a+h)-g(a)}{h}\right]-\lim_{h\to 0} g(a)\left[\frac{f(a+h)-f(a)}{h}\right]$$

$= f(a)g'(a)-g(a)f'(a)$
$= 2\times 2-(-1)\times 1 = 5$

6. (b) For $f(x)$ to be continuous at $x=0$

$$f(0) = \lim_{x\to 0} f(x)$$

$$= \lim_{x\to 0}\frac{\ln(1+ax)-\ln(1-bx)}{x}$$

$$\left[\text{Using } \lim_{x\to 0}\frac{\ln(1+x)}{x}=1\right]$$

$= a+b$

7. (b) $\lim\limits_{n\to\infty}\left(\dfrac{1}{1-n^2}+\dfrac{2}{1-n^2}+\ldots+\dfrac{n}{1-n^2}\right)$

$$= \lim_{n\to\infty}\frac{1+2+3+\ldots+n}{1-n^2}$$

$$= \lim_{n\to\infty}\frac{\Sigma n}{1-n^2} = \lim_{n\to\infty}\frac{n(n+1)}{2(1-n^2)}$$

$$= \lim_{n\to\infty}\frac{1+1/n}{2\left[\frac{1}{n^2}-1\right]} = -1/2$$

8. (d) The given function can be restated as

$$f(x) = \begin{cases}\dfrac{\sin[x]}{[x]}, & \text{if } x\in(-\infty,0)\cup[1,\infty]\\ 0\ , & \text{if } x\in[0,1)\end{cases}$$

$$\therefore\ \lim_{x\to 0^-} f(x) = \lim_{h\to 0}\frac{\sin[-h]}{[-h]}$$

$$= \lim_{x\to 0^-}\frac{\sin(-1)}{(-1)} = \sin 1$$

And $\lim\limits_{x\to 0^+} f(x) = \lim\limits_{h\to 0} 0 = 0$

$\because\ \lim\limits_{x\to 0^-} f(x) \neq \lim\limits_{x\to 0^+} f(x)$ as $\sin 1 \neq 0$

$\therefore\ \lim\limits_{x\to 0} f(x)$ does not exist.

9. (a) We have $f: R\to R$, a differentiable function and $f(1)=4$

NOTE THIS STEP

$$\lim_{x\to 1}\int_4^{f(x)}\frac{2t}{x-1}dt = \lim_{x\to 1}\left[\frac{t^2}{x-1}\right]_4^{f(x)}$$

$$= \lim_{x\to 1}\frac{(f(x))^2-16}{x-1}$$

$$= \lim_{x\to 1}\frac{f(x)-4}{x-1}.\lim_{x\to 1}(f(x)+4)$$

$= f'(1).(f(1)+4) = 8f'(1)$ [Using $f(1)=4$]

10. (b) We have $f(x) = [\tan^2 x]$

$\tan x$ is an increasing function

for $-\dfrac{\pi}{4}<x<\dfrac{\pi}{4}$

$\therefore\ \tan\left(-\dfrac{\pi}{4}\right)<\tan x<\tan\left(\dfrac{\pi}{4}\right)$ **NOTE THIS STEP**

$\Rightarrow\ -1<\tan x<1$
$\Rightarrow\ 0<\tan^2 x<1$
$\Rightarrow\ [\tan^2 x]=0$

Hence, $\lim\limits_{x\to 0} f(x) = \lim\limits_{x\to 0}[\tan^2 x] = 0$

Also $f(0)=0$

$\therefore\ f(x)$ is continuous at $x=0$

11. (c) When x is not an integer, both the functions $[x]$ and $\cos\left(\dfrac{2x-1}{2}\right)\pi$ are continuous.

$\therefore\ f(x)$ is continuous on all non integral points.

For $x = n \in I$

$$\lim_{x\to n^-} f(x) = \lim_{x\to n^-} [x]\cos\left(\frac{2x-1}{2}\right)\pi$$

$$= (n-1)\cos\left(\frac{2n-1}{2}\right)\pi = 0$$

$$\lim_{x\to n^+} f(x) = \lim_{x\to n^+} [x]\cos\left(\frac{2x-1}{2}\right)\pi$$

$$= n\cos\left(\frac{2n-1}{2}\right)\pi = 0$$

Also $f(n) = n\cos\frac{(2n-1)\pi}{2} = 0$

$\therefore$ f is continuous at all integral pts as well.
Thus, f is continuous everywhere.

12. (b) We have $= \lim_{n\to\infty}\frac{1}{n}\sum_{r=1}^{2n}\frac{r}{\sqrt{n^2+r^2}}$

$$= \lim_{n\to\infty}\frac{1}{n}\sum_{r=1}^{2n}\frac{r}{n\sqrt{1+(r/n)^2}}$$

$$= \int_0^2 \frac{x}{\sqrt{1+x^2}}dx \quad \left[\because \lim_{n\to\infty}\frac{1}{r}\sum_{r=0}^{a_n} f\left(\frac{r}{n}\right) = \int_0^a f(x)dx\right]$$

$$= \left[\sqrt{1+x^2}\right]_0^2 = \sqrt{5}-1$$

13. (d) We have $f(x) = [x]^2 - [x^2]$
At $x = 0$,

$$\text{L.H.L.} = \lim_{h\to 0} f(-h) = \lim_{h\to 0}[-h]^2 - [(-h)^2]$$

$$= \lim_{h\to 0} f(-1)^2 - [h^2] = \lim_{h\to 0} 1 - 0 = 1$$

$$\text{R.H.L.} = \lim_{h\to 0} f(h) = \lim_{h\to 0}[h]^2 - [h^2]$$

$$= \lim_{h\to 0} 0 - 0 = 0$$

$\because$ L.H.L. $\neq$ R.H.L.
$\therefore$ $f(x)$ is not continuous at $x = 0$.

At $x = 1$

$$\text{L.H.L.} = \lim_{h\to 0} f(1-h) = \lim_{h\to 0}[1-h]^2 - [(1-h)^2]$$

$$= \lim_{h\to 0} 0 - 0 = 0$$

$$\text{R.H.L.} = \lim_{h\to 0} f(1+h) = \lim_{h\to 0}[1+h]^2 - [(1+h)^2]$$

$$= \lim_{h\to 0} 1 - 1 = 0$$

$f(1) = [1]^2 - [1^2] = 1 - 1 = 0$
$\therefore$ L.H.L. = R.H.L. = f(1)
$\therefore$ f (x) is continuous at x = 1.
Clearly at other integral pts f (x) is not continuous.

14. (d) We have $|x| = \begin{cases} -x & \text{if } x < 0 \\ x & \text{if } x \geq 0 \end{cases}$

Also, $|x^2 - 3x + 2| = |(x-1)(x-2)|$

$$= \begin{cases} (1-x)(2-x) & \text{if } x < 1 \\ (x-1)(2-x) & \text{if } 1 \leq x < 2 \\ (x-1)(x-2) & \text{if } x \geq 2 \end{cases}$$

As $\cos(-\theta) = \cos\theta \Rightarrow \cos|x| = \cos x$

$\therefore$ Given function can be written as

$$f(x) = \begin{cases} (x^2-1)(x-1)(x-2) + \cos x & \text{if } x \leq 1 \\ -(x^2-1)(x-1)(x-2) + \cos x & \text{if } 1 \leq x < 2 \\ (x^2-1)(x-1)(x-2) + \cos x & \text{if } x \geq 2 \end{cases}$$

This function is differentiable at all points except possibly at x = 1 and x = 2.

$$Lf'(1) = \left\{\frac{d}{dx}[(x^2-1)(x-1)(x-2) + \cos x]\right\}_{x=1}$$

$$= -\sin 1$$

$$Rf'(1) = \left\{\frac{d}{dx}[-(x^2-1)(x-1)(x-2) + \cos x]\right\}_{x=1}$$

$$= -\sin 1$$

$\because$ Lf' (1) = Rf' (1)
$\therefore$ f is differentiable at x = 1.

$$Lf'(2) = \left\{\frac{d}{dx}[-(x^2-1)(x-1)(x-2) + \cos x]\right\}_{x=2}$$

$$= -3 - \sin 2$$

$$Rf'(2) = \left\{\frac{d}{dx}((x^2-1)(x-1)(x-2) + \cos x)\right\}_{x=2}$$

$$= 3 - \sin 2$$

Lf' (2) $\neq$ Rf' (2)
$\therefore$ f is not differentiable at x = 2.

15. (c) $\lim_{x\to 0}\frac{x\tan 2x - 2x\tan x}{4\sin^4 x}$

$$= \lim_{x\to 0}\frac{x}{4\sin^4 x}\left[\frac{2\tan x}{1-\tan^2 x} - 2\tan x\right]$$

$$= \lim_{x\to 0}\frac{2x\tan x}{4\sin^4 x}\left[\frac{1-1+\tan^2 x}{1-\tan^2 x}\right]$$

$$= \lim_{x\to 0}\frac{2x\tan^3 x}{4\sin x^4(1-\tan^2 x)}$$

$$= \frac{1}{2}\lim_{x\to 0}\frac{x}{\sin x}.\frac{1}{\cos^3 x}.\frac{1}{1-\tan^2 x}$$

$$= \frac{1}{2}.1.\frac{1}{1^3}.\frac{1}{1-0} = \frac{1}{2}$$

ALTERNATE SOLUTION

$$\lim_{x\to 0}\frac{x\tan 2x-2x\tan x}{(1-\cos 2x)^2}$$

$$=\lim_{x\to 0}\frac{x\left\{2x+\frac{8x^3}{3}+\frac{64x^5}{15}+...\right\}-2x\left\{x+\frac{x^3}{3}+\frac{2x^5}{15}+...\right\}}{4\sin^4 x}$$

$$=\lim_{x\to 0}\frac{x^4\left\{\frac{8}{3}-\frac{2}{3}+\text{terms containing higher positive powers of } x\right\}}{4\sin^4 x}$$

$$=\frac{1}{4}.2=\frac{1}{2}$$

16. **(c)** For $x\in R$,

$$\lim_{x\to\infty}\left(\frac{x-3}{x+2}\right)^x=\lim_{x\to\infty}\left\{\left[1-\frac{5}{x+2}\right]^{\frac{-(x+2)}{5}}\right\}^{\frac{-5x}{x+2}}$$

$$=e^{\lim_{x\to\infty}\frac{-5}{1+\frac{2}{x}}}=e^{-5}$$

17. **(b)** $\lim_{x\to 0}\frac{\sin(\pi\cos^2 x)}{x^2}=\lim_{x\to 0}\frac{\sin(\pi-\pi\sin^2 x)}{x^2}$

$[\sin(\pi-\theta)=\sin\theta]$

$$=\lim_{x\to 0}\frac{\sin(\pi\sin^2 x)}{\pi\sin^2 x}\times\frac{(\pi\sin^2 x)}{x^2}=\pi$$

18. **(a)** At $\underset{\text{at } x=k}{LHD}=\lim_{h\to 0}\frac{f(k)-f(k-h)}{h}$ (k = integer)

$$=\lim_{h\to 0}\frac{[k]\sin k\pi-[k-h]\sin(k-h)\pi}{h}$$

$$=\lim_{h\to 0}\frac{-(k-1)\sin(k-h)\pi}{h}\qquad[\because \sin k\pi=0]$$

$$=\lim_{h\to 0}\frac{-(k-1)\sin(k\pi-h\pi)}{h}$$

$[\sin(k\pi-\theta)=(-1)^{k-1}\sin\theta]$

$$=\lim_{h\to 0}\frac{-(k-1)(-1)^{k-1}\sin h\pi}{h\pi}\times\pi$$

$$=\pi(k-1)(-1)^k$$

19. **(d)** $f(x)=\max.\{x,x^3\}$

$$=\begin{cases}x\ ; & x<-1\\ x^3; & -1\le x\le 0\\ x\ ; & 0\le x\le 1\\ x^3; & x\ge 1\end{cases}$$

$$\therefore\quad f'(x)=\begin{cases}1\ ; & x<-1\\ 3x^2; & -1\le x\le 0\\ 1\ ; & 0\le x\le 1\\ 3x^2; & x\ge 1\end{cases}$$

Clearly f is not differentiable at -1, 0 and 1.

ALTERNATE SOLUTION

KEY CONCEPT

A continuous function $f(x)$ is not differentiable at $x=a$ if graphically it takes a sharp turn at $x=a$.

Graph of $f(x)=\max\{x,x^3\}$ is as shown with solid lines.

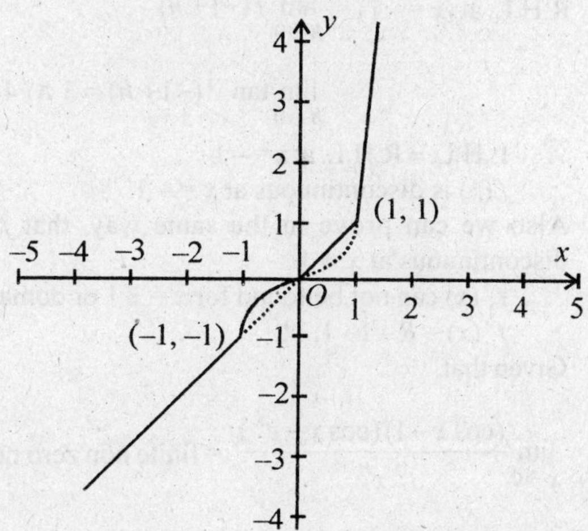

From graph of $f(x)$ at $x=-1$, 0, 1, we have sharp turns.

$\therefore\quad f(x)$ is not differentiable at $x=-1$, 0, 1.

20. **(d)** Let us test each of four options.

(a) $f(x)=\cos|x|+|x|=\begin{cases}\cos x-x, & x<0\\ \cos x+x, & x\ge 0\end{cases}$

$$f'(x)=\begin{cases}-\sin x-1, & x<0\\ -\sin x+1, & x\ge 0\end{cases}$$

At $x=0$, $LD=-1$, $RD=1$

$\therefore$ Not differentiable

(b) $f(x)=\cos|x|-|x|=\begin{cases}\cos x+x, & x<0\\ \cos x-x, & x\ge 0\end{cases}$

$\therefore$ Not differentiable at $x=0$

(c) $f(x)=\sin|x|+|x|=\begin{cases}-\sin x-x, & x<0\\ \sin x-x, & x\ge 0\end{cases}$

$\therefore$ Not differentiable at $x=0$

(d) $f(x)=\sin|x|-|x|=\begin{cases}-\sin x+x, & x<0\\ \sin x-x, & x\ge 0\end{cases}$

$$f'(x)=\begin{cases}-\cos x+1, & x<0\\ \cos x-1, & x\ge 0\end{cases}$$

At $x=0$, $LD=0$, $RD=0$

$\therefore$ f is differentiable at $x=0$.

21. (d) The given function is

$$f(x)=\begin{cases}\tan^{-1}x & \text{if } |x|\le 1\\ \frac{1}{2}(|x|-1) & \text{if } |x|>1\end{cases}$$

$$\Rightarrow f(x)=\begin{cases}\frac{1}{2}(-x-1) & \text{if } x<-1\\ \tan^{-1}x & \text{if } -1\le x\le 1\\ \frac{1}{2}(x-1) & \text{if } x>1\end{cases}$$

Clearly L.H.L. at $(x=-1)=\lim_{h\to 0} f(-1-h)=0$

R.H.L. at $(x=-1)=\lim_{h\to 0} f(-1+h)$

$=\lim_{h\to 0}\tan^{-1}(-1+h)=3\pi/4$

$\therefore$ L.H.L. $\ne$ R.H.L. at $x=-1$

$\therefore$ $f(x)$ is discontinuous at $x=-1$

Also we can prove in the same way, that $f(x)$ is discontinuous at $x=1$

$\therefore$ $f'(x)$ can not be found for $x=\pm 1$ or domain of $f'(x)=R-\{-1,1\}$

22. (c) Given that,

$$\lim_{x\to 0}\frac{(\cos x-1)(\cos x-e^x)}{x^n}=\text{finite non zero number}$$

$$=\lim_{x\to 0}\frac{(1-\cos x)(1+\cos x)(e^x-\cos x)}{x^n(1+\cos x)}$$

$$=\lim_{x\to 0}\left(\frac{\sin^2 x}{x^2}\right)\cdot\left(\frac{e^x-\cos x}{x^{n-2}}\right)\cdot\left(\frac{1}{1+\cos x}\right)$$

$$=\lim_{x\to 0}1^2.\frac{e^x-\cos x}{x^{n-2}}.\frac{1}{2}$$

$$=\frac{1}{2}\lim_{x\to 0}\frac{e^x-\sin x}{(x-2)x^{n-3}} \quad \text{[using L' Jospital's rule]}$$

For this limit to be finite, $n-3=0$

$\Rightarrow$ $n=3$

23. (c) Given that $f;R\to R$ such that $f(1)=3$ and $f'(1)=6$

Then $\lim_{x\to 0}\left[\frac{f(1+x)}{f(1)}\right]^{1/x}$

$$=e^{\lim_{x\to 0}\frac{1}{x}[\log f(1+x)-\log f(1)]}$$

$$=e^{\lim_{x\to 0}\frac{\frac{1}{f(1+x)}f'(1+x)}{1}} \quad \text{[Using L' Hospital rule]}$$

$$=e^{\frac{f'(1)}{f(1)}}=e^{6/3}=e^2$$

24. (d) We are given that

$$\lim_{x\to 0}\frac{[(a-n)nx-\tan x]\sin nx}{x^2}=0$$

where n is non zero real number

$$\Rightarrow \lim_{x\to 0} n.\frac{\sin nx}{nx}\left[\left\{(a-n)n-\frac{\tan x}{x}\right\}\right]=0$$

$\Rightarrow$ $1.n[(a-n)n-1]=0$

$\Rightarrow$ $a=\frac{1}{n}+n$

25. (d) Let $L=\lim_{h\to 0}\frac{f(2h+2+h^2)-f(2)}{f(h-h^2+1)-f(1)}$ $\left[\frac{0}{0}\text{form}\right]$

$\therefore$ Applying L Hospital's rule, we get

$$L=\lim_{h\to 0}\frac{f'(2h+2+h^2).(2+2h)}{f'(h-h^2+1).(1-2h)}$$

$$=\frac{f'(2).2}{f'(1).1}=\frac{6\times 2}{4\times 1}=3.$$

26. (c) Let $L=\lim_{x\to 0}\frac{f(x^2)-f(x)}{f(x)-f(0)}$ $\left[\because f'(a)>0, f \text{ being strictly increasing}\right]$

Using L.H. Rule, we get

$$L=\lim_{x\to 0}\frac{f'(x^2).2x-f'(x)}{f'(x)}$$

$$=\lim_{x\to 0}\frac{f'(x^2).2x}{f'(x)}-1$$

$=0-1=-1$

27. (a) Given function is $y=||x|-1|$

or $y=\begin{cases}-|x|+1 & \text{if } |x|<1\\ |x|-1 & \text{if } |x|\ge 1\end{cases}$

$$=\begin{cases}-|x|+1 & \text{if } -1<x<1\\ |x|-1 & \text{if } x\le -1 \text{ or } x\ge 1\end{cases}$$

$$=\begin{cases}-x-1 & \text{if } x\le -1\\ x+1 & \text{if } -1<x<0\\ -x+1 & \text{if } 0\le x<1\\ x-1 & \text{if } x\ge 1\end{cases}$$

Here $Ly'(-1)=-1$ and $Ry'(-1)=1$

$Ly'(0)=1$ and $Ry'(0)=-1$ and

$Ly'(1)=-1$ and $Ry'(1)=1$

$\Rightarrow$ y is not differentiable at $x=-1, 0, 1$.

ALTERNATE SOLUTION

Graph of $y=||x|-1|$ is as follows :

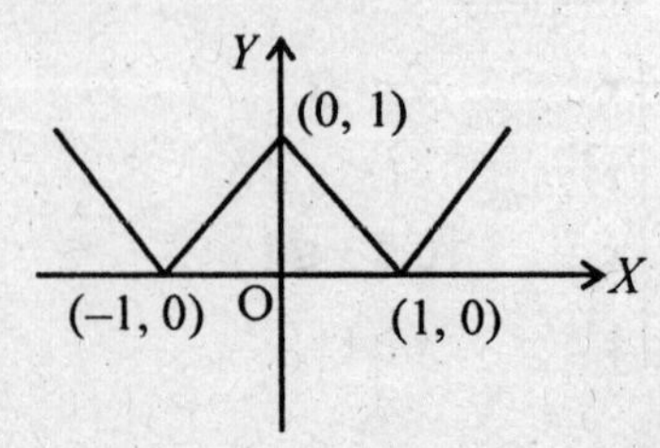

The graph has sharp turnings at $x = -1, 0$ and 1; and hence not differentiable at $x = -1, 0, 1$.

28. **(b)** Given that $f(x)$ is a continuous and differentiable function and

$$f\left(\frac{1}{x}\right) = 0, x = n, n \in I$$

$$\therefore \quad f(0^+) = f\left(\frac{1}{\infty}\right) = 0$$

Since R.H.L. = 0,

$\therefore \quad f(0) = 0$ for $f(x)$ to be continuous.

Also $f'(0) = \lim_{h\to 0} \frac{f(h) - f(0)}{h - 0} = \lim_{h\to 0} \frac{f(h)}{h} = 0$

$= 0$ [Using $f(0) = 0$ and $f(0^+) = 0$]

Hence $f(0) = 0, f'(0) = 0$

29. **(c)** $\lim_{x\to 0} [(\sin x)^{1/x} + (1/x)^{\sin x}]$

$$= \lim_{x\to 0} (\sin x)^{1/x} + \lim_{x\to 0}\left(\frac{1}{x}\right)^{\sin x}$$

$$= 0 + e^{\lim_{x\to 0} \sin x \log\left(\frac{1}{x}\right)}$$

[$\because |\sin x| < 1$ when $x \to 0$]

$$= e^{\lim_{x\to 0} \frac{-\log x}{\operatorname{cosec} x}} = e^{\lim_{x\to 0} \frac{-1/x}{-\operatorname{cosec} x \cot x}}$$

[Using L' Hospital rule]

$$= e^{\lim_{x\to 0} \frac{\sin x}{x} . \tan x} = e^0 = 1$$

30. **(a)** Given that $f(x)$ is differentiable on $(0, \infty)$ with

$$f(1) = 1 \text{ and } \lim_{t\to x} \frac{t^2 f(x) - x^2 f(t)}{t - x} = 1 \text{ for each } x > 0$$

$$\Rightarrow \lim_{t\to x} \frac{2t\, f(x) - x^2 f'(t)}{1} = 1$$

[Using L' Hospital rule]

$$\Rightarrow 2xf(x) - x^2 f'(x) = 1$$

$$\Rightarrow f'(x) - \frac{2}{x} f(x) = -\frac{1}{x^2}$$

[Linear differential equation]

Integrating factor

NOTE THIS STEP

$$e^{\int -\frac{2}{x} dx} = e^{-2\log x} = e^{\log 1/x^2} = \frac{1}{x^2}$$

$$\therefore \quad \text{Solution is } f(x) \times \frac{1}{x^2} = \int\left(-\frac{1}{x^2}\right) \times \frac{1}{x^2}\, dx$$

$$\Rightarrow \frac{f(x)}{x^2} = \frac{1}{3x^3} + C \quad \Rightarrow f(x) = Cx^2 + \frac{1}{3x}$$

Also $f(1) = 1$

$$\Rightarrow \quad 1 = C + \frac{1}{3} \quad \Rightarrow \quad C = 2/3$$

$$\therefore \quad f(x) = \frac{2}{3}x^2 + \frac{1}{3x}$$

31. **(a)** **KEY CONCEPT**

$$\frac{d}{dx}\left[\int_{g(x)}^{h(x)} f(t)\, dt\right] = f(h(x))\, h'(x) - f(g(x)).g'(x)$$

$$= \frac{f(2) \times 2 \times 2 \times 1}{2 \times \frac{\pi}{4}} = \frac{8}{\pi} f(2)$$

$$\text{Let } L = \lim_{x\to \frac{\pi}{4}} \frac{\int_2^{\sec^2 x} f(t)dt}{x^2 - \frac{\pi^2}{16}} \quad \left[\frac{0}{0} \text{ form}\right]$$

On applying L' Hospital's rule, we get

$$L = \lim_{x\to \frac{\pi}{4}} \frac{\frac{d}{dx}\left[\int_2^{\sec^2 x} f(t)dt\right]}{\frac{d}{dx}\left(x^2 - \frac{\pi^2}{16}\right)}$$

$$L = \lim_{x\to \frac{\pi}{4}} \frac{f(\sec^2 x).2\sec^2 x \tan x}{2x}$$

32. **(c)** As per question,

p = left hand derivative of $|x-1|$ at $x = 1 \Rightarrow p = -1$

Also $\lim_{x\to 1+} g(x) = p$

Where $g(x) = \dfrac{(x-1)^n}{\log \cos^m (x-1)}$, $0 < x < 2$,

m, n are integers, $m \neq 0, n > 0$

$\therefore$ we get,

$$\lim_{x\to 1^+} \frac{(x-1)^n}{\log \cos^m (x-1)} = -1 \Rightarrow \lim_{h\to 0} \frac{h^n}{\log \cos^m h} = -1$$

$$\Rightarrow \lim_{h\to 0} \frac{h^n}{m(\log \cos h)} = -1 \text{ [Using L' Hospital's rule]}$$

$$\Rightarrow \lim_{h\to 0} \frac{n h^{n-1} \cos h}{m(-\sin h)} = -1 \text{ [Using L' Hospital's rule]}$$

$$\Rightarrow \lim_{h\to 0} \frac{n h^{n-2} \cos h}{m\left(\frac{\sin h}{h}\right)} = 1 \Rightarrow n = 2 \text{ and } m = 2$$

33. (d) $\lim_{x\to 0}[1+x\ell n(1+b^2)]^{\frac{1}{x}} = 2b\sin^2\theta$

$\Rightarrow e^{\lim_{x\to 0}\frac{1}{x}\ell n[1+x\ell n(1+b^2)]} = 2b\sin^2\theta$

$\Rightarrow e^{\lim_{x\to 0}\frac{\ell n[1+x\ell n(1+b^2)]}{x\ell n(1+b^2)}\times \ell n(1+b^2)} = 2b\sin^2\theta$

$\Rightarrow e^{\ell n(1+b^2)} = 2b\sin^2\theta$

$\Rightarrow 1+b^2 = 2b\sin^2\theta$

$\Rightarrow 2\sin^2\theta = b+\frac{1}{b}$

We know that $2\sin^2\theta \le 2$ and $b+\frac{1}{b}\ge 2$ for $b>0$

$\therefore 2\sin^2\theta = b+\frac{1}{b} = 2 \Rightarrow \sin^2\theta = 1$

As $\theta \in (-\pi, \pi]$,

$\therefore \theta = \pm\frac{\pi}{2}$

34. (b) Given

$$\lim_{x\to\infty}\left(\frac{x^2+x+1}{x+1}-ax-b\right) = 4$$

$$\Rightarrow \lim_{x\to\infty}\frac{x^2+x+1-ax^2-ax-bx-b}{x+1} = 4$$

$$\Rightarrow \lim_{x\to\infty}\frac{(1-a)x^2+(1-a-b)x+(1-b)}{x+1} = 4$$

For this limit to be finite $1-a=0 \Rightarrow a=1$ then given limit reduces to

$$\lim_{x\to\infty}\frac{-bx+(1-b)}{x+1} = 4 \Rightarrow \lim_{x\to\infty}\frac{-b+\frac{(1-b)}{x}}{1+\frac{1}{x}} = 4$$

$\Rightarrow -b=4$ or $b=-4$

Hence $a=1$, $b=-4$.

35. (b) We have $f'(0^+) = \lim_{h\to 0}\frac{f(0+h)-f(0)}{h}$

$$= \lim_{h\to 0}\frac{h^2\left|\cos\frac{\pi}{h}\right|}{h} = \lim_{h\to 0} h\left|\cos\frac{\pi}{h}\right|$$

$= 0\times$ some finite value $= 0$

Also,

$$f'(0^-) = \lim_{h\to 0}\frac{f(0-h)-f(0)}{-h} = \lim_{h\to 0}\frac{h^2\left|\cos\frac{\pi}{-h}\right|}{-h}$$

$= \lim_{h\to 0} -h\left|\cos\frac{\pi}{h}\right| = 0\times$ some finite value

$= 0$

$\because f'(0^+) = f'(0^-) \Rightarrow f$ is differentiable at $x=0$

Now $f'(2^+) = \lim_{h\to 0}\frac{f(2+h)-f(2)}{h}$

$$= \lim_{h\to 0}\frac{(2+h)^2\left|\cos\frac{\pi}{2+h}\right| - 4\left|\cos\frac{\pi}{2}\right|}{h}$$

$$= \lim_{h\to 0}\frac{(2+h)^2\left(\cos\frac{\pi}{2+h}\right)}{h}$$

$$= \lim_{h\to 0}\frac{(2+h)^2}{h}\sin\left(\frac{\pi}{2}-\frac{\pi}{2+h}\right)$$

$$= \lim_{h\to 0}\frac{(2+h)^2}{h}\sin\left(\frac{\pi h}{2(2+h)}\right)$$

$$= \lim_{h\to 0}\frac{(2+h)^2}{h}\times\frac{\sin\left(\frac{\pi h}{2(2+h)}\right)}{\left(\frac{\pi h}{2(2+h)}\right)}\times\frac{\pi h}{2(2+h)} = \pi$$

Also $f'(2^-) = \lim_{h\to 0}\frac{f(2-h)-f(2)}{-h}$

$$= \lim_{h\to 0}\frac{(2-h)^2\left|\cos\left(\frac{\pi}{2-h}\right)\right| - 0}{-h}$$

$$= \lim_{h\to 0}\frac{-(2-h)^2\cos\left(\frac{\pi}{2-h}\right)}{-h}$$

$$= \lim_{h\to 0}\frac{(2-h)^2\sin\left(\frac{\pi}{2}-\frac{\pi}{2-h}\right)}{h}$$

$$= \lim_{h\to 0}\frac{(2-h)^2}{h}\times\frac{\sin\left(\frac{-\pi h}{2(2-h)}\right)}{\left(\frac{-\pi h}{2(2-h)}\right)}\times\left(\frac{-\pi h}{2(2-h)}\right) = -\pi$$

As $f'(2^+)\ne f'(2^-) \Rightarrow f$ is not differentiable at $x=2$.

36. (b) The given equation is

$$(\sqrt[3]{1+a}-1)x^2+(\sqrt{1+a}-1)x+(\sqrt[6]{1+a}-1) = 0$$

Let $a+1=y$, then equation reduces to

$$(y^{1/3}-1)x^2+(y^{1/2}-1)x+(y^{1/6}-1) = 0$$

Dividing both sides by $y-1$, we get

$$\left(\frac{y^{1/3}-1}{y-1}\right)x^2+\left(\frac{y^{1/2}-1}{y-1}\right)x+\left(\frac{y^{1/6}-1}{y-1}\right) = 0$$

Taking limit as $y \to 1$ i.e. $a \to 0$ on both sides we get

$$\frac{1}{3}x^2+\frac{1}{2}x+\frac{1}{6}=0$$

$$\Rightarrow 2x^2+3x+1=0$$

$$\Rightarrow x=-1, -\frac{1}{2} \text{ (roots of the equation)}$$

Thus $\lim_{a\to 0^+} \alpha(a)=-1, \lim_{a\to 0^+} \beta(a)=-\frac{1}{2}$

(D) MCQ's WITH ONE OR MORE THAN ONE CORRECT :

1. **(a, b, d)** Given that $x+|y|=2y$

If $y<0$ then $x-y=2y$

$\Rightarrow y=x/3 \Rightarrow x<0$

If $y=0$ then $x=0$. If $y>0$ then $x+y=2y$

$\Rightarrow y=x \Rightarrow x>0$

Thus we can define $f(x)=y=\begin{cases} x/3 , & x<0 \\ x , & x\ge 0 \end{cases}$

Continuity at $x=0$

$LL=\lim_{h\to 0} f(0-h)=\lim_{h\to 0}(-h/3)=0$

$RL=\lim_{h\to 0} f(0+h)=\lim_{h\to 0} h=0$

$f(0)=0$

As $LL=RL=f(0)$

$\therefore$ $f(x)$ is continuous at $x=0$

Differentiability at $x=0$

$Lf'=1/3$; $Rf'=1$

As $Lf'\ne Rf' \Rightarrow f(x)$ is not differentiable at $x=0$

But for $x<0$, $\frac{dy}{dx}=\frac{1}{3}$.

2. **(b)** We have $f(x)=x(\sqrt{x}-\sqrt{x+1})$

Let us check differentiability of $f(x)$ at $x=0$

$$Lf'(0)=\lim_{h\to 0}\frac{(0-h)[\sqrt{0-h}-\sqrt{0-h+1}]-0}{-h}$$

$$=\lim_{h\to 0}[\sqrt{-h}-\sqrt{-h+1}]$$

$$=0-\sqrt{1}=-1$$

$$Rf'(0)=\lim_{h\to 0}\frac{(0+h)[\sqrt{0+h}-\sqrt{0+h+1}]-0}{h}$$

$$=\lim_{h\to 0}\sqrt{h}-\sqrt{h+1}=-1$$

Since $Lf'(0)=Rf'(0)$

$\therefore$ f is differentiable at $x=0$.

3. **(b,d,e)** We have, $f(x)=1+|\sin x|$

We know that $g(x)=\sin x$ is continuous everywhere on R, also $h(x)=1$ is continuous everywhere on R

$\therefore$ $hog(x)$ must be continuous everywhere.

i.e., $1+|\sin x|$ is continuous everywhere.

ALTERNATE SOLUTION 1

For any real number a,

R.H.L. (at $x=a$) $=\lim_{h\to 0}[1+|\sin(a+h)|]=1+|\sin a|$

L.H.D. (at $x=a$) $=\lim_{h\to 0}[1+|\sin(a-h)|]=1+|\sin a|$

$f(a)=1+|\sin a|$

$\therefore$ R.H.L. (at $x=a$) = L.H.D. (at $x=a$) $=f(a)$, for every real no. a.

$\therefore$ $f(x)$ is continuous everywhere.

Differentiability :

We have $f(x)=\begin{cases} 1-\sin x, & \sin x<0 \\ 1+\sin x, & \sin x\ge 0 \end{cases}$

Clearly $f(x)$ is differentiable at any non critical point i.e. at all real numbers except possibly at critical pts $x=n\pi$; $n=0,\pm 1,\pm 2,....$

Let us check the differentiability of $f(x)$ at $x=n\pi$.

We have $f(n\pi)=1+|\sin n\pi|=1+|0|=1$

$$\text{Now, } Rf'(n\pi)=\lim_{h\to 0}\frac{1+|\sin(n\pi+h)|-1}{h}$$

$$=\lim_{h\to 0}\frac{|\pm\sin h|}{h}=\lim_{h\to 0}\frac{\sin h}{h}=1$$

$$\text{and } Lf'(n\pi)=\lim_{h\to 0}\frac{1+|\sin(n\pi-h)|-1}{-h}$$

$$=\lim_{h\to 0}\frac{|\pm\sin h|}{-h}=\lim_{h\to 0}\frac{\sin h}{-h}=-1$$

Since $Rf'(n\pi)\ne Lf'(n\pi)$.

$\therefore$ $f(x)$ is not differentiable at $x=n\pi$; $n=0,\pm 1,\pm 2,\pm 3,.....$

ALTERNATE SOLUTION 2

The graph of $f(x)=1+|\sin x|$ is as shown in the fig.

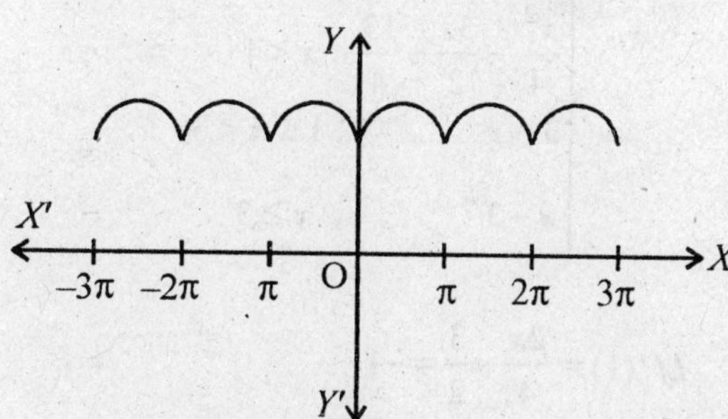

From graph it is clear that function is continuous every where but not differentiable at integral multiples of π ($\because$ at these points curve has sharp turnings)

4. **(a,b,d)** We have, for $-1\le x\le 1$

$\Rightarrow 0\le x\sin\pi x\le 1/2$

$\therefore$ $f(x)=[x\sin\pi x]=0$

Also $x\sin\pi x$ becomes negative and numerically less than 1 when x is slightly greater than 1 and so by definition of $[x]$,

$f(x)=[x\sin\pi x]=-1$ when $1<x<1+h$

Thus $f(x)$ is constant and equal to 0 in the closed interval $[-1, 1]$ and so $f(x)$ is continuous and differentiable in the open interval $(-1, 1)$.

At $x = 1, f(x)$ is clearly discontinuous, since $f(1-0) = 0$ and $f(1+0) = -1$ and $f(x)$ is non-differentiable at $x = 1$.

5. **(a)** The given function is,

$$f(x) = \frac{x}{1+|x|} = \begin{cases} \dfrac{x}{1-x}, & x < 0 \\ \dfrac{x}{1+x}, & x \geq 0 \end{cases}$$

For $x < 0$,

$$f'(x) = \frac{1(1-x)-(-1)x}{(1-x)^2} = \frac{1}{(1-x)^2},$$

For $x > 0$,

$$f'(x) = \frac{1(1+x)-1(x)}{(1-x)^2} = \frac{1}{(1+x)^2},$$

For $x = 0$,

$$Lf'(0) = \lim_{h\to 0} \frac{f(0-h)-f(0)}{-h} = \lim_{h\to 0} \frac{\frac{-h}{1+h}-0}{-h} = 1$$

$$Rf'(0) = \lim_{h\to 0} \frac{f(0+h)-f(0)}{h} = \lim_{h\to 0} \frac{\frac{h}{1+h}-0}{h} = 1$$

$Lf'(0) = Rf'(0)$

$\Rightarrow$ f is differentiable at $x = 0$

Hence f is differentiable in $(-\infty, \infty)$.

6. **(a,b,c)** $f(x) = \begin{cases} |x-3| & , x \geq 1 \\ \dfrac{x^2}{4} - \dfrac{3x}{2} + \dfrac{13}{4} & , x < 1 \end{cases}$

$$= \begin{cases} \dfrac{x^2}{4} - \dfrac{3x}{2} + \dfrac{13}{4} & , x < 1 \\ 3-x & , 1 \leq x < 3 \\ x-3 & , x \geq 3 \end{cases}$$

$$Lf'(1) = \frac{2x}{4} - \frac{3}{2} = -1$$

$Rf'(1) = -1$. Thus $Lf'(1) = Rf'(1)$

$\therefore$ f is differentiable at $x = 1$ and hence continuous at $x = 1$.

Again

$Lf'(3) = -1$ and $Rf'(3) = 1$

$\Rightarrow$ $Lf'(3) \neq Rf'(3)$

$\Rightarrow$ f is not differentiable at $x = 3$

Let us now check the continuity at $x = 3$

$$\text{L.H.L.} = \lim_{h\to 0} f(3-h) = \lim_{h\to 0} [3-(3-h)] = 0$$

$$\text{R.H.L.} = \lim_{h\to 0} f(3+h) = \lim_{h\to 0} [3+h-3] = 0$$

$f(3) = 0$

$\therefore$ f is continuous at $x = 3$

7. **(b)** We have, $f(x) = \dfrac{x}{2} - 1$

$$\therefore \quad [f(x)] = \left[\frac{x}{2} - 1\right] = -1, \ 0 \leq x < 2$$

$\tan[f(x)] = \tan(-1), 0 \leq x < 2$

$= 0, 2 \leq x \leq \pi$

$\therefore$ The function $\tan[f(x)]$ is discontinuous at $x = 2$.

Also the function $\dfrac{1}{f(x)} = \dfrac{1}{\frac{x}{2}-1} = \dfrac{2}{x-2}$ is discontinuous at $x = 2$.

Thus both the given functions $\tan[f(x)]$ as well as $\dfrac{1}{f(x)}$ are discontinuous on the interval $[0, \pi]$.

Also $f^{-1}(x) = y$

$\Rightarrow x = f(y) = \dfrac{y}{2} - 1$

$\Leftrightarrow y = 2(x+1)$

$\therefore$ $f^{-1}(x) = 2(x+1)$ is continuous on $[0, \pi]$

8. **(d)** $\displaystyle\lim_{x\to 0} \frac{\sqrt{\frac{1}{2}(1-\cos 2x)}}{x}$

$$= \lim_{x\to 0} \frac{\sqrt{\frac{1}{2}.2\sin^2 x)}}{x} = \lim_{x\to 0} \frac{|\sin x|}{x}$$

$$\therefore \quad \text{L.H.L.} = \lim_{h\to 0} \frac{|\sin(0-h)|}{0-h} = \lim_{h\to 0} \frac{|-\sin h|}{-h}$$

$$= \lim_{h\to 0} \frac{\sin h}{-h} = -1$$

$$\text{R.H.L.} = \lim_{h\to 0} \frac{|\sin(0+h)|}{0+h} = \lim_{h\to 0} \frac{\sin h}{h} = 1$$

As L.H.L. $\neq$ R.H.L.

$\therefore$ The given limit does not exist.

9. **(b,c)** On $(0, \pi)$

(a) $\tan x = f(x)$

We know that $\tan x$ is discontinuous at $x = \pi/2$

(b) $f(x) = \displaystyle\int_0^x t \sin\left(\frac{1}{t}\right) dt$

NOTE THIS STEP

$\Rightarrow f'(x) = x \sin\left(\dfrac{1}{x}\right)$ which exists on $(0, \pi)$

$\therefore f(x)$, being differentiable, is continuous on $(0, \pi)$.

(c) $f(x) = \begin{cases} 1 & , 0 < x \leq 3\pi/4 \\ 2\sin\dfrac{2x}{9} & , 3\pi/4 < x < \pi \end{cases}$

Clearly $f(x)$ is continuous on $(0, \pi)$ except possibly at $x = \dfrac{3\pi}{4}$, where,

$$\text{L.H.L.} = \lim_{h \to 0} f\left(\frac{3\pi}{4} - h\right) = \lim_{x \to 0} 1 = 1$$

$$\text{R.H.L.} = \lim_{h \to 0} f\left(\frac{3\pi}{4} + h\right) = \lim_{x \to 0} 2\sin\frac{2}{9}\left(\frac{3\pi}{4} + h\right)$$

$$= \lim_{h \to 0} 2\sin\left(\frac{\pi}{6} + \frac{2h}{9}\right) = 2\sin\frac{\pi}{6} = 2.\frac{1}{2} = 1$$

Also $f\left(\frac{3\pi}{4}\right) = 1$

As L.H.L. = R.H.L. = $f\left(\frac{3\pi}{4}\right)$

$\therefore f(x)$ is continuous on $(0, \pi)$

(d) $f(x) = \begin{cases} x\sin x & , 0 < x \le \pi/2 \\ \frac{\pi}{2}\sin(\pi + x) & , \frac{\pi}{2} < x < \pi \end{cases}$

Here $f(x)$ will be continuous on $(0, \pi)$ if it is continuous at $x = \pi/2$.

At $x = \pi/2$,

$$\text{L.H.L.} = \lim_{h \to 0} f\left(\frac{\pi}{2} - h\right)$$

$$= \lim_{h \to 0}\left(\frac{\pi}{2} - h\right)\sin\left(\frac{\pi}{2} - h\right)$$

$$= \frac{\pi}{2}\sin\frac{\pi}{2} = \frac{\pi}{2}$$

$$\text{R.H.L.} = \lim_{h \to 0} f\left(\frac{\pi}{2} + h\right) = \lim_{h \to 0}\frac{\pi}{2}\sin\left(\pi + \frac{\pi}{2} + h\right)$$

$$= \frac{\pi}{2}\sin\left(\pi + \frac{\pi}{2}\right) = \frac{-\pi}{2}\sin\frac{\pi}{2} = -\frac{\pi}{2}$$

As L.H.L. $\neq$ R.H.L.

$\therefore$ $f(x)$ is not continuous on $(0, \pi)$.

10. **(b,c,d)** $f(x) = \begin{cases} 0, & x < 0 \\ x^2, & x \ge 0 \end{cases}$

$\therefore \quad f'(x) = \begin{cases} 0, & x < 0 \\ 2x, & x \ge 0 \end{cases}$

which exists $\forall\, x$ except possibly at $x = 0$.

At $x = 0$, $Lf' = 0 = Rf'$

$\Rightarrow f$ is differentiable.

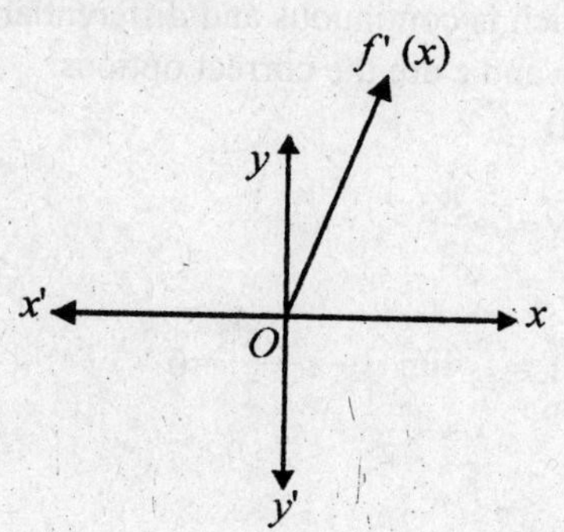

From graph of $f'(x)$, it is clear that $f'(x)$ is continuous but not differentiable at $x = 0$.

11. **(a, b)** We have $g(x) = \begin{cases} x^2\sin\left(\frac{1}{x}\right), & x \neq 0 \\ 0 & , x = 0 \end{cases}$

If $x \neq 0$, $g'(x) = x^2\cos(1/x)\left(-\frac{1}{x^2}\right) + 2x\sin\frac{1}{x}$

$$= -\cos\left(\frac{1}{x}\right) + 2x\sin\left(\frac{1}{x}\right)$$

which exists for $\forall x \neq 0$.

If $x = 0$,

$$g'(0) = \lim_{x \to 0}\frac{g(x) - g(0)}{x - 0}$$

$$= \lim_{x \to 0}\frac{x^2\sin(1/x) - 0}{x - 0}$$

$$= \lim_{x \to 0} x\sin\left(\frac{1}{x}\right) = 0$$

$$\Rightarrow \quad g'(x) = \begin{cases} -\cos\left(\frac{1}{x}\right) + 2x\sin\frac{1}{x}, & x \neq 0 \\ 0, & x = 0 \end{cases}$$

At $x = 0$, $\cos\left(\frac{1}{x}\right)$ is not continuous, therefore $g'(x)$ is not continuous at $x = \mathbf{0}$.

At $x = 0$

$$Lf' = \lim_{x \to 0}\frac{0 - (-x)\sin\left(-\frac{1}{x}\right)}{x}$$

$$= -\sin\left(\frac{1}{x}\right)$$

which does not exist.

$\therefore$ f is not differentiable at $x = 0$.

12. **(a, c)**

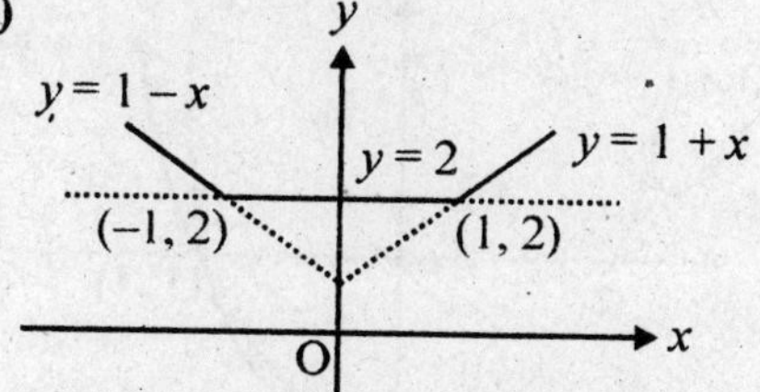

From graph it is clear that $f(x)$ is continuous every where and also differentiable everywhere except at $x = 1$ and -1.

13. **(a, c, d)** From the figure it is clear that

$$h(x) = \begin{cases} x, & \text{if } x \le 0 \\ x^2, & \text{if } 0 < x < 1 \\ x, & \text{if } x \ge 1 \end{cases}$$

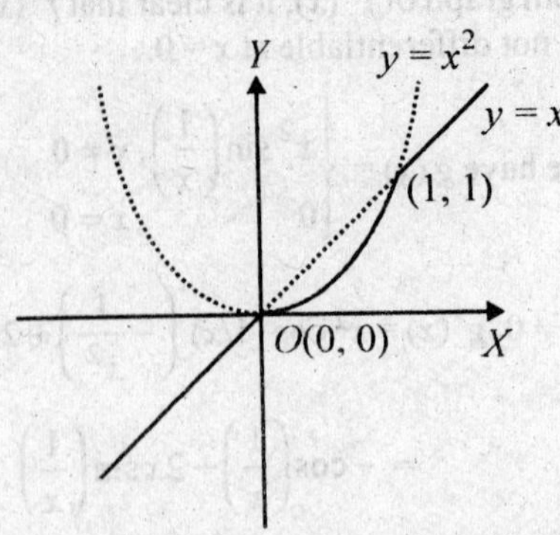

From the graph it is clear that h is continuous for all $x \in R$, $h'(x)=1$ for all $x>1$ and h is not differentiable at $x=0$ and 1.

14. **(d)** L.H.L. $= \lim_{x\to 1^-} \frac{\sqrt{1-\cos[2(x-1)]}}{x-1}$

$= \lim_{x\to 1^-} \frac{\sqrt{2\sin^2(x-1)}}{x-1} = \sqrt{2}.\lim_{x\to 1^-} \frac{\sqrt{\sin^2(x-1)}}{x-1}$

$= \sqrt{2}\lim_{x\to 1^-} \frac{|\sin(x-1)|}{x-1} = \sqrt{2}.\lim_{h\to 0} \frac{|\sin(-h)|}{-h}$

$= \sqrt{2}\lim_{h\to 0} \frac{\sin h}{-h} = -\sqrt{2}$

Again,

R.H.L. $= \lim_{x\to 1^+} \frac{\sqrt{1-\cos(x-1)}}{x-1} = \lim_{x\to 1^+} \sqrt{2}\frac{|\sin(x-1)|}{x-1}$

Put $x = 1+h$, $h>0$ for $x\to 1^+$, $h\to 0$.

$= \lim_{h\to 0} \sqrt{2}\frac{|\sin h|}{h} = \lim_{h\to 0}\sqrt{2}\frac{\sin h}{h} = \sqrt{2}$

L.H.L. $\neq$ R.H.L. Therefore $\lim_{x\to 1} f(x)$ does not exist.

15. **(a, d)** From graph, $f(x)$ is continuous everywhere but not differentiable at $x=1$.

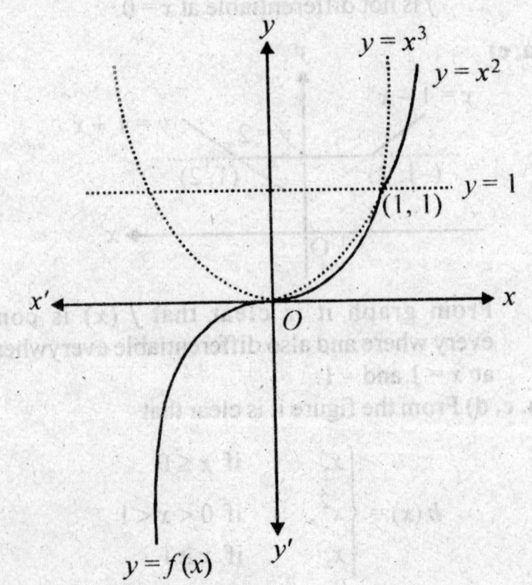

16. **(a,c)** Given that $L = \lim_{x\to 0} \frac{a-\sqrt{a^2-x^2}-\frac{x^2}{4}}{x^4}, a>0$ and L is finite.

Now $L = \lim_{x\to 0} \frac{\frac{x}{\sqrt{a^2-x^2}}-\frac{x}{2}}{4x^3}$

(Using L'Hospital's rule)

$= \lim_{x\to 0} \frac{\frac{1}{\sqrt{a^2-x^2}}-\frac{1}{2}}{4x^2}$

$\because$ L is finite, limiting value of numerator should be zero which is so when $\frac{1}{\sqrt{a^2}}-\frac{1}{2}=0$

i.e. $a=2$ $(\because a>0)$

Applying L 'Hospital's rule again, we get

$L = \lim_{x\to 0} \frac{\frac{x}{(a^2-x^2)^{3/2}}}{8x}$

$= \lim_{x\to 0} \frac{1}{8(a^2-x^2)^{3/2}}$

$= \frac{1}{8\times a^3} = \frac{1}{8\times 8}$ (using $a=2$)

$= \frac{1}{64}$

17. **(b, c)** $\because f(x+y)=f(x)+f(y)\ \forall\ x, y \in R$

$\therefore$ Putting $x=y=0$, we get

$f(0)=0$

Also $f'(x) = \lim_{h\to 0} \frac{f(x+h)-f(x)}{h}$

$= \lim_{h\to 0} \frac{f(h)}{h} = f'(0) = k$ (say)

$\Rightarrow f(x)=kx+c$

But $f(0)=0 \Rightarrow c=0$

$\therefore f(x)=kx$

Which is continuous and differentiable $\forall\ x \in R$.

$\therefore$ b and c are the correct options.

18. **(a, b, c, d)**

At $x = -\frac{\pi}{2}$

LHL $= \lim_{x\to -\frac{\pi}{2}^{-1}} -x-\frac{\pi}{2} = 0$

$$\text{RHL} = \lim_{x\to -\frac{\pi}{2}^+} -\cos x = 0 \text{ and } f\left(\frac{-\pi}{2}\right) = 0$$

$$\therefore \text{LHL} = \text{RHL} = f\left(\frac{-\pi}{2}\right)$$

$\Rightarrow f(x)$ is continuous at $x = -\dfrac{\pi}{2}$

Also at $x = 0$

$Lf'(0) = \sin 0 = 0;\ Rf'(0) = 1 - 0 = 1$

$\because Lf'(0) \neq Rf'(0)$

$\Rightarrow f$ is not differentiable at $x = 0$

At $x = 1$

$Lf'(1) = R'f(1) \Rightarrow f$ is differentiable at $x = 1$.

At $x = \dfrac{-3}{2}$, $f(x) = -\cos x$ which is differentiable.

$\therefore$ All four options are correct.

19. (b, d)

We have $f(x) = \begin{cases} a_n + \sin \pi x, & x \in [2n, 2n+1] \\ b_n + \cos \pi x, & x \in (2n-1, 2n) \end{cases}$

As f is continuous for all n

$\therefore$ At $x = 2n$, LHL = RHL = $f(2n)$

$\Rightarrow b_n + \cos 2\pi n = a_n + \sin 2\pi n = a_n + \sin 2\pi n$

$\Rightarrow b_n + 1 = a_n \Rightarrow a_n - b_n = 1$

$\therefore b$ is correct.

Also at $x = 2n+1$, LHL = RHL = $f(2n+1)$

$$\Rightarrow \lim_{h\to 0} a_n + \sin \pi(2n+1-h)$$

$$= \lim_{h\to 0} b_{n+1} + \cos \pi(2n+1-h) = a_n + \sin(2n+1)\pi$$

$\Rightarrow a_n = b_{n+1} - 1 = a_n \Rightarrow a_n - b_{n+1} = -1$

$\therefore c$ is incorrect

$\Rightarrow a_{n-1} - b_n = -1$

$\therefore d$ is correct.

20. (b, d)

$$\lim_{n\to\infty} \frac{1^a + 2^a + - - - + n^a}{(n+1)^{a-1}[(na+1)+(na+2)+ - - - +(na+n)]} = \frac{1}{60}$$

$$\Rightarrow \lim_{n\to\infty} \frac{n^a}{(n+1)^{a-1}} \frac{\left(\frac{1}{n}\right)^a + \left(\frac{2}{n}\right)^a + - - - + \left(\frac{n}{n}\right)^a}{n^2 a + \frac{n(n+1)}{2}} = \frac{1}{60}$$

$$\Rightarrow \lim_{n\to\infty} \frac{n^{a-1}}{(n+1)^{a-1}} \frac{\frac{1}{n}\sum_{r=1}^{n}\left(\frac{r}{n}\right)^a}{a + \frac{1}{2}\left(1+\frac{1}{n}\right)} = \frac{1}{60}$$

$$\Rightarrow \lim_{n\to\infty} \left(\frac{1}{1+\frac{1}{n}}\right)^{a-1} \frac{\frac{1}{n}\sum_{r=1}^{n}\left(\frac{r}{n}\right)^a}{a + \frac{1}{2}\left(1+\frac{1}{n}\right)} = \frac{1}{60}$$

$$\Rightarrow \frac{\int_0^1 x^a dx}{a + \frac{1}{2}} = \frac{1}{60} \Rightarrow \frac{\left[x^{a+1}\right]_0^1}{(a+1)\left(a+\frac{1}{2}\right)} = \frac{1}{60}$$

$$\Rightarrow \frac{1}{(a+1)\left(a+\frac{1}{2}\right)} = \frac{1}{60}$$

$$\Rightarrow 2a^2 + 3a - 119 = 0 \Rightarrow a = 7 \text{ or } -\frac{17}{2}$$

(E) SUBJECTIVE PROBLEMS :

1. $$\lim_{x\to a} \frac{\sqrt{a+2x} - \sqrt{3x}}{\sqrt{3a+x} - 2\sqrt{x}}$$

$$= \lim_{x\to a} \frac{(\sqrt{a+2x} - \sqrt{3x})(\sqrt{a+2x} + \sqrt{3x})(\sqrt{3a+x} + 2\sqrt{x})}{(\sqrt{3a+x} - 2\sqrt{x})(\sqrt{3a+x} + 2\sqrt{x})(\sqrt{a+2x} + \sqrt{3x})}$$

$$= \lim_{x\to a} \frac{(a+2x-3x)(\sqrt{3a+x} + 2\sqrt{x})}{(3a+x-4x)(\sqrt{a+2x} + \sqrt{3x})}$$

$$= \lim_{x\to a} \frac{(a-x)(\sqrt{3a+x} + 2\sqrt{x})}{3(a-x)(\sqrt{a+2x} + \sqrt{3x})}$$

$$= \lim_{x\to a} \frac{(\sqrt{3a+x} + 2\sqrt{x})}{3(\sqrt{a+2x} + \sqrt{3x})} = \frac{\sqrt{3a+a} + 2\sqrt{a}}{3(\sqrt{a+2a} + \sqrt{3a})}$$

$$= \frac{4\sqrt{a}}{3\times 2\sqrt{3a}} = \frac{2}{3\sqrt{3}}$$

NOTE : The given limit is of the form $\frac{0}{0}$. Hence limit of the function can also be find out by using L' Hospital's Rule.

2. $$f(x) = \int \frac{2\sin x - \sin 2x}{x^3} dx,\ x \neq 0$$

$$\therefore \quad f'(x) = \frac{2\sin x - \sin 2x}{x^3},\ x \neq 0$$

$$\therefore \quad \lim_{x\to 0} f'(x) = \lim_{x\to 0} \frac{2\sin x - \sin 2x}{x^3}$$

$$= \lim_{x\to 0} \frac{2\sin x(1-\cos x)(1+\cos x)}{x^3(1+\cos x)}$$

$$= \lim_{x\to 0} 2.\frac{\sin^3 x}{x^3}.\frac{1}{1+\cos x}$$

$$= 2\times(1)^3 \times \frac{1}{2} = 1$$

3. $\lim_{h\to 0}\frac{(a+h)^2\sin(a+h)-a^2\sin a}{h}$

$= \lim_{h\to 0}\frac{a^2[\sin(a+h)-\sin a]+2ah\sin(a+h)+h^2\sin(a+h)}{h}$

$= \lim_{h\to 0}\frac{a^2\left[2\cos\left(a+\frac{h}{2}\right)\sin\frac{h}{2}\right]}{2\times\frac{h}{2}} + 2a\sin(a+h)$

$+ h\sin(a+h)$

$= a^2\cos a + 2a\sin a$

4. As f(x) is continuous at x = 0, we have

LHL = RHL = f(0)

$\Rightarrow \lim_{h\to 0} f(0-h) = \lim_{h\to 0} f(0+h) = f(0)$

$\Rightarrow f(0)+\lim_{h\to 0} f(-h) = f(0)+\lim_{h\to 0} f(h) = f(0)$

[Using the given property $f(x+y)=f(x)+f(y)$]

$\Rightarrow \lim_{h\to 0} f(-h) = \lim_{h\to 0} f(h) = 0$ ... (1)

Now let x = a be any arbitrary point then at x = a,

LHL = $\lim_{h\to 0} f(a-h) = \lim_{h\to 0}[f(a)+f(-h)]$

[Using, $f(x+y)=f(x)+f(y)$]

$= f(a)+\lim_{h\to 0} f(-h) = f(a)$ [using eq^n (1)]

Similarly, R.H.L. = $\lim_{h\to a} f(a+h) = f(a)$

Thus, we get

$\lim_{h\to 0} f(a-h) = \lim_{h\to 0} f(a+h) = f(a)$

$\Rightarrow$ f is continuous at x = a. But a is any arbitrary point

$\therefore$ f is continuous $\forall x \in R$.

5. $\lim_{x\to 0}\frac{2^x-1}{\sqrt{1+x}-1} = \lim_{x\to 0}\frac{2^x-1}{\sqrt{1+x}-1}\times\frac{\sqrt{1+x}+1}{\sqrt{1+x}+1}$

$= \lim_{x\to 0}\frac{(2^x-1)(\sqrt{1+x}+1)}{1+x-1}$

$= \lim_{x\to 0}\frac{2^x-1}{x}.\lim_{x\to 0}(\sqrt{1+x}+1)$

$= \ln 2.(1+1) = 2\ln 2.$

6. Given function is,

$$f(x)=\begin{cases}\dfrac{\sin(a+1)x+\sin x}{x}, & x<0\\ c, & x=0\\ \dfrac{\sqrt{x+bx^2}-\sqrt{x}}{bx\sqrt{x}}, & x>0\end{cases}$$

Given that it is continuous at $x = 0$, then

$\lim_{x\to 0^-} f(x) = \lim_{x\to 0^+} f(x) = f(0)$

Consider $\lim_{x\to 0^-} f(x) = c$

$\Rightarrow \lim_{h\to 0} f(0-h) = c$

$\Rightarrow \lim_{h\to 0}\frac{\sin(a+1)(-h)+\sin(-h)}{-h} = c$

$\Rightarrow \lim_{h\to 0}\frac{-[\sin(a+1)h+\sin h]}{-h} = c$

$\Rightarrow \lim_{h\to 0}\frac{2\sin\left(\frac{ah+h+h}{2}\right)\cos\left(\frac{ah+h-h}{2}\right)}{h} = c$

$\Rightarrow \lim_{h\to 0}\frac{2\sin\left(\frac{a}{2}+1\right)h}{\left(\frac{a}{2}+1\right)h}\times\left(\frac{a}{2}+1\right)\cos\left(\frac{ah}{2}\right) = c$

$\Rightarrow 2.1\left(\frac{a}{2}+1\right).1 = c$

$\Rightarrow a+2 = c$ ... (1)

Consider $\lim_{x\to 0^+} f(x) = f(0)$

$\Rightarrow \lim_{h\to 0} f(0+h) = c$

$\Rightarrow \lim_{h\to 0}\frac{\sqrt{h+bh^2}-\sqrt{h}}{bh\sqrt{h}} = c$

$\Rightarrow \lim_{h\to 0}\frac{(\sqrt{1+bh}-1)\sqrt{h}}{bh\sqrt{h}} = c$

$\Rightarrow \lim_{h\to 0}\frac{(\sqrt{1+bh}-1)(\sqrt{1+bh}+1)}{bh(\sqrt{1+bh}+1)} = c$

$\Rightarrow \lim_{h\to 0}\frac{1+bh-1}{bh(\sqrt{1+bh}+1)} = c$

$\Rightarrow \lim_{h\to 0}\frac{1}{\sqrt{1+bh}+1} = c$ [Provided $b \neq 0$]

$\Rightarrow \frac{1}{2} = c$ ... (2)

From equation (1)

$\Rightarrow a = c-2 = \frac{1}{2}-2 = \frac{-3}{2}$

Thus we get $a = \frac{-3}{2}$, $c = \frac{1}{2}$, b is any real number except 0.

7. $f(x)=\begin{cases}1+x, & 0\le x\le 2\\ 3-x, & 2<x\le 3\end{cases}$

$f(f(x))=\begin{cases}1+f(x), & 0\le f(x)\le 2\\ 3-f(x), & 2<f(x)\le 3\end{cases}$

Now, $0\le x\le 2 \Rightarrow 1\le f(x)\le 3$

$2<x\le 3 \Rightarrow 0\le f(x)<1$

$\Rightarrow\ 0\le x\le 1 \Rightarrow 1\le f(x)\le 2$

$1<x\le 2 \Rightarrow 2<f(x)\le 3$

$2<x\le 3 \Rightarrow 0\le f(x)<1$

$g(x)=\begin{cases}2+x, & 0\le x\le 1\\ 2-x, & 1<x\le 2\\ 4-x, & 2<x\le 3\end{cases}$

At $x=1$

RHL $=\lim_{h\to 0} g(1+h)=\lim_{h\to 0} 2-(1+h)=1$

$g(1)=3$

$\therefore$ Discontinuous at $x=1$

At $x=2$

RHL $=\lim_{h\to 0} g(2+h)=\lim_{h\to 0} 4-(2+h)=2$

$\therefore$ Discontinuous at $x=2$.

ALTERNATE SOLUTION

Graph of $f(f(x))$ is

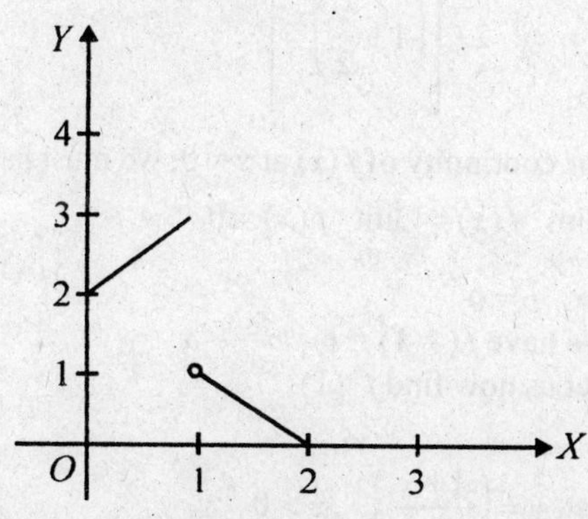

Clearly form graph $f(f(x))$ is discontinuous at $x=1$ and 2.

8. We have $f(x)=\dfrac{x^2}{2}, 0\le x<1$

$=2x^2-3x+\dfrac{3}{2}, 1\le x\le 2$

Here $f(x)$ is continuous everywhere except possibly at $x=1$

$\Rightarrow$ At $x=1$,

$Lf'=\dfrac{2}{2}\times 1=1;\ Rf'=4\times 1-3=1$

$\Rightarrow$ f is differentiable and hence continuous at $x=1$

$\therefore$ $f(x)$ is continuous on $[0, 2]$

$f'(x)=x,\ 0\le x<1$

$=4x-3,\ 1\le x\le 2$

At $x=1$,

$\lim_{x\to 1^-} f'(x)=\lim_{h\to 0} f'(1-h)=\lim_{h\to 0}(1-h)=1$

$\lim_{x\to 1^+} f'(x)=\lim_{h\to 0} f'(1+h)=\lim_{h\to 0} 4(1+h)-3=1$

$f'(1)=4-3=1$

$\therefore$ f' is continuous at $x=1$

$\therefore$ f' is continuous on $[0, 2]$

$f''(x)=\begin{cases}2, & 0\le x<1\\ 4, & 1\le x\le 2\end{cases}$

Clearly $f''(x)$ is discontinuous at $x=1$,

$\therefore$ $f''(x)$ is discontinuous on $[0, 2]$.

9. Given $f(x)=x^3-x^2+x+1$

$\therefore\ f'(x)=3x^2-2x+1=3\left(x^2-\dfrac{2}{3}x+\dfrac{1}{3}\right)$

$=3\left[\left(x-\dfrac{1}{3}\right)^2-\dfrac{1}{9}+\dfrac{1}{3}\right]$

$=3\left[\left(x-\dfrac{1}{3}\right)^2+\dfrac{2}{9}\right]>0\ \forall\, x\in R.$

Hence $f(x)$ is an increasing function of x for all real values of x.

Now max $[f(t): 0\le t\le x]$ means the greatest value of $f(t)$ in $0\le t\le x$ which is obtained at $t=x$, since $f(t)$ is increasing for all t.

$\therefore$ max. $[f(t): 0\le t\le x]=x^3-x^2+x+1$

Hence the function g is defined as follows :

$g(x)=x^3-x^2+x+1$ when $0\le x\le 1$

$=3-x$ when $1<x\le 2$

Now it is sufficient to discuss the continuity and differentiability of $g(x)$ at $x=1$. Since for all other values of x, $g(x)$ is clearly continuous and differentiable, being a polynomial function of x.

We have, $g(1)=2$

$g(1-0)=\lim_{h\to 0}[(1-h)^3-(1-h)^2+(1-h)+1]=2$

$g(1+0)=\lim_{h\to 0}[3-(1+h)]=2$

Hence $g(x)$ is continuous at $x=1$

Now,

$Lg'(1)=\lim_{h\to 0}\dfrac{[(1-h)^3-(1-h)^2+(1-h)+1]-2}{-h}$

$=\lim_{h\to 0}\dfrac{1-3h+3h^2-h^3-1+2h-h^2+1-h+1-2}{-h}$

$=\lim_{h\to 0}\dfrac{-2h+2h^2-h^3}{-h}=\lim_{h\to 0}[2-2h+h^2]=2$

$Rg'(1)=\lim_{h\to 0}\dfrac{[3-(1+h)-2]}{h}=\lim_{h\to 0}\dfrac{-h}{h}=-1$

Since $Lg'(1)\ne Rg'(1)$, the function $g(x)$ is not differentiable at $x=1$

Hence $g(x)$ is continuous on $(0, 2)$. It is also differentiable on $(0, 2)$ except at $x=1$.

10. We have $f(x) = -1,\ -2 \le x \le 0$
$= x - 1,\ 0 < x \le 2$
and $g(x) = f(|x| + |f(x)|$
Hence $g(x)$ involves $|x|$ and $|x-1|$ or $|-1| = 1$
Therefore we should divide the given interval $(-2, 2)$ into the following intervals.

I_1	I_2	I_3
$[-2, 2] = [-2, 0)$	$[0, 1)$	$[1, 2]$
$x = -ve$	$+ ve$	$+ ve$
$\|x\| = -x$	x	x
$f(x) = -1$	$x - 1$	$x - 1$
$f(\|x\|) = -1$	$= x - 1$	$= x - 1$
$\|f(x)\| = \|-1\| = 1$	$\|x - 1\| = -(x-1)$	$\|x - 1\| = x - 1$

$\therefore$ Using above we get
$g(x) = f|x| + |f(x)|$
$= -1 + 1 = 0$ in I_1
$= x - 1 - (x-1) = 0$ in I_2
$= x - 1 + x - 1 = 2(x-1)$ in I_3
Hence g (x) is defined as follows :

$$g(x) = \begin{cases} 0, & -2 \le x < 1 \\ 2(x-1), & 1 \le x \le 2 \end{cases}$$

$Lg'(1) = 0;\ Rg'(1) = 2$ (not equal)
Hence $g(x)$ is not differentiable at $x = 1$.

11. Given that $f(x)$ is a continuous function, and $g(x)$ is a discontinuous function, then for some arbitrary real number a, we must have

$$\lim_{x \to a} f(x) = f(a) \quad ...(1)$$

and $$\lim_{x \to a} g(x) \ne g(a) \quad ...(2)$$

Now, $\lim_{x \to a} [f(x) + g(x)]$

$$= \lim_{x \to a} f(x) + \lim_{x \to a} g(x)$$

$$\ne f(a) + g(a) \quad \text{[Using (1) and (2)]}$$

$\Rightarrow f(x) + g(x)$ is discontinuous.

ALTERNATE SOLUTION

Let $h(x) = f(x) + g(x)$ be continuous.
Then, $g(x) = h(x) - f(x)$
Now, $h(x)$ and $f(x)$ both are continuous functions.
$\therefore\ h(x) - f(x)$ must also be continuous. But it is a contradiction as given that $g(x)$ is discontinuous. Therefore our assumption that $f(x) + g(x)$ a continous function is wrong and hence $f(x) + g(x)$ is discontinuous.

12. Given that $f(x)$ is a function satisfying

$$f(-x) = f(x), \forall x \in R \quad ...(1)$$

Also $f'(0)$ exists
$\Rightarrow f'(0) = Rf'(0) = Lf'(0)$
Now, $Rf'(0) = f'(0)$

$$\Rightarrow \lim_{h \to 0} \frac{f(0+h) - f(0)}{h} = f'(0)$$

$$\Rightarrow \lim_{h \to 0} \frac{f(h) - f(0)}{h} = f'(0)$$

Again $Lf'(0) = f'(0)$ ...(2)

$$\Rightarrow \lim_{h \to 0} \frac{f(0-h) - f(0)}{-h} = f'(0)$$

$$\Rightarrow \lim_{h \to 0} \frac{f(-h) - f(0)}{-h} = f'(0)$$

$$\Rightarrow \lim_{h \to 0} \frac{f(h) - f(0)}{h} = f'(0) \quad ...(3)$$

[Using eq. (1)]

From equations (2) and (3), we get
$\Rightarrow f'(0) = 0$

13. Let $g(x) = ax + b$

$$\text{Then } f(x) = \begin{cases} ax + b, & x \le 0 \\ \left(\dfrac{1+x}{2+x}\right)^{1/x}, & x > 0 \end{cases}$$

Since $f(x)$ is continuous, it is continuous at $x = 0$

Now, $\lim_{x \to 0^-} f(x) = \lim_{x \to 0^-} (ax + b) = b$

and $\lim_{x \to 0^+} f(x) = \lim_{x \to 0^+} \left(\dfrac{1+x}{2+x}\right)^{\frac{1}{x}}$

$$= \lim_{x \to 0^+} \frac{(1+x)^{\frac{1}{x}}}{2^{\frac{1}{x}}\left[\left(1 + \frac{x}{2}\right)^{\frac{2}{x}}\right]^{\frac{1}{2}}} = \frac{e}{e^{\frac{1}{2}}} \cdot \lim_{x \to 0^+} \frac{1}{2^{\frac{1}{x}}} = 0$$

For continuity of $f(x)$ at $x = 0$, we must have

$$\lim_{x \to 0^-} f(x) = \lim_{x \to 0^+} f(x) = 0$$

$\Rightarrow b = 0$
We have $f(-1) = b - a = -a$ $(\because b = 0)$
Let us now find $f'(1)$

$$\text{Let } y = \left(\frac{1+x}{2+x}\right)^{\frac{1}{2}}, x > 0$$

$$\text{Then, } \log y = \frac{1}{x}[\log(1+x) - \log(2+x)]$$

$$\Rightarrow \frac{1}{y}\frac{dy}{dx} = -\frac{1}{x^2}[\log(1+x) - \log(2+x)] + \frac{1}{x}\left[\frac{1}{x+1} - \frac{1}{2+x}\right]$$

$$\Rightarrow \frac{dy}{dx} = \left(\frac{1+x}{2-x}\right)^{\frac{1}{x}}\left[-\frac{1}{x^2}\{\log(1+x) - \log(2+x)\} + \frac{1}{x}\left\{\frac{1}{1+x} - \frac{1}{2+x}\right\}\right]$$

Now, $f'(1) = \left.\frac{dy}{dx}\right|_{x=1} = \left(\frac{1+1}{2+1}\right)^{1/1}$

$$\left[-(\log 2 - \log 3) + \left(\frac{1}{2} - \frac{1}{3}\right)\right]$$

$$= \frac{2}{3}\left[\log 3 - \log 2 + \frac{1}{6}\right] = \frac{2}{3}\log\frac{3}{2} + \frac{1}{9}$$

Since, $f'(1) = f(-1) \quad \Rightarrow \frac{2}{3}\log\frac{3}{2} + \frac{1}{9} = -a$

$\therefore$ The required function is

$$f(x) = \begin{cases} \left(-\frac{2}{3}\log\frac{3}{2} + \frac{1}{9}\right)x, & x \le 0 \\ \left[\frac{1+x}{2+x}\right]^{\frac{1}{x}}, & x > 0 \end{cases}$$

14. Given that,

$$f(x) = \begin{cases} x + a\sqrt{2}\sin x & , 0 \le x < \frac{\pi}{4} \\ 2x\cot x + b & , \frac{\pi}{4} \le x \le \frac{\pi}{2} \\ a\cos 2x - b\sin x & , \frac{\pi}{2} < x \le \pi \end{cases}$$

is continuous for $0 \le x \le \pi$.

$\therefore$ $f(x)$ must be continuous at $x = \frac{\pi}{4}$ and $x = \frac{\pi}{2}$

$$\lim_{x \to \left(\frac{\pi}{4}\right)^-} f(x) = f\left(\frac{\pi}{4}\right)$$

$$\Rightarrow \lim_{h \to 0} f\left(\frac{\pi}{4} - h\right) = \frac{2\pi}{4}\cot\frac{\pi}{4} + b$$

$$\Rightarrow \lim_{h \to 0}\left(\frac{\pi}{4} - h\right) + a\sqrt{2}\sin\left(\frac{\pi}{4} - h\right) = \frac{\pi}{2} + b$$

$$\Rightarrow \frac{\pi}{4} + a = \frac{\pi}{2} + b$$

$$\Rightarrow a - b = \frac{\pi}{4} \qquad ...(1)$$

Also, $\lim_{x \to \left(\frac{\pi}{2}\right)^+} f(x) = f\left(\frac{\pi}{2}\right)$

$$\Rightarrow \lim_{h \to 0} f\left(\frac{\pi}{2} + h\right) = 2.\frac{\pi}{2}\cot\frac{\pi}{2} + b$$

$$\Rightarrow \lim_{h \to 0} a\cos 2\left(\frac{\pi}{2} + h\right) - b\sin\left(\frac{\pi}{2} + h\right) = b$$

$$\Rightarrow a\cos\pi - b\sin\frac{\pi}{2} = b$$

$$\Rightarrow -a - b = b$$

$$\Rightarrow a + 2b = 0 \qquad ...(2)$$

Solving (1) and (2), we get

$$a = \frac{\pi}{6} \text{ and } b = \frac{-\pi}{12}.$$

15. We have, $[x] + |1 - x|, -1 \le x \le 3$

NOTE THIS STEP

$$\text{or } y = \begin{cases} -1 + 1 - x & , -1 \le x < 0 \\ 0 + 1 - x & , 0 \le x < 1 \\ 1 - 1 + x & , 1 \le x < 2 \\ 2 - 1 + x & , 2 \le x < 3 \\ 3 - 1 + x & , x = 3 \end{cases}$$

$$\text{or } y = \begin{cases} -x, & -1 \le x < 0 \\ 1 - x, & 0 \le x < 1 \\ x, & 1 \le x < 2 \\ 1 + x, & 2 \le x < 3 \\ 2 + x, & x = 3 \end{cases}$$

From graph we can say that given functions is not differentiable at $x = 0, 1, 2, 3$.

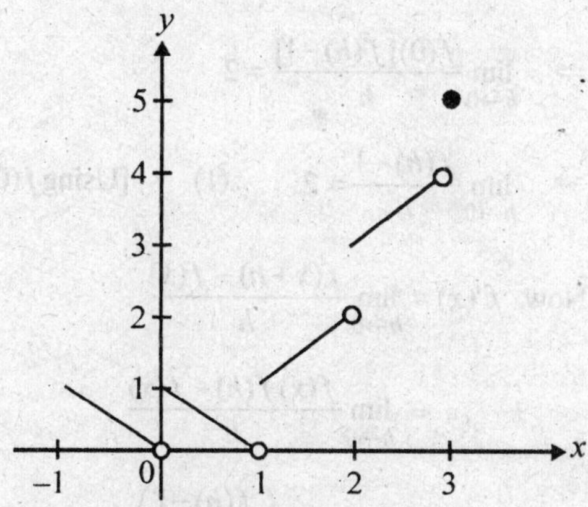

16. We are given that,

$$f(x) = \begin{cases} \frac{1 - \cos 4x}{x^2} & , x < 0 \\ a & , x = 0 \\ \frac{\sqrt{x}}{\sqrt{16 + \sqrt{x}} - 4} & , x > 0 \end{cases}$$

Here L.H.L at $(x = 0)$

$$= \lim_{h \to 0} \frac{1 - \cos 4(0 - h)}{(0 - h)^2} = \lim_{h \to 0} \frac{1 - \cos 4h}{h^2}$$

$$= \lim_{h \to 0} \frac{2\sin^2 2h}{4h^2}.4 = 8$$

R.H.L at $(x=0)$

$$= \lim_{h\to 0} \frac{\sqrt{0+h}}{\sqrt{16+\sqrt{0+h}}-4} = \lim_{h\to 0} \frac{\sqrt{h}(\sqrt{16+h}+4)}{16+\sqrt{h}-16}$$

$$= \lim_{h\to 0} \sqrt{16+\sqrt{h}}+4 = \sqrt{16}+4 = 8$$

For continuity of function $f(x)$, we must have

L.H.L. = R.H.L = f(0)

$\Rightarrow f(0)=8 \Rightarrow a=8$

17. We are given

$f(x+y)=f(x)\,f(y),\ \forall\ x, y \in R$

$f(x) \neq 0$, for any x

f is differentiable at $x=0$

$f'(0)=2$

To prove that $f'(x)=2f(x),\ \forall\ x \in R$ and to find $f(x)$.

We have for $x=y=0$

$f(0+0)=f(0)f(0)$

$\Rightarrow f(0)=[f(0)]^2$

$\Rightarrow f(0)=1$

Again $f'(0) \neq 2$

$$\Rightarrow \lim_{h\to 0} \frac{f(0+h)-f(0)}{h}=2$$

$$\Rightarrow \lim_{h\to 0} \frac{f(0)\,f(h)-f(0)}{h}=2$$

$$\Rightarrow \lim_{h\to 0} \frac{f(0)[f(h)-1]}{h}=2$$

$$\Rightarrow \lim_{h\to 0} \frac{f(h)-1}{h}=2 \quad ...(1) \quad [\text{Using } f(0)=1]$$

Now, $f'(x)=\lim_{h\to 0} \frac{f(x+h)-f(x)}{h}$

$$= \lim_{h\to 0} \frac{f(x)\,f(h)-f(x)}{h}$$

$$= \lim_{h\to 0} f(x)\left(\frac{f(h)-1}{h}\right)$$

$$= f(x) \lim_{h\to 0}\left[\frac{f(h)-1}{h}\right]$$

$= f(x).2$ [Using eq. (1)]

$= 2f(x)$

Also, $\frac{f'(x)}{f(x)}=2$

Integrating on both sides with respect to x, we get

$\log|f(x)| = 2x + C$

At $x=0$, $\log f(0)=C$

$\Rightarrow C=\log 1=0$

$\therefore \log|f(x)|=2x$

$\Rightarrow f(x)=e^{2x}$

18. $$\lim_{x\to 0}\left\{\tan\left(\frac{\pi}{4}\right)+x\right\}^{\frac{1}{x}} = e^{\lim_{x\to 0}\log\left\{\tan\left(\frac{\pi}{4}+x\right)\right\}^{\frac{1}{x}}}$$

$$\left[\text{Using } \lim_{x\to a} f(x) = e^{\lim_{x\to 0}\log f(x)}\right]$$

$$= e^{\lim_{x\to 0}\frac{\log\tan\left(\frac{\pi}{4}+x\right)}{x}} \quad \left[\frac{0}{0}\text{ form}\right]$$

$$= e^{\lim_{x\to 0}\left[\frac{\sec^2\left(\frac{\pi}{4}+x\right)}{\tan\left(\frac{\pi}{4}\right)+x}\right]} \quad [\text{Using } L'\text{ Hospital's rule}]$$

$$= e^{\frac{2}{1}} = e^2$$

19. Given that,

$$f(x)=\begin{cases}(1+|\sin x|)^{\frac{a}{|\sin x|}} & \frac{-\pi}{6}<x<0 \\ b & x=0 \\ e^{\frac{\tan 2x}{\tan 3x}} & 0<x<\frac{\pi}{6}\end{cases}$$

is continuous at $x=0$

$$\therefore \lim_{h\to 0} f(0-h)=f(0)=\lim_{h\to 0} f(0+h)$$

We have,

$$\lim_{h\to 0} f(0-h)=\lim_{h\to 0}[1+|\sin(-h)|]^{\frac{a}{|\sin(-h)|}}$$

$$= \lim_{h\to 0}[1+\sin h]^{\frac{a}{\sin h}}$$

$$e^{\lim_{h\to 0}\frac{a}{\sin h}\log(1+\sin h)} = e^a$$

and $f(0)=b$

$\therefore e^a=b$ ...(1)

Also $$\lim_{h\to 0} f(0+h)=\lim_{h\to 0} e^{\frac{\tan 2h}{\tan 3h}}$$

$$= e^{\lim_{h\to 0}\frac{\tan 2h}{2h}\times\frac{3h}{\tan 3h}\times\frac{2}{3}}$$

$$= e^{\frac{2}{3}}$$

$\therefore e^{\frac{2}{3}}=b$ ...(2)

From (1) and (2)

$$e^a=b=e^{\frac{2}{3}}$$

$$\Rightarrow a=\frac{2}{3} \text{ and } b=e^{\frac{2}{3}}$$

20. $f\left(\frac{x+y}{2}\right)=\frac{f(x)+f(y)}{2}$ ...(1)

Putting $y=0$ and $f(0)=1$ in (1), we get

$$f\left(\frac{x}{2}\right)=\frac{1}{2}[f(x)+1]$$

$$\therefore \quad f(x)=2f\left(\frac{x}{2}\right)-1 \quad ...(2)$$

Now, $f'(x)=\underset{h\to 0}{Lt}\frac{f(x+h)-f(x)}{h}$

$$=\underset{h\to 0}{Lt}\frac{1}{h}\left[\frac{f(2x)+f(2h)}{2}-f(x)\right], \text{ by (1)}$$

$$=\underset{h\to 0}{Lt}\frac{1}{h}\left[\frac{(2f(x)-1)+(2f(h)-1)}{2}-f(x)\right], \text{ by (2)}$$

$$=\underset{h\to 0}{Lt}\frac{1}{h}[f(h)-1]$$

$$=\underset{h\to 0}{Lt}\frac{f(h)-f(0)}{h}=f'(0)=-1$$

Hence $f'(x)=-1$, integrating, we get $f(x)=-x+c$. Putting $x=0$, we get $f(0)=c=1$ by (1) $\therefore f(x)=1-x$

$f(2)=1-2=-1$

21. Given that

$$f(x)=\begin{cases} xe^{-\left[\frac{1}{|x|}+\frac{1}{x}\right]}, & x\neq 0\\ 0, & x=0\end{cases}$$

$$=\begin{cases} xe^{-\left[-\frac{1}{x}+\frac{1}{x}\right]} & x<0\\ 0 & x=0\\ xe^{-\left[\frac{1}{x}+\frac{1}{x}\right]} & x>0\end{cases}$$

$$\Rightarrow f(x)=\begin{cases} x, & \text{if } x<0\\ 0, & \text{if } x=0\\ xe^{\frac{-2}{x}}, & \text{if } x>0\end{cases}$$

Continuity at $x=0$

LHL at $(x=0)=\lim_{h\to 0}f(0-h)=\lim_{h\to 0}f(-h)$

$$=\lim_{h\to 0}(-h)=0$$

RHL at $(x=0)=\lim_{h\to 0}f(0+h)=\lim_{h\to 0}f(h)$

$$=\lim_{h\to 0}he^{\frac{-2}{h}}=0 \quad \left[\because \lim_{x\to\infty}e^{-x}=0\right]$$

Also $f(0)=0$.

Thus LHL = RHL = $f(0)$

$\Rightarrow$ $f(x)$ is continuous at $x=0$

Differentiability at $x=0$

$$Lf'(0)=\lim_{h\to 0}\frac{f(0)-f(0-h)}{h}=\lim_{h\to 0}\frac{0-(-h)}{h}$$

$$=\lim_{h\to 0}1=1$$

$$Rf'(0)=\lim_{h\to 0}\frac{f(0+h)-f(0)}{h}=\lim_{h\to 0}\frac{he^{\frac{-2}{h}}-0}{h}$$

$$=\lim_{h\to 0}e^{\frac{-2}{h}}=0$$

Thus $Lf'(0)\neq Rf'(0)$

$\Rightarrow f$ is not differentiable at $x=0$

22. By the given definition it is clear that the function f is continuous and differentiable at all points except possible at $x=1$ and $x=2$.

Continuity at $x=1$

L.H.L. $=\lim_{h\to 0}[1-(1-h)]=\lim_{h\to 0}h=0$

R.H.L. $=\lim_{h\to 0}[1-(1+h)][2-(1+h)]$

$$=\lim_{h\to 0}\{-h(1-h)\}=0$$

Also, $f(1)=0$

$\therefore$ L.H.L. = R.H.L. $=f(1)=0$

Therefore, f is continuous at $x=1$

Now, differentiability at $x=1$

$$Lf'(1)=\lim_{h\to 0}\frac{f(1-h)-f(1)}{-h}, h>0$$

$$=\lim_{h\to 0}\frac{(1-(1-h)-0}{-h}=\lim_{h\to 0}\left(\frac{h}{-h}\right)=-1$$

and $Rf'(1)=\lim_{h\to 0}\frac{f((1+h)-f(1)}{h}$

$$=\lim_{h\to 0}\frac{\{1-(1-h)\}\{2-(1-h)\}-0}{h}$$

$$=\lim_{h\to 0}\frac{-h(1-h)}{h}=\lim_{h\to 0}(h-1)=-1$$

Since $Lf'(1)=Rf'(1)$

Hence, f is differentiable at $x=1$

Continuous at $x=2$

L.H.L. $=\lim_{h\to 0}[1-(2-h)][2-(2-h)]$

$$\lim_{h\to 0}\{(-1+h)\}\{h\}=0$$

and R.H.L. $=\lim_{h\to 0}[3-(2+h)]=\lim_{h\to 0}(1-h)=1$

Since L.H.L. $\neq$ R.H.L., therefore f is not continuous at $x=2$. As such f cannot be differentiable at $x=2$. Hence f is continuous and differentiable at all points except at $x=2$.

23. Given that, $F(x)=\int_0^x f(t)dt$

NOTE THIS STEP

$\therefore \quad F'(x)=f(x).1-f(0).0$ [Using Leibnitz theorem]

$\Rightarrow \; F'(x)=f(x)$ ... (1), $\forall\, x\geq 0$

Also $F(0)=\int_0^0 f(t)dt=0$

But given that $f(x)\leq cF(x),\ \forall x\geq 0$

$\therefore$ We get $f(0)\leq cF(0)=0$

$\therefore \quad f(0)\leq 0$ ....(2)

But ATQ $f(x)$ is non- negative continuous function on $[0,\infty)$

$\therefore \quad f(x)\geq 0$

$\therefore \quad f(0)\geq 0$ ....(3)

$\therefore$ From (2) and (3) $f(0)=0$

Again $f(x)\leq cF(x)\forall\, x\geq 0$, we get

$$f(x)-cF(x)\leq 0$$

$\Rightarrow \; F'(x)-cF(x)\leq 0, \forall\, x\geq 0$ [Using equation (1)]

$$e^{cx}F'(x)-ce^{-cx}F(x)\leq 0$$

[Multiplying both sides by e^{-cx} (I.F.) and keeping in mind that $e^{-cx}>0, \forall\, x$]

$\Rightarrow \; \frac{d}{dx}[e^{-cx}F(x)]\leq 0$

$\Rightarrow \; g(x)=e^{-cx}F(x)$ is a decreasing function on $[0,\infty)$.

That is $g(x)\leq g(0)$ for all $x\geq 0$

But $g(0)=F(0)=0$

$\therefore \quad g(x)\leq 0,\ \forall\, x\geq 0$

$\Rightarrow \; e^{-cx}F(x)\leq 0,\ \forall\, x\geq 0$

$\Rightarrow \; F(x)\leq 0,\ \forall\, x\geq 0$

$\therefore \quad f(x)\leq cF(x)\leq 0,\ \forall\, x\geq 0$

[$\therefore c>0$ and using $f(x)\leq cF(x)$]

$\Rightarrow \; f(x)\leq 0, \forall\, x\geq 0$

But given $f(x)\geq 0$

$\Rightarrow \; f(x)=0,\ \forall\, x\geq 0$.

24. (I) g is continuous at α and

$f(x)-f(\alpha)=g(x)(x-\alpha), \forall\, x\in R$

$\Rightarrow$ Since g is continuous at $x=\alpha$

and $g(x)=\dfrac{f(x)-f(\alpha)}{x-\alpha}$

We should have, $\lim\limits_{x\to\alpha} g(x)=g(\alpha)$

$\Rightarrow \; \lim\limits_{x\to\alpha}\dfrac{f(x)-f(\alpha)}{x-\alpha}=g(\alpha)\Rightarrow f'(\alpha)=g(\alpha)$

$\Rightarrow$ $f'(\alpha)$ exists and is equal to $g(\alpha)$.

(II) $\because$ $f(x)$ is differentiable at $x=\alpha$

$\therefore \quad \lim\limits_{x\to\alpha}\dfrac{f(x)-f(\alpha)}{x-\alpha}=f'(\alpha)$

exists and is finite.

Let us define,

$$g(x)=\begin{cases}\dfrac{f(x)-f(\alpha)}{x-\alpha}, & x\neq\alpha\\ f'(\alpha), & x=\alpha\end{cases}$$

Then, $f(x)-f(\alpha)=(x-\alpha)\,g(x),\ \forall\ x\neq\alpha$.

Now for continuity of $g(x)$ at $x=\alpha$

$$\lim_{x\to\alpha} g(x)=\lim_{x\to\alpha}\frac{f(x)-f(\alpha)}{x-\alpha}=f'(\alpha)=g(\alpha)$$

$\therefore \quad g$ is continuous at $x=\alpha$.

25. Given that

$$f(x)=\begin{cases}x+a, & \text{if } x<0\\ |x-1|, & \text{if } x\geq 0\end{cases}=\begin{cases}x+a, & \text{if } x<0\\ 1-x, & \text{if } 0\leq x<1\\ x-1, & \text{if } x\geq 1\end{cases}$$

and $g(x)=\begin{cases}(x+1), & \text{if } x<0\\ (x-1)^2+b, & \text{if } x\geq 0\end{cases}$

where $a,b\geq 0$

Then $(gof)(x)=g[f(x)]$

NOTE THIS STEP

$$=\begin{cases}f(x)+1, & \text{if } f(x)<0\\ [f(x)-1]^2+b, & \text{if } f(x)\geq 0\end{cases}$$

(Using definition of $g(x)$

Now, $f(x)<0$ when $x+a<0$ i.e. $x<-a$

$f(x)=0$ when $x=-a$ or $x=1$

$f(x)>0$ when $-a<x<1$ or $x>1$

$$g(f(x))=\begin{cases}f(x)+1, & \text{if } x<-a\\ [f(x)-1]^2+b, & \text{if } x=-a \text{ or } x=1\\ [f(x)-1]^2+b, & \text{if } -a<x<0\\ [f(x)-1]^2+b, & \text{if } 0\leq x<1\\ [f(x)-1]^2+b, & \text{if } x>1\end{cases}$$

[Keeping in mind that $x=0$ and 1 are also the breaking points because of definition of $f(x)$]

$$\therefore g[f(x)]=\begin{cases}x+a+1, & \text{if } x<-a\\ (x+a-1)^2+b, & \text{if } -a\leq x<0\\ (1+x)-1)^2+b, & \text{if } 0\leq x\leq 1\\ (x-1-1)^2+b, & \text{if } x>1\end{cases}$$

Substituting the value of $f(x)$ under different conditions).

$$\therefore \quad g[f(x)]=\begin{cases} x+a+1, & \text{if } x<-a \\ (x+a-1)^2+b, & \text{if } -a\le x<0=F(x)(\text{say}) \\ x^2+b, & \text{if } 0\le x\le 1 \\ (x-2)^2+b, & \text{if } x>1 \end{cases}$$

Now given that $gof(x)\equiv F(x)$ is continuous for all real numbers, therefore it will be continuous at $-a$

$\Rightarrow$ L.H.S = R.H.L = $f(-a)$

$$\lim_{h\to 0} F(-a-h)=\lim_{h\to 0} F(-a+h)=f(-a)$$

Now, $\lim_{h\to 0} F(-a-h)=\lim_{h\to 0}(-a-h+a+1)=1$

$$\lim_{h\to 0} F(-a+h)=\lim_{h\to 0}(-a+h+a-1)^2+b=1+b$$

$F(-a)=1+b$

Thus we should have $1=1+b \Rightarrow b=0$.

Again for continuity at $x=0$

L.H.L = $f(0)$

$$\Rightarrow \lim_{h\to 0} f(0-h)=f(0)$$

$$\Rightarrow \lim_{h\to 0} f(-h+a-1)^2+b=b$$

$\Rightarrow (a-1)^2=0 \Rightarrow a=1$

For $a=1$ and $b=0$, gof becomes

$$gof(x)=\begin{cases} x+2, & x<-1 \\ x^2, & -1\le x\le 1 \\ (x-2)^2 & x>1 \end{cases}$$

Now to check differentiability of $gof(x)$ at $x=0$

We see, $gof(x)=x^2=F(x)$

$\Rightarrow F'(x)=2x$ which exists clearly at $x=0$.

gof is differentiable at $x=0$

26. Given that $f:[-2a,2a]\to R$

f is an odd function.

Lf' at $x=a$ is 0.

$$\Rightarrow \lim_{h\to 0}\frac{f(a-h)-f(a)}{-h}=0$$

$$\Rightarrow \lim_{h\to 0}\frac{f(a-h)-f(a)}{h}=0 \qquad \text{...(1)}$$

To find Lf' at $x=-a$ which is given by

$$\lim_{h\to 0}\frac{f(-a-h)-f(-a)}{-h}=\lim_{h\to 0}\frac{-f(a+h)+f(a)}{-h}$$

$$[\because f(-x)=-f(x)]$$

$$=\lim_{h\to 0}\frac{f(a+h)-f(a)}{h}$$

Again for $x\in[a,2a]$

$f(x)=f(2a-x)$

$\therefore \quad f(a+h)=f(2a-a-h)=f(a-h)$

Substiting this values in last expression we get

$$Lf'(-a)=\lim_{h\to 0}\frac{f(a-h)-f(a)}{h}$$

$$=0 \qquad \text{[Using equation (1)]}$$

Hence $Lf'(-a)=0$

27. To find,

$$\lim_{n\to\infty}\left[(n+1)\frac{2}{\pi}\cos^{-1}\left(\frac{1}{n}\right)-n\right]$$

$$=\lim_{n\to\infty} n\left[\left(1+\frac{1}{n}\right)\frac{2}{\pi}\cos^{-1}\left(\frac{1}{n}\right)-1\right]$$

$$=\lim_{n\to\infty} nf\left(\frac{1}{n}\right)$$

where $f(x)=\left[(1+x)\frac{2}{\pi}\cos^{-1}x-1\right]$ such that

$$f(0)=\left[(1+0)\frac{2}{\pi}\cos^{-1}0-1\right]$$

$$=\frac{2}{\pi}\cdot\frac{\pi}{2}-1=0$$

$\therefore$ Using given relation as $\lim_{n\to\infty} nf\left(\frac{1}{n}\right)=f'(0)$

then given limit becomes

$$=f'(0)=\frac{d}{dx}\left[(1+x)\frac{2}{\pi}\cos^{-1}x-1\right]\Bigg|_{x=0}$$

$$=\frac{2}{\pi}\left[\cos^{-1}x-\frac{1-x}{\sqrt{1-x^2}}\right]\Bigg|_{x=0}$$

$$=\frac{2}{\pi}\left[\frac{\pi}{2}-1\right]=1-\frac{2}{\pi}=\frac{\pi-2}{\pi}$$

28. Given that, $f(x)$ is differentiable at $x=0$.

Hence, $f(x)$ will also be continuous at $x=0$

$$\Rightarrow \lim_{h\to 0} f(0+h)=f(0) \Rightarrow \lim_{h\to 0}\frac{e^{\frac{ah}{2}}-1}{h}=\frac{1}{2}$$

$$\Rightarrow \lim_{h\to 0}\frac{e^{\frac{ah}{2}}-1}{\frac{ah}{2}}\times\frac{a}{2}=\frac{1}{2} \Rightarrow a=1$$

Also differentiability of $f(x)$ at $x=0$, gives

$Lf'(0)=Rf'(0)$

$$\Rightarrow \lim_{h\to 0}\frac{f(0-h)-f(0)}{-h}=\lim_{h\to 0}\frac{f(0+h)-f(0)}{h}$$

$$\Rightarrow \lim_{h\to 0}\frac{b\sin^{-1}\left(\frac{c-h}{2}\right)-\frac{1}{2}}{-h}=\lim_{h\to 0}\frac{e^{\frac{ah}{2}}-1-\frac{1}{2}}{h}$$

$$=\lim_{h\to 0}\frac{2e^{\frac{ah}{2}}-2-h}{2h^2}\quad \left[\text{form }\frac{0}{0}\right]$$

$$\lim_{h\to 0}\frac{\frac{b}{\sqrt{1-\left(\frac{c-h}{2}\right)^2}}\cdot\left(-\frac{1}{2}\right)}{-1}\quad \text{[Using L Hospital's rule]}$$

$$=\lim_{h\to 0}\frac{2.e^{\frac{ah}{2}}.\frac{a}{2}-1}{4h}=\lim_{h\to 0}\frac{e^{\frac{h}{2}}-1}{8\left(\frac{h}{2}\right)}\quad \text{[Putting } a=1]$$

$$\Rightarrow \frac{b}{\sqrt{1-\frac{c^2}{4}}}=\frac{1}{8}$$

$$\Rightarrow 4b=\sqrt{1-\frac{c^2}{4}}$$

$$\Rightarrow 16b^2=\frac{4-c^2}{4}$$

$\Rightarrow 64\,b^2=4-c^2$ Hence proved.

29. Given that,

$f(x-y)=f(x).\,g(y)-f(y)\,.\,g(x)$ ...(i)

$g(x-y)=g(x).\,g(y)+f(x)f(y)$ ...(ii)

In eqn. (i), putting $x=y$, we get

$f(0)=f(x)\,g(x)-f(x)\,g(x)$

$\Rightarrow f(0)=0$

Putting $y=0$, in eqn. (i), we get

$f(x)=f(x)\,g(0)-f(0)\,g(x)$

$\Rightarrow f(x)=f(x)\,g(0)$ [using $f(0)=0$]

$\Rightarrow g(0)=1$

Putting $x=y$ in eqn. (ii), we get

$g(0)=g(x)\,g(x)+f(x)f(x)$

$\Rightarrow 1=[g(x)]^2+[f(x)]^2$ [using $g(0)=1$]

$\Rightarrow [g(x)]^2=1-[f(x)]^2$ ...(iii)

Clearly $g(x)$ will be differentiable only if $f(x)$ is differentiable.

$\therefore$ First we will check the differentiability of $f(x)$

Given that $Rf'(0)$ exists

i.e., $\lim_{h\to 0}\frac{f(0+h)-f(0)}{h}$ exists

i.e., $\lim_{h\to 0}\frac{f(0)g(-h)-f(-h)g(0)}{h}$ exists

i.e., $\lim_{h\to 0}\frac{-f(-h)}{h}$ exists (using $f(0)=0$ and $g(0)=1$)

Which can be written as,

$$\lim_{h\to 0}\frac{f(0)-f(-h)}{-h}=Lf'(0)$$

$\Rightarrow Lf'(0)=Rf'(0)$

$\therefore$ f is differentiable, at $x=0$

Differentiating equation (iii), we get

$2g(x).\,g'(x)=-2f(x).f'(x)$

For $x=0$

$\Rightarrow g(0).\,g'(0)=-f(0)f'(0)$

$\Rightarrow g'(0)=0$ [Using $f(0)=0$ and $g(0)=1$]

F MATCH THE FOLLOWING :

1. (A) $\sin(\pi[x])=0,\ \forall\ x\in R$

$\therefore$ Differentiable everywhere.

$\therefore$ (A) $\rightarrow$ (p)

(B) $\sin(\pi(x-[x]))=f(x)$

We know that

$$x-[x]\begin{cases}=x, \text{if } 0\le x<1\\ =x-1, \text{if } 1\le x<2\\ =x-2, \text{if } 2\le x<3\end{cases}$$

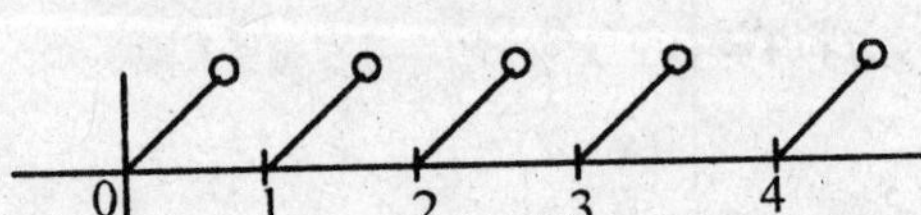

It's graph is, as shown in figure which is discontinuous at $\forall\ x\in z$. Clearly $x-[x]$ and hence $\sin(\pi(x-[x]))$ is not differentiable $\forall\ x\in z$

(B) $\rightarrow$ r

2. (A) $y=x\,|\,x\,|=\begin{cases}-x^2 & \text{if } x<0\\ x^2 & \text{if } x\ge 0\end{cases}$

Graph is as follows :

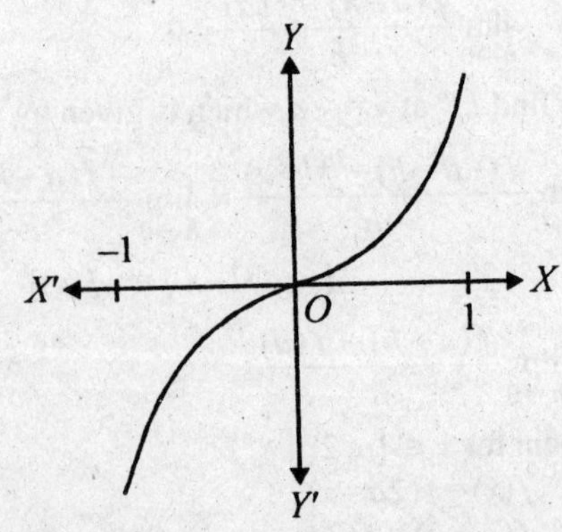

From graph $y = x|x|$ is continuous in $(-1, 1)$ **(p)**
differentiable in $(-1, 1)$ **(q)**
Strictly increasing in $(-1, 1)$. **(r)**

(B) $y = \sqrt{|x|} = \begin{cases} \sqrt{-x} & \text{if } x < 0 \\ \sqrt{x} & \text{if } x \geq 0 \end{cases}$

$\Rightarrow y^2 = -x, x < 0$
{where y can take only + ve values}

and $y^2 = x,\ x \geq 0$

$\therefore$ Graph is as follows :

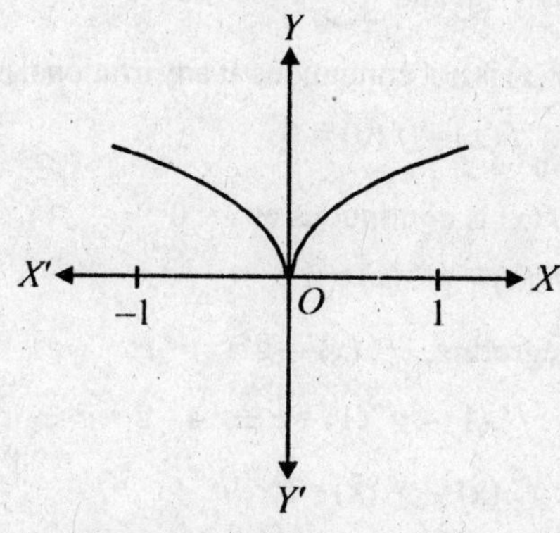

From graph $y = \sqrt{|x|}$ is continuous in $(-1, 1)$ **(*p*)**
not differentiable at $x = 0$ **(*s*)**

(C) **NOTE THIS STEP**

$$y = x + [x] = \begin{cases} - & - & - \\ x-1, & -1 \leq x < 0 \\ x, & 0 \leq x < 1 \\ x+1, & 1 \leq x < 2 \\ - & - & - \end{cases}$$

$\therefore$ Graph of $y = x + [x]$ is as follows :

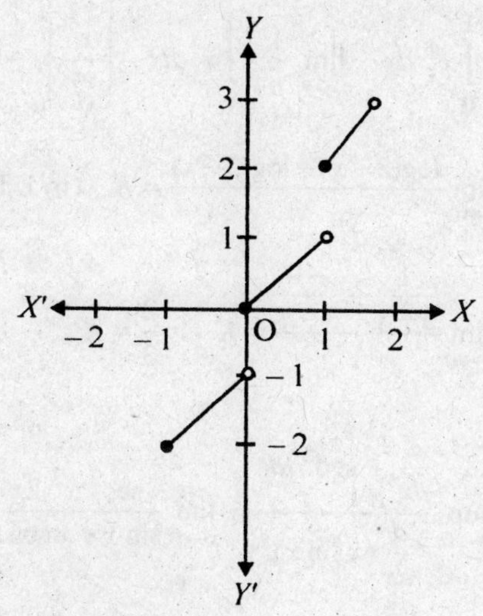

From graph, $y = x + [x]$ is neither continuous, nor differentiable at $x = 0$ and hence in $(-1, 1)$. **(*s*)**
Also it is strictly increasing in $(-1, 1)$ **(*r*)**

(D) $y = |x-1| + |x+1| = \begin{cases} -2x, & x < -1 \\ 2, & -1 \leq x < 1 \\ 2x, & x \geq 1 \end{cases}$

Graph of function is as follows :

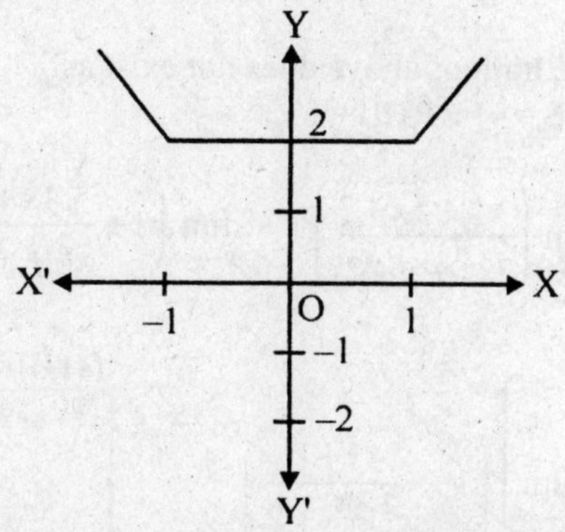

From graph, $y = f(x)$ is continuous **(p)** and differentiable **(q)** in $(-1, 1)$ but not strictly increasing in $(-1, 1)$.

(I) INTEGER VALUE CORRECT TYPE :

1. 6

$$6\int_1^x f(t)dt = 3xf(x) - x^3$$

Differentiating, we get $6f(x) = 3f(x) + 3xf'(x) - 3x^2$

$$\Rightarrow f'(x) - \frac{1}{x}f(x) = x$$

$$\text{I.F.} = \frac{1}{x}$$

$\therefore$ Solution is $f(x).\ \frac{1}{x} = \int 1.dx = x + c$

$\therefore f(x) = x^2 + cx$

But $f(1) = 2 \Rightarrow c = 1 \therefore f(x) = x^2 + x$

Hence $f(2) = 4 + 2 = 6$

Note : Putting $x = 1$ in given integral equation, we get

$$f(1) = \frac{1}{3} \text{ while given } f(1) = 2.$$

$\therefore$ Data given in the question is inconsistent.

Section-B JEE Main/ AIEEE

1. (d) $\lim \dfrac{\sqrt{1-\cos 2x}}{\sqrt{2}x} \Rightarrow \lim \dfrac{\sqrt{1-(1-2\sin^2 x)}}{\sqrt{2}x}$;

$\lim\limits_{x\to 0} \dfrac{\sqrt{2\sin^2 x}}{\sqrt{2}x} \Rightarrow \lim\limits_{x\to 0} \dfrac{|\sin x|}{x}$

The limit of above does not exist as LHS $=-1 \neq$ RHL $=1$

2. (a) $\lim\limits_{x\to\infty}\left(\dfrac{x^2+5x+3}{x^2+x+2}\right)^x = \lim\limits_{x\to\infty}\left(1+\dfrac{4x+1}{x^2+x+2}\right)^x$

$$= \lim_{x\to\infty}\left[\left(1+\frac{4x+1}{x^2+x+2}\right)^{\frac{x^2+x+2}{4x+1}}\right]^{\frac{(4x+1)x}{x^2+x+2}}$$

$$= e^{\lim\limits_{x\to\infty}\frac{4x^2+x}{x^2+x+2}}\left[\because \lim_{x\to\infty}(1+\lambda x)^{\frac{1}{x}} = e^{\lambda}\right]$$

$$= e^{\lim\limits_{x\to\infty}\frac{4+\frac{1}{x}}{1+\frac{1}{x}+\frac{2}{x^2}}} = e^4$$

3. (c) Apply L H Rule

We have, $\lim\limits_{x\to 2}\dfrac{xf(2)-2f(x)}{x-2}\ \left(\dfrac{0}{0}\right)$

$= \lim\limits_{x\to 2} f(2)-2f'(x) = f(2)-2f'(2)$

$=4-2\times 4=-4.$

4. (a) We have $\lim\limits_{n\to\infty}\dfrac{1^p+2^p+....+n^p}{n^{p+1}}$;

$$\lim_{n\to\infty}\sum_{r=1}^{n}\frac{r^p}{n^p\cdot n} = \int_0^1 x^p dx = \left[\frac{x^{p+1}}{p+1}\right]_0^1 = \frac{1}{p+1}$$

5. (d) Since $\lim\limits_{x\to 0}[x]$ does not exist, hence the required limit does not exist.

6. (a) $\lim\limits_{x\to 1}\dfrac{\sqrt{f(x)}-1}{\sqrt{x}-1}\ \left(\dfrac{0}{0}\right)$ form using L' Hospital's rule

$$= \lim_{x\to 1}\frac{\frac{1}{2\sqrt{f(x)}}f'(x)}{1/2\sqrt{x}} = \frac{f'(1)}{\sqrt{f(1)}} = \frac{2}{1} = 2.$$

7. (b) **Let a is a rational number** other than 0, in [–5, 5], then

$f(a)=a$ and $\lim\limits_{x\to a} f(x)=-a$

[As in the immediate neighbourhood of a rational number, we find irrational numbers]

$\therefore f(x)$ is not continuous at any rational number

If a is irrational number, then

$f(a)=-a$ and $\lim\limits_{x\to a} f(x)=a$

$\therefore f(x)$ is not continuous at any irrational number clearly

$\lim\limits_{x\to 0} f(x)=f(0)=0$

$\therefore f(x)$ is continuous at $x=0$

8. (d) $\because f''(x)-g''(x)=0$

Integrating, $f'(x)-g'(x)=c$;

$\Rightarrow f'(1)-g'(1)=c \Rightarrow 4-2=c \Rightarrow c=2.$

$\therefore f'(x)-g'(x)=2$;

Integrating, $f(x)-g(x)=2x+c_1$

$\Rightarrow f(2)-g(2)=4+c_1 \Rightarrow 9-3=4+c_1$;

$\Rightarrow c_1=2 \quad \therefore f(x)-g(x)=2x+2$

At $x=3/2, f(x)-g(x)=3+2=5.$

9. (c) $f(x+y)=f(x)\times f(y)$

Differentiate with respect to x, treating y as constant

$f'(x+y)=f'(x)f(y)$

Putting $x=0$ and $y=x$, we get $f'(x)=f'(0)f(x)$;

$\Rightarrow f'(5)=3f(5)=3\times 2=6.$

10. (a) The given expression can be written as

$$= \lim_{n\to\infty}\frac{1}{n}\sum_{r=1}^{n}\left(\frac{r}{n}\right)^4 - \lim_{n\to\infty}\frac{1}{n}.\lim_{n\to\infty}\frac{1}{n}\left(\frac{r}{n}\right)^3$$

$$= \int_0^1 x^4\,dx - \lim_{n\to\infty}\frac{1}{n}\times\int_0^1 x^3\,dx = \left[\frac{x^5}{5}\right]_0^1 - 0 = \frac{1}{5}$$

11. (d) $\lim\limits_{x\to 0}\dfrac{\log(3+x)-\log(3-x)}{x}=K$ (by L'Hospital rule)

$$\lim_{x\to 0}\frac{\frac{1}{3+x}-\frac{-1}{3-x}}{1}=K \quad \therefore \frac{2}{3}=K$$

12. (d) $$\lim_{x\to 0}\frac{\frac{d}{dx}\int_0^{x^2}\sec^2 t\,dt}{\frac{d}{dx}(x\sin x)} = \lim_{x\to 0}\frac{\sec^2 x^2.2x}{\sin x + x\cos x}$$

(by L' Hospital rule)

$$\lim_{x\to 0}\frac{2\sec^2 x^2}{\left(\frac{\sin x}{x}+\cos x\right)}=\frac{2\times 1}{1+1}=1$$

13. **(b)** $\lim_{x\to a}\frac{f(a)g'(x)-g(a)f'(x)}{g'(x)-f'(x)}=4$

(By L' Hospital rule)

$\lim_{x\to a}\frac{k\,g'(x)-k\,f'(x)}{g'(x)-f'(x)}=4 \qquad \therefore k=4.$

14. **(d)** $\lim_{x\to\frac{\pi}{2}}\frac{\tan\left(\frac{\pi}{4}-\frac{x}{2}\right).(1-\sin x)}{(\pi-2x)^3}$

Let $x=\frac{\pi}{2}+y; y\to 0$

$$=\lim_{y\to 0}\frac{\tan\left(-\frac{y}{2}\right).(1-\cos y)}{(-2y)^3}=\lim_{y\to 0}\frac{-\tan\frac{y}{2}2\sin^2\frac{y}{2}}{(-8).\frac{y^3}{8}.8}$$

$$=\lim_{y\to 0}\frac{1}{32}\frac{\tan\frac{y}{2}}{\left(\frac{y}{2}\right)}.\left[\frac{\sin y/2}{y/2}\right]^2=\frac{1}{32}$$

15. **(c)** $f(0)=0;\ f(x)=xe^{-\left(\frac{1}{|x|}+\frac{1}{x}\right)}$

R.H.L. $\lim_{h\to 0}(0+h)e^{-2/h}=\lim_{h\to 0}\frac{h}{e^{2/h}}=0$

L.H.L. $\lim_{h\to 0}(0-h)e^{-\left(\frac{1}{h}-\frac{1}{h}\right)}=0$

therefore, $f(x)$ is continuous.

R.H.D $=\lim_{h\to 0}\frac{(0+h)e^{-\left(\frac{1}{h}+\frac{1}{h}\right)}-0}{h}=0$

L.H.D. $=\lim_{h\to 0}\frac{(0-h)e^{-\left(\frac{1}{h}-\frac{1}{h}\right)}-0}{-h}=1$

therefore, L.H.D. $\neq$ R.H.D.

$f(x)$ is not differentiable at $x=0$.

16. **(b)** We know that $\lim_{x\to\infty}(1+x)^{\frac{1}{x}}=e$

$$\therefore \lim_{x\to\infty}\left(1+\frac{a}{x}+\frac{b}{x^2}\right)^{2x}=e^2$$

$$\Rightarrow \lim_{x\to\infty}\left[\left(1+\frac{a}{x}+\frac{b}{x^2}\right)^{\left(\frac{1}{\frac{a}{x}+\frac{b}{x^2}}\right)}\right]^{2x\left(\frac{a}{x}+\frac{b}{x^2}\right)}=e^2$$

$$\Rightarrow e^{\lim_{x\to\infty}2\left[a+\frac{b}{x}\right]}=e^2\Rightarrow e^{2a}=e^2\ \Rightarrow a=1 \text{ and } b\in R$$

ALTERNATE SOLUTION

$$\lim_{x\to\infty}\left(1+\frac{a}{x}+\frac{b}{x^2}\right)^{2x}=e^2\Rightarrow e^{\lim_{x\to\infty}\left(1+\frac{a}{x}+\frac{b}{x^2}-1\right)2x}=e^2$$

$$\Rightarrow\lim_{x\to\infty}\left(2a+\frac{2b}{x}\right)=2$$

$$\Rightarrow 2a+0=2, b\in\mathbf{R}\ \Rightarrow a=1,\ b\in\mathbf{R}$$

17. **(c)** $f(x)=\frac{1-\tan x}{4x-\pi}$ is continuous in $\left[0,\frac{\pi}{2}\right]$

$$\therefore f\left(\frac{\pi}{4}\right)=\lim_{x\to\frac{\pi}{4}}f(x)=\lim_{x\to\frac{\pi^+}{4}}f(x)=\lim_{h\to 0}f\left(\frac{\pi}{4}+h\right)$$

$$=\lim_{h\to 0}\frac{1-\tan\left(\frac{\pi}{4}+h\right)}{4\left(\frac{\pi}{4}+h\right)-\pi},h>0=\lim_{h\to 0}\frac{1-\frac{1+\tan h}{1-\tan h}}{4h}$$

$$=\lim_{h\to 0}\frac{-2}{1-\tanh}.\frac{\tan h}{4h}=\frac{-2}{4}=-\frac{1}{2}$$

ALTERNATE SOLUTION

$\because f(x)$ is continuous at $x=\frac{\pi}{4}$

$$\therefore f\left(\frac{\pi}{4}\right)=\lim_{x\to\frac{\pi}{4}}f(x)=\lim_{x\to\frac{\pi}{4}}\frac{1-\tan x}{4x-\pi}$$

$$=\lim_{x\to\frac{\pi}{4}}\frac{-\sec^2 x}{4}\quad\text{[using L' Hospital's rule]}$$

$$=\frac{-\sec^2\frac{\pi}{4}}{4}=\frac{-2}{4}=\frac{-1}{2}$$

18. **(d)** $\lim_{n\to\infty}\left[\frac{1}{n^2}\sec^2\frac{1}{n^2}+\frac{2}{n^2}\sec^2\frac{4}{n^2}+\frac{3}{n^2}\sec^2\frac{9}{n^2}+....+\frac{1}{n}\sec^2 1\right]$

is equal to

$$\lim_{n\to\infty}\frac{r}{n^2}\sec^2\frac{r^2}{n^2}=\lim_{n\to\infty}\frac{1}{n}.\frac{r}{n}\sec^2\frac{r^2}{n^2}$$

$\Rightarrow$ Given limit is equal to value of integral

$$\int_0^1 x\sec^2 x^2 dx$$

or $\frac{1}{2}\int_0^1 2x\sec\ x^2 dx = \frac{1}{2}\int_0^1 \sec^2 t dt$ [put $x^2 = t$]

$= \frac{1}{2}(\tan t)_0^1 = \frac{1}{2}\tan 1.$

19. **(a)** Given limit $= \lim_{x\to\alpha} \frac{1-\cos a(x-\alpha)(x-\beta)}{(x-\alpha)^2}$

$$= \lim_{x\to\alpha} \frac{2\sin^2\left(a\frac{(x-\alpha)(x-\beta)}{2}\right)}{(x-\alpha)^2}$$

$$= \lim_{x\to\alpha} \frac{2}{(x-\alpha)^2} \times \frac{\sin^2\left(a\frac{(x-\alpha)(x-\beta)}{2}\right)}{\frac{a^2(x-\alpha)^2(x-\beta)^2}{4}}$$

$$\times \frac{a^2(x-\alpha)^2(x-\beta)^2}{4}$$

$$= \frac{a^2(\alpha-\beta)^2}{2}.$$

20. **(c)** $f'(1) = \lim_{h\to 0} \frac{f(1+h)-f(1)}{h}$;

As function is differentiable so it is continuous as it is given that $\lim_{h\to 0}\frac{f(1+h)}{h} = 5$ and hence $f(1)=0$

Hence $f'(1) = \lim_{h\to 0}\frac{f(1+h)}{h} = 5$

21. **(a)** As $f(1) = -2$ & $f'(x) \geq 2\ \forall\ x \in [1, 6]$

Applying Lagrange's mean value theorem

$\frac{f(6)-f(1)}{5} = f'(c) \geq 2 \Rightarrow f(6) \geq 10 + f(1)$

$\Rightarrow\quad f(6) \geq 10 - 2 \Rightarrow f(6) \geq 8.$

22. **(b)** $f'(x) = \lim_{h\to 0}\frac{f(x+h)-f(x)}{h}$

$|f'(x)| = \lim_{h\to 0}\left|\frac{f(x+h)-f(x)}{h}\right| \leq \lim_{h\to 0}\left|\frac{(h)^2}{h}\right|$

$\Rightarrow\quad |f'(x)| \leq 0 \Rightarrow f'(x) = 0$

$\Rightarrow\quad f(x) =$ constant

As $f(0) = 0$

$\Rightarrow\quad f(1) = 0.$

23. **(a)** $f(x) = \min\{x+1, |x|+1\} \Rightarrow f(x) = x+1\ \forall\ x \in R$

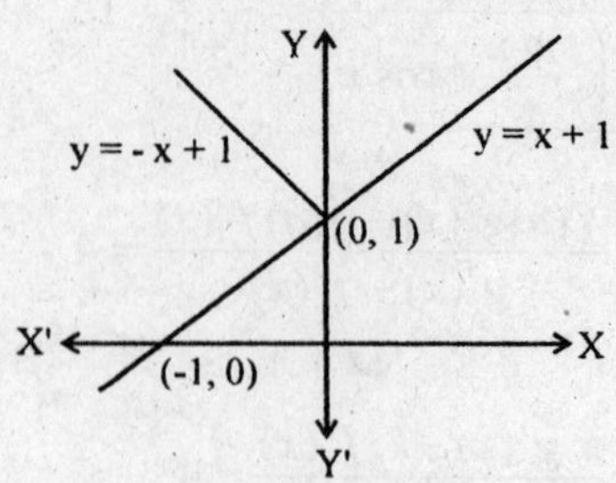

Hence, $f(x)$ is differentiable everywhere for all $x \in R$.

24. **(b)** Given, $f(x) = \frac{1}{x} - \frac{2}{e^{2x}-1} \Rightarrow f(0) = \lim_{x\to 0}\frac{1}{x} - \frac{2}{e^{2x}-1}$

$= \lim_{x\to 0}\frac{(e^{2x}-1)-2x}{x(e^{2x}-1)}$ $\left[\frac{0}{0}\text{ form}\right]$

$\therefore$ using, L'Hospital rule

$f(0) = \lim_{x\to 0}\frac{4e^{2x}}{2(xe^{2x}2+e^{2x}.1)+e^{2x}.2}$

$= \lim_{x\to 0}\frac{4e^{2x}}{4xe^{2x}+2e^{2x}+2e^{2x}}$ $\left[\frac{0}{0}\text{ form}\right]$

$= \lim_{x\to 0}\frac{4e^{2x}}{4(xe^{2x}+e^{2x})} = \frac{4.e^0}{4(0+e^0)} = 1$

25. **(c)** We have $f(x) = \begin{cases}(x-1)\sin\left(\frac{1}{x-1}\right), & if\ x \neq 1\\ 0 & , if\ x = 1\end{cases}$

$Rf'(1) = \lim_{h\to 0}\frac{f(1+h)-f(1)}{h}$

$= \lim_{h\to 0}\frac{h\sin\frac{1}{h}-0}{h} = \lim_{h\to 0}\sin\frac{1}{h}$

= a finite number

Let this finite number be l

$L\,f'(1) = \lim_{h\to 0}\frac{f(1-h)-f(1)}{-h}$

$= \lim_{h\to 0}\frac{-h\sin\left(\frac{1}{-h}\right)}{-h}$

$= \lim_{h\to 0}\sin\left(\frac{1}{-h}\right) = -\lim_{h\to 0}\sin\left(\frac{1}{h}\right)$

$= -$ (a finite number) $= -l$

Thus $Rf'(1) \neq Lf'(1)$

$\therefore$ f is not differentiable at x = 1

Also, $f'(0) = \sin\frac{1}{(x-1)} - \frac{x-1}{(x-1)^2}\cos\left(\frac{1}{x-1}\right)\Bigg]_{x=0}$

$= -\sin 1 + \cos 1$

$\therefore$ f is differentiable at x = 0

26. **(d)** f(x) is a positive increasing function.

$\therefore\ 0 < f(x) < f(2x) < f(3x)$

$$\Rightarrow 0 < 1 < \frac{f(2x)}{f(x)} < \frac{f(3x)}{f(x)}$$

$$\Rightarrow \lim_{x\to\infty} 1 \le \lim_{x\to\infty}\frac{f(2x)}{f(x)} \le \lim_{x\to\infty}\frac{f(3x)}{f(x)}$$

By Sandwich Theorem.

$$\Rightarrow \lim_{x\to\infty}\frac{f(2x)}{f(x)} = 1$$

27. **(d)** $\lim_{x\to 2}\frac{\sqrt{1-\cos\{2(x-2)\}}}{x-2} = \lim_{x\to 2}\frac{\sqrt{2}\,|\sin(x-2)|}{x-2}$

$$\text{L.H.L.} = \lim_{x\to 2^-}\frac{\sqrt{2}\sin(x-2)}{(x-2)} = -\sqrt{2}$$

$$\text{R.H.L.} = \lim_{x\to 2^+}\frac{\sqrt{2}\sin(x-2)}{(x-2)} = \sqrt{2}$$

Thus L.H.L. $\neq$ R.H.L.

Hence, $\lim_{x\to 2}\frac{\sqrt{1-\cos\{2(x-2)\}}}{x-2}$ does not exist.

28. **(b)** $L.H.L = \lim_{x\to 0^-} f(x) = \lim_{h\to 0}\frac{\sin\{(p+1)(-h)\} - \sin(-h)}{-h}$

$$= \lim_{h\to 0}\frac{-\sin(p+1)h}{-h} + \frac{\sin(-h)}{-h} = p+1+1 = p+2$$

$$R.H.L. = \lim_{x\to 0^+} f(x) = \lim_{h\to 0}\frac{\sqrt{1+h}-1}{h}$$

$$= \lim_{h\to 0}\frac{1}{(\sqrt{1+h}+1)} = \frac{1}{2}$$

and $f(0) = q \Rightarrow p = -\frac{3}{2}, q = \frac{1}{2}$

29. **(c)** $f(x) = \begin{cases} x\sin(1/x), & x \neq 0 \\ 0, & x = 0 \end{cases}$ at $x = 0$

$$\text{LHL} = \lim_{h\to 0^-}\left\{-h\sin\left(-\frac{1}{h}\right)\right\}$$

= 0 × a finite quantity betwen – 1 and 1

= 0

$$\text{RHL} = \lim_{h\to 0^+} h\sin\frac{1}{h} = 0$$

Also, $f(0) = 0$

Thus LHL = RHL = $f(0)$

$\therefore$ $f(x)$ is continuous on R.

$f_2(x)$ is not continuous at $x = 0$

30. **(d)** $\lim_{x\to 5}\frac{(f(x))^2 - 9}{\sqrt{|x-5|}} = 0$

$\lim_{x\to 5}[(f(x))^2 - 9] = 0 \Rightarrow \lim_{x\to 5} f(x) = 3$

31. **(c)** $\lim_{x\to a}\frac{x^2 f(a) - a^2 f(x)}{x-a}$

$$= \lim_{x\to a}\frac{2xf(a) - a^2 f'(x)}{1}$$

$$= 2af(a) - a^2 f'(a)$$

32. **(a)** Let $f(x) = [x]\cos\left(\frac{2x-1}{2}\right)$

Doubtful points are $x = n,\ n \in I$

$$\text{L.H.L} = \lim_{x\to n^-}[x]\cos\left(\frac{2x-1}{2}\right)\pi$$

$$= (n-1)\cos\left(\frac{2n-1}{2}\right)\pi = 0$$

($\because$ [x] is the greatest integer function)

$$\text{R.H.L} = \lim_{x\to n^+}[x]\cos\left(\frac{2x-1}{2}\right)\pi = n\cos\left(\frac{2n-1}{2}\right)\pi = 0$$

Now, value of the function at $x = n$ is

$f(n) = 0$

Since, L.H.L = R.H.L. = $f(n)$

$\therefore$ $f(x) = [x]\cos\left(\frac{2x-1}{2}\right)$ is continuous for every real x.

33. **(c)** $f(x) = |x-2| = \begin{cases} x-2, & x-2 \ge 0 \\ 2-x, & x-2 \le 0 \end{cases}$

$$= \begin{cases} x-2, & x \ge 2 \\ 2-x, & x \le 2 \end{cases}$$

Similarly, $f(x) = |x-5| = \begin{cases} x-5, & x \ge 5 \\ 5-x, & x \le 5 \end{cases}$

$\therefore$ $f(x) = |x-2| + |x-5| = \{x-2+5-x = 3,\ 2 \le x \le 5\}$

Thus $f(x) = 3,\ 2 \le x \le 5$

$f'(x) = 0,\ 2 < x < 5$

$f'(4) = 0$ $\therefore$ statement 1 is true.

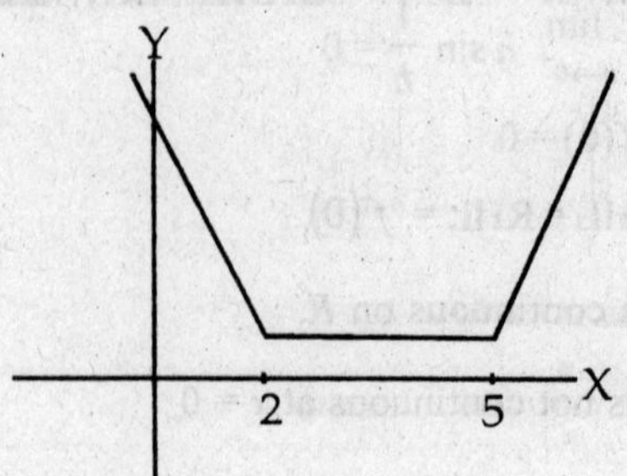

$\because f(2) = 0 + |2 - 5| = 3$ and $f(5) = |5 - 2| + 0 = 3$

$\therefore$ statement-2 is also true and a correct explanation for statement 1.

34. (d) Multiply and divide by x in the given expression, we get

$$\lim_{x\to 0} \frac{(1-\cos 2x)}{x^2} \frac{(3+\cos x)}{1} \cdot \frac{x}{\tan 4x}$$

$$= \lim_{x\to 0} \frac{2\sin^2 x}{x^2} \cdot \frac{3+\cos x}{1} \cdot \frac{x}{\tan 4x}$$

$$= 2 \lim_{x\to 0} \frac{\sin^2 x}{x^2} \cdot \lim_{x\to 0} 3 + \cos x \cdot \lim_{x\to 0} \frac{x}{\tan 4x}$$

$$= 2.4 \frac{1}{4} \lim_{x\to 0} \frac{4x}{\tan 4x} = 2.4.\frac{1}{4} = 2$$

Chapter 12

Differentiation

Section-A JEE Advanced/ IIT-JEE

A FILL IN THE BLANKS :

1. If $y = f\left(\frac{2x-1}{x^2+1}\right)$ and $f'(x) = \sin x^2$, then $\frac{dy}{dx} =$ .... *(1982 - 2 Marks)*

2. If $f_r(x)$, $g_r(x)$, $h_r(x)$, $r = 1, 2, 3$ are polynomials in x such that $f_r(a) = g_r(a) = h_r(a)$, $r = 1, 2, 3$ and $F(x) = \begin{vmatrix} f_1(x) & f_2(x) & f_3(x) \\ g_1(x) & g_2(x) & g_3(x) \\ h_1(x) & h_2(x) & h_3(x) \end{vmatrix}$ then $F'(x)$ at $x = a$ is .... *(1985 - 2 Marks)*

3. If $f(x) = \log_x(\ln x)$, then $f'(x)$ at $x = e$ is .... *(1985 - 2 Marks)*

4. The derivative of $\sec^{-1}\left(\frac{1}{2x^2-1}\right)$ with respect to $\sqrt{1-x^2}$ at $x = \frac{1}{2}$ is .... *(1986 - 2 Marks)*

5. If $f(x) = |x-2|$ and $g(x) = f[f(x)]$, then $g'(x) =$ .... for $x > 20$ *(1990 - 2 Marks)*

6. If $xe^{xy} = y + \sin^2 x$, then at $x = 0$, $\frac{dy}{dx} =$ .... *(1996 - 1 Mark)*

B TRUE / FALSE :

1. The derivative of an even function is always an odd function. *(1983 - 1 Mark)*

C MCQ's WITH ONE CORRECT ANSWER :

1. If $y^2 = P(x)$, a polynomial of degree 3, then $2\frac{d}{dx}\left(y^3 \frac{d^2y}{dx^2}\right)$ equals *(1988 - 2 Marks)*
   (a) $P'''(x) + P'(x)$ (b) $P''(x)\,P'''(x)$
   (c) $P(x)\,P'''(x)$ (d) a constant

2. Let $f(x)$ be a quadratic expression which is positive for all the real values of x. If $g(x) = f(x) + f'(x) + f''(x)$, then for any real x, *(1990 - 2 Marks)*
   (a) $g(x) < 0$ (b) $g(x) > 0$
   (c) $g(x) = 0$ (d) $g(x) \geq 0$

3. If $y = (\sin x)^{\tan x}$, then $\frac{dy}{dx}$ is equal to *(1994)*
   (a) $(\sin x)^{\tan x}(1 + \sec^2 x \log \sin x)$
   (b) $\tan x (\sin x)^{\tan x - 1}.\cos x$
   (c) $(\sin x)^{\tan x} \sec^2 x \log \sin x$
   (d) $\tan x (\sin x)^{\tan x - 1}$

4. Let $f(x) = \begin{vmatrix} x^3 & \sin x & \cos x \\ 6 & -1 & 0 \\ p & p^2 & p^3 \end{vmatrix}$ where p is a constant. Then $\frac{d^3}{dx^3}(f(x))$ at $x = 0$ is *(1997C - 2 Marks)*
   (a) p (b) $p + p^2$
   (c) $p + p^3$ (d) independent of p

5. If $x^2 + y^2 = 1$ then *(2000)*
   (a) $yy'' - 2(y')^2 + 1 = 0$ (b) $yy'' + (y')^2 + 1 = 0$
   (c) $yy'' + (y')^2 - 1 = 0$ (d) $yy'' + 2(y')^2 + 1 = 0$

6. Let $f: (0, \infty) \to R$ and $F(x) = \int_0^x f(t)dt$. If $F(x^2) = x^2(1+x)$, then $f(4)$ equals *(2001S)*
   (a) 5/4 (b) 7
   (c) 4 (d) 2

7. If y is a function of x and $\log(x+y) - 2xy = 0$, then the value of $y'(0)$ is equal to *(2004S)*
   (a) 1 (b) –1
   (c) 2 (d) 0

8. If $f(x)$ is a twice differentiable function and given that $f(1) = 1; f(2) = 4, f(3) = 9$, then *(2005S)*
   (a) $f''(x) = 2$ for $\forall\ x \in (1, 3)$
   (b) $f''(x) = f'(x) = 5$ for some $x \in (2, 3)$
   (c) $f''(x) = 3$ for $\forall\ x \in (2, 3)$
   (d) $f''(x) = 2$ for some $x \in (1, 3)$

9. $\frac{d^2x}{dy^2}$ equals *(2007 -3 marks)*
   (a) $\left(\frac{d^2y}{dx^2}\right)^{-1}$ (b) $-\left(\frac{d^2y}{dx^2}\right)^{-1}\left(\frac{dy}{dx}\right)^{-3}$
   (c) $\left(\frac{d^2y}{dx^2}\right)\left(\frac{dy}{dx}\right)^{-2}$ (d) $-\left(\frac{d^2y}{dx^2}\right)\left(\frac{dy}{dx}\right)^{-3}$

10. Let $g(x) = \log f(x)$ where $f(x)$ is twice differentible positive function on $(0, \infty)$ such that $f(x+1) = x f(x)$. Then, for $N = 1, 2, 3, ...$ *(2008)*

$$g''\left(N+\frac{1}{2}\right) - g''\left(\frac{1}{2}\right) =$$

(a) $-4\left\{1+\frac{1}{9}+\frac{1}{25}+.....+\frac{1}{(2N-1)^2}\right\}$

(b) $4\left\{1+\frac{1}{9}+\frac{1}{25}+.....+\frac{1}{(2N-1)^2}\right\}$

(c) $-4\left\{1+\frac{1}{9}+\frac{1}{25}+.....+\frac{1}{(2N+1)^2}\right\}$

(d) $4\left\{1+\frac{1}{9}+\frac{1}{25}+.....+\frac{1}{(2N+1)^2}\right\}$

E SUBJECTIVE PROBLEMS :

1. Find the derivative of $\sin(x^2+1)$ with respect to x from first principle. *(1978)*

2. Find the derivative of

$$f(x) = \begin{cases} \frac{x-1}{2x^2-7x+5} & \text{when } x \neq 1 \\ -\frac{1}{3} & \text{when } x = 1 \end{cases}$$

at $x = 1$ *(1979)*

3. Given $y = \frac{5x}{3\sqrt{(1-x)^2}} + \cos^2(2x+1)$; Find $\frac{dy}{dx}$. *(1980)*

4. Let $y = e^{x\sin x^3} + (\tan x)^x$. Find $\frac{dy}{dx}$ *(1981 - 2 Marks)*

5. Let f be a twice differentiable function such that

$f''(x) = -f(x)$, and $f'(x) = g(x)$, $h(x) = [f(x)]^2 + [g(x)]^2$

Find $h(10)$ if $h(5) = 11$ *(1982 - 3 Marks)*

6. If α be a repeated root of a quadratic equation $f(x) = 0$ and $A(x)$, $B(x)$ and $C(x)$ be polynomials of degree 3, 4 and 5 respectively, then show that $\begin{vmatrix} A(x) & B(x) & C(x) \\ A(\alpha) & B(\alpha) & C(\alpha) \\ A'(\alpha) & B'(\alpha) & C'(\alpha) \end{vmatrix}$ is divisible by $f(x)$, where prime denotes the derivatives. *(1984 - 4 Marks)*

7. Find the derivative with respect to x of the function

$$(\log_{\cos x} \sin x)(\log_{\sin x} \cos x)^{-1} + \sin^{-1}\left(\frac{2x}{1+x^2}\right) \text{ at } x = \frac{\pi}{4}$$

(1984 - 4 Marks)

8. If $x = \sec\theta - \cos\theta$ and $y = \sec^n\theta - \cos^n\theta$, then show that $(x^2+4)\left(\frac{dy}{dx}\right)^2 = n^2(y^2+4)$ *(1989 - 2 Marks)*

9. Find $\frac{dy}{dx}$ at $x = -1$, when

$$(\sin y)^{\sin\left(\frac{\pi}{2}x\right)} + \frac{\sqrt{3}}{2}\sec^{-1}(2x) + 2^x \tan(\ln(x+2)) = 0$$

(1991 - 4 Marks)

10. If $y = \frac{ax^2}{(x-a)(x-b)(x-c)} + \frac{bx}{(x-b)(x-c)} + \frac{c}{x-c} + 1$,

prove that $\frac{y'}{y} = \frac{1}{x}\left(\frac{a}{a-x} + \frac{b}{b-x} + \frac{c}{c-x}\right)$. *(1998 - 8 Marks)*

H ASSERTION & REASON TYPE QUESTIONS :

1. Let $f(x) = 2 + \cos x$ for all real x.

Statement - 1 : For each real t, there exists a point c in $[t, t+\pi]$ such that $f'(c) = 0$ because

Statement - 2 : $f(t) = f(t+2\pi)$ for each real t. *(2007 -3 marks)*

(a) Statement-1 is True, Statement-2 is True; Statement-2 is a correct explanation for Statement-1

(b) Statement-1 is True, Statement-2 is True; Statement-2 is NOT a correct explanation for Statement-1

(c) Statement-1 is True, Statement-2 is False

(d) Statement-1 is False, Statement-2 is True.

2. Let f and g be real valued functions defined on interval $(-1, 1)$ such that $g''(x)$ is continuous, $g(0) \neq 0$. $g'(0) = 0$, $g''(0) \neq 0$, and $f(x) = g(x)\sin x$

Statement - 1 : $\lim_{x\to 0}[g(x)\cot x - g(0)\operatorname{cosec} x] = f''(0)$

and

Statement - 2 : $f'(0) = g(0)$ *(2008)*

(a) Statement - 1 is True, Statement - 2 is True; Statement - 2 is a correct explanation for Statement - 1

(b) Statement - 1 is True, Statement - 2 is True; Statement - 2 is NOT a correct explaination for Statement - 1

(c) Statement - 1 is True, Statement - 2 is False

(d) Statement - 1 is False, Statement - 2 is True

I INTEGER VALUE CORRECT TYPE :

1. If the function $f(x) = x^3 + e^{\frac{x}{2}}$ and $g(x) = f^{-1}(x)$, then the value of $g'(1)$ is *(2009)*

2. Let $f(\theta) = \sin\left(\tan^{-1}\left(\frac{\sin\theta}{\sqrt{\cos 2\theta}}\right)\right)$, where $-\frac{\pi}{4} < \theta < \frac{\pi}{4}$.

Then the value of $\frac{d}{d(\tan\theta)}(f(\theta))$ is *(2011)*

Section-B JEE Main/ AIEEE

1. If $y = (x + \sqrt{1+x^2})^n$, then $(1+x^2)\frac{d^2y}{dx^2} + x\frac{dy}{dx}$ is **[2002]**

(a) n^2y (b) $-n^2y$
(c) $-y$ (d) $2x^2y$

2. If $f(y) = e^y$, $g(y) = y$; $y > 0$ and

$F(t) = \int_0^t f(t-y)g(y)\,dy$, then **[2003]**

(a) $F(t) = te^{-t}$ (b) $F(t) = 1 - te^{-t}(1+t)$
(c) $F(t) = e^t - (1+t)$ (d) $F(t) = te^t$.

3. If $f(x) = x^n$, then the value of **[2003]**

$f(1) - \frac{f'(1)}{1!} + \frac{f''(1)}{2!} - \frac{f'''(1)}{3!} + \text{.........} \frac{(-1)^n f^n(1)}{n!}$ is

(a) 1 (b) 2^n
(c) $2^n - 1$ (d) 0.

4. Let $f(x)$ be a polynomial function of second degree. If $f(1) = f(-1)$ and a, b, c are in A. P, then $f'(a), f'(b), f'(c)$ are in **[2003]**

(a) Arithmetic -Geometric Progression
(b) A.P
(c) G.P
(d) H.P.

5. If $x = e^{y+e^{y+..to\,\infty}}, x > 0$, then $\frac{dy}{dx}$ is **[2004]**

(a) $\frac{1+x}{x}$ (b) $\frac{1}{x}$
(c) $\frac{1-x}{x}$ (d) $\frac{x}{1+x}$

6. The value of a for which the sum of the squares of the roots of the equation $x^2 - (a-2)x - a - 1 = 0$ assume the least value is **[2005]**

(a) 1 (b) 0
(c) 3 (d) 2

7. If the roots of the equation $x^2 - bx + c = 0$ be two consecutive integers, then $b^2 - 4c$ equals **[2005]**

(a) -2 (b) 3
(c) 2 (d) 1

8. Let $f: R \to R$ be a differentiable function having $f(2) = 6$, $f'(2) = \left(\frac{1}{48}\right)$. Then $\lim_{x\to 2} \int_6^{f(x)} \frac{4t^3}{x-2}\,dt$ equals **[2005]**

(a) 24 (b) 36
(c) 12 (d) 18

9. The set of points where $f(x) = \frac{x}{1+|x|}$ is differentiable is

(a) $(-\infty,0)\cup(0,\infty)$ (b) $(-\infty,-1)\cup(-1,\infty)$
(c) $(-\infty,\infty)$ (d) $(0,\infty)$ **[2006]**

10. If $x^m.y^n = (x+y)^{m+n}$, then $\frac{dy}{dx}$ is

(a) $\frac{y}{x}$ (b) $\frac{x+y}{xy}$ **[2006]**
(c) xy (d) $\frac{x}{y}$

11. Let y be an implicit function of x defined by $x^{2x} - 2x^x \cot y - 1 = 0$. Then $y'(1)$ equals **[2009]**

(a) 1 (b) $\log 2$
(c) $-\log 2$ (d) -1

12. Let $f: (-1, 1) \to \mathbf{R}$ be a differentiable function with $f(0) = -1$ and $f'(0) = 1$. Let $g(x) = [f(2f(x) + 2)]^2$. Then $g'(0) =$ **[2010]**

(a) -4 (b) 0
(c) -2 (d) 4

13. $\frac{d^2x}{dy^2}$ equals : **[2011]**

(a) $-\left(\frac{d^2y}{dx^2}\right)^{-1}\left(\frac{dy}{dx}\right)^{-3}$ (b) $\left(\frac{d^2y}{dx^2}\right)\left(\frac{dy}{dx}\right)^{-2}$

(c) $-\left(\frac{d^2y}{dx^2}\right)\left(\frac{dy}{dx}\right)^{-3}$ (d) $\left(\frac{d^2y}{dx^2}\right)^{-1}$

14. If $y = \sec(\tan^{-1}x)$, then $\frac{dy}{dx}$ at $x = 1$ is equal to :

(a) $\frac{1}{\sqrt{2}}$ (b) $\frac{1}{2}$ **[JEE M 2013]**
(c) 1 (d) $\sqrt{2}$

Solutions & Explanations

Section-A : JEE Advanced/ IIT-JEE

A 1. $\frac{2+2x-2x^2}{(x^2+1)^2}\sin\left(\frac{2x-1}{x^2+1}\right)^2$ 2. zero 3. $1/e$ 4. 4 5. 1 6. 1

B 1. T

C 1. (c) 2. (b) 3. (a) 4. (d) 5. (b) 6. (c) 7. (a)
8. (d) 9. (d) 10. (a)

E 1. $2x\cos(x^2+1)$ 2. $\frac{-2}{9}$

3. $\frac{5}{3}\cdot\frac{1}{(1-x)^2}2\sin(4x+2)$, if $x<1$; $-\frac{5}{3}\cdot\frac{1}{(1-x)^2}-2\sin(4x+2)$, if $x>1$

4. $e^{x\sin x^3}\left[\sin x^3+3x^3\cos x^3\right]+(\tan x)^x\left[\frac{2x}{\sin 2x}+\log\tan x\right]$ 5. 11 7. $-8\log_2 e+\frac{32}{16+\pi^2}$

9. 0

H 1. (b) 2. (a)

I 1. (2) 2. (1)

Section-B : JEE Main/ AIEEE

1. (a) 2. (c) 3. (d) 4. (b) 5. (c) 6. (a) 7. (d)
8. (d) 9. (c) 10. (a) 11. (d) 12. (a) 13. (c) 14. (a)

Section-A JEE Advanced/ IIT-JEE

A FILL IN THE BLANKS :

1. $\frac{2+2x-2x^2}{(x^2+1)^2}\sin\left(\frac{2x-1}{x^2+1}\right)^2$

Given: $y=f\left(\frac{2x-1}{x^2+1}\right)$; $f'(x)=\sin x^2$

$$\therefore\ \frac{dy}{dx}=f'\left(\frac{2x-1}{x^2+1}\right).\frac{d}{dx}\left(\frac{2x-1}{x^2+1}\right)$$

$$=\sin\left(\frac{2x-1}{x^2+1}\right)^2.\frac{2(x^2+1)-2x(2x-1)}{(x^2+1)^2}$$

$$=\frac{2+2x-2x^2}{(x^2+1)^2}\sin\left(\frac{2x-1}{x^2+1}\right)^2$$

2. Given that $F(x)=\begin{vmatrix}f_1(x) & f_2(x) & f_3(x)\\ g_1(x) & g_2(x) & g_3(x)\\ h_1(x) & h_2(x) & h_3(x)\end{vmatrix}$ ...(1)

Where $f_r(x), g_r(x), h_r(x), r=1,2,3$, are polynominals in x and hence differentiable and

$f_r(a)=g_r(a)=h_r(a), r=1,2,3$ ...(2)

Differentiating eq. (1) with respect to x, we get

$$F'(x)=\begin{vmatrix}f_1'(x) & f_2'(x) & f_3'(x)\\ g_1(x) & g_2(x) & g_3(x)\\ h_1(x) & h_2(x) & h_3(x)\end{vmatrix}$$

$$+\begin{vmatrix}f_1(x) & f_2(x) & f_3(x)\\ g_1'(x) & g_2'(x) & g_3'(x)\\ h_1(x) & h_2(x) & h_3(x)\end{vmatrix}+\begin{vmatrix}f_1(x) & f_2(x) & f_3(x)\\ g_1(x) & g_2(x) & g_3(x)\\ h_1'(x) & h_2'(x) & h_3'(x)\end{vmatrix}$$

$$\therefore \quad F'(a)=\begin{vmatrix} f_1'(a) & f_2'(a) & f_3'(a) \\ g_1(a) & g_2(a) & g_3(a) \\ h_1(a) & h_2(a) & h_3(a) \end{vmatrix}$$

$$+\begin{vmatrix} f_1(a) & f_2(a) & f_3(a) \\ g_1'(a) & g_2'(a) & g_3'(a) \\ h_1(a) & h_2(a) & h_3(a) \end{vmatrix}+\begin{vmatrix} f_1(a) & f_2(a) & f_3(a) \\ g_1(a) & g_2(a) & g_3(a) \\ h_1'(a) & h_2'(a) & h_3'(a) \end{vmatrix}$$

$F'(a)=D_1+D_2+D_3$

Using eq. (2) we get $D_1=D_2=D_3=0$ [By the property of determinants that $D=0$ if two rows in D are identical]

$\therefore \quad F'(a)=0.$

3. Given that

$$f(x)=\log_x(ln\, x)=\frac{\log_e(\log_e x)}{(\log_e x)}$$

$$f'(x)=\frac{\frac{1}{\log_e x}\times\frac{1}{x}\times\log_e x-\frac{1}{x}\log_e(\log_e x)}{(\log_e x)^2}$$

$$=\frac{\frac{1}{x}[1-\log_e(\log_e x)]}{(\log_x x)^2}$$

$$f'(e)=\frac{\frac{1}{e}[1-\log_e(\log_e e)]}{(\log_e e)^2}=\frac{\frac{1}{e}[1-\log_e 1]}{(1)^2}$$

$$=\frac{1}{e}(1-0)=\frac{1}{e}.$$

4. Let $u=\sec^{-1}\left(\frac{1}{2x^2-1}\right); v=\sqrt{1-x^2}$

Then to find $\left.\frac{du}{dv}\right|_{x=1/2}$, we have

$u=\cos^{-1}(2x^2-1)=2\cos^{-1}x$

$\therefore \quad \frac{du}{dx}=\frac{-2}{\sqrt{1-x^2}}$ and $\quad v=\sqrt{1-x^2}$

$\therefore \quad \frac{dv}{dx}=\frac{-x}{\sqrt{1-x^2}} \quad \therefore \quad \frac{du}{dv}=\frac{\frac{-2}{\sqrt{1-x^2}}}{\frac{-x}{\sqrt{1-x^2}}}=\frac{2}{x}$

$\therefore \quad \left.\frac{du}{dv}\right|_{x=\frac{1}{2}}=4$

5. $f(x)=|x-2|$

$\Rightarrow \; g(x)=f(f(x))=|f(x)-2|$ as $x>20$

$=||x-2|-2|=|x-2-2|$ as $x>20$

$=|x-4|=x-4$ as $x>20$

$\therefore \quad g'(x)=1$

6. Given : $xe^{xy}=y+\sin^2 x$

Differentiating both sides w. r.to x, we get

$$e^{xy}.1+xe^{xy}\left(y+x\frac{dy}{dx}\right)=\frac{dy}{dx}+2\sin x\cos x$$

$$\text{Put } x=0 \Rightarrow 1+0=\frac{dy}{dx}+0 \Rightarrow \frac{dy}{dx}=1$$

B TRUE / FALSE :

1. Consider $\phi(x)=\frac{f(x)+f(-x)}{2}$, which is an even function

Now, $\psi(x)=\phi'(x)=\frac{f'(x)-f'(-x)}{2}$.

$\psi(-x)=\frac{f'(-x)-f'(x)}{2}=-\psi(x) \;\therefore\; \psi$ is odd.

C MCQ's WITH ONE CORRECT ANSWER :

1. (c) We have $y^2=P(x)$, ...(1)

where $P(x)$ is a polynomial of degree 3 and hence thrice differentiable. Differentiating (1) w.r. to x, we get

$$2y\frac{dy}{dx}=P'(x) \quad ...(2)$$

Again differentiating with respect to x, we get

$$2\left(\frac{dy}{dx}\right)^2+2y\frac{d^2y}{dx^2}=P''(x)$$

$$\Rightarrow \frac{[P'(x)]^2}{2y^2}+2y\frac{d^2y}{dx^2}=P''(x) \quad \text{[Using (2)]}$$

$$\Rightarrow 4y^3\frac{d^2y}{dx^2}=2y^2P''(x)-[P'(x)]^2$$

$$\Rightarrow 4y^3\frac{d^2y}{dx^2}=2P(x)P''(x)-[P'(x)]^2 \quad \text{[Using (1)]}$$

$$\Rightarrow 2y^3\frac{d^2y}{dx^2}=P(x)P''(x)-\frac{1}{2}[P'(x)]^2$$

Again differentiating w.r. to x, we get $2\frac{d}{dx}\left(y^3\frac{d^2y}{dx^2}\right)$

$=P'''(x)P(x)+P''(x)P'(x)-P'(x)P''(x)$

$=P'''(x)P(x)$

2. (b) Let $f(x)=ax^2+bx+c$

As given that $f(x)>0, \;\forall\; x\in R$

$\therefore \quad a>0$ and $D<0$

$\Rightarrow \; a>0$ and $b^2-4ac<0$ .....(1)

Now, $g(x)=f(x)+f'(x)+f''(x)$

$=ax^2+bx+c+2ax+b+2a$

$=ax^2+(2a+b)x+(2a+b+c)$

Here, $D \;=(2a+b)^2-4a(2a+b+c)$

$=4a^2+b^2+4ab-8a^2-4ab-4ac$

$=b^2-4a^2-4ac=-4a^2+b^2-4ac$

$=(-ve)+(-ve)=-ve$ [Using eq. (1)]

Also $a>0$ from (1),

$\therefore \quad g(x)>0, \;\forall\; x\in R$

3. (a) $y=(\sin x)^{\tan x} \Rightarrow \log y=\tan x.\log\sin x$

Differentiating w.r.t. x,

$$\frac{1}{y}\frac{dy}{dx}=\sec^2 x\log\sin x+\tan x.\frac{1}{\sin x}.\cos x$$

$$\Rightarrow \frac{dy}{dx}=(\sin x)^{\tan x}[1+\sec^2 x\log\sin x]$$

4. (d) $f'(x)=\begin{vmatrix} 3x^2 & \cos x & \sin x \\ 6 & -1 & 0 \\ p & p^2 & p^3 \end{vmatrix}$

$$\Rightarrow f''(x)=\begin{vmatrix} 6x & -\sin x & -\cos x \\ 6 & -1 & 0 \\ p & p^2 & p^3 \end{vmatrix}$$

$$\Rightarrow f'''(x)=\begin{vmatrix} 6 & -\cos x & \sin x \\ 6 & -1 & 0 \\ p & p^2 & p^3 \end{vmatrix}$$

$$\therefore \quad f'''(0)=\begin{vmatrix} 6 & -1 & 0 \\ 6 & -1 & 0 \\ p & p^2 & p^3 \end{vmatrix}=0 \qquad (\because R_1\equiv R_2)$$

= independent of p.

5. (b) $x^2+y^2=1 \Rightarrow 2x+2yy'=0 \Rightarrow x+yy'=0$

$$\Rightarrow 1+yy''+(y')^2=0 \Rightarrow yy''+(y')^2+1=0$$

6. (c) $F(x)=\int_0^x f(t)dt$ and $F(x^2)=x^2(1+x)$

$F'(x)=f(x)$ ...... (1)

But $F'(x^2).2x=2x+3x^2$

$$\Rightarrow F'(x^2)=\left(\frac{2+3x}{2}\right) \Rightarrow f(x^2)=\frac{2+3x}{2}$$

$$\Rightarrow f(4)=\frac{2+3\times 2}{2}=\frac{8}{2}=4$$

7. (a) $\log(x+y)=2xy$ when $x=0$ then $y=1$

Differentiating w.r.t. x

$$\frac{1}{x+y}\left[1+\frac{dy}{dx}\right]=2y+\frac{2xdy}{dx}$$

$$\Rightarrow \frac{dy}{dx}=\frac{\frac{1}{x+y}-2y}{2x-\frac{1}{x+y}} \Rightarrow y'(0)=\frac{1-2}{0-1}=1$$

8. (d) Let us consider the function $g(x)=f(x)-x^2$ so that

$g(1)=f(1)-1^2=1-1=0$

$g(2)=f(2)-2^2=4-4=0$

$g(3)=f(3)-3^2=9-9=0$

Since $f(x)$ is twice differentiable we can say $g(x)$ is continuous and differentiable everywhere and

$g(1)=g(2)=g(3)=0$

NOTE THIS STEP:

$\therefore$ By Rolle's theorem, $g'(c)=0$ for some $c\in(1,2)$

and $g'(d)=0$ for some $d\in(2,3)$

Again by Rolle's theorem,

$$g''(e)=0 \text{ for some } e\in(c,d) \Rightarrow e\in(1,3)$$

$$\Rightarrow f''(e)-2=0 \text{ or } f''(e)=2 \text{ for some } x\in(1,3)$$

$f''(x)=2$ for some $x\in(1,3)$

9. (d) $\frac{d^2x}{d^2y}=\frac{d}{dy}\left(\frac{dx}{dy}\right)=\frac{d}{dx}\left(\frac{dx}{dy}\right)\times\frac{dx}{dy}$

$$=\left\{\frac{d}{dx}\left[\frac{1}{\left(\frac{dy}{dx}\right)}\right]\right\}\times\frac{1}{\frac{dy}{dx}}$$

$$=-\frac{1}{\left(\frac{dy}{dx}\right)^2}\times\frac{d^2y}{dx^2}\times\frac{1}{\left(\frac{dy}{dx}\right)}=-\left(\frac{dy}{dx}\right)^{-3}\frac{d^2y}{dx^2}$$

10. (a) Given that $g(x)=\log f(x) \Rightarrow g(x+1)=\log f(x+1)$

$\Rightarrow g(x+1)=\log x f(x) \quad [\because f(x+1)=xf(x)]$

$\Rightarrow g(x+1)=\log x+\log f(x) \Rightarrow g(x+1)-g(x)=\log(x)$

$$\Rightarrow g'(x+1)-g'(x)=\frac{1}{x}$$

$$\Rightarrow g''(x+1)-g''(x)=-\frac{1}{x^2}$$

Putting, $x=x-\frac{1}{2}$, we get

$$\Rightarrow g''\left(x+\frac{1}{2}\right)-g''\left(x-\frac{1}{2}\right)=-\frac{1}{\left(x-\frac{1}{2}\right)^2}=\frac{-2^2}{(2x-1)^2}$$

Putting $x=1,2,3$ ...., N we get

$$g''\left(\frac{3}{2}\right)-g''\left(\frac{1}{2}\right)=-\frac{2^2}{1^2} \qquad ...(1)$$

$$g''\left(\frac{5}{2}\right)-g''\left(\frac{3}{2}\right)=\frac{-2^2}{3^2} \qquad ...(2)$$

$$g''\left(\frac{7}{2}\right)-g''\left(\frac{5}{2}\right)=\frac{-2^2}{5^2} \qquad ...(3)$$

$$g''\left(N+\frac{1}{2}\right)-g''\left(N-\frac{1}{2}\right)=-\frac{2^2}{(2N-1)^2} \qquad ...(N)$$

Adding all the above equations, we get

$$g''\left(N+\frac{1}{2}\right)-g''\left(\frac{1}{2}\right)$$

$$=-4\left[1+\frac{1}{3^2}+\frac{1}{5^2}+.....+\frac{1}{(2N-1)^2}\right]$$

(E) SUBJECTIVE PROBLEMS :

1. Let $f(x) = \sin(x^2+1)$ then

$$f'(x) = \lim_{\delta x \to 0} \frac{\sin[(x+\delta x)^2+1] - \sin[x^2+1]}{\delta x}$$

$$\Rightarrow f'(x) = \lim_{\delta x \to 0} 2\cos\left(\frac{x^2+(\delta x)^2+2x\delta x+1+x^2+1}{2}\right) \cdot \frac{\sin\left(\frac{x^2+(\delta x)^2+2x\delta x+1-x^2-1}{2}\right)}{\delta x}$$

$$= \lim_{\delta x \to 0} \frac{2\cos\left[x^2+1+x\delta x+\frac{(\delta x)^2}{2}\right]\sin\left[\frac{(\delta x)^2+2x\delta x}{2}\right]}{\delta x\left[\frac{\delta x+2x}{2}\right]} \times \left(\frac{\delta x+2x}{2}\right)$$

$$= 2\cos(x^2+1) \lim_{\delta x \to 0} \frac{\sin\left[\frac{(\delta x)^2+2x\delta x}{2}\right]}{\left[\frac{(\delta x)^2+2x\delta x}{2}\right]} \times \left(\frac{\delta x+2x}{2}\right)$$

$$= 2\cos(x^2+1) \times 1 \times \frac{2x}{2} = 2x\cos(x^2+1)$$

2. $$f(x) = \begin{cases} \dfrac{x-1}{2x^2-7x+5}, & x \neq 1 \\ -\dfrac{1}{3}, & x = 1 \end{cases}$$

$$\therefore \quad f'(x)|_{x=1} = \lim_{h \to 0} \frac{f(1+h)-f(1)}{h}$$

$$= \lim_{h \to 0} \left[\frac{\frac{1+h-1}{2(1+h)^2-7(1+h)+5}+\frac{1}{3}}{h}\right]$$

$$= \lim_{h \to 0} \frac{\frac{h}{2h^2-3h}+\frac{1}{3}}{h}$$

$$= \lim_{h \to 0} \frac{\frac{1}{2h-3}+\frac{1}{3}}{h} = \lim_{h \to 0} \frac{2h}{3h(2h-3)}$$

$$= \lim_{h \to 0} \frac{2}{3(2h-3)}$$

$= -2/9$

3. We have,

$$y = \frac{5x}{3|1-x|} + \cos^2(2x+1)$$

(Clearly y is not defined at $x = 1$)

$$\Rightarrow \quad y = \begin{cases} \dfrac{5x}{3(1-x)} + \cos^2(2x+1), & x < 1 \\ \dfrac{5x}{3(x-1)} + \cos^2(2x+1), & x > 1 \end{cases}$$

$$\Rightarrow \quad \frac{dy}{dx} = \begin{cases} \dfrac{5}{3}\left(\dfrac{(1-x)-x(-1)}{(1-x)^2}\right) - 2\sin(4x+2), & x < 1 \\ \dfrac{5}{3}\left(\dfrac{(x-1)-x}{(x-1)^2}\right) - 2\sin(4x+2), & x > 1 \end{cases}$$

$$\text{or} \quad \frac{dy}{dx} = \begin{cases} \dfrac{5}{3}\dfrac{1}{(1-x)^2} - 2\sin(4x+2), & x < 1 \\ -\dfrac{5}{3}\dfrac{1}{(x-1)^2} - 2\sin(4x+2), & x > 1 \end{cases}$$

4. We are given $y = e^{x\sin x^3} + (\tan x)^x$

Here y is the sum of two functions and in the second function base as well as power are functions of x. Therefore we will use logarithmic differentiation here.

Let $y = u + v$

where $u = e^{x\sin x^3}$ ... (1)

and $v = (\tan x)^x$ ...(2)

$$\therefore \quad \frac{dy}{dx} = \frac{du}{dx} + \frac{dv}{dx} \quad ...(3)$$

Differentiating (1) with respect to x, we get

$$\frac{du}{dx} = e^{x\sin x^3} \cdot \frac{d}{dx}(x\sin x^3)$$

$$= e^{x\sin x^3} \cdot [3x^3 \cdot \cos x^3 + \sin x^3]$$

Taking log on both sides on equ^n (2), we get

$$\log v = x \log \tan x$$

Differentiating the above with respect to x, we get

$$\frac{1}{v}\frac{dv}{dx} = x \cdot \frac{1}{\tan x} \cdot \sec^2 x + 1 \cdot \log \tan x$$

$$\therefore \quad \frac{dv}{dx} = (\tan x)^x \left[\frac{2x}{\sin 2x} + \log \tan x\right]$$

Substituting the value of $\frac{du}{dx}$ *and* $\frac{dv}{dx}$ in eq^n (3), we get

$$\frac{dy}{dx} = e^{x\sin x^3}[\sin x^3 + 3x^3 \cos x^3] + (\tan x)^x \left[\frac{2x}{\sin 2x} + \log \tan x\right]$$

5. Given that f is twice differentiable such that

$f''(x) = -f(x)$ and $f'(x) = g(x)$

$h(x) = [f(x)]^2 + [g(x)]^2$

To find $h(10)$ when $h(5) = 11$.

Consider $h'(x) = 2ff' + 2gg' = 2f(x)g(x) + 2g(x)f''(x)$

$[\because\ g(x) = f'(x) \Rightarrow g'(x) = f''(x)]$

$= 2f(x)g(x) + 2g(x)(-f(x))$

$= 2f(x)g(x) - 2f(x)g(x) = 0$

$\therefore\ h'(x) = 0, \forall x$

$\Rightarrow$ h is a constant function

$\therefore\ h(5) = 11$

$\Rightarrow h(10) = 11$.

6. Let $F(x) = \begin{vmatrix} A(x) & B(x) & C(x) \\ A(\alpha) & B(\alpha) & C(\alpha) \\ A'(\alpha) & B'(\alpha) & C'(\alpha) \end{vmatrix}$

Given that α is a repeated root of quadratic equation $f(x) = 0$

$\therefore$ We must have $f(x) = k(x-\alpha)^2$; where k is a non- zero real no.

If we put $x = \alpha$ on both sides of eq. (1); we get

$$F(\alpha) = \begin{vmatrix} A(\alpha) & B(\alpha) & C(\alpha) \\ A(\alpha) & B(\alpha) & C(\alpha) \\ A'(\alpha) & B'(\alpha) & C'(\alpha) \end{vmatrix} = 0$$

$[\because R_1$ and R_2 are identical$]$

$\therefore\ F(\alpha) = 0$

Hence $(x-\alpha)$ is a factor of $F(x)$

Differentiating eq. (1) w.r. to x, we get

$$F'(x) = \begin{vmatrix} A'(x) & B'(x) & C'(x) \\ A(\alpha) & B(\alpha) & C(\alpha) \\ A'(\alpha) & B'(\alpha) & C'(\alpha) \end{vmatrix}$$

NOTE THIS STEP : Putting $x = \alpha$ on both sides, we get

$$F'(\alpha) = \begin{vmatrix} A'(\alpha) & B'(\alpha) & C'(\alpha) \\ A(\alpha) & B(\alpha) & C(\alpha) \\ A'(\alpha) & B'(\alpha) & C'(\alpha) \end{vmatrix} = 0$$

[as R_1 and R_3 are identical]

$\Rightarrow$ $(x-\alpha)$ is a factor of $F'(x)$ also. Or we can say $(x-\alpha)^2$ is a factor of $F(x)$.

$\Rightarrow$ $F(x)$ is divisible by $f(x)$.

7. Given that,

$$f(x) = (\log_{\cos x}\sin x)(\log_{\sin x}\cos x)^{-1} + \sin^{-1}\left(\frac{2x}{1+x^2}\right)$$

$$= \frac{\log_{\cos x}\sin x}{\log_{\sin x}\cos x} + \sin^{-1}\left(\frac{2x}{1+x^2}\right)$$

$$= \left(\frac{\log\sin x}{\log\cos x}\right)^2 + \sin^{-1}\left(\frac{2x}{1+x^2}\right)$$

$= u + v$ (let)

$$f'(x) = \frac{du}{dx} + \frac{dv}{dx} \quad ...(1)$$

and $f'(x)]_{x=\pi/4} = \left.\frac{du}{dx}\right]_{x=\pi/4} + \left.\frac{dv}{dx}\right]_{x=\pi/4}$

Now we have,

$$u = \left(\frac{\log\sin x}{\log\cos x}\right)^2$$

Differentiating both sides w.r. to x, we get

$$\frac{du}{dx} = 2\left(\frac{\log\sin x}{\log\cos x}\right)\left[\frac{\cot x\log\cos x + \tan x\log\sin x}{(\log\cos x)^2}\right]$$

$$\therefore\ \left.\frac{du}{dx}\right]_{x=\pi/4} = 2\left(\frac{\log\sin\pi/4}{\log\cos\pi/4}\right)\left[\frac{\cot\pi/4\log\cos\pi/4 + \tan\pi/4\log\sin\pi/4}{(\log\cos\pi/4)^2}\right]$$

$$\left.\frac{du}{dx}\right]_{x=\pi/4} = -8\log_2 e \quad ...(2)$$

Also, $v = \sin^{-1}\left(\frac{2x}{1+x^2}\right)$

Putting $x = \tan$, we get

$$v = \sin^{-1}\left(\frac{2\tan\theta}{1+\tan^2\theta}\right) = \sin^{-1}\sin 2\theta$$

$v = 2\theta$

$v = 2\tan^{-1}x$ $\quad [x = \tan\theta \Rightarrow \theta = \tan^{-1}x]$

$$\therefore\ \frac{dv}{dx} = \frac{2}{1+x^2}$$

$$\left.\frac{dv}{dx}\right]_{x=\pi/4} = \frac{2}{1+\frac{\pi^2}{16}} = \frac{32}{16+\pi^2} \quad ...(3)$$

Substituting the values of $\left.\frac{du}{dx}\right]_{x=\pi/4}$ and $\left.\frac{dv}{dx}\right]_{x=\pi/4}$

From eq.'s (2) and (3) in eq. (1), we get

$$f'(x)]_{x=\pi/4} = -8\log_2 e + \frac{32}{16+\pi^2}$$

8. We have, $x = \sec\theta - \cos\theta$,

$y = \sec^n\theta - \cos^n\theta$

$$\Rightarrow \frac{dx}{d\theta} = \sec\theta\tan\theta + \sin\theta$$

$= \sec\theta\tan\theta + \tan\theta\cos\theta$

$= \tan\theta(\sec\theta + \cos\theta)$

and $\frac{dy}{d\theta} = n\sec^{n-1}\theta\sec\theta\tan\theta - n\cos^{n-1}\theta(-\sin\theta)$

$= n\sec^n\theta\tan\theta + n\tan\theta\cos^n\theta$

$= n\tan\theta(\sec^n\theta + \cos^n\theta)$

$$\therefore\ \frac{dy}{dx} = \frac{\frac{dy}{d\theta}}{\frac{dx}{d\theta}} = \frac{n\tan\theta(\sec^n\theta + \cos^n\theta)}{\tan\theta(\sec\theta + \cos\theta)}$$

or $\dfrac{dy}{dx}=\dfrac{n(\sec^n\theta+\cos^n\theta)}{(\sec\theta+\cos\theta)}$ ...(1)

Also $x^2+4=(\sec\theta-\cos\theta)^2+4$
$=\sec^2\theta+\cos^2\theta-2\sec\theta\cos\theta+4$
$=\sec^2\theta+\cos^2\theta+2$
$=(\sec\theta+\cos\theta)^2$ ...(2)

and $y^2+4=\sec^n\theta-\cos^n\theta)^2+4$
$=\sec^{2n}\theta+\cos^{2n}\theta-2\sec^n\theta\cos^n\theta+4$
$=\sec^{2n}\theta+\cos^{2n}\theta+2$
$=(\sec^n\theta+\cos^n\theta)^2$ ...(3)

Now we have to prove

$$(x^2+4)\left(\frac{dy}{dx}\right)^2=n^2(y^2+4)$$

$$\text{LHS}=(\sec\theta+\cos\theta)^2.\frac{n^2(\sec^n\theta+\cos^n\theta)^2}{(\sec\theta+\cos\theta)^2}$$ [Using (1) and (2)]

$=n^2(\sec^n\theta+\cos^n\theta)^2$
$=n^2(y^2+4)$ [From eq. (3)]
$=$ RHS

9. We have given the function

$$(\sin y)^{\sin\left(\frac{\pi x}{2}\right)}+\frac{\sqrt{3}}{2}\sec^{-1}(2x)+2^x\tan[\ln(x+2)]=0$$ ... (1)

For $x=-1$, we have

$$(\sin y)^{\sin\left(-\frac{\pi}{2}\right)}+\frac{\sqrt{3}}{2}\sec^{-1}(-2)+2^{-1}\tan[\ln(-1+2)]=0$$

$$\Rightarrow\ (\sin y)^{-1}+\frac{\sqrt{3}}{2}\left(\frac{2\pi}{3}\right)+\frac{1}{2}\tan 0=0$$

$$\Rightarrow\ \frac{1}{\sin y}=-\frac{\pi}{\sqrt{3}}$$

$$\Rightarrow\ \sin y=-\frac{\sqrt{3}}{\pi},\ \text{when } x=-1$$ ....(2)

Now Let $u=(\sin y)^{\sin\left(\frac{\pi x}{2}\right)}$

Taking ln on both sides; we get

$$\ln u=\sin\left(\frac{\pi x}{2}\right)\ln\sin y$$

Differentiating both sides with respect to x, we get

NOTE THIS STEP:

$$\frac{1}{u}\frac{du}{dx}=\frac{\pi}{2}\cos\left(\frac{\pi x}{2}\right)\ln\sin y+\cot y\frac{dy}{dx}\sin\left(\frac{\pi x}{2}\right)$$

$$\Rightarrow\ \frac{du}{dx}=(\sin y)^{\sin\left(\frac{\pi x}{2}\right)}\times\left[\frac{\pi}{2}\cos\left(\frac{\pi x}{2}\right)\ln\sin y+\sin\left(\frac{\pi x}{2}\right)\cot y\frac{dy}{dx}\right]$$ ...(3)

Now differentiating eq. (1), we get

$$\frac{d}{dx}\left[(\sin y)^{\sin\left(\frac{\pi x}{2}\right)}\right]+\frac{\sqrt{3}}{2}\frac{1}{2x\sqrt{4x^2-1}}.2+2^x(\ln 2)\tan[(\ln(x+2)]+2^x\sec^2[\ln(x+2)]\frac{1}{x+2}=0$$

$$\Rightarrow\ (\sin y)^{\sin\left(\frac{\pi x}{2}\right)}\left[\frac{\pi}{2}\cos\left(\frac{\pi x}{2}\right)\ln\sin y+\sin\left(\frac{\pi x}{2}\right)\cot y\frac{dy}{dx}\right]$$
$$+\frac{\sqrt{3}}{2x\sqrt{4x^2-1}}+2^x\ln 2\tan(\ln(x+2))+\frac{2^x\sec^2[\ln(x+2)]}{x+2}=0$$

At $x=-1$ and $\sin y=-\dfrac{\sqrt{3}}{\pi}$, we get

$$\Rightarrow\ \left(-\frac{\sqrt{3}}{\pi}\right)^{-1}\left[0-(-1)\sqrt{\frac{\pi^2}{3}-1}\left(\frac{dy}{dx}\right)_{x=-1}\right]+\frac{\sqrt{3}}{-2\sqrt{3}}+0+2^{-1}=0$$

$$\Rightarrow\ -\frac{\pi}{\sqrt{3}\sqrt{3}}\sqrt{\pi^2-3}\left(\frac{dy}{dx}\right)_{x=-1}-\frac{1}{2}+\frac{1}{2}=0\Rightarrow\left.\frac{dy}{dx}\right]_{x=-1}=0$$

10. $$y=\frac{ax^2}{(x-a)(x-b)(x-c)}+\frac{bx}{(x-b)(x-c)}+\frac{c}{x-c}+1$$

$$=\frac{ax^2}{(x-a)(x-b)(x-c)}+\frac{bx}{(x-b)(x-c)}+\frac{x}{x-c}$$

$$=\frac{ax^2}{(x-a)(x-b)(x-c)}+\left(\frac{b}{x-b}+1\right)\frac{x}{x-c}$$

$$=\frac{ax^2}{(x-a)(x-b)(x-c)}+\frac{x^2}{(x-b)(x-c)}$$

$$=\left(\frac{a}{x-a}+1\right)\frac{x^2}{(x-b)(x-c)}=\frac{x^3}{(x-a)(x-b)(x-c)}$$

$$\Rightarrow\ \log y=3\log x-\log(x-a)-\log(x-b)-\log(x-c)$$

$$\Rightarrow\ \frac{y'}{y}=\frac{3}{x}-\frac{1}{x-a}-\frac{1}{x-b}-\frac{1}{x-c}$$

$$=\left(\frac{1}{x}-\frac{1}{x-a}\right)+\left(\frac{1}{x}-\frac{1}{x-b}\right)+\left(\frac{1}{x}-\frac{1}{x-c}\right)$$

$$=\frac{a}{x(a-x)}+\frac{b}{x(b-x)}+\frac{c}{x(c-x)}$$

$$=\frac{1}{x}\left[\frac{a}{a-x}+\frac{b}{b-x}+\frac{c}{c-x}\right]$$

(H) ASSERTION & REASON TYPE QUESTIONS :

1. (b) Given that $f(x) = 2 + \cos x$ which is continuous and differentiable every where.

Also $f'(x) = -\sin x$

$\therefore \; f'(x) = 0 \Rightarrow x = n\pi$

$\Rightarrow$ There exists $c \in [t, t+\pi]$ for $t \in R$

Such that $f'(c) = 0$

$\therefore$ Statements-1 is true.

Also $f(x)$ being periodic of period 2π, statement-2 is true, but statement-2 is not a correct explanation of statement-1.

2. (a) We have $f(x) = g(x) \sin x$

$\Rightarrow f'(x) = g'(x)\sin x + g(x)\cos x$

$\Rightarrow f'(0) = g'(0) \times 0 + g(0) = g(0) \quad [\because g'(0) = 0]$

$\therefore$ Statement 2 is correct.

$$\text{Also } f''(0) = \lim_{x\to 0} \frac{f'(x) - f'(0)}{x}$$

$$= \lim_{x\to 0} \frac{g(x)\cos x + g'(x)\sin x - g(0)}{x}$$

$$= \lim_{x\to 0} \frac{g(x)\cos x - g(0)}{x} + \lim_{x\to 0} \frac{g'(x)\sin x}{x}$$

$$= \lim_{x\to 0} \frac{g(x)\cos x - g(0)}{x \times \frac{\sin x}{x}} + \lim_{x\to 0} g'(x)$$

$$= \lim_{x\to 0} \frac{g(x)\cos x - g(0)}{\sin x} + g'(0)$$

$$= \lim_{x\to 0} [g(x)\cot(x) - g(0)\operatorname{cosec} x] + 0$$

$$= \lim_{x\to 0} [g(x)\cot x - g(0)\operatorname{cosec} x]$$

$\therefore$ Statement 1 is also true and is a correct explanation for statement 2.

(I) INTEGER VALUE CORRECT TYPE :

1. (2) Given that $f(x) = x^3 + e^{x/2}$

and $g(x) = f^{-1}(x)$

then we should have $gof(x) = x$

$\Rightarrow g(f(x)) = x$

$\Rightarrow g(x^3 + e^{x/2}) = x$

Differentiating both sides with respect to x, we get

$$g'(x^3 + e^{x/2}).\left(3x^2 + e^{x/2}.\frac{1}{2}\right) = 1$$

$$\Rightarrow g'(x^3 + e^{x/2}) = \frac{1}{3x^2 + e^{x/2}.\frac{1}{2}}$$

For $x = 0$, we get $g'(1) = \frac{1}{1/2} = 2$

2. (1)

$$f(\theta) = \sin\left(\tan^{-1}\left(\frac{\sin\theta}{\sqrt{\cos 2\theta}}\right)\right)$$

$$= \sin\left[\sin^{-1}\left(\frac{\sin\theta}{\sqrt{\sin^2\theta + \cos 2\theta}}\right)\right]$$

$$\left[\because \tan^{-1}\frac{x}{y} = \sin^{-1}\frac{x}{\sqrt{x^2+y^2}}\right]$$

$$= \sin\left[\sin^{-1}\left(\frac{\sin\theta}{\cos\theta}\right)\right] = \tan\theta$$

$$\therefore \frac{df(\theta)}{d\tan\theta} = 1.$$

Section-B JEE Main/ AIEEE

1. (a) $y = (x + \sqrt{1+x^2})^n$

$$\frac{dy}{dx} = n(x+\sqrt{1+x^2})^{n-1}\left(1 + \frac{1}{2}(1+x^2)^{-1/2}.2x\right);$$

$$\frac{dy}{dx} = n(x+\sqrt{1+x^2})^{n-1}\frac{(\sqrt{1+x^2}+x)}{\sqrt{1+x^2}}$$

$$= \frac{n(\sqrt{1+x^2}+x)^n}{\sqrt{1+x^2}}$$

or $\sqrt{1+x^2}\frac{dy}{dx} = ny$ or $\sqrt{1+x^2}\, y_1 = ny$

$(y_1 = \frac{dy}{dx})$ Squaring, $(1+x^2)y_1^2 = n^2y^2$

Differentiating, $(1+x^2)2y_1y_2 + y_1^2.2x = n^2.2yy_1$

or $(1+x^2)y_2 + xy_1 = n^2y$

2. (c) $F(t)=\int_0^t f(t-y)g(y)dy$

$=\int_0^t e^{t-y}ydy=e^t\int_0^t e^{-y}ydy$

$=e^t\left[-ye^{-y}-e^{-y}\right]_0^t=-e^t\left[ye^{-y}+e^{-y}\right]_0^t$

$=-e^t\left[te^{-t}+e^{-t}-0-1\right]=-e^t\left[\frac{t+1-e^t}{e^t}\right]$

$=e^t-(1+t)$

3. (d) $f(x)=x^n\Rightarrow f(1)=1$

$f'(x)=nx^{n-1}\Rightarrow f'(1)=n$

$f''(x)=n(n-1)x^{n-2}\Rightarrow f''(1)=n(n-1)$

....

.... $f^n(x)=n!\Rightarrow f^n(1)=n!$

$=1-\frac{n}{1!}+\frac{n(n-1)}{2!}-\frac{n(n-1)(n-2)}{3!}+....+(-1)^n\frac{n!}{n!}$

$={}^nC_0-{}^nC_1+{}^nC_2-{}^nC_3+......+(-1)^n\,{}^nC_n=0$

4. (b) $f(x)=ax^2+bx+c$

$f(1)=f(-1)\Rightarrow a+b+c=a-b+c$ or $b=0$

$\therefore f(x)=ax^2+c$ or $f'(x)=2ax$

Now $f'(a); f'(b)$; and $f'(c)$ are $2a(a); 2a(b); 2a(c)$ i.e. $2a^2, 2ab, 2ac$.

$\Rightarrow$ If a,b,c are in A.P. then $f'(a); f'(b)$ and $f'(c)$ are also in A.P.

5. (c) $x=e^{y+ey+....\infty}\Rightarrow x=e^{y+x}$.

Taking log.

$\log x=y+x\Rightarrow\frac{1}{x}=\frac{dy}{dx}+1$

$\therefore\frac{dy}{dx}=\frac{1}{x}-1=\frac{1-x}{x}$

6. (a) $x^2-(a-2)x-a-1=0$

$\Rightarrow\ \alpha+\beta=a-2;\ \alpha\beta=-(a+1)$

$\alpha^2+\beta^2=(\alpha+\beta)^2-2\alpha\beta=a^2-2a+6=(a-1)^2+5$

For min. value of $\alpha^2+\beta^2$ where α is an integer

$\Rightarrow\ a=1.$

7. (d) Let α, $\alpha+1$ be roots

Then $\alpha+\alpha+1=b=$ sum of roots $\alpha(\alpha+1)=c$ = product of roots

$\therefore\ b^2-4c=(2\alpha+1)^2-4\alpha(\alpha+1)=1.$

8. (d) $\lim_{x\to 2}\int_0^{f(x)}\frac{4t^3}{x-2}dt=\lim_{x\to 0}\frac{\int_0^{f(x)}4t^3\,dt}{x-2}$

Applying L Hospital rule

$\lim_{x\to 2}\frac{[4f(x)^3f'(x)]}{1}=4(f(2))^3f'(2)=4\times 6^3\times\frac{1}{48}=18$

9. (c) $f(x)=\begin{cases}\frac{x}{1-x}, & x<0\\ \frac{x}{1+x}, & x\ge 0\end{cases}$

$\Rightarrow\ f'(x)=\begin{cases}\frac{x}{(1-x)^2}, & x<0\\ \frac{x}{(1+x)^2}, & x\ge 0\end{cases}$

$\therefore\ f'(x)$ exist at everywhere.

10. (a) $x^m.y^n=(x+y)^{m+n}$

$\Rightarrow\ mlnx+nlny=(m+n)ln(x+y)$

Differentiating both sides.

$\therefore\ \frac{m}{x}+\frac{n}{y}\frac{dy}{dx}=\frac{m+n}{x+y}\left(1+\frac{dy}{dx}\right)$

$\Rightarrow\ \left(\frac{m}{x}-\frac{m+n}{x+y}\right)=\left(\frac{m+n}{x+y}-\frac{n}{y}\right)\frac{dy}{dx}$

$\Rightarrow\ \frac{my-nx}{x(x+y)}=\left(\frac{my-nx}{y(x+y)}\right)\frac{dy}{dx}\Rightarrow\ \frac{dy}{dx}=\frac{y}{x}$

11. (d) $x^{2x}-2x^x\cot y-1=0$

$\Rightarrow\ 2\cot y=x^x-x^{-x}$

$\Rightarrow\ 2\cot y=u-\frac{1}{u}$ where $u=x^x$

Differentiating both sides with respect to x, we get

$\Rightarrow\ -2\text{cosec}^2y\frac{dy}{dx}=\left(1+\frac{1}{u^2}\right)\frac{du}{dx}$

where $u=x^x\Rightarrow\log u=x\log x$

$\Rightarrow\ \frac{1}{u}\frac{du}{dx}=1+\log x\ \Rightarrow\ \frac{du}{dx}=x^x(1+\log x)$

$\therefore$ We get

$$-2\operatorname{cosec}^2 y\,\frac{dy}{dx} = (1+x^{-2x}).x^x(1+\log x)$$

$$\Rightarrow \frac{dy}{dx} = \frac{\left(x^x + x^{-x}\right)(1+\log x)}{-2(1+\cot^2 y)} \quad ....(i)$$

Now when $x = 1$, $x^{2x} - 2x^x \cot y - 1 = 0$, gives

$1 - 2\cot y - 1 = 0 \Rightarrow \quad \cot y = 0$

$\therefore$ From equation (i), at $x = 1$ and $\cot y = 0$, we get

$$y'(1) = \frac{(1+1)(1+0)}{-2(1+0)} = -1$$

12. **(a)** $g'(x) = 2\left(f(2f(x)+2)\right)\left(\frac{d}{dx}\left(f(2f(x)+2)\right)\right)$

$$= 2f(2f(x)+2)f'(2f(x))+2).(2f'(x))$$

$$\Rightarrow g'(0) = 2f(2f(0)+2).f'(2f(0)+2).2f'(0) = 4f(0)(f'(0))^2$$

$$= 4(-1)(1)^2 = -4$$

13. **(c)** $\frac{d^2x}{dy^2} = \frac{d}{dy}\left(\frac{dx}{dy}\right) = \frac{d}{dx}\left(\frac{dx}{dy}\right)\frac{dx}{dy}$

$$= \frac{d}{dx}\left(\frac{1}{dy/dx}\right)\frac{dx}{dy} = -\frac{1}{\left(\frac{dy}{dx}\right)^2}.\frac{d^2y}{dx^2}.\frac{1}{\frac{dy}{dx}} = -\frac{1}{\left(\frac{dy}{dx}\right)^3}\frac{d^2y}{dx^2}$$

14. **(a)** Let $y = \sec(\tan^{-1} x)$ and $\tan^{-1} x = \theta$.

$\Rightarrow x = \tan\theta$

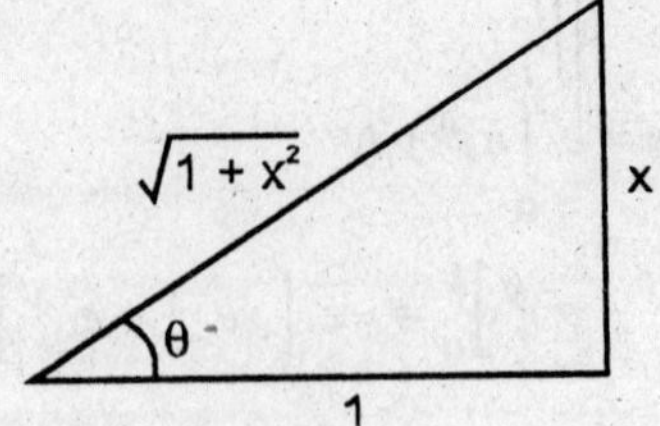

Thus, we have $y = \sec\theta$

$\Rightarrow y = \sqrt{1+x^2}$

$(\because \sec^2\theta = 1 + \tan^2\theta)$

$$\Rightarrow \frac{dy}{dx} = \frac{1}{2\sqrt{1+x^2}}\cdot 2x$$

At $x = 1$,

$$\frac{dy}{dx} = \frac{1}{\sqrt{2}}.$$

❐ ❐

Chapter 13 Properties of Triangle

Section-A JEE Advanced/ IIT-JEE

A FILL IN THE BLANKS :

1. In a ΔABC, $\angle A = 90°$ and AD is an altitude. Complete the relation

$$\frac{BD}{BA} = \frac{AB}{(....)}$$ **(1980)**

2. ABC is a triangle, P is a point on AB, and Q is point on AC such that $\angle AQP = \angle ABC$. Complete the relation

$$\frac{\text{area of } \Delta APQ}{\text{area of } \Delta ABC} = \frac{(...)}{AC^2}$$ **(1980)**

3. ABC is a triangle with $\angle B$ greater than $\angle C$. D and E are points on BC such that AD is perpendicular to BC and AE is the bisector of angle A. Complete the relation **(1980)**

$$\angle DAE = \frac{1}{2}[(\ ) - \angle C]$$

4. The set of all real numbers a such that $a^2 + 2a$, $2a + 3$ and $a^2 + 3a + 8$ are the sides of a triangle is .... **(1985 - 2 Marks)**

5. In a triangle ABC, if cot A, cot B, cot C are in A.P., then a^2, b^2, c^2, are in .... progression. **(1985 - 2 Marks)**

6. A polygon of nine sides, each of length 2, is inscribed in a circle. The radius of the circle is .... **(1987 - 2 Marks)**

7. If the angles of a triangle are 30° and 45° and the included side is $(\sqrt{3}+1)$ cms, then the area of the triangle is .... **(1988 - 2 Marks)**

8. If in a triangle ABC, $\frac{2\cos A}{a} + \frac{\cos B}{b} + \frac{2\cos C}{c} = \frac{a}{bc} + \frac{b}{ca}$, then the value of the angle A is .... degrees. **(1993 - 2 Marks)**

9. In a triangle ABC, AD is the altitude from A. Given $b > c$, $\angle C = 23°$ and $AD = \frac{abc}{b^2 - c^2}$ then $\angle B =$ .... **(1994 - 2 Marks)**

10. A circle is inscribed in an equilateral triangle of side a. The area of any square inscribed in this circle is .... **(1994 - 2 Marks)**

11. In a triangle ABC, $a : b : c = 4 : 5 : 6$. The ratio of the radius of the circumcircle to that of the incircle is .... **(1996 - 1 Mark)**

C MCQ's WITH ONE CORRECT ANSWER :

1. If the bisector of the angle P of a triangle PQR meets QR in S, then **(1979)**
   (a) $QS = SR$ (b) $QS : SR = PR : PQ$
   (c) $QS : SR = PQ : PR$ (d) None of these

2. From the top of a light-house 60 metres high with its base at the sea-level, the angle of depression of a boat is 15°. The distance of the boat from the foot of the light house is **(1983 - 1 Mark)**
   (a) $\left(\frac{\sqrt{3}-1}{\sqrt{3}+1}\right)$ 60 metres (b) $\left(\frac{\sqrt{3}+1}{\sqrt{3}-1}\right)$ 60 metres
   (c) $\left(\frac{\sqrt{3}+1}{\sqrt{3}-1}\right)^2$ metres (d) none of these

3. In a triangle ABC, angle A is greater than angle B. If the measures of angles A and B satisfy the equation $3\sin x - 4\sin^3 x - k = 0$, $0 < k < 1$, then the measure of angle C is **(1990 - 2 Marks)**
   (a) $\frac{\pi}{3}$ (b) $\frac{\pi}{2}$
   (c) $\frac{2\pi}{3}$ (d) $\frac{5\pi}{6}$

4. If the lengths of the sides of triangle are 3, 5, 7 then the largest angle of the triangle is **(1994)**
   (a) $\frac{\pi}{2}$ (b) $\frac{5\pi}{6}$
   (c) $\frac{2\pi}{3}$ (d) $\frac{3\pi}{4}$

5. In a triangle ABC, $\angle B = \frac{\pi}{3}$ and $\angle C = \frac{\pi}{4}$. Let D divide BC internally in the ratio 1 : 3 then $\frac{\sin \angle BAD}{\sin \angle CAD}$ is equal to **(1995S)**
   (a) $\frac{1}{\sqrt{6}}$ (b) $\frac{1}{3}$
   (c) $\frac{1}{\sqrt{3}}$ (d) $\sqrt{\frac{2}{3}}$

6. In a triangle ABC, $2ac\sin\frac{1}{2}(A-B+C)=$ *(2000S)*
(a) $a^2+b^2-c^2$ (b) $c^2+a^2-b^2$
(c) $b^2-c^2-a^2$ (d) $c^2-a^2-b^2$

7. In a triangle ABC, let $\angle C=\frac{\pi}{2}$. If r is the inradius and R is the circumradius of the triangle, then $2(r+R)$ is equal to *(2000S)*
(a) $a+b$ (b) $b+c$ (c) $c+a$ (d) $a+b+c$

8. A pole stands vertically inside a triangular park ΔABC. If the angle of elevation of the top of the pole from each corner of the park is same, then in ΔABC the foot of the pole is at the *(2000S)*
(a) centroid (b) circumcentre
(c) incentre (d) orthocentre

9. A man from the top of a 100 metres high tower sees a car moving towards the tower at an angle of depression of 30°. After some time, the angle of depression becomes 60°. The distance (in metres) travelled by the car during this time is *(2001S)*
(a) $100\sqrt{3}$ (b) $200\sqrt{3}/3$
(c) $100\sqrt{3}/3$ (d) $200\sqrt{3}$

10. Which of the following pieces of data does NOT uniquely determine an acute-angled triangle ABC (R being the radius of the circumcircle)? *(2002S)*
(a) a, $\sin A$, $\sin B$ (b) a, b, c
(c) a, $\sin B$, R (d) a, $\sin A$, R

11. If the angles of a triangle are in the ratio 4 : 1 : 1, then the ratio of the longest side to the perimeter is *(2003S)*
(a) $\sqrt{3}:(2+\sqrt{3})$ (b) $1:6$
(c) $1:2+\sqrt{3}$ (d) $2:3$

12. The sides of a triangle are in the ratio $1:\sqrt{3}:2$, then the angles of the triangle are in the ratio *(2004S)*
(a) $1:3:5$ (b) $2:3:4$ (c) $3:2:1$ (d) $1:2:3$

13. In an equilateral triangle, 3 coins of radii 1 unit each are kept so that they touch each other and also the sides of the triangle. Area of the triangle is *(2005S)*

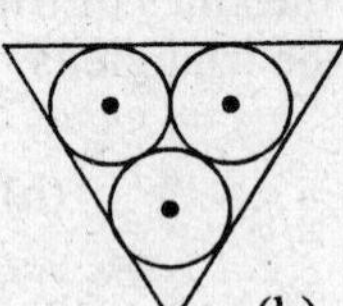

(a) $4+2\sqrt{3}$ (b) $6+4\sqrt{3}$
(c) $12+\frac{7\sqrt{3}}{4}$ (d) $3+\frac{7\sqrt{3}}{4}$

14. In a triangle ABC, a, b, c are the lengths of its sides and A, B, C are the angles of triangle ABC. The correct relation is given by *(2005S)*
(a) $(b-c)\sin\left(\frac{B-C}{2}\right)=a\cos\frac{A}{2}$
(b) $(b-c)\cos\left(\frac{A}{2}\right)=a\sin\frac{B-C}{2}$
(c) $(b+c)\sin\left(\frac{B+C}{2}\right)=a\cos\frac{A}{2}$
(d) $(b-c)\cos\left(\frac{A}{2}\right)=2a\sin\frac{B+C}{2}$

15. One angle of an isosceles Δ is 120° and radius of its incircle $=\sqrt{3}$. Then the area of the triangle in sq. units is *(2006 - 3M, –1)*
(a) $7+12\sqrt{3}$ (b) $12-7\sqrt{3}$
(c) $12+7\sqrt{3}$ (d) 4π

16. Let $ABCD$ be a quadrilateral with area 18, with side AB parallel to the side CD and $2AB = CD$. Let AD be perpendicular to AB and CD. If a circle is drawn inside the quadrilateral $ABCD$ touching all the sides, then its radius is *(2007 - 3 marks)*
(a) 3 (b) 2
(c) $\frac{3}{2}$ (d) 1

17. If the angles A, B and C of a triangle are in an arithmetic progression and if a, b and c denote the lengths of the sides opposite to A, B and C respectively, then the value of the expression $\frac{a}{c}\sin 2C+\frac{c}{a}\sin 2A$ is *(2010)*
(a) $\frac{1}{2}$ (b) $\frac{\sqrt{3}}{2}$
(c) 1 (d) $\sqrt{3}$

18. Let PQR be a triangle of area Δ with $a=2$, $b=\frac{7}{2}$ and $c=\frac{5}{2}$; where a, b, and c are the lengths of the sides of the triangle opposite to the angles at P, Q and R respectively. Then $\frac{2\sin P-\sin 2P}{2\sin P+\sin 2P}$ equals. *(2012)*
(a) $\frac{3}{4\Delta}$ (b) $\frac{45}{4\Delta}$ (c) $\left(\frac{3}{4\Delta}\right)^2$ (d) $\left(\frac{45}{4\Delta}\right)^2$

D MCQ's WITH ONE OR MORE THAN ONE CORRECT :

1. There exists a triangle ABC satisfying the conditions *(1986 - 2 Marks)*
(a) $b\sin A=a, A<\pi/2$
(b) $b\sin A>a, A>\pi/2$
(c) $b\sin A>a, A<\pi/2$
(d) $b\sin A<a, A<\pi/2, b>a$
(e) $b\sin A<a, A>\pi/2, b=a$

2. In a triangle, the lengths of the two larger sides are 10 and 9, respectively. If the angles are in A P. Then the length of the third side can be *(1987 - 2 Marks)*
(a) $5-\sqrt{6}$ (b) $3\sqrt{3}$
(c) 5 (d) $5+\sqrt{6}$ (e) none

3. If in a triangle PQR, $\sin P$, $\sin Q$, $\sin R$ are in A.P., then *(1998 - 2 Marks)*
(a) the altitudes are in A.P. (b) the altitudes are in H.P.
(c) the medians are in G.P. (d) the medians are in A.P.

4. Let $A_0A_1A_2A_3A_4A_5$ be a regular hexagon inscribed in a circle of unit radius. Then the product of the lengths of the line segments A_0A_1, A_0A_2 and A_0A_4 is *(1998 - 2 Marks)*
(a) $\frac{3}{4}$ (b) $3\sqrt{3}$
(c) 3 (d) $\frac{3\sqrt{3}}{2}$

5. In ΔABC, internal angle bisector of $\angle A$ meets side BC in D. $DE \perp AD$ meets AC in E and AB in F. Then **(2006 - 5M, –1)**

(a) AE is HM of b & c (b) $AD = \frac{2bc}{b+c}\cos\frac{A}{2}$

(c) $EF = \frac{4bc}{b+c}\sin\frac{A}{2}$ (d) ΔAEF is isosceles

6. Let ABC be a triangle such that $\angle ACB = \frac{\pi}{6}$ and let a, b and c denote the lengths of the sides opposite to A, B and C respectively. The value(s) of x for which $a = x^2 + x + 1$, $b = x^2 - 1$ and $c = 2x + 1$ is (are) **(2010)**

(a) $-(2+\sqrt{3})$ (b) $1+\sqrt{3}$

(c) $2+\sqrt{3}$ (d) $4\sqrt{3}$

7. In a triangle PQR, P is the largest angle and $\cos P = \frac{1}{3}$. Further the incircle of the triangle touches the sides PQ, QR and RP at N, L and M respectively, such that the lengths of PN, QL and RM are consecutive even integers. Then possible length(s) of the side(s) of the triangle is (are) **(JEE Adv. 2013)**

(a) 16 (b) 18

(c) 24 (d) 22

E SUBJECTIVE PROBLEMS :

1. A triangle ABC has sides $AB = AC = 5$ cm and $BC = 6$ cm Triangle $A'B'C'$ is the reflection of the triangle ABC in a line parallel to AB placed at a distance 2 cm from AB, outside the triangle ABC. Triangle $A''B''C''$ is the reflection of the triangle $A'B'C'$ in a line parallel to $B'C'$ placed at a distance of 2 cm from $B'C'$ outside the triangle $A'B'C'$. Find the distance between A and A''. **(1978)**

2. (a) If a circle is inscribed in a right angled triangle ABC with the right angle at B, show that the diameter of the circle is equal to $AB + BC - AC$.

(b) If a triangle is inscribed in a circle, then the product of any two sides of the triangle is equal to the product of the diameter and the perpendicular distance of the third side from the opposite vertex. Prove the above statement. **(1979)**

3. (a) A balloon is observed simultaneously from three points A, B and C on a straight road directly beneath it. The angular elevation at B is twice that at A and the angular elevation at C is thrice that at A. If the distance between A and B is a and the distance between B and C is b, find the height of the balloon in terms of a and b.

(b) Find the area of the smaller part of a disc of radius 10 cm, cut off by a chord AB which subtends an angle of $22\frac{1}{2}°$ at the circumference. **(1979)**

4. ABC is a triangle. D is the middle point of BC. If AD is perpendicular to AC, then prove that **(1980)**

$$\cos A\,\cos C = \frac{2(c^2 - a^2)}{3ac}$$

5. ABC is a triangle with $AB = AC$. D is any point on the side BC. E and F are points on the side AB and AC, respectively, such that DE is parallel to AC, and DF is parallel to AB. Prove that

$$DF + FA + AE + ED = AB + AC$$ **(1980)**

6. (i) PQ is a vertical tower. P is the foot and Q is the top of the tower. A, B, C are three points in the horizontal plane through P. The angles of elevation of Q from A, B, C are equal, and each is equal to θ. The sides of the triangle ABC are a, b, c; and the area of the triangle ABC is Δ. Show that the height of the tower is $\frac{abc\tan\theta}{4\Delta}$.

(ii) AB is a vertical pole. The end A is on the level ground. C is the middle point of AB. P is a point on the level ground. The portion CB subtends an angle β at P. If $AP = n\,AB$, then show that $\tan\beta = \frac{n}{2n^2+1}$ **(1980)**

7. Let the angles A, B, C of a triangle ABC be in A.P. and let $b : c = \sqrt{3} : \sqrt{2}$. Find the angle A. **(1981 - 2 Marks)**

8. A vertical pole stands at a point Q on a horizontal ground. A and B are points on the ground, d meters apart. The pole subtends angles α and β at A and B respectively. AB subtends an angle γ at Q. Find the height of the pole. **(1982 - 3 Marks)**

9. Four ships A, B, C and D are at sea in the following relative positions : B is on the straight line segment AC, B is due North of D and D is due west of C. The distance between B and D is 2 km. $\angle BDA = 40°$, $\angle BCD = 25°$. What is the distance between A and D? [Take $\sin 25° = 0.423$] **(1983 - 3 Marks)**

10. The ex-radii r_1, r_2, r_3 of ΔABC are in H.P. Show that its sides a, b, c are in A.P. **(1983 - 3 Marks)**

11. For a triangle ABC it is given that $\cos A + \cos B + \cos C = \frac{3}{2}$. Prove that the triangle is equilateral. **(1984 - 4 Marks)**

12. With usual notation, if in a triangle ABC;

$\frac{b+c}{11} = \frac{c+a}{12} = \frac{a+b}{13}$ then prove that $\frac{\cos A}{7} = \frac{\cos B}{19} = \frac{\cos C}{25}$. **(1984 - 4 Marks)**

13. A ladder rests against a wall at an angle α to the horizontal. Its foot is pulled away from the wall through a distance a, so that it slides a distance b down the wall making an angle β with the horizontal. Show that $a = b\tan\frac{1}{2}(\alpha+\beta)$ **(1985 - 5 Marks)**

14. In a triangle ABC, the median to the side BC is of length $\frac{1}{\sqrt{11-6\sqrt{3}}}$ **(1985 - 5 Marks)** and it divides the angle A into angles 30° and 45°. Find the length of the side BC.

15. If in a triangle ABC, $\cos A\cos B + \sin A\sin B\sin C = 1$, Show that $a : b : c = 1 : 1 : \sqrt{2}$ **(1986 - 5 Marks)**

16. A man observes a tower AB of height h from a point P on the ground. He moves a distance d towards the foot of the tower and finds that the angle of elevation has doubled. He further moves a distance $\frac{3}{4}d$ in the same direction and finds that the angle of elevation is three times that at P. Prove that $36h^2 = 35d^2$. **(1986 - 5 Marks)**

17. A 2 metre long object is fired vertically upwards from the mid point of two locations A and B, 8 metres apart. The speed of the object after t seconds is given by $\frac{ds}{dt} = (2t + 1)$ metres per second. Let α and β be the angles subtended by the object at A and B, respectively after one and two seconds. Find the value of $\cos(\alpha - \beta)$. **(1987 - 3 Marks)**

18. A sign-post in the form of an isosceles triangle ABC is mounted on a pole of height h fixed to the ground. The base BC of the triangle is parallel to the ground. A man standing on the ground at a distance d from the sign-post finds that the top vertex A of the triangle subtends an angle β and either of the other two vertices subtends the same angle α at his feet. Find the area of the triangle. **(1988 - 5 Marks)**

19. ABC is a triangular park with $AB = AC = 100$ m. A television tower stands at the midpoint of BC. The angles of elevation of the top of the tower at A, B, C are 45°, 60°, 60°, respectively. Find the height of the tower. **(1989 - 5 Marks)**

20. A vertical tower PQ stands at a point P. Points A and B are located to the South and East of P respectively. M is the mid point of AB. PAM is an equilateral triangle; and N is the foot of the perpendicular from P on AB. Let $AN = 20$ metres and the angle of elevation of the top of the tower at N is $\tan^{-1}(2)$. Determine the height of the tower and the angles of elevation of the top of the tower at A and B. **(1990 - 4 Marks)**

21. The sides of a triangle are three consecutive natural numbers and its largest angle is twice the smallest one. Determine the sides of the triangle. **(1991 - 4 Marks)**

22. In a triangle of base a the ratio of the other two sides is $r (< 1)$. Show that the altitude of the triangle is less than of equal to $\frac{ar}{1-r^2}$ **(1991 - 4 Marks)**

23. A man notices two objects in a straight line due west. After walking a distance c due north he orserves that the objects subtend an angle α at his eye; and, after walking a further distance $2c$ due north, an angle β. Show that the distance between the objects is $\frac{8c}{3\cot\beta - \cot\alpha}$; the height of the man is being ignored. **(1991 - 4 Marks)**

24. Three circles touch the one another externally. The tangent at their point of contact meet at a point whose distance from a point of contanct is 4. Find the ratio of the product of the radii to the sum of the radii of the circles. **(1992 - 4 Marks)**

25. An observer at O notices that the angle of elevation of the top of a tower is 30°. The line joining O to the base of the tower makes an angle of $\tan^{-1}(1/\sqrt{2})$ with the North and is inclined Eastwards. The observer travels a distance of 300 meters towards the North to a point A and finds the tower to his East. The angle of elevation of the top of the tower at A is ϕ, Find ϕ and the height of the tower **(1993 - 5 Marks)**

26. A tower AB leans towards west making an angle α with the vertical. The angular elevation of B, the topmost point of the tower is β as observed from a point C due west of A at a distance d from A. If the angular elevation of B from a point D due east of C at a distance $2d$ from C is γ, then prove that $2\tan\alpha = -\cot\beta + \cot\gamma$. **(1994 - 4 Marks)**

27. Let $A_1, A_2, ..., A_n$ be the vertices of an n-sided regular polygon such that $\frac{1}{A_1A_2} = \frac{1}{A_1A_3} + \frac{1}{A_1A_4}$, Find the value of n. **(1994 - 4 Marks)**

28. Consider the following statements concerning a triangle ABC **(1994 - 5 Marks)**
    (i) The sides a, b, c and area Δ are rational.
    (ii) $a, \tan\frac{B}{2}, \tan\frac{C}{2}$ are rational.
    (iii) $a, \sin A, \sin B, \sin C$ are rational.
    Prove that (i) $\Rightarrow$ (ii) $\Rightarrow$ (iii) $\Rightarrow$ (i)

29. A semicircular arc AB of length $2L$ and a vertical tower PQ are situated in the same vertical plane. The feet A and B of the arch and the base Q of the tower are at the same horizontal level, with B between A and Q. A man at A finds the tower hidden from his view due to the arch. He starts crawling up the arch and just sees the topmost point P of the tower after covering a distance $\frac{L}{2}$ along the arch. He crawls further to the topmost point of the arch and notes the angle of elevation of P to be θ. Compute the height of the tower in terms of L and θ. **(1997C - 5 Marks)**

30. Let A, B, C be three angles such that $A = \frac{\pi}{4}$ and $\tan B \tan C = p$. Find all possible values of p such that A, B, C are the angles of triangle. **(1997C - 5 Marks)**

31. A bird flies in a circle on a horizontal plane. An observer stands at a point on the ground. Suppose 60° and 30° are the maximum and the minimum angles of elevation of the bird and that they occur when the bird is at the points P and Q respectively on its path. Let θ be the angle of elevation of the bird when it is a point on the arc of the circle exactly midway between P and Q. Find the numerical value of $\tan^2\theta$. (Assume that the observer is not inside the vertical projection of the path of the bird.) **(1998 - 8 Marks)**

32. Prove that a triangle ABC is equilateral if and only if $\tan A + \tan B + \tan C = 3\sqrt{3}$. **(1998 - 8 Marks)**

33. Let ABC be a triangle having O and I as its circumcenter and in centre respectively. If R and r are the circumradius and the inradius, respectively, then prove that $(IO)^2 = R^2 - 2Rr$. Further show that the triangle BIO is a right-angled triangle if and only if b is arithmetic mean of a and c. **(1999 - 10 Marks)**

34. Let ABC be a triangle with incentre I and inradius r. Let D,E,F be the feet of the perpendiculars from I to the sides BC, CA and AB respectively. If r_1, r_2 and r_3 are the radii of circles inscribed in the quadrilaterals $AFIE$, $BDIF$ and $CEID$ respecitvely, prove that

$$\frac{r_1}{r-r_1}+\frac{r_2}{r-r_2}+\frac{r_3}{r-r_3}=\frac{r_1 r_2 r_3}{(r-r_1)(r-r_2)(r-r_3)}.$$

(2000 - 7 Marks)

35. If Δ is the area of a triangle with side lengths a, b, c, then show that $\Delta \le \frac{1}{4}\sqrt{(a+b+c)\,abc}$. Also show that the equality occurs in the above inequality if and only if $a=b=c$. ***(2001 - 6 Marks)***

36. If I_n is the area of n sided regular polygon inscribed in a circle of unit radius and O_n be the area of the polygon circumscribing the given circle, prove that

$$I_n=\frac{O_n}{2}\left(1+\sqrt{1-\left(\frac{2I_n}{n}\right)^2}\right).$$

(2003 - 4 Marks)

I INTEGER VALUE CORRECT TYPE :

1. Let ABC and ABC' be two non-congruent triangles with sides $AB = 4$, $AC = AC' = 2\sqrt{2}$ and angle $B = 30°$. The absolute value of the difference between the areas of these triangles is ***(2009)***

2. Consider a triangle ABC and let a, b and c denote the lengths of the sides opposite to vertices A, B and C respectively. Suppose a=6, b=10 and the area of the triangle is $15\sqrt{3}$, if $\angle ACB$ is obtuse and if r denotes the radius of the incircle of the triangle, then r^2 is equal to ***(2010)***

Section-B JEE Main/ AIEEE

1. The sides of a triangle are $3x+4y$, $4x+3y$ and $5x+5y$ where x, $y > 0$ then the triangle is **[2002]**

(a) right angled (b) obtuse angled
(c) equilateral (d) none of these

2. In a triangle with sides $a, b, c, r_1 > r_2 > r_3$ (which are the ex-radii) then **[2002]**

(a) $a>b>c$ (b) $a<b<c$
(c) $a>b$ and $b<c$ (d) $a<b$ and $b>c$

3. The sum of the radii of inscribed and circumscribed circles for an n sided regular polygon of side a , is **[2003]**

(a) $\frac{a}{4}\cot\left(\frac{\pi}{2n}\right)$ (b) $a\cot\left(\frac{\pi}{n}\right)$
(c) $\frac{a}{2}\cot\left(\frac{\pi}{2n}\right)$ (d) $a\cot\left(\frac{\pi}{2n}\right)$.

4. In a triangle ABC, medians AD and BE are drawn. If AD = 4, $\angle DAB=\frac{\pi}{6}$ and $\angle ABE=\frac{\pi}{3}$, then the area of the ΔABC is

(a) $\frac{64}{3}$ (b) $\frac{8}{3}$ (c) $\frac{16}{3}$ (d) $\frac{32}{3\sqrt{3}}$ **[2003]**

5. If in a ΔABC $a\cos^2\left(\frac{C}{2}\right)+c\cos^2\left(\frac{A}{2}\right)=\frac{3b}{2}$, then the sides a, b and c **[2003]**

(a) satisfy $a+b=c$ (b) are in A.P
(c) are in G.P (d) are in H.P.

6. The sides of a triangle are $\sin\alpha$, $\cos\alpha$ and $\sqrt{1+\sin\alpha\cos\alpha}$ for some $0<\alpha<\frac{\pi}{2}$. Then the greatest angle of the triangle is **[2004]**

(a) 150° (b) 90°
(c) 120° (d) 60°

7. A person standing on the bank of a river observes that the angle of elevation of the top of a tree on the opposite bank of the river is 60° and when he retires 40 meters away from the tree the angle of elevation becomes 30°. The breadth of the river is **[2004]**

(a) 60 m (b) 30 m (c) 40 m (d) 20 m

8. In a triangle ABC, let $\angle C=\frac{\pi}{2}$. If r is the inradius and R is the circumradius of the triangle ABC, then 2 $(r+R)$ equals **[2005]**

(a) $b+c$ (b) $a+b$ (c) $a+b+c$ (d) $c+a$

9. If in a ΔABC, the altitudes from the vertices A, B, C on opposite sides are in H.P, then $\sin A$, $\sin B$, $\sin C$ are in **[2005]**

(a) G.P. (b) A. P. (c) A.P-G.P. (d) H. P

10. A tower stands at the centre of a circular park. A and B are two points on the boundary of the park such that AB (= a) subtends an angle of 60° at the foot of the tower, and the angle of elevation of the top of the tower from A or B is 30°. The height of the tower is **[2007]**

(a) $a/\sqrt{3}$ (b) $a\sqrt{3}$ (c) $2a/\sqrt{3}$ (d) $2a\sqrt{3}$.

11. AB is a vertical pole with B at the ground level and A at the top. A man finds that the angle of elevation of the point A from a certain point C on the ground is 60°. He moves away from the pole along the line BC to a point D such that CD=7 m. From D the angle of elevation of the point A is 45°. Then the height of the pole is **[2008]**

(a) $\frac{7\sqrt{3}}{2}\frac{1}{\sqrt{3}-1}m$ (b) $\frac{7\sqrt{3}}{2}(\sqrt{3}+1)m$
(c) $\frac{7\sqrt{3}}{2}(\sqrt{3}-1)m$ (d) $\frac{7\sqrt{3}}{2}\frac{1}{\sqrt{3}+1}m$

12. For a regular polygon, let r and R be the radii of the inscribed and the circumscribed circles. A ***false*** statement among the following is **[2010]**

(a) There is a regular polygon with $\frac{r}{R}=\frac{1}{\sqrt{2}}$
(b) There is a regular polygon with $\frac{r}{R}=\frac{2}{3}$
(c) There is a regular polygon with $\frac{r}{R}=\frac{\sqrt{3}}{2}$
(d) There is a regular polygon with $\frac{r}{R}=\frac{1}{2}$

Solutions & Explanations

Section-A : JEE Advanced/ IIT-JEE

A 1. BC 2. AP^2 3. $\angle B$ 4. $(5, \infty)$ 5. arithmetic 6. $\text{cosec}\,\frac{\pi}{9}$

7. $\frac{\sqrt{3}+1}{2}$ sq. units 8. 90° 9. 113° 10. $\frac{a^2}{6}$ sq. units 11. 16:7

C 1. (c) 2. (b) 3. (c) 4. (c) 5. (a) 6. (b)
7. (a) 8. (b) 9. (b) 10. (d) 11. (a) 12. (d)
13. (b) 14. (b) 15. (c) 16. (b) 17. (d) 18. (c)

D 1. (a, d) 2. (a, d) 3. (b) 4. (c) 5. (a,b,c,d) 6. (b)
7. (b, d)

E 1. $8\sqrt{\frac{17}{5}}$ 3. (a) $\frac{a}{2b}\sqrt{(a+b)(3b-a)}$ (b) 3.91 sq. cm 7. 75°

8. $\frac{d}{\sqrt{\cot^2\alpha+\cot^2\beta-\cot\alpha\cot\beta\cot\gamma}}$ 9. 4.28 km 14. 2 units 17. $5/\sqrt{26}$

18. $(d\tan\beta-h)\sqrt{h^2\cot^2\alpha-d^2}$ 19. $50\sqrt{3}\ m$ 20. $40\sqrt{3}\ m$, 60°, 45° 21. 4, 5, 6

24. 16:1 25. 45°, $150\sqrt{2}\ m$ 27. 7 29. $\frac{2L}{\pi}\left[\frac{\sqrt{2}-\cot\theta}{1-\cot\theta}\right]$

30. $p\in(-\infty, 0]\cup[3+2\sqrt{2}, \infty)$ 31. $\frac{3}{5}$

I 1. 4 2. 3

Section-B : JEE Main/ AIEEE

1. (b) 2. (a) 3. (c) 4. (d) 5. (b) 6. (c)
7. (d) 8. (b) 9. (b) 10. (a) 11. (b) 12. (b)

Section-A JEE Advanced/ IIT-JEE

A FILL IN THE BLANKS :

1. We know that altitude from right vertex to hypotenuse in right angled triangle divides it into two triangles each being similar to the original triangle.

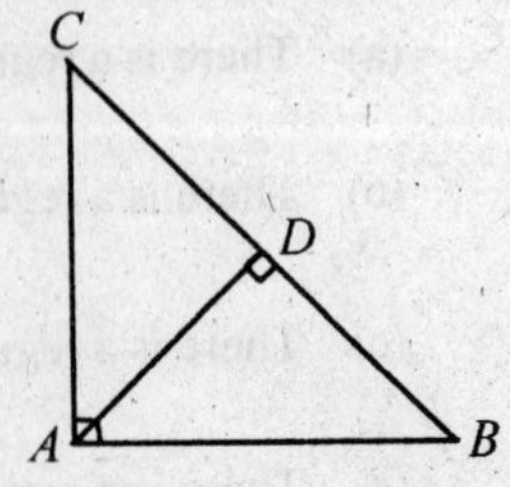

$\therefore\quad \Delta BDA \sim \Delta BAC$

$\Rightarrow\quad \frac{BD}{BA}=\frac{AB}{BC}$

2. In ΔAPQ and ΔACB

$\angle A=\angle A$ (common)

$\angle AQP=\angle ABC$ (given)

$\therefore\quad \Delta APQ \sim \Delta ACB$ fig

is with (iii) (by AA similarity)

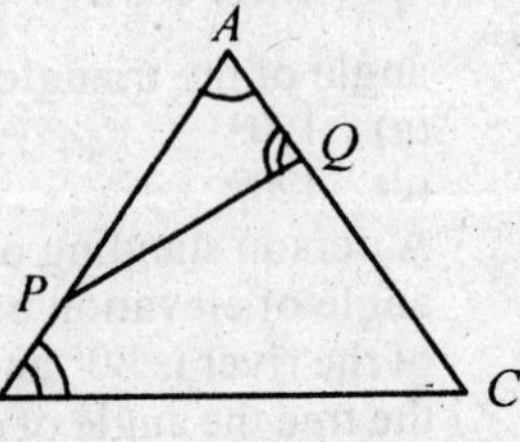

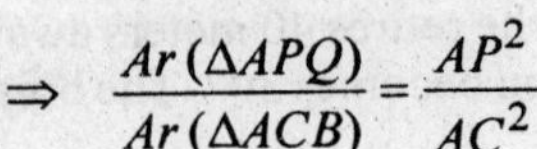

$\Rightarrow\quad \frac{Ar(\Delta APQ)}{Ar(\Delta ACB)}=\frac{AP^2}{AC^2}$

3. We have $\angle BAE = \angle CAE$ (given)
and $\angle ADB = \angle ADC = 90°$ (given)

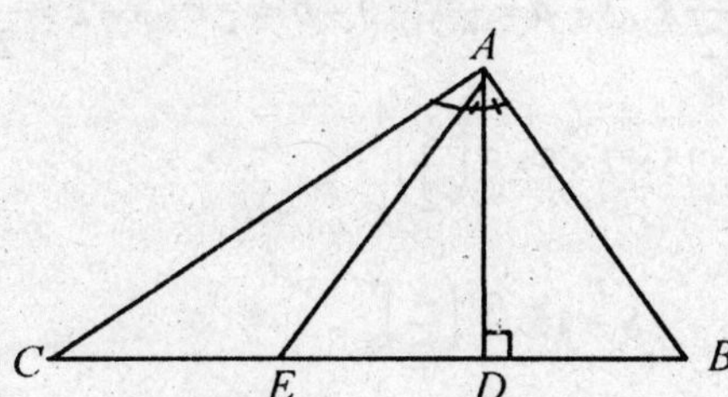

Now $\angle DAE = \angle BAE - \angle BAD$
$= \angle CAE - (90° - \angle B)$
$= (\angle CAD - \angle DAE) - 90° + \angle B$
$= (90° - \angle C) - \angle DAE - 90° + \angle B$
$\Rightarrow\ 2\angle DAE = \angle B - \angle C$
$\Rightarrow\ \angle DAE = \frac{1}{2}(\angle B - \angle C)$

4. If $a^2 + 2a,\ 2a + 3,\ a^2 + 3a + 8$ are sides of a Δ then sum of any two sides is greater than the third side.
Let $x = a^2 + 2a$; $y = 2a + 3$; $z = a^2 + 3a + 8$
Then $x + y > z \Rightarrow a^2 + 4a + 3 > a^2 + 3a + 8$
$\Rightarrow\ a > 5$ .... (1)
$y + z > x$
$\Rightarrow\ a^2 + 5a + 11 > a^2 + 2a \Rightarrow 3a > -11$
$\Rightarrow\ a > -11/3$
$z + x > y$
$\Rightarrow\ 2a^2 + 5a + 8 > 2a + 3$
$\Rightarrow\ 2a^2 + 3a + 5 > 0$
Here coeff. of $a^2 > 0$ and $D = 9 - 40 = -$ve
$\therefore$ it is true for all values of a. Therefore, identity.
Combining (1) and (2), we get $a > 5$.
$\therefore\ a \in (5, \infty)$

5. $\cot A,\ \cot B,\ \cot C$ are in A.P.
$\Rightarrow\ \cot B - \cot A = \cot C - \cot B$
$\Rightarrow\ \dfrac{\sin(A-B)}{\sin A \sin B} = \dfrac{\sin(B-C)}{\sin B \sin C}$
$\Rightarrow\ \sin(A-B)\sin(A+B) = \sin(B+C)\sin(B-C)$
$\Rightarrow\ \sin^2 A - \sin^2 B = \sin^2 B - \sin^2 C$
$\Rightarrow\ a^2 - b^2 = b^2 - c^2 \Rightarrow a^2, b^2, c^2$ are in A.P.

6. Let $AB = 2$ units be one of the sides of the polygon.
Then $\angle AOB = 2\pi/9$ where O is the centre of circle.

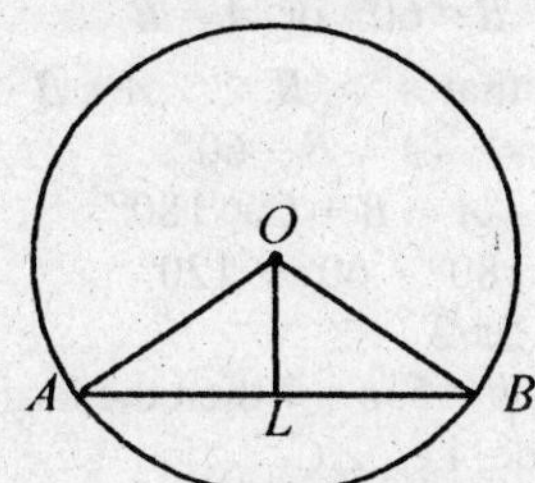

If $OL \perp AB$, then $AL = 1$ and $\angle AOL = \pi/9$
$\therefore$ Radius of the circle
$= OA = AL \operatorname{cosec} \pi/9 = \operatorname{cosec} \pi/9$.

7. $\angle C = 180° - (30° + 45°)$
$= 105°$

NOTE THIS STEP : By Sine Law in ΔABC,

$$\frac{a}{\sin 30°} = \frac{b}{\sin 45°} = \frac{\sqrt{3}+1}{\sin 105°}$$

$$\Rightarrow \frac{a}{1/2} = \frac{b}{1/\sqrt{2}} = \frac{\sqrt{3}+1}{\frac{\sqrt{3}}{2}\cdot\frac{1}{2} + \frac{1}{2}\cdot\frac{1}{\sqrt{2}}}$$

$$\Rightarrow \frac{a}{\sqrt{2}} = \frac{b}{2} = 1 \Rightarrow a = \sqrt{2},\ b = 2$$

$$\therefore\ \Delta = \frac{1}{2}ab\sin C = \frac{1}{2}.\sqrt{2}\,.\,\sqrt{2} \times \sin(105°)$$

$$= \sqrt{2}\left(\frac{\sqrt{3}+1}{2\sqrt{2}}\right) = \frac{\sqrt{3}+1}{2} \text{ sq. units}$$

ALTERNATE SOLUTION :

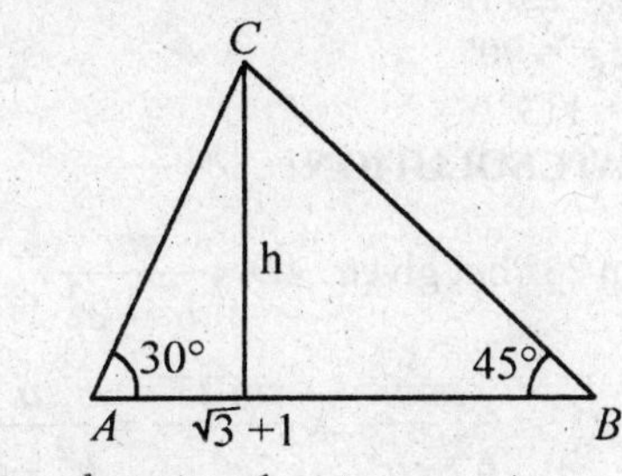

$$\tan 30° = \frac{h}{\sqrt{3}+1-h} = \frac{1}{\sqrt{3}}$$

$\Rightarrow\ h\sqrt{3} + h = \sqrt{3} + 1 \Rightarrow h = 1$

$$\therefore\ \text{Area} = \frac{1}{2} \times (\sqrt{3}+1) \times 1 = \frac{\sqrt{3}+1}{2} \text{ sq. units}$$

8. In ΔABC, we have

$$\frac{2\cos A}{a} + \frac{\cos B}{b} + \frac{2\cos C}{c} = \frac{a}{bc} + \frac{b}{ca}$$

$$\Rightarrow \frac{2\left(\frac{b^2+c^2-a^2}{2bc}\right)}{a} + \frac{\frac{a^2+c^2-b^2}{2ac}}{b} + \frac{2\left(\frac{b^2+a^2-c^2}{2ab}\right)}{c} = \frac{a^2+b^2}{abc}$$

$\Rightarrow\ 2b^2 + 2c^2 - 2a^2 + a^2 + c^2 - b^2 + 2a^2 + 2b^2 - 2c^2 = 2a^2 + 2b^2$
$\Rightarrow\ b^2 + c^2 = a^2$
$\Rightarrow\ \Delta$ is right $\angle$ed' at $A \Rightarrow \angle A = 90°$

9. Given that $AD \perp BC$, $\angle C = 23°$, $b > c$
and $AD = \dfrac{abc}{b^2 - c^2}$; to find $\angle B$

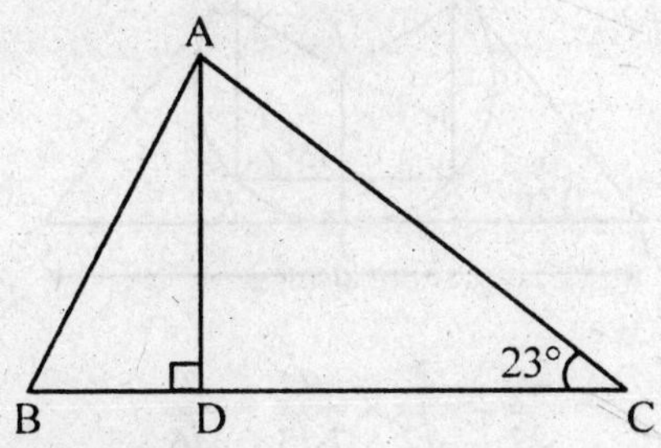

$\therefore$ Let us consider

NOTE THIS STEP : $\cos B = \dfrac{a^2 + c^2 - b^2}{2ac}$

$\Rightarrow \quad \cos B = \dfrac{a^2}{2ac} - \left(\dfrac{b^2 - c^2}{abc}\right) \cdot \dfrac{b}{2}$

$\Rightarrow \quad \cos B = \dfrac{a}{2c} - \dfrac{b}{2AD}$

$\Rightarrow \quad \cos B = \dfrac{\sin A}{2\sin C} - \dfrac{1}{2\sin C}$

$\Rightarrow \quad \cos B = \dfrac{\sin A - 1}{2\sin 23^\circ} \qquad (\because \angle C = 23^\circ)$

$\Rightarrow \quad 2 \sin 23^\circ \cos B = \sin A - 1$

$\Rightarrow \quad \sin(B + 23^\circ) + \sin(23^\circ - B) = \sin[180^\circ - (23^\circ + B)] - 1$

$\Rightarrow \quad \sin(B - 23^\circ) = 1$

$\Rightarrow \quad B - 23^\circ = 90^\circ$

$\Rightarrow \quad \angle B = 113^\circ$

ALTERNATE SOLUTION :

$AD = b \sin 23^\circ$ but given $AD = \dfrac{abc}{b^2 - c^2}$

$\Rightarrow \quad b \sin 23^\circ = \dfrac{abc}{b^2 - c^2} \Rightarrow \dfrac{\sin 23^\circ}{c} = \dfrac{a}{b^2 - c^2} \quad ....(1)$

Also $\quad \dfrac{\sin 23^\circ}{c} = \sin A \quad ....(2)$

(by sine law)

From (1) and (2)

$\dfrac{a}{b^2 - c^2} = \dfrac{\sin A}{a} \Rightarrow \dfrac{a^2}{b^2 - c^2} = \dfrac{\sin^2 A}{\sin^2 B - \sin^2 C} = \sin A$

$\Rightarrow \quad \sin(B + C)\sin(B - C) = \sin A$

$\Rightarrow \quad \sin(B - C) = 1 \Rightarrow B - C = 90^\circ$

$\Rightarrow \quad B = 90^\circ + 23^\circ = 113^\circ$

10. Given that ABC is an equilateral Δ, of side a, r is the radius of circle inscribed in it.

$\therefore$ In ΔIBL

$\tan 30^\circ = \dfrac{r}{a/2} \Rightarrow r = \dfrac{a}{2\sqrt{3}}$

If $PQRS$ is the square inscribed in circle of radius r, then side of square $= 2(r \cos 45^\circ)$

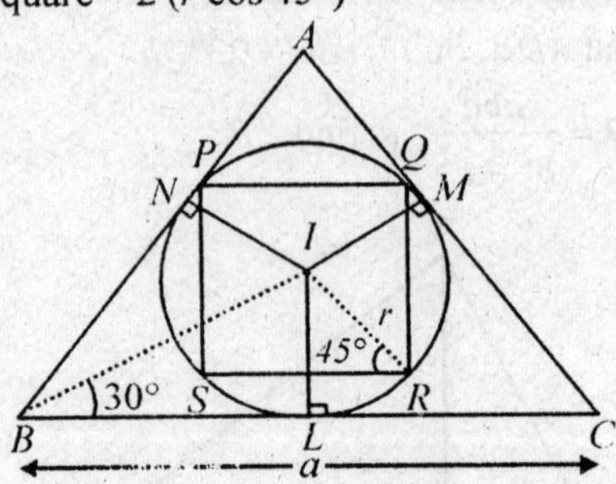

$= \dfrac{2r}{\sqrt{2}} = \dfrac{2}{\sqrt{2}} \times \dfrac{a}{2\sqrt{3}} = \dfrac{a}{\sqrt{6}}$

$\therefore$ area of square $= a^2/6$ sq. units.

11. Let $a = 4k$, $b = 5k$, $c = 6k$

$s = \dfrac{15}{2}k$, $s - a = \dfrac{7}{2}k$, $s - b = \dfrac{5}{2}k$, $s - c = \dfrac{3}{2}k$

$S^2 = 15 \times 7 \times 5 \times 3 \left(\dfrac{k}{2}\right)^4$

$\therefore \quad S = 15\sqrt{7}\left(\dfrac{k}{2}\right)^2$

$r = \dfrac{S}{s} = 15\sqrt{7}\left(\dfrac{k}{2}\right)^4 \div \dfrac{15}{2}k = \sqrt{7}\dfrac{k}{2} \quad R = \dfrac{abc}{4S}$

NOTE THIS STEP :

$= \dfrac{4.5.6k^3}{4.15\sqrt{7}k^2/4} = \dfrac{8}{\sqrt{7}}k.$

$\therefore \quad \dfrac{R}{r} = \dfrac{8}{\sqrt{7}} \div \dfrac{\sqrt{7}}{2} = \dfrac{16}{7}$

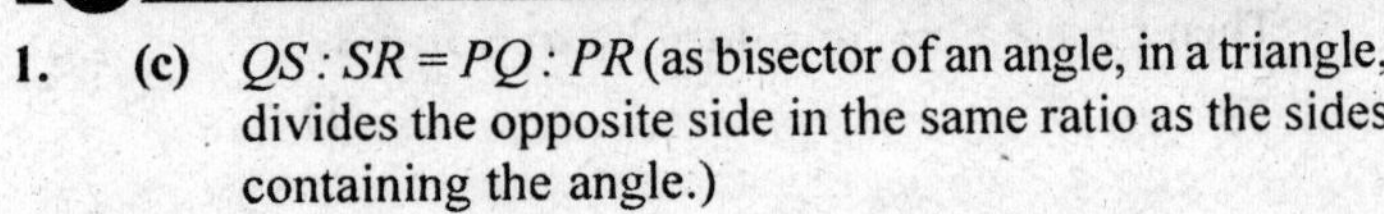

C MCQ's WITH ONE CORRECT ANSWER

1. **(c)** $QS : SR = PQ : PR$ (as bisector of an angle, in a triangle, divides the opposite side in the same ratio as the sides containing the angle.)

2. **(b)** $\tan 15^\circ = \dfrac{60}{x}$

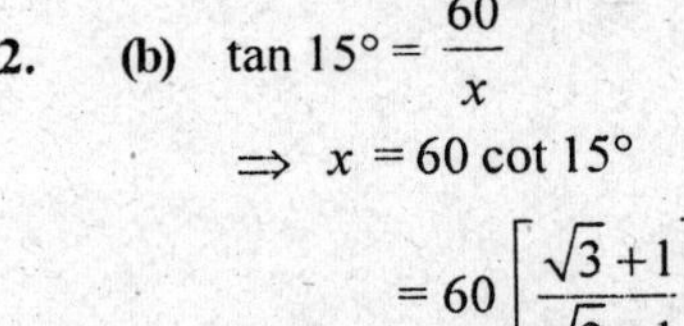

$\Rightarrow x = 60 \cot 15^\circ$

$= 60\left[\dfrac{\sqrt{3}+1}{\sqrt{3}-1}\right]$

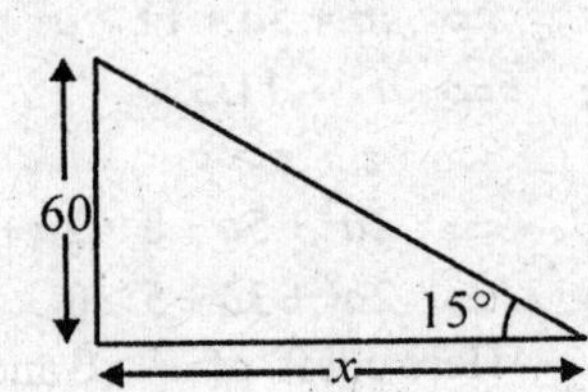

3. **(c)** Given that $A > B$

and $3 \sin x - 4 \sin^3 x - k = 0$, $\quad 0 < k < 1$

$\Rightarrow \quad \sin 3x = k$

As A and B satisfy above eq. (given)

$\therefore \quad \sin 3A = k$, $\sin 3B = k$

$\Rightarrow \quad \sin 3A - \sin 3B = 0$

$\Rightarrow \quad 2\cos\dfrac{3A + 3B}{2} \sin\dfrac{3A - 3B}{2} = 0$

$\Rightarrow \quad \cos\left(\dfrac{3A + 3B}{2}\right) = 0$ or $\sin\left(\dfrac{3A - 3B}{2}\right) = 0$

$\Rightarrow \quad \dfrac{3A + 3B}{2} = 90^\circ$ or $\dfrac{3A - 3B}{2} = 0$

$\Rightarrow \quad A + B = 60^\circ$ or $A = B$

But given that $A > B$, $\therefore A \neq B$

Thus, $\quad A + B = 60^\circ$

But $\quad A + B + C = 180^\circ$

$\therefore \quad C = 180^\circ - 60^\circ = 120^\circ$

$\therefore \quad C = 2\pi/3$

4. **(c)** Let $a = 3$, $b = 5$, $c = 7$ then the largest $\angle$ is opp. to the longest side, i.e., $\angle C$

$\therefore \quad \cos C = \dfrac{a^2 + b^2 - c^2}{2ab} = \dfrac{9 + 25 - 49}{2 \times 3 \times 5} = \dfrac{-1}{2}$

$\Rightarrow \quad C = 2\pi/3$

5. (a)

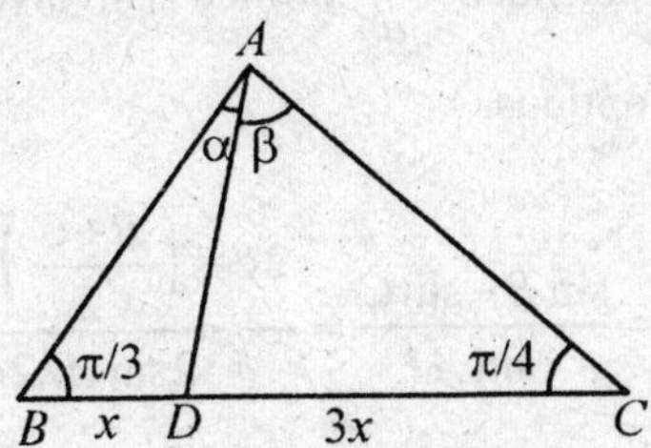

In ΔABD, applying Sine law we get

$$\frac{AD}{\sin \pi/3} = \frac{x}{\sin \alpha}$$

$$\Rightarrow \quad AD = \frac{\sqrt{3}x}{2\sin\alpha} \quad ....(1)$$

In ΔACD, applying Sine law, we get

$$\frac{AD}{\sin \pi/4} = \frac{3x}{\sin\beta}$$

$$\Rightarrow \quad AD = \frac{3x}{\sqrt{2}\sin\beta} \quad ....(2)$$

From (i) and (ii) $\frac{\sqrt{3}x}{2\sin\alpha} = \frac{3x}{\sqrt{2}\sin\beta}$

$$\Rightarrow \quad \frac{\sin\alpha}{\sin\beta} = \frac{1}{\sqrt{6}}$$

6. (b) We know that $A + B + C = 180°$

$$\Rightarrow \quad A + C - B = 180° - 2B$$

Now $2ac \sin\left[\frac{1}{2}(A - B + C)\right] = 2ac \sin(90° - B)$

$$= 2ac \cos B = \frac{2ac\,(a^2 + c^2 - b^2)}{2ac} = a^2 + c^2 - b^2$$

7. (a) Here

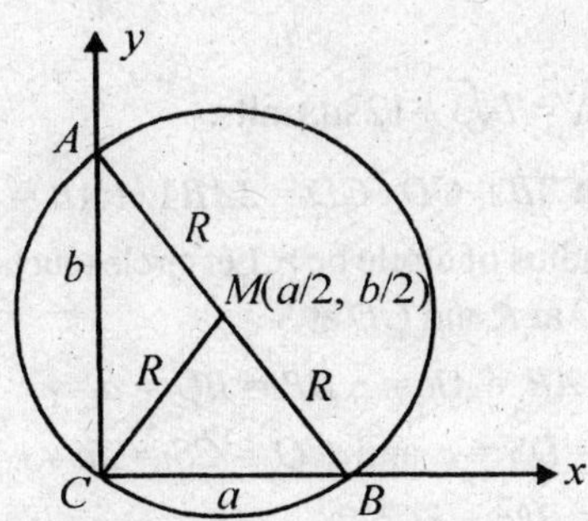

$$R^2 = MC^2 = \frac{1}{4}\left(a^2 + b^2\right) = \frac{1}{4}c^2$$

(by distance from origin) (pythagorous theorem)

$$\Rightarrow \quad R = \frac{c}{2}$$

Next $r = (s - c)\tan(C/2) = (s - c)\tan \pi/4 = s - c$

Thus, $2(r + R) = 2r + 2R = 2s - 2c + c = a + b$

8. (b)

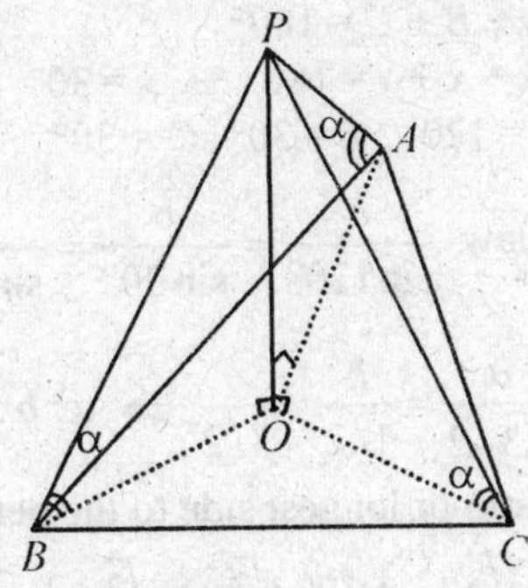

Let OP = Pole,

$\angle PAO = \angle PBO = \angle PCQ = \alpha$

$$\frac{OP}{OB} = \tan\alpha \Rightarrow OB = OP\cot\alpha \quad .... (1)$$

Similarly $\quad OA = OP \cot\alpha \quad .... (2)$

Similarly $\quad OC = OP \cot\alpha \quad .... (3)$

From (1), (2) and (3), $OA = OB = OC \Rightarrow O$ is the point of circumcentre of the triangle ABC.

9. (b)

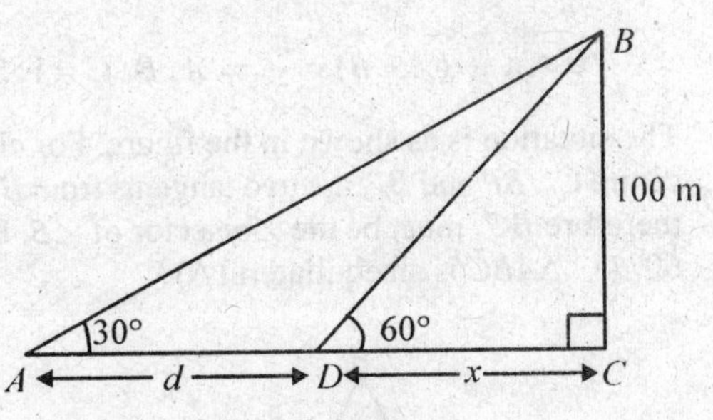

In ΔCBD, $\tan 60° = \frac{100}{x}$

$$\Rightarrow \quad x = \frac{100}{\sqrt{3}}$$

In ΔACB, $\tan 30° = \frac{100}{x + d}$

$x + d = 100\sqrt{3}$.

$$\Rightarrow \quad d = 100\sqrt{3} - \frac{100}{\sqrt{3}} = \frac{200}{\sqrt{3}} = \frac{200\sqrt{3}}{3}m$$

10. (d) We know by Sine law in ΔABC as

$$\frac{a}{\sin A} = \frac{b}{\sin B} = \frac{c}{\sin(\pi - A - B)} = 2R$$

$$\Rightarrow \quad \frac{a}{\sin A} = \frac{b}{\sin B} = \frac{c}{\sin(A + B)} = 2R$$

(a) If we know a, $\sin A$, $\sin B$ we can find b, c; values of $\angle$'s A, B and C all.

(b) Using a b, c we can find $\angle A$, $\angle B$, $\angle C$ using cosine law.

(c) a, $\sin B$, R are given then $\sin A$, b and hence $\sin(A + B)$ and then C can be found.

(d) If we know a, $\sin A$, R then we know only the ratio $\frac{b}{\sin B} = \frac{c}{\sin(A + B)}$; we can not determine the values of b, c, $\sin B$, $\sin C$ seperately.

$\therefore$ Δ can not be determined in this case.

11. (a) Given that $A : B : C = 4 : 1 : 1$

Let $A = 4x,\ B = x,\ C = x$

But $A + B + C = 180°$

$\Rightarrow\ 4x + x + x = 180° \Rightarrow x = 30°$

$\therefore\ A = 120°,\ B = 30°,\ C = 30°$

By sine law, $\dfrac{a}{\sin 120°} = \dfrac{b}{\sin 30°} = \dfrac{c}{\sin 30°}$

$\Rightarrow\ \dfrac{a}{\sqrt{3}/2} = \dfrac{b}{1/2} = \dfrac{c}{1/2} \Rightarrow a : b : c = \sqrt{3} : 1 : 1$

$\therefore$ Ratio of longest side to the perimeter

$= \sqrt{3} : 1 + 1 + \sqrt{3} = \sqrt{3} : 2 + \sqrt{3}$

12. (d) Sides are in the Ratio $1 : \sqrt{3} : 2$

Let $a = k,\ b = \sqrt{3}\,k$ and $c = 2k$

$\cos A = \dfrac{b^2 + c^2 - a^2}{2bc} = \dfrac{\sqrt{3}}{2} \Rightarrow A = \dfrac{\pi}{6}$

$\cos B = \dfrac{c^2 + a^2 - b^2}{2ac} = \dfrac{1}{2} \Rightarrow B = \dfrac{\pi}{3}$

$\Rightarrow\ C = \pi - (A + B) = \dfrac{\pi}{2} \Rightarrow A : B : C = 1 : 2 : 3$

13. (b) The situation is as shown in the figure. For circle with centre C_2, BP and BP' are two tangents from B to circle, therefore BC_2 must be the $\angle$bisector of $\angle B$. But $\angle B = 60°$ ($\because$ ΔABC is an equilateral Δ)

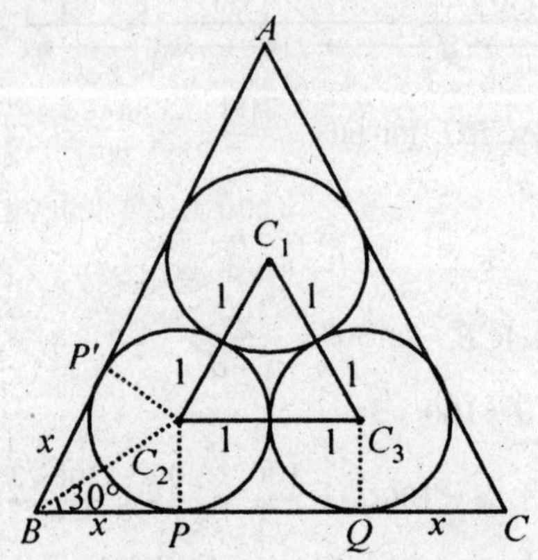

$\therefore\ \angle C_2BP = 30°$

$\therefore\ \tan 30° = \dfrac{1}{x}$

$\Rightarrow\ x = \sqrt{3}$

$\therefore\ BC = BP + PQ + QC$

$= x + 2 + x = 2 + 2\sqrt{3}$

$\therefore$ Area of $\Delta ABC = \dfrac{\sqrt{3}}{4} \times (2 + 2\sqrt{3})^2$

$= \sqrt{3}(1 + \sqrt{3})^2$

$= \sqrt{3}(1 + 3 + 2\sqrt{3})$

$= 4\sqrt{3} + 6$ sq. units.

14. (b) Let us consider $\dfrac{b-c}{a}$, which is involved in each of the these options.

$$\frac{b-c}{a} = \frac{\sin B - \sin C}{\sin A} = \frac{2\cos\left(\frac{B+C}{2}\right)\sin\left(\frac{B-C}{2}\right)}{2\sin A/2 \cos A/2}$$

$$= \frac{\sin A/2\, 2\sin\left(\frac{B-C}{2}\right)}{\sin A/2 \cos A/2} = \frac{\sin\left(\frac{B-C}{2}\right)}{\cos A/2}$$

$\therefore\ (b - c)\cos A/2 = a\sin\left(\dfrac{B-C}{2}\right)$

15. (c) By Sine law in ΔABC

$\dfrac{x}{\sin 30°} = \dfrac{a}{\sin 120°} \Rightarrow a = x\sqrt{3}$

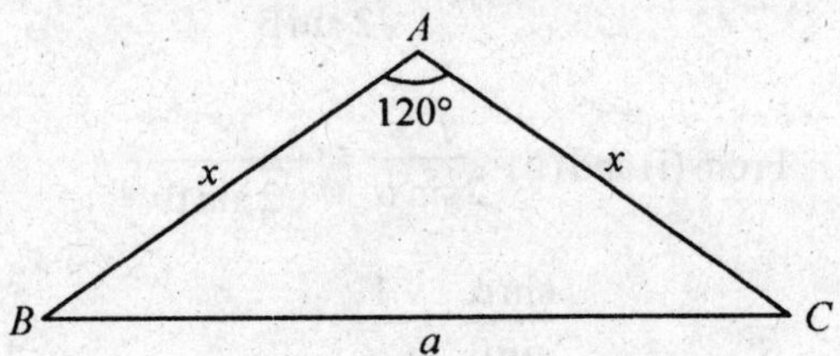

$\therefore\ \Delta = \dfrac{1}{2} \times x \times x \sin 120° = \dfrac{\sqrt{3}}{4}x^2$

Also, $\sqrt{3} = \dfrac{\Delta}{s} \Rightarrow \dfrac{(2x+a)}{2}\sqrt{3} = \dfrac{\sqrt{3}}{4}x^2$

$\Rightarrow\ x = 2(2 + \sqrt{3}) \Rightarrow \Delta = \dfrac{\sqrt{3}}{4} \times 4(4 + 3 + 4\sqrt{3})$

$\Rightarrow\ \Delta = 7\sqrt{3} + 12$ sq. units.

16. (b) Given $AB \parallel CD$, $CD = 2AB$ Let $AB = a$ then $CD = 2a$

Let radius of circle be r. Let circle touches AB at P, BC at Q, AD at R and CD at S.

Then $AR = AP = r$, $BP = BQ = a - r$

$DR = DS = r$ and $CQ = CS = 2a - r$ In ΔBEC

$BC^2 = BE^2 + EC^2$

$\Rightarrow\ (a - r + 2a - r)^2 = (2r)^2 + (a)^2$

$\Rightarrow\ 9a^2 + 4r^2 - 12ar = 4r^2 + a^2$

$\Rightarrow\ a = \dfrac{3}{2}r$ ....(1)

Also Ar (quad. $ABCD$) = 18

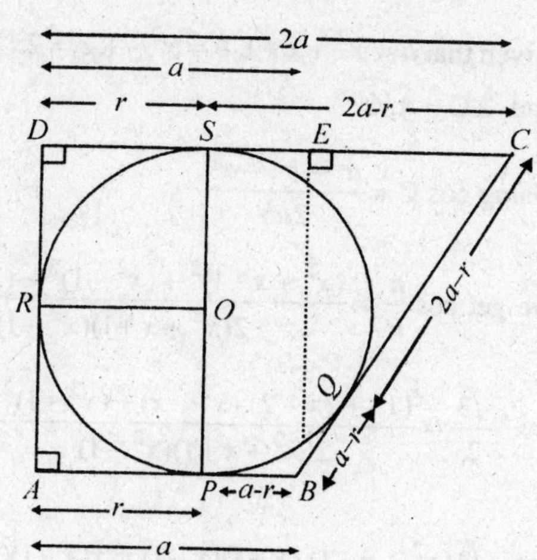

$\Rightarrow \quad a \times 2r + \frac{1}{2} \times a \times 2r = 18$

$\Rightarrow \quad ar = 6 \Rightarrow \quad \frac{3r^2}{2} = 6$ (using equation (1))

$\Rightarrow \quad r^2 = 4 \Rightarrow \quad r = 2$

17. **(d)** $\because$ A, B, C are in A.P.

$\therefore$ $B = 60°$

then $\frac{a}{c}\sin 2C + \frac{c}{a}\sin 2A$

$= \frac{K\sin A}{K\sin C}.2\sin C\cos C + \frac{K\sin C}{K\sin A}.\sin A\cos A$

$= 2(\sin A\cos C + \cos A\sin C) = 2\sin(A+C) = 2\sin B$

$= 2 \times \frac{\sqrt{3}}{2} = \sqrt{3}$

18. **(c)** We have

$\frac{2\sin P - \sin 2P}{2\sin P + \sin 2P} = \frac{2\sin P - 2\sin P\cos P}{2\sin P + 2\sin P\cos P} = \frac{1-\cos P}{1-\sin P}$

$= \frac{2\sin^2\frac{P}{2}}{2\cos^2\frac{P}{2}} = \tan^2\frac{P}{2} = \frac{(s-b)(s-c)}{s(s-a)}$

where $s = \frac{a+b+c}{2}$

$= \frac{(s-b)^2(s-c)^2}{s(s-a)(s-b)(s-c)} = \frac{(a+c-b)^2(a+b-c)^2}{16.\Delta^2}$

$= \frac{1\times 9}{16\Delta^2} = \left(\frac{3}{4\Delta}\right)^2$

D MCQ's WITH ONE OR MORE THAN ONE CORRECT :

1. **(a, d)** We have

$\frac{\sin A}{a} = \frac{\sin B}{b} \Rightarrow a\sin B = b\sin A$

(a) $b\sin A = a \Rightarrow a\sin B = a$

$\Rightarrow \quad \sin B = 1 \Rightarrow B = \pi/2$

Since $A < \pi/2$, the ΔABC is possible.

(b) $b\sin A > a \Rightarrow a\sin B > a \Rightarrow \sin B > 1$

Which is impossible. Hence the possibility (b) is ruled out.

Similarly (c) can be shown to be impossible.

(d) $b\sin A < a \Rightarrow a\sin B < a \Rightarrow \sin B < 1$ so value of $\angle B$ exists.

Now, $b > a \Rightarrow B > A$. Since $A < \pi/2$

The ΔABC is possible when $B >$ or $< \pi/2$.

(e) Since $b = a$, we have $B = A$. But $A > \pi/2$

$\therefore$ $B > \pi/2$. But this is not possible for any triangle.

2. **(a, d)** Since the angles of triangle are in A.P., Let

$\angle A = x - d,\ \angle B = x,\ \angle C = x + d$

Then by $\angle$ sum property of triangle, we have

$\angle A + \angle B + \angle C = 180°$

$\Rightarrow \quad x - d + x + x + d = 180°$

$\Rightarrow \quad 3x = 180°$

$\Rightarrow \quad x = 60°$

$\therefore \quad \angle B = 60°$

Using cosine formula $\cos B = \frac{a^2 + c^2 - b^2}{2ac}$

We get $\cos 60° = \frac{100 + c^2 - 81}{2 \times 10 \times c}$

$\Rightarrow \quad \frac{1}{2} = \frac{19 + c^2}{2 \times 10c} \Rightarrow c^2 - 10c + 19 = 0$

$\Rightarrow \quad c = \frac{10 \pm \sqrt{100-76}}{2} \Rightarrow c = 5 \pm \sqrt{6}$

Given that $a = 10$, $b = 9$ are the longer sides

$\therefore$ $c = 5 \pm \sqrt{6}$ (both the values are less than 9 and 10),

$\therefore$ $c = 5 + \sqrt{6}$ or $5 - \sqrt{6}$ both are possible.

3. **(b)** In ΔPQR let d_1, d_2, d_3 be the altitudes on QR, RP and PQ respectively.

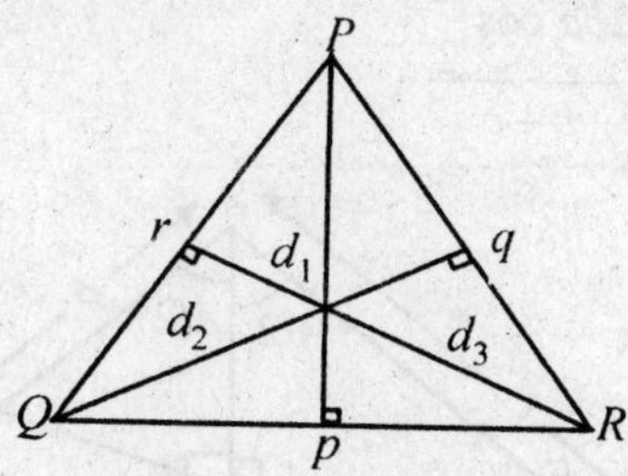

Then, Ar $(\Delta PQR) = \Delta = \frac{1}{2}pd_1 = \frac{1}{2}qd_2 = \frac{1}{2}rd_3$

$\Rightarrow \quad d_1 = \frac{2\Delta}{p}, d_2 = \frac{2\Delta}{q}, d_3 = \frac{2\Delta}{r}$

$\Rightarrow \quad d_1 = \frac{2\Delta}{k\sin P}, d_2 = \frac{2\Delta}{k\sin Q}, d_3 = \frac{2\Delta}{k\sin R}$

[Using sine law]

$\Rightarrow$ d_1, d_2, d_3 are in H.P. (As given that $\sin P$, $\sin Q$, $\sin R$ are in A.P.)

4. **(c)** Given that $A_0A_1A_2A_3A_4A_5$ is a regular hexagon inscribed in a circle of radius 1.

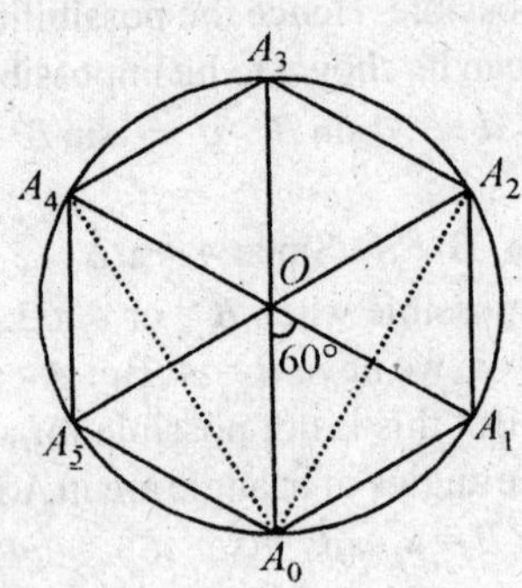

$\angle A_0OA_1 = \frac{360°}{6} = 60°$

But in ΔOA_0A_1, $OA_0 = OA_1 = 1$

$\therefore$ $\angle OA_0A_1 = \angle OA_1A_0 = 60°$

$\therefore$ ΔOA_0A_1 is an equilateral Δ

$\therefore$ $A_0A_1 = 1 = A_1A_2 = A_2A_3 = A_3A_4 = A_4A_5 = A_5A_0$

$\angle A_0A_1A_2 = 60° + 60° = 120°$

In $\Delta A_0A_1A_2$, using cosine law we get

$$\cos 120° = \frac{(A_0A_1)^2 + (A_1A_2)^2 - (A_0A_2)^2}{2(A_0A_1)(A_1A_2)}$$

$$\Rightarrow \quad -\frac{1}{2} = \frac{1+1-(A_0A_2)^2}{2 \times 1 \times 1} \Rightarrow A_0A_2 = \sqrt{3}$$

Thus by symmetry $A_0A_4 = \sqrt{3}$

$\therefore$ $A_0A_1 . A_0A_2 . A_0A_4 = 1. \sqrt{3} . \sqrt{3} = 3$

$\therefore$ (c) is the correct option.

5. **(a,b,c,d)** By simple geometry in ΔAFE, $AF = AE$

$\therefore$ ΔAFE is an isosceles Δ.

Now Ar (ΔABC) = Ar (ΔABD) + Ar (ΔADC)

$$\Rightarrow \quad \frac{1}{2}bc\sin A = \frac{1}{2}cAD\sin\frac{A}{2} + \frac{1}{2}bAD\sin\frac{A}{2}$$

$$\Rightarrow \quad AD = \frac{2bc\cos\frac{A}{2}}{b+c}$$

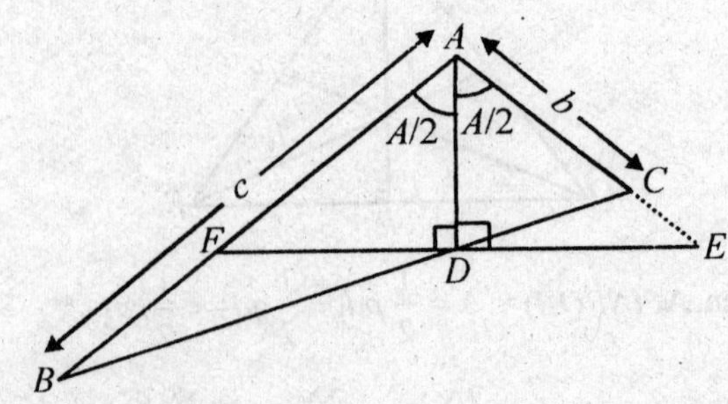

Also $AD = AE\cos\frac{A}{2}$

$$\Rightarrow \quad AE = \frac{2bc}{b+c} = \text{HM of } b \text{ and } c.$$

Again $EF = 2DE = 2.\,AD\tan\frac{A}{2} = \frac{4bc\sin\frac{A}{2}}{b+c}$

6. **(b)** Given that $a = x^2 + x + 1$, $b = x^2 - 1$, $c = 2x + 1$ and $\angle C = \pi/6$

Using $\cos C = \frac{a^2 + b^2 - c^2}{2ab}$

we get $\cos\frac{\pi}{6} = \frac{(x^2+x+1)^2 + (x^2-1)^2 - (2x+1)^2}{2(x^2+x+1)(x^2-1)}$

$$\Rightarrow \quad \frac{\sqrt{3}}{2} = \frac{(x^2+3x+2)(x^2-x) + (x^2-1)^2}{2(x^2+x+1)(x^2-1)}$$

$$\Rightarrow \sqrt{3}(x^2+x+1)(x+1)(x-1) = x(x-1)(x+1)(x+2) + (x+1)^2(x-1)^2$$

$$\Rightarrow (x+1)(x-1)\left[\sqrt{3}(x^2+x+1) - x(x+2) - (x+1)(x-1)\right] = 0$$

$$\Rightarrow (x+1)(x-1)\left[\left(\sqrt{3}-2\right)x^2 + \left(\sqrt{3}-2\right)x + \left(\sqrt{3}+1\right)\right] = 0$$

$$\Rightarrow x = -1, 1, \left(\sqrt{3}+1\right), -\left(\sqrt{3}+2\right)$$

Now for $x = -1$ and 1, $b = 0$ which is not possible

Also for $x = -\left(2+\sqrt{3}\right)$, $c = -4-2\sqrt{3}+1 < 0$ which is not possible

$\therefore$ $x = \sqrt{3}+1$

7. **(b, d)** Let PN = x, QL = x + 2, RM = x + 4 where x is an even integer.

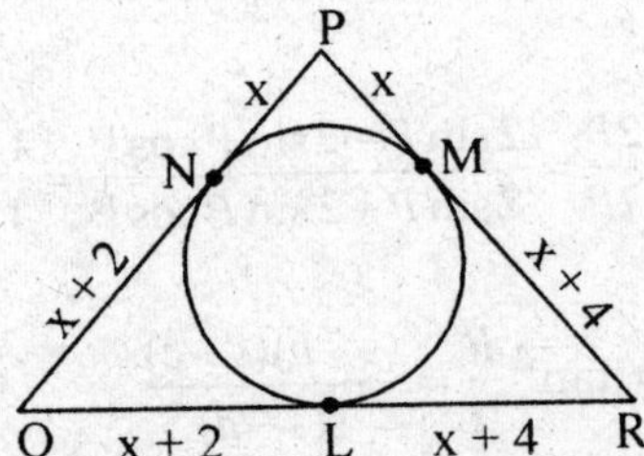

Then PM = PN = x, QN = QL = x + 2 and RL = RM = x + 4

So that PQ = 2x + 2, QR = 2x + 6, PR = 2x + 4

Now $\cos P = \frac{1}{3}$

$$\Rightarrow \frac{PQ^2 + PR^2 - QR^2}{2PQ.PR} = \frac{1}{3}$$

$$\Rightarrow \frac{(2x+2)^2 + (2x+4)^2 - (2x+6)^2}{2.(2x+2).(2x+4)} = \frac{1}{3}$$

$\Rightarrow 3[(x+1)^2 + (x+2)^2 - (x+3)^2] = 2(x+1)(x+2)$

$\Rightarrow 3(x^2-4) = 2(x+1)(x+2)$

$\Rightarrow 3x - 6 = 2x + 2$

$\Rightarrow x = 8$

$\therefore$ PQ = 18, QR = 22, PR = 20

(E) SUBJECTIVE PROBLEMS :

1. Let L be the line parallel to side AB of ΔABC, at a distance of 2 cm from AB, in which the first reflection $\Delta A'B'C'$ is obtained. Let L' be the second line parallel to $B'C'$, at a distance of 2 cm from $B'C'$, in which reflection of $\Delta A'B'C'$ is taken as $\Delta A''B''C''$.
In figure, size of $\Delta A''B''C''$ is same to the size of $\Delta A'B'C'$.

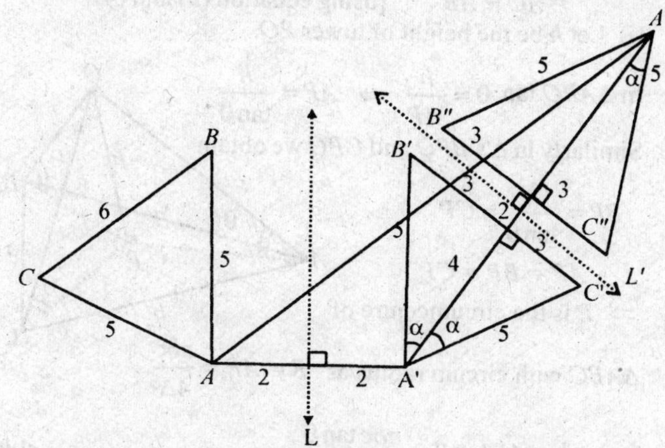

From figure $AA' = 4$ cm and $A'A'' = 12$ cm. So to find AA'' it suffices to know $\angle AA'A''$, clearly

$$\angle AA'A'' = 90^\circ + \alpha \text{ where } \sin\alpha = \frac{3}{5}$$

$\Rightarrow \cos\angle AA'A'' = \cos(90^\circ + \alpha) = -\sin\alpha = -\frac{3}{5}$ and hence

$$AA'' = \sqrt{(AA')^2 + (A'A'')^2 - 2AA' \times A'A''.\cos(90^\circ + \alpha)}$$

$$= \sqrt{16 + 144 + 96 \times \frac{3}{5}}$$

$$= \sqrt{\frac{1088}{5}} = 8\sqrt{\frac{17}{5}} \text{cm}$$

2. (a) Inradius of the circle is given by

$$r = (s-b)\tan\frac{B}{2} = \left(\frac{a+b+c}{2} - b\right)\tan\frac{\pi}{4} = \frac{a+c-b}{2}$$

$2r = a + c - b \Rightarrow$ Diameter $= BC + AB - AC$

(b) Given a ΔABC in which $AD \perp BC$, AE is diameter of circumcircle of ΔABC.

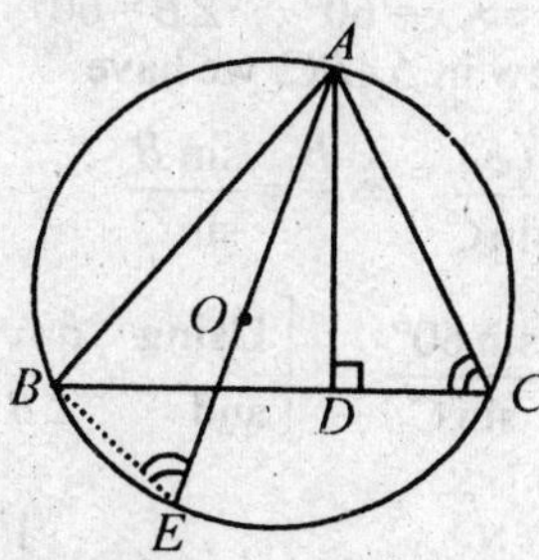

To prove :
$AB \times AC = AE \times AD$
Construction : Join BE

Proof : $\angle ABE = 90^\circ$ ($\angle$ in a semi circle)
Now in Δ's ABE and ADC
$\angle ABE = \angle ADC$ (each 90°)
$\angle AEB = \angle ACD$ ($\angle$'s in the same segment)
$\therefore \Delta ABE \sim \Delta ADC$ (by AA similarity)

$$\Rightarrow \frac{AB}{AD} = \frac{AE}{AC}$$

$\Rightarrow AB \times AC = AD \times AE$ (Proved)

3. By exterior angle theorem, in the adjacent fig.

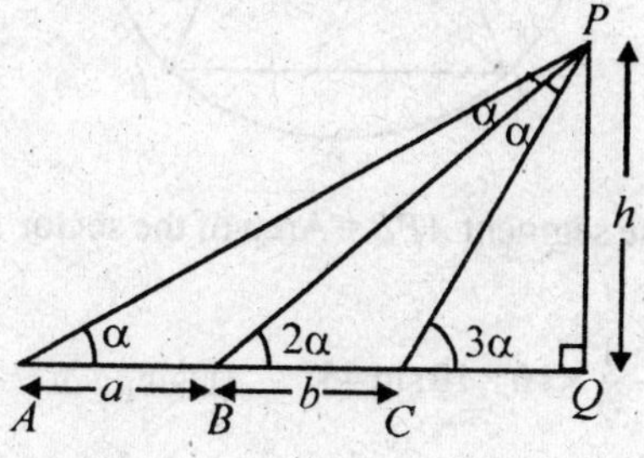

$\angle APB = \angle BPC = \alpha$
Also in ΔABP, $\angle BAP = \angle APB = \alpha$
$\Rightarrow AB = PB = a$
Applying sine law in ΔPBC, we get

$$\Rightarrow \frac{a}{\sin(180^\circ - 3\alpha)} = \frac{b}{\sin\alpha} = \frac{PC}{\sin 2\alpha} \quad ....(1)$$

$$\Rightarrow \frac{a}{3\sin\alpha - 4\sin^3\alpha} = \frac{b}{\sin\alpha} = \frac{PC}{2\sin\alpha\cos\alpha}$$

$$\Rightarrow \frac{a}{3 - 4\sin^2\alpha} = \frac{b}{1} = \frac{PC}{2\cos\alpha}$$

$$\Rightarrow 3 - 4\sin^2\alpha = \frac{a}{b} \Rightarrow \sin^2\alpha = \frac{3b-a}{4b}$$

$$\Rightarrow \cos^2\alpha = \frac{b+a}{4b}$$

$$\Rightarrow \cos\alpha = \frac{1}{2}\sqrt{\frac{b+a}{b}}$$

Also $PC = 2b\cos\alpha = \sqrt{b(a+b)}$
Now in ΔPCQ

$$\sin 3\alpha = \frac{h}{PC} \Rightarrow h = PC.\left(\frac{a\sin\alpha}{b}\right) \quad \text{[Using eqn. (1)]}$$

$$\Rightarrow h = \sqrt{b(a+b)}.\frac{a}{b}.\sqrt{\frac{(3b-a)}{4b}}$$

$$\Rightarrow h = \frac{a}{2b}.\sqrt{(a+b)(3b-a)}$$

(b) $\because \angle ACB = 22\frac{1}{2}^\circ$

$\therefore \quad \angle AOB = 45^\circ$

$r = 10$ cm

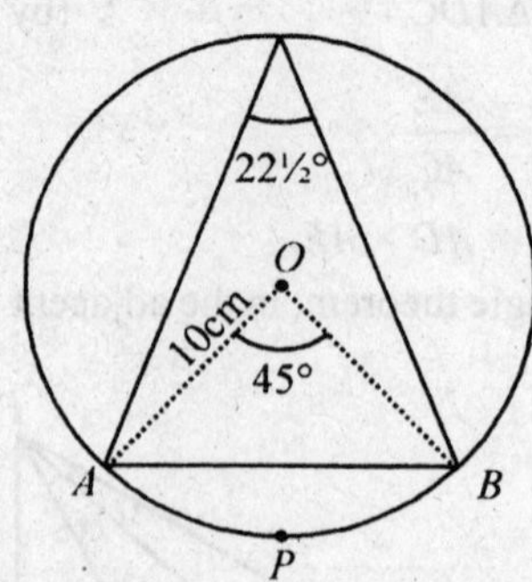

Area of the segment APB = Area of the sector AOB – area of ΔAOB

$$= \frac{1}{8}\pi r^2 - \frac{1}{2} \times 10 \times 10 \sin 45^\circ \quad \left(\text{Using, } \Delta = \frac{1}{2} bc \sin A\right)$$

$$= \frac{3.14 \times 100}{8} - \frac{50}{\sqrt{2}} = 3.91 \text{ sq. cm.}$$

4.

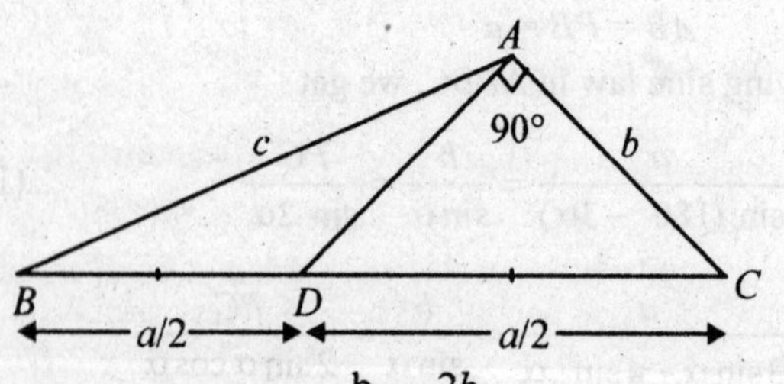

In ΔACD, $\cos C = \frac{b}{a/2} = \frac{2b}{a}$ ....(1)

In ΔABC, $\cos C = \frac{a^2 + b^2 - c^2}{2ab}$ ....(2)

From (1) and (2),

$$\frac{2b}{a} = \frac{a^2 + b^2 - c^2}{2ab} \Rightarrow b^2 = \frac{1}{3}(a^2 - c^2) \quad \text{....(3)}$$

Also $\quad \cos A = \frac{b^2 + c^2 - a^2}{2bc}$

$$\therefore \quad \cos A \cos C = \frac{b^2 + c^2 - a^2}{2bc} \times \frac{2b}{a} = \frac{b^2 + c^2 - a^2}{ac}$$

$$= \frac{\frac{1}{3}(a^2 - c^2) + (c^2 - a^2)}{ac} = \frac{2(c^2 - a^2)}{3ac}$$

5. Given that $AB = AC$

$\therefore \quad \angle 1 = \angle 2$ ....(1)

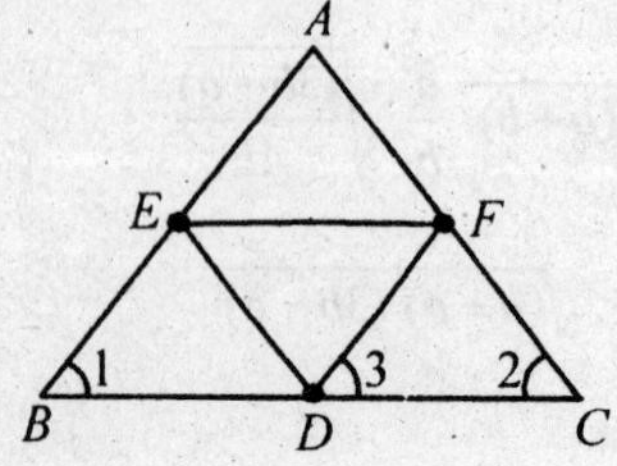

But $AB \parallel DF$ (given) and BC is transversal

$\therefore \quad \angle 1 = \angle 3$ ....(2)

From equation (1) and (2)

$\angle 2 = \angle 3$

$\Rightarrow \quad DF = CF$ ....(3)

Similaraly we can prove

$DE = BE$

Now, $DF + FA + AE + ED = CF + FA + AE + BE$

$= AC + AB$ [using equation (3) and (4)]

6. (i) Let h be the height of tower PQ.

In ΔAPQ $\tan\theta = \frac{h}{AP} \Rightarrow AP = \frac{h}{\tan\theta}$

Similarly in Δ's BPQ and CPQ we obtain

$$BP = \frac{h}{\tan\theta} = CP$$

$\therefore \quad AP = BP = CP$

$\Rightarrow$ P is the circumcentre of

ΔABC with circum radius as $R = AP = \frac{abc}{4\Delta}$

$$\therefore \quad h = AP \tan\theta = \frac{abc \tan\theta}{4\Delta}$$

(ii) Given $AP = AB \times n$

$$\Rightarrow \quad \frac{AB}{AP} = \frac{1}{n} = \tan\theta$$

$\therefore$

$$\tan\theta = \frac{1}{n}$$

Also $\tan(\theta - \beta) = \frac{AC}{AP} = \frac{1}{2}\frac{AB}{AP} = \frac{1}{2n}$

$$\Rightarrow \quad \frac{\tan\theta - \tan\beta}{1 + \tan\theta\tan\beta} = \frac{1}{2n} \Rightarrow \frac{\frac{1}{n} - \tan\beta}{1 + \frac{1}{n}\tan\beta} = \frac{1}{2n}$$

$$\Rightarrow \quad 2n - 2n^2 \tan\beta = n + \tan\beta$$

$$\Rightarrow \quad (2n^2 + 1)\tan\beta = n \Rightarrow \tan\beta = \frac{n}{2n^2 + 1}$$

7. As the angles A, B, C of ΔABC are in AP

$\therefore$ Let $A = x - d$, $B = x$, $C = x + d$

But $A + B + C = 180^\circ$ ($\angle$ Sum prop. of Δ)

$\therefore \quad x - d + x + x + d = 180^\circ$

$\Rightarrow 3x = 180^\circ \Rightarrow x = 60^\circ \therefore \angle B = 60^\circ$

Now by sine law in ΔABC, we have

$$\frac{b}{\sin B} = \frac{c}{\sin C} \Rightarrow \frac{b}{c} = \frac{\sin B}{\sin C}$$

$$\Rightarrow \quad \frac{\sqrt{3}}{\sqrt{2}} = \frac{\sin 60^\circ}{\sin C} \quad \left[\begin{array}{l}\text{Using } b : c = \sqrt{3} : \sqrt{2} \\ \text{and } \angle B = 60^\circ\end{array}\right]$$

$$\Rightarrow \quad \frac{\sqrt{3}}{\sqrt{2}} = \frac{\sqrt{3}}{2\sin C} \Rightarrow \sin C = \frac{1}{\sqrt{2}} = \sin 45^\circ$$

$\therefore \quad \angle C = 45^\circ \Rightarrow \angle A = 180^\circ - (\angle B + \angle C)$

$= 180^\circ - (60^\circ + 45^\circ) = 75^\circ$

8. Let ht. of pole PQ be h.

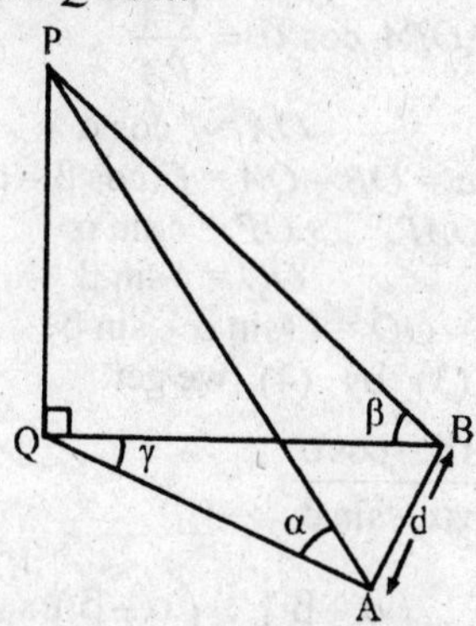

In ΔAPQ, $\tan\alpha = \dfrac{h}{AQ}$

$\Rightarrow \quad AQ = \dfrac{h}{\tan\alpha}$ ....(1)

In ΔBPQ, $\tan\beta = \dfrac{h}{BQ} \Rightarrow BQ = \dfrac{h}{\tan\beta}$ ....(2)

In ΔABQ, $\cos\gamma = \dfrac{AQ^2 + BQ^2 - AB^2}{2AQ\cdot BQ}$

$\therefore \quad \cos\gamma = \dfrac{h^2\cot^2\alpha + h^2\cot^2\beta - d^2}{2h^2\cot\alpha\cot\beta}$

$\therefore \quad -2h^2\cot\alpha\cot\beta\cos\gamma + h^2\cot^2\alpha + h^2\cot^2\beta = d^2$

$\Rightarrow \quad h = \dfrac{d}{\sqrt{\cot^2\alpha + \cot^2\beta - 2\cot\alpha\cot\beta\cos\gamma}}$

9. According to question figure is as follows

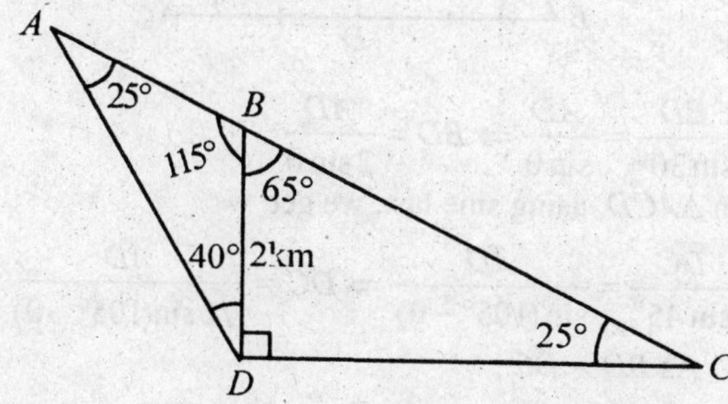

Here, $\angle BDC = 90°, BD = 2$ km

$\angle BDA = 40°, \Rightarrow \angle ADC = 130°,$

$\therefore \quad \angle DAC = 180° - (25° + 130°) = 25°$

From the figure, in ΔABD, using sine law

$\dfrac{AD}{\sin 115°} = \dfrac{BD}{\sin 25°} \Rightarrow$

$AD = \dfrac{2\sin(90° + 25°)}{\sin 25°} = \dfrac{2\cos 25°}{\sin 25°}$

$\Rightarrow AD = 2\cot 25°$

$= 2\sqrt{\dfrac{1}{\sin^2 25°} - 1} = 2\sqrt{\dfrac{1}{(0.423)^2} - 1} = 4.28$ km

10. KEY CONCEPT :

Ex-radii of a ΔABC are $r_1 = \dfrac{\Delta}{s-a}, r_2 = \dfrac{\Delta}{s-b}$

$r_3 = \dfrac{\Delta}{s-c}$ As r_1, r_2, r_3 are in HP. $\therefore \dfrac{1}{r_1}, \dfrac{1}{r_2}, \dfrac{1}{r_3}$ are in AP

$\Rightarrow \dfrac{s-a}{\Delta}, \dfrac{s-b}{\Delta}, \dfrac{s-c}{\Delta}$ are in AP

$\Rightarrow s-a, s-b, s-c$ are in AP

$\Rightarrow -a, -b, -c$ are in AP.

$\Rightarrow a, b, c$ are in A.P.

11. Given that in ΔABC

$\cos A + \cos B + \cos C = 3/2$

$\Rightarrow \dfrac{b^2+c^2-a^2}{2bc} + \dfrac{a^2+c^2-b^2}{2ac} + \dfrac{a^2+b^2-c^2}{2ab} = \dfrac{3}{2}$

$ab^2 + ac^2 - a^3 + a^2b + bc^2 - b^3 + ac^2 + b^2c - c^3 = 3abc$

$\Rightarrow ab^2 + ac^2 + bc^2 + ba^2 + ca^2 + cb^2 - 6abc = a^3 + b^3 + c^3 - 3abc$

$\Rightarrow a(b-c)^2 + b(c-a)^2 + c(a-b)^2 = \left(\dfrac{a+b+c}{2}\right)[(a-b)^2 + (b-c)^2 + (c-a)^2]$

$\Rightarrow (a+b-c)(a-b)^2 + (b+c-a)(b-c)^2 + (c+a-b)(c-a)^2 = 0$ ....(1)

As we know that $\begin{array}{l} a+b>c \\ b+c>a \\ c+a>b \end{array}$ $\left\{\begin{array}{l}\text{Sum of any two sides} \\ \text{of a } \Delta \text{ is greater than} \\ \text{the third side}\end{array}\right.$

$\therefore$ Each part on the LHS of eq. (1) has +ve coeff. multiplied by perfect square, each must be seperately zero

$\therefore \quad a-b=0 \; ; \; b-c=0 \; ; \; c-a=0 \Rightarrow \quad a=b=c$

Hence Δ is an equilateral Δ.

ALTERNATE SOLUTION :

Given that $\cos A + \cos B + \cos C = \dfrac{3}{2}$ in ΔABC

$\Rightarrow 2\cos\dfrac{A+B}{2}\cos\dfrac{A-B}{2} = \dfrac{3}{2} - \cos C$

$\Rightarrow 2\sin\dfrac{C}{2}\cos\dfrac{A-B}{2} = \dfrac{3-2\cos C}{2}$

$\Rightarrow 2\sin\dfrac{C}{2}\cos\dfrac{A-B}{2} = \dfrac{3-2(1-2\sin^2 C/2)}{2}$

$\Rightarrow \cos\left(\dfrac{A-B}{2}\right) = \dfrac{1+4\sin^2 C/2}{4\sin C/2}$

$\Rightarrow \cos\left(\dfrac{A-B}{2}\right) = \dfrac{1+4\sin^2 C/2 - 4\sin C/2 + 4\sin C/2}{4\sin C/2}$

$\Rightarrow \cos\left(\dfrac{A-B}{2}\right) = \dfrac{(1-2\sin C/2)^2}{4\sin C/2} + 1$

Which is possible only when

$1 - 2\sin C/2 = 0 \Rightarrow \sin C/2 = 1/2$

$\Rightarrow \quad C/2 = 30° \Rightarrow C = 60°$

Also then $\cos\dfrac{A-B}{2} = 1 \Rightarrow \dfrac{A-B}{2} = 0 \Rightarrow A - B = 0$ ....(1)

and $\quad A + B = 180° - 60° = 120°$ A + B = 120° ....(2)

From (1) and (2) $A = B = 60°$

Thus we get $A = B = C = 60°$ $\therefore \Delta ABC$ is an equilateral Δ.

12. Given that, in ΔABC,

$$\frac{b+c}{11}=\frac{c+a}{12}=\frac{a+b}{13}$$

where a, b, c are the lengths of sides BC, CA and AB respectively.

Let $\frac{b+c}{11}=\frac{c+a}{12}=\frac{a+b}{13}=k$

$\Rightarrow$ $b+c=11k$ ....(1)

$c+a=12k$ ....(2)

$a+b=13k$ ....(3)

Adding the above three eqs. we get

$2(a+b+c)=36k$

$\Rightarrow$ $a+b+c=18k$ ....(4)

Solving each of (1), (2) and (3) with (4), we get

Now, $\cos A=\frac{b^2+c^2-a^2}{2bc}$

$$=\frac{36k^2+25k^2-49k^2}{2\times 6k\times 5k}=\frac{12k^2}{60k^2}=\frac{1}{5}$$

$$\cos B=\frac{c^2+a^2-b^2}{2ca}=\frac{25k^2+49k^2-36k^2}{2\times 5k\times 7k}$$

$$=\frac{38k^2}{70k^2}=\frac{19}{35}$$

$$\cos C=\frac{a^2+b^2-c^2}{2ab}=\frac{49k^2+36k^2-25k^2}{2\times 7k\times 6k}$$

$$=\frac{60k^2}{84k^2}=\frac{5}{7}$$

$\therefore$ $\frac{\cos A}{1/5}=\frac{\cos B}{19/35}=\frac{\cos C}{5/7}$

$\Rightarrow$ $\frac{\cos A}{7/35}=\frac{\cos B}{19/35}=\frac{\cos C}{25/35}$

$\Rightarrow$ $\frac{\cos A}{7}=\frac{\cos B}{19}=\frac{\cos C}{25}$

13. Let ℓ be the length of the ladder, then

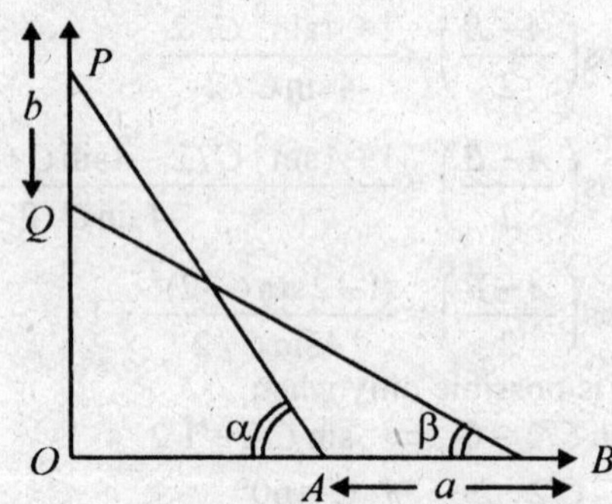

In ΔOQB, $\cos\beta=\frac{OB}{BQ}$

$\Rightarrow$ $OB=\ell\cos\beta$ ....(1)

Similarly in ΔOPA, $\cos\alpha=\frac{OA}{PA}$

$\Rightarrow$ $OA=\ell\cos\alpha$ ....(2)

Now $a=OB-OA=\ell(\cos\beta-\cos\alpha)$ ....(3)

Also from ΔOAP, $OP=\ell\sin\alpha$

and in OQB ; $OQ=\ell\sin\beta$

$\therefore$ $b=OP-OQ=\ell(\sin\alpha-\sin\beta)$ ....(4)

Dividing eq. (3) by (4) we get

$$\frac{a}{b}=\frac{\cos\beta-\cos\alpha}{\sin\alpha-\sin\beta}$$

$$\Rightarrow \frac{a}{b}=\frac{2\sin\left(\frac{\alpha+\beta}{2}\right)\sin\left(\frac{\alpha-\beta}{2}\right)}{2\cos\left(\frac{\alpha+\beta}{2}\right)-\sin\left(\frac{\alpha-\beta}{2}\right)}$$

$$\Rightarrow \frac{a}{b}=\tan\left(\frac{\alpha+\beta}{2}\right)$$

Thus, $a=b\tan\left(\frac{\alpha+\beta}{2}\right)$ is proved.

14. Let AD be the median in ΔABC.

Let $\angle B=\theta$ then $\angle C=105°-\theta$

In ΔABD, using sine law, we get

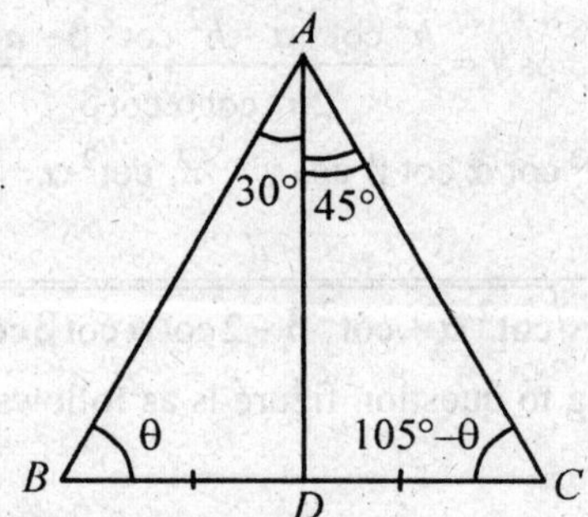

$$\frac{BD}{\sin 30°}=\frac{AD}{\sin\theta}\Rightarrow BD=\frac{AD}{2\sin\theta}$$

In ΔACD, using sine law, we get

$$\frac{DC}{\sin 45°}=\frac{AD}{\sin(105°-\theta)}\Rightarrow DC=\frac{AD}{\sqrt{2}\sin(105°-\theta)}$$

As $BD=DC$

$$\Rightarrow \frac{AD}{2\sin\theta}=\frac{AD}{\sqrt{2}\sin(105°-\theta)}$$

$$\Rightarrow \sin(90°+15°-\theta)=\sqrt{2}\sin\theta$$

$$\Rightarrow \cos 15°\cos\theta+\sin 15°\sin\theta=\sqrt{2}\sin\theta$$

$$\Rightarrow \cot\theta=\frac{\sqrt{2}-\sin 15°}{\cos 15°}=\frac{5-\sqrt{3}}{\sqrt{3}+1}=3\sqrt{3}-4$$

$$\Rightarrow \operatorname{cosec}\theta=\sqrt{1+\cot^2\theta}=\sqrt{1+27+16-24\sqrt{3}}$$

$$\Rightarrow \operatorname{cosec}\theta=2\sqrt{11-6\sqrt{3}}$$

$$\therefore\ BD=\frac{AD}{2\sin\theta}=\frac{1}{\sqrt{11-6\sqrt{3}}}\times\frac{2\sqrt{11-6\sqrt{3}}}{2}=1$$

$\therefore$ $BC=2BD=2$ units

15. We are given that in ΔABC $\cos A \cos B + \sin A \sin B \sin C = 1$

$\Rightarrow \sin A \sin B \sin C = 1 - \cos A \cos B$

$$\Rightarrow \sin C = \frac{1-\cos A \cos B}{\sin A \sin B}$$

$\Rightarrow \frac{1-\cos A\cos B}{\sin A \sin B} \le 1$ $\quad [\because \sin C \le 1]$

$\Rightarrow 1 - \cos A \cos B \le \sin A \sin B$

$\Rightarrow 1 \le \cos A \cos B + \sin A \sin B$

$\Rightarrow 1 \le \cos(A-B)$

$\Rightarrow 1 \le \cos(A-B)$

But we know $\cos(A-B) \le 1$

$\therefore$ We must have $\cos(A-B) = 1$

$\Rightarrow A - B = 0$

$\Rightarrow A = B$

$\therefore \cos A \cos A + \sin A \sin A \sin C = 1$ [For $A = B$]

$\Rightarrow \cos^2 A + \sin^2 A \sin C = 1$

$\Rightarrow \sin^2 A \sin C = 1 - \cos^2 A$

$\Rightarrow \sin^2 A \sin C = \sin^2 A$

$\Rightarrow \sin^2 A(\sin C - 1) = 0$

$\Rightarrow \sin A = 0$ or $\sin C = 1$

The only possibility is $\sin C = 1 \Rightarrow C = \pi/2$

$\therefore A + B = \pi/2$

But $A = B \Rightarrow A = B = \pi/4$

$\therefore$ By Sine law in ΔABC,

$$\frac{a}{\sin A} = \frac{b}{\sin B} = \frac{c}{\sin C}$$

$$\Rightarrow \frac{a}{\sin 45^\circ} = \frac{b}{\sin 45^\circ} = \frac{c}{\sin 90^\circ}$$

$$\Rightarrow \frac{a}{1/\sqrt{2}} = \frac{b}{1/\sqrt{2}} = \frac{c}{1}$$

$$\Rightarrow \frac{a}{1} = \frac{b}{1} = \frac{1}{\sqrt{2}} \Rightarrow a:b:c = 1:1:\sqrt{2}$$

Hence proved the result.

16.

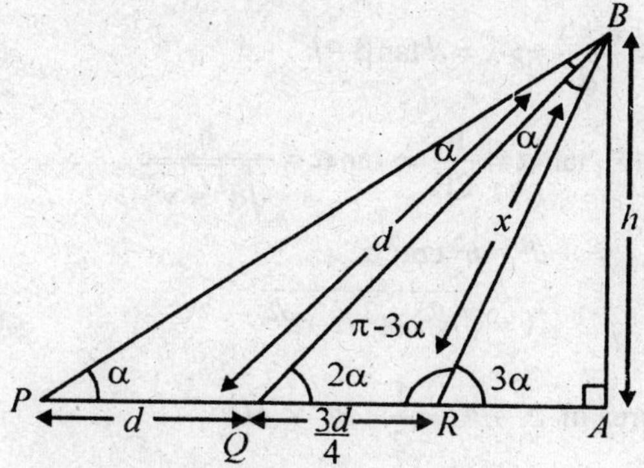

Let $RB = x$

$\angle BQR$ is ext. $\angle$ of ΔPBQ

$\therefore \angle PBQ = 2\alpha - \alpha = \alpha$

Now in ΔPBQ, $\angle PBQ = \angle QPB$

$\Rightarrow PQ = QB = d$

Also $\angle BRA$ is ext. $\angle$ of ΔBQR

$\therefore \angle QBR = 3\alpha - 2\alpha = \alpha$

And $\angle BRQ = \pi - 3\alpha$ (linear pair)

Now in ΔBQR, by applying Sine Law, we get

$$\frac{d}{\sin(\pi-3\alpha)} = \frac{3d/4}{\sin\alpha} = \frac{x}{\sin 2\alpha}$$

$$\Rightarrow \frac{d}{\sin 3\alpha} = \frac{3d}{4\sin\alpha} = \frac{x}{\sin 2\alpha}$$

$$\Rightarrow \frac{d}{3\sin\alpha - 4\sin^3\alpha} = \frac{3d}{4\sin\alpha} = \frac{x}{2\sin\alpha\cos\alpha}$$

$$\Rightarrow \frac{d}{3-4\sin^2\alpha} = \frac{3d}{4} = \frac{x}{2\cos\alpha} \quad ....(1)$$

(I) (II) (III)

From eq. (1), I = II

$$\Rightarrow \frac{d}{3-4\sin^2\alpha} = \frac{3d}{4} \Rightarrow 4 = 9 - 12\sin^2\alpha$$

$$\Rightarrow \sin^2\alpha = \frac{5}{12} \Rightarrow \cos^2\alpha = 7/12$$

Also from eq. (1) using (II) and (III), we have

$$\frac{3d}{4} = \frac{x}{2\cos\alpha} \Rightarrow 4x^2 = 9d^2\cos^2\alpha$$

$$x^2 = \frac{9d^2}{4} \cdot \frac{7}{12} = \frac{21}{16}d^2 \quad ....(3)$$

Again from ΔABR, we have $\sin 3\alpha = \frac{h}{x}$

$$\Rightarrow 3\sin\alpha - 4\sin^3\alpha = \frac{h}{x} \Rightarrow \sin\alpha(3-4\sin^2\alpha) = \frac{h}{x}$$

$$\Rightarrow \sin\alpha\left[3 - 4\times\frac{5}{12}\right] = \frac{h}{x} \quad \left(\text{using } \sin^2\alpha = \frac{5}{12}\right)$$

$$\Rightarrow \frac{4}{3}\sin\alpha = h/x$$

Squaring both sides, we get

$$\frac{16}{9}\sin^2\alpha = \frac{h^2}{x^2} \quad \frac{16}{9}\times\frac{5}{12} = \frac{h^2}{x^2}$$

(again using $\sin^2\alpha = \frac{5}{12}$)

$$\Rightarrow h^2 = \frac{4\times 5}{9\times 3}x^2 \Rightarrow h^2 = \frac{20}{27}\times\frac{21}{16}d^2$$

[using value of x^2 from eq. (3)]

$$\Rightarrow h^2 = \frac{35}{36}d^2 \Rightarrow 36h^2 = 35d^2.$$

Which was to be proved.

17.

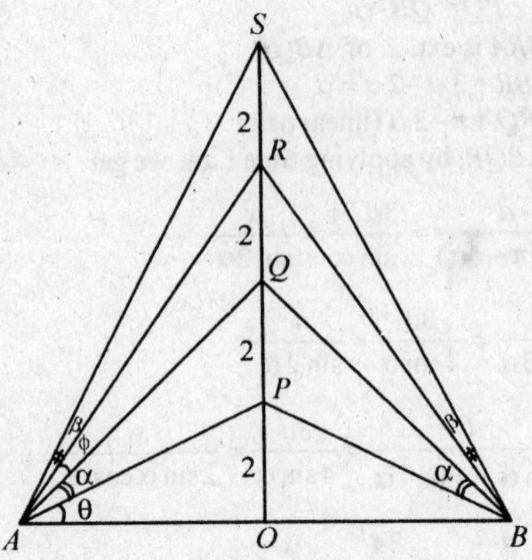

Let O be the mid point of $AB = 8$ m

$\therefore \quad OA = OB = 4$ m

Also $OP = 2\ m$ is the initial position of the object which is $2m$ long. Also we are given, $\frac{ds}{dt} = 2t + 1$

Integrating we get, $\quad s = t^2 + t + k$

(where s is the distance of top of object from O)

when $\quad t = 0,\ s = OP = 2$

$\therefore \quad k = 2$

$\therefore \quad s = t^2 + t + 2 \qquad ....(1)$

For $t = 1$, $s = 4 = OQ \therefore PQ = 2$ where PQ is the position of object after 1 sec.

For $t = 2$, $s = 8 = OS$ but $RS = 2$ where RS is the position of the object after 2 sec.

$\therefore \quad OR = OS - RS = 6$

Also, $OQ = 4$

$\therefore \quad QR = OR - OQ = 6 - 4 = 2$

$\therefore \quad OP = PQ = QR = RS = 2$

As per the condition of the question, PQ and RS, the position of the object at $t = 1$ and $t = 2$ subtend angles α and β at A and B respectively.

Now let $\angle PAO = \theta$ so that $\tan\theta = \frac{2}{4} = \frac{1}{2}$

Also $\tan(\alpha + \theta) = \frac{4}{4} = 1$

Now, $\tan\alpha = \tan[(\alpha + \theta) - \theta]$

$$= \frac{\tan(\alpha + \theta) - \tan\theta}{1 + \tan(\alpha + \theta)\tan\theta} = \frac{1 - \frac{1}{2}}{1 + \frac{1}{2}} = \frac{1}{3}$$

$\therefore \quad \sin\alpha = \frac{1}{\sqrt{10}}$ and $\cos\alpha = \frac{3}{\sqrt{10}} \qquad ....(2)$

Similarly taking $\angle RAQ = \phi$ so that

$\tan\beta = \tan[(\theta + \alpha + \phi + \beta) - (\theta + \alpha + \beta)]$

$$= \frac{\tan(\theta + \alpha + \phi + \beta) - \tan(\theta + \alpha + \beta)}{1 + \tan(\theta + \alpha + \phi + \beta).\tan(\theta + \alpha + \beta)}$$

$$= \frac{\frac{8}{4} - \frac{6}{4}}{1 + \frac{8}{4} \times \frac{6}{4}} = \frac{2 - 3/2}{1 + 2(3/2)} = \frac{1}{8}$$

$\therefore \quad \sin\beta = \frac{1}{\sqrt{65}}$ and $\cos\beta = \frac{8}{\sqrt{65}} \qquad ....(3)$

$\therefore \quad \cos(\alpha - \beta) = \cos\alpha\cos\beta + \sin\alpha\sin\beta$

Using equations (2) and (3) we get

$$\cos(\alpha - \beta) = \frac{3}{\sqrt{10}} \cdot \frac{8}{\sqrt{65}} + \frac{1}{\sqrt{10}} \cdot \frac{1}{\sqrt{65}}$$

$$= \frac{25}{5\sqrt{2}\sqrt{13}} = \frac{5}{\sqrt{26}}$$

18. Let ABC be the isosceles triangular sign board with BC horizontal. DE be the pole of height h. Let the man be standing at P such that $PE = d$

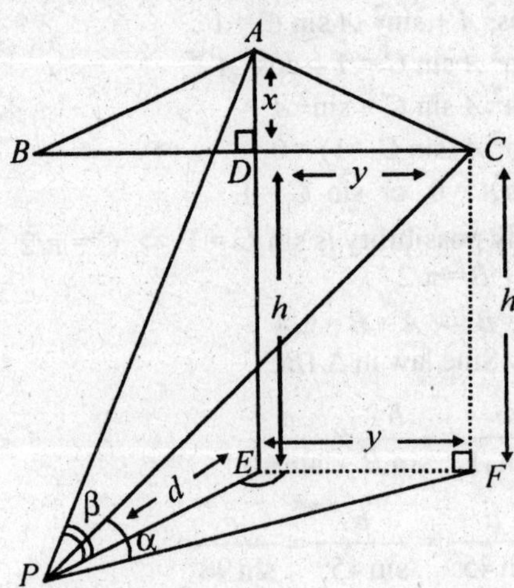

Also ATQ, $\angle APE = \beta$

$\angle CPF = \alpha$

Let $AD = x$ be altitude of ΔABC.

As ΔABC is isosceles with $AB = AC$

$\therefore \quad D$ is mid pt of BC.

Hence $BC = 2y$.

Now in ΔAPE,

$\tan\beta = \frac{h + x}{d} \Rightarrow x = d\tan\beta - h \qquad ....(1)$

In ΔCPF, $\tan\alpha = \frac{h}{PF} \Rightarrow \tan\alpha = \frac{h}{\sqrt{d^2 + y^2}}$

$\Rightarrow \quad y^2 + d^2 = h^2\cot^2\alpha$

$\Rightarrow \quad y = \sqrt{h^2\cot^2\alpha - d^2} \qquad ....(2)$

Now area of $\Delta ABC = \frac{1}{2} \times BC \times AD$

$$= \frac{1}{2} \times 2y \times x = xy$$

$$= (d\tan\beta - h)\sqrt{h^2\cot^2\alpha - d^2}$$

[Using (1) and (2)]

19. Let ABC be the triangular region with $AB = AC = 100\ m$ Let M be the mid pt of BC at which tower LM stands.
As ΔABC is isosceles and M is mid pt. of BC

$\therefore \quad AM \perp BC$.

Let $LM = h$ be the ht. of tower.

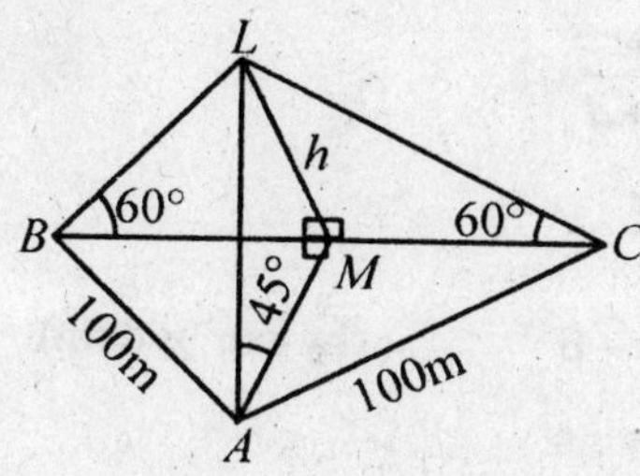

In ΔALM, $\tan 45° = \frac{LM}{MA} \Rightarrow LM = MA$

$\therefore \quad MA = h$

Also in ΔBLM, $\tan 60° = \frac{LM}{BM}$

$\Rightarrow \quad \sqrt{3} = \frac{h}{BM} \Rightarrow BM = \frac{h}{\sqrt{3}}$

Now in $rt\ \Delta AMB$, we have,

$AB^2 = AM^2 + BM^2$

$\Rightarrow \quad (100)^2 = h^2 + \frac{h^2}{3}$

$\Rightarrow \quad \frac{4h^2}{3} = 10000$

$\Rightarrow \quad h = 50\sqrt{3}$ m.

20. Let $PQ = h$
As A and B are located to the south and east of P respectively,
$\therefore \angle APB = 90°$. M is mid pt of AB. PAM is an equilateralΔ

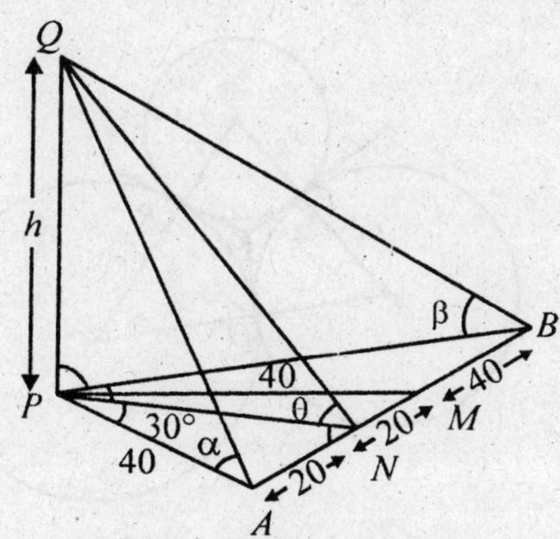

$\therefore \quad \angle APM = 60°$:

Also $PN \perp AB$, therefore $AN = NM = 20$ m

$\Rightarrow \quad AP = 40$ m

Let angles of elevation of top of the tower from A, N and B be α, θ and β respectively. ATQ, $\tan\theta = 2$

In ΔPQN $\quad \tan\theta = \frac{PQ}{PN}$

$\Rightarrow \quad 2 = \frac{h}{PN} \Rightarrow PN = h/2$ ....(1)

Also in ΔAPM, $\angle APM = 60°$ (being equilateral Δ) and PN is altitude $\therefore \angle APN = 30°$ (as in equilateral Δ altitude bisects the vertical angle.

$\therefore \quad$ In ΔAPN $\tan \angle APN = \frac{AN}{PN}$

$\Rightarrow \quad \tan 30° = \frac{20}{h/2}$ [Using eq. (1)]

$\Rightarrow \quad \frac{h}{2\sqrt{3}} = 20 \Rightarrow h = 40\sqrt{3}$ m.

In ΔAPQ $\tan\alpha = \frac{h}{AP} \Rightarrow \tan\alpha = \frac{40\sqrt{3}}{40} = \sqrt{3}$

$\Rightarrow \alpha = 60°$ Also in ΔABQ $\tan\beta = h/PB$ But in $rt\ \Delta PNB$

$PB = \sqrt{PN^2 + NB^2} = \sqrt{(20\sqrt{3})^2 + (60)^2}$

$\therefore \quad PB = \sqrt{1200 + 3600} = \sqrt{4800} = 40\sqrt{3}$

$\therefore \quad \tan\beta = \frac{40\sqrt{3}}{40\sqrt{3}} \Rightarrow \tan\beta = 1 \Rightarrow \beta = 45°$

Thus $h = 40\sqrt{3}m$; $\angle$'s of elevation are $60°, 45°$.

21. Let the sides of Δ be $n, n+1, n+2$ where $n \in N$.
Let $a = n, b = n+1, c = n+2$

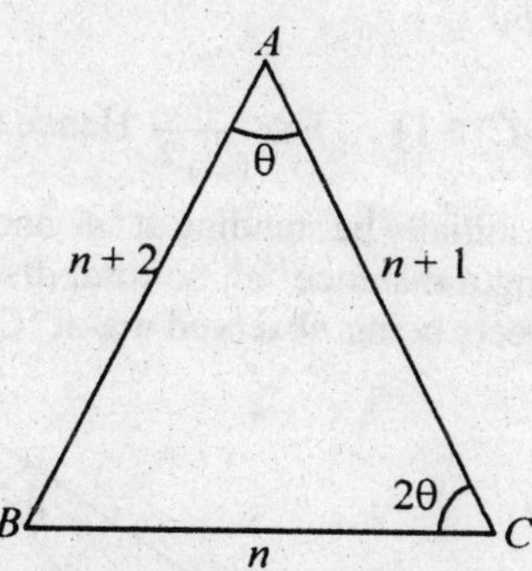

Let the smallest, angle $\angle A = \theta$ then the greatest $\angle C = 2\theta$
In ΔABC by applying Sine Law we get,

$\frac{\sin\theta}{n} = \frac{\sin 2\theta}{n+2}$

$\Rightarrow \quad \frac{\sin\theta}{n} = \frac{2\sin\theta\cos\theta}{n+2} \Rightarrow \frac{1}{n} = \frac{2\cos\theta}{n+2}$ (as $\sin\theta \neq 0$)

$\Rightarrow \quad \cos\theta = \frac{n+2}{2n}$ ....(1)

In ΔABC by Cosine Law, we get

$\cos\theta = \frac{(n+1)^2 + (n+2)^2 - n^2}{2(n+1)(n+2)}$ ....(2)

Comparing the values of $\cos\theta$ from (1) and (2), we get

$$\frac{(n+1)^2 + (n+2)^2 - n^2}{2(n+1)(n+2)} = \frac{n+2}{2n}$$

$\Rightarrow \quad (n+2)^2 (n+1) = n(n+2)^2 + n(n+1)^2 - n^3$

$\Rightarrow \quad n(n+2)^2 + (n+2)^2 = n(n+2)^2 + n(n+1)^2 - n^3$

$\Rightarrow \quad n^2 + 4n + 4 = n^3 + 2n^2 + n - n^3$

$\Rightarrow \quad n^2 - 3n - 4 = 0 \Rightarrow (n+1)(n-4) = 0$

$\Rightarrow \quad n = 4$ (as $n \neq -1$)

$\therefore \quad$ Sides of Δ are $4, 4+1, 4+2$, i.e. 4, 5, 6.

22. Given that, In ΔABC, base = a

and $\frac{c}{b}=r$

To find altitude, h.

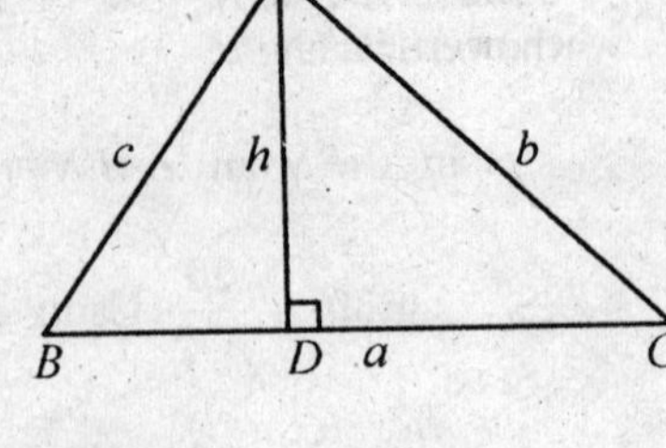

We have, in ΔABD,

$$h=c\sin B=\frac{c\,a\sin B}{a}$$

$$=\frac{c\,k\sin A\sin B}{k\sin A}=\frac{c\,k\sin A\sin B}{\sin(B+C)}$$

$$=\frac{c\sin A\sin B\sin(B-C)}{\sin(B+C)\sin(B-C)}=\frac{c\sin A\sin B\sin(B-C)}{\sin^2 B-\sin^2 C}$$

$$=\frac{c\cdot\frac{a}{k}\cdot\frac{b}{k}\cdot\sin(B-C)}{\frac{b^2}{k^2}-\frac{c^2}{k^2}}=\frac{abc\sin(B-C)}{b^2-c^2}$$

$$=\frac{a\left(\frac{c}{b}\right)\sin(B-C)}{1-\left(\frac{c}{b}\right)^2}=\frac{ar\sin(B-C)}{1-r^2}\le\frac{ar}{1-r^2}$$

$[\because \sin(B-C)\le 1]\ \therefore\ h\le\frac{ar}{1-r^2}$ Hence Proved.

23. Let the man initially be standing at 'A' and 'B' be the position after walking a distance 'c', so total distance becomes $2c$ and the objects being observed are at 'C' and 'D'.

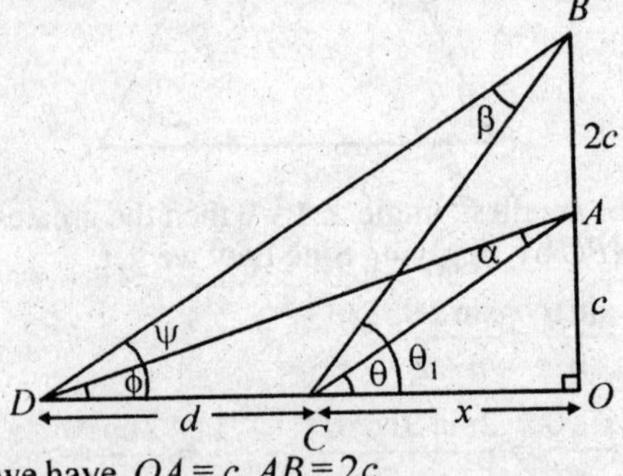

Now we have $OA=c$, $AB=2c$

Let $CO=x$ and $CD=d$

Let $\angle CAD=\alpha$ and $\angle CBD=\beta$

$\angle ACO=\theta$ and $\angle ADC=\phi$

$\angle BCD=\psi$ and $\angle BCO=\theta_1$

In ΔACO, $\tan\theta=\frac{AO}{CO}\Rightarrow\tan\theta=\frac{c}{x}$ ....(1)

In ΔADO, $\tan\phi=\frac{c}{x+d}$ ....(2)

Now, $\theta=\alpha+\phi$ (Using ext. $\angle$ thm.)

$$\Rightarrow\ \alpha=\theta-\phi\ \Rightarrow\ \tan\alpha=\tan(\theta-\phi)\Rightarrow=\frac{\tan\theta-\tan\phi}{1+\tan\theta\tan\phi}$$

$$=\frac{\frac{c}{x}-\frac{c}{x+d}}{1+\frac{c}{x}\cdot\frac{c}{x+d}}$$ (Using equations (1) and (2)

$$\Rightarrow\ \tan\alpha=\frac{cx+cd-cx}{x^2+dx+c^2}$$

$\Rightarrow\ x^2+c^2+xd=cd\cot\alpha$ ....(3)

Again in ΔADO

$\tan\psi=\frac{3c}{x+d}$ ....(4)

$\tan\theta_1=\frac{3c}{x}$ ....(5)

But $\theta_1=\psi+\beta$ (by ext. $\angle$ thrm)

$\Rightarrow\ \beta=\theta_1-\psi$

$$\Rightarrow\ \tan\beta=\tan(\theta_1-\psi)=\frac{\tan\theta_1-\tan\psi}{1+\tan\theta_1\tan\psi}$$

$$\Rightarrow\ \tan\beta=\frac{\frac{3c}{x}-\frac{3c}{x+d}}{1+\frac{3c}{x}\cdot\frac{3c}{x+d}}$$ [Using (4) and (5)]

$$\Rightarrow\ \tan\beta=\frac{3cd}{x^2+xd+9c^2}$$

$\Rightarrow\ x^2+xd+9c^2=3cd\cot\beta$ ....(6)

From (3) and (6), we get

$8c^2=3cd\cot\beta-cd\cot\alpha$

$\Rightarrow\ d=\frac{8c}{3\cot\beta-\cot\alpha}$ Hence Proved.

24. Let us consider three circles with centres at A, B and C and with radii r_1, r_2 and r_3 respectively which touch each other externally at P, Q and R. Let the common tangents at P, Q and R meet each other at O. Then $OP=OQ=OR=4$ (given) (lengths of tangents from a pt to a circle are equal).

Also $OP\perp AB$, $OQ\perp AC$, $OR\perp BC$.

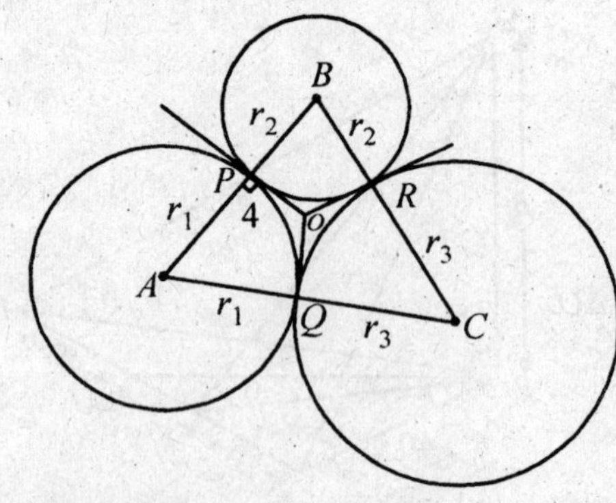

$\Rightarrow$ O is the incentre of the ΔABC

Thus for ΔABC

$$s=\frac{(r_1+r_2)+(r_2+r_3)+(r_3+r_1)}{2}$$

i.e. $s=(r_1+r_2+r_3)$

$\therefore\ \Delta=\sqrt{(r_1+r_2+r_3)\cdot r_1\cdot r_2\cdot r_3}$ (Heron's formula)

Now $r=\frac{\Delta}{s}$

NOTE THIS STEP :

$$\Rightarrow \quad 4=\frac{\sqrt{(r_1+r_2+r_3)r_1\,r_2\,r_3}}{r_1+r_2+r_3}$$

$$\Rightarrow \quad 4=\frac{\sqrt{r_1\,r_2\,r_3}}{\sqrt{r_1+r_2+r_3}}$$

$$\Rightarrow \quad \frac{r_1\cdot r_2\cdot r_3}{r_1+r_2+r_3}=\frac{16}{1}$$

$$\Rightarrow \quad r_1.\,r_2.\,r_3 : r_1+r_2+r_3=16:1$$

25. Let PQ be the tower of height h. A is in the north of O and P is towards east of A.

$\therefore \quad \angle OAP = 90°$

$\angle QOP = 30°$

$\angle QAP = \phi$

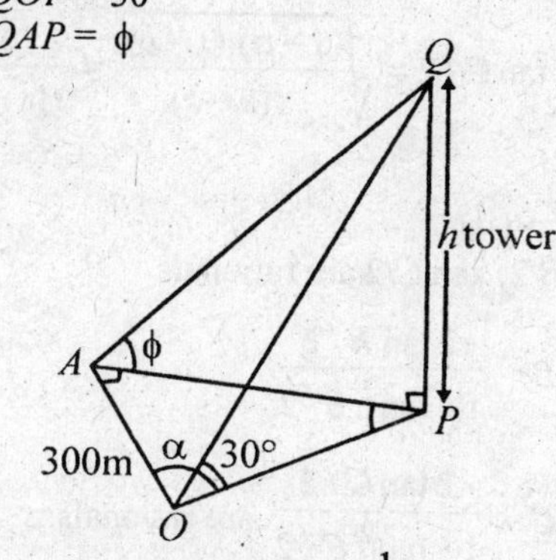

$\angle AOP = \alpha$ s.t. $\tan\alpha = \frac{1}{\sqrt{2}}$

Now in ΔOPQ, $\tan 30° = \frac{h}{OP} \Rightarrow OP = h\sqrt{3}$ ....(1)

In ΔAPQ, $\tan\phi = \frac{h}{AP} \Rightarrow AP = h\cot\phi$ ....(2)

Given that,

$$\tan\alpha = \frac{1}{\sqrt{2}} \Rightarrow \sin\alpha = \frac{\tan\alpha}{\sqrt{1+\tan^2\alpha}} = \frac{1}{3}$$

Now in ΔAOP, $\sin\alpha = \frac{AP}{OP} \Rightarrow \frac{1}{\sqrt{3}} = \frac{h\cot\phi}{h\sqrt{3}}$

[Using (1) and (2)]

$\Rightarrow \quad \cot\phi = 1$

$\Rightarrow \quad \phi = 45°.$

Again in ΔOAP, using Pythagoras thm, we get

$OP^2 = OA^2 + AP^2$

$\Rightarrow \quad 3h^2 = 90000 + h^2\cot^2 45°$

$\Rightarrow \quad h = 150\sqrt{2}\text{m}$

26. Let AB be the tower leaning towards west making an angle α with vertical

At C, $\angle$ of elevation of B is β and at D the

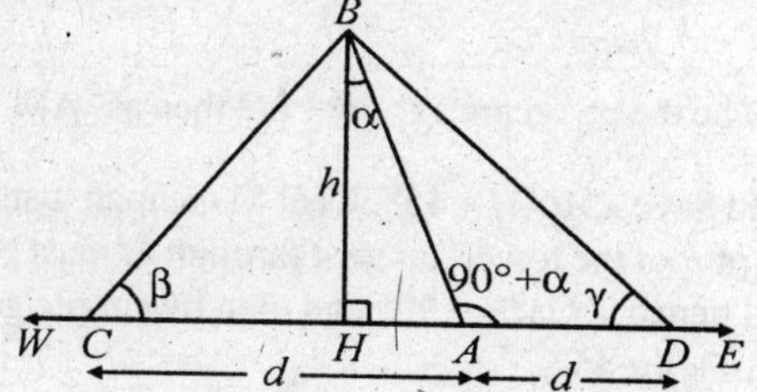

$\angle$of elevation of B is γ

$CA = AD = d$

when in ΔABH

$$\Rightarrow \quad \tan\alpha = \frac{AH}{h} \Rightarrow AH = h\tan\alpha \quad ....(1)$$

In ΔBCH, $\tan\beta = \frac{h}{CH} \Rightarrow CH = h\cot\beta$

$\Rightarrow \quad d - AH = h\cot\beta$

$\Rightarrow \quad d = h(\tan\alpha + \cot\beta)$ ....(2)

(Using eq^n (1))

In ΔBDH, $\tan\gamma = \frac{BH}{HD} \Rightarrow \frac{h}{AH+d}$

$\Rightarrow \quad AH + d = h\cot\gamma$

$\Rightarrow \quad d = h(\cot\gamma - \tan\alpha)$ ....(3)

(Using eq^n (1))

Comparing the values of d from (2) and (3), we get

$h(\tan\alpha + \cot\beta) = h(\cot\gamma - \tan\alpha)$

$\Rightarrow \quad 2\tan\alpha = \cot\gamma - \cot\beta$

Hence Proved.

ALTERNATE SOLUTION :

KEY CONCEPT :

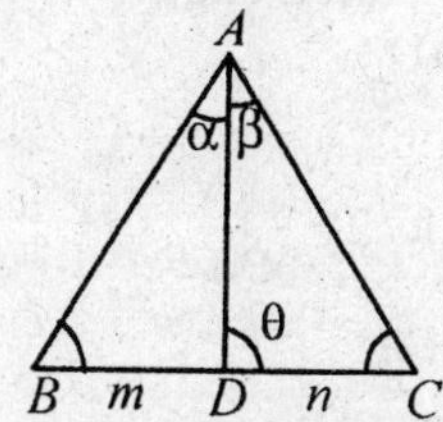

$m:n$ theorem: In ΔABC where point D divides BC in the ratio $m:n$. and $\angle ADC = \theta$

(i) $(m+n)\cot\theta = n\cot B - m\cot C$

(ii) $(m+n)\cot\theta = m\cot\alpha - n\cot\beta$

In ΔBCD, A divides CD in the ratio 1 : 1 where base $\angle$'s are β and γ and $\angle BAD = 90° + \alpha$

$\therefore$ By applying $m:n$ theorem we get

$(1+1)\cot(90°+\alpha) = 1.\cot\beta - 1.\cot\gamma$

$\Rightarrow \quad -2\tan\alpha = \cot\beta - \cot\gamma$

$\Rightarrow \quad 2\tan\alpha = \cot\gamma - \cot\beta$

Hence Proved.

27. Let a be the side of n sided regular polygon

$A_1A_2A_3A_4.....A_n$

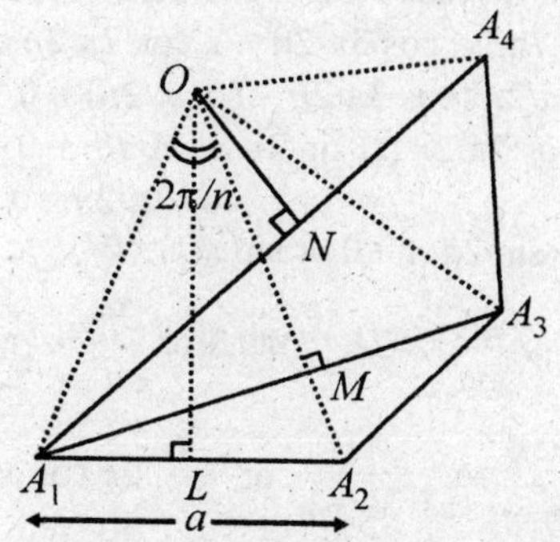

$\therefore$ $\angle$Subtended by each side at centre will be

$$=\frac{2\pi}{n}$$

Let $OL \perp A_1A_2$

Then $\angle OLA_1 = 90°$, $\angle A_1OL = \pi/n$ $\quad (\because OA_1 = OA_2)$

$\therefore$ In ΔOA_1L, $\sin A_1OL = \frac{A_1L}{OA_1}$

$\Rightarrow$ $\sin\pi/n = \frac{a/2}{OA_1}$

$\Rightarrow$ $OA_1 = \frac{a/2}{\sin\pi/n}$ ....(1)

Again by geometry it can be proved that $OM \perp A_1A_3$

In ΔOA_1M, $\sin 2\pi/n = \frac{A_1M}{OA_1}$

$\Rightarrow$ $A_1M = OA_1 \sin 2\pi/n$

$\Rightarrow$ $A_1A_3 = \frac{2a\sin 2\pi/n}{2\sin\pi/n}$ [Using eq^n (1)]

Also if $ON \perp A_1A_4$, then ON bisects angle

$\angle A_1OA_4 = 3(2\pi/n)$

$\therefore$ $\angle A_1ON = 3\pi/n$

$\therefore$ In ΔOA_1N, $\sin 3\pi/2 = \frac{A_1N}{OA_1}$

$\Rightarrow$ $A_1N = OA_1 \sin 3\pi/n$

$\Rightarrow$ $A_1N_4 = 2A_1N = \frac{2a\sin 3\pi/n}{2\sin\pi/n}$

But given that

$$\frac{1}{A_1A_2} = \frac{1}{A_1A_3} + \frac{1}{A_1A_4}$$

$\Rightarrow$ $\frac{1}{a} = \frac{1}{\frac{a\sin 2\pi/n}{\sin\pi/n}} + \frac{1}{\frac{a\sin 3\pi/n}{\sin\pi/n}}$

$\Rightarrow$ $\sin 3\pi/n \sin 2\pi/n$
$= (\sin 3\pi/n + \sin 2\pi/n)\sin\pi/n$

$\Rightarrow$ $2\sin 3\pi/n \sin 2\pi/n$
$= 2\sin 3\pi/n \sin\pi/n + 2\sin 2\pi/n \sin\pi/n$

$\Rightarrow$ $\cos\pi/n - \cos 5\pi/n$
$= \cos 2\pi/n - \cos 4\pi/n + \cos\pi/n - \cos 3\pi/n$

$\Rightarrow$ $\cos 2\pi/n + \cos 5\pi/n = \cos 4\pi/n + \cos 3\pi/n$

$\Rightarrow$ $2\cos 7\pi/2n \cos 3\pi/2n = 2\cos 7\pi/2n \cos\pi/2n$

$\Rightarrow$ $\cos 7\pi/2n (\cos 3\pi/2n - \cos\pi/2n) = 0$

$\Rightarrow$ $\cos 7\pi/2n \,.\, 2\sin 2\pi/n \sin\pi/n = 0$

$\Rightarrow$ $\cos 7\pi/2n = 0$

or $\sin 2\pi/n = 0$ or $\sin\pi/n = 0$

$\Rightarrow$ $\frac{7\pi}{2n} = (2k+1)\pi/2$ or $\frac{2\pi}{n} = k\pi$ or $\pi/n = p\pi$

$\Rightarrow$ $n = \frac{7}{2k+1}$ or $n = 2/k$ or $n = 1/k$

But n should be a +ve integer being no. of sides and $n \geq 4$ (four vertices being considered in the question)

$\therefore$ the only possibility is $n = \frac{7}{2k+1}$ for $k = 0$

$\therefore$ $n = 7$

28. (I) a, b c and Δ are rational.

$\Rightarrow$ $s = \frac{a+b+c}{2}$ is also rational

$\Rightarrow$ $\tan B/2 = \sqrt{\frac{(s-a)(s-c)}{s(s-b)}} = \frac{\Delta}{s(s-b)}$ is also rational

and $\tan C/2 = \sqrt{\frac{(s-a)(s-b)}{s(s-c)}} = \frac{\Delta}{s(s-c)}$ is also rational

Hence (I) $\Rightarrow$ (II).

(II) a, $\tan B/2$, $\tan C/2$ are rational.

$\Rightarrow$ $\sin B = \frac{2\tan B/2}{1+\tan^2 B/2}$

and $\sin C = \frac{2\tan C/2}{1+\tan^2 C/2}$ are rationals.

and $\tan A/2 = \tan\left[90° - \left(\frac{B}{2}+\frac{C}{2}\right)\right] = \cot\left(\frac{B}{2}+\frac{C}{2}\right)$

$= \frac{1}{\tan\left(\frac{B}{2}+\frac{C}{2}\right)} = \frac{1-\tan B/2\tan C/2}{\tan B/2+\tan C/2}$ is rational

$\therefore$ $\sin A = \frac{2\tan A/2}{1+\tan^2 A/2}$ is rational.

Hence (II) $\Rightarrow$ (III)

(III) a, $\sin A$, $\sin B$, $\sin C$ are rational.

But $\frac{a}{\sin A} = 2R$

$\Rightarrow$ R is rational

$\therefore$ $b = 2R\sin B$, $c = 2R\sin C$ are rationals.

$\therefore$ $\Delta = \frac{1}{2}bc\sin A$ is rational

Hence (III) $\Rightarrow$ (I).

29. Let $AMNB$ be the semicircular arc of length $2L$, and PQ be the vertical tower so that A, B, Q are in the same horizontal line.

Let M be the pt. on arc s.t. $AB = L/2$ then as AM $= \frac{1}{4}AB$ we should have $\angle AOM = 45°$, As at M the man just sees the top most pt P of the tower, tangent through M must pass through P and hence $\angle OMP = 90°$ and then by simple geometry we get $\angle PMY = 45°$.

Also N is the top most pt. of arc AB, hence ON must be vertical.

$\therefore \quad ON = r = XQ$

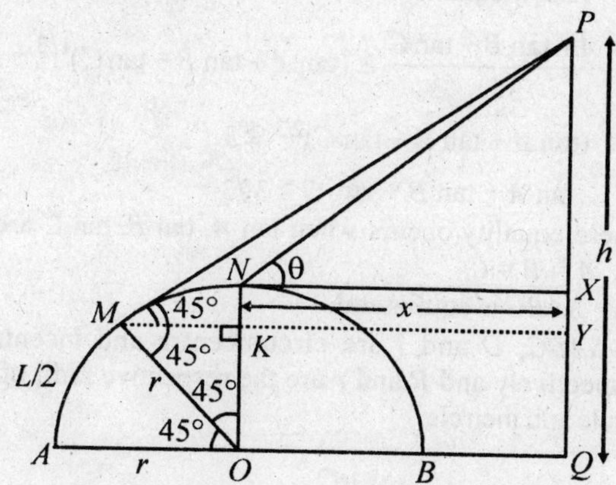

$\therefore \quad PX = h - r$, where h is ht of tower PQ, and $OK = YQ = OM \cos 45° = r/\sqrt{2}$

Similarly, $MK = OM \sin 45° = r\sqrt{2}$

$\therefore \quad$ In $\Delta\, PMY$ we get $\tan 45° = \dfrac{PY}{MY}$

$\Rightarrow \quad PY = MY$

$\Rightarrow \quad h - QY = MK + KY$

$\Rightarrow \quad h - OK = \dfrac{r}{\sqrt{2}} + x$

$\Rightarrow \quad h - \dfrac{r}{\sqrt{2}} + \dfrac{r}{\sqrt{2}} + x$

$\Rightarrow \quad x = h - r\sqrt{2} \quad \text{....(1)}$

In Δ PNX, $\quad \tan\theta = \dfrac{PX}{NX}$

$\Rightarrow \quad \tan\theta = \dfrac{(h-r)}{x}$

$\Rightarrow \quad x = (h-r)\cot\theta \quad \text{....(2)}$

Comparing the values of x from (1) and (2), we get

$$h - r\sqrt{2} = h\cot\theta - r\cot\theta$$

$$h = \frac{r(\sqrt{2} - \cot\theta)}{(1 - \cot\theta)}$$

But arc length $= 2L = \pi r \Rightarrow r = \dfrac{2L}{\pi}$

$$\therefore \quad h = \frac{2L}{\pi}\left[\frac{\sqrt{2} - \cot\theta}{1 - \cot\theta}\right]$$

30. Given that A, B, C, are three $\angle$'s of a Δ therefore $A + B + C = \pi$

also $A = \pi/4 \Rightarrow B + C = 3\pi/4$

$\Rightarrow 0 < B, C < 3\pi/4$

Now, $\tan B \tan C = p$

$$\Rightarrow \quad \frac{\sin B \sin C}{\cos B \cos C} = \frac{p}{1}$$

Applying componendo and dividendo, we get

$$\frac{\sin B \sin C + \cos B \cos C}{\cos B \cos C - \sin B \sin C} = \frac{1+p}{1-p}$$

$$\Rightarrow \quad \frac{\cos(B-C)}{\cos(B+C)} = \frac{1+p}{1-p}$$

$$\Rightarrow \quad \cos(B-C) = \frac{1+p}{1-p}\cdot\left(-\frac{1}{\sqrt{2}}\right) \quad \text{....(1)}$$

$[\because \; B + C = 3\pi/4]$

Now, as B and C can vary from 0 to $3\pi/4$

$\therefore \quad 0 \le B - C < 3\pi/4$

$$\Rightarrow \quad -\frac{1}{\sqrt{2}} < \cos(B-C) \le 1$$

From eq^n (1) substituting the value of cos $(B - C)$, we get

$$-\frac{1}{\sqrt{2}} < \frac{1+p}{\sqrt{2}(p-1)} \le 1$$

$$\Rightarrow \quad -\frac{1}{\sqrt{2}} < \frac{1+p}{\sqrt{2}(p-1)} \text{ and } \frac{1+p}{\sqrt{2}(p-1)} \le 1$$

$$\Rightarrow \quad 0 < 1 + \frac{p+1}{p-1} \text{ and } \frac{(p+1) - \sqrt{2}\,(p-1)}{\sqrt{2}(p-1)} \le 0$$

$$\Rightarrow \quad \frac{2p}{p-1} > 0 \text{ and } \frac{p + 1 - \sqrt{2}\,p + \sqrt{2}}{\sqrt{2}(p-1)} \le 0$$

$$\Rightarrow \quad p(p-1) > 0 \text{ and } \frac{(1-\sqrt{2})\,p + (\sqrt{2}+1)}{(p-1)} \le 0$$

$$\Rightarrow \quad p \in (-\infty, 0) \cup (1, \infty), \text{ and } \frac{-p + (\sqrt{2}+1)^2}{(p-1)} \le 0$$

$$\Rightarrow \quad [p - (3 + 2\sqrt{2})][p - 1] \ge 0$$

+ − +

$-\infty \quad 1 \quad 3+2\sqrt{2} \quad \infty$

$$p \in (-\infty, 1] \cup [3 + 2\sqrt{2}, \infty)$$

Combining the two cases, we get

$$p \in (-\infty, 0) \cup [3 + 2\sqrt{2}, \infty)$$

31. Let A, B and C be the projections of the pts. P, Q and M on the ground.

ATQ, $\angle POA = 60°$

$\angle QOB = 30°$

$\angle MOC = \theta$

Let h be the ht of circle from ground, then

$AP = CM = BQ = h$

Let $OA = x$ and $AB = d$ (diameter of the projection of the circle on ground with C_1 as centre).

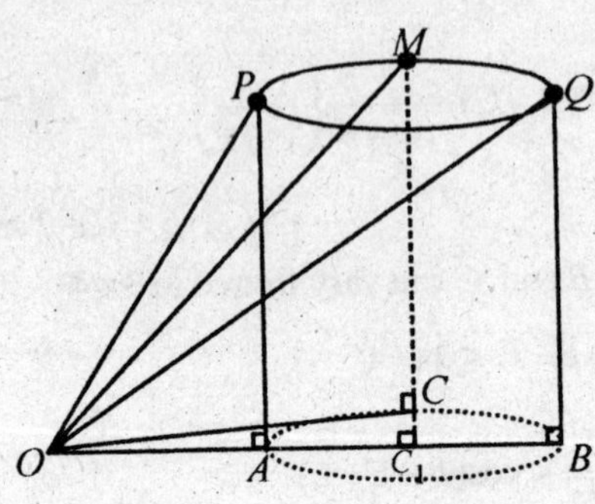

Now in ΔPOA, $\tan 60° = \frac{h}{x} \Rightarrow x = \frac{h}{\sqrt{3}}$ ....(1)

In ΔQBO, $\tan 30° = \frac{h}{x+d} \Rightarrow x + d = h\sqrt{3}$

$\Rightarrow$ $d = h\sqrt{3} - \frac{h}{\sqrt{3}} = \frac{2h}{\sqrt{3}}$ ....(2)

In ΔOMC, $\tan\theta = \frac{h}{OC}$

$$\Rightarrow \tan^2\theta = \frac{h^2}{OC^2} = \frac{h^2}{OC_1^2 + C_1C^2} = \frac{h^2}{\left(x+\frac{d}{2}\right)^2 + \left(\frac{d}{2}\right)^2}$$

$$= \frac{h^2}{\left(\frac{h}{\sqrt{3}} + \frac{h}{\sqrt{3}}\right)^2 + \left(\frac{h}{\sqrt{3}}\right)^2} \text{ [Using (1) and (2)]}$$

$$= \frac{h^2}{\frac{4h^2}{3} + \frac{h^2}{3}} = \frac{3}{5}$$

32. Let ABC is an equilateral Δ then

$A = B = C = 60°$

$\Rightarrow \tan A + \tan B + \tan C = 3\sqrt{3}$

Conversely, suppose

$\tan A + \tan B + \tan C = 3\sqrt{3}$ ....(1)

Now using A.M. $\geq$ G.M. (equality occurs when no's are equal)

For $\tan A$, $\tan B$, $\tan C$, we get

$$\frac{\tan A + \tan B + \tan C}{3} \geq (\tan A \tan B \tan C)^{1/3}$$

NOTE THIS STEP :

But in any ΔABC, we know that

$\tan A + \tan B + \tan C = \tan A \tan B \tan C$

$\therefore$ Last inequality becomes

$$\frac{\tan A + \tan B + \tan C}{3} \geq (\tan A + \tan B + \tan C)^{1/3}$$

$\Rightarrow (\tan A + \tan B + \tan C)^{2/3} \geq 3$

$\Rightarrow \tan A + \tan B + \tan C \geq 3\sqrt{3}$

where equality occurs when $\tan A$, $\tan B$, $\tan C$ are equal, i.e., $A = B = C$

$\Rightarrow$ ΔABC is equilateral.

33. In ΔABC, O and I are circumcentre and incentre of Δ respectively and R and r are the respective radii of circum circle and incircle.

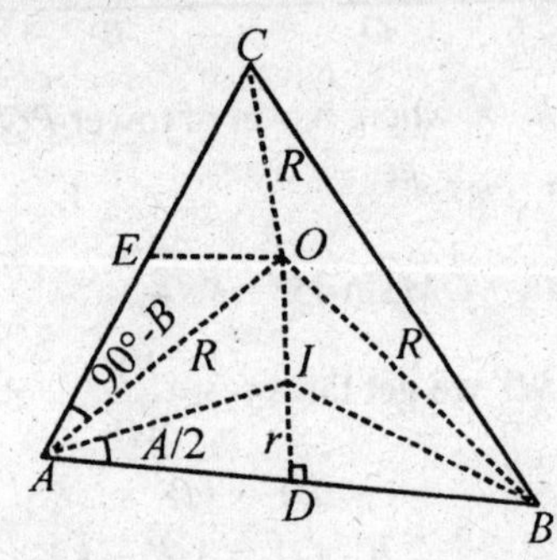

To prove

$(IO)^2 = R^2 - 2Rr$

First of all we will find IO. Using cosine law in ΔAOI

$$\cos\angle OAI = \frac{OA^2 + AI^2 - OI^2}{2.OA.AI} \quad ....(1)$$

where $OA = R$

In ΔAID, $\sin A/2 = \frac{r}{AI}$

$AI = \frac{r}{\sin A/2}$

$= 4R \sin B/2 \sin C/2$

[Using $r = 4R \sin A/2 \sin B/2 \sin C/2$]

Also, $\angle OAI = \angle IAE - \angle OAE$

$= \frac{A}{2} - (90° - \angle AOE)$

$= \frac{A}{2} - 90° + \frac{1}{2}\angle AOC$

$= \frac{A}{2} - 90° + \frac{1}{2}.2B$

$\left(\begin{array}{l}\because O \text{ is circumcentre} \\ \therefore \angle AOC = 2\angle B\end{array}\right)$

$= \frac{A}{2} + B - \frac{A+B+C}{2}$

$= \frac{B-C}{2}$

Substituting all these values in equation (1) we get

$$\cos\frac{B-C}{2}=\frac{R^2+16R^2\sin^2 B/2\sin^2 C/2-OI^2}{2.R.4R\sin B/2\sin C/2}$$

$\Rightarrow \quad OI^2 = R^2 + 16R^2\sin^2 B/2\sin^2 C/2$

$\quad - 8R^2\sin B/2\sin C/2\cos\frac{B-C}{2}$

$= R^2\left[1+\sin B/2\sin C/2\left\{2\sin B/2\sin C/2-\cos\frac{B-C}{2}\right\}\right]$

$= R^2[1+8\sin B/2\sin C/2\{2\sin B/2\sin C/2-\cos\frac{B}{2}\cos\frac{C}{2}$
$-\sin B/2\sin C/2\}]$

$= R^2[1+8\sin B/2\sin C/2\{\sin B/2\sin C/2-\cos B/2\cos C/2]$

$= R^2\left[1-8\sin B/2\sin C/2\cos\frac{B+C}{2}\right]$

$= R^2[1-8\sin A/2\sin B/2\sin C/2]$

$= R^2-2R.4R\sin A/2\sin B/2\sin C/2$

$= R^2-2Rr.$ Hence Proved.

Again if ΔOIB is right $\angle ed\ \Delta$ then

$\Rightarrow \quad OB^2 = OI^2 + IB^2$

$\Rightarrow \quad R^2 = R^2 - 2Rr + \frac{r^2}{\sin^2 B/2}$

NOTE THIS STEP:

$\left[\because \text{ In } \Delta IBD \sin B/2 = \frac{r}{IB}\right]$

$\Rightarrow \quad 2R\sin^2 B/2 = r$

$\Rightarrow \quad 2\frac{abc}{4\Delta}\cdot\frac{(s-a)(s-c)}{ac}=\frac{\Delta}{s}$

$\Rightarrow \quad b(s-a)(s-c) = 2(s-a)(s-b)(s-c)$

$\Rightarrow \quad b = 2s-2b \Rightarrow b = a+c-b$

$\Rightarrow \quad a+c = 2b \Leftrightarrow a, b, c$ are in A.P.

$\Rightarrow$ b is A.M. between a and c. Hence Proved.

34. Let $MN = r_3 = MP = MQ$, $ID = r$

$\Rightarrow \quad IP = r - r_3$

Clearly IP and IQ are tangents to circle with centre M.

$\therefore \quad IM$ must be the $\angle$ bisector of $\angle PIQ$

$\therefore \quad \angle PIM = \angle QIM = \theta_1$

Also from ΔIPM, $\tan\theta_1 = \frac{r_3}{r-r_3} = \frac{MP}{IP}$

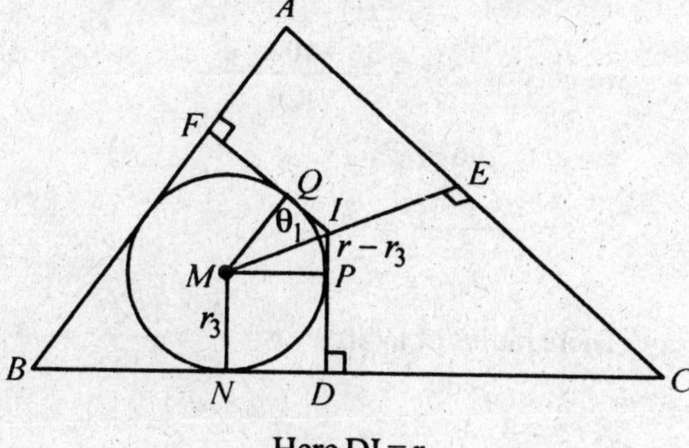

Here DI = r

Similarly, in other quadrilaterals, we get

$$\tan\theta_2 = \frac{r_2}{r-r_2} \text{ and } \tan\theta_3 = \frac{r_1}{r-r_1}$$

Also $2\theta_1+2\theta_2+2\theta_3 = 2\pi \Rightarrow \theta_1+\theta_2+\theta_3 = \pi$

$\Rightarrow \tan\theta_1 + \tan\theta_2 + \tan\theta_3 = \tan\theta_1 . \tan\theta_2 . \tan\theta_3$

NOTE THIS STEP:

$$\Rightarrow \quad \frac{r_1}{r-r_1}+\frac{r_2}{r-r_2}+\frac{r_3}{r-r_3}=\frac{r_1r_2r_3}{(r-r_1)(r-r_2)(r-r_3)}$$

35. We know, $\Delta = \sqrt{s(s-a)(s-b)(s-c)}$

$$= \sqrt{\frac{s}{8}(b+c-a)(c+a-b)(a+b-c)}$$

Since sum of two sides is always greater than third side;

$\therefore \quad b+c-a, c+a-b, a+b-c > 0$

$\Rightarrow \quad (s-a)(s-b)(s-c) > 0$

Let $s-a = x, s-b = y, s-c = z$

Now, $x+y = 2s-a-b = c$

Similary, $\quad y+z = a$

and $\quad z+x = b$

Since $AM \geq GM$

$\Rightarrow \quad \frac{x+y}{2} \geq \sqrt{xy} \Rightarrow 2\sqrt{xy} \leq a \quad \frac{y+z}{2} \geq \sqrt{yz} \Rightarrow 2\sqrt{yz} \leq b$

$\frac{z+x}{2} \geq \sqrt{zx} \Rightarrow 2\sqrt{zx} \leq c \therefore 8xyz \leq abc$

$\Rightarrow \quad (s-a)(s-b)(s-c) \leq \frac{1}{8}abc$

$\Rightarrow \quad s(s-a)(s-b)(s-c) \leq \frac{sabc}{8}$

$\Rightarrow \quad \leq \frac{1}{16}(a+b+c)abc \Rightarrow \Delta \leq \frac{1}{4}\sqrt{abc(a+b+c)}$

and equality holds when $x = y = z \Rightarrow a = b = c$

36. Let OAB be one triangle out of n of a n sided polygon inscribed in a circle of radius 1.

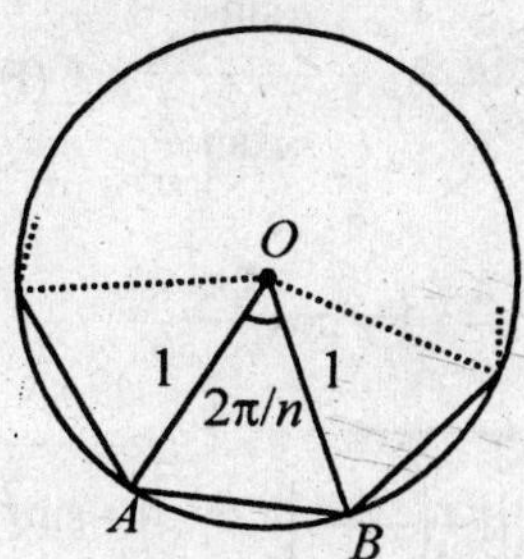

Then $\quad \angle AOB = \frac{2\pi}{n}$

$OA = OB = 1$

$\therefore$ Using Area of isosceles Δ with vertical $\angle\theta$ and equal sides as $r = \frac{1}{2}r^2\sin\theta = \frac{1}{2}\sin\frac{2\pi}{n}$

$\therefore \quad I_n = \frac{n}{2}\sin\frac{2\pi}{n}$ ......(1)

Further consider the n sided polygon subscribing on the circle.

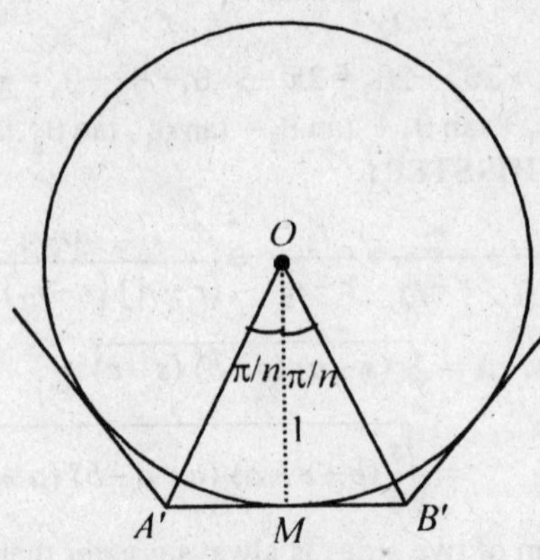

$A'MB'$ is the tangent of the circle at M.

$\Rightarrow \quad A'MB' \perp OM$

$\Rightarrow \quad A'MO$ is right angled triangle, right angle at M.

$$A'M = \tan\frac{\pi}{n}$$

$$\text{Area of } \Delta A'MO = \frac{1}{2}\times 1\times \tan\frac{\pi}{n}$$

$$\therefore \quad \text{Area of } \Delta A'B'O = \tan\frac{\pi}{n}$$

So, $O_n = n\tan\frac{\pi}{n}$ ....(2)

Now, we have to prove

$$I_n = \frac{O_n}{2}\left(1+\sqrt{1-\left(\frac{2I_n}{n}\right)^2}\right)$$

or $$\frac{2I_n}{O_n}-1=\sqrt{1-\left(\frac{2I_n}{n}\right)^2}$$

$$\text{LHS} = \frac{2I_n}{O_n}-1 = \frac{n\sin\frac{2\pi}{n}}{n\tan\frac{\pi}{n}}-1 \text{ (From (1) and (2))}$$

$$= 2\cos^2\frac{\pi}{n}-1=\cos\frac{2\pi}{n}$$

$$\text{RHS} = \sqrt{1-\left(\frac{2I_n}{n}\right)^2} = \sqrt{1-\sin^2\frac{2\pi}{n}} \text{ (From (1))}$$

$$= \cos\left(\frac{2\pi}{n}\right)$$ Hence Proved.

I INTEGER VALUE CORRECT TYPE :

1. Let $\angle ACC' = \theta$

then $\angle AC'C = \theta \; (\because \; AC = AC')$

and $\angle AC'B = 180 - \theta$.

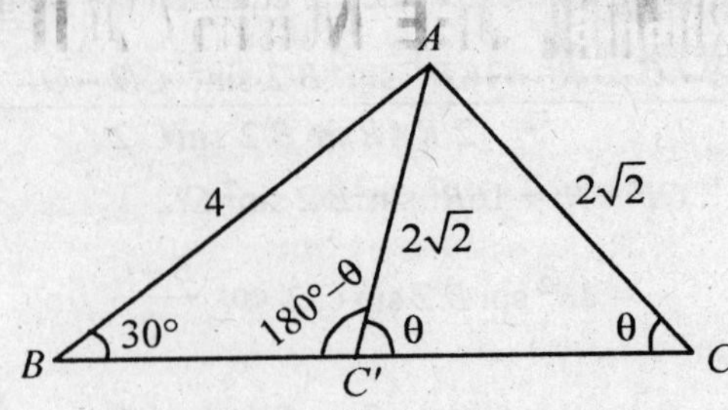

Applying sine law in $\Delta ABC'$, we get

$$\frac{4}{\sin(180-\theta)} = \frac{2\sqrt{2}}{\sin 30^\circ}$$

$$\Rightarrow \quad \sin\theta = \frac{1}{\sqrt{2}} \Rightarrow \quad \theta = 45^\circ$$

$$\therefore \quad \angle CAC' = 90^\circ$$

So, the required area $= ar(\Delta ABC) - ar(\Delta ABC')$

$$= ar(\Delta ACC') \quad = \frac{1}{2}\times AC\times AC'$$

$$= \frac{1}{2}\times 2\sqrt{2}\times 2\sqrt{2} = 4 \text{ sq. units.}$$

2.

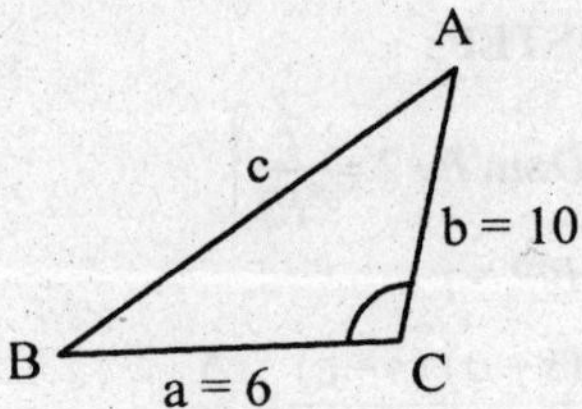

We know that area of $\Delta = \frac{1}{2}ab\sin C$

$$\Rightarrow \frac{1}{2}\times 6\times 10\times \sin C = 15\sqrt{3}$$

$$\Rightarrow \sin C = \frac{\sqrt{3}}{2}$$

$\Rightarrow C = 120^\circ$ as C is obtuse angle.

Now using $\cos C = \frac{a^2+b^2-c^2}{2ab}$,

we get $\cos 120^\circ = \frac{36+100-c^2}{120}$

$\Rightarrow c^2 = 196$ or $c = 14$

$$\therefore \; s = \frac{a+b+c}{2} = 15$$

There radius of incircle, $r = \frac{\Delta}{s} = \frac{15\sqrt{3}}{15} = \sqrt{3}$

$\therefore r^2 = 3$

Section-B JEE Main/ AIEEE

1. **(b)** Let $a = 3x + 4y$, $b = 4x + 3y$ and $c = 5x + 5y$
as $x, y > 0$, $c = 5x + 5y$ is the largest side
$\therefore$ C is the largest angle . Now

$$\cos C = \frac{(3x+4y)^2 + (4x+3y)^3 - (5x+5y)^2}{2(3x+4y)(4x+3y)}$$

$$= \frac{-2xy}{2(3x+4y)(4x+3y)} < 0$$

$\therefore$ C is obtuse angle $\Rightarrow \Delta ABC$ is obtuse angled

2. **(a)** $r_1 > r_2 > r_3 \Rightarrow \frac{\Delta}{s-a} > \frac{\Delta}{s-b} > \frac{\Delta}{s-c}$;
$\Rightarrow s-a < s-b < s-c \Rightarrow -a < -b < -c \Rightarrow a > b > c$

3. **(c)** $\tan\left(\frac{\pi}{n}\right) = \frac{a}{2r}$; $\sin\left(\frac{\pi}{n}\right) = \frac{a}{2R}$

$$r + R = \frac{a}{2}\left[\cot\frac{\pi}{n} + \operatorname{cosec}\frac{\pi}{n}\right]$$

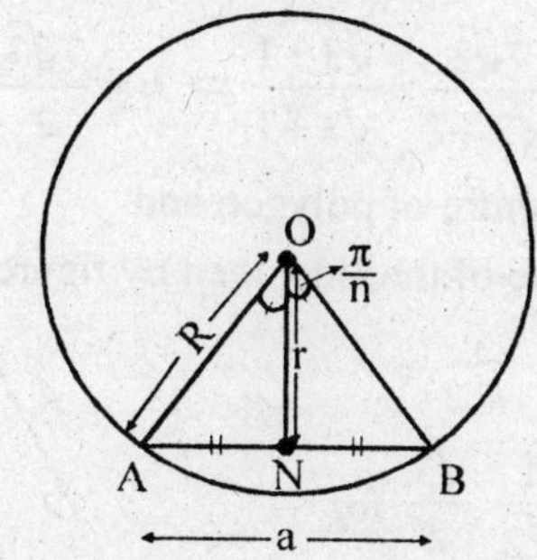

$$= \frac{a}{2}\left[\frac{\cos\frac{\pi}{n}+1}{\sin\frac{\pi}{n}}\right] = \frac{a}{2}\left[\frac{2\cos^2\frac{\pi}{2n}}{2\sin\frac{\pi}{2n}\cos\frac{\pi}{2n}}\right] = \frac{a}{2}\cot\frac{\pi}{2\pi}$$

4. **(d)**

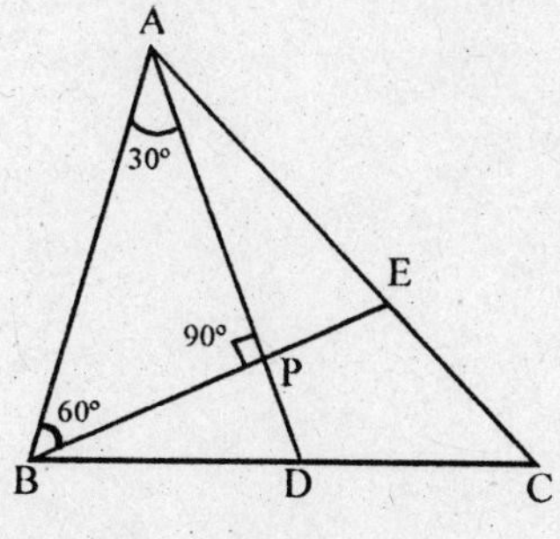

$AP = \frac{2}{3}AD = \frac{8}{3}$; $PD = \frac{4}{3}$; Let $PB = x$

$\tan 60^o = \frac{8/3}{x}$ or $x = \frac{8}{3\sqrt{3}}$

Area of $\Delta ABD = \frac{1}{2} \times 4 \times \frac{8}{3\sqrt{3}} = \frac{16}{3\sqrt{3}}$

$\therefore$ Area of $\Delta ABC = 2 \times \frac{16}{3\sqrt{3}} = \frac{32}{3\sqrt{3}}$

[$\because$ Median of a Δ divides it into two Δ's of equal area.]

5. **(b)** If $a\cos^2\left(\frac{C}{2}\right) + c\cos^2\left(\frac{A}{2}\right) = \frac{3b}{2}$

$$a[\cos C + 1] + c[\cos A + 1] = 3b$$
$$(a+c) + (a\cos C + c\cos B) = 3b$$
$$a + c + b = 3b \text{ or } a + c = 2b \text{ or } a, b, c \text{ are in } A.P.$$

6. **(c)** Let $a = \sin\alpha, b = \cos\alpha$ and $c = \sqrt{1 + \sin\alpha\cos\alpha}$
Clearly a and $b < 1$ but $c > 1$ as $\sin\alpha > 0$ and $\cos\alpha >$ 0
$\therefore c$ is the greatest side and greatest angle is C

$$\therefore \cos C = \frac{a^2 + b^2 - c^2}{2ab}$$

$$= \frac{\sin^2\alpha + \cos^2\alpha - 1 - \sin\alpha\cos\alpha}{2\sin\alpha\cos\alpha} = -\frac{1}{2}$$

$\therefore C = 120°$

7. **(d)**

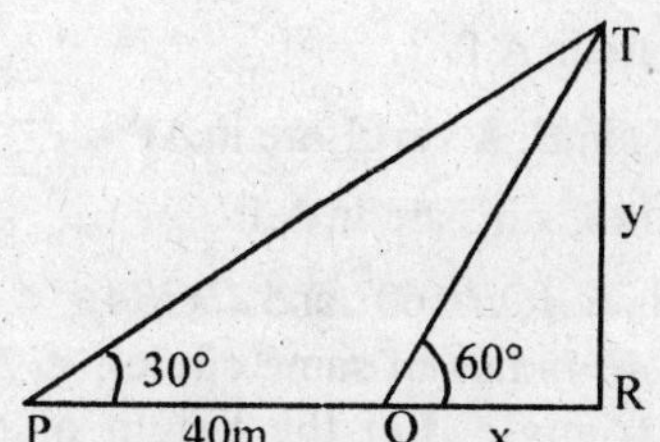

From the figure

$\tan 60° = \frac{y}{x} \Rightarrow y = \sqrt{3}x$ ....(1)

$\tan 30° = \frac{y}{x+40} \Rightarrow y = \frac{x+40}{\sqrt{3}}$ ....(2)

From (1) and (2), $\sqrt{3}x = \frac{x+40}{\sqrt{3}} \Rightarrow x = 20m$

8. **(b)** We know that for the circle circumscribing a right triangle, hypotenutse is the diameter As $\angle C = 90°$

$$\therefore 2R = c \Rightarrow R = \frac{c}{2} \text{ also } r = \frac{\Delta}{s} = \frac{\frac{1}{2} \times a \times b}{\frac{a+b+c}{2}}$$

$\Rightarrow r = \frac{ab}{a+b+c}$ $\qquad \therefore 2r + 2R = \frac{2ab}{a+b+c} + c$

$$= \frac{2ab + ac + bc + c^2}{a+b+c} = \frac{2ab + ac + bc + a^2b^2}{a+b+c}$$

$$\left(\because c^2 = a^2 + b^2\right)$$

$$= \frac{(a+b)^2 + (a+b)c}{a+b+c} = (a+b)$$

ALTERNATE SOLUTION :

We know by sinc rule $\frac{c}{\sin C} = 2R \Rightarrow c = 2R\sin C$

$\Rightarrow c = 2R$ $\quad (\because \angle C = 90^\circ)$

Also $\tan\frac{C}{2} = \frac{r}{s-c}$

$\Rightarrow \tan\frac{\pi}{4} = \frac{r}{s-c}$ $\quad (\because \angle C = 90^\circ)$

$\Rightarrow r = s - c = \frac{a+b-c}{2}$

$\Rightarrow 2r + c = a + b \Rightarrow 2r + 2R = a + b$ (using $c = 2R$)

9. (b) $\Delta = \frac{1}{2}p_1 a = \frac{1}{2}p_2 b = \frac{1}{2}p_3 b$

p_1, p_2, p_3, are in H.P.

$\Rightarrow \frac{2\Delta}{a}, \frac{2\Delta}{b}, \frac{2\Delta}{c}$ are in H.P. $\Rightarrow \frac{1}{a}, \frac{1}{b}, \frac{1}{c}$ are in H.P.

$\Rightarrow$ a, b, c are in A.P.

$\Rightarrow K\sin A, K\sin B, K\sin C$ are in AP

$\Rightarrow$ $\sin A, \sin B, \sin C$ are in A.P.

10. (a) In the ΔAOB, $\angle AOB = 60^\circ$, and $\angle OBA = \angle OAB$ (since $OA = OB = AB$ radius of same circle). $\therefore$ ΔAOB is a equilateral triangle. Let the height of tower is h m. Given distance between two points A & B lie on boundary of circular park, subtends an angle of 60° at the foot of the tower is AB i.e. $AB = a$. A tower OC stands at the centre of a circular park. Angle of elevation of the top of the tower from A and B is 30°.

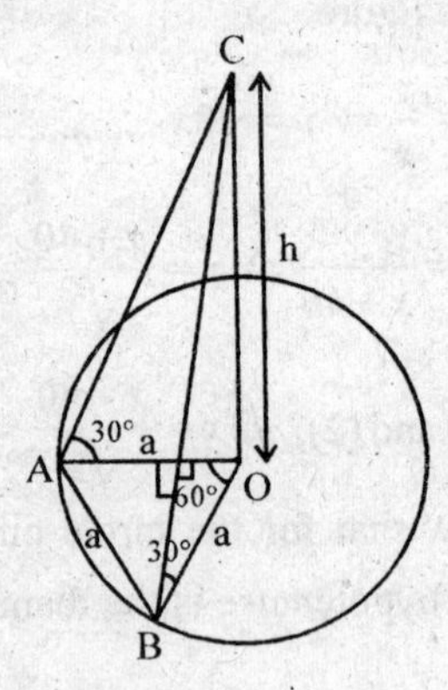

In ΔOAC $\tan 30^\circ = \frac{h}{a} \Rightarrow \frac{1}{\sqrt{3}} = \frac{h}{a} \Rightarrow h = \frac{a}{\sqrt{3}}$

11. (b) In ΔABC

$\frac{h}{x} = \tan 60^\circ = \sqrt{3}$

$\Rightarrow x = \frac{h}{\sqrt{3}}$

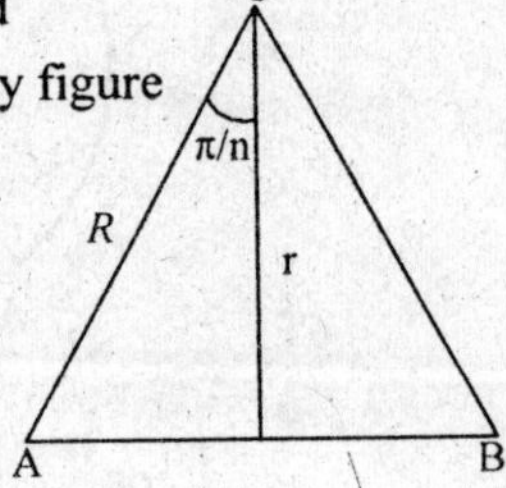

In ΔABD $\frac{h}{x+7} = \tan 45^\circ = 1$

$\Rightarrow h = x + 7 \Rightarrow h - \frac{h}{\sqrt{3}} = 7$

$\Rightarrow h = \frac{7\sqrt{3}}{\sqrt{3}-1} \times \frac{\sqrt{3}+1}{\sqrt{3}+1} \Rightarrow h = \frac{7\sqrt{3}}{2}(\sqrt{3}+1)m$

12. (b) If O is centre of polygon and AB is one of the side, then by figure

$\cos\frac{\pi}{n} = \frac{r}{R}$

$\Rightarrow \frac{r}{R} = \frac{1}{2}, \frac{1}{\sqrt{2}}, \frac{\sqrt{3}}{2}$ for $n = 3, 4, 6$ respectively.

Chapter 14 Inverse Trigonometry

Section-A JEE Advanced/ IIT-JEE

A FILL IN THE BLANKS :

1. Let a, b, c be positive real numbers. Let

$$\theta = \tan^{-1}\sqrt{\frac{a(a+b+c)}{bc}} + \tan^{-1}\sqrt{\frac{b(a+b+c)}{ca}} + \tan^{-1}\sqrt{\frac{c(a+b+c)}{ab}}.$$

Then $\tan\theta =$ ________ *(1981 - 2 Marks)*

2. The numerical value of $\tan\left\{2\tan^{-1}\left(\frac{1}{5}\right)-\frac{\pi}{4}\right\}$ is equal to ________ *(1984 - 2 Marks)*

3. The greater of the two angles $A = 2\tan^{-1}(2\sqrt{2}-1)$ and $B = 3\sin^{-1}(1/3)+\sin^{-1}(3/5)$ is ________. *(1989 - 2 Marks)*

C MCQ's WITH ONE CORRECT ANSWER :

1. The value of $\tan\left[\cos^{-1}\left(\frac{4}{5}\right)+\tan^{-1}\left(\frac{2}{3}\right)\right]$ is *(1983 - 1 Mark)*
(a) $\frac{6}{17}$ (b) $\frac{7}{16}$
(c) $\frac{16}{7}$ (d) none of these

2. If we consider only the principle values of the inverse trigonometric functions then the value of
$\tan\left(\cos^{-1}\frac{1}{5\sqrt{2}}-\sin^{-1}\frac{4}{\sqrt{17}}\right)$ is *(1994)*
(a) $\frac{\sqrt{29}}{3}$ (b) $\frac{29}{3}$
(c) $\frac{\sqrt{3}}{29}$ (d) $\frac{3}{29}$

3. The number of real solutions of
$\tan^{-1}\sqrt{x(x+1)}+\sin^{-1}\sqrt{x^2+x+1}=\pi/2$ is *(1999 - 2 Marks)*
(a) zero (b) one
(c) two (d) infinite

4. If $\sin^{-1}\left(x-\frac{x^2}{2}+\frac{x^3}{4}-\ldots\right)+\cos^{-1}\left(x^2-\frac{x^4}{2}+\frac{x^6}{4}-\ldots\right)=\frac{\pi}{2}$
for $0<|x|<\sqrt{2}$, then x equals *(2001S)*
(a) 1/2 (b) 1
(c) –1/2 (d) –1

5. The value of x for which $\sin(\cot^{-1}(1+x)) = \cos(\tan^{-1}x)$ is *(2004S)*
(a) 1/2 (b) 1
(c) 0 (d) –1/2

6. If $0 < x < 1$, then
$\sqrt{1+x^2}\,[\{x\cos(\cot^{-1}x)+\sin(\cot^{-1}x)\}^2-1]^{1/2} =$ *(2008)*
(a) $\frac{x}{\sqrt{1+x^2}}$ (b) x
(c) $x\sqrt{1+x^2}$ (d) $\sqrt{1+x^2}$

7. The value of $\cot\left(\sum_{n=1}^{23}\cot^{-1}\left(1+\sum_{k=1}^{n}2k\right)\right)$ is *(JEE Adv. 2013)*
(a) $\frac{23}{25}$ (b) $\frac{25}{23}$
(c) $\frac{23}{24}$ (d) $\frac{24}{23}$

D MCQ's WITH ONE OR MORE THAN ONE CORRECT :

1. The principal value of $\sin^{-1}\left(\sin\frac{2\pi}{3}\right)$ is *(1986 - 2 Marks)*
(a) $-\frac{2\pi}{3}$ (b) $\frac{2\pi}{3}$
(c) $\frac{4\pi}{3}$ (d) none

E SUBJECTIVE PROBLEMS :

1. Find the value of : $\cos(2\cos^{-1}x+\sin^{-1}x)$ at $x=\frac{1}{5}$, where $0\le\cos^{-1}x\le\pi$ and $-\pi/2\le\sin^{-1}x\le\pi/2$. *(1981 - 2 Marks)*

2. Find all the solution of $4\cos^2 x\sin x-2\sin^2 x=3\sin x$ *(1983 - 2 Marks)*

3. Prove that $\cos\tan^{-1}\sin\cot^{-1}x=\sqrt{\frac{x^2+1}{x^2+2}}$. *(2002 - 5 Marks)*

F MATCH THE FOLLOWING :

DIRECTIONS (1 & 2) : *Each question contains statements given in two columns, which have to be matched. The statements in Column-I are labelled A, B, C and D, while the statements in Column-II are labelled p, q, r, s and t. Any given statement in Column-I can have correct matching with ONE OR MORE statement(s) in Column-II. The appropriate bubbles corresponding to the answers to these questions have to be darkened as illustrated in the following example :*

If the correct matches are A-p, s and t; B-q and r; C-p and q; and D-s then the correct darkening of bubbles will look like the given.

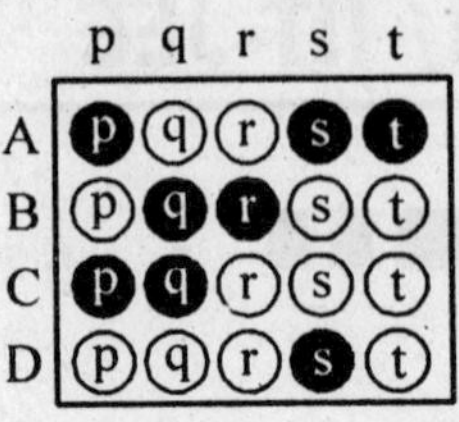

1. Match the following **(2006 - 6M)**

Column I	**Column II**
(A) $\sum_{i=1}^{\infty} \tan^{-1}\left(\frac{1}{2i^2}\right) = t$, then tan t =	(p) 1
(B) Sides a, b, c of a triangle ABC are in AP and $\cos\theta_1 = \frac{a}{b+c}, \cos\theta_2 = \frac{b}{a+c}, \cos\theta_3 = \frac{c}{a+b}$, then $\tan^2\left(\frac{\theta_1}{2}\right) + \tan^2\left(\frac{\theta_3}{2}\right) =$	(q) $\frac{\sqrt{5}}{3}$
(C) A line is perpendicular to $x + 2y + 2z = 0$ and passes through (0, 1, 0). The perpendicular distance of this line from the origin is	(r) $\frac{2}{3}$

2. Let (x, y) be such that $\sin^{-1}(ax) + \cos^{-1}(y) + \cos^{-1}(bxy) = \frac{\pi}{2}$. **(2007)**

Match the statements in Column I with statements in Column II and indicate your answer by darkening the appropriate bubbles in the 4 × 4 matrix given in the ORS.

Column I	**Column II**
(A) If $a = 1$ and $b = 0$, then (x, y)	(p) lies on the circle $x^2 + y^2 = 1$
(B) If $a = 1$ and $b = 1$, they (x, y)	(q) lies on $(x^2 - 1)(y^2 - 1) = 0$
(C) If $a = 1$ and $b = 2$, then (x, y)	(r) lies on $y = x$
(D) If $a = 2$ and $b = 2$, then (x, y)	(s) lies on $(4x^2 - 1)(y^2 - 1) = 0$

DIRECTIONS (Q. 3) : *Following questions. Each question has matching lists. The codes for the lists have choices (a), (b), (c) and (d) out of which ONLY ONE is correct.*

3. Match List I with List II and select the correct answer using the code given below the lists : ***(JEE Adv. 2013)***

List I	**List II**
P. $\left(\frac{1}{y^2}\left(\frac{\cos(\tan^{-1} y) + y\sin(\tan^{-1} y)}{\cot(\sin^{-1} y) + \tan(\sin^{-1} y)}\right)^2 + y^4\right)^{1/2}$ takes value	1. $\frac{1}{2}\sqrt{\frac{5}{3}}$
Q. If $\cos x + \cos y + \cos z = 0 = \sin x + \sin y + \sin z$ then possible value of $\cos\frac{x-y}{2}$ is	2. $\sqrt{2}$

R. If $\cos\left(\frac{\pi}{4}-x\right)\cos 2x + \sin x \sin 2\sec x = \cos x \sin 2x \sec x + \cos\left(\frac{\pi}{4}+x\right)\cos 2x$ then possible value of sec$x$ is — 3. $\frac{1}{2}$

S. If $\cot\left(\sin^{-1}\sqrt{1-x^2}\right)=\sin\left(\tan^{-1}\left(x\sqrt{6}\right)\right), x \neq 0,$ then possible value of x is — 4. 1

Codes:

	P	Q	R	S
(a)	4	3	1	2
(b)	4	3	2	1
(c)	3	4	2	1
(d)	3	4	1	2

Section-B JEE Main/ AIEEE

1. $\cot^{-1}(\sqrt{\cos\alpha})-\tan^{-1}(\sqrt{\cos\alpha})=x$, then $\sin x =$ **[2002]**

(a) $\tan^2\left(\frac{\alpha}{2}\right)$ (b) $\cot^2\left(\frac{\alpha}{2}\right)$

(c) $\tan\alpha$ (d) $\cot\left(\frac{\alpha}{2}\right)$

2. The trigonometric equation $\sin^{-1}x = 2\sin^{-1}a$ has a solution for **[2003]**

(a) $|a| \geq \frac{1}{\sqrt{2}}$ (b) $\frac{1}{2}<|a|<\frac{1}{\sqrt{2}}$

(c) all real values of a (d) $|a|<\frac{1}{2}$

3. If $\cos^{-1}x-\cos^{-1}\frac{y}{2}=\alpha$, then $4x^2-4xy\cos\alpha+y^2$ is equal to **[2005]**

(a) $2\sin 2\alpha$ (b) 4

(c) $4\sin^2\alpha$ (d) $-4\sin^2\alpha$

4. If $\sin^{-1}\left(\frac{x}{5}\right)+\text{cosec}^{-1}\left(\frac{5}{4}\right)=\frac{\pi}{2}$, then the values of x is **[2007]**

(a) 4 (b) 5

(c) 1 (d) 3.

5. The value of $\cot\left(\text{cosec}^{-1}\frac{5}{3}+\tan^{-1}\frac{2}{3}\right)$ is

(a) $\frac{6}{17}$ (b) $\frac{3}{17}$

(c) $\frac{4}{17}$ (d) $\frac{5}{17}$

6. If x, y, z are in A.P. and $\tan^{-1}x$, $\tan^{-1}y$ and $\tan^{-1}z$ are also in A.P., then **[JEE M 2013]**

(a) $x=y=z$ (b) $2x=3y=6z$

(c) $6x=3y=2z$ (d) $6x=4y=3z$

Solutions & Explanations

Section-A : JEE Advanced/ IIT-JEE

A 1. π 2. $\frac{-7}{17}$ 3. A

C 1. (d) 2. (d) 3. (c) 4. (b) 5. (d) 6. (c) 7. (b)

D 1. (d)

E 1. $\frac{-2\sqrt{6}}{5}$ 2. $x = n\pi,\ n\pi + (-1)^n \frac{\pi}{10}, n\pi + (-1)^n\left(\frac{-3\pi}{10}\right)$ where $n \in N$

F 1. (A)-(p), (B)-(r), (C)-(q) 2. (A)-p, (B)-q, (C)-p, (D)-s 3. (b)

Section-B : JEE Main/ AIEEE

1. (a) 2. None 3. (c) 4. (d) 5. (a) 6. (a)

Section-A JEE Advanced/ IIT-JEE

A FILL IN THE BLANKS :

1. Let $a+b+c=u$, then

$$\theta = \tan^{-1}\sqrt{\frac{au}{bc}} + \tan^{-1}\sqrt{\frac{bu}{ca}} + \tan^{-1}\sqrt{\frac{cu}{ab}}$$

Now we know that

$$\tan^{-1}x + \tan^{-1}y = \pi + \tan^{-1}\left(\frac{x+y}{1-xy}\right) \textit{ when } xy > 1$$

$$\sqrt{\frac{au}{bc}} \times \sqrt{\frac{bu}{ca}} = \frac{u}{c} = \frac{a+b+c}{c} > 1; a, b, c$$

being +ve real no's.

$\therefore$ We get

$$\theta = \pi + \tan^{-1}\left[\frac{\sqrt{\frac{au}{bc}} + \sqrt{\frac{bu}{ca}}}{1 - \sqrt{\frac{au}{bc}}\sqrt{\frac{bu}{ca}}}\right] + \tan^{-1}\sqrt{\frac{cu}{ab}}$$

$$\theta = \pi + \tan^{-1}\left[\frac{\frac{a+b}{\sqrt{abc}}\sqrt{u}}{1-\frac{u}{c}}\right] + \tan^{-1}\sqrt{\frac{cu}{ab}}$$

$$\theta = \pi + \tan^{-1}\left[\frac{(u-c)\sqrt{u}}{\sqrt{abc}} \times \frac{c}{-(u-c)}\right] + \tan^{-1}\sqrt{\frac{cu}{ab}}$$

$$\theta = \pi - \tan^{-1}\sqrt{\frac{uc}{ab}} + \tan^{-1}\sqrt{\frac{cu}{ab}}$$

[Using $\tan^{-1}(-x) = -\tan^{-1}x$] $= \pi$

2. $\tan\left(2\tan^{-1}\frac{1}{5} - \frac{\pi}{4}\right)$

$$= \tan\left[\tan^{-1}\left(\frac{2/5}{1-(1/5)^2}\right) - \tan^{-1}(1)\right]$$

$$= \tan\left[\tan^{-1}\left(\frac{5}{12}\right) - \tan^{-1}(1)\right] = \tan\left[\tan^{-1}\left(\frac{5/12-1}{1+5/12}\right)\right]$$

$$= \tan(\tan^{-1}(-7/17)) = -7/17$$

3. We have

$$A = 2\tan^{-1}(2\sqrt{2}-1) = 2\tan^{-1}(2\times 1.414 - 1)$$

$$= 2\tan^{-1}(1.828) > 2\tan^{-1}\sqrt{3} = 2\pi/3$$

$\Rightarrow A > 2\pi/3$ ...(1)

Also $B = 3\sin^{-1}(1/3) + \sin^{-1}(3/5)$

$$= \sin^{-1}\left[3\times\frac{1}{3} - 4\times\frac{1}{27}\right] + \sin^{-1}(3/5)$$

$$= \sin^{-1}\frac{23}{27} + \sin^{-1}(0.6)$$

$$= \sin^{-1}(0.852) + \sin^{-1}(0.6)$$

$$< \sin^{-1}(\sqrt{3}/2) + \sin^{-1}(\sqrt{3}/2) = 2.\pi/3$$

$\Rightarrow B < 2\pi/3$ ....(2)

From (1) and (2) we conclude $A > B$.

C MCQ's WITH ONE CORRECT ANSWER :

1. (d) $\tan\left[\cos^{-1}\left(\frac{4}{5}\right)+\tan^{-1}\left(\frac{2}{3}\right)\right]$

$= \tan\left[\tan^{-1}\frac{3}{4}+\tan^{-1}\frac{2}{3}\right]$

$= \tan\left[\tan^{-1}\left(\frac{3/4+2/3}{1-3/4\times 2/3}\right)\right]=\frac{17}{12}\times\frac{12}{6}=\frac{17}{6}$

2. (d) $\tan\left[\cos^{-1}\frac{1}{5\sqrt{2}}-\sin^{-1}\frac{4}{\sqrt{17}}\right]$

$=$ $\tan\,[\tan^{-1}7-\tan^{-1}4]$

$=$ $\tan\left(\tan^{-1}\left(\frac{3}{29}\right)\right)=\frac{3}{29}$

3. (c) $\tan^{-1}\sqrt{[x(x+1)]}=\pi/2-\sin^{-1}\sqrt{(x^2+x+1)}$

$\Rightarrow\ \tan^{-1}\sqrt{x(x+1)}=\cos^{-1}\sqrt{x^2+x+1}$

$\Rightarrow\ \cos^{-1}\frac{1}{\sqrt{x^2+x+1}}=\cos^{-1}\sqrt{x^2+x+1}$

$\Rightarrow\ x^2+x+1=1\Rightarrow x(x+1)=0$

$\Rightarrow\ x=0,-1$ are the only real solutions.

ALTERNATE SOLUTION:

For $\tan^{-1}\sqrt{x(x+1)}$ to be defined

$x(x+1)\ge 0$ ....(1)

For $\sin^{-1}\sqrt{x^2+x+1}$ to be defined

$x^2+x+1\ge 0$ ....(2)

and $\sqrt{x^2+x+1}\le 1$ ($\because\ \sin^{-1}x$ is defined for $|x|\le 1$)

$\Rightarrow\ x^2+x+1\le 1.$

$\Rightarrow\ x^2+x\le 0$ ....(3)

From (1) and (3) $x(x+1)=0\Rightarrow\ x=0,-1$

For both these values, given equation is satisfied, therefore $x=0$ and $x=-1$ are the possible solutions.

4. (b) $\sin^{-1}\left(x-\frac{x^2}{2}+\frac{x^3}{3}...\right)+\cos^{-1}\left(x^2-\frac{x^4}{2}+\frac{x^6}{4}...\right)=\frac{\pi}{2}$

$\Rightarrow\ \cos^{-1}\left(x^2-\frac{x^4}{2}+\frac{x^6}{4}...\right)=\frac{\pi}{2}-\sin^{-1}\left(x-\frac{x^2}{2}+\frac{x^3}{4}...\right)$

$\Rightarrow\ \cos^{-1}\left(x^2-\frac{x^4}{2}+\frac{x^6}{4}...\right)=\cos^{-1}\left(x-\frac{x^2}{2}+\frac{x^3}{4}...\right)$

$\Rightarrow\ x^2-\frac{x^4}{2}+\frac{x^6}{4}...=x-\frac{x^2}{2}+\frac{x^3}{4}...$

On both sides we have G.P. of infinite terms.

$\therefore\ \frac{x^2}{1-\left(\frac{-x^2}{2}\right)}=\frac{x}{1-\left(-\frac{x}{2}\right)}\Rightarrow\frac{2x^2}{2+x^2}=\frac{2x}{2+x}$

$\Rightarrow\ 2x+x^3=2x^2+x^3\Rightarrow\ x(x-1)=0$

$\Rightarrow\ x=0,1$ but $0<|x|<\sqrt{2}\Rightarrow x=1.$

5. (d) $\sin\left[\cot^{-1}(1+x)\right]=\cos(\tan^{-1}x)$

$\Rightarrow\ \sin\left[\sin^{-1}\left(\frac{1}{\sqrt{1+(1+x)^2}}\right)\right]=\cos\left[\cos^{-1}\left(\frac{1}{\sqrt{1+x^2}}\right)\right]$

$\Rightarrow\ \frac{1}{\sqrt{1+(1+x)^2}}=\frac{1}{\sqrt{1+x^2}}$

$\Rightarrow\ 1+1+2x+x^2=1+x^2\Rightarrow 2x+1=0\Rightarrow x=-\frac{1}{2}$

6. (c) $\sqrt{1+x^2}\left[\{x\cos(\cot^{-1}x)+\sin(\cot^{-1}x)\}^2-1\right]^{\frac{1}{2}}$

$=\sqrt{1+x^2}\left[\left\{x\cos\left(\cos^{-1}\left(\frac{x}{\sqrt{1+x^2}}\right)\right)+\sin\left(\sin^{-1}\left(\frac{1}{\sqrt{1+x^2}}\right)\right)\right\}^2-1\right]^{\frac{1}{2}}$

$=\sqrt{1+x^2}\left[\left\{x.\frac{x}{\sqrt{1+x^2}}+\frac{1}{\sqrt{1+x^2}}\right\}^2-1\right]$

$=\sqrt{1+x^2}\left[\left(\sqrt{1+x^2}\right)^2-1\right]=x\sqrt{1+x^2}$

7. (b) $\cot^{-1}\left(1+\sum_{k=1}^{n}2k\right)=\cot^{-1}[1+n(n+1)]$

$=\tan^{-1}\left[\frac{(n+1)-n}{1+(n+1)n}\right]=\tan^{-1}(n+1)-\tan^{-1}n$

$\therefore\ \sum_{n=1}^{23}[\tan^{-1}(n+1)-\tan^{-1}n]=\tan^{-1}24-\tan^{-1}1=\tan^{-1}\frac{23}{25}$

$\therefore\ \cot\left[\sum_{n=1}^{23}\cot^{-1}\left(1+\sum_{k=1}^{n}2k\right)\right]=\cot\left[\tan^{-1}\frac{23}{25}\right]=\frac{25}{23}$

D MCQ's WITH ONE OR MORE THAN ONE CORRECT :

1. (d) The principal value of $\sin^{-1}\left(\sin\frac{2\pi}{3}\right)=\sin^{-1}\left(\sin\left(\pi-\frac{\pi}{3}\right)\right)$

$=\sin^{-1}(\sin\pi/3)=\pi/3\ \therefore$ (d) is the correct answer.

E SUBJECTIVE PROBLEMS :

1. We have $\cos(2\cos^{-1}x+\sin^{-1}x)$

$=\cos(\cos^{-1}x+\cos^{-1}x+\sin^{-1}x)$

$=\cos(\cos^{-1}x+\pi/2)$ {Using $\cos^{-1}x+\sin^{-1}x=\pi/2$}

$=-\sin(\cos^{-1}x)$

$=-\sqrt{1-\cos^2(\cos^{-1}x)}=-\sqrt{1-[\cos(\cos^{-1}x)]^2}$

$=-\sqrt{1-x^2}$

$=-\sqrt{1-1/25}$ [for $x=1/5$]

$=-\frac{\sqrt{24}}{5}=\frac{-2\sqrt{6}}{5}$

2. Given eq. is,

$4\cos^2x\sin x-2\sin^2x=3\sin x$

$\Rightarrow 4\cos^2x\sin x-2\sin^2x-3\sin x=0$

$\Rightarrow 4(1-\sin^2x)\sin x-2\sin^2x-3\sin x=0$

$\Rightarrow \sin x[4\sin^2x+2\sin x-1]=0$

$\Rightarrow$ either $\sin x=0$ or $4\sin^2x+2\sin x-1=0$

If $\sin x=0\Rightarrow x=n\pi$

$\Rightarrow$ If $4\sin^2x+2\sin x-1=0$

$\sin x=\frac{-1\pm\sqrt{5}}{4}$

If $\sin x=\frac{-1\pm\sqrt{5}}{4}=\sin 18°=\sin\frac{\pi}{10}$

then $x=nx+(-1)^4\frac{\pi}{10}$

If $\sin x=-\left(\frac{\sqrt{5}+1}{4}\right)=\sin(-54°)=\sin\left(\frac{-3\pi}{10}\right)$

then $x=n\pi+(-1)^n\left(\frac{-3\pi}{10}\right)$

Hence, $x=n\pi,\ n\pi+(-1)^n\frac{\pi}{10}$ or $n\pi+(-1)^n\left(\frac{-3\pi}{10}\right)$

where n is some integer

3. To prove that $\cos\tan^{-1}\sin\cot^{-1}x=\sqrt{\frac{x^2+1}{x^2+2}}$.

L.H.S. $=\cos[\tan^{-1}(\sin(\cot^{-1}x))]$

$=\cos[\tan^{-1}(\sin(\sin^{-1}\frac{1}{\sqrt{1+x^2}}))]$ if $x>0$

and $\cos[\tan^{-1}(\sin(\pi-\sin^{-1}\frac{1}{\sqrt{1+x^2}}))]$ if $x<0$

In each case,

$=\cos\left[\tan^{-1}\frac{1}{\sqrt{1+x^2}}\right]=\cos[\cos^{-1}\sqrt{\frac{1+x^2}{2+x^2}}]$

$=\sqrt{\frac{x^2+1}{x^2+2}}=R.H.S.$ *Hence Proved.*

F MATCH THE FOLLOWING :

1. **(A) – (p); (B) – (r); (C) – (q)**

(A) $t=\sum_{i=1}^{\infty}\tan^{-1}\left(\frac{1}{2i^2}\right)$

$=\sum_{i=1}^{\infty}\tan^{-1}\left[\frac{(2i+1)-(2i-1)}{1+4i^2-1}\right]$

$=\sum_{i=1}^{\infty}[\tan^{-1}(2i+1)-\tan^{-1}(2i-1)]$

$t=\tan^{-1}3-\tan^{-1}1+\tan^{-1}5-\tan^{-1}3$
$+......+\tan^{-1}(2n+1)-\tan^{-1}(2n-1)+......\infty$

$\Rightarrow t=\lim_{n\to\infty}[\tan^{-1}(2n+1)-\tan^{-1}1]$

$=\lim_{n\to\infty}\tan^{-1}\left[\frac{2n}{1+(2n+1)}\right]=\lim_{n\to\infty}\tan^{-1}\left[\frac{1}{1+1/n}\right]$

$\Rightarrow t=\tan^{-1}(1)=\frac{\pi}{4}\Rightarrow\tan t=1,$ (A)$\to$(p)

(B) $\because a,b,c$ are in AP $\Rightarrow 2b=a+c$

$\cos\theta_1=\frac{a}{b+c}$

$\Rightarrow \frac{1-\tan^2\theta_1/2}{1+\tan^2\theta_1/2}=\frac{a}{b+c}$

$\Rightarrow \tan^2\frac{\theta_1}{2}=\frac{b+c-a}{b+c+a}$

Similarly, $\tan^2\frac{\theta_3}{2}=\frac{a+b-c}{a+b+c}$

$\Rightarrow \tan^2\frac{\theta_1}{2}+\tan^2\frac{\theta_3}{2}=\frac{2b}{a+b+c}=\frac{2b}{3b}=\frac{2}{3}$, (B)$\to$(r)

(C) Equation of line through (0, 1, 0) and perpendicular to $x+2y+2z=0$ is

$\frac{x}{1}=\frac{y-1}{2}=\frac{z}{2}=\lambda$

For some value of λ, the foot of perpendicular from origin to line is $(\lambda, 2\lambda+1, 2\lambda)$

Dr's of this $\perp$ from origin are $\lambda, 2\lambda+1, 2\lambda$

$\therefore 1.\lambda+2(2\lambda+1)+2.2\lambda=0\Rightarrow\lambda=-\frac{2}{9}$

$\therefore$ Foot of perpendicular is $\left(-\frac{2}{9},\frac{5}{9},-\frac{4}{9}\right)$

$\therefore$ Required distance

$=\sqrt{\frac{4}{81}+\frac{25}{81}+\frac{16}{81}}=\sqrt{\frac{45}{81}}=\frac{\sqrt{5}}{3}$ (C)$\to$(q)

2. (A) - p, (B) - q, (C)- p , (D) - s

$$\sin^{-1}(ax) + \cos^{-1} y + \cos^{-1}(bxy) = \frac{\pi}{2}$$

$\Rightarrow \cos^{-1} y + \cos^{-1}(bxy) = \frac{\pi}{2} - \sin^{-1}(ax) = \cos^{-1}(ax)$

Let $\cos^{-1} y = \alpha, \cos^{-1}(bxy) = \beta, \cos^{-1}(ax) = \gamma$

$\Rightarrow y = \cos\alpha, bxy = \cos\beta, ax = \cos\gamma$

$\therefore$ We get $\alpha + \beta = \gamma$ and $\cos\beta = bxy$

$\Rightarrow \cos(\gamma - \alpha) = bxy$

$\Rightarrow \cos y \cos\alpha + \sin\gamma \sin\alpha = bxy$

$\Rightarrow axy + \sin\gamma\sin\alpha = bxy$

$\Rightarrow (a-b)xy = -\sin\alpha\sin\gamma$

$\Rightarrow (a-b)^2 x^2y^2 = -\sin^2\alpha\sin^2\gamma$

$= (1-\cos^2\alpha)(1-\cos^2\gamma)$

$\Rightarrow (a-b)^2 x^2y^2 = (1-a^2x^2)(1-y^2)$ ....(1)

(A) For $a = 1, b = 0$, equation (1) reduces to
$x^2y^2 = (1-x^2)(1-y^2) \Rightarrow x^2 + y^2 = 1$

(B) For $a = 1, b = 1$ equation (1) becomes
$(1-x^2)(1-y^2) = 0$
$\Rightarrow (x^2-1)(y^2-1) = 0$

(C) For $a = 1, b = 2$ equation (1) reduces to
$x^2y^2 = (1-x^2)(1-y^2)$
$\Rightarrow x^2 + y^2 = 1$

(D) For $a = 2, b = 2$ equation (1) reduces to
$0 = (1-4x^2)(1-y^2)$
$\Rightarrow (4x^2-1)(y^2-1) = 0$

3. (b)

(P) $\left[\frac{1}{y^2}\left(\frac{\cos(\tan^{-1} y) + y\sin(\tan^{-1} y)}{\cot(\sin^{-1} y) + \tan(\sin^{-1} y)}\right)^2 + y^4\right]^{\frac{1}{2}}$

$$= \left[\frac{1}{y^2}\left(\frac{\cos\left(\cos^{-1}\frac{1}{\sqrt{1+y^2}}\right) + y\sin\left(\sin^{-1}\frac{y}{\sqrt{1+y^2}}\right)}{\cot\left(\cot^{-1}\frac{\sqrt{1-y^2}}{y}\right) + \tan\left(\tan^{-1}\frac{y}{\sqrt{1-y^2}}\right)}\right)^2 + y^4\right]^{\frac{1}{2}}$$

$$= \left[\frac{1}{y^2}\left(\frac{\sqrt{1+y^2}}{\frac{1}{y(\sqrt{1-y^2})}}\right)^2 + y^4\right]^{\frac{1}{2}}$$

$= (1 - y^4 + y^4)^{\frac{1}{2}} = 1 \quad \therefore \quad (P) \to (4)$

(Q) We have $\cos x + \cos y = -\cos z$
$\sin x + \sin y = -\sin z$
Squaring and adding we get
$(\cos x + \cos y)^2 + (\sin x + \sin y)^2 = \cos^2 z + \sin^2 z$

$\Rightarrow \quad 2 + 2\cos(x-y) = 1$

$\Rightarrow \quad 4\cos^2\frac{x-y}{2} = 1$ or $\cos\frac{x-y}{2} = \frac{+1}{2}$

$\therefore \quad Q \to (3)$

(R) We have

$\cos\left(\frac{\pi}{4} - x\right)\cos 2x + \sin x \sin 2x \sec x$

$= \cos x \sin 2x \sec x + \cos\left(\frac{\pi}{4} + x\right)\cos 2x$

$\Rightarrow \cos 2x\left[\cos\left(\frac{\pi}{4} - x\right) - \cos\left(\frac{\pi}{4} + x\right)\right]$

$= \sin 2x \sec x (\cos x - \sin x)$

$\Rightarrow 2\sin\frac{\pi}{4}\sin x\cos 2x = 2\sin x(\cos x - \sin x)$

$\Rightarrow 2\sin x\left[\frac{1}{\sqrt{2}}(\cos^2 x - \sin^2 x) - (\cos x - \sin x)\right] = 0$

$\Rightarrow 2\sin x(\cos x - \sin x)\left(\frac{\cos x + \sin x}{\sqrt{2}} - 1\right) = 0$

$\Rightarrow \sin x = 0$ or $\tan x = 1$ or $\cos\left(x - \frac{\pi}{4}\right) = 1$

$\Rightarrow x = 0, \frac{x}{4}$

$\Rightarrow \sec x = 1$ or $\sqrt{2}$

$\therefore (R) \to (2)$

(S) $\cos\left(\sin^{-1}\sqrt{1-x^2}\right) = \sin\left(\tan^{-1} x\sqrt{6}\right)$

$\Rightarrow \frac{x}{\sqrt{1-x^2}} = \frac{x\sqrt{6}}{\sqrt{1+6x^2}} \Rightarrow x = \pm\frac{5}{2\sqrt{3}}$

$\therefore (S) \to (1)$

Hence $(P) \to (4), (Q) \to (3), (R) \to (2), (S) \to (1)$

Section-B JEE Main/ AIEEE

1. (a) $\cot^{-1}(\sqrt{\cos\alpha}) - \tan^{-1}(\sqrt{\cos\alpha}) = x$

$$\tan^{-1}\left(\frac{1}{\sqrt{\cos\alpha}}\right) - \tan^{-1}(\sqrt{\cos\alpha}) = x$$

$$\Rightarrow \tan^{-1}\frac{\frac{1}{\sqrt{\cos\alpha}} - \sqrt{\cos\alpha}}{1+\frac{1}{\sqrt{\cos\alpha}}.\sqrt{\cos\alpha}} = x$$

$$\Rightarrow \tan^{-1}\frac{1-\cos\alpha}{2\sqrt{\cos\alpha}} = x$$

$$\Rightarrow \tan x = \frac{1-\cos\alpha}{2\sqrt{\cos\alpha}} \text{ OR } \cot x = \frac{2\sqrt{\cos\alpha}}{1-\cos\alpha}$$

$$\Rightarrow \sin x = \frac{1-\cos\alpha}{1+\cos\alpha} = \frac{1-(1-2\sin^2\alpha/2)}{1+2\cos^2\alpha/2-1}$$

or $\sin x = \tan^2\frac{\alpha}{2}$

2. (None) $\sin^{-1}x = 2\sin^{-1}a$

$$-\frac{\pi}{2} \le \sin^{-1}x \le \frac{\pi}{2};\ \therefore -\frac{\pi}{2} \le 2\sin^{-1}a \le \frac{\pi}{2}$$

$$-\frac{\pi}{4} \le \sin^{-1}a \le \frac{\pi}{4} \text{ or } \frac{-1}{\sqrt{2}} \le a \le \frac{1}{\sqrt{2}}$$

$$\therefore |a| \le \frac{1}{\sqrt{2}}$$

Out of given four options none is absolutely correct

3. (c) $\cos^{-1}x - \cos^{-1}\frac{y}{2} = \alpha$

$$\cos^{-1}\left(\frac{xy}{2} + \sqrt{(1-x^2)\left(1-\frac{y^2}{4}\right)}\right) = \alpha$$

$$\cos^{-1}\left(\frac{xy+\sqrt{4-y^2-4x^2+x^2y^2}}{2}\right) = \alpha$$

$$\Rightarrow 4-y^2-4x^2+x^2y^2$$

$$= 4\cos^2\alpha + x^2y^2 - 4xy\cos\alpha$$

$$\Rightarrow 4x^2+y^2-4xy\cos\alpha = 4\sin^2\alpha.$$

4. (d) $\sin^{-1}\left(\frac{x}{5}\right) + \text{cosec}^{-1}\left(\frac{5}{4}\right) = \frac{\pi}{2}$

$$\Rightarrow \sin^{-1}\left(\frac{x}{5}\right) = \frac{\pi}{2} - \text{cosec}^{-1}\left(\frac{5}{4}\right)$$

$$\Rightarrow \sin^{-1}\left(\frac{x}{5}\right) = \frac{\pi}{2} - \sin^{-1}\left(\frac{4}{5}\right)$$

$[\because \sin^{-1}x + \cos^{-1}x = \pi/2]$

$$\Rightarrow \sin^{-1}\left(\frac{x}{5}\right) = \cos^{-1}\left(\frac{4}{5}\right)$$

$$\Rightarrow \sin^{-1}\frac{x}{5} = \sin^{-1}\frac{3}{5} \Rightarrow \frac{x}{5} = \frac{3}{5} \Rightarrow x = 3$$

5. (a) $\cot\left(\text{cosec}^{-1}\frac{5}{3} + \tan^{-1}\frac{2}{3}\right) = \cot\left[\tan^{-1}\frac{3}{4} + \tan^{-1}\frac{2}{3}\right]$

$$= \cot\left[\tan^{-1}\left(\frac{\frac{3}{4}+\frac{2}{3}}{1-\frac{3}{4}\times\frac{2}{3}}\right)\right] = \cot\left[\tan^{-1}\frac{17}{6}\right]$$

$$= \cot\left(\cot^{-1}\frac{6}{17}\right) = \frac{6}{17}$$

6. (a) Since, x, y, z are in A.P.

$\therefore\ 2y = x+z$

Also, we have

$2\tan^{-1}y = \tan^{-1}x + \tan^{-1}(z)$

$$\Rightarrow \tan^{-1}\left(\frac{2y}{1-y^2}\right) = \tan^{-1}\left(\frac{x+z}{1-xz}\right)$$

$$\Rightarrow \frac{x+z}{1-y^2} = \frac{x+z}{1-xz}\quad (\because\ 2y = x+z)$$

$$\Rightarrow y^2 = xz \text{ or } x+z=0 \Rightarrow x=y=z$$

Chapter 15 Matrices and Determinants

Section-A JEE Advanced/ IIT-JEE

(A) FILL IN THE BLANKS :

1. Let $p\lambda^4+q\lambda^3+r\lambda^2+s\lambda+t=\begin{vmatrix}\lambda^2+3\lambda & \lambda-1 & \lambda+3\\ \lambda+1 & -2\lambda & \lambda-4\\ \lambda-3 & \lambda+4 & 3\lambda\end{vmatrix}$ be an identity in λ, where p, q, r, s and t are constants. Then, the value of t is .... **(1981 - 2 Marks)**

2. The solution set of the equation $\begin{vmatrix}1 & 4 & 20\\ 1 & -2 & 5\\ 1 & 2x & 5x^2\end{vmatrix}=0$ is .... **(1981 - 2 Marks)**

3. A determinant is chosen at random from the set of all determinants of order 2 with elements 0 or 1 only. The probability that the value of determinant chosen is positive is .... **(1982 - 2 Marks)**

4. Given that $x=-9$ is a root of $\begin{vmatrix}x & 3 & 7\\ 2 & x & 2\\ 7 & 6 & x\end{vmatrix}=0$ the other two roots are .... and .... **(1983 - 2 Marks)**

5. The system of equations

$\lambda x+y+z=0$

$-x+\lambda y+z=0$

$-x-y+\lambda z=0$

Will have a non-zero solution if real values of λ are given by .... **(1984 - 2 Marks)**

6. The value of the determinant $\begin{vmatrix}1 & a & a^2-bc\\ 1 & b & b^2-ca\\ 1 & c & c^2-ab\end{vmatrix}$ is .... **(1988 - 2 Marks)**

7. For positive numbers x, y and z, the numerical value of the determinant $\begin{vmatrix}1 & \log_x y & \log_x z\\ \log_y x & 1 & \log_y z\\ \log_z x & \log_z y & 1\end{vmatrix}$ is .... **(1993 - 2 Marks)**

(B) TRUE / FALSE :

1. The determinants **(1983 - 1 Mark)**

$\begin{vmatrix}1 & a & bc\\ 1 & b & ca\\ 1 & c & ab\end{vmatrix}$ and $\begin{vmatrix}1 & a & a^2\\ 1 & b & b^2\\ 1 & c & c^2\end{vmatrix}$ are not identically equal.

2. If $\begin{vmatrix}x_1 & y_1 & 1\\ x_2 & y_2 & 1\\ x_3 & y_3 & 1\end{vmatrix}=\begin{vmatrix}a_1 & b_1 & 1\\ a_2 & b_2 & 1\\ a_3 & b_3 & 1\end{vmatrix}$

then the two triangles with vertices (x_1,y_1), (x_2,y_2), (x_3,y_3), and (a_1,b_1), (a_2,b_2), (a_3,b_3) must be congruent. **(1985 - 1 Mark)**

(C) MCQ's WITH ONE CORRECT ANSWER :

1. Consider the set A of all determinants of order 3 with entries 0 or 1 only. Let B be the subset of A consisting of all determinants with value 1. Let C be the subset of A consisting of all determinants with value –1. Then **(1981 - 2 Marks)**
   (a) C is empty
   (b) B has as many elements as C
   (c) $A=B\cup C$
   (d) B has twice as many elements as elements as C

2. If $\omega\,(\neq 1)$ is a cube root of unity, then

$\begin{vmatrix}1 & 1+i+\omega^2 & \omega^2\\ 1-i & -1 & \omega^2-1\\ -i & -i+\omega-1 & -1\end{vmatrix}=$ **(1995S)**

   (a) 0 (b) 1
   (c) i (d) ω

3. Let a, b, c be the real numbers. Then following system of equations in x, y and z **(1995S)**

$\frac{x^2}{a^2}+\frac{y^2}{b^2}-\frac{z^2}{c^2}=1,\ \frac{x^2}{a^2}-\frac{y^2}{b^2}+\frac{z^2}{c^2}=1,$

$-\frac{x^2}{a^2}+\frac{y^2}{b^2}+\frac{z^2}{c^2}=1$ has

   (a) no solution (b) unique solution
   (c) infinitely many solutions (d) finitely many solutions

4. If A and B are square matrices of equal degree, then which one is correct among the followings? *(1995S)*
(a) $A+B=B+A$ (b) $A+B=A-B$
(c) $A-B=B-A$ (d) $AB=BA$

5. The determinant $\begin{vmatrix} xp+y & x & y \\ yp+z & y & z \\ 0 & xp+y & yp+z \end{vmatrix}=0$ if *(1997C - 2 Marks)*
(a) x, y, z are in A.P (b) x, y, z are in G.P.
(c) x, y, z are in H.P (d) xy, yz, zx are in A.P.

6. The parameter, on which the value of the determinant
$$\begin{vmatrix} 1 & a & a^2 \\ \cos(p-d)x & \cos px & \cos(p+d)x \\ \sin(p-d)x & \sin px & \sin(p+d)x \end{vmatrix}$$ does not depend upon is *(1997 - 2 Marks)*
(a) a (b) p
(c) d (d) x

7. If $f(x)=\begin{vmatrix} 1 & x & x+1 \\ 2x & x(x-1) & (x+1)x \\ 3x(x-1) & x(x-1)(x-2) & (x+1)x(x-1) \end{vmatrix}$ then $f(100)$ is equal to *(1999 - 2 Marks)*
(a) 0 (b) 1
(c) 100 (d) –100

8. If the system of equations
$x-ky-z=0$, $kx-y-z=0$, $x+y-z=0$ has a non-zero solution, then the possible values of k are *(2000S)*
(a) –1, 2 (b) 1, 2
(c) 0, 1 (d) –1, 1

9. Let $\omega=-\frac{1}{2}+i\frac{\sqrt{3}}{2}$. Then the value of the determinant
$$\begin{vmatrix} 1 & 1 & 1 \\ 1 & -1-\omega^2 & \omega^2 \\ 1 & \omega^2 & \omega^4 \end{vmatrix}$$ is *(2002S)*
(a) 3ω (b) $3\omega(\omega-1)$
(c) $3\omega^2$ (d) $3\omega(1-\omega)$

10. The number of values of k for which the system of equations $(k+1)x+8y=4k$; $kx+(k+3)y=3k-1$ has infinitely many solutions is *(2002S)*
(a) 0 (b) 1
(c) 2 (d) infinite

11. If $A=\begin{bmatrix} \alpha & 0 \\ 1 & 1 \end{bmatrix}$ and $B=\begin{bmatrix} 1 & 0 \\ 5 & 1 \end{bmatrix}$, then value of α for which $A^2=B$, is *(2003S)*
(a) 1 (b) –1
(c) 4 (d) no real values

12. If the system of equations $x+ay=0$, $az+y=0$ and $ax+z=0$ has infinite solutions, then the value of a is *(2003S)*
(a) –1 (b) 1
(c) 0 (d) no real values

13. Given $2x-y+2z=2$, $x-2y+z=-4$, $x+y+\lambda z=4$ then the value of λ such that the given system of equation has NO solution, is *(2004S)*
(a) 3 (b) 1
(c) 0 (d) –3

14. If $A=\begin{bmatrix} \alpha & 2 \\ 2 & \alpha \end{bmatrix}$ and $|A^3|=125$ then the value of α is *(2004S)*
(a) ± 1 (b) ± 2
(c) ± 3 (d) ± 5

15. $A=\begin{bmatrix} 1 & 0 & 0 \\ 0 & 1 & 1 \\ 0 & -2 & 4 \end{bmatrix}$ and $I=\begin{bmatrix} 1 & 0 & 0 \\ 0 & 1 & 0 \\ 0 & 0 & 1 \end{bmatrix}$ and $A^{-1}=\left[\frac{1}{6}(A^2+cA+dI)\right]$, then the value of c and d are *(2005S)*
(a) (–6, –11) (b) (6, 11)
(c) (–6, 11) (d) (6, –11)

16. If $P=\begin{bmatrix} \frac{\sqrt{3}}{2} & \frac{1}{2} \\ -\frac{1}{2} & \frac{\sqrt{3}}{2} \end{bmatrix}$ and $A=\begin{bmatrix} 1 & 1 \\ 0 & 1 \end{bmatrix}$ and $Q=PAP^T$ and $x=P^TQ^{2005}P$ then x is equal to *(2005S)*
(a) $\begin{bmatrix} 1 & 2005 \\ 0 & 1 \end{bmatrix}$
(b) $\begin{bmatrix} 4+2005\sqrt{3} & 6015 \\ 2005 & 4-2005\sqrt{3} \end{bmatrix}$
(c) $\frac{1}{4}\begin{bmatrix} 2+\sqrt{3} & 1 \\ -1 & 2-\sqrt{3} \end{bmatrix}$
(d) $\frac{1}{4}\begin{bmatrix} 2005 & 2-\sqrt{3} \\ 2+\sqrt{3} & 2005 \end{bmatrix}$

17. Consider three points
$P = (-\sin(\beta-\alpha), -\cos\beta)$, $Q = (\cos(\beta-\alpha), \sin\beta)$ and
$R = (\cos(\beta-\alpha+\theta), \sin(\beta-\theta))$, where $0 < \alpha, \beta, \theta < \frac{\pi}{4}$.
Then, *(2008)*
(a) P lies on the line segment RQ
(b) Q lies on the line segment PR
(c) R lies on the line segment QP
(d) P, Q, R are non-collinear

18. The number of 3 × 3 matrices A whose entries are either 0 or 1 and for which the system $A\begin{bmatrix} x \\ y \\ z \end{bmatrix} = \begin{bmatrix} 1 \\ 0 \\ 0 \end{bmatrix}$ has exactly two distinct solutions, is *(2010)*
(a) 0 (b) 2^9-1
(c) 168 (d) 2

19. Let $\omega \neq 1$ be a cube root of unity and S be the set of all non-singular matrices of the form $\begin{bmatrix} 1 & a & b \\ \omega & 1 & c \\ \omega^2 & \omega & 1 \end{bmatrix}$ where each of a, b and c is either ω or ω^2. Then the number of distinct matrices in the set S is *(2011)*
(a) 2 (b) 6 (c) 4 (d) 8

20. Let $P = [a_{ij}]$ be a 3 × 3 matrix and let $Q = [b_{ij}]$, where $b_{ij} = 2^{i+j} a_{ij}$ for $1 \leq i, j \leq 3$. If the determinant of P is 2, then the determinant of the matrix Q is *(2012)*
(a) 2^{10} (b) 2^{11}
(c) 2^{12} (d) 2^{13}

21. If P is a 3 × 3 matrix such that $P^T = 2P + I$, where P^T is the transpose of P and I is the 3 × 3 identity matrix, then there exists a column matrix $X = \begin{bmatrix} x \\ y \\ z \end{bmatrix} \neq \begin{bmatrix} 0 \\ 0 \\ 0 \end{bmatrix}$ such that *(2012)*
(a) $PX = \begin{bmatrix} 0 \\ 0 \\ 0 \end{bmatrix}$ (b) $PX = X$
(c) $PX = 2X$ (d) $PX = -X$

(D) MCQ's WITH ONE OR MORE THAN ONE CORRECT :

1. The determinant $\begin{vmatrix} a & b & a\alpha+b \\ b & c & b\alpha+c \\ a\alpha+b & b\alpha+c & 0 \end{vmatrix}$ is equal to zero, if *(1986 - 2 Marks)*
(a) a, b, c are in A. P.
(b) a, b, c are in G. P.
(c) a, b, c are in H. P.
(d) α is a root of the equation $ax^2 + bx + c = 0$
(e) $(x - \alpha)$ is a factor of $ax^2 + 2bx + c$.

2. If $\begin{vmatrix} 6i & -3i & 1 \\ 4 & 3i & -1 \\ 20 & 3 & i \end{vmatrix} = x + iy$, then *(1998 - 2 Marks)*
(a) $x = 3, y = 1$ (b) $x = 1, y = 3$
(c) $x = 0, y = 3$ (d) $x = 0, y = 0$

3. Let M and N be two 3 × 3 non-singular skew- symmetric matrices such that MN = NM. If P^T denotes the transpose of P, then $M^2N^2(M^TN)^{-1}(MN^{-1})^T$ is equal to *(2011)*
(a) M^2 (b) $-N^2$ (c) $-M^2$ (d) MN

4. If the adjoint of a 3 × 3 matrix P is $\begin{bmatrix} 1 & 4 & 4 \\ 2 & 1 & 7 \\ 1 & 1 & 3 \end{bmatrix}$, then the possible value(s) of the determinant of P is (are)
(a) –2 (b) –1 *(2012)*
(c) 1 (d) 2

5. For 3 × 3 matrices M and N, which of the following statement(s) is (are) **NOT** correct? *(JEE Adv. 2013)*
(a) N^TMN is symmetric or skew symmetric, according as M is symmetric or skew symmetric
(b) MN – NM is skew symmetric for all symmetric matrices M and N
(c) MN is symmetric for all symmetric matrices M and N
(d) (adj M) (adj N) = adj (MN) for all invertible matrices M and N

6. Let ω be a complex cube root of unity with $\omega \neq 1$ and $P = [p_{ij}]$ be a $n \times n$ matrix with $p_{ij} = \omega^{i+j}$. Then $p^2 \neq 0$, when n = *(JEE Adv. 2013)*
(a) 57 (b) 55
(c) 58 (d) 56

(E) SUBJECTIVE PROBLEMS :

1. For what value of k do the following system of equations possess a non trivial (i.e., not all zero) solution over the set of rationals Q?
$x + ky + 3z = 0$
$3x + ky - 2z = 0$
$2x + 3y - 4z = 0$
For that value of k, find all the solutions for the system. *(1979)*

2. Let a, b, c be positive and not all equal. Show that the value of the determinant $\begin{vmatrix} a & b & c \\ b & c & a \\ c & a & b \end{vmatrix}$ is negative. *(1981 - 4 Marks)*

3. Without expanding a determinant at any stage, show that
$$\begin{vmatrix} x^2+x & x+1 & x-2 \\ 2x^2+3x-1 & 3x & 3x-3 \\ x^2+2x+3 & 2x-1 & 2x-1 \end{vmatrix} = xA + B,$$ where A and B are determinants of order 3 not involving x. *(1982 - 5 Marks)*

4. Show that the system of equations $3x-y+4z=3$

$$x+2y-3z=-2$$
$$6x+5y+\lambda z=-3$$

(1983 - 3 Marks)

has at least one solution for any real number $\lambda \neq -5$. Find the set of solutions if $\lambda=-5$

5. Show that

$$\begin{vmatrix} {}^xC_r & {}^xC_{r+1} & {}^xC_{r+2} \\ {}^yC_r & {}^yC_{r+1} & {}^yC_{r+2} \\ {}^zC_r & {}^zC_{r+1} & {}^zC_{r+2} \end{vmatrix} = \begin{vmatrix} {}^xC_r & {}^{x+1}C_{r+1} & {}^{x+2}C_{r+2} \\ {}^yC_r & {}^{y+1}C_{r+1} & {}^{y+2}C_{r+2} \\ {}^zC_r & {}^{z+1}C_{r+1} & {}^{z+2}C_{r+2} \end{vmatrix}$$

(1985 - 2 Marks)

6. Consider the system of linear equations in x, y, z :

$(\sin 3\theta)\ x-y+z=0$

$(\cos 2\theta)\ x+4y+3z=0$

$2x+7y+7z=0$

Find the values of θ for which this system has nontrivial solutions. *(1986 - 5 Marks)*

7. Let $\Delta a = \begin{vmatrix} a-1 & n & 6 \\ (a-1)^2 & 2n^2 & 4n-2 \\ (a-1)^3 & 3n^3 & 3n^2-3n \end{vmatrix}$.

Show that $\sum_{a=1}^{n} \Delta a = c$, a constant. *(1989 - 5 Marks)*

8. Let the three digit numbers $A\,28$, $3B9$, and $62\,C$, where A, B, and C are integers between 0 and 9, be divisible by a fixed integer k. Show that the determinant $\begin{vmatrix} A & 3 & 6 \\ 8 & 9 & C \\ 2 & B & 2 \end{vmatrix}$ is divisible by k. *(1990 - 4 Marks)*

9. If $a \neq p, b \neq q, c \neq r$ and $\begin{vmatrix} p & b & c \\ a & q & c \\ a & b & r \end{vmatrix} = 0$. Then find the value of $\dfrac{p}{p-a}+\dfrac{q}{q-b}+\dfrac{r}{r-c}$ *(1991 - 4 Marks)*

10. For a fixed positive integer n, if *(1992 - 4 Marks)*

$$D = \begin{vmatrix} n! & (n+1)! & (n+2)! \\ (n+1)! & (n+2)! & (n+3)! \\ (n+2)! & (n+3)! & (n+4)! \end{vmatrix}$$

then show that $\left[\dfrac{D}{(n!)^3}-4\right]$ is divisible by n.

11. Let λ and α be real. Find the set of all values of λ for which the system of linear equations

$\lambda x+(\sin\alpha)y+(\cos\alpha)z=0$, $x+(\cos\alpha)y+(\sin\alpha)z=0$,

$-x+(\sin\alpha)y-(\cos\alpha)z=0$

has a non-trivial solution. For $\lambda=1$, find all values of α. *(1993 - 5 Marks)*

12. For all values of A, B, C and P, Q, R show that *(1994 - 4 Marks)*

$$\begin{vmatrix} \cos(A-P) & \cos(A-Q) & \cos(A-R) \\ \cos(B-P) & \cos(B-Q) & \cos(B-R) \\ \cos(C-P) & \cos(C-Q) & \cos(C-R) \end{vmatrix} = 0$$

13. Let $a>0$, $d>0$. Find the value of the determinant *(1996 - 5 Marks)*

$$\begin{vmatrix} \dfrac{1}{a} & \dfrac{1}{a(a+d)} & \dfrac{1}{(a+d)(a+2d)} \\ \dfrac{1}{(a+d)} & \dfrac{1}{(a+d)(a+2d)} & \dfrac{1}{(a+2d)(a+3d)} \\ \dfrac{1}{(a+2d)} & \dfrac{1}{(a+2d)(a+3d)} & \dfrac{1}{(a+3d)(a+4d)} \end{vmatrix}$$

14. Find the value of the determinant $\begin{vmatrix} bc & ca & ab \\ p & q & r \\ 1 & 1 & 1 \end{vmatrix}$ where a, b and c are respectively the pth, q^{th} and r^{th} terms of a harmonic progression. *(1997C - 2 Marks)*

15. Prove that for all values of θ,

$$\begin{vmatrix} \sin\theta & \cos\theta & \sin 2\theta \\ \sin\left(\theta+\dfrac{2\pi}{3}\right) & \cos\left(\theta+\dfrac{2\pi}{3}\right) & \sin\left(2\theta+\dfrac{4\pi}{3}\right) \\ \sin\left(\theta-\dfrac{2\pi}{3}\right) & \cos\left(\theta-\dfrac{2\pi}{3}\right) & \sin\left(2\theta-\dfrac{4\pi}{3}\right) \end{vmatrix} = 0$$

(2000 - 3 Marks)

16. If matrix $A = \begin{bmatrix} a & b & c \\ b & c & a \\ c & a & b \end{bmatrix}$ where a, b, c are real positive numbers, $abc=1$ and $A^TA=I$, then find the value of $a^3+b^3+c^3$. *(2003 - 2 Marks)*

17. If M is a 3×3 matrix, where $\det M=1$ and $MM^T=I$, where 'I' is an identity matrix, prove that $\det(M-I)=0$. *(2004 - 2 Marks)*

18. If $A = \begin{bmatrix} a & 1 & 0 \\ 1 & b & d \\ 1 & b & c \end{bmatrix}, B = \begin{bmatrix} a & 1 & 1 \\ 0 & d & c \\ f & g & h \end{bmatrix}, U = \begin{bmatrix} f \\ g \\ h \end{bmatrix}, V = \begin{bmatrix} a^2 \\ 0 \\ 0 \end{bmatrix}, X = \begin{bmatrix} x \\ y \\ z \end{bmatrix}$

and $AX=U$ has infinitely many solutions, prove that $BX=V$ has no unique solution. Also show that if $afd \neq 0$, then $BX=V$ has no solution. *(2004 - 4 Marks)*

(F) MATCH THE FOLLOWING :

Each question contains statements given in two columns, which have to be matched. The statements in Column-I are labelled A, B, C and D, while the statements in Column-II are labelled p, q, r, s and t. Any given statement in Column-I can have correct matching with ONE OR MORE statement(s) in Column-II. The appropriate bubbles corresponding to the answers to these questions have to be darkened as illustrated in the following example :
If the correct matches are A-p, s and t; B-q and r; C-p and q; and D-s then the correct darkening of bubbles will look like the given.

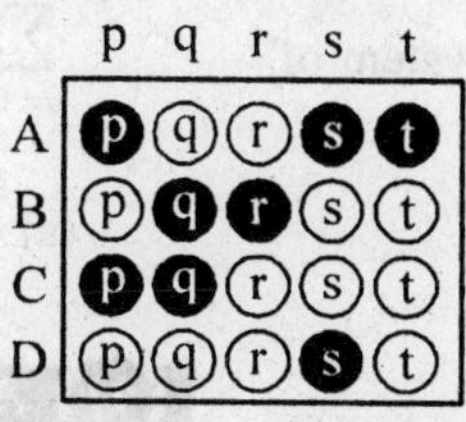

1. Consider the lines given by
$L_1 : x + 3y - 5 = 0;\ L_2 : 3x - ky - 1 = 0;\ L_3 : 5x + 2y - 12 = 0$
Match the Statements / Expressions in **Column I** with the Statements / Expressions in **Column II** and indicate your answer by darkening the appropriate bubbles in the 4 × 4 matrix given in the ORS. ***(2008)***

Column I | ***Column II***

(A) L_1, L_2, L_3 are concurrent, if | (p) $k = -9$

(B) One of L_1, L_2, L_3 is parallel to at least one of the other two, if | (q) $k = -\frac{6}{5}$

(C) L_1, L_2, L_3 from a triangle, if | (r) $k = \frac{5}{6}$

(D) L_1, L_2, L_3 do not form a triangle, if | (s) $k = 5$

2. Match the Statements/Expressions in **Column I** with the Statements / Expressions in **Column II** and indicate your answer by darkening the appropriate bubbles in the 4 × 4 matrix given in the ORS. ***(2008)***

Column I | ***Column II***

(A) The minimum value of $\frac{x^2+2x+4}{x+2}$ is | (p) 0

(B) Let A and B be 3 × 3 matrices of real numbers, where A is symmetric, B is skew-symmetric, and $(A+B)(A-B) = (A-B)(A+B)$. If $(AB)^t = (-1)^k AB$, where $(AB)^t$ is the transpose of the matrix AB, then the possible values of k are | (q) 1

(C) Let $a = \log_3 \log_3 2$. An integer k satisfying $1 < 2^{(-k+3^{-a})} < 2$, must be less than | (r) 2

(D) If $\sin\theta = \cos\phi$, then the possible values of $\frac{1}{\pi}\left(\theta \pm \phi - \frac{\pi}{2}\right)$ are | (s) 3

(G) COMPREHENSION BASED QUESTIONS :

Passage 1

Let $A = \begin{bmatrix} 1 & 0 & 0 \\ 2 & 1 & 0 \\ 3 & 2 & 1 \end{bmatrix}$, and U_1, U_2 and U_3 are columns of a 3 × 3 matrix U. If column matrices U_1, U_2 and U_3 satisfying

$AU_1 = \begin{bmatrix} 1 \\ 0 \\ 0 \end{bmatrix}, AU_2 = \begin{bmatrix} 2 \\ 3 \\ 0 \end{bmatrix}, AU_3 = \begin{bmatrix} 2 \\ 3 \\ 1 \end{bmatrix}$ evaluate as directed in the following questions.

1. The value $|U|$ is ***(2006 - 5M, –2)***

(a) 3 (b) –3 (c) $\frac{3}{2}$ (d) 2

2. The sum of the elements of the matrix U^{-1} is ***(2006 - 5M, –2)***

(a) –1 (b) 0 (c) 1 (d) 3

3. The value of $[3\ 2\ 0]\ U \begin{bmatrix} 3 \\ 2 \\ 0 \end{bmatrix}$ is ***(2006 - 5M, –2)***

(a) 5 (b) $\frac{5}{2}$ (c) 4 (d) $\frac{3}{2}$

Passage 2

Let $\mathscr{A}$ be the set of all 3× 3 symmetric matrices all of whose entries are either 0 or 1. Five of these entries are 1 and four of them are 0.

4. The number of matrices in $\mathscr{A}$ is *(2009)*

(a) 12 (b) 6
(c) 9 (d) 3

5. The number of matrices A in $\mathscr{A}$ for which the system of linear equations

$$A\begin{bmatrix}x\\y\\z\end{bmatrix}=\begin{bmatrix}1\\0\\0\end{bmatrix}$$

has a unique solution, is *(2009)*

(a) less than 4 (b) at least 4 but less than 7
(c) at least 7 but less than 10 (d) at least 10

6. The number of matrices A in $\mathscr{A}$ for which the system of linear equations

$$A\begin{bmatrix}x\\y\\z\end{bmatrix}=\begin{bmatrix}1\\0\\0\end{bmatrix}$$

is inconsistent, is *(2009)*

(a) 0 (b) more than 2
(c) 2 (d) 1

Passage 3

Let p be an odd prime number and T_p be the following set of 2 × 2 matrices :

$$T_P=\left\{A=\begin{bmatrix}a & b\\c & a\end{bmatrix}: a,b,c\in\{0,1,2,\ldots,p-1\}\right\}$$ *(2010)*

7. The number of A in T_p such that A is either symmetric or skew-symmetric or both, and det(A) divisible by p is

(a) $(p-1)^2$ (b) $2(p-1)$
(c) $(p-1)^2+1$ (d) $2p-1$

8. The number of A in T_p such that the trace of A is not divisible by p but det (A) is divisible by p is

[**Note:** The trace of a matrix is the sum of its diagonal entries.]

(a) $(p-1)(p^2-p+1)$ (b) $p^3-(p-1)^2$
(c) $(p-1)^2$ (d) $(p-1)(p^2-2)$

9. The number of A in T_p such that det (A) is not divisible by p is

(a) $2p^2$ (b) p^3-5p
(c) p^3-3p (d) p^3-p^2

Passage 4

(2011)

Let a, b and c be three real numbers satisfying

$$[a\,b\,c]\begin{bmatrix}1 & 9 & 7\\8 & 2 & 7\\7 & 3 & 7\end{bmatrix}=[0\,0\,0] \quad \text{...(E)}$$

10. If the point $P(a, b, c)$, with reference to (E), lies on the plane $2x+y+z=1$, then the value of $7a+b+c$ is

(a) 0 (b) 12 (c) 7 (d) 6

11. Let ω be a solution of $x^3-1=0$ with Im (ω) > 0, if $a = 2$ with b and c satisfying (E), then the value of $\frac{3}{\omega^a}+\frac{1}{\omega^b}+\frac{3}{\omega^c}$ is equal to

(a) −2 (b) 2 (c) 3 (d) −3

12. Let $b = 6$, with a and c satisfying (E). If α and β are the roots of the quadratic equation $ax^2+bx+c=0$, then $\sum_{n=0}^{\infty}\left(\frac{1}{\alpha}+\frac{1}{\beta}\right)^n$ is

(a) 6 (b) 7 (c) $\frac{6}{7}$ (d) ∞

(H) ASSERTION & REASON TYPE QUESTIONS :

1. Consider the system of equations

$$x-2y+3z=-1$$
$$-x+y-2z=k$$
$$x-3y+4z=1$$

STATEMENT - 1 : The system of equations has no solution for $k \neq 3$ and

STATEMENT-2 : The determinant $\begin{vmatrix}1 & 3 & -1\\-1 & -2 & k\\1 & 4 & 1\end{vmatrix}\neq 0$, for $k \neq 3$. *(2008)*

(a) STATEMENT - 1 is True, STATEMENT - 2 is True; STATEMENT - 2 is a correct explanation for STATEMENT - 1

(b) STATEMENT - 1 is True, STATEMENT - 2 is True; STATEMENT - 2 is **NOT** a correct explaination for STATEMENT - 1

(c) STATEMENT - 1 is True, STATEMENT - 2 is False

(d) STATEMENT - 1 is False, STATEMENT - 2 is True

(I) INTEGER VALUE CORRECT TYPE :

1. Let ω be the complex number $\cos\frac{2\pi}{3}+i\sin\frac{2\pi}{3}$. Then the number of distinct complex numbers z satisfying

$$\begin{vmatrix}z+1 & \omega & \omega^2\\\omega & z+\omega^2 & 1\\\omega^2 & 1 & z+\omega\end{vmatrix}=0$$ is equal to *(2010)*

2. Let k be a positive real number and let

$$A=\begin{bmatrix}2k-1 & 2\sqrt{k} & 2\sqrt{k}\\2\sqrt{k} & 1 & -2k\\-2\sqrt{k} & 2k & -1\end{bmatrix} \text{ and } B=\begin{bmatrix}0 & 2k-1 & \sqrt{k}\\1-2k & 0 & 2\sqrt{k}\\-\sqrt{k} & -2\sqrt{k} & 0\end{bmatrix}$$

If det (adj A) + det (adj B) = 10^6. then [k] is equal to

[**Note** : adj M denotes the adjoint of square matrix M and $[k]$ denotes the largest integer less than or equal k. *(2010)*

3. Let M be a 3 × 3 matrix satisfying

$$M\begin{bmatrix}0\\1\\0\end{bmatrix}=\begin{bmatrix}-1\\2\\3\end{bmatrix},\ M\begin{bmatrix}1\\-1\\0\end{bmatrix}=\begin{bmatrix}1\\1\\-1\end{bmatrix}, \text{ and } M\begin{bmatrix}1\\1\\1\end{bmatrix}=\begin{bmatrix}0\\0\\12\end{bmatrix}$$. Then the sum of the diagonal entries of M is *(2011)*

Section-B JEE Main/ AIEEE

1. If $a > 0$ and discriminant of $ax^2+2bx+c$ is –ve, then $\begin{vmatrix} a & b & ax+b \\ b & c & bx+c \\ ax+b & bx+c & 0 \end{vmatrix}$ is equal to **[2002]**

(a) +ve (b) $(ac-b^2)(ax^2+2bx+c)$
(c) –ve (d) 0

2. If the system of linear equations **[2003]**

$x+2ay+az=0$; $x+3by+bz=0$; $x+4cy+cz=0$;

has a non - zero solution, then a, b, c.

(a) satisfy $a+2b+3c=0$ (b) are in A.P
(c) are in G.P (d) are in H.P.

3. If $1, \omega, \omega^2$ are the cube roots of unity, then

$$\Delta = \begin{vmatrix} 1 & \omega^n & \omega^{2n} \\ \omega^n & \omega^{2n} & 1 \\ \omega^{2n} & 1 & \omega^n \end{vmatrix}$$ is equal to **[2003]**

(a) ω^2 (b) 0
(c) 1 (d) ω

4. If $A = \begin{bmatrix} a & b \\ b & a \end{bmatrix}$ and $A^2 = \begin{bmatrix} \alpha & \beta \\ \beta & \alpha \end{bmatrix}$, then **[2003]**

(a) $\alpha = 2ab, \beta = a^2+b^2$

(b) $\alpha = a^2+b^2, \beta = ab$

(c) $\alpha = a^2+b^2, \beta = 2ab$

(d) $\alpha = a^2+b^2, \beta = a^2-b^2$.

5. Let $A = \begin{pmatrix} 0 & 0 & -1 \\ 0 & -1 & 0 \\ -1 & 0 & 0 \end{pmatrix}$. The only correct statement about the matrix A is **[2004]**

(a) $A^2 = I$

(b) $A = (-1)\,I$, where I is a unit matrix

(c) A^{-1} does not exist

(d) A is a zero matrix

6. Let $A = \begin{pmatrix} 1 & -1 & 1 \\ 2 & 1 & -3 \\ 1 & 1 & 1 \end{pmatrix}$ and $10\,B = \begin{pmatrix} 4 & 2 & 2 \\ -5 & 0 & \alpha \\ 1 & -2 & 3 \end{pmatrix}$. If B is the inverseof matrix A, then α is **[2004]**

(a) 5 (b) –1
(c) 2 (d) –2

7. If $a_1, a_2, a_3, ..., a_n, ...$ are in G.P., then the value of the determinant **[2004]**

$$\begin{vmatrix} \log a_n & \log a_{n+1} & \log a_{n+2} \\ \log a_{n+3} & \log a_{n+4} & \log a_{n+5} \\ \log a_{n+6} & \log a_{n+7} & \log a_{n+8} \end{vmatrix}, \text{ is}$$

(a) –2 (b) 1
(c) 2 (d) 0

8. If $A^2 - A + I = 0$, then the inverse of A is **[2005]**

(a) $A+I$ (b) A
(c) $A-I$ (d) $I-A$

9. The system of equations

$\alpha x + y + z = \alpha - 1$
$x + \alpha y + z = \alpha - 1$
$x + y + \alpha z = \alpha - 1$

has infinite solutions, if α is **[2005]**

(a) –2 (b) either – 2 or 1
(c) not – 2 (d) 1

10. If $a^2 + b^2 + c^2 = -2$ and **[2005]**

$$f(x) = \begin{vmatrix} 1+a^2x & (1+b^2)x & (1+c^2)x \\ (1+a^2)x & 1+b^2x & (1+c^2)x \\ (1+a^2)x & (1+b^2)x & 1+c^2x \end{vmatrix},$$

then f (x) is a polynomial of degree

(a) 1 (b) 0
(c) 3 (d) 2

11. If $a_1, a_2, a_3, ..., a_n, ...$ are in G. P., then the determinant

$$\Delta = \begin{vmatrix} \log a_n & \log a_{n+1} & \log a_{n+2} \\ \log a_{n+3} & \log a_{n+4} & \log a_{n+5} \\ \log a_{n+6} & \log a_{n+7} & \log a_{n+8} \end{vmatrix}$$

is equal to **[2005]**

(a) 1 (b) 0
(c) 4 (d) 2

12. If A and B are square matrices of size $n \times n$ such that $A^2 - B^2 = (A-B)(A+B)$, then which of the following will be always true? **[2006]**

(a) $A = B$

(b) $AB = BA$

(c) either of A or B is a zero matrix

(d) either of A or B is identity matrix

13. Let $A = \begin{pmatrix} 1 & 2 \\ 3 & 4 \end{pmatrix}$ and $B = \begin{pmatrix} a & 0 \\ 0 & b \end{pmatrix}$, $a, b \in N$. Then **[2006]**

(a) there cannot exist any B such that AB = BA
(b) there exist more then one but finite number of B's such that AB = BA
(c) there exists exactly one B such that AB = BA
(d) there exist infinitely many B's such that AB = BA

14. If $D=\begin{vmatrix}1 & 1 & 1\\ 1 & 1+x & 1\\ 1 & 1 & 1+y\end{vmatrix}$ for $x\neq 0, y\neq 0$, then D is **[2007]**

(a) divisible by x but not y
(b) divisible by y but not x
(c) divisible by neither x nor y
(d) divisible by both x and y

15. Let $A=\begin{vmatrix}5 & 5\alpha & \alpha\\ 0 & \alpha & 5\alpha\\ 0 & 0 & 5\end{vmatrix}$. If $|A^2|=25$, then $|\alpha|$ equals **[2007]**

(a) 1/5 (b) 5
(c) 5^2 (d) 1

DIRECTIONS (*Q. No.* 16) *is Assertion-Reason type questions. Each of these questions contains two statements: Statement-1(Assertion) and Statement-2(Reason). Each of these questions also has four alternative choices, only one of which is the correct answer. You have to select the correct choice.*

16. Let A be a 2 × 2 matrix with real entries. Let I be the 2 × 2 identity matrix. Denote by tr(A), the sum of diagonal entries of a. Assume that $A^2 = I$. **[2008]**
Statement-1 : If $A \neq I$ and $A \neq -I$, then det(A) = –1
Statement-2 : If $A \neq I$ and $A \neq -I$, then tr (A) ≠ 0.
(a) Statement -1 is false, Statement-2 is true
(b) Statement -1 is true, Statement-2 is true; Statement -2 is a correct explanation for Statement-1
(c) Statement -1 is true, Statement-2 is true; Statement -2 is not a correct explanation for Statement-1
(d) Statement -1 is true, Statement-2 is false

17. Let a, b, c be any real numbers. Suppose that there are real numbers x, y, z not all zero such that $x = cy + bz$, $y = az + cx$, and $z = bx + ay$. Then $a^2 + b^2 + c^2 + 2abc$ is equal to **[2008]**
(a) 2 (b) –1
(c) 0 (d) 1

18. Let A be a square matrix all of whose entries are integers. Then which one of the following is true? **[2008]**
(a) If det $A = \pm 1$, then A^{-1} exists but all its entries are not necessarily integers
(b) If det $A \neq \pm 1$, then A^{-1} exists and all its entries are non integers
(c) If det $A = \pm 1$, then A^{-1} exists but all its entries are integers
(d) If det $A = \pm 1$, then A^{-1} need not exists

DIRECTIONS : Each of these questions contains two statements: **Statement-1 (Assertion) and Statement-2 (Reason).**
Each of these questions also has four alternative choices, only one of which is the correct answer. You have to select the correct choice.

19. Let A be a 2 × 2 matrix
Statement -1 : adj (adj A) = A
Statement -2 : |adj A| = |A| **[2009]**
(a) Statement-1 is true, Statement-2 is true. Statement-2 is not a correct explanation for Statement-1.
(b) Statement-1 is true, Statement-2 is false.
(c) Statement -1 is false, Statement-2 is true.
(d) Statement-1 is true, Statement -2 is true. Statement-2 is a correct explanation for Statement-1.

20. Let a, b, c be such that $b(a + c) \neq 0$ if **[2009]**

$$\begin{vmatrix}a & a+1 & a-1\\ -b & b+1 & b-1\\ c & c-1 & c+1\end{vmatrix}+\begin{vmatrix}a+1 & b+1 & c-1\\ a-1 & b-1 & c+1\\ (-1)^{n+2}a & (-1)^{n+1}b & (-1)^n c\end{vmatrix}=0,$$

then the value of n is :
(a) any even integer (b) any odd integer
(c) any integer (d) zero

21. The number of 3 × 3 non-singular matrices, with four entries as 1 and all other entries as 0, is **[2010]**
(a) 5 (b) 6
(c) at least 7 (d) less than 4

22. Let A be a 2 × 2 matrix with non-zero entries and let $A^2 = I$, where I is 2 × 2 identity matrix. Define
Tr(A) = sum of diagonal elements of A and
$|A|$ = determinant of matrix A.
Statement - 1 : Tr(A) = 0.
Statement -2 : $|A| = 1$. **[2010]**
(a) Statement -1 is true, Statement -2 is true ; Statement -2 is **not** a correct explanation for Statement -1.
(b) Statement -1 is true, Statement -2 is false.
(c) Statement -1 is false, Statement -2 is true .
(d) Statement - 1 is true, Statement 2 is true ; Statement -2 is a correct explanation for Statement -1.

23. Consider the system of linear equations ; **[2010]**
$x_1 + 2x_2 + x_3 = 3$
$2x_1 + 3x_2 + x_3 = 3$
$3x_1 + 5x_2 + 2x_3 = 1$
The system has
(a) exactly 3 solutions
(b) a unique solution
(c) no solution
(d) infinite number of solutions

24. The number of values of k for which the linear equations $4x + ky + 2z = 0$, $kx + 4y + z = 0$ and $2x + 2y + z = 0$ possess a non-zero solution is **[2011]**
(a) 2 (b) 1
(c) zero (d) 3

25. Let A and B be two symmetric matrices of order 3.
Statement-1: $A(BA)$ and $(AB)A$ are symmetric matrices.
Statement-2: AB is symmetric matrix if matrix multiplication of A with B is commutative. **[2011]**
(a) Statement-1 is true, Statement-2 is true; Statement-2 is **not a** correct explanation for Statement-1.
(b) Statement-1 is true, Statement-2 is false.
(c) Statement-1 is false, Statement-2 is true.
(d) Statement-1 is true, Statement-2 is true; Statement-2 is a correct explanation for Statement-1.

26. If the trivial solution is the only solution of the system of equations **[2011 RS]**

$x - ky + z = 0$

$kx + 3y - kz = 0$

$3x + y - z = 0$

then the set of all values of k is :

(a) $R - \{2, -3\}$ (b) $R - \{2\}$

(c) $R - \{-3\}$ (d) $\{2, -3\}$

27. **Statement - 1:**
Determinant of a skew-symmetric matrix of order 3 is zero.
Statement - 2 :
For any matrix A, det $(A)^T$ = det (A) and det (– A) = – det (A).
Where det (B) denotes the determinant of matrix B. Then : **[2011 RS]**

(a) Both statements are true
(b) Both statements are false
(c) Statement-1 is false and statement-2 is true
(d) Statement-1 is true and statement-2 is false

28. Let $A = \begin{pmatrix} 1 & 0 & 0 \\ 2 & 1 & 0 \\ 3 & 2 & 1 \end{pmatrix}$. If u_1 and u_2 are column matrices such that $Au_1 = \begin{pmatrix} 1 \\ 0 \\ 0 \end{pmatrix}$ and $Au_2 = \begin{pmatrix} 0 \\ 1 \\ 0 \end{pmatrix}$, then $u_1 + u_2$ is equal to : **[2012]**

(a) $\begin{pmatrix} -1 \\ 1 \\ 0 \end{pmatrix}$ (b) $\begin{pmatrix} -1 \\ 1 \\ -1 \end{pmatrix}$ (c) $\begin{pmatrix} -1 \\ -1 \\ 0 \end{pmatrix}$ (d) $\begin{pmatrix} 1 \\ -1 \\ -1 \end{pmatrix}$

29. Let P and Q be 3×3 matrices $P \neq Q$. If $P^3 = Q^3$ and $P^2Q = Q^2P$ then determinant of $(P^2 + Q^2)$ is equal to : **[2012]**
(a) –2 (b) 1 (c) 0 (d) –1

30. If $P = \begin{bmatrix} 1 & \alpha & 3 \\ 1 & 3 & 3 \\ 2 & 4 & 4 \end{bmatrix}$ is the adjoint of a 3×3 matrix A and $|A| = 4$, then α is equal to : **[JEE M 2013]**
(a) 4 (b) 11
(c) 5 (d) 0

Solutions & Explanations

Section-A : JEE Advanced/ IIT-JEE

A 1. $t = 0$ 2. $x = -1, 2$ 3. 3/16 4. 2, 7 5. $\lambda = 0$ 6. 0 7. 0

B 1. F 2. F

C 1. (b) 2. (a) 3. (d) 4. (a) 5. (b) 6. (b) 7. (a)
8. (d) 9. (b) 10. (b) 11. (d) 12. (a) 13. (b) 14. (c)
15. (c) 16. (a) 17. (d) 18. (a) 19. (a) 20. (d) 21. (d)

D 1. (b, e) 2. (d) 3. (c) 4. (a, d) 5. (c, d) 6. (b, c, d)

E 1. $x = b,\ y = \dfrac{-2b}{15},\ z = \dfrac{2b}{5},\ b \in R$ 4. $x = \dfrac{4-5k}{7}, \dfrac{13k-9}{7}, z = k\ (for\ \lambda = -5)$

6. $n\pi\ or\ n\pi + (-1)^n \pi/6,\ n \in Z$ 9. 2 11. $n\pi\ or\ n\pi + \pi/4$

13. $\dfrac{4d^4}{a(a+d)^2(a+2d)^3(a+3d)^2(a+4d)}$ 14. 0 16. 4

F 1. (A) → s; (B) → p, q; (C) → r; (D) → p, q, s 2. (A) → r; (B) → q, s; (C) → r, s; (D) → p, r

G 1. (d) 2. (b) 3. (a) 4. (a) 5. (b) 6. (b) 7. (d) 8. (c) 9. (d) 10. (d) 11. (a)
12. (b)

H 1. (a)

I 1. 0 2. 4 3. 9

Section-B : JEE Main/ AIEEE

1. (c) 2. (d) 3. (b) 4. (c) 5. (a) 6. (a) 7. (d)
8. (d) 9. (a) 10. (d) 11. (b) 12. (b) 13. (d) 14. (d)
15. (a) 16. (d) 17. (d) 18. (c) 19. (d) 20. (b) 21. (c)
22. (b) 23. (c) 24. (a) 25. (a) 26. (a) 27. (d) 28. (d)
29. (c) 30. (b)

Section-A JEE Advanced/ IIT-JEE

A FILL IN THE BLANKS :

1. $p\lambda^4+q\lambda^3+r\lambda^2+s\lambda+t=\begin{vmatrix}\lambda^3+3\lambda & \lambda-1 & \lambda+3\\ \lambda+1 & -2\lambda & \lambda-4\\ \lambda-3 & \lambda+4 & 3\lambda\end{vmatrix}$

$\Rightarrow\ p\lambda^4+p\lambda^3+r\lambda^2+s\lambda+t$

$=\begin{vmatrix}\lambda^3 & \lambda-1 & \lambda+3\\ \lambda & -2\lambda & \lambda-4\\ \lambda & \lambda+4 & 3\lambda\end{vmatrix}+\begin{vmatrix}3\lambda & \lambda-1 & \lambda+3\\ 1 & -2\lambda & \lambda-4\\ -3 & \lambda+4 & 3\lambda\end{vmatrix}$

$=\lambda\begin{vmatrix}\lambda^2 & \lambda-1 & \lambda+3\\ 1 & -2\lambda & \lambda-4\\ 1 & \lambda+4 & 3\lambda\end{vmatrix}+3\lambda(-6\lambda^2-\lambda^2+4)-$

$(\lambda-1)(3\lambda+3\lambda-12)+(\lambda+3)(\lambda+4-6\lambda)$

$=(\text{multiple of }\lambda)+3\lambda(-7\lambda^2+4)-(\lambda-1)(6\lambda-12)+(\lambda+3)(4-5\lambda)$

$=(\text{multiple of }\lambda)-21\lambda^3-12\lambda-6\lambda^2+12\lambda+6\lambda-12+4\lambda-5\lambda^2+12-15\lambda$

$=(\text{multiple of }\lambda)-21\lambda^3-11\lambda^2+17\lambda$

$\Rightarrow$ constant term (independent of λ) is zero on RHS.

$\therefore$ Equating two sides of identity we get $t=0$.

ALTERNATE SOLUTION

As given equation is an identity in λ, it must be true for all value of λ.

$\therefore$ For $\lambda=0$ also. Putting $\lambda=0$ we get

$t=\begin{vmatrix}0 & -1 & 3\\ 1 & 0 & -4\\ -3 & 4 & 0\end{vmatrix}=0$

2. Given equation is, $\begin{vmatrix}1 & 4 & 20\\ 1 & -2 & 5\\ 1 & 2x & 5x^2\end{vmatrix}=0$

Clearly on expanding the det. we will get a quadratic equation in x.

$\therefore$ It has 2 roots. We observe that R_3 becomes identical to R_1 if $x=2$. thus at $x=2\Rightarrow\Delta=0$

$\therefore x=2$ is a root of given eq.

Similarly R_3 becomes identical to R_2 if $x=-1$.

thus at $x=-1\Rightarrow\Delta=0$

$\therefore x=-1$ is a root of given eq.

Hence equation has roots as -1 and 2.

3. With 0 and 1 as elements there are $2\times2\times2\times2=16$ determinants of order 2×2 out of which only

$\begin{vmatrix}1 & 0\\ 0 & 1\end{vmatrix},\begin{vmatrix}1 & 1\\ 0 & 1\end{vmatrix},\begin{vmatrix}1 & 0\\ 1 & 1\end{vmatrix}$ are the three det.

whose value is +ve. $\therefore$ Req. prob. = 3/16

4. $\begin{vmatrix}x & 3 & 7\\ 2 & x & 2\\ 7 & 6 & x\end{vmatrix}=0$

Operating $R_1\to R_1+R_2+R_3$ we get

$\begin{vmatrix}x+9 & x+9 & x+9\\ 2 & x & 2\\ 7 & 6 & x\end{vmatrix}=0$

$\Rightarrow(x+9)\begin{vmatrix}1 & 1 & 1\\ 2 & x & 2\\ 7 & 6 & x\end{vmatrix}=0$

Operating $C_2\to C_2-C_1,\ C_3\to C_3-C_1$

$\Rightarrow\quad(x+9)\begin{vmatrix}1 & 0 & 0\\ 2 & x-2 & 0\\ 7 & -1 & x-7\end{vmatrix}=0$

Expanding along R_1

$\Rightarrow\quad(x+9)(x-2)(x-7)=0$

$\Rightarrow\quad x=-9,2,7$

$\therefore$ Other roots are 2 and 7.

5. The given homogeneous system of equations will have non zero solution if $D=0$

$\Rightarrow\quad\begin{vmatrix}\lambda & 1 & 1\\ -1 & \lambda & 1\\ -1 & -1 & \lambda\end{vmatrix}=0$

$\Rightarrow\quad\lambda(\lambda^2+1)-1(-\lambda+1)+1(1+\lambda)=0$

$\Rightarrow\quad\lambda^3+3\lambda=0$

$\Rightarrow\quad\lambda(\lambda^2+3)=0$, but $\lambda^2+3\neq0$ for real λ

$\Rightarrow\quad\lambda=0$

6. $\begin{vmatrix}1 & a & a^2-bc\\ 1 & b & b^2-ca\\ 1 & c & c^2-ab\end{vmatrix}$

Operating $R_1\to R_1-R_2;\ R_2\to R_2-R_3$

$\begin{vmatrix}0 & a-b & (a-b)(a+b+c)\\ 0 & b-c & (b-c)(a+b+c)\\ 1 & c & c^2-ab\end{vmatrix}$

$$= (a-b(b-c)\begin{vmatrix} 1 & 1 & a+b+c \\ 0 & 1 & a+b+c \\ 1 & c & c^2-ab \end{vmatrix} = 0$$

As R_1 and R_2 are identical.

ALTERNATE SOLUTION

$$\begin{vmatrix} 1 & a & a^2-bc \\ 1 & b & b^2-ca \\ 1 & c & c^2-ab \end{vmatrix} = \begin{vmatrix} 1 & a & a^2 \\ 1 & b & b^2 \\ 1 & c & c^2 \end{vmatrix} - \begin{vmatrix} a & a & bc \\ b & b & bc \\ c & c & ab \end{vmatrix}$$

$$= \begin{vmatrix} 1 & a & a^2 \\ 1 & b & b^2 \\ 1 & c & c^2 \end{vmatrix} - \frac{1}{abc}\begin{vmatrix} a & a^2 & abc \\ b & b^2 & abc \\ c & c^2 & abc \end{vmatrix}$$

[Multiplying R_1 by a, R_2 by b and R_3 by c]

$$= \begin{vmatrix} 1 & a & a^2 \\ 1 & b & b^2 \\ 1 & c & c^2 \end{vmatrix} - \begin{vmatrix} a & a^2 & 1 \\ b & b^2 & 1 \\ c & c^2 & 1 \end{vmatrix}$$

$C_1 \leftrightarrow C_2$ and $C_2 \leftrightarrow C_3$

$$= \begin{vmatrix} 1 & a & a^2 \\ 1 & b & b^2 \\ 1 & c & c^2 \end{vmatrix} - \begin{vmatrix} 1 & a & a^2 \\ 1 & b & b^2 \\ 1 & c & c^2 \end{vmatrix} = 0$$

7. Given x, y, z and + ve numbers, then value of

$$D = \begin{vmatrix} 1 & \log_x y & \log_x z \\ \log_y x & 1 & \log_y z \\ \log_z x & \log_z y & 1 \end{vmatrix}$$

$$= \begin{vmatrix} 1 & \frac{\log y}{\log x} & \frac{\log z}{\log x} \\ \frac{\log x}{\log y} & 1 & \frac{\log z}{\log y} \\ \frac{\log x}{\log z} & \frac{\log y}{\log z} & 1 \end{vmatrix} \quad \left(\because \log_b a = \frac{\log a}{\log b}\right)$$

Taking $\frac{1}{\log x}, \frac{1}{\log y}$, and $\frac{1}{\log z}$ common from R_1, R_2 and R_3 respectively

$$D = \frac{1}{\log x \log y \log z}\begin{vmatrix} \log x & \log y & \log z \\ \log x & \log y & \log z \\ \log x & \log y & \log z \end{vmatrix} = 0$$

(Using the prop. that det. vanishes if any two rows are identical.

(B) TRUE / FALSE :

1. $$\begin{vmatrix} 1 & a & bc \\ 1 & b & ca \\ 1 & c & ab \end{vmatrix} = \frac{1}{abc}\begin{vmatrix} a & a^2 & abc \\ b & b^2 & abc \\ c & c^2 & abc \end{vmatrix}$$

$$= \begin{vmatrix} a & a^2 & 1 \\ b & b^2 & 1 \\ c & c^2 & 1 \end{vmatrix} = (-1)^2\begin{vmatrix} 1 & a & a^2 \\ 1 & b & b^2 \\ 1 & c & c^2 \end{vmatrix}$$

$[C_1 \Leftrightarrow C_3$ and then $C_2 \Leftrightarrow C_3]$

$\therefore$ Equal. Hence statement is F.

2. (i) $$\begin{vmatrix} x_1 & y_1 & 1 \\ x_2 & y_2 & 1 \\ x_3 & y_3 & 1 \end{vmatrix} = \begin{vmatrix} a_1 & b_1 & 1 \\ a_2 & b_2 & 1 \\ a_3 & b_3 & 1 \end{vmatrix}$$

$$\Rightarrow \frac{1}{2}\begin{vmatrix} x_1 & y_1 & 1 \\ x_2 & y_2 & 1 \\ x_3 & y_3 & 1 \end{vmatrix} = \frac{1}{2}\begin{vmatrix} a_1 & b_1 & 1 \\ a_2 & b_2 & 1 \\ a_3 & b_3 & 1 \end{vmatrix}$$

$Ar(\Delta_1) = Ar(\Delta_2)$

Where Δ_1 is the area of triangle with vertices (x_1, y_1), (x_2, y_2) and (x_3, y_3); and Δ_2 is the area of triangle with ; vertices (a_1, b_1), (a_2, b_2) and (a_3, b_3). But two Δ's of same area may not be congruent.

$\therefore$ Given statement is false.

(C) MCQ's WITH ONE CORRECT ANSWER :

1. (b) For every 'det, with value 1' ($\in B$) we can find a det. with value – 1 by changing the sign of one entry of '1'. Hence there are equal number of elements in B and C.
$\therefore$ (b) is the correct option

2. (a) $$\begin{vmatrix} 1 & 1+i+\omega^2 & \omega^2 \\ 1-i & -1 & \omega^2-1 \\ -i & -i+\omega-1 & -1 \end{vmatrix}$$

Operating $R_1 \to R_1 - R_2 + R_3$

$$= \begin{vmatrix} 0 & 0 & 0 \\ 1-i & -1 & \omega^2-1 \\ -i & -i+\omega+1 & -1 \end{vmatrix} \quad (\because 1+\omega+\omega^2 = 0)$$

$= 0$

3. (d) Let $\frac{x^2}{a^2} = X, \frac{y^2}{b^2} = Y, \frac{z^2}{c^2} = Z$

Then the given system of equations becomes

$X + Y - Z = 1$

$X - Y + Z = 1, \quad -X + Y + Z = 1$

This is the new system of equations
For new system, we have

$$D=\begin{vmatrix}1 & 1 & -1\\1 & -1 & 1\\-1 & 1 & 1\end{vmatrix}=1(-1-1)-1(1+1)-1(1-1)$$

$=-4\neq 0$

$\therefore$ New system of equations has unique solution.

$$D_1=\begin{vmatrix}1 & 1 & -1\\1 & -1 & 1\\1 & 1 & 1\end{vmatrix}=1(-1-1)-1(1-1)-1(1+1)=-4$$

$$D_2=\begin{vmatrix}1 & 1 & -1\\1 & 1 & 1\\-1 & 1 & 1\end{vmatrix}=1(1-1)-1(1+1)-1(1+1)=-4$$

$$D_3=\begin{vmatrix}1 & 1 & 1\\1 & -1 & 1\\-1 & 1 & 1\end{vmatrix}=1(-1-1)-1(1+1)+1(1-1)=-4$$

Now, $X=\frac{D_1}{D}=\frac{-4}{-4}=1,\ Y=\frac{D_2}{D}=\frac{-4}{-4}=1$

$Z=\frac{D_3}{D}=\frac{-4}{-4}=1$

$\Rightarrow x=\pm a,\ y=\pm b,\ z=\pm c$

4. (a) If A and B are square matrices of same degree then matrices A and B can be added or subtracted or multiplied. By algebra of matrices the only correct option is $A+B=B+A$

5. (b) Given $\begin{vmatrix}xp+y & x & y\\yp+z & y & z\\0 & xp+y & yp+z\end{vmatrix}=0$

Operating $C_1-pC_2-C_3$

$$\begin{vmatrix}0 & x & y\\0 & y & z\\-(xp^2+2py+z) & xp+y & yp+z\end{vmatrix}=0$$

$\Rightarrow (xz-y^2)(xp^2+2py+z)=0$
$\Rightarrow xz-y^2=0$
[also for this D = 0 for other quad. factor in p which shows second factor is a perfect square]
$\Rightarrow y^2=xz\Rightarrow x, y, z$ are in G.P.

6. (b) Let $\Delta=\begin{vmatrix}1 & a & a^2\\\cos(p-d)x & \cos px & \cos(p+d)x\\\sin(p-d)x & \sin px & \sin(p+d)x\end{vmatrix}$

Applying $C_1\to C_1+C_3$

$$\Delta=\begin{vmatrix}1+a^2 & a & a^2\\\cos(p-d)x+\cos(p+d)x & \cos px & \cos(p+d)x\\\sin(p-d)x+\sin(p+d)x & \sin px & \sin(p+d)x\end{vmatrix}$$

NOTE THIS STEP

$$\Rightarrow\Delta=\begin{vmatrix}1+a^2 & a & a^2\\2\cos px\cos dx & \cos px & \cos(p+d)x\\2\sin px\cos dx & \sin px & \sin(p+d)x\end{vmatrix}$$

$C_1\to C_1-(2\cos dx)C_2$

$$\Delta=\begin{vmatrix}1+a^2-2a\cos dx & a & a^2\\0 & \cos px & \cos(p+d)x\\0 & \sin px & \sin(p+d)x\end{vmatrix}$$

Expanding along C_1, we get
$\Delta=(1+a^2-2a\cos dx)[\sin(p+d)x\cos px - \sin px\cos(p+d)x]$
$\Rightarrow\Delta=(1+a^2-2a\cos dx)[\sin\{(p+d)x-px\}]$
$\Rightarrow\Delta=(1+a^2-2a\cos dx)[\sin dx]$
which is independent of p.

7. (a)

$$f(x)=\begin{vmatrix}1 & x & x+1\\2x & x(x-1) & (x+1)x\\3x(x-1) & x(x-1)(x-2) & (x+1)x(x-1)\end{vmatrix}$$

$$=\begin{vmatrix}x+1 & x & x+1\\(x+1)x & x(x-1) & (x+1)x\\(x+1)x(x-1) & x(x-1)(x-2) & (x+1)x(x-1)\end{vmatrix}$$

$[C_1\to C_1+C_2]$

$=0$ [$\because C_1$ and C_2 are identical]
which is free of x, so the function is true for all values of x. Therefore, at $x=100$, $f(x)=0$, i.e., $f(100)=0$

8. (d) For the given homogeneous system to have non zero solution determinant of coefficient matrix should be zero; i.e.,

$$=\begin{vmatrix}1 & -k & -1\\k & -1 & -1\\1 & 1 & -1\end{vmatrix}=1(1+1)+k(-k+1)-1(k+1)=0$$

$\Rightarrow 2-k^2+k-k-1=0\Rightarrow k^2=1\Rightarrow k=\pm 1$

9. (b) Given that $\omega=-\frac{1}{2}+i\frac{\sqrt{3}}{2}$

$\omega^2=-\frac{1}{2}-i\frac{\sqrt{3}}{2}$

Also $1+\omega+\omega^2=0$ and $\omega^3=1$

Now given det. is

$$\Delta=\begin{vmatrix}1 & 1 & 1\\ 1 & -1-\omega^2 & \omega^2\\ 1 & \omega^2 & \omega^4\end{vmatrix}=\begin{vmatrix}1 & 1 & 1\\ 1 & \omega & \omega^2\\ 1 & \omega^2 & \omega\end{vmatrix}$$

[Using $\omega=-1-\omega^2$ and $\omega^3=1$]

Operating $C_1\to C_1+C_2+C_3$

$$\Delta=\begin{vmatrix}3 & 1 & 1\\ 0 & \omega & \omega^2\\ 0 & \omega^2 & \omega\end{vmatrix} \quad (\text{as } 1+\omega+\omega^2=0)$$

Expanding along C_1, we get

$3(\omega^2-\omega^4)=3(\omega^2-\omega)=3\omega(\omega-1)$

10. **(b)** For infinitely many solutions the two equations become identical

$$\Rightarrow \frac{k+1}{k}=\frac{8}{k+3}=\frac{4k}{3k-1}\Rightarrow k=1$$

11. **(d)** Given that $A=\begin{bmatrix}\alpha & 0\\ 1 & 1\end{bmatrix}$ and $B=\begin{bmatrix}1 & 0\\ 5 & 1\end{bmatrix}$

and $A^2=B$

$$\Rightarrow \begin{bmatrix}\alpha & 0\\ 1 & 1\end{bmatrix}\begin{bmatrix}\alpha & 0\\ 1 & 1\end{bmatrix}=\begin{bmatrix}1 & 0\\ 5 & 1\end{bmatrix}$$

$$\Rightarrow \begin{bmatrix}\alpha^2 & 0\\ \alpha+1 & 1\end{bmatrix}=\begin{bmatrix}1 & 0\\ 5 & 1\end{bmatrix}$$

$\Rightarrow \alpha^2=1$ and $\alpha+1=5$

$\Rightarrow \alpha=\pm 1$ and $\alpha=4$

$\because$ There is no common value

$\therefore$ There is no real value of α for which $A^2=B$

12. **(a)** The given system is, $x+ay=0$

$az+y=0$

$ax+z=0$

It is system of homogeneous equations therefore, it will have infinite many solutions if determinant of coefficient matrix is zero. i.e.,

$$\begin{vmatrix}1 & a & 0\\ 0 & 1 & a\\ a & 0 & 1\end{vmatrix}=0$$

$\Rightarrow 1(1-0)-a(0-a^2)=0\Rightarrow 1+a^3=0$

$\Rightarrow a^3=-1 \Rightarrow a=-1$

13. **(b)** Since the system has no solution, $\Delta=0$ and any one amongst $\Delta_x, \Delta_y, \Delta_z$ is non zero.

$$\begin{vmatrix}2 & -1 & 2\\ 1 & -2 & 1\\ 1 & 1 & \lambda\end{vmatrix}=0\Rightarrow\lambda=1$$

Also, $\Delta_z=\begin{vmatrix}2 & -1 & 2\\ 1 & -2 & -4\\ 1 & 1 & 4\end{vmatrix}=6\neq 0$

14. **(c)** $A=\begin{bmatrix}\alpha & 2\\ 2 & \alpha\end{bmatrix}$ and $|A^3|=125\Rightarrow |A|^3=125$

Now, $|A|=\alpha^2-4$

$\Rightarrow(\alpha^2-4)^3=125=5^3\Rightarrow\alpha^2-4=5\Rightarrow\alpha=\pm 3$

15. **(c)** We have $A=\begin{bmatrix}1 & 0 & 0\\ 0 & 1 & 1\\ 0 & -2 & 4\end{bmatrix}$; $I=\begin{bmatrix}1 & 0 & 0\\ 0 & 1 & 0\\ 0 & 0 & 1\end{bmatrix}$

and $A^{-1}=\frac{1}{6}(A^2+cA+dI)$

$\Rightarrow 6AA^{-1}=A^3+cA^2+dAI\Rightarrow A^3+cA^2+dA-6I=0$

We find that, $A^2=\begin{bmatrix}1 & 0 & 0\\ 0 & -1 & 5\\ 0 & -10 & 14\end{bmatrix}$

$$A^3=\begin{bmatrix}1 & 0 & 0\\ 0 & -11 & 19\\ 0 & -38 & 46\end{bmatrix}$$

$\therefore A^3+cA^2+dA-6I=O$

$$\Rightarrow \begin{bmatrix}1 & 0 & 0\\ 0 & -11 & 19\\ 0 & -38 & 46\end{bmatrix}+c\begin{bmatrix}1 & 0 & 0\\ 0 & -1 & 5\\ 0 & -10 & 14\end{bmatrix}$$

$$+d\begin{bmatrix}1 & 0 & 0\\ 0 & 1 & 1\\ 0 & -2 & 4\end{bmatrix}-6\begin{bmatrix}1 & 0 & 0\\ 0 & 1 & 0\\ 0 & 0 & 1\end{bmatrix}=O$$

$$\Rightarrow \begin{bmatrix}1+c+d-6 & 0 & 0\\ 0 & -11-c+d-6 & 19+5c+d\\ 0 & -38-10c-2d & 46+14c+4d-6\end{bmatrix}=\begin{bmatrix}0 & 0 & 0\\ 0 & 0 & 0\\ 0 & 0 & 0\end{bmatrix}$$

$\Rightarrow\quad 1+c+d-6=0, -11-c+d-6=0$

$\Rightarrow\quad c+d=5, -c+d=17$

On solving, we get $c=-6$, $d=11$

ALTERNATE SOLUTION (1)

Given $A=\begin{bmatrix}0 & 0 & 0\\ 0 & 1 & 1\\ 0 & -2 & 4\end{bmatrix}$

$\therefore$ Characteristic eqn of above matrix A is given by $|A-\lambda I|=0$

$$\Rightarrow \begin{bmatrix}1-\lambda & 0 & 0\\ 0 & 1-\lambda & 1\\ 0 & -2 & 4-\lambda\end{bmatrix}=0$$

$\Rightarrow\quad (1-\lambda)(4-5\lambda+\lambda^2+2)=0$

$\Rightarrow\quad \lambda^3-6\lambda^2+11\lambda-6=0$

Also by Cayley Hamilton thm (every square matrix satisfies its characteristic equation) we obtain

$A^3-6A^2+11A-6I=0$

Multiplying by A^{-1}, we get

$A^2-6A+11I-6A^{-1}=0$

$\Rightarrow\quad A^{-1}=\frac{1}{6}(A^2-6A+11I)$

Comparing it with given relation,

$A^{-1}=\frac{1}{6}(A^2-cA+dI)$

we get $c=-6$ and $d=11$

ALTERNATE SOLUTION (2)

$$A=\begin{bmatrix}1&0&0\\0&1&1\\0&-2&4\end{bmatrix}\Rightarrow |A|=6$$

$$\text{Adj. }A=\begin{bmatrix}6&0&0\\0&4&-1\\0&2&1\end{bmatrix};\ A^2=\begin{bmatrix}1&0&0\\0&-1&5\\0&-10&14\end{bmatrix}$$

$$\therefore\quad A^{-1}=\frac{Adj A}{|A|}$$

$$=\frac{1}{6}\begin{bmatrix}6&0&0\\0&4&-1\\0&2&1\end{bmatrix}$$

Now given that $A^{-1}=\frac{1}{6}(A^2+cA+dI)$

$$\Rightarrow\frac{1}{6}\begin{bmatrix}6&0&0\\0&4&-1\\0&2&1\end{bmatrix}$$

$$=\frac{1}{6}\left\{\begin{bmatrix}1&0&0\\0&-1&5\\0&-10&14\end{bmatrix}+\begin{bmatrix}c&0&0\\0&c&c\\0&-2c&4c\end{bmatrix}+\begin{bmatrix}d&0&0\\0&d&0\\0&0&d\end{bmatrix}\right\}$$

$$\Rightarrow\begin{bmatrix}6&0&0\\0&4&-1\\0&2&1\end{bmatrix}=\begin{bmatrix}1+c+d&0&0\\0&-1+c+d&5+c\\0&-10-2c&14+4c+d\end{bmatrix}$$

$\Rightarrow\quad 1+c+d=6$ and $-1=5+c$

$\Rightarrow\quad c=-6$ and $d=11$

16. **(a)** Given that, $P=\begin{bmatrix}\frac{\sqrt{3}}{2}&\frac{1}{2}\\-\frac{1}{2}&\frac{\sqrt{3}}{2}\end{bmatrix}$

$A=\begin{bmatrix}1&1\\0&1\end{bmatrix}$ and $Q=PAP^T$ and $X=P^TQ^{2005}P$

We observe that $Q=PAP^T$

$\Rightarrow\quad Q^2=(PAP^T)(PAP^T)$

$=PA(P^TP)AP^T=PA(IA)P^T$

$\therefore\ PA^2P^T$

Proceeding in the same way, we get

$Q^{2005}=PA^{2005}P^T$

Also $A=\begin{bmatrix}1&1\\0&1\end{bmatrix}\Rightarrow A^2=\begin{bmatrix}1&2\\0&1\end{bmatrix}$

and proceeding in the same way

$A^{2005}=\begin{bmatrix}1&2005\\0&1\end{bmatrix}$

Now, $X=P^TQ^{2005}P$

$=P^T(PA^{2005}P^T)P=(P^TP)A^{2005}(P^TP)$

$=IA^{2005}I=A^{2005}=\begin{bmatrix}1&2005\\0&1\end{bmatrix}$

ALTERNATE SOLUTION

Here we observe that

$|P|=|P^T|$

$=\begin{vmatrix}\sqrt{3}/2&1/2\\-1/2&\sqrt{3}/2\end{vmatrix}=\frac{3}{4}+\frac{1}{4}=1\quad(\because |A|=|A^T|)$

and $|A|=\begin{vmatrix}1&1\\0&1\end{vmatrix}=1$

$\therefore\ |Q|=|PAP^T|$

$=|P||A||P^T|\ (\because |AB|=|A||B|)$

$=1.1.1=1$

$\therefore\ |Q^{2005}|=|Q|^{2005}$

$\therefore\ |x|=|P^TQ^{2005}P|$

$=|P^T||Q|^{2005}|P|=1$

and $|x|=1$ only in the case of option (A)

17. **(d)** The given points are

$P(-\sin(\beta-\alpha),-\cos\beta),\ Q(\cos(\beta-\alpha),\sin\beta)$

$R(\cos(\beta-\alpha+\theta),\sin(\beta-\theta))$

Where $0<\alpha,\beta,\theta<\frac{\pi}{4}$

$$\therefore\ \Delta=\begin{vmatrix}1&1&1\\-\sin(\beta-\alpha)&\cos(\beta-\alpha)&\cos(\beta-\alpha+\theta)\\-\cos\beta&\sin\beta&\sin(\beta-\theta)\end{vmatrix}$$

Operating $C_3 - C_1 \sin\theta - C_2 \cos\theta$, we get

$$\Delta = \begin{vmatrix} 1 & 1 & 1-\sin\theta-\cos\theta \\ -\sin(\beta-\alpha) & \cos(\beta-\alpha) & 0 \\ -\cos\beta & \sin\beta & 0 \end{vmatrix}$$

$$= (1-\sin\theta-\cos\theta)[\cos\beta\cos(\beta-\alpha)-\sin\beta\sin(\beta-\alpha)]$$

$$\Rightarrow \Delta = [1-(\sin\theta+\cos\theta)]\cos(2\beta-\alpha)$$

$\because 0<\alpha,\beta,\theta<\frac{\pi}{4}$ $\quad\therefore \sin\theta+\cos\theta \neq 1$

Also $2\beta-\alpha<\frac{\pi}{4} \Rightarrow \cos(2\beta-\alpha)\neq 0$

$\therefore \Delta \neq 0 \Rightarrow$ the three points are non collinear.

18. **(a)** Let $A = \begin{bmatrix} a_1 & b_1 & c_1 \\ a_2 & b_2 & c_2 \\ a_3 & b_3 & c_3 \end{bmatrix}$ where a_i, b_i, c_i have values 0 or 1 for $i = 1, 2, 3$.

Then the given system is equivalent to

$a_1x+b_1y+c_1z=1,$

$a_2x+b_2y+c_2z=0$

$a_3x+b_3y+c_3z=0$

Which represent three distinct planes. But three planes can not intersect at two distinct points, therefore no such system exists.

19. **(a)** For the given matrix to be non singular

$$\begin{vmatrix} 1 & a & b \\ \omega & 1 & c \\ \omega^2 & \omega & 1 \end{vmatrix} \neq 0$$

$\Rightarrow 1-(a+c)\omega+ac\omega^2 \neq 0$

$\Rightarrow (1-a\omega)(1-c\omega)\neq 0$

$\Rightarrow a\neq\omega^2$ and $c\neq\omega^2$ where ω is complex cube root of unity.

As a, b and c are complex cube roots of unity

$\therefore$ a and c can take only one value i.e. ω while b can take 2 values i.e. ω and ω^2.

$\therefore$ Total number of distinct matrices $= 1\times1\times2 = 2$

20. **(d)** We have

$$|Q| = \begin{vmatrix} 2^2a_{11} & 2^3a_{12} & 2^4a_{13} \\ 2^3a_{21} & 2^4a_{22} & 2^5a_{23} \\ 2^4a_{31} & 2^5a_{32} & 2^6a_{33} \end{vmatrix}$$

$$= 2^2.2^3.2^4 \begin{vmatrix} a_{11} & a_{12} & a_{13} \\ 2a_{21} & 2a_{22} & 2a_{23} \\ 2^2a_{31} & 2^2a_{32} & 2^2a_{33} \end{vmatrix}$$

$$= 2^9.2.2^2 \begin{vmatrix} a_{11} & a_{12} & a_{13} \\ a_{21} & a_{22} & a_{23} \\ a_{31} & a_{32} & a_{33} \end{vmatrix}$$

$$= 2^{12}\times|P| = 2^{12}\times 2 = 2^{13}$$

21. **(d)** We have $P^T = 2P + I$

$\Rightarrow P = 2P^T + I \Rightarrow P = 2(2P+I)+I$

$\Rightarrow P = 4P + 3I \Rightarrow P + I = 0$

$\Rightarrow PX + X = 0 \Rightarrow PX = -X$

(D) MCQ's WITH ONE OR MORE THAN ONE CORRECT :

1. **(b, e)** ATQ $\begin{vmatrix} a & b & a\alpha+b \\ b & c & b\alpha+c \\ a\alpha+b & b\alpha+c & 0 \end{vmatrix} = 0$

Operating $C_3 \to C_3 - C_1\alpha - C_2$, we get

$$\begin{vmatrix} a & b & 0 \\ b & c & 0 \\ a\alpha+b & b\alpha+c & -(a\alpha^2+b\alpha+b\alpha+c) \end{vmatrix} = 0$$

$$\Rightarrow (a\alpha^2+2b\alpha+c)\begin{vmatrix} a & b & 0 \\ b & c & 0 \\ a\alpha+b & b\alpha+c & 1 \end{vmatrix} = 0$$

$\Rightarrow (ac-b^2)(a\alpha^2+2b\alpha+c)=0$

$\Rightarrow$ either $ac-b^2=0$ or $a\alpha^2+2b\alpha+c=0$

$\Rightarrow$ either a, b, c are in G.P. or $(x-\alpha)$ is a factor of $ax^2+2bx+c$

$\Rightarrow$ (b) and (e) are the correct answers.

2. **(d)** $\begin{vmatrix} 6i & -3i & 1 \\ 4 & 3i & -1 \\ 20 & 3 & i \end{vmatrix} = x+iy$ (given)

$$\Rightarrow -3i\begin{vmatrix} 6i & 1 & 1 \\ 4 & -1 & -1 \\ 20 & i & i \end{vmatrix} = 0$$

[$\because C_2$ and C_3 are identical]

$\Rightarrow x+iy=0$

$\Rightarrow x=0, y=0$

3. **(c)** [As a skew symmetric matrix of order 3 cannot be non singular, therefore the data given in the question is inconsistent.]

We have

$M^2N^2(M^TN)^{-1}(MN^{-1})^T = M^2N^2N^{-1}(M^T)^{-1}(N^{-1})^T M^T$

$= M^2N(M^T)^{-1}(N^{-1})^TM^T = -M^2NM^{-1}N^{-1}M$

$(\because M^T=-M, N^T=-N$ and $(N^{-1})^T=(N^T)^{-1}$

$= -M(NM)(NM)^{-1}M$ $\quad(\because MN = NM)$

$= -MM = -M^2$

4. **(a, d)**
We know for a third order matrix P,
$|\text{Adj } P| = |P|^2$
Where
$|\text{Adj } P| = 1(3-7) - 4(6-7) + 4(2-1) = 4$
$\therefore \quad |P|^2 = 4 \Rightarrow |P| = 2 \text{ or } -2$

5. **(c, d)**
(a) $(N' M N)' = (MN)'N = N'M'N = N'MN$ or $-N'MN$
According as M is symm. or skew symm. $\therefore$ correct
(b) $(MN - NM)' = (MN)' - (NM)' = N'M' - M'N'$
$= NM - MN = -(MN - NM)$
$\therefore$ It is skew symm. Statement B is also correct.
(c) $(MN)' = N'M' = NM \neq MN$
$\therefore$ Statement C is incorrect
(d) $(adj\, M)(adj\, N) = adj\,(MN)$ is incorrect.

6. **(b, c, d)**

For $n = 3$, $P = \begin{bmatrix} w^2 & w^3 & w^4 \\ w^3 & w^4 & w^5 \\ w^4 & w^5 & w^6 \end{bmatrix}$

and $P^2 = \begin{bmatrix} 0 & 0 & 0 \\ 0 & 0 & 0 \\ 0 & 0 & 0 \end{bmatrix}$

It shows $P^2 = 0$ if n is a multiple of 3.
So for $P^2 \neq 0$, n should not be a multiple of 3 i.e. n can take values 55, 58, 56

(E) SUBJECTIVE PROBLEMS :

1. We should have,

$$\begin{vmatrix} 1 & k & 3 \\ 3 & k & -2 \\ 2 & 3 & -4 \end{vmatrix} = 0$$

$\Rightarrow 1(-4k + 6) - k(-12 + 4) + 3(9 - 2k) = 0$
$\Rightarrow -2k + 33 = 0$
$\Rightarrow k = \dfrac{33}{2}$

Substituting $k = \dfrac{33}{2}$ and putting $x = b$, where $b \in Q$, we get the system as

$33y + 6z = -2b$ ...(1)
$33y - 4z = -6b$ ...(2)
$3y - 4z = -2b$ ...(3)

$(1) - (2) \Rightarrow 10z = 4b \Rightarrow z = \dfrac{2}{5}\, b$

$(1) \Rightarrow 33y = -2b - \dfrac{12b}{5} = -\dfrac{22b}{5} \Rightarrow y = \dfrac{-2b}{15}$

$\therefore$ The solution is $x = b,\ y = \dfrac{-2b}{15},\ z = \dfrac{2b}{5}$

2. The given det, on expanding along R_1, we get

$$\begin{vmatrix} a & b & c \\ b & c & a \\ c & a & b \end{vmatrix} = a(bc - a^2) - b(b^2 - ac) + c(ab - c^2)$$

$= 3abc - a^3 - b^3 - c^3 = -(a^3 + b^3 + c^3 - 3abc)$
$= -(a + b + c)[a^2 + b^2 + c^2 - ab - bc - ca]$
$= -\dfrac{1}{2}(a + b + c)[2a^2 + 2b^2 + 2c^2 - 2ab - 2bc - 2ca]$
$= -\dfrac{1}{2}(a + b + c)[(a - b)^2 + (b - c)^2 + (c - a)^2]$

As $a, b, c > 0$
$\therefore \quad a + b + c > 0$
Also $a \neq b \neq c$
$\therefore \quad (a - b)^2 + (b - c)^2 + (c - a)^2 > 0$
Hence the given determinant is – ve.

3.
$$\begin{vmatrix} x^2 + x & x + 1 & x - 2 \\ 2x^2 + 3x - 1 & 3x & 3x - 3 \\ x^2 + 2x + 3 & 2x - 1 & 2x - 1 \end{vmatrix} = xA + B$$

$$\text{L.H.S.} = \begin{vmatrix} x^2 + x & x + 1 & x - 2 \\ 2x^2 + 3x - 1 & 3x & 3x - 3 \\ x^2 + 2x + 3 & 2x - 1 & 2x - 1 \end{vmatrix}$$

Operation $R_2 \to R_2 - 2R_1$ and $R_3 \to R_3 - R_1$

$$\begin{vmatrix} x^2 + x & x + 1 & x - 2 \\ x - 1 & x - 2 & x + 1 \\ x + 3 & x - 2 & x + 1 \end{vmatrix}$$

$$= \begin{vmatrix} x^2 & x + 1 & x - 2 \\ 0 & x - 2 & x + 1 \\ 0 & x - 2 & x + 1 \end{vmatrix} + \begin{vmatrix} x & x + 1 & x - 2 \\ x - 1 & x - 2 & x + 1 \\ x + 3 & x - 2 & x + 1 \end{vmatrix}$$

$$= 0 + \begin{vmatrix} x & x + 1 & x - 2 \\ x - 1 & x - 2 & x + 1 \\ x + 3 & x - 2 & x + 1 \end{vmatrix}$$

Operating $R_3 \to R_3 - R_2$ and $R_2 \to R_2 - R_1$

$$= \begin{vmatrix} x & x + 1 & x - 2 \\ -1 & -3 & 3 \\ 4 & 0 & 0 \end{vmatrix} = \begin{vmatrix} x & x & x \\ -1 & -3 & 3 \\ 4 & 0 & 0 \end{vmatrix} + \begin{vmatrix} 0 & 1 & -2 \\ -1 & -3 & 3 \\ 4 & 0 & 0 \end{vmatrix}$$

$$= x\begin{vmatrix} 1 & 1 & 1 \\ -1 & -3 & 3 \\ 4 & 0 & 0 \end{vmatrix} + \begin{vmatrix} 0 & 1 & -2 \\ -1 & -3 & 3 \\ 4 & 0 & 0 \end{vmatrix}$$

$= xA + B =$ R.H.S
Hence Proved.

4. The given system of equations is

$3x - y + 4z = 3$

$x + 2y - 3z = -2$

$6x + 5y + \lambda z = -3$

Then $D = \begin{vmatrix} 3 & -1 & 4 \\ 1 & 2 & -3 \\ 6 & 5 & \lambda \end{vmatrix}$

$= 3(2\lambda + 15) + 1(\lambda + 18) + 4(5 - 12) = 7\lambda + 35 = 7(\lambda + 5)$

For unique sol., $D \neq 0$

$\therefore \lambda + 5 \neq 0$ or $\lambda \neq -5$

$\therefore$ System has unique sol. for $\lambda \neq -5$

For $\lambda = -5$, $D = 0$

Then

$$D_1 = \begin{vmatrix} 3 & -1 & 4 \\ -2 & 2 & -3 \\ -3 & 5 & -5 \end{vmatrix}$$

$= 3(-10 + 15) + 1(10 - 9) + 4(-10 + 6)$

$= 15 + 1 - 16 = 0$

$$D_2 = \begin{vmatrix} 3 & 3 & 4 \\ 1 & -2 & -3 \\ 6 & -3 & -5 \end{vmatrix}$$

$= 3(10 - 9) - 3(-5 + 18) + 4(-3 + 12) = 3 - 39 + 36 = 0$

$$D_3 = \begin{vmatrix} 3 & -1 & 3 \\ 1 & 2 & -2 \\ 6 & -5 & -3 \end{vmatrix}$$

$= 3(-6 + 10) + 1(-3 + 12) + 3(5 - 12) = 12 + 9 - 21 = 0$

$D_1 = D_2 = D_3 = 0$

$\therefore$ Infinite many solutions.

Let $z = k$ then equations becomes

$3x - y = 3 - 4k$

$x + 2y = 3k - 2$

On solving we get

$$x = \frac{4 - 5k}{7},\ y = \frac{13k - 9}{7},\ z = k$$

This satisfies the third equation.

5. On L.H.S.= D, applying operations $C_2 \to C_2 + C_1$ and $C_3 \to C_3 + C_2$ and using ${}^nC_r + {}^nC_{r+1} = {}^{n+1}C_{r+1}$, we get

$$D = \begin{vmatrix} {}^xC_r & {}^{x+1}C_{r+1} & {}^{x+1}C_{r+2} \\ {}^yC_r & {}^{y+1}C_{r+1} & {}^{y+1}C_{r+2} \\ {}^zC_r & {}^{z+1}C_{r+1} & {}^{z+1}C_{r+2} \end{vmatrix}$$

Operating $C_3 + C_2$ and using the same result, we get

$$D = \begin{vmatrix} {}^xC_r & {}^{x+1}C_{r+1} & {}^{x+2}C_{r+2} \\ {}^yC_r & {}^{y+1}C_{r+1} & {}^{y+2}C_{r+2} \\ {}^zC_r & {}^{z+1}C_{r+1} & {}^{z+2}C_{r+2} \end{vmatrix} = \text{RHS}$$

Hence proved

6. The system will have a non-trivial solution if

$$\begin{vmatrix} \sin 3\theta & -1 & 1 \\ \cos 2\theta & 4 & 3 \\ 2 & 7 & 7 \end{vmatrix} = 0$$

Expanding along C_1, we get

$\Rightarrow (28 - 21)\sin 3\theta - (-7 - 7)\cos 2\theta + 2(-3 - 4) = 0$

$\Rightarrow 7\sin 3\theta + 14\cos 2\theta - 14 = 0$

$\Rightarrow \sin 3\theta + 2\cos 2\theta - 2 = 0$

$\Rightarrow 3\sin - 4\sin^3\theta + 2(1 - 2\sin^2\theta) - 2 = 0$

$\Rightarrow 4\sin^3\theta + 4\sin^2\theta - 3\sin\theta = 0$

$\Rightarrow \sin\theta(2\sin\theta - 1)(2\sin\theta + 3) = 0$

$\sin\theta = 0$ or $\sin\theta = 1/2$

($\sin\theta = -3/2$ not possible)

$\Rightarrow \theta = n\pi$ or $\theta = n\pi + (-1)^n \pi/6,\ n \in Z.$

7. We have

$$\Delta a = \begin{vmatrix} (a-1) & n & 6 \\ (a-1)^2 & 2n^2 & 4n-2 \\ (a-1)^3 & 3n^3 & 3n^2 - 3n \end{vmatrix}$$

NOTE THIS STEP

Then $$\sum_{a=1}^{n} \Delta a = \begin{vmatrix} (1-1) & n & 6 \\ (1-1)^2 & 2n^2 & 4n-2 \\ (1-1)^3 & 3n^3 & 3n^2 - 3n \end{vmatrix} + \begin{vmatrix} (2-1) & n & 6 \\ (2-1)^2 & 2n^2 & 4n-2 \\ (2-1)^3 & 3n^3 & 3n^2 - 3n \end{vmatrix} + \ldots + \begin{vmatrix} (n-1) & n & 6 \\ (n-1)^2 & 2n^2 & 4n-2 \\ (n-1)^3 & 3n^3 & 3n^2 - 3n \end{vmatrix}$$

$$= \begin{vmatrix} 1+2+3+\ldots+(n-1) & n & 6 \\ 1^2+2^2+3^2+\ldots+(n-1)^2 & 2n^2 & 4n-2 \\ 1^3+2^3+3^3+\ldots+(n-1)^3 & 3n^3 & 3n^2-3n \end{vmatrix}$$

$$= \begin{vmatrix} \frac{n(n-1)}{2} & n & 6 \\ \frac{n(n-1)(2n-1)}{6} & 2n^2 & 4n-2 \\ \left(\frac{n(n-1)}{2}\right)^2 & 3n^3 & 3n^2-3n \end{vmatrix}$$

$$= \frac{n^2(n-1)}{12} \begin{vmatrix} 6 & 1 & 6 \\ 2(2n-1) & 2n & 2(2n-1) \\ 3n(n-1) & 3n^2 & 3n(n-1) \end{vmatrix}$$

(Taking $\frac{n(n-1)}{12}$ common from C_1 and n from C_2)

$= 0$ (as C_1 and C_3 are identical)

Thus, $\sum_{a=1}^{n} \Delta a = 0 \Rightarrow \sum_{a=1}^{n} \Delta a = c$ (a constant)

Where $c = 0$

8. Given that A, B, C are integers between 0 and 9 and the three digit numbers $A28$, $3B9$ and $62C$ are divisible by a fixed integer k.

Now, $D = \begin{vmatrix} A & 3 & 6 \\ 8 & 9 & C \\ 2 & B & 2 \end{vmatrix}$

On operating $R_2 \to R_2 + 10R_3 + 100R_1$, we get

$$= \begin{vmatrix} A & 3 & 6 \\ A28 & 3B9 & 62C \\ 2 & B & 2 \end{vmatrix} = \begin{vmatrix} A & 3 & 6 \\ kn_1 & kn_2 & kn_3 \\ 2 & B & 2 \end{vmatrix}$$

[As per question $A28$, $3B9$ and $62C$ are divisible by k, therefore,

$A28 = kn_1$
$3B9 = kn_2$
$62C = kn_3$]

$$= k\begin{vmatrix} A & 3 & 6 \\ n_1 & n_2 & n_3 \\ 2 & B & 2 \end{vmatrix} = k \times \text{some integral value.}$$

$\Rightarrow D$ is divisible by k.

9. Consider $\begin{vmatrix} p & b & c \\ a & q & c \\ a & b & r \end{vmatrix} = 0$

Operating $R_1 \to R_1 - R_2$ and $R_2 \to R_2 - R_3$ we get

$$\begin{vmatrix} p-a & -(q-b) & c \\ 0 & q-b & c-r \\ a & b & r \end{vmatrix} = 0$$

Taking $(p-q)$, $(q-b)$ and $(r-c)$ common from C_1, C_2 and C_3 resp, we get

$$\Rightarrow (p-a)(q-b)(r-c)\begin{vmatrix} 1 & -1 & 0 \\ 0 & 1 & -1 \\ \frac{a}{p-a} & \frac{b}{q-b} & \frac{r}{r-c} \end{vmatrix} = 0$$

Expanding along R_1

$$\Rightarrow (p-a)(q-b)(r-c)\left[1\left(\frac{r}{r-c}+\frac{b}{q-b}\right)+\frac{a}{p-a}\right] = 0$$

As given that $p \neq a, q \neq b, r \neq c$

$$\therefore \quad \frac{r}{r-c}+\frac{b}{q-b}+\frac{a}{p-a} = 0$$

$$\Rightarrow \quad \frac{r}{r-c}+\frac{q-(q-b)}{q-b}+\frac{p-(p-a)}{p-a} = 0$$

$$\Rightarrow \quad \frac{r}{r-c}+\frac{q}{q-b}-1+\frac{p}{p-a}-1 = 0$$

$$\Rightarrow \quad \frac{p}{p-a}+\frac{q}{q-b}+\frac{r}{r-c} = 2$$

10. $D = \begin{vmatrix} n! & (n+1)! & (n+2)! \\ (n+1)! & (n+2)! & (n+3)! \\ (n+2)! & (n+3)! & (n+4)! \end{vmatrix}$

$$= n!\,(n+1)!\,(n+2)!\begin{vmatrix} 1 & n+1 & (n+2)(n+1) \\ 1 & n+2 & (n+3)(n+2) \\ 1 & n+3 & (n+4)(n+3) \end{vmatrix}$$

Operating $R_2 \to R_2 - R_1$ and $R_3 \to R_3 - R_2$, we get

$$D = (n!)^3(n+1)^2(n+2)\begin{vmatrix} 1 & n+1 & n^2+3n+2 \\ 0 & 1 & 2n+4 \\ 0 & 1 & 2n+6 \end{vmatrix}$$

Operating $R_3 \to R_3 - R_2$

$$D = (n!)^3(n+1)^2(n+2)\begin{vmatrix} 1 & n+1 & n^2+3n+2 \\ 0 & 1 & 2n+4 \\ 0 & 0 & 2 \end{vmatrix}$$

$$= (n!)^3(n+1)^2(n+2)\,1\,[2]$$

$$\Rightarrow \frac{D}{(n!)^3} = 2\,(n+1)^2(n+2)$$

$$\Rightarrow \frac{D}{(n!)^3} - 4 = 2\,(n^3 + 4n^2 + 5n + 2) - 4$$

$$= 2\,(n^3 + 4n^2 + 5n) = 2n\,(n^2 + 4n + 5)$$

$\Rightarrow \frac{D}{(n!)^3} - 4$ is divisible by n.

11. Given that $\lambda, \alpha \in R$ and system of linear equations

$\lambda x + (\sin\alpha)y + (\cos\alpha)z = 0$
$x + (\cos\alpha)y + (\sin\alpha)z = 0$
$-x(\sin\alpha)y - (\cos\alpha)z = 0$

has a non trivial solution, therefore $D = 0$

$$\Rightarrow \begin{vmatrix} \lambda & \sin\alpha & \cos\alpha \\ 1 & \cos\alpha & \sin\alpha \\ -1 & \sin\alpha & -\cos\alpha \end{vmatrix} = 0$$

$\Rightarrow \lambda(-\cos^2\alpha - \sin^2\alpha) - \sin\alpha(-\cos\alpha + \sin\alpha) + \cos\alpha(\sin\alpha + \cos\alpha) = 0$

$\Rightarrow -\lambda + \sin\alpha\cos\alpha - \sin^2\alpha + \sin\alpha\cos\alpha + \cos^2\alpha = 0$

$\Rightarrow \lambda = \cos^2\alpha - \sin^2\alpha + 2\sin\alpha\cos\alpha$

$\Rightarrow \lambda = \cos 2\alpha + \sin 2\alpha$

For $\lambda = 1$, $\cos 2\alpha + \sin 2\alpha = 1$

$$\frac{1}{\sqrt{2}}\cos 2\alpha + \frac{1}{\sqrt{2}}\sin 2\alpha = \frac{1}{\sqrt{2}}$$

$\Rightarrow \cos 2\alpha \cos \pi/4 + \sin 2\alpha \sin \pi/4 = \dfrac{1}{\sqrt{2}}$

$\Rightarrow \cos(2\alpha - \pi/4) = \cos \pi/4$

$\Rightarrow 2\alpha - \pi/4 = 2n\pi \pm \pi/4$

$\Rightarrow 2\alpha = 2\pi r + \pi/4 + \pi/4;\ 2n\pi - \pi/4 + \pi/4$

$\Rightarrow \alpha = n\pi + \pi/4$ or $n\pi$

12. L.H.S.

$$= \begin{vmatrix} \cos A\cos P + \sin A\sin P & \cos(A-Q) & \cos(A-R) \\ \cos B\cos P + \sin B\sin P & \cos(B-Q) & \cos(B-R) \\ \cos C\cos P + \sin C\sin P & \cos(C-Q) & \cos(C-R) \end{vmatrix}$$

$$= \cos P\begin{vmatrix} \cos A & \cos(A-Q) & \cos(A-R) \\ \cos B & \cos(B-Q) & \cos(B-R) \\ \cos C & \cos(C-Q) & \cos(C-R) \end{vmatrix}$$

$$+\sin P\begin{vmatrix} \sin A & \cos(A-Q) & \cos(A-R) \\ \sin B & \cos(B-Q) & \cos(B-R) \\ \sin C & \cos(C-Q) & \cos(C-R) \end{vmatrix}$$

Operating; $C_2 \to C_2 - C_1 (\cos Q)$; $C_3 \to C_3 - C_1 (\cos R)$ on first determinant and $C_2 \to C_2 - (\sin Q)C_1$ and $C_3 \to C_3 - (\sin R)C_1$ on second determinant, we get

$$= \cos P\begin{vmatrix} \cos A & \sin A\sin Q & \sin A\sin R \\ \cos B & \sin B\sin Q & \sin B\sin R \\ \cos C & \sin C\sin Q & \sin C\sin R \end{vmatrix}$$

$$+\sin P\begin{vmatrix} \sin A & \cos A\cos Q & \cos A\cos R \\ \sin B & \cos B\cos Q & \cos B\cos R \\ \sin C & \cos C\cos Q & \cos C\cos R \end{vmatrix}$$

$$= \cos P\sin Q\sin R\begin{vmatrix} \cos A & \sin A & \sin A \\ \cos B & \sin B & \sin B \\ \cos C & \sin C & \sin C \end{vmatrix}$$

$$+ \sin P\cos Q\cos R\begin{vmatrix} \sin A & \cos A & \cos A \\ \sin B & \cos B & \cos B \\ \sin C & \cos C & \cos C \end{vmatrix}$$

$= 0 + 0$

[Both determinants become zero as $C_2 \equiv C_3$]

$= 0 =$ R.H.S.

13. Let us denote the given determinant by Δ. Taking $\dfrac{1}{a(a+d)(a+2d)}$ as common from R_1, $\dfrac{1}{(a+d)(a+2d)(a+3d)}$ from R_2 and $\dfrac{1}{(a+2d)(a+3d)(a+4d)}$ from R_3, we get

$$\Delta = \frac{1}{a(a+d)^2(a+2d)^3(a+3d)^2(a+4d)}\Delta_1$$

where

$$\Delta_1 = \begin{vmatrix} (a+d)(a+2d) & a+2d & a \\ (a+2d)(a+3d) & a+3d & a+d \\ (a+3d)(a+4d) & a+4d & a+2d \end{vmatrix}$$

Applying $R_3 \to R_3 - R_2$ and $R_2 \to R_2 - R_1$, we get

$$\Delta_1 = \begin{vmatrix} (a+d)(a+2d) & a+2d & a \\ (a+2d)(2d) & d & d \\ (a+3d)(2d) & d & d \end{vmatrix}$$

Applying $R_3 \to R_3 - R_2$, we get

$$\Delta_1 = \begin{vmatrix} (a+d)(a+2d) & a+2d & a \\ (a+2d)(2d) & d & d \\ 2d^2 & 0 & 0 \end{vmatrix}$$

Expanding along R_3, we get

$$\Delta_1 = (2d^2)\begin{vmatrix} a+2d & a \\ d & d \end{vmatrix} = (2d)^2 (d)(a + 2d - a) = 4d^4$$

Thus, $\Delta = \dfrac{4d^4}{a(a+d)^2(a+2d)^3(a+3d)^2(a+4d)}$

14. Given that a, b, c are pth, qth and rth terms of an *H.P.*

$\Rightarrow \dfrac{1}{a}, \dfrac{1}{b}, \dfrac{1}{c}$ are pth, qth, rth terms of an *A.P.*

$$\therefore \left.\begin{aligned} \frac{1}{a} &= A+(p-1)D \\ \frac{1}{b} &= A+(q-1)D \\ \frac{1}{c} &= A+(r-1)D \end{aligned}\right\} \quad ....(1)$$

Now given determinant is

$$\Delta = \begin{vmatrix} bc & ca & ab \\ p & q & r \\ 1 & 1 & 1 \end{vmatrix} = abc\begin{vmatrix} \frac{1}{a} & \frac{1}{b} & \frac{1}{c} \\ p & q & r \\ 1 & 1 & 1 \end{vmatrix}$$

Substituting the values of $1/a$, $1/b$, $1/c$ from (1), we get

$$= abc\begin{vmatrix} A+(p-1)D & A+(q-1)D & A+(r-1)D \\ p & q & r \\ 1 & 1 & 1 \end{vmatrix}$$

Operating $R_1 \to R_1 - (A-D)R_3 - DR_2$, we get

$$\Delta = abc\begin{vmatrix} 0 & 0 & 0 \\ p & q & r \\ 1 & 1 & 1 \end{vmatrix} = 0$$

15. $R_2 \to R_2 + R_3$,

$$\begin{vmatrix} \sin\theta & \cos\theta & \sin 2\theta \\ 2\sin\theta\cos\frac{2\pi}{2} & 2\cos\theta\cos\frac{2\pi}{3} & 2\sin 2\theta\cos\frac{4\pi}{3} \\ \sin\left(\theta-\frac{2\pi}{3}\right) & \cos\left(\theta-\frac{2\pi}{3}\right) & \sin\left(2\theta-\frac{4\pi}{3}\right) \end{vmatrix} = 0$$

$$= \begin{vmatrix} \sin\theta & \cos\theta & \sin 2\theta \\ -\sin\theta & -\cos\theta & -\sin 2\theta \\ \sin\left(\theta-\frac{2\pi}{3}\right) & \cos\left(\theta-\frac{2\pi}{3}\right) & \sin\left(2\theta-\frac{4\pi}{3}\right) \end{vmatrix} = 0$$

16. Given that $A^T A = I$

where $A = \begin{bmatrix} a & b & c \\ b & c & a \\ c & a & b \end{bmatrix}$ $\quad \therefore A^T = \begin{bmatrix} a & b & c \\ b & c & a \\ c & a & b \end{bmatrix}$

$\therefore A^T A = I$

$$\Rightarrow \begin{bmatrix} a & b & c \\ b & c & a \\ c & a & b \end{bmatrix}\begin{bmatrix} a & b & c \\ b & c & a \\ c & a & b \end{bmatrix} = \begin{bmatrix} 1 & 0 & 0 \\ 0 & 1 & 0 \\ 0 & 0 & 1 \end{bmatrix}$$

$$\Rightarrow \begin{bmatrix} a^2+b^2+c^2 & ab+bc+ca & ab+bc+ca \\ ab+bc+ca & a^2+b^2+c^2 & ab+bc+ca \\ ab+bc+ca & ab+bc+ca & a^2+b^2+c^2 \end{bmatrix}$$

$$= \begin{bmatrix} 1 & 0 & 0 \\ 0 & 1 & 0 \\ 0 & 0 & 1 \end{bmatrix}$$

$\Rightarrow a^2 + b^2 + c^2 = 1$ ...(1)

and $ab + bc + ca = 0$ ...(2)

$abc = 1$ (given) ...(3)

Now $a^3 + b^3 + c^3 - 3abc$

$= (a+b+c)(a^2+b^2+c^2-ab-bc-ca)$

$\Rightarrow a^3 + b^3 + c^3 = (a+b+c)[1-0] + 3 \times 1$

[Using (1), (2) and (3)]

$\Rightarrow a^3 + b^3 + c^3 = a + b + c + 3$

Now $(a+b+c)^2 = a^2 + b^2 + c^2 + 2(ab + bc + ca)$

$\Rightarrow (a+b+c)^2 = 1$

$\Rightarrow (a+b+c) = 1$

[$\because$ $(a+b+c) \neq -1$ as a, b, c all are + ve numbers.]

$\therefore$ We get

$a^3 + b^3 + c^3 = 4$

ALTERNATE SOLUTION

Given that $A^T A = I$

$\Rightarrow |A^T A| = |A^T||A| = |A||A| = 1$ $\quad [\because |I| = I]$

$\Rightarrow |A|^2 = 1$ ....(1)

From given matrix $A = \begin{bmatrix} a & b & c \\ b & c & a \\ c & a & b \end{bmatrix}$

$$|A| = \begin{vmatrix} a & b & c \\ b & c & a \\ c & a & b \end{vmatrix} = a^3 + b^3 + c^3 - 3abc \quad \text{....(2)}$$

$\therefore (a^3 + b^3 + c^3 - 3abc)^2 = 1$ (From (1) and (2))

$\Rightarrow a^3 + b^3 + c^3 - 3abc = 1$ or -1

But for a^3, b^3, c^3 using $AM \geq GM$

We get $\frac{a^3+b^3+c^3}{3} \geq \sqrt[3]{a^3b^3c^3}$

$\Rightarrow a^3 + b^3 + c^3 - 3abc \geq 0$

$\therefore$ We must have

$a^3 + b^3 + c^3 - 3abc = 1$

$\Rightarrow a^3 + b^3 + c^3 = 1 + 3 \times 1 = 4$ [Using $abc = 1$]

17. We are given that $MM^T = I$ where M is a square matrix of order 3 and det. $M = 1$.

Consider det $(M - I)$ = det $(M - MM^T)$ [Given $MM^T = I$]

$=$ det $[M(I - M^T)]$

$=$ (det M) (det $(I - M^T)$)

$[\because |AB| = |A||B|]$

$= -$(det M) (det $(M^T - I)$)

$= -1$ [det $(M^T - I)$] $\quad [\because$ det $(M) = 1]$

$= -$det $(M - I)$

$[\because$ det $(M^T - I)$ = det $[(M - I)^T]$ = det $(M - I)]$

$\Rightarrow 2$ det $(M - I) = 0 \Rightarrow$ det $(M - I) = 0$

Hence Proved

18. Given that,

$$A = \begin{bmatrix} a & 1 & 0 \\ 1 & b & d \\ 1 & b & c \end{bmatrix}, X = \begin{bmatrix} x \\ y \\ z \end{bmatrix}, U = \begin{bmatrix} f \\ g \\ h \end{bmatrix}$$

and $AX = U$ has infinite many solutions.

$\Rightarrow |A| = 0 = |A_1| = |A_2| = |A_3|$

Now, $|A| = 0$

$$\Rightarrow \begin{vmatrix} a & 1 & 0 \\ 1 & b & d \\ 1 & b & c \end{vmatrix} = a(bc - bd) - 1(c - d) = 0$$

$\Rightarrow (ab - 1)(c - d) = 0$

$\Rightarrow ab = 1$ or $c = d$ ....(1)

And $|A_1| = \begin{vmatrix} f & 1 & 0 \\ g & b & d \\ h & b & c \end{vmatrix} = 0$

$\Rightarrow f(bc - bd) - 1(gc - hd) = 0$

$\Rightarrow fb(c - d) = gc - hd$ ....(2)

$|A_2| = \begin{vmatrix} a & 1 & f \\ 1 & b & g \\ 1 & b & h \end{vmatrix} = 0$

$\Rightarrow a(gc - hd) - f(c - d) = 0 \Rightarrow a(gc - hd) = f(c - d)$

$|A_3| = \begin{vmatrix} a & 1 & f \\ 1 & b & g \\ 1 & b & h \end{vmatrix} = 0$

$\Rightarrow a(bh - bg) - 1(h - g) + f(b - b) = 0$

$\Rightarrow ab(h - g) - (h - g) = 0$

$\Rightarrow ab = 1$ or $h = g$ ....(3)

Taking $c = d \Rightarrow h = g$ and $ab \neq 1$ (from (1), (2) and (3))

Now taking $BX = V$

where $B = \begin{bmatrix} a & 1 & 1 \\ 0 & d & c \\ f & g & h \end{bmatrix}, V = \begin{bmatrix} a^2 \\ 0 \\ 0 \end{bmatrix}$

Then $|B| = \begin{vmatrix} a & 1 & 1 \\ 0 & d & c \\ f & g & h \end{vmatrix} = 0$

[$\because$ In vew of $c = d$ and $g = h$, C_2 and C_3 are identical]

$\Rightarrow BX = V$ has no unique solution

And $|B_1| = \begin{vmatrix} a^2 & 1 & 1 \\ 0 & d & c \\ 0 & g & h \end{vmatrix} = 0 \quad (\because c = d, g = h)$

$|B_2| = \begin{vmatrix} a & a^2 & 1 \\ 0 & 0 & c \\ f & 0 & h \end{vmatrix} = a^2cf = a^2df \quad (\because c = d)$

$|B_3| = \begin{vmatrix} a & 1 & a^2 \\ 0 & d & 0 \\ f & g & 0 \end{vmatrix} = a^2df$

$\Rightarrow$ If $adf \neq 0$ then $|B_2| = |B_3| \neq 0$

Hence no solution exist.

(F) MATCH THE FOLLOWING :

1. The given lines are

$L_1 : x + 3y - 5 = 0$

$L_2 : 3x - ky - 1 = 0$

$L_3 : 5x + 2y - 12 = 0$

(A) Three lines L_1, L_2, L_3 are concurrent if

$\begin{vmatrix} 1 & 3 & 5 \\ 3 & -k & 1 \\ 5 & 2 & 12 \end{vmatrix} = 0 \Rightarrow 13k - 65 = 0 \Rightarrow k = 5$

$\therefore$ (A) $\to$ (s)

(B) For $L_1 \parallel L_2 \Rightarrow \frac{1}{3} = \frac{-3}{k} \Rightarrow k = -9$

and $L_2 \parallel L_3 \Rightarrow \frac{3}{5} = \frac{-k}{2} \Rightarrow k = -\frac{6}{5}$

$\therefore$ (B) $\to$ (p), (q)

(C) Three lines L_1, L_2, L_3 will form a triangle if no two of them are parallel and no three are concurrent

$\therefore k \neq 5, -9, -6/5 \quad \therefore$ (C) $\to$ r

(D) L_1, L_2, L_3 do not form a triangle if either any two of these are parallel or the three are concurrent i.e.

$k = 5, -9, -6/5$

$\therefore$ (D) $\to$ (p), (q), (s)

2. (A) Let $y = \frac{x^2 + 2x + 4}{x + 2}$

$\Rightarrow x^2 + (2 - y)x + (4 - 2y) = 0$

For x to be real, we have

$D \geq 0$

$\Rightarrow (2 - y)^2 - 4(4 - 2y) \geq 0 \Rightarrow y^2 + 4y - 12 \geq 0$

$\Rightarrow (y + 6)(y - 2) \geq 0 \quad \Rightarrow y \leq -6$ or $y \geq 2$

$\therefore y_{\min}$ does not exist and local minimum value of $y = 2$.

$\therefore$ (A) $\to$ r

(B) As A is symmetric and B is skew symmetric matrix, we should have

$A^t = A$ and $B^t = -B$ ...(1)

Also given that

$(A + B)(A - B) = (A - B)(A + B)$

$\Rightarrow A^2 - AB + BA - B^2 = A^2 + AB - BA - B^2$

$\Rightarrow 2BA = 2AB$ or $AB = BA$ ...(2)

Now given that

$(AB)^t = (-1)^k AB$

$\Rightarrow (BA)^t = (-1)^k AB$ (using equation (2))

$\Rightarrow A'B' = (-1)^k AB$

$\Rightarrow -AB = (-1)^k AB$ [using equation(1)]

$\Rightarrow k$ should be an odd number

$\therefore$ (B) $\to$ (q), (s)

(C) Given that $a = \log_3 \log_3 2$

$\Rightarrow \log_3 2 = 3^a \Rightarrow \frac{1}{\log_2 3} = 3^a$ or $\log_2 3 = 3^{-a}$

$\Rightarrow \quad 3 = 2^{(3^{-a})} \quad ...(1)$

Now $1 < 2^{(-k+3^{-a})} < 2 \Rightarrow 1 < 2^{-k}.2^{3^{-a}} < 2$

$\Rightarrow 1 < 2^{-k}.3 < 2$ (using eq (1))

$\Rightarrow \frac{1}{3} < 2^{-k} < \frac{2}{3} \Rightarrow \frac{3}{2} < 2^k < 3$

$\Rightarrow k = 1$

$\therefore k$ is less than 2 and 3

$\therefore$ (C) $\to$ (r), (s).

(D) Given that $\sin\theta = \cos\phi \Rightarrow \cos\left(\frac{\pi}{2} - \theta\right) = \cos\phi$

$\Rightarrow \frac{\pi}{2} - \theta = 2n\pi \pm \phi,\ n \in Z \Rightarrow \theta \pm \phi - \frac{\pi}{2} = -2n\pi$

$\Rightarrow \frac{1}{\pi}\left(\theta \pm \phi - \frac{\pi}{2}\right) = -2n$

$\therefore$ Here possible values of $\frac{1}{\pi}\left(\theta \pm \phi - \frac{\pi}{2}\right)$ are 0 and 2 for n = 0, –1.

$\therefore$ D $\to$ (p), (r).

G COMPREHENSION BASED QUESTIONS :

1. (d) Let $U_1 = \begin{bmatrix} a \\ b \\ c \end{bmatrix}$ then $\begin{bmatrix} 1 & 0 & 0 \\ 2 & 1 & 0 \\ 3 & 2 & 1 \end{bmatrix}\begin{bmatrix} a \\ b \\ c \end{bmatrix} = \begin{bmatrix} 1 \\ 0 \\ 0 \end{bmatrix}$

$\Rightarrow \begin{bmatrix} a \\ 2a+b \\ 3a+2b+c \end{bmatrix} = \begin{bmatrix} 1 \\ 0 \\ 0 \end{bmatrix} \Rightarrow a = 1, b = -2, c = 1$

$\therefore U_1 = \begin{bmatrix} 1 \\ -2 \\ 1 \end{bmatrix}$

Similarly, $U_2 = \begin{bmatrix} 2 \\ -1 \\ -4 \end{bmatrix}, U_3 = \begin{bmatrix} 2 \\ -1 \\ -3 \end{bmatrix}$

$\therefore U = \begin{bmatrix} 1 & 2 & 2 \\ -2 & -1 & -1 \\ 1 & -4 & -3 \end{bmatrix} \Rightarrow |U| = 3$

2. (b) $U^{-1} = \frac{1}{3}\begin{bmatrix} -1 & -2 & 0 \\ -7 & -5 & -3 \\ 9 & 6 & 3 \end{bmatrix}$

$\Rightarrow$ Sum of elements of $U^{-1} = \frac{1}{3}(0) = 0$

3. (a) $[3\,2\,0]\begin{bmatrix} 1 & 2 & 2 \\ -2 & -1 & -1 \\ 1 & -4 & -3 \end{bmatrix}\begin{bmatrix} 3 \\ 2 \\ 0 \end{bmatrix} = [3\,2\,0]\begin{bmatrix} 7 \\ -8 \\ -5 \end{bmatrix} = 5$

4. (a) Each element of set $\mathcal{A}$ is 3 × 3 symmetric matrix with five of its entries as 1 and four of its entries as 0, we can keep in diagonal either 2 zero and one 1 or no zero and three 1 so that the left over zeros and one's are even in number.

Hence taking 2 zeros and one 1 in diagonal the possible cases are $\frac{3!}{2!} \times \frac{3!}{2!} = 9$

and taking 3 ones in diagonal the possible cases are

$1 \times \frac{3!}{2!} = 3$

$\therefore$ Total elements $\mathcal{A}$ can have = 9 + 3 = 12

5. (b) The given system will have unique solution if $|A| \neq 0$ which is so for the matrices.

$\begin{bmatrix} 1 & 0 & 1 \\ 0 & 0 & 1 \\ 1 & 1 & 0 \end{bmatrix}$ and $\begin{bmatrix} 1 & 1 & 0 \\ 1 & 0 & 1 \\ 0 & 1 & 0 \end{bmatrix}$;

$\begin{bmatrix} 0 & 0 & 1 \\ 0 & 1 & 1 \\ 1 & 1 & 0 \end{bmatrix}$ and $\begin{bmatrix} 0 & 1 & 1 \\ 1 & 1 & 0 \\ 1 & 0 & 0 \end{bmatrix}$;

$\begin{bmatrix} 0 & 1 & 1 \\ 1 & 0 & 0 \\ 1 & 0 & 1 \end{bmatrix}$ and $\begin{bmatrix} 0 & 1 & 0 \\ 1 & 0 & 1 \\ 0 & 1 & 1 \end{bmatrix}$

which are 6 in number.

6. (b) For the given system to be inconsistent $|A| = 0$. The matrices for which $|A| = 0$ are

$\begin{bmatrix} 1 & 0 & 0 \\ 0 & 1 & 1 \\ 0 & 1 & 1 \end{bmatrix}, \begin{bmatrix} 1 & 0 & 1 \\ 0 & 1 & 0 \\ 1 & 0 & 1 \end{bmatrix}, \begin{bmatrix} 1 & 1 & 0 \\ 1 & 1 & 0 \\ 0 & 0 & 1 \end{bmatrix}$

(i) (ii) (iii)

and $\begin{bmatrix} 1 & 1 & 1 \\ 1 & 0 & 0 \\ 1 & 0 & 0 \end{bmatrix}, \begin{bmatrix} 0 & 1 & 0 \\ 1 & 1 & 1 \\ 0 & 1 & 0 \end{bmatrix}, \begin{bmatrix} 0 & 0 & 1 \\ 0 & 0 & 1 \\ 1 & 1 & 1 \end{bmatrix}$

(iv) (v) (vi)

On solving $A\begin{bmatrix} x \\ y \\ z \end{bmatrix} = \begin{bmatrix} 1 \\ 0 \\ 0 \end{bmatrix}$

We find for A = (i)

By Cramer's rule $D_1 = 0 = D_2 = D_3$

$\therefore$ infinite many solution

For A = (ii)

By Cramer's rule $D_1 \neq 0$

$\Rightarrow$ no solution i.e. inconsistent.

Similarly we find the system as inconsistent in cases (iii), (v) and (vi).

Hence for four cases system is inconsistent.

7. **(d)** If A is symmetrie then b = c

$\Rightarrow |A| = a^2 - b^2 = (a+b)(a-b)$

Which is divisible by p if $(a + b)$ is divisible by p or $(a - b)$ is divisible by p.

Now $(a + b)$ is divisible by p if (a, b) can take values $(1, p-1), (2, p-2), (3, p-3), ... (p-1, 1)$

$\therefore (p-1)$ ways.

Also $(a - b)$ is divisible by p only when $a - b = 0$ i.e. $a = b$, then (a, b) can take values (0,0) (1,1), (2,2)..... $(p\acute{Y}, p-2)$

$\therefore p$ ways.

If A is skew symmetric, then $a = 0$ and $b = -c$ or $b + c = 0$ which gives $|A| = 0$ when $b^2 \Rightarrow b = 0, c = 0$

But this possibility is already included when A is symmetric and $(a, b) = (0, 0)$.

Again if A is both symmetric and skew symmetric, then clearly A is null matrix which case is already included.

Hence total number of ways $= p + (p-1) = 2p - 1$

8. **(c)** Trace $A = a + a = 2a$ is not divisible by p

$\Rightarrow a$ is not divisible by p

$\Rightarrow a \neq 0$

But $|A|$ is divisible by p

$\Rightarrow a^2 - bc$ is divisible by p

It will be so if on dividing a^2 by p suppose we get $m\frac{l}{p}$

then on dividing bc by p we should get $n\frac{l}{p}$ for some integeral values of m, n and l.

i.e. the remainder should be same in each case, so that

$$\frac{a^2 - bc}{p} = \left(m + \frac{l}{p}\right) - \left(n + \frac{l}{p}\right) = (m - n) = \text{an integer}$$

For this to happen a can take any value from 1 to p–1, also if b takes any value from 1 to p–1 then c should take only that value corresponding to which the remainder is same.

$\therefore$ No. of ways $= (p-1) \times (p-1) \times 1 = (p-1)^2$.

9. **(d)** Total number of matrices

= total number of ways a, b, c can be selected

$= p \times p \times p = p^3$.

Number of ways when det (A) is divisible by p and trace (A) $\neq$ 0 are equal to number of ways det (A) is divisible by p and trace (A) is not divisible by $p = (p-1)^2$

Also number of ways when det (A) is divisible by p and trace A = 0 are the ways when bc is multiple of p

$\Rightarrow b = 0$ or $c = 0$

for $b = 0$, c can take values 0, 1, 2, ..., $p - 1$

For $c = 0$, b can take values 0, 1, 2, ..., $p - 1$

Here $(b, c) = (0, 0)$ is coming twice.

$\therefore$ Total ways of selecting b and $c = p + p - 1$
$= 2p - 1$

$\therefore$ Total number of ways when det (A) is divisible by $p = (p-1)^2 + 2p - 1 = p^2$

Hence the number of ways when det (A) is not divisible by $p = p^3 - p^2$.

10. **(d)** From equation (E), we get

$a + 8b + 7c = 0$
$9a + 2b + 3c = 0$
$a + b + c = 0$

Here $\begin{vmatrix} 1 & 8 & 7 \\ 9 & 2 & 3 \\ 1 & 1 & 1 \end{vmatrix} = 0,$

Therefore system has infinite many solutions.

Solving these, we get

$b = 6a$ and $c = -7a$

Now (a, b, c) lies on $2x + y + z = 1 \Rightarrow b = 6, c = -7$

$\therefore 2a + 6a - 7a = 1 \Rightarrow a = 1$

$\therefore 7a + b + c = 7 + 6 - 7 = 6 \Rightarrow b = 6, c = -7$

11. **(a)** If $a = 2$ then $b = 12$, $c = -14$

$\therefore \frac{3}{\omega^a} + \frac{1}{\omega^b} + \frac{3}{\omega^c} = \frac{3}{\omega^2} + \frac{1}{\omega^{12}} + \frac{3}{\omega^{-14}}$

$= 3\omega + 1 + 3\omega^2 = 1 - 3 = -2$

12. **(b)** If $b = 6$ then $a = 1$, $c = -7$

$\therefore$ Equation becomes $x^2 + 6x - 7 = 0$ or $(x + 7)(x - 1) = 0$ whose roots are 1 and –7.

Let $\alpha = 1$ and $\beta = -7$

$$\therefore \sum_{n=0}^{\infty}\left(\frac{1}{1} - \frac{1}{7}\right)^n = \sum_{n=0}^{\infty}\left(\frac{6}{7}\right)^n = \frac{1}{1 - \frac{6}{7}} = 7$$

(H) ASSERTION & REASON TYPE QUESTIONS :

(a) The given equations are

$x - 2y + 3z = -1$
$-x + y - 2z = k$
$x - 3y + 4z = 1$

Here $D = \begin{vmatrix} 1 & -2 & 3 \\ -1 & 1 & -2 \\ 1 & -3 & 4 \end{vmatrix} = 0$

and $D_2 = \begin{vmatrix} 1 & -1 & 3 \\ -1 & k & -2 \\ 1 & 1 & 4 \end{vmatrix} = k-3 \neq 0$ if $k \neq 3$

$\therefore$ If $k \neq 3$, the system has no solutions.

Hence statement-1 is true and statement-2 is a correct explanation for statement - 1.

INTEGER VALUE CORRECT TYPE :

1. We have $\omega = \cos\frac{2\pi}{3} + i\sin\frac{2\pi}{3} = \frac{-1+i\sqrt{3}}{2}$

$\therefore 1+\omega+\omega^2 = 0$ and $\omega^3 = 1$

Then $\begin{vmatrix} z+1 & \omega & \omega^2 \\ \omega & z+\omega^2 & 1 \\ \omega^2 & 1 & z+\omega \end{vmatrix} = 0$

$C_1 \leftrightarrow C_1 + C_2 + C_3$

$$\Rightarrow \begin{vmatrix} z+1+\omega+\omega^2 & \omega & \omega^2 \\ z+1+\omega+\omega^2 & z+\omega^2 & 1 \\ z+1+\omega+\omega^2 & 1 & z+\omega \end{vmatrix} = 0$$

$$\Rightarrow \begin{vmatrix} 1 & \omega & \omega^2 \\ 1 & z+\omega^2 & 1 \\ 1 & 1 & z+\omega \end{vmatrix} = 0$$

$$\Rightarrow z\left[1\left(z^2+z\omega+z\omega^2+\omega^3-1\right)-\omega\left(z+\omega-1\right)+\omega^2\left(1-z-\omega^2\right)=0\right]$$

$$\Rightarrow z\left[z^2+z\omega+z\omega^2-z\omega-\omega^2+\omega+\omega^2-z\omega^2-\omega^4\right]=0$$

$$\Rightarrow z\left[z^2\right]=0 \Rightarrow z^3=0 \Rightarrow z=0$$

$\therefore$ z = 0 is the only solution.

2. $|A| = \begin{vmatrix} 2k-1 & 2\sqrt{k} & 2\sqrt{k} \\ 2\sqrt{k} & 1 & -2k \\ -2\sqrt{k} & 2k & -1 \end{vmatrix}$

$$= \begin{vmatrix} 2k-1 & 0 & 2\sqrt{k} \\ 2\sqrt{k} & 1+2k & -2k \\ -2\sqrt{k} & 1+2k & -1 \end{vmatrix}, C_2 \to C_2 - C_3$$

$$= \begin{vmatrix} 2k-1 & 0 & 2\sqrt{k} \\ 4\sqrt{k} & 0 & 1-2k \\ -2\sqrt{k} & 1+2k & -1 \end{vmatrix}, R_2 \to R_2 - R_3$$

$$= (1+2k)(8k-4k+4k^2+1)$$

$$= (2k+1)^3$$

Also $|B| = 0$ as B is skew symmetric of odd order.

$\therefore$ $|\text{Adj A}| + |\text{Adj B}| = |A|^2 + |B|^2 = 10^6$

$\Rightarrow (2k+1)^6 = 10^6 \Rightarrow 2k+1 = 10 \Rightarrow k = 4.5$

$\therefore [k] = 4$

3. 9

Let $M = \begin{bmatrix} a_1 & b_1 & c_1 \\ a_2 & b_2 & c_2 \\ a_3 & b_3 & c_3 \end{bmatrix}$

then $\begin{bmatrix} a_1 & b_1 & c_1 \\ a_2 & b_2 & c_2 \\ a_3 & b_3 & c_3 \end{bmatrix}\begin{bmatrix} 0 \\ 1 \\ 0 \end{bmatrix} = \begin{bmatrix} -1 \\ 2 \\ 3 \end{bmatrix} \Rightarrow \begin{matrix} b_1 = -1 \\ b_2 = 2 \\ b_3 = 3 \end{matrix}$

$\begin{bmatrix} a_1 & b_1 & c_1 \\ a_2 & b_2 & c_2 \\ a_3 & c_3 & c_3 \end{bmatrix}\begin{bmatrix} 1 \\ -1 \\ 0 \end{bmatrix} = \begin{bmatrix} 1 \\ 1 \\ -1 \end{bmatrix} \Rightarrow \begin{matrix} a_1 - b_1 = 1 \\ a_2 - b_2 = 1 \\ a_3 - b_3 = -1 \end{matrix}$

$\Rightarrow a_1 = 0, a_2 = 3, a_3 = 2$

and $\begin{bmatrix} a_1 & b_1 & c_1 \\ a_2 & b_2 & c_2 \\ a_3 & b_3 & c_3 \end{bmatrix}\begin{bmatrix} 1 \\ 1 \\ 1 \end{bmatrix} = \begin{bmatrix} 0 \\ 0 \\ 12 \end{bmatrix} \Rightarrow \begin{matrix} a_3 + b_3 + c_3 = 12 \\ \Rightarrow c_3 = 7 \end{matrix}$

$\therefore$ Sum of diagonal elements $= a_1 + b_2 + c_3 = 0 + 2 + 7 = 9$

Section-B JEE Main/ AIEEE

1. (c) We have $\begin{vmatrix} a & b & ax+b \\ b & c & bx+c \\ ax+b & bx+c & 0 \end{vmatrix}$

By $R_3 \to R_3 - (xR_1 + R_2)$;

$$= \begin{vmatrix} a & b & ax+b \\ b & c & bx+c \\ 0 & 0 & -(ax^2+2bx+C) \end{vmatrix}$$

$= (ax^2+2bx+c)(b^2-ac) = (+)(-) = -\text{ve}.$

2. (d) For homogeneous system of equations to have non zero solution, $\Delta = 0$

$$\begin{vmatrix} 1 & 2a & a \\ 1 & 3b & b \\ 1 & 4c & c \end{vmatrix} = 0 \quad C_2 \to C_2 - 2C_3$$

$$\begin{vmatrix} 1 & 0 & a \\ 1 & b & b \\ 1 & 2c & c \end{vmatrix} = 0 \quad R_3 \to R_3 - R_2, R_2 \to R_2 - R_1$$

$$\begin{vmatrix} 1 & 0 & a \\ 0 & b & b-a \\ 0 & 2c-b & c-b \end{vmatrix} = 0$$

$b(c-b) - (b-a)(2c-b) = 0$

On simplification, $\frac{2}{b} = \frac{1}{a} + \frac{1}{c}$

$\therefore a, b, c$ are in Harmonic Progression.

3. (b) $\Delta = \begin{vmatrix} 1 & \omega^n & \omega^{2n} \\ \omega^n & \omega^{2n} & 1 \\ \omega^{2n} & 1 & \omega^n \end{vmatrix}$

$$= 1\left(\omega^{3n} - 1\right) - \omega^n\left(\omega^{2n} - \omega^{2n}\right) + \omega^{2n}\left(\omega^n - \omega^{4n}\right)$$

$= \omega^{3n} - 1 - 0 + \omega^{3n} - \omega^{6n}$

$= 1 - 1 + 1 - 1 = 0 \left[\because \omega^{3n} = 1\right]$

4. (c) $A^2 = \begin{bmatrix} \alpha & \beta \\ \beta & \alpha \end{bmatrix} = \begin{bmatrix} a & b \\ b & a \end{bmatrix}\begin{bmatrix} a & b \\ b & a \end{bmatrix}$

$$= \begin{bmatrix} a^2+b^2 & 2ab \\ 2ab & a^2+b^2 \end{bmatrix}$$

$\alpha = a^2 + b^2; \beta = 2ab$

5. (a) $A = \begin{bmatrix} 0 & 0 & -1 \\ 0 & -1 & 0 \\ -1 & 0 & 0 \end{bmatrix}$

clearly $A \neq 0$. Also $|A| = -1 \neq 0$

$\therefore A^{-1}$ exists, further $(-1)I = \begin{bmatrix} -1 & 0 & 0 \\ 0 & -1 & 0 \\ 0 & 0 & -1 \end{bmatrix} \neq A$

Also $A^2 = \begin{bmatrix} 0 & 0 & -1 \\ 0 & -1 & 0 \\ -1 & 0 & 0 \end{bmatrix}\begin{bmatrix} 0 & 0 & -1 \\ 0 & -1 & 0 \\ -1 & 0 & 0 \end{bmatrix}$

$$= \begin{bmatrix} 1 & 0 & 0 \\ 0 & 1 & 0 \\ 0 & 0 & 1 \end{bmatrix} = I$$

6. (a) Given that $10B = \begin{bmatrix} 4 & 2 & 2 \\ -5 & 0 & \alpha \\ 1 & -2 & 3 \end{bmatrix}$

$$\Rightarrow B = \frac{1}{10}\begin{bmatrix} 4 & 2 & 2 \\ -5 & 0 & \alpha \\ 1 & -2 & 3 \end{bmatrix}$$

Also since, $B = A^{-1} \Rightarrow AB = I$

$$\Rightarrow \frac{1}{10}\begin{bmatrix} 1 & -1 & 1 \\ 2 & 1 & -3 \\ 1 & 1 & 1 \end{bmatrix}\begin{bmatrix} 4 & 2 & 2 \\ -5 & 0 & \alpha \\ 1 & -2 & 3 \end{bmatrix} = \begin{bmatrix} 1 & 0 & 0 \\ 0 & 1 & 0 \\ 0 & 0 & 1 \end{bmatrix}$$

$$\Rightarrow \frac{1}{10}\begin{bmatrix} 10 & 0 & 5-2 \\ 0 & 10 & -5+\alpha \\ 0 & 0 & 5+\alpha \end{bmatrix} = \begin{bmatrix} 1 & 0 & 0 \\ 0 & 1 & 0 \\ 0 & 0 & 1 \end{bmatrix}$$

$$\Rightarrow \frac{5-\alpha}{10} = 0 \Rightarrow \alpha = 5$$

7. (d) Let r be the common ratio, then

$$\begin{vmatrix} \log a_n & \log a_{n+1} & \log a_{n+2} \\ \log a_{n+3} & \log a_{n+4} & \log a_{n+5} \\ \log a_{n+6} & \log a_{n+7} & \log a_{n+8} \end{vmatrix}$$

$$= \begin{vmatrix} \log a_1 r^{n-1} & \log a_1 r^n & \log a_1 r^{n+1} \\ \log a_1 r^{n+2} & \log a_1 r^{n+3} & \log a_1 r^{n+4} \\ \log a_1 r^{n+5} & \log a_1 r^{n+6} & \log a_1 r^{n+7} \end{vmatrix}$$

$$= \begin{vmatrix} \log a_1+(n-1)\log r & \log a_1+n\log r & \log a_1+(n+1)\log r \\ \log a_1+(n+2)\log r & \log a_1+(n+3)\log r & \log a_1+(n+4)\log r \\ \log a_1+(n+5)\log r & \log a_1+(n+6)\log r & \log a_2+(n+7)\log r \end{vmatrix}$$

$$=0 \left[\text{Apply } c_2 \to c_2 - \frac{1}{2}c_1 - \frac{1}{2}c_3\right]$$

8. (d) Given $A^2 - A + I = 0$

$A^{-1}A^2 - A^{-1}A + A^{-1}.I = A^{-1}.0$

(Multiplying A^{-1} on both sides)

$\Rightarrow A - 1 + A^{-1} = 0$ or $A^{-1} = 1 - A$.

9. (a) $\alpha x + y + z = \alpha - 1$

$x + \alpha y + z = \alpha - 1;$

$x + y + z\alpha = \alpha - 1$

$$\Delta = \begin{vmatrix} \alpha & 1 & 1 \\ 1 & \alpha & 1 \\ 1 & 1 & \alpha \end{vmatrix}$$

$= \alpha(\alpha^2 - 1) - 1(\alpha - 1) + 1(1 - \alpha)$

$= \alpha(\alpha - 1)(\alpha + 1) - 1(\alpha - 1) - 1(\alpha - 1)$

For infinite solutions, $\Delta = 0$

$\Rightarrow (\alpha - 1)[\alpha^2 + \alpha - 1 - 1] = 0$

$\Rightarrow (\alpha - 1)[\alpha^2 + \alpha - 2] = 0$

$\Rightarrow \alpha = -2, 1;$

But $\alpha \neq 1$.

$\therefore \alpha = -2$

10. (d) Applying, $C_1 \to C_1 + C_2 + C_3$ we get

$$f(x) = \begin{vmatrix} 1+(a^2+b^2+c^2+2)x & (1+b^2)x & (1+c^2)x \\ 1+(a^2+b^2+c^2+2)x & 1+b^2x & (1+c^2x) \\ 1+(a^2+b^2+c^2+2)x & (1+b^2)x & 1+c^2x \end{vmatrix}$$

$$= \begin{vmatrix} 1 & (1+b^2)x & (1+c^2)x \\ 1 & 1+b^2x & (1+c^2x) \\ 1 & (1+b^2)x & 1+c^2x \end{vmatrix}$$

[As given that $a^2 + b^2 + c^2 = -2$]

$\therefore a^2 + b^2 + c^2 + 2 = 0$

Applying $R_1 \to R_1 - R_2,\ R_2 \to R_2 - R_3$

$$\therefore f(x) = \begin{vmatrix} 0 & x-1 & 0 \\ 0 & 1-x & x-1 \\ 1 & (1+b^2)x & 1+c^2x \end{vmatrix}$$

$f(x) = (x-1)^2$ Hence degree = 2.

11. (b) $\because a_1, a_2, a_3, \ldots\ldots$ are in G.P.

$\therefore$ Using $a_n = ar^{n-1}$, we get the given determinant, as

$$\begin{vmatrix} \log ar^{n-1} & \log ar^n & \log ar^{n+1} \\ \log ar^{n+2} & \log ar^{n+3} & \log ar^{n+4} \\ \log ar^{n+5} & \log ar^{n+6} & \log ar^{n+7} \end{vmatrix}$$

Operating $C_3 - C_2$ and $C_2 - C_1$ and using

$\log m - \log n = \log \frac{m}{n}$ we get

$$= \begin{vmatrix} \log ar^{n-1} & \log r & \log r \\ \log ar^{n+2} & \log r & \log r \\ \log ar^{n+5} & \log r & \log r \end{vmatrix}$$

$= 0$ (two columns being identical)

12. (b) $A^2 - B^2 = (A - B)(A + B)$

$A^2 - B^2 = A^2 + AB - BA - B^2 \Rightarrow AB = BA$

13. (d) $A = \begin{bmatrix} 1 & 2 \\ 3 & 4 \end{bmatrix}$ $B = \begin{bmatrix} a & 0 \\ 0 & b \end{bmatrix}$

$$AB = \begin{bmatrix} a & 2b \\ 3a & 4b \end{bmatrix}$$

$$BA = \begin{bmatrix} a & 0 \\ 0 & b \end{bmatrix}\begin{bmatrix} 1 & 2 \\ 3 & 4 \end{bmatrix} = \begin{bmatrix} a & 2a \\ 3b & 4b \end{bmatrix}$$

Hence, $AB = BA$ only when $a = b$

$\therefore$ There can be infinitely many B's for which $AB = BA$

14. (d) Given, $D = \begin{vmatrix} 1 & 1 & 1 \\ 1 & 1+x & 1 \\ 1 & 1 & 1+y \end{vmatrix}$

Apply $R_2 \to R_2 - R_1$ and $R \to R_3 - R_1$

$$\therefore D = \begin{vmatrix} 1 & 1 & 1 \\ 0 & x & 0 \\ 0 & 0 & y \end{vmatrix} = xy$$

Hence, D is divisible by both x and y

15. (a) $|A^2| = 25$

$\Rightarrow |A|^2 = 25$

$\Rightarrow (25\alpha)^2 = 25$

$\Rightarrow |\alpha| = \frac{1}{5}$

16. **(d)** Let $A=\begin{bmatrix} a & b \\ c & d \end{bmatrix}$ then $A^2 = I$

$\Rightarrow a^2+bc=1 \quad ab+bd=0$

$ac+cd=0 \quad bc+d^2=1$

From these four relations,

$a^2+bc=bc+d^2 \Rightarrow a^2=d^2$

and $b(a+d)=0=c(a+d) \Rightarrow a=-d$

We can take $a=1, b=0, c=0, d=-1$ as one possible set of values, then $A=\begin{bmatrix} 1 & 0 \\ 0 & -1 \end{bmatrix}$

Clearly $A \neq I$ and $A \neq -I$ and $\det A=-1$

$\therefore$ Statement 1 is true.

Also if $A \neq I$ then $tr(A)=0$

$\therefore$ Statement 2 is false.

17. **(d)** The given equations are

$-x+cy+bz=0$

$cx-y+az=0$

$bx+ay-z=0$

$\because x, y, z$ are not all zero

$\therefore$ The above system should not have unique (zero) solution

$$\Rightarrow \Delta=0 \Rightarrow \begin{vmatrix} -1 & c & b \\ c & -1 & a \\ b & a & -1 \end{vmatrix}=0$$

$\Rightarrow -1(1-a^2)-c(-c-ab)+b(ac+b)=0$

$\Rightarrow -1+a^2+b^2+c^2+2abc=0$

$\Rightarrow a^2+b^2+c^2+2abc=1$

18. **(c)** $\because$ All entries of square matrix A are integers, therefore all cofactors should also be integers.

If $\det A=\pm 1$ then A^{-1} exists. Also all entries of A^{-1} are integers.

19. **(d)** We know that $|adj\,(adj\,A)|=|\text{adj }A|^{2-1}$

$=|A|^{2-1}=|A|$

$\therefore$ Both the statements are true but statement-2 and statement -2 is a correct explantion for statement-1 .

20. **(b)**
$$\begin{vmatrix} a & a+1 & a-1 \\ -b & b+1 & b-1 \\ c & c-1 & c+1 \end{vmatrix}+\begin{vmatrix} a+1 & b+1 & c-1 \\ a-1 & b-1 & c+1 \\ (-1)^{n+2}a & (-1)^{n+1}b & (-1)^n c \end{vmatrix}=0$$

$$\Rightarrow \begin{vmatrix} a & a+1 & a-1 \\ -b & b+1 & b-1 \\ c & c-1 & c+1 \end{vmatrix}+\begin{vmatrix} a+1 & a-1 & (-1)^{n+2}a \\ b+1 & b-1 & (-1)^{n+1}b \\ c-1 & c+1 & (-1)^n c \end{vmatrix}=0$$

(Taking transpose of second determinant)

$C_1 \Leftrightarrow C_3$

$$\Rightarrow \begin{vmatrix} a & a+1 & a-1 \\ -b & b+1 & b-1 \\ c & c-1 & c+1 \end{vmatrix}-\begin{vmatrix} (-1)^{n+2}a & a-1 & a+1 \\ (-1)^{n+2}(-b) & b-1 & b+1 \\ (-1)^{n+2}c & c+1 & c-1 \end{vmatrix}=0$$

$C_2 \Leftrightarrow C_3$

$$\Rightarrow \begin{vmatrix} a & a+1 & a-1 \\ -b & b+1 & b-1 \\ c & c-1 & c+1 \end{vmatrix}+(-1)^{n+2}\begin{vmatrix} a & a+1 & a-1 \\ -b & b+1 & b-1 \\ c & c-1 & c+1 \end{vmatrix}=0$$

$$\Rightarrow \left[1+(-1)^{n+2}\right]\begin{vmatrix} a & a+1 & a-1 \\ -b & b+1 & b-1 \\ c & c-1 & c+1 \end{vmatrix}=0$$

$C_2-C_1,\ C_3-C_1$

$$\Rightarrow \left[1+(-1)^{n+2}\right]\begin{vmatrix} a & 1 & -1 \\ -b & 2b+1 & 2b-1 \\ c & -1 & 1 \end{vmatrix}=0$$

R_1+R_3

$$\Rightarrow \left[1+(-1)^{n+2}\right]\begin{vmatrix} a+c & 0 & 0 \\ -b & 2b+1 & 2b-1 \\ c & -1 & 1 \end{vmatrix}=0$$

$\Rightarrow [1+(-1)^{n+2}](a+c)(2b+1+2b-1)=0$

$\Rightarrow 4b(a+c)[1+(-1)^{n+2}]=0$

$\Rightarrow 1+(-1)^{n+2}=0$ as $b(a+c)\neq 0$

$\Rightarrow$ n should be an odd integer.

21. **(c)** $\begin{bmatrix} 1 & \ldots & \ldots \\ \ldots & 1 & \ldots \\ \ldots & \ldots & 1 \end{bmatrix}$ are 6 non-singular matrices because 6 blanks will be filled by 5 zeros and 1 one.

Similarly, $\begin{bmatrix} \ldots & \ldots & 1 \\ \ldots & 1 & \ldots \\ 1 & \ldots & \ldots \end{bmatrix}$ are 6 non-singular matrices.

So, required cases are more than 7, non-singular 3×3 matrices.

22. **(b)** Let $A=\begin{pmatrix} a & b \\ c & d \end{pmatrix}$ where a, b, c, d $\neq 0$

$$A^2=\begin{pmatrix} a & b \\ c & d \end{pmatrix}\begin{pmatrix} a & b \\ c & d \end{pmatrix}$$

$$\Rightarrow A^2=\begin{pmatrix} a^2+bc & ab+bd \\ ac+cd & bc+d^2 \end{pmatrix}$$

$\Rightarrow a^2+bc=1, bc+d^2=1$

$ab+bd=ac+cd=0$

$c \neq 0$ and $b \neq 0 \Rightarrow a+d=0 \Rightarrow \text{Tr}(A)=0$

$|A|=ad-bc=-a^2-bc=-1$

23. (c) $D=\begin{vmatrix}1 & 2 & 1\\ 2 & 3 & 1\\ 3 & 5 & 2\end{vmatrix}=0 \qquad D_1=\begin{vmatrix}3 & 2 & 1\\ 3 & 3 & 1\\ 1 & 5 & 2\end{vmatrix}\neq 0$

$\Rightarrow$ Given system, does not have any solution.

$\Rightarrow$ No solution

24. (a) $\Delta=0$

$$\Rightarrow \begin{vmatrix}4 & k & 2\\ k & 4 & 1\\ 2 & 2 & 1\end{vmatrix}=0$$

$\Rightarrow 4(4-2)-k(k-2)+2(2k-8)=0$

$\Rightarrow 8-k^2+2k+4k-16=0$

$k^2-6k+8=0$

$\Rightarrow (k-4)(k-2)=0,\ k=4,2$

25. (a) $\because$ A′ = A

B′ = B

Now (A(BA))′ = (BA)′A′

= (A′B′)A′ = (AB)A = A(BA)

Similarly ((AB)A)′ = (AB)A

So, A(BA) and (AB)A are symmetric matrices.

Again (AB)′ = B′A′ = BA

Now if BA = AB, then AB is symmetric matrix.

26. (a) $x-ky+z=0$

$kx+3y-kz=0$

$3x+y-z=0$

The given system of equations will have non trivial solution, if

$$\begin{vmatrix}1 & -k & 1\\ k & 3 & -k\\ 3 & 1 & -1\end{vmatrix}=0$$

$\Rightarrow 1(-3+k)+k(-k+3k)+1(k-9)=0$

$\Rightarrow k-3+2k^2+k-9=0$

$\Rightarrow k^2+k-6=0$

$\Rightarrow k=-3, k=2$

So the equation will have only trivial solution, when $k \in R-\{2,-3\}$

27. (d) **Statement-1 :** Determinant of skew symmetric matrix of odd order is zero.

Statement-2 : $\det\left(A^T\right)=\det(A)$.

det (– A) = – (– 1)^n det (A).

where A is a $n \times n$ order matrix.

28. (d) Let $Au_1=\begin{pmatrix}1\\0\\0\end{pmatrix}$ and $Au_2=\begin{pmatrix}0\\1\\0\end{pmatrix}$

Then, $Au_1+Au_2=\begin{pmatrix}1\\0\\0\end{pmatrix}+\begin{pmatrix}0\\1\\0\end{pmatrix}$

$\Rightarrow \quad A(u_1+u_2)=\begin{pmatrix}1\\1\\0\end{pmatrix}$ ...(1)

Also, $A=\begin{pmatrix}1 & 0 & 0\\ 2 & 1 & 0\\ 3 & 2 & 1\end{pmatrix}$

$\Rightarrow \quad |A|=1(1)-0(2)+0(4-3)=1$

We know,

$$A^{-1}=\frac{1}{|A|}adjA \Rightarrow A^{-1}=adj(A) \quad (\because |A|=1)$$

Now, from equation (1), we have

$$u_1+u_2=A^{-1}\begin{pmatrix}1\\1\\0\end{pmatrix} \quad \Rightarrow \quad u_1+u_2=A^{-1}$$

$$=\begin{bmatrix}1 & 0 & 0\\ -2 & 1 & 0\\ 1 & -2 & 1\end{bmatrix}\begin{pmatrix}1\\1\\0\end{pmatrix}=\begin{bmatrix}1\\-1\\-1\end{bmatrix}$$

29. (c) Given $P^3=Q^3$ ...(1)

and $P^2Q=Q^2P$ ...(2)

Subtracting (1) and (2), we get

$P^3-P^2Q=Q^3-Q^2P$

$\Rightarrow \quad P^2(P-Q)+Q^2(P-Q)=0$

$\Rightarrow \quad (P^2+Q^2)(P-Q)=0$

$\Rightarrow \quad |P^2+Q^2|=0$ as P ≠ Q

30. (b) $|P|=1(12-12)-\alpha(4-6)+3(4-6)=2\alpha-6$

Now, adj A = P $\Rightarrow$ |adj A| = |P|

$\Rightarrow |A|^2=|P|$

$\Rightarrow |P|=16$

$\Rightarrow 2\alpha-6=16$

$\Rightarrow \alpha=11$

Chapter 16 Applications of Derivatives

Section-A JEE Advanced/ IIT-JEE

(A) FILL IN THE BLANKS :

1. The larger of cos (ln θ) and ln (cos θ) if $e^{-\pi/2} < \theta < \frac{\pi}{2}$ is .... **(1983 - 1 Mark)**

2. The function $y = 2x^2 - \ln|x|$ is monotonically increasing for values of $x (\neq 0)$ satisfying the inequalities .... and monotonically decreasing for values of x satisfying the inequalities .... **(1983 - 2 Marks)**

3. The set of all x for which $ln(1+x) \leq x$ is equal to .... **(1987 - 2 Marks)**

4. Let P be a variable point on the ellipse $\frac{x^2}{a^2}+\frac{y^2}{b^2}=1$ with foci F_1 and F_2. If A is the area of the triangle PF_1F_2 then the maximum value of A is .... **(1994 - 2 Marks)**

5 Let C be the curve $y^3 - 3xy + 2 = 0$. If H is the set of points on the curve C where the tangent is horizontal and V is the set of the point on the curve C where the tangent is vertical then H=........... and V= .... **(1994 - 2 Marks)**

(B) TRUE / FALSE :

1. If $x - r$ is a factor of the polynomial $f(x) = a_n x^4 + ... + a_0$, repeated m times $(1 < m \leq n)$, then r is a root of $f'(x) = 0$ repeated m times. **(1983 - 1 Mark)**

2. For $0 < a < x$, the minimum value of the function $\log_a x + \log_x a$ is 2. **(1984 - 1 Mark)**

(C) MCQ's WITH ONE CORRECT ANSWER :

1. If $a + b + c = 0$, then the quadratic equation $3ax^2 + 2bx + c = 0$ has **(1983 - 1 Mark)**
   (a) at least one root in [0, 1]
   (b) one root in [2, 3] and the other in [–2, –1]
   (c) imaginary roots
   (d) none of these

2. AB is a diameter of a circle and C is any point on the circumference of the circle. Then **(1983 - 1 Mark)**
   (a) the area of ΔABC is maximum when it is isosceles
   (b) the area of ΔABC is minimum when it is isosceles
   (c) the perimeter of ΔABC is minimum when it is isosceles
   (d) none of these

3. The normal to the curve $x = a(\cos\theta + \theta\sin\theta)$, $y = a(\sin\theta - \theta\cos\theta)$ at any point 'θ' is such that **(1983 - 1 Mark)**
   (a) it makes a constant angle with the x– axis
   (b) it passes through the origin
   (c) it is at a constant distance from the origin
   (d) none of these

4. If $y = a\ln x + bx^2 + x$ has its extremum values at $x = -1$ and $x = 2$, then **(1983 - 1 Mark)**
   (a) $a = 2, b = -1$ (b) $a = 2, b = -\frac{1}{2}$
   (c) $a = -2, b = \frac{1}{2}$ (d) none of these

5. Which one of the following curves cut the parabola $y^2 = 4ax$ at right angles? **(1994)**
   (a) $x^2 + y^2 = a^2$ (b) $y = e^{-x/2a}$
   (c) $y = ax$ (d) $x^2 = 4ay$

6. The function defined by $f(x) = (x+2)e^{-x}$ is **(1994)**
   (a) decreasing for all x
   (b) decreasing in (–∞, –1) and increasing in (–1, ∞)
   (c) increasing for all x
   (d) decreasing in (–1, ∞) and increasing in (–∞, –1)

7. The function $f(x) = \frac{\ln(\pi + x)}{\ln(e + x)}$ is **(1995S)**
   (a) increasing on (0, ∞)
   (b) decreasing on (0, ∞)
   (c) increasing on (0, π/e), decreasing on (π/e, ∞)
   (d) decreasing on 0, π/e), increasing on (π/e, ∞)

8. On the interval [0, 1] the function $x^{25}(1-x)^{75}$ takes its maximum value at the point **(1995S)**
   (a) 0 (b) $\frac{1}{4}$
   (c) $\frac{1}{2}$ (d) $\frac{1}{3}$

9. The slope of the tangent to a curve $y = f(x)$ at $[x, f(x)]$ is $2x + 1$. If the curve passes through the point (1, 2), then the area bounded by the curve, the x-axis and the line $x = 1$ is **(1995S)**
   (a) $\frac{5}{6}$ (b) $\frac{6}{5}$
   (c) $\frac{1}{6}$ (d) 6

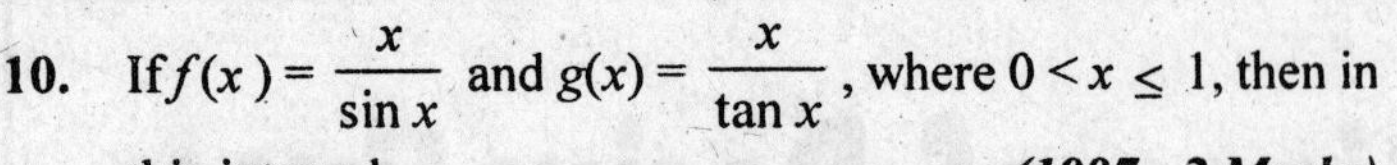

10. If $f(x) = \dfrac{x}{\sin x}$ and $g(x) = \dfrac{x}{\tan x}$, where $0 < x \le 1$, then in this interval *(1997 - 2 Marks)*
(a) both $f(x)$ and $g(x)$ are increasing functions
(b) both $f(x)$ and $g(x)$ are decreasing functions
(c) $f(x)$ is an increasing function
(d) $g(x)$ is an increasing function.

11. The function $f(x) = \sin^4 x + \cos^4 x$ increases if *(1999 - 2 Marks)*
(a) $0 < x < \pi/8$ (b) $\pi/4 < x < 3\pi/8$
(c) $3\pi/8 < x < 5\pi/8$ (d) $5\pi/8 < x < 3\pi/4$

12. Consider the following statments in S and R *(2000S)*
S: Both $\sin x$ and $\cos x$ are decreasing functions in the interval $\left(\dfrac{\pi}{2}, \pi\right)$
R: If a differentiable function decreases in an interval (a, b), then its derivative also decreases in (a, b).
Which of the following is true ?
(a) Both S and R are wrong
(b) Both S and R are correct, but R is not the correct explanation of S
(c) S is correct and R is the correct explanation for S
(d) S is correct and R is wrong

13. Let $f(x) = \int e^x (x-1)(x-2)dx$. Then f decreases in the interval *(2000S)*
(a) $(-\infty, -2)$ (b) $(-2, -1)$
(c) $(1, 2)$ (d) $(2, +\infty)$

14. If the normal to the curve $y = f(x)$ at the point (3,4) makes an angle $\dfrac{3\pi}{4}$ with the positive x-axis, then $f'(3) =$ *(2000S)*
(a) -1 (b) $-\dfrac{3}{4}$
(c) $\dfrac{4}{3}$ (d) 1

15. Let $f(x) = \begin{cases} |x|, & \text{for } 0 < |x| \le 2 \\ 1, & \text{for } x = 0 \end{cases}$ then at $x = 0$, f has *(2000S)*
(a) a local maximum (b) no local maximum
(c) a local minimum (d) no extremum

16. For all $x \in (0,1)$ *(2000S)*
(a) $e^x < 1 + x$ (b) $\log_e(1+x) < x$
(c) $\sin x > x$ (d) $\log_e x > x$

17. If $f(x) = xe^{x(1-x)}$, then $f(x)$ is *(2001S)*
(a) increasing on $[-1/2, 1]$ (b) decreasing on R
(c) increasing on R (d) decreasing on $[-1/2, 1]$

18. The triangle formed by the tangent to the curve $f(x) = x^2 + bx - b$ at the point (1, 1) and the coordinate axes, lies in the first quadrant. If its area is 2, then the value of b is *(2001S)*
(a) -1 (b) 3
(c) -3 (d) 1

19. Let $f(x) = (1 + b^2)x^2 + 2bx + 1$ and let $m(b)$ be the minimum value of $f(x)$. As b varies, the range of $m(b)$ is *(2001S)*
(a) [0, 1] (b) (0, 1/2]
(c) [1/2, 1] (d) (0, 1]

20. The length of a longest interval in which the function $3 \sin x - 4 \sin^3 x$ is increasing, is *(2002S)*
(a) $\dfrac{\pi}{3}$ (b) $\dfrac{\pi}{2}$
(c) $\dfrac{3\pi}{2}$ (d) π

21. The point(s) on the curve $y^3 + 3x^2 = 12y$ where the tangent is vertical, is (are) *(2002S)*
(a) $\left(\pm\dfrac{4}{\sqrt{3}}, -2\right)$ (b) $\left(\pm\sqrt{\dfrac{11}{3}}, 1\right)$
(c) (0,0) (d) $\left(\pm\dfrac{4}{\sqrt{3}}, 2\right)$

22. In [0,1] Lagranges Mean Value theorem is NOT applicable to *(2003S)*
(a) $f(x) = \begin{cases} \dfrac{1}{2} - x & x < \dfrac{1}{2} \\ \left(\dfrac{1}{2} - x\right)^2 & x \ge \dfrac{1}{2} \end{cases}$
(b) $f(x) = \begin{cases} \dfrac{\sin x}{x}, & x \ne 0 \\ 1, & x = 0 \end{cases}$
(c) $f(x) = x|x|$
(d) $f(x) = |x|$

23. Tangent is drawn to ellipse
$\dfrac{x^2}{27} + y^2 = 1$ at $\left(3\sqrt{3}\cos\theta, \sin\theta\right)$ (where $\theta \in (0, \pi/2)$).
Then the value of θ such that sum of intercepts on axes made by this tangent is minimum, is *(2003S)*
(a) $\pi/3$ (b) $\pi/6$
(c) $\pi/8$ (d) $\pi/4$

24. If $f(x) = x^3 + bx^2 + cx + d$ and $0 < b^2 < c$, then in $(-\infty, \infty)$ *(2004S)*
(a) $f(x)$ is a strictly increasing function
(b) $f(x)$ has a local maxima
(c) $f(x)$ is a strictly decreasing function
(d) $f(x)$ is bounded

25. If $f(x) = x^\alpha \log x$ and $f(0) = 0$, then the value of α for which Rolle's theorem can be applied in [0, 1] is *(2004S)*
(a) -2 (b) -1
(c) 0 (d) 1/2

26. If $P(x)$ is a polynomial of degree less than or equal to 2 and S is the set of all such polynomials so that $P(0) = 0$, $P(1) = 1$ and $P'(x) > 0 \ \forall\, x \in [0, 1]$, then *(2005S)*

(a) $S = \phi$
(b) $S = ax + (1-a)x^2 \ \forall \ a \in (0, 2)$
(c) $S = ax + (1-a)x^2 \ \forall \ a \in (0, \infty)$
(d) $S = ax + (1-a)x^2 \ \forall \ a \in (0, 1)$

27. The tangent to the curve $y = e^x$ drawn at the point (c, e^c) intersects the line joining the points $(c-1, e^{c-1})$ and $(c+1, e^{c+1})$ ***(2007 -3 marks)***

(a) on the left of $x = c$ (b) on the right of $x = c$
(c) at no point (d) at all points

28. Consider the two curves $C_1 : y^2 = 4x$, $C_2 : x^2 + y^2 - 6x + 1 = 0$. Then, ***(2008)***

(a) C_1 and C_2 touch each other only at one point.
(b) C_1 and C_2 touch each other exactly at two points
(c) C_1 and C_2 intersect (but do not touch) at exactly two points
(d) C_1 and C_2 neither intersect nor touch each other

29. The total number of local maxima and local minima of the function $f(x) = \begin{cases} (2+x)^3, & -3 < x \le -1 \\ x^{2/3}, & -1 < x < 2 \end{cases}$ is ***(2008)***

(a) 0 (b) 1
(c) 2 (d) 3

30. Let the function $g : (-\infty, \infty) \to \left(-\frac{\pi}{2}, \frac{\pi}{2}\right)$ be given by

$g(u) = 2\tan^{-1}(e^u) - \frac{\pi}{2}$. Then, g is ***(2008)***

(a) even and is strictly increasing in $(0, \infty)$
(b) odd and is strictly decreasing in $(-\infty, \infty)$
(c) odd and is strictly increasing in $(-\infty, \infty)$
(d) neither even nor odd, but is strictly increasing in $(-\infty, \infty)$

(D) MCQ's WITH ONE OR MORE THAN ONE CORRECT :

1. Let $P(x) = a_0 + a_1x^2 + a_2x^4 + ... + a_nx^{2n}$ be a polynomial in a real variable x with

$0 < a_0 < a_1 < a_2 < ..... < a_n$. The function $P(x)$ has

(a) neither a maximum nor a minimum ***(1986 - 2 Marks)***
(b) only one maximum
(c) only one minimum
(d) only one maximum and only one minimum
(e) none of these.

2. If the line $ax + by + c = 0$ is a normal to the curve $xy = 1$, then ***(1986 - 2 Marks)***

(a) $a > 0, b > 0$ (b) $a > 0, b < 0$
(c) $a < 0, b > 0$ (d) $a < 0, b < 0$
(e) none of these.

3. The smallest positive root of the equation, $\tan x - x = 0$ lies in ***(1987 - 2 Marks)***

(a) $\left(0, \frac{\pi}{2}\right)$ (b) $\left(\frac{\pi}{2}, \pi\right)$

(c) $\left(\pi, \frac{3\pi}{2}\right)$ (d) $\left(\frac{3\pi}{2}, 2\pi\right)$

(e) None of these

4. Let f and g be increasing and decreasing functions, respectively from $[0, \infty)$ to $[0, \infty)$. Let $h(x) = f(g(x))$. If $h(0) = 0$, then $h(x) - h(1)$ is ***(1987 - 2 Marks)***

(a) always zero (b) always negative
(c) always positive (d) strictly increasing
(e) None of these.

5. If $f(x) = \begin{cases} 3x^2+12x-1, & -1 \le x \le 2 \\ 37-x & 2 < x \le 3 \end{cases}$ then: ***(1993 - 2 Marks)***

(a) $f(x)$ is increasing on $[-1, 2]$
(b) $f(x)$ is continues on $[-1, 3]$
(c) $f'(2)$ does not exist
(d) $f(x)$ has the maximum value at $x = 2$

6. Let $h(x) = f(x) - (f(x))^2 + (f(x))^3$ for every real number x. Then ***(1998 - 2 Marks)***

(a) h is increasing whenever f is increasing
(b) h is increasing whenever f is decreasing
(c) h is decreasing whenever f is decreasing
(d) nothing can be said in general.

7. If $f(x) = \frac{x^2 - 1}{x^2 + 1}$, for every real number x, then the minimum value of f ***(1998 - 2 Marks)***

(a) does not exist because f is unbounded
(b) is not attained even though f is bounded
(c) is equal to 1
(d) is equal to -1

8. The number of values of x where the function

$f(x) = \cos x + \cos(\sqrt{2}x)$ attains its maximum is ***(1998 - 2 Marks)***

(a) 0 (b) 1
(c) 2 (d) infinite

9. The function $f(x) = \int_{-1}^{x} t(e^t - 1)(t-1)(t-2)^3 (t-3)^5 \, dt$ has a local minimum at $x =$ ***(1999 - 3 Marks)***

(a) 0 (b) 1
(c) 2 (d) 3

10. $f(x)$ is cubic polynomial with $f(2) = 18$ and $f(1) = -1$. Also $f(x)$ has local maxima at $x = -1$ and $f'(x)$ has local minima at $x = 0$, then ***(2006 - 5M, -1)***

(a) the distance between $(-1, 2)$ and $(a f(a))$, where $x = a$ is the point of local minima is $2\sqrt{5}$
(b) $f(x)$ is increasing for $x \in [1, 2\sqrt{5}]$
(c) $f(x)$ has local minima at $x = 1$
(d) the value of $f(0) = 15$

11. Let $f(x) = \begin{cases} e^x, & 0 \le x \le 1 \\ 2 - e^{x-1}, & 1 < x \le 2 \\ x - e, & 2 < x \le 3 \end{cases}$ and $g(x) = \int_0^x f(t)\, dt, x \in [1, 3]$

then $g(x)$ has ***(2006 - 5M, -1)***

(a) local maxima at $x = 1 + \ln 2$ and local minima at $x = e$
(b) local maxima at $x = 1$ and local minima at $x = 2$
(c) no local maxima
(d) no local minima

12. For the function

$$f(x) = x\cos\frac{1}{x},\ x \geq 1, \qquad \textbf{(2009)}$$

(a) for at least one x in the interval $[1, \infty), f(x+2)-f(x)<2$

(b) $\lim_{x\to\infty} f'(x) = 1$

(c) for all x in the interval $[1, \infty), f(x+2)-f(x)>2$

(d) $f'(x)$ is strictly decreasing in the interval $[1, \infty)$

13. If $f(x) = \int_0^x e^{t^2}(t-2)(t-3)dt$ for all $x \in (0,\infty)$, then ***(2012)***

(a) f has a local maximum at $x = 2$

(b) f is decreasing on (2, 3)

(c) there exists some $c \in (0, \infty)$, such that $f''(c) = 0$

(d) f has a local minimum at $x = 3$

14. A rectangular sheet of fixed perimeter with sides having their lengths in the ratio 8 : 15 is converted into an open rectangular box by folding after removing squares of equal area from all four corners. If the total area of removed squares is 100, the resulting box has maximum volume. Then the lengths of the sides of the rectangular sheet are ***(JEE Adv. 2013)***

(a) 24 (b) 32

(c) 45 (d) 60

E SUBJECTIVE PROBLEMS :

1. Prove that the minimum value of $\frac{(a+x)(b+x)}{(c+x)}$, $a, b > c, x > -c$ is $(\sqrt{a-c}+\sqrt{b-c})^2$. ***(1979)***

2. Let x and y be two real variables such that $x > 0$ and $xy = 1$. Find the minimum value of $x+y$. ***(1981 - 2 Marks)***

3. For all x in [0, 1], let the second derivative $f''(x)$ of a function $f(x)$ exist and satisfy $|f''(x)| < 1$. If $f(0) = f(1)$, then show that $|f'(x)| < 1$ for all x in [0, 1]. ***(1981 - 4 Marks)***

4. Use the function $f(x) = x^{1/x}, x > 0$. to determine the bigger of the two numbers e^π and π^e ***(1981 - 4 Marks)***

5. If $f(x)$ and $g(x)$ are differentiable function for $0 \leq x \leq 1$ such that $f(0) = 2, g(0) = 0, f(1) = 6; g(1) = 2$, then show that there exist c satisfying $0 < c < 1$ and $f'(c) = 2g'(c)$. ***(1982 - 2 Marks)***

6. Find the shortest distance of the point $(0, c)$ from the parabola $y = x^2$ where $0 \leq c \leq 5$. ***(1982 - 2 Marks)***

7. If $ax^2 + \frac{b}{x} \geq c$ for all positive x where $a > 0$ and $b > 0$ show that $27ab^2 \geq 4c^3$. ***(1982 - 2 Marks)***

8. Show that $1 + x\ln(x+\sqrt{x^2+1}) \geq \sqrt{1+x^2}$ for all $x \geq 0$ ***(1983 - 2 Marks)***

9. A swimmer S is in the sea at a distance d km from the closest point A on a straight shore. The house of the swimmer is on the shore at a distance L km from A. He can swim at a speed of U km/hr and walk at a speed of V km/hr ($V > U$). At what point on the shore should he land so that he reaches his house in the shortest possible time ? ***(1983 - 2 Marks)***

10. Find the coordinates of the point on the curve $y = \frac{x}{1+x^2}$ where the tangent to the curve has the greatest slope. ***(1984 - 4 Marks)***

11. Find all the tangents to the curve $y = \cos(x+y), -2\pi \leq x \leq 2\pi$, that are parallel to the line $x + 2y = 0$. ***(1985 - 5 Marks)***

12. Let $f(x) = \sin^3 x + \lambda\sin^2 x, \ -\frac{\pi}{2} < x < \frac{\pi}{2}$. Find the intervals in which λ should lie in order that $f(x)$ has exactly one minimum and exactly one maximum. ***(1985 - 5 Marks)***

13. Let $A(p^2, -p)$, $B(q^2, q)$, $C(r^2, -r)$ be the vertices of the triangle ABC. A parallelogram $AFDE$ is drawn with vertices D, E and F on the line segments BC, CA and AB respectively. Using calculus, show that maximum area of such a parallelogram is

$$\frac{1}{4}(p+q)(q+r)(p-r) \qquad \textbf{(1986 - 5 Marks)}$$

14. Find the point on the curve $4x^2 + a^2y^2 = 4a^2$, $4 < a^2 < 8$ that is farthest from the point $(0, -2)$. ***(1987 - 4 Marks)***

15. Investigate for maxima and minima the function ***(1988 - 5 Marks)***

$$f(x) = \int_1^x [2(t-1)(t-2)^3 + 3(t-1)^2(t-2)^2]dt$$

16. Find all maxima and minima of the function

$$y = x(x-1)^2, 0 \leq x \leq 2 \qquad \textbf{(1989 - 5 Marks)}$$

Also determine the area bounded by the curve $y = x(x-1)^2$, the y-axis and the line $y = 2$.

17. Show that $2\sin x + \tan x \geq 3x$ where $0 \leq x < \frac{\pi}{2}$. ***(1990 - 4 Marks)***

18. A point P is given on the circumference of a circle of radius r. Chord QR is parallel to the tangent at P. Determine the maximum possible area of the triangle PQR. ***(1990 - 4 Marks)***

19. A window of perimeter P (including the base of the arch) is in the form of a rectangle surmounded by a semi circle. The semi- circular portion is fitted with coloured glass while the rectangular part is fitted with clear glass transmits three times as much light per square meter as the coloured glass does.

What is the ratio for the sides of the rectangle so that the window transmits the maximum light? ***(1991 - 4 Marks)***

20. A cubic $f(x)$ vanishes at $x = 2$ and has relative minimum / maximum at $x = -1$ and $x = \frac{1}{3}$ if $\int_{-1}^{1} f\,dx = \frac{14}{3}$, find the cubic $f(x)$. ***(1992 - 4 Marks)***

21. What normal to the curve $y = x^2$ forms the shortest chord? ***(1992 - 6 Marks)***

22. Find the equation of the normal to the curve

$y = (1+x)^y + \sin^{-1}(\sin^2 x)$ at $x = 0$ ***(1993 - 3 Marks)***

23. Let $f(x) = \begin{cases} -x^3 + \dfrac{(b^3 - b^2 + b - 1)}{(b^2 + 3b + 2)}, & 0 \le x < 1 \\ 2x - 3, & 1 \le x \le 3 \end{cases}$ ***(1993 - 5 Marks)***

Find all possible real values of b such that $f(x)$ has the smallest value at $x = 1$.

24. The curve $y = ax^3 + bx^2 + cx + 5$, touches the x-axis at $P(-2, 0)$ and cuts the y axis at a point Q, where its gradient is 3. Find a, b, c. ***(1994 - 5 Marks)***

25. The circle $x^2 + y^2 = 1$ cuts the x-axis at P and Q. Another circle with centre at Q and variable radius intersects the first circle at R above the x-axis and the line segment PQ at S. Find the maximum area of the triangle QSR. ***(1994 - 5 Marks)***

26. Let (h, k) be a fixed point, where $h > 0, k > 0$. A straight line passing through this point cuts the positive direction of the coordinate axes at the points P and Q. Find the minimum area of the triangle OPQ, O being the origin. ***(1995 - 5 Marks)***

27. A curve $y = f(x)$ passes through the point $P(1,1)$. The normal to the curve at P is $a(y-1) + (x-1) = 0$ If the slope of the tangent at any point on the curve is proportional to the ordinate of the point, determine the equation of the curve. Also obtain the area bounded by the y-axis, the curve and the normal to the curve at P. ***(1996 - 5 Marks)***

28. Determine the points of maxima and minima of the function

$f(x) = \frac{1}{8} \ln x - bx + x^2, x > 0$, where $b \ge 0$ is a constant. ***(1996 - 5 Marks)***

29. Let $f(x) = \begin{cases} xe^{ax}, & x \le 0 \\ x + ax^2 - x^3, & x > 0 \end{cases}$ ***(1996 - 3 Marks)***

Where a is a positive constant. Find the interval in which $f'(x)$ is increasing.

30. Let $a + b = 4$, where $a < 2$, and let $g(x)$ be a differentiable function.

If $\frac{dg}{dx} > 0$ for all x, prove that $\int_0^a g(x)\,dx + \int_0^b g(x)dx$ increases as $(b - a)$ increases. ***(1997 - 5 Marks)***

31. Suppose $f(x)$ is a function satisfying the following conditions ***(1998 - 8 Marks)***

(a) $f(0) = 2, f(1) = 1$,

(b) f has a minimum value at $x = 5/2$, and

(c) for all x,

$$f'(x) = \begin{vmatrix} 2ax & 2ax-1 & 2ax+b+1 \\ b & b+1 & -1 \\ 2(ax+b) & 2ax+2b+1 & 2ax+b \end{vmatrix}.$$

where a, b are some constants. Determine the constants a, b and the function $f(x)$.

32. A curve C has the property that if the tangent drawn at any point P on C meets the co-ordinate axes at A and B, then P is the mid-point of AB. The curve passes through the point (1, 1). Determine the equation of the curve. ***(1998 - 8 Marks)***

33. Suppose $p(x) = a_0 + a_1x + a_2x^2 + ... + a_nx^n$. If $|p(x)| \le |e^{x-1} - 1|$ for all $x \ge 0$, prove that

$|a_1 + 2a_2 + ... + na_n| \le 1$. ***(2000 - 5 Marks)***

34. Let $-1 \le p \le 1$. Show that the equation $4x^3 - 3x - p = 0$ has a unique root in the interval[1/2, 1] and identify it. ***(2001 - 5 Marks)***

35. Find a point on the curve $x^2 + 2y^2 = 6$ whose distance from the line $x + y = 7$, is minimum. ***(2003 - 2 Marks)***

36. Using the relation $2(1 - \cos x) < x^2, x \ne 0$ or otherwise, prove that $\sin(\tan x) \ge x, \forall\, x \in \left[0, \frac{\pi}{4}\right]$ ***(2003 - 4 Marks)***

37. If the function f: [0,4] $\rightarrow R$ is differentiable then show that

(i) For $a, b \in (0,4)$, $(f(4))^2 - (f(0))^2 = 8f'(a)f(b)$

(ii) $\int_0^4 f(t)dt = 2[\alpha f(\alpha^2) + \beta f(\beta^2)]\ \forall\ 0 < \alpha, \beta < 2$

(2003 - 4 Marks)

38. If $P(1) = 0$ and $\frac{dP(x)}{dx} > P(x)$ for all $x \ge 1$ then prove that $P(x) > 0$ for all $x > 1$. ***(2003 - 4 Marks)***

39. Using Rolle's theorem, prove that there is at least one root in $(45^{1/100}, 46)$ of the polynomial

$P(x) = 51x^{101} - 2323(x)^{100} - 45x + 1035$. ***(2004 - 2 Marks)***

40. Prove that for $x \in \left[0, \frac{\pi}{2}\right]$, $\sin x + 2x \ge \frac{3x(x+1)}{\pi}$. Explain the identity if any used in the proof. ***(2004 - 4 Marks)***

41. If $|f(x_1) - f(x_2)| < (x_1 - x_2)^2$, for all $x_1, x_2 \in R$. Find the equation of tengent to the cuve $y = f(x)$ at the point (1, 2). ***(2005 - 2 Marks)***

42. If $p(x)$ be a polynomial of degree 3 satisfying $p(-1) = 10, p(1) = -6$ and $p(x)$ has maxima at $x = -1$ and $p'(x)$ has minima at $x = 1$. Find the distance between the local maxima and local minima of the curve. ***(2005 - 4 Marks)***

43. For a twice differentiable function $f(x)$, $g(x)$ is defined as $g(x) = (f'(x)^2 + f''(x))\, f(x)$ on $[a, e]$. If for $a < b < c < d < e, f(a) = 0, f(b) = 2, f(c) = -1, f(d) = 2, f(e) = 0$ then find the minimum number of zeros of $g(x)$. ***(2006 - 6M)***

F MATCH THE FOLLOWING :

DIRECTIONS (Q. 1) : *Each question contains statements given in two columns, which have to be matched. The statements in Column-I are labelled A, B, C and D, while the statements in Column-II are labelled p, q, r, s and t. Any given statement in Column-I can have correct matching with ONE OR MORE statement(s) in Column-II. The appropriate bubbles corresponding to the answers to these questions have to be darkened as illustrated in the following example :*
If the correct matches are A-p, s and t; B-q and r; C-p and q; and D-s then the correct darkening of bubbles will look like the given.

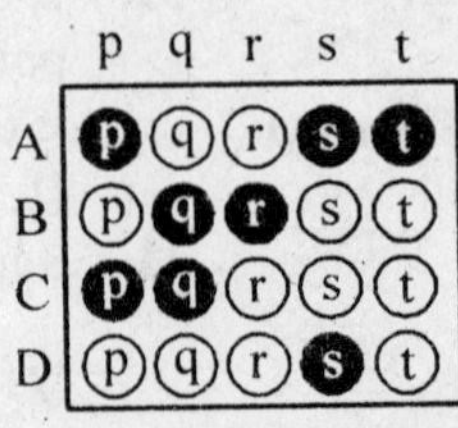

1. In this questions there are entries in columns I and II. Each entry in column I is related to exactly one entry in column II. Write the correct letter from column II against the entry number in column I in your answer book.

Let the functions defined in column I have domain $\left(-\frac{\pi}{2}, \frac{\pi}{2}\right)$ **(1992 - 2 Marks)**

Column I	Column II
(A) $x + \sin x$	(p) increasing
(B) $\sec x$	(q) decreasing
	(r) neither increasing nor decreasing

G COMPREHENSION BASED QUESTIONS :

Passage - 1

If a continuous function f defined on the real line R, assumes positive and negative values in R then the equation $f(x) = 0$ has a root in R. For example, if it is known that a continuous function f on R is positive at some point and its minimum value is negative then the equation $f(x) = 0$ has a root in R.

Consider $f(x) = ke^x - x$ for all real x where k is a real constant.

1. The line $y = x$ meets $y = ke^x$ for $k \le 0$ at **(2007 - 4 marks)**
   (a) no point (b) one point
   (c) two points (d) more than two points

2. The positive value of k for which $ke^x - x = 0$ has only one root is **(2007 - 4 marks)**
   (a) $\frac{1}{e}$ (b) 1
   (c) e (d) $\log_e 2$

3. For $k > 0$, the set of all values of k for which $ke^x - x = 0$ has two distinct roots is **(2007 - 4 marks)**
   (a) $\left(0, \frac{1}{e}\right)$ (b) $\left(\frac{1}{e}, 1\right)$
   (c) $\left(\frac{1}{e}, \infty\right)$ (d) $(0, 1)$

Passage - 2

Let $f(x) = (1-x)^2 \sin^2 x + x^2$ for all $x \in IR$ and let

$$g(x) = \int_1^x \left(\frac{2(t-1)}{t+1} - \ln t\right) f(t)\,dt \text{ for all } x \in (1, \infty).$$ **(2012)**

4. Consider the statements:
   P : There exists some $x \in$ R such that $f(x) + 2x = 2(1 + x^2)$
   Q : There exists some $x \in$ R such that $2f(x) + 1 = 2x(1 + x)$
   Then
   (a) both P and Q are true
   (b) P is true and Q is false
   (c) P is false and Q is true
   (d) both P and Q are false

5. Which of the following is true?
   (a) g is increasing on $(1, \infty)$
   (b) g is decreasing on $(1, \infty)$
   (c) g is increasing on $(1, 2)$ and decreasing on $(2, \infty)$
   (d) g is decreasing on $(1, 2)$ and increasing on $(2, \infty)$

Passage - 3

Let f : [0, 1] → $\mathbb{R}$ (the set of all real numbers) be a function. Suppose the function f is twice differentiable, $f(0) = f(1) = 0$ and satisfies $f''(x) - 2f'(x) + f(x) \ge e^x$, $x \in [0, 1]$.

6. Which of the following is true for $0 < x < 1$? **(JEE Adv. 2013)**
   (a) $0 < f(x) < \infty$ (b) $-\frac{1}{2} < f(x) < \frac{1}{2}$
   (c) $-\frac{1}{4} < f(x) < 1$ (d) $-\infty < f(x) < 0$

7. If the function $e^{-x} f(x)$ assumes its minimum in the interval $[0, 1]$ at $x = \frac{1}{4}$, which of the following is true?

(a) $f'(x) < f(x), \frac{1}{4} < x < \frac{3}{4}$ *(JEE Adv. 2013)*

(b) $f'(x) > f(x), 0 < x < \frac{1}{4}$

(c) $f'(x) < f(x), 0 < x < \frac{1}{4}$

(d) $f'(x) < f(x), \frac{3}{4} < x < 1$

I INTEGER VALUE CORRECT TYPE :

1. The maximum value of the function $f(x) = 2x^3 - 15x^2 + 36x - 48$ on the set $A = \{x \mid x^2 + 20 \leq 9x\}$ is *(2009)*

2. Let $p(x)$ be a polynomial of degree 4 having extremum at $x = 1, 2$ and $\lim_{x \to 0}\left(1 + \frac{p(x)}{x^2}\right) = 2$.
Then the value of $p(2)$ is *(2009)*

3. Let f be a real-valued differentiable function on **R** (the set of all real numbers) such that $f(1) = 1$. If the y-intercept of the tangent at any point $P(x, y)$ on the curve $y = f(x)$ is equal to the cube of the abscissa of P, then find the value of $f(-3)$ *(2010)*

4. Let f be a function defined on **R** (the set of all real numbers) such that $f'(x) = 2010(x-2009)(x-2010)^2(x-2011)^3(x-2012)^4$ for all $x \in \mathbf{R}$.
If g is a function defined on **R** with values in the interval $(0, \infty)$ such that
$f(x) = ln(g(x))$, for all $x \in R$
then the number of points in R at which g has a local maximum is *(2010)*

5. let $f: IR \to IR$ be defined as $f(x) = |x| + |x^2 - 1|$. The total number of points at which f attains either a local maximum or a local minimum is *(2012)*

6. Let $p(x)$ be a real polynomial of least degree which has a local maximum at $x = 1$ and a local minimum at $x = 3$. If $p(1) = 6$ and $p(3) = 2$, then $p'(0)$ is *(2012)*

7. A vertical line passing through the point (h, 0) intersects the ellipse $\frac{x^2}{4} + \frac{y^2}{3} = 1$ at the points P and Q. Let the tangents to the ellipse at P and Q meet at the point R. If $\Delta(h)$ = area of the triangle PQR, $\Delta_1 = \max_{1/2 \leq h \leq 1} \Delta(h)$ and $\Delta_2 = \min_{1/2 \leq h \leq 1} \Delta(h)$, then $\frac{8}{\sqrt{5}}\Delta_1 - 8\Delta_2 =$ *(JEE Adv. 2013)*

Section-B JEE Main/ AIEEE

1. The maximum distance from origin of a point on the curve $x = a \sin t - b \sin\left(\frac{at}{b}\right)$
$y = a \cos t - b \cos\left(\frac{at}{b}\right)$, both $a, b > 0$ is

(a) $a - b$ (b) $a + b$ **[2002]**

(c) $\sqrt{a^2 + b^2}$ (d) $\sqrt{a^2 - b^2}$

2. If $2a + 3b + 6c = 0$, $(a, b, c \in R)$ then the quadratic equation $ax^2 + bx + c = 0$ has **[2002]**

(a) at least one root in [0, 1]

(b) at least one root in [2, 3]

(c) at least one root in [4, 5]

(d) none of these

3. If the function $f(x) = 2x^3 - 9ax^2 + 12a^2x + 1$, where $a > 0$, attains its maximum and minimum at p and q respectively such that $p^2 = q$, then a equals **[2003]**

(a) $\frac{1}{2}$ (b) 3

(c) 1 (d) 2

4. A point on the parabola $y^2 = 18x$ at which the ordinate increases at twice the rate of the abscissa is **[2004]**

(a) $\left(\frac{9}{8}, \frac{9}{2}\right)$ (b) $(2, -4)$

(c) $\left(\frac{-9}{8}, \frac{9}{2}\right)$ (d) $(2, 4)$

5. A function $y = f(x)$ has a second order derivative $f''(x) = 6(x-1)$. If its graph passes through the point (2,1) and at that point the tangent to the graph is $y = 3x - 5$, then the function is **[2004]**

(a) $(x+1)^2$ (b) $(x-1)^3$

(c) $(x+1)^3$ (d) $(x-1)^2$

6. The normal to the curve $x = a(1 + \cos\theta), y = a \sin\theta$ at 'θ' always passes through the fixed point **[2004]**

(a) (a, a) (b) $(0, a)$

(c) $(0, 0)$ (d) $(a, 0)$

7. If $2a+3b+6c=0$, then at least one root of the equation $ax^2+bx+c=0$ lies in the interval **[2004]**
(a) $(1,3)$ (b) $(1,2)$
(c) $(2,3)$ (d) $(0,1)$

8. Area of the greatest rectangle that can be inscribed in the ellipse $\frac{x^2}{a^2}+\frac{y^2}{b^2}=1$ is **[2005]**
(a) $2ab$ (b) ab
(c) $\sqrt{ab}$ (d) $\frac{a}{b}$

9. The normal to the curve **[2005]**
$x=a(\cos\theta+\theta\sin\theta), y=a(\sin\theta-\theta\cos\theta)$ at any point θ' is such that
(a) it passes through the origin
(b) it makes an angle $\frac{\pi}{2}+\theta$ with the x- axis
(c) it passes through $\left(a\frac{\pi}{2},-a\right)$
(d) It is at a constant distance from the origin

10. A spherical iron ball 10 cm in radius is coated with a layer of ice of uniform thickness that melts at a rate of 50 cm^3/min. When the thickness of ice is 5 cm, then the rate at which the thickness of ice decreases is **[2005]**
(a) $\frac{1}{36\pi}$ cm/min. (b) $\frac{1}{18\pi}$ cm/min.
(c) $\frac{1}{54\pi}$ cm/min. (d) $\frac{5}{6\pi}$ cm/min

11. If the equation $a_n x^n + a_{n-1}x^{n-1} + ... + a_1 x = 0$
$a_1 \neq 0, n \geq 2$, has a positive root $x=\alpha$, then the equation
$na_n x^{n-1} + (n-1)a_{n-1}x^{n-2} + ... + a_1 = 0$ has a positive root, which is **[2005]**
(a) greater than α
(b) smaller than α
(c) greater than or equal to α
(d) equal to α

12. The function $f(x)=\frac{x}{2}+\frac{2}{x}$ has a local minimum at **[2006]**
(a) $x=2$ (b) $x=-2$
(c) $x=0$ (d) $x=1$

13. A triangular park is enclosed on two sides by a fence and on the third side by a straight river bank. The two sides having fence are of same length x. The maximum area enclosed by the park is **[2006]**
(a) $\frac{3}{2}x^2$ (b) $\sqrt{\frac{x^3}{8}}$
(c) $\frac{1}{2}x^2$ (d) πx^2

14. A value of c for which conclusion of Mean Value Theorem holds for the function $f(x)=\log_e x$ on the interval [1, 3] is **[2007]**
(a) $\log_3 e$ (b) $\log_e 3$
(c) $2\log_3 e$ (d) $\frac{1}{2}\log_3 e$

15. The function $f(x)=\tan^{-1}(\sin x+\cos x)$ is an increasing function in **[2007]**
(a) $\left(0,\frac{\pi}{2}\right)$ (b) $\left(-\frac{\pi}{2},\frac{\pi}{2}\right)$
(c) $\left(\frac{\pi}{4},\frac{\pi}{2}\right)$ (d) $\left(-\frac{\pi}{2},\frac{\pi}{4}\right)$

16. If p and q are positive real numbers such that $p^2+q^2=1$, then the maximum value of $(p+q)$ is **[2007]**
(a) $\frac{1}{2}$ (b) $\frac{1}{\sqrt{2}}$
(c) $\sqrt{2}$ (d) 2.

17. Suppose the cubic x^3-px+q has three distinct real roots where $p>0$ and $q>0$. Then which one of the following holds? **[2008]**
(a) The cubic has minima at $\sqrt{\frac{p}{3}}$ and maxima at $-\sqrt{\frac{p}{3}}$
(b) The cubic has minima at $-\sqrt{\frac{p}{3}}$ and maxima at $\sqrt{\frac{p}{3}}$
(c) The cubic has minima at both $\sqrt{\frac{p}{3}}$ and $-\sqrt{\frac{p}{3}}$
(d) The cubic has maxima at both $\sqrt{\frac{p}{3}}$ and $-\sqrt{\frac{p}{3}}$

18. How many real solutions does the equation $x^7+14x^5+16x^3+30x-560=0$ have? **[2008]**
(a) 7 (b) 1 (c) 3 (d) 5

19. Let $f(x)=x|x|$ and $g(x)=\sin x$.
Statement-1 : gof is differentiable at $x=0$ and its derivative is continuous at that point.
Statement-2 : gof is twice differentiable at $x=0$. **[2009]**
(a) Statement-1 is true, Statement-2 is true; Statement-2 is not a correct explanation for Statement-1.
(b) Statement-1 is true, Statement-2 is false.
(c) Statement-1 is false, Statement-2 is true.
(d) Statement-1 is true, Statement-2 is true; Statement-2 is a correct explanation for Statement-1.

20. Given $P(x) = x^4 + ax^3 + bx^2 + cx + d$ such that $x = 0$ is the only real root of $P'(x) = 0$. If $P(-1) < P(1)$, then in the interval $[-1, 1]$: **[2009]**
(a) $P(-1)$ is not minimum but $P(1)$ is the maximum of P
(b) $P(-1)$ is the minimum but $P(1)$ is not the maximum of P
(c) Neither $P(-1)$ is the minimum nor $P(1)$ is the maximum of P
(d) $P(-1)$ is the minimum and $P(1)$ is the maximum of P

21. The equation of the tangent to the curve $y = x + \frac{4}{x^2}$, that is parallel to the x-axis, is **[2010]**
(a) $y=1$ (b) $y=2$
(c) $y=3$ (d) $y=0$

22. Let $f: \mathbf{R} \to \mathbf{R}$ be defined by
$$f(x) = \begin{cases} k-2x, & if \ x \le -1 \\ 2x+3, & if \ x > -1 \end{cases}$$
If f has a local minimum at $x = -1$, then a possible value of k is **[2010]**
(a) 0 (b) $-\frac{1}{2}$ (c) -1 (d) 1

23. Let $f: \mathbf{R} \to \mathbf{R}$ be a continuous function defined by
$$f(x) = \frac{1}{e^x + 2e^{-x}}$$ **[2010]**

Statement -1 : $f(c) = \frac{1}{3}$, for some $c \in \mathbf{R}$.

Statement -2 : $0 < f(x) \le \frac{1}{2\sqrt{2}}$, for all $x \in \mathbf{R}$

(a) Statement -1 is true, Statement -2 is true ; Statement -2 is **not** a correct explanation for Statement -1.
(b) Statement -1 is true, Statement -2 is false.
(c) Statement -1 is false, Statement -2 is true .
(d) Statement - 1 is true, Statement 2 is true ; Statement -2 is a correct explanation for Statement -1.

24. The shortest distance between line $y - x = 1$ and curve $x = y^2$ is **[2011]**
(a) $\frac{3\sqrt{2}}{8}$ (b) $\frac{8}{3\sqrt{2}}$
(c) $\frac{4}{\sqrt{3}}$ (d) $\frac{\sqrt{3}}{4}$

25. For $x \in \left(0, \frac{5\pi}{2}\right)$, define $f(x) = \int_0^x \sqrt{t} \sin t \, dt$. Then f has **[2011]**
(a) local minimum at π and 2π
(b) local minimum at π and local maximum at 2π
(c) local maximum at π and local minimum at 2π
(d) local maximum at π and 2π

26. The curve that passes through the point (2, 3), and has the property that the segment of any tangent to it lying between the coordinate axes is bisected by the point of contact is given by : **[2011 RS]**
(a) $2y - 3x = 0$ (b) $y = \frac{6}{x}$
(c) $x^2 + y^2 = 13$ (d) $\left(\frac{x}{2}\right)^2 + \left(\frac{y}{3}\right)^2 = 2$

27. Let f be a function defined by -
$$f(x) = \begin{cases} \frac{\tan x}{x}, & x \ne 0 \\ 1, & x = 0 \end{cases}$$
Statement - 1 : $x = 0$ is point of minima of f

Statement - 2 : $f'(0) = 0$. **[2011 RS]**
(a) Statement-1 is true, statement-2 is true; statement-2 is a correct explanation for statement-1.
(b) Statement-1 is true, statement-2 is true; statement-2 is NOT a correct explanation for statement-1.
(c) Statement-1 is true, statement-2 is false.
(d) Statement-1 is false, statement-2 is true.

28. A spherical balloon is filled with 4500π cubic meters of helium gas. If a leak in the balloon causes the gas to escape at the rate of 72π cubic meters per minute, then the rate (in meters per minute) at which the radius of the balloon decreases 49 minutes after the leakage began is : **[2012]**
(a) $\frac{9}{7}$ (b) $\frac{7}{9}$ (c) $\frac{2}{9}$ (d) $\frac{9}{2}$

29. Let $a, b \in R$ be such that the function f given by $f(x) = \ln|x| + bx^2 + ax, x \ne 0$ has extreme values at $x = -1$ and $x = 2$
Statement-1 : f has local maximum at $x = -1$ and at $x = 2$.

Statement-2 : $a = \frac{1}{2}$ and $b = \frac{-1}{4}$ **[2012]**
(a) Statement-1 is false, Statement-2 is true.
(b) Statement-1 is true, statement-2 is true; statement-2 is a correct explanation for Statement-1.
(c) Statement-1 is true, statement-2 is true; statement-2 is **not** a correct explanation for Statement-1.
(d) Statement-1 is true, statement-2 is false.

30. A line is drawn through the point (1,2) to meet the coordinate axes at P and Q such that it forms a triangle OPQ, where O is the origin. If the area of the triangle OPQ is least, then the slope of the line PQ is : **[2012]**
(a) $-\frac{1}{4}$ (b) -4 (c) -2 (d) $-\frac{1}{2}$

31. The intercepts on x-axis made by tangents to the curve, $y = \int_0^x |t| \, dt, x \in R$, which are parallel to the line $y = 2x$, are equal to : **[JEE M 2013]**
(a) ± 1 (b) ± 2 (c) ± 3 (d) ± 4

Solutions & Explanations

Section-A : JEE Advanced/ IIT-JEE

A **1.** $\cos(\ln \theta)$ **2.** $x \in \left(-\frac{1}{2},\ 0\right) \cup \left(\frac{1}{2},\ \infty\right);\ \left(-\infty,\ -\frac{1}{2}\right) \cup \left(0,\ \frac{1}{2}\right)$ **3.** $x \geq 0$ **4.** abe

5. ϕ, $\{(1, 1)\}$

B **1.** F **2.** F

C **1.** (a) **2.** (a) **3.** (c) **4.** (b) **5.** (d) **6.** (d)
7. (b) **8.** (b) **9.** (a) **10.** (c) **11.** (b) **12.** (d)
13. (c) **14.** (d) **15.** (d) **16.** (b) **17.** (a) **18.** (c)
19. (d) **20.** (a) **21.** (d) **22.** (a) **23.** (c) **24.** (a)
25. (d) **26.** (b) **27.** (a) **28.** (b) **29.** (c) **30.** ([illegible])

D **1.** (c) **2.** (b, c) **3.** (c) **4.** (a) **5.** (a, b, c) **6.** (a, c)
7. (d) **8.** (b) **9.** (b, d) **10.** (b, c) **11.** (a, b) **12.** (b, c, d)
13. (a, b, c, d) **14.** (a, c)

E **2.** 2 **4.** e^{π} **6.** $\sqrt{c-\frac{1}{4}}$ **9.** $\frac{ud}{\sqrt{v^2-u^2}}$ **10.** (0, 0)

11. $2x+4y-\pi=0$
$2x+4y+3\pi=0$ **12.** $\lambda \in \left(-\frac{3}{2},\ 0\right) \cup \left(0,\ \frac{3}{2}\right)$ **14.** (0, 2)

15. f is min at $x=\frac{7}{5}$
and max. at $x=1$ **16.** $\frac{10}{3}$ sq. units **18.** $\frac{3\sqrt{3}}{4}r^2$

19. $6+\pi:6$ **20.** x^3+x^2-x+2 **21.** $x+\sqrt{2}y=\sqrt{2}$ or $x-\sqrt{2}y=-\sqrt{2}$

22. $x+y=1$ **23.** $b \in (-2, -1) \cup (1, \infty)$ **24.** $a=-\frac{1}{2},\ b=\frac{-3}{4},\ c=3$

25. $\frac{4\sqrt{3}}{9}$ sq. units **26.** 2 kh **27.** $y=e^{a(x-1)}$; 1 sq. unit

28. min at $x=\frac{1}{4}(b+\sqrt{b^2-1})$, max at $x=\frac{1}{4}(b-\sqrt{b^2-1})$ **29.** $\left(\frac{-2}{a},\ \frac{a}{3}\right)$

31. $a=\frac{1}{4},\ b=\frac{-5}{4},\ c=2$, $f(x)=\frac{1}{4}x^2-\frac{5}{4}x+2$ **32.** $xy=1$

35. (2, 1) **41.** $y=2$ **42.** $4\sqrt{65}$ **43.** 6

F **1.** (A) $\rightarrow$ p (B) $\rightarrow$ r

G **1.** (c) **2.** (a) **3.** (a) **4.** (c) **5.** (b) **6.** (d) **7.** (c)

I **1.** (7) **2.** (0) **3.** (9) **4.** (1) **5.** (5) **6.** (9) **7.** (9)

Section-B : JEE Main/ AIEEE

1. (b) **2.** (a) **3.** (d) **4.** (a) **5.** (b) **6.** (d) **7.** (d) **8.** (a) **9.** (d) **10.** (b) **11.** (b) **12.** (a)
13. (c) **14.** (c) **15.** (d) **16.** (c) **17.** (a) **18.** (b) **19.** (b) **20.** (a) **21.** (c) **22.** (c) **23.** (d)
24. (a) **25.** (c) **26.** (b) **27.** (b) **28.** (c) **29.** (b) **30.** (c) **31.** (a)

Section-A JEE Advanced/ IIT-JEE

(A) FILL IN THE BLANKS :

1. We have $e^{-\pi/2} < \theta < \pi/2$

$\Rightarrow \quad -\frac{\pi}{2} < \ln\theta < \ln \pi/2$

$\Rightarrow \quad \cos(-\pi/2) < \cos(\ln\theta) < \cos(\ln \pi/2)$

[$\because$ cos is increasing in IV quad]

$\Rightarrow \quad \cos(\ln\theta) > 0$ ....(1)

Also $-1 \le \cos\theta \le 1 \,\forall\, \theta$

$\therefore \quad -1 \le \ln(\cos\theta) \le 0 \,\forall\, 0 < \cos\theta \le 1$

$\Rightarrow \quad \ln(\cos\theta) \le 0$ ....(2)

From (1) and (2) we get, $\cos(\ln\theta) > \ln(\cos\theta)$

$\therefore \quad \cos(\ln\theta)$ is larger.

2. $y = 2x^2 - \ln|x|$

$\frac{dy}{dx} = 4x - \frac{1}{x} = \frac{(2x+1)(2x-1)}{x}$

Critical points are 0, 1/2, –1/2

Clearly $f(x)$ is increasing on $\left(-\frac{1}{2}, 0\right) \cup \left(\frac{1}{2}, \infty\right)$ and

$f(x)$ is decreasing on $\left(-\infty, -\frac{1}{2}\right) \cup \left(0, \frac{1}{2}\right)$.

3. Let $f(x) = \log(1+x) - x$ for $x > -1$

$f'(x) = \frac{1}{1+x} - 1 = \frac{-x}{1+x}$

We observe that,

$f'(x) > 0$ if $-1 < x < 0$ and $f'(x) < 0$ if $x > 0$

Therefore f increases in $(-1, 0)$ and decreases in $(0, \infty)$.

Also $f(0) = \log 1 - 0 = 0$

$\therefore \quad x \ge 0 \Rightarrow f(x) \le f(0)$

$\Rightarrow \quad \log(1+x) - x \le 0 \Rightarrow \log(1+x) \le x$

Thus we get, $\log(1+x) \le x, \forall\, x \ge 0$

4. Let $P(a\cos\theta, b\sin\theta)$ be any point on the ellipse $\frac{x^2}{a^2} + \frac{y^2}{b^2} = 1$ with foci $F_1(ae, 0)$ and $F_2(-ae, 0)$

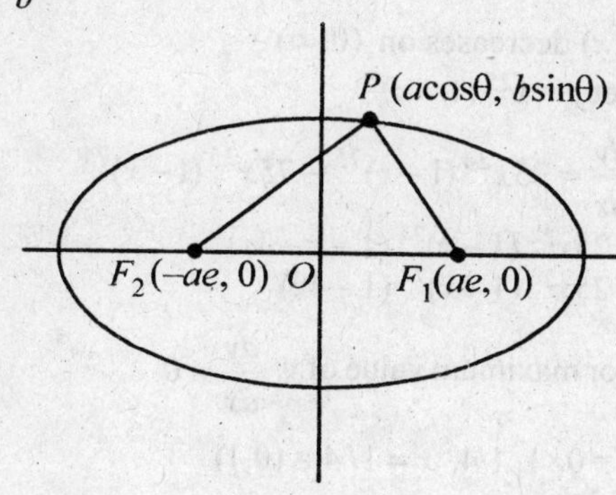

Then area of ΔPF_1F_2 is given by

$$A = \frac{1}{2}\begin{vmatrix} a\cos\theta & b\sin\theta & 1 \\ ae & 0 & 1 \\ -ae & 0 & 1 \end{vmatrix}$$

$$= \frac{1}{2}|-b\sin\theta(ae + ae)| = abe|\sin\theta|$$

$\because \quad |\sin\pi| \le 1$

$\therefore \quad A_{max} = abe$

5. The given curve is $C: y^3 - 3xy + 2 = 0$

Differentiating it with respect to x, we get

$$3y^2\frac{dy}{dx} - 3x\frac{dy}{dx} - 3y = 0$$

$\Rightarrow \quad \frac{dy}{dx} = \frac{y}{-x + y^2}$

$\therefore \quad$ Slope of tangent to C at point (x_1, y_1) is

$\frac{dy}{dx} = \frac{y_1}{-x_1 + y_1^2}$

For horizontal tangent, $\frac{dy}{dx} = 0 \Rightarrow y_1 = 0$

For $y_1 = 0$ in C, we get no value of x_1

$\therefore \quad$ There is no point on C at which tangent is horizontal

$\therefore \quad H = \phi$

For vertical tangent $\frac{dy}{dx} = \frac{1}{0}$

$\Rightarrow \quad -x_1 + y_1^2 = 0 \Rightarrow x_1 = y_1^2$

From C, $y_1^3 - 3y_1^3 + 2 = 0$

$\Rightarrow \quad y_1^3 = 1 \Rightarrow y_1 = 1 \Rightarrow x_1 = 1$

$\therefore \quad$ There is only one point (1, 1) at which vertical tangent can be drawn

$\therefore \quad V = \{(1, 1)\}$

(B) TRUE / FALSE :

1. If $(x - r)$ is a factor of $f(x)$ repeated m times then $f'(x)$ is a polynomial with $(x - r)$ as factor repeated $(m - 1)$ times.

$\therefore \quad$ Statement is false.

2. Given that $0 < a < x$.

Let $f(x) = \log_a x + \log_x a = \log_a x + \frac{1}{\log_a x}$

Consider $g(y) = y + \frac{1}{y}$, where $\log_a x = y$

$\because \quad y + \frac{1}{y} = \left(\sqrt{y} - \frac{1}{\sqrt{y}}\right)^2 + 2 \ge 2$

But equality holds when $y = 1$

$\Rightarrow x = a$ which is not possible.

$\therefore \quad y+\dfrac{1}{y}>2 \Rightarrow g_{\min}$ cannot be 2.

$\therefore \quad f_{\min}$ cannot be 2.

$\therefore$ Statement is false.

C. MCQ's WITH ONE CORRECT ANSWER :

1. **(a)** Consider the function $f(x) = ax^3 + bx^2 + cx$ on $[0, 1]$ then being a polynomial. It is continuous on $[0, 1]$, differentiable on $(0, 1)$ and

$f(0)=f(1)=0$ [as given $a + b + c = 0$]

$\therefore$ By Rolle's theorem $\exists\ x \in (0,1)$ such that

$$f'(x) = 0 \Rightarrow 3ax^2 + 2bx + c = 0$$

Thus equation $3ax^2 + 2bx + c = 0$ has at least one root in $[0, 1]$.

2. **(a)** Area of ΔABC, $A = \dfrac{1}{2}x\sqrt{d^2 - x^2}$

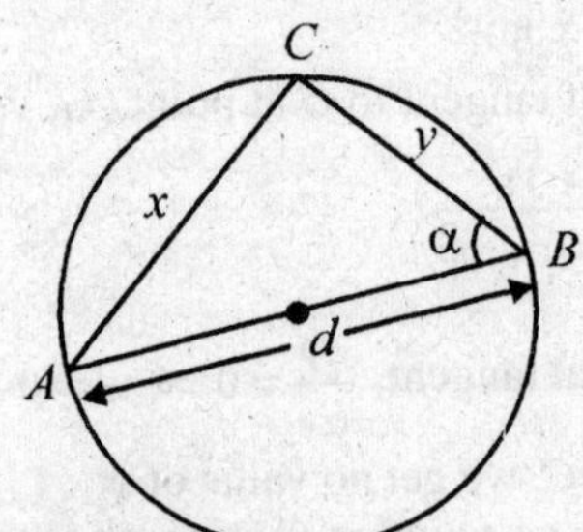

For max/min, $\dfrac{dA}{dx} = 0$

$$\Rightarrow \frac{1}{2}\sqrt{d^2 - x^2} + \frac{1}{2}x\left(\frac{-2x}{2\sqrt{d^2 - x^2}}\right) = 0$$

$$\Rightarrow \frac{d^2 - x^2 - x^2}{2\sqrt{d^2 - x^2}} = 0 \Rightarrow x = d/\sqrt{2}$$

$$\frac{d^2A}{dx^2} = \frac{(-4x).2\sqrt{d^2 - x^2} - 2\dfrac{1}{2\sqrt{d^2 - x^2}}(-2x).(d^2 - 2x^2)}{4(d^2 - x^2)}$$

At $x = d/\sqrt{2}$, $\dfrac{d^2A}{dx^2} = \dfrac{\dfrac{-8d}{\sqrt{2}}.\dfrac{d}{\sqrt{2}} + 0}{4(d^2 - d^2/2)} = -ve$

$\therefore$ A is max at $x = d/\sqrt{2}$ then $y = d/\sqrt{2}$

$\therefore$ Δ is isosceles for max. area.

3. **(c)** $\dfrac{dx}{d\theta} = a(-\sin\theta + \sin\theta + \theta\cos\theta) = a\theta\cos\theta \quad \ldots(1)$

$\dfrac{dy}{d\theta} = a(\cos\theta - \cos\theta + \theta\sin\theta) = a\theta\sin\theta \quad \ldots(2)$

Dividing (2) by (1), we get

$\dfrac{dy}{dx} = \tan\theta$ (slope of tangent)

$\therefore$ Slope of normal $= -\cot\theta$

$\therefore$ Equation of normal is

$$y - a(\sin\theta - \theta\cos\theta) = -\frac{\cos\theta}{\sin\theta}(x - a(\cos\theta + \theta\sin\theta))$$

$\Rightarrow y\sin\theta - a\sin^2\theta + a\sin\theta\cos\theta$
$= -x\cos\theta + a\cos^2\theta + a\theta\sin\theta\cos\theta$

$\Rightarrow x\cos\theta + y\sin\theta = a$

As θ varies inclination is not constant.

$\therefore$ (a) is not correct.

Clearly does not pass through (0, 0).

It's distance from origin $= \left|\dfrac{a}{\sqrt{\cos^2\theta + \sin^2\theta}}\right| = a$

which is constant

4. **(b)** $y = a\ln x + bx^2 + x$
has its extremum values at $x = -1$ and 2

$\therefore \dfrac{dy}{dx} = 0$ at $x = -1$ and 2

$\Rightarrow \dfrac{a}{x} + 2bx + 1 = 0$ or $2bx^2 + x + a = 0$

has -1 and 2 as its roots.

$\therefore$ $2b - 1 + a = 0 \quad \ldots(1)$

$8b + 2 + a = 0 \quad \ldots(2)$

Solving (1) and (2) we get $a = 2$, $b = -1/2$.

5. **(c)** For $y^2 = 4ax$, y-axis is tangent at (0, 0), while for $x^2 = 4ay$, x-axis is tangent at (0, 0). Thus the two curves cut each other at right angles.

6. **(d)** $f'(x) = -(x+2)e^{-x} = -(x+1)e^{-x} = 0$

$\Rightarrow x = -1$

For $x \in (-\infty, -1)$, $f'(x) > 0$ and for $x \in (-1, \infty)$, $f'(x) < 0$

$\therefore$ $f(x)$ is increasing on $(-\infty, -1)$ and decreasing on $(-1, -\infty)$.

7. **(b)** We have $f(x) = \dfrac{\ln(\pi + x)}{\ln(e + x)}$

$$\therefore \quad f'(x) = \frac{\left(\dfrac{1}{\pi + x}\right)\ln(e + x) - \dfrac{1}{(e + x)}\ln(\pi + x)}{[\ln(e + x)]^2}$$

$$= \frac{(e + x)\ln(e + x) - (\pi + x)\ln(\pi + x)}{(e + x)(\pi + x)(\ln(e + x))^2}$$

< 0 on $(0, \infty)$ since $1 < e < \pi$

$\therefore$ $f(x)$ decreases on $(0, \infty)$.

8. **(b)** Let $y = x^{25}(1 - x)^{75}$

$\Rightarrow \dfrac{dy}{dx} = 25x^{24}(1 - x)^{75} - 75x^{25}(1 - x)^{74}$

$= 25x^{24}(1 - x)^{74}(1 - x - 3x)$

$= 25x^{24}(1 - x)^{74}(1 - 4x)$

For maximum value of y, $\dfrac{dy}{dx} = 0$

$\Rightarrow x = 0, 1, 1/4$, $x = 1/4 \in (0,1)$

Also at $x = 0, y = 0$, at $x = 1, y = 0$, and at $x = 1/4, y > 0$

$\therefore$ Max. value of y occurs at $x = 1/4$

9. **(a)** Slope of tangent at $(x, f(x))$ is $2x + 1$

$\Rightarrow f'(x) = 2x + 1 \Rightarrow f(x) = x^2 + x + c$

Also the curve passes through (1, 2)

$\therefore f(1) = 2$

$\Rightarrow 2 = 1 + 1 + c \Rightarrow c = 0, \therefore f(x) = x^2 + x$

$\therefore$ Required area $= \int_0^1 (x^2 + x)\,dx$

$$= \left(\frac{x^3}{3} + \frac{x^2}{2}\right)_0^1 = \frac{1}{3} + \frac{1}{2} = \frac{5}{6}$$

10. **(c)** We have $f(x) = \dfrac{x}{\sin x}, 0 < x \le 1$

$$\Rightarrow f'(x) = \frac{\sin x - x\cos x}{\sin^2 x}$$

where $\sin^2 x$ is always +ve, when $0 < x \le 1$. But to check Nr., we again let

$h(x) = \sin x - x\cos x$

$\Rightarrow h'(x) = x\sin x > 0$ for $0 < x \le 1$

$\Rightarrow h(x)$ is increasing

$\Rightarrow h(0) < h(x)$, when $0 < x \le 1$

$\Rightarrow 0 < \sin x - x\cos x$, when $0 < x \le 1$

$\Rightarrow \sin x - x\cos x > 0$, when $0 < x \le 1$

$\Rightarrow f'(x) > 0,\ x \in (0, 1]$

$\Rightarrow f(x)$ is increasing on (0, 1]

Again $g(x) = \dfrac{x}{\tan x}$

$$\Rightarrow g'(x) = \frac{\tan x - x\sec^2 x}{\tan^2 x}, \text{ when } 0 < x \le 1$$

Here $\tan^2 x > 0$ But to check Nr. we consider

$p(x) = \tan x - x\sec^2 x$

$p'(x) = \sec^2 x - \sec^2 x - x.2\sec x.\sec x\tan x$

$\Rightarrow p'(x) = -2x\sec^2 x\tan x < 0$ for $0 < x \le 1$

$\Rightarrow p(x)$ is decreasing, when $0 < x \le 1$

$\Rightarrow p(0) > p(x) \Rightarrow 0 > \tan x - x\sec^2 x$

$\therefore g'(x) < 0$

Hence $g(x)$ is decreasing when $0 < x \le 1$.

11. **(b)** We are given $f(x) = \sin^4 x + \cos^4 x$

$\Rightarrow f'(x) = 4\sin^3 x\cos x - 4\cos^3 x\sin x$

$= -4\sin x\cos x(\cos^2 x - \sin^2 x)$

$= -2.\sin 2x\cos 2x = -\sin 4x$

Now for $f(x)$ to be increasing function

$f'(x) > 0 \Rightarrow -\sin 4x > 0 \Rightarrow \sin 4x < 0$

$\Rightarrow \pi < 4x < 2\pi \Rightarrow \dfrac{\pi}{4} < x < \dfrac{\pi}{2}$

Since, If $f(x)$ increasing on $(\pi/4, \pi/2)$

$$\frac{\pi}{4} = \frac{4\pi}{8} > \frac{3\pi}{8}$$

It will increasing on $(\pi/4, 3\pi/8)$.

12. **(d)** From graph it is clear that both $\sin x$ and $\cos x$ in the interval $(\pi/2, \pi)$ are decreasing function.

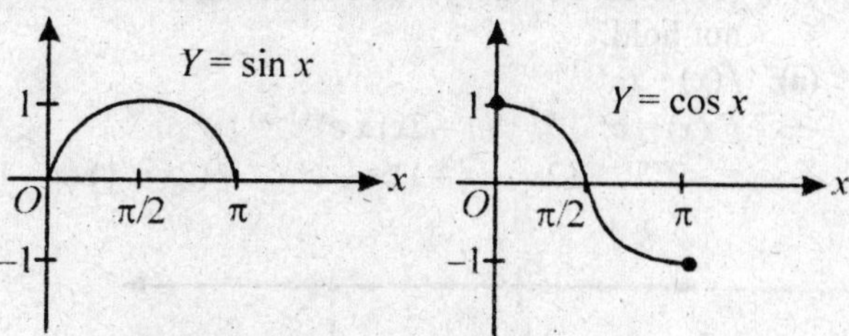

$\therefore$ S is correct.

To disprove R let us consider the counter example :

$f(x) = \sin x$ on $(0, \pi/2)$

so that $f'(x) = \cos x$

Again from graph it is clear that $f(x)$ is increasing on $(0, \pi/2)$ but $f'(x)$ is decreasing on $(0, \pi/2)$

$\therefore$ R is wrong.

13. **(c)** $f(x) = \int e^x (x-1)(x-2)\,dx$

For decreasing function, $f'(x) < 0$

$\Rightarrow e^x(x-1)(x-2) < 0 \Rightarrow (x-1)(x-2) < 0$

$\Rightarrow 1 < x < 2, \quad \because e^x > 0\ \forall x \in R$

14. **(d)** Slope of tangent $y = f(x)$ is $\dfrac{dy}{dx} = f'(x)_{(3,4)}$

Therefore, slope of normal

$$= -\frac{1}{f'(x)_{(3,4)}} = -\frac{1}{f'(3)}$$

but $-\dfrac{1}{f'(3)} = \tan\left(\dfrac{3\pi}{4}\right)$ (given)

or $-\dfrac{1}{f'(3)} = \tan\left(\dfrac{\pi}{2} + \dfrac{\pi}{4}\right) = -1$

$\Rightarrow f'(3) = 1$

15. **(d)** It is clear from figure that at $x = 0$, $f(x)$ is not differentiable.

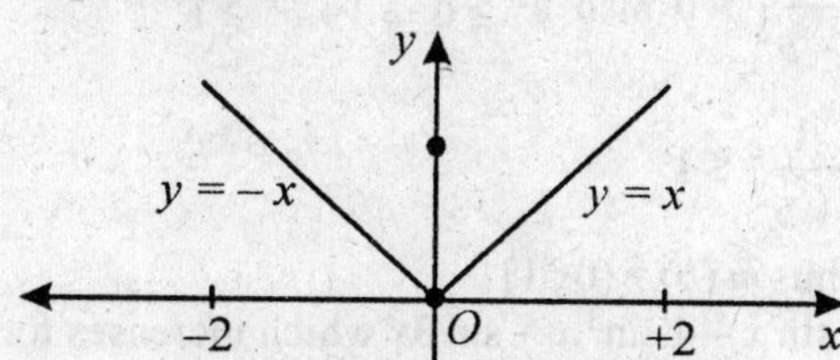

$\Rightarrow f(x)$ has neither maximum nor minimum at $x = 0$.

16. **(b)** Let $f(x) = e^x - 1 - x$ then $f'(x) = e^x - 1 > 0$ for $x \in (0, 1)$

$\therefore$ $f(x)$ is an increasing function.

$\therefore$ $f(x) > f(0), \forall x \in (0, 1)$

$\Rightarrow e^x - 1 - x > 0 \Rightarrow e^x > 1 + x$

$\therefore$ (a) does not hold.

(b) Let $g(x) = \log(1+x) - x$

then $g'(x) = \frac{1}{1+x} - 1 = -\frac{x}{1+x} < 0, \forall x \in (0,1)$

$\therefore$ $g(x)$ is decreasing on (0, 1) $\therefore x > 0$

$\Rightarrow g(x) < g(0)$

$\Rightarrow \log(1+x) - x < 0 \Rightarrow \log(1+x) < x$

$\therefore$ (b) holds. Similarly it can be shown that (c) and (d) do not hold.

17. (a) $f(x) = xe^{x(1-x)}$

$\Rightarrow f'(x) = e^{x(1-x)} + (1-2x)\, x\, e^{x(1-x)}$

$= -e^{x(1-x)}(2x^2 - x - 1) = -e^{x(1-x)}(2x+1)(x-1)$

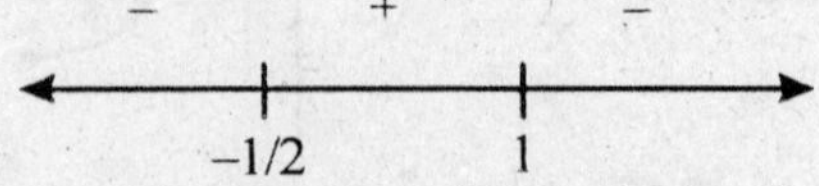

$\therefore$ $f(x)$ is increasing on $[-1/2, 1]$

18. (c) Tangent to $y = x^2 + bx - b$ at (1, 1) is

$y - 1 = (2+b)(x-1)$

$\Rightarrow (b+2)x - y = b + 1$

x-intercept $= \frac{b+1}{b+2}$ and y-intercept $= -(b+1)$

Given $Ar(\Delta) = 2$

$\Rightarrow \frac{1}{2}\left(\frac{b+1}{b+2}\right)[-(b+1)] = 2$

$\Rightarrow b^2 + 2b + 1 = -4(b+2) \Rightarrow b^2 + 6b + 9 = 0$

$\Rightarrow (b+3)^2 = 0 \Rightarrow b = -3$

19. (d) $f(x) = (1+b^2)x^2 + 2bx + 1$

It is a quadratic expression with coeff. of $x^2 = 1 + b^2 > 0$.

$\therefore$ $f(x)$ represents an upward parabola whose min value is $\frac{-D}{4a}$, D being the discriminant.

$\therefore \quad m(b) = -\frac{4b^2 - 4(1+b^2)}{4(1+b^2)}$

$\Rightarrow m(b) = \frac{1}{1+b^2}$

For range of m (b) :

$\frac{1}{1+b^2} > 0$ also $b^2 \ge 0 \Rightarrow 1 + b^2 \ge 1$

$\Rightarrow \frac{1}{1+b^2} \le 1$

Thus $m(b) = (0, 1]$

20. (a) $3\sin x - 4\sin^3 x = \sin 3x$ which increases for

$3x \in \left(-\frac{\pi}{2}, \frac{\pi}{2}\right) \Rightarrow x \in \left(-\frac{\pi}{6}, \frac{\pi}{6}\right)$ whose length is $\frac{\pi}{3}$.

21. (d) The given curve is $y^3 + 3x^2 = 12y$

$\Rightarrow 3y^2 \frac{dy}{dx} + 6x = 12\frac{dy}{dx} \Rightarrow \frac{dy}{dx} = \frac{2x}{4-y^2}$

For vertical tangents $\frac{dy}{dx} = \frac{1}{0} \Rightarrow 4 - y^2 = 0$

$\Rightarrow y = \pm 2$

For y = 2, $x^2 = \frac{24-8}{3} = \frac{16}{3}$

$\Rightarrow x = \pm 4\sqrt{3}$

For $y = -2$, $x^2 = \frac{-24+8}{3} = -ve$ (not possible)

$\therefore$ Req. points are $(\pm 4/\sqrt{3}, 2)$.

22. (a) There is only one function in option (a) whose critical point $\frac{1}{2} \in (0,1)$ for the rest of the parts critical point $0 \notin (0, 1)$. It can be easily seen that functions in options (b), (c) and (d) are continuous on [0, 1] and differentiable in (0, 1).

Now for $f(x) = \begin{cases} \left(\frac{1}{2} - x\right), & x < 1/2 \\ \left(\frac{1}{2} - x\right)^2, & x \ge 1/2 \end{cases}$

Here $f'\left(\frac{1^-}{2}\right) = -1$ and

$f'(1/2^+) = -2\left(\frac{1}{2} - \frac{1}{2}\right) = 0$

$\therefore \quad f'\left(\frac{1^-}{2}\right) \ne f'(1/2^+)$

$\therefore$ f is not differentiable at $1/2 \in (0,1)$

$\therefore$ LMV is not applicable for this function in [0, 1]

23. (c) Equation of tangent to the ellipse $\frac{x^2}{27} + y^2 = 1$ at $(3\sqrt{3}\cos\theta, \sin\theta), \theta \in (0, \pi/2)$ is

$\frac{\sqrt{3}\, x\cos\theta}{9} + y.\sin\theta = 1$

$\therefore$ Intercept on x-axis $= \frac{9}{\sqrt{3}\cos\theta}$;

Intercept on y-axis $= \frac{1}{\sin\theta}$

$\therefore$ Sum of intercepts $= S = 3\sqrt{3}\sec\theta + \operatorname{cosec}\theta$

For min. value of S, $\frac{dS}{d\theta} = 0$

$\Rightarrow 3\sqrt{3}\sec\theta\tan\theta - \operatorname{cosec}\theta\cot\theta = 0$

$\Rightarrow \frac{3\sqrt{3}\sin\theta}{\cos^2\theta} - \frac{\cos\theta}{\sin^2\theta} = 0 \Rightarrow 3\sqrt{3}\sin^3\theta - \cos^3\theta = 0$

$\Rightarrow \tan^3\theta = \frac{1}{3\sqrt{3}} = \left(\frac{1}{\sqrt{3}}\right)^3$

$\Rightarrow \tan\theta = \frac{1}{\sqrt{3}} = \tan\pi/6 \Rightarrow \theta = \pi/6$

24. (a) $f(x)=x^3+bx^2+cx+d, 0<b^2<c$

$f'(x)=3x^2+2bx+c$

Discriminant $=4b^2-12c=4(b^2-3c)<0$

$\therefore \quad f'(x)>0\ \forall\ x\in R$

$\Rightarrow \ f(x)$ is strictly increasing $\forall\ x\in R$

25. (d) For Rolle's theorem in $[a, b]$

$f(a)=f(b)$, In $[0,1] \Rightarrow f(0)=f(1)=0$

$\because$ The function has to be continuous in $[0, 1]$

$\Rightarrow \ f(0)=\lim_{x\to 0^+} f(x)=0 \Rightarrow \lim_{x\to 0} x^{\alpha}\log x=0$

$\Rightarrow \ \lim_{x\to 0}\frac{\log x}{x^{-\alpha}}=0$

Applying L' Hospital's Rule

$\lim_{x\to 0}\frac{1/x}{-\alpha x^{-\alpha-1}}=0 \Rightarrow \lim_{x\to 0}\frac{-x^{\alpha}}{\alpha}=0 \Rightarrow \alpha>0$

26. (b) Let the polynominal be $P(x)=ax^2+bx+c$

Given $P(0)=0$ and $P(1)=1 \Rightarrow c=0$ and $a+b=1$

$\Rightarrow a=1-b$

$\therefore \quad P(x)=(1-b)x^2+bx$

$\Rightarrow \ P'(x)=2(1-b)x+b$

Given $P'(x)>0, \forall\ x\in[0,1]$

$\Rightarrow \ 2(1-b)x+b>0$

$\Rightarrow$ When $x=0$, $b>0$ and when $x=1$, $b<2$

$\Rightarrow \ 0<b<2$

$\therefore \quad S=\left\{(1-a)x^2+ax, a\in(0,2)\right\}$

27. (a) The equation of tangent to the curve $y=e^x$ at (c, e^c) is

$$y-e^c=e^c(x-c) \qquad \ldots(1)$$

and equation of line joining $(c-1, e^{c-1})$ and $(c+1, e^{c+1})$ is

$$y-e^{c-1}=\frac{e^{c+1}-e^{c-1}}{(c+1)-(c-1)}[x-(c-1)]$$

$$\Rightarrow \ y-e^{c-1}=\frac{e^c(e-e^{-1})}{2}[x-c+1] \qquad \ldots(2)$$

Subtracting equation (1) from (2), we get

$$e^c-e^{c-1}=e^c(x-c)\left[\frac{e-e^{-1}-2}{2}\right]+e^c\left(\frac{e-e^{-1}}{2}\right)$$

$$\Rightarrow \ x-c=\frac{\left[1-e^{-1}-\left(\frac{e-e^{-1}}{2}\right)\right]}{\frac{e-e^{-1}-2}{2}}=\frac{2-e-e^{-1}}{e-e^{-1}-2}$$

$$=\frac{e+e^{-1}-2}{2-(e-e^{-1})}.$$

$$=\frac{\frac{e+e^{-1}}{2}-1}{1-\frac{e-e^{-1}}{2}}=\frac{+ve}{-ve}=-ve$$

$\Rightarrow \ x-c<0 \Rightarrow x<c$

$\therefore$ The two lines meet on the left of line $x=c$.

28. (b) The given curves, are

$C_1: y^2=4x \quad \ldots(1)$ and $C_2: x^2+y^2-6x+1=0 \ldots(2)$

Solving (1) and (2) we get

$x^2+4x-6x+1=0 \Rightarrow x=1$ and $\Rightarrow y=2$ or -2

$\therefore$ Points of intersection of the two curves are (1, 2) and (1, –2).

For C_1, $\frac{dy}{dx}=\frac{2}{y}$

$\therefore \left(\frac{dy}{dx}\right)_{(1,2)}=1=m_1$ and $\left(\frac{dy}{dx}\right)_{(1,-2)}=-1=m_1{}'$

For C_2, $\frac{dy}{dx}=\frac{3-x}{y} \ \therefore \left(\frac{dy}{dx}\right)_{(1,2)}=1=m_2$

and $\left(\frac{dy}{dx}\right)_{(1,-2)}=-1=m_2{}'$

$\because m_1=m_2$ and $m_1{}'=m_2{}'$

$\therefore C_1$ and C_2 touch each other at two points.

29. (c)

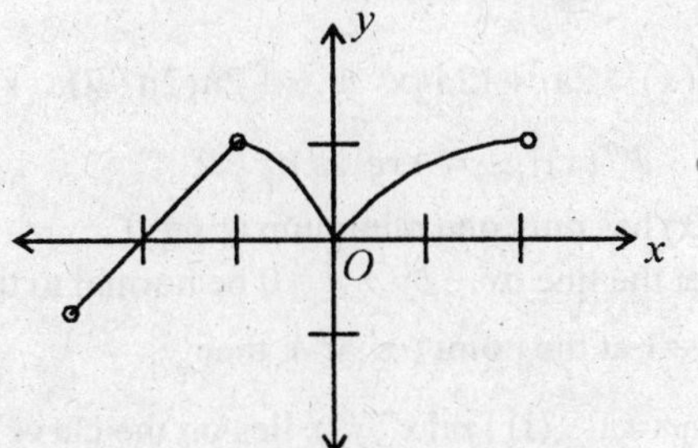

The given function is

$$f(x)=\begin{cases}(2+x)^3, & -3<x\le -1\\ x^{2/3}, & -1<x-2\end{cases}$$

The graph of $y=f(x)$ is as shown in the figure. From graph, clearly, there is one local maximum (at $x=-1$) and one local minima (at $x=0$)

$\therefore$ total number of local maxima or minima = 2.

30. (c) Given that $g(u)=2\tan^{-1}(e^u)-\frac{\pi}{2}$

$$\therefore g(-u)=2\tan^{-1}(e^{-u})-\frac{\pi}{2}=2\tan^{-1}\left(\frac{1}{e^u}\right)-\frac{\pi}{2}$$

$$=2\cot^{-1}(e^u)-\frac{\pi}{2}=2\left[\frac{\pi}{2}-\tan^{-1}(e^u)\right]-\frac{\pi}{2}$$

$$=\pi-2\tan^{-1}(e^u)-\frac{\pi}{2}=\frac{\pi}{2}-2\tan^{-1}(e^u)$$

$=-g(u) \qquad \therefore g$ is an odd function.

Also $g'(u)=\frac{2e^u}{1+e^{2u}}>0,\ \forall\ u\in(-\infty,\infty)$

$\therefore g$ is strictly increasing on $(-\infty,\infty)$.

D MCQ's WITH ONE OR MORE THAN ONE CORRECT :

1. **(c)** The given polynomial is

$$p(x) = a_0 + a_1x^2 + a_2x^4 + \ldots + a_nx^{2n}, x \in R$$

and $0 < a_0 < a_1 < a_2 < \ldots < a_n$

Here we observe that all coefficients of different powers of x, i.e., $a_0, a_1, a_2, \ldots, a_n$ are positive.

Also only even powers of x are involved.

$\therefore$ $P(x)$ can not have any max. value.

Moreover $P(x)$ is minimum i.e., a_0, when $x = 0$.

$\therefore$ $P(x)$ has only one minimum.

ALTERNATE SOLUTION 1

We have $P'(x) = 2a_1x + 4a_2x^3 + \ldots + 2na_nx^{2n-1}$

clearly $P(x)$ increases for all $x > 0$ and decreases for all $x < 0$.

$\therefore$ $P'(x)$ has no. max. value and min. value at $x = 0$.

ALTERNATE SOLUTION 2

We have $P'(x) = 2a_1x + 4a_2x^3 + \ldots + 2na_nx^{2n-1}$

$$P'(x) = 0 \Rightarrow x = 0$$

$$P''(x) = 2a_1 + 12a_2x^2 + \ldots + 2n(2n-1)a_nx^{2n-2}$$

$\therefore$ $P''(x)|_{x=0} = +ve$ as $a_1 > 0$

$\therefore$ $P(x)$ has only one minimum at $x = 0$.

2. **(b, c)** Let the line $ax + by + c = 0$ be normal to the curve $xy = 1$ at the point (x', y'), then

$x'y' = 1 \ldots(1)$ [$pt(x', y')$ lies on the curve]

Also differentiating the curve $xy = 1$ with respect to x we get

$$y + x\frac{dy}{dx} = 0 \Rightarrow \frac{dy}{dx} = -\frac{y}{x}$$

$$\Rightarrow \left.\frac{dy}{dx}\right)_{(x',y')} = \frac{-y'}{x'}$$

$\therefore$ Slope of normal $= \frac{x'}{y'}$

Also equation of normal suggests, slope of normai

$$= \frac{-a}{b}$$

$\therefore$ We must have,

$$\frac{x'}{y'} = -\frac{a}{b} \quad \ldots(2)$$

Now from eq. (1), $x'y' > 0 \Rightarrow x', y'$ are of same sign

$$\Rightarrow \frac{x'}{y'} = +ve \Rightarrow -\frac{a}{b} = +ve \Rightarrow \frac{a}{b} = -ve$$

$\Rightarrow$ a and b are of opposite sign.

$\Rightarrow$ either $a < 0$ and $b > 0$ or $a > 0$ and $b < 0$.

3. **(c)**

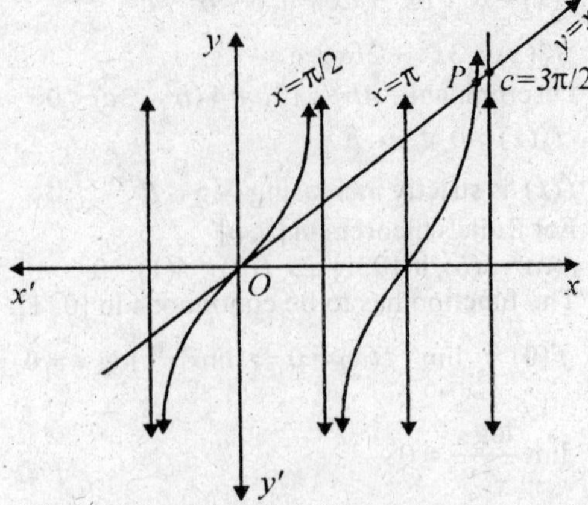

It is clear from the graph that the curves $y = \tan x$ and $y = x$ intersect at P in $(\pi, 3\pi/2)$.

Thus the smallest +ve root of $\tan x - x = 0$ lies in $(\pi, 3\pi/2)$.

4. **(a)** Since g is decreasing in $[0, \infty)$

$\therefore$ For $x \geq y$, $\quad g(x) \leq g(y)$ .... (1)

Also $g(x), g(y) \in [0, \infty)$ and f is increasing from $[0, \infty)$ to $[0, \infty)$.

$\therefore$ For $g(x), g(y) \in [0, \infty)$

such that $g(x) \leq g(y)$

$\Rightarrow$ $f(g(x)) \leq f(g(y))$, where $x \geq y$

$\Rightarrow$ $h(x) \leq h(y)$

$\Rightarrow$ h is decreasing function from $[0, \infty)$ to $[0, \infty)$

$\therefore$ $h(x) \leq h(0)$, $\forall\ x \geq 0$

But $h(0) = 0$ (given)

$\therefore$ $h(x) \leq 0 \forall\ x \geq 0$ ... (2)

Also $h(x) \geq 0 \forall\ x \geq 0$ ... (3)

[as $h(x) \in [0, \infty)$]

From (2) and (3), we get $h(x) = 0$, $\forall\ x \geq 0$

Hence, $h(x) - h(1) = 0 - 0 = 0 \ \forall\ x \geq 0$

5. **(a, b, c)** We are given that

$$f(x) = \begin{cases} 3x^2 + 2x - 1, & -1 \leq x \leq 2 \\ 37 - x, & 2 < x \leq 3 \end{cases}$$

Then on $[-1, 2]$, $f'(x) = 6x + 12$

For $-1 \leq x \leq 2, -6 \leq 6x \leq 12$

$\Rightarrow$ $6 \leq 6x + 12 \leq 24$

$\Rightarrow$ $f'(x) > 0, \forall x \in [-1, 2]$

$\therefore$ f is increasing on $[-1, 2]$

Also $f(x)$ being polynomial for $x \in [-1, 2) \cup (2, 3]$

$f(x)$ is cont. on $[-1, 3]$ except possibly at

At $x = 2$,

$$\text{LHL} = \lim_{h \to 0} f(2-h) = \lim_{h \to 0} 3(2-h)^2 + 12(2-h) - 1 = 35$$

$$\text{RHL} = \lim_{h \to 0} f(2+h) = \lim_{h \to 0} 37 - (2+h) = 35$$

and $f(2) = 3.2^2 + 12.2 - 1 = 35$

LHL = RHL = $f(2)$

$\Rightarrow$ $f(x)$ is continuous at $x = 2$

Hence $f(x)$ is continuous on $[-1, 3]$

Again at $x = 2$

$$RD = \lim_{h\to 0}\frac{f(2+h)-f(2)}{h} = \lim_{h\to 0}\frac{37-(2+h)-35}{h} = 1$$

$$LD = \lim_{h\to 0}\frac{f(2)-f(2-h)}{h}$$

$$= \lim_{h\to 0}\frac{35-3(2-h)^2-12(2-h)+1}{h}$$

$$= \lim_{h\to 0}\frac{-3h^2+24h}{h} = 24$$

As $LD \neq RD$

$\therefore$ $f'(2)$ does not exist. Hence $f(x)$ can not have max. value at $x = 2$.

6. **(a, c)** We have

$$h'(x) = f'(x)[1-2f(x)+3(f(x))^2]$$

$$= 3f'(x)\left[(f(x))^2-\frac{2}{3}f(x)+\frac{1}{3}\right]$$

$$= 3f'(x)[\{f(x)-1/3\}^2+2/9]$$

Note that $h'(x) < 0$ whenever $f'(x) < 0$ and $h'(x) > 0$ whenever $f'(x) > 0$, thus, $h(x)$ increases (decreases) whenever $f(x)$ increases (decreases).

7. **(d)** We have, $f(x) = \dfrac{x^2-1}{x^2+1}$

$$\Rightarrow f'(x) = \frac{4x}{(x^2+1)(x^2+1)^2}$$

For max/min $f'(x) = 0 \Rightarrow x = 0$.

$$f''(x) = \frac{4(1-3x^2)}{(x^2+1)^3},\ f''(x)|_{x=0} = +ve$$

$\Rightarrow$ f is min at $x = 0$

$\therefore$ Min value of $f(x)$ is $\dfrac{0-1}{0+1} = -1$

ALTERNATE SOLUTION

$$f(x) = \frac{x^2-1}{x^2+1} = \frac{(x^2+1)-2}{x^2+1} = 1-\frac{2}{x^2+1}$$

For $f(x)$ to be min $\dfrac{2}{x^2+1}$ should be max, which is so if x^2+1 is min.

And x^2+1 is min at $x = 0$.

$$\therefore\ f_{\min} = \frac{0-1}{0+1} = -1$$

8. **(b)** The maximum value of $f(x) = \cos x + \cos(\sqrt{2x})$ is 2 which occurs at $x = 0$. Also, there is no value of x for which this value will be attained again.

9. **(b,d)** $\dfrac{dy}{dx} = f'(x) \Rightarrow x(e^x-1)(x-1)(x-2)^3(x-3)^5 = 0$

Critical points are 0, 1, 2, 3. Consider change of sign of $\dfrac{dy}{dx}$ at $x = 3$.

$x < 3$, $\dfrac{dy}{dx} = -$ ve and $x > 3$, $\dfrac{dy}{dx} = +$ve

Change is from –ve to +ve, hence minimum at $x = 3$. Again minimum and maximum occur alternately.

$\therefore$ 2nd minimum is at $x = 1$

10. **(b, c)** Let $f(x) = ax^3 + bx^2 + cx + d$

Then, $f(2) = 18 \Rightarrow 8a+4b+2c+d = 18$ ...(1)

$f(1) = -1 \Rightarrow a+b+c+d = -1$ ...(2)

$f(x)$ has local max. at $x = -1$

$\Rightarrow 3a-2b+c = 0$ ...(3)

$f'(x)$ has local min. at $x = 0 \Rightarrow b = 0$ ...(4)

Solving (1), (2), (3) and (4), we get

$$f(x) = \frac{1}{4}(19x^3-57x+34) \Rightarrow f(0) = \frac{17}{2}$$

Also $f'(x) = \dfrac{57}{4}(x^2-1) > 0, \forall x > 1$

Also $f'(x) = 0 \Rightarrow x = 1, -1$

$f''(-1) < 0, f''(1) > 0 \Rightarrow x = -1$ is a point of local max. and $x = 1$ is a point of local min. Distance between $(-1, 2)$ and $(1, f(1))$, i.e. $(1, -1)$ is $\sqrt{13} \neq 2\sqrt{5}$

11. **(a, b)** $\because$ $g(x) = \int_0^x f(t)\,dt$

$$\Rightarrow g'(x) = f(x) = \begin{cases} e^x, & 0 \le x \le 1 \\ 2-e^{x-1}, & 1 < x \le 2 \\ x-e, & 2 < x \le 3 \end{cases}$$

$\therefore$ $g'(x) = 0 \Rightarrow e^{x-1} = 2$ or $x-e = 0$

$\Rightarrow x-1 = \log 2$ or $x = e$

$\Rightarrow x = 1+\ln 2$ or e

$$g''(x) = \begin{cases} e^x, & 0 \le x \le 1 \\ -e^{x-1}, & 1 < x \le 2 \\ 1, & 2 < x \le 3 \end{cases}$$

NOTE THIS STEP

$\therefore$ $g''(1+\ln 2) = -2$ and $g''(e) = 1 \Rightarrow g(x)$ has local max. at $x = 1+\ln 2$ and local min. at $x = e$.

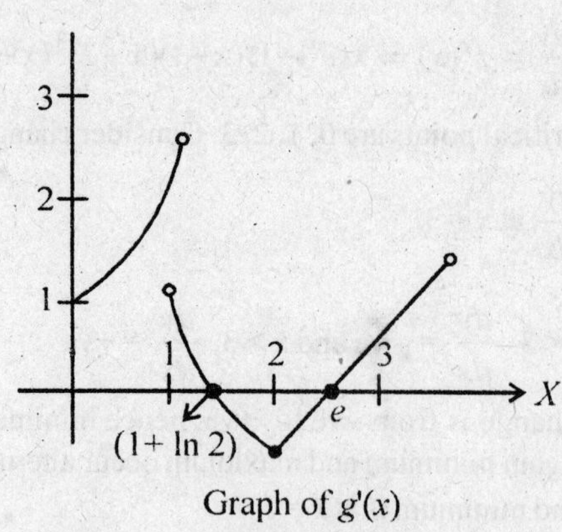

Graph of $g'(x)$

Also graph of $g'(x)$ suggests, $g(x)$ has local max. at $x = 1$ and local min. at $x = 2$

12. (b, c, d) We have, $f(x) = x\cos\frac{1}{x},\ x \ge 1$

$\therefore\ f'(x) = \cos\frac{1}{x} + \frac{1}{x}\sin\frac{1}{x}$

$\lim_{x\to\infty} f'(x) = \cos 0 + (0) \times$ (some finite value)

$\Rightarrow\ \lim_{x\to\infty} f'(x) = 1$

Also $f''(x) = \frac{1}{x^2}\sin\frac{1}{x} - \frac{1}{x^2}\sin\frac{1}{x} - \frac{1}{x^3}\cos\frac{1}{x}$

$\Rightarrow\ f''(x) = \frac{-1}{x^3}\cos\frac{1}{x} < 0,\ \forall x \in [1, \infty)$

$\Rightarrow\ f'(x)$ is strictly decreasing in $[1, \infty)$

$\therefore\ f'(x) > \lim_{x\to\infty} f'(x)$

$\Rightarrow\ \frac{f(x+2) - f(x)}{(x+2) - x} > 1$

$\Rightarrow\ f(x+2) - f(x) > 2$

13. (a, b, c, d)

We have $f(x) = \int_0^x e^{t^2}(t-2)(t-3)dt$

$\Rightarrow f'(x) = e^{x^2}.(x-2)(x-3)$

$f'(x) = 0 \Rightarrow x = 2, 3$

$f''(x) = e^{x^2}.2x(x^2 - 5x + 6) + e^{x^2}(2x - 5)$

$f''(2) = -$ve and $f''(3) = +$ve

$\therefore x = 2$ is a point of local maxima
and $x = 3$ is a point of local minima
Also or $x \in (2, 3)\ f'(x) < 0$
$\Rightarrow f$ is decreasing on $(2, 3)$
Also we observe
$f''(0) < 0$ and $f''(1) > 0$
$\therefore$ There exists some $C \in (0, 1)$ such that $f''(C) = 0$
$\therefore$ All the options are correct.

14. (a, c) Let $L = 8x$, $B = 15x$ and y be the length of square cut off from each corner. Then volume of box
$= (8x - 2y)(15x - 2y)y$
$V = 120x^2y - 46xy^2 + 4y^3$

$\frac{dV}{dy} = 120x^2 - 92xy + 12y^2$

Now $\frac{dV}{dy} = 0$ at $y = 5$ for maximum value of V.

$\Rightarrow\ [30x^2 - 23xy + 3y^2]_{y=5} = 0$
$\Rightarrow\ 6x^2 - 23x + 15 = 0$

$\Rightarrow\ x = 3, \frac{5}{6}$

For $x = 3$, sides are 45 and 24.

E SUBJECTIVE PROBLEMS :

1. $f(x) = \frac{(a+x)(b+x)}{(c+x)}, a, b > c, x > -c$

$= \frac{(a-c+x+c)(b-c+x+c)}{x+c}$

$= \frac{(a-c)(b-c)}{x+c} + (x+c) + a + b - 2c$

$\Rightarrow f'(x) = \frac{-(a-c)(b-c)}{(x+c)^2} + 1$

$\therefore\ f'(x) = 0 \Rightarrow x = -c \pm \sqrt{(a-c)(b-c)}$

$\Rightarrow\ x = -c + \sqrt{(a-c)(b-c)}$ [+ve sign is taken $\because x > -c$]

Also $f''(x) = \frac{2(a-c)(b-c)}{(x+c)^3} > 0$ for $a, b > c$ and $x > -c$

$\therefore\ f(x)$ is least at $x = -c + \sqrt{(a-c)(b-c)}$

$\therefore\ f_{\min} = \frac{(a-c)(b-c)}{\sqrt{(a-c)(b-c)}} + \sqrt{(a-c)(b-c)}$

$+ (a-c) + (b-c)$

$= (a-c) + (b-c) + 2\sqrt{(a-c)(b-c)}$

$= (\sqrt{a-c} + \sqrt{b-c})^2$

2. Given that x and y are two real variables such that $x > 0$ and $xy = 1$.
To find the minimum value of $x + y$.
Let $S = x + y$

$\Rightarrow\ S = x + \frac{1}{x}$ (using $xy = 1$)

$\therefore\ \frac{dS}{dx} = 1 - \frac{1}{x^2}$

For minimum value of $S, \frac{dS}{dx} = 0$

$\Rightarrow\ 1 - \frac{1}{x^2} = 0 \Rightarrow x = \pm 1$

But $x > 0$, $\therefore x = 1$

Now $\frac{d^2S}{dx^2} = \frac{2}{x^3}$

$\Rightarrow \left.\frac{d^2S}{dx^2}\right|_{x=1} = 2 = +ve$

$\therefore$ S is minimum when $x = 1$

$\therefore$ $S_{\min} = 1 + \frac{1}{1} = 2$

3. We are given that

$x \in [0,1], |f''(x)| < 1$ and $f(0) = f(1)$

To prove that $|f'(x)| < 1, \forall x \in [0,1]$

Here $f(x)$ is continuous on [0, 1], differentiable on (0, 1) and $f(0) = f(1)$

$\therefore$ By Rolle's thm.,

$\exists c \in (0,1)$ such that $f'(c) = 0$ ... (1)

Now there may be three cases for $x \in [0,1]$

(i) $x = c$ (ii) $x > c$ (iii) $x < c$

Case I : For $x = c$.

If $x = c$ then $f'(x) = 0 < 1$ [from (1)]

Hence the result $|f'(x)| < 1$ is obtained in this case.

Case II : For $x > c$

Consider the interval $[c, x]$.

As $f'(x)$ is continuous on $[c, x]$ and differentiable on (c, x)

$\therefore$ By LMV $f''(\alpha) = \frac{f'(x) - f'(c)}{x - c}$ where $\alpha \in (c, x)$

$\Rightarrow f'(x) = (x-c) f''(\alpha)$ $[\because f'(c) = 0]$

Now, $x, c \in [0,1]$ and $x > c$

$\therefore x - c < 1$ .... (i)

also $|f''(x)| < 1, \forall x$ (given)

$\therefore |f''(\alpha)| < 1$ .... (ii)

Combining (i) and (ii), $(x-c)|f''(\propto)| < 1$

$\therefore |f'(x)| < 1$. Hence the result in this case.

Case III : For $x < c$

Consder the interval $[x, c]$.

As $f'(x)$ is continuous on $[x, c]$ and differentiable on (x, c)

$\therefore$ By LMV for $\beta \in (x, c)$

$f''(\beta) = \frac{f'(c) - f'(x)}{c - x}$

$\Rightarrow f'(x) = -(c-x) f''(\beta)$

[Using $f'(c) = 0$]

$\therefore |f'(x)| = |(c-x) f''(\beta)|$

as $x, c \in [0,1]$ and $x < c$

$\therefore 0 < c - x < 1$ also $|f''(\beta)| < 1$ as $|f''(x)| < 1, \forall x$

$\therefore |(c-x) f''(\beta)| < 1$

$\therefore |f''(x)| < 1$ hence the result in this case.

Combining all the three cases we get

$|f'(x)| < 1, \forall x \in [0,1]$

4. $f(x) = x^{1/x}, \quad x > 0$

Let $y = x^{1/x}$

$\Rightarrow \log y = \frac{1}{x} \log x$

Differentiating w.r.t. x we get

$\frac{1}{y}\frac{dy}{dx} = \frac{\frac{1}{x}.x - 1.\log x}{x^2} \Rightarrow \frac{dy}{dx} = \frac{y(1 - \log x)}{x^2}$

For max/min value put $\frac{dy}{dx} = 0$

$\Rightarrow \frac{y(1-\log x)}{x^2} = 0 \Rightarrow \log x = 1 \Rightarrow x = e$

Also, $\frac{d^2y}{dx^2} = \frac{\left(\frac{dy}{dx}(1 - \log x) - \frac{1}{x} y\right)x^2 - 2xy(1 - \log x)}{x^4}$

$\left.\frac{d^2y}{dx^2}\right|_{x=e} = \left(\frac{-xy}{x^4}\right)_{x=e}$

$\left[\text{Using } \frac{dy}{dx} = 0, 1 - \log x = 0 \text{ at } x = e\right]$

$= \frac{-e^{1/e}}{e^3} = -ve$

$\therefore$ y is max at $x = e$

$\therefore$ $e^{1/e}$ is the max. value of $f(x)$.

$\therefore$ $x^{1/x} < e^{1/e}, \forall x$

$\therefore$ Put $x = \pi$, we get, $\pi^{1/\pi} < e^{1/e}$

$\Rightarrow$ Raising to the power πe on both sides we get

$\pi^e < e^\pi$ or $e^\pi > \pi^e$

5. Given that $f(x)$ and $g(x)$ are differentiable for $x \in [0,1]$ such that

$f(0) = 2; f(1) = 6, \; g(0) = 0; g(1) = 2$

To show that $\exists c \in (0,1)$ such that $f'(c) = 2g'(c)$

Let us consider $h(x) = f(x) - 2g(x)$

Then $h(x)$ is continuous on [0, 1] and differentiable on (0, 1)

Also $h(0) = f(0) - 2g(0) = 2 - 2 \times 0 = 2$

$h(1) = f(1) - 2g(1) = 6 - 2 \times 2 = 2$

$\therefore$ $h(0) = h(1)$

$\therefore$ All the conditions of Rolle's theorem are satisfied for $h(x)$ on [0, 1]

$\therefore$ $\exists c \in (0,1)$ such that $h'(c) = 0$

$\Rightarrow f'(c) - 2g'(c) = 0 \Rightarrow f'(c) = 2g'(c)$

6. $(0, c), y = x^2, 0 \le c \le 5$.
Any point on parabola is (x, x^2)
Distance between (x, x^2) and (0, 1) is

$$D = \sqrt{x^2 + (x^2 - c)^2}$$

To minimum D we consider

$$D^2 = x^4 - (2c-1)x^2 + c^2$$

$$= \left(x^2 - \frac{2c-1}{2}\right)^2 + c - \frac{1}{4}$$

which is minimum when $x^2 - \frac{2c-1}{2} = 0 \Rightarrow x^2 = \frac{2c-1}{2}$

$$\Rightarrow D_{\min} = \sqrt{c - \frac{1}{4}}$$

7. Given $ax^2 + \frac{b}{x} \ge c$ ...(1)

$\forall x > 0, a > 0, b > 0$

To show that $27ab^2 \ge 4c^3$.
Let us consider the function $f(x) = ax^2 + b/x$

then $f'(x) = 2ax - \frac{b}{x^2} = 0$

$\Rightarrow x^3 = b/2a \Rightarrow x = (b/2a)^{1/3}$

$\therefore \quad f''(x) = 2a + \frac{2b}{x^3}$

$\Rightarrow \quad f''\left(\left(\frac{b}{2a}\right)^{1/3}\right) = 2a + \frac{2b}{b} \times 2a = 6a > 0$

$\therefore \quad f$ is minimum at $x = \left(\frac{b}{2a}\right)^{1/3}$

As (1) is true $\forall\ x$

$\therefore \quad$ so is for $x = \left(\frac{b}{2a}\right)^{1/3}$

$$\Rightarrow a\left(\frac{b}{2a}\right)^{2/3} + \frac{b}{(b/2a)^{1/3}} \ge c$$

$$\Rightarrow \frac{a\left(\frac{b}{2a}\right) + b}{(b/2a)^{1/3}} \ge c \quad \Rightarrow \quad \frac{3b}{2}\left(\frac{2a}{b}\right)^{1/3} \ge c$$

As a, b are +ve, cubing both sides we get

$$\frac{27b^3}{8} \cdot \frac{2a}{b} \ge c^3 \Rightarrow 27ab^2 \ge 4c^3 \quad \text{Hence proved.}$$

8. To show

$$1 + x\ln(x + \sqrt{x^2+1}) \ge \sqrt{1+x^2} \text{ for } x \ge 0.$$

Consider $f(x) = 1 + x\ln(x + \sqrt{x^2+1}) - \sqrt{1+x^2})$

Here, $f'(x) = \ln(x + \sqrt{x^2+1}) + \frac{x}{x + \sqrt{x^2+1}}\left[1 + \frac{x}{\sqrt{x^2+1}}\right] - \frac{x}{\sqrt{1+x^2}}$

$$= \ln(x + \sqrt{x^2+1})$$

As $x + \sqrt{x^2+1} \ge 1$ for $x \ge 1$

$\therefore \quad \ln(x + \sqrt{x^2+1}) \ge 0$

$\therefore \quad f'(x) \ge 0, \forall\ x \ge 0$

Hence $f(x)$ is increasing function.

Now for $x \ge 0 \Rightarrow f(x) \ge f(0)$

$\Rightarrow \quad 1 + x\ln(x + \sqrt{x^2+1}) - \sqrt{1+x^2} \ge 0$

$\Rightarrow \quad 1 + x\ln(x + \sqrt{x^2+1}) \ge \sqrt{1+x^2}$

9. Let the house of the swimmer be at B.
$\therefore \quad d(AB) = L$ km.
Let the swimmer land at C, on the shore and let $d(AC) = x$ km

$\therefore \quad d(SC) = \sqrt{x^2 + d^2}$ and $d(CB) = L - x$

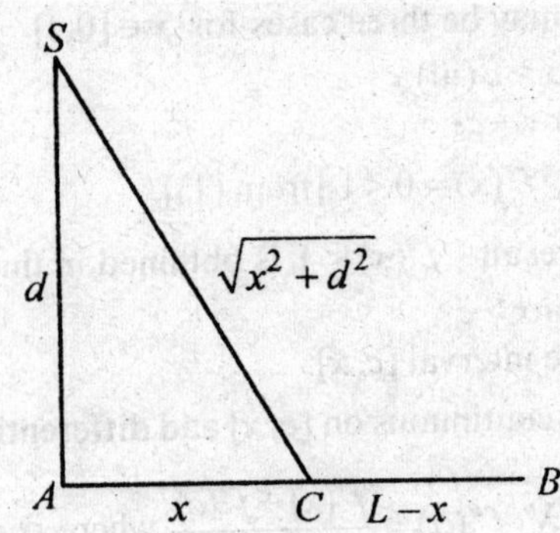

Now using time $= \frac{\text{distance}}{\text{speed}}$

Time from S to B = time S to C + time from C to B

$$\therefore \quad T = \frac{\sqrt{x^2+d^2}}{u} + \frac{L-x}{v}$$

Hence we take, $f(x) = \frac{1}{u}\sqrt{x^2+d^2} + \frac{L}{v} - \frac{x}{v}$

$$\Rightarrow \quad f'(x) = \frac{1}{u}\frac{1.2x}{2\sqrt{x^2+d^2}} + 0 - \frac{1}{v}$$

For either maximum or minimum, $f'(x) = 0$

$$\Rightarrow f'(x) = \frac{x}{u\sqrt{x^2+d^2}} - \frac{1}{v} = 0$$

$$\Rightarrow \quad v^2x^2 = u^2(x^2+d^2) \Rightarrow x^2(v^2-u^2) = u^2d^2$$

$$\Rightarrow \quad x^2 = \frac{u^2d^2}{v^2-u^2}$$

$$\Rightarrow \quad x = \frac{ud}{\sqrt{v^2-u^2}} \quad (\text{as } x \ne -\text{ve})$$

Now, $f''(x)=\frac{1}{u}\left[\frac{\sqrt{x^2+d^2}.1-x.\frac{2x}{2\sqrt{x^2+d^2}}}{(\sqrt{x^2+d^2})^2}\right]$

$=\frac{1}{u}\left[\frac{d^2}{\sqrt{x^2+d^2}.(x^2+d^2)}\right]=\frac{d^2}{u(x^2+d^2)^{3/2}}>0,\forall x$

$\therefore$ f has minimum at $x=\frac{ud}{\sqrt{v^2-u^2}}$

10. Equation of the curve is given by

$y=\frac{x}{1+x^2}$ ...(1)

Differentiating with respect to x, we get

$\frac{dy}{dx}=\frac{1+x^2-x(2x)}{(1+x^2)^2}=\frac{1-x^2}{(1+x^2)^2}$

Again let $f(x)=\frac{1-x^2}{(1+x^2)^2}=\frac{dy}{dx}$

Now, $f'(x)=\frac{(1+x^2)^2(-2x)-(1-x^2)2(1+x^2)(2x)}{(1+x^2)^4}$

$=\frac{(1+x^2)(-2x)-(1-x^2)2.2x}{(1+x^2)^3}=\frac{x(2x^2-6)}{(1+x^2)^3}$

For the greatest value of slope, we have

$f'(x)=\frac{x(2x^2-6)}{(1+x^2)^3}=0 \Rightarrow x=0,\pm\sqrt{3}$

Again we find,

$f''(x)=\frac{12x^2(3-x^2)}{(1+x^2)^4}-\frac{6(1-x^2)}{(1+x^2)^3}$

$\therefore$ $f''(0)=-6$ and $f''(\pm\sqrt{3})=\frac{3}{16}$

Thus, second order derivative at $x=0$ is negative and second order derivative at $x=\pm\sqrt{3}$ is positive.

Therefore, the tangent to the curve has maximum slope at (0, 0).

11. Equation of given curve $y=\cos(x+y)$, $-2\pi\le x\le 2\pi$

Differentiating with respect to x,

$\frac{dy}{dx}=-\sin(x+y).\left[1+\frac{dy}{dx}\right]$

$\Rightarrow$ $[1+\sin(x+y)]\frac{dy}{dx}=-\sin(x+y)$

$\Rightarrow$ $\frac{dy}{dx}=-\frac{\sin(x+y)}{1+\sin(x+y)}$ ...(1)

Since the tangent to given curve is parallel to $x+2y=0$

$\therefore$ $\frac{-\sin(x+y)}{1+\sin(x+y)}=-\frac{1}{2}$ [For parallel line $m_1=m_2$]

$\Rightarrow$ $2\sin(x+y)=1+\sin(x+y)$

$\Rightarrow$ $\sin(x+y)=1$

Thus, $\cos(x+y)=0$

Using equation of curve and above result, we get, $y=0$

$\Rightarrow$ $\sin x=1$

$\Rightarrow$ $x=n\pi+(-1)^n\pi/2, n\in Z$

$\Rightarrow$ $x=\pi/2, -3\pi/2$

which belong to the interval $[-2\pi, 2\pi]$

Thus the points on curve at which tangents are parallel to given line are $(\pi/2, 0)$ and $(-3\pi/2, 0)$

The equation of tangent at $(\pi/2, 0)$ is

$y-0=-\frac{1}{2}(x-\pi/2)$

$\Rightarrow$ $2y=-x+\pi/2$

$\Rightarrow$ $2x+4y-\pi=0$

The equation of tangent at $(-3\pi/2, 0)$ is

$y-0=-\frac{1}{2}(x+3\pi/2)$

$\Rightarrow$ $2y=-x-3\pi/2$

$\Rightarrow$ $2x+4y+3\pi=0$

Thus the required equations of tangents are

$2x+4y-\pi=0$ and $2x+4y+3\pi=0$.

12. The given function is,

$f(x)=\sin^3 x+\lambda\sin^2 x$ for $-\pi/2<x<\pi/2$

$\therefore$ $f'(x)=3\sin^2 x\cos x+2\lambda\sin x\cos x$

$=\frac{1}{2}\sin 2x(3\sin x+2\lambda)$

So, from $f'(x)=0$, we get $x=0$

or $3\sin x+2\lambda=0$

Also, $f''(x)=\cos 2x(3\sin x+2\lambda)+\frac{3}{2}\sin 2x\cos x$

Therefore, for $\lambda=\frac{-3}{2}\sin x$, we have

$f''(x)=3\sin x\cos^2 x=-2\lambda\cos^2 x$

Now, if $0<x<\pi/2$, then $-3/2<\lambda<0$ and therefore $f''(x)>0$.

$\Rightarrow$ $f(x)$ has one minimum for this value of λ.

Also for $x=0$, we have $f''(0)=2\lambda<0$, That is $f(x)$ has a maximum at $x=0$

Again if $-\pi/2<x<0$, then $0<\lambda<3/2$ and therefore $f''(x)=-2\lambda\cos^2 x<0$.

So that $f(x)$ has a maximum.

Also for $x=0, f''(a)=2\lambda>0$ so that $f(x)$ has a minimum.

Thus, for exactly one maximum and minimum value of $f(x)$, λ must lie in the interval

$-3/2<\lambda<0$ or $0<\lambda<3/2$

i.e., $\lambda\in(-3/2,0)\cup(0,3/2)$.

13. ATQ, $AEDF$ is a parallelogram.

Let $AF=x=ED$

and $AE=y=FD$

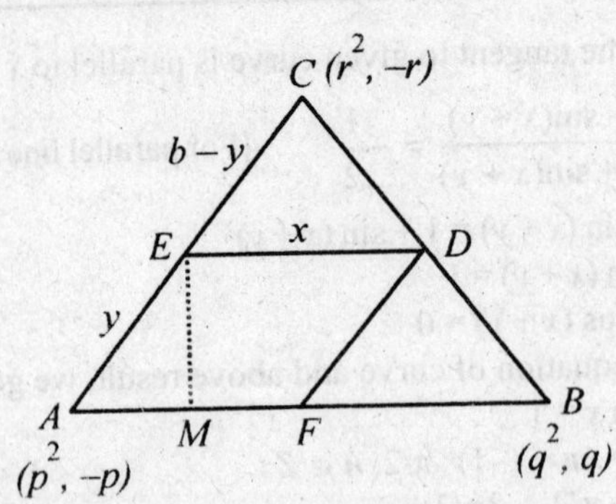

Then from similar Δ's *CED* and *CAB*, we have

$$\frac{CE}{ED}=\frac{CA}{AB} \text{ or } \frac{b-y}{x}=\frac{b}{c} \quad ...(1)$$

Let *P* denote the area of $||^{gm}$ *AFDE*.

$$\text{Then } P = 2\Delta AEF = 2.\frac{1}{2}xy\sin A$$

$$= y.\frac{c}{b}(b-y)\sin A \quad \text{[using eq. (1)]}$$

$$\Rightarrow \quad P=\frac{c}{b}\sin A(by-y^2)$$

$$\therefore \quad \frac{dP}{dy}=\frac{c}{b}\sin A(b-2y)=0$$

Above gives $y=\frac{b}{2}$ and $\frac{d^2P}{dy^2}=\frac{c}{b}\sin A(-2)=-ve$

Hence area is maximum when $y = b/2$ and its value is

$$P=\frac{b}{2}.\frac{c}{b}\left(b-\frac{b}{2}\right)\sin A=\frac{1}{4}bc\sin A=\frac{1}{2}.ar(\Delta ABC)$$

$$\text{Now, } ar(\Delta ABC)=\frac{1}{2}\begin{vmatrix} p^2 & -p & 1 \\ q^2 & q & 1 \\ r^2 & -r & 1 \end{vmatrix}$$

Operating, $R_2 \to R_2 - R_1$ and $R_3 \to R_3 - R_1$, we get

$$=\frac{1}{2}\begin{vmatrix} p^2 & -p & 1 \\ q^2-p^2 & q+p & 0 \\ r^2-p^2 & p-r & 0 \end{vmatrix}$$

$$=\frac{1}{2}(q+p)(p-r)\begin{vmatrix} q-p & 1 \\ -(r+p) & 1 \end{vmatrix}$$

$$=\frac{1}{2}(q+p)(p-r)(q+r)$$

Thus max. area of $||^{gm}$ *AFDE* is

$$=\frac{1}{2}ar(\Delta ABC)$$

$$=\frac{1}{4}(p+q)(q+r)(p-r)$$

Hence Proved.

14. The equation of given curve can be expressed as

$$\frac{x^2}{a^2}+\frac{y^2}{4}=1 \text{ where } 4<a^2<8$$

Clearly it is the question of an ellipse

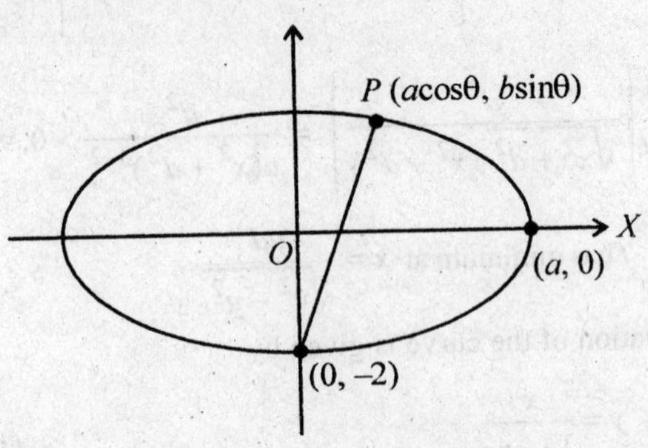

Let us consider a point *P* ($a\cos\theta$, $2\sin\theta$) on the ellipse.
Let the distance of *P* ($a\cos\theta$, $2\sin\theta$) from (0, – 2) is *L*.
Then, $L^2 = (a\cos\theta - 0)^2 + (2\sin\theta + 2)^2$

⇒ Differentiating with respect to θ, we have

$$\frac{d(L^2)}{d\theta}=\cos\theta[-2a^2\sin\theta+8\sin\theta+8]$$

For max. or min. value of *L* we should have

$$\frac{d(L^2)}{d\theta}=0$$

$\Rightarrow \cos\theta[-2a^2\sin\theta+8\sin\theta+8]=0$

$\Rightarrow$ Either $\cos\theta=0$

or $(8-2a^2)\sin\theta+8=0$

$$\Rightarrow \quad \theta=\frac{\pi}{2} \text{ or } \sin\theta=\frac{4}{a^2-4}$$

Since $a^2<8 \Rightarrow a^2-4<4$

$$\Rightarrow \quad \frac{4}{a^2-4}>1 \Rightarrow \sin\theta>1 \text{ which is not possible}$$

$$\text{Also } \frac{d^2(L^2)}{d\theta^2}=\cos\theta[-2a^2\cos\theta+8\cos\theta]$$

$$+(-\sin\theta)[-2a^2\sin\theta+8\sin\theta+8]$$

$$\text{At } \theta=\frac{\pi}{2}, \frac{d^2(L^2)}{d\theta^2}=0-[16-2a^2]=2(a^2-8)<0$$

as $a^2<8$

∴ *L* is max. at θ = π/2 and the farthest point is (0, 2).

15. We have,

$$f(x)=\int_1^x[2(t-1)(t-2)^3+(t-1)^2 3(t-2)^2]dt$$

Then using the theorem,

$$\frac{d}{dx}\left[\int_{\phi(x)}^{\psi(x)}g(t)dt\right]=g[\psi(x)]\psi'(x)-g[\phi(x)]\phi'(x)$$

We get,

$$f'(x)=2(x-1)(x-2)^3+3(x-1)^2(x-2)^2$$
$$=(x-1)(x-2)^2(2x-4+3x-3)$$
$$=(x-1)(x-2)^2(5x-7)$$

For extreme values $f'(x) = 0 \Rightarrow x = 1, 2, 7/5$

Now, $f''(x) = (x-2)^2(5x-7) + 2(x-1)(x-2)(5x-7) + 5(x-1)(x-2)^2$

At $x = 1$,

$f''(x) = 1(-2) = -2 < 0$

$\therefore$ f is max. at $x = 1$

At $x = 2$ $f''(x) = 0$

$\therefore$ f is neither maximum nor minimum at $x = 2$.

At $x = 7/5$

$$f''(x) = 5\left(\frac{7}{5}-1\right)\left(\frac{7}{5}-2\right)^2 = 5\times\frac{2}{5}\times\frac{9}{25} = \frac{18}{25} > 0$$

$\therefore$ $f(x)$ is minimum at $x = 7/5$.

16. We have $y = x(x-1)^2,\ 0 \le x \le 2$

$$\frac{dy}{dx} = (x-1)^2 + 2x(x-1) = (x-1)(3x-1)$$

For max. or min. $\frac{dy}{dx} = 0$

$\Rightarrow$ $(x-1)(3x-1) = 0$

$\Rightarrow$ $x = 1, 1/3$

$$\frac{d^2y}{dx^2} = 3x - 1 + 3(x-1) = 6x - 4$$

At $x = 1$, $\frac{d^2y}{dx^2} = 2(+ve)$ $\therefore$ y is min. at $x = 1$

At $x = 1/3$, $\frac{d^2y}{dx^2} = -2(-ve)$ $\therefore$ y is max. at $x = 1/3$

$\therefore$ Max value of y is $= \frac{1}{3}\left(\frac{1}{3}-1\right)^2 = \frac{4}{27}$

Min value of y is $= 1(1-1)^2 = 0$

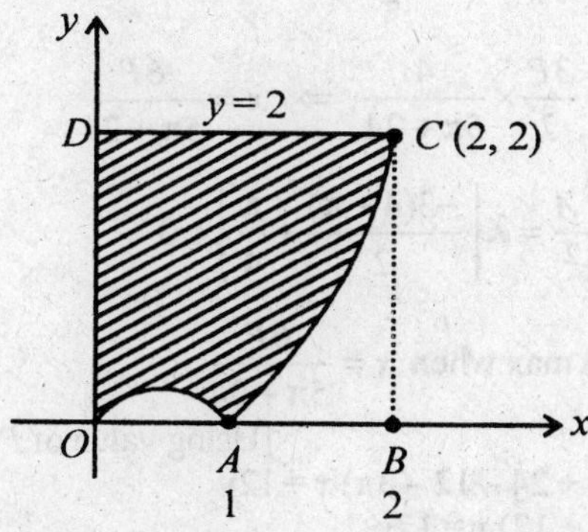

Now the curve cuts the axis x at (0, 0) and (1, 0). When x increases from 1 to 2, y also increases and is +ve.

When $y = 2$, $x(x-1)^2 = 2$

$\Rightarrow$ $x^3 - 2x^2 + x - 2 = 0$

$\Rightarrow$ $(x-2)(x^2+1) = 0$

$\Rightarrow$ $x = 2$

Using max./min. values of y and points of intersection with x-axis, we get the curve as in figure and shaded area is the required area.

$\therefore$ The required area

$$= \text{Area of square } OBCD - \int_0^2 y\,dx$$

$$= 2\times 2 - \int_0^2 x(x-1)^2\,dx$$

$$= 4 - \left[\left(x\frac{(x-1)^3}{3}\right)_0^2 - \frac{1}{3}\int_0^2 (x-1)^3.1\,dx\right]$$

$$= 4 - \left[\frac{x}{3}(x-1)^3 - \frac{(x-1)^4}{12}\right]_0^2$$

$$= 4 - \left[\frac{2}{3} - \frac{1}{12} + \frac{1}{12}\right] = 4 - \frac{2}{3} = \frac{10}{3}\text{ sq.units.}$$

17. Let $f(x) = 2\sin x + \tan x - 3x$

$\therefore$ $f'(x) = 2\cos x + \sec^2 x - 3$

$= (\sec^2 x - 1) - 2(1 - \cos x)$

$$= \tan^2 x - 4\sin^2 x/2 = \frac{\sin^2 x}{\cos^2 x} - 4\sin^2 x/2$$

$$= \frac{4\sin^2 x/2\cos^2 x/2}{\cos^2 x} - 4\sin^2 x/2$$

$$= 4\sin^2 x/2\left[\frac{\cos^2 x/2}{\cos^2 x} - 1\right] \quad ...(i)$$

As $\cos x$ is decreasing function

$\therefore$ $x/2 \le x$ $\Rightarrow$ $\cos x/2 \ge \cos x$

$\Rightarrow$ $\cos^2 x/2 \ge \cos^2 x$ (as both are +ve)

$\Rightarrow$ $\frac{\cos^2 x/2}{\cos^2 x} \ge 1$ $\Rightarrow$ $\frac{\cos^2 x/2}{\cos^2 x} - 1 \ge 0$

$\therefore$ $4\sin^2 x/2\left[\frac{\cos^2 x/2}{\cos^2 x} - 1\right] \ge 0$

$\therefore$ From eq. (1), we get $f'(x) \ge 0$

$\Rightarrow$ $f(x)$ is an increasing function on $(0, \pi/2)$ for $x \ge 0$

$\Rightarrow$ $f(x) \ge f(0)$

$\Rightarrow$ $2\sin x + \tan x - 3x \ge 0$

$\Rightarrow$ $2\sin x + \tan x \ge 3x$. Hence Proved

ALTERNATE SOLUTION

Let $f(x) = 2\sin x + \tan x - 3x$ on $0 \le x < \pi/2$

then $f'(x) = 2\cos x + \sec^2 x - 3$

and $f''(x) = -2\sin x + 2\sec^2 x\tan x$

$= 2\sin x[\sec^3 x - 1]$

for $0 \le x < \pi/2$ $f''(x) \ge 0$

$\Rightarrow$ $f'(x)$ is an increasing function on $0 \le x < \pi/2$.

$\therefore$ For $x \ge 0$, $\Rightarrow$ $f'(x) \ge f'(0)$

$\Rightarrow$ $f'(x) \ge 0$ *or* $0 \le x < \pi/2$

$\Rightarrow$ $f(x)$ is an increasing function on $0 \le x < \pi/2$

$\therefore$ For $x \ge 0$, $f(x) \ge f(0)$

$\Rightarrow$ $2\sin x + \tan x - 3x \ge 0,\ 0 \le x < \pi/2$

$\Rightarrow$ $2\sin x + \tan x \ge 3x,\ 0 \le x < \pi/2$ Hence proved

18. As $QR \parallel XY$ diameter through P is $\perp$ QR.

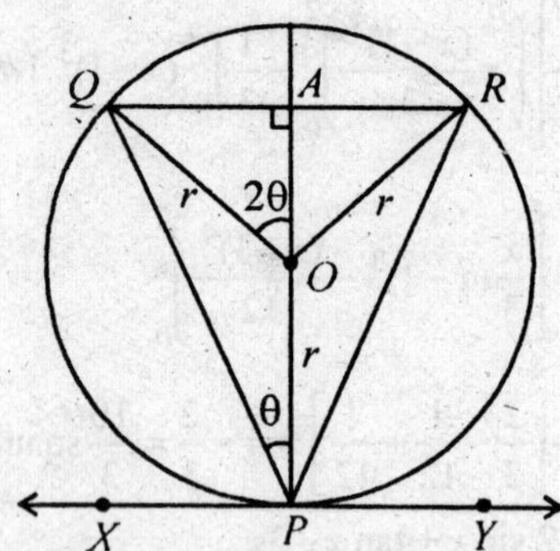

Now area of ΔPQR is given by

$$A = \frac{1}{2} QR.AP$$

But $QR = 2.QA = 2r \sin 2\theta$
and $PA = OA + OP = r\cos 2\theta + r$

$$\therefore \quad A = \frac{1}{2}.2r\sin 2\theta.(r + r\cos 2\theta)$$

$$= r^2.\,2\sin\theta\cos\theta\,.\,2\cos^2\theta = 4r^2\sin\theta\cos^3\theta$$

For max. value of area, $\frac{dA}{d\theta} = 0$

$\Rightarrow$ $4r^2[\cos^4\theta - 3\sin^2\theta\cos^2\theta] = 0$
$\Rightarrow$ $\cos^2\theta(\cos^2\theta - 3\sin^2\theta) = 0$

$\Rightarrow$ $\tan\theta = \frac{1}{\sqrt{3}}$ $\Rightarrow$ $\theta = 30°$

Also $\frac{d^2A}{d\theta^2} = 4r^2[-4\cos^3\theta\sin\theta - 6\sin\theta\cos^3\theta$

$$+ 6\sin^3\theta\cos\theta]$$

$$= 4r^2[-10\sin\theta\cos^3\theta + 6\sin^3\theta\cos\theta]$$

$$\left.\frac{d^2A}{d\theta^2}\right|_{\theta=30°} = 4r^2\left[-10.\frac{1}{2}.\frac{3\sqrt{3}}{8} + 6.\frac{1}{8}.\frac{\sqrt{3}}{2}\right]$$

$$= 4r^2\left[\frac{-15\sqrt{3}}{8} + \frac{3\sqrt{3}}{8}\right] = 4r^2\left(\frac{-12\sqrt{3}}{8}\right) = -ve$$

$\therefore$ A is maximum at $\theta = 30°$
And $A_{max} = 4r^2\sin 30°\cos^3 30°$

$$= 4r^2 \times \frac{1}{2} \times \frac{3\sqrt{3}}{8} = \frac{3\sqrt{3}}{4}r^2$$

19. Let $ABCEDA$ be the window as shown in the figure and let
$AB = x$ m
$BC = y$ m

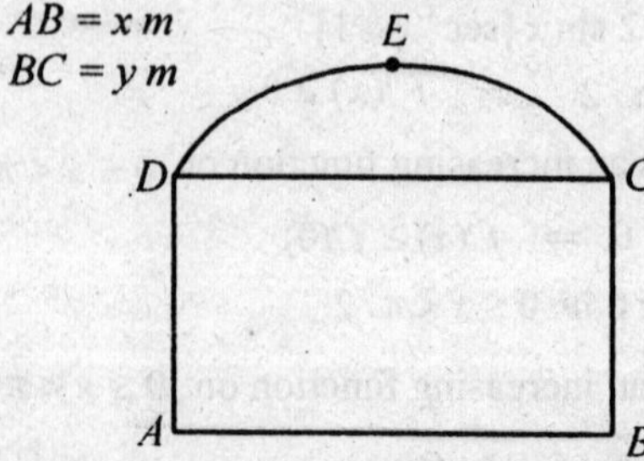

Then its perimeter including the base DC of arch

$$= \left(2x + 2y + \frac{\pi x}{2}\right) m$$

$$\therefore \quad P = \left(2 + \frac{\pi}{2}\right)x + 2y \quad \text{...(1)}$$

Now, area of rectangle $ABCD = xy$

and area of arch $DCED = \frac{\pi}{2}\left(\frac{x}{2}\right)^2$

Let λ be the light transmitted by coloured glass per sq. m. Then 3λ will be the light transmitted by clear glass per sq. m. Hence the area of light transmitted

$$= 3\lambda(xy) + \lambda\left[\frac{\pi}{2}\left(\frac{x}{2}\right)^2\right]$$

$$\Rightarrow \quad A = \lambda\left[3xy + \frac{\pi x^2}{8}\right] \quad \text{...... (2)}$$

Substituting the value of y from (1) in (2), we get

$$A = \lambda\left[3x\frac{1}{2}\left[P - \left(\frac{4+\pi}{2}\right)x\right] + \frac{\pi x^2}{8}\right]$$

$$= \lambda\left[\frac{3Px}{2} - \frac{3(4+\pi)}{4}x^2 + \frac{\pi x^2}{8}\right]$$

$$\therefore \quad \frac{dA}{dx} = \lambda\left[\frac{3P}{2} - \frac{3(4+\pi)}{2}x + \frac{\pi x}{4}\right]$$

For A to be maximum $\frac{dA}{dx} = 01$

$$\Rightarrow \quad x = \frac{\frac{3P}{2}}{\frac{-\pi}{4} + \left(\frac{12+3\pi}{2}\right)}$$

$$\Rightarrow \quad x = \frac{3P}{2} \times \frac{4}{5\pi + 24} \Rightarrow x = \frac{6P}{5\pi + 24}$$

Also $\frac{d^2A}{dx^2} = \lambda\left[\frac{-3(4+\pi)}{2} + \frac{\pi}{4}\right] < 0$

$\therefore$ A is max when $x = \frac{6P}{5\pi + 24}$

[Using value of P from (1)]

$\Rightarrow$ $(5\pi + 24 - 12 - 3\pi)x = 12y$
$\Rightarrow$ $(2\pi + 12)x = 12y$

$\Rightarrow$ $\frac{y}{x} = \frac{\pi + 6}{6}$

$\therefore$ The required ratio of breadth to length of the rectangle $= 6 + \pi : 6$

20. Let $f(x) = ax^3 + bx^2 + cx + d$
ATQ, $f(x)$ vanishes at $x = -2$
$\Rightarrow -8a + 4b - 2c + d = 0$ ...(1)
$f'(x) = 3ax^2 + 2bx + c$
Againg ATQ, $f(x)$ has relative max./min at

$$x = -1 \text{ and } x = \frac{1}{3}$$

$\Rightarrow$ $f'(-1) = 0 = f'(1/3)$

$\Rightarrow \quad 3a-2b+c=0$ ...(2)

and $a+2b+3c=0$ ...(3)

Also, $\int_{-1}^{1} f(x)=\frac{14}{3}$

$\Rightarrow \quad \left(\frac{ax^4}{4}+\frac{bx^3}{3}+\frac{cx^2}{2}+dx\right)_{-1}^{1}=\frac{14}{3}$

$\Rightarrow \quad \left[\frac{a}{4}+\frac{b}{3}+\frac{c}{2}+d\right]-\left[\frac{a}{4}-\frac{b}{3}+\frac{c}{2}-d\right]=\frac{14}{3}$

$\Rightarrow \quad \frac{b}{3}+d=\frac{7}{3}$

$\Rightarrow \quad b+3d=7$ ...(4)

From (1), (2), (3), (4) on solving, we get

$a=1, b=1, c=-1, d=2$

$\therefore$ The required cubic is x^3+x^2-x+2.

21. The given curve is $y=x^2$ ...(1)

Consider any point $A\ (t, t^2)$ on (1) at which normal chord drawn is shortest.

Then eq. of normal to (1) at $A\ (t, t^2)$ is

$$y-t^2=-\frac{1}{\left(\frac{dy}{dx}\right)_{(t,t^2)}}(x-t) \quad \left[\text{where } \frac{dy}{dx}=2x \text{ from (1)}\right]$$

$$y-t^2=-\frac{1}{2t}(x-t)$$

$\Rightarrow \quad x+2ty=t+2t^3$ ...(2)

This normal meets the curve again at point B which can be obtained by solving (1) and (2) as follows :

Putting $y=x^2$ in (2), we get

$2t\,x^2+x-(t+2t^3)=0,$

$D=1+8t\,(t+2t^3)$

$=1+8t^2+16t^4=(1+4t^2)^2$

$$\therefore \quad x=\frac{-1+1+4t^2}{4t}, \frac{-1-1-4t^2}{4t}=t, -\frac{1}{2t}-t$$

$$\therefore \quad y=t^2, t^2+\frac{1}{4t^2}+1$$

Thus, $B\left(-t-\frac{1}{2t}, t^2+\frac{1}{4t^2}+1\right)$

$\therefore$ Length of normal chord

$$AB=\sqrt{\left(2t+\frac{1}{2t}\right)^2+\left(\frac{1}{4t^2}+1\right)^2}$$

Consider $Z=AB^2=\left(2t+\frac{1}{2t}\right)^2+\left(\frac{1}{4t^2}+1\right)^2$

$$\Rightarrow \quad Z=\frac{1}{16t^4}+\frac{3}{4t^2}+3+4t^2$$

For shortest chord, we have to minimize Z, and for that $\frac{dZ}{dt}=0$

$$\Rightarrow \quad -\frac{1}{4t^5}-\frac{3}{2t^3}+8t=0$$

$\Rightarrow \quad -1-6t^2+32t^6=0$

$\Rightarrow \quad 32\,(t^2)^3-6t^2-1=0 \Rightarrow (2t^2-1)(16t^4+8t^2+1)=0$

$\Rightarrow \quad t^2=\frac{1}{2}$ (leaving –ve values of t^2)

$\Rightarrow \quad t=\frac{1}{\sqrt{2}}, -\frac{1}{\sqrt{2}},$

$$\frac{d^2Z}{dt^2}=\frac{5}{4t^6}+\frac{9}{2t^4}+8$$

$$\left.\frac{d^2Z}{dt^2}\right|_{t=\frac{1}{\sqrt{2}}}=+ve \text{ also } \left.\frac{d^2Z}{dt^2}\right|_{t=-\frac{1}{\sqrt{2}}}=+ve$$

$\therefore$ Z is minimum at $t=\frac{1}{\sqrt{2}}$ or $-\frac{1}{\sqrt{2}}$

For $t=\frac{1}{\sqrt{2}}$ normal chord is (from (2)) $x+\sqrt{2}y=\sqrt{2}$

For $t=-\frac{1}{\sqrt{2}}$ normal chord is $x-\sqrt{2}y=-\sqrt{2}$

22. The given curve is $y=(1+x)^y+\sin^{-1}(\sin^2 x)$

Here at $x=0, y=(1+0)^y+\sin^{-1}(0) \Rightarrow y=1$

$\therefore$ Point at which normal has been drawn is (0, 1).

For slope of normal we need to find dy/dx, and for that we consider the curve as

$$y=u+v \Rightarrow \frac{dy}{dx}=\frac{du}{dx}+\frac{dv}{dx}$$

where $u=(1+x)^y$ ...(i)

and $v=\sin^{-1}(\sin^2 x)$ ...(ii)

Taking log on both sides of equation (i) we get

$\log u=y\log(1+x)$

$$\Rightarrow \quad \frac{1}{u}\frac{du}{dx}=\frac{y}{1+x}+\log(1+x).\frac{dy}{dx}$$

$$\Rightarrow \quad \frac{du}{dx}=(1+x)^y\left[\frac{y}{1+x}+\log(1+x)\frac{dy}{dx}\right]$$

Also $v=\sin^{-1}(\sin^2 x)$

$$\Rightarrow \quad \frac{dv}{dx}=\frac{1}{\sqrt{1-\sin^4 x}}.2\sin x\cos x$$

$$\Rightarrow \quad \frac{dv}{dx}=\frac{2\sin x}{\sqrt{1+\sin^2 x}}.$$

Thus, we get,

$$\frac{dy}{dx}=(1+x)^y\left[\frac{y}{1+x}+\log(1+x)\frac{dy}{dx}\right]+\frac{2\sin x}{\sqrt{1+\sin^2 x}}$$

$$\Rightarrow \quad [1-(1+x)^y\log(1+x)]\frac{dy}{dx}$$

$$=y(1+x)^{y-1}+\frac{2\sin x}{\sqrt{1+\sin^2 x}}$$

$$\Rightarrow \frac{dy}{dx} = \frac{y(1+x)^{y-1} + \frac{2\sin x}{\sqrt{1+\sin^2 x}}}{1-(1+x)^y \log(1+x)}$$

$$\left.\frac{dy}{dx}\right|_{(0,1)} = 1, \quad \therefore \text{ Slope of normal} = -1$$

$\therefore$ Equation of normal to given curve at (0, 1) is
$y - 1 = -1(x - 0)$
$\Rightarrow x + y = 1.$

23. We have, $f(x) = \begin{cases} -x^3 + \dfrac{b^3 - b^2 + b - 1}{b^3 + 3b + 2}, & 0 \le x < 1 \\ 2x - 3, & 1 \le x \le 3 \end{cases}$

We can see from definition of the function, that
$f(1) = 2(1) - 3 = -1$
Also $f(x)$ is increasing on $[1, 3]$, $f'(x)$ being $2 > 0$.
$\therefore$ $f(1) = -1$ is the smallest value of $f(x)$
Again $f'(x) = -3x^2$ for $x \in [0, 1]$ such that $f'(x) < 0$
$\Rightarrow$ $f(x)$ is decreasing on $[0, 1]$
$\therefore$ For fixed value of b, its smallest occur when $x \to 1$

i.e., $\lim\limits_{h\to 0} f(1-h) = \lim\limits_{h\to 0} -(1-h)^3 + \dfrac{b^3 - b^2 + b - 1}{b^2 + 3b + 2}$

$$= -1 + \frac{b^3 - b^2 + b - 1}{b^2 + 3b + 2}$$

As given that the smallest value of $f(x)$ occur at $x = 1$
$\therefore$ Any other smallest value $\ge f(1)$

$$\Rightarrow -1 + \frac{b^3 - b^2 + b - 1}{b^2 + 3b + 2} \ge -1$$

$$\Rightarrow \frac{b^3 - b^2 + b - 1}{b^2 + 3b + 2} \ge 0 \Rightarrow \frac{(b^2+1)(b-1)}{(b+2)(b+1)} \ge 0$$

$\Rightarrow (b-1)(b+1)(b+2) \ge 0$

– + – +
←——|——|——|——→
−2 −1 1

$\Rightarrow b \in (-2, -1) \cup (1, \infty).$

24. Given that $y = ax^3 + bx^2 + cx + 5$ touches the x-axis at $P(-2, 0)$

$\Rightarrow \left(\dfrac{dy}{dx}\right)_{x=-2} = 0$ and $P(-2, 0)$ lies on curve
$\Rightarrow 3ax^2 + 2bx + c]_{x=-2} = 0$
$\Rightarrow 12a - 4b + c = 0$ ...(1)
and $-8a + 4b - 2c + 5 = 0$ ...(2)
[$\because$ (–2, 0) lies on curve]
Also the curve cuts the y-axis at Q
$\therefore$ For $x = 0, y = 5$ $\therefore Q(0, 5)$
At Q gradient of the curve is 3

$\Rightarrow \left.\dfrac{dy}{dx}\right|_{x=0} = 3 \Rightarrow 3x^2 + 2bx + c]_{x=0} = 3$
$\Rightarrow c = 3$ ...(3)
Solving (1), (2) and (3), we get
$a = -1/2, b = -3/4$ and $c = 3$.

25. The given circle is $x^2 + y^2 = 1$ ...(1)
which intersect x-axis at $P(-1, 0)$ and $Q(1, 0)$.
Let radius of circle with centre at $Q(1, 0)$ be r, where r is variable.
Then equation of this circle is,
$(x-1)^2 + y^2 = r^2$ ...(2)

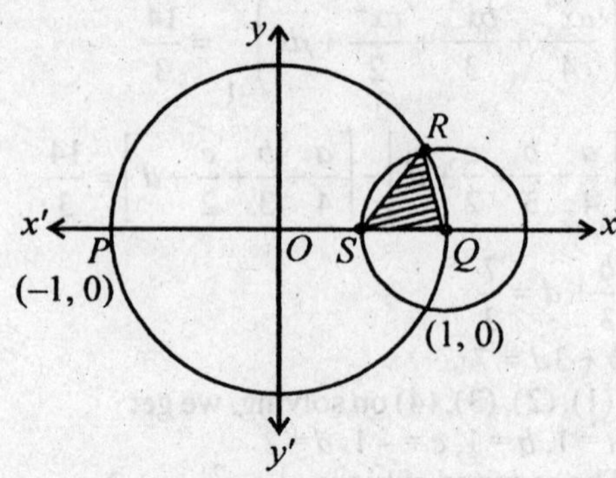

Subtracting (1) from (2) we get
$(x-1)^2 - x^2 = (r^2 - 1)$

$$\Rightarrow -2x + 1 = r^2 - 1 \Rightarrow x = 1 - \frac{r^2}{2}$$

Substituting this value of x in (2), we get

$$\frac{r^4}{4} + y^2 = r^2 \quad \Rightarrow \quad y = \pm r\sqrt{1 - \frac{r^2}{4}}$$

$\therefore$ $R\left(1 - \dfrac{r^2}{2}, r\sqrt{1 - \dfrac{r^2}{4}}\right)$ point being above x-axis.

$\therefore$ Area of $\Delta QRS = \dfrac{1}{2} SQ \times$ ordinate of point R

$$\Rightarrow A = \frac{1}{2} \times r \times r\sqrt{1 - \frac{r^2}{4}}$$

A will be max. if A^2 is max.

$$A^2 = \frac{r^4}{4}\left(1 - \frac{r^2}{4}\right) = \frac{r^4}{4} - \frac{r^6}{16}$$

Differentiating A^2 w.r. to r, we get

$$\frac{dA^2}{dr} = r^3 - \frac{3}{8}r^5$$

For A^2 to be max. $\dfrac{dA^2}{dr} = 0$

$$\Rightarrow r^3\left(1 - \frac{3}{8}r^2\right) = 0 \Rightarrow r = \frac{2\sqrt{2}}{\sqrt{3}}$$

$$\frac{d^2(A^2)}{dr^2} = 3r^2 - \frac{15}{8}r^4$$

$$\Rightarrow \left.\frac{d^2(A^2)}{dr^2}\right|_{r^2=\frac{8}{3}} = 3 \times \frac{8}{3} - \frac{15}{8} \times \frac{64}{9} = -ve$$

$\therefore$ A^2 and hence A is max. when, $r = \dfrac{2\sqrt{2}}{\sqrt{3}}$

$\therefore$ Max. area $= \sqrt{\frac{1}{4}\left(\frac{2\sqrt{2}}{\sqrt{3}}\right)^4 - \frac{1}{16}\left(\frac{2\sqrt{2}}{\sqrt{3}}\right)^6}$

$= \sqrt{\frac{1}{4}\times\frac{64}{9} - \frac{1}{16}\times\frac{512}{27}} = \sqrt{\frac{16}{9} - \frac{32}{27}}$

$= \frac{4}{3\sqrt{3}} = \frac{4\sqrt{3}}{9}$ sq. units.

26. Let the given line be $\frac{x}{a} + \frac{y}{b} = 1$, so that it makes an intercept of a units on x-axis and b units on y-axis. As it passes through the fixed point (h, k), therefore we must have

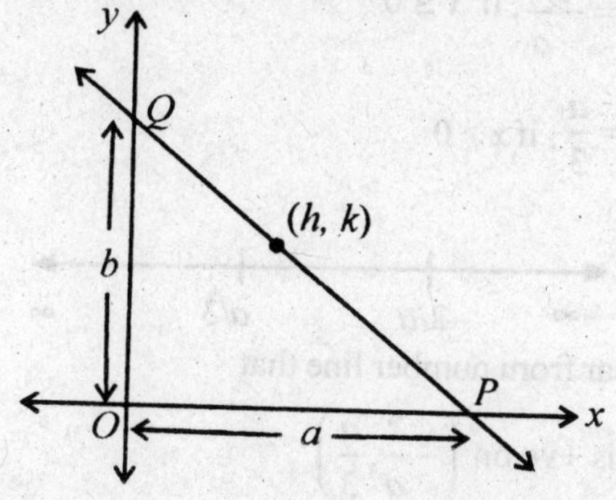

$\Rightarrow \frac{k}{b} = 1 - \frac{h}{a} \quad \Rightarrow \quad b = \frac{ak}{a-h}$ ...(1)

Now Area of $\Delta OPQ = A = \frac{1}{2}ab$

$\therefore \quad A = \frac{1}{2}a\left(\frac{ak}{a-h}\right)$ [using (1)]

or $\quad A = \frac{k}{2}\left[\frac{a^2}{a-h}\right]$

For min. value of A, $\frac{dA}{da} = 0$

$\Rightarrow \frac{k}{2}\left[\frac{2a(a-h) - a^2}{(a-h)^2}\right] = 0$

$\Rightarrow \frac{k}{2}\left[\frac{a^2 - 2ah}{(a-h)^2}\right] = 0 \Rightarrow a = 2h$

Also, $\frac{d^2A}{da^2} = \frac{(2a-2h)(a-h)^2 - 2(a-h)(-1)(a^2 - 2ah)}{(a-h)^4}$

$\therefore \left.\frac{d^2A}{da^2}\right|_{a=2h} = \frac{(2h^3 + 2h)(0)}{h^4} = \frac{2}{h} > 0, [\because h > 0]$

$\therefore$ A is min. when $a = 2h$

$\therefore \quad A_{min} = \frac{k}{2}\left[\frac{4h^2}{h}\right] = 2kh$

27. The normal to the curve at P is
$a(y-1) + (x-1) = 0$

First we consider the case when $a \neq 0$

Slope of normal at $P(1, 1)$ is $= -\frac{1}{a}$

$\Rightarrow$ Slope of the tangent at (1, 1) is $= a$

$\Rightarrow \left(\frac{dy}{dx}\right)_{(1,1)} = a$ ...(1)

But we are given that

$\frac{dy}{dx} \propto y \Rightarrow \frac{dy}{dx} = ky \Rightarrow \frac{dy}{y} = kdx$

$\Rightarrow \log|y| = kx + C \Rightarrow |y| = e^{kx+c} = e^c.e^{kx}$

$\Rightarrow y = \pm e^c e^{kx} \Rightarrow y = Ae^{kx}$

Where A is constant. As the curve passes through (1, 1)

$\therefore \quad 1 = Ae^k \quad \Rightarrow A = e^{-k}$

$\therefore \quad y = e^{k(x-1)} \quad \Rightarrow \frac{dy}{dx} = ke^{k(x-1)}$

$\Rightarrow \left(\frac{dy}{dx}\right)_{(1,1)} = k$

From (1) and (2), $\left(\frac{dy}{dx}\right)_{1,1} = a = k$

$\therefore \quad y = e^{a(x-1)}$ which is the required curve.

Now the area bounded by the curve, y-axis and normal to curve at (1, 1) is as shown the shaded region in the fig.

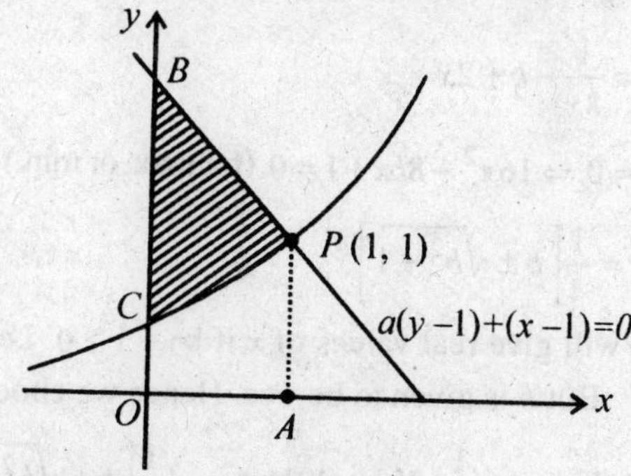

$\therefore$ Req. area $= ar(PBC)$
$= ar(OAPBCO) - ar(OAPCO)$

$= \int_0^1 y_{\text{normal}}\,dx - \int_0^1 y_{\text{curve}}\,dx$

$= \int_0^1\left(-\frac{1}{a}(x-1) + 1\right)dx - \int_0^1 e^{a(x-1)}dx$

$= \left[-\frac{1}{2a}(x-1)^2 + x\right]_0^1 - \left[\frac{1}{a}e^{a(x-1)}\right]_0^1$

$= 1 + \frac{1}{2a} - \frac{1}{a} + \frac{1}{a}e^{-a} = 1 + \frac{1}{a}e^{-a} - \frac{1}{2a}$

Now we consider the case when $a = 0$. Then normal at (1, 1) becomes $x - 1 = 0$ which is parallel to y-axis, therefore tangent at (1, 1) should be parallel to x-axis. Thus

$$\left(\frac{dy}{dx}\right)_{(1,1)} = 0 \qquad \ldots(3)$$

Since $\frac{dy}{dx} \propto y$ gives $y = e^{k(x-1)}$

(as in $a \neq 0$ case)

$$\Rightarrow \frac{dy}{dx} = ke^{k(x-1)}$$

$$\left(\frac{dy}{dx}\right)_{(1,1)} = k \qquad \ldots(4)$$

From (3) and (4), we get $k = 0$ and required curve becomes $y = 1$

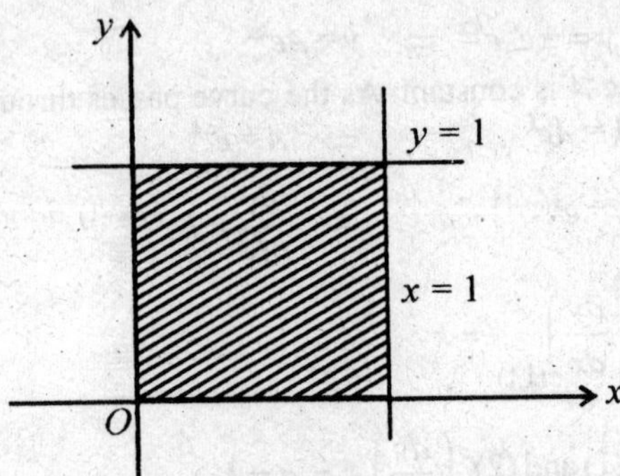

In this case the required area = shaded area in fig. = 1 sq. unit.

28. $f(x) = \frac{1}{8}\ell n x - bx + x^2, x > 0, b \geq 0$

$$f'(x) = \frac{1}{8x} - b + 2x \qquad \ldots(1)$$

$f'(x) = 0 \Rightarrow 16x^2 - 8bx + 1 = 0$ (for max. or min.)

$$\therefore \quad x = \frac{1}{4}\left[b \pm \sqrt{b^2 - 1}\right] \qquad \ldots(2)$$

Above will give real values of x if $b^2 - 1 \geq 0$ i.e. $b \geq 1$ or $b \leq -1$. But b is given to be +ve. Hence we choose $b \geq 1$

If $b = 1$ then $x = \frac{1}{4}$; If $b > 1$ then $x = \frac{1}{4}\left[b \pm \sqrt{b^2 - 1}\right]$

$$f''(x) = -\frac{1}{8x^2} + 2 = \frac{16x^2 - 1}{8x^2}$$

Its sign will depend on N^r, $16x^2 - 1$ as $8x^2$ is +ve. We shall consider its sign for $x = \frac{1}{4}$ and $x = \frac{1}{4}\left[b \pm \sqrt{b^2 - 1}\right]$

$f''(x) = 0$ at $x = 1/4$

$\therefore$ Neither max. nor min. as $f''(x) = 0$

N^r of $f''(x) = 16x^2 - 1 = [b + \sqrt{b^2 - 1}]^2 - 1$
$= +$ve for $b > 1$ $\quad \therefore$ Minima

or N^r of $f''(x) = (b - \sqrt{b^2 - 1})^2 - 1$
$= -$ve for $b > 1$ $\quad \therefore$ Maxima

29. Given that, $f(x) = \begin{cases} xe^{ax}, & x \leq 0 \\ x + ax^2 - x^3, & x > 0 \end{cases}$

Differentiating both sides, we have

$$f'(x) = \begin{cases} axe^{ax} + e^{ax}, & x \leq 0 \\ 1 + 2ax - 3x^2, & x > 0 \end{cases}$$

Again differentiating both sides, we have

$$f''(x) = \begin{cases} 2ae^{ax} + a^2 x e^{ax}; & x \leq 0 \\ 2a - 6x; & x > 0 \end{cases}$$

For critical points, we put $f''(x) = 0$

$\Rightarrow \quad x = -\frac{2}{a}$, if $x \leq 0$

$= \frac{a}{3}$, if $x > 0$

– + –

$-\infty \quad -2/a \quad a/3 \quad \infty$

It is clear from number line that

$f''(x)$ is +ve on $\left(-\frac{2}{a}, \frac{a}{3}\right)$

$\Rightarrow \quad f'(x)$ increases on $\left(-\frac{2}{a}, \frac{a}{3}\right)$

30. Let $b - a = t$, where $a + b = 4$

$\Rightarrow \quad a = \frac{4-t}{2}$ and $b = \frac{t+4}{2}$

as given $a < 2$ and $b > 2 \Rightarrow t > 0$

Now $\int_0^a g(x)\,dx + \int_0^b g(x)\,dx$

$= \int_0^{\frac{4-t}{2}} g(x)\,dx + \int_0^{\frac{4+t}{2}} g(x)\,dx = \phi(t)$ [say]

$\Rightarrow \phi'(t) = g\left(\frac{4-t}{2}\right)\left(-\frac{1}{2}\right) + g\left(\frac{4+t}{2}\right)\left(\frac{1}{2}\right)$ **NOTE THIS STEP**

$\left[\text{Using } \frac{d}{dx}\left[\int_{u(x)}^{v(x)} f(t)\,dt\right] = f[v(x)].v'(x) - f[u(x)].u'(x)\right]$

$$= \frac{1}{2}\left[g\left(\frac{4+t}{2}\right) - g\left(\frac{4-t}{2}\right)\right]$$

Since $g(x)$ is an increasing function (given)

$\therefore$ for $x_1 > x_2 \Rightarrow g(x_1) > g(x_2)$

Here we have $\left(\frac{4+t}{2}\right) > \left(\frac{4-t}{2}\right)$

$$\Rightarrow g\left(\frac{4+t}{2}\right) > g\left(\frac{4-t}{2}\right)$$

$$\Rightarrow \phi'(t) = \frac{1}{2}\left[g\frac{(4+t)}{2} - g\frac{(4-t)}{2}\right] > 0 \Rightarrow \phi'(t) > 0$$

Hence $\phi(t)$ increase as t increases.

$\Rightarrow \int_0^a g(x)dx + \int_0^b g(x)dx$ increases as $(b-a)$ increases.

31. Applying $R_3 \to R_3 - R_1 - 2R_2$ we get

$$f'(x) = \begin{vmatrix} 2ax & 2ax-a & 2ax+b+1 \\ b & b+1 & -1 \\ 0 & 0 & 1 \end{vmatrix}$$

$$= \begin{vmatrix} 2ax & 2ax-1 \\ b & b+1 \end{vmatrix} = \begin{vmatrix} 2ax & -1 \\ b & 1 \end{vmatrix}$$

[Using $C_2 \to C_2 - C_1$]

$\Rightarrow f'(x) = 2ax + b$

Integrating, we get , $f(x) = ax^2 + bx + C$

where C is an arbitrary constant. Since f has a maximum at $x = 5/2$,

$f'(5/2) = 0 \Rightarrow 5a + b = 0$ ... (1)

Also $f(0) = 2 \Rightarrow C = 2$

and $f(1) = 1 \Rightarrow a + b + c = 1$

$\therefore \quad a + b = -1$ ... (2)

Solving (1) and (2) for a, b we get,

$a = 1/4, b = -5/4$

Thus, $f(x) = \frac{1}{4}x^2 - \frac{5}{4}x + 2.$

32. Equation of the tangent at point (x, y) on the curve is

$$Y - y = \frac{dy}{dx}(X - x)$$

This meets axes in

$A\left(x - y\frac{dx}{dy'} 0\right)$ and $B\left(0, y - x\frac{dy}{dx}\right)$

Mid-point of AB is $\left(\frac{1}{2}\left(x - y\frac{dx}{dy}\right), \frac{1}{2}\left(y - x\frac{dy}{dx}\right)\right)$

We are given

$\frac{1}{2}\left(x - y\frac{dx}{dy}\right) = x$ and $\frac{1}{2}\left(y - x\frac{dy}{dx}\right) = ya$

$\Rightarrow x\frac{dy}{dx} = -y \Rightarrow \frac{dy}{y} = -\frac{dx}{x}$

Intergrating both sides,

$$\int\frac{dy}{y} = -\int\frac{dy}{x} \Rightarrow \log y = -\log x + c$$

Put $x = 1, y = 1$,

$\Rightarrow \log 1 = -\log 1 + c \Rightarrow c = 0$

$\Rightarrow \log y + \log x = 0 \Rightarrow \log yx = 0 \Rightarrow yx = e^0 = 1$

Which is a rectangular hyperbola.

33. Given that,

$p(x) = a_0 + a_1 x + a_2 x^2 + \ldots + a_n x^n \ldots (1)$

and $|p(x)| \le |e^{x-1} - 1|, \forall x \ge 0$ ... (2)

To prove that,

$|a_1 + 2a_2 + \ldots + n a_n| \le 1$

It can be clearly seen that in order to prove the result it is sufficient to prove that $|p'(1)| \le 1$

We know that,

$$|p'(1)| = \lim_{h\to 0}\left|\frac{p(1+h) - p(1)}{h}\right|$$

$$\le \lim_{h\to 0}\frac{|p(1+h)| + |p(1)|}{|h|}$$ **NOTE THIS STEP**

[Using $|x - y| \le |x| + |y|$]

But $|p(1)| \le |e^0 - 1|$

[Using equation (2) for $x = 1$]

$\Rightarrow |p(1)| \le 0$

But being absolute value, $|p(1)| \ge 0$.

Thus we must have $|p(1)| = 0$

Also $|p(1+h)| \le |e^h - 1|$

(Using eq^n (2) for $x = 1 + h$)

Thus $|p'(1)| \le \lim_{h\to 0}\frac{|e^h - 1|}{|h|} = 1$

or $|p'(1)| \le 1$

$\Rightarrow |a_1 + 2a_2 + \ldots.. + na^n| \le 1$

34. Given that $-1 \le p \le 1$.

Consider $f(x) = 4x^3 - 3x - p = 0$

Now, $f(1/2) = \frac{1}{2} - \frac{3}{2} - p = -1 - p \le 0$ as $(-1 \le p)$

Also $f(1) = 4 - 3 - p = 1 - p \ge 0$ as $(p \le 1)$

$\therefore$ $f(x)$ has at least one real root between $[1/2, 1]$.

Also $f'(x) = 12x^2 - 3 > 0$ on $[1/2, 1]$

$\Rightarrow$ f is increasing on $[1/2, 1]$

$\Rightarrow$ f has only one real root between $[1/2, 1]$

To find the root, we observe $f(x)$ contains $4x^3 - 3x$ which is multipe angle formula of $\cos 3\theta$ if we put $x = \cos\theta$.

$\therefore$ Let the req. root be $\cos\theta$ then,

$4\cos^3\theta - 3\cos\theta - p = 0$

$\Rightarrow \cos 3\theta = p \Rightarrow 3\theta = \cos^{-1} p$

$\Rightarrow \theta = \frac{1}{3}\cos^{-1}(p)$

$\therefore$ Root is $\cos\left(\frac{1}{3}\cos^{-1}(p)\right)$.

35. The given curve is $\frac{x^2}{6} + \frac{y^2}{3} = 1$ (an ellipse)

Any parametric point on it is $P(\sqrt{6}\cos\theta, \sqrt{3}\sin\theta)$.

Its distance from line $x + y = 7$ is given by

$$D = \frac{\sqrt{6}\cos\theta + \sqrt{3}\sin\theta - 7}{\sqrt{2}}$$

For min. value of D, $\frac{dD}{d\theta} = 0$

$\Rightarrow\quad -\sqrt{6}\sin\theta+\sqrt{3}\cos\theta=0$

$\Rightarrow\quad \tan\theta = 1/\sqrt{2}$

$\Rightarrow\quad \cos\theta=\dfrac{\sqrt{2}}{\sqrt{3}}$ and $\sin\theta=\dfrac{1}{\sqrt{3}}$

$\therefore$ Required point P is (2, 1)

ALTERNATE SOLUTION :

Any point on given ellipse $\dfrac{x^2}{6}+\dfrac{y^2}{3}=1$ is

$P(\sqrt{6}\cos\theta, \sqrt{3}\sin\theta)$

Distance of P from $x + y = 7$ will be minimum if measured along the normal of ellipse at P, and then $PN \perp$ given line.

$\Rightarrow$ (slope of normal at P) × (slope of given line) = – 1

$$\Rightarrow\quad \left(-\frac{dx}{dy}\right)_{(\sqrt{6}\cos\theta,\sqrt{3}\sin\theta)}\times(-1)=-1$$

$$\Rightarrow\quad \left(\frac{2y}{x}\right)_{(\sqrt{6}\cos\theta,\sqrt{3}\sin\theta)}=1$$

$$\Rightarrow\quad \frac{2\sqrt{3}\sin\theta}{\sqrt{6}\cos\theta}=1 \Rightarrow \tan\theta=\frac{1}{\sqrt{2}}\Rightarrow\cos\theta=\frac{\sqrt{2}}{\sqrt{3}}$$

$$\sin\theta=\frac{1}{\sqrt{3}}$$

$\therefore$ Required point is (2, 1).

36. Given that $2(1-\cos x)<x^2,\ x\neq 0$

To prove $\sin(\tan x)\geq x,\ x\in[0,\pi/4)$.

Let us consider $f(x)=\sin(\tan x)-x$

$\Rightarrow\quad f'(x)=\cos(\tan x)\sec^2 x-1$

$$=\frac{\cos(\tan x)-\cos^2 x}{\cos^2 x}$$

As given $2(1-\cos x)<x^2,\ x\neq 0$

$$\Rightarrow\quad \cos x>1-\frac{x^2}{2}$$

Similary, $\cos(\tan x)>1-\dfrac{\tan^2 x}{2}$

$$\therefore\quad f'(x)>\frac{1-\frac{1}{2}\tan^2 x-\cos^2 x}{\cos^2 x}$$

$$=\frac{\sin^2 x\left[1-\frac{1}{2\cos^2 x}\right]}{\cos^2 x}$$

$$=\frac{\sin^2 x(\cos 2x)}{2\cos^4 x}>0,\ \forall\, x\in[0,\pi/4)$$

$\therefore\quad f'(x)>0\Rightarrow f(x)$ is an increasing function.

$\therefore$ For $x\in[0,\pi/4)$,

$x\geq 0\Rightarrow f(x)\geq f(0)$

$\Rightarrow\quad \sin(\tan x)-x\geq\sin(\tan 0)-0$

$\Rightarrow\quad \sin(\tan x)-x\geq 0$

$\Rightarrow\quad \sin(\tan x)\geq x$

Hence proved.

37. Given that f is a differentiable function on [0, 4]

$\therefore$ It will be continuous on [0, 4]

$\therefore$ By Lagrange's mean value theorem, we get

NOTE THIS STEP

$$\frac{f(4)-f(0)}{4-0}=f'(a),\text{ for } a\in(0,4)\quad\ldots(1)$$

Again since f is continuous on [0, 4] by intermediate mean value theorem, we get

$$\frac{f(4)+f(0)}{2}=f(b)\text{ for } b\in(0,4)\quad\ldots(2)$$

[If $f(x)$ is continuous on $[\alpha,\beta]$ then $\exists\,\mu\in(\alpha,\beta)$ such that $f(\mu)=\dfrac{f(\alpha)+f(\beta)}{2}$]

Multiplying (1) and (2) we get

$$\frac{[f(4)]^2-[f(0)]^2}{8}=f'(a)f(b);\ a,b\in(0,4)$$

or $[f(4)]^2-[f(0)]^2=8f'(a)f(b)$

Hence Proved.

(ii) To prove

$$\int_0^4 f(t)\,dt=2[\alpha f(\alpha^2)+\beta f(\beta^2)]\ \forall\ 0<\alpha,\beta<2$$

Let $I=\int_0^4 f(t)\,dt$

Let $t=u^2$ also $t\to 0\Rightarrow u\to 0$

$\Rightarrow dt=2u\,du$ as $t\to 4\Rightarrow u\to 2$

$$\therefore\quad \int_0^4 f(t)\,dt=\int_0^2 f(u^2).2u\,du\quad\ldots(1)$$

Consider, $F(x)=\int_0^x f(u^2).2u\,du$

Then clearly $F(x)$ is differentiable and hence continuous on [0,2]

By LMV theorem, we get some, $\mu\in(0,2)$

such that $F'(\mu)=\dfrac{F(2)-F(0)}{2-0}$

$$\Rightarrow\quad f(\mu^2).2\mu=\frac{\int_0^2 f(u^2)2u\,du}{2}\quad\ldots(2)$$ **NOTE THIS STEP**

Again by intermediate mean value theorem,

$\exists\ \alpha,\beta$ such that $0<\alpha<\mu<\beta<2$

$\Rightarrow\quad F'(\mu)=\dfrac{F'(\alpha)+F'(\beta)}{2}$, as f is continuous on [0, 2]

$\Rightarrow$ F is continuous on [0, 2]

$$\Rightarrow \quad f(\mu^2).2\mu = \frac{f(\alpha^2)2\alpha + f(\beta^2).2\beta}{2}$$

$$\Rightarrow \quad f(\mu^2).2\mu = \alpha f(\alpha^2) + \beta f(\beta^2) \quad \ldots(3)$$

From (2) and (3) we get

$$\int_0^2 f(u^2)2u\,du = 2[\alpha\, f(\alpha^2) + \beta\, f(\beta^2)]$$

where $0 < \alpha, \beta < 2$

$$\int_0^4 f(t)\,dt = 2[\alpha\, f(\alpha^2) + \beta\, f(\beta^2)]$$

where $0 < \alpha, \beta < 2$ (Using eq^n (1))

Hence Proved.

38. We are given that,

$$\frac{dP(x)}{dx} > P(x), \forall, x \geq 1 \text{ and } P(1) = 0$$

$$\Rightarrow \quad \frac{dP(x)}{dx} - P(x) > 0$$

Multiplying by e^{-x}, we get,

$$e^{-x}\frac{dP(x)}{dx} - e^{-x}P(x) > 0$$

$$\Rightarrow \quad \frac{d}{dx}\left[e^{-x}P(x)\right] > 0$$

$\Rightarrow$ $e^{-x}P(x)$ is an increasing function.

$\therefore$ $\forall x > 1, e^{-x}P(x) > e^{-1}P(1) = 0$ [Using $P(1) = 0$]

$\Rightarrow$ $e^{-x}P(x) > 0, \forall x > 1$

$\Rightarrow$ $P(x) > 0, \forall x > 1$ $[\because e^{-x} > 0]$

39. We are given,

$$P(x) = 51x^{101} - 2323x^{100} - 45x + 1035$$

To show that at least one root of $P(x)$ lies in $(45^{1/100}, 46)$, using Rolle's theorem, we consider antiderivative of $P(x)$

i.e. $F(x) = \dfrac{x^{102}}{2} - \dfrac{2323x^{101}}{101} - \dfrac{45x^2}{2} + 1035x$

Then being a polynominal function $F(x)$ is continuous and differentiable.

$$\text{Now, } F(45^{1/100}) = \frac{(45)^{\frac{102}{100}}}{2} - \frac{2323(45)^{\frac{101}{100}}}{101} - \frac{45.(45)^{\frac{2}{100}}}{2} + 1035(45)^{\frac{1}{100}}$$

$$= \frac{45}{2}(45)^{\frac{2}{100}} - 23 \times 45(45)^{\frac{1}{100}} - \frac{45.(45)^{\frac{2}{100}}}{2} + 1035(45)^{\frac{1}{100}} = 0$$

$$\text{And } F(46) = \frac{(46)^{102}}{2} - \frac{2323(46)^{101}}{101} - \frac{45(46)^2}{2} + 1035(46)$$

$$= 23(46)^{101} - 23(46)^{101} - 23 \times 45 \times 46 + 1035 \times 46 = 0$$

$\therefore$ $F(45^{\frac{1}{100}}) = F(46) = 0$

$\therefore$ Rolle's theorem is applicable.

Hence, there must exist at least one root of $F'(x) = 0$

i.e. $P(x) = 0$ in the interval $\left(45^{\frac{1}{100}}, 46\right)$

40. Let us consider,

$$f(x) = \sin x + 2x - \frac{3x(x+1)}{\pi}$$

$$\Rightarrow \quad f'(x) = \cos x + 2 - \frac{3}{\pi}(2x+1)$$

$$\Rightarrow \quad f''(x) = -\sin x - \frac{6}{\pi} < 0, \forall x \in [0, \pi/2]$$

$\Rightarrow$ $f'(x)$ is a decreasing function. .... (1)

Also $f'(0) = 3 - \dfrac{3}{\pi} > 0$ ... (2)

and $f'(\pi/2) = 2 - \dfrac{3}{\pi}(\pi+1) = -1 - \dfrac{3}{\pi} < 0$ ...(3)

Equation (1), (2) and (3) shows that.

$\Rightarrow$ There exists a certain value of $x \in [0, \pi/2]$ for which $f'(x) = 0$ and this point must be a point of maximum for $f(x)$ since the sign of $f'(x)$ changes from +ve to –ve.

Also we can see that $f(0) = 0$ and

$$f\left(\frac{\pi}{2}\right) = \pi + 1 - \frac{3}{2}\left(\frac{\pi}{2} + 1\right) = \frac{\pi}{4} - \frac{1}{2} > 0$$

Let $x = p$ be the point at which the max. of $f(x)$ occurs [**NOTE** There will be only one max. point in $[0, \pi/2]$. Since $f'(x) = 0$ is only once in the interval.]

Consider, $x \in [0, p]$

$\Rightarrow$ $f'(x) > 0 \Rightarrow f(x)$ is an increasing function.

$\Rightarrow$ $f(0) \leq f(x)$ [as $0 \leq x$]

$\Rightarrow$ $f(x) \geq 0$ ..... (4)

Also for $x \in [p, \pi/2]$

$\Rightarrow$ $f'(x) < 0 \Rightarrow f(x)$ is decreasing function.

$\Rightarrow$ for $x < \pi/2, f(x) > f(\pi/2) > 0$ ..... (5)

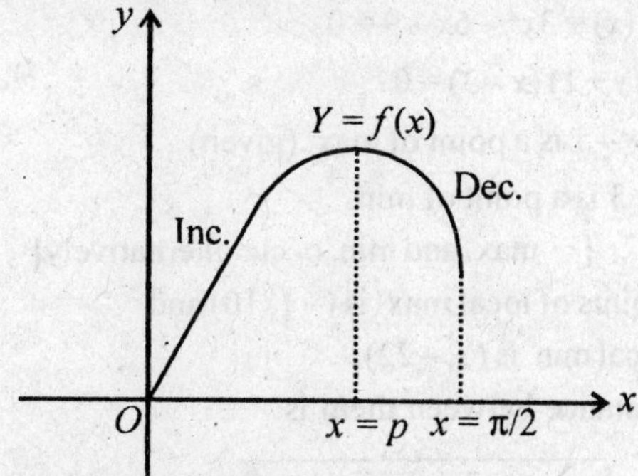

Hence from (4) and (5) we conclude that

$$f(x) \geq 0, \forall x \in [0, \pi/2].$$

41. Given that,

$$|f(x_1)-f(x_2)|<(x_1-x_2)^2,\ x_1,x_2 \in R$$

Let $x_1 = x + h$ and $x_2 = x$ then we get

$|f(x+h)-f(x)|<h^2$

$\Rightarrow |f(x+h)-f(x)|<|h|^2$

$\Rightarrow \left|\frac{f(x+h)-f(x)}{h}\right|<|h|$

Taking limit as $h \to 0$ on both sides, we get

$$\lim_{h\to 0}\left|\frac{f(x+h)-f(x)}{h}\right|<\delta \text{ (a small +ve number)}$$

$\Rightarrow |f'(x)|<\delta \Rightarrow f'(x)=0$

$\Rightarrow f(x)$ is a constant function. Let $f(x) = k$ i.e., $y = k$

As $f(x)$ passes through $(1, 2) \Rightarrow y=2$

$\therefore$ Equation of tangent at $(1, 2)$ is,

$y-2=0\,(x-1)$ i.e. $y=2$

42. Let $p(x) = ax^3 + bx^2 + cx + d$

$p(-1)=10$

$\Rightarrow -a+b-c+d=10$ ...... (i)

$p(1)=-6$

$\Rightarrow a+b+c+d=-6$ ...... (ii)

$p(x)$ has max. at $x=-1$

$\therefore p'(-1)=0$

$\Rightarrow 3a-2b+c=0$ ...... (iii)

$p'(x)$ has min. at $x=1$

$\therefore p''(1)=0$

$\Rightarrow 6a+2b=0$ ...... (iv)

Solving (i), (ii), (iii) and (iv), we get

From (iv), $b=-3a$

From (iii), $3a+6a+c=0 \Rightarrow c=-9a$

From (ii), $a-3a-9a+d=-6 \Rightarrow d=11a-6$

From (i), $-a-3a+9a+11a-6=10$

$\Rightarrow 16a=16 \Rightarrow a=1$

$\Rightarrow b=-3, c=-9, d=5$

$\therefore p(x)=x^3-3x^2-9x+5$

$\Rightarrow p'(x)=3x^2-6x-9=0$

$\Rightarrow 3(x+1)(x-3)=0$

$\Rightarrow x=-1$ is a point of max. (given)

and $x=3$ is a point of min.

[$\because$ max. and min. occur alternatively]

$\therefore$ points of local max. is $(-1, 10)$ and local min. is $(3, -22)$.

And distance between them is

$$=\sqrt{[3-(-1)^2]+(-22-10)^2}$$

$$=\sqrt{16+1024}=\sqrt{1040}=4\sqrt{65}$$

43. $g(x)=(f'(x))^2+f''(x)f(x)=\frac{d}{dx}(f(x)f'(x))$

Let $h(x)=f(x)f'(x)$

Then, $f(x)=0$ has four roots namely a, α, β, e where $b<\alpha<c$ and $c<\beta<d$.

And $f'(x)=0$ at three points k_1, k_2, k_3 where $a<k_1<\alpha, \alpha<k_2<\beta, \beta<k_3<e$

[$\because$ Between any two roots of a polynomial function $f(x)=0$ there lies atleast one root of $f'(x)=0$]

$\therefore$ There are atleast 7 roots of $f(x).f'(x)=0$

$\Rightarrow$ There are atleast 6 roots of $\frac{d}{dx}(f(x)f'(x))=0$

i.e. of $g(x)=0$

F MATCH THE FOLLOWING :

1. (A) $f(x)=x+\sin x$ on $(-\pi/2, \pi/2)$

$f'(x)=1+\cos x$

As $0\le\cos x\le 1$ for $x\in(-\pi/2, \pi/2)$

$\therefore f'(x)>0$ on $(-\pi/2, \pi/2)$

(A) $\to$ p

(B) $f(x)=\sec x \Rightarrow f'(x)=\sec x\tan x$.

Clearly $f'(x)<0$ in $(-\pi/2, 0)$ and $f'(x)>0$ in $(0, \pi/2)$

$\therefore$ On $\left(-\frac{\pi}{2},\frac{\pi}{2}\right)$ $f(x)$ is neither increasing non decreasing.

(B) $\to$ r

G COMPREHENSION BASED QUESTIONS :

1. **(c)** For $k=0$, line $y=x$ meets $y=0$, i.e., x-axis only at one point.

For $k<0$, $y=ke^x$ meets $y=x$ only once as shown in the graph.

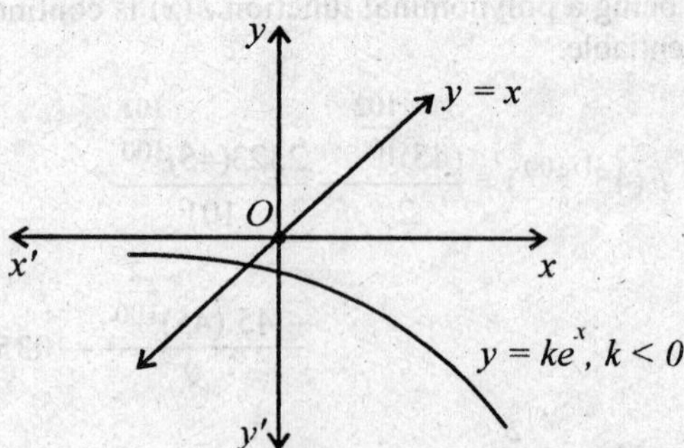

2. **(a)** Let $f(x)=ke^x-x$

Now for $f(x)=0$ to have only one root means the line $y=x$ must be tangential to the curve $y=ke^x$.

Let it be so at (x_1, y_1) then

$$\left(\frac{dy}{dx}\right)_{\text{curve 1}}=\left(\frac{dy}{dx}\right)_{\text{curve 2}} \Rightarrow 1=ke^{x_1}$$

$\Rightarrow \quad e^{x_1} = \frac{1}{k}$ also $y_1 = ke^{x_1}$ and $y_1 = x_1$

$\Rightarrow \quad x_1 = 1 \Rightarrow 1 = ke \Rightarrow k = 1/e$

3. **(a)** $\because$ For $y = x$ to be tangent to the curve $y = ke^x$, $k = 1/e$

$\therefore$ For $y = ke^x$ to meet $y = x$ at two points we should have $k < \frac{1}{e} \Rightarrow k \in \left(0, \frac{1}{e}\right)$ as $k > 0$.

4. **(c)** For the statement P

$f(x) + 2x = 2(1 + x^2)$

$\Rightarrow (1-x)^2 \sin^2 x + x^2 + 2x = 2(1+x^2)$

$\Rightarrow (1-x)^2 \sin^2 x = x^2 - 2x + 1 + 1$

$\Rightarrow (1-x)^2 \sin^2 x = (1-x)^2 + 1$

$\Rightarrow (1-x)^2 \cos^2 x = -1$

Which is not possible for any real value of x.

$\therefore$ P is not true.

Also let $H(x) = 2f(x) + 1 - 2x(1 + x)$

$H(0) = 2f(0) + 1 - 0 = 1$

and $H(1) = 2f(1) + 1 - 4 = -3$

$\Rightarrow H(x)$ has a solution in (0, 1)

$\therefore$ Q is true.

5. **(b)** We have $g(x) = \int_1^x \left(\frac{2(t-1)}{t+1} - \ln t\right) f(t)dt,\ x \in (1, \infty)$

$\therefore g'(x) = \left[\frac{2(x-1)}{x+1} - \ln x\right] f(x)$

$\Rightarrow$ Here $f(x) > 0,\ \forall x \in (1, \infty)$

Also let $h(x) = \frac{2(x-1)}{x+1} - \ln x$

$h'(x) = \frac{4}{(x+1)^2} - \frac{1}{x} = \frac{-(x-1)^2}{(x+1)^2 x} < 0, x \in (1, \infty)$

$\therefore$ $h(x)$ is decreasing function.

$\therefore$ For $x > 1$

$h(x) < h(1)$

$\Rightarrow h(x) < 0\ \forall\ x > 1$

$\therefore g'(x) < 0\ \forall\ x \in (1, \infty)$

$\therefore$ $g(x)$ is decreasing on $(1, \infty)$.

6. **(d)** We have $f''(x) - 2f'(x) + f(x) \geq e^x$

$\Rightarrow [f''(x) - f'(x)] - [f'(x) - f(x)] \geq e^x$

$\Rightarrow \left[e^{-x} f''(x) - e^{-x} f'(x)\right] - \left[e^{-x} f'(x) - e^{-x} f(x)\right] \geq 1$

$\Rightarrow \frac{d}{dx}\left[e^{-x} f'(x)\right] - \frac{d}{dx}\left[e^{-x} f(x)\right] \geq 1$

$\Rightarrow \frac{d}{dx}\left[e^{-x} f'(x) - e^{-x} f(x)\right] \geq 1$

$\Rightarrow \frac{d}{dx}\left[\frac{d}{dx}\left(e^{-x} f(x)\right)\right] \geq 1$

Let $g(x) = e^{-x} f(x)$

Then we have $g''(x) \geq 1 > 0$

So g is concave upward.

Also $g(0) = g(1) = 0$

$g(x) < 0, \forall\ x \in (0, 1)$

$\Rightarrow e^{-x} f(x) < 0$

$\Rightarrow f(x) < 0, \forall\ x \in (0, 1)$

7. **(c)** $g(x) = e^{-x} f(x)$

$\Rightarrow g'(x) = e^{-x} f'(x) - e^{-x} f(x)$

$= e^{-x}(f'(x) - f(x))$

As $x = \frac{1}{4}$ is point of local minima in [0, 1]

$\therefore g'(x) < 0$ for $x \in \left(0, \frac{1}{4}\right)$

and $g'(x) > 0$ for $x \in \left(\frac{1}{4}, 1\right)$

$\therefore$ In $\left(0, \frac{1}{4}\right)$, $g'(x) < 0$

$\Rightarrow e^{-x}(f'(x) - f(x)) < 0$

$\Rightarrow f'(x) < f(x)$

I INTEGER VALUE CORRECT TYPE :

1. **(7)** The given function is $f(x) = 2x^3 - 15x^2 + 36x - 48$

and $A = \{x \mid x^2 + 20 \leq 9x\}$

$\Rightarrow \quad A = \{x \mid x^2 - 9x + 20 \leq 0\}$

$\Rightarrow \quad A = \{x \mid (x-4)(x-5) \leq 0\}$

$\Rightarrow \quad A = [4, 5]$

Also $f'(x) = 6x^2 - 30x + 36 = 6(x^2 - 5x + 6)$

$= 6(x-2)(x-3)$

Clearly $\forall\ x \in A,\ f'(x) > 0$

$\therefore$ f is strictly increasing function on A.

$\therefore$ Maximum value of f on A

$= f(5) = 2 \times 5^3 - 15 \times 5^2 + 36 \times 5 - 48$

$= 250 - 375 + 180 - 48 = 430 - 423 = 7.$

2. **(0)** Let $p(x) = ax^4 + bx^3 + cx^2 + dx + e$

Now $\lim_{x \to 0}\left[1 + \frac{p(x)}{x^2}\right] = 2$

$\Rightarrow \quad \lim_{x \to 0} \frac{p(x)}{x^2} = 1$ ...(1)

$\Rightarrow \quad p(0) = 0 \Rightarrow e = 0$

Applying L'Hospital's rule to eq^n (1), we get

$\lim_{x \to 0} \frac{p'(x)}{2x} = 1 \Rightarrow p'(0) = 0$

$\Rightarrow \quad d = 0$

Again applying L 'Hospital's rule, we get

$\lim_{x \to 0} \frac{p''(x)}{2} = 1 \Rightarrow p''(0) = 2$

$\Rightarrow$ 2 c = 2 or c = 1

$\therefore \quad p(x) = ax^4 + bx^3 + x^2$

$\Rightarrow \quad p'(x) = 4ax^3 + 3bx^2 + 2x$

As $p(x)$ has extremum at $x=1$ and 2

$\therefore \quad p'(1)=0$ and $p'(2)=0$

$\Rightarrow \quad 4a+3b+2=0$ ...(i)

$\Rightarrow \quad 32a+12b+4=0$ or $8a+3b+1=0$ ...(ii)

Solving eq's (i) and (ii) we get $a=\frac{1}{4}$ and $b=-1$

$\therefore \quad p(x)=\frac{1}{4}x^4-x^3+x^2$

So, that $p(2)=\frac{16}{4}-8+4=0$

3. **(9)** The equation of tangent to the curve $y=f(x)$ at the point P (x,y) is

$$\frac{Y-y}{X-x}=\frac{dy}{dx} \text{ or } (X-x)\frac{dy}{dx}-(Y-y)=0$$

$$\Rightarrow X\frac{dy}{dx}-Y=x\frac{dy}{dx}-y$$

Its y-intercept $=y-x\frac{dy}{dx}=x^3$

$$\Rightarrow \frac{dy}{dx}-\frac{y}{x}=-x^2$$

$$\text{I.F.}=e^{\int -\frac{1}{x}dx}=\frac{1}{x}$$

$$\therefore y.\frac{1}{x}=\int -x^2\frac{1}{x}dx=\frac{-x^2}{2}+C$$

$$y=\frac{-x^3}{2}+Cx$$

As $f(1)=1 \Rightarrow At\ x=1, y=1$

$$\therefore 1=\frac{-1}{2}+C \Rightarrow C=3/2$$

$$\therefore y=-\frac{x^3}{2}+\frac{3x}{2}$$

At $x=-3$, $y=\frac{27}{2}-\frac{9}{2}=9$

$\therefore f(-3)=9$.

4. **(1)** We have ,

$f'(x)=2010(x-2009)(x-2010)^2(x-2011)^3(x-2012)^4$

As $f(x)=ln\ g(x)$

$\Rightarrow g(x)=e^{f(x)} \Rightarrow g'(x)=e^{f(x)}.f'(x)$

For max/min, $g'(x)=0 \Rightarrow f'(x)=0$

Out of two points one should be a point of maxima and other that of minima.

$\therefore$ There is only one point of local maxima.

5. **(5)** We have $f(x)=|x|+|x^2-1|$

$$=\begin{cases} -x+x^2-1, & x<-1 \\ -x-x^2+1, & -1\le x\le 0 \\ x-x^2+1, & 0<x<1 \\ x^2+x-1, & x\ge 1 \end{cases}$$

From graph of the function, we observe there are 5 points of local maximum or local minimum.

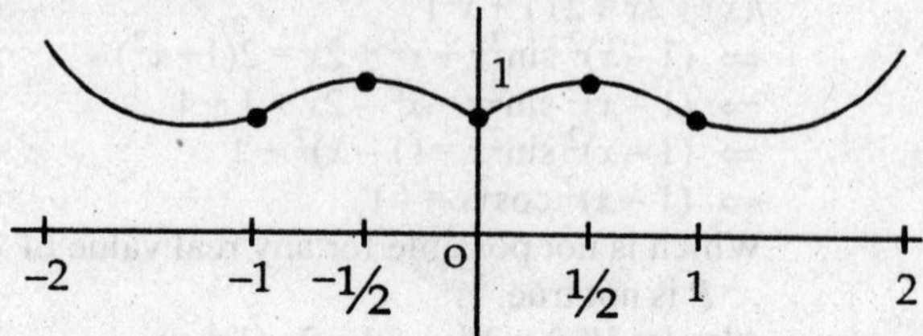

Alternative

We have $f'(x)=\begin{cases} 2x-1, & x<-1 \\ -2x-1, & -1\le x\le 0 \\ -2x+1, & 0<x<1 \\ 2x+1, & x>1 \end{cases}$

Critical pts are $\frac{1}{2},\frac{-1}{2},-1,0$ and 1

$$\xleftarrow{\quad -ve \quad | \quad +ve \quad | \quad -ve \quad | \quad +ve \quad | \quad -ve \quad | \quad +ve \quad}$$

$-1 \quad -\frac{1}{2} \quad 0 \quad \frac{1}{2} \quad 1$

We observe at five points $f'(x)$ changes its sign

$\therefore$ There are 5 points of local maximum or local minimum.

6. **(9)** $\because p(x)$ has a local maximum at $x=1$ and a local minimum at $x=3$ and $p(x)$ is a real polynomial of least degree

$\therefore$ Let $p'(x)=k(x-1)(x-3)=k(x^2-4x+3)$

$$\Rightarrow p(x)=k\left(\frac{x^3}{3}-2x^2+3x\right)+C$$

Given $p(1)=6$ and $p(3)=2$

$$\Rightarrow \frac{4}{3}k+C=6 \text{ and } 0+C=2 \Rightarrow k=3$$

$\therefore p'(x)=3(x-1)(x-3) \Rightarrow p'(0)=9$

7. **(9)** Vertical line $x=h$, meets the ellipse $\frac{x^2}{4}+\frac{y^2}{3}=1$ at

$P\left(h,\frac{\sqrt{3}}{2}\sqrt{4-h^2}\right)$ and $Q\left(h,\frac{-\sqrt{3}}{2}\sqrt{4-h^2}\right)$

By symmetry, tangents at P and Q will meet each other at x-axis.

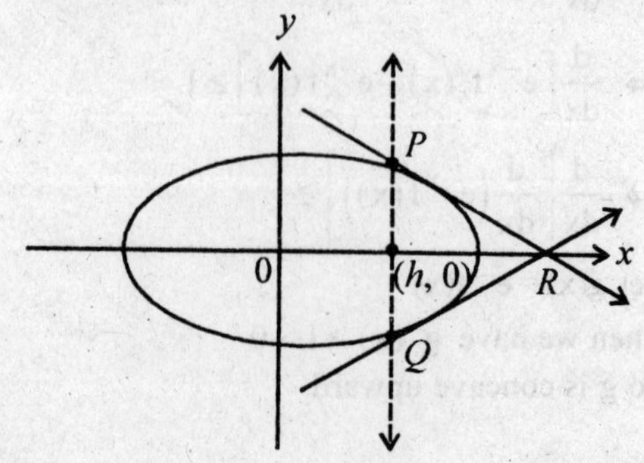

Tangent at P is $\frac{xh}{4}+\frac{y\sqrt{3}}{6}\sqrt{4-h^2}=1$

which meets x-axis at $R\left(\frac{4}{h}, 0\right)$

Area of $\Delta PQR=\frac{1}{2}\times\sqrt{3}\sqrt{4-h^2}\times\left(\frac{4}{h}-h\right)$

i.e., $\Delta(h)=\frac{\sqrt{3}}{2}\frac{(4-h^2)^{3/2}}{h}$

$$\frac{d\Delta}{dh}=-\sqrt{3}\left[\frac{\sqrt{4-h^2}(h^2+2)}{h^2}\right]<0$$

$\therefore$ $\Delta(h)$ is a decreasing function.

$\therefore \frac{1}{2}\le h\le 1 \Rightarrow \Delta_{max}=\Delta\left(\frac{1}{2}\right)$ and $\Delta_{min}=\Delta(1)$

$$\therefore \Delta_1=\frac{\sqrt{3}}{2}\frac{\left(4-\frac{1}{4}\right)^{3/2}}{\frac{1}{2}}=\frac{45}{8}\sqrt{5}$$

$$\Delta_2=\frac{\sqrt{3}}{2}\frac{3\sqrt{3}}{1}=\frac{9}{2}$$

$$\therefore \frac{8}{\sqrt{5}}\Delta_1-8\Delta_2=45-36=9$$

Section-B JEE Main/ AIEEE

1. (b) Distance of origin from $(x, y)=\sqrt{x^2+y^2}$

$=\sqrt{a^2+b^2-2ab\cos\left(t-\frac{at}{b}\right)}$;

$\le\sqrt{a^2+b^2+2ab}\quad\left[\left\{\cos\left(t-\frac{at}{b}\right)\right\}_{min}=-1\right]$

$=a+b$

$\therefore$ Maximum distance from origin $=a+b$

2. (a) Let $f(x)=\frac{ax^3}{3}+\frac{bx^2}{2}+cx \Rightarrow f(0)=0$ and $f(1)$

$=\frac{a}{3}+\frac{b}{2}+c=\frac{2a+3b+6c}{6}=0$

Also $f(x)$ is continuous and differentiable in [0, 1] and [0, 1[. So by Rolle's theorem, $f'(x)=0$.

i.e $ax^2+bx+c=0$ has at least one root in [0, 1].

3. (d) $f(x)=2x^3-9ax^2+12a^2x+1$

$f'(x)=6x^2-18ax+12a^2$; $f''(x)=12x-18a$

For max. or min.

$6x^2-18ax+12a^2=0 \Rightarrow x^2-3ax+2a^2=0$

$\Rightarrow x=a$ or $x=2a$. At $x=a$ max. and at $x=2a$ min

$\therefore p=a$ and $q=2a$

As per question $p^2=q$

$\therefore a^2=2a \Rightarrow a=2$ or $a=0$

but $a>0$, therefore, $a=2$.

4. (a) $y^2=18x \Rightarrow 2y\frac{dy}{dx}=18 \Rightarrow \frac{dy}{dx}=\frac{9}{y}$

Given $\frac{dy}{dx}=2 \Rightarrow \frac{9}{y}=2 \Rightarrow y=\frac{9}{2}$

Putting in $y^2=18x \Rightarrow x=\frac{9}{8}$

$\therefore$ Required point is $\left(\frac{9}{8},\frac{9}{2}\right)$

5. (b) $f''(x)=6(x-1)$. Inegrating, we get

$f'(x)=3x^2-6x+c$

Slope at (2, 1) $=f'(2)=c=3$

[$\because$ slope of tangent at (2,1) is 3]

$\therefore f'(x)=3x^2-6x+3=3(x-1)^2$

Inegrating again, we get $f(x)=(x-1)^3+D$

The curve passes through (2, 1)

$\Rightarrow 1=(2-1)^3+D \Rightarrow D=0$

$\therefore f(x)=(x-1)^3$

6. (d) $\frac{dx}{d\theta}=-a\sin\theta$ and $\frac{dy}{d\theta}=a\cos\theta$

$\therefore \frac{dy}{dx}=-\cot\theta.$

$\therefore$ The slope of the normal at $\theta=\tan\theta$

$\therefore$ The equation of the normal at θ is

$y-a\sin\theta=\tan\theta(x-a-a\cos\theta)$

$\Rightarrow y\cos\theta-a\sin\theta\cos\theta=x\sin\theta-a\sin\theta-a\sin\theta\cos\theta$

$\Rightarrow x\sin\theta-y\cos\theta=a\sin\theta$

$\Rightarrow y=(x-a)\tan\theta$

which always passes through $(a, 0)$.

7. (d) Let us define a function

$$f(x)=\frac{ax^3}{3}+\frac{bx^2}{2}+cx$$

Being polynomial, it is continuous and differentiable, also,

$f(0)=0$ and $f(1)=\frac{a}{3}+\frac{b}{2}+c$

$\Rightarrow f(1)=\frac{2a+3b+6c}{6}=0$ (given)

$\therefore f(0)=f(1)$

$\therefore f(x)$ satisfies all conditions of Rolle's theorem therefore $f'(x)=0$ has a root in (0, 1)

i.e. $ax^2+bx+c=0$ has at lease one root in (0, 1)

8. (a) Area of rectangle $ABCD=2a\cos\theta$ $(2b\sin\theta)=2ab\sin 2\theta$

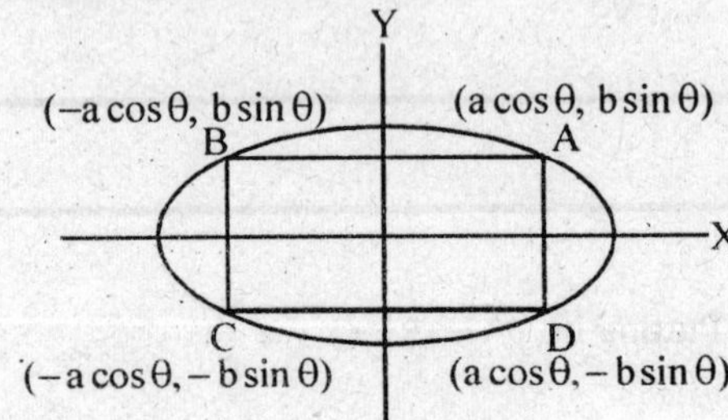

$\Rightarrow$ Area of greatest rectangle is equal to $2ab$ When $\sin 2\theta=1$.

9. (d) $x=a(\cos\theta+\theta\sin\theta)$

$\Rightarrow \frac{dx}{d\theta}=a(-\sin\theta+\sin\theta+\theta\cos\theta)$

$\Rightarrow \frac{dx}{d\theta}=a\theta\cos\theta$ .....(1)

$y=a(\sin\theta-\theta\cos\theta)$

$\frac{dy}{d\theta}=a[\cos\theta-\cos\theta+\theta\sin\theta]$

$\Rightarrow \frac{dy}{d\theta}=a\theta\sin\theta$ .....(2)

From equations (1) and (2) we get

$\frac{dy}{dx}=\tan\theta \Rightarrow$ Slope of normal $=-\cot\theta$

Equation of normal at 'θ' is $y-a(\sin\theta-\theta\cos\theta)$ $=-\cot\theta\,(x-a(\cos\theta+\theta\sin\theta))$

$\Rightarrow y\sin\theta-a\sin^2\theta+a\,\theta\cos\theta\sin\theta$ $=-x\cos\theta+a\cos^2\theta+a\theta\sin\theta\cos\theta$

$\Rightarrow x\cos\theta+y\sin\theta=a$

Clearly this is an equation of straight line which is at a constant distance 'a' from origin.

10. (b) Given that

$\frac{dv}{dt}=50\text{ cm}^3/\text{min} \Rightarrow \frac{d}{dt}\left(\frac{4}{3}\pi r^3\right)=50$

$\Rightarrow 4\pi r^2\frac{dr}{dt}=50$

$\Rightarrow \frac{dr}{dt}=\frac{50}{4\pi(15)^2}=\frac{1}{18\pi}$ cm/min (here $r=10+5$)

11. (b) Let $f(x)=a_nx^n+a_{n-1}x^{n-1}+\text{..........}+a_1x=0$

The other given equation,

$na_nx^{n-1}+(n-1)\,a_{n-1}x^{n-2}+....+a_1=0=f'(x)$

Given $a_1\neq 0 \Rightarrow f(0)=0$

Again $f(x)$ has root α, $\Rightarrow f(\alpha)=0$

$\therefore \quad f(0)=f(\alpha)$

$\therefore$ By Roll's theorem $f'(x)=0$ has root between $(0,\alpha)$

Hence $f'(x)$ has a positive root smaller than α.

12. (a) $\frac{x}{2}+\frac{2}{x}$ is of the form $y+\frac{1}{y}$ where $y+\frac{1}{y}\geq 2$ and equality holds for $y=1$

$\therefore$ Min value of function occurs at $\frac{x}{2}=1$ i.e., at $x=2$

ALTERNATE SOLUTION :

$f(x)=\frac{x}{2}+\frac{2}{x} \Rightarrow f'(x)=\frac{1}{2}-\frac{2}{x^2}=0$

$\Rightarrow x^2=4$ or $x=2,-2;\quad f''(x)=\frac{4}{x^3}$

$f''(x)]_{x=2}=+ve \Rightarrow f(x)$ has local min at $x=2$.

13. (c) Area $=\frac{1}{2}x^2\sin\theta$

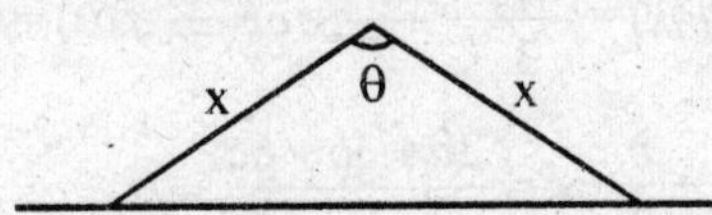

Maximum value of sinθ is 1 at $\theta=\frac{\pi}{2}$

$A_{max}=\frac{1}{2}x^2\left(at\ \sin\theta=1,\ \theta=\frac{\pi}{2}\right)$

14. (c) Using Lagrange's Mean Value Theorem
Let $f(x)$ be a function defined on $[a, b]$

then, $f'(c)=\frac{f(b)-f(a)}{b-a}$ ....(i)

$c\in[a, b]$

$\therefore$ Given $f(x)=\log_e x \therefore f'(x)=\frac{1}{x}$

$\therefore$ equation (i) become

$\frac{1}{c}=\frac{f(3)-f(1)}{3-1}$

$\Rightarrow \frac{1}{c}=\frac{\log_e 3-\log_e 1}{2}=\frac{\log_e 3}{2}$

$\Rightarrow c=\frac{2}{\log_e 3} \Rightarrow c=2\log_3 e$

15. (d) Given $f(x) = \tan^{-1}(\sin x + \cos x)$

$$f'(x) = \frac{1}{1+(\sin x + \cos x)^2}.(\cos x - \sin x)$$

$$= \frac{\sqrt{2}.\left(\frac{1}{\sqrt{2}}\cos x - \frac{1}{\sqrt{2}}\sin x\right)}{1+(\sin x+\cos x)^2}$$

$$= \frac{\left(\cos\frac{\pi}{4}.\cos x - \sin\frac{\pi}{4}.\sin x\right)}{1+(\sin x+\cos x)^2}$$

$$\therefore \quad f'(x) = \frac{\sqrt{2}\cos\left(x+\frac{\pi}{4}\right)}{1+(\sin x+\cos x)^2}$$

if $f'(x) > O$ then $f(x)$ is increasing function.

Hence $f(x)$ is increasing, if $-\frac{\pi}{2} < x+\frac{\pi}{4} < \frac{\pi}{2}$

$$\Rightarrow \quad -\frac{3\pi}{4} < x < \frac{\pi}{4}$$

Hence, $f(x)$ is increasing when $n \in \left(-\frac{\pi}{2}, \frac{\pi}{4}\right)$

16. (c) Given that $p^2 + q^2 = 1 \quad \therefore\ p = \cos\theta$ and $q = \sin\theta$

Then $p + q = \cos\theta + \sin\theta$

We know that

$$-\sqrt{a^2+b^2} \le a\cos\theta + b\sin\theta \le \sqrt{a^2+b^2}$$

$$\therefore -\sqrt{2} \le \cos\theta + \sin\theta \le \sqrt{2}$$

Hence max. value of $p + q$ is $\sqrt{2}$

ALTERNATE SOLUTION :

Since, p and q are positive real numbers

$p^2 + q^2 = 1$ (Given)

Using AM $\ge$ GM

$$\therefore \left(\frac{p+q}{2}\right)^2 \ge \sqrt{(pq)^2} \quad = \frac{p^2+q^2+2pq}{4} \ge pq$$

$$\frac{1+2pq}{4} \ge pq \qquad 1+2pq \ge 4pq$$

$$1 \ge 2pq \qquad \text{or,} \qquad 2pq \le 1$$

$$pq \le \frac{1}{2} \qquad \text{or,} \qquad pq \le \frac{1}{2}$$

Now, $(p+q)^2 = p^2 + q^2 + 2pq$

$$\Rightarrow (p+q)^2 \le 1 + 2 \times \frac{1}{2} \Rightarrow p + q \le \sqrt{2}$$

17. (a) Let $y = x^3 - px + q \Rightarrow \frac{dy}{dx} = 3x^2 - p$

For $\frac{dy}{dx} = 0 \Rightarrow 3x^2 - p = 0 \Rightarrow x = \pm\sqrt{\frac{p}{3}}$

$$\frac{d^2y}{dx^2} = 6x$$

$$\left.\frac{d^2y}{dx^2}\right|_{x=\sqrt{\frac{p}{3}}} = +ve \text{ and } \left.\frac{d^2y}{dx^2}\right|_{x=-\sqrt{\frac{p}{3}}} = -ve$$

$\therefore y$ has minima at $x = \sqrt{\frac{p}{3}}$ and maxima at $x = -\sqrt{\frac{p}{3}}$

18. (b) Let $f(x) = x^7 + 14x^5 + 16x^3 + 30x - 560$

$\Rightarrow f'(x) = 7x^6 + 70x^4 + 48x^2 + 30 > 0, \forall\ x \in R$

$\Rightarrow f$ is an increasing function on R

Also $\lim\limits_{x\to\infty} f(x) = \infty$ and $\lim\limits_{x\to-\infty} f(x) = -\infty$

$\Rightarrow$ The curve $y = f(x)$ crosses x-axis only once.

$\therefore f(x) = 0$ has exactly one real root.

19. (b) Given that $f(x) = x\,|\,x\,|$ and $g(x) = \sin x$

So that

$gof(x) = g(f(x)) = g(x\,|\,x\,|)$

$= \sin x\,|\,x\,|$

$$= \begin{cases} \sin(-x^2), & \text{if } x < 0 \\ \sin(x^2), & \text{if } x \ge 0 \end{cases}$$

$$= \begin{cases} -\sin x^2, & \text{if } x < 0 \\ \sin x^2, & \text{if } x \ge 0 \end{cases}$$

$$\therefore (gof)'(x) = \begin{cases} -2x\cos x^2, & \text{if } x < 0 \\ 2x\cos x^2, & \text{if } x \ge 0 \end{cases}$$

Here we observe

$L(gof)'(0) = 0 = R(gof)'(0)$

$\Rightarrow$ gof is differentiable at $x = 0$

and $(gof)'$ is continuous at $x = 0$

$$\text{Now } (gof)''(x) = \begin{cases} -2\cos x^2 + 4x^2\sin x^2, & x < 0 \\ 2\cos x^2 - 4x^2\sin x^2, & x \ge 0 \end{cases}$$

Here

$L(gof)''(0) = -2$ and $R(gof)''(0) = 2$

$\because$ $L(gof)''(0) \ne R(gof)''(0)$

$\Rightarrow$ $gof(x)$ is not twice differentiable at $x = 0$.

$\therefore$ Statement - 1 is true but statement -2 is false.

20. (a) We have $P(x) = x^4 + ax^3 + bx^2 + cx + d$

$\Rightarrow$ $P'(x) = 4x^3 + 3ax^2 + 2bx + c$

But $P'(0) = 0 \Rightarrow c = 0$

$\therefore$ $P(x) = x^4 + ax^3 + bx^2 + d$

As given that $P(-1) < P(a)$

$\Rightarrow$ $1 - a + b + d \ < \ 1 + a + b + d$

$\Rightarrow$ $a > 0$

Now $P'(x) = 4x^3 + 3ax^2 + 2bx$

$= x(4x^2 + 3ax + 2b)$

As $P'(x) = 0$, there is only one solution $x = 0$, therefore $4x^2 + 3ax + 2b = 0$ should not have any real roots i.e. $D < 0$

$$\Rightarrow 9a^2 - 32b < 0 \qquad \Rightarrow \qquad b > \frac{9a^2}{32} > 0$$

Hence $a, b > 0 \Rightarrow P'(x) = 4x^3 + 3ax^2 + 2bx > 0$

$\forall x > 0$.

$\therefore$ $P(x)$ is an increasing function on (0,1)

$\therefore$ $P(0) < P(a)$

Similarly we can prove $P(x)$ is decreasing on $(-1, 0)$

$\therefore$ $P(-1) > P(0)$

So we can conclude that

Max $P(x) = P(1)$ and Min $P(x) = P(0)$

$\Rightarrow$ $P(-1)$ is not minimum but $P(1)$ is the maximum of P.

21. (c) Since tangent is parallel to x-axis,

$\therefore \frac{dy}{dx} = 0 \Rightarrow 1 - \frac{8}{x^3} = 0 \Rightarrow x = 2 \Rightarrow y = 3$

Equation of tangent is $y - 3 = 0(x-2) \Rightarrow y = 3$

22. (c)

$$f(x) = \begin{cases} k-2x, & \text{if } x \le -1 \\ 2x+3, & \text{if } x > -1 \end{cases}$$

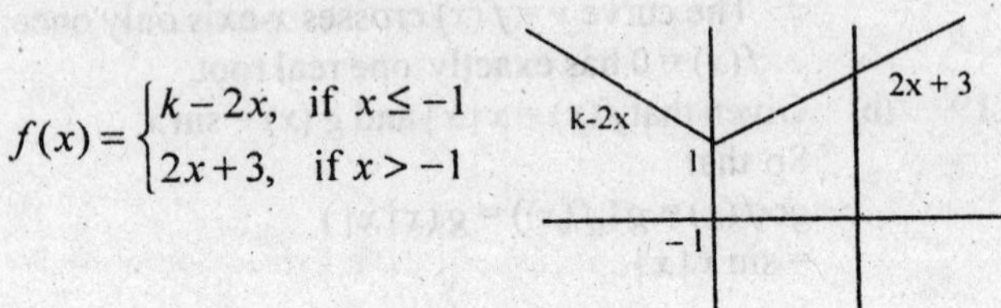

This is true where k = – 1

23. (d) $f(x) = \frac{1}{e^x + 2e^{-x}} = \frac{e^x}{e^{2x}+2}$

$$f'(x) = \frac{(e^{2x}+2)e^x - 2e^{2x}.e^x}{(e^{2x}+2)^2}$$

$f'(x) = 0 \Rightarrow e^{2x} + 2 = 2e^{2x}$

$e^{2x} = 2 \Rightarrow e^x = \sqrt{2}$

maximum $f(x) = \frac{\sqrt{2}}{4} = \frac{1}{2\sqrt{2}}$

$0 < f(x) \le \frac{1}{2\sqrt{2}} \quad \forall x \in R$

Since $0 < \frac{1}{3} < \frac{1}{2\sqrt{2}} \Rightarrow$ for some $c \in R$

$f(c) = \frac{1}{3}$

24. (a) Shortest distance between two curve occurred along the common normal

Slope of normal to $y^2 = x$ at point $P(t^2, t)$ is $-2t$ and slope of line $y - x = 1$ is 1.

As they are perpendicular to each other

$\therefore (-2t) = -1$

$\Rightarrow t = \frac{1}{2}$

$\therefore P\left(\frac{1}{4}, \frac{1}{2}\right)$

and shortest distance $= \left|\frac{\frac{1}{2} - \frac{1}{4} - 1}{\sqrt{2}}\right|$

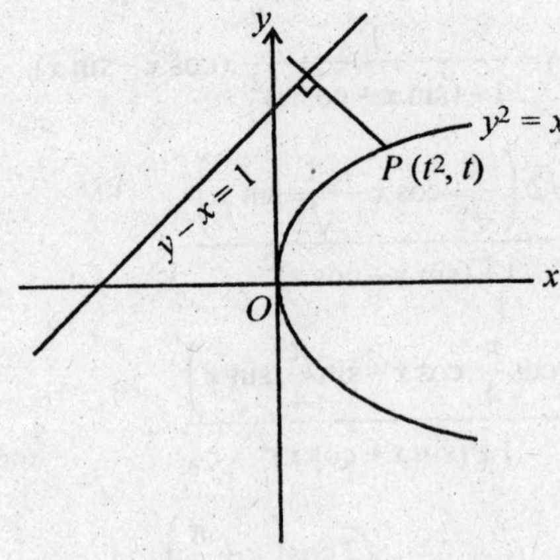

So shortest distance between them is $\frac{3\sqrt{2}}{8}$

25. (c) $f'(x) = \sqrt{x}\sin x$, $f'(x) = 0$

$\Rightarrow x = 0$ or $\sin x = 0$

$\Rightarrow x = 2\pi, \pi \in \left(0, \frac{5\pi}{2}\right)$

$f''(x) = \sqrt{x}\cos x + \frac{1}{2\sqrt{x}}\sin x$

$= \frac{1}{2\sqrt{x}}(2x\cos x + \sin x)$

At $x = \pi$, $f''(x) < 0$

Hence, local maxima at $x = \pi$

At $x = 2\pi$, $f''(x) > 0$

Hence local minima at $x = 2\pi$

26. (b) $Y - y = \frac{dy}{dx}(X - x)$

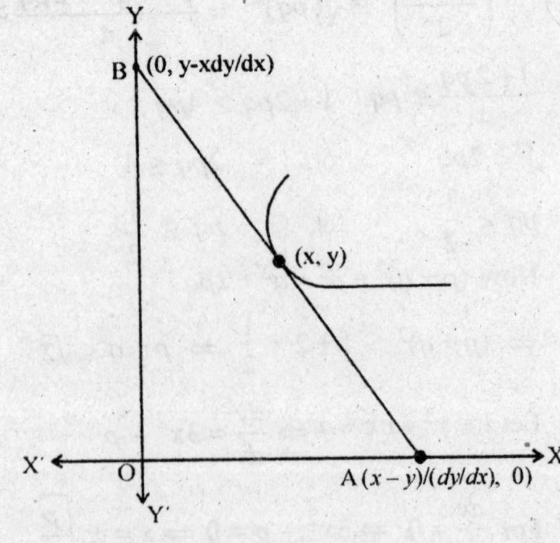

$X\text{-intercept} = x - \frac{y}{dy/dx}$

$Y\text{-intercept} = y - \frac{xdy}{dx}$

According to given statement

$x - \frac{y}{\frac{dy}{dx}} = 2x$ and $y - \frac{xdy}{dx} = 2y$

$\frac{-y}{\frac{dy}{dx}} = x$ and $\frac{-xdy}{dx} = y$

$\Rightarrow \frac{dx}{x} + \frac{dy}{y} = 0$

$\ell ny = -\ell nc + \ell nc$

$y = \frac{c}{x}$

Since the above line passes through the point (2,3)

$\therefore \quad c = 6$

Hence $y = \frac{6}{x}$ is the required equation.

27. **(b)** $f(x) = \begin{cases} \frac{\tan x}{x}, & x \neq 0 \\ 1, & x = 0 \end{cases}$

In right neighbourhood of '0'

$\tan x > x, \frac{\tan x}{x} > 1$

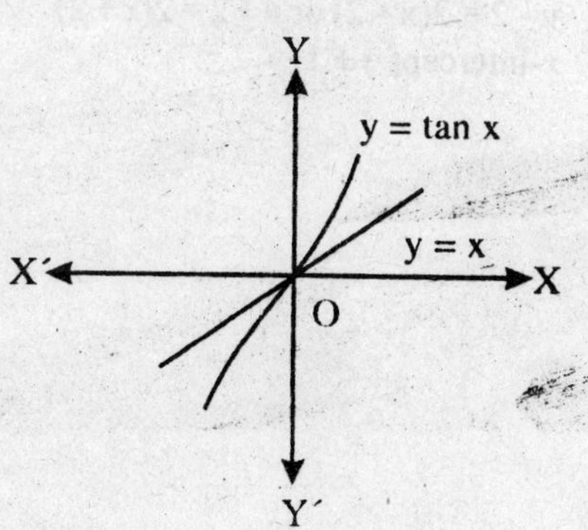

In left neighbourhood of '0'

$\tan x < x, \frac{\tan x}{x} > 1$ as $(\tan x < 0)$

at $x = 0$, $f(x) = 1$

$\Rightarrow \quad x = 0$ is the point of minima

So Statement 1 is true. Statement 2 obvious.

28. **(c)** Volume of spherical balloon $= V = \frac{4}{3}\pi r^3$

$\Rightarrow \quad 4500\,\pi = \frac{4\pi r^3}{3}$ ($\because$ Given, volume $= 4500\pi m^3$)

Differentiating both the sides, w.r.t 't' we get,

$\frac{dV}{dt} = 4\pi r^2 \left(\frac{dr}{dt}\right)$

Now, it is given that $\frac{dV}{dt} = 72\pi$

$\therefore$ After 49 min, Volume $= (4500 - 49 \times 72)\pi$
$= (4500 - 3528)\pi$
$= 972\,\pi\, m^3$

$\Rightarrow \quad V = 972\,\pi\, m^3 \qquad \therefore\ 972\pi = \frac{4}{3}\pi r^3$

$\Rightarrow \quad r^3 = 3 \times 243 = 3 \times 3^5 = 3^6 = (3^2)^3 \Rightarrow r = 9$

Also, we have $\frac{dV}{dt} = 72\pi$

$\therefore \quad 72\pi = 4\pi \times 9 \times 9 \left(\frac{dr}{dt}\right) \Rightarrow \frac{dr}{dt} = \left(\frac{2}{9}\right)$

29. **(b)** Given, $f(x) = ln|x| + bx^2 + ax$

$\therefore \quad f'(x) = \frac{1}{x} + 2bx + a$

At $x = -1$, $f'(x) = -1 - 2b + a = 0$

$\Rightarrow \quad a - 2b = 1$ ...(i)

At $x = 2$, $f'(x) = \frac{1}{2} + 4b + a = 0$

$\Rightarrow \quad a + 4b = -\frac{1}{2}$ ...(ii)

On solving (i) and (ii) we get $a = \frac{1}{2}, b = -\frac{1}{4}$

Thus, $f'(x) = \frac{1}{x} - \frac{x}{2} + \frac{1}{2} = \frac{2 - x^2 + x}{2x}$

$= \frac{-x^2 + x + 2}{2x} = \frac{-(x^2 - x - 2)}{2x} = \frac{-(x+1)(x-2)}{2x}$

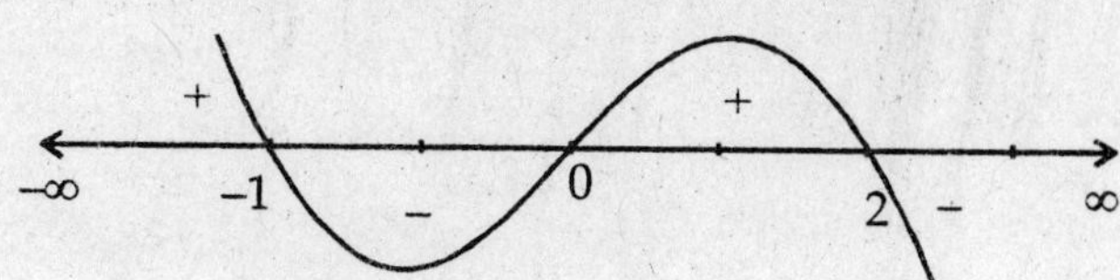

So maxima at $x = -1, 2$

Hence both the statements are true and statement 2 is a correct explanation for 1.

30. **(c)** Equation of a line passing through (x_1, y_1) having slope m is given by $y - y_1 = m\,(x - x_1)$

Since the line PQ is passing through (1,2) therefore its equation is

$(y - 2) = m\,(x - 1)$

where m is the slope of the line PQ.

Now, point $P\,(x, 0)$ will also satisfy the equation of PQ

$\therefore \quad y - 2 = m\,(x - 1)$

$\Rightarrow \quad 0 - 2 = m\,(x - 1)$

$\Rightarrow \quad -2 = m(x-1) \quad \Rightarrow \quad x-1 = \frac{-2}{m}$

$\Rightarrow \quad x = \frac{-2}{m}+1$

Also, $OP = \sqrt{(x-0)^2+(0-0)^2} = x = \frac{-2}{m}+1$

Similarly, point $Q(0,y)$ will satisfy equation of PQ

$\therefore \quad y-2 = m(x-1)$

$\Rightarrow \quad y-2 = m(-1) \Rightarrow \quad y = 2-m$ and $OQ = y = 2-m$

Area of $\Delta POQ = \frac{1}{2}(OP)(OQ) = \frac{1}{2}\left(1-\frac{2}{m}\right)(2-m)$

$(\because \text{Area of } \Delta = \frac{1}{2} \times \text{base} \times \text{height})$

$= \frac{1}{2}\left[2-m-\frac{4}{m}+2\right] = \frac{1}{2}\left[4-\left(m+\frac{4}{m}\right)\right]$

$= 2-\frac{m}{2}-\frac{2}{m}$

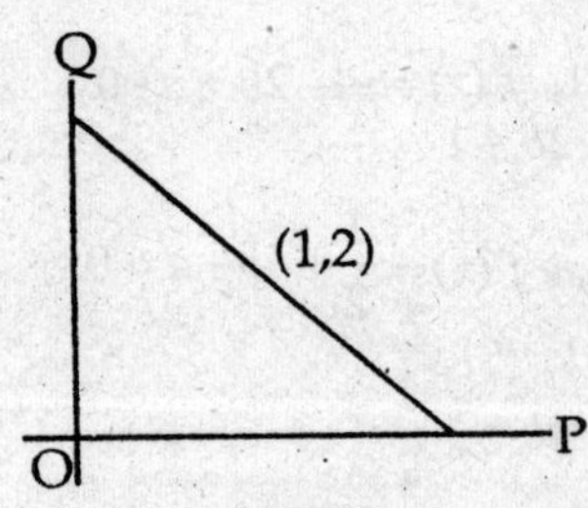

Let Area $= f(m) = 2-\frac{m}{2}-\frac{2}{m}$

Now, $f'(m) = \frac{-1}{2}+\frac{2}{m^2}$

Put $f'(m) = 0 \quad \Rightarrow m^2 = 4 \Rightarrow m = \pm 2$

Now, $f''(m) = \frac{-4}{m^3}$

$f''(m)\big|_{m=2} = -\frac{1}{2} < 0$

$f''(m)\big|_{m=-2} = \frac{1}{2} > 0$

Area will be least at $m = -2$

Hence, slope of PQ is -2.

31. (a) Since, $y = \int_0^x |t|\,dt,\ x \in R$

therefore $\frac{dy}{dx} = |x|$

But from $y = 2x$, $\frac{dy}{dx} = 2$

$\Rightarrow \quad |x| = 2 \quad \Rightarrow \quad x = \pm 2$

Points $y = \int_0^{\pm 2} |t|\,dt = \pm 2$

$\therefore$ equation of tangent is

$y-2 = 2(x-2)$ or $y+2 = 2(x+2)$

$\Rightarrow$ x-intercept $= \pm 1$.

❑ ❑

Chapter 17 Indefinite Integrals

Section-A JEE Advanced/ IIT-JEE

A. FILL IN THE BLANKS :

1. If $\int \frac{4e^x + 6e^{-x}}{9e^x - 4e^{-x}} dx = Ax + B \log(9e^{2x} - 4) + C$, then $A =$ ...., $B =$ .... and $C =$ .... **(1990 - 2 Marks)**

C. MCQ's WITH ONE CORRECT ANSWER :

1. The value of the integral $\int \frac{\cos^3 x + \cos^5 x}{\sin^2 x + \sin^4 x} dx$ is **(1995S)**
   (a) $\sin x - 6\tan^{-1}(\sin x) + c$
   (b) $\sin x - 2(\sin x)^{-1} + c$
   (c) $\sin x - 2(\sin x)^{-1} - 6\tan^{-1}(\sin x) + c$
   (d) $\sin x - 2(\sin x)^{-1} + 5\tan^{-1}(\sin x) + c$

2. If $\int_{\sin x}^{1} t^2 f(t)\, dt = 1 - \sin x$, then $f\left(\frac{1}{\sqrt{3}}\right)$ is **(2005S)**
   (a) $\frac{1}{3}$ (b) $\frac{1}{\sqrt{3}}$
   (c) 3 (d) $\sqrt{3}$

3. $\int \frac{x^2 - 1}{x^3\sqrt{2x^4 - 2x^2 + 1}} dx =$ **(2006 - 3M, –1)**
   (a) $\frac{\sqrt{2x^4 - 2x^2 + 1}}{x^2} + c$ (b) $\frac{\sqrt{2x^4 - 2x^2 + 1}}{x^3} + c$
   (c) $\frac{\sqrt{2x^4 - 2x^2 + 1}}{x} + c$ (d) $\frac{\sqrt{2x^4 - 2x^2 + 1}}{2x^2} + c$

4. Let $I = \int \frac{e^x}{e^{4x} + e^{2x} + 1} dx$, $J = \int \frac{e^{-x}}{e^{-4x} + e^{-2x} + 1} dx$. Then, for an arbitrary constant C, the value of J – I equals **(2008)**
   (a) $\frac{1}{2}\log\left(\frac{e^{4x} - e^{2x} + 1}{e^{4x} + e^{2x} + 1}\right) + C$ (b) $\frac{1}{2}\log\left(\frac{e^{2x} + e^x + 1}{e^{2x} - e^x + 1}\right) + C$
   (c) $\frac{1}{2}\log\left(\frac{e^{2x} - e^x + 1}{e^{2x} + e^x + 1}\right) + C$ (d) $\frac{1}{2}\log\left(\frac{e^{4x} + e^{2x} + 1}{e^{4x} - e^{2x} + 1}\right) + C$

5. The integral $\int \frac{\sec^2 x}{(\sec x + \tan x)^{\frac{9}{2}}} dx$ equals (for some arbitrary constant K) **(2012)**
   (a) $-\frac{1}{(\sec x + \tan x)^{\frac{11}{2}}}\left\{\frac{1}{11} - \frac{1}{7}(\sec x + \tan x)^2\right\} + K$
   (b) $\frac{1}{(\sec x + \tan x)^{\frac{11}{2}}}\left\{\frac{1}{11} - \frac{1}{7}(\sec x + \tan x)^2\right\} + K$
   (c) $-\frac{1}{(\sec x + \tan x)^{\frac{11}{2}}}\left\{\frac{1}{11} + \frac{1}{7}(\sec x + \tan x)^2\right\} + K$
   (d) $\frac{1}{(\sec x + \tan x)^{\frac{11}{2}}}\left\{\frac{1}{11} + \frac{1}{7}(\sec x + \tan x)^2\right\} + K$

E. SUBJECTIVE PROBLEMS :

1. Evaluate $\int \frac{\sin x}{\sin x - \cos x} dx$ **(1978)**
2. Evaluate $\int \frac{x^2 dx}{(a + bx)^2}$ **(1979)**
3. Evaluate the following integrals : **(1980)**
   (a) $\int \sqrt{1 + \sin\left(\frac{1}{2}x\right)}\, dx$ (b) $\int \frac{x^2 dx}{\sqrt{1 - x}}$
4. Evaluate $\int (e^{\log x} + \sin x) \cos x\, dx$. **(1981 - 2 Marks)**
5. Evaluate : $\int \frac{(x - 1)e^x}{(x + 1)^3} dx$ **(1983 - 2 Marks)**
6. Evaluate the following $\int \frac{dx}{x^2(x^4 + 1)^{3/4}}$ **(1984 - 2 Marks)**
7. Evaluate the following $\int \sqrt{\frac{1 - \sqrt{x}}{1 + \sqrt{x}}} dx$ **(1985 - 2½ Marks)**
8. Evaluate : $\int \frac{\sin^{-1}\sqrt{x} - \cos^{-1}\sqrt{x}}{\sin^{-1}\sqrt{x} + \cos^{-1}\sqrt{x}} dx$, **(1986 - 2½ Marks)**
9. Evaluate : $\int \left[\frac{(\cos 2x)^{1/2}}{\sin x}\right] dx$ **(1987 - 6 Marks)**

10. Evaluate $\int(\sqrt{\tan x}+\sqrt{\cot x})dx$ *(1989 - 3 Marks)*

11. Find the indefinite integral $\int\left(\frac{1}{\sqrt[3]{x}+\sqrt[4]{4}}+\frac{\ln(1+\sqrt[6]{x})}{\sqrt[3]{x}+\sqrt{x}}\right)dx$ *(1992 - 4 Marks)*

12. Find the indefinite integral $\int\cos 2\theta \ln\left(\frac{\cos\theta+\sin\theta}{\cos\theta-\sin\theta}\right)d\theta$ *(1994 - 5 Marks)*

13. Evaluate $\int\frac{(x+1)}{x(1+xe^x)^2}dx.$ *(1996 - 2 Marks)*

14. Integrate the following : $\int\left(\frac{1-\sqrt{x}}{1+\sqrt{x}}\right)^{1/2}\frac{dx}{x}$ *(1997C - 3 Marks)*

15. Integrate $\int\frac{x^3+3x+2}{(x^2+1)^2(x+1)}dx.$ *(1999 - 5 Marks)*

16. Evaluate $\int\sin^{-1}\left(\frac{2x+2}{\sqrt{4x^2+8x+13}}\right)dx.$ *(2001 - 5 Marks)*

17. For any natural number m, evaluate

$\int(x^{3m}+x^{2m}+x^m)(2x^{2m}+3x^m+6)^{1/m}dx, x>0.$

(2002 - 5 Marks)

H. ASSERTION & REASON TYPE QUESTIONS :

1. Let $F(x)$ be an indefinite integral of $\sin^2 x$.
STATEMENT-1 : The function $F(x)$ satisfies $F(x+\pi)=F(x)$ for all real x. **because**
STATEMENT-2 : $\sin^2(x+\pi)=\sin^2 x$ for all real x.
(2007 -3 marks)
(a) Statement-1 is True, statement-2 is True; Statement-2 is a correct explanation for Statement-1.
(b) Statement-1 is True, Statement-2 is True; Statement-2 is NOT a correct explanation for Statement-1
(c) Statement-1 is True, Statement-2 is False
(d) Statement-1 is False, Statement-2 is True.

Section-B JEE Main/ AIEEE

1. If $\int\frac{\sin x}{\sin(x-\alpha)}dx = Ax+B\log\sin(x-\alpha),+C$, then value of (A, B) is **[2004]**
(a) $(-\cos\alpha, \sin\alpha)$ (b) $(\cos\alpha, \sin\alpha)$
(c) $(-\sin\alpha, \cos\alpha)$ (d) $(\sin\alpha, \cos\alpha)$

2. $\int\frac{dx}{\cos x-\sin x}$ is equal to **[2004]**
(a) $\frac{1}{\sqrt{2}}\log\left|\tan\left(\frac{x}{2}+\frac{3\pi}{8}\right)\right|+C$
(b) $\frac{1}{\sqrt{2}}\log\left|\cot\left(\frac{x}{2}\right)\right|+C$
(c) $\frac{1}{\sqrt{2}}\log\left|\tan\left(\frac{x}{2}-\frac{3\pi}{8}\right)\right|+C$
(d) $\frac{1}{\sqrt{2}}\log\left|\tan\left(\frac{x}{2}-\frac{\pi}{8}\right)\right|+C$

3. $\int\left\{\frac{(\log x-1)}{1+(\log x)^2}\right\}^2dx$ is equal to **[2005]**
(a) $\frac{\log x}{(\log x)^2+1}+C$ (b) $\frac{x}{x^2+1}+C$
(c) $\frac{xe^x}{1+x^2}+C$ (d) $\frac{x}{(\log x)^2+1}+C$

4. $\int\frac{dx}{\cos x+\sqrt{3}\sin x}$ equals **[2007]**
(a) $\log\tan\left(\frac{x}{2}+\frac{\pi}{12}\right)+C$
(b) $\log\tan\left(\frac{x}{2}-\frac{\pi}{12}\right)+C$
(c) $\frac{1}{2}\log\tan\left(\frac{x}{2}+\frac{\pi}{12}\right)+C$
(d) $\frac{1}{2}\log\tan\left(\frac{x}{2}-\frac{\pi}{12}\right)+C$

5. The value of $\sqrt{2}\int\frac{\sin x dx}{\sin\left(x-\frac{\pi}{4}\right)}$ is **[2008]**
(a) $x+\log|\cos\left(x-\frac{\pi}{4}\right)|+c$
(b) $x-\log|\sin\left(x-\frac{\pi}{4}\right)|+c$
(c) $x+\log|\sin\left(x-\frac{\pi}{4}\right)|+c$
(d) $x-\log|\cos\left(x-\frac{\pi}{4}\right)|+c$

6. If the $\int \frac{5\tan x}{\tan x - 2} dx = x + a\ln|\sin x - 2\cos x| + k$, then a is equal to : **[2012]**
(a) -1 (b) -2 (c) 1 (d) 2

7. If $\int f(x)dx = \psi(x)$, then $\int x^5 f(x^3)dx$ is equal to **[JEE M 2013]**
(a) $\frac{1}{3}\left[x^3\psi(x^3) - \int x^2\psi(x^3)dx\right] + C$
(b) $\frac{1}{3}x^3\psi(x^3) - 3\int x^3\psi(x^3)dx + C$
(c) $\frac{1}{3}x^3\psi(x^3) - \int x^2\psi(x^3)dx + C$
(d) $\frac{1}{3}\left[x^3\psi(x^3) - \int x^3\psi(x^3)dx\right] + C$

Solutions & Explanations

Section-A : JEE Advanced/ IIT-JEE

A 1. $\frac{-3}{2}, \frac{35}{36}$, any real value

C 1. (c) 2. (c) 3. (d) 4. (c) 5. (c)

E 1. $\frac{1}{2}\log|\sin x - \cos x| + \frac{x}{2} + C$ 2. $\frac{1}{b^3}\left[a + bx - 2a\log|a+bx| - \frac{a^2}{a+bx}\right] + C$

3. (a) $\pm 4\left[\sin\frac{x}{4} - \cos\frac{x}{4}\right] + C$ (b) $-2\left[\frac{(1-x)^{5/2}}{5} - \frac{2(1-x)^{3/2}}{3} + \sqrt{1-x}\right] + C$

4. $x\sin x + \cos x - \frac{1}{4}\cos 2x + C$ 5. $\frac{e^x}{(x+1)^2} + C$ 6. $-\left(1 + \frac{1}{x^4}\right)^{\frac{1}{4}} + C$

7. $-2\sqrt{1-x} + \cos^{-1}\sqrt{x} + \sqrt{x}\sqrt{1-x} + C$ 8. $\frac{2}{\pi}\left[\sqrt{x - x^2} - (1-2x)\sin^{-1}\sqrt{x}\right] - x + C$

9. $\frac{1}{\sqrt{2}}\log\left[\frac{\sqrt{2} + \sqrt{1 - \tan^2 x}}{\sqrt{2} - \sqrt{1 - \tan^2 x}}\right] - \log(\cot x + \sqrt{\cot^2 x - 1}) + C$

10. $\sqrt{2}\tan^{-1}\left(\frac{\sqrt{\tan x} - \sqrt{\cot x}}{\sqrt{2}}\right) + c$

11. $\frac{3}{2}x^{2/3} - \frac{12}{7}x^{7/12} + 2x^{1/2} - \frac{12}{5}x^{5/12} + 3x^{1/3} - 4x^{1/4} + 6x^{1/6} - 12x^{1/12}$
$+12\log|x^{1/2} + 1| + 6\left[\left\{\frac{(1+x^{1/6})^3}{3} - \frac{3}{2}(1+x^{1/6})^2 + 3(1+x^{1/6})\right\}\right.$
$\ln(1+x^{1/6}) - \left\{\frac{(1+x^{1/6})^3}{9} - \frac{3}{4}(1+x^{1/6})^2 + 3(1+x^{1/6})\right\}$
$\left. + \frac{(\ln(1+x^{1/6}))^2}{2}\right] + C$

12. $\frac{\sin 2\theta}{2}\ln\left(\frac{\cos\theta + \sin\theta}{\cos\theta - \sin\theta}\right) - \frac{1}{2}\ln\sec 2\theta + C$

13. $\log\left(\frac{1 + xe^x}{xe^x}\right) - \frac{1}{1 + xe^x} + C$

14. $2\cos^{-1}\sqrt{x} - 2\log\left(\frac{1 + \sqrt{1-x}}{\sqrt{x}}\right) + C$

15. $-\frac{1}{2}\log|x+1| + \frac{1}{4}\log(x^2+1) + \frac{3}{2}\tan^{-1}x + \frac{x}{1+x^2} + C$

16. $(x+1)\tan^{-1}\left(\frac{2x+2}{3}\right) - \frac{3}{4}\log(4x^2 + 8x + 13) + C$ 17. $\frac{1}{6}\frac{(2x^{3m} + 3x^{2m} + 6x^m)^{\frac{m+1}{m}}}{m+1} + C$

H 1. (d)

Section-B : JEE Main/ AIEEE

1. (b) 2. (a) 3. (d) 4. (c) 5. (c) 6. (d) 7. (c)

Section-A JEE Advanced/ IIT-JEE

A FILL IN THE BLANKS :

1. Given that

$$\int \frac{4e^x+6e^{-x}}{9e^x-4e^{-x}}dx = Ax + B\log(9e^{2x}-4)+C$$

Let $I=\int \frac{4e^x+6e^{-x}}{9e^x-4e^{-x}}dx$

Let $4e^x+6e^{-x} = L(9e^x-4e^{-x})+M(9e^x+4e^{-x})$

NOTE THIS STEP :

Comparing coefficients of e^x and e^{-x} on both sides,

We get

$9L+9M=4$ ...(1)

$-4L+4M=6$ ...(2)

Solving (1) and (2), we get

$$L=\frac{-19}{36}, M=\frac{35}{36}$$

$$\therefore\quad I=\int \frac{\frac{-19}{36}(9e^x-4e^{-x})+\frac{35}{36}(9e^x+4e^{-x})}{(9e^x-4e^{-x})}dx$$

$$=\frac{-19}{36}\int 1.dx+\frac{35}{36}\int\frac{9e^x+4e^{-x}}{9e^x-4e^{-x}}dx$$

$$=\frac{-19}{36}x+\frac{35}{36}\log|9e^x-4e^{-x}|+C$$

$$=\frac{-19}{36}x+\frac{35}{36}\log\left(\frac{9e^{2x}-4}{e^x}\right)+C$$

$$=\frac{-19}{36}x+\frac{35}{36}\log|9e^{2x}-4|-\frac{35}{36}x+C$$

$$=\frac{-54}{36}x+\frac{35}{36}\log|9e^{2x}-4|+C$$

$$=\frac{-3}{2}x+\frac{35}{36}\log|9e^{2x}-4|+C$$

Comparing above with given value of I, we get $A=-3/2$, $B=35/36$, C can have any real value.

ALTERNATE SOLUTION : We have

$$\int \frac{4e^x+6e^{-x}}{9e^x-4e^{-x}}dx = Ax+B\,ln(9e^{2x}-4)+C$$

$$\Rightarrow\quad \frac{d}{dx}[Ax+B\ln(9e^{2x}-4)+C]=\frac{4e^x+6e^{-x}}{9e^x-4e^{-x}}$$

$$\Rightarrow\quad A+\frac{18Be^x}{9e^x-4e^{-x}}=\frac{4e^x+6e^{-x}}{9e^x-4e^{-x}}$$

$$\Rightarrow\quad \frac{(9A+18B)e^x-4Ae^{-x}}{9e^x-4e^{-x}}=\frac{4e^x+6e^{-x}}{9e^x-4e^{-x}}$$

$$\Rightarrow\quad 9A+18B=4; -4A=6$$

$$\Rightarrow\quad A=\frac{-3}{2}$$

$B=\left(4+\frac{27}{2}\right)\frac{1}{18}=\frac{35}{36}$; C can have any real value.

C MCQ's WITH ONE CORRECT ANSWER :

1. (c) Let $I=\int\frac{\cos^3x+\cos^5x}{\sin^2x+\sin^4x}dx$

$$=\int\frac{(\cos^2x+\cos^4x)\cos x}{\sin^2x(1+\sin^2x)}dx$$

$$=\int\frac{[1-\sin^2x+(1-\sin^2x)^2]\cos x}{\sin^2x(1+\sin^2x)}dx$$

$$=\int\frac{(2-3\sin^2x+\sin^4x)\cos x}{\sin^2x(1+\sin^2x)}dx$$

Put $\sin x = t \Rightarrow \cos x\,dx = dt$

$$I=\int\frac{2-3t^2+t^4}{t^4+t^2}dt,\ I=\int\left(1+\frac{2}{t^2}-\frac{6}{t^2+1}\right)dt$$

$$=t-\frac{2}{t}-6\tan^{-1}(t)+C$$

$$=\sin x-2(\sin x)^{-1}-6\tan^{-1}(\sin x)+C$$

2. (c) Given that $\int_{\sin x}^{1}t^2f(t)\,dt=1-\sin x$

$$\Rightarrow\quad \frac{d}{dx}\int_{\sin x}^{1}t^2f(t)\,dt=\frac{d}{dx}(1-\sin x)$$

$$\Rightarrow\quad -\sin^2x\,f(\sin x).\cos x=-\cos x$$

$$\Rightarrow\quad f(\sin x)=\frac{1}{\sin^2x}$$

$$\Rightarrow\quad f(1/\sqrt{3})=\frac{1}{(1/\sqrt{3})^2}=3$$

3. (d) $\int\frac{x^2-1}{x^3\sqrt{2x^4-2x^2+1}}dx=\int\frac{x^2-1}{x^5\sqrt{2-\frac{2}{x^2}+\frac{1}{x^4}}}dx$

$$=\frac{1}{4}\int\frac{\frac{4}{x^3}-\frac{4}{x^5}}{\sqrt{2-\frac{2}{x^2}+\frac{1}{x^4}}}dx$$

Put $2-\frac{2}{x^2}+\frac{1}{x^4}=t \Rightarrow \left(\frac{4}{x^3}-\frac{4}{x^5}\right)dx=dt$

$$= \frac{1}{4}\int \frac{dt}{\sqrt{t}} = \frac{2\sqrt{t}}{4} + C = \frac{\sqrt{2 - \frac{2}{x^2} + \frac{1}{x^4}}}{2} + C$$

$$= \frac{\sqrt{2x^4 - 2x^2 + 1}}{2x^2} + C$$

4. **(c)** Given $I = \int \frac{e^x}{e^{4x} + e^{2x} + 1} dx,$

$$J = \int \frac{e^{-x}}{e^{-4x} + e^{-2x} + 1} dx = \int \frac{e^{3x}}{e^{4x} + e^{2x} + 1} dx$$

$\therefore$ $$\text{J} - \text{I} = \int \frac{e^x(e^{2x} - 1)}{e^{4x} + e^{2x} + 1} dx$$

Let $e^x = t \Rightarrow e^x dx = dt$

$\therefore$ $$\text{J} - \text{I} = \int \frac{t^2 - 1}{t^4 + t^2 + 1} dt = \int \frac{1 - \frac{1}{t^2}}{t^2 + 1 + \frac{1}{t^2}} dt$$

Let $t - \frac{1}{t} = u \Rightarrow \left(1 - \frac{1}{t^2}\right) dt = du$

$\therefore$ $$\text{J} - \text{I} = \int \frac{du}{u^2 - 1} = \frac{1}{2}\log\left|\frac{u-1}{u+1}\right| + C$$

$$= \frac{1}{2}\log\left|\frac{\frac{t^2+1}{t} - 1}{\frac{t^2+1}{t} + 1}\right| + C = \frac{1}{2}\log\left|\frac{e^{2x} - e^x + 1}{e^{2x} + e^x + 1}\right| + C$$

5. **(c)** $I = \int \frac{\sec^2 x}{(\sec x + \tan x)^{9/2}} dx$

Let $\sec x + \tan x = t \Rightarrow \sec x - \tan x = \frac{1}{t}$

$\Rightarrow$ $\sec x = \frac{1}{2}\left(t + \frac{1}{t}\right)$

Also $\sec x(\sec x + \tan x)\, dx = dt$

$\Rightarrow$ $\sec x\, dx = \frac{dt}{t}$

$\therefore$ $$I = \frac{1}{2}\int \frac{\left(t + \frac{1}{t}\right)dt}{t^{9/2}.t} = \frac{1}{2}\int \left(t^{-9/2} + t^{-13/2}\right)dt$$

$$= \frac{1}{2}\left[\frac{t^{-9/2+1}}{-\frac{9}{2}+1} + \frac{t^{-13/2+1}}{-\frac{13}{2}+1}\right] + K$$

$$= \frac{-1}{7}t^{-7/2} - \frac{1}{11}t^{-11/2} + K$$

$$= -\frac{1}{7t^{7/2}} - \frac{1}{11t^{11/2}} + K = -\frac{1}{t^{11/2}}\left(\frac{1}{11} + \frac{t^2}{7}\right) + K$$

$$= \frac{-1}{(\sec x + \tan x)^{11/2}}\left\{\frac{1}{11} + \frac{1}{7}(\sec x + \tan x)^2\right\} + K$$

E SUBJECTIVE PROBLEMS :

1. $I = \int \frac{\sin x}{\sin x - \cos x} dx = \frac{1}{2}\int \frac{2\sin x}{\sin x - \cos x} dx$

$$= \frac{1}{2}\int \frac{\sin x + \cos x + \sin x - \cos x}{\sin x - \cos x} dx$$

NOTE THIS STEP

$$= \frac{1}{2}\int \frac{\cos x + \sin x}{\sin x - \cos x} dx + \frac{1}{2}\int \frac{\sin x - \cos x}{\sin x - \cos x} dx$$

$$= \frac{1}{2}\log|\sin x - \cos x| + \frac{x}{2} + C$$

2. Let $I = \int \frac{x^2 dx}{(a + bx)^2}$

Let $a + bx = t \Rightarrow x = \left(\frac{t-a}{b}\right) \Rightarrow dx = \frac{dt}{b}$

$\therefore$ $$I = \frac{1}{b^3}\int \frac{(t-a)^2}{t^2} dt = \frac{1}{b^3}\int \frac{t^2 - 2at + a^2}{t^2} dt$$

$$= \frac{1}{b^3}\int \left(1 - \frac{2a}{t} + \frac{a^2}{t^2}\right) dt = \frac{1}{b^3}\left[t - 2a\log|t| - \frac{a^2}{t}\right] + C$$

$$= \frac{1}{b^3}\left[a + bx - 2a\log|a + bx| - \frac{a^2}{a + bx}\right] + C$$

3. **(a)** $\int \sqrt{1 + \sin\frac{x}{2}}\, dx = \int \sqrt{\sin^2\frac{x}{4} + \cos^2\frac{x}{4} + 2\sin\frac{x}{4}\cos\frac{x}{4}}\, dx$

$$= \pm\left[\int \left(\sin\frac{x}{4} + \cos\frac{x}{4}\right) dx\right]$$

$$= \pm\left[\frac{-\cos x/4}{1/4} + \frac{\sin x/4}{1/4}\right] + C = \pm 4\left[\sin\frac{x}{4} - \cos\frac{x}{4}\right] + C$$

(b) $\int \frac{x^2}{\sqrt{1-x}} dx$

Let $1 - x = t^2 \Rightarrow dx = -\, 2t\, dt$

$\therefore$ Integral becomes

$$= \int \frac{(1-t^2)^2}{t}(-2t)dt = -2\int (t^4 - 2t^2 + 1)\, dt$$

$$= -2\left[\frac{t^5}{5} - \frac{2t^3}{3} + t\right] + C$$

$$= -2\left[\frac{(1-x)^{5/2}}{5} - \frac{2(1-x)^{3/2}}{3} + \sqrt{1-x}\right] + C$$

4. To evaluate $\int (e^{\log x} + \sin x)\cos x\, dx$

$= \int (x + \sin x)\cos x\, dx$ [Using $e^{\log x} = x$]

$= \int x\cos x + \frac{1}{2}\int \sin 2x\, dx$

$= x\sin x - \int \sin x\, dx + \frac{1}{2}\left(\frac{-\cos 2x}{2}\right)$

$= x\sin x + \cos x - \frac{1}{4}\cos 2x + C$

5. $I = \int \frac{(x-1)e^x}{(x+1)^3}dx = \int \frac{(x+1-2)e^x}{(x+1)^3}dx$

NOTE THIS STEP:

$= \int \left[\frac{1}{(x+1)^2} - \frac{2}{(x+1)^3}\right]e^x\, dx$

It is of the form, $\int |f(x) + f'(x)|\, e^x dx$

where $f(x) = \frac{1}{(x+1)^2}$ and $f'(x) = \frac{-2}{(x+1)^3}$

$\therefore$ Integral $= e^x f(x)$ $\therefore$ $I = \frac{e^x}{(x+1)^2} + C$

6. Let $I = \int \frac{dx}{x^3 x^2\left(1+\frac{1}{x^4}\right)^{3/4}}$

Put $1 + \frac{1}{x^4} = t \Rightarrow \frac{-4}{x^5}dx = dt \Rightarrow \frac{dx}{x^5} = -\frac{dt}{4}$

$\therefore$ $I = \int \frac{-dt}{4t^{3/4}} = \frac{-1}{4}\left(\frac{t^{-3/4+1}}{\frac{-3}{4}+1}\right) + C$

$= -t^{1/4} + C = -\left(1 + \frac{1}{x^4}\right)^{1/4} + C$

7. $I = \int \sqrt{\frac{1-\sqrt{x}}{1+\sqrt{x}}}\, dx$

Put $x = \cos^2\theta \Rightarrow dx = -2\cos\theta\sin\theta\, d\theta$

$\therefore$ $I = -\int \sqrt{\frac{1-\cos\theta}{1+\cos\theta}}.2\sin\theta\cos\theta\, d\theta$

$= -\int \frac{\sin\theta/2}{\cos\theta/2}.2.2\sin(\theta/2)\cos(\theta/2)\cos\theta\, d\theta$

$= -2\int (1-\cos\theta)\cos\theta\, d\theta$

$= -2\int (\cos\theta - \cos^2\theta)\, d\theta$

$= -2\int \left(\cos\theta - \frac{1+\cos 2\theta}{2}\right)d\theta$

$= -2\left[\sin\theta - \frac{1}{2}\left(\theta + \frac{\sin 2\theta}{2}\right) + C\right]$

$= -2\sqrt{1-x} + \left[\cos^{-1}\sqrt{x} + \sqrt{x}\sqrt{1-x}\right] + C$

[Using $\sin\theta = \sqrt{1-x}$]

$= -2\sqrt{1-x} + \cos^{-1}\sqrt{x} + \sqrt{x}\sqrt{1-x} + C$

8. Let $I = \int \frac{\sin^{-1}\sqrt{x} - \cos^{-1}\sqrt{x}}{\sin^{-1}\sqrt{x} + \cos^{-1}\sqrt{x}}dx$

We know that, $\sin^{-1}\sqrt{x} + \cos^{-1}\sqrt{x} = \pi/2$ ...(1)

also $\cos^{-1}\sqrt{x} = \frac{\pi}{2} - \sin^{-1}x$ ...(2)

Using (1) and (2), we get

$I = \int \frac{\sin^{-1}\sqrt{x} - (\pi/2 - \sin^{-1}\sqrt{x})}{\pi/2}dx$

$= \frac{2}{\pi}\int (2\sin^{-1}\sqrt{x} - \pi/2)\, dx = \frac{4}{\pi}\int \sin^{-1}\sqrt{x}\, dx - \int 1\, dx$

Let $x = \sin^2\theta \Rightarrow dx = 2\sin\theta\cos\theta\, d\theta$

$= \frac{4}{\pi}\int \sin^{-1}(\sin\theta).2\sin\theta\cos\theta\, d\theta - x$

$= \frac{4}{\pi}\int \theta\sin 2\theta d\theta - x = \frac{4}{\pi}\left[\frac{-\theta\cos 2\theta}{2} + \int 1.\frac{\cos 2\theta}{2}d\theta\right] - x + C$

[Integrating by parts]

$= \frac{4}{\pi}\left[\frac{-\theta\cos 2\theta}{2} + \frac{\sin 2\theta}{4}\right] - x + C$

$= \frac{4}{4\times\pi}[-2\theta(1-2\sin^2\theta) + 2\sin\theta\sqrt{1-\sin^2\theta}] + C - x$

$= \frac{1}{\pi}[-2(\sin^{-1}\sqrt{x})(1-2x) + 2\sqrt{x}\sqrt{1-x}] + C - x$

$= \frac{2}{\pi}[\sqrt{x-x^2} - (1-2x)\sin^{-1}\sqrt{x}] - x + C$

9. $I = \int \frac{\sqrt{\cos 2x}}{\sin x}dx = \int \frac{\sqrt{\cos^2 x - \sin^2 x}}{\sin^2 x}dx = \int \sqrt{\cot^2 x - 1}\, dx$

Let $\cot x = \sec\theta \Rightarrow -\text{cosec}^2 x\, dx = \sec\theta\tan\theta\, d\theta$

We get, $I = \int \sqrt{\sec^2\theta - 1}.\frac{\sec\theta\tan\theta}{-(1+\sec^2\theta)}d\theta$

$= -\int \frac{\sec\theta.\tan^2\theta}{1+\sec^2\theta}d\theta = -\int \frac{\sin^2\theta}{\cos\theta + \cos^3\theta}d\theta$

$= -\int \frac{1-\cos^2\theta}{\cos\theta(1+\cos^2\theta)}d\theta = -\int \frac{(1+\cos^2\theta) - 2\cos^2\theta}{\cos\theta(1+\cos^2\theta)}d\theta$

$$= -\int \sec\theta d\theta + 2\int \frac{\cos\theta}{1+\cos^2\theta} d\theta$$

$$= -\log|\sec\theta + \tan\theta| + 2\int \frac{\cos\theta}{2-\sin^2\theta} d\theta$$

$$= -\log|\sec\theta + \tan\theta| + 2.\frac{1}{2\sqrt{2}}\log\left|\frac{\sqrt{2}+\sin\theta}{\sqrt{2}-\sin\theta}\right| + C$$

$$= -\log\left|\cot x + \sqrt{\cot^2 x - 1}\right| + \frac{1}{\sqrt{2}}\log\left|\frac{\sqrt{2}+\sqrt{1-\tan^2 x}}{\sqrt{2}-\sqrt{1-\tan^2 x}}\right| + C$$

10. Let $I = \int (\sqrt{\tan x} + \sqrt{\cot x})dx$

Put $\tan x = t^2 \Rightarrow \sec^2 x\, dx = 2t\, dt$

$$\Rightarrow dx = \frac{2t}{1+\tan^2 x} dt \Rightarrow dx = \frac{2t}{1+t^4} dt$$

$$\therefore \quad I = \int\left(t + \frac{1}{t}\right).\frac{2t}{t^4+1} dt = \int \frac{2(t^2+1)}{t^4+1} dt = 2\int \frac{1+\frac{1}{t^2}}{t^2+\frac{1}{t^2}} dt$$

Put $t - \frac{1}{t} = u$

$$\Rightarrow \left(1+\frac{1}{t^2}\right)dt = du, \text{ also } t^2 + \frac{1}{t^2} = u^2 + 2$$

$$\therefore \quad I = 2\int \frac{du}{u^2+2} = 2.\frac{1}{\sqrt{2}}\tan^{-1}\left(\frac{u}{\sqrt{2}}\right) + C$$

$$= \sqrt{2}\tan^{-1}\left(\frac{t-\frac{1}{t}}{\sqrt{2}}\right) + C = \sqrt{2}\tan^{-1}\left(\frac{\sqrt{\tan x}-\sqrt{\cot x}}{\sqrt{2}}\right) + C$$

ALTERNATE SOLUTION :

$$I = \int (\sqrt{\tan x} + \sqrt{\cot x})dx = \int \frac{\sqrt{\sin x}}{\sqrt{\cos x}} + \frac{\sqrt{\cos x}}{\sqrt{\sin x}} dx$$

$$= \int \frac{\sin x + \cos x}{\sqrt{\sin x \cos x}} dx = \sqrt{2}\int \frac{\sin x + \cos x}{\sqrt{\sin 2x}}$$

Put $\sin x - \cos x = t \Rightarrow (\cos x + \sin x)\, dx = dt$

also $(\sin x - \cos x)^2 = t^2 \Rightarrow 1 - \sin 2x = t^2$

$$\Rightarrow \sin 2x = 1 - t^2 \qquad \because I = \sqrt{2}\int \frac{dt}{\sqrt{1-t^2}}$$

$$= \sqrt{2}\sin^{-1} t + C = \sqrt{2}\sin^{-1}(\sin x - \cos x) + C$$

11. Let $I = \int\left(\frac{1}{\sqrt[3]{x}+\sqrt[4]{x}} + \frac{\ln(1+\sqrt[6]{x})}{\sqrt[3]{x}+\sqrt{x}}\right)dx$

$$= \int \frac{1}{\sqrt[3]{x}+\sqrt[4]{x}} dx + \int \frac{\ln(1+\sqrt[6]{x})}{\sqrt[3]{x}+\sqrt{x}} dx$$

$I = I_1 + I_2$ ...(1)

where $I_1 = \int \frac{1}{\sqrt[3]{x}+\sqrt[4]{x}} dx$

Let $x = y^{12}$ so that $dx = 12\, y^{11}\, dy$

$$\therefore \quad I_1 = \int \frac{12y^{11}}{y^4+y^3} dy = 12\int \frac{y^8}{1+y} dy$$

$$= 12\int\left(y^7 - y^6 + y^5 - y^4 + y^3 - y^2 + y - 1 + \frac{1}{y+1}\right)dy$$

$$= 12\left[\frac{y^8}{8} - \frac{y^7}{7} + \frac{y^6}{6} - \frac{y^5}{5} + \frac{y^4}{4} - \frac{y^3}{3} + \frac{y^2}{2} - y + \log|y+1|\right] + C_1$$

$$= \frac{3}{2}x^{2/3} - \frac{12}{7}x^{7/12} + 2x^{1/2} - \frac{12}{5}x^{5/12} + 3x^{1/3}$$

$$-4x^{1/4} + 6x^{1/6} - 12x^{1/12} + 12\log|x^{1/12} + 1| + C_1 \quad ...(2)$$

$$I_2 = \int \frac{\ln(1+(x)^{1/6})}{(x)^{1/3}+(x)^{1/2}} dx$$

Let $x = z^6$ so that $dx = 6z^5\, dz$

$$= \int \frac{\ln(1+z)}{z^2+z^3}.6z^5 dz = \int \frac{6z^3\ln(z+1)}{z+1} dz$$

Put $z + 1 = t \qquad \Rightarrow dz = dt$

$$\therefore \quad I_2 = \int \frac{6(t-1)^3 \ln t}{t} dt = 6\int (t^2 - 3t + 3 - \frac{1}{t})\ln t\, dt$$

$$= 6\left[\int (t^2 - 3t + 3)\ln t\, dt - \int \frac{1}{t}\ln t\, dt\right]$$

$$= 6\left[\left(\frac{t^3}{3} - \frac{3t^2}{2} + 3t\right)\ln t - \int\left(\frac{t^3}{3} - \frac{3t^2}{2} + 3t\right).\frac{1}{t}dt - \frac{(\ln t)^2}{2}\right]$$

$$= 6\left[\left(\frac{t^3}{3} - \frac{3t^2}{2} + 3t\right)\ln t - \int\left(\frac{t^2}{3} - \frac{3}{2}t + 3\right)dt - \frac{(\ln t)^2}{2}\right]$$

$$= 6\left[\left(\frac{t^3}{3} - \frac{3t^2}{2} + 3t\right)\ln t - \left(\frac{t^3}{9} - \frac{3t^2}{4} + 3t\right) - \frac{(\ln t)^2}{2}\right] + C_2$$

$$= 6\left[\left\{\frac{(1+x^{1/6})^3}{3} - \frac{3}{2}(1+x^{1/6})^2 + 3(1+x^{1/6})\right\}\right.$$

$$\ln(1+x^{1/6}) - \left\{\frac{(1+x^{1/6})^3}{9}\right.$$

$$\left.\left. - \frac{3}{4}(1+x^{1/6})^2 + 3(1+x^{1/6})\right\} + \frac{[\ln(1+x^{1/6})^2]}{2}\right] + C_2 \quad ...(3)$$

Thus we get the value of I on substituting the values of I_1 and I_2 from (2) and (3) in equation (1).

12. Let $I = \int \cos 2\theta \ln\left(\frac{\cos\theta + \sin\theta}{\cos\theta - \sin\theta}\right) d\theta$

Now we observe that

$$\frac{d}{d\theta}\left\{\ln\left(\frac{\cos\theta + \sin\theta}{\cos\theta - \sin\theta}\right)\right\}$$

$$= \frac{d}{d\theta}[\ln(\cos\theta + \sin\theta) - \ln(\cos\theta - \sin\theta)]$$

$$= \frac{\cos\theta - \sin\theta}{\cos\theta + \sin\theta} - \frac{-\cos\theta - \sin\theta}{\cos\theta - \sin\theta}$$

$$= \frac{\cos\theta - \sin\theta}{\cos\theta + \sin\theta} + \frac{\cos\theta + \sin\theta}{\cos\theta - \sin\theta}$$

$$= \frac{\cos^2\theta + \sin^2\theta - 2\sin\theta\cos\theta + \cos^2\theta + \sin^2\theta + 2\sin\theta\cos\theta}{\cos^2\theta - \sin^2\theta} = \frac{2}{\cos 2\theta}$$

$\therefore$ Integrating I with respect to θ, by parts we get

$$I = \frac{\sin 2\theta}{2}\ln\left(\frac{\cos\theta + \sin\theta}{\cos\theta - \sin\theta}\right) - \int \frac{\sin 2\theta}{2}\cdot\frac{2}{\cos 2\theta} d\theta$$

$$= \frac{\sin 2\theta}{2}\ln\left(\frac{\cos\theta + \sin\theta}{\cos\theta - \sin\theta}\right) - \int \tan 2\theta \, d\theta$$

$$= \frac{\sin 2\theta}{2}\ln\left(\frac{\cos\theta + \sin\theta}{\cos\theta - \sin\theta}\right) - \frac{1}{2}\ln\sec 2\theta + C$$

13. $I = \int \frac{(x+1)}{x(1+xe^x)^2} dx = \int \frac{e^x(x+1)}{xe^x(1+xe^x)^2} dx$

Put $1 + xe^x = t$

$(xe^x + e^x)\, dx = dt$

$$\therefore \quad I = \int \frac{dt}{(t-1)t^2} = \int\left(\frac{1}{1-t} + \frac{1}{t} + \frac{1}{t^2}\right) dt$$

$$= -\log|1-t| + \log|t| - \frac{1}{t} + C$$

$$= \log\left|\frac{t}{1-t}\right| - \frac{1}{t} + C = \log\left|\frac{1+xe^x}{-xe^x}\right| - \frac{1}{1+xe^x} + C$$

$$= \log\left(\frac{1+xe^x}{xe^x}\right) - \frac{1}{1+xe^x} + C$$

14. Let $I = \int \sqrt{\frac{1-\sqrt{x}}{1+\sqrt{x}}}\cdot\frac{dx}{x}$

Let $\sqrt{x} = \cos\theta$

$\Rightarrow x = \cos^2\theta \Rightarrow dx = -2\cos\theta\sin\theta\, d\theta$

$$\therefore \quad I = \int\left[\sqrt{\frac{1-\cos\theta}{1+\cos\theta}}\cdot\left(\frac{-2\sin\theta\cos\theta}{\cos^2\theta}\right)\right] d\theta$$

$$= \int \sqrt{\frac{2\sin^2\theta/2}{2\cos^2\theta/2}}\cdot\frac{-2.2\sin\theta/2\cos\theta/2}{\cos\theta} d\theta$$

$$= -2\int \frac{2\sin^2\theta/2}{\cos\theta} d\theta = -2\int \frac{(1-\cos\theta)}{\cos\theta} d\theta$$

$$= 2\int (1 - \sec\theta) d\theta$$

$$= 2[\theta - \log|\sec\theta + \tan\theta|] + C$$

$$= 2\left|\cos^{-1}\sqrt{x} - \log\left|\frac{1}{\sqrt{x}} + \sqrt{\frac{1}{x} - 1}\right|\right| + C$$

$$= 2\cos^{-1}\sqrt{x} - 2\log\left(\frac{1+\sqrt{1-x}}{\sqrt{x}}\right) + C$$

15. $I = \int \frac{x^3 + 3x + 2}{(x^2+1)^2(x+1)} dx.$

$$\frac{x^3 + 3x + 2}{(x^2+1)^2(x+1)} = \frac{A}{x+1} + \frac{Bx + C}{x^2+1} + \frac{Dx + E}{(x^2+1)^2}$$

Comparing and solving, we get,

$$A = \frac{1}{2}, B = \frac{1}{2}, C = \frac{1}{2}, D = 0, E = 2.$$

$$\therefore \quad I = -\frac{1}{2}\int \frac{1}{x+1} dx + \frac{1}{2}\int \frac{x+1}{x^2+1} dx + 2\int \frac{dx}{(x^2+1)^2}$$

$$= -\frac{1}{2}\log|x+1| + \frac{1}{4}\log(x^2+1) + \frac{1}{2}\tan^{-1}x + 2I_1 + C$$

where $I_1 = \int \frac{dx}{(x^2+1)^2}$, putting $x = \tan\theta$,

$$I_1 = \int \frac{\sec^2\theta}{\sec^4\theta} d\theta = \int (\cos^2\theta)\, d\theta = \frac{1}{2}\int (1 + \cos 2\theta) d\theta$$

$$= \frac{1}{2}(\theta + \frac{1}{2}\sin 2\theta) = \frac{1}{2}\tan^{-1}x + \frac{1}{4}\cdot\frac{2x}{1+x^2}$$

$$\therefore \quad I = -\frac{1}{2}\log|x+1| + \frac{1}{4}\log(x^2+1) + \frac{3}{2}\tan^{-1}x + \frac{x}{1+x^2} + C$$

where C is constant of integration.

16. $I = \int \sin^{-1}\left(\frac{2x+2}{\sqrt{4x^2+8x+13}}\right)dx$

$= \int \sin^{-1}\left[\frac{x+1}{\sqrt{x^2+2x+\frac{13}{4}}}\right]dx$

$= \int \sin^{-1}\left[\frac{x+1}{\sqrt{(x+1)^2)+(3/2)^2}}\right]dx$

Put $x+1 = 3/2 \tan\theta$, $dx = \frac{3}{2}\sec^2 d\theta$

NOTE STEP:

$\therefore \quad I = \int \sin^{-1}\left[\frac{\left(\frac{3}{2}\tan\theta\right)}{\sqrt{\frac{9}{4}\tan^2\theta+\frac{9}{4}}}\right]\frac{3}{2}\sec^2\theta\, d\theta$

$= \frac{3}{2}\int \sin^{-1}\left[\frac{\sin\theta}{\cos\theta}\frac{\cos\theta}{1}\right]\sec^2\theta d\theta$

$= \frac{3}{2}\int \theta\sec^2\theta\, d\theta = \frac{3}{2}\left[\theta\tan\theta - \int \tan\theta\, d\theta\right]$

$= \frac{3}{2}[\theta\tan\theta - \log|\sec\theta|] + C$

$I = \frac{3}{2}\left[\frac{2}{3}(x+1)\tan^{-1}\right]\left[\frac{2}{3}(x+1)\right]$

$- \log\sqrt{1+\frac{4}{9}(x+1)^2}] + C$

$= (x+1)\tan^{-1}\left(\frac{2x+2}{3}\right) - \frac{3}{4}\log(9+4x^2+8x+4)$

$+ \frac{3}{4}\log 9 + C$

$= (x+1)\tan^{-1}\left(\frac{2x+2}{3}\right) - \frac{3}{4}\log(4x^2+8x+13) + C$

17. $I \quad = \int (x^{3m}+x^{2m}+x^m)(2x^{2m}+3x^m+6)^{1/m}dx$

$= \int (x^{3m}+x^{2m}+x^m)\left[\frac{2x^{3m}+3x^{2m}+6x^m}{x^m}\right]^{1/m}dx$

$= \int \left(\frac{x^{3m}+x^{2m}+x^m}{x}\right)(2x^{3m}+3x^{2m}+6x^m)^{1/m}dx$

Put $2x^{3m}+3x^{2m}+6x^m = y$

NOTE THIS STEP:

$\therefore \quad I = \frac{1}{6m}\int y^{1/m}dy = \frac{1}{6m}\left(\frac{y^{1/m+1}}{1/m+1}\right) + C$

$= \frac{1}{6}\left(\frac{y^{\frac{m+1}{m}}}{m+1}\right) + C$

$= \frac{1}{6}\frac{(2x^{3m}+3x^{2m}+6x^m)^{\frac{m+1}{m}}}{m+1} + C$

(H) ASSERTION & REASON TYPE QUESTIONS :

1. **(d)** $F(x) = \int \sin^2 x\, dx = \frac{1}{2}\int (1-\cos 2x)dx$

$= \frac{1}{4}(2x - \sin 2x) + C$

Now, $F(x+\pi) = \frac{1}{4}(2x+2\pi-\sin(2x+2\pi)) + C$

$= \frac{1}{4}[2x+2\pi-\sin 2x] + C \neq F(x)$

$\therefore$ Statement –1 is false.

Also $\sin^2(x+\pi) = \sin^2 x, \forall x \in R$

$\therefore$ Statement –2 is true.

Section-B JEE Main/ AIEEE

1. **(b)** $\int \frac{\sin x}{\sin(x-\alpha)}dx = \int \frac{\sin(x-\alpha+\alpha)}{\sin(x-\alpha)}dx$

$= \int \frac{\sin(x-\alpha)\cos\alpha + \cos(x-\alpha)\sin\alpha}{\sin(x-\alpha)}dx$

$= \int \{\cos\alpha + \sin\alpha\cot(x-\alpha)\}dx$

$= (\cos\alpha)x + (\sin\alpha)\log\sin(x-\alpha) + C$

$\therefore A = \cos\alpha, B = \sin\alpha$

2. **(a)** $\int \frac{dx}{\cos x - \sin x} = \int \frac{dx}{\sqrt{2}\cos\left(x+\frac{\pi}{4}\right)}$

$= \frac{1}{\sqrt{2}}\int \sec\left(x+\frac{\pi}{4}\right)dx$

$= \frac{1}{\sqrt{2}}\log\left|\tan\left(\frac{\pi}{4}+\frac{x}{2}+\frac{\pi}{8}\right)\right| + C$

$\left[\because \int \sec x\, dx = \log\left|\tan\left(\frac{\pi}{4}+\frac{x}{2}\right)\right|\right]$

$= \frac{1}{\sqrt{2}}\log\left|\tan\left(\frac{x}{2}+\frac{3\pi}{8}\right)\right| + C$

3. (d) $\int \frac{(\log x - 1)^2}{(1+(\log x)^2)^2}dx = \int \frac{1+(\log x)^2 - 2\log x}{\left[1+(\log x)^2\right]^2}dx$

$= \int\left[\frac{1}{(1+(\log x)^2)} - \frac{2\log x}{(1+(\log x)^2)^2}\right]dx$

$= \int\left[\frac{e^t}{1+t^2} - \frac{2t\,e^t}{(1+t^2)^2}\right]dt$ put $\log x = t \Rightarrow dx = e^t\, dt$

$= \int e^t\left[\frac{1}{1+t^2} - \frac{2t}{(1+t^2)^2}\right]dt$

$\left[\text{Which is of the form } \int e^x\left(f(x)+f'(x)\right)dx\right]$

$= \frac{e^t}{1+t^2} + c = \frac{x}{1+(\log x)^2} + c$

4. (c) $I = \int \frac{dx}{\cos x + \sqrt{3}\sin x} \Rightarrow I = \int \frac{dx}{2\left[\frac{1}{2}\cos x + \frac{\sqrt{3}}{2}\sin x\right]}$

$= \frac{1}{2}\int \frac{dx}{\left[\sin\frac{\pi}{6}\cos x + \cos\frac{\pi}{6}\sin x\right]} = \frac{1}{2}.\int \frac{dx}{\sin\left(x+\frac{\pi}{6}\right)}$

$\Rightarrow \quad I = \frac{1}{2}.\int \operatorname{cosec}\left(x+\frac{\pi}{6}\right)dx$

But we know that

$\int \operatorname{cosec} x\, dx = \log|(\tan x/2)| + C$

$\therefore \quad I = \frac{1}{2}.\log\tan\left(\frac{x}{2}+\frac{\pi}{12}\right) + C$

5. (c) Let $I = \sqrt{2}\int \frac{\sin x dx}{\sin\left(x-\frac{\pi}{4}\right)}$ put $x - \frac{\pi}{4} = t$

$\Rightarrow \quad dx = dt \Rightarrow I = \sqrt{2}\int \frac{\sin\left(t+\frac{\pi}{4}\right)}{\sin t}dt$

$= \frac{\sqrt{2}}{\sqrt{2}}\int\left(\frac{\sin t + \cos t}{\sin t}\right)dt$

$\Rightarrow \quad I = \int(1+\cot t)dt = t + \log|\sin t| + c_1$

$= x - \frac{\pi}{4} + \log\left|\sin\left(x-\frac{\pi}{4}\right)\right| + c_1$

$= x + \log\left|\sin\left(x-\frac{\pi}{4}\right)\right| + c$ (where $c = c_1 - \frac{\pi}{4}$)

6. (d) $\int \frac{5\tan x}{\tan x - 2}dx = \int \frac{5\frac{\sin x}{\cos x}}{\frac{\sin x}{\cos x} - 2}dx$

$= \int\left(\frac{5\sin x}{\cos x} \times \frac{\cos x}{\sin x - 2\cos x}\right)dx$

$= \int \frac{5\sin x\, dx}{\sin x - 2\cos x}$

$= \int\left(\frac{4\sin x + \sin x + 2\cos x - 2\cos x}{\sin x - 2\cos x}\right)dx$

$= \int \frac{(\sin x - 2\cos x) + (4\sin x + 2\cos x)}{\sin x - 2\cos x}dx$

$= \int \frac{(\sin x - 2\cos x) + 2(\cos x + 2\sin x)}{(\sin x - 2\cos x)}dx$

$= \int \frac{\sin x - 2\cos x}{\sin x - 2\cos x}dx + 2\int\left(\frac{\cos x + 2\sin x}{\sin x - 2\cos x}\right)dx$

$= \int dx + 2\int \frac{\cos x + 2\sin x}{\sin x - 2\cos x}dx$

$= I_1 + I_2$ where $I_1 = \int dx$ and $I_2 = 2\int \frac{\cos x + 2\sin x}{\sin x - 2\cos x}dx$

put $\sin x - 2\cos x = t$

$\Rightarrow \quad (\cos x + 2\sin x)\, dx = dt$

$\therefore \; I_2 = 2\int \frac{dt}{t} = 2\ln t + C = 2\ln(\sin x - 2\cos x) + C$

Hence, $I_1 + I_2 = \int dx + 2\ln(\sin x - 2\cos x) + c$

$= x + 2\ln|(\sin x - 2\cos x)| + k \Rightarrow a = 2$

7. (c) Let $\int f(x)dx = \psi(x)$

Let $I = \int x^5 f(x^3)dx$

put $x^3 = t$

$\Rightarrow \; 3x^2 dx = dt$

$I = \frac{1}{3}\int 3\cdot x^2\cdot x^3\cdot f(x^3)\cdot dx$

$= \frac{1}{3}\int t f(t)dt = \frac{1}{3}\left[t\int f(t)dt - \int f(t)dt\right]$

$= \frac{1}{3}\left[t\psi(t) - \int \psi(t)dt\right]$

$= \frac{1}{3}\left[x^3\psi(x^3) - 3\int x^2\psi(x^3)dx\right] + C$

$= \frac{1}{3}x^3\psi(x^3) - \int x^2\psi(x^3)dx + C$

Chapter 18 Definite Integrals and Applications of Integrals

Section-A JEE Advanced/ IIT-JEE

A FILL IN THE BLANKS :

1. $f(x)=\begin{vmatrix} \sec x & \cos x & \sec^2 x+\cot x \operatorname{cosec} x \\ \cos^2 x & \cos^2 x & \operatorname{cosec}^2 x \\ 1 & \cos^2 x & \cos^2 x \end{vmatrix}$.

Then $\int_0^{\pi/2} f(x)\,dx=$ .... **(1987 - 2 Marks)**

2. The integral $\int_0^{1.5} [x^2]dx$, **(1988 - 2 Marks)**

Where [ ] denotes the greatest integer function, equals ....

3. The value of $\int_{-2}^{2} |1-x^2|\,dx$ is....... **(1989 - 2 Marks)**

4. The value of $\int_{\pi/4}^{3\pi/4} \frac{\phi}{1+\sin\phi}\,d\phi$ is .... **(1993 - 2 Marks)**

5. The value of $\int_2^3 \frac{\sqrt{x}}{\sqrt{5-x}+\sqrt{x}}\,dx$ is .... **(1994 - 2 Marks)**

6. If for nonzero x, $af(x)+bf\left(\frac{1}{x}\right)=\frac{1}{x}-5$ where $a \neq b$, then

$\int_1^2 f(x)\,dx=$....... **(1996 - 2 Marks)**

7. For $n>0$, $\int_0^{2\pi} \frac{x\sin^{2n} x}{\sin^{2n} x+\cos^{2n} x}\,dx=$ ...... **(1996 - 1 Mark)**

8. The value of $\int_1^{e^{37}} \frac{\pi\sin(\pi\, ln\, x)}{x}\,dx$ is .... **(1997 - 2 Marks)**

9. Let $\frac{d}{dx}F(x)=\frac{e^{\sin x}}{x}$, $x>0$. If $\int_1^4 \frac{2e^{\sin x^2}}{x}\,dx=F(k)-F(1)$ then one of the possible values of k is ....

(1997 - 2 Marks)

B TRUE / FALSE :

1. The value of the integral $\int_0^{2a} \left[\frac{f(x)}{\{f(x)+f(2a-x)\}}\right]dx$ is equal to a. **(1988 - 1 Mark)**

C MCQ's WITH ONE CORRECT ANSWER :

1. The value of the definite integral $\int_0^1 (1+e^{-x^2})\,dx$ is

(a) -1
(b) 2 **(1981 - 2 Marks)**
(c) $1+e^{-1}$
(d) none of these

2. Let a, b, c be non-zero real numbers such that

$$\int_0^1 (1+\cos^8 x)(ax^2+bx+c)\,dx=\int_0^2 (1+\cos^8 x)(ax^2+bx+c)\,dx.$$

Then the quadratic equation $ax^2+bx+c=0$ has **(1981 - 2 Marks)**

(a) no root in (0, 2)
(b) at least one root in (0, 2)
(c) a double root in (0, 2)
(d) two imaginary roots

3. The area bounded by the curves $y=f(x)$, the x-axis and the ordinates $x=1$ and $x=b$ is $(b-1)\sin(3b+4)$. Then $f(x)$ is

(a) $(x-1)\cos(3x+4)$ **(1982 - 2 Marks)**
(b) $\sin(3x+4)$
(c) $\sin(3x+4)+3(x-1)\cos(3x+4)$
(d) none of these

4. The value of the integral $\int_0^{\pi/2} \frac{\sqrt{\cot x}}{\sqrt{\cot x}+\sqrt{\tan x}}\,dx$ is

(a) $\pi/4$
(b) $\pi/2$ **(1983 - 1 Mark)**
(c) π
(d) none of these

5. For any integer n the integral —

$\int_0^{\pi} e^{\cos^2 x}\cos^3(2n+1)x\,dx$ has the value **(1985 - 2 Marks)**

(a) π
(b) 1
(c) 0
(d) none of these

6. Let $f: R \to R$ and $g: R \to R$ be continous functions. Then the value of the integral

$\int_{-\pi/2}^{\pi/2} [f(x)+f(-x)]\,[g(x)-g(-x)]\,dx$ is **(1990 - 2 Marks)**

(a) π
(b) 1
(c) -1
(d) 0

7. The value of $\int_0^{\pi/2} \frac{dx}{1+\tan^3 x}$ is **(1993 - 1 Marks)**

(a) 0
(b) 1
(c) $\pi/2$
(d) $\pi/4$

8. If $f(x) = A\sin\left(\frac{\pi x}{2}\right) + B$, $f'\left(\frac{1}{2}\right) = \sqrt{2}$ and $\int_0^1 f(x)dx = \frac{2A}{\pi}$, then constants A and B are **(1995S)**

(a) $\frac{\pi}{2}$ and $\frac{\pi}{2}$ (b) $\frac{2}{\pi}$ and $\frac{3}{\pi}$

(c) 0 and $\frac{-4}{\pi}$ (d) $\frac{4}{\pi}$ and 0

9. The value of $\int_\pi^{2\pi} [2\,Sinx]\,dx$ where [.] represents the greatest integer function is **(1995S)**

(a) $\frac{-5\pi}{3}$ (b) $-\pi$ (c) $\frac{5\pi}{3}$ (d) -2π

10. Let f be a positive function.

Let $I_1 = \int_{1-k}^{k} xf[x(1-x)]dx$, $I_2 = \int_{1-k}^{k} f[x(1-x)]dx$,

where $2k-1>0$. Then $\frac{I_1}{I_2}$ is **(1997C - 2 Marks)**

(a) 2 (b) k (c) $\frac{1}{2}$ (d) 1

11. If $g(x) = \int_0^x \cos^4 t\,dt$, then $g(x+\pi)$ equals **(1997 - 2 Marks)**

(a) $g(x)+g(\pi)$ (b) $g(x)-g(\pi)$

(c) $g(x)\,g(\pi)$ (d) $\frac{g(x)}{g(\pi)}$

12. $\int_{\pi/4}^{3\pi/4} \frac{dx}{1+\cos x}$ is equal to **(1999 - 2 Marks)**

(a) 2 (b) -2 (c) 1/2 (d) $-1/2$

13. If for a real number y, $[y]$ is the greatest integer less than or equal to y, then the value of the integral $\int_{\pi/2}^{3\pi/2} [2\sin x]\,dx$ is **(1999 - 2 Marks)**

(a) $-\pi$ (b) 0 (c) $-\pi/2$ (d) $\pi/2$

14. Let $g(x) = \int_0^x f(t)dt$, where f is such that

$\frac{1}{2} \le f(t) \le 1$, for $t \in [0,1]$ and $0 \le f(t) \le \frac{1}{2}$, for $t \in [1,2]$.

Then g(2) satisfies the inequality **(2000S)**

(a) $-\frac{3}{2} \le g(2) < \frac{1}{2}$ (b) $0 \le g(2) < 2$

(c) $\frac{3}{2} < g(2) \le \frac{5}{2}$ (d) $2 < g(2) < 4$

15. If $f(x) = \begin{cases} e^{\cos x}\sin x, & \text{for } |x| \le 2 \\ 2, & \text{otherwise,} \end{cases}$ then $\int_{-2}^{3} f(x)dx =$ **(2000S)**

(a) 0 (b) 1 (c) 2 (d) 3

16. The value of the integral $\int_{e^{-1}}^{e^2} \left|\frac{\log_e x}{x}\right| dx$ is: **(2000S)**

(a) 3/2 (b) 5/2 (c) 3 (d) 5

17. The value of $\int_{-\pi}^{\pi} \frac{\cos^2 x}{1+a^x} dx$, $a > 0$, is **(2001S)**

(a) π (b) $a\pi$ (c) $\pi/2$ (d) 2π

18. The area bounded by the curves $y = |x| - 1$ and $y = -|x| + 1$ is **(2002S)**

(a) 1 (b) 2 (c) $2\sqrt{2}$ (d) 4

19. Let $f(x) = \int_1^x \sqrt{2-t^2}\,dt$. Then the real roots of the equation $x^2 - f'(x) = 0$ are **(2002S)**

(a) ± 1 (b) $\pm\frac{1}{\sqrt{2}}$ (c) $\pm\frac{1}{2}$ (d) 0 and 1

20. Let $T > 0$ be a fixed real number. Suppose f is a continuous function such that for all $x \in R$, $f(x+T) = f(x)$.

If $I = \int_0^T f(x)dx$ then the value of $\int_3^{3+3T} f(2x)dx$ is **(2002S)**

(a) $3/2I$ (b) $2I$ (c) $3I$ (d) $6I$

21. The integral $\int_{-1/2}^{1/2} \left([x] + \ell n\left(\frac{1+x}{1-x}\right)\right) dx$ equal to **(2002S)**

(a) $-\frac{1}{2}$ (b) 0 (c) 1 (d) $2\ell n\left(\frac{1}{2}\right)$

22. If $I(m,n) = \int_0^1 t^m(1+t)^n\,dt$, then the expression for $I(m,n)$ in terms of $I(m+1, n-1)$ is **(2003S)**

(a) $\frac{2^n}{m+1} - \frac{n}{m+1} I(m+1, n-1)$

(b) $\frac{n}{m+1} I(m+1, n-1)$

(c) $\frac{2^n}{m+1} + \frac{n}{m+1} I(m+1, n-1)$

(d) $\frac{m}{n+1} I(m+1, n-1)$

23. If $f(x) = \int_{x^2}^{x^2+1} e^{-t^2}\,dt$, then $f(x)$ increases in **(2003S)**

(a) $(-2, 2)$ (b) no value of x

(c) $(0, \infty)$ (d) $(-\infty, 0)$

24. The area bounded by the curves $y=\sqrt{x}$, $2y+3=x$ and x-axis in the 1st quadrant is *(2003S)*
(a) 9 (b) 27/4 (c) 36 (d) 18

25. If $f(x)$ is differentiable and $\int_0^{t^2} xf(x)dx=\frac{2}{5}t^5$, then $f\left(\frac{4}{25}\right)$ equals *(2004S)*
(a) 2/5 (b) –5/2 (c) 1 (d) 5/2

26. The value of the integral $\int_0^1\sqrt{\frac{1-x}{1+x}}dx$ is *(2004S)*
(a) $\frac{\pi}{2}+1$ (b) $\frac{\pi}{2}-1$ (c) – 1 (d) 1

27. The area enclosed between the curves $y=ax^2$ and $x=ay^2$ $(a>0)$ is 1 sq. unit, then the value of a is *(2004S)*
(a) $1/\sqrt{3}$ (b) 1/2 (c) 1 (d) 1/3

28. $\int_{-2}^{0}\{x^3+3x^2+3x+3+(x+1)\cos(x+1)\}\,dx$ is equal to *(2005S)*
(a) –4 (b) 0 (c) 4 (d) 6

29. The area bounded by the parabolas $y=(x+1)^2$ and $y=(x-1)^2$ and the line $y=1/4$ is *(2005S)*
(a) 4 sq. units (b) 1/6 sq. units
(c) 4/3 sq. units (d) 1/3 sq. units

30. The area of the region between the curves $y=\sqrt{\frac{1+\sin x}{\cos x}}$ and $y=\sqrt{\frac{1-\sin x}{\cos x}}$ bounded by the lines $x=0$ and $x=\frac{\pi}{4}$ is *(2008)*
(a) $\int_0^{\sqrt{2}-1}\frac{t}{(1+t^2)\sqrt{1-t^2}}dt$ (b) $\int_0^{\sqrt{2}-1}\frac{4t}{(1+t^2)\sqrt{1-t^2}}dt$
(c) $\int_0^{\sqrt{2}+1}\frac{4t}{(1+t^2)\sqrt{1-t^2}}dt$ (d) $\int_0^{\sqrt{2}+1}\frac{t}{(1+t^2)\sqrt{1-t^2}}dt$

31. Let f be a non-negative function defined on the interval $[0, 1]$. If $\int_0^x\sqrt{1-(f'(t))^2}\,dt=\int_0^x f(t)\,dt,\quad 0\le x\le 1,$ and $f(0)=0$, then *(2009)*
(a) $f\left(\frac{1}{2}\right)<\frac{1}{2}$ and $f\left(\frac{1}{3}\right)>\frac{1}{3}$
(b) $f\left(\frac{1}{2}\right)>\frac{1}{2}$ and $f\left(\frac{1}{3}\right)>\frac{1}{3}$
(c) $f\left(\frac{1}{2}\right)<\frac{1}{2}$ and $f\left(\frac{1}{3}\right)<\frac{1}{3}$
(d) $f\left(\frac{1}{2}\right)>\frac{1}{2}$ and $f\left(\frac{1}{3}\right)<\frac{1}{3}$

32. The value of $\lim_{x\to 0}\frac{1}{x^3}\int_0^x\frac{t\,ln(1+t)}{t^4+4}dt$ is *(2010)*
(a) 0 (b) $\frac{1}{12}$ (c) $\frac{1}{24}$ (d) $\frac{1}{64}$

33. Let f be a real-valued function defined on the interval $(-1, 1)$ such that $e^{-x}f(x)=2+\int_0^x\sqrt{t^4+1}\,dt$, for all $x\in(-1,1)$, and let f^{-1} be the inverse function of f. Then $(f^{-1})'(2)$ is equal to *(2010)*
(a) 1 (b) $\frac{1}{3}$ (c) $\frac{1}{2}$ (d) $\frac{1}{e}$

34. The value of $\int_{\sqrt{\ell n2}}^{\sqrt{\ell n3}}\frac{x\sin x^2}{\sin x^2+\sin(\ell n6-x^2)}dx$ is *(2011)*
(a) $\frac{1}{4}\ell n\frac{3}{2}$ (b) $\frac{1}{2}\ell n\frac{3}{2}$ (c) $\ell n\frac{3}{2}$ (d) $\frac{1}{6}\ell n\frac{3}{2}$

35. Let the straight line $x=b$ divide the area enclosed by $y=(1-x)^2$, $y=0$, and $x=0$ into two parts R_1 $(0\le x\le b)$ and R_2 $(b\le x\le 1)$ such that $R_1-R_2=\frac{1}{4}$. Then b equals *(2011)*
(a) $\frac{3}{4}$ (b) $\frac{1}{2}$ (c) $\frac{1}{3}$ (d) $\frac{1}{4}$

36. Let $f:[-1, 2]\to[0, \infty)$ be a continuous function such that $f(x)=f(1-x)$ for all $x\in[-1, 2]$

Let $R_1=\int_{-1}^{2}xf(x)dx$, and R_2 be the area of the region bounded by $y=f(x)$, $x=-1$, $x=2$, and the x-axis. Then *(2011)*
(a) $R_1=2R_2$ (b) $R_1=3R_2$
(c) $2R_1=R_2$ (d) $3R_1=R_2$

37. The value of the integral $\int_{-\pi/2}^{\pi/2}\left(x^2+\ln\frac{\pi+x}{\pi-x}\right)\cos x\,dx$ is *(2012)*
(a) 0 (b) $\frac{\pi^2}{2}-4$ (c) $\frac{\pi^2}{2}+4$ (d) $\frac{\pi^2}{2}$

38. The area enclosed by the curves $y=\sin x+\cos x$ and $y=|\cos x-\sin x|$ over the interval $\left[0,\frac{\pi}{2}\right]$ is *(JEE Adv. 2013)*
(a) $4(\sqrt{2}-1)$ (b) $2\sqrt{2}(\sqrt{2}-1)$
(c) $2(\sqrt{2}+1)$ (d) $2\sqrt{2}(\sqrt{2}+1)$

39. Let $f:\left[\frac{1}{2},1\right]\to R$ (the set of all real number) be a positive, non-constant and differentiable function such that

$f'(x) < 2f(x)$ and $f\left(\frac{1}{2}\right)=1$. Then the value of $\int_{1/2}^{1} f(x)\,dx$ lies in the interval **(JEE Adv. 2013)**

(a) $(2e-1, 2e)$ (b) $(e-1, 2e-1)$

(c) $\left(\frac{e-1}{2}, e-1\right)$ (d) $\left(0, \frac{e-1}{2}\right)$

D MCQ's WITH ONE OR MORE THAN ONE CORRECT :

1. If $\int_0^x f(t)\,dt = x + \int_x^1 t\,f(t)\,dt$, then the value of $f(1)$ is **(1998 - 2 Marks)**

(a) 1/2 (b) 0 (c) 1 (d) –1/2

2. Let $f(x) = x - [x]$, for every real number x, where $[x]$ is the integral part of x. Then $\int_{-1}^{1} f(x)\,dx$ is **(1998 - 2 Marks)**

(a) 1 (b) 2 (c) 0 (d) 1/2

3. For which of the following values of m, is the area of the region bounded by the curve $y = x - x^2$ and the line $y = mx$ equals 9/2? **(1999 - 3 Marks)**

(a) –4 (b) –2 (c) 2 (d) 4

4. Let $f(x)$ be a non-constant twice differentiable function definied on $(-\infty, \infty)$ such that $f(x) = f(1-x)$ and $f'\left(\frac{1}{4}\right) = 0$. Then, **(2008)**

(a) $f''(x)$ vanishes at least twice on [0, 1]

(b) $f'\left(\frac{1}{2}\right) = 0$

(c) $\int_{-1/2}^{1/2} f\left(x+\frac{1}{2}\right)\sin x\,dx = 0$

(d) $\int_0^{1/2} f(t)\,e^{\sin \pi t}\,dt = \int_{1/2}^{1} f(1-t)\,e^{\sin \pi t}\,dt$

5. Area of the region bounded by the curve $y = e^x$ and lines $x = 0$ and $y = e$ is **(2009)**

(a) $e-1$ (b) $\int_1^e \ln(e+1-y)\,dy$

(c) $e - \int_0^1 e^x\,dx$ (d) $\int_1^e \ln y\,dy$

6. If $I_n = \int_{-\pi}^{\pi} \frac{\sin nx}{(1+\pi^x)\sin x}dx$ $n = 0, 1, 2, ...,$ then **(2009)**

(a) $I_n = I_{n+2}$ (b) $\sum_{m=1}^{10} I_{2m+1} = 10\pi$

(c) $\sum_{m=1}^{10} I_{2m} = 0$ (d) $I_n = I_{n+1}$

7. The value(s) of $\int_0^1 \frac{x^4(1-x)^4}{1+x^2}dx$ is (are) **(2010)**

(a) $\frac{22}{7} - \pi$ (b) $\frac{2}{105}$

(c) 0 (d) $\frac{71}{15} - \frac{3\pi}{2}$

8. Let f be a real-valued function defined on the interval $(0, \infty)$ by $f(x) = \ln x + \int_0^x \sqrt{1+\sin t}\,dt$. Then which of the following statement(s) is (are) true? **(2010)**

(a) $f''(x)$ exists for all $x \in (0,\infty)$

(b) $f'(x)$ exists for all $x \in (0,\infty)$ and f' is continuous on $(0,\infty)$, but not differentiable on $(0,\infty)$

(c) there exists $\alpha > 1$ such that $|f'(x)| < |f(x)|$ for all $x \in (\alpha, \infty)$

(d) there exists $\beta > 0$ such that $|f(x)| + |f'(x)| \le \beta$ for all $x \in (0, \infty)$

9. Let S be the area of the region enclosed by $y = e^{-x^2}$, $y = 0, x = 0$ and $x = 1$; then **(2012)**

(a) $S \ge \frac{1}{e}$ (b) $S \ge 1 - \frac{1}{e}$

(c) $S \le \frac{1}{4}\left(1+\frac{1}{\sqrt{e}}\right)$ (d) $S \le \frac{1}{\sqrt{2}} + \frac{1}{\sqrt{e}}\left(1-\frac{1}{\sqrt{2}}\right)$

E SUBJECTIVE PROBLEMS :

1. Find the area bounded by the curve $x^2 = 4y$ and the straight line $x = 4y - 2$. **(1981 - 4 Marks)**

2. Show that : $\lim_{n\to\infty}\left(\frac{1}{n+1} + \frac{1}{n+2} + ... + \frac{1}{6n}\right) = \log 6$ **(1981 - 2 Marks)**

3. Evaluate $\int_0^1 (tx+1-x)^n\,dx$, where n is a positive integer and t is a parameter independent of x. Hence show that $\int_0^1 x^k(1-x)^{n-k}\,dx = [{}^nC_k\,(n+1)]^{-1}$ for k=0, 1, ..., n. **(1981 - 4 Marks)**

4. Show that $\int_0^{\pi} xf(\sin x)\,dx = \frac{\pi}{2}\int_0^{\pi} f(\sin x)\,dx$. **(1982 - 2 Marks)**

5. Find the value of $\int_{-1}^{3/2} |x \sin \pi x|\,dx$ **(1982 - 3 Marks)**

6. For any real t, $x = \frac{e^t + e^{-t}}{2}$, $y = \frac{e^t - e^{-t}}{2}$ is a point on the hyperbola $x^2 - y^2 = 1$. Show that the area bounded by the hyperbola and the lines joining its centre to the points corresponding to t_1 and $-t_1$ is t_1. **(1982 - 3 Marks)**

7. Evaluate : $\int_0^{\pi/4} \frac{\sin x+\cos x}{9+16\sin 2x}dx$ **(1983 - 3 Marks)**

8. Find the area bounded by the x-axis, part of the curve $y=\left(1+\frac{8}{x^2}\right)$ and the ordinates at $x = 2$ and $x = 4$. If the ordinate at $x = a$ divides the area into two equal parts, find a. **(1983 - 3 Marks)**

9. Evaluate the following $\int_0^{\frac{1}{2}} \frac{x\sin^{-1}x}{\sqrt{1-x^2}}dx$ **(1984 - 2 Marks)**

10. Find the area of the region bounded by the x-axis and the curves defined by **(1984 - 4 Marks)**

$y = \tan x,\ -\frac{\pi}{3}\le x\le\frac{\pi}{3};\ y = \cot x,\ \frac{\pi}{6}\le x\le\frac{3\pi}{2}$

11. Given a function $f(x)$ such that **(1984 - 4 Marks)**
(i) it is integrable over every interval on the real line and
(ii) $f(t+x)=f(x)$, for every x and a real t, then show that the integral $\int_a^{a+t} f(x)\,dx$ is independent of a .

12. Evaluate the following : $\int_0^{\pi/2} \frac{x\sin x\cos x}{\cos^4 x+\sin^4 x}dx$ **(1985 - 2½ Marks)**

13. Sketch the region bounded by the curves $y=\sqrt{5-x^2}$ and $y=|x-1|$ and find its area. **(1985 - 5 Marks)**

14. Find the area bounded by the curves $x^2 + y^2 = 4$, $x^2 = -\sqrt{2}y$ and $x = y$. **(1986 - 5 Marks)**

15. Evaluate : $\int_0^{\pi} \frac{x\,dx}{1+\cos\alpha\sin x}$, $0<\alpha<\pi$ **(1986 - 2½ Marks)**

16. Find the area bounded by the curves, $x^2 + y^2 = 25$, $4y = |4 - x^2|$ and $x = 0$ above the x-axis. **(1987 - 6 Marks)**

17. Find the area of the region bounded by the curve C : $y = \tan x$, tangent drawn to C at $x=\frac{\pi}{4}$ and the x-axis. **(1988 - 5 Marks)**

18. Evaluate $\int_0^1 \log[\sqrt{1-x}+\sqrt{1+x}]dx$ **(1988 - 5 Marks)**

19. If f and g are continuous function on $[0, a]$ satisfying $f(x) = f(a - x)$ and $g(x) + g(a - x) = 2$, then show that $\int_0^a f(x)g(x)dx = \int_0^a f(x)dx$ **(1989 - 4 Marks)**

20. Show that $\int_0^{\pi/2} f(\sin 2x)\sin x\,dx=\sqrt{2}\int_0^{\pi/4} f(\cos 2x)\cos x\,dx$ **(1990 - 4 Marks)**

21. Prove that for any positive integer k,
$\frac{\sin 2kx}{\sin x}=2[\cos x+\cos 3x+\ldots\ldots\ldots\ldots+\cos(2k-1)x]$

Hence prove that $\int_0^{\pi/2}\sin 2kx\cot x\,dx=\frac{\pi}{2}$ **(1990 - 4 Marks)**

22. Compute the area of the region bounded by the curves $y = ex\ln x$ and $y=\frac{\ln x}{ex}$ where $\ln e = 1$. **(1990 - 4 Marks)**

23. Sketch the curves and identify the region bounded by $x=\frac{1}{2}$, $x = 2$, $y = \ln x$ and $y = 2^x$. Find the area of this region. **(1991 - 4 Marks)**

24. If 'f' is a continous function with $\int_0^x f(t)\,dt\to\infty$ as $|x|\to\infty$, then show that every line $y = mx$

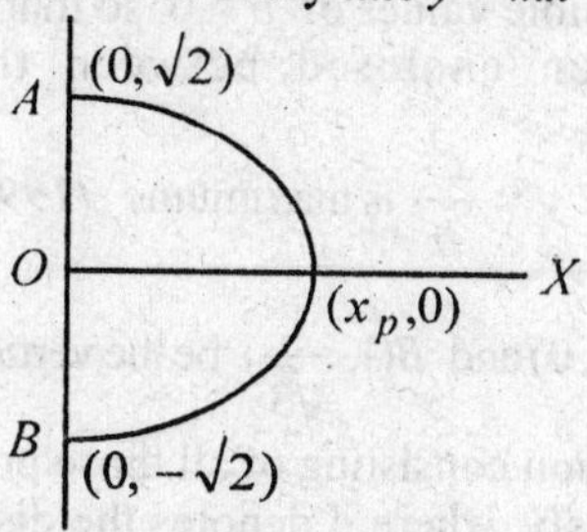

intersects the curve $y^2+\int_0^x f(t)\,dt=2!$ **(1991 - 4 Marks)**

25. Evaluate $\int_0^{\pi}\frac{x\sin 2x\sin\left(\frac{\pi}{2}\cos x\right)}{2x-\pi}dx$ **(1991 - 4 Marks)**

26. Sketch the region bounded by the curves $y = x^2$ and $y=\frac{2}{1+x^2}$. Find the area. **(1992 - 4 Marks)**

27. Determine a positive integer $n\le 5$, such that
$\int_0^1 e^x(x-1)^n dx=16-6e$ **(1992 - 4 Marks)**

28. Evalute $\int_2^3 \frac{2x^5+x^4-2x^3+2x^2+1}{(x^2+1)(x^4-1)}dx.$ **(1993 - 5 Marks)**

29. Show that $\int_0^{n\pi+v}|\sin x|\,dx=2n+1-\cos v$ where n is a positive integer and $0\le v<\pi$. **(1994 - 4 Marks)**

30. In what ratio does the x-axis divide the area of the region bounded by the parabolas $y = 4x - x^2$ and $y = x^2 - x$? **(1994 - 5 Marks)**

31. Let $I_m=\int_0^{\pi}\frac{1-\cos mx}{1-\cos x}dx$. Use mathematical induction to prove that $I_m = m\pi$, $m = 0, 1, 2, \ldots\ldots$ **(1995 - 5 Marks)**

32. Evaluate the definite integral :
$\int_{-1/\sqrt{3}}^{1/\sqrt{3}}\left(\frac{x^4}{1-x^4}\right)\cos^{-1}\left(\frac{2x}{1+x^2}\right)dx$ **(1995 - 5 Marks)**

33. Consider a square with vertices at (1, 1), (– 1, 1), (– 1, – 1) and (1, – 1). Let S be the region consisting of all points inside the square which are nearer to the origin than to any edge. Sketch the region S and find its area. **(1995 - 5 Marks)**

34. Let A_n be the area bounded by the curve $y = (\tan x)^n$ and the lines $x = 0, y = 0$ and $x = \dfrac{\pi}{4}$. Prove that for $n > 2$,

$A_n + A_{n-2} = \dfrac{1}{n-1}$ and deduce $\dfrac{1}{2n+2} < A_n < \dfrac{1}{2n-2}$.

(1996 - 3 Marks)

35. Integrate the following : $\int_0^{\pi/4} \ln(1+\tan x)dx$

(1997C - 2 Marks)

36. Find all the possible values of $b > 0$, so that the area of the bounded region enclosed between the parabolas $y = x - bx^2$ and $y = \dfrac{x^2}{b}$ is maximum. **(1997C - 5 Marks)**

37. Let $O(0, 0)$, $A(2, 0)$ and $B(1, \dfrac{1}{\sqrt{3}})$ be the vertices of a triangle.

Let R be the region consisting of all those points P inside ΔOAB which satisfy, where d denotes the distance from the point to the corresponding line. Sketch the region R and find its area. **(1997C - 5 Marks)**

38. Determine the value of $\int_{-\pi}^{\pi} \dfrac{2x(1+\sin x)}{1+\cos^2 x} dx$.

(1997 - 5 Marks)

39. Let $f(x)$ = Maximum $\{x^2, (1-x)^2, 2x(1-x)\}$, where $0 \le x \le 1$. Determine the area of the region bounded by the curves $y = f(x)$, x-axis, $x = 0$ and $x = 1$. **(1997 - 5 Marks)**

40. Prove that $\int_0^1 \tan^{-1}\left(\dfrac{1}{1-x+x^2}\right) dx = 2\int_0^1 \tan^{-1} x\, dx$.

Hence or otherwise, evaluate the integral

$\int_0^1 \tan^{-1}(1-x+x^2)\, dx$. **(1998 - 8 Marks)**

41. Let C_1 and C_2 be the graphs of the functions $y = x^2$ and $y = 2x$, $0 \le x \le 1$ respectively. Let C_3 be the graph of a function $y = f(x)$, $0 \le x \le 1$, $f(0) = 0$. For a point P on C_1, let the lines through P, parallel to the axes, meet C_2 and C_3 at Q and R respectively (see figure.) If for every position of P (on C_1), the areas of the shaded regions OPQ and ORP are equal, determine the function $f(x)$. **(1998 - 8 Marks)**

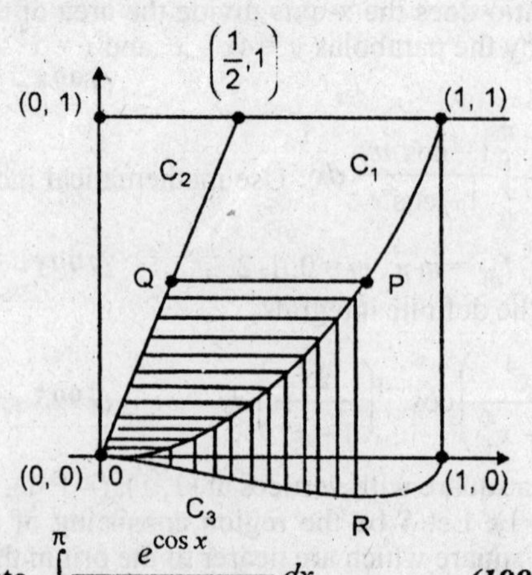

42. Integrate $\int_0^{\pi} \dfrac{e^{\cos x}}{e^{\cos x} + e^{-\cos x}} dx$. **(1999 - 5 Marks)**

43. Let $f(x)$ be a continuous function given by

$f(x) = \begin{cases} 2x, & |x| \le 1 \\ x^2 + ax + b, & |x| > 1 \end{cases}$ **(1999 - 10 Marks)**

Find the area of the region in the third quadrant bounded by the curves $x = -2y^2$ and $y = f(x)$ lying on the left of the line $8x + 1 = 0$.

44. For $x > 0$, let $f(x) = \int_e^x \dfrac{\ln t}{1+t} dt$. Find the function

$f(x) + f\left(\dfrac{1}{x}\right)$ and show that $f(e) + f\left(\dfrac{1}{e}\right) = \dfrac{1}{2}$.

Here, $\ln t = \log_e t$. **(2000 - 5 Marks)**

45. Let $b \ne 0$ and for $j = 0, 1, 2, \ldots, n$, let S_j be the area of the region bounded by the y–axis and the curve $xe^{ay} = \sin by$, $\dfrac{j\pi}{b} \le y \le \dfrac{(j+1)\pi}{b}$. Show that $S_0, S_1, S_2, \ldots, S_n$ are in geometric progression. Also, find their sum for $a = -1$ and $b = \pi$. **(2001 - 5 Marks)**

46. Find the area of the region bounded by the curves $y = x^2$, $y = |2 - x^2|$ and $y = 2$, which lies to the right of the line $x = 1$. **(2002 - 5 Marks)**

47. If f is an even function then prove that **(2003 - 2 Marks)**

$\int_0^{\pi/2} f(\cos 2x)\cos x\, dx = \sqrt{2}\int_0^{\pi/4} f(\sin 2x)\cos x\, dx$.

48. If $y(x) = \int_{\pi^2/16}^{x^2} \dfrac{\cos x \cos\sqrt{\theta}}{1+\sin^2\sqrt{\theta}} d\theta$, then find $\dfrac{dy}{dx}$ at $x = \pi$

(2004 - 2 Marks)

49. Find the value of $\int_{-\pi/3}^{\pi/3} \dfrac{\pi + 4x^3}{2 - \cos\left(|x| + \dfrac{\pi}{3}\right)} dx$

(2004 - 4 Marks)

50. Evaluate $\int_0^{\pi} e^{|\cos x|}\left(2\sin\left(\dfrac{1}{2}\cos x\right) + 3\cos\left(\dfrac{1}{2}\cos x\right)\right)\sin x\, dx$

(2005 - 2 Marks)

51. Find the area bounded by the curves $x^2 = y$, $x^2 = -y$ and $y^2 = 4x - 3$. **(2005 - 4 Marks)**

52. $f(x)$ is a differentiable function and $g(x)$ is a double differentiable function such that $|f(x)| \le 1$ and $f'(x) = g(x)$. If $f^2(0) + g^2(0) = 9$. Prove that there exists some $c \in (-3, 3)$ such that $g(c).g''(c) < 0$. **(2005 - 6 Marks)**

53. If $\begin{bmatrix} 4a^2 & 4a & 1 \\ 4b^2 & 4b & 1 \\ 4c^2 & 4c & 1 \end{bmatrix}\begin{bmatrix} f(-1) \\ f(1) \\ f(2) \end{bmatrix} = \begin{bmatrix} 3a^2+3a \\ 3b^2+3b \\ 3c^2+3c \end{bmatrix}$, $f(x)$ is a quadratic function and its maximum value occurs at a point V. A is a point of intersection of $y = f(x)$ with x-axis and point B is such that chord AB subtends a right angle at V. Find the area enclosed by $f(x)$ and chord AB. **(2005 - 6 Marks)**

54. The value of $5050\dfrac{\int_0^1 (1-x^{50})^{100} dx}{\int_0^1 (1-x^{50})^{101} dx}$ is. **(2006 - 6M)**

(F) MATCH THE FOLLOWING :

Each question contains statements given in two columns, which have to be matched. The statements in Column-I are labelled A, B, C and D, while the statements in Column-II are labelled p, q, r, s and t. Any given statement in Column-I can have correct matching with ONE OR MORE statement(s) in Column-II. The appropriate bubbles corresponding to the answers to these questions have to be darkened as illustrated in the following example :
If the correct matches are A-p, s and t; B-q and r; C-p and q; and D-s then the correct darkening of bubbles will look like the given.

	p	q	r	s	t
A	●	○	○	●	●
B	○	●	●	○	○
C	●	●	○	○	○
D	○	○	○	●	○

1. Match the following : **(2006 - 6M)**

Column I

(A) $\int_0^{\pi/2} (\sin x)^{\cos x}\left(\cos x \cot x - \log(\sin x)^{\sin x}\right)dx$

(B) Area bounded by $-4y^2 = x$ and $x - 1 = -5y^2$

(C) Cosine of the angle of intersection of curves $y = 3^{x-1}\log x$ and $y = x^x - 1$ is

(D) Let $\frac{dy}{dx} = \frac{6}{x+y}$ where $y(0) = 0$ then value of y when $x + y = 6$ is

Column II

(p) 1

(q) 0

(r) 6 ln 2

(s) $\frac{4}{3}$

2. Match the integrals in **Column I** with the values in **Column II** and indicate your answer by darkening the appropriate bubbles in the 4 × 4 matrix given in the *ORS*. **(2007 - 6 marks)**

Column I

(A) $\int_{-1}^{1}\frac{dx}{1+x^2}$

(B) $\int_0^1 \frac{dx}{\sqrt{1-x^2}}$

(C) $\int_2^3 \frac{dx}{1-x^2}$

(D) $\int_1^2 \frac{dx}{x\sqrt{x^2-1}}$

Column II

(p) $\frac{1}{2}\log\left(\frac{2}{3}\right)$

(q) $2\log\left(\frac{2}{3}\right)$

(r) $\frac{\pi}{3}$

(s) $\frac{\pi}{2}$

(G) COMPREHENSION BASED QUESTIONS :

PASSAGE-1

Let the definite integral be defined by the formula $\int_a^b f(x)dx = \frac{b-a}{2}(f(a)+f(b))$. For more accurate result for $c \in (a, b)$, we can use $\int_a^b f(x)dx = \int_a^c f(x)dx + \int_c^b f(x)dx = F(c)$ so that for $c = \frac{a+b}{2}$, we get $\int_a^b f(x)dx = \frac{b-a}{4}(f(a)+f(b)+2f(c))$.

1. $\int_0^{\pi/2} \sin x\, dx =$ **(2006 - 5M, –2)**

(a) $\frac{\pi}{8}(1+\sqrt{2})$ (b) $\frac{\pi}{4}(1+\sqrt{2})$

(c) $\frac{\pi}{8\sqrt{2}}$ (d) $\frac{\pi}{4\sqrt{2}}$

2. If $\lim_{x\to a}\frac{\int_a^x f(x)dx - \left(\frac{x-a}{2}\right)(f(x)+f(a))}{(x-a)^3} = 0$, then $f(x)$ is of maximum degree **(2006 - 5M, –2)**

(a) 4 (b) 3

(c) 2 (d) 1

3. If $f''(x) < 0 \ \forall\, x \in (a, b)$ and c is a point such that $a < c < b$, and $(c, f(c))$ is the point lying on the curve for which $F(c)$ is maximum, then $f'(c)$ is equal to **(2006 - 5M, –2)**

(a) $\frac{f(b)-f(a)}{b-a}$ (b) $\frac{2(f(b)-f(a))}{b-a}$

(c) $\frac{2f(b)-f(a)}{2b-a}$ (d) 0

PASSAGE-2

Consider the functions defined implicitly by the equation $y^3 - 3y + x = 0$ on various intervals in the real line. If $x \in (-\infty, -2) \cup (2, \infty)$, the equation implicitly defines a unique real valued differentiable function $y = f(x)$. If $x \in (-2, 2)$, the equation implicitly defines a unique real valued differentiable function $y = g(x)$ satisfying $g(0) = 0$.

4. If $f(-10\sqrt{2}) = 2\sqrt{2}$, then $f''(-10\sqrt{2}) =$ *(2008)*

(a) $\frac{4\sqrt{2}}{7^3 3^2}$ (b) $-\frac{4\sqrt{2}}{7^3 3^2}$

(c) $\frac{4\sqrt{2}}{7^3 3}$ (d) $-\frac{4\sqrt{2}}{7^3 3}$

5. The area of the region bounded by the curve $y = f(x)$, the x-axis, and the lines $x = a$ and $x = b$, where $-\infty < a < b < -2$, is *(2008)*

(a) $\int_a^b \frac{x}{3\left((f(x))^2 - 1\right)} dx + bf(b) - af(a)$

(b) $-\int_a^b \frac{x}{3\left((f(x))^2 - 1\right)} dx + bf(b) - af(a)$

(c) $\int_a^b \frac{x}{3\left((f(x))^2 - 1\right)} dx - bf(b) + af(a)$

(d) $-\int_a^b \frac{x}{3\left((f(x))^2 - 1\right)} dx - bf(b) + af(a)$

6. $\int_{-1}^{1} g'(x)\, dx =$ *(2008)*

(a) $2g(-1)$ (b) 0

(c) $-2g(1)$ (d) $2g(1)$

PASSAGE-3

Consider the function $f: (-\infty, \infty) \to (-\infty, \infty)$ defined by

$$f(x) = \frac{x^2 - ax + 1}{x^2 + ax + 1}, 0 < a < 2.$$

7. Which of the following is true? *(2008)*

(a) $(2+a)^2 f''(1) + (2-a)^2 f''(-1) = 0$

(b) $(2-a)^2 f''(1) - (2+a)^2 f''(-1) = 0$

(c) $f'(1) f'(-1) = (2-a)^2$

(d) $f'(1) f'(-1) = -(2+a)^2$

8. Which of the following is true? *(2008)*

(a) $f(x)$ is decreasing on $(-1, 1)$ and has a local minimum at $x = 1$

(b) $f(x)$ is increasing on $(-1, 1)$ and has a local minimum at $x = 1$

(c) $f(x)$ is increasing on $(-1, 1)$ but has neither a local maximum nor a local minimum at $x = 1$

(d) $f(x)$ is decreasing on $(-1, 1)$ but has neither a local maximum nor a local minimum at $x = 1$

9. Let $g(x) = \int_0^{e^x} \frac{f'(t)}{1+t^2} dt$. Which of the following is true? *(2008)*

(a) $g'(x)$ is positive on $(-\infty, 0)$ and negative on $(0, \infty)$

(b) $g'(x)$ is negative on $(-\infty, 0)$ and positive on $(0, \infty)$

(c) $g'(x)$ changes sign on both $(-\infty, 0)$ and $(0, \infty)$

(d) $g'(x)$ does not change sign on $(-\infty, \infty)$

PASSAGE-4

Consider the polynomial *(2010)*

$f(x) = 1 + 2x + 3x^2 + 4x^3$.

Let s be the sum of all distinct real roots of $f(x)$ and let $t = |s|$.

10. The real numbers lies in the interval

(a) $\left(-\frac{1}{4}, 0\right)$ (b) $\left(-11, -\frac{3}{4}\right)$

(c) $\left(-\frac{3}{4}, -\frac{1}{2}\right)$ (d) $\left(0, \frac{1}{4}\right)$

11. The area bounded by the curve $y = f(x)$ and the lines $x = 0$, $y = 0$ and $x = t$, lies in the interval

(a) $\left(\frac{3}{4}, 3\right)$ (b) $\left(\frac{21}{64}, \frac{11}{16}\right)$

(c) $(9, 10)$ (d) $\left(0, \frac{21}{64}\right)$

12. The function $f'(x)$ is

(a) increasing in $\left(-t, -\frac{1}{4}\right)$ and decreasing in $\left(-\frac{1}{4}, t\right)$

(b) decreasing in $\left(-t, -\frac{1}{4}\right)$ and increasing in $\left(-\frac{1}{4}, t\right)$

(c) increasing in $(-t, t)$

(d) decreasing in $(-t, t)$

I INTEGER VALUE CORRECT TYPE :

1. Let $f: \mathbf{R} \to \mathbf{R}$ be a continuous function which satisfies

$$f(x) = \int_0^x f(t)\, dt.$$

Then the value of $f(\ln 5)$ is *(2009)*

2. For any real number x, let $[x]$ denote the largest integer less than or equal to x. Let f be a real valued function defined on the interval $[-10, 10]$ by

$$f(x) = \begin{cases} x - [x] & \text{if } [x] \text{ is odd,} \\ 1 + [x] - x & \text{if } [x] \text{ is even} \end{cases}$$

Then the value of $\frac{\pi^2}{10} \int_{-10}^{10} f(x) \cos \pi x\, dx$ is *(2010)*

Section-B JEE Main/ AIEEE

1. $\int_0^{10\pi} |\sin x|\, dx$ is [2002]

(a) 20 (b) 8 (c) 10 (d) 18

2. $I_n = \int_0^{\pi/4} \tan^n x\, dx$ then $\lim_{n\to\infty} n[I_n + I_{n+2}]$ equals [2002]

(a) $\frac{1}{2}$ (b) 1 (c) ∞ (d) zero

3. $\int_0^2 [x^2]dx$ is [2002]

(a) $2-\sqrt{2}$ (b) $2+\sqrt{2}$
(c) $\sqrt{2}-1$ (d) $-\sqrt{2}-\sqrt{3}+5$

4. $\int_{-\pi}^{\pi} \frac{2x(1+\sin x)}{1+\cos^2 x} dx$ is [2002]

(a) $\frac{\pi^2}{4}$ (b) π^2 (c) zero (d) $\frac{\pi}{2}$

5. If $y=f(x)$ makes +ve intercept of 2 and 0 unit on x and y axes and encloses an area of 3/4 square unit with the axes then $\int_0^2 xf'(x)dx$ is [2002]

(a) 3/2 (b) 1 (c) 5/4 (d) –3/4

6. The area bounded by the curves $y = \ln x$, $y = \ln |x|$, $y = |\ln x|$ and $y = |\ln |x||$ is [2002]

(a) 4sq. units (b) 6 sq. units
(c) 10 sq. units (d) none of these

7. The area of the region bounded by the curves $y=|x-1|$ *and* $y=3-|x|$ is [2003]

(a) 6 sq. units (b) 2 sq. units
(c) 3 sq. units (d) 4 sq. units.

8. If $f(a+b-x)=f(x)$ then $\int_a^b xf(x)dx$ is equal to [2003]

(a) $\frac{a+b}{2}\int_a^b f(a+b+x)dx$ (b) $\frac{a+b}{2}\int_a^b f(b-x)dx$

(c) $\frac{a+b}{2}\int_a^b f(x)dx$ (d) $\frac{b-a}{2}\int_a^b f(x)dx$.

9. Let $f(x)$ be a function satisfying $f'(x)=f(x)$ with $f(0)=1$ and $g(x)$ be a function that satisfies $f(x) + g(x) = x^2$. Then the value of the integral $\int_0^1 f(x)\,g(x)dx$, is [2003]

(a) $e+\frac{e^2}{2}+\frac{5}{2}$ (b) $e-\frac{e^2}{2}-\frac{5}{2}$

(c) $e+\frac{e^2}{2}-\frac{3}{2}$ (d) $e-\frac{e^2}{2}-\frac{3}{2}$.

10. The value of the integral $I=\int_0^1 x(1-x)^n dx$ is [2003]

(a) $\frac{1}{n+1}+\frac{1}{n+2}$ (b) $\frac{1}{n+1}$

(c) $\frac{1}{n+2}$ (d) $\frac{1}{n+1}-\frac{1}{n+2}$.

11. $\lim_{n\to\infty}\sum_{r=1}^{n}\frac{1}{n}e^{\frac{r}{n}}$ is [2004]

(a) $e+1$ (b) $e-1$ (c) $1-e$ (d) e

12. The value of $\int_{-2}^{3} |1-x^2|dx$ is [2004]

(a) $\frac{1}{3}$ (b) $\frac{14}{3}$ (c) $\frac{7}{3}$ (d) $\frac{28}{3}$

13. The value of $I=\int_0^{\pi/2}\frac{(\sin x+\cos x)^2}{\sqrt{1+\sin 2x}}dx$ is [2004]

(a) 3 (b) 1 (c) 2 (d) 0

14. If $\int_0^{\pi} xf(\sin x)dx = A\int_0^{\pi/2} f(\sin x)dx$, then A is [2004]

(a) 2π (b) π (c) $\frac{\pi}{4}$ (d) 0

15. If $f(x)=\frac{e^x}{1+e^x}$, $I_1=\int_{f(-a)}^{f(a)} xg\{x(1-x)\}dx$ and $I_2=\int_{f(-a)}^{f(a)} g\{x(1-x)\}dx$, then the value of $\frac{I_2}{I_1}$ is [2004]

(a) 1 (b) –3 (c) –1 (d) 2

16. The area of the region bounded by the curves $y=|x-2|, x=1, x=3$ and the x-axis is [2004]

(a) 4 (b) 2 (c) 3 (d) 1

17. If $I_1=\int_0^1 2^{x^2}dx$, $I_2=\int_0^1 2^{x^3}dx$, $I_3=\int_1^2 2^{x^2}dx$ and $I_4=\int_1^2 2^{x^3}dx$ then [2005]

(a) $I_2>I_1$ (b) $I_1>I_2$
(c) $I_3=I_4$ (d) $I_3>I_4$

18. The area enclosed between the curve $y=\log_e(x+e)$ and the coordinate axes is [2005]

(a) 1 (b) 2 (c) 3 (d) 4

19. The parabolas $y^2=4x$ and $x^2=4y$ divide the square region bounded by the lines $x=4, y=4$ and the coordinate axes. If S_1, S_2, S_3 are respectively the areas of these parts numbered from top to bottom; then $S_1 : S_2 : S_3$ is [2005]

(a) $1:2:1$ (b) $1:2:3$ (c) $2:1:2$ (d) $1:1:1$

20. Let $f(x)$ be a non – negative continuous function such that the area bounded by the curve $y = f(x)$, x - axis and the ordinates $x = \frac{\pi}{4}$ and $x = \beta > \frac{\pi}{4}$ is **[2005]**

$\left(\beta\sin\beta + \frac{\pi}{4}\cos\beta + \sqrt{2}\beta\right)$. Then $f\left(\frac{\pi}{2}\right)$ is

(a) $\left(\frac{\pi}{4}+\sqrt{2}-1\right)$ (b) $\left(\frac{\pi}{4}-\sqrt{2}+1\right)$

(c) $\left(1-\frac{\pi}{4}-\sqrt{2}\right)$ (d) $\left(1-\frac{\pi}{4}+\sqrt{2}\right)$

21. The value of $\int_{-\pi}^{\pi}\frac{\cos^2 x}{1+a^x}dx$, a > 0, is **[2005]**

(a) a π (b) $\frac{\pi}{2}$ (c) $\frac{\pi}{a}$ (d) 2π

22. The value of integral, : $\int_3^6\frac{\sqrt{x}}{\sqrt{9-x}+\sqrt{x}}dx$ is

(a) $\frac{1}{2}$ (b) $\frac{3}{2}$ (c) 2 (d) 1

23. $\int_0^{\pi} xf(\sin x)dx$ is equal to **[2006]**

(a) $\pi\int_0^{\pi} f(\cos x)dx$ (b) $\pi\int_0^{\pi} f(\sin x)dx$

(c) $\frac{\pi}{2}\int_0^{\pi/2} f(\sin x)dx$ (d) $\pi\int_0^{\pi/2} f(\cos x)dx$

24. $\int_{\frac{3\pi}{2}}^{-\frac{\pi}{2}}[(x+\pi)^3+\cos^2(x+3\pi)]dx$ is equal to **[2006]**

(a) $\frac{\pi^4}{32}$ (b) $\frac{\pi^4}{32}+\frac{\pi}{2}$ (c) $\frac{\pi}{2}$ (d) $\frac{\pi}{4}-1$

25. The value of $\int_1^a [x]f'(x)dx$, $a > 1$ where $[x]$ denotes the greatest integer not exceeding x is **[2006]**

(a) $af(a)-\{f(1)+f(2)+.............f([a])\}$

(b) $[a]f(a)-\{f(1)+f(2)+............f([a])\}$

(c) $[a]f([a])-\{f(1)+f(2)+............f(a)\}$

(d) $af([a])-\{f(1)+f(2)+............f(a)\}$

26. Let $F(x) = f(x)+f\left(\frac{1}{x}\right)$, where $f(x) = \int_l^x\frac{\log t}{1+t}dt$, Then $F(e)$ equals **[2007]**

(a) 1 (b) 2 (c) 1/2 (d) 0

27. The solution for x of the equation $\int_{\sqrt{2}}^{x}\frac{dt}{t\sqrt{t^2-1}} = \frac{\pi}{2}$ is **[2007]**

(a) $\frac{\sqrt{3}}{2}$ (b) $2\sqrt{2}$ (c) 2 (d) None

28. The area enclosed between the curves $y^2 = x$ and $y = |x|$ is **[2007]**

(a) 1/6 (b) 1/3 (c) 2/3 (d) 1

29. Let $I = \int_0^1\frac{\sin x}{\sqrt{x}}dx$ and $J = \int_0^1\frac{\cos x}{\sqrt{x}}dx$. Then which one of the following is true?

(a) $I > \frac{2}{3}$ and $J > 2$ (b) $I < \frac{2}{3}$ and $J < 2$

(c) $I < \frac{2}{3}$ and $J > 2$ (d) $I > \frac{2}{3}$ and $J < 2$

30. The area of the plane region bounded by the curves $x + 2y^2 = 0$ and $x + 3y^2 = 1$ is equal to **[2008]**

(a) $\frac{5}{3}$ (b) $\frac{1}{3}$ (c) $\frac{2}{3}$ (d) $\frac{4}{3}$

31. The area of the region bounded by the parabola $(y-2)^2 = x-1$, the tangent of the parabola at the point (2, 3) and the x-axis is: **[2009]**

(a) 6 (b) 9 (c) 12 (d) 3

32. $\int_0^{\pi}[\cot x]\,dx$, where [.] denotes the greatest integer function, is equal to : **[2009]**

(a) 1 (b) –1 (c) $-\frac{\pi}{2}$ (d) $\frac{\pi}{2}$

33. The area bounded by the curves $y = \cos x$ and $y = \sin x$ between the ordinates $x = 0$ and $x = \frac{3\pi}{2}$ is **[2010]**

(a) $4\sqrt{2}+2$ (b) $4\sqrt{2}-1$

(c) $4\sqrt{2}+1$ (d) $4\sqrt{2}-2$

34. Let $p(x)$ be a function defined on **R** such that $p'(x) = p'(1-x)$, for all $x \in [0, 1]$, $p(0) = 1$ and $p(1) = 41$. Then $\int_0^1 p(x)\,dx$ equals **[2010]**

(a) 21 (b) 41 (c) 42 (d) $\sqrt{41}$

35. The value of $\int_0^1\frac{8\log(1+x)}{1+x^2}dx$ is **[2011]**

(a) $\frac{\pi}{8}\log 2$ (b) $\frac{\pi}{2}\log 2$

(c) log 2 (d) π log 2

36. The area of the region enclosed by the curves $y = x$, $x = e$, $y = \frac{1}{x}$ and the positive x-axis is **[2011]**

(a) 1 square unit (b) $\frac{3}{2}$ square units

(c) $\frac{5}{2}$ square units (d) $\frac{1}{2}$ square unit

37. Let [.] denote the greatest integer function then the value of $\int_0^{1.5} x\left[x^2\right]dx$ is :. **[2011 RS]**

(a) 0 (b) $\frac{3}{2}$ (c) $\frac{3}{4}$ (d) $\frac{5}{4}$

38. The area bounded by the curves $y^2 = 4x$ and $x^2 = 4y$ is: **[2011 RS]**

(a) $\frac{32}{3}$ sq units (b) $\frac{16}{3}$ sq units

(c) $\frac{8}{3}$ sq. units (d) 0 sq. units

39. The area between the parabolas $x^2 = \frac{y}{4}$ and $x^2 = 9y$ and the straight line $y = 2$ is : **[2012]**

(a) $20\sqrt{2}$ (b) $\frac{10\sqrt{2}}{3}$ (c) $\frac{20\sqrt{2}}{3}$ (d) $10\sqrt{2}$

40. If $g(x) = \int_0^x \cos 4t\,dt$, then $g(x+\pi)$ equals **[2012]**

(a) $\frac{g(x)}{g(\pi)}$ (b) $g(x) + g(\pi)$

(c) $g(x) - g(\pi)$ (d) $g(x) . g(\pi)$

41. **Statement-1 :** The value of the integral $\int_{\pi/6}^{\pi/3} \frac{dx}{1+\sqrt{\tan x}}$ is equal to $\pi/6$ **[JEE M 2013]**

Statement-2 : $\int_a^b f(x)dx = \int_a^b f(a+b-x)dx.$

(a) Statement-1 is true; Statement-2 is true; Statement-2 is a correct explanation for Statement-1.
(b) Statement-1 is true; Statement-2 is true; Statement-2 is not a correct explanation for Statement-1.
(c) Statement-1 is true; Statement-2 is false.
(d) Statement-1 is false; Statement-2 is true.

42. The area (in square units) bounded by the curves $y = \sqrt{x}$, $2y - x + 3 = 0$, x-axis, and lying in the first quadrant is : **[JEE M 2013]**

(a) 9 (b) 36 (c) 18 (d) $\frac{27}{4}$

Solutions & Explanations

Section-A : JEE Advanced/ IIT-JEE

A **1.** $-\left(\frac{15\pi+32}{60}\right)$ **2.** $2-\sqrt{2}$ **3.** 4 **4.** $\pi(\sqrt{2}-1)$ **5.** $\frac{1}{2}$

6. $\frac{1}{a^2-b^2}\left[a(\log 2-5)+\frac{7b}{2}\right]$ **7.** π^2 **8.** 2 **9.** 16

B **1.** T

C **1.** (d) **2.** (b) **3.** (c) **4.** (a) **5.** (c) **6.** (d)
7. (d) **8.** (d) **9.** (a) **10.** (c) **11.** (a) **12.** (a)
13. (c) **14.** (b) **15.** (c) **16.** (b) **17.** (c) **18.** (b)
19. (a) **20.** (c) **21.** (a) **22.** (a) **23.** (d) **24.** (d)
25. (a) **26.** (b) **27.** (a) **28.** (c) **29.** (d) **30.** (b)
31. (c) **32.** (b) **33.** (b) **34.** (a) **35.** (b) **36.** (c)
37. (b) **38.** (b) **39.** (d)

D **1.** (a) **2.** (a) **3.** (b, d) **4.** (a, b, c, d) **5.** (b, c, d) **6.** (a, b, c)
7. (a) **8.** (b, c) **9.** (a, b, d)

E **1.** $\frac{9}{8}$ sq. units **3.** $\frac{t^{n+1}-1}{(t-1)(n+1)}$ **5.** $\frac{3}{\pi}+\frac{1}{\pi^2}$ **7.** $\frac{1}{20}\log 3$ **8.** $a = 2\sqrt{2}$

9. $\frac{6-\pi\sqrt{3}}{12}$ **10.** $\log\frac{3}{2}$ sq. units **12.** $\frac{\pi^2}{16}$ **13.** $\frac{5\pi-2}{4}$ sq. units

14. $\pi+\frac{1}{3}$ sq. units **15.** $\frac{\pi\alpha}{\sin\alpha}$ **16.** $4+25\sin^{-1}\frac{4}{5}$ **17.** $\frac{1}{2}\left[\log 2-\frac{1}{2}\right]$ sq. units

18. $\frac{1}{2}\left[\log 2+\frac{\pi}{2}-1\right]$ **22.** $\frac{e^2-5}{4e}$ **23.** $\frac{4-\sqrt{2}}{\log 2}-\frac{5}{2}\log 2+\frac{3}{2}$ **25.** $\frac{8}{\pi^2}$

26. $\left(\pi-\frac{2}{3}\right)$ sq. units **27.** $n = 3$ **28.** $\frac{1}{2}\log 6-\frac{1}{10}$ **29.** $2n+1-\cos\gamma$ **30.** 121 : 4

32. $\frac{\pi}{12}\left[\pi+3\log_e(2+\sqrt{3})-4\sqrt{3}\right]$ 33. $\frac{16\sqrt{2}-20}{3}$ 35. $\frac{\pi}{8}\ln 2$ 36. $b=1$ 37. $2-\sqrt{3}$

38. π^2 39. $\frac{17}{27}$ sq. units 40. $\log 2$ 41. $f(x)=x^3-x^2$ 42. $\frac{\pi}{2}$ 43. $\frac{257}{192}$ sq. units

45. $\frac{\pi(1+e)}{1+\pi^2}\left(\frac{e^{n+1}-1}{e-1}\right)$ 46. $\left(\frac{20}{3}-4\sqrt{2}\right)$ sq.units 48. 2π

49. $\frac{4\pi}{\sqrt{3}}\left[\tan^{-1}3-\frac{\pi}{4}\right]$ 50. $\frac{24}{5}\left[e\cos\left(\frac{1}{2}\right)+\frac{1}{2}e\sin\left(\frac{1}{2}\right)-1\right]$ 51. $\frac{1}{3}$ sq. units 53. $\frac{125}{3}$ sq.units

54. 5051

F 1. (A) - p ; (B) -s ; (C) - p ; (D) - r 2. (A) - s ; (B) -s ; (C) - p ; (D) - r

G 1. (a) 2. (d) 3. (b) 4. (b) 5. (a) 6. (d)
7. (a) 8. (a) 9. (b) 10. (c) 11. (a) 12. (b)

I 1. 0 2. 4

Section-B : JEE Main/ AIEEE

1. (a)	2. (b)	3. (d)	4. (b)	5. (d)	6. (a)
7. (d)	8. (c)	9. (d)	10. (d)	11. (b)	12. (d)
13. (c)	14. (b)	15. (d)	16. (d)	17. (b)	18. (a)
19. (d)	20. (d)	21. (b)	22. (b)	23. (d)	24. (c)
25. (b)	26. (c)	27. (d)	28. (a)	29. (b)	30. (d)
31. (b)	32. (c)	33. (d)	34. (a)	35. (d)	36. (b)
37. (c)	38. (b)	39. (c)	40. (b, c)	41. (d)	42. (a)

Section-A JEE Advanced/ IIT-JEE

(A) FILL IN THE BLANKS :

1. Given that,

$$f(x)=\begin{vmatrix}\sec x & \cos x & \sec^2 x+\cot x\,\mathrm{cosec}\,x\\ \cos^2 x & \cos^2 x & \mathrm{cosec}^2 x\\ 1 & \cos^2 x & \cos^2 x\end{vmatrix}$$

Operating $R_1-\sec x.R_3$,

$$=\begin{vmatrix}0 & 0 & \sec^2 x+\cot x\,\mathrm{cosec}\,x-\cos x\\ \cos^2 x & \cos^2 x & \mathrm{cosec}^2 x\\ 1 & \cos^2 x & \cos^2 x\end{vmatrix}$$

Expanding along R_1, we get

$=(\sec^2 x+\cot x\,\mathrm{cosec}\,x-\cos x)(\cos^4 x-\cos^2 x)$

$=\left(\frac{1}{\cos^2 x}+\frac{\cos x}{\sin^2 x}-\cos x\right)\cos^2 x(\cos^2 x-1)$

$=-\sin^2 x-\cos^5 x$

$\therefore \int_0^{\pi/2} f(x)dx=-\int_0^{\pi/2}(\sin^2 x+\cos^5 x)dx$

Using

$\int_0^{\pi/2}\sin^n x\,dx=\int_0^{\pi/2}\cos^n x\,dx=\frac{(n-1)(n-3)\ldots 2\, or\, 1}{(n)(n-2)\ldots ?}$

NOTE THIS STEP : Multiply the above by $\pi/2$ when n is even. We get

$=-\left[\frac{1}{2}.\frac{\pi}{2}+\frac{4}{5}.\frac{2}{3}\right]=-\left[\frac{\pi}{4}+\frac{8}{15}\right]=-\left(\frac{15\pi+32}{60}\right)$

2. $\int_0^{1.5}[x^2]dx$,

We have $0<x<1.5 \Rightarrow 0<x^2<2.25$

$\therefore [x^2]=0,\ 0<x^2<1=1,\ 1\le x^2<2$

NOTE THIS STEP

$=2,\ 2\le x^2<(1.5)^2$

or $[x^2]=0,\ 0<x<1=1,\ 1\le x<\sqrt{2}=2,\sqrt{2}\le x<1.5$

$\therefore I=\int_0^{1.5}[x^2]dx=\int_0^1 0\,dx+\int_1^{\sqrt{2}}1dx+\int_{\sqrt{2}}^{1.5}2dx$

$=0+[x]_1^{\sqrt{2}}+[2x]_{\sqrt{2}}^{1.5}$

$=\sqrt{2}-1+3-2\sqrt{2}=2-\sqrt{2}$

3. Let $I=\int_{-2}^{2}|1-x^2|dx=2\int_0^2|1-x^2|\,dx$

$\left[\because \int_{-a}^{a}f(x)dx=2\int_0^a f(x)dx \text{ if } f \text{ is an even function}\right]$

$=2\int_0^1(1-x^2)dx+2\int_1^2(x^2-1)dx$

$=2\left[x-\frac{x^3}{3}\right]_0^1+2\left[\frac{x^3}{3}-x\right]_1^2=\frac{4}{3}+\frac{8}{3}=\frac{12}{3}=4$

4. We have, $I=\int_{\pi/4}^{3\pi/4}\frac{\phi}{1+\sin\phi}d\phi$ ...(1)

$\Rightarrow\quad I=\int_{\pi/4}^{3\pi/4}\frac{\pi-\phi}{1+\sin(\pi-\phi)}d\phi$

$\left[\text{Using}\int_a^b f(x)dx=\int_a^b f(a+b-x)dx\right]$

$\Rightarrow\quad I=\int_{\pi/4}^{3\pi/4}\frac{\pi-\phi}{1+\sin\phi}d\phi$ ...(2)

Adding (1) and (2), we get

$$2I=\int_{\pi/4}^{3\pi/4}\frac{\pi}{1+\sin\phi}d\phi$$

$$=\pi\int_{\pi/4}^{3\pi/4}\frac{1-\sin\phi}{1-\sin^2\phi}d\phi=\pi\int_{\pi/4}^{3\pi/4}\frac{1-\sin\phi}{\cos^2\phi}d\phi$$

$$=\pi\int_{\pi/4}^{3\pi/4}(\sec^2\phi-\sec\phi\tan\phi)\,d\phi$$

$$=\pi[\tan\phi-\sec\phi]_{\pi/4}^{3\pi/4}$$

$$=\pi[\tan 3\pi/4-\sec 3\pi/4-\tan\pi/4+\sec\pi/4]$$

$$=2\pi(\sqrt{2}-1)\quad\Rightarrow\quad I=\pi(\sqrt{2}-1)$$

5. Let I $=\int_2^3\frac{\sqrt{x}}{\sqrt{5-x}+\sqrt{x}}dx$ ...(1)

$I=\int_2^3\frac{\sqrt{5-x}}{\sqrt{x}+\sqrt{5-x}}dx$ ...(2)

$\left[\text{Using}\int_a^b f(x)dx=\int_a^b f(a+b-x)dx\right]$

Adding (1) and (2), we get

$$2I=\int_2^3\frac{\sqrt{x}+\sqrt{5-x}}{\sqrt{5-x}+\sqrt{x}}dx$$

$$\Rightarrow\quad I=\frac{1}{2}\int_2^3 1dx=\frac{1}{2}(3-2)=\frac{1}{2}$$

6. $af(x)+bf\left(\frac{1}{x}\right)=\frac{1}{x}-5$ ...(1)

Integrating both sides within the limits 1 to 2, we get

$a\int_1^2 f(x)dx+b\int_1^2 f\left(\frac{1}{x}\right)dx=[\log x-5x]_1^2=\log 2-5$ ...(2)

Replacing x by $\frac{1}{x}$ in (1), we get

$$af\left(\frac{1}{x}\right)+bf(x)=x-5$$

Integrating both sides within the limits 1 to 2, we get

$a\int_1^2 f\left(\frac{1}{x}\right)dx+b\int_1^2 f(x)dx=\left[\frac{x^2}{2}-5x\right]_1^2=-\frac{7}{2}$ ...(3)

Eliminate $\int_1^2 f\left(\frac{1}{x}\right)$ between (2) and (3) by multiplying (2) by a and (3) by b and subtracting

$$\therefore\quad (a^2-b^2)\int_1^2 f(x)dx=a(\log 2-5)+b.\frac{7}{2}$$

$$\therefore\quad \int_1^2 f(x)dx=\frac{1}{(a^2-b^2)}\left[a(\log 2-5)+\frac{7b}{2}\right]$$

7. Let $I=\int_0^{2\pi}\frac{x\sin^{2n}x}{\sin^{2n}x+\cos^{2n}x}dx$

$\Rightarrow\quad I=\int_0^{2\pi}\frac{(2\pi-x)\sin^{2n}(\pi-x)}{\sin^{2n}(2\pi-x)+\cos^{2n}(2\pi-x)}dx$

$\left[\text{Using}\int_0^a f(x)dx=\int_0^a f(a-x)dx\right]$

$I=\int_0^{2\pi}\frac{(2\pi-x)\sin^{2n}x}{\sin^{2n}x+\cos^{2n}x}dx$ ...(2)

Adding (1) and (2) we get

$$2I=\int_0^{2\pi}\frac{2\pi\sin^{2n}x}{\sin^{2n}x+\cos^{2n}x}dx$$

$$\Rightarrow\quad I=\pi\int_0^{2\pi}\frac{\sin^{2n}x}{\sin^{2n}x+\cos^{2n}x}dx$$

$$\Rightarrow\quad I=2\pi\int_0^{\pi}\frac{\sin^{2n}x}{\sin^{2n}x+\cos^{2n}x}dx$$

NOTE THIS STEP

$\left[\text{Using}\int_0^{2a}f(x)dx=2\int_0^a(x)dx\text{ if }f(2a-x)=f(x)\right]$

$\Rightarrow\quad I=4\pi\int_0^{\pi/2}\frac{\sin^{2n}x}{\sin^{2n}x+\cos^{2n}x}dx$ ...(3)

[Using above property again]

$\Rightarrow\quad I=4\pi\int_0^{\pi/2}\frac{\cos^{2n}x}{\cos^{2n}x+\sin^{2n}x}dx$ ....(4)

$\left[\text{Using}\int_0^a f(x)dx=\int_0^a(a-x)dx\right]$

Adding (3) and (4) we get

$$2I=4\pi\int_0^{\pi/2}1.dx=4\pi\left(\frac{\pi}{2}-0\right)=2\pi^2$$

$$\Rightarrow\quad I=\pi^2$$

8. Let $I=\int_1^{e^{37}}\frac{\pi\sin(\pi\ln x)}{x}dx$

Let $\pi\ln x=t$

$\Rightarrow\quad \frac{\pi}{x}dx=dt$ also as $x\to 1, t\to 0, x\to e^{37}, t\to 37\pi$

$$\therefore\quad I=\int_0^{37\pi}\sin t\,dt=[-\cos t]_0^{37\pi}=-\cos 37\pi+1$$

$$=-(-1)+1=2$$

9. $\int_1^4 \frac{2e^{\sin x^2}}{x} dx = F(k) - F(1) = [F(x)]_1^k$

Put $x^2 = t$

$\therefore$ $2x dx = dt$; At $x = 1, t = 1$ and at $x = 4, t = 16$

$\therefore$ $I = \int_1^{16} \frac{e^{\sin t}}{t} dt = F[(t)]_1^{16}$ $\therefore$ $k = 16.$

B TRUE / FALSE :

1. Let $I = \int_0^{2a} \frac{f(x)}{f(x)+f(2a-x)} dx$ ...(1)

$= \int_0^{2a} \frac{f(2a-x)}{f(2a-x)+f[2a-(2a-x)]} dx$

[Using $\int_0^a f(x)dx = \int_0^a f(a-x)dx$]

$I = \int_0^{2a} \frac{f(2a-x)}{f(2a-x)+f(x)}$ ...(2)

Adding (1) and (2), we get

$2I = \int_0^{2a} \frac{f(x)+f(2a-x)}{f(x)+f(2a-x)} dx = \int_0^{2a} 1.dx$

$= [x]_0^{2a} = 2a \Rightarrow I = a$

$\therefore$ The given statement is true.

C MCQ's WITH ONE CORRECT ANSWER :

1. (d) $\int_0^1 (1+e^{-x^2})dx = \int_0^1 \left(1+1-\frac{x^2}{1!}+\frac{x^4}{2!}-\frac{x^6}{3!}+...\infty\right)dx$

$= \left[2x - \frac{x^3}{3.1!} + \frac{x^5}{5.2!} - \frac{x^7}{7.3!} + ...\infty\right]_0^1$

$= \left[2 - \frac{1}{3.1!} + \frac{1}{5.2!} - \frac{1}{7.3!} + ...\infty\right]$

2. (b) Given $\int_0^1 (1+\cos^8 x)(ax^2+bx+c)\,dx$

$= \int_0^2 (1+\cos^8 x)(ax^2+bx+c)\,dx$

$= \int_0^1 (1+\cos^8 x)(ax^2+bx+c)\,dx$

$+ \int_1^2 (1+\cos^8 x)(ax^2+bx+c)\,dx$

$\Rightarrow$ $\int_1^2 (1+\cos^8 x)(ax^2+bx+c)\,dx = 0$

Now we know that if $\int_\alpha^\beta f(x)dx = 0$ then it means that $f(x)$ is + ve on some part of (α, β) and – ve on other part of (α, β).

But here $1 + \cos^8 x$ is always + ve,

$\therefore$ $ax^2 + bx + c$ is + ve on some part of [1 , 2] and – ve on other part [1, 2]

$\therefore$ $ax^2 + bx + c = 0$ has at least one root in (1, 2).

$\Rightarrow$ $ax^2 + bx + c = 0$ has at least one root in (0,2).

3. (c) $ATQ \int_1^b f(x)dx = (b-1)\sin(3b+4)$

Differentiating both sides w.r.t b, we get

$f(b) = 3(b-1)\cos(3b+4) + \sin(3b+4)$

$\Rightarrow$ $f(x) = 3(x-1)\cos(3x+4) + \sin(3x+4)$

4. (a) $I \int_0^{\pi/2} \frac{\sqrt{\cot x}}{\sqrt{\cot x}+\sqrt{\tan x}} dx$ ...(1)

$= \int_0^{\pi/2} \frac{\sqrt{\cot(\pi/2-x)}}{\sqrt{\cot(\pi/2-x)}+\sqrt{\tan(\pi/2-x)}} dx$

$I = \int_0^{\pi/2} \frac{\sqrt{\tan x}}{\sqrt{\tan x}+\sqrt{\cot x}} = dx$ ...(2)

Adding (1) and (2) we get

$2I = \int_0^{\pi/2} \frac{\sqrt{\cot x}+\sqrt{\tan x}}{\sqrt{\cot x}+\sqrt{\tan x}} = dx = \int_0^{\pi/2} 1.dx$

$= (x)_0^{\pi/2} = \pi/2$

$\therefore$ $I = \pi/4$

5. (c) $I = \int_0^\pi e^{\cos^2 x} \cos^3(2n+1)x dx, n \in Z$ ...(1)

$= \int_0^\pi e^{\cos^2(\pi-x)} \cos^3[(2n+1)(\pi-x)]dx$

Using $\int_0^a f(x)dx = \int_0^a f(a-x)dx$

$\therefore$ $I = \int_0^\pi e^{\cos^2 x} \cos^3[(2n+1)\pi - (2n+1)x]dx$

$I = \int_0^\pi (-e^{\cos^2 x} \cos^3)(2n+1)x\,dx$ ...(2)

Adding (1) and (2) we get

$2I = 0 \Rightarrow I = 0$

6. (d) We have,

$I = \int_{-\pi/2}^{\pi/2} \{f(x)+f(-x)\}\{g(x)-g(-x)\}dx$

Let $F(x) = (f(x)+f(-x))(g(x)-g(-x))$

then $F(-x) = (f(-x)+f(x))(g(-x)-g(x))$

$= -[f(x)+f(-x)][g(x)-g(-x)]$

$= -F(x)$

$\therefore$ $F(x)$ is an odd function, $\therefore$ we get $I = 0$

7. (d) Let $I = \int_0^{\frac{\pi}{2}} \frac{dx}{1+\tan^3 x} = \int_0^{\frac{\pi}{2}} \frac{\cos^3 x}{\sin^3 x + \cos^3 x} dx$ ...(1)

$I = \int_0^{\frac{\pi}{2}} \frac{\cos^3\left(\frac{\pi}{2}-x\right)}{\sin^3\left(\frac{\pi}{2}-x\right)+\cos^3\left(\frac{\pi}{2}-x\right)} dx$

$$= \int_0^{\frac{\pi}{2}} \frac{\sin^3 x}{\cos^3 x + \sin^3 x} dx \quad \ldots (2)$$

Adding (1) and (2) we get

$$2I = \int_0^{\frac{\pi}{2}} \frac{\cos^3 x + \sin^3 x}{\sin^3 x + \cos^3 x} dx = \int_0^{\frac{\pi}{2}} 1 dx = \frac{\pi}{2}$$

$$\therefore \quad I = \frac{\pi}{4}$$

8. **(d)** $f(x) = A\sin(\pi x/2) + B$

$$\Rightarrow f'(x) = \frac{A\pi}{2}\cos\left(\frac{\pi x}{2}\right) \Rightarrow f'\left(\frac{1}{2}\right) = \frac{A\pi}{2}\cos\frac{\pi}{4} = \sqrt{2}$$

$$\Rightarrow A = 4/\pi \text{ and } \int_0^1 f(x)dx = \frac{2A}{\pi}$$

$$\Rightarrow \int_0^1 \left[A\sin\left(\frac{\pi x}{2}\right) + B\right] dx = \frac{2A}{\pi}$$

$$\Rightarrow \left| -\frac{2A}{\pi}\cos\left(\frac{\pi x}{2}\right) + Bx \right|_0^1 = \frac{2A}{\pi}$$

$$\Rightarrow B + \frac{2A}{\pi} = \frac{2A}{\pi} \Rightarrow B = 0$$

9. **(a)** Let $I = \int_{\pi}^{2\pi} [2\sin x] dx$

$$\pi \le x < 7\pi/6 \Rightarrow -1 \le 2\sin x < 0 \Rightarrow [2\sin x] = -1$$

$$7\pi/6 \le x < 11\pi/6 \Rightarrow -2 \le 2\sin x < -1$$

$$\Rightarrow [2\sin x] = -1$$

NOTE THIS STEP :

$$\therefore \quad I = \int_{\pi}^{7\pi/6} -1dx + \int_{7\pi/6}^{11\pi/6} -2dx + \int_{11\pi/6}^{2\pi} -1dx$$

$$= \left(-\frac{7\pi}{6} + \pi\right) + 2\left(-\frac{11\pi}{6} + \frac{7\pi}{6}\right) + \left(-2\pi + \frac{11\pi}{6}\right)$$

$$= -\frac{\pi}{6} - \frac{8\pi}{6} - \frac{\pi}{6} = -\frac{10\pi}{6} = \frac{-5\pi}{3}$$

10. **(c)** Given that for f to be a positive function

$$I_1 = \int_{1-k}^{k} xf(x(1-x))dx; \quad I_2 = \int_{1-k}^{k} f[x(1-x)]dx$$

$$\text{Here } I_1 = \int_{1-k}^{k} xf[x(1-x)]dx \quad \ldots (1)$$

$$= \int_{1-k}^{k} (1-x)f[(1-x)x]dx \quad \ldots (2)$$

$$\left[\text{Using the prop.} \int_a^b f(x)\,dx = \int_a^b f(a+b-x)dx\right]$$

Adding (1) and (2) we get,

$$2I_1 = \int_{1-k}^{k} f[x(1-x)]dx = I_2 \Rightarrow \frac{I_1}{I_2} = \frac{1}{2}$$

11. **(a)** Given that $g(x) = \int_0^x \cos^4 t\, dt$

$$\therefore g(x+\pi) = \int_0^{x+\pi} \cos^4 t\, dt$$

$$= \int_0^{\pi} \cos^4 t\, dt + \int_{\pi}^{x+\pi} \cos^4 t\, dt$$

$$g(x+\pi) = g(\pi) + I, \text{ where } I = \int_{\pi}^{x+\pi} \cos^4 t\, dt$$

Put $t = \pi + y, dt = dy$

also as $t \to \pi, y \to 0$

as $t \to x + \pi, y \to x$

$$\therefore \quad I = \int_0^x \cos^4(\pi + y)\, dy$$

$$= \int_0^x \cos^4 y\, dy = \int_0^x \cos^4 t\, dt = g(x)$$

$$\therefore \quad g(x+\pi) = g(\pi) + g(x)$$

12. **(a)** We have

$$I = \int_{\pi/4}^{3\pi/4} \frac{dx}{1+\cos x} \quad \ldots (1)$$

$$= \int_{\pi/4}^{3\pi/4} \frac{dx}{1+\cos(\pi - x)}$$

$$\left[\text{Using the prop.} \int_a^b f(x)dx = \int_a^b (f(a+b-x)\,dx\right]$$

$$= \int_{\pi/4}^{3\pi/4} \frac{dx}{1-\cos x} \quad \ldots (2)$$

Adding (1) and (2), we get

$$2I = \int_{\pi/4}^{3\pi/4} \left(\frac{1}{1+\cos x} + \frac{1}{1-\cos x}\right) dx$$

$$= \int_{\pi/4}^{3\pi/4} 2\text{cosec}^2 x\, dx = 2(-\cot x)_{\pi/4}^{3\pi/4}$$

$$= -2[\cot 3\pi/4 - \cot \pi/4] = -2(-1-1) = 4$$

$$\Rightarrow \quad I = 2$$

13. (c) In the range $\frac{\pi}{2}$ to $\frac{3\pi}{2}$, we have to find the value of $[2\sin x]$.

$$[2\sin x]=\begin{cases} 2 & \text{if } x=\pi/2 \\ 1 & \text{if } \frac{\pi}{2}<x\le\frac{5\pi}{6} \\ 0 & \text{if } \frac{5\pi}{6}<x\le\pi \\ -1 & \text{if } \pi<x\le\frac{7\pi}{6} \\ -2 & \text{if } \frac{7\pi}{6}<x\le\frac{3\pi}{2} \end{cases}$$

NOTE THIS STEP:
Thus

$$I=\int_{\pi/2}^{5\pi/6}1.dx+\int_{5\pi/6}^{\pi}0.dx+\int_{\pi}^{7\pi/6}(-1)dx+\int_{7\pi/6}^{3\pi/2}(-2)dx$$

or $I=\left[\frac{5\pi}{6}-\frac{\pi}{2}\right]+0-1\left[\frac{7\pi}{6}-\pi\right]-2\left[\frac{3\pi}{2}-\frac{7\pi}{6}\right]$

$$=\frac{2\pi}{6}-\frac{\pi}{6}-\frac{4\pi}{6}=\frac{-3\pi}{6}=\frac{-\pi}{2}$$

14. (b) $g(x)=\int_0^x f(t)dt$

$$\Rightarrow\quad g(2)=\int_0^2 f(t)dt=\int_0^1 f(t)dt+\int_1^2 f(t)dt$$

Now, $\frac{1}{2}\le f(t)\le 1$ for $t\in[0,1]$

We get $\int_0^1 \frac{1}{2}dt\le\int_0^1 f(t)dt\le\int_0^1 1dt$

(applying line integral on inequality)

$$\Rightarrow\quad \frac{1}{2}\le\int_0^1 f(t)dt\le 1 \qquad \ldots(1)$$

Again, $0\le f(t)\le\frac{1}{2}$ for $t\in[1,2]$

We get $\int_1^2 0dt\le\int_1^2 f(t)dt\le\int_1^2\frac{1}{2}dt$

NOTE THIS STEP:
(applying line integral on inequality)

$$\Rightarrow\quad 0\le\int_1^2 f(t)dt\le\frac{1}{2} \qquad \ldots(2)$$

From (1) and (2), we get

$$\frac{1}{2}\le\int_0^1 f(t)dt+\int_1^2 f(t)dt\le\frac{3}{2}$$

or $\frac{1}{2}\le g(2)\le\frac{3}{2}$

$\Rightarrow\quad 0\le g(2)\le 2$ is the most appropriate solution.

15. (c) If $f(x)=\begin{Bmatrix} e^{\cos x}\sin x & for\,|x|\le 2 \\ 2 & \text{otherwise}\end{Bmatrix}$

$$=\begin{Bmatrix} e^{\cos x}\sin x & for -2\le x\le 2 \\ 2 & \text{otherwise}\end{Bmatrix}$$

$$\int_{-2}^{3} f(x)dx=\int_{-2}^{2} f(x)dx+\int_{2}^{3} f(x)dx$$

$$=\int_{-2}^{2} e^{\cos x}\sin x dx+\int_2^3 2dx=0+2[x]_2^3$$

[$\because$ $e^{\cos x}\sin x$ is an odd function.]

$$=2[3-2]=2$$

$$\therefore\quad \int_{-2}^{3} f(x)dx=2$$

16. (b) Let $I=\int_{e^{-1}}^{e^2}\left|\frac{\log_e x}{x}\right|dx$

We know that for $\frac{1}{e}<x<1, \log_e x<0$ and hence

$\frac{\log_e x}{x}<0$

and for $1<x<e^2$, $\log x>0$ and hence $\frac{\log_e x}{x}>0$

$$\therefore\quad I=\int_{1/e}^{1}\left[-\frac{\log_e x}{x}\right]dx+\int_1^{e^2}\frac{\log_e x}{x}dx$$

$$=-\frac{1}{2}\left[(\log_e x)^2\right]_{1/e}^{1}+\frac{1}{2}\left[(\log_e x)^2\right]_1^{e^2}$$

$$=\frac{1}{2}+2=\frac{5}{2}.$$

17. (c) $I=\int_{-\pi}^{\pi}\frac{\cos^2 x}{1+a^x}dx \qquad .....(1)$

Put x = – y then dx = – dy

$$I=\int_{\pi}^{-\pi}\frac{\cos^2 y}{1+a^{-y}}dy=\int_{-\pi}^{\pi}\frac{a^y\cos^2 y}{1+a^y}dy$$

$$I=\int_{-\pi}^{\pi}\frac{a^x\cos^2 x}{1+a^x}dx \quad \left[\because\int_a^b f(y)dy=\int_a^b f(x)dx\right]$$

.....(2)

But $[x]=\begin{cases}-1, & \text{if } -1\le x<0\\ 0, & \text{if } 0\le x<1\\ 1, & \text{if } x=1\end{cases}$

$\therefore \quad \int_{-1}^{1}[x]dx=\int_{-1}^{0}(-1)dx+\int_{0}^{1}0\,dx=-x]_{-1}^{0}+0=-1$

Thus, putting value in equation (1) we get $\int_{-1}^{1}f(x)dx=1$

3. **(b, d)**

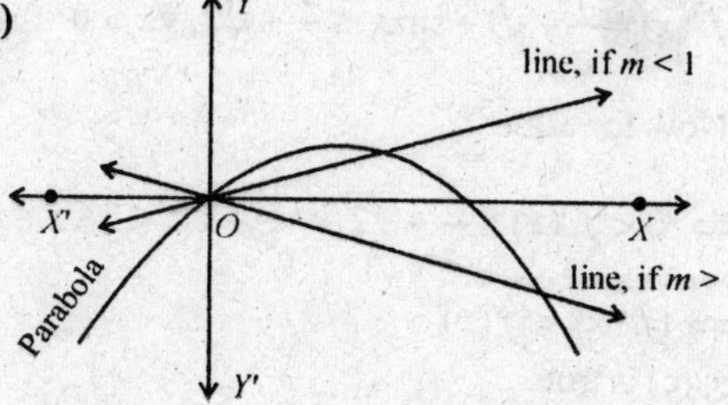

The two curves meet at

$mx=x-x^2$ or $x^2=x(1-m)$, $\quad\therefore\ x=0,1-m$

$\int(y_1-y_2)dx=\int(x-x^2-mx)dx$ (P, L)

Clearly m < 1 or m > 1, but m ≠ 1

Now, $\left[(1-m)\frac{x^2}{2}-\frac{x^3}{3}\right]_0^{1-m}=\frac{9}{2}$, if $m<1$

or $(1-m)^3=27$, $\therefore\ m=-2$

But if m > 1 then 1 – m is – ive, then $\left[(1-m)\frac{x^2}{2}-\frac{x^3}{3}\right]_{1-m}^{0}=\frac{9}{2}$,

$\therefore\quad (1-m)^3=-27$, or $1-m=-3$, $\therefore$ m = 4.

4. **(a, b, c, d)**

$\therefore f(x)$ is a non constant twice differentiable function such that $f(x)=f(1-x)$

$\Rightarrow f'(x)=-f'(1-x)$ ...(1)

For $x=\frac{1}{2}$, we get $f'\left(\frac{1}{2}\right)=-f'\left(1-\frac{1}{2}\right)$

$\Rightarrow\ f'\left(\frac{1}{2}\right)+f'\left(\frac{1}{2}\right)=0\ \Rightarrow\ f'\left(\frac{1}{2}\right)=0$

$\Rightarrow$ (b) is correct

For $x=\frac{1}{4}$, we get $f'\left(\frac{1}{4}\right)=-f'\left(\frac{3}{4}\right)$

but given that $f'\left(\frac{1}{4}\right)=0 \quad \therefore\ f'\left(\frac{1}{4}\right)=f'\left(\frac{3}{4}\right)=0$

Hence, $f'(x)$ satisfies all conditions of Rolle's theorem for $x\in\left[\frac{1}{4},\frac{1}{2}\right]$ and $\left[\frac{1}{2},\frac{3}{4}\right]$. So there exists at least one point $C_1\in\left(\frac{1}{4},\frac{1}{2}\right)$ and at least one point $C_2\in\left(\frac{1}{2},\frac{3}{4}\right)$. Such that $f''(C_1)=0$ and $f''(C_2)=0$

$\therefore f''(x)$ varishes at least twice on [0, 1] $\Rightarrow$ (a) is correct.

Also using $f(x)=f(1-x)$

$\Rightarrow\ f\left(x+\frac{1}{2}\right)=f\left(1-x-\frac{1}{2}\right)=f\left(-x+\frac{1}{2}\right)$

$\Rightarrow\ f\left(x+\frac{1}{2}\right)$ is an even function.

$\Rightarrow \sin x.\ f\left(x+\frac{1}{2}\right)$ is an odd function.

$\Rightarrow\ \int_{-1/2}^{1/2}f\left(x+\frac{1}{2}\right)\sin x\,dx=0$, $\therefore$ (c) is correct.

5. **(b, c, d)**

The area bounded by the curve $y=e^x$ and lines $x=0$ and $y=e$ is as shown in the graph.

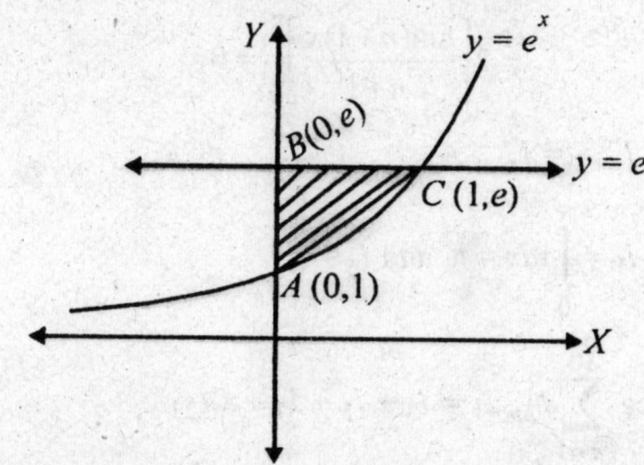

Required area $=\int_0^1(e-e^x)dx=[ex]_0^1-\int_0^1 e^x dx$

$=e-\int_0^1 e^x dx.=1$

Also required area

$=\int_0^e x\,dy$ (where $e^x=y\Rightarrow x=\ln y$)

$=\int_1^e \ln y\,dy$

$=\int_1^e \ln(e+1-y)dy$ $\left[\text{Using the property } \int_a^b f(x)dx=\int_a^b f(a+b-x)dx\right]$

6. **(a, b, c)** We have

$$I_n=\int_{-\pi}^{\pi}\frac{\sin nx}{(1+\pi^x)\sin x}dx \quad ....(1)$$

$$\Rightarrow\quad I_n=\int_{-\pi}^{\pi}\frac{\sin n(-x)}{(1+\pi^{-x})\sin(-x)}dx$$

$\left[\text{Using } \int_a^b f(x)dx=\int_a^b f(a+b-x)dx\right]$

$$\Rightarrow\quad I_n=\int_{-\pi}^{\pi}\frac{\pi^x\sin nx}{(1+\pi^x)\sin x}dx \quad ...(2)$$

Adding equation (1) and (2), we get

$$2I_n = \int_{-\pi}^{\pi} \frac{\sin nx}{\sin x} dx = 2\int_0^{\pi} \frac{\sin nx}{\sin x} dx$$

[as integrand is an even function]

$$\Rightarrow \quad I_n = \int_0^{\pi} \frac{\sin nx}{\sin x} dx$$

Now $I_{n+2} - I_n = \int_0^{\pi} \frac{\sin(n+2)x - \sin nx}{\sin x} dx$

$$= \int_0^{\pi} \frac{2\cos(n+1)x \sin x}{\sin x} dx = 2\int_0^{\pi} \cos(n+1)x\, dx$$

$$= 2\left[\frac{\sin(n+1)x}{n+1}\right]_0^{\pi} = 0$$

$\therefore \quad I_{n+2} = I_n$

Also $I_1 = \int_0^{\pi} 1dx = \pi$ and $I_0 = 0$

Hence $\sum_{m=1}^{10} I_{2m+1} = I_3 + I_5 + I_7 + ... I_{21}$

$= 10\, I$ (using $I_{n+2} = I_n$) $= 10\,\pi$

and $\sum_{m=1}^{10} I_{2m} = I_2 + I_4 + I_6 + ... + I_{20}$

$= 20 \times I_0$ (using $I_{n+2} = I_n$)

$= 20 \times 0 = 0$

7. **(a)** $\int_0^1 \frac{x^4(1-x)^4}{1+x^2} dx = \int_0^1 \left(x^6 - 4x^5 + 5x^4 - 4x^2 + 4 - \frac{4}{1+x^2}\right) dx$

$$= \left[\frac{x^7}{7} - \frac{2x^6}{3} + x^5 - \frac{4x^3}{3} + 4x - 4\tan^{-1} x\right]_0^1$$

$$= \frac{1}{7} - \frac{2}{3} + 1 - \frac{4}{3} + 4 - 4.\frac{\pi}{4} = \frac{22}{7} - \pi$$

8. **(b, c)** We have

$$f(x) = lnx + \int_0^x \sqrt{1+\sin t}\, dt$$

$\Rightarrow f'(x) = \frac{1}{x} + \sqrt{1+\sin x}$ which exists $\forall x \in (0, \infty)$

and $f'(x)$ has finite value $\forall x \in (0, \infty)$, *so* $f'(x)$ is continuous

Also $f''(x) = -\frac{1}{x^2} + \frac{\cos x}{2\sqrt{1+\sin x}}$

Which does not exist at the points where

$\sin x = -1$ like $x = \frac{3\pi}{2}, \frac{7\pi}{2}, ...$

$\therefore f'(x)$ is not differentiable.

$\therefore$ (a) is false but (b) is true

Now $\sqrt{1+\sin t} \geq 0 \Rightarrow \int_0^x \sqrt{1+\sin t}\, dt \geq 0 \forall x \in (0, \infty)$

And $lnx > 0 \,\forall x \in (1, \infty) \Rightarrow f(x) > 0 \,\forall x \in (1, \infty)$

For $x \geq e^3$

$$f(x) = ln\, x + \int_0^x \sqrt{1+\sin t}\, dt \geq 3$$

$$f'(x) = \frac{1}{x} + \sqrt{1+\sin x} \leq \frac{1}{x} + \sqrt{2}, \forall x > 0$$

Now for $x \geq e^3$

$$\Rightarrow 0 < f'(x) \leq \frac{1}{x} + \sqrt{2} < \frac{1}{e^3} + \sqrt{2} < 3 \;\forall x \in (e^3, \infty)$$

$\Rightarrow |f'(x)| < |f(x)|$

$\therefore$ (c) is true.

Also $\lim_{x \to \infty} f(x) = \infty$

$\therefore |f(x)| + |f'(x)|$ is not bounded.

$\therefore$ (d) is wrong.

9. **(a, b, d)** First of all let us draw a rough sketch of $y = e^{-x}$.
At $x = 0, y = 1$
and at $x = 1, y = 1/e$

Also $\frac{dy}{dx} = -2xe^{-x^2} < 0 \;\forall\; x \in (0,1)$

$\therefore y = e^{-x^2}$ is decreasing on (0, 1)

Hence its graph is as shown in figure given below

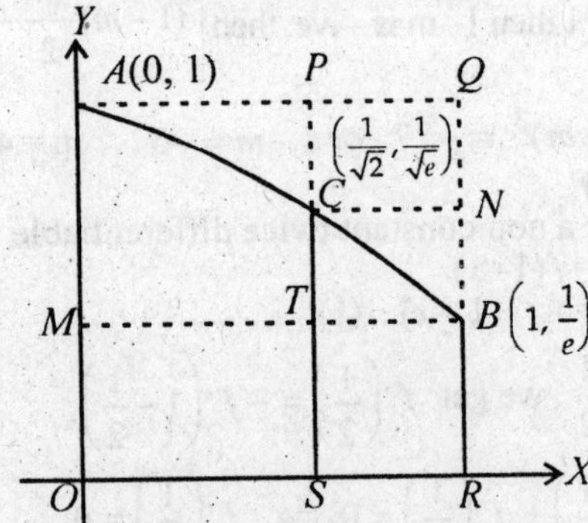

Now, S = area exclosed by curve = ABRO

and area of rectangle ORBM $= \frac{1}{e}$

Clearly $S > \frac{1}{e}$ $\quad\therefore$ A is true.

Also $x^2 < x \;\;\forall\, x \in [0,1]$

$\Rightarrow -x^2 > -x \Rightarrow e^{-x^2} \geq e^{-x} \;\forall\; x \in [0,1]$

$$\Rightarrow \int_0^1 e^{-x^2} dx > \int_0^1 e^{-x} dx = 1 - \frac{1}{e}$$

$\Rightarrow S > 1 - \frac{1}{e}$ $\quad\therefore$ (b) is true.

Now S < area of rectangle APSO + area of rectangle CSRN

$\Rightarrow \quad S < \frac{1}{\sqrt{2}} \times 1 + \left(1 - \frac{1}{\sqrt{2}}\right)\frac{1}{\sqrt{e}}$

$\therefore \quad S < \frac{1}{\sqrt{2}} + \frac{1}{\sqrt{e}}\left(1 - \frac{1}{\sqrt{2}}\right)$ $\quad \because$ (d) is true

Also as $\frac{1}{4}\left(1 + \frac{1}{\sqrt{e}}\right) < 1 - \frac{1}{e}$ $\quad \therefore$ *(c)* is incorrect.

(E) SUBJECTIVE PROBLEMS :

1. To find the area bounded by

$x^2 = 4y$ ...(1)

which is an upward parabola with vertex at (0, 0).

and $x - 4y = -2$ or $\frac{x}{-2} + \frac{y}{1/2} = 1a$ ...(2)

which is a st. line with its intercepts as –2 and 1/2 on axes. For Pt's of intersection of (1) and (2) putting value of 4y fom (2) in (1) we get

$x^2 = x + 2 \Rightarrow x^2 - x - 2 = 0 \Rightarrow (x-2)(x+1) = 0$

$\Rightarrow \quad x = 2, -1 \quad \Rightarrow \quad y = 1, 1/4$

$\therefore \quad A(-1, 1/4) B(2, 1).$

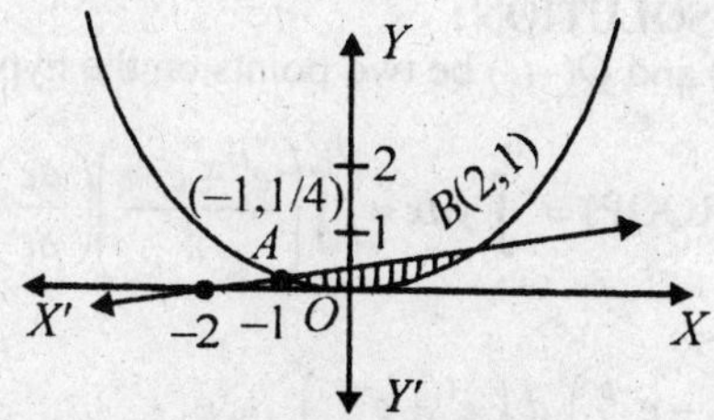

Shaded region in the fig is the req area.

$\therefore$ Required area $= \int_{-1}^{2}(y_{line} - y_{parabola})dx$

$= \int_{-1}^{2}\left(\frac{x+2}{4} - \frac{x^2}{4}\right)dx = \frac{1}{4}\left[\frac{x^2}{2} + 2x - \frac{x^3}{3}\right]_{-1}^{2}$

$= \frac{1}{4}\left[\left(2 + 4 - \frac{8}{3}\right) - \left(\frac{1}{2} - 2 + \frac{1}{3}\right)\right] = 9/8$ sq. units

2. We know that in integration as a limit sum

$$\int_0^1 f(x)dx = \lim_{n\to\infty}\frac{1}{n}\sum_{r=1}^{n} f(r/n)$$

NOTE THIS STEP :

Similarly the given series can be written as

$$\lim_{n\to\infty}\left(\frac{1}{n+1} + \frac{1}{n+2} + \ldots + \frac{1}{6n}\right) = \lim_{n\to\infty}\sum_{r=1}^{5n}\frac{1}{n+r}$$

$$= \lim_{n\to\infty}\frac{1}{n}\sum_{r=1}^{5n}\frac{1}{1+\frac{r}{n}}$$

$$= \int_0^5 \frac{1}{1+x}dx = [\log|1+x|]_0^5 = \log 6 - \log 1 = \log 6$$

3. $\frac{t^{n+1} - 1}{(t-1)(n+1)} = \int_0^1 (tx + 1 - x)^n dx = \int_0^1 [(t-1)x + 1]^n dx$

$= \left|\frac{[t-1]x + 1^{n+1}}{(t-1)(n+1)}\right|_0^1 = \frac{1}{n+1}\left[\frac{t^{n+1}}{t-1} - \frac{1}{t-1}\right]$

$\int_0^1 (tx + 1 - x)^n dx = \frac{t^{n+1} - 1}{(t-1)(n+1)}$ ...(1)

Deduction, $\int_0^1 x^k (1-x)^{n-k} dx = {}^nC_k (n+1)^{-1}$

$k = 0, 1, 2, \ldots, n.$

Consider, $[tx + 1(1-x)]^n = \sum_{k=0}^{n} {}^nC_k (tx)^k (1-x)^{n-k}$

[Using Binomal thm.]

$= \sum_{k=0}^{n} [{}^nC_k x^k (1-x)^{n-k}] t^k$

Integrating both sides from 0 to 1, with respect to x we get.

$\int_0^1 [tx + (1-x)]^n dx = \sum_{k=0}^{n} t^{kn} c_k \int_0^1 x^k (1-x)^{n-k} dx$

$\Rightarrow \quad \frac{t^{n+1} - 1}{(t-1)(n+1)} = \sum_{k=0}^{n} {}^nC_k t^k \left\{\int_0^1 x^k (1-x)^{n-k} dx\right\}$

[Using eq.(1) for LHS]

$\Rightarrow \quad \sum_{k=0}^{n} {}^nC_k t^k \left\{\int_0^1 x^k (1-x)^{n-k} dx\right\}$

$= \frac{1}{n+1}[1 + t + t^2 + t^3 + \ldots + t^n]$ [Using GP]

Equating the coefficiects of t^k on both sides we get

${}^nC_k \int_0^1 x^k (1-x)^{n-k} dx = \frac{1}{n+1}$

NOTE THIS STEP :

$\Rightarrow \quad \int_0^1 x^k (1-x)^{n-k} dx = \frac{1}{{}^nC_k (n+1)}$

$= [{}^nC_k (n+1)]^{-1}$ for $k = 0, 1, 2, \ldots n$

4. Let $I = \int_0^{\pi} x f(\sin x) dx$ ...(1)

$\Rightarrow \quad I = \int_0^{\pi} (\pi - x) f(\sin x) dx$

Adding (1) and (2), we get,

$$2I = \int_0^{\pi} \pi f(\sin x)dx$$

$$I = \frac{\pi}{2}\int_0^{\pi} f(\sin x)dx \qquad \text{Hence Proved.}$$

5. $\displaystyle\int_{-1}^{3/2} |x\sin\pi x|\,dx$

For $-1 \le x < 0 \Rightarrow -\pi < px < 0 \Rightarrow \sin\pi x < 0$

$\Rightarrow x\sin\pi x > 0$

For $1 < x < 3/2 \Rightarrow \pi < \pi x < 3\pi/2 \Rightarrow \sin\pi x < 0$

$\Rightarrow x\sin\pi x < 0$

$$\therefore \int_{-1}^{3/2} |x\sin\pi x|\,dx = \int_{-1}^{1} x\sin\pi x\,dx + \int_{1}^{3/2}(-x\sin\pi x)ax$$

$$= 2\int_0^1 x\sin\pi x\,dx - \int_1^{3/2} x\sin\pi x\,dx$$

$$= 2\left[\frac{-x\cos\pi x}{\pi} + \frac{\sin\pi x}{\pi^2}\right]_0^1 - \left[\frac{-x\cos\pi x}{\pi} + \frac{\sin\pi x}{\pi^2}\right]_1^{3/2}$$

$$= 2\left[\left(\frac{-\cos\pi}{\pi} + 0\right) - (0+0)\right]$$

$$-\left[\left(\frac{-3/2\cos 3\pi/2}{\pi} + \frac{\sin 3\pi/2}{\pi^2}\right) - \left(\frac{-\cos\pi}{\pi} + \frac{\sin\pi}{\pi^2}\right)\right]$$

$$= 2\left[\frac{1}{\pi}\right] - \left[-\frac{1}{\pi^2} - \frac{1}{\pi}\right] = \frac{2}{\pi} + \frac{1}{\pi^2} + \frac{1}{\pi} = \frac{3}{\pi} + \frac{1}{\pi^2}$$

6. It is a point on hyperbola $x^2 - y^2 = 1$.

Then equation of line joining pts t_1 and $-t_1$ that is

$$\left(\frac{e^{t_1}+e^{-t_1}}{2}, \frac{e^{t_1}+e^{-t_1}}{2}\right) \text{ and } \left(\frac{e^{-t_1}+e^{t_1}}{2}, \frac{e^{-t_1}-e^{t_1}}{2}\right) \text{ is}$$

$$y - \frac{e^{t_1}-e^{-t_1}}{2} = \frac{-2\left(\frac{e^{t_1}-e^{-t_1}}{2}\right)}{0}\left(x - \frac{e^{t_1}+e^{-t_1}}{2}\right)$$

$$\Rightarrow x = \frac{e^{t_1}+e^{-t_1}}{2}$$

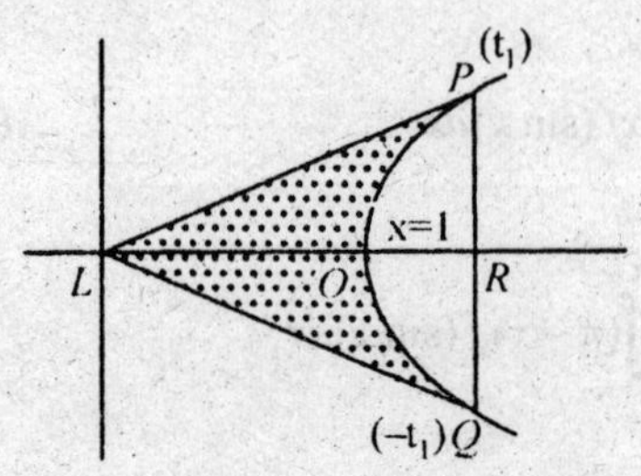

$$\therefore \text{ Area (PRQOP)} = 2\int_1^{\frac{e^{t_1}+e^{-t_1}}{2}} \sqrt{x^2-1}\,dx$$

$$= 2\left[\frac{x}{2}\sqrt{x^2-1} - \frac{1}{2}\log|x + \sqrt{x^2-1}|\right]_1^{\frac{e^{t_1}+e^{-t_1}}{2}}$$

$$= \left(\frac{e^{t_1}+e^{-t_1}}{2}\right)\left(\frac{e^{t_1}-e^{-t_1}}{2}\right) - \log\left|\frac{e^{t_1}+e^{-t_1}}{2} + \frac{e^{t_1}-e^{-t_1}}{2}\right|$$

$$= \frac{e^{2t_1}-e^{-2t_1}}{4} - \log e^{t_1} = \frac{e^{2t_1}-e^{-2t_1}}{4} - t_1 \qquad \ldots(1)$$

$$\text{Ar }\Delta LPQ = 2\times\frac{1}{2}\left(\frac{e^{t_1}+e^{-t_1}}{2}\right)\left(\frac{e^{t_1}-e^{-t_1}}{2}\right)$$

$$= \left(\frac{e^{2t_1}-e^{-2t_1}}{4}\right) \qquad \ldots(2)$$

$\therefore$ Required area = Ar $\Delta LPQ - Ar(POQRP)$

$= t_1$(using (1) and (2))

ALTERNATE SOLUTION :

Let $P(t_1)$ and $Q(-t_1)$ be two points on the hyperbola.

$$\text{Area (PRQOP)} = \int_{-t_1}^{t_1} y\,dx = \int_{-t_1}^{t_1}\left(\frac{e^t+e^{-t}}{2}\right)\left(\frac{dx}{dt}\right)dt$$

$$= \int_{-t_1}^{t_1}\left(\frac{e^t-e^{-t}}{2}\right)\frac{d}{dt}\left(\frac{e^t+e^{-t}}{2}\right).dt$$

$$= \int_{-t_1}^{t_1}\left(\frac{e^t-e^{-t}}{2}\right)dt = \int_{-t_1}^{t_1}\frac{e^{2t}+e^{-2t}-2}{4}dt$$

$$= \left[\frac{e2t}{8} - \frac{e^{-2t}}{8} - \frac{2t}{4}\right]_{-t_1}^{t_1} = \frac{2}{8}(e^{2t_1} - e^{-2t_1} - 4t_1)$$

$$= \frac{e^{2t_1}-e^{-2t_1}}{4} - t_1 \qquad \ldots(1)$$

$$\text{Area of }\Delta LPR = \frac{1}{2}LR\times PQ = LR\times PR$$

$$= \frac{e^{t_1}+e^{-t_1}}{2}\times\frac{e^{t_1}-e^{-t_1}}{2} = \frac{e^{2t_1}-e^{-2t_1}}{4} \qquad \ldots(2)$$

$\therefore$ The required area $= Ar(\Delta LPQ) - Ar(PRQOP)$

$$= \frac{e^{2t_1}-e^{-2t_1}}{4} - \frac{e^{2t_1}-e^{-2t_1}}{4} + t_1 = t_1$$

7. $$I = \int_0^{\pi/4}\frac{\sin x + \cos x}{9 + 16\sin 2x}dx$$

Let $\sin x - \cos x = t$

as $x \to 0, t \to -1$

as $x \to \pi/4, t \to 0$

$\Rightarrow \quad (\cos x + \sin x)dx = dt$

Also, $t^2 = 1 - \sin 2x \Rightarrow \sin 2x = 1 - t^2$

$$I = \int_{-1}^{0} \frac{dt}{9 + 16(1 - t^2)} = \int_{-1}^{0} \frac{dt}{25 - 16t^2}$$

$$= \frac{1}{16}\int_{-1}^{0} \frac{dt}{\left(\frac{5}{4}\right)^2 - t^2} = \frac{1}{16} \cdot \frac{1}{2 \cdot \frac{5}{4}} \log \left[\left|\frac{\frac{5}{4} + t}{\frac{5}{4} - t}\right|\right]_{-1}^{0}$$

$$= \frac{1}{40}\left[\log 1 - \log \frac{1}{9}\right] = \frac{\log 9}{40} = \frac{2\log 3}{40} = \frac{1}{20}\log 3$$

8. $y = 1 + \dfrac{8}{x^2}$

Req. area $= \int_2^4 y dx = \int_2^4 \left(1 + \frac{8}{x^2}\right)dx = \left[x - \frac{8}{x}\right]_2^4 = 4$

If $x = 4a$ bisects the area then we have

$$\int_2^a \left(1 + \frac{8}{x^2}\right)dx = \left[x - \frac{8}{x}\right]_2^a = \left[a - \frac{8}{a} - 2 + 4\right] = \frac{4}{2}$$

$$\Rightarrow \quad a - \frac{8}{a} = 0 \Rightarrow a^2 = 0 \Rightarrow \quad a = \pm 2\sqrt{2}$$

Since $2 < a < 4 \qquad \therefore a = 2\sqrt{2}$

9. Let $I = \int_0^{1/2} \dfrac{x \sin^{-1} x}{\sqrt{1 - x^2}} dx$

Put $x = \sin\theta \Rightarrow dx = \cos\theta \, d\theta$

Also when $x = 0, \theta = 0$

and when $x = 1/2, \theta = \pi/6$

Thus, $I = \int_0^{\pi/6} \dfrac{\sin\theta \sin^{-1}(\sin\theta)}{\sqrt{1 - \sin^2\theta}} \cos\theta \, d\theta$

$$\Rightarrow \quad I = \int_0^{\pi/6} \theta \sin\theta \, d\theta$$

Intergrating the above by parts, we get

$$I = [\theta(-\cos\theta)]_0^{\pi/6} + \int_0^{\pi/6} 1.\cos\theta d\theta$$

$$= [-\theta\cos\theta + \sin\theta]_0^{\pi/6} = \frac{-\pi}{6} \cdot \frac{\sqrt{3}}{2} + \frac{1}{2} = \frac{6 - \pi\sqrt{3}}{12}$$

10. To find the area bold by x - axis and curves

$y = \tan x, -\pi/3 \le x \le \pi/3 \qquad \ldots(1)$

and $y = \cot x, \pi/6 \le x \le 3\pi/2 \qquad \ldots(2)$

The curves intersect at P, where $\tan x = \cot x$, which is satisfied at $x = \pi/4$ within the given domain of x.

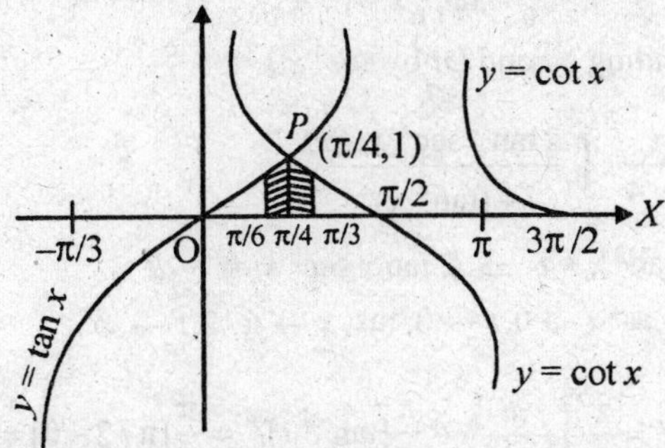

The required area is shaded area

$$A = \int_{\pi/6}^{\pi/4} \tan x \, dx + \int_{\pi/4}^{\pi/3} \cot x \, dx$$

$$= [\log \sec x]_{\pi/6}^{\pi/4} + [\log \sin x]_{\pi/4}^{\pi/3}$$

$$= \left(\log\sqrt{2} - \log\frac{2}{\sqrt{3}}\right) + \left(\log\frac{\sqrt{3}}{2} - \log\frac{1}{\sqrt{2}}\right)$$

$$= 2\left(\log\sqrt{2} \cdot \frac{\sqrt{3}}{2}\right) = 2\log\sqrt{\frac{3}{2}} = \log 3/2 \text{ sq. units}$$

11. Let $\int f(x)dx = F(x) + c$

Then $F'(x) = f(x) \qquad \ldots(1)$

Now $I = \int_a^{a+t} f(x)dx = F(a+t) - F(a)$

$\therefore \quad \dfrac{dI}{da} = F'(a+t) - F(a) = f(a+t) - f(a)$

[Using eq. (1)]

$= f(a) - f(a) \quad$ [Using given condition]

$= 0$

This shows that I is independent of a.

12. Let $I = \int_0^{\pi/2} \dfrac{x \sin x \cos x}{\cos^4 x + \sin^4 x} dx \qquad \ldots(1)$

$$I = \int_0^{\pi/2} \frac{(\pi/2 - x)\sin(\pi/2 - x)\cos(\pi/2 - x)}{\cos^4(\pi/2 - x) + \sin^4(\pi/2 - x)} dx$$

[Using $\int_0^a f(x)dx = \int_0^a f(a - x)dx$]

$$\Rightarrow \quad I = \int_0^{\pi/2} \frac{(\pi/2 - x)\sin x \cos x}{\sin^4 x + \cos^4 x} dx \qquad \ldots(2)$$

Adding (1) and (2), we get

$$2I = \frac{\pi}{2}\int_0^{\pi/2} \frac{\sin x \cos x}{\sin^4 x + \cos^4 x} dx$$

$$\Rightarrow\ I = \frac{\pi}{4}\int_0^{\pi/2}\frac{\sec^2 x\tan x}{\tan^4 x+1}dx$$

(Dividing Nr and Dr by $\cos^4 x$)

$$= \frac{\pi}{2\times 4}\int_0^{\pi/2}\frac{2\tan x\sec^2 x\,dx}{1+(\tan^2 x)^2}$$

Put $\tan^2 x = t \Rightarrow 2\tan x\sec^2 x\,dx = dt$

Also as $x\to 0, t\to 0$; as $x\to\pi/2, t\to\infty$

$$\therefore\ I = \frac{\pi}{8}\int_0^{\infty}\frac{dt}{1+t^2} = \frac{\pi}{8}[\tan^{-1}t]_0^{\infty} = \frac{\pi}{8}[\pi/2-0] = \pi^2/16$$

13. The given curves are

$$y = \sqrt{5-x^2} \quad \ldots(1)$$

$$y = |x-1| \quad \ldots(2)$$

We can clearly see that (on squaring both sides of (1)) eq. (1) represents a circle. But as y is + ve sq. root, $\therefore$ (1) represents upper half of circle with centre (0, 0) and radius $\sqrt{5}$.

Eq. (2) represents the curve

$$y = \begin{cases} -x+1 & \text{if } x<1 \\ x-1 & \text{if } x\ge 1\end{cases}$$

Graph of these curves are as shown in figure with point of intersection of $y=\sqrt{5-x^2}$ and $y=-x+1$ as $A(-1,2)$ and of $y=\sqrt{5-x^2}$ and $y=x-1$ as $C(2,1)$.

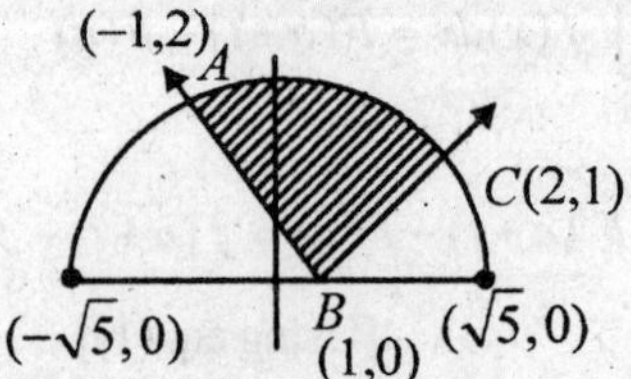

The required area = Shaded area

$$= \int_{-1}^{2}(y_{(1)}-y_{(2)})dx = \int_{-1}^{2}\sqrt{5-x^2}\,dx - \int_{-1}^{2}|x-1|\,dx$$

$$= \left[\frac{x}{2}\sqrt{5-x^2}+\frac{5}{2}\sin^{-1}\left(\frac{x}{\sqrt{5}}\right)\right]_{-1}^{2} - \int_{-1}^{1}\{-(x-1)\}dx - \int_{1}^{2}(x-1)dx$$

$$= \left(\frac{2}{2}\sqrt{5-4}+\frac{5}{4}\sin^{-1}\frac{2}{\sqrt{5}}\right)-\left(\frac{-1}{2}\sqrt{5-1}+\frac{5}{2}\sin^{-1}\left(\frac{-1}{\sqrt{5}}\right)\right)$$

$$-\left(\frac{-x^2}{2}+x\right)_{-1}^{1}-\left(\frac{x^2}{2}-x\right)_{1}^{2}$$

$$= 1+\frac{5}{2}\sin^{-1}\frac{2}{\sqrt{5}}+1+\frac{5}{2}\sin^{-1}\left(\frac{1}{\sqrt{5}}\right)$$

$$-\left[\left(\frac{-1}{2}+1\right)-\left(\frac{-1}{2}-1\right)\right]-\left[(2-2)-\left(\frac{1}{2}-1\right)\right]$$

$$= 2+\frac{5}{2}\left[\sin^{-1}\frac{2}{\sqrt{5}}+\sin^{-1}\frac{1}{\sqrt{5}}\right]-2-\frac{1}{2}$$

$$= \frac{5}{2}\left[\sin^{-1}\frac{2}{\sqrt{5}}+\cos^{-1}\frac{2}{\sqrt{5}}\right]-\frac{1}{2} = \frac{5}{2}\left(\frac{\pi}{2}\right)-\frac{1}{2}$$

$$= \frac{5\pi-2}{4} \text{ square units.}$$

14. The given curves are

$$x^2+y^2=4 \quad \ldots(1)$$

$$x^2 = -\sqrt{2}y \quad \ldots(2)$$

$$x = y \quad \ldots(3)$$

It is clear from the equations of curves that eq. (1) represents a circle with centre (0, 0) and radius 2. (2) represents a downward parabola with vertex at origin.

(3) represents *a* st. line, through origin making an angle of 45° with +ve direction of x- axis.

To find the points of intersection of (1) and (2), substitute the value of x^2 from (2) in (1), we get

$$y^2-\sqrt{2}y-4=0$$

$$\Rightarrow\ y = \frac{\sqrt{2}\pm\sqrt{2+16}}{2} = \frac{\sqrt{2}\pm 3\sqrt{2}}{2}$$

$$= \frac{4\sqrt{2}}{2} \text{ or } \frac{-2\sqrt{2}}{2} = 2\sqrt{2} \text{ or } -\sqrt{2}$$

$$\therefore\ x^2 = -\sqrt{2}(-\sqrt{2})$$

[rejecting +ve value of y as x^2 can not be – ve]

$$\Rightarrow\ x^2 = 2 \Rightarrow x = \pm\sqrt{2}$$

$\therefore$ Points of intersection of (1) and (2) are $(\sqrt{2},-\sqrt{2})$, $(-\sqrt{2},-\sqrt{2})$

Also for pt of intersection of (1) and (3), solving eq. (1) and (3), we get

$$2x^2 = 4 \Rightarrow x^2 = 2 \Rightarrow x = \pm\sqrt{2},\ \therefore\ y = \pm\sqrt{2}$$

$\therefore$ Points of intersection are $(\sqrt{2},-\sqrt{2}),(-\sqrt{2},-\sqrt{2})$.

Thus we observe all the three curves pass through the same point $(-\sqrt{2},-\sqrt{2})$.

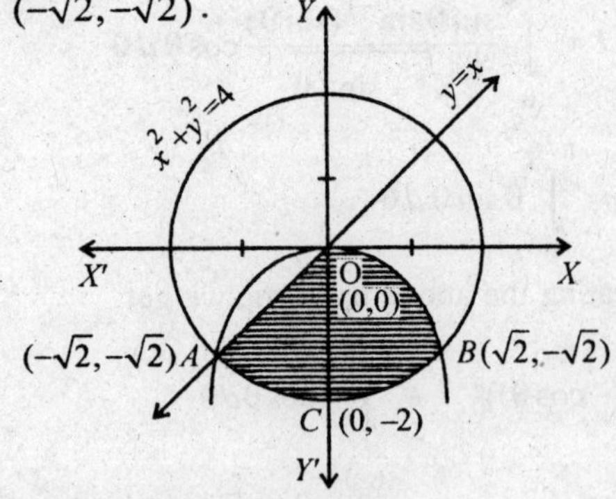

Now required area = shaded area

$$= \int_{-\sqrt{2}}^{0}(y_C-y_L)dx + \int_{0}^{\sqrt{2}}(y_C-y_P)dx$$

NOTE THIS STEP :

$$=\left|\int_{-\sqrt{2}}^{\sqrt{2}} y_C dx\right|-\left|\int_{-\sqrt{2}}^{0} y_L dx\right|-\left|\int_{0}^{\sqrt{2}} y_P dx\right|$$

$$=2\left|\int_{0}^{\sqrt{2}} \sqrt{4-x^2}\,dx\right|-\left|\int_{-\sqrt{2}}^{0} x dx\right|-\left|\int_{0}^{\sqrt{2}}\left(-\frac{x^2}{\sqrt{2}}\right)dx\right|$$

$$=2\left[\frac{x}{2}\sqrt{4-x^2}+\frac{4}{2}\sin^{-1}\frac{x}{2}\right]_0^{\sqrt{2}}-\left|\left[\frac{x^2}{2}\right]_{-\sqrt{2}}^{0}\right|-\left|\left[\frac{-x^3}{3\sqrt{2}}\right]_0^{\sqrt{2}}\right|$$

$$=2\left[\frac{\sqrt{2}}{2}\sqrt{4-2}+2\sin^{-1}\left(\frac{\sqrt{2}}{2}\right)\right]-\left|\frac{-2}{2}\right|-\left|\frac{-2\sqrt{2}}{3\sqrt{2}}\right|$$

$$=2\left[1+2.\frac{\pi}{4}\right]-1-\frac{2}{3}=\pi+\frac{1}{3}$$

15. Let $I=\int_0^{\pi}\frac{x\,dx}{1+\cos\alpha\sin x}$ ...(1)

$$I=\int_0^{\pi}\frac{(\pi-x)dx}{1+\cos\alpha(\sin(\pi-x))}$$

[Using $\int_0^a f(x)dx=\int_0^a f(a-x)dx$]

$\therefore\quad I=\int_0^{\pi}\frac{(\pi-x)dx}{1+\cos\alpha\sin x}$ ...(2)

Adding (1) and (2), we get

$$2I=\int_0^{\pi}\frac{x+\pi-x}{1+\cos\alpha\sin x}dx=\int_0^{\pi}\frac{\pi}{1+\cos\alpha\sin x}dx$$

$$\therefore\quad I=\frac{\pi}{2}\int_0^{\pi}\frac{1}{1+\cos\alpha\sin x}dx=\frac{\pi}{2}.2\int_0^{\pi/2}\frac{1}{1+\cos\alpha\sin x}dx$$

$$=\pi\int_0^{\pi/2}\frac{1}{1+\cos\alpha.\dfrac{2\tan x/2}{1+\tan^2 x/2}}dx$$

$$=\pi\int_0^{\pi/2}\frac{\sec^2}{1+\tan^2 x/2+2\cos\alpha\tan x/2}dx$$

Put $\tan x/2=t$

$\frac{1}{2}\sec^2\frac{x}{2}dx=dt \Rightarrow \sec^2 x/2dx=2dt$

Also when $x\to 0, t\to 0$ as $x\to\pi 2, t\to 1$

$$\therefore\quad I=\pi\int_0^1\frac{2dt}{t^2+(2\cos\alpha)t+1}$$

$$=2\pi\int_0^1\frac{dt}{(t+\cos\alpha)^2+1-\cos^2\alpha}$$

$$=2\pi\int_0^1\frac{dt}{(t+\cos\alpha)^2+\sin^2\alpha}$$

$$=2\pi.\frac{1}{\sin\alpha}\left[\tan^{-1}\left(\frac{t+\cos\alpha}{\sin\alpha}\right)\right]_0^1$$

$$=\frac{2\pi}{\sin\alpha}\left[\tan^{-1}\left(\frac{1+\cos\alpha}{\sin\alpha}\right)-\tan^{-1}\left(\frac{\cos\alpha}{\sin\alpha}\right)\right]$$

$$=\frac{2\pi}{\sin\alpha}\left[\tan^{-1}\left(\frac{2\cos^2\alpha/2}{2\sin\alpha/2\cos\alpha/2}\right)-\tan^{-1}(\cot\alpha)\right]$$

$$=\frac{2\pi}{\sin\alpha}\left[\tan^{-1}(\cot\alpha/2)-\tan^{-1}(\cot\alpha)\right]$$

$$=\frac{2\pi}{\sin\alpha}\left[\tan^{-1}(\tan^{-1}(\pi/2-\alpha/2))-\tan^{-1}(\tan(\pi/2-\alpha))\right]$$

$$=\frac{2\pi}{\sin\alpha}\left[\frac{\pi}{2}-\frac{\alpha}{2}-\frac{\pi}{2}+\alpha\right]=\frac{2\pi}{\sin\alpha}\left[\frac{\alpha}{2}\right]=\frac{\pi\alpha}{\sin\alpha}$$

16. We have to find the area bounded by the curves

$x^2+y^2=25$ ...(1)

$4y=|4-x^2|$ ...(2)

$x=0$ ...(3)

and above x-axis.

Now, $4y=|4x-x^2|=\begin{cases}4-x^2, \text{if } x^2<4\\ x^2-4, \text{if } x^2\geq 4\end{cases}$

$$4y=\begin{cases}4-x^2, \text{if } -2<x<2\\ x^2-4, \text{if } x\geq 2 \text{ or } \leq -2\end{cases}$$

Thus we have three curves

(I) Circle $x^2+y^2=25$

(II) P_1: Parabola, $x^2=-4(y-1), -2\leq x\leq 2$

(III) P_2: Parabola, $x^2=4(y+1), x\geq 2 \text{ or } x\leq -2$

(I) and (II) intersect at $-4y+4+y^2=25$

or $(y-2)^2=5^2 \therefore y-2=\pm 5$

$y=7, y=-3$

$y=-3$, 7 are rejected since.

$y=-3$ is below x-axis and

$y=7$ gives imaginary value of x. So, (I) and (II) do not intersect but II intersects x-axis at (2, 0) and (–2, 0).

(I) and (III) intersect at

$4y+4+y^2=25$ or $(y+2)^2=5^2$

$\therefore\quad y+2=\pm 5 \qquad \therefore\quad y=3,-7.$

$y=-7$ is rejected, $y=3$ gives the points above x-axis. When $y=3$, $x=\pm 4$. Hence the points of intersection of (I) and (III) are (4, 3) and (– 4, 3). Thus we have the shape of the curve as given in figure.

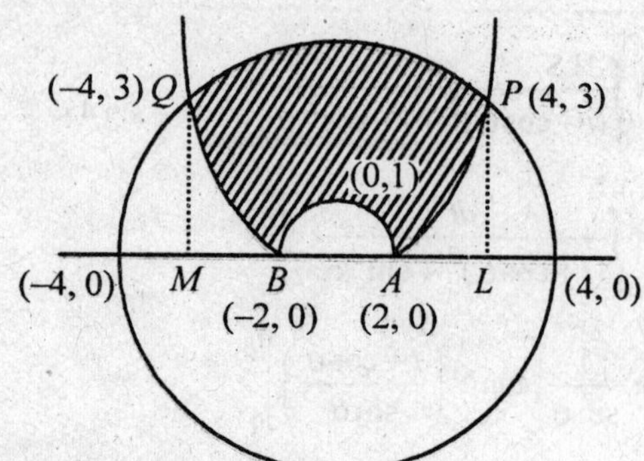

Required area is

$$= 2\left[\int_0^4 y_{circle}dx - \int_0^2 y_{P_1}dx - \int_2^4 y_{P_2}dx\right]$$

$$= 2\left[\int_0^4 \sqrt{25-x^2}\,dx - \frac{1}{4}\int_0^2 (4-x^2)dx - \frac{1}{4}\int_2^4 (x^2-4)dx\right]$$

$$= 2\left[\left[\frac{x}{2}\sqrt{25-x^2} + \frac{25}{2}\sin^{-1}\frac{x}{5}\right]_0^4 - \frac{1}{4}\left(4x - \frac{x^3}{3}\right)_0^2 - \frac{1}{4}\left(\frac{x^3}{3} - 4x\right)_2^4\right]$$

$$= 2\left[6 + \frac{25}{2}\sin^{-1}\frac{4}{5} - \frac{4}{3} - \frac{4}{3} - \frac{4}{3}\right]$$

$$= 12 + 25\sin^{-1}\frac{4}{5} - 8 = 4 + 25\sin^{-1}\frac{4}{5}$$

17. The given curve is $y = \tan x$ ...(1)

Let P be the point on (1) where $x = \pi/4$

$\therefore \quad y = \tan\pi/4 = 1$

i.e. co-ordinates of P are $(\pi/4, 1)$

$\therefore$ Equation of tangent at P is

$y - 1 = 2(x - \pi/4)$

or $y = 2x + 1 - \pi/2$ ...(2)

The graph of (1) and (2) are as shown in the figure.

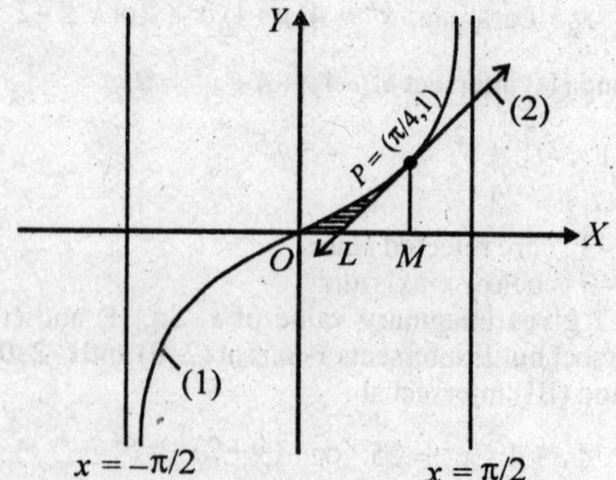

Tangent (2) meets x-axis at, $L\left(\frac{\pi-2}{4}, 0\right)$

Now the required area = shaded area

= Area $OPMO - Ar(\Delta PLM)$

$$= \int_0^{\pi/4} \tan x\,dx - \frac{1}{2}(OM - OL)PM$$

$$= [\log\sec x]_0^{\pi/4} - \frac{1}{2}\left\{\frac{\pi}{4} - \frac{\pi-2}{4}\right\}.1 = \frac{1}{2}\left[\log 2 - \frac{1}{2}\right] \text{ sq.units.}$$

18. Let $I = \int_0^1 1.\log[\sqrt{1-x} + \sqrt{1+x}]\,dx$

Intergrating by parts, we get

$$I = [x\log(\sqrt{1-x} + \sqrt{1+x})]_0^1$$

$$- \int_0^1 x.\frac{1}{\sqrt{1-x}+\sqrt{1+x}}.\left[\frac{-1}{2\sqrt{1-x}} + \frac{1}{2\sqrt{1+x}}\right]dx$$

$$= \log\sqrt{2} - \int_0^1 x\frac{(\sqrt{1+x} - \sqrt{1-x})}{(\sqrt{1+x}+\sqrt{1-x})(\sqrt{1+x}-\sqrt{1-x})} . \frac{(\sqrt{1-x}-\sqrt{1+x})}{2\sqrt{1-x^2}}dx$$

$$= \frac{1}{2}\log 2 + \frac{1}{2}\int_0^1 \frac{x(\sqrt{1+x}-\sqrt{1-x})^2}{(1+x-1+x)\sqrt{1-x^2}}dx$$

$$= \frac{1}{2}\log 2 + \frac{1}{2}\int_0^1 \frac{1+x+1-x-2\sqrt{1-x^2}}{2\sqrt{1-x^2}}dx$$

$$= \frac{1}{2}\log 2 + \frac{1}{2}\int_0^1 \frac{1}{\sqrt{1-x^2}}dx - \frac{1}{2}\int_0^1 1dx$$

$$= \frac{1}{2}\left[\log 2 + \left(\sin^{-1}x\right)_0^1 - (x)_0^1\right] = \frac{1}{2}[\log 2 + \pi/2 - 1]$$

ALTERNATE SOLUTION:

$$I = \int_0^1 \log(\sqrt{1-x} + \sqrt{1+x})dx$$

We can also find the integration by putting $x = \cos\theta$

19. Let $I = \int_0^a f(x)g(x)dx = \int_0^a f(a-x)g(a-x)dx$

[Using the prop. $\int_0^a f(x)dx = \int_0^a f(a-x)dx$]

$$= \int_0^a f(x)(2 - g(x))dx$$

As given that $f(a-x) = f(x)$ and $g(a-x) + g(x) = 2$

$$= 2\int_0^a f(x)dx - \int_0^a f(x)g(x)dx$$

$\therefore \quad I = 2\int_0^a f(x)dx - I$

$\Rightarrow \quad 2I = 2\int_0^a f(x)dx \Rightarrow I = \int_0^a f(x)dx$

Hence the result.

20. We have, $I = \int_0^{\pi/2} f(\sin 2x)\cos x dx$ ...(1)

$I = \int_0^{\pi/2} f(\sin 2x)\sin x dx$ ...(2)

[Using property $\int_0^a f(x)dx = \int_0^a f(a-x)dx$]

Adding (1) and (2), we get

$2I = \int_0^{\pi/2} f(\sin 2x)(\cos x + \sin x)dx$

$\Rightarrow \quad 2I = 2\int_0^{\pi/4} f(\sin 2x)(\sin x + \cos x)dx$

[Using the property,

$\int_0^{2a} f(x)dx = 2\int_0^a f(x)dx$ when $f(2a-x) = f(x)$]

$\Rightarrow \quad I = \int_0^{\pi/4} f(\sin 2x)(\sin x + \cos x)dx$

$= \sqrt{2}\int_0^{\pi/4} f(\sin 2x)\sin(\pi/4 + x)dx$

$= \sqrt{2}\int_0^{\pi/4} f\left[\sin\left(2\left(\frac{\pi}{4} - x\right)\right)\right]\sin(\pi/4 + \pi/4 - x)dx$

$\left[\text{Using the property } \int_0^a f(x)dx = \int_0^a f(a-x)dx\right]$

$= \sqrt{2}\int_0^{\pi/4} f(\cos 2x)\cos x dx$ Hence Proved.

21. To prove : $\frac{\sin 2kx}{\sin x} = 2[\cos x + \cos 3x + ... + \cos(2k-1)x]$

It is equivalent to prove that

$\sin 2kx = 2\sin x\cos x + 2\cos 3x\sin x + ...$

$+ 2\cos(2k-1)x\sin x$

Now, R.H.S.

$= (\sin 2x) + (\sin 4x - \sin 2x) + (\sin 6x - \sin 4x) + ...$

$+ (\sin 2kx - \sin(2k-2)x)$

$= \sin 2kx$ = L.H.S. Hence Proved.

Now $\int_0^{\pi/2} \sin 2kx \cot x dx = \int_0^{\pi/2} \frac{\sin 2kx}{\sin x}.\cos x dx$

$= \int_0^{\pi/2} 2(\cos x + \cos 3x + ... + \cos(2k-1)x)\cos x dx$

[Using the identity proved above]

$= \int_0^{\pi/2} [2\cos^2 x + 2\cos 3x\cos x + 2\cos 5x\cos x + ...$

$+ 2\cos(2k-1)x\cos x]dx$

$= \int_0^{\pi/2} [(1+\cos 2x) + (\cos 4x + \cos 2x)$

$+ (\cos 6x + \cos 4x) + ...$

$+ \{(\cos 2kx) + \cos(2k-2)x\}]dx$

$= \int_0^{\pi/2} 1 + 2[\cos 2x + \cos 4x + \cos 6x + ...$

$+ \cos(2k-2)x] + \cos 2kx\, dx$

$= \left[x + 2\left\{\frac{\sin 2x}{2} + \frac{\sin 4x}{4} + \frac{\sin 6x}{6} + ...\right.\right.$

$\left.\left.+ \frac{\sin(2k-2)x}{2k-2}\right\} + \frac{\sin 2kx}{2k}\right]_0^{\pi/2}$

$= \pi/2$ $[\because \sin n\pi = 0, \forall n \in N]$

Hence Proved

22. The given curves are

$y = ex\log_e x$ ...(1)

and $y = \frac{\log_e x}{ex}$ ...(2)

The two curves intersect where $ex\log x = \frac{\log x}{ex}$

$\Rightarrow \left(ex - \frac{1}{ex}\right)\log x = 0 \Rightarrow x = \frac{1}{e}$ or $x = 1$

At $x = 1/e$ or $ex = 1$, $\log x = -\log e = -1, y = -1$

So that $\left(\frac{1}{e}, -1\right)$ is one point of intersection and at $x = 1$,

$\log 1 = 0 \therefore y = 0$

$\therefore$ (1, 0) is the other common point of intersection of the curves. Now in between these two points, $\frac{1}{e} < x < 1$

or $\log\left(\frac{1}{e}\right) < \log x < \log 1$, or $-1 < \log x < 0$

i.e. $\log x$ is – ve, throughout

$\therefore \quad y_1 = ex \log_e x, y_2 = \frac{\log_e x}{ex}$

Clearly under the condition stated above $y_1 < y_2$ both being –ve in the interval $\frac{1}{e} < x < 1$.

The rough sketch of the two curves is as shown in fig. and shaded area is the required area.

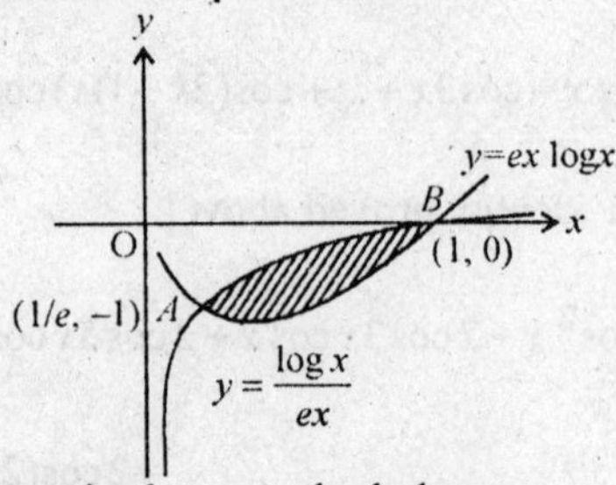

$\therefore$ The required area = shaded area

$$= \left|\int_{1/e}^{1} (y_1 - y_2)\,dx\right| = \left|\int_{1/e}^{1}\left[ex\log x - \frac{\log x}{ex}\right]dx\right|$$

$$= \left|e\int_{1/e}^{1} x\log x - \frac{1}{e}\int_{1/e}^{1}\frac{\log x}{x}dx\right|$$

$$= \left|e\left[\frac{x^2}{2}\log x - \frac{x^2}{4}\right]_{1/e}^{1} - \frac{1}{e}\left[\frac{(\log x)^2}{2}\right]_{1/e}^{1}\right|$$

$$= \left|e\left[\left(-\frac{1}{4}\right) - \left(-\frac{1}{2e^2} - \frac{1}{4e^2}\right)\right] - \frac{1}{e}\left[0 - \frac{1}{2}\right]\right|$$

$$= \left|e\left[-\frac{1}{4} + \frac{3}{4e^2}\right] + \frac{1}{2e}\right| = \left|\frac{-e}{4} + \frac{3}{4e} + \frac{1}{2e}\right|$$

$$= \left|\frac{5}{4e} - \frac{e}{4}\right| = \left|\frac{5-e^2}{4e}\right| = \frac{e^2-5}{4e}$$

23. The given curves are

$x = \frac{1}{2}$ ... (1), $x = 2$...(2), $y = \ln x$...(3), $y = 2^x$...(4)

Clearly (1) and (2) represent straight lines parallel to y-axis at distances 1/2 and 2 units from it, respectively. Line $x = \frac{1}{2}$ meets (3) at $\left(\frac{1}{2}, -\ln 2\right)$ and (4) at $\left(\frac{1}{2}, \sqrt{2}\right)$. Line $x = 2$ meets (3) at (2, ln 2) and (4) at (2, 4).

The graph of curves are as shown in the figure.

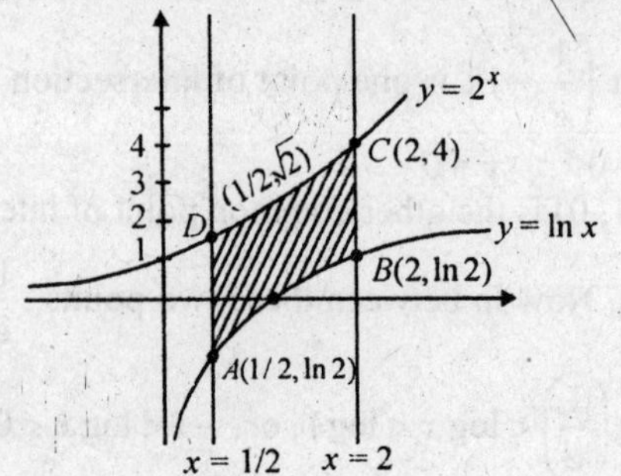

Required area = ABCDA

$$= \int_{1/2}^{1} (-\ln x)dx + \int_{1/2}^{2} 2^x\,dx - \int_{1}^{2}\ln x\,dx$$

$$= \int_{1/2}^{2} 2^x\,dx - \int_{1/2}^{2}\ln x\,dx = \int_{1/2}^{2}(2^x - \ln x)\,dx$$

$$= \left[\frac{2^x}{\log 2} - (x\log x - x)\right]_{1/2}^{2}$$

$$= \left(\frac{4}{\log 2} - 2\log 2 + 2\right) - \left(\frac{\sqrt{2}}{\log 2} + \frac{1}{2}\log 2 - \frac{1}{2}\right)$$

$$= \left(\frac{4-\sqrt{2}}{\log 2} - \frac{5}{2}\log 2 + \frac{3}{2}\right)$$

24. We are given that f is a continuous function and

$$\int_0^x f(t)dt \to \infty \text{ as } |x| \to \infty$$

To show that every line $y = mx$ intersects the curve

$$y^2 + \int_0^x f(t)dt = 2.$$

If possible, let $y = mx$ intersects the given curve, then Substituting $y = mx$ in the equation of the curve we get

$$m^2x^2 + \int_0^x f(t)dt = 2 \quad \text{...........(1)}$$

Consider $F(x) = m^2x^2 + \int_0^x f(t)dt - 2$

Then $F(x)$ is a continuous function as $f(x)$ is given to be continuous.

Also $F(x) \to \infty$ as $|x| \to \infty$

But $F(0) = -2$

Thus $F(0) = -$ve and F(b) = +ve where b is some value of x, and $F(x)$ is continuous.

Therefore $F(x) = 0$ for some value of $x \in (0, b)$ or eq. (1) is solvable for x.

Hence $y = mx$ intersects the given curve.

25. Let $I = \int_0^{\pi} \frac{x\sin(2x)\sin\left(\frac{\pi}{2}\cos x\right)}{2x-\pi}dx$

Consider, $2x - \pi = y$ n$\Rightarrow$ $dx = \frac{dy}{2}$

Also, $x = \left(\frac{\pi}{2} + \frac{y}{2}\right)$

When $x \to 0, y \to -\pi$ when $x \to \pi, y \to \pi$

$\therefore$ We get

$$I = \int_{-\pi}^{\pi} \frac{\left(\frac{\pi+y}{2}\right)\sin(\pi+y)\sin\left[\frac{\pi}{2}\cos\left(\frac{\pi}{2}+\frac{y}{2}\right)\right]}{y}\cdot\frac{dy}{2}$$

$$= \frac{1}{4}\int_{-\pi}^{\pi}\left(\frac{\pi}{y}+1\right)(-\sin y)\sin\left(\frac{-\pi}{2}\sin\frac{y}{2}\right)dy$$

$$= \frac{\pi}{4}\int_{-\pi}^{\pi}\frac{\sin y\sin(\pi/2\sin y/2)}{y}dy$$

$$+\frac{1}{4}\int_{-\pi}^{\pi}\sin y\sin\left(\frac{\pi}{2}\sin\frac{y}{2}\right)dy$$

$$= 0+\frac{2}{4}\int_{0}^{\pi}\sin y\sin(\pi/2\sin y/2)\,dy$$

[Using $\int_{-a}^{a} f(x)\,dx = 0$ if f is odd function

$= 2\int_{0}^{a} f(x)\,dx$ if f *is an even function*]

$$\therefore I = \frac{1}{2}\int_{0}^{\pi} 2\sin y/2\cos y/2\sin(\pi/2\sin y/2)\,dy$$

Let $\sin y/2 = u \Rightarrow \frac{1}{2}\cos y/2\,dy = du$

$\Rightarrow \cos y/2\,dy = 2du$

Also as $y\to 0, u\to 0$ and as $y\to\pi, u\to 1$

$$\therefore I = \int_{0}^{1} 2u\sin\left(\frac{\pi u}{2}\right)du$$

$$= \left[2u\frac{-\cos\frac{\pi u}{2}}{\pi/2}\right]_0^1 + \int_0^1 2.\frac{2}{\pi}\cos\left(\frac{\pi u}{2}\right)du$$

$$= 0+\frac{4}{\pi}\left.\frac{\sin\left(\frac{\pi u}{2}\right)}{\pi/2}\right|_0^1 = \frac{8}{\pi^2}\left(\sin\frac{\pi}{2}-0\right) = \frac{8}{\pi^2}$$

26. The given curves are $y = x^2$ and $y = \frac{2}{1+x^2}$. Here $y = x^2$ is upward parabola with vertex at origin.

Also, $y = \frac{2}{1+x^2}$ is a curve symm. with respect to y-axis.

At $x = 0, y = 2$,

$\frac{dy}{dx} = \frac{-4x}{(1+x^2)^2} < 0$ for $x > 0$

$\therefore$ Curve is decreasing on $(0,\infty)$

Moreover $\frac{dy}{dx} = 0$ at $x = 0$

$\Rightarrow$ At (0,2) tangent to curve is parallel to x – axis.

As $x\to\infty$, $y\to 0$

$\therefore$ y = 0 is asymptote of the given curve.

For the given curves, point of intersection : solving their equations we get x = 1, y = 1, i.e., (1,1).

Thus the graph of two curves is as follows:

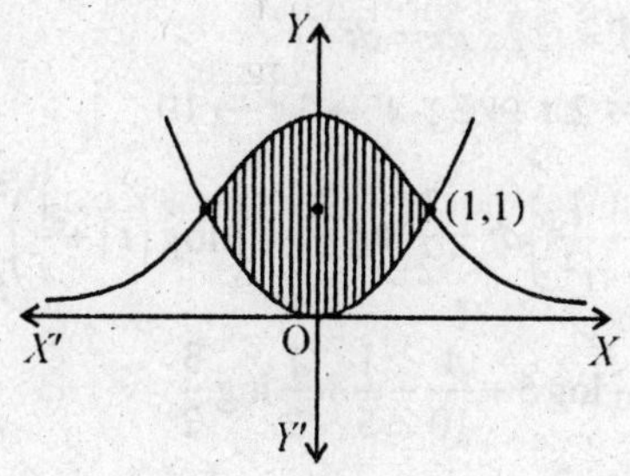

$$\therefore \text{The required area} = 2\int_0^1\left(\frac{2}{1+x^2}-x^2\right)dx$$

$$= \left(4\tan^{-1}x-\frac{2x^3}{3}\right)_0^1 = 4.\frac{\pi}{4}-\frac{2}{3} = \pi-\frac{2}{3}\text{ sq. units.}$$

27. Given that $\int_0^1 e^x(x-1)^n\,dx = 16-6e$

where $n\in N$ and $n\le 5$

To find the value of n.

Let $I_n = \int_0^1 e^x(x-1)^n\,dx$

$$= [(x-1)^n e^x]_0^1 - \int_0^1 n(x-1)^{n-1}e^x\,dx$$

$$= -(-1)^n - \int_0^1 n(x-1)^{n-1}e^x\,dx$$

$$I_n = (-1)^{n+1} - n\,I_{n-1} \qquad ....(1)$$

Also, $I_1 = \int_0^1 e^x(x-1)\,dx$

$$= [e^x(x-1)]_0^1 - \int_0^1 e^x\,dx = -(-1)-(e^x)_0^1$$

$$= -(e-1) = 2-e$$

Using eq. (1), $I_2 = (-1)^3 - 2I_1 = -1-2(2-e) = 2e-5$

Similarly, $I_3 = (-1)^4 - 3I_2 = 1-3(2e-5) = 16-6e$

$\therefore n = 3$

28. $$I = \int_2^3 \frac{2x^5+x^4-2x^3+2x^2+1}{(x^2+1)(x^4-1)}dx$$

$$= \int_2^3 \frac{2x^5-2x^3+x^4+2x^2+1}{(x^2+1)^2(x^2-1)}dx$$

$$= \int_2^3 \frac{2x^3(x^2-1)+(x^2+1)^2}{(x^2+1)^2(x^2-1)}dx$$

$$= \int_2^3 \frac{2x^3}{(x^2+1)^2} + \int_2^3 \frac{1}{x^2-1}dx$$

$$= \int_2^3 \frac{x^2.2x}{(x^2+1)^2} + \left[\frac{1}{2}\log\frac{x-1}{x+1}\right]_2^3$$

$$= \int_5^{10}\frac{t-1}{t^2}dt + \frac{1}{2}\left(\log\frac{2}{4}-\log\frac{1}{3}\right)$$

Put $x^2+1=t,\ 2x\,dx=dt$

when $x\to 2, t\to 5$, $x\to 3, t\to 10$

$$=\int_5^{10}\left(\frac{1}{t}-\frac{1}{t^2}\right)dt+\frac{1}{2}\log\frac{3}{2}=\left(\log|t|+\frac{1}{t}\right)_5^{10}+\frac{1}{2}\log\frac{3}{2}$$

$$=\log 10-\log 5+\frac{1}{10}-\frac{1}{5}+\frac{1}{2}\log\frac{3}{2}$$

$$=\log 2+\left(-\frac{1}{10}\right)+\frac{1}{2}\log\frac{3}{2}=\frac{1}{2}\left[2\log 2+\log\frac{3}{2}\right]-\frac{1}{10}$$

$$=\frac{1}{2}\log 6-\frac{1}{10}$$

29. To prove that

$$\int_0^{n\pi+v}|\sin x|\,dx=2n+1-\cos v$$

Let $I=\int_0^{n\pi+v}|\sin x|\,dx$

$$=\int_0^{v}|\sin x|\,dx+\int_v^{n\pi+v}|\sin x|\,dx$$

Now we know that $|\sin x|$ is a periodic function of period π, So using the property.

$$=\int_a^{a+nT}f(x)dx=n\int_0^{T}f(x)dx$$

where $n\in I$ and $f(x)$ is a periodic function of period T

We get, $I=\int_0^{v}\sin x\,dx+n\int_0^{\pi}\sin x\,dx$

$[\because\ |\sin x|=\sin x$ for $0\le x\le v]$

$$=(-\cos x)_0^v+n(-\cos x)_0^\pi=-\cos v+1+n(1+1)$$

$=2n+1-\cos v=$ R.H.S.

30. The given equations of parabola are

$y=4x-x^2$ or $(x-2)^2=-(y-4)$ ...... (1)

and $y=x^2-x$ or $\left(x-\frac{1}{2}\right)^2=\left(y+\frac{1}{4}\right)$ ...... (2)

Solving the equations of two parabolas we get their points of intersection as $O(0,0), A\left(\frac{5}{2},\frac{15}{4}\right)$

Here the area below $x-$axis,

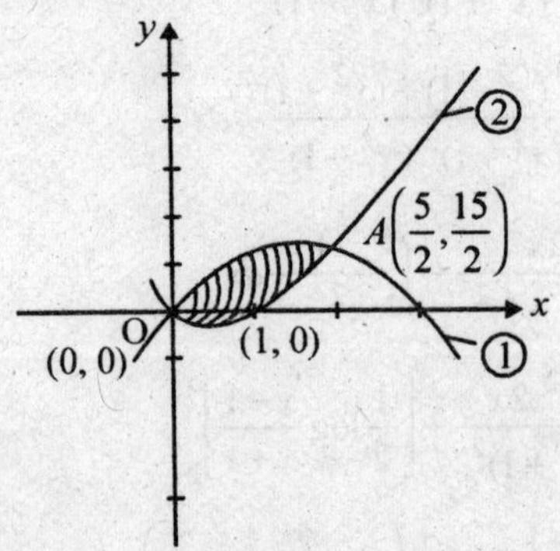

$$A_1=\int_0^1(-y_2)\,dx=\int_0^1(x-x^2)dx$$

$$=\left(\frac{x^2}{2}-\frac{x^3}{3}\right)_0^1=\frac{1}{2}-\frac{1}{3}=\frac{1}{6}\text{ sq. units.}$$

Area above $x-$axis,

$$A_2=\int_0^{5/2}y_1\,dx-\int_1^{5/2}y_2\,dx$$

$$=\int_0^{5/2}(4x-x^2)\,dx-\int_1^{5/2}(x^2-x)\,dx$$

$$=\left(2x^2-\frac{x^3}{3}\right)_0^{5/2}-\left(\frac{x^3}{3}-\frac{x^2}{2}\right)_1^{5/2}$$

$$=\left(\frac{25}{2}-\frac{125}{24}\right)-\left[\left(\frac{125}{24}-\frac{25}{8}\right)-\left(\frac{1}{3}-\frac{1}{2}\right)\right]$$

$$=\frac{25}{2}-\frac{125}{12}+\frac{25}{8}-\frac{1}{6}=\frac{300-250+75-4}{24}=\frac{121}{24}$$

$\therefore$ Ratio of areas above $x-$axis and below $x-$axis.

$$A_2:A_1=\frac{121}{24}:\frac{1}{6}=\frac{121}{4}=121:4$$

31. Given $I_m=\int_0^{\pi}\frac{1-\cos mx}{1-\cos x}dx$

To prove: $I_m=m\pi,\ m=0,1,2,......$

For m = 0

$$I_0=\int_0^{\pi}\frac{1-\cos 0}{1-\cos x}dx=\int_0^{\pi}\frac{1-1}{1-\cos x}dx=0$$

$\because$ Result is true for m = 0

For m = 1,

$$I_1=\int_0^{\pi}\frac{1-\cos x}{1-\cos x}dx=\int_0^{\pi}1.dx$$

$$(x)_0^\pi=\pi-0=\pi$$

$\therefore$ Result is true for $m=1$

Let the result be true for $m\le k$ i.e. $I_k=k\pi$ ....(1)

Consider $I_{k+1}=\int_0^{\pi}\frac{1-\cos(k+1)x}{1-\cos x}dx$

Now, $1-\cos(k+1)x$

$=1-\cos kx\cos x+\sin kx\sin x$

$=1+\cos kx\cos x+\sin kx\sin x-2\cos kx\cos x$

$=1+\cos(k-1)x-2\cos kx\cos x$

$=2-(1-\cos(k-1)x)-2\cos kx\cos x$

$=2(1-\cos kx\cos x)-(1-\cos(k-1)x)$

$=2-2\cos kx+2\cos kx-2\cos kx\cos x-[1-\cos(k-1)x]$

$=2(1-\cos kx)+2\cos kx(1-\cos x)-(1-\cos(k-1)x)$

$$\therefore\ I_{k+1}=\int_0^{\pi}\frac{2(1-\cos kx)+2\cos kx(1-\cos x)-(1-\cos(k-1)x)}{1-\cos x}dx$$

$$=2\int_0^{\pi}\frac{1-\cos kx}{1-\cos x}dx+2\int_0^{\pi}\cos kx\,dx-\int_0^{\pi}\frac{1-\cos(k-1)x}{1-\cos x}dx$$

$$= 2I_k + 2\left(\frac{\sin kx}{k}\right)_0^{\pi} - I_{k-1}$$

$$= 2(k\pi) + 2(0) - (k-1)\pi \qquad \text{[Using (i)]}$$

$$= 2k\pi - k\pi + \pi = (k+1)\pi$$

Thus result is true for $m = k + 1$ as well. Therefore by the principle of mathematical induction, given statement is true for all $m = 0, 1, 2, \ldots\ldots\ldots\ldots$

32. Let $I = \int_{-1/\sqrt{3}}^{1/\sqrt{3}} \frac{x^4}{1-x^4} \cos^{-1}\left(\frac{2x}{1+x^2}\right) dx$

Put $x = -y$, so that $dx = -dy$ and

$$I = \int_{1/\sqrt{3}}^{-1/\sqrt{3}} \frac{y^4}{1-y^4} \cos^{-1}\left(\frac{-2y}{1+y^2}\right)(-1)dy$$

But $\cos^{-1}(-x) = \pi - \cos^{-1} x$ for $-1 \le x \le 1$,

$$I = \int_{-1/\sqrt{3}}^{1/\sqrt{3}} \frac{y^4}{1-y^4}\left[\pi - \cos^{-1}\left(\frac{2y}{1+y^2}\right)\right] dy$$

$$\pi = \int_{-1/\sqrt{3}}^{1/\sqrt{3}} \frac{x^4}{1-x^4} dx - \int_{-1/\sqrt{3}}^{1/\sqrt{3}} \frac{x^4}{1-x^4} \cos^{-1}\left(\frac{2x}{1+x^2}\right) dx$$

$$I = \pi \int_{-1/\sqrt{3}}^{1/\sqrt{3}} \frac{x^4}{1-x^4} dx - I$$

[It is immaterial which variable we use in evaluating a definite integral.]

$$\Rightarrow 2I = \pi \int_{-1/\sqrt{3}}^{1/\sqrt{3}} \frac{x^4}{1-x^4} dx = \pi \int_{-1/\sqrt{3}}^{1/\sqrt{3}} \left[-1 + \frac{1}{1-x^4}\right] dx$$

$$= -\pi \int_{-1/\sqrt{3}}^{1/\sqrt{3}} dx + \pi \int_{-1/\sqrt{3}}^{1/\sqrt{3}} \frac{dx}{1-x^4} = -\pi\left(\frac{1}{\sqrt{3}} + \frac{1}{\sqrt{3}}\right) + \pi I_1$$

where $I_1 = \pi \int_{-1/\sqrt{3}}^{1/\sqrt{3}} \frac{dx}{1-x^4} = 2\int_0^{1/\sqrt{3}} \frac{dx}{1-x^4}$

[$\because$ the integrand is an even function]

$$= \int_0^{1/\sqrt{3}} \left(\frac{1}{1-x^2} + \frac{1}{1+x^2}\right) dx$$

$$= \int_0^{1/\sqrt{3}} \left[\frac{1}{2}\frac{1}{1-x} + \frac{1}{2}\frac{1}{1+x} + \frac{1}{1+x^2}\right] dx$$

[resolving the integrand into partial fractions]

$$= \left[\left(-\frac{1}{2}\log_e|1-x| + \frac{1}{2}\log_e|1+x| + \tan^{-1} x\right)\right]_0^{1/\sqrt{3}}$$

$$= \frac{1}{2}\log_e\left|\frac{1+1/\sqrt{3}}{1-1/\sqrt{3}}\right| + \tan^{-1}(1/\sqrt{3}) = \frac{1}{2}\log_e\left|\frac{\sqrt{3}+1}{\sqrt{3}-1}\right| + \frac{\pi}{6}$$

$$= \frac{1}{2}\log_e \frac{(\sqrt{3}+1)^2}{3-1} + \frac{\pi}{6} = \frac{1}{2}\log_e(2+\sqrt{3}) + \frac{\pi}{6}$$

Then, $2I = \frac{-2\pi}{\sqrt{3}} + \frac{\pi}{2}\log_e(2+\sqrt{3}) + \frac{\pi^2}{6}$

$$= \frac{\pi}{6}[\pi + 3\log_e(2+\sqrt{3}) - 4\sqrt{3}]$$

$$\Rightarrow \quad I = \frac{\pi}{12}[\pi + 3\log_e(2+\sqrt{3}) - 4\sqrt{3}]$$

ALTERNATE SOLUTION :

Let $I = \int_{-1/\sqrt{3}}^{1/\sqrt{3}} \left(\frac{x^4}{1-x^4}\right) \cos^{-1}\left(\frac{2x}{1+x^2}\right) dx$

We know that $\sin^{-1}\left(\frac{2x}{1+x^2}\right) = 2\tan^{-1} x$

Also $\sin^{-1} y + \cos^{-1} y = \frac{\pi}{2}$

$\therefore$ We get $\frac{\pi}{2} - \cos^{-1}\left(\frac{2x}{1+x^2}\right) = 2\tan^{-1} x$

$$\Rightarrow \cos^{-1}\left(\frac{2x}{1+x^2}\right) = \frac{\pi}{2} - 2\tan^{-1} x$$

$$\therefore I = \int_{-1/\sqrt{3}}^{1/\sqrt{3}} \left(\frac{x^4}{1-x^4}\right)\left[\frac{\pi}{2} - 2\tan^{-1} x\right] dx$$

$$= \frac{\pi}{2}\int_{-1/\sqrt{3}}^{1/\sqrt{3}} \frac{x^4}{1-x^4} dx - 2\int_{-1/\sqrt{3}}^{1/\sqrt{3}} \frac{x^4 \tan^{-1} x}{1-x^4} dx$$

$$= 2.\frac{\pi}{2}\int_0^{1/\sqrt{3}} \frac{x^4}{1-x^4} dx - 2\times 0$$

$= \left[\text{Using } \int_{-a}^{a} f(x)dx = 2\int_0^a f(x)dx \text{ if } f \text{ is even}\right.$

$= 0$ if f is odd

$$= \pi\int_0^{1/\sqrt{3}} \frac{x^4}{1-x^4} dx$$

$$\therefore I = -\pi\int_0^{1/\sqrt{3}} \frac{(1-x^4)-1}{1-x^4} dx$$

$$= -\pi\int_0^{1/\sqrt{3}} 1 - \frac{1}{1-x^4} dx = -\pi\int_0^{1/\sqrt{3}} \left[1 - \frac{1}{2}\left(\frac{1}{1-x^2} + \frac{1}{1+x^2}\right)\right] dx$$

$$= -\pi\left[x - \frac{1}{2}\left(\frac{1}{2}\log\left|\frac{1+x}{1+x}\right| + \tan^{-1} x\right)\right]_0^{1/\sqrt{3}}$$

$$= -\pi\left[\frac{1}{\sqrt{3}} - \frac{1}{2}\left(\frac{1}{2}\log\left|\frac{1+1/\sqrt{3}}{1-1/\sqrt{3}}\right| - \tan^{-1}\frac{1}{\sqrt{3}}\right) - 0\right]$$

$$= -\pi\left[\frac{1}{\sqrt{3}} - \frac{1}{4}\log\left(\frac{\sqrt{3}+1}{\sqrt{3}-1}\right) - \frac{\pi}{12}\right]$$

$$= \pi\left[\frac{\pi}{12} + \frac{1}{4}\log(2+\sqrt{3}) - \frac{\sqrt{3}}{3}\right]$$

$$= \frac{\pi}{12}[\pi + 3\log(2+\sqrt{3}) - 4\sqrt{3}]$$

33. Let us consider any point $P(x, y)$ inside the square such that its distance from origin $\leq$ its distance from any of the edges say AD

$\therefore OP \leq PM$ or $\sqrt{(x^2+y^2)} < 1-x$

or $\quad y^2 \leq -2\left(x-\frac{1}{2}\right)$ .... (1)

Above represents all points within and on the parabola 1. If we consider the edges BC then $OP < PN$ will imply

$y^2 \leq 2\left(x+\frac{1}{2}\right)$ .... (2)

Similarly if we consider the edges AB and CD, we will have

$x^2 \leq -2\left(y-\frac{1}{2}\right)$ .... (3)

$x^2 \leq 2\left(y+\frac{1}{2}\right)$ .... (4)

Hence S consists of the region bounded by four parabolas meeting the axes at

$\left(\pm\frac{1}{2}, 0\right)$ and $\left(0, \pm\frac{1}{2}\right)$

The point L is intersection of P_1 and P_3 given by (1) and (3).

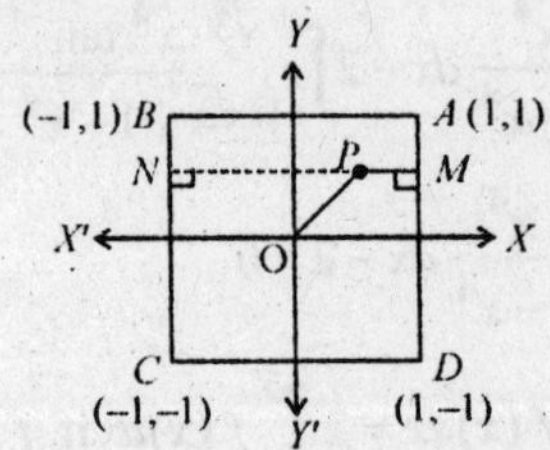

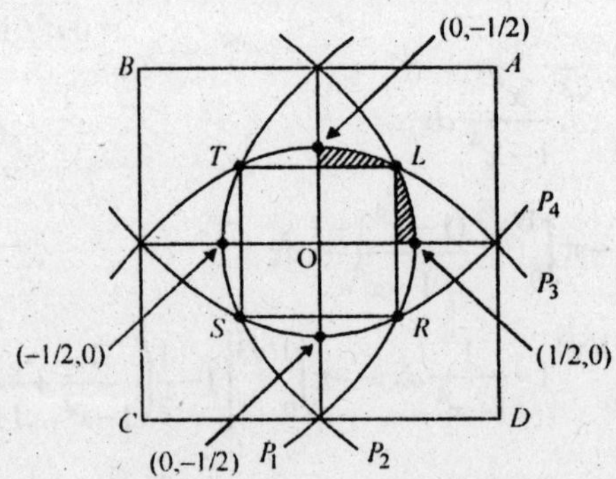

$y^2 - x^2 = -2(x-y) = 2(y-x)$

$\therefore \quad y-x=0 \therefore y=x$

$\therefore \quad x^2+2x-1=0$

$(x+1)^2=2$

$\therefore \quad x=\sqrt{2}-1$ as x is +ve

$\therefore \quad L$ is $(\sqrt{2}-1, \sqrt{2}-1)$

$\therefore$ Total area $= 4\left[\text{square of side}(\sqrt{2}-1)+2\int_{\sqrt{2}-1}^{1/2} y\,dx\right]$

$= 4\left\{(\sqrt{2}-1)^2+2\int_{\sqrt{2}-1}^{1/2}\sqrt{(1-2x)}\,dx\right\}$

$= 4\left[3-2\sqrt{2}-\frac{2}{2}\cdot\frac{2}{3}\{(1-2x)^{3/2}\}_{\sqrt{2}-1}^{1/2}\right]$

$= 4\left[3-2\sqrt{2}-\frac{2}{3}\{0-(1-2\sqrt{2}+2)^{3/2}\right]$

$= 4\left[3-2\sqrt{2}+\frac{2}{3}(3-2\sqrt{2})^{3/2}\right]$

$= 4(3-2\sqrt{2})\left[1+\frac{2}{3}\sqrt{(3-2\sqrt{2})}\right]$

$= 4(3-2\sqrt{2})\left[1+\frac{2}{3}(\sqrt{2}-1)\right]$

$= \frac{4}{3}(3-2\sqrt{2})(1+2\sqrt{2}) = \frac{4}{3}[(4\sqrt{2}-5)] = \frac{16\sqrt{2}-20}{3}$

34. We have

$A_n = \int_0^{\pi/4}(\tan x)^n\,dx$

Since $0 < \tan x < 1$, when $0 < x < \pi/4$, we have

$0 < (\tan x)^{n+1} < (\tan x)^n$ for each $n \in N$

$\Rightarrow \int_0^{\pi/4}(\tan x)^{n+1}dx < \int_0^{\pi/4}(\tan x)^n dx$

$\Rightarrow A_{n+1} < A_n$

Now, for $n > 2$

$A_n + A_{n+2} = \int_0^{\pi/4}[(\tan x)^n + (\tan x)^{n+2}]\,dx$

$= \int_0^{\pi/4}(\tan x)^n + (1+\tan^2 x)\,dx$

$= \int_0^{\pi/4}(\tan x)^n + (\sec^2 x)\,dx$

$= \left[\frac{1}{(n+1)}(\tan x)^{n+1}\right]_0^{\pi/4}$

$\left[\because \int [f(x)]^n f'(x)dx = \frac{[f(x)]^{n+1}}{n+1}\right]$

$= \frac{1}{(n+1)}(1-0)$

Since $A_{n+2} < A_{n+1} < A_n$. we get

$A_n + A_{n+2} < 2A_n$

$\Rightarrow \frac{1}{n+1} < 2A_n \Rightarrow \frac{1}{2n+2} < A_n$ ....(1)

Also for $n > 2, A_n + A_n < A_n + A_{n-2} = \frac{1}{n-1}$

$\Rightarrow 2A_n < \frac{1}{n-1}$

$\Rightarrow A_n < \frac{1}{2n-2}$ ....(2)

Combining (1) and (2) we get

$\frac{1}{2n+2} < A_n < \frac{1}{2n-2}$ *Hence Proved.*

35. Let, $I = \int_0^{\pi/4} \ln(1+\tan x)dx$ ....(1)

$= \int_0^{\pi/4} \ln(1+\tan(\pi/4 - x))dx$

$\left[\text{Using} \int_0^a f(x)dx = \int_0^a f(a-x)\,dx\right]$

$= \int_0^{\pi/4} \ln\left[1+\frac{1-\tan x}{1+\tan x}\right]dx$

$= \int_0^{\pi/4} \ln\left(\frac{2}{1+\tan x}\right)dx$

$I = \int_0^{\pi/4} [\ln 2 - \ln(1+\tan x)]dx$ ....(2)

Adding equation (1) and (2) we get

$2I = \int_0^{\pi/4} \ln 2dx = \ln 2[x]_0^{\pi/4} = \ln 2.\left[\frac{\pi}{4}\right] \Rightarrow I = \frac{\pi}{8}\ln 2$

36. The given curves are

$y = x - bx^2$ ....(1)

and $y = x^2/b$ ....(2)

$\Rightarrow \left(y - \frac{1}{4b}\right) = -b\left(x - \frac{1}{2b}\right)^2$ and $x^2 = by$

Here clearly first curve is a downward parabola with vertex at $\left(\frac{1}{2b}, \frac{1}{4b}\right)$ and meeting x – axis at (0, 0) and $(1/b, 0)$

while second is an upward parabola with vertex at (0, 0).

Solving (1) and (2) we get the intersection points (0, 0) and $\left(\frac{b}{1+b^2}, \frac{b}{(1+b^2)^2}\right)$ of two curves.

Hence, the graph of given curves is as below

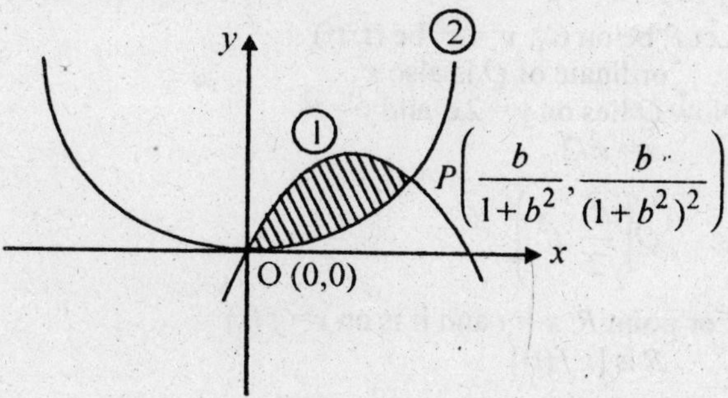

Shaded portion represents the required area, given by

$A = \int_0^{\frac{b}{1+b^2}} \left(x - bx^2 - \frac{x^2}{b}\right)dx$

$\Rightarrow A = \left(\frac{x^2}{2} - \frac{bx^3}{3} - \frac{x^3}{3b}\right)_0^{\frac{b}{1+b^2}} = \frac{x^2}{2} - \frac{b^2+1}{b}.\frac{x^3}{3}$

$= \frac{b^2}{2(1+b^2)^2} - \frac{b^2}{3(1+b^2)^2} = \frac{b^2}{6(1+b^2)^2}$

For area to be max/min we should have $\frac{dA}{db} = 0$

$\Rightarrow \frac{1}{6}[2b(1+b^2)^2 - 2(1+b^2).2b.b^2] = 0$

$2b(1+b^2)(1+b^2-2b^2) = 0$

$\Rightarrow 1 - b^2 = 0 \Rightarrow b = 1, -1$

but given that $b > 0$

$\therefore b = 1$

37. OAB is given Δ. Consider OI the $\angle$ bisector of $\angle BOA$. We know any point on OI must be at equal distances from OB and OA.

Thus for $d(P, OA) \le \min[d(P, OB), d(P, AB)]$, point P must lie under OI and similarly under AI.

Thus within or on $\Delta\, OIA$

$\therefore$ Req, area $= Ar(\Delta OIA)$

Now, $\tan\angle BOA = \frac{1/\sqrt{3}}{1} = \frac{1}{\sqrt{3}}$

$\Rightarrow \angle BOA = 30^\circ$

$\therefore \angle IOA = 15^\circ$

Similarly, $\angle IAO = 15^\circ$

$\therefore$ Eq^n of OI is $y = (\tan 15)x$ ....(1)

Eq^n of IA is $y = (-\tan 15)(x-2)$ ....(2)

with point of intersection $I(1, \tan 15)$

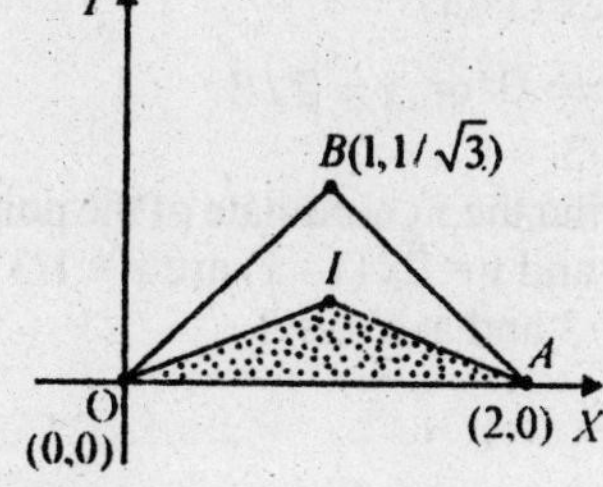

$\therefore$ Required area

$= \int_0^1 (\tan 15)xdx + \int_1^2 \{(-\tan 15)(x-2)\}dx$

$= \tan 15\left[\left(\frac{x^2}{2}\right)_0^1 + \left(2x - \frac{x^2}{2}\right)_1^2\right]$

$= \tan 15\left[\left(\frac{1}{2}\right) + 4 - \frac{4}{2} - 2 + \frac{1}{2}\right]$

$= \tan 15[1] = \tan 15^\circ = 2 - \sqrt{3}$ sq. units.

38. $\int_{-\pi}^{\pi}\frac{2x(1+\sin x)}{1+\cos^2 x}dx = I$ (say)

or $I = \int_{-\pi}^{\pi}\frac{2x}{1+\cos^2 x}dx + \int_{-\pi}^{\pi}\frac{2x\sin x}{1+\cos^2 x}dx$

$I = 0 + 2\int_0^{\pi}\frac{2x\sin x}{1+\cos^2 x}dx \quad \left[\because \frac{2x}{1+\cos^2 x} \text{ is an odd function}\right]$

or $I = 4\int_0^{\pi}\frac{x\sin x}{1+\cos^2 x}dx$ .....(1)

or $I = 4\int_0^{\pi}\frac{(\pi - x)\sin(\pi - x)}{1+\{\cos(\pi - x)\}^2}dx$

$= 4\int_0^{\pi}\frac{(\pi - x)\sin x}{1+\cos^2 x}dx$

or $I = 4\pi\int_0^{\pi}\frac{\sin x}{1+\cos^2 x}dx - \int_0^{\pi}\frac{x\sin x}{1+\cos^2 x}dx$

or $I = 4\pi\int_0^{\pi}\frac{\sin x}{1+\cos^2 x}dx - 1$ [from (1)]

or $2I = 4\pi\int_0^{\pi}\frac{\sin x}{1+\cos^2 x}dx$

Putting $\cos x = t$, $-\sin x\, dx = dt$

When $x \to 0, t \to 1$ and when $x \to \pi, t \to -1$

$\Rightarrow I = 2\pi\int_1^{-1}\frac{-dt}{1+r^2} = 2\pi\int_{-1}^{1}\frac{dt}{1+t^2} = 4\pi\int_0^1\frac{dt}{1+t^2}$

$\Rightarrow I = 4\pi\left(\tan^{-1} t\right)_0^1 = 4\pi\{\tan^{-1}(1) - \tan^{-1}(0)\}$

$\Rightarrow I = 4\pi\left\{\frac{\pi}{4} - 0\right\} = \pi^2$

39. We draw the graph of $y = x^2$, $y = (1-x)^2$ and $y = 2x(1-x)$ in figure.

Let us find the point of intersection of $y = x^2$ and $y = 2x(1-x)$

The x – coordinate of the point of intersection satisfies the equation $x^2 = 2x(1-x)$.

$\Rightarrow 3x^2 = 2x \Rightarrow 0$ or $x = 2/3$

$\therefore$ At B, $x = 2/3$

Similarly, we find the x coordinate of the points of intersection of $y = (1-x)^2$ and $y = 2x(1-x)$ are $x = 1/3$ and $x = 1$

$\therefore$ At A, $x = 1/3$ and at C $x = 1$

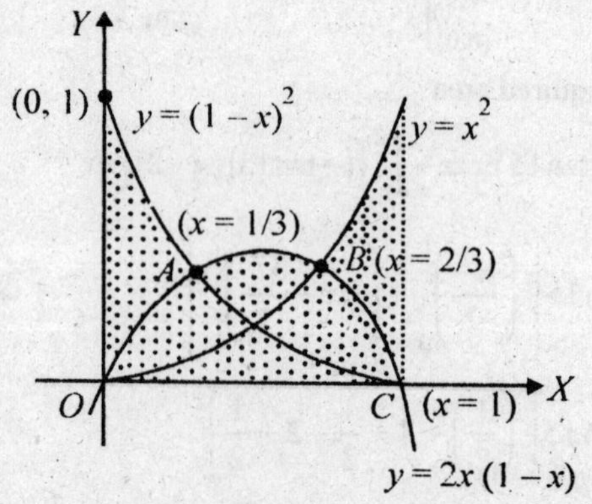

From the figure it is clear that

$$f(x) = \begin{cases} (1-x)^2 & \text{for } 0 \le x \le 1/3 \\ 2x(1-x) & \text{for } 1/3 \le x \le 2/3 \\ x^2 & \text{for } 2/3 \le x \le 1 \end{cases}$$

NOTE THIS STEP:

The required area A is given by

$A = \int_0^1 f(x)dx$

$= \int_0^{1/3}(1-x)^2 dx + \int_{1/3}^{2/3} 2x(1-x)dx + \int_{2/3}^{1} x^2 dx$

$= \left[-\frac{1}{3}(1-x)^3\right]_0^{1/3} + \left[x^2 - \frac{2x^2}{3}\right]_{1/3}^{2/3} + \left[\frac{1}{3}x^3\right]_{2/3}^{1}$

$= -\frac{1}{3}\left(\frac{2}{3}\right)^3 + \frac{1}{3}\left(\frac{2}{3}\right)^2 - \frac{2}{3}\left(\frac{2}{3}\right)^3 - \left(\frac{1}{3}\right)^2 + \frac{2}{3}\left(\frac{1}{3}\right)^3$

$+\frac{1}{3}(1) - \frac{1}{3}\left(\frac{2}{3}\right)^3 = \frac{17}{27}$ sq. units

40. $\therefore I = \int_0^1 y dx = \int_0^1 \tan^{-1} x dx - \int_0^1 \tan^{-1}(x-1)dx$

$= \int_0^1 \tan^{-1} x dx - \int_0^1 \tan^{-1}\{(1-x)-1\}$

$\left[\text{Using} \int_0^a f(x)dx = \int_0^a f(a-x)dx\right]$

$= \int_0^1 \tan^{-1} x\, dx - \int_0^1 (-\tan^{-1} x)\, dx = 2\int_0^1 \tan^{-1} x dx$ (Proved)

$= 2\left[x\tan^{-1} dx - \frac{1}{2}\log(1+x^2)\right]_0^1$

$= \frac{\pi}{2} - \log 2$ ...... (1)

Now, $\int_0^1 \tan^{-1}(1-x+x^2)dx$

$= \int_0^1 \cot^{-1}\frac{1}{1-x+x^2}dx = \int_0^1\left(\frac{\pi}{2} - \tan^{-1}\frac{1}{1-x+x^2}\right)dx$

$= \left[\frac{\pi}{2}x\right]_0^1 - I = \frac{\pi}{2} - \left(\frac{\pi}{2} - \log 2\right) = \log 2$ by (1)

41. $f(x) = x^3 - x^2$

Let P be on C_1, $y = x^2$ be (t, t^2)

$\therefore$ ordinate of Q is also t^2.

Now Q lies on $y = 2x$, and $y = t^2$

$\therefore$ $x = t^2/2$

$\therefore$ $Q\left(\frac{t^2}{2}, t^2\right)$

For point R, $x = t$ and it is on $y = f(x)$

$\therefore$ R is $[t, f(t)]$

Area $OPQ = \int_0^{t^2} \underset{C_1\ \ C_2}{(x_1 - x_2)} dy = \int_0^{t^2} \left(\sqrt{y} - \frac{y}{2}\right) dy$

$= \frac{2}{3}t^3 - \frac{t^4}{4}$ ...(1)

Area $OPR = \int_{0_{C_1}}^{t} ydx + \left|\int_{0_{C_3}}^{t} ydx\right|$

$= \int_0^t x^2 dx + \left|\int_0^t f(x)dx\right| = \frac{t^3}{3} + \left|\int_0^t f(x)dx\right|$ ....(2)

Equating (1) and (2), we get,

$\frac{t^3}{3} - \frac{t^4}{4}\left|\int_0^t f(x)dx\right|$

Differentiating both sides,we get,

$t^2 - t^3 = -f(t)$

$\therefore \quad f(t) = x^3 - x^2.$

42. $I = \int_0^{\pi} \frac{e^{\cos x}}{e^{\cos x} + e^{-\cos x}} dx$

$\Rightarrow I = \int_0^{\pi} \frac{e^{\cos(\pi - x)}}{e^{\cos(\pi-x)} + e^{-\cos(\pi-x)}}$

$\Rightarrow I = \int_0^{\pi} \frac{e^{-\cos x}}{e^{-\cos x} + e^{\cos x}}$

Adding, $2I = \int_0^{\pi} dx = \pi \Rightarrow I = \pi/2$

43. $f(x) = \begin{cases} x^2 + ax + b; & x < -1 \\ 2x & ; -1 \le x \le 1 \\ x^2 + ax + b; & x > 1 \end{cases}$

$\because f(x)$ is continuous at $x = -1$ and $x = 1$

$\therefore (-1)^2 + a(-1) + b = -2$

and $2 = (1)^2 + a.1 + b$

i.e. $a - b = 3$

and $a + b = 1$

On solving we get $a = 2, b = -1$

$\therefore \quad f(x) = \begin{cases} x^2 + 2x - 1; & x < -1 \\ 2x & ; -1 \le x \le 1 \\ x^2 + 2x - 1; & x > 1 \end{cases}$

Given curves are

$y = f(x), x = -2y^2$ and $8x + 1 = 0$

Solving $x = -2y^2$, $y = x^2 + 2x - 1$ $(x < -1)$ we get $x = -2$

Also $y = 2x$, $x = -2y^2$ meet at (0, 0)

$y = 2x$ and $x = -1/8$ meet at $\left(-\frac{1}{8}, \frac{-1}{4}\right)$

The required area is the shaded region in the figure.

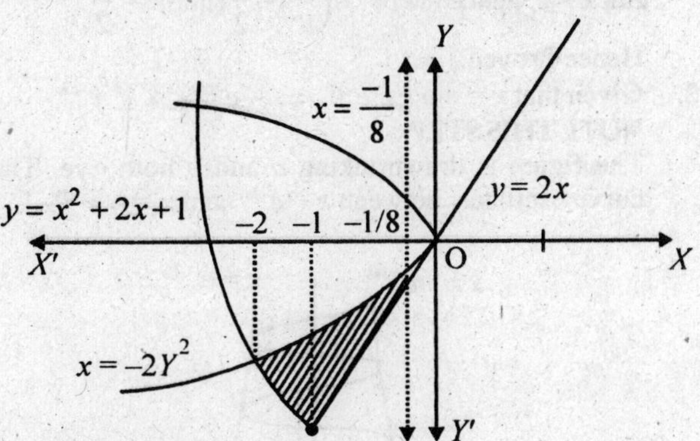

$\therefore$ Required area

NOTE THIS STEP :

$= \int_{-2}^{-1}\left[\sqrt{\frac{-x}{2}} - (x^2 + 2x - 1)\right]dx + \int_{-1}^{-1/8}\left[\sqrt{\frac{-x}{2}} - 2x\right]dx$

$= \left[\frac{1}{\sqrt{2}}\frac{2(-x)^{3/2}}{3} - \frac{x^3}{3} - x^2 + x\right]_{-2}^{-1}$

$+ \left[\frac{1}{\sqrt{2}}\frac{2(-x)^{3/2}}{3} - x^2\right]_{-1}^{-1/8}$

$= \left(\frac{\sqrt{2}}{3} + \frac{1}{3} - 1 - 1\right) - \left(\frac{4}{3} + \frac{8}{3} - 4 - 2\right)$

$+ \left(\frac{\sqrt{2}}{3}.\frac{1}{16\sqrt{2}} - \frac{1}{64}\right) - \left(\frac{\sqrt{2}}{3} - 1\right)$

$= \left(\frac{\sqrt{2} - 5}{3}\right) - \left(\frac{4 + 8 - 18}{3}\right) + \left(\frac{4 - 3}{192}\right) - \left(\frac{\sqrt{2} - 3}{3}\right)$

$= \frac{257}{192}$ sq. units

44. $f(x) = \int_1^x \frac{\ln t}{1 + t} dt$ for $x > 0$ (given)

Now $f\left(\frac{1}{x}\right) = \int_1^{1/x} \frac{\ln t}{1 + t} dt$: Put $t = \frac{1}{u}$, so that

$dt = -\frac{1}{u^2} du$

Therefore $f\left(\frac{1}{x}\right) = \int_1^x \frac{\ln(1/u)}{1 + \frac{1}{u}} . \frac{(-1)}{u^2} du$

$= \int_1^x \frac{\ln u}{u(u+1)} du = \int_1^x \frac{\ln t}{t(t+1)} dt$

Now, $f(x) + f\left(\frac{1}{x}\right) = \int_1^x \frac{\ln t}{1 + t} dt + \int_1^x \frac{\ln t}{t(1+t)} dt$

$= \int_1^x \frac{(1+t)\ln t}{t(1+t)} dt = \int_1^x \frac{\ln t}{t} dt = \frac{1}{2}(\ln t)^2]_1^x = \frac{1}{2}(\ln x)^2$

Put $x = e$, hence $f(e) + f\left(\frac{1}{e}\right) = \frac{1}{2}(\ln e)^2 = \frac{1}{2}$

Hence Proved.

45. Given that $x = \sin by.\, e^{-ay} \Rightarrow -e^{-ay} \le x \le e^{-ay}$

NOTE THIS STEP :

The figure is drawn taking a and b both +ve. The given curve oscillates between $x = e^{-ay}$ and $x = -e^{-ay}$

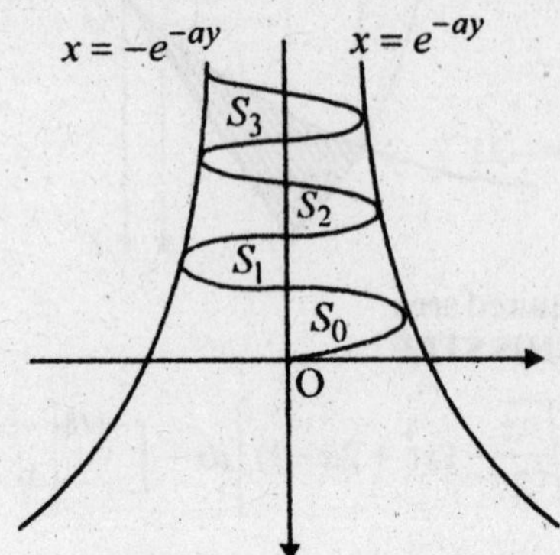

Clearly, $S_j = \int_{\frac{j\pi}{b}}^{\frac{(j+1)\pi}{b}} \sin by.e^{-ay}.dy$

Integrating by parts, $I = \int \sin by.e^{-ay}.dy$

We get $I = -\frac{e^{-ay}}{a^2+b^2}(a\sin by + b\cos by)$

$$\text{So, } S_j = \left| -\frac{1}{a^2+b^2}\left[e^{-a\frac{(j+1)\pi}{b}}\{a\sin(j+1)\pi + b\cos(j+1)\pi - e^{\frac{-aj\pi}{b}}(a\sin j\pi + b\cos j\pi)\right] \right|$$

$$\Rightarrow S_j = \left| -\frac{1}{a^2+b^2}\left[e^{-\frac{a}{b}(j+1)\pi} b(-1)^{j+1} - e^{-\frac{a}{b}j\pi} b(-1)^j \right] \right|$$

$$= \left| \frac{b.(-1)^j e^{-\frac{a}{b}j\pi}}{a^2+b^2}\left(e^{-\frac{a}{b}\pi}+1\right) \right| = b.\frac{e^{-\frac{a}{b}j\pi}}{a^2+b^2}\left(e^{-\frac{a}{b}\pi}+1\right)$$

$$\text{Now, } \frac{S_j}{S_{j-1}} = \frac{e^{-\frac{a}{b}j\pi}}{e^{-\frac{a}{b}(j-1)\pi}} = e^{-\frac{a}{b}\pi} = \text{constant}$$

$\Rightarrow S_o, S_1, S_2, \ldots\ldots\ldots S_j$ form a G.P.

For $a = -1$ and $b = \pi$ $\quad S_j = \frac{\pi e^j}{(1+\pi^2)}(1+e)$

$$\Rightarrow \sum_{j=0}^{n} S_j = \frac{\pi(1+e)}{(1+\pi^2)}.\frac{(e^{(n+1)}-1)}{(e-1)}.$$

46. The given curves are $y = x^2$

which is an upward parabola with vertex at (0, 0)

$y = |2 - x^2|$

$$\text{or } y = \begin{cases} 2 - x^2 \text{ if } & -\sqrt{2} \le x \le \sqrt{2} \\ x^2 - 2 \text{ if } x < -\sqrt{2} \text{ or } x > \sqrt{2} \end{cases}$$

or $x^2 = -(y-2);\ -\sqrt{2} < x < \sqrt{2}$ ....(2)

a downward parabola with vertex at (0, 2)

$x^2 = y + 2;\ x < -\sqrt{2},\ x > \sqrt{2}$ ....(3)

An upward parabola with vertex at (0, – 2)

$y = 2$ ....(4)

A straight line parallel to x – axis

$x = 1$ ....(5)

A straight line parallel to y – axis

The graph of these curves is as follows.

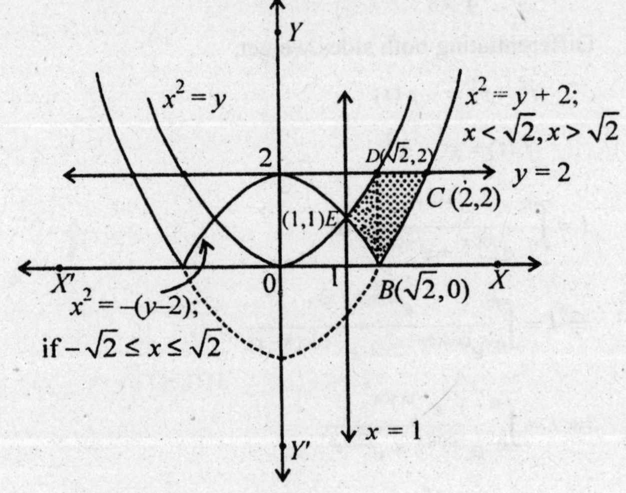

$\therefore$ Required area = $BCDEB$

$$= \int_1^{\sqrt{2}} [Y_{(1)} - Y_{(2)}]dx + \int_{\sqrt{2}}^{2} [Y_{(4)} - Y_{(3)}]dx \quad \text{.........(1)}$$

$$= \int_1^{\sqrt{2}} [x^2 - (2 - x^2)]\,dx + \int_{\sqrt{2}}^{2} [2 - (x^2 - 2)]dx$$

$$= \int_1^{\sqrt{2}} (2x^2 - 2)\,dx + \int_{\sqrt{2}}^{2} (4 - x^2)dx$$

$$= \left[\frac{2x^3}{3} - 2x\right]_1^{\sqrt{2}} + \left[4x - \frac{x^3}{3}\right]_{\sqrt{2}}^{2}$$

$$= \left(\frac{4\sqrt{2}}{3} - 2\sqrt{2} - \frac{2}{3} + 2\right) + \left(8 - \frac{8}{3} - 4\sqrt{2} + \frac{2\sqrt{2}}{3}\right)$$

$$= -\frac{2}{3}\sqrt{2} + \frac{4}{3} + \frac{16}{3} - \frac{10\sqrt{2}}{3}$$

$$= \frac{20 - 12\sqrt{2}}{3} = \left(\frac{20}{3} - 4\sqrt{2}\right) \text{ sq. units.}$$

47. Given that $f(x)$ is an even function, then to prove

$$\int_0^{\pi/2} f(\cos 2x)\cos x\,dx = \sqrt{2}\int_0^{\pi/4} f(\sin 2x)\cos x\,dx$$

Let $I=\int_0^{\pi/2} f(\cos 2x)\cos x\,dx$ ....(1)

$$=\int_0^{\pi/2} f\left[\cos 2\left(\frac{\pi}{2}-x\right)\right]\cos\left(\frac{\pi}{2}-x\right)dx$$

$$\left[\text{Using}\int_0^a f(x)dx=\int_0^a f(a-x)dx\right]$$

$$=\int_0^{\pi/2} f(-\cos 2x)\sin x\,dx$$

$$I=\int_0^{\pi/2} f(\cos 2x)\sin x\,dx \quad ....(2)$$

[As f is an even function]

Adding two values of I in (1) and (2) we get

$$2I=\int_0^{\pi/2} f(\cos 2x)(\sin x+\cos x)dx$$

$$\Rightarrow I=\frac{\sqrt{2}}{2}\int_0^{\pi/2} f(\cos 2x)\left[\frac{1}{\sqrt{2}}\sin x+\frac{1}{\sqrt{2}}\cos x\right]dx$$

$$I=\frac{\sqrt{2}}{2}\int_0^{\pi/2} f(\cos 2x)\cos(x-\pi/4)dx$$

Let $x-\pi/4=t$ so that $dx=dt$

as $x\to 0,\ t\to -\pi/4$

and as $x\to\pi/4,\quad t\to\pi/2-\pi/4=\pi/4$

$$\therefore\quad I=\frac{1}{\sqrt{2}}\int_{-\pi/4}^{\pi/4} f[\cos 2(t+\pi/4)]\cos t\,dt$$

$$=\frac{1}{\sqrt{2}}\int_{-\pi/4}^{\pi/4} f[-\sin 2t]\cos t\,dt$$

$$=\frac{1}{\sqrt{2}}\int_{-\pi/4}^{\pi/4} f(\sin 2t)\cos t\,dt$$

[$\because f$ is an even function]

$$=\frac{2}{\sqrt{2}}\int_0^{\pi/4} f(\sin 2t)\cos t\,dt$$

$$=\sqrt{2}\int_0^{\pi/4} f(\sin 2x)\cos x\,dx$$

R.H.S. *Hence proved.*

48. We have,

$$y(x)=\int_{\pi^2/16}^{x^2}\frac{\cos x\cos\sqrt{\theta}}{1+\sin^2\sqrt{\theta}}d\theta$$

$$=\cos x\int_{\pi^2/16}^{x^2}\frac{\cos\sqrt{\theta}}{1+\sin^2\sqrt{\theta}}d\theta$$

[$\because\ \cos x$ is independent of θ]

$$\Rightarrow\frac{dy}{dx}=-\sin x\int_{\pi^2/16}^{x^2}\left[\frac{\cos\sqrt{\theta}}{1+\sin^2\sqrt{\theta}}\right]d\theta$$

$$+\cos x\frac{d}{dx}\left[\int_{\pi^2/16}^{x^2}\frac{\cos\sqrt{\theta}}{1+\sin^2\sqrt{\theta}}d\theta\right]$$

$$=-\sin x\int_{\frac{\pi^2}{16}}^{x^2}\frac{\cos\sqrt{\theta}}{1+\sin^2\sqrt{\theta}}d\theta$$

$$+\cos x\left[\int_{\pi^2/16}^{x^2}\frac{\cos x}{1+\sin^2 x}.2x-0\right]\ \text{(Using Leibnitz thm.)}$$

$$\Rightarrow\left.\frac{dy}{dx}\right|_{x=\pi}=-\sin x\int_{\pi^2/16}^{x^2}\frac{\cos\sqrt{\theta}}{1+\sin^2\sqrt{\theta}}.d\theta+\frac{(\cos^2\pi).2\pi}{1+\sin^2\pi}$$

$=0+2\pi=2\pi$

49. Let $I=\int_{-\pi/3}^{\pi/3}\frac{\pi+4x^3}{2-\cos\left(|x|+\frac{\pi}{3}\right)}dx$

$$=\int_{-\pi/3}^{\pi/3}\frac{\pi}{2-\cos\left(|x|+\frac{\pi}{3}\right)}dx+\int_{-\pi/3}^{\pi/3}\frac{4x^3}{2-\cos\left(|x|+\frac{\pi}{3}\right)}dx$$

The second integral becomes zero integrand being an odd function of x.

$$=2\pi\int_0^{\pi/3}\frac{dx}{2-\cos\left(x+\frac{\pi}{3}\right)}$$

{ using the prop. of even function and also $|x|=x$ for $0\le x\le\pi/3$ }

Let $x+\pi/3=y\Rightarrow dx=dy$

also as $x\to 0,\ y\to\pi/3$ as $x\to\pi/3,\ y\to 2\pi/3$

$\therefore$ The given integral becomes

$$=2\pi\int_{\pi/3}^{2\pi/3}\frac{dy}{2-\cos y}=2\pi\int_{\pi/3}^{2\pi/3}\frac{dy}{2-\frac{1-\tan^2 y/2}{1+\tan^2 y/2}}$$

$$=2\pi\int_{\pi/3}^{2\pi/3}\frac{\sec^2 y/2}{3\tan^2 y/2+1}dy$$

$$=\frac{2\pi}{3}\int_{\pi/3}^{2\pi/3}\frac{\sec^2 y/2}{\tan^2 y/2+(1/\sqrt{3})^2}dy$$

$$=\frac{4\pi\sqrt{3}}{3}\left[\tan^{-1}(\sqrt{3}\tan y/\sqrt{2})\right]_{\pi/3}^{2\pi/3}$$

$$=\frac{4\pi}{3}\left[\tan^{-1}3-\tan^{-1}1\right]=\frac{4\pi}{\sqrt{3}}\left[\tan^{-1}3-\pi/4\right]$$

50. Let

$$I=\int_0^{\pi}e^{|\cos x|}\left[2\sin\left(\frac{1}{2}\cos x\right)+3\cos\left(\frac{1}{2}\cos x\right)\right]\sin x\,dx$$

$$=\int_0^{\pi}e^{|\cos x|}2\sin\left(\frac{1}{2}\cos x\right)\sin x\,dx$$

$$+\int_0^{\pi}e^{|\cos x|}3\cos\left(\frac{1}{2}\cos x\right)\sin x\,dx$$

$=I_1+I_2$

Now using the property that

$$\int_0^{2a} f(x)dx = 2\int_0^a f(x)dx \text{ if } f(2a-x)=f(x)$$
$$= 0 \text{ if } f(2a-x) = -f(x)$$

We get, $I_1 = 0$

and $I_2 = 2\int_0^{\pi/2} e^{|\cos x|} 3\cos\left(\frac{1}{2}\cos x\right)\sin x\, dx$

$$= 6\int_0^{\pi/2} e^{\cos x}\cos\left(\frac{1}{2}\cos x\right)\sin x\, dx$$

Put $\cos x = t \Rightarrow -\sin x\, dx = dt,$

$\therefore \quad I_2 = 6\int_0^1 e^t \cos t/2\, dt$

Integrating by parts, we get

$$I_2 = 6[(e^t \cos t/2)_0^1 + \frac{1}{2}\int_0^1 e^t \sin t/2\, dt]$$

$$= 6\left\{e\cos(1/2) - 1 + \frac{1}{2}\left\{(e^t \sin t/2)_0^1 - \frac{1}{2}\int_0^1 e^t \cos t/2\, dt\right\}\right\}$$

$$I_2 = 6\left[e\cos\left(\frac{1}{2}\right) - 1 + \frac{1}{2}\left\{e\sin\left(\frac{1}{2}\right) - \frac{1}{2}\cdot\frac{1}{6}I_2\right\}\right]$$

$$I_2 = 6\left[e\cos(1/2) - 1 + \frac{1}{2}(e\sin t/2) - \frac{1}{24}I_2\right]$$

$$I_2 + \frac{1}{4}I_2 = 6\left[e\cos(1/2) + \frac{1}{2}e\sin(1/2) - 1\right]$$

$$\frac{5I_2}{4} = 6\left[e\cos(1/2) + \frac{1}{2}e\sin(1/2) - 1\right]$$

$$\Rightarrow I_2 = \frac{24}{5}\left[e\cos(1/2 + \frac{1}{2}e\sin\left(\frac{1}{2}\right) - 1\right]$$

51. The given curves are, $x^2 = y$ ....(i)

$x^2 = -y$ ....(ii)

$y^2 = 4x - 3$ ....(iii)

Clearly point of intersection of (i) and (ii) is (0, 0). For point of intersection of (i) and (iii), solving them as follows

$$x^4 - 4x + 3 = 0$$

$$(x-1)(x^3 + x^2 + x - 3) = 0$$

or $(x-1)^2(x^2 + 2x + 3) = 0$

$\Rightarrow x = 1$ and then $y = 1$

$\therefore$ Req. point is (1, 1). Similarly point of intersection of (ii) and (iii) is (1, – 1). The graph of three curves is as follows:

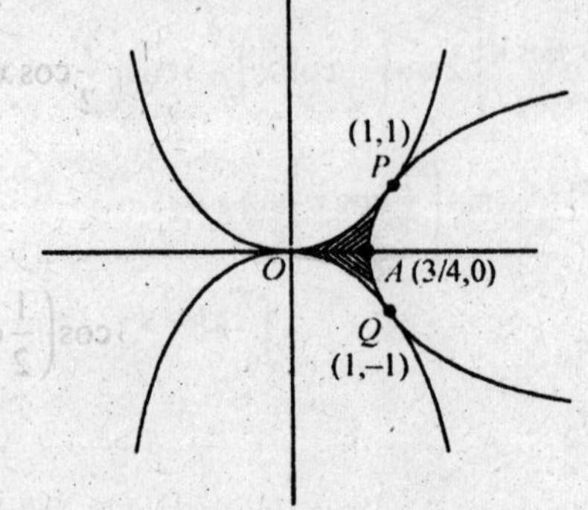

We also observe that at $x = 1$ and $y = 1$

$\frac{dy}{dx}$ for (i) and (iii) is same and hence the two curves touch each other at (1, 1).

Same is the case with (ii) and (iii) at (1, –1).

Required area = Shaded region in figure = 2 (*Ar OPA*)

$$= 2\left[\int_0^1 x^2 dx - \int_{3/4}^1 \sqrt{4x-3}\, dx\right]$$

$$= 2\left[\left(\frac{x^3}{3}\right)_0^1 - \left(\frac{2(4x-3)^{3/2}}{4\times 3}\right)_{3/4}^1\right] = 2\left[\frac{1}{3} - \frac{1}{6}\right]$$

$$= 2\times\frac{1}{6} = \frac{1}{3} \text{ sq. units}$$

52. Given that $f(x)$ is a differentiable function such that $f'(x) = g(x)$, then

$$\int_0^3 g(x)\, dx = \int_0^3 f'(x)dx = [f(x)]_0^3 = f(3) - f(0)$$

But $|f(x)| < 1 \Rightarrow -1 < f(x) < 1, \forall x \in R$

$\therefore \quad f(3) = f(0) \in (-1, 1)$

Similarly

$$\int_{-3}^0 g(x)dx = \int_{-3}^0 f'(x)dx = [f(0) - f(3)] \in (-2, 2)$$

Also given $[f(0)]^2 + [g(0)]^2 = 9$

$\Rightarrow [g(0)]^2 = 9 - [f(0)]^2$

$\Rightarrow |g(0)|^2 > 9 - 1 \quad [\because |f(x)| < 1]$

$\Rightarrow |g(0)| > 2\sqrt{2} \Rightarrow g(0) > 2\sqrt{2}$

or $\quad g(0) < -2\sqrt{2}$

First let us consider $g(0) > 2\sqrt{2}$

Let us suppose that $g''(x)$ be positive for all $x \in (-3, 3)$.

Then $g''(x) > 0 \Rightarrow$ the curve $y = g(x)$ is open upwards.

Now one of the two situations are possible.

(i) g(*x*) is increasing

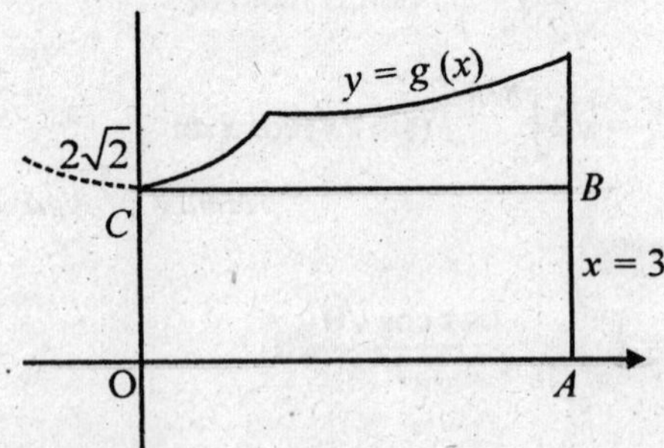

$\therefore \left|\int_0^3 g(x)dx\right| >$ area of rect. *OABC*

i.e. $\left|\int_0^3 g(x)dx\right| > 6\sqrt{2} > 2$

a contradiction as $\int_0^1 g(x)dx \in (-2, 2)$

$\therefore$ at least at one of the point $c \in (-3, 3)$, $g''(x) < 0$. But $g(x) > 0$ on $(-3, 3)$

Hence $g(x)\,g''(x) < 0$ at some $x \in (-3, 3)$.

(ii) $g(x)$ is decreasing

$\therefore \left|\int_{-3}^{0} g(x)dx\right| >$ area of rect. $OABC$

i.e. $\left|\int_{-3}^{0} g(x)dx\right| > 3.2\sqrt{2} = 6\sqrt{2} > 2$

a contradiction as $\int_{-3}^{0} g(x)dx \in (-2, 2)$

$\therefore$ at least at one of point $c \in (-3, 3)\; g''(x)$ should be – ve. But $g(x) > 0$ on $(-3, 3)$.

Hence $g(x)\,g''(x) < 0$ at some $x \in (-3, 3)$.

Secondly let us consider $g(0) < -2\sqrt{2}$.

Let us suppose that $g''(x)$ be – ve on $(-3, 3)$.

then $g''(x) < 0 \Rightarrow$ the curve $y = g(x)$ is open downward.

Again one of the two situations are possible

(i) $g(x)$ is decreasing then

$\left|\int_{0}^{3} g(x)dx\right| >$ Ar or rect. $OABC = 3.2\sqrt{2} = 6\sqrt{2} > 2$

a contradiction as $\int_{0}^{3} g(x)dx \in (-2, 2)$

$\therefore$ At least at one of the point $c \in (-3, 3)$, $g''(x)$ is + ve. But $g(x) < 0$ on $(-3, 3)$.

Hence $g(x)\,g''(x) < 0$ for some $x \in (-3, 3)$.

(ii) $g(x)$ is increasing then

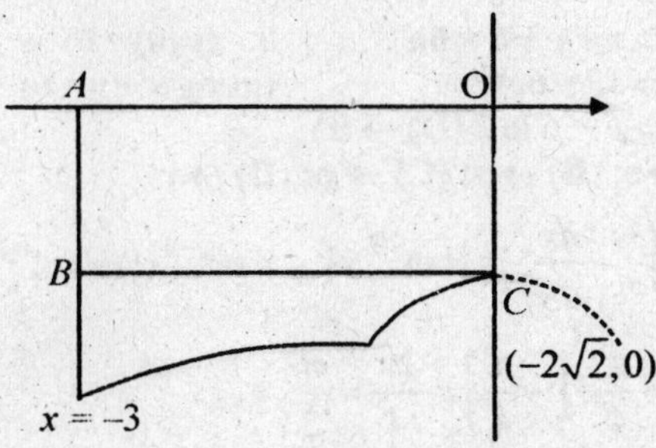

$\left|\int_{-3}^{0} g(x)dx\right| >$ Ar of rect. $OABC = 3.2\sqrt{2} = 6\sqrt{2} > 2$

a contradiction as as $\int_{-3}^{0} g(x)dx \in (-2, 2)$

$\therefore$ At least at one of the point $c \in (-3, 3)$ $g''(x)$ is + ve. But $g(x) < 0$ on $(-3, 3)$.

Hence $g(x)\,g''(x) < 0$ for some $x \in (-3, 3)$.

Combining all the cases, discussed aboe, we can conclude that at least at one point in $(-3, 3)$, $g(x)\,g''(x) < 0$.

53. We have, $\begin{bmatrix} 4a^2 & 4a & 1 \\ 4b^2 & 4b & 1 \\ 4c^2 & 4c & 1 \end{bmatrix}\begin{bmatrix} f(-1) \\ f(1) \\ f(2) \end{bmatrix} = \begin{bmatrix} 3a^2+3a \\ 3b^2+3b \\ 3c^2+3c \end{bmatrix}$

$\Rightarrow 4a^2 f(-1) + 4af(1) + f(2) = 3a^2 + 3a$

$4b^2 f(-1) + 4bf(1) + f(2) = 3b^2 + 3a$

$4c^2 f(-1) + 4cf(1) + f(2) = 3c^2 + 3c$

Consider the equation

$4x^2 f(-1) + 4xf(1) + f(2) = 3x^2 + 3x$

or $[4f(-1) - 3]x^2 + [4f(1) - 3]x + f(2) = 0$

Then clearly this eqn. is satisfied by $x = a, b, c$

A quadratic eqn. satisfied by more than two values of x means it is an identity and hence

$4f(-1) - 3 = 0 \Rightarrow f(-1) = 3/4$

$4f(1) - 3 = 0 \quad f(1) = 3/4$

$f(2) = 0 \quad f(2) = 0$

Let $f(x) = px^2 + qx + r$ [$f(x)$ being a quadratic eqn.]

$f(-1) = \frac{3}{4} \Rightarrow p - q + r = \frac{3}{4}$

$f(1) = \frac{3}{4} \Rightarrow p + q + r = \frac{3}{4}$

$f(2) = 0 \Rightarrow 4p + 2q + r = 0$

Solving the above we get $q = 0,\ p = -\frac{1}{4}, r = 1$

$\therefore \quad f(x) = -\frac{1}{4}x^2 + 1$

It's maximum value occur at $f'(x) = 0$

i.e., $x = 0$ then $f(x) = 1$, $\therefore V(0, 1)$

Let $A(-2, 0)$ be the point where curve meet x–axis.

Let B be the point $\left(h. \frac{4-h^2}{4}\right)$

As $\angle AVB = 90^0$, $m_{AV} \times m_{BV} = -1$

$\Rightarrow \left(\frac{0-1}{-2-1}\right) \times \left(\frac{\frac{4-h^2}{4} - 1}{h - 0}\right) = -1$

$\Rightarrow \frac{1}{2} \times \left(\frac{-h}{4}\right) = -1 \Rightarrow h = 8$

$\therefore \quad B(8, -15)$

Equation of chord AB is

$$y+15=\frac{0-(-15)}{-2-8}(x-8) \Rightarrow y+15=-\frac{3}{2}(x-8)$$

$$\Rightarrow 2y+30=-3x+24 \Rightarrow 3y+2y+6=0$$

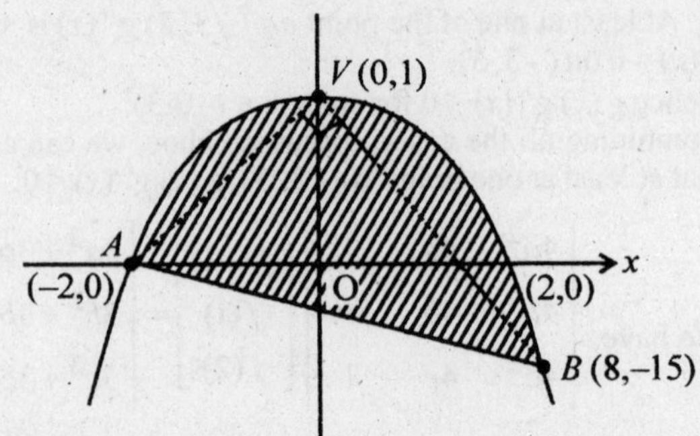

Required area is the area of shaded region given by

$$=\int_{-2}^{2}\left(-\frac{x^2}{4}+1\right)dx+\int_{-2}^{8}\left\{-\left(\frac{-6-3x}{2}\right)\right\}dx-\int_{2}^{8}\left\{-\left(-\frac{x^2}{4}+1\right)\right\}dx$$

$$=2\int_{0}^{2}\left(-\frac{x^2}{4}+1\right)dx+\frac{1}{2}\int_{-2}^{8}(6+3x)\,dx+\frac{1}{4}\int_{2}^{8}\left(-x^2+4\right)dx$$

$$=2\left[\left(\frac{-x^3}{12}+x\right)_0^2\right]+\frac{1}{2}\left[6x+\frac{3x^2}{2}\right]_{-2}^{8}+\frac{1}{4}\left[\frac{-x^3}{3}+4x\right]_2^8$$

$$=2\left[\frac{-8}{12}+2\right]+\frac{1}{2}[(48+3\times 32)-(-12+6)]$$

$$+\left[\frac{1}{4}\left(\frac{-512}{3}+32\right)-\left(\frac{-8}{3}+8\right)\right]$$

$$=2\left[\frac{4}{3}\right]+\frac{1}{2}[150]+\frac{1}{4}\left[\frac{-432}{3}\right]=\frac{125}{3}\text{ sq. units.}$$

54. Let $I=\int_0^1(1-x^{50})^{100}\,dx$ and $I'=\int_0^1(1-x^{50})^{101}\,dx$

Then, $I'=\int_0^1 1.(1-x^{50})^{101}\,dx$

$$=\left[x(1-x^{50})^{101}\right]_0^1+101\int_0^1 50x^{50}(1-x^{50})^{100}\,dx$$

$$=+5050\int_0^1 x^{50}(1-x^{50})^{100}\,dx$$

$$-I'=+5050\int_0^1 -x^{50}(1-x^{50})^{100}\,dx$$

$$\Rightarrow 5050\,I-I'=5050\int_0^1(1-x^{50})^{100}\,dx$$

$$+5050\int_0^1\left[-x^{50}(1-x^{50})^{100}\right]dx$$

$$\Rightarrow 5050\int_0^1(1-x^{50})^{101}\,dx=5050\,I'$$

$$\Rightarrow 5050\,I=5051\,I' \Rightarrow 5050\,\frac{I}{I'}=5051$$

(F) MATCH THE FOLLOWING :

1. **(A) → p, (B) → s, (C) → p, (D) → r**

(A) $\int_0^{\pi/2}(\sin x)^{\cos x}(\cos x\cot x-\log(\sin x)^{\sin x})dx$

$=\int_0^1 du$ where $(\sin x)^{\cos x}=u=1$

(A) → (p)

(B)

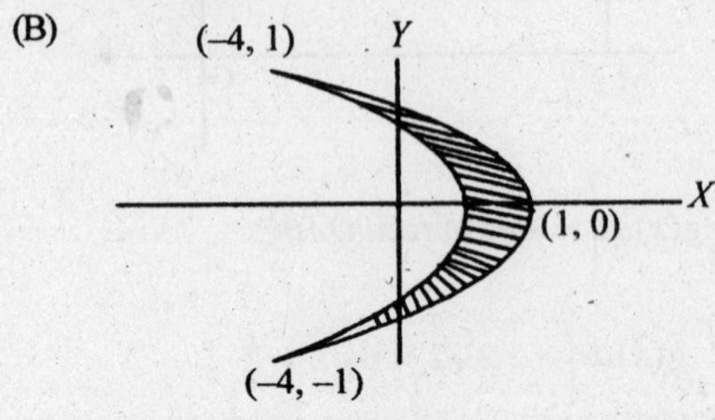

Solving $y^2=-\frac{1}{4}x$ and $y^2=-\frac{1}{5}(x-1)$, we get intersection points as $(-4,\pm 1)$

∴ Required area

$$=\int_{-1}^{1}[(1-5y^2)+4y^2]dy=2\int_0^1(1-y^2)dy=\frac{4}{3},$$

(B) → (s)

(C) By inspection, the point of intersection of two curves $y=3^{x-1}\log x$ and $y=x^x-1$ is (1, 0)

For first curve $\frac{dy}{dx}=\frac{3^{x-1}}{x}+3^{x-1}\log 3\log x$

$$\Rightarrow\left(\frac{dy}{dx}\right)_{(1,0)}=1=m_1$$

For second curve $\frac{dy}{dx}=x^x(1+\log x)$

$$\Rightarrow\left(\frac{dy}{dx}\right)_{(1,0)}=1=m_2$$

∵ $m_1=m_2$ ⇒ Two curves touch each other

⇒ Angle between them is 0°

∴ $\cos\theta=1$,

(C) → (p)

(D) $\frac{dy}{dx}=\frac{6}{x+y} \Rightarrow \frac{dx}{dy}-\frac{1}{6}x=\frac{y}{6}$

I.F. $=e^{-y/6}$

⇒ Solution is $xe^{-y/6}=-ye^{-y/6}-6e^{-y/6}+c$

$\Rightarrow x+y+6=ce^{y/6}$

$\Rightarrow x+y+6=6e^{y/6}$ ∴ $(y(0)=0)$

$\Rightarrow 12=6e^{y/6}$ (using $x+y=6$)

$\Rightarrow y=6\ln 2$ (D) → (r)

2. **(A) → s; (B) → s; (C) → p; (D) → r**

(A) $\int_{-1}^{1}\frac{dx}{1+x^2}=[\tan^{-1}x]_{-1}^{1}=\tan^{-1}(1)-\tan^{-1}(-1)$

$$=\frac{\pi}{4}-\left(-\frac{\pi}{4}\right)=\frac{2\pi}{4}=\frac{\pi}{2}$$

(B) $\int_0^1 \frac{dx}{\sqrt{1-x^2}} = \left[\sin^{-1}x\right]_0^1 = \sin^{-1}(1) - \sin^{-1}(0)$

$= \frac{\pi}{2} - 0 = \frac{\pi}{2}$

(C) $\int_2^3 \frac{dx}{1-x^2} = \left[\frac{1}{2}\log\left|\frac{1+x}{1-x}\right|\right]_2^3 = \frac{1}{2}[\log 2 - \log 3]$

$= \frac{1}{2}\log 2/3$

(D) $\int_1^2 \frac{dx}{x\sqrt{x^2-1}} = \left[\sec^{-1}x\right]_1^2 = \sec^{-1}2 - \sec^{-1}1$

$= \frac{\pi}{3} - 0 = \frac{\pi}{3}$

G COMPREHENSION BASED QUESTIONS :

1. (a) $\int_0^{\pi/2} \sin x\,dx = \frac{\left(\frac{\pi}{2}-0\right)}{4}\left(\sin 0 + \sin\frac{\pi}{2} + 2\sin\frac{\pi}{4}\right)$

$= \frac{\pi}{8}(1+\sqrt{2})$

2. (d) $\lim_{x\to a} \frac{\int_a^x f(x)dx - \left(\frac{x-a}{2}\right)(f(x)+f(a))}{(x-a)^3} = 0$

$\lim_{h\to 0} \frac{\int_a^{a+h} f(x)dx - \frac{h}{2}(f(a+h)+f(a))}{h^3} = 0$

$\Rightarrow \lim_{h\to 0} \frac{f(a+h) - \frac{1}{2}[f(a)+f(a+h)] - \frac{h}{2}(f'(a+h))}{3h^2} = 0$

[Using L'Hospital rule]

$\Rightarrow \lim_{h\to 0} \frac{\frac{1}{2}f(a+h) - \frac{1}{2}f(a) - \frac{h}{2}f'(a+h)}{3h^2} = 0$

$\Rightarrow \lim_{h\to 0} \frac{\frac{1}{2}f'(a+h) - \frac{1}{2}f'(a+h) - \frac{h}{2}f''(a+h)}{6h} = 0$

[Using L'Hospital rule]

$\Rightarrow \lim_{h\to 0} -\frac{f''(a+h)}{12} = 0 \Rightarrow f''(x) = 0,\ \forall a \in R$

$\Rightarrow f(x)$ must be of max. degree 1.

3. (b) $f''(x) < 0,\ \forall x \in (a, b)$, for $c \in (a, b)$

$F(c) = \frac{c-a}{2}(f(a)+f(c)) + \frac{b-c}{2}(f(b)+f(c))$

$= \frac{b-a}{2}f(c) + \frac{c-a}{2}f(a) + \frac{b-c}{2}f(b)$

$\Rightarrow F'(c) = \frac{b-a}{2}f'(c) + \frac{1}{2}f(a) - \frac{1}{2}f(b)$

$= \frac{1}{2}[(b-a)f'(a) - f(b)]$

$F''(c) = \frac{1}{2}(b-a)f''(c) < 0$

$[\because f''(x) < 0, \forall x \in (a,b) \text{ and } b > a]$

$\therefore$ $F(c)$ is max. at the point $(c, f(c))$ where $F'(c) = 0$

$\Rightarrow f'(c) = \frac{1}{2}\left(\frac{f(b)-f(a)}{b-a}\right)$.

(For 4-6). Given the implicit function $y^3 - 3y + x = 0$

For $x \in (-\infty, -2) \cup (2, \infty)$ it is $y = f(x)$ real valued differentiable function.

and for $x \in (-2, 2)$ it is $y = g(x)$ real valued differentiable function.

4. (b) We have $y^3 - 3y + x = 0 \Rightarrow 3y^2\frac{dy}{dx} - 3\frac{dy}{dx} + 1 = 0$

$\Rightarrow \frac{dy}{dx} = \frac{1}{3(1-y^2)}$ or $f'(x) = \frac{1}{3\left[1-[f(x)]^2\right]}$

Also $3y^2\frac{d^2y}{dx^2} + 6y\left(\frac{dy}{dx}\right)^2 - 3\frac{d^2y}{dx^2} = 0$

$\Rightarrow \frac{d^2y}{dx^2} = \frac{2y}{1-y^2}\left(\frac{dy}{dx}\right)^2$

$\Rightarrow f''(x) = \frac{2f(x)}{9\left[1-[f(x)]^2\right]^3}$

$\therefore\ f''(-10\sqrt{2}) = \frac{2\times 2\sqrt{2}}{9[1-8]^3} = \frac{-4\sqrt{2}}{3^2\times 7^3}$

5. (a) For $x < -2$

we have, $3y - y^3 < -2 \Rightarrow y^3 - 3y - 2 > 0$

$\Rightarrow (y+1)^2(y-2) > 0 \Rightarrow y > 2\ \forall\ x < -2$

$\Rightarrow f(x)$ is positive $\forall\ x < -2$

Hence required area $= \int_a^b f(x)dx = \int_a^b 1.f(x)dx$

$= x f(x)]_a^b - \int_a^b x f'(x)dx$

$= b f(b) - a f(a) - \int_a^b \frac{x.1}{3\left[1-(f(x))^2\right]}dx$

$= \int_a^b \frac{x}{3\left[(f(x))^2 - 1\right]} + b f(b) - a f(a)$

6. (d) For $y = g(x)$, we have $y^3 - 3y + x = 0$

$\Rightarrow [g(x)]^3 - 3[g(x)] + x = 0$ ...(1)

Putting $x = -x$, we get

$\Rightarrow [g(-x)]^3 - 3[g(-x)] - x = 0$ ...(2)

Adding equations (1) and (2) we get

$[g(x)]^3 + [g(-x)]^3 - 3\{[g(x)] + [g(-x)]\} = 0 \frac{n!}{r!(n-r)!}$

$\Rightarrow [g(x) + g(-x)]$

$\left[(g(x))^2 + (g(-x))^2 - g(x)g(-x) - 3\right] = 0$

For $g(0) = 0$, we should have

$g(x) + g(-x) = 0$

[$\because$ From other factor we get $g(0) = \pm\sqrt{3}$]

$\Rightarrow g(x)$ is an odd function

$\therefore \int_{-1}^{1} g'(x)dx = [g(x)]_{-1}^{1} = g(1) - g(-1)$

$= g(1) + g(1) = 2\,g(1).$

7. (a) We have $f(x) = \frac{x^2 - ax + 1}{x^2 + ax + 1}$; $0 < a < 2$

$\Rightarrow f'(x) = \frac{2a(x^2-1)}{(x^2+ax+1)^2}$

$\Rightarrow (x^2+ax+1)^2 f'(x) = 2a(x^2-1)$

$\Rightarrow (x^2+ax+1)^2 f''(x) + 2(x^2+ax+1)$

$(2x+a)f'(x) = 4ax$ ...(1)

Putting $x = -1$ in equation (1), we get

$(2-a^2)f''(-1) = -4a$ ...(2)

Putting $x = 1$ in equation (1), we get

$(2+a)^2 f''(1) = 4a$ ...(3)

Adding equations (2) and (3), we get

$(2+a)^2 f''(1) + (2-a)^2 f''(-1) = 0$

8. (a) We have $f'(x) = \frac{2a(x^2-1)}{(x^2+ax+1)^2}$

$f'(x) = 0 \Rightarrow x = -1, 1$ are the critical points.

$\longleftarrow + \;\; -1 \;\; - \;\; 1 \;\; + \longrightarrow$

$\therefore f(x)$ is decreasing on $(-1, 1)$

Also using equation (1), $f''(-1) = \frac{-4a}{(2-a)^2} < 0$

and $f''(1) = \frac{4a}{(2+a)^2} > 0$

$\therefore x = -1$ is a point of local maximum and $x = 1$ is a point of local minimum.

9. (b) $g(x) = \int_0^{e^x} \frac{f'(t)}{1+t^2}dt \Rightarrow g'(x) = \frac{f'(e^x)}{1+e^{2x}}$

$= \frac{2a(e^{2x}-1)e^x}{(e^{2x}+ae^x+1)^2(1+e^{2x})} = \frac{2ae^x}{1+e^{2x}} \cdot \frac{e^{2x}-1}{(e^{2x}+ae^x+1)^2}$

Now $g'(x) > 0$ for $e^{2x} - 1 > 0 \Rightarrow x > 0$

and $g'(x) < 0$ for $e^{2x} - 1 < 0 \Rightarrow x < 0$

$\therefore g'(x)$ is negative on $(-\infty, 0)$ and positive on $(0, \infty)$

10 (c) $f(x) = 4x^3 + 3x^2 + 2x + 1$

$\because$ $f(x)$ is a cubic polynomial

$\therefore$ It has at least one real root.

Also $f'(x) = 12x^2 + 6x + 2 = 2\left(6x^2 + 3x + 1\right) > 0 \; \forall \; x \in R$

$\therefore$ $f(x)$ is strictly increasing function

$\Rightarrow$ There is only one real root of $f(x) = 0$

Also $f(-1/2) = 1 - 1 + \frac{3}{4} - \frac{1}{2} > 0$

and $f(-3/4) = 1 - \frac{3}{2} + \frac{27}{16} - \frac{27}{16} < 0$

$\therefore$ Real root lies between $-\frac{3}{4}$ and $-\frac{1}{2}$ and hence

$s \in \left(-\frac{3}{4}, -\frac{1}{2}\right)$

11. (a) $y = f(x)$, $x = 0$, $y = 0$ and $x = t$ bounds the area as shown in the figure

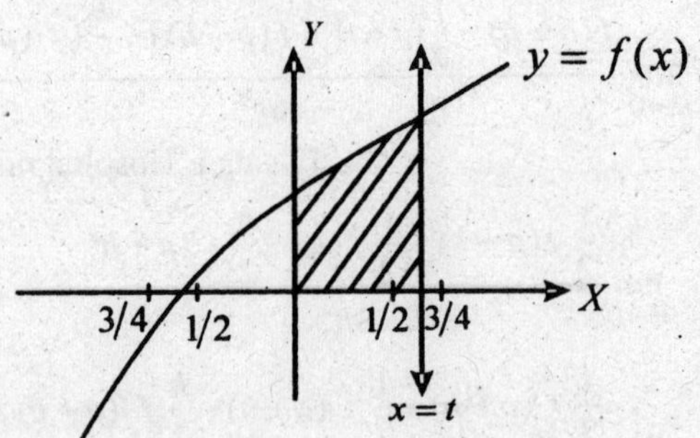

$\therefore$ Required area is given by

$A = \int_0^t dx = \int_0^t \left(4x^3 + 3x^2 + 2x + 1\right)dx$

$= t^4 + t^3 + t^2 + t = t\left(t+1\right)\left(t^2+1\right)$

New $\frac{1}{2} < t < \frac{3}{4}$; $\frac{3}{2} < t + 1 < \frac{7}{4}$; $\frac{5}{4} < t^2 + 1 < \frac{25}{16}$.

$\therefore \frac{1}{2} \times \frac{3}{2} \times \frac{5}{4} < A < \frac{3}{4} \times \frac{7}{4} \times \frac{25}{16}$

or $A \in \left(\frac{15}{16}, \frac{525}{256}\right) \subset \left(\frac{3}{4}, 3\right)$

12. **(b)** $f'(x) = 2(6x^2 + 3x + 1)$

$f''(x) = 6(4x + 1)$

Critical point $x = -1/4$

$\therefore$ decreasing in $\left(-t, -\frac{1}{4}\right)$

$\longleftarrow t \quad -\ \frac{1}{4}\ + \quad t \longrightarrow$

and increasing in $\left(-\frac{1}{4}, t\right)$

I INTEGER VALUE CORRECT TYPE :

1. Given that $f(x) = \int_0^x f(t)dt$

Clearly $f(0) = 0$. Also $f'(x) = f(x)$

$\Rightarrow \quad \frac{f'(x)}{f(x)} = 1$

Integrating both sides with respect to x, we get

$\int \frac{f'(x)}{f(x)} dx = \int 1\, dx$

$\Rightarrow \quad \ln f(x) = x + \ln C \quad \Rightarrow f(x) = Ce^x$

Now $f(0) = 0 \quad \Rightarrow \quad Ce^x = 0 \Rightarrow C = 0$

$\therefore \quad f(x) = 0\ \forall\ x \Rightarrow \qquad f(\ln 5) = 0$

2. Given function is $f(x) = \begin{cases} x - [x] & \text{if } [x] \text{ is odd} \\ 1 + [x] - x & \text{if } [x] \text{ is even} \end{cases}$

The graph of this function is as below

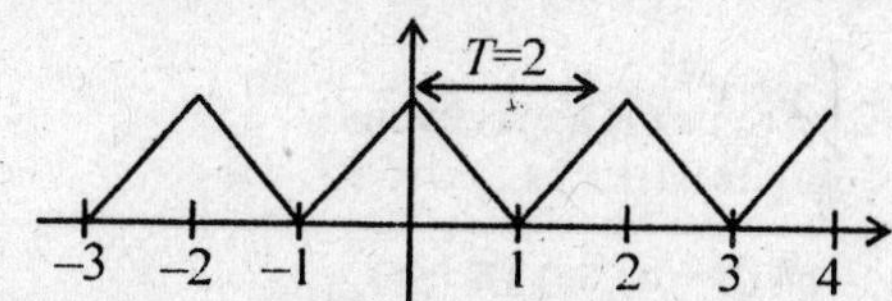

Clearly $f(x)$ is periodic with period 2

Also $\cos \pi x$ is periodic with period 2

$\therefore \ f(x) \cos \pi x$ is periodic with period 2

$\therefore \ I = \frac{\pi^2}{10} \int_{-10}^{10} f(x) \cos \pi x\, dx$

$= \frac{\pi^2}{10} \times 10 \int_0^2 f(x) \cos \pi x\, dx$

$= \pi^2 \left[\int_0^1 (1 - x) \cos \pi x\, dx + \int_1^2 (x - 1) \cos \pi x\, dx \right]$

$= \pi^2 \left[\left\{ (1 - x) \frac{\sin \pi x}{\pi} \Big|_0^1 + \int_0^1 \frac{\sin \pi x}{\pi} dx \right\} + \right.$

$\left. \left\{ (x - 1) \frac{\sin \pi x}{\pi} \Big|_1^2 - \int_1^2 \frac{\sin \pi x}{\pi} dx \right\} \right]$

$= \pi^2 \left[\left(-\frac{1}{\pi^2} \cos \pi x \right) \Big|_0^1 - \left(-\frac{1}{\pi^2} \cos \pi x \right)^2 \right]$

$= [(-\cos \pi + \cos 0) - (-\cos 2\pi + \cos \pi)] = [2 + 2] = 4$

Section-B JEE Main/ AIEEE

1. **(a)** $I = \int_0^{10\pi} |\sin x|\, dx = 10 \int_0^{\pi} |\sin x|\, dx = 10 \int_0^{\pi} \sin x\, dx$

[$\because |\sin x|$ is periodic with period π and $\sin x > 0$ if $0 < x < \pi$]

$I = 20 \int_0^{\pi/2} \sin x\, dx = 20[-\cos x]_0^{\pi/2} = 20$

2. **(b)** $I_n + I_{n+2} = \int_0^{\pi/4} \tan^n x (1 + \tan^2 x) dx$

$= \int_0^{\pi/4} \tan^n x \sec^2 x\, dx = \left[\frac{\tan^{n+1} x}{n+1} \right]_0^{\pi/4}$

$= \frac{1 - 0}{n+1} = \frac{1}{n+1}$

$\therefore \ I_n + I_{n+2} = \frac{1}{n+1} \Rightarrow \lim_{n \to \infty} n [I_n + I_{n+2}]$

$= \lim_{n \to \infty} n . \frac{1}{n+1} = \lim_{n \to \infty} \frac{n}{n+1} = \lim_{n \to \infty} \frac{n}{n\left(1 + \frac{1}{n}\right)} = 1$

3. **(d)** $\int_0^2 [x^2] dx = \int_0^1 [x^2] dx + \int_1^{\sqrt{2}} [x^2] dx + \int_{\sqrt{2}}^{\sqrt{3}} [x^2] + \int_{\sqrt{3}}^{2} [x^2] dx$

$= \int_0^1 0 dx + \int_1^{\sqrt{2}} 1 dx + \int_{\sqrt{2}}^{\sqrt{3}} 2 dx + \int_{\sqrt{3}}^{2} 3 dx$

$= [x]_1^{\sqrt{2}} + [2x]_{\sqrt{2}}^{\sqrt{3}} + [3x]_{\sqrt{3}}^{2}$

$= \sqrt{2} - 1 + 2\sqrt{3} - 2\sqrt{2} + 6 - 3\sqrt{3} = 5 - \sqrt{3} - \sqrt{2}$

4. **(b)** $\int_{-\pi}^{\pi} \frac{2x(1 + \sin x)}{1 + \cos^2 x} dx$

$= \int_{-\pi}^{\pi} \frac{2x\, dx}{1 + \cos^2 x} + 2 \int_{-\pi}^{\pi} \frac{x \sin x}{1 + \cos^2 x} dx$

$= 0 + 4 \int_0^{\pi} \frac{x \sin x\, dx}{1 + \cos^2 x}; \qquad \left[\because \int_{-a}^{a} f(x) dx = 0 \right]$

if f(x) is odd

$= 2\int_0^a f(x)\,dx$ if $f(x)$ is even.

$$I = 4\int_0^{\pi} \frac{(\pi - x)\sin(\pi - x)}{1+\cos^2(\pi - x)}dx$$

$$I = 4\int_0^{\pi} \frac{(\pi - x)\sin x}{1+\cos^2 x}dx$$

$$\Rightarrow I = 4\pi\int_0^{\pi} \frac{\sin x\,dx}{1+\cos^2 x} - 4\int \frac{x\sin x\,dx}{1+\cos^2 x}$$

$$\Rightarrow 2I = 4\pi\int_0^{\pi} \frac{\sin x}{1+\cos^2 x}dx$$

put $\cos x = t \Rightarrow -\sin x\,dx = dt$

$$\therefore I = -2\pi\int_1^{-1} \frac{1}{1+t^2}dt = 2\pi\int_{-1}^{1} \frac{1}{1+t^2}dt$$

$$= 2\pi\left[\tan^{-1}t\right]_{-1}^{1} = 2\pi\left[\tan^{-1}1 - \tan^{-1}(-1)\right]$$

$$= 2\pi\left[\frac{\pi}{4} - \left(\frac{-\pi}{4}\right)\right] = 2\pi.\frac{\pi}{2} = \pi^2$$

5. **(d)** We have $\int_0^2 f(x)dx = \frac{3}{4}$; Now,

$$\int_0^2 xf'(x)dx = x\int_0^2 f'(x)dx - \int_0^2 f(x)dx$$

$$= [x\,f(x)]_0^2 - \frac{3}{4} = 2f(2) - \frac{3}{4}$$

$$= 0 - \frac{3}{4}\ (\because f(2) = 0) = -\frac{3}{4}.$$

6. **(a)** First we draw each curve as separate graph

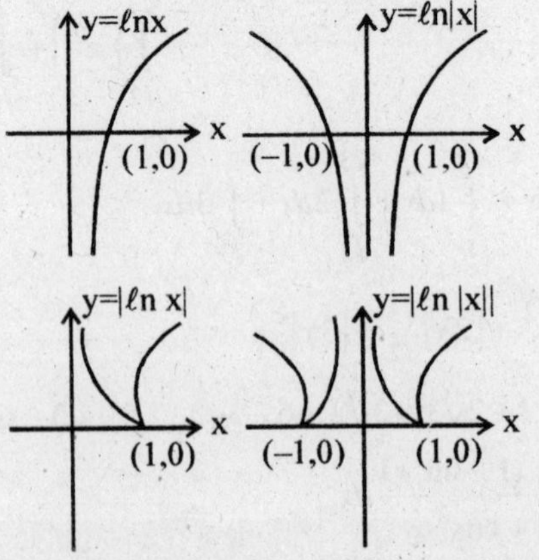

NOTE : Graph of $y = |f(x)|$ can be obtained from the graph of the curve $y = f(x)$ by drawing the mirror image of the portion of the graph below x-axis, with respect to x-axis.

Clearly the bounded area is as shown in the following figure.

Required area $= 4\int_0^1 (-\ell n x)dx$

$$= -4[x\ \ell n\ x - x]_0^1 = 4 \text{ sq. units}$$

7. **(d)**

$$A = \int_{-1}^{0} \{(3+x) - (-x+1)\}\,dx +$$

$$\int_0^1 \{(3-x) - (-x+1)\}\,dx + \int_1^2 \{(3-x) - (x-1)\}\,dx$$

$$= \int_{-1}^{0}(2+2x)dx + \int_0^1 2dx + \int_1^2 (4-2x)dx$$

$$= \left[2x - x^2\right]_{-1}^{0} + [2x]_0^1 + \left[4x - x^2\right]_1^2$$

$$= 0 - (-2+1) + (2-0) + (8-4) - (4-1)$$

$$= 1 + 2 + 4 - 3 = 4 \text{ sq. units}$$

8. **(c)** $I = \int_a^b xf(x)dx = \int_a^b (a+b-x)f(a+b-x)dx$

$$= (a+b)\int_a^b f(a+b-x)dx - \int_a^b xf(a+b-x)dx$$

$$= (a+b)\int_a^b f(x)\,dx - \int_a^b xf(x)dx$$

[$\because$ given that $f(a+b-x) = f(x)$]

$$2I = (a+b)\int_a^b f(x)dx \Rightarrow I = \frac{(a+b)}{2}\int_a^b f(x)dx$$

9. **(d)** Given $f'(x) = f(x) \Rightarrow \frac{f'(x)}{f(x)} = 1$

Integrating $\log f(x) = x + c \Rightarrow f(x) = e^{x+c}$

$f(0) = 1 \Rightarrow f(x) = e^x$

$$\therefore \int_0^1 f(x)g(x)dx = \int_0^1 e^x(x^2 - e^x)dx$$

$$= \int_0^1 x^2e^x dx - \int_0^1 e^{2x}dx$$

$$= \left[x^2 e^x\right]_0^1 - 2\left[xe^x - e^x\right]_0^1 - \frac{1}{2}\left[e^{2x}\right]_0^1$$

$$= e - \left[\frac{e^2}{2} - \frac{1}{2}\right] - 2[e - e + 1] = e - \frac{e^2}{2} - \frac{3}{2}$$

10. **(d)** $I = \int_0^1 x(1-x)^n dx = \int_0^1 (1-x)(1-1+x)^n dx$

$$= \int_0^1 (1-x)x^n dx = \left[\frac{x^{n+1}}{n+1} - \frac{x^{n+2}}{n+2}\right]_0^1 = \frac{1}{n+1} - \frac{1}{n+2}$$

11. **(b)** $\lim_{n\to\infty} \sum_{r=1}^{n} \frac{1}{n} e^{\frac{r}{n}}$ [Using definite integrals as limit of sum]

$$= \int_0^1 e^x dx = e - 1$$

12. **(d)** $\int_{-2}^{3} |1 - x^2| dx = \int_{-2}^{3} |x^2 - 1| dx$

Now $|x^2 - 1| = \begin{cases} x^2 - 1 & if \quad x \le -1 \\ 1 - x^2 & if \quad -1 \le x \le 1 \\ x^2 - 1 & if \quad x \ge 1 \end{cases}$

$\therefore$ Integral is $\int_{-2}^{-1} (x^2 - 1)dx + \int_{-1}^{1} (1 - x^2)dx + \int_{1}^{3} (x^2 - 1)dx$

$$\left[\frac{x^3}{3} - x\right]_{-2}^{-1} + \left[x - \frac{x^3}{3}\right]_{-1}^{1} + \left[\frac{x^3}{3} - x\right]_{1}^{3}$$

$$= \left(-\frac{1}{3} + 1\right) - \left(-\frac{8}{3} + 2\right) + \left(2 - \frac{2}{3}\right) + \left(\frac{27}{3} - 3\right) - \left(\frac{1}{3} - 1\right)$$

$$= \frac{2}{3} + \frac{2}{3} + \frac{4}{3} + 6 + \frac{2}{3} = \frac{28}{3}$$

13. **(c)** $I = \int_0^{\frac{\pi}{2}} \frac{(\sin x + \cos x)^2}{\sqrt{1 + \sin 2x}} dx$

We know $[(\sin x + \cos x)^2 = 1 + \sin 2x]$, so

$$I = \int_0^{\frac{\pi}{2}} \frac{(\sin x + \cos x)^2}{(\sin x + \cos x)} dx = \int_0^{\frac{\pi}{2}} (\sin x + \cos x) dx$$

$$\left[\because \sin x + \cos x > 0 \; if \; 0 < x < \frac{\pi}{2}\right]$$

or $I = [-\cos x + \sin x]_0^{\frac{\pi}{2}} = 2$

14. **(b)** Let $I = \int_0^{\pi} xf(\sin x)dx = \int_0^{\pi} (\pi - x)f(\sin x)dx$

$$\therefore 2I = \pi \int_2^{\pi} f(\sin x)dx = \pi.2 \int_0^{\frac{\pi}{2}} f(\sin x)dx$$

$$\therefore I = \pi \int_0^{\frac{\pi}{2}} f(\sin x)dx \Rightarrow A = \pi$$

15. **(d)** $f(x) = \frac{e^x}{1 + e^x} \Rightarrow f(-x) = \frac{e^{-x}}{1 + e^{-x}} = \frac{1}{e^x + 1}$

$\therefore f(x) + f(-x) = 1 \; \forall \; x$

Now $I_1 = \int_{f(-a)}^{f(a)} xg\{x(1-x)\}dx$

$$= \int_{f(-a)}^{f(a)} (1-x)g\{x(1-x)\}dx$$

$$\left[\text{using} \int_a^b f(x)\,dx\; a = \int_a^b f(a + b - x)\,dx\right]$$

$$= I_2 - I_1 \Rightarrow 2I_1 = I_2 \quad \therefore \frac{I_2}{I_1} = 2$$

16. **(d)** The required area is shown by shaded region

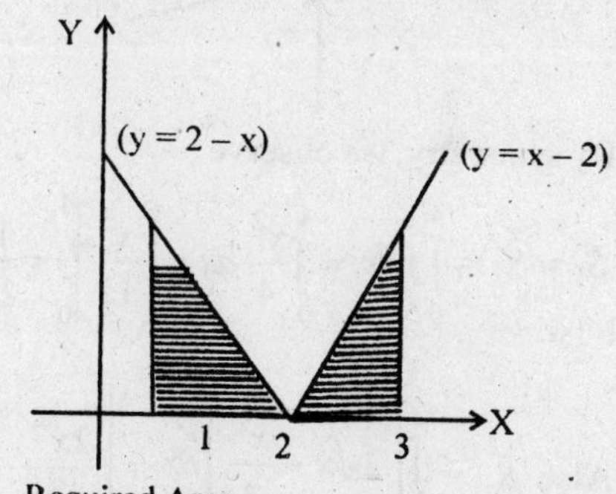

Required Area

$$A = \int_1^3 |x - 2| dx = 2\int_2^3 (x - 2)dx = 2\left[\frac{x^2}{2} - 2x\right]_2^3 = 1$$

17. **(b)** $I_1 = \int_0^1 2^{x^2} dx,\; I_2 = \int_0^1 2^{x^3} dx,$

$$= I_3 = \int_0^1 2^{x^2} dx,\; I_4 = \int_0^1 2^{x^3} dx \;\forall 0 < x < 1,\; x^2 > x^3$$

$$\Rightarrow \int_0^1 2^{x^2} dx > \int_0^1 2^{x^3} dx \Rightarrow I_1 > I_2$$

18. (a) The graph of the curve $y = \log_e(x+e)$ is as shown in the fig.

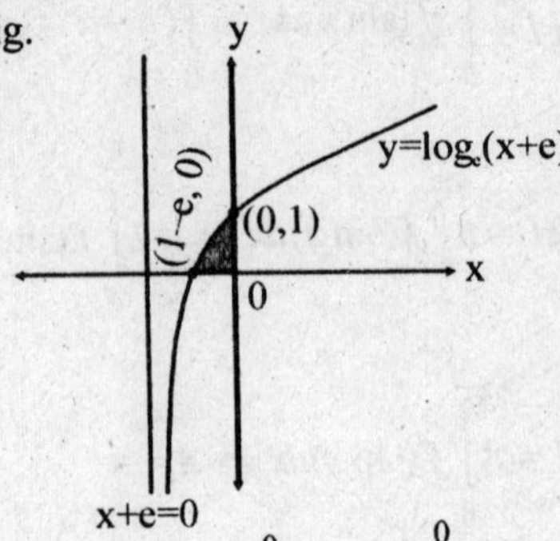

Required area $A = \int_{1-e}^{0} y dx = \int_{1-e}^{0} \log_e(x+e)dx$

put $x+e=t \Rightarrow dx = dt$ also At $x = 1-e, t = 1$

At $x=0, t=e$ $\quad \therefore A = \int_{1}^{e} \log_e t dt = [t\log_e t - t]_1^e$

$e - e - 0 + 1 = 1$

Hence the required area is 1 square unit.

19. (d) Intersection points of $x^2 = 4y$ and $y^2 = 4x$ are (0, 0) and (4, 4). The graph is as shown in the figure.

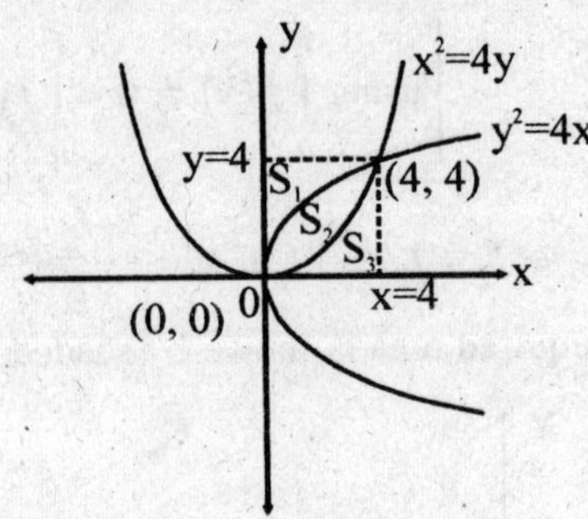

By symmetry, we observe

$$S_1 = S_3 = \int_0^4 y dx = \int_0^4 \frac{x^2}{4} dx = \left[\frac{x^3}{12}\right]_0^4 = \frac{16}{3} \text{ sq. units}$$

$$\text{Also } S_2 = \int_0^4 \left(2\sqrt{x} - \frac{x^2}{4}\right) dx = \left[\frac{2x^{\frac{3}{2}}}{\frac{3}{2}} - \frac{x^3}{12}\right]_0^4$$

$$= \frac{4}{3}\times 8 - \frac{16}{3} = \frac{16}{3} \text{ sq. units}$$

$\therefore S_1 : S_2 : S_3 = 1:1:1$

20. (d) Given that $\int_{\pi/4}^{\beta} f(x)dx = \beta\sin\beta + \frac{\pi}{4}\cos\beta + \sqrt{2}\beta$

Differentiating w. r. t β

$$f(\beta) = \beta\cos\beta + \sin\beta - \frac{\pi}{4}\sin\beta + \sqrt{2}$$

$$f\left(\frac{\pi}{2}\right) = \left(1 - \frac{\pi}{4}\right)\sin\frac{\pi}{2} + \sqrt{2} = 1 - \frac{\pi}{4} + \sqrt{2}$$

21. (b) Let $I = \int_{-\pi}^{\pi} \frac{\cos^2 x}{1+a^x} dx$ ....(1)

$$= \int_{-\pi}^{\pi} \frac{\cos^2(-x)}{1+a^{-x}} dx$$

$$\left[\text{Using} \int_a^b f(x)dx = \int_a^b f(a+b-x)dx\right]$$

$$= \int_{-\pi}^{\pi} \frac{\cos^2 x}{1+a^x} dx \quad ....(2)$$

Adding equations (1) and (2) we get

$$2I = \int_{-\pi}^{\pi} \cos^2 x\left(\frac{1+a^x}{1+a^x}\right)dx = \int_{-\pi}^{\pi} \cos^2 x\, dx$$

$$= 2\int_0^{\pi} \cos^2 x\, dx = 2\times 2\int_0^{\frac{\pi}{2}} \cos^2 x\, dx = 4\int_0^{\frac{\pi}{2}} \sin^2 x\, dx$$

$$\Rightarrow I = 2\int_0^{\frac{\pi}{2}} \sin^2 x\, dx = 2\int_0^{\frac{\pi}{2}} \left(1 - \cos^2 x\right)dx$$

$$\Rightarrow I = 2\int_0^{\frac{\pi}{2}} dx - 2\int_0^{\frac{\pi}{2}} \cos^2 x\, dx$$

$$\Rightarrow I + I = 2\left(\frac{\pi}{2}\right) = \pi \Rightarrow I = \frac{\pi}{2}$$

22. (b) $I = \int_3^6 \frac{\sqrt{x}}{\sqrt{9-x}+\sqrt{x}} dx$ ...(1)

$$I = \int_3^6 \frac{\sqrt{9-x}}{\sqrt{9-x}+\sqrt{x}} dx \quad ...(2)$$

[using $\int_a^b f(x)dx = \int_a^b f(a+b-x)dx$]

Adding equation (1) and (2)

$$2I = \int_3^6 dx = 3 \Rightarrow I = \frac{3}{2}$$

23. (d) $I = \int_0^{\pi} x f(\sin x)dx = \int_0^{\pi} (\pi - x) f(\sin x)dx$

$$= \pi\int_0^{\pi} f(\sin x)dx - I$$

$$\Rightarrow 2I = \pi\int_0^{\pi} f(\sin x)dx$$

$$I=\frac{\pi}{2}\int_0^{\pi} f(\sin x)dx=\pi\int_0^{\pi/2} f(\sin x)dx$$

$$=\pi\int_0^{\pi/2} f(\cos x)dx$$

24. **(c)** $I=\int_{-\frac{3\pi}{2}}^{-\frac{\pi}{2}}[(x+\pi)^3+\cos^2(x+3\pi)]dx$

Put $x+\pi=t$

$$I=\int_{-\frac{\pi}{2}}^{\frac{\pi}{2}}[t^3+\cos^2 t)dt=2\int_{-\frac{\pi}{2}}^{\frac{\pi}{2}}\cos^2 t dt$$

[using the property of even and odd function]

$$=\int_0^{\frac{\pi}{2}}(1+\cos 2t)dt=\frac{\pi}{2}+0$$

25. **(b)** Let $a=k+h$ where k is an integer such that $[a]=k$ and $0\le h<1$

$$\therefore \int_1^a [x]f'(x)dx=\int_1^2 1f'(x)dx+\int_2^3 2f'(x)dx+...$$

$$\int_{k-1}^{k}(k-1)dx+\int_k^{k+h} kf'(x)dx$$

$\{f(2)-f(1)\}+2\{f(3)-f(2)\}+3\{f(4)-f(3)\}$
$+........+(k-1)\{f(k)-f(k-1)\}$
$+k\{f(k+h)-f(k)\}$
$=-f(1)-f(2)-f(3)\,.........-f(k)+kf(k+h)$
$=[a]f(a)-\{f(1)+f(2)+f(3)+..........f([a])\}$

26. **(c)** Given $f(x)=f(x)+f\left(\frac{1}{x}\right)$, where $f(x)=\int_1^x\frac{\log t}{1+t}dt$

$$\therefore F(e)=f(e)+f\left(\frac{1}{e}\right)$$

$$\Rightarrow F(e)=\int_1^e\frac{\log t}{1+t}dt+\int_1^{1/e}\frac{\log t}{1+t}dt \quad ....(A)$$

Now for solving, $I=\int_1^{1/e}\frac{\log t}{1+t}dt$

$\therefore$ Put $\frac{1}{t}=z\Rightarrow -\frac{1}{t^2}dt=dz \Rightarrow dt=-\frac{dz}{z^2}$

and limit for $t=1\Rightarrow z=1$ and for $t=1/e\Rightarrow z=e$

$$\therefore I=\int_1^e\frac{\log\left(\frac{1}{z}\right)}{1+\frac{1}{z}}\left(-\frac{dz}{z^2}\right)=\int_1^e\frac{(\log 1-\log z).z}{z+1}\left(-\frac{dz}{z^2}\right)$$

$$=\int_1^e-\frac{\log z}{(z+1)}\left(-\frac{dz}{z}\right) \qquad [\because \log 1=0]$$

$$=\int_1^e\frac{\log z}{z(z+1)}dz$$

$$\therefore I=\int_1^e\frac{\log t}{t(t+1)}dt$$

[By property $\int_a^b f(t)dt=\int_a^b f(x)dx$]

Equation (A) becomes

$$F(e)=\int_1^e\frac{\log t}{1+t}dt+\int_1^e\frac{\log t}{t(1+t)}dt$$

$$=\int_1^e\frac{t.\log t+\log t}{t(1+t)}dt=\int_1^e\frac{(\log t)(t+1)}{t(1+t)}dt$$

$$\Rightarrow F(e)=\int_1^e\frac{\log t}{t}dt$$

Let $\log t=x$ $\therefore \frac{1}{t}dt=dx$

[for limit $t=1, x=0$ and $t=e, x=\log e=1$]

$$\therefore F(e)=\int_0^1 x\,dx \quad F(e)=\left[\frac{x^2}{2}\right]_0^1 \Rightarrow F(e)=\frac{1}{2}$$

27. **(d)** $\int_{\sqrt{2}}^{x}\frac{dt}{t\sqrt{t^2-1}}=\frac{\pi}{2}$

$$\therefore \left[\sec^{-1}t\right]_{\sqrt{2}}^{x}=\frac{\pi}{2} \qquad \left[\because \int\frac{dx}{x\sqrt{x^2-1}}=\sec^{-1}x\right]$$

$$\Rightarrow \sec^{-1}x-\sec^{-1}\sqrt{2}=\frac{\pi}{2}$$

$$\Rightarrow \sec^{-1}x-\frac{\pi}{4}=\frac{\pi}{2}\Rightarrow \sec^{-1}x=\frac{\pi}{2}+\frac{\pi}{4}$$

$$\Rightarrow \sec^{-1}x=\frac{3\pi}{4}\Rightarrow x=\sec\frac{3\pi}{4}\Rightarrow x=-\sqrt{2}$$

28. **(a)** The area enclosed between the curves $y^2=x$ and $y=|x|$

From the figure, area lies between $y^2=x$ and $y=x$

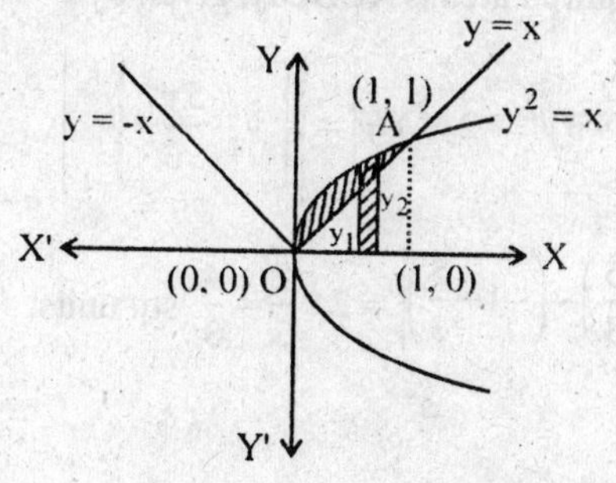

$\therefore$ Required area $= \int_0^1 (y_2 - y_1)dx$

$= \int_0^1 (\sqrt{x} - x)dx = \left[\frac{x^{3/2}}{3/2} - \frac{x^2}{2}\right]_0^1$

$\therefore$ Required area $= \frac{2}{3}\left[x^{3/2}\right]_0^1 - \frac{1}{2}\left[x^2\right]_0^1 = \frac{2}{3} - \frac{1}{2} = \frac{1}{6}$

29. (b) We know that $\frac{\sin x}{x} < 1$, for $x \in (0, 1)$

$\Rightarrow \frac{\sin x}{\sqrt{x}} < \sqrt{x}$ on $x \in (0, 1)$

$\Rightarrow \int_0^1 \frac{\sin x}{\sqrt{x}} dx < \int_0^1 \sqrt{x} dx = \left[\frac{2x^{3/2}}{3}\right]_0^1$

$\Rightarrow \int_0^1 \frac{\sin x}{\sqrt{x}} dx < \frac{2}{3} \Rightarrow I < \frac{2}{3}$ Also $\frac{\cos x}{\sqrt{x}} < \frac{1}{\sqrt{x}}$ for $x \in (0, 1)$

$\Rightarrow \int_0^1 \frac{\cos x}{\sqrt{x}} dx < \int_0^1 x^{-1/2} dx = \left[2\sqrt{x}\right]_0^1 = 2$

$\Rightarrow \int_0^1 \frac{\cos x}{\sqrt{x}} dx < 2 \Rightarrow J < 2$

30. (d) $x + 2y^2 = 0 \Rightarrow y^2 = -\frac{x}{2}$

[Left handed parabola with vertex at (0, 0)]

$x + 3y^2 = 1 \Rightarrow y^2 = -\frac{1}{3}(x-1)$

[Left handed parabola with vertex at (1, 0)]

Solving the two equations we get the points of intersection as (–2, 1), (–2, –1)

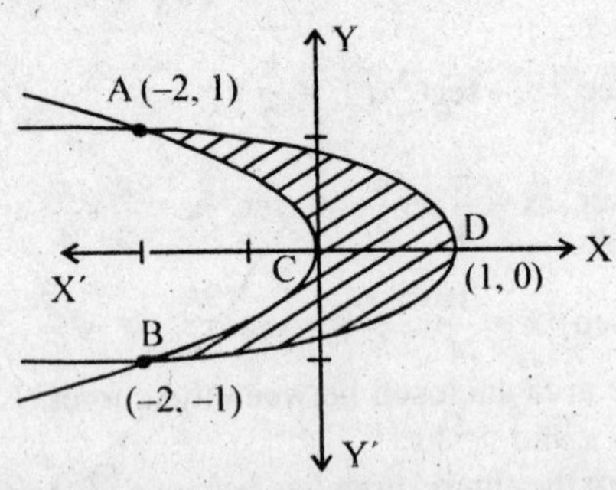

The required area is ACBDA, given by

$= \left|\int_{-1}^{1} (1 - 3y^2 - 2y^2)dy\right| = \left|\left[y - \frac{5y^3}{3}\right]_{-1}^{1}\right|$

$= \left|\left(1 - \frac{5}{3}\right) - \left(-1 + \frac{5}{3}\right)\right| = 2 \times \frac{2}{3} = \frac{4}{3}$ sq. units.

31. (b) The given parabola is $(y-2)^2 = x - 1$

Vertex (1, 2) and it meets x–axis at (5, 0)

Also it gives $y^2 - 4y - x + 5 = 0$

So, that equation of tangent to the parabola at (2, 3) is

$y.3 - 2(y+3) - \frac{1}{2}(x+2) + 5 = 0$

or $x - 2y + 4 = 0$

which meets x-axis at (– 4, 0).

In the figure shaded area is the required area.

Let us draw PD perpendicular to y – axis.

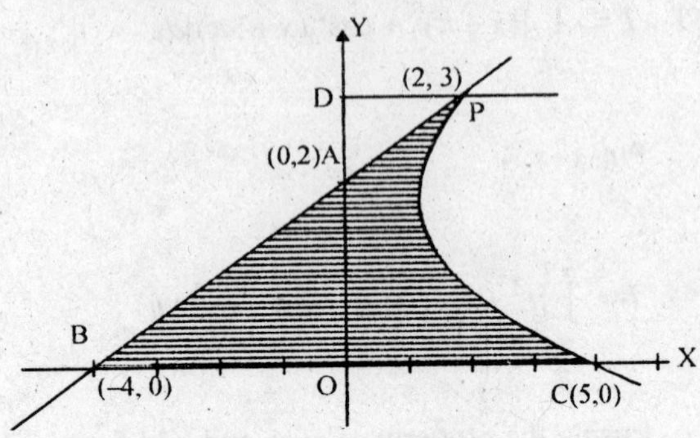

Then required area = Ar ΔBOA + Ar (OCPD) – Ar (ΔAPD)

$= \frac{1}{2} \times 4 \times 2 + \int_0^3 x dy - \frac{1}{2} \times 2 \times 1$

$= 3 + \int_0^3 (y-2)^2 + 1\, dy = 3 + \left[\frac{(y-2)^3}{3} + y\right]_0^3$

$= 3 + \left[\frac{1}{3} + 3 + \frac{8}{3}\right] = 3 + 6 = 9$ Sq. units

32. (c) Let $I = \int_0^{\pi} [\cot x]\, dx$ ....(1)

$= \int_0^{\pi} [\cot(\pi - x)]\, dx = \int_0^{\pi} [-\cot x]\, dx$ ....(2)

Adding two values of I in eq^n s (1) & (2),

We get

$2I = \int_0^{\pi} ([\cot x] + [-\cot x])\, dx$

$= \int_0^{\pi} (-1) dx$

$[\because [x] + [-x] = -1$, if $x \notin z$ and $[x] + [-x] = 0$, if $x \in z]$

$= [-x]_0^{\pi} = -\pi$

$\Rightarrow I = -\frac{\pi}{2}$

33. (d)

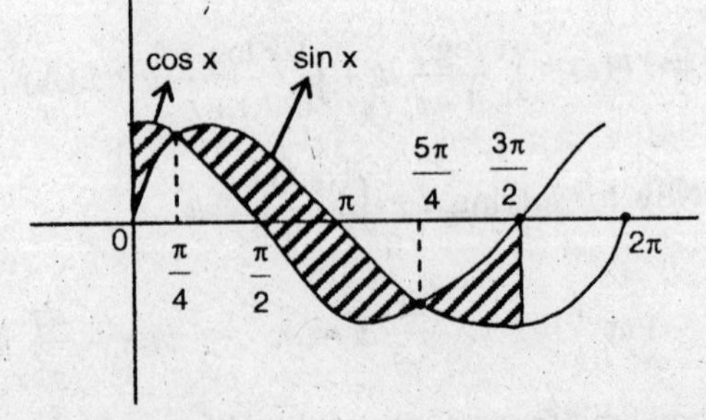

Area above x-axis = Area below x-axis

$\therefore$ Required area

$$= 2\left[\int_0^{\frac{\pi}{4}}(\cos x - \sin x)dx + \int_{\frac{\pi}{4}}^{\frac{\pi}{2}}(\sin x - \cos x)dx + \int_{\frac{\pi}{2}}^{\pi}\sin x\, dx\right]$$

$$= 4\sqrt{2} - 2$$

34. (a) $p'(x) = p'(1-x)$

$\Rightarrow p(x) = -p(1-x) + c$

at $x = 0$

$p(0) = -p(1) + c \Rightarrow 42 = c$

Now, $p(x) = -p(1-x) + 42 \Rightarrow p(x) + p(1-x) = 42$

$\Rightarrow I = \int_0^1 p(x)dx$ ...(i)

$\Rightarrow I = \int_0^1 p(1-x)dx$ ...(ii)

on adding (i) and (ii), $2I = \int_0^1 (42)\, dx \Rightarrow I = 21$

35. (d) $I = \int_0^1 \frac{8\log(1+x)}{1+x^2}dx$

Put $x = \tan\theta$, $\therefore \frac{dx}{d\theta} = \sec^2\theta \Rightarrow dx = \sec^2\theta d\theta$

$\therefore I = 8\int_0^{\pi/4}\frac{\log(1+\tan\theta)}{1+\tan^2\theta}.\sec^2\theta d\theta$

$I = 8\int_0^{\pi/4}\log(1+\tan\theta)d\theta$ ...(i)

$= 8\int_0^{\pi/4}\log\left[1+\tan\left(\frac{\pi}{4}-\theta\right)\right]d\theta$

$= 8\int_0^{\pi/4}\log\left[1+\frac{1-\tan\theta}{1+\tan\theta}\right]d\theta$

$= 8\int_0^{\pi/4}\log\left[\frac{2}{1+\tan\theta}\right]d\theta$

$= 8\int_0^{\pi/4}[\log 2 - \log(1+\tan\theta)]d\theta$

$I = 8.(\log 2)[x]_0^{\pi/4} - 8\int_0^{\pi/4}\log(1+\tan\theta)d\theta$

$I = 8.\frac{\pi}{4}.\log 2 - I$ [From equation (i)]

$\Rightarrow 2I = 2\pi\log 2$

$\therefore I = \pi\log 2$

36. (b) Area of required region $AOBC$

$= \int_0^1 x dx + \int_1^e \frac{1}{x}dx$

$= \frac{1}{2} + 1 = \frac{3}{2}$ sq. units

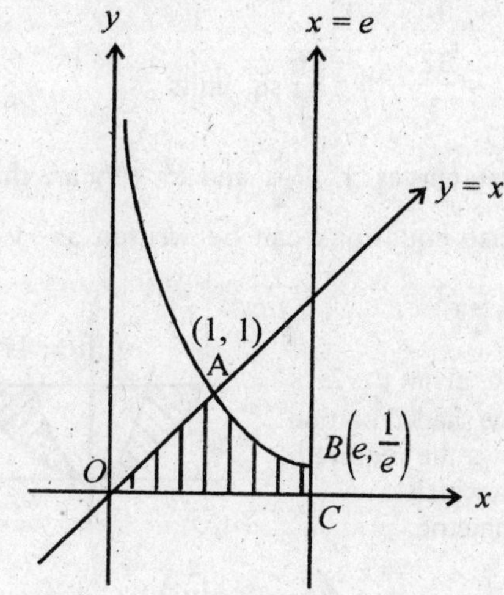

37. (c) $\int_0^{1.5} x\left[x^2\right]dx = \int_0^1 x[x^2]dx + \int_1^{\sqrt{2}} x\left[x^2\right]dx + \int_{\sqrt{2}}^{1.5} x\left[x^2\right]dx$

$= \int_0^1 x.0\, dx + \int_1^{\sqrt{2}} x dx + \int_{\sqrt{2}}^{1.5} 2x dx$

$= 0 + \left[\frac{x^2}{2}\right]_1^{\sqrt{2}} + \left[x^2\right]_{\sqrt{2}}^{1.5}$

$= \frac{1}{2}(2-1) + (2.25-2)$

$= \frac{1}{2} + 0.25$

$= \frac{1}{2} + \frac{1}{4} = \frac{3}{4}$

38. (b)

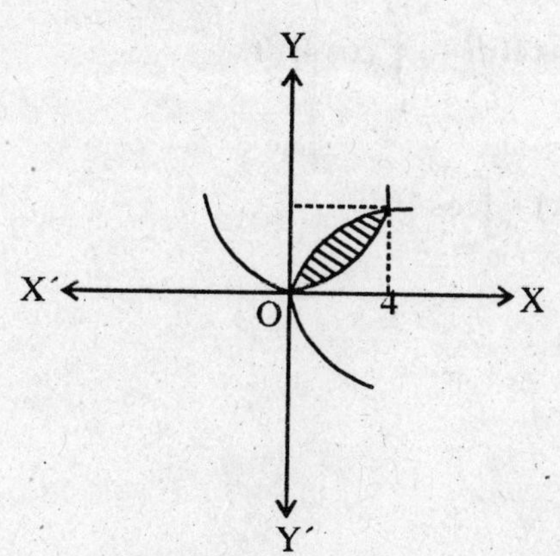

$$\text{Area} = \int_0^4 \left(2\sqrt{x} - \frac{x^2}{4}\right)dx$$

$$= \left[2\left(\frac{x^{3/2}}{3/2}\right) - \frac{x^3}{12}\right]_0^4$$

$$= \frac{4}{3} \times 8 - \frac{64}{12}$$

$$= \frac{32}{3} - 16 = \frac{16}{3} \text{ sq. units}$$

39. **(c)** Given curves $x^2 = \frac{y}{4}$ and $x^2 = 9y$ are the parabolas whose equations can be written as $y = 4x^2$ and $y = \frac{1}{9}x^2$.

Also, given $y = 2$.
Now, shaded portion shows the required area which is symmetric.

$y = 4x^2$ $y = \frac{1}{9}x^2$ $y = 2$ Y X

$$\therefore \quad \text{Area} = 2\int_0^2 \left(\sqrt{9y} - \sqrt{\frac{y}{4}}\right)dy$$

$$\text{Area} = 2\int_0^2 \left(3\sqrt{y} - \frac{\sqrt{y}}{2}\right)dy$$

$$= 2\left[\frac{2}{3} \times 3.y^{\frac{3}{2}} - \frac{1}{2} \times \frac{2}{3}.y^{\frac{3}{2}}\right]_0^2$$

$$= 2\left[2y^{\frac{3}{2}} - \frac{1}{3}y^{\frac{3}{2}}\right] = 2 \times \frac{5}{3}y^{\frac{3}{2}}\Bigg|_0^2$$

$$= 2.\frac{5}{3}2\sqrt{2} = \frac{20\sqrt{2}}{3}$$

40. **(b, c)** $g(x+\pi) = \int_0^{x+\pi} \cos 4t\, dt$

$$= \int_0^{\pi} \cos 4t\, dt + \int_{\pi}^{\pi+x} \cos 4t\, dt$$

$$= g(\pi) + \int_0^{x} \cos 4t\, dt$$

Putting $t = \pi + y$ in second integral, we get

$$\int_x^{\pi+x} \cos 4t\, dt = \int_0^{\pi} \cos 4t\, dt$$

$= g(\pi) + g(x) = g(x) - g(\pi)$

$\therefore g(\pi) = 0)$

41. **(d)** Let $I = \int_{\pi/6}^{\pi/3} \frac{dx}{1+\sqrt{\tan x}}$

$$= \int_{\pi/6}^{\pi/3} \frac{dx}{1+\sqrt{\tan\left(\frac{\pi}{2}-x\right)}} = \int_{\pi/6}^{\pi/3} \frac{\sqrt{\tan x}\, dx}{1+\sqrt{\tan x}} \quad \text{...(i)}$$

Also, Given, I

$$= \int_{\pi/6}^{\pi/3} \frac{\sqrt{\tan x}\, dx}{1+\sqrt{\tan x}} \quad \text{...(2)}$$

By adding (1) and (2), we get

$$2I = \int_{\pi/6}^{\pi/3} dx$$

$$\Rightarrow \quad I = \frac{1}{2}\left[\frac{\pi}{3} - \frac{\pi}{6}\right] = \frac{\pi}{12}, \text{ statement-1 is false}$$

$$\int_a^b f(x)dx = \int_a^b f(a+b-x)\,dx$$

It is fundamental property.

42. **(a)** Given curves are

$y = \sqrt{x}$ ...(1)

and $2y - x + 3 = 0$ ...(2)

On solving both we get $y = -1, 3$

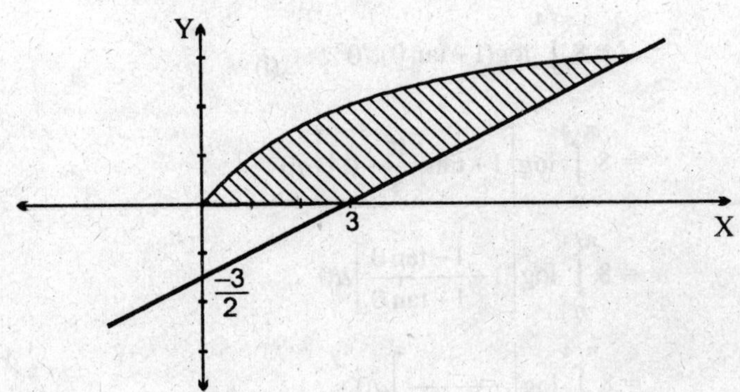

$$\text{Required area} = \int_0^3 \left\{(2y+3) - y^2\right\}dy$$

$$= y^2 + 3y - \frac{y^3}{3}\Bigg|_0^3 = 9.$$

Chapter 19 Differential Equations

Section-A JEE Advanced/ IIT-JEE

(A) FILL IN THE BLANKS :

1. A spherical rain drop evaporates at a rate proportional to its surface area at any instant t. The differential equation giving the rate of change of the radius of the rain drop is .... *(1997C – 2 Marks)*

(C) MCQ's WITH ONE CORRECT ANSWER :

1. A solution of the differential equation *(1999 – 2 Marks)*

$\left(\frac{dy}{dx}\right)^2 - x\frac{dy}{dx} + y = 0$ is

(a) $y=2$ (b) $y=2x$
(c) $y=2x-4$ (d) $y=2x^2-4$

2. If $x^2+y^2=1$, then *(2000S)*

(a) $yy''-2(y')^2+1=0$ (b) $yy''+(y')^2+1=0$
(c) $yy''+(y')^2-1=0$ (d) $yy''+2(y')^2+1=0$

3. If $y(t)$ is a solution of $(1+t)\frac{dy}{dt}-ty=1$ and $y(0)=-1$, then $y(1)$ is equal to *(2003S)*

(a) $-1/2$ (b) $e+1/2$
(c) $e-1/2$ (d) $1/2$

4. If $y=y(x)$ and $\frac{2+\sin x}{y+1}\left(\frac{dy}{dx}\right)=-\cos x, y(0)=1,$

then $y\left(\frac{\pi}{2}\right)$ equals *(2004S)*

(a) 1/3 (b) 2/3
(c) –1/3 (d) 1

5. If $y=y(x)$ and it follows the relation $x\cos y+y\cos x=\pi$ then $y''(0)=$ *(2005S)*

(a) 1 (b) –1
(c) $\pi-1$ (d) $-\pi$

6. The solution of primitive integral equation $(x^2+y^2)\,dy=xy\,dx$ is $y=y(x)$. If $y(1)=1$ and $(x_0)=e$, then x_0 is equal to *(2005S)*

(a) $\sqrt{2(e^2-1)}$ (b) $\sqrt{2(e^2+1)}$
(c) $\sqrt{3}\,e$ (d) $\sqrt{\frac{e^2+1}{2}}$

7. For the primitive integral equation $ydx+y^2dy=x\,dy$; $x\in R, y>0, y=y(x), y(1)=1$, then $y(-3)$ is *(2005S)*

(a) 3 (b) 2
(c) 1 (d) 5

8. The differential equation $\frac{dy}{dx}=\frac{\sqrt{1-y^2}}{y}$ determines a family of circles with *(2005S)*

(a) variable radii and a fixed centre at (0, 1)
(b) variable radii and a fixed centre at (0, –1)
(c) fixed radius 1 and variable centres along the x–axis.
(d) fixed radius 1 and variable centres along the y–axis.

(D) MCQ's WITH ONE OR MORE THAN ONE CORRECT :

1. The order of the differential equation whose general solution is given by *(1998 – 2 Marks)*
$y=(C_1+C_2)\cos(x+C_3)-C_4e^{x+C_5}$, where C_1, C_2, C_3, C_4, C_5, are arbitrary constants, is

(a) 5 (b) 4
(c) 3 (d) 2

2. The differential equation representing the family of curves $y^2=2c\left(x+\sqrt{c}\right)$, where c is a positive parameter, is of *(1999 – 3 Marks)*

(a) order 1 (b) order 2
(c) degree 3 (d) degree 4

3. A curve $y=f(x)$ passes through (1, 1) and at $P(x, y)$, tangent cuts the x–axis and y–axis at A and B respectively such that BP : AP = 3 : 1, then *(2006 – 5M, –1)*

(a) equation of curve is $xy'-3y=0$
(b) normal at (1, 1) is $x+3y=4$
(c) curve passes through (2, 1/8)
(d) equation of curve is $xy'+3y=0$

4. If $y(x)$ satisfies the differential equation $y'-y\tan x = 2x\sec x$ and $y(0)=0$, then *(2012)*

(a) $y\left(\frac{\pi}{4}\right)=\frac{\pi^2}{8\sqrt{2}}$ (b) $y'\left(\frac{\pi}{4}\right)=\frac{\pi^2}{18}$

(c) $y\left(\frac{\pi}{3}\right)=\frac{\pi^2}{9}$ (d) $y'\left(\frac{\pi}{3}\right)=\frac{4\pi}{3}+\frac{2\pi^2}{3\sqrt{3}}$

5. A curve passes through the point $\left(1,\frac{\pi}{6}\right)$. Let the slope of the curve at each point (x, y) be $\frac{y}{x}+\sec\left(\frac{y}{x}\right), x>0$. Then the equation of the curve is *(JEE Adv. 2013)*

(a) $\sin\left(\frac{y}{x}\right)=\log x+\frac{1}{2}$ (b) $\operatorname{cosec}\left(\frac{y}{x}\right)=\log x+2$

(c) $\sec\left(\frac{2y}{x}\right)=\log x+2$ (d) $\cos\left(\frac{2y}{x}\right)=\log x+\frac{1}{2}$

(E) SUBJECTIVE PROBLEMS :

1. If $(a + bx)\, e^{y/x} = x$, then prove that $x^3 \frac{d^2y}{dx^2} = \left(x\frac{dy}{dx} - y\right)^2$

(1983 – 3 Marks)

2. A normal is drawn at a point $P(x, y)$ of a curve. It meets the x–axis at Q. If PQ is of constant length k, then show that the differential equation describing such curves is

$$y\frac{dy}{dx} = \pm\sqrt{k^2 - y^2}$$ *(1994 – 5 Marks)*

Find the equation of such a curve passing through $(0, k)$.

3. Let $y = f(x)$ be a curve passing through (1, 1) such that the triangle formed by the coordinate axes and the tangent at any point of the curve lies in the first quadrant and has area 2. Form the differential equation and determine all such possible curves. *(1995 – 5 Marks)*

4. Determine the equation of the curve passing through the origin, in the form $y = f(x)$, which satisfies the differential equation $\frac{dy}{dx} = \sin(10x + 6y)$. *(1996 – 5 Marks)*

5. A and B are two separate reservoirs of water, Capacity of reservoir A is double the capacity of reservoir B. Both the reservoirs are filled completely with water, their inlets are closed and then the water is released simultaneously from both the reservoirs. The rate of flow of water out of each reservoir at any instant of time is proportional to the quantity of water in the reservoir at the time. One hour after the water is released, the quantity of water in reservoir A is $1\frac{1}{2}$ times the quantity of water in reservoir B. After how many hours do both the reservoirs have the same quantity of water?

(1997C – 5 Marks)

6. Let $u(x)$ and $v(x)$ satisfy the differential equation $\frac{du}{dx} + p(x)\,u = f(x)$ and $\frac{dv}{dx} + p(x)\,v = g(x)$, where $p(x)$ $f(x)$ and $g(x)$ are continuous functions. If $u(x_1) > v(x_1)$ for some x_1 and $f(x) > g(x)$ for all $x > x_1$, prove that any point (x, y) where $x > x_1$, does not satisfy the equatons $y = u(x)$ and $y = v(x)$.

(1997 – 5 Marks)

7. A curve passing through the point (1, 1) has the property that the perpendicular distance of the origin from the normal at any point P of the curve is equal to the distance of P from the x–axis. Determine the equation of the curve.

(1999 – 10 Marks)

8. A country has a food deficit of 10%. Its population grows continously at a rate of 3% per year. Its annual food production every year is 4% more than that of the last year. Assuming that the average food requirement per person remains constant, prove that the country will become self–sufficient in food after n years, where n is the smallest integer bigger than or equal to $\frac{\ln 10 - \ln 9}{\ln(1.04) - 0.03}$. *(2000 – 10 Marks)*

9. A hemispherical tank of radius 2 metres is initially full of water and has an outlet of 12 cm^2 cross–sectional area at the bottom. The outlet is opened at some instant. The flow through the outlet is according to the law $v(t) = 0.6\sqrt{2gh(t)}$, where $v(t)$ and $h(t)$ are respectively the velocity of the flow through the outlet and the height of water level above the outlet at time t, and g is the acceleration due to gravity. Find the time it takes to empty the tank. (Hint : Form a differential equation by relating the decrease of water level to the outflow). *(2001 – 10 Marks)*

10. A right circular cone with radius R and height H contains a liquid which evaporates at a rate proportional to its surface area in contact with air (proportionality constant = $k > 0$). Find the time after which the cone is empty.

(2003 – 4 Marks)

11. A curve 'C' passes through (2, 0) and the slope at (x, y) as $\frac{(x+1)^2 + (y-3)}{x+1}$. Find the equation of the curve. Find the area bounded by curve and x–axis in fourth quadrant.

(2004 – 4 Marks)

12. If length of tangent at any point on the curve $y = f(x)$ intecepted between the point and the x–axis is of length 1. Find the equation of the curve. *(2005 – 4 Marks)*

(F) MATCH THE FOLLOWING :

Each question contains statements given in two columns, which have to be matched. The statements in Column-I are labelled A, B, C and D, while the statements in Column-II are labelled p, q, r, s and t. Any given statement in Column-I can have correct matching with ONE OR MORE statement(s) in Column-II. The appropriate bubbles corresponding to the answers to these questions have to be darkened as illustrated in the following example :

If the correct matches are A-p, s and t; B-q and r; C-p and q; and D-s then the correct darkening of bubbles will look like the given.

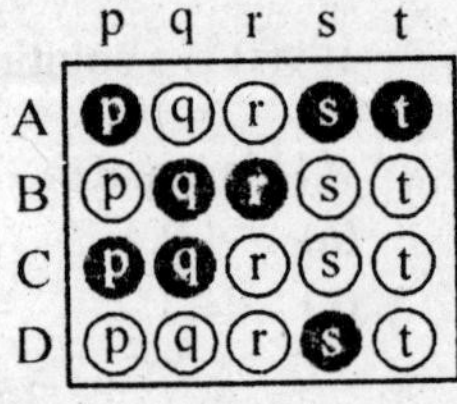

1. Match the statements/expressions in **Column I** with the open intervals in **Column II**. *(2009)*

Column I		Column II
(A) Interval contained in the domain of definition of non-zero solutions of the differential equation $(x-3)^2 y' + y = 0$	(p)	$\left(-\frac{\pi}{2}, \frac{\pi}{2}\right)$
(B) Interval containing the value of the integral $\int_1^5 (x-1)(x-2)(x-3)(x-4)(x-5)dx$	(q)	$\left(0, \frac{\pi}{2}\right)$
(C) Interval in which at least one of the points of local maximum of $\cos^2 x + \sin x$ lies	(r)	$\left(\frac{\pi}{8}, \frac{5\pi}{4}\right)$
(D) Interval in which $\tan^{-1}(\sin x + \cos x)$ is increasing	(s)	$\left(0, \frac{\pi}{8}\right)$
	(t)	$(-\pi, \pi)$

(H) ASSERTION & REASON TYPE QUESTIONS :

1. Let a solution $y = y(x)$ of the differential equation $x\sqrt{x^2-1}\,dy - y\sqrt{y^2-1}\,dx = 0$ satisfy $y(2) = \frac{2}{\sqrt{3}}$.

STATEMENT-1 : $y(x) = \sec\left(\sec^{-1}x - \frac{\pi}{6}\right)$ and

STATEMENT-2 : $y(x)$ is given by $\frac{1}{y} = \frac{2\sqrt{3}}{x} - \sqrt{1-\frac{1}{x^2}}$ *(2008)*

(a) STATEMENT - 1 is True, STATEMENT - 2 is True; STATEMENT - 2 is a correct explanation for STATEMENT - 1
(b) STATEMENT - 1 is True, STATEMENT - 2 is True; STATEMENT - 2 is NOT a correct explanation for STATEMENT - 1
(c) STATEMENT - 1 is True, STATEMENT - 2 is False
(d) STATEMENT - 1 is False, STATEMENT - 2 is True

(I) INTEGER VALUE CORRECT TYPE :

1. Let $y'(x) + y(x)g'(x) = g(x), g'(x), y(0) = 0, x \in R$, where $f'(x)$ denotes $\frac{df(x)}{dx}$ and $g(x)$ is a given non-constant differentiable function on R with $g(0) = g(2) = 0$. Then the value of $y(2)$ is *(2011)*

Section-B JEE Main/ AIEEE

1. The order and degree of the differential equation $\left(1+3\frac{dy}{dx}\right)^{2/3} = 4\frac{d^3y}{dx^3}$ are **[2002]**
(a) $(1, \frac{2}{3})$ (b) $(3, 1)$
(c) $(3, 3)$ (d) $(1, 2)$

2. The solution of the equation $\frac{d^2y}{dx^2} = e^{-2x}$ **[2002]**
(a) $\frac{e^{-2x}}{4}$ (b) $\frac{e^{-2x}}{4} + cx + d$
(c) $\frac{1}{4}e^{-2x} + cx^2 + d$ (d) $\frac{1}{4}e^{-4x} + cx + d$

3. The degree and order of the differential equation of the family of all parabolas whose axis is x - axis, are respectively.
(a) 2, 3 (b) 2, 1 **[2003]**
(c) 1, 2 (d) 3, 2.

4. The solution of the differential equation $(1+y^2) + (x - e^{\tan^{-1}y})\frac{dy}{dx} = 0$, is **[2003]**
(a) $xe^{2\tan^{-1}y} = e^{\tan^{-1}y} + k$
(b) $(x-2) = ke^{2\tan^{-1}y}$
(c) $2xe^{\tan^{-1}y} = e^{2\tan^{-1}y} + k$
(d) $xe^{\tan^{-1}y} = \tan^{-1}y + k$

5. The - differential equation for the family of circle $x^2 + y^2 - 2ay = 0$, where a is an arbitrary constant is **[2004]**
(a) $(x^2+y^2)y' = 2xy$ (b) $2(x^2+y^2)y' = xy$
(c) $(x^2-y^2)y' = 2xy$ (d) $2(x^2-y^2)y' = xy$

6. Solution of the differential equation $ydx + (x + x^2y)dy = 0$ is **[2004]**
(a) $\log y = Cx$ (b) $-\frac{1}{xy} + \log y = C$
(c) $\frac{1}{xy} + \log y = C$ (d) $-\frac{1}{xy} = C$

7. The differential equation representing the family of curves $y^2 = 2c(x+\sqrt{c})$, where $c > 0$, is a parameter, is of order and degree as follows : **[2005]**
(a) order 1, degree 2 (b) order 1, degree 1
(c) order 1, degree 3 (d) order 2, degree 2

8. If $x\frac{dy}{dx} = y(\log y - \log x + 1)$, then the solution of the equation is **[2005]**
(a) $y\log\left(\frac{x}{y}\right) = cx$ (b) $x\log\left(\frac{y}{x}\right) = cy$
(c) $\log\left(\frac{y}{x}\right) = cx$ (d) $\log\left(\frac{x}{y}\right) = cy$

9. The differential equation whose solution is $Ax^2 + By^2 = 1$ where A and B are arbitrary constants is of
(a) second order and second degree **[2006]**
(b) first order and second degree
(c) first order and first degree
(d) second order and first degree

10. The differential equation of all circles passing through the origin and having their centres on the x-axis is **[2007]**
(a) $y^2 = x^2 + 2xy\frac{dy}{dx}$ (b) $y^2 = x^2 - 2xy\frac{dy}{dx}$
(c) $x^2 = y^2 + xy\frac{dy}{dx}$ (d) $x^2 = y^2 + 3xy\frac{dy}{dx}$.

11. The soluton of the differential equation $\frac{dy}{dy} = \frac{x+y}{x}$ satisfying the condition $y(1) = 1$ is
(a) $y = \ln x + x$ (b) $y = x\ln x + x^2$ **[2008]**
(c) $y = xe^{(x-1)}$ (d) $y = x\ln x + x$

12. The differential equation which represents the family of curves $y = c_1 e^{c_2 x}$, where c_1, and c_2 are arbitrary constants, is [2009]
(a) $y'' = y'y$ (b) $yy'' = y'$
(c) $yy'' = (y')^2$ (d) $y' = y^2$

13. Solution of the differential equation
$\cos x\, dy = y(\sin x - y)\, dx,\ 0 < x < \frac{\pi}{2}$ is [2010]
(a) $y \sec x = \tan x + c$ (b) $y \tan x = \sec x + c$
(c) $\tan x = (\sec x + c)\, y$ (d) $\sec x = (\tan x + c)\, y$

14. If $\frac{dy}{dx} = y + 3 > 0$ and $y(0) = 2$, then $y(\ln 2)$ is equal to : [2011]
(a) 5 (b) 13
(c) −2 (d) 7

15. Let I be the purchase value of an equipment and $V(t)$ be the value after it has been used for t years. The value $V(t)$ depreciates at a rate given by differential equation $\frac{dV(t)}{dt} = -k(T - t)$, where $k > 0$ is a constant and T is the total life in years of the equipment. Then the scrap value $V(T)$ of the equipment is [2011]
(a) $I - \frac{kT^2}{2}$ (b) $I - \frac{k(T-t)^2}{2}$
(c) e^{-kT} (d) $T^2 - \frac{1}{k}$

16. Consider the differential equation
$y^2 dx + \left(x - \frac{1}{y}\right) dy = 0$. If $y(1) = 1$,
then x is given by : [2011 RS]
(a) $4 - \frac{2}{y} - \frac{e^{\frac{1}{y}}}{e}$ (b) $3 - \frac{1}{y} + \frac{e^{\frac{1}{y}}}{e}$
(c) $1 + \frac{1}{y} - \frac{e^{\frac{1}{y}}}{e}$ (d) $1 - \frac{1}{y} + \frac{e^{\frac{1}{y}}}{e}$

17. The population $p(t)$ at time t of a certain mouse species satisfies the differential equation $\frac{dp(t)}{dt} = 0.5\, p(t) - 450$. If $p(0) = 850$, then the time at which the population becomes zero is : [2012]
(a) $2 \ln 18$ (b) $\ln 9$ (c) $\frac{1}{2} \ln 18$ (d) $\ln 18$

18. At present, a firm is manufacturing 2000 items. It is estimated that the rate of change of production P w.r.t. additional number of workers x is given by $\frac{dP}{dx} = 100 - 12\sqrt{x}$. If the firm employs 25 more workers, then the new level of production of items is
(a) 2500 (b) 3000 [JEE M 2013]
(c) 3500 (d) 4500

Solutions & Explanations

Section-A : JEE Advanced/ IIT-JEE

A 1. $\frac{dr}{dt} = -k$

C 1. (c) 2. (b) 3. (a) 4. (a) 5. (c) 6. (c) 7. (a)
8. (c)

D 1. (c) 2. (a, c) 3. (c, d) 4. (a, d) 5. (a)

E 3. $x + y = 2$ and $xy = 1,\ x, y > 0$ 4. $y = \frac{1}{3}\left[\tan^{-1}\left(\frac{5\tan 4x}{4 - 3\tan 4x}\right) - 5x\right]$

5. $T = \left(\frac{\log 2}{\log \frac{4}{3}}\right)$ 7. $x^2 + y^2 - 2x = 0,\ x - 1 = 0$ 9. $\frac{14\pi}{27\sqrt{g}}(10)^5$ units

10. H/k 11. $\frac{4}{3}$ sq. units 12. $\log\left|\frac{1 - \sqrt{1 - y^2}}{y}\right| + \sqrt{1 - y^2} = \pm x + c$

F 1. (A) → p,q,r,s,t; (B) → p,t; (C) → p,q,r,t; (D) → s

H 1. (c)

I 1. 0

Section-B : JEE Main/ AIEEE

1. (c) 2. (b) 3. (d) 4. (c) 5. (c) 6. (b) 7. (c) 8. (c) 9. (d) 10. (a) 11. (d) 12. (c)

Section-A JEE Advanced/ IIT-JEE

A FILL IN THE BLANKS :

1. If S denotes the surface area and V volume of the rain drop

then $ATQ\ \frac{dV}{dt} \propto -S$

$\Rightarrow \quad \frac{dV}{dt} = -k\,S$ where $k > 0$

('–'ve sign shows V decreases with time.)

$\Rightarrow \quad \frac{d}{dt}\left[\frac{4}{3}\pi r^3\right] = -k(4\pi r^2)$

$\Rightarrow \quad 4\pi r^2 \frac{dr}{dt} = -k(4\pi r^2) \Rightarrow \quad \frac{dr}{dt} = -k$

C MCQ's WITH ONE CORRECT ANSWER :

1. (c) 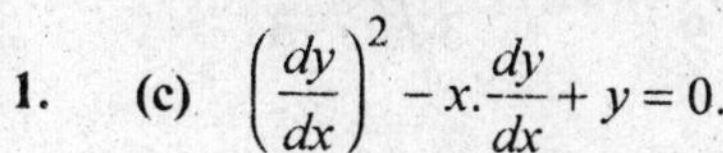 $\left(\frac{dy}{dx}\right)^2 - x.\frac{dy}{dx} + y = 0.$

By actual verification we find that the choice (c), i.e. $y = 2x - 4$ satisfies the given differential equation.

2. (b) Given $x^2 + y^2 = 1$. Differentiating w.r.t. x, we get $2x + 2yy' = 0$ or $x + yy' = 0$. Again differentiating w.r.t. x,

we get $1 + y'y' + yy'' = 0$ or $1 + (y')^2 + yy'' = 0$

3. (a) The given differential equation is

$$\frac{dy}{dt} - \frac{t}{1+t}y = \frac{1}{1+t}$$

I.F. $= e^{-\int \frac{t}{1+t}dt} = e^{-\int\left(1-\frac{1}{1+t}\right)dt} = e^{-(t-\log(1+t))}$

$= e^{-t}.e^{\log(1+t)} = (1+t)e^{-t}$ ∴ Solution is

$$y.e^{-t}(1+t) = \int \frac{1}{(1+t)} e^{-t}(1+t)\,dt + C$$

$\Rightarrow \quad y.e^{-t}(1+t) = -e^{-t} + C \Rightarrow \quad y = -\frac{1}{1+t} + \frac{Ce^t}{1+t}$

Given that $y(0) = -1 \Rightarrow \quad -1 = -1 + C \Rightarrow C = 0$

$\therefore\ y = -\frac{1}{1+t} \quad \therefore \quad y(1) = -\frac{1}{1+1} = -\frac{1}{2}$

4. (a) $\frac{dy}{dx}\left(\frac{2+\sin x}{1+y}\right) = -\cos x, y(0) = 1$

$\Rightarrow \quad \frac{dy}{(1+y)} = \frac{-\cos x}{2+\sin x}dx$

Integrating both sides

$\Rightarrow \quad \ln(1+y) = -\ln(2+\sin x) + C$

Put $x = 0$ and $y = 1 \Rightarrow \ln(2) = -\ln 2 + C \Rightarrow C = \ln 4$

Put $x = \frac{\pi}{2}$ $\ln(1+y) = -\ln 3 + \ln 4 = \ln\frac{4}{3} \Rightarrow \quad y = \frac{1}{3}$

5. (c) Given that $y = y(x)$

and $x\cos y + y\cos x = \pi$ ...(1)

For $x = 0$ in (1) we get $y = \pi$

Differentiating (1) with respect to x, we get

$-x\sin y.y' + \cos y + y'\cos x - y\sin x = 0$

$\Rightarrow \quad y' = \frac{y\sin x - \cos y}{\cos x - x\sin y}$ ...(2)

$\Rightarrow \quad y'(0) = 1$ (Using $y(0) = \pi$)

Differentiating (2) with respect to x, we get

$$y'' = \frac{(y'\sin x + y\cos x + \sin y.y')(\cos x - x\sin y) - (-\sin x - \sin y - x\cos y y')(y\sin x - \cos y)}{(\cos x - x\sin y)^2}$$

$\Rightarrow \quad y''(0) = \frac{\pi(1) - 1}{1} = \pi - 1$

6. (c) The given D.E. is $(x^2 + y^2)dy = xy\,dx$ s.t. $y(1) = 1$ and $y(x_0) = e$

The given eq^n can be written as

$$\frac{dy}{dx} = \frac{xy}{x^2 + y^2}$$

Put $y = vx$, $\therefore\ v + x\frac{dv}{dx} = \frac{v}{1+v^2}$

$\Rightarrow \quad x\frac{dv}{dx} = \frac{-v^3}{1+v^2} \Rightarrow \int \frac{1+v^2}{v^3}dv + \int \frac{dx}{x} = 0$

$\Rightarrow \quad -\frac{1}{2v^2} + \log|v| + \log|x| = C$

$\Rightarrow \quad \log y = C + \frac{x^2}{2y^2}$ (using $v = y/x$)

Also, $y(1) = 1 \Rightarrow \log 1 = C + \frac{1}{2} \Rightarrow C = -\frac{1}{2}$

$\therefore \quad \log y = \frac{x^2 - y^2}{2y^2}$, But given $y(x_0) = e$

$\Rightarrow \quad \log e = \frac{x_0^2 - e^2}{2e^2} \Rightarrow \quad x_0^2 = 3e^2 \Rightarrow \quad x_0 = \sqrt{3}e$

7. (a) The given eq^n is

$$ydx + y^2dy = x\,dy\ ; x \in R, y > 0, y(1) = 1$$

$\Rightarrow \quad \frac{ydx - xdy}{y^2} + dy = 0 \Rightarrow \quad \frac{d}{dx}\left(\frac{x}{y}\right) + dy = 0$

On integrating, we get

$\frac{x}{y} + y = C \quad y(1) = 1. \Rightarrow 1 + 1 = C \Rightarrow C = 2$

$\therefore \quad \frac{x}{y} + y = 2$

Now to find $y(-3)$, putting $x=-3$ in above eq^n we get

$$-\frac{3}{y}+y=2 \Rightarrow y^2-2y-3=0 \Rightarrow y=3,-1$$

But given that $y>0$, $\therefore$ $y=3$

8. **(c)** $\frac{dy}{dx}=\frac{\sqrt{1-y^2}}{y} \Rightarrow \frac{-2y}{\sqrt{1-y^2}}dy+2dx=0$

$$\Rightarrow 2\sqrt{1-y^2}+2x=2c \Rightarrow \sqrt{1-y^2}+x=c$$

$$\Rightarrow (x-c)^2+y^2=1$$

which is a circle of fixed radius 1 and variable centre $(c, 0)$ lying on x-axis.

D MCQ's WITH ONE OR MORE THAN ONE CORRECT :

1. **(c)** The given solution of D.E. is

$$y=(c_1+c_2)\cos(x+c_3)=c_4e^{x+c_5}$$

$$=(c_1+c_2)\cos(x+c_3)-c_4.e^{c_5}.e^x$$

$$=A\cos(x+c_3)-Be^x$$

[Taking $c_1+c_2=A, c_4e^{c_5}=B$]

Thus, there are actually three arbitrary constants and hence this differential equation should be of order 3.

2. **(a, c)** $2yy_1=2c \Rightarrow c=yy_1$

Eliminating, c, we get,

$$y^2=2yy_1(x+\sqrt{yy_1}) \text{ or } (y-2xy_1)^2=4yy_1^3$$

It involves only Ist order derivative, its order is 1 but its degree is 3 as y_1^3 is there.

3. **(c, d)** Tangent to the curve $y=f(x)$ at (x, y) is

$$Y-y=\frac{dy}{dx}(X-x)$$

$$\therefore \quad A\left(\frac{x\frac{dy}{dx}-y}{\frac{dy}{dx}},0\right);\ B\left(0,-x\frac{dy}{dx}+y\right)$$

$$\because \quad BP:PA=3:1 \Rightarrow x=\frac{\dfrac{3\left(x\frac{dy}{dx}-y\right)}{\frac{dy}{dx}}+1\times 0}{4}$$

$$\Rightarrow x\frac{dy}{dx}+3y=0 \Rightarrow \int\frac{dy}{y}=\int -3\frac{dx}{x}$$

$$\Rightarrow \log y=-3\log x+\log c \Rightarrow y=\frac{c}{x^3}$$

As curve passes through (1, 1), $c=1$

$\therefore$ curve is $x^3y=1$, which also passes through $\left(2,\frac{1}{8}\right)$.

4. **(a, d)**
The given differential equation is

$$\frac{dy}{dx}-y\tan x=2x\sec x,$$

$$\text{I.F.}=e^{-\int\tan x dx}=\cos x$$

$$\therefore \quad y.\cos x=\int 2x\,dx=x^2+c \quad y(0)=0 \Rightarrow c=0$$

$$\therefore y=x^2\sec x$$

Now at $x=\frac{\pi}{4}$, $y=\frac{\pi^2}{16}\times\sqrt{2}=\frac{\pi^2}{8\sqrt{2}}$

At $x=\frac{\pi}{3}$, $y=\frac{\pi^2}{9}\times 2=\frac{2\pi^2}{9}$

At $x=\frac{\pi}{4}$, $y'=\frac{2\pi}{4}\times\sqrt{2}+\frac{\pi^2}{8\sqrt{2}}\times 1=\frac{\pi^2}{8\sqrt{2}}+\frac{\pi}{\sqrt{2}}$

At $x=\frac{\pi}{3}$, $y'=\frac{2\pi}{3}\times 2+\frac{2\pi^2}{9}\times\sqrt{3}=\frac{2\pi^2}{3\sqrt{3}}+\frac{4\pi}{3}$

5. **(a)** $\frac{dy}{dx}=\frac{y}{x}+\sec\frac{y}{x}$

Putting $y=vx$ and $\frac{dy}{dx}=v+x\frac{dv}{dx}$, we get $x\frac{dv}{dx}=\sec v$

or $\int\cos v\,dv=\int\frac{dx}{x}$

$\Rightarrow \sin v=\log x+c \qquad (\because x>0)$

$\Rightarrow \sin\frac{y}{x}=\log x+c$

It passes through $\left(1,\frac{\pi}{6}\right) \Rightarrow C=\frac{1}{2}$

$\therefore \quad \sin\frac{y}{x}=\log x+\frac{1}{2}$

E SUBJECTIVE PROBLEMS :

1. **(a)** $(a+bx)e^{\frac{y}{x}}=x$

$$\Rightarrow e^{\frac{y}{x}}=\frac{x}{a+bx} \qquad ...(1)$$

Diff. w.r. to x, we get

$$e^{\frac{y}{x}}\frac{\left[(x)\frac{dy}{dx}-y\right]}{x^2}=\frac{a+bx-bx}{(a+bx)^2}$$

$$\text{or} \quad \left(x\frac{dy}{dx}-y\right)e^{\frac{y}{x}}=\frac{ax^2}{(a+bx)^2} \qquad ...(2)$$

From (1) using $e^{\frac{y}{x}}=\frac{x}{a+bx}$, we get

$$\left(x\frac{dy}{dx}-y\right)=\frac{ax}{a+bx} \qquad ...(3)$$

Differentiating (3) w. r. to x, we get

$$x\frac{d^2y}{dx^2}+\frac{dy}{dx}-\frac{dy}{dx}=\frac{(a+bx)a-axb}{(a+bx)^2}$$

or $\quad x\frac{d^2y}{dx^2}=\frac{a^2}{(a+bx)^2}$

$$\Rightarrow \quad x^3\frac{d^2y}{dx^2}=\left(\frac{ax}{a+bx}\right)^2 \qquad ...(4)$$

Comparing (3) and (4) we get

$$x^3\frac{d^2y}{dx^2}=\left(x\frac{dy}{dx}-y\right)^2$$

2. The length of normla PQ to any curve $y=f(x)$ is given by

$$y\sqrt{1+\left(\frac{dy}{dx}\right)^2}$$

According to question length of $PQ=k$

$$\Rightarrow \quad \left(y\frac{dy}{dx}\right)^2+y^2=k^2$$

$$\Rightarrow \quad y\frac{dy}{dx}=\pm\sqrt{k^2-y^2}$$

which is the required differential equation of given curve. On solving this D.E. we get the eq^n of curve as follows

$$\int\frac{ydy}{\sqrt{k^2-y^2}}=\int\pm dx$$

$$-\frac{1}{2}.2\sqrt{k^2-y^2}=\pm x+C$$

$$-\sqrt{k^2-y^2}=\pm x+C$$

As it passes through (0, k) we get $C=0$

$\therefore$ Eq^n of curve is

$$-\sqrt{k^2-y^2}=\pm x \quad\text{or}\quad x^2+y^2=k^2$$

3. Equation of the tangent to the curve $y=f(x)$ at point (x, y) is

$$Y-y=f'(x)(X-x) \qquad ...(1)$$

The line (1) meets X-axis at $P\left(x-\frac{y}{f'(x)},0\right)$ and Y-axis in $Q(0, y-xf'(x))$

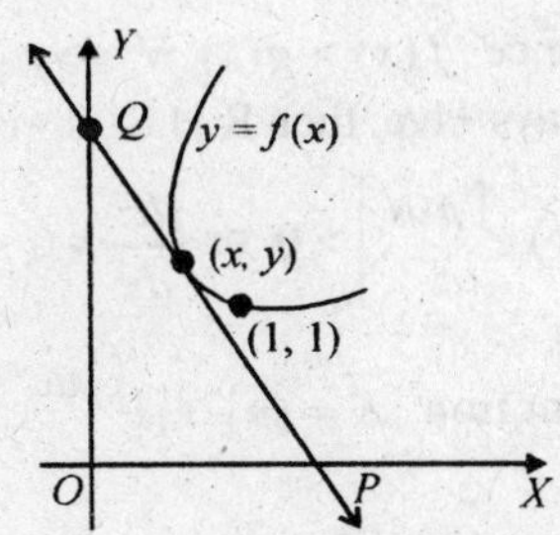

Area of triangle OPQ is

$$=\frac{1}{2}(OP)(OQ)=\frac{1}{2}\left(x-\frac{y}{f'(x)}\right)(y-xf'(x))$$

$$=-\frac{(y-xf'(x))^2}{2f'(x)}$$

We are given that area of $\Delta OPQ=2$

$$\Rightarrow \quad \frac{-(y-xf'(x))^2}{2f'(x)}=2$$

$$\Rightarrow \quad (y-xf'(x))^2+4f'(x)=0$$

$$\Rightarrow \quad (y-px)^2+4p=0 \qquad ...(2)$$

where $\quad p=f'(x)=\frac{dy}{dx}$

Since $OQ>0, y-xf'(x)>0$. Also note that

$$p=f'(x)<0$$

We can write (2) as

$$y-px=2\sqrt{-p}$$

$$\Rightarrow \quad y=px+2\sqrt{-p} \qquad ...(3)$$

Differentiating (3) with respect to x, we get

$$\frac{dy}{dx}=p+\frac{dp}{dx}x+2\left(\frac{1}{2}\right)(-p)^{-1/2}(-1)\frac{dp}{dx}$$

$$\Rightarrow \quad \frac{dp}{dx}x-(-p)^{-1/2}\frac{dp}{dx}=0$$

$$\Rightarrow \quad \frac{dp}{dx}[x-(-p)^{-1/2}]=0$$

$$\Rightarrow \quad \frac{dp}{dx}=0 \text{ or } x=(-p)^{-1/2}$$

If $\frac{dp}{dx}=0$, then $p=c$ where $c<0$ [$\because p<0$]

Putting this value in (3) we get

$$y=cx+2\sqrt{-c} \qquad ...(4)$$

This curve will pass through (1, 1) if

$$1=c+2\sqrt{-c} \Rightarrow -c-2\sqrt{-c}+1=0$$

$$\Rightarrow \quad (\sqrt{-c}-1)^2=0$$

or $\quad \sqrt{-c}=1 \Rightarrow -c=1$ or $c=-1$

Putting the value of c in (4), we get

$$y=-x+2, \text{ or } x+y=2$$

Next, putting $x=(-p)^{-1/2}$ or $-p=x^{-2}$ in (3) we get

$$y=\frac{-x}{x^2}+2\left(\frac{1}{x}\right)=\frac{1}{x}$$

$$\Rightarrow \quad xy=1(x>0, y>0)$$

Thus, the two required curves are $x+y=2$ and $xy=1, (x>0, y>0)$.

4. Put $10x+6y=v$

$\therefore \quad 10+6\dfrac{dy}{dx}=\dfrac{dv}{dx} \quad \therefore \quad \dfrac{dv}{dx}-10=6\sin v$

$\Rightarrow \quad \dfrac{dv}{6\sin v+10}=dx$ or $\dfrac{dv}{12\sin\frac{v}{2}\cos\frac{v}{2}+10}=dx$

Divide numerator and denomenator by $\cos^2\left(\frac{v}{2}\right)$ and put

$\tan\left(\frac{v}{2}\right)=t$

$\therefore \quad \dfrac{2dt}{12t+10(1+t^2)}=dx$ or $\dfrac{dt}{5t^2+6t+5}=dx$

$\dfrac{dt}{\left(t+\frac{3}{5}\right)^2+\left(\frac{4}{5}\right)^2}=5dx$

or $\dfrac{5}{4}\tan^{-1}\dfrac{5t+3}{4}=5x+5c$ or $\tan^{-1}\dfrac{5t+3}{4}=4x+c$

At origin $x=0, y=0$

$\therefore \quad v=0, \therefore t=\tan\frac{v}{2}=0$

Hence, from above

$\tan^{-1}\dfrac{3}{4}=c \Rightarrow \tan^{-1}\dfrac{5t+3}{4}-\tan^{-1}\dfrac{3}{4}=4x$

or $\dfrac{\frac{5t+3}{4}-\frac{3}{4}}{1+\frac{5t+3}{4}\frac{3}{4}}=\tan 4x$ or $\dfrac{20t}{25+15t}=\tan 4x$

or $4t=(5+3t)\tan 4x$ or $t(4-3\tan 4x)=5\tan 4x$

or $\tan\dfrac{v}{2}=\dfrac{5\tan 4x}{4-3\tan 4x}$

or $\tan(5x+3y)=\dfrac{5\tan 4x}{4-3\tan 4x}$

or $5x+3y=\tan^{-1}\left(\dfrac{5\tan 4x}{4-3\tan 4x}\right)$

or $y=\dfrac{1}{3}\left(\tan^{-1}\left(\dfrac{5\tan 4x}{4-3\tan 4x}\right)-5x\right)$

5. Let at any instant t, x be the vol. of water in the reserver A and y of that in B.

Then $\dfrac{dx}{dt}\propto x \Rightarrow \dfrac{dx}{dt}=k_1x$

and $\dfrac{dy}{dt}\propto y \Rightarrow \dfrac{dy}{dt}=k_2y$

$\Rightarrow \quad \dfrac{dx}{x}=k_1dt \Rightarrow \log x=k_1t+C_1$

and $\dfrac{dy}{y}=k_2dt \Rightarrow \log y=k_2t+C_2$

$\Rightarrow \quad x=e^{k_1t}.e^{C_1}$ ...(1)

and $y=e^{k_2t}.e^{C_2}$ ...(2)

Now at $t=0, x=2y$ i.e., $\dfrac{x}{y}=2$

$\therefore$ From (1) and (2) we get

$\dfrac{e^{C_1}}{e^{C_2}}=2$ ...(3)

Also at $t=1, x=\frac{3}{2}y$, i.e., $\dfrac{x}{y}=\dfrac{3}{2}$

$\Rightarrow \quad \dfrac{e^{k_1}}{e^{k_2}}\dfrac{e^{C_1}}{e^{C_2}}=\dfrac{3}{2} \Rightarrow e^{k_1-k_2}=\dfrac{3}{4}$

Let at $t=T, x=y$, i.e. $\dfrac{x}{y}=1$

then $\dfrac{e^{k_1T}}{e^{k_2T}}\dfrac{e^{C_1}}{e^{C_2}}=1 \Rightarrow e^{(k_1-k_2)T}.2=1$

$\Rightarrow \quad \left(\dfrac{3}{4}\right)^T=\dfrac{1}{2}$

Taking log on both sides, we get

$T\log\left(\dfrac{3}{4}\right)=\log\left(\dfrac{1}{2}\right)$

$\Rightarrow . \ T=\dfrac{-\log 2}{-\log\frac{4}{3}} \Rightarrow T=\left(\dfrac{\log 2}{\log\frac{4}{3}}\right)$

6. (i) $y=u(x)$ and $y=v(x)$ are solutions of given differential equations.

(ii) $u(x_1)>v(x_1)$ for some x_1

(iii) $f(x)>g(x), \forall x>x_1$

$\dfrac{du}{dx}+p(x)u=f(x)$

$\therefore \quad \dfrac{d}{dx}\left[u.e^{\int p\,dx}\right]=f(x)e^{\int p\,dx}$

NOTE THIS STEP:

Similarly, $\dfrac{d}{dx}\left[v.e^{\int p\,dx}\right]=g(x)e^{\int p\,dx}$

Subtracting, $\dfrac{d}{dx}\left[(u-v).e^{\int p\,dx}\right]=[f(x)-g(x)]e^{\int p\,dx}$

From above since $f(x)>g(x), \forall x>x_1$ and exponential function is always +ive, then R.H.S. is +ive.

$\therefore \quad \dfrac{d}{dx}\left[(u-v).e^{\int p\,dx}\right]>0$ or $\dfrac{dF}{dx}>0$

Hence the function $F=(u-v)e^{\int pdx}$ is an increasing function.

Again $u(x_1) > v(x_1)$ for some x_1

$\therefore$ $F = (u-v)e^{\int pdx}$ is +ive at $x = x_1$

$\Rightarrow$ $F = (u-v)e^{\int pdx}$ is +ive $\forall\, x > x_1$

(F being increasing function)

$\therefore$ $u(x) > v(x), \forall\, x > x_1$

$\therefore$ Hence there is no point (x, y) such that $x > x_1$ which can satisfy the equations.

$y = u(x)$ and $y = v(x)$.

7. Equation of normal is,

$$\frac{dx}{dy}(X-x) + Y - y = 0$$

$$\therefore \quad \frac{\left|x\frac{dx}{dy} + y\right|}{\sqrt{1+\left(\frac{dx}{dy}\right)^2}} = |y|$$

$$\Rightarrow x^2\left(\frac{dx}{dy}\right)^2 + y^2 + 2xy\frac{dx}{dy} = y^2 + y^2\left(\frac{dx}{dy}\right)^2$$

$$\Rightarrow \left(\frac{dx}{dy}\right) = 0, \text{ or } \frac{dx}{dy} = -\frac{2xy}{y^2 - x^2}$$

If $\frac{dx}{dy} = 0$, then $x = c$, when $x = 1, y = 1, c = 1$.

$$\therefore \quad x = 1 \qquad ...(1)$$

When $\frac{dx}{dy} = \frac{2xy}{y^2 - x^2}$ (homogeneous)

Putting $x = vy$

$$\Rightarrow \frac{dx}{dy} = v + y\frac{dv}{dy}$$

$$\therefore \quad v + y\frac{dv}{dy} = \frac{2v}{1-v^2}$$

$$\Rightarrow \quad y\frac{dv}{dy} = \frac{2v}{1-v^2} - v = \frac{2v - v + v^3}{1-v^2} = \frac{v+v^3}{1-v^2}$$

$$\Rightarrow \frac{(1-v^2)dv}{v(1+v^2)} = \frac{dy}{y} \Rightarrow \left(\frac{1}{v} - \frac{2v}{(1+v^2)}\right)dv = \frac{dy}{y}$$

$$\Rightarrow \log v - \log(1+v^2) = \log y + \log c$$

$$\text{or} \quad \frac{v}{1+v^2} = cy \Rightarrow \frac{xy}{x^2+y^2} = cy$$

$$\Rightarrow \frac{x}{x^2+y^2} = c,$$

$x = 1, y = 1$

gives $c = \frac{1}{2}$

$\therefore$ Solution is $x^2 + y^2 - 2x = 0$ ...(2)

Hence the solutions are,

$$x^2 + y^2 - 2x = 0,\ x - 1 = 0.$$

8. Let X_0 be initial population of the country and Y_0 be its initial food production. Let the average consumption be a units. Therefore, food required initially aX_0. It is given

$$Y_0 = aX_0\left(\frac{90}{100}\right) = 0.9aX_0 \qquad ...(i)$$

Let X be the population of the country in year t.

Then $\frac{dX}{dt}$ = rate of change of population

$$= \frac{3}{100}X = 0.03X$$

$\frac{dX}{X} = 0.03\,dt$; Integrating, $\int \frac{dX}{X} = \int 0.03\,dt$

$\Rightarrow$ $\log X = 0.03t + c$ $\Rightarrow$ $X = A.e^{0.03t}$

At $t = 0, X = X_0$, thus $X_0 = A, X = X_0e^{0.03t}$

Let Y be the food production in year t.

Then $Y = Y_0\left(1 + \frac{4}{100}\right)^t = 0.9aX_0(1.04)^t$

(since $Y = 0.9aX_0$ from (i))

Food consumption in the year t is $aX_0e^{0.03t}$

Again $Y - aX \geq 0$ (given)

$$\Rightarrow \quad 0.9a(1.04)^t \geq aX_0e^{0.03t} \Rightarrow \frac{(1.04)^t}{e^{0.03t}} \geq \frac{1}{0.9} = \frac{10}{9}$$

Taking log on both sides,

$$t[\ln(1.04) - 0.03] \geq \ln 10 - \ln 9$$

$$\Rightarrow \quad t \geq \frac{\ln 10 - \ln 9}{\ln(1.04) - 0.03}$$

Thus, the least integral value of the year n, when the country becomes self-sufficient, is the smallest integer greater than or equal to $\frac{\ln 10 - \ln 9}{\ln(1.04) - 0.03}$.

9. Let the water level be at a height h after time t, and water level falls by dh in time dt and the corresponding volume of water gone out be dV.

$\Rightarrow$ $|dV| = |\pi r^2 dh|$ ($\because$ dh is very small)

NOTE THIS STEP :

$\Rightarrow$ $\frac{dV}{dt} = -\pi r^2\frac{dh}{dt}$ ($\because$ as t increases, h decreases)

Now, velocity of water, $v = \frac{3}{5}\sqrt{2gh}$

Rate of flow of water $= Av\,(A = 12\,\text{cm}^2)$

$$\Rightarrow \quad \frac{dV}{dt} = \left(\frac{3}{5}\sqrt{2gh}A\right) = -\pi r^2\frac{dh}{dt}$$

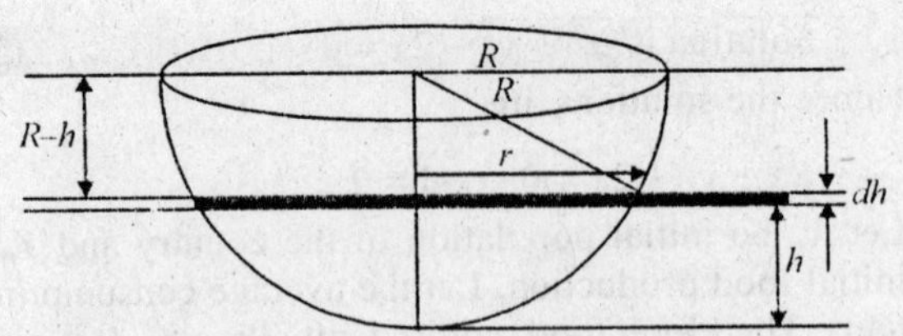

Also from figure,

$$R^2 = (R-h)^2 + r^2$$

$$\Rightarrow r^2 = 2hR - h^2$$

So, $\frac{3}{5}\sqrt{2g}.\sqrt{h}A = -\pi(2hR - h^2).\frac{dh}{dt}$

$$\Rightarrow \frac{2hR - h^2}{\sqrt{h}} dh = -\frac{3}{5\pi}\sqrt{2g}.A.dt$$

Integrating,

$$\int_R^0 (2R\sqrt{h} - h^{3/2})dh = -\frac{3\sqrt{2}g}{5\pi}.A.\int_0^T dt$$

$$\Rightarrow T = \frac{5\pi}{3A\sqrt{2g}}\left(2R.\frac{h^{\frac{3}{2}}}{\frac{3}{2}} - .\frac{h^{\frac{5}{2}}}{\frac{5}{2}}\right)_R^0$$

$$= \frac{5\pi}{3A\sqrt{2g}}\left(-\frac{2}{5}R^{5/2} + \frac{4R}{3}.R^{3/2}\right)$$

$$= \frac{5\pi}{3A\sqrt{2g}}.\frac{14}{15}R^{5/2}$$

$$= \frac{56\pi}{9A\sqrt{g}}(10)^5 = \frac{56\pi}{9\times 12\sqrt{g}}(10)^5 = \frac{14\pi}{27\sqrt{g}}(10)^5 \text{ units.}$$

10. Let at time t, r and h be the radius and height of cone of water.

$\therefore$ At time t, surface area of liquid in contact with air $= \pi r^2$

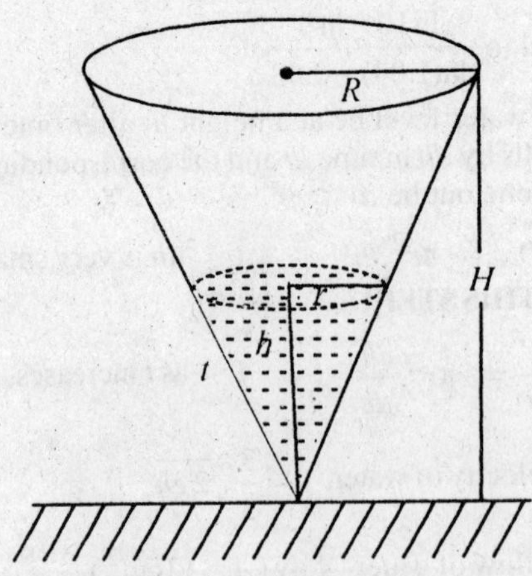

$ATQ \quad -\frac{dV}{dt} \propto \pi r^2$

[$\because$ '–'ve sign shows that V decreases with time.]

$$\Rightarrow \frac{dV}{dt} = -k\pi r^2 \Rightarrow \frac{d}{dt}\left[\frac{1}{3}\pi r^2 h\right] = -k\pi r^2$$

$$\Rightarrow \frac{1}{3}\pi\frac{d}{dt}[r^2 h] = -k\pi r^2$$

But from figure $\frac{r}{h} = \frac{R}{H}$ [Using similarity of Δ's]

$$\Rightarrow h = \frac{rH}{R}$$

$\therefore$ We get, $\frac{1}{3}\frac{d}{dt}\left[r^2.\frac{rH}{R}\right] = -kr^2$

$$\Rightarrow \frac{r^2 H}{R}\frac{dr}{dt} = -kr^2$$

$$\Rightarrow \frac{dr}{dt} = -\frac{kR}{H} \Rightarrow r = \frac{-kR}{H}t + C$$

But at $t = 0, r = R \Rightarrow R = 0 + C \Rightarrow C = R$

$$\therefore \quad r = \frac{-kRt}{H} + R$$

Now let the time at which cone is empty be T then at T, $r = 0$ (no liquid is left)

$$\therefore \quad 0 = \frac{-kRT}{H} + R \Rightarrow T = \frac{H}{k}$$

11. According to question

slope of curve C at $(x, y) = \frac{(x+1)^2 + (y-3)}{(x+1)}$

$$\Rightarrow \frac{dy}{dx} = (x+1) + \frac{y-3}{x+1}$$

$$\Rightarrow \frac{dy}{dx} - \left(\frac{1}{x+1}\right)y = x + 1 - \frac{3}{x+1}$$

$$\text{I.F.} = e^{-\int\frac{1}{x+1}dx} = e^{-\log(x+1)} = \frac{1}{x+1}$$

$\therefore$ Sol^n is $y\frac{1}{x+1} = \int\left[1 - \frac{3}{(x+1)^2}\right]dx$

$$\frac{y}{x+1} = x + \frac{3}{x+1} + C$$

$$y = x(x+1) + 3 + C(x+1) \quad ...(1)$$

As the curve passes through (2, 0)

$\therefore \quad 0 = 2.3 + 3 + C.3$

$\Rightarrow \quad C = -3$

$\therefore$ Eq^n. (1) becomes

$$y = x(x+1) + 3 - 3x - 3$$

$$y = x^2 - 2x \quad ...(2)$$

which is the required eq^n of curve.

This can be written as $(x-1)^2 = (y+1)$

[Upward parabola with vertex at (1, –1), meeting x-axis at (0, 0) and (2, 0)]

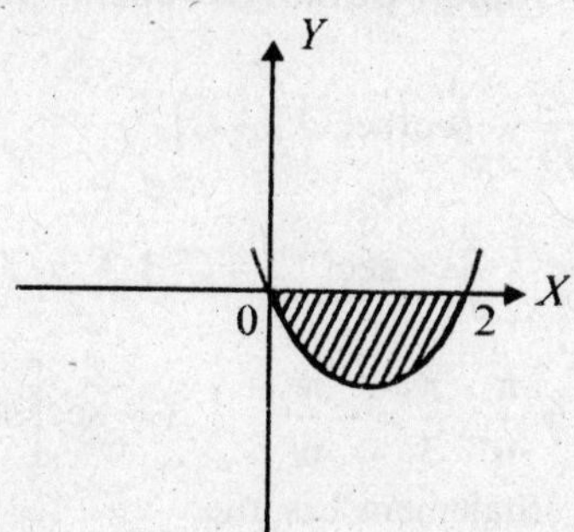

Area bounded by curve and x-axis in fourth quadrant is as shaded region in fig. given by

$$A=\left|\int_0^2 y\,dx\right|=\left|\int_0^2 (x^2-2x)\,dx\right|=\left[\frac{x^3}{3}-x^2\right]_0^2$$

$$=\left|\frac{8}{3}-4\right|=\frac{4}{3}\text{ sq. units.}$$

12. We know that length of tangent to curve $y=f(x)$ is given by

$$\left|\frac{y\sqrt{1+\left(\frac{dy}{dx}\right)^2}}{\left(\frac{dy}{dx}\right)}\right|$$

NOTE THIS STEP :

As per question $\left|\frac{y\sqrt{1+\left(\frac{dy}{dx}\right)^2}}{\left(\frac{dy}{dx}\right)}\right|=1$

$\Rightarrow\quad y^2\left(1+\left(\frac{dy}{dx}\right)^2\right)=\left(\frac{dy}{dx}\right)^2$

$\Rightarrow\quad \left(\frac{dy}{dx}\right)^2=\frac{y^2}{1-y^2}\quad\Rightarrow\quad \frac{dy}{dx}=\pm\frac{y}{\sqrt{1-y^2}}$

$\Rightarrow\quad \int\frac{\sqrt{1-y^2}}{y}dy=\int\pm dx$

$\Rightarrow$ Put $y=\sin\theta$

so that $dy=\cos\theta\,d\theta$

$\Rightarrow\quad \int\frac{\cos\theta}{\sin\theta}\cos\theta\,d\theta=\pm x+c$

$\Rightarrow\quad \int(\operatorname{cosec}\theta-\sin\theta)d\theta=\pm x+c$

$\Rightarrow\quad \log|\operatorname{cosec}\theta-\cot\theta|+\cos\theta=\pm x+c$

$\Rightarrow\quad \log\left|\frac{1-\sqrt{1-y^2}}{y}\right|+\sqrt{1-y^2}=\pm x+c$

(F) MATCH THE FOLLOWING :

1. ***A-p, q, r, s, t; B-p, t; C-p, q, r, t; D-s***

(A) $(x-3)^2y'+y=0$

$\Rightarrow\quad (x-3)^2\frac{dy}{dx}=-y$

$\Rightarrow\quad \int\left(-\frac{1}{y}\right)dy=\int\frac{1}{(x-3)^2}dx$

or $\quad \log|y|=\frac{1}{x-3}+\log c,\ x\neq 3$

$\Rightarrow\quad \log\left(\frac{y}{c}\right)=\frac{1}{x-3},x\neq 3$

$\Rightarrow\quad \frac{y}{c}=e^{\frac{1}{x-3}}$ or $y=ce^{\frac{1}{x-3}},\ x\neq 3$

$\therefore$ The solution set is $(-\infty,\infty)-\{3\}$

The interval $\left(-\frac{\pi}{2},\frac{\pi}{2}\right),\left(0,\frac{\pi}{2}\right),\left(\frac{\pi}{8},\frac{5\pi}{4}\right),\left(0,\frac{\pi}{8}\right)$ and $(-\pi,\pi)$ contained in the domain

$\therefore$ ***(A) → p, q, r, s, t***

(B) $\int_1^5(x-1)(x-2)(x-3)(x-4)(x-5)dx$

Let $(x-3)=t\Rightarrow dx=dt$

Also when $x\to 1, t\to -2$

and when $x\to 5, t\to 2$

$\therefore$ Integral becomes

$\int_{-2}^2(t+2)(t+1)t(t-1)(t-2)dt$

$=\int_{-2}^2 t(t^2-1)(t^2-4)dt=0$

as integrand is an odd function.

O is contained by $\left(-\frac{\pi}{2},\frac{\pi}{2}\right)$ and $(-\pi,\pi)$

$\therefore$ ***(B) → p, t.***

(C) Let $f(x)=\cos^2x+\sin x$

$\Rightarrow\quad f'(x)=-2\sin x\cos x+\cos x$

For critical point $f'(x)=0$

$\Rightarrow\quad \sin x=\frac{1}{2}$ or $\cos x=0$

$\Rightarrow\quad x=\frac{\pi}{6},\frac{5\pi}{6},$ or $x=\frac{\pi}{2},-\frac{\pi}{2}$

Now $f''(x)=-2\cos 2x-\sin x$

$f''(x)|_{x=\pi/6}=-ve\quad f''(x)|_{x=5\pi/6}=-ve$

$f''(x)|_{x=\pi/2}=+ve$ and $f''(x)|_{x=-\pi/2}=+ve$

$\therefore\quad \frac{\pi}{6}$ and $\frac{5\pi}{6}$ are the points of local maxima.

Clearly all the intervals given in column II except $\left(0, \frac{\pi}{8}\right)$ contain at least one point of local maxima.

$\therefore$ **(C) → p, q, r, t**

(D) Let $f(x) = \tan^{-1}(\sin x + \cos x)$

$$= \tan^{-1}\left[\sqrt{2}\sin\left(x + \frac{\pi}{4}\right)\right]$$

$$f'(x) = \frac{1}{1 + 2\sin^2\left(x + \frac{\pi}{4}\right)} . \sqrt{2}\cos\left(x + \frac{\pi}{4}\right)$$

For $f(x)$ to be an increasing function,

$f'(x) > 0$

$$\Rightarrow \cos\left(x + \frac{\pi}{4}\right) > 0 \Rightarrow -\frac{\pi}{2} < x + \frac{\pi}{4} < \frac{\pi}{2}$$

$$\Rightarrow -\frac{3\pi}{4} < x < \frac{\pi}{4}$$

Clearly only $\left(0, \frac{\pi}{8}\right) \subset \left(-\frac{3\pi}{4}, \frac{\pi}{4}\right)$

$\therefore$ **(D) → s.**

(H) ASSERTION & REASON TYPE QUESTIONS :

1. **(c)** The given differential equation is

$$x\sqrt{x^2 - 1}\,dy - y\sqrt{y^2 - 1}\,dx = 0$$

$$\Rightarrow \int \frac{dy}{y\sqrt{y^2 - 1}} = \int \frac{dx}{x\sqrt{x^2 - 1}}$$

$$\Rightarrow \sec^{-1} y = \sec^{-1} x + C$$

$$\Rightarrow y = \sec\left[\sec^{-1} x + C\right] \qquad \because y(2) = \frac{2}{\sqrt{3}}$$

$$\Rightarrow \frac{2}{\sqrt{3}} = \sec\left(\sec^{-1} 2 + C\right)$$

$$\Rightarrow \sec^{-1}\frac{2}{\sqrt{3}} - \sec^{-1} 2 = C$$

$$\Rightarrow C = \frac{\pi}{6} - \frac{\pi}{3} = -\frac{\pi}{6} \qquad \therefore y = \sec\left[\sec^{-1} x - \frac{\pi}{6}\right]$$

$\Rightarrow$ Statement 1 is true.

Also $\frac{1}{y} = \cos\left[\cos^{-1}\frac{1}{x} - \frac{\pi}{6}\right]$

$$= \cos\left(\cos^{-1}\frac{1}{x}\right)\cos\frac{\pi}{6} + \sin\left(\cos^{-1}\frac{1}{x}\right)\sin\frac{\pi}{6}$$

$$\Rightarrow \frac{1}{y} = \frac{\sqrt{3}}{2x} + \frac{1}{2}\sqrt{1 - \frac{1}{x^2}} \quad \therefore \text{Statement 2 is false.}$$

(I) INTEGER VALUE CORRECT TYPE :

1. **0**

The given equation is

$$\frac{dy}{dx} + g'(x)y = g(x)\,g'(x)$$

$$\text{I.F.} = e^{\int g'(x)dx} = e^{g(x)}$$

$\therefore$ Solution is $y.e^{g(x)} = \int e^{g(x)} g(x).g'(x)dx$

put g (x) = t so that g′(x) dx = dt

$$= \int e^t t\,dt = e^t(t - 1) + c$$

$\therefore$ $y.e^{g(x)} = e^{g(x)}[g(x) - 1] + c$

As y (0) = 0 and g (0) = 0

$\therefore$ C = 1

$\therefore$ $y.e^{g(x)} = e^{g(x)}[g(x) - 1] + 1$

As g (2) = 0, putting $x = 2$ we get

$y(2).e^{g(2)} = e^{g(2)}[g(2) - 1] + 1$

$\Rightarrow$ y (2) = 0

Section-B JEE Main/ AIEEE

1. **(c)** $\left(1 + 3\frac{dy}{dx}\right)^2 = \left(\frac{4d^3y}{dx^3}\right)^3 \Rightarrow \left(1 + 3\frac{dy}{dx}\right)^2 = 16\left(\frac{d^3y}{dx^3}\right)^3$

2. **(b)** $\frac{d^2y}{dx^2} = e^{-2x};\ \frac{dy}{dx} = \frac{e^{-2x}}{-2} + c;\ y = \frac{e^{-2x}}{4} + cx + d$

3. **(c)** $y^2 = 4a(x - h),\ 2yy_1 = 4a \Rightarrow yy_1 = 2a$

Differentiating, $\Rightarrow y_1^2 + yy_2 = 0$

Degree = 1, order = 2.

4. **(c)** $(1 + y^2) + (x - e^{\tan^{-1} y})\frac{dy}{dx} = 0$

$$\Rightarrow \frac{dx}{dy} + \frac{x}{(1 + y^2)} = \frac{e^{\tan^{-1} y}}{(1 + y^2)}$$

$$I.F = e^{\int \frac{1}{(1+y^2)}dy} = e^{\tan^{-1} y}$$

$$x(e^{\tan^{-1} y}) = \int \frac{e^{\tan^{-1} y}}{1 + y^2} e^{\tan^{-1} y}\,dy$$

$$x(e^{\tan^{-1} y}) = \frac{e^{2\tan^{-1} y}}{2} + C$$

$$\therefore 2xe^{\tan^{-1} y} = e^{2\tan^{-1} y} + k$$

5. (c) $x^2 + y^2 - 2ay = 0$ ....(1)

Differentiate,

$$2x + 2y\frac{dy}{dx} - 2a\frac{dy}{dx} = 0 \Rightarrow a = \frac{x + yy'}{y'}$$

Put in (1), $x^2 + y^2 - 2\left(\frac{x + yy'}{y'}\right)y = 0$

$$\Rightarrow (x^2 + y^2)y' - 2xy - 2y^2y' = 0$$

$$\Rightarrow (x^2 - y^2)y' = 2xy$$

6. (b) $ydx + (x + x^2y)dy = 0$

$$\Rightarrow \frac{dx}{dy} = -\frac{x}{y} - x^2 \Rightarrow \frac{dx}{dy} + \frac{x}{y} = -x^2,$$

It is Bernoullis form. Divide by x^2

$$x^{-2}\frac{dx}{dy} + x^{-1}\left(\frac{1}{y}\right) = -1.$$

put $x^{-1} = t, -x^{-2}\frac{dx}{dy} = \frac{dt}{dy}$

We get, $-\frac{dt}{dy} + t\left(\frac{1}{y}\right) = -1 \Rightarrow \frac{dt}{dy} - \left(\frac{1}{y}\right)t = 1$

It is linear in t.

Integrating factor $= e^{\int -\frac{1}{y}dy} = e^{-\log y} = y^{-1}$

$\therefore$ Solution is $t(y^{-1}) = \int (y^{-1})dy + c$

$$\Rightarrow \frac{1}{x}.\frac{1}{y} = \log y + c \Rightarrow \log y - \frac{1}{xy} = c$$

7. (c) $y^2 = 2c(x + \sqrt{c})$ .... (i)

$2yy' = 2c.1$ or $yy' = c$ .... (ii)

$$\Rightarrow y^2 = 2yy'(x + \sqrt{yy'})$$

[On putting value of c from (ii) in (i)]

On simplifying, we get

$(y - 2xy')^2 = 4yy'^3$ .... (iii)

Hence equation (iii) is of order 1 and degree 3.

8. (c) $\frac{xdy}{dx} = y(\log y - \log x + 1)$

$$\frac{dy}{dx} = \frac{y}{x}\left(\log\left(\frac{y}{x}\right) + 1\right)$$

Put $y = vx$

$$\frac{dy}{dx} = v + \frac{xdv}{dx} \Rightarrow v + \frac{xdv}{dx} = v(\log v + 1)$$

$$\frac{xdv}{dx} = v\log v \Rightarrow \frac{dv}{v\log v} = \frac{dx}{x}$$

Put $\log v = z$

$$\frac{1}{v}dv = dz \Rightarrow \frac{dz}{z} = \frac{dx}{x}$$

$\ln z = \ln x + \ln c$

$x = cx$ or $\log v = cx$ or $\log\left(\frac{y}{x}\right) = cx.$

9. (d) $Ax^2 + By^2 = 1$ ....(1)

$$Ax + By\frac{dy}{dx} = 0 \quad ....(2)$$

$$A + By\frac{d^2y}{dx^2} + B\left(\frac{dy}{dx}\right)^2 = 0 \quad ....(3)$$

From (2) and (3)

$$x\left\{-By\frac{d^2y}{dx^2} - B\left(\frac{dy}{dx}\right)^2\right\} + By\frac{dy}{dx} = 0$$

Dividing both sides by –B, we get

$$\Rightarrow xy\frac{d^2y}{dx^2} + x\left(\frac{dy}{dx}\right)^2 - y\frac{dy}{dx} = 0$$

Which is a DE of order 2 and degree 1.

10. (a) General equation of circles passing through origin and having their centres on the x-axis is

$x^2 + y^2 + 2gx = 0$ ...(i)

On differentiating w.r.t x, we get

$$2x + 2y.\frac{dy}{dx} + 2g = 0 \Rightarrow g = -\left(x + y\frac{dy}{dx}\right)$$

$\therefore$ equation (i) be

$$x^2 + y^2 + 2\left\{-\left(x + y\frac{dy}{dx}\right)\right\}.x = 0$$

$$\Rightarrow x^2 + y^2 - 2x^2 - 2x\frac{dy}{dx}.y = 0$$

$$\Rightarrow y^2 = x^2 + 2xy\frac{dy}{dx}$$

11. (d) $\frac{dy}{dx} = \frac{x + y}{x} = 1 + \frac{y}{x}$

Putting $y = vx$ and $\frac{dy}{dx} = v + x\frac{dv}{dx}$

we get

$$v + x\frac{dv}{dx} = 1 + v \Rightarrow \int\frac{dx}{x} = \int dv$$

$\Rightarrow v = \ln x + c \Rightarrow y = x\ln x + cx$

As $y(1) = 1$

$\therefore c = 1$ So solution is $y = x\ln x + x$

12. (c) We have $y = c_1e^{c_2x}$

$$\Rightarrow y' = c_1c_2e^{c_2x} = c_2y$$

$$\Rightarrow \frac{y'}{y} = c_2 \Rightarrow \frac{y''y - (y')^2}{y^2} = 0 \Rightarrow y''y = (y')^2$$

13. (d) $\cos x\, dy = y(\sin x - y)\, dx$

$$\frac{dy}{dx} = y\tan x - y^2 \sec x$$

$$\frac{1}{y^2}\frac{dy}{dx} - \frac{1}{y}\tan x = -\sec x \quad ...(i)$$

Let $\frac{1}{y} = t \Rightarrow -\frac{1}{y^2}\frac{dy}{dx} = \frac{dt}{dx}$

From equation (i)

$$-\frac{dt}{dx} - t\tan x = -\sec x \Rightarrow \frac{dt}{dx} + (\tan x)t = \sec x$$

$$\text{I.F.} = e^{\int \tan x\, dx} = (e)^{\log|\sec x|} \sec x$$

Solution : $t(\text{I.F}) = \int(\text{I.F})\sec x\, dx \Rightarrow \frac{1}{y}\sec x = \tan x + c$

14. (d) $\frac{dy}{dx} = y + 3 \Rightarrow \int \frac{dy}{y+3} = \int dx$

$\Rightarrow \ell n|y+3| = x + c$

Since $y(0) = 2$, $\therefore \ell n 5 = c$

$\Rightarrow \ell n|y+3| = x + \ell n 5$

When $x = \ell n 2$, then $\ell n|y+3| = \ell n 2 + \ell n 5$

$\Rightarrow \ell n|y+3| = \ell n 10$

$\therefore y + 3 = \pm 10 \Rightarrow y = 7, -13$

15. (a) $\frac{dV(t)}{dt} = -k(T - t)$

$$\Rightarrow \int dVt = -k\int (T-t)dt$$

$$V(t) = \frac{k(T-t)^2}{2} + c$$

$$V(0) = I \Rightarrow I = \frac{KT^2}{2} + C \Rightarrow C = I - \frac{KT^2}{2}$$

$$\therefore V(T) = 0 + C = I - \frac{KT^2}{2}$$

16. (c) $\frac{dx}{dy} + \frac{x}{y^2} = \frac{1}{y^3}$

$$\text{I.F.} = e^{\int \frac{1}{y^2}dy} = e^{-\frac{1}{y}}$$

So $x.e^{-\frac{1}{y}} = \int \frac{1}{y^3} e^{-\frac{1}{y}} dy$

Let $\frac{-1}{y} = t \Rightarrow \frac{1}{y^2}dy = dt$

$$\Rightarrow I = -\int te^t dt = e^t - te^t = e^{-\frac{1}{y}} + \frac{1}{y}e^{-\frac{1}{y}} + c$$

$$\Rightarrow xe^{-\frac{1}{y}} = e^{-\frac{1}{y}} + \frac{1}{y}e^{-\frac{1}{y}} + c$$

$$\Rightarrow x = 1 + \frac{1}{y} + c.e^{1/y}$$

Since $y(1) = 1$

$$\therefore c = -\frac{1}{e}$$

$$\Rightarrow x = 1 + \frac{1}{y} - \frac{1}{e}.e^{1/y}$$

17. (a) Given differential equation is

$$\frac{dp(t)}{dt} = 0.5p(t) - 450$$

$$\Rightarrow \frac{dp(t)}{dt} = \frac{1}{2}p(t) - 450$$

$$\Rightarrow \frac{dp(t)}{dt} = \frac{p(t) - 900}{2}$$

$$\Rightarrow 2\frac{dp(t)}{dt} = -[900 - p(t)]$$

$$\Rightarrow 2\frac{dp(t)}{900 - p(t)} = -dt$$

Integrate both the side, we get

$$-2\int \frac{dp(t)}{900 - p(t)} = \int dt$$

Let $900 - p(t) = u$

$\Rightarrow -dp(t) = du$

$\therefore$ We have, $2\int \frac{du}{u} = \int dt \Rightarrow 2\, ln\, u = t + c$

$\Rightarrow 2ln\,[900 - p(t)] = t + c$

when $t = 0, p(0) = 850$

$2\, ln(50) = c$

$$\Rightarrow 2\left[ln\left(\frac{900 - p(t)}{50}\right)\right] = t$$

$$\Rightarrow 900 - p(t) = 50e^{\frac{t}{2}}$$

$$\Rightarrow p(t) = 900 - 50e^{\frac{t}{2}}$$

let $p(t_1) = 0$

$0 = 900 - 50e^{\frac{t_1}{2}}$ $\therefore t_1 = 2ln\, 18$

18. (c) Given, Rate of change is $\frac{dP}{dx} = 100 - 12\sqrt{x}$

$\Rightarrow dP = (100 - 12\sqrt{x})\, dx$

By integrating

$$\int dP = \int (100 - 12\sqrt{x})dx$$

$P = 100x - 8x^{3/2} + C$

Given, when $x = 0$ then $P = 2000$

$\Rightarrow C = 2000$

Now when $x = 25$ then

$P = 100 \times 25 - 8 \times (25)^{3/2} + 2000$

$= 4500 - 1000$

$\Rightarrow P = 3500$

Chapter 20 Vector Algebra and Three Dimensional Geometry

Section-A JEE Advanced/ IIT-JEE

A FILL IN THE BLANKS :

1. Let $\vec{A}$, $\vec{B}$, $\vec{C}$ be vectors of length 3, 4, 5 respectively. Let $\vec{A}$ be perpendicular to $\vec{B}+\vec{C}, \vec{B}$ to $\vec{C}+\vec{A}$ and $\vec{C}$ *to* $\vec{A}+\vec{B}$. Then the length of vector $\vec{A}+\vec{B}+\vec{C}$ is .... **(1981 - 2 Marks)**

2. The unit vector perpendicular to the plane determined by $P(1,-1,2)$, $Q(2,0,-1)$ and $R(0,2,1)$ is .... **(1983 - 1 Mark)**

3. The area of the triangle whose vertices are $A(1,-1,2)$, $B(2,1,-1)$, $C(3,-1,2)$ is .... **(1983 - 1 Mark)**

4. A, B, C and D, are four points in a plane with position vectors ***a, b, c*** and ***d*** respectively such that $(a-d)(b-c)=(b-d)(c-a)=0$ **(1984 - 2 Marks)** The point D, then, is the .... of the triangle ABC.

5. If $\begin{vmatrix} a & a^2 & 1+a^3 \\ b & b^2 & 1+b^3 \\ c & c^2 & 1+c^3 \end{vmatrix}=0$ and the vectors $\vec{A}=(1,a,a^2)$, $\vec{B}=(1,b,b^2)$, $\vec{C}=(1,c,c^2)$, are non-coplanar, then the product $abc=$ .... **(1985 - 2 Marks)**

6. If $\vec{A}\,\vec{B}\,\vec{C}$ are three non-coplanar vectors, then –
$$\frac{\vec{A}\cdot\vec{B}\times\vec{C}}{\vec{C}\times\vec{A}\cdot\vec{B}}+\frac{\vec{B}\cdot\vec{A}\times\vec{C}}{\vec{C}\cdot\vec{A}\times\vec{B}}= ....$$ **(1985 - 2 Marks)**

7. If $\vec{A}=(1,1,1)$, $\vec{C}=(0,1,-1)$ are given vectors, then a vector B satisfying the equations $\vec{A}\times\vec{B}=\vec{C}$ and $\vec{A}\cdot\vec{B}=3$ ...... **(1985 - 2 Marks)**

8. If the vectors $a\hat{i}+\hat{j}+\hat{k}$, $\hat{i}+b\hat{j}+\hat{k}$ and $\hat{i}+\hat{j}+c\hat{k}$ $(a\neq b\neq c\neq 1)$ are coplanar, then the value of $\frac{1}{(1-a)}+\frac{1}{(1-b)}+\frac{1}{(1-c)}=$ .... **(1987 - 2 Marks)**

9. Let $b=4\hat{i}+3\hat{j}$ and $\vec{c}$ be two vectors perpendicular to each other in the *xy*-plane. All vectors in the same plane having projections 1 and 2 along $\vec{b}$ and $\vec{c}$, respectively, are given by .... **(1987 - 2 Marks)**

10. The components of a vector $\vec{a}$ along and perpendicular to a non-zero vector $\vec{b}$ are.......and ....respectively. **(1988 - 2 Marks)**

11. Given that $\vec{a}=(1,1,1)$, $\vec{c}=(0,1,-1)$, $\vec{a}.\vec{b}=3$ and $\vec{a}\times\vec{b}=\vec{c}$, then $\vec{b}=$ .... **(1991 - 2 Marks)**

12. A unit vector coplanar with $\vec{i}+\vec{j}+2\vec{k}$ and $\vec{i}+2\vec{j}+\vec{k}$ and perpendicular to $\vec{i}+\vec{j}+\vec{k}$ is ....**(1992 - 2 Marks)**

13. A unit vector perpendicular to the plane determined by the points $P(1,-1,2)$ $Q(2,0,-1)$ and $R(0,2,1)$ is .... **(1994 - 2 Marks)**

14. A nonzero vector $\vec{a}$ is parallel to the line of intersection of the plane determined by the vectors $\hat{i},\hat{i}+\hat{j}$ and the plane determined by the vectors $\hat{i}-\hat{j}$, $\hat{i}+\hat{k}$. The angle between $\vec{a}$ and the vector $\hat{i}-2\hat{j}+2\hat{k}$ is .... **(1996 - 2 Marks)**

15. If $\vec{b}$ and $\vec{c}$ are any two non-collinear unit vectors and $\vec{a}$ is any vector, then $(\vec{a}.\vec{b})\vec{b}+(\vec{a}.\vec{c})\vec{c}+\frac{\vec{a}.(\vec{b}\times\vec{c})}{|\vec{b}\times\vec{c}|}(\vec{b}\times\vec{c})=$ .... **(1996 - 2 Marks)**

16. Let $\vec{a}$, $\vec{b}$ and $\vec{c}$ be three vectors having magnitudes 1, 1 and 2 respectively. If $\vec{a}\times(\vec{a}\times\vec{c})+\vec{b}=\vec{0}$, then the acute angle between $\vec{a}$ and $\vec{c}$ is ...... **(1997C - 2 Marks)**

17. Let ***OA = a, OB*** = 10 ***a*** + 2***b*** and ***OC = b*** where O, A and C are non-collinear points. Let p denote the area of the quadrilateral $OABC$, and let q denote the area of the parallelogram with OA and OC as adjacent sides. If $p=kq$, then $k=$ .... **(1997 - 2 Marks)**

B TRUE / FALSE :

1. Let $\vec{A}, \vec{B}$ and $\vec{C}$ be unit vectors suppose that $\vec{A}.\vec{B}=\vec{A}.\vec{C}=0$, and that the angle between $\vec{B}$ and $\vec{C}$ is $\pi/6$. Then $\vec{A}=\pm 2(\vec{B}\times\vec{C})$. **(1981 - 2 Marks)**

2. If $X.A=0, X.B=0, X.C=0$ for some non-zero vector X, then $[A\,B\,C]=0$ **(1983 - 1 Mark)**

3. The points with position vectors $a+b, a-b$, and $a+kb$ are collinear for all real values of k. *(1984 - 1 Mark)*

4. For any three vectors $\vec{a}, \vec{b}$, and $\vec{c}$,

$(\vec{a}-\vec{b}).(\vec{b}-\vec{c})\times(\vec{c}-\vec{a}) = 2\vec{a}.\vec{b}\times\vec{c}$. *(1989 - 1 Mark)*

C MCQ's WITH ONE CORRECT ANSWER :

1. The scalar $\vec{A}.(\vec{B}+\vec{C})\times(\vec{A}+\vec{B}+\vec{C})$ equals : *(1981 - 2 Marks)*

(a) 0 (b) $[\vec{A}\ \vec{B}\ \vec{C}]+[\vec{B}\ \vec{C}\ \vec{A}]$

(c) $[\vec{A}\ \vec{B}\ \vec{C}]$ (d) None of these

2. For non-zero vectors $\vec{a},\vec{b},\vec{c}$, $|(\vec{a}\times\vec{b}).\vec{c}| = |\vec{a}||\vec{b}||\vec{c}|$ holds if and only if *(1982 - 2 Marks)*

(a) $\vec{a}.\vec{b}=0,\ \vec{b}.\vec{c}=0$ (b) $\vec{b}.\vec{c}=0,\ \vec{c}.\vec{a}=0$

(c) $\vec{c}.\vec{a}=0\ \vec{a}.\vec{b}=0$ (d) $\vec{a}.\vec{b}=\vec{b}.\vec{c}=\vec{c}.\vec{a}=0$

3. The volume of the parallelopiped whose sides are given by $\overrightarrow{OA}=2i-2j$, $\overrightarrow{OB}=i+j-k$, $\overrightarrow{OC}=3i-k$, is *(1983 - 1 Mark)*

(a) $\frac{4}{13}$ (b) 4

(c) $\frac{2}{7}$ (d) none of these

4. The points with position vectors $60i+3j$, $40i-8j$, $ai-52j$ are collinear if *(1983 - 1 Mark)*

(a) $a=-40$ (b) $a=40$

(c) $a=20$ (d) none of these

5. Let $\vec{a}, \vec{b}, \vec{c}$, be three non-coplanar vectors and $\vec{p}, \vec{q}, \vec{r}$, are vectors defined by the relations $\vec{p}=\frac{\vec{b}\times\vec{c}}{[\vec{a}\,\vec{b}\,\vec{c}]}$, $\vec{q}=\frac{\vec{c}\times\vec{a}}{[\vec{a}\,\vec{b}\,\vec{c}]}$, $\vec{r}=\frac{\vec{a}\times\vec{b}}{[\vec{a}\,\vec{b}\,\vec{c}]}$ then the value of the expression $(\vec{a}+\vec{b}).\vec{p}+(\vec{b}+\vec{c}).\vec{q}+(\vec{c}+\vec{a}).\vec{r}$ is equal to *(1988 - 2 Marks)*

(a) 0 (b) 1 (c) 2 (d) 3.

6. Let a, b, c be distinct non-negative numbers. If the vectors $a\hat{i}+a\hat{j}+c\hat{k}, \hat{i}+\hat{k}$ and $c\hat{i}+c\hat{j}+b\hat{k}$ lie in a plane, then c is *(1993 - 1 Marks)*

(a) the Arithmetic Mean of a and b
(b) the Geometric Mean of a and b
(c) the harmonic Mean of a and b
(d) equal to zero

7. Let $\vec{p}$ and $\vec{q}$ be the position vectors of P and Q respectively, with respect to O and $|\vec{p}|=p, |\vec{q}|=q$. The points R and S divide PQ internally and externally in the ratio 2 : 3 respectively. If OR and OS are perpendicular then *(1994)*

(a) $9q^2=4q^2$ (b) $4p^2=9q^2$

(c) $9p=4q$ (d) $4p=9q$

8. Let α, β, γ be distinct real numbers. The points with position vectors $\alpha\hat{i}+\beta\hat{j}+\gamma\hat{k}, \beta\hat{i}+\gamma\hat{j}+\alpha\hat{k}, \gamma\hat{i}+\alpha\hat{j}+\beta\hat{k}$ *(1994)*

(a) are collinear
(b) form an equilateral triangle
(c) form a scalene triangle
(d) form a right angled triangle

9. Let $\vec{a}=\hat{i}-\hat{j}, \vec{b}=\hat{j}-\hat{k}, \vec{c}=\hat{k}-\hat{i}$. If $\vec{d}$ is a unit vector such that $\vec{a}.\vec{d}=0=[\vec{b}\,\vec{c}\,\vec{d}]$, then $\vec{d}$ equals *(1995S)*

(a) $\pm\frac{\hat{i}+\hat{j}-2\hat{k}}{\sqrt{6}}$ (b) $\pm\frac{\hat{i}+\hat{j}-\hat{k}}{\sqrt{3}}$

(c) $\pm\frac{\hat{i}+\hat{j}+\hat{k}}{\sqrt{3}}$ (d) $\pm\hat{k}$

10. If $\vec{a},\vec{b},\vec{c}$ are non coplanar unit vectors such that $\vec{a}\times(\vec{b}\times\vec{c})=\frac{(\vec{b}+\vec{c})}{\sqrt{2}}$, then the angle between $\vec{a}$ and $\vec{b}$ is *(1995S)*

(a) $\frac{3\pi}{4}$ (b) $\frac{\pi}{4}$ (c) $\pi/2$ (d) π

11. Let $\vec{u}, \vec{v}$ and $\vec{w}$ be vectors such that $\vec{u}+\vec{v}+\vec{w}=0$. If $|\vec{u}|=3, |\vec{v}|=4$ and $|\vec{w}|=5$, then $\vec{u}.\vec{v}+\vec{v}.\vec{w}+\vec{w}.\vec{u}$ is *(1995S)*

(a) 47 (b) -25 (c) 0 (d) 25

12. If $\vec{a}, \vec{b}$ and $\vec{c}$ are three non coplanar vectors, then *(1995S)* $(\vec{a}+\vec{b}+\vec{c}).[(\vec{a}+\vec{b})\times(\vec{a}+\vec{c})]$ equals

(a) 0 (b) $[\vec{a}\,\vec{b}\,\vec{c}]$

(c) $2[\vec{a}\,\vec{b}\,\vec{c}]$ (d) $-[\vec{a}\,\vec{b}\,\vec{c}]$

13. If $\vec{p}, \vec{q}, \vec{r}$ be three mutually perpendicular vectors of the same magnitude. If a vector $\vec{x}$ satisfies the equation

$$\vec{p}\times\left((\vec{x}-\vec{q})\times\vec{p}\right)+\vec{q}\times\left((\vec{x}-\vec{r})\times\vec{q}\right)+\vec{r}\times\left((\vec{x}-\vec{p})\times\vec{r}\right)=\vec{0}$$

then $\vec{x}$ is given by *(1997C - 2 Marks)*

(a) $\frac{1}{2}(\vec{p}+\vec{q}-2\vec{r})$ (b) $\frac{1}{2}(\vec{p}+\vec{q}+\vec{r})$

(c) $\frac{1}{3}(\vec{p}+\vec{q}+\vec{r})$ (d) $\frac{1}{3}(2\vec{p}+\vec{q}-\vec{r})$

14. Let $\boldsymbol{a}=\boldsymbol{2i}+\boldsymbol{j}-\boldsymbol{2k}$ and $\boldsymbol{b}=\boldsymbol{i}+\boldsymbol{j}$. If c is a vector such that $\boldsymbol{a}$.

$c = |c|, |c - a| = 2\sqrt{2}$ and the angle between $(a \times b)$ and c is $30°$, then $|(a \times b) \times c| =$ **(1999 - 2 Marks)**

(a) 2/3 (b) 3/2
(c) 2 (d) 3

15. Let $a = 2i + j + k$, $b = i + 2j - k$ and a unit vector c be coplanar. If c is perpendicular to a, then $c =$ **(1999 - 2 Marks)**

(a) $\frac{1}{\sqrt{2}}(-j+k)$ (b) $\frac{1}{\sqrt{3}}(-i-j-k)$

(c) $\frac{1}{\sqrt{5}}(i-2j)$ (d) $\frac{1}{\sqrt{3}}(i-j-k)$

16. If the vectors $\vec{a}, \vec{b}$ and $\vec{c}$ form the sides BC, CA and AB respectively of a triangle ABC, then **(2000S)**

(a) $\vec{a}.\vec{b} + \vec{b}.\vec{c} + \vec{c}.\vec{a} = 0$ (b) $\vec{a} \times \vec{b} = \vec{b} \times \vec{c} = \vec{c} \times \vec{a}$

(c) $\vec{a}.\vec{b} = \vec{b}.\vec{c} = \vec{c}.\vec{a}$ (d) $\vec{a} \times \vec{b} + \vec{b} \times \vec{c} + \vec{c} \times \vec{a} = \vec{0}$

17. Let the vectors $\vec{a}, \vec{b}, \vec{c}$ and $\vec{d}$ be such that $\left(\vec{a} \times \vec{b}\right) \times \left(\vec{c} \times \vec{d}\right) = 0$. Let P_1 and P_2 be planes determined by the pairs of vectors $\vec{a}, \vec{b}$ and $\vec{c}, \vec{d}$ respectively. Then the angle between P_1 and P_2 is **(2000S)**

(a) 0 (b) $\frac{\pi}{4}$

(c) $\frac{\pi}{3}$ (d) $\frac{\pi}{2}$

18. If $\vec{a}, \vec{b}$ and $\vec{c}$ are unit coplanar vectors, then the scalar triple product $\left[2\vec{a} - \vec{b}, 2\vec{b} - \vec{c}, 2\vec{c} - \vec{a}\right] =$ **(2000S)**

(a) 0 (b) 1 (c) $-\sqrt{3}$ (d) $\sqrt{3}$

19. Let $\vec{a} = \vec{i} - \vec{k}$, $\vec{b} = x\vec{i} + \vec{j} + (1-x)\vec{k}$ and $\vec{c} = y\vec{i} + x\vec{j} + (1+x-y)\vec{k}$. Then $\left[\vec{a}\,\vec{b}\,\vec{c}\right]$ depends on **(2001S)**

(a) only x (b) only y
(c) Neither x Nor y (d) both x and y

20. If $\vec{a}, \vec{b}$ and $\vec{c}$ are unit vectors, then $\left|\vec{a}-\vec{b}\right|^2 + \left|\vec{b}-\vec{c}\right|^2 + \left|\vec{c}-\vec{a}\right|^2$ does NOT exceed **(2001S)**

(a) 4 (b) 9 (c) 8 (d) 6

21. If $\vec{a}$ and $\vec{b}$ are two unit vectors such that $\vec{a} + 2\vec{b}$ and $5\vec{a} - 4\vec{b}$ are perpendicular to each other then the angle between $\vec{a}$ and $\vec{b}$ is **(2002S)**

(a) 45° (b) 60°

(c) $\cos^{-1}\left(\frac{1}{3}\right)$ (d) $\cos^{-1}\left(\frac{2}{7}\right)$

22. Let $\vec{V} = 2\vec{i} + \vec{j} - \vec{k}$ and $\vec{W} = \vec{i} + 3\vec{k}$. If $\vec{U}$ is a unit vector, then the maximum value of the scalar triple product $|\vec{U}\vec{V}\vec{W}|$ is

(a) -1 (b) $\sqrt{10} + \sqrt{6}$ **(2002S)**

(c) $\sqrt{59}$ (d) $\sqrt{60}$

23. The value of k such that $\frac{x-4}{1} = \frac{y-2}{1} = \frac{z-k}{2}$ lies in the plane $2x - 4y + z = 7$, is **(2003S)**

(a) 7 (b) -7
(c) no real value (d) 4

24. The value of 'a' so that the volume of parallelopiped formed by $\hat{i} + a\hat{j} + \hat{k}$, $\hat{j} + a\hat{k}$ and $a\hat{i} + \hat{k}$ becomes minimum is **(2003S)**

(a) -3 (b) 3 (c) $1/\sqrt{3}$ (d) $\sqrt{3}$

25. If $\vec{a} = (\hat{i} + \hat{j} + \hat{k})$, $\vec{a}.\vec{b} = 1$ and $\vec{a} \times \vec{b} = \hat{j} - \hat{k}$, then $\vec{b}$ is

(a) $\hat{i} - \hat{j} + \hat{k}$ (b) $2\hat{j} - \hat{k}$ **(2004S)**

(c) $\hat{i}$ (d) $2\hat{i}$

26. If the lines $\frac{x-1}{2} = \frac{y+1}{3} = \frac{z-1}{4}$ and $\frac{x-3}{1} = \frac{y-k}{2} = \frac{z}{1}$ intersect, then the value of k is **(2004S)**

(a) 3/2 (b) 9/2 (c) $-2/9$ (d) $-3/2$

27. The unit vector which is orthogonal to the vector $3\hat{i} + 2\hat{j} + 6\hat{k}$ and is coplanar with the vectors $2\hat{i} + \hat{j} + \hat{k}$ and $\hat{i} - \hat{j} + \hat{k}$ is **(2004S)**

(a) $\frac{2\hat{i} - 6\hat{j} + \hat{k}}{\sqrt{41}}$ (b) $\frac{2\hat{i} - 3\hat{j}}{\sqrt{13}}$

(c) $\frac{3\hat{i} - \hat{k}}{\sqrt{10}}$ (d) $\frac{4\hat{i} + 3\hat{j} - 3\hat{k}}{\sqrt{34}}$

28. A variable plane at a distance of the one unit from the origin cuts the coordinates axes at A, B and C. If the centroid $D(x, y, z)$ of triangle ABC satisfies the relation $\frac{1}{x^2} + \frac{1}{y^2} + \frac{1}{z^2} = k$, then the value k is **(2005S)**

(a) 3 (b) 1 (c) $\frac{1}{3}$ (d) 9

29. If $\vec{a}, \vec{b}, \vec{c}$ are three non-zero, non-coplanar vectors and

$$\vec{b}_1 = \vec{b} - \frac{\vec{b}\cdot\vec{a}}{|\vec{a}|^2}\vec{a}, \quad \vec{b}_2 = \vec{b} + \frac{\vec{b}\cdot\vec{a}}{|\vec{a}|^2}\vec{a},$$

$$\vec{c}_1 = \vec{c} - \frac{\vec{c}\cdot\vec{a}}{|\vec{a}|^2}\vec{a} + \frac{\vec{b}\cdot\vec{c}}{|\vec{c}|^2}\vec{b}_1, \quad \vec{c}_2 = \vec{c} - \frac{\vec{c}\cdot\vec{a}}{|\vec{a}|^2}\vec{a} - \frac{\vec{b}_1\cdot\vec{c}}{|\vec{b}_1|^2}\vec{b}_1,$$

$$\vec{c}_3 = \vec{c} - \frac{\vec{c}\cdot\vec{a}}{|\vec{c}|^2}\vec{a} + \frac{\vec{b}\cdot\vec{c}}{|\vec{c}|^2}\vec{b}_1, \quad \vec{c}_4 = \vec{c} - \frac{\vec{c}.\vec{a}}{|\vec{c}|^2}\vec{a} = \frac{\vec{b}.\vec{c}}{|\vec{b}|^2}\vec{b}_1,$$

then the set of orthogonal vectors is (2005S)

(a) $(\vec{a}, \vec{b_1}, \vec{c_3})$ (b) $(\vec{a}, \vec{b_1}, \vec{c_2})$

(c) $(\vec{a}, \vec{b_1}, \vec{c_1})$ (d) $(\vec{a}, \vec{b_2}, \vec{c_2})$

30. A plane which is perpendicular to two planes $2x-2y+z=0$ and $x-y+2z=4$, passes through $(1, -2, 1)$. The distance of the plane from the point $(1, 2, 2)$ is (2006 - 3M, –1)

(a) 0 (b) 1 (c) $\sqrt{2}$ (d) $2\sqrt{2}$

31. Let $\vec{a}=\hat{i}+2\hat{j}+\hat{k}, \vec{b}=\hat{i}-\hat{j}+\hat{k}$ and $\vec{c}=\hat{i}+\hat{j}-\hat{k}$. A vector in the plane of $\vec{a}$ and $\vec{b}$ whose projection on $\vec{c}$ is $\frac{1}{\sqrt{3}}$, is (2006 - 3M, –1)

(a) $4\hat{i}-\hat{j}+4\hat{k}$ (b) $3\hat{i}+\hat{j}-3\hat{k}$

(c) $2\hat{i}+\hat{j}-2\hat{k}$ (d) $4\hat{i}+\hat{j}-4\hat{k}$

32. The number of distinct real values of λ, for which the vectors $-\lambda^2\hat{i}+\hat{j}+\hat{k}$, $\hat{i}-\lambda^2\hat{j}+\hat{k}$ and $\hat{i}+\hat{j}-\lambda^2\hat{k}$ are coplanar, is (2007 -3 marks)

(a) zero (b) one (c) two (d) three

33. Let $\vec{a}, \vec{b}, \vec{c}$ be unit vectors such that $\vec{a}+\vec{b}+\vec{c}=\vec{0}$. Which one of the following is correct ? (2007 -3 marks)

(a) $\vec{a}\times\vec{b}=\vec{b}\times\vec{c}=\vec{c}\times\vec{a}=\vec{0}$

(b) $\vec{a}\times\vec{b}=\vec{b}\times\vec{c}=\vec{c}\times\vec{a}\neq\vec{0}$

(c) $\vec{a}\times\vec{b}=\vec{b}\times\vec{c}=\vec{a}\times\vec{c}\neq\vec{0}$

(d) $\vec{a}\times\vec{b}, \vec{b}\times\vec{c}, \vec{c}\times\vec{a}$ are muturally perpendicular

34. The edges of a parallelopiped are of unit length and are parallel to non-coplanar unit vectors $\hat{a}, \hat{b}, \hat{c}$ such that $\hat{a}.\hat{b}=\hat{b}.\hat{c}=\hat{c}.\hat{a}=\frac{1}{2}$. Then, the volume of the parallelopiped is (2008)

(a) $\frac{1}{\sqrt{2}}$ (b) $\frac{1}{2\sqrt{2}}$ (c) $\frac{\sqrt{3}}{2}$ (d) $\frac{1}{\sqrt{3}}$

35. Let two non-collinear unit vectors $\hat{a}$ and $\hat{b}$ form an acute angle. A point P moves so that at any time t the position vector $\overrightarrow{OP}$ (where O is the origin) is given by $\hat{a}\cos t+\hat{b}\sin t$. When P is farthest from origin O, let M be the length of $\overrightarrow{OP}$ and $\hat{u}$ be the unit vector along $\overrightarrow{OP}$. Then, (2008)

(a) $\hat{u}=\frac{\hat{a}+\hat{b}}{|\hat{a}+\hat{b}|}$ and $M=(1+\hat{a}\cdot\hat{b})^{1/2}$

(b) $\hat{u}=\frac{\hat{a}-\hat{b}}{|\hat{a}-\hat{b}|}$ and $M=(1+\hat{a}\cdot\hat{b})^{1/2}$

(c) $\hat{u}=\frac{\hat{a}+\hat{b}}{|\hat{a}+\hat{b}|}$ and $M=(1+2\hat{a}\cdot\hat{b})^{1/2}$

(d) $\hat{u}=\frac{\hat{a}-\hat{b}}{|\hat{a}-\hat{b}|}$ and $M=(1+2\hat{a}\cdot\hat{b})^{1/2}$

36. Let $P(3, 2, 6)$ be a point in space and Q be a point on the line

$$\vec{r}=(\hat{i}-\hat{j}+2\hat{k})+\mu(-3\hat{i}+\hat{j}+5\hat{k})$$

Then the value of μ for which the vector $\overrightarrow{PQ}$ is parallel to the plane $x-4y+3z=1$ is (2009)

(a) $\frac{1}{4}$ (b) $-\frac{1}{4}$ (c) $\frac{1}{8}$ (d) $-\frac{1}{8}$

37. If $\vec{a}, \vec{b}, \vec{c}$ and $\vec{d}$ are unit vectors such that

$(\vec{a}\times\vec{b})\cdot(\vec{c}\times\vec{d})=1$ and $\vec{a}\cdot\vec{c}=\frac{1}{2}$, then (2009)

(a) $\vec{a}, \vec{b}, \vec{c}$ are non-coplanar

(b) $\vec{b}, \vec{c}, \vec{d}$ are non-coplanar

(c) $\vec{b}, \vec{d}$ are non-parallel

(d) $\vec{a}, \vec{d}$ are parallel and $\vec{b}, \vec{c}$ are parallel

38. A line with positive direction cosines passes through the point $P(2, -1, 2)$ and makes equal angles with the coordinate axes. The line meets the plane

$$2x+y+z=9$$

at point Q. The length of the line segment PQ equals (2009)

(a) 1 (b) $\sqrt{2}$ (c) $\sqrt{3}$ (d) 2

39. Let P, Q, R and S be the points on the plane with position vectors $-2\hat{i}-\hat{j}, 4\hat{i}, 3\hat{i}+3\hat{j}$ and $-3\hat{i}+2\hat{j}$ respectively. The quadrilateral PQRS must be a (2010)

(a) parallelogram, which is neither a rhombus nor a rectangle

(b) square

(c) rectangle, but not a square

(d) rhombus, but not a square

40. Equation of the plane containing the straight line $\frac{x}{2}=\frac{y}{3}=\frac{z}{4}$ and perpendicular to the plane containing the straight lines $\frac{x}{3}=\frac{y}{4}=\frac{z}{2}$ and $\frac{x}{4}=\frac{y}{2}=\frac{z}{3}$ is (2010)

(a) $x+2y-2z=0$ (b) $3x+2y-2z=0$

(c) $x-2y+z=0$ (d) $5x+2y-4z=0$

41. If the distance of the point $P(1, -2, 1)$ from the plane $x+2y-2z=\alpha$, where $\alpha>0$, is 5, then the foot of the perpendicular from P to the plane is (2010)

(a) $\left(\frac{8}{3}, \frac{4}{3}, -\frac{7}{3}\right)$ (b) $\left(\frac{4}{3}, -\frac{4}{3}, \frac{1}{3}\right)$

(c) $\left(\frac{1}{3}, \frac{2}{3}, \frac{10}{3}\right)$ (d) $\left(\frac{2}{3}, -\frac{1}{3}, \frac{5}{2}\right)$

42. Two adjacent sides of a parallelogram ABCD are given by

$$\overrightarrow{AB}=2\hat{i}+10\hat{j}+11\hat{k} \text{ and } \overrightarrow{AD}=\hat{i}+2\hat{j}+2\hat{k}$$

The side AD is rotated by an acute angle α in the plane of the parallelogram so that AD becomes AD'. If AD' makes a right angle with the side AB, then the cosine of the angle α

is given by **(2010)**

(a) $\frac{8}{9}$ (b) $\frac{\sqrt{17}}{9}$ (c) $\frac{1}{9}$ (d) $\frac{4\sqrt{5}}{9}$

43. Let $\vec{a}=\hat{i}+\hat{j}+\hat{k}$, $\vec{b}=\hat{i}-\hat{j}+\hat{k}$ and $\vec{c}=\hat{i}-\hat{j}-\hat{k}$ be three vectors. A vector $\vec{v}$ in the plane of $\vec{a}$ and $\vec{b}$, whose projection on $\vec{c}$ is $\frac{1}{\sqrt{3}}$, is given by **(2011)**

(a) $\hat{i}-3\hat{j}+3\hat{k}$ (b) $-3\hat{i}-3\hat{j}-\hat{k}$

(c) $3\hat{i}-\hat{j}+3\hat{k}$ (d) $\hat{i}+3\hat{j}-3\hat{k}$

44. The point P is the intersection of the straight line joining the points $Q(2, 3, 5)$ and $R(1, -1, 4)$ with the plane $5x-4y-z=1$. If S is the foot of the perpendicular drawn from the point $T(2, 1, 4)$ to QR, then the length of the line segment PS is **(2012)**

(a) $\frac{1}{\sqrt{2}}$ (b) $\sqrt{2}$ (c) 2 (d) $2\sqrt{2}$

45. The equation of a plane passing through the line of intersection of the planes $x+2y+3z=2$ and $x-y+z=3$ and at a distance $\frac{2}{\sqrt{3}}$ from the point $(3, 1, -1)$ is **(2012)**

(a) $5x-11y+z=17$ (b) $\sqrt{2}x+y=3\sqrt{2}-1$

(c) $x+y+z=\sqrt{3}$ (d) $x-\sqrt{2}y=1-\sqrt{2}$

46. If $\vec{a}$ and $\vec{b}$ are vectors such that $|\vec{a}+\vec{b}|=\sqrt{29}$ and $\vec{a}\times(2\hat{i}+3\hat{j}+4\hat{k})=(2\hat{i}+3\hat{j}+4\hat{k})\times\vec{b}$, then a possible value of $(\vec{a}+\vec{b}).(-7\hat{i}+2\hat{j}+3\hat{k})$ is **(2012)**

(a) 0 (b) 3 (c) 4 (d) 8

D MCQ's WITH ONE OR MORE THAN ONE CORRECT :

1. Let $\vec{a}=a_1i+a_2j+a_3k$, $\vec{b}=b_1i+b_2j+b_3k$ and $\vec{c}=c_1i+c_2j+c_3k$ be three non-zero vectors such that $\vec{c}$ is a unit vector perpendicular to both the vectors $\vec{a}$ and $\vec{b}$. If the angle between $\vec{a}$ and $\vec{b}$ is $\frac{\pi}{6}$, then $\begin{vmatrix} a_1 & a_2 & a_3 \\ b_1 & b_2 & b_3 \\ c_1 & c_2 & c_3 \end{vmatrix}^2$ is equal to **(1986 - 2 Marks)**

(a) 0

(b) 1

(c) $\frac{1}{4}(a_1^2+a_2^2+a_3^2)(b_1^2+b_2^2+b_3^2)$

(d) $\frac{3}{4}(a_1^2+a_2^2+a_3^2)(b_1^2+b_2^2+b_3^2)(c_1^2+c_2^2+c_3^2)$

2. The number of vectors of unit length perpendicular to vectors $\vec{a}=(1, 1, 0)$ and $\vec{b}=(0, 1, 1)$ is **(1987 - 2 Marks)**

(a) one (b) two (c) three (d) infinite

(e) None of these.

3. Let $\vec{a}=2\hat{i}-\hat{j}+\hat{k}, \vec{b}=\hat{i}+2\hat{j}-\hat{k}$ and $\vec{c}=\hat{i}+\hat{j}-2\hat{k}-2\hat{k}$ be three vectors. A vector in the plane of $\vec{b}$ and $\vec{c}$, whose projection on $\vec{a}$ is of magnitude $\sqrt{2/3}$, is : **(1993 - 2 Marks)**

(a) $2\hat{i}+3\hat{j}-3\hat{k}$ (b) $2\hat{i}+3\hat{j}+3\hat{k}$

(c) $-2\hat{i}-\hat{j}+5\hat{k}$ (d) $2\hat{i}+\hat{j}+5\hat{k}$

4. The vector $\frac{1}{3}(2\hat{i}-2\hat{j}+\hat{k})$ is **(1994)**

(a) a unit vecotr

(b) makes an angle $\frac{\pi}{3}$ with the vector $(2\hat{i}-4\hat{j}+3\hat{k})$

(c) parallel to the vector $\left(-\hat{i}+\hat{j}-\frac{1}{2}\hat{k}\right)$

(d) perpendicular to the vector $3\hat{i}+2\hat{j}-2\hat{k}$

5. If $\boldsymbol{a}=\boldsymbol{i}+\boldsymbol{j}+\boldsymbol{k}, \vec{b}=4\boldsymbol{i}+3\boldsymbol{j}+4\boldsymbol{k}$ and $\boldsymbol{c}=\boldsymbol{i}+\alpha\boldsymbol{j}+\beta\boldsymbol{k}$ are linearly dependent vectors and $|c|=\sqrt{3}$, then **(1998 - 2 Marks)**

(a) $\alpha=1, \beta=-1$ (b) $\alpha=1, \beta=\pm1$

(c) $\alpha=-1, \beta=\pm1$ (d) $\alpha=\pm1, \beta=1$

6. For three vectors $\boldsymbol{u}, \boldsymbol{v}, \boldsymbol{w}$ which of the following expression is not equal to any of the remaining three? **(1998 - 2 Marks)**

(a) $\boldsymbol{u}\cdot(\boldsymbol{v}\times\boldsymbol{w})$ (b) $(\boldsymbol{v}\times\boldsymbol{w})\cdot\boldsymbol{u}$

(c) $\boldsymbol{v}\cdot(\boldsymbol{u}\times\boldsymbol{w})$ (d) $(\boldsymbol{u}\times\boldsymbol{v})\cdot\boldsymbol{w}$

7. Which of the following expressions are meaningful? **(1998 - 2 Marks)**

(a) $\boldsymbol{u}(\boldsymbol{v}\times\boldsymbol{w})$ (b) $(\boldsymbol{u}\cdot\boldsymbol{v})\cdot\boldsymbol{w}$

(c) $(\boldsymbol{u}\cdot\boldsymbol{v})\boldsymbol{w}$ (d) $\boldsymbol{u}\times(\boldsymbol{v}\cdot\boldsymbol{w})$

8. Let $\boldsymbol{a}$ and $\boldsymbol{b}$ be two non-collinear unit vectors. If $\boldsymbol{u}=\boldsymbol{a}-(\boldsymbol{a}\,.\,\boldsymbol{b})\,\boldsymbol{b}$ and $\boldsymbol{v}=\boldsymbol{a}\times\boldsymbol{b}$, then $|\boldsymbol{v}|$ is **(1999 - 3 Marks)**

(a) $|\boldsymbol{u}|$ (b) $|\boldsymbol{u}|+|\boldsymbol{u}\,.\,\boldsymbol{a}|$

(c) $|\boldsymbol{u}|+|\boldsymbol{u}\,.\,\boldsymbol{b}|$ (d) $|\boldsymbol{u}|+\boldsymbol{u}.(\boldsymbol{a}+\boldsymbol{b})$

9. Let $\vec{A}$ be vector parallel to line of intersection of planes P_1 and P_2. Plane P_1 is parallel to the vectors $2\hat{j}+3\hat{k}$ and $4\hat{j}-3\hat{k}$ and that P_2 is parallel to $\hat{j}-\hat{k}$ and $3\hat{i}+3\hat{j}$, then the angle between vector $\vec{A}$ and a given vector $2\hat{i}+\hat{j}-2\hat{k}$ is **(2006 - 5M, –1)**

(a) $\frac{\pi}{2}$ (b) $\frac{\pi}{4}$ (c) $\frac{\pi}{6}$ (d) $\frac{3\pi}{4}$

10. The vector (s) which is/are coplanar with vectors $\hat{i}+\hat{j}+2\hat{k}$ and $\hat{i}+2\hat{j}+\hat{k}$, and perpendicular to the vector $\hat{i}+\hat{j}+\hat{k}$ is/are **(2011)**

(a) $\hat{j}-\hat{k}$ (b) $-\hat{i}+\hat{j}$ (c) $\hat{i}-\hat{j}$ (d) $-\hat{j}+\hat{k}$

11. If the straight lines $\frac{x-1}{2}=\frac{y+1}{k}=\frac{z}{2}$ and $\frac{x+1}{5}=\frac{y+1}{2}=\frac{z}{k}$ are coplanar, then the plane (s) containing these two lines is (are) **(2012)**

(a) $y+2z=-1$ (b) $y+z=-1$
(c) $y-z=-1$ (d) $y-2z=-1$

12. A line l passing through the origin is perpendicular to the lines *(JEE Adv. 2013)*

$$l_1:(3+t)\hat{i}+(-1+2t)\hat{j}+(4+2t)\hat{k}, \quad -\infty<t<\infty$$

$$l_2:(3+2s)\hat{i}+(3+2s)\hat{j}+(2+s)\hat{k}, \quad -\infty<s<\infty$$

Then, the coordinate(s) of the point(s) on l_2 at a distance of $\sqrt{17}$ from the point of intersection of l and l_1 is (are)

(a) $\left(\frac{7}{3},\frac{7}{3},\frac{5}{3}\right)$ (b) $(-1,-1,0)$

(c) $(1,1,1)$ (d) $\left(\frac{7}{9},\frac{7}{9},\frac{8}{9}\right)$

13. Two lines $L_1: x=5, \frac{y}{3-\alpha}=\frac{z}{-2}$ and $L_2: x=\alpha, \frac{y}{-1}=\frac{z}{2-\alpha}$ are coplanar. Then α can take value(s) *(JEE Adv. 2013)*
(a) 1 (b) 2 (c) 3 (d) 4

(E) SUBJECTIVE PROBLEMS :

1. From a point O inside a triangle ABC, perpendiculars OD, OE, OF are drawn to the sides BC, CA, AB respectively. Prove that the perpendiculars from A, B, C to the sides EF, FD, DE are concurrent. *(1978)*

2. $A_1, A_2, \ldots\ldots\ldots\ldots\ldots\ldots A_n$ are the vertices of a regular plane polygon with n sides and O is its centre. Show that $\sum_{i=1}^{n-1}(OA_i \times OA_{i+1}) = (1-n)(OA_2 \times OA_1)$ *(1982 - 2 Marks)*

3. Find all values of λ such that $x, y, z, \neq (0,0,0)$ and $(\vec{i}+\vec{j}+3\vec{k})x+(3\vec{i}-3\vec{j}+\vec{k})y+(-4\vec{i}+5\vec{j})z$ $=\lambda(x\vec{i}\times\vec{j}\ y+\vec{k}\ z)$ where $\vec{i}$, $\vec{j}$, $\vec{k}$ are unit vectors along the coordinate axes. *(1982 - 3 Marks)*

4. If c be a given non-zero scalar, and $\vec{A}$ and $\vec{B}$ be given non-zero vectors such that $\vec{A}\perp\vec{B}$ find the vector $\vec{X}$ which satisfies the equations $\vec{A}\cdot\vec{X}=c$ and $\vec{A}\times\vec{X}=\vec{B}$. *(1983 - 3 Marks)*

5. A vector $\vec{A}$ has components A_1, A_2, A_3 in a right -handed rectangular Cartesian coordinate system $oxyz$. The coordinate system is rotated about the x-axis through an angle $\frac{\pi}{2}$. Find the components of A in the new coordinate system, in terms of A_1, A_2, A_3. *(1983 - 2 Marks)*

6. The position vectors of the points A, B, C and D are $3\hat{i}-2\hat{j}-\hat{k}, 2\hat{i}+3\hat{j}-4\hat{k}, -\hat{i}+\hat{j}+2\hat{k}$ and $4\hat{i}+5\hat{j}+\lambda\hat{k}$, respectively. If the points A, B, C and D lie on a plane, find the value of λ. *(1986 - 2½ Marks)*

7. If A, B, C, D are any four points in space, prove that – *(1987 - 2 Marks)*

$\left|\overrightarrow{AB}\times\overrightarrow{CD}+\overrightarrow{BC}\times\overrightarrow{AD}+\overrightarrow{CA}\times\overrightarrow{BD}\right|=4$ (area of triangle ABC)

8. Let $OA\ CB$ be a parallelogram with O at the origin and OC a diagonal. Let D be the midpoint of OA. Using vector methods prove that BD and CO intersect in the same ratio. Determine this ratio. *(1988 - 3 Marks)*

9. If vectors $\vec{a}, \vec{b}, \vec{c}$ are coplanar, show that

$$\begin{vmatrix} \vec{a} & \vec{b} & \vec{c} \\ \vec{a}.\ \vec{a} & \vec{a}.\ \vec{b} & \vec{a}.\ \vec{c} \\ \vec{b}.\ \vec{a} & \vec{b}.\ \vec{b} & \vec{b}.\ \vec{c} \end{vmatrix} = \vec{0}$$ *(1989 - 2 Marks)*

10. In a triangle OAB, E is the midpoint of BO and D is a point on AB such that $AD : DB = 2 : 1$. If OD and AE intersect at P, determine the ratio $OP : PD$ using vector methods. *(1989 - 4 Marks)*

11. Let $\vec{A}=2\vec{i}+\vec{k}$, $\vec{B}=\vec{i}+\vec{j}+\vec{k}$, and $\vec{C}=4\vec{i}-3\vec{j}+7\vec{k}$. Determine a vector $\vec{R}$. Satisfying $\vec{R}\times\vec{B}=\vec{C}\times\vec{B}$ and $\vec{R}\cdot\vec{A}=0$ *(1990 - 3 Marks)*

12. Determine the value of 'c' so that for all real x, the vector $cx\hat{i}-6\hat{j}-3\hat{k}$ and $x\hat{i}+2\hat{j}+2cx\hat{k}$ make an obtuse angle with each other. *(1991 - 4 Marks)*

13. In a triangle ABC, D and E are points on BC and AC respectively, such that $BD = 2\ DC$ and $AE = 3EC$. Let P be the point of intersection of AD and BE. Find BP/PE using vector methods. *(1993 - 5 Marks)*

14. If the vectors $\vec{b}, \vec{c}, \vec{d}$, are not coplanar, then prove that the vector

$(\vec{a}\times\vec{b})\times(\vec{c}\times\vec{d})+(\vec{a}\times\vec{c})\times(\vec{d}\times\vec{b})+(\vec{a}\times\vec{d})\times(\vec{b}\times\vec{c})$ is parallel to $\vec{a}$. *(1994 - 4 Marks)*

15. The position vectors of the vertices A, B and C of a tetrahedron $ABCD$ are $\hat{i}+\hat{j}+\hat{k}$, $\hat{i}$ and $3\hat{i}$, respectively. The altitude from vertex D to the opposite face ABC meets the median line through A of the triangle ABC at a point E. If the length of the side AD is 4 and the volume of the tetrahedron is $\frac{2\sqrt{2}}{3}$, find the position vector of the point E for all its possible positions. *(1996 - 5 Marks)*

16. Let $\vec{a}$, $\vec{b}$ and $\vec{c}$ be non coplanar unit vectors, equally inclined to one another at an angle θ. If $\vec{a}\times\vec{b}+\vec{b}\times\vec{c}=p\vec{a}+q\vec{b}+r\vec{c}$, find scalars p, q and r in terms of θ. *(1997C - 5 Marks)*

17. If **A**, **B** and **C** are vectors such that $|B|=|C|$. Prove that $[(A+B)\times(A+C)]\times(B\times C)(B+C)=0$. *(1997 - 5 Marks)*

18. Prove, by vector methods or otherwise, that the point of intersection of the diagonals of a trapezium lies on the line passing through the mid-points of the parallel sides. (You may assume that the trapezium is not a parallelogram.) *(1998 - 8 Marks)*

19. For any two vectors u and v, prove that *(1998 - 8 Marks)*
(a) $(u\cdot v)^2+|u\times v|^2=|u|^2|v|^2$ and
(b) $(1+|u|^2)(1+|v|^2)=(1-u\cdot v)^2+|u+v+(u\times v)|^2$.

20. Let $\boldsymbol{u}$ and $\boldsymbol{v}$ be unit vectors. If $\boldsymbol{w}$ is a vector such that $\boldsymbol{w}+(\boldsymbol{w}\times\boldsymbol{u})=\boldsymbol{v}$, then prove that $|(\boldsymbol{u}\times\boldsymbol{v})\cdot\boldsymbol{w}|\le 1/2$ and that the equality holds if and only if $\boldsymbol{u}$ is perpendicular to $\boldsymbol{v}$. ***(1999 - 10 Marks)***

21. Show, by vector methods, that the angular bisectors of a triangle are concurrent and find an expression for the position vector of the point of concurrency in terms of the position vectors of the vertices. ***(2001 - 5 Marks)***

22. Find 3–dimensional vectors $\vec{v}_1, \vec{v}_2, \vec{v}_3$ satisfying

$\vec{v}_1.\vec{v}_1 = 4, \vec{v}_1.\vec{v}_2 = -2, \vec{v}_1.\vec{v}_3 = 6, \vec{v}_2.\vec{v}_2 = 2, \vec{v}_2.\vec{v}_3 = -5, \vec{v}_3.\vec{v}_3 = 29$ ***(2001 - 5 Marks)***

23. Let $\vec{A}(t) = f_1(t)\hat{i} + f_2(t)\hat{j}$ and

$\vec{B}(t) = g_1(t)\hat{i} + g_2(t)\hat{j}, t \in [0,1],$

where f_1, f_2, g_1, g_2 are continuous functions. If $\vec{A}(t)$ and $\vec{B}(t)$ are nonzero vectors for all t and $\vec{A}(0) = 2\hat{i} + 3\hat{j}$, $\vec{A}(1) = 6\hat{i} + 2\hat{j}$, $\vec{B}(0) = 3\hat{i} + 2\hat{j}$ and $\vec{B}(1) = 2\hat{i} + 6\hat{j}$. Then show that $\vec{A}(t)$ and $\vec{B}(t)$ are parallel for some t. ***(2001 - 5 Marks)***

24. Let V be the volume of the parallelopiped formed by the vectors $\vec{a} = a_1\hat{i} + a_2\hat{j} + a_3\hat{k}$, $\vec{b} = b_1\hat{i} + b_2\hat{j} + b_3\hat{k}$, $\vec{c} = c_1\hat{i} + c_2\hat{j} + c_3\hat{k}$. If a_r, b_r, c_r, where $r = 1, 2, 3$, are non-negative real numbers and $\sum_{r=1}^{3}(a_r + b_r + c_r) = 3L$, show that $V \le L^3$. ***(2002 - 5 Marks)***

25. (i) Find the equation of the plane passing through the points (2, 1, 0), (5, 0, 1) and (4, 1, 1).

(ii) If P is the point (2, 1, 6) then find the point Q such that PQ is perpendicular to the plane in (i) and the mid point of PQ lies on it. ***(2003 - 4 Marks)***

26. If $\vec{u}, \vec{v}, \vec{w}$, are three non-coplanar unit vectors and α, β, γ are the angles between $\vec{u}$ and $\vec{v}$ and $\vec{w}$, $\vec{w}$ and $\vec{u}$ respectively and $\vec{x}, \vec{y}, \vec{z}$ are unit vectors along the bisectors of the angles α, β, γ respectively. Prove that

$$[\vec{x}\times\vec{y} \;\; \vec{y}\times\vec{z} \;\; \vec{z}\times\vec{x}] = \frac{1}{16}[\vec{u}\;\vec{v}\;\vec{w}]^2 \sec^2\frac{\alpha}{2}\sec^2\frac{\beta}{2}\sec^2\frac{\gamma}{2}.$$

(2003 - 4 Marks)

27. If $\vec{a}, \vec{b}, \vec{c}$ and $\vec{d}$ are distinct vectors such that $\vec{a}\times\vec{c} = \vec{b}\times\vec{d}$ and $\vec{a}\times\vec{b} = \vec{c}\times\vec{d}$. Prove that

$(\vec{a}-\vec{d}).(\vec{b}-\vec{c}) \ne 0$ i.e. $\vec{a}.\vec{b} + \vec{d}.\vec{c} \ne \vec{d}.\vec{b} + \vec{a}.\vec{c}$ ***(2004 - 2 Marks)***

28. Find the equation of plane passing through (1, 1, 1) & parallel to the lines L_1, L_2 having direction ratios (1,0, –1), (1, –1,0). Find the volume of tetrahedron formed by origin and the points where these planes intersect the coordinate axes. ***(2004 - 2 Marks)***

29. A parallelopiped 'S' has base points A, B, C and D and upper face points A', B', C' and D'. This parallelopiped is compressed by upper face $A'B'C'D'$ to form a new parallelopiped 'T' having upper face points A'', B'', C'' and D''. Volume of parallelopiped T is 90 percent of the volume of parallelopiped S. Prove that the locus of 'A''', is a plane. ***(2004 - 2 Marks)***

30. P_1 and P_2 are planes passing through origin. L_1 and L_2 are two line on P_1 and P_2 respectively such that their intersection is origin. Show that there exists points A, B, C, whose permutation A', B', C' can be chosen such that (i) A is on L_1, B on P_1 but not on L_1 and C not on P_1 (ii) A' is on L_2, B' on P_2 but not on L_2 and C' not on P_2 ***(2004 - 4 Marks)***

31. Find the equation of the plane containing the line $2x - y + z - 3 = 0, 3x + y + z = 5$ and at a distance of $\frac{1}{\sqrt{6}}$ from the point (2, 1, –1). ***(2005 - 2 Marks)***

32. If the incident ray on a surface is along the unit vector $\hat{v}$, the reflected ray is along the unit vector $\hat{w}$ and the normal is along unit vector $\hat{a}$ outwards. Express $\hat{w}$ in terms of $\hat{a}$ and $\hat{v}$. ***(2005 - 4 Marks)***

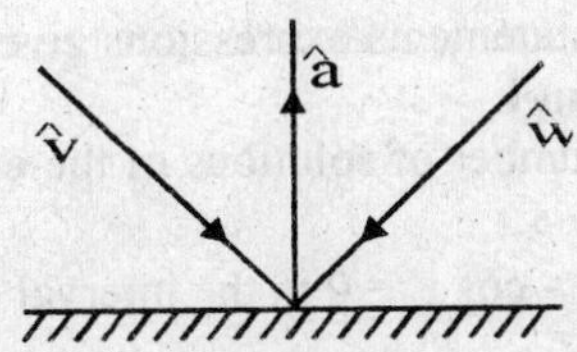

(F) MATCH THE FOLLOWING :

DIRECTIONS (for Qs. 1-6) *Each question contains statements given in two columns, which have to be matched. The statements in Column-I are labelled A, B, C and D, while the statements in Column-II are labelled p, q, r, s and t. Any given statement in Column-I can have correct matching with ONE OR MORE statement(s) in Column-II. The appropriate bubbles corresponding to the answers to these questions have to be darkened as illustrated in the following example :*

If the correct matches are A-p, s and t; B-q and r; C-p and q; and D-s then the correct darkening of bubbles will look like the given.

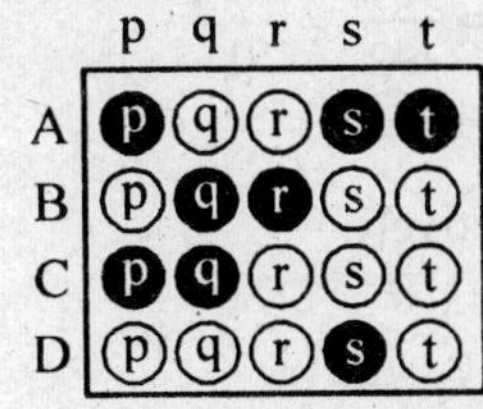

1. Match the following : *(2006 - 6M)*

(A) Two rays $x + y = |a|$ and $ax - y = 1$ intersects each other in the first quadrant in the interval $a \in (a_0, \infty)$, the value of a_0 is — (p) 2

(B) Point (α, β, γ) lies on the plane $x + y + z = 2$. — (q) $\frac{4}{3}$

Let $\vec{a} = \alpha\hat{i} + \beta\hat{j} + \gamma\hat{k},\ \hat{k}\times(\hat{k}\times\vec{a}) = 0$, then $\gamma =$

(C) $\left|\int_0^1 (1-y^2)\,dy\right| + \left|\int_1^0 (y^2-1)\,dy\right|$ — (r) $\left|\int_0^1 \sqrt{1-x}\,dx\right| + \left|\int_{-1}^0 \sqrt{1+x}\,dx\right|$

(D) If $\sin A \sin B \sin C + \cos A \cos B = 1$, then the value of $\sin C =$ — (s) 1

2. Consider the following linear equations

$$ax + by + cz = 0; \quad bx + cy + az = 0; \quad cx + ay + bz = 0$$

Match the conditions/expressions in **Column I** with statements in **Column II** and indicate your answer by darkening the appropriate bubbles in the 4 × 4 matrix given in the *ORS*. *(2007)*

Column I	Column II
(A) $a + b + c \neq 0$ and $a^2 + b^2 + c^2 = ab + bc + ca$	(p) the equations represent planes meeting only at asingle point
(B) $a + b + c = 0$ and $a^2 + b^2 + c^2 \neq ab + bc + ca$	(q) the equations represent the line $x = y = z$.
(C) $a + b + c \neq 0$ and $a^2 + b^2 + c^2 \neq ab + bc + ca$	(r) the equations represent identical planes.
(D) $a + b + c = 0$ and $a^2 + b^2 + c^2 = ab + bc + ca$	(s) the equations represent the whole of the three dimensional space.

3. Match the statements / expressions given in **Column-I** with the values given in **Column-II.** *(2009)*

Column-I	Column-II
(A) Root(s) of the equation $2\sin^2\theta + \sin^2 2\theta = 2$	(p) $\frac{\pi}{6}$
(B) Points of discontinuity of the unction $f(x) = \left[\frac{6x}{\pi}\right]\cos\left[\frac{3x}{\pi}\right]$, f where $[y]$ denotes the largest integer less than or equal tc y	(q) $\frac{\pi}{4}$
(C) Volume of the parallelopiped with its edges represented by the vectors $\hat{i}+\hat{j},\ \hat{i}+2\hat{j}$ and $\hat{i}+\hat{j}+\pi\hat{k}$	(r) $\frac{\pi}{3}$
(D) Angle between vector $\vec{a}$ and $\vec{b}$ where $\vec{a}$, $\vec{b}$ and $\vec{c}$ are unit vectors satisfying $\vec{a}+\vec{b}+\sqrt{3}\,\vec{c} = \vec{0}$	(s) $\frac{\pi}{2}$
	(t) π

4. Match the statements/expressions given in **Column-I** with the values given in **Column-II.** *(2009)*

Column-I	Column-II
(A) The number of solutions of the equation $x e^{\sin x} - \cos x = 0$ in the interval $\left(0, \frac{\pi}{2}\right)$	(p) 1
(B) Value(s) of k for which the planes $kx + 4y + z = 0$, $4x + ky + 2z = 0$ and $2x + 2y + z = 0$ intersect in a straight line	(q) 2
(C) Value(s) of k for which $\|x-1\| + \|x-2\| + \|x+1\|\ \| + \|x+2\| = 4k$ has integer solution(s)	(r) 3
(D) If $y' = y + 1$ and $y(0) = 1$, then value(s) of $y(\ln 2)$	(s) 4
	(t) 5

5. Match the statement in **Column-1** with the values in **Column -II** *(2010)*

Column – I	Column – II
(A) A line from the origin meets the lines $\frac{x-2}{1} = \frac{y-1}{-2} = \frac{z+1}{1}$ and $\frac{x-\frac{8}{3}}{2} = \frac{y+3}{-1} = \frac{z-1}{1}$ at P and Q respectively. If length $PQ = d$, then d^2 is	(p) –4

(B) The values of x satisfying
$\tan^{-1}(x+3)-\tan^{-1}(x-3)=\sin^{-1}\left(\frac{3}{5}\right)$ are (q) 0

(C) Non-zero vectors $\vec{a}, \vec{b}$ and $\vec{c}$ satisfy $\vec{a}.\vec{b}=0$.
$(\vec{b}-\vec{a}).(\vec{b}+\vec{c})=0$ and $2|\vec{b}+\vec{c}|=|\vec{b}-\vec{a}|$.
If $\vec{a}=\mu\vec{b}+4\vec{c}$, then the possible values of μ are (r) 4

(D) Let f be the function on $[-\pi, \pi]$ given by $f(0)=9$
and $f(x)=\sin\left(\frac{9x}{2}\right)\Big/\sin\left(\frac{x}{2}\right)$ *for* $x \neq 0$ (s) 5
The value of $\frac{2}{\pi}\int_{-\pi}^{\pi} f(x)dx$ is
(t) 6

6. Match the statements given in Column-I with the values given in Column-II. *(2011)*

Column-I **Column-II**

(A) If $\vec{a}=\hat{j}+\sqrt{3}\hat{k}, \vec{b}=-\hat{j}+\sqrt{3}\hat{k}$ and $\vec{c}=2\sqrt{3}\hat{k}$ form a triangle, then the internal angle of the triangle between $\vec{a}$ and $\vec{b}$ is (p) $\frac{\pi}{6}$

(B) If $\int_a^b (f(x)-3x)\,dx = a^2-b^2$, then the value of $f\left(\frac{\pi}{6}\right)$ is (q) $\frac{2\pi}{3}$

(C) The value of $\frac{\pi^2}{\ell n 3}\int_{7/6}^{5/6} \sec(\pi x)\,dx$ is (r) $\frac{\pi}{3}$

(D) The maximum value of $\left|Arg\left(\frac{1}{1-z}\right)\right|$ for $|z|=1, z\neq 1$ is given by (s) π

(t) $\frac{\pi}{2}$

DIRECTIONS (Q. 7 & 8) : *Following questions. Each question has matching lists. The codes for the lists have choices (a), (b), (c) and (d) out of which ONLY ONE is correct.*

7. Match List I with List II and select the correct answer using the code given below the lists : *(JEE Adv. 2013)*

List I **List II**

P. Volume of parallelepiped determined by vectors $\vec{a}, \vec{b}$ and $\vec{c}$ is 2. Then the volume of the parallelepiped determined by vectors $2(\vec{a}\times\vec{b}), 3(\vec{b}\times\vec{c})$ and $2(\vec{c}\times\vec{a})$ is — 1. 100

Q. Volume of parallelepiped determined by vectors $\vec{a}, \vec{b}$ and $\vec{c}$ is 5. Then the volume of the parallelepiped determined by vectors $3(\vec{a}+\vec{b}), 3(\vec{b}+\vec{c})$ and $2(\vec{c}+\vec{a})$ is — 2. 30

R. Area of a triangle with adjacent sides determined by vectors $\vec{a}$ and $\vec{b}$ is 20. Then the area of the triangle with adjacent sides determined by vectors $(2\vec{a}+3\vec{b})$ and $(\vec{a}-\vec{b})$ is — 3. 24

S. Area of a parallelogram with adjacent sides determined by vectors $\vec{a}$ and $\vec{b}$ is 30. Then the area of the parallelogram with adjacent sides determined by vectors $(\vec{a}+\vec{b})$ and $\vec{a}$ is — 4. 60

Codes:

	P	Q	R	S
(a)	4	2	3	1
(b)	2	3	1	4
(c)	3	4	1	2
(d)	1	4	3	2

8. Consider the lines $L_1 : \frac{x-1}{2} = \frac{y}{-1} = \frac{z+3}{1}, L_2 \frac{x-4}{1} = \frac{y+3}{1} = \frac{z+3}{2}$ and the planes $P_1 : 7x + y + 2z = 3$, $P_2 = 3x + 5y - 6z = 4$. Let $ax + by + cz = d$ be the equation of the plane passing through the point of intersection of lines L_1 and L_2, and perpendicular to planes P_1 and P_2.
Match List I with List II and select the correct answer using the code given below the lists : **(JEE Adv. 2013)**

	List I		List II
P.	$a=$	1.	13
Q.	$b=$	2.	-3
R.	$c=$	3.	1
S.	$d=$	4.	-2

Codes:

	P	Q	R	S
(a)	3	2	4	1
(b)	1	3	4	2
(c)	3	2	1	4
(d)	2	4	1	3

G COMPREHENSION BASED QUESTIONS :

Consider the lines

$$L_1 : \frac{x+1}{3} = \frac{y+2}{1} = \frac{z+1}{2} \quad L_2 : \frac{x-2}{1} = \frac{y+2}{2} = \frac{z-3}{3}$$

1. The unit vector perpendicular to both L_1 and L_2 is **(2008)**
   (a) $\frac{-\hat{i}+7\hat{j}+7\hat{k}}{\sqrt{99}}$ (b) $\frac{-\hat{i}-7\hat{j}+5\hat{k}}{5\sqrt{3}}$
   (c) $\frac{-\hat{i}+7\hat{j}+5\hat{k}}{5\sqrt{3}}$ (d) $\frac{7\hat{i}-7\hat{j}-\hat{k}}{\sqrt{99}}$

2. The shortest distance between L_1 and L_2 is **(2008)**
   (a) 0 (b) $\frac{17}{\sqrt{3}}$
   (c) $\frac{41}{5\sqrt{3}}$ (d) $\frac{17}{5\sqrt{3}}$

3. The distance of the point (1, 1, 1) from the plane passing through the point (–1, –2, –1) and whose normal is perpendicular to both the lines L_1 and L_2 is **(2008)**
   (a) $\frac{2}{\sqrt{75}}$ (b) $\frac{7}{\sqrt{75}}$
   (c) $\frac{13}{\sqrt{75}}$ (d) $\frac{23}{\sqrt{75}}$

H ASSERTION & REASON TYPE QUESTIONS :

1. Consider the planes $3x - 6y - 2z = 15$ and $2x + y - 2z = 5$.
STATEMENT-1 : The parametric equations of the line of intersection of the given planes are $x = 3 + 14t, y = 1 + 2t, z = 15t$. **because**

STATEMENT-2 : The vector $14\hat{i} + 2\hat{j} + 15\hat{k}$ is parallel to the line of intersection of given planes. ***(2007 -3 marks)***
   (a) Statement-1 is True, Statement-2 is True; Statement-2 is a correct explanation for Statement-1
   (b) Statement-1 is True, Statement-2 is True; Statement-2 is NOT a correct explanation for Statement-1
   (c) Statement-1 is True, Statement-2 is False
   (d) Statement-1 is False, Statement-2 is True.

2. Let the vectors $\overrightarrow{PQ}, \overrightarrow{QR}, \overrightarrow{RS}, \overrightarrow{ST}, \overrightarrow{TU}$ and $\overrightarrow{UP}$ represent the sides of a regular hexagon.
STATEMENT-1 : $\overrightarrow{PQ} \times \left(\overrightarrow{RS} + \overrightarrow{ST}\right) \neq \overrightarrow{0}$. **because**
STATEMENT-2 : $\overrightarrow{PQ} \times \overrightarrow{RS} = \overrightarrow{0}$ and $\overrightarrow{PQ} \times \overrightarrow{ST} \neq \overrightarrow{0}$.
(2007 -3 marks)
   (a) Statement-1 is True, statement-2 is True; Statement-2 is a correct explanation for Statement-1.
   (b) Statement-1 is True, Statement-2 is True; Statement-2 is NOT a correct explanation for Statement-1
   (c) Statement-1 is True, Statement-2 is False
   (d) Statement-1 is False, Statement-2 is True.

3. Consider three planes
$P_1 : x - y + z = 1$ $\quad P_2 : x + y - z = 1$
$P_3 : x - 3y + 3z = 2$
Let L_1, L_2, L_3 be the lines of intersection of the planes P_2 and P_3, P_3 and P_1, P_1 and P_2, respectively.
STATEMENT - 1 : At least two of the lines L_1, L_2 and L_3 are non-parallel and
STATEMENT - 2 : The three planes doe not have a common point. **(2008)**
   (A) STATEMENT - 1 is True, STATEMENT - 2 is True; STATEMENT - 2 is a correct explanation for STATEMENT - 1
   (B) STATEMENT - 1 is True, STATEMENT - 2 is True; STATEMENT - 2 is **NOT** a correct explanation for STATEMENT - 1
   (C) STATEMENT - 1 is True, STATEMENT - 2 is False
   (D) STATEMENT - 1 is False, STATEMENT - 2 is True

I INTEGER VALUE CORRECT TYPE :

1. If $\vec{a}$ and $\vec{b}$ are vectors in space given by $\vec{a} = \frac{\hat{i}-2\hat{j}}{\sqrt{5}}$ and $\vec{b} = \frac{2\hat{i}+\hat{j}+3\hat{k}}{\sqrt{14}}$, then find the value of $\left(2\vec{a}+\vec{b}\right)$. $\left[\left(\vec{a}\times\vec{b}\right)\times\left(\vec{a}-2\vec{b}\right)\right]$. **(2010)**

2. If the distance between the plane $Ax - 2y + z = d$ and the plane containing the lines $\frac{x-1}{2} = \frac{y-2}{3} = \frac{z-3}{4}$ and $\frac{x-2}{3} = \frac{y-3}{4} = \frac{z-4}{5}$ is $\sqrt{6}$, then find $|d|$. *(2010)*

3. Let $\vec{a} = -\hat{i} - \hat{k}, \vec{b} = -\hat{i} + \hat{j}$ and $\vec{c} = \hat{i} + 2\hat{j} + 3\hat{k}$ be three given vectors. If $\vec{r}$ is a vector such that $\vec{r} \times \vec{b} = \vec{c} \times \vec{b}$ and $\vec{r}.\vec{a} = 0$, then the value of $\vec{r}.\vec{b}$ is *(2011)*

4. If $\vec{a}, \vec{b}$ and $\vec{c}$ are unit vectors satisfying *(2012)* $|\vec{a} - \vec{b}|^2 + |\vec{b} - \vec{c}|^2 + |\vec{c} - \vec{a}|^2 = 9$, then $|2\vec{a} + 5\vec{b} + 5\vec{c}|$ is

5. Consider the set of eight vectors $V = \{a\hat{i} + b\hat{j} + c\hat{k} : a, b, c \in \{-1, 1\}\}$. Three non-coplanar vectors can be chosen from V in 2^p ways. Then p is *(JEE Adv. 2013)*

6. A pack contains n cards numbered from 1 to n. Two consecutive numbered cards are removed from the pack and the sum of the numbers on the remaining cards is 1224. If the smaller of the numbers on the removed cards is k, then $k - 20 =$ *(JEE Adv. 2013)*

Section-B JEE Main/ AIEEE

1. A plane which passes through the point (3, 2, 0) and the line $\frac{x-4}{1} = \frac{y-7}{5} = \frac{z-4}{4}$ is **[2002]**
(a) $x - y + z = 1$ (b) $x + y + z = 5$
(c) $x + 2y - z = 1$ (d) $2x - y + z = 5$

2. If $|\vec{a}| = 4, |\vec{b}| = 2$ and the angle between $\vec{a}$ and $\vec{b}$ is $\pi/6$ then $(\vec{a} \times \vec{b})^2$ is equal to **[2002]**
(a) 48 (b) 16
(c) $\vec{a}$ (d) none of these

3. If $\vec{a}, \vec{b}, \vec{c}$ are vectors such that $[\vec{a}\ \vec{b}\ \vec{c}] = 4$ then $[\vec{a} \times \vec{b}\ \ \vec{b} \times \vec{c}\ \ \vec{c} \times \vec{a}] =$ **[2002]**
(a) 16 (b) 64 (c) 4 (d) 8

4. If $\vec{a}, \vec{b}, \vec{c}$ are vectors show that $\vec{a} + \vec{b} + \vec{c} = 0$ and $|\vec{a}| = 7, |\vec{b}| = 5, |\vec{c}| = 3$ then angle between vector $\vec{b}$ and $\vec{c}$ is **[2002]**
(a) 60° (b) 30° (c) 45° (d) 90°

5. If $|\vec{a}| = 5, |\vec{b}| = 4, |\vec{c}| = 3$ thus what will be the value of $|\vec{a}.\vec{b} + \vec{b}.\vec{c} + \vec{c}.\vec{a}|$, given that $\vec{a} + \vec{b} + \vec{c} = 0$ **[2002]**
(a) 25 (b) 50 (c) –25 (d) –50

6. If the vectors $\vec{c}, \vec{a} = x\hat{i} + y\hat{j} + z\hat{k}$ and $\hat{b} = \hat{j}$ are such that $\vec{a}, \vec{c}$ and $\vec{b}$ form a right handed system then $\vec{c}$ is : **[2002]**
(a) $z\hat{i} - x\hat{k}$ (b) $\vec{0}$
(c) $y\hat{j}$ (d) $-z\hat{i} + x\hat{k}$

7. $\vec{a} = 3\hat{i} - 5\hat{j}$ and $\vec{b} = 6\hat{i} + 3\hat{j}$ are two vectors and $\vec{c}$ is a vector such that $\vec{c} = \vec{a} \times \vec{b}$ then $|\vec{a}| : |\vec{b}| : |\vec{c}|$
(a) $\sqrt{34} : \sqrt{45} : \sqrt{39}$ (b) $\sqrt{34} : \sqrt{45} : 39$ **[2002]**
(c) $34 : 39 : 45$ (d) $39 : 35 : 34$

8. If $\vec{a} \times \vec{b} = \vec{b} \times \vec{c} = \vec{c} \times \vec{a}$ then $\vec{a} + \vec{b} + \vec{c} =$ **[2002]**
(a) abc (b) –1 (c) 0 (d) 2

9. The d.r. of normal to the plane through (1, 0, 0), (0, 1, 0) which makes an angle $\pi/4$ with plane $x + y = 3$ are **[2002]**
(a) $1, \sqrt{2}, 1$ (b) $1, 1, \sqrt{2}$
(c) $1, 1, 2$ (d) $\sqrt{2}, 1, 1$

10. Let $\vec{u} = \hat{i} + \hat{j}, \vec{v} = \hat{i} - \hat{j}$ *and* $\vec{w} = \hat{i} + 2\hat{j} + 3\hat{k}$. If $\hat{n}$ is a unit vector such that $\vec{u}.\hat{n} = 0$ and $\vec{v}.\hat{n} = 0$, then $|\vec{w}.\hat{n}|$ is equal to **[2003]**
(a) 3 (b) 0 (c) 1 (d) 2.

11. A particle acted on by constant forces $4\hat{i} + \hat{j} - 3\hat{k}$ and $3\hat{i} + \hat{j} - \hat{k}$ is displaced from the point $\hat{i} + 2\hat{j} - 3\hat{k}$ to the point $5\hat{i} + 4\hat{j} + \hat{k}$. The total work done by the forces is
(a) 50 units (b) 20 units **[2003]**
(c) 30 units (d) 40 units.

12. The vectors $\overrightarrow{AB} = 3\hat{i} + 4\hat{k}$ & $\overrightarrow{AC} = 5\hat{i} - 2\hat{j} + 4\hat{k}$ are the sides of a triangle ABC. The length of the median through A is **[2003]**
(a) $\sqrt{288}$ (b) $\sqrt{18}$ (c) $\sqrt{72}$ (d) $\sqrt{33}$

13. The shortest distance from the plane $12x + 4y + 3z = 327$ to the sphere $x^2 + y^2 + z^2 + 4x - 2y - 6z = 155$ is
(a) 39 (b) 26 (c) $11\frac{4}{13}$ (d) 13

14. The two lines $x = ay + b, z = cy + d$ and $x = a'y + b', z = c'y + d'$ will be perpendicular, if and only if **[2003]**
(a) $aa' + cc' + 1 = 0$
(b) $aa' + bb' + cc' + 1 = 0$
(c) $aa' + bb' + cc' = 0$
(d) $(a + a')(b + b') + (c + c') = 0$.

15. The lines $\frac{x-2}{1}=\frac{y-3}{1}=\frac{z-4}{-k}$ and $\frac{x-1}{k}=\frac{y-4}{1}=\frac{z-5}{1}$ are coplanar if **[2003]**

(a) $k=3$ or -2 (b) $k=0$ or -1
(c) $k=1$ or -1 (d) $k=0$ or -3.

16. $\vec{a},\vec{b},\vec{c}$ are 3 vectors, such that $\vec{a}+\vec{b}+\vec{c}=0$, $|\vec{a}|=1$ $|\vec{b}|=2, |\vec{c}|=3$, then $\vec{a}.\vec{b}+\vec{b}.\vec{c}+\vec{c}.\vec{a}$ is equal to **[2003]**

(a) 1 (b) 0 (c) -7 (d) 7

17. The radius of the circle in which the sphere $x^2+y^2+z^2+2x-2y-4z-19=0$ is cut by the plane $x+2y+2z+7=0$ is **[2003]**

(a) 4 (b) 1 (c) 2 (d) 3

18. A tetrahedron has vertices at O(0, 0, 0), A(1, 2, 1) B(2, 1, 3) and C(–1, 1, 2). Then the angle between the faces OAB and ABC will be **[2003]**

(a) 90° (b) $\cos^{-1}\left(\frac{19}{35}\right)$
(c) $\cos^{-1}\left(\frac{17}{31}\right)$ (d) 30°

19. If $\begin{vmatrix} a & a^2 & 1+a^3 \\ b & b^2 & 1+b^3 \\ c & c^2 & 1+c^3 \end{vmatrix}=0$ and vectors $(1,a,a^2)$, $(1,b,b^2)$ and $(1,c,c^2)$ are non- coplanar, then the product abc equals **[2003]**

(a) 0 (b) 2 (c) -1 (d) 1

20. Consider points A, B, C and D with position vectors $7\hat{i}-4\hat{j}+7\hat{k}, \hat{i}-6\hat{j}+10\hat{k}, -\hat{i}-3\hat{j}+4\hat{k}$ and $5\hat{i}-\hat{j}+5\hat{k}$ respectively. Then ABCD is a **[2003]**

(a) parallelogram but not a rhombus
(b) square
(c) rhombus
(d) rectangle.

21. If $\vec{u},\vec{v}$ *and* $\vec{w}$ are three non- coplanar vectors, then $(\vec{u}+\vec{v}-\vec{w}).(\vec{u}-\vec{v})\times(\vec{v}-\vec{w})$ equals **[2003]**

(a) $3\vec{u}.\vec{v}\times\vec{w}$ (b) 0 (c) $\vec{u}.\vec{v}\times\vec{w}$ (d) $\vec{u}.\vec{w}\times\vec{v}$

22. Two system of rectangular axes have the same origin. If a plane cuts them at distances a,b,c and a',b',c' from the origin then **[2003]**

(a) $\frac{1}{a^2}+\frac{1}{b^2}+\frac{1}{c^2}-\frac{1}{a'^2}-\frac{1}{b'^2}-\frac{1}{c'^2}=0$

(b) $\frac{1}{a^2}+\frac{1}{b^2}+\frac{1}{c^2}+\frac{1}{a'^2}+\frac{1}{b'^2}+\frac{1}{c'^2}=0$

(c) $\frac{1}{a^2}+\frac{1}{b^2}-\frac{1}{c^2}+\frac{1}{a'^2}+\frac{1}{b'^2}-\frac{1}{c'^2}=0$

(d) $\frac{1}{a^2}-\frac{1}{b^2}-\frac{1}{c^2}+\frac{1}{a'^2}-\frac{1}{b'^2}-\frac{1}{c'^2}=0$.

23. Distance between two parallel planes $2x+y+2z=8$ and $4x+2y+4z+5=0$ is **[2004]**

(a) $\frac{9}{2}$ (b) $\frac{5}{2}$ (c) $\frac{7}{2}$ (d) $\frac{3}{2}$

24. A line with direction cosines proportional to 2, 1, 2 meets each of the lines $x=y+a=z$ and $x+a=2y=2z$. The co-ordinates of each of the points of intersection are given by **[2004]**

(a) $(2a,3a,3a),(2a,a,a)$ (b) $(3a,2a,3a),(a,a,a)$
(c) $(3a,2a,3a),(a,a,2a)$ (d) $(3a,3a,3a),(a,a,a)$

25. If the straight lines **[2004]**

$x=1+s, y=-3-\lambda s, z=1+\lambda s$ and $x=\frac{t}{2}, y=1+t, z=2-t$,

with parameters s and t respectively, are co-planar, then λ equals.

(a) 0 (b) -1 (c) $-\frac{1}{2}$ (d) -2

26. The intersection of the spheres $x^2+y^2+z^2+7x-2y-z=13$ and $x^2+y^2+z^2-3x+3y+4z=8$ is the same as the intersection of one of the sphere and the plane **[2004]**

(a) $2x-y-z=1$ (b) $x-2y-z=1$
(c) $x-y-2z=1$ (d) $x-y-z=1$

27. Let $\vec{a},\vec{b}$ and $\vec{c}$ be three non-zero vectors such that no two of these are collinear. If the vector $\vec{a}+2\vec{b}$ is collinear with $\vec{c}$ and $\vec{b}+3\vec{c}$ is collinear with $\vec{a}$ (λ being some non-zero scalar) then $\vec{a}+2\vec{b}+6\vec{c}$ equals **[2004]**

(a) 0 (b) $\lambda\vec{b}$ (c) $\lambda\vec{c}$ (d) $\lambda\vec{a}$

28. A particles is acted upon by constant forces $4\hat{i}+\hat{j}-3\hat{k}$ and $3\hat{i}+\hat{j}-\hat{k}$ which displace it from a point $\hat{i}+2\hat{j}+3\hat{k}$ to the point $5\hat{i}+4\hat{j}+\hat{k}$. The work done in standard units by the forces is given by **[2004]**

(a) 15 (b) 30 (c) 25 (d) 40

29. If $\vec{a}$, $\vec{b}$, $\vec{c}$ are non-coplanar vectors and λ is a real number, then the vectors $\vec{a}+2\vec{b}+3\vec{c}$, $\lambda\vec{b}+4\vec{c}$ and $(2\lambda-1)\vec{c}$ are non coplanar for **[2004]**

(a) no value of λ
(b) all except one value of λ
(c) all except two values of λ
(d) all values of λ

30. Let $\bar{u}, \bar{v}, \bar{w}$ be such that $|\bar{u}|=1, |\bar{v}|=2, |\bar{w}|=3$. If the projection $\bar{v}$ along $\bar{u}$ is equal to that of $\bar{w}$ along $\bar{u}$ and $\bar{v}$, $\bar{w}$ are perpendicular to each other then $|\bar{u}-\bar{v}+\bar{w}|$ equals **[2004]**

(a) 14 (b) $\sqrt{7}$ (c) $\sqrt{14}$ (d) 2

31. Let $\bar{a}, \bar{b}$ and $\bar{c}$ be non-zero vectors such that $(\bar{a}\times\bar{b})\times\bar{c} = \frac{1}{3}|\bar{b}||\bar{c}|\bar{a}$. If θ is the acute angle between the vectors $\bar{b}$ and $\bar{c}$, then $\sin\theta$ equals **[2004]**

(a) $\frac{2\sqrt{2}}{3}$ (b) $\frac{\sqrt{2}}{3}$ (c) $\frac{2}{3}$ (d) $\frac{1}{3}$

32. If C is the mid point of AB and P is any point outside AB, then **[2005]**

(a) $\overrightarrow{PA}+\overrightarrow{PB}=2\overrightarrow{PC}$ (b) $\overrightarrow{PA}+\overrightarrow{PB}=\overrightarrow{PC}$

(c) $\overrightarrow{PA}+\overrightarrow{PB}+2\overrightarrow{PC}=\vec{0}$ (d) $\overrightarrow{PA}+\overrightarrow{PB}+\overrightarrow{PC}=\vec{0}$

33. If the angel θ between the line $\frac{x+1}{1}=\frac{y-1}{2}=\frac{z-2}{2}$ and the plane $2x-y+\sqrt{\lambda}\,z+4=0$ is such that $\sin\theta=\frac{1}{3}$ then the value of λ is **[2005]**

(a) $\frac{5}{3}$ (b) $\frac{-3}{5}$ (c) $\frac{3}{4}$ (d) $\frac{-4}{3}$

34. The angle between the lines $2x=3y=-z$ and $6x=-y=-4z$ is **[2005]**

(a) 0° (b) 90° (c) 45° (d) 30°

35. If the plane $2ax-3ay+4az+6=0$ passes through the midpoint of the line joining the centres of the spheres $x^2+y^2+z^2+6x-8y-2z=13$ and $x^2+y^2+z^2-10x+4y-2z=8$ then a equals **[2005]**

(a) -1 (b) 1 (c) -2 (d) 2

36. The distance between the line $\vec{r}=2\hat{i}-2\hat{j}+3\hat{k}+\lambda(i-j+4k)$ and the plane $\vec{r}.(\hat{i}+5\hat{j}+\hat{k})=5$ is **[2005]**

(a) $\frac{10}{9}$ (b) $\frac{10}{3\sqrt{3}}$ (c) $\frac{3}{10}$ (d) $\frac{10}{3}$

37. For any vector $\vec{a}$, the value of $(\vec{a}\times\hat{i})^2+(\vec{a}\times\hat{j})^2+(\vec{a}\times\hat{k})^2$ is equal to **[2005]**

(a) $3\vec{a}^2$ (b) $\vec{a}^2$ (c) $2\vec{a}^2$ (d) $4\vec{a}^2$

38. If non zero numbers a, b, c are in H.P., then the straight line $\frac{x}{a}+\frac{y}{b}+\frac{1}{c}=0$ always passes through a fixed point. That point is **[2005]**

(a) $(-1,2)$ (b) $(-1,-2)$ (c) $(1,-2)$ (d) $\left(1,-\frac{1}{2}\right)$

39. Let a, b and c be distinct non- negative numbers. If the vectors $a\hat{i}+a\hat{j}+c\hat{k}$, $\hat{i}+\hat{k}$ and $c\hat{i}+c\hat{j}+b\hat{k}$ lie in a plane, then c is **[2005]**

(a) the Geometric Mean of a and b
(b) the Arithmetic Mean of a and b
(c) equal to zero
(d) the Harmonic Mean of a and b

40. If $\vec{a},\vec{b},\vec{c}$ are non coplanar vectors and λ is a real number then $[\lambda(\vec{a}+\vec{b})\ \lambda^2\vec{b}\ \lambda\vec{c}]=[\vec{a}\ \ \vec{b}+\vec{c}\ \ \vec{b}]$ for **[2005]**

(a) exactly one value of λ
(b) no value of λ
(c) exactly three values of λ
(d) exactly two values of λ

41. Let $\vec{a}=\hat{i}-\hat{k}$, $\vec{b}=x\hat{i}+\hat{j}+(1-x)\hat{k}$ and $\vec{c}=y\hat{i}+x\hat{j}+(1+x-y)\hat{k}$. Then $[\vec{a},\vec{b},\vec{c}]$ depends on **[2005]**

(a) only y (b) only x
(c) both x and y (d) neither x nor y

42. The plane $x+2y-z=4$ cuts the sphere $x^2+y^2+z^2-x+z-2=0$ in a circle of radius **[2005]**

(a) 3 (b) 1 (c) 2 (d) $\sqrt{2}$

43. If $(\bar{a}\times\bar{b})\times\bar{c}=\bar{a}\times(\bar{b}\times\bar{c})$ where $\bar{a},\bar{b}$ and $\bar{c}$ are any three vectors such that $\bar{a}.\bar{b}\neq 0$, $\bar{b}.\bar{c}\neq 0$ then $\bar{a}$ and $\bar{c}$ are **[2006]**

(a) inclined at an angle of $\frac{\pi}{3}$ between them
(b) inclined at an angle of $\frac{\pi}{6}$ between them
(c) perpendicular
(d) parallel

44. The values of a, for which points A, B, C with position vectors $2\hat{i}-\hat{j}+\hat{k}$, $\hat{i}-3\hat{j}-5\hat{k}$ and $a\hat{i}-3\hat{j}+\hat{k}$ respectively are the vertices of a right angled triangle with $C=\frac{\pi}{2}$ are **[2006]**

(a) 2 and 1 (b) -2 and -1
(c) -2 and 1 (d) 2 and -1

45. The two lines $x=ay+b$, $z=cy+d$; and $x=a'y+b'$, $z=c'y+d'$ are perpendicular to each other if **[2006]**

(a) $aa'+cc'=-1$ (b) $aa'+cc'=1$

(c) $\frac{a}{a'}+\frac{c}{c'}=-1$ (d) $\frac{a}{a'}+\frac{c}{c'}=1$

46. The image of the point (–1, 3, 4) in the plane $x-2y=0$ is

(a) $\left(-\frac{17}{3},-\frac{19}{3},4\right)$ (b) (15,11,4) **[2006]**

(c) $\left(-\frac{17}{3},-\frac{19}{3},1\right)$ (d) None of these

47. If a line makes an angle of $\pi/4$ with the positive directions of each of x- axis and y- axis, then the angle that the line makes with the positive direction of the z-axis is **[2007]**

(a) $\frac{\pi}{4}$ (b) $\frac{\pi}{2}$ (c) $\frac{\pi}{6}$ (d) $\frac{\pi}{3}$

48. If $\hat{u}$ and $\hat{v}$ are unit vectors and θ is the acute angle between them, then $2\hat{u}\times 3\hat{v}$ is a unit vector for **[2007]**

(a) no value of θ
(b) exactly one value of θ
(c) exactly two values of θ
(d) more than two values of θ

49. If (2, 3, 5) is one end of a diameter of the sphere $x^2+y^2+z^2-6x-12y-2z+20=0$, then the cooordinates of the other end of the diameter are **[2007]**

(a) (4, 3, 5) (b) (4, 3, –3)
(c) (4, 9, –3) (d) (4, –3, 3).

50. Let $\vec{a}=\hat{i}+\hat{j}+\hat{k}, \vec{b}=\hat{i}-\hat{j}+2\hat{k}$ and $\vec{c}=x\hat{i}+(x-2)\hat{j}-\hat{k}$. If the vectors $\vec{c}$ lies in the plane of $\vec{a}$ and $\vec{b}$, then x equals **[2007]**

(a) –4 (b) –2 (c) 0 (d) 1.

51. Let L be the line of intersection of the planes $2x+3y+z=1$ and $x+3y+2z=2$. If L makes an angle α with the positive x-axis, then cos α equals **[2007]**

(a) 1 (b) $\frac{1}{\sqrt{2}}$ (c) $\frac{1}{\sqrt{3}}$ (d) $\frac{1}{2}$.

52. The vector $\vec{a}=\alpha\hat{i}+2\hat{j}+\beta\hat{k}$ lies in the plane of the vectors $\vec{b}=\hat{i}+\hat{j}$ and $\vec{c}=\hat{j}+\hat{k}$ and bisects the angle between $\vec{b}$ and $\vec{c}$. Then which one of the following gives possible values of α and β? **[2008]**

(a) $\alpha=2, \beta=2$ (b) $\alpha=1, \beta=2$
(c) $\alpha=2, \beta=1$ (d) $\alpha=1, \beta=1$

53. The non-zero vectors are $\vec{a}$, $\vec{b}$ and $\vec{c}$ are related by $\vec{a}=8\vec{b}$ and $\vec{c}=-7\vec{b}$. Then the angle between $\vec{a}$ and $\vec{c}$ is **[2008]**

(a) 0 (b) $\frac{\pi}{4}$ (c) $\frac{\pi}{2}$ (d) π

54. The line passing through the points (5, 1, a) and (3, b, 1) crosses the yz-plane at the point $\left(0,\frac{17}{2},\frac{-13}{2}\right)$. Then

(a) $a=2, b=8$ (b) $a=4, b=6$
(c) $a=6, b=4$ (d) $a=8, b=2$

55. If the straight lines $\frac{x-1}{k}=\frac{y-2}{2}=\frac{z-3}{3}$ and $\frac{x-2}{3}=\frac{y-3}{k}=\frac{z-1}{2}$ intersect at a point, then the integer k is equal to **[2008]**

(a) –5 (b) 5 (c) 2 (d) –2

56. Let the line $\frac{x-2}{3}=\frac{y-1}{-5}=\frac{z+2}{2}$ lie in the plane $x+3y-\alpha z+\beta=0$. Then (α, β) equals **[2009]**

(a) (–6, 7) (b) (5, –15)
(c) (–5, 5) (d) (6, –17)

57. The projections of a vector on the three coordinate axis are 6, –3, 2 respectively. The direction cosines of the vector are : **[2009]**

(a) $\frac{6}{5},\frac{-3}{5},\frac{2}{5}$ (b) $\frac{6}{7},\frac{-3}{7},\frac{2}{7}$

(c) $\frac{-6}{7},\frac{-3}{7},\frac{2}{7}$ (d) 6, –3, 2

58. If $\vec{u},\vec{v},\vec{w}$ are non-coplanar vectors and p, q are real numbers, then the equality $[3\vec{u}\ p\vec{v}\ p\vec{w}]-[p\vec{v}\ \vec{w}\ q\vec{u}]-[2\vec{w}\ q\vec{v}\ q\vec{u}]=0$ holds for : **[2009]**

(a) exactly two values of (p, q)
(b) more than two but not all values of (p, q)
(c) all values of (p, q)
(d) exactly one value of (p, q)

59. Let $\vec{a}=\hat{j}-\hat{k}$ and $\vec{c}=\hat{i}-\hat{j}-\hat{k}$. Then the vector $\vec{b}$ satisfying $\vec{a}\times\vec{b}+\vec{c}=\vec{0}$ and $\vec{a}.\vec{b}=3$ **[2010]**

(a) $2\hat{i}-\hat{j}+2\hat{k}$ (b) $\hat{i}-\hat{j}-2\hat{k}$

(c) $\hat{i}+\hat{j}-2\hat{k}$ (d) $-\hat{i}+\hat{j}-2\hat{k}$

60. If the vectors $\vec{a}=\hat{i}-\hat{j}+2\hat{k}$, $\vec{b}=2\hat{i}+4\hat{j}+\hat{k}$ and $\vec{c}=\lambda\hat{i}+\hat{j}+\mu\hat{k}$ are mutually orthogonal, then $(\lambda,\mu)=$ **[2010]**

(a) (2, –3) (b) (–2, 3) (c) (3, –2) (d) (–3, 2)

61. **Statement -1 :** The point A(3, 1, 6) is the mirror image of the point B(1, 3, 4) in the plane $x-y+z=5$.
Statement -2: The plane $x-y+z=5$ bisects the line segment joining A(3, 1, 6) and B(1, 3, 4). **[2010]**

(a) Statement -1 is true, Statement -2 is true ; Statement -2 is **not** a correct explanation for Statement -1.
(b) Statement -1 is true, Statement -2 is false.
(c) Statement -1 is false, Statement -2 is true .
(d) Statement - 1 is true, Statement 2 is true ; Statement -2 is a correct explanation for Statement -1.

62. A line AB in three-dimensional space makes angles 45° and 120° with the positive x-axis and the positive y-axis respectively. If AB makes an acute angle θ with the positive z-axis, then θ equals **[2010]**

(a) 45° (b) 60° (c) 75° (d) 30°

63. If the angle between the line $x = \frac{y-1}{2} = \frac{z-3}{\lambda}$ and the plane $x + 2y + 3z = 4$ is $\cos^{-1}\left(\sqrt{\frac{5}{14}}\right)$, then λ equals [2011]

(a) $\frac{3}{2}$ (b) $\frac{2}{5}$ (c) $\frac{5}{3}$ (d) $\frac{2}{3}$

64. If $\vec{a} = \frac{1}{\sqrt{10}}(3\hat{i} + \hat{k})$ and $\vec{b} = \frac{1}{7}(2\hat{i} + 3\hat{j} - 6\hat{k})$, then the value of $(2\vec{a} - \vec{b})[(\vec{a} \times \vec{b}) \times (\vec{a} + 2\vec{b})]$ is [2011]

(a) –3 (b) 5 (c) 3 (d) –5

65. The vectors $\vec{a}$ and $\vec{b}$ are not perpendicular and $\vec{c}$ and $\vec{d}$ are two vectors satisfying $\vec{b} \times \vec{c} = \vec{b} \times \vec{d}$ and $\vec{a}.\vec{d} = 0$. Then the vector $\vec{d}$ is equal to [2011]

(a) $\vec{c} + \left(\frac{\vec{a}.\vec{c}}{\vec{a}.\vec{b}}\right)\vec{b}$ (b) $\vec{b} + \left(\frac{\vec{b}.\vec{c}}{\vec{a}.\vec{b}}\right)\vec{c}$

(c) $\vec{c} - \left(\frac{\vec{a}.\vec{c}}{\vec{a}.\vec{b}}\right)\vec{b}$ (d) $\vec{b} - \left(\frac{\vec{b}.\vec{c}}{\vec{a}.\vec{b}}\right)\vec{c}$

66. **Statement-1:** The point $A(1, 0, 7)$) is the mirror image of the point $B(1, 6, 3)$ in the line : $\frac{x}{1} = \frac{y-1}{2} = \frac{z-2}{3}$ [2011]

Statement-2: The line $\frac{x}{1} = \frac{y-1}{2} = \frac{z-2}{3}$ bisects the line segment joining $A(1, 0, 7)$ and $B(1, 6, 3)$.

(a) Statement-1 is true, Statement-2 is true; Statement-2 is **not a** correct explanation for Statement-1.
(b) Statement-1 is true, Statement-2 is false.
(c) Statement-1 is false, Statement-2 is true.
(d) Statement-1 is true, Statement-2 is true; Statement-2 is a correct explanation for Statement-1.

67. If the vector $p\hat{i} + \hat{j} + \hat{k}$, $\hat{i} + q\hat{j} + \hat{k}$ and $\hat{i} + \hat{j} + r\hat{k}$ $(p \neq q \neq r \neq 1)$ are coplanar, then the value of $pqr - (p+q+r)$ is [2011 RS]

(a) 2 (b) 0 (c) –1 (d) –2

68. The distance of the point (1, – 5, 9) from the plane $x - y + z = 5$ measured along a straight $x = y = z$ is [2011 RS]

(a) $10\sqrt{3}$ (b) $5\sqrt{3}$ (c) $3\sqrt{10}$ (d) $3\sqrt{5}$

69. Let $\vec{a}, \vec{b}, \vec{c}$ be three non-zero vectors which are pairwise non-collinear. If $\vec{a} + 3\vec{b}$ is collinear with $\vec{c}$ and $\vec{b} + 2\vec{c}$ is collinear with $\vec{a}$, then $\vec{a} + 3\vec{b} + 6\vec{c}$ is : [2011 RS]

(a) $\vec{a}$ (b) $\vec{c}$ (c) $\vec{0}$ (d) $\vec{a} + \vec{c}$

70. The length of the perpendicular drawn from the point $(3, -1, 11)$ to the line $\frac{x}{2} = \frac{y-2}{3} = \frac{z-3}{4}$ is : [2011 RS]

(a) $\sqrt{29}$ (b) $\sqrt{33}$ (c) $\sqrt{53}$ (d) $\sqrt{66}$

71. Let $\vec{a}$ and $\vec{b}$ be two unit vectors. If the vectors $\vec{c} = \hat{a} + 2\hat{b}$ and $\vec{d} = 5\hat{a} - 4\hat{b}$ are perpendicular to each other, then the angle between $\hat{a}$ and $\hat{b}$ is : [2012]

(a) $\frac{\pi}{6}$ (b) $\frac{\pi}{2}$ (c) $\frac{\pi}{3}$ (d) $\frac{\pi}{4}$

72. A equation of a plane parallel to the plane $x - 2y + 2z - 5 = 0$ and at a unit distance from the origin is : [2012]

(a) $x - 2y + 2z - 3 = 0$ (b) $x - 2y + 2z + 1 = 0$
(c) $x - 2y + 2z - 1 = 0$ (d) $x - 2y + 2z + 5 = 0$

73. If the line $\frac{x-1}{2} = \frac{y+1}{3} = \frac{z-1}{4}$ and $\frac{x-3}{1} = \frac{y-k}{2} = \frac{z}{1}$ intersect, then k is equal to: [2012]

(a) –1 (b) $\frac{2}{9}$ (c) $\frac{9}{2}$ (d) 0

74. Let $ABCD$ be a parallelogram such that $\overrightarrow{AB} = \vec{q}, \overrightarrow{AD} = \vec{p}$ and $\angle BAD$ be an acute angle. If $\vec{r}$ is the vector that coincide with the altitude directed from the vertex B to the side AD, then $\vec{r}$ is given by : [2012]

(a) $\vec{r} = 3\vec{q} - \frac{3(\vec{p}.\vec{q})}{(\vec{p}.\vec{p})}\vec{p}$ (b) $\vec{r} = -\vec{q} + \frac{(\vec{p}.\vec{q})}{(\vec{p}.\vec{p})}\vec{p}$

(c) $\vec{r} = \vec{q} - \frac{(\vec{p}.\vec{q})}{(\vec{p}.\vec{p})}\vec{p}$ (d) $\vec{r} = -3\vec{q} - \frac{3(\vec{p}.\vec{q})}{(\vec{p}.\vec{p})}\vec{p}$

75. Distance between two parallel planes $2x + y + 2z = 8$ and $4x + 2y + 4z + 5 = 0$ is [JEE M 2013]

(a) $\frac{3}{2}$ (b) $\frac{5}{2}$ (c) $\frac{7}{2}$ (d) $\frac{9}{2}$

76. If the lines $\frac{x-2}{1} = \frac{y-3}{1} = \frac{z-4}{-k}$ and $\frac{x-1}{k} = \frac{y-4}{2} = \frac{z-5}{1}$ are coplanar, then k can have [JEE M 2013]

(a) any value (b) exactly one value
(c) exactly two values (d) exactly three values

77. If the vectors $\overrightarrow{AB} = 3\hat{i} + 4\hat{k}$ and $\overrightarrow{AC} = 5\hat{i} - 2\hat{j} + 4\hat{k}$ are the sides of a triangle ABC, then the length of the median through A is [JEE M 2013]

(a) $\sqrt{18}$ (b) $\sqrt{72}$ (c) $\sqrt{33}$ (d) $\sqrt{45}$

Solutions & Explanations

Section-A : JEE Advanced/ IIT-JEE

A **1.** $5\sqrt{2}$ **2.** $\pm\dfrac{2\hat{i}+\hat{j}+\hat{k}}{\sqrt{6}}$ **3.** $\sqrt{13}$ **4.** orthocentre **5.** –1 **6.** 0

7. $\dfrac{5}{3}\hat{i}+\dfrac{2}{3}\hat{j}+\dfrac{2}{3}\hat{k}$ **8.** 1 **9.** $2\hat{i}-\hat{j}$ **10.** $\left(\dfrac{\vec{a}\cdot\vec{b}}{|\vec{b}|^2}\right)\vec{b}\ ,\ \vec{a}-\left(\dfrac{\vec{a}\cdot\vec{b}}{|\vec{b}|^2}\right)\vec{b}$

11. $\dfrac{5\hat{i}+2\hat{j}+2\hat{k}}{3}$ **12.** $\dfrac{\hat{j}-\hat{k}}{\sqrt{2}}$ or $\dfrac{-\hat{j}+\hat{k}}{\sqrt{2}}$ **13.** $-\left(\dfrac{2\hat{i}+\hat{j}+\hat{k}}{\sqrt{6}}\right)$ **14.** $\dfrac{\pi}{4}$ or $\dfrac{3\pi}{4}$ **15.** $\vec{a}$

16. $\dfrac{\pi}{6}$ **17.** 6

B **1.** T **2.** T **3.** T **4.** F

C **1.** (a) **2.** (d) **3.** (d) **4.** (a) **5.** (d) **6.** (b) **7.** (a)
8. (b) **9.** (a) **10.** (a) **11.** (b) **12.** (d) **13.** (b) **14.** (b)
15. (a) **16.** (b) **17.** (a) **18.** (a) **19.** (c) **20.** (b) **21.** (b)
22. (c) **23.** (a) **24.** (c) **25.** (c) **26.** (b) **27.** (c) **28.** (d)
29. (b) **30.** (d) **31.** (a) **32.** (c) **33.** (b) **34.** (a) **35.** (a)
36. (a) **37.** (c) **38.** (c) **39.** (a) **40.** (c) **41.** (a) **42.** (b)
43. (c) **44.** (a) **45.** (a) **46.** (c)

D **1.** (c) **2.** (b) **3.** (a, c) **4.** (a, c, d) **5.** (d) **6.** (c)
7. (a, c) **8.** (a, c) **9.** (b, d) **10.** (a, d) **11.** (b, c) **12.** (b, d) **13.** (a, d)

E **3.** $\lambda=0,-1$ **4.** $\dfrac{c}{|\vec{A}|^2}\vec{A}-\dfrac{1}{|\vec{A}|^2}(\vec{A}\times\vec{B})$ **5.** $A_2\hat{i}-A_1\hat{j}+A_3\hat{k}$ **6.** $\dfrac{146}{17}$

11. $-\hat{i}-8\hat{j}+2\hat{k}$ **12.** $-\dfrac{4}{3}<c<0$ **13.** $8:3$ **15.** $(-1, 3, 3)$ or $(3, -1, -1)$

16. $p=\dfrac{1}{\sqrt{1+2\cos\theta}},\ q=\dfrac{-2\cos\theta}{\sqrt{1+2\cos\theta}},\ r=\dfrac{1}{\sqrt{1+2\cos\theta}}$

22. $\vec{v_1}=2\hat{i}\ ;\ \vec{v_2}=-\hat{i}\pm\hat{j}\ ;\vec{v_3}=3\hat{i}\pm2\hat{j}\pm4\hat{k}$ are some possible values. **25.** (i) $x+y-2z=3$ (ii) $Q(6, 5, -2)$

28. $x+y+z=3\ ;\ \dfrac{9}{2}$ cubic units **31.** $62x+29y+19z-105=0$ **32.** $\hat{\omega}=\hat{v}-2(\hat{a}\cdot\hat{v})\hat{a}$

F **1.** (A) → s; (B) → p; (C) → r, q; (D) → s **2.** (A) → r; (B) → q; (C) → p; (D) → s
3. (A) → q, s; (B) → p, r, s,t; (C) → t; (D) → r **4.** (A) → p; (B) → q, s; (C) → q, r, s, t; (D) → r
5. (A) → t; (B) → p, r; (C) → q, s; (D) → r **6.** (A) → q; (B) → p; (C) → s; (D) → t
7. (c) **8.** (a)

G **1.** (b) **2.** (d) **3.** (c)

H **1.** (d) **2.** (c) **3.** (d)

I **1.** 5 **2.** 6 **3.** 9 **4.** 3 **5.** 5 **6.** 5

Section-B : JEE Main/ AIEEE

1. (a)	2. (b)	3. (a)	4. (a)	5. (a)	6. (a)	7. (b)
8. (c)	9. (b)	10. (a)	11. (d)	12. (d)	13. (d)	14. (a)
15. (d)	16. (c)	17. (d)	18. (b)	19. (c)	20. (none)	21. (c)
22. (a)	23. (c)	24. (b)	25. (d)	26. (a)	27. (c)	28. (d)
29. (c)	30. (c)	31. (a)	32. (a)	33. (a)	34. (b)	35. (c)
36. (b)	37. (c)	38. (c)	39. (a)	40. (b)	41. (d)	42. (b)
43. (d)	44. (a)	45. (a)	46. (none)	47. (b)	48. (b)	49. (c)
50. (b)	51. (c)	52. (d)	53. (d)	54. (c)	55. (a)	56. (a)
57. (b)	58. (d)	59. (d)	60. (d)	61. (a)	62. (b)	63. (d)
64. (d)	65. (c)	66. (a)	67. (d)	68. (a)	69. (c)	70. (c)
71. (c)	72. (a)	73. (c)	74. (b)	75. (c)	76. (c)	77. (c)

Section-A JEE Advanced/ IIT-JEE

(A) FILL IN THE BLANKS :

1. Given that $|\vec{A}|=3$; $|\vec{B}|=4$; $|\vec{C}|=5$

$\vec{A}\perp(\vec{B}+\vec{C}) \Rightarrow \vec{A}.(\vec{B}+\vec{C})=0 \Rightarrow \vec{A}.\vec{B}+\vec{A}.\vec{C}=0$ ...(1)

$\vec{B}\perp(\vec{C}+\vec{A}) \Rightarrow \vec{B}.(\vec{C}+\vec{A})=0 \Rightarrow \vec{B}.\vec{C}+\vec{B}.\vec{A}=0$ ...(2)

$\vec{C}\perp(\vec{A}+\vec{B}) \Rightarrow \vec{C}.(\vec{A}+\vec{B})=0 \Rightarrow \vec{C}.\vec{A}+\vec{C}.\vec{B}=0$ ...(3)

Adding (1), (2) and (3) we get

$2\left(\vec{A}\,\vec{B}+\vec{B}\,\vec{C}+\vec{C}\,\vec{A}\right)=0$ ...(4)

Now, $|\vec{A}+\vec{B}+\vec{C}|^2=\left(\vec{A}+\vec{B}+\vec{C}\right).\left(\vec{A}+\vec{B}+\vec{C}\right)$

$=\vec{A}.\vec{A}+\vec{B}.\vec{B}+\vec{C}.\vec{C}+2\vec{A}.\vec{B}+2\vec{B}.\vec{C}+2\vec{C}.\vec{A}$

$=|\vec{A}|^2+|\vec{B}|^2+|\vec{C}|^2+2\left(\vec{A}.\vec{B}+\vec{B}.\vec{C}+\vec{C}.\vec{A}\right)$

$=9+16+25+0$ (using equation 4)

$=50$ $\qquad\therefore |\vec{A}+\vec{B}+\vec{C}|=5\sqrt{2}$

2. Required unit vector, $\hat{n}=\pm\dfrac{\overrightarrow{PQ}\times\overrightarrow{PR}}{|\overrightarrow{PQ}\times\overrightarrow{PR}|}$

$\overrightarrow{PQ}=\hat{i}+\hat{j}-3\hat{k}$; $\overrightarrow{PR}=-\hat{i}+3\hat{j}-\hat{k}$

$$\therefore \overrightarrow{PQ}\times\overrightarrow{PR}=\begin{vmatrix}\hat{i} & \hat{j} & \hat{k}\\ 1 & 1 & -3\\ -1 & 3 & -1\end{vmatrix}$$

$=(-1+9)\hat{i}+(3+1)\hat{j}+(3+1)\hat{k}\;=8\hat{i}+4\hat{j}+4\hat{k}$

$|\overrightarrow{PQ}\times\overrightarrow{PR}|=\sqrt{64+16+16}=\sqrt{96}=4\sqrt{6}$

$$\hat{n}=\pm\left(\frac{8\hat{i}+4\hat{j}+4\hat{k}}{4\sqrt{6}}\right)=\pm\left(\frac{2\hat{i}+\hat{j}+\hat{k}}{\sqrt{6}}\right)$$

3. Area of $\Delta ABC=\frac{1}{2}|\overrightarrow{BA}\times\overrightarrow{BC}|$

$\overrightarrow{BA}=-\hat{i}-2\hat{j}+3\hat{k}$, $\overrightarrow{BC}=\hat{i}-2\hat{j}+3\hat{k}$

$$\therefore \Delta=\frac{1}{2}\begin{vmatrix}\hat{i} & \hat{j} & \hat{k}\\ -1 & -2 & 3\\ 1 & -2 & 3\end{vmatrix}=\frac{1}{2}|6\hat{j}+4\hat{k}|$$

$=|3\hat{j}+2\hat{k}|=\sqrt{9+4}=\sqrt{13}$

4. Given that $\vec{a}$, $\vec{b}$, $\vec{c}$, $\vec{d}$ are position vectors of points A, B, C and D respectively, such that

$(\vec{a}-\vec{d}).(\vec{b}-\vec{c})=(\vec{b}-\vec{d}).(\vec{c}-\vec{a})=0$

$\Rightarrow \overrightarrow{DA}.\overrightarrow{CB}=\overrightarrow{DB}.\overrightarrow{AC}=0$

$\Rightarrow \overrightarrow{DA}\perp\overrightarrow{CB}$ and $\overrightarrow{DB}\perp\overrightarrow{AC}$

Clearly D is orthocentre of $\Delta\,ABC$

5. Given that $\begin{vmatrix}a & a^2 & 1+a^3\\ b & b^2 & 1+b^3\\ c & c^2 & 1+c^3\end{vmatrix}=0$

$$\Rightarrow\begin{vmatrix}a & a^2 & 1\\ b & b^2 & 1\\ c & c^2 & 1\end{vmatrix}+abc\begin{vmatrix}1 & a & a^2\\ 1 & b & b^2\\ 1 & c & c^2\end{vmatrix}=0$$

Operating $C_2\leftrightarrow C_3$ and then $C_1\leftrightarrow C_2$ in first determinant

$$\Rightarrow\begin{vmatrix}1 & a & a^2\\ 1 & b & b^2\\ 1 & c & c^2\end{vmatrix}+abc\begin{vmatrix}1 & a & a^2\\ 1 & b & b^2\\ 1 & c & c^2\end{vmatrix}=0$$

$$\Rightarrow (1+abc)\begin{vmatrix} 1 & a & a^2 \\ 1 & b & b^2 \\ 1 & c & c^2 \end{vmatrix} = 0$$

$$\Rightarrow \text{either } 1 + abc = 0 \text{ or } \begin{vmatrix} 1 & a & a^2 \\ 1 & b & b^2 \\ 1 & c & c^2 \end{vmatrix} = 0$$

Also given that the vectors $\vec{A}, \vec{B}, \vec{C}$ are noncoplanar

i.e., $\left[\vec{A}\vec{B}\vec{C}\right] \neq 0$

where $\vec{A} = \hat{i} + a\hat{j} + a^2\hat{k}$

$\vec{B} = \hat{i} + b\hat{j} + b^2\hat{k}$, $\vec{C} = \hat{i} + c\hat{j} + c^2\hat{k}$

$$\Rightarrow \begin{vmatrix} 1 & a & a^2 \\ 1 & b & b^2 \\ 1 & c & c^2 \end{vmatrix} \neq 0$$

$\therefore$ We must have $1 + abc = 0 \Rightarrow abc = -1$

6. As given that $\vec{A}, \vec{B}, \vec{C}$ are three noncoplanar vectors, therefore, $\left[\vec{A}\vec{B}\vec{C}\right] \neq 0$

Also by the property of scalar triple product we have

$\vec{A}.(\vec{B}\times\vec{C}) = \left[\vec{A}\vec{B}\vec{C}\right]$, $\vec{B}.(\vec{A}\times\vec{C}) = -\left[\vec{A}\vec{B}\vec{C}\right]$

$\vec{C}\times(\vec{A}.\vec{B}) = \left[\vec{A}\vec{B}\vec{C}\right]$, $\vec{C}.(\vec{A}\times\vec{B}) = \left[\vec{A}\vec{B}\vec{C}\right]$

$$\therefore \frac{\vec{A}.(\vec{B}\times\vec{C})}{(\vec{C}\times\vec{A}).\vec{B}} + \frac{\vec{B}.(\vec{A}\times\vec{C})}{(\vec{C}\times\vec{A}).\vec{B}} = \frac{\left[\vec{A}\vec{B}\vec{C}\right]}{\left[\vec{A}\vec{B}\vec{C}\right]} + \frac{-\left[\vec{A}\vec{B}\vec{C}\right]}{\left[\vec{A}\vec{B}\vec{C}\right]} = 0$$

7. Given $\vec{A} = \hat{i} + \hat{j} + \hat{k}$, $\vec{C} = \hat{j} - \hat{k}$

Let $\vec{B} = x\hat{i} + y\hat{j} + z\hat{k}$

$$ATQ,\ \vec{A}\times\vec{B} = \vec{C} \Rightarrow \begin{vmatrix} \hat{i} & \hat{j} & \hat{k} \\ 1 & 1 & 1 \\ x & y & z \end{vmatrix} = \hat{j} - \hat{k}$$

$$\Rightarrow (z-y)\,\hat{i} + (x-z)\,\hat{j} + (y-x)\,\hat{k} = \hat{j} - \hat{k}$$

$$\left.\begin{matrix} z - y = 0 \\ \Rightarrow\ x - z = 1 \\ y - x = -1 \end{matrix}\ \Rightarrow\ \begin{matrix} \\ y = z \\ x = 1 + z \end{matrix}\right\} \quad ...(1)$$

Also, $\vec{A}.\vec{B} = 3 \Rightarrow x + y + z = 3$ ...(2)

Using equations (1) and (2) we get

$1 + z + z + z = 3$

$\Rightarrow z = 2/3 \Rightarrow y = 2/3, x = 5/3$

$$\therefore \vec{B} = \frac{5}{3}\hat{i} + \frac{2}{3}\hat{j} + \frac{2}{3}\hat{k}$$

8. Given that the vectors $\hat{u} = a\hat{i} + \hat{j} + \hat{k}$, $\hat{v} = \hat{i} + b\hat{j} + \hat{k}$ and $\hat{w} = \hat{i} + \hat{j} + c\hat{k}$ where $a \neq b \neq c \neq 1$ are coplanar

$$\therefore \left[\vec{u}\,\vec{v}\,\vec{w}\right] = 0 \Rightarrow \begin{vmatrix} a & 1 & 1 \\ 1 & b & 1 \\ 1 & 1 & c \end{vmatrix} = 0$$

Oeprating $C_1 - C_2$, $C_2 - C_3$

$$\begin{vmatrix} a-1 & 0 & 1 \\ 1-b & b-1 & 1 \\ 0 & 1-c & c \end{vmatrix} = 0$$

Taking $(1 - a)$, $(1 - b)$, $(1 - c)$ common from R_1, R_2 and R_3 respectively.

$$\Rightarrow (1-a)(1-b)(1-c)\begin{vmatrix} -1 & 0 & \frac{1}{1-a} \\ 1 & -1 & \frac{1}{1-b} \\ 0 & 1 & \frac{c}{1-c} \end{vmatrix} = 0$$

$$\Rightarrow (1-a)(1-b)(1-c)\left[-\left\{\frac{-c}{1-c} - \frac{1}{1-b}\right\} + \frac{1}{1-a}(1-0)\right] = 0$$

$$\Rightarrow (1-a)(1-b)(1-c)\left[\frac{1}{1-a} + \frac{1}{1-b} + \frac{c}{1-c}\right] = 0$$

$$\Rightarrow (1-a)(1-b)(1-c)\left[\frac{1}{1-a} + \frac{1}{1-b} - \frac{(1-c)-1}{1-c}\right] = 0$$

$$\Rightarrow (1-a)(1-b)(1-c)\left[\frac{1}{1-a} + \frac{1}{1-b} + \frac{1}{1-c} - 1\right] = 0$$

But $a \neq b \neq c \neq 1$ (given)

$$\therefore \frac{1}{1-a} + \frac{1}{1-b} + \frac{1}{1-c} - 1 = 0$$

$$\Rightarrow \frac{1}{1-a} + \frac{1}{1-b} + \frac{1}{1-c} = 1$$

9. Let $\vec{c} = \alpha\hat{i} + \beta\hat{j}$

As $\hat{b} \perp \hat{c}$ (given) $\quad \therefore \vec{b}.\vec{c} = 0$

$\Rightarrow (4\hat{i} + 3\hat{j}).(\alpha\hat{i} + \beta\hat{j}) = 0 \Rightarrow 4\alpha + 3\beta = 0$

$$\Rightarrow \alpha = -\frac{3\beta}{4} \Rightarrow \frac{\alpha}{+3} = \frac{\beta}{-4} = \lambda$$

$\Rightarrow \alpha = +3\lambda, \beta = -4\lambda$ ...(1)

Now, let $\vec{a} = x\hat{i} + y\hat{j}$ be the required vectors.

Then as per question

$$\text{Projection of } \vec{a} \text{ along } \vec{b} = \frac{\vec{a}.\vec{b}}{|\vec{b}|} \Rightarrow \frac{4x+3y}{\sqrt{4^2+3^2}} = 1$$

$\Rightarrow 4x + 3y = 5$ ...(2)

Also, projection of $\vec{a}$ along $\vec{c} = 2$

$$\Rightarrow \frac{\vec{a}.\vec{c}}{|\vec{c}|} = 2 \Rightarrow \frac{\alpha x + \beta y}{\sqrt{\alpha^2 + \beta^2}} = 2 \Rightarrow \frac{3\lambda x - 4\lambda y}{\sqrt{(3\lambda)^2 + (-4\lambda)^2}} = 2$$

$\Rightarrow 3\lambda x - 4\lambda y = 10\lambda$

$\Rightarrow 3x - 4y = 10$ ...(3)

Solving (2) and (3), we get $x = 2, y = -1$

$\therefore$ The required vector is $2\hat{i} - \hat{j}$

10. Component of $\vec{a}$ along $\vec{b} = \overline{OD} = OA\cos\theta.\hat{b}$

$$= \left(\frac{\vec{a}.\vec{b}}{|\vec{b}|}\right)\frac{\vec{b}}{|\vec{b}|} = \left(\frac{\vec{a}.\vec{b}}{|\vec{b}|^2}\right)\vec{b}$$

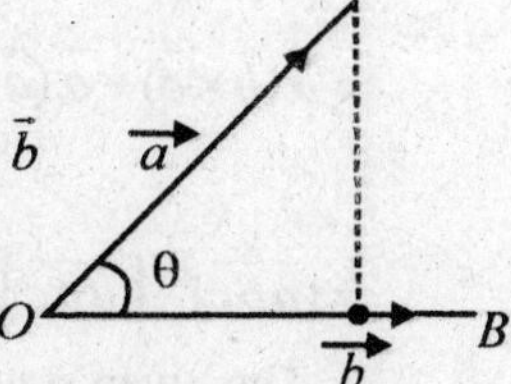

Component of $\vec{a}$ perpendicular to $\vec{b}$

$$= \overline{DA} = \vec{a} - \overline{OD} = \vec{a} - \left(\frac{\vec{a}.\vec{b}}{|\vec{b}|^2}\right)\vec{b}$$

11. See the solution to Q-7

12. Let $x\hat{i} + y\hat{j} + z\hat{k}$ be a unit vector, coplanar with $\hat{i} + \hat{j} + 2\hat{k}$ and $\hat{i} + 2\hat{j} + \hat{k}$ and also perpendicular to $\hat{i} + \hat{j} + \hat{k}$

Then, $\begin{vmatrix} x & y & z \\ 1 & 1 & 2 \\ 1 & 2 & 1 \end{vmatrix} = 0$

$\Rightarrow -3x + y + z = 0$ .....(i)

and $x + y + z = 0$ .....(ii)

Solving the above by cross multiplication method, we get

$$\frac{x}{0} = \frac{y}{4} = \frac{z}{-4} \text{ or } \frac{x}{0} = \frac{y}{1} = \frac{z}{-1} = \lambda(say)$$

$\Rightarrow x = 0, y = \lambda, z = -\lambda$

As $x\hat{i} + y\hat{j} + z\hat{k}$ is a unit vector, therefore

$$0 + \lambda^2 + \lambda^2 = 1 \Rightarrow \lambda^2 = \frac{1}{2} \Rightarrow \lambda = \pm\frac{1}{\sqrt{2}}$$

$\therefore$ The required vector is $\frac{\hat{j} - \hat{k}}{\sqrt{2}}$ or $\frac{-\hat{j} + \hat{k}}{\sqrt{2}}$

13. We have position vectors of points $P(\hat{i} - \hat{j} + 2\hat{k})$, $Q(2\hat{i} - \hat{k})$, $R(2\hat{j} + \hat{k})$

$\therefore \overline{QP} = (\hat{i} - \hat{j} + 2\hat{k}) - (2\hat{i} - \hat{k}) = -\hat{i} - \hat{j} + 3\hat{k}$

$\therefore \overline{QR} = 2\hat{j} + \hat{k} - 2\hat{i} + \hat{k} = -2\hat{i} + 2\hat{j} + 2\hat{k}$

Now any vector perpendicular to the plane formed by pts PQR is given by $\overline{QP} \times \overline{QR}$

$$= \begin{vmatrix} \hat{i} & \hat{j} & \hat{k} \\ -1 & -1 & 3 \\ -2 & 2 & 2 \end{vmatrix} = -8\hat{i} - 4\hat{j} - 4\hat{k}$$

$\therefore$ Unit vector normal to plane $= \pm\left(\frac{-8\hat{i} - 4\hat{j} - 4\hat{k}}{\sqrt{64 + 16 + 16}}\right)$

$$= \pm\left(\frac{2\hat{i} + \hat{j} + \hat{k}}{\sqrt{6}}\right)$$

14. Eq^n of plane containing vectors $\hat{i}$ and $\hat{i} + \hat{j}$ is

$$\left[\vec{r} - \hat{i}\ \hat{i}\ \hat{i} + \hat{j}\right] = 0 \Rightarrow \begin{vmatrix} x-1 & y & z \\ 1 & 0 & 0 \\ 1 & 1 & 0 \end{vmatrix} = 0$$

$\Rightarrow z = 0$ ....(1)

Similarly, eq^n of plane containing vectors $\hat{i} - \hat{j}$ and $\hat{i} + \hat{k}$ is

$[\vec{r} - (\hat{i} - \hat{j})\ \hat{i} - \hat{j}\ \hat{i} + \hat{k}] = 0$

$$\Rightarrow \begin{vmatrix} x-1 & y+1 & z \\ 1 & -1 & 0 \\ 1 & 0 & 1 \end{vmatrix} = 0$$

$\Rightarrow (x-1)(-1-0) - (y+1)(1-0) + z(0+1) = 0$

$\Rightarrow -x + 1 - y - 1 + z = 0$

$\Rightarrow x + y - z = 0$ ....(2)

Let $\vec{a} = a_1\hat{i} + a_2\hat{j} + a_3\hat{k}$

Since $\vec{a}$ is parallel to (1) and (2)

$a_3 = 0$ and $a_1 + a_2 - a_3 = 0 \Rightarrow a_1 = -a_2, a_3 = 0$

$\therefore$ *a vector in direction of* $\vec{a}$ is $\hat{i} - \hat{j}$

Now if θ is the angle between $\vec{a}$ and $\hat{i} - 2\hat{j} + 2\hat{k}$ then

$$\cos\theta = \pm\frac{1.1 + (-1)(-2)}{\sqrt{1+1}\sqrt{1+4+4}} = \pm\frac{3}{\sqrt{2}.3}$$

$\Rightarrow \cos\theta = \pm\frac{1}{\sqrt{2}} \Rightarrow \theta = \pi/4$ or $3\pi/4$

15. Let us consider $\vec{b} = \hat{i}$ and $\vec{c} = \hat{j}$ then $\vec{b} \times \vec{c} = \hat{k}$

Let $\vec{a} = x\hat{i} + y\hat{j} + z\hat{k}$

Then, $(\vec{a}.\vec{b})\vec{b} + (\vec{a}.\vec{c})\vec{c} + \frac{\vec{a}.\vec{b} \times \vec{c}}{|\vec{b} \times \vec{c}|}(\vec{b} \times \vec{c}) = x\hat{i} + y\hat{j} + z\hat{k} = \vec{a}$

ALTERNATE SOLUTION :

Since $\vec{b}$ and $\vec{c}$ are non collinear therefore $\vec{b} \times \vec{c}$ is normal to their plane. Thus $\vec{b}$, $\vec{c}$, $\vec{b} \times \vec{c}$ are three non coplanar vectors

$\therefore$ Any vector $\vec{a}$ can be experessed in terms of these three non coplanar vectors as follows :

$$\vec{a} = x\vec{b} + y\vec{c} + z(\vec{b} \times \vec{c})$$

where x, y, z are projections of $\vec{a}$ in the directions of $\vec{b}$, $\vec{c}$ and $\vec{b} \times \vec{c}$ respectively.

$$\therefore x = \frac{\vec{a} \cdot \vec{b}}{|\vec{b}|} = \vec{a} \cdot \vec{b},\ y = \frac{\vec{a} \cdot \vec{c}}{|\vec{c}|} = \vec{a} \cdot \vec{c},\ z = \frac{\vec{a} \cdot (\vec{b} \times \vec{c})}{|\vec{b} \times \vec{c}|}$$

Hence $\vec{a} = (\vec{a} \cdot \vec{b})\vec{b} + (\vec{a} \cdot \vec{c})\vec{c} + \left(\frac{\vec{a} \cdot (\vec{b} \times \vec{c})}{|\vec{b} \times \vec{c}|}\right)(\vec{b} \times \vec{c})$

16. $\vec{a} \times (\vec{a} \times \vec{c}) + \vec{b} = 0 \Rightarrow (\vec{a}.\vec{c})\vec{a} - (\vec{a}.\vec{a})\vec{c} + \vec{b} = 0$

NOTE THIS STEP :

$\Rightarrow (2\cos\theta)\vec{a} - \vec{c} + \vec{b} = 0$

(Using $|\vec{a}| = 1, |\vec{b}| = 1, |\vec{c}| = 2$)

$\Rightarrow (2\cos\theta\ \vec{a} - \vec{c})^2 = (-\vec{b})^2$

$\Rightarrow 4\cos^2\theta.\ |\vec{a}|^2 + |\vec{c}|^2 - 2.2\cos\theta.\ \vec{a}.\vec{c} = |\vec{b}|^2$

$\Rightarrow 4\cos^2\theta + 4 - 8\cos\theta.\cos\theta = 1$

$\Rightarrow 4\cos^2\theta - 8\cos^2\theta + 4 = 1$

$\Rightarrow 4\cos^2\theta = 3 \Rightarrow \cos\theta = \pm\sqrt{3}/2$

For θ to be acute, $\cos\theta = \sqrt{3}/2 \Rightarrow \theta = \pi/6$

17. q = area of parallelogram with $\overline{OA}$ and $\overline{OC}$ as adjacent sides $= \left|\overline{OA} \times \overline{OC}\right| = \left|\vec{a} \times \vec{b}\right|$

and p = area of quadrilateral $OABC$

$= \frac{1}{2}\left|\overline{OA} \times \overline{OB}\right| + \frac{1}{2}\left|\overline{OB} \times \overline{OC}\right|$

$= \frac{1}{2}\left|\vec{a} \times \left(10\vec{a} + 2\vec{b}\right)\right| + \frac{1}{2}\left|\left(10\vec{a} + 2\vec{b}\right) \times \vec{b}\right|$

$= \left|\vec{a} \times \vec{b}\right| + 5\left|\vec{a} \times \vec{b}\right| = 6\left|\vec{a} \times \vec{b}\right| \quad \therefore\ p = 6q \Rightarrow k = 6$

(B) TRUE / FALSE :

1. $\vec{A}, \vec{B}, \vec{C}$ are three unit vectors such that

$\vec{A}.\vec{B} = \vec{A}.\vec{C} = 0$ ...(1)

and angle between $\vec{B}$ and $\vec{C}$ is $\pi/6$.

Now eq. (1) shows that $\vec{A}$ is perpendicular to both $\vec{B}$ and $\vec{C}$.

$\therefore \vec{B} \times \vec{C} \parallel \vec{A} \Rightarrow \vec{B} \times \vec{C} = \lambda\vec{A}$ where λ is any scalar.

$\Rightarrow |\vec{B} \times \vec{C}| = |\lambda\vec{A}| \Rightarrow \sin\pi/6 = \pm\lambda$

(as $\pi/6$ is the angle between $\vec{B}$ & $\vec{C}$)

$\Rightarrow \lambda = \pm\frac{1}{2} \Rightarrow \vec{B} \times \vec{C} = \pm\frac{1}{2}\vec{A} \Rightarrow \vec{A} = \pm 2\left(\vec{B} \times \vec{C}\right)$

$\therefore$ Given statement is true.

2. $\vec{X}.\vec{A} = 0 \Rightarrow$ either $\vec{A} = 0$ or $\vec{X} \perp \vec{A}$

$\vec{X}.\vec{B} = 0 \Rightarrow$ either $\vec{B} = 0$ or $\vec{X} \perp \vec{B}$

$\vec{X}.\vec{C} = 0 \Rightarrow$ either $\vec{C} = 0$ or $\vec{X} \perp \vec{C}$

In any of three cases,

if $\vec{A}$ or $\vec{B}$ or $\vec{C} = 0 \Rightarrow \left[\vec{A}\vec{B}\vec{C}\right] = 0$

Otherwise if $\vec{X} \perp \vec{A}, \vec{X} \perp \vec{B}, \vec{X} \perp \vec{C}$ then $\vec{A}, \vec{B}, \vec{C}$ are coplanar $\Rightarrow \left[\vec{A}\vec{B}\vec{C}\right] = 0$

$\therefore$ Given statement is true.

3. Let position vectors of pts A, B and C be $\vec{a} + \vec{b}$, $\vec{a} - \vec{b}$ and $\vec{a} + k\vec{b}$ respectively.

Then, $\overline{AB}$ = p.v. of B – p.v of $A = (\vec{a} - \vec{b}) - (\vec{a} + \vec{b}) = -2\vec{b}$

Similarly, $\overline{BC} = \vec{a} + k\vec{b} - \vec{a} + \vec{b} = (k+1)\vec{b}$

Clearly $\overline{AB} \parallel \overline{BC}\ \ \forall k \in R$

$\Rightarrow A, B, C$ are collinear $\forall k \in R$

$\therefore$ Statement is true.

4. For any three vectors $\vec{a}, \vec{b}$ and $\vec{c}$, we have

L.H.S. $= (\vec{a} - \vec{b}).(\vec{b} - \vec{c}) \times (\vec{c} - \vec{a})$

$= (\vec{a} - \vec{b}).(\vec{b} \times \vec{c} - \vec{b} \times \vec{a} - \vec{c} \times \vec{c} + \vec{c} \times \vec{a})$

$= (\vec{a} - \vec{b}).(\vec{b} \times \vec{c} + \vec{a} \times \vec{b} + \vec{c} \times \vec{a})$

$= \vec{a}.(\vec{b} \times \vec{c}) + \vec{a}.(\vec{a} \times \vec{b}) + \vec{a}.(\vec{c} \times \vec{a})$

$\qquad - \vec{b}.(\vec{b} \times \vec{c}) - \vec{b}.(\vec{a} \times \vec{b}) - \vec{b}.(\vec{c} \times \vec{a})$

$= [\vec{a}\vec{b}\vec{c}] - [\vec{a}\vec{b}\vec{c}] = 0 \neq$ R.H.S.

$\therefore$ The given statement is false.

(C) MCQ's WITH ONE CORRECT ANSWER :

1. (a) $\vec{A}.(\vec{B} + \vec{C}) \times (\vec{A} + \vec{B} + \vec{C})$

$= \vec{A}.[\vec{B} \times \vec{A} + \vec{B} \times \vec{B} + \vec{B} \times \vec{C} + \vec{C} \times \vec{A} + \vec{C} \times \vec{B} + \vec{C} \times \vec{C}]$

$= \vec{A}.\vec{B} \times \vec{A} + \vec{A}.\vec{B} \times \vec{C} + \vec{A}.\vec{C} \times \vec{A} + \vec{A}.\vec{C} \times \vec{B}$

(Using $\vec{a} \times \vec{a} = 0$)

$= 0 + [\vec{A}\vec{B}\vec{C}] + 0 + [\vec{A}\vec{C}\vec{B}]$

(as $\left[\vec{a}\vec{b}\vec{c}\right] = 0$ if any two vector are equal out of $\vec{a}, \vec{b}, \vec{c}$)

$= \left[\vec{A}\vec{B}\vec{C}\right] - \left[\vec{A}\vec{B}\vec{C}\right]$ [Using $[\vec{a}\vec{b}\vec{c}] = -[\vec{a}\vec{c}\vec{b}]$]

$= 0$

2. (d) $|(\vec{a} \times \vec{b}).\vec{c}| = |\vec{a}||\vec{b}||\vec{c}|$

$\Rightarrow |\hat{a}||\hat{b}|\sin\theta\,\hat{n}.\vec{c}| = |\hat{a}||\hat{b}||\vec{c}|$

where θ is angle between $\vec{a}$ and $\vec{b}$.

$\Rightarrow \quad |\hat{a}||\hat{b}||\hat{c}|\ |\sin\theta\cos\alpha|$

$= |\hat{a}||\hat{b}||\hat{c}|$ where α is angle between $\vec{c}$ and $\hat{n}$.

$\Rightarrow \quad |\sin\theta|\,|\cos\alpha| = 1 \Rightarrow \theta = \pi/2$ and $\alpha = 0$

$\Rightarrow \quad \vec{a} \perp \vec{b}$ and $\vec{c} \parallel \vec{n} \Rightarrow \vec{a}.\vec{b} = \vec{b}.\vec{c} = \vec{c}.\vec{a} = 0$

3. (d) Vol. of parallelopiped $= \left[\vec{a}\,\vec{b}\,\vec{c}\right]$

$= \begin{vmatrix} 2 & -2 & 0 \\ 1 & 1 & -1 \\ 3 & 0 & -1 \end{vmatrix} = 2(-1) + 2(-1+3) = 2$

4. (a) Three pts A, B, C are collinear if $\overline{AB} \parallel \overline{AC}$

$\overline{AB} = -20\hat{i} - 11\hat{j}$; $\overline{AC} = (a-60)\hat{i} - 55\hat{j}$

$\overline{AB} \parallel \overline{AC} \Rightarrow \frac{a-60}{-20} = \frac{-55}{-11} \Rightarrow a = -40$

5. (d) Given that $\vec{a}, \vec{b}, \vec{c}$ are non coplanar

$\therefore \left[\vec{a}\vec{b}\vec{c}\right] \neq 0$

Also $\vec{p} = \frac{\vec{b} \times \vec{c}}{[\vec{a}\,\vec{b}\,\vec{c}]}, \ \vec{q} = \frac{\vec{c} \times \vec{a}}{[\vec{a}\,\vec{b}\,\vec{c}]}, \ \vec{r} = \frac{\vec{a} \times \vec{b}}{[\vec{a}\,\vec{b}\,\vec{c}]}$ ..(1)

Now, $(\vec{a}+\vec{b}).\vec{p}+(\vec{b}+\vec{c}).\vec{q}+(\vec{c}+\vec{a}).\vec{r}$

$$=(\vec{a}+\vec{b}).\frac{\vec{b}\times\vec{c}}{[\vec{a}\vec{b}\vec{c}]}+(\vec{b}+\vec{c}).\frac{\vec{c}\times\vec{a}}{[\vec{a}\vec{b}\vec{c}]}+(\vec{c}+\vec{a}).\frac{\vec{a}\times\vec{b}}{[\vec{a}\vec{b}\vec{c}]}$$

$$=\frac{\vec{a}.\vec{b}\times\vec{c}}{[\vec{a}\,\vec{b}\,\vec{c}]}+\frac{\vec{b}.\vec{c}\times\vec{a}}{[\vec{a}\,\vec{b}\,\vec{c}]}+\frac{\vec{c}.\vec{a}\times\vec{b}}{[\vec{a}\,\vec{b}\,\vec{c}]}$$

[Using $\vec{b}.\vec{b}\times\vec{c}=\vec{c}.\vec{c}\times\vec{a}=\vec{a}.\vec{a}\times\vec{b}=0$]

$$=\frac{[\vec{a}\,\vec{b}\,\vec{c}]}{[\vec{a}\,\vec{b}\,\vec{c}]}+\frac{[\vec{a}\,\vec{b}\,\vec{c}]}{[\vec{a}\,\vec{b}\,\vec{c}]}+\frac{[\vec{a}\,\vec{b}\,\vec{c}]}{[\vec{a}\,\vec{b}\,\vec{c}]}=1+1+1=3$$

6. (b) a, b, c are distinct non negative numbers and the vectors $a\hat{i}+a\hat{j}+c\hat{k}$, $\hat{i}+\hat{k}$ and $c\hat{i}+c\hat{j}+b\hat{k}$ are coplanar.

$$\therefore\begin{vmatrix}a&a&c\\1&0&1\\c&c&b\end{vmatrix}=0\Rightarrow\begin{vmatrix}a&a&c-a\\1&0&0\\c&c&b-c\end{vmatrix}$$

Operating $C_3\to C_3-C_1$
Expanding along R_2, we get

$$-\begin{vmatrix}a&c-a\\c&b-c\end{vmatrix}=c(c-a)-a(b-c)=0$$

$\Rightarrow c^2-ac-ab+ac=0$
$\Rightarrow c^2=ab\Rightarrow$ a, c, b are in G.P.
$\therefore$ c is the G.M. of a and b.

7. (a) We have $\overrightarrow{OR}=\dfrac{3\vec{p}+2\vec{q}}{3+2}=\dfrac{1}{2}(3\vec{p}+2\vec{q})$ [Internal divison]

and $\overrightarrow{OS}=\dfrac{3\vec{p}-2\vec{q}}{3-2}=3\vec{p}-2\vec{q}$ [external divison]

Given $\overrightarrow{OR}\perp\overrightarrow{OS}\Rightarrow\overrightarrow{OR}.\overrightarrow{OS}=0$

$\Rightarrow\ \dfrac{1}{5}[3\vec{p}+2\vec{q}].(3\vec{p}-2\vec{q})=0$

$\Rightarrow\ 9|\vec{p}|^2=4|\vec{q}|^2\ \Rightarrow\ 9p^2=4q^2$

8. (b) Let the given position vectors be of point A, B and C respectively, then

$|\overrightarrow{AB}|=\sqrt{(\beta-\alpha)^2+(\gamma-\beta)^2+(\alpha-\gamma)^2}$

$|\overrightarrow{BC}|=\sqrt{(\gamma-\beta)^2+(\alpha-\gamma)^2+(\alpha-\beta)^2}$

$|\overrightarrow{CA}|=\sqrt{(\alpha-\gamma)^2+(\beta-\alpha)^2+(\gamma-\beta)^2}$

$\because\ |\overrightarrow{AB}|=|\overrightarrow{BC}|=|\overrightarrow{CA}|$

$\Rightarrow\ \Delta ABC$ is an equilateral Δ.

9. (a) Let $\vec{d}=x\hat{i}+y\hat{j}+z\hat{k}$

where $x^2+y^2+z^2=1$ ....(1)

($\vec{d}$ being unit vector)

$\therefore\ \vec{a}.\vec{d}=0$

$\Rightarrow x-y=0\Rightarrow x=y$ ...(2)

$[\vec{b}\,\vec{c}\,\vec{d}]=0$

$$\Rightarrow\begin{vmatrix}0&1&-1\\-1&0&1\\x&y&z\end{vmatrix}=0$$

$\Rightarrow x+y+z=0$
$\Rightarrow 2x+z=0$ (using (2))
$\Rightarrow z=-2x$ ...(3)
From (1), (2) and (3)

$x^2+x^2+4x^2=1\Rightarrow x=\pm\dfrac{1}{\sqrt{6}}$

$$\therefore d=\pm\left(\frac{1}{\sqrt{6}}\vec{i}+\frac{1}{\sqrt{6}}\vec{j}-\frac{2}{\sqrt{6}}\vec{k}\right)=\pm\left(\frac{\vec{i}+\vec{j}-2\vec{k}}{\sqrt{6}}\right)$$

10. (a) Since $\vec{a}\times(\vec{b}\times\vec{c})=\dfrac{\vec{b}+\vec{c}}{\sqrt{2}}$

$\therefore(\vec{a}.\vec{c})\vec{b}-(\vec{a}.\vec{b})\vec{c}=\dfrac{1}{\sqrt{2}}\vec{b}+\dfrac{1}{\sqrt{2}}\vec{c}\ \Rightarrow\ \vec{a}.\vec{c}=\dfrac{1}{\sqrt{2}}$

[$\because\ \vec{b}$ and $\vec{c}$ are non-coplanar]

and $\vec{a}.\vec{b}=-\dfrac{1}{\sqrt{2}}\Rightarrow\cos\theta=-\dfrac{1}{\sqrt{2}}$

$\Rightarrow\cos\dfrac{3\pi}{4}=\cos\theta\Rightarrow\theta=3\pi/4$

11. (b) $\because\vec{u}+\vec{v}+\vec{w}=0\quad\therefore\ |\vec{u}+\vec{v}+\vec{w}|^2=0$

$\Rightarrow|\vec{u}|^2+|\vec{v}|^2+|\vec{w}|^2+2(\vec{u}.\vec{v}+\vec{v}.\vec{w}+\vec{w}.\vec{u})=0$

$\Rightarrow 9+16+25+2(\vec{u}.\vec{v}+\vec{v}.\vec{w}+\vec{w}.\vec{u})=0$

$\Rightarrow(\vec{u}.\vec{v}+\vec{v}.\vec{w}+\vec{w}.\vec{u})=-25$

12. (d) $(\vec{a}+\vec{b}+\vec{c}).[(\vec{a}+\vec{b})\times(\vec{a}+\vec{c})]$

$=(\vec{a}+\vec{b}+\vec{c}).[\vec{a}\times\vec{a}+\vec{a}\times\vec{c}+\vec{b}\times\vec{a}+\vec{b}\times\vec{c}]$

$=(\vec{a}+\vec{b}+\vec{c}).[\vec{a}\times\vec{c}+\vec{b}\times\vec{a}+\vec{b}\times\vec{c}]\quad[\because\ \vec{a}\times\vec{a}=0]$

$=\vec{a}.\vec{a}\times\vec{c}+\vec{a}.\vec{b}\times\vec{a}+\vec{a}.\vec{b}\times\vec{c}+\vec{b}.\vec{a}\times\vec{c}$
$+\vec{b}.\vec{b}\times\vec{a}+\vec{b}.\vec{b}\times\vec{c}+\vec{c}.\vec{a}\times\vec{c}+\vec{c}.\vec{b}\times\vec{a}+\vec{c}.\vec{b}\times\vec{c}$

$=[\vec{a}\vec{b}\vec{c}]-[\vec{a}\vec{b}\vec{c}]-[\vec{a}\vec{b}\vec{c}]=-[\vec{a}\vec{b}\vec{c}]$

13. (b) As $\vec{p},\vec{q}$ and $\vec{r}$ are three mutually perpendicular vectors of same magnitude, so let us consider

$\vec{p}=a\hat{i},\vec{q}=a\hat{j},\vec{r}=a\hat{k}$

Also, let $\vec{x}=x_1\hat{i}+y_1\hat{j}+z_1\hat{k}$

Given that $\vec{x}$ satisfies the eq^n

$\vec{p}\times[(\vec{x}-\vec{q})\times\vec{p}]+\vec{q}\times[(\vec{x}-\vec{r})\times\vec{q}]+\vec{r}\times[(\vec{x}-\vec{p})\times\vec{r}]=0$ ...(1)

Now $\vec{p}\times[(\vec{x}-\vec{q})\times\vec{p}]=\vec{p}\times[\vec{x}\times\vec{p}-\vec{q}\times\vec{p}]$

$=\vec{p}\times(\vec{x}\times\vec{p})-\vec{p}\times(\vec{q}\times\vec{p})$

$= (\vec{p}.\vec{p})\vec{x} - (\vec{p}.\vec{x})\vec{p} - (\vec{p}.\vec{p})\vec{q} + (\vec{p}.\vec{q})\vec{p}$

$= a^2\vec{x} - a^2 x_1\hat{i} - a^3\hat{j} + 0$

Similarly, $\vec{q}\times[(\vec{x}-\vec{r})\times\vec{q}]$

$= (\vec{q}.\vec{q})\vec{x} - (\vec{q}.\vec{x})\vec{q} - (\vec{q}.\vec{q})\vec{r} + (\vec{q}.\vec{r})\vec{q}$

$= a^2\vec{x} - a^2 y_1\hat{j} - a^3\hat{k}$

and $\vec{r}\times[(\vec{x}-\vec{p})\times\vec{r}] = (\vec{r}.\vec{r})\vec{x} - (\vec{r}.\vec{x})\vec{r} - (\vec{r}.\vec{r})\vec{p} + (\vec{r}.\vec{p})\vec{r}$

$= a^2\vec{x} - a^2 z_1\hat{k} - a^3\hat{i}$

Substituting these values in eq^n (1) we get

$3a^2\vec{x} - a^2(x_1\hat{i} + y_1\hat{j} + z_1\hat{k}) - a^2(a\hat{i} + a\hat{j} + a\hat{k}) = 0$

$\Rightarrow 3a^2\vec{x} - a^2\vec{x} - a^2(\vec{p}+\vec{q}+\vec{r}) = \vec{0}$

$\Rightarrow 2a^2\vec{x} = (\vec{p}+\vec{q}+\vec{r})a^2 \Rightarrow \vec{x} = \frac{1}{2}(\vec{p}+\vec{q}+\vec{r})$

14. **(b)** $|(\vec{a}\times\vec{b})\times\vec{c}| = |\vec{a}\times\vec{b}||\vec{c}|\sin 30^\circ$

$= \frac{1}{2}|\vec{a}\times\vec{b}||\vec{c}|$ ...(1)

We have, $\vec{a} = 2\hat{i} + \hat{j} - 2\hat{k}$ and $\vec{b} = \hat{i} + \hat{j}$

$\Rightarrow \vec{a}\times\vec{b} = 2\hat{i} - 2\hat{j} + \hat{k} \Rightarrow |\vec{a}\times\vec{b}| = \sqrt{9} = 3$

Also given $|\vec{c}-\vec{a}| = 2\sqrt{2}$

$\Rightarrow |\vec{c}-\vec{a}|^2 = 8 \Rightarrow (\vec{c}-\vec{a}).(\vec{c}-\vec{a}) = 8$

$\Rightarrow |\vec{c}|^2 - \vec{c}.\vec{a} - \vec{a}.\vec{c} + |\vec{a}|^2 = 8$

As $|\vec{a}| = 3$ and $\vec{a}.\vec{c} = |\vec{c}|$, we get

$|\vec{c}|^2 - 2|\vec{c}| + 1 = 0$

$(|\vec{c}|-1)^2 = 0 \Rightarrow |\vec{c}| = 1$

Substituting values of $|\vec{a}\times\vec{b}|$ and $|\vec{c}|$ in (1), we get

$|(\vec{a}\times\vec{b})\times\vec{c}| = \frac{1}{2}\times 3\times 1 = \frac{3}{2}$

15. **(a)** As c is coplanar with *a* and *b*, we take,

$c = \alpha a + \beta b$ ...(1)

where α, β are scalars.

As c is perpendicular to a, $c.a = 0$

$\therefore$ From (1) we get, $0 = \alpha\, a.a + \beta\, b.a$

$\Rightarrow 0 = \alpha(6) + \beta(2+2-1) = 3(2\alpha+\beta)$

$\Rightarrow \beta = -2\alpha.$

Thus, $c = \alpha(a-2b) = \alpha(-3j+3k) = 3\alpha(-j+k)$

$\Rightarrow |c|^2 = 9\alpha^2(1+1) = 18\alpha^2 \Rightarrow 1 = 18\alpha^2$

$\Rightarrow \alpha = \pm\frac{1}{3\sqrt{2}}$ $\therefore c = \pm\frac{1}{\sqrt{2}}(-j+k).$

Thus, we may take $c = \frac{1}{\sqrt{2}}(-j+k).$

16. **(b)** Given $\vec{a}+\vec{b}+\vec{c} = 0$ (by triangle law)

$\therefore \vec{a}\times(\vec{a}+\vec{b}+\vec{c}) = \vec{a}\times\vec{0} = \vec{0}$

$\Rightarrow \vec{a}\times\vec{a} + \vec{a}\times\vec{b} + \vec{a}\times\vec{c} = \vec{0}$

$\Rightarrow \vec{a}\times\vec{b} = \vec{c}\times\vec{a}$ $[\because \vec{a}\times\vec{a} = 0]$

Similarly, $\vec{a}\times\vec{b} = \vec{b}\times\vec{c}$;

Therefore $\vec{a}\times\vec{b} = \vec{b}\times\vec{c} = \vec{c}\times\vec{a}.$

17. **(a)** Given that $\vec{a},\vec{b},\vec{c},\vec{d}$ are vectors such that

$(\vec{a}\times\vec{b})\times(\vec{c}\times\vec{d}) = 0$ ...(1)

P_1 is the plane determined by vectors $\vec{a}$ and $\vec{b}$

$\therefore$ Normal vectors $\vec{n_1}$ to P_1 will be given by $\vec{n_1} = \vec{a}\times\vec{b}$

Similarly, P_2 is the plane determined by vectors $\vec{c}$ and $\vec{d}$

$\therefore$ Normal vectors $\vec{n_2}$ to P_2 will be given by $\vec{n_2} = \vec{c}\times\vec{d}$

Substituting the values of $\vec{n_1}$ and $\vec{n_2}$ in eq^n (1)

We get, $\vec{n_1}\times\vec{n_2} = 0 \Rightarrow \vec{n_1} \parallel \vec{n_2}$

and hence the planes will also be parallel to each other. Thus angle between the planes = 0.

18. **(a)** $\vec{a},\vec{b},\vec{c}$ are unit coplanar vectors, $\overrightarrow{2a}-\vec{b}, \overrightarrow{2b}-\vec{c}$ and $\overrightarrow{2c}-\vec{a}$ are also coplanar vectors. being linear combination of $\vec{a},\vec{b}$ and $\vec{c}$.

Thus, $[\overrightarrow{2a}-\vec{b}\ \ \overrightarrow{2b}-\vec{c}\ \ \overrightarrow{2c}-\vec{a}] = 0$

ALTERNATE SOLUTION

$\because$ $\vec{a},\vec{b},\vec{c}$ are unit coplanar vectors

$\therefore$ $[\vec{a}\vec{b}\vec{c}] = 0$

Also $[2\vec{a}-\vec{b}\ \ 2\vec{b}-\vec{c}\ \ 2\vec{c}-\vec{a}]$

$= (2\vec{a}-\vec{b}).(2\vec{b}-\vec{c})\times(2\vec{c}-\vec{a})$

$= (2\vec{a}-\vec{b}).[4\vec{b}\times\vec{c} - 2\vec{b}\times\vec{a} - 2\vec{c}\times\vec{c} + \vec{c}\times\vec{a})$

$= (2\vec{a}-\vec{b}).[4\vec{b}\times\vec{c} + 2\vec{a}\times\vec{b} + \vec{c}\times\vec{a}]$

$= 8\vec{a}.\vec{b}\times\vec{c} - \vec{b}.\vec{c}\times\vec{a}$

(using $\vec{a}.\vec{a}\times\vec{b} = \vec{a}.\vec{c}\times\vec{a} = \vec{b}.\vec{b}\times\vec{c} = \vec{b}.\vec{a}\times\vec{b} = 0$)

$= 8[\vec{a}\vec{b}\vec{c}] - [\vec{a}\vec{b}\vec{c}] = 7[\vec{a}\vec{b}\vec{c}] = 0$ as $[\vec{a}\vec{b}\vec{c}] = 0$

19. **(c)** $\vec{a} = \hat{i} - \hat{k}, \vec{b} = x\hat{i} + \hat{j} + (1-x)\hat{k},$

$\vec{c} = y\hat{i} + x\hat{j} + (1+x-y)\hat{k}$

$$[\vec{a}\vec{b}\vec{c}] = \begin{vmatrix} 1 & 0 & -1 \\ x & 1 & 1-x \\ y & x & 1+x-y \end{vmatrix}$$

$= 1(1+x-y-x+x^2) - 1(x^2-y) = 1$

$\therefore$ Depends neither on *x* nor on *y*.

20. **(b)** $\hat{a},\hat{b},\hat{c}$ are units vectors.

$\therefore$ $\hat{a}.\hat{a} = \hat{b}.\hat{b} = \hat{c}.\hat{c} = 1$

Now, $x = |\hat{a}-\hat{b}|^2 + |\hat{b}-\hat{c}|^2 + |\hat{c}-\hat{a}|^2$

$= \hat{a}.\hat{a} + \hat{b}.\hat{b} - 2\hat{a}.\hat{b} + \hat{b}.\hat{b} + \hat{c}.\hat{c} - 2\hat{b}.\hat{c} + \hat{c}.\hat{c} + \hat{a}.\hat{a} - 2\hat{c}.\hat{a}$

$= 6 - 2(\hat{a}.\hat{b} + \hat{b}.\hat{c} + \hat{c}.\hat{a})$ ...(1)

Also

$\Rightarrow |\hat{a}+\hat{b}+\hat{c}| \geq 0 \Rightarrow |\hat{a}+\hat{b}+\hat{c}|^2 \geq 0$

$\Rightarrow \hat{a}.\hat{a} + \hat{b}.\hat{b} + \hat{c}.\hat{c} + 2(\hat{a}.\hat{b} + \hat{b}.\hat{c} + \hat{c}.\hat{a}) \geq 0$

$\Rightarrow 3 + 2(\hat{a}.\hat{b} + \hat{b}.\hat{c} + \hat{c}.\hat{a}) \geq 0$

$\Rightarrow 2(\hat{a}.\hat{b} + \hat{b}.\hat{c} + \hat{c}.\hat{a}) \geq -3$

$\Rightarrow -2(\hat{a}.\hat{b} + \hat{b}.\hat{c} + \hat{c}.\hat{a}) \leq 3$

$\Rightarrow 6 - 2(\hat{a}.\hat{b} + \hat{b}.\hat{c} + \hat{c}.\hat{a}) \leq 9$ ...(2)

From (1) and (2), $x \leq 9$

$\therefore$ x does not exceed 9

21. **(b)** Given that $\vec{a}$ and $\vec{b}$ are two unit vectors

$\therefore |\vec{a}| = 1$ and $|\vec{b}| = 1$

Also, given that $(\vec{a}+2\vec{b}) \perp (5\vec{a}-4\vec{b})$

$\Rightarrow (\vec{a}+2\vec{b}).(5\vec{a}-4\vec{b}) = 0$

$\Rightarrow 5|\vec{a}|^2 - 8|\vec{b}|^2 - 4\vec{a}.\vec{b} + 10\vec{b}.\vec{a} = 0$

$\Rightarrow 5 - 8 + 6\vec{a}.\vec{b} = 0 \Rightarrow 6|\vec{a}||\vec{b}|\cos\theta = 3$

[where θ is the angle between $\vec{a}$ and $\vec{b}$]

$\Rightarrow \cos\theta = \frac{1}{2} \Rightarrow \theta = 60°$

22. **(c)** Given that $\vec{v} = 2\hat{i} + \hat{j} - \hat{k}$ and $\vec{w} = \hat{i} + 3\hat{k}$ and u is a unit vector

$\therefore |\vec{u}| = 1$

Now, $[\vec{u}\,\vec{v}\,\vec{w}] = \vec{u}.(\vec{v} \times \vec{w})$

$= \vec{u}.(2\hat{i} + \hat{j} - \hat{k}) \times (\hat{i} + 3\hat{k})$

$= \vec{u}.(3\hat{i} - 7\hat{j} - \hat{k}) = \sqrt{3^2 + 7^2 + 1^2}\cos\theta$

which is max. when $\cos\theta = 1$

$\therefore$ Max. value of $[\vec{u}\,\vec{v}\,\vec{w}] = \sqrt{59}$

23. **(a)** As the line $\frac{x-4}{1} = \frac{y-2}{1} = \frac{z-k}{2}$ lies in the plane $2x - 4y + z = 7$, the point (4, 2, k) through which line passes must also lie on the given plane and hence $2 \times 4 - 4 \times 2 + k = 7 \Rightarrow k = 7$

24. **(c)** Vol. of paralleopiped formed by $\vec{u} = \hat{i} + a\hat{j} + \hat{k}, \vec{v} = \hat{j} + a\hat{k}, \hat{w} = a\hat{i} + \hat{k}$ is

$$V = [\vec{u}\,\vec{v}\,\vec{w}] = \begin{vmatrix} 1 & a & 1 \\ 0 & 1 & a \\ a & 0 & 1 \end{vmatrix}$$

$= 1(1-0) - a(0-a^2) + 1(0-a) = 1 + a^3 - a$

For V to be min $\frac{dV}{da} = 0$

$\Rightarrow 3a^2 - 1 = 0 \Rightarrow a = \pm\frac{1}{\sqrt{3}}$

25. **(c)** $\because (\vec{a} \times \vec{b}) \times \vec{a} = (\vec{a}.\vec{a})\vec{b} - (\vec{a}.\vec{b})\vec{a}$

$\therefore (\hat{j} - \hat{k}) \times (\hat{i} + \hat{j} + \hat{k}) = (\sqrt{3})^2(\vec{b}) - (\hat{i} + \hat{j} + \hat{k})$

$\Rightarrow 3\vec{b} = 3\hat{i} \Rightarrow \vec{b} = \hat{i}$

26. **(b)** $\frac{x-1}{2} = \frac{y+1}{3} = \frac{z-1}{4} = \lambda$

$\Rightarrow x = 2\lambda + 1, y = 3\lambda - 1$ and $z = 4\lambda + 1$

$\frac{x-3}{1} = \frac{y-k}{2} = \frac{z}{1} = \mu$

$\Rightarrow x = 3 + \mu, y = k + 2\mu$ and $z = \mu$

Since above lines intersect

$\Rightarrow 2\lambda + 1 = 3 + \mu$ ...(1)

$3\lambda - 1 = 2\mu + k$ ...(2)

$\mu = 4\lambda + 1$ ...(3)

Solving (1) and (3) and putting the value of λ and μ in (2)

we get $k = \frac{9}{2}$.

27. **(c)** Any vector coplanar to $\vec{a}$ and $\vec{b}$ can be written as $\vec{r} = \vec{a} + \lambda\vec{b}$

NOTE THIS STEP :

$\vec{r} = (1+2\lambda)\hat{i} + (-1+\lambda)\hat{j} + (1+\lambda)\hat{k}$

Since $\vec{r}$ is orthogonal to $5\hat{i} + 2\hat{j} + 6\hat{k}$

$\Rightarrow 5(1+2\lambda) + 2(-1+\lambda) + 6(1+\lambda) = 0$

$\Rightarrow 9 + 18\lambda = 0 \Rightarrow \lambda = -\frac{1}{2}$

$\therefore \vec{r}$ is $3\hat{j} - \hat{k}$

Since $\hat{r}$ is a unit vector

$\therefore \hat{r} = \frac{3\hat{j} - \hat{k}}{\sqrt{10}}$

28. **(d)** Let the eq^n of variable plane be $\frac{x}{a} + \frac{y}{b} + \frac{z}{c} = 1$ which meets the axes at A (a, 0, 0), B (0, b, 0) and C (0. 0, c).

$\therefore$ Centroid of ΔABC is $\left(\frac{a}{3}, \frac{b}{3}, \frac{c}{3}\right)$

and it satisfies the relation

$\frac{1}{x^2} + \frac{1}{y^2} + \frac{1}{z^2} = k \Rightarrow \frac{9}{a^2} + \frac{9}{b^2} + \frac{9}{c^2} = k$

$\Rightarrow \frac{1}{a^2} + \frac{1}{b^2} + \frac{1}{z^2} = \frac{k}{9}$ ...(1)

Also given that the distance of plane $\frac{x}{a} + \frac{y}{b} + \frac{z}{c} = 1$ from (0, 0, 0) is 1 unit.

$$\Rightarrow \frac{1}{\sqrt{\frac{1}{a^2}+\frac{1}{b^2}+\frac{1}{c^2}}}=1 \Rightarrow \frac{1}{a^2}+\frac{1}{b^2}+\frac{1}{c^2}=1 \quad ...(2)$$

From (1) and (2), we get $\frac{k}{9}=1$ i.e. $k=9$

29. **(b)** We observe that

$$\vec{a}.\vec{b}_1=\vec{a}.\vec{b}-\left(\frac{\vec{b}.\vec{a}}{|\vec{a}|^2}\right)\vec{a}.\vec{a}=\vec{a}.\vec{b}-\vec{a}.\vec{b}=0$$

$$\vec{a}.\vec{c}_2=\vec{a}.\left(\vec{c}-\frac{\vec{c}.\vec{a}}{|\vec{a}|^2}\vec{a}-\frac{\vec{c}.\vec{b}_1}{|\vec{b}_1|^2}\vec{b}_1\right)$$

$$=\vec{a}.\vec{c}-\vec{c}.\frac{\vec{a}.\vec{c}}{|\vec{a}|^2}|\vec{a}|^2-\frac{\vec{c}.\vec{b}_1}{|\vec{b}_1|^2}(\vec{a}.\vec{b}_1)$$

$$=\vec{a}.\vec{c}-\vec{a}.\vec{c}-0$$

$$=0 \qquad [\because \vec{a}.\vec{b}_1=0]$$

And $\vec{b}_1.\vec{c}_2=\vec{b}_1.\left(\vec{c}-\frac{\vec{c}.\vec{a}}{|\vec{a}|^2}\vec{a}-\frac{\vec{c}.\vec{b}_1}{|\vec{b}_1|^2}\vec{b}_1\right)$

$$=\vec{b}_1.\vec{c}-\frac{(\vec{c}.\vec{a})(\vec{b}_1.\vec{a})}{|\vec{a}|^2}-\frac{\vec{c}.\vec{b}_1}{|\vec{b}_1|^2}\vec{b}_1.\vec{b}_1$$

$$=\vec{b}_1.\vec{c}-0-\vec{b}_1.\vec{c} \qquad (\text{Using } \vec{b}_1.\vec{a}=0)$$

$=0$

Hence $\vec{a}.\vec{b}_1=\vec{a}.\vec{c}_2=\vec{b}_1.\vec{c}_2=0$

$\Rightarrow$ $(\vec{a},\vec{b},\vec{c}_2)$ is a set of orthogonal vectors.

30. **(d)** The equation of plane through the point (1, –2, 1) and perpendicular to the planes $2x-2y+z=0$ and $x-y+2z=4$ is given by

$$\begin{vmatrix} x-1 & y+2 & z-1 \\ 2 & -2 & 1 \\ 1 & -1 & 2 \end{vmatrix}=0$$

$\Rightarrow$ $x+y+1=0$

It's distance from the point (1, 2, 2) is

$$\left|\frac{1+2+1}{\sqrt{2}}\right|=2\sqrt{2}.$$

31. **(a)** A vector in the plane of $\vec{a}$ and $\vec{b}$ is

$$\vec{u}=\vec{a}+\lambda\vec{b}=(1+\lambda)\hat{i}+(2-\lambda)\hat{j}+(1+\lambda)\hat{k}$$

Projection of $\vec{u}$ on $\vec{c}=\frac{1}{\sqrt{3}} \Rightarrow \frac{\vec{u}.\vec{c}}{|\vec{c}|}=\frac{1}{\sqrt{3}}$

$\Rightarrow$ $\vec{u}.\vec{c}=1 \Rightarrow |1+\lambda+2-\lambda-1-\lambda|=1$

$\Rightarrow$ $|2-\lambda|=1 \Rightarrow \lambda=1$ or 3

$\Rightarrow$ $\vec{u}=2\hat{i}+\hat{j}+2\hat{k}$ or $4\hat{i}-\hat{j}+4\hat{k}$

32. **(c)** We know that three vector are coplanar if their scalar triple product is zero.

$$\begin{vmatrix} -\lambda^2 & 1 & 1 \\ 1 & -\lambda^2 & 1 \\ 1 & 1 & -\lambda^2 \end{vmatrix}=0$$

$R_1 \to R_1+R_2+R_3$

$$\Rightarrow \begin{vmatrix} 2-\lambda^2 & 2-\lambda^2 & 2-\lambda^2 \\ 1 & -\lambda^2 & 1 \\ 1 & 1 & -\lambda^2 \end{vmatrix}=0$$

$$\Rightarrow (2-\lambda^2)\begin{vmatrix} 1 & 1 & 1 \\ 1 & -\lambda^2 & 1 \\ 1 & 1 & -\lambda^2 \end{vmatrix}=0$$

$$\Rightarrow (2-\lambda^2)\begin{vmatrix} 1 & 1 & 1 \\ 0 & -(1+\lambda^2) & 0 \\ 0 & 0 & -(1+\lambda^2) \end{vmatrix}=0$$

(R_2-R_1, R_3-R_1)

$\Rightarrow$ $(2-\lambda^2)(1+\lambda^2)^2=0 \Rightarrow \lambda=\pm\sqrt{2}$

$\therefore$ Two real solutions.

33. **(b)** Since, $\vec{a}+\vec{b}+\vec{c}=0$ and $\vec{a},\vec{b},\vec{c}$ are unit vectors, therefore $\vec{a},\vec{b},\vec{c}$ form an equilateral triangle.

$\Rightarrow$ $\vec{a}\times(\vec{a}+\vec{b}+\vec{c})=\vec{0}$

$\Rightarrow$ $\vec{a}\times\vec{a}+\vec{a}\times\vec{b}+\vec{a}\times\vec{c}=\vec{0} \Rightarrow \vec{a}\times\vec{b}=\vec{c}\times\vec{a}$

Similarly, $\vec{b}\times\vec{c}=\vec{c}\times\vec{a}$

$\therefore$ $\vec{a}\times\vec{b}=\vec{b}\times\vec{c}=\vec{c}\times\vec{a}$

Also since $\vec{a},\vec{b},\vec{c}$ are non parallel (these form an equilateral Δ).

$\therefore$ $\vec{a}\times\vec{b}=\vec{b}\times\vec{c}=\vec{c}\times\vec{a}\neq\vec{0}$

34. **(a)** We know that the volume of a parallelopipe with coterminus edges as the vectors $\vec{a}$, $\vec{b}$, $\vec{c}$ is given by

$$V=\left[\vec{a}\ \vec{b}\ \vec{c}\right]^2=\begin{vmatrix} \vec{a}.\vec{a} & \vec{a}.\vec{b} & \vec{a}.\vec{c} \\ \vec{b}.\vec{a} & \vec{b}.\vec{b} & \vec{b}.\vec{c} \\ \vec{c}.\vec{a} & \vec{c}.\vec{b} & \vec{c}.\vec{c} \end{vmatrix}$$

$$\Rightarrow V^2=\begin{vmatrix} 1 & 1/2 & 1/2 \\ 1/2 & 1 & 1/2 \\ 1/2 & 1/2 & 1 \end{vmatrix}=\frac{1}{2} \Rightarrow V=\frac{1}{\sqrt{2}}$$

35. **(a)** Given $\overrightarrow{OP}=\hat{a}\cos t+\hat{b}\sin t$

$$\Rightarrow |\overrightarrow{OP}|^2=\cos^2 t+\sin^2 t+2\hat{a}.\hat{b}\sin t\cos t$$

$$\Rightarrow |\overrightarrow{OP}|^2=1+\hat{a}.\hat{b}\sin 2t\le 1+\hat{a}.\hat{b} \quad (\text{Max. at } t=\frac{\pi}{4})$$

$$\therefore |\overrightarrow{OP}|_{\max}=\sqrt{1+\hat{a}.\hat{b}}$$

Also $\hat{u} = |\widehat{OP}|_{max}$

Maximum occurs at $t = \frac{\pi}{4}$

$\therefore\ |\overline{OP}|_{max} = \frac{\hat{a}+\hat{b}}{\sqrt{2}}$ $\quad \therefore\ |\hat{O}P|_{max} = \frac{\hat{a}+\hat{b}}{|\hat{a}+\hat{b}|}$

Hence $\hat{u} = \frac{\hat{a}+\hat{b}}{|\hat{a}+\hat{b}|}$ and $M = \sqrt{1+\hat{a}.\hat{b}}$

36. **(a)** Given that $P(3, 2, 6)$ is a point in space and Q is a point on line

$\vec{r} = (\hat{i} - \hat{j} + 2\hat{k}) + \mu(-3\hat{i} + \hat{j} + 5\hat{k})$

or $\frac{x-1}{-3} = \frac{y+1}{1} = \frac{z-2}{5} = \mu$

Let coordinates of Q be $(-3\mu+1, \mu-1, 5\mu+2)$

$\therefore$ d.r's of $\overrightarrow{PQ} = -3\mu-2,\ \mu-3,\ 5\mu-4$

As $\overrightarrow{PQ}$ is parallel to the plane $x-4y+3z=1$

$\therefore\ 1.(-3\mu-2)-4.(\mu-3)+3.(5\mu-4)=0$

$\Rightarrow\ 8\mu = 2$ or $\mu = \frac{1}{4}$

37. **(c)** $\vec{a}, \vec{b}, \vec{c}$ and $\vec{d}$ are unit vectors,

Let $\vec{a}\times\vec{b} = (\sin\alpha)\overrightarrow{n_1}$ and $\vec{c}\times\vec{d} = (\sin\beta)\overrightarrow{n_2}$

then $(\vec{a}\times\vec{b}).(\vec{c}\times\vec{d}) = 1$

$\Rightarrow\ (\sin\alpha)\overrightarrow{n_1}.(\sin\beta)\overrightarrow{n_2} = 1$

$\Rightarrow\ \sin\alpha\sin\beta\ \overrightarrow{n_1}.\overrightarrow{n_2} = 1 \Rightarrow \sin\alpha\sin\beta\cos\gamma = 1$

where γ is the angle between $\overrightarrow{n_1}$ and $\overrightarrow{n_2}$.

$\alpha = \frac{\pi}{2}, \beta = \frac{\pi}{2}$ and $\gamma = 0°$

Now $\gamma = 0° \Rightarrow \vec{a}\times\vec{b} \parallel \vec{c}\times\vec{d}$

Let $\vec{a}\times\vec{b} = \lambda(\vec{c}\times\vec{d})$

$\Rightarrow\ (\vec{a}\times\vec{b}).\vec{c} = \lambda(\vec{c}\times\vec{d}).\vec{c} = 0$

and $(\vec{a}\times\vec{b}).\vec{d} = \lambda(\vec{c}\times\vec{d}).\vec{d} = 0$

$\therefore\ \vec{a}, \vec{b}, \vec{c}$ are coplanar and $\vec{a}, \vec{b}, \vec{d}$ are coplanar

$\Rightarrow\ \vec{a}, \vec{b}, \vec{c}, \vec{d}$ are coplanar

Also $\alpha = 90°$

$\Rightarrow\ \vec{a}\perp\vec{b}$ and $\beta = 90° \quad \Rightarrow\ \vec{c}\perp\vec{d}$

But angle between $\vec{a}$ and $\vec{c}$ is $\pi/3$ ($\because\ \vec{a}.\vec{c} = \frac{1}{2}$)

So, angle between $\vec{b}$ and $\vec{d}$ should also be $\pi/3$.

Hence $\vec{b}$ and $\vec{d}$ are non parallel.

38. **(c)** The line has +ve and equal direction cosines, these are $\frac{1}{\sqrt{3}}, \frac{1}{\sqrt{3}}, \frac{1}{\sqrt{3}}$ or direction ratios are 1, 1, 1. Also the lines passes through $P(2, -1, 2)$.

$\therefore$ Equation of line is

$$\frac{x-2}{1} = \frac{y+1}{1} = \frac{z-2}{1} = \lambda \text{ (say)}$$

Let $Q(\lambda+2, \lambda-1, \lambda+2)$ be a point on this line where it meets the plane

$$2x+y+z=9$$

Then Q must satisfy the eq^n of plane

i.e. $2(\lambda+2)+\lambda-1+\lambda+2 = 9 \Rightarrow \lambda = 1$

$\therefore$ Q has coordintes (3, 0, 3)

Hence the length of line segments PQ

$= \sqrt{(2-3)^2+(-1-0)^2+(2-3)^2} = \sqrt{3}$

39. **(a)** We have $\overrightarrow{PQ} = 6\hat{i}+\hat{j}, \overrightarrow{QR} = -\hat{i}+3\hat{j}, \overrightarrow{SR} = 6\hat{i}+\hat{j},$ $\overrightarrow{PS} = -\hat{i}+3\hat{j}$

$\Rightarrow \overrightarrow{PQ} = \overrightarrow{SR}; \overrightarrow{QR} = \overrightarrow{PS}$ and $\overrightarrow{PQ}.\overrightarrow{PS} \neq 0$

Also $|\overrightarrow{PQ}| \neq |\overrightarrow{QR}|$

$\Rightarrow$ PQRS is a parallelogram but neither a rhombus nor a rectangle.

40. **(c)** Plane containing two lines $\frac{x}{3} = \frac{y}{4} = \frac{z}{2}$ and $\frac{x}{4} = \frac{y}{2} = \frac{z}{3}$ is given by

$$\begin{vmatrix} x & y & z \\ 3 & 4 & 2 \\ 4 & 2 & 3 \end{vmatrix} = 0 \Rightarrow 8x - y - 10z = 0$$

Now equation of plane containing the line $\frac{x}{2} = \frac{y}{3} = \frac{z}{4}$ and perpendicular to the plane $8x - y - 10z = 0$ is

$$\begin{vmatrix} x & y & z \\ 2 & 3 & 4 \\ 8 & -1 & -10 \end{vmatrix} = 0$$

$\Rightarrow -26x + 52y - 26z = 0$ *or* $x - 2y + z = 0$

41. **(a)** As perpendicular distance of $x+2y-2z = \alpha$ from the point $(1, -2, 1)$ is 5

$\therefore \left|\frac{1-4-2-\alpha}{3}\right| = 5$

$\Rightarrow \frac{-5-\alpha}{3} = 5$ or -5

$\Rightarrow \alpha = -20$ or 10

As $\alpha > 0 \Rightarrow \alpha = 10$

P(1, –2, 1)

N

$x+2y-2z=\alpha$

$\therefore$ Plane becomes $x+2y-2z-10=0$

Equation of PN is $\frac{x-1}{1} = \frac{y+2}{2} = \frac{z-1}{-2} = \lambda$

For some value of λ, $N(\lambda+1, 2\lambda-2, -2\lambda+1)$

It lies on $x+2y-2z-10=0$

$\therefore \lambda+1+4\lambda-4+4\lambda-2 = 10$

$\Rightarrow 9\lambda = 15 \Rightarrow \lambda = 5/3$

$\therefore N\left(\frac{2}{3}, \frac{4}{3}, \frac{-7}{3}\right)$

42. (b)

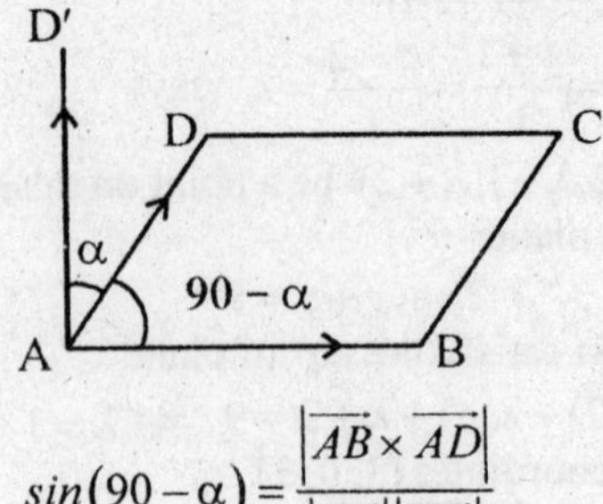

$$sin(90-\alpha)=\frac{|\overrightarrow{AB}\times\overrightarrow{AD}|}{|\overrightarrow{AB}||\overrightarrow{AD}|}$$

Where, $\overrightarrow{AB}\times\overrightarrow{AD}=\begin{vmatrix}\hat{i} & \hat{j} & \hat{k}\\ 2 & 10 & 11\\ -1 & 2 & 2\end{vmatrix}=-2\hat{i}-15\hat{j}+14\hat{k}$

$\therefore\ |\overrightarrow{AB}\times\overrightarrow{AD}|=\sqrt{4+225+196}=\sqrt{425}$

$|\overrightarrow{AB}|=\sqrt{4+100+121}=\sqrt{225}=15$

$|\overrightarrow{AD}|=\sqrt{1+4+4}=3$

$\therefore\ \sin(90-\alpha)=\frac{\sqrt{425}}{15\times 3}=\frac{\sqrt{17}}{9}\Rightarrow\cos\alpha=\frac{\sqrt{17}}{9}$

43. (c) As $\vec{v}$ lies in the plane of $\vec{a}$ and $\vec{b}$

$\therefore\ \vec{v}=\lambda\vec{a}+\mu\vec{b}$

$\Rightarrow\vec{v}=(\lambda+\mu)\ \hat{i}+(\lambda-\mu)\hat{j}+(\lambda+\mu)\hat{k}$

$\because$ Projection of $\vec{v}$ on $\vec{c}$ is $\frac{1}{\sqrt{3}}$

$\therefore\ \frac{\vec{v}.\vec{c}}{|\vec{c}|}=\frac{1}{\sqrt{3}}$

$\Rightarrow\ \frac{(\lambda+\mu)-(\lambda-\mu)-(\lambda+\mu)}{\sqrt{3}}=\frac{1}{\sqrt{3}}$

$\Rightarrow\mu-\lambda=1$ or $\mu=\lambda+1$

$\Rightarrow\vec{v}=(2\lambda+1)\hat{i}-\hat{j}+(2\lambda+1)\hat{k}$

For $\lambda=1,\ \vec{v}=3\hat{i}-\hat{j}+3\hat{k}.$

44. (a) Equation of st. line joining Q (2, 3, 5) and $R(1,-1,4)$ is

$$\frac{x-2}{-1}=\frac{y-3}{-4}=\frac{z-5}{1}=\lambda$$

Let $P(-\lambda+2,-4\lambda+3,-\lambda+5)$

As P lies on $5x-4y-z=1$

$\therefore\quad -5\lambda+10+16\lambda-12+\lambda-5=1$

$\Rightarrow\quad 12\lambda=8\Rightarrow\lambda=\frac{2}{3}\qquad\therefore\ P=\left(\frac{4}{3},\frac{1}{3},\frac{13}{3}\right)$

Now let point S on QR be

$(-\mu+2,-4\mu+3,-\mu+5)$

$\because$ S is the foot of perpendicular drawn from $T(2,1,4)$ to QR, where dr's of ST are $\mu, 4\mu-2, \mu-1$ and dr's of QR are $-1,-4,-1$

$\therefore\quad -\mu-16\mu+8-\mu+1=0$

$\Rightarrow\quad 18\mu=9\Rightarrow\mu=\frac{1}{2}$

$\therefore\quad S=\left(\frac{3}{2},1,\frac{9}{2}\right)$

$\therefore$ Distance between P and S

$$=\sqrt{\left(\frac{4}{3}-\frac{3}{2}\right)^2+\left(\frac{1}{3}-1\right)^2+\left(\frac{13}{3}-\frac{9}{2}\right)^2}$$

$$=\sqrt{\frac{1}{36}+\frac{4}{9}+\frac{1}{36}}=\frac{1}{\sqrt{2}}$$

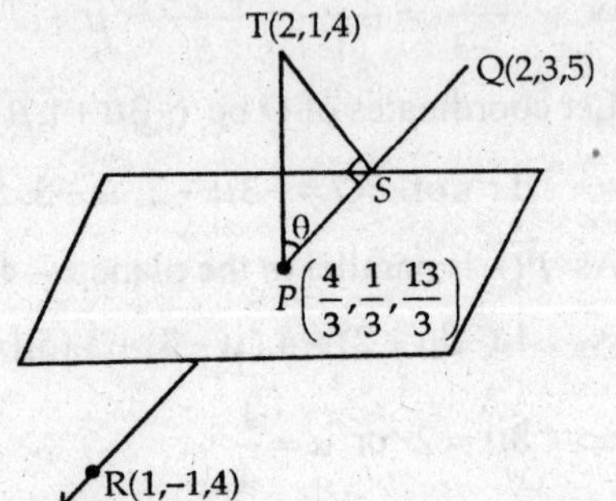

ALTERNATIVE

As in previous method P is $\left(\frac{4}{3},\frac{1}{3},\frac{13}{3}\right)$

$\therefore$ dr's of PT are $2, 2, -1$

Also dr's of QR are $1, 4, 1$

$\therefore$ If θ be the angle between PT and QR, then

$$\cos\theta=\frac{2\times 1+2\times 4-1\times 1}{\sqrt{4+4+1}\sqrt{1+16+1}}=\frac{1}{\sqrt{2}}\Rightarrow\theta=45°$$

$\Rightarrow PS=TS=PT\cos 45°=\frac{1}{\sqrt{2}}\qquad(\because PT=1)$

45. (a) The plane passing through the intersection line of given planes is

$(x+2y+3z-2)+\lambda(x-y+z-3)=0$

or $\quad(1+\lambda)x+(2-\lambda)y+(3+\lambda)z+(-2-3\lambda)=0$

Its distance from the point (3, 1, –1) is $\frac{2}{\sqrt{3}}$

$$\therefore\ \left|\frac{3(1+\lambda)+1(2-\lambda)-1(3+\lambda)+(-2-3\lambda)}{\sqrt{(1+\lambda)^2+(2-\lambda)^2+(3+\lambda)^2}}\right|=\frac{2}{\sqrt{3}}$$

$$\Rightarrow\left|\frac{-2\lambda}{\sqrt{3\lambda^2+4\lambda+14}}\right|=\frac{2}{\sqrt{3}}$$

$\Rightarrow 3\lambda^2+4\lambda+14=3\lambda^2\ \Rightarrow\lambda=-\frac{7}{2}$

$\therefore$ Required equation of plane is

$(x+2y+3z-2)-\frac{7}{2}\ (x-y+z-3)=0$

or $\quad 5x-11y+z=17$

46. **(c)** Given that $\vec{a}\times(2\hat{i}+3\hat{j}+4\hat{k})=(2\hat{i}+3\hat{j}+4\hat{k})\times\vec{b}$

$\Rightarrow (\vec{a}+\vec{b})\times(2\hat{i}+3\hat{j}+4\hat{k})=\vec{0}$

But neither $\vec{a}+\vec{b}$ nor $2\hat{i}+3\hat{j}+4\hat{k}$ is a null vector

$\therefore (\vec{a}+\vec{b}) \,||\, (2\hat{i}+3\hat{j}+4\hat{k}) \Rightarrow \vec{a}+\vec{b}=\lambda(2\hat{i}+3\hat{j}+4\hat{k})$

Also given $|\vec{a}+\vec{b}|=\sqrt{29} \Rightarrow \lambda=\pm 1$

$\therefore \vec{a}+\vec{b}=\pm(2\hat{i}+3\hat{j}+4\hat{k})$

$\therefore (\vec{a}+\vec{b}).(-7\hat{i}+2\hat{i}+3\hat{k})=\pm 4$

D MCQ's WITH ONE OR MORE THAN ONE CORRECT :

1. **(c)** We are given that, $\vec{a}=a_1\hat{i}+a_2\hat{j}+a_3\hat{k}$

$\vec{b}=b_1\hat{i}+b_2\hat{j}+b_3\hat{k}$

$\vec{c}=c_1\hat{i}+c_2\hat{j}+c_3\hat{k}$

Then $\begin{vmatrix} a_1 & a_2 & a_3 \\ b_1 & b_2 & b_3 \\ c_1 & c_2 & c_3 \end{vmatrix}^2=[\vec{a}\vec{b}\vec{c}]^2=(\vec{a}\times\vec{b}.\vec{c})^2$

$=(|\vec{a}\times\vec{b}|.1\cos 0^\circ)^2$ $\left[\begin{array}{l}\vec{c}\text{ is a unit vector, } \therefore |\vec{c}|=1 \\ \text{Also } \vec{c} \text{ is } \perp \text{ to } \vec{a} \text{ as well} \\ \text{as to } \vec{b}, \therefore \vec{c}\perp(\vec{a}\times\vec{b})\end{array}\right.$

$=(|\vec{a}\times\vec{b}|)^2$

$=\left(|\vec{a}||\vec{b}|.\sin\frac{\pi}{6}\right)^2$

[$\because$ angle between $\vec{a}$ and $\vec{b}$ is $\frac{\pi}{6}$]

$=\left(\frac{1}{2}\sqrt{a_1^2+a_2^2+a_3^2}\sqrt{b_1^2+b_2^2+b_3^2}\right)^2$

$=\frac{1}{4}\left(a_1^2+a_2^2+a_3^2\right)\left(b_1^2+b_2^2+b_3^2\right)$

2. **(b)** Let (x, y, z) be the unit vector perpendicular to $\vec{a}=(1,1,0)$ and $\vec{b}=(0,1,1)$

then $x+y=0$ ...(1)

$y+z=0$ ...(2)

and $x^2+y^2+z^2=1$ ...(3)

From eq. (1) $x=-y$

From eq. (2) $z=-y$

Substituting these values in eq. (3), we have

$y^2+y^2+y^2=1$

$\Rightarrow 3y^2=1 \Rightarrow y=\pm\frac{1}{\sqrt{3}}$

As there are two possible values of y, there are two values of required vectors.

ALTERNATE SOLUTION

We know that if $\hat{n}$ is $\perp$ to $\vec{a}$ as well as $\vec{b}$ then

$\hat{n}=\frac{\vec{a}\times\vec{b}}{|\vec{a}\times\vec{b}|}$ or $\frac{\vec{b}\times\vec{a}}{|\vec{b}\times\vec{a}|}$

as $\vec{a}\times\vec{b}$ and $\vec{b}\times\vec{a}$ represent two vectors in opp. directions.

$\therefore$ We have two possible values of $\hat{n}$

3. **(a, c)** We have

$a=2\hat{i}-\hat{j}+\hat{k}, b=\hat{i}+2\hat{j}-\hat{k}, c=\hat{i}+\hat{j}-2\hat{k}$

Any vector in the plane of $\vec{b}$ and $\vec{c}$ is

$\vec{u}=\vec{b}+\lambda\vec{c}$

$=(\hat{i}+2\hat{j}-\hat{k})+\lambda(\hat{i}+\hat{j}-2\hat{k})$

$=(1+\lambda)\hat{i}+(2+\lambda)\hat{j}-(1+2\lambda)\hat{k}$

ATQ, magnitude of projection of $\vec{u}$ on $\vec{a}$ is $\sqrt{\frac{2}{3}}$

$\Rightarrow \sqrt{\frac{2}{3}}=\left|\frac{\vec{u}.\vec{a}}{|\vec{a}|}\right| \Rightarrow \sqrt{\frac{2}{3}}=\left|\frac{2(1+\lambda)-(2+\lambda)-(1+2\lambda)}{\sqrt{6}}\right|$

$\Rightarrow |-\lambda-1|=2 \Rightarrow \lambda+1=2$ or $\lambda+1=-2$

$\Rightarrow \lambda=1$ or $\lambda=-3$

$\therefore$ The required vector is either,

$2\hat{i}+3\hat{j}-3\hat{k}$ or $-2\hat{i}-\hat{j}+5\hat{k}$

4. **(a, c, d)** $|\vec{a}|^2=\frac{1}{9}(4+4+1)=1 \Rightarrow |\vec{a}|=1$

Let $\vec{b}=2\hat{i}-4\hat{j}+\hat{3}$ then

$\cos\theta=\frac{\vec{a}.\vec{b}}{|\vec{a}||\vec{b}|}=\frac{5}{\sqrt{29}} \Rightarrow \theta\neq\frac{\pi}{3}$

Let $\vec{c}=-\hat{i}+\hat{j}-\frac{1}{2}\hat{k}=\frac{-3}{2}\hat{a} \Rightarrow \vec{c}\,||\,\vec{a}$

Let $\vec{d}=3\hat{i}+2\hat{j}+2\hat{k}$ then $\vec{a}.\vec{d}=0 \Rightarrow \vec{a}\perp\vec{d}$

5. **(d)** Given that, $\vec{a}=\hat{i}+\hat{j}+\hat{k}, \vec{b}=4\hat{i}+3\hat{j}+4\hat{k}$ and $\vec{c}=\hat{i}+\alpha\hat{j}+\beta\hat{k}$ are linearly dependent,

NOTE THIS STEP :

$\therefore$ $\vec{c}=l\vec{a}+m\vec{b}$ for some scalars l and m not all zeros.

$\hat{i}+\alpha\hat{j}+\beta\hat{k}=(l+4m)\hat{i}+(l+3m)\hat{j}+(l+4m)\hat{k}$

$\Rightarrow l+4m=1$ ...(1)

$l+3m=\alpha$ ...(2)

$l+4m=\beta$ ...(3)

From (1) and (3) we have, $\beta=1$

Also given that $|\vec{c}|=\sqrt{3} \Rightarrow 1+\alpha^2+\beta^2=3$

Substituting the value of β we get $\alpha^2=1$

$\Rightarrow \alpha=\pm 1$

6. (c) $[\vec{u}\,\vec{v}\,\vec{w}] = [\vec{v}\,\vec{w}\,\vec{u}] = [\vec{w}\,\vec{u}\,\vec{v}]$
I II IV

but $[\vec{v}\,\vec{u}\,\vec{w}] = -[\vec{u}\,\vec{v}\,\vec{w}]$
III

7. (a, c) As dot product of two vectors gives a scalar quantity.

8. (a, c) We have

$\vec{v} = \vec{a}\times\vec{b} = |\vec{a}||\vec{b}|\sin\theta\hat{n} = \sin\theta\hat{n}$

[$\because \vec{a}$ and $\vec{b}$ are unit vectors.]

$\therefore \quad |\vec{v}| = \sin\theta$

Now, $\vec{u} = \vec{a} - (\vec{a}.\vec{b})\vec{b}$

$= \vec{a} - \vec{b}\cos\theta \qquad$ (where $\vec{a}.\vec{b} = \cos\theta$)

$\therefore \quad |\vec{u}|^2 = |\vec{a} - \vec{b}\cos\theta|^2$

$= 1 + \cos^2\theta - 2\cos\theta.\cos\theta$

$= 1 - \cos^2\theta = \sin^2\theta = |v|^2 \Rightarrow |\vec{u}| = |\vec{v}|$

Also, $\vec{u}.\vec{b} = \vec{a}.\vec{b} - (\vec{a}.\vec{b})(\vec{b}.\vec{b}) = \vec{a}.\vec{b} - \vec{a}.\vec{b} = 0 \therefore |\vec{u}.\vec{b}| = 0$

$\therefore \quad |\vec{v}| = |\vec{u}| + |\vec{u}.\vec{b}|$ is also correct

9. (b, d) Normal to plane P_1 is

$\vec{n}_1 = (2\hat{i} + 3\hat{k})\times(4\hat{j} - 3\hat{k}) = -18\hat{i}$

Normal to plane P_2 is

$\vec{n}_2 = (\hat{j} - \hat{k})\times(3\hat{i} + 3\hat{j}) = 3\hat{i} - 3\hat{j} - 3\hat{k}$

$\therefore \quad \vec{A}$ is parallel to $\pm(\vec{n}_1\times\vec{n}_2) = \pm(-54\hat{j} + 54\hat{k})$

Now, angle between $\vec{A}$ and $2\hat{i} + \hat{j} - 2\hat{k}$ is given by

$$\cos\theta = \pm\frac{(-54\hat{j} + 54\hat{k}).(2\hat{i} + \hat{j} - 2\hat{k})}{54\sqrt{2}.3} = \pm\frac{1}{\sqrt{2}}$$

$$\theta = \frac{\pi}{4} \text{ or } \frac{3\pi}{4}$$

10. (a,d) Let $\vec{a} = \hat{i} + \hat{j} + 2\hat{k},\ \vec{b} = \hat{i} + 2\hat{j} + \hat{k},\ \vec{c} = \hat{i} + \hat{j} + \hat{k}$

$\because$ Required vector is coplanar with $\vec{a}$ and $\vec{b}$

$\therefore \vec{r} = \lambda\vec{a} + \mu\vec{b}$

or $\vec{r} = (\lambda + \mu)\,\hat{i} + (\lambda + 2\mu)\,\hat{j} + (2\lambda + \mu)\,\hat{k}$

As $\vec{r}\perp\vec{c} \Rightarrow \vec{r}.\vec{c} = 0$

$\Rightarrow \lambda + \mu + \lambda + 2\mu + 2\lambda + \mu = 0 \Rightarrow \lambda + \mu = 0 \Rightarrow \lambda = -\mu$

$\therefore \vec{r} = \mu(\hat{j} - \hat{k})$

For $\mu = 1$, we get $\vec{r} = \hat{j} - \hat{k}$

and for $\mu = -1$, we get $\vec{r} = -\hat{j} + \hat{k}$

$\therefore$ a and d are the correct options.

11. (b, c) For given lines to be coplanar, we should have

$$\begin{vmatrix} 2 & 0 & 0 \\ 2 & k & 2 \\ 5 & 2 & k \end{vmatrix} = 0 \Rightarrow k = \pm 2$$

For $k = 2$, obviously the plane $y + 1 = z$ is common in both lines.

For $k = -2$, the plane is given by

$$\begin{vmatrix} x-1 & y+1 & z \\ 2 & -2 & 2 \\ 5 & 2 & -2 \end{vmatrix} = 0 \Rightarrow y + z + 1 = 0$$

12. (b, d) The given lines are

$$\ell_1 : \frac{x-3}{1} = \frac{y+1}{2} = \frac{z-4}{2} = t$$

$$\ell_2 : \frac{x-3}{2} = \frac{y-3}{2} = \frac{z-2}{1} = s$$

Let direction ratios of ℓ be a, b, c then as $\ell\perp\ell_1$ and ℓ_2

$\therefore \quad a + 2b + 2c = 0$
$2a + 2b + c = 0$

$$\Rightarrow \quad \frac{a}{-2} = \frac{b}{3} = \frac{c}{-2}$$

$$\therefore \quad \ell : \frac{x}{2} = \frac{y}{-3} = \frac{z}{2} = \lambda$$

Any point on ℓ_1 is $(t + 3, 2t - 1, 2t + 4)$ and any point on ℓ is $(2\lambda, -3\lambda, 2\lambda)$

$\therefore$ For Intersection point P of ℓ and ℓ_1

$t + 3 = 2\lambda,\ 2t - 1 = -3\lambda,\ 2t + 4 = 2\lambda$

$\Rightarrow \quad t = -1,\ \lambda = 1 \qquad \therefore P(2, -3, 2)$

Any point Q on ℓ_2 is $(2s + 3, 2s + 3, s + 2)$

As per question $PQ = \sqrt{17}$

$\Rightarrow \quad (2s + 1)^2 + (2s + 6)^2 + s^2 = 17$

$\Rightarrow \quad 9s^2 + 28s + 20 = 0$

$$\Rightarrow \quad s = -2, \frac{-10}{9}$$

$\therefore$ Point Q can be $(-1, -1, 0)$ and $\left(\frac{7}{9}, \frac{7}{9}, \frac{8}{9}\right)$

13. (a, d) $L_1 : \frac{x-5}{0} = \frac{y}{3-\alpha} = \frac{z}{-2}$

$$L_2 : \frac{x-\alpha}{0} = \frac{y}{-1} = \frac{z}{2-\alpha}$$

As L_1, L_2 are coplanar, therefore

$$\begin{vmatrix} 5-\alpha & 0 & 0 \\ 0 & 3-\alpha & -2 \\ 0 & -1 & 2-\alpha \end{vmatrix} = 0$$

$\Rightarrow (5-\alpha)[6 - 5\alpha + \alpha^2 - 2] = 0$

$\Rightarrow (5-\alpha)(\alpha - 1)(\alpha - 4) = 0 \Rightarrow \alpha = 1, 4, 5.$

E SUBJECTIVE PROBLEMS :

1. Let with respect to O, position vectors of points A, B, C, D, E, F be $\vec{a}, \vec{b}, \vec{c}, \vec{d}, \vec{e}, \vec{f}$.

Let perpendiculars from A to EF and from B to DF meet each other at H. Let position vector of H be $\vec{r}$. we join CH. In order to prove the statement given in question, it is sufficient to prove that CH is perpendicular to DE.

Now, as $OD \perp BC \Rightarrow \vec{d}.(\vec{b} - \vec{c}) = 0$

$\Rightarrow \vec{d}.\vec{b} = \vec{d}.\vec{c}$ ...(1)

As $OE \perp AC \Rightarrow \vec{e}.(\vec{c}-\vec{a}) = 0 \Rightarrow \vec{e}.\vec{c} = \vec{e}.\vec{a}$ ...(2)

As $OF \perp AB \Rightarrow \vec{f}.(\vec{a}-\vec{b}) = 0 \Rightarrow \vec{f}.\vec{a} = \vec{f}.\vec{b}$ ...(3)

Also $AH \perp EF \Rightarrow (\vec{r}-\vec{a}).(\vec{e}-\vec{f}) = 0$

$\Rightarrow \vec{r}.\vec{e} - \vec{r}.\vec{f} - \vec{a}.\vec{e} + \vec{a}.\vec{f} = 0$ ...(4)

and $BH \perp FD \Rightarrow (\vec{r}-\vec{b}).(\vec{f}-\vec{d}) = 0$

$\Rightarrow \vec{r}.\vec{f} - \vec{r}.\vec{d} - \vec{b}.\vec{f} + \vec{b}.\vec{d} = 0$ ...(5)

Adding (4) and (5), we get

$\vec{r}.\vec{e} - \vec{a}.\vec{e} + \vec{a}.\vec{f} - \vec{r}.\vec{d} - \vec{b}.\vec{f} + \vec{b}.\vec{d} = 0$

$\Rightarrow \vec{r}.(\vec{e}-\vec{d}) - \vec{e}.\vec{c} + \vec{d}.\vec{c} = 0$

(using (1), (2) and (3))

$\Rightarrow \vec{r}.(\vec{e}-\vec{d}) - \vec{c}.(\vec{e}-\vec{d}) = 0 \Rightarrow (\vec{r}-\vec{c}).(\vec{e}-\vec{d}) = 0$

$\Rightarrow \overrightarrow{CH}.\overrightarrow{ED} = 0 \Rightarrow CH \perp ED$ Hence Proved.

2. $\overrightarrow{OA_1}, \overrightarrow{OA_2}, ..., \overrightarrow{OA_n}$ all vectors are of same magnitude, say 'a' and angle between any two consecutive vector is same that is $\frac{2\pi}{n}$ radians. Let $\hat{p}$ be the unit vectors $\perp$ to the plane of the polygon.

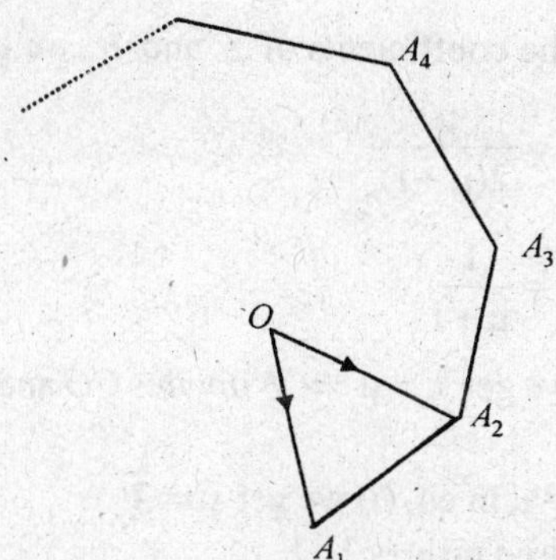

$\therefore \overrightarrow{OA_1} \times \overrightarrow{OA_2} = a^2 \sin\frac{2\pi}{n}\hat{p}$ ...(i)

Now, $\sum_{i=1}^{n-1} \overrightarrow{OA_i} \times \overrightarrow{OA_{i+1}} = \sum_{i=1}^{n-1} a^2 \sin\frac{2\pi}{n}\hat{p}$

$= (n-1)a^2 \sin\frac{2\pi}{n}\hat{p} \quad = -(n-1)[\overrightarrow{OA_2} \times \overrightarrow{OA_1}]$

[using eq^n. Czx(1)]

$= (1-n)[\overrightarrow{OA_2} \times \overrightarrow{OA_1}] = R.H.S$

3. $(\hat{i}+\hat{j}+3\hat{k})x + (3\hat{i}-3\hat{j}+\hat{k})y + (-4\hat{i}+5\hat{j})z$

$= \lambda(x\hat{i}+y\hat{j}+z\hat{k})$

$\Rightarrow \quad x+3y-4z = \lambda x \Rightarrow (1-\lambda)x+3y-4z = 0$

$x-3y+5z = \lambda y$

$x-(3+\lambda)y+5z = 0$

$3x+y+0z = \lambda z$

$3x+y-\lambda z = 0$

All the above three equations are satisfied for x, y, z not all zero if

$$\begin{vmatrix} 1-\lambda & 3 & -4 \\ 1 & -(3+\lambda) & 5 \\ 3 & 1 & -\lambda \end{vmatrix} = 0$$

$\Rightarrow \quad (1-\lambda)[3\lambda+\lambda^2-5] - 3[-\lambda-15] - 4[1+9+3\lambda] = 0$

$\Rightarrow \quad \lambda^3 + 2\lambda^2 + \lambda = 0 \Rightarrow \lambda(\lambda+1)^2 = 0 \Rightarrow \lambda = 0, -1.$

4. We are given that

$\vec{A} \perp \vec{B} \Rightarrow \vec{A}.\vec{B} = 0$ ...(1)

Also $\vec{A} \times \vec{X} = \vec{B} \Rightarrow \vec{X} \perp \vec{B} \Rightarrow \vec{X}.\vec{B} = 0$ ...(2)

Now, consider $[\vec{X}\ \vec{A}\ \vec{A}\times\vec{B}]$

$= \vec{X}.\vec{A}\times(\vec{A}\times\vec{B}) = \vec{X}.[(\vec{A}.\vec{B})\vec{A} - (\vec{A}.\vec{A})\vec{B}]$

NOTE THIS STEP :

$= (\vec{A}.\vec{A})(\vec{X}.\vec{B})$

$= 0$ [Using (1) and (2)]

$\Rightarrow \quad \vec{X}, \vec{A}, \vec{A}\times\vec{B}$ are coplanar.

So $\vec{X}$ can be represented as a linear combination of $\vec{A}$ and $\vec{A}\times\vec{B}$.

Let us define $\vec{X} = l\vec{A} + m(\vec{A}\times\vec{B})$ where l and m are scalar quantities.

Now, $\quad \vec{A}.\vec{X} = C \Rightarrow \vec{A}.(l\vec{A} + m(\vec{A}\times\vec{B})) = C$

$\Rightarrow \quad l\vec{A}.\vec{A} + m\vec{A}.\vec{A}\times\vec{B} = C \Rightarrow l|\vec{A}|^2 + m(0) = C$

$\Rightarrow \quad l = \frac{C}{|\vec{A}|^2}$

Also, $\vec{A}\times\vec{X} = \vec{B}$

$\Rightarrow \quad \vec{A}\times(l\vec{A} + m\vec{A}\times\vec{B}) = \vec{B}$

$\Rightarrow \quad l(\vec{A}\times\vec{A}) + m\vec{A}\times(\vec{A}\times\vec{B}) = \vec{B}$

$\Rightarrow \quad 0 - m|\vec{A}|^2\vec{B} = \vec{B} \Rightarrow m = -\frac{1}{|\vec{A}|^2}$

$\therefore \quad \vec{X} = \frac{C}{|\vec{A}|^2}\vec{A} - \frac{1}{|\vec{A}|^2}(\vec{A}\times\vec{B})$

5. Since vector $\vec{A}$ has components A_1, A_2, A_3, in the coordinate system $OXYZ$,

$\therefore \quad \vec{A} = \hat{i}A_1 + \hat{j}A_2 + \hat{k}A_3$

When given system is rotated through $\frac{\pi}{2}$. the new x-axis is along old y-axis and new y-axis is along the old negative x-axis z remains same as before.

Hence the components of A in the new system are $A_2, -A_1, A_3$.

$\therefore \vec{A}$ becomes $A_2\hat{i} - A_1\hat{j} + A_3\hat{k}$.

6. Then $\overrightarrow{AB} = -\hat{i} - 5\hat{j} - 3\hat{k}$, $\overrightarrow{AC} = -4\hat{i} + 3\hat{j} + 3\hat{k}$

$\overrightarrow{AD} = \hat{i} + 7\hat{j} + (1-\lambda)\hat{k}$

We know that A, B, C, D lie in a plane if $\overrightarrow{AB}, \overrightarrow{AC}, \overrightarrow{AD}$ are coplanar i.e. $\left[\overrightarrow{AB}\,\overrightarrow{AC}\,\overrightarrow{AD}\right] = 0$

$$\Rightarrow \begin{vmatrix} -1 & 5 & -3 \\ -4 & 3 & 3 \\ 1 & 7 & 1-\lambda \end{vmatrix} = 0$$

$$\Rightarrow \quad -1(3-3\lambda-21) - 5(-4+4\lambda-3) - 3(-28-3) = 0$$

$$\Rightarrow \quad 3\lambda + 18 - 20\lambda + 35 + 93 = 0 \Rightarrow 17\lambda = 146 \Rightarrow \lambda = \frac{146}{17}$$

7. Let the position vectors of points A, B, C, D be a, b, c, and d respectively with respect to some origin O.

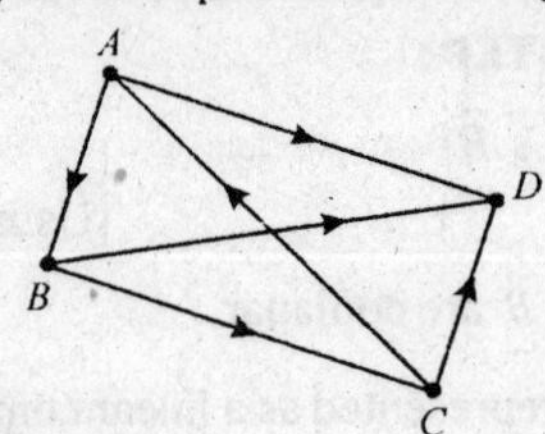

Then, $\overrightarrow{AB} = \vec{b} - \vec{a}$, $\quad \overrightarrow{AD} = \vec{d} - \vec{a}$,

$\overrightarrow{BC} = \vec{c} - \vec{b}$, $\quad \overrightarrow{BD} = \vec{d} - \vec{b}$,

$\overrightarrow{CD} = \vec{d} - \vec{c}$, $\quad \overrightarrow{CA} = \vec{a} - \vec{c}$

Now, $|\overrightarrow{AB} \times \overrightarrow{CD} + \overrightarrow{BC} \times \overrightarrow{AD} + \overrightarrow{CA} \times \overrightarrow{BD}|$

$= |(\vec{b}-\vec{a}) \times (\vec{d}-\vec{c}) + (\vec{c}-\vec{b}) \times (\vec{d}-\vec{a}) + (\vec{a}-\vec{c}) \times (\vec{d}-\vec{b})|$

$= |\vec{b} \times \vec{d} - \vec{a} \times \vec{d} - \vec{b} \times \vec{c} + \vec{a} \times \vec{c} + \vec{c} \times \vec{d} - \vec{c} \times \vec{a}$

$\quad -\vec{b} \times \vec{d} + \vec{b} \times \vec{a} + \vec{a} \times \vec{d} - \vec{a} \times \vec{b} - \vec{c} \times \vec{d} + \vec{c} \times \vec{b}|$

$= |-\vec{b} \times \vec{c} + \vec{a} \times \vec{c} - \vec{c} \times \vec{a} + \vec{b} \times \vec{a} - \vec{a} \times \vec{b} + \vec{c} \times \vec{b}|$

$= 2|\vec{b} \times \vec{a} + \vec{c} \times \vec{b} + \vec{a} \times \vec{c}|$ ...(1)

Also Area of ΔABC is

$= \frac{1}{2}|\overrightarrow{BC} \times \overrightarrow{BA}| = \frac{1}{2}|(\vec{c}-\vec{b}) \times (\vec{a}-\vec{b})|$

$= \frac{1}{2}|(\vec{c} \times \vec{a} - \vec{c} \times \vec{b} - \vec{b} \times \vec{a} + \vec{b} \times \vec{b}|$

$= \frac{1}{2}|-\vec{b} \times \vec{a} - \vec{c} \times \vec{b} - \vec{a} \times \vec{c}| = \frac{1}{2}|\vec{b} \times \vec{a} + \vec{c} \times \vec{b} + \vec{a} \times \vec{c}|$

$\Rightarrow \quad 2Ar(\Delta ABC) = |\vec{b} \times \vec{a} + \vec{c} \times \vec{b} + \vec{a} \times \vec{c}|$ ...(2)

From (1) and (2), we get

$|\overrightarrow{AB} \times \overrightarrow{CD} + \overrightarrow{BC} \times \overrightarrow{AD} + \overrightarrow{CA} \times \overrightarrow{BD}|$

$= 2(2Ar(\Delta ABC))$

$= 4Ar(\Delta ABC)$ Hence Proved.

8. $OACB$ is a parallelogram with O as origin. Let with respect to O position vectors of A and B be $\vec{a}$ and $\vec{b}$ respectively. Then p.v. of C is $\vec{a} + \vec{b}$.

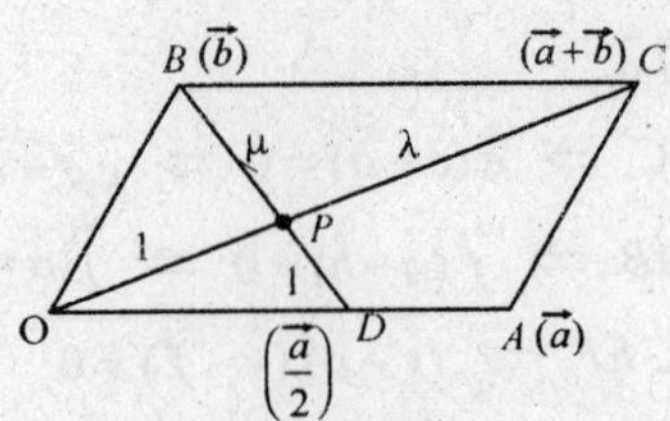

Also D is mid pt. of OA, therefore position vector of D is $\frac{\vec{a}}{2}$.

CO and BD intersect each other at P.

Let P divides CO in the ratio $\lambda : 1$ and BD in the ratio $\mu : 1$

Then by section theorem,

position vector of pt. P dividing CO in ratio

$$\lambda : 1 \quad = \frac{\lambda \times 0 + 1 \times (\vec{a}+\vec{b})}{\lambda + 1} = \frac{(\vec{a}+\vec{b})}{\lambda+1} \qquad ...(1)$$

And position vector of pt. P dividing BD in the ratio $\mu : 1$ is

$$= \frac{\mu\left(\frac{\vec{a}}{2}\right) + 1(\vec{b})}{\mu + 1} = \frac{\mu\vec{a} + 2\vec{b}}{2(\mu+1)} \qquad ...(2)$$

As (1) and (2) represent the position vector of same point, we should have

$$\frac{\vec{a}+\vec{b}}{\lambda+1} = \frac{\mu\vec{a} + 2\vec{b}}{2(\mu+1)}$$

Equating the coefficients of $\vec{a}$ and $\vec{b}$, we get

$$\frac{1}{\lambda+1} = \frac{\mu}{2(\mu+1)} \qquad ...(i)$$

$$\frac{1}{\lambda+1} = \frac{1}{\mu+1} \qquad ...(ii)$$

From (ii) we get $\lambda = \mu \Rightarrow P$ divides CO and BD in the same ratio.

Putting $\lambda = \mu$ in eq. (i) we get $\mu = 2$

Thus required ratio is 2 : 1.

9. Given that $\vec{a}, \vec{b}, \vec{c}$ are three coplanar vectors.

$\therefore$ There exists scalars x, y, z, not all zero, such that

$$x\vec{a} + y\vec{b} + z\vec{c} = \vec{0} \qquad ...(1)$$

Taking dot product of $\vec{a}$ and (1), we get

$$x\vec{a}.\vec{a} + y\vec{a}.\vec{b} + z\,\vec{a}.\vec{c} = \vec{0} \qquad ...(2)$$

Again taking dot product of $\vec{b}$ and (1), we get

$$x\vec{b}.\vec{a} + y\vec{b}.\vec{b} + z\,\vec{b}.\vec{c} = \vec{0} \qquad ...(3)$$

Now equations (1), (2), (3) form a homogeneous system of equations, where x, y, z are not all zero.

$\therefore$ system must have non trivial solution and for this, determinant of coefficient matrix should be zero

i.e. $\begin{vmatrix} \vec{a} & \vec{b} & \vec{c} \\ \vec{a}.\vec{a} & \vec{a}.\vec{b} & \vec{a}.\vec{c} \\ \vec{b}.\vec{a} & \vec{b}.\vec{b} & \vec{b}.\vec{c} \end{vmatrix} = 0$ Hence Proved.

10. With O as origin let $\vec{a}$ and $\vec{b}$ be the position vectors of A and B respectively.

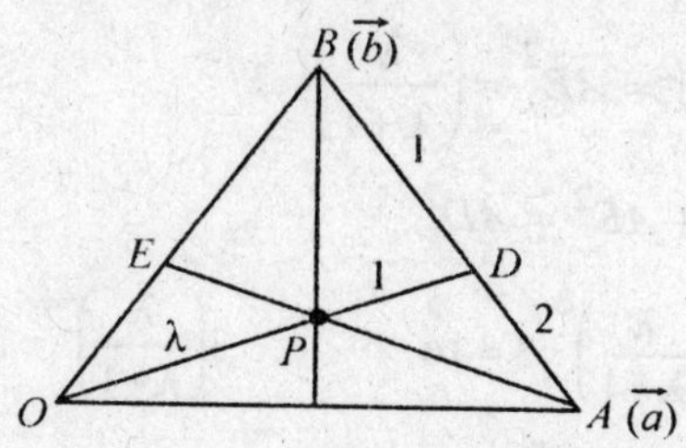

Then the position vector of E, the mid point of OB is $\frac{\vec{b}}{2}$.

Again since $AD : DB = 2 : 1$, the position vector of D is

$$\frac{1.\vec{a}+2\vec{b}}{1+2} = \frac{\vec{a}+2\vec{b}}{3}$$

∴ Equation of OD is

$$\vec{r} = t\left(\frac{\vec{a}+2\vec{b}}{3}\right) \quad ...(1)$$

and Equation of AE is

$$\vec{r} = \vec{a} + s\left(\frac{\vec{b}}{2} - \vec{a}\right) \quad ...(2)$$

If OD and AE intersect at P, then we will have identical values of $\vec{r}$. Hence comparing the coefficients of $\vec{a}$ and $\vec{b}$, we get

$$\frac{t}{3} = 1 - s \text{ and } \frac{2t}{3} = \frac{s}{2} \Rightarrow t = \frac{3}{5} \text{ and } s = \frac{4}{5}$$

Putting value of t in eq. (1) we get position vector of point of intersection P as

$$\frac{\vec{a}+2\vec{b}}{5} \quad ...(3)$$

Now if P divides OD in the ratio $\lambda : 1$, then p.v. of P is

$$\frac{\lambda\left(\frac{\vec{a}+2\vec{b}}{3}\right)+1.0}{\lambda+1} = \frac{\lambda}{3(\lambda+1)}(\vec{a}+2\vec{b}) \quad ...(4)$$

From (3) and (4) we get

$$\frac{\lambda}{3(\lambda+1)} = \frac{1}{5} \Rightarrow 5\lambda = 3\lambda + 3 \Rightarrow \lambda = 3/2$$

∴ $OP : PD = 3 : 2$

11. We are given that $\vec{A} = 2\hat{i} + \hat{k}, \vec{B} = \hat{i} + \hat{j} + \hat{k}$ and $\vec{C} = 4\hat{i} - 3\hat{j} + 7\hat{k}$ and to determine a vector $\vec{R}$ such that $\vec{R}\times\vec{B} = \vec{C}\times\vec{B}$ and $\vec{R}.\vec{A} = 0$

Let $\vec{R} = x\hat{i} + y\hat{j} + z\hat{k}$

Then $\vec{R}\times\vec{B} = \vec{C}\times\vec{B}$

$$\Rightarrow \begin{vmatrix} \hat{i} & \hat{j} & \hat{k} \\ x & y & z \\ 1 & 1 & 1 \end{vmatrix} = \begin{vmatrix} \hat{i} & \hat{j} & \hat{k} \\ 4 & -3 & 7 \\ 1 & 1 & 1 \end{vmatrix}$$

$$\Rightarrow (y-z)\hat{i} - (x-z)\hat{j} + (x-y)\hat{k} = 10\hat{i} - 11\hat{j} + 7\hat{k}$$

$\Rightarrow y - z = -10 \quad ...(1)$

$z - x = -11 \quad ...(2)$

$x - y = 7 \quad ...(3)$

Also $\vec{R}.\vec{A} = 0$

$\Rightarrow 2x + z = 0 \quad ...(4)$

Substituting $y = x - 7$ and $z = -2x$ from (3) and (4) respectively in eq. (1) we get

$$x - 7 + 2x = -10 \Rightarrow 3x = -3$$

$\Rightarrow x = -1,\ y = -8$ and $z = 2$

∴ $\vec{R} = -\hat{i} - 8\hat{j} + 2\hat{k}$

12. We have, $\vec{a} = cx\hat{i} - 6\hat{j} + 3\hat{k},\ \vec{b} = x\hat{i} - 2\hat{j} + 2cx\hat{k}$

Now we know that $\vec{a}.\vec{b} = |\vec{a}||\vec{b}|\cos\theta$

As angle between $\vec{a}$ and $\vec{b}$ is obtuse, therefore

$\cos\theta < 0 \Rightarrow \vec{a}.\vec{b} < 0$

$\Rightarrow cx^2 - 12 + 6cx < 0 \Rightarrow -cx^2 - 6cx + 12 > 0, \forall x \in R$

$\Rightarrow -c > 0$ and $D < 0 \Rightarrow c < 0$ and $36c^2 + 48c < 0$

$\Rightarrow c < 0$ and $c(3c+4) < 0 \Rightarrow c < 0$ and $(3c+4) > 0$

$\Rightarrow c < 0$ and $c > -4/3 \Rightarrow -4/3 < c < 0$

13. Let $\vec{a}, \vec{b}, \vec{c}$, be the position vectors of pt A, B and C respectively with respect to some origin.

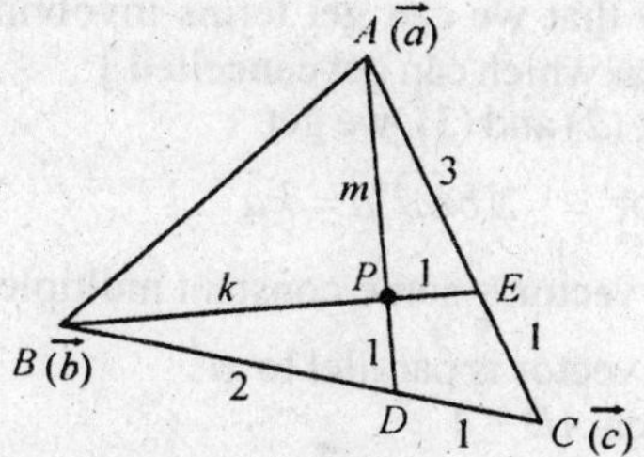

ATQ, D divides BC in the ratio 2 : 1 and E divides AC in the ratio 3 : 1.

∴ position vector of D is $\frac{\vec{b}+2\vec{c}}{3}$ and position vector of E is $\frac{\vec{a}+3\vec{c}}{4}$

Let pt. of intersection P of AD and BE divides BE in the ratio $k : 1$ and AD in the ratio $m : 1$, then position vectors of P in these two cases are

$$\frac{\vec{b}+k\left(\frac{\vec{a}+3\vec{c}}{4}\right)}{k+1} \text{ and } \frac{\vec{a}+m\left(\frac{\vec{b}+2\vec{c}}{3}\right)}{m+1} \text{ respectively.}$$

Equating the position vectors of P in two cases we get

$$\frac{k}{4(k+1)}\vec{a} + \frac{1}{k+1}\vec{b} + \frac{3k}{4(k+1)}\vec{c}$$

$$= \frac{1}{m+1}\vec{a} + \frac{m}{3(m+1)}\vec{b} + \frac{2m}{3(m+1)}\vec{c}$$

$$\Rightarrow \frac{k}{4(k+1)} = \frac{1}{m+1} \quad ...(1)$$

$$\frac{1}{k+1} = \frac{m}{3(m+1)} \quad ...(2)$$

$$\frac{3k}{4(k+1)} = \frac{2m}{3(m+1)} \quad \ldots(3)$$

Dividing (3) by (2) we get

$\frac{3k}{4} = 2 \Rightarrow k = \frac{8}{3} \Rightarrow$ the req. ratio is 8 : 3.

14. Given that $\vec{b}, \vec{c}, \vec{d}$ are not coplanar $\quad \therefore \ [\vec{b}, \vec{c}, \vec{d}] \neq 0$

Consider, $(\vec{a} \times \vec{b}) \times (\vec{c} \times \vec{d}) + (\vec{a} \times \vec{c}) \times (\vec{d} \times \vec{b})$
$+ (\vec{a} \times \vec{d}) \times (\vec{b} \times \vec{c})$

Here, $(\vec{a} \times \vec{b}) \times (\vec{c} \times \vec{d}) = -(\vec{c} \times \vec{d}) \times (\vec{a} \times \vec{b})$

$$= -(\vec{c} \times \vec{d}.\vec{b})\vec{a} + (\vec{c} \times \vec{d}.\vec{a})\vec{b}$$

$$= \left[\vec{a}\,\vec{c}\,\vec{d}\right]\vec{b} - \left[\vec{b}\,\vec{c}\,\vec{d}\right]\vec{a} \quad \ldots(1)$$

$(\vec{a} \times \vec{c}) \times (\vec{d} \times \vec{b}) = -(\vec{d} \times \vec{b}) \times (\vec{a} \times \vec{c})$

$$= -(\vec{d} \times \vec{b}.\vec{c})\vec{a} + (\vec{d} \times \vec{b}.\vec{a})\vec{c}$$

$$= [\vec{a}\,\vec{d}\,\vec{b}]\vec{c} - [\vec{c}\,\vec{d}\,\vec{b}]\vec{a} \quad \ldots(2)$$

$(\vec{a} \times \vec{d}) \times (\vec{b} \times \vec{c}) = \left((\vec{a} \times \vec{d})\vec{c}\right)\vec{b} - (\vec{a} \times \vec{d}.\vec{b})\vec{c}$

$$= -[\vec{a}\,\vec{c}\,\vec{d}]\vec{b} - [\vec{a}\,\vec{d}\,\vec{b}]\vec{c} \quad \ldots(3)$$

[NOTE : Here we have tried to write the given expression in such a way that we can get terms involving $\vec{a}$ and other terms similar which can get cancelled.]

Adding (1), (2) and (3), we get

given vector $= -2[\vec{b}\,\vec{c}\,\vec{d}]\vec{a} = k\vec{a}$

$\Rightarrow$ given vector = some constant multiple of $\vec{a}$

$\Rightarrow$ given vector is parallel to $\vec{a}$.

15. We are given $AD = 4$

Volume of tetrahedron $= \frac{2\sqrt{2}}{3}$

$\Rightarrow \frac{1}{3} Ar(\Delta ABC) p = \frac{2\sqrt{2}}{3}$

$\therefore \quad \frac{1}{2} |\overrightarrow{BA} \times \overrightarrow{BC}| \, p = 2\sqrt{2}$

$\frac{1}{2} |(\hat{j} + \hat{k}) \times 2\hat{i}| \, p = 2\sqrt{2}$ or $|\hat{j} - \hat{k}| \, p = 2\sqrt{2}$

or $\quad \sqrt{2} p = 2\sqrt{2} \quad \therefore p = 2$

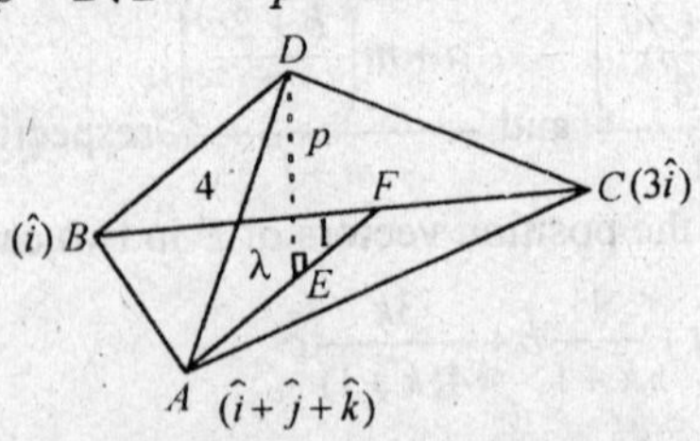

We have to find the P.V. of point E. Let it divides median AF in the ratio $\lambda : 1$

$\therefore \quad$ P. V. of E is $\frac{\lambda.2\hat{i} + (\hat{i} + \hat{j} + \hat{k})}{\lambda + 1} \quad \ldots(2)$

$\therefore \quad \overline{AE} =$ P.V. of $E -$ P.V. of $A = \frac{\lambda}{\lambda+1}(\hat{i} - \hat{j} - \hat{k})$

$\therefore \quad |\overline{AE}|^2 = \overline{AE}^2 = \left(\frac{\lambda}{\lambda+1}\right)^2 .3 \quad \ldots(3)$

Now, $p^2 + AE^2 = AD^2$

or $\quad 4 + \left(\frac{\lambda}{\lambda+1}\right)^2 .3 = 16 \quad \therefore \quad 3\left(\frac{\lambda}{\lambda+1}\right)^2 = 12$

or $\quad \left(\frac{\lambda}{\lambda+1}\right) = \pm 2$

$\lambda = \pm(2\lambda + 2) \qquad \therefore \ \lambda = -2$ or $-2/3$

Putting the value of λ in (2) we get the P.V. of possible positions of E as $(-1, 3, 3)$ or $(3, -1, -1)$

16. Given that $\vec{a}, \vec{b}, \vec{c}$ are three unit vectors inclined at an angle with each other. Also a, b, c are non coplanar, therefore

$[\vec{a}\vec{b}\vec{c}] \neq 0$

Also given that

$$\vec{a} \times \vec{b} + \vec{b} \times \vec{c} = \overrightarrow{pa} + q\vec{b} + r\vec{c}$$

Taking dot product on both sides with $\vec{a}$, we get

$$p + q\cos\theta + r\cos\theta = [\vec{a}\vec{b}\vec{c}] \quad \ldots(1)$$

Similarly, taking dot product on both sides with $\vec{b}$ and $\vec{c}$, we get respectively

$$p\cos\theta + q + r\cos\theta = 0 \quad \ldots(2)$$

and $\quad p\cos\theta + q\cos\theta + r = [\vec{a}\vec{b}\vec{c}] \quad \ldots(3)$

Adding the above three $eq^n s$, we get

$$p + q + r = \frac{2[\vec{a}\vec{b}\vec{c}]}{2\cos\theta + 1} \quad \ldots(4)$$

Multiplying (4) by $\cos\theta$ and subtracting (1), we get

$$p(\cos\theta - 1) = \frac{2[\vec{a}\vec{b}\vec{c}]\cos\theta}{2\cos\theta + 1} - [\vec{a}\vec{b}\vec{c}]$$

or $\quad p(\cos\theta - 1) = \frac{-[\vec{a}\vec{b}\vec{c}]}{2\cos\theta + 1}$

$\Rightarrow \quad p = \frac{[\vec{a}\vec{b}\vec{c}]}{(1 - \cos\theta)(1 + 2\cos\theta)}$

Similarly, $(4) \times \cos\theta - (2)$ gives,

$$q(\cos\theta - 1) = \frac{2[\vec{a}\vec{b}\vec{c}]\cos\theta}{2\cos\theta + 1}$$

$\Rightarrow \quad q = \frac{-2[\vec{a}\vec{b}\vec{c}]\cos\theta}{(1 + 2\cos\theta)(1 - \cos\theta)}$

and $(4) \times \cos\theta - (3)$ gives,

$$r(\cos\theta - 1) = \frac{2[\vec{a}\vec{b}\vec{c}]\cos\theta}{2\cos\theta + 1} - [\vec{a}\vec{b}\vec{c}]$$

$\Rightarrow \quad r = \frac{[\vec{a}\vec{b}\vec{c}]}{(1 - \cos\theta)(1 + 2\cos\theta)}$

But we have to find p, q, r in terms of θ only. So let us find the value of $[\vec{a}\vec{b}\vec{c}]$

We know that

$$[\vec{a}\vec{b}\vec{c}]^2 = \begin{vmatrix} \vec{a}.\vec{a} & \vec{a}.\vec{b} & \vec{a}.\vec{c} \\ \vec{b}.\vec{a} & \vec{b}.\vec{b} & \vec{b}.\vec{c} \\ \vec{c}.\vec{a} & \vec{c}.\vec{b} & \vec{c}.\vec{c} \end{vmatrix} = \begin{vmatrix} 1 & \cos\theta & \cos\theta \\ \cos\theta & 1 & \cos\theta \\ \cos\theta & \cos\theta & 1 \end{vmatrix}$$

On operating $C_1 + C_2 + C_3$ we get

$$\begin{vmatrix} 1+2\cos\theta & \cos\theta & \cos\theta \\ 1+2\cos\theta & 1 & \cos\theta \\ 1+2\cos\theta & \cos\theta & 1 \end{vmatrix} = (1+2\cos\theta)\begin{vmatrix} 1 & \cos\theta & \cos\theta \\ 1 & 1 & \cos\theta \\ 1 & \cos\theta & 1 \end{vmatrix}$$

Operating $R_1 - R_2$ and $R_2 - R_3$, we get

$$= (1+2\cos\theta)\begin{vmatrix} 0 & \cos\theta-1 & 0 \\ 0 & 1-\cos\theta & \cos\theta-1 \\ 1 & \cos\theta & 1 \end{vmatrix}$$

$= (1+2\cos\theta)(1-\cos\theta)^2$ (Expanding along C_1)

$\therefore\ [\vec{a}\vec{b}\vec{c}] = (1-\cos\theta)\sqrt{1+2\cos\theta}$

Thus we get,

$$p = \frac{1}{\sqrt{1+2\cos\theta}}, q = \frac{-2\cos\theta}{\sqrt{1+2\cos\theta}}, r = \frac{1}{\sqrt{1+2\cos\theta}}$$

17. We have, $(\vec{A}+\vec{B})\times(\vec{A}+\vec{C})$

$= \vec{A}\times\vec{A} + \vec{B}\times\vec{A} + \vec{A}\times\vec{C} + \vec{B}\times\vec{C}$

$= \vec{B}\times\vec{A} + \vec{A}\times\vec{C} + \vec{B}\times\vec{C}$ $\quad [\because \vec{A}\times\vec{A} = 0]$

Thus, $[(\vec{A}+\vec{B})\times(\vec{A}+\vec{C})]\times(\vec{B}\times\vec{C})$

$= [\vec{B}\times\vec{A} + \vec{A}\times\vec{C} + \vec{B}\times\vec{C}]\times(\vec{B}\times\vec{C})$

$= (\vec{B}\times\vec{A})\times(\vec{B}\times\vec{C}) + (\vec{A}\times\vec{C})\times(\vec{B}\times\vec{C})$ $\quad [\because x\times x = 0]$

$= \{(\vec{B}\times\vec{A}).\vec{C}\}\vec{B} - \{(\vec{B}\times\vec{A}).\vec{B}\}\vec{C}$

$+ \{(\vec{A}\times\vec{C}).\vec{C}\}\vec{B} - \{(\vec{A}\times\vec{C}).\vec{B}\}\vec{C}$

$[\because\ (a\times b)\times c = (a.c)b - (b.c)a]$

$= [\vec{B}\vec{A}\vec{C}]\vec{B} - [\vec{A}\vec{C}\vec{B}]C$

$[\because$ [A B C]=0 if any two of A, B, C are equal.]

$= [\vec{A}\vec{C}\vec{B}]\{\vec{B} - \vec{C}\}$

Thus, LHS of the given expression

$= [\vec{A}\vec{C}\vec{B}]\{(\vec{B}-\vec{C}).(\vec{B}+\vec{C})\}$

$= [\vec{A}\vec{C}\vec{B}]\{|\vec{B}|^2 - |\vec{C}|^2\} = 0$ $\quad [\because |B| = |C|]$

18. The P.Vs. of the points A, B, C, D are

$A(\vec{O}), B(\vec{b}), D(\vec{d}), C(\vec{d}+t\vec{b})$

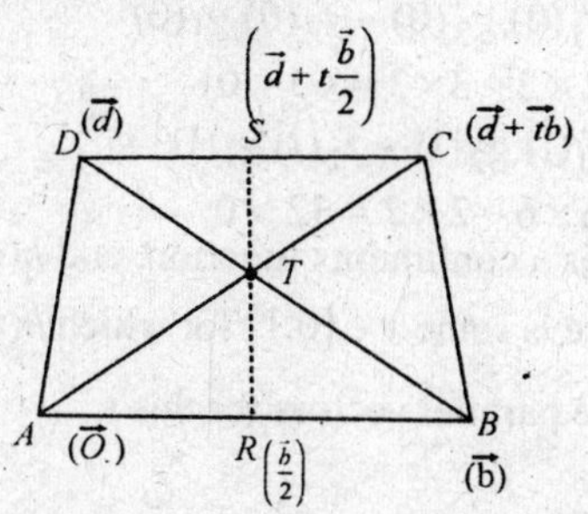

Equations of AC and BD are

$r = \lambda(d+tb)$ and $r = (1-\mu)b + \mu d$

For point of intersection say T compare the coefficients

$\lambda = \mu,\ t\lambda = 1-\mu = 1-\lambda$ or $(t+1)\lambda = 1$

$$\therefore\quad \lambda = \frac{1}{t+1} = \mu$$

$$\therefore\quad T \text{ is } \frac{d+tb}{t+1} \qquad ...(1)$$

Let R and S be mid-points of parallel sides AB and DC then

R is $\frac{b}{2}$ and S is $d + t\frac{b}{2}$.

Equation of RS by $r = a + s(b-a)$ is

$$r = \frac{b}{2} + s\left[d + (t-1)\frac{b}{2}\right]$$

The point (1) will lie on above if,

$$\frac{d+tb}{1+t} = \frac{b}{2} + s\left[d + (t-1)\frac{b}{2}\right]$$

Comparing the coefficients, we get

$$\frac{t}{1+t} = \frac{1}{2} + s\frac{(t-1)}{2} \text{ and } \frac{t}{(1+t)} = s,$$

$$\therefore\quad \frac{t}{1+t} = \frac{1}{2} + \frac{1}{1+t}.\frac{(t-1)}{2} = \frac{2t}{2(1+t)} = \frac{t}{1+t}$$

which is true . Hence proved.

19. (a) We have, $\vec{u}.\vec{v} = |\vec{u}||\vec{v}|\cos\theta$

and $\vec{u}\times\vec{v} = |\vec{u}||\vec{v}|\sin\theta\,\hat{n}$

Where θ is the angle between $\vec{u}$ and $\vec{v}$ and $\hat{n}$ is a unit vector perpendicular to both $\vec{u}, \vec{v}$ and is such that $\vec{u}, \vec{v}, \hat{n}$ form a right handed system.

Thus, $|\vec{u}.\vec{v}|^2 = |\vec{u}|^2|\vec{v}|^2\cos^2\theta$

and $|\vec{u}\times\vec{v}|^2 = |\vec{u}|^2|\vec{v}|^2\sin^2\theta\,\hat{n}.\hat{n} = |\vec{u}|^2|\vec{v}|^2\sin^2\theta$

$\therefore\quad |u.v|^2 + |u\times v|^2 = |u|^2|v|^2(\cos^2\theta + \sin^2\theta)$

$= |u|^2|v|^2$

(b) Let $|u| = a$, $|v| = b$, $u\times v = ab\sin\theta\,\hat{n}$, where $\hat{n}$ is perpendicular to both u and v, $|a|^2 = a^2$

L.H.S. $= (1+a^2)(1+b^2)$

R.H.S. $= (1 - ab\cos\theta)^2 + (u+v)^2\times(u\times v)^2$

$+ 2(u+v).ab\sin\theta\,\hat{n}$

$= 1 + a^2b^2\cos^2\theta - 2ab\cos\theta + a^2$

$+ b^2 + 2ab\cos\theta + a^2b^2\sin^2\theta.1 + 0$

as $\hat{n}$ is $\perp$ to both u and v.

$= 1 + a^2b^2(\cos^2\theta + \sin^2\theta) + a^2 + b^2$

$= 1 + a^2 + b^2 + a^2b^2 = (1+a^2)(1+b^2)$

20. $[\vec{u}\,\vec{v}\,\vec{w}] = (\vec{u}\times\vec{v}).(\vec{v}-\vec{w}\times\vec{u}) = (\vec{u}\times\vec{v}).(\vec{u}\times\vec{w})$

$$= \begin{vmatrix} \vec{u}.\vec{u} & \vec{u}.\vec{w} \\ \vec{v}.\vec{u} & \vec{v}.\vec{w} \end{vmatrix}$$

Now, $\vec{u}.\,\vec{u} = 1$

$\vec{u}.\vec{w} = \vec{u}.(\vec{v}-\vec{w}\times\vec{u}) = \vec{u}.\vec{v} - [\vec{u}\,\vec{w}\,\vec{u}] = \vec{u}.\vec{v}$

$\vec{v}.\vec{w} = \vec{v}.(\vec{v}-\vec{w}\times\vec{u}) = 1 - [\vec{v}\,\vec{w}\,\vec{u}] = 1 - [\vec{u}\,\vec{v}\,\vec{w}]$

$$\therefore \quad [\vec{u}\,\vec{v}\,\vec{w}] = \begin{vmatrix} 1 & \cos\theta \\ \cos\theta & 1-(\vec{u}\,\vec{v}\,\vec{w}) \end{vmatrix}$$

(θ is angle between $\vec{u}$ and $\vec{v}$)

$$= 1 - [\vec{u}\,\vec{v}\,\vec{w}] - \cos^2\theta$$

$$\therefore \quad [\vec{u}\,\vec{v}\,\vec{w}] = \frac{1}{2}\sin^2\theta \le \frac{1}{2}$$

Equality holds when

$\sin^2\theta = 1$ i.e., $\theta = \pi/2 \therefore \vec{u} \perp \vec{v}$.

21. Let $\vec{a}, \vec{b}, \vec{c}$ be the position vectors by A, B, and C respectively,

Let AB, BE and CF be the bisectors of $\angle A$, $\angle B$, and $\angle C$ respectively.

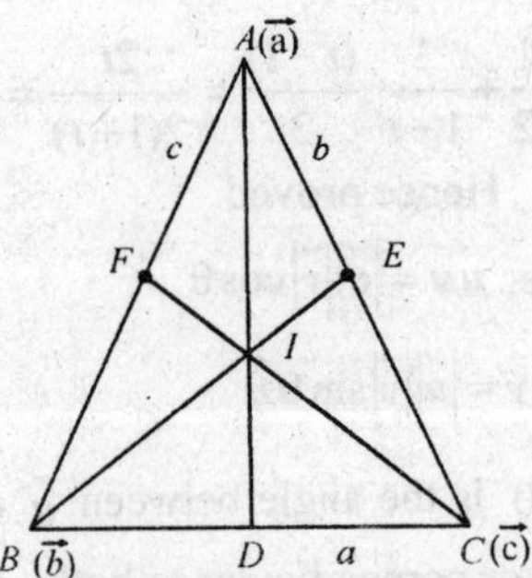

a, b, c are the lengths of sides BC, CA and AB respectively. Now we know by angle bisector thm that AD divides, BC in the ratio

$$BD : DC = AB : AC = c : b.$$

$\therefore$ The position vector of D is $\vec{d} = \dfrac{b\vec{b}+c\vec{c}}{b+c}$

Let I be the point of intersection of BE and AD. Then in ΔABD, BI is bisector of $\angle B$.

$\therefore$ DI : IA = BD : BA

$$\text{But } \frac{BD}{DC} = \frac{c}{b} \Rightarrow \frac{BD}{BD+DC} = \frac{c}{c+b}$$

$$\Rightarrow \frac{BD}{BC} = \frac{c}{c+b} \Rightarrow BD = \frac{ac}{b+c}$$

$$\therefore DI : IA = \frac{ac}{b+c} : c = a : (b+c)$$

$$\therefore \text{P.V. of } I = \frac{\vec{a}.\,a + \vec{d}(b+c)}{a+b+c}$$

$$= \frac{a\,\vec{a} + \left(\dfrac{b\vec{b}+c\vec{c}}{b+c}\right)(b+c)}{a+b+c} = \frac{a\vec{a}+b\vec{b}+c\vec{c}}{a+b+c}$$

As p.v. of I is symm. in $\vec{a}, \vec{b}, \vec{c}$ and a, b, c.

$\therefore$ It must lie on CF as well.

[We can also see that p.v. of intersection of AD and CF is also $\dfrac{a\vec{a}+b\vec{b}+c\vec{c}}{a+b+c}$]

Above prove that all the $\angle$ bisectors pass through I, i.e., these are concurrent.

22. Given data is insufficient to uniquely determine the three vectors as there are only 6 equations involving 9 variables.

$\therefore$ We can obtain infinitely many set of three vectors, $\vec{v}_1, \vec{v}_2, \vec{v}_3$, satisfying these conditions.

From the given data, we get

$$\vec{v}_1 . \vec{v}_1 = 4 \Rightarrow |\vec{v}_1| = 2$$

$$\vec{v}_2 . \vec{v}_2 = 2 \Rightarrow |\vec{v}_2| = \sqrt{2}$$

$$\vec{v}_3 . \vec{v}_3 = 29 \Rightarrow |\vec{v}_3| = \sqrt{29}$$

Also $\vec{v}_1 . \vec{v}_2 = -2$

$\Rightarrow |\vec{v}_1|\,|\vec{v}_2| \cos\theta = -2$

[where θ is the angle between $\vec{v}_1$ and $\vec{v}_2$]

$$\Rightarrow \quad \cos\theta = \frac{-1}{\sqrt{2}} \Rightarrow \theta = 135°$$

Now since any two vectors are always coplanar, let us suppose that $\vec{v}_1$ and $\vec{v}_2$ are in x-y plane.

Let $\vec{v}_1$ is along the positive direction of x-axis

then $\vec{v}_1 = 2\hat{i}$. $\quad [\because |\vec{v}_1| = 2]$

As $\vec{v}_2$ makes an angle 135° with $\vec{v}_1$ and lines in x-y plane,

Also keeping in mind $|\vec{v}_2| = \sqrt{2}$, we obtain $\vec{v}_2 = -\hat{i} \pm \hat{j}$

Again let $\vec{v}_3 = \alpha\hat{i} + \beta\hat{j} + \gamma\hat{k}$

$\because \quad \vec{v}_3 . \vec{v}_1 = 6 \Rightarrow 2\alpha = 6 \Rightarrow \alpha = 3$

and $\vec{v}_3 . \vec{v}_2 = -5 \Rightarrow -\alpha \pm \beta = -5 \Rightarrow \beta = \pm 2$

Also $|\vec{v}_3| = \sqrt{29} \Rightarrow \alpha^2 + \beta^2 + \gamma^2 = 29 \Rightarrow \gamma = \pm 4$

Hence $\vec{v}_3 = 3\hat{i} \pm 2\hat{j} \pm 4\hat{k}$

Thus, $\vec{v}_1 = 2\hat{i}$; $\vec{v}_2 = -\hat{i} \pm \hat{j}$; $\vec{v}_3 = 3\hat{i} \pm 2\hat{j} \pm 4\hat{k}$ are some possible answers.

23. $\vec{A}(t)$ is parallel to $\vec{B}(t)$ for some $t \in [0,1]$ if and only if

$$\frac{f_1(t)}{g_1(t)} = \frac{f_2(t)}{g_2(t)} \text{ for some } t \in [0,1]$$

or $f_1(t).g_2(t) = f_2(t).g_1(t)$ for some $t \in [0,1]$

Let $h(t) = f_1(t).g_2(t) - f_2(t).g_1(t)$

$h(0) = f_1(0).g_2(0) - f_2(0).g_1(0)$

$= 2\times 2 - 3\times 3 = -5 < 0$

$h(1) = f_1(1).g_2(1) - f_2(1).g_1(1)$

$= 6\times 6 - 2\times 2 = 32 > 0$

Since h is a continuous function, and $h(0).h(1) < 0$

$\Rightarrow$ There is some $t \in [0,1]$ for which $h(t) = 0$ i.e., $\vec{A}(t)$ and $\vec{B}(t)$ are parallel vectors for this t.

24. Given that $\vec{a} = a_1\hat{i} + a_2\hat{j} + a_3\hat{k}$

$\vec{b} = b_1\hat{i} + b_2\hat{j} + b_3\hat{k}$

$\vec{c} = c_1\hat{i} + c_2\hat{j} + c_3\hat{k}$

where $a_r, b_r, c_r, r = 1,2,3$ are all non negative real numbers.

Also $\sum_{r=1}^{3}(a_r + b_r + c_r) = 3L$

To prove $V \le L^3$ where V is vol. of parallelopiped formed by the vectors $\vec{a}, \vec{b}$, and $\vec{c}$

$\therefore$ We have $V = [\vec{a}\ \vec{b}\ \vec{c}\,] = \begin{vmatrix} a_1 & a_2 & a_3 \\ b_1 & b_2 & b_3 \\ c_1 & c_2 & c_3 \end{vmatrix}$

$\Rightarrow V = (a_1b_2c_3 + a_2b_3c_1 + a_3b_1c_2)$
$\quad - (a_1b_3c_2 + a_2b_1c_3 + a_3b_2c_1)$ ....(1)

Now we know that $AM \ge GM$

$\therefore \frac{(a_1+b_1+c_1)+(a_2+b_2+c_2)+(a_3+b_3+c_3)}{3}$

$\ge [(a_1+b_1+c_1)(a_2+b_2+c_2)(a_3+b_3+c_3)]^{1/3}$

$\Rightarrow \frac{3L}{3} \ge [(a_1+b_1+c_1)(a_2+b_2+c_2)(a_3+b_3+c_3)]^{1/3}$

$\Rightarrow L^3 \ge (a_1+b_1+c_1)(a_2+b_2+c_2)(a_3+b_3+c_3)$

$\Rightarrow L^3 \ge a_1b_2c_3 + a_2b_3c_1 + a_3b_1c_2 + 24$ more such terms

$L^3 \ge a_1b_2c_3 + a_2b_3c_1 + a_3b_1c_2$ $[\because a_r, b_r, c_r \ge 0 \text{ for } r = 1,2,3]$

$L^3 \ge (a_1b_2c_3 + a_2b_3c_1 + a_3b_1c_2) - (a_1b_3c_2 + a_2b_1c_3 + a_3b_2c_1)$ [same reason]

$L^3 \ge V$ from (1) Hence Proved.

25. (i) We know that eq^n of plane passing through three points $(x_1, y_1, z_1), (x_2, y_2, z_2)$ and (x_3, y_3, z_3) is

$$\begin{vmatrix} x-x_1 & y-y_1 & z-z_1 \\ x_2-x_1 & y_2-y_1 & z_2-z_1 \\ x_3-x_1 & y_3-y_1 & z_3-z_1 \end{vmatrix} = 0$$

$\Rightarrow$ Plane passing through (2, 1, 0), (5, 0, 1) and (4, 1, 1) is

$$\begin{vmatrix} x-2 & y-1 & z-0 \\ 5-2 & 0-1 & 1-0 \\ 4-2 & 1-1 & 1-0 \end{vmatrix} = 0$$

$$\Rightarrow \begin{vmatrix} x-2 & y-1 & z \\ 3 & -1 & 1 \\ 2 & 0 & 1 \end{vmatrix} = 0$$

$\Rightarrow (x-2)(-1-0) - (y-1)(3-2) + z(0-(-2)) = 0$

$\Rightarrow -x+2-y+1+2z = 0 \Rightarrow x+y-2z = 3$

(ii) ATQ, we have to find a pt. Q such that PQ is $\perp$ to the plane $x+y-2z=3$ ....(1)

And mid pt. of PQ lies on the plane, (Clearly we have to find image of pt. P with respect to plane).

Let Q be (α, β, γ)

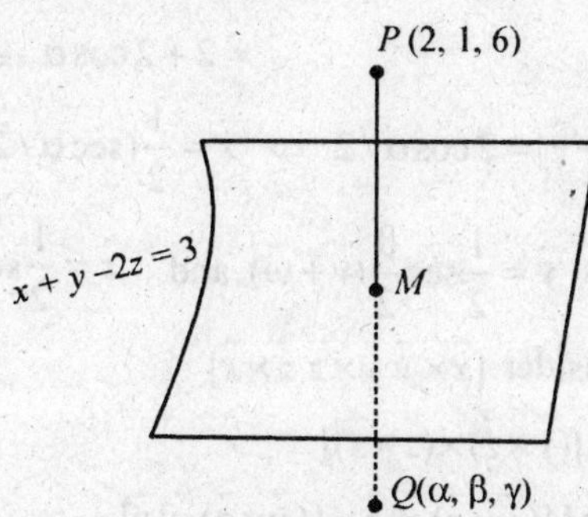

Eq^n of PM passing through $P(2, 1, 6)$ and $\perp$ to plane $x+y-2z=3$, is given by

$\frac{x-2}{1} = \frac{y-1}{1} = \frac{z-6}{-2} = \lambda$

For some value of λ, $Q(\alpha, \beta, \gamma)$ lies on PM

$\therefore \quad \frac{\alpha-2}{1} = \frac{\beta-1}{1} = \frac{\gamma-6}{-2} = \lambda$

$\Rightarrow \quad \alpha = \lambda+2, \beta = \lambda+1, \gamma = -2\lambda+6$

$\therefore$ Mid. pt. of PQ

i.e. $M\left(\frac{2+\lambda+2}{2}, \frac{1+\lambda+1}{2}, \frac{6-2\lambda+6}{2}\right)$

$= \left(\frac{\lambda+4}{2}, \frac{\lambda+2}{2}, \frac{12-2\lambda}{2}\right)$

But M lies on plane (1)

$\therefore \quad \frac{\lambda+4}{2} + \frac{\lambda+2}{2} - 12 - 2\lambda = 3$

$\Rightarrow \lambda+4+\lambda+2-24+4\lambda = 6 \Rightarrow 6\lambda = 24 \Rightarrow \lambda = 4$

$\therefore \quad Q(4+2, 4+1, -8+6) = (6, 5, -2)$

26. Given that $\vec{u}, \vec{v}, \vec{\omega}$ are three non coplanar unit vectors. Angle between $\vec{u}$ and $\vec{v}$ is α, between $\vec{v}$ and $\vec{\omega}$ is β and between $\vec{\omega}$ and $\vec{u}$ it is γ. In fig. $\overrightarrow{OA}$ and $\overrightarrow{OB}$ represent $\vec{u}$ and $\vec{v}$. Let P be a pt. on angle bisector of $\angle AOB$ such that $OAPB$ is a parallelogram.

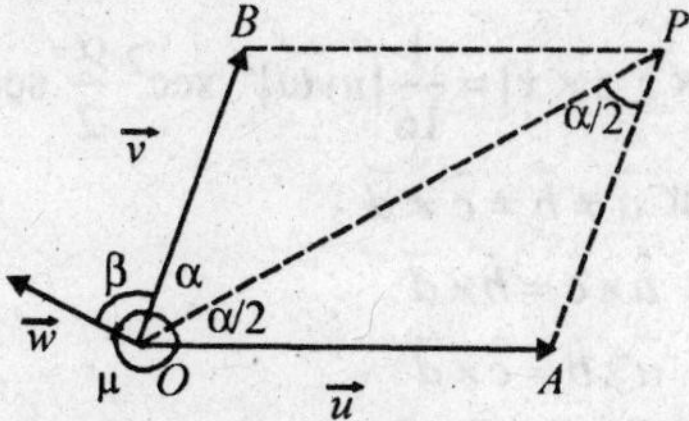

Also $\angle POA = \angle BOP = \alpha/2$

$\therefore \quad \angle APO = \angle BOP = \alpha/2$ (alt. int. $\angle$'s)

$\therefore$ In $\Delta OAP, OA = AP$

a unit vector in the direction of $\overrightarrow{OP}$

$\therefore \quad \overrightarrow{OP} = \overrightarrow{OA} + \overrightarrow{AP} = \vec{u} + \vec{v}$

$\therefore$ A unit vector in the direction of

$\overrightarrow{OP} = \frac{\vec{u}+\vec{v}}{|\vec{u}+\vec{v}|}$ i.e. $\vec{x} = \frac{\vec{u}+\vec{v}}{|\vec{u}+\vec{v}|}$

But $|\vec{u}+\vec{v}|^2=(\vec{u}+\vec{v}).(\vec{u}+\vec{v}) = 1+1+2\vec{u}.\vec{v}$ $[\because |\vec{u}|=|\vec{v}|=1]$

$= 2+2\cos\alpha = 4\cos^2\alpha/2$

$\therefore \quad |\vec{u}+\vec{v}| = 2\cos\alpha/2 \Rightarrow \vec{x} = \frac{1}{2}(\sec\alpha/2)(\vec{u}+\vec{v})$

Similarly, $\vec{y} = \frac{1}{2}\sec\frac{\beta}{2}(\vec{v}+\vec{\omega})$ and $\vec{z} = \frac{1}{2}\sec\frac{\gamma}{2}(\vec{\omega}+\vec{u})$

Now consider $[\vec{x}\times\vec{y}\ \vec{y}\times\vec{z}\ \vec{z}\times\vec{x}]$

$=(\vec{x}\times\vec{y}).[(\vec{y}\times\vec{z})\times(\vec{z}\times\vec{x})]$

$=(\vec{x}\times\vec{y}).[\{(\vec{y}\times\vec{z}).\vec{x}\}\vec{z}-\{(\vec{y}\times\vec{z}).\vec{z}\}\vec{x}]$

[Using def^n of vector triple product.]

$=(\vec{x}\times\vec{y}).[[\vec{x}\vec{y}\vec{z}]\vec{z}-0]$ $\quad [\therefore [\vec{y}\vec{z}\vec{z}]=0]$

$=[\vec{x}\vec{y}\vec{z}][\vec{x}\vec{y}\vec{z}] = [\vec{x}\vec{y}\vec{z}]^2$ ...(1)

Also $[\vec{x}\vec{y}\vec{z}] = \left[\frac{1}{2}\left(\sec\frac{\alpha}{2}\right)(\vec{u}+\vec{v})\frac{1}{2}\sec\beta/2\right.$

$\left.(\vec{v}+\vec{\omega})\frac{1}{2}\sec\gamma/2(\vec{\omega}+\vec{u})\right]$

$=\frac{1}{8}\sec\alpha/2\sec\beta/2\sec\gamma/2[\vec{u}+\vec{v}\ \vec{v}+\vec{\omega}\ \vec{\omega}+\vec{u}]$

$=\frac{1}{8}\sec\alpha/2\sec\beta/2\sec\gamma/2[(\vec{u}+\vec{v}).\{(\vec{v}+\vec{\omega})\times(\vec{\omega}+\vec{u})\}]$

$=\frac{1}{8}\sec\alpha/2\sec\beta/2\sec\gamma/2[(\vec{u}+\vec{v}).(\vec{v}\times\vec{\omega}+\vec{v}\times\vec{u}+\vec{\omega}\times\vec{u})]$

$=\frac{1}{8}\sec\alpha/2\sec\beta/2\sec\gamma/2[\vec{u}.(\vec{v}\times\vec{\omega})+\vec{v}.(\vec{\omega}\times\vec{u})]$

($\because [\vec{a}\vec{b}\vec{c}]=0$ when ever any two vectors are same)

$=\frac{1}{8}(\sec\alpha/2\sec\beta/2\sec\gamma/2)2[\vec{u}\vec{v}\vec{\omega}]$

$=\frac{1}{4}(\sec\alpha/2\sec\beta/2\sec\gamma/2)[\vec{u}\vec{v}\vec{\omega}]$

$\therefore [\vec{x}\vec{y}\vec{z}]^2 = \frac{1}{16}[\vec{u}\vec{v}\vec{\omega}]^2\sec^2\alpha/2\sec^2\beta/2\ \sec^2\gamma/2]$ ...(2)

From (1) and (2),

$[\vec{x}\times\vec{y}\ \vec{y}\times\vec{z}\ \vec{z}\times\vec{x}] = \frac{1}{16}[\vec{u}\vec{v}\vec{\omega}]^2\ \sec^2\frac{\alpha}{2}.\sec^2\frac{\beta}{2}\sec^2\frac{\gamma}{2}$

27. Given that $\vec{a}\neq\vec{b}\neq\vec{c}\neq\vec{d}$

Such that $\vec{a}\times\vec{c}=\vec{b}\times\vec{d}$ ...(1)

$\vec{a}\times\vec{b}=\vec{c}\times\vec{d}$ ...(2)

To prove $(\vec{a}-\vec{d}).(\vec{b}-\vec{c})\neq 0$

Subtracting eq^n (2) from (1) we get

$\vec{a}\times(\vec{c}-\vec{b})=(\vec{b}-\vec{c})\times\vec{d}$

$\Rightarrow \vec{a}\times(\vec{c}-\vec{b})=\vec{d}\times(\vec{c}-\vec{b})$

$\Rightarrow \vec{a}\times(\vec{c}-\vec{b})-\vec{d}\times(\vec{c}-\vec{b})=0$

$\Rightarrow (\vec{a}-\vec{d})\times(\vec{c}-\vec{b})=0 \Rightarrow (\vec{a}-\vec{d})\,||\,(\vec{c}-\vec{b})$

$[\because \vec{a}-\vec{d}\neq 0, \vec{c}-\vec{b}\neq 0$ as all distinct $]$

$\Rightarrow$ Angle between $\vec{a}-\vec{d}$ and $\vec{c}-\vec{b}$ is either 0 or 180°.

$\Rightarrow (\vec{a}-\vec{d}).(\vec{c}-\vec{b})=|\vec{a}-\vec{d}||\vec{c}-\vec{b}|\cos 0\,[\text{or}\cos 180°]\neq 0$

as *a, b, c, d* all are different. Hence Proved.

28. $\because$ The plane is parallel to the lines L_1 and L_2 with direction ratios as (1, 0, –1) and (1, –1, 0) $\therefore$

A vector perpendicular to L_1 and L_2 will be parallel to the normal $(\vec{n})$ to the plane.

$\therefore \vec{n} = \begin{vmatrix}\hat{i} & \hat{j} & \hat{k}\\ 1 & 0 & -1\\ 1 & -1 & 0\end{vmatrix} = -\hat{i}-\hat{j}-\hat{k}$

$\therefore$ Eqn. of plane through (1, 1, 1) and having normal vector $\vec{n}=-\hat{i}-\hat{j}-\hat{k}$ is given by

$(\vec{r}-\vec{a}).\vec{n}=0$

$\Rightarrow -1(x-1)-1(y-1)-1(z-1)=0 \Rightarrow x+y+z=3$

$\Rightarrow \frac{x}{3}+\frac{y}{3}+\frac{z}{3}=1$ ...(1)

Now the pts where this plane meets the axes are $A(3,0,0), B(0,3,0), C(0,0,3)$

$\therefore$ Vol.of tetrahedron *OABC*

$=\frac{1}{6}\times$ Area of base $\times$ altitude

$=\frac{1}{6}\times \text{Ar}(\Delta ABC)\times$ length of $\perp^{\text{lar}}$ (0,0,0) to plane (1)

$=\frac{1}{6}\times\frac{1}{2}\left[\frac{\sqrt{3}}{4}\times|\overline{AB}|^2\right]\times\left[\left|\frac{-3}{\sqrt{1+1+1}}\right|\right]$

(Note that ΔABC is an equilateral Δ here.)

$=\frac{1}{12}\times\frac{\sqrt{3}}{4}\times(3\sqrt{2})^2\times\sqrt{3} = \frac{3\times 18}{48}=\frac{9}{2}$ cubic units.

29. ATQ 'S' is the parallelopiped with base points *A, B, C* and *D* and upper face points A', B', C' and D'. Let its vol. be V_S.

By compressing it by upper face A', B', C', D', a new parallelopiped 'T' is formed whose upper face pts are now A'', B'', C'' and D''. Let its vol. be V_T.

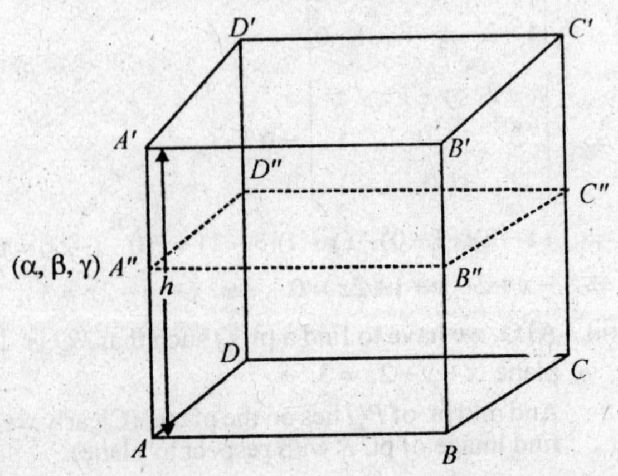

Let h be the height of original parallelopiped S.

Then $V_S = (ar\ ABCD) \times h$ ...(1)

Let equation of plane $ABCD$ be

$ax + by + cz + d = 0$ and $A''(\alpha, \beta, \gamma)$

Then height of new parallelopipe T is the length of perpendicular from A'' to $ABCD$

i.e. $\dfrac{a\alpha + b\beta + c\gamma + d}{\sqrt{a^2 + b^2 + c^2}}$

$\therefore \quad V_T = (ar\ ABCD) \times \dfrac{a\alpha + b\beta + c\gamma + d}{\sqrt{a^2 + b^2 + c^2}}$ ...(2)

But given that,

$V_T = \dfrac{90}{100} V_s$ ...(3)

From (1), (2) and (3) we get,

$$\frac{a\alpha + b\beta + c\gamma + d}{\sqrt{a^2 + b^2 + c^2}} = 0.9h$$

$\Rightarrow \quad a\alpha + b\beta + c\gamma + (d - 0.9h\sqrt{a^2 + b^2 + c^2}) = 0$

$\therefore$ Locus of $A''(\alpha, \beta, \gamma)$ is

$ax + by + cz + (d - 0.9h\sqrt{a^2 + b^2 + c^2}) = 0$

which is a plane parallel to $ABCD$. Hence proved.

30. Following fig. shows the possible situation for planes P_1 and P_2 and the lines L_1 and L_2

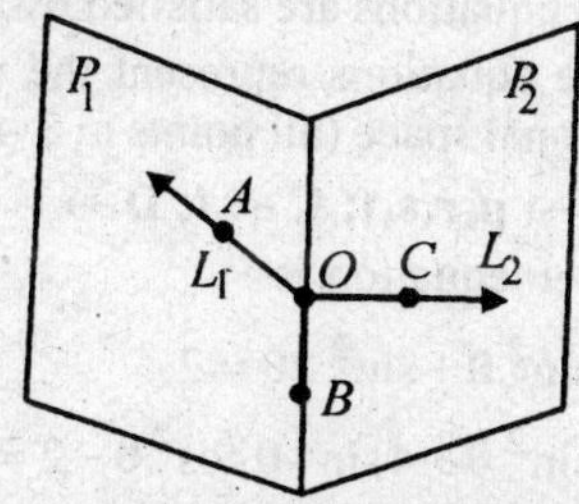

Now if we choose pts A, B, C as follows.

A on L_1, B on the line of intersection of P_1 and P_2 but other than origin and C on L_2 again other than origin then we can consider

A corresponds to one of A', B', C' and

B corresponds to one of the remaining of A', B', C' and

C corresponds to third of A', B', C' e.g.

$A' \equiv C; B' \equiv B; C' \equiv A$

Hence one permutation of $[ABC]$ is $[CBA]$. Hence Proved.

31. The given line is

$2x - y + z - 3 = 0 = 3x + y + z - 5$

Which is intersection line of two planes

$2x - y + z - 3 = 0$ ...(i)

and $3x + y + z - 5 = 0$ ...(ii)

Any plane containing this line will be the plane passing through the intersection of two planes (i) and (ii).

Thus the plane containing given line can be written as

$(2x - y + z - 3) + \lambda(3x + y + z - 5) = 0$

$\Rightarrow \quad (3\lambda + 2)x + (\lambda - 1)y + (\lambda + 1)z + (-5\lambda - 3) = 0$

As its distance from the pt. $(2, 1, -1)$ is $\dfrac{1}{\sqrt{6}}$

$$\therefore \quad \left| \frac{(3\lambda + 2)2 + (\lambda - 1)1 + (\lambda + 1)(-1) + (-5\lambda - 3)}{\sqrt{(3\lambda + 2)^2 + (\lambda - 1)^2 + (\lambda + 1)^2}} \right| = \frac{1}{\sqrt{6}}$$

$$\Rightarrow \quad \left| \frac{\lambda - 1}{\sqrt{11\lambda^2 + 12\lambda + 6}} \right| = \frac{1}{\sqrt{6}}$$

Squaring both sides, we get

$$\frac{(\lambda - 1)^2}{11\lambda^2 + 12\lambda + 6} = \frac{1}{6}$$

$\Rightarrow \quad 6\lambda^2 - 12\lambda + 6 - 11\lambda^2 - 12\lambda - 6 = 0$

$\Rightarrow \quad 5\lambda^2 + 24\lambda = 0 \Rightarrow \lambda(5\lambda + 24) = 0$

$\Rightarrow \quad \lambda = 0$ or $-24/5$

$\therefore$ The required equations of planes are

$2x - y + z - 3 = 0$

and $\left[3\left(\dfrac{-24}{5}\right) + 2\right]x + \left[-\dfrac{24}{5} - 1\right]y + \left[-\dfrac{24}{5} + 1\right]z - 5\left(\dfrac{-24}{5}\right) - 3 = 0$

or $\quad -62x - 29y - 19z + 105 = 0$

or $\quad 62x + 29y + 19z - 105 = 0$

32. Given that incident ray is along $\hat{v}$, reflected ray is along $\hat{w}$ and normal is along $\hat{a}$, outwards. The given figure can be redrawn as shown.

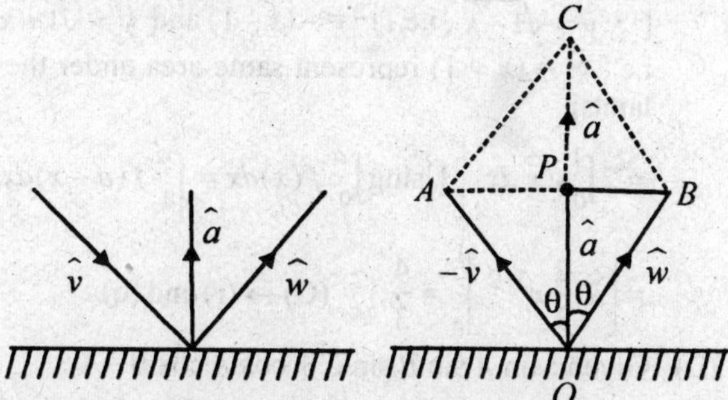

We know that incident ray, reflected ray and normal lie in a plane, and angle of incidence = angle of reflection.

Therefore $\hat{a}$ will be along the angle bisector of $\hat{w}$ and $-\hat{v}$,

i.e., $\hat{a} = \dfrac{\hat{w} + (-\hat{v})}{|\hat{w} - \hat{v}|}$ ...(1)

NOTE THIS STEP:

[$\because$ Angle bisector will along a vector dividing in same ratio as the ratio of the sides forming that angle.]

But $\hat{a}$ is a unit vector.

Where $|\hat{w} - \hat{v}| = OC = 2OP = 2|\hat{w}|\cos\theta = 2\cos\theta$

Substituting this value in equation (1) we get

$$\hat{a} = \frac{\hat{w} - \hat{v}}{2\cos\theta}$$

$\therefore \quad \hat{w} = \hat{v} + (2\cos\theta)\hat{a} = \hat{v} - 2(\hat{a}.\hat{v})\hat{a} \quad [\because \hat{a}.\hat{v} = -\cos\theta]$

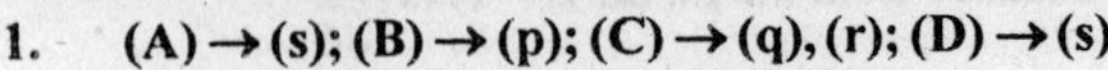

1. (A)→(s); (B)→(p); (C)→(q), (r); (D)→(s)

(A) On solving the given equations $x+y=|a|$ and $ax-y=1$, we get

$$x=\frac{1+|a|}{a+1} \text{ and } y=\frac{a|a|-1}{a+1}$$

∵ Rays intersect each other in I quad.

$\therefore x, y>0 \Rightarrow a+1>0$ and $a|a|-1>0 \Rightarrow a>1$

$\therefore a_0=1$ (A)→(s)

(B) (α, β, γ) lies on the plane $x+y+z=2$

$\Rightarrow \alpha+\beta+\gamma=2$

Also $\hat{k}\times(\hat{k}\times\vec{a})=(\hat{k}.\vec{a})\hat{k}-(\hat{k}.\hat{k})\vec{a}$

$\Rightarrow \gamma\hat{k}-\alpha\hat{i}-\beta\hat{j}-\gamma\hat{k}=0 \Rightarrow \alpha\hat{i}+\beta\hat{j}=0$

$\Rightarrow \alpha=0=\beta \Rightarrow \gamma=2$ $\quad(\because \alpha+\beta+\gamma=2)$

(B)→(p)

(C) $\left|\int_0^1(1-y^2)dy\right|+\left|\int_0^1(y^2-1)dy\right|$

$=2\int_0^1(1-y^2)dy=\frac{4}{3}$

Also, $\left|\int_0^1\sqrt{1-x}\,dx\right|+\int_{-1}^0\sqrt{1+x}\,dx=2\int_0^1\sqrt{1-x}\,dx$

NOTE THIS STEP :

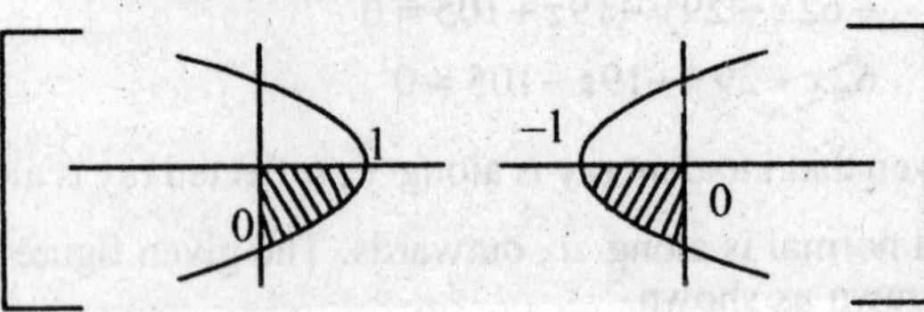

[∵ $y=\sqrt{1-x}$, i.e., $y^2=-(x-1)$ and $y=\sqrt{1+x}$ i.e., $y^2=(x+1)$ represent same area under the given limits]

$$=2\int_0^1\sqrt{x}\,dx\left[\text{Using}\int_0^a f(x)dx=\int_0^a f(a-x)dx\right]$$

$$=\left[2.\frac{2}{3}x^{3/2}\right]_0^1=\frac{4}{3}, \quad \text{(C)}\to\text{(r) and (q)}$$

(D) Given : $\sin A\sin B\sin C+\cos A\cos B=1$

But $\sin A\sin B\sin C+\cos A\cos B\le\sin A\sin B+\cos A\cos B=\cos(A-B)$

$\Rightarrow \cos(A-B)\ge 1 \Rightarrow \cos(A-B)=1$

$\Rightarrow A-B=0 \Rightarrow A=B$

∴ Given relation becomes $\sin^2 A\sin C+\cos^2 A=1$

$\Rightarrow \sin C=1,$

(D)→(s)

2. (A)→r ; (B)→q; (C)→p ; (D)→s

Here we have, the determinant of the coefficient matrix of given equation, as

$$\Delta=\begin{vmatrix}a & b & c\\ b & c & a\\ c & a & b\end{vmatrix}=-(a+b+c)(a^2+b^2+c^2-ab-bc-ca)$$

$$=-\frac{1}{2}(a+b+c)[(a-b)^2+(b-c)^2+(c-a)^2]$$

(A) $a+b+c\ne 0$ and $a^2+b^2+c^2-ab-bc-ca=0$

$\Rightarrow (a-b)^2+(b-c)^2+(c-a)^2=0$

$\Rightarrow a=b=c$ (but $\ne 0$ as $a+b+c\ne 0$)

This equation represent identical planes.

(B) $a+b+c=0$ and $a^2+b^2+c^2-ab-bc-ca\ne 0$

$\Rightarrow \Delta=0$ and a, b, c are not all equal.

∴ All equations are not identical but have infinite many solutions.

$\therefore ax+by=(a+b)z$ (using $a+b+c=0$)

and $bx+cy=(b+c)z$

$\Rightarrow (b^2-ac)y=(b^2-ac)z \quad \Rightarrow y=z$

$\Rightarrow ax+by+cy=0 \Rightarrow ax=ay \Rightarrow x=y$

$\Rightarrow x=y=z$

∴ The equations represent the line $x=y=z$

(C) $a+b+c\ne 0$ and $a^2+b^2+c^2-ab-bc-ca\ne 0$

$\Rightarrow \Delta\ne 0 \Rightarrow$ Equations have only trivial solution i.e., $x=y=z=0$

∴ the equations represents the three planes meeting at a single point namely origin.

(D) $a+b+c=0$ and $a^2+b^2+c^2-ab-bc-ca=0$

$\Rightarrow a=b=c$ and $\Delta=0$

$\Rightarrow a=b=c=0$

⇒ All equations are satisfied by all x, y, and z.

⇒ The equations represent the whole of the three dimensional space (all points in 3–D)

3. **A → q,s; B → p,r,s,t; C → t; D → r**

(A) The given equation is

$$2\sin^2\theta+\sin^2 2\theta=2$$

$\Rightarrow 2\sin^2\theta+4\sin^2\theta\cos^2\theta-2=0$

$\Rightarrow \sin^2\theta+2\sin^2\theta(1-\sin^2\theta)-1=0$

$\Rightarrow 2\sin^4\theta-3\sin^2\theta+1=0$

$\Rightarrow 2\sin^4\theta-2\sin^2\theta-\sin^2\theta+1=0$

$\Rightarrow 2\sin^2\theta(\sin^2\theta-1)-1(\sin^2\theta-1)=0$

$\Rightarrow (\sin^2\theta-1)(2\sin^2\theta-1)=0$

$\Rightarrow \sin^2\theta=1$ or $\sin^2\theta=\frac{1}{2}$

$\Rightarrow \sin^2\theta=\sin^2\frac{\pi}{2}$ or $\sin^2\theta=\sin^2\frac{\pi}{4}$

$\Rightarrow \theta=n\pi\pm\frac{\pi}{2}$ or $n\pi\pm\frac{\pi}{4} \Rightarrow \theta=\frac{\pi}{2}$ or $\frac{\pi}{4}$

(B) We know that $[x]$ is discontinuous at all integral values, therefore $\left[\frac{6x}{\pi}\right]$ is discontinuous at $x=\frac{\pi}{6},\frac{\pi}{3},\frac{\pi}{2}$ and π. Also $\cos\left[\frac{3x}{\pi}\right]\ne 0$ for any of these values of x.

$\therefore \quad \left[\frac{6x}{\pi}\right]\cos\left[\frac{3x}{\pi}\right]$ is discontinuous at $x = \frac{\pi}{6}, \frac{\pi}{3}, \frac{\pi}{2}$ and π.

(C) We know that the volume of a parallelopipe with coterminus edges as $\vec{a}, \vec{b}$ and $\vec{c}$ is given by $[\vec{a}\ \vec{b}\ \vec{c}]$

$\therefore$ The required volume is $= \vec{a}.\vec{b}\times\vec{c}$

$$= \begin{vmatrix} 1 & 1 & 0 \\ 1 & 2 & 0 \\ 1 & 1 & \pi \end{vmatrix} = \pi$$

(D) We have $\vec{a}+\vec{b} = -\sqrt{3}\vec{c} \Rightarrow |\vec{a}+\vec{b}|^2 = 3|\vec{c}|^2$

$\Rightarrow (\vec{a}+\vec{b}).(\vec{a}+\vec{b}) = 3\vec{c}.\vec{c}$

$\Rightarrow \vec{a}.\vec{a}+\vec{b}.\vec{b}+2\vec{a}.\vec{b} = 3\vec{c}.\vec{c} \Rightarrow 1+1+2\cos\theta = 3$

(where θ is the angle between $\vec{a}$ and $\vec{b}$)

$\Rightarrow \cos\theta = \frac{1}{2} \Rightarrow \theta = \frac{\pi}{3}$

4. **A → p; B → q,s; C → q,r,s,t; D → r**

(A) For the solution of $xe^{\sin x} - \cos x = 0$ in $\left(0, \frac{\pi}{2}\right)$

Let us consider two functions

$y = xe^{\sin x}$ and $y = \cos x$

The range of $y = xe^{\sin x}$ is $\left(0, \frac{\pi e}{2}\right)$, also it is an increasing function on $\left(0, \frac{\pi}{2}\right)$. Their graph are as shown in the figure below :

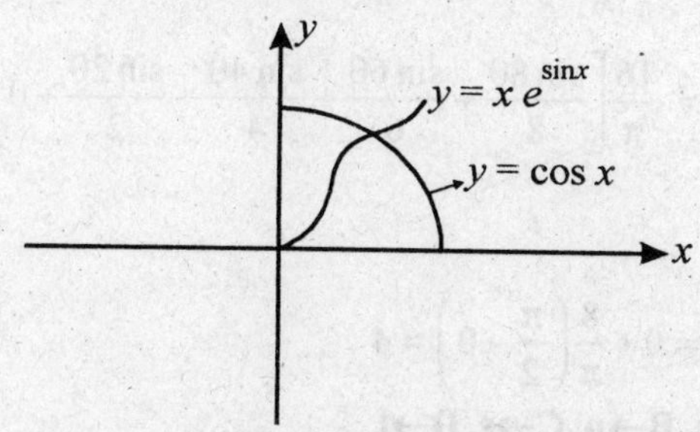

Clearly the two curves meet only at one point, therefore the given equation has only one solution in $\left(0, \frac{\pi}{2}\right)$.

(B) Three given planes are

$kx + 4y + z = 0$

$4x + ky + 2z = 0$

$2x + 2y + z = 0$

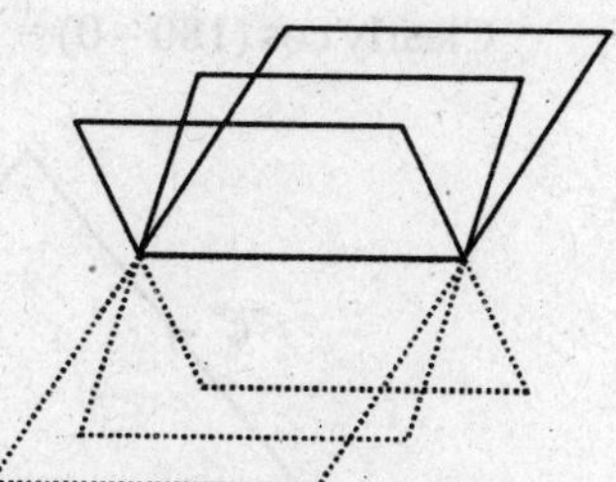

Clearly all the planes pass through (0,0,0).

$\therefore$ Their line of intersection also pass through (0, 0, 0)

Let a, b, c, be the direction ratios of required line, then we should have

$ka + 4b + c = 0$

$4a + kb + 2c = 0$

$2a + 2b + c = 0$

For the required line to exist the above system of equations in *a, b, c,* should have non trivial solution i.e.

$$\begin{vmatrix} k & 4 & 1 \\ 4 & k & 2 \\ 2 & 2 & 1 \end{vmatrix} = 0$$

$\Rightarrow k(k-4) - 4(4-4) + 1(8-2k) = 0$

$\Rightarrow k^2 - 6k + 8 = 0 \Rightarrow (k-2)(k-4) = 0$

$\Rightarrow k = 2$ or 4

(C) We have $f(x) = |x-1| + |x-2| + |x+1| + |x+2|$

$$= \begin{cases} -4x, & x \le -2 \\ -2x+4, & -2 < x \le -1 \\ 6, & -1 < x \le 1 \\ 2x+4, & 1 < x \le 2 \\ 4x, & x \ge 2 \end{cases}$$

The graph of the above function is as given below

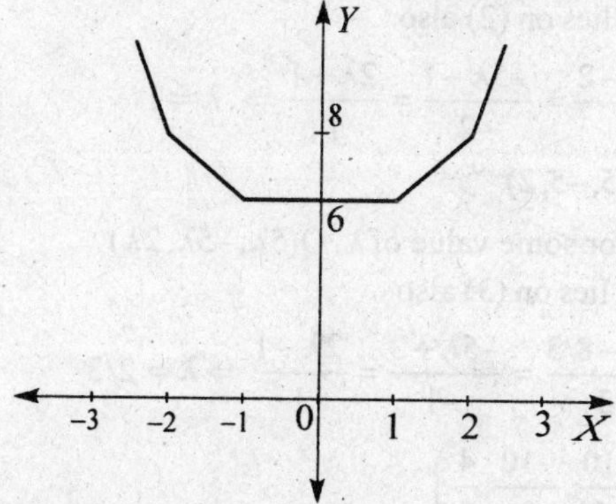

Clearly, from graph, $f(x) \ge 6$

$\Rightarrow 4k \ge 6 \Rightarrow k \ge \frac{3}{2}$

$\therefore k = 2, 3, 4, 5, 6, ...$

(D) Given that

$\frac{dy}{dx} = y + 1$ and $y(0) = 1$

$\Rightarrow \int \frac{dy}{y+1} = \int dx \Rightarrow \ln|y+1| = x + c$

At $x = 0, y = 1 \Rightarrow c = \ln 2$

$\therefore \ln|y+1| = x + \ln 2 \Rightarrow y + 1 = 2e^x \Rightarrow y = 2e^x - 1$

$\therefore y(\ln 2) = 2e^{\ln 2} - 1 = 2\times 2 - 1 = 3$

5. **(A) → t ; (B) → p,r ; (C) → q,s ; (D) → r**

Let the line through origin be $\frac{x}{a} = \frac{y}{b} = \frac{z}{c}$ —(1)

then as it intersects

$\frac{x-2}{1} = \frac{y-1}{-2} = \frac{z+1}{1}$ —(2)

and $\dfrac{x-8/3}{2}=\dfrac{y+3}{-1}=\dfrac{z-1}{1}$ —(3)

at P and Q, shortest distance of (1) with (2) and (3) should be zero.

$\therefore$ Using $\begin{vmatrix} x_2-x_1 & y_2-y_1 & z_2-z_1 \\ a_1 & b_1 & c_1 \\ a_2 & b_2 & c_2 \end{vmatrix}=0$

we get $\begin{vmatrix} 2 & 1 & -1 \\ a & b & c \\ 1 & -2 & 1 \end{vmatrix}=0 \Rightarrow a+3b+5c=0$ —(4)

and $\begin{vmatrix} 8/3 & -3 & 1 \\ a & b & c \\ 2 & -1 & 1 \end{vmatrix}=0 \Rightarrow 3a+b-5c=0$ —(5)

Solving (4) and (5), we get

$$\frac{a}{-15-5}=\frac{b}{15+5}=\frac{c}{1-9} \text{ or } \frac{a}{5}=\frac{b}{-5}=\frac{c}{2}$$

Hence equation (1) becomes $\dfrac{x}{5}=\dfrac{y}{-5}=\dfrac{z}{2}=\lambda$

For some value of λ, $P(5\lambda,-5\lambda,2\lambda)$ which lies on (2) also

$$\therefore \frac{5\lambda-2}{1}=\frac{-5\lambda-1}{-2}=\frac{2\lambda+1}{1} \Rightarrow \lambda=1$$

$\therefore$ P (5, –5, 2)

Also for some value of λ, $Q(5\lambda,-5\lambda,2\lambda)$ which lies on (3) also

$$\therefore \frac{5\lambda-8/3}{3}=\frac{-5\lambda+3}{-1}=\frac{2\lambda-1}{1} \Rightarrow \lambda=2/3$$

$$\therefore Q\left(\frac{10}{3},\frac{-10}{3},\frac{4}{3}\right)$$

Hence $d^2=PQ^2=\left(\dfrac{25}{9}+\dfrac{25}{9}+\dfrac{4}{9}\right)=6$

(B) $\tan^{-1}(x+3)-\tan^{-1}(x-3)=\tan^{-1}\dfrac{3}{4}$

$$\Rightarrow \tan^{-1}\left(\frac{x+3-x+3}{1+x^2-9}\right)=\tan^{-1}\left(\frac{3}{4}\right), x^2-9\geq -1$$

$$\Rightarrow \frac{6}{x^2-8}=\frac{3}{4} \Rightarrow x^2=16 \text{ or } x=4,-4$$

(C) We have $\vec{c}=\dfrac{\vec{a}-\mu\vec{b}}{4}$

Then $(\vec{b}-\vec{a}).(\vec{b}+\vec{c})=0$

$$\Rightarrow (\vec{b}-\vec{a}).\left(\vec{b}+\frac{\vec{a}-\mu\vec{b}}{4}\right)=0$$

$$\Rightarrow (\vec{b}-\vec{a}).\left(\frac{4-\mu}{4}\vec{b}+\frac{\vec{a}}{4}\right)=0$$

$$\Rightarrow \frac{4-\mu}{4}b^2-\frac{a^2}{4}=0$$

$$\Rightarrow (4-\mu)b^2-a^2=0 \quad \text{—(1)}$$

Also, $2^2\left|\dfrac{4-\mu}{4}\vec{b}+\dfrac{\vec{a}}{4}\right|^2=|\vec{b}-\vec{a}|^2$

$$\Rightarrow (4-\mu)^2b^2+a^2=4b^2+4a^2$$

$$\Rightarrow \left[(4-\mu)^2-4\right]b^2=3a^2 \quad \text{—(2)}$$

From (1) and (2), we get

$$\frac{(4-\mu)^2-4}{4-\mu}=\frac{3}{1}$$

$$\Rightarrow \mu^2-8\mu+12=12-3\mu \Rightarrow \mu^2-5\mu=0$$

$$\Rightarrow \mu=0 \text{ or } 5$$

(D) $I=\dfrac{2}{\pi}\displaystyle\int_{-\pi}^{\pi}\frac{\sin(9x/2)}{\sin(x/2)}dx=\frac{2}{\pi}\times 2\int_0^{\pi}\frac{\sin 9x/2}{\sin x/2}dx$

Let $\dfrac{x}{2}=0 \Rightarrow dx=2d\theta$

Also at $x=0, \theta=0$ and at $x=x, \theta=\pi/2$

$$\therefore I=\frac{8}{\pi}\int_0^{\pi/2}\frac{\sin 9\theta}{\sin\theta}d\theta$$

$$I=\frac{8}{\pi}\int_0^{\pi/2}\left[\frac{\sin 9\theta-\sin 7\theta}{\sin\theta}+\frac{(\sin 7\theta-\sin 5\theta)}{\sin\theta}+\frac{(\sin 5\theta-\sin 3\theta)}{\sin\theta}+\frac{(\sin 3\theta-\sin\theta)}{\sin\theta}+\frac{\sin\theta}{\sin\theta}\right]d\theta$$

$$=\frac{16}{\pi}\int_0^{\pi/2}(\cos 8\theta+\cos 6\theta+\cos 4\theta+\cos 2\theta+1)d\theta+\frac{8}{\pi}\int_0^{\pi/2}d\theta$$

$$=\frac{16}{\pi}\left[\frac{\sin 8\theta}{8}+\frac{\sin 6\theta}{6}+\frac{\sin 4\theta}{4}+\frac{\sin 2\theta}{2}+\theta\right]_0^{\pi/2}+\frac{8}{\pi}(\theta)_0^{\pi/2}$$

$$=0+\frac{8}{\pi}\left(\frac{\pi}{2}-0\right)=4$$

6. **A→q, B→p, C→s, D→t**

As $\vec{a}+\vec{b}=\vec{c}$

$\therefore$ The figure is as shown.

Clearly $\cos(180-\theta)=\dfrac{\vec{a}.\vec{b}}{|\vec{a}||\vec{b}|}=\dfrac{1}{2}$

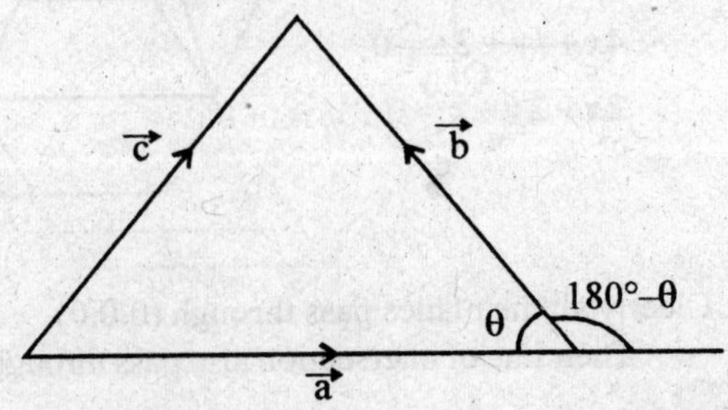

$\Rightarrow \cos\theta = -\frac{1}{2} \Rightarrow \theta = \frac{2\pi}{3}$

$\Rightarrow$ **A → q**

$\int_a^b (f(x) - 3x)\,dx = a^2 - b^2$

$\Rightarrow \int_a^b f(x)\,dx + \frac{3}{2}[-b^2 + a^2] = a^2 - b^2$

$\Rightarrow \int_a^b f(x)\,dx = -\frac{1}{2}(a^2 - b^2)$

$\Rightarrow \frac{d}{db}\left[\int_a^b f(x)\,dx\right] = b \Rightarrow f(b) = b \Rightarrow f\left(\frac{\pi}{6}\right) = \frac{\pi}{6}$

$\Rightarrow$ **B → p**

$\frac{\pi^2}{\ell n3}\int_{7/6}^{5/6} \sec(\pi x)\,dx = \frac{\pi^2}{\pi \ell n3}[\ell n|\sec \pi x + \tan \pi x|]_{7/6}^{5/6}$

$= \frac{\pi}{\ell n3}\left[\ell n|\sec\frac{5\pi}{6} + \tan\frac{5\pi}{6} - \ell n|\sec\frac{7\pi}{6} + \tan\frac{7\pi}{6}|\right]$

$= \frac{\pi}{\ell n3}\left[\ell n\left|-\frac{2}{\sqrt{3}} - \frac{1}{\sqrt{3}}\right| - \ell n\left|-\frac{2}{\sqrt{3}} + \frac{1}{\sqrt{3}}\right|\right] = \frac{\pi}{\ell n3}\ell n3 = \pi$

$\therefore$ **C → s**

For $|z| = 1$ and $z \neq 1$. Let $z = e^{i\theta}$

Then $1 - z = 1 - \cos\theta - i\sin\theta = 2\sin^2\frac{\theta}{2} - 2i\sin\frac{\theta}{2}\cos\frac{\theta}{2}$

or $1 - z = 2\sin\frac{\theta}{2}\left[\sin\frac{\theta}{2} - i\cos\frac{\theta}{2}\right]$

$\therefore \frac{1}{1-z} = \frac{1}{2}\left[1 + i\cot\frac{\theta}{2}\right]$

Here real part of $\frac{1}{1-z}$ is always $\frac{1}{2}$

$\therefore$ Locus of $\frac{1}{1-z}$ is $x = \frac{1}{2}$

For which max $\left|Arg\left(\frac{1}{1-z}\right)\right|$ is max. value of ϕ i.e. $\frac{\pi}{2}$.

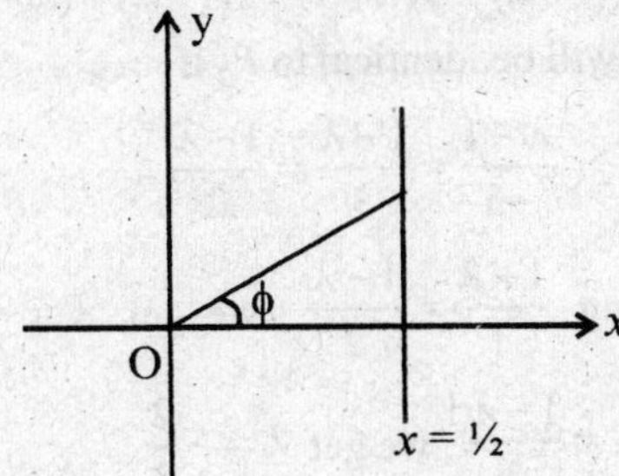

Clearly max. $\left|Arg\left(\frac{1}{1-z}\right)\right|$ approaches to $\frac{\pi}{2}$ but will not be attained.

$\therefore$ **D → t.**

7. (c) (P) $[\vec{a}\ \vec{b}\ \vec{c}] = 2$

$\therefore \left[2(\vec{a}\times\vec{b})\ \ 3(\vec{b}\times\vec{c})\ \ \vec{c}\times\vec{a}\right]$

$= 6\left[\vec{a}\times\vec{b}\ \ \vec{b}\times\vec{c}\ \ \vec{c}\times\vec{a}\right]$

$= 6[\vec{a}\ \vec{b}\ \vec{c}]^2 = 6 \times 4 = 24$

$\therefore$ (P) → (3)

(Q) $[\vec{a}\ \vec{b}\ \vec{c}] = 5$

$\therefore \left[3(\vec{a}+\vec{b})\ \ \vec{b}+\vec{c}\ \ 2(\vec{c}+\vec{a})\right]$

$= 6\left[\vec{a}+\vec{b}\ \ \vec{b}+\vec{c}\ \ \vec{c}+\vec{a}\right]$

$= 6\times 2[\vec{a}\ \vec{b}\ \vec{c}] = 6\times 2\times 5 = 60$

$\therefore$ (Q) → (4)

(R) $\frac{1}{2}|\vec{a}\times\vec{b}| = 20 \Rightarrow |\vec{a}\times\vec{b}| = 40$

$\therefore \frac{1}{2}\left|(2\vec{a}+3\vec{b})\times(\vec{a}-\vec{b})\right|$

$= \frac{1}{2}\left|-2\vec{a}\times\vec{b} + 3\vec{b}\times\vec{a}\right|$

$= \frac{1}{2}\times 5|\vec{a}\times\vec{b}| = \frac{5}{2}\times 40 = 100$

$\therefore$ (R) → (1)

(S) $|\vec{a}\times\vec{b}| = 30$

$\therefore \left|(\vec{a}+\vec{b})\times\vec{a}\right| = \left|(\vec{b}\times\vec{a})\right| = 30$

$\therefore$ (s) → (2)

8. (a) Any point on L_1 is $(2\lambda + 1, -\lambda, \lambda - 3)$ and that on L_2 is $(\mu + 4, \mu - 3, 2\mu - 3)$

For point of intersection of L_1 and L_2

$2\lambda + 1 = \mu + 4, -\lambda = \mu - 3, \lambda - 3 = 2\mu - 3$

$\Rightarrow \quad \lambda = 2, \mu = 1$

$\therefore$ Intersection point of L_1 and L_2 is (5, –2, – 1)

$\because$ $ax + by + cz = d$ is perpendicular to p_1 & p_2

$\therefore$ $7a + b + 2c = 0$ and $3a + 5b - 6c = 0$

$\Rightarrow \frac{a}{-16} = \frac{b}{48} = \frac{c}{32} \Rightarrow \frac{a}{1} = \frac{b}{-3} = \frac{c}{-2}$

$\therefore$ Equation of plane is $x - 3y - 2z = d$

As it passes through (5, –2, –1)

$\therefore$ $5 + 6 + 2 = d = 13$

$\therefore$ $a = 1, b = -3, c = -2, d = 13$

or (P) → (3)

(Q) → (2)

(R) → (4)

(S) → (1)

COMPREHENSION BASED QUESTIONS :

1. (b) Vector in the direction of $L_1 = \overrightarrow{n_1} = 3\hat{i} + \hat{j} + 2\hat{k}$

Vector in the direction of $L_2 = \overrightarrow{n_2} = \hat{i} + 2\hat{j} + 3\hat{k}$

$\therefore$ Vector perpendicular to both L_1 and L_2

$= \vec{n_1} \times \vec{n_2} = \begin{vmatrix} \hat{i} & \hat{j} & \hat{k} \\ 3 & 1 & 2 \\ 1 & 2 & 3 \end{vmatrix} = -\hat{i} - 7\hat{j} + 5\hat{k}$

$\therefore$ Required unit vector

$= \hat{n} = \dfrac{-\hat{i}-7\hat{j}+5\hat{k}}{\sqrt{1+49+25}} = \dfrac{-\hat{i}-7\hat{j}+5\hat{k}}{5\sqrt{3}}$

2. **(d)** The shortest distance between L_1 and L_2 is

$= \dfrac{(\vec{a}_2 - \vec{a}_1).\vec{b}_1 \times \vec{b}_2}{|\vec{b}_1 \times \vec{b}_2|} = (\vec{a}_2 - \vec{a}_1).\hat{n}$

where $a_1 = -\hat{i} - 2\hat{j} - \hat{k}$

$a_2 = 2\hat{i} - 2\hat{j} + 3\hat{k}$

$\therefore \vec{a}_2 - \vec{a}_1 = 3\hat{i} + 4\hat{k} \quad \therefore (\vec{a}_2 - \vec{a}_1).\hat{n}$

$\therefore (3\hat{i}+4\hat{k}).\left(\dfrac{-\hat{i}-7\hat{j}+5\hat{k}}{5\sqrt{3}}\right) = \dfrac{-3+20}{5\sqrt{3}} = \dfrac{17}{5\sqrt{3}}$

3. **(c)** The plane passing through (–1, –2, –1) and having normal along $\vec{n}$ is

$-1(x+1) - 7(y+2) + 5(z+1) = 0$

or $x + 7y - 5z + 10 = 0$

$\therefore$ Distance of point (1, 1, 1) from the above plane is

$= \dfrac{1+7\times1-5\times1+10}{\sqrt{1+49+25}} = \dfrac{13}{\sqrt{75}}$

(H) ASSERTION & REASON TYPE QUESTIONS :

1. **(d)** The line of intersection of given plane is

$3x - 6y - 2z - 15 = 0 = 2x + y - 2z - 5$

For $z = 0$, we obtain $x = 3$ and $y = -1$

$\therefore$ Line passes through (3, –1, 0).

Let *a, b, c* be the d'rs of line of intersection, then

$3a - 6b - 2c = 0$

$2a + b - 2c = 0$

Solving the above equations using cross product method, we get $a : b : c = 14 : 2 : 15$

$\therefore$ Eqn. of line is

$\dfrac{x-3}{14} = \dfrac{y+1}{2} = \dfrac{z}{15} = t$

whose parametric form is

$x = 3 + 14t, y = 1 + 2t, z = 15t$

$\therefore$ Statement-I is false (value of *y* is not matching).

Since dr's of line intersection of given planes are 14,2,15

$\therefore$ $14\hat{i} + 2\hat{j} + 15\hat{k}$ is parallel to this line.

$\therefore$ Statement 2 is true.

2. **(c)** $\overrightarrow{PQ} \times (\overrightarrow{RS} + \overrightarrow{ST}) = \overrightarrow{PQ} \times \overrightarrow{RT}$

$= |\overrightarrow{PQ}| \times |\overrightarrow{RT}| \sin 150 \hat{n} \neq 0$

$\neq 0$

Statement-1 is true.

Also, $\overrightarrow{PQ} \times \overrightarrow{RS} = |\overrightarrow{PQ}| \times |\overrightarrow{RS}| \sin 120° \times \hat{n}_1 \neq 0$

And $\overrightarrow{PQ} \times \overrightarrow{ST} = \overrightarrow{PQ} | \times | \overrightarrow{ST} | \sin 180° \times \hat{n}_2 = 0$

$\therefore$ Statement-2 is false.

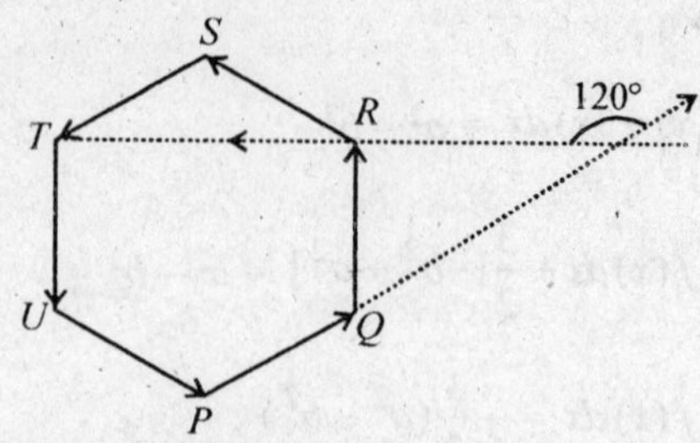

3. **(d)** The given planes are

$P_1 : x - y + z = 1$ ...(1)

$P_2 : x + y - z = -1$ ...(2)

$P_3 : x - 3y + 3z = 2$ ...(3)

Line L_1 is intersection of planes P_2 and P_3.

$\therefore L_1$ is parallel to the vector

$= \begin{vmatrix} \hat{i} & \hat{j} & \hat{k} \\ 1 & 1 & -1 \\ 1 & -3 & 3 \end{vmatrix} = -4\hat{j} - 4\hat{k}$

Line L_2 is intersection of P_3 and P_1

$\therefore L_2$ is parallel to the vector

$= \begin{vmatrix} \hat{i} & \hat{j} & \hat{k} \\ 1 & -1 & 1 \\ 1 & -3 & 3 \end{vmatrix} = -2\hat{j} - 2\hat{k}$

Line L_3 is intersection of P_1 and P_2

$\therefore L_3$ is parallel to the vector

$= \begin{vmatrix} \hat{i} & \hat{j} & \hat{k} \\ 1 & -1 & 1 \\ 1 & 1 & -1 \end{vmatrix} = 2\hat{j} + 2\hat{k}$

Clearly lines L_1, L_2 and L_3 are parallel to each other.

$\therefore$ Statement–1 is False

Also family of planes passing through the intersection of P_1 and P_2 is $P_1 + \lambda P_2 = 0$. If plane P_3 is represented by $P_1 + \lambda P_2 = 0$ for some value of λ then the three planes pass through the same point.

Here $P_1 + \lambda P_2 = 0$

$\Rightarrow x(1+\lambda) + y(\lambda - 1) + z(1-\lambda) + (\lambda - 1) = 0$

This will be identical to P_3 if

$\dfrac{1+\lambda}{1} = \dfrac{\lambda-1}{-3} = \dfrac{1-\lambda}{3} = \dfrac{1-\lambda}{2}$ ...(1)

Taking $\dfrac{1+\lambda}{1} = \dfrac{1-\lambda}{2}$, we get $\lambda = -\dfrac{1}{3}$ and taking $\dfrac{1+\lambda}{1} = \dfrac{1-\lambda}{3}$, we get $\lambda = -\dfrac{2}{3}$.

$\therefore$ There is no value of λ which satisfies eq (1).

$\therefore$ The three planes do not have a common point.

$\Rightarrow$ Statement 2 is true.

$\therefore$ (d) is the correct option.

1 INTEGER VALUE CORRECT TYPE :

1. We have $\bar{a}=\frac{\hat{i}-2j}{\sqrt{5}}, \bar{b}=\frac{2\hat{i}+j+3\hat{k}}{\sqrt{14}}$

Clearly $|\vec{a}|=1$ $|\vec{b}|=1$ and $\vec{a}.\vec{b}=0$

$$(2\vec{a}+\vec{b}).\left[(\vec{a}\times\vec{b})\times(\vec{a}-2\vec{b})\right]$$

$$=-(2\vec{a}+\vec{b}).\left[(\vec{a}-2\vec{b})\times(\vec{a}\times\vec{b})\right]$$

$$=-(2\vec{a}+\vec{b}).\left[((\vec{a}-2\vec{b}).\vec{b})\vec{a}-((\vec{a}-2\vec{b}).\vec{a})\vec{b}\right]$$

$$=-(2\vec{a}+\vec{b}).\left[\left(\vec{a}.\vec{b}-2|\vec{b}|^2\right)\vec{a}-\left(|\vec{a}|^2-2\vec{b}.\vec{a}\right)\vec{b}\right]$$

$$=-(2\vec{a}+\vec{b}).\left[-2\vec{a}-\vec{b}\right]$$

$$=(2\vec{a}+\vec{b}).(2\vec{a}+\vec{b})=4|\vec{a}|^2+|\vec{b}|^2\ (\because \vec{a}.\vec{b}=0)$$

$$=4+1=5.$$

2. The equation of plane containing the lines $\frac{x-1}{2}=\frac{y-2}{3}=\frac{z-3}{4}$ and $\frac{x-2}{3}=\frac{y-3}{4}=\frac{z-4}{5}$ is

$$\begin{vmatrix} x-1 & y-2 & z-3 \\ 2 & 3 & 4 \\ 3 & 4 & 5 \end{vmatrix}=0 \Rightarrow x-2y+z=0$$

$\therefore$ Distance between $x-2y+z=0$ and $Ax-2y+z=d$

= Perpendicular distance between parallel planes $(\because A=1)$

$$=\frac{|d|}{\sqrt{6}}=\sqrt{6} \Rightarrow |d|=6.$$

3. (9) We have $\vec{r}\times\vec{b}-\vec{c}\times\vec{b}=\vec{0}$

$\Rightarrow (\vec{r}-\vec{c})\times\vec{b}=\vec{0} \Rightarrow \vec{r}-\vec{c}\parallel\vec{b}$

Let $\vec{r}-\vec{c}=\lambda\vec{b}$ or $\vec{r}=\vec{c}+\lambda\vec{b}$

$\Rightarrow \vec{r}=\hat{i}+2\hat{j}+3\hat{k}-\lambda\hat{i}+\lambda\hat{j}=(1-\lambda)i+(2+\lambda)\hat{j}+3\hat{k}$

$\vec{r}.\vec{a}=0 \Rightarrow -1+\lambda-3=0 \Rightarrow \lambda=4$

$\therefore \vec{r}=-3\hat{i}+6\hat{j}+3\hat{k}$

$\therefore \vec{r}.\vec{b}=3+6=9$

4. (3) $\vec{a},\vec{b},\vec{c}$ are units vectors such that

$$|\vec{a}-\vec{b}|^2+|\vec{b}-\vec{c}|^2+|\vec{c}-\vec{a}|^2=9$$

$$\Rightarrow 2\left(|\vec{a}|^2+|\vec{b}|^2+|\vec{c}|^2\right)-2(\vec{a}.\vec{b}+\vec{b}.\vec{c}+\vec{c}.\vec{a})=9$$

$$\Rightarrow \vec{a}.\vec{b}+\vec{b}.\vec{c}+\vec{c}.\vec{a}=\frac{-3}{2}$$

Also $|\vec{a}+\vec{b}+\vec{c}|^2=|\vec{a}|^2+|\vec{b}|^2+|\vec{c}|^2+2(\vec{a}.\vec{b}+\vec{b}.\vec{c}+\vec{c}.\vec{a})$

$$=1+1+1+2\times\left(-\frac{3}{2}\right)=0$$

$$\Rightarrow \vec{a}+\vec{b}+\vec{c}=0 \Rightarrow (\vec{b}+\vec{c})=-\vec{a}$$

$$\therefore |2\vec{a}+5(\vec{b}+\vec{c})|=|2\vec{a}-5\vec{a}|=|-3\vec{a}|=3$$

5. (5) Given 8 vectors are
(1, 1, 1), (–1, –1, –1); (–1, 1, 1), (1, –1, –1); (1, –1, 1), (–1, 1, –1); (1, 1, –1), (–1, –1, 1)
These are 4 diagonals of a cube and their opposites.
For 3 non coplanar vectors first we select 3 groups of diagonals and its opposite in ${}^4C_3$ ways. Then one vector from each group can be selected in 2 × 2 × 2 ways.

$\therefore$ Total ways $={}^4C_3\times2\times2\times2=32=2^5$

$\therefore$ $p=5$

6. (5) Let $k, k+1$ be removed from pack.

$\therefore (1+2+3+...+n)-(k+k+1)=1224$

$$\frac{n(n+1)}{2}-2k=1225$$

$$k=\frac{n(n+1)-2450}{4}$$

for $n=50,\ k=25$

$\therefore$ $k-20=5$

Section-B JEE Main/ AIEEE

1. (a) As the point (3, 2, 0) lies on the given line $\frac{x-4}{1}=\frac{y-7}{5}=\frac{z-4}{4}$

$\therefore$ There can be infinite many planes passing through this line. But here out of the four options only first option is satisfied by the coordinates of both the points (3, 2, 0) and (4, 7, 4)

$\therefore$ $x-y+z=1$ is the required plane.

2. (b) We have, $\vec{a}\cdot\vec{b}=|\vec{a}||\vec{b}|\cos\frac{\pi}{6}$

$$=4\times2\times\frac{\sqrt{3}}{2}=4\sqrt{3}.$$

Now, $(\vec{a}\times\vec{b})^2+(\vec{a}\cdot\vec{b})^2=a^2b^2$;

$$\Rightarrow (\vec{a}\times\vec{b})^2+48=16\times4 \Rightarrow (\vec{a}\times\vec{b})^2=16$$

3. (a) We have, $[\vec{a}\times\vec{b}\ \ \vec{b}\times\vec{c}\ \ \vec{c}\times\vec{a}]$

$$=(\vec{a}\times\vec{b}).\left\{(\vec{b}\times\vec{c})\times(\vec{c}\times\vec{a})\right\}$$

$$=(\vec{a}\times\vec{b}).\left\{(\vec{m}\cdot\vec{a})\vec{c}-(\vec{m}\cdot\vec{c})\vec{a}\right\}$$

(where $\vec{m} = \vec{b} \times \vec{c}$)

$= \{(\vec{a} \times \vec{b}).\vec{c}\}.\{(\vec{a} \cdot (\vec{b} \times \vec{c})\} = [\vec{a}\ \vec{b}\ \vec{c}]^2 = 4^2 = 16 \cdot$

4. (a) $\vec{a} + \vec{b} + \vec{c} = 0 \Rightarrow \vec{b} + \vec{c} = -\vec{a}$

$\Rightarrow (\vec{b} + \vec{c})^2 = (\vec{a})^2 = 5^2 + 3^2 + 2\vec{b} \cdot \vec{c} = 7^2$

$\Rightarrow 2|\vec{b}||\vec{c}|\cos\theta = 49 - 34 = 15;$

$\Rightarrow 2 \times 5 \times 3\cos\theta = 15;$

$\Rightarrow \cos\theta = 1/2; \Rightarrow \theta = \frac{\pi}{3} = 60°$

5. (a) We have, $\vec{a} + \vec{b} + \vec{c} = \vec{0} \Rightarrow (\vec{a} + \vec{b} + \vec{c})^2 = 0$

$\Rightarrow |\vec{a}|^2 + |\vec{b}|^2 + |\vec{c}|^2 + 2(\vec{a} \cdot \vec{b} + \vec{b} \cdot \vec{c} + \vec{c} \cdot \vec{a}) = 0$

$\Rightarrow 25 + 16 + 9 + 2(\vec{a} \cdot \vec{b} + \vec{b} \cdot \vec{c} + \vec{c} \cdot \vec{a}) = 0$

$\Rightarrow (\vec{a} \cdot \vec{b} + \vec{b} \cdot \vec{c} + \vec{c} \cdot \vec{a}) = -25 \cdot$

$\therefore |\vec{a} \cdot \vec{b} + \vec{b} \cdot \vec{c} + \vec{c} \cdot \vec{a}| = 25 \cdot$

6. (a) Since $\vec{a}, \vec{c}, \vec{b}$ form a right handed system,

$$\therefore \quad \vec{c} = \vec{b} \times \vec{a} = \begin{vmatrix} \hat{i} & \hat{j} & \hat{k} \\ 0 & 1 & 0 \\ x & y & z \end{vmatrix} = z\hat{i} - x\hat{k}$$

7. (b) We have $\vec{a} \times \vec{b} = 39\vec{k} = \vec{c}$

Also $|\vec{a}| = \sqrt{34}, |\vec{b}| = \sqrt{45}, |\vec{c}| = 39;$

$\therefore |\vec{a}| : |\vec{b}| : |\vec{c}| = \sqrt{34} : \sqrt{45} : 39 \cdot$

$\vec{a} + \vec{b} + \vec{c} = \vec{r}$. Then

$(\vec{a} + \vec{b} + \vec{c}) = \vec{a} \times \vec{r} \Rightarrow 0 + \vec{a} \times \vec{b} + \vec{a} \times \vec{c} = \vec{a} \times \vec{r}$

$\times \vec{b} - \vec{c} \times \vec{a} = \vec{a} \times \vec{r} \Rightarrow \vec{a} \times \vec{r} = \vec{0}$

...milarly $\vec{b} \times \vec{r} = \vec{0}$ & $\vec{c} \times \vec{r} = \vec{0}$

...bove three conditions will be satisfied for non-zero ...ectors if and only if $\vec{r} = \vec{0}$

9. (b) Equation of plane through (1, 0, 0) is

$a(x-1) + by + cz = 0$ ...(i)

(i) passes through (0, 1, 0).

$-a + b = 0 \Rightarrow b = a$; Also, cos 45°

$= \frac{a+a}{\sqrt{2(2a^2+c^2)}} \Rightarrow 2a = \sqrt{2a^2+c^2} \Rightarrow 2a^2 = c^2$

$\Rightarrow c = \sqrt{2}a \cdot$

So d.r of normal are a, a $\sqrt{2}a$ i.e. 1, 1, $\sqrt{2}$.

10. (a) since $\vec{n}$ is perpendicular $\vec{u}$ and $\vec{v}$, $\vec{n} = \frac{\vec{u} \times \vec{v}}{|\vec{u}||\vec{v}|}$

$$\hat{n} = \frac{\begin{vmatrix} i & j & k \\ 1 & 1 & 0 \\ 1 & -1 & 0 \end{vmatrix}}{\sqrt{2} \times \sqrt{2}} = \frac{-2\hat{k}}{2} = -\hat{k}$$

$|\vec{\omega}.\hat{n}| = |(i + 2j + 3k).(-k)| = |-3| = 3$

11. (d) $\vec{F} + \vec{F}_1 + \vec{F}_2 = 7i + 2j - 4k$

$\vec{d} = P.V\ of\ \vec{B} - P.V\ of\ \vec{A} = 4i + 2j - 2k$

$W = \vec{F}.\vec{d} = 28 + 4 + 8 = 40\ unit$

12. (d)

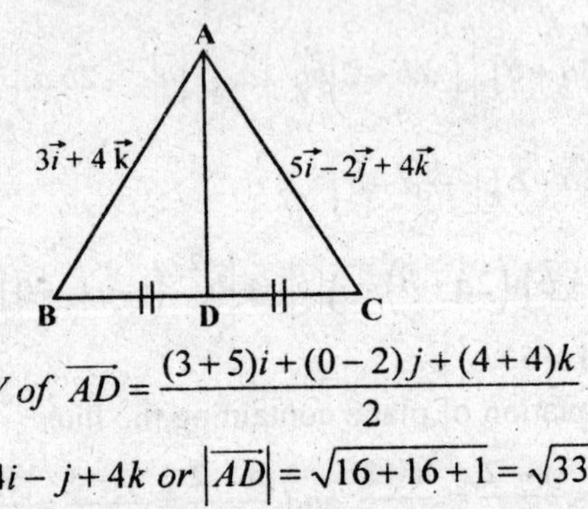

$P.V\ of\ \overline{AD} = \frac{(3+5)i + (0-2)j + (4+4)k}{2}$

$= 4i - j + 4k$ or $|\overline{AD}| = \sqrt{16+16+1} = \sqrt{33}$

13. (d) Shortest distance = perpendicular distance between the plane and sphere = distance of plane from centre of sphere – radius

$= \left|\frac{-2\times12 + 4\times1 + 3\times3 - 327}{\sqrt{144+9+16}}\right| - \sqrt{4+1+9+155}$

$= 26 - 13 = 13$

14. (a) $\frac{x-b}{a} = \frac{y}{1} = \frac{z-d}{c}; \frac{x-b'}{a'} = \frac{y}{1} = \frac{z-d'}{c'}$

For perpendicularity of lines $aa' + 1 + cc' = 0$

15. (d) $\begin{vmatrix} x_2 - x_1 & y_2 - y_1 & z_2 - z_1 \\ l_1 & m_1 & n_1 \\ l_2 & m_2 & n_2 \end{vmatrix} = 0$

$$\therefore \begin{vmatrix} 1 & -1 & -1 \\ 1 & 1 & -k \\ k & 2 & 1 \end{vmatrix} = 0 \Rightarrow \begin{vmatrix} 0 & 0 & -1 \\ 2 & 1+k & -k \\ k+2 & 1 & 1 \end{vmatrix} = 0$$

$k^2 + 3k = 0 \Rightarrow k(k+3) = 0$ or $k = 0$ or -3

16. (c) $\vec{a} + \vec{b} + \vec{c} = 0 \Rightarrow (\vec{a} + \vec{b} + \vec{c}).(\vec{a} + \vec{b} + \vec{c}) = 0$

$|\vec{a}|^2 + |\vec{b}|^2 + |\vec{c}|^2 + 2(\vec{a}.\vec{b} + \vec{b}.\vec{c} + \vec{c}.\vec{a}) = 0$

$\vec{a}.\vec{b} + \vec{b}.\vec{c} + \vec{c}.\vec{a} = \frac{-1-4-9}{2} = -7$

17. (d)

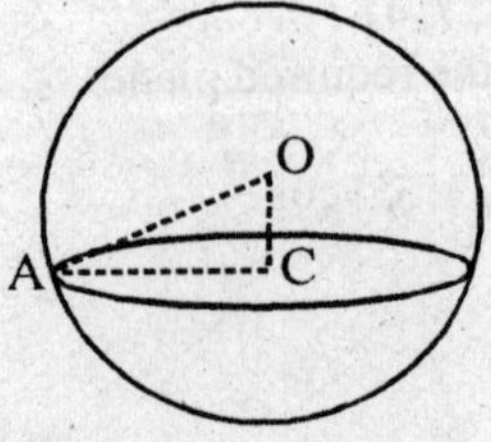

Centre of sphere = (–1, 1, 2)

Radius of sphere $\sqrt{1+1+4+19}=5$

Perpendicular distance from centre to the plane

$$OC = d = \left|\frac{-1+2+4+7}{\sqrt{1+4+4}}\right| = \frac{12}{3} = 4.$$

$$AC^2 = AO^2 - OC^2 = 5^2 - 4^2 = 9 \Rightarrow AC = 3$$

18. **(b)** Vector perpendicular to the face OAB

$$= \overrightarrow{OA} \times \overrightarrow{OB} = \begin{vmatrix} \hat{i} & \hat{j} & \hat{k} \\ 1 & 2 & 1 \\ 2 & 1 & 3 \end{vmatrix} = 5\hat{i} - \hat{j} - 3\hat{k}$$

Vector perpendicular to the face ABC

$$= \overrightarrow{AB} \times \overrightarrow{AC} = \begin{vmatrix} \hat{i} & \hat{j} & \hat{k} \\ 1 & -1 & 2 \\ -2 & -1 & 1 \end{vmatrix} = \hat{i} - 5\hat{j} - 3\hat{k}$$

Angle between the faces = angle between their normals

$$\cos\theta = \left|\frac{5+5+9}{\sqrt{35}\sqrt{35}}\right| = \frac{19}{35} \text{ or } \theta = \cos^{-1}\left(\frac{19}{35}\right)$$

19. **(c)**

$$\begin{vmatrix} a & a^2 & 1+a^3 \\ b & b^2 & 1+b^3 \\ c & c^2 & 1+c^3 \end{vmatrix} = 0 \Rightarrow \begin{vmatrix} a & a^2 & 1 \\ b & b^2 & 1 \\ c & c^2 & 1 \end{vmatrix} + \begin{vmatrix} a & a^2 & a^3 \\ b & b^2 & b^3 \\ c & c^2 & c^3 \end{vmatrix} = 0$$

$$\Rightarrow (1+abc)\begin{vmatrix} 1 & a & a^2 \\ 1 & b & b^2 \\ 1 & c & c^2 \end{vmatrix} = 0$$

$$As \begin{vmatrix} 1 & a & a^2 \\ 1 & b & b^2 \\ 1 & c & c^2 \end{vmatrix} \neq 0 \; (given\ condition) \quad \therefore abc = -1$$

20. $A = (7,-4,7), B = (1,-6,10), C = (-1,-3,4)$ and $D = (5,-1,5)$

$$AB = \sqrt{(7-1)^2 + (-4+6)^2 + (7-10)^2}$$

$$= \sqrt{36+4+9} = 7$$

Similarly $BC = 7, CD = \sqrt{41}, DA = \sqrt{17}$

∴ None of the options is satisfied

21. **(c)** $(\vec{u}+\vec{v}-\vec{w}).(\vec{u}\times\vec{v}-\vec{u}\times\vec{w}-\vec{v}\times\vec{v}+\vec{v}\times\vec{w})$

$= (\vec{u}+\vec{v}-\vec{w}).(\vec{u}\times\vec{v}-\vec{u}\times\vec{w}+\vec{v}\times\vec{w}) = \vec{u}.(\vec{u}\times\vec{v})$
$-\vec{u}.(\vec{u}\times\vec{w}) + \vec{u}.(\vec{v}\times\vec{w}) + \vec{v}.(\vec{u}\times\vec{v}) - \vec{v}.(\vec{u}\times\vec{w})$
$+\vec{v}.(\vec{v}\times\vec{w}) - \vec{w}.(\vec{u}\times\vec{v}) + \vec{w}.(\vec{u}\times\vec{w}) - \vec{w}.(\vec{u}\times\vec{w})$
$= \vec{u}.(\vec{v}\times\vec{w}) - \vec{v}.(\vec{u}\times\vec{w}) - \vec{w}.(\vec{u}\times\vec{v})$
$= [\vec{u}\vec{v}\vec{w}] + [\vec{v}\vec{w}\vec{u}] - [\vec{w}\vec{u}\vec{v}] = \vec{u}.(\vec{v}\times\vec{w})$

22. **(a)** Eq. of planes be $\frac{x}{a}+\frac{y}{b}+\frac{z}{c}=1$ & $\frac{x}{a'}+\frac{y}{b'}+\frac{z}{c'}=1$

($\perp r$ distance on plane from origin is same.)

$$\left|\frac{-1}{\sqrt{\frac{1}{a^2}+\frac{1}{b^2}+\frac{1}{c^2}}}\right| = \left|\frac{-1}{\sqrt{\frac{1}{a'^2}+\frac{1}{b'^2}+\frac{1}{c'^2}}}\right|$$

$$\frac{1}{a^2}+\frac{1}{b^2}+\frac{1}{c^2}-\frac{1}{a'^2}-\frac{1}{b'^2}-\frac{1}{c'^2}=0$$

23. **(c)** The planes are $2x+y+2z-8=0$. ...(1)

and $4x+2y+4z+5=0$

or $2x+y+2z+\frac{5}{2}=0$ ...(2)

∴ Distance between (1) and (2)

$$= \left|\frac{\frac{5}{2}+8}{\sqrt{2^2+1^2+2^2}}\right| = \left|\frac{21}{2\sqrt{9}}\right| = \frac{7}{2}$$

24. **(b)** Let a point on the line $x = y+a = z$ is $(\lambda, \lambda-a, \lambda)$ and a point on the line $x+a = 2y = 2z$ is $\left(\mu-a, \frac{\mu}{2}, \frac{\mu}{2}\right)$, then direction ratio of the line joining these points are $\lambda-\mu+a, \lambda-a-\frac{\mu}{2}, \lambda-\frac{\mu}{2}$

If it respresents the required line, then

$$\frac{\lambda-\mu+a}{2} = \frac{\lambda-a-\frac{\mu}{2}}{1} = \frac{\lambda-\frac{\mu}{2}}{2}$$

on solving we get $\lambda = 3a, \mu = 2a$

∴ The required points of intersection are (3*a*, 3*a*-*a*,3*a*) and $\left(2a-a, \frac{2a}{2}, \frac{2a}{2}\right)$

or $(3a, 2a, 3a)$ and (a, a, a)

25. **(d)** The given lines are

$$x-1 = \frac{y+3}{-\lambda} = \frac{z-1}{\lambda} = s \text{ ....(1)}$$

and $2x = y-1 = \frac{z-2}{-1} = t$ ....(2)

The lines are coplanar, if

$$\begin{vmatrix} 0-(-1) & -1-3 & -2-(-1) \\ 1 & -\lambda & \lambda \\ \frac{1}{2} & 1 & -1 \end{vmatrix} = 0$$

$$c_2 \to c_2 + c_3; \begin{vmatrix} 1 & -5 & -1 \\ 1 & 0 & \lambda \\ \frac{1}{2} & 0 & -1 \end{vmatrix} = 0$$

$$\Rightarrow 5\left(-1-\frac{\lambda}{2}\right) = 0 \Rightarrow \lambda = -2$$

26. (a) The equations of spheres are

$S_1: x^2+y^2+z^2+7x-2y-z-13=0$ *and*

$S_2: x^2+y^2+z^2-3x+3y+4z-8=0$

Their plane of intersection is

$S_1-S_2=0 \Rightarrow 10x-5y-5z-5=0$

$\Rightarrow 2x-y-z=1$

27. (c) Let $\vec{a}+2\vec{b}=t\vec{c}$ and $\vec{b}+3\vec{c}=s\,\vec{a}$, where t and s are scalars. Adding, we get

$\vec{a}+3\vec{b}+3\vec{c}=t\vec{c}+s\vec{a} \Rightarrow \vec{a}+2\vec{b}+6\vec{c}=t\vec{c}+s\vec{a}-\vec{b}+3\vec{c}$

$=t\vec{c}+(\vec{b}+3\vec{c})-\vec{b}+3\vec{c}=(t+6)\vec{c}$

$\left[\text{using } s\,\vec{a}=\vec{b}+3\vec{c}\right]$

$=\lambda\vec{c}$, where $\lambda=t+6$

28. (d) Resultant of forces $\vec{F}=7\hat{i}+2\hat{j}-4\hat{k}$

Disptacement $\vec{d}=4\hat{i}+2\hat{j}-2\hat{k}$

$\therefore$ Work done $=\vec{F}.\vec{d}=28+4+8=40$

29. (c) Vectors $\vec{a}+2\vec{b}+3\vec{c}, \lambda\vec{b}+4\vec{c}$, and $(2\lambda-1)\vec{c}$ are

coplanar if $\begin{vmatrix} 1 & 2 & 3 \\ 0 & \lambda & 4 \\ 0 & 0 & 2\lambda-1 \end{vmatrix}=0$

$\Rightarrow \lambda(2\lambda-1)=0 \Rightarrow \lambda=0$ or $\frac{1}{2}$

$\therefore$ Forces are noncoplanar for all λ, except $\lambda=0, \frac{1}{2}$

30. (c) Projection of $\vec{v}$ along $\vec{u}$ $=\frac{\vec{v}.\vec{u}}{|\vec{u}|}=\frac{\vec{v}.\vec{u}}{2}$

projection of $\vec{w}$ along $\vec{u}$ $=\frac{\vec{w}.\vec{u}}{|\vec{u}|}=\frac{\vec{w}.\vec{u}}{2}$

Given $\frac{\vec{v}.\vec{u}}{2}=\frac{\vec{w}.\vec{u}}{2}$ ....(1)

Also , $\vec{v}.\vec{w}=0$ ....(2)

Now $|\vec{u}-\vec{v}+\vec{w}|^2$

$=|\vec{u}|^2+|\vec{v}|^2+|\vec{w}|^2-2\vec{u}.\vec{v}-2\vec{v}.\vec{w}+2\vec{u}.\vec{w}$

$=1+4+9+0$ [From (1) and (2)] $=14$

$\therefore |\vec{u}-\vec{v}+\vec{w}|=\sqrt{14}$

31. (a) Given $(\vec{a}\times\vec{b})\times\vec{c}=\frac{1}{3}|\vec{b}||\vec{c}|\,\vec{a}$

Clearly $\vec{a}$ and $\vec{b}$ are noncollinear

$\Rightarrow (\vec{a}.\vec{c})\vec{b}-(\vec{b}.\vec{c})\vec{a}=\frac{1}{3}|\vec{b}||\vec{c}|\,\vec{a}$

$\therefore\ \vec{a}.\vec{c}=0$ and $-\vec{b}.\vec{c}=\frac{1}{3}|\vec{b}||\vec{c}| \Rightarrow \cos\theta=\frac{-1}{3}$

$\therefore \sin\theta=\sqrt{1-\frac{1}{9}}=\frac{2\sqrt{2}}{3}$

[θ is acute angle between $\vec{b}$ and $\vec{c}$]

32. (a) $\overrightarrow{PA}+\overrightarrow{AP}=0$ and $\overrightarrow{PC}+\overrightarrow{CP}=0$

$\Rightarrow \overrightarrow{PA}+\overrightarrow{AC}+\overrightarrow{CP}=0$ and $\overrightarrow{PB}+\overrightarrow{BC}+\overrightarrow{CP}=0$

Adding, we get $\overrightarrow{PA}+\overrightarrow{PB}+\overrightarrow{AC}+\overrightarrow{BC}+2\overrightarrow{CP}=0.$

Since $\overrightarrow{AC}=-\overrightarrow{BC}$ & $\overrightarrow{CP}=-\overrightarrow{PC}$

$\Rightarrow \overrightarrow{PA}+\overrightarrow{PB}-2\overrightarrow{PC}=0.$

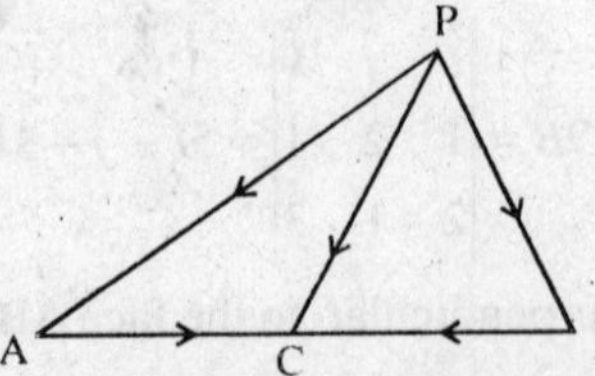

33. (a) If θ is the angle between line and plane then $\left(\frac{\pi}{2}-\theta\right)$ is the angle between line and normal to plane given by

$\cos\left(\frac{\pi}{2}-\theta\right)=\frac{(\hat{i}+2\hat{j}+2\hat{k}).(2\hat{i}-\hat{j}+\sqrt{\lambda}\hat{k})}{3\sqrt{4+1+\lambda}}$

$\cos\left(\frac{\pi}{2}-\theta\right)=\frac{2-2+2\sqrt{\lambda}}{3\times\sqrt{5+\lambda}}$

$\Rightarrow \sin\theta=\frac{2\sqrt{\lambda}}{3\sqrt{5+\lambda}}=\frac{1}{3} \Rightarrow 4\lambda=5+\lambda$

$\Rightarrow \lambda=\frac{5}{3}.$

34. (b) The given lines are $2x=3y=-z$

or $\frac{x}{3}=\frac{y}{2}=\frac{z}{-6}$ [Dividing by 6]

and $6x=-y=-4z$

or $\frac{x}{2}=\frac{y}{-12}=\frac{z}{-3}$ [Dividing by 12]

$\therefore$ Angle between two lines is

$\cos\theta=\frac{3.2+2.(-12)+(-6).(-3)}{\sqrt{3^2+2^2+(-6)^2}\sqrt{2^2+(-12)^2+(-3)^2}}$

$=\frac{6-24+18}{\sqrt{49}\sqrt{157}}=0 \Rightarrow \theta=90^\circ$

35. (c) Centers of given spheres are $(-3, 4, 1)$ and $(5, -2, 1)$. Mid point of centres is $(1, 1, 1)$.
Satisfying this in the equation of plane, we get

$2a-3a+4a+6=0 \Rightarrow a=-2.$

36. (b) A point on line is $(2, -2, 3)$ its perpendicular distance from the plane $x+5y+z-5=0$ is

$=\left|\frac{2-10+3-5}{\sqrt{1+25+1}}\right|=\frac{10}{3\sqrt{3}}$

37. (c) Let $\vec{a}=x\vec{i}+y\vec{j}+z\vec{k}$

$\vec{a}\times\vec{i}=z\vec{j}-y\vec{k} \Rightarrow (\vec{a}\times\vec{i})^2=y^2+z^2$

Similarly, $(\vec{a}\times\vec{j})^2 = x^2+z^2$ and $(\vec{a}\times\vec{k})^2 = x^2+y^2$

$\Rightarrow (\vec{a}\times\vec{i})^2+(\vec{a}\times\vec{j})^2+(\vec{a}\times\vec{k})^2$

$= 2(x^2+y^2+z^2) = 2\vec{a}^2$

38. (c) a, b, c are in H.P. $\Rightarrow \frac{1}{a},\frac{1}{b},\frac{1}{c}$ are in A.P.

$\Rightarrow \frac{2}{b}=\frac{1}{a}+\frac{1}{c} \Rightarrow \frac{1}{a}-\frac{2}{b}+\frac{1}{c}=0$

$\therefore \frac{x}{a}+\frac{y}{a}+\frac{1}{c}=0$ passes through (1, –2)

39. (a) Vector $a\vec{i}+a\vec{j}+c\vec{k}$, $\vec{i}+\vec{k}$ and $c\vec{i}+c\vec{j}+b\vec{k}$ are coplanar

$$\begin{vmatrix} a & a & c \\ 1 & 0 & 1 \\ c & c & b \end{vmatrix} = 0 \Rightarrow c^2 = ab \Rightarrow c = \sqrt{ab}$$

$\therefore$ c is G.M. of a and b.

40. (b) Let us consider

$\vec{a} = a_1\hat{i}+a_2\hat{j}+a_3\hat{k}$; $\vec{b} = b_1\hat{i}+b_2\hat{j}+b_3\hat{k}$]

$\vec{c} = c_1\hat{i}+c_2\hat{j}+c_3\hat{k}$

then as per question

$$\left[\lambda(\vec{a}+\vec{b})\ \lambda^2\vec{b}\ \lambda\vec{c}\right] = \left[\vec{a}\ \vec{b}+\vec{c}\ \vec{b}\right]$$

$$\Rightarrow \begin{vmatrix} \lambda(a_1+b_1) & \lambda(a_2+b_2) & \lambda(a_3+b_3) \\ \lambda^2 b_1 & \lambda^2 b_2 & \lambda b_3 \\ \lambda c_1 & \lambda c_2 & \lambda c_3 \end{vmatrix}$$

$$= \begin{vmatrix} a_1 & a_2 & a_3 \\ b_1+c_1 & b_2+c_2 & b_3+c_3 \\ b_1 & b_2 & b_3 \end{vmatrix}$$

$$\Rightarrow \lambda^4 \begin{vmatrix} a_1+b_1 & a_2 b_2 & a_3+b_3 \\ b_1 & b_2 & b_3 \\ c_1 & c_2 & c_3 \end{vmatrix}$$

$$= \begin{vmatrix} a_1 & a_2 & a_3 \\ b_1+c_1 & b_2+c_2 & b_3+c_3 \\ b_1 & b_2 & b_3 \end{vmatrix}$$

$R_1 - R_2 \qquad R_2 - R_3$

$$\Rightarrow \lambda^4 \begin{vmatrix} a_1 & a_2 & a_3 \\ b_1 & b_2 & b_3 \\ c_1 & c_2 & c_3 \end{vmatrix} = \begin{vmatrix} a_1 & a_2 & a_3 \\ c_1 & c_2 & c_3 \\ b_1 & b_2 & b_3 \end{vmatrix} \Rightarrow \lambda^4 = -1$$

Hence λ has no real values.

41. (d) $\vec{a} = \hat{i}-\hat{k}$, $\vec{b} = x\hat{i}+\hat{j}+(1-x)\hat{k}$ and

$\vec{c} = y\hat{i}+x\hat{j}+(1+x-y)\hat{k}$

$$[\vec{a}\ \vec{b}\ \vec{c}] = \vec{a}.\vec{b}\times\vec{c} = \begin{vmatrix} 1 & 0 & -1 \\ x & 1 & 1-x \\ y & x & 1+x-y \end{vmatrix}$$

$= 1\left[1+x-y-x+x^2\right]-\left[-x^2-y\right]$

$= 1-y+x^2-x^2+y = 1$

Hence $\left[\vec{a}\ \vec{b}\ \vec{c}\right]$ is independent of x and y both.

42. (b) Perpendicular distance of centre $\left(\frac{1}{2},0,-\frac{1}{2}\right)$

from $x+2y-2=4$ is given by

$$\frac{\left|\frac{1}{2}+\frac{1}{2}-4\right|}{\sqrt{6}} = \sqrt{\frac{3}{2}}$$

$\sqrt{\frac{3}{2}}$ $\sqrt{\frac{5}{2}}$ $\left(\frac{1}{2}, 0, -\frac{1}{2}\right)$

radius of sphere

$= \sqrt{\frac{1}{4}+\frac{1}{4}+2} = \sqrt{\frac{5}{2}}$

$\therefore$ radius of circle $= \sqrt{\frac{5}{2}-\frac{3}{2}} = 1$.

43. (d) $(\vec{a}\times\vec{b})\times\vec{c} = \vec{a}\times(\vec{b}\times\vec{c})$, $\vec{a}.\vec{b}\neq 0$, $\vec{b}.\vec{c}\neq 0$

$\Rightarrow (\vec{a}.\vec{c}).\vec{b}-(\vec{b}.\vec{c})\vec{a} = (\vec{a}.\vec{c}).\vec{b}-(\vec{a}.\vec{b}).\vec{c}$

$\Rightarrow (\vec{a}.\vec{b}).\vec{c} = (\vec{b}.\vec{c})\vec{a} \Rightarrow \vec{a}\parallel\vec{c}$.

44. (a) $\overrightarrow{CA} = (2-a)\hat{i}+2\hat{j}$;

$\overrightarrow{CB} = (1-a)\hat{i}-6\hat{k}$

$\overrightarrow{CA}.\overrightarrow{CB} = 0 \Rightarrow (2-a)(1-a) = 0$

$\Rightarrow a = 2, 1$

45. (a) Equation of lines $\frac{x-b}{a} = \frac{y}{1} = \frac{z-d}{c}$

$\frac{x-b'}{a'} = \frac{y}{1} = \frac{z-d'}{c'}$

Line are perpendicular $\Rightarrow aa'+1+cc' = 0$

46. (none) If (α, β, γ) be the image, then mid point of (α, β, γ) and (–1, 3, 4) must lie on $x-2y=0$

$\therefore \frac{\alpha-1}{2}-2\left(\frac{\beta+3}{2}\right) = 0$

$\therefore \alpha-1-2\beta-6=0 \Rightarrow \alpha-2\beta = 7$ ..... (1)

Also line joining (α, β, γ) and (–1, 3, 4) should be parallel to the normal of the plane $x-2y=0$

$\therefore \frac{\alpha+1}{1} = \frac{\beta-3}{-2} = \frac{\gamma-4}{0} = \lambda$

$\Rightarrow \alpha = \lambda-1, \beta = -2\lambda+3, \gamma = 4$ ..... (2)

From (1) and (2)

$\alpha = \frac{9}{5}, \beta = -\frac{13}{5}, \gamma = 4$

None of the option matches.

47. (b) Let the angle of line makes with the positive direction of z-axis is α direction cosines of line with the +ve directions of x-axis, y-axis, and z-axis is l, m, n respectively.

$\therefore l = \cos\frac{\pi}{4}, m = \cos\frac{\pi}{4}, n = \cos\alpha$

as we know that, $l^2+m^2+n^2 = 1$

$\therefore\ \cos^2\frac{\pi}{4}+\cos^2\frac{\pi}{4}+\cos^2\alpha=1$

$\Rightarrow\ \frac{1}{2}+\frac{1}{2}+\cos^2\alpha=1$

$\Rightarrow\ \cos^2\alpha=0\Rightarrow\ \alpha=\frac{\pi}{2}$

Hence, angle with positive direction of the z-axis is $\frac{\pi}{2}$

48. (b) Given $|2\hat{u}\times 3\hat{v}|=1$ and θ is acute angle between $\hat{u}$ and $\hat{v}$, $|\hat{u}|=1, |\hat{v}|=1\Rightarrow 6|\hat{u}||\hat{v}||\sin\theta|=1$

$\Rightarrow\ 6|\sin\theta|=1\ \Rightarrow\ \sin\theta=\frac{1}{6}$

Hence, there is exactly one value of θ for which $2\hat{u}\times 3\hat{v}$ is a unit vector.

49. (c) For given sphere centre is (3, 6, 1)
Coordinates of one end of diameter of the sphere are (2, 3, 5). Let the coordinates of the other end of diameter are (α,β,γ)

$\therefore\ \frac{\alpha+2}{2}=3,\ \frac{\beta+3}{2}=6,\ \frac{\gamma+5}{2}=1$

$\Rightarrow\ \alpha=4,\ \beta=9$ and $\gamma=-3$

$\therefore$ Coordinate of other end of diameter are (4, 9, –3)

50. (b) Given $\vec{a}=\hat{i}+\hat{j}+\hat{k}$, $\vec{b}=\hat{i}-\hat{j}+2\hat{k}$ and $\vec{c}=x\hat{i}+(x-2)\hat{j}-\hat{k}$

If $\vec{c}$ lies in the plane of $\vec{a}$ and $\vec{b}$, then $[\vec{a}\ \vec{b}\ \vec{c}]=0$

i.e. $\begin{vmatrix}1 & 1 & 1\\ 1 & -1 & 2\\ x & (x-2) & -1\end{vmatrix}=0$

$\Rightarrow\ 1[1-2(x-2)]-1[-1-2x]+1[x-2+x]=0$
$\Rightarrow\ 1-2x+4+1+2x+2x-2=0$
$\Rightarrow\ 2x=-4\Rightarrow\ x=-2$

51. (c) Let the direction cosines of line L be l, m, n, then
$2l+3m+n=0$ ....(i)
and $l+3m+2n=0$ ....(ii)
on solving equation (i) and (ii), we get

$\frac{l}{6-3}=\frac{m}{1-4}=\frac{n}{6-3}\ \Rightarrow\ \frac{l}{3}=\frac{m}{-3}=\frac{n}{3}$

Now $\frac{l}{3}=\frac{m}{-3}=\frac{n}{3}=\frac{\sqrt{l^2+m^2+n^2}}{\sqrt{3^2+(-3)^2+3^2}}$

$\because\ l^2+m^2+n^2=1$

$\therefore\ \frac{l}{3}=\frac{m}{-3}=\frac{n}{3}=\frac{1}{\sqrt{27}}$

$\Rightarrow\ l=\frac{3}{\sqrt{27}}=\frac{1}{\sqrt{3}}, m=-\frac{1}{\sqrt{3}}, n=\frac{1}{\sqrt{3}}$

Line L, makes an angle α with +ve x-axis

$\therefore\ l=\cos\alpha\ \Rightarrow\ \cos\alpha=\frac{1}{\sqrt{3}}$

52. (d) $\because\ \vec{a}$ lies in the plane of $\vec{b}$ and $\vec{c}$

$\therefore\ \vec{a}=\vec{b}+\lambda\vec{c}$

$\Rightarrow\alpha\hat{i}+2\hat{j}+\beta\hat{k}=\hat{i}+\hat{j}+\lambda(\hat{j}+\hat{k})$
$\Rightarrow\alpha=1, 2=1+\lambda, \beta=\lambda\Rightarrow\alpha=1,\beta=1$

ALTERNATE SOLUTION

$\because\ \vec{a}$ bisects the angle between $\vec{b}$ and $\vec{c}$.

$\therefore\ \vec{a}=\lambda(\hat{b}+\hat{c})\ \Rightarrow\alpha\hat{i}+2\hat{j}+\beta\hat{k}=\frac{\lambda(\hat{i}+2\hat{j}+\hat{k})}{\sqrt{2}}$

$\Rightarrow\alpha=\frac{\lambda}{\sqrt{2}},\ \lambda=\sqrt{2},\ \beta=\frac{\lambda}{\sqrt{2}}\Rightarrow\alpha=\beta=1$

53. (d) Clearly $\vec{a}=-\frac{8}{7}\vec{c}$

$\Rightarrow\vec{a}\parallel\vec{c}$ and are opposite in direction

$\therefore$ Angle between $\vec{a}$ and $\vec{c}$ is π.

ALTERNATE SOLUTION

$\cos\theta=\frac{\vec{a}.\vec{c}}{|\vec{a}||\vec{c}|}=\frac{8\vec{b}\cdot(-7\vec{b})}{8|\vec{b}|7|\vec{b}|}=-1\Rightarrow\theta=\pi$

54. (c) Equation of line through (5, 1, a) and (3, b, 1) is $\frac{x-5}{-2}=\frac{y-1}{b-1}=\frac{z-a}{1-a}=\lambda$

$\therefore$ Any point on this line is a
$[-2\lambda+5,(b-1)\lambda+1,(1-a)\lambda+a]$
It crosses yz plane where $-2\lambda+5=0$

$\lambda=\frac{5}{2}\quad\therefore\ \left(0,(b-1)\frac{5}{2}+1,(1-a)\frac{5}{2}+a\right)=\left(0,\frac{17}{2},\frac{-13}{2}\right)$

$\Rightarrow(b-1)\frac{5}{2}+1=\frac{17}{2}$ and $(1-a)\frac{5}{2}+a=-\frac{13}{2}$

$\Rightarrow b=4$ and $a=6$

55. (a) The two lines intersect if shortest distance between them is zero i.e.

$\frac{(\vec{a}_2-\vec{a}_1)\cdot\vec{b}_1\times\vec{b}_2}{|\vec{b}_1\times\vec{b}_2|}=0\ \Rightarrow(\vec{a}_2-\vec{a}_1)\cdot\vec{b}_1\times\vec{b}_2=0$

where $\vec{a}_1=\hat{i}+2\hat{j}+3\hat{k}$, $\vec{b}_1=k\hat{i}+2\hat{j}+3\hat{k}$

$\vec{a}_2=2\hat{i}+3\hat{j}+\hat{k}$, $\vec{b}_2=3\hat{i}+k\hat{j}+2\hat{k}$

$\Rightarrow\begin{vmatrix}1 & 1 & -2\\ k & 2 & 3\\ 3 & k & 2\end{vmatrix}=0$

$\Rightarrow 1(4-3k)-1(2k-9)-2(k^2-6)=0$

$\Rightarrow -2k^2-5k+25=0\Rightarrow k=-5$ or $\frac{5}{2}$

$\because k$ is an integer, therefore $k=-5$

56. (a) $\because$ The line $\frac{x-2}{3}=\frac{y-1}{-5}=\frac{z+2}{2}$ lie in the plane $x+3y-\alpha z+\beta=0$

$\therefore$ Pt (2, 1, –2) lies on the plane
i.e. $2+3+2\alpha+\beta=0$

or $2\alpha+\beta+5=0$ ....(i)

Also normal to plane will be perpendicular to line,

$\therefore\ 3\times 1-5\times 3+2\times(-\alpha)=0$

$\Rightarrow\ \alpha=-6$

From equation (i) then, $\beta = 7$

$\therefore$ $(\alpha, \beta) = (-6, 7)$

57. (b) Let $P(x_1, y_1, z_1)$ and $Q(x_2, y_2, z_2)$ be the initial and final points of the vector whose projections on the three coordiante axes are 6, – 3, 2 then

$x_2 - x_1 = 6$; $y_2 - y_1 = -3$; $z_2 - z_1 = 2$

So that direction ratios of $\overrightarrow{PQ}$ are 6, – 3, 2

$\therefore$ Direction cosines of $\overrightarrow{PQ}$ are

$$\frac{6}{\sqrt{6^2+(-3)^2+2^2}}, \frac{-3}{\sqrt{6^2+(-3)^2+2^2}}, \frac{2}{\sqrt{6^2+(-3)^2+2^2}} = \frac{6}{7}, \frac{-3}{7}, \frac{2}{7}$$

58. (d) $[3\vec{u}\ p\vec{v}\ p\vec{\omega}] - [p\vec{v}\ \vec{\omega}\ q\vec{u}] - [2\vec{\omega}\ q\vec{v}\ q\vec{u}] = 0$

$\Rightarrow (3p^2 - pq + 2q^2)[\vec{u}\ \vec{v}\ \vec{\omega}] = 0$

$\Rightarrow 3p^2 - pq + 2q^2 = 0 \quad (\because [\vec{u}\ \vec{v}\ \vec{\omega}] \neq 0)$

$\Rightarrow 2p^2 + p^2 - pq + \frac{q^2}{4} + \frac{7q^2}{4} = 0$

$\Rightarrow 2p^2 + \left(p - \frac{q}{2}\right)^2 + \frac{7}{4}q^2 = 0$

$\Rightarrow p = 0, q = 0, p = \frac{q}{2} \Rightarrow p = 0, q = 0$

$\therefore$ Exactly one value of (p, q)

59. (d) $\vec{c} = \vec{b} \times \vec{a} \Rightarrow \vec{b}.\vec{c} = \vec{b}.(\vec{b} \times \vec{a}) \Rightarrow \vec{b}.\vec{c} = 0$

$\Rightarrow \left(b_1\hat{i} + b_2\hat{j} + b_3\hat{k}\right).\left(\hat{i} - \hat{j} - \hat{k}\right) = 0,$

where $\vec{b} = b_1\hat{i} + b_2\hat{j} + b_3\hat{k}$

$b_1 - b_2 - b_3 = 0$ ...(i)

and $\vec{a}.\vec{b} = 3 \Rightarrow (\hat{j} - \hat{k}).(b_1\hat{i} + b_2\hat{j} + b_3\hat{k}) = 3$

$\Rightarrow b_2 - b_3 = 3$

From equation (i)

$b_1 = b_2 + b_3 = (3 + b_3) + b_3 = 3 + 2b_3$

$\vec{b} = (3 + 2b_3)\hat{i} + (3 + b_3)\hat{j} + b_3\hat{k}$

From the option given, it is clear that b_3 equal to either 2 or –2.

If $b_3 = 2$ then $\vec{b} = 7\hat{i} + 5\hat{j} + 2\hat{k}$ which is not possible

If $b_3 = -2$, then $\vec{b} = -\hat{i} + \hat{j} - 2\hat{k}$

60. (d) Since, $\vec{a}, \vec{b}$ and $\vec{c}$ are mutually orthogonal

$\therefore \vec{a}.\vec{b} = 0,\ \vec{b}.\vec{c} = 0,\ \vec{c}.\vec{a} = 0$

$\Rightarrow 2\lambda + 4 + \mu = 0$ ...(i)

$\lambda - 1 + 2\mu = 0$ ...(ii)

On solving (i) and (ii), we get $\lambda = -3, \mu = 2$

61. (a) A(3, 1, 6); B = (1, 3, 4)

Mid-point of AB = (2, 2, 5) lies on the plane.

and d.r's of AB = (2, –2, 2)

d.r's of normal to plane = (1, –1, 1).

Direction ratio of AB and normal to the plane are proportional therefore,

AB is perpendicular to the plane

$\therefore$ A is image of B

Statement-2 is correct but it is not correct explanation.

62. (b) Direction cosines of the line :

$\ell = \cos 45° = \frac{1}{\sqrt{2}}$, $m = \cos 120° = \frac{-1}{2}$, $n = \cos\theta$

where θ is the angle, which line makes with positive z-axis.

Now $\ell^2 + m^2 + n^2 = 1$

$\Rightarrow \frac{1}{2} + \frac{1}{4} + \cos^2\theta = 1,\ \cos^2\theta = \frac{1}{4}$

$\Rightarrow \cos\theta = \frac{1}{2}$ (θ being acute)

$\Rightarrow \theta = \frac{\pi}{3}$

63. (d) If θ be the angle between the given line and plane, then

$$\sin\theta = \frac{1\times1 + 2\times2 + \lambda\times3}{\sqrt{1^2+2^2+\lambda^2}.\sqrt{1^2+2^2+3^2}} = \frac{5+3\lambda}{\sqrt{14}.\sqrt{5+\lambda^2}}$$

But it is given that $\theta = \cos^{-1}\sqrt{\frac{5}{14}} \Rightarrow \sin\theta = \frac{3}{\sqrt{14}}$

$\therefore \frac{5+3\lambda}{\sqrt{14}\sqrt{5+\lambda^2}} = \frac{3}{\sqrt{14}} \Rightarrow \lambda = \frac{2}{3}$

64. (d) We have $\vec{a}\cdot\vec{b} = 0,\ \vec{a}\cdot\vec{a} = 1,\ \vec{b}\cdot\vec{b} = 1$

$(2\vec{a} - \vec{b})\cdot\left[(\vec{a}\times\vec{b})\times(\vec{a} + 2\vec{b})\right]$

$= (2\vec{a} - \vec{b})\cdot\left[\{\vec{a}\cdot(\vec{a} + 2\vec{b})\}\vec{b} - \{\vec{b}\cdot(\vec{a} + 2\vec{b})\vec{a}\}\right]$

$= (2\vec{a} - \vec{b})\cdot\left[(\vec{a}\cdot\vec{a} + 2\vec{a}\cdot\vec{b})\vec{b} - (\vec{a}\cdot\vec{b} + 2\vec{b}\cdot\vec{b})\vec{a}\right]$

$= (2\vec{a} - \vec{b})\cdot\left[\vec{b} - 2\vec{a}\right] = 4\vec{a}\cdot\vec{b} - \vec{b}\cdot\vec{b} - 4\vec{a}\cdot\vec{a} = -5$

65. (c) $\vec{a}.\vec{b} \neq 0,\ \vec{a}.\vec{d} = 0$

Now, $\vec{b}\times\vec{c} = \vec{b}\times\vec{d}$

$\Rightarrow \vec{a}\times(\vec{b}\times\vec{c}) = \vec{a}\times(\vec{b}\times\vec{d})$

$\Rightarrow (\vec{a}.\vec{c})\vec{b} - (\vec{a}.\vec{b})\vec{c} = (\vec{a}.\vec{d})\vec{b} - (\vec{a}.\vec{b})\vec{d}$

$\Rightarrow (\vec{a}.\vec{b})\vec{d} = -(\vec{a}.\vec{c})\vec{b} + (\vec{a}.\vec{b})\vec{c}$

$\vec{d} = \vec{c} - \left(\frac{\vec{a}.\vec{c}}{\vec{a}.\vec{b}}\right)\vec{b}$

66. (a) The direction ratios of the line segment joining points A(1, 0, 7) and B(1, 6, 3) are 0, 6, – 4.

The direction ratios of the given line are 1, 2, 3.

Clearly $1\times0 + 2\times6 + 3\times(-4) = 0$

So, the given line is perpendicular to line AB.

Also , the mid point of A and B is (1, 3, 5) which lies on the given line.

So, the image of B in the given line is A, because the given line is the perpendicular bisector of line segment joining points A and B, But statement-2 is not a correct explanation for statement-1.

67. (d) The given vectors are collinear if

$$\begin{vmatrix} p & 1 & 1 \\ 1 & q & 1 \\ 1 & 1 & r \end{vmatrix} = 0$$

$\Rightarrow \quad p(qr-1)+1(1-r)+1(1-q)=0$

$\Rightarrow \quad pqr-p+1-r+1-q=0 \Rightarrow pqr-(p+q+r)=-2$

68. (a) Equation of line through $P(1,-5,9)$ and parallel to the plane $x=y=z$ is

$$\frac{x-1}{1}=\frac{y+5}{1}=\frac{z-9}{1}=\lambda(say)$$

$Q=(x=1+\lambda, y=-5+\lambda, z=9+\lambda)$

Given plane $x-y+z=5$

$\therefore \quad 1+\lambda+5-\lambda+9+\lambda=5 \Rightarrow \lambda=-10$

$\therefore \quad Q=(-9,-15,-1)$

$\therefore \quad PQ=\sqrt{(1+9)^2+(15-5)^2+(9+1)^2}$

$=\sqrt{300}=10\sqrt{3}$

69. (c) $\vec{a}+3\vec{b}=\lambda\vec{c}$ ....(i)

$\vec{b}+2\vec{c}=\mu\vec{a}$ ....(ii)

On solving equation (i) and (ii)

$(1+3\mu)\vec{a}-(\lambda+6)\vec{c}=0$

As $\vec{a}$ and $\vec{c}$ are non collinear,

$\therefore \quad 1+3\mu=0$ and $\lambda+6=0$

From (1), $\vec{a}+3\vec{b}+6\vec{c}=\vec{0}$

70. (c) Let feet of perpendicular is $(2\alpha, 3\alpha+2, 4\alpha+3)$

$\Rightarrow$ Direction ratio of the $\perp$ line is $2\alpha-3, 3\alpha+3, 4\alpha-8$.

$\Rightarrow$ Direction ratio of the line 2, 3, 4 are

$\Rightarrow \quad 2(2\alpha-3)+3(3\alpha+3)+4(4\alpha-8)=0$

$\Rightarrow \quad 29\alpha-29=0 \Rightarrow \alpha=1$

$\Rightarrow$ Feet of $\perp$ is (2, 5, 7)

$\Rightarrow$ Length $\perp$ is $\sqrt{1^2+6^2+4^2}=\sqrt{53}$

71. (c) Let $\vec{c}=\hat{a}+2\hat{b}$ and $\vec{d}=5\hat{a}-4\hat{b}$

Since $\vec{c}$ and $\vec{d}$ are perpendicular to each other

$\therefore \quad \vec{c}.\vec{d}=0 \Rightarrow (\hat{a}+2\hat{b}).(5\hat{a}-4\hat{b})=0$

$\Rightarrow \quad 5+6\hat{a}.\hat{b}-8=0 \quad (\because \hat{a}.\hat{a}=1)$

$\Rightarrow \quad \hat{a}.\hat{b}=\frac{1}{2} \Rightarrow \theta=\frac{\pi}{3}$

72. (a) Given equation of a plane is $x-2y+2z-5=0$

So, Equation of parallel plane is given by $x-2y+2z+d=0$

Now, it is given that distance from origin to the parallel plane is 1.

$\therefore \quad \left|\frac{d}{\sqrt{1^2+2^2+2^2}}\right|=1 \Rightarrow d=\pm 3$

So equation of required plane $x-2y+2z\pm 3=0$

73. (c) Given lines are $\frac{x-1}{2}=\frac{y+1}{3}=\frac{z-1}{4}$

and $\frac{x-3}{1}=\frac{y-k}{2}=\frac{z}{1}$

These lines will intersect if shortest distance between them = 0

i.e. $\left[\vec{a}-\vec{c}, \vec{b}, \vec{d}\right]=0$

where $\vec{a}(1,-1,1)$, $\vec{b}(2,3,4)$, $\vec{c}(3,k,0)$; and $\vec{d}(1,2,1)$

Now, $\vec{a}-\vec{c}=(3-1, k+1, 0-1)=(2, k+1, -1)$

$$\Rightarrow \begin{vmatrix} 2 & k+1 & -1 \\ 2 & 3 & 4 \\ 1 & 2 & 1 \end{vmatrix}=0$$

$\Rightarrow \quad 2(3-8)-k+1(2-4)-1(4-3)=0$

$\Rightarrow \quad 2(-5)-(k+1)(-2)-1(1)=0$

$\Rightarrow \quad -10+2k+2-1=0 \Rightarrow k=\frac{9}{2}$

74. (b) Let $ABCD$ be a parallelogram such that $\overrightarrow{AB}=\vec{q}$, $\overrightarrow{AD}=\vec{p}$ and $\angle BAD$ be an acute angle.

We have

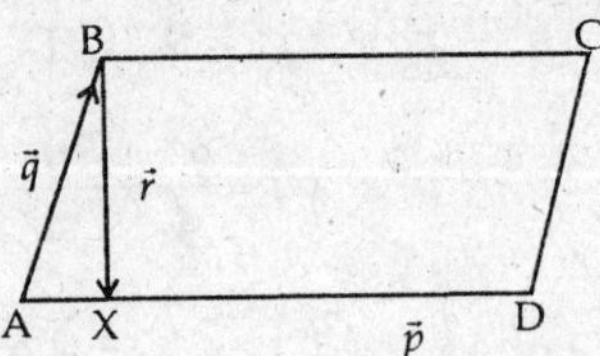

$$\overrightarrow{AX}=\left(\frac{\vec{p}.\vec{q}}{|\vec{p}|}\right)\left(\frac{\vec{p}}{|\vec{p}|}\right)=\frac{\vec{p}.\vec{q}}{|\vec{p}|^2}\vec{p}$$

Let $\vec{r}=\overrightarrow{BX}=\overrightarrow{BA}+\overrightarrow{AX}=-\vec{q}+\frac{\vec{p}.\vec{q}}{|\vec{p}|^2}\vec{p}$

75. (c) $2x+y+2z-8=0$ ...(Plane 1)

$2x+y+2z+\frac{5}{2}=0$ ...(Plane 2)

Distance between Plane 1 and 2

$$=\left|\frac{-8-\frac{5}{2}}{\sqrt{2^2+1^2+2^2}}\right|=\left|\frac{-21}{6}\right|=\frac{7}{2}$$

76. (c) Given lines will be coplanar

If $\begin{vmatrix} -1 & 1 & 1 \\ 1 & 1 & -k \\ k & 2 & 1 \end{vmatrix}=0$

$\Rightarrow \quad -1(1+2k)-(1+k^2)+1(2-k)=0$

$\Rightarrow \quad k=0,-3$

77. (c) We have, $\overrightarrow{AB}+\overrightarrow{BC}+\overrightarrow{CA}=0 \Rightarrow \overrightarrow{BC}=\overrightarrow{AC}-\overrightarrow{AB}$

Now, $\overrightarrow{BM}=\frac{\overrightarrow{AC}-\overrightarrow{AB}}{2}$

$\left(\because \overrightarrow{BM}=\frac{\overrightarrow{BC}}{2}\right)$

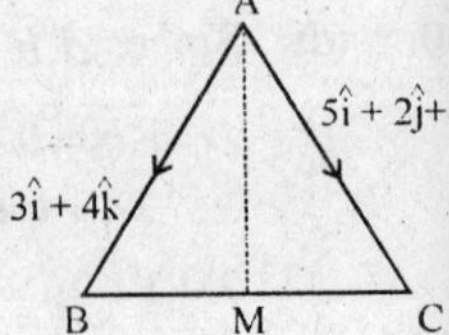

Also, we have

$\overrightarrow{AB}+\overrightarrow{BM}+\overrightarrow{MA}=0 \Rightarrow \overrightarrow{AB}+\frac{\overrightarrow{AC}-\overrightarrow{AB}}{2}=\overrightarrow{AM}$

$\Rightarrow \overrightarrow{AM}=\frac{\overrightarrow{AB}+\overrightarrow{AC}}{2}=4\hat{i}-\hat{j}+4\hat{k} \Rightarrow \left|\overrightarrow{AM}\right|=\sqrt{33}$

Probability

Section-A JEE Advanced/ IIT-JEE

A FILL IN THE BLANKS :

For a biased die the probabilities for the different faces to turn up are given below :

Face	1	2	3	4	5	6
Prob.	0.1	0.32	0.21	0.15	0.05	0.17

This die is tossed and you are told that either face 1 or face 2 has turned up. Then the probability that it is face 1 is .... *(1981 - 2 Marks)*

$P(A \cup B) = P(A \cap B)$ if and only if the relation between $P(A)$ and $P(B)$ is .... *(1985 - 2 Marks)*

A box contains 100 tickets numbered 1, 2, ..., 100. Two tickets are chosen at random. It is given that the maximum number on the two chosen tickets is not more than 10. The minimum number on them is 5 with probability .... *(1985 - 2 Marks)*

If $\frac{1+3p}{3}$, $\frac{1-p}{4}$ and $\frac{1-2p}{2}$ are the probabilities of three mutually exclusive events, then the set of all values of p is .... *(1986 - 2 Marks)*

Urn A contains 6 red and 4 black balls and urn B contains 4 red and 6 black balls. One ball is drawn at random from urn A and placed in urn B. Then one ball is drawn at random from urn B and placed in urn A. If one ball is now drawn at random from urn A, the probability that it is found to be red is .... *(1988 - 2 Marks)*

A pair of fair dice is rolled together till a sum of either 5 or 7 is obtained. Then the probability that 5 comes before 7 is .... *(1989 - 2 Marks)*

Let A and B be two events such that $P(A) = 0.3$ and $P(A \cup B) = 0.8$. If A and B are independent events then $P(B)$ = ... *(1990 - 2 Marks)*

If the mean and the variance of a binomial variate X are 2 and 1 respectively, then the probability that X takes a value greater than one is equal to .... *(1991 - 2 Marks)*

Three faces of a fair die are yellow, two faces red and one blue. The die is tossed three times. The probability that the colours, yellow, red and blue, appear in the first, second and the third tosses respectively is ....*(1992 - 2 Marks)*

If two events A and B are such that $P(A^c) = 0.3$, $P(B) = 0.4$ and $P(A \cap B^c) = 0.5$, then $P(B/(A \cup B^c)]$ = ... *(1994 - 2 Marks)*

Three numbers are chosen at random without replacement from {1, 2, ..., 10}. The probability that the minimum of the chosen numbers is 3, or their maximum is 7, is .... *(1997C - 2 Marks)*

B TRUE / FALSE :

1. If the letters of the word "Assassin" are written down at random in a row, the probability that no two S's occur together is 1/35 *(1983 - 1 Mark)*
2. If the probability for A to fail in an examination is 0.2 and that for B is 0.3, then the probability that either A or B fails is 0.5. *(1989 - 1 Mark)*

C MCQ's WITH ONE CORRECT ANSWER :

1. Two fair dice are tossed. Let x be the event that the first die shows an even number and y be the event that the second die shows an odd number. The two events x and y are : *(1979)*
   (a) Mutually exclusive
   (b) Independent and mutually exclusive
   (c) Dependent
   (d) None of these.
2. Two events A and B have probabilities 0.25 and 0.50 respectively. The probability that both A and B occur simultaneously is 0.14. Then the probability that neither A nor B occurs is *(1980)*
   (a) 0.39 (b) 0.25 (c) 0.11 (d) none of these
3. The probability that an event A happens in one trial of an experiment is 0.4. Three independent trials of the experiment are performed. The probability that the event A happens at least once is *(1980)*
   (a) 0.936 (b) 0.784 (c) 0.904 (d) none of these
4. If A and B are two events such that $P(A) > 0$, and $P(B) \neq 1$, then $P\left(\frac{\bar{A}}{\bar{B}}\right)$ is equal to *(1982 - 2 Marks)*
   (a) $1-P(\frac{A}{B})$ (b) $1-P(\frac{\bar{A}}{B})$
   (c) $\frac{1-P(A \cup B)}{P(\bar{B})}$ (d) $\frac{P(\bar{A})}{P(\bar{B})}$
   (Here $\bar{A}$ and $\bar{B}$ are complements of A and B respectively).
5. Fifteen coupons are numbered 1, 2 ..... 15, respectively. Seven coupons are selected at random one at a time with replacement. The probability that the largest number appearing on a selected coupon is 9, is *(1983 - 1 Mark)*
   (a) $\left(\frac{9}{16}\right)^6$ (b) $\left(\frac{8}{15}\right)^7$ (c) $\left(\frac{3}{5}\right)^7$ (d) none of these
6. Three identical dice are rolled. The probability that the same number will appear on each of them is *(1984 - 2 Marks)*
   (a) 1/6 (b) 1/36 (c) 1/18 (d) 3/28

7. A box contains 24 identical balls of which 12 are white and 12 are black. The balls are drawn at random from the box one at a time with replacement. The probability that a white ball is drawn for the 4th time on the 7th draw is *(1984 - 2 Marks)*
(a) 5/64 (b) 27/32 (c) 5/32 (d) 1/2

8. One hundred identical coins, each with probability, p, of showing up heads are tossed once. If $0 < p < 1$ and the probabilitity of heads showing on 50 coins is equal to that of heads showing on 51 coins, then the value of p is *(1988 - 2 Marks)*
(a) 1/2 (b) 49/101 (c) 50/101 (d) 51/101.

9. India plays two matches each with West Indies and Australia. In any match the probabilities of India getting, points 0, 1 and 2 are 0.45, 0.05 and 0.50 respectively. Assuming that the outcomes are independent, the probability of India getting at least 7 points is *(1992 - 2 Marks)*
(a) 0.8750 (b) 0.0875 (c) 0.0625 (d) 0.0250

10. An unbiased die with faces marked 1, 2, 3, 4, 5 and 6 is rolled four times. Out of four face values obtained, the probability that the minimum face value is not less than 2 and the maximum face value is not greater than 5, is then: *(1993 - 1 Mark)*
(a) 16/81 (b) 1/81 (c) 80/81 (d) 65/81

11. Let A, B, C be three mutually independent events. Consider the two statements S_1 and S_2
S_1 : A and $B \cup C$ are independent
S_2 : A and $B \cap C$ are independent
Then, *(1994)*
(a) Both S_1 and S_2 are true
(b) Only S_1 is true
(c) Only S_2 is true
(d) Neither S_1 nor S_2 is true

12. The probability of India winning a test match against west Indies is 1/2. Assuming independence from match to match the probability that in a 5 match series India's second win occurs at third test is *(1995S)*
(a) 1/8 (b) 1/4 (c) 1/2 (d) 2/3

13. Three of the six vertices of a regular hexagon are chosen at random. The probability that the triangle with three vertices is equilateral, equals *(1995S)*
(a) 1/2 (b) 1/5 (c) 1/10 (d) 1/20

14. For the three events A, B, and C, P (exactly one of the events A or B occurs) = P (exactly one of the two events B or C occurs) = P(exactly one of the events C or A occurs) = p and P (all the three events occur simultaneously) = p^2, where $0 < p < 1/2$. Then the probability of at least one of the three events A, B and C occurring is *(1996 - 2 Marks)*
(a) $\frac{3p+2p^2}{2}$ (b) $\frac{p+3p^2}{4}$
(c) $\frac{p+3p^2}{2}$ (d) $\frac{3p+2p^2}{4}$

15. If the integers m and n are chosen at random from 1 to 100, then the probability that a number of the form $7^m + 7^n$ is divisible by 5 equals *(1999 - 2 Marks)*
(a) 1/4 (b) 1/7 (c) 1/8 (d) 1/49

16. Two numbers are selected randomly from the set $S = \{1, 2, 3, 4, 5, 6\}$ without replacement one by one. The probability that minimum of the two numbers is less than 4 is *(2003S)*
(a) 1/15 (b) 14/15 (c) 1/5 (d) 4/5

17. If $P(B) = \frac{3}{4}, P(A \cap B \cap \bar{C}) = \frac{1}{3}$ and *(2003S)*
$P(\bar{A} \cap B \cap \bar{C}) = \frac{1}{3}$, then $P(B \cap C)$ is
(a) 1/12 (b) 1/6 (c) 1/15 (d) 1/9

18. If three distinct numbers are chosen randomly from the first 100 natural numbers, then the probability that all three of them are divisible by both 2 and 3 is *(2004S)*
(a) 4/25 (b) 4/35 (c) 4/33 (d) 4/1155

19. A six faced fair dice is thrown until 1 comes, then the probability that 1 comes in even no. of trials is *(2005S)*
(a) 5/11 (b) 5/6 (c) 6/11 (d) 1/6

20. One Indian and four American men and their wives are to be seated randomly around a circular table. Then the conditional probability that the Indian man is seated adjacent to his wife given that each American man is seated adjacent to his wife is *(2007 -3 marks)*
(a) $\frac{1}{2}$ (b) $\frac{1}{3}$ (c) $\frac{2}{5}$ (d) $\frac{1}{5}$

21. Let E^c denote the complement of an event E. Let E, F, G be pairwise independent events with $P(G) > 0$ and $P(E \cap F \cap G) = 0$. Then $P(E^c \cap F^c \mid G)$ equals *(2007-3 marks)*
(a) $P(E^c) + P(F^c)$ (b) $P(E^c) - P(F^c)$
(c) $P(E^c) - P(F)$ (d) $P(E) - P(F^c)$

22. An experiment has 10 equally likely outcomes. Let A and B be non-empty events of the experiment. If A consists of 4 outcomes, the number of outcomes that B must have so that A and B are independent, is *(2008)*
(a) 2, 4 or 8 (b) 3, 6 or 9 (c) 4 or 8 (d) 5 or 10

23. Let ω be a complex cube root of unity with $\omega \neq 1$. A fair die is thrown three times. If r_1, r_2 and r_3 are the numbers obtained on the die, then the probability that $\omega^{r_1} + \omega^{r_2} + \omega^{r_3} = 0$ is *(2010)*
(a) $\frac{1}{18}$ (b) $\frac{1}{9}$ (c) $\frac{2}{9}$ (d) $\frac{1}{36}$

24. A signal which can be green or red with probability $\frac{4}{5}$ and $\frac{1}{5}$ respectively, is received by station A and then transmitted to station B. The probability of each station receiving the signal correctly is $\frac{3}{4}$. If the signal received at station B is green, then the probability that the original signal was green is *(2010)*
(a) $\frac{3}{5}$ (b) $\frac{6}{7}$ (c) $\frac{20}{23}$ (d) $\frac{9}{20}$

25. Four fair dice D_1, D_2, D_3 and D_4; each having six faces numbered 1, 2, 3, 4, 5 and 6 are rolled simultaneously. The probability that D_4 shows a number appearing on one of D_1, D_2 and D_3 is *(2012)*
(a) $\frac{91}{216}$ (b) $\frac{108}{216}$ (c) $\frac{125}{216}$ (d) $\frac{127}{216}$

D MCQ's WITH ONE OR MORE THAN ONE CORRECT :

1. If M and N are any two events, the probability that exactly one of them occurs is **(1984 - 3 Marks)**
   (a) $P(M)+P(N)-2P(M\cap N)$
   (b) $P(M)+P(N)-P(M\cap N)$
   (c) $P(M^c)+P(N^c)-2P(M^c\cap N^c)$
   (d) $P(M\cap N^c)+P(M^c\cap N)$
2. A student appears for tests I, II and III. The student is successful if he passes either in tests I and II or tests I and III. The probabilities of the student passing in tests I, II and III are p, q and $\frac{1}{2}$ respectively. If the probability that the student is successful is $\frac{1}{2}$, then **(1986 - 2 Marks)**
   (a) $p=q=1$ (b) $p=q=\frac{1}{2}$
   (c) $p=1, q=0$ (d) $p=1, q=\frac{1}{2}$
   (e) none of these
3. The probability that at least one of the events A and B occurs is 0.6. If A and B occur simultaneously with probability 0.2, then $P(\bar{A})+P(\bar{B})$ is **(1987 - 2 Marks)**
   (a) 0.4 (b) 0.8 (c) 1.2 (d) 1.4
   (e) none
   (Here $\bar{A}$ and $\bar{B}$ are complements of A and B, respectively).
4. For two given events A and B, $P(A\cap B)$ **(1988 - 2 Marks)**
   (a) not less than $P(A)+P(B)-1$
   (b) not greater than $P(A)+P(B)$
   (c) equal to $P(A)+P(B)-P(A\cup B)$
   (d) equal to $P(A)+P(B)+P(A\cup B)$
5. If E and F are independent events such that $0<P(E)<1$ and $0<P(F)<1$, then **(1989 - 2 Marks)**
   (a) E and F are mutually exclusive
   (b) E and F^c (the complement of the event F) are independent
   (c) E^c and F^c are independent
   (d) $P(E|F)+P(E^c|F)=1.$
6. For any two events A and B in a sample space **(1991 - 2 Marks)**
   (a) $P(A/B)\geq\frac{P(A)+P(B)-1}{P(B)}, P(B)\neq 0$ is always true
   (b) $P(A\cap\bar{B})=P(A)-P(A\cap B)$ does not hold
   (c) $P(A\cup B)=1-P(\bar{A})P(\bar{B})$, if A and B are independent
   (d) $P(A\cup B)=1-P(\bar{A})P(\bar{B})$, if A and B are disjoint.
7. E and F are two independent events. The probability that both E and F happen is 1/12 and the probability that neither E nor F happens is 1/2. Then, **(1993 - 2 Marks)**
   (a) $P(E)=1/3, P(F)=1/4$
   (b) $P(E)=1/2, P(F)=1/6$
   (c) $P(E)=1/6, P(F)=1/2$
   (d) $P(E)=1/4, P(F)=1/3$
8. Let $0<P(A)<1, 0<P(B)<1$ and $P(A\cup B)=P(A)+P(B)-P(A)P(B)$ then **(1995S)**
   (a) $P(B/A)=P(B)-P(A)$
   (b) $P(A'-B')=P(A')-P(B')$
   (c) $P(A\cup B)'=P(A')P(B')$
   (d) $P(A/B)=P(A)$
9. If from each of the three boxes containing 3 white and 1 black, 2 white and 2 black, 1 white and 3 black balls, one ball is drawn at random, then the probability that 2 white and 1 black ball will be drawn is **(1998 - 2 Marks)**
   (a) 13/32 (b) 1/4 (c) 1/32 (d) 3/16
10. If $\bar{E}$ and $\bar{F}$ are the complementary events of events E and F respectively and if $0<P(F)<1$, then **(1998 - 2 Marks)**
    (a) $P(E/F)+P(\bar{E}/F)=1$
    (b) $P(E/F)+P(E/\bar{F})=1$
    (c) $P(\bar{E}/F)+P(E/\bar{F})=1$
    (d) $P(E/\bar{F})+P(\bar{E}/\bar{F})=1$
11. There are four machines and it is known that exactly two of them are faulty. They are tested, one by one, in a random order till both the faulty machines are identified. Then the probability that only two tests are needed is **(1998 - 2 Marks)**
    (a) 1/3 (b) 1/6 (c) 1/2 (d) 1/4
12. If E and F are events with $P(E)\leq P(F)$ and $P(E\cap F)>0$, then **(1998 - 2 Marks)**
    (a) occurrence of $E\Rightarrow$ occurrence of F
    (b) occurrence of $F\Rightarrow$ occurrence of E
    (c) non-occurrence of $E\Rightarrow$ non-occurrence of F
    (d) none of the above implications holds
13. A fair coin is tossed repeatedly. If the tail appears on first four tosses, then the probability of the head appearing on the fifth toss equals **(1998 - 2 Marks)**
    (a) 1/2 (b) 1/32 (c) 31/32 (d) 1/5
14. Seven white balls and three black balls are randomly placed in a row. The probability that no two black balls are placed adjacently equals **(1998 - 2 Marks)**
    (a) 1/2 (b) 7/15 (c) 2/15 (d) 1/3
15. The probabilities that a student passes in Mathematics, Physics and Chemistry are m, p and c, respectively. Of these subjects, the student has a 75% chance of passing in at least one, a 50% chance of passing in at least two, and a 40% chance of passing in exactly two. Which of the following relations are true? **(1999 - 3 Marks)**
    (a) $p+m+c=19/20$ (b) $p+m+c=27/20$
    (c) $pmc=1/10$ (d) $pmc=1/4$
16. Let E and F be two independent events. The probability that exactly one of them occurs is $\frac{11}{25}$ and the probability of none of them occurring is $\frac{2}{25}$. If $P(T)$ denotes the probability of occurrence of the event T, then **(2011)**
    (a) $P(E)=\frac{4}{5}, P(F)=\frac{3}{5}$ (b) $P(E)=\frac{1}{5}, P(F)=\frac{2}{5}$
    (c) $P(E)=\frac{2}{5}, P(F)=\frac{1}{5}$ (d) $P(E)=\frac{3}{5}, P(F)=\frac{4}{5}$

17. A ship is fitted with three engines E_1, E_2 and E_3. The engines function independently of each other with respective probabilities $\frac{1}{2}, \frac{1}{4}$ and $\frac{1}{4}$. For the ship to be operational at least two of its engines must function. Let X denote the event that the ship is operational and let X_1, X_2 and X_3 denote respectively the events that the engines E_1, E_2 and E_3 are functioning. Which of the following is(are) true ? *(2012)*

(a) $P\left[X_1^c \mid X\right] = \frac{3}{16}$

(b) P [Exactly two engines of the ship are functioning $\left|X\right] = \frac{7}{8}$

(c) $P\left[X \mid X_2\right] = \frac{5}{16}$ (d) $P\left[X \mid X_1\right] = \frac{7}{16}$

18. Let X and Y be two events such that $P(X \mid Y) = \frac{1}{2}$, $P(Y/X) = \frac{1}{3}$ and $P(X \cap Y) = \frac{1}{6}$. Which of the following is (are) correct ? *(2012)*

(a) $P(X \cup Y) = \frac{2}{3}$

(b) X and Y are independent

(c) X and Y are not independent

(d) $P(X^c \cap Y) = \frac{1}{3}$

19. Four persons independently solve a certain problem correctly with probabilities $\frac{1}{2}, \frac{3}{4}, \frac{1}{4}, \frac{1}{8}$. Then the probability that the problem is solved correctly by at least one of them is *(JEE Adv. 2013)*

(a) $\frac{235}{256}$ (b) $\frac{21}{256}$ (c) $\frac{3}{256}$ (d) $\frac{253}{256}$

(E) SUBJECTIVE PROBLEMS :

1. Balls are drawn one-by-one without replacement from a box containing 2 black, 4 white and 3 red balls till all the balls are drawn. Find the probability that the balls drawn are in the order 2 black, 4 white and 3 red. *(1978)*

2. Six boys and six girls sit in a row randomly. Find the probability that
   (i) the six girls sit together
   (ii) the boys and girls sit alternately. *(1979)*

3. An anti-aircraft gun can take a maximum of four shots at an enemy plane moving away from it. The probabilities of hitting the plane at the first, second, third and fourth shot are 0.4, 0.3, 0.2 and 0.1 respectively. What is the probability that the gun hits the plane ? *(1981 - 2 Marks)*

4. A and B are two candidates seeking admission in IIT. The probability that A is selected is 0.5 and the probability that both A and B are selected is atmost 0.3. Is it possible that the probability of B getting selected is 0.9 ? *(1982 - 2 Marks)*

5. Cards are drawn one by one at random from a well-shuffled full pack of 52 playing cards until 2 aces are obtained for the first time. If N is the number of cards required to be drawn, then show that $P_r\{N = n\} = \frac{(n-1)(52-n)(51-n)}{50 \times 49 \times 17 \times 13}$ where $2 \le n \le 50$ *(1983 - 3 Marks)*

6. A, B, C are events such that *(1983 - 2 Marks)*
$P(A) = 0.3, P(B) = 0.4, P(C) = 0.8$
$P(AB) = 0.08, P(AC) = 0.28;\quad P(ABC) = 0.09$
If $P(A \cup B \cup C) \ge 0.75$, then show that $P(BC)$ lies in the interval $0.23 \le x \le 0.48$

7. A and B are two independent events. The probability that both A and B occur is 1/6 and the probability that neither of them occurs is 1/3. Find the probability of the occurrence of A. *(1984 - 2 Marks)*

8. In a certain city only two newspapers A and B are published, it is known that 25% of the city population reads A and 20% reads B while 8% reads both A and B. It is also known that 30% of those who read A but not B look into advertisements and 40% of those who read B but not A look into advertisements while 50% of those who read both A and B look into advertisements. What is the percentage of the population that reads an advertisement? *(1984 - 4 Marks)*

9. In a multiple-choice question there are four alternative answers, of which one or more are correct. A candidate will get marks in the question only if he ticks the correct answers. The candidate decides to tick the answers at random, if he is allowed upto three chances to answer the questions, find the probability that he will get marks in the questions. *(1985 - 5 Marks)*

10. A lot contains 20 articles. The probability that the lot contains exactly 2 defective articles is 0.4 and the probability that the lot contains exactly 3 defective articles is 0.6. Articles are drawn from the lot at random one by one without replacement and are tested till all defective articles are found. What is the probability that the testing procedure ends at the twelth testing. *(1986 - 5 Marks)*

11. A man takes a step forward with probability 0.4 and backwards with probability 0.6 Find the probability that at the end of eleven steps he is one step away from the starting point. *(1987 - 3 Marks)*

12. An urn contains 2 white and 2 black balls. A ball is drawn at random. If it is white it is not replaced into the urn. Otherwise It is replaced along with another ball of the same colour. The process is repeated. Find the probability that the third ball drawn is black. *(1987 - 4 Marks)*

13. A box contains 2 fifty paise coins, 5 twenty five paise coins and a certain fixed number N (≥ 2) of ten and five paise coins. Five coins are taken out of the box at random. Find the probability that the total value of these 5 coins is less than one rupee and fifty paise.
(1988 - 3 Marks)

14. Suppose the probability for A to win a game against B is 0.4. If A has an option of playing either a "best of 3 games" or a "best of 5 games" match against B, which option should be choose so that the probability of his winning the match is higher? (No game ends in a draw). *(1989 - 5 Marks)*

15. A is a set containing n elements. A subset P of A is chosen at random. The set A is reconstructed by replacing the elements of P. A subset Q of A is again chosen at random. Find the probability that P and Q have no common elements. *(1990 - 5 Marks)*

16. In a test an examine either guesses or copies or knows the answer to a multiple choice question with four choices. The probability that he make a guess is 1/3 and the probability that he copies the answer is 1/6. The probability that his answer is correct given that he copied it, is 1/8. Find the probability that he knew the answer to the question given that he correctly answered it. *(1991 - 4 Marks)*

17. A lot contains 50 defective and 50 non defective bulbs. Two bulbs are drawn at random, one at a time, with replacement. The events A, B, C are defined as *(1992 - 6 Marks)*
A = (the first bulb is defective)
B = (the second bulb is non-defective)
C = (the two bulbs are both defective or both non defective)
Determine whether
(i) A, B, C are pairwise independent
(ii) A, B, C are independent

18. Numbers are selected at random, one at a time, from the two-digit numbers 00, 01, 02....., 99 with replacement. An event E occurs if only if the product of the two digits of a selected number is 18. If four numbers are selected, find probability that the event E occurs at least 3 times. *(1993 - 5 Marks)*

19. An unbiased coin is tossed. If the result is a head, a pair of unbiased dice is rolled and the number obtained by adding the numbers on the two faces is noted. If the result is a tail, a card from a well shuffled pack of eleven cards numbered 2, 3, 4,.....12 is picked and the number on the card is noted. What is the probability that the noted number is either 7 or 8? *(1994 - 5 Marks)*

20. In how many ways three girls and nine boys can be seated in two vans, each having numbered seats, 3 in the front and 4 at the back? How many seating arrangements are possible if 3 girls should sit together in a back row on adjacent seats? Now, if all the seating arrangements are equally likely, what is the probability of 3 girls sitting together in a back row on adjacent seats? *(1996 - 5 Marks)*

21. Sixteen players S_1, S_2 .... S_{16} play in a tournament. They are divided into eight pairs at random. From each pair a winner is decided on the basis of a game played between the two players of the pair. Assume that all the players are of equal strength. *(1997C - 5 Marks)*
(a) Find the probability that the player S_1 is among the eight winners.
(b) Find the probability that exactly one of the two players S_1 and S_2 is among the eight winners.

22. If p and q are chosen randomly from the set {1, 2, 3, 4, 5, 6, 7, 8, 9, 10}, with replacement, determine the probability that the roots of the equation $x^2 + px + q = 0$ are real. *(1997 - 5 Marks)*

23. Three players, A, B and C, toss a coin cyclically in that order (that is A, B, C, A, B, C, A, B, ...) till a head shows. Let p be the probability that the coin shows a head. Let α, β and γ be, respectively, the probabilities that A, B and C gets the first head. Prove that $\beta = (1 - p)\alpha$. Determine α, β and γ (in terms of p). *(1998 - 8 Marks)*

24. Eight players $P_1, P_2, ...P_8$ play a knock-out tournament. It is known that whenever the players P_i and P_j play, the player P_i will win if $i < j$. Assuming that the players are paired at random in each round, what is the probability that the player P_4 reaches the final? *(1999 - 10 Marks)*

25. A coin has probability p of showing head when tossed. It is tossed n times. Let p_n denote the probability that no two (or more) consecutive heads occur. Prove that $p_1=1$, $p_2=1-p^2$ and $p_n=(1-p).\ p_{n-1} + p(1-p)\,p_{n-2}$ for all $n \geq 3$. *(2000 - 5 Marks)*

26. An urn contains m white and n black balls. A ball is drawn at random and is put back into the urn along with k additional balls of the same colour as that of the ball drawn. A ball is again drawn at random. What is the probability that the ball drawn now is white? *(2001 - 5 Marks)*

27. An unbiased die, with faces numbered 1, 2, 3, 4, 5, 6, is thrown n times and the list of n numbers showing up is noted. What is the probability that, among the numbers 1, 2, 3, 4, 5, 6, only three numbers appear in this list? *(2001 - 5 Marks)*

28. A box contains N coins, m of which are fair and the rest are biased. The probability of getting a head when a fair coin is tossed is 1/2, while it is 2/3 when a biased coin is tossed. A coin is drawn from the box at random and is tossed twice. The first time it shows head and the second time it shows tail. What is the probability that the coin drawn is fair? *(2002 - 5 Marks)*

29. For a student to qualify, he must pass at least two out of three exams. The probability that he will pass the 1st exam is p. If he fails in one of the exams then the probability of his passing in the next exam is $\frac{p}{2}$ otherwise it remains the same. Find the probability that he will qualify. *(2003 - 2 Marks)*

30. A is targeting to B, B and C are targeting to A. Probability of hitting the target by A, B and C are $\frac{2}{3}, \frac{1}{2}$ and $\frac{1}{3}$ respectively. If A is hit then find the probability that B hits the target and C does not. *(2003 - 2 Marks)*

31. A and B are two independent events. C is event in which exactly one of A or B occurs. Prove that
$P(C) \geq P(A \cup B)P(\bar{A} \cap \bar{B})$ *(2004 - 2 Marks)*

32. A box contains 12 red and 6 white balls. Balls are drawn from the box one at a time without replacement. If in 6 draws there are at least 4 white balls, find the probability that exactly one white is drawn in the next two draws. (binomial coefficients can be left as such) *(2004 - 4 Marks)*

33. A person goes to office either by car, scooter, bus or train, the probability of which being $\frac{1}{7}, \frac{3}{7}, \frac{2}{7}$ and $\frac{1}{7}$ respectively. Probability that he reaches office late, if he takes car, scooter, bus or train is $\frac{2}{9}, \frac{1}{9}, \frac{4}{9}$ and $\frac{1}{9}$ respectively. Given that he reached office in time, then what is the probability that he travelled by a car. *(2005 - 2 Marks)*

(G) COMPREHENSION BASED QUESTIONS :

Passage - 1

There are n urns, each of these contain $n + 1$ balls. The ith urn contains i white balls and $(n + 1 - i)$ red balls. Let u_i be the event of selecting ith urn, $i = 1, 2, 3 \ldots\ldots\ldots, n$ and w the event of getting a white ball.

1. If $P(u_i) \propto i$, where $i = 1, 2, 3, \ldots\ldots, n$, then $\lim_{n \to \infty} P(w) =$ *(2006 - 5M, –2)*

   (a) 1 (b) 2/3 (c) 3/4 (d) 1/4

2. If $P(u_i) = c$, (a constant) then $P(u_n/w) =$ *(2006 - 5M, –2)*

   (a) $\frac{2}{n+1}$ (b) $\frac{1}{n+1}$ (c) $\frac{n}{n+1}$ (d) $\frac{1}{2}$

3. Let $P(u_i) = \frac{1}{n}$, if n is even and E denotes the event of choosing even numbered urn, then the value of $P(w / E)$ is *(2006 - 5M, –2)*

   (a) $\frac{n+2}{2n+1}$ (b) $\frac{n+2}{2(n+1)}$ (c) $\frac{n}{n+1}$ (d) $\frac{1}{n+1}$

Passage - 2

A fair die is tossed repeatedly until a six is obtained. Let X denote the number of tosses required. *(2009)*

4. The probability that $X = 3$ equals

   (a) $\frac{25}{216}$ (b) $\frac{25}{36}$ (c) $\frac{5}{36}$ (d) $\frac{125}{216}$

5. The probability that $X \geq 3$ equals

   (a) $\frac{125}{216}$ (b) $\frac{25}{36}$ (c) $\frac{5}{36}$ (d) $\frac{25}{216}$

6. The conditional probability that $X \geq 6$ given $X > 3$ equals

   (a) $\frac{125}{216}$ (b) $\frac{25}{216}$ (c) $\frac{5}{36}$ (d) $\frac{25}{36}$

Passage - 3

Let U_1 and U_2 be two urns such that U_1 contains 3 white and 2 red balls, and U_2 contains only 1 white ball. A fair coin is tossed. If head appears then 1 ball is drawn at random from U_1 and put into U_2. However, if tail appears then 2 balls are drawn at random from U_1 and put into U_2. Now 1 ball is drawn at random from U_2. *(2011)*

7. The probability of the drawn ball from U_2 being white is

   (a) $\frac{13}{30}$ (b) $\frac{23}{30}$ (c) $\frac{19}{30}$ (d) $\frac{11}{30}$

8. Given that the drawn ball from U_2 is white, the probability that head appeared on the coin is

   (a) $\frac{17}{23}$ (b) $\frac{11}{23}$ (c) $\frac{15}{23}$ (d) $\frac{12}{23}$

Passage - 4

A box B_1 contains 1 white ball, 3 red balls and 2 black balls. Another box B_2 contains 2 white balls, 3 red balls and 4 black balls. A third box B_3 contains 3 white balls, 4 red balls and 5 black balls.

9. If 1 ball is drawn from each of the boxes B_1, B_2 and B_3, the probability that all 3 drawn balls are of the same colour is *(JEE Adv. 2013)*

   (a) $\frac{82}{648}$ (b) $\frac{90}{648}$ (c) $\frac{558}{648}$ (d) $\frac{566}{648}$

10. If 2 balls are drawn (without replacement) from a randomly selected box and one of the balls is white and the other ball is red, the probability that these 2 balls are drawn from box B_2 is

    (a) $\frac{116}{181}$ (b) $\frac{126}{181}$ (c) $\frac{65}{181}$ (d) $\frac{55}{181}$

(H) ASSERTION & REASON TYPE QUESTIONS :

1. Let $H_1, H_2, \ldots, H_n$ be mutually exclusive and exhaustive events with $P(H_i) > 0$, $i = 1, 2, \ldots, n$. Let E be any other event with $0 < P(E) < 1$.

   STATEMENT-1 :
   $P(H_i \mid E) > P(E \mid H_i) . P(H_i)$ for $i = 1, 2, \ldots, n$ **because**

   STATEMENT-2 : $\sum_{i=1}^{n} P(H_i) = 1$. *(2007 -3 marks)*

   (a) Statement-1 is True, statement-2 is True; Statement-2 is a correct explanation for Statement-1.
   (b) Statement-1 is True, Statement-2 is True; Statement-2 is NOT a correct explanation for Statement-1
   (c) Statement-1 is True, Statement-2 is False
   (d) Statement-1 is False, Statement-2 is True.

2. Consider the system of equations $ax + by = 0$; $cx + dy = 0$, where $a, b, c, d \in \{0, 1\}$

   STATEMENT - 1 : The probability that the system of equations has a unique solution is $\frac{3}{8}$.

   and

   STATEMENT - 2 : The probability that the system of equations has a solution is 1. *(2008)*

   (a) STATEMENT - 1 is True, STATEMENT - 2 is True; STATEMENT - 2 is a correct explanation for STATEMENT - 1
   (b) STATEMENT - 1 is True, STATEMENT - 2 is True; STATEMENT - 2 **is NOT** a correct explaination for STATEMENT - 1
   (c) STATEMENT - 1 is True, STATEMENT - 2 is False
   (d) STATEMENT - 1 is False, STATEMENT - 2 is True

(I) INTEGER VALUE CORRECT TYPE :

1. Of the three independent events E_1, E_2 and E_3, the probability that only E_1 occurs is α, only E_2 occurs is β and only E_3 occurs is γ. Let the probability p that none of events E_1, E_2 or E_3 occurs satisfy the equations $(\alpha\ 2\beta)p = \alpha\beta$ and $(\beta - 3\gamma)p = 2\beta\gamma$. All the given probabilities are assumed to lie in the interval (0, 1). *(JEE Adv. 2013)*

   Then $\frac{\text{Probability of occurrence of } E_1}{\text{Probability of occurrence of } E_3}$

Section-B JEE Main/ AIEEE

1. A problem in mathematics is given to three students A, B, C and their respective probability of solving the problem is $\frac{1}{2}, \frac{1}{3}$ and $\frac{1}{4}$. Probability that the problem is solved is **[2002]**

(a) $\frac{3}{4}$ (b) $\frac{1}{2}$ (c) $\frac{2}{3}$ (d) $\frac{1}{3}$

2. A and B are events such that $P(A \cup B)=3/4$, $P(A \cap B)=1/4$, $P(\overline{A})=2/3$ then $P(\overline{A} \cap B)$ is **[2002]**

(a) 5/12 (b) 3/8 (c) 5/8 (d) 1/4

3. A dice is tossed 5 times. Getting an odd number is considered a success. Then the variance of distribution of success is **[2002]**

(a) 8/3 (b) 3/8 (c) 4/5 (d) 5/4

4. The mean and variance of a random variable X having binomial distribution are 4 and 2 respectively, then $P(X=1)$ is **[2003]**

(a) $\frac{1}{4}$ (b) $\frac{1}{32}$ (c) $\frac{1}{16}$ (d) $\frac{1}{8}$

5. Events A, B, C are mutually exclusive events such that $P(A)=\frac{3x+1}{3}$, $P(B)=\frac{1-x}{4}$ and $P(C)=\frac{1-2x}{2}$ The set of possible values of x are in the interval. **[2003]**

(a) $[0, 1]$ (b) $\left[\frac{1}{3}, \frac{1}{2}\right]$ (c) $\left[\frac{1}{3}, \frac{2}{3}\right]$ (d) $\left[\frac{1}{3}, \frac{13}{3}\right]$

6. Five horses are in a race. Mr. A selects two of the horses at random and bets on them. The probability that Mr. A selected the winning horse is **[2003]**

(a) $\frac{2}{5}$ (b) $\frac{4}{5}$ (c) $\frac{3}{5}$ (d) $\frac{1}{5}$

7. The probability that A speaks truth is $\frac{4}{5}$, while the probability for B is $\frac{3}{4}$. The probability that they contradict each other when asked to speak on a fact is **[2004]**

(a) $\frac{4}{5}$ (b) $\frac{1}{5}$ (c) $\frac{7}{20}$ (d) $\frac{3}{20}$

8. A random variable X has the probability distribution:

X:	1	2	3	4	5	6	7	8
p(X):	0.2	0.2	0.1	0.1	0.2	0.1	0.1	0.1

For the events $E=\{X \text{ is a prime number}\}$ and $F=\{X<4\}$, the $P(E \cup F)$ is **[2004]**

(a) 0.50 (b) 0.77 (c) 0.35 (d) 0.87

9. The mean and the variance of a binomial distribution are 4 and 2 respectively. Then the probability of 2 successes is **[2004]**

(a) $\frac{28}{256}$ (b) $\frac{219}{256}$ (c) $\frac{128}{256}$ (d) $\frac{37}{256}$

10. Three houses are available in a locality. Three persons apply for the houses. Each applies for one house without consulting others. The probability that all the three apply for the same house is **[2005]**

(a) $\frac{2}{9}$ (b) $\frac{1}{9}$ (c) $\frac{8}{9}$ (d) $\frac{7}{9}$

11. A random variable X has Poisson distribution with mean 2. Then $P(X>1.5)$ equals **[2005]**

(a) $\frac{2}{e^2}$ (b) 0 (c) $1-\frac{3}{e^2}$ (d) $\frac{3}{e^2}$

12. Let A and B be two events such that $P(\overline{A \cup B})=\frac{1}{6}$, $P(A \cap B)=\frac{1}{4}$ and $P(\overline{A})=\frac{1}{4}$, where $\overline{A}$ stands for complement of event A. Then events A and B are

(a) equally likely and mutually exclusive **[2005]**
(b) equally likely but not independent
(c) independent but not equally likely
(d) mutually exclusive and independent

13. The resultant R of two forces acting on a particle is at right angles to one of them and its magnitude is one third of the other force. The ratio of larger force to smaller one is: **[2005]**

(a) $2:1$ (b) $3:\sqrt{2}$ (c) $3:2$ (d) $3:2\sqrt{2}$

14. At a telephone enquiry system the number of phone cells regarding relevant enquiry follow Poisson distribution with an average of 5 phone calls during 10 minute time intervals. The probability that there is at the most one phone call during a 10-minute time period is **[2006]**

(a) $\frac{6}{5^e}$ (b) $\frac{5}{6}$ (c) $\frac{6}{55}$ (d) $\frac{6}{e^5}$

15. Two aeroplanes I and II bomb a target in succession. The probabilities of I and II scoring a hit correctly are 0.3 and 0.2, respectively. The second plane will bomb only if the first misses the target. The probability that the target is hit by the second plane is **[2007]**

(a) 0.2 (b) 0.7 (c) 0.06 (d) 0.14.

16. A pair of fair dice is thrown independently three times. The probability of getting a score of exactly 9 twice is **[2007]**

(a) 8/729 (b) 8/243 (c) 1/729 (d) 8/9.

17. It is given that the events A and B are such that $P(A)=\frac{1}{4}, P(A|B)=\frac{1}{2}$ and $P(B|A)=\frac{2}{3}$. Then $P(B)$ is **[2008]**

(a) $\frac{1}{6}$ (b) $\frac{1}{3}$ (c) $\frac{2}{3}$ (d) $\frac{1}{2}$

18. A die is thrown. Let A be the event that the number obtained is greater than 3. Let B be the event that the number obtained is less than 5. Then $P(A \cup B)$ is **[2008]**

(a) $\frac{3}{5}$ (b) 0 (c) 1 (d) $\frac{2}{5}$

19. In a binomial distribution $B\left(n, p=\frac{1}{4}\right)$, if the probability of at least one success is greater than or equal to $\frac{9}{10}$, then n is greater than: **[2009]**

(a) $\frac{1}{\log_{10} 4+\log_{10} 3}$ (b) $\frac{9}{\log_{10} 4-\log_{10} 3}$

(c) $\frac{4}{\log_{10} 4-\log_{10} 3}$ (d) $\frac{1}{\log_{10} 4-\log_{10} 3}$

20. One ticket is selected at random from 50 tickets numbered 00,01,02,...,49. Then the probability that the sum of the digits on the selected ticket is 8, given that the product of these digits is zero, equals: **[2009]**

(a) $\frac{1}{7}$ (b) $\frac{5}{14}$ (c) $\frac{1}{50}$ (d) $\frac{1}{14}$

21. Four numbers are chosen at random (without replacement) from the set $\{1, 2, 3, ...20\}$. **[2010]**

Statement -1: The probability that the chosen numbers when arranged in some order will form an AP is $\frac{1}{85}$.

Statement -2 : If the four chosen numbers form an AP, then the set of all possible values of common difference is $(\pm1, \pm2, \pm3, \pm4, \pm5)$.

(a) Statement -1 is true, Statement -2 is true ; Statement -2 is **not** a correct explanation for Statement -1
(b) Statement -1 is true, Statment -2 is false
(c) Statement -1 is false, Statment -2 is true.
(d) Statement -1 is true, Statement -2 is true ; Statement -2 is a correct explanation for Statement -1.

22. An urn contains nine balls of which three are red, four are blue and two are green. Three balls are drawn at random without replacement from the urn. The probability that the three balls have different colours is **[2010]**

(a) $\frac{2}{7}$ (b) $\frac{1}{21}$ (c) $\frac{2}{23}$ (d) $\frac{1}{3}$

23. Consider 5 independent Bernoulli's trials each with probability of success p. If the probability of at least one failure is greater than or equal to $\frac{31}{32}$, then p lies in the interval **[2011]**

(a) $\left(\frac{3}{4}, \frac{11}{12}\right]$ (b) $\left[0, \frac{1}{2}\right]$

(c) $\left(\frac{11}{12}, 1\right]$ (d) $\left(\frac{1}{2}, \frac{3}{4}\right]$

24. If C and D are two events such that $C \subset D$ and $P(D) \neq 0$, then the correct statement among the following is **[2011]**

(a) $P(C|D) \geq P(C)$ (b) $P(C|D) < P(C)$

(c) $P(C|D) = \frac{P(D)}{P(C)}$ (d) $P(C|D) = P(C)$

25. Let A, B, C, be pairwise independent events with $P(C) > 0$ and $P(A \cap B \cap C)=0$. Then $P\left(A^c \cap B^c / C\right)$. **[2011 RS]**

(a) $P\left(B^c\right)-P\left(B^c\right)$ (b) $P\left(A^c\right)+P\left(B^c\right)$

(c) $P\left(A^c\right)-P\left(B^c\right)$ (d) $P\left(A^c\right)-P(B)$

26. Three numbers are chosen at random without replacement from $\{1,2,3,..8\}$. The probability that their minimum is 3, given that their maximum is 6, is : **[2012]**

(a) $\frac{3}{8}$ (b) $\frac{1}{5}$ (c) $\frac{1}{4}$ (d) $\frac{2}{5}$

27. A multiple choice examination has 5 questions. Each question has three alternative answers of which exactly one is correct. The probability that a student will get 4 or more correct answers just by guessing is: **[JEE M 2013]**

(a) $\frac{17}{3^5}$ (b) $\frac{13}{3^5}$ (c) $\frac{11}{3^5}$ (d) $\frac{10}{3^5}$

Solutions & Explanations

Section-A : JEE Advanced/ IIT-JEE

A 1. $\frac{5}{21}$ 2. $P(A)=P(B)$ 3. 1/9 4. $\frac{1}{3}\le p\le\frac{1}{2}$ 5. 32/55
6. 2/5 7. 5/7 8. 11/16 9. 1/36 10. 1/4 11. 11/40

B 1. F 2. F

C 1. (d) 2. (a) 3. (b) 4. (c) 5. (c) 6. (b) 7. (c) 8. (d) 9. (b)
10. (a) 11. (a) 12. (b) 13. (c) 14. (a) 15. (a) 16. (d) 17. (a) 18. (d)
19. (a) 20. (c) 21. (c) 22. (d) 23. (c) 24. (c) 25. (a)

D 1. (a,c,d) 2. (c) 3. (c) 4. (a, b, c) 5. (b, c, d) 6. (a, c) 7. (a, d) 8. (c, d) 9. (a)
10. (a, d) 11. (b) 12. (d) 13. (a) 14. (b) 15. (b, c) 16. (a, d) 17. (b, d) 18. (a, b)
19. (a)

E 1. $\frac{1}{1260}$ 2. (i) $\frac{1}{132}$ (ii) $\frac{1}{462}$ 3. 0.6976 4. No

7. $\frac{1}{2}$ and $\frac{1}{3}$ or $\frac{1}{3}$ and $\frac{1}{2}$ 8. 13.9% 9. 1/5 10. 99/1900

11. 0.37 12. 23/30 13. $1-\frac{10(N+2)}{^{N+7}C_5}$ 14. best of three games 15. $\left(\frac{3}{4}\right)^n$

16. 24/29 17. A, B, C are pairwise independent but A, B, C are dependent. 18. $\frac{97}{(25)^4}$

19. 0.2436 20. 7(13)!, 12!, $\frac{1}{91}$ 21.(a) $\frac{1}{2}$ (b) $\frac{8}{15}$ 22. 0.62

23. $\alpha=\frac{p}{1-(1-p)^3}$, $\beta=\frac{(1-p)p}{1-(1-p)^3}$, $\gamma=\frac{p(1-p)^2}{1-(1-p)^3}$ 24. $\frac{4}{35}$ 26. $\frac{m}{m+n}$

27. $\frac{^6C_3[3^n-3(2^n)+3]}{6^n}$ 28. $\frac{9m}{m+8N}$ 29. $2p^2-p^3$ 30. $\frac{1}{2}$

32. $\frac{^{10}C_1\times{}^2C_1}{^{12}C_2}\times\frac{^{12}C_2\times{}^6C_4}{^{18}C_6}+\frac{^{11}C_1\times{}^1C_1}{^{12}C_2}\times\frac{^{12}C_1\times{}^6C_5}{^{18}C_6}$ 33. $\frac{1}{7}$

G 1. (b) 2. (a) 3. (b) 4. (a) 5. (b) 6. (d) 7. (b) 8. (d) 9. (a) 10. (d)
H 1. (d) 2. (b)
I 1. 6

Section-B : JEE Main/ AIEEE

1. (a) 2. (a) 3. (d) 4. (b) 5. (b) 6. (a) 7. (c) 8. (b) 9. (a)
10. (b) 11. (c) 12. (c) 13. (d) 14. (d) 15. (d) 16. (b) 17. (b) 18. (c)
19. (d) 20. (d) 21. (b) 22. (a) 23. (b) 24. (a) 25. (d) 26. (b) 27. (c)

Section-A JEE Advanced/ IIT-JEE

(A) FILL IN THE BLANKS :

1. Let $E_1 \equiv$ face 1 has turned up.
$E_2 \equiv$ face 1 or 2 has turned up.
By the given data
$P(E_2) = 0.1 + 0.32 = 0.42, \quad P(E_1 \cap E_2) = P(E_1) = 0.1$
Given that E_2 has happened and we have to find the probability of happening of E_1.
$\therefore$ By conditional probability theorem, we have

$$P(E_1 / E_2) = \frac{P(E_1 \cap E_2)}{P(E_2)} = \frac{0.1}{0.42} = \frac{10}{42} = \frac{5}{21}.$$

2. Given that $P(A \cup B) = P(A \cap B)$
$\Rightarrow P(A) + P(B) - P(A \cap B) = P(A \cap B)$
$\Rightarrow [P(A) - P(A \cap B)] + [P(B) - P(A \cap B)] = 0$
But $P(A) - P(A \cap B), P(B) - P(A \cap B) \geq 0$
$[\because P(A \cap B) \leq P(A), P(B)]$
$\Rightarrow P(A) - P(A \cap B) = 0$ and $P(B) - P(A \cap B) = 0$
[$\because$ Sum of two non-negative no's can be zero only when these no's are zeros]
$\Rightarrow P(A) = P(B) = P(A \cap B)$
Which is the required relationship.

3. Let A be the event that max. number on the two chosen tickets in not more than 10, and B is the event that min. number on them is 5. We have to find $P(B / A)$.

We know that $P(B / A) = \frac{P(B \cap A)}{P(A)}$

Total ways to select two tickets out of 100 = ${}^{100}C_2$.
Number of ways favourable to A
= number of ways of selecting any 2 numbers from 1 to 10
$= {}^{10}C_2 = 45$
$A \cap B$ contains one number 5 and other greater than 5 and ≤ 10
So ways favourable to $A \cap B = {}^5C_1 = 5$

Therefore, $P(A) = \frac{45}{{}^{100}C_2}$ and $P(B \cap A) = \frac{5}{{}^{100}C_2}$

Thus, $P(B/A) = \frac{5 / {}^{100}C_2}{45 / {}^{100}C_2} = \frac{5}{45} = \frac{1}{9}$

4. Let $P(A) = \frac{1+3p}{3}, P(B) = \frac{1-p}{4}, P(C) = \frac{1-2p}{2}$
As A, B and C are three mutually exclusive events
$\therefore P(A) + P(B) + P(C) \leq 1$

$\Rightarrow \frac{1+3p}{3} + \frac{1-p}{4} + \frac{1-2p}{2} \leq 1$

$\Rightarrow 4 + 12p + 3 - 3p + 6 - 12p \leq 12$

$\Rightarrow 3p \geq 1 \Rightarrow p \geq 1/3$ ... (i)

Also $0 \leq P(A) \leq 1 \Rightarrow 0 \leq \frac{1+3p}{3} \leq 1$

$\Rightarrow 0 \leq 1 + 3p \leq 3$

$\Rightarrow -\frac{1}{3} \leq p \leq \frac{2}{3}$ ... (ii)

$0 \leq P(B) \leq 1 \Rightarrow 0 \leq \frac{1-p}{4} \leq 1$

$\Rightarrow 0 \leq 1 - p \leq 4$

$\Rightarrow -3 \leq p \leq 1$ ... (iii)

$0 \leq P(C) \leq 1 \Rightarrow 0 \leq \frac{1-2p}{2} \leq 1 \Rightarrow -\frac{1}{2} \leq p \leq \frac{1}{2}$ ... (iv)

Combining (i), (ii), (iii) and (iv), we get $\frac{1}{3} \leq p \leq \frac{1}{2}$

5. There may be following cases:
Case I : Red from A to B and red from B to A then prob. of drawing a red ball from A is

$$= \frac{6}{10} \times \frac{5}{11} \times \frac{6}{10} = \frac{180}{1100} = \frac{18}{110}$$

Case II : Red from A to B and black from B to A then prob. of drawing a red from A is

$$= \frac{6}{10} \times \frac{6}{11} \times \frac{5}{10} = \frac{180}{1100} = \frac{18}{110}$$

Case III : Black from A to B and red from B to A then prob. of drawing red from A is

$$= \frac{4}{10} \times \frac{4}{11} \times \frac{7}{10} = \frac{56}{550}$$

Case IV : Black from A to B and black from B to A then prob. of drawing red from A is

$$= \frac{4}{10} \times \frac{7}{11} \times \frac{6}{10} = \frac{168}{1100} = \frac{84}{550}$$

$\therefore$ The required prob. is

$$= \frac{18}{110} + \frac{18}{110} + \frac{56}{550} + \frac{84}{550} = \frac{90 + 90 + 56 + 84}{550}$$

$$= \frac{320}{550} = \frac{32}{55}$$

6. Probability of getting a sum of $5 = \frac{4}{36} = \frac{1}{9} = P(A)$ as favourable cases are $\{(1, 4), (4, 1), (2, 3), (3, 2)\}$
Similarly favourable cases of getting a sum of 7 are $\{(1, 6), (6, 1), (2, 5), (5, 2), (3, 4), (4, 3)\} = 6$

$\therefore$ Prob. of getting a sum of $7 = \frac{6}{36} = \frac{1}{6}$

$\therefore$ Prob. of getting a sum of 5 or 7

$= \frac{1}{6} + \frac{1}{9} = \frac{5}{18}$ [as events are mutually exclusive.]

$\therefore$ Prob of getting neither a sum of 5 nor of 7

$$= 1 - \frac{5}{18} = \frac{13}{18}$$

Now we get a sum of 5 before a sum of 7 if either we get a sum of 5 in first chance or we get neither a sum of 5 nor of 7 in first chance and a sum of 5 in second chance and so on. Therefore the required prob. is

$$= \frac{1}{9} + \frac{13}{18} \times \frac{1}{9} + \frac{13}{18} \times \frac{13}{18} \times \frac{1}{9} + \ldots \infty$$

$$= \frac{1/9}{1 - 13/18} = \frac{1}{9} \times \frac{18}{5} = \frac{2}{5}$$

7. $P(A \cup B) = 0.8$

$\Rightarrow P(A \cup B) = P(A) + P(B) - P(A \cap B)$

$\Rightarrow P(A \cup B) = P(A) + P(B) - P(A)\,P(B)$

[As A and B are independent events]

$\Rightarrow 0.8 = 0.3 + P(B) - 0.3\,P(B)$

$\Rightarrow 0.5 = 0.7\,P(B) \Rightarrow P(B) = 5/7$

8. KEY CONCEPT:
For a binomial distribution,

$$P(X = r) = {}^nC_r\, p^r q^{n-r}$$

where, n = no. of trials
p = prob. of success in one trial
q = prob. of failure in one trial
r = no. of successes.

For a binomial distribution, we know,
mean = np and variance = npq

$\therefore$ ATQ, $np = 2$; $npq = 1 \Rightarrow q = 1/2$

$\Rightarrow p = 1/2$ and $n = 4$

$\therefore P(X > 1) = P(X = 2) + P(X = 3) + P(X = 4)$

$= 1 - P(X = 0) - P(X = 1)$

$$= 1 - {}^4C_0\left(\frac{1}{2}\right)^0\left(\frac{1}{2}\right)^4 - {}^4C_1\left(\frac{1}{2}\right)^1\left(\frac{1}{2}\right)^3 = 1 - \frac{1}{16} - \frac{4}{16} = \frac{11}{16}$$

9. Sample space = {Y, Y, Y, R, R, B} where Y stands for yellow colour, R for red and B for blue.
Prob. that the colours yellow, red and blue appear in the first second, and third tosses respectively

$$= \frac{3}{6} \times \frac{2}{6} \times \frac{1}{6} = \frac{1}{36}$$

10. Given that $P(A^c) = 0.3$, $P(B) = 0.4$ and $P(A \cap B^c) = 0.5$

then $P[B/(A \cup B^c)] = \dfrac{P[B \cap (A \cup B^c)]}{P(A \cup B^c)}$

$$= \frac{P((B \cap A) \cup (B \cap B^c))}{P(A \cup B^c)} = \frac{P(A \cap B)}{P(A) + P(B^c) - P(A \cap B^c)}$$

$$= \frac{P(A) - P(A \cap B^c)}{1 - P(A^c) + 1 - P(B) - P(A \cap B^c)}$$

NOTE THIS STEP:

$$= \frac{1 - 0.3 - 0.5}{1 - 0.3 + 1 - 0.4 - 0.5} = \frac{0.2}{0.8} = \frac{1}{4}$$

11. Let $E_1 \equiv$ getting min no. 3, $E_2 \equiv$ getting max no. 7
Then $P(E_1) = P$ (getting one no. 3 and other 2 from no's 4 to 10)

$$= \frac{{}^1C_1 \times {}^7C_2}{{}^{10}C_3} = \frac{7}{40}$$

Similarly $P(E_2) = P$ (getting one no. 7 and other 2 from no's 1 to 6)

$$= \frac{{}^1C_1 \times {}^6C_2}{{}^{10}C_3} = \frac{1}{8}$$

and $P(E_1 \cap E_2) = P$ (getting one no. 3, second no. 7 and third from no's 4 to 6)

$$= \frac{{}^1C_1 \times {}^1C_1 \times {}^3C_1}{{}^{10}C_3} = \frac{1}{40}$$

$\therefore P(E_1 \cup E_2) = P(E_1) + P(E_2) - P(E_1 \cap E_2)$

$$= \frac{7}{40} + \frac{1}{8} - \frac{1}{40} = \frac{7 + 5 - 1}{40} = \frac{11}{40}$$

B TRUE / FALSE :

1. Let E be the event "No two S's occur together".

A, A, I, N can be arranged in $\dfrac{4!}{2!} = 12$ ways

$-A-A-I-N-$ Creating 5 places for 4 S. Out of 5 places 4 can be selected in ${}^5C_4 = 5$ ways.

$\therefore$ No two S's occur together in $= 12 \times 5 = 60$ ways

Total no.of arranging all letters of word 'assassin'

$$= \frac{8!}{4!2!} = 840$$

$\therefore$ Req. prob. $= \dfrac{60}{840} = \dfrac{1}{14}$ $\therefore$ Statement is False.

2. $P(A \cup B) = P(A) + P(B) - P(A \cap B)$

$= P(A) + P(B) - P(A)\,P(B)$

[$\because$ A and B are independent events]

$= 0.2 + 0.3 - 0.2 \times 0.3$

$= 0.5 - 0.06 = 0.44 \neq 0.5$

$\therefore$ The statement is false.

C MCQ's WITH ONE CORRECT ANSWER :

1. (d) The two events can happen simultaneously e.g., (2, 3)
$\therefore$ not mutually exclusive.
Also are not dependent on each other.

2. (a) $P(A \cup B) = P(A) + P(B) - P(A \cap B)$

$= 0.25 + 0.50 - 0.14 = 0.61$

$\therefore P(A' \cap B') = P((A \cup B)') = 1 - P(A \cup B)$

$= 1 - 0.61 = 0.39$

3. (b) $p = 0.4, n = 3, P(X \geq 1) = ? \Rightarrow q = 0.6$

$\therefore P(X \geq 1) = 1 - P(X = 0)$

$= 1 - {}^3C_0(0.4)^0(0.6)^3$

$= 1 - 0.216 = 0.784$

4. (c) $$P(\bar{A}/\bar{B}) = \frac{P(\bar{A} \cap \bar{B})}{P(\bar{B})} = \frac{P(\overline{A \cup B})}{P(\bar{B})} = \frac{1 - P(A \cup B)}{P(\bar{B})}$$

5. (c) $n = 7$
Prob. of getting any no. out 1, 2, 3, ... 9 is $p = 9/15$

$\therefore q = 6/5$

$P(x = 7) = {}^7C_7 p^7 q^0$ [Binomial distribution]

$$= \left(\frac{9}{15}\right)^7 = \left(\frac{3}{5}\right)^7$$

6. (b) Favourable cases = 6; {(1, 1, 1), (2, 2, 2),...(6, 6, 6)}

Total cases $= 6 \times 6 \times 6 = 216$ $\therefore$ Req. prob. $= \dfrac{6}{216} = \dfrac{1}{36}$

7. (c) Prob. of a getting a white ball in a single draw

$$= p = \frac{12}{24} = \frac{1}{2}$$

Prob. of getting a white ball 4th time in the 7th draw
$= P$ (getting 3 W in 6 draws)
$\times P$ (getting W ball at 7th draw)

$$= {}^6C_3\left(\frac{1}{2}\right)^6 \cdot \frac{1}{2} = \frac{5}{32}$$

8. (d) Prob. of one coin showing head = p

$\therefore$ Prob of one coin showing tail = $1-p$

ATQ coin is tossed 100 times and prob. of 50 coins showing head = prob of 51 coins showing head.

Using binomial prob. distribution

$$P(X=r)={}^nC_r p^r q^{n-r},$$

NOTE THIS STEP :

we get

$${}^{100}C_{50}\,p^{50}(1-p)^{50}={}^{100}C_{51}\,p^{51}(1-p)^{49}$$

$$\Rightarrow \frac{1-p}{p}=\frac{{}^{100}C_{51}}{{}^{100}C_{50}}=\frac{50!50!}{51!49!}=\frac{50}{51}\Rightarrow 51-51p=50p$$

$$\Rightarrow 101p=51\Rightarrow p=\frac{51}{101}$$

9. (b) P (at least 7 pts) = P (7pts) + P (8 pts)

[$\because$ At most 8 pts can be scored.]

Now 7 pts can be scored by scoring 2 pts in 3 matches and 1 pt. in one match, similarly 8 pts can be scored by scoring 2 pts in each of the 4 matches.

$\therefore$ Req. prob.

$={}^4C_1\times[P(2\text{ pts})]^3P(1\text{pt})+[P(2\text{ pts})]^4$

$=4(0.5)^3\times0.05+(0.50)^4=0.0250+0.0625=0.0875$

10. (a) The min. face value is not less than 2 and max. face value is not greater than 5 if we get any of the numbers 2, 3, 4, 5, while total possible out comes are 1, 2, 3, 4, 5 and 6.

$\therefore$ In one thrown of die, prob. of getting any no.

Out of 2, 3, 4 and 5 is $=\frac{4}{6}=\frac{2}{3}$

If the die is rolled four times, then all these events being independent, the required prob. is

$$\left(\frac{2}{3}\right)^4=\frac{16}{81}$$

11. (a) $P[A\cap(B\cup C)]=P[(A\cap B)\cup(A\cap C)]$

$=P(A\cap B)+P(A\cap C)-P(A\cap B\cap C)$

$=P(A)\,P(B)+P(A)\,P(C)-P(A)\,P(B)P(C)$

$=P(A)[P(B)+P(C)-P(B\cap C)]=P(A)\,P(B\cup C)$

$\therefore$ S_1 is true.

$P(A\cap(B\cap C))=P(A)\,P(B)\,P(C)=P(A)\,P(B\cap C)$

$\therefore$ S_2 is also true.

12. (b) Given that P (India wins) = p = 1/2

$\therefore$ P (India loses) = p' = 1/2

Out of 5 matches india's second win occurs at third test

$\Rightarrow$ India wins third test and simultaneously it has won one match from first two and lost the other.

$\therefore$ Required prob. = $P(LWW)+P(WLW)$

$$=\left(\frac{1}{2}\right)^3+\left(\frac{1}{2}\right)^3=\frac{1}{4}$$

13. (c) Out of 6 vertices 3 can be chosen in ${}^6C_3$ ways.

Δ will be equilateral if it is ΔACE or ΔBDF (2 ways)

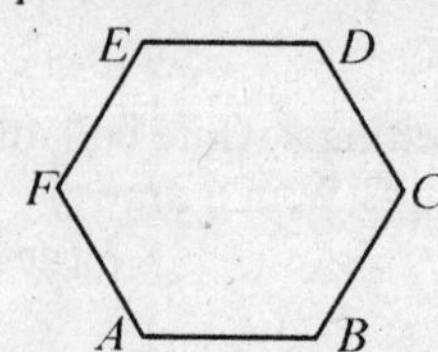

$\therefore$ Required prob. $=\frac{2}{{}^6C_3}=\frac{2}{20}=\frac{1}{10}$

14. (a) We know that P (exactly one of A or B occurs)

$=P(A)+P(B)-2P(A\cap B)$.

Therefore, $P(A)+P(B)-2P(A\cap B)=p$ ...(1)

Similarly, $P(B)+P(C)-2P(B\cap C)=p$ ...(2)

and $P(C)+P(A)-2P(C\cap A)=p$ ...(3)

Adding (1), (2) and (3) we get

$$2[P(A)+P(B)+P(C)-P(A\cap B)-P(B\cap C)-P(C\cap A)]=3p$$

$$\Rightarrow P(A)+P(B)+P(C)-P(A\cap B)-P(B\cap C)-P(C\cap A)=3p/2 \quad ...(4)$$

We are also given that,

$P(A\cap B\cap C)=p^2$ ...(5)

Now, P (at least one of A, B and C)

$=P(A)+P(B)+P(C)-P(A\cap B)-P(B\cap C)-P(C\cap A)+P(A\cap B\cap C)$

$=\frac{3p}{2}+p^2$ [using (4) and (5)]

$=\frac{3p+2p^2}{2}$

15. (a) We know that,

$7^1=7, 7^2=49, 7^3=343, 7^4=2401, 7^5=16807$

$\therefore$ 7^k (where $k\in Z$), results in a number whose unit's digit is 7 or 9 or 3 or 1.

Now, 7^m+7^n will be divisible by 5 if unit's place digit of resulting number is 5 or 0 clearly it can never be 5. But it can be 0 if we consider values of m and n such that the sum of unit's place digits become 0. And this can be done by choosing

$m=1,5,9,\ldots97$ and correspondingly $n=3,7,11,\ldots99$ } (25 options each) [7 + 3 = 10]

$m=2,6,10,\ldots\ldots98$ and $n=4,8,12,\ldots\ldots100$ } (25 options each) [9 + 1 = 10]

case I:

Thus m can be chosen in 25 ways and n can be chosen in 25 ways

case II:

m can be chosen in 25 ways and n can be chosen in 25 ways

$\therefore$ Total no. of selections of m, n so that 7^m+7^n is divisible by 5 = (25 × 25 + 25 × 25) × 2

Note we can interchange values of m and n.

Also no. of total possible selections of m and n out of 100 = 100 × 100

$\therefore$ Req. prob. $=\frac{2(25\times25+25\times25)}{100\times100}=\frac{1}{4}$.

16. (d) The minimum of two numbers will be less than 4 if at least one of the numbers is less than 4.

$\therefore$ P (at least one no. is < 4),

$=1-P$ (both the no's are ≥ 4)

$=1-\frac{3}{6}\times\frac{2}{5}=1-\frac{6}{30}=1-\frac{1}{5}=4/5$

17. (a) Given that $P(B) = 3/4$, $P(A \cap B \cap \bar{C}) = 1/3$

$P(\bar{A} \cap B \cap \bar{C}) = 1/3$

From venn diagram, we see

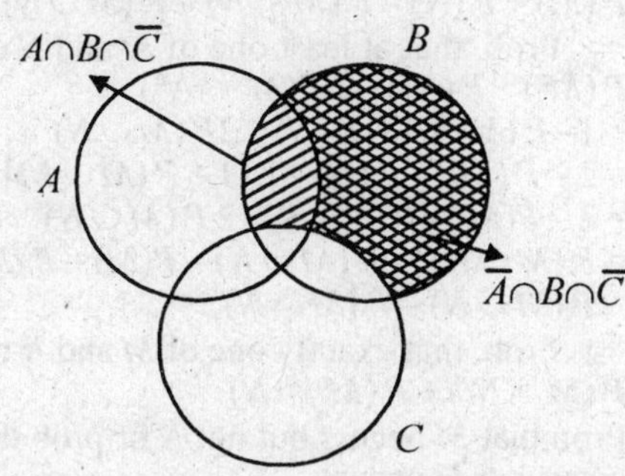

$B \cap C \equiv B - (A \cap B \cap \bar{C}) - (\bar{A} \cap B \cap \bar{C})$

NOTE THIS STEP :

$\Rightarrow P(B \cap C) = P(B) - P(A \cap B \cap \bar{C}) - P(\bar{A} \cap B \cap \bar{C})$

$\Rightarrow P(B \cap C) = \frac{3}{4} - \frac{1}{3} - \frac{1}{3} = \frac{9-4-4}{12} = \frac{1}{12}$

18. (d) If a no. is to be divisible by both 2 and 3. It should be divisible by their L.C.M.

$\therefore$ L.C.M. of (2 and 3) = 6

$\therefore$ Numbers are = 6, 12, 18 ... 96.

Total numbers are = 16

$\therefore$ Probability $= \frac{^{16}C_3}{^{100}C_3} = \frac{4}{1155}$

19. (a) In single throw of a dice, probability of getting 1 is $= \frac{1}{6}$

and prob. of not getting 1 is $\frac{5}{6}$.

Then getting 1 in even no. of chances = getting 1 in 2nd chance or in 4th chance or in 6th chance and so on

$\therefore$ Req. Prob.

$= \frac{5}{6} \times \frac{1}{6} + \left(\frac{5}{6}\right)^3 \times \frac{1}{6} + \left(\frac{5}{6}\right)^5 \times \frac{1}{6} + \ldots \infty$

$= \frac{5}{36}\left[\frac{1}{1-\frac{25}{36}}\right] = \frac{5}{36} \times \frac{36}{11} = \frac{5}{11}$

20. (c) Let $E_1 \equiv$ The Indian man is seated adjacent to his wife.

$E_2 \equiv$ Each American man is seated adjacent to his wife.

Then $P(E_1 / E_2) = \frac{P(E_1 \cap E_2)}{P(E_2)}$

Now $E_1 \cap E_2 \equiv$ All men are seated adjacent to their wives.

$\therefore$ We can consider the 5 couples as single-single objects which can be arranged in a circle in 4! ways. But for each couple, husband and wife can interchange their places in 2! ways.

$\therefore$ Number of ways when all men are seated adjacent to their wives $= 4! \times (2!)^5$

Also in all 10 persons can be seated in a circle in 9! ways.

$\therefore \quad P(E_1 \cap E_2) = \frac{4! \times (2!)^5}{9!}$

Similarly if each American man is seated adjacent to his wife, considering each American couple as single object and Indian woman and man as seperate objects there are 6 different objects which can be arranged in a circle in 5! ways. Also for each American couple, husband and wife can interchange their places in 2 ! ways.

So the number of ways in which each American man is seated adjacent to his wife.

$= 5! \times (2!)^4 \therefore P(E_2) = \frac{5! \times (2!)^4}{9!}$

So $P(E_1 / E_2) = \frac{(4! \times (2!)^5)/9!}{(5! \times (2!)^4)/9!} = \frac{2}{5}$

21. (c) $P(E^c \cap F^c / G) = P(E \cup F)^c / G)$

$1 - P(E \cup F / G)$

$= 1 - P(E/G) - P(F/G) + P(E \cap F/G)$

$= 1 - P(E) - P(F) + O$

($\because$ E, F, G are pairwise independent and $P(E \cap F \cap G) = 0$

$\Rightarrow P(E).P(F) = 0$ as $P(G) > 0$)

$= P(E^c) - P(F)$

22. (d) We have n (S) = 10

$n(A) = 4$

Let $n(B) = x$ and $n(A \cap B) = y$

Then for A and B to be independent events

$P(A \cap B) = P(A) P(B)$

$\Rightarrow \frac{y}{10} = \frac{4}{10} \times \frac{x}{10} \Rightarrow x = \frac{5}{2} y$

$\Rightarrow$ y can be 2 or 4 so that $x = 5$ or 10

$\therefore n(B) = 5$ or 10

23. (c) If ω is a complex cube root of unity then, we know that

$\omega^{3m} + \omega^{3n+1} + \omega^{3p+2} = 0$

Where m, n, p are integers.

$\therefore r_1, r_2, r_3$ should be of the form $3m$, $3n + 1$ and $3p + 2$ taken in any order. As r_1, r_2, r_3 are the numbers obtained on die, these can take any value from 1 to 6.

$\therefore$ m can take values 1 or 2

n can take values 0 or 1

p can take values 0 or 1

$\therefore$ Number of ways of selecting r_1, r_2, r_3

$= {}^2C_1 \times {}^2C_1 \times {}^2C_1 \times 3!$.

Also the total number of ways of getting r_1, r_2, r_3 on die $= 6 \times 6 \times 6$

$\therefore$ Required probability $= \frac{{}^2C_1 \times {}^2C_1 \times {}^2C_1 \times 3!}{6 \times 6 \times 6} = \frac{2}{9}$

24. **(c)** Let G ≡ oreginal signal is green ⇒ P(G) = 4/5

E_1 ≡ A receives the signal correctly $P(E_1) = 3/4$

E_2 ≡ B receives the signal correctly $P(E_2) = 3/4$

E ≡ Signal received by B is green.

Then E can happen in the following ways

Original Signal	Received at A	Received at B
Red ⟶	Red ⟶	Green
Red ⟶	Green ⟶	Green
Green ⟶	Green ⟶	Green
Green ⟶	Red ⟶	Green

$$\therefore P(E) =)\left(\bar{G} \cap E_1 \cap \bar{E}_2\right) + \left(\bar{G} \cap E_1 \cap \bar{E}_2\right)$$

$$+ P\left(G \cap E_1 \cap E_2\right) + P\left(G \cap \bar{E}_1 \cap \bar{E}_2\right)$$

$$= \frac{1}{5}\times\frac{3}{4}\times\frac{1}{4} + \frac{1}{5}\times\frac{3}{4}\times\frac{1}{4} + \frac{4}{5}\times\frac{3}{4}\times\frac{3}{4} + \frac{4}{5}\times\frac{1}{4}\times\frac{1}{4}$$

$$= \frac{3+3+36+4}{80} = \frac{46}{80} = \frac{23}{40}$$

$$\therefore P(G/E) = \frac{P(G \cap E)}{P(E)} = \frac{P(G \cap E_1 \cap E_2) + P(G \cap \bar{E}_1 \cap \bar{E}_2)}{P(E)}$$

$$= \frac{\frac{4}{5}\times\frac{3}{4}\times\frac{3}{4} + \frac{4}{5}\times\frac{1}{4}\times\frac{1}{4}}{23/40} = \frac{40/80}{23/40} = \frac{20}{23}$$

25. **(a)** D_4 can show a number appearing on one of D_1, D_2 and D_3 in the following cases.

Case I : D_4 shows a number which is shown by only one of D_1, D_2 and D_3.

D_4 shows a number in ${}^6C_1$ ways.

One out of D_1, D_2 and D_3 can be selected in ${}^3C_1$ ways.

The selected die shows the same number as on D_4 in one way and rest two dice show the different number in 5 ways each.

∴ Number of ways to happen case I

$= {}^6C_1 \times {}^3C_1 \times 1 \times 5 \times 5 = 450$

Case II : D_4 shows a number which is shown by only two of D_1, D_2 and D_3.

As discussed in case I, it can happen in the following number of ways

$= {}^6C_1 \times {}^3C_2 \times 1 \times 1 \times 5 = 90$

Case III : D_4 shows a number which is shown by all three dice D_1, D_2 and D_3.

Number of ways it can be done

$= {}^6C_1 \times {}^3C_3 \times 1 \times 1 \times 1 = 6$

∴ Total number of favourable ways

$= 450 + 90 + 6 = 546$

Also total ways $= 6 \times 6 \times 6 \times 6$

$$\therefore \text{Required Probability} = \frac{546}{6\times6\times6\times6} = \frac{91}{216}$$

D MCQ's WITH ONE OR MORE THAN ONE CORRECT :

1. **(a, c, d)** Given that M and N are any two events. To check the probability that exactly one of them occurs. We check all the options one by one.

(a) $P(M) + P(N) - 2P(M \cap N)$
$= [P(M) + P(N) - P(M \cap N)] - P(M \cap N)$
$= P(M \cup N) - P(M \cap N)$
⇒ Prob. that exactly one of M and N occurs.

(b) $P(M) + P(N) - P(M \cap N) = P(M \cup N)$
⇒ Prob. that at least one of M and N occurs.

(c) $P(M^c) + P(N^c) - 2P(M^c \cap N^c)$
$= 1 - P(M) + 1 - P(N) - 2P(M \cup N)^c$
$= 2 - P(M) - P(N) - 2[1 - P(M \cup N)]$
$= 2 - P(M) - P(N) - 2 + 2P(M \cup N)$
$= P(M \cup N) + P(M \cup N) - P(M) - P(N)$
$= P(M \cup N) - P(M \cap N)$
⇒ Prob. that exactly one of M and N occurs.

(d) $P(M \cap N^c) + P(M^c \cap N)$
⇒ Prob that M occurs but not N or prob that M does not occur but N occurs.
⇒ Prob. that exactly one of M and N occurs.

Thus we can conclude that (a), (c) and (d) are the correct options.

2. **(c)** Let A, B, C be the events that the student passes test I, II, III respectively.

Then, ATQ ; $P(A) = p$; $P(B) = q$; $P(C) = \frac{1}{2}$

Now the student is successful if A and B happen or A and C happen or A, B and C happen.

ATQ, $P(AB\bar{C}) + P(AC\bar{B}) + P(ABC) = \frac{1}{2}$

$$\Rightarrow pq\left(1 - \frac{1}{2}\right) + p.\frac{1}{2}.(1-q) + p.q.\frac{1}{2} = \frac{1}{2}$$

$$\Rightarrow \frac{1}{2}pq + \frac{1}{2}p - \frac{1}{2}pq + \frac{1}{2}pq = \frac{1}{2}$$

$\Rightarrow p + pq = 1 \Rightarrow p(1+q) = 1$

which holds for $p = 1$ and $q = 0$.

3. **(c)** Given that $P(A \cup B) = 0.6$; $P(A \cap B) = 0.2$

$\therefore P(\bar{A}) + P(\bar{B}) = 1 - P(A) + 1 - P(B)$
$= 2 - (P(A) + P(B))$
$= 2 - [P(A \cup B) + P(A \cap B)]$.
$= 2 - [0.6 + 0.2] = 2 - 0.8 = 1.2$

4. **(a, b, c)** We know that,

$P(A \cap B) = P(A) + P(B) - P(A \cup B)$ ...(1)

Also $P(A \cup B) \le 1$

$\Rightarrow -P(A \cup B) \ge -1$ ...(2)

$\therefore P(A \cap B) \ge P(A) + P(B) - 1$ [Using (1) and (2)]

∴ (a) is true. Again $P(A \cup B) \ge 0$

$\Rightarrow -P(A \cup B) \le 0$ ...(3)

$\Rightarrow P(A \cap B) \le P(A) + P(B)$ [Using (1) and (3)]

∴ (b) is also correct.

From (1) (c) is true and (d) is not correct.

5. **(b, c, d)** Since E and F are independent

$\therefore P(E \cap F) = P(E).P(F)$ ...(1)

Now, $P(E \cap F^c) = P(E) - P(E \cap F)$
$= P(E) - P(E)P(F)$ [Using (1)]
$= P(E)[1 - P(F)] = P(E)P(F^c)$

∴ E and F^c are independent.

Again $P(E^c \cap F^c) = P(E \cup F)^c = 1 - P(E \cup F)$
$= 1 - P(E) - P(F) + P(E \cap F)$
$= 1 - P(E) - P(F) + P(E)P(F)$
$= ((1 - P(E))(1 - P(F)) = P(E^c)P(F^c)$

$\therefore$ E^c and F^c are independent.
Also $P(E/F) + P(E^c/F)$

$$= \frac{P(E \cap F)}{P(F)} + \frac{P(E^c \cap F)}{P(F)} = \frac{P(E)P(F) + P(E^c)P(F)}{P(F)}$$

$$= \frac{P(F)(P(E) + P(E^c))}{P(F)} = 1$$

6. **(a, c)** For any two events A and B

(a) $P(A/B) = \frac{P(A \cap B)}{P(B)}$

Now we know $P(A \cup B) \le 1$

$P(A) + P(B) - P(A \cap B) \le 1$

$\Rightarrow P(A \cap B) \ge P(A) + P(B) - 1$

$\Rightarrow \frac{P(A \cap B)}{P(B)} \ge \frac{P(A) + P(B) - 1}{P(B)}$ $\left[\begin{array}{l}\text{As } P(B) \ne 0 \\ \therefore P(B) > 0\end{array}\right]$

$\Rightarrow P(A/B) \ge \frac{P(A) + P(B) - 1}{P(B)}$ $\therefore$ (a) is correct statement.

(b)

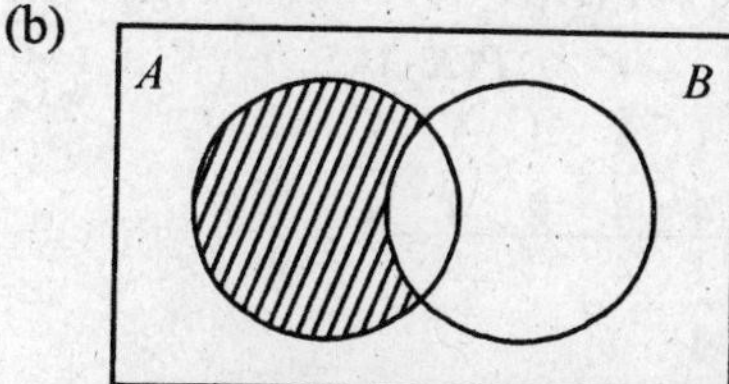

From venn diagram we can clearly conclude that

$P(A \cap \bar{B}) = P(A) - P(A \cap B)$

$\therefore$ (b) is incorrect statement.

(c) $P(A \cup B) = P(A) + P(B) - P(A \cap B)$

$= 1 - P(\bar{A}) + 1 - P(\bar{B}) - P(A)P(B)$

[$\because$ A & B are independent events]

$= 2 - P(\bar{A}) - P(\bar{B}) - [1 - P(\bar{A})][1 - P(\bar{B})]$

$= 2 - P(\bar{A}) - P(\bar{B}) - 1 + P(\bar{A}) + P(\bar{B}) - P(\bar{A})P(\bar{B})]$

$= 1 - P(\bar{A})P(\bar{B})$ $\therefore$ (c) is the correct statement.

(d) For disjoint events $P(A \cup B) = P(A) + P(B)$

$\therefore$ (d) is the incorrect statement.

7. **(a, d)** Let $P(E) = x$ and $P(F) = y$

ATQ, $P(E \cap F) = \frac{1}{12}$

As E and F are independent events

$\therefore$ $P(E \cap F) = P(E)P(F)$

$\Rightarrow \frac{1}{12} = xy \Rightarrow xy = \frac{1}{12}$ ...(1)

Also $P(\bar{E} \cap \bar{F}) = P(\overline{E \cup F}) = 1 - P(E \cup F)$

$\Rightarrow \frac{1}{2} = 1 - [P(E) + P(F) - P(E)P(F)]$

$\Rightarrow x + y - xy = \frac{1}{2} \Rightarrow x + y = \frac{7}{12}$ ...(2)

Solving (1) and (2) we get

either $x = \frac{1}{3}$ and $y = \frac{1}{4}$ or $x = \frac{1}{4}$ and $y = \frac{1}{3}$

$\therefore$ (a) and (d) are the correct options.

8. **(c, d)**
$P(A \cup B)' = 1 - P(A \cup B)$
$= 1 - [P(A) + P(B) - P(A \cap B)]$
$= 1 - P(A) - P(B) + P(A)P(B)$
$= P(A')P(B')$

Also $P(A \cup B) = P(A) + P(B) - P(A)P(B)$

$\Rightarrow P(A \cap B) = P(A) + P(B)$

$\therefore$ $P(A/B) = \frac{P(A \cap B)}{P(B)} = \frac{P(A)P(B)}{P(B)} = P(A)$

9. **(a)** P(2 white and 1 black)
$= P(W_1 W_2 B_3 \text{ or } W_1 B_2 W_3 \text{ or } B_1 W_2 W_3)$
$= P(W_1 W_2 B_3) + P(W_1 B_2 W_3) + P(B_1 W_2 W_3)$
$= P(W_1)P(W_2)P(B_3) + P(W_1)P(B_2)P(W_3) + P(B_1)P(W_2)P(W_3)$

$$= \frac{3}{4}.\frac{2}{4}.\frac{3}{4} + \frac{3}{4}.\frac{2}{4}.\frac{1}{4} + \frac{1}{4}.\frac{2}{4}.\frac{1}{4} = \frac{1}{32}(9 + 3 + 1) = \frac{13}{32}$$

10. **(a, d)** We have,

(a) $P(E/F) + P(\bar{E}/F) = \frac{P(E \cap F)}{P(F)} + \frac{P(\bar{E} \cap F)}{P(F)}$

$$= \frac{P(E \cap F) + P(\bar{E} \cap F)}{P(F)} = \frac{P(F)}{P(F)} = 1$$

$\therefore$ (a) holds.

Also

(b) $P(E/F) + P(E/\bar{F}) = \frac{P(E \cap F)}{P(F)} + \frac{P(E \cap \bar{F})}{P(\bar{F})}$

$$= \frac{P(E \cap F)[1 - P(F)] + P(F)[P(E \cap \bar{F})]}{P(F)P(\bar{F})}$$

$$= \frac{P(E \cap F) + P(F)[P(E \cap \bar{F}) - P(E \cap F)]}{P(F)P(\bar{F})} \ne 1$$

$\therefore$ (b) does not hold. Similarly we can show that (c) does not hold but (d) holds.

11. **(b)** The probability that only two tests are needed = (probability that the second machine tested is faulty given the first machine tested is faulty).

$$= \frac{2}{4} \times \frac{1}{3} = \frac{1}{6}$$

12. **(d)** Given that $P(E) \le P(F)$ and $P(E \cap F) > 0$. It doesn't necessarily mean that E is the subset of F.

$\therefore$ The choices (a), (b), (c) do not hold in general. Hence (d) is the right choice here.

13. **(a)** The event that the fifth toss results in a head is independent of the event that the first four tosses result in tails.

$\therefore$ Probability of the required event = 1/2.

14. **(b)** The no. of ways of placing 3 black balls without any restrition is ${}^{10}C_3$. Now the no. of ways in which no two black balls put together is equal to the no of ways of choosing 3 places marked out of eight places.

$- W - W - W - W - W - W - W -$

This can be done is ${}^{8}C_3$ ways. Thus, probability of the required event

$$= \frac{{}^{8}C_3}{{}^{10}C_3} = \frac{8 \times 7 \times 6}{10 \times 9 \times 8} = \frac{7}{15}.$$

$\therefore$ (b) is the correct option.

15. **(b, c)** According to the problem,

$m+p+c-mp-mc-pc+mpc=3/4$ ...(1)

$mp(1-c)+mc(1-p)+pc(1-m)=2/5$

or $mp+mc+pc-3mpc=2/5$ ...(2)

Also $mp+pc+mc-2mpc=1/2$ ...(3)

(2) and (3) $\Rightarrow mpc=\frac{1}{2}-\frac{2}{5}=\frac{1}{10}$

$\therefore \quad mp+mc+pc=\frac{2}{5}+\frac{3}{10}=\frac{7}{10}$

$\therefore \quad m+p+c=\frac{3}{4}+\frac{7}{10}-\frac{1}{10}=\frac{15+14-2}{20}=\frac{27}{20}$

16. **(a,d)** $\because$ E and F are independent events

$\therefore P(E\cap F)=P(E).P(F)$ ...(1)

Given that

$P(E\cap\bar{F})+P(\bar{E}\cap F)=\frac{11}{25}$

$\Rightarrow P(E)P(\bar{F})+P(\bar{E})P(F)=\frac{11}{25}$

$\Rightarrow P(E)(1-P(F))+(1-P(E))P(F)=\frac{11}{25}$

$\Rightarrow P(E)-P(E)P(F)+P(F)-P(E)P(F)=\frac{11}{25}$

$\Rightarrow P(E)+P(F)-2P(E).P(F)=\frac{11}{25}$ ...(2)

and $P(\bar{E}\cap\bar{F})=\frac{2}{25}\Rightarrow P(\bar{E})\ P(\bar{F})=\frac{2}{25}$

$\Rightarrow [1-P(E)][1-P(F)]=\frac{2}{25}$

$\Rightarrow 1-P(E)-P(F)+P(E)P(F)=\frac{2}{25}$ ...(3)

Adding equation (2) and (3) we get

$1-P(E)P(F)=\frac{13}{25}$ or $P(E)P(F)=\frac{12}{25}$ ...(4)

Using the result in equation (2) we get

$P(E)+P(F)=\frac{35}{25}$ ...(5)

Solving (4) and (5) we get

$P(E)=\frac{3}{5}$ and $P(F)=\frac{4}{5}$ or $P(E)=\frac{4}{5}$ and $P(F)=\frac{3}{5}$

$\therefore$ (a) and (d) are the correct options.

17. **(b,d)**

We have $P(X_1)=\frac{1}{2}$, $P(X_2)=\frac{1}{4}$, $P(X_3)=\frac{1}{4}$

$P(X)=P$(at least 2 engines are functioning)

$=P(X_1\cap X_2\cap X_3^C)+P(X_1\cap X_2^C\cap X_3)$

$+P(X_1^C\cap X_2\cap X_3)+P(X_1\cap X_2\cap X_3)$

$=\frac{1}{2}\times\frac{1}{4}\times\frac{3}{4}+\frac{1}{2}\times\frac{3}{4}\times\frac{1}{4}+\frac{1}{2}\times\frac{1}{4}\times\frac{1}{4}+\frac{1}{2}\times\frac{1}{4}\times\frac{1}{4}=\frac{1}{4}$

(a) $P(X_1^C/X)=\frac{P(X_1^C\cap X)}{P(X)}=\frac{P(X_1^C\cap X_2\cap X_3)}{P(X)}$

$=\frac{\frac{1}{2}\times\frac{1}{4}\times\frac{1}{4}}{\frac{1}{4}}=\frac{1}{8}$

(b) P [Exactly two engines are functioning $/X$]

$=\frac{P[(\text{Exactly two engines are functioning})\cap X]}{P(X)}$

$=\frac{P(X_1^C\cap X_2\cap X_3)+P(X_1\cap X_2^C\cap X_3)+P(X_1\cap X_2\cap X_3^C)}{P(X)}$

$=\frac{\frac{1}{2}\times\frac{1}{4}\times\frac{1}{4}+\frac{1}{2}\times\frac{3}{4}\times\frac{1}{4}+\frac{1}{2}\times\frac{1}{4}\times\frac{3}{4}}{\frac{1}{4}}=\frac{7}{8}$

(c) $P(X/X_2)=\frac{P(X\cap X_2)}{P(X_2)}$

$=\frac{P(X_1\cap X_2\cap X_3)+P(X_1^C\cap X_2\cap X_3)+P(X_1\cap X_2\cap X_3^C)}{P(X_2)}$

$=\frac{\frac{1}{2}\times\frac{1}{4}\times\frac{1}{4}+\frac{1}{2}\times\frac{1}{4}\times\frac{1}{4}+\frac{1}{2}\times\frac{1}{4}\times\frac{3}{4}}{\frac{1}{4}}=\frac{5}{8}$

(d) $P(X/X_1)=\frac{P(X\cap X_1)}{P(X_1)}$

$=\frac{P(X_1\cap X_2\cap X_3)+P(X_1\cap X_2^C\cap X_3)+P(X_1\cap X_2\cap X_3^C)}{P(X_1)}$

$=\frac{\frac{1}{2}\times\frac{1}{4}\times\frac{1}{4}+\frac{1}{2}\times\frac{3}{4}\times\frac{1}{4}+\frac{1}{2}\times\frac{1}{4}\times\frac{3}{4}}{\frac{1}{2}}=\frac{7}{16}$

18. **(a, b)**

We know $P(X/Y)=\frac{P(X\cap Y)}{P(Y)}$

$\Rightarrow \frac{1}{2}=\frac{1/6}{P(Y)}\Rightarrow P(Y)=\frac{1}{3}$

Similarly, $P(Y/X)=\frac{P(X\cap Y)}{P(X)}$

$\Rightarrow \frac{1}{3}=\frac{1/6}{P(X)}\Rightarrow P(X)=\frac{1}{2}$

$\therefore \quad P(X\cup Y)=P(X)+P(Y)-P(X\cap Y)=\frac{1}{2}+\frac{1}{3}-\frac{1}{6}=\frac{2}{3}$

Also $P(X\cap Y)=P(X)P(Y)$

$\Rightarrow$ X and Y are independent events.

$\therefore$ X^C and Y are also independent events.

$\therefore\ P(X^C\cap Y)=P(X^C)\times P(Y)=\frac{1}{2}\times\frac{1}{3}=\frac{1}{6}$

19. **(a)** P (atleast one of them solves the problem)

$=1-P$ (none of them solves it)

$=1-\frac{1}{2}\times\frac{1}{4}\times\frac{3}{4}\times\frac{7}{8}=1-\frac{21}{256}=\frac{235}{256}$

(E) SUBJECTIVE PROBLEMS :

1. To draw 2 black, 4 white and 3 red balls in order is same as arranging two black balls at first 2 places, 4 white at next 4 places, (3rd to 6th place) and 3 red at still next 3 places (7th to 9th place), i.e., $B_1B_2W_1W_2W_3W_4R_1R_2R_3$, which can be done in $2! \times 4! \times 3!$ ways. And total ways of arranging all $2+4+3 = 9$ balls is 9!

$\therefore$ Required probability $= \dfrac{2! \times 4! \times 3!}{9!} = \dfrac{1}{1260}$

2. (i) 6 boys and 6 girls sit in a row randomly.
Total ways of their seating = 12!
No. of ways in which all the 6 girls sit together = $6! \times 7!$ (considering all 6 girls as one person)
$\therefore$ Probability of all girls sitting together

$$= \frac{6! \times 7!}{12!} = \frac{720}{12 \times 11 \times 10 \times 9 \times 8} = \frac{1}{132}$$

(ii) Staring with boy, boys can sit in 6! ways leaving one place between every two boys and one a last.
B_B_B_B_B_B_
These left over places can be occupied by girls in 6! ways.
$\therefore$ If we start with boys. no. of ways of seating boys and girls alternately $= 6! \times 6!$
In the similar manner, if we start with girl, no. of ways of seating boys and girls alternately
$= 6! \times 6!$
G_G_G_G_G_G_
Thus total ways of alternate seating arrangements
$= 6! \times 6! + 6! \times 6!$
$= 2 \times 6! \times 6!$
$\therefore$ Probability of making alternate seating arrangement for 6 boys and 6 girls

$$= \frac{2 \times 6! \times 6!}{12!} = \frac{2 \times 720}{12 \times 11 \times 10 \times 9 \times 8 \times 7} = \frac{1}{462}$$

3. (a) Let us define the events as :
$E_1 \equiv$ First shot hits the target plane,
$E_2 \equiv$ Second shot hits the target plane
$E_3 \equiv$ third shot hits the target plane,
$E_4 \equiv$ fourth shot hits the target plane
then ATQ, $P(E_1) = 0.4;\ P(E_2) = 0.3;$
$P(E_3) = 0.2;\ P(E_4) = 0.1$
$\Rightarrow P(\bar{E}_1) = 1 - 0.4 = 0.6;\ P(\bar{E}_2) = 1 - 0.3 = 0.7$
$P(\bar{E}_3) = 1 - 0.2 = 0.8;\ P(\bar{E}_4) = 1 - 0.1 = 0.9$
(where $\bar{E}_1$ denotes not happening of E_1)
Now the gun hits the plane if at least one of the four shots hit the plane.
Also, P (at least one shot hits the plane).
$= 1 - P$ (none of the shots hits the plane)
$= 1 - P(\bar{E}_1 \cap \bar{E}_2 \cap \bar{E}_3 \cap \bar{E}_4)$
$= 1 - P(\bar{E}_1).P(\bar{E}_2).P(\bar{E}_3).P(\bar{E}_4)$
[Using multiplication thm for independent events]
$= 1 - 0.6 \times 0.7 \times 0.8 \times 0.9 = 1 - 0.3024 = 0.6976$

4. Let A denote the event that the candidate A is selected and B the event that B is selected. It is given that
$P(A) = 0.5$ ...(1)
$P(A \cap B) \le 0.3$ ...(2)
Now, $P(A) + P(B) - P(A \cap B) = P(A \cup B) \le 1$
or $0.5 + P(B) - P(A \cap B) \le 1$ [Using (1)]
or $P(B) \le 0.5 + P(A \cap B) \le 0.5 + 0.3$ [Using (2)]
or $P(B) \le 0.8 \therefore P(B)$ can not be 0.9

5. We must have one ace in $(n-1)$ attempts and one ace in the nth attempt. The probability of drawing one ace in first $(n-1)$ attempts is $\dfrac{{}^4C_1 \times {}^{48}C_{n-2}}{{}^{52}C_{n-1}}$ and other one ace in the nth attempt is, $\dfrac{{}^3C_1}{[52-(n-1)]} = \dfrac{3}{53-n}$
Hence the required probability,

$$= \frac{4.48!}{(n-2)!(50-n)!} \times \frac{(n-1)!(53-n)}{52!} \times \frac{3}{53-n}$$
$$= \frac{(n-1)(52-n)(51-n)}{50.49.17.13}$$

6. Given that
$P(A) = 0.3, P(B) = 0.4, P(C) = 0.8$
$P(AB) = 0.08, P(AC) = 0.28, P(ABC) = 0.09$
$P(A \cup B \cup C) \ge 0.75$
To find $P(BC) = x$ (say)
Now we know,
$P(A \cup B \cup C) = P(A) + P(B) + P(C)$
$- P(AB) - P(BC) - P(CA) + P(ABC)$
$\Rightarrow P(A \cup B \cup C) = 0.3 + 0.4 + 0.8$
$-0.08 - x - 0.28 + 0.09 = 1.23 - x$
Also we have,
$P(A \cup B \cup C) \ge 0.75$ and $P(A \cup B \cup C) \le 1$
$\therefore\ 0.75 \le P(A \cup B \cup C) \le 1$
$\Rightarrow\ 0.75 \le 1.23 - x \le 1$
$\Rightarrow\ 0.23 \le x \le 0.48$

7. Given that A and B are independent events
$\therefore\ P(A \cap B) = P(A).P(B)$ ...(1)
Also given that $P(A \cap B) = \dfrac{1}{6}$ ...(2)
and $P(\bar{A} \cap \bar{B}) = \dfrac{1}{3}$ ...(3)
also $P(\bar{A} \cap \bar{B}) = 1 - P(A \cup B)$
$\Rightarrow P(\bar{A} \cap \bar{B}) = 1 - P(A) - P(B) + P(A \cap B)$
$\Rightarrow \dfrac{1}{3} = 1 - P(A) - P(B) + \dfrac{1}{6}$
$\Rightarrow P(A) + P(B) = \dfrac{5}{6}$ ...(4)
From (1) and (2) we get
$P(A).P(B) = 1/6$ ...(5)
Let $P(A) = x$ and $P(B) = y$ then eq's (4) and (5) become
$x + y = \dfrac{5}{6},\ xy = \dfrac{1}{6}$
$\Rightarrow\ x - y = \pm\sqrt{(x+y)^2 - 4xy}$
$= \pm\sqrt{25/36 - 4/6} = \pm 1/6$
$\therefore$ We get $x = 1/2$ and $y = 1/3$
or $x = 1/3$ and $y = 1/2$
Thus $P(A) = 1/2$ and $P(B) = 1/3$
or $P(A) = 1/3$ and $P(B) = 1/2$.

8. **KEY CONCEPT :**
(Total prob. theorem) If $E_1, E_2, E_3 \ldots E_n$ are mutually exclusive and exhaustive events and E is an event which can take place in conjunction with any one of E_i then
$P(E)=P(E_1)P(E/E_1)+P(E_2)P(E/E_2)+\ldots+P(E_n)P(E/E_n)$
Let $P(A)$ denotes the prob. of people reading newspaper A and $P(B)$ that of people reading newspaper B

Then, $P(A)=\frac{25}{100}=0.25$

$P(B)=\frac{20}{100}=0.20,\quad P(AB)=\frac{8}{100}=0.08$

Prob. of people reading the newspaper A but not $B=P(AB^c)$
$=P(A)-P(AB)=0.25-0.08=0.17$
Similarly,
$P(A^cB)=P(B)-P(AB)=0.20-0.08=0.12$
Let E be the event that a person reads an advertisement.
Therefore

ATQ, $P(E/AB^c)=\frac{30}{100};\ P(E/A^cB)=\frac{40}{100}$

$P(E/AB)=\frac{50}{100}$

$\therefore$ By total prob. theorem (as AB^c, A^cB and AB are mutually exclusive)
$P(E)=P(E/AB^c)\,P(AB^c)+P(E/A^cB)\,P(A^cB)+P(E/AB).P(AB)$

$=\frac{30}{100}\times0.17+\frac{40}{100}\times0.12+\frac{50}{100}\times0.08$
$=0.051+0.048+0.04$
$=0.139.$

Thus the population that reads an advertisement is 13.9%.

9. The total number of ways of ticking the answers in any one attempt $=2^4-1=15$.
The student is taking chance at ticking the correct answer, It is reasonable to assume that in order to derive maximum benefit, the three solutions which he submit must be all different.
$\therefore$ n = total no. of ways $={}^{15}C_3$
m = the no. of ways in which the correct solution is excluded $={}^{14}C_3$

Hence the required probability $=1-\frac{{}^{14}C_3}{{}^{15}C_3}=1-\frac{4}{5}=\frac{1}{5}$

ALTERNATE SOLUTION :
The candidate may tick one or more of the alternatives. As each alternative may or may not be chosen, the total number of exhaustive possibilites are $2^4-1=15$.
Therefore the prob. that the questions are correctly answered by condidate is $\frac{1}{15}$.
As such the candidate may be correct on the first, second or third chance. As these events are mutually exclusive, the total probability will be given by

$=\frac{1}{15}+\frac{14}{15}\times\frac{1}{14}+\frac{14}{15}\times\frac{13}{14}\times\frac{1}{13}=\frac{1}{15}+\frac{1}{15}+\frac{1}{15}=\frac{3}{15}=\frac{1}{5}$

Thus the probability that the candidate gets marks in the question is $\frac{1}{5}$.

10. Let A_1 be the event that the lot contains 2 defective articles and A_2 the event that the lot contains 3 defective articles. Also let A be the event that the testing procedure ends at the twelth testing. Then according to the question :
$P(A_1)=0.4$ and $P(A_2)=0.6$
Since $0<P(A_1)<1, 0<P(A_2)<1$, and $P(A_1)+P(A_2)=1$
$\therefore$ The events A_1, A_2 form a partition of the sample space.
Hence by the theorem of total probability for compound events, we have
NOTE THIS STEP :
$P(A)=P(A_1)\,P(A/A_1)+P(A_2)\,P(A/A_2)$ ...(1)
Here $P(A/A_1)$ is the probability of the event the testing procedure ends at the twelth testing when the lot contains 2 defective articles. This is possible when out of 20 articles, first 11 draws must contain 10 non defective and 1 defective article and 12th draw must give a defective article.

$\therefore\quad P(A/A_1)=\frac{{}^{18}C_{10}\times{}^{2}C_1}{{}^{20}C_{11}}\times\frac{1}{9}=\frac{11}{190}$

Similarly, $P(A/A_2)=\frac{{}^{17}C_9\times{}^{3}C_1}{{}^{20}C_{11}}\times\frac{1}{9}=\frac{11}{228}$

Now substituting the values of $P(A/A_1)$ and $P(A/A_2)$ in eq. (1), we get

$P(A)=0.4\times\frac{11}{190}+0.6\times\frac{11}{228}=\frac{11}{475}+\frac{11}{380}=\frac{99}{1900}$

11. Since the man is one step away from starting point means that either
(i) man has taken 6 steps forward and 5 steps backward.
or (ii) man has taken 5 steps forward and 6 steps backward.
Taking movement 1 step forward as success and 1 step backward as failure.
$\therefore$ p = Probability of success = 0.4
and q = Probability of failure = 0.6
$\therefore$ Required probability $=P(X=6 \text{ or } X=5)$
$=P(X=6)+P(X=5)$
$={}^{11}C_6p^6q^5+{}^{11}C_5p^5q^6$
$={}^{11}C_5(p^6q^5+p^5q^6)={}^{11}C_5(p+q)(p^5q^5)$
$=\frac{11.10.9.8.7}{1.2.3.4.5}(0.4+0.6)(0.4\times0.6)^5$
$=462\times1\times(0.24)^5=0.37$
Hence the required prob. = 0.37

12. (a) There are following four possible ways of drawing first two balls
(i) Both the first and the second balls drawn are white.
(ii) The first ball drawn is white and the second ball drawn is black.
(iii) The first ball drawn is black and the second ball drawn is white.
(iv) Both the first and the second balls drawn are black. Let us define events (i), (ii), (iii) and (iv) by E_1, E_2, E_3 and E_4 respectively. Also let E denotes the event that the third ball drawn is black.

Then, $P(E_1)=\frac{2}{4}\times\frac{1}{3}=\frac{1}{6},\quad P(E_2)=\frac{2}{4}\times\frac{2}{3}=\frac{1}{3}$

$P(E_3)=\frac{2}{4}\times\frac{2}{5}=\frac{1}{5},\quad P(E_4)=\frac{2}{4}\times\frac{3}{5}=\frac{3}{10}$

Also $P(E/E_1)=1$, since when the event E, has already happend i.e. the first two balls drawn are both white, they are not replaced and so there are left 2 black balls in the urn so that the probability that the third ball drawn in this case is black $=\frac{2}{2}=1$.

Again $P(E/E_2) = \frac{3}{4}$, since when the event E_2 has already happened there are 3 black and one white balls in the urn. So in this case the probability that the third ball drawn is black $= \frac{3}{4}$.

Similarly, $P(E/E_3) = \frac{3}{4}$. and $P(E/E_4) = \frac{2}{3}$

Now by thm of total prob. for compound events, we have

$$P(E) = P(E_1)P(E/E_1) + P(E_2)P(E/E_2) + P(E_3)P(E/E_3) + P(E_4)P(E/E_4)$$

$$= \frac{1}{6}\times 1 + \frac{1}{3}\times\frac{3}{4} + \frac{1}{5}\times\frac{3}{4} + \frac{3}{10}\times\frac{2}{3} = \frac{1}{6} + \frac{1}{4} + \frac{3}{20} + \frac{1}{5} = \frac{23}{30}$$

13. Here the total number of coins is $N + 7$. Therefore the total number of ways of choosing 5 coins out of $N + 7$ is ${}^{N+7}C_5$.
Let E denotes the event that the sum of the values of the coins is less than one rupee and fifty paise.
Then E' denotes the event that the total value of the five coins is equal to or more than one rupee and fifty paise.

NOTE THIS STEP :

The number of cases favourable to E' is

$$= {}^2C_1 \times {}^5C_4 \times {}^NC_0 + {}^2C_2 \times {}^5C_3 \times {}^NC_0 + {}^2C_2 \times {}^5C_2 \times {}^NC_1$$
$$= 2 \times 5 + 10 + 10N = 10(N+2)$$

$$\therefore \quad P(E') = \frac{10(N+2)}{{}^{N+7}C_5}$$

$$\Rightarrow \quad P(E) = 1 - P(E') = 1 - \frac{10(N+2)}{{}^{N+7}C_5}$$

. The probability p_1 (say) of winning the best of three games is = the prob. of winning two games + the prob. of winning three games.

$= {}^3C_2\,(0.6)\,(0.4)^2 + {}^3C_3\,(0.4)^3$ [Using Binomial distribution]

Similarly the probability of winning the best five games is p_2 (say) = the prob. of winning three games + the prob. of winning four games + the prob. of winning 5 games

$$= {}^5C_3(0.6)^2\,(0.4)^3 + {}^5C_4(0.6)(0.4)^4 + {}^5C_5(0.4)^5$$

We have $p_1 = 0.288 + 0.064 = 0.352$

and $p_2 = 0.2304 + 0.0768 + 0.01024 = 0.31744$

As $p_1 > p_2$

$\therefore$ A must choose the first offer i.e. best of three games.

Set A has n elements.

$\therefore$ Number of subsets of $A = 2^n$.

$\therefore$ Each one of P and Q can be selected in 2^n ways.

Hence total no. of ways of selecting P and $Q = 2^n \cdot 2^n = 4^n$.

Let P contains r elements, where r varies from 0 to n, Then, P can be chosen in ${}^nC_r$ ways. Now as $P \cap Q = \phi$, Q can be chosen from the set of all subsets of set consisting of remaining $(n - r)$ elements. This can be done in 2^{n-r} ways.

$\therefore$ P and Q can be chosen in ${}^nC_r.2^{n-r}$ ways.

But, r can vary from 0 to n

$\therefore$ total number of disjoint sets P and Q are

$$= \sum_{r=0}^{n} {}^nC_r 2^{n-r} = (1+2)^n = 3^n$$

NOTE THIS STEP :

$$\therefore \quad \text{Required probability} = \frac{3^n}{4^n} = \left(\frac{3}{4}\right)^n$$

ALTERNATE SOLUTION :

Let $A = \{a_1, a_2, a_3, \ldots, a_n\}$

For each a_i, $1 \le i \le n$, there aries 4 cases

(i) $a_i \in P$ and $a_i \in Q$ (ii) $a_i \notin P$ and $a_i \in Q$

(iii) $a_i \in P$ and $a_i \notin Q$ (iv) $a_i \notin P$ and $a_i \notin Q$

$\therefore$ Total no. of ways of choosing P and Q is 4^n. Here case (i) is not favourable as $P \cap Q = \phi$

$\therefore$ For each element there are 3 favourable cases and hence total no. of favourable cases $= 3^n$.

Hence prob. $(P \cap Q = \phi) = \frac{3^n}{4^n} = \left(\frac{3}{4}\right)^n$.

ALTERNATE SOLUTION :

The set P be the empty set, or one element set or two elements set .... or n elements set. Then the set Q will be chosen from amongst the remaining n elements or $(n - 1)$ elements or $(n - 2)$ elements .... or no elements. Now if P is the empty set then prob. of its choosing is ${}^nC_0/2^n$, if it is one element set then prob. of its choosing is ${}^nC_1/2^n$, and so on. When the set P consisting of r elements is chosen from A, then the prob. of choosing the set Q from amongst the remaining $n - r$ elements is $2^{n-r}/2^n$. Hence the prob. that P and Q have no common elements is given by

$$\sum_{r=0}^{n} \frac{{}^nC_r}{2^n}.\frac{2^{n-r}}{2^n} = \frac{1}{4^n}\sum_{r=0}^{n} {}^nC_r 2^{n-r}$$

$$= \frac{1}{4^n}(1+2)^n \text{ (Using Binomial thm.)} = \frac{3^n}{4^n} = \left(\frac{3}{4}\right)^n$$

16. **KEY CONCEPT :**

Baye's theorem : If $E_1, E_2, E_3, \ldots, E_n$ are mutually exclusive and exhaustive events and E is an event which takes place in conjunction with any one of E_i then the probability of the event E_i happening when the event E has taken place is given by

$$P(E_i/E) = \frac{P(E_i)P(E/E_i)}{\sum_{i=1}^{n} P(E_i)P(E/E_i)}$$

Let us define the events :

$A_1 \equiv$ the examinee guesses the answer,

$A_2 \equiv$ the examinee copies the answer

$A_3 \equiv$ the examinee knows the answer,

$A \equiv$ the examinee answers correctly.

$$\text{ATQ}, P(A_1) = \frac{1}{3}; P(A_2) = \frac{1}{6}$$

As any one happens out of A_1, A_2, A_3, these are mutually exclusive and exhaustive events.

$$\therefore \quad P(A_1) + P(A_2) + P(A_3) = 1$$

$$\Rightarrow \quad P(A_3) = 1 - \frac{1}{3} - \frac{1}{6} = \frac{6-2-1}{6} = \frac{3}{6} = \frac{1}{2}$$

Also we have, $P(A/A_1) = \frac{1}{4}$

[$\because$ out of 4 choices only one is correct.] $P(A/A_2) = \frac{1}{8}$ (given) $P(A/A_3) = 1$

[If examinee knows the ans., it is correct. i.e. true event]

To find $P(A_3/A)$. By Baye's thm, $P(A_3/A)$

$$= \frac{P(A/A_3)P(A_3)}{P(A/A_1)P(A_1) + P(A/A_2)P(A_2) + P(A/A_3)P(A_3)}$$

$$= \frac{1.\frac{1}{2}}{\frac{1}{4}.\frac{1}{3} + \frac{1}{8}.\frac{1}{6} + 1.\frac{1}{2}} = \frac{1/2}{\frac{29}{48}} = \frac{1}{2}\times\frac{48}{29} = \frac{24}{29}$$

17. Let X = defective and Y = non defective. Then all possible outcomes are $\{XX, XY, YX, YY\}$

Also $P(XX)=\frac{50}{100}\times\frac{50}{100}=\frac{1}{4}$,

$P(XY)=\frac{50}{100}\times\frac{50}{100}=\frac{1}{4}$ $P(YX)=\frac{50}{100}\times\frac{50}{100}=\frac{1}{4}$,

$P(YY)=\frac{50}{100}\times\frac{50}{100}=\frac{1}{4}$

Here, $A = XX\cup XY; B = XY\cup YY; C = XX\cup YY$

$\therefore\ P(A)=P(XX)+P(XY)=\frac{1}{4}+\frac{1}{4}=\frac{1}{2}$

$\therefore\ P(B)=P(XY)+P(YX)=\frac{1}{4}+\frac{1}{4}=\frac{1}{2}$

$P(C)=P(XX)+P(YY)=\frac{1}{4}+\frac{1}{4}=\frac{1}{2}$

Now, $P(AB)=P(XY)=\frac{1}{4}=P(A).P(B)$

$\therefore$ A and B are independent events.

$P(BC)=P(YX)=\frac{1}{4}=P(B).P(C)$

$\therefore$ B and C are independent events.

$P(CA)=P(XX)=\frac{1}{4}=P(C).P(A)$

$\therefore$ C and A are independent events.

$P(ABC)=0$ (impossible event)

$\neq P(A)P(B)P(C)$

$\therefore$ A, B, C are dependent events.

Thus we can conclude that A, B, C are pairwise independent but A, B, C are dependent events.

18. The given numbers are 00, 01, 02, ..., 99. These are total 100 numbers, out of which the numbers, the product of whose digits is 18, are 29, 36, 63 and 92.

$\therefore\ p=P(E)=\frac{4}{100}=\frac{1}{25}\Rightarrow q=1-p=\frac{24}{25}$

From Binomial distribution

P(E occuring at least 3 times) = P(E occuring 3 times) + P(E occuring 4 times)

$^4C_3p^3q+{}^4C_4p^4=4\times\left(\frac{1}{25}\right)^3\left(\frac{24}{25}\right)+\left(\frac{1}{25}\right)^4=\frac{97}{(25)^4}$

19. $E_1\equiv$ number noted is 7, $E_2\equiv$ number notes is 8,

$H\equiv$ getting head on coin, $T\equiv$ getting tail on coin.

Then by total probability theorem,

$P(E_1)=P(H)P(E_1/H)+P(T)P(E_1/T)$

and $P(E_2)=P(H)P(E_2/H)+P(T)P(E_2/T)$

where $P(H)=\frac{1}{2};P(T)=\frac{1}{2}$

$P(E_1/H)$ = prob. of getting a sum of 7 on two dice. Here favourable cases are

$\{(1,6),(6,1),(2,5),(5,2),(3,4),(4,3)\}$

$\therefore\ P(E_1/H)=\frac{6}{36}=\frac{1}{6}$

Also $P(E_1/T)$ = prob. of getting '7' numbered card out of 11 cards $=\frac{1}{11}$.

$P(E_2/H)$ = Prob. of getting a sum of 8 on two dice. Here favourable cases are $\{(2,6)(6,2)(4,4),(5,3),(3,5)\}$

$\therefore\ P(E_2/H)=\frac{5}{36}$

$P(E_2/T)$ = prob. of getting '8' numbered card out of 11 cards $=\frac{1}{11}$

$\therefore\ P(E_1)=\frac{1}{2}\times\frac{1}{6}+\frac{1}{2}\times\frac{1}{11}=\frac{1}{12}+\frac{1}{22}=\frac{11+6}{132}=\frac{17}{132}$

$P(E_2)=\frac{1}{2}\times\frac{5}{36}+\frac{1}{2}\times\frac{1}{11}=\frac{1}{2}\left[\frac{55+36}{396}\right]=\frac{91}{792}$

Now E_1 and E_2 are mutually exclusive events therefore

$P(E_1 \text{ or } E_2)=P(E_1)+P(E_2)=\frac{17}{132}+\frac{91}{792}$

$=\frac{102+91}{792}=\frac{193}{792}=0.2436.$

20. We have 14 seats in two vans. And there are 9 boys and 3 girls. The no. of ways of arranging 12 people on 14 seats without restriction is

$^{14}P_{12}=\frac{14!}{2!}=7(13!)$

Now the no. of ways of choosing back seats is 2. And the no. of ways of arranging 3 girls on adjacent seats is 2 (3!). And the no. of ways of arranging 9 boys on the remaining 11 seats is $^{11}P_9$

Therefore, the required number of ways

$=2.(2.3!).\ ^{11}P_9=\frac{4.3!.11!}{2!}=12!$

Hence, the probability of the required event

$=\frac{12!}{7.13!}=\frac{1}{91}$

21. (a) Prob. of S_1 to be among the eight winners = (prob. of S_1 being in a pair) × (prob. of S_1 winning in the group).

$=1\times\frac{1}{2}$ [$\because$ S_1 is definitely in a group i.e. certain event]

$=\frac{1}{2}$

(b) If S_1 and S_2 are in the same pair then exactly one wins. If S_1 and S_2 are in two pairs seperately then exactly one of S_1 and S_2 will be among the eight winners if S_1 wins and S_2 loses or S_1 loses and S_2 wins.

Now the prob. of S_1, S_2 being in the same pair and one wins.

= (prob. of S_1, S_2 being in the same pair) × (prob. of any one winning in the pair)

And the prob. of S_1, S_2 being in the same pair

$=\frac{n(E)}{n(S)}$, where n (S) = the no. of ways in which 16 persons can be divided in 8 pairs ; n (E) = the no. of ways in which S_1, S_2 are in same pair or 14 persons can be divided into 7 pairs.

$\therefore\ n(E)=\frac{14!}{(2!)^7.7!}$ and $n(S)=\frac{16!}{(2!)^8.8!}$

$\therefore$ Prob. of S_1 and S_2 being in the same pair

$$=\frac{\frac{14!}{(2!)^7.7!}}{\frac{16!}{(2!)^8.8!}}=\frac{2!.8}{16.15}=\frac{1}{15}$$

The prob. of any one winning in the pair of $S_1, S_2 = P$ (Certain event) = 1

$\therefore$ The pair of S_1, S_2 being in two pairs seperately and any one of S_1, S_2 wins.

= the prob. of S_1, S_2 being in two pairs seperately and S_1 wins, S_2 loses + the prob. of S_1, S_2 being in two pairs seperately and S_1 loses, S_2 wins.

$$=\left[1-\frac{\frac{14!}{(2!)^7.7!}}{\frac{16!}{(2!)^8.8!}}\right]\times\frac{1}{2}\times\frac{1}{2}+\left[1-\frac{\frac{14!}{(2!)^7.7!}}{\frac{16!}{(2!)^8.8!}}\right]\times\frac{1}{2}\times\frac{1}{2}$$

$$=2\times\frac{\frac{16-14!\times16}{(2!)^8.8!}}{\frac{16!}{(2!)^8.8!}}\times\frac{1}{4}=\frac{1}{2}\times\frac{14\times14!}{15\times14!}=\frac{7}{15}$$

$\therefore$ Required prob. $=\frac{1}{15}+\frac{7}{15}=\frac{8}{15}$

22. The required probability = 1 – (probability of the event that the roots of $x^2 + px + q = 0$ are non-real if and only if $p^2 - 4q < 0$ i.e. if $p^2 < 4q$.

We enumerate the possible values of p and q, for which this can happen in the following table.

q	p	Number of pairs of p,q
1	1	1
2	1,2	2
3	1,2,3	3
4	1,2,3	3
5	1,2,3,4	4
6	1,2,3,4	4
7	1,2,3,4,5	5
8	1,2,3,4,5	5
9	1,2,3,4,5	5
10	1,2,3,4,5,6	6

Thus, the number of possible pairs = 38. Also, the total number of possible pairs is $10 \times 10 = 100$.

$\therefore$ The required probability

$$=1-\frac{38}{100}=1-0.38=0.62$$

23. Given that p is the prob. that coin shows a head then $1-p$ will be the prob. that coin shows a tail.

Now, $\alpha = P$ (A gets the 1st head in 1st try) + P (A gets the 1st head in 2nd try) + ...

$\Rightarrow$ $\alpha = P(H) + P(T)\,P(T)\,P(T)\,P(H) + P(T)\,P(T)\,P(T)\,P(T)\,P(T)\,P(T)\,P(H)$

$= p + (1-p)^3 p + (1-p)^6 p + ...$

$= p\,[1 + (1-p)^3 (1-p)^6 + ...]$

$$=\frac{p}{1-(1-p)^3} \quad ...(i)$$

NOTE THIS STEP:

Similarly $\beta = P$ (B gets the 1st head in 1st try) + P (B gets the 1st head in 2nd try) + ...

$= P(T)\,P(H) + P(T)\,P(T)\,P(T)\,P(T)\,P(H) + ...$

$= (1-p)\,p + (1-p)^4 p + ...$

$$=\frac{(1-p)\,p}{1-(1-p)^3} \quad ...(ii)$$

From (i) and (ii) we get $\beta = (1-p)\,\alpha$

Also (i) and (ii) give expresion for α and β in terms of p.

Also $\alpha + \beta + \gamma = 1$ (exhaustive events and mutually exclusive events)

$\Rightarrow$ $\gamma = 1 - \alpha - \beta = 1 - \alpha - (1-p)\,\alpha$

$$=1-(2-p)\,\alpha=1-(2-p)\frac{p}{1-(1-p)^3}$$

$$=\frac{1-(1-p)^3-(2p-p^2)}{1-(1-p)^3}$$

$$=\frac{1-1+p^3+3p(1-p)-2p+p^2}{1-(1-p)^3}$$

$$=\frac{p^3-2p^2+p}{1-(1-p)^3}=\frac{p(p^2-2p+1)}{1-(1-p)^3}=\frac{p(1-p)^2}{1-(1-p)^3}$$

24. The number of ways in which $P_1, P_2, ..., P_8$ can be paired in four pairs

$$=\frac{1}{4!}\times{}^8C_2\times{}^6C_2\times{}^4C_2\times{}^2C_2=105$$

Now, at least two players certainly reach the second round in between P_1, P_2 and P_3. And P_4 can reach in final if exactly two players play against each other in between P_1, P_2, P_3 and remaining player will play against one of the players from P_5, P_6, P_7, P_8 and P_4 plays against one of the remaining three from P_5, P_6, P_7, P_8.

This can be possible in ${}^3C_2 \times {}^4C_1 \times {}^3C_1 = 36$ ways

$\therefore$ Prob. that P_4 and exactly one of $P_5 ... P_8$ reach second round $=\frac{36}{105}=\frac{12}{35}$

If P_1, P_i, P_4 and P_j where $i = 2$ or 3 and $j = 5$ or 6 or 7 reach the second round then they can be paired in 2 pairs in

$$\frac{1}{2!}\times{}^4C_2\times{}^2C_2=3 \text{ ways}$$

But P_4 will reach the final if P_1 plays against P_i and P_4 plays against P_j.

Hence the prob. that P_4 reach the final round from the second $=\frac{1}{3}$.

$\therefore$ prob. that P_4 reach the final is $\frac{12}{35}\times\frac{1}{3}=\frac{4}{35}$

25. Given that the probability of showing head by a coin when tossed = p

$\therefore$ Prob. of coin showing a tail = $1-p$

Now p_n = prob. that no two or more consecutive heads occur when tossed n times.

$\therefore$ p_1 = prob. of getting one or more or no head = prob. of H or T = 1

Also p_2 = prob. of getting one H or no H

$= P(HT) + P(TH) + P(TT)$

$= p(1-p) + p(1-p) p + (1-p)(1-p)$

$= 1 - p^2$, For $n \geq 3$

p_n = prob. that no two or more consecutive heads occur when tossed n times.

$= P$ (last out come is T) P (no two or more consecutive heads in $(n-1)$ throw) + P (last out come is H) P (($n-1$)th throw results in a T) P (no two or more consecutive heads in $(n-2)$ n throws)

$= (1-p)\,P_{n-1} + p\,(1-p)\,p_{n-2}$

Hence Proved.

26. Let W_1 (B_1) be the event that a white (a back) ball is drawn in the first draw and let W be the event that a white ball is drawn in the second draw. Then

$P(W) = P(B_1)\,.\,P(W/B_1) + P(W_1)\,.\,P(W/W_1)$

$$= \frac{n}{m+n}\cdot\frac{m}{m+n+k} + \frac{m}{m+n}\cdot\frac{m+k}{m+n+k}$$

$$= \frac{m(n+m+k)}{(m+n)(m+n+k)} = \frac{m}{m+n}$$

27. The total no. of outcomes $= 6^n$

We can choose three numbers out of 6 in ${}^6C_3$ ways. By using three numbers out of 6 we can get 3^n sequences of length n. But these include sequences of length n which use exactly two numbers and exactly one number.

The number of n-sequences which use exactly two numbers $= {}^3C_2\,[2^n - 1^n - 1^n] = 3\,(2^n - 2)$ and the number of n sequences which are exactly one number $= ({}^3C_1)\,(1^n) = 3$.

Thus, the number of sequences, which use exactly three numbers

$= {}^6C_3\,[3^n - 3\,(2^n - 2) - 3] = {}^6C_3[3^n - 3(2^n) + 3]$

$\therefore$ Probability of the required event,

$$= \frac{{}^6C_3[3^n - 3(2^n) + 3]}{6^n}$$

28. Let E_1 be the event that the coin drawn is fair and E_2 be the event that the coin drawn is biased.

$\therefore\ P(E_1) = \dfrac{m}{N}$ and $P(E_2) = \dfrac{N-m}{N}$

A is the event that on tossing the coin the head appears first and then appears tail.

$\therefore\ P(A) = P(E_1 \cap A) + P(E_2 \cap A)$

$= P(E_1)\,P(A/E_1) + P(E_2)\,P(A/E_2)$

$$= \frac{m}{N}\left(\frac{1}{2}\right)^2 + \left(\frac{N-m}{N}\right)\left(\frac{2}{3}\right)\left(\frac{1}{3}\right) \quad \ldots(1)$$

We have to find the probability that A has happened because of E_1

$$\therefore\ P(E_1/A) = \frac{P(E_1 \cap A)}{P(A)}$$

$$= \frac{\frac{m}{N}\left(\frac{1}{2}\right)^2}{\frac{m}{N}\left(\frac{1}{2}\right)^2 + \frac{N-m}{N}\left(\frac{2}{3}\right)\left(\frac{1}{3}\right)} \quad \text{(by (1))}$$

$$= \frac{m/4}{m/4 + \frac{2(N-m)}{9}} = \frac{9m}{m+8N}$$

29. Let us consider

$E_1 \equiv$ event of passing I exam.

$E_2 \equiv$ event of passing II exam.

$E_3 \equiv$ event of passing III exam.

Then a student can qualify in anyone of following ways

1. He passes first and second exam.
2. He passes first, fails in second but passes third exam.
3. He fails in first, passes second and third exam.

$\therefore$ Required probability

$= P(E_1)\,P(E_2/E_1) + P(E_1)\,P(\bar{E}_2/E_1)\,P(E_3/\bar{E}_2) + P(\bar{E}_1)\,P(E_2/\bar{E}_1)\,P(E_3/E_2)$

[as an event is dependent on previous one]

$$= p.p + p.(1-p)\,.\,\frac{p}{2} + (1-p).\frac{p}{2}.p$$

$$= p^2 + \frac{p^2}{2} - \frac{p^3}{2} + \frac{p^2}{2} - \frac{p^3}{2} = 2p^2 - p^3$$

30. Let us consider the events

$E_1 \equiv A$ hits B Then $P(E_1) = 2/3$

$E_2 \equiv B$ hits A $P(E_2) = 1/2$

$E_3 \equiv C$ hits A $P(E_3) = 1/3$

$E \equiv A$ is hit

$P(E) = P(E_2 \cup E_3) = 1 - P(\bar{E}_2 \cap \bar{E}_3)$

$$= 1 - P(\bar{E}_2)\,.\,P(\bar{E}_3) = 1 - \frac{1}{2}\cdot\frac{2}{3} = \frac{2}{3}$$

To find $P(E_2 \cap \bar{E}_3 / E)$

$$= \frac{P(E_2 \cap \bar{E}_3)}{P(E)}$$ [$\because P(E_2 \cap \bar{E}_3 \cap E) = P(E_2 \cap \bar{E}_3)$ i.e., B hits A and A is hit $= B$ hits A]

$$= \frac{P(E_2).P(\bar{E}_3)}{P(E)} = \frac{1/2 \times 2/3}{2/3} = \frac{1}{2}$$

31. Given that A and B are two independent events. C is the event in which exactly of A or B occurs.

Let $P(A) = x,\ P(B) = y$

then $P(C) = P(A \cap \bar{B}) + P(\bar{A} \cap B)$

$= P(A)P(\bar{B}) + P(\bar{A})P(B)$

[$\because$ If A and B are independent so are 'A and $\bar{B}$' and '$\bar{A}$ and B'.]

$\Rightarrow\ P(C) = x(1-y) + y(1-x)$ ...(1)

Now consider, $P(A \cup B)\,P(\bar{A} \cap \bar{B})$

$= [P(A) + P(B) - P(A)\,P(B)]\,[P(\bar{A})P(\bar{B})]$

$= (x + y - xy)\,(1-x)\,(1-y)$

$= (x+y)\,(1-x)\,(1-y) - xy(1-x)\,(1-y) \le (x+y)\,(1-x)\,(1-y)$

[$\because x, y \in (0,1)$]

$= x(1-x)(1-y) + y(1-x)\,(1-y)$

$= x(1-y) + y(1-x) - x^2(1-y) - y^2(1-x) \le x(1-y) + y(1-x)3$

$= P(C)$ [Using eq^n (1)]

Thus $P(C) \ge P(A \cup B)P(\bar{A} \cap \bar{B})$ is proved.

32. Let us define the following events

$A \equiv$ 4 white balls are drawn in first six draws

$B \equiv$ 5 white balls are drawn in first six draws

$C \equiv$ 6 white balls are drawn in first six draws

$E \equiv$ exactly one white ball is drawn in next two draws (i.e. one white and one red)

Then $P(E) = P(E/A)\,P(A) + P(E/B)\,P(B) + P(E/C)\,P(C)$

But $P(E/C) = 0$ [As there are only 6 white balls in the bag.]

$P(E) = P(E/A)\,P(A) + P(E/B)\,P(B)$

$$= \frac{{}^{10}C_1 \times {}^2C_1}{{}^{12}C_2}\cdot\frac{{}^{12}C_2 \times {}^6C_4}{{}^{18}C_6} + \frac{{}^{11}C_1 \times {}^1C_1}{{}^{12}C_2}\cdot\frac{{}^{12}C_1 \times {}^6C_5}{{}^{18}C_6}$$

33. Let us define the following events
$C \equiv$ person goes by car,
$S \equiv$ person goes by scooter,
$B \equiv$ person goes by bus,
$T \equiv$ person goes by train,
$L \equiv$ person reaches late
Then we are given in the question

$$P(C)=\frac{1}{7}; P(S)=\frac{3}{7}; P(B)=\frac{2}{7}; P(T)=\frac{1}{7}$$

$$P(L/C)=\frac{2}{9}; P(L/S)=\frac{1}{9}; P(L/B)=\frac{4}{9}; P(L/T)=\frac{1}{9}$$

To find the prob. $P(C/\bar{L})$ [$\because$ reaches in time $\equiv$ not late]
Using Baye's theorem

$$P(C/\bar{L}) = \frac{P(\bar{L}/C)P(C)}{P(\bar{L}/C)P(C)+P(\bar{L}/S)P(S)+P(\bar{L}/B)P(B)+A(\bar{L}/T)P(T)} \quad \text{... (i)}$$

Now, $P(\bar{L}/C)=1-\frac{2}{9}=\frac{7}{9};\ P(\bar{L}/S)=1-\frac{1}{9}=\frac{8}{9}$

$P(\bar{L}/B)=1-\frac{4}{9}=\frac{5}{9};\ P(\bar{L}/T)=1-\frac{1}{9}=\frac{8}{9}$

Substituting these values in eqn. (i) we get

$$P(C/\bar{L})=\frac{\frac{7}{9}\times\frac{1}{7}}{\frac{7}{9}\times\frac{1}{7}+\frac{8}{9}\times\frac{3}{7}+\frac{5}{9}\times\frac{2}{7}+\frac{8}{9}\times\frac{1}{7}}$$

$$=\frac{7}{7+24+10+8}=\frac{7}{49}=\frac{1}{7}$$

G COMPREHENSION BASED QUESTIONS :

$P(u_i)\propto i \Rightarrow P(u_i)=ki$, But $\Sigma P(u_i)=1$

$$\Rightarrow \Sigma ki=1 \Rightarrow k\Sigma i=1 \Rightarrow k=\frac{2}{n(n+1)}$$

$$\Rightarrow \quad P(u_i)=\frac{2i}{n(n+1)}$$

By total prob. theorem

$$P(w)=\sum_{i=1}^{n} P(u_i)P(w/u_i) = \sum_{i=1}^{n}\frac{2i}{n(n+1)}\times\frac{i}{n+1}$$

$$=\frac{2}{n(n+1)^2}\cdot\frac{n(n+1)(2n+1)}{6}=\frac{2n+1}{3n+3}$$

$$\therefore \lim_{n\to\infty} P(w)=\lim_{n\to\infty}\frac{2n+1}{3n+3}=\lim_{n\to\infty}\frac{2+1/n}{3+3/n}=\frac{2}{3}$$

$P(u_i)=c$
Using Baye's thm.

$$P(u_n/w)=\frac{P(w/u_n)P(u_n)}{\sum_{i=1}^{n}P(w/u_i)P(u_i)}$$

$$=\frac{c\times\frac{n}{n+1}}{c\left[\frac{1}{n+1}+\frac{2}{n+1}+\ldots+\frac{n}{n+1}\right]}=\frac{n}{n+1}\times\frac{n+1}{\frac{n(n+1)}{2}}=\frac{2}{n+1}$$

3. $P(w/E)=\frac{P(w\cap E)}{P(E)}$

NOTE THIS STEP :

$$=\frac{\frac{1}{n}\times\frac{2}{n+1}+\frac{1}{n}\times\frac{4}{n+1}+\frac{1}{n}\times\frac{6}{n+1}+\ldots+\frac{1}{n}\times\frac{n}{n+1}}{\frac{1}{n}+\frac{1}{n}+\ldots+\frac{1}{n}\left(\frac{n}{2}\text{ times}\right)}$$

$$=\frac{\frac{2}{n(n+1)}\left[1+2+3\ldots+\frac{n}{2}\right]}{\frac{1}{n}\times\frac{n}{2}} \quad (n \text{ being even})$$

$$=\frac{4}{n(n+1)}\left[\frac{\frac{n}{2}\left(\frac{n}{2}+1\right)}{2}\right]=\frac{n+2}{2(n+1)}$$

4. $P(X=3)$ = (probability of not a six in first chance) $\times$ (probability of not a six in second chance) $\times$ (probability of a six in third chance)

$$=\frac{5}{6}\times\frac{5}{6}\times\frac{1}{6}=\frac{25}{216}$$

5. $P(X\ge 3)=1-(X<3)$

$$=1-[P(X=1)+P(X=2)]$$

$$=1-\left[\frac{1}{6}+\frac{5}{6}\times\frac{1}{6}\right]=1-\frac{11}{36}=\frac{25}{36}$$

ALTERNATE SOLUTION :

$$P(X\ge 3)=P(X=3)+P(X=4)+P(X=5)+\ldots\ldots\infty$$

$$=\frac{5}{6}\times\frac{5}{6}\times\frac{1}{6}+\frac{5}{6}\times\frac{5}{6}\times\frac{5}{6}\times\frac{1}{6}+\frac{5}{6}\times\frac{5}{6}\times\frac{5}{6}\times\frac{5}{6}\times\frac{1}{6}+\ldots\ldots\infty$$

$$=\frac{25}{216}\left[1+\frac{5}{6}+\left(\frac{5}{6}\right)^2+\ldots\infty\right]=\frac{25}{216}\times\frac{1}{1-\frac{5}{6}}=\frac{25}{216}\times\frac{6}{1}=\frac{25}{36}$$

6. Let us define the events
$A\equiv X\ge 6$ and $B\equiv X>3$
so that $A\cap B\equiv X\ge 6\equiv A$

Now $P(A)=\left(\frac{5}{6}\right)^5\times\frac{1}{6}+\left(\frac{5}{6}\right)^6\times\frac{1}{6}+\ldots\infty$

$$=\left(\frac{5}{6}\right)^5\times\frac{1}{6}\left[1+\frac{5}{6}+\left(\frac{5}{6}\right)^2+\ldots\infty\right]=\left(\frac{5}{6}\right)^5\times\frac{1}{6}\times\frac{1}{1-\frac{5}{6}}=\left(\frac{5}{6}\right)^5$$

and $P(B)=\left(\frac{5}{6}\right)^3\times\frac{1}{6}+\left(\frac{5}{6}\right)^4\times\frac{1}{6}+\ldots\infty=\left(\frac{5}{6}\right)^3$

$$\therefore \quad P(A/B)=\frac{P(A\cap B)}{P(B)}=\frac{P(A)}{P(B)}=\frac{\left(\frac{5}{6}\right)^5}{\left(\frac{5}{6}\right)^3}=\frac{25}{36}$$

7. (b) P (white) = P (H ∩ white) + P (T ∩ white)
= P (H) P (white/H) + P (T) P (white/T)

$$=\frac{1}{2}\left\{\frac{3}{5}\times 1+\frac{2}{5}\times\frac{1}{2}\right\}+\frac{1}{2}$$

$$\times\left\{\frac{^3C_2}{^5C_2}\times 1+\frac{^2C_2}{^5C_2}\times\frac{1}{3}+\frac{^3C_1.^2C_1}{^5C_2}\times\frac{2}{3}\right\}$$

$$=\frac{1}{2}\times\frac{8}{10}+\frac{1}{2}\times\left(\frac{3}{10}+\frac{1}{30}+\frac{12}{30}\right)=\frac{4}{10}+\frac{11}{30}=\frac{23}{30}$$

8. (d) $P(H/\text{white})=\frac{P(H\cap\text{white})}{P(\text{white})}=\frac{\frac{1}{2}\left[\frac{3}{5}\times 1+\frac{2}{5}\times\frac{1}{2}\right]}{\frac{23}{30}}$

$$=\frac{\frac{4}{10}}{\frac{23}{30}}=\frac{12}{23}$$

9. (a) Probability that all balls are of same colour
= P (all red) + P (all white) + P (all black)

$$=\frac{3}{6}\times\frac{3}{9}\times\frac{4}{12}+\frac{1}{6}\times\frac{2}{9}\times\frac{3}{12}+\frac{2}{6}\times\frac{4}{9}\times\frac{5}{12}=\frac{82}{648}$$

10. (d) B_1 [1W, 3R, 2B] B_2 [2W, 3R, 4B] B_3 [3W, 4R, 5B]

Let E_1, E_2, E_3 be the events that bag B_1, B_2 and B_3 is selected respectively.
Let E be the event that one white and one red ball is selected.
Then by baye's theorem,

$$P(E_2\setminus E)=\frac{P(E\setminus E_2)P(E_2)}{P(E\setminus E_1)P(E_1)+P(E\setminus E_2)P(E_2)+P(E\setminus E_3)P(E_3)}$$

$$=\frac{\frac{2\times 3}{9C_2}}{\frac{1\times 3}{^6C_2}+\frac{2\times 3}{^9C_2}+\frac{3\times 4}{^{12}C_2}}=\frac{55}{181}$$

(H) ASSERTION & REASON TYPE QUESTIONS :

1. We know $P(H_i/E)=\frac{P(H_i\cap E)}{P(E)}=\frac{P(E/H_i)P(H_i)}{P(E)}$

$\Rightarrow P(H_i/E)P(E)=P(E/H_i)P(H_i)\Rightarrow P(E)=\frac{P(E/H_i)P(H_i)}{P(H_i/E)}$

Now given that $0<P(E)<1\Rightarrow\quad 0<\frac{P(E/H_i)P(H_i)}{P(H_i/E)}<1$

$\Rightarrow P(E/H_i)P(H_i)<P(H_i/E)$ But if $P(H_i\cap E)=0$ then $P(H_i/E)=P(E/H_i)=0$
Then $P(E/H_i)P(H_i)<P(H_i/E)$ is not true.
∴ Statement -1 is not always true.
Also as $H_1, H_2, \ldots H_n$ are mutually exclusive and exhaustive events, therefore $\sum_{i=1}^{n}P(H_i)=1$. ∴Statement -2 is true.

2. The given system of equastions is
$ax+by=0\ cx+dy=0$ where $a, b, c, d\in\{0, 1\}$
For the system to have unique solution,

$\frac{a}{c}\neq\frac{b}{d}$ i.e. $\begin{vmatrix}a & b\\ c & d\end{vmatrix}\neq 0$

This is so in each of the following cases -

$$\begin{vmatrix}1&1\\1&0\end{vmatrix},\begin{vmatrix}1&1\\0&1\end{vmatrix},\begin{vmatrix}0&1\\1&1\end{vmatrix},\begin{vmatrix}1&0\\1&1\end{vmatrix},\begin{vmatrix}1&0\\0&1\end{vmatrix},\begin{vmatrix}0&1\\1&0\end{vmatrix}$$

∴ Favourable cases for the system to have unique solution = 6.

Also total possible cases for $\begin{vmatrix}a & b\\ c & d\end{vmatrix}=2^4=16$

(∵ each entry can be either 0 or 1)

∴ Probability of the given system to have unique solution $=\frac{6}{16}=\frac{3}{8}$ ∵ Statment –1 is true.

∵ Homogeneous system of equations always has a solution (Trival solution $x=0, y=0$)

∴ The probability that the system of equations has a solution is 1.

Hence the statement-2 is true but is not a correct explanation of statement- 1.

(I) INTEGER VALUE CORRECT TYPE :

1. 6 Let $P(E_1)=x$; $P(E_2)=y$, $P(E_3)=z$
$P(\text{only } E_1)=x(1-y)(1-z)=\alpha$
$P(\text{only } E_2)=(1-x)y(1-z)=\beta$
$P(\text{only } E_3)=(1-x)(1-y)z=\gamma$
$P(\text{none})=(1-x)(1-y)(1-z)=p.$
Now given $(\alpha-2\beta)p=\alpha\beta\Rightarrow x=2y$
and $(\beta-3r)p=2\beta r$

$\Rightarrow y=3z\ \therefore x=6z$ Hence $\frac{P(E_1)}{P(E_3)}=\frac{x}{z}=6$

Section-B JEE Main/ AIEEE

1. (a) $P(E_1)=\frac{1}{2}, P(E_2)=\frac{1}{3}$ and $P(E_3)=\frac{1}{4}$;

$$P(E_1UE_2UE_3)=1-P(\bar{E}_1)P(\bar{E}_2)P(\bar{E}_3)$$

$$=1-\left(1-\frac{1}{2}\right)\left(1-\frac{1}{3}\right)\left(1-\frac{1}{4}\right)$$

$$=1-\frac{1}{2}\times\frac{2}{3}\times\frac{3}{4}=\frac{3}{4}$$

2. (a) $P(A\cup B)=P(A)+P(B)-P(A\cap B)$;

$$\Rightarrow\frac{3}{4}=1-P(\overline{A})+P(B)-\frac{1}{4}$$

$$\Rightarrow 1=1-\frac{2}{3}+P(B)\Rightarrow P(B)=\frac{2}{3};$$

Now, $P(\overline{A}\cap B)=P(B)-P(A\cap B)=\frac{2}{3}-\frac{1}{4}=\frac{5}{12}$.

3. (d) The event follows binomial distribution with $n=5, p=3/6=1/2$.
$q=1-p=1/2$.; ∴ Variance $=npq=5/4$.

4. (b) $\left.\begin{array}{l} np=4 \\ npq=2 \end{array}\right\} \Rightarrow q=\frac{1}{2}, p=\frac{1}{2}, n=8$

$$p(X=1)={}^8C_1\left(\frac{1}{2}\right)\left(\frac{1}{2}\right)^7=8.\frac{1}{2^8}=\frac{1}{2^5}=\frac{1}{32}$$

5. (b) $P(A)=\frac{3x+1}{3},\ P(B)=\frac{1-x}{4},\ P(C)=\frac{1-2x}{2}$

$\because$ For any event $E, 0 \le P(E) \le 1$

$$\Rightarrow 0 \le \frac{3x+1}{3} \le 1,\ 0 \le \frac{1-x}{4} \le 1 \text{ and } 0 \le \frac{1-2x}{2} \le 1$$

$$\Rightarrow -1 \le 3x \le 2, -3 \le x \le 1 \text{ and } -1 \le 2x \le 1$$

$$\Rightarrow -\frac{1}{3} \le x \le \frac{2}{3} \le -3 \le x \le 1, \text{ and } -\frac{1}{2} \le x \le \frac{1}{2}$$

Also for mutually exclusive events A, B, C,

$$P(A \cup B \cup C)=P(A)+P(B)+P(C)$$

$$\Rightarrow P(A \cup B \cup C)=\frac{3x+1}{3}+\frac{1-x}{4}+\frac{1-2x}{2}$$

$$\therefore\ 0 \le \frac{1+3x}{3}+\frac{1-x}{4}+\frac{1-2x}{2} \le 1$$

$$0 \le 13-3x \le 12 \Rightarrow 1 \le 3x \le 13 \Rightarrow \frac{1}{3} \le x \le \frac{13}{3}$$

Considering all inequations, we get

$$\max\left\{-\frac{1}{3},-3,-\frac{1}{2},\frac{1}{3}\right\} \le x \le \min\left\{\frac{2}{3},1,\frac{1}{2},\frac{13}{3}\right\}$$

$$\frac{1}{3} \le x \le \frac{1}{2} \Rightarrow x \in \left[\frac{1}{3},\frac{1}{2}\right]$$

6. (a) Let 5 horses are H_1, H_2, H_3, H_4 and H_5. Selected pair of horses will be one of the 10 pairs (i.e.; ${}^5C_2$): $H_1 H_2$, $H_1 H_3$, $H_1 H_4$, $H_1 H_5$, H_2H_3, $H_2 H_4$, $H_2 H_5$, $H_3 H_4$, $H_3 H_5$ and $H_4 H_5$.
Any horse can win the race in 4 ways.
For example : Horses H_2 win the race in 4 ways $H_1 H_2$, H_2H_3, H_2H_4 and H_2H_5.

Hence required probability $=\frac{4}{10}=\frac{2}{5}$

7. (c) A and B will contradict each other if one speaks truth and other false . So , the required

$$\text{Probability }=\frac{4}{5}\left(1-\frac{3}{4}\right)+\left(1-\frac{4}{5}\right)\frac{3}{4}$$

$$=\frac{4}{5}\times\frac{1}{4}+\frac{1}{5}\times\frac{3}{4}=\frac{7}{20}$$

8. (b) $P(E)=P(2 \text{ or } 3 \text{ or } 5 \text{ or } 7)$

$=0.23+0.12+0.20+0.07=0.62$

$P(F)=P(1 \text{ or } 2 \text{ or } 3)\ =0.15+0.23+0.12=0.50$

$P(E \cap F)=P(2 \text{ or } 3)=0.23+0.12=0.35$

$\therefore P(EUF)=P(E)+P(F)-P(E \cap F)$

$=0.62+0.50-0.35=0.77$

9. (a) mean $= np=4$ and variance $= npq=2$

$\therefore p=q=\frac{1}{2}$ and $n=8$

$$\therefore P(2 \text{ success})={}^8C_2\left(\frac{1}{2}\right)^6\left(\frac{1}{2}\right)^2=\frac{28}{2^8}=\frac{28}{256}$$

10. (b) For a particular house being selected

Probability $=\frac{1}{3}$

P (all the persons apply for the same house)

$$=\left(\frac{1}{3}\times\frac{1}{3}\times\frac{1}{3}\right)3=\frac{1}{9}.$$

11. (c) According to Poission distribution, prob. of getting k successes is

$$P(x=k)=e^{-\lambda}\frac{\lambda^k}{k!}\ P(x \ge 2)=1-P(x=0)-P(x=1)$$

$$=1-e^{-\lambda}-e^{-\lambda}\left(\frac{\lambda}{1!}\right)=1-\frac{3}{e^2}.$$

12. (c) $P(\overline{A \cup B})=\frac{1}{6}, P(A \cap B)=\frac{1}{4}$ and $P(\overline{A})=\frac{1}{4}$

$\Rightarrow P(A \cup B)=\frac{5}{6}\ P(A)=\frac{3}{4}$

Also $\Rightarrow P(A \cup B)=P(A)+P(B)-P(A \cap B)$

$\Rightarrow P(B)=\frac{5}{6}-\frac{3}{4}+\frac{1}{4}=\frac{1}{3}$

$\Rightarrow P(A)\ P(B)=\frac{3}{4}\cdot\frac{1}{3}=\frac{1}{4}=P(A \cap B)$

Hence A and B are independent but not equally likely.

13. (d) $F'=3F\cos\theta$ and $F=3F\sin\theta$

$\Rightarrow F'=2\sqrt{2}\,F \Rightarrow F:F'::\ 3:2\sqrt{2}.$

14. (d) $P(X=r)=\frac{e^{-m}m^r}{r!}$

P (at most 1 phone call)

$=P(X \le 1)=P(X=0)+P(X=1)$

$=e^{-5}+5\times e^{-5}=\frac{6}{e^5}$

15. (d) Given : Probability of aeroplane I, scoring a target correctly i.e., $P(I)=0.3$ probability of scoring a target correctly by aeroplane II, i.e. $P(II)=0.2$

$\therefore P(\overline{I})=1-0.3=0.7$ $\therefore$ The required probability

$=P(\overline{I} \cap II)=P(\overline{I}).P(II)=0.7\times0.2=0.14$

16. (b) A pair of fair dice is thrown, the sample space $S=(1,1)$, $(1,2)(1,3)\ldots\ldots=36$
Possibility of getting 9 are (5, 4), (4, 5), (6, 3), (3, 6)
$\therefore$ Possibility of getting score 9 in a single throw

$=\frac{4}{36}=\frac{1}{9}$

$\therefore$ Probability of getting score 9 exactly twice

$$={}^3C_2\times\left(\frac{1}{9}\right)^2\cdot\left(1-\frac{1}{9}\right)=\frac{3!}{2!}\times\frac{1}{9}\times\frac{1}{9}\times\frac{8}{9}$$

$$=\frac{3.2!}{2!}\times\frac{1}{9}\times\frac{1}{9}\times\frac{8}{9}=\frac{8}{243}$$

17. **(b)** $P(A)=1/4, P(A/B)=\frac{1}{2}, P(B/A)=2/3$

By conditional probability,
$P(A\cap B)=P(A)P(B/A)=P(B)P(A/B)$

$\Rightarrow \frac{1}{4}\times\frac{2}{3}=P(B)\times\frac{1}{2} \Rightarrow P(B)=\frac{1}{3}$

18. **(c)** $A\equiv$ number is greater than 3

$\Rightarrow P(A)=\frac{3}{6}=\frac{1}{2}$

$B\equiv$ number is less than 5 $\Rightarrow P(B)=\frac{4}{6}=\frac{2}{3}$

$A\cap B\equiv$ number is greater than 3 but less than 5.

$\Rightarrow P(A\cap B)=\frac{1}{6}$

$\therefore P(A\cup B)=P(A)+P(B)-P(A\cap B)$

$=\frac{1}{2}+\frac{2}{3}-\frac{1}{6}=\frac{3+4-1}{6}=1$

19. **(d)** We have

$P(x\geq 1)\geq\frac{9}{10} \Rightarrow 1-P(x=0)\geq\frac{9}{10}$

$\Rightarrow \quad 1-{}^nC_0\left(\frac{1}{4}\right)^0\left(\frac{3}{4}\right)^n\geq\frac{9}{10}$

$\Rightarrow \quad 1-\frac{9}{10}\geq\left(\frac{3}{4}\right)^n \Rightarrow \left(\frac{3}{4}\right)^n\leq\left(\frac{1}{10}\right)$

Taking log to the base 3/4, on both sides, we get

$n\log_{3/4}\left(\frac{3}{4}\right)\geq\log_{3/4}\left(\frac{1}{10}\right)$

$\Rightarrow \quad n\geq-\log_{3/4}10=\frac{-\log_{10}10}{\log_{10}\left(\frac{3}{4}\right)}=\frac{-1}{\log_{10}3-\log_{10}4}$

$\Rightarrow \quad n\geq\frac{1}{\log_{10}4-\log_{10}3}$

20. **(d)** Let $A\equiv$ Sum of the digits is 8

$B\equiv$ Product of the digits is 0

Then A = {08, 17, 26, 35, 44}

B = {00, 01, 02, 03, 04, 05, 06, 07, 08, 09, 10, 20, 30, 40,}

$A\cap B=\{08\}$

$\therefore \quad P(A/B)=\frac{P(A\cap B)}{P(B)}=\frac{1/50}{14/50}=\frac{1}{14}$

21. **(b)** $n(S)={}^{20}C_4$

Statement-1:

common difference is 1; total number of cases = 17
common difference is 2; total number of cases = 14
common difference is 3; total number of cases = 11
common difference is 4; total number of cases = 8
common difference is 5; total number of cases = 5
common difference is 6; total number of cases = 2

Prob. $=\frac{17+14+11+8+5+2}{{}^{20}C_4}=\frac{1}{85}$

Statement -2 is false, because common difference can be 6 also.

22. **(a)** $n(S)={}^9C_3,\ n(E)={}^3C_1\times{}^4C_1\times{}^2C_1$

Probability $=\frac{3\times4\times2}{{}^9C_3}=\frac{24\times3!}{9!}\times6!=\frac{24\times6}{9\times8\times7}=\frac{2}{7}$

23. **(b)** p (at least one failure) $\geq\frac{31}{32}$

$\Rightarrow 1-p$ (no failure) $\geq\frac{31}{32}$

$\Rightarrow 1-p^5\geq\frac{31}{32} \Rightarrow p^5\leq\frac{1}{32} \Rightarrow p\leq\frac{1}{2}$

But $p\geq0$,

Hence p lies in the interval $\left[0,\frac{1}{2}\right]$.

24. **(a)** In this case

$P\left(\frac{C}{D}\right)=\frac{P(C\cap D)}{P(D)}=\frac{P(C)}{P(D)}$

Where, $0\leq P(D)\leq1$, hence $P\left(\frac{C}{D}\right)\geq P(C)$

25. **(d)** $P(A^c\cap B^c/C)=\frac{P\left(\left(A^c\cap B^c\right)\cap C\right)}{P(C)}=\frac{P\left((A\cup B)^c\cap C\right)}{P(C)}$

$=\frac{P((1-A\cup B)\cap C)}{P(C)}=\frac{P((1-A-B+A\cap B)\cap C)}{P(C)}$

$=\frac{P(C)-P(A\cap C)-P(B\cap C)+P(A\cap B\cap C)}{P(C)}$

$=\frac{P(C)-P(A).P(C)-P(B)P(C)+0}{P(C)}$

$=1-P(A)-P(B)=P\left(A^c\right)-P(B)$

26. **(b)** Given sample space = {1,2,3,.....,8}

Let Event

A : Maximum of three numbers is 6.

B : Minimum of three numbers is 3.

This is the case of conditional probability

We have to find P (minimum) is 3 when it is given that P (maximum) is 6.

$\therefore \quad P\left(\frac{B}{A}\right)=\frac{P(B\cap A)}{P(A)}=\frac{{}^2C_1/{}^8C_3}{{}^5C_2/{}^8C_3}=\frac{{}^2C_1}{{}^5C_2}=\frac{2}{10}=\frac{1}{5}$

27. **(c)** $p=P$ (correct answer), $q=P$ (wrong answer)

$\Rightarrow p=\frac{1}{3}, q=\frac{2}{3}, n=5$

By using Binomial distribution

Required probability $={}^5C_4\left(\frac{1}{3}\right)^4\cdot\frac{2}{3}+{}^5C_5\left(\frac{1}{3}\right)^5$

$=5\cdot\frac{2}{3^5}+\frac{1}{3^5}=\frac{11}{3^5}$

Miscellaneous (Sets, Relations, Statistics)

Section-A JEE Advanced/ IIT-JEE

A FILL IN THE BLANKS :

1. For real numbers x and y, we write $x * y$ if $x - y + \sqrt{2}$ is an irrational number. Then, the relation* is an equivalence relation. **(1981 - 2 Marks)**
2. A variable takes value x with frequency ${}^{n+x-1}C_x$, $x = 0, 1, 2, \ldots n$. The mode of the variable is................. **(1982 - 2 Marks)**

B TRUE / FALSE :

1. For real numbers x and y, we write $x * y$ if $x - y + \sqrt{2}$ is an irrational number. Then, the relation* is an equivalence relation. **(1981 - 2 Marks)**

C MCQ's WITH ONE CORRECT ANSWER :

1. If X and Y are two sets, then $X \cap (X \cup Y)^c$ equals. **(1979)**
   (a) X (b) Y
   (c) ϕ (d) None of these.
2. The expression $\dfrac{12}{3+\sqrt{5}+2\sqrt{2}}$ is equal to **(1980)**
   (a) $1-\sqrt{5}+\sqrt{2}+\sqrt{10}$ (b) $1+\sqrt{5}+\sqrt{2}-\sqrt{10}$
   (c) $1+\sqrt{5}-\sqrt{2}+\sqrt{10}$ (d) $1-\sqrt{5}-\sqrt{2}+\sqrt{10}$
3. Select the correct alternative in each of the following. Indicate your choice by the appropriate letter only.
   Let S be the standard deviation of n observations. Each of the n observations is multiplied by a constant c. Then the standard deviation of the resulting number is **(1980)**
   (a) s (b) $c\,s$
   (c) $s\sqrt{c}$ (d) none of these
4. The standard deviation of 17 numbers is zero. Then **(1980)**
   (a) the numbers are in geometric progression with common ratio not equal to one.
   (b) eight numbers are positive, eight are negative and one is zero.
   (c) either (a) or (b) (d) none of these
5. Consider any set of 201 observations $x1, x2, \ldots x200, x201$. It is given that $x1 < x2 < \ldots < x200 < x201$. Then the mean deviation of this set of observations about a point k is minimum when k equals **(1981 - 2 Marks)**
   (a) $(x_1 + x_2 + \ldots + x_{200} + x_{201})/201$
   (b) x_1
   (c) x_{101}
   (d) x_{201}
6. If $x1, x2, \ldots\ldots\ldots\ldots, xn$ are any real numbers and n is any postive integer, then **(1982 - 2 Marks)**
   (a) $n\sum_{i=1}^{n} x_i^2 < \left(\sum_{i=1}^{n} x_i\right)^2$ (b) $\sum_{i=1}^{n} x_i^2 \ge \left(\sum_{i=1}^{n} x_i\right)^2$
   (c) $\sum_{i=1}^{n} x_i^2 \ge n\left(\sum_{i=1}^{n} x_i\right)^2$ (d) none of these
7. Let S={1, 2, 3, 4} . The total number of unordered pairs of disjoint subsets of S is equal to **(2010)**
   (a) 25 (b) 34
   (c) 42 (d) 41
8. Let $P = \{\theta : \sin\theta - \cos\theta = \sqrt{2}\cos\theta\}$ and $Q = \{\theta : \sin\theta + \cos\theta = \sqrt{2}\sin\theta\}$ be two sets. Then **(2011)**
   (a) $P \subset Q$ and $Q - P \ne \varnothing$
   (b) $Q \not\subset P$
   (c) $P \not\subset Q$
   (d) $P = Q$

D MCQ's WITH ONE OR MORE THAN ONE CORRECT :

1. In a college of 300 students every student reads 5 newspapers and every newspaper is read by 60 students. The number of newpapers is **(1998 - 2 Marks)**
   (a) at least 30 (b) at most 20
   (c) exactly 25 (d) none of these
2. Let $S_1, S_2, \ldots\ldots$ be squares such that for each $n \ge 1$, the length of a side of S_n equals the length of a diagonal of S_{n+1}. If the length of a side of S_1 is 10 cm, then for which of the following values of n is the area of S_n less than 1 sq. cm? **(1999 - 3 Marks)**
   (a) 7 (b) 8
   (c) 9 (d) 10

E SUBJECTIVE PROBLEMS :

1. An investigator interviewed 100 students to determine their preferences for the three drinks : milk (M), coffee (C) and tea (T). He reported the following : 10 students had all the three drinks M, C and T; 20 had M and C; 30 had C and T; 25 had M and T; 12 had M only; 5 had C only; and 8 had T only. Using a Venn diagram find how many did not take any of the three drinks. **(1978)**
2. (a) Construct a triangle with base 9 cm and altitude 4 cm, the ratio of the other two sides being 2 : 1.
   (b) Construct a triangle in which the sum of the three sides is 15 cm with base angles 60° and 45°. Justify your steps. **(1979)**

3. A tent is made in the form of a frustrum A of a right circular cone surmounted by another right circular cone B. The diameter of the ends of the frustrum A are 8 m and 4 m, its height is 3 m and the height of the cone B is 2 m. Find the area of the canvas required. ***(1979)***

4. In calculating the mean and variance of 10 readings, a student wrongly used the figure 52 for the correct figure of 25. He obtained the mean and variance as 45.0 and 16.0 respectively. Determine the correct mean and variance. ***(1979)***

5. The diameter PQ of a semicircle is 6 cm. Construct a square $ABCD$ with points A, B on the circumference, and the side CD on the diameter PQ. Describe briefly the method of construction. ***(1980)***

6. C and D are any two points on the same side of a line L. Show how to find a point P on the line L such that PC and PD are equally inclined to the line L. Justify your steps. ***(1980)***

7. (i) $DABC$ is a regular tetrahedron. The length of each edge of the tetrahedron is 2a. M is the middle point of the altitude DN. Prove that MA, MB, MC are mutually perpendicular.
   (ii) There is circular plot of land of diameter 24 metres. At the centre of the plot, a pit in the form of a frustrum of a cone is dug up. The depth of the pit is 3 meters. The diameters of the top and the bottom of the pit are 4 metres and 2 metres respectively. The earth dug out from the pit is spread uniformly over the remaining area of the plot. Find the height to which the level of the plot rises. ***(1980)***

8. (i) AB is a diameter of a circle, CD is a chord parallel to AB, and $2CD = AB$. The tangent at B meets the line AC produced at E. Prove that $AE = 2AB$.
   (ii) If $x > y > 0$, then show that the expression
   $$(\sqrt{2}\,(2x+\sqrt{x^2-y^2}\,)\,(\sqrt{x-\sqrt{x^2-y})})$$
   can be simplified to $\sqrt{(x+y)^3}-\sqrt{(x-y)^3}$ ***(1980)***

9. (i) Set A has 3 elements, and set B has 6 elements. What can be the minimum number of elements in the set $A \cup B$? ***(1980)***
   (ii) P, Q, R are subsets of a set A. Is the following equality true ?
   $R \times (P^c \cup Q^c)^c = (R \times P) \cap (R \times Q)$? ***(1980)***
   (iii) For any two subset X and Y of a set A define
   $X \circ Y = (X^c \cap Y) \cup (X \cap Y^c)$
   Then for any three subsets X, Y and Z of the set A, is the following equality true.
   $(X \circ Y) \circ Z = X \circ (Y \circ Z)$? ***(1980)***

10. Suppose $A_1, A_2, \ldots\ldots A_{30}$ are thirty sets each with five elements and $B_1, B_2, \ldots\ldots B_n$ are n sets each with three elements. Let $\overset{30}{\underset{i=1}{U}} A_i = \overset{n}{\underset{j=1}{U}} B_j = S$. Assume that each element of S belongs to exactly ten of the Ai' s and to exactly nine of the Bj' s. Find n. ***(1981 - 2 Marks)***

11. The mean square deviations of a set of observations $x_1, x_2, \ldots, x_n$ about a points c is defined to be $\frac{1}{n}\sum_{i=1}^{n}(x_1-c)^2$.
    The mean suqare deviations about –1 and +1 of a set of observatons are 7 and 3 respectively. Find the standard deviation of this set of observations. ***(1981 - 2 Marks)***

12. The marks obtained by 40 students are grouped in a frequency table in class intervals of 10 maks each. The mean and the variance obtained from this distribution are found to be 40 and 49 respectively. It was later discovered that two observations belonging to the class interval (21–30) were included in the class interval (31–40) by mistake. Find the mean and the variance after correcting the error. ***(1982 - 3 Marks)***

13. A relation R on the set of complex numbers is defined by $z_1\ R\ z_2$ if and only if $\frac{z_1-z_2}{z_1+z_2}$ is real show that R is an equivalence relation. ***(1982 - 2 Marks)***

F MATCH THE FOLLOWING

Each question contains statements given in two columns, which have to be matched. The statements in Column-I are labelled A, B, C and D, while the statements in Column-II are labelled p, q, r, s and t. Any given statement in Column-I can have correct matching with ONE OR MORE statement(s) in Column-II. The appropriate bubbles corresponding to the answers to these questions have to be darkened as illustrated in the following example :

If the correct matches are A-p, s and t; B-q and r; C-p and q; and D-s then the correct darkening of bubbles will look like the given.

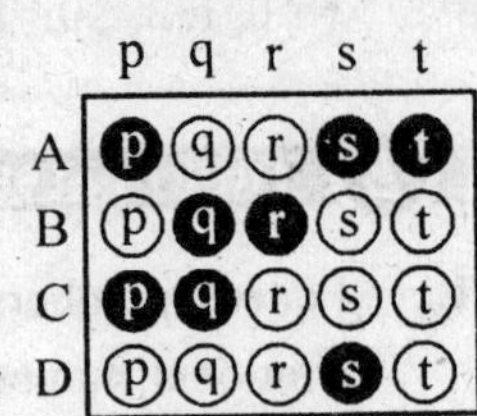

1. Match the statements given in Column-I with the intervals/ union of intervals given in Column-II. ***(2011)***

Column-I	Column-II
(A) The set $\{\text{Re}\left(\frac{2iz}{1-z^2}\right)$: z is a complex number, $\lvert z\rvert = 1\ z \neq \pm 1\}$ is	(p) $(-\infty,-1)\cup(1,\infty)$
(B) The domain of the function $f(x)=\sin^{-1}\left(\frac{8(3)^{x-2}}{1-3^{2(x-1)}}\right)$ is	(q) $(-\infty,-0)\cup(0,\infty)$
(C) If $f(\theta)=\begin{vmatrix}1 & \tan\theta & 1\\ -\tan\theta & 1 & \tan\theta\\ -1 & -\tan\theta & 1\end{vmatrix}$, then the set $\left\{f(\theta):0\le\theta<\frac{\pi}{2}\right\}$ is	(r) $[2,\infty)$
	(s) $(-\infty,-1]\cup[1,\infty)$
(D) If $f(x)=x^{3/2}(3x-10), x\ge 0$ then $f(x)$ is increasing in	(t) $(-\infty,0]\cup[2,\infty)$

I INTEGER VALUE CORRECT TYPE :

1. The value of
$$6+\log_{\frac{3}{2}}\left(\frac{1}{3\sqrt{2}}\sqrt{4-\frac{1}{3\sqrt{2}}\sqrt{4-\frac{1}{3\sqrt{2}}\sqrt{4-\frac{1}{3\sqrt{2}}\ldots}}}\right)$$ is
(2012)

Section-B JEE Main/ AIEEE

1. In a class of 100 students there are 70 boys whose average marks in a subject are 75. If the average marks of the complete class is 72, then what is the average of the girls?
(a) 73 (b) 65 **[2002]**
(c) 68 (d) 74

2. The sum of two forces is 18 N and resultant whose direction is at right angles to the smaller force is 12 N. The magnitude of the two forces are **[2002]**
(a) 13, 5 (b) 12, 6
(c) 14, 4 (d) 11, 7

3. A bead of weight w can slide on smooth circular wire in a vertical plane. The bead is attached by a light thread to the highest point of the wire and in equilibrium, the thread is taut and make an angle θ with the vertical then tension of the thread and reaction of the wire on the bead are
(a) $T = w\cos\theta$ $R = w\tan\theta$ **[2002]**
(b) $T = 2w\cos\theta$ $R = w$
(c) $T = w$ $R = w\sin\theta$
(d) $T = w\sin\theta$ $R = w\cot\theta$

4. The median of a set of 9 distinct observations is 20.5. If each of the largest 4 observations of the set is increased by 2, then the median of the new set **[2003]**
(a) remains the same as that of the original set
(b) is increased by 2
(c) is decreased by 2
(d) is two times the original median.

5. A couple is of moment $\vec{G}$ and the force forming the couple is $\vec{P}$. If $\vec{P}$ is turned through a right angle the moment of the couple thus formed is $\vec{H}$. If instead, the force $\vec{P}$ are turned through an angle α, then the moment of couple becomes **[2003]**
(a) $\vec{H}\sin\alpha - \vec{G}\cos\alpha$ (b) $\vec{G}\sin\alpha - \vec{H}\cos\alpha$
(c) $\vec{H}\sin\alpha + \vec{G}\cos\alpha$ (d) $\vec{G}\sin\alpha + \vec{H}\cos\alpha$.

6. The resultant of forces $\vec{P}$ and $\vec{Q}$ is $\vec{R}$. If $\vec{Q}$ is doubled then $\vec{R}$ is doubled. If the direction of $\vec{Q}$ is reversed, then $\vec{R}$ is again doubled. Then $P^2 : Q^2 : R^2$ is
(a) 2 : 3 : 1 (b) 3 : 1 : 1 **[2003]**
(c) 2 : 3 : 2 (d) 1 : 2 : 3.

7. A body travels a distance s in t seconds. It starts from rest and ends at rest. In the first part of the journey, it moves with constant acceleration f and in the second part with constant retardation r. The value of t is given by
(a) $\sqrt{2s\left(\frac{1}{f}+\frac{1}{r}\right)}$ (b) $2s\left(\frac{1}{f}+\frac{1}{r}\right)$ **[2003]**
(c) $\frac{2s}{\frac{1}{f}+\frac{1}{r}}$ (d) $\sqrt{2s(f+r)}$

8. Two stones are projected from the top of a cliff h metres high, with the same speed u, so as to hit the ground at the same spot. If one of the stones is projected horizontally and the other is projected horizontally and the other is projected at an angle θ to the horizontal then $\tan\theta$ equals
(a) $u\sqrt{\frac{2}{gh}}$ (b) $\sqrt{\frac{2u}{gh}}$ **[2003]**
(c) $2g\sqrt{\frac{u}{h}}$ (d) $2h\sqrt{\frac{u}{g}}$

9. Two particles start simultaneously from the same point and move along two straight lines, one with uniform velocity $\vec{u}$ and the other from rest with uniform acceleration $\vec{f}$. Let α be the angle between their directions of motion. The relative velocity of the second particle w.r.t. the first is least after a time **[2003]**
(a) $\frac{u\cos\alpha}{f}$ (b) $\frac{u\sin\alpha}{f}$
(c) $\frac{f\cos\alpha}{u}$ (d) $u\sin\alpha$

10. The upper $\frac{3}{4}$ th portion of a vertical pole subtends an angle $\tan^{-1}\frac{3}{5}$ at a point in the horizontal plane through its foot and at a distance 40 m from the foot. A possible height of the vertical pole is **[2003]**
(a) 80 *m* (b) 20 *m*
(c) 40 *m* (d) 60 *m*.

11. Let R_1 and R_2 respectively be the maximum ranges up and down an inclined plane and R be the maximum range on the horizontal plane. Then R_1, R, R_2 are in **[2003]**
(a) H.P (b) A.G.P
(c) A.P (d) G.P.

12. In an experiment with 15 observations on *x*, the following results were available: **[2003]**
$$\Sigma x^2 = 2830,\ \Sigma x = 170.$$
One observation that was 20 was found to be wrong and was replaced by the correct value 30. The corrected variance is **[2003]**
(a) 8.33 (b) 78.00
(c) 188.66 (d) 177.33

13. Let R = $\{(1,3),(4,2),(2,4),(2,3),(3,1)\}$ be a relation on the set $A = \{1, 2, 3, 4\}$.. The relation R is **[2004]**
(a) reflexive (b) transitive
(c) not symmetric (d) a function

14. Consider the following statements :
(A) Mode can be computed from histogram
(B) Median is not independent of change of scale
(C) Variance is independent of change of origin and scale.
Which of these is / are correct ? **[2004]**
(a) (A), (B) and (C) (b) only (B)
(c) only (A) and (B) (d) only (A)

15. In a series of 2 n observations, half of them equal a and remaining half equal $-a$. If the standard deviation of the observations is 2, then $|a|$ equals. **[2004]**
(a) $\frac{\sqrt{2}}{n}$ (b) $\sqrt{2}$
(c) 2 (d) $\frac{1}{n}$

16. With two forces acting at point, the maximum affect is obtained when their resultant is 4N. If they act at right angles, then their resultant is 3N. Then the forces are **[2004]**
(a) $\left(2+\frac{1}{2}\sqrt{3}\right)N$ and $\left(2-\frac{1}{2}\sqrt{3}\right)N$
(b) $\left(2+\sqrt{3}\right)N$ and $\left(2-\sqrt{3}\right)N$
(c) $\left(2+\frac{1}{2}\sqrt{2}\right)N$ and $\left(2-\frac{1}{2}\sqrt{2}\right)N$
(d) $\left(2+\sqrt{2}\right)N$ and $\left(2-\sqrt{2}\right)N$

17. In a right angle ΔABC, $\angle A = 90°$ and sides a, b, c are respectively, 5 cm, 4 cm and 3 cm. If a force $\vec{F}$ has moments 0, 9 and 16 in N cm. units respectively about vertices A, B and C, then magnitude of $\vec{F}$ is **[2004]**
(a) 9 (b) 4
(c) 5 (d) 3

18. Three forces $\vec{P}, \vec{Q}$ and $\vec{R}$ acting along IA, IB and IC, where I is the incentre of a ΔABC are in equilibrium. Then $\vec{P}:\vec{Q}:\vec{R}$ is **[2004]**
(a) $\operatorname{cosec}\frac{A}{2}:\operatorname{cosec}\frac{B}{2}:\operatorname{cosec}\frac{C}{2}$
(b) $\sin\frac{A}{2}:\sin\frac{B}{2}:\sin\frac{C}{2}$
(c) $\sec\frac{A}{2}:\sec\frac{B}{2}:\sec\frac{C}{2}$
(d) $\cos\frac{A}{2}:\cos\frac{B}{2}:\cos\frac{C}{2}$

19. A paticle moves towards east from a point A to a point B at the rate of 4 km/h and then towards north from B to C at the rate of 5km/hr. If $AB = 12$ km and $BC = 5$ km, then its average speed for its journey from A to C and resultant average velocity direct from A to C are respectively **[2004]**
(a) $\frac{13}{9}$ km/h and $\frac{17}{9}$ km/h
(b) $\frac{13}{4}$ km/h and $\frac{17}{4}$ km/h
(c) $\frac{17}{9}$ km/h and $\frac{13}{9}$ km/h
(d) $\frac{17}{4}$ km/h and $\frac{13}{4}$ km/h

20. A velocity $\frac{1}{4}$ m/s is resolved into two components along OA and OB making angles 30° and 45° respectively with the given velocity. Then the component along OB is **[2004]**
(a) $\frac{1}{8}(\sqrt{6}-\sqrt{2})$m/s (b) $\frac{1}{4}(\sqrt{3}-1)$ m/s
(c) $\frac{1}{4}$m/s (d) $\frac{1}{8}$m/s

21. If t_1 and t_2 are the times of flight of two particles having the same initial velocity u and range R on the horizontal, then $t_1^2+t_2^2$ is equal to **[2004]**
(a) 1 (b) $4u^2/g^2$
(c) $u^2/2g$ (d) u^2/g

22. Let $R = \{(3, 3), (6, 6), (9, 9), (12, 12), (6, 12), (3, 9), (3, 12), (3, 6)\}$ be a relation on the set $A = \{3, 6, 9, 12\}$. The relation is **[2005]**
(a) reflexive and transitive only
(b) reflexive only
(c) an equivalence relation
(d) reflexive and symmetric only

23. ABC is a triangle. Forces $\vec{P}$, $\vec{Q}$, $\vec{R}$ acting along IA, IB, and IC respectively are in equilibrium, where I is the incentre of ΔABC. Then $P:Q:R$ is **[2005]**
(a) $\sin A : \sin B : \sin C$
(b) $\sin\frac{A}{2}:\sin\frac{B}{2}:\sin\frac{C}{2}$
(c) $\cos\frac{A}{2}:\cos\frac{B}{2}:\cos\frac{C}{2}$
(d) $\cos A : \cos B : \cos C$

24. If in a frequency distribution, the mean and median are 21 and 22 respectively, then its mode is approximately
(a) 22.0 (b) 20.5 **[2005]**
(c) 25.5 (d) 24.0

25. A lizard, at an initial distance of 21 cm behind an insect, moves from rest with an acceleration of $2\ cm/s^2$ and pursues the insect which is crawling uniformly along a straight line at a speed of 20 cm/s. Then the lizard will catch the insect after **[2005]**
(a) $20\,s$ (b) $1\,s$
(c) $21\,s$ (d) $24\,s$

26. Two points A and B move from rest along a straight line with constant acceleration f and f' respectively. If A takes m sec. more than B and describes 'n' units more than B in acquiring the same speed then [2005]

(a) $(f-f')m^2 = ff'n$ (b) $(f+f')m^2 = ff'n$

(c) $\frac{1}{2}(f+f')m = ff'n^2$ (d) $(f'-f)n = \frac{1}{2}ff'm^2$

27. A and B are two like parallel forces. A couple of moment H lies in the plane of A and B and is contained with them. The resultant of A and B after combining is displaced through a distance

(a) $\frac{2H}{A-B}$ (b) $\frac{H}{A+B}$ [2005]

(c) $\frac{H}{2(A+B)}$ (d) $\frac{H}{A-B}$

28. Let $x_1, x_2, ... x_n$ be n observations such that $\sum x_i^2 = 400$ and $\sum x_i = 80$. Then the possible value of n among the following is [2005]

(a) 15 (b) 18

(c) 9 (d) 12

29. A particle is projected from a point O with velocity u at an angle of 60° with the horizontal. When it is moving in a direction at right angles to its direction at O, its velocity then is given by [2005]

(a) $\frac{u}{3}$ (b) $\frac{u}{2}$

(c) $\frac{2u}{3}$ (d) $\frac{u}{\sqrt{3}}$

30. ABC is a triangle, right angled at A. The resultant of the forces acting along $\overline{AB}, \overline{BC}$ with magnitudes $\frac{1}{AB}$ and $\frac{1}{AC}$ respectively is the force along $\overline{AD}$, where D is the foot of the perpdicular from A onto BC. The magnitude of the resultant is [2006]

(a) $\frac{AB^2 + AC^2}{(AB)^2(AC)^2}$ (b) $\frac{(AB)(AC)}{AB + AC}$

(c) $\frac{1}{AB} + \frac{1}{AC}$ (d) $\frac{1}{AD}$

31. Let W denote the words in the English dictionary. Define the relation R by $R = \{(x, y) \in W \times W|$ the words x and y have at least one letter in common.$\}$ Then R is [2006]

(a) not reflexive, symmetric and transitive

(b) relexive, symmetric and not transitive

(c) reflexive, symmetric and transitive

(d) reflexive, not symmetric and transitive

32. Suppose a population A has 100 observations 101, 102, ..., 200 and another population B has 100 obsevrations 151, 152, ... 250. If V_A and V_B represent the variances of the two populations, respectively then $\frac{V_A}{V_B}$ is

(a) 1 (b) $\frac{9}{4}$ [2006]

(c) $\frac{4}{9}$ (d) $\frac{2}{3}$

33. A particle has two velocities of equal magnitude inclined to each other at an angle θ. If one of them is halved, the angle between the other and the original resultant velocity is bisected by the new resultant. Then θ is [2006]

(a) 90° (b) 120°

(c) 45° (d) 60°

34. A body falling from rest under gravity passes a certain point P. It was at a distance of 400 m from P, 4s prior to passing through P. If $g = 10 m/s^2$, then the height above the point P from where the body began to fall is [2006]

(a) 720 m (b) 900 m

(c) 320 m (d) 680 m

35. The resultant of two forces Pn and $3n$ is a force of $7n$. If the direction of $3n$ force were reversed, the resultant would be $\sqrt{19}\,n$. The value of P is [2007]

(a) $3\,n$ (b) $4\,n$

(c) $5\,n$ (d) $6\,n$.

36. A particle just clears a wall of height b at a distance a and strikes the ground at a distance c from the point of projection. The angle of projection is [2007]

(a) $\tan^{-1}\frac{bc}{a(c-a)}$ (b) $\tan^{-1}\frac{bc}{a}$

(c) $\tan^{-1}\frac{b}{ac}$ (d) 45°.

37. The average marks of boys in class is 52 and that of girls is 42. The average marks of boys and girls combined is 50. The percentage of boys in the class is [2007]

(a) 80 (b) 60

(c) 40 (d) 20.

38. A body weighing 13 kg is suspended by two strings 5m and 12m long, their other ends being fastened to the extremities of a rod 13m long. If the rod be so held that the body hangs immediately below the middle point, then tensions in the strings are [2007]

(a) 5 kg and 12 kg (b) 5 kg and 13 kg

(c) 12 kg and 13 kg (d) 5 kg and 5 kg

39. The mean of the numbers a, b, 8, 5, 10 is 6 and the variance is 6.80. Then which one of the following gives possible values of a and b ? [2008]

(a) $a=0, b=7$ (b) $a=5, b=2$

(c) $a=1, b=6$ (d) $a=3, b=4$

DIRECTIONS (for Qs. 40 to 43): Each of these questions contains two statements: Statement-1(Assertion) and Statement-2(Reason). Each of these questions also has four alternative choices, only one of which is the correct answer. You have to select the correct choice.

40. Let p be the statement "x is an irrational number", q be the statement "y is a transcendental number", and r be the statement " x is a rational number if f y is a transcendental number". [2008]

Statement-1 : r is equivalent to either q or p

Statement-2 : r is equivalent to $\sim(p \leftrightarrow \sim q)$.

(a) Statement -1 is false, Statement-2 is true
(b) Statement -1 is true, Statement-2 is true; Statement -2 is a correct explanation for Statement-1
(c) Statement -1 is true, Statement-2 is true; Statement -2 is not a correct explanation for Statement-1
(d) Statement -1 is true, Statement-2 is false

41. The statement $p \to (q \to p)$ is equivalent to **[2008]**
(a) $p \to (p \to q)$ (b) $p \to (p \vee q)$
(c) $p \to (p \wedge q)$ (d) $p \to (p \leftrightarrow q)$

42. **Statement-1 :** $\sim (p \leftrightarrow \sim q)$ is equivalent to $p \leftrightarrow q$.
Statement-2 : $\sim (p \leftrightarrow \sim q)$ is a tautology **[2009]**
(a) Statement-1 is true, Statement-2 is true; Statement-2 is not a correct explanation for Statement-1.
(b) Statement-1 is true, Statement-2 is false.
(c) Statement-1 is false, Statement-2 is true.
(d) Statement-1 is true, Statement-2 is true, Statement-2 is a correct explanation for statement -1

43. **Statement-1 :** The variance of first n even natural numbers is $\frac{n^2-1}{4}$.
Statement-2 : The sum of first n natural numbers is $\frac{n(n+1)}{2}$ and the sum of squares of first n natural numbers is $\frac{n(n+1)(2n+1)}{6}$. **[2009]**
(a) Statement-1 is true, Statement-2 is true Statement-2 is not a correct explanation for Statement-1.
(b) Statement-1 is true, Statement-2 is false.
(c) Statement-1 is false, Statement-2 is true.
(d) Statement-1 is true, Statement-2 is true. Statement-2 is a correct explanation for Statement-1.

44. If A, B and C are three sets such that $A \cap B = A \cap C$ and $A \cup B = A \cup C$, then **[2009]**
(a) $A = C$ (b) $B = C$
(c) $A \cap B = \phi$ (d) $A = B$

45. If the mean deviation of the numbers 1, 1 + d, 1 + 2d, ... 1 + 100d from their mean is 255, then d is equal to : **[2009]**
(a) 20.0 (b) 10.1
(c) 20.2 (d) 10.0

46. Let S be a non-empty subset of R. Consider the following statement :
P : There is a rational number $x \in$ S such that $x > 0$.
Which of the following statements is the negation of the statement P ? **[2010]**
(a) There is no rational number $x \in$ S such than $x \le 0$.
(b) Every rational number $x \in$ S satisfies $x \le 0$.
(c) $x \in$ S and $x \le 0 \Rightarrow x$ is not rational.
(d) There is a rational number $x \in$ S such that $x \le 0$.

47. Consider the following relations:
R = $\{(x, y) \mid x, y$ are real numbers and $x = wy$ for some rational number $w\}$;
$S = \{\left(\frac{m}{n}, \frac{p}{q}\right) \mid m, n, p$ and q are integers such that $n, q \neq 0$ and $qm = pn\}$.
Then **[2010]**
(a) Neither R nor S is an equivalence relation
(b) S is an equivalence relation but R is not an equivalence relation
(c) R and S both are equivalence relations
(d) R is an equivalence relation but S is not an equivalence relation

48. For two data sets, each of size 5, the variances are given to be 4 and 5 and the corresponding means are given to be 2 and 4, respectively. The variance of the combined data set is **[2010]**
(a) $\frac{11}{2}$ (b) 6
(c) $\frac{13}{2}$ (d) $\frac{5}{2}$

49. Let R be the set of real numbers.
Statement-1: $A = \{(x, y) \in R \times R : y - x$ is an integer$\}$ is an equivalence relation on R.
Statement-2: $B = \{(x, y) \in R \times R : x = \alpha y$ for some rational number $\alpha\}$ is an equivalence relation on R. **[2011]**
(a) Statement-1 is true, Statement-2 is true; Statement-2 is not a correct explanation for Statement-1.
(b) Statement-1 is true, Statement-2 is false.
(c) Statement-1 is false, Statement-2 is true.
(d) Statement-1 is true, Statement-2 is true; Statement-2 is a correct explanation for Statement-1.

50. Consider the following statements **[2011]**
P : Suman is brilliant
Q : Suman is rich
R : Suman is honest
The negation of the statement "Suman is brilliant and dishonest if and only if Suman is rich" can be expressed as
(a) $\sim (Q \leftrightarrow (P \wedge \sim R))$ (b) $\sim Q \leftrightarrow \sim P \wedge R$
(c) $\sim (P \wedge \sim R) \leftrightarrow Q$ (d) $\sim P \wedge (Q \leftrightarrow \sim R)$

51. If the mean deviation about the median of the numbers a, $2a$,.......,$50a$ is 50, then $|a|$ equals **[2011]**
(a) 3 (b) 4
(c) 5 (d) 2

52. A scientist is weighing each of 30 fishes. Their mean weight worked out is 30 gm and a standarion deviation of 2 gm. Later, it was found that the measuring scale was misaligned and always under reported every fish weight by 2 gm. The correct mean and standard deviation (in gm) of fishes are respectively : **[2011 RS]**
(a) 32,2 (b) 32,4
(c) 28,2 (d) 28,4

53. The only statement among the following that is a tautology is **[2011 RS]**
(a) $A \wedge (A \vee B)$ (b) $A \vee (A \wedge B)$
(c) $[A \wedge (A \to B)] \to B$ (d) $B \to [A \wedge (A \to B)]$

54. Consider the following relation R on the set of real square matrices of order 3.
$R = \{(A, B) \mid A = P^{-1}$ BP for some invertible matirx $P\}$

Statement-1 : R is equivalence relation.

Statement-2 : For any two invertible 3×3 matrices M and N, $(MN)^{-1} = N^{-1}M^{-1}$. **[2011 RS]**

(a) Statement-1 is true, statement-2 is true and statement-2 is a correct explanation for statement-1.

(b) Statement-1 is true, statement-2 is true; statement-2 is not a correct explanation for statement-1.

(c) Statement-1 is true, stement-2 is false.

(d) Statement-1 is false, statement-2 is true.

55. The negation of the statement **[2012]**

"If I become a teacher, then I will open a school", is :

(a) I will become a teacher and I will not open a school.

(b) Either I will not become a teacher or I will not open a school.

(c) Neither I will become a teacher nor I will open a school.

(d) I will not become a teacher or I will open a school.

56. Let $x_1, x_2, ..., x_n$ be n observations, and let $\bar{x}$ be their arithmetic mean and σ^2 be the variance. **[2012]**

Statement-1 : Variance of $2x_1, 2x_2, ..., 2x_n$ is $4\sigma^2$.

Statement-2 : Arithmetic mean $2x_1, 2x_2, ..., 2x_n$ is $4\bar{x}$.

(a) Statement-1 is false, Statement-2 is true.

(b) Statement-1 is true, statement-2 is true; statement-2 is a correct explanation for Statement-1.

(c) Statement-1 is true, statement-2 is true; statement-2 is ***not*** a correct explanation for Statement-1.

(d) Statement-1 is true, statement-2 is false.

57. Let $X = \{1,2,3,4,5\}$. The number of different ordered pairs (Y,Z) that can formed such that $Y \subseteq X$, $Z \subseteq X$ and $Y \cap Z$ is empty is : **[2012]**

(a) 5^2 (b) 3^5

(c) 2^5 (d) 5^3

58. Let A and B two sets containing 2 elements and 4 elements respectively. The number of subsets of A × B having 3 or more elements is **[JEE M 2013]**

(a) 256 (b) 220

(c) 219 (d) 211

59. Consider

Statement-1 : $(p \wedge \sim q) \wedge (\sim p \wedge q)$ is a fallacy.

Statement-2 : $(p \to q) \leftrightarrow (\sim q \to \sim p)$ is a tautology. **[JEE M 2013]**

(a) Statement-1 is true; Statement-2 is true; Statement-2 is a correct explanation for Statement-1.

(b) Statement-1 is true; Statement-2 is true; Statement-2 is not a correct explanation for Statement-I.

(c) Statement-1 is true; Statement-2 is false.

(d) Statement-1 is false; Statement-2 is true.

60. All the students of a class performed poorly in Mathematics. The teacher decided to give grace marks of 10 to each of the students. Which of the following statistical measures will not change even after the grace marks were given ? **[JEE M 2013]**

(a) mean (b) median

(c) mode (d) variance

Solutions & Explanations

Section-A : JEE Advanced/ IIT-JEE

A 1. False

B 1. F

C 1. (c) 2. (b) 3. (b) 4. (c) 5. (c) 6. (d)
7. (d) 8. (d)

D 1. (c) 2. (b, c, d)

E 1. 20 3. $2(3\sqrt{13} + 2\sqrt{2})\pi m^2$ 4. Correct mean 42.3 ; Correct variance 43. 81

7. (ii) 5cm 9. (i) 6 (ii) True (iii) Yes 10. 45 11. $\sqrt{3}$

12. new mean = 39.5, new variance = 49.25

F 1. A-s; B-t; C-r; D-r

I 1. 4

Section-B : JEE Main/ AIEEE

1. (b)	2. (a)	3. (b)	4. (a)	5. (c)	6. (c)
7. (a)	8. (a)	9. (a)	10. (c)	11. (a)	12. (b)
13. (c)	14. (c)	15. (c)	16. (c)	17. (c)	18. (d)
19. (d)	20. (a)	21. (b)	22. (a)	23. (c)	24. (d)
25. (c)	26. (d)	27. (b)	28. (b)	29. (d)	30. (d)
31. (b)	32. (a)	33. (b)	34. (a)	35. (c)	36. (a)
37. (a)	38. (a)	39. (d)	40. (None)	41. (d)	42. (b)
43. (c)	44. (b)	45. (b)	46. (b)	47. (b)	48. (a)
49. (b)	50. (a)	51. (b)	52. (a)	53. (c)	54. (b)
55. (a)	56. (d)	57. (b)	58. (c)	59. (b)	60. (d)

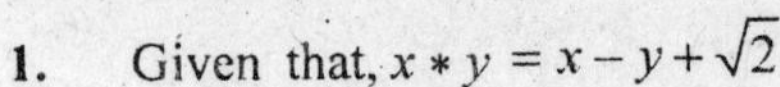

Section-A JEE Advanced/ IIT-JEE

A FILL IN THE BLANKS :

1. Given that, $x * y = x - y + \sqrt{2}$

Consider $x = 2\sqrt{2}$, $y = \sqrt{2}$

then $x * y = 2\sqrt{2} - \sqrt{2} + \sqrt{2} = 3\sqrt{2}$ (irrational)

and $y * x = \sqrt{2} - 2\sqrt{2} + \sqrt{2} = 0$ (rational)

$\therefore x * y \neq y * x$

Hence $*$ is not symm. $\Rightarrow *$ is not an equivalence relation

2. Frequency for variable x is ${}^{n+x-1}C_x$

where $x = 0, 1, 2, ...n.$

Mode is the variable for which freq. is max.

Now, ${}^nC_r$ is max for $r = n/2$, if n is even

$r = \frac{n+1}{2}$ if n is odd.

If $n + x - 1$ is even then for max value of ${}^{n+x-1}C_x$,

$x = \frac{n+x-1}{2} \Rightarrow x = n-1$, $\therefore$ freq ${}^{2n-2}C_{n-1}$

If $n + x - 1$ is odd then for max value of ${}^{n+x-1}C_x$

$x = \frac{n+x-1+1}{2} \Rightarrow x = n$, $\therefore$ freq ${}^{2n-1}C_n$

But we know ${}^{2n-1}C_n = \frac{2n-1}{n}\ {}^{2n-2}C_{n-1}$ **NOTE THIS STEP**

i.e., ${}^{2n-1}C_n > {}^{2n-2}C_{n-1}$

$\therefore$ Mode should be n.

B TRUE / FALSE :

1. Given that, $x * y = x - y + \sqrt{2}$

Consider $x = 2\sqrt{2}$, $y = \sqrt{2}$

then $x * y = 2\sqrt{2} - \sqrt{2} + \sqrt{2} = 3\sqrt{2}$ (irrational)

and $y * x = \sqrt{2} - 2\sqrt{2} + \sqrt{2} = 0$ (rational)

$\therefore x * y \neq y * x$

Hence $*$ is not symm. $\Rightarrow *$ is not an equivalence relation

C MCQ's WITH ONE CORRECT ANSWER :

1. (c) $X \cap (X \cup Y)^c = X \cap (X^c \cap Y^c) = (X \cap X^c) \cap Y^c$

$= \phi \cap Y^c = \phi$

2. (b) $\frac{12}{(3+\sqrt{5})+2\sqrt{2}} = \frac{12}{(3+\sqrt{5})+2\sqrt{2}} \times \frac{(3+\sqrt{5})-2\sqrt{2}}{(3+\sqrt{5})-2\sqrt{2}}$

$= \frac{12[3+\sqrt{5}-2\sqrt{2}]}{(3+\sqrt{5})^2-(2\sqrt{2})^2} = \frac{12[3+\sqrt{5}-2\sqrt{2}]}{9+5+6\sqrt{5}-8}$

$= \frac{12[3+\sqrt{5}-2\sqrt{2}]}{6(\sqrt{5}+1)} \times \frac{\sqrt{5}-1}{\sqrt{5}-1}$

$= \frac{2[3\sqrt{5}+5-2\sqrt{10}-3-\sqrt{5}+2\sqrt{2}]}{5-1}$

$= \frac{2+2\sqrt{5}+2\sqrt{2}-2\sqrt{10}}{2} = 1+\sqrt{2}+\sqrt{5}-\sqrt{10}$

3. (b) If each of n observations is multiplied by a constant C, the standard deviation also gets multiplied by C.

4. (c) If s. d. = 0, statements like (a) and (b) can not be given.

5. (c) Given that $x_1 < x_2 < x_3 < .... < x_{201}$

$\therefore$ Median of the given observation $= \frac{201+1}{2}$ th items

$= 101\text{th item} = x_{101}$

Now, deviations will be minimum if taken from the median

$\therefore$ Mean deviation will be min if $k = x_{101}$.

6. (d) If any of the inequations hold, it must hold for any real numbers $x_1, x_2, ... x_n$ and any $n \in N$.

$\therefore$ let $x_1 = 1, x_2 = 2, x_3 = 3$; $n = 3$ then we can check none of the inequalities (a), (b) or (c) are satisfied.

7. (d) $S = \{1, 2, 3, 4\}$

Let P and Q be disjoint subsets of S

Now for any element $a \in s$, following cases are possible

$a \in P$ and $a \notin Q$, $a \notin P$ and $a \in Q$, $a \notin P$ and $a \notin Q$

$\Rightarrow$ For every element there are three option

$\therefore$ Total options $= 3^4 = 81$

Here $P \neq Q$ except when $P = Q = \phi$

$\therefore$ 80 ordered pairs (P, Q) are there for which $P \neq Q$.

Hence total number of unordered pairs of disjoint

subsets $= \frac{80}{2} + 1 = 41$

8. (d) $P = \{\theta : \sin\theta - \cos\theta = \sqrt{2}\cos\theta\}$

$\sin\theta = (\sqrt{2}+1)\cos\theta$, $\tan\theta = \sqrt{2}+1$

$Q = \{\theta : \sin\theta + \cos\theta = \sqrt{2}\sin\theta\}$

$\cos\theta = (\sqrt{2}-1)\sin\theta$ or $\tan\theta = \sqrt{2}+1$

$\therefore P = Q$

D MCQ's WITH ONE OR MORE THAN ONE CORRECT :

1. (c) Let n be the number of newspapers which are read. Then $60\,n = (300)(5) \Rightarrow n = 25$

2. (b, c, d)

Given side of square S_1 = 10 cm

As it is same as diagonal of S_2.

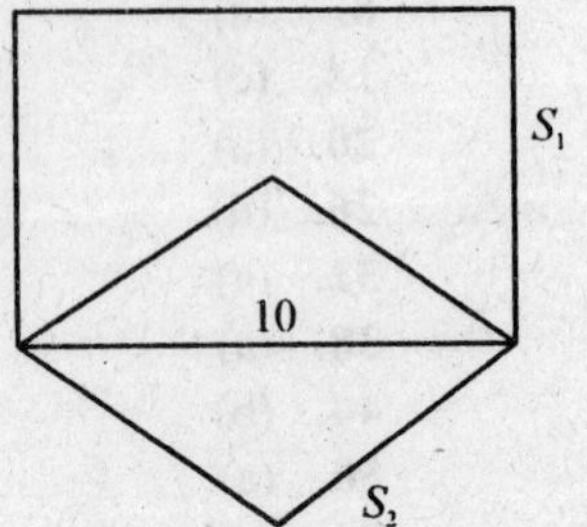

$\therefore$ Side of square $S_2 = \left(\frac{10}{\sqrt{2}}\right)$ cm

Similarly, side of square $S_3 = \frac{10}{(\sqrt{2})^2}$ cm

In the same manner, side of square

$S_n = \frac{10}{(\sqrt{2})^{n-1}}$ cm

$\therefore$ Area of square

$$S_n = \left[\frac{10}{(\sqrt{2})^{n-1}}\right]^2 = \left[\frac{10}{(\sqrt{2})^{n-1}}\right] < 1$$

$\Rightarrow 100 < (2)^{n-1} \Rightarrow 2^{n-1} > 100$

$\Rightarrow n - 1 \geq 7 \Rightarrow n \geq 8$

E SUBJECTIVE PROBLEMS :

1. We have

$n(U) = 100$, where U stands for universal set

$n(M \cap C \cap T) = 10$; $n(M \cap C) = 20$;.

$n(C \cap T) = 30$; $n(M \cap T) = 25$;

n (M only) = 12; n (only C) = 5; n (only T) = 8

Filling all the entries we obtain the Venn diagram as shown below:

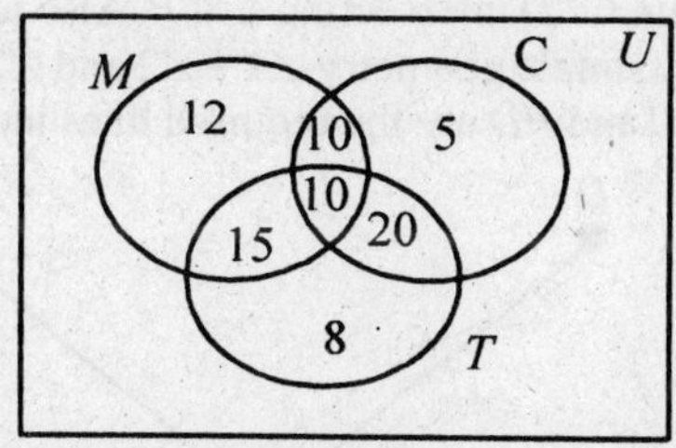

$\therefore$ $n(M \cap C \cup T) = 12 + 10 + 5 + 15 + 10 + 20 + 8 = 80$

$\therefore$ $n(M \cup C \cup T)' = 100 - 80 = 20$

2. (a) To construct a Δ with base = 9 cm, altitude = 4 cm and ratio of the other two sides as 2 : 1.

Steps of Construction :

1. Draw $BC = 9$ cm
2. Divide BC internally at P and externally at Q in the ratio 2 : 1.

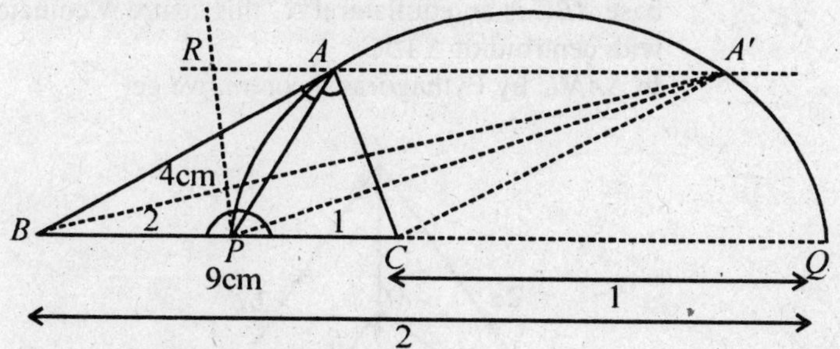

3. Draw a semicircle on PQ.
4. Draw a line $\parallel BQ$ at a distance of 4 cm from intersecting semicircle at A and A'.
5. ABC and $A'BC$ are the required Δ's.

(b) To construct a Δ with perimeter = 15 cm, base angles 60° and 45°.

Steps of Construction :

1. Draw $PQ = 15$ cm
2. At P draw $\angle RPQ = 60°$ and $\angle XPQ = \frac{60°}{2} = 30°$

and at Q draw $\angle SQP = 45°$ and $\angle YQP = \frac{45°}{2} = 22\frac{1}{2}°$

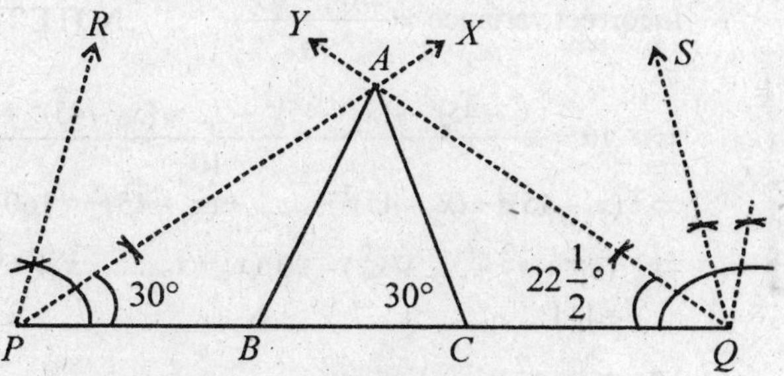

3. PX and QY meet each other at A.
4. Through A draw $AB \parallel PR$ and $AC \parallel QS$.
5. ABC is the required Δ.

Justification: $\because$ $AB \parallel PR$ and PA transversal

$\therefore$ $\angle PAB = \angle RPA = \frac{1}{2} \times 60° = 30°$

$\angle APB = \angle BAP = 30° \Rightarrow AB = PB$

Similarly $AC = CQ$

$\therefore$ $AB + BC + CA = PB + BC + CQ = 15$ cm

Also $\angle ABC = \angle RPB = 60°$ and $\angle ACB = \angle SQS = 45°$

3. Slant height of frustum A

$= \sqrt{(4-2)^2 + 3^2} = \sqrt{13}$

$\therefore$ Curved surface area of frustum

$= \pi(4+2)\sqrt{13} = 6\sqrt{13}\pi\ m^2$

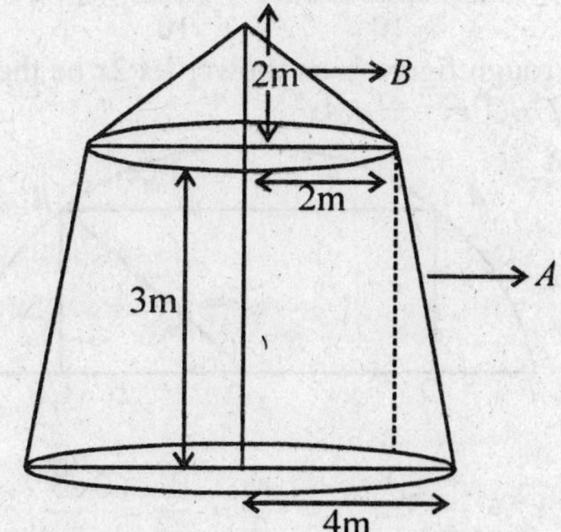

Also slant height of cone $B = \sqrt{2^2 + 2^2} = 2\sqrt{2}$

$\therefore$ Curved surface area of cone

$= \pi \times 2 \times 2\sqrt{2} = 4\sqrt{2}\pi\ m^2$

$\therefore$ Area of canvas required $= 6\sqrt{13}\pi + 4\sqrt{2}\pi$

$= 2\left(3\sqrt{13} + 2\sqrt{2}\right)\pi$ m^2.

4. Let the remaining 9 readings be $x_1, x_2, x_3 ..... x_9$ and tenth is taken by the student as 52.

$\therefore$ Incorrect mean $= \dfrac{x_1 + x_2 + ... + x_9 + 52}{10} = 45$

$\Rightarrow x_1 + x_2 + ... + x_9 = 450 - 52 = 398$ ....(1)

$\therefore$ Correct mean $= \dfrac{x_1 + x_2 + ... + x_9 + 25}{10} = \dfrac{398 + 25}{10}$

$\Rightarrow$ Correct mean = 42.3

Incorrect variance $= \dfrac{\Sigma(x_i - \bar{x})^2}{n}$ **NOTE THIS STEP**

$\Rightarrow 16 = \dfrac{(x_1 - 45)^2 + (x_2 - 45)^2 + ... + (x_9 - 45)^2 + (52 - 45)^2}{10}$

$\Rightarrow (x_1 - 45)^2 + (x_2 - 45)^2 + ... + (x_9 - 45)^2 = 160 - 49 = 111$

$\Rightarrow (x_1^2 + x_2^2 + ... + x_9^2) - 90 (x_1 + x_2 ..... + x_9) + 9 \times (45)^2 = 111$

$\Rightarrow x_1^2 + x_2^2 + ... + x_9^2 = 111 + 90 \times 398 - 9 \times (45)^2 = 17706$ ...(2)

Correct variance $= \dfrac{\Sigma(x_i - 42.3)^2}{10}$

$= \dfrac{(x_1 - 42.3)^2 + (x_2 - 42.3)^2 + ... + (x_9 - 42.3)^2 + (25 - 42.3)^2}{10}$

$= \dfrac{(x_1^2 + x_2^2 + ... + x_9^2) - 84.6(x_1 + x_2 + ... + x_9) + 9 \times (42.3)^2 + 299.29}{10}$

$= \dfrac{17706 - 84.6 \times 398 + 16103.61 + 299.29}{10}$

$= \dfrac{34108.9 - 33670.8}{10} = \dfrac{438.1}{10} = 43.81$

5. The rough figure is as shown, let $2x$ be the side of square. In ΔOBC $r^2 = x^2 + 4x^2$

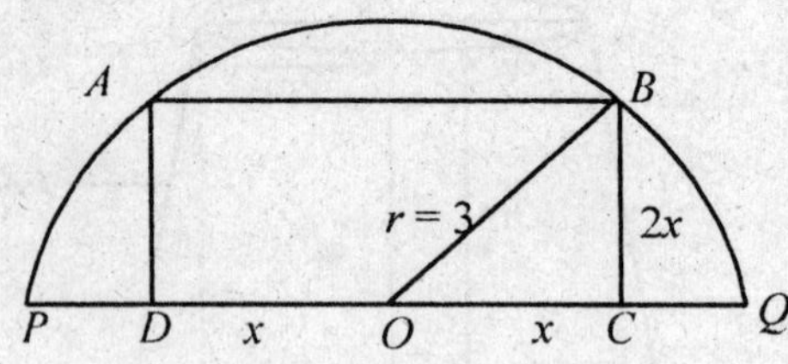

$\Rightarrow x^2 = \dfrac{r^2}{5} \Rightarrow x = \dfrac{r}{\sqrt{5}} = \dfrac{3}{\sqrt{5}} = \dfrac{3\sqrt{5}}{5}$

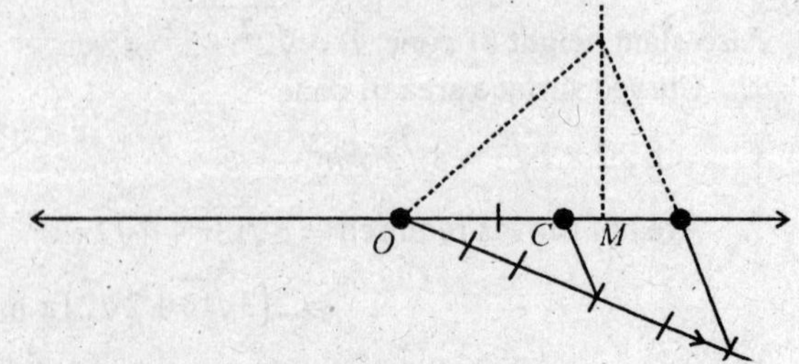

Now on number line we locate M such that $OM = \sqrt{5}$ Divide OM in five equal parts and take a point C on it such that $OC : CM = 3 : 2$. So, that $OC = \frac{3}{5}\sqrt{5}$

Now, draw a line segment $PQ = 6$ cm whose mid-point is O.

From this cut a line segment OC $= \dfrac{3\sqrt{5}}{5}$ and $OD = \dfrac{3\sqrt{5}}{5}$ on opposite sides of O.

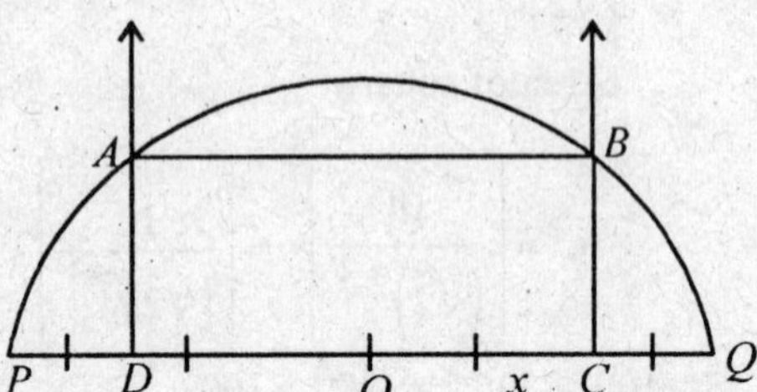

At C and B draw perpendiculars to PQ. With O as centre and OP as radius, draw a semicircle intersecting the perpendiculars draw at C & D at A and P respectively. Join AB. $ABCD$ is the required square.

6. Here we are given two points C and D on the same side of line L. To find a point P on L such that PC and PD are equally inclined to L.

Steps of Construction :

1. From C draw $CE \perp L$ and produce it to C' such that $EC' = EC$.
2. Join $C'D$ intersecting L at P. Also join CP.
3. By simple geometry $\angle 1 = \angle 2$ and $\angle 2 = \angle 3 \Rightarrow \angle 1 = \angle 3$

$\therefore$ PC and PD are the required lines inclined equally to L.

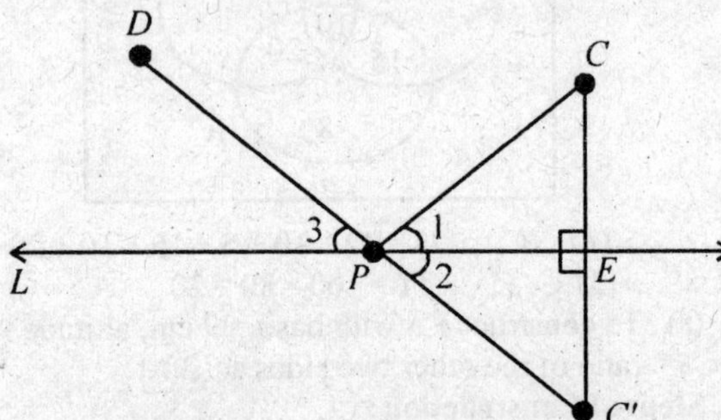

7. (i) Let $DABC$ is the regular tetrahedron with base ABC, and each edge being $2a$. Let DN be the altitudes of tetrahedron then N should be the centre of base. As base ABC is an equilateral Δ., this centre N coincides with centroid of ΔABC.

In $\Delta AN'C$ by Pythagoras theorem, we get

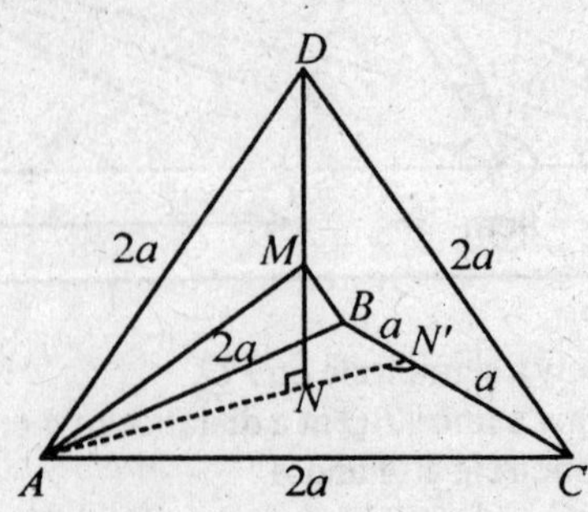

$$AN' = \sqrt{(2a)^2 - a^2} = a\sqrt{3}$$

$$\Rightarrow \quad AN = \frac{2}{3}AN' = \frac{2a}{\sqrt{3}}$$

In ΔDAN,

$$\angle AND = 90°, AD = 2a, AN = \frac{2a}{\sqrt{3}}$$

$\therefore$ By Pythagoras theorem

$$DN = \sqrt{(2a)^2 - \left(\frac{2a}{\sqrt{3}}\right)^2} = 2a\frac{\sqrt{2}}{3}$$

As M is mid-point of DN. $\therefore MN = a\frac{\sqrt{2}}{3}$

Now in ΔAMN,

$$MA^2 = MN^2 + AN^2 = \frac{2a^2}{3} + \frac{4a^2}{3} = 2a^2$$

Similarly, we can prove that

$$MB^2 = MC^2 = 2a^2$$

Now,

$\therefore$ $MA^2 + MB^2 = 4a^2 = AB^2$

$\therefore$ MA and MB are at right angles to each other. Similarly we can prove that MA and MC and MB and MC are perpendicular to each other.

Hence, MA, MB, MC are mutually perpendicular.

(ii) Volume of earth dug out from pit = Vol. of frustum.

$$= \frac{1}{3}\pi(2^2 + 1^2 + 2\times 1)\times 3m^3 = 7\pi m^3$$

Area over which this earth is to be spread

$$= \pi(12^2 - 2^2) = 140\pi m^2$$

$\therefore$ Height upto which the level of plot rises

$$= \frac{7\pi}{140\pi} = \frac{1}{20}m = 5cm$$

8. (i) $\because$ $CD = \frac{1}{2}AB$

$\therefore$ $OC = CD = OD$ (each radii)

$\therefore$ ΔOCD is an equilateral Δ.

$\therefore$ $\angle COD = 60° = \angle OCD = \angle ODC$

$$\Rightarrow \quad \angle CBD = \frac{1}{2}\angle COD = 30° \text{ [*]}$$

[* $\because$ The angle which an arc of a circle subtends at the centre is double that which it subtends at any point on the remaining part of the circumference]

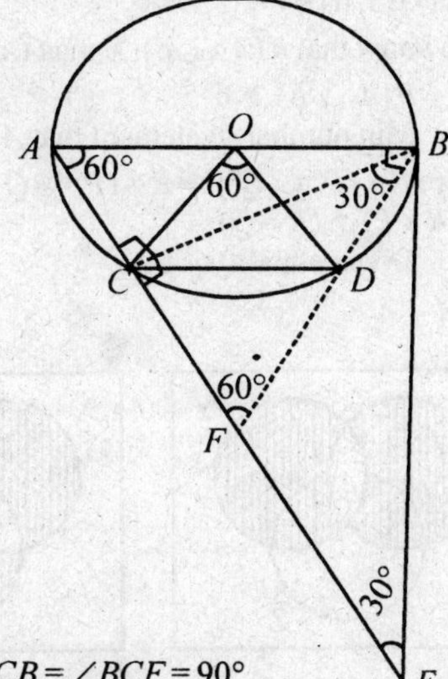

Also, $\angle ACB = \angle BCF = 90°$

$\therefore$ In ΔBCF, $\angle BFC = 60°$

Now $AB \parallel CD$ and OC is transversal

$\therefore$ $\angle AOC = \angle OCD = 60°$

In ΔAOC, $\angle AOC = 60°$ and $OA = OC$

$\therefore$ $\angle OAC = 60° = \angle OCA$

Similarly, $\angle OBD = 60°$

So, ΔABF is an equilateral Δ.

$\therefore$ $FB = AB$

Also $\angle BFE = 180 - 60° = 120°$ and in right ΔABE, $\angle AEB = 30°$

$\therefore$ In ΔBEF by angle sum property $\angle EBF = 30°$

Now $\angle AEB = \angle EBF \Rightarrow \angle FEB = \angle EBF \Rightarrow FE = BF$

But $FB = AB = AF$

$\therefore$ $EF = AB$

Now, $AE = AF + EF = AB + AB = 2AB$

(ii) Given $x > y > 0$

Then $\sqrt{2}[2x + \sqrt{x^2 - y^2})]\left(\sqrt{x - \sqrt{x^2 - y^2}}\right)$

$$= \sqrt{2}[\sqrt{(x + x + \sqrt{x^2 - y^2})\left(x - \sqrt{x^2 - y^2}\right)}]$$

$$\sqrt{2x + \sqrt{x^2 - y^2}}$$

$$= \sqrt{2}\sqrt{[x^2 - x\sqrt{x^2 - y^2} + y^2][2x + \sqrt{x^2 - y^2}]}$$

$$= \sqrt{2}\sqrt{2x^3 + 2xy^2 - 2x^2\sqrt{x^2 - y^2} + x^2\sqrt{x^2 - y^2} - x^3 + xy^2 + y^2\sqrt{x^2 - y^2}}$$

$$= \sqrt{2}\sqrt{x^3 + 3xy^2 - (x^2 - y^2)^{3/2}}$$

$$= \sqrt{2(x^3 + 3xy^2) - 2(x + y)^{3/2}(x - y)^{3/2}}$$

Now let $a = x + y$ and $b = x - y$

then $a^3 + b^3 = (x + y)^3 + (x - y)^3 = 2(x^3 + 3xy^2)$

and $a^{3/2} b^{3/2} = (x^2 - y^2)^{3/2}$

So, we get

$$= \sqrt{a^3 + b^3 - 2a^{3/2}b^{3/2}} = \sqrt{(a^{3/2} - b^{3/2})^2}$$

$$= a^{3/2} - b^{3/2} = (x + y)^{3/2} - (x - y)^{3/2}$$

$$= \sqrt{(x + y)^3} - \sqrt{(x - y)^3}$$

9. (i) $n(A) = 3$, $n(B) = 6$

We know that $n(A \cup B) \geq \max(n(A), n(B))$

$\Rightarrow \quad n(A \cup B) \geq 6$

$\therefore$ Min number of element that $A \cup B$ can have is 6.

(ii) Here $R \times (P^c \cup Q^c)^c = R \times (P \cap Q)$
$= (R \times P) \cap (R \times Q)$

$\therefore$ Given equality is true.

(iii) Yes

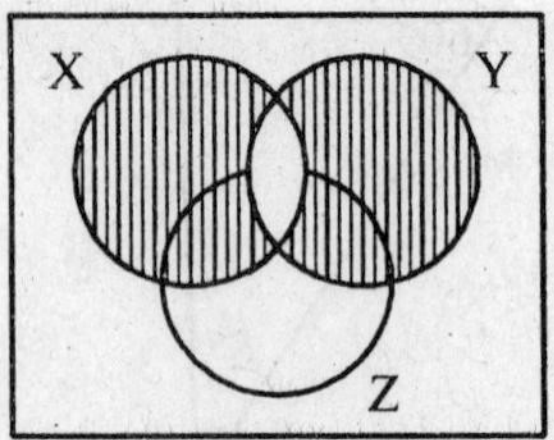

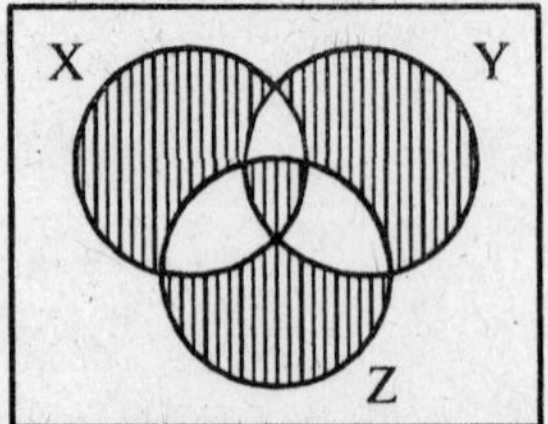

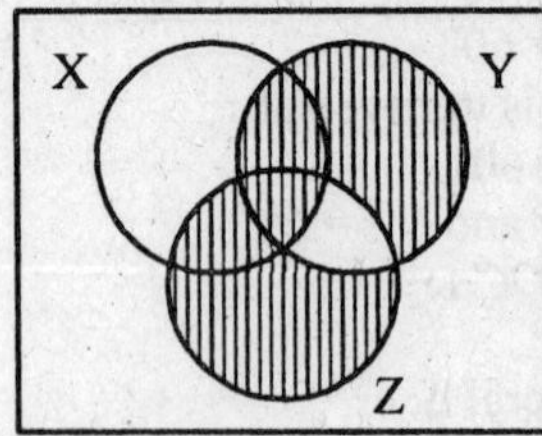

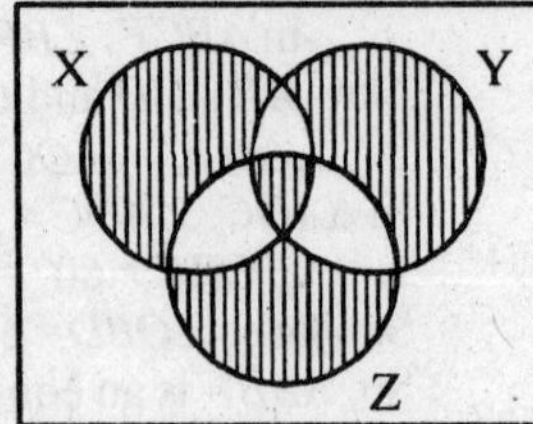

From (1) and (2) $(X \circ Y) \circ Z = X \circ (Y \circ Z)$

10. We are given that

$$\bigcup_{i=1}^{30} A_i = \bigcup_{j=1}^{n} B_j = S \quad ....(1)$$

Each $A_i'^s$ contain 5 elements, so $\bigcup_{i=1}^{30} A_i$ contains $5 \times 30 = 150$ elements (with repetition) out of which each element is repeated 10 times, (as given that each element of S belongs to 10 $A_i'^s$)

$\therefore$ Number of different elements in $\bigcup_{i=1}^{30} A_i$ is $= \frac{150}{10} = 15$

$\therefore$ From eqn (1) we can say S contains 15 elements...(2)

Again each $B_j'^s$ contains 3 elements, so $\bigcup_{j=1}^{n} B_j$ contains $3 \times n = 3n$ elements (with repetition), out of which each element is repeated 9 times (as each element of S belongs to 9 $B_j'^s$)

$\therefore$ No. of different elements in $\bigcup_{j=1}^{n} B_j = \frac{3n}{9} = \frac{n}{3}$

$\therefore$ From eqn (1) we can say S contain $n/3$ elements....(3)

From (2) and (3) we get $\frac{n}{3} = 15 \Rightarrow n = 45$

11. Given that : Mean square deviation for the observations $x_1, x_2, ... x_n$, about a point. c is given by $\frac{1}{n}\sum_{i=1}^{n}(x_i - c)^2$.

Also given that mean square deviations about – 1 and + 1 are 7 and 3 for a particular set of observations.

$$\Rightarrow \frac{1}{n}\sum_{i=1}^{n}(x_i+1)^2 = 7 \text{ and } \frac{1}{n}\sum_{i=1}^{n}(x_i-1)^2 = 3$$

$$\Rightarrow \sum_{i=1}^{n}(x_i^2 + 2x_i + 1) = 7n \text{ and } \sum_{i=1}^{n}(x_i^2 - 2x_i + 1) = 3n$$

$$\Rightarrow \sum x_i^2 + 2\sum x_i + n = 7n \text{ and}$$ **NOTE THIS STEP**

$$\sum x_i^2 - 2\sum x_i + n = 3n$$

$$\Rightarrow \sum x_i^2 + 2\sum x_i = 6n \quad ...(1)$$

$$\text{and } \sum x_i^2 - 2\sum x_i = 2n \quad ...(2)$$

Subtracting (2) from (1), we get

$$4\sum x_i = 4n \Rightarrow \frac{\sum x_i}{n} = 1 \Rightarrow \bar{x} = 1$$

Now standard deviation for same set of observations

$$= \sqrt{\frac{1}{n}\sum_{i=1}^{n}(x_i - \bar{x})^2} = \sqrt{\frac{1}{n}\sum_{i=1}^{n}(x_i - 1)^2} = \sqrt{3}$$

{using the given value}

12. $n = 40$, $\bar{x} = 40$, Var. = 49

$$\frac{\sum f_i x_i}{40} = \bar{x} = 40 \Rightarrow \sum f_i x_i = 1600 \quad ....(1)$$

$$\text{Also Var.} = 49 = \frac{1}{40}\sum f_i(x_i - 40)^2$$

$$\Rightarrow 49 = \frac{1}{40}\sum f_i x_i^2 - 2\sum f_i x_i + 40\sum f_i$$

$$\Rightarrow 49 = \frac{1}{40}\sum f_i x_i^2 - 2 \times 1600 + 40 \times 40$$

$$\Rightarrow \frac{1}{40}\sum f_i x_i^2 = 1649 \quad ...(2)$$

Let 21 – 30 and 31 – 40 denote the kth and $(k+1)^{th}$ class intervals respectively.

Then if before correction f_k and f_{k+1} are the frequencies of these intervals then after correction (2 observations are shifted from 31 – 40 to 21 – 30), frequency of kth intervals becomes $f_k + 2$ and frequency of $(k+1)^{th}$ interval becomes $f_{k+1} - 2$.

Then, we get.

$$\Rightarrow \bar{x}_{new} = \frac{1}{40}\sum_{i=1}^{40} f_i x_i + \frac{2}{40}(x_k - x_{k+1})$$

$$= \frac{1}{40}\sum_{i=1}^{40} f_i x_i + \frac{1}{20}(-10) = 40 - 0.5 = 39.5$$

$$Var_{new} = \frac{1}{40}\left[\sum_{\substack{i=1\\ i\neq k,k+1}}^{40} f_i(x_i-39.5)^2 + f_k(x_k-39.5)^2 + f_{k+1}(x_{k+1}-39.5)^2\right]$$

$$= \frac{1}{40}\left[\sum_{\substack{i=1\\ i\neq k,k+1}}^{40} (f_i x_i^2 - 79 f_i x_i + 39.5)^2 f_i)\right]$$

$$+\frac{1}{40}\left[f_k x_k^2 - 79 f_k x_k + (39.5)^2 f_k + f_{k+1}x_k^2 + 1 - 79 f_{k+1}x_{k+1} + (39.5)^2 f_{k+1}\right]$$

$$= \frac{1}{40}\sum_{i=1}^{40} f_i x_i^2 - 79.\frac{1}{40}\sum_{i=1}^{40} f_i x_i + (39.5)^2 \frac{1}{40}\sum f_i$$

$= 1649 - 3160 + 1560.25 = 49.25$

[Using eqn. (1) and (2)]

13. Given that $z_1 R z_2$ iff $\dfrac{z_1 - z_2}{z_1 + z_2}$ is real.

To show that R is an equivalence relation.

Reflexivity : For $z_1 = z_2 = z$ (say)

$\dfrac{z_1 - z_2}{z_1 + z_2} = \dfrac{z - z}{z + z} = 0$ which is real

$\therefore\ z R z\ \forall\ z$ $\qquad\therefore$ R is reflexive.

Symmetric : Let $z_1 R z_2 \Rightarrow \dfrac{z_1 - z_2}{z_1 + z_2}$ is real

$\Rightarrow -\left(\dfrac{z_1 - z_2}{z_1 + z_2}\right)$ is also real

$\Rightarrow \dfrac{z_2 - z_1}{z_2 + z_1}$ is real $\Rightarrow z_2 R z_1 \Rightarrow$ R is symmetric.

Transitivity : Let $z_1 R z_2$ and $z_2 R z_3$

$\Rightarrow \dfrac{z_1 - z_2}{z_1 + z_2}$ is real and $\dfrac{z_2 - z_3}{z_2 + z_3}$ is also real

Now, $\dfrac{z_1 - z_2}{z_1 + z_2}$ is real $\Rightarrow I_m\left(\dfrac{z_1 - z_2}{z_1 + z_2}\right) = 0$

$\Rightarrow I_m\left(\dfrac{(x_1 - x_2) + i(y_1 - y_2)}{(x_1 + x_2) + i(y_1 + y_2)}\right) = 0$

$\Rightarrow I^m((x_1 - x_2) + i(y_1 - y_2))((x_1 + x_2) - i(y_1 + y_2)) = 0$

$\Rightarrow (x_1 + x_2)(y_1 - y_2) - (x_1 - x_2)(y_1 + y_2) = 0$

$\Rightarrow x_2 y_1 - x_1 y_2 = 0$

$\Rightarrow \dfrac{x_1}{y_1} = \dfrac{x_2}{y_2}$ ...(1)

Similarly, $I_m\left(\dfrac{z_2 - z_3}{z_2 + z_3}\right) = 0 \Rightarrow \dfrac{x_2}{y_2} = \dfrac{x_3}{y_3}$ ...(2)

From (1) and (2) we get $\dfrac{x_1}{y_1} = \dfrac{x_3}{y_3}$

$\Rightarrow I_m\left(\dfrac{z_1 - z_3}{z_1 + z_3}\right) = 0 \Rightarrow \dfrac{z_1 - z_3}{z_1 + z_3}$ is real

$\Rightarrow z_1 R z_3 \therefore R$ is transitive.

Thus R is reflexive, symmetric and transitive.

Hence R is an equivalence relation.

F MATCH THE FOLLOWING :

1. **A → (s), B → (t), C → (r) D → (r)**

(A) As $|z| = 1$ and $z \neq \pm 1$

$\therefore$ Let $z = e^{i\theta}$ where $\theta \neq n\pi$

Then $\text{Re}\left(\dfrac{2iz}{1 - z^2}\right) = \text{Re}\left[\dfrac{2i(\cos\theta + i\sin\theta)}{1 - \cos 2\theta - i\sin 2\theta}\right]$

$= \text{Re}\left[\dfrac{i(\cos\theta + i\sin\theta)}{\sin\theta(\cos(\frac{\pi}{2} - \theta) - i\sin(\frac{\pi}{2} - \theta)}\right]$

$= \text{Re}\left(\dfrac{i}{\sin\theta}\left[\cos\dfrac{\pi}{2} + i\sin\dfrac{\pi}{2}\right]\right) = \text{Re}\left(\dfrac{-1}{\sin\theta}\right) = \text{Re}(-\text{cosec}\theta)$

as $+\ \text{cosec}\theta \leq -1$ or $\text{cosec}\theta \geq 1$

$\therefore -\text{cosec}\theta \geq 1$ or $-\text{cosec}\theta \leq -1$

$\therefore$ Required set is $(-\infty, -1] \cup [1, \infty)$

$\therefore$ A → s

Alternative method

Let $z = x + iy$ where and $x^2 + y^2 = 1$ and $x \neq \pm 1$

Then $\text{Re}\left(\dfrac{2iz}{1 - z^2}\right) = \text{Re}\left(\dfrac{2i(x + iy)}{1 - (x^2 - y^2 + 2ixy)}\right)$

$= \text{Re}\left(\dfrac{-2y + 2ix}{1 - x^2 + y^2 - 2ixy}\right) = \text{Re}\left(\dfrac{-2y + 2ix}{2y(y - ix)}\right)$

$= \text{Re}\left(\dfrac{-1}{y}\right) = \dfrac{-1}{y}$

where $-1 \leq y \leq | \Rightarrow \dfrac{-1}{y} \geq 1$ of $\dfrac{-1}{y} \leq -1$

$\therefore \text{Re}\left(\dfrac{2iz}{1 - z^2}\right) \in (-\infty, -1] \cup [1, \infty)$

(B) For the domain of $f(x) = \sin^{-1}\left(\dfrac{8(3)^{x-2}}{1 - 3^{2(x-1)}}\right)$

We should have

$-1 \leq \left(\dfrac{8(3)^{x-2}}{1 - 3^{2(x-1)}}\right) \leq 1 \Rightarrow -1 \leq \dfrac{8.3^x}{9 - 3^{2x}} \leq 1$

Let $3^x = t$ then $-1 \leq \dfrac{8t}{9 - t^2} \leq 1$

$\Rightarrow \quad \frac{8t}{9-t^2} \geq -1$ and $\frac{8t}{9-t^2} \leq 1$

$\Rightarrow \quad \frac{8t+9-t^2}{9-t^2} \geq 0$ and $\frac{8t-9+t^2}{9-t^2} \leq 0$

$\Rightarrow \quad \frac{t^2-8t-9}{t^2-9} \geq 0$ and $\frac{t^2+8t-9}{t^2-9} \geq 0$

$\Rightarrow \quad \frac{(t-9)(t+1)}{(t-3)(t+3)} \geq 0$ and $\frac{(t+9)(t-1)}{(t+3)(t-3)} \geq 0$

Also $t = 3^x$ can not be – ve

$\therefore \; t \in (0,3) \cup [9,\infty)$ and $t \in (0,1] \cup [3,\infty)$

$\Rightarrow x \in (-\infty,1) \cup [2,\infty)$ and $x \in (-\infty,0] \cup (1,\infty)$

Combining the two, we get

$x \in (-\infty,0] \cup [2,\infty)$

$\therefore$ **B → t**

(C) $f(\theta) = \begin{vmatrix} 1 & \tan\theta & 1 \\ -\tan\theta & 1 & \tan\theta \\ -1 & -\tan\theta & 1 \end{vmatrix}$

$R_1 + R_3$

$= \begin{vmatrix} 0 & 0 & 2 \\ -\tan\theta & 1 & \tan\theta \\ -1 & -\tan\theta & 1 \end{vmatrix}$

$= 2(1+\tan^2\theta) = 2\sec^2\theta \in [2,\infty)$ for $0 \leq \theta < \frac{\pi}{2}$

$\therefore$ **C → r**

(D) $f(x) = x^{3/2}(3x-10), x \geq 0$

$\therefore f'(x) = \frac{3}{2}\sqrt{x}(3x-10) + 3x\sqrt{x}$

For f(x) to be increasing $f'(x) \geq 0$

$\Rightarrow 3\sqrt{x}[3x-10+2x] \geq 0$

$\Rightarrow \sqrt{x}(5x-10) \geq 0$ but $x \geq 0 \Rightarrow x \geq 2$

$\therefore f(x)$ is incresing on $[2,\infty)$

$\therefore$ **D → r.**

(I) INTEGER VALUE CORRECT TYPE :

1. (4) Let $\sqrt{4-\frac{1}{3\sqrt{2}}\sqrt{4-\frac{1}{3\sqrt{2}}\sqrt{4-\frac{1}{3\sqrt{2}}\ldots\ldots}}} = y$

$\Rightarrow 4-\frac{1}{3\sqrt{2}}\sqrt{4-\frac{1}{3\sqrt{2}}\sqrt{4-\frac{1}{3\sqrt{2}}\ldots\ldots}} = y^2$

$\Rightarrow 4-\frac{y}{3\sqrt{2}} = y^2 \Rightarrow y^2 + \frac{y}{3\sqrt{2}} - 4 = 0$

$\Rightarrow 3\sqrt{2}y^2 + y - 12\sqrt{2} = 0$

$\Rightarrow y = \frac{-1+\sqrt{1+288}}{6\sqrt{2}}$

(rejecting -ve value as y is a +ve square)

$\Rightarrow y = \frac{-1+17}{6\sqrt{2}} = \frac{8}{3\sqrt{2}}$

$\therefore$ The required value is

$6 + \log_{3/2}\left(\frac{1}{3\sqrt{2}} \times \frac{8}{3\sqrt{2}}\right) = 6 + \log_{3/2}\frac{4}{9}$

$= 6 + \frac{2\log_{2/3}}{\log_{3/2}} = 6 - 2 = 4$

Section-B JEE Main/ AIEEE

1. (b) Total student = 100;

for 70 stds. total marks = 75 × 70 = 5250

$\Rightarrow$ Total marks of girls = 7200 – 5250 = 1950

Average of girls = $\frac{1950}{30} = 65$

2. (a) Given $P + Q = 18$ ....(1)

$P^2 + Q^2 + 2PQ\cos\alpha = 144$ ......(2)

$\tan 90° = \frac{Q\sin\alpha}{P+Q\cos\alpha}$

$\Rightarrow \quad P + Q\cos\alpha = 0$ ......(3)

From (2) and (3),

$Q^2 - P^2 = 144 \Rightarrow (Q-P)(Q+P) = 144$

$\therefore \quad Q - P = \frac{144}{18} = 8$

From (1), On solving, we get $Q = 13, P = 5$

3. (b) $\angle TQW = 180 - \theta$; $\angle RQW = 2\theta$; $\angle RQT = 180 - \theta$

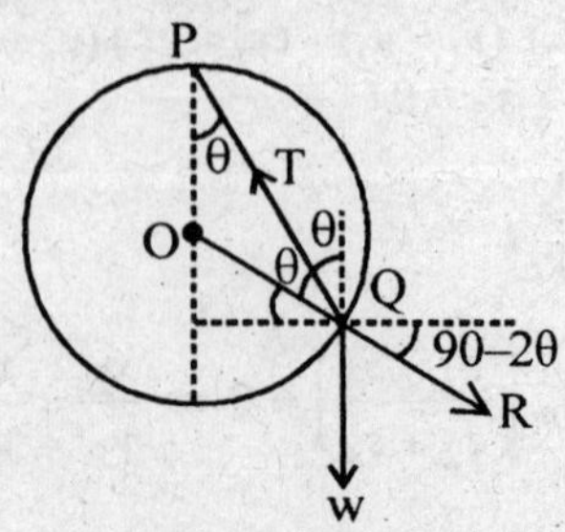

Applying Lami's theorem at Q.

$$\frac{T}{\sin 2\theta}=\frac{R}{\sin(180-\theta)}=\frac{W}{\sin(180-\theta)}$$

$\Rightarrow$ $R = W$ and $T = 2W\cos\theta$

4. (a) $n = 9$ then median term $=\left(\frac{9+1}{2}\right)^{th} = 5^{th}$ term. Last four observations are increased by 2. The median is 5th observation which is remaining unchanged.

$\therefore$ there will be no change in median.

5. (c) $\vec{G}=\vec{r}\times\vec{p}$; $|\vec{G}| = rp\sin\theta$

$|\vec{H}| = rp\cos\theta \left[\because \sin(90^o+\theta)=\cos\theta\right]$

$G = rp\sin\theta$.......(1) $H = rp\cos\theta$............(2)

$x = rp\sin(\theta+\alpha)$......(3)

From (1), (2) & (3), $x = \vec{G}\cos\alpha + \vec{H}\sin\alpha$.

6. (c) $R^2 = P^2+Q^2+2PQ\cos\theta$ ....(1)

$4R^2 = P^2+4Q^2+4PQ\cos\theta$ ....(2)

$4R^2 = P^2+Q^2-2PQ\cos\theta$ ....(3)

On (1) + (3), $5R^2 = 2P^2+2Q^2$ ....(4)

On (3) × 2 + (2), $12R^2 = 3P^2+6Q^2$ .....(5)

$2P^2+2Q^2-5R^2=0$ .....(6)

$3P^2+6Q^2-12R^2=0$ .....(7)

$$\frac{P^2}{-24+30}=\frac{Q^2}{24-15}=\frac{R^2}{12-6}$$

$$\frac{P^2}{6}=\frac{Q^2}{9}=\frac{R^2}{6} \text{ or } P^2:Q^2:R^2 = 2:3:2$$

7. (a) Let the body travels from A to B with constant acceleration t and from B to C with constant retardation r.

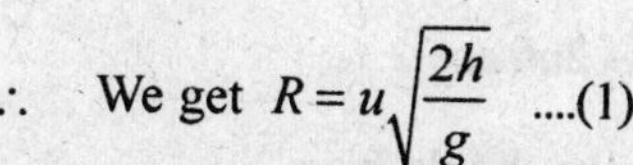

If $AB = x$, $BC = y$, time taken from A to $B = t_1$ and time taken from B to $C = t_2$, then $s = x + y$ and $t = t_1 + t_2$

For the motion from A to B

$$v^2 = u^2+2fs \Rightarrow v^2 = 2fx \ (\because u=0)$$

$\Rightarrow$ $x = \frac{v^2}{2f}$ ....(1)

and $v = u + ft$ $\Rightarrow v = ft_1$

$\Rightarrow$ $t_1 = \frac{v}{f}$ ...(2)

For the motion from B to C

$$v^2 = u^2+2fs$$

$\Rightarrow$ $0 = v^2 - 2ry$

$\Rightarrow$ $y = \frac{v^2}{2r}$ ...(3)

and $v = u + ft \Rightarrow 0 = v - rt_2$

$\Rightarrow$ $t_2 = \frac{v}{r}$

Adding equations (1) and (3), we get

$$x+y=\frac{v^2}{2}\left[\frac{1}{f}+\frac{1}{r}\right]=s$$

Adding equations (2) and (4), we get

$$t_1+t_2 = v\left[\frac{1}{f}+\frac{1}{r}\right]=t$$

$$\therefore \quad \frac{t^2}{2s}=\frac{v^2\left[\frac{1}{f}+\frac{1}{r}\right]^2}{2\times\frac{v^2}{2}\left(\frac{1}{f}+\frac{1}{r}\right)}=\frac{1}{f}+\frac{1}{r}$$

$$\Rightarrow \quad t=\sqrt{2s\left(\frac{1}{f}+\frac{1}{r}\right)}$$

8. (a) For the stone projected horizontally, for horizontal motion, using distance

= speed × time $\Rightarrow R = ut$

and for vertical motion

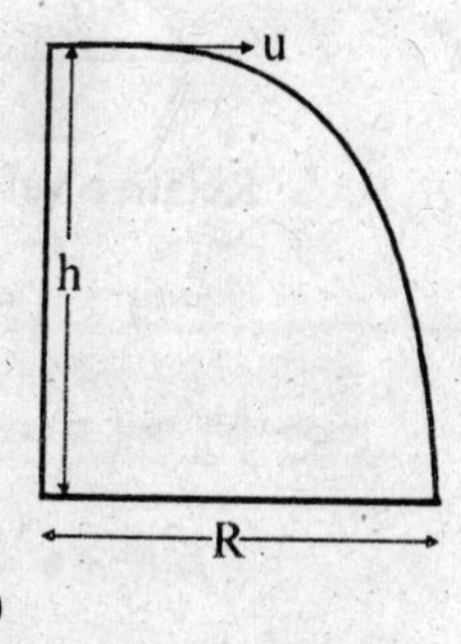

$$h = 0\times t + \frac{1}{2}gt^2$$

$$\Rightarrow t=\sqrt{\frac{2h}{g}}$$

$\therefore$ We get $R = u\sqrt{\frac{2h}{g}}$ ....(1)

For the stone projected at an angle θ, for horizontal and vertical motions, we have

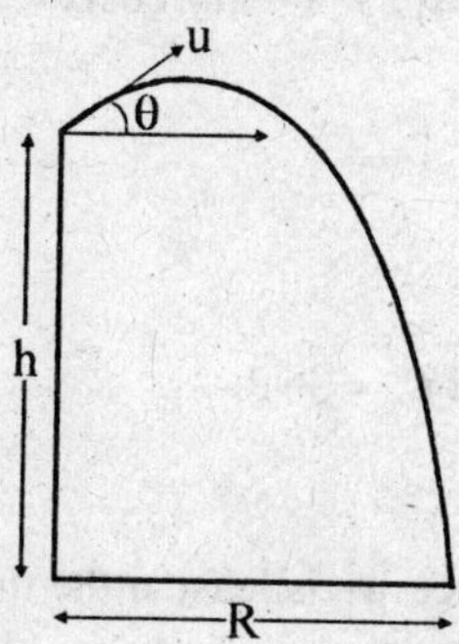

$R = u\cos\theta \times t$ ....(2)

and $h = -u\sin\theta \times t + \frac{1}{2}gt^2$ ....(3)

From (1) and (2) we get

$\Rightarrow u\sqrt{\frac{2h}{g}} = u\cos\theta \times t \Rightarrow t = \frac{1}{\cos\theta}\sqrt{\frac{2h}{g}}$

Substituting this value of t in eq (3) we get

$$h = -\frac{u\sin\theta}{\cos\theta}\sqrt{\frac{2h}{g}} + \frac{1}{2}g\left[\frac{2h}{g\cos^2\theta}\right]$$

$$h = -u\sqrt{\frac{2h}{g}}\tan\theta + h\sec^2\theta$$

$$h = -u\sqrt{\frac{2h}{g}}\tan\theta + h\tan^2\theta + h$$

$$\tan^2\theta - u\sqrt{\frac{2}{hg}}\tan\theta = 0; \therefore \tan\theta = u\sqrt{\frac{2}{hg}}$$

9. (a) We can consider the two velocities as $\vec{v}_1 = u\hat{i}$

and $\vec{v}_2 = (ft\cos\alpha)\hat{i} + (ft\sin\alpha)\hat{j}$

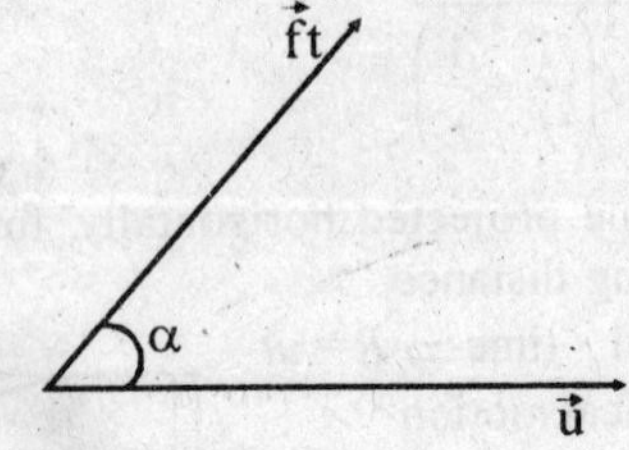

$\therefore$ Relative velocity of second with respect to first

$\vec{v} = \vec{v}_2 - \vec{v}_1 = (ft\cos\alpha - u)\hat{i} + ft\sin\alpha\ \hat{j}$

$\Rightarrow |\vec{v}|^2 = (ft\cos\alpha - u)^2 + (ft\sin\alpha)^2$

$= f^2t^2 + u^2 - 2uft\cos\alpha$

For $|\vec{v}|$ to be min we should have

$$\frac{d|v|^2}{dt} = 0 \Rightarrow 2f^2t - 2uf\cos\alpha = 0$$

$$\Rightarrow t = \frac{u\cos\alpha}{f}$$

Also $\frac{d^2|v|^2}{dt^2} = 2f^2 = +ve$

$\therefore |v|^2$ and hence $|v|$ is least at the time $\frac{u\cos\alpha}{f}$

10. (c)

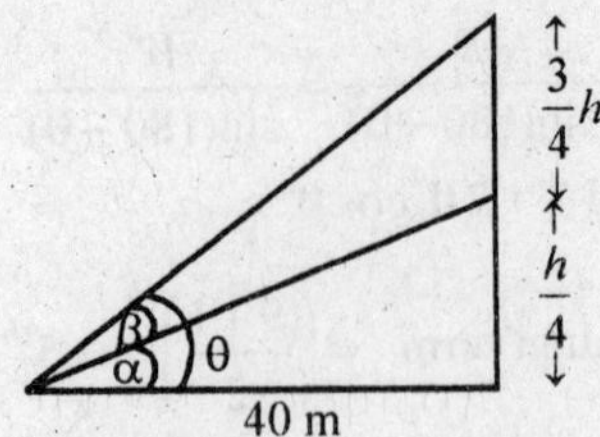

$\theta = \alpha + \beta, \beta = \tan^{-1}\left(\frac{3}{5}\right)$

or $\beta = \theta - \alpha$

$$\Rightarrow \tan\beta = \frac{\tan\theta - \tan\alpha}{1 + \tan\theta.\tan\alpha} \text{ or } \frac{3}{5} = \frac{\frac{h}{40} - \frac{h}{160}}{1 + \frac{h}{40}.\frac{h}{160}}$$

$h^2 - 200h + 6400 = 0 \Rightarrow h = 40$ or 160 metre

$\therefore$ possible height $= 40$ metre

11. (a) Let β be the inclination of the plane to the horizontal and u be the velocity of projection of the projectile

$$R_1 = \frac{u^2}{g(1+\sin\beta)} \text{ and } R_2 = \frac{u^2}{g(1-\sin\beta)}$$

$$\frac{1}{R_1} + \frac{1}{R_2} = \frac{2g}{u^2} \text{ or } \frac{1}{R_1} + \frac{1}{R_2} = \frac{2}{R}\left[\because R = \frac{u^2}{g}\right]$$

$\therefore R_1, R, R_2$ *are in H.P.*

12. (b) $\Sigma x = 170, \Sigma x^2 = 2830$ increase in $\Sigma x = 10$, then

$\Sigma x' = 170 + 10 = 180$

Increase in $\Sigma x^2 = 900 - 400 = 500$ then

$\Sigma x'^2 = 2830 + 500 = 3330$

Variance $= \frac{1}{n}\Sigma x'^2 - \left(\frac{1}{n}\Sigma x'\right)^2$

$= \frac{1}{15} \times 3330 - \left(\frac{1}{15} \times 180\right)^2 = 222 - 144 = 78.$

13. (c) $\because (1, 1) \notin R \Rightarrow R$ is not reflexive $(2,3) \in R$ but $(3,2) \notin R$

$\therefore$ R is not symmetric

14. (c) Only first (A) and second (B) statements are correct.

15. (c) Clearly mean $A = 0$

Standard deviation $\sigma = \sqrt{\frac{\sum(x-A)^2}{2n}}$

$$2 = \sqrt{\frac{(a-0)^2 + (a-0)^2 + ...(0-a)^2 + ...}{2n}} = \sqrt{\frac{a^2.2n}{2n}} = |a|$$

Hence $|a| = 2$

16. (c) Let forces be P and Q. then $P+Q=4$ ....(1)

and $P^2+Q^2=3^2$ ....(2)

Solving we get the forces

$$\left(2+\frac{\sqrt{2}}{2}\right)N \text{ and } \left(2-\frac{\sqrt{2}}{2}\right)N$$

17. (c) Since, the moment about A is zero, hence $\vec{F}$ passes through A. Taking A as origin. Let the line of action of force $\vec{F}$ be $y = mx$.(see *figure*)

Moment about $B = \frac{3m}{\sqrt{1+m^2}}|\vec{F}|=9$ ....(1)

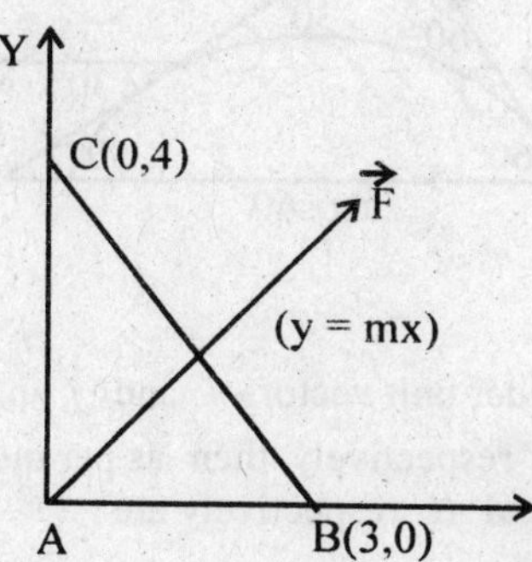

Moment about $C = \frac{4}{\sqrt{1+m^2}}|\vec{F}|=16$ ....(2)

Dividing (1) by (2), we get $m=\frac{3}{4} \Rightarrow |\vec{F}|=5N$.

18. (d) IA, IB, IC are bisectors of the angles A, B and C as I is incentre of ΔABC.

Now $\angle BIC = 180 - \frac{B}{2} - \frac{C}{2} = 90° + \frac{A}{2}$ *etc.*

Applying Lami's theorem at I

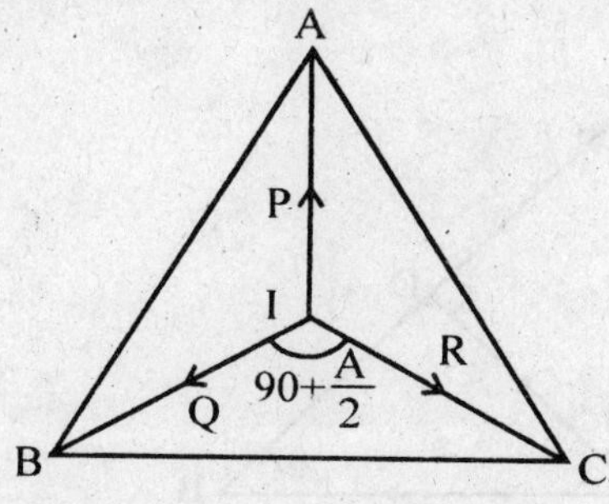

$$\frac{P}{Sin(90°+\frac{A}{2})} = \frac{Q}{\sin\left(90°+\frac{B}{2}\right)} = \frac{R}{\sin\left(90°+\frac{C}{2}\right)}$$

$$\Rightarrow \quad P:Q:R = \cos\frac{A}{2}:\cos\frac{B}{2}:\cos\frac{C}{2}$$

19. (d) Time taken by the particle in complete journey

$$T = \frac{12}{4}+\frac{5}{5} = 4\ hr.$$

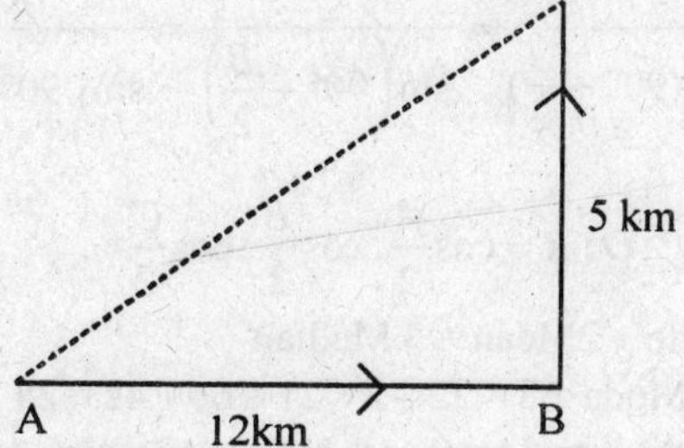

$\therefore$ Average speed $= \frac{12+5}{4} = \frac{17}{4}$

Average velocity $= \sqrt{\frac{12^2+5^2}{4}} = \frac{13}{4}$

[using vector addition]

20. (a) If $v = \frac{1}{4}$, component along OB

$$= \frac{v\sin 30°}{\sin(45°+30°)} = \frac{\frac{1}{4}\times\frac{1}{2}}{\frac{\sqrt{3}+1}{2\sqrt{2}}} = \frac{\sqrt{6}-\sqrt{2}}{8}$$

21. (b) For same horizontal range the angles of projection must be α and $\frac{\pi}{2}-\alpha$

$$\therefore t_1 = \frac{2u\sin\alpha}{g} \text{ and } t_2 = \frac{2u\sin\left(\frac{\pi}{2}-\alpha\right)}{g} = \frac{2u\cos\alpha}{g}$$

$$\therefore t_1^2+t_2^2 = \frac{4u^2}{g^2}$$

22. (a) Reflexive and transitive only.

e.g. (3, 3), (6, 6), (9, 9), (12, 12) [Reflexive]
(3, 6), (6, 12), (3, 12) [Transitive].

$(3, 6) \in R$ but $(6, 3) \notin R$ [non symmetric]

23. (c) IA, IB, IC are bisectors of the angles A, B and C as I is incentre of ΔABC.

Now $\angle BIC = 180 - \frac{B}{2} - \frac{C}{2} = 90° + \frac{A}{2}$ etc.

Applying Lami's theorem at I

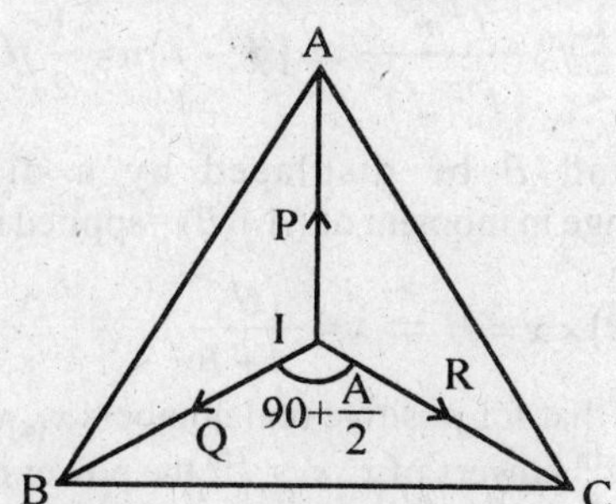

$$\frac{P}{Sin(90°+\frac{A}{2})}=\frac{Q}{\sin\left(90°+\frac{B}{2}\right)}=\frac{R}{\sin\left(90°+\frac{C}{2}\right)}$$

$$\Rightarrow P:Q:R=\cos\frac{A}{2}:\cos\frac{B}{2}:\cos\frac{C}{2}$$

24. (d) Mode + 2Mean = 3 Median

$\Rightarrow$ Mode $=3\times 22-2\times 21=66-42=24$.

25. (c) Let the lizard catches the insect after time t then distance covered by lizard = 21cm + distance covered by insect

$$\Rightarrow \frac{1}{2}ft^2=4\times t+21$$

$$\Rightarrow \frac{1}{2}\times 2\times t^2=20\times t+21$$

$$\Rightarrow t^2-20t-21=0$$

$$\Rightarrow t=21\text{ sec}$$

26. (d)

A —(u=0, f)— t+m — s+n —→ v

B —(u=0, f')— t — s —→ v

As per question if point *B* moves s distance in *t* time then point *A* moves $(s+n)$ distance in time $(t+m)$ after which both have same velocity *v*.

Then using equation $v=u+at$ we get

$$v=f(t+m)=f't\Rightarrow t=\frac{f\,m}{f'-f} \quad ....(1)$$

Using equation $v^2=u^2+2$, as we get

$$v^2=2f(s+n)=2f's \quad \Rightarrow s=\frac{f\,n}{f'-f} \quad ....(2)$$

Also for point B using the eqn $s=ut+\frac{1}{2}at^2$, we get

$$s=\frac{1}{2}f't^2$$

Substituting values of t and s from equations (1) and (2) in the above relation, we get

$$\frac{f\,n}{f'-f}=\frac{1}{2}f'\frac{f^2m^2}{(f'-f)^2}\Rightarrow (f'-f)n=\frac{1}{2}ff'm^2$$

27. (b) Let *A* and *B* be displaced by a distance *x* then Change in moment of $(A+B)$ =applied moments

$$\Rightarrow (A+B)\times x=H\Rightarrow x=\frac{H}{A+B}$$

28. (b) We know that for positive real numbers $x_1, x_2, ..., x_n$, A.M. of kth powers of $x'_i s \geq k^{th}$ the power of A.M. of $x'_i s$

$$\Rightarrow \frac{\sum x_1^2}{n}\geq\left(\frac{\sum x_1}{n}\right)^2 \Rightarrow \frac{400}{n}\geq\left(\frac{80}{n}\right)^2$$

$\Rightarrow n\geq 16$. So only possible value for $n=18$

29. (d) $u\cos 60°=v\cos 30°$

(as horizontal component of velocity remains the same)

$$\Rightarrow u\cdot\frac{1}{2}=v\cdot\frac{\sqrt{3}}{2} \quad\text{or}\quad v=\frac{1}{\sqrt{3}}u$$

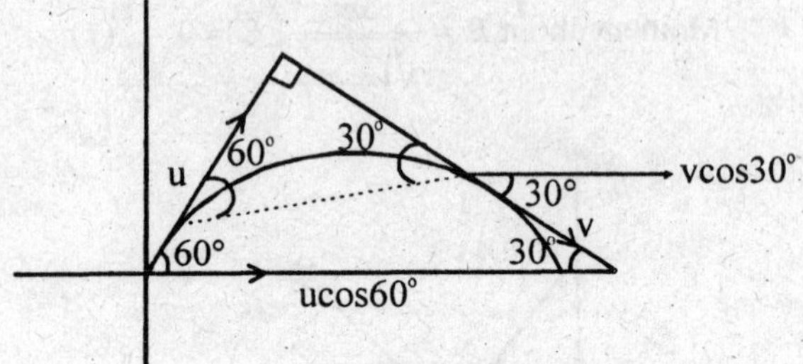

30. (d) If we consider unit vectors $\hat{i}$ and $\hat{j}$ in the direction *AB* and *AC* respectively, then as per quesiton, forces along *AB* and *AC* respectively are

$$\left(\frac{1}{AB}\right)\hat{i} \text{ and } \left(\frac{1}{AC}\right)\hat{j}$$

$\therefore$ Their resultant along $AD=\left(\frac{1}{AB}\right)i+\left(\frac{1}{AC}\right)j$

$\therefore$ Magnitude of resultant is

$$=\sqrt{\left(\frac{1}{AB}\right)^2+\left(\frac{1}{AC}\right)^2}=\sqrt{\frac{AC^2+AB^2}{AB^2+AC^2}}$$

$$=\frac{BC}{AB.AC}$$

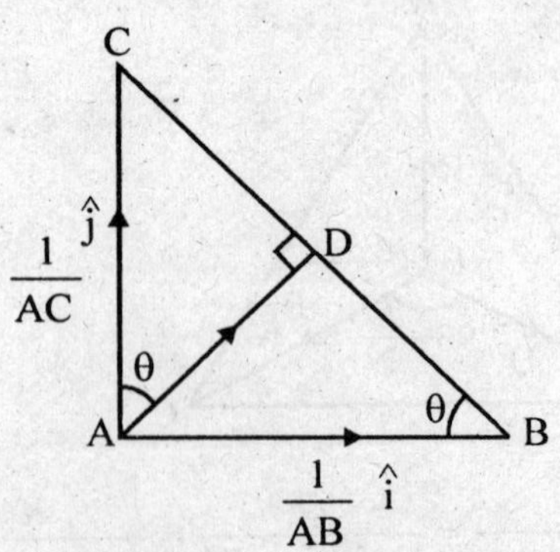

But from figure $\Delta ABC\sim\Delta DBA$

$$\Rightarrow \frac{BC}{AB}=\frac{AC}{AD}\Rightarrow\frac{BC}{AB\times AC}=\frac{1}{AD}$$

$\therefore$ The required magnitude of resultant becomes $\frac{1}{AD}$.

31. **(b)** Clearly $(x, x) \in R \forall x \in W$. So R is relexive.

Let $(x, y) \in R$, then $(y, x) \in R$ as x and y have at least one letter in common. So, R is symmetric.

But R is not transitive for example

Let x = INDIA, y = BOMBAY and z = JOKER

then $(x, y) \in R$ (A is common) and $(y, z) \in R$ (O is common) but $(x, z) \notin R$. (as no letter is common)

32. **(a)** 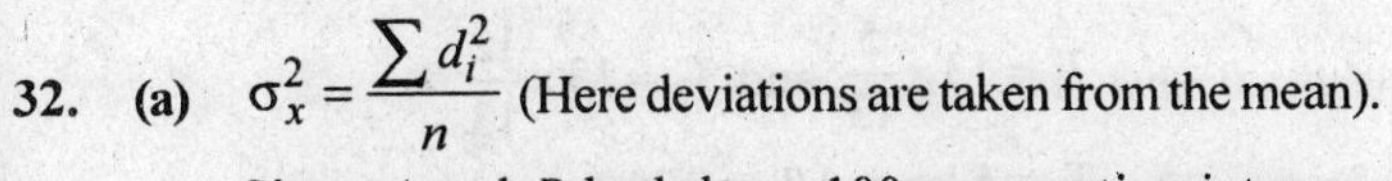

$\sigma_x^2 = \frac{\sum d_i^2}{n}$ (Here deviations are taken from the mean).

Since A and B both have 100 consecutive integers, therefore both have same standard deviation and hence the variance. $\therefore \frac{V_A}{V_B} = 1$

(As $\sum d_i^2$ is same in both the cases)

33. **(b)** For two velocities u and u at an angle θ to each other the resultant is given by

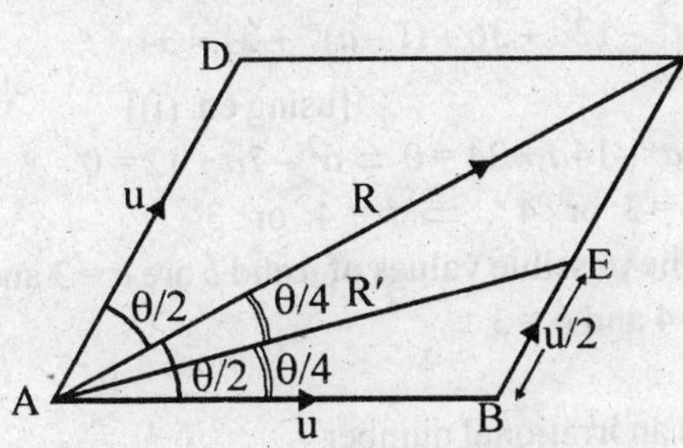

$$R^2 = u^2 + u^2 + 2u^2 \cos\theta = 2u^2 (1 + \cos\theta)$$

$$\Rightarrow R^2 = 4u^2 \cos^2 \theta/2 \text{ or } R = 2u \cos\frac{\theta}{2}$$

Now in second case, the new resultant AE (i.e., R') bisects $\angle CAB$, therefore using angle bisector theorem in ΔABC, we get

$$\frac{AB}{AC} = \frac{BE}{EC} \Rightarrow \frac{u}{R} = \frac{u/2}{u/2} \Rightarrow R = u \Rightarrow 2u\cos\frac{\theta}{2} = u$$

$$\Rightarrow \cos\frac{\theta}{2} = \frac{1}{2} = \cos 60° \Rightarrow \frac{\theta}{2} = 60° \text{ or } \theta = 120°$$

ALTERNATE SOLUTION

$$\tan\frac{\theta}{4} = \frac{\frac{u}{2}\sin\theta}{u + \frac{u}{2}\cos\theta}$$

$$\Rightarrow \sin\frac{\theta}{4} + \frac{1}{2}\sin\frac{\theta}{4}\cos\theta = \frac{1}{2}\sin\theta\cos\frac{\theta}{4}$$

$$\therefore 2\sin\frac{\theta}{4} = \sin\frac{3\theta}{4} = 3\sin\frac{\theta}{4} - 4\sin^3\frac{\theta}{4}$$

$$\therefore \sin^2\frac{\theta}{4} = \frac{1}{4} \Rightarrow \frac{\theta}{4} = 30° \text{ or } \theta = 120°$$

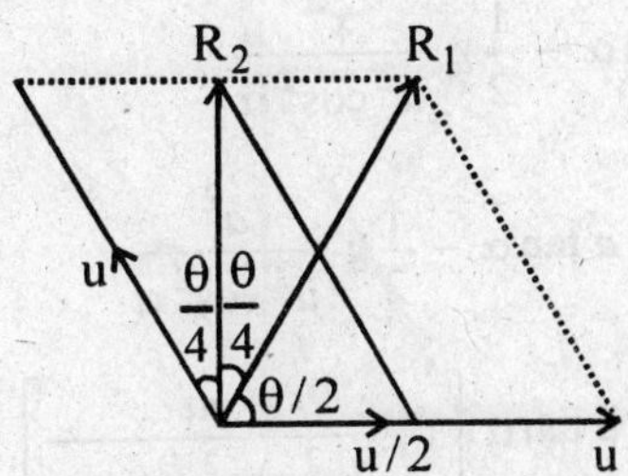

34. **(a)** Using $h = \frac{1}{2}gt^2$ and $h + 400 = \frac{1}{2}g(t + 4)^2$

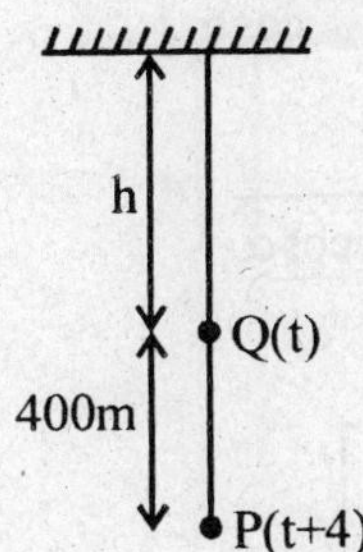

Subtracting, we get $400 = 8g + 4gt$

$\Rightarrow t = 8$ sec

$\therefore h = \frac{1}{2} \times 10 \times 64 = 320m$

$\therefore$ Desired height $= 320 + 400 = 720\ m$

35. **(c)** Given : Force $P = Pn$, $Q = 3n$, resultant $R = 7n$ & $P' = Pn$, $Q' = (-3)n$, $R' = \sqrt{19}\ n$

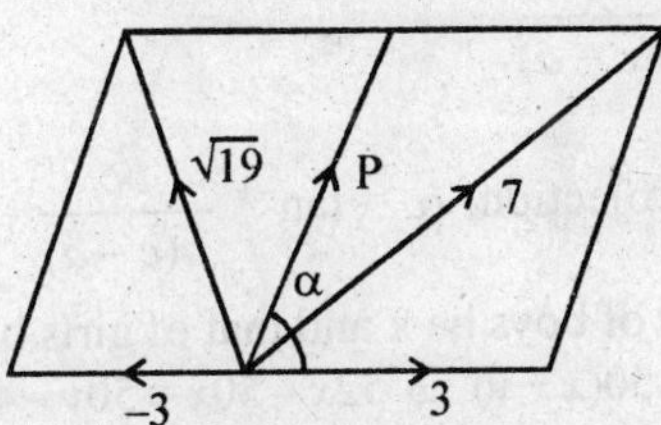

We know that $R^2 = P^2 + Q^2 + 2PQ\cos\alpha$

$\Rightarrow (7)^2 = P^2 + (3)^2 + 2 \times P \times 3\cos\alpha$

$\Rightarrow 49 = P^2 + 9 + 6P\cos\alpha$

$\Rightarrow 40 = P^2 + 6P\cos\alpha$ .....(i)

and $\left(\sqrt{19}\right)^2 = P^2 + (-3)^2 + 2P \times -3\cos\alpha$

$\Rightarrow 19 = P^2 + 9 - 6P\cos\alpha$

$\Rightarrow 10 = P^2 - 6P\cos\alpha$ .....(ii)

Adding (i) and (ii) $50 = 2P^2$

$\Rightarrow P^2 = 25 \Rightarrow P = 5n$.

36. **(a)** Let B be the top of the wall whose coordinates will be (a, b). Range $(R) = c$

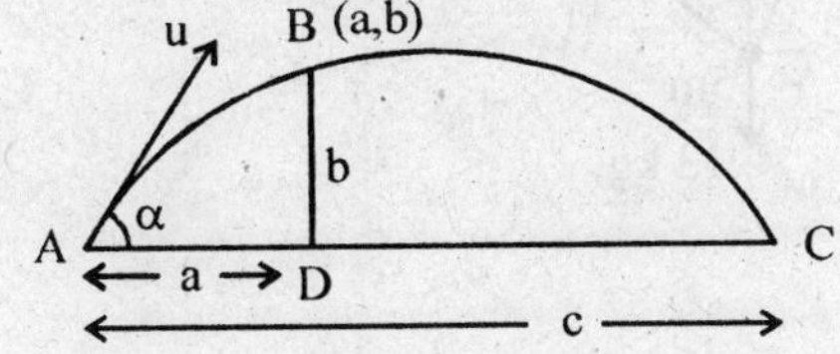

B lies on the trajectory

$$y = x\tan\alpha - \frac{1}{2}g\frac{x^2}{u^2\cos^2\alpha}$$

$$\Rightarrow \quad b = a\tan\alpha - \frac{1}{2}g\frac{a^2}{u^2\cos^2\alpha}$$

$$\Rightarrow \quad b = a\tan\alpha\left[1 - \frac{ga}{2u^2\cos^2\alpha\tan\alpha}\right]$$

$$= a\tan\alpha\left[1 - \frac{a}{\frac{2u^2}{g}\cos^2\alpha.\frac{\sin\alpha}{\cos\alpha}}\right]$$

$$= a\tan\alpha\left[1 - \frac{a}{\frac{u^2.2\sin\alpha\cos\alpha}{g}}\right]$$

$$= a\tan\alpha\left[1 - \frac{a}{\frac{u^2\sin 2\alpha}{g}}\right]$$

$$= a\tan\alpha\left[1 - \frac{a}{R}\right] \qquad \left(\because R = \frac{u^2\sin^2\alpha}{g}\right)$$

$$\Rightarrow \quad b = a\tan\alpha\left[1 - \frac{a}{c}\right] \Rightarrow b = a\tan\alpha.\left(\frac{c-a}{c}\right)$$

$$\Rightarrow \quad \tan\alpha = \frac{bc}{a(c-a)}$$

The angle of projection, $\alpha = \tan^{-1}\dfrac{bc}{a(c-a)}$

37. **(a)** Let the number of boys be x and that of girls be y.

$\Rightarrow 52x + 42y = 50(x+y) \Rightarrow 52x - 50x = 50y - 42y$

$\Rightarrow 2x = 8y \Rightarrow \dfrac{x}{y} = \dfrac{4}{1}$ and $\dfrac{x}{x+y} = \dfrac{4}{5}$

Required % of boys $= \dfrac{x}{x+y}\times 100 = \dfrac{4}{5}\times 100 = 80\ \%$

38. **(a)**

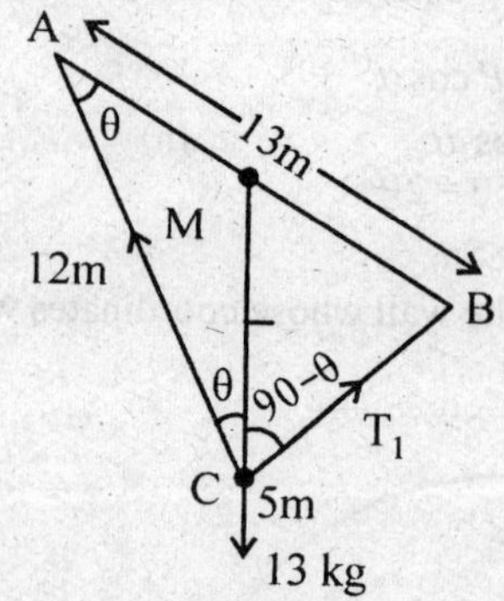

$\because\ 13^2 = 5^2 + 12^2 \Rightarrow AB^2 = AC^2 + BC^2$

$\Rightarrow \angle ACB = 90°$

$Q\ m$ is mid point of the hypotenuse AB, therefore $MA = MB = MC \Rightarrow \angle A = \angle ACM = \theta$

Applying Lami's theorem at C, we get

$$\frac{T_1}{\sin(180-\theta)} = \frac{T_2}{\sin(90+\theta)} = \frac{13kg}{\sin 90°}$$

$\Rightarrow \quad T_1 = 13\sin\theta$ and $T_2 = 13\cos\theta$

$\Rightarrow \quad T_1 = 13\times\dfrac{5}{13}$ and $T_2 = 13\times\dfrac{12}{13}$

$\Rightarrow \quad T_1 = 5$ kg and $T_2 = 12$ kg

39. **(d)** Mean of $a, b, 8, 5, 10$ is 6

$$\Rightarrow \quad \frac{a+b+8+5+10}{5} = 10 \Rightarrow a+b = 7 \quad ...(i)$$

Variance of $a, b, 8, 5, 10$ is 6.80

$$\Rightarrow \quad \frac{(a-6)^2 + (b-6)^2 + (8-6)^2 + (5-6)^2 + (10-6)^2}{5} = 6.80$$

$$\Rightarrow \quad a^2 - 12a + 36 + (1-a)^2 + 21 = 34$$

[using eq. (i)]

$\Rightarrow \quad 2a^2 - 14a + 24 = 0 \Rightarrow a^2 - 7a + 12 = 0$

$\Rightarrow \quad a = 3$ or $4 \quad \Rightarrow b = 4$ or 3

$\therefore$ The possible values of a and b are $a = 3$ and $b = 4$ or, $a = 4$ and $b = 3$

40. **(None)**

p : x is an irrational number

q : y is a transcendental number

r : x is a rational number iff y is a transcendental number.

clearly $r : \sim p \leftrightarrow q$

Let us use truth table to check the equivalence of 'r' and 'q or p'; 'r' and $\sim(p \leftrightarrow \sim q)$

				1	2		3
p	q	~p	~q	~p ↔q	q or p	p↔~q	~(p↔~q)
T	T	F	F	F	T	F	T
T	F	F	T	T	T	T	F
F	T	T	F	T	T	T	F
F	F	T	T	F	F	F	T

From columns (1), (2) and (3), we observe, none of the these statements are equivalent to each other.

$\therefore$ Statement 1as well as statement 2 both are false.

$\therefore$ None of the options is correct.

41. **(d)** Let us make the truth table for the given statements, as follows :

p	q	p∨q	q→p	p →(q→p)	p →(p∨q)
T	T	T	T	T	T
T	F	T	T	T	T
F	T	T	F	T	T
F	F	F	T	T	T

From table we observe

$p \to (q \to p)$ is equivalent to $p \to (p \vee q)$

42. (b) The truth table for the logical statements, involved in statement 1, is as follows :

p	q	$\sim q$	$p \leftrightarrow \sim q$	$\sim(p \leftrightarrow \sim q)$	$p \leftrightarrow q$
T	T	F	F	T	T
T	F	T	T	F	F
F	T	F	T	F	F
F	F	T	F	T	T

We observe the columns for $\sim(p \leftrightarrow \sim q)$ and $p \leftrightarrow q$ are identical, therefore

$\sim(p \leftrightarrow \sim q)$ is equivalent to $p \leftrightarrow q$

But $\sim(p \leftrightarrow \sim q)$ is not a tautology as all entries in its column are not T.

$\therefore$ Statement-1 is true but statement-2 is false.

43. (c) For the numbers 2, 4, 6, 8, ..., $2n$

$$\bar{x} = \frac{2[n(n+1)]}{2n} = (n+1)$$

$$\text{And } Var = \frac{\Sigma(x-\bar{x})^2}{2n} = \frac{\Sigma x^2}{n} - (\bar{x})^2$$

$$= \frac{4\Sigma n^2}{n} - (n+1)^2 = \frac{4n(n+1)(2n+1)}{6n} - (n+1)^2$$

$$= \frac{2(2n+1)(n+1)}{3} - (n+1)^2 = (n+1)\left[\frac{4n+2-3n-3}{3}\right]$$

$$= \frac{(n+1)(n-1)}{3} = \frac{n^2-1}{3}$$

$\therefore$ Statement-1 is false. Clearly, statement - 2 is true.

44. (b) Let $x \in A$ and $x \in B \Leftrightarrow x \in A \cup B$

$\Leftrightarrow x \in A \cup C \quad (\because A \cup B = A \cup C)$

$\Leftrightarrow x \in C$

$\therefore \quad B = C.$

Let $x \in A$ and $x \in B \Leftrightarrow x \in A \cap B$

$\Leftrightarrow x \in A \cap C \quad (\because A \cap B = A \cap C)$

$\Leftrightarrow x \in C$

$\therefore B = C$

45. (b) Mean $= \dfrac{101 + d(1+2+3+......+100)}{101}$

$$= 1 + \frac{d \times 100 \times 101}{101 \times 2} = 1 + 50\,d$$

$\because$ Mean deviation from the mean = 255

$$\Rightarrow \frac{1}{101}[|1-(1+50d)| + |(1+d)-(1+50d)| + |(1+2d) - (1+50d)| + ... + |(1+100d)-(1+50d)|] = 255$$

$$\Rightarrow \quad 2d[1+2+3+...+50] = 101 \times 255$$

$$\Rightarrow \quad 2d \times \frac{50 \times 51}{2} = 101 \times 255$$

$$\Rightarrow \quad d = \frac{101 \times 255}{50 \times 51} = 10.1$$

46. (b) P : there is a rational number $x \in S$ such that $x > 0$

$\sim$P : Every rational number $x \in S$ satisfies $x \le 0$

47. (b) x Ry need not implies yRx

$\therefore$ R is not symmetric and hence not an equivalence relation.

$$S : \frac{m}{n} s \frac{p}{q}$$

Given $qm = pn \Rightarrow \dfrac{p}{q} = \dfrac{m}{n}$

$\therefore \dfrac{m}{n} s \dfrac{m}{n}$ (reflexive) $\dfrac{m}{n} s \dfrac{p}{q} \Rightarrow \dfrac{p}{q} s \dfrac{m}{n}$ (symmetric)

$$\frac{m}{n} s \frac{p}{q}, \frac{p}{q} s \frac{r}{s} \Rightarrow qm = pn, ps = rq$$

$\Rightarrow \dfrac{p}{q} = \dfrac{m}{n} = \dfrac{r}{s} \Rightarrow$ ms = rn (transitive).

S is an equivalence relation.

48. (a) $\sigma_x^2 = 4, \sigma_y^2 = 5, x = 2, y = 4$

$$\frac{1}{5}\sum x_i^2 - (2)^2 = 4; \ \frac{1}{5}\sum y_i^2 - (4)^2 = 5$$

$$\sum x_i^2 = 40; \sum y_i^2 = 105 \Rightarrow \sum\left(x_i^2 + y_i^2\right) = 145$$

$$\Rightarrow \sum(x_i + y_i) = 5(2) + 5(4) = 30$$

Variance of combined data

$$= \frac{1}{10}\sum\left(x_i^2 + y_i^2\right) - \left(\frac{1}{10}\sum(x_i + y_i)\right)^2$$

$$= \frac{145}{10} - 9 = \frac{11}{2}$$

49. (b) $x - y$ is an integer.

$\because x - x = 0$ is an integer $\Rightarrow A$ is reflexive.

Let $x - y$ is an integer

$\Rightarrow y - x$ is an integer

$\Rightarrow A$ is symmetric

Let $x - y, y - z$ are integers

$\Rightarrow x - y + y - z$ is also an integer

$\Rightarrow x - z$ is an integer

$\Rightarrow A$ is transitive

$\therefore A$ is an equivalence relation.

Hence statement 1 is true.

Also B can be considered as

xBy if $\dfrac{x}{y} = \alpha$, a rational number

$\because \dfrac{x}{x} = 1$ is a rational number

$\Rightarrow B$ is reflexive

But $\frac{x}{y} = \alpha$, a rational number need not imply $\frac{y}{x} = \frac{1}{\alpha}$, a rational number because

$\frac{0}{1}$ is rational $\Rightarrow \frac{1}{0}$ is not rational

$\therefore$ B is not an equivalence relation.

0. (a) Suman is brilliant and dishonest if and only if Suman is rich is expressed as

$Q \leftrightarrow (P \wedge \sim R)$

Negation of it will be $\sim (Q \leftrightarrow (P \wedge \sim R))$

51. (b) Median is the mean of 25th and 26th observation

$$\therefore M = \frac{25a + 26a}{2} = 25.5a$$

$$M.D(M) = \frac{\sum |x_i - M|}{N}$$

$$\Rightarrow 50 = \frac{1}{50}[2 \times |a| \times (0.5 + 1.5 + 2.5 + ...24.5)]$$

$$\Rightarrow 2500 = 2|a| \times \frac{25}{2}(25) \Rightarrow |a| = 4$$

52. (a) Correct mean = observed mean + 2 = 30 + 2 = 32

Correct S. D. = observed S.D. = 2

3. (c)

A	B	A ∨ B	A ∧ B	A ∧ (A ∨ B)	A ∨ (A ∧ B)	A → B	A ∧ (A → B)	[A ∧ (A → B) → B]	[B → [A ∧ (A → B)]]
T	F	T	F	T	T	F	F	T	T
F	T	T	F	F	F	T	F	T	F
T	T	T	T	T	T	T	T	T	T
F	F	F	F	F	F	T	F	T	T

4. (b) **For reflexive :** $(A, A) \in R$

$A = P^{-1}AP$ is true,

For $P = I$, which is an invertible matrix.

$\therefore$ R is reflexive.

For symmetry : As $(A, B) \in R$ for matrix P

$$A = P^{-1}BP$$

$$\Rightarrow PAP^{-1} = B$$

$$\Rightarrow B = PAP^{-1}$$

$$\Rightarrow B = (P^{-1})^{-1} A (P^{-1})$$

$\therefore$ (B, A) $\in$ R for matrix P^{-1}

$\therefore$ R is symmetric.

For transitivity

$A = P^{-1}BP$ and $B = P^{-1}CP$

$$\Rightarrow A = P^{-1}(P^{-1}CP)P$$

$$\Rightarrow A = (P^{-1})^2 CP^2$$

$$\Rightarrow A = (P^2)^{-1} C(P^2)$$

$\therefore$ $(A, C) \in R$ for matrix P^2

$\therefore$ R is transitive.

So R is equivalence

. (a) Let p : I become a teacher.

q : I will open a school

Negation of $p \to q$ is $\sim (p \to q) = p \wedge \sim q$

i.e. I will become a teacher and I will not open a school.

56. (d) **KEY CONCEPT :** If each observation is multiplied by k, mean gets multiplied by k and variance gets multiplied by k^2. Hence the new mean should be $2\overline{x}$ and new variance should be $k^2\sigma^2$.

So statement-1 is true and statement-2 is false.

57. (b) Let $X = \{1,2,3,4,5\}$

Total no. of elements = 5

Each element has 3 options. Either set Y or set Z or none. ($\because Y \cap Z = \phi$)

So, number of ordered pairs = 3^5

58. (c) Given

$n(A) = 2, n(B) = 4, n(A \times B) = 8$

Required number of subsets =

$${}^8C_3 + {}^8C_4 + ... + {}^8C_8 = 2^8 - {}^8C_0 - {}^8C_1 - {}^8C_2 = 256 - 1 - 8 - 28 = 219$$

59. (b) **Statement-2 :** $(p \to q) \leftrightarrow (\sim q \to \sim p)$

$\equiv (p \to q) \leftrightarrow (p \to q)$

which is always true.

So statement 2 is true

Statement-1 : $(p \wedge \sim q) \wedge (\sim p \wedge q) = p \wedge \sim q \wedge \sim p \wedge q$

$= p \wedge \sim p \wedge \sim q \wedge q = f \wedge f = f$

So statement-1 is true

60. (d) If initially all marks were x_i then $\sigma_1^2 = \frac{\sum_i (x_i - \overline{x})^2}{N}$

Now each is increased by 10

$$\sigma_1^2 = \frac{\sum_i [(x_i + 10) - (\overline{x} + 10)]^2}{N} = \frac{\sum_i (x_i - \overline{x})^2}{N} = \sigma_1^2$$

Hence, variance will not change even after the grace marks were given.

❑ ❑